LEXI-COMP'S

Laboratory Test Handbook

C·O·N·C·I·S·E

with
Disease Index

D0939607

David S. Jacobs, MD
Wayne R. DeMott, MD
Dwight K. Oxley, MD

Laboratory Test Handbook

C·O·N·C·I·S·E

with
Disease Index

3rd Edition

David S. Jacobs, MD, FACP, FCAP
Co-Editor-in-Chief
President, Pathologists Chartered
Consultant in Pathology and Laboratory Medicine
Overland Park, Kansas

Dwight K. Oxley, MD, FCAP
Co-Editor-in-Chief
Consultant in Pathology and Laboratory Medicine
Wichita, Kansas

Wayne R. DeMott, MD, FCAP
Consultant in Pathology and Laboratory Medicine
Shawnee Mission, Kansas

LEXI-COMP, INC
Hudson (Cleveland), OH

Dedication

To Judy

To Patricia

To Lee

With love and thanks

The *Concise III* edition of the *Laboratory Test Handbook* is intended to serve as a useful reference at the time of publication, and not as a complete laboratory testing resource. No warranty is provided that the information herein is in every way accurate or complete. The explosion of information in many directions, in multiple scientific disciplines, with advances in laboratory techniques, and continuing evolution of knowledge requires constant scholarship. The publication covers common, as well as some esoteric testing procedures. The authors, editors, reviewers, contributors, and publishers cannot be responsible for the continued currency of the information or for any errors or omissions in this book or for any consequences arising therefrom, given the certainty of human error and the continuing dynamic changes in science. Readers are advised that decisions regarding diagnosis and treatment must be based on the independent judgment of the clinician. The editors or authors are not responsible for any inaccuracy of quotation or for any false or misleading implication that may arise due to the text. Sources widely understood to be reliable have been used to try to provide information. For brevity, footnotes and many references have been deleted from *Laboratory Test Handbook Concise III*. The reader is encouraged to consult the *Laboratory Test Handbook*, 5th edition for further footnote and reference listings documenting information presented in *Concise III* and to examine other sources of information including internet web sites, as well as other printed or electronic materials.

This manual was produced using the Pathfinder™ Program –
a complete publishing service of Lexi-Comp Inc.

1100 Terex Road
Hudson, Ohio 44236
(330) 650-6506

ISBN 1-59195-080-5

TABLE OF CONTENTS

FOREWORD

The timely application of information is important, if not critical, to success and survival in today's world. The medical community strives in a challenging environment to provide quality patient care that is timely and applicable. Improved diagnostic and therapeutic modalities must be applied within a challenging environment. The *Laboratory Test Handbook Concise III (LTH Concise III)* assists the clinician by providing a compact source of data relevant to most available clinical laboratory tests.

This handbook, produced by Lexi-Comp, Inc, its editors, and contributing authors, is the 3rd edition of *LTH Concise*. It is an abridged, now updated and more portable version of the established comprehensive *Laboratory Test Handbook*, 5th edition *(LTH-5)*. As with the previous edition, clinical laboratory test information is presented conveniently (using a consistent and compact format) in order to facilitate timely application by those working or training in the healthcare professions. Multiple aspects of each test are presented including name, synonyms, available related procedures, patient preparation, specimen requirements, reference intervals, and interpretive considerations (including appropriate application, limitations, contraindications, and additional information). Discussion of methodology is often truncated or deleted.

Test listing is alphabetical and includes some common synonyms with referral by page number to the appropriate location.

Apropos to the tragic events of September 11, 2001, biological and chemical warfare are addressed. The 2004 volume has been updated to provide presently needed essential information and references. These address West Nile virus, SARS, newborn screening by tandem mass spectrometry, newly developing tests such as B-type natriuretic peptide, serum myeloperoxidase, and CD40, and other needs which have developed since distribution of the 2002 printing. Many monographs have been updated, including HER-2/*neu*, and other new entities added.

The *LTH Concise III* format, as with its earlier edition, does not include footnotes. Some text, tables, and illustrations (present in *LTH-5*) have been deleted. Documentation of text comments, in particular new or recent findings, will often be found in the references that accompany each test entry. For credit and recognition of original sources, the reader is referred by extension to the literature citations present in the 2001 *LTH-5*.

If a test is no longer of clinical value (and is not commonly performed), interpretive sections may be truncated with an indication of current status, usage may be discouraged, or the test listing may be deleted. Omission of a test from this book is not to be considered implied discouragement of its use. A large number of valuable but infrequently requested tests and assays are available nationally.

Current and updated information may also be found in the electronic handheld version of *LTH Concise* and in *LTH, 5th Edition* on CD-ROM and on-line. Please visit the Lexi-Comp web site at www.lexi.com for information on these products.

The reader is urged to consult the references and to discuss reference ranges, if indicated, with physicians or other scientists responsible for laboratories that provide patient test results.

The important area of drugs and their effect on laboratory tests requires current and comprehensive coverage. We recommend Young DS, *Effects of Drugs on Clinical Laboratory Tests*, 5th ed, Washington, DC: AACC Press, American Association of Chemistry, 2000, to provide encyclopedic specific data, references, and understanding of this area.

Lexi-Comp, Inc, its authors, and editors are sincerely hopeful that the *LTH Concise III* will benefit its readership in their goal to provide quality, timely, and applicable patient care.

The contributions of current and previous authors is acknowledged and applauded with gratitude. Limitations of space do not allow documentation of many worthy contributions. For more complete recognition, the reader may consult *LTH-5*.

ACKNOWLEDGMENTS

The editors of the *Laboratory Test Handbook Concise III* wish to thank several individuals whose concerted efforts have made this book possible. Mr Robert Kerscher, the publisher and president of Lexi-Comp Inc, is the person who foresaw the usefulness of such a book and provided the resources required for its production. His dedication to the project and his support and development of the many unique and innovative features included in the book (eg, format, internal cross-references, comprehensive indexing, and Disease Index) contribute substantially to the content and usefulness of the book.

Barbara Kerscher, production manager, has made essential administrative contributions, including the facilitation of revisions, new entries, and overall coordination among the numerous authors and contributors who have contributed to this edition. Despite many changes in format and content, Ms Kerscher has exhibited outstanding patience, intelligence, flexibility, and extraordinary good humor. Assisting her have been other members of the Lexi-Comp staff: David C. Marcus, computer programming design; Matthew C. Kerscher, product manager; Kathy Smith, RN, and Robin Farabee, production assistants; Alexandra J. Hart, composition specialist; and Tracey Reinecke, graphics artist.

We thank the authors and contributors who have made this volume possible. We hope that they have shared with us the sense and pride of purpose which we have seen develop as the edition evolved.

I. Cori Baill, MD and Jonathan Todd Jacobs, MD have each provided much appreciated clinical insight for several monographs.

Finally, we hope our readers find this handbook a practical source of useful information to support good patient care.

ACKNOWLEDGMENTS *(Continued)*

Sincere appreciation and many thanks for their contributions are extended to the following *Laboratory Test Handbook*, 5th edition authors and contributors:

Christopher D. Ackley, MD, PhD

Uri S. Alon, MD

Leland B. Baskin, MD, MS, FCAP

Jennifer A. Brainard, MD

Leigh Ann Cahill, CT(ASCP)

Marilyn K. Davis-Cansler, BS SCT (ASCP) CMIAC

John Foxworth, PharmD

Charles W. Gorodetzky, MD, PhD

Harold J. Grady, PhD

Daniel R. Hinthorn, MD, FACP

Rebecca Horvat, PhD

Daniel H. Jacobs, MD

Michael Laposata, MD, PhD

Phillip A. Munoz, MD

Eugene S. Olsowka, MD, PhD

Mary Ann Pedigo CT(ASCP)

Diane L. Persons, MD

Frederick V. Plapp, MD, PhD

Jasbir Singh, PhD

Barry S. Skikne, MBBCh, FACP, FCP(SA)

Karen Stephens, PhD

Jonathan F. Tait, MD, PhD

Ossama Tawfik, MD, PhD

The editors and authors acknowledge and are thankful for the use of powerful data processing resources embodied in the Lexi-Comp Pathfinder™ Program, the Medline® System, and for the computer-based publishing expertise of Lexi-Comp, Inc. Clinical Laboratory-related information is also available in other titles that form "Lexi-Comp's Clinical Reference Laboratory". In particular, the *Poisoning and Toxicology Handbook, Clinician's Guide to Laboratory Medicine Series, Infectious Diseases Handbook,* and *Drug Information Handbook* are recommended.

Publisher's Acknowledgment and Dedication

The *Laboratory Test Handbook with Key Word Index* is the direct result of over 20 years of dedicated service by David S. Jacobs, MD and Wayne R. DeMott, MD.

Drs Jacobs and DeMott have provided clear vision and extraordinary commitment, establishing higher standards for each edition. The *Laboratory Test Handbook* was originally the effort of four community-based pathologists offering their insights on tests routinely available in a modern clinical laboratory. This 5th edition represents the culmination of contributions of over 50 medical and editorial professionals reflecting state-of-the-art offerings available in contemporary medical practice.

ABOUT THE AUTHORS

The transformation of *Laboratory Test Handbook*, 5th edition, to *Laboratory Test Handbook Concise III*, is attributable to the following:

David S. Jacobs, MD, FACP

Dr Jacobs was Director of Laboratories at Providence Medical Center in Kansas City, Kansas for 29 years, leaving that position in 1994. He served as a member of the institutional Credentials Committee for 20 years. During Dr Jacobs' tenure as Director, then Medical Director of the School of Medical Technology, 173 medical technologists graduated from 1965 to 1990.

He remains Clinical Professor at the University of Kansas and the University of Missouri-Kansas City Schools of Medicine. He is a Fellow of the American College of Physicians, the College of American Pathologists, and the American Society of Clinical Pathologists. He is a member of the International Academy of Pathology, the AMA, and the Kansas Medical Society. He was a member of the Editorial Board of *Kansas Medicine* and was Book Review Editor of that journal. Dr Jacobs is past President of the Kansas Society of Pathologists, the Kansas City Society of Pathologists, and of the Medical Staff of the Providence Medical Center (formerly Providence-St Margaret Health Center). He held a position on the Board of Directors of the Wyandotte County Medical Society for several years, and for much of his career served as an inspector for both the College of American Pathologists and for the American Association of Blood Banks.

He received his premedical education at the University of Michigan. Entering the UM Medical School in the "Letters and Medicine" program, Dr Jacobs remained at Michigan as a general rotating intern, then did his residency in Anatomic and Clinical Pathology at the same institution. Following service in the U.S. Army Medical Corps as a pathologist with 13 months residency, he returned to Ann Arbor and completed the pathology program under Drs A. J. French and M. R. Abell. He then enjoyed additional professional experience in clinical pathology in Chicago with Drs Israel Davidsohn, Douglas Huestis, and Norbert Tietz. Dr Jacobs is certified in both Anatomic and Clinical Pathology by the American Board of Pathology.

Dr Jacobs' special interests include general pathology, particularly surgical pathology, interpretation of clinical laboratory tests, and transfusion medicine. His publications address topics in surgical pathology and the clinical relevance of laboratory testing. He is active as a consultant in pathology and laboratory medicine.

Wayne R. DeMott, MD

Dr DeMott is a graduate of the University of Oregon Medical School (1959). He completed his internship at Madigan General Hospital outside Tacoma, Washington. After service in the U.S. Air Force as a General Medical Officer, he completed the pathology residency program at the University of California San Francisco Medical Center (1967). He was certified in Anatomic and Clinical Pathology by the American Board of Pathology in 1968.

As a practicing pathologist at Providence Medical Center in Kansas City, Kansas from 1968-1994, Dr DeMott gave special attention to the areas of Hematology and Coagulation. He participated in research activities involving separation of white cells from peripheral blood, measurement of zinc levels in peripheral blood leukocytes utilizing atomic absorption spectrophotometry, and separation of neutrophil enzymes by isoelectric focusing. He was an active contributor to the Providence Medical Center (formerly Providence-St Margaret Health Center) School of Medical Technology, is a past president of the Medical Staff of Providence Medical Center, and was chairman or a member of several of the hospital professional committees, including the Cancer Committee. He directed the Cancer Conference for many years.

ABOUT THE AUTHORS *(Continued)*

Dr DeMott is a Fellow of the American Society of Clinical Pathologists and College of American Pathologists, and is a member of the American Association for Clinical Chemistry, American Medical Association, Kansas Society of Pathologists, and Kansas City Society of Pathologists.

Dwight K. Oxley, MD

Dr Oxley received his MD degree from the University of Kansas and took his Anatomic Pathology residency at that institution; his Clinical Pathology residency was at the U.S. Naval Hospital, St Albans, New York. He took a fellowship in Nuclear Medicine at Johns Hopkins University. From 1988 to 2003, he was the Medical Director of Pathology at Wesley Medical Center, Wichita, Kansas. Prior to 1988 he was a community pathologist in Kansas City, Missouri (St Joseph Hospital) and, before that in Rancho Mirage, California (Eisenhower Medical Center).

Dr Oxley is a Trustee of the American Board of Pathology and served as President of the Board in 1999. He has been a member of the Executive Committee of the American Board of Medical Specialties, and is a member of the CAP Foundation Board of Directors. He has served on the editorial boards of the *American Journal of Clinical Pathology*, the *Archives of Pathology and Laboratory Medicine*, and *Clinica Chimica Acta*. While practicing in Kansas City he was active in the teaching of the sophomore pathology course at the University of Kansas, and he has an appointment as a Clinical Professor of Pathology at that institution.

Dr Oxley continues his major interest in clinical pathology, particularly the clinical relevance of laboratory test results. He is a Fellow of the College of American Pathologists and the American Society of Clinical Pathologists. He is a member of the American Pathology Foundation, the American Association for the History of Medicine, the Kansas Society of Pathologists, the American Medical Association, the Kansas Medical Society, and the Medical Society of Sedgwick County.

Bradly D. Clark, MD

Dr Clark is a fellow in cytopathology at VCU Medical Center in Richmond, Virginia. He completed his residency in Anatomic Pathology at the National Cancer Institute's Laboratory of Pathology in Bethesda, Maryland.

Uttam Garg, PhD, DABCC, DABFT, FACB

Dr Garg received his PhD from the Postgraduate Institute of Medical Education and Research, India. He did his postdoctoral training in Cell Biology and Pharmacology at the New York Medical College. He received additional training in Clinical Chemistry and Toxicology at the University of Minnesota Medical School. Before joining the staff at Children's Mercy Hospital in Kansas City, Missouri as Associate Professor and Director of Clinical Chemistry and Toxicology, he was Assistant Professor at the University of Minnesota Medical School. He also served the New York University Medical Center as Research Assistant Professor.

Dr Garg is a recipient of a number of awards and honors including Major General Amir Chand Gold Medal for graduate excellence, American Heart Association Fellowship, FIDIA Research Foundation Award, and Outstanding Scientific Achievements Award by a young investigator by the American Association for Clinical Chemistry. Dr Garg is a Fellow of the Academy of Clinical Biochemistry. In 2003, Dr Garg was Chairman of the Midwest Section of the American Association of Clinical Chemistry. He is on the Board of Directors of the American Board of Clinical Chemistry and the National Registry of Certified Chemists. He serves as a member of the toxicology subcommittee of the American Board of Clinical Chemistry and is the Chairman of the Chemistry Exam Committee of the National Registry of Certified Chemists. He is also an advisor to the area committee on clinical chemistry and toxicology of the National Committee for Clinical Laboratory Standards.

He is certified in Clinical Chemistry by the American Board of Clinical Chemistry and the American Society of Clinical Pathology. He is also certified in Toxicological Chemistry by the American Board of Clinical Chemistry and the National Registry of Certified Chemists, and in Forensic Toxicology by the American Board of Forensic Toxicology. He has published over 70 peer-reviewed research articles, which have been cited in over 1600 publications. He has also published 7 book chapters. He has been an invited reviewer for several journals including *Clinical Chemistry*, *Clinica Chimica Acta*, *Journal of Laboratory and Clinical Medicine*, *Life Sciences*, and *Neuroscience*. His research interests include method developments in clinical laboratory diagnosis.

Susan H. Hsu, PhD

Dr Hsu received her postdoctoral training from 1970-1974 in the Immunogenetics Laboratories, Division of Medical Genetics of the Johns Hopkins Medical Institution (JHMI) at Baltimore, Maryland. She has been the Assistant and subsequently the Codirector of the Immunogenetics Lab from between 1976-1986 and associate Professor in the Department of Medicine at the JHMI from 1983-1986. Since 1987 she has assumed the position of the Scientific Director and the Director of the HLA/Molecular Genetics Department of the American Red Cross, Blood Services, Penn-Jersey Region in Philadelphia, Pennsylvania.

Geralyn M. Meny, MD

Dr Meny received her BS degree in medical technology from the University of Kentucky. She received certification as a Medical Technologist and a Specialist in Blood Banking from the American Society of Clinical Pathologists. Dr Meny received her MD degree from the University of Texas Southwestern Medical Center at Dallas in 1990. She completed a pathology residency at Parkland Memorial Hospital in Dallas in 1995. She is board certified in Anatomic and Clinical Pathology and Blood Banking/Transfusion Medicine by the American Board of Pathology.

Dr Meny is currently Medical Director at the American Red Cross, Penn-Jersey Region, Philadelphia, Pennsylvania.

Celeste N. Powers, MD, PhD

Dr Celeste N. Powers received her PhD in Microbiology and Immunology from Baylor College of Medicine, Houston, Texas and her MD from The University of Texas Medical School at Houston. She also did her residency training in Anatomic and Clinical Pathology at UTMSH. Dr Powers completed fellowship training in Surgical and Cytopathology at the Medical College of Virginia in Richmond, Virginia. She is board certified in Anatomic and Clinical Pathology and has Added Qualifications in Cytopathology from the American Board of Pathology. Dr Powers has played an active role in several national pathology societies, including American Society of Clinical Pathologists, American Society of Cytopathology and the United States and Canadian Academy of Pathology. She has authored numerous articles and book chapters, including a book on the fine needle aspiration biopsy of the head and neck.

Dr Powers is currently Professor and Chair of the Division of Surgical and Cytopathology as well as Director of Anatomic Pathology Services for the Department of Pathology, Virginia Commonwealth University Health System, Medical College of Virginia Hospitals.

ABOUT THE AUTHORS *(Continued)*

Charlotte E. Shideler, PhD

Dr Shideler earned her PhD in biochemistry from the University of Arkansas for Medical Sciences in Little Rock, Arkansas and received postdoctoral training in clinical chemistry at the University of Virginia Medical Center in Charlottesville, Virginia. Dr Shideler is certified in Clinical Chemistry by the American Board of Clinical Chemistry and National Registry in Clinical Chemistry and is employed by Wesley Pathology Consultants in Wichita, Kansas.

Dr Shideler is a member of the American Association for Clinical Chemistry, American Chemical Society, American Society of Clinical Pathologists, Association of Clinical Scientists, Clinical Laboratory Management Association, and American Association for the Advancement of Science.

HOW TO USE THIS HANDBOOK

The *Laboratory Test Handbook Concise III* consists of clinical laboratory test listings arranged alphabetically and covering Anatomic Pathology, Chemistry, Clinical Microscopy, Coagulation, Cytogenetics, Cytopathology, Hematology, Immunology, Infectious Diseases, Molecular Genetic Testing, Molecular Oncology, Therapeutic Drug Monitoring, Toxicology/Drugs of Abuse, Trace Elements, Transfusion Medicine (Blood Bank), and Urinalysis. Located at the beginning of the book and entitled, Maximizing the Information From Laboratory Tests - The Ulysses Syndrome is a discussion of matters pertaining to laboratory accuracy, statistics, and "normal range". Also presented initially are brief introductions to the major clinical laboratory disciplines, including importantly a section devoted to Specimen Collection. The laboratory tests are cross-referenced with synonyms referring the user to the actual test name.

Each individual test listing is arranged in a consistent format providing specific types of information. The fields of information include the following. The **test name**; **related information** which lists other tests that may be of interest and the page number in which such tests can be found; **synonyms** or other common names for a test are noted; topics or procedures which are not exact synonyms but have similar instructions or require similar consideration are referred to under the **applies to** heading; tests **replaced by** a current procedure may be noted; a definition of procedures included within the named test is given under **test includes**; an **abstract** or overview of the test is often provided; patient **preparation** includes patient care considerations prior to the collection of specimen or performance of a test; **aftercare** includes patient care considerations following the collection of a specimen or performance of a procedure; the specific **specimen** required, the **container, sampling time**, specific **collection** instructions, specimen **storage instructions, causes for rejection** of the specimen by the laboratory, **turnaround time** when relevant, and **special instructions** indicating additional pertinent considerations relating to the specimen are listed; a discussion of basic information relevant to the clinical application of the test, including **reference** (or normal) **range, critical values**, and **possible panic ranges**, specific **use** of the test, **limitations** of the test method(s), specific test **methodology** where appropriate, **contraindications** to the test, and **additional information** which may contribute to the interpretation or utilization of the test are given.

References

References are included at the end of each test listing and may refer to specific literature quotations, specific points of information, opinions, and may provide access to needed additional information. Footnotes and a more complete reference base are available in the *Laboratory Test Handbook*, 5th edition and its updated electronic versions. Such sources may be consulted as an extension of citations presented in the *Concise III* edition.

The importance of the clinicopathologic conference as a teaching tool continues to be supported by the Case Records of the Massachusetts General Hospital in the *New England Journal of Medicine*. We have inserted some of these exercises as references in the *Laboratory Test Handbook Concise III*.

Internet Web Sites

This edition of the *Concise III* includes citations to internet web sites for those seeking additional information. There are numerous citations to special interest web sites, many of which are excellent. Various governmental agencies maintain web sites, such as the Centers for Disease Control and Prevention (ww.cdc.gov) and the Food and Drug Administration (www.fda.gov). There is growing interest in on-line medical textbooks (eg, emedicine.com) which are edited by established investigators and are usually fairly complete.

The editors point out that these sites are undergoing frequent revisions. Thus, we cannot be responsible for the quality of information on each site at any given

HOW TO USE THIS HANDBOOK *(Continued)*

time. Information at each web site should be compared with other sites for verification.

Acronyms and Abbreviations Glossary

This glossary provides a useful listing of many acronyms and abbreviations commonly associated with laboratory medicine. We offer this glossary not as an exhaustive authoritative list, but more as a guide to assist in interpretation of frequently used terminology.

Disease Index

The Disease Index is not intended in any way to suggest patterns of physicians' orders, nor is it complete. Rather, it is the intent of the authors and editors to make information easier to find and utilize in order to support better patient care.

The Disease Index provides a reference to test names based on a diagnostic property, disease entity, organ system, or syndrome for which the test may be useful. It provides lists of specific tests. Some may support possible clinical diagnoses or help to rule out other diagnostic possibilities. The index is further refined by using symbols: the (••) symbol indicates a test regarded as essential in the diagnosis of the disease; the (•) symbol denotes a test that is frequently used in the diagnosis or management of the disease. Other tests, listed without a symbol, are those that may or may not be useful, depending on particular clinical circumstances. **A negative laboratory test result can be, and frequently is, highly relevant in the practice of medicine.**

Clinical diagnosis is determined following history, physical examination, and usual laboratory investigation with selected additional tests. Complete blood count (CBC) with differential, urinalysis, and a basic chemistry profile are not only good medicine, they are in fact cost effective. Thus, these basic tests are excluded from much of the Disease Index.

A few of the entities in the Disease Index include a very brief explanation of the process or additional information which may be helpful.

Diagnoses with *International Classification of Disease—Ninth Revision—Clinical Modification* (ICD-9-CM) codes are indicated within the [ ] symbol.

MAXIMIZING THE INFORMATION FROM LABORATORY TESTS — THE ULYSSES SYNDROME

TESTS IN SEARCH OF A DISEASE

Dwight K. Oxley, MD
Uttam Garg, PhD
Eugene S. Olsowka, MD, PhD

Appropriate use of laboratory testing is so difficult that it has been compared with the obstacles encountered by the Greek hero Ulysses on his return from the Trojan War.[1] In this metaphor, the patient, like Ulysses, is put at risk, here because of misleading results from tests that should not, in a more perfect world, have been ordered. To protect the patient from misinterpreted or erroneously ordered tests, four subtasks are relevant: how a "reference range" is established, why "normal range" is a misnomer, how to assess the analytical error involved in testing, and how to measure the usefulness of a test in actual practice.

The "Reference Range" Problem and Biologic Variation

Every test result from an accredited laboratory is accompanied by a "reference range." For certain tests, the reference range has been determined with reasonable accuracy and is well known; examples include hemoglobin, serum calcium, serum sodium, and leukocyte count. Even with these basic tests, however, the concept of a population-wide reference range is problematic. For clinical decision-making, the population-wide data for blood hemoglobin should be partitioned into one reference range for women and another for men. In fact, population-wide data distributions for many analytes are partitioned into subsets based on age, gender, and other patient attributes (eg, pregnancy).

The term "normal range" has been frequently used in the past and is not recommended for clinical practice because it is potentially misleading; the reference population used to define the reference range is selected for specific attributes rather than "normalcy." **Reference interval** is the preferred term for a statement which comprises lower and upper boundary values against which a patient result will be compared. A test result from an accredited laboratory is always reported with such a reference interval. Results outside of this interval are often flagged by the laboratory and interpreted by physicians as being significantly different from some accepted benchmark. Consider the example total serum calcium, an analyte in which concentration is under tight homeostatic control. The population-based "reference range" for such an analyte is approximately the same as the reference interval reported by most laboratories: 8.6-10.0 mg/dL. However, if the patient is a newborn full-term infant, the appropriate interval is quite different, viz 7.6-10.4 mg/dL. And if the patient has a low concentration of serum proteins, the reference interval must be adjusted by a formula based on the protein value. Patient age and serum protein status are only two of the many variables which influence the reference interval and therefore the interpretation of this test result. Other variables include gender, ethnicity, hydration status, lean body mass, pregnancy, posture, diet, exercise, current medications, how the specimen was obtained, and how the specimen was handled and stored before reaching the analytical system. **Preanalytical variability** is the term which includes all of these effects.

Another important source of variability in test results is that due to normal physiologic fluctuations (**circadian rhythms**). Consider the example of serum iron. An individual's serum iron is highest in the morning. The result from a specimen obtained at 2 PM may be as much as 50% less than the value obtained at 8 AM.[2] Some analytes have fluctuations with periods that are measured in days or weeks. **Biologic variability** is the term which refers to this source of variation. A

MAXIMIZING THE INFORMATION FROM LABORATORY TESTS — THE ULYSSES SYNDROME *(Continued)*

compilation of biologic variability is available.[3] Some examples of the within-day biologic variability of common analytes are presented in the table.

Total and Analytical Variation for Serum Tests on Specimens Obtained at 8 AM and 2 PM

Constituent	Mean	Total Variation (%)	Analytical Variation (%)
Potassium (mmol/L)	4.4	7.1	2.8
Calcium (mg/dL)	10.8	3.2	2.7
Chloride (mmol/L)	102	3.8	3.4
Phosphate (mg/dL)	3.8	10.7	2.4
Urea N (mg/dL)	14	22.5	2.5
Creatinine (mg/dL)	1.0	14.5	6.3
Uric acid (mg/dL)	5.6	11.5	2.6
Iron (μg/dL)	116	36.6	3.4
Cholesterol (mg/dL)	193	14.8	5.7
Albumin (g/dL)	4.5	5.5	3.9
Total protein (g/dL)	7.3	4.8	1.7
Total lipids (g/L)	5.3	25.0	3.6
Aspartate aminotransferase (units/L)	9	25	6
Alanine aminotransferase (units/L)	6	56	17
Acid phosphatase (units/L)	3	15	8
Alkaline phosphatase (units/L)	63	20	3
Lactate dehydrogenase (units/L)	195	16	12

11 male subjects, age 21-27 years, studied at 8 AM, 11 AM, and 2 PM.

Adapted from Burtis CA and Ashwood ER, eds, *Tietz Textbook of Clinical Chemistry*, 3rd ed, Philadelphia, PA: WB Saunders Co, 1999, 59.

The effects of both preanalytical and biologic variability on **plasma glucose** assays have been extensively studied. Fasting plasma glucose has a reference interval often stated as 60-109 mg/dL. But, as with calcium, this interval does not apply to all age groups. Much more important than the upper boundary of 109 mg/dL is the **clinical decision limit** of 126 mg/dL.[4] This clinical decision limit applies only to the diagnosis of diabetes mellitus. The point for emphasis is that a clinical decision limit is conceptually different from a reference range. A contingency affecting the clinical decision limit for the fasting plasma glucose reference interval arises from biologic variability: as revealed by the work of Troisi et al,[5] fasting plasma glucose levels are lower in the afternoon than in the morning. Since the criterion of 126 mg/dL was based on morning specimens, applying the 126 mg/dL clinical decision limit to patients whose blood is obtained in the afternoon will miss half of the cases of undiagnosed diabetes mellitus. One way to deal with this is to use a different clinical decision limit, viz 114 mg/dL, for patients whose specimens are obtained in the afternoon; another is to use only morning specimens.

Some of the most common sources of preanalytical variability are summarized below.

1. **Pregnancy**. Many of the analytes affected by pregnancy are listed in the following table.[6] An encyclopedic source also is available.[7]

Mean Serum and Plasma Laboratory Findings During Normal Pregnancy, Expressed as a Percentage of the Nonpregnant Mean[1]

Analyte	Time of Gestation					
	12 wk	28 wk	32 wk	36 wk	Term	1 d pp[2]
Bicarbonate	85	85	85	85	81	88
Urea nitrogen	77	63	63	63	77	72
Creatinine	71	71	74	79	81	74
Fasting glucose	98	94	94	91	94	94
Bilirubin, unconjugated	56	56	67	67	78	78
Albumin	93	78	78	78	78	71
Protein	92	83	83	83	83	77
Uric acid	68	79	92	106	120	135
Calcium	98	94	94	95	97	94
Free ionized calcium	99	102	101	102	102	100
Parathyroid hormone, intact					140	
Vitamin 1,25-$(OH)_2D_3$					400	
Phosphate	108	99	97	103	96	106
Magnesium	92	90	87	87	87	86
Alkaline phosphatase	90	131	203	274	347	284
Creatine kinase	87	86	86	90	135	257
α_1-Antitrypsin	129	169	174	189	191	187
Transferrin	105	145	160	160	170	139
Cholesterol	100	132	144	148	156	138
HDL cholesterol	121	121	119	127	130	116
LDL cholesterol	80	118	118	150	146	121
Fasting triglycerides	141	244	300	356	349	328
Iron	112	82	94	94	94	82
Iron-binding capacity	95	129	139	142	144	128
Transferrin saturation	136	68	68	76	64	56
Ferritin	81	33	33	37	59	81
Thyroxine	103	102	107	99	100	92
Triiodothyronine	100	121	121	116	121	95
Free thyroxine	98	71	72	62	74	80
Thyroxine-binding globulin	114	177	155	155	182	150
Thyroid-stimulating hormone	111	106	122	111	139	111
Cortisol	111	284	301	292	309	238
Aldosterone					1500	
Prolactin					800	
Hemoglobin	95	89	90	93	96	89
Hematocrit	94	89	91	94	97	91
Leukocyte count	144	167	167	165	240	222
Fibrinogen	119	132	154	157	165	161

[1]The values are the means in pregnant subjects expressed as a percentage of the means in nonpregnant controls. (Most of the values are from Lockitch G, ed, *Handbook of Diagnostic Biochemistry and Hematology in Normal Pregnancy*, Boca Raton, FL: CRC Press, 1993.)

[2]pp = postpartum.

Adapted from Burtis CA and Ashwood ER, eds, *Tietz Textbook of Clinical Chemistry*, 3rd ed, Philadelphia, PA: WB Saunders Co, 1999, 1741.

2. **Exercise**. Individuals who exercise regularly are apt to experience a moderate or marked increase in high density lipoprotein cholesterol (HDLC) and slight elevations of urea nitrogen (BUN), and lactate dehydrogenase (LD). Immediately after strenuous exercise there are often elevations of lactate, creatine kinase (including the MB fraction), aldolase, alanine aminotransferase (ALT), aspartate aminotransferase (AST), phosphorus, acid phosphatase, creatinine, uric acid, haptoglobin, uric acid, transferrin, catecholamines, and leukocyte count; decreases may be seen in albumin, iron, and sodium. Marathon

MAXIMIZING THE INFORMATION FROM LABORATORY TESTS — THE ULYSSES SYNDROME *(Continued)*

runners may develop marked hyponatremia during a race and may present emergently with hyponatremic encephalopathy.[8,9]

3. **Neonatal Period and Childhood**. Reference intervals appropriate for adults often are not valid for babies and children. Pediatric reference values are available in standard textbooks.[10,11]

4. **Older Individuals**. This is a relatively new field of investigation. A compilation of geriatric reference values is available.[12]

5. **Body Weight**. Most persons who are "overweight" have increased body fat (obesity) while a much smaller subset has increased lean body mass. Obesity itself has subdivisions based on the distribution of the excess adipose tissue (central vs peripheral). The effects of obesity as a preanalytical variable are not well understood. For example, obese individuals have elevated C-reactive protein levels but we do not know whether this is a marker for systemic inflammation or a preanalytical variable.[13]

6. **Posture** affects a number of test results. When blood is drawn in the upright position, the following analytes are described as increased: total protein, albumin, calcium, hemoglobin and hematocrit, renin, catecholamines (urine), alkaline phosphatase, cholesterol, alanine aminotransferase (ALT), and iron. As a person moves from the recumbent to the upright position, plasma water enters the interstitial fluid resulting in higher plasma concentrations of substances which do not readily pass the capillary endothelium.

7. **Diet** is a major source of preanalytic variability. For many analytes, such variation is controlled by obtaining specimens in the fasting state after 2 weeks on a diet of stable composition. Some of the more common tests for which an overnight fast is recommended are plasma glucose, lipids, iron, iron-binding capacity, vitamin B_{12}, folate, insulin, and gastrin.

    Blood specimens obtained shortly after a meal are problematic for many reasons. Postprandial turbidity interferes with many analytes, especially bilirubin, total protein, uric acid, and urea nitrogen (BUN). Postprandial potassium, triglyceride, and alkaline phosphatase values are increased, while phosphorus may be decreased. Diets high in protein can result in increased urea nitrogen (BUN), ammonia, and uric acid. Diets high in purines increase uric acid levels. The serotonin metabolite, 5-hydroxyindoleacetic acid (5-HIAA), is increased by diets rich in bananas, pineapples, tomatoes, and avocados. Caffeine and theophylline elevate catecholamines.

8. **Ethanol**. The ingestion of ethyl alcohol is followed acutely by increases in the serum levels of uric acid, lactate, gamma glutamyl transferase, triglycerides, and aspartate aminotransferase (AST). Long-term alcohol abuse is associated with increases in bilirubin, alkaline phosphatase, and the ratio AST:alanine aminotransferase (ALT). See also footnote 14.

9. **Oral Contraceptives and Estrogens**. These agents increase thyroxine-binding globulin, alpha$_1$-antitrypsin, iron, triglycerides, alanine aminotransferase, and gamma glutamyl transferase. Albumin may be decreased. See footnote 14.

10. **Other Drugs**. The potential effects of other drugs on laboratory tests are enormous. Current encyclopedic references are very helpful in practice.[14,15]

11. **Hemolysis** is a cause for specimen rejection in most nonemergent situations. Hemolysis causes increases in lactate dehydrogenase, bilirubin, potassium, aspartate aminotransferase (AST), creatine kinase (CK), alanine aminotransferase (ALT), and magnesium. Hemolysis

invalidates the results of most coagulation tests and can mask hemolyzing antibodies in the antibody screen and crossmatch.

12. **Sampling Problems**. It is prudent to avoid, when possible, obtaining blood specimens from an extremity which is also an intravenous infusion site. Such samples are apt to have artefactually lowered values due to dilutional effects; this is a major problem in the interpretation of electrolytes, glucose, urea nitrogen (BUN), glucose, and coagulation tests.

Tourniquets cause increases in potassium and lactate, and a decrease in pH. **Capillary whole blood samples** should be reported as such and accompanied by reference values appropriate for this matrix.

Arterial samples should be reported as such and accompanied by reference values appropriate for this matrix.

Inappropriate blood and urine collection containers, if unrecognized, can lead to misleading results.

Specimens containing **clots** should be rejected. When serum is in prolonged contact with clots in venipuncture tubes, glucose decreases, while lactate dehydrogenase, potassium, and iron increase.

Other Specimen Mishandling. Exposure to sunlight causes increases in leukocyte count, platelet count, and erythrocyte sedimentation rate, while bilirubin is decreased. Any delay beyond ~5 minutes in analyzing blood gases causes significant errors.

13. **Circadian Rhythms:** Circadian (approximately 24-hour) rhythms have implications for physiology, measurement of many laboratory tests, drug excretion (eg, salicylates, sulfonamides), and responses to therapy. Levels fluctuating very significantly during the 24-hour cycle include cortisol (which has different normals for 8 AM and 8 PM), growth hormone, serum acid phosphatase, aldosterone (high 6 AM to 3 PM), transferrin (maximum 4 PM to 8 PM), ACTH, serum iron, serum creatinine (7 PM values 130% of 7 AM concentration), eosinophils (low in afternoon), lymphocytes (maximum in early AM), WBC (maximum in early AM), leukocyte function, and urine urobilinogen (maximum excretion in afternoon). Urinary excretion of potassium, LH, FSH, TSH, testosterone, and some less commonly ordered hormones have some diurnal variation. Parathyroid hormone is best drawn at 8 AM.

Triglyceride is higher in the afternoon, as is phosphate, BUN, and the hematocrit. Bilirubin falls in the PM, but overnight fasting itself causes bilirubin to increase.

The waves which characterize circadian rhythms may be square shaped or may occur as a series of pulses. The latter pattern is seen with plasma cortisol concentration, which begins to increase during sleep. A large difference exists between this level and that found in the evening. Still another pattern is a single daily pulse such as occurs with growth hormone secretion.

The magnitude of the effect of circadian rhythms is greater than is generally recognized. Although only 10% variation exists for plasma potassium concentration, urinary potassium excretion can vary fivefold during the day.

Some hormone secretion cycles are longer (infradian) – eg, the menstrual cycle.

An extraordinary compilation of the effects of preanalytical variables is available.[16]

MAXIMIZING THE INFORMATION FROM LABORATORY TESTS — THE ULYSSES SYNDROME (Continued)

Probabilities of Test Results Within Reference Range on Multiple Panels

A reference range for a particular analyte generally covers 95% of healthy or normal results. It means that 5% of results from healthy individuals will fall outside this limit. Thus, the probability of a healthy person having two test results within reference range is 0.95 x 0.95 = 0.9025 or 90.25%. This phenomenon will progress as multiple tests are ordered (see table below).

Probability That a Healthy Individual Will Have All Test Results Within Reference Range

Number of Tests Ordered	Probability (%)
1	95.0
2	90.2
5	77.4
10	59.9
20	35.8

Analytical Error and the Coefficient of Variation

Imprecision, also known as **random error**, refers to the distribution of results when an assay is performed repeatedly on the same specimen. Imprecision may be measured as within-run, within-day, day-to-day, or over any other time interval. In general, the longer the time interval, the larger the random error. When the distribution of repeated results is approximately Gaussian in distribution (ie, a bell-shaped curve), imprecision is expressed numerically as the **coefficient of variation (CV)**. **The CV is the standard deviation of the results divided by the mean, and expressed as a percentage**. A small CV implies good precision, while a large CV implies relatively poor precision. All laboratories can, and should, calculate the within-run and day-to-day imprecision of their assays and make the results available to interested clients.

Accuracy refers to how close a reported test result is to the true value. Proficiency testing results, both external and internal, provide a measure of accuracy. External proficiency testing programs often provide results that are obtained by, or linked to, definitive or reference methods. The difference between a particular clinical result and the result obtained by the definitive (or reference) method is proportional to the **bias** or **systematic error**. It is important to note that even reference methods can have bias when comparing results. As an example, reference methods may give a different concentration of a particular analyte when compared to gravimetrically prepared specimens. Ionicity, protein concentrations, and other environmental factors are known matrix effects in proficiency testing. Matrix effects are inherent in any analytical system. Laboratories should choose methods that minimize matrix effects.

While random error is relatively easy to calculate, and systematic error is difficult, what really counts is the combination of the two, or **total error**. Methods of calculating total error, and its components, can be found in the following section of this chapter as well as in standard references.[17] A laboratory director should furnish information on total error performance to interested physicians and other scientists.

Decision Analysis — Predictive Value Theory — What Does This Result Mean?

The ability of a laboratory test to identify a particular disease is quantified by two measurements: **sensitivity** and **specificity**.

Sensitivity is the frequency of a positive (abnormal) test result among all patients with a particular disease ("positivity in disease" is a commonly used descriptive that is easy to remember). In other words sensitivity of a test indicates its ability to generate a true positive (TP), not false negative (FN), result when a diseased individual is tested. It is calculated by testing a population of patients who have been found to have a particular disease by some gold standard method. Mathematically it is expressed as:

$$\text{Sensitivity} = \frac{TP}{TP + FN} \times 100\%$$

Specificity is the frequency of a negative (normal) test result among all persons who do not have disease ("negativity in health" is a commonly used descriptive that is easy to remember). In other words specificity of a test indicates its ability to generate a true negative (TN), not false positive (FP), result when a healthy individual is tested. It is calculated by testing a group of persons who do not have a particular disease in question. Mathematically it is expressed as:

$$\text{Specificity} = \frac{TN}{TN + FP} \times 100\%$$

Example: A method is evaluated for its sensitivity and specificity for disease XYZ. 100 healthy individuals and 100 individuals with disease XYZ are tested with this test. The following results are obtained:

Test Result	DISEASE XYZ Present	Absent	Total
Positive	95 (TP)	10 (FP)	105
Negative	5 (FN)	90 (TN)	95
Total	100 (TP + FN)	100 (FP + TN)	200

Sensitivity and specificity of the test are:

$$\text{Sensitivity} = \frac{95}{95 + 5} \times 100\% = 95\%$$

$$\text{Specificity} = \frac{90}{90 + 10} \times 100\% = 90\%$$

Other statistics, which are used to assess the diagnostic value of a test, are **predictive value** and **efficiency** of a test. Predictive value of a positive or negative result relates to the number of correct positive or negative results respectively. Efficiency relates to the total number of correct results. Mathematically these terms are defined as:

$$\text{Predictive value (positive result)} = \frac{TP}{TP + FP} \times 100\%$$

$$\text{Predictive value (negative result)} = \frac{TN}{TN + FN} \times 100\%$$

$$\text{Efficiency} = \frac{TP + TN}{TP + FP + TN + FN} \times 100\%$$

What will be the predictive value of the above test for the diagnosis of disease XYZ? To answer that question we must know one more statistic: the prevalence of disease XYZ in the population under study. Illustrated below is the performance of the test where the prevalence of XYZ is 50%. In actual practice, what we really need to know is how predictive a positive (or negative) test result is for the presence (or absence) of disease XYZ in a particular patient. To determine the predictive value we calculate as follows:

MAXIMIZING THE INFORMATION FROM LABORATORY TESTS — THE ULYSSES SYNDROME (Continued)

| | DISEASE XYZ (PREVALENCE 50%) | | |
Test Result	Present	Absent	Total
Positive	95	10	105
Negative	5	90	95
Total	100	100	200

PV positive = 95/105 = 90.4%
PV negative = 90/95 = 94.7%

The **predictive value (PV)** of a positive test result is the frequency of disease XYZ among all persons with a positive (abnormal) test result. As illustrated above, the predictive value of a positive test in this situation is ~90%; this means that 90% of persons with the disease will have a positive test, and 10% will have a negative test (false negative). The PV of a negative test result is the frequency of disease XYZ absence among all persons with a negative test result. Approximately 95% of persons free of disease XYZ will have a negative (normal) result, while ~5% of such persons will have a false positive result.

The following table illustrates the effect of disease prevalence on predictive values.

| | DISEASE XYZ (PREVALENCE 10%) | | |
Test Result	Present	Absent	Total
Positive	95	90	185
Negative	5	810	815
Total	100	900	1000

PV positive = 95/185 = 51.3%
PV negative = 810/815 = 99.3%

A point for emphasis is the marked decrease in the predictive value of a positive test when the prevalence changes from high to low. In the low prevalence situation only about 50% of the positive tests are true positives, while in the high prevalence situation ~90% of positive results are true positives. In many clinical scenarios, the disease prevalence may be ~1%, or even lower; in such situations false positives may outnumber true positives by factors of 100 or 1000. A corollary of this is that, to function as a good screening test – a test applied across an entire population of asymptomatic persons – test sensitivity must approach 100%.

Receiver Operating Characteristic (ROC) Curves

The statistics of sensitivity, specificity, and predictive values can be formatted in many ways. One of the more popular ways is the construction of receiver operating characteristic (ROC) curves. Whereas the tables shown allow one to determine the predictive values at one decision limit, ROC curve analysis portrays, in graphic fashion, the trade-off between true positives and false positives.

ROC curves were developed during World War II. Radar receiver operators detected blips that may have represented hostile aircraft. Many a receptor squadron scrambled on the basis of such blips to find a flock of wild geese instead of hostile aircraft. It was recognized that there was a trade-off between sensitivity and specificity. As sensitivity increased so did the false-positive interpretations.

The principle of ROC states that for any given test system, the false-positive rate (1-specificity) is a function of sensitivity. As sensitivity increases, the false-positive rate increases linearly – up to a point. After this point, the false-positive rate rises disproportionately high. Thus, there is a limit to the sensitivity of any test system. When the limit is exceeded, the false-positive rate may become unacceptable. A laboratory test becomes unacceptable when desired sensitivity cannot be attained because of a large number of false positives. The limitation of the ROC curve analysis is that it depends on a gold standard (independent criterion) for the presence of clinical disease. The analysis will not be very good if the gold standard is a poor one. ROC analysis should be used in conjunction with clinical decision analysis by clinicians who have an understanding of the benefits and risks of the therapeutic and diagnostic procedures that may follow from the results of a particular laboratory test.

Figure 1 presents a prototypic ROC curve. The dashed diagonal line represents a test with a sensitivity of 50% and a specificity of 50% – equivalent to a coin toss. A good test will have an ROC curve that lies to the left of the dashed diagonal; the farther to the left, the better the combination of sensitivity and specificity and the better the performance in practice. The curve denoted as a nonideal test performs only slightly better than a coin toss, and the curve denoted ideal is typical of many clinically useful tests. The curve denoted "Brand X" is typical of tests which, while perhaps adequate for limited purposes, do not perform well and, over time, are replaced by better ones.

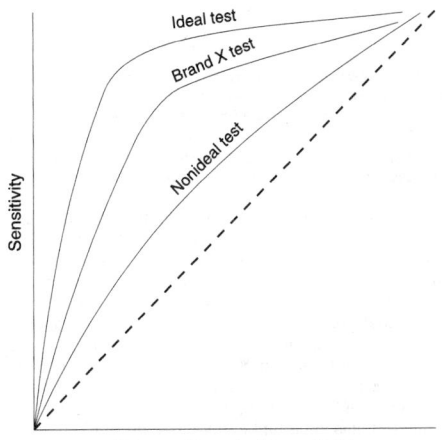

Figure 1. Typical receiver operating curve

A laboratory director must decide what numerical **clinical decision limit** (cutoff value) is best for a given test. The selection of such a decision limit involves an inevitable trade-off between sensitivity and specificity. Thus, selection of a very high (or stringent) limit will usually reduce sensitivity and increase specificity – more false negatives and fewer false positives. A less stringent decision limit has the opposite effects – increased sensitivity and decreased specificity.

Figure 2[18] illustrates two ROC curves, one for prostate-specific antigen (PSA) and one for prostatic acid phosphatase (PAP) in the detection of prostate carcinoma.

Note that the true-positive rate is on the vertical axis and the false-positive rate is on the horizontal axis. Moving upward on the vertical axis, and moving to the right on the horizontal axis, corresponds to lowering the cutoff value associated with a positive test. Diagnostic accuracy is approximately proportional to the area under the ROC curve. The fact that the PSA curve lies to the left of the PAP curve means that the area under the PSA curve is greater than the area under the PAP curve, and, therefore, that the PSA test performs better.

MAXIMIZING THE INFORMATION FROM LABORATORY TESTS — THE ULYSSES SYNDROME *(Continued)*

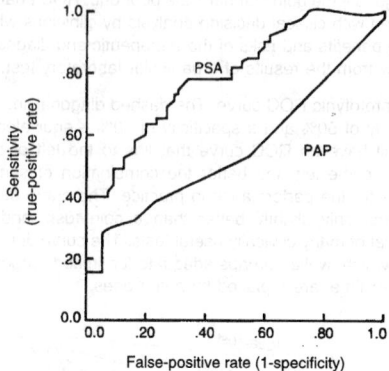

Figure 2. Receiver operating characteristic curves of prostatic acid phosphatase (PAP) and prostate-specific antigen (PSA) assays for patients with benign hyperplasia and carcinoma of prostate. Because the PSA assay curve is above the PAP assay curve at all points, the better assay for the patients tested is the PSA assay.

Adapted from *Tietz Textbook of Clinical Chemistry*, 3rd ed, Burtis CA and Ashwood ER, eds, Philadelphia, PA: WB Saunders Co, 1999, 314.

Another example of ROC analysis is in the evaluation of new tests. Bender et al[19] have compared two tests for fetal lung maturity: the conventional leci-thin:sphingomyelin ratio (L:S ratio) and a fluorescence polarization immuno-assay, the surfactant:albumin ratio (S:A ratio). They reported that, between 29 and 37 weeks gestation, a wide range of values is observed at every interval, and this indicates that gestational age accounts for only a small fraction of the total variation in test results for both tests. They found a wide divergence between the L:S ratio results and S:A ratio results, which suggests either a nonlinear relationship between the two tests, or that one (or both) test(s) is/are affected by nonsurfactant substances in amniotic fluid. Their ROC curves are presented in Figure 3. It is important to note that the ROC curve for the S:A ratio (heavy line) lies to the left of the ROC curve for the L:S test; stated another way, the area under the S:A curve (0.869) is slightly larger than the area under the L:S curve (0.847). The clinically important implication from the Figure 3 is that there is no statistically significant difference between the area under the two ROC curves. Stated another way, once the result from one of the tests is known, determining the other result appears to provide no additional information.

Concurrent Disease Processes

While most physicians of a certain age were indoctrinated in medical school with the goal of finding one grand unifying diagnosis to explain all of a patient's symptoms, signs, laboratory results and imaging findings, the reality is that many people have multiple disease processes at once. Quite obviously, concurrent diseases will also modify test results and alter the significance of certain refer-ence intervals. A useful listing of how these concurrent diseases affect test results is available.[20]

Ulysses Reaches Home at Last

Sources of variation, preanalytical and biologic, and sources of analytic error are so numerous that, at the current state-of-the-art, no test result should be inter-preted without consideration of the entire clinical picture. Even with computer-based decision support systems,[21] test interpretation remains a

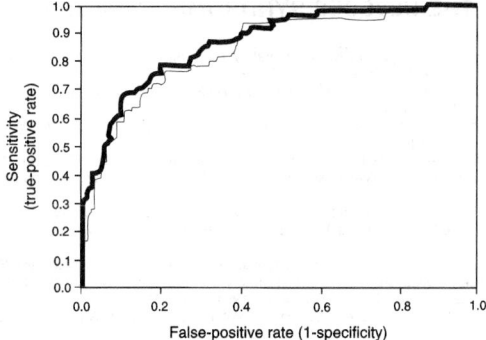

Figure 3. ROC curves comparing L/S (thin line) and S/A (heavy line). The area defined by the L/S curve is 0.847 (SE 0.0438); that for the S/A curve, 0.869 (SE 0.0406). The Z score for difference in areas was 0.729 (P >0.05).

Adapted from Bender TM, Stone LR, and Amenta JS, "Diagnostic Power of Lecithin/Sphingomyelin Ratio and Fluorescence Polarization Assays for Respiratory Distress Syndrome Compared by Relative Operating Characteristic Curves," *Clin Chemistry*, 1994, 40(4):543.

complex process, and patients are still at risk from unintended consequences of well-intentioned diagnostic evaluations. We began this chapter by referring to an analogy between the Greek hero Ulysses and the patient at risk. We conclude by turning to the lessons taught in Homer's masterpiece. Recall that Ulysses finally reached his home in Ithaca only after overcoming a host of obstacles and injuries, many of which could have been predicted and, therefore, avoided. During his long arduous journey of self-discovery, Ulysses received help from a number of people, but most significantly from the beautiful Phaeacian princess, Nausicaä. The astute clinician models an error-avoidance strategy after the wily Ulysses, and adopts a skeptical attitude toward test results, especially results which do not fit clinically. A successful outcome will be even more likely if the clinician's pathologist follows the model of Nausicaä, who showed Ulysses how to resist the temptation of the moment and stay the homeward course.

Footnotes

1. Rang M, "The Ulysses Syndrome," *Can Med Assoc J*, 1972, 106(2):122-3.
2. Young DS and Bermes EW, "Specimen Collecting and Processing: Sources of Biological Variation," *Tietz Textbook of Clinical Chemistry*, 3rd ed, Burtis CA and Ashwood ER, eds, Philadelphia, PA: WB Saunders Co, 1999, 42-72.
3. Fraser CG, "Biological Variation in Clinical Chemistry," *Am J Clin Pathol*, 1992, 116(9):916-23.
4. "Report of the Expert Committee on the Diagnosis and Classification of Diabetes Mellitus," *Diabetes Care*, 1997, 20(7):1183-97.
5. Troisi RJ, Cowie CC, and Harris MI, "Diurnal Variation in Fasting Plasma Glucose: Implications for Diagnosis of Diabetes Patients Examined in the Afternoon," *JAMA*, 2000, 284(24):3157-9.
6. Ashwood ER, "Clinical Chemistry of Pregnancy," *Tietz Textbook of Clinical Chemistry*, 3rd ed, Burtis CA and Ashwood ER, eds, Philadelphia, PA: WB Saunders Co, 1999, 1736-75.
7. Ramsay MM, James DK, Weiner CP, et al, *Normal Values in Pregnancy*, London, England: WB Saunders Co, 1996.
8. Ayus JC, Varon J, and Arieff AI, "Hyponatremia, Cerebral Edema, and Noncardiogenic Pulmonary Edema in Marathon Runners," *Ann Intern Med*, 2000, 132(9):711-4.
9. Foran SE, Lewandrowski KB, and Kratz A, "Effects of Exercise on Laboratory Test Results," *Lab Med*, 2003, 34(10):736-42.
10. Soldin SJ, Brugnara C, and Hicks JM, eds, *Pediatric Reference Ranges*, 3rd ed, Washington, DC: AACC Press, 1999.
11. Nicholson JF and Pesce MA, "Reference Ranges for Laboratory Tests and Procedures," *Nelson Textbook of Pediatrics*, 16th ed, Behrman RE, Kliegman RM, and Jenson HB, eds, Philadelphia, PA: WB Saunders Co, 2000, 2181-229.
12. Faulkner WR, ed, *Geriatric Clinical Chemistry Reference Values*, Washington, DC: AACC Press, 1994.
13. Visser M, Bouter LM, McQuillan GM, et al, "Elevated C-Reactive Protein Levels in Overweight and Obese Adults," *JAMA*, 1999, 282(22):2131-5.

MAXIMIZING THE INFORMATION FROM LABORATORY TESTS — THE ULYSSES SYNDROME *(Continued)*

14. Young DS, *Effects of Drugs on Clinical Laboratory Tests*, 5th ed, Volume One: Listing by Test, Washington, DC: AACC Press, 2000.
15. Young DS, *Effects of Drugs on Clinical Laboratory Tests*, 5th ed, Volume Two: Listing by Drug, Washington, DC: AACC Press, 2000.
16. Young DS, *Effects of Preanalytical Variables on Clinical Laboratory Tests*, 2nd ed, Washington, DC: AACC Press, 1997.
17. Koch DD and Peters T, "Selection and Evaluation of Methods," *Tietz Textbook of Clinical Chemistry*, 3rd ed, Burtis CA and Ashwood ER, eds, Philadelphia, PA: WB Saunders Co, 1999, 320-35.
18. Shultz EK, "Selection and Interpretation of Laboratory Procedures," *Tietz Textbook of Clinical Chemistry*, 3rd ed, Burtis CA and Ashwood ER, eds, Philadelphia, PA: WB Saunders Co, 1999, 314.
19. Bender TM, Stone LR, and Amenta JS, "Diagnostic Power of Lecithin/Sphingomyelin Ratio and Fluorescence Polarization Assays for Respiratory Distress Syndrome Compared by Relative Operating Characteristic Curves," *Clin Chem*, 1994, 40(4):541-5.
20. Friedman RB, Effects of Disease in Clinical Laboratory Tests, 3rd ed, Washington, DC: AACC Press, 1997.
21. van Wijk MA, van der Lei J, Mooseveld M, et al, "Assessment of Decision Support for Blood Test Ordering in Primary Care," *Ann Intern Med*, 2001, 134(4):274-81.

SPECIMEN COLLECTION

Eugene S. Olsowka, MD, PhD

Uttam Garg, PhD

Proper specimen collection is pivotal for provision of meaningful clinical laboratory information. **The laboratory must have an optimum, properly labeled specimen**. This section includes general information pertaining to the collection of laboratory specimens. Specimen collection information specific for each individual test is provided with the detailed discussion of the test within the individual monographs. A discussion of the special requirements for collection of specimens for detection of **drugs of abuse and therapeutic drug levels** is presented in the introduction to Therapeutic Drug Monitoring *on page 57* and Toxicology/ Drugs of Abuse *on page 63*, and a typical **chain-of-custody** form is illustrated in the Toxicology/Drugs of Abuse Introduction *on page 76*. The unique considerations required for the collection of specimens for **trace element** testing are discussed in the Introduction to Trace Elements *on page 77*. **Transfusion Service** needs are addressed in that introduction *on page 78*. Special requirements exist for **coagulation** testing, as described in that introduction *on page 29*.

Overview and Regulatory Considerations

Every healthcare employee, from nurse to housekeeper, has some (albeit small) risk of exposure to HIV and other viral agents such as hepatitis B, hepatitis C, and Jakob-Creutzfeldt agent: See monograph, Blood and Fluid Precautions, Specimen Collection *on page 271*. Exposure management should include preexposure education and immediate postexposure care and counseling. Chemoprophylaxis, despite its questionable effectiveness, is widely used. Each year, the CDC estimates that approximately 384,000 healthcare workers in the United States are exposed to infectious body fluids. Exposure to tuberculosis by healthcare workers is also significant. An understanding of the appropriate procedures, responsibilities, and risks inherent in the collection and handling of patient specimens is necessary and is required by Occupational Safety and Health Administration (OSHA) regulations.

The Occupational Safety and Health Administration published its "Final Rule on Occupational Exposure to Bloodborne Pathogens" in the Federal Register on December 6, 1991. The rule was revised on January 18, 2001 to include the Needlestick Safety and Prevention Act. The rule requires preparation of a plan to reduce needlesticks and other percutaneous injuries resulting in exposure to infectious material. OSHA has chosen to follow the Centers for Disease Control (CDC) definition of universal precautions. The Final Rule provides full legal force to universal precautions and requires employers and employees to treat blood and certain body fluids as if they were infectious. The Final Rule mandates that healthcare workers must avoid parenteral contact and must avoid splattering blood or other potentially infectious material on their skin, hair, eyes, mouth, mucous membranes, or on their personal clothing. Hazard abatement strategies must be used to protect workers. Such plans typically include, but are not limited to, the following:

- safe handling of sharp items ("sharps") and disposal of such into puncture resistant containers
- gloves required for employees handling items soiled with blood or equipment contaminated by blood or other body fluids
- provisions of protective clothing when more extensive contact with blood or body fluids may be anticipated (eg, surgery, autopsy, or deliveries)
- resuscitation equipment to reduce necessity for mouth to mouth resuscitation
- restriction of HIV-1 or hepatitis B-exposed employees to noninvasive procedures

SPECIMEN COLLECTION *(Continued)*

Housekeeping protocols: OSHA requires that all bins, cans, and similar recep-tacles, intended for reuse which have a reasonable likelihood for becoming contaminated, be inspected and decontaminated immediately or as soon as feasible upon visible contamination and on a regularly scheduled basis. Broken glass that may be contaminated must not be picked up directly with the hands. Mechanical means (eg, brush, dust pan, tongs, or forceps) must be used. Broken glass must be placed in a proper sharps container.

Pre-exposure and postexposure protocols: OSHA's Final Rule includes the provision that employees, who are exposed to contamination, be offered hepa-titis B vaccine at no cost to the employee. Employees may decline; however, a declination form must be signed. The employee must be offered free vaccine if he/she changes his/her mind. Vaccination to prevent the transmission of hepa-titis B virus (HBV) in the healthcare setting is widely regarded as sound practice. In the event of exposure, a confidential medical evaluation and follow-up must be offered at no cost to the employee. Follow-up must include collection and testing of blood from the source individual for HBV and HIV if permitted by state law if a blood sample is available.

The employee follow-up must also include appropriate postexposure prophy-laxis, counseling, and evaluation of reported illnesses. The employee has the right to decline baseline blood collection and/or testing. If the employee gives consent for the collection but not the testing, the sample must be preserved for 90 days in the event that the employee changes his/her mind within that time. Confidentiality related to blood testing must be ensured. **The employer does not have the right to know the results** of the testing of either the source individual or the exposed employee.

Communication of Hazards

Communication regarding the dangers of bloodborne infections through the use of labels, signs, information, and education is required. Storage locations (eg, refrigerators and freezers, waste containers) that are used to store, dispose of, transport, or ship blood or other potentially infectious materials require labels. The label background must be red or bright orange with the biohazard design and the word biohazard in a contrasting color. The label must be part of the container or affixed to the container by permanent means.

Education provided by a qualified and knowledgeable instructor is mandated. The sessions for employees must include:

- accessible copies of the regulation
- general epidemiology of bloodborne diseases
- modes of bloodborne pathogen transmission
- an explanation of the exposure control plan and a means to obtain copies of the written plan
- an explanation of the tasks and activities that may involve exposure
- the use of exposure prevention methods and their limitations (eg, engi-neering controls, work practices, personal protective equipment)
- information on the types, proper use, location, removal, handling, decontamination, and disposal of personal protective equipment)
- an explanation of the basis for selection of personal protective equip-ment
- information on the HBV vaccine, including information on its efficacy, safety, and method of administration and the benefits of being vacci-nated (ie, the employee must understand that the vaccine and vaccina-tion will be offered free of charge)
- information on the appropriate actions to take and persons to contact in an emergency involving exposure to blood or other potentially infec-tious materials
- an explanation of the procedure to follow if an exposure incident occurs, including the method of reporting the incident

- information on the postexposure evaluation and follow-up that the employer is required to provide for the employee following an exposure incident
- an explanation of the signs, labels, and color coding
- an interactive question-and-answer period

Record Keeping

The OSHA Final Rule requires that the employer maintain both education and medical records. Medical records must be kept confidential and be maintained for the duration of employment plus 30 years. They must contain a copy of the employee's HBV vaccination status and postexposure incident information. The training record must contain dates of training, the content or summary of training sessions, and the names, qualifications, and job titles of trainers. Education records must be maintained for 3 years from the date the program was given.

OSHA has the authority to conduct inspections without notice. Penalties for cited violation may be assessed as follows.

Patient Satisfaction Issues

The Joint Commission on Accreditation of Healthcare Organizations (JCAHO) has recognized patient satisfaction issues as an important indicator of quality and requires its members to collect data regarding patient satisfaction. The phlebotomy service is typically the major segment of the laboratory that has direct patient contact., although in the current climate of healthcare, phlebotomy services are sometimes taken over by nursing staff. The patient's level of satisfaction with the laboratory often revolves around the phlebotomy experience. Recent articles have begun to address patient satisfaction issues. Factors that are associated with patient dissatisfaction include discourteous treatment, large bruises (>25 mm), more discomfort than expected, a wait of longer than 30 minutes, and a bruise of any size. The factor most strongly associated with patient dissatisfaction is discourteous treatment. It is interesting that the factor that can be controlled best has the most profound impact on the patient's level of satisfaction.

References

Bailey EM, "Exposure to Bloodborne Pathogens," *Am J Nurs*, 1998, 98(3):67-8.

Dale JC and Ruby SG, "Specimen Collection Volumes for Laboratory Tests: A College of American Pathologists Study of 140 Laboratories," *Arch Pathol Lab Med*, 2003, 127(2):162-8.

Fraser VJ and Powderly WG, "Risks of HIV Infection in the Health Care Setting," *Annu Rev Med*, 1995, 46:203-11.

Holland NT, Smith MT, Eskenazi B, et al, "Biological Sample Collection and Processing for Molecular Epidemiological Studies," *Mutat Res*, 2003, 543(3):217-34.

Klosinski DD, "Collecting Specimens From the Elderly Patient," *Lab Med*, 1997, 28(8):518-20.

McGovern PM, Kochevar LK, Vesley D, et al, "Laboratory Professionals' Compliance With Universal Precautions," *Lab Med*, 1997, 28(11):725-30.

"Occupation Exposure to Bloodborne Pathogens: Needlestick and Other Sharps Injuries; Final Rule," *Federal Register*, 2001, 66:5317-25.

Osburn EHS, Papadakis MA, and Gerberding JL, "Occupational Exposures to Body Fluids Among Medical Students: A Seven Year Longitudinal Study," *Ann Intern Med*, 1999, 130(1):45-51.

Web Sites

http:/www.osha.gov

ANATOMIC PATHOLOGY

Phillip A. Munoz, MD
David S. Jacobs, MD

Contributors:
Ossoma W. Tawfik, MD, PhD
Marilyn K. Davis-Cansler, BS SCT (ASCP) CMIAC
Uri S. Alon, MD
Dwight K. Oxley, MD
Wayne R. DeMott, MD
Diane L. Persons, MD

The specialty of anatomic pathology is deeply rooted in the history of modern medicine. Its foundation is based on astute observations of anatomical changes induced by disease at the gross and microscopic levels. For the pathologist, the journey to mastering this discipline begins at the autopsy table, at which the sum of a lifetime of disease is dissected, classified, and placed in a perspective appropriate for the case at hand. The autopsy is the ultimate quality control procedure in the care of patients. It remains the validator or the invalidator of the results of many of the more recent diagnostic modalities.

For most pathologists, the journey continues in the surgical pathology laboratory, where tissue diagnoses based on gross, microscopic, and sometimes ultrastructural studies form critical junctures for the clinical decision-making process. With increasing emphasis on minimally invasive procedures, the journey is carried into the cytology laboratory, where minute fragments of tissue and individual cells are morphologically interrogated to yield clues addressing the underlying disease process. Such is a thumbnail history of diagnostic anatomic pathology.

Today, a morphologic diagnosis is often just the first step to complete a pathologic evaluation. With increasing frequency, we are called on to augment morphologic diagnoses with critical descriptors of phenotype, genotype, and fundamental pathology at the molecular level. Faced with the challenge of completing this task in an appropriate, cost-effective, and timely fashion, we are compelled to continually update our fund of knowledge and convey our collective experience in a venue that meets the needs of a broad range of clinical specialties.

We are indebted to the many contributors who have shaped this chapter through its previous editions. We especially recognize the contribution of the late Dr John G. Gruhn, a true scholar, physician, and gentleman.

References

Mills SE, Carter D, Greenson JK, et al, *Sternberg's Diagnostic Surgical Pathology*, 4th ed, Volume II, Philadelphia, PA: Lippincott Williams & Wilkins, 2004.

CHEMISTRY

David S. Jacobs, MD

Uttam Garg, PhD

Dwight K. Oxley, MD

Charlotte Shideler, PhD

Wayne R. DeMott, MD

Harold J. Grady, PhD

Eugene S. Olsowka, MD, PhD

Leland Baskin, MD

Uri Alon, MD

Sridevi Devaraj, MD

Major advances in instrumentation have resulted in the migration of tests formerly performed in hematology and immunology sections to a central chemistry laboratory. Most analytical techniques are performed by automated analyzers, which have large test menus, rapid turnaround time, and high through-put. Such contemporary systems are coupled to laboratory information systems (LIS) which track and record the patient sample from order entry to the final user. All control values are stored, and detailed control histories are available in various interactive formats.

Point-of-care testing (POCT) is increasingly used in hospitals, clinics, and homes. Such instruments, analyze whole blood, saliva, or cutaneous interstitial fluid. POCT is summarized in a monograph.

Some terminology has been changed to meet current standards of acceptance. The ambiguous term "normal interval" has been replaced with "reference interval," a term which can be used for any population, provided a description of that population is furnished.

A review of many of the factors involved in the decision as to whether or not a result is within the "normal" or "reference" range for a given patient is found in the chapter entitled Maximizing the Information From Laboratory Tests – the Ulysses Syndrome on page 11.

It is hazardous to interpret a set of test results without clinical information, and even more hazardous to interpret a single test result without knowing the clinical context. The effects of drugs are complex.[1] An encyclopedic reference is highly recommended.[2] Therefore, we stress relationships between clinical context and test results. The revisions and new entries in this handbook reflect the rapid increase in knowledge and expertise of laboratory medicine and the dedication of laboratory scientists to implementation of contemporary advances.

The editors and authors wish to acknowledge and express appreciation to those who have written the Chemistry chapter in prior editions. We especially note the contributions of the late Dr Paul R. Finley, leading author of Chemistry in the 3rd edition of the *Laboratory Test Handbook*, 1994. Dr Finley was an outstanding physician and clinical chemist.

Footnotes

1. Kailajärvi M, Takala T, Grönroos P, et al, "Reminders of Drug Effects on Laboratory Test Results," *Clin Chem*, 2000, 46(9):1395-1400.
2. Young DS, *Effects of Drugs on Clinical Laboratory Tests*, 5th ed, Volume 1: Listing by Test, Washington, DC: AACC Press, American Association of Clinical Chemistry, 2000, Section 3, 258-61.

References

Burtis C and Ashwood E, *Tietz Textbook of Clinical Chemistry*, 3rd ed, Philadelphia, PA: WB Saunders Co, 1999.
Gilbert-Barness E and Barness LA, *Metabolic Diseases*, Natick, MA: Eaton Publishing Co, 2000.
Larsen PR, Kronenberg HM, Melmed S, et al, *Williams Textbook of Endocrinology*, 10th ed, Philadelphia, PA: WB Saunders Co, 2002.

CLINICAL MICROSCOPY

Wayne R. DeMott, MD
David S. Jacobs, MD

Use of the microscope to study all manner of materials (in particular body fluids and tissues) of human/animal origin is one of the oldest disciplines of medical/patient investigation. The microscope has played, and continues to play, a central role in the evolution of medical science.

COAGULATION

Elizabeth M. Van Cott, MD

Michael Laposata, MD, PhD

Hemostasis requires a balance between procoagulant and anticoagulant pathways. Normal hemostasis involves the formation of blood clots to stop bleeding from injured vessels, with natural anticoagulant and fibrinolytic systems that limit clot formation to sites of injury to prevent excessive clot formation. Fibrinolysis (degradation of clots) also helps restore normal blood flow by removing the clot once it is no longer needed. An imbalance between the procoagulant and anticoagulant systems can result in disorders of hemostasis, which may be hereditary or acquired. Defects in clot formation, due to insufficient quantity or function of platelets or coagulation factors, can lead to bleeding disorders. Excessive fibrinolysis can also lead to bleeding disorders (eg, in rare hereditary disorders). Defects in the natural anticoagulant systems or, less commonly defective fibrinolysis, lead to hypercoagulability. New evidence suggests that elevated levels of coagulation factors may also be associated with hypercoagulability. A growing menu of laboratory tests can be used to provide a specific diagnosis in affected patients. Several new therapeutic anticoagulants are also commonly in use, in particular danaparoid, hirudin, and argatroban. The Coagulation Laboratory can offer new tests to monitor these anticoagulants.

CYTOGENETICS

Diane L. Persons, MD

Contributor: Wayne R. DeMott, MD

Chromosomes, as they appear in a metaphase spread, consist of tightly coiled DNA and protein. A karyotype is the somatic chromosomal complement of an individual or species. For the human, a normal karyotype consists of 46 chromosomes including 22 pairs of autosomes, identical in males and females, and one pair of sex chromosomes, XX in the female and XY in the male. The primary constriction of a chromosome, the centromere, divides the chromosome into an upper (short) arm and lower (long) arm designated "p" (petite) and "q", respectively. The chromosomes are aligned in a standard sequence on the basis of size, centromere location, and banding pattern. The karyotype is one of the basic tools of the cytogeneticist.

Chromosomal abnormalities are defined according to a uniform system, The International System for Cytogenetic Nomenclature. An example of the abbreviations provided in this system include the use of a plus sign to signify gain of a chromosome: +21 is interpreted as gain of one copy of chromosome 21. A translocation is designated by "t" and is defined as the exchange of chromosomal material between two or more nonhomologous chromosomes. A list of some of the most commonly utilized terms is provided in the following table.

Partial Listing of Symbols and Abbreviated Terms

del	Deletion
der	Derivative chromosome
dmin	Double minute
hsr	Homogeneously staining region
i	Isochromosome
ins	Insertion
inv	Inversion
mar	Marker chromosome
mat	Maternal origin
minus (-)	Loss of
p	Short arm
pat	Paternal origin
plus (+)	Gain of
q	Long arm
r	Ring chromosome
t	Translocation

From Mitelman F, *An International System of Human Cytogenetic Nomenclature*, 1995.

The greatest advance in the field of cytogenetics over the last decade has been the development of molecular cytogenetic techniques. Traditionally, cytogenetic data have been obtained through the direct microscopic analysis of chromosomes displaying characteristic bands with Giemsa, quinacrine, or other staining methods from cells arrested in metaphase. Progress in the study of the organization and function of nucleic acid sequences has made it possible to gain information about the chromosomes in an interphase or terminally differentiated cell by means of specially developed chromosome-specific probes and fluorescence *in situ* hybridization (FISH). These probes are labeled with fluorescent dyes, either directly or indirectly. Different combinations of haptenated probes (eg, labeled with biotin, digoxigenin, or dinitrophenol) and different fluorophores [green (fluorescein), red (rhodamine or Texas Red), and blue (AMCA or Cascade Blue)] allows visualization of three or more separate chromosomal DNA sequences concurrently. Specific DNA target sequences in individual cells in tissue sections, single-cell, or chromosome preparations can be detected with FISH. With this approach, DNA sequences can be mapped on specific chromosomes; repositioning of sequences between chromosomes or within a particular chromosome, as a result of a chromosomal rearrangement, can be determined; small rearrangements, not detectable with standard karyotypic analysis, can be uncovered; breakpoints using probes for defined DNA sequences can be detected and characterized; and numerical chromosomal abnormalities can be ascertained in interphase and/or metaphase cells. With respect to the latter, cell culture can be omitted, resulting in an expedited diagnosis. The newly developed techniques of spectral karyotyping and multicolor FISH identify all 24 human chromosomes simultaneously, greatly facilitating recognition of chromosome aberrations.

Another central area of diagnostic testing that has profited from advances in cytogenetic technology is cancer genetics. Cancer is an acquired genetic disorder resulting from loss of normal regulation of cell growth. This loss is manifested in a number of different ways, including alteration of the normal chromosomal complement. Cytogenetic analysis of both benign and malignant neoplasms has resulted in the definition of characteristic chromosomal anomalies which serve as important, if not essential, diagnostic aids. Identification of the aberrant chromosomal bands has provided a basis for molecular approaches in establishing the definitive genes affected and the associated consequences of these gene alterations. Moreover, many of these abnormalities are important prognostically.

Dr Julia Bridge is acknowledged with appreciation for her significant contributions to this chapter in the previous edition.

CYTOPATHOLOGY

Bradly D. Clark, MD

Celeste N. Powers, MD, PhD

Leigh Ann Cahill, CT(ASCP)

Mary Ann Pedigo, CT(ASCP)

David S. Jacobs, MD

Cytopathology is the study of alterations within individual cells reflective of changes within their environment. Examination of such alterations at the cellular, as well as molecular level, allows the diagnosis of a wide range of benign, preneoplastic and malignant conditions. Like all morphologic studies, clinical and radiologic findings are of inestimable value in the accurate diagnosis of cytologic specimens. In addition, proper specimen procurement is absolutely essential for reliable interpretation.

Cytopathology can be divided into two major areas: exfoliative and aspiration cytopathology. The most well known and widely used cytopathology test is the Pap smear. Developed by Dr George Papanicolaou in the 1940s, the implementation of the Pap smear as a screening technique for cervical cancer has resulted in the dramatic decline of cervical cancer over the past 50 years. Current advances include The Bethesda System, a classification scheme developed by gynecologists, pathologists, and oncologists; development of better sampling and preparatory devices; and computer-assisted screening of Pap smears.

Screening techniques have also been developed for other body sites such as lung (sputum), bladder (urine), effusions, and cerebrospinal fluids and have established the utility of the Cytopathology Laboratory in the interpretation of nongynecologic specimens. Advances in endoscopic procedures have allowed more extensive evaluations of the respiratory, gastrointestinal, and urinary tracts by cytologic methods. The advent of fine needle aspiration (FNA) techniques utilized for both superficial or palpable masses and deep lesions requiring radiologic guidance has brought the Cytopathology Laboratory into the forefront of diagnostic pathology.

In addition to light microscopy, numerous ancillary techniques can be used to evaluate cellular material. While electron microscopy, immunocytochemistry, and flow/laser cytometry are proven methods, newer molecular diagnostic techniques are being evaluated for diagnostic and prognostic reliability.

Absolutely Essential Information

To facilitate the correct processing of a specimen for interpretation by cytopathologists, certain necessary documentation must accompany the sample. Such documentation includes patient demographics, relevant history, clinical impression and prior abnormal pathology results, as well as referring physician information for report distribution. In addition, the submitting physician must designate an ICD-9-CM (International Classification of Disease, 9th revision, Clinical Modification) code or diagnostic narrative for each specimen. ICD-9 codes are extremely important for all types of samples and absolutely necessary to differentiate between screening and diagnostic Pap smears, as currently required by Medicare and other payers. Since Medicare may not pay for a particular test, such as annual screening Pap smear, patients need to sign an ABN (Advanced Beneficiary Notice) which documents their agreement to pay for a test if it is denied.

ICD-9 codes designated by the ordering physician are required to identify the reason (signs/symptoms/history) the test is performed. The CPT (Current Procedural Terminology) coding is used by the laboratory for reporting medical services performed, with each specimen being assigned one or more of these codes. CPT codes indicate the method of slide evaluations (manual, computer assisted, automated screening, etc) as well as the type of preparation of the sample (monolayer, smear, cytospin, etc) The list of CPT codes is maintained by the AMA for annual review/revision.

Cytology laboratories are inspected by a variety of enforcing agencies and must comply with the government mandated CLIA '88 (Clinical Laboratory Improvement Amendments of 1988) Final Rules. CLIA '88 sets standards for laboratory personnel, quality assurance, and controls workload limits. Proficiency testing of all individuals screening or interpreting gynecologic cytology is proposed. In an attempt to detect errors and improve performance, CLIA '88 requires that each cytology laboratory must: 1) review at least 10% of gynecologic cases interpreted as negative; 2) compare all malignant and premalignant gynecologic reports with corresponding histopathology reports to determine the cause of any discrepancies; 3) and for each patient with a current diagnosis of a high grade lesion or above, the laboratory must perform a retrospective review of all negative Pap smears within the previous 5 years. If any significant discrepancies are found that would affect current patient care, the laboratory must issue an amended report to the patient's physician.

In an attempt to standardize reporting terminology and better communicate with clinicians, the Bethesda System (TBS) for reporting cervical/vaginal cytology was developed at the National Cancer Institute in 1988 and further modified in both 1991 and 2001. The format of TBS includes a statement of specimen adequacy, a classification of the Pap smear based on the worst type of cellular change present, and a descriptive interpretation/result.

The authors express appreciation to prior authors of this chapter, including Melanie J. Castelli, MD, Russell M. Fiorella, MD, Patricia G. Tweeddale, EdD, CT(ASCP), and others.

References

Atkinson BF, *Atlas of Diagnostic Cytopathology*, Second Edition, Philadelphia, PA: WB Saunders Co, 2003.

Cibas ES and Ducatman BS, *Cytology: Diagnostic Principles and Clinical Correlates*, Philadelphia, PA: WB Saunders Co, 2003.

Gray W, *Diagnostic Cytopathology*, 2nd ed, New York, NY: Churchill-Livingstone, 2003.

Geisinger KR, Stanley MW, Raab SS, et al, *Modern Cytopathology*, New York, NY: Churchill-Livingston, 2003.

Solomon D, et al, "The 2001, Bethesda System: Terminology for Reporting Results of Cervical Cytology," *JAMA*, 2002, 287():2114-9.

Web Sites

American Society for Colposcopy and Cervical Pathology (ASCCP) Bethesda 2001 Consensus Guideline web site: http://www.asccp.org/consensus.shtml.

NCI Bethesda System 201 web site: http://bethesda.2001.cancer.gov.

HEMATOLOGY

Wayne R. DeMott, MD

Barry S. Skikne, MD

Hematology involves the study of blood, bone marrow, and components of the reticuloendothelial system as found in a number of discreet and diffuse organs and systems including liver, spleen, gastrointestinal tract, and lymph nodes. Physiologic and biochemical processes that affect the quantity, quality, and function of the cellular components of blood (erythrocytes, leukocytes, and platelets) are an integral part of this discipline of medicine.

The genetic and molecular biologic basis of hematologic disorders is being established upon previous largely morphologic foundations. Understanding of the hemoglobinopathies (including the thalassemias), red cell enzyme deficiencies, and red cell membrane/structural protein abnormalities, is continually enhanced. Paroxysmal nocturnal hemoglobinuria (PNH), relates to the complement system through glycosyl-phosphatidylinositol (GPI) anchored proteins and to the gene PIG-A (for phosphatidylinositol glycan-class A). PNH provides substantial insight into molecular mechanisms of disease. A new generation of analytic techniques is being applied to the analysis of molecular biologic and cell-based mechanisms of hematologic disease. These new methods complement and, in many cases, supplant the earlier generation of morphologic-based observations and concepts. Dissection of lymphocyte function and pathology at the molecular level is of particular importance to a variety of human afflictions. Older analytic techniques such as electrophoresis of nucleic acid fragments with Southern (DNA) or Northern (RNA) hybridization, and restriction fragment length polymorphism (RFLP) analyses are now often pre-empted by more efficient and ever more commonly employed, fluorescence *in situ* hybridization (FISH), polymerase chain reaction, flow cytometry, and gene array based testing. With continuing budgetary restraints, the ratio of cost to patient benefit remains an area of serious concern.

Table of Unit Equivalency

Procedure	Conventional Unit	SI Equivalent
Red blood cell count	$10^6/\mu L$ ($10^6/mm^3$)	$10^{12}/L$
White blood cell count	$10^3/\mu L$ ($10^3/mm^3$)	$10^9/L$
Platelet count	$10^3/\mu L$ ($10^3/mm^3$)	$10^9/L$
Reticulocyte count	% (or .../mm³)	% (or ... x $10^9/L$)
Hemoglobin	g/dL	g/L
Mean cell volume	fL	fL
Mean cell hemoglobin	pg	pg
Mean cell hemoglobin concentration	g/dL	g/L
Mean cell diameter	μm	μm
Plasma hemoglobin	mg/dL	mg/L
Vitamin B_{12}, serum	pg/mL	pmol/L
Folate, serum	ng/mL	nmol/L

References

Bain BJ, *Hemoglobinopathy Diagnosis*, Malden, MA: Blackwell Science, Ltd, 2001.

Berliner N, "Use of Molecular Techniques in the Analysis of Hematologic Diseases," *Hematology: Basic Principles and Practice*, 3rd ed, Chapter 160, Hoffman R, Benz EJ, Shattil SJ, et al, eds, New York, NY: Churchill Livingstone, 2001, 2511-9.

Bessler M and Atkinson JP, "Paroxysmal Nocturnal Hemoglobinuria," *The Molecular Basis of Blood Diseases*, 3rd ed, Part III, Chapter 17, Stamatoyannopoulos G, Majerus PW, Perlmutter RM, et al, eds, Philadelphia PA: WB Saunders Company, 2001, 564-77.

Le Beau MM and Larson RA, "Cytogenetics and Neoplasia," *Hematology: Basic Principles and Practice*, 3rd ed, Chapter 48, Hoffman R, Benz EJ, Shattil SJ, et al, eds, New York, NY: Churchill Livingstone, 2000, 848-70.

Paraskevas F, "Clinical Flow Cytometry," *Wintrobe's Clinical Hematology*, 10th ed, Chapter 4, Lee RG, Foerster J, Lukens, J, et al, eds, Baltimore, MD: Williams & Wilkins, 1999, 56-71.

Provan D and Gribben J, eds, foreword by Perutz MF, *Molecular Haematology*, Malden, MA: Blackwell Science, Ltd, 2001.

HEMATOLOGY — ANEMIA FLOWCHART

The flowchart that follows is intended as an aid to the diagnosis of anemia. The flowchart is not meant to replace a standard history and physical. It is intended as a teaching tool to develop patterns for test ordering in the evaluation of anemia. Cost-effectiveness is a prime concern. Initial evaluation is based on results of the CBC usually produced by automated blood counters. This output includes red blood cell count, MCV, and RDW (see Complete Blood Count *on page 442* and the tables in that listing). The method of Bessman has been adapted for this beginning classification. In real life, some conditions may not fit the flowchart precisely. The reticulocyte count (section C) and the reticulocyte production index (RPI) play important roles. See Reticulocyte Count *on page 1156* entry for more information.

It will become apparent that the flowchart often ends with the diagnosis of iron deficiency anemia. This should not be surprising since iron deficiency anemia is the most common cause of decreased hemoglobin, especially in the outpatient setting. Normal values for many tests will vary between laboratories. Use reference ranges (normal values) from your own laboratory when appropriate. This flowchart is not intended to cover every hematologic possibility. There may not be a direct flow from one category to another in all possible circumstances. In order to follow the pathway of testing in an individual patient, follow the alphabetical designations.

References

Bessman JD, Gilmer PR Jr, and Gardner FH, "Improved Classification of Anemias by MCV and RDW," *Am J Clin Pathol*, 1983, 80(3):322-6.

Geaghan SM, "Hematologic Values and Appearances in the Healthy Fetus, Neonate, and Child," *Clin Lab Med*, 1999, 19(1):1-37.

Goodnough LT, Skikne B, and Brugnara C, "Erythropoietin, Iron, and Erythropoiesis," *Blood*, 2000, 96(3):823-33.

HEMATOLOGY *(Continued)*

Is the patient anemic?

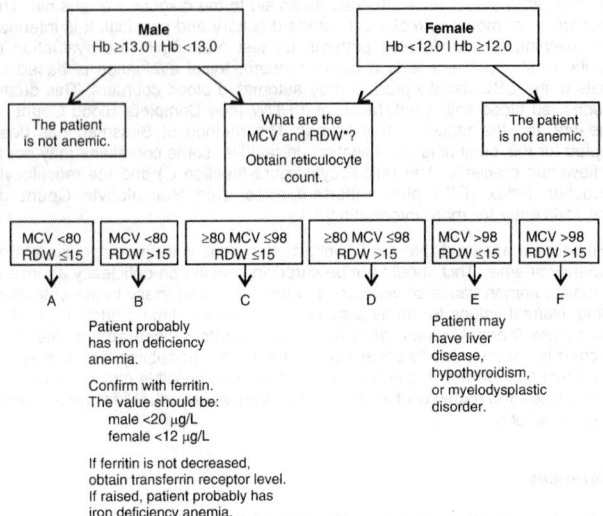

Male		Female
Hb ≥13.0 I Hb <13.0		Hb <12.0 I Hb ≥12.0

The patient is not anemic.

What are the MCV and RDW*? Obtain reticulocyte count.

The patient is not anemic.

MCV <80 RDW ≤15	MCV <80 RDW >15	≥80 MCV ≤98 RDW ≤15	≥80 MCV ≤98 RDW >15	MCV >98 RDW ≤15	MCV >98 RDW >15
A	B	C	D	E	F

B: Patient probably has iron deficiency anemia.

Confirm with ferritin. The value should be:
male <20 μg/L
female <12 μg/L

If ferritin is not decreased, obtain transferrin receptor level. If raised, patient probably has iron deficiency anemia.

E: Patient may have liver disease, hypothyroidism, or myelodysplastic disorder.

Although helpful in a number of instances, the RDW is not sufficiently sensitive/specific to make clear decisions in these evaluations.

Note: Hb levels may be higher at elevated altitudes.

A
What is the RBC?

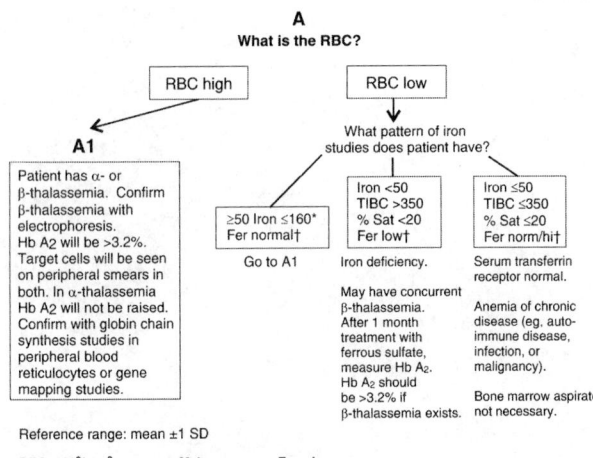

RBC high

RBC low

What pattern of iron studies does patient have?

A1

Patient has α- or β-thalassemia. Confirm β-thalassemia with electrophoresis. Hb A$_2$ will be >3.2%. Target cells will be seen on peripheral smears in both. In α-thalassemia Hb A$_2$ will not be raised. Confirm with globin chain synthesis studies in peripheral blood reticulocytes or gene mapping studies.

≥50 Iron ≤160* Fer normal†	Iron <50 TIBC >350 % Sat <20 Fer low†	Iron ≤50 TIBC ≤350 % Sat ≤20 Fer norm/hi†

Go to A1

Iron deficiency.

May have concurrent β-thalassemia. After 1 month treatment with ferrous sulfate, measure Hb A$_2$. Hb A$_2$ should be >3.2% if β-thalassemia exists.

Serum transferrin receptor normal.

Anemia of chronic disease (eg. auto-immune disease, infection, or malignancy).

Bone marrow aspirate not necessary.

Reference range: mean ±1 SD

RBC x 10^6/mm^3	Male	Female
15-17 y	5.0 ±0.4	4.5 ±0.3
18-64 y	5.0 ±0.4	4.4 ±0.4
65-74 y	4.8 ±0.5	4.5 ±0.4

*Adult males, values are slightly (5% to 10%) lower for adult females.
†Ferritin, reference range:
adult male 20-250 ng/mL
adult female
<40 y 12-122 ng/mL
>40 y 12-250 ng/mL

B
Does red cell morphology demonstrate schistocytes?

Yes

B1

Consider red cell fragmentation:

A) Microangiopathic hemolytic anemia
  1. DIC (look for underlying disease eg, adenocarcinoma, gram-positive or -negative sepsis, etc)
  2. TTP / hemolytic uremic syndrome (many schistocytes, thrombo-Cytopenia, transient neurological signs, raised LDH, low haptoglobin)

B) Vasculitis (eg, SLE, other connective tissue diseases, Rocky Mountain spotted fever, etc)

C) Giant cavernous hemangioma

D) Abnormal heart valves

E) Eclampsia / preeclampsia, HELLP syndrome

F) March hemoglobinuria

G) Severe burns

H) Others

No

Is the patient black?

Yes

Is sickle cell test positive?

Yes

Patient has sickle-beta thalassemia.

Confirm with electrophoresis.

No

Patient probably has iron deficiency anemia.

Confirm with ferritin or serum transferrin receptor. The ferritin value should be <12 ng/mL or serum transferrin receptor should be raised.

No

Patient probably has iron deficiency anemia.

Confirm with ferritin or serum transferrin receptor The value should be <12 ng/mL.

C
What is the uncorrected reticulocyte count?

Obtain single correction reticulocyte count (reticulocyte index) (S)

S = reticulocyte count x (patient's hematocrit / 0.45)

Double correction reticulocyte count or reticulocyte production index (RPI) is calculated by dividing the single correction reticulocyte count(s) by the maturation index (T).

Hct	T
0.40 - 0.50	1
0.30 - 0.40	1.5
0.20 - 0.30	2.0
0.10 - 0.20	2.5

Reticulocyte production index = U

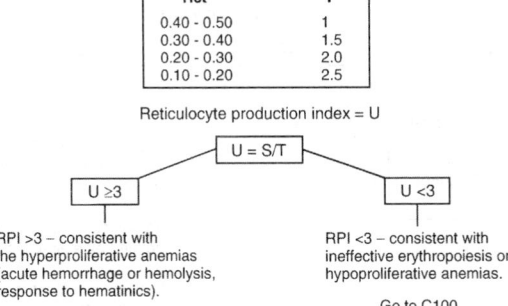

$U = S/T$

$U \geq 3$

$U < 3$

RPI >3 – consistent with the hyperproliferative anemias (acute hemorrhage or hemolysis, response to hematinics).

Go to C1

RPI <3 – consistent with ineffective erythropoiesis or hypoproliferative anemias.

Go to C100

HEMATOLOGY *(Continued)*

C1

Raised RPI ≥3

Examine peripheral blood smear for abnormal morphology.

Obtain:
Bilirubin
LDH
Serum free hemoglobin
Urine hemoglobin
Urine hemosiderin
Haptoglobin

Haptoglobin >40 mg/dL	Haptoglobin <30 mg/dL*

Patient probably has/had an acute hemorrhage or is responding to hematinic.

Patient should be evaluated for external or internal bleeding. Stools checked for occult blood.

Probably has hemolytic anemia. Bilirubin is usually between 1.0 and 5.0 mg/dL. Most should be indirect bilirubin.
Serum LDH is raised. The LDH isoezymes will show a flipped 1:2 pattern similar to that of an acute myocardial infarction.

If intravascular hemolysis is present, hemoglobinemia, hemoglobinuria, and hemosiderinuria may be present. A bone marrow aspirate (usually not indicated) will reveal erythroid hyperplasia with fairly normal maturation.

*Haptoglobin levels are decreased during pregnancy and with estrogen medication. With acute inflammation, haptoglobin, as an acute phase reactant may increase, masking a decrease relating to concurrent hemolysis (see also Haptoglobin, Serum).

C2

Does the patient have any of the following morphological abnormalities?

C3) Sickle cells
C4) Hb C crystals
C5) Target cells
C6) Spherocytes
C7) Elliptocytes
C8) Acanthocytes
C9) Stomatocytes
C10) Schistocytes
C11) Malaria / *Babesia*
C12) None of the above

C3	C4	C5

C3

Do hemoglobin electrophoresis. The patient has sickle cell disease (SS) or is doubly heterozygous (SC, SD, SO, S-thal, etc).

C4

The patient probably has homozygous Hb C disease. Confirm with electro-phoresis.

C5

Four possibilities:

1) Liver disease...elevated enzymes, (AST, ALT, LDH, GGT, Alk Phos), decreased albumin, increased PT, etc.

2) Hemoglobin C trait or disease, confirm by electrophoresis.

3) Thalassemia, confirm with elevated A2 (>3.2%) by electrophoresis.

4) Postsplenectomy

C6

Is the antiglobulin test, direct (Coombs test, direct) positive?

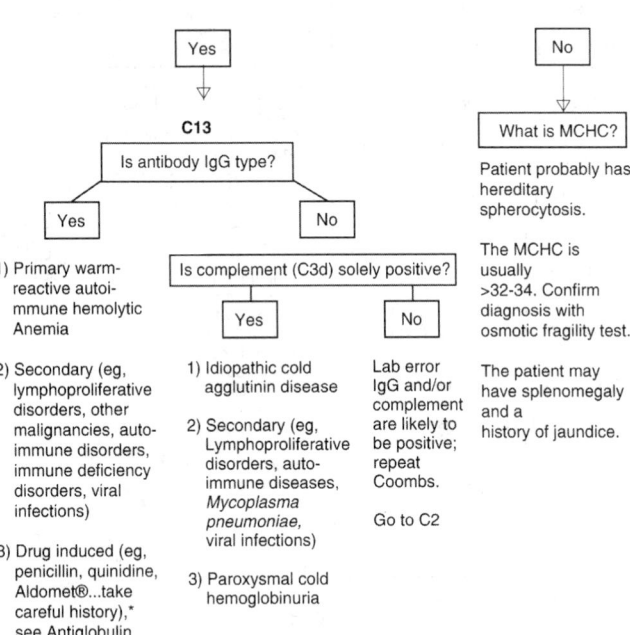

Yes	No

Yes branch → C13

C13

Is antibody IgG type?

Yes	No

Yes:

1) Primary warm-reactive autoimmune hemolytic Anemia

2) Secondary (eg, lymphoproliferative disorders, other malignancies, autoimmune disorders, immune deficiency disorders, viral infections)

3) Drug induced (eg, penicillin, quinidine, Aldomet®...take careful history),* see Antiglobulin Test, Indirect, and Direct

No: Is complement (C3d) solely positive?

Yes	No

Yes:

1) Idiopathic cold agglutinin disease

2) Secondary (eg, Lymphoproliferative disorders, autoimmune diseases, *Mycoplasma pneumoniae*, viral infections)

3) Paroxysmal cold hemoglobinuria

No: Lab error IgG and/or complement are likely to be positive; repeat Coombs.

Go to C2

No branch → What is MCHC?

Patient probably has hereditary spherocytosis.

The MCHC is usually >32-34. Confirm diagnosis with osmotic fragility test.

The patient may have splenomegaly and a history of jaundice.

---

*Young DS, *Effects of Drugs on Clinical Laboratory Tests*, 5th ed, Volume One: Listing by Test, Washington, DC: AACC Press, 2000, Section 3, 220-1.

C7	C8	C9
Repeat peripheral smear on another occasion (ensure that results are repeatable). Assure that the result is not an artifact. If results are repeatable then the diagnosis is hereditary elliptocytosis (very rarely of clinical significance).	Due to: 1) Severe liver disease. See Liver Disease: Laboratory Assessment, Overview 2) Congenital abetalipoproteinemia 3) Postsplenectomy	Patient has hereditary stomatocytosis, or changes may be an artifact.

HEMATOLOGY *(Continued)*

C10

Causes of red cell fragmentation:

A) Microangiopathic hemolytic anemia.

    1) DIC* (look for underlying disease eg, adenocarcinoma, gram-positive or -negative sepsis, etc)

    2) TTP* (many schistocytes, thrombocytopenia, renal changes, transient neurological signs)

    3) Hemolytic uremic syndrome*

B) Vasculitis (eg, SLE, other connective tissue diseases,* Rocky Mountain spotted fever, etc)

C) Prosthetic / abnormal heart valves

D) Eclampsia / preeclampsia, HELLP syndrome*

E) Malignant hypertension

F) March hemoglobinuria*

G) Severe burns

*See listing in Key Word Index.

C12

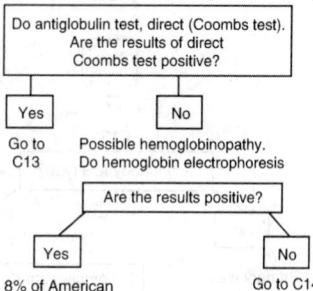

Do antiglobulin test, direct (Coombs test). Are the results of direct Coombs test positive?

Yes → Go to C13

No → Possible hemoglobinopathy. Do hemoglobin electrophoresis

Are the results positive?

Yes → 8% of American blacks have sickle trait (AS). 0.15% of American blacks have sickle cell disease (SS). Hemoglobin C is 10 times less likely than hemoglobin S. Numerous hemoglobin variants exist, most of them quite rare.

No → Go to C14

C11

The patient has red blood cell inclusions (eg, malaria, *Babesia*, ehrlichiosis). Confirm with thick smear, serological tests.

C14

Do pyruvate kinase (PK) and G-6-PD enzyme screens.

Is PK present?

Is G-6-PD present?

PK (Yes) G-6-PD (No)	PK (No) G-6-PD (Yes)	PK (Yes) G-6-PD (Yes)
The patient has G-6-PD deficiency. False-negatives can occur if measured at time of a high reticulocyte count associated with a hemolytic episode.	Patient has PK deficiency	Other enzyme deficiencies could be the cause of the hemolysis. Send sample of blood to a reference lab for screening. The chance of such deficiency is 1 in 300,000.

C15

Repeat history and physical, examine for splenomegaly.

C25

What are vitamin B$_{12}$, folate, and RBC

B$_{12}$ >250 pg/mL Folate <2 RBC folate <200	B$_{12}$ <250 pg/mL Folate >2 RBC folate >200	B$_{12}$ <250 pg/mL Folate <2 RBC folate <200	Any other combination?
Patient has folate deficiency.	The patient has B$_{12}$ deficiency.	The patient has combined B$_{12}$ and folate deficiencies.	Do bone marrow aspirate and biopsy and evaluate microscopically.
Homocysteine level raised, methylmalonic acid level normal.	Homocysteine and methylmalonic acid levels raised.	The bone marrow aspirate will show hyperplastic, megaloblastic changes.	A) Myelophthisic state - this shows replacement of the normal hematopoietic elements of the bone marrow by leukemia, metastic carcinoma, TB, fibrosis, etc.
The bone marrow aspirate will show hyperplastic, megaloblastic changes.	Obtain parietal cell antibody, intrinsic factor blocking antibodies.		B) Hypoplasia/aplasia - bone marrow shows decreased erythrocytic precursors only (pure red cell aplasia) or panhypoplasia of marrow elements.
	Schilling test, parts 1 & 2 may distinguish gastric from ileal problem.		C) Endocrinopathy - slight hypoplasia due to Hashimoto thyroiditis, Addison disease, etc. Do appropriate endocrine tests; see Key Word Index.
	The bone marrow aspirate will show hyperplastic, megaloblastic changes.		D) Normal marrow - obtain heme consult.

C100

What pattern of iron studies does patient have?

Iron <50 TIBC >350 % Sat <20 Fer low	Iron <50 TIBC ≤250 % Sat ≤20 Fer norm/hi	Iron >160 TIBC <250 % Sat >20 Fer norm/hi	≥50 Iron ≤160 Fer norm/hi	Any other combination
The patient has iron deficiency, raised serum transferrin receptor level.	Anemia of chronic disease (inflammatory disease, infection or malignancy, RE iron block).	The patient has sideroblastic anemia, either congenital, or acquired (lead poisoning, alcohol, isoniazid, myelodysplasia, etc).	**A1** Patient has beta thalassemia. Confirm with electrophoresis. Hb A will be >3.2%.	What is patient's serum creatinine?
A bone marrow is not necessary, but would show absence of stainable iron.	Normal serum transferrin receptor level; inappropriately reduced erythropoietin level.	Bone marrow may show normal/increased erythroid precursors, ringed sideroblasts on doing iron stain.	Target cells will be seen on peripheral smear.	Creatinine ≤1.8 → Go to C25 / Creatinine >2.0 → Chronic renal failure
	Bone marrow would show the presence of iron stores.	**Note:** Obtain cytogenetic studies on the bone marrow sample.		Obtain erythropoietin level. Bone marrow may show mild hypoplasia due to lack of erythropoietin. Bone marrow is not usually required.

HEMATOLOGY *(Continued)*

D

Normocytic, heterogeneous anemia is present.
Possible early iron deficiency, sideroblastic or megaloblastic
anemia or mixed deficiency is present.

What are ferritin and RBC folate / vitamin B_{12} values?

Ferritin low RBC folate <200 Vitamin B_{12} <250	Ferritin normal RBC folate <200 Vitamin B_{12} <250	Ferritin normal RBC folate >200 Vitamin B_{12} >250

Mixed iron and megaloblastic anemia are present.

Obtain serum transferrin receptor, homocysteine, and methylmalonic acid levels if necessary.

Megaloblastic anemia is present.

Obtain homocysteine and methylmalonic acid levels if necessary.

Sideroblastic anemia may be present. Confirm with observation of ringed sideroblasts in bone marrow.

or

If patient is black, also consider hemoglobinopathy (eg, SS or SC hemoglobinopathy).

or

Myeloproliferative disorder may be present. A leukoerythroblastic reaction should be present (nucleated RBCs and immature myeloid series). Diagnosis confirmed on bone marrow examination.

F

**Do antiglobulin test, direct (Coombs test, direct).
Are results of direct Coombs test positive?**

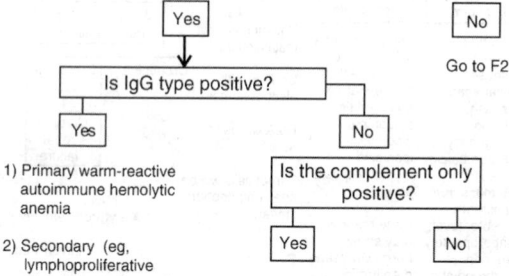

Yes	No

No → Go to F2

Is IgG type positive?

Yes	No

Yes:

1) Primary warm-reactive autoimmune hemolytic anemia

2) Secondary (eg, lymphoproliferative disorders, malignancies, autoimmune disorders, immune deficiency disorders, viral infections)

3) Drug induced (eg, penicillin, quinidine, Aldomet®... take careful history)

No → Is the complement only positive?

Yes	No

Yes:

1) Idiopathic cold agglutinin disease

2) Secondary (eg, lymphoproliferative disorders, autoimmune disease, *Mycoplasma*, pneumonia, viral infections

3) Paroxysmal cold hemoglobinuria

No:

Lab error; IgG and/or complement should be positive: repeat Coombs.

Go to F2

F2

Macrocytic, heterogeneous anemia

What pattern of folate / vitamin B$_{12}$ studies does patient have?

>250 B$_{12}$ <1000 Folate <2 RBC folate <200	B$_{12}$ <250 Folate >2 RBC folate >200	B$_{12}$ <250 RBC folate <200	Normal B$_{12}$ >250 Folate >2 RBC folate >200
The patient has folate deficiency.*	The patient has B$_{12}$ deficiency.*	The patient has both B$_{12}$ and folate deficiencies.*	Ensure no liver disease, alcoholism, raised reticulocyte count (accelerated erythropoiesis), hypothyroidism.
Measure homocysteine and methylmalonic acid, LDH, indirect bilirubin.	Measure homocysteine and methylmalonic acid, LDH, indirect bilirubin.	Measure homocysteine and methylmalonic acid, LDH, indirect bilirubin.	Do bone marrow aspirate with chromosomes and biopsy.
Homocysteine raised confirmed folate deficiency.	Both raised - confirmed vitamin B$_{12}$ deficiency.	Both raised - confirmed vitamin B$_{12}$ deficiency or mixed vitamin B$_{12}$ deficiency and folate deficiency.	If the diagnosis is not apparent (eg, myelodysplastic syndrome) then repeat CBC and begin work-up again.
Bone marrow may not be required.	Obtain intrinsic factor and parietal cell antibody studies to help confirm/exclude pernicious anemia. A Shilling test part 1 and 2 may help to confirm the presence of pernicious anemia.	Obtain intrinsic factor and parietal cell antibody studies to help confirm/exclude pernicious anemia. A Shilling test part 1 and 2 may help to confirm the presence of pernicious anemia.	
	Bone marrow may not be required.	Bone marrow may not be required.	

*A deficiency state may be present in a patient, but this may not necessarily be the cause of the anemia. Confirmatory studies are thus required.

HEMATOLOGY (Continued)

CLUSTER OF DIFFERENTIATION (CD) ANTIGENS
Human Leukocyte Cell Surface (Membrane) Markers

CD #	Synonyms	Cell Type	Identity	Molecular Size of Antigen	Function / Clinical Application
1	T6; Leu-6; OKT6		HLA class I-like molecule	45-kDa	Expressed as complex with β-2 microglobulin on cortical immature thymocytes and on Langerhans' cells
2	T11; Leu-6; OKT11	Pan-T cells	Sheep red blood cell receptor	50-kDa	Involved in T-cell activation; a single chain glycoprotein member of Ig superfamily that transduces signal to T cells
3	T3; Leu-4; OKT3	Pan-T cells	Multicomponent; five polypeptides	16-28 kDa each	Associated with T-cell receptor, T-lymphocyte activation
4	T4; Leu-3a; OKT4	Helper T cells; monocytes; macrophages		59-kDa	A single chain glycoprotein member of Ig superfamily involved in T-cell activation; receptor for HLA class II; receptor for HIV
7	3AI; Leu-9			40-kDa	Member of Ig superfamily; present on T cells, NK cells
8	T8; Leu-2a; OKT8	Suppressor T cells; some monocytes; macrophages	Homo-or heterodimer with α and β chains	32-kDa	Receptor for HLA class I antigens; member of Ig superfamily; involved in T-cell activation
10	T5; CALLA (common acute lymphoblastic leukemia antigen)	Early B and T cells; germinal center cells; neutrophils	Integrin glycoprotein	100-kDa	Single chain glycoprotein; peptide cleavage (endopeptidase)
11a	LFA-1		α-integrin glycoprotein complexed with β-integrin CD18	180-kDa	Receptor for ICAM-1, -2, and -3 adherence molecules
11b	Mac-1; OKM-1; CR3; Leu-15	Monocytes; neutrophils; NK cells	α-integrin glycoprotein complexed with β-integrin CD18	165-kDa	Receptor for ICAM-1 and fibrinogen complement 3b receptor
11c	CR4; Leu-M5; BLY6	Monocytes; activated lymphs; granulocytes	α-integrin glycoprotein complexed with β-integrin CD18	150-kDa	Receptor for ICAM-1 and fibrinogen complement 3b receptor
14	Mo2; MY4; Leu-M3	Monocytes; immature granulocytes	Glycoprotein linked to phosphatidyl-inositol	55-kDa	Lipopolysaccharide receptor

CLUSTER OF DIFFERENTIATION (CD) ANTIGENS
Human Leukocyte Cell Surface (Membrane) Markers (continued)

CD #	Synonyms	Cell Type	Identity	Molecular Size of Antigen	Function / Clinical Application
16	FcrRIII	NK cells; macrophages; granulocytes	Glycoprotein	60-kDa	IgG receptor, low affinity antibody-dependent cellular toxicity
18	LFA-1; Mac-1	Leukocytes	β integrin glycoprotein complexed with α-integrins CD 11a, b, or c	95-kDa	
19	B4; Leu-12	B cells; follicular dentritic cells	Single chain glycoprotein	95-kDa	B-lymphocyte activation
20	B1; Bp35; Leu-16; L26	B cells	Single chain protein	35-kDa	Modulates B-cell activation (Ca^{++} channel involved)
21	CR2	B cells	Single chain complement regulatory protein	145-kDa	B-lymphocyte activation; complement 3 receptor; Epstein-Barr virus receptor; amplified signal from B-cell receptor
22	FCeRII; Bgp135; Leu-14	B cells	Ig superfamily		B-cell activation receptor for IgE
23	FCeRII; B6; Leu-20	Activated B cells; follicular dentritic cells; eosinophils; platelets	Lectin like	45-kDa	Receptor for IgE
25	TAC antigen; IL-2R α-chain	Activated T and B cells; macrophages	Glycoprotein	55-kDa	Low affinity receptor for IL-2
28	TP44	T-cell subset activated B cells	Ig superfamily	44-kD; homodimer	Costimulatory for CD80 binding results in T-cell activation receptor for B7
29	VLA	Numerous	β-integrin complexed with α-integrin chains	130-kDa	
32	FcγRII	Numerous	Single chain protein	40-kDa	IgG receptor
34	MY10	T cells; stem cells	Single chain glycoprotein	110-kDa	?adhesion molecule; ?marker for HSC
35	CRI; C36 receptor	B cells; NK cells; PMNs; monocytes; red blood cells		190-250-kDa	Binds immune complexes
38	OKT10; T10; CRI; Leu-17	Activated lymphocytes; immature B cells	Single chain glycoprotein		Complement 3b receptor

HEMATOLOGY *(Continued)*

CLUSTER OF DIFFERENTIATION (CD) ANTIGENS
Human Leukocyte Cell Surface (Membrane) Markers *(continued)*

CD #	Synonyms	Cell Type	Identity	Molecular Size of Antigen	Function / Clinical Application
40	gp50	B cells; macrophages; follicular dendritic cells	Single chain polypeptide homologue of tumor necrosis factor receptor, Fas	48-kDa	Growth factor receptor; B-cell activation; involved in B cell apoptosis
43	Leukosialin; leukophorin	Lymphocytes NK; cells monocytes	Glycoprotein	95-135-kDa	Leukocyte activation; cell surface mucin
44	Pgp-1; Hermes receptor type III	Lymphocytes; NK cells; monocytes; red blood cells	Single chain glycoprotein	80-95-kDa	Leukocyte homing; cell migration; adhesion to hyalutonic acid with implication to tumor cell metastasis
45	Leukocyte common antigen; T200 isoforms: CD45RA, CD45RB, CD45RO	Leukocytes	Single chain glycoprotein with differentially expressed isoforms	Multiple isoforms; 180-220-kDa; 220-kDa form; 220, 205, 190-kDa forms; 180-kDa form	B cells; monocytes CD8 SP; T cells; naive T cells; B cells; monocytes; macrophages; subsets of memory T cells; monocytes; thymocytes; activated T cells; memory T cells
46	MCP	Numerous	Dimeric glycoprotein	60-kDa	Cofactor, cleavage of complement 36 and 46
50	ICAM-3	Leukocytes	Single chain glycoprotein; Ig superfamily adhesion molecule	120-kDa	Adhesion molecule binds CD11/CD18
54	ICAM-1	Activated lymphocytes; endothelial cells	Ig superfamily adhesion molecule	90-kDa	Binds CD11/CD18; ligand for LFA-1, rhinoviruses
55	DAF	Many	Single chain glycoprotein	70-kDa	Inactivates complement factors C3 and C5
56	Leu-19	NK cells	Single chain glycoprotein	140-200-kDa	Adhesion molecule
59	Protectin	Many	Glycoprotein	20-kDa	Inhibits production of membrane attack complex of complement
62 E, L, P	Selectins E, L, P; ELAM-1; LAM-1	E: activated endothelium; L: neutrophils/monocytes and NK cells	Single chain glycoprotein	E: 110-kDa L: 76-kDa P: 140-kDa	Adhesion; P-selectin mediates adhesion of platelets

CLUSTER OF DIFFERENTIATION (CD) ANTIGENS
Human Leukocyte Cell Surface (Membrane) Markers *(continued)*

CD #	Synonyms	Cell Type	Identity	Molecular Size of Antigen	Function / Clinical Application
64	FcγRI	Monocytes/macrophages	Single chain glycoprotein	75-kDa	High affinity IgG receptor mediates phagocytosis; ab-dependent cell cytotoxicity; cytokine release
71	Transferrin receptor	Most activated cells	Homodimeric glycoprotein	95-kDa	Receptor for iron uptake protein
73	Ecto 5′-nucleotidase	Lymphocyte subsets			Regulates uptake of nucleotides
74	HLA class II		Single chain glycoprotein	35-kDa	Antigen presentation
79 a and b	Igα; Igβ	B cells			Part of B-cell antigen receptor
80	B7-1	Activated B cells; monocytes; macrophages; dendritic cells	Ig superfamily	60-kDa	Ligand for CD28
86	B7-2	Activated lymphocytes; monocytes; macrophages; dendritic cells	Ig superfamily		Ligand for CD152
87	Urokinase/plasminogen receptor (uPA-R)	Activated T cells; neutrophils; macrophages	Phosphatidyl-inositol linked single chain glycoprotein anchored by GPI linkage	60-kDa	uPA-R and receptors for plasminogen focus plasmin on surface of tumor (or inflammatory cells) to assist metastatic invasion (and aid diapedesis at inflammatory sites)
95	APO-1; Fas	Activated T and B cells; NK cells	Glycoprotein; member of TNF receptor family	45-kDa	Mediates apoptosis; delivers apoptotic signal on binding Fas ligand; sequence of 65 aas (intracytoplasmic) form the "death domain" (promotes the death signal)
115	M-CSF receptor; c-fms product	Macrophages	Single chain glycoprotein	150-kDa	Growth factor receptor
115w	GM CSF	Myeloid precursors	Single chain glycoprotein	80-kDa	Growth factor receptor
117	c-kit, stem cell factor receptor	Hemopoietic cells	Single chain glycoprotein	145-kDa	Growth factor receptor
117w	IFN-γ receptor	Many	Single chain glycoprotein	90-kDa	Growth factor receptor

HEMATOLOGY *(Continued)*

CLUSTER OF DIFFERENTIATION (CD) ANTIGENS
Human Leukocyte Cell Surface (Membrane) Markers *(continued)*

CD #	Synonyms	Cell Type	Identity	Molecular Size of Antigen	Function / Clinical Application
120a	Tumor necrosis; factor receptor I; TNFR I	Many	Member NGFR/TNFR superfamily	50-kDa	Cytokine receptor; involved in septic shock
120b	Tumor necrosis; factor receptor II; TNFR II	Many	Member NGFR/TNFR superfamily	75-kDa	Cytokine receptor; induces nitric oxide (NO); activates neutrophils
121a	IL-1 receptor type 1	Activated T and B cells; macrophages	Glycoproteins		Cytokine receptor
121b	IL-1 receptor type 2	Activated T and B cells; macrophages	Glycoproteins		Cytokine receptor
122	IL-2 receptor	B cells; activated macrophages	Glycoproteins	75-kDa	IL-2 receptor, β-chain
123	IL-3 receptor	B cells; hematopoietic stem cells			Binds IL-3; affects early hemopoietic (pluripotential stem cell stage) cell differentiation
124	IL-4 receptor	Many, including B cells, T cells, hemopoietic precursors, marrow stroma, fibroblasts, mast cells, others			IL-4 binding; high affinity switch factor for IgE
125-129	IL-5-9 receptor	Widespread distribution			Multiple cytokine stimulatory functions; signal transduction functions
152	CTLA-4; Ly 56	Activated T cells	Ig superfamily		Negative signals; immune response
154	CD40 ligand, gp39	Activated T cells			Receptor for CD40

131-166 — new CD antigens from the 6th International Workshop on Human Leukocyte Differentiation Antigens

IMMUNOLOGY

Wayne R. DeMott, MD

David S. Jacobs, MD

Fred V. Plapp, MD, PhD

Dwight K. Oxley, MD

Susan H. Hsu, PhD

Contributor: Eugene S. Olsowka, MD, PhD

The science of clinical and laboratory immunology includes its cellular and humoral constituents. Its mechanisms have been addressed in recent literature.[1,2,3]

The focus is on laboratory tests pertinent to diseases which are known, or believed, to be of immune pathogenesis, including connective tissue/autoimmune/rheumatic diseases, vasculitides, organ-specific autoimmune diseases, allergic disorders, transplantation immunology, and immunodeficiency syndromes. Clinical immunology, defined broadly, includes study of acquired immunodeficiency syndrome (AIDS), some neoplastic entities including especially lymphoreticular malignant diseases, immunity related to infections, immunogenetics, and immune-based therapy.[3]

Serology has classically involved antibodies and reactions to exogenous antigens. The traditional approaches used to investigate serologic responses to disease were labor-intensive, as well as insensitive and nonspecific. Automation of serologic tests, broad availability of monoclonal antibody reagents, and dramatic improvements in flow cytometry instrumentation and software have combined to expand the readily available capabilities for immunologic evaluation in many clinical laboratories. Generally, diagnosis of disease by immunoassay involves either the detection of specific antibody or specific antigen. Detection of autoimmune antibodies may require interpretation of specific patterns.

The sensitivity of agar gel diffusion methods is substantially less than that of electroimmunodiffusion. Enzyme immunoassay (EIA), radioimmunoassay (RIA), fluorometric assay, and chemiluminescent assay provide still greater sensitivity. Polymerase chain reaction (PCR) is addressed in a separate monograph in this book. DNA methods have rapidly progressed.

The tests for antibodies may detect all classes of immunoglobulins or may be specific for IgG, IgM, or IgA. Detection of autoimmune antibodies may require interpretation of specific patterns. Current immunologic tests include useful assays that may improve diagnosis of specific diseases. We have tried to provide information to orchestrate application of immunologic assays, and to include specifics of each test. Potential problems and applications are briefly addressed.

The authors express appreciation to Dr David F. Keren, who provided numerous additions and enhancements to the Immunology and Serology test listings in the 3rd edition of the *Laboratory Test Handbook*.

Footnotes

1. Schwartz RS, "Advances in Immunology - A New Series of Review Articles," *N Engl J Med*, 2000, 343(1):61-2.
2. Mackay IR and Rosen FS, "Advances in Immunology - Autoimmune Diseases," *N Engl J Med*, 2001, 345(5):340-50.
3. Nakamura RM, Keren DF, Bylund DJ, eds, *Clinical and Laboratory Evaluation of Human Autoimmune Diseases*, Chicago, IL: Amercian Society for Clinical Pathology, ASCP Press, 2002.

INFECTIOUS DISEASES

Rebecca T. Horvat, PhD

David S. Jacobs, MD

Dwight Oxley, MD

Bradly D. Clark, MD

Daniel R. Hinthorn, MD

Although there have been tremendous gains in the treatment and prevention of infectious diseases during the last few centuries, around the world infectious diseases still remains one of the major causes of morbidity and mortality faced by humans. Diseases due to bacterial, viral, and parasitic organisms remain the leading cause of death globally. According to WHO statistics, one of every three deaths around the world in 1997 resulted from an infectious disease.

Tests performed by the diagnostic laboratory are often used to identify an infectious process. A critical step in obtaining a useful laboratory report is the collection and transport of the specimen. The material sent for culture or assay should be collected from the appropriate site with minimal contamination and should be submitted in the appropriate collection container. All specimens should be labeled properly to avoid incorrect results from affecting patient care. Finally, specimens should be promptly transported to the laboratory to prevent overgrowth of organisms or loss of organism viability (especially viral cultures).

A rapid result for diagnosis of infectious disease is essential to both patient management and hospital and public infection control. The Gram stain still remains one of the most rapid results available. However, other rapid tests are gradually being introduced such as antigen and antibody detection methods that have increased sensitivity. A new area of rapid diagnosis in infectious disease is the use of molecular technology. This technology permits rapid detection of fastidious and nonculturable infectious agents. Presently, a complete culture result is usually not available until 48 hours after submission. Certain fastidious organisms, mycobacteria, and fungal cultures require longer incubations, and results require as long as 4-6 weeks. Molecular methods have shortened this time.

The Disease Index now includes a topic, Biological Warfare, intended to lead the reader to some useful references and introductory information.

As with any progressive work, the contents of this chapter have been based on the scholarship of previous authors. The current authors wish to acknowledge Drs David F. Keren, Larry D. Gray, Christopher J. Papasian, and Bernard L. Kasten for their contributions in previous editions.

References

Koneman EW, Allen SD, Janda WM, et al, *Color Atlas and Textbook of Diagnostic Microbiology*, 5th ed, Philadelphia, PA: Lippincott Williams & Wilkins, 1997.

Murray PR, Baron EJ, Pfaller MA, et al, *Manual of Clinical Microbiology, 8th ed,*, Washington, DC: American Society of Microbiology, 2003.

MOLECULAR GENETIC TESTING

Karen Stephens, PhD

Jonathan F. Tait, MD

David S. Jacobs, MD

Uttam Garg, PhD

The first DNA-based genetic tests became available in the mid 1980s. Since then, the field has expanded rapidly, and today there are clinical tests available for hundreds of diseases[1]. The number of tests will continue to increase as information from the Human Genome Project is translated into new diagnostic procedures. DNA-based genetic testing is also moving out of specialized genetics clinics and is becoming more common in general clinical practice.

The new DNA-based genetic tests represent a genuine advance, but they also pose new challenges for patients, for the physicians providing them to patients, and for the laboratories offering these test services. On the one hand, news reports of genetic breakthroughs can create unrealistic expectations for the availability or utility of new genetic tests. On the other hand, both public and professional groups have raised various concerns about the quality, appropriateness, and potential misuse of genetic testing. Although most evidence to date indicates that molecular-genetic testing is accurate and has a low frequency of adverse events,[2] and does not lead to discrimination in health insurance coverage decisions,[3] some caution is appropriate, particularly for certain tests that are either controversial or have major impact on life events. This introduction provides some general guidance on factors to consider for effective and responsible use of genetic tests.

Approach to Testing

Molecular-genetic tests are highly heterogeneous; some can be ordered and used as part of a panel of general laboratory tests (eg, testing for genetic thrombophilia in a patient presenting with venous thrombosis), while others require careful pretest preparation and post-test follow-up for patient and family (eg, tests for Huntington disease or hereditary risk for breast and ovarian cancer).

A general approach is as follows.

Pretest Evaluation and Counseling

- Evaluate the indication for testing - diagnostic versus presymptomatic versus carrier testing. The nature of testing and counseling procedures can often differ when the same laboratory test is used for different clinical indications.

- Assess need for specialized genetic counseling. Directories of specialized genetics clinics are available online.[1]

- Assess clinical and psychosocial appropriateness of testing.

- Address confidentiality concerns. Patients may be concerned about disclosure of information to relatives, or its impact on insurability or employability.

- For some presymptomatic tests, it may be appropriate to first confirm the presence of a detectable mutation in an affected relative.

MOLECULAR GENETIC TESTING *(Continued)*

- Obtain informed consent. Explain the nature of the test procedure, the expected outcomes, the risks, benefits, and alternatives. Specific items to cover vary greatly from test to test, but can include:
    - clinical utility of results
    - alternatives to testing (ie, risk calculation for carrier status, diagnosis on clinical grounds instead of diagnostic testing)
    - sensitivity, specificity, cost, and turnaround time
    - possible psychological risks and impact on family members
    - irrevocable nature of results
    - possible risks to insurance and employment

Post-test Counseling and Follow-up

This will also vary greatly from test to test, but factors to consider include:

- How and when to give results. Tests with a major impact on life events should usually be given in person, and presence of a support person may be helpful.
- Psychological and emotional impact of test results.
- Clinical follow-up and referral required for the disease or predisposition in question.
- Counseling of at-risk family members. Particularly for a family newly diagnosed with a genetic disease, counseling or referral of relatives is important. Some genetic diseases such as hemochromatosis are highly treatable, and other relatives may be spared by timely diagnosis. Reproductive decisions may also be affected. At the same time, one does not want to encourage testing for its own sake; each family member needs to decide what is best for himself or herself.
- Confidentiality concerns. Patients may consider some test results especially sensitive, and may wish to apply a higher standard of confidentiality than is normally the case for healthcare information.

Special Situations and Ethical Issues

For tests that have major impact on medical care or life planning, some clinical circumstances call for special care before testing is performed:

- Adoptees. There may be pressure from the adoptive parents or the adoption agency to determine if the child carries a disease present or suspected in the birth parents. In evaluating these requests, one should ask whether testing serves the best interests of the child. For example, presymptomatic testing for an untreatable adult-onset disease may provide no clinical benefit to the adoptee. Guidelines for genetic testing during adoption have been provided recently.[4]
- Testing of children for adult-onset diseases. Testing for adult-onset diseases that are currently not treatable or preventable may provide no benefit to a child, and may be harmful if it leads to discriminatory treatment in school, family, or community. If a test result primarily affects adult decisions such as choice of career, insurance coverage, marriage, or reproductive options, testing may be best postponed until a child can make his or her own decision as an adult.[5]
- Persons with impaired ability to give consent. Caution should be exercised in performing predictive testing on people such as the mentally handicapped, critically ill patients, or prisoners.
- Persons at 25% risk for an autosomal dominant disease. The person requesting testing should be aware that a positive result will reveal the status of the parent at 50% risk, and be encouraged to reach a prior understanding with the parent about how testing will be done and who will have access to the information.

Alternatives to Testing

Because of the technical limitations, complexity or expense of some molecular-genetic tests, it is appropriate to consider alternatives to testing. Depending on the circumstances, one or more of the following alternatives might be most appropriate for a given individual or family. Discussion and choice of alternatives is often well handled by specialized genetics clinics.

- Information and reassurance. This may be helpful for some people who perceive their risk of a genetic disease to be unrealistically high, or who have unrealistic expectations about the utility of a given genetic test result. This situation commonly occurs with many of the current tests that detect genetic predisposition to a variety of cancers; these tests are generally applicable only to a small subset of people with very strong family histories of cancer, and clinically proven surveillance or preventive strategies based on test results are not necessarily available.

- Diagnosis on clinical grounds or by other test methods. In some instances, other laboratory methods provide a better means for initial diagnosis. For example, screening for thalassemia carriers is still best done with a combination of hemoglobin electrophoresis and quantitation, red-cell indices, and special staining for intraerythrocytic inclusions. DNA-based tests are reserved for confirmation or for prenatal diagnosis. Similarly, conventional clinical chemistry testing provides the best initial screening test for hereditary hemochromatosis.

- Risk estimates. For families with known genetic diseases, it is often possible to estimate an individual's risk of carrying the mutant gene based on simple pedigree analysis. For people without a known family history, their risk of carrying a mutant gene can often be estimated from the population frequency of that gene in their particular ethnic group.

  For some forms of cancer, empirical risk estimates can be provided based on analysis of a patient's known risk factors. An example would be analysis of the risk of breast cancer.[6]

- DNA banking. This simple, low-cost alternative is often useful in the setting of terminally ill patients with undiagnosed diseases. DNA can be safely stored for years, thus preserving the option of future genetic diagnosis for an individual or family. For example, patients terminally ill with cancer of suspected genetic origin can store their DNA to allow future diagnosis for their at-risk relatives.

Electronic References on Genetic Diseases and Genetic Testing

The GeneTests web site provides a directory of laboratories providing clinical or research-based testing for genetic diseases, and directories of clinics providing specialized services for genetic diseases.[1] GeneClinics provides a series of brief reviews on many genetic diseases.[7] Online Mendelian Inheritance in Man (OMIM) provides an encyclopedic reference source on genetic diseases of all kinds.[8] The American Society of Human Genetics provides a series of policy statements on ethical, legal and social issues related to genetic testing.[9]

Footnotes

1. Anonymous, *GeneTests: Directory of Genetic Testing Laboratories* [database online] Children's Health System, 1999, available at: http://www.genetests.org/
2. Hofgartner WT and Tait JF, "Frequency of Problems During Clinical Molecular-Genetic Testing," *Am J Clin Pathol*, 1999; 112:14-21.
3. Hall MA and Rich SS, "Laws Restricting Health Insurers' Use of Genetic Information: Impact on Genetic Discrimination," *Am J Hum Genet*, 2000, 66:293-307.
4. Anonymous, "Genetic Testing in Adoption," The American Society of Human Genetics Social Issues Committee and The American College of Medical Genetics Social, Ethical, and Legal Issues Committee, *Am J Hum Genet*, 2000, 66:761-7.
5. "Points to Consider: Ethical, Legal, and Psychosocial Implications of Genetic Testing in Children and Adolescents," American Society of Human Genetics Board of Directors, American College of Medical Genetics Board of Directors, *Am J Hum Genet*, 1995, 57:1233-41.

MOLECULAR GENETIC TESTING *(Continued)*

6. Armstrong K, Eisen A, and Weber B, "Assessing the Risk of Breast Cancer," *N Engl J Med*, 2000, 342:564-71.
7. Anonymous, *GeneClinics: Medical Genetics Knowledge Base* [database online] 1995, available at: http://www.geneclinics.org/ (updated weekly).
8. Anonymous, *Online Mendelian Inheritance in Man, OMIM (TM)*, McKusick-Nathans Institute for Genetic Medicine, Johns Hopkins University (Baltimore, MD) and National Center for Biotechnology Information, National Library of Medicine (Bethesda, MD), 2000, available at: http://www.ncbi.nlm.nih.gov/omim/
9. American Society of Human Genetics, *Society Policy Statements and Reports*, available at: http://www.faseb.org/genetics/ashg/policy/pol-00.htm

MOLECULAR ONCOLOGY

Diane L. Persons, MD

Wayne R. DeMott, MD

David S. Jacobs, MD

Contributor: Uttam Garg, PhD

Research that is defining the molecular blueprints of malignancy has uncovered a bewildering array of gene-based mechanisms. Involvement of proto-oncogenes, oncogenes, tumor suppressor genes, molecular signaling mechanisms, and control of apoptosis includes, importantly, gene expression/ transcription of effector proteins, and their interaction and collaboration as multimeric complexes. Malignant transformation usually requires activation of multiple proto-oncogenes and involvement of tumor suppressor genes (the multiple hit hypothesis). The activation of proto-oncogenes and oncogenes may be the result of point mutation, deletion, recombination, insertional mutation, chromosomal translocation, gene amplification, and/or abnormal expression of oncogene-encoded cell growth factors.[1] The pathway to malignant transformation includes interwoven viral DNA and damaged DNA resultant from environmental carcinogens. DNA-related molecular testing procedures have evolved in tandem with implications to the diagnosis of malignant neoplasia, but in particular with clinical applications in therapy, monitoring of therapy, and prognosis.

The genetic control and containment of cell maturation and proliferation is subverted and deranged with the onset of tumorigenesis. Tissue growth and molding during organogenesis is dependent upon orderly maturation of cells and also upon their timely elimination (demise). As noted above, a number of gene-based processes are involved in malignant transformation. Proto-oncogenes are derived from normal components of the genome, but may be activated to oncogenes. The viral oncogenes *abl, erb-B, ets, mos, myc, myb, H-ras, K-ras* and *Sis* include four with proto-oncogenes that may have a role in human malignancy (*myc, ras, abl* and *erb-B*). Proto-oncogenes encode proteins involved in normal growth and differentiation that can be activated (eg, by mutation or other genetic mechanism) to oncogenes which can then be activated to transform cells.[2]

By negative regulation of transcription factors, by inhibition of apoptotic cell death, or by other mechanisms, tumor suppressor genes (eg, retinoblastoma (*RB*), and *p53* genes) lead to the development of malignancy. The proto-oncogene *bcl-2* is considered to act as a suppressor of programmed cell death (direct action in the prevention of apoptosis). *RB* and *p53* genes both encode nuclear phosphoproteins, the degree of phosphorylation fluctuating with the cell cycle. The phosphoprotein pRB exerts growth inhibition only in its hypophosphorylated form. In some tumors, presence of mutated *p53* is indicative of short survival and of resistance to chemotherapy.[3] Thus, the clinical interest in demonstrating expression/overexpression of *p53*.

The test listings in this chapter deal largely with molecular oncology and are certainly not comprehensive. A common thread is the often uncertain application and utility that these costly analyses may have in clinical environments. In screening situations, cost savings may be claimed, not only in terms of human concern and suffering but also in the identification of individuals who may benefit from preventive measures. The expense of some procedures may detract from compliance with recommended screening strategies, even when significant benefit is evident. Illustrative is the current status of screening for hereditary nonpolyposis colorectal cancer (HNPCC). A study from Finland concludes (on the basis of a 15-year trial in families with HNPCC) that screening by colonoscopy at 3-year intervals more than halves the risk of colorectal cancer, prevents death, and decreases overall mortality by about 65%.[4] Over half (57%) of members of HNPCC families, however, were found to have declined genetic testing.[5] Hopefully, new technology (eg, microarray profiling)[6,7,8] may improve (lower) the cost:benefit ratio.

MOLECULAR ONCOLOGY *(Continued)*

Footnotes

1. Fitzgerald PJ, *From Demons and Evil Spirits to Cancer Genes: The Development of Concepts Concerning the Causes of Cancer and Carcinogenesis,* Washington, DC: The American Registry of Pathology, 2000, 207-40.
2. Kimmelman A, Bafico A, and Aaronson SA, "Oncogenes and Signal Transduction," *The Molecular Basis of Cancer,* 2nd ed, Chapter 7, Mendelsohn J, Howley PM, Israel MA, et al, eds, Philadelphia, PA: WB Saunders Co, 2001, 115-33.
3. Soussi T, "The p53 Tumor Suppressor Gene: From Molecular Biology to Clinical Investigation," *Ann N Y Acad Sci,* 2000, 910:121-37.
4. Jarvinen HJ, Aarnio M, Mustonen H, et al, "Controlled 15-year Trial on Screening for Colorectal Cancer in Families With Hereditary Nonpolyposis Colorectal Cancer," *Gastroenterology,* 2000, 118(5):829-34.
5. Lerman C, Hughes C, Trock BJ, et al, "Genetic Testing in Families With Hereditary Nonpolyposis Colon Cancer," *JAMA,* 1999, 281(17):1618-22.
6. Rosewald A, Wright G, Chan W, et al, "The Use of Molecular Profiling to Predict Survival After Chemotherapy for Diffuse Large-B-Cell Lymphoma," *N Engl J Med,* 2002, 346(18):1937-47.
7. Su A, Welsh J, Sapinoso L, et al, "Molecular Classification of Human Carcinomas by Use of Gene Expression Signatures," *Cancer Res,* 2001, 7388-93.
8. Woeste S, "Molecular Profiling for Cancer," *Lab Med,* 2003, 771-4.

THERAPEUTIC DRUG MONITORING

Uttam Garg, PhD

David S. Jacobs, MD

Harold J. Grady, PhD

John Foxworth, PharmD

Charles W. Gorodetzky, MD, PhD

Contributors:
Eugene S. Olsowka, MD, PhD
Daniel H. Jacobs, MD

Optimal dosing of many drugs can be achieved by measurement of blood drug levels (therapeutic drug monitoring - TDM). This strategy requires that three prerequisites be satisfied.

1. Therapeutic and dose-related toxic effects are initiated through interaction of the drug with specific receptors on the cells of the target issue.

2. Therapeutic/toxic effects are proportional to drug concentration at the receptor site, which is represented by the free or unbound concentration at that site.

3. Concentration of free drug at the receptor site is directly proportional to the free drug concentration in the serum and, in most cases, to the total drug concentration.

Exceptions to the relationship concerning total drug concentration occur when the drug is highly protein bound. Changes in the fraction bound are produced by various physiological or pathological processes.

Although many drugs are safely and effectively administered without TDM, it is useful to monitor blood levels when one or more of the following conditions apply:

-   The drug has a narrow therapeutic range.
-   The drug exhibits large intra- or interindividual variation.
-   The drug does not produce the desired therapeutic effect or produces toxicity when empiric dosing is used.
-   Concurrent disease alters drug utilization.
-   Noncompliance is suspected.
-   Drug interactions may have taken place.
-   Medicolegal verification is desired.
-   Bioavailability of the drug is suspect.
-   The therapeutic or toxic effect cannot be easily determined by clinical observation.
-   There is a change in patient's physiology.

The availability of a rapid and accurate assay method is necessary for TDM. For the most part, current laboratory methods meet these requirements. Current methods usually employ **immunoassay techniques (IA)**. Enzyme immunoassays are most frequently used. Sensitivity of immunoassays has been tremendously increased by use of **fluorescent** and **chemiluminescent labels**. A few current methods require extraction and **high performance liquid chromatography (HPLC) or gas chromatography (GC)**. Methods must be validated with respect to accuracy, precision, specificity, limit of detection, linear dynamic range, and robustness. The concept of robustness refers to the method's stability in the presence of changes in components of the system, such as reagent lot. Also important is the practice of measuring and reporting values for active metabolites.

THERAPEUTIC DRUG MONITORING *(Continued)*

It is common practice to use TDM to monitor many of the drugs in the following classes:

- antibiotics
- antiasthmatics
- anticonvulsants
- antipsychotics

- antidepressants
- antiarrhythmics
- immunosuppressants
- analgesics

The serum drug concentration achieved is a consequence of a wide variety of processes. The study of the interrelationships of such processes and their consequences is **pharmacokinetics**. An appreciation of some pharmacokinetic principles helps in the interpretation of serum drug levels. The aim of TDM is to achieve a steady-state drug concentration within the therapeutic range. This is achieved by controlling a dosage regimen which consists of two parts, the **dose** (amount of drug given per dose) and the **dose interval** (how often this amount is given). When starting (or changing) a dosage regimen, it is necessary to wait a period of time to allow for establishment of a stable concentration (steady-state = C). To do so without the use of a loading dose usually requires about five to six half-lives ($t_{1/2}$). The **elimination half-life** is the time required for the serum level to change from one value to half that value. Increasing the dose rate will increase the steady-state concentration while increasing the dose interval will increase the **peak/trough ratio**. The **peak** value is the highest value obtained after a given dose, and the **trough** is the lowest value which usually occurs just before the next dose. Blood samples for TDM are normally drawn at either peak or trough or both. For proper interpretation of TDM values, the sampling time and the time of the last dose must be known and recorded on the order and the report. Without this information, appropriate reference ranges cannot be chosen. **Trough** samples (used for most drugs) are usually drawn just before the next dose. **Peak** samples are drawn at various times depending upon the route of administration or the formulation of the oral dose. For most drugs, for intravenous administration, peak samples are drawn 30-60 minutes after completion of the infusion. For intramuscular administration, the peak time is 2-4 hours after dosing. For oral dosing, 2-3 hours following administration can usually be used for peak sampling except for sustained-release products. Exceptions to the use of only trough samples are many drugs, including aminoglycoside antibiotics and theophylline, which are sampled at both peak and trough.

Any of the following factors may influence the serum level achieved after a given dosage regimen:

- patient age
- genetic variability
- disease processes
- compliance
- absorption
- distribution

- metabolism
- excretion
- drug tolerance
- toxicity
- use of multiple drugs

In the drug listings in this chapter, therapeutic ranges are given when TDM is applicable. Separate ranges apply to peak and trough values and are so labeled. The following pharmacokinetic parameters are listed when applicable:

- $T_{1/2}$ (elimination half-life)
- Vd (volume of distribution)

- PB (protein binding)

where Vd = dose/concentration

In the text of the individual monographs, special items concerning toxicity, sampling time, drug-drug interactions, and other helpful clinical and laboratory data are presented. Tables summarizing some of the above information for several classes of drugs can be found in this introduction *on page 60*. Three other important kinetic parameters can be calculated from the above: the **elimination rate constant** (K), the **clearance** (Cl), and the **dose** (D) required to produce a chosen **steady-state concentration** (C_{SS}).

- $K = 0.693/T_{1/2}$
- $Cl = K \times Vd$

- $Css = D/(Cl \times dosage\ interval)$

A number of drugs are highly protein bound (80% to 90%) by serum albumin and an alpha globulin. It is the "free", or unbound, concentration of the drug that is in equilibrium with intracellular free drug. It determines the pharmacological effect, the rate of liver metabolism, and the amount of parent drug presented to the kidney for excretion. Decreases in binding can occur because of competition for protein binding sites by bilirubin and metabolites accumulating secondary to renal insufficiency. Increase in free drug will result, but increased liver metabolism and/or renal excretion will reasonably soon bring concentrations back down to near the previous level. The total drug concentration will, however, decrease and may be deceptively low. Increasing the drug level by changing dosage regimen under such circumstances could cause toxicity. Analysis of free or unbound drug level would be the ideal solution, and is available in some laboratories for a number of drugs. Somewhat labor-intensive and expensive, it is not a common practice. In situations not involving changes in protein binding, total drug concentration is useful and almost always proportional to the free value. Certain classes of drugs may require specialized knowledge concerning drug interactions and level of efficacy of the individual members of that group, as illustrated by the following discussion of antiepileptics. Antiepileptic drugs (AED) are titrated clinically to prevent seizures and other adverse effects. Some patients require drug levels outside the classical therapeutic range. Thus, some dosing may be empiric, rather than based on the AED level. Empiric dosage may be based on therapeutic response rather than patient weight. With drugs of this class, changes in serum levels may not always be related to efficacy or toxicity. For example, phenytoin and carbamazepine may mutually lower the level of the other drug without decrease in efficacy.

It is also true that undesirable side effects can occur with drug levels in the therapeutic range. Both desktop and large automated laboratory systems are available for TDM analysis; essentially all provide acceptable accuracy and precision.

References

Bates DW, "Improving the Use of Therapeutic Drug Monitoring," *Ther Drug Monit*, 1998, 20(5):550-5.

Bowers L, Shaw L, et al, "Therapeutic Drug Monitoring Conference," *Clin Chem*, 1998, 44:369-436.

Burtis C and Ashwood E, *Tietz Textbook of Clinical Chemistry*, 3rd ed, Philadelphia, PA: WB Saunders Co, 1999.

Hardman JG and Limbird LE, eds, *Goodman and Gilman's, The Pharmacological Basis of Therapeutics*, 10th ed, New York, NY: McGraw-Hill, 2001.

Kaplan L, Pesce A, and Kazmierczak S, *Clinical Chemistry: Theory, Analysis, and Correlation*, 4th ed, St Louis, MO: Mosby, 2003.

Lacy CF, Armstrong LL, Goldman MP, et al, *Drug Information Handbook*, 12th ed, Hudson, OH: Lexi-Comp Inc, 2004.

Soldin SJ and Steele BW, "Mini Review: Therapeutic Drug Monitoring in Pediatrics," *Clin Biochem*, 2000, 33:333-5.

Young DS, *Effects of Drugs on Clinical Laboratory Tests*, 5th ed, Volume 1: Listing by Test, Washington, DC: AACC Press, American Association of Clinical Chemistry, 2000.

THERAPEUTIC DRUG MONITORING *(Continued)*

Table A. Class of Drug — Anticonvulsants

Drug	Therapeutic Range	Toxic Level	Half-Life	Time to Sample	Protein Binding (%)	Active Metabolites	Route of Excretion	Influence on Hepatic Drug Metabolizing Enzymes
Carbamazepine	8-12 μg/mL	>12 μg/mL	15-40 h	3-8 d	60-80	10,11-N-epoxide	Hepatic	Inducer
Clonazepam	10-50 ng/mL	>100 μg/mL	20-60 h	5-10 d	80-90	7-Amino		None
Ethosuximide	40-100 μg/mL	>150 μg/mL	25-70 h	10-13 d	0-5		Hepatic	
Felbamate	20-100 μg/mL		20-23 h	5-14 d	20-25			Inducer and inhibitor
Gabapentin	1-2 μg/mL		5-7 h		<3			None
Lamotrigine	2-4 μg/mL		24-30 h		50-60			None
Mephenytoin	25-40 μg/mL	>50 μg/mL	8 h	2 d	20-50	5-Ethyl, 5-phenyl-hydantoin		
Phenobarbital	20-40μg/mL	>40 μg/mL	50-140 h	20 d	40-50		Hepatic	Inducer
Phenytoin	10-20 μg/mL	>25 μg/mL	20-40 h		85-95			Inducer
Primidone	5-12 μg/mL	>12 μg/mL	4-12 h	5 d	0-20	Phenobarbital		Inducer
Valproic acid	50-100 μg/mL	>200 μg/mL	8-15 h	4 d	85-95		Renal	Inhibitor

See also the table comparing anticonvulsants in the listing, Antiepileptic Drugs, New, Overview *on page*

Table B. Class of Drug — Antibiotics

Drug	Therapeutic Range*	Toxic Level	Half-Life	Time to Sample (after starting)	Protein Binding %	Route of Excretion
Amikacin	P: 15-30 µg/mL†; T: 5-8 µg/mL	T: >8 µg/mL	2-3 h	15 h	4	Renal
Chloramphenicol	P: 10-25 µg/mL; T: <5 µg/mL	P: >25 µg/mL	1.5-5 h‡	10-15 h‡	50-60	Renal§
Gentamicin	P: 4-10 µg/mL†; T: <2 µg/mL	T: >2 µg/mL	2-3 h	15 h	10	Renal
Tobramycin	P: 4-10 µg/mL†; T: <2 µg/mL	T: >2 µg/mL	2-3 h	15 h	10	Renal
Vancomycin¶	P: 20-40 µg/mL; T: 5-10 µg/mL	P: >80 µg/mL	4 h	24 h	55	Renal

P = peak, T = trough.

*Dependent upon site of infection and individual MIC of drug. See individual drugs.

†Higher peak levels will be attained with once daily dosing.

‡Varies substantially with age.

§Hepatic inactivation very important.

¶Routine monitoring of serum vancomycin levels is necessary.

THERAPEUTIC DRUG MONITORING (Continued)

Table C. Class of Drug — Cardiac Drugs

Drug	Therapeutic Range	Toxic Level	Half-Life	Time to Sample	Protein Binding %	Active Metabolites	Route of Excretion	Major Drug Interactions
Digitoxin	18-35 ng/mL	>35 ng/mL	150-250 h		90-95	Digoxin	Hepatic	
Digoxin	0.8-2.0 ng/mL	>2.5 ng/mL	20-60 h	5 d	20-25		Renal	Quinidine
Disopyramide	2.8-3.2 µg/mL	>7.0 µg/mL	4-10 h	30 h	30-70	N-desisopropyl disopyramide	Renal	
Lidocaine	1.5-5.0 µg/mL	>6.0µg/mL	1.5-2 h	12 h	60-80	MEGX	Hepatic	Phenobarbital
Procainamide	4-10 µg/mL	>12 µg/mL	2-6 h	20 h	10-20	N-acetylprocainamide	Renal	
Propranolol	50-100 ng/mL	>1000 ng/mL	4-6 h	30 h	90-95	4-Hydroxy-propranolol	Hepatic	
Quinidine	2-5 µg/mL	>8 µg/mL	6-8 h	35 h	70-90	3-Hydroxy-quinidine	Hepatic	Digitalis

TOXICOLOGY / DRUGS OF ABUSE

Uttam Garg, PhD

David S. Jacobs, MD

Eugene S. Olsowka, MD, PhD

Harold J. Grady, PhD

Leland B. Baskin, MD

Jasbir Singh, PhD

Contributor: Wayne R. DeMott, MD

Toxicology

Modern toxicology is employed to resolve environmental, clinical, and forensic problems. Most medical laboratory toxicology concerns the identification of substances involved in acute or chronic poisoning of man. The laboratory is frequently asked to perform a comprehensive drug screen, often when little or no clue from the history and/or physical examination is available concerning the toxin. The term "comprehensive" is a relative one, since no hospital laboratory can truly screen for all possible toxic substances. Two widely used systems, a thin-layer chromatographic method called Toxi-Lab®, and an automated high performance liquid chromatographic instrument called Remedi®, can screen for several hundred drugs. In recent years, mass-spectrometry (gas and liquid) is also being widely used. Most often, they are supplemented with immunoassays for drugs of abuse, acetaminophen, salicylate, and tricyclic antidepressants. Quantitative measurement, usually carried out on serum/plasma, is most often used for the following drugs and toxins:

- acetaminophen
- salicylate
- carboxyhemoglobin (whole blood)
- digoxin (usually from TDM)
- ethanol, methanol, isopropanol, ethylene glycol
- lithium
- methemoglobin
- iron
- theophylline (usually from TDM)
- phenytoin (usually from TDM)

Of the other anticonvulsants, carbamazepine, pentobarbital, primidone, phenobarbital, and valproic acid are most commonly monitored, although others may also be measured. Toxicology laboratories should consider provision of cholinesterase assays, due to increased concern of chemical attack.

The laboratory can be most helpful when it has all the information available concerning possible substances involved. It should be noted that in overdose situations, the usual values for elimination rates (half-lives) may not apply, because when enzyme systems become saturated typical first-order kinetics are no longer valid.

Drugs of Abuse

The drugs or drug classes most commonly listed as drugs of abuse include:

- amphetamine/methamphetamine
- barbiturates
- benzodiazepines
- cannabinoids (marijuana or THC)
- cocaine
- opiates (heroin, morphine, codeine)
- phencyclidine (PCP)
- methadone
- methaqualone
- propoxyphene

TOXICOLOGY / DRUGS OF ABUSE *(Continued)*

These drugs are measured under one of two circumstances. One is the overdose situation in which the analysis is treated as any toxicological sample. The second involves testing for the presence of a drug in clinically well persons. Most of the following discussion applies to the latter situation. Analysis for such drugs in clinically-well subjects frequently involves two sequential tests, the first a screening procedure, and the second a confirmatory test performed only on positive screens. The screening test must have good sensitivity but may lack some specificity, while the confirmatory test must be both sensitive and specific and involve a different chemical principle. When used strictly for clinical purposes, the screening test result may be used without confirmation if an occasional false-positive will do no serious harm. However, when medicolegal or forensic application is a possibility, confirmation is essential. It is also extremely important for forensic applications that the sample be accompanied by a chain-of-custody document which will assure the integrity of the sample through the process of collection, delivery, receipt, and analysis (see the Toxicology/ Drugs of Abuse Introduction *on page 76* for a Chain-of-Custody form). Most often, samples without such a document lack forensic value, regardless of the quality of the analysis. The sample for analysis of drugs of abuse is urine, because of the ease of collection and because concentrations of drugs and metabolites are usually higher than in serum/plasma or saliva. Laboratory reports for detection of drugs of abuse in well persons are generally not quantitative and are usually expressed as "positive" or "negative." For each drug, a predetermined threshold or cutoff value has been agreed upon by scientific and regulatory groups. Results equal to the cutoff or above are considered positive. Thus, a report of "negative" does not necessarily mean absence of the drug, but rather a result less than the cutoff. Cutoff values for a given drug are often different for the screening evaluation than for the confirmatory test.

A federal agency is responsible for regulation of laboratories involved in analysis of drugs of abuse for federal employees and for companies operating under federal control. It is the Substance Abuse and Mental Health Services Administration (SAMHSA). This agency has set up strict guidelines for sample collection, handling, measurement, and reporting of drugs of abuse in urine. This aspect of federal control of drugs-of-abuse testing was formerly under the National Institute for Drug Abuse (NIDA), which continues to perform other functions in this area. SAMHSA certifies laboratories following a rigorous proficiency testing and inspection procedure. Only SAMHSA-certified laboratories may perform drugs-of-abuse testing for federal agencies or for firms contracting with federal agencies. Only five drugs are on the SAMHSA panel: amphetamine/methamphetamine, cannabinoids, cocaine, opiates, and phencyclidine. Other drugs may be measured by SAMHSA-certified laboratories but are not part of the certification. The College of American Pathologists (CAP), under CAP-Forensic Urine Drug Testing program, also accredits laboratories for toxicology and drugs-of-abuse testing, using similar proficiency testing and inspection procedures. A majority of the laboratories screening for drugs of abuse employ immunoassay methodology, which has adequate sensitivity and, in most cases, reasonable specificity. The amphetamine/methamphetamine class produces the most problems with false-positives, caused by interference from over-the-counter antiallergy and anticold medications. They, of course, give negative confirmatory tests. Confirmatory testing is frequently done by gas chromatography/mass spectrometry (GC/MS), which is clearly the method of choice and is considered the "gold standard."

Urine collection procedures for drugs-of-abuse testing should incorporate certain checks and precautions to preserve sample integrity. The collection room should not have warm water available, and the stool water should be colored with a dye. If the specimen is to be used for forensic purposes, witnessed voiding is preferred. The temperature within 4 minutes of collection should be between 90°F and 99°F. Later measurement of pH should be between 5 and 9, creatinine ≥ 5 mg/dL, and the specific gravity (refractometer) ≥ 1.003. Some laboratories also test for adulterants such as nitrates, chromium, bleach, etc. Any unusual colors, odors, or physical appearance should be noted.

References

Baselt RC, *Disposition of Toxic Drugs and Chemicals in Man*, 6th ed, Foster City, CA: Biomedical Publications, 2002.

Burtis CA and Ashwood ER, *Tietz Textbook of Clinical Chemistry*, 3rd ed, Philadelphia, PA: WB Saunders Co, 1999.

Ellenhorn K, Schonwald S, et al, *Ellenhorn's Medical Toxicology*, 2nd ed, Baltimore, MD: Williams & Wilkins, 1997.

Kaplan LA and Pesce AJ, *Clinical Chemistry: Theory Analysis and Correlation*, 3rd ed, St. Louis, MO: Mosby, 1996.

Klaasen CD, *Casarette and Doull's Toxicology*, 5th ed, New York, NY: McGraw Hill, 1996.

Leikin J and Paloucek F, *Poisoning and Toxicology Handbook*, Hudson, OH: Lexi-Comp Inc, 2002.

Sipes G, McQueen C, and Gandolfi A, *Comprehensive Toxicology*, 7 volumes, New York, NY: Elsevier Science Inc, 1997.

Wu AH, McKay C, Broussard LA, et al, "National Academy of Clinical Biochemistry Laboratory Medicine Practice Guidelines: Recommendations for the Use of Laboratory Tests to Support Poisoned Patients Who Present to the Emergency Department," *Clin Chem*, 2003, 49(3):357-79.

Web Sites

www.aapcc.org
www.health.org
www.samhsa.gov
www.soft-tox.org

TOXICOLOGY / DRUGS OF ABUSE *(Continued)*

American Association of Poison Control Centers
U.S. Poison Control Center Members
Updated September 2003
(certified centers are written in *italics*)

ALABAMA

Alabama Poison Center
2503 Phoenix Drive
Tuscaloosa, AL 35405
Emergency Phone: (800) 222-1222

Regional Poison Control Center
Children's Hospital
1600 7th Avenue South
Birmingham, AL 35233
Emergency Phone: (800) 222-1222

ALASKA

Oregon Poison Center
Oregon Health Sciences University
3181 SW Sam Jackson Park Road, CB550
Portland, OR 97201
Emergency Phone: (800) 222-1222

ARIZONA

Arizona Poison & Drug Info Center
Arizona Health Sciences Center, Room 1156
1501 North Campbell Avenue
Tucson, AZ 85724
Emergency Phone: (800) 222-1222

Banner Poison Control Center
Good Samaritan Regional Medical Center
1111 East McDowell
Phoenix, AZ 85006
Emergency Phone: (800) 222-1222

ARKANSAS

Arkansas Poison & Drug Information Center
College of Pharmacy, University of Arkansas for Medical Sciences
4301 West Markham, Mail Slot 522-2
Little Rock, AR 72205
Emergency Phone: (800) 222-1222
TDD/TTY: (800) 641-3805

CALIFORNIA

California Poison Control System – Fresno/Madera Division
Children's Hospital Central California
9300 Valley Children's Place, MB 15
Madera, CA 93638-8762
Emergency Phone: (800) 222-1222
TDD/TTY: (800) 972-3323

California Poison Control System – Sacramento Division
UC Davis Medical Center
2315 Stockton Boulevard
Sacramento, CA 95817
Emergency Phone: (800) 222-1222
TDD/TTY: (800) 972-3323

California Poison Control System – San Francisco Division
UCSF Box 1369
San Francisco, CA 94143-1369
Emergency Phone: (800) 222-1222
TDD/TTY: (800) 972-3323

California Poison Control System – San Diego Division
University of California, San Diego, Medical Center
200 West Arbor Drive
San Diego, CA 92103-8925
Emergency Phone: (800) 222-1222
TDD/TTY: (800) 972-3323

COLORADO

Rocky Mountain Poison & Drug Center
777 Bannock Street
Mail Code 0180
Denver, CO 80204-4028
Emergency Phone: (800) 222-1222
TDD/TTY: (303) 739-1127

CONNECTICUT

Connecticut Poison Control Center
University of Connecticut Health Center
263 Farmington Avenue
Farmington, CT 06030-5365
Emergency Phone: (800) 222-1222

DELAWARE

The Poison Control Center
Children's Hospital of Philadelphia
34th & Civic Center Blvd
Philadelphia, PA 19104-4303
Emergency Phone: (800) 222-1222
TDD/TTY: (215) 590-8789

DISTRICT OF COLUMBIA

National Capital Poison Center
3201 New Mexico Avenue, NW, Suite 310
Washington, DC 20016
Emergency Phone: (800) 222-1222
TDD/TTY: (800) 222-1222

FLORIDA

Florida Poison Information Center – Jacksonville
655 West Eighth Street
Jacksonville, FL 32209
Emergency Phone: (800) 222-1222
TDD/TTY: (800) 222-1222; (800) 282-3171 (FL only)

Florida Poison Information Center – Miami
University of Miami, Dept of Pediatrics
PO Box 016960 (R-131)
Miami, FL 33101
Emergency Phone: (800) 222-1222

Florida Poison Information Center – Tampa
Tampa General Hospital
PO Box 1289
Tampa, FL 33601
Emergency Phone: (800) 222-1222

TOXICOLOGY / DRUGS OF ABUSE *(Continued)*

GEORGIA

Georgia Poison Center
Hughes Spalding Children's Hospital
Grady Health System
80 Jesse Hill Jr Drive, SE
PO Box 26066
Atlanta, GA 30335-3801
Emergency Phone: (800) 222-1222
TDD/TTY: (404) 616-9287 (TDD)

HAWAII

Rocky Mountain Poison & Drug Center
777 Bannock Street
Mail Code 0180
Denver, CO 80204-4028
Emergency Phone: (800) 222-1222
TDD/TTY: (303) 739-1127

IDAHO

Rocky Mountain Poison & Drug Center
777 Bannock Street
Mail Code 0180
Denver, CO 80204-4028
Emergency Phone: (800) 222-1222
TDD/TTY: (303) 739-1127

INDIANA

Indiana Poison Center
Methodist Hospital
Clarian Health Partners
I-65 at 21st Street
Indianapolis, IN 46206-1367
Emergency Phone: (800) 222-1222
TDD/TTY: (317) 962-2336 (TTY)

ILLINOIS

Illinois Poison Center
222 South Riverside Plaza, Suite 1900
Chicago, IL 60606
Emergency Phone: (800) 222-1222
TDD/TTY: (312) 906-6185

IOWA

Iowa Statewide Poison Control Center
St Luke's Regional Medical Center
2910 Hamilton Boulevard Lower A
Sioux City, IA 51104
Emergency Phone: (800) 222-1222

KANSAS

Mid-America Poison Control Center
University of Kansas Medical Center
3901 Rainbow Blvd, Room B-400
Kansas City, KS 66160-7231
Emergency Phone: (800) 222-1222
TDD/TTY: (913) 588-6639 (TDD)

KENTUCKY

Kentucky Regional Poison Center
Medical Towers South, Suite 847
234 East Gray Street
Louisville, KY 40202
Emergency Phone: (800) 222-1222

LOUISIANA

Louisiana Drug and Poison Information Center
University of Louisiana at Monroe
College of Pharmacy, Sugar Hall
Monroe, LA 71209-6430
Emergency Phone: (800) 222-1222

MAINE

Northern New England Poison Center
22 Bramhall Street
Portland, ME 04102
Emergency Phone: (800) 222-1222
TDD/TTY: (877) 299-4447 (ME only); (207) 871-2879

MARYLAND

Maryland Poison Center
University of MD at Baltimore
School of Pharmacy
20 North Pine Street, PH 772
Baltimore, MD 21201
Emergency Phone: (800) 222-1222
TDD/TTY: (410) 706-1858 (TDD)

National Capital Poison Center
3201 New Mexico Avenue, NW, Suite 310
Washington, DC 20016
Emergency Phone: (800) 222-1222
TDD/TTY: (800) 222-1222

MASSACHUSETTS

Regional Center for Poison Control and Prevention Serving
Massachusetts and Rhode Island
300 Longwood Avenue
Boston, MA 02115
Emergency Phone: (800) 222-1222
TDD/TTY: (888) 244-5313

MICHIGAN

Children's Hospital of Michigan Regional Poison Control Center
4160 John R. Harper Professional Office Bldg, Suite 616
Detroit, MI 48201
Emergency Phone: (800) 222-1222
TDD/TTY: (800) 356-3232 (TDD)

DeVos Children's Hospital Regional Poison Center
1300 Michigan, NE, Suite 203
Grand Rapids, MI 49503
Emergency Phone: (800) 222-1222
TDD/TTY: (800) 222-1222

TOXICOLOGY / DRUGS OF ABUSE *(Continued)*

MINNESOTA

Hennepin Regional Poison Center
Hennepin County Medical Center
701 Park Avenue
Minneapolis, MN 55415
Emergency Phone: (800) 222-1222
TDD/TTY: (800) 222-1222; (612) 904-4691 (TTY)

MISSISSIPPI

Mississippi Regional Poison Control Center
University of Mississippi Medical Center
2500 North State Street
Jackson, MS 39216
Emergency Phone: (800) 222-1222

MISSOURI

Missouri Regional Poison Center
7980 Clayton Rd, Suite 200
St Louis, MO 63117
Emergency Phone: (800) 222-1222
TDD/TTY: (314) 612-5705

MONTANA

Rocky Mountain Poison & Drug Center
777 Bannock Street
Mail Code 0180
Denver, CO 80204-4028
Emergency Phone: (800) 222-1222
TDD/TTY: (303) 739-1127

NEBRASKA

Nebraska Regional Poison Center
8200 Dodge Street
Omaha, NE 68114
Emergency Phone: (800) 222-1222

NEVADA

Oregon Poison Center
Oregon Health Sciences University
3181 SW Sam Jackson Park Road, CB550
Portland, OR 97201
Emergency Phone: (800) 222-1222

Rocky Mountain Poison & Drug Center
777 Bannock Street
Mail Code 0180
Denver, CO 80204-4028
Emergency Phone: (800) 222-1222
TDD/TTY: (303) 739-1127

NEW HAMPSHIRE

New Hampshire Poison Information Center
Dartmouth-Hitchcock Medical Center
One Medical Center Drive
Lebanon, NH 03756
Emergency Phone: (800) 222-1222

NEW JERSEY

New Jersey Poison Information and Education System
located at University of Medicine and Dentistry at New Jersey
65 Bergen Street
Newark, NJ 07107-3001
Emergency Phone: (800) 222-1222
TDD/TTY: (973) 926-8008

NEW MEXICO

New Mexico Poison & Drug Info Center
Health Science Center Library, Room 130
University of New Mexico
Albuquerque, NM 87131-1076
Emergency Phone: (800) 222-1222

NEW YORK

Central New York Poison Center
750 East Adams Street
Syracuse, NY 13210
Emergency Phone: (800) 222-1222

Finger Lakes Regional Poison & Drug Information Center
University of Rochester Medical Center
601 Elmwood Avenue, Box 321
Rochester, NY 14642
Emergency Phone: (800) 222-1222
TDD/TTY: (585) 273-3854 (TTY)

Long Island Regional Poison and Drug Information Center
Winthrop University Hospital
259 First Street
Mineola, NY 11501
Emergency Phone: (800) 222-1222
TDD/TTY: (516) 924-8811 (Suffolk); (516) 747-3323 (Nassau)

New York City Poison Control Center
NYC Bureau of Labs
455 First Avenue, Room 123, Box 81
New York, NY 10016
Emergency Phone: (800) 222-1222
TDD/TTY: (212) 689-9014 (TDD)

Western New York Poison Center
Children's Hospital of Buffalo
219 Bryant Street
Buffalo, NY 14222
Emergency Phone: (800) 222-1222

NORTH CAROLINA

Carolinas Poison Center
Carolinas Medical Center
5000 Airport Center Parkway, Suite B
Charlotte, NC 28208
Emergency Phone: (800) 222-1222

NORTH DAKOTA

Hennepin Regional Poison Center
Hennepin County Medical Center
701 Park Avenue
Minneapolis, MN 55415
Emergency Phone: (800) 222-1222
TDD/TTY: (800) 222-1222; (612) 904-4691 (TTY)

TOXICOLOGY / DRUGS OF ABUSE *(Continued)*

OHIO

Central Ohio Poison Center
700 Children's Drive, Room L032
Columbus, OH 43205
Emergency Phone: (800) 222-1222
TDD/TTY: (614) 228-2272 (TTY)

Cincinnati Drug & Poison Information Center
3333 Burnet Avenue
Vernon Place – 3rd Floor
Cincinnati, OH 45229-9004
Emergency Phone: (800) 222-1222
TDD/TTY: (800) 253-7955

Greater Cleveland Poison Control Center
11100 Euclid Avenue
Cleveland, OH 44106-6010
Emergency Phone: (800) 222-1222

OKLAHOMA

Oklahoma Poison Control Center
Children's Hospital at OU Medical Center
940 NE 13th Street, Room 3510
Oklahoma City, OK 73104
Emergency Phone: (800) 222-1222
TDD/TTY: (405) 271-1122

OREGON

Oregon Poison Center
Oregon Health Sciences University
3181 SW Sam Jackson Park Road, CB550
Portland, OR 97201
Emergency Phone: (800) 222-1222

PENNSYLVANIA

Pittsburgh Poison Center
Children's Hospital of Pittsburgh
3705 Fifth Avenue
Pittsburgh, PA 15213
Emergency Phone: (800) 222-1222

The Poison Control Center
Children's Hospital of Philadelphia
34th & Civic Center Blvd
Philadelphia, PA 19104-4303
Emergency Phone: (800) 222-1222
TDD/TTY: (215) 590-8789

PUERTO RICO

San Jorge Children's Hospital Poison Center
Calle San Jorge #252
Santurce, Puerto Rico 00912
Emergency Phone: (800) 222-1222

RHODE ISLAND

**Regional Center for Poison Control and Prevention
Serving Massachusetts and Rhode Island**
300 Longwood Avenue
Boston, MA 02115
Emergency Phone: (800) 222-1222
TDD/TTY: (888) 244-5313

SOUTH CAROLINA

Palmetto Poison Center
College of Pharmacy
University of South Carolina
Columbia, SC 29208
Emergency Phone: (800) 222-1222

SOUTH DAKOTA

Hennepin Regional Poison Center
Hennepin County Medical Center
701 Park Avenue
Minneapolis, MN 55415
Emergency Phone: (800) 222-1222
TDD/TTY: (800) 222-1222; (612) 904-4691 (TTY)

TENNESSEE

Middle Tennessee Poison Center
501 Oxford House
1161 21st Avenue South
Nashville, TN 37232-4632
Emergency Phone: (800) 222-1222
TDD/TTY: (615) 936-2047 (TDD)

Southern Poison Center
University of Tennessee
875 Monroe Avenue, Suite 104
Memphis, TN 38163
Emergency Phone: (800) 222-1222

TEXAS

Central Texas Poison Center
Scott and White Memorial Hospital
2401 South 31st Street
Temple, TX 76508
Emergency Phone: (800) 222-1222

North Texas Poison Center
at Parkland Memorial Hospital
5201 Harry Hines Blvd
Dallas, TX 75235
Emergency Phone: (800) 222-1222

Southeast Texas Poison Center
The University of Texas Medical Branch
3.112 Trauma Building
Galveston, TX 77555-1175
Emergency Phone: (800) 222-1222

South Texas Poison Center
The University of Texas Health Science Center – San Antonio
Department of Surgery, Mail Code 7849
7703 Floyd Curl Drive
San Antonio, TX 78229-3900
Emergency Phone: (800) 222-1222

Texas Panhandle Poison Center
1501 South Coulter
Amarillo, TX 79106
Emergency Phone: (800) 222-1222

West Texas Regional Poison Center
Thomason Hospital
4815 Alameda Avenue
El Paso, TX 79905
Emergency Phone: (800) 222-1222

TOXICOLOGY / DRUGS OF ABUSE *(Continued)*

UTAH

Utah Poison Control Center
585 Komas Drive, Suite 200
Salt Lake City, UT 84108
Emergency Phone: (800) 222-1222

VERMONT

Northern New England Poison Center
22 Bramhall Street
Portland, ME 04102
Emergency Phone: (800) 222-1222
TDD/TTY: (877) 299-4447 (ME only); (207) 871-2879

VIRGINIA

Blue Ridge Poison Center
Jefferson Park Place
1222 Jefferson Park Ave
Charlottesville, VA 22903
Emergency Phone: (800) 222-1222

National Capital Poison Center
3201 New Mexico Avenue, NW, Suite 310
Washington, DC 20016
Emergency Phone: (800) 222-1222
TDD/TTY: (800) 222-1222

Virginia Poison Center
Medical College of Virginia Hospitals
Virginia Commonwealth University Health System
PO Box 980522
Richmond, VA 23298-0522
Emergency Phone: (800) 222-1222

WASHINGTON

Washington Poison Center
155 NE 100th Street, Suite 400
Seattle, WA 98125-8011
Emergency Phone: (800) 222-1222
TDD/TTY: (206) 517-2394 (TDD); (800) 572-0638 (TDD WA only)

WEST VIRGINIA

West Virginia Poison Center
3110 MacCorkle Ave, SE
Charleston, WV 25304
Emergency Phone: (800) 222-1222
TDD/TTY: (304) 388-9698

WISCONSIN

Children's Hospital of Wisconsin Poison Center
PO Box 1997, Mail Station 677A
Milwaukee, WI 53201-1997
Emergency Phone: (800) 222-1222
TDD/TTY: (414) 266-2542

WYOMING

Nebraska Regional Poison Center
8200 Dodge Street
Omaha, NE 68114
Emergency Phone: (800) 222-1222

ANIMAL POISON CONTROL CENTER

ASPCA
Animal Poison Control Center
1717 South Philo Road, Suite 36
Urbana, IL 61802
Emergency Phone: (888) 426-4435

TOXICOLOGY / DRUGS OF ABUSE (Continued)

CHAIN-OF-CUSTODY FORM

External chain-of-custody forms vary significantly from laboratory to laboratory, depending on the legal requirements. A typical form covers the following aspects. The following form is a combination requisition and external chain-of-custody form. The strip at the bottom peels off and is used to seal the specimen. The number on the strip appears on all copies of the form and serves to positively associate the form with the sample.

ACCOUNT NAME AND ADDRESS	LABORATORY NAME	TOXICOLOGY
		LIS NO.
		ACC NO.

COLLECTOR / EMPLOYMENT REPRESENTATIVE IDENTIFICATION | **ADDRESS** | MEDICAL REVIEW OFFICER'S NAME & ADDRESS

MEDICATIONS WITHIN LAST 30 DAYS

REASON FOR TESTING

1 ☐ PRE-EMPLOYMENT
2 ☐ FOLLOW-UP
3 ☐ REASONABLE CAUSE
4 ☐ RETURN ON DUTY
5 ☐ POST-ACCIDENT
6 ☐ RANDOM
7 ☐ OTHER (SPECIFY)_____

PROFILES
DRUGS OF ABUSE SCREEN
DRUGS OF ABUSE SCREEN (with confirmation)
BLOOD ALCOHOL
URINE ALCOHOL
OTHER (specify):

TEMPERATURE (to be read within 4 minutes of collection)

TEMPERATURE WITHIN RANGE (90☐ to 100☐F / 32☐ to 38☐C)　☐ Yes　☐ No

TO BE COMPLETED BY COLLECTOR

NAME _____ COLLECTION FACILITY _____

I CERTIFY THAT THE SPECIMEN IDENTIFIED ON THIS FORM IS THE SPECIMEN PRESENTED TO ME BY THE DONOR SIGNING THIS FORM, AND THAT THE SPECIMEN BEARS AN IDENTIFICATION NUMBER IDENTICAL TO THE NUMBER BELOW, AND THAT IT HAS BEEN COLLECTED, LABELED, AND SEALED WITH THE SECURITY LABEL.

_____ _____ _____
SIGNATURE　　　　　　　　　　DATE / TIME　　　　　　PHONE

TO BE COMPLETED BY DONOR

I _____ HEREBY CONSENT TO HAVE A SPECIMEN OF MY URINE AND/OR BLOOD TAKEN, AND I UNDERSTAND THAT IT WILL BE USED FOR DRUG
PRINT NAME
ANALYSIS BY THE LABORATORY. THE RESULTS OF THE TESTS ON MY SPECIMEN WILL THEN BE MADE AVAILABLE TO THE ABOVE NAMED COMPANY/EMPLOYER FOR EMPLOYMENT EVALUATION ONLY.

IN ADDITION, I HEREBY ACKNOWLEDGE THAT THE SPECIMEN LABELED WITH THE IDENTIFICATION NUMBER BELOW IS MY OWN, AND THE SPECIMEN WAS LABELED AND SEALED IN MY PRESENCE.

IN ADDITION, I ACKNOWLEDGE THAT THE FOLLOWING PRESCRIPTION AND/OR NONPRESCRIPTION MEDICATIONS ARE THE ONLY MEDICATIONS I HAVE TAKEN WITHIN THE LAST 7 DAYS:

I HEREBY RELEASE ALL TESTING FACILITIES, THE ABOVE NAMED EMPLOYER/COMPANY, THEIR EMPLOYEES, AGENTS, AND REPRESENTATIVES FROM ANY AND AL LIABILITY ARISING FROM THE
RELEASE OF THE INFORMATION DISCOVERED FROM MY TEST.

_____ _____
SIGNATURE OF DONOR　　　　　　　　DATE / TIME

CHAIN OF CUSTODY: TO BE INITIATED BY THE COLLECTOR AND COMPLETED AS NECESSARY THEREAFTER

DATE MO. DAY YR.	SPECIMEN RELEASED BY	SPECIMEN RECEIVED BY	PURPOSE OF CHANGE
/ /	DONOR - NO SIGNATURE	Signature Name	PROVIDE SPECIMEN FOR TESTING
/ /	Signature Name	Signature Name	
/ /	Signature Name	Signature Name	
/ /	Signature Name	Signature Name	
/ /	Signature Name	Signature Name	
/ /	Signature Name	Signature Name	
/ /	Signature Name	Signature Name	

LAB USE ONLY

SEAL INTACT	YES	NO	VOLUME	SPECIMEN IS	☐ ACCEPTED ☐ NOT SUITABLE FOR ANALYSIS
COMMENTS					

Peel and place over CAP　　　Specimen ID Number: 123456

TRACE ELEMENTS

Eugene S. Olsowka, MD, PhD

Uttam Garg, PhD

Knowledge of trace elements in human toxicity, nutrition, and trace element-related disease states has recently made substantial advances. Generally required in very small amounts, all trace elements are toxic if given in excessive quantities.

Many essential trace elements have specific binding proteins, and all bind nonspecifically to various serum proteins. Knowledge of such binding characteristics is often essential to properly interpret trace element analyses.

Blood Collection Methods for Trace Elements

Since at least 1971, reports have appeared detailing trace metal contamination or alteration of blood and serum samples by blood collection needles, syringes, and vacuum tubes. Problems have included the leaching of chromium or manganese from metal needles; the contamination of the sample by the glass or the rubber parts of syringes or by the rubber stopper; or the adsorption with time of selenium or lead onto ordinary glass blood tubes, leading to falsely low levels for these elements.

Because of such problems, the gold-standard methods that evolved are:

- draw the sample through a plastic catheter preplaced in the vein
- use a syringe (acid-leached, all plastic) that allowed centrifugation in the syringe, or transfer of the blood to a plastic centrifuge tube, and
- transfer or store serum or blood sample in a special acid-leached plastic vial

Although these methods provide reliable results, they are cumbersome for clinical practice. Due to availability of "trace metal" certified tubes, syringes, needles, and other accessories, alternate methods have been developed.

Specifically, if using a vacuum device for drawing blood, use trace metal-free certified blood collection tubes. Powder-free gloves should be worn during blood collection. If several tubes of blood are to be drawn together, draw all trace metal tubes first so as not to contaminate the needle by puncture of the ordinary rubber stoppers, as they are heavily contaminated by several trace metals.

Not reliable are ordinary "red top" clot tubes or the use of needles with metal hubs or syringes with rubber plungers. Directions for obtaining blood samples are briefly summarized for each specific test.

Acknowledgment is given for helpful advice from Phillip H. Stoltenberg, MD, for review of the entries for serum copper and urine copper; and to H. Ray Adams, PhD[1] and Edward D. Harris, PhD,[2] for review of the entire chapter in a prior edition. We are grateful to Dr Glen Willie, the author of earlier editions, for his superlative efforts.

Footnotes

1. H. Ray Adams, PhD, Chief of General Chemistry and Toxicology, Department of Pathology, Texas A&M University Health Science Center College of Medicine, Scott & White Clinic, and Memorial Hospital, Temple, Texas.
2. Edward D. Harris, PhD, Professor of Biochemistry and Biophysics, Texas A&M University, Bryan-College Station, Texas.

References

Jackson MJ, "Diagnosis and Detection of Deficiencies of Micronutrients: Minerals," *Br Med Bull*, 1999, 55(3):634-42.

Moyer TP, Mussmann GV, and Nixon DE, "Blood-Collection Device for Trace and Ultra-Trace Metal Specimens Evaluated," *Clin Chem*, 1991, 37:709-14.

Subramanian KS, "Storage and Preservation of Blood and Urine for Trace Element Analysis," *Biol Trace Element Res*, 1995, 49:187-210.

Young VR, "Trace Element Biology: The Knowledge Base and Its Application for the Nutrition of Individuals and Populations," *J Nutr*, 2003, 133(5 Suppl 1):1581S-7S.

TRANSFUSION MEDICINE
(BLOOD BANK)

Geralyn Meny, MD

David S. Jacobs, MD

The close of the 20st century has witnessed a rush to eliminate transfusion-transmissible disease – a process which was spurred by the introduction of the human immunodeficiency virus (HIV) into the blood supply and which continues today. The following screening tests have been implemented as a means to reduce the risk of transfusion-transmitted disease.

Test	Year Introduced
Syphilis	~1939
Hepatitis B surface antigen	1971
Anti-CMV	1981
Anti-HIV	1985
Antihepatitis B core	1986
ALT[1]	1986
Anti-HTLV I	1989
Anti-HCV	1990
Anti-HIV 1/2	1992
Anti-HTLV I/II	1995
HIV p24 antigen[1]	1996
HIV-1 genomic nucleic acid test	1999
HCV genomic nucleic acid test	1999
West Nile virus genomic nucleic acid test (IND)[2]	2003

[1]No longer required.

[2]Investigational use, however, utilized by most blood collection organizations.

While reducing the risk of transfusion-transmitted infectious disease has been the focus of transfusion medicine over the past two decades, current risk from transfusion is largely due to noninfectious hazards. Transfusion of ABO incompatible blood is reported to occur in 1 out of every 38,000 red cell units and to cause 12-13 deaths per year in the United States. Factors associated with transfusion errors include lack of automation, insufficient staffing or patient monitoring, inadequate education, confusing labeling, poor communication, and complex processes used under urgent conditions. The transfusion safety officer, who identifies, solves, and monitors unsafe practices, and new technology, including wireless handheld portable digital assistants, advanced bar coding, radiofrequency identification, and imbedded chip technology are measures some institutions are implementing to further improve patient safety. (Dzik WH, Corwin H, Goodnough LT, et al, "Patient Safety and Blood Transfusion: New Solution," *Trans Med Rev*, 2003, 17(3):169-80.)

It is also important to remember the blood donor. Apheresis and blood donation must continue to be safe procedures. For example, donors should be monitored for 15-20 minutes postdonation for signs or symptoms of untoward effects before release. Physicians and surgeons, re-examining their usage of blood, have gradually come to realize that blood transfusion and the maintenance of arbitrary hemoglobin levels are not as important as they were once thought to be. Informed consent for transfusion should be obtained. It is desirable that patients understand why transfusions of blood and components may be necessary, the risks involved (in appropriate perspective), and ways of reducing the risks. A few states (California, New Jersey, and Pennsylvania) require by law informed consent from patients planning to undergo a medical procedure which may involve a transfusion.

Obviously, such information should be available to the patient well before the anticipated event. Many blood services now provide informative brochures about blood transfusion, its risks, and the alternatives. These materials can be placed in clinics and physicians' offices to provide a basis for informed consent for blood transfusion.

Preparation for Transfusion

1. **Sample Collection.** Ask the patient his or her name and check this against the information on the identification band. At the patient's bedside, label the sample tube with two unique forms of patient identification, such as the patient's full name and hospital number; also include date and identity of the collector. Other information may be locally required. The same identification must be placed on a requisition form. If the Blood Bank uses a special transfusion identification system, collect the sample and label it appropriately.

   If an identification plate is used, information on the patient's identification plate must agree fully with that on the wristband or the sample should not be drawn. Adjust procedures for emergencies or disasters with great care to avoid the increased likelihood of mix-ups which are inherent in such situations.

   - The patient's identification band should remain attached to an accessible part of the patient's body and should not be removed except by established procedure. If the band has been removed for an unauthorized reason, the identification procedure must be repeated and another sample obtained.
   - Any identification discrepancies or problems must be corrected promptly.

   Clerical error is the most common cause of blood incompatibility and transfusion reactions which follow. Full and correct labeling **at the bedside** is essential for all samples of blood drawn for the Blood Bank. The Blood Bank must not be permitted to accept unlabeled tubes or those with unattached or loose identification, no matter who presents them. Specimen identification requirements are outlined in a listing bearing that name in the specimen collection chapter.

2. **Issuing Blood.** Unless there is some good reason to do otherwise, the oldest unit crossmatched for a patient is dispensed first. But if a patient has autologous and/or directed as well as allogeneic (homologous) units, issue them in this order: first autologous, then directed, and last allogeneic.

   The selected unit is removed from the refrigerator. It is carefully inspected for clots, hemolysis, or discoloration. Any unit of abnormal appearance is quarantined.

   Identifying information on a transfusion request form and blood bag tag should be matched according to established procedures prior to issuing blood. Information may include the following:

   - patient's name, hospital number, other identifying data
   - patient's ABO and Rh type
   - donor blood ABO and Rh type
   - donor number, expiration date, component being issued
   - crossmatch result (if applicable)
   - result of antibody screen

   Any discrepancies must be resolved before blood is released. Date and time of issue must be documented.

   With some exceptions (emergencies, major surgery), only one unit of blood at a time should be issued for a patient receiving elective transfusions. Otherwise, the second unit may remain at room temperature for an extended period.

TRANSFUSION MEDICINE (BLOOD BANK) *(Continued)*

Personnel should never be allowed to pick up two units for two or more different patients at the same time.

Administration of Blood and Blood Components

1. Except in urgent situations, avoid starting transfusions at night for three reasons. First, hospital staffing is thinner at night, making it harder to keep a close watch on patients being transfused. Second, patients need sleep. Third, any untoward reaction will awaken the patient and create turmoil on the ward.

2. If possible, before picking up the blood from the Blood Bank, an intravenous infusion of isotonic saline using a Y-type blood-recipient set having a standard clot filter should be started. A needle of 19-gauge or bigger should be used. Smaller needles and special sets may be necessary for infants and children. The saline should be run at a slow drip. With an already existing I.V., the needle gauge and the I.V. site should be checked. The I.V. tubing should be changed, if necessary. If solutions other than isotonic saline have been running, a saline flush is indicated.

    Do not use any other solution or medications which may cause decreased survival of stored blood. Ringer's, for example, may cause formation of small clots that block the needle. Hypotonic solutions, particularly those containing dextrose, may cause lysis. For similar reasons of pharmacologic incompatibility, no medications should be added to stored blood or to tubing containing blood, unless it is flushed with saline before and after the medication.

    Check other I.V. medications that may have to be given and try to reschedule them before or after the blood transfusion. If necessary, other I.V.s can be given into another extremity.

3. The intended recipient must be correctly identified.

    *   All information on the transfusion form must be compared with that on the patient's identification band or bands, and on the donor blood unit. Make sure that everything conforms.

4. The blood unit must be inspected for hemolysis, discoloration, or other abnormal appearance. In case of anything out of the ordinary, the Blood Bank must be consulted before the transfusion is begun.

5. Vital signs, including temperature, must be recorded.

6. Do not attempt to vent plastic blood containers.

7. Once the blood unit is obtained from the Blood Bank, it should be mixed thoroughly by repeatedly inverting the bag, and started at once. The blood must be returned to the Blood Bank without delay if anything interferes with starting the transfusion.

8. **Do not store blood**, even temporarily, in conventional refrigerators on nursing stations, in operating rooms, emergency departments, or anywhere other than in blood storage units that are specially designated and continuously monitored. To do so violates all accreditation standards and federal regulations.

9. Hang the blood.
    *   Mix gently.
    *   Carefully insert the plastic cannula of the infusion set to avoid puncturing the wall of the bag.
    *   In the case of CPDA-1 red blood cells, lower the unit and allow 50-100 mL of saline to run into the bag. This allows easier and faster infusion. Additive unit red blood cells do not require this step.

10. Special blood filters may be indicated for some patients. See Filters for Blood *on page 588.*

11. The time the blood is started should be recorded.

12. Remain with and observe the patient for at least the first 5 minutes. Take vital signs again at 15 minutes and at the end of the transfusion. The patient should be checked frequently.

13. Infusion generally should not exceed 2 hours for one unit of RBCs. If there is danger of pulmonary edema or congestive heart failure, or if there is a possible immunologic problem or a history of reactions, the time may be extended to 4 hours. If circumstances require even slower transfusion, it may be better to ask the Blood Bank to divide the unit in half and provide it as two transfusions.

14. After the transfusion, dispose of the empty blood bags as per established procedure. If necessary, the residual small amounts of blood can be very useful in the event of reaction.

Release of Blood Set Up for Transfusion, Then Not Used

1. Blood issued for transfusion and then not used can be used for another patient as long as:
   - the unit has not been entered or punctured
   - the unit (red cells or whole blood) have not been allowed to warm above 10°C or cool below 1°C
   - records indicate that the blood has been reissued and that it has been inspected prior to reissue
   - At least one sealed segment of integral donor tubing has remained attached to the container

2. When blood has been set up for a patient but not used, most hospitals must limit the time it can be held for that patient. Otherwise, there is a likelihood the blood would be held indefinitely and would outdate, because someone forgot to notify the Blood Bank that the patient no longer needed a transfusion. The time limit may vary according to the needs of the institution, but is usually 24-48 hours, or 24 hours after an indicated surgical procedure.

   The foregoing policy should be automatic, but not absolute. Clinicians must have the option of requesting that units be held longer for specific clinical indications, provided those indications do not lead to increased adverse effects of transfusion. The availability of a validated computer crossmatch system allows better utilization of available inventory for patients who have no clinically significant antibodies detected by antibody screening and history review. For such patients, a postcrossmatch time limitation will become a moot point.

   Another exception may be when antigen-negative units have been obtained (often with great difficulty) for a patient with a particular blood group antibody. The likelihood of a continuing need will dictate that such units be reserved for the patient.

3. Patients receiving a series of transfusions are at particular risk of forming blood group antibodies that could cause future transfusion reactions. Because such antibodies can form quickly and unpredictably, standards require that a crossmatch sample be valid only for 3 days, after which a new one must be obtained. This is particularly important in patients who have either been pregnant or had transfusions within the past 3 months. The rule may be waived for those who have not had either such event, but the Blood Bank seldom is provided with such information.

   Both the foregoing rules (ie, the 1-day hold of blood and the 3-day hold of crossmatch samples) are best written into standard policies so as to form part of the operating routine. Requests for exceptions should be the responsibility of the clinical physician and should be referred to the Blood Bank physician for evaluation.

Transfusions in Trauma and Other Emergencies

Transfusions may be needed at once in an emergency, in which delay may result in loss of life. In such a case, somewhat different procedures are essential and some steps, such as the crossmatch, may have to be dispensed within the

TRANSFUSION MEDICINE (BLOOD BANK) *(Continued)*

interest of time. All abbreviated procedures increase the risk to the patient, but to varying degrees.

In this discussion, understand that "blood" usually means red blood cells (previously also known as "packed cells"). Whole blood may be the preferred transfusion medium in acute blood loss or in hypovolemic shock, but it is seldom presently available. Experience has shown that emergencies and massive surgical procedures can be effectively treated with red blood cells and appropriate plasma expanders.

The following procedures are suggested for different periods of time available.

1. **No sample, blood needed now.** Give type O Rh-negative red cells (packed cells) without delay and a blood sample requested. If O Rh-negative blood is in short supply, the physician should be informed and O Rh-positive red cells (packed cells) issued. Better a live, immunized patient than a dead one without antibodies. As soon as the patient's own blood type is known, type-specific units can be issued.

2. **Blood sample provided, blood needed in 5-10 minutes.** Give blood of the patient's own type without crossmatch. If any blood is needed before the typing is finished, use two units of O Rh-negative red cells (packed cells) immediately, with type-specific to follow.

   In both the foregoing situations, it is important for all persons involved to understand that **blood issued in this manner may turn out to be incompatible with the recipient.** The patient's physician must be notified as soon as possible if any test (eg, antibody screen or crossmatch) subsequently becomes positive or reactive.

3. **Blood sample provided, blood needed in 15-30 minutes.** It may be possible to complete the antibody screen (and crossmatches) within this time, depending on the technique used and number of units issued. If the antibody screen is completed and no clinically significant antibodies detected, chances are results on the crossmatches will be the same.

4. **Afterwards.** Complete all serologic testing, including the antibody screen and all crossmatches so that the patient's record will be complete. Documentation must be received on whose authority this departure from normal transfusion protocol was performed. Many hospitals require an Emergency Release form to document the original request.

5. **Additional guidelines.** Blood Bank and Emergency Department should develop policies and procedures together, so that surprises do not emerge during emergencies. Such topics include the importance of sample pickup, identification, and the principles of blood selection. The emergency department should be notified of times of blood shortage.

   A well-designed emergency identification system is vital, since the patient's identity may be unknown and since some of the usual precautions cannot be used. The system must also be applicable to the multiple trauma situation (imagine dealing with 4 or 5 people from an accident, all with the same surname) and to the mass casualty disaster.

   If red top tubes are used, remember that blood samples do not clot immediately, and there may be difficulty with fibrin shreds in incompletely clotted samples.

   The community's supply of O Rh-negative blood should not be exhausted on a single massive trauma case. It is reasonable to switch to Rh-positive early, particularly in the case of a male or female of nonchildbearing age, in whom the effects of immunization are more manageable. This, too, can be part of an understanding between the

Blood Bank and the Emergency Department. The Blood Bank physician should always make sure that such decisions are appropriately documented with the names of the deciding parties.

See listing, Uncrossmatched Blood, Emergency *on page 1281.*

Guidelines for Outpatient Transfusion

With some minor exceptions, mostly concerning the timing of the blood sample and crossmatch, outpatient transfusions are handled as are those for inpatients. The precautions are the same, and if the hospital clinic staff is not accustomed to dealing with transfusions, it must be trained to do so.

1. **Informed consent.** The patient should give informed consent. In 1995, the Joint Commission on Accreditation of Healthcare Organizations added the requirement for blood transfusion informed consent to its standards for hospital accreditation. Follow state or local guidelines as to the frequency of required informed consent.

2. **Pretransfusion testing.** Positive identification procedures must be followed for outpatients. There must be no doubt about the identity of the patient and the sample. Alternative methods for identification have been developed, including photographic identification such as a driver's license. If the patient has been pregnant or transfused with allogeneic red cells in the preceding 3 months, a sample shall be obtained within 3 days of the scheduled transfusion.

3. **Quantity to be transfused.** Since most of these patients have chronic conditions, and many are elderly, it may not be possible or desirable to infuse their units rapidly. For these reasons, it is seldom practical to give more than two units in 1 day.

4. **Observation.** The patient should be observed continuously for the first 15 minutes, frequently thereafter. After the transfusion is completed, the patient should remain under observation for at least 30 minutes. If the transfusion is performed in a clinic, it is better that a relative or friend be available to accompany the patient home.

   Staff should report any untoward reaction at once to the clinic physician or the patient's physician (see Transfusion Reaction Work-up *on page 1269* and Risks of Transfusion *on page 1166*). In such a case, the transfusion should be stopped and a saline drip allowed to continue. The Blood Bank should be notified promptly.

5. **Instructions.** It is potentially useful to have a set of printed instructions or an informative brochure to provide to patients.

6. **Follow-up.** The patient's physician should inform the Blood Bank of any delayed untoward reactions to the transfusion, or of any evidence of transfusion-associated infectious disease. It may be helpful to see that the physician gets a copy of the *Circular of Information for the Use of Human Blood and Blood Components*, prepared jointly by the American Association of Blood Banks, America's Blood Centers, and American Red Cross, which should be obtainable from the community blood center or hospital transfusion service.

Quality Assurance

Transfusion services and blood centers strive for quality improvement as a means to improve patient outcome. Physicians practicing in hospitals and using blood services are likely to find their transfusion practices monitored by an institutional Transfusion Review Committee. This is a requirement of the Joint Commission on Accreditation of Health Care Organizations (JCAHO). The commission requires detailed records of transfusion practices, including indications, single-unit transfusions, transfusion, transfusion reactions, and clinical effectiveness. Other accrediting agencies, such as the American Association of Blood Banks (AABB) and the College of American Pathologists (CAP), sponsor inspection and accreditation programs. Centers for Medicare and Medicaid Services (CMS), formerly HCFA, through its CLIA '88 regulation, requires laboratories

TRANSFUSION MEDICINE (BLOOD BANK) *(Continued)*

to follow a quality assurance program, which both CAP and AABB provide. The Food and Drug Administration (FDA), as noted in 21 CFR 211, requires an independent quality assurance unit with responsibility for the overall quality of the finished "product", so called. Development and use of quality and operational systems, guided by a quality program, is changing the approach from error detection to prevention, to improve the safety and adequacy of the blood supply.

Current concern over adverse effects, cost, and potential litigious climate of transfusion medicine all justify an important role for quality assurance procedures. The physician must not only strive to become educated about transfusion medicine and follow established transfusion criteria, but is urged to document in every patient's clinical records the rationale for using blood or its components in the clinically existing circumstances.

The authors and editors express appreciation to
two outstanding authors of this chapter in earlier editions,
Dr Douglas W. Huestis and Dr David C. Jenkins.

Summary Chart of Blood Components

Component	Major Indications	Action	Not Indicated for –	Special Precautions	Hazards	Rate of Infusion
Whole blood	Symptomatic anemia with large volume deficit	Restoration of oxygen-carrying capacity, restoration of blood volume	Condition responsive to specific component	Must be ABO-identical Labile coagulation factors deteriorate within 24 hours after collection	Infectious diseases; graft-vs-host disease; septic/toxic, allergic, febrile reactions; circulatory overload	For massive loss, fast as patient can tolerate
Red blood cells; red blood cells, adenine-saline added	Symptomatic anemia	Restoration of oxygen-carrying capacity	Pharmacologically treatable anemia Coagulation deficiency	Must be ABO-compatible	Infectious diseases; graft-vs-host disease; septic/toxic, allergic, febrile reactions	As patient can tolerate but less than 4 hours
Red blood cells, leukocytes removed	Symptomatic anemia, febrile reactions from leukocyte antibodies	Restoration of oxygen-carrying capacity	Pharmacologically treatable anemia Coagulation deficiency	Must be ABO-compatible	Infectious diseases; graft-vs-host disease; septic/toxic, allergic reaction (unless plasma also removed, eg, by washing)	As patient can tolerate but less than 4 hours
Fresh frozen plasma	Deficit of labile and stable plasma coagulation factors and TTP	Source of labile and nonlabile plasma factors	Condition responsive to volume replacement	Should be ABO-compatible	Infectious diseases; allergic reactions; circulatory overload	Less than 4 hours
Liquid plasma, plasma, and thawed plasma	Deficit of stable coagulation factors	Source of nonlabile factors	Deficit of labile coagulation factors or volume replacement	Should be ABO-compatible	Infectious diseases, allergic reactions	Less than 4 hours
Cryoprecipitated AHF	Hemophilia A,[1] von Willebrand disease,[1] hypofibrinogenemia, factor XIII deficiency	Provides factor VIII, fibrinogen, vWF, factor XIII	Conditions not deficient in contained factors	Frequent repeat doses may be necessary	Infectious diseases, allergic reactions	Less than 4 hours
Platelets; platelets, pheresis	Bleeding from thrombocytopenia or platelet function abnormality	Improves hemostasis	Plasma coagulation deficits and some conditions with rapid platelet destruction (eg, ITP)	Should not use microaggregate filters	Infectious diseases; graft-vs-host disease; septic/toxic, allergic, febrile reactions	Less than 4 hours
Granulocytes, pheresis	Neutropenia with infection	Provides granulocytes	Infection responsive to antibiotics	Must be ABO-compatible, do not use microaggregate filters	Infectious diseases; graft-vs-host disease; allergic, febrile reactions	One apheresis unit over over 2- to 4-hour period – closely observe for reactions

[1]Use when virus-inactivated concentrates are not available.

Modified from *Circular of Information for the Use of Human Blood and Blood Components*, American Association of Blood Banks, America's Blood Centers, and American Red Cross, 2002.

TRANSFUSION MEDICINE (BLOOD BANK) *(Continued)*

References

American Association of Blood Banks, *Standards for Hematopoietic Progenitor Cell Services*, 3rd ed, Bethesda, MD: American Association of Blood Banks Press, 2002.

AuBuchon JP, "Optimizing the Cost-Effectiveness of Quality Assurance in Transfusion Medicine," *Arch Pathol Lab Med*, 1999, 123(7):603-6.

Belanger AC, "Joint Commission on Accreditation of Healthcare Organizations' Expectations for Transfusion Medicine in Healthcare Organizations," *Arch Pathol Lab Med*, 1999, 123(6):472-4.

Blumberg N, "The Costs and Consequences of Management Fads and Politically Driven Regulatory Oversight: The Case of Blood Transfusion," *Arch Pathol Lab M ed*, 1999, 123(7):580-4.

Brecher ME, *Technical Manual*, 14th ed, Bethesda, MD: American Association of Blood Banks Press, 2002.

Carmel R and Shulman IA, "Blood Transfusion in Medically Treatable Chronic Anemia. Pernicious Anemia as a Model for Transfusion Overuse," *Arch Pathol Lab Med*, 1989, 113(9):995-7.

Code of Federal Regulations, Title 21 CFR Parts 200-299, 600-799, Washington, DC: U.S. Government Printing Office, 1998.

Fridey J, ed, *Standards for Blood Banks and Transfusion Services*, 22nd ed, Bethesda, MD: American Association of Blood Banks, 2003.

Joint Commission Resources, Inc, *Comprehensive Accreditation Manual for Hospitals: The Official Handbook (CAMH)*, Oakbrook Terrace, IL, 2004.

Goldman RL, "The Reliability of Peer Assessments of Quality of Care," *JAMA*, 1992, 267(7):958-60.

Goodnough LT and Audet AM, "Utilization Review for Red Cell Transfusion. Are We Just Going Through the Motions?" *Arch Pathol Lab Med*, 1996, 120(9):802-3.

Gustafson M, "Blood Safety: The Food and Drug Administration's Role," *Arch Pathol Lab Med*, 1999, 123(6):475-7.

Hamlin WB, "Requirements for Accreditation by the College of American Pathologists Laboratory Accreditation Program," *Arch Pathol Lab Med*, 1999, 123(6):465-7.

Hanson M, "The Ps and Qs of Quality Systems," *Arch Pathol Lab Med*, 1999, 123(7):576-9.

Lam HT, Schweitzer SO, Petz L, et al, "Are Retrospective Peer-Review Transfusion Monitoring Systems Effective in Reducing Red Blood Cell Utilization?" *Arch Pathol Lab Med*, 1996, 120(9):810-16.

Marconi M, Almini D, Pizzi N, et al, "Quality Assurance of Clinical Transfusion Practice by Implementation of the Privilege of Blood Prescriptions and Computerized Prospective Audit of Blood Requests," *Transfus Med*, 1996, 6(1):11-9.

McCurdy K and Gregory K, *Blood Bank Regulations: A to Z*, 2nd ed, Bethesda, MD: American Association of Blood Banks, 1999.

Otter J and Cooper ES, "What Do the Accreditation Organizations Expect? American Association of Blood Banks," *Arch Pathol Lab Med*, 1999, 123(6):468-71.

Sazama K, "College of American Pathologists Conference XXXIII on Transfusion Medicine Performance Improvement," *Arch Pathol Lab Med*, 1999, 123(8):680-1.

Sherman LA, "Outcomes in Transfusion," *Arch Pathol Lab Med*, 1999, 123(7):599-602.

Shulman IR, Saxena S, Ramer L, et al, "Assessing Blood Administering Practices," *Arch Pathol Lab Med*, 1999, 123(7):595-8.

Simon TL, Dzik WH, Snyder EL, et al, *Rossi's Principles of Transfusion Medicine*, 3rd ed, Philadelphia, PA: Lippincott Williams & Wilkins, 2002.

Smith DM and Otter J, "Performance Improvement in a Hospital Transfusion Service. The American Association of Blood Banks' Quality Systems Approach," *Arch Pathol Lab Med*, 1999, 123(7):585-91.

Toy P, "Guiding the Decision to Transfuse: Interventions That Do and Do Not Work," *Arch Pathol Lab Med*, 1999, 123(7):592-4.

Wanamaker V, "Health Care Financing Administration/Clinical Laboratory Improvement Amendments of 1988," *Arch Pathol Lab Med*, 1999, 123(6):4 78-81.

Web Sites

Hematopoietic Progenitor Cells

American Association of Blood Banks: www.aabb.org
Bone Marrow Donors Worldwide: www.bmdw.org
Cord Blood Donor Foundation: www.cordblooddonor.org
International Society for Hematotherapy and Graft Engineering: www.ishage.org
The Blood and Marrow Transplant Newsletter Web Site: www.bmtnews.org
The National Marrow Donor Program (NMDP): www.marrow.org
TransWeb: www.transweb.org

Recalls / Withdrawals / Safety Issues

Center for Biologics Evaluation and Research: www.fda.gov/cber
The Pennsylvania Society of Health-System Pharmacists: www.pshp.org

Transfusion Medicine (General)

American Association of Blood Banks: www.aabb.org
American Red Cross: www.redcross.org
America's Blood Centers: www.americasblood.org
International Council for Commonality in Blood Banking Automation, Inc: www.iccbba.com
National Institutes of Health: www.nih.org

URINALYSIS

David S. Jacobs, MD

Uri Alon, MD

Contributor: Wayne R. DeMott, MD

Analysis of urine dates to ancient times. Clinical microscopy has been practiced since the mid-17th century. Thus, many of the tests presented are among the most enduring in medicine. Other more recently developed tests (ie, certain reagent strip screening procedures) yield significant clinical information rapidly and relatively inexpensively.

Urine color has been of interest for centuries. Its color is determined by its concentration, the presence of drugs, exogenous and endogenous compounds, and its pH. **Colorless** urine may be normal or secondary to diuretic use, high fluid intake, diabetes insipidus, or diabetes mellitus. **Cloudy or hazy** urine may reflect the presence of phosphates, pyuria, bacteruria, chyluria, or radiographic dye. On oxidation, development of a **black** color is evidence for alkaptonuria.[1] Increased indican may cause the urine to blacken on standing. **Dark** urine is the second most common sign of acute intermittent porphyria; urine in porphyria has been described as port wine in color. Very rarely, dark urine may indicate the presence of malignant melanoma. **Green** urine may be produced by indigo carmine, methylene blue, phenol, and in some cases of iodochlorhydroxyquin (clioquinol)-induced subacute myelo-opticoneuropathy. Other causes of green urine are reported as *Pseudomonas* bacteremia, urinary bile pigments, amitriptyline hydrochloride or methocarbamol ingestion, and breath freshener abuse.[2] **Red** urine was described elegantly by Berman[3] and is described further in the listing, Blood, Urine. Chlorpromazine and haloperidol may cause pink, red, or red-brown discoloration. **Red to brown to tea** colored plasma and urine indicate hemoglobin or myoglobin; clear plasma with red urine may occur in congenital erythropoietic porphyria and cutanea tarda porphyria. **Red brown to brown black** urine may be caused by phenacetin, quinine, or methemoglobin, and **brown-black** urine can be caused by methemoglobin, homogentisic acid, or melanin. **Purple** urine, after standing, may be due to porphyrins.[4] The plastic urine bag may discolor **purple** in the presence of the indican produced by *Providencia* or *Klebsiella* species.[5] **Yellow** to **orange** urine may contain bile. Other causes of darker yellow to orange urine include increased concentration of urine or the presence of riboflavin, quinacrine (Atabrine®), rifampin (Rifadin®, Rimactane®), phenazopyridine (Pyridium®), or salicylazosulfapyridine (Azulfidine®). Color and appearance of urine are outlined well by Fuller, Threatte, and Henry.[6] As in some examples above, **drugs** may cause unusual urine colors.[7,8]

Urine volume is not *per se* a laboratory test, although it is measured in 24-hour or other timed urine collections. It is relevant for nephrolithiasis.

The editors express appreciation to Dr Glen Willie for his contributions to an earlier edition.

Footnotes

1. Gaines JJ, "The Pathology of Alkaptonuric Ochronosis," *Hum Pathol*, 1989, 20:40-6.
2. Norfleet RG, "Green Urine," *JAMA*, 1982, 247:29.
3. Berman LB, "When the Urine Is Red," *JAMA*, 1977, 237:2753-4.
4. Nolan CR, McKinney TD, and Forland M, "Urinalysis and Renal Function Tests," *Internal Medicine*, 5th ed, Chapter 102, Stein JH, ed, St Louis, MO: Mosby, 1998, 742-56.
5. Dealler SF, Belfield PW, Belford M, et al, "Purple Urine Bags," *J Urol*, 1989, 142:769-70.
6. Fuller CE, Threatte GA, and Henry JB, "Basic Examination of Urine," *Clinical Diagnosis and Management by Laboratory Methods*, 20th ed, Chapter 18, Philadelphia, PA: WB Saunders Co, 2001, 367-402.
7. Lubran MM, "Effect of Drugs on Clinical Laboratory Tests," *Clinical Pathology in the Elderly*, Chapter 24, Rochman H, ed, 1988, 193-6.
8. Young DS, *Effect of Drugs on Clinical Laboratory Tests*, 5th ed, Volume 1: Listing by Test, Washington, DC: AACC Press, 2000. 24, Rochman H, ed, 1988, 193-6.

ALPHABETICAL LISTING OF TESTS

- **A₁AT** *see* Alpha₁-Antitrypsin, Serum *on page 134*
- **βA4 Peptide** *see* Apolipoprotein E, Plasma *on page 204*
- **Aα Fragment** *see* Hypercoagulation Panel *on page 758*
- **Abciximab** *see* CD40 Ligand (Soluble), Serum or Plasma *on page 353*
- **ABC Proteins** *see* Inherited Diseases of Metabolism and Cell Structure *on page 792*
- **Abdominal Mass Aspiration** *see* Fine Needle Aspiration, Deep and Superficial Masses *on page 590*
- **A Beta₍₁₋₄₂₎** *see* Cerebrospinal Fluid and Plasma β-Amyloid₍₁₋₄₂₎ *on page 359*
- **ABGs** *see* Blood Gases and pH, Arterial *on page 275*
- **ABO Group and Rh Type** *see* Pretransfusion Testing *on page 1088*
- **Acalix®** *see* Diltiazem, Serum or Plasma *on page 515*
- **Accuracy of Test Results** *see page 11*
- **ACE** *see* Angiotensin Converting Enzyme, Serum *on page 159*
- **Acetaminophen** *see* Liver Disease: Laboratory Assessment, Overview *on page 869*
- **Acetaminophen Hepatotoxicity** *see* Aspartate Aminotransferase, Serum *on page 216*

Acetaminophen, Serum

Related Information
Alanine Aminotransferase, Serum *on page 116*
Aspartate Aminotransferase, Serum *on page 216*
Liver Disease: Laboratory Assessment, Overview *on page 869*

Synonyms Anacin-3®; Datril®; Paracetamol; Tempra®; Tylenol®; Ty-Pap; Uni-Ace®

Applies to Acetanilide; Phenacetin

Abstract Acetaminophen is an analgesic-antipyretic available singly or in combination with a number of other medications such as codeine, acetylsalicylic acid, and caffeine. It is the most frequently ingested medication in the U.S., and is frequently seen in the deliberate overdose situation. There is an intermediate metabolite, N-acetyl-p-benzoquinoneimine which is inactivated by cysteine and mercapturic acid conjugation. Glutathione is required for formation of cysteine and mercapturic acid conjugates. In overdose or prolonged use of acetaminophen, glutathione stores are depleted. Such depletion results in free N-acetyl-p-benzoquinoneimine causing liver damage. In addition to suicidal overdose, acetaminophen-related hepatic necrosis is recognized in a "therapeutic misadventure" scenario. The alcohol-acetaminophen syndrome, in which an alcoholic uses acetaminophen in doses exceeding those recommended (4 g/24 hours), is characterized by a direct toxic reaction. It may be the most common form of acute liver failure. The prognosis depends on the amount ingested and the time of presentation.

Specimen Serum

Container Red top tube

Sampling Time If ingestion time is known, the sample should be drawn 4 hours after ingestion.

Reference Interval Acetaminophen, serum: 10-30 µg/mL (SI: 66-169 µmol/L)

Critical Values Toxic: >150 µg/mL (SI: >990 µmol/L) (4 hours postingestion); >50 µg/mL (SI: >330 µmol/L) (12 hours postingestion). See nomogram.

Use Evaluation of possible toxicity or therapeutic monitoring

Methodology Immunoassay, gas chromatography (GC)

Additional Information
- Half-life: adults: 1-3 hours; neonates: 2-5 hours
- Volume of distribution: 0.95 L/kg
- Protein binding: 20% to 50%

Acetaminophen is an analgesic and antipyretic agent used for headache, fever, and relief of pain. Acetaminophen is the analgesic/antipyretic of choice in children 13 years of age or younger due to the association of aspirin with possible development of Reye syndrome. Children are less prone to acetaminophen toxicity due to smaller contribution of the P450 system.

Acetaminophen is rapidly absorbed from the GI tract. Peak plasma concentrations are reached in 30-60 minutes after a therapeutic dose. However, following overdose, the peak plasma concentration may not be reached for 4 hours or more.

In an acute overdose situation, the Rumack nomogram will help to determine necessity of acetylcysteine treatment and likelihood of hepatotoxicity. The first sample for drug level should not be drawn until at least 4 hours after ingestion. This allows for complete absorption. Using the Rumack nomogram, if the concentration of acetaminophen at 4 hours postingestion, is above the broken line, an entire course of the antidote, acetylcysteine (Mucomyst®), is necessary. The nomogram is not useful in chronic ingestion. Standard therapy of acetaminophen poisoning is a 72-hour protocol of oral N-acetylcysteine administration. Recently, it has been suggested that this protocol may be excessive for many episodes of acute acetaminophen poisoning. Hepatotoxicity from such poisoning is preventable with intravenous administration of N-acetylcysteine for 20 hours. Like oral therapy, intravenous treatment should be initiated within 8-10 hours of ingestion. Increased prothrombin time (PT) and transaminases may signal hepatic necrosis.

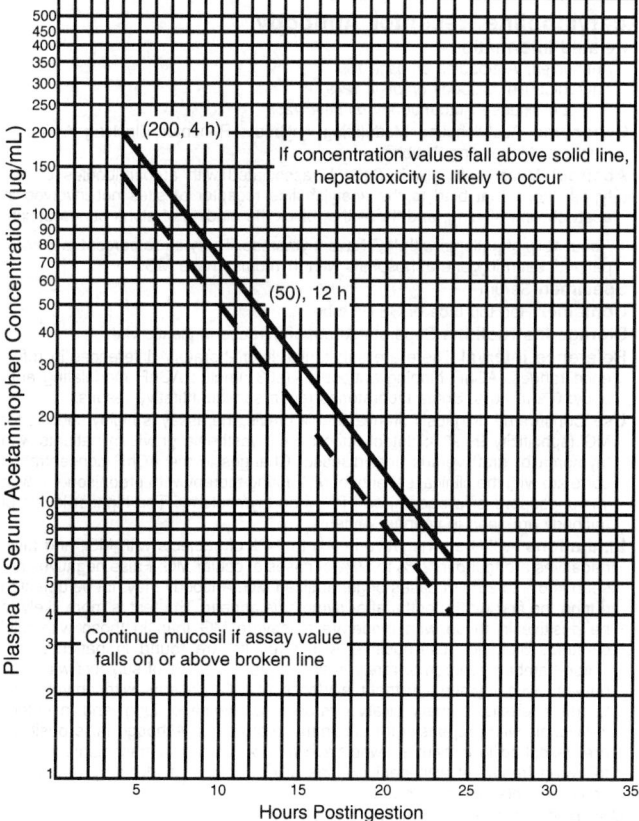

From Rumack BH and Matthews H, "Acetaminophen Poisoning and Toxicity," *Pediatrics,* 1975, 55:871-6, with permission.

References
Isbister G, Whyte I, and Dawson A, "Pediatric Acetaminophen Poisoning," *Arch Pediatr Adolesc Med,* 2001, 155(3):417-9.
Jones AL, "Mechanism of Action and Value of N-acetylcystine in the Treatment of Early and Late Acetaminophen Poisoning: A Critical Review," *J Toxicol Clin Toxicol,* 1998, 36(4):277-85.
(Continued)

Acetaminophen, Serum *(Continued)*

Rumack BH and Matthew H, "Acetaminophen Poisoning and Toxicity," *Pediatrics*, 1975, 55(6):871-6.

Salgia AD and Kosnik SD, "When Acetaminophen Use Becomes Toxic. Treating Acute Accidental and Intentional Overdose," *Postgrad Med*, 1999, 105(4):81-90.

Yip L and Dart RC, "A 20-Hour Treatment for Acute Acetaminophen Overdose," *N Engl J Med*, 2003, 348(24):2471-2.

♦ **Acetanilide** *see* Acetaminophen, Serum *on page 90*

♦ **Acetest**® *see* Ketones, Urine *on page 817*

♦ **Acetoacetate** *see* Ketone Bodies, Blood *on page 816*

♦ **Acetoacetic Acid, Urine** *see* Ketones, Urine *on page 817*

♦ **Acetone** *see* Ketone Bodies, Blood *on page 816*

♦ **Acetone** *see* Volatile Screen, Blood or Urine *on page 1320*

♦ **Acetone, Semiquantitative, Urine** *see* Ketones, Urine *on page 817*

♦ **Acetylcholine Modulating Antibody** *see* Acetylcholine Receptor Antibody *on page 92*

Acetylcholine Receptor Antibody

Related Information
Antinuclear Antibodies *on page 189*
Thyroglobulin Antibody *on page 1249*
Thyroperoxidase Autoantibody *on page 1253*

Synonyms Acetylcholine Modulating Antibody; Receptor Blocking Antibody; Receptor Modulating Antibody

Abstract Myasthenia gravis (MG) is associated with autoantibodies of two types: those that bind to the acetylcholine receptor at sites not involved in acetylcholine binding, and those that block the binding of alpha-bungarotoxin.

About 15% of patients with MG have thymomas. Thirty-three percent to 50% of patients seen in general hospitals with thymomas have MG.

Specimen Serum

Container Red top tube or SST™ tube

Storage Instructions Separate serum and freeze in plastic vial

Reference Interval There are substantial interlaboratory differences in reference ranges. AChR binding antibody: <0.03 nmol/L; AChR modulating antibody: 0% to 20%. Some laboratories report semiquantitative results.

Use Confirm the diagnosis of myasthenia gravis; the assay is highly specific for MG (specificity 99.9%); detect subclinical myasthenia gravis in patients with thymoma or graft versus host disease. Changes in the AChR concentration correlate with the clinical severity of MG during therapy with prednisone or with immunosuppressive agents, and following thymectomy. The highest levels of antibody are seen in younger patients.

Limitations AChR are negative in 7% to 34% of subjects with MG, and false negatives are found in 21% to 50% of cases of ocular MG. False negatives are seen in 6% to 25% of cases of generalized MG. Antibody may not be detected during the first 6-12 months after symptoms appear. The test is more likely to be positive in those with moderate to severe MG than in those with mild disease. Biologic false-positive AChR results are found in patients with Eaton-Lambert syndrome, rarely in first-degree relatives of subjects with MG, patients with thymoma without evidence of MG, patients with amyotrophic lateral sclerosis, primary biliary cirrhosis, carcinoma of lung, and in elderly individuals with propensity for autoimmune diseases. Although false positives are described in individuals who have had bone marrow transplantation and following treatment with penicillamine, clinical signs of MG may develop in some patients in those groups.

Contraindications Recent radioactive scan

Methodology The binding-antibody assay provides relatively high sensitivity and specificity. It is a radioimmunoassay (RIA) procedure. Tests done with human acetylcholine-receptor antigen are more sensitive than those performed with rat or ape antigen. Other assays include those for blocking antibody and tests for modulating antibody.

Additional Information The antibodies in patients with MG are believed to block acetylcholine-binding sites, damage the postsynaptic membrane, and accelerate receptor degradation. Other tests for MG are available.

References

Shimosato Y and Mukai K, "Tumors of the Mediastinum," *Atlas of Tumor Pathology*, Washington DC: Armed Forces Institute of Pathology, American Registry of Pathology, 1997.

Siao P and Zuckerberg LR, "A 69-Year Old Man With Myasthenia Gravis and a Mediastinal Mass," Case Records of the Massachusetts General Hospital, Case 15-2000, Scully RE, Mark EJ, McNeely WF, et al, eds, *N Engl J Med*, 2000, 342(20):1508-14.

Spickett G, *Oxford Handbook of Clinical Immunology*, New York, NY: Oxford Press, 1999.

AcetylcholinesteraseE, Amniotic Fluid

Related Information

AcetylcholinesteraseE, Red Cell and Serum *on page 93*

Alpha$_1$-Fetoprotein, Amniotic Fluid *on page 135*

Abstract AcetylcholinesteraseE (AChE) is an isoenzyme not present in normal amniotic fluid (AF).

Specimen Amniotic fluid

Sampling Time Amniotic fluid AChE is independent of gestational age.

Causes for Rejection Fetal blood in the specimen invalidates results; AChE is present in fetal blood.

Reference Interval Absent

Use AF AChE is one of the tests performed when screening tests (maternal serum alpha fetoprotein, chorionic gonadotropin, and unconjugated estriol) are abnormal and ultrasonography results do not resolve the issue. In this setting, AF AChE is used with amniotic fluid alpha-fetoprotein (AF AFP) to detect open neural tube and open ventral wall defects. AChE has high specificity. Its sensitivity for anencephaly is 97%, and for open spina bifida 99%. Visible AChE bands are found about 95% of the instances of gastroschisis but much less frequently with omphalocele.

AChE is not detected in cases of congenital nephrosis, a disease consistently associated with increased AF AFP. AF AChE and AFP both are associated with fetal death.

Limitations Both AFP and AChE are likely to be found in increased concentrations following substantial fetal hemorrhage into amniotic fluid.

Methodology Polyacrylamide gel electrophoresis followed by an inhibitor of AChE. While normal amniotic fluid includes a single cholinesterase, that from pregnancies with open neural tube and abdominal wall defects includes a more rapidly migrating second cholinesterase.

Additional Information The ratio between AChE and pseudocholinesterase densities permits differentiation between open neural tube and open ventral wall defects. Typically, AChE:pseudocholinesterase ratio is <0.10 in ventral wall defects and >0.15 in open neural tube defects.

Neural tissue is the source of amniotic fluid AChE until day 28 of fetal development. AChE persists in amniotic fluid until about the 11th week of fetal development. The presence of amniotic fluid AChE throughout pregnancy indicates that the neural tube is not closed.

References

Baskin LB, "Pregnancy and Prenatal Testing," *The Handbook of Clinical Pathology*, 2nd ed, Chapter 21, McKenna RW and Keffer JH, eds, Chicago, IL: American Society of Clinical Pathologists, 2000, 281-92.

Haddow JE and Palomaki GE, "Biochemical Markers of Fetal Disorders in Maternal Serum and Amniotic Fluid," *Medicine of the Fetus and Mother*, 2nd ed, Chapter 40, Reece EA and Hobbins JC, eds, Philadelphia, PA: Lippincott-Raven, 1999, 689-706.

Acetylcholinesterase, Red Cell and Serum

Related Information

AcetylcholinesteraseE, Amniotic Fluid *on page 93*

Dibucaine Number, Serum or Plasma *on page 510*

Organophosphate Pesticides, Urine, Blood, or Serum *on page 975*

Pseudocholinesterase, Serum *on page 1122*

Synonyms Cholinesterase, Erythrocytic; Cholinesterase I; Cholinesterase, True; Erythrocyte Cholinesterase

Applies to Carbamate Toxicity; Organophosphate Toxicity; Sarin Exposure; Succinylcholine Sensitivity

Abstract The **red cell** enzyme (**true cholinesterase**) is specific for the substrate acetylcholine. The **serum** enzyme (**pseudocholinesterase**) hydrolyzes other choline esters.

Specimen Red blood cells

(Continued)

Acetylcholinesterase, Red Cell and Serum *(Continued)*

Container Green top (heparin) tube or heparinized capillary tubes

Storage Instructions Stable at 4°C to 25°C for 1 week.

Reference Interval Not well established, varies with method, age, sex, and use of oral contraceptives. Typical value: 30-50 units/g Hb.

Use Red cell cholinesterase is measured to diagnose organophosphate and carbamate toxicity and to detect atypical forms of the enzyme. Cholinesterase is irreversibly inhibited by organophosphate insecticides and reversibly inhibited by carbamate insecticides. Serum or plasma pseudocholinesterase is commonly used to measure acute toxicity, while erythrocyte levels are better for chronic exposure. (Serum level returns to normal before red cell levels.) Half-life of serum pseudocholinesterase is ~8 days, whereas that of acetylcholinesterase in red blood cells is 3 months. Persons with an atypical form of the enzyme (with low enzyme activity) exhibit prolonged apnea following the use of certain suxamethonium-type muscle relaxants in anesthesia (succinylcholine sensitivity - AA phenotype). These atypical forms may be detected by the use of fluoride or dibucaine inhibition.

The assay was used to evaluate sarin toxicity (*vide infra*).

Limitations Values decrease as erythrocytes age. Values are higher in younger red blood cells and reticulocytosis, and may mask the effect of acetylcholinesterase inhibition. Activity in red blood cells may not always provide a good index of intoxication with acetylcholine inhibitors.

Pseudocholinesterase in serum is the indicated test for succinylcholine sensitivity.

Methodology Methods are based on determination of the rate of hydrolysis of an ester catalyzed by the enzyme acetylcholinesterase and include colorimetry, fluorometry, spectrophotometry based systems.

Additional Information Cholinesterase activity is low at birth and higher in adult males than females. Because of the many constituent amino acids, many molecular variants are possible. The RBC level is **increased** in hemolytic states such as the thalassemias, spherocytosis, hemoglobin SS, and acquired hemolytic anemias. It is **decreased** in paroxysmal nocturnal hemoglobinuria and in relapse of megaloblastic anemia and it returns to normal with therapy. It is not widely regarded as useful as a test for paroxysmal nocturnal hemoglobinuria.

In patients poisoned by systemic insecticides (eg, organophosphates or carbamates), both RBC acetylcholinesterase and plasma cholinesterase are usually inhibited. The effect on the serum enzyme is more marked, however, and serum levels are usually utilized in diagnosis and assessment of recovery. Recovery is best determined by looking for a plateau in erythrocyte cholinesterase activity. Toxic potential may vary, plasma versus red cell cholinesterase, such that in some cases erythrocyte levels may be needed for diagnosis and/or monitoring. If there is suspicion that a decrease in cholinesterase activity may not relate to the inhibitor effect of an organophosphate, then red cell level of acetylcholinesterase should be obtained. If both serum and RBC levels are significantly decreased, findings are those of exogenous toxic effect.

AchE activity use was studied following a terrorist attack in Tokyo subways in which terrorists used sarin. Systemic poisoning was apparently less likely to evolve when pupil size was normal on arrival. Miosis was perceived as a more sensitive index of exposure to sarin vapor then RBC AchE.

References

Carlock LL and Chen WL, "Regulating and Assessing Risks of Cholinesterase-Inhibiting Pesticides," *J Toxicol Environ Health*, 1999, 2(2):105-60.

Mason HJ, "The Recovery of Plasma Cholinesterase and Erythrocyte Acetylcholinesterase Activity in Workers After Overexposure to Dichlorvos," *Occp Med (Lond)*, 2000, 50(5):543-7.

Worek F, Mast U, Kiderlen D, et al, "Improved Determination of Acetylcholinesterase Activity in Human Whole Blood," *Clin Chim Acta*, 1999, 288(1-2):73-90.

♦ **6-0-Acetyl Morphine** *see* Morphine, Urine *on page 921*

♦ **Ac-Globulin (Factor V)** *see* Coagulation Factor Assays *on page 418*

♦ **Acid-Base Status** *see* pCO$_2$, Blood *on page 1008*

♦ **Acid-Base Status Evaluation** *see* Carbon Dioxide, Total, Blood *on page 339*

♦ **Acid Elution for Fetal Hemoglobin** *see* Kleihauer-Betke *on page 822*

♦ **Acid Elution Test** *see* Rosette Test for Fetomaternal Hemorrhage *on page 1172*

Acid-Fast Stain, Routine or Modified

Related Information

Fine Needle Aspiration Culture *on page 589*
Mycobacteria by DNA Probe *on page 928*
Mycobacterial Culture, Biopsy or Body Fluid *on page 929*
Mycobacterial Culture, Cerebrospinal Fluid *on page 931*
Mycobacterial Culture, Cutaneous and Subcutaneous Tissue *on page 932*
Mycobacterial Culture, Sputum *on page 933*
Mycobacterial Culture, Urine *on page 935*

Synonyms AFB Smear; Hank's Stain for *Nocardia* Species; Modified Acid-Fast Stain; *Mycobacterium* Smear; *Nocardia* Smear; TB Smear

Applies to Auramine-Rhodamine Stain; Fluorochrome Stain; Kinyoun Stain; Ziehl-Neelsen Stain

Test Includes Acid-fast stain and culture are usually ordered together.

Abstract Acid-fast bacilli are surrounded by a waxy cell wall that is resistant to destaining by acid alcohol. Heat (classic Ziehl-Neelsen), prolonged exposure, or detergent (Tergitol™ Kinyoun method) is required to allow carbol-fuchsin stain to penetrate the cell wall. Once stained, acid-fast bacteria resist decolorization with acid alcohol. Auramine O, a fluorochrome stain that binds to mycolic acids, is more sensitive than the Ziehl-Neelsen or Kinyoun methods, especially when low numbers of acid-fast bacilli are present. The acid-fast smear is inexpensive, has high specificity (<100%), but it has low sensitivity.

Nocardia are aerobic bacteria, although they produce a fungus-like mycelium. *Nocardia* species are acid-fast when stained by the modified acid-fast stain. *Actinomyces* and *Streptomyces* species, which may be microscopically similar to *Nocardia* on Gram stain, are negative with the modified acid-fast stain.

Patient Preparation Same as for mycobacteria culture of given site

Specimen The appropriate specimen for an acid-fast smear is the same as for culture. Such specimens should be cultured as well.

Causes for Rejection Specimen received on a dry swab

Reference Interval No acid-fast organisms

Use Determine the presence of mycobacteria; establish the etiology of maduromycosis

Limitations Cultures are more sensitive than smears. Acid-fast stains are not specific for *M. tuberculosis*; other species in the genus *Mycobacterium* will stain acid-fast, and other organisms will occasionally stain acid-fast (eg, *Nocardia* species, *Rhodococcus equi*). *Nocardia* species do not always stain acid-fast by this method and thus the presence of branching, gram-positive bacilli on Gram stain may have greater sensitivity.

Methodology Acid-fast stain of concentrated or unconcentrated specimen (Ziehl-Neelsen, Kinyoun, or fluorochrome stain). Positive smears are reported in a quantitated fashion. Modified acid-fast stain is usually the Kinyoun stain followed by light decolorization with 3% acid alcohol.

Additional Information The sensitivity and specificity of acid-fast staining is approximately 87% and 99.8%, respectively.

Molecular target amplification techniques are currently being used with significant success for rapid detection and identification of *Mycobacterium tuberculosis* in clinical specimens. Certain laboratories may use these techniques as an adjunct to AFB smears. See Mycobacteria by DNA Probe *on page 928*.

Occasionally there are requests for stat AFB stains. Performance without prior digestion and concentration compromises sensitivity. Discontinuing AFB or respiratory isolation on the basis of a negative stat result is inappropriate. Identification of positive AFB smear 12-18 hours earlier than routine smears is useful for controlling exposure of healthcare workers.

Infections with *Nocardia* species resemble other more common diseases. Because therapy differs, it is important to establish a definitive diagnosis, preferably by culture. The diagnosis of nocardiosis should be considered in unexplained cavitary lung disease, granulomatous lung disease of established cause not responsive to appropriate therapy, brain abscess particularly in the (Continued)

Acid-Fast Stain, Routine or Modified *(Continued)*

presence of cavitary lung disease, alveolar proteinosis, with mycetoma, and in any patient in whom a disseminated granulomatous disease is considered.

References

Behr MA, Warren SA, Salamon H, et al, "Transmission of *Mycobacterium tuberculosis* From Patients Smear-Negative for Acid-Fast Bacilli," *Lancet*, 1999, 353(9151):444-9.

Krane JF and Renshaw AA, "Relative Value and Cost-Effectiveness of Cultures and Special Stains in Fine Needle Aspirates of the Lung," *Acta Cytol*, 1998, 42(2):305-11.

Olson ES, Simpson AJ, Norton AJ, et al, "Not Everything Acid-Fast Is *Mycobacterium tuberculosis*," *J Clin Pathol*, 1998, 51(7):535-6.

Sridhar MS, Sharma S, Reddy MK, et al, "Clinicomicrobiological Review of *Nocardia keratitis*," *Cornea*, 1998, 17(1):17-22.

Acid-Fast Stain, Modified, Parasites

Related Information

Acid-Fast Stain, Routine or Modified *on page 95*
Cryptosporidium Antigen Detection by EIA *on page 483*
Cryptosporidium Direct Staining Procedures *on page 484*
Ova and Parasites, Direct Exam *on page 985*

Test Includes Modified acid-fast stain on concentrated specimens. Traditional concentration and staining methods used for ova and parasite exams are inadequate for detection of oocysts of *Cyclospora*, *Cryptosporidium*, and *Microsporidium*.

Abstract *Cryptosporidium* and *Microsporidium* species are protozoan pathogens of humans. These pathogens are obligate intracellular parasites. Microsporidia is usually associated with immunocompromised hosts. *Cyclospora* and *Cryptosporidium* are causes of chronic diarrhea and are responsible for water-borne and food-borne outbreaks. The small size and poor staining properties may result in the under-reporting of these infections. For accurate detection, a modified acid-fast stain will usually detect these organisms in stool specimens of infected patients.

Specimen Stool

Causes for Rejection Specimen received on a swab

Reference Interval No acid-fast organisms seen

Use Determine the presence or absence of the gastrointestinal parasites *Cryptosporidium parvum*, *Cyclospora cayetanensis*, and *Microsporidium* species (eg, *Enterocytozoon bieneusi*)

Limitations The critical step in this procedure is the destaining reagent, since over-destaining will make the organisms difficult to detect. Formed stool may have artifact material that can be confused with oocysts.

Methodology Kinyoun stain followed by light decolorization with 1% to 3% H_2SO_4

Additional Information *Cryptosporidium* and *Cyclospora* can cause diarrheal disease in humans. Infection is usually associated with food and water. *Cryptosporidium* infection is associated with traveling, and person-to-person transmission. *Cyclospora cayetanensis* has been associated with prolonged diarrheal disease that can last for 7-8 weeks. It is found in developing countries including Guatemala, Mexico, Peru, and the Caribbean. Microsporidia infection occurs more frequently in immunocompromised hosts, but is being recognized as a cause of diarrheal disease in the immunocompetent individual as well.

See Ova and Parasites, Direct Exam *on page 985*.

References

Connor BA, Reidy J, and Soave R, "Cyclosporiasis: Clinical and Histopathologic Correlates," *Clin Infect Dis*, 1999, 28(6):1216-22.

Curry A and Smith HV, "Emerging Pathogens: *Isospora*, *Cyclospora*, and Microsporidia," *Parasitology*, 1998, 117(Suppl):S143-59.

Herwalt BL, "*Cyclospora caytanesis*: A Review, Focusing on the Outbreaks of Cyclosporiasis in the 1990s," *Clin Infect Dis*, 2000, 31(4):1040-57.

Kotler DP and Orenstein JM, "Clinical Syndromes Associated With Microsporidiosis," *Adv Parasitol*, 1998, 40:321-49.

Osterholm MT, "Lessons Learned Again: Cyclosporiasis and Raspberries," *Ann Intern Med*, 1999, 130(3):134-5.

Reisner BS and Spring J, "Evaluation of a Combined Acid-Fast Trichrome Stain for Detection of Microsporidia and *Cryptosporidium parvum*," *Arch Pathol Lab Med*, 2000, 124(5):777-9.

♦ **Acidified Serum Test for PNH** *see* Ham Test *on page 665*

Acid Phosphatase, Plasma

Related Information

Prostate Specific Antigen, Serum *on page 1098*

Synonyms *o*-Phosphoric-Monoester Phosphohydrolase; PAP; Phosphatase, Acid; Prostatic Acid Phosphatase

Applies to Tartrate-Resistant Acid Phosphatase

Abstract Several acid phosphatases are present in serum, including prostatic and osseous sources.

Patient Preparation Do not obtain specimen immediately after rectal examination, prostate biopsy, TUR, or prostatic massage. Fasting specimen is preferred, as lipemia may interfere.

Specimen Serum or plasma

Container Lavender top (EDTA) tube is preferred; red top tube may be acceptable.

Collection The test should be performed as soon as possible.

Storage Instructions Check with laboratory.

Reference Interval Method dependent; the following values are representative.
- Male: enzymatic, total: 2-12 units/L; enzymatic, prostatic: 0.2-3.5 units/L; RIA, prostatic: 2.5-3.7 ng/mL
- Female: enzymatic, total: 0.3-9.2 units/L; prostatic: 0-0.8 units/L

Use Serum prostate specific antigen (PSA) has replaced serum acid phosphatase in the detection, staging, and monitoring of prostate carcinoma. PSA is much more sensitive and specific than acid phosphatase.

Limitations Acid phosphatase may be **increased** in prostate diseases other than carcinoma (eg, infarct, prostatitis, benign hyperplasia). Increased serum PAP with normal serum PSA may provide indication of significant extraprostatic (nonprostatic) disease and has been rarely elevated in other cancers not metastatic to bone. Moderate elevations of total acid phosphatase have been observed also with malignant invasion of bone from nonprostatic primaries, as well as with myelocytic leukemia, Gaucher disease, and Niemann-Pick disease.

The nonspecificity of acid phosphatase can be partially mitigated by:
- using selective inhibitors (eg, tartrate)
- using selective substrates (eg, thymolphthalein, alpha naphthyl phosphate).

Prostatic acid phosphatase is inhibited by tartrate. Tartrate-resistant acid phosphatase is a bone resorption marker. Substrates like thymolphthalein are more sensitive to prostatic acid phosphatase than to acid phosphatase originating elsewhere.

In males, approximately half of the normal total acid phosphatase is of prostatic origin. High serum bilirubin (>2.0 mg/dL) almost totally interferes with determination of serum tartrate-resistant acid phosphatase.

Additional Information The acid phosphatases are a family of genetically distinct isoenzymes. Erythrocytic and lysosomal forms show widespread distribution in most cells. Prostatic and macrophagic forms have more limited expression and distribution. Erythrocytic and macrophagic forms comprise the tartrate-resistant group and are linked with miscellaneous disorders such as osteolysis, Gaucher disease, and hairy cell leukemia. Exacerbations and remissions of adenocarcinoma of prostate are not correlated with acid phosphatase levels. See Prostate Specific Antigen, Serum *on page 1098*.

References

Nakasato YR, Janckila AJ, Halleen JM, et al, "Clinical Significance of Immunoassays for Type-5 Tartrate-Resistant Acid Phosphatase," *Clin Chem*, 1999, 45(12):2150-7.

Wada N, Ishii S, Ikeda T, et al, "Serum Tartrate Resistant Acid Phosphatase as a Potential Marker of Bone Metastasis From Breast Cancer," *Anticancer Res*, 1999, 19(5C):4515-21.

Actinomyces Culture

Related Information

Acid-Fast Stain, Routine or Modified *on page 95*
Bacterial Culture, Anaerobes *on page 231*
Fine Needle Aspiration, Deep and Superficial Masses *on page 590*
Nocardia Culture *on page 964*

Test Includes Anaerobic culture for *Actinomyces* species and direct microscopic examination of Gram stain for sulfur granules and gram-positive branching bacilli

Abstract *Actinomyces* species are fastidious anaerobic organisms. It is, therefore, essential that the specimen be placed into the appropriate anaerobic transport tube and delivered to the laboratory as quickly as possible. Actinomycosis is a chronic progressive suppurative disease characterized by the formation of multiple abscesses, draining sinuses, and dense fibrosis. Classic presentations include cervicofacial, thoracic, abdominal, and pelvic infections.

Patient Preparation Cleanse the skin around the opening of a draining sinus or lesion and obtain the specimen from as deep within the lesion as possible.

Specimen Exudate, material from draining sinus, tissue

Container Anaerobic specimen transport medium or needle and syringe

Collection All air should be expelled from syringes before transport or transfer. Swabs should be transported in anaerobic transport medium.

Storage Instructions Specimens should be transported immediately to the laboratory and processed as soon as possible.

Causes for Rejection Specimens exposed to air, specimens which have been refrigerated or have an excessive delay in transit, specimens from sites which have anaerobic bacteria as normal flora

Turnaround Time Cultures with no growth are reported after 14 days.

Special Instructions Inform the laboratory that actinomycosis is clinically suspected.

Use Establish the etiology of suppurative/granulomatous fibrosing disease, of a chronic draining sinus, and/or of fever of unknown origin (FUO), particularly in immunocompromised patients

Limitations *Actinomyces* species are relatively slow growing and will often fail to grow in the period in which most laboratories incubate routine cultures. Additionally, even when incubated appropriately, recovery of *Actinomyces* species may be hindered by overgrowth with obligate and facultative anaerobic bacteria.

Methodology Anaerobic culture including thioglycolate broth media

Additional Information In tissue, *Actinomyces* species produce chronic suppuration. Microscopic examination of material from abscesses and/or sinuses often reveals tangled masses of filamentous elements and granules called **sulfur granules**. The presence of such granules is highly suggestive of *Actinomyces* infection. The differential diagnosis between actinomycosis and nocardiosis is relevant because treatment of each is different.

Actinomycetes are not acid-fast by the modified acid-fast stain used for *Nocardia* species.

References

Belmont MJ, Behar PM, and Wax MK, "Atypical Presentations of Actinomycosis", *Head Neck*, 1999, 21(3):264-8.

Lippes J, "Pelvic *Actinomyces*: A Review and Preliminary Look at Prevalence," *Am J Obstet Gynecol*, 1999, 180(2 Pt 1):265-9.

Smego RA Jr and Foglia G, "Actinomycosis," *Clin Infect Dis*, 1998, 26(6):1255-61.

Zitsch 3rd RP and Bothwell M, "Actinomycosis: A Potential Complication of Head and Neck Surgery," *Am J Otolaryngol*, 1999, 20(4):260-2.

Activated Clotting Time

Related Information

Activated Partial Thromboplastin Time *on page 100*
Heparin Antifactor Xa Assay *on page 693*
Heparin Neutralization *on page 697*
Platelet Count *on page 1050*
Point-of-Care Testing *on page 1065*

Synonyms ACT; Activated Coagulation Time

Applies to Heparin

Abstract The ACT is a bedside clotting test that is most useful for monitoring high-dose heparin anticoagulation.

Specimen Whole blood

Container One tube containing an activator of coagulation, such as celite (diatomaceous earth), kaolin, or glass particles. For methods that use cartridges rather than tubes, whole blood may be collected into a plastic syringe or tube and then immediately transferred into the cartridge.

Collection Routine venipuncture. Do not collect from a line that contains heparin. Some tubes require approximately 10 vigorous shakes to disperse the activator; other tubes require gentle mixing. Perform test immediately.

Storage Instructions Specimen cannot be stored; test is performed immediately after collection.

Turnaround Time Minutes

Reference Interval Reference range varies considerably depending on the method; it usually falls somewhere within 70-180 seconds. With cardiopulmonary bypass heparinization, the goal is to exceed 400-500 seconds (commonly >480 seconds), depending on the method, representing a mean heparin level of approximately 4-5 units/mL. For other indications, the ACT goal is typically lower than it is for cardiopulmonary bypass. The ACT goal can also vary depending on the test method. For example, Hemochron® ACT measurements tend to be higher than HemoTec® ACT measurements, although this is not always the case. A HemoTec® ACT >275-300 seconds or a Hemochron® ACT >350 seconds has been recommended for coronary angioplasty.

Use Monitor high-dose heparin anticoagulation, such as during cardiopulmonary bypass surgery. May also be used when an immediate measure of heparin anticoagulation is required at the bedside, such as with extracorporeal membrane oxygenation (ECMO), hemodialysis, cardiac catheterization, and vascular surgery.

Limitations The ACT is less precise than the PTT, and lacks high correlation with the PTT or with heparin antifactor Xa levels. The ACT is influenced by a number of variables, including platelet count, platelet function, lupus anticoagulants, factor deficiencies, ambient temperature, hypothermia, and hemodilution. The various methods are not standardized, and therefore, results from different methods are not interchangeable. Aprotinin prolongs celite-based ACTs but generally not kaolin-based ACTs. Thus, celite-based ACTs may overestimate the amount of heparin anticoagulation when aprotinin is present. However, very high doses of aprotinin, such as following a large initial bolus, may prolong kaolin-based ACTs.

Methodology Whole blood is collected into a tube containing an activator of coagulation, such as celite (diatomaceous earth), kaolin, or glass particles. These activate the intrinsic pathway of coagulation, causing the blood to clot. The tubes are placed into a specialized coagulation analyzer (eg, Hemochron®, International Technidyne; Actalyke®, Helena Laboratories), which measures the time it takes for the blood to clot. Clot formation can be detected by the mobility of a magnet inside the blood test tube. As the instrument rolls the test tube, the magnet rolls along the bottom. When the clot forms, the clot pulls the magnet away from a magnetic detector. With other methods, whole blood is placed into a specialized cartridge that contains an activator of coagulation, such as celite, kaolin or silica, and the clotting time is measured (eg, HemoTec®, Medtronics Inc; i-STAT, i-STAT Corp; GEM PCL, Instrumentation Laboratories). The Medtronics instrument detects clotting by a plunger that moves through the blood sample. When a clot forms, the clot resists the plunger. With the i-STAT instrument, clotting is detected indirectly by the presence of a substrate for thrombin in the cartridge. The substrate resembles the site on fibrinogen that thrombin normally cleaves to form fibrin clot. Thrombin is generated as coagulation is activated, and it cleaves the substrate, releasing an electroactive compound that is detected amperometrically. With GEM PCL cartridges, clot formation is monitored as the instrument draws the blood back and forth across a light detection window. Once the blood clots, the blood no longer flows across the window.

Additional Information With high doses of heparin, the PTT cannot be used to monitor heparin therapy because the PTT is unclottable. The ACT or heparin antifactor Xa levels are used instead of the PTT in such situations. Rarely, ACTs are collected in anticoagulated tubes and the test is performed in a central laboratory.

(Continued)

Activated Clotting Time *(Continued)*

References

Ferguson JJ, "Conventional Antithrombotic Approaches," *Am Heart J*, 1995, 130(3 Pt 2):651-7.

Hirsh J, Warkentin TE, Raschke R, et al, "Heparin and Low-Molecular Weight Heparin. Mechanisms of Action, Pharmacokinetics, Dosing Considerations, Monitoring, Efficacy, and Safety," *Chest*, 1998, 114(5 Suppl):489-510.

Olson JD, Arkin CF, Brandt JT, et al, "College of American Pathologists Conference XXXI on Laboratory Monitoring of Anticoagulant Therapy. Laboratory Monitoring of Unfractionated Heparin Therapy," *Arch Pathol Lab Med*, 1998, 122(9):782-98.

♦ **Activated Coagulation Time** *see* Activated Clotting Time *on page 98*

Activated Partial Thromboplastin Time

Related Information

Activated Clotting Time *on page 98*

Antiphospholipid Antibody (Lupus Anticoagulant and/or Anticardiolipin Antibody) *on page 193*

Coagulation Factor Assays *on page 418*

Cryoprecipitate *on page 481*

Disseminated Intravascular Coagulation Screen *on page 517*

Factor Inhibitors *on page 566*

Fibrinogen *on page 583*

Heparin Antifactor Xa Assay *on page 693*

Heparin Neutralization *on page 697*

High-Molecular Weight Kininogen *on page 731*

Mixing Studies (Coagulation) *on page 918*

Point-of-Care Testing *on page 1065*

Prekallikrein *on page 1085*

Prothrombin Time *on page 1116*

Thrombin Time *on page 1246*

Warfarin, Serum or Plasma *on page 1325*

Synonyms APTT; aPTT; Partial Thromboplastin Time; PTT

Applies to Argatroban; Common Pathway; Extrinsic Pathway; Heparin; Heparin Resistance; Hirudin; Intrinsic Pathway

Abstract The activated partial thromboplastin time (PTT) measures the clotting time from the activation of factor XII, through the formation of fibrin clot (see figure). This measures the integrity of the intrinsic and common pathways of coagulation, whereas the prothrombin time (PT) measures the integrity of the extrinsic and common pathways of coagulation. PTT prolongations are caused by either factor deficiencies (especially of factors VIII, IX, XI, and/or XII), or inhibitors (most commonly, lupus anticoagulants, or therapeutic anticoagulants such as heparin, hirudin, or argatroban).

Specimen Plasma

Container One blue top (citrate) tube; 3.2% citrate tubes are now recommended instead of 3.8% citrate tubes.

Sampling Time See Additional Information for sampling times with the various anticoagulant therapies.

Collection Routine venipuncture. If multiple tests are being drawn, draw blue top tubes after any red top tubes but before any lavender top (EDTA), green top (heparin), or gray top (oxalate/fluoride) tubes. Recent data suggest that an initial discard tube is not necessary. Immediately invert tube gently at least 4 times to mix. Tubes must be appropriately filled. Deliver tubes immediately to the laboratory, otherwise factor VIII may degrade thereby falsely raising the PTT, or falsely low values may occur in heparinized samples as platelets release platelet factor 4 (PF4) which neutralizes heparin.

Specimens drawn from a heparinized line are easily contaminated with heparin, even when the initial volume drawn is discarded. Therefore, coagulation tests are best drawn directly from a peripheral vein, avoiding the arm in which heparin, hirudin, or argatroban is being infused (if relevant).

Storage Instructions Separate plasma from cells as soon as possible, preferably within 1 hour if the PTT is used to monitor heparin, otherwise, PF4 released from platelets neutralizes heparin and can falsely lower the PTT value. To minimize the amount of PF4 in specimens, laboratories should ensure that the plasma contains <10 x 10^9/L platelets. With or without heparin, plasma may be stored on ice for up to 4 hours, otherwise, store frozen.

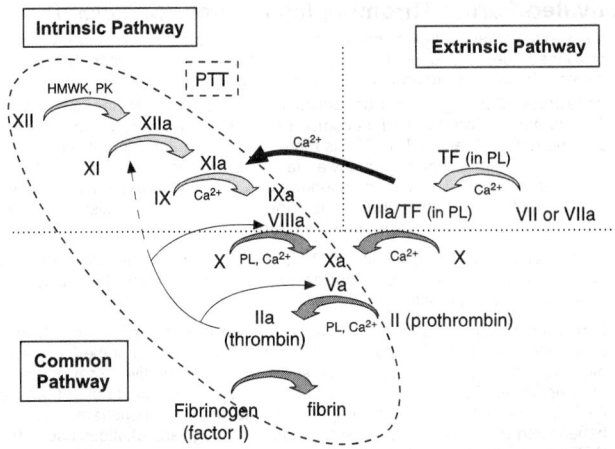

The PTT measures the clotting time from factor XII through fibrin formation (intrinsic and common pathways of coagulation). The PT measures the clotting time from factor VII through fibrin formation (extrinsic and common pathways of coagulation). The intrinsic pathway is activated when factor XII binds to a negatively charged "foreign" surface exposed to the blood, with sequential activation of factor XI, then IX, then X, followed by II, and finally fibrinogen is converted to fibrin. Factors V and VIII and phospholipid serve as cofactors. Many steps also require calcium. It is now believed that in vivo, coagulation is primarily initiated through the extrinsic pathway, upon exposure of blood to tissue factor (TF) at sites of tissue injury. In this model of coagulation, the ability to activate factor IX (by TF/VIIa) and factor XI (by thrombin) without factor XII indicates that factor XII, prekallikrein, and HMWK of the intrinsic pathway are not needed in normal procoagulant pathways. This is consistent with the observation that deficiencies of the latter three factors are not associated with bleeding symptoms, whereas, deficiencies of the other factors may cause a bleeding tendency.

Key:
TF = tissue factor (a transmembrane protein; thus, it is associated with phospholipid in vivo).
PK = prekallikrein.
HMWK = high molecular weight kininogen.
PL = phospholipid.
Ca^{2+} = calcium.

Adapted from Van Cott EM and Laposata M, "Coagulation, Fibrinolysis and Hypercoagulation," *Clinical Diagnosis and Management by Laboratory Methods*, 20th ed, Henry JG, ed, Philadelphia, PA: WB Saunders Co, 2001, 642-59.

Causes for Rejection Specimen received more than 4 hours after collection, tubes not filled, clotted specimens, visible hemolysis

Turnaround Time Less than 1 day; often less than 1 hour if requested stat. The PT and PTT are the most readily available coagulation tests.

Reference Interval Varies significantly among different reagent-instrument combinations. The approximate lower limit of normal is 20-25 seconds; the approximate upper limit of normal is 32-39 seconds. Newborns normally have prolonged PTTs in comparison with adults. The PTT is up to 55 seconds at birth, and the PTT gradually decreases into the adult normal range by the age of 6 months. However, newborns and infants do not normally experience bleeding, because a balance between procoagulants and natural anticoagulants is maintained.

Critical Values >100-150 seconds (varies depending on reagent-instrument combination and laboratory policies)

Use To screen the integrity of the intrinsic pathway of coagulation (factors VIII, IX, XI, and XII) and to a lesser extent the common pathway (fibrinogen and factors II, V, and X). May detect lupus anticoagulants, but the PTT should not
(Continued)

Activated Partial Thromboplastin Time *(Continued)*

be used to screen for lupus anticoagulants because the PTT may or may not be prolonged (depending on the reagents). Also used to monitor therapeutic heparin, hirudin, or argatroban anticoagulation.

Limitations With single factor deficiencies, the deficient factor has to be below 15% to 45% before the PTT becomes prolonged, depending on the reagent and the deficient factor. The PTT is more sensitive to intrinsic pathway factor deficiencies than to common pathway factor deficiencies. With multiple factor deficiencies, the PTT becomes prolonged with less severe decreases in factor levels. Factor VIII elevations shorten the PTT. Factor VIII elevations are common because they occur during acute phase reactions.

Factors VII and XIII do not affect the PTT. The PT can screen for factor VII (and common pathway factor) deficiencies and a specific factor XIII assay can screen for factor XIII deficiencies.

Lupus anticoagulants and deficiencies of certain factors (eg, factor XII) may prolong the baseline PTT and/or accentuate the prolongation of the PTT when heparin is added. Therefore, in these situations, an alternative assay, such as the Heparin Antifactor Xa Assay *on page 693*, should be used rather than the PTT to monitor heparin. If the heparin anti-Xa assay demonstrates that the heparinized PTT is not affected by the lupus anticoagulant, cautious use of the PTT may be considered in that patient.

With very high doses of heparin, as used in cardiac bypass surgery, the PTT is unclottable (>150 seconds) and therefore not useful. The activated coagulation time (ACT) is typically used instead in such situations.

Methodology PTT reagent (phospholipid with an intrinsic pathway activator such as silica, celite, kaolin, ellagic acid) and calcium are added to patient plasma, and the time until clot formation is measured in seconds. Phospholipid in the PTT assay is called "partial thromboplastin" because tissue factor is not present. Tissue factor is present with phospholipid in (complete) thromboplastin reagents that are used for PT assays. Tissue factor activates the extrinsic pathway of coagulation, which is not measured in PTT assays. Phospholipid and calcium are required cofactors in the coagulation cascade. Citrate in the blue top tube prevents clotting by chelating calcium. When the PTT test is ready to be performed, excess calcium is added to overcome citrate.

More recently point-of-care PTT test methods have become available which use a single drop of whole blood, and these methods are undergoing evaluation.

Additional Information To determine the etiology of an unexplained PTT prolongation, the first step is usually to determine if heparin contamination is the cause (see Heparin Neutralization *on page 697*). If this demonstrates that heparin is not present, a mixing study is usually the next step (see Mixing Studies (Coagulation) *on page 918*). Mixing studies can predict whether the cause of the PTT prolongation is a factor deficiency or an inhibitor.

Factor deficiencies that prolong PTT: The PTT is more sensitive to deficiencies of the intrinsic pathway (factors VIII, IX, XI, XII, prekallikrein, HMWK) than it is to deficiencies of the common pathway (fibrinogen, and factors II, V, and X). If a mixing study suggests a factor deficiency, assays for factors VIII, IX, XI, and XII can be performed. If the PT is also prolonged, assays for fibrinogen and factors II, V, VII, and X can also be performed. Prekallikrein and HMWK are often not assayed because deficiencies of these two factors are rare and do not cause bleeding, despite causing a prolonged PTT. Factor XII deficiencies also do not cause bleeding, but factor XII deficiencies are relatively common. If the factor assays are all normal, lupus anticoagulant tests can be considered, because occasionally the mixing study will not detect the presence of a lupus anticoagulant.

The effects of hereditary or acquired factor deficiencies on PT and PTT are shown in Tables 1 and 2 in Coagulation Factor Assays *on page 418*.

Inhibitors that prolong PTT: Inhibitors are usually antibodies (lupus anticoagulants or specific factor inhibitors) or anticoagulants such as heparin, hirudin, or argatroban. Lupus anticoagulants bind to phospholipid and interfere with phospholipid's role as an essential cofactor in the coagulation cascade, thereby prolonging various clotting times such as the PTT. Despite the PTT prolongation, lupus anticoagulants are associated with thrombosis rather than bleeding. Specific factor inhibitors are antibodies directed against a specific coagulation factor, such as a factor VIII inhibitor. PTT mixing studies have a characteristic pattern when a factor VIII inhibitor is present. In such cases, factor VIII inhibitor tests should be performed (see Mixing Studies (Coagulation) *on page 918* and Factor Inhibitors *on page 566*). When a PTT mixing study suggests an inhibitor other than a factor VIII inhibitor is present, lupus anticoagulant tests may be performed, as lupus anticoagulants are by far the most common inhibitor. If the lupus anticoagulant tests are negative, factor assays may be performed as described above. If one factor is significantly decreased, specific factor inhibitor assays may be performed to determine if there is an inhibitor against that factor (see Factor Inhibitors). Specific factor inhibitors are rare.

Acquired causes of PTT prolongations are much more common than hereditary causes, especially among inpatients (see following list). The liver synthesizes all of the coagulation factors. Therefore, with liver disease, multiple factor deficiencies can develop which prolong the PT earlier and more than the PTT. Coumadin® or vitamin K deficiency impairs the function of factors II, VII, IX, and X, leading to PT and eventually PTT prolongations. In disseminated intravascular coagulation (DIC), multiple factor deficiencies may arise due to activation and consumption of factors, prolonging the PT more often than the PTT. Heparin inhibits activated factors II, X, IX, XI, XII, and kallikrein by enhancing antithrombin activity. Hirudin and argatroban inhibit only activated factor II (thrombin).

Causes of PTT Prolongations
Hereditary:
- Deficiency of factor VIII, IX, XI, XII, prekallikrein, or HMWK *(PT is normal)*
- Deficiency of fibrinogen or factor II, V, or X *(PT is also prolonged)*

Acquired:
- Lupus anticoagulants *(PT usually normal)*
- Heparin *(PT less affected than PTT, PT may be normal)*
- Hirudin or argatroban *(PT usually also prolonged)*
- Liver dysfunction *(PT affected earlier and more than PTT)*
- Vitamin K deficiency *(PT affected earlier and more than PTT)*
- Coumadin® *(PT affected earlier and more than PTT)*
- Disseminated intravascular coagulation (DIC) *(PT affected earlier and more than PTT)*
- Specific factor inhibitors *(PT normal except in the rare cases of an inhibitor against fibrinogen, factor II, V, or X)*

Monitoring heparin: Low-dose, subcutaneous, prophylactic unfractionated heparin (eg, 5000 units two or three times daily) is typically not monitored with coagulation tests. Platelet counts should be followed to ensure that if heparin-induced thrombocytopenia develops, the diagnosis will be made promptly. These low levels of unfractionated heparin usually do not affect the PTT. Full-dose, therapeutic levels of unfractionated heparin should be monitored, and the platelet count also followed. The PTT is the most commonly used assay for unfractionated heparin monitoring because it is inexpensive, automated, and usually available 24 hours a day. The therapeutic range is the PTT range that corresponds to an antifactor Xa level of 0.3-0.7 units/mL. Each laboratory determines its own therapeutic range, but it is often a PTT range that is about 1.5-2.5 times the mean of normal PTT. Therapeutic levels of heparin are most often administered as an initial intravenous bolus followed by a continuous intravenous infusion. The PTT is measured every 6 hours during the first day of unfractionated heparin therapy and 6 hours after any dosage change. If the PTT is therapeutic, it can be checked once daily while patients are on heparin. A less common approach is to administer therapeutic unfractionated heparin doses subcutaneously twice daily, drawing the PTT 6 hours after injection. Peak levels are reached 2-4 hours after subcutaneous injection, although this is variable. If patients on unfractionated heparin are started on Coumadin,® heparin is continued until the INR is therapeutic for 2 days. With some PT reagents, heparin can prolong the PT (and therefore the INR) to some extent. Conversely, the PTT can be prolonged somewhat by Coumadin.® Low-molecular weight heparin (LMWH) usually does not significantly prolong the PTT, therefore, the PTT is not used to monitor LMWH. Antifactor Xa assays can be used to monitor LMWH, when indicated.
(Continued)

Activated Partial Thromboplastin Time (Continued)

Heparin resistance is a condition in which the PTT does not prolong as much as expected despite high doses of heparin. This is commonly due to an acute phase reaction, because many acute phase reactant proteins bind and neutralize heparin. Additionally, factor VIII becomes elevated during acute phase reactions, which shorten the PTT. Rarely, heparin resistance is due to antithrombin deficiency. Mild decreases of antithrombin commonly occur as a result of heparin therapy, but mild decreases do not cause significant heparin resistance. Thus, if a patient has heparin resistance, indices of an acute phase reaction may be ordered (eg, fibrinogen, factor VIII), and a heparin assay (antifactor Xa assay) may be helpful.

Monitoring hirudin (lepirudin, Refludan®): Hirudin is a direct thrombin inhibitor that is commonly used as an anticoagulant for the treatment of thrombosis in patients with heparin-induced thrombocytopenia. Hirudin treatment should be monitored with the PTT. The usual therapeutic dose of hirudin in patients with normal kidney function is 0.4 mg/kg intravenous bolus followed by 0.15 mg/kg/hour continuous intravenous infusion. The dose has to be significantly reduced when the creatinine is >1.6 mg/dL. The PTT is performed 4 hours after starting hirudin and 4 hours after any dosage change. If the PTT is in the desired therapeutic range (1.5-2.5 times mean of normal PTT), the PTT can be checked once daily while on hirudin. **Note:** See Coagulation Factor Assays on page 418 for the use of chromogenic factor X assays to monitor Coumadin® in patients receiving hirudin or argatroban.

Some potentially important new anticoagulants that may not require laboratory monitoring are undergoing evaluation. While low-molecular-weight heparins are replacing unfractionated heparin in many clinical situations (in part because they are less likely to cause immunologic thrombocytopenia), a "next generation" heparin-derived anticoagulant, fondaparinux, has been synthesized and undergone initial clinical evaluation. Fondaparinux is a synthetic pentasaccharide (the minimal antithrombin-binding unit of heparin), binds only to antithrombin and is a specific inhibitor of factor Xa. Anticoagulants that are direct thrombin inhibitors (and act independently of other plasma proteins) include hirudin, argatroban, melagatran, and ximelagatran (a chemically modified form of melagatran that is more readily absorbed resulting in a new oral anticoagulants).

References

Andrew M, Paes B, and Johnston M, "Development of the Hemostatic System in the Neonate and Young Infant," Am J Pediatr Hematol Oncol, 1990, 12(1)95-104.

Bajaj SP and Joist JH, "New Insights Into How Blood Clots: Implications for the Use of APTT and PT as Coagulation Screening Tests and in Monitoring of Anticoagulant Therapy," Semin Thromb Hemost, 1999, 25(4):407-18.

Fischer KG and Sutor AH, "Hirudin," Semin Thromb Hemost, 2002, 28(5):403-89.

Gottfried EL and Adachi MM, "Prothrombin Time and Activated Partial Thromboplastin Time Can Be Performed on the First Tube," Am J Clin Pathol, 1997, 107(6):681-3.

Hirsh J, Anand SS, Halperin JL, et al, "Guide to Anticoagulant Therapy: Heparin. A Statement for Healthcare Professionals From the American Heart Association," Circulation, 2001, 103(24):2994-3018.

Kitchen S and Preston FE, "Standardization and Control of Heparin Therapy," Advanced Laboratory Methods in Haematology, Part 6, Chapter 13, Rowan RM, van Assendelft OW, and Preston FE, eds, London, UK: Arnold, 2002, 293-315.

National Committee for Clinical Laboratory Standards (NCCLS), "Collection, Transport, and Processing of Blood Specimens for Coagulation Testing and General Performance of Coagulation Assays: Approved Guideline 3rd edition," NCCLS document H21-A3, NCCLS, 940 West Valley Road, Wayne, Pennsylvania 19087, USA, 1998.

Shapiro S, "Treating Thrombosis in the 21st Century," N Engl J Med, 2003, 349(18):1762-4.

♦ **Activated Protein C Resistance** see Activated Protein C Resistance and the Factor V Leiden Mutation on page 104

♦ **Activated Protein C Resistance** see Hypercoagulation Panel on page 758

Activated Protein C Resistance and the Factor V Leiden Mutation

Related Information

Activated Partial Thromboplastin Time on page 100
Antithrombin on page 198
Hypercoagulation Panel on page 758
Polymerase Chain Reaction on page 1069

Protein C *on page 1101*

Protein S *on page 1110*

Synonyms Activated Protein C Resistance; APC; Protein C Resistance, Activated

Applies to Factor V Leiden Mutation

Abstract Resistance to activated protein C (APC) is a condition which leads to a hypercoagulable state with an increased risk for venous thrombosis. The effect of exogenous APC on patient's clotting time [usually activated partial thromboplastin time (PTT)] is used to detect presence of resistance to APC (as occurs in individuals with the factor V Leiden mutation). A few laboratories might use clotting times other than the PTT. DNA-based assays can be used to directly detect the presence of the factor V Leiden mutation.

Specimen Plasma (for clotting time-based screening assay) and whole blood (for DNA-based confirmatory assay)

Container Blue top (sodium citrate) tube for screening assay. Container varies with laboratory for DNA-based assay (blue top, yellow top, or lavender top).

Collection Routine venipuncture. If multiple tests are being drawn, draw blue top tubes after any red top tubes but before any lavender top (EDTA), green top (heparin), or gray top (oxalate/fluoride) tubes. Immediately invert tube gently at least 4 times to mix. Tubes must be appropriately filled. Deliver tubes immediately to the laboratory.

Storage Instructions Clotting-time based assay: Store plasma at 4°C or room temperature if testing is performed within 4 hours; otherwise, store frozen until testing. According to one manufacturer, storage up to 24 hours without freezing is acceptable (Coatest APC Resistance V by Chromogenix). **DNA-based assay:** Store whole blood at 4°C or at room temperature.

Causes for Rejection Tube not full, specimen clotted, specimen received more than 4-24 hours after collection

Turnaround Time Clotting-time based assay: usually less than 1 day. DNA-based assay: several days (depending on how often test batches are performed).

Special Instructions Do not centrifuge or freeze whole blood specimen for DNA-based assay.

Reference Interval APC prolongs the PTT usually more than twofold in controls (normal persons) and less than twofold in affected individuals.

Some laboratories report the result as a normalized ratio, which is the result of the activated protein C resistance assay for the patient, divided by the result for normal pooled plasma.

Use The clotting-time assay identifies individuals who have resistance to APC. The DNA test identifies factor V Leiden as the cause of the APC resistance. Activated protein C resistance is the most prevalent hereditary predisposition to venous thrombosis. It is present in 5% of the general Caucasian population and is less common or rare in other ethnic groups. It accounts for 20% of unselected patients with a first deep vein thrombosis and 50% of familial cases of thrombosis. The vast majority of cases are due to the factor V Leiden mutation, which renders factor V resistant to degradation by activated protein C, resulting in an increased risk for venous thrombosis.

Limitations Lupus anticoagulants, hirudin, or argatroban may cause inaccurate results in the commonly used PTT clotting-time based assay but do not affect DNA-based tests. Various alternative assays are not affected by lupus anticoagulants.

Methodology Clotting time (usually PTT)-based. Test provides a measure of the APC-dependent prolongation of the clotting time (PTT), in essence, of the ability of APC to act as an anticoagulant. The specimen is first diluted 1:5 in factor V deficient plasma. An activated partial thromboplastin time (PTT) is performed on the diluted specimen, in the presence and absence of exogenously supplied activated protein C. Exogenous activated protein C degrades the patient's factors Va and VIIIa, thereby prolonging the PTT. The ratio of the PTT with activated protein C divided by the baseline PTT is calculated. Normal individuals usually have a ratio >2.0, whereas individuals with factor V Leiden usually have a ratio <2.0 because their mutated factor Va resists activated protein C degradation (each laboratory determines its own reference range). If the result is abnormal, a DNA-based assay (eg, polymerase chain reaction

(Continued)

Activated Protein C Resistance and the Factor V Leiden Mutation *(Continued)*

(PCR)-based assay) should be performed to determine if the patient has the factor V Leiden mutation, which confers activated protein C resistance. DNA-based methods allow precise determination of heterozygosity and homozygosity for the mutation.

The sensitivity and specificity of the PTT-based assay for detection of factor V Leiden mutation approach 100%. Some laboratories do not include the dilution into factor V deficient plasma described above, in which case the sensitivity and specificity are reduced for detecting the factor V Leiden mutation. In addition, patients with an abnormal baseline PTT (usually including patients receiving Coumadin® or heparin) cannot be tested without the dilution step. Whether or not the test without the dilution step provides information regarding hypercoagulability is currently under investigation.

Additional Information The anticoagulant action of activated protein C normally involves degradation of activated factors V and VIII by proteolytic cleavage at specific arginine residues, thereby inhibiting coagulation. Individuals with factor V Leiden have a mutation at one of the arginine cleavage sites in factor V, such that factor V resists degradation by activated protein C. The factor V Leiden mutation is a point mutation in which the guanine at nucleotide position 1691 is replaced by an adenine, resulting in substitution of arginine with glutamine at amino acid residue 506. One, and possibly two, additional factor V mutations at another arginine cleavage site are a very rare cause of activated protein C resistance, and other factor V mutations are also under investigation. Mutations in the factor VIII gene causing resistance to activated protein C are theoretically possible but have not yet been described. Using the normalized ratio reduces intra- and interlaboratory variability in the assay. However, it has not improved the ability of the assay to distinguish activated protein C resistance from normal.

See Hypercoagulation Panel *on page 758.*

References

Jorquera JI, Montoro JM, Fernandez MA, et al, "Modified Test for Activated Protein C Resistance," *Lancet*, 1994, 344:1162-3.

Mann KG and Kalafatis M, "Factor V: A Combination of Dr Jekyll and Mr Hyde," *Blood*, 2003, 101(1):20-30.

Oh D, Kim SH, Kang MS, et al, "Acquired Activated Protein C Resistance, High Tissue Factor Expression, and Hyperhomocysteinemia in Systemic Lupus Erythematosus," *Am J Hematol*, 2003, 72(2):103-8.

Svensson PJ and Dahlbäck B, "Resistance to Activated Protein C as a Basis for Venous Thrombosis," *N Engl J Med*, 1994, 330(8):517-22.

Van Cott EM and Laposata M, "Laboratory Evaluation of Hypercoagulable States," *Hematol Oncol Clin NA*, 1998, 12:1141-66.

- ♦ **Acute Phase Reactant** *see* Alpha₁-Antitrypsin, Serum *on page 134*
- ♦ **Acute Phase Reactant** *see* C-Reactive Protein, Serum *on page 467*
- ♦ **Acute Phase Reactant** *see* Fibrinogen *on page 583*
- ♦ **Acute Phase Reactant** *see* Plasminogen *on page 1042*
- ♦ **Acute Phase Reactant** *see* Plasminogen Activator Inhibitor 1 *on page 1043*
- ♦ **Acute Phase Reactant** *see* von Willebrand Factor *on page 1321*
- ♦ **Acylcarnitines** *see* Newborn Screening by Tandem Mass Spectrometry (MS/MS) *on page 957*
- ♦ **Acylcholine Acylhydrolase** *see* Pseudocholinesterase, Serum *on page 1122*

AD7c Neural Thread Protein, CSF or Urine

Related Information

Apolipoprotein E, Plasma *on page 204*

Cerebrospinal Fluid and Plasma β-Amyloid$_{(1-42)}$ *on page 359*

Applies to Neuronal Thread Proteins

Abstract AD7c-neural thread protein (AD7c-NTP) is present in certain neurons. It is increased in the brains and CSF of patients with Alzheimer disease (AD). The protein in CSF is 41-kD in size, and its concentration in CSF correlates with the severity of dementia. The same protein is excreted in the urine, and preliminary studies report a higher concentration of AD7c-NTP in urine of AD patients compared with non-AD controls.

Specimen Cerebrospinal fluid or urine

Collection Collect CSF by lumbar puncture; the CSF specimen should be clear and free of blood. Use a first morning, midstream urine collection.

Storage Instructions Freeze at -20°C or colder until analysis.

Reference Interval CSF: ≤2.0 ng/mL; urine: ≤1.5 ng/mL

Use The test is a presumptive diagnostic aid that can be used with other relevant clinical information to diagnose AD.

Limitations Both the CSF and urinary tests are relatively new and only a few published clinical studies are presently available.

Methodology Enzyme-linked sandwich immunoassay (ELISA), microparticle enzyme immunoasay (MEIA)

Additional Information The test in CSF is reported to have sensitivities of 83% in patients with probable/possible AD and 89% in patients with early AD. The specificities of the CSF test in non-AD dementia controls and normal controls are 94% and 89%, respectively. Preliminary studies suggest that the urine test is approximately 80% to 85% sensitive and 91% specific. However, these levels of sensitivity and specificity have not yet been independently confirmed.

References

Ghanbari HA, Ghanbari K, Beheshti I, et al, "Biochemical Assay for AD7c-NTP in Urine as an Alzheimer's Disease Marker," *J Clin Lab Anal*, 1998, 12(5):285-8.

Ghanbari HA, Ghanbari K, Munzar M, et al, "Specificity of AD7c-NTP as a Biochemical Marker for Alzheimer's Disease," *J Contemp Neurol*, 1998, 4A:2-6.

Mulder C, Scheltens P, Visser JJ, et al, "Genetic and Biochemical Markers for Alzheimer's Disease: Recent Developments," *Ann Clin Biochem*, 2000, 37(Pt 5):593-607.

Internet Web Sites

www.nymox.com

♦ **AD7c-Neuronal Thread Protein** *see* Cerebrospinal Fluid and Plasma β-Amyloid(1-42) *on page 359*

♦ **Adam** *see* 3,4 Methylenedioxymethamphetamine, Urine *on page 907*

♦ **Adapin®** *see* Antidepressants, Cyclic, Serum or Plasma *on page 171*

♦ **Adapin®** *see* Doxepin, Serum or Plasma *on page 524*

Adenosine Deaminase, CSF, Pleural Fluid, Pericardial Fluid, Peritoneal Fluid

Related Information

Mycobacterial Culture, Biopsy or Body Fluid *on page 929*

Mycobacterial Culture, Cerebrospinal Fluid *on page 931*

Abstract Adenosine deaminase (ADA) has a broad tissue distribution and exists in at least three isoforms (ADA_1, ADA_{1+CP}, and ADA_2). It appears to be involved in the proliferation and differentiation of lymphocytes associated with the immune response.

Specimen Cerebrospinal fluid, pleural fluid, pericardial fluid, or peritoneal fluid

Storage Instructions Centrifuge specimen at ambient temperature. Store supernatant at -20°C until analysis.

Causes for Rejection Specimen unfrozen

Reference Interval

- CSF: <6 units/L
- Pleural fluid: ≤47 units/L
- Pericardial fluid: ≤50 units/L
- Peritoneal fluid: ≤32 units/L

Use Elevated body fluid adenosine deaminase (ADA) activity has been used as an indirect marker for diagnosing tuberculous meningitis (CSF), tuberculous pleural effusion (pleural fluid), tuberculous pericarditis (pericardial fluid), and tuberculous peritonitis (peritoneal or ascitic fluid). Values are higher in tuberculosis than in carcinomatosis.

Limitations There is a high frequency of false-negative pleural fluid ADA levels. The authors of a recent publication concluded that pleural fluid ADA testing is inadequate for diagnostic or therapeutic decisions.

Methodology Kinetic spectrophotometric

Additional Information Polymerase chain reaction (PCR) methods for detection of mycobacterial tuberculosis DNA in body fluids are available and have the potential for more effective diagnosis of tuberculous meningitis/effusions. (Continued)

Adenosine Deaminase, CSF, Pleural Fluid, Pericardial Fluid, Peritoneal Fluid (Continued)

CSF: A recent study showed that an ADA cutoff ≥6 units/L was 90.9% sensitive and 94% specific in detection of tuberculous meningitis. However, significant overlap was observed between patients with tuberculous meningitis and those with cryptococcal meningitis or acute bacterial meningitis. Improved differentiation among these groups was achieved by measurement of the ADA_2 isoenzyme.

Pleural fluid: 253 of 254 patients (99.6%) with known tuberculous pleural effusions had pleural fluid ADA activities >47 units/L. Most ADA activity in these patients was due to the ADA_2 isoenzyme.

Pericardial fluid: An ADA cutoff of 50 units/L was 100% sensitive and 83% specific for the diagnosis of tuberculous effusion in a study that included subjects with tuberculous pericarditis, malignant pericardial effusion, uremic pericarditis, purulent pericarditis, and no pericardial disease. This same study found no correlation between serum and pericardial ADA activity.

Peritoneal (ascites) fluid: Voigt et al, using an ADA cutoff of 32.3 units/L, conducted retrospective and prospective studies and reported, respectively, sensitivities of 95% and 100% and specificities of 98% and 96% in distinguishing patients with tuberculous peritonitis from patients with ascites of other causes.

A discussant in a 1998 Case Records of the Massachusetts General Hospital exercise reported a high sensitivity and excellent specificity of this analyte for detection of *M. tuberculosis* in pleural or peritoneal fluid, and noted that some authorities predicted that the current gold standard for diagnosis (a peritoneal biopsy) may eventually be replaced by this test.

References

Eintracht S, Silber E, Sonnenberg P, et al, "Analysis of Adenosine Deaminase Isoenzyme-2 (ADA2) in Cerebrospinal Fluid in the Diagnosis of Tuberculosis Meningitis," *J Neurol Neurosurg Psychiatry*, 2000, 69(1):137-8.

Garcia-Zamalloa A, Ruiz-Irastorza G, Aguayo FJ, et al, "Pseudochylothorax. Report of 2 Cases and Review of the Literature," *Medicine*, 1999, 78(3):200-7.

Sheets EE and Smith RN, "A 31-Year-Old Woman With a Pleural Effusion, Ascites, and Persistent Fever Spikes," Case Records of the Massachusetts General Hospital, Case 3-1998, Scully RE, Mark EJ, McNeely WF, et al, eds, *N Engl J Med*, 1998, 338(4):248-54.

Adenosine Deaminase, Erythrocyte

Abstract Adenosine deaminase (ADA) catalyzes the deamination of adenosine and deoxyadenosine to inosine and deoxyinosine. It is deficient in certain forms of severe combined immunodeficiency (SCID).

Specimen Whole blood

Container Lavender top (EDTA) tube

Storage Instructions Stable at 4°C for 1 week.

Special Instructions Specimens must be sent to a specialized referral laboratory.

Reference Interval 0.8-1.3 units/g Hb

Use Activities are decreased in patients with ADA-deficient SCID. Usually these patients present at birth with severe lymphopenia (absolute lymphocyte count <500/mm^3) and develop repeated infections and failure to thrive.

Limitations Recent transfusion may obscure the presence of abnormality.

Methodology Spectrophotometric, kinetic assay; fluorometric

Additional Information ADA deficiency is due to an autosomal-recessive, single-gene defect that is responsible for approximately 15% to 25% of SCID cases. Over 50 mutations have been identified in the ADA gene. ADA deficiency causes an accumulation of cytotoxic purine intermediates with subsequent T- and B-cell damage and lethal loss of immune protection. This disorder was the first human genetic disease to be treated by gene transfer techniques with limited success. Enzyme replacement with polyethylene-glycol (PEG) modified ADA has been shown to be effective and safe, but expensive.

References

Buckley RH, "Combined B- and T-Cell Diseases," *Nelson Textbook of Pediatrics*, 16th ed, Chapter 126, Behrman RE, Kliegman RM, and Jenson HB, eds, Philadelphia, PA: WB Saunders Co, 2000, 601-6.

Hershfield MS, "Adenosine Deaminase Deficiency: Clinical Expression, Molecular Basis, and Therapy," *Semin Hematol*, 1998, 35(4):291-8.

Adenovirus Culture and Serology

Related Information

Viral Culture *on page 1307*

Virus, Direct Detection by Fluorescent Antibody *on page 1311*

Test Includes Culture or serology for adenovirus only; adenovirus is usually detected in a routine/general virus culture.

Abstract The name "adenovirus" is based on viral isolation from adenoids. The virus is best known for its propensity to infect the upper airway, where it may cause an exudative pharyngitis similar to that of group A *Streptococcus*. It is the most common cause of epidemic conjunctivitis. A definitive diagnosis of adenovirus infection depends on the isolation of the virus from a patient specimen, demonstration of adenovirus antigen, or detection of an increased antibody titer (greater than fourfold) during the course of the illness.

Specimen Midstream urine, stool or rectal swabs, nasopharyngeal secretions, conjunctival exudates, throat swab or tissue, cerebrospinal fluid for viral culture; serum for serology

Container Culture requires a sterile container and swabs should be placed into cold viral transport medium. A red top tube is used to collect serum for antibody detection.

Collection Acute and convalescent sera drawn 2-4 weeks apart

Storage Instructions Keep culture specimens cold and moist. Adenoviruses are more stable than most other viruses; however, specimens should not be stored or refrigerated for long periods of time. Specimens should be delivered immediately to the clinical laboratory.

Turnaround Time Variable (1-14 days) depending upon culture method used and amount of virus in the specimen; serology may take 1-2 weeks.

Special Instructions Acute and convalescent sera must be tested simultaneously. Tests should, therefore, not be performed unless both specimens are received.

Reference Interval No virus isolated from culture. A fourfold or greater increase in antibody titer in paired sera is indicative of a recent virus infection. Expected value single specimen: <1:8.

Use Aid in the diagnosis of disease caused by adenovirus (eg, conjunctivitis, cystitis, gastroenteritis, pneumonia, and pharyngoconjunctivitis)

Limitations Adenovirus infections are relatively common and circulating antibody may be long-lasting. Consequently, titers on unpaired sera are essentially uninterpretable.

Methodology Conventional culture: Inoculation of specimen into cell cultures and observation for characteristic cytopathic effect (CPE), and identification by fluorescent monoclonal antibody.

Rapid culture: Specimens are centrifuged onto cell cultures and incubated for 2-5 days. Fluorescein-labeled monoclonal antibodies are applied to the infected cells to detect viral antigens that are expressed in the membranes of the cells. Characteristic fluorescent foci indicate the presence of virus.

Serology: Complement fixation (CF), enzyme-linked immunosorbent assay (ELISA), hemagglutination inhibition (HAI), serum neutralization

Additional Information Adenoviruses are spread directly by oral transmission or infectious aerosols. Infections with adenoviruses occur throughout the year, especially in people who are grouped together (eg, those in schools, day care centers, nursing home facilities, and hospitals). Adenoviruses can cause severe respiratory infections in children and immunocompromised adults, which can mimic pertussis and which are sometimes fatal. They cause ocular infections in both children and adults. These include pharyngoconjunctival fever, usually acquired in the summer, sporadic or epidemic keratoconjunctivitis, and acute hemorrhagic conjunctivitis. Types 40 and 41 are common causes of viral diarrhea in children. However, culture of type 40 and 41 is more difficult than culture of other adenoviruses and may not be isolated. Serology to detect adenovirus antibodies can be performed using complement fixation, ELISA, HAI, or serum neutralization. HAI and serum neutralization are not widely available. Serologic evidence of adenovirus, and even isolation of an

(Continued)

Adenovirus Culture and Serology *(Continued)*

adenovirus from a patient, may be coincidental rather than the cause of the patient's present complaints.

References

Chien JW and Johnson JL, "Viral Pneumonias. Infection in the Immunocompromised Host," *Postgrad Med*, 2000, 107(2):67-80.

Ginsberg HS, "The Life and Times of Adenoviruses," *Adv Virus Res*, 1999, 54:1-13.

King JC Jr, "Community Respiratory Viruses in Individuals With Human Immunodeficiency Virus Infection," *Am J Med*, 1997, 102(3A):19-24.

Munoz FM, Piedra PA, and Demmler GJ, "Disseminated Adenovirus Disease in Immunocompromised and Immunocompetent Children," *Clin Infect Dis*, 1998, 27(5):1194-200.

Wadell G, Allard A, and Hierholzer JC, "Adenoviruses," *Manual of Clinical Microbiology*, 7th ed, Murray PR, Baron EJ, Pfaller MA, et al, eds, Washington, DC: ASM Press, American Society of Microbiology, 1999, 970-82.

♦ **ADH** *see* Antidiuretic Hormone, Plasma *on page 172*

Adrenal Cortex: Laboratory Assessment Overview

Related Information

Adrenocorticotropic Hormone, Plasma *on page 114*
Aldosterone, Serum or Plasma *on page 122*
Aldosterone, Urine *on page 124*
Corticotropin-Releasing Hormone Stimulation Test *on page 455*
Corticotropin Stimulation Test (Rapid) *on page 456*
Cortisol, Free, Urine *on page 459*
Cortisol, Serum or Plasma *on page 460*
Follicle Stimulating Hormone, Serum, Plasma, or Urine *on page 609*
21-Hydroxylase Antibodies, Serum *on page 755*
17-Hydroxyprogesterone, Whole Blood, Serum, or Plasma *on page 755*
Insulin Tolerance Test *on page 804*
Metyrapone Stimulation Test, Serum *on page 910*
Potassium, Serum or Plasma *on page 1078*
Potassium, Urine *on page 1080*
Renin Activity, Plasma *on page 1149*
Testosterone, Total and Free, Serum or Plasma *on page 1238*
Urinary Cortisol/Creatinine Increment *on page 1292*

Applies to Dexamethasone Suppression Test; Loperamide Inhibition Test; Naloxone Stimulation Test

Abstract Most endocrine disorders of the adrenal cortex reflect an excess or deficiency of either cortisol or aldosterone, or of one of their secretagogues. The interpretation of an isolated serum cortisol or aldosterone value is usually impossible because hormone secretion fluctuates over a broad and unpredictable range, both in patients with adrenal disease and in acutely ill patients with normal adrenals. For diagnosis, patients require specialized tests which evaluate the dynamics of the negative feedback systems that control the secretion of cortisol and aldosterone.

The primary stimulants of the adrenal cortex are corticotropin (ACTH) and angiotensin II.

Use

CORTISOL-RELATED DISORDERS

Cushing Syndrome (CS)

The biochemical hallmark of CS is increased cortisol secretion, most readily identified by the 24-hour **urine free cortisol** (see Cortisol, Free, Urine *on page 459*). **Serum cortisol** values, especially when drawn at midnight, are often, but not invariably, elevated. Cortisol values drawn at 8 AM are resistant to suppression by a 1 mg dose of dexamethasone; for some patients a multiple-dose dexamethasone suppression protocol is required (see Cortisol, Serum or Plasma *on page 460*). Once CS has been biochemically confirmed, the underlying cause must be identified from among the following.

1. Corticotropin (ACTH)-secreting pituitary adenoma (**Cushing disease - CD**) - 65% to 70% of CS cases
2. **Adrenal adenoma** or **carcinoma** - 15% to 20% of CS cases
3. The **ectopic corticotropin syndrome** (ECS) - 10% to 15% of CS cases. The most common nonpituitary corticotropin-producing neoplasm is small cell carcinoma of lung.

4. **Pseudo-Cushing syndrome**
5. **Ectopic corticotropin-releasing hormone (CRH) syndrome**
6. **Food-dependent CS**
7. **Iatrogenic** (therapy in asthma and rheumatoid arthritis)

Diagnoses 1 and 3 are collectively referred to as **corticotropin-dependent CS**. Patients typically have elevated serum ACTH (see Adrenocorticotropic Hormone, Plasma *on page 114*) and hyperplastic adrenals. Specialized tests include Corticotropin-Releasing Hormone Stimulation Test *on page 455*, in which patients with CD have an exaggerated cortisol and ACTH response; Metyrapone Stimulation Test, Serum *on page 910*, in which patients with CD will have a marked increase in serum 11-deoxycortisol and ACTH; and the **high-dose dexamethasone suppression test** in which approximately 60% to 70% of patients with CD will have a diagnostic decrease in urine free cortisol.

The **ectopic corticotropin syndrome (ECS)**, with elevated cortisol and ACTH levels, occurs in patients with various neuroendocrine neoplasms. Classically, the very elevated ACTH levels are resistant to **high-dose dexamethasone suppression testing**, and the **metyrapone stimulation test** demonstrates no significant increase in serum 11-deoxycortisol. Small cell lung carcinoma accounts for 75% to 80% of ECS cases, and these patients usually present with an acute syndrome (hypokalemia, edema, glucose intolerance, and hypertension) that is not a diagnostic problem. The other 20% to 25% of patients have carcinoid tumors, or other neuroendocrine neoplasms, and a clinical presentation much like CD. When the differential is ECS vs CD, a very effective procedure is to measure the basal and the CRH-stimulated serum ACTH gradient (ratio) between the inferior petrosal sinuses and a peripheral vein; patients with CD have a twofold or higher basal gradient, and a threefold or higher stimulated gradient. (See Corticotropin-Releasing Hormone Stimulation Test *on page 455*.) Other useful procedures are the Metyrapone Stimulation Test, Serum *on page 910*, the low-dose and high-dose dexamethasone suppression test, or a protocol that combines metyrapone and dexamethasone testing.

Diagnoses 2, 5, and 6 are collectively referred to as **corticotropin-independent CS**. Patients with adrenal adenoma or carcinoma have elevated serum cortisol and low or absent serum ACTH. These patients classically show no suppression of cortisol secretion in the **high-dose dexamethasone suppression test**.

Food-dependent CS is a very rare entity in which the diurnal rhythm of cortisol secretion is inverted: low in the early morning, with marked increases following food ingestion. These patients have low or absent serum ACTH and macronodular adrenal hyperplasia.

The **ectopic CRH syndrome** clinically resembles the ectopic corticotropin syndrome. Serum ACTH is elevated and, in typical cases, suppressed in the high-dose dexamethasone suppression test. The anatomic substrate is usually a bronchial carcinoid.

Pseudo-CS is a heterogeneous grouping which includes patients with one or more of the following: depression, chronic alcoholism, severe obesity, eating disorder, hypertension, and hirsutism. The biochemical diagnosis of pseudo-CS, an exercise in exclusion, is facilitated by one or more of the following tests:
- midnight serum cortisol (values >7.5 µg/dL (SI: >207 nmol/L) suggest CD)
- sequential CRH-dexamethasone stimulation test (pseudo-CS patients have markedly decreased cortisol after dexamethasone, and CS patients have a markedly increased response to CRH)
- naloxone stimulation test (CS patients release less ACTH and cortisol in response to the opioid antagonist, naloxone)
- loperamide inhibition test (pseudo-CS patients have a marked decrease in serum cortisol following loperamide hydrochloride, an opiate agonist)
- insulin tolerance test (serum cortisol increases in pseudo-CS, but not in CS)

Adrenal Insufficiency (AI) refers to any disorder in which the adrenal secretion of cortisol is insufficient for metabolic needs. AI is a potentially life-threatening condition. Successful diagnosis and management requires both clinical and laboratory information. Reliance on only one category of information can lead to disastrous consequences. Anatomically, AI may be primary
(Continued)

Adrenal Cortex: Laboratory Assessment Overview
(Continued)

(disease in the adrenals) or secondary (disease in the pituitary or hypothalamus). The clinical onset may be abrupt (acute AI) or gradual (chronic AI).

Chronic primary AI (Addison disease) refers to destruction of the adrenals by neoplasm, infection, autoimmune disease, or other causes. Patients have increased **serum ACTH** and hyperpigmentation of the skin. 21-hydroxylase autoantibodies are markers for autoimmune Addison disease; see 21-Hydroxylase Antibodies, Serum *on page 755*.

Congenital adrenal hyperplasia is most commonly due to 21-hydroxylase deficiency. These autosomal recessive diseases are included here since adrenocortical insufficiency is present in some. See 17-Hydroxyprogesterone, Whole Blood, Serum, or Plasma *on page 755* and the Disease Index.

Acute primary AI is usually due to adrenal hemorrhage and/or necrosis; the differential diagnosis includes meningococcemia, anticoagulant therapy, thrombosis due to a thrombophilic state, and (very rarely) a procoagulant defect. The **serum cortisol** is low and the **ACTH** high.

Chronic secondary AI is caused by cessation of long-term glucocorticoid therapy, neoplasms involving the pituitary or hypothalamus, surgical procedures in the pituitary region, and involvement of the pituitary by inflammatory disease. Findings are similar to those in chronic primary AI, except that serum ACTH is low and there is no hyperpigmentation.

Acute secondary AI is usually caused by hemorrhage or necrosis involving the pituitary/hypothalamic region. The differential diagnosis includes anticoagulant therapy, thrombosis due to a thrombophilic state, surgical procedures in the pituitary or hypothalamus, and (very rarely) a procoagulant defect. The serum cortisol and ACTH are low.

First-line testing: The diagnosis of AI in an unstressed, noncritically ill patient is confirmed if the **basal morning serum cortisol** is <5.0 µg/dL (SI: 140 nmol/L). If the AI is primary, **serum ACTH** is >100 pg/mL (SI: 22 pmol/L). A random serum cortisol >20 µg/dL (SI: >550 nmol/L) excludes the diagnosis of AI, **but only in an "unstressed" patient.**

Increased secretion of cortisol is a normal response to the metabolic requirements imposed by sepsis, trauma, systemic neoplasm, and surgical procedures (collectively referred to as stress). In the acutely ill patient, Cooper and Stewart use a random serum cortisol with the following interpretive guidelines: if the random cortisol is >34 µg/dL (>938 nmol/L), AI is unlikely; if the random cortisol is <15 µg/dL (414 nmol/L), AI is likely. Between these two values, a dynamic test is needed. A similar approach, recommended by Arit and Allolio, uses a random assay of cortisol with a result >25 µg/dL (>700 nmol/L) interpreted as excluding AI. If AI is not so excluded, ACTH is measured in the same specimen as the cortisol (since hydrocortisone therapy has usually been started in such patients) and dynamic testing is conducted. A point for emphasis is that authorities differ on the interpretation of test results.

Dynamic tests: The most commonly used dynamic test to evaluate AI is the Corticotropin Stimulation Test (Rapid) *on page 456*. There are two protocols: a high-dose protocol (using a pharmacologic dose), and a low-dose protocol (using a physiologic dose). The high-dose test correctly identifies patients with primary AI, but false-negative results are a problem in patients with secondary AI. The low-dose test is reported to be more sensitive, but its use remains controversial. Additional dynamic tests include the Metyrapone Stimulation Test, Serum *on page 910* and the Insulin Tolerance Test *on page 804*.

A recently described test for AI is the Urinary Cortisol/Creatinine Increment *on page 1292*. This procedure, which is noninvasive and performed at home, uses sleep as a stimulus for ACTH release.

The **insulin tolerance test (ITT)** is the gold standard for evaluation of the hypothalamic-pituitary axis (HPA). In this procedure, insulin-induced hypoglycemia stimulates hypothalamic corticotropin-releasing hormone (CRH), and this produces increases in ACTH and cortisol.

Metyrapone is a testing agent which metabolically inhibits the synthesis of cortisol at the 11-deoxycortisol step. When this agent is given to a person whose HPA is functionally intact, the serum ACTH and 11-deoxycortisol increase, and the serum cortisol decreases. An impaired response to the metyrapone stimulation test is seen in patients with pituitary or hypothalamic disease, as well as some cases of Cushing syndrome. When testing patients for secondary AI, the ITT is less sensitive than the metyrapone stimulation test.

ALDOSTERONE-RELATED DISORDERS

Primary Hyperaldosteronism (PH) (Conn syndrome) refers to a group of disorders which share characteristic features: hypertension, hypokalemia, urine potassium >30 mmol/day, metabolic alkalosis, elevated serum aldosterone, and low serum renin. Despite such unifying features, PH has significant clinical and biochemical diversity. It is important to avoid basing a diagnostic conclusion on one or two findings. In PH, increased aldosterone secretion is autonomous and at least partially independent of the renin-angiotensin system. Helpful diagnostic algorithms are found in references by Demers and Stewart. PH includes the following differential diagnoses:

- aldosterone-producing adenoma (APA) of the adrenal cortex (65%)
- idiopathic adrenal hyperplasia (IAH) (30%)
- glucocorticoid-remediable aldosteronism (GRA) (rare)
- adrenal carcinoma (very rare)
- ectopic aldosterone-secreting tumor (very rare)

Biochemical confirmation of diagnoses 1, 2, and 4 may be obtained from several different dynamic tests, each of which involves manipulating the renin-angiotensin system and using serum aldosterone as the end point. Maneuvers that normally increase renin (eg, upright posture, sodium restriction, furosemide diuresis, or infusion of angiotensin II) fail to increase aldosterone in patients with PH. Maneuvers that normally decrease renin (eg, supine posture, saline infusion) fail to suppress aldosterone in PH. The **ratio of serum aldosterone to plasma renin** is >50 in PH patients. A ratio >30 deserves further evaluation.

Patients with PH, after 3 days of a high sodium diet (urine Na >250 mmol/day), have 24-hour urine aldosterone values >14 μg/day (SI: >38.9 nmol/day).

As many as 20% of total APA cases have a normal aldosterone response to angiotensin II, but otherwise are typical. The biochemical findings of PH can be closely simulated in patients who are taking thiazide diuretics. Confounding test results also occur in patients taking angiotensin-converting enzyme inhibitors and calcium channel blockers. It is recommended that antihypertensive medications be discontinued for at least 2, and preferably 4, weeks before biochemical testing for PH.

IAH is poorly defined. Classical patients may have findings similar to classical APA, but others (78% in one recent series) have increased aldosterone in response to angiotensin II. Many patients with IAH may actually have essential hypertension.

The diagnosis and treatment of **GRA**, an autosomal dominant inherited condition, requires recognition that the hyperaldosteronism is produced by ACTH and responds to treatment with dexamethasone.

Secondary hyperaldosteronism is not a disease entity but a physiological adaption to a variety of disease processes which have decreased "effective" blood volume as a common denominator. Examples include diseases associated with edema formation: congestive heart failure, cirrhosis, and nephrotic syndrome. In addition, secondary hyperaldosteronism occurs in patients with renal artery stenosis and renin-secreting tumors, possibly reflecting situations in which blood volume "seems" reduced at the level of the juxtaglomerular apparatus and afferent arteriole, even though actual blood volume is normal. In none of these diseases is aldosterone the primary abnormality. Moreover, the clinical and biochemical alterations, apart from elevated aldosterone, are completely different from PH.

References

Arlt W and Allolio B, "Adrenal Insufficiency," *Lancet*, 2003, 361:1881-93.

Cooper MS and Stewart PM, "Corticosteroid Insufficiency in Acutely Ill Patients," *N Engl J Med*, 2003, 348(8):727-34.

Ganguly A, "Primary Aldosteronism," *N Engl J Med*, 1998, 339(25):1828-34.

(Continued)

Adrenal Cortex: Laboratory Assessment Overview
(Continued)

Henzen C, Suter A, Lerch E, et al, "Suppression and Recovery of Adrenal Response After Short-Term, High-Dose Glucocorticoid Treatment," *Lancet*, 2000, 355(9203):542-5.

Jacobs DS, DeMott WR, Oxley DK, et al, *Laboratory Test Handbook*, 5th ed, Hudson, OH: Lexi-Comp Inc, 2001.

Levine LS and DiGeorge AM, "Disorders of the Adrenal Glands," *Nelson Textbook of Pediatrics*, 16th ed, Behrman RE, Kliegman RM, and Jenson HB, eds, Philadelphia, PA: WB Saunders Co, 2000, 1722-43.

Oelkers W, "Adrenal Insufficiency," *N Engl J Med*, 1996, 335(16):1206-12.

Orth DN and Kovacs WJ, "The Adrenal Cortex," *Williams Textbook of Endocrinology*, 9th ed, Wilson JD, Foster DW, Kronenberg HM, et al, eds, Philadelphia, PA: WB Saunders Co, 1998, 517-664.

Stewart PM, "Mineralocorticoid Hypertension," *Lancet*, 1999, 353(9161):1341-7.

Streeten DH, Anderson GH Jr, and Bonaventura MM, "The Potential for Serious Consequences From Misinterpreting Normal Responses to the Rapid Adrenocorticotropin Test," *J Clin Endocrinol Metab*, 1996, 81(1):285-90.

♦ **Adrenal Mass Aspiration** *see* Fine Needle Aspiration, Deep and Superficial Masses *on page 590*

Adrenocorticotropic Hormone, Plasma
Related Information
Adrenal Cortex: Laboratory Assessment Overview *on page 110*
Corticotropin-Releasing Hormone Stimulation Test *on page 455*
Corticotropin Stimulation Test (Rapid) *on page 456*
Cortisol, Free, Urine *on page 459*
Cortisol, Serum or Plasma *on page 460*
Growth Hormone, Serum *on page 662*
Metyrapone Stimulation Test, Serum *on page 910*
Testosterone, Total and Free, Serum or Plasma *on page 1238*
Urinary Cortisol/Creatinine Increment *on page 1292*

Synonyms ACTH; Corticotropin

Applies to Dexamethasone Suppression

Abstract Hypothalamic corticotropin-releasing hormone (CRH) stimulates the secretion of ACTH which, in turn, stimulates the secretion from the adrenal cortex of cortisol, adrenal androgens, and mineralocorticoids.

Specimen Plasma

Container Use chilled syringe. Use two lavender top (EDTA) tubes previously cooled in ice. (Check with laboratory for appropriate container.)

Sampling Time ACTH normally has diurnal variation. Peak values occur in the morning. Normal secretion, as well as that in Cushing disease, is pulsatile and may require multiple samples. For sequential follow-up, ACTH should always be drawn at the same time each day.

Collection Samples for demonstration of the normal circadian rhythm should be drawn between 6 AM and 10 AM and between 9 PM and midnight. Simultaneously obtained cortisol levels may be helpful. Transport specimen **immediately** to the laboratory following collection.

Storage Instructions Separate plasma in refrigerated centrifuge and freeze immediately. Store frozen at -70°C in plastic tubes. Aprotinin (Trasylol®) 500 kU/mL should be added for long-term storage.

Reference Interval
- Cord blood: 50-570 pg/mL (SI: 11-125 pmol/L)
- Newborns: 10-185 pg/mL (SI: 2.2-41 pmol/L)
- Adults, 8 AM: <120 pg/mL (SI: <26.0 pmol/L)
- Adults, 8 AM (ICMA): 10-60 pg/mL (SI: 2.2-13.2 pmol/L)

Use Evaluate the etiology of Cushing syndrome; differentiate pituitary from other causes of corticosteroid excess and deficiency syndromes (ie, distinguish corticotropin-dependent from corticotropin-independent Cushing syndrome); evaluate ectopic ACTH production by neoplasms; examine results of transsphenoidal surgery; follow up patients after bilateral adrenalectomy; for diagnosis of Nelson syndrome (a tumor of the anterior pituitary gland with skin pigmentation following bilateral adrenalectomy.) Evaluate secondary hypopituitarism (target gland failures secondary to lack of stimulation by a pituitary hormone like ACTH). Increased ACTH concentrations in a subject with hypocortisolism signal primary adrenocortical insufficiency. A suppressed level in a patient with Cushing syndrome or hypercortisolism is consistent with cortisol-producing adrenocortical adenoma or carcinoma, with primary adreno-cortical micronodular hyperplasia, or with use of exogenous corticosteroids. A normal to increased concentration in a subject with Cushing syndrome is in keeping with ACTH-dependent Cushing syndrome, caused by ACTH production from an adenoma of the anterior pituitary or by an ectopic source of ACTH [eg, small cell carcinoma of lung, carcinoid (especially those of lung or thymus)]. Cushing syndrome rarely coexists with gastrinoma, multiple endocrine neoplasia syndrome, and Zollinger-Ellison syndrome. Evaluate hirsutism.

Plasma ACTH is also used, in conjunction with cortisol, to evaluate possible adrenal insufficiency (AI) (see Adrenal Cortex: Laboratory Assessment Overview *on page 110*). In the setting of suspected AI, if plasma cortisol is <6.0 µg/dL (<165 nmol/L), then an ACTH >100 pg/mL suggests that the AI is primary and not due to pituitary failure (see Arit and Allolio).

Clinical Diagnosis of Cushing Syndrome

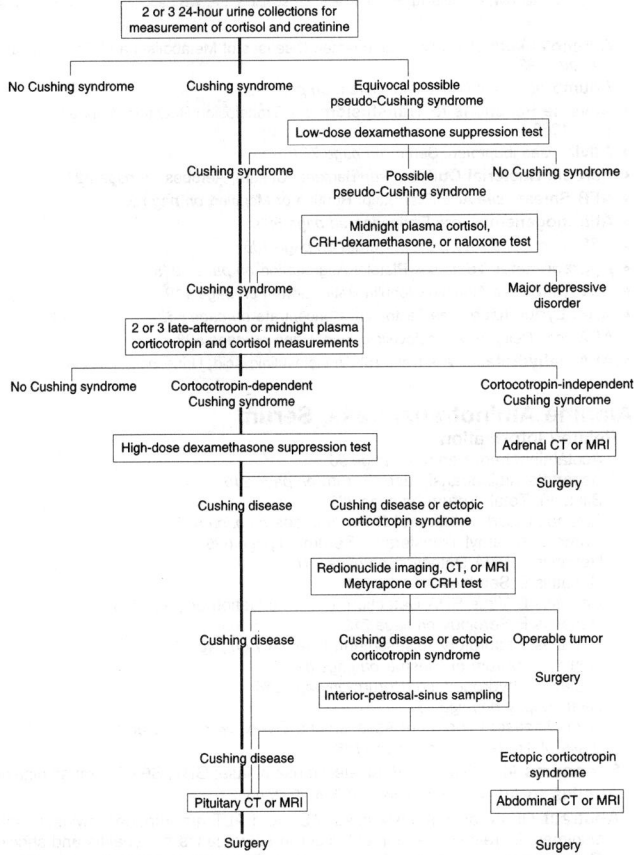

An approach to the diagnosis of Cushing syndrome and its cause. The heavy line indicates the diagnostic path for the majority of patients who have Cushing disease. CT denotes computed tomography and MRI magnetic resonance imaging.

Adapted from Orth DN, "Cushing's Syndrome," *N Engl J Med*, 1995, 332(12):795.

Limitations A single determination may be within normal limits in patients with either excessive production (Cushing disease) or borderline deficiency. ACTH level is affected by stress, which may obscure the normal diurnal and pulsatile changes. ACTH level must be correlated with cortisol levels.
(Continued)

Adrenocorticotropic Hormone, Plasma *(Continued)*

Other useful tests include low-dose dexamethasone suppression test, high-dose dexamethasone suppression tests, metyrapone stimulation test, and petrosal venous sinus catheterization (see below and Adrenal Cortex: Laboratory Assessment Overview *on page 110*).

Methodology Radioimmunoassay (RIA) after separatory step and immunoradiometric assay (IRMA), immunochemiluminometric assay (ICMA).

Additional Information See Adrenal Cortex: Laboratory Assessment Overview *on page 110*. See previous page for an algorithm for the diagnosis of Cushing syndrome.

References

Arit W and Allolio B, "Adrenal Insufficiency," *Lancet*, 2003, 361:1881-93.

Demers LM and Whitley RJ, "Function of the Adrenal Cortex," *Tietz Textbook of Clinical Chemistry*, 3rd ed, Burtis CA and Ashwood ER, eds, Philadelphia, PA: WB Saunders Co, 1999, 1530-69.

Orth DN and Kovacs WJ, "The Adrenal Cortex," *Williams Textbook of Endocriology*, 9th ed, Wilson JD, Foster DW, Kronenberg HM, et al, eds, Philadelphia, PA: WB Saunders Co, 1998, 517-664.

- ♦ **Adrenoleukodystrophy** *see* Inherited Diseases of Metabolism and Cell Structure *on page 792*
- ♦ **Adumbran®** *see* Oxazepam, Serum *on page 990*
- ♦ **Adverse Reactions to Transfusion** *see* Transfusion Reaction Work-up *on page 1269*
- ♦ **Advil®** *see* Ibuprofen, Serum *on page 764*
- ♦ **Aerobic Bacterial Culture** *see* Bacterial Culture, Aerobes *on page 229*
- ♦ **AFB Smear** *see* Acid-Fast Stain, Routine or Modified *on page 95*
- ♦ **Afibrinogenemia** *see* Fibrinogen *on page 583*
- ♦ **AFP** *see* Alpha₁-Fetoprotein, Serum *on page 136*
- ♦ **Aggregometer Test** *see* Platelet Aggregation *on page 1045*
- ♦ **A:G Ratio** *see* Albumin:Globulin Ratio, Serum *on page 119*
- ♦ **AHF, Lyophilized** *see* Factor VIII Concentrate *on page 563*
- ♦ **ALA** *see* Delta (5)-Aminolevulinic Acid, Urine *on page 508*
- ♦ **ALA Dehydratase** *see* Delta (5)-Aminolevulinic Acid, Urine *on page 508*

Alanine Aminotransferase, Serum

Related Information

Acetaminophen, Serum *on page 90*
Aspartate Aminotransferase, Serum *on page 216*
Bilirubin, Total, Serum *on page 265*
Ethanol, Blood, Urine, and Other Sources *on page 558*
Gamma-Glutamyl Transferase, Serum *on page 629*
Hepatitis B DNA Detection *on page 701*
Hepatitis B Serology *on page 702*
Hepatitis C Virus RNA Detection and Quantitation *on page 705*
Hepatitis E Serology *on page 712*
Hepatitis: Laboratory Assessment, Overview *on page 713*
Isoniazid, Serum or Plasma *on page 813*
Lactate Dehydrogenase, Serum *on page 825*
Liver Biopsy *on page 864*
Liver Disease: Laboratory Assessment, Overview *on page 869*
Risks of Transfusion *on page 1166*

Synonyms ALT; Glutamic Pyruvate Transaminase; GPT; SGPT; Transaminase

Applies to Aminotransferases; AST:ALT Ratio

Abstract Of the aminotransferases, AST and ALT are important, widely used enzymes. Increases over tenfold occur in some cases of hepatitis and shock. The greatest amount of ALT is in the liver. See table in listing Bilirubin, Total, Serum *on page 265* and Table 2 in Liver Biopsy *on page 864*.

Unexplained increases of the aminotransferases (ALT, AST) can be followed by noninvasive serologic investigations; see Liver Disease: Laboratory Assessment, Overview *on page 869*.

Patient Preparation Strenuous exercise can cause elevation.

Specimen Serum; plasma may be used.

Container Red top tube; green top (heparin) tube for plasma

Reference Interval Slightly increased intervals in infancy compared to adult normal interval.

- Male: 10-40 units/L (SI: 0.17-0.68 μkat/L)
- Female: 8-35 units/L (SI: 0.14-0.6 μkat/L)

Use A liver function test, ALT is more sensitive for detection of hepatocyte injury than for biliary obstruction. ALT is more specific for liver injury than is AST (SGOT). Hepatic steatosis/steatohepatitis may be the most common etiology of slight to moderate enzyme increases. Other causes of chronic increases of aminotransferase concentrations include alcohol abuse, medications, chronic hepatitis B and C, hepatic fibrosis, cirrhosis, autoimmune hepatitis, hemochromatosis, Wilson disease in subjects 40 years of age and younger, and alpha$_1$-antitrypsin deficiency. AST and ALT levels are elevated before signs or symptoms of viral hepatitis appear. Increased with AST in Reye syndrome. Screening test for hepatitis; acute hepatitis A, B, C, D, or E; Epstein-Barr virus or cytomegalovirus can be confirmed serologically or by other means: see individual listings. Other viruses for possible consideration include herpes and adenovirus. Negative serological findings in the presence of hepatitis-like chemical abnormalities may suggest acute drug-induced hepatitis, an impression supported by resolution after removal of the offending agent. The combination of increased AST and ALT with negative hepatitis markers occurs in a number of other entities, including infectious mononucleosis. Correlating ALT levels with liver cell necrosis in hepatitis C virus - associated cirrhosis, ALT has been proposed as a marker for risk of development of hepatocellular carcinoma in this patient group.

Both AST and ALT are increased after liver cell injury but ALT elevations are more specific for liver cell damage. Increased AST and triglycerides are the most reliable markers for hepatic steatosis. ALT equals or exceeds AST level in nonalcoholic steatohepatitis. AST is increased after myocardial infarction but ALT usually is not. As liver function tests, the transaminases play a role in evaluation of inborn errors of metabolism.

Causes of chronic elevations not deriving from the liver include celiac sprue and inherited and acquired disorders of striated muscle.

Limitations Grossly hemolyzed samples can generate somewhat spurious results. The activity in red cells is six times that of serum. Elevations are found in trauma to striated muscle, rhabdomyolysis, polymyositis, and dermatomyositis, but CK (CK-MM fraction) is increased in such patients and is preferable for evaluation of diseases of skeletal muscle. ALT is less sensitive than is AST to alcoholic liver disease and can even be within normal interval in the presence of severe alcohol abuse. Increased ALT is found with obesity. Although ALT is used to select hepatitis C virus-infected subjects for therapy and liver biopsy, high visit-to-visit variability is reported; ALT is not consistently elevated with hepatitis C. Thyroid disease can cause moderate increases. ALT is characterized by diurnal variation, day-to-day variation, and can be affected by exercise.

Drugs: A few agents can cause a decrease by some methods (eg, metronidazole). A large number of drugs can cause an increase, including acyclovir; allopurinol; antibiotics including synthetic penicillins, ciprofloxacin, nitrofurantoin, ketoconazole, fluconazole, and isoniazid; carbamazepine; carbenicillin; cefoxitin; chloramphenicol; dicumarol; diethylstilbestrol; doxorubicin; erythromycin; esterified estrogens; flutamide; furosemide; gentamicin; inhibitors of hydroxymethylglutaryl-coenzyme A reductase including simvastatin, pravastatin, lovastatin, and atorvastatin; ibuprofen; interleukin-2; mefenamic acid; meprobamate; methotrexate; methyldopa; methyltestosterone; naproxen; nonsteroidal anti-inflammatory drugs such as acetaminophen and aminosalicylic acid; phenobarbital; phenothiazine; phenytoin; progesterone; propranolol; pyrazinamide; rifampin; sulfa drugs; sulfonylureas for hypoglycemia (eg, glipizide); thiazides; ticarcillin; tolbutamide; trimethoprim; troleandomycin; valproic acid; zidovudine; and many others.

Certain **over-the-counter drugs, herbs including Chinese herbs, and alternative and homeopathic treatments** may cause elevations.

Drugs/substances of abuse causing increases include anabolic steroids, cocaine, "ecstasy" (see 3,4 Methylenedioxymethamphetamine, Urine *on* (Continued)

117

Alanine Aminotransferase, Serum *(Continued)*

page 907), "angel dust" (see Phencyclidine, Qualitative, Urine *on page 1019*), glues and solvents containing toluene, trichlorethylene, and chloroform.

Methodology Spectrophotometry by rate assay

Additional Information Among entities in which AST and ALT increases occur are therapeutic applications of bovine or porcine heparin. LD (LDH) abnormality with elevation of hepatic fractions has also been reported.

In children with acute lymphoblastic leukemia, high ALT activity at diagnosis is associated with rapidly progressive ALL.

Acetaminophen hepatotoxicity may be potentiated in alcoholics, the alcohol-acetaminophen syndrome, in which coagulopathy and extremely abnormal ALT and AST are found. The ALT and AST, above 9000 units/L, distinguish the alcohol-acetaminophen syndrome from alcoholic or viral hepatitis. Such levels are found with overdose as well.

AST:ALT ratios are highest in alcoholic liver disease, typically at least 2:1, but are often above unity in nonalcoholic cirrhosis. Of ratios >3:1, >96% of patients have alcoholic liver disease. AST > ALT is reported with typhoid fever. AST:ALT ratios are commonly 0.5-0.8 with acute and chronic viral hepatitis. (See Aspartate Aminotransferase, Serum *on page 216*.)

Elevations of ALT and AST <5 times normal, if confirmed, may initially be followed by prothrombin time; albumin; CBC with platelet count; serology for hepatitis A, B, and C; serum Fe; TIBC; and ferritin. If such serologic tests are negative, additional studies to be considered include ultrasound, ANA, smooth muscle antibody, ceruloplasmin, and alpha₁ antitrypsin. Abnormal results may lead to liver biopsy.

References

American Gastroenterological Association, "American Gastroenterological Association Medical Position Statement: Evaluation of Liver Chemistry Tests," *Gastroenterology*, 2002, 123(4):1364-6.

Berasain C, Betes M, Panizo A, et al, "Pathological and Virological Findings in Patients With Persistent Hypertransaminasaemia of Unknown Aetiology," *Gut*, 2000, 47(3)429-35.

Green RM and Flamm S, American Gastroenterological Association, "AGA Technical Review on the Evaluation of Liver Chemistry Tests," *Gastroenterology*, 2002, 123(4):1367-84.

Mathiesen UL, Franzen LE, Fryden A, et al, "The Clinical Significance of Slightly to Moderately Increased Transaminase Values in Asymptomatic Patients," *Scand J Gastroenterol*, 1999, 34(1):85-91.

Pratt DS and Kaplan MM, "Evaluation of Abnormal Liver-Enzyme Results in Asymptomatic Patients," *N Engl J Med*, 2000, 342(17):1266-71.

Young DS, *Effects of Drugs on Clinical Laboratory Tests*, 5th ed, Volume 1: Listing by Test, Washington, DC: AACC Press, American Association of Clinical Chemistry, 2000, Section 3, 8-23.

♦ **Al, Bone** *see* Aluminum, Bone and Bone Biopsy *on page 139*

Albumin and Plasma Protein Fraction for Infusion

Related Information

Albumin, Serum *on page 120*
Fibrinogen *on page 583*
Prothrombin Time *on page 1116*

Applies to Normal Serum Albumin (Human); Plasma Protein Fraction (Human) (PPF)

Abstract A commercially prepared derivative of donor plasma, human albumin is 96% pure. The manufacturing process and heat inactivation for 10 hours at 60°C destroys viruses. Albumin is available as either 5% or 25% (weight/volume). For volume expansion alone, hydroxyethyl starch or electrolyte solutions are much cheaper and often suitable. **Plasma protein fraction (PPF)** is 83% albumin.

Patient Preparation Dosage and administration: Give a 500 mL dose (10-20 mL/kg in children) rapidly for shock. Hypotension may be seen at rapid rates of administration with PPF (see Limitations). In the absence of shock, administer at a rate of 1-2 mL/minute.

Aftercare Follow blood pressure after rapid administration.

Use A volume expander used mostly for replacement of colloid in emergencies such as burns, pancreatitis, shock due to trauma, hemorrhage, or surgery; adult respiratory distress syndrome; with removal of ascitic fluid and for hypotension related to hemodialysis. Used as standard replacement in therapeutic plasma exchange. The 25% albumin is used in patients who are not dehydrated, but it may be used with large volumes of normal saline or lactated Ringer's solution.

Limitations Does not contain clotting factors; brief retention; not effective for nutritional objectives. **Side effects and hazards**: Fast administration can cause fluid overload, especially with the hyperosmotic 25 g/dL albumin. Although albumin is concentrated about fivefold in 25% albumin, the electrolytes are not. When diluted with water or dextrose in water, fluid electrolytes become proportionally low. Large volumes of such fluid infused rapidly may lead to hyponatremia and cerebral swelling.

Hemolysis may occur when sterile water is used to dilute 25% albumin.

Additional saline must be given to patients who are dehydrated. PPF may contain Hageman factor fragments that can cause hypotension on rapid infusion (>10 mL/minute). PPF may also lead to anaphylaxis in IgA-deficient patients. It is expensive. There are periodic shortages.

The use of colloids vs crystalloids in critically ill patients was recently reviewed. There was no evidence in the 31 trials reviewed which met the inclusion criteria that albumin reduced the risk of death. There was a strong suggestion that it may increase the risk of death. A 2001 editorial by Cook and Guyatt also deserves careful reading.

Contraindications Cardiac failure with congestion; not suitable for nutritional support

Additional Information Albumin is not "salt-poor". Sodium content is 100-160 mmol/L; PPF has 130-160 mmol/L. No filter is needed during infusion. No antibodies are present. In plasma exchange, replacement can usually be part albumin and part isotonic saline.

References

Anderson P, Bunn F, Lefebvre C, et al, "Human Albumin Solution for Resuscitation and Volume Expansion in Critically Ill Patients," *Cochrane Database Syst Rev*, 2002, 1:CD001208.

Cook D and Guyatt G, "Colloid Use for Fluid Resuscitation: Evidence and Spin," *Ann Intern Med*, 2001, 135(3):205-8.

Kravath RE, "More on Dangerous Dilution of 25% Albumin," *N Engl J Med*, 1998, 339(9):634.

Trissel LA and Manasse HR, "More on Dangerous Dilution of 25% Albumin," *N Engl J Med*, 1998, 339(9):634-5.

Triulzi DJ, *Blood Transfusion Therapy. A Physician's Handbook*, 6th ed, Bethesda, MD: American Association of Blood Banks Press, 1999, 45-7.

Albumin:Globulin Ratio, Serum

Related Information

Synonyms A:G Ratio

Test Includes Total protein, albumin, A:G ratio

Abstract The A:G ratio is a calculation derived from total protein and albumin measurements. Principally of historic interest, it was used to detect abnormalities in serum protein composition. Information from the A:G ratio alone is rarely useful in contemporary practice.

Reference Interval Generally, ratios ≥1 are expected. Interpretation is confounded, however, by different analytical methods. A:G ratios are much higher by capillary zone electrophoresis than by agarose gel electrophoresis.

Use Low A:G ratio is found in cirrhosis and other liver diseases, chronic glomerulonephritis and nephrotic syndromes, myeloma, macroglobulinemia of Waldenström, sarcoidosis and other granulomatous diseases, connective tissue diseases, severe infections, cachexia, burns, ulcerative colitis and other chronic inflammatory states. Elevated ratios are usually clinically insignificant.

Limitations The A:G ratio lacks specificity. Electrophoresis provides considerably more information.

(Continued)

Albumin:Globulin Ratio, Serum *(Continued)*

Methodology Electrophoretic or chemically determined albumin divided by [total protein minus albumin]

Additional Information Total protein minus albumin equals globulins. Albumin divided by globulins equals the ratio.

References

Katzmann JA, Clark R, Sanders E, et al, "Prospective Study of Serum Protein Capillary Zone Electrophoresis and Immunotyping of Monoclonal Proteins by Immunosubtraction," *Am J Clin Pathol*, 1998, 110(4):503-9.

Albumin, Serum

Related Information

Albumin:Globulin Ratio, Serum *on page 119*
Anion Gap, Serum, Plasma, or Urine *on page 160*
C-Reactive Protein, Serum *on page 467*
Liver Disease: Laboratory Assessment, Overview *on page 869*
Protein Electrophoresis, Capillary Zone *on page 1103*
Protein Electrophoresis, Serum *on page 1104*
Protein, Quantitative, Urine *on page 1108*
Protein, Total, Serum *on page 1114*
Zinc, Serum or Plasma *on page 1340*

Applies to Albumin:Globulin Ratio

Abstract Albumin is the most abundant protein in human plasma, constituting 55% to 65% of total protein. Serum albumin is synthesized in the liver. It is lower in liver disease, malnutrition, malabsorption (decreased synthesis), and in renal disease (increased loss in urine in diseases including nephrotic syndrome, chronic glomerulonephritis, diabetes mellitus).

Specimen Serum

Container Red top tube

Reference Interval 0-1 year: 2.9-5.5 g/dL (SI: 29-55 g/L); 1-31 years: 3.5-4.8 g/dL (SI: 35-50 g/L) with A:G ratio >1. After age 40, the interval decreases. By serum protein electrophoresis in agarose gel, the interval is 3.1-4.6 g/dL. Capillary zone electrophoresis provides 3.7-5.2.

Possible Panic Range <1.5 g/dL (SI: <15 g/L)

Use Evaluate nutritional status, blood oncotic pressure, liver disease, renal disease with proteinuria, outcome of hemodialysis, and other chronic illnesses. Hypoalbuminemia is reported as an independent risk factor in older subjects for mortality. Decreased admission serum albumin in geriatric patients is a predictor of longer hospital stay and less satisfactory outcome. Low serum albumin in preoperative patients is a negative predictor of 30-day complications, mortality, and morbidity. Use of C-reactive protein as well provides somewhat better sensitivity and specificity. Among patients presenting with colorectal adenocarcinoma, prognostic indicators include pretreatment serum albumin as well as Dukes' stage, histopathologic differentiation, and patient age.

High albumin indicates dehydration. Look for increase in hemoglobin, hematocrit in such patients.

Low albumin is found with use of intravenous fluids, rapid hydration, overhydration; hemodilution by inappropriate secretion of antidiuretic hormone, cirrhosis, other liver disease, including chronic alcoholism; in pregnancy and with oral contraceptive use; many chronic diseases, including the nephrotic syndromes, neoplasia, protein-losing enteropathies (including Crohn disease and ulcerative colitis), peptic ulcer, thyroid disease, burns, severe skin disease, prolonged immobilization, heart failure, chronic catabolic entities such as autoimmune diseases and other chronic catabolic states.

Starvation, malabsorption, or malnutrition: In the absence of intravenous fluid therapy and in patients without liver or renal disease, low albumin may be regarded as an indication of inadequate body protein reserves. Serum albumin has a half-life of about 18-20 days. Its half-life is decreased in patients with catabolic states: infection and with protein loss through the kidneys (eg, nephrosis), gastrointestinal tract, and skin (eg, burns). Its prognostic application is most useful in patients with weight loss, anorexia, surgical therapy, hemorrhage, and infection. Total iron binding capacity <240 µg/dL (SI: <43 μmol/L) and/or low transferrin levels would support an impression of inadequate protein reserves. In **severe malnutrition**, albumin may decrease to <2.5 g/dL (SI: <25 g/L).

Albumin levels ≤2.0-2.5 g/dL (SI: ≤20-25 g/L) may be the cause of edema (eg, nephrotic syndrome, protein-losing enteropathies).

Albumin, prealbumin, and transferrin are regarded as "negative" acute phase reactants (ie, these proteins decrease with acute inflammatory/infectious processes).

Hypoalbuminemia can mask the presence of increased anion gap.

Limitations Albumin levels decrease (<0.5 g/dL) in patients in supine position.

Drugs: Drugs which may cause decrease of albumin include acetaminophen (in severe poisoning), amiodarone, estrogen/progestin therapy in postmenopausal women, interleukin-2, oral contraceptives, phenytoin, prednisone, valproic acid and other agents. Drugs which may cause increase of albumin include other anticonvulsants, furosemide, phenobarbital, prednisolone, and others.

Methodology Bromcresol green (BCG) and bromcresol purple (BCP) measure some alpha-globulins, and therefore provide slightly higher values than does serum protein electrophoresis. Albumin can also be measured by turbidimetry or nephelometry.

Additional Information Twenty-four hour urine collection to measure protein loss is helpful in work-up of some patients with hypoalbuminemia.

Other tests useful in assessment of nutritional status include serum prealbumin, retinal binding protein, transferrin, iron, absolute lymphocyte count, and vitamin B_{12}/folate levels.

Globulin is generally provided as a calculation by the laboratory: total protein minus albumin = globulin. Globulins are measured by serum protein electrophoresis or immunofixation. Quantitative IgA, IgM, and IgG are also more precise.

References
Carfray A, Patel K, Whitaker P, et al, "Albumin as an Outcome Measure in Haemodialysis in Patients: The Effect of Variation in Assay Method," *Nephrol Dial Transplant*, 2000, 15(11):1819-22.

Offringa M, "Excess Mortality After Human Albumin Administration in Critically Ill Patients. Clinical and Pathophysiological Evidence Suggests Albumin Is Harmful," *BMJ*, 1998, 317(7153):223-4.

Orth SR and Ritz E, "The Nephrotic Syndrome," *N Engl J Med*, 1998, 338(17):1202-11.

Sullivan DH, Sun S, and Walls RC, "Protein-Energy Undernutrition Among Elderly Hospitalized Patients: A Prospective Study," *JAMA*, 1999, 281(21):2013-9.

Young DS, *Effects of Drugs on Clinical Laboratory Tests*, 5th ed, Volume 1: Listing by Test, Washington, DC: AACC Press, American Association of Clinical Chemistry, 2000, Section 3, 23-32.

♦ **Albumin, Urine** *see* Protein, Semiquantitative, Urine *on page 1113*

♦ **Alcohol** *see* Ethanol, Blood, Urine, and Other Sources *on page 558*

♦ **ALD** *see* Aldolase, Plasma or Serum *on page 121*

Aldolase, Plasma or Serum

Related Information

Creatine Kinase, Serum *on page 470*
Duchenne/Becker Muscular Dystrophy DNA Detection *on page 526*
Muscle Biopsy *on page 927*
Myoglobin, Blood, Serum, or Plasma *on page 940*

Synonyms ALD; Fructose Biphosphate Aldolase

Abstract Very high elevations of this enzyme occur in certain diseases of skeletal muscle such as muscular dystrophy and dermatomyositis but not in diseases of neurogenic origin (eg, polio, multiple sclerosis).

Specimen Serum or plasma; plasma is preferred over serum because of possible release of platelet enzyme during clotting.

Container Red top tube, SST™ tube, green top (heparin) tube

Storage Instructions Separate serum and freeze immediately. May be stored at -20°C until analysis. The addition of boric acid will stabilize aldolase.

Causes for Rejection Hemolysis (red cells contain aldolase)

Reference Interval
- 0-2 years: <16.3 units/L
- 3-16 years: <8.3 units/L

(Continued)

Aldolase, Plasma or Serum *(Continued)*

- 17 years and older: 7.4 units/L

Use Clinical applications are very limited because of nonspecificity. Very high levels are found in progressive Duchenne muscular dystrophy (MD), with the highest concentrations found early. Elevations occur in carriers of MD, in limb-girdle dystrophy and other dystrophies, in dermatomyositis, polymyositis, and trichinosis, but not in neurogenic atrophies (eg, myasthenia gravis, multiple sclerosis, or poliomyelitis).

Limitations As muscle mass diminishes, aldolase decreases. Serum aldolase elevation is not specific for muscle disease and is also seen in patients with hepatitis, pancreatitis, acute myocardial infarct, and many other diseases. Assay of creatine kinase (CK) is preferred for evaluation of muscle disease. It is more specific for skeletal muscle degeneration. AST and LD also reflect damage to muscle, but lack specificity.

Methodology Ultraviolet, kinetic, coupled enzymatic

References

Mayo Medical Laboratories, *2001 Test Catalogue*, Rochester, MN, 46.

Moss DW and Henderson AR, "Clinical Enzymology," *Tietz Textbook of Clinical Chemistry*, 3rd ed, Burtis CA and Ashwood ER, eds, Philadelphia, PA: WB Saunders Co, 1999, 666-8.

Aldosterone, Serum or Plasma

Related Information

Adrenal Cortex: Laboratory Assessment Overview *on page 110*
Aldosterone, Urine *on page 124*
Potassium, Serum or Plasma *on page 1078*
Potassium, Urine *on page 1080*
Renin Activity, Plasma *on page 1149*

Applies to Renin

Abstract Aldosterone is a mineralocorticoid hormone produced in the zona glomerulosa of the adrenal cortex under complex control by the renin-angiotensin system. Primary hyperaldosteronism (PH) is characterized by hypokalemia, hypersecretion of aldosterone, hypertension, and suppressed plasma renin activity. PH is found in 1% to 2% of unselected subjects with hypertension.

Patient Preparation Preanalytic variables which must be controlled are sodium balance, posture, blood pressure medications, and time of day. Hypokalemia, if present, should be corrected before additional testing. Blood pressure medications and diuretics should be discontinued for **at least** 2 weeks (4-6 weeks is preferable) before additional testing is undertaken.

In the most common scenario, a patient with hypertension and increased urine potassium (>30 mmol/L) will be evaluated with a **screening protocol** in which the patient fasts overnight and a blood specimen for plasma renin and aldosterone is obtained the next day after 4 hours in the upright position. Some clinicians augment this protocol by giving furosemide, 40-80 mg, orally or intravenously on arising.

A patient whose plasma **aldosterone (ng/mL):renin (ng/mL/hour) ratio** is >20 on the screening evaluation can then be further evaluated by a **sodium loading protocol**, consisting of 1) a fasting, early morning blood specimen for aldosterone and renin before arising; 2) spending 2 hours in the upright position; and 3) 2 L of isotonic saline intravenously over 4 hours, at the completion of which another blood specimen for aldosterone and renin is obtained. A variation in this protocol is provided by Ganguly.

The **captopril protocol** is useful for patients who cannot tolerate the sodium loading protocol. Captopril is an antihypertensive agent which inhibits the conversion of angiotensin I to angiotensin II, thus, removing the physiologic stimulus to aldosterone secretion. The patient is given 25 mg (adult dose) of captopril by mouth; blood specimens for aldosterone are obtained just before, and 2 hours after, the drug is taken. This test, however, has fallen short of expectations.

Specimen Serum or plasma, peripheral blood and if indicated, percutaneous catheterization for sampling adrenal vein aldosterone and cortisol

Container Lavender top (EDTA) tube (if also to be used for renin). For aldosterone alone, use heparin, EDTA, citrate, or red top tube; consult laboratory during the assay.

Sampling Time The timing of sampling (morning) and the posture of the patient before sampling (upright) require standardization; *vide supra.*

Collection Specify time and exact source of specimen.

Storage Instructions Transport at once to the laboratory on ice. Freeze serum or plasma in a plastic vial as soon as possible after sampling.

Reference Interval Method dependent. See table.

	Serum Aldosterone ng/dL (SI: pmol/L)	Plasma Renin ng/mL/h	Ratio Aldosterone (ng/dL): Renin (ng/mL/h)
Screening protocol			
Normal subjects			<20*
1-10 y	3.5-124.0† (96.9-3434.8)†		
>11 y	1.0-21.0† (27.7-581.7)†		
18-39 y		2.9-24.0†	
>40 y		2.9-10.8†	
Sodium loading protocol			
Normal subjects (all ages)	<5 (SI: <140)‡		
18-39 y		<0.6-4.3†	
≥40 y		<0.6-3.0†	
Primary hyperaldosteronism (PH)	>10 (>280)‡	Very low, often undetectable	
Captopril protocol			
Normal subjects	<10 (<280)‡		
	<15 (<420)*		

*Orth DN and Kovacs WJ, "The Adrenal Cortex," *William's Textbook of Endocrinology*, 9th ed, Philadelphia, PA: WB Saunders Co, 1998, 596.

†Mayo Medical Laboratories, *2001 Test Catalogue*, Rochester, MN, 47, 457.

‡Demers LM and Whitley RJ, "Function of the Adrenal Cortex," *Tietz Textbook of Clinical Chemistry*, 3rd ed, Burtis CA and Ashwood ER, eds, Philadelphia, PA: WB Saunders Co, 1999, 1530-69.

Critical Values Ratio of plasma aldosterone to renin >30-50; a ratio >30 deserves further evaluation. Montori et al and Kaplan have recently addressed problems in use of the aldosterone:renin ratio.

Use The key to the diagnosis of PH is demonstrating increased aldosterone (blood and/or urine) at the same time that plasma renin is low.

Hypertensive patients with PH have the potential of being cured by adrenalectomy, and therefore extensive testing is warranted. The **screening protocol** (above) may be done after the patient has been off all antihypertensive and diuretic medications for 2 weeks, and after hypokalemia, if present, has been corrected. **Note:** Many patients with PH have serum potassium >3.5 mmol/L. **If 4.0 mmol/L is used as the decision threshold, nearly all patients with PH will be identified, but with the usual trade-off in false positives.** This screen takes advantage of the fact that the upright posture stimulates renin release; the augmented protocol adds sodium depletion - an additional renin stimulant.

In the **sodium loading protocol** and the **captopril protocol** the physiologic secretion of renin is strongly suppressed; hypertensives without PH have a proportionate decrease in aldosterone.

Methodology Two-site immunoradiometric assays, chemiluminescence immunoassay (CIA), enzyme-linked immunosorbent assay (ELISA), radioimmunoassay (RIA)

(Continued)

Aldosterone, Serum or Plasma *(Continued)*

Additional Information The principal pitfall to be avoided is performing tests for aldosterone on a patient with secondary hyperaldosteronism (see Adrenal Cortex: Laboratory Assessment Overview *on page 110*) and misinterpreting the elevated aldosterone as indicative of PH. Secondary hyperaldosteronism is a common physiologic response to several disease processes which have the common denominators of decreased effective renal plasma flow or contracted plasma volume. (Examples include congestive heart failure, cirrhosis, and the nephrotic syndrome.) It is important to recall that patients with renal artery stenosis also have secondary hyperaldosteronism, reflecting the pathophysiology of their hypertension. The combination of elevated renin and aldosterone is also encountered in Bartter syndrome and the Gitelman syndrome. Patients on thiazide diuretics will have laboratory testing results strikingly similar to patients with PH.

References

Bornstein SR, Stratakis CA, and Chrousos GP, "Adrenocortical Tumors: Recent Advances in Basic Concepts and Clinical Management," *Ann Intern Med*, 1999, 130(9):759-71.

Conn JW, "Presidential Address II. Primary Aldosteronism, a New Clinical Syndrome," *J Lab Clin Med*, 1955, 45:3-17.

Demers LM and Whitley RJ, "Function of the Adrenal Cortex," *Tietz Textbook of Clinical Chemistry*, 3rd ed, Burtis CA and Ashwood ER, eds, Philadelphia, PA: WB Saunders Co, 1999, 1530-69.

Ganguly A, "Primary Aldosteronism," *N Engl J Med*, 1998, 339(25):1828-34.

Ghose RP, Hall PM, and Bravo EL, "Medical Management of Aldosterone-Producing Adenomas," *Ann Intern Med*, 1999, 131(2):105-8.

Kaplan NM, "Caution About the Overdiagnosis of Primary Aldosteronism," *Mayo Clin Proc*, 2001, 76(9):875-6.

Montori VM, Schwartz GL, Chapman AB, et al, "Validity of the Aldosterone-Renin Ratio Used to Screen for Primary Aldosteronism," *Mayo Clin Proc*, 2001, 76(9):877-82.

Aldosterone, Urine

Related Information

Adrenal Cortex: Laboratory Assessment Overview *on page 110*
Aldosterone, Serum or Plasma *on page 122*
Potassium, Serum or Plasma *on page 1078*
Potassium, Urine *on page 1080*
Renin Activity, Plasma *on page 1149*
Sodium, Urine *on page 1213*
Urine Collection, 24-Hour *on page 1295*

Applies to Renin

Abstract See Aldosterone, Serum or Plasma *on page 122*.

Patient Preparation Diuretics, antihypertensive drugs, cyclic progestogens, estrogens, and licorice should be terminated for at least 2 weeks (and preferably 4-6 weeks) prior to testing. Patient should be on a diet containing 135 mmol (3 g) sodium/day for at least 2 weeks and preferably 30 days prior to testing. No recent radioactive scans. Potassium deficiencies should be corrected before specimen is collected.

Specimen 24-hour urine

Container Refrigerated during collection

Collection Boric acid preservative, 50% acetic acid (25 mL/24-hour specimen) added to reach pH 2-4. Consult reference laboratory.

Storage Instructions Freeze

Reference Interval There are significant interlaboratory differences.

- 0-30 days: 0.7-11.0 µg/24 hours (SI: 1.94-30.5 nmol/day)
- 1-11 months: 0.7-22.0 µg/24 hours (SI: 1.94-61.0 nmol/day)
- >12 months: 2.0-16.0 µg/24 hours (SI: 5.5-44.3 nmol/day)

Note: Measure also the **sodium** in the urine specimen. A person with normal renin-angiotensin dynamics on a high sodium diet (urine sodium >200 mEq/24 hours), should have urine aldosterone <10 µg/24 hours (SI: <27.7 nmol/day). Urinary potassium and creatinine may also be needed.

Use Urine aldosterone measurements are useful in the diagnosis of primary hyperaldosteronism (PH). The key finding in PH is elevated aldosterone (serum and/or urine) simultaneous with low plasma renin. See discussions on Aldosterone, Serum or Plasma *on page 122* and Adrenal Cortex: Laboratory Assessment Overview *on page 110*.

Methodology Radioimmunoassay (RIA) following acid hydrolysis, enzyme-linked immunosorbent assay (ELISA)

References
Desai SP and Isa-Pratt S, *Clinician's Guide to Laboratory Medicine*, Cleveland, OH: Lexi-Comp, 2000, Chapter 17.

Ganguly A, "Primary Aldosteronism," *N Engl J Med*, 1998, 339(25):1828-34.

Mayo Medical Laboratories, *2001 Test Catalogue*, Rochester, MN, 47.

Orth DN and Kovacs WJ, "The Adrenal Cortex," *Williams Textbook of Endocrinology*, 9th ed, Philadelphia, PA: WB Saunders Co, 1998, 517-664.

Weber KT, "Aldosterone and Spironolactone in Heart Failure," *N Engl J Med*, 1999, 341(10):753-4.

♦ **Aliseum**®   see Diazepam, Serum *on page 510*

Alkaline Phosphatase, Heat Stable, Serum
Related Information
Alkaline Phosphatase Isoenzymes, Serum *on page 125*
Alkaline Phosphatase, Serum *on page 127*
Calcium, Serum *on page 329*
Gamma-Glutamyl Transferase, Serum *on page 629*
Hydroxyproline, Total, Urine *on page 757*
Liver Disease: Laboratory Assessment, Overview *on page 869*
N-Telopeptides, Urine *on page 967*
Osteocalcin, Serum or Plasma *on page 983*

Applies to Nagao Isoenzyme; Regan Isoenzyme

Test Includes Total alkaline phosphatase and heat stable alkaline phosphatase as a percent of total.

Abstract The placental isoenzyme is very heat stable, as is the Regan isoenzyme (a placental-like fetal form occurring in some cancers). Liver isoenzyme is more stable than the bone form. In hepatobiliary disease, GGT is often helpful and more definitive than this fractionation of alkaline phosphatase. GGT is increased with hepatobiliary disease. Assays for bone-specific alkaline phosphatase are also available.

Patient Preparation Patient should be fasting.

Specimen Serum

Container Red top tube or SST™ tube

Storage Instructions Refrigerate

Causes for Rejection Hemolysis

Reference Interval In nonpregnant subjects, heating at 55°C for 15 minutes resulting in percent residual activity >25% favors hepatic origin; <10% favors bone origin. The addage "bone burns, liver lives" makes the differentiation easier to remember. If >90% of stability, probably a placental form is present.

Use Differentiate liver and bone diseases in patients with increased alkaline phosphatase. High alkaline phosphatase of bone origin is found with Paget disease of bone, in which very high concentrations may occur. High levels may be found with osteogenic sarcoma and increases occur in hyperparathyroidism, rickets, osteomalacia, Fanconi syndrome, fracture healing, and with physiologic bone growth in childhood and adolescence. Levels are normal in uncomplicated osteoporosis. **Preferred method for this purpose is separation of isoenzyme by electrophoresis.** See Alkaline Phosphatase Isoenzymes, Serum *on page 125*.

Additional Information Heat stable alkaline phosphatase provides an alternative to alkaline phosphatase electrophoresis. Postmenopausal females generally have slightly elevated total alkaline phosphatase and a low percentage of heat stable fraction, indicating osseous origin.

References
Farley JR, Hall SL, Herring S, et al, "Reference Standards for Quantification of Skeletal Alkaline Phosphatase Activity in Serum by Heat Inactivation and Lectin Precipitation," *Clin Chem*, 1993, 39(9):1878-84.

Watts NB, "Clinical Utility of Biochemical Markers of Bone Remodeling," *Clin Chem*, 1999, 45(8B):1359-68.

Alkaline Phosphatase Isoenzymes, Serum
Related Information
Alkaline Phosphatase, Heat Stable, Serum *on page 125*
Alkaline Phosphatase, Serum *on page 127*
Calcium, Serum *on page 329*
Hydroxyproline, Total, Urine *on page 757*
Liver Disease: Laboratory Assessment, Overview *on page 869*
N-Telopeptides, Urine *on page 967*
(Continued)

Alkaline Phosphatase Isoenzymes, Serum *(Continued)*

Osteocalcin, Serum or Plasma *on page 983*
Pyridinolines (Pyridinoline and Deoxypyridinoline), Urine *on page 1126*

Applies to Bone Alkaline Phosphatase; Nagao Isoenzyme; Regan Isoenzyme

Test Includes May include combinations of heat and/or L-phenylalanine inactivation with or without electrophoretic differentiation.

Abstract Isoelectric focusing, use of selective inhibitors and enzyme immunometric techniques allows identification of the principle isoenzymes of alkaline phosphatase, which include those from bone, liver, intestine, and placenta.

Patient Preparation Patient should be fasting.

Specimen Serum

Container Red top tube or SST™ tube

Causes for Rejection Hemolysis

Special Instructions Send 1 mL frozen serum in plastic container on dry ice to reference laboratory.

Use Evaluate contribution of liver, bone, placental, and Regan isoenzymes to total alkaline phosphatase. Bone fraction is increased in Paget disease of bone. In a chemistry panel, marked isolated increase of alkaline phosphatase in a nonpregnant, older patient who has no healing fracture, with other tests within normal range, is likeliest to indicate Paget disease of bone. Osteoblastic tumor can also cause increased alkaline phosphatase. Other causes of increased serum alkaline phosphatase concentrations of bone origin include hyperparathyroidism, rickets, and osteomalacia. Monitor bone mineral density response to hormonal replacement therapy and treatment of osteoporosis. Aid in detection in bone metastases from prostate and breast carcinoma. Useful as a marker of bone formation - serum bone alkaline phosphatase isoenzymes may be a predictor of effectiveness of growth hormone therapy in children with growth hormone deficiency. If gamma-glutamyl transferase is elevated, the source of the elevated ALP is most likely the liver.

Limitations In the presence of liver disease, the specificity of alkaline phosphatase measurements may be improved if necessary by measuring bone alkaline phosphatase.

Additional Information In most cases, elevation in serum total alkaline phosphatase (T-ALP) is reasonably well defined on the basis of other already established clinical-pathologic findings. LD (LDH) isoenzyme fractionation or serum gamma-glutamyl transferase activity frequently may help to define the clinical problem sufficiently. In a minority of patients, elevation of T-ALP resists explanation. Here, application of ALP isoenzyme studies may indicate whether T-ALP is increased on the basis of contributions from liver, bone, intestinal, placental, endothelial cell, or pathologic (tumor markers Regan and Nagao) fractions.

Total liver and bone ALP are increased in hyperthyroid patients. B-ALP is most commonly and significantly increased. I-ALP is not elevated in the hyperthyroid state.

T-ALP may be elevated in rheumatic diseases (30% to 50% of cases) (eg, rheumatoid arthritis and ankylosing spondylitis). Osteoarthritis and inactive RA are nearly always associated with normal T-ALP. A few cases of RA have increase in liver AP. Increase in T-ALP and in bone fraction has been shown to correlate with disease activity and the number of involved joints.

Cobalamin (vitamin B_{12}) deficient patients have reduced bone ALP. The degree of megaloblastic anemia has been found to correlate with the decrease in enzyme level. T-ALP level, however, is usually within normal range in B_{12} deficient patients.

Newer applications for alkaline phosphatase isoenzymes include monitoring the reduction of bone turnover after alendronate therapy in postmenopausal osteoporotic women or after hormonal replacement therapy. Skeletal alkaline phosphatase may help to provide information for staging prostate cancer and breast cancer. Isoforms of bone alkaline phosphatase may provide information on bone turnover as well as bone turnover within specific bone compartments. Immunoassays for bone alkaline phosphatase (skeletal alkaline phosphatase) are available. Bone alkaline phosphatase, like osteocalcin, is a marker of bone formation. Bone alkaline phosphatase derives from osteoblasts. Since it is cleared by the liver, its concentrations may be elevated with diseases of the liver.

A number of ALP isoenzymes have been described (rarely) in association with carcinoma. They are most commonly seen with hepatocellular cancer or carcinoma metastatic to liver. They include Regan, Magoo, Regan variant, Kashahara, fetal intestinal, and Timperley types. The Regan isoenzyme, which is similar to placental ALP, is seen in 1% to 3% of carcinomas (varying in primary site of origin) metastatic to liver.

References

Kress BC, Mizrahi IA, Armour KW, et al, "Use of Bone Alkaline Phosphatase to Monitor Alendronate Therapy in Individual Postmenopausal Osteoporotic Women," *Clin Chem*, 1999 45(7):1009-17.

Lorente JA, Valenzuela H, Morote J, et al, "Serum Bone Alkaline Phosphatase Levels Enhance the Clinical Utility of Prostate Specific Antigen in the Staging of Newly Diagnosed Prostate Cancer Patients," *Eur J Nucl Med*, 1999, 26(6):625-32.

Watts NB, "Clinical Utility of Biochemical Markers of Bone Remodeling," *Clin Chem*, 1999, 45(8 Pt 2):1359-68.

Wolff JM, Ittel TH, Borchers H, et al, "Metastatic Workup of Patients With Prostate Cancer Employing Alkaline Phosphatase and Skeletal Alkaline Phosphatase," *Anticancer Res*, 1999, 19(4A):2653-5.

Alkaline Phosphatase, Serum

Related Information

Alkaline Phosphatase, Heat Stable, Serum *on page 125*
Alkaline Phosphatase Isoenzymes, Serum *on page 125*
Antimitochondrial Antibody *on page 183*
Aspartate Aminotransferase, Serum *on page 216*
Bilirubin, Total, Serum *on page 265*
Carbohydrate-Deficient Transferrin, Serum *on page 338*
Ethanol, Blood, Urine, and Other Sources *on page 558*
Gamma-Glutamyl Transferase, Serum *on page 629*
Hepatitis B Serology *on page 702*
Hepatitis C Virus RNA Detection and Quantitation *on page 705*
Hepatitis: Laboratory Assessment, Overview *on page 713*
Hydroxyproline, Total, Urine *on page 757*
Immunoglobulin M *on page 779*
Kidney Stone Analysis *on page 820*
Leucine Aminopeptidase (LAP), Serum and Urine *on page 844*
Liver Biopsy *on page 864*
Liver Disease: Laboratory Assessment, Overview *on page 869*
Liver/Kidney Microsomal Type 1 Antibodies *on page 873*
5' Nucleotidase, Serum *on page 968*
Osteocalcin, Serum or Plasma *on page 983*
Smooth Muscle Antibody *on page 1207*

Abstract Eighty percent of serum alkaline phosphatase (ALP) activity normally originates from liver and bone. Other sources include intestine, kidney, and placenta. Synthesized by biliary epithelium, ALP is excreted in bile. Serum total ALP level provides a useful but nonspecific indication of liver or bone disease. Increased in cholestatic, infiltrative and inflammatory liver disease, ALP is increased with obstructive biliary processes, even small secondary bile duct obstruction and, thus, may be elevated when bilirubin is normal due to compensatory bilirubin excretion by the rest of the liver. ALP may be helpful in localized obstructive problems such as hepatic metastases. With biliary tract obstruction, the rise in ALP parallels increases in serum bilirubin. Increased synthesis of ALP can take place even without increases in bilirubin. Heating serum at 56°C causes significant inactivation of ALP of bone origin, but electrophoresis (see Alkaline Phosphatase Isoenzymes, Serum *on page 125*) or determination of serum gamma-glutamyl transferase are better tests for the determination of the source of elevated ALP. GGT is more readily available. See the table in Bilirubin, Total, Serum *on page 265*.

Patient Preparation Patient should be fasting.

Specimen Serum

Container Red top tube or capillary tube

Storage Instructions Refrigerate. Serum alkaline phosphatase increases slowly with storage. Increases of 5% to 10% can be expected after less than 4 hours storage at 4°C. For this reason, it is best to analyze on the day of collection.

(Continued)

Alkaline Phosphatase, Serum *(Continued)*

Reference Interval Method dependent. Normal values are higher for pediatric patients and in pregnancy. Levels are two to three or more the adult interval in children and are increased in puberty compared to adult interval. During episodes of very rapid growth, levels as high as 1000 units/L may be normal. The high level of ALP in childhood reflects osteoblastic activity of bone growth. Postpuberty, serum ALP is mostly of liver origin. Adult normal interval is approximately 50-120 units/L (IFCC reference method at 37°C). ALP is up to two times adult upper limit in pregnancy. Values in adult males are slightly higher than in adult females. With menopause and after, values in women increase, are similar to or higher than those in men, and are higher than in younger subjects.

Use Causes of **high alkaline phosphatase** include nonfasting specimen, bone growth, healing fracture, acromegaly, osteogenic sarcoma, liver or bone metastases, leukemia, myelofibrosis, mastocytosis, and rarely myeloma. Alkaline phosphatase is used as a tumor marker.

In rickets and osteomalacia, serum calcium and phosphorus are low to normal, and alkaline phosphatase may be normal or increased. Hypervitaminosis D may cause elevations in alkaline phosphatase, as may vitamin D malabsorption (eg, celiac sprue).

In Paget disease of bone, there is often isolated elevation of ALP, some among the highest levels seen.

Hyperthyroidism, by its effects upon bone, may elevate alkaline phosphatase.

Hyperparathyroidism, in some patients; pseudohyperparathyroidism.

Chronic alcohol ingestion (in chronic alcoholism, alkaline phosphatase may be normal or increased, but often with high AST (SGOT) and/or high bilirubin and especially with high GGT; MCV may be high).

Biliary obstruction (eg, tenfold increase may be seen with carcinoma of head of pancreas, choledocholithiasis); cholestasis; GGT also high. Cholecystitis with cholangitis: in most patients with cholecystitis and cholangitis who do not have a common duct stone, alkaline phosphatase is within normal limits or only slightly increased. Sclerosing cholangitis (eg, with ulcerative colitis), although 3% of cases of symptomatic sclerosing cholangitis may have normal serum ALP.

Cirrhosis, especially in primary biliary cirrhosis, in which fivefold or more increases are seen; antimitochondrial antibodies needed for evaluation of primary biliary cirrhosis. GGT and 5′ nucleotidase serum levels parallel those of ALP, while AST and ALT are normal to slightly increased in primary biliary cirrhosis. Bilirubin is usually normal early. Liver biopsy is used to confirm the diagnosis and evaluate stage of disease.

Infiltrative/granulomatous liver diseases (eg, sarcoid, TB, amyloidosis, metastatic tumor, abscess). When ALP is >1000 units/L with hepatobiliary source confirmed with increased liver fraction of ALP or with high GGT and bilirubin <1.0 mg/dL, suggested diagnoses include infiltrative/granulomatous liver disease including sarcoidosis, mycoses, tuberculosis, and lymphoma. In early primary biliary cirrhosis and primary sclerosing cholangitis, there is also increased ALP with normal bilirubin.

Autoimmune cholangiopathy includes features of primary biliary cirrhosis or primary sclerosing cholangitis, pruritus with high ALP without antimitochondrial antibodies.

With primary or metastatic tumor in the liver, there may be a marked increase in alkaline phosphatase and GGT. Other relevant laboratory markers for metastatic carcinoma include GGT and CEA.

Gilbert syndrome; postoperative cholestasis, pancreatitis, carcinoma of pancreas, cystic fibrosis.

Hepatitis: Moderate increases in alkaline phosphatase occur in viral hepatitis, but greater elevations of the transaminases (AST (SGOT), ALT (SGPT)) are usually found. Increased in typhoid fever, with AST > ALT, in patients with jaundice and encephalopathy.

Hepatic steatosis (fatty metamorphosis) (moderate increase occurs in acute fatty liver).

Diabetes mellitus, diabetic hepatic steatosis.

Sepsis and certain viral diseases, including infectious mononucleosis, cytomegalovirus infections, and AIDS. *Clostridium difficile* colitis in AIDS patients can also be a cause of unexplained elevation of serum alkaline phosphatase.

Healing infarcts of lung and occasionally in other organs, including kidney. Other situations in which angiofibroplasia occurs, such as healing in a large decubitus ulcer.

Tumors, especially renal cell carcinoma; neoplastic ectopic production (Regan, Nagao isoenzymes); lymphoma. Paraneoplastic serum alkaline phosphatase elevation in renal cell carcinoma patients portends an unfavorable prognosis. Additional paraneoplastic syndromes further worsen the prognosis. The return of serum alkaline phosphatase to normal does not guarantee cure.

Fanconi syndrome, familial hyperphosphatasemia, idiopathic.

Peptic ulcer, erosion; intestinal strangulation or obstruction, or ulcerative lesion; steatorrhea, malabsorption (from bone, secondary to vitamin D deficiency); ulcerative colitis with pericholangitis, other erosive lesions of colon.

Congestive heart failure, parenteral hyperalimentation of glucose, intravenous albumin administration.

Benign familial increase of alkaline phosphatase.

Transient hyperphosphatasemia: Very high (sometimes in thousands) levels of alkaline phosphatase, mainly in children under 5 years of age. In these patients there is no sign of bone or liver disease. More than 400 cases have been reported. Excessive diagnostic procedures should be avoided in these patients.

Drugs - aminoglycosides, amiodarone, anticonvulsants, aspirin (prolonged use), chlorpromazine, erythromycin, esterified estrogens, isoniazid, levothyroxine, lithium, oral hypoglycemic agents, methyltestosterone, oral contraceptives, phenothiazines, and any drug producing hypersensitivity or toxic cholestasis. Many commonly and uncommonly used drugs may elevate alkaline phosphatase, and tenfold increases may be seen with drug cholestasis.

Causes of **low alkaline phosphatase** are said to include hypothyroidism, but most hypothyroid patients have normal alkaline phosphatase.

Some cases of Wilson disease.

Pernicious anemia - in very few patients.

Hypophosphatasia: Very low ALP values are found in the presence of normocalcemia or hypocalcemia. This diagnosis may be confirmed by quantitation of urinary phosphoethanolamine.

Malnutrition has been reported to relate to low values, but in practice, diseases causing malnutrition relate often to high ALP (eg, disseminated neoplasia).

Some **drugs** (alendronate, clofibrate, estrogens in postmenopausal women on estrogen replacement therapy, theophylline, and other agents) may lower serum ALP activity.

Limitations Normal intervals are dependent upon methodology, age, sex, and pregnancy status. Used alone, alkaline phosphatase may be misleading. Elevations occur, usually 2-4 hours after a fatty meal, especially in people who are Lewis positive secretors of blood type O or B. Standing of blood specimen before analysis; up to 30% increase with storage of serum.

Methodology End-point, kinetic spectrophotometric or fluorescent procedures. More recent techniques utilize chromogenic substrates (eg, methylumbelliferyl phosphate) and improved buffer systems with resultant increased sensitivity. Specific immunoassay for bone alkaline phosphatase is now available.

Additional Information There are distinctive forms of ALP in the placenta and small intestine; hepatic, renal, and osteoblast (bone) ALP are similar molecules.

An electrophoretically slow moving isoenzyme with high relative mass may occur in some patients with bile duct obstruction and hepatic metastases and may result in false elevation of CK-MB.
(Continued)

Alkaline Phosphatase, Serum *(Continued)*

To confirm biliary abnormality, additional useful tests include GGT, bilirubin, and occasionally 5'-nucleotidase. They are elevated in hepatobiliary disease, not in uncomplicated bone disease. In the presence of liver disease, the specificity of alkaline phosphatase measurements is improved by measuring bone alkaline phosphatase isoenzyme levels. In most other clinical situations, total serum alkaline phosphatase levels appear to provide sufficient clinical information. Serum alkaline phosphatase levels may serve as an indicator of liver function after hepatectomy but may not reflect morphological regeneration of the liver.

Marked decline of the high ALP of pregnancy is seen with placental insufficiency and imminent fetal demise.

A characteristic of acute liver failure in Wilson disease is the combination of a very high bilirubin, >30 mg/dL (>513 µmol/L) with decreased serum alkaline phosphatase activity. The ratio of ALP to bilirubin <2.0 is relatively distinctive.

See the following diagram.

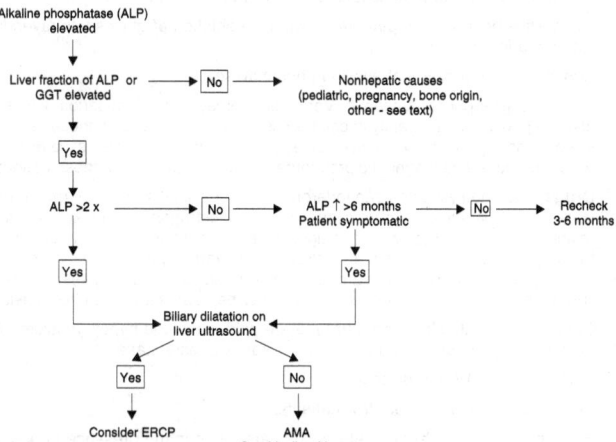

Basic approach to patient with elevated alkaline phosphatase (ALP). ERCP = endoscopic retrograde cholangiopancreatography, GGT = γ-glutamyltransferase. Other relevant tasks include bilirubin, AST, ALT, antimitochondrial antibody, IgM, prothrombin time, and carbohydrate-deficient transferrin.
Modified from Kamath PS, "Clinical Approach to the Patient With Abnormal Liver Test Results," *Mayo Clin Proc*, 1996, 71:1089-95.

In the presence of jaundice, see the table in Bilirubin, Total, Serum *on page 265.*

References

Kamath PS, "Clinical Approach to the Patient With Abnormal Liver Test Results," *Mayo Clin Proc*, 1996, 71(11):1089-95.

Pratt DS and Kaplan MM, "Evaluation of Abnormal Liver-Enzyme Results in Asymptomatic Patients," *N Engl J Med*, 2000, 342(17):1266-71.

Young DS, *Effects of Drugs on Clinical Laboratory Tests*, 5th ed, Volume 1: Listing by Test, Washington, DC: AACC Press, American Association of Clinical Chemistry, 2000, Section 3, 36-48.

♦ **Alkalosis** *see* Chloride, Urine *on page 394*

♦ **Alkylating Agents** *see* Chemical Warfare Agents *on page 382*

♦ **Allegron®** *see* Nortriptyline, Serum *on page 966*

♦ **Allen Test** *see* Arterial Blood Collection *on page 211*

♦ **Allen Test** *see* Blood Gases and pH, Arterial *on page 275*

Allergen Specific IgE Antibody

Related Information

Immunoglobulin E *on page 773*

Synonyms IgE Allergen Specific; Radioallergosorbent Test (RAST®)

Applies to Immunocapture; Latex Sensitization; Phadenzyme RAST®; Pharmacia CAP; RAST®

Test Includes *Alternaria tenuis*, Bermuda grass, cat epithelium, common ragweed, *Dermatophagoides farinae*, dog epithelium, egg white, English plantain, house dust, maple, oak, timothy, or specific mini panels of grasses, foods, animal danders, etc, one or multiple allergens from a large library (available commercially) can be chosen for analysis dependent upon clinical guidance.

Abstract Radiolabeled anti-IgE is used to detect binding of patient's IgE to specific allergens present on a paper disk (radioallergosorbent test-RAST). A number of commercial variations are available, some nonradiolabeled (eg, with enzyme-labeled detection system such as used by Phadenzyme RAST®). Absolute specific IgE levels (reported in mass units) are measured by the recently introduced UniCAP™ and Pharmacia CAP systems.

Patient Preparation No radioisotopes administered 24 hours prior to collection of specimen (not applicable to nonisotope-based methods)

Specimen Serum; plasma may be used including heparinized plasma but fibrin, if formed, may cause high interassay variation and increased nonspecific binding.

Container Red top tube

Storage Instructions Separate serum and refrigerate. It is best to let blood clot and retract overnight at 4°C. Store at -20°C.

Turnaround Time Test is commonly performed by a reference laboratory.

Reference Interval Each allergen is scored from 0-4, 0 meaning no IgE detected, 1 meaning a borderline result, and 2-4 increasing IgE antibody against the allergen

Use Detect possible allergic responses to various substances in the environment such as animals, antibiotics, foods, grasses, house dust, mites, insects, insulin, molds, smuts, trees, and weeds. Evaluate hay fever, extrinsic asthma, atopic eczema, respiratory allergy.

Radioallergosorbent (RAST®) test, or its equivalent, is indicated when:
- specific allergic sensitivity is needed to allow immunotherapy ("desensitization shots") to be initiated
- testing for food or chemical sensitivity, in which skin testing is unreliable
- there is a history of severe allergic reaction to skin testing
- testing infants
- evaluating patients who refuse skin tests or who are unable to have them because of dermatopathic conditions
- immunotherapy or other therapeutic measures based on skin testing results have not led to a satisfactory remission of symptoms

Limitations RAST® results should be interpreted in the context of all available clinical and laboratory findings. False-negative results are possible and may reflect the timing of the blood sample relative to a previous adverse reaction. There may be overlap of IgE values between atopic and nonatopic conditions. High levels of total IgE (>3000 IU/mL as may be seen due to parasitic infestation) may cause nonspecific binding and, thus, false-positive RAST® results. Total quantitative IgE level must usually be ordered separately. When the level of total IgE is less than the geometric mean of IgE reference range, RAST® results may not have clinical significance.

Contraindications Recently administered radioisotopes will interfere with the radioisotope based forms of this test, causing spurious results.

RAST® is contraindicated when:
- all skin tests are negative
- the patient has only mild symptoms or can be successfully treated with medication and avoidance
- IgE levels are <10 IU/mL unless there is strong clinical suggestion of allergic disease
- patients have successfully responded to immunotherapy
- evaluating non-IgE mediated disease, such as certain drug and food reactions

(Continued)

131

Allergen Specific IgE Antibody *(Continued)*

Methodology Radioallergosorbent (RAST®) test (a radioimmunoassay). In this procedure specific allergen is adsorbed on a paper disk. Immunospecific IgE, if present in the test (patient's) serum will bind to the disk. Detection is effected by radiolabeled anti-IgE. Different scoring systems comparing test results with the absolute binding of a negative control are in use. Enzyme immunoassay and immunofluorometric-based systems have been developed. Older methods will likely be replaced by recently introduced quantitative IgE assays that quantitate specific IgE antibodies in mass units (ie, UniCAP and Pharmacia CAP systems).

Additional Information IgE is elevated 4-30 times normal in various diseases, among which atopic disorders and parasitic infections are most prominent. The principal limitation of this test is the wide and overlapping range of IgE values between atopic and nonatopic disease states. A positive value is usually meaningful; a negative value is equivocal. RAST® test has value with patients who do not respond to environmental control or conservative medical management and in whom skin tests are contraindicated.

Numerous reports comparing skin testing and RAST® have accumulated in the literature, generally to assess which method has the better sensitivity/specificity. The results appear to vary with the allergen, that is, the relative performance of the two different methods, skin prick test (SPT) vs allergen specific IgE level, is allergen-dependent. This finding has provided the stimulus for expanded allergen specific comparison studies in the decade of the eighties.

With the advent of specific IgE quantitative assays (UniCAP and Pharmacia CAP systems), a new round of comparison studies to evaluate the clinical significance of such quantitation is underway. Applications in the area of predicting development of atopic disorders are also important (eg, egg-specific IgE levels in the first year of life vs subsequent aeroallergen sensitization/asthma, mite sensitization in early infancy vs subsequent asthma, and many other applications). A comparison of skin prick test (SPT), Pharmacia CAP RAST®, and intradermal skin tests in cat allergy concluded that the latter "added little to the diagnostic evaluation." Skin prick testing, however, has not been considered "obsolete" for a number (many of them clinical) reasons.

Allergen-specific IgE testing has a large and growing number of applications. A sampling of some more unusual areas includes latex sensitization, fish hypersensitivity, venom allergy, cow's milk protein, and toxoplasmosis (IgE immunocapture method).

References

Grote M, Mahler V, Spitzauer S, et al, "*in situ* Localization of Latex Allergens in 3 Different Brands of Latex Gloves by Means of Immunogold Field Emission Scanning and Transmission Electron Microscopy," *J Allergy Clin Immunol*, 2000, 105(3):561-9.

Hamilton RG and Adkinson NF Jr, "Immunological Tests for Diagnosis and Management of Human Allergic Disease: Total and Allergen-Specific IgE and Allergen-Specific IgG," Rose NR, de Macario EC, Folds JD, et al, eds, *Manual of Clinical Laboratory Immunology*, 1997, 881-92.

Paganelli R, Ansotegui IJ, Sastre J, et al, "Specific IgE Antibodies in the Diagnosis of Atopic Disease. Clinical Evaluation of a New *In Vitro* Test System, UniCAP, in Six European Allergy Clinics" *Allergy*, 1998, 53(8):763-8.

Sainte-Laudy J, Sabbah A, Droute M, et al, "Diagnosis of Venom Allergy by Flow Cytometry. Correlation With Clinical History, Skin Tests, Specific IgE, Histamine and Leukotriene C4 Release," *Clin Exp Allergy*, 2000, 30(8):1166-71.

Sicherer SH and Sampson HA, "Cow's Milk Protein-Specific IgE Concentrations in Two Age Groups of Milk-Allergic Children and in Children Achieving Clinical Tolerance," *Clin Exp Allergy*, 1999, 29(4):507-12.

Villena I, Aubert D, Brodard V, et al, "Detection of Specific Immunoglobulin E During Maternal, Fetal, and Congenital Toxoplasmosis," *J Clin Microbiol*, 1999, 37(11):3487-90.

Wood RA, Phipatanakul W, Hamilton RG, et al, "A Comparison of Skin Prick Tests, Intradermal Skin Tests, and RASTs in the Diagnosis of Cat Allergy," *J Allergy Clin Immunol*, 1999, 103(5 Pt 1):773-9.

Yunginger JW, Ahlstedt S, Eggleston PA, et al, "Quantitative IgE Antibody Assays in Allergic Diseases," *J Allergy Clin Immunol*, 2000, 105(6 Pt 1):1077-84.

Alpha₁-Acid Glycoprotein, Serum

Related Information

Haptoglobin, Serum *on page 667*

Synonyms Orosomucoid

Abstract Alpha₁-acid glycoprotein (AAG) is an acute phase reactant, appearing in plasma within 12 hours of injury and peaking at 3-5 days. AAG binds with certain basic drugs and hormones, in some cases rendering them ineffective. Markedly increased serum levels are found in inflammation, pregnancy, and certain malignancies.

Patient Preparation Patient should be fasting.

Specimen Serum

Container Red top tube

Storage Instructions Refrigerate (4°C) for up to 72 hours. After 72 hours, the specimen should be stored frozen at -20°C and thawed only once prior to analysis.

Causes for Rejection Gross hemolysis, lipemia

Reference Interval

- Infants: birth: 12-56 mg/dL, 1 month: 24-93 mg/dL, 3 months: 27-107 mg/dL, 6 months: 35-149 mg/dL
- Children 3-16 years: 43-308 mg/dL
- Adults 20-60 years: 50-120 mg/dL

Use Serum AAG is increased by acute-phase inflammatory conditions and glucocorticoids; it is decreased in nephrotic syndrome, protein-losing enteropathy, and estrogen therapy. In patients with suspected hemolysis, measurement of AAG has been suggested as an aid in interpretation of low haptoglobin levels (see Haptoglobin, Serum *on page 667*), since both proteins are affected similarly by inflammation, glucocorticoids, and estrogens.

Methodology Immunonephelometry, immunoturbidimetry, enzyme-linked immunosorbent assay (ELISA), radioimmunoassay (RIA), and radial immunodiffusion (RID)

References

Dati F, Schumann G, Thomas L, et al, "Consensus of a Group of Professional Societies and Diagnostic Companies on Guidelines for Interim Reference Ranges for 14 Proteins in Serum Based on the Standardization Against the IFCC/BCR/CAP Reference Material (CRM 470). International Federation of Clinical Chemistry. Community Bureau of Reference of the Commission of the European Communities. College of American Pathologists," *Eur J Clin Chem Clin Biochem*, 1996, 34(6):517-20.

Johnson AM, Rohlfs EM, and Silverman LM, "Proteins," *Tietz Textbook of Clinical Chemistry*, 3rd ed, Chapter 20, Burtis CA and Ashwood ER, eds, Philadelphia, PA: WB Saunders Co, 1999, 477-540.

Alpha₁-Antitrypsin Phenotyping

Related Information

Alpha₁-Antitrypsin, Serum *on page 134*
Aspartate Aminotransferase, Serum *on page 216*
Liver Biopsy *on page 864*
Liver Disease: Laboratory Assessment, Overview *on page 869*
Protein Electrophoresis, Serum *on page 1104*

Synonyms α_1-Antiprotease Phenotype; α_1-AT Phenotype

Applies to Protease Inhibitors

Test Includes Serum trypsin inhibitory capacity

Abstract Alpha₁-antitrypsin (α_1-AT) deficiency, characterized by varying levels of severity, is the most common genetic cause of liver disease in the pediatric population. Affected patients are detected by lack or diminution of the alpha₁ band on serum protein electrophoresis, abnormal migration of the alpha₁ band or by decreased levels determined immunochemically.

Hereditary α_1-AT deficiency accounts for 1% to 2% of emphysema cases.

Patient Preparation Fasting is preferred.

Specimen Serum

Container Red top tube, lavender top (EDTA) tube for molecular analysis

Sampling Time Misleading results can follow sampling during acute illness; *vide infra*.

Storage Instructions Separate serum and refrigerate or freeze.

Reference Interval Interpretation usually accompanies report; phenotypes are designated. Pi•MM phenotype is normal; Pi•MZ is heterozygous, intermediate
(Continued)

Alpha₁-Antitrypsin Phenotyping *(Continued)*

deficient; and Pi•ZZ is homozygous, severely deficient. Over 70 alleles are described. Biosynthesis of α₁-AT is controlled at the Pi locus by a pair of genes. There is codominant expression. The phenotype is "Pi" for protease inhibitor. Z and S denote mutant proteins. A null-null state occurs as well. In a dysfunctional type, α₁-AT is found in normal amounts but does not function normally.

Use Alpha₁-AT phenotyping provides definitive analysis of hereditary α₁-AT deficiency, which is associated with chronic obstructive pulmonary disease (COPD), hepatic cirrhosis, and hepatoma. α₁-AT deficiency is a cause of neonatal cholestasis. Cholestasis with neonatal hepatitis is found in a minority of neonates with α₁-AT deficiency. Such neonatal hepatitis leads to cirrhosis.

Limitations α₁-AT is a positive acute phase reactant. Therefore, patients with α₁-AT deficiency who suffer from bronchitis, pneumonia, or similar respiratory inflammation may have misleadingly normal levels during acute illness. After the acute phase of illness has passed, repeat determinations often reveal the baseline α₁-AT level which is indicative of heterozygous phenotypic deficiency.

Serum α₁-AT may be increased in patients during normal pregnancy, chronic pulmonary diseases, hereditary angioneurotic edema, gastric diseases, liver diseases, pancreatitis, diabetes, carcinomas, renal diseases, and rheumatic diseases and may be decreased in patients with severe protein loss or with improper storage of specimen.

Methodology Isoelectric focusing in a narrow range pH gradient, crossed immunoelectrophoresis. Use of high-resolution electrophoresis which would detect the slower electrophoretic migration of the Z and S variants is preferred over quantification of α₁-AT by nephelometry as a screen for this deficiency. Further, a high-resolution electrophoretic system will detect heterozygotes, which could lead to important family studies of potentially deficient first-degree relatives who may benefit from therapy.

Additional Information An M null genotype will have the MM phenotype associated with low serum α₁-AT.

References

Cuvelier A, Muir JF, Hellot MF, et al, "Distribution of α₁-Antitrypsin Alleles in Patients With Bronchiectasis," *Chest*, 2000, 117(2):415-9.

Jacobs DS, DeMott WR, Oxley DK, et al, *Laboratory Test Handbook*, 5th ed, Hudson, OH: Lexi-Comp Inc, 2001.

Snider GL, "Pulmonary Disease in Alpha₁-Antitrypsin Deficiency," *Ann Intern Med*, 1989, 111(12):957-9.

Wulfsberg EA, Hoffmann DE, and Cohen MM, "Alpha₁-Antitrypsin Deficiency. Impact of Genetic Discovery on Medicine and Society," *JAMA*, 1994, 271(3):217-22.

Yang P, Tremaine WJ, Meyer RL, et al, "Alpha₁-Antitrypsin Deficiency and Inflammatory Bowel Diseases," *Mayo Clin Proc*, 2000, 75:450-5.

Alpha₁-Antitrypsin, Serum

Related Information

Alpha₁-Antitrypsin Phenotyping *on page 133*
Bilirubin, Total, Serum *on page 265*
C-Reactive Protein, Serum *on page 467*
Liver Biopsy *on page 864*
Liver Disease: Laboratory Assessment, Overview *on page 869*
Protein Electrophoresis, Serum *on page 1104*

Synonyms A₁AT; α₁-Antiprotease; α₁-AT

Applies to Acute Phase Reactant; Protease Inhibitors

Abstract Deficiency of α₁-antitrypsin (α₁-AT) may present as emphysema, classically in persons younger than 40. It is the most common genetic cause of liver disease in the pediatric population. α₁-antitrypsin is the most abundant proteinase inhibitor (Pi) in plasma. α₁-AT is an acute phase reactant.

Specimen Serum. Prenatal diagnosis is possible.

Container Red top tube, lavender top (EDTA) tube for molecular analysis

Reference Interval 110-140 mg/dL, method dependent. Levels are normally low at birth but rise soon thereafter.

Critical Values Levels <70 mg/dL (SI: <0.70 g/L) are likely to correlate with homozygous deficiency; subjects having levels <125 mg/dL (SI: <1.40 g/L) should be phenotyped.

Use Detect hereditary decreases in the production of alpha₁-antitrypsin (α₁-AT) (see Alpha₁-Antitrypsin Phenotyping *on page 133*). Decreased or nearly absent levels of α_1-AT are important in chronic obstructive lung disease and in liver disease. An increased prevalence of non-MM phenotypes is found with cirrhosis, chronic liver disease, and with hepatocellular carcinoma. Chronic liver disease in a child should raise consideration of α_1-AT deficiency or Wilson disease. α_1-AT deficiency is a leading cause of childhood liver disease. However, only a minority of children with Pi•ZZ phenotype develop evidence of liver disease by adolescence.

Limitations α_1-AT may be elevated into normal range in heterozygous deficient patients during concurrent infection, pregnancy, estrogen therapy, steroid therapy, cancer, and during postoperative periods. Homozygous deficient patients will not show such elevation. Normal α_1-AT levels may occur in patients with liver disease who are heterozygotes. In normals, pregnancy and contraceptive medication may elevate levels. α_1-AT is often elevated in inflammatory states (eg, rheumatoid arthritis, bacterial infection, vasculitis, neoplasia).

Contraindications If CRP is elevated, retest α_1-AT in 10-14 days.

Methodology Nephelometry, radial immunodiffusion (RID), molecular analysis by polymerase chain reaction (PCR), isoelectric focusing for phenotyping

Additional Information This assay should be performed when alpha₁ globulin in serum protein electrophoresis is decreased, when two bands are seen in the alpha₁ region, when the alpha₁ region is obscured by alpha₁ lipoprotein and especially on clinical indications. Heterozygous patients (Pi•MZ phenotype) or homozygous patients with PiSS exhibit α_1-AT levels which are commonly about 60% of normal, but most of these patients do not have clinically important liver disease. In one report, the P variant in conjunction with Z allele has been attributed to the progression of liver disease in one patient. One of the P variants, P$_{lowell}$ mutation, results in alpha₁-AT that is rapidly degraded by hepatocytes. Homozygous (Pi•ZZ phenotype) patients exhibit activity levels of about 10% to 18% of normal. Phenotyping is desirable on patients with low values and on all patients being worked up for α_1-AT-deficient liver disease. Most pathologic is homozygous state ZZ. Early emphysema exacerbated by smoking develops in PiZZ individuals with A₁AT concentrations <10%. However, this is a very uncommon cause of chronic obstructive pulmonary disease. Neutrophil elastase, inhibited in pulmonary parenchyma by A₁AT, leads to the emphysema of A₁AT deficiency. An M null genotype will have phenotype as MM but a low serum level.

References

Barnes PJ, "Chronic Obstructive Pulmonary Disease," *N Engl J Med*, 2000, 343(4):269-80.

Perlmutter DH, "Alpha-1-Antitrypsin Deficiency: Biochemistry and Clinical Manifestations," *Ann Med*, 1996, 28(5):385-94.

Yang P, Tremaine WJ, Meyer RL, et al, "α_1-Antitrypsin Deficiency and Inflammatory Bowel Diseases," *Mayo Clin Proc*, 2000, 75:450-5.

Alpha₁-Fetoprotein, Amniotic Fluid

Related Information

AcetylcholinesteraseE, Amniotic Fluid *on page 93*

Acetylcholinesterase, Red Cell and Serum *on page 93*

Alpha₁-Fetoprotein, Serum *on page 136*

Amniotic Fluid, Chromosome and Genetic Abnormality Analysis *on page 152*

Cystic Fibrosis DNA Detection *on page 491*

Duchenne/Becker Muscular Dystrophy DNA Detection *on page 526*

Fluorescence *in situ* Hybridization *on page 602*

Kleihauer-Betke *on page 822*

Applies to Amniotic Fluid Acetylcholinesterase

Test Includes Cytogenetic studies can be initiated from the same specimen.

Abstract Amniotic fluid AFP (AF-AFP) testing is done following positive maternal AFP screening, but may also be done when the family history is positive for neural tube defect. Prediction of neural tube defects can be projected much more accurately by amniotic fluid AFP than by serum screening.

Specimen Amniotic fluid

Container Sterile syringe
(Continued)

Alpha₁-Fetoprotein, Amniotic Fluid (Continued)

Collection Although the optimal time to collect amniotic fluid for AFP is between the 16th and 18th weeks of gestation, reference intervals are generally established for 14-25 weeks. Provide gestational age to the laboratory. If the amniotic fluid is traumatic (bloody), a maternal blood specimen should also be submitted. One or two drops of fetal blood in amniotic fluid can give false-positive results.

Storage Instructions Refrigerate

Causes for Rejection Sample determined not to be amniotic fluid; contamination of amniotic fluid with maternal or fetal blood; recently administered radioisotopes if RIA is used for assay. Urine urea nitrogen (UUN) may be used to distinguish maternal urine from amniotic fluid, since UUN of maternal urine is >100 g/day (that of normal amniotic fluid is much less).

Special Instructions Amniotic fluid for AFP probably should not be collected before the 14th week.

Reference Interval Reference intervals are stratified by weeks of gestation, decreasing with increasing maturity. **Reference intervals are supplied by the laboratory performing the assay.** Assay results are expressed as multiples of the median (MoM) and >0.5 MoM and <2.5 MoM. Most authorities, however, regard MoM >2.0 as abnormal until proven otherwise. The curves for unaffected and open spina bifida pregnancies, during the second trimester, slightly overlap. MoM is **not** corrected for maternal race, maternal weight, and maternal insulin-dependent diabetes mellitus. Amniotic fluid AFP peak differs from that of maternal serum.

Possible Panic Range Median AF-AFP is ~7 MoM in open spina bifida and ~20 MoM in anencephaly.

Use An elevated maternal serum AFP usually is followed by an ultrasound exam, after which amniotic fluid may be obtained. If this is done, the amniotic fluid is usually analyzed for AFP and acetylcholinesterase.

Limitations Contamination with fetal blood will produce marked elevations of AFP, but not acetylcholine.

Methodology Immunoassay, solid-phase and enzyme-labeled monoclonal antibody directed to different epitopes

Additional Information See Amniotic Fluid, Chromosome and Genetic Abnormality Analysis *on page 152.*

References

Ashwood ER, "Clinical Chemistry of Pregnancy," *Tietz Textbook of Clinical Chemistry*, 3rd ed, Burtis CA and Ashwood ER, eds, Philadelphia, PA: WB Saunders Co, 1999, 1736-75.

Haddow JE and Palomaki GE, "Biochemical Markers of Fetal Disorders in Maternal Serum and Amniotic Fluid," *Medicine of the Fetus and Mother*, 2nd ed, Reece EA and Hobbins JC, eds, Philadelphia, PA: Lippincott-Raven Publishers, 1999, 689-706.

Alpha₁-Fetoprotein, Serum

Related Information

Acetylcholinesterase, Red Cell and Serum *on page 93*
Alpha₁-Fetoprotein, Amniotic Fluid *on page 135*
Amniotic Fluid, Chromosome and Genetic Abnormality Analysis *on page 152*
Body Fluid Chemical Analysis *on page 291*
CA 19-9, Serum *on page 321*
Carcinoembryonic Antigen, Serum *on page 342*
Chorionic Gonadotropin, Human, Serum and Urine *on page 397*
Cyst Fluid Cytology *on page 490*
Fanconi Anemia, Chromosome Breakage Study *on page 569*
Inhibin A, Serum *on page 799*
Pregnancy-Associated Protein A, Serum *on page 1082*

Synonyms AFP

Applies to Triple Test

Abstract Alpha₁-fetoprotein (AFP) is increased in hepatic disorders attended by hepatocyte regenerative activity, in hepatocellular carcinoma, and in nonseminomatous germ cell tumors.

Maternal serum AFP (MSAFP) is used as a prenatal screen for two classes of fetal abnormality:

1. elevated MSAFP is associated with neural tube defects, abdominal wall defects, and other abnormalities
2. low MSAFP is associated with fetal chromosome abnormalities, trisomy 21 (Down syndrome), and trisomy 18

Specimen Serum; can also be done on cerebrospinal fluid (CSF)

Container Plain red top tube; not SST™ if intended for maternal screening.

Sampling Time The optimal time to draw **maternal serum** for prenatal screening is at 16-18 weeks of gestation. Maternal serum can be collected between 15 and 21 weeks. AFP loses sensitivity when sampled earlier than 15 weeks. Check with laboratory.

One of the most common reasons for an abnormal result is uncorrected gestational age. Thus, confirmation of gestational age with ultrasound is helpful and recalculation is necessary if the original gestational age is inaccurate.

Storage Instructions Freeze

Special Instructions Maternal serum screening programs usually require maternal birth date, first day of last menstrual period, gestational age by calculation and gestation by ultrasound and by physical examination, expected date of delivery, maternal weight, race, and diabetic status.

Reference Interval

Tumor marker:

Children:

0-1 month: male: 0.6-16,387 ng/mL; female: 0.6-18,964 ng/mL (values are very high at birth)

1-12 months: male: 0.6-28.3 ng/mL; female: 0.6-77.0 ng/mL

1-3 years: male: 0.6-7.9 ng/mL; female: 0.6-11.1 ng/mL

4-6 years: male: 0.6-5.6 ng/mL; female: 0.6-4.2 ng/mL

7-12 years: male: 0.6-3.7 ng/mL; female: 0.6-5.6 ng/mL

13-18 years: male: 0.6-3.9 ng/mL; female: 0.6-4.2 ng/mL

Adults:

- nonpregnant: <15 ng/mL
- pregnant: ≤500 ng/mL in 3rd trimester

Prenatal screen: Results are usually expressed as multiples of the median (MoM), with the conventional reference interval of 0.5-2.5 MoM. Lenke et al suggest 0.5-2.0 MoM. The MoM result is often adjusted for gestational age, maternal weight, diabetes mellitus, maternal race, and the number of fetuses. Physicians advising patients must be familiar with the practices of the laboratory performing the test.

Use

Tumor marker: AFP is used to detect and monitor the treatment of **hepatocellular carcinoma**, a disease in which serum levels of AFP are elevated, often dramatically. AFP >1000 ng/mL is a negative prognostic factor in patients with hepatocellular carcinoma. AFP is often combined with human chorionic gonadotropin (hCG) to stage and monitor nonseminomatous **germ cell tumors**, especially testicular tumors. In yolk sac tumors (endodermal sinus tumor), AFP is elevated and hCG normal; in choriocarcinoma, hCG is elevated while AFP is normal. In embryonal carcinoma, both are elevated. Patients with seminoma have normal AFP, but hCG may be elevated in the subset containing trophoblastic elements.

Prenatal screening: Many screening programs use the **triple test** which measures AFP, hCG, and unconjugated estriol (uE₃) in maternal serum. When the maternal serum screen finds elevated AFP, attention is directed to the possibility of neural tube defect, abdominal wall defect, or other anatomic malformation (see table and Yankowitz et al). Follow-up often includes ultrasound and measurement of amniotic fluid AFP and acetylcholine. When the maternal serum screen finds low AFP, attention is directed toward the possibility of a chromosome abnormality, such as Down syndrome or trisomy 18; a number of diagnostic approaches may then be undertaken. (In Down syndrome, the classical pattern is low AFP, low uE₃, and elevated hCG). See following table.

(Continued)

Alpha₁-Fetoprotein, Serum *(Continued)*

Maternal Serum: Abnormalities of Gestation Summary

	AFP	hCG	uE₃
Abortion, spontaneous or impending		Low	Low
Anencephaly	Very high		Very low
Atresia, esophagus, duodenum	High		
Encephalocoele	High		
Fetal blood contamination	High		
Fetal demise	Low	Low	Low
Gastroschisis	High		
Hemolytic disease of the newborn	High		
Herpesviral fetal liver necrosis	High		
Hydrocephalus	High		
Molar gestation	Undetectable	Very high	Very low
Molar gestation, partial	Low to normal	Very high	Low to normal
Multiple gestation	High	High	High
Myelomeningocoele	High		
Omphalocoele	High		
Preeclampsia		High	
Sulfatase deficiency (fetal)			Very low
Trisomy 13	High		
Trisomy 18	Low	Low	Very low
Trisomy 21 (Down syndrome)	Low	High	Low

Replacement of unconjugated estriol (uE₃) with inhibin A in the multiple marker screening test has been advocated for Down syndrome. See also Inhibin A, Serum and PAPP-A, Serum.

Modified from Ashwood ER, "Clinical Chemistry of Pregnancy" in *Tietz Textbook of Clinical Chemistry* 3rd ed, Burtis CA and Ashwood ER, eds, Philadelphia, PA: WB Saunders Co, 1999.

Limitations Serum AFP is elevated in some cases of nonmalignant liver disease (eg, massive hepatic necrosis, hepatitis, cirrhosis). AFP levels are usually <150 µg/L in such nonmalignant diseases.

False positives and false negatives occur. Abnormal results in maternal serum screening programs are never conclusive, and must be followed by more specific procedures (eg, high-resolution ultrasound, amniocentesis, chorionic villus sampling, cordocentesis, etc).

Contraindications Recently administered radioisotopes may interfere with RIA testing.

Methodology Enzyme immunoassay (EIA), monoclonal immunofluorescent assay, immunochemiluminescence assay, radioimmunoassay (RIA)

References

Baliff JP and Mooney RA, "New Developments in Prenatal Screening for Down Syndrome," *Am J Clin Pathol*, 2003, 120(Suppl 1):S14-24.

Chan DW and Sell S, "Tumor Markers," *Tietz Textbook of Clinical Chemistry*, 3rd ed, Burtis CA and Ashwood ER, eds, Philadelphia, PA: WB Saunders Co, 1999, 722-49.

Goldstein NS, Blue DE, Hankin R, et al, "Serum Alpha-Fetoprotein Levels in Patients With Chronic Hepatitis C," *Am J Clin Pathol*, 1999, 111(6):811-16.

Haddow JE and Palomaki GE, "Biochemical Markers of Fetal Disorders in Maternal Serum and Amniotic Fluid," *Medicine of the Fetus and Mother*, 2nd ed, Chapter 40, Reece EA and Hobbins JC, eds, Philadelphia, PA: Lippincott-Raven, 1999, 689-706.

Palomaki GE and Haddow JE, "Prenatal Screening for Open Neural-Tube Defects in Maine," *N Engl J Med*, 1999, 340(13)1049-50.

Shen BY, Li HW, Regimbeau JM, et al, "Recurrence After Resection of Hepatocellular Carcinoma," *Hepatobiliary Pancreat Dis Int*, 2002, 1(3):401-5.

Soldin SJ, Brugnara C, Gunter KC, et al, *Pediatric Reference Ranges*, 2nd ed, Washington, DC: AACC Press, 1997, 9.

Wenstrom KD, Owen J, Chu DC, et al, "Maternal Serum Alpha-Fetoprotein and Dimeric Inhibin A Detect Aneuploidies Other Than Down Syndrome," *Am J Obstet Gynecol*, 1998, 179(4):966-70.

Yankowitz J and Williamson RA, "Abnormalities of Alpha-Fetoprotein and Other Biochemical Tests," *High Risk Pregnancy - Management Options*, James DK, et al, eds, London: WB Saunders Co, 1999, 153-70.

Yaron Y, Cherry M, Kramer RL, et al, "Second-Trimester Maternal Serum Marker Screening: Maternal Serum Alpha-Fetoprotein, Beta-Human Chorionic Gonadotropin, Estriol, and Their Various Combinations as Predictors of Pregnancy Outcome," *Am J Obstet Gynecol*, 1999, 181(4):968-74.

♦ **Alpha₁ Lipoprotein Cholesterol** *see* High Density Lipoprotein Cholesterol, Serum *on page 729*

♦ **Alpha₂-Antiplasmin** *see* Antiplasmin *on page 196*

Alpha₂-Macroglobulin, Serum

Abstract α_2-Macroglobulin (AMG) is synthesized primarily by hepatic parenchymal cells. The loss of AMG in the urine is prevented by its large size. The plasma concentration of AMG increases in nephrotic syndrome due to volume shifts and a compensatory increase in synthesis in response to albumin loss in urine. AMG is not an acute phase reactant. Levels are decreased in proteolytic diseases such as acute pancreatitis and peptic ulcer.

Patient Preparation Patient should be fasting.

Specimen Serum

Container Red top tube

Storage Instructions Refrigerate (4°C) for up to 72 hours. After 72 hours, the specimen should be stored frozen at -20°C and thawed only once prior to analysis.

Causes for Rejection Gross hemolysis, lipemia

Reference Interval The reference interval in adults 20-60 years of age is 130-300 mg/dL. Since estrogen increases AMG synthesis, levels in women of childbearing age are higher than in age-matched men. Children and infants have levels higher than adults. Reference intervals for infants are as follows: birth: 125-392 mg/dL, 1 month: 217-439 mg/dL, 3 months: 261-466 mg/dL, 6 months: 269-503 mg/dL. In subjects 6 months to 15 years, results range from 250-640 mg/dL, peaking at 2-4 years of age.

Use Measurement of AMG may be used to evaluate patients with nephrotic syndrome and patients with proteolytic conditions such as pancreatitis, or to further characterize an electrophoretic change in the α_2-zone.

Limitations Estrogen therapy, exercise, pregnancy, and any number of disease states (eg, diabetes mellitus, hepatitis, cirrhosis of the liver, α_1-antitrypsin deficiency, cerebral infarction, amyloidosis) may result in increased AMG serum levels. Decreased concentrations have been observed following surgery.

Methodology Immunonephelometry, immunoturbidimetry, radial immunodiffusion (RID)

Additional Information AMG is one of the proteins which forms a complex with prostate-specific antigen (PSA). However, most PSA assays do not detect the PSA-AMG complex. Pretreatment concentrations are decreased in patients with advanced carcinoma of prostate.

References

Dati F, Schumann G, Thomas L, et al, "Consensus of a Group of Professional Societies and Diagnostic Companies on Guidelines for Interim Reference Ranges for 14 Proteins in Serum Based on the Standardization Against the IFCC/BCR/CAP Reference Material (CRM 470). International Federation of Clinical Chemistry. Community Bureau of Reference of the Commission of the European Communities. College of American Pathologists," *Eur J Clin Chem Clin Biochem*, 1996, 34(6):517-20.

Johnson AM, Rohlfs EM, and Silverman LM, "Proteins," *Tietz Textbook of Clinical Chemistry*, 3rd ed, Chapter 20, Burtis CA and Ashwood ER, eds, Philadelphia, PA: WB Saunders Co, 1999, 477-540.

Wu YY, Delgado RM, Sunderland T, et al, "Detection of a Deletion Polymorphism of the Human α_2-Macroglobulin Gene (A2M-2) by a Semiautomated PCR-Single-Stranded Conformational Polymorphism Method," *Clin Chem*, 1999, 45(9):1572-3.

♦ **Alpha Fetoprotein** *see* Immunoperoxidase Procedures *on page 780*

♦ **5-Alpha Reductase** *see* Dihydrotestosterone, Serum *on page 514*

♦ **Alpha Tocopherol** *see* Vitamin E, Serum or Plasma *on page 1319*

♦ **Al, Serum** *see* Aluminum, Serum or Urine *on page 141*

♦ **ALT** *see* Alanine Aminotransferase, Serum *on page 116*

♦ **ALT:AST Ratio** *see* Liver Disease: Laboratory Assessment, Overview *on page 869*

♦ **Alternate Site Testing** *see* Point-of-Care Testing *on page 1065*

Aluminum, Bone and Bone Biopsy

Related Information

Aluminum, Serum or Urine *on page 141*
Calcium, Serum *on page 329*
Deferoxamine Infusion Test *on page 504*
Histopathology *on page 733*
(Continued)

Aluminum, Bone and Bone Biopsy *(Continued)*

Osteocalcin, Serum or Plasma *on page 983*
Parathyroid Hormone, Serum *on page 1001*
Vitamin D, Serum *on page 1318*

Synonyms Al, Bone

Applies to Bone Biopsy; Bone Histomorphometry; Histomorphometry

Test Includes Aluminum measured on anterior iliac crest bone biopsy specimen

Abstract Aluminum (Al), one of the most common elements in the earth's crust, is a potentially toxic metal, interfering with bone mineralization in individuals with dialysis-dependent renal failure. The human body is protected from Al accumulation in normal individuals because the gut is rather impervious to Al, and almost all of the Al that is absorbed is excreted. Long-term hemodialysis with Al contaminants in the dialysate and long-term parenteral nutritional support with trace Al contamination in parenteral solutions (especially in those individuals with chronic renal insufficiency) are the two major causes of Al-related osteodystrophy and are causes of progressive dementia.

Patient Preparation Tetracycline and Declomycin® fluoresce differently under ultraviolet light and can be separately distinguished under the microscope so as to indicate the amount of bone formed between the two tetracycline labels. Tetracycline and Declomycin® should be taken between meals, and all antacids, calcium supplements, and phosphate binders should be avoided on the days when these labels are taken. Days 1, 2 (2 days): tetracycline 500 mg twice daily, midmorning and midafternoon; days 3-12: no tetracycline; days 13, 14, 15, 16 (4 days): Declomycin® 300 mg twice daily, midmorning and midafternoon; days 17, 18 (2 days): no tetracycline; day 19, 20, or 21: do bone biopsy. The dates and times of labels and of biopsy should be recorded and submitted with the specimen to facilitate interpretation.

Specimen The patient's skeleton is prelabeled with tetracycline (see Patient Preparation). The timing of the labels and the day of the biopsy are critical to standardize the time between tetracycline labels (which fluoresce) and the day of biopsy. The biopsy specimen is a **0.8 cm diameter core of bone** taken under local or general anesthesia full thickness from the anterior iliac crest in a standardized fashion, with a hollow bone biopsy instrument so as to obtain both layers of cancellous bone as well as internal trabecular bone.

Container Acid-washed plastic vial containing 95% ethanol. (**Do not fix the specimen in formalin.**) **Avoid decalcification of the bone biopsy specimen.**

Special Instructions Bone Al is usually measured in concert with bone histomorphometry and is performed by prearrangement with a reference laboratory specializing in bone histomorphometry. These instructions assume bone histomorphometry will also be performed in parallel with bone Al determination.

Reference Interval Bone Al: <15 µg/g dry weight (see figure).

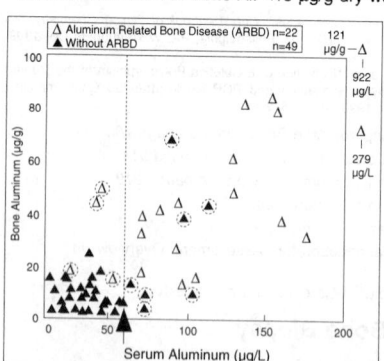

Relationship between serum aluminum, bone aluminum, and aluminum-related bone disease (ARBD). The sensitivity and specificity of a serum aluminum value of 60 µg/L (↑) in the detection of ARBD is illustrated. ▲ without ARBD, △ with ARBD as diagnosed by histochemistry, histology, and bulk analysis; ⊛ false-positive; ⊘ false negatives.

Bone Al relates to total body burden of Al, whereas **serum Al** may be only recently elevated in heavy exposure and may not correlate with Al-related bone disease (ARBD). Conversely, individuals with industrial exposure to Al might have normal serum Al concentrations.

Histomorphometry normal values:
- trabecular bone: 3% nonmineralized, 97% mineralized
- osteoid covers 25% bone surface, lined with osteoblasts
- osteoclasts 4% of trabecular surface

Variations from normal histomorphometry are interpreted by the pathologist as compatible with pure osteitis fibrosa, pure osteomalacia, aplastic bone disease, or mixed bone disease. Bone resorption and turnover may be determined. The measured distance between the two tetracycline labels allows calculation of bone formation rate, which is reduced in Al-related bone disease. Many feel that an Al stain, however, is most specific for Al-related bone disease if Al is detected on the mineralization front, blocking calcium deposition and resulting in osteomalacia. The rate of bone formation is inversely related to the amount of Al present.

Use Diagnose or confirm ARBD in patients with renal failure (with or without dialysis) or in those receiving parenteral nutrition. Aluminum bone disease among parenteral nutrition patients has become much less of a clinical problem, as nutritional products for intravenous use have been improved. Parenteral nutrition solutions prepared with purified amino acids rather than from casein hydrolysate have diminished the incidence of ARBD. Intravenous albumin products may still be a significant source of Al.

Limitations Histomorphometry does not always correlate with total bone Al content. Secondary hyperparathyroidism provides relative protection from clinical Al bone disease, despite the presence of substantial Al stored in bone.

Contraindications Bone biopsy should be done with caution in the presence of a coagulopathy; history of allergy to tetracyclines precludes tetracycline labeling.

Methodology Atomic absorption (AA), graphite furnace flameless atomic absorption with Zeeman background correction, inductively-coupled plasma-mass spectroscopy (ICP-MS)

Additional Information Aluminum interferes with normal bone formation by several mechanisms, including direct reduction of osteoblast function and population; reduction of parathyroid hormone release down-regulating bone turnover and in the presence of citrate, direct inhibition of calcium phosphate crystal growth. Bone serves as a store of Al, and Al from bone can be released back to the blood and other tissues during stress, illness, hyperthyroidism, or failed renal transplant, precipitating Al encephalopathy.

Stability of serum Al concentrations in one study of dialysis patients has remained constant.

References

Duarte ME, Peixoto AL, Pacheco AS, et al, "The Spectrum of Bone Disease in 200 Chronic Hemodialysis Patients: A Correlation Between Clinical, Biochemical and Histological Findings," *Rev Paul Med*, 1998, 116(5):1790-7.

Fernández Martín JL, Canteros A, and Serrano M, "Prevention of Aluminum Exposure Through Dialysis Fluids. Analysis of Changes in the Last 8 Years," *Nephrol Dial Transplant*, 1998, 13(Suppl 3):78-81.

Hruska KA and Teitelbaum SL, "Renal Osteodystrophy," *N Engl J Med*, 1995, 333(3):166-74.

Aluminum, Serum or Urine

Related Information

Aluminum, Bone and Bone Biopsy *on page 139*
Calcium, Serum *on page 329*
Deferoxamine Infusion Test *on page 504*
Red Blood Cell Indices *on page 1136*

Synonyms Al, Serum

Abstract Aluminum (Al), a lightweight, ubiquitous, silvery metal, is never found alone but it commonly occurs as Al silicate or as a silicate of Al mixed with other metals. Bauxite, an impure hydrated Al oxide, is the commercial source of Al and its compounds. The potential toxicity of Al includes microcytic hypochromic (Continued)

Aluminum, Serum or Urine *(Continued)*

anemia and its role in progressive dementia and osteodystrophy. There is evidence that Al may play a role in Alzheimer disease.

Specimen Serum, dialysis fluid, urine, cerebrospinal fluid

Container Serum collected in metal-free tubes; urine in acid-washed container.

Causes for Rejection Use of ordinary collection tubes or containers

Special Instructions The patient should take no Al-containing antacids or medicines (eg, Amphojel®, Basaljel®, Gelusil®, Maalox®, Mylanta®, Sucralfate, etc) for 24 hours prior to obtaining specimens.

Reference Interval Serum (normal patient): 0-6 ng/mL (SI: 0-0.22 µmol/L) (may vary with laboratory); serum (dialysis patients): up to 40 ng/mL (SI: <1.48 µmol/L) without apparent acute effects; urine: 0-32 ng/day (SI: 0-1.2 µmol/day); dialysate: <0.01 mg/L (AAMI standards). As Al exposure of dialysis patients by medications has fallen in recent years, there is a tendency to move the lower limit of normal down toward 10 ng/mL (SI: 0.38 µmol/L) or 12 ng/mL, even for dialysis patients.

Critical Values Serum: >100 ng/mL (SI: >3.7 µmol/L) possible CNS toxicity, >200 ng/mL (SI: >7.4 µmol/L) probable multisystem toxicity

Use Monitor patients for prior and ongoing exposure to Al. Patients at risk include:

- infants on parenteral fluids, particularly parenteral nutrition
- burn patients through administration of intravenous albumin, particularly with coexisting renal failure
- adult and pediatric patients with chronic renal failure, who accumulate Al readily from medications and dialysate
- adult parenteral nutrition patients (less so, recently)
- persons with industrial exposure

Monitor dialysate and water to prepare dialysate fluid.

Limitations Serum levels rise and fall after each dose of Al-containing phosphate binder or sulfacrate. Al antacids, Al hydroxide, lithium, and sulfacrate can cause an increase.

Methodology Atomic absorption (AA), inductively-coupled plasma atomic emission spectrometry, graphite furnace atomic absorption spectroscopy with Zeeman correction (GFAAS); *vide infra*

Additional Information Signs and symptoms of possible chronic intoxication include:

- encephalopathy (stuttering, gait disturbance, myoclonic jerks, seizures, coma, abnormal EEG)
- osteomalacia or aplastic bone disease (associated with painful spontaneous fractures, hypercalcemia, tumorous calcinosis)
- proximal myopathy
- increased risk of infection
- increased left ventricular mass and decreased myocardial function
- microcytic anemia
- with very high levels, sudden death

The gastrointestinal tract is relatively impervious to Al, absorption normally being only about 2%. Aluminum is absorbed by a mechanism related to that for calcium. Gastric acidity, organic acids, and oral citrate favor absorption, and H_2-blockers reduce absorption. As is true for several trace elements, transferrin is the primary protein binder and carrier for Al in the plasma, where 80% is protein bound and 20% is free or complexed to small molecules such as citrate. Peripheral cells appear to take up Al from transferrin rather than from citrate. Serum Al correlates with encephalopathy; red cell Al correlates with microcytic anemia; and bone Al correlates with Al bone disease. Prolonged intravenous feeding of preterm infants with solutions containing Al is associated with impaired neurologic development.

References

Bishop NJ, Morley R, Day JP, et al, "Aluminum Neurotoxicity in Preterm Infants Receiving Intravenous Feeding Solutions," *N Engl J Med*, 1997, 336(22):1557-61.

Di Paolo N, Masti A, Comparini IB, et al, "Uremia, Dialysis and Aluminum," *Int J Artif Organs*, 1997, 20(10):547-52.

Hachinski V, "Aluminum Exposure and Risk of Alzheimer Disease," *Arch Neurol*, 1998, 55(5):742.

♦ **Alupram®** *see* Diazepam, Serum *on page 510*

♦ **Alurate®** *see* Barbiturates, Quantitative, Serum or Plasma *on page 248*

♦ **Alveolar Echinococcosis** *see* Echinococcosis Diagnostic Procedures *on page 530*

♦ **AMA** *see* Antimitochondrial Antibody *on page 183*

♦ **Amazin®** *see* Chlorpromazine, Serum *on page 395*

♦ **Amethopterin** *see* Methotrexate, Serum or Plasma *on page 905*

♦ **Amikacin, Serum** *see* Aminoglycosides, Serum *on page 147*

♦ **Aminoacidopathies** *see* Amino Acids, Urine *on page 145*

Amino Acids, Plasma

Related Information
Amino Acids, Urine *on page 145*
Ammonia, Plasma *on page 150*
Newborn Screen for Phenylketonuria *on page 954*
Phenylalanine, Blood *on page 1022*

Synonyms Inborn Errors of Metabolism Screen; Metabolic Screen for Amino Acids

Applies to Biotinidase; Carnitine; Organic Acids

Test Includes Screening and quantitation of amino acids

Abstract A variety of inherited metabolic disorders result in aminoacidemia/aminoaciduria. Typically, the abnormal amino acids concentration exceeds 3-4 times normal, but may be >10 times normal.

Patient Preparation Infants: 4-hour fast; children and adults: 12-hour fast. Protein intake does not affect diurnal variation, but it influences absolute concentrations of amino acids in blood or urine.

Specimen Plasma

Container Green top (heparin) tube

Reference Interval Varies significantly with age. Established by each laboratory. See following table as a guide. Age specific reference intervals are also published.

Amino Acids in Plasma
(µmol/L)

Amino Acid	Men	Women	Adolescents	Children
Alanine	146-494	218-474	242-594	120-600
α-Aminobutyrate	15-35	7-35	8-36	12-43
Arginine	28-96	28-108	1-81	12-112
Asparagine	32-92	26-74	34-94	15-83
Aspartic Acid	2-9	3-6	3-15	1-17
Citrulline	19-47	10-58	19-52	8-47
Cystine	24-54	31-49	36-58	23-68
Glutamic Acid	6-62	6-38	17-69	14-78
Glutamine	466-798	340-696	457-857	333-809
Glycine	147-299	100-384	166-330	107-343
Histidine	72-108	68-104	68-108	47-135
Isoleucine	46-90	39-67	34-106	6-122
Leucine	113-205	98-142	86-206	30-246
Lysine	135-243	119-203	116-276	66-270
Methionine	213-37	214-30	213-41	223-43
Ornithine	55-135	36-96	47-195	20-136
Phenylalanine	46-74	42-62	34-86	26-98
Proline	97-297	112-220	58-324	40-332
Serine	89-165	78-166	92-196	70-194
Taurine	27-95	18-66	20-90	20-120
Threonine	92-180	93-197	102-246	40-204
Tryptophan	25-65	17-53	—	12-69
Tyrosine	37-77	26-78	35-107	19-119
Valine	179-335	172-248	155-343	132-480

(Continued)

Amino Acids, Plasma *(Continued)*

Sampling Time Plasma amino acids concentrations are higher after a high protein meal. Therefore, fasting samples are preferred, **except** in hyperammonemia screening. Postprandial ammonia samples are preferred since elevations are maximum in the fed state. Marked circadian rhythms are exhibited with values highest in afternoon and lowest in early morning.

Collection Routine venipuncture

Storage Instructions Centrifuge. Transfer plasma to plastic vial and freeze within 1 hour of collection. Stable for 1 week at -20°C. For longer periods, deproteinize sample and store at -70°C.

Congenital Disorders of Amino Acid Metabolism

Name	Enzyme or Metabolic Pathways	Clinical Findings	Laboratory Findings
Classic phenylketonuria	Phenylalanine hydroxylase	Mental retardation, psychiatric dysfunction	Plasma phenylalanine >15 mg/dL
Benign phenylalaninemia	Phenylalanine hydroxylase	Asymptomatic	Increased plasma phenylalamine
Malignant phenylalaninemia	Dihydropteridine reductase	Mental retardation, psychiatric dysfunction	Increased plasma phenylalanine
Malignant phenylalaninemia	GTP cyclohydrolase	Mental retardation, psychiatric dysfunction	Increased plasma phenylalanine
Hereditary tyrosinemia	p-hydroxy phenyl acetic acid hydroxylase	Hepatic cirrhosis, renal tubular dysfunction	Increased plasma tyrosine
Alkaptonuria	Homogentisic acid oxidase	Ochronosis, arthritis	Increased urinary homogentisic acid
Histidinemia	Histidine-ammonia lyase	Hearing and speech defect	Increased plasma and urine histidine
Branched-chain aminoacidemia	Branched-chain amino acid oxidase	Seizures, ketosis, mental retardation	Increased urine and plasma branched-chain amino acids
Homocystinuria	Cystathionine synthase	Mental retardation, thromboembolism	Increased plasma and urine homocystine and methionine
Cystathioninuria	Cystathionase	Asymptomatic	Increased urine cystathionine
Cystinuria	Renal transport system for cystine and dibasic amino acids	Cystine stones	Increased urine cystine and dibasic amino acids
Hyperglycinemias			
Ketotic form	Propionyl-CoA-carboxylase	Ketosis, neutropenia, mental retardation	Increased urine and plasma glycine and propionic acid
Nonketotic form	Glycine decarboxylase	Developmental retardation	Increased urine and plasma glycine
Urea cycle abnormalities	Carbamoylphosphate synthase, ornithine-carbamoyltransferase, citrulline aspartate lyase, argininosuccinate arginine-lyase	Developmental retardation, vomiting, lethargy, seizures, hepatomegaly	Increased urine and plasma ammonia, glutamine, citrulline, and arginosuccinate
Glycinuria	Renal transport system for glycine and amino acids	Asymptomatic	Increased urine glycine, proline, and hydroxyproline
Hartnup disease	Renal transport system for neutral amino acids	Ataxia, retardation	Increased urine neutral amino acids
Fanconi syndrome	General renal transport deficiency	Acidosis and rickets	General aminoaciduria, glycinemia, phosphaturia

Use Screen for inborn errors of metabolism of amino acids (eg, investigation of metabolic acidosis, ketosis, hyperammonemia, developmental impairment, failure to thrive)

Limitations For some of these entities, investigation should include urine organic acids, amino acids, blood ammonia, biotinidase, and carnitine.

Methodology Plasma screen by single dimension thin-layer chromatography (TLC), amino acid analyzer (ion-exchange chromatography), gas chromatography (GC), and high performance liquid chromatography (HPLC)

Additional Information

Cystinuria has an autosomal recessive mode of inheritance, is a disorder of amino acid transport involving renal tubules/GI tract and should be suspect in cases of urinary stone disease. It is characterized by the formation of radiopaque urinary stones and by the presence of characteristic hexagonal crystals in the urine. Dibasic amino acids are increased in urine (see Cystine, Urine *on page 494*).

Lysinuric protein intolerance is another disorder of membrane transport. As in some cases of cystinuria, cationic amino acids (lysine, arginine and ornithine) are involved. Lysine is present in large amounts in the urine but is normal or decreased in plasma. Patients have poor appetite, fail to thrive, develop hepatosplenomegaly, hypotonia, sparse hair, osteoporosis, mental retardation, and a variety of other problems. In **Hartnup disease**, there is impaired neutral amino acid transport involving the kidneys and small intestine. It is characterized clinically by pellagra-like features, mental retardation and/or psychotic behavior, intermittent ataxia and is inherited as an autosomal recessive. A comprehensive review of these and other amino acidurias is provided in the text edited by Scriver et al. Congenital disorders of amino acids are summarized in the table.

The usual approach is to screen urine for amino acids. However, urinary levels are variable: plasma measurements usually have better predictive values. Quantitative tests for plasma amino acids are available. In almost all cases in which an amino acid is elevated in blood, it will also be elevated in urine.

CSF amino acids are useful in the diagnosis of nonketotic hyperglycinemia.

When urea cycle defects are suspected, plasma quantitative amino acids, urine orotic acids and organic acids should be assessed. See previous table.

References

Burlina AB, Bonafé L, and Zacchello F, "Clinical and Biochemical Approach to the Neonate With a Suspected Inborn Error of Amino Acid and Organic Acid Metabolism," *Semin Perinatol*, 1999, 23(2):162-73.

Hancock WS and Harding DR, "Review of Separation Conditions," *CRC Handbook of HPLC for the Separation of Amino Acids, Peptides, and Proteins*, Volume I, Hancock WS, ed, Boca Raton, FL: CRC Press, 1984, 235-62.

Lindor NM and Karnes PS, "Laboratory Medicine and Pathology: Initial Assessment of Infants and Children With Suspected Inborn Errors of Metabolism," *Mayo Clin Proc*, 1995, 70(10):987-8.

Scriver CR, Kaufman S, Eisensmith RC, et al, "Amino Acids," Part 5, *The Metabolic and Molecular Basis of Inherited Disease*, 7th ed, Scriver CR, Beaudet AL, Sly WS, et al, eds, New York, NY: McGraw-Hill Inc, 1995, 1015-368.

Shih VE, "Amino Acid Analysis," *Physician's Guide to the Laboratory Diagnosis of Metabolic Diseases*, Blau N, Duran M, and Blaskovics, eds, London: Chapman and Hall, 1996, 1-29.

Slocum RH and Cummings JG, "Amino Acid Analysis of Physiological Samples," *Techniques in Diagnostic Human Biochemical Genetics*, Hommes FA, ed, New York, NY: Wiley-Liss, 1991, 87-126.

Soldin SJ, Brugnara C, Hicks JM, "Pediatric Reference Ranges," Washington, DC: AACC Press, American Association of Clinical Chemistry, 1999, 11-9.

Amino Acids, Urine

Related Information

Amino Acids, Plasma *on page 143*
Ammonia, Plasma *on page 150*
Cystine, Urine *on page 494*
Newborn Screening Tests for Galactosemia *on page 960*
pH, Blood *on page 1018*
Phenylalanine, Blood *on page 1022*
Phosphorus, Serum *on page 1031*
Potassium, Serum or Plasma *on page 1078*
Vitamin D, Serum *on page 1318*
(Continued)

Amino Acids, Urine (Continued)

Synonyms Metabolic Screen for Amino Acids

Applies to Aminoacidopathies; Beta-Amino Isobutyrate; Cystathionine; Cystine; Glycine; Homocyst(e)ine; Hydroxyproline; Isoleucine; Leucine; Methionine; Organic Acids; Ornithine; Phenylalanine; Proline; Tryptophan; Tyrosine; Valine

Test Includes Urine amino acid screen may include alanine, arginine, citrulline, glutamine, beta-amino isobutyrate, cystathionine, cystine, glycine, homocyst(e)ine, hydroxyproline, isoleucine, leucine, methionine, ornithine, phenylalanine, proline, tryptophan, tyrosine, and valine.

Abstract Aminoacidurias may be primary or secondary; the former being due to inherited enzyme defect or inborn error of metabolism. Secondary aminoaciduria is due to organ failure, such as liver or renal failure.

Amino acids filtered through the renal glomerulus are normally almost entirely reabsorbed. In Fanconi syndrome they are not, in spite of the presence of normal plasma concentrations.

Patient Preparation Amphetamines, norepinephrine, levodopa, some antibiotics (particularly penicillins and cephalosporins), methyldopa, levodopa, and polythiazide have been reported to interfere chemically with this test. Amino acid concentrations in urine are physiologically increased by aspirin, bismuth, hydrocortisone, insulin, lead poisoning, triamcinolone, and valproic acid.

Specimen Urine

Collection Random specimen acceptable, no preservative. Morning urine preferred.

Storage Instructions Refrigerate; freeze for long-term storage.

Causes for Rejection Specific gravity of the urine must be ≥1.010

Reference Interval Subjective interpretation based on comparison of patient, normal, and control urines of comparable age. Interpretation is age dependent. Urinary amino acids separated by thin-layer chromatography (TLC) are not usually quantitated. If any amino acids appear to be in high concentration compared to normals or controls, quantitation can be carried out by separate methodology, such as amino acid analyzer (ion-exchange chromatography), gas chromatography (GC), and high performance liquid chromatography (HPLC). Significant differences exist between laboratories.

Use This urine screen is used in investigation of failure to thrive, acidosis, hypokalemia, hypophosphatemia, and abnormalities of vitamin D metabolism. It is a screen for "inborn errors of metabolism" of amino acids, Fanconi syndrome, Wilson disease, and Lowe syndrome.

Methodology Screening: thin-layer chromatography (TLC); quantitation: amino acid analyzer (ion-exchange chromatography), gas chromatography (GC), and high performance liquid chromatography (HPLC)

Additional Information Excretion of certain amino acids is increased in several specific **aminoacidurias**, such as phenylketonuria and maple syrup urine disease. Dibasic amino acids (lysine, ornithine, arginine, cystine) are increased in cystinuria. **Aminoacidopathies**, with ninhydrin-positive urinary amino acids, include phenylketonuria, cystinuria, nonketotic hyperglycinemia and homocystinuria. Their investigation includes urine organic acids and amino acids, plasma amino acids, blood NH_3^+, biotinidase, and carnitine. Aminoaciduria may also be seen in a variety of other disorders, including viral hepatitis, multiple myeloma, rickets, hyperparathyroidism, and chronic renal failure. A positive test should be followed up with quantitation on a 24-hour collection. See table in listing Amino Acids, Plasma *on page 143*.

The Fanconi syndrome (FS), characterized by dysfunction of the proximal renal tubule, is caused by a variety of hereditary and acquired diseases. The former include primary idiopathic FS, cystinosis, Lowe's syndrome, tyrosinemia type 1, galactosemia, hereditary fructose intolerance, glycogen storage disease, Wilson disease, and other entities. Acquired FS may be caused by aminoglycosides, outdated tetracyclines, cephalothins, valproic acid, streptozotocin, 6-mercaptopurine, azathioprine, and cis-platinum; toxins including heavy metals, toluene sniffing and paraquat. It may be found with nephrotic syndrome, myeloma, amyloidosis, antitubular basement membrane antibody, renal vein thrombosis, cancer, and following renal transplantation.

References

Burlina AB, Bonafé L, and Zacchello F, "Clinical and Biochemical Approach to the Neonate With a Suspected Inborn Error of Aino Acid and Organic Acid Metabolism," *Semin Perinatol*, 1999, 23(2):162-73.

Lindor NM and Karnes PS, "Initial Assessment of Infants and Children With Suspected Inborn Errors of Metabolism," *Mayo Clin Proc*, 1995, 70(10):987-8.

Soldin SJ, Brugnara C, Hicks JM, "Pediatric Reference Ranges," Washington, DC: AACC Press, American Association of Clinical Chemistry, 1999, 20-30.

Aminoglycosides, Serum

Related Information

Beta$_2$-Microglobulin, Serum or Urine *on page 254*

Magnesium, Serum *on page 885*

Vancomycin, Serum *on page 1298*

Applies to Amikacin, Serum; Gentamicin, Serum; Tobramycin, Serum

Abstract Aminoglycoside antibiotics are used primarily in combination with other drugs (often beta-lactams or vancomycin) in order to treat infections caused by aerobic gram-negative bacilli. They are used to treat susceptible nontuberculous mycobacterial infections. Amikacin is an alternative drug for nocardiosis. When used in combination with penicillins and some of their derivatives, aminoglycosides may have synergistic bactericidal activity against gram-positive cocci such as *Staphylococcus aureus* and *Enterococcus faecalis*, as well as some gram-negative aerobic rods, such as *Pseudomonas aeruginosa*. High-level resistance of *Enterococcus* species to aminoglycosides is not rare. Since aminoglycoside transport through cell walls depends on oxygen (among other things), they are not active against anaerobic organisms.

Specimen Serum

Container Red top tube

Sampling Time Peak: it is acceptable to administer the drug intravenously over 30-60 minutes and either obtain a peak level right away, or alternatively, 30 minutes after the infusion is complete. Generally, peak sample is drawn after 30 minutes to allow drug distribution. **Trough** levels should be drawn immediately prior to the next dose. When the drug is administered intramuscularly, specimens may be drawn 60-90 minutes later (**peak**) and just prior to the next dose (**trough**).

Storage Instructions Separate within 1 hour of collection and refrigerate or freeze until assayed. Must be frozen if a β-lactam antibiotic is also present, because of potential inactivation of aminoglycosides.

Reference Interval

- Amikacin: peak: 15-30 µg/mL, trough: 5-8 µg/mL
- Gentamicin: peak: 4-10 µg/mL, trough: 1-2 µg/mL
- Tobramycin: peak: 4-10 µg/mL, trough: 1-2 µg/mL

Possible Panic Range

- Amikacin: toxic: peak: >35 µg/mL (SI: >74 µmol/L), trough: >8 µg/mL (SI: >17 µmol/L)
- Gentamicin or tobramycin: peak: >12 µg/mL (>25 µmol/L), trough: >2 µg/mL (SI: >4 µmol/L),

Methodology Immunoassay

Additional Information

- Half-life: 0.5 to >24 hours (longer with decreased renal function)
- Volume of distribution: 0.4-1.3 L/kg (adults)
- Protein binding: minimal

Aminoglycosides are cleared by the kidney (no metabolism) and may accumulate in renal proximal tubular cells. Nephrotoxicity may relate to the length of time that trough levels exceed 2 µg/mL (SI: >4 µmol/L) for gentamicin and tobramycin, and 8 µg/mL (SI: >17 µmol/L) for amikacin. Creatinine levels should be monitored every 2-3 days as an indicator of impending renal toxicity, or as a sign of a decline in renal function independent of the aminoglycosides, but signaling a need for reduction in dosage of the aminoglycoside. These drugs can potentially be nephrotoxic, against which the best defense is careful dosing and monitoring. Other factors (eg, very ill patients at risk for organ dysfunction, hypotension, nephrotoxic drugs including pressor agents, contrast dye, amphotericin B, etc) may play a role as well. Aminoglycosides are cleared by hemodialysis as well.

Aminoglycosides may also cause irreversible ototoxicity.

(Continued)

Aminoglycosides, Serum *(Continued)*

References
Edson RS and Terrell CL, "The Aminoglycosides," *Mayo Clin Proc*, 1999, 74(5):519-28.

Gonzalez LS 3rd and Spencer JP, "Aminoglycosides: A Practical Review," *Am Fam Phys*, 1998, 58(8):1811-20.

Henry NK, Hoecker JL, and Rhodes KH, "Antimicrobial Therapy for INfants and Children: Guidelines for the Inpatient and Outpatient Practice of Pediatric Infectious Diseases," *Mayo Clin Proc*, 2000, 75(1):86-97.

Jacobs DS, DeMott WR, Oxley DK, et al, *Laboratory Test Handbook*, 5th ed, Hudson, OH: Lexi-Comp Inc, 2001.

Schutze GE, Lowry JA, and Kearns GL, "Monitoring of Aminoglycoside Serum Concentrations," *Pediatr Infect Dis J*, 2000, 19(5):489-90.

♦ **Aminolevulinic Acid** *see* Delta (5)-Aminolevulinic Acid, Urine *on page 508*

♦ **5-Aminolevulinic Acid** *see* Delta (5)-Aminolevulinic Acid, Urine *on page 508*

♦ **Aminophylline** *see* Theophylline, Serum *on page 1243*

♦ **Aminotransferases** *see* Alanine Aminotransferase, Serum *on page 116*

♦ **Aminotransferases** *see* Aspartate Aminotransferase, Serum *on page 216*

Amiodarone, Serum

Related Information
Digoxin, Serum *on page 512*
Lidocaine, Serum or Plasma *on page 850*
Procainamide, Serum *on page 1092*
Quinidine, Serum *on page 1129*

Synonyms Cordarone®; Pacerone®

Test Includes Desethylamiodarone

Abstract Amiodarone is an antiarrhythmic characterized by substantial toxicity and very prolonged half-life. Some of its adverse effects are potentially fatal. It is used for atrial fibrillation and life-threatening recurrent ventricular arrhythmias which have not responded to alternative therapy, and in patients at risk for sudden death. Amiodarone is effective for converting atrial fibrillation to sinus rhythm in a wide range of patients. Aggravation of arrhythmia has usually taken place in subjects with hypokalemia or those receiving another antiarrhythmic agent.

Specimen Serum

Container Red top tube; do not draw a sample in a gel-containing tube.

Sampling Time Although steady-state plasma concentrations are reached in 50-300 days, sample can be drawn 18 hours after the last dose. Time to peak serum concentrations is 4-7 hours after an oral dose.

Reference Interval 1.0-2.5 µg/mL (SI: 1.6-3.9 µmol/L) (parent); desethyl metabolite is active and is present in equal concentration to parent drug.

Possible Panic Range Adverse effects at >3.5 µg/mL (SI: >5.6 µmol/L) (parent) and >7.0 µg/mL (SI: >9.8 µmol/L) (both)

Use Compliance and toxicity assessment; used for rhythm abnormalities

Limitations Life-threatening side effects and management difficulties occur, but unlike other antiarrhythmic agents amiodarone has not been shown to increase mortality. Wide interpatient variability in dose concentration relationships limits usefulness of serum concentrations.

Contraindications Amiodarone decreases the hepatic enzyme activity needed to metabolize many drugs. This results in a decrease in clearance and therefore an increase in concentration, which may result in toxicity. Serum concentrations and pharmacologic effects of the following drugs may be increased by amiodarone: cyclosporine, digitalis, flecainide, lidocaine, phenytoin, procainamide, quinidine, theophylline, and warfarin type oral anticoagulants.

Methodology High performance liquid chromatography (HPLC)

Additional Information
- Half-life: 250-1200 hours or more
- Volume of distribution: 20-100 L/kg
- Protein binding: 95% to 97%

The bioavailability of Cordarone® is approximately 50%, but has varied between 35% and 65% in various studies. Up to 5% to 10% of patients on the drug develop hypothyroidism or hyperthyroidism. It may cause increased thyroxine levels and decreased T_3. TSH levels may be useful for diagnosis of thyroid complications. Amiodarone thyrotoxicosis occurs more frequently in areas of low iodine intake. Potassium or magnesium deficiency should be corrected. AST and ALT should be monitored on a regular basis, because side effects of amiodarone include liver complications, and it is cleared by hepatic metabolism. Increased aminotransferase and alkaline phosphatase levels are found in 25% of subjects.

Because pulmonary toxicity is the most serious noncardiac complication, chest radiographs are recommended. Pulmonary toxicity has been fatal about 10% of the time.

Amiodarone can potentiate effects of warfarin, elevating prothrombin time. It can elevate serum digoxin level and concentrations of other antiarrhythmic drugs including quinidine, procainamide, mexiletine, and propafenone.

In patients with shock-resistant out-of-hospital ventricular fibrillation, amiodarone leads to substantially higher rates of survival as compared with lidocaine (Dorian et al).

References
Daniels GH, "Amiodarone-Induced Thyrotoxicosis," *J Clin Endocrinol Metab*, 2001, 86(1):3-8.

Dorian P, Cass D, Schwartz B, et al, "Amiodarone as Compared With Lidocaine for Shock-Resistant Ventricular Fibrillation," *N Engl J Med*, 2002, 346(12):884-90.

Letelier LM, Udol K, Ena J, et al, "Effectiveness of Amiodarone for Conversion of Atrial Fibrillation to Sinus Rhythm: A Meta-analysis," *Arch Intern Med*, 2003, 163(7):777-85.

Loh KC, "Amiodarone-Induced Thyroid Disorders: A Clinical Review," *Postgrad Med J*, 2000, 76(893):133-40.

Roy D, Talajic M, Dorian P, et al, "Amiodarone to Prevent Recurrence of Atrial Fibrillation. Canadian Trial of Atrial Fibrillation Investigators," *N Engl J Med*, 2000, 342(13):913-20.

Stevenson WG and Stevenson LW, "Atrial Fibrillation in Heart Failure," *N Engl J Med*, 1999, 341(12):910-1.

♦ **Amitriptyline and Nortriptyline** *see* Antidepressants, Cyclic, Serum or Plasma *on page 171*

Amitriptyline, Serum or Plasma
Related Information
Antidepressants, Cyclic, Serum or Plasma *on page 171*
Nortriptyline, Serum *on page 966*

Synonyms Elavil®; Endep®; Etrafon®; Limbitrol®; Triavil®

Applies to Pamelor®

Test Includes Nortriptyline levels

Abstract Amitriptyline is a tricyclic antidepressant indicated for the relief of symptoms of endogenous depression. Its mechanism of action is through inhibition of reuptake of serotonin and norepinephrine. Nortriptyline is a major active metabolite. Sometimes it is used in combination with benzodiazepines to treat depression with anxiety. In children and adolescents, tricyclic antidepressants have been used for the treatment of enuresis, obsessive compulsive disorder (OCD), and attention deficit hyperactivity disorder (ADHD), and in patients with less compelling evidence of anxiety and tic disorders and as-yet unproven efficacy in major depression.

Specimen Serum or plasma

Container Red top tube, green top (heparin) tube

Sampling Time Trough levels at steady-state (100-200 hours)

Causes for Rejection Specimen collected in gel tube

Reference Interval Amitriptyline: 80-200 ng/mL (SI: 289-722 nmol/L); nortriptyline: 50-150 ng/mL (SI: 190-570 nmol/L); combined: 120-250 ng/mL

Critical Values >300 ng/mL (SI: >1080 nmol/L); amitriptyline + nortriptyline: >500 ng/mL

Possible Panic Range >1000 ng/mL

Use Therapeutic monitoring and toxicity assessment

Limitations Some data indicate no definite correlation of concentration and clinical outcome and/or severity of side effects. Severity of overdose is better correlated with an EKG finding of QRS widening.

Methodology Immunoassay, high performance liquid chromatography (HPLC), gas chromatography (GC). Immunoassays generally provide semiquantitative total tricyclic antidepressant concentrations.

Additional Information
- Half-life: 20-40 hours
- Volume of distribution: 10-36 L/kg
- Protein binding: 85% to 95%

(Continued)

Amitriptyline, Serum or Plasma (Continued)

Anticholinergic side effects are common with this drug. They are not severe and may diminish with continued therapy, or can be treated with other pharmacologic and nonpharmacologic therapies. Anticholinergic side effects are more troublesome in the elderly. Sedation may also decrease with continued use. Amitriptyline can lower the seizure threshold and cause orthostasis and arrhythmias. Its cardiovascular effects are more common in patients with underlying cardiovascular disorders. Due to side effects of tricyclic antidepressants, low doses of the drugs have been tried. A recent meta-analysis on effects and side effects of low-dose tricyclic antidepressants has been published (Furukawa et al).

Tricyclic antidepressants should be avoided in pregnant and lactating women because the safety of such drugs has not been established in these subjects. Geriatric patients are especially susceptible to orthostasis, urinary retention, constipation, and sedation. Tricyclic antidepressants are commonly seen in overdose situations. Cardiovascular, anticholinergic, and central nervous system toxicities can be lethal. Use of tricyclic antidepressants is contraindicated in patients receiving monoamine oxidase inhibitors and in patients with narrow angle glaucoma.

References

Barbui C and Hotopf M, "Amitriptyline vs the Rest: Still the Leading Antidepressant After 40 Years of Randomised Controlled Trials," Br J Psychiatry, 2001, 178:129-44.

Daly JM and Wilens T, "The Use of Tricyclic Antidepressants in Children and Adolescents," Pediatr Clin North Am, 1998, 45(5):1123-35.

Furukawa TA, McGuire H, and Barbui C, "Meta-analysis of Effects and Side Effects of Low Dosage Tricyclic Antidepressants in Depression: Systematic Review," BMJ, 2002, 325(7371):991.

Geller B, Reising D, Leonard HL, et al, "Critical Review of Tricyclic Antidepressant Use in Children and Adolescents," J Am Acad Child Adolesc Psychiatry, 1999, 38(5):513-6.

Linder MW and Keck PE Jr, "Standards of Laboratory Practice: Antidepressant Drug Monitoring. National Academy of Clinical Biochemistry," Clin Chem, 1998, 44(5):1073-84.

♦ **Ammonia, Cerebrospinal Fluid** see Ammonia, Plasma on page 150

Ammonia, Plasma

Related Information

Amino Acids, Plasma on page 143
Cerebrospinal Fluid Glutamine on page 363
Insulin, Serum on page 803
Lactic Acid, Whole Blood or Plasma on page 827

Synonyms NH_3, Blood

Applies to Ammonia, Cerebrospinal Fluid

Abstract May be useful for diagnosis of Reye syndrome, urea cycle metabolic abnormalities, organic acidurias, hepatic encephalopathy, and in monitoring patients on hyperalimentation therapy.

Patient Preparation Patient should avoid smoking prior to sampling. Avoid clenching of fist.

Specimen Plasma

Container Green top (sodium or lithium heparin) tube or lavender top (EDTA) tube. One author, however, suggests that heparin will produce false low results.

Collection Tube must be filled completely and kept tightly stoppered at all times. Avoid hemolysis, which increases plasma ammonia. Specimen must be placed on ice immediately and rotated, then centrifuged at 4°C. Plasma should be very promptly separated from the cells. Test must be performed within 20 minutes of the venipuncture, or the plasma frozen immediately. Concentration rapidly increases on standing. Never freeze whole blood.

Plasma Ammonia

Age	µg N/dL	SI: µmol N/L
Neonates	90-150	64-107
<2 wk	79-129	56-92
Children	29-70	21-50
Adults	15-45	11-32

Note: Values are somewhat higher in capillary blood.

Storage Instructions In separated plasma, ammonia is stable for several days at -70°C.

Causes for Rejection Specimen not received on ice within 20 minutes of collection

Reference Interval Reference intervals vary among laboratories. See table for approximate intervals. Ammonia level in cerebrospinal fluid is about 33% to 50% of that in arterial blood.

Use Ammonia is elevated in liver disease; urinary tract infection with distention and stasis; Reye syndrome; urea cycle disorders; HHH (hyperornithinemia, hyperammonemia-homocitrullinuria) syndrome; certain organic disorders such as propionic, methylmalonic, and isovaleric acidemias; some normal neonates (usually returning to normal in 48 hours); total parenteral nutrition; ureterosigmoidostomy; gastrointestinal bleed; and sodium valproate therapy. Ammonia determination is indicated in neonates with neurological deterioration, subjects with lethargy and/or emesis not explained, and in patients with possible encephalopathy.

Ammonia is used in the diagnosis of **urea cycle enzyme deficiencies** (to be considered in any neonate with unexplained nausea, vomiting, or neurological deterioration appearing after the first feeding). Investigation includes plasma amino acids, urine orotic and organic acids.

In **Reye syndrome** there are marked elevations of ammonia, AST, ALT, and prothrombin time, while bilirubin is normal. Glucose should be monitored to anticipate hypoglycemia. Acid-base and osmolal status are monitored to anticipate encephalopathy. Liver biopsy is recommended to confirm the diagnosis and exclude other metabolic and toxic liver disorders.

Limitations The correlation between blood ammonia levels and **hepatic coma** is poor. Ammonia determinations are not reliable predictors of impending hepatic coma. Ammonia levels are not always high in all patients with urea cycle disorders. High protein diet may cause increased levels. Ammonia levels may also be elevated with **gastrointestinal hemorrhage**. If portal hypertension develops with cirrhosis, hepatic blood flow is altered, leading to elevated blood ammonia levels.

Methodology Enzymatic assay, spectrophotometric endpoint, ammonia-selective electrode

Additional Information Metabolic acidosis with ketosis are found in the **organic acidemias**, in which hyperammonemia is found. Plasma amino acids, urine organic acids, and amino acids are indicated with NH_3, biotinidase, and carnitine.

Recently, hyperinsulinaemic hypoglycaemia with persistent hyperammonia has been described. This disorder is not associated with any of the abnormalities of amino acids or organic acids observed in urea cycle enzyme defects, and is thought to be due to mutations in glutamate dehydrogenase.

References

Glaser B, "Hyperinsulinism of the Newborn," *Semin Perinatol*, 2000, 24(2):150-63.

Kitaura J, Miki Y, Kato H, et al, "Hyperinsulinaemic Hypoglycaemia Associated With Persistent Hyperammonaemia," *Eur J Pediatr*, 1999, 158(5):410-3.

Lindor NM and Karnes PS, "Initial Assessment of Infants and Children With Suspected Inborn Errors of Metabolism," *Mayo Clin Proc*, 1995, 70(10):987-8.

Miga DE and Roth KS, "Hyperammonemia: The Silent Killer," *South Med J*, 1993, 86(7):742-7.

Stanley CA, Lieu YK, Hsu BY, et al, "Hyperinsulinism and Hyperammonemia in Infants With Regulatory Mutations of the Glutamate Dehydrogenase Gene," *N Engl J Med*, 1998, 338(19):1352-7.

Treem WR, "Inherited and Acquired Syndromes of Hyperammonemia and Encephalopathy in Children," *Semin Liver Dis*, 1994, 14(3):236-58.

♦ **Amniotic Fluid Acetylcholinesterase** *see* Alpha$_1$-Fetoprotein, Amniotic Fluid *on page 135*

♦ **Amniotic Fluid Analysis for Erythroblastosis Fetalis** *see* Bilirubin, Amniotic Fluid, Delta A450 *on page 261*

♦ **Amniotic Fluid Analysis for Hemolytic Disease of the Newborn** *see* Bilirubin, Amniotic Fluid, Delta A450 *on page 261*

♦ **Amniotic Fluid Bilirubin** *see* Bilirubin, Amniotic Fluid, Delta A450 *on page 261*

Amniotic Fluid, Chromosome and Genetic Abnormality Analysis

Related Information

Alpha₁-Fetoprotein, Amniotic Fluid *on page 135*

Alpha₁-Fetoprotein, Serum *on page 136*

Beta-Hexosaminidase, Serum, White Blood Cells *on page 255*

Bilirubin, Amniotic Fluid, Delta A450 *on page 261*

Chorionic Gonadotropin, Human, Serum and Urine *on page 397*

Chorionic Villus Sampling, Chromosome and Genetic Abnormality Analysis *on page 400*

Chromosome Analysis, Blood *on page 406*

Chromosome Analysis, Products of Conception *on page 412*

Creatinine, Amniotic Fluid *on page 472*

Cystic Fibrosis DNA Detection *on page 491*

Duchenne/Becker Muscular Dystrophy DNA Detection *on page 526*

Fluorescence *in situ* Hybridization *on page 602*

Fragile X Syndrome DNA Test *on page 611*

Inherited Diseases of Metabolism and Cell Structure *on page 792*

Inhibin A, Serum *on page 799*

Lecithin:Sphingomyelin Ratio, Amniotic Fluid *on page 836*

Mucopolysaccharides, Urine *on page 922*

Phosphatidylglycerol, Amniotic Fluid *on page 1030*

Polymerase Chain Reaction *on page 1069*

Pregnancy-Associated Protein A, Serum *on page 1082*

Pulmonary Surfactant, Amniotic Fluid *on page 1124*

Test Includes Chromosomal complement of fetal cells in amniotic fluid are examined for determination of abnormalities.

Abstract A karyotype is the somatic chromosomal complement of an individual or species. For the human, a normal karyotype consists of 46 chromosomes aligned in a standard sequence on the basis of size, centromere location, and banding pattern. Amniocentesis is performed to obtain cells of fetal origin in the amniotic fluid for culturing and chromosomal analysis. Examination of the chromosomes by banding techniques can reveal numerical and/or structural abnormalities.

Patient Preparation The patient should be placed on her abdomen for ~20 minutes prior to the amniocentesis. Traditional amniocentesis is performed around 15-16 weeks gestation and early amniocentesis may be performed before 15 weeks (11-14 weeks). Ultrasound studies (to verify fetal life, detect multiple gestation, confirm fetal age, localize placenta, and detect fetal/uterine/adnexal abnormality) are usually carried out.

Specimen Amniotic fluid

Container Sterile container

Sampling Time At or after 16 weeks gestation

Collection An optimum quantity of 20 mL should be obtained by amniocentesis, using strict aseptic technique. Pertinent medical findings should accompany the request, including maternal age, gestational age by sonography, reason for study, relevant history, medication history, transfusion history, note of any viral infection, number of pregnancies and miscarriages, and suspected diagnosis. In the case of twins or triplets, amniotic fluid must be collected separately from each amniotic sac.

Storage Instructions Specimen should be transported to the laboratory at room temperature and under sterile conditions as quickly as possible.

Causes for Rejection Specimen frozen or clotted (due to excess contamination with blood)

Turnaround Time 1-2.5 weeks may be needed.

Reference Interval Forty-six chromosomes to include 22 sets of normal autosomal chromosomes and one set of normal sex chromosomes (XX for female; XY for male). Interpretative information is usually included.

Use Prenatal detection of chromosomal abnormalities, especially Down syndrome, in groups of pregnant women at risk. Such groups include women age 35 years or older, previous child with a chromosomal abnormality or multiple congenital abnormalities, three or more previous spontaneous abortions, familial history of a chromosomal abnormality, or known carrier of an

X-linked disorder. At the same time that amniotic fluid is collected for chromosomal analysis, additional sample can be obtained for testing of inherited metabolic disorders (enzyme deficiency analyses on cultured cells), molecular genetic disorders, or neural tube defects (alpha$_1$-fetoprotein).

Limitations Failure of cells to grow in culture and/or contamination precludes complete analysis (fluorescence *in situ* hybridization may be of use if this occurs). Overall culture success rate has been reported as 97% with a fetal loss (within 4 weeks of the amniocentesis) of 1.2%. Higher fetal loss and increased incidence of musculoskeletal foot deformities have been described in early amniocentesis (11-12 weeks) compared to midtrimester (14-16 weeks) amniocentesis.

Contraindications Environment lacking capability in ultrasonography, genetic counseling, amniocentesis, amniotic fluid culturing, and chromosomal analysis techniques

Additional Information At least 0.5% of newborns are born with a chromosomal abnormality. Among these, the most common is trisomy 21 or Down syndrome. It affects approximately 1 in 800 newborns and is a major cause of mental retardation. The incidence is higher in children born to mothers 35 years of age and older. The risk of obstetric complications for amniocentesis is <0.5%. Because the risk of having a child with a chromosomal abnormality for a mother older than 35 years of age is greater than the risk of the procedure for amniocentesis or chorionic villus sampling, maternal age is an indication for prenatal testing. Rapid detection (24 hours) of the most common chromosome abnormalities (trisomy 21, 18, and 13 and sex chromosome aneusomy) can also be accomplished using fluorescence *in situ* hybridization (FISH) on interphase nuclei of uncultured amniocytes. See Fluorescence *in situ* Hybridization on page 602. In addition to numeric chromosome abnormalities, structural chromosome rearrangements are also detected by cytogenetic analysis.

Birth defects and genetic disorders are encountered in ~3% of liveborn infants (less than $1/3$ of which are the result of a chromosomal abnormality). Prenatal diagnosis is possible for more than 1000 inherited diseases, including inborn errors of metabolism.

References

DiLiberti JH, Greenstein MA, and Rosengren SS, "Prenatal Diagnosis," *Pediatr Rev*, 1992, 13(9):334-42.

Pergament E, Chen PX, Thangavelu M, et al, "The Clinical Application of Interphase FISH in Prenatal Diagnosis," *Prenat Diagn*, 2000, 20(3):215-20.

Warburton D, "*De novo* Balanced Chromosome Rearrangements and Extra Marker Chromosomes Identified at Prenatal Diagnosis: Clinical Significance and Distribution of Breakpoints," *Am J Hum Genet*, 1991, 49(5):995-1013.

♦ **Amniotic Fluid Creatinine** *see* Creatinine, Amniotic Fluid *on page 472*

♦ **Amniotic Fluid Glucose** *see* Body Fluid Glucose *on page 294*

♦ **Amniotic Fluid Lamellar Bodies** *see* Lamellar Bodies, Amniotic Fluid *on page 830*

♦ **Amniotic Fluid Lecithin:Sphingomyelin Ratio** *see* Lecithin:Sphingomyelin Ratio, Amniotic Fluid *on page 836*

♦ **Amniotic Fluid Phosphatidylglycerol** *see* Phosphatidylglycerol, Amniotic Fluid *on page 1030*

♦ **Amniotic Fluid Pulmonary Surfactant** *see* Pulmonary Surfactant, Amniotic Fluid *on page 1124*

♦ **Amniotic Fluid Spectral Analysis** *see* Bilirubin, Amniotic Fluid, Delta A450 *on page 261*

♦ **Amobarb** *see* Barbiturates, Qualitative, Urine *on page 248*

♦ **Amobarbital** *see* Barbiturates, Quantitative, Serum or Plasma *on page 248*

♦ **Amoxapine and 8-Hydroxyamoxapine** *see* Antidepressants, Cyclic, Serum or Plasma *on page 171*

Amoxapine, Serum or Plasma

Related Information

Antidepressants, Cyclic, Serum or Plasma *on page 171*
Fluoxetine, Serum or Plasma *on page 604*

Synonyms Asendin®; Demolox®; Moxadil®; Omnipres®

Applies to Loxapine

Test Includes 8-OH-amoxapine

(Continued)

Amoxapine, Serum or Plasma *(Continued)*

Abstract Amoxapine is a second generation antidepressant. It acts by blocking reuptake of norepinephrine in a more selective manner than the tricyclic antidepressants. Amoxapine should be reserved as second-line therapy for patients unresponsive to standard tricyclic antidepressants. Amoxapine use is contraindicated with concomitant use of monoamine oxidase inhibitors.

Specimen Serum or plasma

Container Red top tube or green top (heparin) tube

Sampling Time Trough level at steady-state. Time to steady-state is 35-50 hours.

Reference Interval Amoxapine 20-100 ng/mL (SI: 64-319 nmol/L); 8-OH-amoxapine 150-300 ng/mL (SI: 478-956 nmol/L); both 200-400 ng/mL (SI: 637-1275 nmol/L)

Critical Values >600 ng/mL (SI: >1913 nmol/L)

Possible Panic Range ≥1000 ng/mL

Use Toxicity assessment

Methodology High performance liquid chromatography (HPLC), gas chromatography (GC). Amoxapine and its metabolite can be qualitatively detected by thin layer chromatography (TLC).

Additional Information
- Half-life: 8-15 hours
- Volume of distribution: 1.0-1.2 L/kg
- Protein binding: 80% to 90%; highly bound to tissues also

Amoxapine is completely and quickly absorbed. The peak concentration occurs 1-4 hours after an oral dose, with a bioavailability of 46% to 82%.

Anticholinergic side effects are common with this drug. They are not severe and may diminish with continued therapy or can be treated with other pharmacologic and nonpharmacologic therapies. Amoxapine can lower the seizure threshold and cause orthostasis and arrhythmias. Its cardiovascular effects are more common in patients with underlying cardiovascular disorders.

Drug interactions are common with the tricyclic antidepressants. Concomitant treatment with cimetidine, fluoxetine, and antipsychotics produce unexpectedly high concentrations of amoxapine. Enzyme inducers (eg, phenytoin, chloral hydrate, smoking, and the barbiturates) will decrease amoxapine concentrations. Monoamine oxidase inhibitors (MAOIs) and thyroid hormones potentiate the toxicity of cyclics. Hyperthermia, delirium, convulsions, coma, and fatalities have occurred with the combination of MAOIs and cyclics.

Signs and symptoms of acute overdose include grand mal convulsions, photosensitivity, insomnia, hyperprolactinemia, cognitive dysfunction, nystagmus, acidosis, coma, supraventricular arrhythmias, hematuria, incomplete right bundle-branch block, renal failure (acute), myoglobinuria, and hematuria. Neurotoxic effects may be permanent.

References

Fenton J, "Amoxapine," *The Laboratory and the Poisoned Patient*, Washington, DC: AACC Press, American Association of Clinical Chemistry, 1998, 50-3.

Lacy CF, Armstrong LL, Goldman MP, et al, *Drug Information Handbook*, 12th ed, Hudson, OH: Lexi-Comp Inc, 2004.

Mazzola CD, Miron S, and Jenkins AJ, "Loxapine Intoxication: Case Report and Literature Review," *J Anal Toxicol*, 2000, 24(7):638-41.

Merigian KS, Browning RG, and Leeper KV, "Successful Treatment of Amoxapine-Induced Refractory Status Epilepticus With Propofol," *Acad Emerg Med*, 1995, 2(2):128-33.

♦ **AMP, Cyclic, Plasma** *see* Cyclic AMP, Plasma *on page 486*

♦ **AMP, Cyclic, Urine** *see* Cyclic AMP, Urine *on page 486*

Amphetamine, Qualitative, Urine

Related Information

Methamphetamine, Qualitative, Urine *on page 902*

Synonyms Bennies; Crystal; Dexedrine®; Dexies; Ferndex®; Speed; Uppers

Test Includes Amphetamine, methamphetamine

Abstract Amphetamine is a central nervous system stimulant for the management of severe obesity, hyperkinetic syndromes, and narcolepsy. Due to euphoric effects, it has a high potential for abuse. The use of amphetamine for weight reduction has been curtailed significantly. It is a DEA schedule II drug.

Specimen Random urine

Collection If forensic, observe precautions (see the Introduction *on page 63*).

Storage Instructions Refrigerate

Turnaround Time Usually 1-2 hours for screen if done in-house. Confirmation, 1-3 days.

Special Instructions If forensic, use Chain-of-Custody form. See the Chain-of-Custody form in the Introduction *on page 63*.

Reference Interval Negative (less than cutoff)

Critical Values Substance Abuse and Mental Health Services Administration (SAMHSA) cutoff: screen: 1000 ng/mL; confirmation: 500 ng/mL. For methamphetamine, a positive report requires methamphetamine ≥500 ng/mL and amphetamine (a metabolite) ≥200 ng/mL in the same sample.

Use Drug abuse evaluation; toxicity assessment; patient compliance; therapeutic drug monitoring

Limitations Some over-the-counter cold and antiallergy medications may cross react in certain immunoassay screens; confirmation by a different, more sensitive and specific method (eg, GC/MS) is necessary.

Methodology Screen: immunoassay; confirmation: gas chromatography/mass spectrometry (GC/MS)

Additional Information

- Half-life: 10-20 hours (usual urine pH); 5-10 hours (acidic urine)
- Volume of distribution: 3-4 L/kg
- Protein binding: 10% to 40%
- Urine detection time for amphetamine and methamphetamine: 2-4 days

Once in the Department of Defense, stimulants, particularly amphetamines, were used in "GO/NO GO" packs as countermeasures to fatigue and extended flight operations. There is no specific antidote for amphetamine overdose and the treatment is generally supportive. Tolerance to the drug is developed after repeated use.

In interpretation of methamphetamine and amphetamine concentrations, positive test knowledge of legitimate and illicit sources is important. Amphetamines are sometimes prescribed as weight-reducing medicines. Many substances (amphetaminil, benzphetamine, clobenzorex, deprenyl, dimethylamphetamine, ethylamphetamine, famprofazone, fencamine, fenethylline, fenproporex, furfenorex, mefenorex, mesocarb, and prenylamine) which are available as prescription drugs, are metabolized in the body to methamphetamine or amphetamine. Knowledge of d and l enantiomers and their distinction is many times helpful to distinguish between legitimate and illegal use.

Due to the observed collection and improved adulteration control, use of oral fluid as a matrix for drugs of abuse testing is becoming increasing popular.

References

Musshoff F, "Illegal or Legitimate Use? Precursor Compounds to Amphetamine and Methamphetamine," *Drug Metab Rev*, 2000, 32(1):15-44.

Schepers RJ, Oyler JM, Joseph RE Jr, et al, "Methamphetamine and Amphetamine Pharmacokinetics in Oral Fluid and Plasma After Controlled Oral Methamphetamine Administration to Human Volunteers," *Clin Chem*, 2003, 49(1):121-32.

Waksman J, Taylor RN, Bodor GS, et al, "Acute Myocardial Infarction Associated With Amphetamine Use," *Mayo Clin Proc*, 2001, 76(3):323-6.

Williams RH, Erickson T, and Broussard LA, "Evaluating Sympathomimetic Intoxication in an Emergency Setting," *Lab Med*, 2000, 31(9):497-507.

♦ **Amphiphysin** *see* Antineuronal Nuclear Antibody, Type 1 (Anti-Hu) *on page 185*

♦ **Amsterdam Criteria** *see* Colon Cancer, Hereditary Nonpolyposis Type *on page 432*

♦ **Amylase, Body Fluid** *see* Body Fluid Amylase *on page 287*

♦ **Amylase:Creatinine Clearance Ratio** *see* Amylase, Urine *on page 157*

Amylase, Serum

Related Information

Amylase, Urine *on page 157*
Bile Fluid Examination *on page 259*
Bilirubin, Total, Serum *on page 265*
Body Fluid Amylase *on page 287*
Calcium, Serum *on page 329*
C-Reactive Protein, Serum *on page 467*
(Continued)

Amylase, Serum *(Continued)*

Lipase, Serum *on page 851*
Triglycerides, Serum or Plasma *on page 1275*

Synonyms 1,4-α-D Glucanohydrolase, Serum

Applies to Interleukin-6; Trypsinogen Activation Peptide

Abstract Serum amylase is usually a sensitive and useful diagnostic method in those patients with acute pancreatitis who present within hours of the onset of pain. **Serum lipase** assay provides somewhat better sensitivity and specificity and is optimally used with amylase determination. Simultaneous determination of both is widely recommended for evaluation of abdominal pain.

Specimen Serum, plasma; amylase can also be determined from body fluids.

Container Red top tube, green top (heparin) tube

Sampling Time Usually elevated within 12 hours of onset of pancreatitis, serum amylase increases persist for 3-4 days. Since the biologic half-life of amylase is shorter than that of lipase, delay in laboratory evaluation may lead to a normal amylase result with increased lipase. For detection of blunt injury to the pancreas, determination more than 3 hours following trauma is advocated. Serum amylase should be drawn 4 hours following endoscopic sphincterotomy to assess postprocedure pancreatitis.

Collection Anticoagulants other than heparin diminish amylase activity

Storage Instructions Amylase is stable for 1 week at 25°C and 2 months at 4°C.

Special Instructions Dilution of lipemic sera may cause amylase values to increase.

Reference Interval Pancreatic amylase: 11-54 units/L (immunoassay); interval is method dependent. Children up to 2 years of age have virtually no pancreatic amylase.

Use Useful in diagnosis of acute pancreatitis; desirable to support the clinical significance of elevated serum lipase. About 80% of subjects with acute pancreatitis have increased serum amylase within the first 24 hours. Both amylase and lipase assays are recommended in organophosphate poisoning. Causes of **high serum pancreatic enzymes** include acute pancreatitis, chronic pancreatitis, pancreatic pseudocyst, pancreatic ascites, pancreatic abscess, neoplasm in or adjacent to pancreas, trauma to pancreas, and common duct stones.

Limitations Poor specificity and limited sensitivity (~80%). Nonpancreatic causes of hyperamylasemia include inflammatory salivary lesions (eg, mumps), perforated peptic ulcer, intestinal obstruction and infarction, afferent loop syndrome, biliary tract disease, hepatic cirrhosis, aortic aneurysm, peritonitis, acute appendicitis, cerebral trauma, burns and traumatic shock, the postoperative state, diabetic ketoacidosis, and extrapancreatic carcinomas (especially of esophagus, lung, ovary). The term "salivary amylase" includes other nonpancreatic sources of the enzyme. Serum amylase is cleared by renal excretion. Urinary amylase increases often persist longer than do those of serum. Serum amylase may increase up to three times the upper limit of normal in renal failure without diagnostic significance. In such cases, urine amylase is normal or low.

Many **drugs** cause increased amylase levels, including those which cause spasm of the sphincter of Oddi (eg, cholinergics, bethanechol, codeine, fentanyl, meperidine, morphine and other narcotics, pentazocine) and those which may cause pancreatitis (eg, aminosalicylic acid, amoxapine, azathioprine, chlorthalidone, cimetidine, clozapine, diazoxide, dideoxyinosine, felbamate, fluvastatin, glucocorticoids, hydantoin derivatives, hydrochlorothiazide, hydroflumethiazide, isoniazid, mercaptopurine, minocycline, mirtazapine, pegaspargase, penicillamine, sulfamethoxazole, sulfisoxazole). Some drugs may cause parotitis (eg, phenylbutazone, potassium iodide, procyclidine). Other drugs which may cause increased levels include cisplatin, thiazides, and valproic acid.

Macroamylase is amylase complexed to immunoglobulin and it occurs in normal as well as abnormal subjects. Such individuals have slightly increased serum amylase levels which do not fluctuate, and normal (or low) urine amylase. See Amylase, Urine *on page 157.*

Methodology Amyloclastic, saccharogenic, chromolytic; up to 200 methods exist. Monoclonal antibody techniques are in use.

Additional Information Although C-reactive protein is elevated in many cases of acute pancreatitis, its concentration is also increased in substantial numbers of instances of nonpancreatic acute abdomen. Other tests recommended for acute pancreatitis include interleukin-6 and trypsinogen activation peptide.

Pancreatitis occurs in children as well as in adults. A rare cause is pinworm infestation with involvement of the pancreatic duct.

Isoenzymes of amylase exist: pancreatic and salivary type, which can be separated by polyacrylamide gel or agarose film electrophoresis, isoelectric focusing, ion exchange chromatography, plant isoamylase inhibitors, and by monoclonal antibody technique on a centrifugal analyzer. Amylase isoenzymes are separated in few laboratories. Where available, the procedure is moderately expensive. Separations of amylase into its P and S isoenzymes adds little to the diagnosis of pancreatitis.

References

Benkow KJ and Winter HS, "A 15-Year-Old Girl With Abdominal Pain and Bloody Stools," Case Records of the Massachusetts General Hospital, Case 2-1999, Scully RE, Mark EJ, McNeely WF, et al, eds, *N Engl J Med*, 1999, 340(3):215-21.

Frank B and Gottlieb K, "Amylase Normal, Lipase Elevated: Is it Pancreatitis?" *Am J Gastroenterol*, 1999, 94(2):463-9.

Lankisch PG, Burchard-Reckert S, and Lehnick D, "Underestimation of Acute Pancreatitis: Patients With Only a Small Increase in Amylase/Lipase Levels Can Also Have or Develop Severe Acute Pancreatitis," *Gut*, 1999, 44(4):542-4.

Mayo Medical Laboratories, *2001 Test Catalogue*, Rochester, MN, 67.

Moss DW and Henderson AR, "Clinical Enzymology," *Tietz Textbook of Clinical Chemistry*, 3rd ed, Burtis CA and Ashwood ER, eds, Philadelphia, PA: WB Saunders Co, 1999, 689-98.

Steinberg W and Tenner S, "Acute Pancreatitis," *N Engl J Med*, 1994, 330(17):1198-210.

Torrens JK and McWhinney PH, "Acute Pancreatitis. Normal Serum Amylase Does Not Exclude Severe Acute Pancreatitis," *BMJ*, 1998, 316(7149):1982-3.

Uretsky G, Goldschmiedt M, and James K, "Childhood Pancreatitis," *Am Fam Phys*, 1999, 59(9):2507-12.

Amylase, Urine

Related Information

Amylase, Serum *on page 155*
Body Fluid Amylase *on page 287*
Lipase, Serum *on page 851*

Synonyms 1,4-α-D Glucanohydrolase, Urine

Applies to Amylase:Creatinine Clearance Ratio; Trypsin, Immunoreactive

Abstract Urine amylase is elevated early in acute pancreatitis. However, diagnostic utility of the combination of serum amylase and lipase is superior.

Patient Preparation Fasting from 10 PM to 6 AM is recommended before a 24-hour urine collection.

Specimen 2-hour urine specimen is preferred, no preservative

Collection Collect timed specimen. Instruct the patient to void at the beginning of the collection period and discard the specimen. Collect all urine including the final specimen voided at the end of the collection period. Centrifugation to provide optically clear specimen is desirable.

Storage Instructions Keep refrigerated.

Special Instructions Requisition should include date and time collection started, date and time collection finished.

Reference Interval Method dependent

Methodology Maltopentose, other methods also available

Additional Information The diagnostic specificity of amylase assays can be improved by measuring the **amylase/creatinine** clearance ratio (ACCR). The ACCR is simply the urinary clearance of amylase divided by the urinary clearance of creatinine:

[urine amylase (units/L) x serum creatinine (mg/L) / serum amylase (units/L) x urine creatinine (mg/L)] x 100

The ACCR reference interval is ~2.5%. Increased values are obtained in patients with acute pancreatitis, but similar increases are seen in most of the nonpancreatic causes of increased serum amylase (see Amylase, Serum *on page 155*). **Macroamylasemia** is a benign condition in which patients have slight-to-moderate elevations of serum amylase with low-to-normal urine amylase. In macroamylasemia, the ACCR is usually <2%.
(Continued)

Amylase, Urine (Continued)

Some patients with pancreatitis have very high triglyceride levels. **Immunoreactive trypsin** has not been widely available, but a urine trypsinogen-2 test strip has recently been described. Further evaluation is needed.

References
Hedström J, Svens E, Kenkimaki P, et al, "Evaluation of a New Urinary Amylase Test Strip in the Diagnosis of Acute Pancreatitis," *Scand J Clin Lab Invest*, 1998, 58(8):611-6.

Kylänpää-Bäck ML, Kemppainen E, Puolakkainen P, et al, "Reliable Screening for Acute Pancreatitis With Rapid Urine Trypsinogen-2 Test Strip," *Br J Surg*, 2000, 87:49-52.

Steinberg W and Tenner S, "Acute Pancreatitis," *N Engl J Med*, 1994, 330(17):1198-210.

♦ **β-Amyloid₄₂** *see* Cerebrospinal Fluid and Plasma β-Amyloid$_{(1-42)}$ *on page 359*

♦ **Amyloid A, Serum** *see* C-Reactive Protein, Serum *on page 467*

♦ **Amyloid Precursor Protein (APP), CSF** *see* Cerebrospinal Fluid and Plasma β-Amyloid$_{(1-42)}$ *on page 359*

♦ **Amytal®** *see* Barbiturates, Quantitative, Serum or Plasma *on page 248*

♦ **ANA** *see* Antinuclear Antibodies *on page 189*

♦ **Anacin®** *see* Salicylate, Serum or Plasma *on page 1176*

♦ **Anacin-3®** *see* Acetaminophen, Serum *on page 90*

♦ **ANCA** *see* Antineutrophil Cytoplasmic Antibody *on page 187*

♦ **Ancillary Testing** *see* Point-of-Care Testing *on page 1065*

♦ **Ancobon®** *see* Flucytosine, Serum *on page 600*

♦ **Androstenedione** *see* Dehydroepiandrosterone and Dehydroepiandrosterone Sulfate, Serum or Plasma *on page 506*

Androstenedione, Serum

Related Information

Chorionic Villus Sampling, Chromosome and Genetic Abnormality Analysis *on page 400*

Cortisol, Serum or Plasma *on page 460*

Dehydroepiandrosterone and Dehydroepiandrosterone Sulfate, Serum or Plasma *on page 506*

17-Hydroxyprogesterone, Whole Blood, Serum, or Plasma *on page 755*

Testosterone, Total and Free, Serum or Plasma *on page 1238*

Abstract Androstenedione (AS) is synthesized in the adrenal cortex. AS enters a complex metabolic matrix with pathways leading to both **estrogens** and **androgens**. In tissues rich in the aromatase enzymes (eg, fat and liver), AS is a prohormone for estradiol and estrone. AS also is metabolized to testosterone. The regulatory mechanisms controlling these pathways are poorly understood. AS and dehydroepiandrosterone (DHEA) are the predominant androgens in females.

Patient Preparation Fasting morning specimen is preferred. Collect 1 week before or after menstrual period.

Specimen Serum

Container Red top tube

Storage Instructions Freeze serum.

Reference Interval Variation exists between laboratories. Some representative intervals are given in the following table.

Androstenedione, Serum

Age	Male		Female	
	ng/dL	nmol/L	ng/dL	nmol/L
1-5 mo	5-45	0.2-1.6	5-35	0.2-1.2
1-9 y	5-55	0.2-1.9	5-45	0.2-1.6
10-17 y	10-100	0.3-3.5	25-200	0.9-7.0
Adults	50-250	1.7-8.7	50-250	1.7-8.7

Values are higher in pregnancy and highest at delivery. A marked diurnal variation exists, with a peak around 7 AM and a nadir around 4 PM. Levels rise sharply after puberty to peak at about 20 years of age. An abrupt decline occurs after menopause. Concentrations are lower following bilateral oophorectomy.

Use AS is elevated in virilizing adrenal tumors, the Stein-Leventhal syndrome, ovarian stromal hyperplasia, congenital adrenal hyperplasia, Cushing syndrome, and idiopathic hirsutism. Values >1000 ng/dL (SI: >34.9 nmol/L) suggest an adrenal tumor. See Adrenal Cortex: Laboratory Assessment Overview *on page 110.*

AS is one of the analytes which may be assessed in the prenatal diagnosis of congenital adrenal hyperplasia. Other tests include 17-hydroxyprogesterone, testosterone, 21-deoxycortisol, and HLA typing. Early diagnosis can be made with molecular genetic testing of chorionic villus biopsies.

Limitations There is poor correlation of plasma levels with clinical severity.

Methodology Radioimmunoassay (RIA)

Additional Information AS is marketed as a "dietary supplement" (often called simply "andro") in the U.S. and is widely perceived in the media as an anabolic substance which promotes muscle growth and other androgenic end points. The accuracy of this perception is strongly challenged by the results of a multicenter research study (Broeder et al, 2000).

References

Broeder CE, Quindry J, Brittingham K, et al, "The Andro Project: Physiological and Hormonal Influences of Androstenedione Supplementation in Men 35 to 65 Years Old Participating in a High-Intensity Resistance Training Program," *Arch Intern Med*, 2000, 160(20):3093-104.

Catlin DH, Leder BZ, Ahrens B, et al, "Trace Contamination of Over-the-Counter Androstenedione and Positive Urine Test Results for a Nandrolone Metabolite," *JAMA*, 2000, 284(20):2618-21.

Levine LS and Pang S, "Prenatal Diagnosis and Treatment of Congenital Adrenal Hyperplasia," *J Pediatr Endocrinol*, 1994, 7(3):193-200.

Soldin SJ, Brugnara C, Gunter KC, et al, "Pediatric Reference Ranges," 3rd ed, Washington, DC: AACC Press, American Association of Clinical Chemistry, 1999, 30.

♦ **Androsterone** *see* 17-Ketosteroids Fractionation, Urine *on page 818*

♦ **Anemia Flowchart** *see page 35*

♦ **Anestacon®** *see* Lidocaine, Serum or Plasma *on page 850*

♦ **Angel Dust** *see* Phencyclidine, Qualitative, Urine *on page 1019*

♦ **Angelman Syndrome** *see* Gene Rearrangement for Leukemia and Lymphoma *on page 633*

♦ **Angilol®** *see* Propranolol, Serum *on page 1096*

♦ **Angiotensin** *see* Renin Activity, Plasma *on page 1149*

♦ **Angiotensin Converting Enzyme, CSF** *see* Angiotensin Converting Enzyme, Serum *on page 159*

Angiotensin Converting Enzyme, Serum

Synonyms ACE; Angiotensin-I-Converting Enzyme

Applies to Angiotensin Converting Enzyme, CSF; Cerebrospinal Fluid Angiotensin Converting Enzyme

Abstract Angiotensin-I-converting enzyme (ACE) is especially known for its generation of the octapeptide, angiotensin II, by releasing the dipeptide histidyl-leucine from angiotensin I. The major site of normal ACE production is the pulmonary endothelial cells.

In an appropriate clinical context, increased concentrations of ACE support a diagnosis of sarcoidosis, but the diagnostic value of the test is controversial. Its most established role is as a marker of disease activity in subjects whose diagnosis is established.

Patient Preparation Angiotensin converting enzyme inhibiting drugs cause decreased ACE values.

Specimen Serum

Container Red top tube or SST™ tube; EDTA inhibits ACE

Storage Instructions Separate serum (or plasma) immediately. Stable 1 week at 4°C, 6 months at -20°C.

Angiotensin Converting Enzyme, Serum

Genotype	ACE Reference Interval (units/L)
II	4.6-30.6
ID	10.0-47.6
DD	17.9-64.3

(Continued)

Angiotensin Converting Enzyme, Serum *(Continued)*

Reference Interval The reference interval is approximately 15-70 units/L, though large interindividual variations exist. Reference intervals related to genotype may improve the diagnostic sensitivity of the test in acute sarcoidosis. See table on previous page.

Reference intervals for children and adolescents may be up to 50% greater than specimens from individuals 20 years of age and older.

Use Results are elevated in sarcoidosis, more often when the disease is active and are of value in assessing the response of sarcoidosis to corticosteroid therapy. A marked decrease is found in some patients on prednisone. It also is used in investigation for Gaucher disease and may be useful for monitoring noncompliance with ACE inhibitor treatment. ACE is thought to be produced by epithelioid cells and macrophages; elevations are found in a variety of granulomatous diseases.

Limitations ACE lacks specificity and sensitivity for diagnosis of sarcoidosis. Elevations have been reported in about 35% to 91% of cases of sarcoidosis (see reference by Jordan et al for entry to the somewhat older literature on this subject). ACE levels are less likely to be increased with chronic sarcoidosis. Elevations have been found in patients with diabetes mellitus, Gaucher disease and leprosy. Twenty-five percent of 86 patients with acute histoplasmosis had elevated levels. Increased levels have been observed in many other diseases. ACE is physiologically decreased by administration of captopril, enalapril, and lisinopril.

Methodology Spectrofluorometric or radioimmunoassay (RIA), spectrophotometric utilizing synthetic substrates

Additional Information Other abnormalities found in some sarcoidosis patients may include elevations of serum alkaline phosphatase, calcium, gamma globulin with polyclonal gammopathy, and hypercalciuria. Serum angiotensin converting enzyme is not usually elevated in cases of active tuberculosis or Hodgkin lymphoma. The diagnosis of sarcoidosis is a histopathologic/clinical complex. Lymph nodes, liver, skin, and lung, especially transbronchial lung biopsies, are often useful. Cultures and special stains to rule out mycobacterial and fungal infection are needed, and polarizing microscopy is utilized to identify crystalline material in granulomas in tissue sections. (Noncaseating granulomas must be proven not to be caused by histoplasmosis or other microbiologic entities.) Berylliosis is a very rare cause of such granulomas.

References

Jordan DR, Anderson RL, Nerad JA, et al, "The Diagnosis of Sarcoidosis," *Can J Ophthalmol*, 1988, 23(5):203-7.

Newman LS, Rose CS, and Maier LA, "Sarcoidosis," *N Engl J Med*, 1997, 336(17):1224-34.

Sharma P, Smith I, Maguire G, et al, "Clinical Value of ACE Genotyping in Diagnosis of Sarcoidosis," *Lancet*, 1997, 349(9065):1602-3.

Singer DR, Missouris CG, and Jeffery S, "Angiotensin-Converting Enzyme Gene Polymorphism. What to Do About All the Confusion?" *Circulation*, 1996, 94(3):236-9.

Smith CC, Mandel J, and Bush B, "Less Is More," *N Engl J Med*, 2001, 344(14):1079-82.

♦ **Angiotensin-I-Converting Enzyme** *see* Angiotensin Converting Enzyme, Serum *on page 159*

Anion Gap, Serum, Plasma, or Urine

Related Information

Bicarbonate, Blood *on page 258*
Chloride, Serum, Plasma, or Blood *on page 391*
Chloride, Urine *on page 394*
Electrolyte Panel, Serum *on page 532*
Ethylene Glycol, Serum or Plasma *on page 561*
Ibuprofen, Serum *on page 764*
Ketone Bodies, Blood *on page 816*
Ketones, Urine *on page 817*
Lactic Acid, Whole Blood or Plasma *on page 827*
Osmolality, Calculated, Serum or Plasma *on page 976*
Osmolality, Serum *on page 978*
pH, Blood *on page 1018*
Phosphorus, Serum *on page 1031*
Salicylate, Serum or Plasma *on page 1176*

Sodium, Serum or Plasma *on page 1210*
Sodium, Urine *on page 1213*
Volatile Screen, Blood or Urine *on page 1320*

Applies to Anion Gap, Urine

Test Includes A calculation from sodium, potassium, HCO_3^-, and chloride to ascertain quantities of unmeasured cations and anions

Abstract The anion gap is useful in evaluation of patients with acid-base abnormalities. It is based on the principle of electroneutrality, which requires that the sum of anions and cations be equal in blood or other body fluids. This calculation (anion gap) is an estimate of unmeasured anions. Most cases of increased anion gap are caused by lactic acidosis or ketoacidosis.

Aftercare Patient urinary output may be relevant in cases of increased anion gap.

Specimen Serum, plasma, urine

Container Red top tube or green top (heparin) tube, plastic urine container

Reference Interval Using ion-selective electrode technology: 3-11 mmol/L; considerable variation has been shown among instruments, and each laboratory should establish or verify its own reference interval. The reference interval by flame photometry is approximately 8-16 mmol/L using the following:

$$[Na^+] - [HCO_3^- + Cl^-]$$

If K is included with the cations, the reference interval is 10-20 mmol/L.

Use The major clinical application of the anion gap is in the differential diagnosis of metabolic acidosis. Within the laboratory, the anion gap is often used as an internal check.

A marked elevation of anion gap suggests **metabolic acidosis**. Increased anion gaps are found in states such as renal failure and toxic ingestions. A result >30 mmol/L is commonly secondary to **lactic acidosis** or **ketoacidosis** but can be caused also by rhabdomyolysis or nonketotic hyperglycemic coma. See Lactic Acid, Whole Blood or Plasma *on page 827*.

Acid-base abnormalities may coexist (eg, loss of acid by vomiting and increased acid production from ketoacidosis of diabetes mellitus). See table.

Anion Gap in Metabolic Acid-Base Disorders

Mechanism	Anion Gap	Osmotic Gap	Chloride	Potassium	Other
Metabolic Acidosis					
Increased acid	↑	N[1]	N	↑	↑ Lactate or production ketones = anion gap
Acid precursor ingestion	↑	↑[2]	N	↑	Measure methanol, ethylene glycol, and salicylates, depending on history
Decreased acid excretion	↑	N	N	↑	Severe renal failure also present; usually low urine output
Increased base excretion	N	N	↑	↑ or ↓	Urine anion gap ↑ with stool losses; ↓ with renal tubular acidosis
Metabolic Alkalosis					
Dehydration	N	N	↓	↓	Urine chloride, sodium usually undetectable
Vomiting	N	N	↓	↓	Urine chloride undetectable; urine sodium usually normal
Base ingestion	N	N	↓	↓	Very high urine anion gap
Primary mineralocorticoid excess	N	N	↓	↓	Urine chloride, sodium usually measurable

↑Indicates increased; N indicates normal; ↓ indicates decreased.

[1]Osmotic gap may be slightly high in patients with ketoacidosis.

[2]Osmotic gap is normal with salicylates.

Adapted from Dufour DR, "Laboratory Recognition and Testing in Acid-Base Disorders," *Lab Med*, 1999, 30(12):776-81.

(Continued)

Anion Gap, Serum, Plasma, or Urine *(Continued)*

Limitations A spurious increase may follow excessive exposure of the sample to room air as well as underfilling the Vacutainer® tube. All metabolic abnormalities are not detected by abnormal gaps (eg, isopropanol ingestion is accompanied by a normal gap, but ketone bodies are positive). There are a number of causes of normal anion gap acidosis associated with hyperchloremia. The anion gap is unsuitable as a quick screen for lactic acidosis due to its lack of sensitivity. In fact, hyperlactatemia is in the differential diagnosis of normal anion gap acidosis. The anion gap should not replace assays for lactate, ketone bodies, or osmolality. In one study, only 71% of patients with an anion gap of 20-29 mmol/L could be proven to have an organic acidosis. In critically ill newborns, however, a recent study found anion gap >16 mmol/L to be highly predictive of lactic acidosis and <8 mmol/L highly predictive of absence of lactic acidosis.

Methodology Most commonly, the anion gap is calculated from the electrolyte measurements as:

$Na^+ - (Cl^- + HCO_3^-)$

Less often, the following is used:

$(Na^+ + K^+) - (Cl^- + HCO_3)$

Additional Information The existence of a "gap" reflects the fact that some anions and cations are not measured in routine practice. Unmeasured cations include Ca^{2+} and Mg^{2+}. Unmeasured anions include protein, PO_4^{3-}, SO_4^{2-}, and organic acids. Organic acids include lactate and ketoacids as well as others.

High anion gaps are caused by elevated concentrations of unmeasured anions. **When the anion gap is high and pH is low**, possible causes include uremia, ketoacidosis, lactic acidosis, salicylate toxicity, methanol toxicity, ethylene glycol toxicity, paraldehyde toxicity, or toluene toxicity. Toluene exposure by glue sniffing can cause severe high anion gap metabolic acidosis, which can convert to hyperchloremic acidosis. Metabolic acidoses with profoundly elevated anion gaps appear to be due to multifactorial causes, including renal failure, rhabdomyolysis, nonketotic hyperglycemic hyperosmolar syndrome, marked hyperphosphatemia, hemoconcentration, and identified and unidentified organic metabolic acidosis. Abnormal anion gaps due to uremia are usually seen only when the creatinine is >4.0 mg/dL (SI: >354 µmol/L). Uremic acidosis is rare without hyperphosphatemia. **When both the anion gap and pH are high**, the cause could be due to extracellular volume contraction, massive transfusion (with renal failure and/or volume contraction), carbenicillin or penicillin (in large doses), or salts of organic acids such as citrate. Common mnemonic "MUDPILES" is used to remember conditions causing increased anion gap. These are **M**ethanol toxicity, **U**remia of renal failure, **D**iabetes mellitus, **P**araldehyde toxicity, **I**soniazid/Iron toxicity, **L**actic acidosis, **E**thylene glycol toxicity, **S**alicylate toxicity.

Low anion gaps are caused by retained unmeasured anions. The most common causes are hypoalbuminemia (eg, in nephrosis or cirrhosis), dilution, hypernatremia, very marked hypercalcemia, very severe hypermagnesemia, IgG myeloma, and polyclonal gamma globulinemia. Hyperviscosity, lithium toxicity, and bromism also have been associated with low anion gaps. Decreased anion gap with spurious hyperchloremia and with hyponatremia is reported in hyperlipidemia. Dilution of extracellular fluid may cause a decreased gap. The finding of a **low anion gap** is perceived as an unreliable diagnostic finding but should be strongly considered as an **indication of laboratory error.** Use of an adjusted anion gap in cases of hypoalbuminemia has been suggested.

Normal anion gaps occur with **metabolic acidosis**. Causes include diarrhea, renal tubular acidosis, hyperalimentation, ureteroileostomy, ureterosigmoidostomy, external drainage of pancreaticobiliary fluids, and administration of NH_4Cl and other drugs.

The urinary anion gap (Na + K - Cl) is sometimes used in the diagnosis of hyperchloremic metabolic acidosis and evaluation of renal potassium wasting. Marked increase occurs following ingestion of baking soda (sodium bicarbonate) and accumulation of citrate following massive transfusions, with metabolic alkalosis. As for its role in diagnosis of metabolic acidosis, the urinary anion gap has been suggested as a replacement for the uncommonly measured urinary ammonium concentration. Recent studies, however, give varying assessments with respect to its accuracy and usefulness for this purpose.

The Fencl-Stewart Method for estimating the portion of base excess in plasma due to unmeasured anions (BEua) has been shown to be superior to the anion gap, the standard base excess calculation, and the determination of plasma lactate in predicting mortality in a population of pediatric intensive care patients. The calculation is derived from the standard base excess (BE net) derived from bicarbonate levels (determined in arterial blood using blood gas instrumentation) and plasma levels of sodium, chloride, and albumin. It is given below:

$$BE\ ua = BE\ net - \{ [0.3 \times (Na - 140)] + [102 - (Cl \times 140/Na)] + [3.4 \times (4.5 - albumin)] \}$$

References

Balasubramanyan N, Havens PL, and Hoffman GM, "Unmeasured Anions Identified by the Fencl-Stewart Method Predict Mortality Better Than Base Excess, Anion Gap, and Lactate in Patients in the Pediatric Intensive Care Unit," *Crit Care Med*, 1999, 27(8):1577-81.

DuFour DR, "Laboratory Recognition and Testing in Acid-Base Disorders," *Lab Med*, 1999, 30(12):776-81.

Lorenz JM, Kleinman LI, Markarian K, et al, "Serum Anion Gap in the Differential Diagnosis of Metabolic Acidosis in Critically Ill Newborns," *J Pediatr*, 1999, 135(6):751-5.

Uribarri J, Oh MS, and Carroll HJ, "D-Lactic Acidosis. A Review of Clinical Presentation, Biochemical Features, and Pathophysiologic Mechanisms," *Medicine (Baltimore)*, 1998, 77(2):73-82.

♦ **Anion Gap, Urine** *see* Anion Gap, Serum, Plasma, or Urine *on page 160*

♦ **ANNA-1** *see* Antineuronal Nuclear Antibody, Type 1 (Anti-Hu) *on page 185*

♦ **ANNA-2** *see* Antineuronal Nuclear Antibody, Type 2 (Anti-Ri) *on page 185*

♦ **Annexin V (Annexin A5)** *see* Apoptosis Assays *on page 205*

Anthrax Detection

Related Information

Bacterial Culture, Blood *on page 232*
Fine Needle Aspiration Culture *on page 589*

Test Includes Aerobic culture and Gram stain of specimen

Abstract Anthrax is caused by the bacteria *Bacillus anthracis*, and is one of the most serious biological agents that may be used as a weapon. The natural reservoir for this gram-positive nonmotile, nonhemolytic *Bacillus* is the soil. It exists in an infected host as a vegetative bacillus, and in the environment as a spore. Spores can remain viable for decades, and are the usual infective form. Bacterial culture for aerobic organisms will provide proper growth conditions for *Bacillus anthracis*; however, it is important to notify the laboratory if this organism is suspected.

Specimen Blood, sputum, cerebrospinal fluid, pleural fluid, stool, rectal swab, or cutaneous lesion fluid may be cultured; *vide infra*.

Container Sterile container or swab

Storage Instructions If transportation of swabs requires more than 1 hour, then store at 2°C to 8°C.

Causes for Rejection Nasal swabs and environmental samples (ie, powder) should not be tested by a routine clinical laboratory. Nasal swabs should **not** be used routinely to determine diagnosis or therapy.

Turnaround Time Negative results are typically reported as follows: blood and CSF: 5-7 days; sputum and wounds: 2 days. Preliminary morphologic information for positive cultures is usually available within 24 hours. Suspicious isolates should be forwarded to a state or local public health laboratory for complete identification. Analytic test time for rapid-cycle real-time PCR assays may be only 20-30 minutes.

Use Detect the presence of *Bacillus anthracis* in clinical specimens from humans, grazing animals including cattle, goats, and sheep.

Limitations Patients with inhalation anthrax become ill and often die rapidly. Sputum cultures are rarely positive.

Methodology Inoculation of sheep's blood agar, incubation of media at 35°C in ambient or CO_2-enhanced atmospheric conditions. Cultures are nonhemolytic.

Blood, impression smears from a lesion, or rarely, cerebrospinal fluid reveal short chains of large, encapsulated gram-positive bacilli.
(Continued)

Anthrax Detection *(Continued)*

Enzyme-linked immunosorbent assay (ELISA) and polymerase chain reaction (PCR) are available at reference laboratories. Immunohistochemical staining methods are available. Rapid-cycle real-time PCR for anthrax has recently been described. LightCycler PCR instrumentation permits anthrax detection following autoclaving.

Additional Information Anthrax is a zoonotic infection that may present as cutaneous, pulmonary, or gastrointestinal disease. Although rare, human anthrax does occur throughout the world, including the United States, and has usually been acquired by contact with contaminated herbivores or animal products such as hides, wool, or other fractions. Means of transmission include contact with inoculation of minor lesions, meat ingestion, handling, or inhalation.

Because of its potential application as an agent of biological warfare, laboratories throughout the U.S. should have procedures in place to make a preliminary identification and to notify the proper authorities if a suspicious isolate is recovered. It is recommended that laboratories utilize biosafety level 2 facilities and practices for handling specimens and cultures for *Bacillus anthracis*.

References

Dixon TC, Meselson JG, Guillemin J, et al, "Anthrax," *N Engl J Med*, 1999, 341(11):815-26.

Eachempati SR, Flomenbaum N, and Barie PS, "Biological Warfare: Current Concerns for the Health Care Provider," *J Trauma*, 2002, 52(1):179-86.

Espy MJ, Uhl JR, Sloan LM, et al, "Detection of Vaccinia Virus, Herpes Simplex Virus, Varicella-Zoster Virus, and *Bacillus anthracis* DNA by LightCycler Polymerase Chain Reaction After Autoclaving: Implications for Biosafety of Bioterrorism Agents," *Mayo Clin Proc*, 2002, 77(7):624-8.

Grinberg LM, Abramova FA, Yampolskaya OV, et al, "Quantitative Pathology of Inhalational Anthrax I: Quantitative Microscopic Findings," *Mod Pathol*, 2001, 14(5):482-95.

Henderson DA, Inglesby TV, and O'Toole T, *Bioterrorism: Guidelines for Medical and Public Health Management*, Chicago, IL: AMA Press, 2002.

Inglesby TV, O'Toole T, Henderson DA, et al, "Anthrax as a Biological Weapon, 2002 - Updated Recommendations for Management," *JAMA*, 2002, 287(17):2236-52.

McGovern TW and Norton SA, "Recognition and Management of Anthrax," *N Engl J Med*, 2002, 346(12):943-5.

Meyer RF and Morse SA, "Bioterrorism Preparedness for the Public Health and Medical Communities," *Mayo Clin Proc*, 2002, 77(7):619-21.

Moser R Jr, White GL, Lewis-Younger CR, et al, "Preparing for Expected Bioterrorism Attacks," *Military Medicine*, 2001, 166(5):369-74.

Swartz MN, "Recognition and Management of Anthrax - An Update," *N Engl J Med*, 2001, 345(22):1621-6.

Uhl JR, Bell CA, Sloan LM, et al, "Application of Rapid-Cycle Real-Time Polymerase Chain Reaction for the Detection of Microbial Pathogens: The Mayo-Roche Rapid Anthrax Test," *Mayo Clin Proc*, 2002, 77(7):673-80.

Varkey P, Poland GA, Cockerill FR III, et al, "Confronting Bioterrorism: Physicians on the Front Line," *Mayo Clin Proc*, 2002, 77(7):661-2.

Internet Web Sites

www.asm.org
www.bt.cdc.gov
www.bt.cdc.gov/labissues/index.asp
www.cdc.gov/mmwr/preview/mmwrhtml/mm5041a1.htm
www.cdc.gov/mmwr/preview/mmwrhtml/mm5042a1.htm
www.cdc.gov/mmwr/preview/mmwrhtml/mm5043a1.htm
www.cdc.gov/mmwr/preview/mmwrhtml/mm5044a1.htm
www.idsociety.org/bt/toc.htm

♦ **Antiactin Antibodies** *see* Smooth Muscle Antibody *on page 1207*

♦ **Anti-ASGPR** *see* Smooth Muscle Antibody *on page 1207*

Antibodies to IgA

Related Information

Frozen Red Blood Cells *on page 614*
Immunoglobulin A *on page 770*
Plasma, Fresh Frozen *on page 1039*
Plasma, Frozen, Donor Retested *on page 1041*
Platelet Transfusion *on page 1058*
Red Blood Cells *on page 1139*
Red Blood Cells, Washed *on page 1143*
Risks of Transfusion *on page 1166*
Transfusion Reaction Work-up *on page 1269*

Synonyms Anti-immunoglobulin A; IgA Antibodies

Abstract While 1 in 600 individuals are IgA deficient, IgA anaphylactic transfusion reactions occur infrequently, in approximately 1 in 20,000 to 47,000 transfusions. Antibodies to IgA may be found in sera of individuals who have anaphylactic reactions. Most, but not all, of these individuals are IgA deficient. The frequency of antibodies in IgA-deficient individuals has been reported to be from 20% to 40%. Anti-IgA may be detected in 20% to 25% of individuals with common variable immune deficiency. IgA antibodies can also be detected in 2% to 59% of normal individuals. IgA-deficient individuals who have developed anti-A in the context of an anaphylactic reaction can only accept plasma fractions prepared from donors completely lacking IgA.

Patient Preparation If possible, delay transfusion in patients who have had an anaphylactic reaction until a pretransfusion sample can be tested for the presence of antibodies to IgA. If antibodies are present, transfuse with IgA-deficient or autologous blood components. Frozen, thawed-deglycerolized red blood cells, washed red blood cells, or washed platelets may be used. Consult manufacturer's package insert prior to infusion of any plasma derivative (eg, immunoglobulin, factor concentrate) and use only those noted to be IgA deficient.

Aftercare Signs and symptoms may occur up to as long as 1 hour after transfusion and are related to the cutaneous, respiratory, gastrointestinal, and cardiovascular systems. For anaphylactic reactions, the transfusion must be stopped and I.V. line kept open with normal saline. Save blood component, notify Blood Bank, and perform transfusion reaction work-up (see Transfusion Reaction Work-up *on page 1269*).

Specimen Serum

Container Red top tube or SST™ tube

Sampling Time Ideally, utilize specimen obtained prior to transfusion of blood component or derivative implicated in the anaphylactic reaction.

Turnaround Time Results of testing for antibody to IgA may not be available prior to the decision to transfuse again, as this test is sent to specialized reference laboratories.

Reference Interval Anti-IgA: negative

Use Evaluate anaphylactic adverse effect of transfusion

Limitations Pre-existing antibodies to other serum proteins may exist which also cause anaphylaxis. Thus, a negative test for antibodies to IgA does not mean that the patient may not suffer another anaphylactic reaction.

Detection of antibodies to IgA in a nontransfused individual by passive hemagglutination assays does not reliably predict risk of an anaphylactic transfusion reaction.

Additional Information Exposure to blood fractions containing IgA in IgA-deficient individuals may induce IgG or IgE antibodies, which can cause anaphylactic reactions. Immunoglobulin preparations may contain some IgA and lead to serious reactions.

IgA **red cell** autoantibodies are also recognized. Sometimes referred to as a "Coombs-negative hemolytic anemia," they may be idiopathic or associated with neoplasms.

References

Bardill B, Mengis C, Tschopp M, et al, "Severe IgA-Mediated Autoimmune Haemolytic Anaemia in a 48-Year-Old Woman," *Eu J Haematol*, 2003, 70(1):60-3.

Baroti Toth C, Kramer J, Pinter J, et al, "IgA Content of Washed Red Blood Cell Concentrates," *Vox Sang*, 1998, 74(1):13-4.

de Albuquerque Campos R, Sato MN, da Silva Duarte AJ, "IgG Anti-IgA Subclasses in Common Variable Immunodeficiency and Association With Severe Adverse Reactions to Intravenous Immunoglobulin Therapy," *J Clin Immunol*, 2000, 20(1):77-82.

Lilic D and Sewell WA, "IgA Deficiency: What We Should - or Should Not - Be Doing," *J Clin Pathol*, 2001, 54(5):337-8.

Mollison PL, Engelfriet CP, Contreras M, et al, "Some Unfavorable Effects of Transfusion," *Blood Transfusion in Clinical Medicine*, 10th ed, Chapter 15, Oxford, UK: Blackwell Scientific Publications, 1997, 487-508.

Shimada E, Tadokoro K, Watanabe Y, et al, "Anaphylactic Transfusion Reactions in Haptoglobin-Deficient Patients With IgE and IgG Haptoglobin Antibodies," *Transfusion*, 2002, 42(6):766-73.

Vamvakas EC and Pineda AA, "Allergic and Anaphylactic Reactions," *Transfusion Reactions*, Popovsky MA, ed, Bethesda, MD: American Association of Blood Banks Press, 2001, 83-127.

Antibody Detection/Identification, Red Cell

Related Information

Antibody Titer *on page 167*
(Continued)

Antibody Detection/Identification, Red Cell *(Continued)*

Antiglobulin Test, Indirect *on page 177*
Hemolytic Disease of the Newborn, Antibody Identification *on page 690*
Pretransfusion Testing *on page 1088*
Transfusion Reaction Work-up *on page 1269*

Synonyms Irregular Antibody Detection/Identification

Applies to Autoimmune Hemolytic Anemia Work-up; Hemolytic Disease of the Newborn, Investigation

Test Includes Absorption/elution procedures to distinguish between auto- and alloantibodies; antibody titration studies; red cell antigen typing

Abstract The incidence of immunization to red cell antigens has been determined to be in the order of 1% to 1.5% in the population at large. It is mandatory, therefore, that pretransfusion testing includes an antibody screen. The finding of a positive antibody screen requires further testing with an extended panel of phenotyped red cells to establish the identity of the antibody.

Specimen Blood

Container One red top tube and one lavender top (EDTA) tube is usually sufficient (complex cases may require additional specimen).

Causes for Rejection Gross hemolysis, sample collected in wrong tube (eg, serum separator tube), improper labeling

Special Instructions Provide Blood Bank with diagnosis, medications, and history of prior pregnancy and transfusions.

Use The object of antibody detection is to detect and subsequently identify clinically important red cell antibodies. Thus, antibody screening is a routine pretransfusion test. Antibody detection and identification also play an important role in the investigation protocols for hemolytic disease of the newborn (HDN), hemolytic anemia, and transfusion reactions.

Limitations Antibody directed to low incidence or "private" antigens, which may not be represented on the testing cell panel, will not be detected. On some occasions, antibody levels are too low for detection.

Contraindications A blood sample taken shortly after massive transfusion or exchange transfusion will not be representative of the patient's blood. In these cases, a limited pretransfusion testing protocol may be implemented.

Methodology Patients' sera are screened for unexpected red cell antibodies by testing against an abbreviated red cell panel selected to contain all the clinically important antigens. The tests are carried out by a variety of techniques with the expectation that all clinically significant antibodies will be detected. Antibody identification is performed in the same way but with extended panels of fully phenotyped red cells.

In cases of HDN and hemolytic anemia, testing may require elution of antibody from autologous or test red cells. The presence of autoantibody may mask the serologic expression of underlying alloantibody. In these cases, absorption procedures may be necessary to remove the autoantibody. Autoabsorption and elution studies may require the provision of further blood samples from the patient.

Alternative methods to traditional tube testing have been devised. Some methods fail to detect both IgM and IgG antibodies. Information may not be provided regarding temperature or phase of antibody reactivity. More commonly used alternative methods include solid-phase red cell adherence and gel column agglutination technology.

Additional Information Determination of antibody specificity permits the appropriate selection of donor blood for compatibility tests.

It is customary to screen for unexpected antibodies associated with HDN early in pregnancy. All positive screen results require antibody identification to permit assessment of clinical significance. Titration of antibody levels throughout the course of the pregnancy may indicate the likelihood of HDN. Similarly, antibody characterization may assist in the diagnosis of autoimmune hemolytic anemia.

See Anemia (Hemolytic) in the Disease Index.

References

Issitt PD and Anstee DJ, *Applied Blood Group Serology*, 4th ed, Durham, NC: Montgomery Scientific Publications, 1998, 877-905, 1045-95.

Brecher ME, *Technical Manual*, 14th ed, Bethesda, MD: American Association of Blood Banks Press, 2002, 253-69.

Bunker ML, Thomas CL, and Geyer SJ, "Optimizing Pretransfusion Antibody Detection and Identification: A Parallel, Blinded Comparison of Tube PEG, Solid-Phase and Automated Methods," *Transfusion*, 2001, 41(5):621-6.

Weisbach V, Ziener A, Zimmermann R, et al, "Comparison of the Performance of Four Microtube Column Agglutination Systems in the Detection of Red Cell Alloantibodies," *Transfusion*, 1999, 39(10):1045-50.

Antibody Titer

Related Information

Antibody Detection/Identification, Red Cell *on page 165*
Prenatal Screen, Immunohematology *on page 1086*

Applies to Hemolytic Disease of the Newborn, Antibody Titer

Test Includes Known positive antibody screen and identification

Specimen Serum

Container Red top tube

Storage Instructions Keep serum frozen to permit subsequent parallel titrations.

Causes for Rejection Gross hemolysis, sample collected in a serum separator tube, improper labeling

Possible Panic Range Increase of titration value over previous level (at least two dilutions)

Use Antibody titration is an attempt to quantitate antibody by testing serial twofold dilutions against a selected red cell. Results are expressed as the reciprocal of the highest dilution showing serologic activity. Comparative titrations can demonstrate a change in antibody levels over time. This information may be valuable in prenatal studies of Rh-negative mothers with anti-D or other antibody associated with hemolytic disease of the newborn (HDN). Decisions to perform invasive procedures may be based to some extent on maternal antibody titration results. However, maternal titers are being replaced by Doppler ultrasound to detect fetal anemia in cases where there is a history of an affected fetus or infant.

Limitations Manual titration methods are inherently subjective. It is important that successive studies of the same patient be performed in the same laboratory by a standardized technique with the red cells from the same donor, if possible; if not, then with cells of the same phenotype. The use of semiautomated pipettes is highly advised.

Contraindications Antibody screening test is negative or antibody identification indicates that the detected antibody is not associated with HDN; or paternal red cell studies show that father lacks the offending antigen.

Methodology Serial (usually doubling) dilutions of the serum are tested against red cells of the appropriate phenotype by a technique that demonstrates activity of the antibody. The end-point is recorded as the reciprocal of the highest dilution expressing activity.

Additional Information Although an indirect antiglobulin titration end-point of 16-32 or higher has been considered significant for anti-D, it does not follow that this result is significant in the case of other alloantibody specificities. Failure to demonstrate a rising titer of maternal alloantibody through the course of pregnancy does not necessarily preclude HDN.

References

Brecher ME, *Technical Manual*, 14th ed, Bethesda, MD: American Association of Blood Banks Press, 2002, 497-515.

Moise KJ Jr, "Management of Rhesus Alloimmunization in Pregnancy," *Obstet Gynecol*, 2002, 100(3):600-11.

♦ **Antibody to Double-Stranded DNA** *see* Anti-DNA *on page 173*
♦ **Antibody to Native DNA** *see* Anti-DNA *on page 173*

Anticardiolipin Antibody

Related Information

Antinuclear Antibodies *on page 189*
Antiphospholipid Antibody (Lupus Anticoagulant and/or Anticardiolipin Antibody) *on page 193*
Factor Inhibitors *on page 566*
Hypercoagulation Panel *on page 758*
Prothrombin Time *on page 1116*
RPR *on page 1174*
Sjögren Antibodies *on page 1199*
(Continued)

Anticardiolipin Antibody *(Continued)*

VDRL, Serum or Cerebrospinal Fluid *on page 1303*

Applies to β2-Glycoprotein 1; Lupus Anticoagulant

Test Includes Detection of IgG, IgM, and IgA antibody to the phospholipid, cardiolipin

Abstract Anticardiolipin antibodies (ACA) belong to the antiphospholipid antibody group of proteins and may have anticoagulant activity (similar to that of lupus anticoagulants). The **antiphospholipid antibody syndrome** is characterized by recurrent clinical events: noninflammatory thrombosis of small or large arteries and/or veins and fetal loss, with demonstrable antiphospholipid antibodies (anticardiolipin antibody or lupus anticoagulant). They are autoantibodies found in subjects with systemic lupus erythematosus (SLE) and related entities, lupus-like diseases, infectious diseases, and drug reactions. The syndrome is primary if SLE is not present. When patients have SLE and also have antiphospholipid antibodies (with corresponding clinical features) cases are considered as secondary antiphospholipid antibody syndrome.

Specimen Serum

Container Red top tube

Storage Instructions Repeated freeze-thaw cycles alters stability of anticardiolipin antibodies.

Reference Interval Negative

Use Differential diagnosis of recurrent thromboses, lupus-like syndromes, false-positive VDRL or RPR, recurrent fetal loss, and rarely, severe hemorrhage

Limitations Anticardiolipin levels by enzyme-linked immunosorbent assay (ELISA) are associated with poor reproducibility.

Methodology Enzyme-linked immunosorbent assay (ELISA) for IgG or IgM anticardiolipin antibody

Additional Information Antibody to cardiolipin (the diphosphatidyl glycerol component of many phospholipid membranes) is at least partially cross reactive with the reagin antibody of syphilis and the lupus anticoagulant. The commonality between these diseases is antibody involving phosphate groups (lupus-DNA; VDRL-phospholipid of cardiolipin; and ACA - antibody directed against phospholipids of the coagulation system).

Binding of ACA to phospholipid is mediated by β_2-glycoprotein 1 (also known as apolipoprotein H), a serum protein with anticoagulant properties that may inhibit thrombin generation. Upon binding of β_2-glycoprotein 1 to phospholipid/cardiolipin, resultant change in structural configuration appears to expose new epitopes that stimulate formation of antiphospholipid antibodies. Structural variants (mutants) with functional significance have been described.

Neoepitopes, however, are not universally accepted as the antigenic targets. Pathophysiologic considerations of the recently proposed "catastrophic antiphospholipid syndrome" (CAPS), an accelerated form of primarily microvascular thrombosis as a severe manifestation of the antiphospholipid syndrome and with multiorgan involvement, focus on the role of endothelial cells in promoting a procoagulant state. A sufficient "density" of proteins bound to phospholipid surfaces are required, representing a threshold for antibody binding. Endothelial cells, platelets, and monocytes are "activated" and adhesion molecules are upregulated.

ACA is associated with a host of clinical and laboratory abnormalities. Abnormal tests may include thrombocytopenia, reactive VDRL or RPR, SS-A/Ro antibodies, and prolonged activated partial thromboplastin time (APTT) (lupus anticoagulant). Clinically, patients have lupus-like symptoms, often "ANA negative," recurrent venous and arterial thromboses, recurrent fetal loss (usually more than two episodes for a strong association), mitral valve endocarditis, chorea, and epilepsy. The entire constellation represents the antiphospholipid antibody syndrome. The association between thrombosis and recurrent fetal loss and ACA in patients with a prolonged APTT is especially strong in patients in whom ACA is not induced by infection or medication. IgG anticardiolipin is more likely to be influenced by disease activity than is IgM anticardiolipin. Plasmapheresis along with anticoagulant therapy may be used in symptomatic cases. Lupus anticoagulant and anticardiolipin antibodies are found together in about 70% of patients with antiphospholipid antibody syndrome. LA is found in about 20% to 40% and ACA is found in some 45% of subjects with SLE. An increased incidence of antiphospholipid antibodies has been found among patients with monoclonal gammopathy of undetermined significance.

Retinopathy has been shown to be associated with anticardiolipin antibody and with central nervous system lupus. Presence of ACA does not appear to affect the international normalized ratio. A recently developed ELISA system to detect complement-fixing ability of ACA with reported sensitivity of 78% and specificity of 84% (relative to the occurrence of thrombotic events) is possibly a marker for thrombotic manifestations of the antiphospholipid syndrome. High titers of cerebrospinal fluid IgG-ACA have been shown to occur in symptomatic cerebral lupus patients with evidence of intrathecal synthesis. The presence of ACA in patients with chronic hepatitis C virus (HCV) infection is reportedly significantly higher than in subjects with other inflammatory diseases of the liver (eg, chronic hepatitis B virus infection, primary biliary cirrhosis). The frequency of thrombotic complications was similar in ACA-positive and ACA-negative patients with chronic HCV infection. Sera from all but one ACA-positive HCV patient was negative for phospholipid-dependent anti-β_2-glycoprotein 1 antibodies. Thus, while ACA commonly occurs in patients with chronic HCV infection, it may be without clinical import.

References
Gushiken FC, Arnett FC, and Thiagarajan P, "Primary Antiphospholipid Antibody Syndrome With Mutations in the Phospholipid Binding Domain of β_2-Glycoprotein I," *Am J Hematol*, 2000, 65(2):160-5.

Lai NS and Lan JL, "Evaluation of Cerebrospinal Anticardiolipin Antibodies in Lupus Patients With Neuropsychiatric Manifestations," *Lupus*, 2000, 9(5):353-7.

Male C, Mitchell L, Julian J, et al, "Acquired Activated Protein C Resistance Is Associated With Lupus Anticoagulants and Thrombotic Events in Pediatric Patients With Systemic Lupus Erythematosus," *Blood*, 2001, 97(4):844-9.

Mant MJ, Stang L, and Etches WS, "Warfarin Monitoring in Patients With Anticardiolipin Antibodies, but Without Lupus Anticoagulants," *Thromb Res*, 2000, 99(5):477-82.

Tanikawa K and Sata M, "High Prevalence of Anticardiolipin Antibodies in Hepatitis C Virus Infection: Lack of Effects on Thrombocytopenia and Thrombotic Complications," *J Gastroenterol*, 2000, 35(4):272-7.

Triplett DA and Asherson RA, "Pathophysiology of the Catastrophic Antiphospholipid Syndrome (CAPS)," *Am J Hematol*, 2000, 65(2):154-9.

Ushiyama O, Ushiyama K, Koarada S, et al, "Retinal Disease in Patients With Systemic Lupus Erythematosus," *Ann Rheum Dis*, 2000, 59(9):705-8.

♦ **Anti-CCP** *see* Anticyclic Citrullinated Peptide Antibody *on page 169*

♦ **Anticitrullinated Peptide Antibodies** *see* Anticyclic Citrullinated Peptide Antibody *on page 169*

♦ **Anticoagulants, Oral** *see* Warfarin, Serum or Plasma *on page 1325*

Anticyclic Citrullinated Peptide Antibody
Related Information
Anti-DNA *on page 173*
Antinuclear Antibodies *on page 189*
Antiphospholipid Antibody (Lupus Anticoagulant and/or Anticardiolipin Antibody) *on page 193*
Jo-1 Antibody *on page 815*
Rheumatoid Factor, Serum or Body Fluid *on page 1161*
Sjögren Antibodies *on page 1199*
Smith (Sm) and Ribonucleoprotein (RNP) Antibodies *on page 1206*

Synonyms Anti-CCP; Anticitrullinated Peptide Antibodies; CCP Antibodies

Applies to Filaggrin

Abstract Anti-CCP antibodies have high specificity for rheumatoid arthritis (RA), especially in disease of recent onset. Demonstration of anti-CCP antibodies may be of use as a diagnostic test in early RA.

Specimen Serum

Container Red top tube

Reference Interval 1-39 units; cutoff value of 50 or 60 units (populations dependent) has been recommended (this level contributes to high specificity).

Use Diagnosis of rheumatoid arthritis, especially early stage disease. Test has high level of specificity (97%) but only moderate sensitivity for RA. When anti-CCP and rheumatoid factor (RF) tests are used in conjunction, specificity is increased (>99%). Anti-CCP has been used in the evaluation of palindromic rheumatism.

(Continued)

Anticyclic Citrullinated Peptide Antibody *(Continued)*

Methodology Enzyme-linked immunosorbent assay (ELISA) with intraassay coefficient of variation of 5% to 13%

Additional Information Serologic support for the diagnosis of RA is dependent on the RF test (see Rheumatoid Factor, Serum or Body Fluid *on page 1161*) which, however, lacks specificity. Citrulline is an amino acid of the protein filaggrin which has origin from profilaggrin (a component of some keratohyalin granules) during cell differentiation. With citrullination (conversion of peptidyl-arginine to peptidyl-citrulline) there is induction of autoantibody which can be measured by ELISA utilizing cyclic citrullinated peptides as antigens. Anti-CCP antibodies have high specificity for RA in recent-onset arthritis and in patients who also have demonstrable RF (IgG, IgA, and IgM).

Children with juvenile idiopathic arthritis may have anti-CCP antibodies but these are found largely in a subset of patients who have polyarticular IgM-RF.

References

Bas S, Genevay S, Meyer O, et al, "Anticyclic Citrullinated Peptide Antibodies, IgM and IgA Rheumatoid Factors in the Diagnosis and Prognosis of Rheumatoid Arthritis," *Rheumatology*, 2003, 42(5):677-80.

Bizzaro N, Mazzanti G, Tonutti E, et al, "Diagnostic Accuracy of the Anti-Citrulline Antibody Assay for Rheumatoid Arthritis," *Clin Chem*, 2001, 47(6):1089-93.

Jansen LMA, Van Schaardenburg D, Van Der Horst-Bruinsma IE, et al, "The Predictive Value of Anticyclic Citrullinated Peptide Antibodies in Early Arthritis," *J Rheumatol*, 2003, 30(8):1691-5.

Rantapää-Dahlqvist S, de Jong BAW, Berglin E, et al, "Antibodies Against Cyclic Citrullinated Peptide and IgA Rheumatoid Factor Predict the Development of Rheumatoid Arthritis," *Arth Rheum*, 2003, 48(10):2741-9.

Salvador G, Gomez A, Vinas O, et al, "Prevalence and Clinical Significance of Anticyclic Citrullinated Peptide and Antikeratin Antibodies in Palindromic Rheumatism. An Abortive Form of Rheumatoid Arthritis?" *Rheumatology*, 2003, 42(8):972-5.

Schellekens GA, Visser H, De Jong BAW, et al, "The Diagnostic Properties of Rheumatoid Arthritis Antibodies Recognizing a Cyclic Citrullinated Peptide," *Arth Rheum*, 2000, 43(1):155-63.

Van Rossum M, Van Soesberger R, De Kort S, et al, "Anticyclic Citrullinated Peptide (Anti-CCP) Antibodies in Children With Juvenile Idiopathic Arthritis," *J Rheumatol*, 2003, 30(4):825-8.

Antideoxyribonuclease-B Titer, Serum

Related Information

Antistreptolysin O Titer, Serum *on page 197*

Bacterial Culture, Throat, and Antigen Detection Testing for Group A Streptococci *on page 245*

Group A *Streptococcus* Screen, Rapid *on page 659*

Sedimentation Rate, Erythrocyte *on page 1181*

Streptozyme *on page 1225*

Synonyms Anti-DNase-B Titer; Antistreptococcal DNase-B Titer; Streptodornase

Abstract Detection of an immune response to extracellular products of *S. pyogenes*, such as DNase B, is useful in demonstration of evidence of streptococcal infections in patients without documentation of recent infection, who present with nonsuppurative sequelae such as rheumatic fever or glomerulonephritis.

Specimen Serum

Container Red top tube

Reference Interval Children: preschool: ≤60 units; school: ≤170 units; adults: ≤85 units; a rise in titer of two or more dilution increments between acute and convalescent sera is significant.

Use Document recent streptococcal infection (eg, to evaluate possible rheumatic fever).

Limitations Normal ranges may vary in different populations; test must use a pure source of DNase-B to ensure specificity.

Contraindications Not valid in patients with hemorrhagic pancreatitis

Methodology Colorimetry based on hydrolysis of DNA

Additional Information DNase-B is antigenically the most consistent of four streptococcal DNases. The presence of antibodies to streptococcal DNase is an indicator of recent infection, especially if a rise in titer can be documented. This test has advantages over the ASO test: It is more sensitive for streptococcal pyoderma, it is not as subject to false positives due to liver disease, reagents are not likely to oxidize, and it is not affected by the site of infection. It is positive, like ASO, in about 80% to 85% of patients with streptococcal infections. Application of both DNase B antibodies and ASO detects about 95% of streptococcal infection.

References

Bisno AL, "Acute Pharyngitis," *N Engl J Med*, 2001, 344(3):205-11.

Cunningham MW, "Pathogenesis of Group A Streptococcal Infections," *Clin Microbiol Rev*, 2000, 13(3):470-511.

Efstratiou A, "Group A Streptococci in the 1990s," *J Antimicro Chemother*, 2000, 45(Suppl):3-12.

Olivier C, "Rheumatic Fever - Is It Still a Problem?" *J Antimicro Chemother*, 2000, 45(Suppl):13-21.

Antidepressants, Cyclic, Serum or Plasma

Related Information

Amitriptyline, Serum or Plasma *on page 149*
Amoxapine, Serum or Plasma *on page 153*
Doxepin, Serum or Plasma *on page 524*
Fluoxetine, Serum or Plasma *on page 604*
Imipramine, Serum or Plasma *on page 767*
Maprotiline, Serum *on page 892*
Nortriptyline, Serum *on page 966*
Sertraline, Serum *on page 1191*
Trazodone, Serum or Plasma *on page 1272*

Synonyms CAD; Cyclic Antidepressants

Applies to Adapin®; Amitriptyline and Nortriptyline; Amoxapine and 8-Hydroxy-amoxapine; Aventyl®; Desipramine; Doxepin and Nordoxepine; Etrafon®; Fluoxetine and Norfluoxetine; Imipramine and Desipramine; Maprotiline; Norpramin®; Nortriptyline; Pamelor®; Pertofrane®; Presamine®; Protriptyline; Sinequan®; TCA; Tetracyclic Antidepressants; Tofranil®; Trazodone; Tricyclic Antidepressants; Vivactil®

Abstract Drugs in this class are widely used as antidepressants. They are frequently involved in suicidal ingestion and responsible for a large percentage of drug-related deaths. The central nervous system and cardiovascular systems are primarily affected in toxicity of the cyclic antidepressants. Drowsiness, seizures, coma, and cardiac dysrhythmias occur. Hypoventilation and anticholinergic findings are seen.

Specimen Serum or plasma

Container Red top tube, green top (heparin) tube; avoid serum separator tubes as drugs tend to bind to gel.

Sampling Time Steady-state specimen after 1 week of dose schedule; draw specimen 12 hours after the last dose.

Storage Instructions Remove serum within 2 hours of drawing. Samples are stable for 24 hours at room temperature, 4 weeks at 4°C, and more than 1 year at -20°C.

Causes for Rejection Specimen collected in gel tube

Special Instructions Order individual drug level or tricyclic overdose screen

Reference Interval See table.

Cyclic Antidepressants Therapeutic and Toxic Levels

Drug / Active Metabolite	Range (ng/mL)	Toxic Level (ng/mL)	Half-life (h) (Drug / Metabolite)
Amitriptyline + nortriptyline	120-250	>500	20-40/20-90
Amoxapine + 8-hydroxyamoxapine	200-400	>600	8-15/25-40
Desipramine	75-300	>500	20-90
Doxepin + nordoxepine	150-250	>500	10-25/35-55
Fluoxetine + norfluoxetine	300-1200	>2000	24-72/170-360
Imipramine + desipramine	100-300	>500	5-25/20-90
Maprotiline	200-400	>1000	25-30
Nortriptyline	50-150	>500	20-60
Protriptyline	70-250	>500	60-90
Trazodone	800-1600	>5000	6-15

Use Concentrations play a role in therapeutic monitoring. Blood levels may be useful for diagnosis of toxicity but not in projection of severity of exposure. (Continued)

Antidepressants, Cyclic, Serum or Plasma (Continued)

Contraindications Patient taking more than one cyclic antidepressant, patient taking phenothiazines or monoamine oxidase inhibitors

Methodology Immunoassay, gas chromatography (GC), high performance liquid chromatography (HPLC)

Additional Information Cyclic antidepressants (CAs) are metabolized to secondary active compounds. African-Americans usually have 50% higher blood levels than whites for same dose schedule. The most important of the more serious or toxic effects of CAs is cardiotoxicity. Arrhythmias and conduction defects with precipitation of congestive heart failure and possibly myocardial infarction are common at combined levels >1000 ng/mL. Widening of the QRS interval to >100 msec is highly suggestive of a tricyclic antidepressant overdose. Total tricyclic concentrations reflect the severity of overdose; however, the better clinical correlation is with an EKG finding of QRS widening.

References

Geller B, Reising D, Leonard HL, et al, "Critical Review of Tricyclic Antidepressant Use in Children and Adolescents," *J Am Acad Child Adolesc Psychiatry*, 1999, 38(5):513-6.

Hulten BA, Adams R, Askenasi R, et al, "Predicting Severity of Tricyclic Antidepressant Overdose," *J Toxicol Clin Toxicol*, 1992, 30(12):161-70.

Williams JW Jr, Mulrow CD, Chiquette E, et al, "A Systematic Review of Newer Pharmacotherapies for Depression in Adults: Evidence Report Summary," *Ann Intern Med*, 2000, 132(9):743-56.

Antidiuretic Hormone, Plasma

Related Information

Concentration Test, Urine *on page 446*
Methadone, Serum or Urine *on page 900*
Osmolality, Calculated, Serum or Plasma *on page 976*
Osmolality, Serum *on page 978*
Osmolality, Urine *on page 979*
Sodium, Serum or Plasma *on page 1210*
Specific Gravity, Urine *on page 1216*

Synonyms ADH; Arginine[8]-Vasopressin; Arginine-Vasopressin; AVP; Vasopressin

Test Includes Serum osmolality, urine osmolality, serum sodium

Abstract Antidiuretic hormone (ADH), synthesized in the hypothalamus and stored and released by the posterior pituitary, initiates a series of events resulting in water reabsorption in the distal renal tubule. Synthesis and release are regulated by hypothalamic osmoreceptors that respond to changes in osmolality and baroreceptors that respond to changes in blood volume.

Deficiency of, or lack of normal response to, ADH results in polyuria, increased serum osmolality, hypernatremia, and decreased urine osmolality. Excess production or secretion of ADH results in oliguria, decreased serum osmolality, hyponatremia, and increased urine osmolality.

Patient Preparation Patient should avoid substances that influence ADH secretion (eg, nicotine, alcohol, caffeine, diuretics). Fasting, water deprivation, or water loading may be required, depending upon diagnostic test protocol in use.

Specimen Plasma

Container Prechilled lavender top (EDTA) tube(s)

Storage Instructions Centrifuge immediately in a refrigerated (4°C) centrifuge at sufficient speed and time to produce platelet-poor plasma (eg, 3600 g for 20 minutes). Remove the plasma into a plastic transport tube, and freeze at -20°C until analysis. Transport to referral laboratory frozen.

Causes for Rejection Specimen not received frozen, recently administered radioisotopes

Special Instructions Following collection, place the tube in ice and deliver to the laboratory for immediate processing.

Reference Interval Basal values in normally hydrated individuals: 0.5-2.0 ng/L (SI, 0.5-1.9 pmol/L). In a study of 203 children, age 1 day to 18 years, basal levels averaged 1.1 ±0.6 ng/L (SI: 1.0 ±0.6 pmol/L) with no significant differences between males and females and no age correlation within the study population. This same study reported basal levels of 1.0 ±0.5 ng/L (SI: 0.9 ±0.5 pmol/L) in 16 adult controls. Results are best interpreted with simultaneously determined plasma osmolalities. See table.

Antidiuretic Hormone, Plasma

ADH ng/L (SI: pmol/L)	Osmolality (mOsm/kg)
<1.5 (<1.4)	270-280
<2.5 (<2.3)	280-285
1-5 (0.9-4.6)	285-290
2-7 (1.9-6.5)	290-295
4-12 (3.7-11.1)	295-300

Use Abnormal ADH function is usually assessed by serum and urine osmolality and serum sodium determinations. Plasma ADH measurements are occasionally useful in the differential diagnosis of syndrome of inappropriate secretion of ADH (SIADH), diabetes insipidus (DI), chronic hyponatremia, and psychogenic water intoxication.

Overnight water deprivation protocols are useful for differentiating polyuric patients with DI from other causes of polyuria (eg, psychogenic polydipsia) and for characterizing DI as neurogenic (cranial or central) or nephrogenic (failure of kidneys to respond to ADH). After water deprivation, patients with DI have high plasma osmolalities, low urine osmolalities, and either lower than expected (neurogenic) or high (nephrogenic) plasma ADH levels. Patients with psychogenic polydipsia respond similarly to normal individuals, having plasma osmolalities within the normal range, appropriately concentrated urines, and ADH levels that correlate predictably with the plasma osmolality. Saline infusion test protocols are sometimes used when overnight water deprivation studies are inconclusive.

Limitations Assays are complex to perform and lack sensitivity and specificity. Consequently, it is important to evaluate plasma osmolality levels concurrently with ADH levels. Since ADH is contained in platelets, plasma contaminated with platelets will result in overestimation of plasma ADH.

Methodology Radioimmunoassay following extraction of ADH from plasma

Additional Information Idiopathic DI is the most common type, but a variety of causes of central DI exist, including trauma, autosomal dominant type of inherited disease, and pregnancy. Tumors of the hypothalamus or pituitary which can cause DI include craniopharyngioma, ependymona, germinoma, pinealoma, leukemic infiltrates, some tumors of the anterior pituitary, and metastases. Langerhans cell histiocytosis, sarcoidosis, and tuberculosis may cause DI.

SIADH is characterized by hyponatremia, low plasma osmolality, and increased plasma ADH. Water loading protocols, though hazardous, are occasionally useful for diagnosis of difficult cases of SIADH.

References

Crawford GA and Gyory AZ, "Measuring Arginine Vasopressin in Children and Babies," *Clin Chem*, 1990, 36(9):1689.

Demers LM, "Pituitary Function," *Tietz Textbook of Clinical Chemistry*, 3rd ed, Chapter 41, Burtis CA and Ashwood ER, eds, Philadelphia, PA: WB Saunders Co, 1999, 1470-95.

Kluge M, Riedl S, Erhart-Hofmann B, et al, "Improved Extraction Procedure and RIA for Determination of Arginine⁸-Vasopressin in Plasma: Role of Premeasurement Sample Treatment and Reference Values in Children," *Clin Chem*, 1999, 45(1):98-103.

Anti-DNA

Related Information

Antinuclear Antibodies *on page 189*
Complement C3, Serum *on page 434*
Complement, Total, Serum or Body Fluid *on page 441*
Sjögren Antibodies *on page 1199*
Smith (Sm) and Ribonucleoprotein (RNP) Antibodies *on page 1206*
Topoisomerase I Antibody *on page 1261*

Synonyms Antibody to Double-Stranded DNA; Antibody to Native DNA; Anti-Double-Stranded DNA; Anti-ds-DNA; DNA Antibody; n-DNA

Applies to ss-DNA

Replaces Anti-ss-DNA; LE Cell Preparation

Test Includes Titers on positive specimens

(Continued)

Anti-DNA (Continued)

Abstract IgG autoantibodies to ds-DNA are found characteristically in subjects with systemic lupus erythematosus (SLE) and only rarely with other connective tissue diseases. The other autoantibodies relatively specific for SLE include anti-Sm, which provides less sensitivity. Sjögren antibodies SS-A/Ro and SS-B/La may be helpful. Antibodies to ds-DNA are found in 60% to 83% of patients with SLE at some time during their disease course; thus, the absence of anti-ds-DNA antibody does not exclude the diagnosis of SLE. Results must be interpreted with other clinical and laboratory observations.

Specimen Serum

Container Red top tube

Storage Instructions Refrigerate immediately. Store at 4°C for up to 72 hours or at -20°C or colder without freezing and thawing indefinitely.

Reference Interval Normal: low levels of antibody or none (units and reference range will depend on laboratory and methodology). Most normal individuals have IgM antibodies to single-stranded DNA (ss-DNA). Such antibodies bear only low affinity for DNA. The specificity of anti-ss-DNA for SLE is substantially less than that of ds-DNA. Anti-ss-DNA is no longer considered useful for clinical diagnosis. IgG antibodies to double-stranded DNA are less common.

One reference laboratory has for IgG ds-DNA antibody:

- negative: <25 IU
- borderline positive: 25-30 IU
- positive: 31-200 IU
- strongly positive: >200 IU

Critical Values Higher values are more specific for active SLE.

Use Confirmatory test for systemic lupus erythematosus (SLE); monitor clinical course and response to treatment. Associated with lupus nephritis, anti-DNA antibodies are rare in drug-induced LE.

Limitations False-positive tests due to antibodies against histones have been reported with use of the *Crithidia luciliae* substrate assay, but are rare using calf thymus DNA.

Wide day-to-day variation in control sera ≥20% occurs.

Serum antibodies to single-stranded DNA are found in SLE, drug-induced SLE, and other entities.

Methodology Most tests are reactive to B DNA, the right-handed form of ds-DNA. Methods include enzyme-linked immunosorbent assay (ELISA), indirect fluorescent antibody (IFA) using *Crithidia luciliae* substrate, radioimmunoassay (RIA), and Farr assay.

Additional Information The ANA may be positive in a number of diseases, some of which have clinical features similar to those of SLE. With ANA reactivity, additional tests can help establish the diagnosis of SLE: tests for autoantibodies to ds-DNA (anti-ds-DNA), to Smith antigen (anti-Sm), and to Ro (SS-A).

Anti-DNA antibody, in practice, refers to antibodies that bind specifically to double stranded DNA. Anti-ds-DNA antibodies may be detected by a radioimmunoassay technique known as Farr assay. The Farr assay is more specific for the diagnosis of SLE than the more commonly used enzyme immunoassays and is the assay which is most likely to predict flares, especially exacerbations of glomerulonephritis, and/or to predict response to therapy. Anti-ds-DNA antibodies have specificity for SLE of nearly 100%. High antibody concentrations are more specific for SLE than are low concentrations just above the upper limit of normal (4 IU/mL). Between 50% and 85% of SLE patients have high titer antibodies. Weakly positive anti-ds-DNA antibodies occur infrequently in other autoimmune disease and in healthy persons.

In addition to diagnosis of SLE, measurement of anti-ds-DNA antibodies is helpful in assessing prognosis and monitoring disease activity in context of the entire clinical picture. Renal disease is more common in patients with high antibody levels. An increase in anti-ds-DNA concentration often precedes flares of disease activity in SLE patients. Because of this association, many physicians measure anti-DNA levels serially. The disease facets most likely to be heralded are flares of glomerulonephritis, vasculitis, or both.

Other tests relevant to SLE and additional to the complements are CBC including platelet count, ESR, urinalysis with urine sediment microscopy, urinary protein excretion, and creatinine. Falling concentrations of C3 and C4 can also predict exacerbation. Rising ESR, falling WBC counts, increasing urinary protein excretion, or observation of microscopic hematuria may also measure disease activity in SLE. Creatinine and creatinine clearance are relatively insensitive to deterioration of glomerular filtration rate in lupus nephritis.

References

Hahn BH, "Antibodies to DNA," *Mechanisms of Disease*, Epstein FH, ed, *N Engl J Med*, 1998, 338(19):1359-68.

Homburger HA, "Advances in the Diagnosis and Laboratory Evaluation of Systemic Rheumatic Diseases Other Than Rheumatoid Arthritis," *Clinical and Laboratory Evaluation of Human Autoimmune Diseases*, Chapter 11, Nakamura RA, Keren DF, and Bylund DJ, eds, Chicago, IL: American Society for Clinical Pathology, 2002, 153-64.

Kavanaugh A, Tomar R, Reveille J, et al, "Guidelines for Clinical Use of the Antinuclear Antibody Test and Tests for Specific Autoantibodies to Nuclear Antigens," *Arch Pathol Lab Med*, 2000, 124(1):71-81.

Moder KG, "Use and Interpretation of Rheumatologic Tests: A Guide for Clinicians," *Mayo Clin Proc*, 1996, 71(4):391-6.

♦ **Anti-DNase-B Titer**   *see* Antideoxyribonuclease-B Titer, Serum *on page 170*

♦ **Anti-Double-Stranded DNA**   *see* Anti-DNA *on page 173*

♦ **Anti-ds-DNA**   *see* Anti-DNA *on page 173*

Antiepileptic Drugs in Current Use

Drug	Indication	Reference Range	Adverse Effect
Carbamazepine	SPS, CPS, 2 GTC	8-12 µg/mL	Diplopia, dizziness, idiosyncratic aplastic anemia, rash, hyponatremia, osteoporosis
Diazepam	Acute seizures	0.2-1.0 µg/mL	Hypotension, respiratory depression, sedation, tolerance
Clonazepam	Myoclonic, atonic, 1 GTC	10-50 ng/mL	Hypotension, respiratory depression, sedation, tolerance
Ethosuximide	Absence seizures	40-100 µg/mL	Sedation, GI distress
Felbamate	SPS, CPS, 1 and 2 GTC, atonic, absence	20-100 µg/mL	Dizziness, headache, idiosyncratic hepatic failure or aplastic anemia, insomnia, weight loss
Gabapentin	SPS, CPS, 2 GTC	1-2 µg/mL	Fatigue, transient GI distress
Lamotrigine	SPS, CPS, 1 and 2 GTC	2-4 µg/mL	Dizziness, headache, rash
Levetiracetam	SPS, CPS, 1 and 2 GTC	5-45 µg/mL	Somnolence, coordination difficulties
Oxacarbazepine	SPS, CPS, 1 and 2 GTC	10-30 µg/mL as metabolite	Dizziness, diplopia, ataxia, hyponatremia
Phenobarbital	SPS, CPS, 1 and 2 GTC	15-40 µg/mL	Cognitive effects, respiratory depression, sedation
Phenytoin	SPS, CPS, 1 and 2 GTC	10-20 µg/mL	Ataxia, gingival hyperplasia, hirsutism, lymphadenopathy, nystagmus, osteoporosis
Primidone	SPS, CPS, 1 and 2 GTC	5-12 µg/mL	Sedation, depression, dizziness
Tiagabine	SPS, CPS, 1 and 2 GTC	5-35 ng/mL trough	GI distress, cognitive effects
Topiramate	SPS, CPS, 1 and 2 GTC	2-5 µg/mL	Impaired memory, weight loss, word-finding difficulty
Vigabatrin	SPS, CPS, 2 GTC, infantile spasms	20-160 µg/mL	Visual field defects, psychiatric symptoms, somnolence, fatigue
Zonisamide	SPS, CPS, 1 and 2 GTC	15-30 µg/mL	Somnolence, dizziness, agitation, difficulty concentrating, weight loss

CPS = complex partial seizure; GI = gastrointestinal; 1 GTC = primary generalized tonic-clonic seizure; 2 GTC = secondary generalized seizure; SPS - simple partial seizure.

Adapted from Sirven JI, "Acute and Chronic Seizures in Patients Older Than 60 Years," *Mayo Clin Proc*, 2001, 76:175-83.

Antiepileptic Drugs Overview

Related Information

Phenobarbital, Serum or Plasma *on page 1021*

Applies to Felbamate; Felbatol®; Gabapentin; Gabitril®; Keppra®; Lamictal®; Lamotrigine; Levetiracetam; Neurontin®; Oxcarbazepine; Sabril®; Sabrilex®; Tiagabine; Topamax®; Topiramate; Trileptal®; Vigabatrin; Zonegran™; Zonisamide Excegran®

Abstract A number of new antiepileptic drugs have reached the market in the past 10 years. Very little hard information exists regarding desirable therapeutic plasma ranges of such drugs. Some may become candidates for therapeutic monitoring. Such monitoring may be useful in assessing compliance or in evaluating circumstances in which therapeutic or toxic effects are difficult to interpret. Many investigators feel that these drugs are not candidates for routine monitoring at this time due to lack of solid information on therapeutic ranges and dose-related side effects.

Specimen Usually plasma or serum

Container Green top (heparin) tube or red top tube

Storage Instructions Usually stored and shipped frozen.

Turnaround Time 2-4 days for reference laboratories

Reference Interval Use range given by the laboratory that analyzes the sample.

Use When assay is appropriate, to follow therapy and assess toxicity and compliance

Limitations Sufficient data from large studies is not yet available for these drugs.

Methodology Usually immunoassays, gas chromatography (GC), high performance liquid chromatography (HPLC), or tandem mass-spectrometry

Additional Information The previous table lists some currently used antiepileptic drugs, indications, reference ranges, and adverse effects.

Tidwell and Swims reviewed the newer antiepileptic drugs with emphasis on FDA-approved indications, mechanisms of action, adverse effects, and pharmacokinetics.

References

Cramer JA, Fisher R, Ben-Menachem E, et al, "New Antiepileptic Drugs: Comparison of Key Clinical Trials," *Epilepsia*, 1999, 40(5):590-600.

Dichter MA and Brodie MJ, "New Antiepileptic Drugs," *N Engl J Med*, 1996, 334(24):1583-90.

Tatum WO, Galvez R, Benbadis S, et al, "New Antiepileptic Drugs: Into the New Millenium," *Arch Fam Med*, 2000, 9(10):1135-41.

Tidwell A and Swims M, "Review of the Newer Antiepileptic Drugs," *Am J Manag Care*, 2003, 9(3):253-76.

Tomson T and Johannessen SI, "Therapeutic Monitoring of the New Antiepileptic Drugs," *Eur J Clin Pharmacol*, 2000, 55(10):697-705.

♦ **Anti-F Actin** *see* Smooth Muscle Antibody *on page 1207*

♦ **Antifactor Xa Assay** *see* Heparin Antifactor Xa Assay *on page 693*

♦ **Antifreeze** *see* Ethylene Glycol, Serum or Plasma *on page 561*

♦ **Antiglobulin-Augmented Lymphotoxicity Assay** *see* Platelet Transfusion *on page 1058*

Antiglobulin Test, Direct

Related Information

Anemia Flowchart *on page 35*
Cord Blood Antibody Screen *on page 453*
Hemosiderin Stain, Urine *on page 692*
Pretransfusion Testing *on page 1088*
Rh$_o$(D) Immune Globulin (Human) *on page 1164*
Transfusion Reaction Work-up *on page 1269*

Synonyms Antihuman Globulin; Coombs Test, Direct; DAT; Direct Antiglobulin Test; Direct Coombs Test

Applies to Antiglobulin Test, Direct, Complement; Sensitization

Test Includes Direct antiglobulin testing with monospecific antihuman globulin reagents, red cell elution, and antibody identification

Abstract The direct antiglobulin test (DAT) utilizes reagent antihuman globulin to detect nonagglutinating antibodies bound *in vivo* to the surface of red cells. The test may be performed with polyspecific reagent that will detect red cell bound IgG and complement components (C3) or with monospecific anti-IgG/anticomplement reagents.

Specimen Blood (cord blood for the investigation of HDN)

Container One red top tube (cord blood) for HDN investigation; one lavender top (EDTA) tube in other cases. The DAT should be performed on an EDTA (lavender) top tube to avoid artifactual complement activation (see Limitations). Therefore, do not request a red top tube for non-HDN investigations.

Causes for Rejection Gross hemolysis, sample not in EDTA tube (other than HDN testing), improper labeling

Special Instructions Provide diagnosis, transfusion history, obstetric history, and medication list to the Blood Bank.

Reference Interval Negative

Use Polyspecific antiglobulin reagent detects red cell bound immunoglobulin (especially IgG) as well as complement components. The DAT is commonly utilized in the investigation of antibody-induced hemolysis (eg, HDN, autoimmune hemolytic anemia, and transfusion reaction).

Limitations The finding of red cell bound IgG or complement components is not invariably associated with *in vivo* hemolysis. A small percentage of normal blood donors present with positive direct antiglobulin tests. The most frequent cause of a positive DAT in nonhemolysing patients is associated with drug-induced antibodies. Whole blood samples should not be refrigerated before testing. At low temperature, cold autoantibodies present in many sera can activate complement. Artifactual complement activation can be avoided by the exclusive use of EDTA specimens for direct antiglobulin testing.

Methodology Red cells washed free of extraneous immunoproteins are exposed to antihuman globulin and anticomplement components (C3) as polyspecific or monospecific reagents. Agglutination indicates sensitization. Newer techniques are available for the DAT; however, some methods have reportedly performed less well than conventional tube testing.

Additional Information The direct antiglobulin (Coombs) test will detect red cells sensitized with IgG in cases of HDN and warm antibody hemolytic anemia. The appropriate reagent will detect red cells sensitized with complement C3 components in cases of cold antibody hemolytic anemia. The test is usefully applied in other cases of suspected *in vivo* hemolysis, such as transfusion reactions and drug-induced hemolytic anemia.

Mechanisms of positive DAT include warm autoantibodies, hypergammaglobulinemia, and passively acquired antibodies. Warm autoimmune hemolytic anemias are usually mediated by IgG, less frequently with IgG with C3 and uncommonly with C3 alone.

References

Dittmar K, Procter JL, Cipolone K, et al, "Comparison of DATs Using Agglutination to Gel Column and Affinity Column Procedures," *Transfusion*, 2001, 41(10):1258-62.

Petz LD, "Drug-Induced Autoimmune Hemolytic Anemia," *Transfus Med Rev*, 1993, 7(4):242-54.

Petz LD, "Treatment of Autoimmune Hemolytic Anemias," *Curr Opin Hematol*, 2001, 8(6):411-6.

♦ **Antiglobulin Test, Direct, Complement** *see* Antiglobulin Test, Direct *on page 176*

Antiglobulin Test, Indirect

Related Information

Anemia Flowchart *on page 35*
Antibody Detection/Identification, Red Cell *on page 165*
Hemolytic Disease of the Newborn, Antibody Identification *on page 690*
Prenatal Screen, Immunohematology *on page 1086*
Pretransfusion Testing *on page 1088*

Synonyms Coombs Test, Direct; Indirect Coombs Test

Abstract The indirect antiglobulin test (IAT) was developed to demonstrate the presence of nonagglutinating IgG alloantibodies in serum or plasma.

Specimen Blood

Container Red top tube or lavender top (EDTA) tube

Causes for Rejection Gross hemolysis, sample collected in serum separator tube, improper labeling

Reference Interval Negative
(Continued)

Antiglobulin Test, Indirect *(Continued)*

Use The indirect antiglobulin test (IAT) is a reliable screening procedure for clinically relevant antibodies in patient sera. Effective in detection of alloantibodies (99.6% detection), the IAT is used in antibody screening and identification tests in prenatal evaluations. It is used in investigation of hemolytic conditions, and especially in routine pretransfusion compatibility testing.

Limitations Antibodies directed to low incidence "private" antigens may not be represented on the screening or identification panels and will not be detected. At the time of testing, antibody levels may be below detectable limits.

Methodology This is a two-stage test system. In the first stage, IgG antibody in patient's serum sensitizes test red cell sample (panel cell or donor unit). Red cell bound antibody is detected in the second stage by the use of anti-IgG antiglobulin reagent.

Additional Information The difference between the direct and indirect antiglobulin (Coombs) test is that the direct test is used to demonstrate *in vivo* red cell sensitization with antibody whereas the indirect test is used to demonstrate the presence of antibody in a patient's serum or plasma through *in vitro* sensitization.

Since its introduction in 1945, the indirect antiglobulin test became the most widely used test for detection and identification of irregular IgG antibodies. It is a simple test with high sensitivity capable of detecting as few as 100-200 molecules of IgG per cell.

References

Brecher ME, *Technical Manual*, 14th ed, Bethesda, MD: American Association of Blood Banks Press, 2002, 253-69.

Issitt PD and Anstee DJ, *Applied Blood Group Serology*, 4th ed, Durham, NC: Montgomery Scientific Publications, 1998, 115-22.

Roback JD, Barclay S, and Hillyer CD, "An Automatable Format for Accurate Immunohematology Testing by Flow Cytometry," *Transfusion*, 2003, 43(7):918-27.

♦ **Anti-HCV** *see* Hepatitis C Virus Serology *on page 706*

♦ **Antihemophilic Factor (Factor VIII)** *see* Coagulation Factor Assays *on page 418*

♦ **Antihemophilic Factor (Human)** *see* Factor VIII Concentrate *on page 563*

♦ **Antihistidyl Transfer tRNA Synthetase** *see* Jo-1 Antibody *on page 815*

♦ **Anti-Hu** *see* Antineuronal Nuclear Antibody, Type 1 (Anti-Hu) *on page 185*

♦ **Antihuman Globulin** *see* Antiglobulin Test, Direct *on page 176*

♦ **Anti-immunoglobulin A** *see* Antibodies to IgA *on page 164*

♦ **Anti-Ku** *see* Topoisomerase I Antibody *on page 1261*

♦ **Antiliver Cytosol 1 Antibodies** *see* Liver/Kidney Microsomal Type 1 Antibodies *on page 873*

♦ **Antiliver/Kidney Microsomal Antibodies** *see* Liver/Kidney Microsomal Type 1 Antibodies *on page 873*

♦ **Anti-LKM1** *see* Liver/Kidney Microsomal Type 1 Antibodies *on page 873*

Antimicrobial Susceptibility Testing, Aerobic and Facultatively Anaerobic Bacteria

Related Information

    Aminoglycosides, Serum *on page 147*
    Antimicrobial Susceptibility Testing, Fungi *on page 180*
    Antimicrobial Susceptibility Testing, Mycobacteria *on page 181*
    Antimicrobial Susceptibility Testing, Unusual Isolates/Fastidious Organisms *on page 182*
    Bacterial Culture, Aerobes *on page 229*
    Beta-Lactamase Test *on page 256*
    Serum Bactericidal Test *on page 1192*
    Vancomycin, Serum *on page 1298*

Applies to MIC; Minimal Inhibitory Concentration; Susceptibility Testing

Test Includes Qualitative or quantitative determination of antimicrobial susceptibility of an isolated organism

Abstract The purpose of antimicrobial susceptibility testing is to determine the antibacterial activity of agents against specific pathogens. These susceptibility assays have been standardized for use in clinical laboratories by the National

Committee for Clinical Laboratory Standards (NCCLS). Standards include use of quality control microorganisms to support reliability and use of standard agar and broth media to diminish variability between laboratories.

Specimen Viable pure culture of a rapidly growing aerobic or facultatively anaerobic organism

Turnaround Time Usually 1 day after recovering an organism from a clinical specimen

Reference Interval Results are qualitatively reported as susceptible (S), intermediate (I), or resistant (R) and may include a minimal inhibitory concentration (MIC)

Use Determine antimicrobial susceptibility of organisms involved in infectious processes. The pattern of antibiotic susceptibility is sometimes used to monitor nosocomial infections such as those due to methicillin-resistant *Staphylococcus aureus* and to evaluate or follow the development of resistance to new antimicrobial drugs.

Limitations Interpretive guidelines developed for antimicrobial susceptibility tests are applicable to most nonfastidious, rapidly growing bacteria (*Staphylococcus* species, *Enterococcus* species, *Pseudomonas aeruginosa*, and members of the *Enterobacteriaceae*), and some common fastidious pathogens (eg, *Streptococcus pneumoniae*, *Haemophilus influenzae*, *Neisseria gonorrhoeae*). Interpretive criteria for less common pathogens (eg, *Neisseria meningitidis*, *Corynebacterium* species, *Nocardia* species, rapidly-growing mycobacteria, *Bacillus* species) have not been standardized.

Methodology Disk diffusion (qualitative) broth dilution, microbroth dilution, agar dilution (quantitative), or gradient diffusion (quantitative) or antimicrobial concentrations are selected to correspond to therapeutically relevant levels. Several automated instruments have been developed to perform susceptibility testing of microorganisms.

Additional Information Effective antimicrobial therapy is usually selected with intent to achieve a peak level two to four times the MIC at the site of infection. An antimicrobial level 10 times the MIC is usually sought in urinary tract infections. The "breakpoints" are based on achievable levels of antibiotic in the blood and indicate MICs above which organisms are moderately or very resistant and would not be expected to respond to that antibiotic.

Susceptible: This category implies that this infection may be appropriately treated with the dosage of antimicrobial agent recommended for that type of infection and infecting species, unless otherwise contraindicated.

Intermediate: This category provides a "buffer zone," which should prevent small, uncontrolled, technical factors from causing major discrepancies in interpretations.

Antimicrobial agents that are excreted via the kidneys can usually be used to treat uncomplicated cystitis due to an organism with intermediate susceptibility. For systemic infections, antimicrobials with intermediate activity should be avoided unless used in combination with a nonantagonistic agent with greater activity against the infecting organism; if such an agent does not exist, use maximal nontoxic dosing of the agent with intermediate activity.

Resistant: Strains falling in this category are not inhibited by the usually achievable systemic concentrations of the agent with normal dosage schedules and/or fall in the range in which specific microbial resistance mechanisms are likely (eg, beta-lactamases), and clinical efficacy has not been reliable in treatment studies.

References

Hessen MT and Kaye D, "Principles of Selection and Use of Antibacterial Agents. *In Vitro* Activity and Pharmacology," *Infect Dis Clin North Am*, 2000, 14(2):265-79.

National Committee for Clinical Laboratory Standards, "Performance Standards for Antimicrobial Disk Susceptibility Tests," Approved Standard M2-A7, Villanova, PA: National Committee for Clinical Laboratory Standards, 2000.

National Committee for Clinical Laboratory Standards, "Methods for Dilution Susceptibility Tests for Bacteria That Grow Aerobically." Approved Standard M7-A5, Villanova, PA: National Committee for Clinical Laboratory Standards, 2000.

National Committee for Clinical Laboratory Standards, "Performance Standards for Antimicrobial Testing," Supplemental Tables, NCCLS Document M100-S10 (M7-A5 Aerobic Dilution), Villanova, PA: National Committee for Clinical Laboratory Standards, 2000.

Rubenstein E, "Antimicrobial Resistance - Pharmacological Solutions," *Infection*, 1999, 27(S):32-4.

(Continued)

Antimicrobial Susceptibility Testing, Aerobic and Facultatively Anaerobic Bacteria *(Continued)*

Thompson RL and Wright AJ, "General Principles of Antimicrobial Therapy," *Mayo Clin Proc*, 1998, 73(10):995-1006.

Virk A and Steckelberg JM, "Clinical Aspects of Antimicrobial Resistance," *Mayo Clin Proc*, 2000, 75(2):200-14.

Antimicrobial Susceptibility Testing, Anaerobic Bacteria

Related Information

Antimicrobial Susceptibility Testing, Unusual Isolates/Fastidious Organisms *on page 182*

Bacterial Culture, Anaerobes *on page 231*

Beta-Lactamase Test *on page 256*

Applies to Susceptibility Testing

Abstract Most anaerobic infections are treated empirically due to the time necessary to recover and test anaerobic bacteria. When there is treatment failure, testing of anaerobic bacteria for antimicrobial susceptibility is important in selection of alternative therapy.

Specimen A pure culture of the isolated organism to be tested, prepared by the laboratory

Turnaround Time 2-5 days from time organism is isolated and identified

Special Instructions Appropriateness and scope of anaerobic susceptibility testing in a particular clinical setting is supported by laboratory consultation.

Use Antimicrobial therapy of anaerobic infections is usually empiric because many of these infections are polymicrobic, and because there is an unavoidable delay in time of results. The major reasons for testing are to determine susceptibility to new antimicrobials, to establish institutional or regional susceptibility patterns which may be used to guide empiric therapy, and to assist in management of infection in individual patients. Circumstances such as anaerobic brain abscess, endocarditis, osteomyelitis, and prosthetic device or vascular graft infections are particularly relevant.

Limitations Anaerobic infections are often polymicrobial (involving aerobic and anaerobic organisms), and are associated with tissue necrosis and abscess formation resulting in impaired delivery of antimicrobial agents to the infection site. Methods for recovering and identifying anaerobic bacteria are cumbersome and time consuming, resulting in elapsed periods of nearly a week before susceptibility test results are available. Standardized methods for testing anaerobes are not always reliable and often produce growth patterns that are difficult to interpret.

Methodology Broth microdilution, macrobroth dilution, and E-Test (gradient diffusion), agar dilution technique, beta-lactamase testing

Additional Information At present, routine susceptibility testing of anaerobic isolates is not recommended. Infections involving anaerobes frequently contain mixed flora, and appropriate drainage rather than antimicrobial therapy seems to be the most crucial factor in successful treatment of such infections. Thus, in many laboratories, susceptibility testing is generally performed only on anaerobic isolates from blood, pleural fluid, peritoneal fluid, and CSF. The physician is advised to contact the laboratory regarding the specific antibiotic(s) to be tested and the testing method available.

References

Falagas ME and Siakavellas E, "*Bacteroides, Prevotella,* and *Porphyromonas* species: A Review of Antibiotic Resistance and Therapeutic Options," *Int J Antimicrob Agents*, 2000, 15(1):1-9.

Hecht DW, "Resistance Trends in Anaerobic Bacteria," *Clin Microbiol Newslet*, 2000, 22(6):41-4.

Nguyen MH, Yu VL, Morris AJ, et al, "Antimicrobial Resistance and Clinical Outcome of *Bacteroides,* Bacteremia: Findings of a Multicenter Prospective Observational Trial," *Clin Infect Dis,* 2000, 30(6):870-6.

Virk A and Steckelberg JM, "Clinical Aspects of Antimicrobial Resistance," *Mayo Clin Proc*, 2000, 75(2):200-14.

Wexler HM, Molitoris E, and Molitoris D, "Susceptibility Testing of Anaerobes: Old Problems, New Options?" *Clin Infect Dis*, 1997, 25(Suppl 2):S275-8.

Antimicrobial Susceptibility Testing, Fungi

Related Information

Antimicrobial Susceptibility Testing, Aerobic and Facultatively Anaerobic Bacteria *on page 178*

Fungal Culture, Biopsy or Body Fluid *on page 619*
Fungal Culture, Blood *on page 620*
Fungal Culture, Skin *on page 623*
Fungal Culture, Sputum *on page 624*
Fungus Smear, Stain *on page 626*
Itraconazole, Serum *on page 814*

Applies to Susceptibility Testing

Test Includes Broth dilution or agar dilution testing of antifungal agents. Results may be quantitative or qualitative.

Abstract Fungi have emerged in the last 10 years as important nosocomial pathogens. As the incidence of serious fungal infections has increased a number of new antifungal agents have been introduced, thus the need for *in vitro* antifungal susceptibility tests. Susceptibility testing of yeasts is now standardized, but standard methods and clinical relevance of antifungal susceptibility tests for filamentous fungi have not been determined.

Specimen A pure culture of the isolated organism to be tested, prepared by the laboratory

Special Instructions Consult the laboratory to determine availability and choice of methods.

Use Routine antifungal susceptibility testing is clearly unwarranted. In certain clinical circumstances, it may be appropriate to test *Candida* species. The dimorphic fungi need not be tested since they are virtually always susceptible.

Limitations Although susceptibility testing of yeasts has recently been standardized, the relationship between test results and patient response to therapy has not been established. Susceptibility testing of yeasts against the azoles often produces variable results. Susceptibility of yeasts against amphotericin B is problematic since results fall within a narrow range and are, therefore, difficult to categorize. Susceptibility testing of filamentous fungi has not been standardized and is quite variable; isolates should be sent to a well respected reference laboratory for testing.

Methodology A standardized broth-dilution method is available for susceptibility testing of yeasts.

Additional Information Interpretation of *in vitro* susceptibility data for antifungal drugs has been hindered by the absence of standardized test criteria. Thus, it has been extremely difficult to identify a clear relation between *in vitro* minimal inhibitory concentrations and clinical outcome, which is highly dependent on host factors. The situation appears more readily resolvable for yeast-like than for filamentous fungi, since the former are more easily quantified by standardized microbiologic techniques.

References

Gianinni MA, Pearson T, and Patrick CC, "Fungal Susceptibility Testing," *Pediatr Infect Dis J*, 1999, 18(11):1021-2.

Klepser ME, Lewis RE, and Pfaller MA, "Therapy of *Candida* Infections: Susceptibility Testing, Resistance and Therapeutic Options," *Ann Pharmacother*, 1998, 32(12):1353-61.

National Committee for Clinical Laboratory Standards, "Reference Method for Broth Dilution Antifungal Susceptibility Testing for Yeasts," Approved Standard Document M-27-A, Villanova, PA: National Committee for Clinical Laboratory Standards, 1997.

Pfaller MA, Rex JH, and Rinaldi MG, "Antifungal Susceptibility Testing. Technical Advances and Potential Clinical Applications," *Clin Infect Dis*, 1997, 24(5):776-84.

Sheehan DJ, Hitchcock CA, and Sibley CM, "Current and Emerging Azole Antifungal Agents," *Clin Microbiol Rev*, 1999, 12(1):40-79.

Antimicrobial Susceptibility Testing, Mycobacteria

Related Information

Antimicrobial Susceptibility Testing, Aerobic and Facultatively Anaerobic Bacteria *on page 178*
Mycobacterial Culture, Biopsy or Body Fluid *on page 929*
Mycobacterial Culture, Cutaneous and Subcutaneous Tissue *on page 932*
Mycobacterial Culture, Sputum *on page 933*
Mycobacterial Culture, Urine *on page 935*

Applies to Susceptibility Testing

Test Includes Panel of antimycobacterial agents tested against clinical isolates at appropriate concentrations

Abstract Susceptibility testing of *Mycobacterium tuberculosis* is important in determining long-term therapy. Broth methods are more rapid and correlate well with the proportion method and have replaced it in many laboratories. (Continued)

Antimicrobial Susceptibility Testing, Mycobacteria
(Continued)

Broth-dilution methods are used to test rapidly-growing mycobacteria, but methods and interpretation have not been standardized.

Specimen A pure culture of the isolated organism to be tested, prepared by the laboratory

Turnaround Time 4-6 weeks after organism is isolated and identified

Use Determine the susceptibility of the isolated organism to a panel of antimycobacterial agents

Limitations Methodology and interpretive criteria for *M. tuberculosis* is well standardized. Methodology and interpretive criteria for *Mycobacterium* species other than *M. tuberculosis* have not been standardized; consequently, the distinction between susceptible and resistant isolates is often unclear.

Methodology Disk diffusion, agar containing antibiotic, or broth containing antibiotic

Additional Information Susceptibilities are performed on the first organism isolated from a patient and at 3- to 6-month intervals, if that organism continues to be isolated while the patient is on therapy. Susceptibility tests should be performed in patients with recurrent tuberculosis, as resistant strains are common in recurrent infection.

References
Neville K, Bromberg A, Bromberg R, et al, "The Third Epidemic - Multidrug-Resistant Tuberculosis," *Chest*, 1994, 105(1):45-8.

Pfyffer GE, Bonato DA, Ebrahimzadeh A, et al, "Multicenter Laboratory Validation of Susceptibility Testing of *Mycobacterium tuberculosis* Against Classical Second-Line and Newer Antimicrobial Drugs by Using the Radiometric Bactec® 460 Technique and the Proportion Method With Solid Media," *J Clin Microbiol*, 1999, 37(10):3179-86.

Rusch-Gerdes S, Domehl C, Nardi G, et al, "Multicenter Evaluation of the Mycobacteria Growth Indicator Tube for Testing Susceptibility of *Mycobacterium tuberculosis* to First-Line Drugs," *J Clin Microbiol*, 1999, 37(1):45-8.

Antimicrobial Susceptibility Testing, Unusual Isolates/ Fastidious Organisms

Related Information

Antimicrobial Susceptibility Testing, Aerobic and Facultatively Anaerobic Bacteria *on page 178*
Antimicrobial Susceptibility Testing, Anaerobic Bacteria *on page 180*

Applies to MIC; Minimal Inhibitory Concentration; Susceptibility Testing

Test Includes Panel of antibiotics or single antibiotic selected by physician

Abstract Broth-dilution susceptibility testing is used to determine the minimal inhibitory concentration of one or more antibiotics considered for treatment of an infection due to a fastidious organism. It is especially useful when published studies of antimicrobial susceptibility are not available, or when the patient is unable to take the most commonly prescribed antibiotics.

Specimen A pure culture of the isolated organism to be tested, prepared by the laboratory

Turnaround Time 24-48 hours after isolation of bacterium

Special Instructions An isolate of the organism of interest must be saved at the request of the physician.

Reference Interval Results are qualitatively reported as susceptible (S), intermediate (I), and resistant (R) if interpretive standards are available.

Use Determine minimum inhibitory concentration (MIC) of a given organism to an antimicrobial agent

Limitations The organism may fail to grow in media used for testing. If interpretive criteria do not exist, the MIC or zone size will be reported.

Methodology Broth microdilution or macrodilution technique, disk diffusion, agar diffusion (E-test). Special methods are required for individual species. These generally include addition of supplement to the agar or broth and/or alteration of incubation atmosphere.

Additional Information The terms "fastidious or unusual isolates" refer to organisms that do not grow well on Mueller-Hinton medium or are unusual in that there is insufficient data to document that reliable susceptibility testing can be performed by routine methods. Organisms such as *Haemophilus influenzae* and *Neisseria gonorrhoeae* have developed resistance to beta-lactam antibiotics by plasmid or chromosomal genes mediating the production of beta-lactamase. Other organisms, such as *Streptococcus pneumoniae*, have developed chromosomal gene mediated alteration of the penicillin-binding proteins. Because of the emergence of resistance, empiric therapy with penicillin or ampicillin can no longer be relied upon, and susceptibility testing for such "fastidious" organisms may be necessary. Susceptibility testing of group A streptococci to penicillin is not necessary because the organism remains highly susceptible; however, resistance to erythromycin and tetracycline has been reported, and susceptibility testing may be appropriate if these drugs are used to treat serious infections.

The following organisms generally require special susceptibility testing procedures: *Helicobacter* species, *Corynebacterium* species, *Francisella tularensis*, *Haemophilus influenzae*, *Legionella* species, *Listeria monocytogenes*, *Neisseria meningitidis*, *Neisseria gonorrhoeae*, *Nocardia* species, nonfermentative bacteria, *Streptococcus pneumoniae*, *Streptococcus* species, *Streptococcus* species peridoxal dependent.

References

Fox KK, Knapp JS, Holmes KK, et al, "Antimicrobial Resistance of *Neisseria gonorrhoeae* in the United States, 1988-1994; The Emergence of Decreased Susceptibility to Fluoroquinolones," *J Infect Dis*, 1997, 175(6):1396-1403.

Jones RN and Pfaller MA, "*In Vitro* Activity of Newer Fluoroquinolones for Respiratory Tract Infections and Emerging Patterns of Antimicrobial Resistance: Data from the SENTRY Antimicrobial Surveillance Program," *Clin Infect Dis*, 2000, 31(Suppl 2):S16-23.

Jorgensen JH and Ferraro MJ, "Antimicrobial Susceptibility Testing: Special Needs for Fastidious Organisms and Difficult-to-Detect Resistance Mechanisms," *Clin Infect Dis*, 2000, 30(5):799-808.

Venglarcik JS 3rd, "*Streptococcus pneumoniae* Antimicrobial Susceptibility Testing," *Pediatr Infect Dis J*, 2000, 19(4):329-31.

Antimitochondrial Antibody

Related Information

Alkaline Phosphatase, Serum *on page 127*
Antineutrophil Alloantibody and Autoantibody *on page 186*
Antineutrophil Cytoplasmic Antibody *on page 187*
Aspartate Aminotransferase, Serum *on page 216*
Bilirubin, Total, Serum *on page 265*
Gamma-Glutamyl Transferase, Serum *on page 629*
Immunoglobulin M *on page 779*
Liver Biopsy *on page 864*
Liver Disease: Laboratory Assessment, Overview *on page 869*
Liver/Kidney Microsomal Type 1 Antibodies *on page 873*
Smooth Muscle Antibody *on page 1207*

Synonyms AMA; Mitochondrial Antibody

Abstract Biliary tract diseases which lead to cirrhosis include primary biliary cirrhosis (PBC), primary and secondary sclerosing cholangitis, hepatic diseases associated with chronic inflammatory bowel disease, and duct obstruction. They are characterized by disproportionately increased concentrations of serum alkaline phosphatase. PBC is a chronic progressive autoimmune cholestatic disease in which intrahepatic bile ducts undergo continual destruction with portal inflammation and scarring, leading ultimately to cirrhosis and hepatic failure. Mitochondrial antibodies are found in up to 95% of patients with primary biliary cirrhosis, but may be found in other circumstances as well. They are usually absent in subjects with extrahepatic jaundice.

Specimen Serum

Container Red top tube

Reference Interval ≤1:20 considered nondiagnostic; patients with PBC usually have titers >1:160; patients with other autoimmune disease often have titers between 1:20 and 1:80.

Use Tests for mitochondrial antibody are needed in differential diagnosis of chronic liver disease and to provide confirmatory evidence for diagnosis of PBC. Antimitochondrial antibodies (AMA) have high specificity for PBC when in substantial titer.

Limitations AMA is also found in some instances of cryptogenic cirrhosis and in cases which have been classified as autoimmune hepatitis. AMA is rarely found in patients with extrahepatic biliary obstruction, drug-induced hepatitis, viral hepatitis, alcoholic and other forms of cirrhosis, and hepatic malignancy.
(Continued)

Antimitochondrial Antibody *(Continued)*

There is an incidence of 1% positives in a general hospital population, mostly people with other autoimmune disease. AMA may be found in SLE and other disorders, but usually with lower titers.

AMA titer does not correlate with disease severity.

Methodology Indirect immunofluorescence, enzyme-linked immunosorbent assay (ELISA)

Additional Information PBC is a cholestatic disease found most frequently in women, with an incidence which is highest in the 35- to 60-year age group. The diagnosis of PBC is based upon clinical observations including pruritus and often fatigue, malabsorption of fat soluble vitamins, histopathologic findings on liver biopsy, markedly increased serum alkaline phosphatase activity and cholesterol concentrations, elevated IgM levels, and presence of mitochondrial antibodies. Increases of gamma-glutamyl transferase parallel those of alkaline phosphatase. Aminotransferases are increased only two- to fourfold in PBC. In >90% of patients, the key M2 antigen has been identified as the E_2 component of the pyruvate dehydrogenase complex, a mitochondrial enzyme. Enzyme-linked immunosorbent assays developed using pyruvate, branched-chain ketoacid, and alpha-ketoglutarate dehydrogenase promise to add objectivity to analysis of these antibodies.

An overlap syndrome is recognized in which histopathologic features of autoimmune hepatitis accompany antimitochondrial antibodies.

With pruritus and high serum alkaline phosphatase, autoimmune cholangiopathy resembles primary biliary cirrhosis or primary sclerosing cholangitis but lacks antimitochondrial antibodies.

The following features are useful in the differential diagnosis between PBC and primary sclerosing cholangitis. See table.

Antimitochondrial Antibody

	PBC	PSC
History of ulcerative colitis, Crohn disease	−	++
Antimitochondrial antibody	+++	±
Antinuclear antibody, double-stranded DNA, smooth muscle antibody	+	±
↑ IgM	+++	−
↑ Alkaline phosphatase	++	++
Endoscopic retrograde cholangiopancreatography findings	+	+++
Long indolent course	+	+++

PBC = primary biliary cirrhosis; PSC = primary sclerosing cholangitis; + = present to a 1+ degree; ++ = present to a 2+ degree; +++ = present to a 3+ degree; − = not present; ± = minimal degree of involvement.

Adapted from Ferrell LD, "Liver and Gallbladder Pathology," *The Difficult Diagnosis in Surgical Pathology*, Weidner N, ed, Philadelphia, PA: WB Saunders Co, 1996, 290.

Patients with PBC are likely to have evidence of other autoimmune diseases, including Sjögren syndrome, scleroderma, CREST syndrome, and autoimmune thyroiditis. Subjects with PBC may have reactivity for rheumatoid factor, anti-smooth muscle antibodies, ANA, or thyroid antibodies.

The laboratory tests of prognostic relevance in PBC include serum bilirubin, albumin, and prothrombin time. Serum bilirubin is the most important test for prediction of survival in candidates for liver transplantation in PBC. GGT and C-reactive protein are among tests useful to monitor the course of PBC.

References

Ferrel L, "Liver Pathology: Cirrhosis, Hepatitis, and Primary Liver Tumors: Update and Diagnostic Problems," *Mod Pathol*, 2000, 13(6):679-704.

Galperin C and Gershwin ME, "Immunopathogenesis of Gastrointestinal and Hepatobiliary Disease," *JAMA*, 1997, 278(22):1946-55.

Kaplan MM, "Primary Biliary Cirrhosis," *N Engl J Med*, 1996, 335(21):1570-80.

Krawitt EL, "Autoimmune Hepatitis," *N Engl J Med*, 1996, 334(14):897-903.

Antineuronal Nuclear Antibody, Type 1 (Anti-Hu)

Related Information

Antineuronal Nuclear Antibody, Type 2 (Anti-Ri) *on page 185*

Synonyms ANNA-1; Anti-Hu

Applies to Amphiphysin; CRMP; PCA-1; PCA-2; PCA-Tr

Abstract Anti-Hu is one of several antibodies detected in the serum of patients with neurologic paraneoplastic syndrome. Anti-Hu antibody causes a subacute syndrome of encephalomyeloradiculopathy, sensory neuropathy, or autonomic neuropathy predominantly affecting the gastrointestinal tract.

Specimen Serum or CSF

Container Red top or SST™ tube

Reference Interval Serum: <1:60; CSF: <1:2

Use The presence of anti-Hu antibody suggests that a middle-aged patient with neurologic symptoms has a paraneoplastic syndrome. The underlying cancer is usually a small cell lung carcinoma. These tumors are not infrequently unsuspected but are found with searching in >80% of subjects. Clinically, the rapid onset of dementia without focal cerebral signs, but with cerebellar, brain stem, or peripheral nerve dysfunction, suggests paraneoplastic dementia.

Limitations Anti-Hu is not recommended as a screening test for lung cancer. The absence of antibodies does not exclude a paraneoplastic syndrome or cancer.

Methodology Indirect immunofluorescence on sections of cerebellum, Western blot using recombinant protein, and immunohistochemical staining of brain tissue

Additional Information Anti-Hu antibody, an IgG marker, causes a spectrum of paraneoplastic neurologic disorders. Most frequent are pure sensory, predominantly autonomic, and mixed sensorimotor neuropathies. Some patients may present with gastroparesis or intestinal obstruction. When anti-Hu is found in high titer with encephalitis or sensory neuropathy, small cell lung carcinoma is almost always present, but may be difficult to find. Approximately 15% of patients will have another tumor coexisting with small cell lung carcinoma. These syndromes occur twice as frequently in women as in men. Occasionally, other primaries are responsible (eg, carcinoma of prostate).

Anti-Hu has been detected in children with intestinal dysmotility, cerebellar ataxia, and brainstem encephalitis with and without peripheral neuroblastoma.

Anti-Hu is not detected in the serum or CSF of healthy individuals. It is found in about 5% to 10% of patients with small cell carcinoma of lung, not complicated by a neurological autoimmune disease.

Other paraneoplastic IgG markers recognized by immunofluorescence include ANNA-2, PCA-1, PCA-2, PCA-Tr, amphiphysin, and CRMP.

References

Benyahia B, Liblau R, Merle-Beral H, et al, "Cell-Mediated Autoimmunity in Paraneoplastic Neurological Syndromes With Anti-Hu Antibodies," *Ann Neurol*, 1999, 45(2):162-7.

Galanis E, Frytak S, Rowland KM, et al, "Neuronal Autoantibody Titers in the Course of Small-Cell Lung Carcinoma and Platinum-Associated Neuropathy," *Cancer Immunol Immunother*, 1999, 48(2-3):85-90.

Lucchinetti CF, Kimmel DW, and Lennon VA, "Paraneoplastic and Oncologic Profiles of Patients Seropositive for Type 1 Antineuronal Nuclear Autoantibodies," *Neurology*, 1998, 50(3):652-7.

Posner J, "Anti-Yo and Anti-Hu," *Lab Med*, 1999, 30:770.

Antineuronal Nuclear Antibody, Type 2 (Anti-Ri)

Related Information

Antineuronal Nuclear Antibody, Type 1 (Anti-Hu) *on page 185*

Synonyms ANNA-2; Anti-Ri

Abstract Anti-Ri is the least common of the paraneoplastic autoantibodies.

Specimen Serum or cerebrospinal fluid

Container Red top or SST™ tube for blood

Collection 3 mL serum or 2 mL CSF

Reference Interval Serum: <1:60; CSF: <1:2

Use Detection of anti-Ri antibody identifies an otherwise unexplained neurological disorder as autoimmune and paraneoplastic. A positive result prompts a search for an underlying occult malignancy.

(Continued)

Antineuronal Nuclear Antibody, Type 2 (Anti-Ri)
(Continued)

Limitations A negative result does not rule out a paraneoplastic syndrome or cancer. Anti-Ri antibodies are seldom detected in patients with breast carcinoma who do not have neurological dysfunction. In patients with ovarian carcinoma, the frequency of antineuronal antibodies is greater than the frequency of paraneoplastic syndromes. The presence of antibody does not necessarily lead to appearance of a paraneoplastic neurological syndrome.

Additional Information Anti-Ri antibodies are detected most commonly in postmenopausal women who usually present with signs of midbrain, brain stem, cerebellar and/or spinal cord dysfunction. Ocular opsoclonus-myoclonus may be a prominent symptom.

Most patients have a primary carcinoma of the breast. Small cell lung and gynecological cancer are less frequently associated with this syndrome. Treatment of the cancer can lead to decreased antibody titer and improvement of the neurological disorder.

References

Drlicek M, Bianchi G, Bogliun G, et al, "Antibodies of the Anti-Yo and Anti-Ri Type in the Absence of Paraneoplastic Neurological Syndromes: A Long-Term Survey of Ovarian Cancer Patients," *J Neurol*, 1997, 244(2):85-9.

Hunter SF, Parisi JE, Mastovich SL, et al, "Chronic Progressive Paraneoplastic Syndrome With Prominent Brainstem and Spinal Cord Involvement Associated With Type-2 Antineuronal Nuclear Antibodies (ANNA-2) and Breast Carcinoma," *J Neuropath Exp Neurol*, 1995, 54:464.

Moll JW, Hooijkaas H, van Goorbergh BC, et al, "Systemic and Antineuronal Autoantibodies in Patients With Paraneoplastic Neurological Disease," *J Neurol*, 1996, 243(1):51-6.

Antineutrophil Alloantibody and Autoantibody

Related Information

Antimitochondrial Antibody *on page 183*
Transfusion Reaction Work-up *on page 1269*
White Blood Cell Count *on page 1330*

Synonyms Granulocyte Antibody; Neutrophil Antibody

Applies to Granulocyte Agglutination Test; Granulocyte Immunofluorescence Test; Monoclonal Antibody Specific Immobilization of Granulocyte Antigens; Neutropenia

Test Includes For **autoimmune neutropenia:** neutrophil antibody identification and direct neutrophil testing. For **neonatal alloimmune neutropenia:** maternal antibody identification, crossmatching between maternal serum and paternal neutrophils and maternal and paternal neutrophil typing.

Abstract Neutrophil antibodies, both alloantibodies and autoantibodies, have been associated with a variety of clinical syndromes. **Autoimmune neutropenia** in adults may be idiopathic or may occur secondary to other diseases such as systemic lupus erythematosus. Autoimmune neutropenia in children, most frequently occurring between ages 6 months to 2 years, is relatively benign. Maternal antibodies directed against fetal neutrophil alloantigens cause **neonatal alloimmune neutropenia.** Neutrophil alloantibodies cause neutropenia after stem cell transplantation and refractoriness to granulocyte transfusion. Medications should be considered as a cause of immune-mediated neutropenia. Febrile, nonhemolytic transfusion reactions may be caused by neutrophil antibodies, although they are more frequently caused by HLA antibodies or passive transfer of cytokines. Either neutrophil or HLA antibodies may also induce transfusion-related acute lung injury (TRALI) (see Risks of Transfusion *on page 1166*).

Specimen For autoimmune neutropenia: serum and neutrophils from patient. For neonatal alloimmune neutropenia: maternal serum and neutrophils, paternal neutrophils.

Container Red top tube (serum), lavender top (EDTA) (whole blood - provides neutrophils)

Sampling Time Always check with the laboratory performing testing as frequently cases must be scheduled.

Storage Instructions Serum may be separated and stored frozen, but neutrophils may not survive for long periods of time. Check with testing laboratory prior to collection.

Reference Interval Negative

Limitations Premature infants and patients with certain clinical conditions (eg, chronic myelogenous leukemia) may have depression of granulocyte antigens.

Methodology Granulocyte agglutination test (GAT), granulocyte immunofluorescence test (GIFT), monoclonal antibody specific immobilization of granulocyte antigens (MAIGA). A combination of tests is usually beneficial. PCR techniques have been described.

References

Bux J and Chapman J, "Report on the Second International Granulocyte Serology Workshop," *Transfusion*, 1997, 37(9):977-83.

Bux J, Gehrens G, Jaeger G, et al, "Diagnosis and Clinical Course of Autoimmune Neutropenia in Infancy: Analysis of 240 Cases," *Blood*, 1998, 91(1):181-6.

Hadley A and Soothill P, *Alloimmune Disorders of Pregnancy. Anemia, Thrombocytopenia, and Neutropenia in the Fetus and Newborn*, United Kingdom: Cambridge University Press, 2002, 235-52.

Meyer O, Gaedicke G, and Salama A, "Demonstration of Drug-Dependent Antibodies in Two Patients With Neutropenia and Successful Treatment With Granulocyte Colony-Stimulating Factor," *Transfusion*, 1999, 39(5):527-30.

Antineutrophil Cytoplasmic Antibody

Related Information

Antimitochondrial Antibody *on page 183*
Antinuclear Antibodies *on page 189*
Bronchial Brushings/Washings Cytology *on page 310*
Cryoglobulin, Qualitative, Serum and Plasma *on page 478*
Glomerular Basement Membrane Antibody *on page 638*
Kidney Biopsy *on page 818*
Urinalysis *on page 1289*

Synonyms ANCA

Applies to C-ANCA; Myeloperoxidase Antibody (MPO-ANCA); P-ANCA; Proteinase 3 (PR3)

Abstract Antineutrophil cytoplasmic antibodies (ANCA) are autoantibodies against several cytoplasmic antigens of neutrophils and monocytes. Two different immunofluorescent staining patterns have been identified, cytoplasmic (C-ANCA) and perinuclear (P-ANCA), which reflect different antigenic specificities. Diagnostically, ANCA tests are helpful adjuncts in the differential diagnosis of systemic vasculitides including Wegener granulomatosis (WG), microscopic polyangiitis, renal-limited vasculitis, Churg-Strauss angiitis, drug-induced ANCA-associated vasculitis, and idiopathic crescentic glomerulonephritis (ICGN). ANCA is best utilized in subjects with WG and microscopic polyangiitis with the combination of indirect immunofluorescence of neutrophils and ELISA for proteinase 3 (PR3)-ANCA and myeloperoxidase (MPO) ANCA.

Specimen Serum

Container Red top tube or SST™ tube

Reference Interval Absent. Standard units for PR3-ANCA and MPO-ANCA by ELISA are not available.

Use Diagnostically, ANCA tests are helpful adjuncts in the differential diagnosis of systemic vasculitides, multiple pulmonary nodules, destructive lesions of the upper airways, chronic otitis, and subglottic tracheal stenosis. C-ANCA is most often present in patients with WG, a necrotizing granulomatous inflammatory disease often including giant cells and epithelioid granulomas, involving the upper and/or lower respiratory tract, necrotizing vasculitis of small to medium vessels and necrotizing glomerulonephritis. P-ANCA is present in some patients with WG, but it is less sensitive and specific than the C-ANCA pattern. P-ANCA is most often found in patients with microscopic polyarteritis, crescentic glomerulonephritis, and occasionally inflammatory bowel disease.

Limitations Although C-ANCA is useful in supporting the initial diagnosis, treatment should not be based soley on its presence. A positive or negative ANCA does not obviate the need for a tissue diagnosis in a patient with clinical manifestations suggestive of Wegener granulomatosis or another type of vasculitis.

Heat inactivation of serum leads to false positives.

Samples should be examined for PR3-ANCA and MPO-ANCA by ELISA, especially those negative by immunofluorescence. However, 5% of samples are positive **only** by ELISA.
(Continued)

Antineutrophil Cytoplasmic Antibody *(Continued)*

ANCA-negative entities include leukocytoclastic skin vasculitis, Henoch-Schönlein purpura, classical polyarteritis nodosa, Kawasaki disease, Behçet disease, thromboarteritis obliterans, and giant cell (Takayasu) arteritis. About 10% of subjects with WG or microscopic polyangiitis lack positive assays for ANCA.

Methodology Indirect immunofluorescence using whole buffy coat preparations. It is recommended that immunofluorescent identification be confirmed with enzyme-linked immunosorbent assay (ELISA). Maximum specificity is provided by combined application of both.

Patient's serum is incubated on slides containing ethanol-fixed human neutrophils. If ANCAs are present, they bind to neutrophils and are detected by adding fluorescently-labeled antihuman IgG. Two different immunofluorescent staining patterns have been identified, cytoplasmic (C-ANCA) and perinuclear (P-ANCA), which reflect different antigenic specificities. **Antibodies producing a C-ANCA pattern usually bind to proteinase-3**, a serine protease in the primary granules of neutrophils. **Antibodies producing a P-ANCA pattern recognize one or more positively charged proteins such as myeloperoxidase (MPO)**, elastase, cathepsin G, lactoferrin, and lysozyme. Immunofluorescence titer is recommended when serum is positive to immunofluorescence and negative in ELISA for PR3 and MPO.

Additional Information The different staining patterns have diagnostic significance. **C-ANCA** with PR3 specificity by enzyme-linked immunosorbent assay is detected in 70% to 90% of individuals who have generalized WG. ANCA is detectable in only 65% of patients without active renal disease (limited Wegener).

Evaluation for WG is supported by negative cultures. Bronchial brush and lavage may be useful. Biopsies are pivotal.

P-ANCA is present in 5% to 30% of patients with WG, but it is much less sensitive and specific than the C-ANCA pattern. In most cases of generalized active WG, C-ANCA with PR3 specificity is found; in up to 25%, P-ANCA with MPO specificity is demonstrated. About 60% of patients with microscopic polyangiitis or pauci-immune segmental necrotizing glomerulonephritis have P-ANCA with MPO specificity and 30% have C-ANCA with PR3 specificity. In a few, antiglomerular basement membrane antibodies are seen. WG and microscopic polyangiitis are more common causes of pulmonary-renal syndrome than antiglomerular basement membrane disease. ANA can cause a false-positive P-ANCA. Therefore, an ANA should be ordered on any specimen with a positive P-ANCA, before the test is interpreted as clinically significant. ELISA for PR3-ANCA and MPO-ANCA are recommended as confirmation of P-ANCA specificity.

Anti-MPO antibodies are present in <1% of patients with systemic rheumatic diseases and are much more specific for vasculitis than the P-ANCA fluorescent pattern.

When asthma and eosinophilia are found with vasculitis, Churg-Strauss syndrome must be suspected.

See tables in Glomerular Basement Membrane Antibody *on page 638.*

P-ANCA occurs with some cases of autoimmune diseases, including ulcerative colitis more frequently than with Crohn disease; primary sclerosing cholangitis; autoimmune hepatitis, Goodpasture syndrome, rheumatoid arthritis, and with SLE. Usually, specificities in such cases are other than MPO.

References

Bylund DJ and McCallum RM, "Idiopathic Vasculitis and Antineutrophilic Cytoplasmic Antibodies," *Clinical and Laboratory Evaluation of Human Autoimmune Diseases*, Chapter 15, Nakamura RM, Keren DF, and Bylund SJ, eds, Chicago, IL: ASCP Press, American Society for Clinical Pathology, 2002, 213-31.

Helfgott SM and Smith RN, "A 21-Year-Old Man With Arthritis During Treatment for Hyperthyroidism," Case Records of the Massachusetts General Hospital, Case 21-2002, Scully RE, Mark EJ, McNeely WF, et al, eds, *N Engl J Med*, 2002, 347(2):122-30.

Lim LC, Taylor JG III, Schmitz JL, et al, "Diagnostic Usefulness of Antineutrophil Cytoplasmic Autoantibody Serology: Comparative Evaluation of Commercial Indirect Fluorescent Antibody Kits and Enzyme Immunoassay Kits," *Am J Clin Pathol*, 1999, 111(3):363-9.

O'Sullivan BP, Erickson LA, and Niles JL, "An Eight-Year-Old Girl With Fever, Hemoptysis, and Pulmonary Consolidations," Case Records of the Massachusetts General Hospital, Case 30-2002, Scully RE, Mark EJ, McNeely WF, et al, eds, *N Engl J Med*, 2002, 347(13):1009-17.

Savige J, Gillis D, Benson E, et al, "International Consensus Statement on Testing and Reporting of Antineutrophil Cytoplasmic Antibodies (ANCA)," *Am J Clin Pathol*, 1999, 111(4):507-13.

Antinuclear Antibodies

Related Information

Anticardiolipin Antibody *on page 167*
Anti-DNA *on page 173*
Antiphospholipid Antibody (Lupus Anticoagulant and/or Anticardiolipin Antibody) *on page 193*
Aspartate Aminotransferase, Serum *on page 216*
Centromere/Kinetochore Antibody *on page 354*
Complement C4, Serum *on page 436*
Complement Components, Overview *on page 437*
Complement, Total, Serum or Body Fluid *on page 441*
Factor Inhibitors *on page 566*
Jo-1 Antibody *on page 815*
Kidney Biopsy *on page 818*
Liver Biopsy *on page 864*
Procainamide, Serum *on page 1092*
Rheumatoid Factor, Serum or Body Fluid *on page 1161*
RPR *on page 1174*
Sjögren Antibodies *on page 1199*
Skin Biopsy *on page 1200*
Smith (Sm) and Ribonucleoprotein (RNP) Antibodies *on page 1206*
Smooth Muscle Antibody *on page 1207*
Topoisomerase I Antibody *on page 1261*
VDRL, Serum or Cerebrospinal Fluid *on page 1303*

Synonyms ANA; FANA

Applies to ENA; Extractable Nuclear Antigens; Nucleolar Antibody

Replaces LE Cell Preparation

Test Includes Titers and pattern of nuclear fluorescence on positive samples

Abstract Antinuclear antibodies (ANA) are a central feature of systemic lupus erythematosus (SLE) and related systemic rheumatic diseases, presently called the connective tissue diseases. ANAs are thought to be directly involved in the pathogenesis of these disorders. The antinuclear antibody (ANA) test provides good sensitivity for SLE, and therefore is often used to screen for autoimmune rheumatic diseases. Nearly all patients with active, untreated SLE have a positive ANA, and a negative ANA effectively rules out SLE.

Aftercare If ANA is ≥1:160, further investigation and testing is usually indicated. Recommended secondary tests for confirmation and identification include anti-DNA, RNP, Smith (Sm), and Sjögren antibodies (SS-A/Ro and SS-B/La). Other tests which may be needed include topoisomerase 1 antibody, Scl-70, Jo-1 antibody, and antiphospholipid antibody. Complement studies and urinalysis are desirable. Renal biopsy may be indicated. Anti-ds-DNA is generally the test of choice to monitor disease progression in SLE, especially for lupus nephritis. C3 and C4 are among other tests useful to monitor patients with SLE.

Specimen Serum

Container Red top tube or SST™ tube

Storage Instructions Store at 4°C for up to 72 hours or -20°C or colder without freezing and thawing for an indefinite time period; -70°C used.

Special Instructions A complete drug history is needed to rule out drug-induced lupus.

Reference Interval Negative; initial serum dilution may vary from laboratory to laboratory. A titer of 1:80 is often reported, but a titer ≥1:160 is considered significant by IFA, and <1:40 is considered negative. Cutoff value ≥3 units on enzyme immunoassay has been recommended. Mayo Medical Laboratories has elected to use ≥1 units for EIA as a positive result, and to use ≥3 units for follow-up tests.

Critical Values A titer ≥1:320 has specificity of almost 97% for SLE or a related disorder.

Use An ANA test should be ordered if the physician feels there is a reasonable clinical suspicion of SLE or another systemic rheumatic disease based on the (Continued)

Antinuclear Antibodies *(Continued)*

clinical history, physical findings, and results of other laboratory tests. The indirect immunofluorescence (IFM) is the best screening test for SLE. A large number of diverse conditions in addition to SLE are associated with positive ANA test results, including many chronic inflammatory and infectious diseases and some drugs. Significant test reactivity is usually followed by selected clinically appropriate secondary tests (see Related Information).

Limitations A substantial number of normal healthy persons have a positive ANA. The prevalence of ANAs among healthy individuals varies with sex and age; older persons, particularly women older than 65 years, more commonly have positive results. Titers are usually low and <5% of healthy people have a titer of ≥1:160. Due to the low prevalence of SLE (50 cases per 100,000 persons) in the general population, the majority of persons randomly discovered to have a positive ANA result do not have SLE. Thus, lack of specificity is recognized. (Anti-ds-DNA and anti-Smith are specific for SLE.)

Repeat testing with ANA is not indicated for prediction of disease activity.

Autoimmune diseases in which indirect immunofluorescence microscopy (IFM-ANA) may be negative include polymyositis/dermatomyositis, rheumatoid arthritis, Sjögren syndrome, and scleroderma and a small percentage of patients with SLE. These may be reactive to other tests.

Drugs inducing positive ANA tests and clinical evidence of SLE most commonly are hydralazine and procainamide, but others are associated with ANA positivity and SLE as well.

Methodology Indirect immunofluorescence (IFM) on HEp-2 cell line or animal tissue, sandwich immunoassay (immunometric assay), enzyme immunoassay, fluorescent image analysis

Additional Information Different laboratories may test patient samples at various initial dilutions ranging from 1:40-1:160. Positive samples are then diluted and both the fluorescent pattern and titer are reported.

The overall clinical importance of staining patterns has diminished. Considerable overlap exists between the different fluorescent patterns and rheumatic diseases and more specific autoantibody tests have become available. A useful current description of staining patterns is provided by Homburger, who also points out that different patterns may provide useful directions for further testing.

ANAs and specific autoantibodies are an integral part of the diagnostic criteria of the American College of Rheumatology for systemic lupus erythematosus (SLE). Nearly all patients (95% to 100%) with active, untreated SLE have a positive ANA. A reactive ANA may be found with discoid LE.

American College of Rheumatology (truncated) Classification Criteria for SLE 1982

Malar rash
Discoid rash
Photosensitivity
Oral or nasal ulcers
Arthritis of two or more peripheral joints
Pleuritis or pericarditis
Renal disease with persistent proteinuria (>0.5 g/d or >3+) or cellular casts
Unexplained seizures or psychosis
Hemolytic anemia, leukopenia (WBC <4.0 x 10⁹/L), or lymphopenia on two or more occasions
Positive ANA, anti-ds-DNA antibodies, anti-Sm antibodies, false-positive serologic test for syphilis, or antiphospholipid antibodies

An expansion of this table is available. Raynaud phenomenon, hair loss, FUO, lymphadenopathy and/or splenomegaly, and thromboembolic phenomena are also characteristic of SLE.

Adapted from Bloom BJ and Zukerberg LR, "Case Records of the Massachusetts General Hospital. Weekly Clinicopathological Exercises. Case 14-1999. A 9-Year-Old Girl With Fever and Cervical Lymphadenopathy," Scully RE, Mark EJ, McNeely WF, et al, eds, *N Engl J Med*, 1999, 340(19):1491-7.

Clinical Utility of ANA	Frequency (%) of Positive ANA
Diseases for which ANA is very useful for diagnosis	
SLE	95-100
Systemic sclerosis (scleroderma)	60-80
Diseases for which ANA is somewhat useful for diagnosis	
Sjögren syndrome	40-70
Idiopathic inflammatory myositis (dermatomyositis or polymyositis)	30-80
Diseases for which ANA is useful for monitoring or prognosis	
Juvenile chronic oligoarticular arthritis with uveitis	20-50
Raynaud phenomenon	20-60
Conditions in which ANA is part of the diagnostic criteria	
Drug-induced SLE	~100
Autoimmune hepatic disease	~100
Mixed connective tissue disease (MCTD)	~100
Diseases for which an ANA is not directly useful in diagnosis, but may be relevant to differential diagnosis	
Rheumatoid arthritis	30-50
Multiple sclerosis	25
Idiopathic thrombocytopenic purpura	10-30
Thyroid disease	30-50
Discoid lupus	5-25
Infectious diseases	Varies widely
Malignancies	Varies widely
Patients with silicone breast implants	15-25
Fibromyalgia	15-25
Relatives of patients with autoimmune diseases	5-25
Normal persons with titer >1:40	20-30
Normal persons with titer >1:160	5

Adapted from Kavanaugh A, Tomar R, Reveille J, et al, "Guidelines of Clinical Use of the Antinuclear Antibody Test and Test for Specific Autoantibodies to Nuclear Antigens," *Arch Pathol Lab Med*, 2000, 124(1):71-81.

Patients who have few signs or symptoms suggestive of SLE have a low pretest probability of having the disease. In these patients, a positive ANA test result does little to increase the probability that the patient has SLE. In fact, positive ANA results in such cases can be misleading and precipitate unnecessary testing, erroneous diagnosis, or even inappropriate therapy. Patients classified as "ANA-negative LE" often demonstrate reactivity to Ro (SS-A).

Patients with **scleroderma (systemic sclerosis)** usually present with a distinct set of clinical signs and symptoms. Sixty to 80% of patients have a positive result. A positive ANA is not required for diagnosis, but is supportive. Performed on HEp-2 substrate, the ANA is positive in nearly all patients with scleroderma. A negative result might lead the physician to consider other fibrosing illnesses such as local scleroderma, the CREST syndrome, eosinophilic fasciitis, or scleredema.

Forty to 70% of patients with **Sjögren syndrome** have a positive ANA test result. This finding supports the diagnosis, but is not an absolute requirement. Testing is useful in patients with persistent sicca symptoms or in women who have given birth to a child with congenital heart block.

The ANA is positive in 30% to 70% of patients with **polymyositis** and **dermatomyositis**. A positive result is supportive, but a negative result does not rule out the diagnosis.

A positive ANA result is an integral component of the diagnosis of **drug-induced lupus erythematosus**, **autoimmune hepatitis**, and **mixed connective tissue disease**. Patients must have a positive ANA before these (Continued)

Antinuclear Antibodies *(Continued)*

diagnoses can be made. All studies of drug-induced SLE have included a positive ANA result in the definition of the syndrome. Similarly, criteria for the diagnosis of certain types of autoimmune hepatitis and mixed connective tissue disease (MCTD) dictate that the ANA result be positive for diagnosis. A reactive ANA may be found with autoimmune thyroiditis.

Raynaud phenomenon is diagnosed by physical examination and by specific clinical history. An ANA test does not establish the diagnosis, but may provide prognostic information. Raynaud phenomenon may be associated with several connective tissue diseases including SLE, rheumatoid arthritis, and scleroderma. However, the Raynaud phenomenon is also common among the general population and 80% of these individuals do not develop a systemic rheumatic disease. A positive ANA in a patient with Raynaud phenomenon increases the likelihood of development of a systemic rheumatic disease from 20% to 30%, while a negative result decreases the likelihood to about 7%.

The ANA is not useful in establishing diagnosis of **juvenile chronic arthritis**. However, in children with known disease, a positive ANA result may predict the development of **uveitis**, a serious complication. Approximately 33% of patients with a positive ANA develop uveitis. In such patients, a positive ANA is an indication for screening for uveitis.

An ANA is not necessary for diagnosis of the **antiphospholipid antibody syndrome**. However, approximately half of patients with this syndrome have a positive ANA. Positivity of ANA increases likelihood that the syndrome is secondary to SLE.

Except for subjects with the Raynaud phenomenon, juvenile chronic arthritis, and antiphospholipid antibody syndrome, ANA testing does not provide useful prognostic information. Since the ANA does not provide useful information about prognosis or disease activity, there is no indication for sequentially monitoring the ANA titer. (Serial measurements of anti-ds-DNA are utilized.)

ANA is positive in some individuals with chronic nonviral hepatitis, especially type 1 autoimmune hepatitis, in which antismooth muscle antibodies are also detected. These antibodies may also be found in some instances of primary biliary cirrhosis. In type 2 autoimmune hepatitis, antiliver/kidney microsomal type 1 antibodies are found; ANA and antismooth muscle antibodies are not detected. See table in Smooth Muscle Antibody *on page 1207*. ANA positivity may occur with autoimmune cholangitis.

Reactive ANA may occur with autoimmune thyroid disease, including Graves disease and Hashimoto thyroiditis.

Autoantibodies may be present for many years before the diagnosis of SLE is made.

References

Arbuckle MR, McClain MT, Rubertone MV, et al, "Development of Autoantibodies Before the Clinical Onset of Systemic Lupus Erythematosus," *N Engl J Med*, 2003, 349(16):1526-33.

Homburger HA, "Advances in the Diagnosis and Laboratory Evaluation of Systemic Rheumatic Diseases Other Than Rheumatoid Arthritis," *Clinical and Laboratory Evaluation of Human Autoimmune Diseases*, Chapter 11, Nakamura RA, Keren DF, and Bylund DJ, eds, Chicago, IL: American Society for Clinical Pathology, 2002, 153-64.

Kavanaugh A, Tomar R, Reveille J, et al, "Guidelines of Clinical Use of the Antinuclear Antibody Test and Tests for Specific Autoantibodies to Nuclear Antigens," *Arch Pathol Lab Med*, 2000, 124(1):71-81.

Moder KG, "Use and Interpretation of Rheumatologic Tests: A Guide for Clinicians," *Mayo Clin Proc*, 1996, 71(4):391-6.

Nakabayashi T, Kumagai T, Yamauchi K, et al, "Evaluation of the Automatic Fluorescent Image Analyzer, Image Titer, for Quantitative Analysis of Antinuclear Antibodies," *Am J Clin Pathol*, 2001, 115(3):424-9.

Nakamura RM, and Tan EM, "Clinical and Laboratory Evaluation of Systemic Lupus Erythematosus and Lupus-Related Disorders," *Clinical and Laboratory Evaluation of Human Autoimmune Diseases*, Chapter 9, Nakamura RA, Keren DF, and Bylund DJ, eds, Chicago, IL: American Society for Clinical Pathology, 2002, 111-40.

Reisner BS, DiBlasi J, and Goel N, "Comparison of an Enzyme Immunoassay to an Indirect Fluorescent Immunoassay for the Detection of Antinuclear Antibodies," *Am J Clin Pathol*, 1999, 111(4):503-6.

Antioxidant Concentrations, Plasma

Related Information

Ascorbic Acid, Serum or Plasma *on page 215*

Vitamin A, Serum or Plasma *on page 1314*

Vitamin E, Serum or Plasma *on page 1319*

Applies to Oxygen Radical Absorbance Capacity; Total Antioxidant Capacity; Total Radical Absorbin Parameter

Test Includes Measurement of total antioxidant capacity of plasma

Abstract Total antioxidant capacity: Halliwell and Gutteridge defined antioxidants as any substance that delays or inhibits damage to a target molecule by an oxidant. The balance between pro-oxidant challenge and the presence of antioxidants determines net oxidative stress. Antioxidants include vitamins A, C, and E and many of the carotenoids. Assessment of antioxidant status employs methods to measure total antioxidant capacity.

Patient Preparation Patient must fast for a minimum of 8 hours.

Specimen Plasma

Container Green top (heparin) tube

Collection Draw in chilled tube protected from light. Keep specimen on ice until separation of plasma.

Storage Instructions Store at -20°C for 3 months or -70°C for longer periods of time.

Turnaround Time 6 hours

Reference Interval Measures as units of Trolox

Use Assess total antioxidant deficiency (eg, vitamin C or E); especially useful for water soluble antioxidant status

Methodology All published methods measure the inhibition of an artificially generated species. A free radical species is generated in a solution containing an oxidation target; antioxidants in the added sample quench the target response by interaction with the reactive oxygen species. Published methods differ in the choice of free radical generator used, target, and type of measurement used to detect the oxidized product. One of the most widely adapted techniques is the total peroxyl radical-trapping antioxidant parameter (TRAP). In this assay, the water-soluble compound, 2,2′-azobis(2-amidinopropane hydrochloride) (AAPH), undergoes thermal decomposition to produce peroxyl free radicals. The rate of oxygen uptake is monitored by performing the reaction in an oxygen-electrode chamber. The measurement is calibrated using Trolox, a water-soluble alpha-tocopherol analog. In an effort to improve this assay, a method has been developed for use on a centrifugal analyzer with plasma volumes as low a 3 mL. In this assay, 2,2′-azinobis-(3-ethylbenzothiazoline-6-sulphonic acid) (ABTS) is incubated with metmyoglobin (this acts as a peroxidase) and hydrogen peroxide, resulting in the formation of the long-lived cation, ABTS⁺. The presence of an antioxidant in heparinized plasma reduces this radical cation, the concentration of which is measured by absorbance. The assay is calibrated using Trolox. However, its clinical utility needs to be assessed.

Another widely used method for determining total antioxidant capacity is the ORAC (oxygen radical absorbance capacity) assay based on the procedure described by Cao et al. The method utilizes β-phycoerythrin (β-PE) as an indicator protein and AAPH as a peroxyl radical generator. Under appropriate conditions, the loss of β-PE fluorescence in the presence of reactive species is an index of oxidative damage of the protein. The inhibition by an antioxidant, which is reflected in the protection against the loss of β-PE fluorescence in the ORAC assay, is a measure of its antioxidant capacity.

References

Halliwell B and Gutteridge JM, "Free Radicals and Antioxidants in the Year 2000," *Ann N Y Acad Sci*, 2000, 899:136-47

Kushi LH, Folsom AR, Prineas RJ, et al, "Dietary Antioxidant Vitamins and Death From Coronary Heart Disease in Postmenopausal Women," *N Engl J Med*, 1996, 334(18):1156-62.

Wayner DD, Burton GW, Ingold KU, et al, "The Relative Contribution of Vitamin E, Urate, Asorbate and Proteins to TRAP in Human Blood Plasma," *Biochim Biophys Acta*, 1987, 924(3):408-19.

♦ **Antiphosphatidylserine Antibodies** *see* Antiphospholipid Antibody (Lupus Anticoagulant and/or Anticardiolipin Antibody) *on page 193*

Antiphospholipid Antibody (Lupus Anticoagulant and/or Anticardiolipin Antibody)

Related Information

Activated Partial Thromboplastin Time *on page 100*

(Continued)

Antiphospholipid Antibody (Lupus Anticoagulant and/ or Anticardiolipin Antibody) *(Continued)*

Anticardiolipin Antibody *on page 167*
Antinuclear Antibodies *on page 189*
Chlorpromazine, Serum *on page 395*
Factor Inhibitors *on page 566*
HIV-1/HIV-2 Serology *on page 736*
Hypercoagulation Panel *on page 758*
Mixing Studies (Coagulation) *on page 918*
Platelet Count *on page 1050*
RPR *on page 1174*
VDRL, Serum or Cerebrospinal Fluid *on page 1303*

Synonyms Lupus Inhibitor

Applies to Antiphosphatidylserine Antibodies; Beta-2-Glycoprotein I; Cardiolipin; Circulating Anticoagulant; Lupus Anticoagulant; PTT

Abstract Antiphospholipid antibodies are associated with an increased risk of thrombosis, thrombocytopenia, and recurrent fetal loss.

Specimen Lupus anticoagulant: plasma; anticardiolipin antibody: serum

Container Lupus anticoagulant: blue top (sodium citrate) tube; anticardiolipin antibody: red top tube

Collection Routine venipuncture. If multiple tests are being drawn, draw blue top tubes after any red top tubes but before any lavender top (EDTA), green top (heparin), or gray top (oxalate/fluoride) tubes. Immediately invert gently at least 4 times, mixing thoroughly. Blue top tubes must be appropriately filled. Deliver immediately to the laboratory.

Storage Instructions Separate plasma (or serum) from cells as soon as possible. Plasma (or serum) may be stored on ice for up to 4 hours; otherwise, store frozen. Platelet count must be <10 x 10^9/L in plasma prior to freezing, or false-negative lupus anticoagulant results may occur.

Causes for Rejection Blue top specimen received more than 4 hours after collection; clotted specimen; blue top not filled; patient on hirudin, danaparoid, or argatroban anticoagulation

Turnaround Time Lupus anticoagulant: less than 1 day if negative, longer if positive because confirmatory assays need to be performed; anticardiolipin: often several days because testing is batched.

Special Instructions Notify laboratory if patient is on heparin, including subcutaneous low-dose heparin or low-molecular weight heparin. In some assays, heparin may cause false-positive lupus anticoagulant results. Therefore, heparin must first be removed from the specimen by the laboratory. Other assays contain a heparin neutralizer that tolerates specimens containing up to 1 unit/mL heparin. Results can be interpreted correctly in patients on Coumadin®. Hirudin, danaparoid, or argatroban anticoagulation may cause false-positive results in some assays.

Reference Interval Negative for lupus anticoagulant; <15 units of anticardiolipin antibody

Use If an antiphospholipid antibody is suspected, assays for both lupus anticoagulant and anticardiolipin antibody should be performed. Used to evaluate hypercoagulable states, recurrent miscarriage, thrombocytopenia, or prolonged PTT. Lupus anticoagulants may or may not prolong the PTT.

Limitations Factor VIII inhibitors can cause false-positive lupus anticoagulant tests (a factor VIII assay showing normal factor VIII levels rules out this possibility). The transient presence of antiphospholipid antibodies may accompany infections or drugs; *vide infra.*

Methodology

Anticardiolipin antibody: Enzyme-linked immunosorbent assay (ELISA) using cardiolipin, a phospholipid, as the antigen. Newer ELISA assays are available that test for anti-β_2-glycoprotein I antibodies or antiphosphatidylserine antibodies.

Lupus anticoagulant: To improve sensitivity, two screening tests are suggested. These tests are clotting-time based assays, such as the Russell viper venom time, PTT-based assays, kaolin clotting time, or dilute prothrombin time (tissue thromboplastin inhibition test). Lupus anticoagulants prolong various clotting times in the laboratory because they bind to phospholipid and thereby interfere with the ability of phospholipid to serve its essential cofactor function in the coagulation cascade. Lupus anticoagulant screening assays usually have a low concentration of phospholipid to enhance sensitivity. Any abnormal (prolonged) screening result is repeated after a 1:1 mixture of patient plasma with normal plasma to demonstrate that the clotting time remains prolonged upon mixing. Confirmatory assays are performed if the screening assay remains abnormal after the 1:1 mixture. Confirmatory assays typically demonstrate that upon addition of excess phospholipid, the clotting time shortens toward normal. The "platelet neutralization procedure" is a confirmatory assay in which the source of the excess phospholipid is freeze-thawed platelets. **Note:** The routine PTT may or may not be prolonged, depending on the amount of phospholipid in the reagent. In addition, elevated factor VIII can normalize an otherwise prolonged PTT. PTT-based lupus anticoagulant screening assays have a low concentration of phospholipid to enhance sensitivity. When the PTT is prolonged, a PTT mixing study may be a useful first test. **When lupus anticoagulants are present, the PTT remains prolonged upon mixing with an equal volume of normal plasma.**

Additional Information The two principal types of antiphospholipid antibodies are lupus anticoagulants and anticardiolipin antibodies. They are present in 0% to 5% of the general population and in 12% or more of patients with thrombosis. Antiphospholipid antibodies are acquired autoantibodies directed against phospholipid-protein complexes. These antibodies are associated with an increased risk for arterial or venous thrombosis, thrombocytopenia, and fetal loss. Associations with cardiac valve disease, livedo reticularis, and other features are also recognized. The mechanism of thrombosis is not entirely clear, although a number of mechanisms have been proposed. In a recent prospective study involving individuals with antiphospholipid antibodies, the **incidence of thrombosis** per year was 1% in individuals with no history of thrombosis, 4% in patients with systemic lupus erythematosus, 5.5% in patients with a history of thrombosis, and 6% in individuals with high titer IgG anticardiolipin antibody (>40 units).

The diagnosis of **antiphospholipid antibody syndrome** requires a positive test in the antiphospholipid antibody panel (lupus anticoagulant and/or anticardiolipin antibody) on two separate occasions, at least 6 weeks apart, in the setting of thrombosis, thrombocytopenia, or recurrent miscarriage.

Anticardiolipin antibodies recognize cardiolipin bound to β_2-glycoprotein I. Most lupus anticoagulants recognize phospholipid bound to prothrombin, but others recognize phospholipid bound to β_2-glycoprotein I or other proteins. Rarely, prothrombin levels become decreased as a result of a lupus anticoagulant, and an increased risk for bleeding may develop. As cardiolipin is the antigen used for syphilis screening tests (VDRL, Venereal Disease Research Laboratories; and RPR, rapid plasma reagin), false-positive syphilis tests may occur in patients with anticardiolipin antibodies. Conversely, true syphilis infections can cause positive anticardiolipin antibody test results.

Despite the prolonged clotting times, bleeding is not a typical feature associated with these antibodies. Thrombocytopenia, if present, is usually mild. Patients may have either a lupus anticoagulant or an anticardiolipin antibody or they may have both antibodies. A high percentage of patients with systemic lupus erythematosus (SLE) or related autoimmune diseases have these antibodies. These antibodies may also develop in patients without an underlying disorder. The antibodies can appear transiently in association with certain medications (eg, hydralazine, phenytoin) or infections. The human immunodeficiency virus (HIV) is commonly associated with positive tests for antiphospholipid antibodies. Infection-associated antibodies may not be associated with clinical symptoms of antiphospholipid antibody syndrome, and they tend to recognize phospholipid rather than the phospholipid-protein complexes described above.

Heparin treatment in patients with lupus anticoagulants can be complicated by the fact that lupus anticoagulants may prolong the baseline PTT and/or accentuate the PTT prolongation when heparin is added. As such, heparin may be monitored with antifactor Xa assays. If the antifactor Xa assay demonstrates
(Continued)

Antiphospholipid Antibody (Lupus Anticoagulant and/ or Anticardiolipin Antibody) (Continued)

that the heparinized PTT is not affected by the lupus anticoagulant, cautious use of the PTT may be considered for that patient.

References

Brandt JT, Triplett DA, Alving B, et al, "Criteria for the Diagnosis of Lupus Anticoagulants: An Update," Thromb Haemost, 1995, 74(4):1185-90.

Doig RG, O'Malley CJ, Dauer R, et al, "An Evaluation of 200 Consecutive Patients With Spontaneous or Recurrent Thrombosis for Primary Hypercoagulable States," Am J Clin Pathol, 1994, 102(6):797-801.

Erkan D, Yazici Y, Peterson MG, et al, "A Cross-Sectional Study of Clinical Thrombotic Risk Factors and Preventive Treatments in Antiphospholipid Syndrome," Rheumatology, 2002, 41(8):924-9.

Finazzi G, Brancaccio V, Moia M, et al, "Natural History and Risk Factors for Thrombosis in 360 Patients With Antiphospholipid Antibodies: A Four-Year Prospective Study From the Italian Registry," Am J Med, 1996, 100(5):530-6.

Galli M and Barbui T, "Antiphospholipid Syndrome: Definition and Treatment," Semin Thromb Hemost, 2003, 29(2):195-204.

Ginsburg KS, Liang MH, Newcomer L, et al, "Anticardiolipin Antibodies and the Risk for Ischemic Stroke and Venous Thrombosis," Ann Intern Med, 1992, 117(12):997-1002.

Sugi T and Makino T, "Plasma Contact System, Kallikrein-Kinin System and Antiphospholipid-Protein Antibodies in Thrombosis and Pregnancy," J Reprod Immunol, 2000, 47(2):169-84.

Wilson WA, Gharavi AE, Koike T, et al, "International Consensus Statement on Preliminary Classification Criteria for Definite Antiphospholipid Syndrome," Arthritis Rheum, 1999, 42(7):1309-11.

Antiplasmin

Related Information

Hypercoagulation Panel on page 758

Plasminogen on page 1042

Plasminogen Activator Inhibitor 1 on page 1043

Synonyms Alpha$_2$-Antiplasmin; α_2-Antiplasmin; Plasmin Inhibitor

Abstract Antiplasmin is a major inhibitor of plasmin. Hereditary antiplasmin deficiency is a rare familial bleeding disorder due to excessive fibrinolysis.

Specimen Plasma

Container Blue top (sodium citrate) tube

Collection Routine venipuncture. If multiple tests are being drawn, draw blue top tubes after any red top tubes but before any lavender top (EDTA), green top (heparin), or gray top (oxalate/fluoride) tubes. Immediately invert tube gently at least 4 times to mix. Tubes must be appropriately filled. Deliver tubes immediately to the laboratory.

Storage Instructions Separate plasma from cells as soon as possible; plasma may be stored on ice for up to 4 hours; otherwise store frozen.

Causes for Rejection Specimen received more than 4 hours after collection, tube not filled, clotted specimen

Turnaround Time Several days, because test is often sent out

Special Instructions Specimens for functional assays should not contain fibrinolysis inhibitors (eg, epsilon-aminocaproic acid, aprotinin) or heparin. Elevated α_2-macroglobulin levels >200% may slightly interfere with functional assays.

Reference Interval Approximately 80% to 130% functional; approximately 48-80 mg/dL antigen. Antiplasmin levels (measured by antigen assay) are slightly lower during the first 5 days of life.

Use Not a commonly performed clinical assay. May be considered in patients with strong evidence for a familial bleeding disorder and normal test results for more common bleeding disorders, such as von Willebrand disease.

Methodology

Functional (activity) assays: Excess plasmin is added to patient plasma. Antiplasmin in the patient plasma binds to and inhibits plasmin, forming a plasmin-antiplasmin complex. Residual plasmin then cleaves a chromogenic substrate, releasing a colored compound that can be detected spectrophotometrically. The amount of plasmin detected is inversely proportional to the concentration of antiplasmin in the patient specimen.

Antigen (immunologic) assay by radial immunodiffusion: Plasma is placed in a cylindrical well of an agarose gel. The agarose gel contains an antibody monospecific for antiplasmin. Antiplasmin in the specimen diffuses from the well into the gel where it forms a complex with the antibody, creating a precipitin ring. The size of the ring is proportional to the amount of antiplasmin in the plasma.

Additional Information Plasmin mediates fibrinolysis, and antiplasmin inhibits plasmin. Activated factor XIII cross-links antiplasmin to fibrin, and antiplasmin protects fibrin from plasmin-mediated fibrinolysis. Antiplasmin also binds to plasminogen and may inhibit plasminogen binding to fibrin. Antiplasmin is synthesized in the liver. Acquired causes of decreased antiplasmin include liver disease, thrombolytic therapy, and disseminated intravascular coagulation (DIC). Hereditary deficiencies of antiplasmin are either type I or type II. Type I deficiencies are quantitative, in which both functional and antigen levels are reduced. Type II deficiencies are qualitative, with decreased functional levels but normal or near normal antigen levels.

References
Andrew M, Paes B, Milner R, et al, "Development of the Human Coagulation System in the Full-Term Infant," *Blood*, 1987, 70(1):165-72.

Lijnen HR, Okada K, Matsuo O, et al, "α2-Antiplasmin Gene Deficiency in Mice Is Associated With Enhanced Fibrinolytic Potential Without Overt Bleeding," *Blood*, 1999, 93(7):2274-81.

♦ **α_2-Antiplasmin** *see* Antiplasmin *on page 196*

♦ **Antiplatelet Antibody** *see* Platelet Antibody, Immunohematologic *on page 1049*

♦ **Anti-PM-1** *see* Topoisomerase I Antibody *on page 1261*

♦ **Anti-PM-Scl** *see* Topoisomerase I Antibody *on page 1261*

♦ **α_1-Antiprotease** *see* Alpha$_1$-Antitrypsin, Serum *on page 134*

♦ **α_1-Antiprotease Phenotype** *see* Alpha$_1$-Antitrypsin Phenotyping *on page 133*

♦ **Anti-Ri** *see* Antineuronal Nuclear Antibody, Type 2 (Anti-Ri) *on page 185*

♦ **Antisacer®** *see* Phenytoin, Serum or Plasma *on page 1026*

♦ **Antismooth Muscle Antibody** *see* Smooth Muscle Antibody *on page 1207*

♦ **Anti-sn RNP** *see* Smith (Sm) and Ribonucleoprotein (RNP) Antibodies *on page 1206*

♦ **Antispermatozoal Antibody Test** *see* Infertility Screen *on page 786*

♦ **Anti-ss-DNA** *see* Anti-DNA *on page 173*

♦ **Antistreptococcal DNase-B Titer** *see* Antideoxyribonuclease-B Titer, Serum *on page 170*

Antistreptolysin O Titer, Serum

Related Information

Antideoxyribonuclease-B Titer, Serum *on page 170*

Bacterial Culture, Throat, and Antigen Detection Testing for Group A Streptococci *on page 245*

Group A *Streptococcus* Screen, Rapid *on page 659*

Sedimentation Rate, Erythrocyte *on page 1181*

Streptozyme *on page 1225*

Synonyms ASO

Abstract Detection of elevated immune responses to extracellular products of *S. pyogenes* (eg, streptolysin O), is useful in demonstration of streptococcal infections in patients without documentation of recent infection, who present with nonsuppurative sequelae such as rheumatic fever or glomerulonephritis. While sequential ASO assays are required, a single increased titer of anti-DNase provides evidence of recent streptococcal infection.

Specimen Serum

Container Red top tube

Reference Interval Younger than 2 years of age: usually <50 Todd units; 2-5 years: <100 Todd units; 5-19 years: <166 Todd units; adults: <125 Todd units. A rise in titer of four or more dilution increments between acute and convalescent specimens is considered to be significant regardless of the magnitude of the titer. For a single specimen, ASO titers ≤166 Todd units are considered normal; but higher titers may be "normal" in demographic groups or may be associated with chronic pharyngeal carriage.

Use Used with throat culture, antideoxyribonuclease-B and other investigations to document streptococcal infection

Limitations False-positive ASO titers can be caused by increased levels of serum betalipoprotein produced in liver disease and by contamination of the

(Continued)

Antistreptolysin O Titer, Serum *(Continued)*

serum with *Bacillus cereus* and *Pseudomonas* species. ASO is not sensitive for the diagnosis of streptococcal pyoderma. Test is subject to false positives due to technical and clinical circumstances.

Methodology Nephelometry (the nephelometric assay may provide optimal sensitivity), hemolysis inhibition, latex agglutination (LA)

Additional Information Streptolysin is a hemolysin produced by group A streptococci. In an infected individual streptolysin O acts as a protein antigen, and the patient mounts an antibody response. A rise in titer begins about 1 week after infection and peaks 2-4 weeks later. In the absence of complications or reinfection, the ASO titer will usually fall to preinfection levels within 6-12 months. A marked rise in titer or a persistently elevated ASO titer indicates a recent or current infection with beta-hemolytic group A *Streptococcus* (*Streptococcus pyogenes*). Elevated titers are seen in 80% to 85% of patients with acute rheumatic fever and in up to 95% of patients with acute glomerulonephritis, but in patients with glomerulonephritis responses are variable.

References

Aviles RJ, Ramakrishna G, Mohr DN, et al, "Poststreptococcal Reactive Arthritis in Adults: A Case Series," *Mayo Clin Proc*, 2000, 75(2):144-7.

Cunningham MW, "Pathogenesis of Group A Streptococcal Infections," *Clin Microbiol Rev*, 2000, 13(3):470-511.

Efstratiou A, "Group A Streptococci in the 1990s," *J Antimicro Chemother*, 2000, 45(Suppl):3-12.

Kaplan EL, Rothermel CD, and Johnson DR, "Antistreptolysin O and Antideoxyribonuclease B Titers: Normal Values for Children Ages 2 to 12 in the United States," *Pediatrics*, 1998, 101(1):86-8.

Olivier C, "Rheumatic Fever - Is It Still a Problem?" *J Antimicro Chemother*, 2000, 45(Suppl):13-21.

♦ **Antisynthetases** *see* Jo-1 Antibody *on page 815*

Antithrombin

Related Information

Activated Protein C Resistance and the Factor V Leiden Mutation *on page 104*
Heparin Antifactor Xa Assay *on page 693*
Hypercoagulation Panel *on page 758*
Protein C *on page 1101*
Protein S *on page 1110*

Applies to Heparin; Heparin Cofactor II; Heparin Resistance

Replaces Antithrombin III Assay

Abstract A deficiency of antithrombin, a natural anticoagulant protein, leads to a hypercoagulable state with an increased risk for venous thrombosis. Acquired antithrombin deficiencies are more common than hereditary deficiencies.

Specimen Plasma

Container One blue top (sodium citrate) tube

Collection Routine venipuncture. If multiple tests are being drawn, draw blue top tubes after any red top tubes but before any lavender top (EDTA), green top (heparin), or gray top (oxalate/fluoride) tubes. Immediately invert tube gently at least 4 times to mix. Tubes must be appropriately filled. Deliver tubes immediately to the laboratory.

Storage Instructions Separate plasma from cells as soon as possible. Plasma may be stored on ice for up to 4 hours, otherwise, store frozen.

Causes for Rejection Specimen received more than 4 hours after collection, tubes not filled, clotted specimens

Turnaround Time 2-4 hours; longer if testing is batched

Reference Interval Results are often reported as a percent of the amount expected in normal plasma. By definition, the mean value in normal plasma is 100%. The reference range is approximately 80% to 130%.

Results may also be reported in mg/dL; reference range is approximately 17-39 mg/dL (SI: 170-390 mg/L).

At birth, antithrombin levels average 63% (range 39% to 87%) of adult levels. Antithrombin increases to adult values within 6 months. Spontaneous thromboses do not develop in normal infants because a balance between procoagulants and inhibitors is maintained.

Use A functional assay should be performed first, because both type I and type II hereditary antithrombin deficiencies will be detected. If the result of the functional assay is decreased, the antigenic assay is needed to differentiate between type I and type II deficiencies. If the antigen assay is performed without the functional assay, type II deficiencies will not be detected (see Additional Information).

Antithrombin levels are **increased** with Coumadin® (possibly)

Antithrombin levels are **decreased** with heparin, liver disease, thrombosis (eg, pulmonary embolism, acute myocardial infarction, thrombophlebitis), disseminated intravascular coagulation (DIC), surgery, nephrotic syndrome, oral contraceptives, pregnancy (possibly)

Limitations

Chromogenic (functional) assays: Heparin cofactor II, another natural thrombin inhibitor, produces falsely elevated levels of antithrombin in some thrombin-based assays. One commercially available kit uses heparin that has been treated with a bacterial enzyme (chondroitinase) such that the heparin no longer enhances heparin cofactor II activity, and heparin cofactor II interference is thus essentially eliminated. In factor Xa-based assays, high levels of heparin cofactor II will not lead to an overestimate of antithrombin, because heparin cofactor II does not inhibit factor Xa.

Hirudin or argatroban anticoagulation can interfere with thrombin-based assays.

Antigenic assays: If used without the functional assay, type II antithrombin deficiencies will not be detected. An antigenic test is usually performed only if the functional test result is decreased, to determine if the patient has type I or type II deficiency (see Additional Information).

Methodology Functional (activity) and antigenic (immunologic) tests are available. Functional assays are usually chromogenic. To perform the chromogenic test, heparin and thrombin are added to the patient's plasma. Antithrombin in the patient's plasma will bind and inhibit thrombin. A chromogenic substrate that resembles thrombin's natural substrate is then added. Any unbound thrombin will cleave the substrate, liberating a chromogenic substance that can be measured spectrophotometrically. The amount detected is inversely proportional to the amount of antithrombin in the patient. Factor Xa-based methods are also available; these are similar in principle to the thrombin-based assays described above, except that factor Xa is used instead of thrombin. Antigenic (immunologic) assay: A commonly used automated method involves latex particles coated with antibodies directed against antithrombin. In the presence of antithrombin, the latex particles form aggregates that absorb light passing through the specimen. The amount of light absorbance is directly related to the amount of antithrombin in the specimen.

Additional Information Antithrombin is a natural inhibitor of thrombin as well as factors Xa, IXa, XIa, XIIa, and kallikrein. The activity of antithrombin is greatly accelerated by interaction with the glycosaminoglycans, heparan sulfate or heparin. Heparan sulfate is naturally located *in vivo* on the endothelial cell surface. Antithrombin deficiency is present in 0.17% of the general population. It accounts for 1.1% of unselected patients with venous thrombosis and up to 5% of patients younger than age 70 years with thrombosis. Over 127 mutations in the antithrombin gene are known to cause hereditary antithrombin deficiency. Individuals heterozygous for antithrombin deficiency have a fivefold increased risk for venous thrombosis. Homozygous deficiencies are incompatible with life, except for patients with type II deficiency due to heparin-binding mutations. Heterozygotes generally have antithrombin levels between 45% to 75%, although levels as high as 78% have been observed. The risk for thrombosis is further increased in the presence of a second risk factor. The age at onset of thrombosis is usually between 10-50 years (peak 15-35 years) in heterozygous individuals. The risk of arterial thrombosis remains uncertain, but 2% of individuals developed arterial thrombosis in one study.

Decreased antithrombin also arises from acquired conditions, such as:

- decreased hepatic synthesis from liver disease or L-asparaginase treatment
- consumption from thrombosis, DIC (disseminated intravascular coagulation) or surgery

(Continued)

Antithrombin (Continued)

- increased clearance from full-dose heparin use
- proteinuria

Mild decreases occasionally result from elevated estrogen levels (eg, pregnancy or oral contraceptive use). Colitis has been associated with low antithrombin levels. If a patient with low antithrombin has any of the conditions listed above, the test should be repeated once the condition is no longer present, if possible. Confirmation of a hereditary antithrombin deficiency may require documenting antithrombin deficiency in a relative. In contrast to protein C and protein S, which are decreased by Coumadin®, antithrombin levels may increase while on Coumadin®. Antithrombin levels in premenopausal women may be somewhat lower than in men, but postmenopausal women have higher levels than men.

Antithrombin deficiencies are quantitative (type I) or qualitative (type II). In type I deficiencies, normal antithrombin molecules are made, but in reduced quantity. In type II deficiencies, normal amounts of antithrombin are made, but the antithrombin is defective. Accordingly, type I deficiencies have decreased antithrombin in both functional and antigenic assays. Type II deficiencies have normal (or near normal) antigenic antithrombin levels, with decreased functional antithrombin. Thus, if only antigenic assays are performed, type II deficiencies will not be detected. Therefore, a functional assay should be used as the initial screening assay. If the result is decreased, an antigenic assay should be performed to determine if the deficiency is type I or type II. According to one study, 0.02% of the general population have type I antithrombin deficiency and 0.15% have type II. The heparin-binding variant, which is one of the mutations that causes type II deficiency, has a low risk of thrombosis in comparison to the other mutations, and it is present in at least 0.01% of the general population. For patients with test results suggesting type II deficiency, a method has been described that tests for the heparin-binding mutation.

Patients with marked decreases in antithrombin may demonstrate heparin resistance, in which very high doses of heparin are required to obtain a therapeutic PTT prolongation. Antithrombin concentrates are available for the treatment of hereditary antithrombin deficiency. The use of antithrombin concentrates for certain acquired antithrombin deficiencies, such as DIC, is under investigation.

See Hypercoagulation Panel *on page 758.*

References

Andrew M, Paes B, Milner R, et al, "Development of the Human Coagulation System in the Full-Term Infant," *Blood*, 1987, 70(1):165-72.

Demers C, Ginsberg JS, Hirsh J, et al, "Thrombosis in Antithrombin III-Deficient Persons: Report of a Large Kindred and Literature Review," *Ann Intern Med*, 1992, 116(9):754-61.

Melissari E, Monte G, Lindo VS, et al, "Congenital Thrombophilia Among Patients With Venous Thromboembolism," *Blood Coagul Fibrinolysis*, 1992, 3(6):749-58.

Minakami H, Yamada H, and Suzuki S, "Gestational Thrombocytopenia and Pregnancy-Induced Antithrombin Deficiency: Progenitors to the Development of the HELLP Syndrome and Acute Fatty Liver of Pregnancy," *Semin Thromb Hemost*, 2002, 28(6):515-8.

Rodeghiero F and Tosetto A, "The Epidemiology of Inherited Thrombophilia: The VITA Project," *Thromb Haemost*, 1999, 78(1):636-40.

Tait RC, Walker ID, Perry DJ, et al, "Prevalence of Antithrombin Deficiency in the Healthy Population," *Br J Haematol*, 1994, 87(1):106-12.

van der Meer FJM, Koster T, Vandenbroucke JP, et al, "The Leiden Thrombophilia Study (LETS)," *Thromb Haemost*, 1997, 78(5):631-5.

Apheresis, Therapeutic

Related Information

Low Density Lipoprotein Cholesterol *on page 874*
Plasma Exchange *on page 1038*
Platelets, Apheresis, Donation *on page 1054*
Sickle Cell Tests *on page 1195*

Synonyms Cytapheresis, Therapeutic; Cytoreduction; Plasmapheresis; Therapeutic Cytapheresis

Applies to Erythrocytapheresis; Granulocytapheresis; Leukapheresis, Therapeutic; Photopheresis; Red Cell Exchange; Stem Cell Collection

Test Includes ABO, Rh, antibody screen, and/or crossmatch if appropriate in the following instances: Red cells must be selected and crossmatched for a red cell exchange. Fresh frozen plasma must be selected and thawed if chosen as a replacement solution.

Abstract Selective removal of a pathologic component of a patient's blood to assist in the treatment of a disease. Separation techniques are varied and include manual (whole blood bags) and automated instruments (centrifugation, membrane filtration, and adsorption). With the numerous automated instruments available, manual techniques are seldom used.

Patient Preparation Excellent venous access is a daily requirement. Indwelling dual- or triple-lumen central venous catheters suitable for apheresis procedures are frequently necessary. Hypotension secondary to hypovolemia will become a concern if the extracorporeal volume exceeds 15% of the patient's blood volume. Patients taking angiotensin-converting-enzyme (ACE) inhibitors may be prone to marked hypotension prior to routine or immunosorption treatment. Such therapy should be removed from their medication if possible. Understanding which drugs being given to the patient will be altered by the procedure is relevant, so that dosage can be adjusted accordingly (withhold administration for 1 hour before and after procedure if possible). Obtain and document informed consent.

Aftercare Monitor patient for signs and symptoms of reduced ionized calcium (tingling, perioral paresthesias, cardiac arrhythmias) and hypotension secondary to hypovolemia. Hypocalcemia may be managed by decreasing the inlet flow rate or administering intravenous calcium.

Causes for Rejection Guidelines for evidence of therapeutic efficacy have been developed by the American Association of Blood Banks and the American Society for Apheresis.

Special Instructions Emergent procedures may include those for patients with thrombotic thrombocytopenic purpura with end-organ dysfunction; presence of monoclonal gammopathy (paraproteinemia) with significant hyperviscosity; and acute leukemia with an extremely high circulating blast count (>100 x 10^9/L) and evidence of leukostasis.

Use During therapeutic plasmapheresis, plasma containing the pathologic substance is removed and replaced with crystalloids, albumin, or fresh frozen plasma. A combination of replacement fluids is often used. Usually 1-1.5 plasma volumes are exchanged. Selective removal of pathologic substances in plasma via adsorption columns (eg, dextran sulfate removal of LDL cholesterol and staphylococcal protein A removal of IgG antibodies) eliminates the need for replacement fluid. Remove abnormal red cells and replace with normal red cells in sickle cell disease with crisis or in hypertransfusion regimens. Reduce platelets in thrombocythemic patients or leukocytes in hyperleukocytic leukemia. Treat cutaneous T-cell lymphoma via photopheresis (separate lymphocytes, add psoralens, and treat the cells with ultraviolet irradiation prior to returning white cells to patient).

Additional Information Policies relating to therapeutic apheresis should be established and indications for treatment and outcomes periodically reviewed by responsible physicians.

References

"Clinical Applications of Therapeutical Apheresis," *J Clin Apheresis*, 2000, 15:1-159 (special issue).
Jones HG and Bandarenko N, "Management of the Therapeutic Apheresis Patient," *Apheresis: Principles and Practice*, 2nd ed, McLeod BC, Price TH, Weinstein R, eds, Bethesda, MD: American Association of Blood Banks Press, 2003, 253-82.
McLeod BC, Price TH, and Weinstein R, "Apheresis Principles and Practice," 2nd ed, Bethesda, MD: American Association of Blood Banks Press, 2003.
Moake JL, "Thrombotic Microangiopathies," *N Engl J Med*, 2002, 347(8):589-600.

(Continued)

Apheresis, Therapeutic (Continued)

Shehata N, Kouroukis C, and Kelton JG, "A Review of Randomized Controlled Trials Using Thera-peutic Apheresis," *Trans Med Rev*, 2002, 16(3):200-9.

♦ **Apo A** *see* Apolipoprotein A-I, Serum *on page 202*

♦ **Apo-A-I** *see* Cholesterol, Total, Serum or Plasma *on page 396*

♦ **Apo-A-I** *see* Lipids, Overview *on page 853*

♦ **Apo B** *see* Apolipoprotein B-100, Serum *on page 203*

♦ **Apo B-100** *see* Cholesterol, Total, Serum or Plasma *on page 396*

♦ **Apo B-100** *see* Lipids, Overview *on page 853*

♦ **APOE** *see* Apolipoprotein E, Plasma *on page 204*

♦ **Apolipoprotein A** *see* Apolipoprotein A-I, Serum *on page 202*

♦ **Apolipoprotein A-I** *see* High Density Lipoprotein Cholesterol, Serum *on page 729*

Apolipoprotein A-I, Serum

Related Information

Apolipoprotein B-100, Serum *on page 203*
Apolipoprotein E, Plasma *on page 204*
Cholesterol, Total, Serum or Plasma *on page 396*
High Density Lipoprotein Cholesterol, Serum *on page 729*
Lipid Panel, Serum *on page 852*
Lipids, Overview *on page 853*
Low Density Lipoprotein Cholesterol *on page 874*
Triglycerides, Serum or Plasma *on page 1275*

Synonyms Apo A; Apolipoprotein A

Abstract Apolipoprotein A-I (apo A-I) is the principal protein associated with the HDL particle. Apo A-I participates in the removal of excess cholesterol from tissues and is the primary lipoprotein in the interstitial space. Similar to HDL, serum apo A-I is a negative risk factor for coronary heart disease (CHD) and stroke.

Patient Preparation Many laboratories require a 9- to 14-hour fast. However, no significant differences in apo-A-I result between fasting and nonfasting subjects have been reported.

Specimen Serum or plasma

Container Red top tube or lavender (EDTA) top tube

Storage Instructions Separate serum or plasma and refrigerate or freeze according to the instructions provided by the individual testing laboratory.

Population Distributions for Apolipoprotein A-I (mg/dL)

Age (y)	Percentile						
	Male						
	5	10	25	50	75	90	95
12-20	107	121	128	142	154	169	191
22-29	102	113	116	135	150	174	184
30-39	104	110	119	134	150	161	176
40-49	103	108	119	132	146	161	171
50-59	102	107	119	131	145	159	172
60-69	100	105	116	132	150	168	183
70-79	92	104	114	128	144	175	202
	Female						
12-20	97	112	127	144	166	179	188
22-29	107	117	128	137	152	177	220
30-39	107	119	132	149	160	182	199
40-49	114	121	134	149	168	191	202
50-59	116	122	135	153	174	198	211
60-69	119	124	137	152	170	193	212
70-79	116	118	140	154	173	186	189

Data for ages 12-20 obtained from African-American adolescents (Rifai et al, 1996) and data for ages 22-79 obtained from white adults (Contois et al, 1996).

Adapted from Rifai N, Bachorik PS, and Albers JS, "Lipids, Lipoproteins and Apolipoproteins," *Tietz Textbook of Clinical Chemistry*, 3rd ed, Burtis CA and Ashwood ER, eds, Philadelphia, PA: WB Saunders Co, 1999, 828.

Reference Interval For this analyte, high coronary risk is associated with low serum levels. See Lipids, Overview *on page 853* and previous table.

Use Apo A-I assays are primarily of use in investigating patients with low HDL cholesterol concentrations (eg, familial apo A-I deficiency, Tangier disease, and LCAT deficiency).

Limitations Apo A-I measurement may provide no advantage over high density lipoprotein cholesterol (HDLC) measurement for prediction of coronary risk: the topic is controversial.

Methodology Analytical methods include immunonephelometry, immunoturbidimetry, enzyme-linked immunosorbent assay (ELISA), radioimmunoassay (RIA), and radical immunodiffusion (RID). Current commercial apo A-I assays are standardized with WHO-IFCC reference materials. Automation has become available for this assay. Addition of Tween 20 has been advocated to enhance diagnostic discrimination for CHD.

References
Bachorik PS, Lovejoy KL, Carroll MD, et al, "Apolipoprotein B and AI Distributions in the United States, 1988-1991: Results of the National Health and Nutrition Examination Survey III (NHANES III)," *Clin Chem*, 1997, 43(12):2364-78.

Contois JH, McNamara JR, Lammi-Keefe CJ, et al, "Reference Intervals for Plasma Apolipoprotein A-I Determined With a Standardized Commercial Immunoturbidimetric Assay: Results From the Framingham Offspring Study," *Clin Chem*, 1996, 42(4):507-14.

♦ **Apolipoprotein B** *see* Apolipoprotein B-100, Serum *on page 203*

Apolipoprotein B-100, Serum

Related Information
Apolipoprotein A-I, Serum *on page 202*
Apolipoprotein E, Plasma *on page 204*
Cholesterol, Total, Serum or Plasma *on page 396*
High Density Lipoprotein Cholesterol, Serum *on page 729*
Lipid Panel, Serum *on page 852*
Lipids, Overview *on page 853*
Lipoprotein (a), Serum *on page 861*
Low Density Lipoprotein Cholesterol *on page 874*
Triglycerides, Serum or Plasma *on page 1275*

Synonyms Apo B; Apolipoprotein B

Abstract Apolipoprotein B-100 (apo B-100) is an important constituent of the following lipoproteins: VLDL, IDL, LDL, and Lp(a). Apo B-100 is synthesized in the liver and secreted into plasma as part of the VLDL particle. Apo B-100 participates in the delivery of cholesterol to the tissues and interacts directly with the LDL receptor. Serum Apo B-100 is a positive risk factor for coronary heart disease. See Lipids, Overview *on page 853*.

Patient Preparation Many laboratories require a 9- to 14-hour fast. However, no significant differences in apo-B-100 results between fasting and nonfasting subjects have been reported.

Specimen Serum or plasma

Container Red top tube or lavender (EDTA) top tube

Storage Instructions Separate serum or plasma and refrigerate or freeze according to the instructions provided by the individual testing laboratory.

Reference Interval See Lipids, Overview *on page 853* and the following table.

Use Apo B-100 measurement is used in the evaluation of risk for coronary heart disease and in the diagnosis of abetalipoproteinemia, hypobetalipoproteinemia, and hyperabetalipoproteinemia.

Limitations Plasma lipoproteins change significantly during cardiac catheterization and measurements shortly thereafter should be avoided.

Methodology Immunonephelometry, immunoturbidimetry, enzyme-linked immunosorbent assay (ELISA), radioimmunoassay (RIA), and radical immunodiffusion (RID). Current commercial apo B-100 assays are standardized with WHO-IFCC reference materials.

Additional Information Unlike the LDL cholesterol calculated by the Friedewald equation, apo B-100 measurement does not require a fasting specimen and is useful for patients with high triglyceride levels.

(Continued)

Apolipoprotein B-100, Serum *(Continued)*

Population Distributions for Apolipoprotein B-100 (mg/dL)

Age (y)	Percentile						
	5	10	25	50	75	90	95
12-20	M: 59 F: 55	M: 61 F: 60	M: 75 F: 74	M: 86 F: 90	M: 108 F: 104	M: 122 F: 121	M: 140 F: 138
22-29	M: 54 F: 59	M: 60 F: 61	M: 68 F: 69	M: 84 F: 80	M: 98 F: 97	M: 107 F: 107	M: 109 F: 114
30-39	M: 57 F: 50	M: 64 F: 57	M: 76 F: 66	M: 91 F: 75	M: 107 F: 88	M: 123 F: 99	M: 133 F: 110
40-49	M: 66 F: 58	M: 73 F: 63	M: 88 F: 73	M: 104 F: 87	M: 118 F: 103	M: 134 F: 119	M: 144 F: 130
50-59	M: 68 F: 66	M: 78 F: 72	M: 90 F: 84	M: 105 F: 99	M: 120 F: 116	M: 134 F: 137	M: 146 F: 150
60-69	M: 69 F: 72	M: 78 F: 79	M: 90 F: 90	M: 106 F: 103	M: 120 F: 120	M: 134 F: 137	M: 145 F: 145
70-79	M: 74 F: 78	M: 77 F: 79	M: 86 F: 92	M: 104 F: 106	M: 123 F: 120	M: 133 F: 136	M: 144 F: 142

Data for ages 12-20 obtained from African-American adolescents (Rifai et al 1996) and data for ages 22-79 obtained from white adults (Contois et al 1996, 43).

Adapted from Rifai N, Bachorik PS, and Albers JS, "Lipids, Lipoproteins and Apolipoproteins," *Tietz Textbook of Clinica Chemistry*, 3rd ed, Burtis CA and Ashwood ER, eds, Philadelphia, PA: WB Saunders Co, 1999, 828.

References

Albers JJ and Marcovina SM, "Standardization of Apolipoprotein B and A-I Measurements," *Clin Chem*, 1989, 35(7):1357-61.

Contois JH, McNamara JR, Lammi-Keefe CJ, et al, "Reference Intervals for Plasma Apolipoprotein A-I Determined With a Standardized Commercial Immunoturbidimetric Assay: Results From the Framingham Offspring Study," *Clin Chem*, 1996, 42(4):507-14.

Apolipoprotein E, Plasma

Related Information

AD7c Neural Thread Protein, CSF or Urine *on page 106*
Apolipoprotein A-I, Serum *on page 202*
Apolipoprotein B-100, Serum *on page 203*
Cerebrospinal Fluid and Plasma β-Amyloid$_{(1-42)}$ *on page 359*
Cerebrospinal Fluid Protein *on page 371*
Cerebrospinal Fluid Protein Electrophoresis *on page 374*
Cholesterol, Total, Serum or Plasma *on page 396*
High Density Lipoprotein Cholesterol, Serum *on page 729*
Lipid Panel, Serum *on page 852*
Lipids, Overview *on page 853*
Lipoprotein (a), Serum *on page 861*
Low Density Lipoprotein Cholesterol *on page 874*
Triglycerides, Serum or Plasma *on page 1275*

Synonyms APOE; Apo E; Apoprotein E

Applies to βA4 Peptide; Clusterin (Apo J); Neurofibrillary Tangles; Tau Protein

Test Includes ε2, ε3, and ε4

Abstract ApoE is a ligand for the LDL and apoE receptors, and has a major role in the clearance of apoB-containing lipoproteins, especially chylomicrons, VLDL, and VLDL remnants. Deficient or defective apoE usually presents as an increase in apoB-containing lipoproteins, especially VLDL and chylomicron-containing remnants. There are multiple molecular forms of apoE. This polymorphism is expressed with the common alleles, ε2, ε3, and ε4. Apoε4 predisposes to coronary artery disease (CAD) and to Alzheimer dementia.

Specimen Plasma

Container Lavender top (EDTA) tube

Storage Instructions Store at 2°C to 8°C. Do not freeze.

Causes for Rejection Hemolysis, heparinized blood collected, specimen frozen

Special Instructions Test will usually be performed by a reference laboratory. Schedule patient sample requirements, handling, and timing with laboratory.

Reference Interval Variation in methodology has resulted in wide variation of reported **mean** values in healthy individuals. A reference interval should be supplied by the laboratory performing the test.

Use This test is used to evaluate abnormal lipid metabolism associated with coronary artery disease, study predisposing factors to the development of Alzheimer disease (AD), and aid in the differential diagnosis of AD.

Methodology Total ApoE: radioimmunoassay (RIA), immunonephelometry, radial immunodiffusion (RID), and electroimmunoassay. ApoE identification (phenotyping) can be performed using polyacrylamide gel isoelectric focusing (PAGE-IEF) or PAGE-IEF of plasma followed by immunoblotting or immunofixation. ApoE genotyping is generally performed by restriction enzyme isoform genotyping (restriction isotyping). "MADGE" (microplate array diagonal gel electrophoresis) should allow lower cost and large scale apoE genotyping.

Additional Information Apolipoprotein E plays a variety of roles in lipid metabolism with increasing evidence of significance in both vascular and central nervous system (Alzheimer) degenerative disease.

ApoE and Vascular Disease

ApoE mediates removal of chylomicron and very low density lipoprotein (VLDL) remnants from plasma by binding these particles to low density lipoprotein (LDL).

The apoε2/ε2 phenotype (apoε3 deficient) is associated with familial dysbetalipoproteinemia in which there is chylomicronemia, increased plasma triglyceride, cholesterol, and β-VLDL. The ε2/ε2 genotype is uncommon (about 1% of individuals) and, as additional risk factors are involved, only 2% to 5% of ε2/ε2 individuals develop type III hyperlipidemia. The ε4 allele is present in 10% to 15% of individuals in the population and, in contrast to the ε3 allele, is over-represented in association with coronary heart disease. The lowest cholesterol levels are associated with ε3/ε3 phenotype. There is compelling evidence that apoE polymorphism is a major determinant of risk for development of atherosclerotic vascular disease and its complications. Demand for apoE genotype and/or phenotype determination is likely to increase, in particular, if therapeutic modulation of the effect of apoε4 is developed.

ApoE and Alzheimer Disease

There have been many reports confirming an association between Alzheimer disease (AD) and apoε4. The allele frequency of apoε4 is increased in patients with late-onset AD and in cases of sporadic AD. A review by Mulder et al of recent developments in markers for AD is available.

References

Davignon J, Gregg RE, and Sing CF, "Apolipoprotein E Polymorphism and Atherosclerosis," *Arteriosclerosis*, 1988, 8(1):1-21.

Jacobs DS, DeMott WR, Oxley DK, et al, *Laboratory Test Handbook*, 5th ed, Hudson, OH: Lexi-Comp Inc, 2001.

Mulder C, Scheltens P, Visser JJ, et al, "Genetic and Biochemical Markers for Alzheimer's Disease: Recent Developments," *Ann Clin Biochem*, 2000, 37(Pt 5):593-607.

Roses AD, "Apolipoprotein E Genotyping in the Differential Diagnosis, Not Prediction, of Alzheimer's Disease," *Ann Neurol*, 1995, 38(1):6-14.

♦ **Apoprotein E** *see* Apolipoprotein E, Plasma *on page 204*

Apoptosis Assays

Related Information

bcl-2 Gene Rearrangement *on page 251*
Flow Cytometry, Overview *on page 592*
p53, Functional Assay/Sequencing *on page 995*
Retinoblastoma Gene DNA Detection *on page 1159*
White Blood Cell Count *on page 1330*

Applies to Annexin V (Annexin A5); Caspase Enzymes; CD95; FAS Receptor; TUNEL Assay

Abstract Apoptosis, a genetically controlled mechanism of cell death distinct from necrosis, has specific morphologic characteristics. The process is of critical import to embryogenesis, lymphocyte development, and thus, immunity, and to autoimmunity, inflammation, neoplasia, and the cellular response to toxins (eg, chemotherapeutic drugs and ionizing radiation). Numerous approaches are available to test for apoptosis and its resultant biologic effects, including morphologic observation, flow cytometry, and molecular analysis. (Continued)

Apoptosis Assays *(Continued)*

Specimen Specimen requirements are method dependent (see Methodology).

Special Instructions Confer with laboratory consultants concerning appropriate application and availability of different assays.

Use Cell, tissue, and molecular-based assays to estimate the characteristics of apoptosis in a broad range of pathologic and developmental processes. Assays for apoptosis can be applied to monitor therapy for malignant disease, to assess apoptotic change in neural cells (as may occur in neurodegenerative disorders), to evaluate transplant rejection, and to evaluate apoptotic change in fibroblast activity in the regulation of tissue repair (ie, as in the control of scar formation). Clinical usefulness is not limited to these examples.

Limitations Methods must be well controlled (and may not have been in some, if not many, instances) in order to avoid producing the process (apoptosis) that one is attempting to study (see reference by Zhang et al).

Methodology Apoptosis is a multistep process (see Additional Information) which has allowed the development and application of a surprising number of laboratory assays, each usually targeting a specific stage and/or event of programmed cell death. Distinct morphologic changes include condensation of nuclear chromatin, shrinkage of cytoplasm, membrane bleb formation, and formation of apoptotic bodies.

Morphologic observation, annexin V (annexin A5) binding, and TUNEL (terminal dUTP nick end labeling) are especially commonly employed.

Apoptosis - identification by morphologic features: In cultured cells there is loss of adhesion to substrate with cell rounding, surface bleb formation, cell shrinkage with cessation of blebbing, membrane deterioration, nuclear matrix degeneration with chromatin condensation, and formation of condensed apoptotic cell bodies.

Quantitative Assessment of Apoptosis Using Flow Cytometry

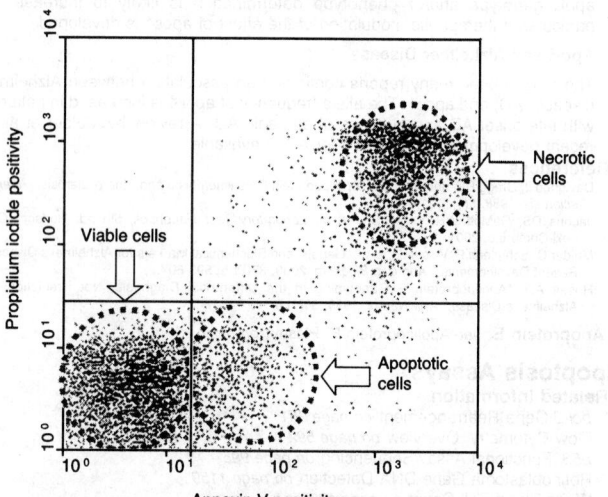

The illustration reflects activated peripheral blood lymphocytes that have been stained with propidium iodide (PI) or annexin V and subjected to fluorescence activated cell sorting. Each dot represents an individual cell. Viable cells stain with neither PI nor annexin V. Apoptotic cells stain with annexin V only, whereas necrotic cells stain with both.

Adapted from Afford S and Randhawa S, "Apoptosis," *Mol Pathol*, 2000, 53(2):55-63.

Criteria for detection of apoptotic leukocytes in peripheral blood smears:

- cell shrinkage
- peripheral condensation of chromatin along the nuclear membrane, nuclear fragmentation, formation of cytoplasmic blebs, membrane-bound apoptotic bodies, and cytoplasmic vacuoles
- persistence of specific granules and plasma membrane

Annexin V (annexin A5) is an anticoagulant protein which can bind to phosphatidyl serine. The latter, usually hidden, is externally exposed as a result of apoptosis. Annexin A5 is then able to bind and can be detected by flow cytometry. Plasma levels of annexin A5 are anticoagulant-related (higher in EDTA vs citrated anticoagulated samples). This variation is under evaluation and may involve the existence (and function) of polymorphisms but in particular the role of calcium in the binding of annexin A5 to negatively-charged phospholipids. See diagram.

The **TUNEL** (*in situ* terminal deoxynucleotidyl transferase d uridine triphosphate nick end labeling) assay and the **INSEL** (*in situ* DNA end labeling) methods are based on the detection of DNA fragmentation. Both techniques reflect endonuclease dependent cleavage of DNA by labeling nick end DNA with a specific complex conjugated to a chromogen or fluorogen.

Leukocyte viability by light scatter: A rapid and simple method for quantifying apoptosis using the Abbott CD4000 hematology analyzer has been proposed. Based on cellular light scatter properties with the demonstration of membrane fragility (as occurs with apoptosis) this method has been reported to correlate well with TUNEL (slide based) and annexin V (by flow cytometry). In addition to multiangle cellular light scatter data the Cell-Dyn 4000 measures red fluorescence emitted by cells that have been labeled with propidium iodide, with generation of a leukocyte viability index.

Additional Information Apoptosis, a term first used in 1972, refers to the process of programmed cell death. It is a critical end stage of cellular homeostasis and is highly regulated. It is a coordinated mechanism of cellular death morphologically distinct from necrosis. Malignancy, autoimmunity and immunodeficiency may develop in relation to abnormalities in the regulation of apoptosis.

Operationally, apoptosis can be conveniently divided into four steps:

- initiation
- monitoring and decision
- execution
- removal

Initially, there is interaction between the cell surface and its environment, mediated importantly by the FAS receptor (CD95). FAS may then activate the second phase of apoptosis, involving the intracellular proteins of the Bcl-2 family. Within this family are molecules that block (Bcl-2, Bcl-x1) and some that promote (BAX, Bcl-xs and BAD) apoptosis. These proteins are situated on the outer mitochondrial surface, act as pore forming units and share protein dimerization sequences. The other Bcl-2 family proteins on the mitochondrial surface represent the decision mechanism and control the fate of the cell (survival or death). A positive death decision results in the release of cytochrome C and other factors from mitochondria into the cell cytoplasm. In the third step (execution), cytochrome C activates the caspase family of cysteine proteinase enzymes that lead to a broad range of intracellular events resulting in the dismantling of the cell. The importance of the caspase family of enzymes was developed by Yuan and Horvitz who showed that the protein Ced-3 was essential for the execution of apoptosis. Ced-3 came to be known as a member of the caspase family. Mutations of Ced-3 that prevented its catalytic activity also prevented apoptosis. Caspase-1, a cytoplasmic protease, activates interleukin-1B, a proinflammatory cytokine, and was originally identified as interleukin-1B converting enzyme (ICE). At present there are 14 known caspase family members in the cascade that modulates apoptosis. DNA is "cut" by the caspases resulting in a characteristic "laddering" pattern of DNA fragments on electrophoretic study, one of the laboratory modalities for investigation of apoptosis. In the final step, phagocytic macrophages, as a result of signaling (by caspases) at the cell surface are attracted and remove the apoptotic dead (Continued)

Apoptosis Assays *(Continued)*

cell body, completing the death process without inciting a true inflammatory reaction.

A fraction of circulating tumor-specific DNA and RNA appear to have origin from apoptotic or necrotic cells. Messenger RNA from apoptotic tumor cells remains stable in serum within apoptotic bodies, protected from degradation by circulating ribonucleases.

References

Badley AD, Pilon AA, Landay A, et al, "Mechanisms of HIV-associated Lymphocyte Apoptosis," *Blood*, 2000, 96(9):2951-64.

Cotran RS, Kumar V, and Collins TC, et al, eds, "Cellular Pathology I: Cell Injury and Cell Death," *Robbins Pathologic Basis of Disease*, 6th ed, Chapter 1, Philadelphia, PA: WB Saunders Company, 1999, 18-25.

Cotter FE, "Laboratory Assessment of Apoptosis," *Clin Lab Haematol*, 1997, 19(4):289-320.

Eichhorst ST and Krammer PH, "Derangement of Apoptosis in Cancer," *Lancet*, 2001, 358:345-6.

Hasselman DO, Rappl G, Tilgen W, et al, "Extracellular Tyrosinase nRNA Within Apoptotic Bodies Is Protected From Degradation in Human Serum," *Clin Chem*, 2001, 47(8):1488-9.

McKenney CA, Romzek MR, and Ziemba SE, "Apoptosis - When Cells Die," *Lab Med*, 1999, 30(12):791.

Méhes G, Witt A, Kubista E, et al, "Circulating Breast Cancer Cells Are Frequently Apoptotic," *Am J Pathol*, 2001, 159(1):17-20.

Mentz F, Baudet S, Maloum K, et al, "Quantification of Apoptosis by the Abbott CD4000 Hematology Analyzer," *Hematol Cell Ther*, 1998, 40(5):183-8.

Shidham VB and Swami VK, "Evaluation of Apoptotic Leukocytes in Peripheral Blood Smears," *Arch Pathol Lab Med*, 2000, 124(9):1291-4.

van Heerde WL, Kenis H, Schoormans S, et al, "The -1C>T Mutation in the *Annexin* A5 Gene Does Not Affect Plasma Levels of Annexin A5," *Blood*, 2003, 101(10):4223-5.

Willingham MC, "Cytochemical Methods for the Detection of Apoptosis," *J Histochem Cytochem*, 1999, 47(9):1101-10.

Zhang X, Chen J, Davis B, et al, "Hoechst 33342 Induces Apoptosis in HL-60 Cells and Inhibits Topoisomerase I *In Vivo*," *Arch Pathol Lab Med*, 1999, 123(10):921-7.

♦ **A-Poxide**® *see* Chlordiazepoxide, Serum *on page 391*

♦ **Aprobarbital** *see* Barbiturates, Quantitative, Serum or Plasma *on page 248*

♦ **Apsolol**® *see* Propranolol, Serum *on page 1096*

Apt-Downey Test

Related Information

Fetal Hemoglobin *on page 581*

Synonyms Fetal Hemoglobin Test in Newborn

Abstract The Apt-Downey test (Apt test) uses alkali denaturation of fetal hemoglobin to determine if blood present in the stool of a newborn is the result of swallowing maternal blood or is due to GI hemorrhage.

Specimen Blood stained diaper, grossly bloody (red) stool, or bloody vomitus or mucus

Container Use clean glass or plastic container for specimen or send blood stained diaper.

Causes for Rejection Specimen is not grossly bloody or there is evidence of melena/coffee ground aspirate (*vide infra*).

Turnaround Time 1-2 hours

Reference Interval Report will provide indication if blood is of maternal or infant origin (adult or fetal hemoglobin).

Use Diagnose swallowed blood syndrome and differentiate this condition from gastrointestinal hemorrhage in the newborn

Limitations The specimen must be grossly bloody, red, not tarry. Test performed in cases of melena or with coffee ground material (denatured blood) may produce a false-positive result as oxyhemoglobin has been converted to hematin and may be falsely read as adult Hb. Visual judgment of color produced by test procedure may lead to error if only a small amount of blood is present. Bilirubin containing meconium and possibly other substances may cause stool color interference. Use of a spectrophotometric-based or high performance liquid chromatography (HPLC) procedure may avoid these problems.

Methodology Dissolved blood (one volume of bloody stool or vomitus mixed with five volumes of water) is treated with 1% NaOH, 1-4 mL of hemolysate (alkali denaturation test). The mixture is then centrifuged at 2000 rpm for 1-2 minutes. Fetal hemoglobin resists elution by alkali, and the solution will remain pink. Maternal blood will be converted to alkaline hematin in 1-2 minutes, and the solution becomes yellow to dark green-brown. Thus, if the supernatant remains pink (indicative of fetal blood), additional clinical investigation must be pursued. The newborn's blood should be tested concurrently as a control to exclude the possibility of adult Hb in the test infant.

Additional Information In the swallowed blood syndrome, blood or bloody stools are passed usually on the second or third day of life. The blood may be swallowed during delivery or may be from a fissure of the mother's nipple. This condition must be differentiated from gastrointestinal hemorrhage of the newborn. The test is based on the fact that the infant's blood contains >60% fetal hemoglobin which is alkali resistant. Swallowed blood of maternal origin contains adult hemoglobin which is converted to brownish alkaline hematin on the addition of alkali. A sensitive and accurate spectrophotometric-based procedure has been developed.

References

Chen D, Wilhite TR, Smith CH, et al, "HPLC Detection of Fetal Blood in Meconium: Improved Sensitivity Compared With Qualitative Methods," *Clin Chem*, 1998, 44(11):2277-80.

Lehman CM and Baron B, "Apt Testing," Q & A, *Lab Med*, 2000, 31(7):365-6.

Ogur G, Gül D, Özen S, et al, "Application of the Apt Test in Prenatal Diagnosis to Evaluate the Fetal Origin of Blood Obtained by Cordocentesis: Results of 30 Pregnancies," *Prenat Diagn*, 1997, 17(9):879-82.

◆ **APTT** *see* Activated Partial Thromboplastin Time *on page 100*

◆ **Argatroban** *see* Activated Partial Thromboplastin Time *on page 100*

◆ **Argatroban** *see* Heparin-Induced Thrombocytopenia *on page 695*

◆ **Argatroban** *see* Mixing Studies (Coagulation) *on page 918*

◆ **Argatroban** *see* Prothrombin Time *on page 1116*

◆ **Arginine** *see* Cystine, Urine *on page 494*

◆ **Arginine⁸-Vasopressin** *see* Antidiuretic Hormone, Plasma *on page 172*

◆ **Arginine-Vasopressin** *see* Antidiuretic Hormone, Plasma *on page 172*

◆ **Arsenate** *see* Arsenic, Urine *on page 211*

Arsenic, Blood

Related Information

Arsenic, Hair, Nails *on page 210*
Arsenic, Urine *on page 211*
Heavy Metal Screen, Blood *on page 668*
Heavy Metal Screen, Urine *on page 669*

Synonyms As

Applies to Hair Analysis; Heavy Metal Screen, Arsenic

Abstract Arsenic is a toxic heavy metal. It exists in various forms. Arsine gas, As^{3+}, and As^{5+} are toxic forms; organic forms are not much less toxic. Arsenic is found in soil and rocks. Major sources of human exposure are arsenic in food resulting from broad use of arsenical insecticides and from drinking water, especially well water. Acute arsenic toxicity follows accidental ingestion, industrial accidents, suicide or homicide. For children, 2 mg/kg body weight can cause lethal arsenic poisoning. Adamsite and Lewisite, war gases of World War I, are arsenic compounds. In certain areas of the United States, fresh water supplies contain up to 1.4 mg/L arsenic, substantially in excess of the acceptable limit of 0.01 mg/L.

Container Trace metal-free certified EDTA tubes

Collection See Blood Collection Methods for Trace Elements in the Trace Elements Introduction *on page 77*.

Causes for Rejection Containers not metal-free

Reference Interval <70 µg/L (SI: 0.93 µmol/L)

Critical Values Poisoning: ≥100 µg/L (SI: ≥1.33 µmol/L)

Use Blood arsenic is for the diagnosis of acute poisoning only; use urine, hair, or nails for chronic poisoning. Acute arsenic poisoning may be signaled by the abrupt onset of vomiting and diarrhea.

Limitations Short half-life in blood, 4-6 hours. **Serum or plasma is the least useful specimen, except in acute poisoning.**

Methodology Inductively-coupled plasma-mass spectrometry (ICP-MS), electrothermal atomic absorption spectrometry (AA)

Additional Information
- Half-life: 4-6 hours
- Volume of distribution: 0.2 L/kg

(Continued)

209

Arsenic, Blood *(Continued)*

Arsenic can be absorbed through the gastrointestinal tract, by inhalation, and by penetration of the skin.

Arsine gas (AsH_3), combining with the globin chain in red cells, causes hemolysis with hemoglobinuria and hematuria. Acute renal failure may cause death.

Arsenic intoxication causes hypotension, tachycardia, conduction blocks, dysrhythmias, changes of mental status, rhabdomyolysis, pulmonary edema, encephalopathy, seizures, neuropathy, hepatic and renal dysfunction, hemolytic anemia, and bone marrow toxicity. Arsenic trioxide has provided remission of acute promyelocytic leukemia. Cell Therapeutics Inc. issued a "Dear Health Care Provider" letter reminding clinicians that QTc prolongation with torsade de pointes arrhythmia and sudden death have been associated with the use of arsenic trioxide.

References

Campbell BG, "Broadsheet Number 48: Mercury, Cadmium and Arsenic: Toxicology and Laboratory Investigation," *Pathology*, 1999, 31(1):17-22.

Graeme KA and Pollack CV, "Heavy Metal Toxicity, Part I: Arsenic and Mercury," *J Emerg Med*, 1998, 16(1):45-56.

Lazo G, Kantarjian H, Estey E, et al, "Use of Arsenic Trioxide (As2O3) in the Treatment of Patients With Acute Promyelocytic Leukemia: The M. D. Anderson Experience," *Cancer*, 2003, 97(9):2218-24.

Ratnaike RN, "Acute and Chronic Arsenic Toxicity," *Postgrad Med J*, 2003, 79(933):391-6.

Internet Web Sites

www.fda.gov/medwatch/safety/2001/Trisenox_letter.pdf

♦ **Arsenic, Gastric Content** *see* Arsenic, Urine *on page 211*

Arsenic, Hair, Nails

Related Information

Arsenic, Blood *on page 209*
Arsenic, Urine *on page 211*
Heavy Metal Screen, Blood *on page 668*
Heavy Metal Screen, Urine *on page 669*

Synonyms As^{3+} (As III); As^{5+} (As V); As, Quantitative

Applies to Hair Analysis; Mee's Lines

Abstract Chronic toxicity manifests cutaneous changes, including alopecia, hyperkeratosis, hyperpigmentation, and tumors including basal and squamous cell carcinoma. See Arsenic, Blood *on page 209*.

Specimen Clean hair or nails, ≥0.5 g

Container Clean envelope or heavy metal-free screw top plastic container

Collection Extreme care is necessary to avoid surface contamination. Hair from the nape of the neck indicates more recent ingestion, while axillary or pubic hair provides evidence of earlier exposure (6-12 months before). Toenails are preferable to fingernails; they are less prone to surface contamination.

Special Instructions Hair should be clean, free of oil and tonic; clip close. Nails should be thoroughly washed, dried, and clipped close to cuticle.

Reference Interval Up to 1 µg/g (SI: 0.13 nmol/g)

Critical Values Values >100 µg/g (SI: >13.4 nmol/g) of hair are considered toxic

Use Diagnose chronic arsenic exposure and intoxication. Complications of chronic exposure include carcinomas of skin; associations with carcinomas of lung and with transitional cell carcinoma of bladder and kidney.

Limitations Hairs and nails do not detect recent exposure.

Methodology Inductively-coupled plasma-mass spectrometry (ICP-MS) is useful in arsenic speciation, electrothermal atomic absorption spectrometry (AA). ICP-MS is useful in arsenic speciation.

Additional Information Chronic exposure to arsenic commonly involves insecticides, industrial sources, or contamination of food, water, soil, or medications. Evidence of chronic arsenic poisoning may be manifested 2-8 weeks after ingestion. A 2003 study shows that arsenic speciation in fingernails correlates better with arsenism than that in hair.

References

Gallagher RE, "Arsenic - New Life for an Old Potion," *N Engl J Med*, 1998, 339(19):1389-91.

Mandal BK, Ogra Y, and Suzuki KT, "Speciation of Arsenic in Human Nail and Hair From Arsenic-Affected Area by HPLC-Inductively Coupled Argon Plasma Mass Spectrometry," *Toxicol Appl Pharmacol*, 2003, 189(2):73-83.

Seidel S, Kreutzer R, Smith D, et al, "Assessment of Commercial Laboratories Performing Hair Mineral Analysis," *JAMA*, 2001, 285(1):67-72.

Arsenic, Urine

Related Information
Arsenic, Blood *on page 209*
Arsenic, Hair, Nails *on page 210*
Heavy Metal Screen, Blood *on page 668*
Heavy Metal Screen, Urine *on page 669*

Synonyms As^{3+} (As III); As^{5+} (As V); As, Quantitative, Urine

Applies to Arsenate; Arsenic, Gastric Content; Arsenite; Dimethylarsine; DMA; MMA; Monomethylarsine

Test Includes Organic forms of arsenic can also be evaluated.

Abstract This toxic heavy metal appears in urine and stools. Its excretion rate in urine is used to determine toxicity. Arsenic exists in various inorganic and organic forms, of which arsine gas, As^{3+}, and As^{5+} are most toxic. As^{3+} and As^{5+} are partially detoxified to monomethylarsine (MMA) and dimethylarsine (DMA) and excreted in urine.

Patient Preparation Patient should avoid seafood for 48 hours before collection is begun. Organic arsenic may be found especially in shellfish, cod, and haddock. Seafood may contain arsenic as high as 10 mg/lb.

Specimen 24-hour urine

Container Acid-washed plastic container, no preservative, no metal cap or insert

Storage Instructions Refrigerate or freeze.

Reference Interval Ranges for urine arsenic levels can be variable among different laboratories. A general guideline is given: normal: <120 µg/L (SI: >1.59 µmol/L); chronic exposure: 100-200 µg/L (SI: 1.3-2.6 µmol/L).

Critical Values Toxic: >850 µg/L (SI: >11.3 µmol/L).

Use Evaluate recent exposure to arsenic, arsenic toxicity

Limitations Spot levels, if normal, may not rule out arsenic poisoning. Seafood, particularly shellfish, can increase urinary As to as much as 2000 µg/L.

Methodology Atomic absorption spectrometry (AA), inductively-coupled plasma-mass spectrometry (ICP-MS). Chromatography before analysis is used to distinguish between organic, nontoxic forms and inorganic, toxic forms.

Additional Information 25 mL acidified gastric washing is acceptable for arsenic analysis; gastric content normally contains no arsenic. Random urine samples are acceptable.

Arsenic is radiopaque. Abdominal x-rays may prove helpful, but usually are not.

References
Graeme KA and Pollack CV, "Heavy Metal Toxicity, Part I: Arsenic and Mercury," *J Emerg Med*, 1998, 16(1):45-56.
Peters GR, McCurdy RF, and Hindmarsh JT, "Environmental Aspects of Arsenic Toxicity," *Crit Rev Clin Lab Sci*, 1996, 33(6):457-93.

♦ **Arsenite** *see* Arsenic, Urine *on page 211*
♦ **Arterial-Ascitic Fluid pH Gradient** *see* Body Fluid pH *on page 295*

Arterial Blood Collection

Related Information
Blood Gases and pH, Arterial *on page 275*
Oxygen Saturation, Blood *on page 991*
pCO$_2$, Blood *on page 1008*
pH, Blood *on page 1018*

Synonyms Arterial Puncture

Applies to Allen Test

Test Includes Brachial, radial, or femoral artery puncture by trained personnel to obtain arterial blood, most frequently for blood gas analysis

Patient Preparation The patient should be resting for 20-30 minutes before collection of the specimen.

Aftercare Direct pressure must be applied to the arterial puncture site and should be maintained for a minimum of 10 minutes. Patients with bleeding tendency due to anticoagulation, platelet deficiency, factor deficiency, or liver disease may bleed excessively and form a hematoma. Such patients should be monitored carefully after the procedure to ascertain that bleeding has been (Continued)

Arterial Blood Collection *(Continued)*

controlled. Arterial spasm preventing aspiration of the specimen and thrombosis of the punctured artery can occur.

Specimen Arterial blood

Container Heparinized syringe with 21- or 23-gauge needle. Alternatively, a 21- or 23-gauge Butterfly® infusion set may be used. Glass syringes are preferred over plastic syringes, as gases can dissolve in plastic.

Collection The experienced arterial puncturist should carefully select an appropriate artery. If the radial artery is used, **Allen's test** to assure collateral circulation to the hand from the ulnar artery is performed: the hand is closed tightly by the patient or by an assistant to form a fist. Pressure is then applied at the wrist, compressing and obstructing both the radial and ulnar arteries. The hand is then opened (but not fully extended), revealing a blanched palm and fingers. The obstructing pressure is next removed from only the ulnar artery while the palm and fingers, including thumb, are observed; they should become flushed within 15 seconds as the blood from the ulnar artery refills the empty capillary bed. If the ulnar artery does not adequately supply the entire hand (a negative Allen test), the radial artery should not be used as a puncture site; an alternate artery should be selected.

Careful preparation of the puncture site is performed with 70% alcohol (isopropanol) using a circular motion working out from the site. Dry with gauze or let it air dry. The artery is stabilized by holding with a finger. Take care not to contaminate the puncture site. The artery is punctured at a 30° angle for the radial artery, 45° angle for the brachial, 45° or 90° angle for the femoral. The bevel of the needle or Butterfly® should be pointed toward the direction of blood flow. The syringe should fill spontaneously. Small bore needles or plastic syringes may require gentle slow suction. Be sure no air bubbles are aspirated into the syringe. After adequate sample volume is obtained, quickly remove the needle and apply pressure. See Aftercare. The detailed technique for arterial puncture is described in NCCLS Standard H11-A2. Place specimen on ice after sealing the needle into a piece of hard rubber or plastic. Deliver to the laboratory within 15 minutes of collection.

Storage Instructions Keep the specimen air tight and water tight in a container of ice.

Causes for Rejection Clots in specimen, specimen left at room temperature for more than 10 minutes (if collected for blood gas analysis)

Special Instructions Provide patient's temperature on the laboratory requisition if specimen is collected for blood gases.

Use Obtain arterial blood for analysis. See Blood Gases and pH, Arterial *on page 275*.

Limitations Arterial puncture should be performed by persons familiar with the procedure and the potential complications. For example, forearm compartment syndrome occurs after percutaneous arterial blood sampling and is usually associated with anticoagulation therapy.

Additional Information The evaluation of alveolar pO_2 is routinely done by assuming that the respiratory gas exchange ratio is equal to 0.8. A study of the respiratory gas exchange ratio in patients undergoing arterial puncture revealed that in ~25% of cases, there is a transient change in alveolar ventilation associated with arterial puncture that may cause a change in the gas exchange ratio and lead to at least a 10 mm Hg error in estimating alveolar pO_2. Evidently, some patients respond to arterial puncture by transient breath holding or by taking rapid shallow breaths. Arterialized capillary blood is addressed in Skin Puncture Blood Collection *on page 1203*.

References

Cinel D, Markwell K, Lee R, et al, "Variability of the Respiratory Gas Exchange Ratio During Arterial Puncture," *Am Rev Respir Dis*, 1991, 143(2):217-8.

National Committee for Clinical Laboratory Standards, "Percutaneous Collection of Arterial Blood for Laboratory Analysis," Approved Standard, H11-A2, 2nd ed, Wayne, PA: National Committee for Clinical Laboratory Standards, 1992.

Pardo M and Burkhardt DH, "Preventing Complications of Arterial Puncture During Central Venous Catheter Insertion," *Crit Care Med*, 2000, 28(7):2677-8.

Safran MR, Bernstein A, and Lesavoy MA, "Forearm Compartment Syndrome Following Brachial Arterial Puncture in Uremia," *Ann Plast Surg*, 1994, 32(5):535-8.

♦ **Arterial Blood Gases** *see* Blood Gases and pH, Arterial *on page 275*

♦ **Arterialized Capillary Blood** *see* Skin Puncture Blood Collection *on page 1203*

♦ **Arterial Puncture** *see* Arterial Blood Collection *on page 211*

Arthropod Identification

Related Information

Babesiosis Serology *on page 225*
Cerebrospinal Fluid Analysis: Overview *on page 355*
Ehrlichiosis Serology *on page 531*
Lyme Disease DNA Detection *on page 878*
Lyme Disease Serology *on page 879*
Q Fever Serology *on page 1128*
Rocky Mountain Spotted Fever Serology *on page 1171*
Tularemia Diagnostic Procedures *on page 1279*

Synonyms Insect Identification

Applies to Ants; Beetles; Centipedes; *Cimex* Identification; Crustaceans; *Dermacentor andersoni*; Entomology; Flea Identification; Flies; Hemipterae; Hymenopterae; *Ixodes dammini* Identification; *Latrodectus hasselti*; Lepidopterae; Lice Identification; Millipedes; Mite Identification; Nits Identification; *Pediculus humanus* Identification; *Phthirus pubis* Identification; *Sarcoptes scabiei* Skin Scrapings Identification; Scorpions; Skin Scrapings for *Sarcoptes scabiei* Identification; Spiders; Tick-Borne Diseases; Tick Identification

Abstract Arthropoda is a phylum which includes *Arachnida* and *Insecta*, among other classes. Species include parasites and vectors. *Arachnida* includes spiders, ticks, mites, and scorpions. *Hymenoptera* include ants, bees, wasps, and hornets.

Tick-borne illnesses include ehrlichiosis, Lyme disease, babesiosis, relapsing fever, tularemia, Rocky Mountain spotted fever, and tick typhus.

Specimen Gross arthropod, skin scrapings

Collection Arthropods (gross) are to be submitted in alcohol (70%) or formaldehyde in tube or container with secure closure. To establish the diagnosis of scabies, skin scrapings may be collected with a scalpel and a drop of mineral oil. The liquid may be examined directly, or alternatively the organism may be teased away from its burrow or papule with a needle or scalpel.

Storage Instructions Maintain specimen at room temperature. Fill the container with preservative as completely as possible to avoid damage to the specimen by air bubbles in the container.

Special Instructions Correct identification of vectors can play a role in clinical diagnosis, especially in instances in which definitive diagnosis may be difficult (eg, Lyme disease). Assistance of individuals trained in entomology in specimen identification may be pivotal.

Use Identify arthropods affecting man; establish the presence of ectoparasite infestation

Limitations Ticks are distinguished by four pairs of legs, but antennae of some insects can be confused with legs.

Methodology Macroscopic evaluation

Additional Information See graphic.

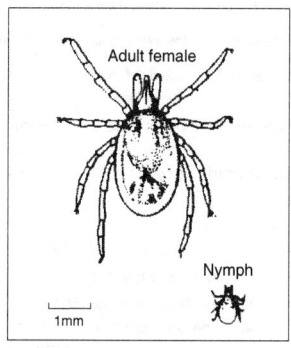

Deer tick (*Ixodes dammini*)

(Continued)

Arthropod Identification *(Continued)*

Lyme disease and **babesiosis** may be transmitted to humans by two species of ticks. *Ixodes scapularis* (or *I. dammini*) has been implicated along the Eastern seaboard, and in Southern and North Central states. *I. pacificus* is the most common vector in Western states. In Europe, **Lyme disease** is most often transmitted to humans by *I. ricinus. Ehrlichia* sp, known to cause human **ehrlichiosis**, is carried by the tick *Amblyomma americanum*. Other diseases in which ticks are vectors include **Colorado tick fever, Rocky Mountain spotted fever, tick typhus, Q fever, tularemia**, and **babesiosis**.

The only vector of louse-borne **relapsing fever** is the body louse. The other variety, endemic relapsing fever, is tick-borne.

Tick paralysis, caused by an engorged, attached tick, presents as ascending paralysis. Prompt recovery follows tick removal (see Cerebrospinal Fluid Analysis: Overview *on page 355*). The differential diagnosis includes myasthenia gravis, botulism, or Guillain-Barré syndrome. *Dermacentor andersoni* (the North American wood tick), *D. variabilis* (the common dog tick), and *Ixodes holocyclus* affect humans. *Dermacentor* ticks can harbor *Rickettsia* and transmit Rocky Mountain spotted fever.

Scabies is a skin disease in which the itch mite, *Sarcoptes scabiei*, bores into the stratum corneum. Mites act as vectors of **Western equine encephalitis, St Louis encephalitis, murine typhus**, and rickettsialpox. Lice are vectors of **typhus, trench fever**, and **epidemic (louse-borne) relapsing fever**. Mosquitoes are vectors of **malaria, filariasis, viral encephalitis, dengue fever**, and **yellow fever**. Deer flies are vectors of **loiasis** and **tularemia**. Black flies are vectors of **onchocerciasis**. Tsetse flies are vectors of **trypanosomiasis**. Sand flies are vectors of **Leishmaniasis bartonellosis** and **sand fly fever**. Cone-nose (reduvid) (triatomid) bugs are vectors of **Chagas disease**. Fleas are vectors of **plague, murine typhus**, *Rickettsia felis, Dypilidium caninum, Hymenolepsis diminuta*, and *Bartonella henselae*.

An Australian study of 750 definite spider bites concludes that most bites have minor effects. Significant clinical problems occurred in 6% of cases (including 37 of 56 "redback" - *Latrodectus hasselti* - widow bites). These caused significant pain for over 24 hours, 11% received antivenom.

An outstanding atlas we recommend provides useful, extensive tables, maps, transmission cycle diagrams, and color plates of human disease as well as of arthropods, including eggs, nymphs, larvae, and adults (Peters, 1992).

References

Falco RC, Fish D, and D'Amico V, "Accuracy of Tick Identification in a Lyme Disease Endemic Area," *JAMA*, 1998, 280(7):602-3.

Felz MW, Smith CD, and Swift TR, "A Six-Year-Old Girl With Tick Paralysis," *N Engl J Med*, 2000, 342(2):90-4.

Fix AD, Strickland GT, and Grant J, "Tick Bites and Lyme Disease in an Endemic Setting: Problematic Use of Serologic Testing and Prophylactic Antibiotic Therapy," *JAMA*, 1998, 279(3):206-10.

Isbister GK and Gray MR, "A Prospective Study of 750 Definite Spider Bites, With Expert Spider Identification," *Q J Med*, 2002, 95(11):723-31.

Peters W, *A Colour Atlas of Arthropods in Clinical Medicine*, London, England: Wolfe Publishing Ltd, 1992.

Schaumburg HH and Herskovitz S, "The Weak Child - A Cautionary Tale," *N Engl J Med*, 2000, 342(2):127-9.

Wu ML, Warren DJ, and Jones VA, "Ticked Off: *Ixodes*," *Arch Pathol Lab Med*, 2000, 124(6):925.

♦ **Arylamidase** *see* Leucine Aminopeptidase (LAP), Serum and Urine *on page 844*

♦ **Arylamidase Naphthylamidase** *see* Leucine Aminopeptidase (LAP), Serum and Urine *on page 844*

♦ **As** *see* Arsenic, Blood *on page 209*

♦ **As³⁺ (As III)** *see* Arsenic, Hair, Nails *on page 210*

♦ **As³⁺ (As III)** *see* Arsenic, Urine *on page 211*

♦ **As⁵⁺ (As V)** *see* Arsenic, Hair, Nails *on page 210*

♦ **As⁵⁺ (As V)** *see* Arsenic, Urine *on page 211*

♦ **ASA** *see* Salicylate, Serum or Plasma *on page 1176*

Asbestos, Lung or Sputum

Related Information

Body Cavity Fluid Cytology *on page 285*
Bronchoalveolar Lavage (BAL) *on page 311*

Test Includes Precise weight of lung required - 5 grams is the standard sample. Lung tissue or sputum is digested, which may require 24 hours.

Abstract Asbestiform minerals include chrysotile, crocidolite, amosite, tremolite, anthophyllite, and actinolite.

Ferruginous bodies are baton-shaped refractile rods onto which iron salts have precipitated. They are strongly associated with asbestosis. Iron stain may be helpful in detection.

Patient Preparation If sputum sample is collected, advise patient not to remove fixative from container.

Specimen Fresh tissue, tissue fixed in appropriate fixative for lung biopsy, single or continuous multiple expectorated sputum samples

Container Wide-neck jar with appropriate fixative for lung biopsy. If fresh tissue is obtained, submit in sterile gauze premoistened with saline, enclosed in a plastic bag. Wide-mouth jar with Saccomanno fixative is used for expectorated sputa.

Collection Lung biopsies collected during procedures should be placed directly into fixative or wrapped in moistened gauze and delivered immediately. Sputa samples can be collected at the patient's bedside or a container with fixative can be sent home with the patient for collection.

Storage Instructions Specimen should be delivered to the Cytology Laboratory as soon as possible. Unfixed tissue should be either refrigerated or processed immediately. Weight of submitted tissue specimen should be recorded prior to processing. Appropriate documentation is needed.

Use It was recognized in 1960 that many subjects with mesothelioma had asbestos exposure.

Asbestos is considered responsible for 4000-6000 annual deaths from carcinoma of lung, including all histopathologic subtypes. A synergistic and multiplicative effect is postulated for smoking and exposure to asbestos.

Additional Information Cases are often the subject of legal proceedings.

References

Colby TV, Koss MN, and Travis WD, "Tumors of the Lower Respiratory Tract," *Atlas of Tumor Pathology*, Third Series, Fascicle 13, Washington, DC: Armed Forces Institute of Pathology, American Registry of Pathology, 1995.

DeMay R, *The Art and Science of Cytopathology-Exfoliative Cytology*, Chicago, IL: ASCP Press, American Society of Clinical Pathologists, 1996.

Hammar SP, "Pleural Diseases," *Pulmonary Pathology Tumors*, Chapter 5, Dar DH, Hammar SP, and Colby TV, eds, New York, NY: Springer-Verlag, 1995, 405-522.

♦ **Aschheim-Zondek Test** *see* Chorionic Gonadotropin, Human, Serum and Urine *on page 397*

♦ **Ascitic Fluid Analysis** *see* Body Fluid Analysis, Cell Count *on page 288*

♦ **Ascitic Fluid Analysis** *see* Body Fluid Chemical Analysis *on page 291*

♦ **Ascitic Fluid Cytology** *see* Body Cavity Fluid Cytology *on page 285*

♦ **Ascitic Fluid/Serum:Total Bilirubin Ratio** *see* Bilirubin, Total, Serum *on page 265*

♦ **Ascorbate** *see* Ascorbic Acid, Serum or Plasma *on page 215*

Ascorbic Acid, Serum or Plasma

Related Information

Antioxidant Concentrations, Plasma *on page 192*
Oxalate, Urine *on page 989*
Urinalysis *on page 1289*

Synonyms Ascorbate; Vitamin C

Abstract Decreased plasma levels indicate nutritional deficiency. An individual on a diet totally deficient in vitamin C will develop scurvy in 60-90 days. Vitamin C has a half-life of about 16 days. If vitamin C ingestion is suddenly stopped, steady state concentrations of 55-60 μmol/L will probably serve to prevent deficiency for a month.

Patient Preparation Patient should be fasting.

Specimen Plasma or leukocytes; uncommonly, urine following a loading test
(Continued)

Ascorbic Acid, Serum or Plasma *(Continued)*

Container Green top (heparin) tube preferred; red top tube, lavender top (EDTA) tube, or gray top (sodium fluoride) tube also acceptable; check with the laboratory.

Collection Draw blood in chilled tube. Keep specimen on ice.

Storage Instructions Keep specimen refrigerated until frozen. Freeze separated plasma. Stable 30 minutes at 25°C. Stable 4 days at -20°C.

Causes for Rejection Specimen not frozen

Reference Interval Plasma: 0.6-2.0 mg/dL (SI: 34-114 µmol/L); leukocytes: 20-50 µg/10³ WBC. Lassitude (fatigue) appears at concentrations <20 µmol/L.

Use Evaluate vitamin C deficiency. Principal clinical findings in scurvy include bleeding gums, petechiae, follicular hyperkeratosis, perifollicular hemorrhages beginning on the lower thighs, muscle aches, easy fatiguability, effusions, and emotional changes. The metabolic product of hypervitaminosis C, oxalate, may lead to oxalate renal calculi. Safe vitamin C intake is reported to be 1 g/day.

Methodology High performance liquid chromatography (HPLC) with coulometric electrochemical detection, acidic 2,4-dinitrophenylhydrazine with photometry. HPLC is a preferred method; spectrophotometric methods tend to overestimate low concentrations.

Additional Information The recommended dietary allowance (RDA), proposed in 1999, is 120 mg/day. The tolerable upper intake is proposed to be <1 g/day. Plasma or serum levels of vitamin C are an adequate measurement of clinical status, although leukocyte levels are superior but more difficult to obtain. Smokers have lower levels than nonsmokers. The disease scurvy and its treatment, from Tierra del Fuego in 1519 to the present, represents a tapestry of naval and world history. The defeat of the French and Spanish at Trafalgar and the successful blockade of the French at Brest may have been supported by the interest in prevention and treatment of scurvy by the British Admiralty, limited as that interest may have been. When the British controlled the Caribbean, their seamen were provided with limes to prevent scurvy, leading to the expression "limey". Vitamin C is a very important antioxidant and its possibly uncertain role in the prevention of atherogenesis is frequently cited.

Vitamin C promotes small intestinal iron absorption 1.5-10-fold.

Vitamin C, not bound to plasma proteins, is dialyzable. Those on dialysis require replacements.

Intake ≥250 mg of vitamin C causes false-negative results on stool guaiac testing.

References

Ausman LM, "Criteria and Recommendations of Vitamin C Intake," *Nutr Rev*, 1999, 57(7):222-4.
Cuppage FE, *James Cook and the Conquest of Scurvy*, Westport, CT: Greenwood Press, 1994.
Levine M, Rumsey SC, Daruwala R, et al, "Criteria and Recommendation for Vitamin C Intake," *JAMA*, 1999, 281(15):1415-23.
Margolis SA and Duewer DL, "Measurement of Ascorbic Acid in Human Plasma and Serum: Stability, Intralaboratory Repeatability, and Interlaboratory Reproducibility," *Clin Chem*, 1996, 42(8 Pt 1):1257-62.

Aspartate Aminotransferase, Serum

Related Information

Synonyms AST; Glutamic Oxaloacetic Transaminase, Serum; GOT; SGOT; Transaminase

Applies to Acetaminophen Hepatotoxicity; Aminotransferases; AST:ALT Ratio; AST:LD Ratio; L-Aspartate-2-Oxoglutarate Aminotransferase

Abstract AST and ALT, widely used enzymes, are increased in diseases of the liver and in many other disease entities. The differential diagnosis of isolated AST increase includes hemolysis, myopathy, and macro-AST as well as alcohol or drug-related liver damage. AST is found in cardiac and skeletal muscle and in lesser amounts in kidneys, brain, lungs, pancreas, spleen, white cells, and erythrocytes. See Table 2 in Liver Biopsy *on page 864* and the table in Liver Disease: Laboratory Assessment, Overview *on page 869*. The aminotransferases (transaminases) are extensively used by virtue of their ease of assay and extensive clinical experience with their concentrations. The sensitivity of AST is widely recognized.

Patient Preparation Avoid strenuous exercise before sampling.

Specimen Serum; plasma may be used.

Container Red top tube; green top (heparin) tube for plasma

Storage Instructions Stable 3 days at 25°C and 1 week at 4°C; fairly stable refrigerated or frozen.

Reference Interval Levels in infancy are two to three times those found in adults. Intervals decrease during childhood years.
- Newborns: 25-75 units/L (SI: 0.43-1.28 µkat/L)
- Infants: 15-60 units/L (SI: 0.26-1.02 µkat/L)
- Adult male: 20-40 units/L (SI: 0.34-0.68 µkat/L)
- Adult female: 15-30 units/L (SI: 0.25-0.51 µkat/L)

Use A wide range of disease entities alters AST (SGOT). When an increased AST is from the liver, it is likely to relate to disease of the **hepatocyte**. Other enzymes, including alkaline phosphatase and gamma-glutamyl transferase (GGT), are more sensitive indicators of **biliary** obstruction.

Causes of low AST: uremia, vitamin B$_6$ deficiency.

Causes of high AST: chronic alcohol ingestion, not limited to overt chronic alcoholism; cirrhosis. In alcoholic hepatitis, AST values usually are <250 units/L. AST is distributed in both cytosol and mitochondria, while serum activity of ALT is predominantly related to cytosol. The damage from alcohol is predominantly to mitochondria; therefore, the increase of AST is greater than that of ALT. Pyridoxine deficiency is commonplace in alcoholism. ALT is more sensitive to pyridoxine deficiency than is AST; thus, the **AST:ALT ratio** is driven higher; it is >2.0 in alcoholic hepatitis. AST is rarely >8 times the upper reference interval in subjects with alcohol abuse. The AST:ALT ratio is generally less helpful in chronic diseases of the liver, but has a role in assessment of the patient with chronic hepatitis C, in whom AST:ALT ratio of one or more provides suggestion of cirrhosis.

Most instances of toxic necrosis lead to ALT as high or higher than those of AST.

In viral hepatitis, look for high **AST:LD ratio**, >3, and very high AST peaking at 500-3000 units/L in acute viral hepatitis (ie, in clinical acute viral hepatitis the transaminases may be increased 10 times or more above their upper limits of normal). See Hepatitis: Laboratory Assessment, Overview *on page 713*

AST increases are found in other types of liver disease, including earlier stages of hemochromatosis and chemical injury. Some instances of cholecystitis cause increased AST. AST increase, not greater than fivefold, is found early in choledocholithiasis. AST can increase 10-fold with cholangitis.
(Continued)

Aspartate Aminotransferase, Serum *(Continued)*

Transaminases increase abruptly and may exceed 10,000 IU/L in ischemia, as in subjects in congestive heart failure with hypotension.

Transaminases can be increased with nonalcoholic steatohepatitis, which often accompanies obesity, diabetes mellitus, jejunoileal bypass, with amiodarone, total parenteral nutrition, and without recognized relationship. ALT may be equal to or more than AST in nonalcoholic steatohepatitis.

Elevated aminotransferase concentrations may be found in asymptomatic patients with Wilson disease, for which the initial screening test is serum ceruloplasmin.

AST and ALT are increased in Reye syndrome. In infectious mononucleosis, LD is commonly considerably higher than AST. Trauma (including head trauma and surgery) and skeletal muscle diseases, including dystrophy, dermatomyositis, trichinosis, polymyositis, and gangrene cause AST increases. Both AST and ALT elevations are found with Duchenne muscular dystrophy. Look for high CK in myositis, with high LD_5 (or isomorphic pattern in some instances of polymyositis) on LD isoenzymes. CK and aldolase are at least as sensitive to diseases of skeletal muscle.

In myocardial infarction (MI) AST peaks about 24 hours after infarct but it is no longer used for this purpose. In acute MI without shock or heart failure, ALT is not apt to increase significantly. AST increases in congestive failure with centrilobular liver congestion, in which high LD_5 is found, and in pericarditis, myocarditis, pancreatitis, and other inflammatory states including Legionnaires' disease. In renal infarction LD is usually high, out of proportion to AST. Lung infarction and other disease entities leading to necrosis including large, necrotic tumors cause increased AST; LD is commonly also increased in such instances. Shock (LD also usually increased); hypothyroidism (LD and/or CK not infrequently increased in myxedema); hemolytic anemias (LD high with increased LD_1) and certain CNS diseases may increase AST. AST may be increased with nonbiliary sepsis. There may be moderate increase with thyroid disease.

Drugs: Some drugs may cause **decreases** in AST, including allopurinol, cyclosporine, progesterone, and others. A large number of commonly used drugs have been reported to **elevate** AST, some of them by particular chemistry methods, but many by pathophysiologic means. They include acetaminophen, aminosalicylic acid, amiodarone, amitriptyline, anabolic steroids, anticonvulsants, ascorbic acid, aspirin, carbamazepine, cephalosporins, chlorambucil, chloroform, chlorothiazide, chlorpromazine, conjugated estrogens, cyclosporine, diclofenac, erythromycin, fluconazole, gentamicin, haloperidol, halothane, hydralazine, ibuprofen, indomethacin, interferon alpha$_2$, isoniazid, isoproterenol, levodopa, lovastatin, mefenamic acid, meprobamate, methotrexate, methyldopa, methyltestosterone, metronidazole (certain methods), naproxen, niacin (large doses), nonsteroidal anti-inflammatory agents, nortriptyline, opiates, oral contraceptives, oxacillin, papaverine, penicillamine, penicillin, phenobarbital, phenothiazines, procainamide, progesterone, propylthiouracil, pyrazinamide, quinidine, rifampin, ritonavir, streptomycin, sulfonamides, tamoxifen, ticarcillin, tobramycin, tocainide, tolbutamide, valproic acid, verapamil, certain over-the-counter medications, herbal/alternative preparations, and others. See Alanine Aminotransferase, Serum *on page 116*.

Acetaminophen hepatotoxicity deserves special mention. In alcoholics, apparently moderate doses of acetaminophen have caused severe hepatotoxicity. Doses of 2.6-16.5 g/24 hours are reported with total bilirubin 1.3-23.9 mg/dL (SI: 22-409 µmol/L), AST 1960-29,700 units/L, and ALT 12,000-12,550 units/L. The characteristic findings include mild to severe coagulopathy and AST greater than ALT by a considerable margin.

Macroenzyme (macro-AST) causing unexplained increase of AST is found with normal levels of CK and ALT. Macro-AST is complexed with an immunoglobulin, thus, not readily cleared.

See the following diagram.

Approach to Patient With Elevated Aspartate Aminotransferase (AST)

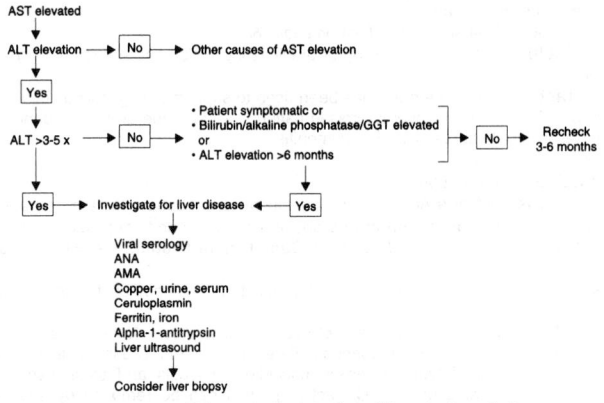

ALT = alanine aminotransferase, AMA = antimitochondrial antibody, ANA = antinuclear antibody.
Modified from Kamath PS, "Clinical Approach to the Patient With Abnormal Liver Test Results," *Mayo Clin Proc,*
1996, 71:1089-95.

Limitations Gross hemolysis causes falsely high values. Specific diagnoses require more specialized studies. Serum enzymes are relatively insensitive to uncomplicated hepatic steatosis. ALT is more specific for liver but both enzymes and others are helpful. Use of several enzymes together is recommended. (See Liver Disease: Laboratory Assessment, Overview *on page 869*).

AST can be falsely low following dialysis. It varies moderately day to day and with exercise, and may be 15% higher in African-American males.

Additional Information Very high values, >500 units/L, usually suggest hepatitis, ischemia, or other types of hepatocellular necrosis but can also be found with large necrotic tumors, congestive failure, and shock. Unexplained AST elevations should first be investigated with ALT and GGT.

AST >3x the upper limit of reference interval supports diagnosis of the **HELLP** syndrome.

Laboratory findings supportive of the diagnosis of **acute liver failure** include high aminotransferases, low glucose concentrations, and evidence of respiratory alkalosis.

References

Green RM and Flamm S, American Gastroenterological Association, "AGA Technical Review on the Evaluation of Liver Chemistry Tests," *Gastroenterology,* 2002, 123(4):1367-84.

Pratt DS and Kaplan MM, "Evaluation of Abnormal Liver-Enzyme Results in Asymptomatic Patients," *N Engl J Med,* 2000, 342(17):1266-71.

Stone JH, "HELLP Syndrome: Hemolysis, Elevated Liver Enzymes, and Low Platelets," *JAMA,* 1998, 280(6):559-62.

Whitehead MW, Hawkes ND, Hainsworth I, et al, "A Prospective Study of the Causes of Notably Raised Aspartate Aminotransferase of Liver Origin," *Gut,* 1999, 45(1):129-33.

Young DS, *Effects of Drugs on Clinical Laboratory Tests,* 5th ed, Volume 1: Listing by Test, Washington, DC: AACC Press, American Association of Clinical Chemistry, 2000, Section 3, 93-110.

Zimmerman HJ, *Hepatotoxicity: The Adverse Effects of Drugs and Other Chemicals on the Liver,* 2nd ed, Baltimore, MD: Lippincott Williams & Wilkins, 1999.

♦ ***Aspergillus niger / fumigatus* Precipitating Antibodies** *see* Hypersensitivity Pneumonitis Serology *on page 761*

Aspergillus Serology

Related Information

Bronchial Brushings/Washings Cytology *on page 310*
Fungal Culture, Biopsy or Body Fluid *on page 619*
Fungal Culture, Blood *on page 620*
Fungal Culture, Sputum *on page 624*
Hypersensitivity Pneumonitis Serology *on page 761*
(Continued)

Aspergillus Serology *(Continued)*

Immunoglobulin E *on page 773*

Lymphocyte Transformation Test *on page 882*

Test Includes Complement fixing or precipitating antibodies specific for *Aspergillus*

Abstract *Aspergillus* serology has been used to support a diagnosis of invasive aspergillosis in immunocompromised patients, and as one of the components in testing for hypersensitivity pneumonitis.

Specimen Serum

Container Red top tube

Special Instructions Acute and convalescent serum specimens are desirable.

Reference Interval Immunodiffusion: no precipitin bands detected; positive: at least one precipitin band detected. **Complement fixation:** titer <1:8 or less than fourfold increase.

Use Confirm the presence of serum precipitating antibodies to *Aspergillus* species

Limitations A negative test does not rule out aspergillosis. Nonspecific precipitin bands could be due to presence of C-reactive protein. Cross reactions may occur in cases of histoplasmosis, coccidioidomycosis and blastomycosis. Bands due to reaction with C-reactive protein can be removed by sodium citrate.

The value of complement fixing antibodies in the diagnosis of pulmonary aspergillosis is not established.

Methodology Immunodiffusion (ID), complement fixation (CF), enzyme-linked immunosorbent assay (ELISA)

Additional Information Aspergillosis immunodiffusion: Sera can be tested against a polyvalent antigen mixture, or a series of species preparations. The greater the number of bands, the greater the likelihood of either a fungus ball or invasive aspergillosis. *Aspergillus* precipitins are seen in 90% of patients with **fungus balls** (aspergillomas), 70% of patients with **allergic bronchopulmonary aspergillosis**, and less often in patients with **invasive aspergillosis**, who usually are immunocompromised, have been on long-term corticosteroids and antibiotics, and often fail to produce detectable antibodies. Invasive pulmonary aspergillosis is a subacute pneumonia, in the periphery of which a thrombosed vessel may be found. Hematogenous dissemination takes place in ~25% of cases.

References

Ho PL and Yuen KY, "Aspergillosis in Bone Marrow Transplant Recipients," *Crit Rev Oncol Hematol*, 2000, 34(1):55-69.

Ledesma D and Pearce WH, Images in Clinical Medicine, "Septic (*Aspergillus*) Embolus," *N Engl J Med*, 2000, 342(14):1015.

Patterson TF, Kirkpatrick WR, White M, et al, "Invasive Aspergillosis. Disease Spectrum, Treatment Practices, and Outcomes. I3 *Aspergillus* Study Group," *Medicine (Baltimore)*, 2000, 79(4):250-60.

Stevens DA, Kan VL, Judson MA, et al, "Practice Guidelines for Diseases Caused by *Aspergillus*. Infectious Diseases Society of America," *Clin Infect Dis*, 2000, 30(4)696-709.

♦ **Atrial Natriuretic Peptide** *see* B-Type Natriuretic Peptide *on page 314*

♦ **Auramine-Rhodamine Stain** *see* Acid-Fast Stain, Routine or Modified *on page 95*

♦ **Auranofin** *see* Gold, Serum *on page 657*

♦ **Aurothioglucose** *see* Gold, Serum *on page 657*

Autohemolysis Test

Related Information

Anemia Flowchart *on page 35*
Glucose-6-Phosphate Dehydrogenase, Blood *on page 641*
Hemosiderin Stain, Urine *on page 692*
Hypertonic Cryohemolysis Test *on page 762*
Osmotic Fragility *on page 980*
Osmotic Fragility, Incubated *on page 982*
Peripheral Blood: Red Blood Cell Morphology *on page 1016*
Pyruvate Kinase Assay, Erythrocytes *on page 1127*
Red Blood Cell Enzyme Deficiency *on page 1134*

Abstract Autohemolysis test measures the degree to which patient's red cells lyse without additives, with glucose, and with ATP. The test has some application to the differential diagnosis of hemolytic states, but is rarely used and is of limited value. The hypertonic cryohemolysis test may be more effective than the autohemolysis test. (See Hypertonic Cryohemolysis Test *on page 762*.)

Specimen Defibrinated sterile blood

Container Sterile syringe

Collection Using sterile technique, 25 mL of blood is drawn and immediately defibrinated by swirling in a bottle which contains glass beads.

Storage Instructions Specimen is taken immediately to the laboratory, blood defibrinated, and tubes prepared for incubation.

Causes for Rejection Specimen hemolyzed, clotted, or more than 5 minutes in transit; specimen contaminated with bacteria (as from a patient with septicemia)

Special Instructions Defibrinated blood is usually collected by laboratory personnel.

Reference Interval Percent red cell lysis at 48 hours: blood alone: 0.2% to 2.0%; blood incubated with glucose: 0.0% to 0.9%; blood incubated with ATP: 0.5% to 2.5%

Use Diagnosis of **hereditary spherocytosis** (HS); detect conditions producing spontaneous hemolysis, particularly hereditary spherocytosis; categorize RBC enzyme deficiencies; evaluate hemolytic anemia

Limitations Large sample of blood required and must be obtained by a trauma-free venipuncture. Test lacks sensitivity and specificity. While this test may still find application in the diagnosis of hereditary spherocytosis, it has largely been supplanted by specific enzyme spot assays for the diagnosis of nonspherocytic congenital hemolytic anemia.

Contraindications Bacteremic patients

Methodology The test must always be run with and compared to a control. Test should be run in duplicate to possibly allow detection of reagent inactivity or bacterial contamination of specimen or reagent. It is difficult to maintain sterility and full ATP activity.

Additional Information The findings in **G6PD deficiency**, **PK deficiency**, and **hereditary spherocytosis** are summarized in the table. **Hereditary elliptocytosis** has normal autohemolysis. **Hereditary pyropoikilocytosis** has

Autohemolysis Test

Condition	Incubation at 37°C for 48 Hours	Incubation + 10% Glucose	Incubation + ATP
Normal	0.2%-2.0%	0.0%-0.9%	0.5%-2.5%
G6PD deficiency	3.0%-5.0%	0.2%-2.0%	0.2%-2.0%
Pyruvate kinase deficiency	12%-16%	12%-16%	0.2%-2.0%
Hereditary spherocytosis	12.0%-15.0%	3.0%-5.0%	3.0%-5.0%

(Continued)

Autohemolysis Test *(Continued)*

increased autohemolysis, not corrected by glucose. **Triosephosphate isomerase deficiency** has an abnormal result which corrects completely with glucose or ATP. The findings in other red cell disorders have been described (Jacobs et al).

References

Cochran DL and Burnside LK, "Detecting and Identifying Hereditary Pyropoikilocytosis," Clinical Pathology Rounds, *Lab Med*, 1999, 30(1):26-9.

Dacie JV and Lewis SM, *Practical Haematology*, 8th ed, New York, NY: Churchill Livingstone, 1995, 222-5.

Gallagher PG, Forget BC, and Lux SE, "Disorders of the Erythrocyte Membrane," *Hematology of Infancy and Childhood*, 5th ed, Chapter 16, Nathan DG and Orkin SH, eds, Philadelphia, PA: WB Saunders Co, 1998, 594-6.

Jacobs DS, DeMott WR, Oxley DK, et al, *Laboratory Test Handbook*, 5th ed, Hudson, OH: Lexi-Comp Inc, 2001, 405-6.

Thornburg CD and Ware RE, "The Utility of the Autohaemolysis Test for Children With Congenital Haemolytic Anaemia," *Clin Lab Haematol*, 2003, 25(1):25-8.

♦ **Autoimmune Hemolytic Anemia Work-up** *see* Antibody Detection/Identification, Red Cell *on page 165*

♦ **Autoimmune Lymphoproliferative Syndrome** *see* CD4/CD8 Enumeration *on page 349*

Autologous Transfusion, Intraoperative Blood Salvage

Related Information

Autologous Transfusion, Preoperative Deposit *on page 223*

Synonyms Blood Salvage, Intraoperative Autologous Transfusion

Applies to Normovolemic Hemodilution; Postoperative Cell Salvage

Abstract A patient should have the option to receive his/her own blood, when possible. The use of autologous blood provides many advantages, including preventing transfusion-transmitted disease and preventing alloimmunization, in addition to supplementing a tenuous blood supply.

Special Instructions Collection and recovery services require the coordinated efforts of a multidiscipline team composed of surgeons, anesthesiologists, transfusion medicine specialists, and personnel trained in the use of the equipment.

Limitations Certain transfusion-transmitted diseases may not be prevented, such as bacterial contamination. The risk of bacterial contamination may be lower than with preoperative autologous donation. This type of transfusion is likely to be more costly than transfusion of allogeneic blood. Air embolism is a potential risk in intraoperative cell salvage.

Methodology Although **preoperative** autologous blood collection is the best-known way of avoiding a transfusion of homologous blood, a patient's own blood may be saved and returned **during** the operation or immediately **afterwards**.

Three procedures are available.

1. **Intraoperative hemodilution (acute normovolemic hemodilution):** Blood is withdrawn into standard blood bags containing anticoagulant after the induction of anesthesia but before the surgical incision is made. Crystalloids or colloids can support the blood volume. Autologous units collected during acute normovolemic hemodilution (ANH) are returned at the conclusion of the surgical procedure (within 8 hours if stored at room temperature and 24 hours if stored in a monitored refrigerator). Recombinant erythropoietin has been used to increase its effectiveness.
   - Patients should be selected for ANH if the following criteria are met: the likelihood of transfusion exceeds 10%; the preoperative hemoglobin is at least 12 g/dL; lack of severe hypertension; lack of infection/bacteremia; and lack of clinically significant coronary, pulmonary, renal, or liver disease. ANH can be considered equivalent to preoperative autologous donation in knee and hip replacement and radical prostatectomy.
   - This procedure may diminish the need for allogeneic transfusions. The blood procured requires no testing, diminishing costs. Since the units removed remain in the operating room, mixup of units is diminished. Risk of bacterial contamination is almost eliminated.

2. **Intraoperative cell salvage:** This procedure describes the technique of collecting and reinfusing blood lost by a patient during surgery, for procedures such as liver transplantation, repair of scoliosis, cardiac and vascular surgery. Many instruments collect shed blood, washing it in normal saline and concentrating the red blood cells. Most of the plasma and platelets are removed. Devices that neither wash nor concentrate blood prior to infusion increase the risk of adverse effects. A specially trained operator is required. Other systems usually involve collection of shed blood into a container from which it is returned to the patient through a filter. Without a wash step, such systems risk transfusion of thromboplastic debris admixed with the collected blood. Other risks include aspiration of tumor cells, the presence of infectious agents, or other constraints. The procedure is contraindicated in the presence of bacterial contamination; the equivalent of at least two units of blood must be recovered to reach cost-effective levels. Its use in the presence of bacterial contamination and malignant disease is a contraindication in the British Committee for Standards in Hematology (BCSH) Guidelines for Cell Salvage.

3. **Postoperative cell salvage:** Blood from mediastinal drainage and from other sterile operative sites is collected and reinfused, with or without washing. Reinfusion must be initiated within 6 hours of collection or the blood must be discarded. Recovered blood is dilute and somewhat hemolyzed. The quantities salvaged are often too small to be cost-effective. The safety and benefit of postoperative cell salvage have been questioned.

References

Duguid JK, "Autologous Blood Transfusion," *Clin Lab Haematol*, 1999, 21(6):371-6.

Goodnough LT, "Acute Normovolemic Hemodilution," *Vox Sang*, 2002, 83(Suppl 1):211-5.

Goodnough LT, Brecher ME, Kanter MH, et al, "Transfusion Medicine. Second of Two Parts - Blood Conservation," *N Engl J Med*, 1999, 340(7):525-33.

Santrach P, *Standards for Perioperative Autologous Blood Collection and Administration*, 1st ed, Bethesda, MD: American Association of Blood Banks, 2001.

"Transfusion Alert: Use of Autologous Blood," National Heart, Lung, and Blood Institute Expert Panel on the Use of Autologous Blood, *Transfusion*, 1995, 35(8):703-11.

Autologous Transfusion, Preoperative Deposit

Related Information

Autologous Transfusion, Intraoperative Blood Salvage *on page 222*
Donation, Blood *on page 521*
Donation, Blood, Directed *on page 523*
Erythropoietin, Serum *on page 551*

Synonyms Autotransfusion; Predeposit Autologous Donation; Transfusion, Autologous

Applies to Cryopreservation; Erythropoietin; Normovolemic Hemodilution

Test Includes ABO group and Rh type of the unit are performed on the donor blood and recipient sample. If the unit will be transfused outside of the collecting facility, tests for hepatitis B surface antigen, anti-HIV-1, anti-HIV-2, anti-HCV, antihepatitis B core, anti-HTLV-I, anti-HTLV-II, nucleic acid testing for detection of HIV-1, HCV, and West Nile virus, and a serologic test for syphilis must be performed on the first unit of blood collected within a 30-day period. Each unit must be labeled "autologous donor" and a special label "for autologous use only" is required. In addition, a "biohazard" label is required if any of the above tests are confirmed positive, or found repeat reactive only and confirmatory testing is not completed. The patient and patient's physician must be informed of any abnormal test results.

Abstract This procedure includes removal of blood or components from a donor/patient for subsequent autologous transfusion. Autologous transfusion alleviates concern about the safety and availability of blood for transfusion. Blood collected from a patient is reserved for that patient for elective surgery. There is no risk of transmission of hepatitis, HIV, or other donor-related infectious diseases, nor reaction to serum proteins or red cell antigens. But risks do exist, both during donation (donor reactions, anemia of donation) and subsequent reinfusion (identification mixup, bacterial contamination, volume overload). Not all patients are suitable candidates.

Categories of autologous programs:
- preoperative phlebotomy (the principal type discussed in this listing)
- immediate preoperative phlebotomy with hemodilution, also known as acute normovolemic hemodilution

(Continued)

Autologous Transfusion, Preoperative Deposit
(Continued)

- intraoperative cell salvage
- postoperative salvage

Note that combinations of the above techniques may be used to limit the patient's exposure to allogeneic transfusion.

Patient Preparation Patients should generally be in good health prior to donation. It has generally been recommended that supplemental iron usually should be prescribed by the patient's primary physician and administered **prior to** collection of the first unit.

Aftercare Continued replacement of iron is important. Increase fluid intake on the days of donation. Immediately after donation of blood, the patient/donor should remain lying down for a few minutes. When moved to a chair **with assistance**, the donor/patient should remain seated for 15-30 minutes and offered a drink (eg, orange juice and often cookies).

Container Each unit must be labeled "autologous donor" and a special label "for autologous use only" is required. In addition, a "biohazard" label is required if any of the above tests are confirmed positive, or found repeat reactive only and confirmatory testing is not completed.

Collection Documented requests for autologous collection must be received from the patient's primary physician. Sufficient numbers of units should ideally be collected prior to surgery to avoid exposure to allogeneic blood. Two-unit collections via red cell apheresis may be an option. Collections should also be scheduled far enough in advance of surgery to avoid anemia, through compensatory erythropoiesis.

The final collection should occur no sooner than 72 hours prior to scheduled surgery.

Adverse donor reactions are 12 times as high as those associated with voluntary donations.

Storage Instructions Keep blood unit(s) in a monitored Blood Bank refrigerator for up to 42 days. For longer storage, blood may be frozen within 5 days of collection if such facilities are available. This option is expensive.

Causes for Rejection Criteria vary. Some criteria which disqualify donors for homologous donation do not apply to autologous donation.

The following are rejection criteria for participation in a preoperative autologous donation program:

- active infection/bacteremia
- significant aortic stenosis
- unstable angina
- uncontrollable seizure disorder
- myocardial infarction or cerebral vascular accident within 6 months of donation
- significant cardiac or pulmonary disease without clearance for donation by treating physician
- high-grade left main coronary artery disease
- cyanotic heart disease

Other possible criteria for rejection of a proposed autologous donor include:

- pregnancy - autologous blood is seldom indicated in an uncomplicated pregnancy. Policies should be developed, nonetheless, for situations in which blood may be needed for the infant or in women with multiple or high-incidence antibodies.
- uncontrolled hypertension

In all cases, the patient's primary physician and the Blood Bank physician share responsibility for acceptance of a patient and collection of the autologous blood.

Special Instructions Usually done by appointment with Blood Bank. Physician should write out a prescription giving the date of the intended surgery, the number and type of components requested and the specified surgical procedure. An order for Type and Screen is desirable to cover unanticipated blood needs that might exceed the amount of autologous blood.

Use Alleviates concerns regarding risk of transfusion-transmitted disease and alloimmunization in selected types of elective surgery. Autologous transfusion may be the only suitable source of compatible blood for patients with extremely rare blood types, patients with antibodies to high incidence antigens, or patients with antibodies to multiple antigens.

Limitations Combined with advances in the safety of allogeneic transfusion, the cost-effectiveness of the procedure has been questioned. Preoperative donations diminish presurgical hemoglobin, which may result in the need for additional transfusion.

Autologous predeposit programs do not alleviate all risks of transfusion. Autologous blood has been issued to the wrong intended recipient. Bacterial contamination of units has not been eliminated. Air embolism has been reported. Up to 90% of such units go unused when intended for such procedures as hysterectomy, vaginal delivery, or prostatic TUR.

Contraindications While the safety of cardiac patients as predeposit autologous donors is supported, others have noted a higher severe reaction rate during autologous donation. A U.S. Supreme Court decision (Bragdon vs Abbott) may make it illegal to deny autologous blood services to individuals protected under the Americans with Disabilities Act (ADA), including those who test positive for HIV.

Methodology Similar to conventional blood donation

Additional Information Some general considerations of a preoperative autologous blood program follow.
- Guidelines are needed to establish or improve a program.
- The patient should take iron for at least 1 week prior to the first donation, particularly if multiple units are needed.
- Units of blood are normally collected at weekly intervals, although if only 2 units are needed they can be collected via apheresis at a single collection in eligible donors.
- Autologous donation stimulates erythropoietin production.
- Use of recombinant human erythropoietin may further assist recovery during autologous donation. Its use may be integrated into a set of guidelines. It is costly.
- Surgery must be appropriately scheduled to permit needed intervals between donations.

References
AuBuchon J, *Guidelines for Blood Utilization Review*, Bethesda, MD: American Association of Blood Banks, 2001, 20-4.

Brecher ME and Goodnough LT, "The Rise and Fall of Preoperative Autologous Donation," *Transfusion*, 2001, 41(12):1459-62.

Duguid JK, "Autologous Blood Transfusion," *Clin Lab Haematol*, 1999, 21(6):371-6.

Fontaine MJ, Winters JL, Moore SB, et al, "Frozen Preoperative Autologous Blood Donation for Heart Transplantation at Mayo Clinic From 1988 to 1999," *Transfusion*, 2003, 43(4):476-80.

Goodnough LT, Brecher ME, Kanter MH, et al, "Transfusion Medicine, Second of Two Parts: Blood Conservation," *N Engl J Med*, 1999, 340(7):525-33.

Klein HG, "Transfusion Safety: Avoid Unnecessary Bloodshed," *Mayo Clin Proc*, 2000, 75(1):5-7.

Nuttall GA, Santrach PJ, Oliver WC Jr, et al, "Possible Guidelines for Autologous Red Blood Cell Donations Before Total Hip Arthroplasty Based on the Surgical Blood Order Equation," *Mayo Clin Proc*, 2000, 75(1):10-7.

Babesiosis Serology

Related Information

Anemia Flowchart *on page 35*
(Continued)

Babesiosis Serology *(Continued)*

Arthropod Identification *on page 213*
Ehrlichiosis Serology *on page 531*
Lyme Disease Serology *on page 879*
Malaria Smear and Tests *on page 888*
Risks of Transfusion *on page 1166*

Synonyms Nantucket Fever Serological Test

Abstract *Babesia microti* and *Babesia divergens* are tick-borne intraerythrocytic protozoans, which can cause symptoms resembling those of *Plasmodium falciparum*. Like malaria, babesiosis causes hemolytic anemia. Asplenic, immunocompromised, and elderly subjects are especially at risk, but immunocompetent persons can develop the disease. These organisms also cause disease in cattle, including Texas fever. Babesiosis is enzootic in Southern New England, Southern New York, Wisconsin, and Minnesota.

Specimen Serum, blood for smear

Container Red top tube

Storage Instructions Refrigerate at 4°C.

Reference Interval Negative

Critical Values Titers >1:128 are considered consistent with infection. A fourfold increase in titer establishes diagnosis.

Use *Babesia* species serological test is used to diagnose babesiosis

Limitations False reactivity may be seen in patients with malaria. Patients exposed to *Babesia* on the West Coast of the U.S. should be tested for antibodies to the *Babesia* WA1 species rather than the *Babesia microti* antigen due to the lack of cross-reactivity between these organisms.

Methodology Indirect immunofluorescent antibody (IFA), serum. **Note:** Intraerythrocytic ring forms and tetrads can be identified in the peripheral blood film. The ring forms resemble those of *P. falciparum* malaria, but the rare tetrad forms are diagnostic of babesiosis.

Additional Information The geographic distribution of babesiosis is worldwide. Babesiosis is transmitted in nature by hard-bodied ticks. The ticks become infected by feeding on various infected vertebrate animals (cattle, deer, moles, and mice). *Babesia* can also be transmitted to humans by blood transfusions. *Babesia divergens* is the most common species reported in Europe while *B. microti* is the agent most frequently identified in the U.S. Recently, two variants of *Babesia* species have been reported in the U.S. One variant is from the states of Washington and California (WA1 type) and the other is from Missouri (MO-1).

Babesiosis is particularly severe in patients who have undergone splenectomy (and who lack the RBC "pitting" function of the spleen). It may be potentially life-threatening in immunosuppressed persons or those of advanced age. Infections in individuals with normal spleens may often be asymptomatic. The Northern deer tick, *Ixodes scapularis* (*Ixodes dammini*), which transmits one of the *Babesia* species (*B. microti*, a rodent parasite) also transmits Lyme disease. Subjects with either disease should be considered for the other. Concurrent babesiosis and ehrlichiosis has been reported.

Pantanowitz et al have recently outlined helpful diagnostic features in the peripheral blood film.

References

Gelfand JA and Callahan MV, "Babesiosis," *Curr Clin Top Infect Dis*, 1998, 18:201-16.

Gorenflot A, Moubri K, Precigout E, et al, "Human Babesiosis," *Ann Trop Med Parasitol*, 1998, 92(4):489-501.

Homer MJ, Aguilar-Delfin I, Telford SR, et al, "Babesiosis," *Clin Microbiol Rev*, 2000, 13(3):451-69.

Javed MZ, Srivastava M, Zhang S, et al, "Concurrent Babesiosis and Ehrlichiosis in an Elderly Host," *Mayo Clin Proc*, 2001, 76:563-5.

Krause PJ, Lepore T, Sikand VK, et al, "Atovaquone and Azithromycin for the Treatment of Babesiosis," *N Engl J Med*, 2000, 343(20):1454-8.

Pantanowitz L, Ballesteros E, and DeGirolami P, "Laboratory Diagnosis of Babesiosis," *Lab Med*, 2001, 32(4):184-7.

Internet Web Sites

www.astdhpphe.org/infect/babesiosis.html
www.cdc.gov/ncidod/dpd/parasites/babesia/default.htm

♦ **Baby Bilirubin** *see* Bilirubin, Neonatal, Serum *on page 263*

Bacteremia Detection, Buffy Coat Micromethod

Related Information

Acid-Fast Stain, Routine or Modified *on page 95*
Bacterial Antigens, Rapid Detection Methods *on page 228*
Bacterial Culture, Blood *on page 232*
Buffy Coat Smear Study of Peripheral Blood *on page 316*
Gram Stain *on page 658*
Histoplasmosis Antibody *on page 734*
Malaria Smear and Tests *on page 888*
Microfilariae, Peripheral Blood Preparation *on page 915*
Peripheral Blood: Differential Leukocyte Count *on page 1010*

Synonyms Buffy Coat Method for Detection of Bacteremia

Abstract A buffy coat smear, stained for microorganisms, is a simple test but lacks sensitivity and specificity.

Specimen Blood

Container Heparinized capillary tubes

Collection Transport immediately to the laboratory for processing. Blood should be cultured concurrently.

Turnaround Time 1-2 hours

Reference Interval Negative

Possible Panic Range Positive

Use An infrequently used aid in the diagnosis of acute bacterial blood infection, bacteremia; can provide rapid, prompt diagnosis of septicemia in preterm infants; assist in the detection and identification of rarely encountered microorganisms that may cause potentially fatal septicemia.

Limitations There have been conflicting reports on the usefulness of this technique. False-positive and false-negative results complicate interpretation.

Methodology Gram stain is applied to smear of buffy coat. Wright stain of buffy coat can show histoplasmosis; Ziehl-Neelsen stain can show mycobacteria.

Additional Information A variety of unusual organisms, sites of infection and clinical circumstances may produce septicemia. The QBC tube (quantitative buffy coat, utilizing tubes precoated with acridine orange) has been considered to have value in the rapid detection of *Wuchereria bancrofti* microfilarial organisms but has low sensitivity and may be unable to provide species identification when utilized for malaria case identification in the field. The ability to detect and diagnose *Mycobacterium avium-intracellulare* and *Cryptococcus neoformans* using both a stain of the buffy coat and culture of the buffy coat from AIDS patients has been studied. Culture of buffy coat was much more effective in early diagnosis of those organisms than examining the stained buffy coat smear. Use of the buffy coat smear was rapid and specific for detection of *Mycobacterium avium* complex infection in AIDS patients, although lacking in sensitivity. Buffy coat smears have been employed in the diagnosis of histoplasmosis in AIDS patients and in the detection of *Malassezia* species deep-line catheter-associated sepsis. In select populations (eg, immunosuppressed patients, hyposplenism, presence of indwelling catheters, especially in premature neonates with signs of illness), intracellular bacteria may be identified on study of routine peripheral blood smears without preparation of buffy coat slides.

An unusual but clinically significant microbial cause of septicemia is *Capnocytophaga canimorsus*. Septicemia followed by fatal Waterhouse-Friderichsen syndrome may progress rapidly following inoculation by animal bite (especially dog bite). The organism is a fastidious, microaerophilic, gram-negative rod and is susceptible to antibiotics, including penicillin.

Study of buffy coat smears may provide detection of organisms prior to development of clinical signs of infection.

In select populations (eg, premature infants), careful examination of a routine peripheral blood smear may disclose bacteria within neutrophils without preparation of buffy coat smears.

References

Damsker B and Bottone EJ, "Mycobacteria and Cryptococci Cultured From the Buffy Coat of AIDS Patients Prior to Symptomatology: A Rationale for Early Therapy," *AIDS Res*, 1986, 2(4):343-8.
Howard MR and Smith RA, "Early Diagnosis of Septicaemia in Preterm Infants From Examination of Peripheral Blood Films," *Clin Lab Haematol*, 1999, 21(5):365-8.

(Continued)

Bacteremia Detection, Buffy Coat Micromethod
(Continued)

Mirza I, Wolk J, Toth L, et al, "Waterhouse-Fredericksen Syndrome Secondary to *Capnocytophaga canimorsus* Septicemia and Demonstration of Bacteremia by Peripheral Blood Smear: A Case Report and Review of the Literature," *Arch Pathol Lab Med*, 2000, 124(6):859-63.

♦ **Bacterial Antigens, CSF** *see* Bacterial Antigens, Rapid Detection Methods *on page 228*

Bacterial Antigens, Rapid Detection Methods
Related Information
Bacteremia Detection, Buffy Coat Micromethod *on page 227*
Bacterial Culture, Cerebrospinal Fluid *on page 236*
Cerebrospinal Fluid Analysis: Overview *on page 355*
Cerebrospinal Fluid Glucose *on page 362*
Cerebrospinal Fluid Protein *on page 371*
Cryptococcal Antigen Titer *on page 482*
Gram Stain *on page 658*
Group A *Streptococcus* Screen, Rapid *on page 659*
Group B *Streptococcus* Screen, Rapid *on page 660*
Mycobacterial Culture, Cerebrospinal Fluid *on page 931*
Viral Culture *on page 1307*

Applies to Bacterial Antigens, CSF; Cerebrospinal Fluid Bacterial Antigen Testing

Test Includes Latex agglutination of antigens of *H. influenzae*, *S. pneumoniae*, *N. meningitidis*. Test may also include testing for group B *Streptococcus* and *E. coli* K1 antigen in neonates.

Abstract Gram staining and culture take precedence over bacterial antigen testing. Rapid bacterial antigen tests on CSF are usually done by latex agglutination. The causative microorganisms in true-positive latex agglutination CSF specimens are generally found in Gram stain as well. It is recommended that **these latex agglutination tests are not intended as a substitute for bacterial culture. Confirmatory diagnosis of bacterial meningitis infection is possible only with appropriate culture procedures.**

Specimen Cerebrospinal fluid

Container Sterile CSF tube

Storage Instructions Do not refrigerate.

Reference Interval Negative

Use Although the test is intended to detect bacterial antigens in CSF for the rapid diagnosis of meningitis, its specific indications remain questionable.

Limitations Antigen detection does not replace Gram stain and culture. The test may be negative in early meningitis. *Staphylococcus aureus* and *Pseudomonas aeruginosa* are not detected by these methods. Most members of the Enterobacteriaceae also fail to react. Antigenic cross-reactions are seen. The sensitivity of commercial antigen detection kits remains imperfect. False-negative results occur with low antigen load. Pneumococcal and *Haemophilus* strains not possessing capsular antigens may not be detected by immunological techniques. False positives cause expense and complications.

Methodology Latex agglutination (LA) of capsular polysaccharide bacterial antigen

Additional Information The CSF white blood cell count and differential with Gram stain are the best predictors of meningitis.

Antigen detection methods should never be substituted for culture and Gram stain. Culture and Gram stain must always have priority when limited quantities of CSF are available.

Bacterial antigens may be detected despite previous antibiotic therapy. Positive latex agglutination in patients with negative cultures caused by prior antimicrobial treatment were not found in several studies.

References
Hussein AS and Shafran SD, "Acute Bacterial Meningitis in Adults. A 12-Year Review," *Medicine (Baltimore)*, 2000, 79(6):360-8.
Mein J and Lum G, "CSF Bacterial Antigen Detection Tests Offer no Advantage Over Gram's Stain in the Diagnosis of Bacterial Meningitis," *Pathology*, 1999, 31(1):67-9.
Perkins MD, Mirrett S, and Reller LB, "Rapid Bacterial Antigen Detection Is Not Clinically Useful," *J Clin Microbiol*, 1995, 33(6):1486-91.

Smith AL, "Bacterial Meningitis," *Pediatr Rev*, 1993, 14(1):11-8.

Sobanski MA, Gray SJ, Cafferkey M, et al, "Meningitis Antigen Detection: Interpretation of Agglutination by Ultrasound-Enhanced Latex Immunoassay," *Br J Biomed Sci*, 1999, 56(4):239-46.

Internet Web Sites

www.cdc.gov/ncidod/dbmd/diseaseinfo/meningococcal_g.htm

♦ **Bacterial Culture, Abscess** *see* Bacterial Culture, Aerobes *on page 229*

♦ **Bacterial Culture, Abscess** *see* Bacterial Culture, Anaerobes *on page 231*

Bacterial Culture, Aerobes

Related Information

Anthrax Detection *on page 163*

Antimicrobial Susceptibility Testing, Aerobic and Facultatively Anaerobic Bacteria *on page 178*

Bacterial Culture, Anaerobes *on page 231*

Bacterial Culture, Blood *on page 232*

Bacterial Culture, Lower Respiratory *on page 241*

Beta-Lactamase Test *on page 256*

Bronchoalveolar Lavage (BAL) *on page 311*

Gram Stain *on page 658*

Synonyms Aerobic Bacterial Culture

Applies to Bacterial Culture, Abscess; Enterobacteriaceae Culture; *Neisseria* sp; *Pseudomonas* spp; *Staphylococcus aureus*

Test Includes Culture of aerobic or facultative anaerobes contributing to an infectious process. Antimicrobial susceptibility testing is commonly needed. Cultures for fungi and mycobacteria should be requested if indicated. Bacterial culture on tissue, abscess material, or sterile fluids should include a Gram stain as well as bacterial cultures.

Abstract Culture for aerobic bacteria utilizes methods capable of detecting obligately aerobic (those incapable of reproducing in the absence of oxygen) and facultatively anaerobic (those capable of reproducing in the presence or absence of oxygen) organisms. Obligately anaerobic organisms (those incapable of reproducing in the presence of oxygen) will not be recovered in aerobic cultures.

The overwhelming majority of bacterial pathogens are facultatively anaerobic organisms (eg, all streptococci, enterococci, staphylococci, members of the family Enterobacteriaceae, *Haemophilus influenzae*, *Pasteurella multocida*, *Vibrio* spp). There are relatively few obligately aerobic bacteria that are human pathogens; those that are commonly encountered include *Pseudomonas* spp and most *Neisseria* spp.

Most abscesses are caused by mixed bacterial growth that includes both aerobic and anaerobic bacteria. Initial antibiotic coverage should be broad-spectrum to cover these possibilities. Routine culture will not detect slow growing microorganisms such as mycobacteria.

A useful overview of aerobic microbiology is that provided by Pezzlo.

Patient Preparation For collection of abscess material or tissue, the site is decontaminated (surgical soap and 70% isopropyl alcohol) to eliminate potentially contaminating aerobic and anaerobic bacteria which colonize many surfaces. Adjacent skin surfaces must not be touched.

Specimen Body site, tissue, or fluid associated with infection; fluid, pus, or other material properly obtained from an abscess

Container Specimens may be aspirated into a syringe and capped; all air should be expelled from the syringe prior to transport. Clinical material may be transferred from the syringe to commercially available vials. Specimens may also be transferred to sterile containers. Specimens in syringes with needles are not acceptable because of concerns about needlestick injuries. Swabs are inferior specimens.

Collection Contamination with normal flora must be avoided. Ideally, pus obtained by needle aspiration through an intact surface, which has been aseptically prepared, is put directly into a transport device or transported directly to the laboratory in the original syringe. Sampling of open lesions is enhanced by deep aspiration using a sterile needle and syringe. Curettings of the base of an open lesion may also provide a good yield. Irrigation should be done with nonbacteriostatic sterile saline. Pulmonary samples are obtained by transtracheal percutaneous needle aspiration by trained physicians or by use of a (Continued)

Bacterial Culture, Aerobes *(Continued)*

special sheathed catheter. If swabs must be used, two should be collected; one for culture and one for Gram stain. Specimens collected and transported in syringes should be transported to the laboratory within 30 minutes of collection.

The portion of the biopsy specimen submitted for culture should be separated from the portion submitted for histopathology by the surgeon or pathologist, utilizing sterile technique. Bedside inoculation of sterile body fluid into blood culture bottles improves sensitivity.

Causes for Rejection Specimens exposed to air, refrigerated specimens, or those having excessive delay in transit have a suboptimal yield; specimens in fixative

Turnaround Time Varies with specimen. **Final** reports of negative results are typically available as follows: sputum and wounds: 2 days; aseptically obtained body fluids: 5 days; tissue and abscess material: 2 days. **Preliminary** information from positive cultures is usually available within 24 hours; speciation and antimicrobial susceptibility is often available 24 hours later, but certain organisms require more time to identify. If actinomycosis is suspected clinically, the specimen may be held for 2 weeks.

Use Recover and identify obligately aerobic and facultatively anaerobic bacteria suspected of causing infections and provide a guide for therapy

Limitations It is often difficult to distinguish contaminating organisms from etiologic agents of infection. Organisms most likely to contaminate specimens include *Corynebacterium* species, coagulase-negative staphylococci, alpha-hemolytic streptococci, *Propionibacterium* species, and *Bacillus* species. These organisms are not invariably contaminants, however, and may be pathogenic in certain settings. Occasionally, organisms may grow slowly, be difficult to isolate in pure culture, or be difficult to identify. Fastidious bacteria may not be recovered despite significant efforts to collect and properly submit a specimen. A Gram stain should be performed on certain specimens (sputum, tissue, abscess material, etc) if sufficient material is obtained, to provide early presumptive information and to help interpret culture results.

Methodology Inoculation of microbiological media, incubation of media at temperatures varying from 25°C to 42°C (usually 35°C) in ambient or CO_2-enhanced atmospheric conditions

Additional Information *Staphylococcus aureus* causes community- and hospital-acquired infections which are increasing in frequency. The emergence of multidrug-resistant strains has made therapy more difficult. These gram-positive cocci cause diseases which include **toxic shock syndrome**, **staphylococcal scalded skin syndrome**, and **food poisoning**. **Metastatic infections** include spread to the skeleton, kidneys, and lungs. *S. aureus* is among the most common pathogens causing **sepsis**.

S. aureus is the most frequent organism in **osteomyelitis**. Other causes include coagulase-negative staphylococci, *Proprionibacterium*, Enterobacteriaceae, *Pseudomonas aeruginosa*, streptococci, anaerobic bacteria, *Salmonella*, *Streptococcus pneumoniae*, *Bartonella henselae*, *Pasteurella multocida*, Eikenella corrodens, *Aspergillus*, *Mycobacterium avium* complex, *Mycobacterium tuberculosis*, *Candida albicans*, *Brucella*, *Coxiella burnetii*, and various fungi.

Pneumonia caused by gram-negative bacilli may be community- or hospital-acquired. The former patient group is composed almost entirely of individuals with chronic diseases (eg, chronic obstructive pulmonary disease, alcoholism, or malignant disease). The second (nosocomial) group is principally found in subjects whose pneumonia is secondary (eg, the postoperative state). Pneumonias caused by gram-negative bacilli are commonly related to aspiration. Blood cultures are useful in diagnosis, especially in those with community-acquired infections. The yield of positive cultures from effusion fluid is about 30%. Bronchoalveolar lavage or bronchial brush techniques are useful.

References

Elliott D, Kufera JA, and Myers RA, "The Microbiology of Necrotizing Soft Tissue Infections," *Am J Surg*, 2000, 179(5):361-6.

Funke G, "Algorithm for Identification of Aerobic Gram-Positive Rods," *Manual of Clinical Microbiology*, 7th ed, Murray PR, Baron EJ, Pfaller MA, et al, eds, Washington, DC: ASM Press, American Society for Microbiology, 1999, 316-18.

Jacobs DS, DeMott WR, Oxley DK, et al, *Laboratory Test Handbook*, 5th ed, Hudson, OH: Lexi-Comp Inc, 2001.

Koneman EW, Allen SD, Janda WM, et al, "The Aerobic Gram-Positive Bacilli," *Color Atlas and Textbook of Diagnostic Microbiology*, 5th ed, New York, NY: Lippincott, 1997, 651-708.

Pezzlo M, "Aerobic Bacteriology," *Essential Procedures for Clinical Microbiology*, Isenberg HD, ed, Washington, DC: ASM Press, American Society for Microbiology, 1998, 37-126.

Stone HH, "Soft Tissue Infections," *Am J Surg*, 2000, 66(2):162-5.

Tuomanen EI, Austrian R, and Masure HR, "Pathogenesis of Pneumococcal Infection," *N Engl J Med*, 1995, 332(19):1280-4.

Bacterial Culture, Anaerobes

Related Information

Applies to Bacterial Culture, Abscess; Body Fluid Culture

Test Includes Culture for anaerobic bacteria; Gram stain is usually performed. Cultures for fungi and mycobacteremia should be requested if indicated.

Abstract Culture for anaerobic bacteria utilizes methods capable of detecting obligate anaerobes (those incapable of reproducing in the absence of oxygen). Facultatively anaerobic (those capable of reproducing in the presence of oxygen) organisms are also recovered in these cultures, but the primary purpose of anaerobic cultures is to recover obligate anaerobes. Anaerobic bacteria most likely to be identified include *Bacteroides* sp, *Prevotella* sp, *Fusobacterium* sp, *Peptostreptococcus* sp, other streptococci, *Gemella morbillorum*, *Staphylococcus saccharolyticus*, *Veillonella* sp, *Actinomyces* sp, *Eubacterium* sp, *Proprionibacterium* sp, *Bifidobacterium* sp, *Lactobacillus* sp, and *Clostridium* sp.

Specimen Abscess, blood, aseptically obtained body fluid (eg, pleural, peritoneal, synovial), wounds, etc are appropriate specimens

Container Capped syringe; biopsy; anaerobic transport media, blood culture media

Collection Swabs usually cause unacceptable exposure of anaerobes to oxygen and have a propensity to dry out. Some anaerobes will be killed by contact with oxygen for only a few seconds. Swabs are considered inferior specimens; if they must be used, two are advised: one for Gram stain, the other for culture.

Causes for Rejection Specimens from sites in which anaerobic bacteria are normal flora (eg, throat, rectal swabs, urine, bronchial washes, cervico-vaginal mucosal swabs, sputums) are unacceptable for anaerobic culture. Specimens that have not been appropriately protected from atmospheric oxygen cannot yield accurate results.

Turnaround Time Negative cultures are typically reported as follows: blood cultures: 5-7 days; aseptically obtained body fluids: 5 days; swab specimen: 2-5 days.

Use Recover and identify obligately anaerobic bacteria suspected of causing infections. Cultures from a variety of clinical settings should be accomplished to provide both aerobic and anaerobic culture (eg, those from biopsies and body fluids, abscesses, wounds, and bites).

Limitations Anaerobic bacterial cultures are intended to recover most common obligately anaerobic bacterial pathogens. Some obligate anaerobes die after very brief exposure to oxygen and are very difficult to recover in culture. Anaerobic bacteria may contaminate clinical specimens that are not collected aseptically; it is often challenging to distinguish contaminants from etiologic agents of infection. Specimens in anaerobic transport containers are suboptimal for fungus culture.

Contraindications It is usually not relevant to seek anaerobes in acute cholecystitis, acute osteomyelitis, acute otitis media, acute sinusitis, appendicitis, bronchitis, cystitis, meningitis, pharyngitis, primary peritonitis, pyelonephritis, or superficial skin lesions.

Methodology Inoculation of microbiological media suitable for recovering anaerobes and incubation of media at 35°C in an atmosphere lacking oxygen. (Continued)

Bacterial Culture, Anaerobes *(Continued)*

Biochemical, gas-liquid chromatography (GLC), DNA hybridization, and RNA homology and sequencing define anaerobic genera.

Additional Information Serious anaerobic infections are often due to mixed flora, which are pathologic synergists. Anaerobes frequently recovered from closed postoperative wound infections include *Bacteroides fragilis*, ~50%; *Prevotella melaninogenica* (previously *Bacteroides melaninogenicus*), ~25%; *Peptostreptococcus prevotii*, ~15%; and *Fusobacterium* species, ~25%. Anaerobes are seldom recovered in pure culture (10% to 15% of cultures). Aerobes and facultative bacteria, when present, are frequently found in lesser numbers than the anaerobes. Anaerobic infection is most commonly associated with operations involving opening or manipulating the bowel or a hollow viscus (eg, appendectomy, cholecystectomy, colectomy, gastrectomy, bile duct exploration, etc). The ratio of anaerobes to facultative species is normally about 10:1 in the mouth, vagina, and sebaceous glands and at least 1000:1 in the colon.

Anaerobic bacteria can be involved in all types of bacterial infections, since these bacteria consist of a major part of the indigenous flora of humans. The most common anaerobe encountered in human infections is *Bacteroides fragilis*. A number of other anaerobes also cause human disease (*Prevotella* sp, *Porphyromonas* sp, *Fusobacterium* sp, *Veillonella* sp). Tetanus and botulism are serious diseases due to toxins produced by anaerobes. See Tetanus Antibody *on page 1240* or Botulism Diagnostic Procedures *on page 302*.

References
Anuradha DE, Saraswathi K, and Gogate A, "Anaerobic Bacteraemia: A Review of 17 Cases," *J Postgrad Med*, 1998, 44(2):63-6.

Brook I and Frazier EH, "Microbiology of Subphrenic Abscesses: A 14-Year Experience," *Am J Surg*, 1999, 65(11):1049-53.

Jacobs DS, DeMott WR, Oxley DK, et al, *Laboratory Test Handbook*, 5th ed, Hudson, OH: Lexi-Comp Inc, 2001.

Walker CK, Workowski KA, Washington AE, et al, "Anaerobes in Pelvic Inflammatory Disease: Implications for the Centers for Disease Control and Prevention Guidelines for Treatment of Sexually Transmitted Diseases," *Clin Infect Dis*, 1999, 28(Suppl 1):S29-36.

Bacterial Culture, Blood

Related Information

Anthrax Detection *on page 163*
Bacteremia Detection, Buffy Coat Micromethod *on page 227*
Bacterial Culture, Aerobes *on page 229*
Bacterial Culture, Anaerobes *on page 231*
Bacterial Culture, Cerebrospinal Fluid *on page 236*
Bacterial Culture, Lower Respiratory *on page 241*
Bacterial Culture, Stool *on page 243*
Brucellosis Culture and Serology *on page 312*
Buffy Coat Smear Study of Peripheral Blood *on page 316*
Entamoeba histolytica Antigen Detection and Serology *on page 538*
Fungal Culture, Blood *on page 620*
Leptospira Culture *on page 842*
Opiates, Qualitative, Urine *on page 974*
Tularemia Diagnostic Procedures *on page 1279*
Viral Culture *on page 1307*
Yersinia enterocolitica Antibody *on page 1337*
Yersinia pestis Diagnostic Procedures *on page 1337*

Test Includes Isolation of both aerobic and anaerobic microorganisms and antimicrobial susceptibility testing on all significant isolates

Abstract A blood culture is one of the most significant procedures that a laboratory performs. Positive results should be promptly called to the ordering physician. New automated blood culture systems provide continuous monitoring, which allows detection of positive cultures 24 hours/day. A major problem of blood cultures is the possibility of contamination with normal skin flora. Other factors relevant to detection of microbial pathogens in blood include the volume of blood cultured, the number of separate cultures, the extent of dilution, the types of media, the devices selected, the presence of unusual or fastidious organisms, and the presence of antibiotics.

Patient Preparation Blood cultures are sometimes contaminated by skin flora. Such contamination can be markedly reduced by careful attention to skin preparation and antisepsis **prior** to collection of the specimen.

After location of the vein by palpation, the venipuncture site should be cleansed with 70% alcohol (isopropyl or ethyl) and then swabbed in a circular motion concentrically from the center outward using tincture of iodine or a povidone iodine solution. **The iodine should be allowed to dry before the venipuncture is undertaken.** If palpation is required during the venipuncture, the glove covering the palpating finger tip should be disinfected. In iodine-sensitive patients, a double alcohol, green soap, or acetone alcohol preparation may be substituted.

Alcoholic chlorhexidine has also been recommended for skin preparation.

Aftercare Iodine used in the skin preparation should be carefully removed from the skin after venipuncture.

Specimen Venous blood. The yield of positives is not increased by culturing arterial blood, even in endocarditis.

Container Bottles of trypticase soy broth or other standard medium

Blood Culture Collection

Clinical Disease Suspected	Culture Recommendation	Rationale
Sepsis, meningitis, osteomyelitis, septic arthritis, bacterial pneumonia	Two to three sets of cultures, each 10-30 mL for adults	Assure sufficient sampling in cases of intermittent or low level bacteremia. Minimize the confusion caused by a positive culture resulting from transient bacteremia or skin contamination.
Fever of unknown origin (eg, occult abscess, empyema, typhoid fever, etc)	Two to three sets of cultures - one from each of two prepared sites, the second or third drawn after a brief time interval (30 minutes). If cultures are negative after 24-48 hours obtain two more sets, preferably prior to an anticipated temperature rise.	The yield after four sets of cultures is minimal. A maximum of three sets per patient per day for 3 consecutive days is recommended.
Endocarditis		
Acute	Obtain two to three blood culture sets within 2 hours, then begin therapy.	95% to 99% of acute endocarditis patients (untreated) will yield a positive in one of the first three cultures.
Subacute	Obtain two to three blood culture sets on day 1, repeat if negative after 24 hours. If still negative or if the patient had prior antibiotic therapy, repeat again.	Adequate sample volume despite low level bacteremia or previous therapy should result in a positive yield.
Immunocompromised host (eg, AIDS)		
Septicemia, fungemia mycobacteremia	Obtain two to three sets of cultures from each of two prepared sites; consider lysis concentration technique to enhance recovery for fungi and mycobacteria.	Low levels of fungemia and mycobacteremia frequently encountered.
Previous antimicrobial therapy		
Septicemia, bacteremia; monitor effect of antimicrobial therapy	Obtain two to three sets of cultures from each of two prepared sites; consider use of antimicrobial removal device (ARD) or increased volume >10 mL/set.	Recovery of organisms is enhanced by dilution, increased sample volume, and removal of inhibiting antimicrobials.

(Continued)

Bacterial Culture, Blood *(Continued)*

Sampling Time Ideally, two to three sets of blood cultures should be collected per febrile episode; collection of each set should be separated by at least 1 hour from the previous specimen. Such intervals provide maximum recovery of microorganisms in patients with intermittent bacteremia, and documentation of persistent bacteremia in patients with intravascular infections (eg, endocarditis, intravenous catheter site infections). If multiple sets must be collected simultaneously, draw two sets initially from separate sites, and collect a third set at least 1 hour later. Although three blood culture sets provide optimal yield, the cost-effectiveness of this approach has been challenged and some individuals propose collection of only two sets per febrile episode.

Collection Blood cultures should be drawn prior to initiation of antimicrobial therapy. If more than one culture is ordered, the specimens should be drawn from separately prepared sites. A syringe and needle, transfer set, or pre-evacuated set of tubes containing culture media may be used to collect blood. Collection tubes should be held below the level of the venipuncture to avoid reflux. A sample volume of 10-30 mL in adults or 1-5 mL in children is collected for each set. The likelihood of recovering a pathogen increases as the volume of blood sampled increases; however, **drawing of more than three blood culture sets per bacteremic episode rarely increases yield**. If a syringe and needle or transfer set is used, the top of the blood culture bottle should also be aseptically prepared. See table on previous page.

Interpretation of results can be enhanced by collecting blood cultures from more than one site and after a time interval (1 hour). Cultures should be taken as early as possible in the course of a febrile episode.

Storage Instructions Specimens collected in tubes with SPS® (sodium polyanetholsulfonate) should be processed without delay. The specimen should be transferred to appropriate culture media to avoid any possible decrease in yield due to storage or prolonged contact with SPS®. Culture bottles from some automated systems can sit at room temperature for several hours.

Turnaround Time Common laboratory procedure is to issue a final negative culture report after 5-7 days. A preliminary positive culture report based upon Gram stain and primary subculture is usually available at 24-72 hours.

Special Instructions The requisition should indicate current antibiotic therapy, clinical diagnosis, and relevant history.

Reference Interval Negative

Critical Values Positive cultures are immediately phoned to the nursing station or physician.

Use Isolate, identify, and determine antimicrobial susceptibility of pathogenic organisms causing bacteremia

Limitations Transient bacteremia caused by brushing teeth, bowel movements, or scratching the skin may cause a positive blood culture, but usually will not cause all three sets to be positive. Blood culture contamination produces false-positive results and subjects patients to the side effects of inappropriate therapy. The most common bacterial contaminants in blood cultures include coagulase-negative staphylococci, *Corynebacterium*, viridans streptococci, and *Bacillus* species.

Prior antibiotic therapy may cause negative blood cultures or delayed growth. Blood cultures from patients suspected of having *Brucella*, tularemia, or *Leptospira* must be requested as special cultures. Consultation with the laboratory for the recovery of these organisms prior to collection of the specimen is recommended. When patients with infective endocarditis have negative bacterial blood cultures, the possibility of a fungal infection should be considered. Yeast often are isolated from routine blood cultures. However, if fungi are specifically suspected, a separate fungal blood culture should be drawn along with each of the routine blood culture specimens. See Fungal Culture, Blood *on page 620* for proper collection of specimen. *Mycobacterium avium-intracellulare* (MAI) is recovered from blood of immunocompromised patients, particularly those with acquired immunodeficiency syndrome (AIDS). Special procedures are required for the recovery of these organisms (ie, lysis filtration concentration or use of a special mycobacteria blood culture medium).

Contraindications Use of a 2% iodine preparation is contraindicated in patients sensitive to iodine. Green soap may be substituted for the iodine or alcohol acetone alone may be used. See Patient Preparation; *vide supra.*

Methodology Aerobic and anaerobic culture in broth media with detection of bacterial growth by a variety of methods. The antimicrobial removal device procedure (ARD) includes use of an adsorbent resin in the aerobic bottle. Resin-containing bottles are also available for several automated detection systems. In the lysis centrifugation procedure, blood is lysed and centrifuged using a Wampole Isolator™ tube or similar method. The sediment is inoculated to media appropriate for growing aerobic and anaerobic bacteria, fungi, and mycobacteria. A method of continuously monitoring media for bacterial growth is available from several commercial sources.

Additional Information Blood culture should be collected from patients with community-acquired pneumonia. The most important single test for diagnosis of infective endocarditis is the blood culture. The diagnosis of bacterial meningitis is accomplished by blood culture as well as culture and examination of cerebrospinal fluid. Most children with bacterial meningitis are initially bacteremic.

Sequential blood cultures in nonendocarditis patients using a 20 mL sample resulted in an 80% positive yield after the first set, a 93% yield after the second set, and a 98% yield after the third set. The volume of blood cultured seems to be more important than the specific culture technique being employed by the laboratory. The isolation of coagulase-negative staphylococci (CNS) poses a critical and difficult clinical dilemma. Although CNS are the most commonly isolated organism from blood cultures, only a few (6.3%) of the isolates represent "true" clinically significant bacteremia. Conversely, CNS are well recognized as a cause of infections involving prosthetic devices, cardiac valves, CSF shunts, dialysis catheters, and indwelling vascular catheters. Ultimately, the physician is responsible for determination of whether an organism is a contaminant or a pathogen. The decision is based on both laboratory and clinical data. Patient data including patient history, physical examination, body temperatures, clinical course, and laboratory data (ie, culture results, white blood cell count, and differential) are relevant. Clinical experience and judgment may play a significant role in resolution of this clinical dilemma. Various sources of contamination include the patient's own skin flora, transient benign bacteremias, and perhaps, disinfection materials.

Recovery of mycobacteria, atypical mycobacteria, and *Legionella* may also be enhanced by lysis filtration.

The use of antimicrobial removal devices (ARD) or resin bottles to attempt to increase the yield of blood cultures drawn from patients on antimicrobial therapy is controversial. Some microorganisms are occasionally not recovered with the use of ARD blood cultures. It is, therefore, advised that at least one culture in a series of three be requested without the use of the ARD bottles. ARD blood cultures are substantially more expensive than routine blood cultures.

References

Chien JW, "Making the Most of Blood Cultures. Tips for Optimal Use of This Time-Honored Test," *Postgrad Med*, 1998, 104(1):119-27.

DesJardin JA, Falagas ME, Ruthazer R, et al, "Clinical Utility of Blood Cultures Drawn From Indwelling Central Venous Catheters in Hospitalized Patients With Cancer," *Ann Intern Med*, 1999, 131(9):641-7.

Edmond MB, Wallace SE, McClish DK, et al, "Nosocomial Bloodstream Infections in United States Hospitals: A Three-Year Analysis," *Clin Infect Dis*, 1999, 29(2):239-44.

Glerant JC, Hellmuth D, Schmit JL, et al, "Utility of Blood Cultures in Community-Acquired Pneumonia Requiring Hospitalization: Influence of Antibiotic Treatment Before Admission," *Respir Med*, 1999, 93(3):208-12.

Isaacman DJ, Shults J, Gross Tk, et al, "Predictors of Bacteremia in Febrile Children 3-36 Months of Age," *Pediatrics*, 2000, 106(5):977-82.

Luna CM, Videla A, Mattera J, et al, "Blood Cultures Have Limited Value in Predicting Severity of Illness and as a Diagnostic Tool in Ventilator-Associated Pneumonia," *Chest*, 1999, 116(4):1075-84.

Pizzo PA, "Fever in Immunocompromised Patients," *N Engl J Med*, 1999, 341(12):893-90.

Waterer GW, Jennings SG, and Wunderink RG, "The Impact of Blood Cultures on Antibiotic Therapy in Pneumococcal Pneumonia," *Chest*, 1999, 116(5):1278-81.

Wheeler AP and Bernard GR, "Treating Patients With Severe Sepsis," *N Engl J Med*, 1999, 340(3):207-14.

Bacterial Culture, Cerebrospinal Fluid

Related Information

Bacterial Antigens, Rapid Detection Methods *on page 228*
Bacterial Culture, Blood *on page 232*
Cerebrospinal Fluid Analysis: Overview *on page 355*
Cerebrospinal Fluid Cytology *on page 361*
Cerebrospinal Fluid Glucose *on page 362*
Cerebrospinal Fluid Lactate *on page 369*
Cerebrospinal Fluid Protein *on page 371*
Enterovirus Polymerase Chain Reaction *on page 540*
Fungal Culture, Cerebrospinal Fluid *on page 621*
Gram Stain *on page 658*
India Ink Preparation *on page 784*
Mycobacterial Culture, Cerebrospinal Fluid *on page 931*
VDRL, Serum or Cerebrospinal Fluid *on page 1303*
Viral Culture *on page 1307*

Test Includes Aerobic culture and Gram stain. Many laboratories inoculate specimens onto broth media that can support growth of anaerobic bacteria.

Abstract The major test to be performed on the CNS for meningitis is the bacteriologic culture. Bacteria commonly isolated include *Neisseria meningitidis*, *Haemophilus influenzae* group B, *Streptococcus pneumoniae*, *Listeria monocytogenes*, *Escherichia coli*, *Streptococcus agalactiae*, *Flavobacterium meningosepticum*; and *Mycobacterium* sp. The gold standard for the diagnosis of bacterial meningitis is the isolation of a bacterium from the cerebrospinal fluid. Cell count, differential, Gram stain, CSF glucose, and CSF protein also are useful. Blood cultures are often positive with meningitis.

Patient Preparation Aseptic preparation of the aspiration site

Specimen Cerebrospinal fluid

Container Sterile CSF tube

Collection Contamination with normal flora from skin or other body surfaces must be avoided. Peripheral blood white cell count and differential are important part of the clinical investigation.

Tubes should be numbered 1, 2, 3 indicating the sequence of collection. Contamination with normal flora from skin or other body surfaces must be avoided. The third tube collected is most suitable for culture. Special requests are needed if mycobacteria are suspected.

Storage Instructions The specimen should be transported immediately to the laboratory. If the specimen cannot be processed immediately, it should be kept at room temperature or placed in an incubator. Refrigeration inhibits viability of certain anaerobic organisms and may prevent the recovery of common fastidious aerobic pathogens.

Turnaround Time Gram stain results are reported within 1 hour. Preliminary culture results are usually available at 24 hours. Identification of pathogens may require 48 hours.

Special Instructions The laboratory should be informed of the specific source of specimen, age of patient, current antibiotic therapy, clinical diagnosis, and time of collection.

Reference Interval No growth

Critical Values Positive Gram stain and growth of bacteria

Use Isolate and identify pathogenic organisms causing meningitis, shunt infection, brain abscess, subdural empyema, cerebral or spinal epidural abscess, bacterial endocarditis with embolism. Gram stain with cultures of CSF in suspected bacterial meningitis are fundamental to appropriate diagnosis and treatment.

Limitations Cultures may be negative in partially treated cases of meningitis. Microorganisms such as *Neisseria meningitidis* and *Haemophilus influenzae* are sensitive to temperature shifts. Refrigeration can inhibit their isolation from the specimen. Gram stains should be interpreted with care. Gram-positive organisms may decolorize (ie, stain gram-negative in partially treated cases).

When cultures are negative in the presence of clinical meningitis, identification of positive blood culture(s) with CSF pleocytosis (leukocyte count ≥10 cells/mL of CSF) can support diagnosis.

Methodology Aerobic bacterial culture

Additional Information Bacterial meningitis remains a diagnostic problem. Symptoms suggestive of the diagnosis are those associated with febrile illness (eg, fever, lethargy, and anorexia); meningeal inflammation giving rise to nausea, vomiting, photophobia, and nuchal rigidity, leading to apathy; and encephalopathy with headache, confusion, and seizures. Stupor, coma, and focal neurologic signs indicate a poor prognosis if present before start of therapy. **Mortality of bacterial meningitis** reaches 30%. Prognosis is worse in the very young, very old, in the presence of sickle cell disease, asplenia, and with endocarditis. The rapid appearance of hemorrhagic eruptions and a shock-like state is indicative of meningococcaemia (Waterhouse-Fridericksen syndrome). Complications occur in survivors despite early diagnosis and appropriate use of antimicrobial drugs. Developmental or neurologic sequelae are found in 33% to 40% of survivors. *Haemophilus influenzae, Neisseria meningitidis,* and *Streptococcus pneumoniae* are the organisms commonly associated with bacterial meningitis. *Streptococcus agalactiae, E. coli,* and *Listeria monocytogenes* can be associated with meningitis in children younger than 3 months of age. Infections of cerebrospinal fluid shunts pose a difficult clinical problem.

**Most Common Pathogens in Patients
With Suspected Bacterial Meningitis
Who Have a Nondiagnostic Gram Stain of CSF**

Group of Patients	Likely Pathogen
Immunocompetent	
<3 mo	*S. agalactiae, E. coli, L. monocytogenes*
3 mo to <18 y	*N. meningitidis, S. pneumoniae,* or *H. influenzae*
18-50 y	*S. pneumoniae* or *N. meningitidis*
>50 y	*S. pneumoniae, L. monocytogenes,* or gram-negative bacilli
With impaired cellular immunity	*L. monocytogenes* or gram-negative bacilli
With head trauma, neurosurgery, or CSF shunt	Staphylococci, gram-negative bacilli, or *S. pneumoniae*

Adapted from Quagliarello VJ and Scheld WM, "Treatment of Bacterial Meningitis," *N Engl J Med,* 1997, 336(10):708-16.

A polymerase chain reaction-based test utilizing a range of bacterial primers to detect conserved regions of the microbial 16S RNA gene in the CSF may allow identification of the pathogen in pretreated, culture-negative cases of bacterial meningitis.

References

Aronin SI, Peduzzi P, and Quagliarello VJ, "Community-Acquired Bacterial Meningitis: Risk Stratification for Adverse Clinical Outcome and Effect of Antibiotic Timing," *Ann Intern Med,* 1998, 129(11):862-9.

Hussein AS and Shafran SD, "Acute Bacterial Meningitis in Adults. A 12-Year Review," *Medicine (Baltimore),* 2000, 79(6):360-8.

"Meningococcal Infections," *Red Book: 2003 Report of the Committee on Infectious Diseases,* 26th ed, Section 3, "Summaries of Infectious Diseases," Pickering LK, Baker CJ, Overturf GD, et al, eds, Am Acad Pediatrics, Elk Grove, IL, 2003, 430-6.

Negrini B, Kelleher KJ, and Wald ER, "Cerebrospinal Fluid Findings in Aseptic Versus Bacterial Meningitis," *Pediatrics,* 2000, 105(2):316-9.

Pollard AJ, Probe G, Trombleg C, et al, "Evaluation of a Diagnostic Polymerase Chain Reaction Assay for *Neisseria meningitidis* in North America and Field Experience During an Outbreak," *Arch Pathol Lab Med,* 2002, 126(10):1209-15.

Rajnik M and Ottolini MG, "Serious Infections of the Central Nervous System: Encephalitis, Meningitis, and Brain Abscess," *Adolesc Med,* 2000, 11(2):401-25.

Bacterial Culture, Conjunctiva

Related Information

Adenovirus Culture and Serology *on page 109*
Bartonella Culture *on page 249*
Bartonella Diagnostic Procedures *on page 250*
Chlamydia Group Serology *on page 385*
Chlamydia trachomatis Culture *on page 385*
Chlamydia trachomatis Direct Antigen Test *on page 387*
(Continued)

Bacterial Culture, Conjunctiva *(Continued)*

Chlamydia trachomatis Nucleic Acid Detection *on page 388*
Fungal Culture, Ocular Infections *on page 622*
Fungus Smear, Stain *on page 626*
Gram Stain *on page 658*
Herpes Simplex Virus Culture and Antigen Detection *on page 721*
Neisseria gonorrhoeae Culture and Smear *on page 945*
Neisseria gonorrhoeae Nucleic Acid Detection *on page 947*
Ocular Cytology *on page 973*
Sjögren Antibodies *on page 1199*
Tularemia Diagnostic Procedures *on page 1279*
Varicella-Zoster Virus Culture and Serology *on page 1300*
Viral Culture *on page 1307*

Test Includes Aerobic bacterial culture and smears (Gram and Giemsa) if specifically requested

Abstract Hyperacute bacterial conjunctivitis is usually related to **gonococcal infection**, is often accompanied by preauricular lymphadenopathy, and may lead to perforation. The leading cause of **red eye** is viral infection, which may accompany or follow an upper respiratory tract infection, and which is highly contagious. Replicating **adenovirus** is found in 95% of patients with that infection 10 days after appearance of symptoms. **Chlamydial conjunctivitis** causes inclusion conjunctivitis, a sexually transmitted disease. Trachoma is caused by *Chlamydia*.

Collection Inoculation of prewarmed plates at the time of collection of the specimen (C-streak) is a useful adjunct to optimal culture yield because of the low numbers of organisms usually present.

Use Isolate and identify potentially pathogenic organisms. Pathogens which may cause bacterial conjunctivitis include *H. influenzae*, *H. aegyptius*, *Staphylococcus* sp, *Streptococcus pneumoniae*, and *Neisseria gonorrhoeae*. *Pseudomonas* is the most common bacterial pathogen in those who wear contact lenses. Viral conjunctivitis may be caused by adenovirus, echovirus, and Coxsackievirus. Causes of keratitis include herpes simplex, adenovirus, *Streptococcus pneumoniae*, *Staphylococcus aureus*, *Pseudomonas*, *Acanthamoeba*, and chemical agents.

Ophthalmia neonatorum, occurring in infants younger than 4 weeks of age, may be caused by chemical irritation (silver nitrate), *Neisseria gonorrhoeae*, or *Pseudomonas*. In the U.S., the most common cause is *Chlamydia trachomatis*. Gram stain and culture are indicated when conjunctivitis develops more than 48 hours after birth.

Limitations The procedure will not detect *Chlamydia*, viruses, fungal agents, or mycobacteria which may cause conjunctivitis and/or keratitis. Scrapings are a more useful specimen than a swab for Gram stain.

Methodology Aerobic culture on blood and chocolate agar, incubation at 37°C with CO_2

Additional Information Eye infections include eyelid infections, blepharitis, dacryocystitis, orbital cellulitis, conjunctivitis, keratitis, endophthalmitis retinitis, and chorioretinitis. Blepharitis may be caused by a variety of organisms, including *Staphylococcus*. **Pinkeye** is usually caused by adenovirus. It presents as bilateral conjunctivitis with a sudden onset. Herpes simplex and varicella-zoster present as periorbital or corneal infections. Nontuberculous mycobacterial keratitis may occur following trauma or surgery accompanied by the use of local corticosteroids.

Giemsa and Gram stains must specifically be requested. If gonorrhea is suspected, a Thayer-Martin plate should be inoculated.

A minority of patients with cat scratch disease develop Perinaud oculoglandular syndrome, characterized by conjunctivitis and preauricular lymphadenopathy. This syndrome may also occur in oculoglandular tularemia, in which the conjunctiva is the portal of entry.

References

Brodovsky SC and Snibson GR, "Corneal and Conjunctival Infections," *Curr Opin Ophthalmol*, 1997, 8(4):2-7.
Bullington RH Jr, Lanier JD, and Font RL, "Nontuberculous Mycobacterial Keratitis. Report of Two Cases and Review of the Literature," *Arch Ophthalmol*, 1992, 110(4):519-24.
Leibowitz HM, "The Red Eye," *N Engl J Med*, 2000, 343(5):345-51.

Morrow GL and Abbott RL, "Conjunctivitis," *Am Fam Phys*, 1998, 57(4):735-46.

Ormerod LD and Dailey JP, "Ocular Manifestations of Cat-Scratch Disease," *Curr Opin Ophthalmol*, 1999, 10(3):209-16.

Bacterial Culture, Genital Specimen

Related Information

Cervical/Vaginal Cytology *on page 376*
Chlamydia trachomatis Culture *on page 385*
Chlamydia trachomatis Direct Antigen Test *on page 387*
Chlamydia trachomatis Nucleic Acid Detection *on page 388*
Endometrial Cytology *on page 536*
Fibronectin, Fetal, Cervicovaginal Secretions *on page 585*
Genital Culture for *Ureaplasma urealyticum on page 635*
Group B *Streptococcus* Screen, Rapid *on page 660*
Herpes Simplex Virus Culture and Antigen Detection *on page 721*
Mycoplasma pneumoniae Culture and Serology *on page 936*
Neisseria gonorrhoeae Culture and Smear *on page 945*
Neisseria gonorrhoeae Nucleic Acid Detection *on page 947*
Polymerase Chain Reaction *on page 1069*
RPR *on page 1174*
Trichomonas Preparation *on page 1273*
VDRL, Serum or Cerebrospinal Fluid *on page 1303*
Viral Culture *on page 1307*

Abstract The most common pathogens recovered from nonpregnant women are *Chlamydia trachomatis* and *Neisseria gonorrhoeae*. These infections are often asymptomatic and are usually not diagnosed before membrane rupture. The organisms most often found with chorioamnionitis (infections of the fetal membranes) and fetal infection following rupture of the membranes are group B streptococci and *E. coli*. Less frequent pathogens requiring consideration include *Enterococcus* sp, *Gardnerella vaginalis*, and anaerobes (especially *Actinomyces*). Additional organisms to be considered include genital *Mycoplasma* and *Listeria monocytogenes*.

Collection

Female: Rayon-tipped swabs are used to collect cervical, vaginal, and vesicle cultures. Small dacron-tipped swabs are used to collect cultures from the urethra for gonococcal and chlamydial organisms. Separate transport media are necessary for recovery of gonococci, *Chlamydia trachomatis* and herpesvirus, and *Mycoplasma* and *Ureaplasma* sp.

Male: Urethral exudate is cultured for *Neisseria gonorrhoeae* and *Chlamydia trachomatis*. When no exudate can be expressed, insertion of a swab (cotton, rayon, or Dacron) into the distal urethra may be required. Appropriate transport media are essential.

Currently, *Neisseria gonorrhoeae* and *Chlamydia trachomatis* are often diagnosed using the more sensitive, nonculture methods. Specimens for these assays should be collected in the specified kit.

Storage Instructions Recovery of *N. gonorrhoeae* is enhanced by cultures taken at the bedside (onto special medium at room temperature) and delivery to the laboratory within 30 minutes.

Special Instructions The laboratory should be informed of the specific source of specimen, current antibiotic therapy, clinical diagnosis, and time of collection.

Critical Values Recovery of *N. gonorrhoeae* and/or beta-hemolytic group B streptococci during pregnancy

Use Cultures are often used in the setting of acute sexually transmitted disease. The role of cultures in chorioamnionitis and postpartum endometritis (PPE) is more controversial. Screening is often recommended for *Chlamydia trachomatis* infection in women.

Limitations *N. gonorrhoeae* is a fastidious organism; if left in the environment, it may not grow in culture at all. *Chlamydia trachomatis* is an intracellular pathogen. Cultures may be negative if the specimen does not contain adequate cells. Infections with *Chlamydia trachomatis* and *N. gonorrhoeae* are often evaluated using molecular assays. It is important to note that, because of inhibitors, these techniques may be insensitive in pregnant women.
(Continued)

Bacterial Culture, Genital Specimen *(Continued)*

Methodology Culture of swabs on agar media that support the growth of *N. gonorrhoeae* (chocolate agar with antibiotics) as well as colonizers such as beta-hemolytic group B streptococci (sheep blood agar). Special transport media and culture systems are required for the identification of genital *Chlamydia* and *Mycoplasma* organisms.

In addition to cultures, nucleic acid amplification, enzyme-linked immunoassays, and direct fluorescent antibody stains may also be used. Nucleic acid amplification techniques are more sensitive for identification of both *Chlamydia trachomatis* and *Neisseria gonorrhoeae* in urethral exudates, endocervical exudates, and urine specimens.

Additional Information Vaginitis is one of the most commonly encountered complaints of female patients. A significant portion appear to be due to specific etiologic agents such as *Candida* species or *Trichomonas vaginalis*. Nonspecific vaginitis, also called bacterial vaginosis, is characterized by an excessive malodorous vaginal discharge associated with a decrease in the number of lactobacilli and an increase in the number of *Gardnerella vaginalis* and other anaerobic bacteria such as *Bacteroides* species, *Prevotella* species, *Peptostreptococcus* species, and *Mobiluncus* species. **Presently, bacterial cultures do not contribute to the diagnosis of bacterial vaginosis.** Minimum diagnostic requirements for bacterial vaginosis include three of the following signs:

- excessive vaginal discharge
- vaginal pH >4.5
- "clue" cells (vaginal epithelial cells covered by small gram-negative rods)
- a fishy amine-like odor in the KOH test (10% KOH added to vaginal discharge)

Candida species are frequently present as normal flora in vagina. A saline wet mount may demonstrate yeast cells or pseudohyphae and may provide rapid diagnostic information. The most common clinical presentation is a characteristic clumpy white cottage cheese appearance with vaginal or vulvar itching. Vaginitis, which frequently complicates pregnancy and diabetes, is seen with broad spectrum antibiotic therapy and in conditions which lower host resistance (eg, AIDS).

Listeria monocytogenes causes uterine infection, chorioamnionitis, placental abscesses, neonatal sepsis, abortion, stillbirth, premature birth, and other diseases.

Infections caused by **group B streptococci** (GBS) (*Streptococcus agalactiae*) can include sepsis, meningitis, pneumonia, and mortality among neonates. GBS infections are the leading cause of serious U.S. neonatal disease.

The high-risk criteria include:

- previous infant with invasive GBS infection
- GBS bacteriuria during the current gestation
- delivery at <37 weeks gestation
- duration of rupture of membranes ≥18 hours
- intrapartum fever ≥100.4°F (≥38°C)

See also Group B *Streptococcus* Screen, Rapid *on page 660* and Fibronectin, Fetal, Cervicovaginal Secretions *on page 585*.

References

Centers for Disease Control and Prevention, "Adoption of Perinatal Group B Streptococcal Disease Prevention Recommendations by Prenatal-Care Providers - Connecticut and Minnesota, 1998," *JAMA*, 2000, 283(18):2384-5.

Cline MK, Bailey-Dorton C, and Cayelli M, "Maternal Infections: Diagnosis and Management," *Prim Care*, 2000, 27(1):13-33.

Goldenberg RL, Hauth JC, and Andrews WW, "Intrauterine Infection and Preterm Delivery," *N Engl J Med*, 2000, 342(20):1500-7.

McGregor JA and French JI, "Bacterial Vaginosis in Pregnancy," *Obstet Gynecol Surv*, 2000, 55(5 Suppl 1):S1-19.

Pimenta J, Catchpole M, Gray M, et al, "Evidence Based Health Policy Report. Screening for Genital Chlamydial Infection," *BMJ*, 2000, 321(7261):629-31.

Schrag SJ, Zywicki S, Farley MM, et al, "Group B Streptococcal Disease in the Era of Intrapartum Antibiotic Prophylaxis," *N Engl J Med*, 2000, 342(1):15-20.

Internet Web Sites

Group B streptococci: www.cdc.gov/groupbstrep/

Bacterial Culture, Lower Respiratory

Related Information

Anthrax Detection *on page 163*
Bacterial Culture, Blood *on page 232*
Bordetella pertussis Serology *on page 301*
Bronchial Brushings/Washings Cytology *on page 310*
Bronchoalveolar Lavage (BAL) *on page 311*
C-Reactive Protein, Serum *on page 467*
Fungal Culture, Blood *on page 620*
Fungal Culture, Sputum *on page 624*
Gram Stain *on page 658*
Hypersensitivity Pneumonitis Serology *on page 761*
Legionella Culture and Serology *on page 838*
Legionella Direct Fluorescent Antigen Smear *on page 839*
Mycobacterial Culture, Sputum *on page 933*
Mycoplasma pneumoniae Culture and Serology *on page 936*
Pneumocystis Immunofluorescence *on page 1063*
Sputum Cytology *on page 1222*
Viral Culture *on page 1307*

Applies to Bronchial Culture; Bronchoalveolar Lavage Culture; Sputum Culture

Test Includes Gram stain, quantitative scoring for specimen adequacy, and inoculation of culture media (acceptable specimens only)

Abstract Laboratory evaluation begins with the microscopic examination of a Gram stain to guide initial therapy and to determine the suitability of the specimen for culture. Even when meticulous technique is used (see below), sputum is a problematic specimen because of contaminating organisms from the mouth and pharyngeal region. Alternative specimen sources, highly effective in recovering organisms which cause lower respiratory infections, include transtracheal aspirates and specimens obtained via bronchoscopy and bronchoalveolar lavage. Depending on the clinical situation, **blood culture may also be very useful in the diagnosis of bacterial pneumonia**. When pleural effusion or empyema exists, it too can be evaluated and cultured if clinically indicated. Defining the bacterial etiology of lower respiratory infections occasionally requires collection of specimens by bronchoscopy/bronchial brushing/bronchoalveolar lavage. Bronchoscopy specimens are always contaminated with oropharyngeal flora and even use of a protected brush does not eliminate problems of contamination.

Patient Preparation The patient should remove dentures, rinse mouth, and gargle with water. (Mouthwashes, which often contain antibacterial agents, should not be used.) The patient should then cough deeply and expectorate sputum into sterile container.

Container Sterile sputum container, sputum trap, sterile tracheal aspirate or bronchoscopy aspirate tube. Bartlett catheters should be submitted in 1 mL of sterile nonbacteriostatic saline in a sterile container. Bronchial washes or bronchoalveolar lavages should be submitted in tightly sealed, sterile containers.

Collection Collect expectorated sputum under direct supervision of nurse or physician. Specimen collected, at time of bronchoscopy, by aspiration or by transtracheal aspiration by a physician skilled in the procedure. The specimen should be transported to the laboratory within 1 hour of collection for processing.

Storage Instructions Refrigerate if the specimen cannot be promptly processed.

Causes for Rejection Microscopy provides an index of the magnitude and character of exudate. Gram stain is commonly utilized. The best expectorated sputum specimens have many PMNs and few to no squamous epithelial cells. Such specimens are likely to include the etiologic agent of the infection in relatively high numbers and comparatively few contaminating organisms from the upper respiratory tract.

Microscopic examination of specimens under oil (1000x total magnification) often provides preliminary morphologic information about the etiologic agent that can be used to guide empiric therapy. Organisms are especially important in Gram stain when phagocytosis is seen. Interpretation of results of bacterial sputum cultures without considering specimen quality (determined by Gram stain) is problematic.

(Continued)

Bacterial Culture, Lower Respiratory *(Continued)*

Specimen rejection criteria are not applied when the suspected organisms include *Mycoplasma pneumoniae*, *Legionella* sp, mycobacteria fungi, and viruses.

Turnaround Time Stat Gram stain results usually are available in 15-30 minutes. Identification of pathogens usually requires at least 48 hours for completion.

Special Instructions The most productive sputum is that which represents bronchial secretions. A plug or cast of the infected bronchus may be evident. Such identified portions are inoculated on selective and nonselective media.

Reference Interval Normal upper respiratory flora. Tracheal aspirate and bronchoscopy specimens can be contaminated with normal oral flora. Transtracheal aspiration should have no growth.

- Bartlett catheters (in 1 mL saline): $<10^3$ cfu/mL is within the expected level of contamination
- Bronchoalveolar lavages (BAL): bacteria: $<10^6$ cfu/mL of original specimen; normal total cell count: $4\text{-}23 \times 10^6$; differential, 95% alveolar macrophages; 3% lymphocytes; 1% polymorphonuclear cells, 0.2% eosinophils
- Bronchial washes: cannot be established; often contaminated heavily with oral flora.

Critical Values The presence or absence of intracellular organisms on Gram staining is important: >7% cells containing intracellular organisms appear to correlate with ventilator-assisted pneumonia.

Use Identify the etiology of pulmonary infections. BAL is useful for diagnosis of opportunistic infections in immunosuppressed subjects. Quantitative culture for respiratory tract pathogens: $>10^3$ to 10^4 bacterial colonies/mL BAL fluid supports the diagnosis of acute bacterial pneumonia. Bronchial brush suspensions may also require colony counts.

Media are selected to support the growth of the most common pathogens causing pneumonia. In adults: *Streptococcus pneumoniae* is the most common cause of community-acquired pneumonia. Other causes include *Haemophilus influenzae*, *Staphylococcus aureus*, many gram-negative rods, *Legionella* spp, and *Moraxella catarrhalis*. In the elderly, *Klebsiella pneumoniae*, and group B streptococci are also included. In critically ill patients, additional possibilities include *Mycoplasma pneumoniae*, *Chlamydia trachomatis*, and *Chlamydia pneumoniae*. In HIV-infected patients, additional possibilities include *Pneumocystis carinii*, cytomegalovirus, *Mycoplasma pneumoniae*, and *Cryptococcus neoformans*. In children, the most common pathogens causing pneumonia are viruses: respiratory syncytial virus and parainfluenza virus type 3.

Limitations Contamination of sputum specimens by organisms normally present in the mouth and pharyngeal area is the principal factor limiting the usefulness of sputum cultures. Therefore, it is imperative that everyone involved in the collection and handling of a sputum specimen exercise meticulous technique. Even when this is done, contamination will occur in some specimens. Contamination with oral pharyngeal secretions causes false-positive aerobic and anaerobic bacterial cultures; quantitative or semi-quantitative cultures on bronchoalveolar lavage and Bartlett catheter specimens circumvent this problem.

Agents such as *Bordetella pertussis*, *Corynebacterium diphtheriae*, *Legionella pneumophila*, *Mycoplasma pneumoniae*, and *Mycobacterium tuberculosis* require special culture for isolation. Clinical suspicion of involvement by such organisms should be communicated to the laboratory.

Methodology Semiquantitative or quantitative aerobic bacterial culture; semiquantitative aerobic and anaerobic bacterial cultures on solid media for protected catheter brushes (PCB) (Bartlett catheter)

Additional Information The abundant normal anaerobic flora of the mouth makes anaerobic cultures of any specimen contaminated by oral secretions (eg, bronchial washes, sputums) essentially useless for defining the anaerobic bacterial etiology of lower respiratory infection. Only specimens that bypass the mouth should be cultured anaerobically. Transtracheal aspirates meet this need, but in most hospital settings, they are rarely, if ever, collected. They have been replaced by bronchoscopically obtained specimens using a PCB. Specimens obtained by PCB are subject to low level contamination with normal oral flora.

Bronchoalveolar lavage (BAL) has become an established procedure for defining the etiology of pulmonary infections. It is particularly useful for recovering opportunistic pathogens (eg, pneumocystis, *Histoplasma capsulatum*, *Candida* spp, *Aspergillus* spp, *Mycobacterium* spp) from immunocompromised individuals and in defining the etiology of nosocomial pneumonia in patients undergoing mechanical ventilation. It has a role for evaluation of *Legionella*, *Nocardia*, *T. gondii*, viruses including CMV, RSV, adenovirus, and herpes simplex virus. BAL is performed with a catheter wedged into a segmented bronchus. Bronchial washings or airway washings are collected with a nonwedged, more proximally positioned scope tip. Bronchial washings, therefore, preferentially sample airways. Lavage is preferred for the diagnosis of *Pneumocystis* pneumonia, which is primarily an alveolar process. Quantitative bacterial cultures of BAL specimens have also proven useful for defining the etiology of acute bacterial pneumonia. Blood cultures are indicated in cases of acute pneumonia.

References

Almirall J, Bolibar I, Vidal J, et al, "Epidemiology of Community-Acquired Pneumonia in Adults: A Population-Based Study," *Eur Respir J*, 2000, 15(4):757-63.

Bartlett JG and Mundy L, "Community-Acquired Pneumonia," *N Engl J Med*, 1995, 333(24):1618-24.

Hindiyeh M and Carroll KC, "Laboratory Diagnosis of Atypical Pneumonia," *Semin Respir Infect*, 2000, 15(2):101-13.

Juven T, Mertsola J, Waris M, et al, "Etiology of Community-Acquired Pneumonia in 254 Hospitalized Children," *Pediatr Infect Dis J*, 2000, 19(4):293-8.

Shuster LT and McDougall JC, "Pneumonia Management Guidelines - Why, How, and Where to Start," *Mayo Clin Proc*, 1998, 73(1):96-7.

Skerrett SJ, "Diagnostic Testing for Community-Acquired Pneumonia," *Clin Chest Med*, 1999, 20(3):531-48.

Bacterial Culture, Stool

Related Information

Bacterial Culture, Blood *on page 232*
Botulism Diagnostic Procedures *on page 302*
Clostridium difficile Toxin Assay and Culture *on page 416*
Cryptosporidium Direct Staining Procedures *on page 484*
Electron Microscopic Examination for Viruses, Stool *on page 533*
Entamoeba histolytica Antigen Detection and Serology *on page 538*
Enterovirus Culture *on page 539*
Fecal Lactoferrin *on page 575*
Methylene Blue Stain, Stool *on page 906*
Microsporidia Diagnostic Procedures *on page 915*
Ova and Parasites, Direct Exam *on page 985*
Rotavirus, Direct Detection *on page 1173*
Viral Culture *on page 1307*
Yersinia enterocolitica Antibody *on page 1337*

Test Includes Culture with selective agar to detect common enteric pathogens; may include Gram stain or methylene blue stain for leukocytes

Abstract Cultures from outpatients and persons hospitalized for fewer than 3 days should be placed on media which can support the most common enteric pathogens (*Salmonella* spp, *Shigella* spp, and *Campylobacter* spp). Clinical information should be supplied if less common organisms are sought (eg, enterohemorrhagic *Escherichia coli* [most commonly serotype O157], *Yersinia enterocolitica*, *Vibrio* spp, *Aeromonas hydrophila*, *Plesiomonas shigelloides*, and *Listeria monocytogenes*). Gastrointestinal disease may be the presenting and often only symptom for many food-borne pathogens (eg, *Salmonella* and *Campylobacter* species). Devastating disease may be seen with some food-borne pathogens (eg, listeriosis, *E. coli* O157:H7, typhoid fever).

Specimen Fresh random stool, rectal swab

Container Stool container, Culturette®

Sampling Time *Escherichia coli* O157:H7, *Salmonella* species, and *Campylobacter jejuni* are most often cultured from stools in summer months.

Collection If stool is collected in a clean bedpan, it must not be contaminated with urine, residual soap, or disinfectants. Swabs of lesions of the rectal wall during proctoscopy or sigmoidoscopy are preferred; *vide infra*.
(Continued)

Bacterial Culture, Stool *(Continued)*

Rectal swab: Insert the swab past the anal sphincter, move the swab circumferentially around the rectum. Allow 15-30 seconds for organisms to adsorb onto the swab. Withdraw swab, place in Culturette® tube, and crush media compartment.

Storage Instructions Fresh specimens should be promptly delivered to the laboratory and processed within 1-2 hours. If specimen transport or processing will be delayed, the specimen may be preserved in modified Cary-Blair or buffered glycerol-saline media and refrigerated.

Causes for Rejection Most laboratories reject specimens from patients hospitalized more than 3 days. This recommendation is based on a study from the College of American Pathologists.

Because of risk to laboratory personnel, specimens sent on a diaper or tissue paper, or specimen contamination of the outside of a transport container may not be acceptable. Specimens containing interfering substances (eg, castor oil, bismuth, Metamucil®, barium), specimens delayed in transit, and those contaminated with urine may not have optimal yield.

Turnaround Time Minimum 48 hours if negative; 72 hours if identification of *Yersinia* is required

Special Instructions The laboratory should be informed of the specific pathogen suspected if not *Salmonella*, *Shigella*, or *Campylobacter*.

Reference Interval Negative for *Campylobacter*, *Salmonella*, and *Shigella*. In endemic areas the isolation of a pathogen may not indicate the only cause of diarrhea.

Use Indications for stool culture include:
- bloody diarrhea
- fever
- tenesmus
- severe or persistent symptoms
- recent travel to a third world country
- known exposure to a bacterial agent
- presence of fecal leukocytes

Limitations An enormous variety of microorganisms have been proven to cause diarrhea, and many others have been associated with diarrhea. Still, even with the most extensive work-ups, a substantial proportion of patients with gastrointestinal disorders fail to yield an etiologic agent. Routine bacterial stool cultures are designed to recover only a few specific organisms; these vary with the laboratory, but usually include *Salmonella*, *Shigella*, *Campylobacter* species, and may include *E. coli* O157:H7. Culture for the last may require special request, unless the stool is bloody. The laboratory should be consulted if other etiologic agents are suspected (eg, *Vibrio* or *Yersinia*), or a more extensive work-up is required. Rectal swab cultures may not be as effective as stool cultures for identifying individuals with small numbers of organisms. *Clostridium difficile* infection is usually diagnosed by toxin assays rather than culture, often supplemented by endoscopy with biopsies.

Contraindications A rectal swab culture is not as effective as a stool culture for detection of the carrier state.

Methodology Aerobic culture on selective media

Additional Information In acute or subacute diarrhea, three common syndromes are recognized: gastroenteritis, enteritis, and colitis (dysenteric syndrome). With colitis, patients have fecal urgency and tenesmus. Stools are frequently small in volume and contain blood, mucus, and leukocytes. External hemorrhoids are common and painful. Diarrhea of small bowel origin is indicated by the passage of few large volume stools. This is due to accumulation of fluid in the large bowel before passage. **Leukocytes** usually indicate bacterial colonic inflammation or ulcerative colitis rather than a specific pathogen; see Methylene Blue Stain, Stool *on page 906.* Bacterial diarrhea may be present in the absence of fecal leukocytes, and fecal leukocytes may be present in the absence of bacterial or parasitic agents (ie, idiopathic inflammatory bowel disease). Although most bacterial diarrhea is transient (1-30 days), cases of persistent symptoms (10 months) have been reported. Infants younger than 1 year of age with a history of blood in the stool, more than 10 stools in 24 hours, and temperature greater than 39°C have a high probability of having bacterial diarrhea. Diarrhea is also a common side effect of long-term antibiotic treatment. Although often associated with *Clostridium difficile*, other bacteria and yeasts have been implicated.

In patients hospitalized more than 3 days, the most common causes of diarrhea are antibiotic administration, enteral feeding, osmotic diarrhea, and *Clostridium difficile* infection. Therefore, evaluation of such patients should begin with assessing these possibilities (see *Clostridium difficile* Toxin Assay and Culture *on page 416*).

In the special case of suspected food poisoning, a variety of diagnostic techniques are available, including latex agglutination and enzyme-linked immunoassays.

Diarrhea is common in patients with the acquired immunodeficiency syndrome (AIDS). It may be caused by classic bacterial pathogens or unusual opportunistic bacteria viral or parasites (eg, *Giardia*, Microsporidia, *Cryptosporidium*, and *Entamoeba histolytica*). Rectal swabs are useful for the diagnosis of *Neisseria gonorrhoeae* and *Chlamydia* infections.

References

Bauer TM, Lalvani A, Fehrenbach J, et al, "Derivation and Validation of Guidelines for Stool Cultures for Enteropathogenic Bacteria Other Than *Clostridium difficile* in Hospitalized Adults," *JAMA*, 2001, 285(3):313-9.

Bishop WP and Ulshen MH, "Bacterial Gastroenteritis," *Pediatr Clin North Am*, 1988, 35(1):69-87.

Jacobs DS, DeMott WR, Oxley DK, et al, *Laboratory Test Handbook*, 5th ed, Hudson, OH: Lexi-Comp Inc, 2001.

Murray BE, "Vancomycin-Resistant Enterococcal Infections," *N Engl J Med*, 2000, 342(10):710-20.

Valenstein P, Pfaller M, and Yungbluth M, "The Use and Abuse of Routine Stool Microbiology: A College of American Pathologists Q-Probes Study of 601 Institutions," *Arch Pathol Lab Med*, 1996, 120(2):206-11.

Bacterial Culture, Throat, and Antigen Detection Testing for Group A Streptococci

Related Information

Antideoxyribonuclease-B Titer, Serum *on page 170*
Antistreptolysin O Titer, Serum *on page 197*
Bacterial Culture, Blood *on page 232*
Chlamydia Group Serology *on page 385*
Corynebacterium diphtheriae Throat Culture *on page 463*
Gram Stain *on page 658*
Group A *Streptococcus* Screen, Rapid *on page 659*
Infectious Mononucleosis Screening Test *on page 785*
Neisseria gonorrhoeae Culture and Smear *on page 945*
Polymerase Chain Reaction *on page 1069*
Streptozyme *on page 1225*

Test Includes Evaluation for group A beta-hemolytic streptococci

Abstract The rate of patient visits to physicians for acute pharyngitis is more than twice that of any other infectious disease. The differential diagnosis of acute pharyngotonsillitis includes a wide range of infectious agents. The most common cause of acute pharyngotonsillitis is viral (adenovirus, Epstein-Barr virus, coxsackie A, adenovirus, and others) but laboratories rarely receive requests to identify these. The most important bacterial causes are group A streptococci, group C beta-hemolytic streptococci, *Neisseria gonorrhoeae*, and *H. influenzae* type b; very rarely *Corynebacterium diphtheriae*, *Chlamydia pneumoniae* and *Candida albicans* cause pharyngitis. *Mycoplasma pneumoniae* may cause acute pharyngitis. *Arcanobacterium haemolyticum* is a rarely diagnosed cause of pharyngotonsillitis which may resemble streptococcal pharyngitis.

Patient Preparation Do not swab throat in cases of acute epiglottitis unless provisions to establish an alternate airway are readily available.

Specimen Throat swab

Container Sterile Culturette®; cotton, dacron, or alginate swabs are acceptable.

Collection The specimen should be transported to the laboratory promptly. Both tonsillar pillars and the oropharynx should be swabbed. Exudates should be swabbed, and the tongue and uvula should be avoided.

Storage Instructions Refrigerate

(Continued)

Bacterial Culture, Throat, and Antigen Detection Testing for Group A Streptococci (Continued)

Turnaround Time Usually 24-48 hours for completion; cultures with no growth are usually reported after 48 hours.

Reference Interval Negative for beta-hemolytic streptococci group A

Use Isolate and identify beta-hemolytic Group A *Streptococcus*, which causes pharyngitis, tonsillitis, and scarlet fever

Limitations Delineation between the acutely infected patient and the asymptomatic carrier with intercurrent viral pharyngitis is not made either by culture nor by rapid antigen detection testing. Most laboratories employ a procedure that is optimized for the growth and identification of group A, beta-hemolytic streptococci only. If any of the other potential pathogens are suspected, the laboratory must be notified so that appropriate procedures will be undertaken.

Methodology Aerobic incubation on sheep's blood agar

Additional Information Group A *Streptococcus* causes 15% to 30% of cases of acute pharyngitis in children and 5% to 10% in adults. It is characterized by the sudden onset of sore throat with pain on swallowing, pharyngeal exudate, tender cervical lymphadenopathy, fever, headache, abdominal pain, nausea, and vomiting. Children younger than age 3 may have coryza and crusting, but usually not exudative pharyngitis. Uncharacteristic features in older subjects include coryza, hoarseness, cough, diarrhea, conjunctivitis, or discrete ulcers.

Antigen detection tests are now widely used to diagnose streptococcal pharyngotonsillitis. Such tests have improved sensitivity and high specificity. They can be performed in a physician's office, with the results available immediately. When the results of an antigen detection test are negative, a confirmatory culture should be performed. Antigen detection tests are less sensitive (between 80% and 90%) than culture. The direct antigen tests, when positive, can be considered equivalent to a positive culture. See also Group A *Streptococcus* Screen, Rapid *on page 659*.

References
Bisno AL, "Acute Pharyngitis," *N Engl J Med*, 2001, 344(3):205-11.

Cunningham MW, "Pathogenesis of Group A Streptococcal Infections," *Clin Microbiol Rev*, 2000, 13(3):470-511.

DiMatteo L, "Managing Streptococcal Pharyngitis: A Review of Clinical Decision-Managing Strategies, Diagnostic Evaluation, and Treatment," *J Am Acad Nurse Pract*, 1999, 11(2):57-62.

Ebell MH, Smith MA, Barry HC, et al, "Does This Patient Have Strep Throat?" *JAMA*, 2000, 284(22):2912-8.

Pichichero ME, "Group A Beta-Hemolytic Streptococcal Infections," *Pediatr Rev*, 1998, 19(9):291-302.

Sela S and Barzilai A, "Why Do We Fail With Penicillin in the Treatment of Group A *Streptococcus* Infections?" *Ann Med*, 1999, 31(5):303-7.

Bacterial Culture, Urine

Related Information

Chlamydia trachomatis Culture *on page 385*
Fungal Culture, Urine *on page 625*
Gram Stain *on page 658*
Kidney Stone Analysis *on page 820*
Leukocyte Esterase, Urine *on page 849*
Mycobacterial Culture, Urine *on page 935*
Nitrite, Urine *on page 962*
Polymerase Chain Reaction *on page 1069*
Urinalysis *on page 1289*
Viral Culture *on page 1307*

Test Includes Aerobic culture of clean catch urine; significant bacterial growth is typically defined as ≥10,000 colony forming units/mL

Abstract Urinary tract infections (UTI) caused by bacteria are common. Clinical presentations include the acute dysuric syndrome, acute and chronic cystitis, and acute and chronic pyelonephritis. Diagnostically challenging are patients who are asymptomatic, or who present with acute abdominal pain simulating appendicitis or cholecystitis. *Escherichia coli* causes ~80% to 90%, and *Staphylococcus saprophyticus* ~10% to 20% of bacterial infections in the urinary tract; rare isolates include various species of *Proteus*, *Klebsiella*, *Enterobacter*, and *Enterococcus*. Many physicians and nurses have been taught to screen for bacterial UTI using the routine dipstick urinalysis (with or without microscopic examination of the sediment). There is, however, recent very substantial controversy on this issue (see Additional Information).

Patient Preparation Meticulous attention to technique and cleansing is necessary.

Clean voided urine (midstream urine): Instruct patient to wash hands thoroughly, wash penis or vulva using downward strokes four times with four soapy sponges, then once with sponge wet with warm water. Urethral meatus and perineum must be washed. Each sponge must be discarded after one use. Urinate about 30 mL (1 ounce) of urine directly into toilet or bedpan - **stop** - position container and take middle portion of urine sample. Screw cap securely on container without touching the inside rim. Apply the completed patient label to the specimen cup. Most patients, with instruction, do better with privacy than with an attendant.

Suprapubic puncture: Aseptic preparation of the aspiration site. Collect the specimen to avoid contamination with normal skin flora. Fluids should be forced prior to collection of the specimen to distend the bladder. Successful suprapubic collection requires a distended bladder; 6-10 hours may be required for the bladder to fill.

Specimen Clean voided urine specimen

Container Sterile plastic urine container or sterile tube

Collection See Patient Preparation.

Clean voided urine: Early morning specimens yield highest bacterial counts from overnight incubation in the bladder. Forced fluids dilute the urine and may cause reduced colony counts. Hair from perineum will contaminate the specimen. The stream from a male may be contaminated by bacteria from beneath the prepuce. Bacteria from vaginal secretions, vulva, or distal urethra may also contaminate the specimen, as may organisms from hands or clothing. Receptacle must be sterile. Provide time and date of urine collection.

Suprapubic puncture: After aseptic preparation of the skin, the specimen is aspirated in a sterile syringe by a physician skilled in the technique.

Storage Instructions Deliver to the laboratory without delay. Storage at room temperature is not acceptable; bacteria multiply at room temperature.

Reference Interval Positivity depends on the technique of specimen collection, the character of the patient group at risk, specimen handling, and perhaps the presence of comorbid conditions.

Clean voided urine: Clinical decision limits vary with the clinical syndrome. For example, ≥10,000 colony-forming units (cfu)/mL is usually a clinically significant finding in a properly collected specimen. However, for patients with the acute dysuric syndrome, as few as 1000 cfu/mL, and as few as 100 cfu/mL in the presence of pyuria may be significant in young sexually active females. In an unselected outpatient male population, it has been suggested that as few as 1000 cfu/mL in voided urine specimens are clinically significant.

Suprapubic puncture: Any growth is clinically significant if proper technique was followed in collecting the specimen.

Use The urine culture is the gold standard for the diagnosis of bacterial UTI. Some physicians screen for UTI by using urinalysis, and proceed with culture only if one or more of the urinalysis results are positive (eg, leukocyte esterase, nitrites, bacteria, or leukocytes on microscopy). See Additional Information.

Limitations Bacteria present in numbers <1000 organisms/mL may not be detected by routine methods. Contamination of urine specimens is reported as 23% from females, 11% from males. Failure to recover aerobic organisms from patients with pyuria or positive Gram stains on urinary sediment may indicate the presence of mycobacteria or anaerobes.

Contraindications Foley catheter tips are consistently contaminated with organisms from the urethra and should not be cultured.

Methodology Quantitative aerobic culture; usually plated at a 0.001 dilution, allowing detection of organisms in enumeration $\geq 10^3$ cfu/mL.

Additional Information The use of urinalysis alone as a screen for UTI is controversial. The dipstick urinalysis may be "cost-effective" in screening **asymptomatic pregnant** patients for bacteriuria. However, a positive culture result could not be accurately predicted from the results of urinalysis in other patients; specifically, 14% to 19% of positive cultures would have been missed. (Continued)

Bacterial Culture, Urine (Continued)

The leukocyte esterase result has a sensitivity of 57% and specificity of 94%; adding the nitrite result increased the sensitivity to 78% and reduced the specificity to 75%. In addition, a semiquantitative microscopic examination has a sensitivity of 83%, and produced a false-positive frequency of 40%. See Nitrite, Urine *on page 962* and Leukocyte Esterase, Urine *on page 849*.

Certain biologic factors also militate against dipstick results as an accurate screen. Urine specimens with bacteria, but free of blood and hemoglobin, have false-positive dipstick results for hematuria.

References

American Academy of Pediatrics, "The Diagnosis, Treatment, and Evaluation of the Initial Urinary Tract Infection in Febrile Infants and Young Children (AC9830)," *Pediatrics*, 2000, 103(4):1-17.

Bartkowski DP, "Recognizing UTIs in Infants and Children. Early Treatment Prevents Permanent Damage," *Postgrad Med*, 2001, 109(1):171-2, 177-81.

Tambyah PA, Halvorson KT, and Maki DG, "A Prospective Study of Pathogenesis of Catheter-Associated Urinary Tract Infections," *Mayo Clin Proc*, 1999, 74(2):131-6.

Internet Web Sites

www.pediatrics.org/cgi/content/full/103/4/e54

- ♦ **Bacteria Screen, Urine** *see* Leukocyte Esterase, Urine *on page 849*
- ♦ **Bacteria Screen, Urine** *see* Nitrite, Urine *on page 962*
- ♦ **Bacteriology Specimen Identification** *see* Specimen Identification Requirements *on page 1217*
- ♦ **Barbita®** *see* Phenobarbital, Serum or Plasma *on page 1021*
- ♦ **Barbiturate Screen, Urine** *see* Barbiturates, Quantitative, Serum or Plasma *on page 248*

Barbiturates, Qualitative, Urine

Synonyms Amobarb; Butalbital; Mephobarb; Pentobarb; Phenobarb; Secobarb

Test Includes Identification and confirmation of barbiturates in urine

Abstract Barbiturates are sedative hypnotics which are also drugs of abuse. Overdose causes CNS depression. The effects of barbiturates are augmented by other CNS depressants such as ethanol and benzodiazepines.

Specimen Random urine

Collection If forensic, observe precautions (see the Introduction *on page 63*).

Storage Instructions Refrigerate specimen.

Special Instructions If forensic, use Chain-of-Custody form. See Introduction *on page 63* for the Chain-of-Custody form.

Reference Interval Less than cutoff

Critical Values Cutoff: screen: 300 ng/mL, confirmation: 50 or 100 ng/mL

Use Urine drugs-of-abuse testing, pre-employment screens, random drug testing.

Limitations Short- and intermediate-acting barbiturates can be detected in urine 24-72 hours following ingestion, longer-acting drugs up to 7 days.

Methodology Screen: immunoassay; confirmation: gas chromatography/mass spectrometry (GC/MS)

Additional Information Barbiturates are nonselective CNS depressants that may be used as sedative-hypnotics or anticonvulsants. They are capable of producing all levels of CNS mood effects from sedation to hypnosis to deep coma and anesthesia. Sensory cortex functions, cerebellar functions, and motor activity are decreased.

References

Coupey SM, "Barbiturates," *Pediatr Rev*, 1997, 18(8):260-4; quiz 265.

Hall BJ and Brodbelt JS, "Determination of Barbiturates by Solid-Phase Microextraction (SPME) and Ion Trap Gas Chromatography-Mass Spectrometry," *J Chromatogr A*, 1997, 777(2):275-82.

Meatherall R, "GC/MS Confirmation of Barbiturates in Blood and Urine," *J Forensic Sci*, 1997, 42(6):1160-70.

Barbiturates, Quantitative, Serum or Plasma

Related Information

Barbiturates, Qualitative, Urine *on page 248*
Ethanol, Blood, Urine, and Other Sources *on page 558*
Methylphenobarbital, Serum *on page 909*
Phenobarbital, Serum or Plasma *on page 1021*

Applies to Alurate®; Amobarbital; Amytal®; Aprobarbital; Barbiturate Screen, Urine; Blue Angels; Butabarbital; Butalbital; Butisol Sodium®; Fiorinal®;

Gemonil®; Lotusate®; Luminal®; Mebaral®; Mephobarbital; Metharbital; Nembutal®; Pentobarbital; Phenobarbital; Red Devils; Secobarbital; Seconal™; Talbutal; Yellow Jackets

Abstract Measurement of barbiturates as a class is usually used for drug-of-abuse testing or as evidence for toxicity. A urine sample is generally used for barbiturate abuse testing. Certain barbiturates (eg, secobarbital, amobarbital, phenobarbital, pentobarbital) are measured in blood for therapeutic drug monitoring purposes. See individual listings for details.

Specimen Plasma or serum

Container Lavender top (EDTA) tube; green top (heparin) tube; red top tube (avoid serum separator tube)

Sampling Time Trough

Reference Interval Therapeutic: short-acting (secobarbital): 1-5 µg/mL (SI: 4.2-21.0 µmol/L); intermediate-acting (amobarbital): 5-15 µg/mL (SI: 22-66 µmol/L); long-acting (phenobarbital): 15-40 µg/mL (SI: 65-172 µmol/L) for seizure control

Critical Values Toxic: short-acting: >10 µg/mL (SI: >43 µmol/L); intermediate-acting: >20 µg/mL (SI: >86 µmol/L); long-acting: >40 µg/mL (SI: >172 µmol/L)

Use Evaluate barbiturate toxicity, drug abuse, therapeutic levels; if barbiturates are suspected in a drug overdose, determination of long-, medium-, or short-acting may influence treatment

Limitations Only barbiturates as a class will be identified by immunoassays.

Methodology Gas chromatography (GC), high performance liquid chromatography (HPLC), immunoassay

Additional Information See individual entries.

References

Alvarez N, "Barbiturates in the Treatment of Epilepsy in People With Intellectual Disability," *J Intellect Disabil Res*, 1998, 42(Suppl 1):16-23.

Coupey SM, "Barbiturates," *Pediatr Rev*, 1997, 18(8):260-4; quiz 265.

Simpson D, Braithwaite RA, Jarvie DR, et al, "Screening for Drugs of Abuse (II): Cannabinoids, Lysergic Acid Diethylamide, Buprenorphine, Methadone, Barbiturates, Benzodiazepines and other Drugs," *Ann Clin Biochem*, 1997, 34(Pt 5):460-510.

Bartonella Culture

Related Information

Bartonella Diagnostic Procedures *on page 250*
Polymerase Chain Reaction *on page 1069*

Abstract *Bartonella* species are small gram-negative rods. These bacteria are responsible for bacillary angiomatosis, bacillary peliosis hepatis, bacteremia, and cat scratch disease (CSD). Bacillary angiomatosis, characterized by subcutaneous nodules and fever, occurs primarily in subjects with AIDS but has been found as well in immunocompetent patients and in recipients of organ transplants. Cat scratch disease can occur in any patient but most cases occur in children or adolescents. The disease is characterized by swollen lymph nodes 2 weeks after a scratch or bite of a cat. Other symptoms such as low grade fever, tiredness, and headache may accompany such lymphadenopathy. Cat scratch disease is more likely to be associated with *B. henselae* infection.

Specimen Blood, tissue from skin, liver, spleen, or lymph node

Collection Collect blood in either EDTA tubes or Isolator™ blood-lysis tubes.

Causes for Rejection Tissue fixed in formalin

Reference Interval Negative

Use Diagnose diseases associated with *Bartonella* infection

Limitations Culture may require more than 7 days of incubation. Recovery of isolates from skin specimens may be overgrown with indigenous bacterial skin flora. *B. henselae* is inhibited by the sodium-polyethylene sulfonate used in many blood culture media.

Methodology Culture on solid agar media for 3-4 weeks

Additional Information *Bartonella* species are considered emerging pathogens. *Bartonella quintana* infections have been detected among the urban homeless population. *Bartonella henselae* is now recognized as the organism associated with cat scratch disease in immunocompetent individuals and bacillary angiomatosis in immunocompromised patients. This pleomorphic organism may be recognized with the Warthin-Starry method in tissue sections and can be labeled by immunocytochemistry. Culture of *Bartonella* species is difficult, (Continued)

Bartonella **Culture** *(Continued)*

time consuming, and costly, but several reference laboratories or large local laboratories will offer specific culture for this organism. Some conventional blood culture systems can also detect *Bartonella* species and an effort can be made when there is clinical suspicion of active infection.

Detection of specific DNA sequences has enhanced epidemiologic investigations. Polymerase chain reaction amplification in arthropods has provided understanding of such zoonotic infections. Molecular detection will likely replace culture as the preferred diagnostic test for *Bartonella* species.

References

Brougui P, Lascola B, Roux V, et al, "Chronic *Bartonella quintana* Bacteremia in Homeless Patients," *N Engl J Med*, 1999, 340(3):184-9.

Maguina C and Gotuzzo E, "Bartonellosis. New and Old," *Infect Dis Clin North Am*, 2000, 14(1):1-22.

Maurin M, Birtles R, and Raoult D, "Current Knowledge of *Bartonella* Species," *Eur J Clin Microbiol Infect Dis*, 1997, 16(7):487-506.

Ohl ME and Spach DH, "*Bartonella quintana* and Urban Trench Fever," *Clin Infect Dis*, 2000, 31(1):131-5.

Spach DH and Koehler JE, "*Bartonella*-Associated Infections," *Infect Dis Clin North Am*, 1998, 12(1):137-55.

Internet Web Sites

www.astdhpphe.org/infect
www.thebody.com/cdc/pdfs/oi_1101.pdf

Bartonella **Diagnostic Procedures**

Related Information

Bartonella Culture *on page 249*
Lymph Node Biopsy *on page 880*
Polymerase Chain Reaction *on page 1069*
Skin Biopsy *on page 1200*

Abstract *Bartonella*, small gram-negative rods, include *B. quintana* and *B. henselae*, which are responsible for bacillary angiomatosis, bacillary peliosis hepatis, bacteremia, and cat scratch disease (CSD). **Bacillary angiomatosis**, characterized by subcutaneous nodules and fever, occurs primarily in subjects with AIDS but has been found as well in immunocompetent patients and in recipients of organ transplants.

Cat scratch disease is found in immunocompetent individuals, 80% of whom are younger than 21 years of age. It is associated with *B. henselae* infection. It presents as an inoculation-site papule in >50% of patients, followed by local lymphadenopathy. Cat scratch or bite is recognized in ~75%. See table.

Diseases Associated With *Bartonella* Species Infection

Cutaneous bacillary angiomatosis
Extracutaneous infection
Bacillary peliosis of the liver (bacillary peliosis hepatitis) and spleen
Fever and bacteremia (*Bartonella* bacteremic syndrome)
Cat scratch disease (associated only with *Bartonella henselae*)
Trench fever (associated only with *B. quintana*)
Endocarditis

Adapted from Maguina C and Gotuzzo E, "Bartonellosis. New and Old," *Infect Dis Clin North Am*, 2000, 14(1):1-22.

Specimen Serology: serum, separated quickly. Tissue.

Collection Serology: paired serum samples are desirable, one drawn 2-4 weeks after the other. Blood cultures are indicated as well.

Causes for Rejection Serology: plasma, severe lipemia, hemolysis

Reference Interval Serology: IgG: <1:64; IgM: <1:16

Critical Values Serology: fourfold increase in titer is meaningful. IgG: ≥1:256; IgM: ≥1:16; provide evidence of recent or present infection.

Use *Bartonella quintana* is a cause of "culture-negative" endocarditis. Most cases of cat scratch disease are diagnosed on the basis of history, physical findings, histopathologic characteristics, and serology.

Limitations Serology: there is cross-reactivity between antibodies to *B. quintana* and *B. henselae*. Additionally, some assays will also cross-react with *Chlamydia* species and *Coxiella burnetii*. Culture of the organism is difficult. See *Bartonella* Culture *on page 249*.

Methodology Indirect fluorescent antibody (IFA), enzyme immunoassay (EIA), DNA amplification from specimens. Polymerase chain reaction-based confirmation of cat scratch disease and *Bartonella henselae* infection can detect organisms in tissue secretions.

Additional Information Although IFA has been very reliable for detecting antibody to *Bartonella*, recently developed EIAs detect both IgG and IgM with a specificity of 92.5% and sensitivity of 89.6%. Biopsies of skin and lymph node lesions include well described features. The pleomorphic bacilli may be recognized (without species specificity) with the Warthin-Starry method in tissue sections and can be labeled by immunocytochemistry. Detection of specific DNA sequences has enhanced epidemiologic investigations and polymerase chain reaction amplification in arthropods has provided understanding of such zoonotic infections. Blood cultures can detect *Bartonella* species and should be obtained when there is a clinical suspicion of active infection.

References

Avidor B, Varon M, Marmor S, et al, "DNA Amplification for the Diagnosis of Cat-Scratch Disease in Small-Quantity Clinical Specimens," *Am J Clin Pathol*, 2001, 115(6):900-9.

Glaser C, Lewis P, and Wong S, "Pet-, Animal-, and Vector-Borne Infections," *Pediatr Rev*, 2000, 21(7):219-32.

Harrison TG and Doshi N, "Serologica Evidence of *Bartonella* spp Infection in the UK," *Epidemiol Infect*, 1999, 123(2):233-40.

Karem KL, "Immune Aspects of *Bartonella*," *Crit Rev Microbiol*, 2000, 26(3):133-45.

Litwin CM, Martins TB, and Hill HR, "Immunologic Response to *Bartonella henselae* as Determined by Enzyme Immunoassay and Western Blot Analysis," *Am J Clin Pathol*, 1997, 108(2):202-9.

Maguina C and Gotuzzo E, "Bartonellosis. New and Old," *Infect Dis Clin North Am*, 2000, 14(1):1-22.

Margolis B, Kuzu I, Herrmann M, et al, "Rapid Polymerase Chain Reaction-Based Confirmation of Cat Scratch Disease and *Bartonella henselae* Infection," *Arch Pathol Lab Med*, 2003, 127(6):706-10.

Not T, Canciane M, Buratti E, et al, "Serologic Response to *Bartonella henselae* in Patients With Cat Scratch Disease and in Sick and Healthy Children," *Acta Paediatr*, 1999, 88(3):284-9.

Internet Web Sites

www.astdhpphe.org/infect/catscratch.html

♦ **Base Excess** *see* Blood Gases and pH, Arterial *on page 275*

♦ **Basement Membrane Zone Antibodies** *see* Skin Biopsy, Immunofluorescence *on page 1203*

♦ **Bay Clor®** *see* Chlorpromazine, Serum *on page 395*

♦ **Baylocaine®** *see* Lidocaine, Serum or Plasma *on page 850*

♦ **Bβ 1-42 Fragment** *see* Hypercoagulation Panel *on page 758*

bcl-2 Gene Rearrangement

Related Information

Apoptosis Assays *on page 205*

Gene Rearrangement for Leukemia and Lymphoma *on page 633*

Lymph Node Biopsy *on page 880*

p53, Functional Assay/Sequencing *on page 995*

Polymerase Chain Reaction *on page 1069*

Applies to Oncogenes

Test Includes Identification of unique DNA bands associated with the *bcl-2* oncogene rearrangement in B-cell lymphomas

Abstract The *bcl-2* oncogene codes for a protein that is located in mitochondria of the cell. The *bcl-2* protein regulates cell death, and when it is overexpressed the cell is resistant to the "natural" death cycle, called apoptosis. Rearrangement of the *bcl-2* gene with the B-cell receptor genes is found in a number of different B-cell lymphomas. This translocation can be identified cytogenetically as t(14;18) in follicular lymphoma (FL), its original relationship, or large cell diffuse lymphomas.

Specimen 0.1 g or more of frozen tissue, specifically from the involved area of the lymph node or tumor. Polymerase chain reaction protocols are available for use with tissue that has been formalin fixed/paraffin embedded.

Container Tissue must be shipped on dry ice or in 95% ethanol.

Collection Tissue must be carefully cut from the surgically removed tumor and contain at least 10% of tumor cells from the involved area.

Storage Instructions Tissue can be stored in a -70°C freezer until shipped.

Causes for Rejection Tissue samples that have thawed during transit cannot be used for DNA analysis.

(Continued)

bcl-2 Gene Rearrangement *(Continued)*

Turnaround Time Results are usually available within 10 days to 2 weeks.

Reference Interval No rearrangement of *bcl-2* is normal.

Use Detect *bcl-2* rearrangement in B-cell lymphomas, in both frozen and paraffin sections. It is used most commonly to distinguish follicular lymphoma from follicular hyperplasia in lymphoid tissue. Reactive follicles are negative for *bcl-2*. The rearrangement is found in FL, large cell diffuse B-cell lymphomas, and undifferentiated lymphomas. Usually a reciprocal translocation with the J_H region on chromosome 14 is involved, thus forming t(14;18). Abnormal expression of *bcl-2* has also been identified in solid malignant neoplasms. In association with HPV, it may play a role in early stages of tumorigenesis of the uterine cervix.

Limitations Rearrangement will not be found if the tissue is not from the involved tumor. Tissue samples that are too small will not yield enough DNA to do an accurate Southern blot analysis. The method is not useful to distinguish follicular lymphomas from other lymphoma cell types.

Methodology DNA is extracted from the clinical sample and digested with restriction enzymes. The digested fragments of DNA are electrophoresed using an agarose gel and then transferred to a nylon membrane (Southern blotting). The DNA on the membrane is hybridized with a labeled DNA probe specific for the *bcl-2* gene. Hybridization of the *bcl-2* probe is detected using autoradiography or enzymatic color development. Clinical samples are always compared to a normal control sample that does not have a *bcl-2* rearrangement. Hybridization to a DNA fragment different from the control sample typically indicates a rearranged *bcl-2* gene. Polymerase chain reaction (PCR) with conventional gel electrophoresis is now commonly used.

In immunohistochemical application, *bcl-2* is best recovered after fixation in B5 or Bouin fixed tissues, unless microwaved.

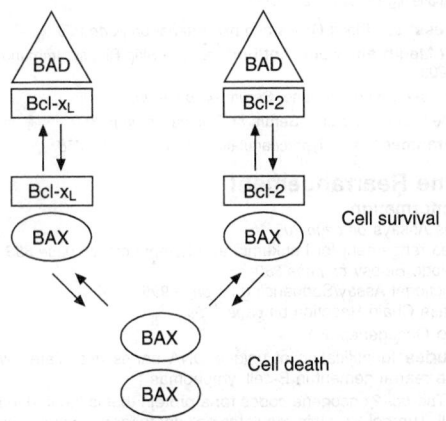

Cell survival

Cell death

≥50% of BAX heterodimerized with
bcl-2 or *bcl-x$_L$* = survival

BAD is a regulator of apoptosis. BAD displaces BAX from *bcl-2*/BAX or *bcl-x$_L$*/BAX heterodimers, allowing more BAX/BAX homodimer formation, which promotes death.

Adapted from Yang E and Korsmeyer SJ, "Molecular Thanatopsis: A Discourse on the BCL-2 Family and Cell Death," *Blood*, 1996, 88(2):386-401.

Additional Information The protein (p26 Bcl-2) coded for by the oncogene, *bcl-2*, acts by suppressing the cell death program (apoptosis). Its role involves control of cell growth. The gene *bcl-x$_L$*, a homolog of *bcl-2*, encodes for p29 Bcl-x$_L$, also an antiapoptotic protein. p26 Bcl-2 and p29 Bcl-x$_L$ localize on the outer mitochondrial membrane, smooth endoplasmic reticulum, and perinu-clear membrane. Their antiapoptotic effect results from inhibition of the mito-chondrial permeability transition with increase in the generation of reactive oxygen species and by blocking the release of cytochrome c into the cyto-plasm. BAX, BAK, and BAD are proapoptotic members of the *bcl-2* family and can form heterodimers with Bcl-2 or Bcl-x$_L$. When the Bcl-2:BAX ratio is low (due to increase in BAX) cell death is caused by anticancer drug therapy. Low levels of BAX may result in a poor response to chemotherapy (see diagram).

Apoptosis occurs in all cells but is especially important in immune and hemato-poietic cells, which have a high cell turnover rate. When the *bcl-2* gene is overexpressed, it will act to prevent apoptosis and may thus render cells resistant to cell death by irradiation and certain chemotherapeutic agents. A translocation between immunoglobulin genes (heavy chain or light chain genes) and *bcl-2* results in the overexpression of *bcl-2* protein and thus the expansion of B cells due to suppression of cell death. This translocation is found in some 85% of patients with follicular lymphoma. It is found in some cases of chronic lymphocytic leukemia, acute lymphoblastic leukemia, and small noncleaved cell lymphoma as well as cases of Hodgkin lymphoma and myeloid neoplasms. The t(14;18) is rarely detected in monocytoid B-cell lymphoma and MALT lymphomas. Some 25% to 35% of cases of FL undergo transformation to a more aggressive diffuse subtype of lymphoma (actuarial risk of histologic transformation, 60% to 70% at 10 years). This change is usually accompanied by the mutation/deletion of the p53 gene (and less commonly by alterations of the *c-myc* gene) in addition to the pre-existent *bcl-2* abnormality. Transformation has also been associated with inactivation of p16 by deletion, mutation or hypermethylation. Overexpression of *bcl-2* is not pathognomic for lymphomas. It is found in 10% of reactive lymph nodes, and in some normal cells (eg, lymphoid and myeloid precursors, medullary thymo-cytes, most T cells, nongerminal center B cells, and plasma cells). It is not expressed in reactive germinal centers.

References

Bagg A and Kallakury BV, "Molecular Pathology of Leukemia and Lymphoma," *Am J Clin Pathol*, 1999, 112(1 Suppl 1):S76-92.

Kroft SH, Domiati-Saad R, Finn WG, et al, "Precursor B-Lymphoblastic Transformation of Grade I Follicle Center Lymphoma," *Am J Clin Pathol*, 2000, 113(3):411-8.

Lones MA, Pinkus GS, Shintaku IP, et al, "*bcl-2* Oncogene Protein Is Preferentially Expressed in Reed-Sternberg Cells in Hodgkin Disease of the Nodular Sclerosis Subtype," *Am J Clin Pathol*, 1994, 102(4):464-7.

Benzodiazepines, Qualitative, Urine

Related Information

(Continued)

Benzodiazepines, Qualitative, Urine *(Continued)*

Synonyms Tranquilizers (Valium®, Librium®, Dalmane®, Tranxene®, Klonopin™, Ativan®, Serax®, Centrax®, Restoril®, Xanax®, Halcion®, Versed®, Doral®, etc)

Applies to Benzodiazepine, Serum and Urine

Test Includes Alprazolam, chlordiazepoxide, clorazepate, diazepam, lorazepam, oxazepam, flurazepam, temazepam, triazolam, clonazepam, midazolam

Abstract Benzodiazepines are sedative-hypnotics, anticonvulsants, and anxiolytics. They are used by more Americans than any other single prescription drug due to their efficacy, safety, low addiction potential, and minimal side effects. These are also commonly abused drugs.

Specimen Random urine

Storage Instructions Refrigerate or freeze if not analyzing immediately

Reference Interval None present unless prescribed. When used as drug-of-abuse screen, negative (less than cutoff).

Critical Values Screening cutoff: 300 ng/mL (generally as oxazepam); Confirmation cutoff: typically 200 ng/mL for low-potency benzodiazepines such as chlordiazepoxide, diazepam, flurazepam, oxazepam, and temazepam; lower cutoff for high-potency benzodiazepines such as alprazolam, clonazepam, lorazepam, midazolam, and triazolam. Cutoffs vary between laboratories.

Use Drug abuse evaluation, toxicity assessment

Methodology Screen: immunoassay; Confirmation: gas chromatography/mass spectrometry (GC/MS)

Additional Information
- Half-life: 3-150 hours, depending on a particular benzodiazepine
- Volume of distribution (for most): 2-5 L/kg
- Protein binding (for most): 90% to 95%

Benzodiazepines are a class of chemically-related central nervous system depressants used as sedative-hypnotics to treat sleep disorders, anxiety, alcohol withdrawal, and seizure disorders. This drug class in low doses can cause sedation, drowsiness, blurred vision, fatigue, mental depression, and loss of coordination. In higher doses or when used chronically, they can cause confusion, slurred speech, hypotension, and diminished reflexes. Chronic use may produce a physical dependence and a withdrawal syndrome which can last for weeks. Benzodiazepine use in the elderly is associated with an increased likelihood of motor vehicle crashes.

Despite their widespread use, abuse of benzodiazepines is relatively infrequent and is more likely to occur in individuals who abuse other drugs or alcohol. Their CNS depressive effect is synergistic with barbiturate or alcohol use. Treatment for benzodiazepine overdose is supportive.

References

Drummer OH, "Benzodiazepines - Effects on Human Performance and Behavior," *Forensic Sci Rev*, 2002, 14:1-14.

Fraser AD, "Use and Abuse of Benzodiazepines," *Ther Drug Monit*, 1998, 20(5):481-9.

Hemmelgarn B, Suissa S, Huang A, et al, "Benzodiazepine Use and the Risk of Motor Vehicle Crash in the Elderly," *JAMA*, 1997, 278(1):27-31.

Segura M, Barbosa J, Torrens M, et al, "Analytical Methodology for the Detection of Benzodiazepine Consumption in Opioid-Dependent Subjects," *J Anal Toxicol*, 2001, 25(2):130-6.

◆ **Benzoyl Cholinesterase** *see* Pseudocholinesterase, Serum *on page 1122*

◆ **Benzoylecgonine** *see* Cocaine (Cocaine Metabolite), Qualitative, Urine or Hair *on page 427*

◆ **Ber-EP₄** *see* Flow Cytometry, Overview *on page 592*

◆ **Berkolol®** *see* Propranolol, Serum *on page 1096*

◆ **Berkomine®** *see* Imipramine, Serum or Plasma *on page 767*

◆ **Berry Spot Test** *see* Inherited Diseases of Metabolism and Cell Structure *on page 792*

◆ **Bespar®** *see* Buspirone, Serum *on page 318*

◆ **Beta-2-Glycoprotein I** *see* Antiphospholipid Antibody (Lupus Anticoagulant and/or Anticardiolipin Antibody) *on page 193*

Beta₂-Microglobulin, Serum or Urine

Related Information

Aminoglycosides, Serum *on page 147*

Cadmium, Blood or Urine *on page 325*
CD4/CD8 Enumeration *on page 349*
Cerebrospinal Fluid Cytology *on page 361*
HIV-1/HIV-2 Serology *on page 736*
Syndecan-1, Serum *on page 1228*
Zidovudine, Serum or Plasma *on page 1339*

Abstract Beta-2 microglobulin (BMG), a low molecular weight serum protein derived from cell membranes, is the light chain component of the class I human leukocyte antigen (HLA) complex. BMG is filtered by the renal glomerulus, and most is reabsorbed by the tubules. Elevated values are found in a large number of diseases, including renal failure (any cause), multiple myeloma, other lymphomas, many neoplasms, chronic inflammation, amyloidosis, and common variable immunodeficiency with granulomatous complications.

Patient Preparation Avoid recent administration of radioisotopes if assay performed by RIA.

Specimen Serum or 24-hour urine

Container Red top tube or SST™ tube, plastic urine container

Storage Instructions Urine BMG is unstable when pH is <5.5.

Reference Interval There are significant interlaboratory differences. One reference laboratory uses the cutoff for serum: <0.27 mg/dL.

A standard text uses, for serum, **average** values:
- Neonates: 0.3 mg/dL
- 0-59 years: 0.19 mg/dL
- 60-69 years: 0.21 mg/dL
- >69 years: 0.24 mg/dL

Urine: <120 µg/day

Use Prognosis assessment in multiple myeloma: The serum BMG reflects tumor size, growth rate, and renal function. A value at presentation <0.4 mg/dL has the best prognosis; and a value >2.0 mg/dL has the worst.

Renal tubular function. In the past BMG was measured in urine to assess proximal tubular function, but is rarely used now for that purpose.

Limitations Increased synthesis of BMG in Crohn disease, hepatitis, sarcoidosis, vasculitis, hyperthyroidism, viral infections, and some malignancies decreases the usefulness of serum levels.

Additional Information Serum BMG predicts response in subjects with low grade lymphoma: at 42 months no patient with a level ≥3.0 mg/L was projected to be in remission. Urinary BMG becomes abnormal before serum creatinine in aminoglycoside nephrotoxicity. BMG is increased in AIDS patients with progressive disease, particularly those with opportunistic infection; it decreases in response to therapy. Its use has been combined with CD4 lymphocyte counts to calculate the probability of an HIV-infected person developing AIDS within the next 3 years.

BMG is reported to delineate a subset of subjects with primary amyloidosis whose outcomes are unfavorable, but it is not useful in such patients as an index of response to therapy.

Pretreatment BMG concentrations >2.9 mg/dL are reported to recognize patients with stage I Philadelphia-positive chronic myelogenous leukemia, on interferon-α-based treatment, who are at greater risk of adverse outcome.

References
Mayo Medical Laboratories, *2000 Test Catalogue*, Rochester, MN, 100.

Painter PC, Cope JY, and Smith JL, *Tietz Textbook of Clinical Chemistry*, 3rd ed, Philadelphia, PA: WB Saunders Co, 1999, 1826.

Rodriguez J, Cortes J, Talpaz M, et al, "Serum β-2 Microglobulin Levels Are a Significant Prognostic Factor in Philadelphia Chromosome-Positive Chronic Myelogenous Leukemia," *Clin Cancer Res*, 2000, 6(1):147-52.

♦ **Beta-Amino Isobutyrate** *see* Amino Acids, Urine *on page 145*
♦ **Beta-Cell Peptides** *see* C-Peptide, Serum *on page 465*
♦ **Beta-Gamma Bridging** *see* Protein Electrophoresis, Serum *on page 1104*

Beta-Hexosaminidase, Serum, White Blood Cells

Synonyms Beta-N-Acetylglucosaminidase; Beta-N-Acetylhexosaminidase; Hexosaminidase; β-Hexosaminidase, Serum; N-Acetyl-β-Hexosaminidase A; NAG; β-NAG A

(Continued)

Beta-Hexosaminidase, Serum, White Blood Cells
(Continued)

Test Includes May include isolation of isoenzyme hexosaminidase A, B, and S. Patients with Tay-Sachs disease lack hexosaminidase A, while B may be normal or elevated.

Abstract The enzyme, beta-hexosaminidase, includes the isoenzymes, hexosaminidase A, hexosaminidase B, and hexosaminidase S. In Tay-Sachs disease, isoenzyme A is missing and B and S may be normal or increased. Deficiency of hexosaminidase A results in more than a 100-fold increase in G_{M2} ganglioside in the neurons of affected children. In Sandhoff disease, even less common, A and B are absent.

Specimen Serum, white blood cells, cultured fibroblasts, amniotic fluid cells

Container Red top tube for serum, yellow top for white blood cells

Storage Instructions Store at -20°C.

Special Instructions Usually sent to reference laboratory.

Reference Interval Hexosaminidase A isoenzyme: noncarriers, 450-600 µmol/hour/mL; Tay-Sachs homozygotes: none present; heterozygotes: 200-300 µmol/hour/mL. If isoenzyme A is <50% of total activity, carrier status is indicated.

Use Evaluation of Tay-Sachs disease and Sandhoff disease

Limitations Total hexosaminidase levels are not useful for evaluation of Tay-Sachs disease. A mutant allele may prevent reliable measurement of enzyme activity when artificial substrates are employed. Serum assays are ambiguous on pregnant females.

Methodology DEAE chromatography followed by photometric or fluorometric assay of appropriate isoenzyme

Additional Information Hexosaminidase A can be measured in serum, cultured fibroblasts or leukocytes. Prenatal diagnosis of Tay-Sachs disease is done with amniocentesis or chorionic villus sampling. If females wish to be screened for carrier status of hexaminidase A deficiency by serum assays, they must be tested prior to pregnancy, as serum assays are ambiguous on pregnant females. More than 50 mutations have been reported in alpha subunit of hexaminidase A gene. DNA-based assays using PCR are available, but the enzyme assay is presently recommended as a general screen. Ultimately, molecular methods will replace the enzyme assay.

References

Gilbert-Barness E and Barness LA, *Metabolic Diseases*, Natick, MA: Eaton Publishing, 2000, 323-90.

Kaback M, Lim-Steele J, Dabholkar D, et al, "Tay-Sachs Disease - Carrier Screening, Prenatal Diagnosis, and the Molecular Era. An International Perspective, 1970 to 1993," *JAMA*, 1993, 270(19):2307-15.

Wappner RS, "Lysosomal Storage Disorders," *Oski's Pediatrics: Principles and Practice*, McMillan JA, DeAngelis CD, Feigin RD, et al, eds, Philadelphia, PA: JB Lippincott Co, 1999, 1863-83.

♦ **Beta-Hydroxybutyrate** *see* C-Peptide, Serum *on page 465*

♦ **Beta-Hydroxybutyrate** *see* Ketone Bodies, Blood *on page 816*

♦ **Beta-Hydroxybutyric Acid, Urine** *see* Ketones, Urine *on page 817*

♦ **11-Beta-Hydroxylase** *see* 17-Hydroxyprogesterone, Whole Blood, Serum, or Plasma *on page 755*

Beta-Lactamase Test

Related Information

Antimicrobial Susceptibility Testing, Aerobic and Facultatively Anaerobic Bacteria *on page 178*

Antimicrobial Susceptibility Testing, Anaerobic Bacteria *on page 180*

Bacterial Culture, Aerobes *on page 229*

Neisseria gonorrhoeae Culture and Smear *on page 945*

Applies to Cephalosporinases; Penicillinases

Test Includes Rapid testing of isolated bacterial colonies for the production of beta lactamase

Abstract Certain bacteria produce enzymes that inactivate beta-lactam antibiotics. Some enzymes can hydrolyze penicillin (penicillinases); others hydrolyze cephalosporins (cephalosporinases). Recently, newer beta lactamases have been identified that have an extended spectrum of antimicrobial activity. These enzymes have been named extended-spectrum beta lactamases (ESBL). In all cases, the detection of enzyme production by bacterial isolates is essential in determination of appropriate therapy.

Specimen Isolated colonies of *Haemophilus influenzae*, *Moraxella* (*Branhamella*) *catarrhalis*, *Neisseria gonorrhoeae*, enterococci, *Staphylococcus aureus*, or gram-negative anaerobic rods including *Bacteroides fragilis*

Use Rapid detection of beta-lactamase production; all clinically significant isolates of *E. coli* and *Klebsiella* species should be tested for the presence of ESBLs.

Limitations Beta-lactamase negative strains may be resistant to penicillins or cephalosporins by other mechanisms.

Methodology Paper disk with a chromogenic cephalosporin reagent. The hydrolysis of the beta-lactam ring results in a color change.

Extended-spectrum beta lactamases (ESBL): Several methods have been suggested by the National Committee for Clinical Laboratory Standards for the screening and detection of ESBLs. The initial screen uses a broth-based method to detect MIC values >1 µg/mL to selected cephalosporins (cefpodoxime, ceftazidime, cefotaxime, ceftriaxone) or aztreonam. The confirmatory test uses a broth-based method with both cefotaxime and ceftazidime alone and in combination with clavulanic acid. Several commercial tests have been developed to screen and detect *E. coli* and *Klebsiella* species for the presence of ESBLs.

Additional Information The beta-lactamase test can provide clinically useful information when used for organisms in which the primary mechanism of resistance to beta-lactam antibiotics is by means of beta-lactamase enzymes, and in which resistance patterns to other antimicrobial agents is predictable. Consequently, beta-lactamase testing is restricted to a few specific circumstances that meet such criteria. *H. influenzae* and *M. catarrhalis* isolates that produce beta lactamase are resistant to ampicillin, but resistance to other agents is predictable (ie, virtually all isolates are presently susceptible to most second and third generation cephalosporins, beta-lactamase inhibitor combinations). *N. gonorrhoeae* isolates that are beta-lactamase positive are resistant to penicillin, but uniformly susceptible to ceftriaxone. Beta-lactamase testing of anaerobic gram-negative bacilli (eg, *Bacteroides*, *Porphyromonas*, and *Prevotella* sp) can be performed using the nitrocefin disk assay. However, beta-lactamase testing of anaerobic bacteria has limited clinical utility since a number of anaerobic bacteria can be resistant to beta-lactam antibiotics in the absence of beta lactamases.

Recently, it has been recognized that changes have occurred in the beta lactamases produced by gram-negative bacteria. These new enzymes have a broader spectrum of activity against several beta-lactam antimicrobial agents. These enzymes are not homogeneous (there are many different types) and do not have predictable responses to beta-lactam antibiotics in the laboratory. More important, patients infected with these ESBL producing infections often do not respond to beta-lactam antibiotics, even when routine susceptibility testing indicates a susceptible phenotype. It is important for the laboratory to properly screen and detect these enzymes to prevent placing patients with serious gram-negative infections at risk.

References

Dennesen PJ, Bonten MJ, and Weinstein RA, "Multiresistant Bacteria as a Hospital Epidemic Problem," *Ann Med*, 1998, 30(2):176-85.

Hand WL, "Current Challenges in Antibiotic Resistance," *Adolesc Med*, 2000, 11(2):427-38.

MacKenzie FM and Gould IM, "Extended Spectrum Beta-Lactamases," *J Infect*, 1998, 36(3):255-8.

National Committee for Clinical Laboratory Standards, "Performance Standards for Antimicrobial Susceptibility Testing; 11th Informational Supplement," NCCLS Document M100-S11, Volume 21(1), Wayne, Pennsylvania, 2001.

Thomson KS and Smith Moland E, "Version 2000: The New Beta-Lactamases of Gram-Negative Bacteria at the Dawn of the New Millennium," *Microbes Infect*, 2000, 2(10):1225-35.

- **Beta Subunit Human Chorionic Gonadotropin, Urine or Serum** *see* Pregnancy Test, Serum or Urine *on page 1084*
- **Beta-Thromboglobulin** *see* Hypercoagulation Panel *on page 758*
- **Beta-Thromboglobulin** *see* Platelet Aggregation *on page 1045*
- **Bethesda Assay** *see* Factor Inhibitors *on page 566*
- **Bethesda System** *see* Cervical/Vaginal Cytology *on page 376*
- **Beutler-Baluda Test** *see* Newborn Screening Tests for Galactosemia *on page 960*
- **Beutler Test** *see* Newborn Screening Tests for Galactosemia *on page 960*
- **BGP** *see* Osteocalcin, Serum or Plasma *on page 983*
- **BH₄ Cofactor Deficiencies** *see* Newborn Screen for Phenylketonuria *on page 954*
- **Bhang** *see* Cannabinoids (Marijuana Metabolites), Qualitative, Urine *on page 335*
- **Biased Statistic** *see page 11*
- **Bicarbonate** *see* Carbon Dioxide, Total, Blood *on page 339*

Bicarbonate, Blood

Related Information

Anion Gap, Serum, Plasma, or Urine *on page 160*
Blood Gases and pH, Arterial *on page 275*
Carbon Dioxide, Total, Blood *on page 339*
Chloride, Serum, Plasma, or Blood *on page 391*
Creatinine, Serum or Plasma *on page 474*
Electrolyte Panel, Serum *on page 532*
Ketone Bodies, Blood *on page 816*
Ketones, Urine *on page 817*
Osmolality, Calculated, Serum or Plasma *on page 976*
pCO₂, Blood *on page 1008*
pH, Blood *on page 1018*
Potassium, Serum or Plasma *on page 1078*
Transfusion Reaction Work-up *on page 1269*
Urea Nitrogen, Serum or Plasma *on page 1284*

Synonyms Carbon Dioxide; CO_2; HCO_3^-

Applies to Henderson-Hasselbalch Equation; TCO_2

Test Includes Test is part of blood gases panel and electrolytes panel

Abstract The two major extracellular anions are chloride and bicarbonate. Bicarbonate (HCO_3^-) makes up about 25 mmol/L of the anions found in normal plasma and is a major contributor to the bicarbonate/carbonic acid plasma-buffering system that maintains acid-base homeostasis. Its estimation is fundamental in evaluation of acid-base and electrolyte status.

Specimen Whole blood, plasma or serum

Container Heparinized blood gas syringe, green top (heparin) tube, red top tube

Reference Interval Newborns and infants: whole blood: 16-24 mmol/L; children and adults: serum or plasma: arterial: 21-28 mmol/L, venous: 22-29 mmol/L

Possible Panic Range <10 mmol/L, >40 mmol/L

Use With anion gap, HCO_3 is used as a preliminary screen for many abnormalities of acid-base balance, but additional studies are needed for some diseases (eg, acute respiratory disorders).

HCO_3^- is **increased** with metabolic alkalosis, respiratory acidosis, and compensated respiratory acidosis. Metabolic alkalosis is usually acute and often is accompanied by hypokalemia. It may be seen with dehydration, use of some diuretics, and with vomiting. Ingestion of baking soda and citrate intoxication from massive transfusions leads to increased urine anion gap. Increased cortisol or aldosterone may cause metabolic alkalosis from urinary losses of hydrogen and potassium ions.

HCO_3^- is **decreased** with metabolic acidosis (eg, low in ketoacidosis) and compensated respiratory alkalosis. Severe metabolic acidemia bears an implication of bicarbonate levels ≤8 mmol/L.

The combination of decreased HCO_3^- with high chloride and normal anion gap, resulting from loss of bicarbonate, occurs with diarrhea and with renal tubular acidosis. Such disorders are called hyperchloremic or nonanion gap metabolic acidosis, but when gastrointestinal HCO_3^- losses are profuse, the anion gap can be raised. Subjects with renal tubular acidosis and hypobicarbonatemia can present with hypokalemia.

Severe acidemia is treated with intravenous sodium bicarbonate, among other measures. To provide it judiciously, HCO_3^- must be followed with other studies.

Marked hypobicarbonatemia is seen in alcoholic ketoacidosis.

Limitations Excessive heparin dilution can falsely lower calculated arterial bicarbonate by as much as 10 mmol/L, and should be avoided by using only enough heparin sufficient to fill the dead space of the syringe. Inadequate filling of red top evacuated collection tubes will decrease apparent serum bicarbonate concentrations.

Methodology HCO_3^- is not necessarily directly measured. Instead, in whole blood specimens, pH and pCO_2 are measured using standard blood gas instrumentation, and the concentration of HCO_3^- ($cHCO_3^-$) is calculated using the Henderson-Hasselbalch equation:

$$pH = 6.103 + \log [HCO_3^- / (0.0306 \times pCO_2)].$$

Additionally, methods for measuring total carbon dioxide (TCO_2) in serum or plasma are widely used to estimate HCO_3^-. See Carbon Dioxide, Total, Blood *on page 339*.

Ion-selective electrode (ISE) and enzymatic methods are in current use.

Additional Information Under normal conditions, HCO_3^- constitutes ~95% of TCO_2. In the vast majority of clinical settings, the calculated HCO_3^- value is comparable to the measured TCO_2 for diagnostic and therapeutic considerations. If only TCO_2 is measured (without pH) it is often used interchangeably with HCO_3^-. Error using this estimate is usually ≤2 mmol/L.

Other laboratory tests used in the evaluation of diabetic ketoacidosis include the other electrolytes, plasma glucose, blood and urine ketone bodies, hematocrit, BUN, pH and osmolality.

In early infancy, higher bicarbonate and lower serum chloride provide help to distinguish infants with pyloric stenosis from those with gastroesophageal reflux. Bicarbonate concentrations ≥29 mmol/L identified infants with pyloric stenosis with sensitivity of 36%, specificity of 99%, and positive predictive value of 99%. Chloride levels ≤98 mmol/L were found in infants with pyloric stenosis with sensitivity of 50%, specificity of 99%, and positive predictive value of 97%.

References

Adrogué HJ and Madias NE, "Management of Life-Threatening Acid-Base Disorders," First of Two Parts, *N Engl J Med*, 1998, 338(1):26-34.

Heusel JW, Siggaard-Andersen O, and Scott MG, "Physiology and Disorders of Water, Electrolyte, and Acid-Base Metabolism," *Tietz Textbook of Clinical Chemistry*, 3rd ed, Chapter 32, Burtis CA and Ashwood ER, eds, Philadelphia, PA: WB Saunders Co, 1999, 1095-124.

Smith GA, Mihalov L, and Shields BJ, "Diagnostic Aids in the Differentiation of Pyloric Stenosis From Severe Gastrointestinal Reflux During Early Infancy," *Am J Emerg Med*, 1999, 17(1):28-31.

Bile Fluid Examination

Related Information

Ova and Parasites, Direct Exam *on page 985*

Synonyms Biliary Drainage Examination; Biliary Sludge Examination; Crystal Examination; Duodenal Drainage Examination

Abstract A variety of bile/duodenal content specimens obtained by tube aspiration, endoscopic retrograde cholangiopancreatography, or direct gallbladder puncture are studied for the presence of cholesterol, calcium bilirubinate, calcium carbonate crystals, bilirubin crystals, or parasites (predominantly *Giardia lamblia*). Crystals may support diagnoses of gallbladder and/or pancreatic disease. **Biliary sludge** is a suspension of precipitates of cholesterol monohydrate crystals or calcium bilirubinate granules in bile, but other calcium salts including calcium carbonate microspheroliths may be relevant as well.

Sludge in most persons resolves without treatment, but other outcomes include correlation with cholecystolithiasis, biliary colic, acute pancreatitis, and acute cholecystitis.

Patient Preparation Specimen is obtained by use of a gastroduodenal endoscopy study, either by direct aspiration or into a trap. Patient must take nothing by mouth after midnight before the test.

Specimen Bile fluid, duodenal drainage specimen (gallbladder bile is preferred over hepatic bile)
(Continued)

Bile Fluid Examination *(Continued)*

Collection The collection may be divided into multiple containers, usually three: "A" bile (yellow - common duct origin), "B" bile (viscous and green or green-brown - gallbladder bile), and "C" bile (lighter color - hepatic bile duct origin). Many physicians send only one specimen, either a "pool" or a collection of largely "B" bile. Such "B" bile (gallbladder origin) may have a volume of 30-50 mL.

Storage Instructions Cholesterol that does not redissolve may crystallize with freezing. Centrifuge the specimen if it cannot be examined immediately; such sediment can be safely frozen. Refrigeration of whole bile samples may be followed by bacterial proliferation.

Reference Interval Gallbladder bile is normally clear and brown. Normal bile should not include any significant number of cholesterol crystals, calcium bilirubinate or bilirubin crystals, parasites, or inflammatory cells.

Use Evaluate cholelithiasis, cholecystitis, and biliary colic in cases in which a high index of suspicion exists, but in which primary tests are negative. Direct microscopy of gallbladder content is more sensitive for detection of sludge than is ultrasonography. To evaluate results of litholytic therapy, cholesterol crystals in bile can identify most cholesterol calculi and calcium bilirubinate granules can be used to identify pigment stones. Bile fluid examination may be used to detect occult microlithiasis in cases of "idiopathic" acute pancreatitis. Biliary sludge has been described with pregnancy, rapid weight loss, critical illness, prolonged fasting, prolonged parenteral nutrition, ceftriaxone or octreotide use, and with transplantation.

Such fluid may be used to establish the presence of giardiasis. See also Ova and Parasites, Direct Exam *on page 985.*

Limitations Specimen pH <4.5 may produce a false-positive bilirubin precipitate which may be confused with calcium bilirubinate.

This diagnostic approach for identification of crystals and diagnosis of cholelithiasis is not widely used and is not found in all major textbooks. It is less clinically applicable than ultrasonography. The sensitivity of transabdominal ultrasonography for sludge is only ~55% while sensitivity of endoscopic ultrasonography is ~96%.

Gallstones are formed only in a minority of subjects with sludge.

Additional Information In studies of gallbladder bile (obtained at cholecystectomy), presence of rhomboid, birefringent cholesterol crystals as an indication of cholesterol gallstones has a sensitivity approaching 90% and a specificity of nearly 100%. Reddish brown bilirubinate crystals alone as predictors of pigment stones have a lower level of sensitivity (about 70%) and specificity (slightly >90%).

Almost 70% of cases of acute pancreatitis are caused by gallstones or alcohol abuse. Microscopic examination of centrifuged duodenal bile in patients recovering from an episode of acute pancreatitis will show cholesterol, bilirubinate, or calcium carbonate microspheroliths in nearly 70% of cases while bile from postalcoholic pancreatitis patients is usually negative for crystals.

Bile cholesterol supersaturation with nearly all patients having cholesterol crystals in their bile has been reported in postcolectomy ulcerative colitis patients. Cholesterol monohydrate predominates in pregnancy, with weight loss, and with octreotide therapy.

A sensitivity of 83% and specificity of 100% for recognition of cholelithiasis has been reported on the basis of microscopic examination of bile samples obtained at endoscopic retrograde cholangiography.

Dahan et al report different figures for sensitivity and specificity but conclude that when ultrasonography and microscopic examination of duodenal bile are negative, the risk of underdiagnosis of cholecystolithiasis is negligible.

While a variety of protozoan, trematode and nematode parasites have been found in duodenal specimens, *Giardia lamblia* is most frequently encountered. In patients clinically suspected of opisthorchiasis, in whom stool specimens are negative for ova, bile microscopy may be of special importance in establishing the diagnosis.

References

Bockus HL, Shay H, Willard JH, et al, "Comparison of Biliary Drainage and Cholecystography in Gallstone Diagnosis: With Especial Reference to Bile Microscopy," *JAMA*, 1931, 96:311-17.

Dahan P, Andant C, Levy P, et al, "Prospective Evaluation of Endoscopic Ultrasonography and Microscopic Examination of Duodenal Bile in the Diagnosis of Cholecystolithiasis in 45 Patients With Normal Conventional Ultrasonography," *Gut*, 1996, 38(2):277-81.

Ko CW, Sekijima JH, and Lee SP, "Biliary Sludge," *Ann Intern Med*, 1999, 130(4 Pt 1):301-11.

- **Bile, Urine** see Bilirubin, Urine *on page 268*
- **Biliary Drainage Examination** see Bile Fluid Examination *on page 259*
- **Biliary Sludge Examination** see Bile Fluid Examination *on page 259*
- **Biliprotein** see Bilirubin, Direct, Serum *on page 262*

Bilirubin, Amniotic Fluid, Delta A450

Related Information

Amniotic Fluid, Chromosome and Genetic Abnormality Analysis *on page 152*
Bilirubin, Neonatal, Serum *on page 263*
Cord Blood Antibody Screen *on page 453*
Hemolytic Disease of the Newborn, Antibody Identification *on page 690*
Newborn Crossmatch and Transfusion *on page 952*
Prenatal Screen, Immunohematology *on page 1086*
Rh Genotype *on page 1162*

Synonyms Amniotic Fluid Analysis for Hemolytic Disease of the Newborn; Amniotic Fluid Bilirubin; Amniotic Fluid Spectral Analysis; Liley Test; OD 450 Method

Replaces Amniotic Fluid Analysis for Erythroblastosis Fetalis

Abstract The amniotic fluid bilirubin level is directly proportional to the degree of fetal anemia in hemolytic disease of the newborn (HDN). The test is performed to monitor the pregnancies of women with atypical antibodies (see Hemolytic Disease of the Newborn, Antibody Identification *on page 690*). Measurement of the 450 nm peak (delta A450) by scanning spectrophotometry estimates the bilirubin level in amniotic fluid, which is directly related to the severity of disease. Each patient's delta A450 numerical result is compared to normative criteria which were originally developed by Liley and later extended by Queenan et al (see the following graphic). A graphic format is usual, and includes four zones, with zone 1 implying an unaffected fetus and zone 4 implying a high risk for intrauterine death from HDN.

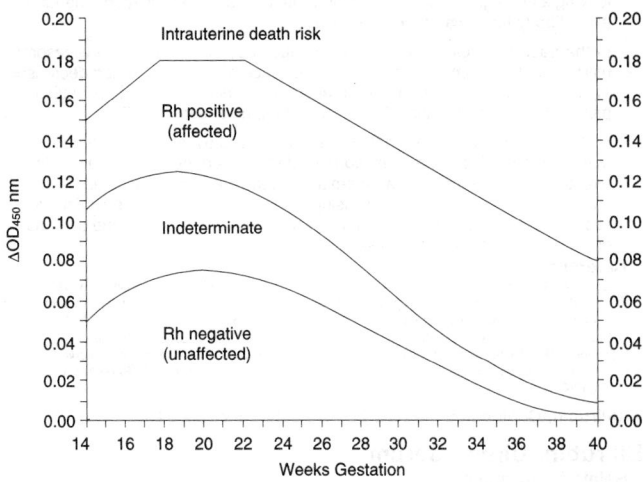

Amniotic fluid ΔOD_{450} management zones

(Continued)

Bilirubin, Amniotic Fluid, Delta A450 *(Continued)*

Aftercare All nonsensitized Rh-negative patients should receive anti-D immuno-globulin within 72 hours after amniocentesis.

Specimen Amniotic fluid

Collection Ultrasound-guided amniocentesis is performed by the physician, usually at 27 weeks or more gestation. Protect the specimen from light for transport to the laboratory. In the case of multiple pregnancy, each amniotic sac should be sampled and analyzed individually.

Storage Instructions Centrifuge the specimen promptly and filter the supernatant in the dark. Store the supernatant protected from light at 4 °C until analysis.

Reference Interval See Use.

Use The reference interval cutoffs for zone 1, implying an unaffected fetus, vary with gestational age (see graphic). These cutoffs apply most specifically to pregnancies complicated by Rh(D) isoimmunization, and to the first such affected pregnancy. In practice, however, the same cutoffs are applied to HDN caused by other red cell antigens.

Limitations Amniocentesis is an invasive procedure and poses potential hazards to the fetus, including enhanced maternal sensitization, amnionitis, or premature rupture of membranes.

Amniotic fluid contaminated with meconium and/or maternal blood can give erroneous results. Maternal urine can be aspirated inadvertently instead of amniotic fluid, but can be conveniently identified with urea nitrogen and creatinine measurements.

The most common antibodies causing HDN are anti-D and anti-K.

In cases of **Kell isoimmunizations** (HDN due to anti-K), delta A450 testing is probably not helpful since fetal anemia is due to erythroid suppression rather than hemolysis. This fact, in addition to the low predictive value of anti-Kell titers in these instances, mandates aggressive surveillance of the Kell positive fetus and, if necessary, fetal umbilical blood determinations to assess disease severity.

Methodology Spectral analysis of centrifuged amniotic fluid is done using a scanning spectrophotometer. The peak at 450 nm corresponds to bilirubin.

Additional Information Monitoring serum antibody titers of anti-D antibodies has been reported to be superior to delta A450 determinations for predicting fetal anemia; however, the titers must be less ≤32 or ≥1000 to be useful, leaving a large gray area. Consultation with clinical experts is recommended for alloantibody titrations other than anti-D.

In the past, most cases of HDN were caused by anti-D. With the widespread utilization of Rh immune globulin, the incidence of anti-D HDN has decreased. This has resulted in a **relative** increase in the incidence of other antibodies, particularly other Rh antibodies and anti-Kell, as causes of HDN.

High amniotic fluid delta-OD450 is usually followed by fetal blood sampling; the management of pregnancies complicated by red cell isoimmunization is directed at evaluation of fetal anemia, then intervention if indicated. Noninvasive fetal RhD genotyping is possible in maternal plasma, beginning in the second trimester. Prenatal diagnosis by molecular methods has the potential to decrease requirements for invasive testing.

References

Queenan JT, Tomai TP, Ural SH, et al, "Deviation in Amniotic Fluid Optical Density at a Wavelength of 450 nm in Rh-Immunized Pregnancies From 14 to 40 Weeks Gestation: A Proposal for Clinical Management," *Am J Obstet Gynecol*, 1993, 168(5):1370-6.

Saade GR, "Noninvasive Testing for Fetal Anemia," *N Engl J Med*, 2000, 342(1):52-3.

Weiner CP, "Fetal Hemolytic Disease," *High Risk Pregnancy Management Options*, 2nd ed, Chapter 22, James DK, Steer PJ, Weiner CP, et al, eds, London: WB Saunders Co, 1999, 343-61.

♦ **Bilirubin, Conjugated** *see Bilirubin, Direct, Serum on page 262*

Bilirubin, Direct, Serum

Related Information

Bilirubin, Neonatal, Serum *on page 263*
Bilirubin, Total, Serum *on page 265*
Bilirubin, Urine *on page 268*

Hepatitis B Serology *on page 702*
Hepatitis C Virus RNA Detection and Quantitation *on page 705*
Hepatitis: Laboratory Assessment, Overview *on page 713*
Liver Disease: Laboratory Assessment, Overview *on page 869*
Urobilinogen, 2-Hour Urine *on page 1296*

Synonyms Bilirubin, Conjugated

Applies to Biliprotein; Conjugated Hyperbilirubinemia; Delta Bilirubin

Abstract The mono- and diconjugated bilirubin and some of the delta bilirubin (covalently bound to albumin) account for the "direct" value. If more than half of total bilirubin is direct-reacting, jaundice is expressed as **"conjugated hyperbilirubinemia,"** which indicates hepatocellular dysfunction or cholestasis.

Specimen Serum

Container Red top tube, red top Microtainer™ for babies

Collection Pediatrics: Blood drawn from a heelstick. If blood is drawn by capillary puncture, avoid excessive squeezing or milking (to avoid hemolysis).

Storage Instructions Store in refrigerator. **Protect from light.**

Reference Interval Newborns: varies with age in days, prematurity vs maturity (see table in listing Bilirubin, Neonatal, Serum *on page 263*); adults: ≤0.4 mg/dL (SI: ≤7 µmol/L)

Use Increased direct bilirubin occurs with hepatobiliary diseases, including both intrahepatic and extrahepatic lesions. Hepatocellular causes of elevation include hepatitis, cirrhosis, and advanced neoplastic states. Increased with cholestatic drug reactions, Dubin-Johnson syndrome, and Rotor syndrome. In the latter two syndromes, the level is usually <5 mg/dL.

Infections (eg, bacterial sepsis, hepatitis B, syphilis, toxoplasmosis, rubella, CMV, herpes), drug-induced cholestasis, and parenteral nutrition-associated cholestasis are among acquired causes of increased direct bilirubin.

Genetic and metabolic disorders which may increase direct bilirubin include galactosemia, tyrosinemia, Niemann-Pick disease, and trisomy 18.

Limitations Cord blood samples may yield elevated values. Visibly hemolyzed samples may yield spurious results by some methods. **Drugs** which may cause increases include acetaminophen, aminosalicylic acid, anabolic steroids (cholestatic syndrome), chlorpromazine, interleukin-2, mephenytoin, nalidixic acid, methyldopa, oral contraceptives, phenothiazine, propylthiouracil, sulfasalazine, and other agents.

Contraindications Measurement of direct bilirubin is usually not necessary when the total bilirubin is <1.2 mg/dL (SI: <21 µmol/L).

Additional Information Direct bilirubin is the water soluble (conjugated) fraction. When conjugated bilirubin is sufficiently increased in serum, bilirubin becomes positive in urine.

Physiologic jaundice, occurring 2-4 days after birth, is due to lack of liver glucuronyl transferase. Physiological jaundice does not produce conjugated (direct) bilirubin levels >1.5-2.0 mg/dL.

When bilirubin is rising in early liver disease, very little delta bilirubin (biliprotein) is present. In resolving liver disease, conjugated fractions may return to normal quickly whereas delta bilirubin may stay elevated for a long time due to long half-life (20 days) of albumin.

References

Iyanagi T, Emi Y, and Ikushiro S, "Biochemical and Molecular Aspects of Genetic Disorders of Bilirubin Metabolism," *Biochim Biophys Acta,* 1998, 1407(3):173-84.

Shiomi S, Habu D, Kuroki T, et al, "Clinical Usefulness of Conjugated Bilirubin Levels in Patients With Acute Liver Diseases," *J Gastroenterol,* 1999, 34(1):88-93.

Young DS, *Effects of Drugs on Clinical Laboratory Tests,* 5th ed, Volume 1: Listing by Test, Washington, DC: AACC Press, American Association of Clinical Chemistry, 2000, Section 3, 139-40.

Bilirubin, Neonatal, Serum

Related Information

Bilirubin, Amniotic Fluid, Delta A450 *on page 261*
Bilirubin, Direct, Serum *on page 262*
Bilirubin, Total, Serum *on page 265*
Cord Blood Antibody Screen *on page 453*
Hemolytic Disease of the Newborn, Antibody Identification *on page 690*
Newborn Crossmatch and Transfusion *on page 952*
(Continued)

Bilirubin, Neonatal, Serum *(Continued)*

Synonyms Baby Bilirubin; Microbilirubin; Total Bilirubin, Neonatal
Applies to Transcutaneous Bilirubinometry
Specimen Serum
Container Microbilirubin tube
Collection Draw blood from heel using capillary pipette. Avoid excessive squeezing (to avoid specimen hemolysis).
Storage Instructions Protect sample from light; bilirubin is photosensitive.
Reference Interval Reference limit depends on whether baby is premature or term, and age in days. See table.

Bilirubin, Neonatal Upper Reference Limit (mg/dL)

Age	Premature	Full-Term
Cord	2.9	2.5
<24 h	8.0	6.0
<48 h	12.0	10.0
3-5 d	15.0	12.0
7 d	15.0	10.0

Note: At 7 days, occasional premature infants may develop kernicterus at 10.0-12.0 mg/dL of bilirubin.

Critical Values Jaundice in the first 24 hours of life is indication for a neonatal bilirubin determination.
Possible Panic Range >15.0 mg/dL (SI: >257 µmol/L) in term infants, 10.0-15.0 mg/dL (SI: 171-257 µmol/L) in premature babies. Premature infants are at greater risk for bilirubin toxicity. In HDN, published indications for exchange transfusion include:

- hematocrit <45%, positive direct antiglobulin test with bilirubin >4 mg/dL in cord blood
- Postnatal rise of bilirubin >1 mg/dL/hour for more than 6 hours
- Progressive anemia with rate of increase of bilirubin >0.5 mg/dL/hour
- Continuing progression of anemia
- Using bilirubin only, bilirubin >15 mg/dL for more than 48 hours
- Other criteria

Use Detection and evaluation of jaundice. Physiologic jaundice is characterized by unconjugated hyperbilirubinemia which peaks by the third or fourth day in full-term neonates, then declines. In premature infants, bilirubin peaks on the fifth to seventh day. Pathologic jaundice usually is found within the first 24 hours; bilirubin rises rapidly, often >5 mg/dL/day.
Limitations If direct spectrophotometric method is used, only total bilirubin is measured with "neonatal bilirubin." Procedure is not utilized for patients older than 10 days of age due to formation of endogenous carotenoids. Ten percent fat emulsion has been reported to interfere with neonatal bilirubin measurement.
Contraindications Spectrometric assay for bilirubin should not be used for infants older than 10 days of age, for whom usual total bilirubin is indicated.
Methodology Spectrophotometric, direct (bichromatic), bilirubin oxidase, diazo-dye binding, transcutaneous bilirubin by reflectance.

Bilirubin measurement by transcutaneous bilirubinometry has become available and may prove to be an adjunctive screening tool which enhances patient care by reducing frequent blood draws.
Additional Information In 1932, Diamond showed that hydrops fetalis and kernicterus were aspects of the same disease, in which fetal and neonatal hemolysis takes place. Kernicterus is caused by deposition of unconjugated bilirubin in the brain and leads to devastating injury. HDN occurs from maternal alloimmunization to RhD, other Rh antibodies including hr'(c), ABO incompatibility (A), and antibodies involving additional blood groups including Kell.

Uncommon causes of neonatal jaundice include galactosemia, sepsis, hepatitis, syphilis, toxoplasmosis, cytomegalovirus, rubella, and G6PD and pyruvate kinase deficiencies.

If bilirubin concentration is >15-17 mg/dL in a term infant at 25-48 hours, phototherapy may be indicated, an individual clinical judgment related to age and bilirubin level. The American Academy of Pediatrics has published a practice parameter which includes management tables, treatment options, and algorithms.

Drugs may displace bilirubin from albumin. It is the so-called "free" form of bilirubin, thus displaced, which crosses the blood-brain barrier and causes neurotoxicity.

Information relevant to this topic is included in Bilirubin, Amniotic Fluid, Delta A450 on page 261 and Hemolytic Disease of the Newborn, Antibody Identification on page 690.

References

American Academy of Pediatrics, "Practice Parameter: Management of Hyperbilirubinemia in the Healthy Term Newborn," *Pediatrics*, 1994, 94(4):558-65.

Bowman JM, "RhD Hemolytic Disease of the Newborn," *N Engl J Med*, 1998, 339(24):1775-7 (editorial).

Bratlid D, "Criteria for Treatment of Neonatal Jaundice," *J Perinatol*, 1996, 16(3 Pt 2):S83-S88.

Dale JC and Hamrick HJ, "Neonatal Bilirubin Testing Practices. Reports From 312 Laboratories Enrolled in the College of American Pathologists Excel Proficiency Testing Program," *Arch Pathol Lab Med*, 2000, 124(10):1425-8.

Dennery PA, Seidman DS, and Stevenson DK, "Neonatal Hyperbilirubinemia," *N Engl J Med*, 2001, 344(8):581-90.

Lo YD, Hjelm YM, Fidler C, et al, "Prenatal Diagnosis of Fetal RhD Status by Molecular Analysis of Maternal Plasma," *N Engl J Med*, 1998, 339(24):1734-8.

Internet Web Sites

www.aap.org

Bilirubin, Total, Serum

Related Information

Acetaminophen, Serum on page 90
Alanine Aminotransferase, Serum on page 116
Alkaline Phosphatase, Serum on page 127
Amylase, Serum on page 155
Anemia Flowchart on page 35
Aspartate Aminotransferase, Serum on page 216
Bilirubin, Direct, Serum on page 262
Bilirubin, Neonatal, Serum on page 263
Bilirubin, Urine on page 268
Carbohydrate-Deficient Transferrin, Serum on page 338
Chlorpromazine, Serum on page 395
Ethanol, Blood, Urine, and Other Sources on page 558
Gamma-Glutamyl Transferase, Serum on page 629
Hepatitis B Serology on page 702
Hepatitis C Virus RNA Detection and Quantitation on page 705
Hepatitis: Laboratory Assessment, Overview on page 713
Isoniazid, Serum or Plasma on page 813
Leucine Aminopeptidase (LAP), Serum and Urine on page 844
Lipase, Serum on page 851
Liver Biopsy on page 864
Liver Disease: Laboratory Assessment, Overview on page 869
Liver/Kidney Microsomal Type 1 Antibodies on page 873
3,4 Methylenedioxymethamphetamine, Urine on page 907
Prothrombin Time on page 1116
Smooth Muscle Antibody on page 1207
Transfusion Reaction Work-up on page 1269
Urobilinogen, 2-Hour Urine on page 1296

Applies to Ascitic Fluid/Serum:Total Bilirubin Ratio; Fasting Bilirubin Test

Abstract Used to monitor diseases of the liver, biliary tract, and hemolytic diseases.

Specimen Serum

Container Red top tube; capillary tube for babies

Collection Pediatrics: Blood drawn from a heelstick. If sample is collected by capillary puncture, excess squeezing should be avoided (to avoid hemolysis and dilution with tissue fluids).

Storage Instructions Protect sample from light.

Causes for Rejection Gross hemolysis

(Continued)

Bilirubin, Total, Serum *(Continued)*

Reference Interval Newborns: see table in listing Bilirubin, Neonatal, Serum *on page 263*; adults: 0.3-1.0 mg/dL (SI: 5-17 µmol/L). Approximately 70% is indirect (unconjugated).

Use

Causes of **High Bilirubin**

- Hepatobiliary disease: hepatitis, cholangitis, cholecystitis, even without common duct calculi; cirrhosis, other types of liver disease (including primary or secondary neoplasia); alcoholism (usually with high AST (SGOT), GGT, MCV, or some combination of these findings); cholestasis (intrahepatic or extrahepatic); infectious mononucleosis (look also for increased LD (LDH), lymphocytosis); Dubin-Johnson syndrome; Gilbert disease (familial hyperbilirubinemia). If >80% of total bilirubin is indirect and total bilirubin is <6.0 mg/dL hemolysis or Gilbert syndrome is suggested.
- Malnutrition, anorexia, or prolonged fasting: 36 hours or more may cause moderate rise.
- Pernicious anemia, hemolytic anemias, erythroblastosis fetalis, other neonatal jaundice, hematoma and following a blood transfusion, especially if several units are given in a short time or with delayed hemolytic transfusion reaction. The major source of bilirubin is hemoglobin catabolism from lysis of red blood cells.
- Pulmonary embolism/infarct
- Congestive heart failure
- **Drugs:** A large number of drugs can cause jaundice by *in vivo* action. Drugs which may cause cholestasis and/or hepatocellular damage include acetaminophen, aminosalicylic acid, anabolic steroids, azathioprine, chlorpromazine, clindamycin, erythromycin, esterified estrogens, gentamicin, indinavir, indomethacin, isoniazid, MAO inhibitors, methyldopa, nortriptyline, oleandomycin, oral contraceptives, penicillin, phenothiazines, procainamide, progesterone, pyrazinamide, sulfonamides, valproic acid, warfarin, drugs of abuse (eg, 3,4 methylenedioxymethamphetamine - MDMA), and many other agents. A few drugs can cause analytical decreases (eg, amikacin, high doses of ascorbic acid, theophylline) and a large number of drugs can cause analytic, physiologic, or pathologic increases.

Limitations Differential diagnosis of liver diseases requires total and direct bilirubin values, as well as other tests. Visibly hemolyzed sera and lipemia can produce erroneous results. Serum alkaline phosphatase can be more sensitive to focal biliary obstruction.

Methodology Diazo reaction photometric method for adults is the most common; differential spectrophotometry for neonates (not useful for infants >10 days old); bilirubin oxidase method is also available.

Additional Information Interpretation of increased bilirubin is greatly enhanced by other chemistry results and selectively, with other studies; see table. In acute viral hepatitis with jaundice, for instance, the transaminases ALT (SGPT) and AST (SGOT) are consistently increased, while an isolated elevation of bilirubin is seen in Gilbert disease. **Obstruction** causes increases in bilirubin and alkaline phosphatase greater than and out of proportion to the transaminases. Gamma-glutamyl transferase is also increased in obstructive jaundice. Amylase and lipase are useful in differential diagnosis of obstructive jaundice. In **intrahepatic cholestasis,** the transaminases are not as increased, relative to bilirubin, as they are in hepatitis.

Nicotinic acid increases the formation of bilirubin in the spleen, leading to a rise in unconjugated bilirubin. This can be used as a test for **Gilbert disease** in which there is a moderate elevation of bilirubin with otherwise unremarkable chemistries. In Gilbert disease, decreased hepatic clearance of unconjugated bilirubin occurs with the criterion of basal total bilirubin >1.2 mg/dL. Although the indirect bilirubin level is increased in normal controls when nicotinic acid is given, the increase is greater in patients with Gilbert disease. The **fasting bilirubin test** can be used to support the diagnosis of constitutional hyperbilirubinemia. It involves fasting for 24 hours with a light breakfast, with only 100 g of sucrose and water allowed. Unconjugated bilirubin increases 1 mg/dL in subjects with Gilbert disease, or 1.5 mg/dL increase of total bilirubin.

Acute liver failure in **Wilson disease** is characterized by very high bilirubin, often >30 mg/dL (513 µmol/L) with decreased alkaline phosphatase (see

Clinical Approach to the Patient With Jaundice

Diagnostic Factors	Hemolytic	Hepatocellular	Intrahepatic Cholestatic	Extrahepatic Cholestatic
			Type of Jaundice	
Symptoms	May be asymptomatic or backache, joint pain	Nausea, vomiting, fever, anorexia	Deep jaundice, dark-colored urine, light-colored stools, pruritus	Deep jaundice, dark-colored urine, light-colored stools, pruritus, cholangitis, biliary colic
Physical findings	Splenomegaly	Tender hepatomegaly, splenomegaly[1]	Tender hepatomegaly	Hepatomegaly, palpable gallbladder
Liver tests				
Bilirubin				
Total	<6 mg/dL	Variable	Variable, may be >30 mg/dL	<30 mg/dL
Direct	<20%	>50%	>50%	>50%
Alanine aminotransferase (ALT)	Normal	>5-fold increase	2- to 5-fold increase	<2- to 3-fold increase; >3- to 5-fold increase with cholangitis
Alkaline phosphatase	Normal	<2- to 3-fold increase	>3- to 5-fold increase	>3- to 5-fold increase
Prothrombin time	Normal	Prolonged	Prolonged	Prolonged
Corrected by vitamin K		No	Variable	Yes
Ultrasonography of liver	No	No	No	Yes
Biliary dilatation				
Endoscopic retrograde cholangiopancreatography	Not necessary	Not necessary	Usually not necessary	Usually necessary

[1]May or may not be present.
Modified from Kamath PS, "Clinical Approach to the Patient With Abnormal Liver Test Results," *Mayo Clin Proc*, 1996, 71:1089-95.

(Continued)

Bilirubin, Total, Serum *(Continued)*

Copper, Serum *on page 448* and Copper, Urine *on page 452*). A ratio of alkaline phosphatase to bilirubin <2.0 is fairly distinctive.

Ascitic fluid/serum:total bilirubin ratio >6.0 supports distinction of exudate (eg, malignancy) from transudate with accuracy of 80%.

References

Elis A, Meisel S, Tishler T, et al, "Ascitic Fluid to Serum Bilirubin Concentration Ratio for the Classification of Transudates or Exudates," *Am J Gastroenterol*, 1998, 93(3):401-3.

Kamath PS, "Clinical Approach to the Patient With Abnormal Liver Test Results," *Mayo Clin Proc*, 1996, 71:1089-95.

Lee WM, "Acute Liver Failure," *N Engl J Med*, 1993, 329(25):1862-72.

Young DS, *Effects of Drugs on Clinical Laboratory Tests*, 5th ed, Volume 1: Listing by Test, Washington, DC: AACC Press, American Association for Clinical Chemistry, 2000, Section 3, 122-38.

Bilirubin, Urine

Related Information

Bilirubin, Direct, Serum *on page 262*
Bilirubin, Total, Serum *on page 265*
Liver Disease: Laboratory Assessment, Overview *on page 869*
Urobilinogen, 2-Hour Urine *on page 1296*

Synonyms Bile, Urine

Applies to Ictotest®; Urobilinogen, Urine

Abstract Since all urine bilirubin is conjugated, its presence is evidence of hepatocellular or biliary disease. Of heritable abnormalities of bilirubin metabolism, urine bilirubin is positive in Dubin-Johnson and Rotor types, but not in Gilbert or Crigler-Najjar disease.

Specimen Random urine

Storage Instructions Test immediately.

Special Instructions The specimen should be tested as soon as possible after voiding.

Reference Interval Negative (absent). (The chemical upper limit of normal, about 0.02 mg/dL is insufficient for detection by usual clinical testing).

Use In hepatocellular and obstructive disease of the biliary tract, urine bilirubin is frequently positive. Urine bilirubin has a sensitivity of 78% and a specificity of 86% as a screen for elevated serum bilirubin; the sensitivity is only 47% and specificity 89% as a screen for any liver function test abnormality. See table.

Differential Diagnosis Using Urine Bilirubin and Urobilinogen Tests

Type of Jaundice	Urine Bilirubin	Urine Urobilinogen
Normal	0	0 - trace
Hepatocellular jaundice (eg, hepatitis, chemical, or drug injury)	↑	↑
Biliary obstruction (extrahepatic and intrahepatic); obstructive jaundice	↑	0
Hemolytic jaundice	0	↑

0 = absent; ↑ = increased.

Limitations False positives are caused by stool contamination and drugs, including Ponstel® (mefenamic acid), Thorazine®, Ormazine® (chlorpromazine), rifampin, and etodolac. False negatives are caused by prolonged standing, vitamin C, and nitrites.

Methodology Diazotization reaction in an acid medium yields a blue to purple color (reagent strips)

Additional Information A more specific and sensitive **tablet** test, such as the Ictotest®, has been recommended when such dip-and-read results are inconclusive. The manufacturer expresses need for proper storage of the reagent. The tablet method is more sensitive than dipsticks. It detects as little as 0.05-0.1 mg bilirubin/100 mL. The test is exquisitely sensitive for bilirubinuria. Rifampin and chlorpromazine metabolites may interfere, and metabolites of mefenamic acid and flufenamic acid may cause false positives.

References
Binder L, Smith D, Kupka T, et al, "Failure of Prediction of Liver Function Test Abnormalities With the Urine Urobilinogen and Urine Bilirubin Assays," *Arch Pathol Lab Med*, 1989, 113(1):73-6.

- ◆ **Binomial Probability Function** *see page 11*
- ◆ **Biocoryl**® *see Procainamide, Serum on page 1092*
- ◆ **Biologic Variability** *see page 11*
- ◆ **Biologic Variation** *see page 11*
- ◆ **Biopsy** *see Histopathology on page 733*
- ◆ **Biopterin Cofactor Deficiency** *see Phenylalanine, Blood on page 1022*
- ◆ **Biotin** *see Lactic Acid, Whole Blood or Plasma on page 827*
- ◆ **Biotinidase** *see Amino Acids, Plasma on page 143*
- ◆ **Biotin-Labeled Red Cell Survival** *see* [51]Cr Red Cell Survival *on page 476*
- ◆ **Biquin**® *see Quinidine, Serum on page 1129*
- ◆ **Bisalbuminemia** *see Protein Electrophoresis, Capillary Zone on page 1103*

Blastomycosis Serology

Related Information

Fungal Culture, Biopsy or Body Fluid *on page 619*
Fungal Culture, Sputum *on page 624*
Fungal Culture, Urine *on page 625*
Fungus Smear, Stain *on page 626*

Abstract Blastomycosis is caused by the dimorphic fungus *Blastomyces dermatitidis*. The disease may present with subacute pneumonia, acute pneumonia, or as disseminated extrapulmonary disease. Extrapulmonary infections can affect the skin, bone, or genitourinary tract (including the prostate), and other sites (eg, CNS). This mold is a natural inhabitant of the soil, and most cases in the United States have occurred around the Great Lakes and Upper Mississippi River.

Specimen Serum

Container Red top tube

Reference Interval Complement fixation: titers <1:8; immunodiffusion: no precipitin band; enzyme immunoassay: titers <1:32

Use Support the diagnosis of infection due to *Blastomyces dermatitidis*. Acute and convalescent titers are helpful

Limitations Failure to demonstrate precipitin antibodies does not rule out blastomycosis. Cross reactions are seen in patients with histoplasmosis and coccidioidomycosis. Skin testing prior to the test may elevate the complement fixation titer. The complement fixation test lacks sensitivity and specificity and gives positive results in <50% of culture proven cases. Complement fixation assays are often cross-reactive. Newer EIA tests for blastomycosis have shown greater sensitivity with no compromise in specificity compared to other tests. EIA for antibody to purified A antigen is 90% sensitive, with some cross reaction with cases of histoplasmosis.

Methodology Immunodiffusion (ID), enzyme immunoassay (EIA), radioimmunoassay (RIA). DNA probes are being developed. The EIA assay uses purified antigen A and is 80% to 100% sensitive.

Additional Information The diagnosis of blastomycosis is established by demonstration of the organisms in smear, tissue sections, or by culture. Skin lesions are found in ~50% of patients, providing access to histopathologic and mycologic diagnosis.

Almost 50% of patients infected with *Blastomyces dermatitidis* have a negative antibody test on initial testing. Several serial samples should be tested to accurately diagnose blastomycosis. A negative serologic result has little value and in no way excludes the existence of blastomycosis. Cross reactions producing lines of partial identity are seen in patients with histoplasmosis and coccidioidomycosis. Repeated testing at 3-week intervals may be needed to secure a diagnosis. After diagnosis is established, falling titers are a good prognostic sign.

References
Chapman SW, Bradsher RW, Campbell GD, et al, "Practice Guidelines for the Management of Patients With Blastomycosis. Infectious Diseases Society of America," *Clin Infect Dis*, 2000, 30(4):679-83.

Lemos LB, Guo M, and Baliga M, "Blastomycosis: Organ Involvement and Etiologic Diagnosis: A Review of 123 Patients From Mississippi," *Ann Diagn Pathol*, 2000, 4(6):391-406.

(Continued)

Blastomycosis Serology *(Continued)*

McCullough MJ, DiSalvo AF, Clemons KV, et al, "Molecular Epidemiology of *Blastomyces dermatitidis*," *Clin Infect Dis*, 2000, 30(2):328-35.

Pappas PG, "Blastomycosis in the Immunocompromised Patient," *Semin Respir Infect*, 1997, 12(3):243-51.

Internet Web Sites
www.cdc.gov/ncidod/dbmd/diseaseinfo/blastomycosis_a.htm

Bleeding Time

Related Information
Platelet Aggregation *on page 1045*
Platelet Count *on page 1050*
von Willebrand Factor *on page 1321*

Applies to Bleeding Time, Duke; Bleeding Time, Ivy; Bleeding Time, Mielke

Abstract The bleeding time is intended to measure platelet function, but it is neither a sensitive nor a specific test. For this reason, its use is declining and at some institutions this test has been eliminated.

Patient Preparation Aspirin prolongs the bleeding time, and therefore, patients should not have taken aspirin or related compounds for at least 1 week prior to testing.

Clinicians may wish to inform the patient that a scar might form as a result of a bleeding time test, particularly if the patient has a history of keloids.

Aftercare A butterfly bandage is placed over the incision and kept in place for 24 hours.

Specimen None - performed at bedside by a coagulation technologist or other trained healthcare professional.

Turnaround Time 30 minutes or less after the coagulation technologist arrives at the bedside

Reference Interval Approximately 1.5-9.5 minutes (shorter in newborns)

Use Its intended use is as a measure of platelet function, but due to its inaccuracies, it is generally not useful.

Limitations Lacks sensitivity and specificity. Platelet counts <100,000/µL, low hematocrit, aspirin, other platelet inhibitory drugs, and certain other medications can prolong the bleeding time. Many variables influence the result, including skin thickness, temperature, blood vessel characteristics, the blade, orientation of the incision (horizontal vs vertical), location of the incision, handedness, and other features.

Methodology A trained healthcare professional makes a small incision on the patient's arm, and every 30 seconds gently blots the blood with filter paper to see if the bleeding has stopped. The filter paper must not touch the wound. Prior to making the cut, a blood pressure cuff is placed on the patient's arm at 40 mm Hg.

Additional Information The bleeding time can be prolonged in von Willebrand disease and other hereditary platelet function disorders, uremia, macroglobulinemia, and a variety of other conditions. However, it is not a reliable test for diagnosis or for predicting bleeding risk. In 1990, an analysis of 862 publications on the bleeding time concluded that the bleeding time is not a useful test, particularly as a preoperative screening test in a patient with a negative bleeding history. More recent publications continue to support this concept.

Historically, the Duke bleeding time was used, in which the earlobe or fingertip was pierced with a lancet. This was later replaced with the Ivy bleeding time, in which a blood pressure cuff was placed on the arm at 40 mm Hg and the forearm was cut with a lancet. This approach was later modified into the template bleeding time (Mielke bleeding time), which attempted to standardize the size and depth of the cut by placing a template on the skin. A spring-loaded blade within the template device creates a cut through a slit in the template. Two such template devices are Surgicutt® (International Technidyne Corp) and Simplate® (Organon Teknika Corp).

References
Brown BA, *Hematology: Principles and Procedures*, 6th ed, Philadelphia, PA: Lea and Febiger, 1993, 267-70.

Rodgers RP and Levin J, "A Critical Reappraisal of the Bleeding Time," *Semin Thromb Hemost*, 1990, 16(1):1-20.

♦ **Bleeding Time, Duke** *see* Bleeding Time *on page 270*

♦ **Bleeding Time, Ivy** *see* Bleeding Time *on page 270*
♦ **Bleeding Time, Mielke** *see* Bleeding Time *on page 270*

Blood and Fluid Precautions, Specimen Collection

Related Information

Arterial Blood Collection *on page 211*
Hepatitis B Antigen Detection *on page 699*
HIV-1/HIV-2 Serology *on page 736*
Phlebotomist Procedures *on page 1028*
Skin Puncture Blood Collection *on page 1203*
Venous Blood Collection *on page 1304*

Synonyms Isolation Patients, Precautions for Specimen Collection; Precautions, Specimen Collection

Abstract Exposure to blood and body fluid-borne pathogens can be significantly reduced by following universal precautions. It is also well known that universal precautions are not strictly followed among all healthcare workers.

Patient Preparation The **Occupational Safety and Health Administration (OSHA)** Final Rule requires that the risk to healthcare workers of accidental exposure to infection be minimized. By careful planning and thoughtful attention to detail, an appropriate and representative specimen can be safely collected. See Overview and Regulatory Considerations, discussed in the Specimen Collection Introduction *on page 23*.

Before entering the isolation room or drawing area:
- Check orders and assemble the equipment needed for this patient.
- Read the isolation sign on the door or patient's chart. It will explain the type of isolation and what you must wear and do. **Follow these directions carefully.**
- Find out if it is necessary to take a tourniquet and/or a plastic holder into the room. Many times these items will be there already.
- Take in the minimum equipment needed: tourniquet; plastic holder; evacuated tube needle; alcohol sponges; evacuated blood collection tubes or blood culture media; glass slides (if a blood smear is to be made).

In the room:
- Ask the patient to state his/her name to confirm patient identification. Check wristband.
- Put on gloves.
- Place paper towels on table and place your equipment on these towels.
- Obtain blood samples in the usual manner, avoiding any unnecessary contact with the patient and the bed.
- After obtaining blood samples, leave tourniquet and plastic holder in room and discard needle in proper container.
- Place several clean paper towels on the table, one on top of the other. If the outside of the tubes is contaminated, follow established laboratory decontamination procedures.
- If blood smears were made, place smears on two clean paper towels. When ready to leave, wrap smears and tubes in the top paper towel and discard the bottom paper towel.
- Label specimens for proper identification (see Specimen Identification Requirements *on page 1217*) as directed by institutional policy. Label specimens for infectious hazards in a distinctive manner as required by institutional policy. Since the implementation of universal blood and body fluid precautions for **all** patients, special labeling for specific patients may be eliminated, depending upon institutional policies and local regulations. In any case, **universal precautions must be observed.**
- Wash hands.
- Bring specimens to the laboratory.

Collection All specimens of blood and body fluids should be put in a well-constructed container with a secure lid to prevent leaking during transport.

Special Instructions

Precautions for laboratories: Universal precautions should be followed at all times. Blood and other body fluids from **all** patients should be considered infective. To supplement universal blood and body fluid precautions, the following precautions are recommended for healthcare workers in clinical laboratories.

(Continued)

271

Blood and Fluid Precautions, Specimen Collection
(Continued)

All persons collecting and processing blood and body fluid specimens should wear gloves. Masks, protective eyewear, and laboratory coats or gowns should be worn if contact with blood or body fluids is anticipated. Gloves should be changed and hands washed after completion of specimen processing.

For routine procedures, such as histologic and pathologic studies or microbiologic culturing, a biological safety cabinet is not necessary. However, biological safety cabinets (class I or II) should be used whenever procedures are conducted that have a high potential for generating droplets. These include activities such as blending, sonicating, and vigorous mixing. Mechanical pipetting devices must be used for manipulating all liquids in the laboratory. **Mouth pipetting must not be done.**

Use of needles and syringes should be limited to situations in which there is no alternative, and the recommendations for preventing injuries with needles outlined under universal precautions must be followed.

Laboratory work surfaces should be decontaminated with an appropriate chemical germicide after a spill of blood or other body fluids and when work activities are completed.

Contaminated materials used in laboratory tests should be decontaminated before reprocessing or be placed in bags and disposed in accordance with institutional policies for disposal of infective waste.

Scientific equipment that has been contaminated with blood or other body fluids should be decontaminated and cleaned before being repaired in the laboratory or transported to the manufacturer.

All persons must wash their hands after completing laboratory activities and should remove personal protective equipment before leaving the laboratory.

Implementation of universal blood and body fluid precautions for **all** patients eliminates the need for warning labels on specimens, since blood and other body fluids from all patients should be considered infective. OSHA rules require "Biohazard" labeling or color coding of containers of regulated waste, refrigerators and freezers containing blood or other potentially infectious material, and containers used to store, transport, or ship such materials.

Additional Information

Universal Precautions: Since medical history and examination cannot reliably identify all patients infected with HIV or other blood-borne pathogens, blood and body fluid precautions should be consistently used for **all** patients. This approach, recommended by CDC and referred to as "universal blood and body fluid precautions" or "universal precautions," must be used in the care of **all** patients as a result of OSHA's Final Rule. (See References.)

All healthcare workers must take precautions to prevent injuries caused by needles, scalpels, and other sharp instruments or devices during procedures; when cleaning used instruments; during disposal of used needles; and when handling sharp instruments after procedures. To prevent needlestick injuries, needles must not be recapped, purposely bent or broken by hand, removed from disposable syringes, or otherwise manipulated by hand. After they are used, disposable syringes and needles, scalpel blades, and other sharp items must be placed in puncture-resistant containers for disposal; the puncture-resistant containers should be located as close as possible to the area of use. Large-bore reusable needles should be placed in a puncture-resistant container for transport to the reprocessing area. Recapping of syringes, if absolutely necessary, may be done by a one-handed method which employs the use of a recapping block.

Although saliva has not been implicated in HIV transmission, to minimize the need for emergency mouth-to-mouth resuscitation, mouthpieces, resuscitation bags, or other ventilation devices should be available for use in areas in which the need for resuscitation is predictable.

Healthcare workers who have exudative lesions or weeping dermatitis should refrain from all direct patient care and from handling patient care equipment until the condition resolves.

Pregnant healthcare workers are not known to be at greater risk of contracting HIV infection than healthcare workers who are not pregnant; however, if a healthcare worker develops HIV infection during pregnancy, the infant is at risk of infection resulting from perinatal transmission. Because of this risk, **pregnant healthcare workers should be especially familiar with and strictly adhere to precautions to minimize the risk of HIV transmission.**

Sterilization and Disinfection: Standard sterilization and disinfection procedures for patient care equipment currently recommended for use in a variety of healthcare settings, including hospitals, medical and dental clinics and offices, hemodialysis centers, emergency care facilities, and long-term nursing care facilities, are adequate to sterilize or disinfect instruments, devices, or other items contaminated with blood or other body fluids from persons infected with blood-borne pathogens including HIV.

Cleaning and Decontaminating Spills of Blood or Other Body Fluids: Chemical germicides that are approved for use as "hospital disinfectants" and are tuberculocidal when used at recommended dilutions can be used to decontaminate spills of blood and other body fluids. Strategies for decontaminating spills of blood and other body fluids in a patient care setting are different than for spills of cultures or other materials in clinical, public health, or research laboratories. In patient care areas, visible material should first be removed and then the area should be decontaminated. With large spills of cultured or concentrated infectious agents in the laboratory, the contaminated area should be flooded with a liquid germicide before cleaning, then decontaminated with fresh germicidal agent. In both settings, gloves should be worn during the cleaning and decontaminating procedures.

HIV is inactivated rapidly after being exposed to commonly used chemical germicides at concentrations that are much lower than used in practice. Embalming fluids (formalin preparations) are similar to the types of chemical germicides that have been tested and found to completely inactivate HIV. Formalin may not rapidly inactivate hepatitis B virus nor quickly kill bacteria. It is a slow-acting antiseptic agent requiring 18 hours or more to kill microorganisms. In addition to commercially available chemical germicides, a solution of sodium hypochlorite (household bleach) prepared daily is an inexpensive and effective germicide. Concentrations ranging from ~500 ppm (1:100 dilution of household bleach) sodium hypochlorite to 5000 ppm (1:10 dilution of household bleach) are effective, depending on the amount of organic material (eg, blood, mucus) present on the surface to be cleaned and disinfected. Disinfecting surfaces in cases of known Jakob-Creutzfeld agent may require full strength bleach. Commercially available chemical germicides may be more compatible with certain medical devices that might be corroded by repeated exposure to sodium hypochlorite.

Housekeeping: Environmental surfaces such as walls and floors are not associated with transmission of infections to patients or healthcare workers. Therefore, extraordinary attempts to disinfect or sterilize such environmental surfaces are unnecessary. Cleaning and removal of soil should be done routinely.

Infective Waste: There is no epidemiologic evidence to suggest that most hospital waste is any more infective than residential waste. Moreover, there is no epidemiologic evidence that hospital waste has caused disease in the community as a result of improper disposal. Therefore, identifying wastes for which special precautions are indicated is largely a matter of judgment about relative risk of disease transmission. The most practical approach to the management of infective waste is to identify those wastes with potential for causing infection during handling and disposal and for which some special precautions appear prudent. Hospital wastes for which special precautions are required include microbiology laboratory waste, pathology waste, blood specimens or blood products, and other potentially infectious material. Any item that has had contact with blood, exudates, or secretions may be potentially infective. Infective waste, in general, should either be incinerated or should be autoclaved before disposal in a sanitary landfill. Bulk blood, suctioned fluids, excretions, and secretions may be carefully poured down a drain connected to a sanitary sewer. Sanitary sewers may also be used to dispose of other infectious wastes capable of being ground and flushed into the sewer.
(Continued)

Blood and Fluid Precautions, Specimen Collection
(Continued)

Survival of HIV in the Environment: The most extensive study on the survival of HIV after drying involved greatly concentrated HIV samples (ie, 10 million tissue culture infectious doses/mL). This concentration is at least 100,000 times greater than that typically found in the blood or serum of patients with HIV infection. HIV was detectable by tissue culture techniques 1-3 days after drying, but the rate of inactivation was rapid. Studies performed at CDC have also shown that drying HIV causes a rapid (within several hours) 1-2 log (90% to 99%) reduction in HIV concentration. In tissue culture fluid, cell-free HIV could be detected up to 15 days at room temperature, up to 11 days at 37°C (98.6°F), and up to 1 day if the HIV was cell-associated. HIV can be isolated from peripheral blood mononuclear cells and plasma for up to 48 hours after sample collection.

Risk to Healthcare Workers of Acquiring Hepatitis or HIV in Healthcare Settings: Comparative risks of needlestick transmission are estimated by the rule of threes: Hepatitis B is transmitted in 30% of exposures, hepatitis C in 3%, and HIV-1 in 0.3%. An estimated 2.7 million people in the U.S. have active hepatitis C infection. The risks associated with occupational mucous membrane and cutaneous exposures are likely to be substantially smaller.

References

Department of Labor, Occupational Safety and Health Administration, "Occupational Exposure to Blood-Borne Pathogens; Final Rule (29 CFR Part 1910.1030)," *Fed Regist*, 1991, 64004-182.

Dillman CM, "Hepatitis C: A Danger to Healthcare Workers," *Nurs Forum*, 1999, 34(2):23-8.

Kim LE, Evanoff BA, Parks RL, et al, "Compliance With Universal Precautions Among Emergency Department Personnel: Implications for Prevention Programs," *Am J Infect Control*, 1999, 27(5):453-5.

Lauer GM and Walker BD, "Hepatitis C Virus Infection," *N Engl J Med*, 2001, 345(1):41-52.

Michalsen A, Delclos GL, Felknor SA, et al, "Compliance With Universal Precautions Among Physicians," *J Occup Environ Med*, 1997, 39(2):130-7.

Perry C and Barnett J, "Principles of Universal Precautions," *Emerg Nurse*, 1998, 6(6):25-8.

Tube Codes

Color	Optimum Volume/ Minimum Volume	Additive
Blue	4.5 mL/4.5 mL	Sodium citrate
Blue/navy	7 mL/3 mL	No additive (for trace metals) Heparin (for trace metals)
Culture (yellow)	8.3 mL/8.3 mL	SPS
FSP (blue)	2 mL/2 mL	Thrombin, trypsin inhibitor
Gray	5 mL/5 mL 7 mL/7 mL	Potassium oxalate, sodium fluoride
Green	10 mL/3.5 mL	Heparin
Lavender	7 mL/2 mL	EDTA
Orange	10 mL/NA	Thrombin
Red	10 mL/NA	None
Red/gray (gel)	10 mL/NA	Inert barrier material; clot activator
Yellow	5 mL/NA	ACD
Yellow/black	7 mL	Thrombin
Pediatric Tubes		
Blue	2.7 mL/2.7 mL	Sodium citrate
Culture (yellow)	3.3 mL/3.3 mL	SPS
Green	2 mL/2 mL	Heparin
Lavender	2 mL/0.6 mL 3 mL/0.9 mL 4 mL/1 mL	EDTA
Red	2 mL/NA 3 mL/NA 4 mL/NA	None

Blood Collection Tube Information

Related Information

Chain-of-Custody Protocol *on page 381*
Phlebotomist Procedures *on page 1028*
Venous Blood Collection *on page 1304*

Synonyms Blood Container Description; Tubes for Blood Collection; Vacutainer® Tube Description

Special Instructions See individual listings throughout this book for particular test requirements.

Additional Information The table describes most commonly used color codes, optimum and minimum volumes required, and additives contained in common vacuum draw tubes. In the last 5-10 years, due to the advent of different anticoagulant combinations and other additives, tube color codes have changed significantly and are becoming increasingly confusing. International Standard Organization (ISO), along with blood collection tube manufacturers is making efforts to standardize color codes. **It is important to be certain that a tube is filled with the prescribed minimum volume in order to avoid spurious results due to an inappropriate anticoagulant to specimen ratio.**

See the introductions for Coagulation *on page 29*, Transfusion Services *on page 78*, Trace Metals *on page 77*, Therapeutic Drug Monitoring *on page 57*, and Toxicology Drugs of Abuse *on page 63* for appropriate specimen requirements.

♦ **Blood Collection, Venous** *see* Venous Blood Collection *on page 1304*

♦ **Blood Container Description** *see* Blood Collection Tube Information *on page 275*

♦ **Blood Culture, *Brucella*** *see* Brucellosis Culture and Serology *on page 312*

Blood Gases and pH, Arterial

Related Information

Anion Gap, Serum, Plasma, or Urine *on page 160*
Arterial Blood Collection *on page 211*
Bicarbonate, Blood *on page 258*
Blood Gases and pH, Capillary *on page 277*
Blood Gases and pH, Umbilical Cord *on page 278*
Blood Gases and pH, Venous *on page 279*
Carbon Dioxide, Total, Blood *on page 339*
Carboxyhemoglobin, Blood *on page 340*
Lactic Acid, Whole Blood or Plasma *on page 827*
Oxygen Saturation, Blood *on page 991*
P_{50}, Blood *on page 993*
pCO_2, Blood *on page 1008*
pH, Blood *on page 1018*
Red Blood Cells *on page 1139*
Uncrossmatched Blood, Emergency *on page 1281*
Whole Blood *on page 1333*

Synonyms ABGs; Arterial Blood Gases; Gases, Arterial

Applies to Allen Test; Base Excess; FiO_2; HCO_3^-; Oxygen Saturation; PaO_2; pCO_2; pH; pO_2; TCO_2

Test Includes Measured results include pH, pCO_2 ($PaCO_2$), and pO_2 (PaO_2). Calculated values include, among others, total carbon dioxide (TCO_2), bicarbonate (HCO_3^-), oxygen saturation, and base excess.

Abstract The measurement in arterial blood of pH, pCO_2, pO_2 and the calculation of HCO_3^-, TCO_2, and O_2 saturation, are used to evaluate oxygen and carbon dioxide exchange, respiratory function, and acid-base balance. Arterial blood is preferred for these determinations due to its superior uniformity throughout the body, but venous pH is extremely similar in most situations and is more easily obtained.

Patient Preparation Patient should be supine, relaxed. The patient's temperature, breathing pattern, and concentration of inspired air (FiO_2) should be recorded. Refer to NCCLS Approved Standard H11-A3 for complete guidelines.

(Continued)

Blood Gases and pH, Arterial (Continued)

Aftercare Observe for bleeding. Pressure must be applied to puncture site for at least 10-15 minutes; longer times are required for anticoagulated patients. See Arterial Blood Collection *on page 211*.

Specimen Whole blood (arterial)

Container Heparinized blood gas syringe (plastic or glass) or via an indwelling arterial line. Gases are capable of dissolving in plastic, which may alter results in some situations. Mahoney et al reported clinically significant increases in pO_2 levels in whole blood stored in iced plastic syringes for 30 minutes. Specimens stored in iced glass syringes did not change significantly.

Collection See Arterial Blood Collection *on page 211*. Very small diameter needles are used. Specimen is drawn into air-free heparinized syringe, then stoppered. The radial artery is frequently used after the Allen test, which assesses the presence of normal collateral circulation. The brachial artery is the second choice. **The specimen should be transported to the laboratory immediately and analysis should be prompt. If testing will be delayed by more than 10-15 minutes, the specimen should be placed in a slurry of ice chips and water.** Mode of oxygen delivery (quantity of therapeutic oxygen or room air) and patient's temperature must be indicated. Rapid changes may occur if collected immediately after exercise. Avoid excessive heparin. Strict anaerobiosis must be maintained.

Storage Instructions Testing should occur **immediately**; therefore, specimens should not require storage. However, if testing must be delayed more than 10-15 minutes, the specimen should be cooled to about 0°C by placing the syringe into a slurry of ice chips and water. Delay of analysis should not exceed 1 hour. Specimens for critical alveolar-arterial oxygen tension or shunt fraction samples must be analyzed immediately (within 10 minutes) to minimize changes in gas tensions. The following *in vitro* changes occur in blood gas parameters: pH decreases by <0.01 pH units/hour at 4°C, pCO_2 increases by about 0.5 mm Hg/hour at 4°C, and pO_2 decreases negligibly (<3 mm Hg/hour) if collected in a glass syringe and stored in an ice water slurry.

Causes for Rejection Specimen **not** received correctly iced, air bubbles or clots in syringe, unsealed/open syringe

Special Instructions Sample obtained just after a change in inspired oxygen concentration (FiO_2) (eg, room air or quantity of therapeutic oxygen delivered) is likely to generate confusing results. Normally, arterial blood gases and pH will achieve steady-state levels within a minute or two after a change in FiO_2 or alveolar ventilation, but in certain disease states (eg, lung disease) the time necessary to achieve equilibrium can be as long as 20-30 minutes.

Reference Interval
- Arterial pH: 7.35-7.45
- TCO_2: 23-29 mmol/L
- pCO_2: 35-45 mm Hg
- pO_2: newborns: 60-70 mm Hg, adults: 80-95 mm Hg
- O_2 saturation: 95% to 99%

Such intervals must be interpreted in light of the FiO_2 and other variables.

Possible Panic Range pH: <7.2, >7.55; pCO_2: <20 mm Hg, >60 mm Hg; pO_2: <40 mm Hg

Use Blood gas and pH testing are done to evaluate oxygen and carbon dioxide gas exchange, respiratory function including hypoxia and acid-base status. They are clinically indicated in a wide variety of medical and surgical situations involving cardiorespiratory, metabolic, and central nervous system disturbances.

The context in which the specimen is drawn is pivotal relevant to its significance (eg, pH 7.10 drawn in the immediate postictal state may be of little consequence, but is ominous in methanol intoxication).

Acidosis and alkalosis are addressed in Anion Gap, Serum, Plasma, or Urine *on page 160*. Refer as well to the Disease Index.

Limitations Arterial puncture is a hazardous procedure and may be extremely difficult in some individuals. Complications of arterial puncture potentially include hematoma, bleeding, arterial occlusion and insufficiency, infection, and, very rarely, gangrene.

Markedly elevated leukocytes and/or platelets in a blood gas specimen will significantly alter blood gas and pH results regardless of specimen transport/storage methods. Point-of-care or *in vivo* methods of analysis are helpful in these situations.

By some instruments, O_2 saturation is calculated from oxyhemoglobin and total hemoglobin. The reported value may be misleading when nonfunctional hemoglobins (CoHB, MetHb, or sulfhemoglobin) are present or other hemoglobins with different dissociation curves are present. Calculations commonly assume body temperature of 37°C.

Variability of results occurs; changes in pO_2 in isolated reports must be interpreted cautiously and in light of data trends, oxygen delivery, and the patient's clinical appearance. Such variation occurs without change in FiO_2 or the patient's clinical status.

Arterial gases and pH are of little value in treatment decisions for carbon monoxide poisoning.

Methodology Selective electrodes measuring pH, pCO_2, and pO_2

Additional Information A **pH value <7.25, without elevation of pCO_2,** may indicate need for a lactate determination.

Potassium leaves the intracellular fluids in acidemia, leading to hyperkalemia.

The **acute respiratory distress syndrome (ARDS)** is usually initially characterized by respiratory alkalosis and hypoxemia.

Assessment of acid-base status of tissues in patients in circulatory failure may be misleading if only arterial blood gas data is available. Adrogué presents data supporting the need for information on **mixed venous** as well as arterial gases in care of critically ill patients.

For a review of definitions and calculated values, see a two-part review in the outstanding *New England Journal of Medicine* and NCCLS Approved Standard C12-A.

References
Adrogué HJ and Madias NE, "Management of Life-Threatening Acid-Base Disorders," First of Two Parts, *N Engl J Med*, 1998, 338(1):26-34, Second of Two Parts, *N Engl J Med*, 1998, 338(2):107-11.

Hood VL and Tannen RL, "Protection of Acid-Base Balance by pH Regulation of Acid Production," *N Engl J Med*, 1998, 339(12):819-26.

Mahoney JJ, Harvey JA, Wong RJ, et al, "Changes in Oxygen Measurements When Whole Blood Is Stored in Iced Plastic or Glass Syringes," *Clin Chem*, 1991, 37(7):1244-8.

National Committee for Clinical Laboratory Standards, *Procedures for the Collection of Arterial Blood Specimens: Approved Standard*, 3rd ed, H11-A3, Wayne, PA: NCCLS, 1999.

Blood Gases and pH, Capillary

Related Information
Bicarbonate, Blood *on page 258*
Blood Gases and pH, Arterial *on page 275*
Blood Gases and pH, Venous *on page 279*
Carbon Dioxide, Total, Blood *on page 339*
Oxygen Saturation, Blood *on page 991*
pCO_2, Blood *on page 1008*
pH, Blood *on page 1018*

Synonyms Capillary Blood Gases

Test Includes Measured results include pH, pCO_2, and pO_2. Calculated values include total carbon dioxide (TCO_2), bicarbonate (HCO_3^-), oxygen saturation, and base excess.

Abstract Although arterial specimens are preferred, capillary pCO_2 and pH determinations are entirely adequate for many purposes and are usually used for monitoring blood gases in neonates or other patients in whom arterial blood collection is not practical. Arterialization of capillary blood by prewarming the puncture site yields a blood specimen similar to arterial blood with results for pH and pCO_2 that match very well those of arterial blood. Capillary pO_2 values do not correlate well with arterial pO_2 values, especially in sick patients with either abnormal peripheral vasoconstriction or vasodilation.

Patient Preparation See Skin Puncture Blood Collection *on page 1203*. The puncture site (eg, lateral exterior side of the foot, tip of finger or toe, earlobe) should be prewarmed to about 42°C to dilate the capillaries and increase arteriolar flow and free bleeding. The puncture should be deep enough to allow
(Continued)

Blood Gases and pH, Capillary *(Continued)*

a free flow of blood. Blood is then collected in heparinized capillary tubes, which should be filled as much as possible, capped and mixed well.

Aftercare Elevate the site above the body and apply direct pressure to the puncture site with sterile gauze until bleeding stops. Bandaids or bandages are generally not applied because of the risk of skin sensitization to tape and the risk of aspiration, should the bandage come loose.

Specimen Whole blood (capillary)

Container Heparinized capillary tube(s) capped tightly with internal mixing flea

Collection See Skin Puncture Blood Collection *on page 1203.* Fill capillary tubes completely excluding any air bubbles, and mix immediately with heparin to avoid clotting.

Storage Instructions Specimen should be analyzed immediately (within 10-15 minutes). If delay in analysis is unavoidable or otherwise anticipated, put specimen in a slurry of ice chips and water.

Causes for Rejection No heparin, sample clotted, specimen not received on ice

Reference Interval pH: 7.35-7.45, pCO_2: 26-41 mm Hg

Possible Panic Range pH: <7.2, >7.55; pCO_2: <20 mm Hg, >60 mm Hg

Use Testing is done to assess and monitor acid-base balance. Although capillary blood is satisfactory for most purposes for pH and pCO_2, the role of capillary pO_2 is limited. Capillary blood sampling is less likely to cause complications than is arterial puncture.

Limitations Arterialized capillary blood specimens should not be used for pO_2 determinations in patients with systolic blood pressure <95 mm Hg, in patients with vasoconstriction, for patients on O_2 therapy, for newborns during the first few hours following birth, or newborns with respiratory distress syndrome. These situations are associated with high likelihoods for venous admixture and erroneously low pO_2 results.

Methodology Specific electrodes for pH, pCO_2, and pO_2

References

Dong SH, Liu HM, Song GW, et al, "Arterialized Capillary Blood Gases and Acid-Base Studies in Normal Individuals From 29 Days to 24 Years of Age," *Am J Dis Child*, 1985, 139(10):1019-22.

McLain BI, Evans J, Dear PR, et al, "Comparison of Capillary and Arterial Blood Gas Measurements in Neonates," *Arch Dis Child*, 1988, 63(7 Spec No):743-7. CA and Ashwood ER, eds, Philadelphia, PA: WB Saunders Co, 1999, 1056-92.

Blood Gases and pH, Umbilical Cord

Synonyms pH; pH, Umbilical Venous Blood Gases (pCO_2 and pO_2); Umbilical Arterial Blood Gases (pCO_2 and pO_2)

Test Includes Measured pH, pCO_2, and pO_2 and calculated total carbon dioxide (TCO_2), bicarbonate (HCO_3^-), O_2 saturation, and base excess

Abstract Umbilical cord blood acid-base analysis provides a means for ruling out asphyxiation in newborns with low Apgar scores. **Asphyxia** is defined as hypoxia with metabolic acidosis.

Patient Preparation Double clamping of the umbilical cord must occur immediately after delivery. Delays in clamping of as little as 20-30 seconds can significantly alter the pCO_2 and pH. The American College of Obstetricians and Gynecologists (ACOG) recommends that cords be clamped for all deliveries regardless of whether cord blood pH and gas analyses are indicated.

Specimen Whole blood (umbilical cord); the umbilical arteries contain blood that is returning from the fetus to the placenta and provides the better cord blood specimen for evaluating fetal/newborn acid-base status. Blood gas and pH values analyzed in umbilical vein specimens can be normal concurrently with significantly abnormal umbilical artery values. ACOG recommends obtaining a specimen from an artery on the chorionic surface of the placenta, if a specimen cannot be obtained from the umbilical artery.

Container Glass or plastic heparinized syringe

Sampling Time As soon as it has been determined that cord blood pH and gases are indicated, sampling from the clamped cord should occur. pH, pCO_2, and pO_2 values have been shown to be stable from umbilical cord segments left at room temperature for up to 60 minutes.

Collection Blood is anaerobically collected into the heparinized syringe from either an umbilical artery (preferable) or umbilical vein from the clamped umbilical cord. Residual air is ejected from the syringe, and the syringe is sealed and transported immediately to the laboratory for analysis.

Storage Instructions Analysis should occur immediately. If delay beyond 10-15 minutes is unavoidable or anticipated, specimen should be placed in a slurry of ice chips and water. Any delay >30 minutes may yield misleading results.

Causes for Rejection Air bubbles in syringe, unsealed/open syringe, improperly iced syringe

Reference Interval The table gives the 5th and 95th percentiles from a study of cord blood specimens from 1015 preterm infants delivered vaginally. The same study examined 3522 term infants delivered vaginally, and the intervals were similar.

	Arterial Cord Blood	Venous Cord Blood
pH	7.14-7.40	7.23-7.46
pCO_2 (mm Hg)	32-69	28-57
pO_2 (mm Hg)	8-33	15-42
HCO_3^- (mmol/L)	16.0-27.1	17.4-25.4
BE (mmol/L)	-7.6 to +1.3	-5.8 to +0.7
O_2 saturation (%)	5-59	14-75

Critical Values Umbilical arterial blood pH <7.00 is consistent with pathologic fetal acidemia that is of sufficient degree to be associated with birth asphyxia or hypoxia capable of subsequent neurologic abnormalities.

Use Severe fetal acidemia is associated with increased perinatal mortality and increased risk for impaired neurodevelopment. An umbilical artery blood pH <7.00 with a metabolic pattern (ie, normal pCO_2, low HCO_3^-, and base deficit) and persistent Apgar scores ≤3 for 5 minutes or longer are consistent with a degree of birth asphyxia or hypoxia capable of subsequent neurologic abnormalities.

Limitations Even with umbilical arterial pH levels below the critical threshold of 7.00, the majority of newborns will be neurologically normal with no apparent morbidity.

Methodology Blood gas instrumentation with specific pH, pCO_2, and pO_2 electrodes

Additional Information ACOG recommends that cord blood pH and gases be done only in cases in which serious abnormalities arise in the delivery process and/or when a problem with the neonate's condition persists beyond the first 5 minutes after birth. While ACOG does not recommend that cord blood pH and gases be done for all deliveries, doubly clamping the umbilical cord immediately after each birth is recommended in the event assessment of fetal acidemia is warranted.

References
ACOG Technical Bulletin Number 216-November 1995 (Replaces No. 127, April 1989), "Umbilical Artery Blood Acid-Base Analysis," *Int J Gynaecol Obstet*, 1996, 52(3):305-10.

Andres RL, Saade G, Gilstrap LC, et al, "Association Between Umbilical Blood Gas Parameters and Neonatal Morbidity and Death in Neonates With Pathologic Fetal Acidemia," *Am J Obstet Gynecol*, 1999, 181(4):867-71.

Arikan GM, Scholz HS, Haeusler MC, et al, "Low Fetal Oxygen Saturation at Birth and Acidosis," *Obstet Gynecol*, 2000, 95(4):565-71.

Blood Gases and pH, Venous

Related Information
Blood Gases and pH, Arterial *on page 275*
Blood Gases and pH, Capillary *on page 277*
Carbon Dioxide, Total, Blood *on page 339*
pH, Blood *on page 1018*
Venous Blood Collection *on page 1304*

Synonyms Venous Blood Gases

Applies to Central Venous Blood

Test Includes Measured results include pH, pCO_2, and pO_2. Calculated values include, among others, total carbon dioxide (TCO_2), bicarbonate (HCO_3^-), oxygen saturation, and base excess.
(Continued)

Blood Gases and pH, Venous *(Continued)*

Abstract Determination of pH and pCO_2 can be done reliably from venous blood in most clinical situations, but measurements involving oxygen are **not** usually useful when done on venous blood.

Patient Preparation The patient should be supine, relaxed.

Specimen Whole blood (venous)

Container Heparinized syringe, green top (heparin) tube

Collection See Venous Blood Collection *on page 1304.* Draw specimen into air-free heparinized syringe or green top vacuum blood collection tube. If a vacuum blood collection tube is used, it must be completely filled and removed from needle before needle is removed from patient's arm. It is best not to use a tourniquet and to avoid hand clenching. Indicate specimen source (ie, venous) and mode of oxygen delivery or room air if applicable on requisition.

Storage Instructions Specimen should be analyzed immediately (within 10-15 minutes). If delay in analysis is unavoidable or otherwise anticipated, place specimen in a slurry of ice chips and water.

Causes for Rejection Specimen **not** received correctly iced, specimen clotted

Reference Interval Venous pH: 7.32-7.43, TCO_2: 23-30 mmol/L, pCO_2: 38-50 mm Hg, pO_2 should be about 40 mm Hg, O_2 saturation should be about 75%.

Possible Panic Range pH: <7.2, >7.55; pCO_2: <20 mm Hg, >60 mm Hg

Use The tests are used to evaluate cellular hypoxia and acid-base balance. A major use is to obtain pH without arterial puncture in infants, children, and adults in whom oxygen measurements are not needed. In many metabolic situations a venous pH is adequate and arterial puncture is unnecessary. Both arterial and central venous blood samples play a role in assessment of acid-base status in subjects in critical hemodynamic compromise. With severe hypoperfusion central venous blood better detects hypercapnia and acidemia. The pO_2, pCO_2, and pH from pulmonary arterial samples correlate with central venous specimens. The value of venous blood gas assays is supported in well-perfused subjects. A strong linear relationship between venous and capillary gas results exists, with the exception of pO_2.

Limitations In hypotensive subjects with severe circulatory failure, Adrogué et al describe substantial differences between mean arterial and central venous pH and pCO_2.

Methodology Specific electrodes for pH, pCO_2, and pO_2

Additional Information The arteriovenous pH difference is usually extremely small (0.01-0.03), except in patients in congestive heart failure and in shock. The differences in pH and pCO_2 widen only slightly with moderate cardiac failure. Total CO_2 values are slightly higher in venous blood than in arterial blood. Arterial blood, however, must be used to accurately measure pO_2 and oxygen saturation.

References

Adrogué HJ and Madias NE, "Management of Life-Threatening Acid-Base Disorders," First of Two Parts, *N Engl J Med*, 1998, 338(1):26-34, Second of Two Parts, *N Engl J Med*, 1998, 338(2):107-11.

Adrogué HJ, Rashad MN, Gorin AB, et al, "Assessing Acid-Base Status in Circulatory Failure: Differences Between Arterial and Central Venous Blood," *N Engl J Med*, 1989, 320(20):1312-6.

McGillivray D, Ducharme FM, Charron Y, et al, "Clinical Decision Making Based on Venous Versus Capillary Blood Gas Values in the Well-Perfused Child," *Ann Emerg Med*, 1999, 34(1):58-63.

Blood, Urine

Related Information

Anemia Flowchart *on page 35*
Antinuclear Antibodies *on page 189*
Glomerular Basement Membrane Antibody *on page 638*
Hemoglobin, Qualitative, Urine *on page 688*
Hemosiderin Stain, Urine *on page 692*
Kidney Biopsy *on page 818*
Kidney Stone Analysis *on page 820*
Myoglobin, Qualitative, Urine *on page 941*
Platelet Count *on page 1050*
Protein, Quantitative, Urine *on page 1108*
Prothrombin Time *on page 1116*
Urinalysis *on page 1289*
Urinary Tract Cytology *on page 1293*

Synonyms Blood, Occult, Urine; Hemoglobin, Urine; Occult Blood, Urine

Applies to Dysmorphic Red Blood Cells; Urine Calcium:Creatinine Ratio; Urine Protein:Creatinine Ratio

Test Includes Dipstick method for occult blood is a part of urinalysis.

Abstract When the urine dipstick is positive for "blood", the differential includes red cells in the urine (hematuria), hemoglobinuria, and myoglobinuria. Microscopy, preferably phase-contrast, of the urine sediment should follow a positive dipstick test.

Specimen Centrifuged sediment from random urine

Storage Instructions Refrigeration if examination is delayed. Immediate examination is best.

Causes for Rejection Standing for more than 2 hours at room temperature may be a reason to consider a specimen unacceptable.

Reference Interval Dipstick: negative; microscopy: 0-2 red cells/hpf

Use Urine sediment microscopy, preferably using phase-contrast technique, should be preformed to evaluate the significance of a positive urine dipstick test for "blood". Microscopy will identify cellular elements and crystals, and assist in the differentiation between hematuria and hemoglobinuria/myoglobinuria. Urine dipsticks are an imperfect tool for the detection of hematuria, with a reported sensitivity of 80% and specificity of 35% using microscopy as the gold standard. The resulting predictive values were 65% for positive results and 68% for negative results. See tables.

Blood, Urine: Summary

Causes of Hemoglobinuria
Hemolysis associated with
red cell enzyme deficiency
microorganisms (eg, malaria, *Bartonella*)
drugs (eg, acetanilid)
chemicals
antibodies
Unstable hemoglobins
March hemoglobinuria
secondary to exercise
Transfusion reactions
incompatible blood
Burns
Crush injury (myoglobinuria)
Poisoning
snake or spide bite
Paroxysmal nocturnal hemoglobinuria
Paroxysmal cold hemoglobinuria
Myoglobin (may be detected as hemoglobin)

(Continued)

Blood, Urine *(Continued)*

Blood, Urine: Summary

Causes of Hematuria	
Renal and ureteral diseases	**Ureteral diseases**
glomerulonephritis	ureterolithiasis
nephrotic syndrome	carcinoma
hemolytic uremic syndrome	**Bladder entities**
vasculitis	exercise
tumor	cystitis
polycystic kidney disease	neoplasm
infarct	tuberculosis
infection	**Prostatic diseases**
trauma	prostatitis
renal vein thrombosis	benign prostatic hyperplasia
Alport syndrome	prostatic carcinoma
thin-GBM disease	**Urethral diseases**
benign hematuria	urethritis
loin-pain hematuria	tumor
stone / hypercalciuria / hyperuricosuria / hyperoxaluria	**Trauma**
Blood diseases	**Drugs**
thrombocytopenia, any cause	Coumadin®
Henoch-Schönlein purpura	heparin
infective endocarditis	salicylate
hemophilias	many others
abnormal hemoglobins	

Limitations Myoglobinuria results in positive dipstick test for "blood". False-positive dipstick tests for "blood" are caused by iodine, bleach (hypochlorite), menstrual blood, and bacterial peroxidases. False-negative dipstick results are caused by vitamin C, proteinuria, nitrites, Pyridium® (phenazopyridine) and Serenium® (ethoxazene hydrochloride), and a prolonged delay before testing (especially if the specimen is not refrigerated).

Methodology Dipstick is based on the pseudoperoxidase activity of heme from hemoglobin or myoglobin. Microscopy (preferably phase-contrast) is used for recognition of red blood cells and red cell casts.

Additional Information The number of red cells in a given volume of urine does not correlate with the degree or significance of urologic pathologic findings. Contemporary long-term anticoagulation protocols infrequently cause hematuria. The use of nonsteroidal anti-inflammatory drugs including aspirin may be associated with hematuria.

Algorithms for diagnosis of hematuria are published for children and adults (see Restrep et al and Lieu et al).

References

Culclasure TF, Bray VJ, and Hasbargen JA, "The Significance of Hematuria in the Anticoagulated Patient," *Arch Intern Med*, 1994, 154(6):649-52.

Lieu TA, Grasmeder HM 3d, and Kaplan BS, "An Approach to the Evaluation and Treatment of Microscopic Hematuria," *Pediatr Clin North Am*, 1991, 38(3):579-92.

Restrepo NC and Carey PO, "Evaluating Hematuria in Adults," *Am Fam Phys*, 1989, 40(2):149-56.

Threatte GA and Henry JB, "Urine and Other Body Fluids," *Clinical Diagnosis and Management by Laboratory Methods*, Henry JB, ed, Philadelphia, PA: WB Saunders Co, 2001, 389.

Blood Volume

Related Information

[51]Cr Red Cell Survival *on page 476*
Erythropoietin, Serum *on page 551*
Hematocrit *on page 674*
Peripheral Blood: Red Blood Cell Morphology *on page 1016*
Phlebotomy, Therapeutic *on page 1029*
Red Blood Cell Indices *on page 1136*

Red Cell Mass *on page 1144*

Applies to Plasma Volume Measurement; Red Cell Volume

Test Includes Total blood volume, red cell mass, and plasma volume, measured or derived and with measured values reported with predicted values for comparison

Abstract Using labeled red cells and/or labeled albumin, this procedure measures either the red cell volume or the plasma volume (or both). The dilution of the label is inversely proportional to the volume of the compartment in which it has been diluted. Useful red cell labels include ^{53}Cr (a stable isotope), carbon monoxide (CO), and four radioactive labels (^{51}Cr, ^{99m}Tc, ^{113m}In, ^{111}In). Plasma labels include both nonradioactive agents (hydroxyethyl starch, indocyanine green [ICG], and Evans blue dye) as well as radioactive materials (^{125}I albumin). These procedures may prove useful in bedside, critical care, or intraoperative situations. Blood volume study may be an invaluable contribution to some clinical situations (eg, polycythemia, acute blood loss) in which determination of Hb, a concentration, or Hct, a fraction, could be misleading.

Patient Preparation Varies with method.

Specimen Whole blood; the CO and ICG methods do not require collection of postlabeling blood samples.

Container Method dependent. For ^{51}Cr-labeled red cell methods, ACD-NIH or Strumia's ACD solution, ratio of one part ACD to five parts blood. EDTA anticoagulated blood may be used, but excess EDTA must be avoided. EDTA causes shrinkage of red cells. The resultant red cell volume is too low unless EDTA is used in a concentration of 1.5 ±0.25 mg/mL of blood. Samples of blood for ^{125}I- or ^{131}I-labeled albumin plasma volume methods may be collected in heparinized syringe.

Sampling Time The laboratory will obtain timed postdose blood samples; the patient must be available for such sampling.

Causes for Rejection Patient with recent radioisotope administration (consult laboratory), patient not available for postdose blood sampling

Turnaround Time Method dependent, varying from 10 minutes to 12 hours; most isotope dilution methods will require 2-6 hours for earliest availability of results, unsatisfactory for many critical care applications.

Reference Interval Normal values are method dependent. Blood volume varies with body habitus, age, sex, weight, and height. There is special correlation with body surface area. As the amount of blood in fat is about 2/35 of lean tissue, the normal value for an obese individual is less than that for a lean person of same weight. See Red Cell Mass *on page 1144* for application of biologic impedance measuring devices to normalizing total RBC volume in relation to body fat content. Careful clinical assessment as to degree of obesity, edema, etc, must be a part of the determination of "normal." Difficulties with determination of predicted (calculated) blood volume parameters (using weight, height, and in particular, body surface area) have limited the use of these studies in some, especially acute situations. See table.

Estimated Blood Volumes

Age	Plasma Volume (mL/kg) (PV)	Red Cell Mass (mL/kg) (RCM)	Total Blood Volume (mL/kg)	
			From PV	From RCM
Newborns	41.3	43.1	82.1	86.1
1-12 mo	46.1	25.5	78.1	72.8
1-3 y	44.4	24.9	73.8	69.1
4-6 y	48.5	25.5	80.0	67.5
7-9 y	52.2	24.3	87.6	67.5
10-12 y	51.9	26.3	87.6	67.4
13-15 y	51.2		88.3	
16-18 y	50.1		90.2	
Adults	40-50	M: 25-30 F: 20-30	PV added to RCM: M: 65-85 F: 60-80	

Note: Values vary widely relating to method and study population.

Adapted from Shinton NK, ed, *CRC Desk Reference for Hematology*, Boca Raton, FL: CRC Press, 1998, 95-6, 717 and Price DC and Ries C, *Nuclear Medicine in Clinical Pediatrics*, Handmaker H and Lowenstein JM, eds, New York, NY: Society of Nuclear Medicine, 1975, 279.

(Continued)

Blood Volume *(Continued)*

Use Differentiate relative from absolute **polycythemia**. Polycythemia may be defined as increased red cells and is usually considered when hemoglobin is 18 g/dL, hematocrit is 52%, and RBC count is 6 million/mm³. These values do not tell whether the red cell mass is increased or the plasma volume is decreased. Relative polycythemias are caused by decreased plasma volumes such as in burns, severe sweating, shock, dehydration, or any other cause of hemoconcentration. Absolute polycythemia occurs when red cell mass is increased. This can be because of increased erythropoietin in secondary polycythemia (appropriate or inappropriate) or occurs spontaneously as in polycythemia vera, one of the myeloproliferative syndromes in which splenomegaly is usually present.

Limitations Any *in vivo* isotope test may affect radioisotope determined blood volume (eg, bone scans, liver scans, brain scans). Check with the laboratory to see if blood volume determination would be valid. A dependable, accurate, "double tag" procedure (simultaneous use of ¹²⁵I albumin and ⁵¹Cr-labeled RBCs) is the preferred method but is technically rigorous and time consuming compared to other clinical laboratory procedures. Methods in which the red cell volume or plasma volume are measured and total blood volume is calculated using the packed cell volume are subject to error.

Contraindications Patient actively bleeding, combative, or with situations leading to loss of capillary vascular integrity or "third space" effects

Methodology Dilution of label (volume = quantity/concentration)

Additional Information Hb, Hct, or RBC count determinations are concentration expressed parameters and may be misleading when the clinical situation requires assessment of the absolute volume of blood or one of its components. Acute shift in body fluids between the intravascular and extravascular spaces as may occur with heart failure, shock, and third space pooling are examples of such misleading circumstances.

There is an increase in plasma volume in the second to last trimester of pregnancy, toxemia of pregnancy, and in uremia. As red cell volume shows proportionally less rise (or may decrease), there is an element of dilutional anemia. On the basis of independent measure of red cells using ⁵³Cr, a stable, nonradioactive isotope and Evans blue determination of plasma volume ("double tag" study), blood volume is decreased in patients with pre-eclampsia (as compared with normotensive subjects) but normal in patients with gestational hypertension.

The plasma volume is decreased by some 2 mL/kg as the result of positive/prolonged bedrest. Blood volume may be as much as 16% higher in the evening than in the morning. Blood volume is decreased in cases of subarachnoid hemorrhage. There is evidence that total blood volume (TBV) decreases with age in healthy men of comparable size and levels of physical activity. In postmenopausal women, TBV is decreased in healthy but sedentary subjects but maintained in physically active females. Hormone replacement therapy (in postmenopausal women) had no effect on TBV.

Runner's anemia (a mechanically induced hemolytic anemia) is characterized by increase in plasma volume in the presence of mild hemolysis.

See Red Cell Mass *on page 1144.*

References
Berlin NI and Lewis SM, "Measurement of Total RBC Volume Relative to Lean Body Mass for Diagnosis of Polycythemia," *Am J Clin Pathol*, 2000, 114(6):922-6.

Dang CV, "Runner's Anemia," *JAMA*, 2001, 286(6):714-6.

Pollycove M and Tono M, "Blood Volume," *Diagnostic Nuclear Medicine*, 3rd ed, Volume 2, Chapter 42, Sandler MP, Coleman RE, Wackers FJT, et al, eds, Baltimore, MD: Lippincott Williams & Wilkins, 1996, 827-34.

Sato K, Karibe H, and Yoshimoto T, "Circulating Blood Volume in Patients With Subarachnoid Hemorrhage," *Acta Neurochir*, 1999, 141(10):1069-73.

Silver HM, Seebeck M, and Carlson R, "Comparison of Total Blood Volume in Normal, Pre-eclamptic, and Nonproteinuric Gestational Hypertensive Pregnancy by Simultaneous Measurement of Red Blood Cell and Plasma Volumes," *Am J Obstet Gynecol*, 1998, 179(1):87-93.

Spivak JL, "Polycythemia Vera: Myths, Mechanisms, and Management," *Blood*, 2002, 100(13):4272-90.

Tefferi A, "Polycythemia Vera: A Comprehensive Review and Clinical Recommendations," *Mayo Clin Proc*, 2003, 78(2):174-94.

- ♦ **Bloom Syndrome** *see* Fanconi Anemia, Chromosome Breakage Study *on page 569*
- ♦ **Blue Angels** *see* Barbiturates, Quantitative, Serum or Plasma *on page 248*
- ♦ **BNP** *see* B-Type Natriuretic Peptide *on page 314*

Body Cavity Fluid Cytology

Related Information

Asbestos, Lung or Sputum *on page 215*
Bacterial Culture, Aerobes *on page 229*
Body Fluid Amylase *on page 287*
Body Fluid Analysis, Cell Count *on page 288*
Body Fluid Chemical Analysis *on page 291*
Body Fluid Glucose *on page 294*
Body Fluid Lactate Dehydrogenase *on page 294*
Body Fluid pH *on page 295*
Bronchoalveolar Lavage (BAL) *on page 311*
CA 27.29, Serum *on page 322*
CA 125, Serum *on page 323*
Carcinoembryonic Antigen, Serum *on page 342*
Cyst Fluid Cytology *on page 490*
Fine Needle Aspiration, Deep and Superficial Masses *on page 590*
Flow Cytometry, Overview *on page 592*
Fungal Culture, Biopsy or Body Fluid *on page 619*
Gene Rearrangement for Leukemia and Lymphoma *on page 633*
Gram Stain *on page 658*
Immunoperoxidase Procedures *on page 780*
Mycobacterial Culture, Biopsy or Body Fluid *on page 929*
Polymerase Chain Reaction *on page 1069*
Synovial Fluid Analysis *on page 1229*
Viral Culture *on page 1307*
Washing Cytology *on page 1326*

Synonyms Effusion Cytology; Fluids Cytology

Applies to Ascitic Fluid Cytology; Culdocentesis; Paracentesis Fluid Cytology; Pericardial Fluid Cytology; Peritoneal Fluid Cytology; Pleural Fluid Cytology; Thoracentesis Fluid Cytology

Test Includes Cytologic evaluation of smears, cytocentrifuge preparations and/or liquid-based monolayer preparations, filter preparations, and cell block preparations when indicated

Abstract The major coelomic cavities are the pleural spaces, the pericardial sac, and the peritoneum. The tunica vaginalis testis exists in males. Cytologic evaluation of body fluids, in conjunction with chemical analysis and clinical profile, can render and/or enhance diagnosis of a variety of benign and malignant conditions. In addition to classical cytologic studies, immunocytochemical analysis on cytocentrifuged specimens and flow cytometric immunophenotyping can enhance diagnosis.

Fluid may be obtained during staging surgical procedures (see Washing Cytology *on page 1326*) or by puncture. The most common cause of fluid accumulations in the body spaces is congestive heart failure. The second most common cause is neoplastic disease.

Patient Preparation Patient should sign informed consent prior to procedure. Puncture site should be carefully cleaned and prepared as for any tap. In cases of suspected malignancy, the smallest gauge needle (22 g) should be used.

Specimen The volume, clarity or opalescence, color, malodor, or viscosity may be relevant.

Container Use clear container to which anticoagulant can be added prior to collection. The optimal amount of fluid for cytology is 200-500 mL. The practice of salvaging large amounts (in excess of 500 mL) of fluid for cytologic examination is not recommended. The best diagnostic aliquot is the last portion that is drawn off, not the first.

Collection Gently agitate the container as fluid is collected in order to mix the heparin with the fluid; fluid may also be collected fresh without anticoagulant and sent to the laboratory in the fresh state immediately. Other tests are usually
(Continued)

Body Cavity Fluid Cytology *(Continued)*

needed as well; see Related Information above. **Venous blood** drawn at the same time may be helpful; comparisons between serum and body fluid protein, LD, glucose, and other tests are often useful. When pleural fluid is sampled, a **pleural biopsy** may provide diagnosis, especially of granulomatous diseases as well as carcinoma.

Storage Instructions Fluid with or without anticoagulant may be stored at 4°C; cells in the fluid can be preserved at this temperature for up to 1 week, without appreciable deterioration of cellular detail. With special techniques, specimens may be kept frozen at -70°C for 1 year without serious loss of cellular details.

Causes for Rejection Prolonged period (over 2 hours) at room temperature; improper container (thoracentesis and paracentesis drainage bags and large syringes are not acceptable containers)

Special Instructions Add 1 mL of heparin per 100 mL of fluid anticipated (each mL of heparin contains 1000 units). The common anticoagulants that can be used are heparin, 5-10 units per mL of fluid to be placed in the collecting vessel; or 3.8% sodium citrate, 1 mL per 10 mL of fluid; or EDTA, 1 mg per 1 mL of fluid. The cells in these fluids do not deteriorate rapidly if refrigerated, and no fixative need be added if the **fluid is refrigerated** within 30 minutes of collection. Provide pertinent clinical information including previous malignancy, drugs, radiation therapy, or history of alcohol abuse.

Use Establish the presence of primary (mesothelioma) or metastatic neoplasms. Aid in the diagnosis of rheumatoid pleuritis; systemic lupus erythematosus; myeloproliferative and lymphoproliferative disorders; viral, fungal, and parasitic infestation; and fistulas involving serous cavities. The presence of neoplastic cells in body fluid usually indicates that the patient has widespread metastases, with the exception of patients with primary pulmonary lesions. Positive peritoneal cytology is associated with advanced disease. Effusion fluid may be submitted for flow cytometric analysis in those cases suspected of myeloproliferative or lymphoproliferative disorder, and gene rearrangement analysis can be useful as well. Pleural biopsy is advocated when >50% of cells in a pleural effusion are lymphocytes. In that situation, the likelihood of neoplastic disease or of tuberculosis is 95%. In addition to lymphoma and TB, sarcoidosis is a cause of pleural fluid lymphocytosis.

Examination of **synovial fluid** from a joint effusion may aid in the diagnosis of metabolic arthritis (gout or pseudogout), rheumatoid arthritis, or traumatic arthritis as well as septic arthritis (gonococcal arthritis). See Synovial Fluid Analysis *on page 1229.*

Limitations Clots may contain diagnostic cells which are available for recovery by preparation of a cell block while routine smears may fail to reveal such cells. Malignant cells cannot be recovered from all fluids from all subjects with malignant disease. Very well differentiated carcinomas may be difficult to distinguish from reactive states.

Ascitic fluid from patients with cirrhosis may contain markedly atypical cells which may be derived from mesothelial cells.

Contraindications Relative contraindications include documented bleeding diathesis and full anticoagulated state.

Additional Information Fluids should be submitted **fresh, unfixed,** and **heparinized** with anticoagulant added before collection to provide well-preserved, representative, diagnostic material. Exfoliated cells deteriorate rapidly in effusions, both in and out of the body. Fixatives, such as formalin and alcohol, or other types of fixatives must not be used since they prevent adherence of the cells to the slides, do not allow cells to flatten out for optimal presentation of cellular details, and hinder quality staining by the Papanicolaou method. Alcohol also causes precipitation of protein which may interfere with cell analysis. If feasible, cell blocks can be prepared from the fluid sediment. Cell blocks may be especially helpful in evaluation of clotted specimens, and for ultrastructural examination, special straining, and immunocytochemistry.

Tumor markers support discrimination between benign and malignant effusions. **Carcinoembryonic antigen** (CEA) is used both as an immunocytochemical marker and as a test available for serum and other body fluids. **CA 125** elevation with negative CEA assay occurs with serous and endometrioid carcinomas of ovary and adenocarcinoma of endometrium and fallopian tube. By contrast, increased fluid CEA with negative CA 125 results are found with mucinous adenocarcinomas of ovary, lungs, gastrointestinal tract (including pancreas), or breast. Both antigens are within normal range with lymphoma, melanoma, and with benign effusions.

Molecular detection of pathogens can be applied (eg, body cavity-based lymphoma and human herpesvirus 8).

Other techniques may also supplement classical cytologic methods (eg, chemistry). See Body Fluid Chemical Analysis on page 291 and Flow Cytometry, Overview on page 592. Davidson et al compared the efficiency of immunophenotyping using flow cytometry with four epithelial markers to other methods.

When fluids are examined in the clinical microscopy laboratory, detection of neoplastic cells cannot be expected to compare with the capabilities of the cytopathology laboratory.

References

Carella R, Deleonardi G, D'Errico A, et al, "Immunohistochemical Panels for Differentiating Epithelial Malignant Mesothelioma From Lung Adenocarcinoma: A Study With Logistic Regression Analysis," Am J Surg Pathol, 2001, 25(1):43-50.

Davidson B, Dong HP, Berner A, et al, "Detection of Malignant Epithelial Cells in Effusions Using Flow Cytometric Immunophenotyping," Am J Clin Pathol, 2002, 118(1):85-92.

DeMay RM, "Fluids," The Art and Science of Cytopathology, Chapter 8, Chicago, IL: ASCP Press, American Society of Clinical Pathologists, 1996, 257-325.

Mullaney BP, Ng VL, Herndier BG, et al, "Comparative Genomic Analyses of Primary Effusion Lymphoma," Arch Pathol Lab Med, 2000, 124(6):824-6.

Nathan NA, Narayan E, Smith MM, et al, "Cell Block Cytology. Improved Preparation and its Efficacy in Diagnostic Cytology," Am J Clin Pathol, 2000, 114(4):599-606.

Body Fluid Amylase

Related Information

Amylase, Serum on page 155
Amylase, Urine on page 157
Body Cavity Fluid Cytology on page 285
Body Fluid Chemical Analysis on page 291
Cyst Fluid Cytology on page 490
Lipase, Serum on page 851

Synonyms Amylase, Body Fluid

Abstract High pleural or peritoneal fluid amylase is associated with pancreatitis (and its complications), rupture or perforation of the esophagus and occasionally with tumors, especially adenocarcinoma of lung and ovary. Pancreatitis is related to left-sided or bilateral pleural effusion. (Right-sided pleural effusions related to abdominal disease are often caused by cirrhosis, subdiaphragmatic abscess, hepatic abscess, and Meig syndrome).

Specimen Body fluid (ie, ascitic fluid, pleural fluid, etc) and simultaneously drawn serum for amylase. Often, fluid for cytopathology is indicated.

Container Clean container, no preservative

Collection Peritoneal fluid may be obtained by peritoneal lavage. Centrifugation is desirable.

Storage Instructions Amylase is fairly stable, at 4°C, at normal levels.

Reference Interval Body fluid (peritoneal or pleural) amylase values are usually compared to simultaneously obtained serum values. No established reference intervals exist for body fluids.

Use Pancreatitis, with or without pseudocyst formation or pancreatic pleural fistula, is the most common cause of a pleural or peritoneal fluid amylase value greater than the serum amylase. Rupture of the esophagus is the second most common cause and malignant effusion is the third. Other causes include pancreatic ascites and pancreatic duct trauma. A defect in the wall of the gastrointestinal tract (eg, perforated peptic ulcer) will allow pancreatic secretion to enter the peritoneal cavity. Similarly, peritoneal fluid amylase elevations may be found in the presence of necrotic bowel.

An ascitic fluid amylase >3 times the serum amylase is strong evidence for disease in the pancreas, including pancreatitis, pancreatic pseudocyst, or trauma.

Limitations In collection of ascitic fluid, the localization of the catheter is likely to affect the chemistry result. Oxalate or citrate depress results. Lipemic sample may contain inhibitors which falsely depress results. Benign ovarian cyst fluids (Continued)

Body Fluid Amylase (Continued)

may have significant amylase activity. In about 10% of instances of pancreatic disease, ascitic fluid, as well as serum amylase, may be within normal limits.

Additional Information Most patients with pancreatic ascites have high peritoneal fluid amylase as well as amylase and lipase elevations in serum. Pancreatitis may present with pleural effusion. Tumors of lung and serous tumors of ovary are especially recognized as causes of hyperamylasemia with pleural effusion.

Dark, prune juice-like peritoneal fluid is described as characteristic of severe, necrotizing pancreatitis.

The presence of bacteria in a foul-smelling peritoneal fluid indicates perforated viscus in the differential diagnosis of pancreatitis and peritonitis.

References

DeMay RM, "Fluids," *The Art and Science of Cytopathology*, Chapter 8, Chicago, IL: ASCP Press, American Society of Clinical Pathologists, 1996, 257-325.

Steinberg W and Tenner S, "Acute Pancreatitis," *N Engl J Med*, 1994, 330(17):1198-210.

Body Fluid Analysis, Cell Count

Related Information

Applies to Ascitic Fluid Analysis; Cell Count Ratio; Cyst Fluid Analysis; Joint Fluid Analysis; Paracentesis Fluid Analysis; Pericardial Fluid Analysis; Peritoneal Fluid Analysis; Peritoneal Lavage, Diagnostic; Pleural Fluid Analysis; Serous Fluid Analysis; Thoracentesis Fluid Analysis

Test Includes Total WBC count and differential; total RBC count; multiple additional tests (eg, protein, sugar, lactate dehydrogenase [LD], amylase) are commonly done on body fluids.

Abstract Benign and/or malignant cells, crystals, bacteria, and various forms of particulate material can be analyzed by cell counting techniques (manual/automated) and by microscopic study of stained slides bearing smears from unprocessed fluid or centrifuged sediment from fluid. Cytocentrifuge preparations, as available from the laboratory, are strongly recommended (in particular for cerebrospinal fluid) in order to obtain the best preservation of cell morphology.

Patient Preparation Aseptic preparation for aspiration

Specimen Body fluid (ie, pleural fluid, synovial fluid, cyst fluid, paracentesis fluid, pericardial fluid, lavage fluid, etc)

Container Glass test tube or large glass container

Collection Add heparin to specimen.

Storage Instructions Specimen should be brought directly to the laboratory after collection. Cell studies should be done within 1 hour of collection. If complement level is measured (synovial fluid), test immediately or freeze specimen at -70°C.

Causes for Rejection Clotted specimen

Special Instructions Commonly cytologic, microbiologic and chemical examinations are also helpful.

Reference Interval See table. Glucose and amylase levels in fluids approximate whole blood levels. A pleural fluid LD to serum LD ratio >0.6 suggests exudate. In pleural fluids, total protein >3.0 g/dL indicates exudate, protein <3.0 g/dL indicates transudate. For peritoneal fluid, the cutoff point is lower, 2.0-2.5.

Pericardial fluids have no established cutoff point in differentiating transudates from exudates.

Expected Normal Findings – Body Fluids

Type of Fluid	Appearance	Amount (mL)	Cells	Glucose	Total Protein
Pleural	Clear, colorless to pale yellow	1-10	<1000/mm^3 <25% polys 0 RBC	Approximate WB glucose	
Peritoneal	Clear, colorless to pale yellow	<100	<500/mm^3 <25% polys <100,000 RBC/mm^3		
Pericardial	Clear, colorless to pale yellow	20-25	<500 WBC/mm^3 <25% polys 0 RBC		
Synovial	No crystals	<4	<200 WBC/mm^3 <25% polys	Blood / synovial difference <10 mg/dL	1.0-3.0 g/dL

Use Evaluate body fluids; differential diagnosis of exudate, transudate

Methodology Cells are usually counted/studied using manual or automated methods. Specimen should be observed macroscopically and on initial smear for presence of fibrin clots, cell aggregates, crystals, etc, that might occlude apertures of automated counters. Stained smears of fluid are prepared and/or sediment is obtained from centrifuged fluid. Wright/Giemsa stains are used for white cell morphology/parasites, Papanicolaou-stained smears for identification and study of tumor cells, Gram stain for bacteria, acid-fast stain for mycobacteria, Gomori methenamine silver (GMS), periodic acid Schiff (PAS), and other stains for fungi.

Optimal processing is achieved with cytocentrifuge concentration and slide preparation, in particular, for cell studies of cerebrospinal fluid.

For best results, fresh unfixed specimens should be processed using specific protocols including washing cells of serous fluids, liquefaction of synovial fluids with hyaluronidase, agitation of clots (to free up cells for study), and use of albumin. A standardized dilution protocol and a "reagent blank" (to detect bacterial contamination of supposedly sterile reagents) are also recommended. Aliquots for culture, cytology, and biochemistry are best received in separate containers.

Additional Information Elevated lymphocytes can be associated with congestive heart failure (pleural fluid), tuberculosis, tumors, lymphomas, lymphatic leukemia, rheumatoid arthritis, and postpneumonic effusions. Elevated polymorphonuclear leukocytes are associated with acute infectious processes. The difference between uninfected ascitic fluid and that of bacterial peritonitis is reported, for example, as WBC count of 122/mm^3 vs 2686/mm^3. PMN count provides the highest sensitivity; its reliability is enhanced with pH (low with peritonitis, <7.35). Elevated eosinophils are associated with tumors, infarcts, systemic lupus erythematosus (SLE), rheumatoid arthritis, rheumatic fever, parasites, postpneumonic effusions, pneumothorax, or may have no clinical significance. The level of eosinophil granule proteins can be determined in body fluids and may serve as a marker for previously lysed eosinophils (see Eosinophil Granule Proteins *on page 545*). Elevated plasma cells can be associated with lymphoma, especially Hodgkin disease or with chronic inflammation. Sometimes atypical plasma cells are seen in the presence of multiple myeloma. A low glucose level with rheumatoid factor supports diagnosis of rheumatoid effusion. Creatinine level distinguishes ascitic collection from tap of an overdistended bladder.

Diagnostic peritoneal lavage (DPL) was introduced in 1965. Classic criteria for a positive DPL have been considered as >100,000 red blood cells (RBC)/mm^3, >500 white blood cells (WBC)/mm^3, serum amylase >175 IU, or particulate material in the lavage return (with use of 1000 mL of lactated Ringer's solution as lavage fluid). DPL has been considered negative if there are <50,000 RBC/mm^3, <100 WBC/mm^3, and amylase activity is <75 IU. With positive findings, laparotomy was usually performed. With the advent of noninvasive imaging (eg, computed tomography, ultrasonography) the role (and use) (Continued)

Body Fluid Analysis, Cell Count *(Continued)*

of DPL has been reduced. When imaging is unavailable or results are inconclusive, DPL studies with new diagnostic criteria may assist in determination of the need for laparotomy. In a series of 429 DPLs performed for the question of peritoneal penetration as a result of gunshot wound to the abdomen with threshold criterion of 10,000 red blood cells/mm³, DPL was sensitive (99%), specific (98%), and accurate (98%). In a study of 250 patients with blunt abdominal trauma, classic criteria were used with addition of WBC ≥ RBC/150 (when DPL was positive for hemoperitoneum) for determination of intestinal injury. Diagnostic sensitivity was 97% with specificity of 99% when DPL was performed within the time frame of 3-18 hours after trauma. A **"cell count ratio"** defined as the ratio between WBC count and RBC count of lavage fluid divided by the ratio of the same counts (WBC/RBC) in peripheral blood was applied to 212 patients who had a positive DPL (by classic criteria). A cell count ratio ≥1 predicted hollow organ perforation with a specificity of 97% and sensitivity of 100%. A low (<1) cell count ratio can contribute to more confident nonoperative management. See graphic.

Management of Blunt Abdominal Injury

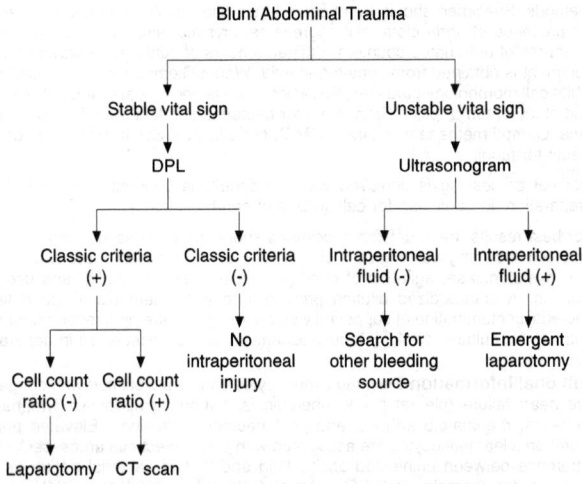

Adapted from Fang JF, Chen RJ, and Lin BC, "Cell Count Ratio: New Criterion of Diagnostic Peritoneal Lavage for Detection of Hollow Organ Perforation," *J Trauma*, 1998, 45(3):540-4.

A common cause of serous and/or bloody effusions is malignancy, usually malignant lymphoma or metastatic carcinoma, and less commonly sarcoma or malignant mesothelioma. The malignancy may be an unexpected discovery on study of stained smears of body cavity fluid. Initially or subsequently, cytocentrifuge slide preparations with Wright-Giemsa stain, provide a basis for study and identification of the malignant cell population.

References

Cornbleet PJ, "Wright-Giemsa Cytology of Body Fluids: Criteria for Identification of Malignant Cells," *Lab Med*, 1998, 29(1):26-31.

Gall JJ, "Laboratory Evaluation of Body Fluids," *Clinical Hematology: Principles, Procedures, Correlations*, 2nd ed, Chapter 30, Stiene-Martin EA, Lotspeich-Steininger CA, and Koepke JA, eds, Philadelphia, PA: Lippincott-Raven, 1998, 400-14.

Jones CD and Cornbleet PJ, "Wright-Giemsa Cytology of Body Fluids," *Lab Med*, 1997, 28(11):713-6.

Kjeldsberg CR and Knight JA, *Body Fluids - Laboratory Examination of Amniotic, Cerebrospinal, Seminal, Serous, and Synovial Fluids*, 3rd ed, Chicago, IL: ASCP Press, American Society of Clinical Pathologists, 1993, 159-253, 265-301.

Body Fluid Chemical Analysis

Related Information

Adenosine Deaminase, CSF, Pleural Fluid, Pericardial Fluid, Peritoneal Fluid *on page 107*
Body Cavity Fluid Cytology *on page 285*
Body Fluid Amylase *on page 287*
Body Fluid Analysis, Cell Count *on page 288*
Body Fluid Glucose *on page 294*
Body Fluid Lactate Dehydrogenase *on page 294*
Body Fluid pH *on page 295*
CA 125, Serum *on page 323*
Carcinoembryonic Antigen, Serum *on page 342*
Cerebrospinal Fluid Analysis: Overview *on page 355*
Fine Needle Aspiration Culture *on page 589*
Fine Needle Aspiration, Deep and Superficial Masses *on page 590*
Fungal Culture, Biopsy or Body Fluid *on page 619*
Gram Stain *on page 658*
Liver Disease: Laboratory Assessment, Overview *on page 869*
Mycobacterial Culture, Biopsy or Body Fluid *on page 929*
Rheumatoid Factor, Serum or Body Fluid *on page 1161*
Synovial Fluid Analysis *on page 1229*
Washing Cytology *on page 1326*

Synonyms Ascitic Fluid Analysis; Fluid, Pericardial; Fluid, Peritoneal; Fluid, Pleural; Paracentesis Fluid Analysis; Pericardial Fluid Analysis; Peritoneal Fluid Analysis; Pleural Fluid Analysis; Thoracentesis Fluid Analysis

Applies to Albumin, Ascites Fluid; CEA, Body Fluid; Chylous Fluid; Cyst Fluid Chemistry; Lactic Acid, Body Fluid; LD, Body Fluid; Protein, Body Fluids; Rheumatoid Factor, Body Fluid; Serum-Ascites Albumin Gradient (Alb$_{s-a}$)

Test Includes Tests commonly helpful in work-up of a fluid include **cell count** and differential, hemoglobin/hematocrit, **glucose, lactate dehydrogenase (LD), albumin**, specific gravity, **amylase** and **pH**. Protein quantitation is occasionally useful for pleural fluid but is less reliable for peritoneal fluid. **Cultures** are very commonly indicated, require sterile specimens, and generally are ordered for routine, anaerobic, TB and sometimes fungi. **Gram and acid-fast smears** are often essential for guiding diagnosis and therapy. **Cytology** and **tumor markers** may be indicated.

Abstract Serous effusions (pleural, peritoneal, and pericardial) are fluids which may be classified either as transudates or exudates. Transudates are produced in systemic diseases (eg, congestive heart failure, hepatic cirrhosis, nephrotic syndrome) due to noninflammatory processes: increased hydrostatic pressure or decreased plasma oncotic pressure, changes at the capillary level. Exudates result from inflammatory processes (eg, infection, esophageal or other hollow viscus rupture, subphrenic or liver abscess, rheumatoid arthritis, pancreatitis, lung embolization or infarct, trauma, systemic LE) that increase capillary permeability or decrease absorption of fluid by the lymphatic system (eg, by lymphatic obstruction). (Fluids caused by malignant diseases are classified as exudates). Laboratory tests are done to determine whether an effusion is a transudate or an exudate. Follow-up testing is done to characterize the cause of fluid accumulation and establish its etiology.

A third type of fluid accumulation, **chylous effusion**, is also recognized. Chylous pleural effusions may be secondary to leakage and/or obstruction of the thoracic duct and are further described below.

When the explanation for a fluid accumulation is unknown, consider cirrhosis, carcinomatosis, or tuberculosis. See Body Cavity Fluid Cytology *on page 285.*

While pleural transudates are apt to be bilateral, pleural exudates (pleuritis) are commonly unilateral.

Specimen Pleural, peritoneal, or pericardial fluid

Container Fluids are often aspirated using syringes anticoagulated with heparin or EDTA. Chemistry testing requires red top tube or green top (heparin) collection tubes. Check with the laboratory for requirements for hematology, microbiology, and cytology testing.

Collection Specimen is obtained surgically by the physician using sterile technique. Since testing of body fluids usually occurs in several laboratory sections, (Continued)

Body Fluid Chemical Analysis *(Continued)*

a common error is to provide the laboratory insufficient quantity of fluid for adequate examinations. 50 mL is desirable, divided into appropriate containers. A simultaneously drawn blood specimen is desirable for appropriate serum chemistry testing: protein and LD, sometimes glucose.

Special Instructions Laboratory must be made aware of the source of the specimen.

Reference Interval Since accumulations of fluid in the body cavities are abnormal, no reference intervals exist.

Use Testing is done to determine whether a fluid is a transudate or an exudate (see table). Some authors publish slightly varying figures.

	Transudate	Exudate
Fluid appearance	Clear, colorless to yellow	Cloudy, variable color (yellow, green, red)
Coaguability	Will not clot	May clot due to presence of fibrinogen
WBC/mm^3	<1000/mm^3	>1000/mm^3
Specific gravity	<1.016	>1.016
Fluid to serum total protein ratio	<0.5	>0.5
Fluid to serum LD ratio	<0.6	>0.6
Glucose	Equal to serum level	Less than or equal to serum levels - especially low in rheumatoid effusion
Cholesterol pleural	<60 mg/dL (SI: <1.55 mmol/L)	>60 mg/dL (SI: >1.55 mmol/L)
Cholesterol peritoneal	<46 mg/dL (SI: <1.19 mmol/L)	>46 mg/dL (SI: >1.19 mmol/L)
Albumin (serum-fluid)	1.6 ±0.5 g/dL	0.6 ±0.4 g/dL

Limitations Postural changes are significant in exudates but not in transudates in pleural effusions.

Additional Information Lung and breast carcinoma are the two most common tumors causing pleural effusion. Exfoliative cytology is often critically important in diagnosing malignant effusions, though sensitivity is low. In addition to cytology, **tumor markers** in body fluids have been used, including CEA, CA 125, cytokeratin 19 fragments (CYFRA 21-1), CA 15-3, hCG, and neuron-specific enolase (NSE). CEA increases occur with many carcinomas primary in the gastrointestinal tract, breast, and lung. Increased CA 125 without elevated CEA is consistent with primary carcinoma of ovary, fallopian tube, or endometrium, but may occur with stage III or IV endometriosis. CEA is commonly negative in Müllerian carcinomas but positive with mucinous cystadenocarcinoma of ovary and adenocarcinoma of endocervix. Normal CEA and CA 125 concentrations in malignant fluids are found with mesothelioma, melanoma, and lymphoma.

Pericardial fluid specimens are often exudates. Significant palliation may be achieved with diagnosis and therapy of some malignant effusions. Common primaries include lung, breast, esophagus, melanoma, and lymphoma. Other causes of pericardial effusions include bacterial infection, TB, coxsackieviruses, AIDS, uremia, myocardial infarction, radiation, hypothyroidism, fluid overload, hypoproteinemia, connective tissue diseases, and trauma. The most common cause of pericardial transudate is congestive heart failure.

In descending order of frequency, the **causes of ascites** (peritoneal fluid accumulation) in the U.S. are cirrhosis, cancer, heart failure, tuberculosis, dialysis, and pancreatic disease. Malignant causes include carcinomas of ovary, breast, stomach, pancreas, liver, colon and rectum, endometrium, cervix, lymphoma, and mesothelioma.

Long-term dialysis is a cause of pleural, pericardial or ascitic effusions.

Other tests that are sometimes helpful include **pH**, especially in chest fluids. **Urea nitrogen (BUN)** is helpful if a question of bladder content versus ascitic fluid exists.

Green fluids may be tested for **bilirubin**; positive results are consistent with perforated intestine, peptic ulcer, or gallbladder.

Bloody fluids, if not caused by traumatic tap, are generally exudates. Trauma, TB, as well as infarct or cancer must be considered. **Hematocrit** of the fluid is useful for diagnosis of hemothorax or hemoperitoneum (eg, trauma).

Chylous fluid appears milky, may appear bloody, yellow or green, contains chylomicrons, and has very high **triglyceride** levels (>110 mg/dL). Fluid triglyceride concentrations <60 mg/dL are inconsistent with chylous effusion. For fluids having triglyceride levels of 60-110 mg/dL, lipoprotein electrophoresis is recommended to identify the presence or absence of chylomicrons. Chylous effusions relate to trauma, surgical procedures, lymphoma, carcinoma, and tuberculosis, and an idiopathic group is recognized as well. The most common cause is lymphoma. **Pseudochylous effusions**, with triglycerides <50 mg/dL, are found with cases of rheumatoid pleuritis, tuberculosis, and myxedema.

High levels of **rheumatoid factor** in a pleural fluid support a diagnosis of rheumatoid effusion, while in SLE, rheumatoid factor titers are apt to be only about 1:40.

Peritoneal and pleural fluid **lactate** levels are increased with infection. In uninfected ascites, ascitic fluid lactate was 15 ±5 mg/dL (SI: 1.7 ±0.6 mmol/L), while in bacterial peritonitis, 14 patients ranged 45 ±37 mg/dL (SI: 5.0 ±4.1 mmol/L). A cutoff >25 mg/dL (SI: >2.8 mmol/L) is suggested. Lactate may also be increased in malignant disease in body fluids.

The **ratio of LD** in serum to that of ascitic fluid is helpful in the differential diagnosis of ascites. Ratios of less than unity occur with malignant disease, while ratios >1 are reported mostly with cirrhosis, when these two groups are compared. However, a few instances of cancer have ratios greater than unity. The LD of the ascitic fluid of cirrhosis is usually <60% that of serum.

Since **albumin** in body fluids is the main determinant of oncotic pressure, the **serum-ascites albumin gradient (Alb$_{s-a}$)** has been used for differentiating transudates and exudates in peritoneal fluid. It is calculated by subtracting the albumin value of the fluid from that of serum and is greater in transudates than exudates. It correlates with portal pressure (ie, the pressure gradient between the portal capillaries and the peritoneal cavity). Gradients ≥1.1 g/dL (SI: 11 g/L) strongly correlate with portal hypertension. Disease states related to high gradients include cirrhosis, alcoholic hepatitis, cardiac failure, massive liver metastases, fulminant hepatic failure, Budd-Chiari syndrome, portal vein thrombosis, veno-occlusive disease, and fatty liver of pregnancy. Constrictive pericarditis is usually associated with high gradients. **Low gradients** are found with peritoneal carcinomatosis and tuberculosis, pancreatic and biliary ascites, nephrotic syndrome, serositis, and bowel obstruction or infarct. Other diseases reported with low gradients include SLE, certain ovarian diseases, and severe hypoalbuminemia.

"High protein ascites," >2.5 g total protein/dL (SI: >25 g/L), is found in 15% to 20% of subjects who have hepatic disease, while a similar fraction of patients with malignant disease have **"low protein ascites."**

Since **secondary peritonitis** often requires surgery, its distinction from spontaneous peritonitis is critical. Secondary peritonitis is characterized by total protein ≥1.0 g/dL, fluid LD greater than upper limit of normal for serum, glucose <50 mg/dL, and polymicrobial infection. See Body Fluid Amylase *on page 287*.

Gram stains for bacteria and Wright stains for white cell count may also be helpful.

References

DeMay RM, "Fluids," *The Art and Science of Cytopathology*, Chapter 8, Chicago, IL: ASCP Press, American Society of Clinical Pathologists, 1996, 257-325.

Miedouge M, Rouzaud P, Salama G, et al, "Evaluation of Seven Tumour Markers in Pleural Fluid for the Diagnosis of Malignant Effusions," *Br J Cancer*, 1999, 81(5):1059-65.

Sheets EE and Smith RN, "A 31-Year-Old Woman With a Pleural Effusion, Ascites, and Persistent Fever Spikes," Case Records of the Massachusetts General Hospital, Case 3-1998, Scully RE, Mark EJ, McNeely WF, et al, eds, *N Engl J Med*, 1998, 338(4):248-54.

(Continued)

Body Fluid Chemical Analysis (Continued)

Wilkes JD, Fidias P, Vaickus L, et al, "Malignancy-Related Pericardial Effusion. 127 Cases From the Roswell Park Cancer Institute," *Cancer*, 1995, 76(8):1377-87.

♦ **Body Fluid Creatinine** *see* Body Fluid pH *on page 295*

♦ **Body Fluid Culture** *see* Bacterial Culture, Anaerobes *on page 231*

♦ **Body Fluid Fungus Culture** *see* Fungal Culture, Biopsy or Body Fluid *on page 619*

♦ **Body Fluid GGT** *see* Gamma-Glutamyl Transferase, Serum *on page 629*

Body Fluid Glucose

Related Information

Body Cavity Fluid Cytology *on page 285*
Body Fluid Chemical Analysis *on page 291*
Body Fluid Lactate Dehydrogenase *on page 294*
Body Fluid pH *on page 295*
Cerebrospinal Fluid Glucose *on page 362*
Glucose, Fasting, Plasma *on page 643*
Washing Cytology *on page 1326*

Synonyms Glucose, Body Fluid

Applies to Amniotic Fluid Glucose

Abstract Decreased body fluid glucose may be found in bacterial infection, tuberculosis, rheumatoid effusion and occasionally with malignant disease.

Specimen Body fluid; simultaneously drawn plasma glucose

Reference Interval Fluid glucose concentration is usually similar to plasma glucose concentration. A value <60 mg/dL, or a value 40 mg/dL less than the simultaneous plasma level is considered decreased.

Use Pleural fluid glucose levels <60 mg/dL indicate often grossly purulent parapneumonic effusion or rheumatoid effusion, and glucose <50 mg/dL (SI: <2.8 mmol/L) characterizes rheumatoid effusion. In contrast, fluid glucose in SLE is usually >60 mg/dL. Pericardial effusions with decreased glucose are reported with malignant disease and with bacterial endocarditis. Ascitic fluid glucose is often decreased in tuberculous peritonitis and with malignant disease but is usually normal with cirrhosis or congestive failure.

Limitations Peritoneal fluid glucose is rarely of clinical value, unlike that of cerebrospinal fluid and pleural fluid.

Additional Information Amniotic fluid glucose when low, is a marker for intrauterine infection in women with preterm labor. Its specificity is 94% to 100%, but it has poor sensitivity. However, in a study comparing several amniotic fluid markers of intra-amniotic infection (glucose, polymorphonuclear leukocytes, Gram stain, and culture), amniotic fluid glucose levels <20 mg/dL were the most sensitive predictors of histologic chorioamnionitis.

References

Odibo AO, Rodis JF, Sanders MM, et al, "Relationship of Amniotic Fluid Markers of Intra-amniotic Infection With Histopathology in Cases of Preterm Labor With Intact Membranes," *J Perinatol*, 1999, 19(6 Pt 1):407-12.

♦ **Body Fluid Identification** *see* Specimen Identification Requirements *on page 1217*

Body Fluid Lactate Dehydrogenase

Related Information

Body Cavity Fluid Cytology *on page 285*
Body Fluid Chemical Analysis *on page 291*
Body Fluid Glucose *on page 294*
Body Fluid pH *on page 295*
Lactate Dehydrogenase, Serum *on page 825*

Synonyms LD, Body Fluid

Abstract Body fluid LD levels are often higher than serum levels in malignant effusions. Elevations are also found in inflammatory states, including tuberculous effusions.

Specimen Body fluid; simultaneously drawn serum

Reference Interval Exudates exhibit pleural fluid to serum LD ratios >0.6, whereas transudates exhibit ratios <0.6.

Use Differential diagnosis and classification of effusions; aid in the differential diagnosis of traumatic tap vs central nervous system (CNS) hemorrhage in newborns. Analysis of serum and body fluid drawn at the same time, for LD and total protein, often provides distinction between transudates and exudates (see table in Body Fluid Chemical Analysis *on page 291*).

Limitations This test has limited usefulness due to its nonspecificity. Like use of total protein, LD fails to classify fluids accurately as transudate or exudate. Though transudates and exudates are often said to be characterized by fluid LD activities <200 units/L and >200 units/L, respectively, assay methodologies give widely varying results, thus, making the fluid to serum ratio a more useful assessment.

Additional Information Lactate dehydrogenase (LD) is a normal component of CSF and is increased in bacterial and viral meningitis. LD_1 and LD_2 are decreased in lavage fluid in pulmonary alveolar proteinosis. With ascitic fluid cholesterol, ascitic fluid LDH is helpful in the differential diagnosis of hepatocellular carcinoma versus ascites from other malignant neoplasms. See Body Fluid Chemical Analysis *on page 291* for more information.

References

Castaldo G, Oriani G, Cimino L, et al, "Total Discrimination of Peritoneal Malignant Ascites From Cirrhosis- and Hepatocarcinoma-Associated Ascites by Assays of Ascitic Cholesterol and Lactate Dehydrogenase," *Clin Chem*, 1994, 40(3):478-83.

Drapkin MS and Mark EJ, "A 38-Year-Old Man With Fever, Cough, and a Pleural Effusion," Case Records of the Massachusetts General Hospital, Case 25-1996, Scully RE, Mark EJ, McNeely WF, et al, eds, *N Engl J Med*, 1996, 335(7):499-505.

♦ **Body Fluid Lipase** *see Lipase, Serum on page 851*

Body Fluid pH

Related Information

Body Cavity Fluid Cytology *on page 285*
Body Fluid Chemical Analysis *on page 291*
Body Fluid Glucose *on page 294*
Body Fluid Lactate Dehydrogenase *on page 294*
Gram Stain *on page 658*
Washing Cytology *on page 1326*

Synonyms pH, Body Fluid

Applies to Arterial-Ascitic Fluid pH Gradient; Body Fluid Creatinine; Creatinine, Ratio; Pericardial Fluid pH; Peritoneal Fluid pH; pH Body Fluid; Pleural Fluid pH; Rheumatoid Effusion; Thoracentesis Fluid pH

Abstract Pleural, peritoneal, or pericardial fluid pH levels have been used where appropriate in the evaluation of pneumonia, rheumatoid pleuritis, malignancy, uremia, empyema, tuberculosis, hemothorax, esophageal rupture, and cirrhosis with spontaneous bacterial peritonitis.

Container Heparinized blood gas syringe or green top (lithium heparin) tube

Collection The usual recommendation is for the sample to be collected anaerobically, directly into a heparinized blood gas syringe or green top (lithium heparin) tube to prevent clotting. "Indirect" collection recently has been shown to be acceptable as well, in which pleural fluid initially is collected into a large (30- to 60-mL) heparinized syringe followed by aliquot transfer to a heparinized blood gas syringe. If the specimen is collected in a syringe, all air should be expelled and needle sealed and capped. The pH should be measured anaerobically without delay.

Storage Instructions Most recommend that the pH be measured anaerobically, keeping the specimen on ice when delay of any kind is unavoidable. A recent study, however, showed that pleural fluid pH preserved anaerobically at room temperature is stable during the first hour following thoracentesis.

Causes for Rejection Purulent or viscous sample

Reference Interval Pleural fluid pH is usually about 7.6.

Use Low pleural fluid pH concentrations (<7.3) are considered abnormal and are usually found in exudates, including empyema, connective tissue diseases, rupture or perforation of esophagus, rheumatoid disease, and tuberculosis. In malignant disease, low pH is suggestive of positive cytopathologic findings and poor prognosis. Levels ≥7.3 are considered a benign finding and are found in transudates.

Methodology Blood gas instrumentation should be used. pH meters and pH paper should not be used.

(Continued)

Body Fluid pH *(Continued)*

Additional Information

Pleural fluid: Low pleural fluid pH levels (<7.2-7.3) identify patients with effusions due to pneumonia, empyema or lung abscess; such conditions require aggressive treatment. Low pleural fluid pH is accompanied by low pleural fluid glucose in these patients.

pH of pleural fluid <6.0 is highly suggestive of rupture of esophagus. Pleural fluid amylase is very helpful for this diagnosis as well (see Body Fluid Amylase *on page 287*).

The pleural fluid pH is often <7.2 and consistently <7.3 in rheumatoid pleural effusion, which is characterized by low glucose, high LD, and high rheumatoid factor. Effusions related to lupus erythematosus generally have pleural fluid pH levels >7.35.

Low pleural fluid pH with negative cytologic examination may occur in a malignant effusion as well as tuberculosis or rheumatoid disease.

A low, transudative, pleural fluid pH (<7.30) accompanied by a pleural fluid creatinine to serum creatinine ratio >1 is consistent with urinothorax, a condition found in some patients with obstructive uropathy.

Peritoneal (ascitic) fluid: The most reliable diagnostic criteria for spontaneous bacterial peritonitis have been reported to include ascitic fluid pH <7.35 (with a mean of 7.24), ascitic fluid PMN >500/mm^3, and an **arterial-ascitic fluid pH gradient** >0.10. Reduction of ascitic fluid pH and in some, but not all, series increments of ascitic fluid PMN counts may be found with peritoneal metastases.

Pericardial fluid: Pericardial fluid pH measurements are not widely used, though decreased (<7.30) levels are observed in rheumatic and purulent disorders, malignancy, uremia, tuberculosis, and other conditions.

References

Burrows CM, Mathews WC, and Colt HG, "Predicting Survival in Patients With Recurrent Symptomatic Malignant Pleural Effusions: An Assessment of the Prognostic Values of Physiologic, Morphologic, and Quality of Life Measures of Extent of Disease," *Chest*, 2000, 117(1):73-8.

DeMay RM, "Fluids," *The Art and Science of Cytopathology*, Chapter 8, Chicago, IL: ASCP Press, 1996, 257-325.

Heffner JE, Nietert PJ, and Barbieri C, "Pleural Fluid pH as a Predictor of Pleurodesis Failure: Analysis of Primary Data," *Chest*, 2000, 117(1):87-95.

Heffner JE, Nietert PJ, and Barbieri C, "Pleural Fluid pH as a Predictor of Survival for Patients With Malignant Pleural Effusions," *Chest*, 2000, 117(1):79-86.

♦ **Body Fluid, Specific Gravity** *see* Specific Gravity, Urine *on page 1216*

♦ **Body Mass Index (BMI)** *see* Leptin, Serum or Plasma *on page 842*

♦ **Body Surface Area** *see* Creatinine Clearance and Urine Creatinine *on page 473*

♦ **Body Weight** *see page 11*

♦ **Bone Alkaline Phosphatase** *see* Alkaline Phosphatase Isoenzymes, Serum *on page 125*

♦ **Bone Biopsy** *see* Aluminum, Bone and Bone Biopsy *on page 139*

♦ **Bone Formation** *see* Osteocalcin, Serum or Plasma *on page 983*

♦ **Bone GLA Protein** *see* Osteocalcin, Serum or Plasma *on page 983*

♦ **Bone Histomorphometry** *see* Aluminum, Bone and Bone Biopsy *on page 139*

Bone Marrow

Related Information

Anemia Flowchart *on page 35*
Breakpoint Cluster Region Rearrangement in CML *on page 304*
Buffy Coat Smear Study of Peripheral Blood *on page 316*
Cerebrospinal Fluid Cytology *on page 361*
Chorionic Villus Sampling, Chromosome and Genetic Abnormality Analysis *on page 400*
Chromosomal Translocations, Molecular Detection *on page 404*
Chromosome Analysis, Blood *on page 406*
Chromosome Analysis, Bone Marrow *on page 407*
Cobalamin, Serum *on page 424*
Complete Blood Count *on page 442*

Test Includes H&E stain, Wright or Wright/Giemsa stain, iron stain, cytochemistry, special histochemistry, and immunohistochemistry stains as indicated for the particular disorders/indications under investigation. Always consider obtaining samples for flow cytometry, gene rearrangement studies, and cytogenetics when obtaining a bone marrow. If these are not initially obtained, it may be necessary to repeat the bone marrow in order to establish the diagnosis. Bone marrow samples in liquid phase (heparinized test tube) can be held prior to sending for special studies, with the decision to send dependent upon findings of an initial inspection of a stained aspirate slide. Bone marrow studies are done in concert with obtaining/examining the peripheral blood smear and obtaining a bone marrow clot section from the aspirated bone marrow material.

Abstract A 20 cc plastic syringe should be used for the **bone marrow aspiration**. Disposable aspiration kits are available. A needle guard is used when obtaining sample from the sternum. **Bone marrow biopsy** is obtained using a specialized needle (eg, Jamshidi needle). Samples are obtained to examine the cellularity of the various normal marrow constituents, for morphologic changes in populations of marrow cells, to identify and study malignant infiltrates, and to determine the presence of infections, cytogenetic abnormalities, and status of iron stores. There are many other applications.

Patient Preparation Explain procedure and pain potential to the patient. Obtain informed consent from the patient or the patient's guardian. Use an adequate quantity of local anesthetic, 1% lidocaine. Systemic control of pain (conscious sedation) may be required (eg, Demerol® 25-50 mg I.V. or oral and Valium® 5-10 mg oral or Versed®). If systemic pain control is administered, provide for monitoring of vital signs, oxygen saturation, and facilities to administer oxygen. The procedure is performed using aseptic technique, typically from the posterior superior iliac crest or the sternum. **Note:** Marrow biopsy is never attempted from the sternum.

Aftercare Apply septic barrier/cover to the needle entry site for 24 hours. Instruct the patient not to bathe for 24 hours. If conscious sedation is utilized, ensure that the patient's condition is fully stable prior to releasing patient from the facility where bone marrow was obtained.

Specimen Bone marrow aspirate is obtained for making coverslip or glass slide preparations. The smearing technique and rapid handling of the aspirated marrow material is crucial to obtain quality preparations, in view of the tendency of the marrow to clot rapidly. Only 1 mL of aspirated marrow is required for the production of slide preparations. Additional aspirated marrow samples may be obtained for flow cytometry, cytogenetic studies, or for obtaining bacterial, fungal, virology cultures (~5 mL in heparinized test tube). Obtain a 1 mL sample for making slides prior to obtaining larger aspiration volumes for the cytogenetic, flow cytometry, and culture samples.

(Continued)

Bone Marrow (Continued)

Container Glass coverslips or slides, pipette for handling/transferring sample to glass coverslip or slide, petri dish for distribution of aspirated material, heparinized tubes for additional studies as noted above; microbiology Isolator™ tubes for culture studies; tubes containing Zenker solution or formalin for clot sections and biopsy samples

Special Instructions Marrow procedures require advance planning/scheduling in order to obtain/coordinate other special studies (eg, flow cytometric studies, cytogenetics, gene rearrangement studies, and in situ hybridization studies).

Reference Interval Normal ranges for infants and adults have been determined (see references by Bain and Geaghan).

Use Evaluation of overall cellularity; morphology of erythroid, myeloid, lymphoid monocytic/macrophage precursors and megakaryocytes; maturation of the precursors of each of these cell lines; and erythroid:myeloid ratio. Identify maturational abnormality or infiltration by cells foreign to the bone marrow. Assess presence of abnormal reticulin or myelofibrosis. Quantify iron status by Perls' stain and establish presence/absence of ring(ed) sideroblasts. Examine for the presence of infectious organisms (eg, histoplasmosis, various mycobacteria, cytomegalovirus, parvovirus inclusions involving erythroid precursors). In general, a bone marrow examination can be omitted in uncomplicated iron deficiency anemia when serum ferritin levels are reduced below 20 µg/L, serum transferrin receptor levels are raised above normal, or the transferrin level/TIBC is raised along with concomitant reduction in the fasting serum iron level <55 mg/dL and resultant transferrin saturation is reduced to <18%. (See the Anemia Flowchart in the Hematology introduction.) Repetitive bone marrow examinations are valuable for assessment of response to therapy in cases of hematological malignancies (eg, acute leukemias, lymphomas, multiple myeloma, as well as nonhematological malignancies, eg, Ewing sarcoma, carcinoma). Evaluation of response to therapy. Fluorescence in situ hybridization (FISH) and polymerase chain reaction (PCR) may be used to determine minimal disease in acute progranulocytic leukemia and the number of Philadelphia-positive metaphases in chronic myelogenous leukemia.

Limitations Bone marrow involvement may be spotty in certain disorders (eg, multiple myeloma, carcinoma). It may be necessary to perform bilateral bone marrow biopsies when evaluating possible marrow involvement by Hodgkin lymphoma as well as by the non-Hodgkin lymphomas. In certain conditions in which the bone marrow is heavily infiltrated by malignant cells or in which there is underlying myelofibrosis, a "dry tap" is not unusual as a result of attempting to obtain a bone marrow aspirate. Under such circumstances, touch imprints of the bone marrow biopsy should be performed on a glass coverslip or slide. When a "dry tap" is encountered, it may be possible to obtain a single cell suspension from a bone marrow biopsy sample for flow cytometry or other special studies. It may be difficult to determine the significance of lymphoid aggregates in the bone marrow, even after performing immunocytochemistry. PCR study may be helpful in such cases. Potentially, repeated biopsies from the same site may result in fibrosis which could be misinterpreted and result in false diagnosis of myelofibrosis.

Contraindications Avoid doing a bone marrow from the sternum in any patient with multiple myeloma, or in a patient with a thoracic aortic aneurysm. Great care should be taken, especially in elderly patients with osteoporosis. **Note:** It is not necessary to correct a low platelet count or coagulopathy prior to performing a bone marrow aspirate/biopsy. Significant bleeding is unusual, even in the presence of a reduced platelet count or coagulopathy. A pressure dressing and compression over the site is usually all that is required.

Methodology A variety of disposable bone marrow aspirate needles and biopsy needles are currently available. Biopsy needles are based on the Jamshidi needle. In general, bone marrow **aspirates** are obtained from the sternum or the posterior or anterior iliac crests. Bone marrow **biopsies** are obtained from the posterior or anterior iliac crests. In infants, the anterior tibia is usually the preferred site.

Additional Information The French American British classification has generally been used for the classification of acute myeloid leukemia, acute lymphoid leukemia, and the myelodysplastic syndromes. Recently, the WHO classification has been proposed and is likely to be universally accepted as an all

encompassing classification of the myeloid and lymphoid malignancies. See
Leukocyte Cytochemistry on page 846 for classification of acute myeloid
leukemias and related conditions, including immunophenotypic and cytogenetic
characteristics.

References

Bain BJ, Clark DM, and Lampert IA, "The Normal Bone Marrow," *Bone Marrow Pathology*, Chapter 1, Cambridge, MA: Blackwell Scientific Publications, 1996, 24-30.

Ben-Ezra J, Hazelgrove K, Ferreira-Gonzalez A, et al, "Can Polymerase Chain Reaction Help Distinguish Benign From Malignant Lymphoid Aggregates in Bone Marrow Aspirates?" *Arch Pathol Lab Med*, 2000, 124(4):511-5.

Braun S, Pantel K, Müller P, et al, "Cytokeratin-Positive Cells in the Bone Marrow and Survival of Patients With Stage I, II, or III Breast Cancer," *N Engl J Med*, 2000, 342(8):525-33.

Dunphy CH, Polski J, Evans HL, et al, "Evaluation of Bone Marrow Specimens With Acute Myelogenous Leukemia for CD34, CD15, CD117, and Myeloperoxidase. Comparison of Flow Cytometric and Enzyme Cytochemical Versus Immunohistochemical Techniques," *Arch Pathol Lab Med*, 2001, 125(8):1063-9.

Foucar K, *Bone Marrow Pathology*, 2nd ed, Chicago, IL: ASCP Press, 2001.

Geaghan SM, "Hematologic Values and Appearances in the Healthy Fetus, Neonate, and Child," *Clinics in Laboratory Medicine: Diagnostic Pediatric Hematology*, Volume 19, Philadelphia, PA: WB Saunders, 1999, 1-37.

Gupta D, Wu SL, and Nguyen AN, "Human Parvovirus B19 in the Bone Marrow With Negative Viral Serologic Results," *Lab Med*, 2001, 32(8):429-31.

Harris NL, Jaffé ES, Diebold J, et al, "World Health Organization Classification of Neoplastic Diseases of the Hematopoietic and Lymphoid Tissues. Report of the Clinical Advisory Committee Meeting, Airlie House, Virginia, November, 1997," *J Clin Oncol*, 1999, 17(12):3835-49.

Manaloor EJ, Neiman RS, Heilman DK, et al, "Immunohistochemistry Can Be Used to Subtype Acute Myeloid Leukemia in Routinely Processed Bone Marrow Biopsy Specimens. Comparison With Flow Cytometry," *Am J Clin Pathol*, 2000, 113(6):814-22.

◆ **Bone Marrow Culture for *Brucella*** *see Brucellosis Culture and Serology on page 312*

◆ **Bone Marrow Fungus Culture** *see Fungal Culture, Biopsy or Body Fluid on page 619*

◆ **Bone Marrow Transplant, Allogeneic** *see Hematopoietic Progenitor Cells, Marrow on page 677*

◆ **Bone Marrow Transplant, Autologous** *see Hematopoietic Progenitor Cells, Marrow on page 677*

◆ **Bone Needle Aspiration Cytology** *see Fine Needle Aspiration, Deep and Superficial Masses on page 590*

◆ **Bone Resorption** *see Osteocalcin, Serum or Plasma on page 983*

Bordetella pertussis Culture and Direct Fluorescent Antibody

Related Information
Bordetella pertussis Serology on page 301

Test Includes Culture for *Bordetella pertussis* and detection directly in nasopharyngeal specimens using fluorescent-labeled antibody specific for organism

Abstract *B. pertussis* is the predominant cause of pertussis (whooping cough) which occurs during a primary infection in unimmunized children. Beginning as a mild respiratory infection, the infection can be recognized within 2 weeks by a distinctive whooping cough that occurs during the paroxysmal stage. Since other infectious agents can also cause a pertussis-like disease (*Chlamydia trachomatis*, adenovirus, and respiratory syncytial virus) it is very important to diagnose pertussis infection quickly in order to prevent further spread. Laboratory diagnosis of pertussis requires both a nasopharyngeal culture, direct detection using fluorescent antibody (DFA) or PCR assays and detection of *B. pertussis*-specific IgG or IgM (especially IgM). For optimal culture of *B. pertussis* from nasopharyngeal specimens, the laboratory must be notified. Routine agar plates do not support the growth of this organism and special media must be prepared or purchased. Even when appropriate culture media are used, positive growth is seen in only 50% of infected patients.

Patient Preparation Patient should not be on antimicrobial therapy prior to the collection of the specimen.

Specimen Nasopharyngeal swab, cough plate optional

Container Flexible calcium alginate swab (Calgiswab®) and Bordet Gengou plate. Transport medium composed of half strength Oxoid charcoal agar CM19 supplemented with 40 µg/mL cephalexin and 10% hemolyzed defibrinated horse blood may be used.

(Continued)

Bordetella pertussis Culture and Direct Fluorescent Antibody *(Continued)*

Collection A swab is passed through the nose gently and into the nasopharynx. Stay near septum and the floor of the nose. Rotate and remove. Recovery of the organism depends on collection of an adequate specimen. Inoculate the plate or transport medium directly at the bedside.

The following procedure optimizes the laboratory diagnosis of pertussis.
- Collect nasopharyngeal specimens in the early stage of illness.
- For swab collected specimens, use a transport medium consisting of half strength Oxoid charcoal agar supplemented with 10% hemolyzed, defibrinated horse blood, and 40 µg/mL cephalexin.
- Inoculate a selective primary plating medium composed of Oxoid charcoal agar, 10% defibrinated horse blood, and 40 µg/mL cephalexin. A nonselective medium without cephalexin may be used in addition to the selective medium.
- Perform direct fluorescent antibody (DFA) tests on appropriately collected nasopharyngeal secretions with *B. pertussis*- and *B. parapertussis*-conjugated antisera to facilitate earlier diagnosis.
- After inoculation of primary plating media, retain swabs in the original transport medium at room temperature. If cultures become overgrown with indigenous bacterial flora or fungi, use swabs to inoculate additional media.
- Identify suspicious isolates with appropriate cultural and biochemical tests. The DFA test performed on growth from isolated colonies is an excellent procedure for confirmatory or definitive identification.

Storage Instructions Do not refrigerate. Transport to the laboratory as soon as possible.

Turnaround Time Growth of *Bordetella pertussis* requires at least 72 hours for detection. Direct detection takes 24-48 hours.

Special Instructions Consult the laboratory prior to collection of the specimen so that the special isolation medium can be obtained. The laboratory should be made aware of the specific request to screen for *Bordetella pertussis* with information relevant to current antibiotic therapy.

Reference Interval No *B. pertussis* or *B. parapertussis* isolated

Use Isolate and identify *B. pertussis*, and *B. parapertussis*; directly detect and identify *B. pertussis* and *B. parapertussis* to establish the diagnosis of whooping cough

Limitations Isolation of *Bordetella* species probably has sensitivity of <50% compared to comprehensive serologic testing or detection by PCR. Direct detection assays are always limited by the adequacy of the sample. Negative direct detection assays do not rule out pertussis. Many laboratories do not keep reagents immediately on hand, due to the low incidence of disease in vaccinated populations.

Contraindications Current antibiotic therapy; lack of clinical symptoms of pertussis; history of vaccination is a relative contraindication

Methodology Culture on selective medium (selective chocolate agar with 10% defibrinated horse blood and 40 µg/mL cephalexin), presumptive confirmation by DFA. Culture after enrichment in transport medium for 48 hours increases yield. Detection by DNA amplification is currently in development and seems to be more sensitive than culture. DFA on nasopharyngeal specimen.

Additional Information Despite extensive vaccination programs, pertussis is still one of the 10 most common causes of death from infectious disease worldwide. Pertussis causes 350,000 deaths annually, the majority occurring in infants. Typical pertussis can be recognized clinically by the distinctive whooping cough that occurs during the paroxysmal stage. Pertussis is highly contagious and even immunized individuals can become transiently colonized, thus, spreading the organism to unvaccinated individuals.

To prevent infections of pertussis it is important to recognize the disease, avoid further spread, and provide pertussis vaccination to children. DFA procedures and PCR assays seem to provide more rapid results and are being increasingly used in the diagnosis of *B. pertussis* infection.

References

Hallander HO, "Microbiological and Serological Diagnosis of Pertussis," *Clin Infect Dis*, 1999, 28(S2):S99-106.

Loeffelholz MJ, Thompson CJ, Long KS, et al, "Comparison of PCR, Culture and Direct Fluores-cent-Antibody Testing for Detection of *Bordetella pertussis*," *J Clin Microbiol*, 1999, 37(9):2872-6.

Smith C and Vyas H, "Early Infantile Pertussis: Increasingly Prevalent and Potentially Fatal," *Eur J Pediatr*, 2000, 159(12):898-900.

Stojanov S, Liese J, and Belohradsky BH, "Hospitalization and Complications in Children Under 2 Years of Age With *Bordetella pertussis* Infection," *Infection*, 2000, 28(2):106-10.

Internet Web Sites
www.astdhpphe.org/infect/per.html
www.cdc.gov/nip/publications/pink/pert.pdf

Bordetella pertussis Serology

Related Information

Bacterial Culture, Lower Respiratory *on page 241*
Bordetella pertussis Culture and Direct Fluorescent Antibody *on page 299*

Test Includes Detect antibodies to *Bordetella pertussis* and/or pertussis toxin

Abstract Pertussis is one of the most common childhood diseases and a major cause of childhood mortality. Serologic testing to document either infection with the bacteria or prior immunization is useful in epidemiology and at times can be used for clinical diagnosis. However, the test is not routinely available. Direct fluorescent antibody (DFA) procedures as well as PCR assays provide more rapid results than does serology and should be used in the diagnosis of *B. pertussis* infection.

Specimen Serum

Container Red top tube

Storage Instructions Refrigerate serum at 4°C.

Reference Interval Less than fourfold rise in titer in paired sera; IgM antibody: negative in unimmunized individual; positive in vaccinated individual.

Use Evaluate acute infection with *Bordetella pertussis*; establish immunity following vaccination for *Bordetella pertussis*

Limitations The clinical role for serologic diagnosis of acute pertussis is severely limited by the time required for seroconversion. There is a lack of association between antibody levels and immunity. Assays are not well standardized.

Methodology Microhemagglutination, complement fixation (CF), toxin neutralization, enzyme-linked immunosorbent assay (ELISA)

Additional Information Serologic testing alone should not be relied upon for clinical confirmation of pertussis. However, patients with acute infection will develop IgG, IgM, and IgA antibodies to febrile agglutinogens. Following vaccination, IgG and IgM antibodies can be demonstrated, except in infants. IgA antibodies do not develop.

References

Smith C and Vyas H, "Early Infantile Pertussis: Increasingly Prevalent and Potentially Fatal," *Eur J Pediatr*, 2000, 159(12):898-900.

Stojanov S, Liese J, and Belohradsky BH, "Hospitalization and Complications in Children Under 2 Years of Age With *Bordetella pertussis* Infection," *Infection*, 2000, 28(2):106-10.

Internet Web Sites
www.cdc.gov/nip/publications/pink/pert.pdf
www.astdhpphe.org/infect/per.html

Borrelia burgdorferi Culture

Related Information

Lyme Disease Serology *on page 879*
Polymerase Chain Reaction *on page 1069*

Abstract *Borrelia burgdorferi* is a corkscrew shaped organism that is the causative agent for Lyme disease. *B. burgdorferi* is transmitted to humans by the bite of an infected deer tick (*Ixodes scapularis*). The number of reported cases of Lyme disease has increased in the U.S. since national surveillance began in 1982. There were nearly 17,000 cases reported to the CDC in 1998. Most of the cases were reported from Northeastern and Mid-Atlantic states and two North-Central States.

Specimen Tick taken from patient; skin, blood, cerebrospinal fluid, or joint fluid

Collection Collect blood in either EDTA or serum tubes.

Causes for Rejection Tissue fixed in formalin

Reference Interval Negative

Limitations Culture may often be negative in patients with the disease; lengthy incubation is required before bacteria can be detected (4-6 weeks). Recovery of isolates from skin specimens may be overgrown with indigenous bacterial

(Continued)

Borrelia burgdorferi Culture *(Continued)*

skin flora. Culture of *B. burgdorferi* is less sensitive than PCR or serology, especially in neuroborreliosis.

Methodology Culture in BSKII medium with a neutral pH. Incubation at 30°C to 37°C. Using a darkfield microscope, monitor for 4-6 weeks.

Additional Information The diagnosis of Lyme disease is primarily based on clinical findings and elevated serology. Clinical signs and symptoms of Lyme disease include one or more of the following: malaise, fever, headache, myalgias, swollen lymph nodes, and a characteristic skin rash, erythema migrans (EM). Culture provides a definitive diagnosis, but culture is often negative in patients without skin manifestations.

References

Eppes SC, Nelson DK, Lewis LL, et al, "Characterization of Lyme Meningitis and Comparison With Viral Meningitis in Children," *Pediatrics*, 1999, 103(5Pt1):957-60.

Nadelman RB and Wormser GP, "Lyme Borreliosis," *Lancet*, 1998, 352(9127):557-65.

Orloski KA, Hayes EB, Campbell GL, et al, "Surveillance for Lyme Disease - United States, 1992-1998," *MMWR Morb Mortal Wkly Report*, 2000, 49(SS3):1-11.

Strle F, Nadelman RB, Cimperman J, et al, "Comparison of Culture-Confirmed Erythema Migrans Caused by *Borrelia burgdorferi sensu stricto* in New York State and by *Borrelia afzelii* in Slovenia," *Ann Intern Med*, 1999, 130(1):32-6.

Internet Web Sites

www.amm.co.uk/pubs/fa_lyme.htm
www.astdhpphe.org/infect/lyme.html
www.cdc.gov/ncidod/dvbid/lyme/index.htm

♦ **Botanical Dietary Supplements** *see* Digoxin, Serum *on page 512*

Botulism Diagnostic Procedures

Related Information

Bacterial Culture, Stool *on page 243*
Organophosphate Pesticides, Urine, Blood, or Serum *on page 975*
Polymerase Chain Reaction *on page 1069*
Tetanus Antibody *on page 1240*

Abstract Botulism is a toxin-mediated disease caused by *Clostridium botulinum*. The anaerobic bacteria produce a potent neurotoxic protein which affects the nervous system without direct bacterial invasion of the CNS. Clinical manifestations include diplopia, dysphonia, dysarthria, and dysphagia with symmetric descending progression of flaccid paralysis, leading to respiratory embarrassment in afebrile patients. Botulism is classified into four categories:

- food-borne
- wound botulism
- infant botulism
- botulism from intestinal colonization of children older than infants

Specimen Vomitus, serum, stool, gastric contents, wound or autopsy tissue; food samples

Container Sterile wide-mouth, leakproof, screw-cap jar; red top tube

Storage Instructions Keep refrigerated at 4°C, except for unopened food samples. Do not freeze.

Turnaround Time 3-7 days

Special Instructions The laboratory should be notified prior to obtaining specimen, in order to prepare for transport of the specimen to the State Health Laboratory or Centers for Disease Control.

Reference Interval No toxin identified, no *Clostridium botulinum* bacteria isolated

Use Diagnose classic food-borne botulism, a neuroparalytic disease; wound botulism (rare), infant botulism, sudden death syndrome, or floppy baby syndrome

Limitations The toxin from *C. botulinum* binds almost irreversibly to individual nerve terminals; thus, specimens can yield false-negative results.

Contraindications Due to the difficulty in performance of the diagnostic test and because of the extensive epidemiological studies initiated upon receipt of the specimen, State Department of Health Laboratories require specific clinical symptomatology for botulism testing. The physician may be asked to submit specific forms before a specimen is submitted to the State Department of Health or CDC. Physicians are encouraged to call the CDC, which provides consultation and services to state and local health departments, at (404) 639-2206. An emergency number is (404) 639-2888.

Methodology Toxin neutralization test in mice, isolation of *Clostridium botulinum* from feces. ELISA may detect toxin on nasal mucosa for 24 hours following inhalation.

Additional Information *C. botulinum* is an obligate anaerobic, spore-forming, gram-positive bacillus that can be recovered from a wide variety of environmental sources including soils. The toxin from this bacteria is up to 100,000 times more toxic than sarin. **Infant botulism** occurs most commonly in the second and third postnatal months and is rarely seen after the sixth month of life. The disease occurs when infants ingest *C. botulinum* spores that germinate and produce botulinum toxin (usually serotypes A or B) within the gastrointestinal tract. The spectrum of disease in infants varies from mild constipation to sudden death.

Food-borne botulism usually develops 12-36 hours after toxin ingestion, commonly from home-canned foods. Initial complaints consist of nausea, dry mouth, and diarrhea followed by evidence of cranial nerve dysfunction. The toxin is destroyed by cooking. Foodborne botulism typically occurs in outbreaks that require identification, then control to prevent further disease.

In **wound botulism**, spores are introduced into a wound in which they germinate and produce toxin. Clinically, wound botulism lacks the prodromal gastrointestinal disorder of foodborne botulism, but is otherwise similar. Persons who inject illicit drugs, such as black-tar heroin, are at increased risk for wound botulism.

The differential diagnosis in individual cases includes Guillain-Barré syndrome, myasthenia gravis, tick paralysis, and intoxication with organophosphates, atropine, carbon monoxide, and aminoglycosides.

Intensive surveillance for botulism in the U.S. is maintained by state health departments and the CDC. Prompt recognition of botulism by physicians is needed. A potential exists for the use of botulism as a bioterrorist weapon. National governments and terrorist organizations have been reported to maintain stockpiles of botulinum toxin. One gram of aerosolized botulinum toxin potentially could kill 1.5 million people or more. Botulinum toxin, aerosolized, would be absorbed in the respiratory tract. The toxin could also be used as a terrorist weapon through food contamination.

Arnon et al have recently discussed the differential diagnosis of botulism.

References

Arnon SS, Schechter R, Inglesby TV, et al, "Botulinum Toxin as a Biological Weapon. Medical and Public Health Management," *JAMA*, 2001, 285(8):1059-70.

Centers for Disease Control and Prevention, "Infant Botulism - New York City, 2001-2002," *JAMA*, 2003, 289(7):834-6.

Franz DR, Jahrling PB, Friedlander AM, et al, "Clinical Recognition and Management of Patients Exposed to Biological Warfare Agents," *JAMA*, 1997, 278(5):399-411.

Henderson DA, Inglesby TV, and O'Toole T, *Bioterrorism: Guidelines for Medical and Public Health Management*, Chicago, IL: AMA Press, 2002.

McMaster P, Piper S, Schell D, et al, "A Taste of Honey," *J Paediart Child Health*, 2000, 36(6):596-7.

Meyer RF and Morse SA, "Bioterrorism Preparedness for the Public Health and Medical Communities," *Mayo Clin Proc*, 2002, 77(7):619-21.

Shapiro RL, Hatheway C, and Becher J, "Botulism Surveillance and Emergency Response. A Public Health Strategy for a Global Challenge," *JAMA*, 1997, 278(5):433-5.

Shapiro RL, Hatheway C, and Swerdlow DL, "Botulism in the United States: A Clinical and Epidemiologic Review," *Ann Intern Med*, 1998, 129(3):221-8.

Varkey P, Poland GA, Cockerill FR III, et al, "Confronting Bioterrorism: Physicians on the Front Line," *Mayo Clin Proc*, 2002, 77(7):661-72.

Internet Web Sites

vm.cfsan.fda.gov/~mow/chap2.html
www.cdc.gov/ncidod/dbmd/diseaseinfo/botulism_g.htm
www.epi.hss.state.ak.us/pubs/botulism/bot_01.htm
www.phppo.cdc.gov/phtn/botulism/alaska/alaska.asp

♦ **Brain Needle Aspiration** *see* Fine Needle Aspiration, Deep and Superficial Masses *on page 590*

♦ **BRCA1** *see* Breast Cancer, Hereditary, BRCA1, BRCA2 *on page 307*

♦ **BRCA2** *see* Breast Cancer, Hereditary, BRCA1, BRCA2 *on page 307*

Breakpoint Cluster Region Rearrangement in CML

Related Information
Bone Marrow *on page 296*
Chromosome Analysis, Blood *on page 406*
Chromosome Analysis, Bone Marrow *on page 407*
Leukocyte Alkaline Phosphatase *on page 845*
Leukocyte Cytochemistry *on page 846*
Polymerase Chain Reaction *on page 1069*

Synonyms *bcr/c-abl* Translocation; Gene Rearrangement, *bcr*

Applies to Chronic Neutrophilic Leukemia; Philadelphia Chromosome

Test Includes DNA detection of chromosomal translocation associated with chronic myelogenous leukemia (CML)

Abstract Chronic myelogenous leukemia (CML) is a myeloproliferative disorder characterized by the transformation of pluripotent hematopoietic stem cells. Ninety percent to 95% of the CML cases have a translocation between chromosome 9 and chromosome 22. This chromosomal translocation results in an abnormal gene rearrangement that can be detected using nucleic acid technology. The translocation involves a rearrangement of the breakpoint cluster region (*bcr*) gene located on chromosome 22 with the *c-abl* oncogene on chromosome 9.

Specimen Blood or bone marrow

Container Blood should be collected in a lavender top (EDTA) Vacutainer® tube; bone marrow should be collected in a syringe with heparin or transferred to a green top (heparin) tube or lavender top (EDTA) tube.

Turnaround Time 1-2 weeks

Reference Interval No rearrangement observed

Use This test is used for the confirmation of CML along with bone marrow examination, cytogenetics, and leukocyte alkaline phosphatase score.

The *bcr/c-abl* rearrangement assay is clinically useful for:
- confirmation of Philadelphia chromosome-positive CML
- diagnosis of Philadelphia-negative CML
- diagnosis and monitoring of CML blast crisis during and after chemotherapy and bone marrow transplantation
- detection of remission and early detection of relapse (for the characterization and monitoring of chronic myelogenous leukemia)

Additional Information Chronic myelogenous leukemia is a malignant clonal disorder of hematopoietic stem cells, characterized by increase in myeloid cells (and in many cases erythroid cells and platelets) in the peripheral blood, and by marked predominantly myeloid hyperplasia in the bone marrow. It affects males and females in a ratio of 1.5:1 with a peak incidence between 40 and 60 years of age and a median age at presentation of 53 years. All ages (including children) may be affected.

CML is characterized by a reciprocal translocation between chromosomes 9 and 22, producing the Philadelphia chromosome. The *c-abl* gene (a cellular oncogene - the Abelson proto-oncogene) is the human homologue of a gene for a murine oncogenic virus, the Abelson murine leukemia virus, that maps to chromosome 9q34. The hybrid *bcr/c-abl* gene is transcribed into an abnormal messenger RNA, which is translated into an abnormal tyrosine kinase of 210,000 molecular weight instead of the normal 145,000 molecular weight protein. More than 90% of patients with CML have the Philadelphia chromosome by cytogenetic analysis as well as rearrangement of *bcr/c-abl*. **Most patients with clinically documented CML that lack the Philadelphia chromosome still have the *bcr/c-abl* rearrangement.** A small number of patients do not have the Philadelphia chromosome as the *bcr/c-abl* rearrangement. During reassessment many of these patients have a myelodysplastic syndrome, usually chronic myelomonocytic leukemia. A very small number of patients with clinical CML remain both Philadelphia chromosome negative and *bcr/c-abl* negative.

Cytogenetically the Philadelphia chromosome has been found in 20% to 25% of adults and 5% of children with acute lymphoblastic leukemia (ALL) and 2% of patients with acute myelogenous leukemia (AML). The Philadelphia chromosome from ALL cases appears similar to CML Philadelphia chromosomes in cytogenetic analysis. However, the two chromosomes result from distinct molecular rearrangements that can be analyzed and detected with DNA analysis. Some ALL Philadelphia chromosomes have been found to be identical to the CML Philadelphia chromosome even at the molecular level. These ALL cases are generally regarded as the blast crisis of CML. Some laboratories perform a DNA amplification assay, polymerase chain reaction (PCR), for detection of this translocation. This may assist in the detection of minimal residual disease.

In addition to the *bcr-abl* hybrid gene-encoded p210 tyrosine kinase, additional fusion proteins (p230 *bcr-abl* and p190 *bcr-abl*) have been described. The fusion protein p230 *bcr-abl* has been linked to the phenotype of chronic neutrophilic leukemia.

Current therapy (for cure) is dependent upon bone marrow transplantation from sibling or volunteer donors. The use of tryphostins appears to be a major advance in therapy.

References

Chopra R, Pu QQ, and Elefanty AG, "Biology of *bcr-abl*," *Blood Rev*, 1999, 13(4):211-29.

Faderl S, Talpaz M, Estrov Z, et al, "The Biology of Chronic Myeloid Leukemia," *N Engl J Med*, 1999, 341(3):164-72.

Franklin IM and Mills K, "Chronic Myeloid Leukemia," *Molecular Haematology*, Chapter 5, Provan D and Gribben J, eds, Malden, MA: Blackwell Science Ltd, 2001, 61-74.

Hansen JA, Gooley TA, Martin PJ, et al, "Bone Marrow Transplants From Unrelated Donors for Patients With Chronic Myeloid Leukemia," *N Engl J Med*, 1998, 338(14):962-8.

Kalidas M, Kantarjian H, and Talpaz M, "Chronic Myelogenous Leukemia," *JAMA*, 2001, 286(8):895-8.

Pane F, Frigeri F, Sindona M, et al, "Neutrophilic-Chronic Myeloid Leukemia: A Distinct Disease With a Specific Molecular Marker (*bcr-abl* With C3/A2 Junction)," *Blood*, 1996, 88(7):2410-4.

Stewart AK and Schuh AC, "White Cells 2: Impact of Understanding the Molecular Basis of Haematological Malignant Disorders on Clinical Practice," *Lancet*, 2000, 355(9213):1447-53.

♦ **Breast Aspiration Cytology** *see* Fine Needle Aspiration, Deep and Superficial Masses *on page 590*

Breast Biopsy

Related Information

Breast Cancer, Hereditary, BRCA1, BRCA2 *on page 307*
CA 15-3, Serum *on page 320*
Carcinoembryonic Antigen, Serum *on page 342*
Cyst Fluid Cytology *on page 490*
Estrogen and Progesterone Receptor Assay *on page 556*
Fine Needle Aspiration, Deep and Superficial Masses *on page 590*
Frozen Section *on page 615*
HER-2/*neu on page 716*
Histopathology *on page 733*
Immunoperoxidase Procedures *on page 780*
p53, Functional Assay/Sequencing *on page 995*
Sentinel Lymph Node Biopsy *on page 1189*

Applies to DNA Ploidy; Gross Cystic Disease Fluid Protein-15 (GCDFP-15); Lumpectomy; Ploidy; S Phase

Abstract Breast biopsies are performed to evaluate the pathologic nature of mammographically detected or palpable abnormalities. Breast lesions may be sampled by fine needle aspiration, skinny needle biopsy, mammotome suction biopsy, and open surgical biopsy.

Container When lesions are initially detected by mammography, the excised tissue is commonly radiographed to confirm the presence of lesional tissue. Such specimens are often submitted in a device with radiopaque grids, permitting the coordinates of the abnormality to be precisely documented for histopathologic correlation.

Use The primary intent of breast biopsy is to determine the presence of cancer and, for open biopsies, evaluate the adequacy of excision. The information derived from the biopsy has profound implications on the subsequent management of malignant disease.
(Continued)

Breast Biopsy (Continued)

Limitations Needle biopsies are susceptible to crush artifact. Evaluation of margins is complicated or rendered impossible by tissue fragmentation.

Additional Information Some benign lesions carry no associated increased risk of malignancy while others (eg, atypical ductal hyperplasia) carry a significantly increased risk for the eventual development of cancer. Malignant tumors include *in situ* carcinoma, infiltrating carcinoma, combinations thereof, and some phylloides tumors. Rarely, stromal tumors and lymphomas are documented. The proximity of malignant tumors to the surgical margins plays a pivotal role in the subsequent management of neoplastic disease.

In cases in which microcalcifications are mammographically detected, histopathologic confirmation of their presence is imperative. A high percentage of such cases demonstrates significant breast disease in fibrous tissue adjacent to benign tissues containing microcalcifications. Calcium oxalate is often rendered inconspicuous by routine tissue processing, but is readily found using polarized microscopy.

The role of frozen section examination for the intraoperative management of breast conservation therapy is controversial. If mastectomy is planned to immediately follow a biopsy diagnosis of malignancy, frozen section examination is indicated. Frozen section exam, touch imprint cytology, or scrape preparation cytology is useful to establish a working diagnosis of malignancy to triage the specimen for ancillary studies (*vide infra*). **Frozen section examination of nonpalpable breast lesions is contraindicated. Frozen section examination to evaluate surgical margins of specimens with grossly discernible tumor is not clearly indicated.** Connolly et al conclude that frozen section evaluation of margins grossly free of tumor has no significant role in intraoperative management and may compromise margin evaluation in permanent section. Margins are best evaluated on intact segmental resections.

Histopathologic classification and grading are important in the final management of epithelial breast cancers. The most commonly used histologic grading system was described by Bloom and Scharff and subsequently modified in the following table.

Nottingham Modification for the Bloom-Richardson System

Tubule Formation
- 1 point: tubules in >75% of tumor
- 2 points: tubules in 10% to 75% of tumor
- 3 points: tubules in <10% of tumor

Nuclear Pleomorphism
- 1 point: nuclei with minimal variation in size and shape
- 2 points: nuclei with moderate variation in size and shape
- 3 points: nuclei with marked variation in size and shape

Mitotic Count (actual counts vary somewhat based on microscope field size)
- 1 point: 0-4/40x field
- 2 points: 6-10/40x field
- 3 points: >11/40x field

Numerous prognostic factors have been studied in the setting of breast cancer. Recently published guidelines detail the current status of prognostic markers in breast cancer. Hormone receptor and HER-2/*neu* are both predictive and prognostic markers in breast cancer. Hormone receptor status is best determined by immunohistochemical techniques using paraffin embedded tissue.

Despite some conflicting data, there is a consensus that ER positivity correlates with longer disease-free and overall survival, at least during the first 5 years. Survival curves tend to merge with longer follow-up. Stage-matched patients with higher expression of ER tend to follow a more favorable course than do patients with low levels of expression. Generally, both ER and PR expression tend to reflect tumor growth rate rather than metastatic potential.

ER expression is highly correlated with response to endocrine therapy. Antiestrogen therapy produces a tumor static response generally in direct proportion to the degree of ER expression, especially when PR is coexpressed.

HER-2/*neu* expression also confers both predictive and prognostic value. Intraductal carcinoma expressing HER-2/*neu* is associated with high nuclear grade and comedocarcinoma architecture. HER-2/*neu* overexpression is correlated with increased local recurrence rate. In cases of infiltrating ductal carcinoma, HER-2/*neu* expression is correlated with decreased disease-free and overall survival, particularly in those with lymph node positive disease. HER-2/*neu* overexpression may also predict patients with increased resistance to cyclophosphamide therapy and enhanced sensitivity to chemotherapeutic regimens containing Adriamycin®.

Flow cytometric determination of DNA ploidy and S-phase fraction (SPF) have gained wide acceptance in the evaluation of breast cancer. Most reference laboratories can obtain DNA ploidy and S-phase determinations from the same tissue sample submitted for hormone receptor assays. These studies can also be obtained from paraffin-embedded tissue. Approximately 66% of breast cancers are aneuploid and 30% to 40% are regarded as high SPF tumors. Several studies have demonstrated shorter disease-free and shorter overall survival in patients with aneuploid tumors or with high SPF. Aneuploidy and high SPF cancers correlate with high nuclear grade and are more commonly hormone receptor-negative.

The risk of local recurrence is dependent on tumor size, adequacy of resection, histologic grade with mitotic count, and presence of extensive intraductal carcinoma. Age and presence of certain phenotypic markers also have a bearing on local recurrence. Local recurrence is significantly reduced with adjuvant radiotherapy.

The risk of regional and distant metastasis is associated with a number of factors. The single most important factor for distant metastasis is the presence of axillary lymph node metastases. Recent advances in the identification and biopsy of sentinel lymph node are hoped to promote the identification of patients at risk for developing distant metastases (see Sentinel Lymph Node Biopsy *on page 1189*).

References

Carter D, *Interpretation of Breast Biopsies*, 4th ed, Baltimore, MD: Lippincott Williams & Wilkins, 2002.

Connolly J and Schnitt S, "Evaluation of Breast Biopsy Specimens in Patients Considered for Treatment by Conservative Surgery and Radiation Therapy for Early Breast Cancer," *Pathol Annu*, 1988, (23 Pt 1):1-23.

Fisher B, Dignam J, Wolmark N, et al, "Lumpectomy and Radiation Therapy for the Treatment of Intraductal Breast Cancer: Findings From National Surgical Adjuvant Breast and Bowel Project B-17," *J Clin Oncol*, 1998. 16(2):441-52.

Fitzgibbons PL, Page DL, Weaver D, et al, "Prognostic Factors in Breast Cancer," College of American Pathologists Consensus Statement 1999, *Arch Pathol Lab Med*, 2000. 124(7):966-78.

Rosen PP, *Rosen's Breast Pathology*, 2nd ed, Baltimore, MD: Lippincott Williams & Wilkins, 2001.

Ross J. and Fletcher J, "HER-2/*neu* (c-*erb*-B2) Gene and Protein in Breast Cancer," *Am J Clin Pathol*, 1999. 122(Suppl 1):S53-S67.

♦ **Breast Cancer** see Estrogen and Progesterone Receptor Assay *on page 556*

Breast Cancer, Hereditary, BRCA1, BRCA2

Related Information

Breast Biopsy *on page 305*
CA 15-3, Serum *on page 320*
p53, Functional Assay/Sequencing *on page 995*
Polymerase Chain Reaction *on page 1069*

Synonyms BRCA1; BRCA2

Applies to CHEK2 Mutation

Test Includes This molecular test involves end-to-end sequencing and/or protein truncation assay to detect carriers of mutations in the gene.

Abstract Mutations in BRCA1 and BRCA2 are characterized by predisposition to breast and ovarian cancers, but such mutations are rare in sporadic carcinoma of breast. About 5% of patients with carcinoma of breast have a pattern of autosomal dominant inheritance. About 20% of families who have evidence of inherited susceptibility to carcinoma of breast have mutations of BRCA1, and another 20% BRCA2 genes, which can exist and be passed to offspring by
(Continued)

Breast Cancer, Hereditary, BRCA1, BRCA2 (Continued)

males as well as females. BRCA1 and BRCA2 are the two major susceptibility genes for breast carcinoma. Complex interactions include patient age, family history, and other factors relevant to consideration for testing or interpretation of results. Of cases of early-onset familial carcinoma of breast, a substantial proportion are related to BRCA1 or BRCA2 mutations. About 10% of cases of carcinoma of ovary are attributed to BRCA mutations.

Patient Preparation Pretest education and counseling should be provided prior to orders for molecular genetic testing.

Specimen Whole blood

Container Blue top (sodium citrate) tube or yellow top (ACD) tube. Struewing et al used fingerstick procedures and collection cards.

Storage Instructions Specimens should be sent to the laboratory **immediately** after collection, preferably by overnight delivery. Specimens should be kept at room temperature or refrigerated, never frozen.

Causes for Rejection Lysed or frozen blood sample

Use This test is indicated for families with a history of autosomal dominant early-onset breast/ovarian cancer (ie, identification of those with high risk). Genetic testing is appropriate when carcinoma of breast develops in younger women, especially in those with positive family history of breast and/or ovarian cancer, two or more family members younger than 50 years of age, male carcinoma of breast, Ashkenaai ancestry, and/or women whose blood relatives bear BRCA1 or BRCA2 mutations. Genetic testing can identify those who carry a mutation bearing serious risk from those who do not. A degree of increased risk for contralateral breast neoplasm is recognized in patients with BRCA1- or BRCA2-associated carcinoma of breast.

Limitations Modifying factors exist, influencing the effect of a BRCA mutation.

Methodology Referral to a center providing specialized genetic counseling for BRCA testing is recommended. Call 1-800-422-6337.

DNA and/or RNA are isolated from the specimen. A protein truncation assay will detect mutations that cause shortened protein products. Regions within the BRCA1 gene are amplified by PCR and sequenced in entirety. A 2003 paper describes the importance of having sequenced the human genome and of array-based comparative genomic hybridization. van de Vijer et al, using microarray analysis, projected disease outcome, but their study was limited to a specific group of patients.

Additional Information The risk for BRCA1/BRCA2 mutation increases with the strength of the family history and with earlier ages of onset (premenopausal). Although the occurrence of breast cancer **in families** may be due to chance alone or shared environment, the inheritance of a gene mutation increasing susceptibility to breast cancer is the most common reason. It is

The BRCA1 and BRCA2 Genes

Characteristic	BRCA1	BRCA2
Underlies what percentage of breast cancer in the population	2% to 3%	2% to 3%
Chromosome location	17	13
Breast cancer risk if woman carries gene and has:		
Strong family history	50% by age 50 70% to 90% by age 70	Same as BRCA1
Limited pedigree or no family history	Uncertain; probably about 50%	Same as BRCA1
Ovarian cancer risk if woman carries gene and has:		
Strong family history	30% by age 50 20% to 45% by age 70	About 15%
Limited pedigree or no family history	Unknown	Unknown
Risk of male breast cancer	Low	6% to 7%

Some published papers provide somewhat different estimates of risk. BRCA genes may behave differently in different women.

Adapted from Muss HB, "Breast Cancer and Differential Diagnosis of Benign Nodules," *Cecil Textbook of Medicine*, 21st ed, Chapter 258, Goldman L and Bennett JC, Philadelphia, PA: WB Saunders Co, 2000, 1373-80.

estimated that 1 in 200 to 1 in 400 women carry BRCA1 mutations. The lifetime risk for breast cancer in a woman who has inherited a mutation in BRCA1 is up to 85% by age 85 as opposed to 11% for women in the general population, and the risk for ovarian cancer is up to 40% or more by age 70 compared to 1% in the general population. (See table.) Once breast cancer has been diagnosed, the risk for carcinoma in the contralateral breast in the presence of hereditary breast cancer is about 60%. Breast and ovarian cancer survival rates are significantly better when diagnosis occurs in an early, localized stage.

Women found to carry the mutation should be encouraged to undergo earlier and more intensive surveillance for breast cancer, which is expected to lead to earlier detection and improved outcome. BRCA1/BRCA2 testing can help rule out family members who, because they do not carry the mutation, do not require increased surveillance.

Most BRCA1-associated carcinomas of breast are negative for estrogen receptors.

CHEK2 was recently discovered. It is a mutation of the cell cycle-check point kinase gene which doubles the risk of carcinoma of breast in females. It increases risk among men by a factor of ten. CHEK2 accounts for ~5% of familial carcinoma of breast.

Men who carry a BRCA1 mutation appear not to be at great risk for breast cancer but male carriers of BRCA1 or BRCA2 are at increased risk for prostate cancer. The risk of prostate carcinoma is ~16% by age 70. Men with BRCA2 are at risk for male carcinoma of breast.

Fanconi anemia proteins (see Fanconi Anemia, Chromosome Breakage Study *on page 569*) appear to function similarly to those of BCRA1 and BCRA2 (important to the processes of DNA repair with suppression of spontaneous chromosomal aberrations).

See Breast Biopsy *on page 305*.

References

Boyd J, Sonoda Y, Federici MG, et al, "Clinicopathologic Features of *BRCA*-Linked and Sporadic Ovarian Cancer," *JAMA*, 2000, 283(17):2060-5.

Frank TS, "Laboratory Determination of Hereditary Susceptibility to Breast and Ovarian Cancer," *Arch Pathol Lab Med*, 1999, 123(11):1023-6.

Haber D, "Prophylactic Oophorectomy to Reduce the Risk of Ovarian and Breast Cancer in Carriers of BRCA Mutations," *N Engl J Med*, 2002, 346(21):1660-2.

Hartge P, "Genes, Hormones, and Pathways to Breast Cancer," *N Engl J Med*, 2003, 348(23):2352-4.

Hedenfalk I, Duggan D, Chen Y, et al, "Gene-Expression Profiles in Hereditary Breast Cancer," *N Engl J Med*, 2001, 344(8):539-48.

Kallioniemi A, "Molecular Signature of Breast Cancer: Predicting the Future," *N Engl J Med*, 2002, 347(25):2067-8.

Kauff ND, Satagopan JM, Robson ME, et al, "Risk-Reducing Salpingo-Oophorectomy in Women With a BRCA1 or BRCA2 Mutation," *N Engl J Med*, 2002, 346(21):1609-15.

King MC, Marks JH, Mandell JB, et al, "Breast and Ovarian Cancer Risks Due to Inherited Mutations in BRCA1 and BRCA2," *Science*, 2003, 302(5645):634-6.

Rebbeck TR, Lynch HT, Neuhausen SL, et al, "Prophylactic Oophorectomy in Carriers of BRCA1 or BRCA2 Mutations," *N Engl J Med*, 2002, 346(21):1616-22.

van de Vijver MJ, He YD, van't Veer LJ, et al, "A Gene-Expression Signature as a Predictor of Survival in Breast Cancer," *N Engl J Med*, 2002, 347(25):1999-2009.

Venkitaraman AR, "A Growing Network of Cancer - Susceptibility Genes," *N Engl J Med*, 2003, 348(19):1917-9.

Wooster R and Weber BL, "Breast and Ovarian Cancer," *N Engl J Med*, 2003, 348(23):2339-47.

Internet Web Sites

www.geneclinics.org/profiles/brca1

cancernet.nci.nih.gov/genesrch.shtml

A description of laboratory methods used is available through the National Auxiliary Publications Service (NAPS) and the Breast Cancer Information Core site (http://research.nhgri.nih.gov/bic/). See NAPS document no. 05401 for 5 pages of supplementary material. Order from NAPS, c/o Microfiche Publications, P.O. box 3513, Grand Central Station, New York, NY 10163-3513. Remit in advance (in US funds only) $11.65 for photocopies or $5 for microfiche. Outside the U.S., add postage ($4.50 for up to 20 pages, $5.50 for over 20 pages, or $1.50 for microfiche. There is a $15 invoicing charge on all orders filled before payment.

♦ **Breast Carcinoma-Associated Antigen** *see* CA 27.29, Serum *on page 322*

♦ **Breast Cyst Fluid Cytology** *see* Cyst Fluid Cytology *on page 490*

♦ **Breast Ductal Lavage** *see* Washing Cytology *on page 1326*

♦ **Breath Hydrogen Analysis** *see* Lactose Tolerance Test *on page 829*

♦ **Breath Hydrogen Analysis** *see* Reducing Substances, Stool *on page 1147*

♦ **Breath Testing for H. pylori** *see Helicobacter pylori* Antigen and Serology *on page 671*

♦ **Britiazim®** *see* Diltiazem, Serum or Plasma *on page 515*

♦ **Bromism** *see* Chloride, Serum, Plasma, or Blood *on page 391*

Bronchial Brushings/Washings Cytology

Related Information
Aspergillus Serology *on page 219*
Bronchoalveolar Lavage (BAL) *on page 311*
Brushings Cytology *on page 314*
Fine Needle Aspiration, Deep and Superficial Masses *on page 590*
Sputum Cytology *on page 1222*
Transbronchial Fine Needle Aspiration *on page 1266*

Test Includes Cytologic evaluation of direct smears, cytocentrifuge (cytospin) preparations, and/or monolayer preparations most commonly; membrane filtration preparations and cell blocks are less often used.

Abstract Bronchial brushings are generally used with bronchial washings for diagnosis of potential pulmonary neoplasms in patients with abnormal imaging findings or persistent respiratory symptoms. Bronchial cytology is useful in the evaluation of patients with abnormal sputum cytology. Visualization of tumor results in excellent diagnostic yield.

Specimen The freshly aspirated washing specimen should be immediately sent to the Cytopathology Laboratory without fixative for processing.

Storage Instructions If a delay in processing must occur, the brushing or washing specimen should be refrigerated.

Causes for Rejection Prolonged period (more than 1 hour) at room temperature

Special Instructions Specify need for special stains for microorganisms.

Use Bronchial cytology is generally used to evaluate patients with the clinical symptoms of prolonged cough, localized wheezing, hemoptysis, or bronchial obstruction, structural deformity, extrinsic compression, as well as those with radiographic abnormalities including a new solitary pulmonary nodule, atelectasis or persistent pulmonary infiltrates. The various types of diagnostic respiratory cytology are complementary to one another. Pleural based lesions or those in the extreme periphery of the pulmonary parenchyma may not be accessible to the bronchoscope. Bronchial brushing is most sensitive for the diagnosis of squamous cell carcinoma and most specific in identifying small cell carcinoma and adenocarcinoma. Bronchial brushing should be used in combination with bronchial washing to maximize diagnostic yield. The combination of cytology with biopsy provides specific diagnosis more often than either alone. Repeat specimens may be helpful in the setting of an initial negative result in the presence of a high degree of clinical suspicion.

Respiratory infections may also be diagnosed using this technique, though bronchoalveolar lavage (BAL) is generally more useful.

Limitations To be considered adequate, the specimen should contain large numbers of well-preserved bronchial epithelial cells and pulmonary macrophages. Sparsely cellular samples or specimens composed predominantly of oral squamous cells are unsatisfactory for evaluation. Obscuring inflammation, blood, or extensive air-drying artifact may also render a specimen unsatisfactory. A specimen is considered less than optimal for evaluation if there is a lack of accompanying clinical information.

Limited extent of sampling is provided by bronchial brushings.

References
Bibbo M, "Bronchial Brushing," *Compendium on Diagnostic Cytology*, 8th ed, Wied GL, Bibbo M, Keebler CM, et al, eds, Chicago, IL: Tutorials of Cytology, 1997, 231-2.

DeMay RM, "The Art and Science of Cytopathology: Exfoliative Cytology," *Respiratory Cytology*, Chapter 7, Chicago, IL: ASCP Press, 1996, 207-56.

Linder J, "Lung Cancer Cytology - Something Old, Something New," *Am J Clin Pathol*, 2000, 114(2):169-71.

Papanicolaou Society Task Force on Standards of Practice, "Guidelines of the Papanicolaou Society of Cytopathology for the Examination of Cytologic Specimens Obtained From the Respiratory Tract," *Diagn Cytopathol*, 1999, 21(1):61-9.

Rabb SS, Oweity T, Hughes JH, et al, "Effect of Clinical History on Diagnostic Accuracy in the Cytologic Interpretation of Bronchial Brush Specimens," *Am J Clin Pathol*, 2000, 114(1):78-83.

Sturgis CD, Nassar DL, D'Antonio JA, et al, "Cytologic Features Useful for Distinguishing Small Cell From Nonsmall Cell Carcinoma in Bronchial Brush and Wash Specimens," Am J Clin Pathol, 2000, 114(2):197-202.

♦ **Bronchial Culture** *see* Bacterial Culture, Lower Respiratory *on page 241*

Bronchoalveolar Lavage (BAL)

Related Information

Amiodarone, Serum *on page 148*
Asbestos, Lung or Sputum *on page 215*
Bacterial Culture, Aerobes *on page 229*
Bacterial Culture, Lower Respiratory *on page 241*
Bronchial Brushings/Washings Cytology *on page 310*
Brushings Cytology *on page 314*
Cytomegalovirus Culture *on page 495*
Cytomegalovirus Cytology *on page 497*
Cytomegalovirus Serology *on page 499*
Fine Needle Aspiration, Deep and Superficial Masses *on page 590*
Fungal Culture, Sputum *on page 624*
Herpes Simplex Virus DNA Detection *on page 723*
Histoplasmosis Antibody *on page 734*
Mycobacterial Culture, Sputum *on page 933*
Pneumocystis carinii Preparation *on page 1062*
Pneumocystis Immunofluorescence *on page 1063*
Sputum Cytology *on page 1222*
Transbronchial Fine Needle Aspiration *on page 1266*
Viral Culture *on page 1307*
Virus, Direct Detection by Fluorescent Antibody *on page 1311*

Abstract Useful in the diagnosis of cancer, opportunistic infections in immuno-compromised patients, culture in those with aspiration pneumonia, pneumonia secondary to chronic disease or to complications (eg, the postoperative state), interstitial lung disease, granulomatous disease (including sarcoid), hypersensitivity pneumonia, drug-induced pulmonary toxicity, asbestosis, pulmonary hemorrhage, hemosiderosis, and evaluation of transplant rejection.

Patient Preparation Informed consent for procedure. Premedication is often given.

Specimen Lavage fluid

Container Sterile, leakproof disposable container

Collection The bronchoscope is wedged in a subsegmental bronchus. 100-300 mL of warm, pyrogen-free, isotonic, sterile solution is infused with recovery of 40% to 60%. Fixative should not be used.

Storage Instructions Specimen should be delivered to the Cytology Laboratory as soon as possible. Refrigeration of the specimen is essential if there is a delay in delivering to the laboratory or delay in specimen processing. Detection of *Pneumocystis carinii* is not compromised by delayed processing; however, it is more difficult to identify in patients who are already receiving treatment for this organism.

Special Instructions Specify need for special stains for microorganisms or cell count. Routine processing in many cytology laboratories includes only Papanicolaou stained smears.

Use Bronchoalveolar lavage (BAL), transbronchial and percutaneous lung fine needle aspirations are investigative methods for lung diseases (masses, cavitary lesions, infiltrates, and other abnormalities) detected by imaging. BAL is useful in the diagnosis of infections of the lung and is now commonly used as a diagnostic procedure in diffuse lung infections in both normal and especially in immunocompromised patients, including pneumonia caused by gram-negative bacilli, and particularly opportunistic infectious organisms such as cytomegalovirus, *P. carinii*, herpes simplex virus (HSV), *Cryptococcus neoformans*, *Candida* spp, *Blastomyces dermatitidis*, *Aspergillus* spp, *Nocardia* spp, *Actinomyces* spp, and atypical mycobacteria. BAL is also useful in the diagnosis and management of sarcoid; fibrosing alveolitis; bronchiolitis obliterans organizing pneumonia; eosinophilic pneumonia; pneumoconioses including silica particles, asbestos, titanium, tantalum, nickel, and chromium lung disease; chronic beryllium disease; alveolar proteinosis; idiopathic pulmonary hemosiderosis; and other interstitial lung disease. The presence of foamy cytoplasm in
(Continued)

Bronchoalveolar Lavage (BAL) *(Continued)*

alveolar macrophages results from accumulation of phospholipids in amiodarone toxicity. With electron microscopy, a diagnosis of Langerhans cell (eosinophilic) granulomatosis can be established. Various neoplastic processes are diagnosed by this method including primary bronchogenic carcinoma, bronchoalveolar carcinoma, metastatic tumors including lymphangitic carcinomatosis, leukemias, and lymphomas. BAL is useful in evaluation of response to therapy and follow up (eg, allergic alveolitis).

Limitations Wide variability in cell type and numbers recovered, particularly in smokers. Standardized volumes and concentrations not yet established. Clark et al have found lung fine needle aspirations to provide a superior method for pulmonary disease, over BAL. However, in particular cases, either method may be diagnostic while the other is not.

Contraindications Severe hypoxemia with impending respiratory failure

Methodology Samples are often shared between cytology and microbiology laboratories.

Additional Information The presence of bacteria phagocytosed by >7% of lavaged cells is thought to bear predictive relevance and may be useful to direct antimicrobial treatment before the results of cultures become available. Detection of lipid-laden macrophages in increased numbers (>40% of cells) correlates well with chronic aspiration pneumonia.

References

Clark BD, Vezza PR, Copeland C, et al, "Diagnostic Sensitivity of Bronchoalveolar Lavage Versus Lung Fine Needle Aspirate," *Mod Pathol*, 2002, 15(12):1259-65.

DeMay RM, *The Art and Science of Cytopathology*, Chicago, IL: ASCP Press, American Society of Clinical Pathologists, 1996, 212.

Faling LJ and Marke J, "A 65-Year-Old woman With a Dry Cough and Pulmonary Nodules," Case Records of the Massachusetts General Hospital, Case 35-1997, Scully RE, Mark EJ, McNeely WF, et al, eds, *N Engl J Med*, 1997, 337(20):1449-58.

Guzman y Rotaeche J and Costabel U, "Bronchoalveolar Lavage in Diagnostic Cytology," *Compendium on Diagnostic Cytology*, 8th ed, Wied GL, Bibbo M, Keebler CM, et al, eds, Chicago, IL: Tutorials of Cytology, 1997, 233-47.

♦ **Bronchoalveolar Lavage Culture** *see* Bacterial Culture, Lower Respiratory *on page 241*

Brucellosis Culture and Serology

Related Information

Bacterial Culture, Blood *on page 232*

Synonyms Undulant Fever Culture and Serology

Applies to Blood Culture, *Brucella*; Bone Marrow Culture for *Brucella*

Test Includes Culture that optimizes the growth of *Brucella* species. Serology will detect *Brucella*-specific agglutinins. Positive serum specimens are titered.

Abstract Brucellosis is a febrile zoonosis caused by several species of *Brucella*. These small, gram-negative coccobacilli are intracellular pathogens that cause a mild disease or a severe septicemic febrile illness in humans. Transmission of brucellosis is through contaminated milk or milk products or by direct contact with infected animals. *Brucella melitensis* usually infects goats and sheep, *Brucella suis* swine, *Brucella abortus* cattle, and *Brucella canis* infects dogs.

Used as a weapon of bioterrorism, brucellosis could be delivered in bomblets or as a dry aerosol, and thus the CDC has placed it in Category B, bacteria with the potential to infect humans by the aerosol route.

Specimen Blood, bone marrow, infected tissues, spleen, liver biopsies; rarely, cerebrospinal fluid, urine, pleural or peritoneal fluid are submitted for *Brucella* culture. Serology requires serum.

Container Ideally, blood should be collected in biphasic blood culture bottles; if these are not available, blood may be collected in routine blood culture bottles or in lysis centrifugation tubes. Specimens other than blood should be collected in a sterile tightly closed container. Serology requires serum collected in red top tubes.

Sampling Time Acute phase serum for serology should be drawn at presentation.

Collection Collect prior to antimicrobial therapy, when possible.

Storage Instructions Specimens should be cultured promptly.

Turnaround Time Blood cultures may be held for up to 6 weeks before reporting out as negative, however, recent studies with newer blood culture instruments report isolation in less than 7 days.

Special Instructions *Brucella* spp pose a significant infection control hazard to laboratorians and require special procedures for recovery. The organisms are highly infectious by aerosol, and can survive in dust for 6 weeks. In order to ensure that proper precautions and methods are used, **the laboratory must be informed when brucellosis is suspected.** Serologic testing requires paired specimens that should be tested together in the same laboratory.

Reference Interval No growth; negative serology; a fourfold rise in titer on paired (acute and convalescent) sera drawn 14-21 days apart is strongly indicative of the diagnosis. Titers of 1:160 are suggestive of active past or present disease. Ninety percent of patients with titers $\geq$1:320 have bacteremia.

Use Investigation of patients with episodic fever, sweating, and malaise. Brucellosis may cause osteomyelitis, spondylitis, arthritis, and/or splenomegaly. The differential diagnosis often includes tuberculosis and may include lymphoma and causes of infective endocarditis.

Limitations Blood cultures for *Brucella* are useful in acute infection but are rarely positive in subacute or chronic brucellosis; bone marrow should always be cultured when subacute or chronic brucellosis is suspected. Cultures may be misinterpreted.

Previous vaccination may have an effect on the antibody titer. Serology must be done utilizing a standard antigen prepared from *B. abortus* strain 1119, which will not detect antibodies to *B. canis*. Serologic testing for *B. canis* infection requires use of *B. canis* or *B. ovis* antigen. Testing for *B. canis* is available in some veterinary laboratories. Prozone phenomena may occur, requiring testing at dilutions of 1:320. Blocking antibodies may interfere at low titers. There are cross reactions with *Proteus* OX-19, *Yersinia enterocolitica*, *Francisella tularensis*, and *Vibrio cholerae* including cholera vaccination, as well as with skin tests for *Brucella*. Subjects who live where *Brucella* infections are endemic may have titers $\geq$160. Therefore, documentation by a fourfold rise in titer is advised.

Methodology Extended incubation in an atmosphere of 5% to 10% CO_2 on media capable of supporting *Brucella* growth. Solid media (usually available in the laboratory) that is capable of supporting growth of *Brucella* spp includes selective buffered charcoal yeast extract agar, Thayer-Martin agar, and chocolate agar containing VCNT. Cultures must be kept for at least 6 weeks, with periodic subculturing.

For culture of blood or other body fluids, a biphasic medium is recommended. Commercial blood culture bottles can be vented and incubated at 35°C to 37°C with subculture to solid media every 4-5 days for 30 days. Lysis centrifugation techniques have also been used with success.

Serology is tested using tube agglutination (most common); slide agglutination, complement fixation (CF), or enzyme-linked immunosorbent assay (ELISA). Radioimmunoassay (RIA) is available.

Additional Information Brucellosis is common worldwide. The organisms infect lung, spleen, liver, central nervous system, bone marrow, testes, the gallbladder, and the skeletal system. They cause vertebral osteomyelitis, large joint infections, and sacroiliitis. Genitourinary infections occur. Endocarditis and CNS infections are rare, but account for most fatalities.

B. melitensis occurs particularly in Russia, the Mediterranean area, Spain, the Arabian Gulf, Indian subcontinent, and parts of Mexico and Central and South America. Brucellosis is rare in the United States with ~100 cases reported annually, primarily in persons who have traveled to endemic areas or have been exposed to animals or laboratory cultures. Clinical suspicion in low prevalence populations is complicated by the fact that human *Brucella* infections have variable incubation periods, insidious or abrupt onset, and no pathognomonic symptoms or signs. Clinical suspicion is often based merely on risk factors such as travel to endemic areas or occupational exposure to animals (eg, butchers, abattoir workers, farmers, dairymen, veterinarians) and laboratory workers.

Serologic diagnosis is often the method of choice for the diagnosis of brucellosis because of the difficulty of culturing the bacteria. *Brucella* agglutinins (Continued)

Brucellosis Culture and Serology *(Continued)*

appear during the second week in acute cases and peak in 3-6 weeks. With newer ELISA assays, IgG and IgM antibodies to *Brucella* are used both for initial diagnosis and for follow-up of the patient. After successful treatment of brucellosis, specific IgG may be present for as long as 1 year. A newly developed immunoassay uses the dipstick method to detect serum antibody to *Brucella* species. This assay was found to have a sensitivity of 83% to 95% and a specificity of 94% when compared to culture and agglutination titers. **Diagnosis of brucellosis ideally requires isolation (blood and/or bone marrow cultures)** or cultures from infected sites, but only about 20% of cases are confirmed.

Histopathologic examination of aspirates or other tissue may reveal granulomas.

References

Alcala L, Munoz P, Rodriguez-Creixems M, et al, "*Brucella* spp Peritonitis," *Am J Med*, 1999, 107(3):300.

Araj GF, "Human Brucellosis: A Classical Infectious Disease With Persistent Diagnostic Challenges," *Clin Lab Sci*, 1999, 12(4):207-12.

Corbel MJ, "Brucellosis: An Overview," *Emerging Infect Dis*, 1997, 3(2):213-21.

Durmaz G, Us T, Aydinli A, et al, "Optimum Detection Times for Bacteria and Yeast Species With the BACTEC 9120 Aerobic Blood Culture System: Evaluation for a 5-Year Period in a Turkish University Hospital," *J Clin Microbiol*, 2003, 41(2):819-21.

Franz DR, Jahrling PB, Friedlander AM, et al, "Clinical Recognition and Management of Patients Exposed to Biological Warfare Agents," *JAMA*, 1997, 278(5):399-411.

Kontoyiannis DP and Versalovic J, "A 64-Year-Old Man With Fever and Gram-Negative Bacteremia," Case Records of the Massachusetts General Hospital. Case 20-2001, Scully RE, Mark EJ, McNeely WF, et al, eds, *N Engl J Med*, 2001, 344(26):2009-14.

Mishal J, Ben-Israel N, Levin Y, et al, "Brucellosis Outbreak: Analysis of Risk Factors and Serologic Screening," *Int J Mol Med*, 1999, 4(6):655-8.

Noble JT and Mark EJ, "A 37-Year-Old Man With Unexplained Fever After a Long Trip Through South America," Case Records of the Massachusetts General Hospital, Case 22-2002, Harris NL, McNeely WF, Shepard J-AO, et al, eds, *N Engl J Med*, 2002, 347(3):200-6.

Ozturk R, Mert A, Kocak F, et al, "The Diagnosis of Brucellosis by Use of BACTEC 9240 Blood Culture System," *Diagn Microbiol Infect Dis*, 2002, 44(2):133-5.

Internet Web Sites

www.cdc.gov/ncidod/dbmd/diseaseinfo/brucellosis_g.htm

www.who.int/inf-fs/en/fact173.html

Brushings Cytology

Related Information

Bronchial Brushings/Washings Cytology *on page 310*

Sputum Cytology *on page 1222*

Abstract Cytologic specimens using endoscopy and the brushing technique can be obtained from the respiratory and gastrointestinal tracts. Additionally, the endoscope has also provided a means of obtaining small tissue biopsies for histologic examination. The general principle in this type of cytology is that the lesion is visible endoscopically and must primarily involve the tissue surface.

Bronchial brushing cytology can be used for diagnosis of granulomatous and infectious processes.

Application of endoscopic retrograde cholangiopancreatography (ERCP) has provided increasing numbers of brush specimens from the biliary tract and pancreatic ducts. Cytologic brushing samples can also be beneficial in the diagnosis of inflammatory disease involving the GI tract, as well as neoplastic lesions.

Use In addition to the differential diagnosis of neoplastic and tumor-like lesions, infectious agents are encountered. These include histoplasmosis, candidiasis, herpes infections, CMV, mycobacterial and *H. pylori* infections, infestation by giardiasis, cryptosporidiosis, and microsporidiosis.

References

Johnston WW, "Cytopathology of the Lung: Diagnostic Applications of Sputum, Bronchial Brushings and Fine Needle Aspiration Biopsy Specimens," *Compendium on Diagnostic Cytology*, 8th ed, Wied GL, Bibbo M, Keebler CM, et al, eds, Chicago, IL: Tutorials of Cytology, 1997, 216-29.

Rumalla A, Baron TH, Leontovich O, et al, "Improved Diagnostic Yield of Endoscopic Biliary Brush Cytology by Digital Image Analysis," *Mayo Clin Proc*, 2001, 76(1):29-33.

B-Type Natriuretic Peptide

Related Information

Albumin, Serum *on page 120*

Synonyms BNP; Brain Natriuretic Peptide; Natriuretic Peptide, Brain

Applies to Atrial Natriuretic Peptide; Ventricular Hypertrophy

Abstract B-type natriuretic peptide (BNP), a 32-amino acid polypeptide, is secreted from the ventricular myocardium in the presence of volume expansion and pressure overload. The natriuretic peptides possess diuretic, natriuretic, and vasodilator effects. They suppress the renin-angiotensin-aldosterone system and the sympathetic nervous system. BNP is used as a diagnostic marker of congestive heart failure (CHF) and its levels correlate with severity of CHF, the degree of left ventricular dysfunction, as well as with prognosis. It is useful in distinguishing between cardiac and noncardiac causes of acute dyspnea. When used in conjunction with other clinical information, it is useful in ruling-in or ruling-out the diagnosis of CHF in patients with acute dyspnea.

Specimen Plasma; whole blood for rapid assay

Container Lavender top (EDTA) tube

Sampling Time BNP may be normal or low early in some individuals with very rapid onset of pulmonary edema.

Storage Instructions If the analysis cannot be performed within a few hours, plasma should be frozen.

Reference Interval Significant variation with age and sex exists. Approximate ranges are given in the following table. BNP concentrations increase with age. They are higher in women.

Critical Values 200-400 pg/mL: likely moderate CHF; >400 pg/mL: likely moderate to severe CHF

Use An association between BNP levels and the severity of CHF exists. BNP can be assayed rapidly at point-of-care settings and can be utilized to quantify functional classes, provide estimation of prognosis and evaluation of therapeutic efficacy for heart failure, as well as to aid in confirmation of the diagnosis of CHF when clinical diagnostic ambiguity exists. With other evaluations, it may be used to rule in or rule out CHF in subjects with acute dyspnea.

Limitations In very acute congestive heart failure or ventricular flow obstruction, BNP may not increase. Elevation in BNP can occur in other diseases. In addition to CHF, they include right heart failure with cor pulmonale, pulmonary hypertension, and acute pulmonary embolism.

Methodology Immunoassay, electrochemiluminescence. One of the commonly used assays, Triage BNP Test from Biosite Inc, is a fluorescence immunoassay for quantitative determination of BNP. The device can perform BNP on plasma or EDTA whole blood samples.

BNP Functional Classification

Functional Class	Symptoms	BNP (pg/mL) 5th-95th Percentile	Median BNP (pg/mL)
I	Cardiac disease, but without resulting limitations of physical activity.	15-499	95
II	More than ordinary activity causes fatigue, palpitation, dyspnea, or anginal pain.	10-1080	222
III	Ordinary physical activity results in fatigue, palpitation, dyspnea, or anginal pain.	38-1300	459
IV	Cardiac disease resulting in inability to carry on any physical activity without discomfort. Symptoms of heart disease or the anginal syndrome may be present even at rest.	147-1300	1006

(Continued)

B-Type Natriuretic Peptide *(Continued)*

Additional Information A-type (atrial) natriuretic peptide derives from atrial myocardium as a response to dilation. C-type is produced by endothelial cells. Levels of the natriuretic peptides relate to stress.

It is estimated that in the U.S., five million people have CHF. A clinical diagnosis, as many as 500,000 new cases are diagnosed annually. The New York Heart Association, based on the severity of the symptoms, classifies heart failure into four functional classes. BNP levels roughly correlate with the classification shown in the truncated table on the previous page.

BNP levels have been compared with other evaluations such as echocardiography, 6-minute walk test, left ventricular ejection fraction, and clinical criteria in 139 CHF patients. Using death as an endpoint, multivariate analysis showed that BNP independently identified patients with the worst prognosis. Furthermore, BNP levels have been used in monitoring therapy for congestive CHF and appear to be an objective marker to assess response to treatment.

References

Baughman KL, "B-Type Natriuretic Peptide - A Window to the Heart," *N Engl J Med*, 2002, 347(3):158-9.

Cowie MR and Mendez GF, "BNP and Congestive Heart Failure," *Prog Cardiovasc Dis*, 2002, 44(4):293-321.

Dao Q, Krishnaswamy P, Kazanegra R, et al, "Utility of B-Type Natriuretic Peptide in the Diagnosis of Congestive Heart Failure in an Urgent-Care Setting," *J Am Coll Cardiol*, 2001, 37(2):379-85.

Maisel AS, Krishnaswamy P, Nowak RM, et al, "Rapid Measurement of B-Type Natriuretic Peptide in the Emergency Diagnosis of Heart Failure," *N Engl J Med*, 2002, 347(3):161-7.

Mayo Reference Services, "B-Type Natriuretic Peptide (BNP) Plasma," Aug 2002.

Shapiro BP, Chen HH, Burnett JC, et al, "Use of Plasma Brain Natriuretic Peptide Concentration to Aid in the Diagnosis of Heart Failure," *Mayo Clin Proc*, 2003, 78(4):481-6.

Vasan RS, Benjamin EJ, Larson MG, et al, "Plasma Natriuretic Peptides for Community Screening for Left Ventricular Hypertrophy and Systolic Dysfunction: The Framingham Heart Study," *JAMA*, 2002, 288(10):1252-9.

Internet Web Sites

www.biosite.com/products/bnp/pdf/pi.pdf

♦ **Bufferin®** *see* Salicylate, Serum or Plasma *on page 1176*
♦ **Buffy Coat Method for Detection of Bacteremia** *see* Bacteremia Detection, Buffy Coat Micromethod *on page 227*

Buffy Coat Smear Study of Peripheral Blood

Related Information

Bacteremia Detection, Buffy Coat Micromethod *on page 227*
Bacterial Culture, Blood *on page 232*
Bone Marrow *on page 296*
Chagas Disease Diagnostic Procedures *on page 381*
Ehrlichiosis Serology *on page 531*
Lymph Node Biopsy *on page 880*
Malaria Smear and Tests *on page 888*
Peripheral Blood: Differential Leukocyte Count *on page 1010*
Polymerase Chain Reaction *on page 1069*
Toxoplasmosis Diagnostic Procedures *on page 1265*

Applies to *Capnocytophaga canimorsus*

Abstract The detection of microorganisms in peripheral blood is enhanced by preparing (with centrifugation) and staining the concentrated white cell fraction ("buffy coat") of blood. The buff or pale tan colored layer contains the nucleated blood cells, while the off-white colored top layer usually corresponds to platelets. PCR assays for detecting bacteria, viruses, and parasites have been developed for use on buffy coat white cells.

Specimen Blood

Container Lavender top (EDTA) tube

Storage Instructions Cannot be stored. Specimen is processed and smears are prepared and studied 1-2 hours after blood is obtained, after which stained smears can be archived.

Use Low cost maneuver to detect uncommon cells or pathogenic organisms in blood. Can be used for detection of abnormal, immature, blast, or malignant white blood cells or other nucleated cell forms. Used to detect leukemic cells, circulating malignant cells, or immature blood cells in cases of marrow myelophthisic processes; may be used in histocytochemical evaluation of leukemias when a "dry tap" of the marrow occurs. May assist in the search for circulating plasma cells (as in cases of multiple myeloma) and in the evaluation of megalo-blastic anemia (identification of megaloblastic nucleated red cells and hyper-segmented neutrophils).

Limitations Preparation of buffy coat smears may distort cells; artifact affects especially fragile cells. When used to detect microorganisms, a variable inci-dence of false-negative results is anticipated.

Methodology Wright/Giemsa type stained smear of buffy coat developed in centrifuged tube or capillary of anticoagulated whole blood.

Additional Information In some cases, erroneous results relate to an uneven distribution of cells in the buffy coat layer. Removal and mixing of the entire buffy coat layer prior to making smears prevents uneven distribution of the leukocytes and may assist in avoiding artifactual distortion of WBC morphology. Unusual microorganisms, tumor cells, and cells with inclusions (eg, *Strongyloides stercoralis* hyperinfection in an immune-suppressed indi-vidual, microfilaria *Borrelia* species spirochetes [cause of relapsing fever], AIDS, Howell-Jolly body-like inclusions, RBC-associated bacillus *Tropheryma whippelii* [causative agent of Whipple disease], *Ehrlichia* neutrophil inclusions in human ehrlichiosis, myeloma, and other circulating malignant cells) may be more readily detected by buffy coat study.

Currently, buffy coat specimens are also obtained to harvest hemopoietic progenitors (stem cells) and for performance of polymerase chain reaction (PCR) analyses.

Buffy coat smears prepared from peripheral blood and bone marrow and stained for specific organisms are still of value. They provide a low cost approach to some difficult diagnostic situations (eg, sputum-negative cases of pulmonary tuberculosis, tuberculosis of inaccessible extrapulmonary sites, septicemia due to unusual microorganisms such as *Capnocytophaga cani-morsus* [which is rapidly progressive and potentially fatal]).

References

Lowsky R, Archer GL, Fyles G, et al, "Brief Report: Diagnosis of Whipple's Disease by Molecular Analysis of Peripheral Blood," *N Engl J Med*, 1994, 331(20):1343-6.

Mirza I, Wolk J, Toth L, et al, "Waterhouse-Fredericksen Syndrome Secondary to *Capnocytophaga carnimorsus* Septicemia and Demonstration of Bacteremia by Peripheral Blood Smear: A Case Report and Review of the Literature," *Arch Pathol Lab Med*, 2000, 124(6):859-63.

Shafer JA, "Preparation of Blood Films for Examination," *Clinical Hematology: Principles, Proce-dures, and Correlations*, 2nd ed, Chapter 3, Stiene-Martin EA, Lotspeich-Steininger CA, and Koepke JA, eds, Philadelphia, PA: Lippincott Raven Publishers, 1998, 26-7.

♦ **Bullets** *see* Chain-of-Custody Protocol *on page 381*

♦ **BUN** *see* Urea Nitrogen, Serum or Plasma *on page 1284*

♦ **BUN:Creatinine Ratio** *see* Urea Nitrogen:Creatinine Ratio *on page 1283*

♦ **Bunya Virus** *see* California Encephalitis Virus Serology *on page 334*

Bupropion, Serum or Plasma

Related Information

Antidepressants, Cyclic, Serum or Plasma *on page 171*

Synonyms Wellbutrin®; Wellbutrin SR®; Wellbutrin XL™; Zyban®

Abstract An antidepressant that decreases the reuptake of serotonin, norepi-nephrine, and dopamine. It is not a monoamine oxidase inhibitor and is not related to tricyclic or tetracyclic antidepressants. It is also used as an aid to smoking cessation and attention-deficit hyperactivity disorder (ADHD).

Specimen Serum or plasma

Container Plain red top tube or lavender top (EDTA) tube

Sampling Time Trough

Storage Instructions Freeze sample.

Turnaround Time 2-4 days

Reference Interval Therapeutic: 50-100 ng/mL

Critical Values Levels >170 ng/mL are associated with seizures.

Possible Panic Range >300 ng/mL

Use Monitor therapy; evaluate possible toxicity

Methodology Gas chromatography (GC), high performance liquid chromatog-raphy (HPLC)

Additional Information

- Half-life: 14 hours
- Volume of distribution: 30-60 L/kg

(Continued)

317

Bupropion, Serum or Plasma *(Continued)*

- Protein binding: 70% to 90%

Bupropion is metabolized to hydroxybupropion, an active metabolite with a half-life of 20 hours. The other two metabolites, threohydrobupropion and erythrohydrobupropion are also active. Use of the drug is contraindicated in patients with seizure disorders. Its use is contraindicated in patients using monoamine oxidase inhibitors. Symptoms of overdose include labored breathing, salivation, ataxia, and convulsions.

References

Britton J and Jarvis MJ, "Bupropion: A New Treatment for Smokers. Nicotine Replacement Treatment Should Also Be Available on the NHS," *BMJ*, 2000, 321(7253):65-6.

Hays JT and Ebbert JO, "Bupropion for the Treatment of Tobacco Dependence: Guidelines for Balancing Risks and Benefits," *CNS Drugs*, 2003, 17(2):71-83.

Holm KJ and Spencer CM, "Bupropion: A Review of its Use in the Management of Smoking Cessation," *Drugs*, 2000, 59(4):1007-24.

Horst WD and Preskorn SH, "Mechanisms of Action and Clinical Characteristics of Three Atypical Antidepressants: Venlafaxine, Nefazodone, Bupropion," *J Affect Disord*, 1998, 51(3):237-54.

♦ **BuSpar®** *see* Buspirone, Serum *on page 318*

Buspirone, Serum

Related Information

Antidepressants, Cyclic, Serum or Plasma *on page 171*
Lithium, Serum *on page 863*

Synonyms Bespar®; BuSpar®

Abstract This is an anxiolytic drug that antagonizes serotonin receptors, without affecting benzodiazepine-GABA receptors. It is also used in the treatment of panic disorder, manic depressive disorder, and obsessive compulsive disorder.

Specimen Serum

Container Red top plain tube; do not use gel separation tubes.

Sampling Time Peak specimen 40-90 minutes postdose; trough specimen just before a dose, following 5 half-lives.

Reference Interval Peak: 100-800 ng/mL; trough: 40-350 ng/mL

Use Monitor therapy

Methodology Gas chromatography (GC), high performance liquid chromatography (HPLC), radioimmunoassay (RIA)

Additional Information

- Half-life: 2-3 hours
- Volume of distribution: 3-8 L/kg
- Protein binding: 95%

Buspirone undergoes extensive first-pass metabolism, with absolute bioavailability of only 4%. One of the major metabolites of buspirone, 1-pyrimidinyl piperazine (1-PP), is pharmacologically active.

Adverse effects of buspirone include bradycardia, dizziness, headache, ataxia, and nausea. There is no specific treatment for buspirone toxicity.

References

Apter JT and Allen LA, "Buspirone: Future Directions," *J Clin Psychopharmacol*, 1999, 19(1):86-93.

Leikin JB and Paloucek FP, *Poisoning and Toxicology Compendium*, Hudson, OH: Lexi-Comp Inc, 1998, 160.

Mahmood I and Sahajwalla C, "Clinical Pharmacokinetics and Pharmacodynamics of Buspirone, an Anxiolytic Drug," *Clin Pharmacokinet*, 1999, 36(4):277-87.

♦ **Butabarbital** *see* Barbiturates, Quantitative, Serum or Plasma *on page 248*

♦ **Butalbital** *see* Barbiturates, Qualitative, Urine *on page 248*

♦ **Butalbital** *see* Barbiturates, Quantitative, Serum or Plasma *on page 248*

♦ **Butisol Sodium®** *see* Barbiturates, Quantitative, Serum or Plasma *on page 248*

C1 Esterase Inhibitor, Serum

Related Information

Complement Components, Overview *on page 437*

Synonyms C1 Inactivator; C1 Inhibitor

Applies to C1q

Abstract C1 esterase inhibitor is a serum alpha$_2$ globulin acute-phase protein and a member of the serpin family of protease inhibitors that is synthesized by hepatocytes, monocytes (of blood), dermal fibroblasts, and vascular endothelium. Its physiologic function is inhibition of the catalytic subunits of the first component of the classic complement pathway (C1r and C1s). Deficiency of C1 esterase inhibitor results in the inappropriate activation of C1 and generation of C2 kinin. The latter molecule increases vascular permeability and is believed to be the mediator of the angioedema observed in patients with C1 esterase inhibitor deficiency. There are resultant recurrent bouts of circumscribed brawny, nonpitting (deep) edema involving variably subcutaneous tissue, gastrointestinal and respiratory tracts. Deficiency of C1 esterase inhibitor (C1EI) is the most common congenital deficiency in the complement system. Acquired C1 esterase inhibitors also occur (eg, in patients with lymphoproliferative diseases and monoclonal gammopathies).

C1 inhibitor plays a central role in regulation of the coagulation and contact (kinin-forming) systems as well as in control of the complement cascade. It inhibits C1r and C1s in the complement system, factor XII and kallikrein in the intrinsic pathway of coagulation, and activates plasminogen to plasmin in the fibrinolytic system.

C1 esterase inhibitor **function** is decreased in **hereditary** types I and II and **acquired** types I and II.

Specimen Plasma

Container Lavender top (EDTA) tube

Storage Instructions Refrigerate

Reference Interval Total: 18-40 mg/dL

Use C1 esterase inhibitor (C1EI) deficiency or dysfunction occurs in acquired and hereditary forms of angioneurotic edema.

Limitations C1EI **antigen levels** may be within normal limits in some patients with the type II inherited form of the disease, and in some patients with types I and II acquired angioedema.

Antigenic concentrations of C1EI protein may be decreased or normal in type II hereditary angioedema and in types I and II acquired angioedema. C1EI **function** is decreased in types I and II hereditary angioedema and the types I and II acquired forms.

Methodology Both functional and antigen assays are recommended. A functional (enzymatic) method should be used that tests the ability of a serum to inhibit the esterolytic activity of a preparation of activated C1 (C1 esterase). Some patients and family members may have dysfunctional C1 inhibitor molecules with normal serum concentration. Thus, selection of inappropriate test methodology may deny therapy for a potentially fatal condition. Radial immunodiffusion (RID) or nephelometry for measurement of antigenic material (may be used to measure concentration if presence of deficiency has been established). A functional (hemolytic) assay has been described and an ELISA method is available. See tables.

	C1 Esterase Inhibitor		
	Antigen	Function	Autoantibody
Type I HAE (85% of HAE patients)	↓	↓	–
Type II HAE	normal or ↓	↓	–
Type I AAE	normal or ↓	↓	–
Type II AAE	normal or ↓	↓	+

Note: HAE = hereditary angioedema; AAE = acquired angioedema.

Modified from Carney DF, "Complement and Autoimmune Diseases," *Clinical and Laboratory Evaluation of Human Autoimmune Diseases*, Chapter 6, Nakamura RA, Keren DF, and Bylund DJ, eds, Chicago, IL: American Society for Clinical Pathology, 2002, 71-84.

Differentiation of Acquired and Hereditary Deficiencies

Type of Deficiency	C1q Levels	C2 and C4 Levels
Inherited	Normal	Decreased
Acquired	Decreased	Decreased

(Continued)

C1 Esterase Inhibitor, Serum *(Continued)*

Additional Information Fatality may occur with involvement of airway/lungs. There may be massive swelling involving skin/subcutaneum. GI symptoms include severe abdominal pain, vomiting, and self-limited intestinal obstruction commonly followed by diarrhea.

The inherited forms are usually detected in the first or second decade of life and have an autosomal dominant pattern of inheritance. All patients with hereditary angioneurotic edema (HAE) are heterozygous for the deficiency with one normal gene controlling the synthesis of normal C1 inhibitor. This genetic basis, with a variable rate of biosynthesis of C1 inhibitor, may explain the variability in symptoms. **The acquired forms** primarily affect adult or elderly patients with autoimmune or lymphoproliferative disorders. Type II AAE relates to antibody activity to C1 inhibitor.

The more common form of **hereditary angioneurotic edema** (85% of cases) is due to an absolute decrease in the synthesis of the C1 esterase inhibitor. The less common form (15% of cases) is due to production of normal quantities of a functionally deficient protein. In both hereditary subtypes, C1 activation proceeds unabated, resulting in normal levels of C1q. (C1q is a subunit of the first component of complement.) Levels of C2 and C4 are decreased because of the uncontrolled activity of C1s. Patients in prolonged remission may have normal C4 levels but with activation of C1 (by incubation of serum at 37°C for 60 minutes) hemolytic (functional) C4 activity will be destroyed (by C1).

Patients with the **acquired form** C1 esterase inhibitor deficiency produce immune complexes that consume large amounts of C1q and C1 esterase inhibitor, resulting in quantitative and functional deficiency of the C1 esterase inhibitor and C1q. C2, C3, and C4 levels are reduced in some patients. Autoantibodies to C1EI which inhibit functional activity may be found in some cases of acquired angioedema (type II acquired angioedema, AAE). They may be detected by an enzyme-linked immunosorbent assay.

References

Bowen B, Hawk JJ, Sibunka S, et al, "A Review of the Reported Defects in the Human C1 Esterase Inhibitor Gene Producing Hereditary Angioedema Including Four New Mutations," *Clin Immunol*, 2001, 98(2):157-63.

Caliezi C, Wuillemin WA, Zeerleder S, et al, "C1 Esterase Inhibitor: An Anti-inflammatory Agent and Its Potential Use in the Treatment of Diseases Other Than Hereditary Angioedema," *Pharmacol Rev*, 2000, 52(1):91-112.

Carney DF, "Complement and Autoimmune Diseases," *Clinical and Laboratory Evaluation of Human Autoimmune Diseases*, Chapter 6, Nakamura RM, Keren DF, and Bylund DJ, eds, Chicago, IL: American Society for Clinical Pathology, ASCP Press, 2002, 71-84.

Cicardi M and Agostoni A, "Hereditary Angioedema," *N Engl J Med*, 1996, 334(25):1666-7.

Farkas H, Fust G, Fekete B, et al, "Eradication of *Helicobacter pylori* and Improvement of Hereditary Angioneurotic Oedema," *Lancet*, 2001, 358(9294):1695-6.

Markovic SN, Inwards DJ, Frigas EA, et al, "Acquired C1 Esterase Inhibitor Deficiency," *Ann Intern Med*, 2000, 132(2):144-50.

♦ **C1 Inactivator** *see* C1 Esterase Inhibitor, Serum *on page 318*

♦ **C1 Inhibitor** *see* C1 Esterase Inhibitor, Serum *on page 318*

♦ **C1q** *see* C1 Esterase Inhibitor, Serum *on page 318*

♦ **C3 Complement** *see* Complement C3, Serum *on page 434*

♦ **C4 Complement** *see* Complement C4, Serum *on page 436*

CA 15-3, Serum

Related Information

Body Fluid Chemical Analysis *on page 291*

Breast Biopsy *on page 305*

Breast Cancer, Hereditary, BRCA1, BRCA2 *on page 307*

CA 125, Serum *on page 323*

Carcinoembryonic Antigen, Serum *on page 342*

Synonyms Cancer Antigen 15-3; Carbohydrate Antigen 15-3; *MUC1*; Polymorphic Epithelial Mucin

Abstract CA 15-3 has been proposed as a biomarker for breast cancer. The substance actually measured in assays labeled CA 15-3 is a glycoprotein product of the *MUC1* gene, variously known as CA 15-3, polymorphic epithelial mucin (PEM), DF3 and *MUC1*. CA 15-3 is the designation most often found in the medical literature. This glycoprotein is overexpressed in many carcinomas, including both adenocarcinomas and squamous carcinomas. Such overexpression often results in increased serum levels of CA 15-3.

Specimen Serum

Container Red top tube

Storage Instructions Refrigerate serum. Stable at 4°C for 2 weeks. For long-term storage, hold frozen at -70°C.

Reference Interval Usually stated as <30 kU/L, but actually method dependent, with cutoff values ranging from <26 kU/L to <39 kU/L.

Use No established usefulness at this time. There is no evidence that CA 15-3 is useful in screening, diagnosis, or staging of breast cancer. Proposals, based largely on theoretical considerations, that CA 15-3 would be useful in patients with breast cancer by 1) detecting relapse, recurrence, and metastasis, and 2) monitoring the response to therapy, have **not** been confirmed. In a prospective study of 664 patients, each with 6 months of follow-up, CA 15-3 values had a positive predictive value of only 27% and a negative predictive value of 91% in the detection of relapse.

Attempts to use multiple CA 15-3 measurements in kinetic models (marker half-life and doubling time) have **not** yet been shown clinically effective.

According to the practice guidelines for breast cancer from the American Society of Clinical Oncology (ASCO), CA 15-3 is **not** recommended for screening, diagnosis, staging, or surveillance following primary treatment. CA 15-3 is also **not** recommended by ASCO as a stand-alone monitor of response to treatment; one potential exception is that in patients who have no readily measurable disease to follow, rising levels of CA 15-3 are suggestive of treatment failure.

Limitations Elevations occur in benign diseases of the breast and liver.

Methodology Radioimmunoassay (RIA), enzyme-linked immunosorbent assay (ELISA), microparticle enzyme immunoassay (MEIA), and chemiluminescent immunoassay (CIA)

References
"Clinical Practice Guidelines for the Use of Tumor Markers in Breast and Colorectal Cancer Adopted on May 17, 1996 by the American Society of Clinical Oncology," *J Clin Oncol*, 1996, 14:2843-77.

Smith TJ, Davidson NE, Schapira DV, et al, "American Society of Clinical Oncology 1998 Update of Recommended Breast Cancer Surveillance Guidelines," *J Clin Oncol*, 1999, 17:1080-2.

CA 19-9, Serum

Related Information
Alpha$_1$-Fetoprotein, Serum *on page 136*
Body Fluid Chemical Analysis *on page 291*
CA 15-3, Serum *on page 320*
CA 125, Serum *on page 323*
Carcinoembryonic Antigen, Serum *on page 342*
Fine Needle Aspiration, Deep and Superficial Masses *on page 590*
Immunoperoxidase Procedures *on page 780*

Synonyms Cancer Antigen 19-9; Carbohydrate Antigen 19-9

Applies to CA 50; CA 242; TAG-72; Tissue Polypeptide Antigen; TPA

Abstract Proposed as a marker for pancreatic carcinoma (and occasionally for colorectal and hepatocellular carcinomas), CA 19-9 can be measured in serum and body fluids and can be localized by immunohistology. The concentration of CA 19-9 in blood and body fluids is strongly influenced by the patient's Lewis blood group phenotype, secretor genotype, and gender. CA 19-9 is synthesized by a wide range of epithelial tissues, including colonic, pancreatic, biliary, gastrointestinal, salivary, and endometrial; however, CA 19-9 is not synthesized in persons who have the Le (a-b-) phenotype.

Specimen Serum

Container Red top tube

Storage Instructions Freeze to ship.

Reference Interval
- Detection, diagnosis, and prognosis of pancreatic cancer. Most commonly reported is the arbitrary cutoff of <37 kU/L, but <35 kU/L and <40 kU/L also are used. When the patient's Lewis blood group genotype and secretor
(Continued)

CA 19-9, Serum *(Continued)*

genotype are taken into account, the cutoff values range from <10.3 kU/L (in *Le/le*, *Se/Se* individuals) to <61.3 kU/L (in *Le/Le*, *se/se* individuals).

- Monitoring the effectiveness of cancer treatment and the potential for recurrence. In this situation, the patient's current value(s) is usually compared with his/her previous value(s), obviating the need to resort to an arbitrary cutoff.

Use According to the 1997 update of the clinical practice guidelines of the American Society of Clinical Oncology (ASCO), CA 19-9 is **not** recommended for "screening, diagnosis, staging, surveillance, or monitoring treatment of patients with colorectal cancer."

Elevated values (>37 kU/L) are reported in 75% to 80% of pancreatic carcinomas, 67% of hepatobiliary carcinomas, and <50% of gastric and hepatocellular carcinomas. The Mayo Clinic reports that CA 19-9 values >200 kU/L in a nonjaundiced patient with a "confirming" CT scan are highly predictive of pancreatic carcinoma. In addition, this group and others use CA 19-9 values to monitor the response to therapy and to predict disease-free survival and median survival following pancreatic resection.

The biologic variability (both intraindividual and interindividual) of CA 19-9 is high. When CA 19-9 values are used to monitor treatment, the current values are compared to previous values from the same patient. Based on data from an asymptomatic white population, sequential values should differ by at least 40% to 50% in order to conclude that a significant change has occurred.

Limitations

Nonspecificity: Marked elevations of CA 19-9 are common in patients with acute liver failure, regardless of etiology. Such high values are believed to reflect hepatocellular regeneration, and they should not be interpreted as a reason to delay liver transplantation in otherwise appropriate candidates. Elevated CA 19-9 values have been reported in a wide variety of benign and malignant processes.

Population distribution: CA 19-9 is absent from the serum (and other body fluids) of individuals who have the Le(a-b-) phenotype. This phenotype is found in ~6% of the U.S. white population and 22% of the U.S. black population. Women have higher CA 19-9 values than men.

Methodology Immunoradiometric assay (IRMA), microparticle enzyme immunoassay (MEIA), and chemiluminescence enzyme immunoassay

References

American Association of Blood Banks, *Technical Manual*, 13th ed, Bethesda, MD: American Association of Blood Banks Press, 1999, 286-7.

"1997 Update of Recommendations for the Use of Tumor Markers in Breast and Colorectal Cancer," Adopted November 7, 1997 by the American Society of Clinical Oncology, *J Clin Oncol*, 1998, 16(2):793-5.

Vestergaard EM, Hein HO, Meyer H, et al, "Reference Values and Biological Variation for Tumor Marker CA 19-9 in Serum for Different Lewis and Secretor Genotypes and Evaluation of Secretor and Lewis Genotyping in a Caucasian Population," *Clin Chem*, 1999; 45:54-61.

CA 27.29, Serum

Related Information

Body Cavity Fluid Cytology *on page 285*

Breast Biopsy *on page 305*

Breast Cancer, Hereditary, BRCA1, BRCA2 *on page 307*

CA 15-3, Serum *on page 320*

Carcinoembryonic Antigen, Serum *on page 342*

Fine Needle Aspiration, Deep and Superficial Masses *on page 590*

Sentinel Lymph Node Biopsy *on page 1189*

Synonyms Breast Carcinoma-Associated Antigen; Cancer Antigen 27.29

Applies to MUC1 Gene

Test Includes Long-term serial monitoring of results

Abstract CA 27.29 is a tumor marker proposed for detection of breast cancer recurrence. It is an antigen defined by a monoclonal antibody (B27.29) specific to the protein core of the breast-cancer-associated mucin encoded by the MUC1 gene.

Specimen Serum

Container Red top tube

Storage Instructions Refrigerate or freeze specimen according to directions of testing laboratory.

Reference Interval ≤37.7 units/mL (SI: ≤37.7 kU/L)

Use CA 27.29 testing is not approved for screening. In addition, according to the American Society of Clinical Oncology, there is insufficient basis to recommend use of this tumor marker for routine breast cancer surveillance. CA 27.29 is proposed for the monitoring of the recurrence of breast carcinoma in patients diagnosed with stage II or III disease.

Methodology Radioimmunoassay (RIA), chemiluminescent immunoassay

Additional Information The CA 27.29 assay, in a multicenter study, was shown to have a sensitivity of 57.7%, specificity of 97.9%, positive predictive value of 83.3%, and negative predictive value of 92.6% for detection of breast cancer recurrence in a group of 166 patients previously diagnosed with either stage II (80.1% of study population) or III (19.9% of study population) disease. A retrospective study of 275 primary breast cancer patients and 83 healthy controls showed CA 27.29 to have better discriminating power than CA 15.3 in differentiating these populations.

References
Chan DW, Beveridge RA, Muss H, et al, "Use of the Truquant BR Radioimmunoassay for Early Detection of Breast Cancer Recurrence in Patients With Stage II and Stage III Disease," *J Clin Oncol*, 1997, 15(6):2322-8.

Gion M, Mione R, Leon AE, et al, "Comparison of the Diagnostic Accuracy of CA 27.29 and CA 15-3 in Primary Breast Cancer," *Clin Chem*, 1999, 45(5):630-7.

Smith TJ, Davidson NE, Schapira DV, et al, "American Society of Clinical Oncology 1998 Update of Recommended Breast Cancer Surveillance Guidelines," *J Clin Oncol*, 1999, 17(3):1080-2.

♦ **CA 50** see CA 19-9, Serum on page 321

♦ **CA-125** see Immunoperoxidase Procedures on page 780

CA 125, Serum

Related Information
Body Cavity Fluid Cytology on page 285
Body Fluid Analysis, Cell Count on page 288
Body Fluid Chemical Analysis on page 291
Breast Cancer, Hereditary, BRCA1, BRCA2 on page 307
CA 15-3, Serum on page 320
CA 19-9, Serum on page 321
Carcinoembryonic Antigen, Serum on page 342
Cyst Fluid Cytology on page 490
Fine Needle Aspiration, Deep and Superficial Masses on page 590
Heterophilic Antibodies on page 727
Immunoperoxidase Procedures on page 780

Synonyms Cancer Antigen 125; Carbohydrate Antigen 125

Applies to Lysophatidic Acid

Abstract CA 125 is a mainstay in the management of patients with ovarian carcinoma, and has potential application in certain other malignancies (endometrium, fallopian tube, pancreas, breast, colon, and lung). CA 125 is present in the serum of normal persons at levels <35 kU/L, and, in normal persons, has a biological half-life in the serum of 4.8 days. In patients with completely resected Stage I and II ovarian carcinoma, the half-life ranges from 5.1-12 days. In patients with more extensive carcinoma, the half-life is often substantially longer.

Specimen Serum

Container Red top tube

Sampling Time Abdominal surgery, by itself, causes a temporary increase in serum CA 125, therefore, no postsurgical specimen should be obtained until **3 weeks after the procedure**. It may be increased during menstruation.

Storage Instructions Refrigerate within 2 hours of collection. Freeze at -20°C for long-term storage.

Reference Interval Most commonly used is <35 kU/L; however, depending on the purpose of testing, other cutoff values and other approaches (ie, half-life and other kinetic measurements) are used (see Use). Concentrations are typically less than 100 units/mL when increased secondary to benign disorders.

Critical Values Values greater than 100 units/mL are seen in malignant diseases, in pregnancy, and with some instances of benign disease of liver and pancreas.

(Continued)

CA 125, Serum *(Continued)*

Use

Monitor the response to treatment and the probability of recurrence of ovarian carcinoma. This is the principal use for CA 125, and serial serum levels are often required in clinical trials. In general, serum levels are proportional to tumor burden, and the extent to which CA 125 values decrease after treatment reflects the effectiveness of the treatment. Some investigators measure the kinetics of serum CA 125 to monitor treatment and reach prognostic conclusions. Rosman et al found a low probability for recurrence in women with a minimum CA 125 <35 kU/L and a half-life of 12 days or less. Buller et al derived an "ideal" exponential regression curve which yielded a serum CA 125 half-life of 10.4 days in patients whose primary tumors were completely removed. When patient values are entered into this model, the results can provide early evidence of treatment failure (within 60 days of surgery) and, thus, lead to treatment modification sooner than would otherwise be possible. Hawkins et al and Hunter et al have found the CA 125 half-life a useful prognostic assessment. Gadducci et al have demonstrated the prognostic value of measuring the CA 125 half-life in early chemotherapy.

CA 125-based standard response criteria in ovarian carcinoma. The World Health Organization (WHO) gold standard for defining an objective response to cancer treatment requires serial measurements of tumor deposits, either directly at surgery or indirectly from a standardized imaging study. This WHO consensus statement (Miller et al, 1981) specifically excludes biochemical measurements from any role in the determination of a response. Rustin et al undertook a study, involving a total of 403 accessible patients, to determine whether serial CA 125 values could be as accurate as the WHO gold standard. These investigators used a subset of 117 patients to derive definitions of a 50% response and a 75% response. They then tested such definitions in a subset of 186 patients and found predictive values that compare favorably with those using the gold standard. Bridgewater et al have tested such criteria in 769 patients and found them equivalent in accuracy, and superior in cost and convenience, to the gold standard. The precise definitions of a 50% response and a 75% response are complex, requiring a minimum of 4 and 3 samples, respectively; they have been incorporated into a computer program, reported to be available from Dr Rustin.

Screen for ovarian carcinoma. A review of twenty-five screening studies (1983-1996), yielded inconclusive results. When CA 125 was used as a stand alone test, and applied across all age groups, the predictive values were too low for an effective screen.

When restricted to asymptomatic postmenopausal women, using a cutoff of 30 kU/L, Jeyarajah et al found an odds ratio of 21.56 in screening for ovarian carcinoma. In this study, elevated CA 125 values were also recorded in patients who, on investigation, had other gynecologic cancers. Although women with elevated CA 125 values were more likely to have a history of breast cancer, breast cancer did not develop more frequently in the women with elevated CA 125 values. Therefore, based on these findings, asymptomatic postmenopausal women with elevated CA 125 values should be investigated for gynecologic cancer.

Another study in **postmenopausal** women (Jacobs et al, 1999) found a positive predictive value of 20% (one ovarian carcinoma surgically confirmed for every five women explored), using a screening cutoff of CA 125 >30 kU and performing transvaginal ultrasound on all patients with elevated values.

Screening programs for ovarian cancer are based on the premise that many ovarian carcinomas begin as cystic lesions which are readily identified by ultrasound scans. It was, therefore, surprising that in a study of >5000 subjects, the removal of persistent benign ovarian cysts was **not** associated with reduced mortality due to carcinoma of the ovary (Crayford et al, 2000).

Other gynecologic malignancies. CA 125 values are often elevated in patients with carcinoma of the the fallopian tube and endometrium. The magnitude of the elevation is generally proportional to total tumor burden. According to Kurihara et al, a CA 125 value <20 kU/L in a postmenopausal woman with endometrial carcinoma indicates a low probability of myometrial invasion and extrauterine spread.

Therapy selection for endometrial carcinoma. One report suggests that women with endometrial carcinoma and a preoperative CA 125 <20 kU/L can be treated more conservatively than patients with higher values.

Monitoring cardiac function. Nagele et al have reported that CA 125 values reflect heart failure status and the response to treatment in patients with congestive heart failure.

Limitations Elevations of CA 125 are also found in carcinomas of the fallopian tube, endometrium, pancreas, lung, breast, prostate, and gastrointestinal tract. Elevated values have also been reported in mesothelioma, primary peritoneal carcinoma, and rhabdomyosarcoma of the uterus. It may be increased with ascites from benign or neoplastic diseases. Benign diseases in which elevations are reported include endometriosis, liver disease, pregnancy, pelvic inflammatory disease, ovarian cysts, tuberculous peritonitis, Meigs syndrome, and pseudo-Meigs syndrome. It is increased with acute and chronic liver diseases, especially in the presence of ascites secondary to cirrhosis.

Methodology Enzyme immunoassay (EIA), radioimmunoassay (RIA), immunoradiometric assay (IRMA), microparticle enzyme immunoassay (MEIA)

Additional Information An assay for lysophosphatidic acid may have better sensitivity than CA 125 in early ovarian cancer.

References

Bidart JM, Thuillier F, Augereau C, et al, "Kinetics of Serum Tumor Marker Concentrations and Usefulness in Clinical Monitoring," *Clin Chem*, 1999, 45:1695-1707.

Bridgewater JA, Nelstrop AE, Rustin GS, et al, "Comparison of Standard and CA-125 Response Criteria in Patients With Epithelial Ovarian Cancer Treated With Platinum or Paclitaxel," *J Clin Oncol*, 1999, 17(2):501-8.

Crayford TJ, Campbell S, Bourne TH, et al, "Benign Ovarian Cysts and Ovarian Cancer: A Cohort Study With Implications for Screening," *Lancet*, 2000, 355(9209):1060-3.

Devarbhavi H, Kaese D, Williams AW, et al, "Cancer Antigen 125 in Patients With Chronic Liver Disease," *Mayo Clin Proc*, 2002, 77(6):538-541.

Gadducci A, Zola P, Landoni FT, et al, "Serum Half-Life of CA 125 During Early Chemotherapy as an Independent Prognostic Variable for Patients With Advanced Epithelial Ovarian Cancer: Results of a Multicentric Italian Study," *Gynecol Oncol*, 1995, 58(1):42-7.

Imai A, Horibe S, Takagi A, et al, "Drastic Elevation of Serum CA 125, CA 72-4, and CA 19-9 Levels During Menses in a Patient With Probable Endometriosis," *Eur J Obstet Gynecol Reprod Biol*, 1998, 78(1):79-81.

Jacobs IJ, Skates SJ, MacDonald N, et al, "Screening for Ovarian Cancer: A Pilot Randomised Controlled Trial," *Lancet*, 1999, 353(9160):1207-10.

Jeyarajah AR, Ind TE, Skates S, et al, "Serum CA 125 Elevation and Risk of Clinical Detection of Cancer in Asymptomatic Postmenopausal Women," *Cancer*, 1999, 85(9):2068-72.

Kurihara T, Mizunuma KT, Obara N, et al, "Determination of Normal Level of Serum CA 125 in Postmenopausal Women as a Tool for Preoperative Evaluation and Postoperative Surveillance of Endometrial Carcinoma," *Gynecol Oncol*, 1998, 69(3):192-6.

Nagele F, Bahlo M, Klapdor R, et al, "CA 125 and its Relation to Cardiac Function," *Am Heart J*, 1999, 137(6):1044-9.

Rustin GJ, Nelstrop AE, McClean P, et al, "Defining Response of Ovarian Carcinoma to Initial Chemotherapy According to Serum CA 125," *J Clin Oncol*, 1996, 14(5):1545-51.

Xu Y, Shen Z, Wiper DW, et al, "Lysophosphatidic Acid as a Potential Biomarker for Ovarian and Other Gynecologic Cancers," *JAMA*, 1998, 280(8):719-23.

♦ **CA 242** see CA 19-9, Serum *on page 321*

♦ **Ca, Blood** see Calcium, Serum *on page 329*

♦ **CAD** see Antidepressants, Cyclic, Serum or Plasma *on page 171*

Cadmium, Blood or Urine

Related Information

Beta₂-Microglobulin, Serum or Urine *on page 254*
Heavy Metal Screen, Blood *on page 668*
Heavy Metal Screen, Urine *on page 669*
Mercury, Blood *on page 897*

Synonyms Cd, Blood; Cd, Urine

Abstract Cadmium and cadmium compounds are highly toxic, particularly to lungs and kidneys. Exposure is predominantly occupational, predominantly from spray painting, using organic-based paints without protective apparatus, mining, and smelting. Acute life-threatening pneumonitis results from inhalation of cadmium fumes.

Specimen 24-hour urine is recommended for chronic exposure
(Continued)

Cadmium, Blood or Urine *(Continued)*

Container Plastic (preferably polycarbonate) urine container; green top (heparin) tube for blood

Storage Instructions Refrigerate urine.

Causes for Rejection Specimen allowed to contact metal or dusts containing metals

Reference Interval
- Urine: nonsmoker: <1 µg/L (SI: <8.9 nmol/L) or <1 µg/g creatinine
- Whole blood: nonsmoker: 0.3-1.2 µg/L (SI: 2.7-10.7 nmol/L), smoker: 0.6-3.9 µg/L (SI: 5.3-34.7 nmol/L)

Possible Panic Range Levels >10 µg/L (SI: >88.97 µmol/L) in whole blood and urine values >15 µg/g creatinine indicate severe exposure.

Use Evaluate cadmium toxicity in industrial exposure to cadmium fumes or cadmium ingestion

Limitations Blood or urine levels of cadmium reflect current exposure only.

Methodology Inductively-coupled plasma (ICP) spectrometry, atomic absorption spectrometry (AA)

Additional Information The major route of absorption of many heavy metals, including cadmium, is by inhalation. Inhalation of cadmium fumes produces an acute chemical pneumonitis which can produce pulmonary edema and respiratory failure. Long-term exposure may lead to emphysema (with decreased alpha$_1$-antitrypsin), pulmonary fibrosis, and osteomalacia.

Acute cadmium toxicity may cause hepatic failure. Chronic cadmium toxicity leads to progressive renal dysfunction with proteinuria of slow onset. With slow excretion and constant exposure, cadmium values increase with age. Body cadmium elimination half-life may be greater than 20 years. When renal tubular toxicity is suspected, urinary β$_2$-microglobulin, lysozyme, and retinol-binding proteins are useful. Chest radiographs to reveal pulmonary fibrosis and bone films to assess osteomalacia are recommended.

References
Baselt RC, *Disposition of Toxic Drugs and Chemicals in Man*, 6th ed, Foster City, CA: Chemical Toxicology Institute, 2002, 147-9.

Choudhury H, Harvey T, Thayer WC, et al, "Urinary Cadmium Elimination as a Biomarker of Exposure for Evaluating a Cadmium Dietary Exposure - Biokinetics Model," *J Toxicol Environ Health A*, 2001, 63(5):321-50.

Moyer TP, "Toxic Metals," *Tietz Textbook of Clinical Chemistry*, Chapter 28, Burtis CA and Ashwood ER, eds, Philadelphia, PA: WB Saunders Co, 1999, 988.

Caffeine, Serum

Related Information

Theophylline, Serum *on page 1243*

Synonyms Coffee Break®; Dexitac®; Durvitan®; Magnum®; Max Alert Magnum®; Mole®; No-Doz®; Pep Back®; Percoffedrinol N®; Percutafeine®; Pick-me-up®; Pro-Plus®; Stay Awake®; Vivarin®

Abstract Caffeine is a methylxanthine structurally related to theophylline. It is a mild CNS stimulant that also produces diuresis, myocardial and respiratory stimulation. An average cup of coffee or cola drink contains 35-100 mg caffeine. It is used clinically with other measures to treat idiopathic apnea of prematurity. It is a metabolite of theophylline in infants. Unlike its metabolism in adults, theophylline in neonates is extensively metabolized to caffeine. However, due to its better tolerance and wider therapeutic index, caffeine treatment is preferred over theophylline. It is also used in combination with analgesics as a migraine remedy.

Specimen Serum

Container Red top tube

Collection Indicate exact time of blood drawn and relationship to last theophylline or caffeine dose on request.

Reference Interval The therapeutic range in the treatment of neonatal apnea is 8-14 µg/mL (SI: 41-72 µmol/L).

Possible Panic Range Toxic: >50 µg/mL (SI: >256 µmol/L); **fatal:** >100 µg/mL (SI: >512 µmol/L)

Methodology High performance liquid chromatography (HPLC), immunoassay, gas chromatography (GC)

Additional Information
- Half-life (adults): 3-5 hours; up to 100 hours in neonates
- Volume of distribution: 0.7 L/kg; 0.8 L/kg in neonates
- Protein binding: 35%

Apnea of prematurity is a common condition in neonates born at less than 37 weeks' gestational age. It affects approximately 90% of premature neonates weighing <1000 g and 25% weighing <2500 g. Caffeine significantly decreases incidence of apnea (Comer et al).

Physical dependence may develop upon repeated use. Withdrawal symptoms include headache and fatigue. Hypertensive patients are more prone to the effects of caffeine.

Epidemiological studies suggest that caffeine increases the risk of spontaneous abortion. Women who consumed 500 mg or more caffeine per day had double the risk of spontaneous abortion as compared to those who consumed <100 mg of caffeine (Cnattingius et al).

References
Cnattinguis S, Signorello LB, Annerén G, et al, "Caffeine Intake and the Risk of First-Trimester Spontaneous Abortion," *N Engl J Med*, 2000, 343(25):1839-45.

Comer AM, Perry CM, and Figgitt DP, "Caffeine Citrate: A Review of Its Use in Apnoea of Prematurity," *Paediatr Drugs*, 2001, 3(1):61-79.

Nurminen ML, Nittynen L, Korpela R, et al, "Coffee, Caffeine, and Blood Pressure: A Critical Review," *Eur J Clin Nutr*, 1999, 53(11):831-9.

Pesce AJ, Rashkin M, and Kotagal U, "Standards of Laboratory Practice: Theophylline and Caffeine Monitoring. National Academy of Clinical Biochemistry," *Clin Chem*, 1998, 44(5):1124-8.

♦ **Calan®** *see* Verapamil, Serum or Plasma *on page 1306*

♦ **Calcitonin** *see* Vitamin D, Serum *on page 1318*

Calcitonin, Serum or Plasma

Related Information
Calcium, Serum *on page 329*
Carcinoembryonic Antigen, Serum *on page 342*
Catecholamines, Fractionation, Plasma *on page 345*
Catecholamines, Fractionation, Urine *on page 347*
Fine Needle Aspiration, Deep and Superficial Masses *on page 590*
Immunoperoxidase Procedures *on page 780*
Multiple Endocrine Neoplasia/Familial Medullary Thyroid Carcinoma *on page 924*
Parathyroid Hormone, Serum *on page 1001*
Phosphorus, Urine *on page 1032*

Synonyms Thyrocalcitonin

Applies to Calcium Stimulation Test; Chromogranin A, Serum; MEN A; MEN B; Pentagastrin Stimulation Test; Sipple Syndrome

Abstract Calcitonin is a hypocalcemic polypeptide secreted by the C cells (parafollicular cells) of the thyroid gland, by tumors of the C cells (medullary carcinoma of thyroid - MCT), as well as by bronchopulmonary and thymic neuroendocrine cells, and some cells of the adrenal medulla.

Patient Preparation Patient should fast overnight.

Specimen Serum or plasma

Container Red top tube or green top (heparin) tube

Collection Avoid hemolysis.

Storage Instructions Collect into chilled tube. Process within 10 minutes of collection. Separate in a refrigerated centrifuge. Separate serum (plasma) into plastic tube and freeze.

Reference Interval <19 pg/mL (SI: <19 ng/L) basal for a sensitive RIA assay or column chromatography. Following stimulation with calcium or pentagastrin, reference intervals are <350 pg/mL in men and <100 pg/mL in women.

Use Calcitonin concentrations are increased in both sporadic and familial types of MCT and in its precursor, C-cell hyperplasia. Calcitonin levels correlate with tumor burden and the degree of tumor differentiation, and are used in postoperative patients with CEA to monitor residual, recurrent, and/or metastatic carcinoma. Recurrent or persistent postoperative elevation reliably indicates tumor persistence or relapse.

Calcitonin is also used in family studies to detect early, subclinical cases of C-cell hyperplasia or MCT.
(Continued)

Calcitonin, Serum or Plasma *(Continued)*

Limitations In many patients with MCT (especially those with familial MCT), the baseline calcitonin may be normal; however, an abnormal large calcitonin response may follow stimulation with calcium or pentagastrin.

Methodology Radioimmunoassay (RIA), immunoradiometric assay (IRMA) or enzyme immunoassay (EIA), chemiluminescent enzyme immunoassay

Additional Information High concentrations of calcitonin occur not only in patients with MCT but are also found in patients with other neuroendocrine tumors, pancreatitis, thyroiditis, renal failure, and pernicious anemia.

Serum chromogranin A may be increased as well with MCT. CEA is also a marker for MCT.

See also Multiple Endocrine Neoplasia/Familial Medullary Thyroid Carcinoma *on page 924.*

References

Fleming JB, Lee JE, Bouvet M, et al, "Surgical Strategy for the Treatment of Medullary Thyroid Carcinoma," *Ann Surg,* 1999, 230(5):697-707.

Guilloteau D, Perdrisot R, Calmettes C, et al, "Diagnosis of Medullary Carcinoma of the Thyroid (MCT) by Calcitonin Assay Using Monoclonal Antibodies: Criteria for the Pentagastrin Stimulation Test in Hereditary MCT," *J Clin Endocrinol Metab,* 1990, 71(4):1064-7.

Heshmati HM, Gharib H, Khosla S, et al, "Genetic Testing in Medullary Thyroid Carcinoma Syndromes: Mutation Types and Clinical Significance," *Mayo Clin Proc,* 1997, 72(5):430-6.

Isomura M, Honda N, Kawada A, et al, "Development of a Highly Sensitive Enzyme Immunoassay for Human Calcitonin Using Solid Phase Coupled With Multiple Antibodies," *Ann Clin Biochem,* 1999, 36(Pt 5):629-35.

Scheuba C, Kaserer K, Weinhausl A, et al, "Is Medullary Thyroid Cancer Predictable? A Prospective Study of 86 Patients With Abnormal Pentagastrin Tests," *Surgery,* 1999, 126(6):1089-95.

♦ **Calcitriol** *see* Vitamin D, Serum *on page 1318*

♦ **Calcium:Creatinine Ratio** *see* Calcium, Urine *on page 332*

♦ **Calcium:Creatinine Ratio, Urinary** *see* Parathyroid Hormone, Serum *on page 1001*

Calcium, Ionized, Serum

Related Information

Calcium, Serum *on page 329*
Calcium, Urine *on page 332*
Kidney Stone Analysis *on page 820*
Parathyroid Hormone-Related Protein, Serum *on page 1000*
Parathyroid Hormone, Serum *on page 1001*

Synonyms Ionized Calcium

Abstract This is the physiologically active portion of serum calcium and represents about 43% of total serum calcium. The serum ionized calcium has a diurnal variation with minimum values at 8 PM and peak values at 10 AM. Binding of free calcium to proteins is pH dependent; alkalosis decreases and acidosis increases ionized calcium.

Patient Preparation Patient should be recumbent for 30 minutes prior to collection.

Calcium, Ionized, Serum

	Reference Interval	
	Conventional (mg/dL)	SI (mmol/L)
Whole blood		
1 d	4.2-5.48	1.05-1.37
2-4 d	4.4-5.68	1.10-1.42
5 d	4.8-5.92	1.20-1.48
Adults	4.6-5.10	1.15-1.27
Plasma		
Adults	4.12-4.92	1.03-1.23
Serum		
1-18 y	4.8-5.52	1.20-1.38
Adults	4.64-5.28	1.16-1.32

Specimen Whole blood (preferred), serum, or plasma

Container Green top (heparin) tube if whole blood or plasma is used; red top tube for serum

Collection Collect anaerobically, leave stoppers in; do not use tourniquet. Heparin syringe is best; 1 unit of heparin/mL of blood lowers ionized calcium 0.01 mmol/L. The use of dry, electrolyte balanced heparin virtually eliminates the heparin interference.

Storage Instructions Store anaerobically. Such specimens can be stored 48 hours at 4°C or 2 hours at room temperature.

Special Instructions Controversy exists over the ideal specimen. Serum and plasma have generally similar values, while whole blood is 1% to 2% higher.

Reference Interval See table.

Critical Values <0.8 mmol/L; hypocalcemia may result in tetany or seizures.

Possible Panic Range <0.70 mmol/L

Use Ionized calcium is increased in the same diseases that produce elevations in total serum calcium, of which the most important are hyperparathyroidism, cancer, and granulomatous diseases (see Calcium, Serum *on page 329*).

In a series of 60 proven cases of primary hyperparathyroidism, increased total calcium was found in only 47, but 59 cases had increased ionized calcium. Ionized calcium is the more helpful test for evaluation of hyperparathyroidism and hypoparathyroidism.

Ionized calcium is measured in patients with renal failure and/or transplantation, in whom problems include secondary hyperparathyroidism. It is indicated for patients with sepsis, with magnesium deficiency, and in pancreatitis. Hypocalcemia occurs following administration of citrate (eg, liver transplantation) or infusion of other fluids during extracorporeal membrane oxygenation or surgery, and measurement of ionized calcium is needed to ascertain balance in dialysis patients.

Ionized calcium is measured in premature infants with hypoproteinemia and acidosis.

Low ionized calcium values are encountered in hypoparathyroidism, vitamin D deficiency and resistance, pseudohypoparathyroidism, and anxiety-related hyperventilation. Low or high in genetic abnormalities of the calcium-sensing receptor.

Methodology Ion-selective electrode (ISE)

References

Glendenning P, Gutteridge DH, Retallack RW, et al, "High Prevalence of Normal Total Calcium and Intact PTH in 60 Patients With Proven Primary Hyperparathyroidism: A Challenge to Current Diagnostic Criteria," *Aust N Z J Med*, 1998, 28(2):173-8.

Nordin BE, Need AG, Hartley TF, et al, "Improved Method for Calculating Calcium Fractions in Plasma: Reference Values and Effect of Menopause," *Clin Chem*, 1989, 35(1):14-7.

Painter PC, Cope JY, and Smith JL, "Reference Information for the Clinical Laboratory," *Tietz Textbook of Clinical Chemistry*, 3rd ed, Philadelphia, PA: WB Saunders Co, 1999, 1804.

♦ **Calcium Oxalate, Urine** *see* Oxalate, Urine *on page 989*
♦ **Calcium Pyrophosphate Dihydrate Crystals** *see* Synovial Fluid Analysis *on page 1229*

Calcium, Serum

Related Information

(Continued)

Calcium, Serum (Continued)

Phosphorus, Serum *on page 1031*
Phosphorus, Urine *on page 1032*
Potassium, Serum or Plasma *on page 1078*
Pyridinolines (Pyridinoline and Deoxypyridinoline), Urine *on page 1126*
Vitamin D, Serum *on page 1318*

Synonyms Ca, Blood; Total Calcium, Serum

Applies to Chloride:Phosphorus Ratio; Parathyroid Hormone-Related Protein

Abstract Total plasma calcium has three components, protein-bound (~47%), ionized (free ~43%), and complexed (~10%). The extent of protein binding varies with protein concentrations and pH. The two most common causes of hypercalcemia are primary hyperparathyroidism (HPT) and malignancy. Hypercalcemia is also a feature of certain granulomatous diseases (eg, sarcoidosis, cat-scratch disease, and lymphoma) which have in common the presence of elevated serum 1,25 dihydroxyvitamin D (1,25[OH]$_2$D).

Specimen Serum

Container Red top tube

Sampling Time Morning, fasting sample is desirable. The diurnal variation has peaks at 5 PM and 4 AM.

Storage Instructions Refrigerate in stoppered vials, not in sample cups.

Causes for Rejection Gross hemolysis

Reference Interval See table.

	mg/dL	mmol/L
Cord	8.2-11.2	2.05-2.80
Premature	6.2-11.0	1.55-2.75
0-10 d	7.6-10.4	1.90-2.60
10 d to 24 mo	9.0-11.0	2.25-2.75
2-12 y	8.8-10.8	2.20-2.70
Adults	8.6-10.0	2.15-2.50
Male (>60 y)	8.8-10.2	2.20-2.55

These intervals apply to persons with normal serum albumin. Although it is rarely necessary, the following formula may be used to adjust total serum calcium for abnormalities of albumin concentration.

adjusted calcium (mmol/L) = measured calcium (mmol/L) + 0.02 (mean normal albumin - measured albumin [g/L])

Critical Values <7.0 mg/dL (SI: <1.75 mmol/L) may lead to tetany and in young children, seizures. Calcium >12.0 mg/dL (SI: >2.99 mmol/L) may induce coma, although most patients tolerate higher levels. More commonly, hypercalcemia leads to polyuria, anorexia, nausea, and constipation.

Possible Panic Range Possibly life-threatening levels: ≤6.0 mg/dL (SI: ≤1.50 mmol/L); **severe hypercalcemia** is defined as ≥14.0 mg/dL (SI: ≥3.5 mmol/L). Extremely high levels may be found with primary parathyroid carcinomas, in patients with malignancies and infants with Williams syndrome.

Use

Causes of High Calcium

- Primary hyperparathyroidism (HPT) - elevated ionized calcium and hypophosphatemia. Hyperparathyroidism may coexist with other endocrine tumors (multiple endocrine adenomatosis syndromes). Individuals with hypercalcemia, elevated intact parathyroid hormone concentration, and normal renal function, with few exceptions, have primary hyperparathyroidism. In a series of 60 cases of proven hyperparathyroidism, increased corrected calcium concentrations were found in only 78%, but 98.3% had elevated ionized calcium. The latter is the more useful test for hyperparathyroidism and hypoparathyroidism.
- Carcinoma, with or without bone metastases.
  - Humoral hypercalcemia of malignancy (HHM), tumor-induced hypercalcemia in patients without bone metastases, is seen especially in primary squamous cell carcinoma of lung, head and neck, but other primary sites include breast, kidney, liver, bladder, and ovary. HHM is usually caused by **parathyroid hormone-related protein** (see Parathyroid

Hormone-Related Protein, Serum *on page 1000*). Rarely, a benign tumor is associated with increased production of parathyroid hormone-related protein.

- **The chloride:phosphorus ratio** is predominantly of value when it is <29 mmol/L, to provide evidence **against** a diagnosis of primary hyperparathyroidism.

- Dehydration is a common cause of slight hypercalcemia.
- Granulomatous diseases, including sarcoidosis
- Chronic hypervitaminosis D; ectopic production of 1,25-dihydroxy vitamin D_3; vitamin A intoxication, isotretinoin (a vitamin A derivative).
- Prolonged immobilization (uncommon), in patient with increased bone turnover (eg, Paget disease of bone, malignancy, children).
- Milk-alkali syndrome: prolonged use of calcium-containing materials and alkali (eg, $CaCO_3$) or other absorbable alkali ulcer remedies with high milk intake (now rare).
- Idiopathic hypercalcemia of infancy, Williams syndrome
- Endocrine: hyperthyroidism; Addison disease; acromegaly; pheochromocytoma (rare cause of hypercalcemia); vasoactive intestinal polypeptide hormone-producing tumor
- Advanced chronic liver disease
- Bacteremia
- Twenty-four hour urinary calcium is increased in HPT, low in **familial hypocalciuric hypercalcemia** (FHH) which is characterized by hypercalcemia and hypocalciuria. An autosomal dominant, it apparently has no complications. Ratio of renal calcium clearance to creatinine clearance <0.01 suggests this genetic disease. The calcium:creatinine clearance ratio is said to discriminate between FHH and hyperparathyroidism.
- Aluminum-induced renal osteodystrophy
- Parenteral nutrition
- Renal insufficiency
- Postrenal transplant
- Drugs: calcium salts, lithium, thiazide/chlorthalidone therapy, other diuretics; antiestrogens and estrogens (rapid increase in patients with breast carcinoma).

Causes of Low Calcium
- Low albumin (*vide supra*: Reference Interval)
- High phosphorus: renal insufficiency, hypoparathyroidism, pseudohypoparathyroidism
- Vitamin D deficiency and resistance
- Osteomalacia (including Milkman fractures)
- Pseudovitamin D deficiency rickets
- Celiac disease and other malabsorption disorders
- Renal tubular acidosis
- Pancreatitis, acute
- Dilutional: Intravenous fluids
- Bacteremia
- Hypomagnesemia
- Anticonvulsants and other common drugs, most by *in vivo* action, can depress calcium. Barbiturates in elderly may cause calcium decrease; other drugs including calcitonin, corticosteroids, gastrin, glucagon, glucose, insulin, magnesium salts, methicillin, and tetracycline in pregnancy.
- A rare genetic disorder can cause hypocalemia due to an abnormality in the calcium-sensing receptor.

Limitations Gross hemolysis falsely elevates results. Patients receiving citrate-containing blood or blood fractions may have increased total serum calcium, despite decreased ionized calcium. In such patients, measurement of ionized calcium is indicated (see Calcium, Ionized, Serum *on page 328*).

Methodology Spectrophotometry; atomic absorption spectrometry (AA) is not used extensively, but remains the reference method.

References
Cundy T and Reid I, "Calcium, Phosphate and Magnesium," *Clinical Biochemistry*, Marshall WJ and Bangert SK, eds, New York, NY: Churchill Livingstone, 1995, 93-4.

Glendenning P, Gutteridge DH, Retallack RW, et al, "High Prevalence of Normal Total Calcium and Intact PTH in 60 Patients With Proven Primary Hyperparathyroidism: A Challenge to Current Diagnostic Criteria," *Aust N Z J Med*, 1998, 28(2):173-8.

(Continued)

Calcium, Serum (Continued)

Glendenning P, Gutteridge DH, and Retallack RW, "Treatment of Primary Hyperparathyroidism," *N Engl J Med*, 2000, 342(13):976-7.

Gunn IR and Wallace JR, "Urine Calcium and Serum Ionized Calcium, Total Calcium and Parathyroid Hormone Concentrations in the Diagnosis of Primary Hyperparathyroidism and Familial Benign Hypercalcaemia," *Ann Clin Biochem*, 1992, 29(Pt 1):52-8.

Irvin GL 3rd and Carneiro DM, "Management Changes in Primary Hyperparathyroidism," *JAMA*, 2000, 284(8):934-6.

Knecht TP, Behling CA, Burton DW, et al, "The Humoral Hypercalcemia of Benignancy. A Newly Appreciated Syndrome," *Am J Clin Pathol*, 1996, 105(4):487-92.

Marx SJ, "Hyperparathyroid and Hypoparathyroid Disorders," *N Engl J Med*, 2000, 343(25):1863-75.

Seymour JF, Gagel RF, Hagemeister FB, et al, "Calcitriol Production in Hypercalcemic and Normocalcemic Patients With non-Hodgkin Lymphoma," *Ann Intern Med*, 1994, 121(9):633-40.

♦ **Calcium Stimulation Test** see Calcitonin, Serum or Plasma *on page 327*

Calcium, Urine

Related Information

Calcium, Serum *on page 329*
Hydroxyproline, Total, Urine *on page 757*
Kidney Stone Analysis *on page 820*
Magnesium, Urine *on page 886*
Osteocalcin, Serum or Plasma *on page 983*
Parathyroid Hormone-Related Protein, Serum *on page 1000*
Parathyroid Hormone, Serum *on page 1001*
Phosphorus, Urine *on page 1032*
Uric Acid, Urine *on page 1287*
Urine Collection, 24-Hour *on page 1295*
Vitamin D, Serum *on page 1318*

Applies to Calcium:Creatinine Ratio

Abstract Urine calcium determinations are used to investigate calcium metabolism. For renal stone studies, see Kidney Stone Analysis *on page 820*.

Specimen 24-hour urine is preferred; random urine is acceptable for calcium:creatinine ratio calculations

Container Plastic urine container or acid-washed glass bottle

Reference Interval Varies with diet; based on average calcium intake of 600-800 mg/24 hours (SI: 15-20 mmol/day): excretion may be 100-250 mg/24 hours (SI: 2.5-6.2 mmol/day). On a diet of 400-800 mg/24 hours of calcium daily (SI: 10-20 mmol/day), others set the upper limit at 200 mg/24 hours of calcium (SI: 5 mmol/day) in a 24-hour urine collection. More than 4 mg/kg is associated with increased prevalence of stone formation. Low calcium diet: <150 mg/24 hours (SI: <3.7 mmol/day) excreted. High calcium diet: 250-300 mg/24 hours (SI: 6.2-7.5 mmol/day) excreted. Hypercalciuria is calcium excretion in excess of 250 mg/24 hours (SI: 6.2 mmol/day) for women, 300 mg/24 hours (SI: 7.5 mmol/day) for men. Calcium excretion, like other laboratory results, must be related to the individual patient. The rate of calcium excretion can also be expressed as a **calcium:creatinine ratio**. In healthy individuals with constant muscle mass, urinary calcium (mg/dL):creatinine (mg/dL) is <0.14 (SI: calcium (mmol/L):creatinine (mmol/L) is <0.40). Values >0.20 (mg/dL) or >0.57 (mmol/L units) suggest hypercalciuria.

In children younger than 6 years of age, urine calcium:creatinine ratio is inversely related to age with the upper limit of normal being 0.8 mg/mg (SI: 2.25) at age 1 month. See graphic.

Critical Values Urine calcium excretion >300 mg/24 hours (7.5 mmol/day) provides an indication for parathyroidectomy in the presence of other criteria for hyperparathyroidism.

Use Low in familial hypocalciuric hypercalcemia, for which urine calcium measurements are mandatory (see Calcium, Serum *on page 329*). Gitelman syndrome, low with thiazide diuretics, vitamin D deficiency, renal osteodystrophy, vitamin D resistant rickets, hypoparathyroidism, pseudohypoparathyroidism and pre-eclampsia.

High in 30% to 80% of instances of primary hyperparathyroidism, but urinary calcium excretion does not consistently, reliably distinguish hyperparathyroidism from other entities. High in about 50% of patients with sarcoidosis. Increased with immobilization, with steroid therapy, with Paget disease of bone, and in primary (idiopathic) hypercalciuria. Increased with entities causing high ultrafiltrable calcium: humoral hypercalcemia of malignancy, some cases of renal tubular acidosis, Fanconi syndrome, Bartter syndrome, increased calcium intake, vitamin D intoxication, hyperthyroidism, diabetes mellitus, acromegaly, glucocorticoid excess, some cases of Crohn disease and ulcerative colitis, myeloma, some instances of leukemia and lymphoma, and carcinoma metastatic to bone. Reported relationship to hematuria in children.

In the fasting state when intestinal and renal components are relatively constant, calcium excretion is used to assess the skeletal component of calcium metabolism. Values >0.16 mg (>0.04 mmol/L)/100 mL of glomerular filterate implies an increase in osteoclastic bone resorption. The following equation is used to calculate calcium excretion in urine.

UCa (mg/100 mL glomerular filterate) = [UCa (mg/dL) x serum creatinine (mg/dL)] / urinary creatinine (mg/dL)

Methodology Spectrophotometry, atomic absorption (AA) spectrometry

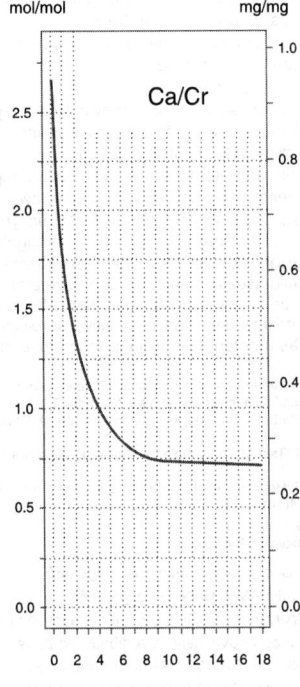

Estimated 95th percentiles for urinary Ca/Cr ratios in relation to age.

References

Endres DB and Rude RK, "Mineral and Bone Metabolism," *Tietz Textbook of Clinical Chemistry*, 3rd ed, Burtis CA and Ashwood ER, eds, Philadelphia, PA: WB Saunders Co, 1999, 1395-457.

Gunn IR and Wallace JR, "Urine Calcium and Serum Ionized Calcium, Total Calcium and Parathyroid Hormone Concentrations in the Diagnosis of Primary Hyperparathyroidism and Familial Benign Hypercalcaemia," *Ann Clin Biochem*, 1992, 29(Pt 1):52-8.

Irvin GL and Carneiro DM, "Management Changes in Primary Hyperparathyroidism," *JAMA*, 2000, 284(8):934-6.

Matos V, Van Melle G, Boulat O, et al, "Urinary Phosphate/Creatinine, Calcium/Creatinine and Magnesium/Creatinine Ratios in Healthy Pediatric Populations," *J Pediatr*, 1997, 131(2):252-7.

Osorio AV and Alon US, "The Relationship Between Urinary Calcium, Sodium, and Potassium Excretion and the Role of Potassium in Treating Idiopathic Hypercalcinuria," *Pediatrics*, 1997, 100(4):675-81.

(Continued)

Calcium, Urine *(Continued)*

Young DS, *Effects of Disease on Clinical Laboratory Tests*, 5th ed, Volume 1: Listing by Test, Washington, DC: AACC Press, American Association of Clinical Chemistry, 2000, Section 3, 149-59.

♦ **Calculus Analysis** *see* Kidney Stone Analysis *on page 820*
♦ **California** *see* California Encephalitis Virus Serology *on page 334*

California Encephalitis Virus Serology
Related Information
Bacterial Culture, Cerebrospinal Fluid *on page 236*
Cerebrospinal Fluid Analysis: Overview *on page 355*
Eastern Equine Encephalitis Virus Serology *on page 529*
Encephalitis Viral Serology *on page 535*
St Louis Encephalitis Virus Serology *on page 1224*
Viral Culture *on page 1307*
Western Equine Encephalitis Virus Serology *on page 1328*

Applies to Bunya Virus; California; Encephalitis Virus; Jamestown Canyon Virus; LaCrosse Virus; Snowshoe Hare Virus

Abstract These agents are arboviruses causing endemic disease.

Specimen Serum, cerebrospinal fluid

Container Red top tube, sterile container

Collection Acute and convalescent sera drawn 10-14 days apart are required.

Reference Interval Less than fourfold increase in titer in paired sera; IgM in CSF is considered diagnostic of CNS infection.

Use Support diagnosis of California encephalitis serogroup virus infection

Limitations Complement fixing antibodies appear slowly

Methodology Indirect fluorescent antibody (IFA), enzyme-linked immunosorbent assay (ELISA), complement fixation (CF), hemagglutination

Additional Information Infections are often asymptomatic. Symptomatic individuals develop a mild aseptic meningitis or a severe encephalitis. Most of the acute illnesses occur in children younger than 15 years of age. Diagnosis depends primarily on the detection of positive serology. Recent studies are investigating PCR as a potential technique to improve the diagnosis.

References
Huang C, Campbell WP, Grady L, et al, " Diagnosis of Jamestown Canyon Encephalitis by Polymerase Chain Reaction," *Clin Infect Dis*, 1999, 28(6):1294-7.
Lundstrom JO, "Mosquito-Borne Viruses in Western Europe: A Review," *J Vector Ecol*, 1999, 24(1):1-39.
McJunkin JE, Khan RR, Tsai TF, "California-La Crosse Encephalitis," *Infect Dis Clin North Am*, 1998, 12(1):83-93.
Rust RS, Thompson WH, Matthews, et al, "La Crosse and Other Forms of California Encephalitis," *J Child Neurol*, 1999, 14(1):1-14.

Internet Web Sites
www.cdc.gov/ncidod/dvbid/arbor/lacfact.htm

♦ **cAMP, Plasma** *see* Cyclic AMP, Plasma *on page 486*
♦ **cAMP, Urine** *see* Cyclic AMP, Urine *on page 486*
♦ **Canale-Smith Syndrome** *see* CD4/CD8 Enumeration *on page 349*
♦ **C-ANCA** *see* Antineutrophil Cytoplasmic Antibody *on page 187*
♦ **Cancer Antigen 15-3** *see* CA 15-3, Serum *on page 320*
♦ **Cancer Antigen 19-9** *see* CA 19-9, Serum *on page 321*
♦ **Cancer Antigen 27.29** *see* CA 27.29, Serum *on page 322*
♦ **Cancer Antigen 125** *see* CA 125, Serum *on page 323*

Candida Antigen Detection and Serology
Related Information
Fungal Culture, Biopsy or Body Fluid *on page 619*
Fungal Culture, Blood *on page 620*

Test Includes Antigen detection is performed with a latex agglutination test. Serology includes a precipitin test by agar gel diffusion.

Abstract Numerous reports have studied the diagnosis of disseminated candidiasis by detection of *Candida* antigens or antibody in serum. Controversies and problems with such diagnostic testing remain; therefore, antigen and serology testing should not be used alone to confirm or rule out disseminated candidiasis.

Specimen Serum

Container Red top tube

Storage Instructions Separate and refrigerate serum at 4°C.

Reference Interval Negative for antigen and/or antibody; a fourfold increase in antibody titer in paired sera drawn 10-14 days apart is usually indicative of acute infection.

Use Diagnosis of candidiasis in immunocompromised patients by detection of *Candida* antigens in serum

Limitations *Candida* antigen tests are insensitive, with a high incidence of false-negative as well as false-positive results. Evaluation of these assays has been hampered by difficulty in establishing the diagnosis of disseminated candidiasis by other clinical and laboratory methods.

Cross-reactive antibody with the latex agglutination test occur in cases of cryptococcosis and tuberculosis. Negative results do not rule out candidiasis. This test is difficult to interpret because precipitins are found in 20% to 30% of the normal population. Very severe cases of vaginitis or mucocutaneous candidiasis can produce positive results. Clinical correlation is needed. The decision to initiate therapy should not be based only on results of such assays.

Methodology Antigen assays: latex agglutination (LA), enzyme-linked immunosorbent assay (ELISA); serology assays: latex agglutination (LA), crossed electrophoresis, immunodiffusion (ID), enzyme-linked immunosorbent assay (ELISA)

Additional Information Detection of disseminated candidiasis is particularly important in immunocompromised patients. Unfortunately, the sensitivity of *Candida* antigen tests is too low to rule out candidiasis. Consequently, a negative result does not preclude the use of empiric antifungal therapy. A positive test is usually reliable but many of these patients also have positive blood cultures. The decision to initiate therapy should not be based only on results of *Candida* serologic assays.

In general, *Candida*-specific antigen tests have limited use in diagnosis. Quantitative tests on sera taken at biweekly intervals are of value in monitoring the progress of infection before and after therapy.

References

Elsayed S, Fitzgerald V, Massey V, et al, "Evaluation of the Candigen Enzyme-Linked Immunosorbent Assay for Quantitative Detection of *Candida* Species Antigen," *Arch Pathol Lab Med*, 2001, 125(3):344-6.

Iwasaki H, Misaki H, Nakamura T, et al, "Surveillance of the Serum *Candida* Antigen Titer for Initiation of Antifungal Therapy After Postremission Chemotherapy in Patients With Acute Leukemia," *Int J Hematol*, 2000, 71(3):266-72.

Knoke M, Bernhardt H, Schulz K, et al, "Funguria and *Candida*-Specific Immunoglobulins in Patients With Systemic Candidiasis," *Mycoses*, 2000, 43(3-4):145-9.

Reiss E and Morrison CJ, "Nonculture Methods for Diagnosis of Disseminated Candidiasis," *Clin Microbiol Rev*, 1993, 6(4):311-23.

Richardson MD and Kokki MH, "New Perspectives in the Diagnosis of Systemic Fungal Infections," *Ann Med*, 1999, 31(5):327-35.

Internet Web Sites

www.cdc.gov/ncidod/dbmd/diseaseinfo/candidiasis_g.htm

Cannabinoids (Marijuana Metabolites), Qualitative, Urine

Related Information

Ethanol, Blood, Urine, and Other Sources *on page 558*

Synonyms Bhang; Cannabis; Carboxy THC; Ganja; Hashish; Hemp; Marijuana; 11-Nor-9-Carboxy-Delta-9-Tetrahydrocannabinol; Pot

Abstract Marijuana is the most common illicit drug used by children and adolescents in the United States. The main active ingredient of marijuana (cannabinoids) is tetrahydrocannabinol (THC). Its behavioral effects include feelings of euphoria, relaxation, altered time perception, lack of concentration, impaired memory, and paranoia. Its major metabolite is 11-carboxy-THC, which is excreted in the urine. Immunoassays are designed to detect this metabolite. In chronic users, the metabolite is detected in urine for several weeks.

Specimen Random urine

Collection For employee screening or forensic purpose, use precautions during collection (see the Introduction *on page 63*).

Causes for Rejection Evidence of urine dilution or alteration

(Continued)

Cannabinoids (Marijuana Metabolites), Qualitative, Urine *(Continued)*

Special Instructions If forensic, use chain-of-custody protocol and form. See Chain-of-Custody Protocol *on page 381* and the Chain-of-Custody form in the Introduction *on page 63.*

Reference Interval Negative (less than cutoff)

Critical Values Substance Abuse and Mental Health Services Administration (SAMHSA) screening cutoff: 50 ng/mL (20 ng/mL in some laboratories); confirmation cutoff: 15 ng/mL for 11-carboxy-THC

Use Drug abuse evaluation, toxicity assessment

Methodology Screen: immunoassay; Confirmation: gas chromatography/mass spectrometry (GC/MS)

Additional Information
- Half-life: 20-40 hours
- Volume of distribution: 4-19 L/kg
- Protein binding: 97%

THC is highly lipophilic and is stored in body fat for a long period of time. It is then released from storage sites slowly over time making drug detection possible for up to 6 weeks in chronic users.

Driving experiments show that marijuana affects a wide range of skills needed for safe driving. Thinking and reflexes are slowed, making it difficult for drivers to respond to sudden unexpected events.

When conventional antiemetic agents fail, the use of tetrahydrocannabinol (THC) to relieve nausea and vomiting associated with cancer chemotherapy has been described. Other uses include control of intraocular pressure, reduction of muscle spasms, and relief from chronic pain. Though available as pure THC, proponents of marijuana use insist that only smoked marijuana leaves are effective. Use of marijuana for medicinal use is currently highly debated.

References
DuPont RL, "Examining the Debate on the Use of Medical Marijuana," *Proc Assoc Am Physicians*, 1999, 111(2):166-72.

Gruber AJ and Pope HG, "Marijuana Use Among Adolescents," *Clin North Am*, 2002, 49(2):389-413.

Martin BR, "Medical Marijuana - Moving Beyond the Smoke," *Lancet*, 2002, 360(9326):4-5.

Rouse BA, "Epidemiology of Illicit and Abused Drugs in the General Population, Emergency Department Drug-Related Episodes, and Arrestees," *Clin Chem*, 1996, 42(8 Pt 2):1330-6.

- **Cannabis** *see* Cannabinoids (Marijuana Metabolites), Qualitative, Urine *on page 335*
- **Capillary Blood Collection** *see* Skin Puncture Blood Collection *on page 1203*
- **Capillary Blood Gases** *see* Blood Gases and pH, Capillary *on page 277*
- **Capillary Electrophoresis** *see* Protein Electrophoresis, Capillary Zone *on page 1103*
- **Capnocytophaga canimorsus** *see* Buffy Coat Smear Study of Peripheral Blood *on page 316*
- **Capnocytophaga canimorsus** *see* Peripheral Blood: Differential Leukocyte Count *on page 1010*
- **Captopril Test** *see* Renin Activity, Plasma *on page 1149*
- **Carbamate Toxicity** *see* Acetylcholinesterase, Red Cell and Serum *on page 93*
- **Carbamate Toxicity** *see* Organophosphate Pesticides, Urine, Blood, or Serum *on page 975*
- **Carbamazepine** *see* Phenytoin, Serum or Plasma *on page 1026*

Carbamazepine-10,11-Epoxide, Serum

Related Information

Carbamazepine, Serum *on page 337*

Synonyms Carbamazepine Metabolite

Test Includes Carbamazepine and carbamazepine-10,11-epoxide

Abstract Carbamazepine-10,11-epoxide is the active metabolite of carbamazepine. Occasional cases of carbamazepine toxicity occur with normal levels of carbamazepine due to accumulation of 10,11-epoxide.

Specimen Serum

Container Red top tube

Reference Interval 0.8-3.2 µg/mL. High level: In patients on chronic carbamazepine therapy, the addition of valpromide or progabide produces clinical toxicity with high levels of metabolite and normal levels of parent compound.

Methodology High performance liquid chromatography (HPLC), fluorescence polarization immunoassay (FPIA)

Additional Information Carbamazepine-induced thrombocytopenia is probably due to carbamazepine-10,11-epoxide. Valproic acid, an anticonvulsant chemically related to valpromide, may increase the epoxide:carbamazepine ratio by eliminating excretion of the epoxide. Since most cases of fatal valproate hepatotoxicity occur in young children on multiple anticonvulsants, the combination of valproate and carbamazepine is not recommended in the susceptible population. Phenytoin may also increase the ratio of epoxide:parent compound.

References

Divanoglou D, Orologas A, Iliadis S, et al, "Pharmacokinetic Behaviour of Carbamazepine and its Main Metabolite-10,11 Epoxide of Carbamazepine in Monotherapy or in Combination With Other Antiepileptic Drugs," *Eur J Neurol*, 1998, 5(4):397-400.

Potter JM and Donnelly A, "Carbamazepine-10,11-Epoxide in Therapeutic Drug Monitoring," *Ther Drug Monit*, 1998, 20(6):652-7.

Shimoyama R, Ohkubo T, and Sugawara K, "Monitoring of Carbamazepine and Carbamazepine 10,11-Epoxide in Breast Milk and Plasma by High Performance Liquid Chromatography," *Ann Clin Biochem*, 2000, 37(Pt 2):210-5.

♦ **Carbamazepine Metabolite** *see* Carbamazepine-10,11-Epoxide, Serum *on page 336*

Carbamazepine, Serum

Related Information

Carbamazepine-10,11-Epoxide, Serum *on page 336*
Phenytoin, Serum or Plasma *on page 1026*
Valproic Acid, Serum or Plasma *on page 1297*
Verapamil, Serum or Plasma *on page 1306*

Synonyms Carbamazepinum; Carbategretal®; Carbatrol®; Carbazep®; CBZ; Epitrol®; Tegretol®; Tegretol® XR

Applies to P450 System

Abstract Carbamazepine is a first-line antiepileptic drug for generalized and partial seizures. It is also used for control of neurogenic pain from trigeminal neuralgia and diabetic neuropathy. It has been successfully used in the treatment of bipolar disease and other psychiatric and neurologic illnesses. Other investigational uses include ethanol withdrawal, restless leg syndrome, psychotic behavior associated with dementia and post-traumatic stress. It has a distinctive pharmacokinetic property of inducing the hepatic enzymes responsible for increase in its own clearance, called "autoinduction."

Patient Preparation Levels should be drawn before next oral dose with patient at steady-state.

Specimen Serum

Container Red top tube

Sampling Time At steady-state concentration. Time to reach steady-state is 3-8 days.

Reference Interval
- Total: 8-12 µg/mL (SI: 34-51 µmol/L); with other anticonvulsants: 4-8 µg/mL
- Free: 0.5-4.0 (SI: 2-17 µmol/L)
- Toxic concentration: total: >15 µg/mL (SI: >64 µmol/L); free: >4 µg/mL (SI: 17 µmol/L)

Use Monitor for compliance, efficacy, or possible toxicity

Limitations See Carbamazepine-10,11-Epoxide, Serum *on page 336*.

Contraindications Carbamazepine should not be used in patients with a history of previous bone marrow depression or known sensitivity to any of the tricyclic compounds, such as amitriptyline, desipramine, imipramine, protriptyline, and nortriptyline. Half-life of warfarin is shortened; coadministration of monoamine oxidase inhibitors is not recommended. It should not be used in most circumstances if the absolute neutrophil count (ANC) is <1500. Caution should be used in pregnancy as it crosses the placenta and may harm the fetus. The elderly may have increased risk of SIADH-like syndrome.
(Continued)

Carbamazepine, Serum *(Continued)*

Methodology Immunoassay, gas chromatography (GC), high performance liquid chromatography (HPLC). Immunoassays estimate both carbamazepine and carbamazepine 10,11-epoxide (an active metabolite).

Additional Information
- Half-life: 15-40 hours
- Volume of distribution: 0.8-1.8 L/kg
- Protein binding: 60% to 80%

Carbamazepine is only commercially available in oral formulations. It is absorbed slowly and is about 80% bioavailable. Plasma concentrations peak at about 6 hours after an oral dose. Carbamazepine is totally cleared hepatically and has one active metabolite, the 10,11-epoxide.

Low level: The most common cause of a low level is noncompliance. The addition of anticonvulsants which induce the P450 system, such as phenytoin, primidone, and phenobarbital, may decrease carbamazepine levels without causing seizures. Because of autoinduction of metabolism, patients in the first 2 months of therapy may have diminishing levels and be at ongoing risk for seizures. **High level:** Drugs which inhibit the P450 system, including isoniazid, fluoxetine, propoxyphene, quetiapine, verapamil, and stiripentol can cause a precipitous rise in carbamazepine levels and clinical toxicity, usually within 48 hours.

In uremia, hepatic dysfunction, and other conditions which affect binding of carbamazepine with albumin, the free carbamazepine concentrations are more useful for dosage adjustment than are total levels.

Patients on carbamazepine are at 5-8 times greater risk of developing aplastic anemia and agranulocytosis than the general population. Leukopenia is the most common hematologic side effect, with incidences as high as 10% being reported. It may be dose-related. In most patients, leukopenia is transient. Nonetheless complete CBC at the baseline should be obtained. If the patient exhibits low WBC or platelet count, he/she should be monitored closely and discontinuation of the drug should be considered if significant bone marrow depression develops.

References

Leucht S, McGrath J, White P, et al, "Carbamazepine Augmentation for Schizophrenia: How Good Is the Evidence?" *J Clin Psychiatry*, 2002, 63(3):218-24.

Lifshitz M, Gavrilov V, and Sofer S, "Signs and Symptoms of Carbamazepine Overdose in Young Children," *Pediatr Emerg Care*, 2000, 16(1):26-7.

Liu H and Delgado MR, "A Comprehensive Study of the Relation Between Serum Concentrations, Concentration Ratios, and Level/Dose Ratios of Carbamazepine and Its Metabolite With Age, Weight, Dose, and Clearances in Epileptic Children," *Epilepsia*, 1994, 35(6):1221-9.

Shen S, Elin RJ, and Soldin SJ, "Characterization of Cross Reactivity by Carbamazepine 10,11-Epoxide With Carbamazepine Assays," *Clin Biochem*, 2001, 34(2):157-8.

Sirven JI, "Antiepileptic Drug Therapy for Adults: When to Initiate and How to Choose," *Mayo Clin Proc*, 2002, 77(12):1367-75.

- **Carbamazepinum** *see Carbamazepine, Serum on page 337*
- **Carbategretal®** *see Carbamazepine, Serum on page 337*
- **Carbatrol®** *see Carbamazepine, Serum on page 337*
- **Carbazep®** *see Carbamazepine, Serum on page 337*
- **Carbohydrate Antigen 15-3** *see CA 15-3, Serum on page 320*
- **Carbohydrate Antigen 19-9** *see CA 19-9, Serum on page 321*
- **Carbohydrate Antigen 125** *see CA 125, Serum on page 323*
- **Carbohydrate-Deficient Glycoprotein Syndrome** *see Carbohydrate-Deficient Transferrin, Serum on page 338*

Carbohydrate-Deficient Transferrin, Serum

Related Information

Alanine Aminotransferase, Serum *on page 116*
Alkaline Phosphatase, Serum *on page 127*
Aspartate Aminotransferase, Serum *on page 216*
Bilirubin, Total, Serum *on page 265*
Ethanol, Blood, Urine, and Other Sources *on page 558*
Folic Acid, RBC *on page 606*
Folic Acid, Serum *on page 606*
Gamma-Glutamyl Transferase, Serum *on page 629*

Liver Biopsy *on page 864*
Liver Disease: Laboratory Assessment, Overview *on page 869*
Osmolality, Serum *on page 978*

Applies to Carbohydrate-Deficient Glycoprotein Syndrome; Congenital Disorders of Glycosylation

Abstract Carbohydrate-deficient transferrin (CDT) is increased in alcoholics, and in patients with a group of autosomal recessive disorders of infancy and childhood, until recently known as carbohydrate-deficient glycoprotein syndrome. Presently designated "congenital disorders of glycosylation" (CDG), there are variable phenotypes. Its characteristics include mental retardation, cerebellar hypoplasia, liver dysfunction, or stroke-like occurrences.

Specimen Serum

Container Red top tube or SST™ tube

Storage Instructions Freeze immediately; ship on dry ice.

Reference Interval Reference intervals vary according to different analytic specificities and various recovery rates of CDT assays. Values are reported either in relative (ie, percentage of total transferrin) or absolute (ie, units/L) terms.

Relative: <2.5%
Absolute:
- children: <30 units/L
- adult male: <20 units/L, <27 units/L
- adult female: <26 units/L, <35 units/L

Values for CDG evaluation are:
- mono-oligosaccharide/di-oligosaccharide: <0.074 units/L
- indeterminate: 0.075-0.109 units/L
- α-oligosaccharide/di-oligosaccharide: <0.022 units/L

Use CDT is used as a marker for excessive alcohol consumption, and for monitoring relapse or progressive abstinence in alcoholics. It is also used to diagnose carbohydrate-deficient glycoprotein syndrome, a central nervous system developmental disorder characterized by muscular weakness and hypotonia.

Limitations False-negative and false-positive CDT results occur in diagnosis of alcohol abuse; thus, the test is not recommended for screening in a general population. CDT may be useful in case findings and appears to be the best single marker for alcoholism. The sensitivity and/or specificity may be affected by female sex, smoking, body mass index, diastolic hypertension, iron stores, and other factors. It is especially insensitive among women who abuse alcohol and is relatively insensitive among those without liver damage. False positives (relevant to alcohol consumption) occur in cases of nonalcoholic liver disease, women on estrogen replacement, and recipients of combined pancreas and kidney transplantation. Sensitivity of 58% and 70% and specificity 82% to 98% for detection of alcohol abuse are reported.

Additional Information The combined use of CDT with gamma-glutamyl transpeptidase (GGT) and mean corpuscular volume (MCV) raises sensitivity above 90% for alcoholism. Total bilirubin, AST:ALT ratio, and alkaline phosphatase are all relevant as well. During abstinence, serum CDT exhibits a half-life of ~2 weeks.

References

Dibbelt L, "Does Trisialo-Transferrin Provide Valuable Information for the Laboratory Diagnosis of Chronically Increased Alcohol Consumption by Determination of Carbohydrate-Deficient Transferrin?" *Clin Chem*, 2000, 46(8 Pt 1):1203-5.

Laposata M, "Assessment of Ethanol Intake," *Am J Clin Pathol*, 1999, 112(4):443-50.

Mayo Reference Services Publication, "Carbohydrate Deficient Transferrin, Serum," *New Test Announcement*, Rochester, MN: Mayo Medical Laboratories, 2000.

♦ **Carbolith®** *see* Lithium, Serum *on page 863*
♦ **Carbon Dioxide** *see* Bicarbonate, Blood *on page 258*

Carbon Dioxide, Total, Blood
Related Information

Bicarbonate, Blood *on page 258*
Blood Gases and pH, Arterial *on page 275*
Blood Gases and pH, Capillary *on page 277*
Blood Gases and pH, Venous *on page 279*
(Continued)

Carbon Dioxide, Total, Blood *(Continued)*

Chloride, Serum, Plasma, or Blood *on page 391*
Electrolyte Panel, Serum *on page 532*
pCO_2, Blood *on page 1008*
pH, Blood *on page 1018*

Synonyms CO_2; CO_2 Content; $ctCO_2$; TCO_2

Applies to Acid-Base Status Evaluation; Bicarbonate; Oximeters, Pulse; Tidal Volume

Abstract Total carbon dioxide (TCO_2) consists primarily of bicarbonate (HCO_3^-), plus small to trace amounts of carbonic acid (H_2CO_3), dissolved CO_2, carbonate ions (CO_3^{2-}), and CO_2 bound amino groups (carbamino compounds). Laboratory measurement of TCO_2 is done to estimate the HCO_3^- concentration, which usually makes up all but ≤2 mmol/L of the TCO_2 concentration. The two terms, TCO_2 and HCO_3^-, are used interchangeably in routine practice. (See Bicarbonate, Blood *on page 258*.)

Specimen Whole blood; plasma or serum

Container Red top tube or green top (heparin) tube

Collection Specimen should be kept tightly closed, as CO_2 will diffuse out, causing erroneously low values. This loss may amount to 6 mmol/hour. Anaerobic conditions are best.

Reference Interval Whole blood - infancy to 2 years: 18-28 mmol/L; 2 years and older: arterial: 23-29 mmol/L; venous: 22-26 mmol/L. Plasma or serum - venous: 23-29 mmol/L.

Possible Panic Range <15 mmol/L, >50 mmol/L

Use TCO_2 is used in the evaluation of acid-base and electrolyte status. High results may represent respiratory acidosis with CO_2 retention (eg, advanced pulmonary emphysema) or metabolic alkalosis (eg, prolonged vomiting). Low values may indicate respiratory alkalosis as in hyperventilation or metabolic acidosis (eg, diabetes with ketoacidosis). Hypercapnia may go unrecognized despite adequate arterial oxygen saturation measured by pulse oximetry.

Limitations Interpretation requires clinical information and evaluation of other laboratory data (eg, electrolytes, blood gases, glucose). Organic acids and nitrate have been reported to interfere (positively) in one TCO_2 (direct ion-specific electrode) method.

Methodology Colorimetry, enzyme assay, pCO_2 electrode ("indirect" electrode assay), or TCO_2 electrode ("direct" ISE). TCO_2 can be calculated from blood gas measurements.

Additional Information Hypercapnia from impaired elimination of CO_2 is due to abnormalities in mechanisms controlling respiratory drive, the muscles of respiration, and the function of the lung. Elimination of CO_2 from the lung involves alveolar ventilation but not dead-space ventilation. Partitioning of these spaces is expressed as a ratio between dead space and total volume per breath: the tidal volume. The tidal volume normally is <0.30. These and other aspects of pulmonary gas exchange, ventilation and their consequences are addressed as the partial pressure of arterial carbon dioxide, $PaCO_2$, a part of arterial blood gases.

References

McLain BI, Evans J, Dear PR, et al, "Comparison of Capillary and Arterial Blood Gas Measurements in Neonates," *Arch Dis Child*, 1988, 63(7 Spec No):743-7.

O'Leary TD and Langton SR, "Calculated Bicarbonate or Total Carbon Dioxide?" *Clin Chem*, 1989, 35(8):1697-700.

Scott MG, Heusel JW, LeGrys VA, et al, "Electrolytes and Blood Gases," *Tietz Textbook of Clinical Chemistry*, 3rd ed, Chapter 31, Burtis CA and Ashwood ER, eds, Philadelphia, PA: WB Saunders Co, 1999, 1056-92.

Weinberger SE, Schwartzstein RM, and Weiss JW, "Hypercapnia," *N Engl J Med*, 1989, 321(18):1223-31.

Zenger M, Brenner M, Hua P, et al, "Measuring Oxygen Uptake and Carbon Dioxide Production in Critically Ill Patients Using a Standard Blood Gas Analyzer," *Crit Care Med*, 1994, 22(5):783-8.

♦ **Carbon Monoxide** *see* Carboxyhemoglobin, Blood *on page 340*

♦ **Carboxyhemoglobin** *see* Methemoglobin, Whole Blood *on page 903*

Carboxyhemoglobin, Blood

Related Information

Blood Gases and pH, Arterial *on page 275*
Cotinine, Serum, Plasma, or Urine *on page 464*

Cyanide, Blood *on page 485*
Methemoglobin, Whole Blood *on page 903*
Nicotine, Serum or Plasma *on page 961*

Synonyms Carbon Monoxide; CO; COHb

Applies to Methylene Chloride; Oximeters, Pulse

Test Includes COHb is sometimes included in Blood Gases, but may be ordered as a separate test.

Abstract A byproduct of incomplete combustion of hydrocarbons, carbon monoxide (CO) is a colorless, tasteless, and odorless gas. It binds tightly to hemoglobin (Hb) to form COHb, reducing oxygen-carrying capacity of blood. The affinity of Hb for CO is 200-250 times that of oxygen. It also binds to myoglobin and cytochrome oxidase. Carbon monoxide poisoning is seen from smoke inhalation, suicide attempt and accidental exposure. It is the most common cause of poisoning death in the U.S., accounting for thousands of emergency department visits and some 800 deaths annually. This test measures hemoglobin-bound carbon monoxide.

Patient Preparation In suspected carbon monoxide poisoning, the specimen should be collected immediately.

Specimen Whole blood, venous or arterial

Container Green top (heparin) tube or lavender top (EDTA) tube, depending upon laboratory methods

Sampling Time Draw before the patient is started on oxygen, if possible.

Collection Keep tube capped

Storage Instructions Refrigerate immediately after collection. Do not remove cap. Carboxyhemoglobin is stable 4 months in a filled, well-capped tube.

Reference Interval:
- **Nonsmoker:** <3%
- **Smoker:** 1-2 packs/day: 4% to 5%, >2 packs/day: 8% to 10%

Carboxyhemoglobin in the **newborn** may run to 10% to 12%. Carbon monoxide is a metabolic product of hemoglobin catabolism. The increased turnover of hemoglobin in the newborn together with decreased efficiency of the infant's respiratory system may and does lead to higher levels of carboxyhemoglobin.

Critical Values Exposure to CO concentrations 80-140 ppm for 1-2 hours can lead to COHb results of 3% to 6%; in some patients, even these levels can precipitate angina and cardiac arrhythmias. Healthy individuals can tolerate levels of 10% without any symptoms. Toxic concentration is 20%; lethal is >50%.

Possible Panic Range Disturbance of judgment, headache, and dizziness occur at 10% to 30%; coma at 50% to 60%; **fatality** occurs at 30% to 60% or more, and rapid death at level of 80%.

Use Determine the extent of carbon monoxide poisoning

Limitations CO binds to cytochrome oxidase, interfering with cellular respiration. Thus, although COHb assays provide information on exposure to CO, they do not always consistently correlate with symptoms or prognosis.

Methodology Spectrophotometry; gas chromatography is used to measure CO.

Additional Information A danger of missed diagnosis of CO intoxication is continued exposure of the patient and others to a toxic environment. The cherry red color of CO poisoning is not consistently seen. CO intoxication may contribute to the risk of myocardial infarction. The half-life of carboxyhemoglobin at room air is ~6 hours. The half-life with 100% O_2 administration, at atmospheric pressure, is 80 minutes. With O_2 at three atmospheres, the half-life is 24 minutes. Use of hyperbaric oxygen in the treatment of carbon monoxide poisoning is more effective than normobaric oxygen therapy (Weaver et al 2002).

References
Ernst A and Zibrak JD, "Carbon Monoxide Poisoning," *N Engl J Med*, 1998, 339(22):1603-8.
Varon J, Marik PE, Fromm RE Jr, et al, "Carbon Monoxide Poisoning: A Review for Clinicians," *J Emerg Med*, 1999, 17(1):87-93.
Weaver LK, "Carbon Monoxide Poisoning," *Crit Care Clin*, 1999, 15(2):297-317.
Weaver LK, Hopkins RO, Chan KJ, et al, "Hyperbaric Oxygen for Acute Carbon Monoxide Poisoning," *N Engl J Med*, 2002, 347(14):1057-67.

♦ **Carboxy THC** *see* Cannabinoids (Marijuana Metabolites), Qualitative, Urine *on page 335*

Carcinoembryonic Antigen, Serum

Related Information

Alpha₁-Fetoprotein, Serum *on page 136*
Body Cavity Fluid Cytology *on page 285*
Body Fluid Analysis, Cell Count *on page 288*
Body Fluid Chemical Analysis *on page 291*
Body Fluid Glucose *on page 294*
Breast Biopsy *on page 305*
CA 15-3, Serum *on page 320*
CA 19-9, Serum *on page 321*
CA 27.29, Serum *on page 322*
CA 125, Serum *on page 323*
Colon Cancer, Hereditary Nonpolyposis Type *on page 432*
Cyst Fluid Cytology *on page 490*
Fine Needle Aspiration, Deep and Superficial Masses *on page 590*
Heterophilic Antibodies *on page 727*
Homovanillic Acid, Urine *on page 743*
Immunoperoxidase Procedures *on page 780*
Liver Disease: Laboratory Assessment, Overview *on page 869*
Occult Blood, Stool *on page 969*

Synonyms CEA

Abstract CEA, an oncofetal glycoprotein antigen, is not organ specific. Abnormalities are found in a range of tumor types, but application is widest with adenocarcinomas of the gastrointestinal tract, especially colorectal tumors.

CEA may be used as an adjunct to staging and as a monitor. Increasing concentrations of CEA may be the first evidence of postoperative tumor progression. Survival rates as high as 20% to 30% in selected patients undergoing resection for isolated hepatic or pulmonary metastases have been described.

CEA is a useful reagent in immunocytochemistry as well.

Specimen Serum or plasma, usually serum; effusion fluid; bronchoalveolar lavage; cerebrospinal fluid

Container Plain red top tube, SST™ tube, or lavender top (EDTA) tube; avoid heparin anticoagulant

Sampling Time Preoperative; ~4 weeks postoperative and subsequently, every 2-3 months in patients with stage II or III disease for 2 years or longer following diagnosis

Storage Instructions Separate serum (or plasma) from cells and refrigerate if assayed within 24 hours. For longer storage, freeze at -20°C.

Special Instructions Transport to reference laboratory on dry ice.

Reference Interval Adults: nonsmoker: <2.5 ng/mL (SI: <2.5 µg/L), smoker: ≤5.0 ng/mL (SI: ≤5.0 µg/L); method-dependent

Use CEA testing has been widely used in the staging and monitoring of patients with **colorectal carcinoma**, but the actual clinical effectiveness is controversial. CEA is not effective as a screen for colorectal carcinoma (or any other tumor). Ahmad et al report that approximately 5% of patients with a Dukes A primary have an elevated CEA, and 25% of patients with Dukes B; even at stage D, the positivity is 65% to 90%. CEA is used for postoperative surveillance; when positive, a CT scan is indicated. Approximately 65% to 79% of patients with recurrent colorectal carcinoma have an elevated CEA at the time of, or before, the recurrence is detected by other means.

According to the recommendations of the American Society of Clinical Oncology (ASCO), CEA may be useful in the following situations in patients with colorectal carcinoma:

- preoperatively, if the result will assist in surgical management or staging
- Postoperatively (every 2-3 months in patients with stage II or III disease) in patients for whom liver resection would be clinically indicated.

ASCO recommendations point out that a preoperative CEA >5 mg/mL is not sufficient to institute adjuvant therapy. In addition, an isolated elevated postoperative value is **not** a sufficient finding to institute adjuvant or systemic therapy.

CEA testing for other carcinomas is controversial. Serum CEA concentrations may be more sensitive for metastases in bone than are bone scans. CEA may also detect other primary carcinomas of entodermal origin, including stomach and pancreas. Significant elevations may be found with primaries of breast and lung. High CEA occurs with medullary carcinoma of thyroid.

Limitations CEA levels are modestly elevated in smokers; patients with inflammation including infections, peptic ulcer, some instances of liver disease, inflammatory bowel disease, and pancreatitis; some patients with hypothyroidism and cirrhosis may have increased concentrations. CA 15-3 is a better marker than CEA in breast cancer. Many negatives occur in patients with early carcinoma. Doubtful cost-effectiveness for many patients. According to the work of Safi et al, approximately 30% of all recurrent colorectal carcinomas do not produce CEA. Increased CEA concentrations are found only in 10% to 20% of subjects with surgically resectable gastric carcinoma.

Contraindications CEA is **not** a screening test for occult cancer.

Additional Information Progressive elevations of CEA may herald tumor recurrence 3-36 months before clinical evidence of metastases. Increase bears an implication of treatment failure or recurrence and may be a signal for second look procedures.

CEA may be measured in effusion fluids. False positives are reported with empyema and complicated parapneumonic effusions.

References

Ahmad S and Ross MI, "Tumor Markers," *Surgery: Basic Science and Clinical Evidence*, Norton JA, et al, eds, New York, NY: Springer-Verlag, 2001, 1619-39.

American Society of Clinical Oncology, "Clinical Practice Guidelines for the Use of Tumor Markers in Breast and Colorectal Cancer," *J Clin Oncol*, 1996, 14(10):2843-77.

Bakalakos EA, Burak WE, Young DC, et al, "Is Carcinoembryonic Antigen Useful in the Follow-up Management of Patients With Colorectal Liver Metastases?" *Am J Surg*, 1999, 177(1):2-6.

Chapman MA, Buckley D, Henson DB, et al, "Preoperative Carcinoembryonic Antigen Is Related to Tumour Stage and Long-Term Survival in Colorectal Cancer," *Br J Cancer*, 1998, 78(10):1346-9.

Compton CC, Fielding LP, Burgart LJ, et al, "Prognostic Factors in Colorectal Cancer. College of American Pathologists Consensus Statement 1999," *Arch Pathol Lab Med*, 2000, 124(7):979-94.

Desch CE, Benson AB, Smith TJ, et al, "Recommended Colorectal Cancer Surveillance Guidelines by the American Society of Clinical Oncology," *J Clin Oncol*, 1999, 17(4):1312-21.

Graham RA, Wang S, Catalano PJ, et al, "Postsurgical Surveillance of Colon Cancer: Preliminary Cost Analysis of Physical Examination, Carcinoembryonic Antigen Testing, Chest X-ray, and Colonoscopy," *Ann Surg*, 1998, 228(1):59-63.

Horie Y, Miura K, Matsui K, et al, "Marked Elevation of Plasma Carcinoembryonic Antigen and Stomach Carcinoma," *Cancer*, 1996, 77(10):1991-7.

Rockall TA and McDonald PJ, "Carcinoembryonic Antigen: Its Value in the Follow-up of Patients With Colorectal Cancer," *Int J Colorectal Dis*, 1999, 14(1):73-7.

Rubins JB, Dunitz J, Rubins HB, et al, "Serum Carcinoembryonic Antigen as an Adjunct to Preoperative Staging of Lung Cancer," *J Thorac Cardiovasc Surg*, 1998, 116(3):412-16.

Safi F and Beyer HG, "The Value of Follow-up After Curative Surgery of Colorectal Carcinoma," *Cancer Detect Prev*, 1993, 17(3):417-24.

♦ **Cardiac Enzymes/Isoenzymes** *see* Cardiac Markers: Laboratory Assessment, Overview *on page 343*

♦ **Cardiac Index** *see* Creatine Kinase MB and Other Isoenzymes, Serum *on page 469*

Cardiac Markers: Laboratory Assessment, Overview

Related Information

C-Reactive Protein, Serum *on page 467*
Creatine Kinase MB and Other Isoenzymes, Serum *on page 469*
Creatine Kinase, Serum *on page 470*
Myoglobin, Blood, Serum, or Plasma *on page 940*
Troponins, Serum *on page 1278*

Applies to Cardiac Enzymes/Isoenzymes; LD Isoenzymes; Myocardial Infarct Panel

Abstract The major criterion for the diagnosis of acute myocardial infarction (AMI) has been redefined as the typical rise and fall of biochemical markers of cardiac muscle necrosis, according to a joint consensus statement published by the American College of Cardiology and the European Society of Cardiology (Alpert, 2000). These markers are: **cardiac troponin I** (cTI), **cardiac troponin T** (cTT), and the **MB fraction of creatine kinase** (CK-MB). In addition to satisfying the biochemical criteria, at least **one** of the following is required for the diagnosis of AMI:

- ischemic symptoms
- development of pathologic Q waves in the EKG
- EKG changes of ischemia (ST segment changes)

(Continued)

Cardiac Markers: Laboratory Assessment, Overview
(Continued)

- a coronary artery intervention (eg, angioplasty)

Moreover, the consensus document states that cTI and cTT are more sensitive than CK-MB in the detection of cardiac muscle necrosis, a conclusion well supported by the recent literature. Myoglobin, the marker elevated earliest after AMI, is acceptable as an early marker, but is less specific than the troponins; if the myoglobin result is positive, confirmatory values should be obtained using one of the troponins or CK-MB.

During the past 30 years, other useful cardiac markers have included aspartate aminotransferase (AST) and isoenzyme 1 of lactate dehydrogenase (LD-1), both of which are less sensitive and less specific than the troponins and CK-MB. The availability of the troponins and CK-MB has rendered these obsolete. According to the consensus document, these markers should no longer be used to diagnose cardiac damage.

Specimen Serum; plasma or anticoagulated whole blood are the specimens of choice for the stat analysis of cardiac markers.

Container Red top tube, green top (lithium heparin) tube

Sampling Time According to the consensus document, blood should be obtained for testing on admission, at 6-9 hours, and, if the first two samples have negative results, again at 12-24 hours.

Collection Blood samples for cardiac markers should be referenced to the time of presentation in the emergency department (eg, "T_0" would indicate arrival).

Turnaround Time Stat cardiac marker testing with a target turnaround time of 1 hour or less is recommended.

Reference Interval Decision limits are method dependent. Each laboratory should evaluate its own methods and experience in concert with cardiologists and with active chart review.

Use Diagnosis of acute infarct of myocardium (acute MI or AMI); evaluation of chest pain. The two most important acute coronary arterial disease entities are the poles of a spectrum which include AMI and unstable angina pectoris. The latter is the cause of more 750,000 annual hospitalizations.

Distinction of unstable angina from non-Q-wave AMI may be difficult. "High-risk unstable angina" has been defined as elevated TnI with negative CK-MB.

Goals of the cardiac markers should include preventing or minimizing loss of myocardium.

Limitations At onset of acute infarction of myocardium, all enzymes and isoenzymes are normal. It is desirable to draw cardiac enzymes at onset for a baseline. CK-MB elevation has been reported during marathon training with increases of total CK. Myositis, various myopathies, rhabdomyolysis, and Reye syndrome are reported to cause elevations of CK-MB. In these situations CK-MB is elevated and remains elevated for days, while the values peak in about 1 day after onset of AMI and then begin to decline. Thus, the test is not entirely specific for AMI. Confirmation by troponin I is widely advocated. The negative predictive value of myoglobin early in AMI is very high, but its role as a marker exists only as a narrow window at presentation and its use may not be appropriate for some institutions.

The system for isoform testing is labor-intensive.

Methodology Immunoassays and immunoinhibition are more rapid, sensitive, and specific than electrophoresis (for CK-MB). Immunochemical CK-2 is commonly used and reported in mass units. It has distinct advantages over electrophoresis, including a short turnaround time (15-20 minutes).

Additional Information Newer tests which may allow for earlier detection of acute myocardial injury include CK-MM and CK-MB isoforms, and quantitation of human ventricular myosin light chains in blood. Assay of troponin I has represented the most important advance. Its combination in serial mode with myoglobin has been advocated as the best combination in one study, but CK-MB is more specific than myoglobin assay for injury of myocardium. Another study has advocated using troponin I as a second-line assay in a subset of higher-risk patients in whom the CK-MB is normal and early exercise testing is not an option.

The literature continues to call attention to the poor sensitivity of the ECG for acute myocardial infarction, especially when the infarct is small. In emergency room patients, the sensitivity of electrocardiograms may be as poor as 50% in evaluation of patients with acute chest pain.

Increased concentrations of troponin T and C-reactive protein are related to long-term (as well as short-term) risk of cardiac death.

Cost-effectiveness has been studied.

References

Alpert JS and Thygesen K, "Myocardial Infarction Redefined - A Consensus Document of The Joint European Society of Cardiology/American College of Cardiology Committee for the Redefinition of Myocardial Infarction," The Joint European Society of Cardiology/American College of Cardiology Committee, J Am Coll Cardiol, 2000, 36(3):959-69.

Caragher TE, Fernandez BB, and Barr LA, "Long-Term Experience With an Accelerated Protocol for Diagnosis of Chest Pain," Arch Pathol Lab Med, 2000, 124(10):1434-9.

Kahn SE, "The Challenge of Evaluating the Patient With Chest Pain," Arch Pathol Lab Med, 2000, 124(10):1418-9.

Keffer JH, "Assessment of Myocardial Injury," The Handbook of Clinical Pathology, 2nd ed, Chapter 17, Chicago, IL: ASCP Press, American Society of Clinical Pathologists, 2000, 235-44.

Kost GJ, Kirk JD, and Omand K, "A Strategy for the Use of Cardiac Injury Markers (Troponin I and T, Creatinine Kinase-MB Mass and Isoforms, and Myoglobin) in the Diagnosis of Acute Myocardial Infarction," Arch Pathol Lab Med, 1998, 122(3):245-51.

Lindahl B, Toss H, Siegbahn A, et al, "Markers of Myocardial Damage and Inflammation in Relation to Long-Term Mortality in Unstable Coronary Artery Disease. FRISC Study Group. Fragmin During Instability in Coronary Artery Disease," N Engl J Med, 2000, 343(16):1139-47.

Mehta RH and Eagle KA, "Missed Diagnoses of Acute Coronary Syndromes in the Emergency Room - Continuing Challenges," N Engl J Med, 2000, 342(16):1207-10.

Polaczyk CA, Kuntz KM, Sacks DB, et al, "Emergency Department Triage Strategies for Acute Chest Pain Using Creatine Kinase-MB and Troponin I Assays: A Cost-Effectiveness Analysis," Ann Intern Med, 1999, 131(12):909-18.

Rader DJ, "Inflammatory Markers of Coronary Risk," N Engl J Med, 2000, 343(16):1179-82.

Wu AH, Apple FS, Gibler WB, et al, "National Academy of Clinical Biochemistry Standards of Laboratory Practice: Recommendations for the Use of Cardiac Markers in Coronary Artery Diseases," Clin Chem, 1999, 45(7):1104-21.

Zaninotto M, Altinier S, Lachin M, et al, "Strategies for the Early Diagnosis of Acute Myocardial Infarction Using Biochemical Markers," Am J Clin Pathol, 1999, 111(3):399-405.

◆ **Cardiac Troponins** see Troponins, Serum on page 1278

◆ **Cardiolipin** see Antiphospholipid Antibody (Lupus Anticoagulant and/or Anticardiolipin Antibody) on page 193

◆ **Cardioquin®** see Quinidine, Serum on page 1129

◆ **Cardioreg®** see Digoxin, Serum on page 512

◆ **Cardizem®** see Diltiazem, Serum or Plasma on page 515

◆ **Carnitine** see Amino Acids, Plasma on page 143

◆ **Carotene** see Vitamin A, Serum or Plasma on page 1314

◆ **Cartia XT™** see Diltiazem, Serum or Plasma on page 515

◆ **Caspase Enzymes** see Apoptosis Assays on page 205

◆ **Casts, Urine** see Urinalysis on page 1289

◆ **Catapres®** see Clonidine, Serum or Plasma on page 415

◆ **Catch 22 Syndrome** see Gene Rearrangement for Leukemia and Lymphoma on page 633

◆ **Catecholamines** see Metanephrines, Urine or Plasma on page 899

Catecholamines, Fractionation, Plasma

Related Information

Catecholamines, Fractionation, Urine on page 347
Homovanillic Acid, Urine on page 743
Metanephrines, Urine or Plasma on page 899
Vanillylmandelic Acid, Urine on page 1299

Synonyms Epinephrine, Norepinephrine, Dopamine; Pressor Amines

Applies to Chromogranin A, Plasma; Clonidine Suppression Test; Dopamine

Test Includes Epinephrine (E) and norepinephrine (N); some laboratories also report dopamine.

Abstract Epinephrine (E), norepinephrine (NE), and dopamine are the catecholamines of interest in diagnostic medicine and are synthesized in the adrenal medulla, brain, and sympathetic nervous system. Pheochromocytoma, when hormonally active, secretes large quantities of E, NE, or both. Early diagnosis of pheochromocytoma is important because it is a surgically curable cause of severe hypertension, and in ~10% of cases the neoplasm is malignant. (Continued)

Catecholamines, Fractionation, Plasma *(Continued)*

The catecholamines are highly labile and have very brief (~2 minute) half-life in the circulation. Their plasma levels are influenced by multiple environmental factors, foods, and drugs; a successful testing protocol must control these preanalytical variables. E, NE, and dopamine undergo complex metabolic changes and are ultimately excreted in the urine, primarily as metanephrines.

Patient Preparation Meticulous patient preparation is recommended. The laboratory should provide a list of drugs and foods which interfere with its technique. At the time of specimen collection, the patient should be relaxed and at rest; emotional stress and physical activity result in elevated plasma catecholamines. Epinephrine and epinephrine-like drugs (eg, Aldomet®, Inderal®) interfere; such drugs should be stopped a week prior to testing.

Clonidine suppression protocol: The patient remains recumbent for the entire 3-hour test period. Testing is conducted in the morning after an overnight fast. After a baseline blood specimen for plasma catecholamines is obtained, clonidine, 4.3 µg/kg body weight is given by mouth with water.

Specimen Plasma

Container Green top (heparin) tube or lavender top (EDTA) tube; check with reference laboratory before obtaining specimen.

Sampling Time Blood specimens are obtained in anticoagulant just before the clonidine is given, and 3 hours later.

Collection Patient should be fasting and without use of tobacco for at least 4 hours. Patient should remain supine in quiet surroundings for at least 30 minutes before specimen collection.

Reference Interval Method-dependent; the following are reference intervals reported by one reference laboratory using high performance liquid chromatography.

Norepinephrine:
- supine: 70-750 pg/mL
- standing: 200-1700 pg/mL

Epinephrine:
- supine: undetectable: 110 pg/mL
- standing: undetectable: 140 pg/mL

Dopamine: <30 pg/mL (does not vary with posture)

Study of ~500 Hypertensive Individuals

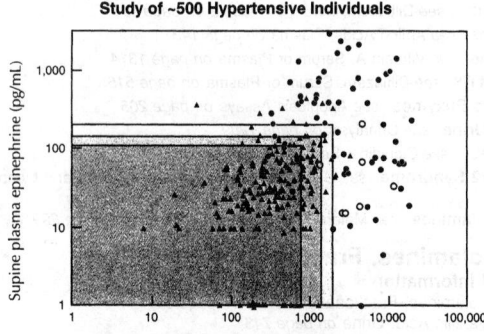

▲ Individuals without pheochromocytoma
o Extra-adrenal pheochromocytomas
● Surgically proven pheochromocytoma

The darkest-shaded zone represents patients with epinephrine (E) values ≤110 pg/mL and norepinephrine (NE) values ≤750 pg/mL. Concentrations outside of these limits result in sensitivities and specificities in the diagnosis of pheochromocytoma as shown in the following chart. Similarly, there is a medium-shaded area with E values ≤140 pg/mL and NE values ≤1400 pg/mL. The next lighter-shaded area has upper limits for E of 200 pg/mL and NE of 2000 pg/mL. See the following for the sensitivity and specificity figures for values falling outside of each of the areas defined above.

Adapted from *2000 Test Catalogue*, Mayo Medical Laboratories, Rochester, MN.

Limits From the Graph and Diagnostic Efficacy
in Diagnosis of Pheochromocytoma

Limits	Sensitivity (%)	Specificity (%)
E: 110 pg/mL NE: 750 pg/mL	93 (66/71)	68 (130/191)
E: 140 pg/mL NE: 1400 pg/mL	90 (64/71)	90 (172/191)
E: 200 pg/mL NE: 2000 pg/mL	85 (60/71)	95 (182/191)

Adapted from Mayo Medical Laboratories, *2000 Test Catalogue*, 123-4 and 526-7.

Use Resting values: Plasma catecholamines are used to diagnose catecholamine-secreting neoplasms, most often adrenal pheochromocytomas. Some of the preanalytic variability is mitigated if specimens are always obtained after 15 minutes in the supine position in a minimally stressful environment. The interpretation of individual patient results is subject to the usual tradeoffs between sensitivity and specificity. Presented on previous page and above are a graph and table containing results from ~500 hypertensive patients seen at the Mayo Clinic. The sensitivity and specificity results are shown, in the table, at three different cutoffs for E and NE.

Clonidine suppression: Clonidine suppresses catecholamines in persons with essential hypertension, but not in persons with pheochromocytoma. Plasma NE should be within the reference interval at 3 hours. Levels above this threshold are consistent with pheochromocytoma.

Limitations There are many causes of elevated plasma catecholamines, including recent surgery, traumatic injury, upright posture, cold, anxiety, pain, clonidine withdrawal, and concurrent acute or chronic illness. Drug-induced changes in catecholamine concentrations are published (Rosano and Whitley).

Methodology High performance liquid chromatography (HPLC) with electrochemical detection is the method of choice.

References
Bravo EL, "Plasma or Urinary Metanephrines for the Diagnosis of Pheochromocytoma? That Is the Question," *Ann Intern Med*, 1996, 125(4):331-2.

Héron E, Chatellier G, Billaud E, et al, "The Urinary Metanephrine-to-Creatinine Ratio for the Diagnosis of Pheochromocytoma," *Ann Intern Med*, 1996, 125(4):300-3.

Jacobs DS, DeMott WR, Oxley DK, et al, *Laboratory Test Handbook*, 5th ed, Hudson, OH: Lexi-Comp Inc, 2001.

Mayo Medical Laboratories, *2000 Test Catalogue*, Rochester, MN, 123-4 and 526-7.

Panholzer TJ, Beyer J, and Lichtwald K, "Coupled-Column Liquid Chromatographic Analysis of Catecholamines, Serotonin, and Metabolites in Human Urine," *Clin Chem*, 1999, 45(2):262-8.

Rosano TG and Whitley RJ, "Catecholamines and Serotonin," *Tietz Textbook of Clinical Chemistry*, 3rd ed, Burtis CA and Ashwood ER, eds, Philadelphia, PA: WB Saunders Co, 1999, 1570-600.

Wassell J, Reed P, Kane J, et al, "Freedom From Drug Interference in New Immunoassays for Urinary Catecholamines and Metanephrines," *Clin Chem*, 1999, 45(12):2216-23.

Catecholamines, Fractionation, Urine

Related Information
Calcitonin, Serum or Plasma *on page 327*
Catecholamines, Fractionation, Plasma *on page 345*
Homovanillic Acid, Urine *on page 743*
Metanephrines, Urine or Plasma *on page 899*
Urine Collection, 24-Hour *on page 1295*
Vanillylmandelic Acid, Urine *on page 1299*

Synonyms Free Catecholamine Fractionation, Urine

Applies to Dopamine, Urine; Epinephrine, Urine; Norepinephrine, Urine

Replaces Total Urinary Catecholamines

Abstract See Catecholamines, Fractionation, Plasma *on page 345*.

Patient Preparation Meticulous preparation is required. The laboratory should provide a list of drugs and foods which interfere with its techniques.

Specimen 24-hour urine

Overnight collections for urinary metanephrines and catecholamines have been advocated for diagnosis of pheochromocytoma.
(Continued)

Catecholamines, Fractionation, Urine *(Continued)*

Container Brown urine container with sufficient acetic acid to keep pH between 2-4. Typically, 25 mL of 50% acetic acid for an adult, and 15 mL of 50% acetic acid for a child younger than 5 years of age.

Storage Instructions Refrigerate during and after collection.

Special Instructions Consult performing laboratory

Reference Interval The following intervals are from a reference laboratory employing HPLC methodology.

Epinephrine:
- <1 year: 0.0-2.5 µg/24 hours
- 1 year: 0.0-3.5 µg/24 hours
- 2-3 years: 0.0-6.0 µg/24 hours
- 4-9 years: 0.2-10.0 µg/24 hours
- 10-15 years: 0.5-20.0 µg/24 hours
- ≥16 years: 0.0-20.0 µg/24 hours

Norepinephrine:
- <1 year: 0-10 µg/24 hours
- 1 year: 1-17 µg/24 hours
- 2-3 years: 4-29 µg/24 hours
- 4-6 years: 8-45 µg/24 hours
- 7-9 years: 13-65 µg/24 hours
- ≥10 years: 15-80 µg/24 hours

Dopamine:
- <1 year: 0-85 µg/24 hours
- 1 year: 10-140 µg/24 hours
- 2-3 years: 40-260 µg/24 hours
- ≥4 years: 65-400 µg/24 hours

Use Measurement of urine catecholamines is useful for the diagnosis of catecholamine-secreting neoplasms. More than 90% of these are adrenal pheochromocytomas; other catecholamine-secreting neoplasms include paragangliomas and neuroblastomas. Presented in the graph and table in the listing Catecholamines, Fractionation, Plasma *on page 345* are results from ~500 hypertensive patients seen at the Mayo Clinic. Note the usual tradeoff between sensitivity and specificity. See diagram.

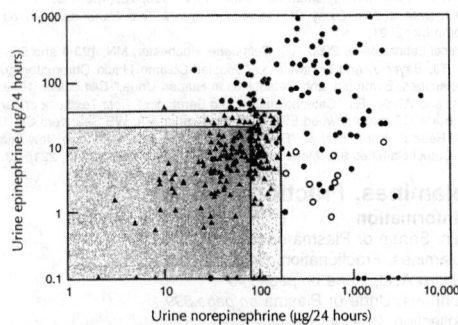

▲ Individuals without pheochromocytoma
○ Extra-adrenal pheochromocytomas
● Surgically proven pheochromocytoma

The darkest-shaded zone represents patients with epinephrine (E) values ≤ 20 µg/24 hours and norepinephrine (NE) values ≤80 µg/24 hours. Concentrations outside of these limits result in sensitivities and specificities in the diagnosis of pheochromocytoma as shown in the following chart. Similarly, the medium-shaded area has upper limits for E of 35 µg/24 hours and NE of 170 µg/24 hours. The next lighter-shaded area has upper limits for E of 200 pg/mL and NE of 2000 pg/mL. See the following for the sensitivity and specificity figures for values falling outside of each of the areas defined above.

Adapted from *2000 Test Catalogue*, Mayo Medical Laboratories, Rochester, MN.

**Limits From the Graph and Their Diagnostic Efficacy
in Diagnosis of Pheochromocytoma**

Limits	Sensitivity (%)	Specificity (%)
E: 20 µg/24 hours NE: 80 µg/24 hours	98 (90/92)	45 (90/200)
E: 35 µg/24 hours NE: 170 µg/24 hours	97 (89/92)	95 (191/200)

Adapted from Mayo Medical Laboratories, *2000 Test Catalogue.*

Limitations There are many causes of elevated plasma catecholamines including recent surgery, traumatic injury, upright posture, cold, anxiety, pain, clonidine withdrawal, and concurrent acute or chronic illness.

Methodology High performance liquid chromatography (HPLC), fluorometry, radioenzymatic assay, radioimmunoassay (RIA)

Additional Information Drugs which increase and decrease urinary catecholamines are published (Rosano and Whitley). New immunoassays are reported to provide advantages from drug interference.

See Catecholamines, Fractionation, Plasma *on page 345* and Metanephrines, Urine or Plasma *on page 899.*

References

Mayo Medical Laboratories, *2000 Test Catalogue*, Rochester, MN, 124.

Panholzer TJ, Beyer J, and Lichtwald K, "Coupled-Column Liquid Chromatographic Analysis of Catecholamines, Serotonin, and Metabolites in Human Urine," *Clin Chem*, 1999, 45(2):262-8.

Payne RB, "Urinary Catecholamine Excretion in Relation to Renal Function," *Ann Clin Biochem*, 2000, 37(Pt 2):228-9.

Rosano TG and Whitley RJ, "Catecholamines and Serotonin," *Tietz Textbook of Clinical Chemistry*, 3rd ed, Chapter 14, Burtis CA and Ashwood ER, eds, Philadelphia, PA: WB Saunders Co, 1999, 1570-600.

Wassell J, Reed P, Kane J, et al, "Freedom From Drug Interference in New Immunoassays for Urinary Catecholamines and Metanephrines," *Clin Chem*, 1999, 45(12):2216-23.

♦ **CBC** *see* Complete Blood Count *on page 442*

♦ **CBZ** *see* Carbamazepine, Serum *on page 337*

♦ **CCCT** *see* Follicle Stimulating Hormone, Serum, Plasma, or Urine *on page 609*

♦ **CCP Antibodies** *see* Anticyclic Citrullinated Peptide Antibody *on page 169*

♦ **CD** *see* Flow Cytometry, Overview *on page 592*

CD4/CD8 Enumeration

Related Information

Beta$_2$-Microglobulin, Serum or Urine *on page 254*
Complete Blood Count *on page 442*
Flow Cytometry, Overview *on page 592*
HIV-1/HIV-2 Serology *on page 736*
Peripheral Blood: Differential Leukocyte Count *on page 1010*
White Blood Cell Count *on page 1330*
Zidovudine, Serum or Plasma *on page 1339*

Applies to Autoimmune Lymphoproliferative Syndrome; Canale-Smith Syndrome; CD4/CD8 Ratio; Lymphocyte CD4 Counts

Test Includes Lymphocyte subpopulation enumeration

Abstract The introduction of flow cytometry has allowed subclassification of lymphocytes on the basis of function and differentiation. Mature T cells in the peripheral blood usually express either the CD4 or CD8 antigen. CD4$^+$ T cells are functionally defined as T helper cells, while CD8$^+$ cells exhibit either suppressor or cytotoxic activity.

Depletion of CD4$^+$ helper/inducer T lymphocytes is one of the most significant surrogate markers to study progression of human immunodeficiency virus disease.

Specimen Whole blood; peripheral blood leukocytes can be harvested from leukopheresis packs or lymphoid tissue by use of Ficoll-Hypaque density gradient centrifugation. Skin puncture specimens may be used in children or adults with difficult venous access.

Container Green top (heparin) tube
(Continued)

CD4/CD8 Enumeration *(Continued)*

Storage Instructions Blood ideally is delivered to the laboratory immediately; however, whole blood may be held for 24 hours at room temperature prior to assay. Do not refrigerate or freeze sample.

Special Instructions With additional processing, polypropylene centrifuge tubes may be used. Polystyrene or polycarbonate should not be used in order to avoid cell adhesion. If RBC contamination occurs (due to clinical low mean corpuscular hemoglobin content), the red cells can be removed by hypotonic saline or ammonium chloride lysis.

Reference Interval See table.

Lymphocyte	Absolute Count (cells/μL)	Relative %
Mature T cells (CD3)	650-3036	65-92
Helper T cells (CD4)	310-2112	31-64
Suppressor T cells (CD8)	80-1353	8-41
CD4:8 ratio	1.0-1.5	

Use Enumeration of CD4 and CD8 cells is useful in the diagnosis and monitoring of patients with human immunodeficiency virus infections.

Limitations Values can be abnormal if patient is taking steroids or other immunosuppressives, has a severe intercurrent illness, or has had recent surgery requiring general anesthesia; patterns of maturation are disordered and inconsistent in lymphomas.

Methodology Flow cytometry (FC)

Additional Information Imbalances or deficiencies in the immune system can result from abnormalities in either the CD4 or CD8 population. In cases of Fas deficiency (result of specific Fas mutations) there is nonmalignant but potentially massive proliferation of T lymphocytes that are both CD4⁻ and CD8⁻. This condition is generally referred to as the autoimmune lymphoproliferative syndrome (ALPS). It has also been referred to as the Canale-Smith syndrome. Fas (apo-1, CD95) represents a cell surface molecule of the tumor necrosis factor (TNF) group. Interaction of Fas with Fas-L (Fas-ligand) as occurs with prolonged lymphocyte activation leads to lymphocyte cell death by apoptosis.

HIV-1 infection causes significant changes in the number of CD4⁺ and CD8⁺ lymphocytes; within 6 months of seroconversion, the CD4 count falls about 30%, while the CD8 count increases by 40%, causing a decrease in the CD4:CD8 ratio. Healthy persons generally have absolute CD4 cell counts of about 1000 cells/μL and a CD4:CD8 ratio >1.0. CD4 counts <400 cells/μL are generally associated with progression to AIDS. The continuing lymphocyte depletion may relate to one or more mechanisms, including direct killing of CD4 cells by HIV, syncytial formation with removal of both HIV+ and HIV- lymphocytes, immune responses to gp120 adsorbed on uninfected CD4 cells, and induction of programmed cell death (apoptosis) by the interaction of gp120 with CD4 (the HIV receptor). With decrease in the absolute number of CD4 cells, the CD8 population is also affected with change from CD8⁺ CD28⁺ to CD8⁺ CD28⁻ phenotype, accelerated apoptosis, and increased turnover with chromosome telomere shortening. Increase in CD4 count from very low levels (CD4 count nadir) to at least 200 cells/mm³ has been reported to be associated with a decreased rate of disease progression.

The increase in CD8⁺ cells is not unique to HIV-1 infection, since many viruses and vaccinations cause a transient increase in the CD8 population. It is important to determine absolute numbers of CD4⁺ and CD8⁺ cells, not just the CD4:CD8 ratio, in order to distinguish HIV-1 from other viral infections. In postsplenectomy patients, the absolute lymphocyte count may increase such that use of the percent of CD4 lymphocytes may be more reliable.

Rare individuals, occasionally with manifestations of immunodeficiency, have unusually low CD4 levels but no evidence of HIV infection. They have been labeled as cases of "idiopathic CD4 lymphocytopenia" (ICL). These patients often have general lymphocytopenia with decrease in CD8 T cells, B cells, and NK cells, as well as CD4 T cells.

Modest variations in CD4, CD8, and CD4:CD8 ratio occur with age, sex, and race. Absolute lymphocyte count (and each derived subset) peak in infancy and decline with age. The Asian population (in particular males) appears to have lower mean percentages of CD3 and CD4 cells, a lower CD4:CD8 ratio and a lower absolute level of CD4 lymphocytes.

References

Cracknell SE, Hinchliffe RF, and Lilleyman JS, "Lymphocyte Subset Counts in Skin Puncture and Venous Blood Compared," *J Clin Pathol*, 1995, 48(12):1137-8.

Drappa J, Vaishnaw AK, Sullivan KE, et al, "Fas Gene Mutations in the Canale-Smith Syndrome, an Inherited Lymphoproliferative Disorder Associated With Autoimmunity," *N Engl J Med*, 1996, 335(22):1643-9.

Freier S, Kerem E, Dranitzki Z, et al, "Hereditary CD4+ T Lymphocytopenia," *Arch Dis Child*, 1998, 78(4):371-2.

Gougeon ML and Montagnier, L, "Programmed Cell Death as a Mechanism of CD4 and CD8 T Cell Deletion in AIDS: Molecular Control and Effect of Highly Active Antiretroviral Therapy," Mechanisms of Cell Death, The Second Annual Conference of the Cell Death Society, *Ann N Y Acad Sci*, New York, NY: New York Academy of Sciences, 1999, 199-209.

Smith DK, Neal JJ, and Holmberg SD, "Unexplained Opportunistic and CD4+ T-Lymphocytopenia Without HIV Infection," *N Engl J Med*, 1993, 328(6):373-9.

Straus SE, Sneller M, Lenardo MJ, et al, "An Inherited Disorder of Lymphocyte Apoptosis: The Autoimmune Lymphoproliferative Syndrome," NIH Conference, *Ann Intern Med*, 1999, 130(7):591-601.

♦ **CD4/CD8 Ratio** *see* CD4/CD8 Enumeration *on page 349*

♦ **CD34** *see* Hematopoietic Progenitor Cells, Peripheral Blood *on page 679*

♦ **CD34 Analysis** *see* Hematopoietic Progenitor Cells, Marrow *on page 677*

CD34⁺ Hematopoietic Stem Cells by Flow Cytometry

Related Information

Flow Cytometry, Overview *on page 592*
Hematopoietic Progenitor Cells, Cord Blood/Placental Blood *on page 676*
Hematopoietic Progenitor Cells, Marrow *on page 677*
Hematopoietic Progenitor Cells, Peripheral Blood *on page 679*
Thrombopoietin, Serum or Plasma *on page 1247*

Applies to Stem Cells

Abstract The CD34 membrane antigen is present on the surface of ~3% of bone marrow hematopoietic precursors or "stem-like" cells and ~0.01% to 0.05% of nucleated cells in the peripheral blood. Administration of hematopoietic growth factors increases circulating (CD34⁺) stem cells in patients and in healthy donors, such that peripheral blood stem cells (PBSC) have largely replaced bone marrow as the source for stem cells in marrow transplantation.

Specimen Peripheral whole blood

Container Lavender top (EDTA) tube

Collection Proper identification of the sample is of critical importance. Working Party Guideline recommends that samples be labeled with patient's full name, a numeric patient identifier, date of birth, patient location, and date and time of collection.

Storage Instructions Total WBC count must be obtained within 6 hours of sample collection. Specimen should be stored at 4°C and processing completed within 12 hours.

Causes for Rejection Presence of blood clots, hemolysis, receipt after 12 hours of sampling

Special Instructions Timing of stem cell harvesting is dependent on the CD34⁺ cell count. Harvesting requires a peripheral blood CD34 count >10 cells/μL (with the goal of obtaining ≥2 x 10⁶ CD34⁺ cells).

Use To assess stem cell (hematopoietic progenitor cell) numbers in peripheral blood, cord blood, and apheresis specimens for potential use in PBSC transplantation

Limitations Electronic white blood cell determinations, used to obtain absolute CD34⁺ counts, must be corrected for presence of nucleated red blood cells; reverse pipetting must be used to avoid error from the introduction of bubbles

Methodology *In vitro* assay of colony-forming units (CFU-GM) to assess hematopoietic potential (numbers of clonogenic-committed granulocytic-monocytic progenitor cells grown *in vitro*). Single or dual platform flow cytometry. Samples should be analyzed immediately post red blood cell lysis (or kept on melting ice until analysis) and analyses must be completed within 1 hour following lysis. The Working Party Guideline specifies reagents for detection of the CD45 antigen (the CD34 receptor), and emphasizes that a sufficient number of (Continued)

CD34⁺ Hematopoietic Stem Cells by Flow Cytometry
(Continued)

events must be analyzed to obtain a statistically valid result (an intra-assay coefficient of variation of 10% requires that a minimum of 100 CD34⁺ events must be collected).

Additional Information CD34⁺ hematopoietic precursor blast-like cells were originally described in 1984. The earliest marrow progenitor cells are CD34⁻ with subsequent switch of this resting stem cell from CD34⁻ to CD34⁺. With external stimulation, prior dormant CD34⁻ cells respond by upregulation of multiple surface receptors (see diagram) and become CD34⁺ cells. They may then become susceptible to DNA damage and development of leukemia. Currently, there is interest (with therapeutic application) in identifying (and potentially separating) a population of benign hematopoietic progenitors at this early stage of maturation.

CD34⁺ Hematopoietic Stem-Cells by Flow Cytometry

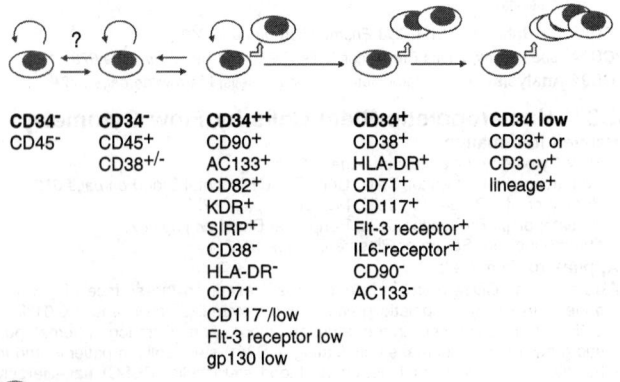

CD34⁻	CD34⁻	CD34⁺⁺	CD34⁺	CD34 low
CD45⁻	CD45⁺	CD90⁺	CD38⁺	CD33⁺ or
	CD38⁺/⁻	AC133⁺	HLA-DR⁺	CD3 cy⁺
		CD82⁺	CD71⁺	lineage⁺
		KDR⁺	CD117⁺	
		SIRP⁺	Flt-3 receptor⁺	
		CD38⁻	IL6-receptor⁺	
		HLA-DR⁻	CD90⁻	
		CD71⁻	AC133⁻	
		CD117⁻/low		
		Flt-3 receptor low		
		gp130 low		

⟳ Self-renewal

⌐ Differentiation capacity

Adapted from Brendel C and Neubauer A, "Characteristics and Analysis of Normal and Leukemic Stem Cells: Current Concepts and Future Directions," *Leukemia*, 2000, 14(10):1711-7.

With the exception of a tiny population of very immature hematopoietic progenitor cells (HPC) that are CD34⁻, the CD34 molecule is unique in that it is expressed at various levels on essentially all HPC. There are <1% of CD34⁺ cells in all HPC preparations and the expression of CD34⁺ varies within the HPC population. A number of different glycosylated CD34⁺ epitopes exist with resultant antigenic diversity.

Factors influencing the recovery of CD34⁺ cells during collection and the significance of rapidity of clearance of CD34⁺ cells from peripheral blood after HSC infusion have been studied. Optimal timing of stem cell collection can be based on rising WBC and platelet counts which provides increased number of CD34⁺ cells in leukapheresis as compared with collection on a fixed day following initiation of the mobilization protocol.

The use of CD34⁺ subsets (eg, CD34⁻38⁻, CD34⁺33⁻, CD34⁺33⁺, CD34⁺41⁺, CD34⁺CD61⁺, CD34⁺CD105⁺) in predicting the engraftment success and quality of marrow transplantation and in following early maturation is under investigation.

The bone marrow transplantation treatment modality has been expanded by the discovery that umbilical cord blood (UCB) can be used as an alternate source for HSC.

CD34 antigen present on some leukemic cell membrane surfaces may have clinical relevance. CD34 expression in acute promyelocytic leukemia is associated with hyperleukocytosis, an immature hypogranular morphology, and a poor clinical outcome.

References
Barnett D, Janossy G, Lubenko A, et al, "Guideline for the Flow Cytometric Enumeration of CD34+ Haematopoietic Stem Cells. Prepared by the CD34+ Haematopoietic Stem Cell Working Party," General Haematology Task Force of the British Committee for Standards in Haematology, *Clin Lab Haematol*, 1999, 21(5):301-8.

Hughes VC, "Cord Blood Transplantation: Hallmarks of the 20th Century," *Lab Med*, 2000, 31(12):672.

Jacobs DS, DeMott WR, Oxley DK, et al, *Laboratory Test Handbook*, 5th ed, Hudson, OH: Lexi-Comp Inc, 2001, 413-5.

Krieger MS, Schiller G, Berenson JR, et al, "Collection of Peripheral Blood Progenitor Cells (PBPC) Based on a Rising WBC and Platelet Count Significantly Increases the Number of CD34+ Cells," *Bone Marrow Transplant*, 1999, 24(1):25-8.

Lee JJ, Cho D, Chung IJ, et al, "CD34 Expression Is Associated With Poor Clinical Outcome in Patients With Acute Promyelocytic Leukemia," *Am J Hematol*, 2003, 73(2):149-53.

Spangrude GJ and Cooper DD, "Paradigm Shifts in Stem-Cell Biology," *Semin Hematol*, 2000, 37(1 Suppl 2):3-10.

Wright-Kanuth MS and Smith LA, "Hematopoietic Stem Cell Transplantation," *Clin Lab Sci*, 2001, 14(2):112-7.

CD40 Ligand (Soluble), Serum or Plasma

Related Information
C-Reactive Protein, Serum *on page 467*
Plasminogen Activator Inhibitor 1 *on page 1043*
Platelet Hyperaggregation *on page 1053*

Applies to Abciximab; Platelets

Abstract CD40 ligand is released from the surfaces of platelets upon platelet activation. An elevated level of soluble CD40 ligand in a patient with unstable coronary artery disease may be associated with an increased risk of cardiovascular events.

Specimen Plasma is preferred. CD40 ligand can be in its soluble form or may be associated with microparticles in the serum. Microparticle surface ligand protein may be included in error in the measurement of soluble CD40 ligand. Essentially all the CD40 ligand in plasma has been reported to be soluble.

Storage Instructions In published studies, storage was 80°C.

Reference Interval CD40 ligand levels >5.0 μg/L have been considered as elevated.

Use Assess risk of adverse cardiovascular events in patients with unstable coronary artery disease; monitor reduction of increased risk relating to elevated CD40 ligand levels in patients receiving abciximab therapy

Limitations This assay and its clinical application are in early evaluation. The measure of soluble CD40 ligand could be widely applied in evaluation of patients with acute coronary syndromes. Soluble CD40 ligand levels do not correlate with markers of inflammation such as C-reactive protein.

Methodology Enzyme-linked immunosorbent assay (ELISA)

Additional Information The CD40-CD40 ligand system is present on leukocytes, endothelial/smooth muscle cells, and on activated platelets. CD40 is homologous to the tumor necrosis factor receptor. In some cells, CD40 interacts with CD154 to initiate inflammatory responses. CD154 may contribute to the inflammatory component of atherosclerosis. With platelet stimulation, a soluble form of CD40 ligand is released to the circulation. Soluble CD40 ligand binds a major platelet integrin, $\alpha IIb\beta 3$ (glycoprotein IIb/IIIa) and is a platelet agonist, important for stability of arterial thrombi. Soluble CD40 ligand is considered to play a role in the development and evolution of acute coronary syndromes.

Initial study has shown that the incidence of death or nonfatal myocardial infarction was higher in placebo treated patients with high levels (>5.0 μg/L) of soluble CD40 ligand. A treatment (abciximab - antiplatelet therapy) group had ≤5.0 μg/L levels of CD40 ligand prior to coronary angioplasty.

References
Conde ID and Kleiman NS, "Soluble CD40 Ligand in Acute Coronary Syndromes," *N Engl J Med*, 2003, 348(25):2575-7.

Freedman JE, "CD40 Ligand - Assessing Risk Instead of Damage?" *N Engl J Med*, 2003, 348(12):1163-5.

(Continued)

CD40 Ligand (Soluble), Serum or Plasma (Continued)

Heeschen C, Dimmeler S, Hamm CW, et al, "Soluble CD40 Ligand in Acute Coronary Syndromes," N Engl J Med, 2003, 348(12):1104-11.

Henn V, Steinback S, Büchner K, et al, "The Inflammatory Action of CD40 Ligand (CD154) Expressed on Activated Human Platelets Is Temporarily Limited by Coexpressed CD40," Blood, 2003, 98(4):1047-54.

- ◆ **CD95** see Apoptosis Assays on page 205
- ◆ **Cd, Blood** see Cadmium, Blood or Urine on page 325
- ◆ **Cd, Urine** see Cadmium, Blood or Urine on page 325
- ◆ **CEA** see Carcinoembryonic Antigen, Serum on page 342
- ◆ **CEA, Body Fluid** see Body Fluid Chemical Analysis on page 291
- ◆ **CellCept™** see Mycophenolic Acid, Serum on page 936
- ◆ **Cell Count, CSF** see Cerebrospinal Fluid Analysis: Overview on page 355
- ◆ **Cell Count Ratio** see Body Fluid Analysis, Cell Count on page 288
- ◆ **Cell Cycle Analysis by Flow Cytometry** see Flow Cytometry, Overview on page 592
- ◆ **Centipedes** see Arthropod Identification on page 213
- ◆ **Centralgine®** see Meperidine, Serum or Urine on page 895
- ◆ **Central Venous Blood** see Blood Gases and pH, Venous on page 279

Centromere/Kinetochore Antibody

Related Information

Antinuclear Antibodies on page 189

Topoisomerase I Antibody on page 1261

Abstract Anticentromere antibodies (ACA), a subset of antinuclear antibodies, are strongly associated with CREST syndrome.

Patients with scleroderma are categorized primarily into two types of disease: limited and diffuse. Patients with the form of limited disease known as CREST syndrome, a scleroderma variant, tend to have a better prognosis than those with diffuse disease. Subcutaneous **c**alcinosis, **R**aynaud phenomenon, **e**sophageal dysfunction, **s**clerodactyly, and **t**elangiectasia characterize the CREST syndrome (3 of 5 must be evident). Autoantibodies may be useful in differentiating these two types. Limited disease is most commonly associated with the anticentromere pattern of ANA staining.

Specimen Serum

Container Red top tube or SST™ tube

Reference Interval IFA: negative; ELISA: negative

Use Anticentromere autoantibodies are a marker for diagnosis of CREST syndrome, a less aggressive form of scleroderma. Association with diffuse scleroderma is less strong, about 10% to 20%. Anticentromere antibodies are rarely found in systemic lupus erythematosus.

Anticentromere pattern is found in up to a quarter of individuals with idiopathic Raynaud phenomenon. In such patients, the presence of this antibody may predict progression to the CREST syndrome.

Limitations May be absent in some patients with CREST syndrome and many with scleroderma. Positive in 15% of healthy controls; there is greater sensitivity for patients with limited cutaneous disease.

Methodology Indirect immunofluorescent antibody (IFA) detected on HEp-2 tissue culture cell substrate, enzyme immunoassay (EIA)

Additional Information Approximately 60% of patients with limited scleroderma have anticentromere antibodies. Diffuse scleroderma is associated with antinuclear antibodies (ANA). The presence of ACA and topoisomerase I is mutually exclusive.

References

Homburger HA, "Advances in the Diagnosis and Laboratory Evaluation of Systemic Rheumatic Diseases Other Than Rheumatoid Arthritis," Clinical and Laboratory Evaluation of Human Autoimmune Diseases, Chapter 11, Nakamura RA, Keren DF, and Bylund DJ, eds, Chicago, IL: American Society for Clinical Pathology, 2002, 153-64.

Kavanaugh A, Tomar R, Reveille J, et al, "Guidelines for Clinical Use of the Antinuclear Antibody Test and Tests for Specific Autoantibodies to Nuclear Antigens," Arch Pathol Lab Med, 2000, 124(1):71-81.

Moder KG, "Use and Interpretation of Rheumatologic Tests: A Guide for Clinicians," Mayo Clin Proc, 1996, 71(4):391-6.

- ◆ **Cephalosporinases** see Beta-Lactamase Test on page 256

♦ **c-*erb*B-2** *see* HER-2/neu *on page 716*

Cerebrospinal Fluid Analysis: Overview

Related Information

Synonyms Cell Count, CSF; CSF Analysis; Spinal Fluid Analysis

Applies to CSF Leukocyte Aggregation Score; Lumbar Puncture

Test Includes Color of supernatant, volume, turbidity, WBC/mm^3, polys/mm^3, lymphs/mm^3, RBC/mm^3, percent of crenated RBC, protein and glucose, VDRL, and other tests as clinically indicated in selected cases. India ink preparation and cryptococcal antigen titer are best not forgotten.

Abstract Bacterial meningitis is a medical emergency and requires immediate evaluation and treatment (ie, with cerebrospinal fluid (CSF) analysis, and other studies). Examination of CSF also contributes to management of other disease entities of the central nervous system including encephalitis, encephalopathy, instances of vasculitis, demyelinating diseases, certain tumors, paraneoplastic entities, polyneuritis, cases of seizure disorders, and confusional states. For diagnosis of meningitis, culture and then Gram stain study have priority over all other testing, when only a small quantity of CSF is available. Cell count with differential deserves the next priority, followed by glucose and protein.

Patient Preparation Aseptic preparation for aspiration

Specimen Cerebrospinal fluid. A sample of peripheral blood for serum should be obtained concurrently if CSF is being studied for presence of oligoclonal bands (eg, as in demyelinating diseases, in particular multiple sclerosis) or when CSF glucose levels are needed as in cases of possible infection (eg, meningitis or encephalitis). Blood culture and plasma glucose should be drawn as well as CBC and differential in cases of possible meningitis.

Container Sterile test tubes from lumbar puncture (LP) tray

Collection Specimens of spinal fluid, and blood for culture, should be obtained prior to initiation of antibiotic treatment of meningitis. Tubes must be labeled with patient's name, date, and labeled with number indicating sequence in which tubes were obtained. Specimen must be delivered to the laboratory **promptly**.

(Continued)

Cerebrospinal Fluid Analysis: Overview *(Continued)*

Storage Instructions Do **not** store; place in the hands of a laboratory technologist. The laboratory should not discard such specimens for several days; need for additional testing may be subsequently recognized.

Special Instructions When a diagnosis of meningitis is considered, culture of other materials as well as culture of CSF may be helpful. Most children with bacterial meningitis are initially bacteremic, and blood cultures are of value. In neonates and small children, urine culture may be positive. Culture and Gram stain of petechiae may provide immediate diagnosis.

Reference Interval

- Premature neonate: <29 cells/mm^3
- Younger than 1 month: <32 cells/mm^3
- 1 month to 1 year: <10 cells/mm^3
- 1-4 years: <8 cells/mm^3
- 5 years to puberty: <5 cells/mm^3
- Adults: 0-5 cells/mm^3, all lymphocytes and monocytes; 0 red blood cells; protein: lumbar: 15-50 mg/dL, cisternal: 15-25 mg/dL, ventricular: 6-15 mg/dL; glucose: 50-80 mg/dL

Possible Panic Range Increased number of cells

Use Evaluate bacterial or viral encephalitis, meningitis, meningoencephalitis, mycobacterial or fungal infection, parasitic infestations, primary or secondary malignancy, leukemia/malignant lymphoma of CNS, trauma, vascular occlusive disease, vasculitis, heredofamilial and/or degenerative processes. **Table 1** outlines findings in selected commonplace entities.

The nucleated blood cell count in the cerebrospinal fluid is the most sensitive CSF test for bacterial meningitis. In patients who have not received antimicrobial agents the ultimate diagnosis of bacterial meningitis is based on results of culture. The cell count and differential count are mandatory, and a Gram stain must be examined promptly in work-up of possible meningitis. Glucose and protein are important in the work-up for possible meningitis. In a study of aseptic versus bacterial meningitis, the mean CSF cell counts in aseptic and bacterial meningitis were respectively 228 cells/mm^3 (range 6-2650) and 4035 cells/mm^3 (range 16-17,650). **Early viral infection** as well as **bacterial meningitis** can elicit neutrophil leukocytosis in blood and CSF. In viral meningitis, a shift to mononuclear predominance often occurs subsequently. In 205 cases of acute **viral** meningitis, the 25th percentile, median, and 75th percentile leukocyte counts (10^6/L) were 37/100/250, with 3%, 33%, and 75%, respectively PMNs. In 217 cases of acute **bacterial** meningitis, leukocyte counts (10^6/L) were 330/1195/4400 with 70%, 86%, and 97% PMNs, respectively. Seasonal curves for viral and bacterial infection go in opposite directions; viral meningitis is a disease of midsummer, while bacterial meningitis is relatively more common in the winter.

The differential diagnosis of a number of entities requires clinical and laboratory correlation. Botulism is characterized by descending paralysis, but both botulism and tick paralysis can mimic the Guillain-Barré syndrome. Guillain-Barré syndrome progresses to peak neurologic impairment within 4 weeks; in tick paralysis, CSF is normal and progression to peak paralysis requires only hours to a few days. See **Table 2** on page 358.

Signals of possible meningeal infection in the newborn: Leukocyte count >30 cells/mm^3 with more than 60% PMNs, CSF protein >100 mg/dL, CSF glucose lower than 40% of the level in blood are findings in bacterial meningitis. CSF WBC >10 cells/mm^3 in very young infants, and 5 cells/mm^3 in older infants and children with >1 PMN/mm^3 is abnormal.

Limitations A traumatic (bloody) tap may make interpretation difficult. Normal CSF may be found early in meningitis.

Contraindications Lumbar puncture (LP) is generally considered unsafe and contraindicated in cases of brain abscess proven by neuroimaging. LP may cause tentorial herniation and death. In cases of suspected or possible abscess, neuroimaging, usually with contrast, should be done to exclude brain abscess prior to LP. CSF findings with brain abscess may be normal, or resemble those of bacterial meningitis.

Table 1. Differential Diagnosis of Initial Cerebrospinal Fluid Findings in Suppurative Diseases of the Central Nervous System and Meninges

Condition	Pressure (mm H$_2$O)	Leukocytes/mm^3	Protein (mg/dL)	Sugar (mg/dL)	Specific Findings
Acute bacterial meningitis	Usually elevated	Several hundred to >60,000; usually a few thousand; occasionally <100 (especially meningococcal or early in disease); PMNs[1] predominate	Usually 100–500, occasionally >1000	<40 in >60% of cases (normal glucose does not rule out bacterial meningitis)	Organism usually seen on smear or culture in >90% of cases
Acute viral encephalitis	Usually elevated	Few to >1000; PMNs early, followed by mononuclear cells	Moderately elevated	Normal to slightly reduced	No organisms seen on smear or culture; oligoclonal bands may be found
Aseptic meningitis	Slightly increased	10 to >3000 (avg 50–500); neutrophils early, quickly replaced by mononuclear cells	Normal to slightly elevated, usually <100 mg/dL	Normal to <40 mg/dL	Enteroviruses cause >80% of cases
Herpetic meningitis		Mononuclear pleocytosis	Slightly elevated	Normal to slightly low	Etiology HSV-2 by viral culture or PCR
Tuberculous infection	Usually elevated; may be low with dynamic block in advanced stages	Usually 25–100, rarely more than 500; lymphocytes predominate, except in early stages when PMNs may account for 80% of the cells	Nearly always elevated, usually 100–200; may be much higher if dynamic block	Usually reduced; <50 in 75% of cases	Acid-fast organisms may be seen on smear of protein coagulum (pellicle) or recovered from inoculated guinea pig or by culture
Cryptococcal infection	Usually elevated; average, 225	Average, 50 (0–800); lymphocytic pleocytosis; diminished or absent inflammatory response may be found in subjects with and without AIDS	Average, 100; usually 20–500	Reduced in >50% of cases; average 30; often higher in patients with concomitant diabetes mellitus	Organisms may be seen in India ink preparation and on culture (Sabouraud's medium); will usually grow on blood agar; may produce alcohol in cerebrospinal fluid from fermentation of glucose; culture is imperative; large volumes of CSF are recommended for culture (5–10 mL).
Syphilis (acute)	Usually elevated	Average, 500; usually lymphocytes; rare PMNs	Average, 100; globulin often high, with abnormal colloidal gold curve	Normal (rarely reduced)	Positive results of reagin test for syphilis; spirochetes not demonstrable by usual techniques of smear or culture
CMV polyradiculopathy			Marked increase		CMV inclusions on cytology examination

[1]Polymorphonuclear leukocytes.

Modified from Feigin RD and Cherry JD, eds, *Textbook of Pediatric Infectious Diseases*, Volume 1, Philadelphia, PA: WB Saunders Co, 1992, 410.

Cerebrospinal Fluid Analysis: Overview *(Continued)*

Table 2. Features of Four Similar Syndromes of Ascending Paralysis

Feature	Tick Paralysis	Guillain-Barré Syndrome	Spinal Cord Lesion	Poliomyelitis
Rate of progression	Hours to days	Days to weeks	Gradual or abrupt	Days to weeks
Babinski sign	Absent	Absent	Present	Absent
Sensory loss	None	Mild	Present	None
Meningeal signs	Absent	Rare	Absent	Present
Fever	Absent	Rare	Absent	Present
Cerebrospinal fluid findings:				
Protein level	Normal	High	Normal or high	High
White cell count/mm^3	<10	<10	Variable	>10
Time to recovery	<24 hours after tick removal	Weeks to months	Variable, depending on cause	Months to years or no recovery (permanent paresis)

Modified from Felz MW, Smith CD, and Swift TR, "Brief Report: A Six-Year-Old Girl With Tick Paralysis," *N Engl J Med*, 2000, 342(2):90-4.

Methodology Manual cell count using hemacytometer. Electronic cell counters lack precision and validity when used to analyze most CSF specimens (background is high relative to total white cell count). Differential cell study using manually prepared smears or cytocentrifuge methods. Bovine albumin, 22%, mixed with spinal fluid specimen is helpful in maintaining the morphologic integrity of smeared cells. For chemical, microbiologic, and other analyses, see appropriate entries.

Additional Information More extensive testing: Cytology, conventional cultures and cultures for mycobacteria, fungi, and viruses, additional chemistry and serologic determinations must usually be ordered separately. Gram stain of punch biopsy or needle aspiration of hemorrhagic skin lesions in meningococcemia can provide rapid diagnosis.

Correction for traumatic tap: If cell counts and protein determinations are performed on CSF and blood obtained at the same time, correction for "bloody tap" can be calculated. All CSF measurements should be made from the same tube. The ratio of RBC count CSF to RBC count blood provides a factor which when multiplied by the blood WBC count or blood protein level indicates the expected level of contribution of these parameters from the blood to the spinal fluid. These contributed WBC or protein values can then be subtracted from the respective values measured in the spinal fluid. For example:

1. RBCs (CSF)/RBCs (blood) x WBCs (blood) or x protein (blood).
2. WBCs (CSF) or protein (CSF) - product calculated in 1 = true CSF WBC or true CSF protein.

If the peripheral blood is normal and traumatic tap occurs, about 1 WBC is added to the CSF for each 700 RBCs that have been transferred into the CSF. RBC contamination from a traumatic tap does not adversely affect the laboratory diagnosis of bacterial meningitis. Prompt centrifugation and analysis of xanthochromic fluid can help to differentiate a traumatic tap from bloody CSF. Clearing of RBCs from tubes 1-4 suggests traumatic tap.

When only a small amount of CSF can be obtained, Gram stain and culture must always have priority over antigen detection testing. **Antigen detection methods, if done at all, cannot replace culture and Gram stain.** Gram stain of spun sediment is positive in 60% to 80% of cases of bacterial meningitis. A threshold ≥50 leukocytes/µL CSF has been proposed as justification for bacterial antigen testing. Another group found a CSF nucleated blood cell count <6/mm^3 provided a criterion for an abbreviated CSF evaluation. Correlation exists between bacterial concentration in CSF and the numbers of PMNs found.

There is substantial mortality and morbidity in subjects with bacterial meningitis. Despite early diagnosis and the use of appropriate therapy, both complications of meningitis and death may occur. As many as 50% of survivors of meningitis have some sequelae.

A "CSF leukocyte aggregation score" (LAS) has been developed and applied to the rapid differentiation of bacterial (potentially fatal) meningitis from viral meningitis (usually benign). The CSF LAS (ratio of cell clusters of more than 3 cells to individual cells in CSF smears) was standardized and showed <5% intra- and interobserver variability. The mean LAS was significantly higher in bacterial meningitis (32.1% with range of 0% to 84%) than in viral meningitis (0% with range of 0% to 16.6%). In children with meningitis of undefined etiology, the mean LAS score was 0% with a range of 0% to 20.7%. Compared with 11 other laboratory tests, the LAS was the best predictor of bacterial meningitis (odds ratio of 1.6:3.7 in a logistic regression model).

White cell pleocytosis is found in only 33% of patients with multiple sclerosis (MS). The white count rarely exceeds 50 cells/mm^3. Most patients with MS (66%) have normal total protein. In contrast, patients with Guillain-Barré syndrome usually have no excess white cells but show elevated CSF protein >40 mg/dL.

Nonsteroidal anti-inflammatory drugs, antibiotics, intravenous immunoglobulins, and monoclonal antibodies against the T_3 receptor are the most common causes of drug-induced aseptic meningitis. Neutrophilic pleocytosis with the remainder of clinical and CSF findings obscures distinction from infectious meningitis but resolution takes place several days after discontinuation of the drug.

References

Attia J, Hatala R, Cook DJ, et al, "Does This Adult Patient Have Acute Meningitis?" *JAMA*, 1999, 282(2):175-81.

Felz MW, Smith CD, and Swift TR, "Brief Report: A Six-Year-Old Girl With Tick Paralysis," *N Engl J Med*, 2000, 342(2):90-4.

Forgacs P, Geyer CA, and Freidberg SR, "Characterization of Chemical Meningitis After Neurological Surgery," *Clin Infect Dis*, 2001, 32(2):179-85.

Gray LD and Fedorko DP, "Laboratory Diagnosis of Bacterial Meningitis," *Clin Microbiol Rev*, 1992, 5(2):130-45.

Mazor SS, McNulty JE, and Roosevelt GE, "Interpretation of Traumatic Lumbar Punctures: Who Can Go Home?" *Pediatrics*, 2003, 111(3):525-8.

Michelow IC, Nicol M, Tiemessen C, et al, "Value of Cerebrospinal Fluid Leukocyte Aggregation in Distinguishing the Causes of Meningitis in Children," *Pediatr Infect Dis J*, 2000, 19(1):66-72.

Moris G and Garcia-Monco JC, "The Challenge of Drug-Induced Aseptic Meningitis," *Arch Intern Med*, 1999, 159(11):1185-94.

Norris CM, Danis PG, and Gardner TD, "Aseptic Meningitis in the Newborn and Young Infant," *Am Fam Phys*, 1999, 59(10):2761-70.

Schaumburg HH and Herskovitz S, "The Weak Child - A Cautionary Tale," *N Engl J Med*, 2000, 342(2):127-9.

Yogev R, "Meningitis," *Pediatric Infectious Diseases*, Hensen HB and Baltimore RD, eds, Philadelphia, PA: WB Saunders Co, 2002, 630-50.

Cerebrospinal Fluid and Plasma β-Amyloid(1-42)

Related Information

AD7c Neural Thread Protein, CSF or Urine *on page 106*
Apolipoprotein E, Plasma *on page 204*
Cerebrospinal Fluid Protein *on page 371*
Cerebrospinal Fluid Protein Electrophoresis *on page 374*
Complement Components, Overview *on page 437*

Synonyms A Beta(1-42); β-Amyloid42; Plasma β-Amyloid(1-42)

Applies to AD7c-Neuronal Thread Protein; Amyloid Precursor Protein (APP), CSF; Creutzfeldt-Jakob Disease; CSF 14-3-3 Protein Assay; Presenilin; Tau Protein

Abstract Progressive deterioration in cognitive function gradually develops in Alzheimer disease (AD). Extracellular senile plaques and intracellular neurofibrillary tangles accumulate in the brains of affected patients in both familial and sporadic forms. Senile plaques are formed by deposition of β-amyloid peptides, which are toxic to neurons. The predominant form of β-amyloid in AD is β-amyloid42 which aggregates more readily than do shorter forms of β-amyloid. CSF levels of β-amyloid(1-42) are decreased in the majority of patients with AD. (Continued)

Cerebrospinal Fluid and Plasma β-Amyloid(1-42)
(Continued)

The explanation for low concentrations of β-amyloid$_{42}$ in CSF in AD, while increased amounts are found in brain, is obscure. The lowest concentrations occur in subjects with early-onset AD and in those with Ae4/e4 genotype.

Patient Preparation See Cerebrospinal Fluid Protein *on page 371*.

Aftercare Patient should be maintained supine for a few hours postpuncture in an attempt to avoid postpuncture headache.

Specimen Cerebrospinal fluid, plasma

Collection Lumbar puncture

Storage Instructions Refrigerate

Causes for Rejection Specimen contaminated with blood

Turnaround Time Currently, β-amyloid$_{(1-42)}$, tau protein, and myelin basic protein levels are available only from a few referral/research laboratories.

Special Instructions Advance preparation (following the recommendations of research/referral laboratory) should be made prior to obtaining specimen (see Limitations).

Reference Interval CSF cutoff >1130 pg/mL

Critical Values CSF concentrations of β-amyloid are decreased in AD to about 50% of those in controls. Alternatively, the CSF ratio of β-amyloid$_{42}$ to β-amyloid$_{40}$ is increased in AD, early onset autosomal dominant forms.

Use The interest in using amyloid precursor protein derivatives as laboratory markers to help confirm a clinical diagnosis of Alzheimer disease arises from evidence that cerebral accumulation of the β-amyloid$_{(1-42)}$ (Aβ42) peptide occurs in many cases of Alzheimer disease. Low levels are not dependent on disease stage.

Limitations Different sizes of amyloid fragments exist in CSF. Measurement of total β-amyloid fragments does not correlate with the presence of AD and has no clinical utility.

Measurement of plasma Aβ40 levels is not useful in diagnosis of AD, but elevated plasma levels of Aβ$_{(1-42)}$ can be detected several years prior to onset of symptoms in incipient AD. Plasma β-amyloid$_{(1-42)}$ concentrations are increased in patients with Down syndrome, but are similar to levels from other mentally retarded individuals.

Currently, this test is essentially unavailable in the United States from major general reference/research laboratories. The CSF 14-3-3 protein assay (for diagnosis of Creutzfeldt-Jakob disease - see below) is performed as a service by the Laboratory of CNS Studies, NINDS, National Institutes of Health, Bethesda, MD (301-496-4821).

Different groups have used different sets of antibodies.

More autopsy correlative studies are needed.

Methodology Immunoblotting, enzyme-linked immunosorbent assay (ELISA)

Additional Information AD usually begins in the seventh to ninth decades of life, but an early-onset form is well recognized, and histopathologic findings of Alzheimer encephalopathy appear in the brains of patients with Down syndrome. Several studies of Aβ42 in CSF have shown a decrease in the level of β-amyloid peptides in patients with AD compared to age-matched normal or neurologic disease control subjects. The probable biological explanation for a decrease in CSF Aβ42 is that the levels decrease as the peptide becomes increasingly insoluble and deposits in senile plaques.

The amyloid gene encodes a large protein called amyloid precursor protein (APP). Sequential proteolytic cleavages of APP by proteases referred to as beta and gamma secretases result in formation of two sizes of β-amyloid peptides (Aβ), containing 40 or 42 amino acids. Mutations in the APP and presenilin genes are responsible for the relatively infrequent autosomal dominant form of early-onset AD. The E4 allele of apolipoprotein E is a strong susceptibility gene for the development of AD in individuals in their 60s and 70s, apparently affecting age of onset. It appears to somehow enhance Aβ42 aggregation and intracellular deposition. Other risk factors are cognitive impairment and family history, including presence of Down syndrome. Elevation of plasma concentration of the 42-residue β-amyloid is found in presymptomatic carriers of mutations which are associated with familial AD.

Combined application with tau protein improves diagnostic classification: high tau levels with low Aβ42 are found in the CSF of most subjects with AD. Abnormal highly phosphorylated forms of the microtubule-associated tau protein are major constituents of the paired helical filaments that form neurofibrillary tangles in Alzheimer disease. Significant increase of CSF tau protein has been found in AD patients of five different national origins. Tau protein in CSF may provide better discrimination of brain dysfunction in those younger than age 70 than in older patients. Traumatic brain injury/environmental insults (with resultant neuronal or axonal injury) may elevate CSF amyloid peptide Aβ42, and correlate with increase in CSF tau protein. There are recent reports of increased CSF tau protein in multiple sclerosis and in normal pressure hydrocephalus. In cases of dementia with Lewy bodies (DLB), the second most common neurodegenerative disease that causes dementia after AD, CSF Aβ42 levels are decreased but CSF tau levels have been reported as normal. These findings in CSF are consistent with studies of DLB brain in which numerous senile plaques but few NFTs are usually found on neuropathologic study.

Decreased CSF β-amyloid$_{(1-42)}$ has been reported in patients with Creutzfeldt-Jakob disease (CJD). Interest has developed in 14-3-3 protein as a possible marker for the *in vivo* diagnosis of CJD. The 14-3-3 proteins are involved in the regulation of protein phosphorylation and the mitogen-activated protein kinase pathway. Apparent false-negative results occur as well as false-positive results as reported in patients with herpes simplex encephalitis, multi-infarct dementia, acute infarction, stroke, subarachnoidal hemorrhage, viral encephalitis, and carcinomatous meningitis. A commercial kit for 14-3-3 protein utilizing chemiluminescence is available (Amersham Buchler); see Limitations (above) concerning a resource performing the 14-3-3 protein assay as a service.

Recently, CSF tau and AD7c-neuronal thread protein (each measured by enzyme-linked immunosorbent assay) have been suggested (Kahle et al) as biomarkers for AD (together with reported specificity of 93% and sensitivity of 63%); their utility remains under investigation. Combined use of CSF Aβ42, tau, and F$_2$-isoprostane (a biomarker of oxidative damage) has been studied (small patient sample, however) and found to have specificity of 89% and sensitivity of 84%.

References

Andreasen N, Hesse C, Davidsson P, et al, "Cerebrospinal Fluid β-Amyloid$_{(1-42)}$ in Alzheimer Disease: Differences Between Early- and Late-Onset Alzheimer Disease and Stability During the Course of Disease," *Arch Neurol*, 1999, 56(6):673-80.

Hulstaert F, Blennow K, Ivanoiu A, et al, "Improved Discrimination of AD Patients Using β-amyloid$_{(1-42)}$ and Tau Levels in CSF," *Neurology*, 1999, 52(8):1555-62.

Jacobs DS, DeMott WR, Oxley DK, et al, *Laboratory Test Handbook*, 5th ed, Hudson, OH: Lexi-Comp Inc, 2001.

Kahle PJ, Jakowec M, Teipel SJ, et al, "Combined Assessment of Tau and Neuronal Thread Protein in Alzheimer's Disease CSF," *Neurology*, 2000, 54(7):1498-504.

Martin JB, "Molecular Basis of the Neurodegenerative Disorders," *N Engl J Med*, 1999, 340(25):1970-8.

Mayeux R, Tang MX, Jacobs DM, et al, "Plasma Amyloid Beta-Peptide 1-42 and Incipient Alzheimer's Disease," *Ann Neurol*, 1999, 46(3):412-6.

Montine TJ, Kaye JA, Montine SK, et al, "Cerebrospinal Fluid Aβ42, Tau, and F$_2$-Isoprostane Concentrations in Patients With Alzheimer Disease, Other Dementias, and in Age-Matched Controls," *Arch Pathol Lab Med*, 2001, 125(4):510-12.

Moussavian M, Potolicchio S, and Jones R, "The 14-3-3 Brain Protein and Transmissible Spongiform Encephalopathy," *N Engl J Med*, 1997, 336(12):873-4.

Otto M, Esselmann H, Schulz-Schaeffer W, et al, "Decreased β-Amyloid$_{1-42}$ in Cerebrospinal Fluid of Patients With Creutzfeldt-Jakob Disease," *Neurology*, 2000, 54(5):1099-102.

Sunderland T, Linker G, Mirza N, et al, "Decreased β-Amyloid$_{(1-42)}$ and Increased Tau Levels in Cerebrospinal Fluid of Patients With Alzheimer's Disease," *JAMA*, 2003, 289(16):2094-103.

♦ **Cerebrospinal Fluid Angiotensin Converting Enzyme** *see* Angiotensin Converting Enzyme, Serum *on page 159*

♦ **Cerebrospinal Fluid Bacterial Antigen Testing** *see* Bacterial Antigens, Rapid Detection Methods *on page 228*

Cerebrospinal Fluid Cytology
Related Information
Bacterial Culture, Cerebrospinal Fluid *on page 236*
Beta$_2$-Microglobulin, Serum or Urine *on page 254*
(Continued)

Cerebrospinal Fluid Cytology (Continued)

Cerebrospinal Fluid Analysis: Overview on page 355
Cerebrospinal Fluid Protein Electrophoresis on page 374
Cryptococcal Antigen Titer on page 482
Cytomegalovirus Cytology on page 497
Flow Cytometry, Overview on page 592
Fungal Culture, Cerebrospinal Fluid on page 621
Immunoperoxidase Procedures on page 780
Mycobacterial Culture, Cerebrospinal Fluid on page 931
Polymerase Chain Reaction on page 1069
Viral Culture on page 1307

Test Includes Depending on the clinical setting, immunohistochemistry, flow cytometry, and other ancillary tests can be performed on cerebrospinal fluid (CSF) specimens.

Abstract Cytologic examination of CSF fluid may be used as a diagnostic test in patients with clinical or neuroimaging evidence of central nervous system (CNS) abnormality.

Specimen CSF specimens should be collected in sterile, leakproof, plastic containers and sent immediately to the Cytopathology Laboratory for processing. The second or third tube collected is usually preferred for cytologic evaluation. The volume of CSF sent to cytopathology is usually 1-3 mL. If malignant disease is suspected, as much as 10 mL has been recommended for adults and older children.

Storage Instructions If a delay in processing must occur, the CSF specimen must be promptly refrigerated. Without immediate processing, degeneration of cells in the spinal fluid begins within 20 minutes at room temperature.

Special Instructions Indicate the type of CSF specimen obtained (lumbar puncture versus shunt/reservoir or other source) as well as all pertinent clinical information.

Use CSF cytology may aid in the diagnosis of neoplasm, infection, trauma, vascular disorders or demyelinating disease. Depending on the clinical impression, CSF cytology may be used in conjunction with microbiologic studies and hematologic evaluation to maximize diagnostic yield. Secondary malignancies, including lymphoma/leukemia and metastases, are much more commonly diagnosed in the CSF than are primary brain tumors. Common sites of origin of tumors metastatic to the brain include lung, breast, gastrointestinal tract including stomach and pancreas, kidney, bladder, prostate, and ovary. CNS metastases are also frequently seen with melanoma and choriocarcinoma. In general, most primary brain tumors are deeply situated in the brain parenchyma and not amenable to diagnosis by exfoliative cytology. Medulloblastoma is the primary brain tumor most commonly diagnosed in the CSF.

Limitations Only processes that exfoliate cells and involve the leptomeninges or are located adjacent to the ventricles can be diagnosed with cerebrospinal fluid cytology. Thus, a negative CSF cytology does not exclude malignancy. A repeat evaluation may be helpful. False negatives remain a problem. Certain procedures including lumbar puncture and myelography as well as chemotherapy and radiation may produce striking reactive changes that may be mistaken for malignancy.

References

DeMay RM, "Cerebrospinal FLuid," The Art and Science of Cytopathology - Exfoliative Cytology, Chapter 11, Chicago, IL: ASCP Press, American Society of Clinical Pathologists, 1996, 427-62.
Rosenthal DL, "Cytopathology of the Central Nervous System," Compendium on Diagnostic Cytology, 8th ed, Wied GL, Bibbo M, Keebler CM, et al, eds, Chicago, IL: Tutorials of Cytology, 1997, 303-9.
Stanton C and Stanley MW, "Cytopathology of Cerebrospinal Fluid and Brain," Atlas of Difficult Diagnoses in Cytopathology, Chapter 8, Atkinson BF and Silverman JF, eds, Philadelphia, PA: WB Saunders Co, 1998, 259-98.

Cerebrospinal Fluid Glucose

Related Information

Bacterial Culture, Cerebrospinal Fluid on page 236
Body Fluid Glucose on page 294
Cerebrospinal Fluid Analysis: Overview on page 355
Cerebrospinal Fluid Lactate on page 369
Cerebrospinal Fluid Protein on page 371
Cerebrospinal Fluid Protein Electrophoresis on page 374

Synonyms CSF Glucose; Glucose, Cerebrospinal Fluid; Spinal Fluid Glucose

Applies to Hypoglycorrhachia; Neuroglycopenia

Abstract For diagnosis of meningitis, culture and Gram staining have priority over all other testing when only a small quantity of cerebrospinal fluid (CSF) is available. Cell count with differential deserve the next priority, followed by glucose and protein.

Patient Preparation Blood (ie, plasma) glucose is needed also. Ideally, it should be drawn 2 hours before the lumbar puncture, the equilibration time.

Specimen Cerebrospinal fluid

Container Clean, sterile CSF tube

Reference Interval 40-70 mg/dL (SI: 2.2-3.9 mmol/L) in fasting patients, should be interpreted with plasma glucose. Values may be somewhat higher in infants and young children, 60-80 mg/dL (SI: 3.4-4.5 mmol/L). CSF glucose should be 60% to 70% of plasma glucose.

Use CSF glucose values may be helpful in the distinction of bacterial versus viral meningitis; the CSF glucose values are low (classically <40 mg/dL [SI: <2.2 mmol/L]) in bacterial and tuberculous meningitis. CSF glucose is generally normal in viral disease, but occasionally it may be low with aseptic meningitis.

Limitations Falsely decreased levels may result from cellular and bacterial utilization of glucose if the test is not performed immediately. Xanthochromic samples may give misleading results. The sensitivity of CSF glucose for bacterial meningitis was only 72% in a series from Minnesota, inferior to the sensitivity of the nucleated blood cell count, but multiple tests are used together. Bloody taps cause falsely increased glucose, since there is more glucose in blood.

Methodology Same procedures as used for blood glucose (eg, glucose oxidase, hexokinase reactions); a reagent strip and a handheld analyzer method have been reported as reliable for use in the bedside determination of CSF glucose.

Additional Information A major textbook of pediatrics points out that acute viral meningitis is often differentiated from acute bacterial meningitis because the latter is characterized by a CSF glucose <30 mg/dL, a CSF glucose:blood glucose ratio <0.2-0.3 as well as a protein >200 mg/dL, a CSF PMN count >1000/mm^3, and an 80% to 90% likelihood of positive Gram stain in an illness often occurring during the winter in a child younger than 2 years of age. **The gold standard for the diagnosis of bacterial meningitis is the culture,** which is fundamental to appropriate diagnosis and treatment. Sarcoidosis and neurosyphilis are reported causes of low CSF glucose. Other very uncommon causes of low CSF glucose include meningeal cysticercosis, trichinosis, and with the chemical meningitis which accompanies intrathecal therapy. Low CSF glucose may also occur in subarachnoid hemorrhage and neoplasia (eg, medulloblastoma). Low CSF glucose may be found in CNS leukemia.

References

Behrman RE, Kliegman RM, and Jenson HB, *Nelson Textbook of Pediatrics*, 16th ed, Philadelphia, PA: WB Saunders Co, 2000, 1793-802.

Olukoga AO, Bolodeoku J, and Donaldson D, "Cerebrospinal Fluid Analysis in Clinical Diagnosis," *J Clin Pathol*, 1997, 50(3):187-92.

Spanos A, Harrell FE Jr, and Durack DT, "Differential Diagnosis of Acute Meningitis. An Analysis of the Predictive Value of Initial Observations," *JAMA*, 1989, 262(19):2700-7.

Cerebrospinal Fluid Glutamine

Related Information

Ammonia, Plasma *on page 150*

Synonyms CSF Glutamine; Glutamine, Spinal Fluid

Abstract CSF glutamine is increased in metabolic diseases causing hyperammonemia (liver disease, urea cycle defects, certain organic acidurias).

Specimen Cerebrospinal fluid

Container Clean, sterile CSF tube

Collection Tube should be labeled with the number indicating the sequence in which tubes were obtained.

Storage Instructions With immediate deproteinization CSF samples may be stored for 9 months at -80°C.

Causes for Rejection Samples contaminated with red blood cells

Special Instructions Specimen must be transported **immediately** to the laboratory.

(Continued)

Cerebrospinal Fluid Glutamine (Continued)

Reference Interval Large intermethod variability. Typical intervals are:
- neonates: 6.74-13.94 mg/dL
- 3 months to 2 years: 5.97-8.57 mg/dL
- 2-10 years: 5.56-7.96 mg/dL
- adults: 6-15 mg/dL

See Methodology.

Use Evaluate encephalopathy, including coma and Reye syndrome

Limitations This test is not used extensively.

Methodology Enzymatic, amino acid analyzer, high performance liquid chromatography (HPLC), capillary-isotachophoresis. Glutamine/glutamate levels vary widely with different methods and with different specimen management protocols.

References

Olukoga AO, Bolodeoku J, and Donaldson D, "Cerebrospinal Fluid Analysis in Clinical Diagnosis," *J Clin Pathol*, 1997, 50(3):187-92.

Painter RC, Cope JY, and Smith JL, "Reference Information for Clinical Laboratory," *Tietz Textbook of Clinical Chemistry*, 3rd ed, Burtis CA and Ashwood ER, eds, 1999, Philadelphia, PA: WB Saunders Co, 1815.

Scriver CR, Kaufman S, Eisensmith RC, et al, "Amino Acids," Part 5, *The Metabolic and Molecular Basis of Inherited Disease*, 7th ed, Scriver CR, Beaudet AL, Sly WS, et al, New York, NY: McGraw-Hill Inc, 1995, 1015-368.

Cerebrospinal Fluid Glycine

Related Information

Amino Acids, Plasma *on page 143*
Amino Acids, Urine *on page 145*

Synonyms CSF Glycine; Glycine, Cerebrospinal Fluid

Applies to Isovaleric Aciduria; Methylmalonic Aciduria; Nonketotic Hyperglycinemia; Propionic Aciduria

Test Includes Cerebrospinal fluid (CSF) and plasma glycine

Abstract Patients with nonketotic hyperglycinemia (NKH) develop rapidly progressive neurological symptoms. Laboratory studies are negative except that glycine is elevated in plasma, urine, and CSF. Increase in CSF:plasma glycine ratio is more diagnostic than absolute CSF or plasma values in the diagnosis of NKH.

Specimen Cerebrospinal fluid; blood sample should also be drawn at the same time for plasma glycine.

Container CSF tube, green top (heparin) tube for blood

Sampling Time Fasting sample is preferred.

Causes for Rejection Traumatic CSF tap (falsely increases glycine values)

Reference Interval
- Plasma: 120-375 µmol/L
- CSF: 3-10 µmol/L
- CSF:plasma ratio: 0.01-0.04

Use CSF and plasma glycine levels are used in the diagnosis of nonketotic hyperglycinemia. The following pathological values have been reported.

Plasma:
- neonatal type: 460-2580 µmol/L
- late-onset type: 340-920 µmol/L

CSF: neonatal type: 33-440 µmol/L

CSF:plasma ratio:
- neonatal type: 0.09-0.25 (diagnostic)
- late-onset type: 0.06-0.10

Other authorities report that a CSF:plasma ration ≥0.08 is diagnostic (Gilbert-Barness et al).

Limitations With traumatic tap, contamination of CSF with blood makes the test invalid.

Methodology Amino acid analyzer (ion-exchange chromatography), high performance liquid chromatography (HPLC)

Additional Information NKH is an autosomal recessive disorder, due to deficiency of the glycine cleavage system. Several mutations have been identified in the glycine cleavage enzyme complex. The disorder is clinically divided into two major types: neonatal and late-onset. Neonatal patients present with lethargy, hypotonia, and myoclonic jerks, generally progressing to apnea and often death. Those who survive such episodes develop intractable seizures and profound mental retardation. A minority of the patients develop symptoms later in life. These patients present with progressive spastic diplegia and optic atrophy. However, these patients do not develop seizures, and their intellectual function remains preserved.

References

Gilbert-Barness E and Barness LA, *Metabolic Diseases*, Natick, MA: Eaton Publishing, 2000, 46-8.

Hamosh A, Johnston MV, and Valle D, "Nonketotic Hyperglycinemia," *The Metabolic and Molecular Basis of Inherited Disease*, 7th ed, Scriver CR, Beaudet AL, Sly WS, et al, eds, New York, NY: McGraw-Hill Inc, 1995, 1337-48.

Tada K, "Disorders of Glycine and Imino Acids," *Physician's Guide to the Laboratory Diagnosis of Metabolic Diseases*, Blau N, Duran M, and Blaskovics, eds, London: Chapman & Hall, 1996, 201-8.

Cerebrospinal Fluid IgG:Albumin Ratio, IgG Index, and IgG Synthesis Rate

Related Information

Cerebrospinal Fluid Analysis: Overview *on page 355*
Cerebrospinal Fluid Immunoglobulin G *on page 367*
Cerebrospinal Fluid Oligoclonal Bands *on page 369*
Cerebrospinal Fluid Protein *on page 371*
Immunofixation Electrophoresis, Serum or Urine *on page 768*

Synonyms CSF IgG/CSF Albumin Ratio; IgG Ratios and IgG Index, Cerebrospinal Fluid

Applies to IgG Synthesis Rate; Oligoclonal Bands, CSF

Test Includes Protein measurements, frequently immunochemical, on CSF and/or serum with determination of ratio or ratios of one protein to another.

Abstract Multiple sclerosis is a relapsing and remitting demyelinating disease of the central nervous system for which there are no specific diagnostic laboratory tests. About 50% of multiple sclerosis patients have elevated CSF protein levels and about 75% have increased gamma globulin. Several tests have been devised to determine if the elevated gamma globulin is the result of intrathecal IgG synthesis. The most useful tests to aid in the diagnosis of multiple sclerosis are detection of oligoclonal bands in the CSF and estimations of the intrathecal synthesis of IgG. The detection of oligoclonal bands is the most helpful of the laboratory tests. Synthesis of IgG in the central nervous system is typically oligoclonal, for reasons that are unknown. Estimates of intrathecal protein synthesis include calculations of IgG index, CSF IgG index, and the IgG synthesis rate. All methods require quantitation of serum and CSF albumin and IgG. The most helpful of the estimates of intrathecal protein synthesis is the IgG index. An abnormal IgG index or IgG synthesis rate (indicative of increased CSF IgG production) is noted in some 90% of cases of clinically definite MS. The IgG synthesis rate is a complex formula which provides no more useful clinical information than the IgG index.

Patient Preparation See Cerebrospinal Fluid Protein *on page 371*.

Aftercare Patient should be maintained supine for a few hours postpuncture in an attempt to avoid postpuncture headache.

Specimen Cerebrospinal fluid and serum; minimum of 0.1-0.5 mL of CSF is required, preferably, 3 mL of CSF and one SST™ tube of blood. Contact referral laboratory for specific specimen requirements.

Container Clean, sterile CSF tube; red top tube or SST™ tube of blood

Collection Tube should be labeled with the number indicating the sequence in which tubes were obtained.

Storage Instructions Store CSF in refrigerator at 4°C.

Causes for Rejection CSF received without concurrently obtained serum, insufficient quantity of CSF, CSF contaminated with blood ("bloody tap")

Special Instructions Before performing lumbar puncture, communicate with laboratory to determine which measurements and ratios are offered or can be obtained on a referral basis. The third tube of routinely obtained three tube set of CSF should be used for these studies. Simultaneous detection of oligoclonal bands by immunofixation or isoelectric focusing is recommended.
(Continued)

Cerebrospinal Fluid IgG:Albumin Ratio, IgG Index, and IgG Synthesis Rate *(Continued)*

Reference Interval Published cutoff data utilizing receiver operator curves for CSF IgG index, CSF IgG:albumin ratio, and CSF IgG synthesis rate (mg/day) listed in the following table are from McMillan et al using rate nephelometry methods with specificity threshold set at 90%. Published information is relevant only to lumbar CSF. Ventricular, cisternal, or cervical CSF may have different reference ranges. Alternative reference ranges from *Tietz Textbook of Clinical Chemistry* include:

- CSF IgG index: <0.7
- CSF IgG:albumin ratio: <0.27
- CSF IgG synthesis rate: -9.9-3.3 mg/day

Values >0.7, 0.27, and 8 mg/day are interpreted as increased. Reference ranges used are dependent upon methodology and ideally are determined by each individual laboratory.

Formula	Specificity	Sensitivity	Cutoff Value	Predictive Value		
				Positive	Negative	
Index	90%	78%	0.7	86	83	96
Ratio	90%	62%	0.22	84	74	87
IgG synthesis rate	90%	47%	15	79	67	78

Cutoff values giving a specificity of 90% with the three formulae using rate nephelometry.

Adapted from McMillan SA, Douglas JP, Droogan AG, et al, "Evaluation of Formulae for CSF IgG Synthesis Using Data Obtained From Two Methods: Importance of Receiver Operator Characteristic Curve Analysis," *J Clin Pathol*, 1996, 49(1):24-8.

Use Oligoclonal bands and estimation of IgG production by the central nervous system are helpful in support of the diagnosis of multiple sclerosis, which is essentially a clinical diagnosis.

Limitations Conditions in which lymphoreticular elements of the CNS produce immunoglobulins may result in oligoclonal IgG synthesis. Such conditions include, but are not limited to, aseptic meningitis, lymphoma, subacute sclerosing panencephalitis (SSPE), sarcoidosis, neurosyphilis, Guillain-Barré syndrome, and cerebral lupus erythematosus. These conditions, however, are either uncommon and may be ruled out clinically or by finding a CSF total protein >100 mg/dL, a leukocyte count >50/µL, or a positive test for neurosyphilis. Oligoclonal bands and IgG index are less commonly positive in childhood cases of multiple sclerosis. Contamination of CSF by blood (even by small amounts) may elevate the IgG index and IgG synthesis rate.

Methodology Wide variety of generally immunochemical based methods (eg, rate nephelometry). See also Cerebrospinal Fluid Immunoglobulin G *on page 367*.

Additional Information Cerebrospinal fluid (CSF) is an ultrafiltrate of plasma lacking the highest molecular weight proteins (ie, IgM and alpha$_2$-macroglobulin). IgG may be synthesized in the CNS and normally comprises <10% of the total CSF protein. Elevated CNS IgG levels can be either a result of local synthesis or diffusion of plasma IgG across an altered blood brain barrier. Patients with demyelinating disorders often have elevated CSF IgG concentrations due to intrathecal synthesis. It is useful to determine the degree of permeability of the blood brain barrier when assessing CSF IgG synthesis, as increased CSF IgG may merely be a function of increased permeability of the blood-brain barrier. Albumin is a protein which is neither synthesized nor metabolized in the CSF compartment. Calculation of the CSF:serum albumin ratio allows one to assess the degree of permeability of the CSF.

The CSF/serum albumin index is calculated as:

CSF/serum albumin index = albumin$_{CSF}$ / albumin$_S$

where albumin$_{CSF}$ is expressed in mg/dL and serum albumin (albumin$_S$) is expressed in g/dL. An index value <9 is consistent with an intact barrier. In approximately 70% of multiple sclerosis cases, CSF albumin concentrations are not elevated.

The CSF IgG:albumin ratio is given as:

$$IgG_{CSF}:albumin_{CSF}$$

Both IgG_{CSF} and $albumin_{CSF}$ are expressed in mg/dL. This ratio is a method of standardizing the CSF IgG relative to albumin. This formula especially estimates CSF IgG correcting for the permeability of the blood brain barrier.

The CSF IgG index is calculated as:

$$(IgG_{CSF} / albumin_{CSF}) / (IgG_{serum} / albumin_{serum})$$

where IgG_{CSF} and $albumin_{CSF}$ are expressed in mg/dL and IgG_S and $albumin_S$ are expressed in g/dL. This ratio further refines the estimation of CSF IgG synthesis by estimating IgG as a function of both serum concentration and permeability of the blood brain barrier to albumin and IgG.

Although these ratios are useful clinically, they do not quantify the CNS IgG production rate. Tourtellotte CSF IgG synthesis rate is calculated as follows:

$$5 \left[\left(CSF\ IgG - \frac{serum\ IgG}{369} \right) - \left(\left(CSF\ alb - \frac{serum\ alb}{230} \right) \times \frac{0.43\ (serum\ IgG)}{serum\ alb} \right) \right]$$

Where alb = albumin

Protein concentrations are expressed in mg/dL. The numbers 369 and 230 are the average normal serum:CSF ratios for IgG and albumin, respectively, 0.43 is the molecular weight ratio of albumin: IgG and 5 is the daily CSF production expressed in dL. The reference interval is -9.9 to +3.3 mg/day. Negative values are considered normal. Multiple sclerosis patients usually have a synthesis rate >8.0. The CSF IgG synthesis rate (Tourtellotte) calculation is more complex but does not provide more clinical information. Therefore, it may not be routinely performed as part of multiple sclerosis panels. In the receiver operator curve study of McMillan et al, the CSF IgG synthesis rate had the lowest sensitivity of any index (47% sensitivity for IgG synthesis rate vs 78% sensitivity for the IgG index and 62% sensitivity for the IgG/albumin ratio).

Up to 20% of patients with neurosyphilis, SSPE, chronic fungal meningitis, and Guillain-Barré syndrome may have some of these protein abnormalities.

Patients with sciatica caused by lumbar disk herniation may leak plasma proteins into the CSF, resulting in increased CSF total protein. Disk patients with paresis may have significant increase in CSF total protein, CSF albumin and IgG, and in the CSF/serum albumin and IgG ratios but not in the IgG index (as compared to patients with lumbar disk herniation but without symptoms).

References

Cavuoti D, Baskin L, and Jialal I, "Detection of Oligoclonal Bands in Cerebrospinal Fluid by Immunofixation Electrophoresis," *Am J Clin Pathol*, 1998, 109(5):585-8.

Johnson AM, Rohlfs EM, and Silverman LM, "Proteins," *Tietz Textbook of Clinical Chemistry*, 3rd ed, Chapter 20, Burtis CA and Ashwood ER, eds, Philadelphia, PA: WB Saunders Co, 1999, 515-7.

McMillan SA, Douglas JP, Droogan AG, et al, "Evaluation of Formulae for CSF IgG Synthesis Using Data Obtained From Two Methods: Importance of Receiver Operator Characteristic Curve Analysis," *J Clin Pathol*, 1996, 49(1):24-8.

Skouen JS, Larsen JL, and Vollset SE, "Cerebrospinal Fluid Protein Concentrations Related to Clinical Findings in Patients With Sciatica Caused by Disk Herniation," *J Spinal Disord*, 1994, 7(1):12-8.

Willis MS and Bai Y, "Confounding Issues in the Diagnosis of Multiple Sclerosis: Lyme Disease Testing," *Lab Med*, 2003, 34(6):467-75.

Cerebrospinal Fluid Immunoglobulin G

Related Information

Cerebrospinal Fluid IgG:Albumin Ratio, IgG Index, and IgG Synthesis Rate *on page 365*
Cerebrospinal Fluid Oligoclonal Bands *on page 369*
Cerebrospinal Fluid Protein *on page 371*
Cerebrospinal Fluid Protein Electrophoresis *on page 374*

Synonyms Immunoglobulin G, Cerebrospinal Fluid

Replaces Colloidal Gold Curve

Abstract The pathogenesis of multiple sclerosis (MS) may involve immunoglobulins. Increased concentrations of immunoglobulins may be found in serum, cerebrospinal fluid (CSF), and brain of subjects with multiple sclerosis.

Patient Preparation See Cerebrospinal Fluid Protein *on page 371*.

(Continued)

Cerebrospinal Fluid Immunoglobulin G *(Continued)*

Aftercare Patient should be maintained supine for a few hours postpuncture in an attempt to avoid postpuncture headache.

Specimen Cerebrospinal fluid; usually at least 0.1-0.5 mL of CSF is required.

Container Clean, sterile CSF tube

Collection Tube should be labeled with the sequence in which tubes were obtained.

Storage Instructions Store in refrigerator.

Causes for Rejection CSF received without concurrently obtained serum, insufficient quantity of CSF, CSF contaminated with blood ("bloody tap")

Special Instructions The third tube of a routinely obtained three tube set of CSF should be used for CSF IgG study.

Reference Interval Normal CSF IgG: 3% to 12% of total CSF protein. IgG in CSF is normally <5 mg/dL. There is no evidence of diurnal variation.

Use Evaluate central nervous system involvement by infection, neoplasm, or primary neurologic disease (in particular, multiple sclerosis) Multiple sclerosis is essentially a clinical diagnosis.

Limitations Normal levels do not exclude disease; clinical correlation is needed. Patients in early phases of multiple sclerosis are those most likely to have normal CSF concentrations of IgG. Free kappa light chains are found less often than are oligoclonal bands, but provide greater specificity for multiple sclerosis than do intact IgG abnormalities.

Methodology Radial immunodiffusion (RID), electroimmunodiffusion, immunofluorometry, immunoprecipitation, immunonephelometry, rate immunonephelometry

Additional Information Cerebrospinal fluid protein is elevated in many conditions which affect the central nervous system primarily or secondarily. In inflammatory or destructive processes in which serum leaks into CSF, both IgG and albumin will be present in the CSF in increased amounts. Since albumin is not, but immunoglobulins are, synthesized in the central nervous system, a relative increase in CSF IgG indicates presence of a process involving the central nervous system primarily, in particular multiple sclerosis.

Clinical evidence of MS includes (but is not limited to) weakness, early fatigue, double vision, numbness and tingling, difficulty with coordination, and dizziness.

While MS patients are maintained on ACTH and/or steroid therapy or during remission, CSF IgG levels decrease but generally remain significantly elevated.

Electrophoresis of CSF may also be enlightening if oligoclonal gamma globulin bands are demonstrated, which also suggests, but is not diagnostic of, multiple sclerosis. See also Cerebrospinal Fluid Oligoclonal Bands *on page 369.*

While CSF laboratory findings alone cannot establish or exclude the diagnosis of MS and CSF total protein or albumin level is normal in most patients, CSF immunoglobulin is usually elevated relative to the other proteins with the implication of intrathecal synthesis. IgG is predominantly increased with excess of IgG lambda and kappa light chains. The IgG level is commonly ratioed to another CSF protein component. The Poser committee diagnostic criteria for MS use CSF laboratory findings to support a diagnosis of MS. A number of IgG ratios and the IgG index have been considered. The IgG index (normal value <0.66) is a "ratio of ratios" and compares CSF with serum parameters as follows:

$$\text{CSF Ig Index} = [\text{IgG}_{CSF}/\text{albumin}_{CSF}] / [\text{IgG}_{serum}/\text{albumin}_{serum}]$$

Tourtellotte's IgG synthesis rate has also been utilized. Quantitative measures of CNS IgG production are said to be less sensitive than isoelectric focusing of CNS protein (for oligoclonal bands). The use of both oligoclonal bands (sensitivity 90%) and IgG index (sensitivity 80%) test results increases the sensitivity to 95%.

References

Andersson M, Alvarez-Cermeno J, Bernardi G, et al, "Cerebrospinal Fluid in the Diagnosis of Multiple Sclerosis: A Consensus Report," *J Neurol Neurosurg Psychiatry,* 1994, 57(8):897-902.

Kyle RA and Katzmann JA, "Immunochemical Characterization of Immunoglobulins," Section C, Kyle RA, section ed, *Manual of Clinical Laboratory Immunology,* 5th ed, Nakamura RM, volume ed, Washington, DC: ASM Press, American Society for Microbiology, 1997, 172-4.

Olek MJ and Dawson DM, "Multiple Sclerosis and Other Inflammatory Demyelinating Diseases of the Central Nervous System," *Neurology in Clinical Practice: The Neurological Disorders*, 3rd ed, Volume 2, Chapter 60, Bradley WG, Daroff RB, Fenichel GM, et al, eds, Boston MA: Butterworth-Heinemann, 2000, 1431-65.

Rudick RA, "Multiple Sclerosis and Related Conditions," *Cecil Textbook of Medicine*, 21st ed, Chapter 482, Goldman L and Bennett JC, eds, Philadelphia, PA: WB Saunders Co, 2000, 2141-49.

Cerebrospinal Fluid Lactate

Related Information

Cerebrospinal Fluid Glucose *on page 362*
Lactic Acid, Whole Blood or Plasma *on page 827*

Synonyms CSF Lactate; Lactate, Cerebrospinal Fluid; Lactic Acid, Cerebrospinal Fluid; Spinal Fluid Lactate

Abstract Interest in CSF lactate is related to increases in bacterial meningitis.

Specimen Cerebrospinal fluid

Container Clean, sterile CSF tube

Storage Instructions Unstable at room temperature

Reference Interval Increased in first 2 weeks of life; adults: 10-22 mg/dL (SI: 1.1-2.4 mmol/L)

Use CSF lactate was proposed as a test to differentiate bacterial from other types of meningitis, but clinical effectiveness has not been established. CSF lactate is also elevated in cerebral infarct, Creutzfeldt-Jacob disease, cerebral hemorrhage, subarachnoid hemorrhage, primary CSF acidosis, malignant hypertension, hepatic encephalopathy, diabetes mellitus, hypoglycemic coma, and in the first 3 days following head injury. An entity of developmental delay with infantile seizures and with depressed CSF glucose and lactate levels is noteworthy, because it responds to a special diet.

Limitations As stated above, the test is nonspecific and of limited utility.

Methodology Enzymatic, gas-liquid chromatography (GLC), amperometric utilizing a lactate-sensitive electrode

References

Cameron PD, Boyce JM, and Ansori BM, "Cerebrospinal Fluid Lactate in Meningitis and Meningo-carcemia," *J Infect*, 1993, 26(3):245-52.

Fishman RA, *Cerebrospinal Fluid in Diseases of the Nervous System*, 2nd ed, Philadelphia, PA: WB Saunders Co, 1992.

Lebel MH, "Meningitis," *Oski's Pediatrics: Principles and Practice*, 3rd ed, McMillan JA, DeAngelis CD, Feigin RD, et al, eds, Philadelphia, PA: JB Lippincott Co, 1999, 413-6.

Sacks DB, "Carbohydrates," *Tietz Textbook of Clincal Chemistry*, 3rd ed, Burtis CA and Ashwood ER, eds, Philadelphia, PA: WB Saunders, 1999, 788.

Stacpoole PW, Bunch ST, Neiberger RE, et al, "The Importance of Cerebrospinal Fluid Lactate in the Evaluation of Congenital Lactic Acidosis," *J Pediatr*, 1999, 134(1):99-102.

♦ **Cerebrospinal Fluid Methotrexate** *see* Methotrexate, Serum or Plasma *on page 905*

Cerebrospinal Fluid Oligoclonal Bands

Related Information

Cerebrospinal Fluid IgG:Albumin Ratio, IgG Index, and IgG Synthesis Rate *on page 365*
Cerebrospinal Fluid Immunoglobulin G *on page 367*
Cerebrospinal Fluid Protein *on page 371*
Cerebrospinal Fluid Protein Electrophoresis *on page 374*

Synonyms Oligoclonal Bands, Cerebrospinal Fluid

Applies to Myelin Basic Protein

Test Includes High-resolution electrophoresis of cerebrospinal fluid (CSF) and serum obtained concurrently

Abstract Oligoclonal bands on CSF electrophoresis are typical of but not pathognomonic for multiple sclerosis.

Patient Preparation Fasting serum specimen is preferred (to avoid hyperlipemia). See Cerebrospinal Fluid Protein *on page 371*.

Aftercare Patient should be maintained supine for a few hours postpuncture in an attempt to avoid postpuncture headache.

Specimen Cerebrospinal fluid (5 mL) and serum obtained concurrently

Container Clean, sterile CSF tube; red top tube or SST™ tube of blood

Storage Instructions Serum and CSF specimens may be kept at room temperature for up to 6 hours or at 4°C to 8°C for up to 24 hours before testing. If there is greater delay, they should be stored at -20°C.

(Continued)

Cerebrospinal Fluid Oligoclonal Bands *(Continued)*

Causes for Rejection CSF contaminated with blood ("bloody tap"), CSF submitted without accompanying serum specimen

Reference Interval Normal CSF has no demonstrable oligoclonal bands.

Use Oligoclonal CSF bands contribute to the diagnosis of inflammatory and autoimmune disease of the CNS. In particular, they are found in 83% to 94% of subjects with clinically definite multiple sclerosis and in 100% of patients with subacute sclerosing panencephalitis (SSPE), and in other degenerative states as well (eg, presenile dementia). SSPE may be the result of long-standing measles infection. With widespread measles vaccination, it has been nearly eradicated.

CSF examination is useful in entities such as neuropsychiatric SLE to exclude infectious meningitis.

Limitations Test has a satisfactorily high level of sensitivity (~90%) for association with multiple sclerosis, but it is not specific. Serum protein electrophoresis must be run concurrently to assure that any CSF bands detected do not have origin in the serum (diffusion of serum bands into the CSF). A few cases of MS may have oligoclonal bands in both serum and CSF; the incidence increases with use of isoelectric focusing.

Although oligoclonal bands have been seen in neuropsychiatric SLE, they lack specificity.

Methodology Thin gel agarose high-resolution electrophoresis, immunofixation electrophoresis, isoelectric focusing; requires concentration of CSF.

Additional Information During CSF protein electrophoresis, IgG in normal spinal fluid migrates as a faint diffuse zone. In demyelinating diseases, IgG migrates as discrete bands called oligoclonal bands. An abnormal result is the finding of two or more oligoclonal bands in the CSF that are not present in a concurrent serum sample. Oligoclonal bands are present in the CSF of 90% of patients with multiple sclerosis. The occurrence of oligoclonal bands in both CSF and serum is seen in CLL, lymphoma, malignancies, autoimmune hepatitis, and viral illnesses.

Oligoclonal bands are not specific for MS but have been described in many other disorders, including subacute sclerosing panencephalitis, syphilis, Jakob-Creutzfeldt disease, encephalitis, Guillain-Barré syndrome, neurosyphilis, stroke, cerebral vasculitis, lupus erythematosus, and neoplasms. Immunoglobulin D oligoclonal bands have been noted in the CSF of some patients with central nervous system tumors. However, in most of these diseases, oligoclonal bands are uncommon.

The Optic Neuritis Treatment Trial has found that lumbar puncture (to detect CSF oligoclonal bands) does not add predictive value (5-year risk) of definite MS in patients who have abnormal magnetic resonance imaging (MRI) at the time of onset of monosymptomatic optic neuritis. Testing for oligoclonal bands, however, may assist assessment of risk for MS in patients with normal brain MRI at the onset of optic neuritis.

At least some MS patients have oligoclonal band patterns with EBNA-1 (Epstein Barr virus nuclear antigen-1) specific comigration. Some oligoclonal bands could be absorbed with EBNA-1 antigens. EBNA-1 may be a target epitope for some MS-associated oligoclonal bands.

Oligoclonal banding was present in 26% of asymptomatic, HIV seropositive men.

A symposium through the December 1997 issue of *Mayo Clinic Proceedings* began with a foreward in the July issue.

Myelin basic protein (MBP) assays have also been applied to the diagnosis, evaluation, and post-therapeutic monitoring of multiple sclerosis. The methods suffer from fragmentation and variation of the myelin antigen(s). Newer assays have increased sensitivity but specificity would be expected to be low.

References

Cole SR, Beck RW, Moke PS, et al, "The Predictive Value of CSF Oligoclonal Banding for MS 5 Years After Optic Neuritis," *Neurology*, 1998, 51(3):885-7.

Fortini AS, Sanders EL, Weinshenker BG, et al, "Cerebrospinal Fluid Oligoclonal Bands in the Diagnosis of Multiple Sclerosis. Isoelectric Focusing With IgG Immunoblotting Compared With

High-Resolution Agarose Gel Electrophoresis and Cerebrospinal Fluid IgG Index," *Am J Clin Pathol*, 2003, 120(5):672-5.

Keren DR, "Optimizing Detection of Oligoclonal Bands in Cerebrospinal Fluid by Use of Isoelectric Focusing With IgG Immunoblotting," *Am J Clin Pathol*, 2003, 120(5):649-51.

Mavra M, Drulovic J, Levic Z, et al, "CNS Tumors: Oligoclonal Immunoglobulin D in Cerebrospinal Fluid and Serum," *Acta Neurol Scand*, 1999, 100(2):117-8.

Ohta M, Ohta K, Ma J, et al, "Clinical and Analytical Evaluation of an Enzyme Immunoassay for Myelin Basic Protein in Cerebrospinal Fluid," *Clin Chem*, 2000, 46(9):1326.

Rodriguez M, "Multiple Sclerosis: Insights Into Molecular Pathogenesis and Therapy," *Mayo Clin Proc*, 1997, 72(7):663-4.

Wekerle H, "Immunology of Multiple Sclerosis," *McAlpine's Multiple Sclerosis*, 3rd ed, Chapter 12, Compston A, Ebers G, Lassmann H, et al, eds, New York, NY: Churchill Livingstone, 1998, 379.

Willis MS and Bai Y, "Confounding Issues in the Diagnosis of Multiple Sclerosis: Lyme Disease Testing," *Lab Med*, 2003, 34(6):467-75.

Cerebrospinal Fluid Protein

Related Information

Bacterial Antigens, Rapid Detection Methods *on page 228*
Bacterial Culture, Cerebrospinal Fluid *on page 236*
Cerebrospinal Fluid Analysis: Overview *on page 355*
Cerebrospinal Fluid and Plasma β-Amyloid$_{(1-42)}$ *on page 359*
Cerebrospinal Fluid Glucose *on page 362*
Cerebrospinal Fluid IgG:Albumin Ratio, IgG Index, and IgG Synthesis Rate *on page 365*
Cerebrospinal Fluid Immunoglobulin G *on page 367*
Cerebrospinal Fluid Lactate *on page 369*
Cerebrospinal Fluid Oligoclonal Bands *on page 369*
Cerebrospinal Fluid Protein Electrophoresis *on page 374*
Fungal Culture, Cerebrospinal Fluid *on page 621*
Mycobacterial Culture, Cerebrospinal Fluid *on page 931*
VDRL, Serum or Cerebrospinal Fluid *on page 1303*
Viral Culture *on page 1307*

Synonyms Protein, Cerebrospinal Fluid

Applies to Prealbumin, CSF; Tau Fraction

Test Includes Culture, Gram stain, cell count with differential, glucose, and protein are usually ordered together to work up possible meningitis.

Abstract Increased CSF protein may indicate presence of infectious/inflammatory, hemorrhagic, neoplastic or demyelinating disease of the CNS (central nervous system). Great increases may be found in acute bacterial meningitis, including tuberculous meningitis. See Cerebrospinal Fluid Analysis: Overview *on page 355*, in which protein concentrations in various entities are outlined.

Patient Preparation Patient should be informed, relaxed, and properly positioned for lumbar puncture. Signed consent form is required. Antiseptic agents must be carefully and properly applied to the proposed puncture site. Local anesthetic (to which patient is not allergic) is usually advised. Lumbar puncture is not without potential complications. If increased intracranial pressure due to a mass lesion (eg, abscess) is suspected, lumbar puncture should be deferred until appropriate neuroimaging study is performed and the issue resolved. Lumbar puncture is not without potential complications.

Aftercare Patient should be maintained supine for a few hours postpuncture in an attempt to avoid postpuncture headache.

Specimen Cerebrospinal fluid, usually at least 0.1-0.5 mL of CSF is required.

Container Clean, sterile CSF tube

Collection Tubes should be labeled with patient's name, hospital number, and date and time of collection.

Storage Instructions Do not store. Must be delivered to clinical laboratory immediately.

Turnaround Time 1-2 hours

Special Instructions Usually three tubes of CSF are collected for count and culture in addition to protein and glucose with collection of 1 mL in each tube labeled #1, #2, #3 in order of collection. **For diagnosis of meningitis, culture and then Gram staining have priority over all other testing**, when only a small quantity of cerebrospinal fluid (CSF) is available. **Cell count with differential deserves the next priority, followed by glucose and total protein.**

Reference Interval Lumbar CSF: 0-1 month: <150 mg/dL (SI: <1.5 g/L); 1-6 months: approximately 30-100 mg/dL (SI: 0.30-1.00 g/L); 6 months and up: approximately 15-45 mg/dL (SI: 0.15-0.50 g/L); elderly adults: 15-60 mg/dL (SI: 0.15-0.50 g/L). (Continued)

Cerebrospinal Fluid Protein (Continued)

Ventricular CSF protein is generally lower, 5-15 mg/dL. Cisternal CSF is midway, 15-25 mg/dL. See tables.

Infant Reference Values

CSF Total Protein Levels, First 3 Months of Life		
Analyzer-Vitros 700 (Ortho, formerly Kodak Ektachem)		
Age	mg/dL	SI (g/L)
1-8 d	26-135	0.26-1.35
8-30 d	26-115	0.26-1.15
1-2 mo	18-86	0.18-0.86
2-3 mo	10-74	0.10-0.74

Adapted from Evans RW, "Complications of Lumbar Puncture," *Neurol Clin N America*, 1998, 16(1):83-105.

Protein Analyte

Protein	Molecular Weight	Plasma Concentration (mg/L)	CSF Concentration (mg/L)	Plasma:CSF Ratio
Prealbumin	61,000	238	17.3	14
Albumin	69,000	36,600	155.0	236
Transferrin	81,000	2040	14.4	142
Ceruloplasmin	152,000	366	1.0	366
IgG	150,000	9870	12.3	802
IgA	150,000	1750	1.3	1346
Alpha$_2$ macroglobulin	798,000	2220	2.0	1111
Fibrinogen	340,000	2964	0.6	4940
IgM	800,000	700	0.6	1167
Beta lipoprotein	2,239,000	3728	0.6	6213

Data from Felgenhauser, 1974 as cited by Fishman RA, "Cerebrospinal Fluid in Diseases of the Nervous System," 2nd ed, Philadelphia, PA: WB Saunders Co, 1992 and Swaiman KF and Ashwal S, *Pediatric Neurology: Principles and Practice*, 3rd ed, St Louis, MO: Mosby, 1999.

Hitachi (turbidimetric method) has been found to give values 21% lower, on average, than the Vitros (copper binding Biuret colorimetry based) method.

Use Increased with bacterial meningitis including tuberculous meningitis; brain abscess, meningovascular syphilis, diabetes mellitus, CVA (including cases in which no hemorrhage has occurred), arachnoiditis, dehydration, some major depressive disorders, some patients with sciatica due to lumbar disk herniation, POEMS syndrome (**P**olyneuropathy, **O**rganomegaly, **E**ndocrinopathy, **M** protein, and **S**kin changes), drug effects, and subarachnoid hemorrhage. Protein is normal or **slightly** high in psychiatric disease. Used for differential diagnosis of multiple sclerosis; encephalomyelitis; other degenerative processes causing neurologic disease, some neoplastic diseases, some cases of myxedema and other instances of endocrine disorders, traumatic tap, and in CSF recovered from below the level of an obstruction of the spinal cord.

Decreased CSF protein falls rapidly during the first few months of life, followed by a low plateau in the 20-30 mg/dL general range from age 6 months to 10 years. CSF protein is decreased with dilution from water intoxication, CSF leak (CSF rhinorrhea or otorrhea, leaks following lumbar puncture, other fistulas), with removal of large volumes of CSF, in some patients with benign intracranial hypertension, in some leukemic subjects, and with hyperthyroidism. Low CSF protein is 10-20 mg/dL.

Limitations Fresh blood in the specimen (traumatic tap) will invalidate the protein result; turbid samples may exhibit a positive interference; hemolyzed or xanthochromic specimens may falsely depress results; diabetics may have high CSF protein; ampicillin, gentamicin, and vancomycin increase the apparent CSF protein in at least some cases (method dependent). Problems in the differential diagnosis of multiple sclerosis, meningitis, and other entities are discussed below.

Additional Information Spinal fluid is an ultrafiltrate of plasma that lacks high molecular weight proteins such as beta lipoprotein, alpha$_2$-macroglobulin, IgM, and polymeric haptoglobins. Thus, most CSF protein is albumin and has origin from plasma. The protein concentration of spinal fluid is <1% of plasma proteins.

Increased Protein	Decreased Protein
Inflammation	Water intoxication
Tumor	Leukemia
Demyelinating disorders	CSF leakage
Subarachnoid hemorrhage	Rhinorrhea, otorrhea
Traumatic tap	Hyperthyroidism
Phenothiazine medications	Pneumoencephalography

Infants have higher protein levels (60-150 mg/dL) due to increased blood brain barrier permeability. Cisternal and ventricular fluids have both lower total protein concentration and leukocyte levels than fluid obtained by lumbar puncture. Ventricular fluid protein level may be <5 mg/dL. Such uneven distribution must be considered when CNS bacterial infection is a possibility.

The significance of elevated CSF protein should be carefully considered, in relation to the clinical findings, in particular if blood is present in the CSF. This could occur at the time of needling the subarachnoid space ("bloody tap") and be clinically misleading, or reflect clinically significant CNS hemorrhage, trauma, vascular anomaly or tumor, and have prime clinical significance.

If the spinal canal is obstructed (eg, by tumor) CSF protein content is increased, apparently the result of absence of reabsorption by arachnoidal villi. High CSF protein levels (several 100 mg/dL) may occur and due to the presence of fibrinogen, the CSF may clot (Froin syndrome).

Patients with elevated CSF protein may require additional analyses (eg, cerebrospinal fluid IgG/albumin index, IgG synthetic rate, and high resolution agarose protein electrophoresis for demonstration of "oligoclonal" bands), in particular if there are clinical findings of multiple sclerosis. (Multiple sclerosis is a clinical diagnosis.) Total protein in CSF is within normal limits in many subjects with MS but may be slightly increased. Protein levels >75 mg/dL bring a diagnosis of MS or other neurologic disease into question.

Protein may be normal or increased in viral meningitis/aseptic meningitis but is usually slightly increased, 50-80 mg/dL. Protein in most cases of viral meningitis is <100 mg/dL. It is usually 100-500 mg/dL in acute bacterial meningitis and is occasionally >1000 mg/dL.

AIDS patients with primary CNS disease or secondary CNS infections may have elevated CSF protein. Studies of asymptomatic, HIV seropositive individuals, however, have shown that pleocytosis or elevated CSF protein occurs in some 50% of such patients, with 12% to 26% showing oligoclonal banding. The percentage of protein abnormalities increases with inclusion of patients with AIDS and AIDS-related complex.

CSF cellular and protein abnormalities occur with 10% to 20% of subjects who have primary syphilis and with 30% to 70% of those with secondary lues.

Lumbar disk herniation with sciatica may be associated with increased CSF total protein (see comments in Cerebrospinal Fluid IgG:Albumin Ratio, IgG Index, and IgG Synthesis Rate *on page 365*).

Methods have been developed for detection of CSF leakage (as occurs most commonly from the subarachnoid space into nasal or aural cavities). The "marker protein" O-sialotransferrin (also called β_2-transferrin, asialotransferrin, or the tau fraction) is detected by high resolution agarose electrophoresis and immunofixation or, most recently, by isoelectric focusing on polyacrylamide gel followed by immunofixation of transferrins and silver staining.

References

Biou D, Benoist JF, Nguyen-Thi C, et al, "Cerebrospinal Fluid Protein Concentrations in Children: Age-related Values in Patients Without Disorders of the Central Nervous System," *Clin Chem*, 2000, 46(3):399-403.

Evans RW, "Complications of Lumbar Puncture," *Neurol Clin*, 1998, 16(1):83-105.

(Continued)

Cerebrospinal Fluid Protein *(Continued)*

Kleine TO and Althaus H, "Detection of Asialo-transferrin and β-trace Protein in Mixtures of Blood Serum and Cerebrospinal Fluid (CSF) as Models for CNS Contamination With Rhinorrhea and Otorrhea," *Clin Chem*, 2000, 46(Suppl 6):A49-50.

Olek MJ and Dawson DW, "Multiple Sclerosis and Other Demyelinating Disease of the Central Nervous System," *Neurology in Clinical Practice: The Neurological Disorders*, 3rd ed, Chapter 60, Bradley WG, Daroff RB, Fenichel GM, et al, eds, Boston, MA: Butterworth Heinemann, 2000, (2):1448-9.

Roelandse FW, van der Zwart N, Didden JH, et al, "Detection of CSF Leakage by Isoelectric Focusing on Polyacrylamide Gel, Direct Immunofixation of Transferrins and Silver Staining," *Clin Chem*, 1998, 44(2):351-2.

Cerebrospinal Fluid Protein Electrophoresis

Related Information

Cerebrospinal Fluid Cytology *on page 361*
Cerebrospinal Fluid Glucose *on page 362*
Cerebrospinal Fluid IgG:Albumin Ratio, IgG Index, and IgG Synthesis Rate *on page 365*
Cerebrospinal Fluid Immunoglobulin G *on page 367*
Cerebrospinal Fluid Oligoclonal Bands *on page 369*
Cerebrospinal Fluid Protein *on page 371*
Immunofixation Electrophoresis, Serum or Urine *on page 768*
Protein Electrophoresis, Serum *on page 1104*
Protein, Total, Serum *on page 1114*

Synonyms Protein Electrophoresis, Spinal Fluid

Applies to Tau Fraction

Test Includes Total protein

Abstract This is among several studies used in evaluation of patients presenting with signs and/or symptoms of multiple sclerosis (MS), but findings are not exclusively diagnostic of this condition.

Patient Preparation See Cerebrospinal Fluid Protein *on page 371*.

Aftercare Patient should be maintained supine for a few hours postpuncture in an attempt to avoid postpuncture headache.

Specimen Cerebrospinal fluid

Container Clean, sterile CSF tube

Collection Tube should be labeled with the number indicating the sequence in which tubes were obtained.

Storage Instructions Store refrigerated.

Causes for Rejection Insufficient quantity of CSF, CSF contaminated with blood ("bloody tap")

Special Instructions The third tube of routinely obtained three tube set of CSF should be used for protein electrophoretic study.

Reference Interval See table.

Total protein	15-60 mg/dL
Prealbumin	1% to 8%
Albumin	50% to 80%
Alpha$_1$ globulin	2% to 8%
Alpha$_2$ globulin	2% to 12%
Beta globulin	8% to 18%
Gamma globulin	3% to 12%

The second of the two transferrin bands is called the tau band.

Use The primary reason to perform protein electrophoresis is to detect oligoclonal bands, which are most often associated with multiple sclerosis. CSF protein electrophoresis can be used to assist in the detection of CSF leakage (with resultant nasal or aural discharge).

Methodology Cellulose acetate, agarose electrophoresis, "high resolution" agarose gel electrophoresis. The latter is the method of choice in the initial detection and study of oligoclonal bands. A procedure for CSF protein separation by capillary isoelectric focusing has recently been developed.

Additional Information Spinal fluid is an ultrafiltrate of plasma that lacks high molecular weight proteins such as beta lipoprotein, alpha$_2$-macroglobulin, IgM, and polymeric haptoglobulins (see table of protein analytes in Cerebrospinal

Fluid Protein *on page 371*). The protein concentration of spinal fluid is <1% of plasma proteins. A normal CSF protein electrophoretic pattern has relatively more prealbumin, less alpha$_2$-globulin, and less gamma globulin than serum from the same individual. Albumin and beta bands appear similar to serum. If the CSF shows presence of a monoclonal band, immunofixation study of patient's serum as well as CSF should be performed (most M-proteins will cross the blood-brain barrier). If a monoclonal band is present in the CSF but none is detected in the serum, presence of central nervous system (CNS) lymphoma or plasmacytoma of the CNS (admittedly rare) should be considered. Cytology study of CSF may be important in establishing a diagnosis.

References

Kyle RA, "Sequence of Testing for Monoclonal Gammopathies," *Arch Pathol Lab Med*, 1999, 123(2):114-8.

Manabe T, Miyamoto H, Inoue K, et al, "Separation of Human Cerebrospinal Fluid Proteins by Capillary Isoelectric Focusing in the Absence of Denaturing Agents," *Electrophoresis*, 1999, 20(18):3677-83.

Zaret D, Morrison N, Gulbranson R, et al, "Immunofixation to Quantify β 2-Transferrin in Cerebrospinal Fluid to Detect Leakage of CSF From Skull Injury," *Clin Chem*, 1992, 38(9):1909-12.

♦ **Cerebyx**® *see* Phenytoin, Serum or Plasma *on page 1026*

Ceruloplasmin, Serum or Plasma

Related Information

Copper, Serum *on page 448*
Copper, Urine *on page 452*
Liver Biopsy *on page 864*
Liver Disease: Laboratory Assessment, Overview *on page 869*

Applies to Transcuprein

Abstract Ceruloplasmin, a copper-binding protein and acute phase reactant, is synthesized in the liver. In Wilson disease (WD) (hepatolenticular degeneration), an autosomal recessive condition, deficient ceruloplasmin synthesis leads to excessive copper deposition in liver, brain, cornea, kidney, and other sites. Diagnosis of WD can sometimes be made on the basis of hypoceruloplasminemia, hypocupremia, and hypercupruria, with abnormalities in liver-related tests.

Patient Preparation Patient should be fasting.

Reference Interval 19 years and older: 22.9-43.1 mg/dL; see table.

Ceruloplasmin

Test	Age	n	mg/dL	n	mg/dL
	Serum				
	0-5 d	73*	50-260	73*	50-260
	1-3 y	51*	240-460	51*	240-460
1.	4-6 y	39*	240-420	39*	240-420
	7-9 y	39*	240-400	39*	240-400
	10-13 y	36	220-360	45	230-430
	14-19 y	46	140-340	66	200-450
	Serum / Plasma				
	1-30 d	35	70-230	36	30-250
	31-365 d	119	140-440	87	140-390
	1-3 y	127	230-510	114	260-490
2.	4-6 y	99	260-510	81	240-490
	7-9 y	75	230-470	84	210-440
	10-12 y	69	190-460	90	190-440
	13-15 y	73	180-450	72	190-420
	16-18 y	49	180-410	73	200-450

*No significant differences were found for males and females. Ranges derived from combined data.

Test 1: Adapted from Lockitch G, Halstead AC, Quigley G, et al, "Age and Sex-Specific Pediatric Reference Intervals: Study, Design, and Methods Illustrated by Measurement of Serum Proteins With the Behring LN Nephelometer," *Clin Chem*, 1988, 34:1618-21.

Test 2: Adapted from Soldin SJ, Gunter KC, Brugnara C, et al, *Pediatric Reference Ranges*, 2nd ed, Washington, DC: AACC Press, 1997.

(Continued)

Ceruloplasmin, Serum or Plasma *(Continued)*

Specimen Serum or plasma

Container Red top tube, SST™ tube, or lavender top (EDTA) tube

Collection Centrifuge immediately. Freeze (dry ice) until assay.

Storage Instructions Separate serum or plasma and freeze.

Use Serum ceruloplasmin is decreased in most (>75%) patients with Wilson disease (WD). However, interpretive caution is essential because ceruloplasmin is an acute phase reactant; multiple specimens may be required. WD is a devastating, but treatable disease. The diagnosis of WD requires consideration of history, physical findings, multiple laboratory tests, and pedigree. Additional tests often include serum and urine copper and copper content of a liver biopsy (*vide infra*).

Ceruloplasmin is **low** in Menkes syndrome, a disease due to defective absorption and utilization of dietary copper and in the occipital horn syndrome.

Ceruloplasmin is **high** in many neoplastic and inflammatory states since it is an acute phase reactant. High levels also occur in pregnancy, with estrogens, and with oral contraceptive use when the agent contains estrogen as well as progesterone. Increased in copper intoxication.

Limitations A normal ceruloplasmin result does not rule out Wilson disease, especially in childhood. Serum and liver copper should often be measured. Discrepancies occur between immunologic and enzymatic assays in serum of patients with Wilson disease.

Additional Information About 70% of serum copper is associated with ceruloplasmin, 7% with transcuprein, 19% with albumin, and 2% with amino acids.

Measurement of copper in a liver biopsy specimen is often essential for the diagnosis of Wilson disease. The histopathologic findings in WD are usually those of chronic hepatitis, and are not specific for WD. Demonstration of failure to incorporate radiolabeled copper into ceruloplasmin is a definitive test for Wilson disease.

Additional information is provided in the entries Copper, Serum *on page 448* and Copper, Urine *on page 452*. The former monograph includes a table relevant to ceruloplasmin as well as copper.

References

Cauza E, Maier-Dobersberger T, Polli C, et al, "Screening for Wilson's Disease on Patients With Liver Diseases by Serum Ceruloplasmin," *J Hepatol*, 1997, 27(2):358-62.

Gahl WA, "Wilson Disease," *Cecil Textbook of Medicine*, 21st ed, Goldman L and Bennett JC, eds, Philadelphia, PA: WB Saunders Co, 2000, 1130-2.

Gilbert-Barness E and Barness L, *Metabolic Diseases: Foundations of Clinical Management, Genetics, and Pathology*, Natick, MA: Eaton Publishing, 2000, 443-55.

Ludwig J, Moyer TP, and Rakela J, "The Liver Biopsy Diagnosis of Wilson's Disease," *Am J Clin Pathol*, 1994, 102(4):443-6.

Soldin SJ, Gunter KC, Brugnara C, et al, *Pediatric Reference Ranges*, 2nd ed, Washington, DC: AACC Press, 1997.

Cervical/Vaginal Cytology

Related Information

Bacterial Culture, Genital Specimen *on page 239*
Chlamydia trachomatis Culture *on page 385*
Chlamydia trachomatis Direct Antigen Test *on page 387*
Cytomegalovirus Culture *on page 495*
Endometrial Cytology *on page 536*
Hormonal Evaluation, Cytologic *on page 744*
Human Papillomavirus (HPV) DNA Tests *on page 749*
Neisseria gonorrhoeae Culture and Smear *on page 945*
Trichomonas Preparation *on page 1273*
Viral Culture *on page 1307*

Synonyms Pap Smear; Pap Test

Applies to Bethesda System; Herpes Smear; Human Papillomaviruses (HPV); Koilocytosis; Vulvar Cytology

Test Includes Cervical scraping smear, vaginal pool smear, lateral vaginal wall smear, direct scraping smear, cytobrush, or monolayer preparation

Abstract The spectrum of abnormality of cervical epithelium includes benign inflammatory processes (often due to bacterial, fungal, parasitic, or viral infections), low-grade squamous intraepithelial lesions (**LSIL**, mild dysplasia, **CIN**

1), high-grade squamous intraepithelial lesions (**HSIL**, moderate and severe dysplasia, **CIS, CIN 2-3**), and carcinoma. About 80% of carcinomas of the uterine cervix are squamous. Most of the remainder are adenocarcinomas and mixed adenosquamous carcinomas. Other entities which may sometimes be detected include adenocarcinoma of endometrium, sarcomas, carcinoids, endocrine carcinomas, and other rare epithelial tumors and metastatic cancers.

A notable reduction in mortality from cervical cancer followed implementation of Pap test screening, but problems in sampling, specimen adequacy, microscopic screening, and interpretation remain. Even with strict federal guidelines, compliance requirements, and inspection by approved agencies, false negatives can occur in the best cytopathology laboratory.

Presently, >80% of those women dying of carcinoma of the cervix have either never had a Pap test or have not had one in the prior 5 years.

Patient Preparation Patients are advised to avoid douches 48-72 hours prior to examination; however, this should not preclude taking of the smear.

Specimen Cervical scrape and endocervical brush samples are recommended in all cases. The optimal sample for hormonal evaluation is a lateral vaginal wall scrape, but aspiration of posterior vaginal fornix fluid (vaginal pool) may be used. Endometrial aspirations are not advised for routine use. For lesions of the vagina or vulva, scrapings made directly from the lesion are most diagnostic.

Container Bottle filled with 95% ethanol, cardboard mailer or plastic slide holder if slides are spray-fixed. **Hair spray should never be used to fix Pap smear slides.** Appropriate vials with collection fluid are used for liquid-based preparations.

Collection Using a graphite pencil, write the patient's name on the frosted end of the glass slide. It is possible to label nonfrosted slides with a diamond point pen. Collection vials for liquid-based preparations should be labeled. The speculum must be introduced **without** lubricant; in certain cases, running the speculum through warm saline will prove helpful prior to insertion into an atrophic, stenotic, or small introitus.

Sampling:

Endocervix: Gentle scrape or brush of endocervical canal

- Scrape - Rotate narrow end of spatula in the cervical os and gently smear onto labeled glass slide and fix immediately.
- Brush - Use a synthetic fiber brush to sample endocervical cells; remove mucous plug if present, before sampling the endocervical canal. Do not rub onto the slide; rather, lightly roll brush over the slide surface.

Ectocervical scrape: With spatula thoroughly scrape the entire ectocervix with emphasis on the squamocolumnar junction (transformation zone). Spread material evenly onto labeled glass slide and fix immediately. Immediate fixation is imperative.

Lateral vaginal wall smear: Scraping from upper lateral one-third of vaginal mucosa. Used for cytohormonal evaluations.

Direct scraping smear: Direct scrape of grossly visible lesion, smeared and fixed as previously described.

Liquid-based preparations require a brush instrument for sample collection. The brush may be disconnected or cut and submerged in the collection fluid provided. The vial is then submitted to the laboratory for processing.

Special Instructions Include pertinent clinical history such as age, LMP, parity, postmenopausal status, surgery, exogenous hormones, history of carcinoma, radiation, chemotherapy, abnormal vaginal bleeding, and history of previous abnormal Pap tests.

Use Diagnose primary or metastatic neoplasms; diagnose cervical dysplasia (cervical intraepithelial neoplasia (CIN)); diagnose genital infections with herpes, *Candida* species, *Trichomonas vaginalis*, cytomegalovirus and *Actinomyces*; aid in the diagnosis of vaginal adenosis, cervicovaginal endometriosis, condyloma, human papillomavirus infection, lymphogranuloma venereum; aid in evaluation of hormonal status.

Presently, the primary concerns should be consistent availability and use of the Pap test for screening with adequate follow-up. The Papanicolaou test is the most widely used cancer screening measure in the U.S.
(Continued)

Cervical/Vaginal Cytology *(Continued)*

Limitations In spite of its great success, the Pap test (gynecologic cytology) should still be considered a screening test. And the interpretation of Pap tests, even though based on well-documented criteria, is a subjective evaluation. Higher areas of disease prevalence are related to higher levels of sensitivity. There are guidelines that establish the adequacy of gynecologic samples. Each report should contain a statement of adequacy.

"Satisfactory for evaluation" (without any quality statements) indicates that the specimen has all of the following:

- appropriate labeling and identifying information
- relevant clinical information
- adequate numbers of well-preserved and well-visualized squamous epithelial cells
- an adequate endocervical/transformation zone component - at least 10 well-preserved endocervical or metaplastic cells (from those patients who have a cervix)

A specimen is **"satisfactory for evaluation"** (but may have **quality statements**) if there are any of the following:

- lack of pertinent clinical patient information (age, date of last menstrual period as a minimum; additional information as appropriate)
- partially obscuring blood, inflammation, thick areas, poor fixation, air-drying artifact, contaminant, etc, that precludes interpretation of approximately 50% to 75% of the epithelial cells.
- absence of an endocervical/transformation zone component as defined above

A specimen is **"unsatisfactory for evaluation..."** if any of the following apply:

- lack of patient identification on the specimen and/or requisition form
- a slide that is broken and cannot be repaired
- scant squamous epithelial component: <8000 squamous cells for conventional smears; <5000 squamous cells for liquid-based preparations
- obscuring blood, inflammation, thick areas, poor fixation, air-drying artifact, contaminant, etc, that precludes interpretation >75% of the epithelial cells.

The "unsatisfactory" designation indicates that the specimen is unreliable for detection of cervical epithelial abnormalities. A diagnosis should not be made.

About two-thirds of false negatives result from sampling error, the remaining third from detection error.

The finding of atypical glandular cells (AGC) is important clinically because the percentages of cases associated with underlying high-grade disease is higher than for ASC-US.

The greatest obstacle to prevention of carcinoma of the cervix in the U.S. is failure to be tested. An Agency for Healthcare Policy and Research Evidence Report/Technology Assessment recognized that "...a large proportion of cervical cancer occurs in women with very limited or no screening" but the assessment "...did not examine programs or policies designed to improve screening compliance."

Pap smears are presently underused.

Contraindications Air drying of smears prior to fixation must not be permitted.

Additional Information Infection with human papillomavirus (HPV) is a necessary factor in the development of cervical neoplasia. Fortunately, most women infected with HPV do not develop significant cervical abnormalities (Walboomers). Infection with HPV is easily spread during sexual intercourse. With an effective immune response, most women clear the infection or significantly reduce the viral load in an average of 8-24 months (Ho, Moscicki, Woodman). Approximately 30 HPV subtypes have been found in the anogenital tract, with 66% of the HPV-associated anogenital neoplasmas involving types 6, 11, 16, and 18. HPV subtypes have been segmented into low-risk and high-risk categories. Low-grade squamous intraepithelial lesions (LGSIL) of the cervix are associated with low-risk virus (eg, types 6 and 11) and such lesions most often are self-limited, are cleared by the immune system, and seldom progress to higher grade lesions. On the other hand, it is estimated that up to 100% of women with high-grade squamous intraepithelial lesions (HGSIL) will test positive for the high-risk type of HPV (eg, types 16 and 18). However, many women may harbor HPV in their lower genital tracts without showing changes on cervical/vaginal cytology or by histopathology on cervical biopsies (Herrero). Furthermore, the cofactors that allow HPV infection to develop into squamous intraepithelial lesions, or worse, are not fully characterized, but a small percentage of infected women do not clear HPV infection and will develop cervical neoplasia.

Cervical/Vaginal Cytology

2001 Bethesda System (Abridged)
SPECIMEN ADEQUACY
Satisfactory for evaluation (note presence/absence of endocervical/transformation zone component)
Unsatisfactory for evaluation...(specify reason)
Specimen rejected/not precessed (specify reason)
Specimen processed and examined but unsatisfactory for evaluation of epithelial abnormality because of (specify reason)
GENERAL CATEGORIZATION (optional)
Negative for intraepithelial lesion or malignancy
Epithelial cell abnormality
Other
INTERPRETATION/RESULT
Negative for Intraepithelial Lesion or Malignancy
Organisms
Trichomonas vaginalis
Fungal organisms morphologically consistent with *Candida* sp
Shift in flora suggestive of bacterial vaginosis
Bacteria morphologically consistent with *Actinomyces* sp
Cellular changes consistent with herpes simplex virus
Other non-neoplastic findings (optional to report; list not comprehensive)
Reactive cellular changes associated with:
inflammation (includes typical repair)
radiation
intrauterine contraceptive device
Glandular cells status posthysterectomy
Atrophy
Epithelial Cell Abnormalities
Squamous cell
Atypical squamous cells (ASC)
of undetermined significance (ASC-US)
cannot exclude HSIL (ASC-H)
Low-grade squamous intraepithelial lesion (LSIL)
encompassing: human papillomavirus/mild dysplasia/cervical
intraepithelial neoplasia (CIN) 1
High-grade squamous intraepithelial lesion (HSIL)
encompassing: moderate and severe dysplasia, carcinoma *in situ*, CIN 2 and CIN 3
Squamous cell carcinoma
Glandular cell
Atypical glandular cells (AGC) (specify endocervical, endometrial, or not otherwise specified)
Atypical glandular cells, favor neoplastic (specify endocervical or not otherwise specified)
Endocervical adenocarcinoma *in situ* (AIS)
Adenocarcinoma
Other (list not comprehensive)
Endometrial cells in a woman ≥40 years of age
AUTOMATED REVIEW AND ANCILLARY TESTING (include as appropriate)
EDUCATIONAL NOTES AND SUGGESTIONS (optional)

A variety of molecular assays exist for the detection and quantitation of HPV. One of these, the Hybrid Capture® 2 (HC2) test, manufactured by Digene Corporation, Inc (Gaitherburg, MD), is the sole test presently (2003) approved

(Continued)

Cervical/Vaginal Cytology *(Continued)*

by the U.S. Food and Drug Administration (FDA) for *in vitro* diagnostic use (Hubbard). The FDA initially approved the HPV DNA test in the year 2000 for testing women who had abnormal Pap test results to determine whether they needed to be referred for further examination. The newest FDA-approved indication allows the test to be used for screening, in conjunction with the Pap test, of women older than 30 years of age for HPV infection (*FDA News*). It is important to emphasize that the HC2 test should be used along with the Pap test, a complete medical history, and an evaluation of other risk factors to help physicians determine what kind of follow-up is necessary. See Human Papillomavirus (HPV) DNA Tests *on page 749*.

Discrepancies occur between cervicovaginal smears and subsequent biopsies, even when done by quality laboratories. SIL may be underestimated in 17.5% of cases. About 15% to 25% of individuals with SIL are reported to have normal cervical/vaginal cytology.

The practitioner and patients should insist on case evaluation by licensed cytotechnologists and board certified pathologists and should be cautious about simply seeking the lowest price. The present Pap smear liability crisis is primarily a threat to women and the public health. Analysis and interpretation of the Pap smear is a sophisticated professional consultation which should be selected with the same care utilized in selection of surgeons or other specialists.

Conferences of leading cytopathologists and gynecologists, held at Bethesda in 1988 and 1991, formulated recommendations for uniform diagnostic terminology for cervical/vaginal cytology. Such recommendations eliminated reporting by classes. These conferences, sponsored by the National Cancer Institute, led to a new classification, the Bethesda System (TBS). See table. TBS provides a means to support communication of cervical/vaginal cytologic diagnoses in a format which is concise, reproducible, and easily understood.

References

Agency for Healthcare Policy and Research, Evidence Report/Technology Assessment No. 5, "Evaluation of Cervical Cytology," *AHCPR*, No. 99-E009, January, 1999.

Austin RM and McLendon WW, "The Papanicolaou Smear: Medicine's Most Successful Cancer Screening Procedure Is Threatened," *JAMA*, 1997, 277(9):754-5.

Bosch FX, Manos MM, Munoz N, et al, "Prevalence of Human Papillomavirus in Cervical Cancer: A Worldwide Perspective. International Biological Study on Cervical Cancer (IBSCC) Study Group," *J Natl Cancer Inst*, 1995, 87(11):796-802.

Cain JM and Howett MK, "Preventing Cervical Cancer," *Science*, 2000, 288(5472):1753-4.

Cannistra SA and Niloff JM, "Cancer of the Uterine Cervix," *N Engl J Med*, 1996, 334(16):1030-8.

Ellerbrock TV, Chiasson MA, Bush TJ, et al, "Incidence of Cervical Squamous Intraepithelial Lesions in HIV-Infected Women," *JAMA*, 2000, 283(8):1031-7.

FDA News, FDA approves expanded use of HPV test. P03-26, March 31, 2003 (www.fda.gov/bbs/topics/news/2003/new00890.html).

Feldman S, "How Often Should We Screen for Cervical Cancer?" *N Engl J Med*, 2003, 349(16):1495-6.

Frable WJ, "Toward a Process Standard in Cytopathology," *Cancer*, 1998, 84(3):127-9.

Godfrey SE, "The Pap Smear, Automated Rescreening, and Negligent Nondisclosure," *Am J Clin Pathol*, 1999, 111(1):14-7.

Goodman A and Wilbur DC, "A 37-Year-Old Woman With Atypical Squamous Cells on a Papanicolaou Smear," Case Records of the Massachusetts General Hospital, Case 32-2003, Scully RE, Mark EJ, McNeely WF, et al, eds, *N Engl J Med*, 2003, 349(16):1555-64.

Herrero R, Hildesheim A, Bratti C, et al, "Population-Based Study of Human Papillomavirus Infection and Cervical Neoplasia in Rural Costa Rica," *J Natl Cancer Inst*, 2000, 92(6):464-74.

Ho GY, Bierman R, Beardsley L, et al, "Natural History of Cervicovaginal Papillomavirus Infection in Young Women," *N Engl J Med*, 1998, 338(7):423-8.

Hubbard RA, "Human Papillomavirus Testing Methods," *Arch Pathol Lab Med*, 2003, 127(8):940-5.

Kiviat N, "Natural History of Cervical Neoplasia: Overview and Update," *Am J Obstet Gynecol*, 1996, 175(4 Pt 2):1099-104.

Moscicki AB, Shiboski S, Broering J, et al, "The Natural History of Human Papillomavirus Infection as Measured by Repeated DNA Testing in Adolescent and Young Women," *J Pediatr*, 1998, 132(2):277-84.

Schiffman MH and Brinton LA, "The Epidemiology of Cervical Carcinogenesis," *Cancer*, 1995, 76(10 Suppl):1888-901.

Sawaya GF, McConnell KJ, Kulasingam SL, et al, "Risk of Cervical Cancer Associated With Extending the Interval Between Cervical-Cancer Screenings," *N Engl J Med*, 2003, 349(16):1501-9.

Walboomers JMM, Jacobs MV, Manos MM, et al, "Human Papillomavirus Is a Necessary Cause of Invasive Cervical Cancer Worldwide," *J Pathol*, 1999, 189(1):12-9.

Woodman CB, Collins S, Winter H, et al, "Natural History of Cervical Human Papillomavirus Infection in Young Women: A Longitudinal Cohort Study," *Lancet*, 2001, 357(9271):1831-6.

Internet Web Sites
www.fda.gov/bbs/topics/news/2003/new00890.html

♦ *CFTR* **Gene Mutation Analysis** *see* Cystic Fibrosis DNA Detection *on page 491*

♦ **CH$_{50}$** *see* Complement C3, Serum *on page 434*

♦ **CH$_{50}$** *see* Complement, Total, Serum or Body Fluid *on page 441*

Chagas Disease Diagnostic Procedures

Related Information

Buffy Coat Smear Study of Peripheral Blood *on page 316*
Malaria Smear and Tests *on page 888*
Risks of Transfusion *on page 1166*

Abstract American trypanosomiasis (Chagas disease), caused by a protozoan, *Trypanosoma cruzi*, is transmitted to humans by bloodsucking triatomine insects. Chagas disease is a life-long infection with the highest incidence found in Brazil, Argentina, Chile, Bolivia, and Venezuela. This disease is found only in the Western hemisphere. Up to 100,000 Latin American immigrants in the U.S. are estimated to be infected with *T. cruzi*. Chronic disease includes myocarditis, cardiomyopathy, and megadisease of esophagus and colon. It can cause placentitis, and maternal transmission to the fetus leads to congenital Chagas disease or abortion. Severe recrudescence may occur with immunosuppression (eg, organ transplantation, AIDS). Laboratory workers may be accidentally infected. Correlation between infectivity and Chagas antibodies is suboptimal.

Specimen Serum

Container Red top tube

Reference Interval Indirect hemagglutination titer: <1:128; complement fixation titer: <1:8; immunoelectrophoresis titer: <1:64

Use Support the clinical diagnosis of Chagas disease, which can present as an acute febrile illness. In endemic areas, serologic testing is needed in blood banks

Limitations False-positive serologic reactions occur in persons with leishmaniasis, malaria, toxoplasmosis, hepatitis, leprosy, syphilis, and collagen diseases. An individual can be serologically negative but still be infectious.

Methodology Indirect fluorescent antibody (IFA) and enzyme-linked immunosorbent assay (ELISA) are considered more sensitive than an indirect hemagglutination test. A radioimmunoprecipitation assay is available.

Stained peripheral blood smears may demonstrate tryptomastigotes of *T. cruzi* in acute cases. A large posterior kinetoplast is visible in the C-shaped organism. Examination of fresh blood specimen, hemoculture, or xenodiagnosis may provide diagnosis.

Additional Information CF shows a high degree of sensitivity in the acute stages of the disease since the CF assay detects IgM. Serologic tests return to normal in a large majority of patients 12-24 months after treatment. IgM antibodies in a newborn baby of a seropositive mother usually is considered evidence of congenital infection.

A serum should test positive by at least two different assays before the results are accepted.

References

Bahia-Oliveira LM, Gomes JA, Cancado JR, et al, "Immunological and Clinical Evaluation of Chagasic Patients Subjected to Chemotherapy During the Acute Phase of *Trypanosoma cruzi* Infection 14-30 Years Ago," *J Infect Dis*, 2000, 182(2):634-8.

Betonico GN, Miranda EO, Silva DA, et al, "Evaluation of a Synthetic Tripeptide as Antigen for Detection of IgM and IgG Antibodies to *Trypanosoma cruzi* in Serum Samples From Patients With Chagas' Disease or Viral Diseases," *Trans R Soc Trop Med Hyg*, 1999, 93(6):603-6.

Center for Disease Control and Prevention, "Chagas Disease After Organ Transplantation - United States, 2001," *JAMA*, 2002, 287(14):1795-6.

Miles MA, "American Trypanosomiasis (Chagas' Disease)," *Manson's Tropical Diseases*, 21st ed, Chapter 74, Cook GC and Zumla A, eds, Philadelphia, PA: WB Saunders Co, 2003, 1325-37.

Saez-Alquezar A, Sabino EC, Salles N, et al, "Serological Confirmation of Chagas' Disease by a Recombinant and Peptide Antigen Line Immunoassay: INNO-LIA Chagas," *J Clin Microbiol*, 2000, 38(2):851-4.

Chain-of-Custody Protocol

Synonyms Chain-of-Evidence Form; Specimen Chain-of-Custody Protocol

Applies to Bullets; Medical Legal Specimens

(Continued)

Chain-of-Custody Protocol *(Continued)*

Abstract A procedure to ensure sample integrity from collection through transport, receipt, sampling, and analysis. It is associated with a Chain-of-Custody form. See Chain-of-Custody form in the Introduction *on page 63*. Similar forms are used (chain-of-evidence) for other forensic materials such as guns, bullets, chemicals, etc.

Specimen Usually urine for drugs-of-abuse-related monitoring; blood for alcohol testing

Container Plastic urine cup with locking lid covered by seal which is signed or initialed (if for drugs of abuse)

Collection See the Introduction *on page 63*.

Causes for Rejection Sample container not sealed or labeled

Special Instructions Form requires signature of sample donor as well as that of the person receiving the sample at the collection site.

Reference Interval Normal: all seals intact and Chain-of-Custody form completed.

Use Maintain sample integrity in the collection, handling, and storage of urine or other samples

Additional Information A written record of specimen transfer from patient, to analyst, to storage and disposal is maintained on all specimens covered by chain-of-custody. All drug screens, blood alcohols, most bullets, or any other tests or objects that have medicolegal significance should be accompanied by Chain-of-Custody and a written release form.

References
"Mandatory Guidelines for Federal Workplace Drug Testing Programs," *Fed Regist*, 1994, 59:29916-31.

Smith ML, Bronner WE, Shimomura ET, et al, "Quality Assurance in Drug Testing Laboratories," *Clin Lab Med*, 1990, 10(3):503-16.

Wu AH, Bristol B, Sexton K, et al, "Adulteration of Urine by Urine Luck," *Clin Chem*, 1999, 45(7):1051-7.

Internet Web Sites
www.health.org/workplace

♦ **Chain-of-Evidence Form** *see* Chain-of-Custody Protocol *on page 381*

♦ **Chan Su** *see* Digoxin, Serum *on page 512*

♦ **CHEK2 Mutation** *see* Breast Cancer, Hereditary, BRCA1, BRCA2 *on page 307*

♦ **Chemical Terrorism** *see* Chemical Warfare Agents *on page 382*

Chemical Warfare Agents

Related Information
Acetylcholinesterase, Red Cell and Serum *on page 93*
Blood Gases and pH, Arterial *on page 275*
Cyanide, Blood *on page 485*
Methemoglobin, Whole Blood *on page 903*
Organophosphate Pesticides, Urine, Blood, or Serum *on page 975*
Pseudocholinesterase, Serum *on page 1122*
Thiocyanate, Serum, Plasma, or Urine *on page 1245*

Synonyms Chemical Terrorism; Warfare Agents, Chemicals

Applies to Alkylating Agents; Chemical Weapons; Cyanide; Cyanogen Chloride; GA; GB; GD; Hydrogen Cyanide; Lewisite; Nitrogen Mustard; Organophosphates; Phosgene; Sarin; Soman; Sulphur Mustard; Tabun; VX

Abstract Chemical warfare agents have been known for centuries. Their use in World War I, Iraqi use against the Kurds, and the Tokyo subway terrorists' attack is well documented. Although military personnel are relatively well aware of chemical warfare agents, most civilian medical communities are not adequately aware or prepared. Due to the September 2001 World Trade Center catastrophe, various anthrax attacks in the United States, and repeated terrorists threats, public and healthcare awareness to chemical and biological weapons has increased.

The Chemical Weapons Convention defines toxic chemicals as "any chemical which, through its chemical effect on living processes, may cause death, temporary loss of performance, or permanent injury to people and animals".

Specimen Blood, pulmonary lavage, and urine depending on type of exposure

Limitations Special assays are not readily available in clinical laboratories.

Methodology Cholinesterase by spectrophotometry; nerve agents by gas-chromatography (GC) with flame ionization, nitrogen-phosphorus or mass spectrometer (MS) detectors and biosensor technology; mustard agents or their metabolites measured directly by GC or GC-MS or by measuring alkylated adducts of hemoglobin or albumin by GC, high performance liquid chromatography (HPLC) or HPLC-MS or immunoassay, sulfur mustard hydrolysis product thiodiglycol by GC-MS; chlorine measurement by electrochemical method, protein measurement in bronchoalveolar lavage; Conway microdiffusion, GC, spectrophotometry for cyanide assay

Additional Information Chemical warfare agents are categorized into four major classes. The table lists various chemical agents, their codes, persistence, physical state and site of action.

1. **Nerve agents**: Their action is similar to organophosphates (ie, they inhibit acetylcholinesterase, resulting in accumulation of acetylcholine and overstimulation of muscarinic and nicotinic receptors). They inhibit both acetylcholinesterase and pseudocholinesterase. See Acetylcholinesterase, Red Cell and Serum *on page 93* and Pseudocholinesterase, Serum *on page 1122*. Examples of these agents include tabun (GA, ethyl N- dimethylphophoroamidocyanidate), sarin (GB or isopropyl methylphosphonofluoridate) and soman (GD or pinacolyl methylphosphonofluoridate), and VX (methylphosphonothioic acid S-(2-(bis(1-methyl-ethyl)amino)ethyl) O-ethyl ester or V).

The clinical symptoms are associated with multiorgan effects. Symptoms include miosis, rhinorrhea, bronchial secretion, diarrhea, bronchospasm, bradycardia, muscle twitching, weakness, and paralysis.

Treatment includes injection of atropine, possibly in combination with pralidoxime chloride (Protopam® chloride; 2-PAMCl) an oxime. Oximes attach to the nerve agent that is inhibiting the cholinesterase and break the agent-enzyme bond to restore the normal activity of the enzyme.

Routine toxicology laboratory testing does not identify nerve agents. Measurement of RBC or serum cholinesterase has been used as an index of severity of toxicity of nerve agents, but with poor correlation. Erythrocyte enzyme activity is more sensitive to acute nerve agent exposure than is plasma enzyme activity. However this approach is not always reliable. In the Tokyo subway sarin attack, miosis and RBC acetylcholinesterase activity were compared; miosis was a more sensitive index to sarin exposure than was RBC acetylcholinesterase activity.

Cholinesterase assay is available on most automated chemistry instruments.

2. **Vesicants or blistering agents**: These agents, on contact with body surfaces, cause burns and blisters. The most susceptible organs are eyes, mucous membranes, lungs, and skin. Examples include HD - sulphur mustard (Yperite), HN - nitrogen mustard, and L - Lewisite (arsenical vesicants may be used in a mixture with HD). Sulphur mustard was used as a weapon in World War I. Mechanism of action includes irreversibly alkylating DNA, RNA, and proteins and, thus, disrupting cell function, causing cell death. Actively dividing cells, such as epithelium and hematopoietic cells, are more severely affected. These agents also cause glutathione depletion, resulting in diminished protective capacity in cells from oxygen-free radicals.

Clinical symptoms include burning skin and eyes, damage of upper respiratory airway mucosa, and gastrointestinal irritation. There may be a latent period of 4-6 hours. Diagnosis is based on history and clinical symptoms. No specific treatment is available. Treatment is supportive and most patients recover. Several agents including antioxidants, anti-inflammatory agents, and sulfur-group donors (glutathione and N-acetylcysteine) are under investigation.

No laboratory tests are available to detect these agents. Leukocytosis correlates roughly with the extent of tissue injury, primarily to skin or pulmonary tissue. A leukocyte count ≤500 is a sign of an unfavorable prognosis.

There is no clinical laboratory test for mustard in blood or tissue. A method for analysis of urine for thiodiglycol, a metabolite of mustard has been described.

3. **Pulmonary intoxicants and irritants**: These agents include perfluroisobutylene (PFIB), phosgene, diphosgene, and chlorine. They cause pulmonary edema and irritation of nose, larynx, pharynx, trachea, and bronchi. (Continued)

Chemical Warfare Agents *(Continued)*

Protein concentration in bronchoalveolar lavage is increased and may be useful as an indicator of the extent of edema.

4. **Cyanides - cellular asphyxiants**: These include cyanide, hydrogen cyanide, and cyanogen chloride. The mechanism of action includes binding of cyanide to iron in cytochrome oxidase, thus, inhibiting intracellular oxygen utilization. After exposure to high concentrations, signs and symptoms include seizures, respiratory and cardiac arrest. Sodium nitrite and sodium thiosulfate are effective antidotes. Due to high LC50 and high volatility, cyanide is not considered a major threat for mass casualities. The French used 4000 tons of cyanide in World War I without significant success. Activated charcoal in a chemical protective mask effectively absorbs cyanide.

Laboratory findings include increased blood cyanide, metabolic acidosis with high concentration of lactic acid, increased anion gap, and increased oxygen content. Blood cyanide levels are generally not available and other laboratory findings are nonspecific.

Chemical Warfare Agents

Chemical Agent	Code	Persistence (Hours) at 20°C to 32°C	State at 20°C	Site and Mechanism of Action
Nerve agents				
Tabun	GA	24-48	Liquid	Cholinesterase inhibition
Sarin	GB	1-24	Liquid	Cholinesterase inhibition
Soman	GD	24-48	Liquid	Cholinesterase inhibition
VX	VX	240-720	Liquid	Cholinesterase inhibition
Vesicants				
Distilled mustard	HD	24-48	Liquid	Blistering, alkylating
Nitrogen mustard	HN-1,2,3	24-72	Liquid	Blistering, alkylating
Lewisite	L	18-36	Liquid	Blistering, irritating
Mustard Lewisite	HL	24-36	Liquid	Blistering, irritating
Pulmonary agents				
Phosgene	CG	0.5-1	Gas	Choking, lung damaging
Diphosgene	DP	0.5-3	Liquid	Choking, lung damaging
Chlorine	CL	0.5-2	Gas	Choking, lung damaging
Cyanides				
Hydrogen cyanide	AC	0.25-0.5	Gas	Cytochrome oxidase inhibition
Cyanogen chloride	CK	0.25-0.5	Gas	Cytochrome oxidase inhibition

References

Brennan RJ, Waeckerle JF, Sharp TW, et al, "Chemical Warfare Agents: Emergency Medical and Emergency Public Health Issues," *Ann Emerg Med*, 1999, 34(2):191-204.

Goozner B, Lutwick LI, and Bourke E, "Chemical Terrorism: A Primer for 2002," *J Assoc Acad Minor Phys*, 2002, 13(1):14-8.

Jortani SA, Snyder JW, and Valdes R Jr, "The Role of the Clinical Laboratory in Managing Chemical or Biological Terrorism," *Clin Chem*, 2000, 46(12):1883-93.

Knudson GB, Elliott TB, and Brook I, "Nuclear, Biological, and Chemical Combined Injuries and Countermeasures on the Battlefield," *Mil Med*, 2002, 167(2 Suppl):95-7.

Lawler A, "Antiterrorism Programs: The Unthinkable Becomes Real for a Horrified World," *Science*, 2001, 293(5538):2182-5.

Richard CF, et al, "Emergency Physicians and biological Terrorism," *Ann Emerg Med*, 1999, 34(2):183-90.

Smith JR, Shih ML, Price EO, et al, "Army Medical Laboratory Telemedicine: Role of Mass Spectrometry in Telediagnosis for Chemical and Biological Defense," *J Appl Toxicol*, 2001, 21(Suppl 1):S35-41.

U.S. Army Medical Research Institute of Chemical Defense, *Medical Management of Chemical Casualties Handbook*, 3rd ed, 2000.

Wetter DC, Daniell WE, and Treser CD, "Hospital Preparedness for Victims of Chemical or Biological Terrorism," *Am J Public Health*, 2001, 91(5):710-6.

♦ **Chemical Weapons** *see* Chemical Warfare Agents *on page 382*

♦ **Chewing** *see* Cocaine (Cocaine Metabolite), Qualitative, Urine or Hair *on page 427*

♦ **Childhood Reference Ranges** *see page 11*

♦ **Chi-Square Distribution** *see page 11*

Chlamydia Group Serology

Related Information

Bacterial Culture, Throat, and Antigen Detection Testing for Group A Strepto-cocci *on page 245*
Chlamydia trachomatis Culture *on page 385*
Chlamydia trachomatis Direct Antigen Test *on page 387*
Chlamydia trachomatis Nucleic Acid Detection *on page 388*
Psittacosis Serology *on page 1123*

Test Includes Detection of antibody titer to *Chlamydia* species

Abstract *Chlamydia* are intracellular bacteria. Three *Chlamydia* species are associated with disease in humans: *C. psittaci*, *C. trachomatis*, and *C. pneumo-niae*. Chronic asymptomatic or persistent infections in humans are common for *C. trachomatis* and *C. pneumoniae*. Because of this, characteristic serology is not useful in the diagnosis of active disease.

Specimen Serum

Container Red top tube

Collection Collect acute phase blood as soon as possible after onset (no later than 1 week). Convalescent blood should be drawn 1-2 weeks after acute (no less than 2 weeks after onset).

Reference Interval Normal individuals without acute disease may have an IgG titer due to past infections. Determination of IgM antibody may indicates acute infection.

Use Evaluation of atypical pneumonia in infants, lymphogranuloma venereum, or psittacosis

Limitations The antigen used in the test is group specific, not species specific. Conjunctivitis, urethritis, and pneumonia of the newborn do not usually induce an antibody response detectable by complement fixation. A very high "back-ground" of immunity in the general population makes interpretation of levels difficult. Due to the low sensitivity, specificity, and predictive value serology is not useful for the diagnosis of active disease.

Methodology Complement fixation (CF), indirect fluorescence antibody (IFA), enzyme immunoassay (EIA)

Additional Information *C. pneumoniae* is a cause of community-acquired pneumonia, causing 5% to 10% of cases. *Chlamydia pneumoniae* also causes infantile pneumonia, bronchitis, and pharyngitis. Recent reports have associ-ated *C. pneumoniae* with atherosclerotic cardiovascular disease using seroepidemiologic studies. However, other reports have been unable to repro-duce these results and found no relationship in large studies between infection with *C. pneumoniae* and heart disease. Further study is required to confirm association between infection with *C. pneumoniae* and coronary heart disease.

Chlamydia trachomatis is a common sexually transmitted bacterial pathogen in the United States. It is gaining increasing recognition as a respiratory pathogen and is also associated with blindness, ophthalmia neonatorum, lymphogranu-loma venereum, pelvic inflammatory disease, ectopic pregnancy, urethritis, epididymitis, and infertility. In a patient being evaluated for trachomatis disease, nonserologic methods (culture or PCR) can be used. For genitourinary infec-tions, serologic test for syphilis and culture for *Neisseria gonorrhoeae* are desirable as well.

References

Dayan L, "*Chlamydia* Detection and Management," *Aust Fam Physician*, 2000, 29(6):522-6.
Persson K and Boman J, "Comparison of Five Serologic Tests for Diagnosis of Acute Infections by *Chlamydia pneumoniae*," *Clin Diagn Lab Immunol*, 2000, 7(5):739-44.
Tompkins LS, Schachter J, Boman J, et al, "Collaborative Multidisciplinary Workshop Report: Detection, Culture, Serology, and Antimicrobial Susceptibility Testing of *Chlamydia pneumoniae*," *J Infect Dis*, 2000, 181:S460-1.
Tuuminen T, Palomaki P, and Paavonen J, "The Use of Serologic Tests for the Diagnosis of Chlamydial Infections," *J Microbiol Methods*, 2000, 42(3):265-79.

Internet Web Sites

www.cdc.gov/std/Chlamydia/STDFact-Chlamydia.htm

♦ *Chlamydia* Smears Cytology *see* Ocular Cytology *on page 973*

Chlamydia trachomatis Culture

Related Information

Bacterial Culture, Genital Specimen *on page 239*
Bacterial Culture, Urine *on page 246*
(Continued)

Chlamydia trachomatis Culture (Continued)

Cervical/Vaginal Cytology *on page 376*
Chlamydia Group Serology *on page 385*
Chlamydia trachomatis Direct Antigen Test *on page 387*
Chlamydia trachomatis Nucleic Acid Detection *on page 388*
Neisseria gonorrhoeae Culture and Smear *on page 945*
Neisseria gonorrhoeae Nucleic Acid Detection *on page 947*
Viral Culture *on page 1307*

Abstract *Chlamydia trachomatis* is an intracellular pathogen and can be cultured in a variety of cell lines. The specificity of *C. trachomatis* culture approaches 100% but the sensitivity is between 70% and 90%. Cell culture is the only test that should be used to establish *C. trachomatis* infection with legal implications (eg, children suspected of being victims of sexual abuse).

Specimen Obtain swab specimens containing columnar epithelial cells of urethra, cervix, rectum, conjunctiva, posterior nasopharynx, or throat.

Container Culturette® (dacron) swabs should be placed in *Chlamydia* transport medium.

Collection Urethra: Remove mucous/pus. The swab should be inserted 2-4 cm into the urethra. Use firm pressure to scrape cells from the mucosal surface. If possible repeat with second swab. Patient should not urinate within 1 hour prior to specimen collection.

Cervix: Remove mucous/pus with a Culturette® and use firm and rotating pressure to obtain specimen with another swab. May be combined with a urethral swab into same transport medium. This two-swab method is highly recommended.

Rectum: Sample anal crypts with a Culturette®.

Conjunctiva: Remove mucous and exudate. Use a Culturette® and firm pressure to scrape away epithelial cells from upper and lower lids.

Posterior nasopharynx or throat: Collect epithelial cells by using a Culturette®.

Storage Instructions Deliver inoculated transport medium **immediately** to the laboratory. **Specimens must be refrigerated** if stored for 2 days. Specimen must be frozen at -70°C if stored more than 2 days.

Turnaround Time Cultures with no growth usually are reported after 7 days. Rapid culture methods require 48 hours or more.

Reference Interval No *Chlamydia trachomatis* isolated

Use Aid in the diagnosis of infections caused by *Chlamydia trachomatis*

Limitations Culture may be negative in presence of *Chlamydia trachomatis* infection. Culture is probably not the gold standard for detection of *C. trachomatis*. The sensitivity of culture is only 70% to 90% because *C. trachomatis* does not always survive transit to the laboratory, and often there is inadequate sampling with (multiple) swabs.

Methodology Inoculation of specimen onto McCoy cell, HeLa-229, or Buffalo green monkey cell culture and subsequent detection of *Chlamydia*-infected cells by monoclonal antibody and immunofluorescence

Additional Information Genital infections due to *C. trachomatis* are the most frequent reportable bacterial sexually transmitted disease in the United States. There are more than 4 million infections reported annually. Many of the cases are asymptomatic or minimally symptomatic. Some will progress to serious infections including pelvic inflammatory disease, ectopic pregnancy, and infertility in women. Infection with *C. trachomatis* during pregnancy places the newborn infant at risk of pneumonia and conjunctivitis.

Direct immunofluorescence techniques and enzyme immunoassays are available to detect *C. trachomatis* in clinical specimens. These methods usually provide reliable results in high-prevalence populations and detect both viable and nonviable organisms, but have for the most part been replaced by nucleic acid-based methods.

Culture should be the test-of-choice in cases of child abuse, ascending pelvic infections, rectal and throat infections, and when a test-for-cure is desired.

References
Dayan L, "*Chlamydia* Detection and Management," *Aust Fam Physician*, 2000, 29(6):522-6.

Guaschino S and DeSeta F, "Update on *Chlamydia trachomatis*," *Ann N Y Acad Sci*, 2000, 900:293-300.

Hammerschlag MR, Ajl S, and Laraque D, "Inappropriate Use of Nonculture Tests for the Detection of *Chlamydia trachomatis* in Suspected Victims of Child Sexual Abuse: A Continuing Problem," *Pediatrics*, 1999, 104(5 Pt 1):1137-9.

Kirchner JT and Emmert DH, "Sexually Transmitted Diseases in Women. *Chlamydia trachomatis* and Herpes Simplex Infections," *Postgrad Med*, 2000, 107(1):55-65.

Internet Web Sites
www.cdc.gov/std/Chlamydia/STDFact-Chlamydia.htm

Chlamydia trachomatis Direct Antigen Test

Related Information
Bacterial Culture, Genital Specimen *on page 239*
Cervical/Vaginal Cytology *on page 376*
Chlamydia Group Serology *on page 385*
Chlamydia trachomatis Culture *on page 385*
Chlamydia trachomatis Nucleic Acid Detection *on page 388*
Neisseria gonorrhoeae Culture and Smear *on page 945*
Neisseria gonorrhoeae Nucleic Acid Detection *on page 947*
Ocular Cytology *on page 973*
Viral Culture *on page 1307*

Abstract Several tests are available for direct detection of *Chlamydia trachomatis*, specific antigen using monoclonal antibodies. The sensitivity of these tests reaches >70% and the specificity 97% to 99%. These tests should **not** be used to establish *C. trachomatis* infection with legal implications (eg, children suspected of being victims of sexual abuse).

Patient Preparation For urogenital specimens, patient should not urinate 1 hour prior to collection.

Specimen Direct smear or urogenital specimen.

Container Single well (8 mm) glass slide, dacron swabs (one large, one small), one cytobrush, methanol fixative (0.5 mL vial). These items are contained in a commonly used direct detection kit (collection pack) known as MicroTrak®.

Collection Endocervical with cytology brush: Nonpregnant women. Use large swab to remove exudate or mucous from endocervix. Insert cytobrush into cervical os past the squamocolumnar junction. Rest 2-3 seconds, rotate brush 360 degrees to gather columnar cells and withdraw brush. Do not touch vaginal walls with brush, and prepare slides immediately by rotating and twisting brush back and forth across center of slide well.

Endocervical with swab: Pregnant women. Use large swab to remove exudate or mucous from exocervix. Insert another large dacron swab until tip is no longer visible, rotate swab 5-10 seconds, and withdraw swab. Do not touch vaginal walls and prepare slides immediately. Firmly roll one side of swab over top half of well. Turn swab over and roll other side over bottom half of slide well.

Urethral: Males. Patient should not urinate 1 hour before sampling. Remove pus or exudate, insert small swab with wire shaft 2-4 cm into penis. Gently rotate swab to dislodge cells, rest swab 2 seconds, withdraw swab, and prepare slide immediately as above.

Rectal: Symptomatic patients only. Use large swab. Insert ~3 cm into anal canal. Move swab from side to side to sample crypts. If fecal contamination occurs, discard swab and obtain another specimen. Prepare slide immediately as above.

Conjunctival: Neonates, symptomatic only. Use large swab to gently remove pus or discharge and discard. If both eyes are sampled, swab less affected eye first. Swab inside of lower, then upper lid, and prepare slide immediately as above.

Nasopharyngeal: Neonates, symptomatic only. Use small swab or nasal aspirator. Collect specimen from posterior nasopharynx using standard collection method. If swab was used, prepare slide immediately. If nasal aspirate was collected, deliver to the laboratory technician or technologist immediately for slide preparation.

All specimens: Allow specimen to air dry. Lay slide flat and flood with methanol fixative. Let entire quantity evaporate. Refold pack without touching fixed specimen.
(Continued)

Chlamydia trachomatis Direct Antigen Test
(Continued)

Storage Instructions Refrigerate slides at 2°C to 8°C or at room temperature (20°C to 30°C) until taken to the laboratory. Slides must be stained within 7 days of collection.

Causes for Rejection Less than 10 columnar or cuboidal epithelial cells on slide

Turnaround Time 24-48 hours

Special Instructions Specify specimen origin. Include all pertinent information, label slide and collection pack.

Reference Interval No *Chlamydia trachomatis* detected

Use Aid in the diagnosis of disease caused by *Chlamydia trachomatis*

Limitations The direct fluorescent antibody procedure is considerably less sensitive than the cell culture procedure and molecular amplification methods. The number of cells on the slide can be too low for diagnosis. It is cumbersome and requires a highly skilled microscopist for proper interpretation.

The direct detection of *Chlamydia* in specimens depends largely on the preparation of the cell smear. Smears that are too thick or lumpy can cause false-positive results. Contamination with red blood cells makes the smears difficult to interpret. Smears with too few cells can cause false-negative results. This test detects only *Chlamydia trachomatis* major outer membrane protein (MOMP). The test does not distinguish between living and dead organisms. Therefore, the test does not necessarily serve as a test-of-cure.

Methodology The *Chlamydia trachomatis* direct test uses fluorescein-conjugated monoclonal antibodies (reactive with all 15 known serotypes of *C. trachomatis*) to detect elementary bodies in clinical smears or an enzyme immunoassay to detect outer membrane protein.

Additional Information In some populations, up to 45% of women who have gonorrheal infection have chlamydial infection as well. (Other sexually transmitted diseases are included in the Disease Index). *Chlamydia trachomatis* has been implicated in neonatal/infantile conjunctivitis and afebrile pneumonia. Thirty-three percent to 50% of babies born vaginally to mothers with chlamydial infection of the cervix will be infected; the majority of these neonates will develop inclusion conjunctivitis and/or a respiratory tract infection that can lead to the distinctive (afebrile) pneumonia syndrome. Conjunctivitis in infected neonates usually occurs between the 5th and 12th day after birth. In neonates born to mothers with premature rupture of the membranes, *C. trachomatis* has been detected, in rare cases, as early as the first day following birth.

References
Chan EL, Brandt K, Stoneham H, et al, "Comparison of the Effectiveness of Polymerase Chain Reaction and Enzyme Immunoassay in Detecting *Chlamydia trachomatis* in Different Female Genitourinary Specimens," *Arch Pathol Lab Med*, 2000, 124(6):840-3.

Dayan L, "*Chlamydia* Detection and Management," *Aust Fam Physician*, 2000, 29(6):522-6.

Guaschino S and DeSeta F, "Update on *Chlamydia trachomatis*," *Ann N Y Acad Sci*, 2000, 900:293-300.

Hammerschlag MR, Ajl S, and Laraque D, "Inappropriate Use of Nonculture Tests for the Detection of *Chlamydia trachomatis* in Suspected Victims of Child Sexual Abuse: A Continuing Problem," *Pediatrics*, 1999, 104(5 Pt 1):1137-9.

Internet Web Sites
www.cdc.gov/std/Chlamydia/STDFact-Chlamydia.htm

Chlamydia trachomatis Nucleic Acid Detection
Related Information
Bacterial Culture, Genital Specimen *on page 239*
Chlamydia Group Serology *on page 385*
Chlamydia trachomatis Culture *on page 385*
Chlamydia trachomatis Direct Antigen Test *on page 387*
Neisseria gonorrhoeae Culture and Smear *on page 945*
Neisseria gonorrhoeae Nucleic Acid Detection *on page 947*
Viral Culture *on page 1307*

Test Includes Detection of *Chlamydia trachomatis* nucleic acid in clinical specimens, either directly or after nucleic acid amplification.

Abstract *Chlamydia trachomatis* is the most common sexually transmitted bacterial infection. In 1998, over 600,000 new cases of genital *Chlamydia* infection was reported to the CDC. *Chlamydia* infections may be asymptomatic in up to 70% of women and 30% of men. Disease and infection is associated with a high rate of tubal pregnancies, pelvic inflammatory disease, and infertility. *Chlamydia* also causes a severe infection of the eye that, if untreated, may lead to blindness. While in the past, *Chlamydia trachomatis* had been principally diagnosed by culture or immunoassay, the commercial availability of molecular assays has altered the detection of this agent. In addition to swab specimens of cervix and urethra, the new assays are sufficiently sensitive to detect organism in the urine of infected males.

Specimen Swabs of an infected site or urine (depending on the assay). Although the rectum and nasopharynx are also sites that may be infected with *Chlamydia*, swabs of these sites have not been approved for use with molecular assays.

Container Special collection and transport kits are an integral component of the assay and must be used with the appropriate commercial detection assay. A commercial kit typically contains a swab and a transport media or device. Contact the laboratory for appropriate collection kit.

Collection Two swabs are provided in the typical commercial kit for use in females. The cervix or endocervix is first cleaned with one swab and the second swab is used to collect the specimen. The swab is inserted into the endocervical canal and rotated to collect epithelial cells from the infected site. The swab is then placed into transport media and sent to the laboratory. In males, the swab is inserted into the urethral meatus and rotated to collect epithelial cells. The swab is then placed into transport media and sent to the laboratory.

Storage Instructions Many of these specimens may be contained at room temperature and sent without freezing to the laboratory. Specimens collected for commercial molecular assays are typically stable for up to 1 week after collection. Contact the laboratory for appropriate instructions.

Causes for Rejection Collection of a specimen from a nonapproved site; some assays cannot detect genomic material if the specimen contains excess blood. Specimens from a suspected sexual abuse case will be rejected.

Turnaround Time Results are usually available within 24 hours.

Reference Interval Negative for *Chlamydia trachomatis*

Use The rapid detection of *C. trachomatis* in clinical specimens. The sensitivity of the test exceeds that for culture and other nonmolecular assays.

Limitations These tests should not be used to establish *C. trachomatis* infection with legal implications (eg, children suspected of being victims of sexual abuse).

Methodology Several molecular assays have been approved or are under review by the Food and Drug Administration. The first PCR test licensed by the FDA in 1993 was for *C. trachomatis*. The ligase chain reaction procedure incorporates the extraction and denaturation of DNA followed by specific amplification through the ligation of two segments of the target. The Gen-Probe® assay detects the presence of ribosomal RNA, which is in vast excess to single copy genes of an organism. The probe incorporates a chemiluminescent chemical which upon activation emits light. Recently, the Gen-Probe® assay has incorporated a transcription-based amplification assay that will amplify the ribosomal RNA before detection. Additionally the ProbeTec ET system has been introduced and uses yet a different method of amplification called strand displacement. It simultaneously amplifies and detects *Chlamydia trachomatis* DNA using fluorescence and chemiluminescence. Various instrumentation has been utilized to make the assays automated.

Additional Information The molecular probe assays provide several advantages over traditional culture assays or antibody-based tests. One of the most significant advantages is the ability to detect nonviable organisms, since only the presence of nucleic acids is necessary. This latter property also minimizes the need to rapidly transfer the specimen to culture media or cells. The sensitivity of molecular assays exceeds that of culture techniques, depending upon the prevalence of disease and the site of collection. The ability to detect organisms in urine provides an additional advantage over alternative methods and may have a significant impact on public health attempts to limit disease spread.
(Continued)

Chlamydia trachomatis Nucleic Acid Detection

(Continued)

References

Bull SS, Jones CA, Granberry-Owens D, et al, "Acceptability and Feasibility of Urine Screening for *Chlamydia* and Gonorrhea in Community Organizations: Perspectives From Denver and St. Louis," *Am J Public Health*, 2000, 90(2):285-6.

Dayan L, "*Chlamydia* Detection and Management," *Aust Fam Physician*, 2000, 29(6):522-6.

Guaschino S and DeSeta F, "Update on *Chlamydia trachomatis*," *Ann N Y Acad Sci*, 2000, 900:293-300.

Keenan GF, "Polymerase Chain Reaction as a Diagnostic Tool," *Adolesc Med*, 1998, 9(1);35-43.

Internet Web Sites

www.cdc.gov/std/Chlamydia/STDFact-Chlamydia.htm

Chloramphenicol, Serum

Synonyms Chloromycetin®; Mychel-S®

Abstract The use of chloramphenicol has greatly diminished in recent years. It is still appropriate to use this agent to treat certain rickettsial infections in selected patients, or in carefully selected penicillin-allergic patients with bacterial meningitis. Its most feared side effect, aplastic anemia, is idiopathic, and is therefore not serum concentration related. Since its action is through inhibition of bacterial protein synthesis, the drug is more bacteriostatic than bactericidal.

Specimen Serum

Container Red top tube

Sampling Time Collect trough level immediately before next dose; peak level 1-2 hours after oral dose, 30 minutes after I.V. dose (time to peak can be variable).

Storage Instructions Freeze processed specimen

Reference Interval Therapeutic: 10-25 µg/mL (SI: 31-77 µmol/L); trough: <5 µg/mL (SI: <15 µmol/L)

Critical Values Toxic: >25 µg/mL (SI: >77 µmol/L)

Use Monitor drug therapy; monitoring for potential toxicity. See Table B in the Therapeutic Drug Monitoring Introduction *on page 61.*

Methodology High performance liquid chromatography (HPLC), gas chromatography (GC), immunoassay

Additional Information

- Half-life (adults): 1.5-5 hours
- Volume of distribution: 0.9 L/kg
- Protein binding: 50% to 60%

Hematologic toxicities of chloramphenicol have limited its use. They can be manifested as concentration-related, reversible bone marrow suppression, or rare, irreversible idiosyncratic aplastic anemia. Concentration-related bone marrow suppression has been associated with peak concentrations >25 µg/mL. Reversible bone marrow suppression begins with reticulocytopenia and a decrease in hemoglobin. With continued use, thrombocytopenia and neutropenia may occur. The more feared aplastic anemia is not concentration related. There may be a genetic predisposition to this adverse effect. There may be an increased incidence of leukemia in survivors of this toxicity. Gray baby syndrome is the other dreaded adverse effect seen with chloramphenicol use. This toxicity generally occurs in infants, but may occur in susceptible adults (those with hepatic dysfunction). In infants, there is a limited ability to conjugate chloramphenicol with glucuronide, which inactivates the parent compound. This syndrome includes vomiting, ashen gray color, abdominal distention, metabolic acidosis, and about a 40% mortality rate. Deaths due to chloramphenicol are associated with serum levels usually >200 µg/mL.

Chloramphenicol is a hepatic enzyme inhibitor. It decreases the clearance of many drugs, including warfarin, tolbutamide, chlorpropamide, and phenytoin. Chloramphenicol clearance is increased by some enzyme inducers including phenobarbital, phenytoin, and rifampin.

References

Hammett-Stabler CA and Johns T, "Laboratory Guidelines for Monitoring of Antimicrobial Drugs," National Academy of Clinical Biochemistry, *Clin Chem*, 1998, 44(5):1129-40.

Kasten MJ, "Clindamycin, Metronidazole, and Chloramphenicol," *Mayo Clin Proc*, 1999, 74(8):825-33.

Wareham DW and Wilson P, "Chloramphenicol in the 21st Century," *Hosp Med*, 2002, 63(3):157-61.

Chlordiazepoxide, Serum

Related Information

Benzodiazepines, Qualitative, Urine *on page 253*
Diazepam, Serum *on page 510*

Synonyms A-Poxide®; Equibral®; Librax®; Libritabs®; Librium®; Methaminodiazepoxide Hydrochloride; Mitran®; Resposan-10®; SK-Lygen®; Smail®; Solium®; Tropium®

Abstract Chlordiazepoxide is a widely prescribed benzodiazepine drug used as a sedative-hypnotic (tranquilizer) prescribed for anxiety and panic. The drug is also used for withdrawal symptoms of acute alcoholism and for preoperative apprehension.

Specimen Serum

Container Red top tube; do not use gel-containing tubes.

Sampling Time Trough levels at steady state (40-60 hours); collect for peak level 4 hours after oral dosing.

Storage Instructions Process immediately; avoid exposure to light; freeze if not analyzed immediately.

Reference Interval Therapeutic: 0.7-1.0 µg/mL (SI: 2.3-3.3 µmol/L)

Critical Values Toxic: >5 µg/mL (SI: >17 µmol/L)

Use Monitor therapeutic drug level for compliance; determine toxic level

Methodology High performance liquid chromatography (HPLC), gas chromatography (GC)

Additional Information

- Half-life: 8-12 hours, longer with renal disease, up to 30-63 hours with cirrhosis
- Volume of distribution: 3 L/kg
- Protein binding: 90% to 95%

Overdosage with the benzodiazepines is frequent, but serious sequelae are rare. However, there is an additive effect when used with other CNS depressants (eg, ethanol and barbiturates).

Benzodiazepines have been associated with anterograde amnesia. Paradoxical reactions, including hyperactive or aggressive behavior have been reported with benzodiazepines, particularly in adolescent/pediatric or psychiatric patients. Chlordiazepoxide lacks analgesic, antidepressant, or antipsychotic properties.

References

Fraser AD, "Use and Abuse of the Benzodiazepines," *Ther Drug Monit*, 1998, 20(5):481-9.
Giri AK and Banerjee S, "Genetic Toxicology of Four Commonly Used Benzodiazepines: A Review," *Mutat Res*, 1996, 340(2-3):93-108.
Iqbal MM, Sobhan T, and Ryals T, "Effects of Commonly Used Benzodiazepines on the Fetus, the Neonate, and the Nursing Infant," *Psychiatr Serv*, 2002, 53(1):39-49.
Saitz R and O'Malley SS, "Pharmacotherapies for Alcohol Abuse. Withdrawal and Treatment," *Med Clin North Am*, 1997, 81(4):881-907.

♦ **Chloride:Phosphorus Ratio** *see* Calcium, Serum *on page 329*

Chloride, Serum, Plasma, or Blood

Related Information

Anion Gap, Serum, Plasma, or Urine *on page 160*
Bicarbonate, Blood *on page 258*
Carbon Dioxide, Total, Blood *on page 339*
Chloride, Urine *on page 394*
Electrolyte Panel, Serum *on page 532*
Point-of-Care Testing *on page 1065*
Sodium, Serum or Plasma *on page 1210*

Synonyms Cl, Serum

Applies to Bromism

Abstract Chloride is a component of the serum electrolyte and metabolic panels.

Specimen Serum, plasma, or whole blood

Container Red top tube or green top (heparin) tube

Collection Pediatrics: Blood drawn from heelstick for capillary.

Storage Instructions Refrigerate

Reference Interval Premature: 95-110 mmol/L; full-term: 96-106 mmol/L; children and adults: 97-107 mmol/L

(Continued)

Chloride, Serum, Plasma, or Blood (Continued)

Possible Panic Range <80 mmol/L (SI: <80 mmol/L), >115 mmol/L (SI: >115 mmol/L)

Use Chloride measurement is useful in the differential diagnosis of acidemias and alkalemias; an important use of chloride is in application of the anion gap. (Consult Anion Gap, Serum, Plasma, or Urine *on page 160* for more information.) Chloride values must be interpreted with results of other electrolytes and acid-base analytes.

Chloride is **increased** in mineralocorticoid deficiency, with ammonium chloride administration, and in cases of hyperchloremic (nongap) metabolic acidosis: excessive infusion of hyperchloremic (normal) saline, diarrhea/GI losses, renal tubular acidosis, pancreatic fistula, and enterovesical fistula.

Chloride is **decreased** with overhydration, congestive failure, syndrome of inappropriate secretion of ADH, vomiting, gastric suction, chronic or compensated respiratory acidosis, Addison disease, salt-losing nephritis, burns, metabolic alkalosis, diabetic ketoacidosis, and in some instances of diuretic therapy.

In differential diagnosis of emesis of uncertain etiology in early infancy, higher bicarbonate and lower chloride (mean 95.7 vs 104 mmol/L) favor pyloric stenosis over gastroesophageal reflux. Serum CL ≤98 mmol signaled the diagnosis of pyloric stenosis with sensitivity of 50%, specificity 99%, positive predictive value 97%.

Methodology Coulometric-amperometric titration, spectrophotometry, mercurimetric titration, and ion-selective electrode (ISE) methods

References

Adrogué HJ and Madias NE, "Hypernatremia," *N Engl J Med*, 2000, 342(20):1493-9.

Adrogué HJ and Madias NE, "Hyponatremia," *N Engl J Med*, 2000, 342(21):1581-9.

Adrogué HJ and Madias NE, "Management of Life-Threatening Acid-Base Disorders," First of Two Parts, *N Engl J Med*, 1998, 338(1):26-34, Second of Two Parts, *N Engl J Med*, 1998, 338(2):107-11.

Batlle DC, Hizon M, Cohen E, et al, "The Use of the Urinary Anion Gap in the Diagnosis of Hyperchloremic Metabolic Acidosis," *N Engl J Med*, 1988, 318(10):594-9.

Koch SM and Taylor RW, "Chloride Ion in Intensive Care Medicine," *Crit Care Med*, 1992, 20(2):227-40.

Smith GA, Mihalov L, and Shields BJ, "Diagnostic Aids in the Differentiation of Pyloric Stenosis From Severe Gastroesophageal Reflux During Early Infancy: the Utility of Serum Bicarbonate and Serum Chloride," *Am J Emerg Med*, 1999, 17(1):28-31.

Chloride, Sweat

Related Information

Cystic Fibrosis DNA Detection *on page 491*

d-Xylose Absorption Test, Serum, Urine *on page 527*

Synonyms Cystic Fibrosis Sweat Test; Iontophoresis; Sweat, Chloride

Applies to Pilocarpine Iontophoresis; *Pseudomonas aeruginosa*, Mucoid; Sodium, Sweat; Trypsin Activity, Stool; Trypsinogen, Immunoreactive

Test Includes Sodium level may also be measured.

Abstract Sweat chloride is used in the diagnostic evaluation of persons, usually infants and small children, with clinical manifestations suggesting possible cystic fibrosis (CF). In this test, forearm sweating is induced by pilocarpine iontophoresis, and the resulting sweat is collected on to filter paper, gauze, or a macroduct tube. The sweat is then assayed for chloride. Individuals contemplating performing or interpreting sweat chloride tests are urged to study the NCCLS Document 34.A.

Specimen Sweat

Reference Interval

Children and adults to age 20 years:
- normal: 0-40 mmol/L
- borderline/indeterminate: 41-60 mmol/L
- consistent with cystic fibrosis: >60 mmol/L

Adults older than 20 years: consistent with cystic fibrosis >70 mmol/L

Note: All values should be interpreted with family history and clinical presentations. Sweat chloride values <40 mmol/L have been documented in patients with genetically proven CF; clinical correlation is necessary.

Critical Values Strongly positive and (with characteristic clinical findings or family history) confirmatory: ≥80 mmol/L.

Use Evaluate possible CF in children with family history of CF, frequent and/or foul stools, diarrhea, malnutrition and failure to thrive, depletion of the fat-soluble vitamins, malabsorption, pancreatic insufficiency, history of meconium ileus, neonatal intestinal obstruction, rectal prolapse, infant celiac disease, chronic sinopulmonary and pulmonary disease, asthma, chronic cough, digital clubbing, salt depletion syndromes, chronic metabolic alkalosis, and *Pseudomonas* bronchitis. Evaluation of young adult males for aspermia and for absence of vas deferens.

Limitations Skin involved by inflammation should not be tested. Elevations have been reported in several other diseases; see table. In some of the diseases listed, sweat chloride and sodium elevations lack the constancy that is found in CF. False low results have been described with edema, hypoproteinemia, and excessive sweating. Results are highly variable in adults, especially women in whom sweat chloride levels vary with the menstrual cycle. Sweat chloride levels in adults must be interpreted cautiously: false positives and false negatives may occur.

The differential diagnosis of an elevated sweat chloride includes several disorders other than CF. See table.

Differential Diagnosis of Elevated Sweat Chloride

Anorexia nervosa
Atopic dermatitis
Autonomic dysfunction
Ecodermal dysplasia
Environmental deprivation
Familial cholestasis
Fucosidosis
Glucose-6-phosphate dehydrogenase deficiency
Glycogen storage disease: type 1
Hypogammaglobulinemia
Klinefelter syndrome
Long-term prostaglandin E1 infusion
Mauriac syndrome
Mucopolysaccharidosis type 1
Nephrogenic diabetes insipidus
Nephrosis
Protein calorie malnutrition
Pseudohypoaldosteronism
Psychosocial failure to thrive
Untreated adrenal insufficiency
Untreated hypothyroidism

Borderline tests must be repeated. Even negative tests should be repeated if the clinical picture suggests cystic fibrosis. U.S. Cystic Fibrosis Foundation recommends repeating all positives on a separate occasion. Sweat chloride test does not identify carriers (heterozygotes). Reliability in the first weeks of life is questionable.

A small fraction of CF patients do not have diagnostic sweat chloride patterns. The measurement of sodium and the determination of the Na:Cl ratio may be useful. An application of evolving technology involves isolation of DNA with amplification utilizing the polymerase chain reaction: see Cystic Fibrosis DNA Detection *on page 491.* Molecular recombinant DNA diagnostic techniques may ultimately be the best way to diagnose the disease and the carrier states.

Contraindications Dermatitis. Do not collect sweat from the palm of the hand, or from any site following excessive sweating such as following high temperature or heavy exercise. Improper placement of pad or electrode can cause skin burn.

Methodology Chloride in sweat from the forearm by pilocarpine-ionotophoresis. Coulometric titration by chloridometer. Measurement by an ion-specific electrode (ISE). If gauze or filter paper method of sweat collection is used, at least
(Continued)

Chloride, Sweat (Continued)

75 mg of sweat, on 2" x 2" stimulated skin, must be collected for test validity. Usual collections, using proper equipment producing 1.5-4 mA, range from 100-400 mg of sweat. Laboratory confirmation of the diagnosis of CF requires two or more sweat tests done on separate days with duplicated samples of sweat weighing >75 mg. If microtube tubing is used for sweat collection, minimal volume should be 15 μL.

Additional Information Other laboratory abnormalities in CF may include culture of mucoid *Pseudomonas aeruginosa*, nonmucoid strains, *Staphylococcus aureus* and *Burkholderia cepacia*; low total protein; prolongation of prothrombin time and liver disease, with abnormalities which may relate to evolving hepatic cirrhosis.

Testing for stool trypsin activity is less reliable than the sweat chloride test and should not be used in its place. Assay of immunoreactive trypsinogen on dried blood spots in CF screening programs is improved when combined with DNA analysis, which provides increased sensitivity and specificity.

Different clinical facets of CF are recognized. Mutations which cause less severe pancreatic disease are known, and mutations associated with mild lung disease are described.

References

Chmiel JF, Drumm ML, Konstan MW, et al, "Pitfall in the Use of Genotype Analysis as the Sole Diagnostic Criterion for Cystic Fibrosis," *Pediatrics*, 1999, 103(4 Pt 1):823-6.

Grody, WW, "Cystic Fibrosis: Molecular Diagnosis, Population Screening, and Public Policy," *Arch Pathol Lab Med*, 1999, 123(11):1041-6.

LeGrys VA, "Sweat Testing for the Diagnosis of Cystic Fibrosis: *Pediatrics*, 1999, 103(4 Pt 1):823-6. Practical Considerations," *J Pediatr*, 1996, 129(6):892-7.

National Committee for Clinical Laboratory Standards (NCCLS), "Sweat Testing: Sample Collection and Quantitative Analysis: Approved Guidelines," *NCCLS Document*, C34-A2, 2000, 20(14):1-40.

Stern RC, "The Diagnosis of Cystic Fibrosis," *N Engl J Med*, 1997, 336(7):487-91.

Internet Web Sites

www.nccls.org

Chloride, Urine

Related Information

Anion Gap, Serum, Plasma, or Urine *on page 160*
Blood Gases and pH, Arterial *on page 275*
Chloride, Serum, Plasma, or Blood *on page 391*
Potassium, Urine *on page 1080*
Sodium, Urine *on page 1213*
Urine Collection, 24-Hour *on page 1295*

Synonyms Cl, Urine; Urine Cl

Applies to Alkalosis

Replaces Electrolytes, Urine

Specimen Timed or random urine

Container No preservative

Reference Interval Infants: 2-10 mmol/24 hours; children: 15-40 mmol/24 hours; adults: 110-250 mmol/24 hours

Use Urinary chloride is used to evaluate acid-base balance, and particularly to distinguish whether or not a case of metabolic alkalosis is chloride-responsive (salt responsive). Chloride depleted patients excrete urine with low chloride, <10 mmol/L. Such patients are chloride-responsive (ie, they respond to chloride sufficient to return body stores to normal).

Endogenous or exogenous corticosteroids produce urine chloride values in excess of 20 mmol/L. Such patients are chloride resistant. The finding of chloride resistant metabolic alkalosis may be a clue to the identification of an ACTH- or aldosterone-producing neoplasm (eg, Cushing syndrome or Conn syndrome). A diagrammatic presentation of the differential diagnosis of hypokalemia includes application of urinary chloride (see Potassium, Urine *on page 1080*).

Methodology Coulometric titration, ion-selective electrode (ISE)

Additional Information Urine chloride is often ordered with sodium and potassium as a timed urine. The **urinary anion gap** $[Na^+ - (Cl^- + HCO_3^-)]$ or $[(Na^+ + K^+) - (Cl^-)]$ may be used in the initial evaluation of hyperchloremic metabolic acidosis, and is discussed in the listing Anion Gap, Serum, Plasma, or Urine *on* *page 160.* In metabolic alkalosis, spot urine chloride in the presence of hypertension can provide important information.

References

Guay-Woodford LM, "Bartter Syndrome: Unraveling the Pathophysiologic Enigma," *Am J Med,* 1998, 105(2):151-61.

Harrington JT and Cohen JJ, "Measurement of Urinary Electrolytes - Indications and Limitations," *N Engl J Med,* 1975, 293:1241-3.

Painter PC, Cope JY, and Smith JL, "Reference Information for the Clinical Laboratory," *Tietz Textbook of Clinical Chemistry,* 3rd ed, Burtis CA and Ashwood ER, eds, Philadelphia, PA: WB Saunders Co, 1999, 1806.

Sherman RA and Eisinger RP, "The Use (and Misuse) of Urinary Sodium and Chloride Measurements," *JAMA,* 1982, 247:3121-4.

♦ **Chloridorrhea, Congenital** *see* pH, Stool *on page 1034*

♦ **Chloromycetin®** *see* Chloramphenicol, Serum *on page 390*

♦ **Chlorpromazine** *see* Phenothiazines, Serum *on page 1021*

Chlorpromazine, Serum

Related Information

Phenothiazines, Serum *on page 1021*

Synonyms Amazin®; Bay Clor®; Dozine®; Hibanil®; Largactil®; Ormazine®; Prozil®; Repazine®; Thorazine®

Applies to Phenothiazines

Abstract This is an aliphatic phenothiazine used as an antipsychotic and sedative agent. The drug is also used for intractable hiccups and to control nausea and vomiting. Other uses include relief of restlessness and apprehension before surgery; treatment of acute intermittent porphyria; adjunct in the treatment of tetanus; combativeness and/or explosive hyperexcitable behavior in children 1-12 years of age; and short-term treatment of hyperactive children.

Specimen Serum

Container Red top tube

Sampling Time Trough (serum) at steady state (150 hours)

Collection Serum

Storage Instructions Refrigerate

Reference Interval Therapeutic: 50-300 ng/mL (SI: 157-942 nmol/L)

Critical Values Toxic: >750 ng/mL (SI: >2355 nmol/L)

Use Screen for chlorpromazine in urine; evaluate possibility of chlorpromazine poisoning or drug toxicity (serum)

Limitations Urine test is not specific for chlorpromazine; it may detect other phenothiazines if present. Correlation between serum concentration and clinical response is poor.

Contraindications The drug should not be used in the presence of nervous system depressants such as alcohol, barbiturates, and narcotics.

Methodology Gas chromatography (GC), high performance liquid chromatography (HPLC)

Additional Information

- Half-life: 30 hours
- Volume of distribution: 20 L/kg
- Protein binding: 95% to 98%

Due to the complex metabolism of this drug, the pharmacokinetics are variable and follow a multiphasic pattern. Attempts to correlate drug levels with clinical responses have not been successful. Chlorpromazine in overdose causes drowsiness, fainting, hypotension or hypertension, tachycardia, tremor, dizziness, hypoglycemia, coma, and other signs and symptoms. Overdose fatalities are relatively rare.

References

Boehme C and Strobel HW, "High Performance Liquid Chromatographic Methods for the Analysis of Haloperidol and Chlorpromazine Metabolism *In Vitro* by Purified Cytochrome P450 Isoforms," *J Chromatogr B Biomed Sci Appl,* 1998, 718(2):259-66.

Friedman NL, "Hiccups: A Treatment Review," *Pharmacotherapy,* 1996, 16(6):986-95.

Lacy CF, Armstrong LL, Goldman MP, et al, *Drug Information Handbook,* 12th ed, Hudson, OH: Lexi-Comp Inc, 2004.

♦ **Cholecalciferol** *see* Vitamin D, Serum *on page 1318*

♦ **Cholesterol-Binding Pancreatic Proteinase** *see* Fecal Pancreatic Elastase 1 *on page 576*

♦ **Cholesterol, Total** *see* Cholesterol, Total, Serum or Plasma *on page 396*

♦ **Cholesterol, Total** *see* High Density Lipoprotein Cholesterol, Serum *on page 729*

♦ **Cholesterol, Total** *see* Lipids, Overview *on page 853*

Cholesterol, Total, Serum or Plasma

Related Information
Apolipoprotein A-I, Serum *on page 202*
Apolipoprotein B-100, Serum *on page 203*
C-Reactive Protein, Serum *on page 467*
Endomysial Antibodies *on page 537*
High Density Lipoprotein Cholesterol, Serum *on page 729*
Homocyst(e)ine, Plasma *on page 741*
Lipid Panel, Serum *on page 852*
Lipids, Overview *on page 853*
Low Density Lipoprotein Cholesterol *on page 874*
Mevalonic Acid, Urine or Amniotic Fluid *on page 911*
Triglycerides, Serum or Plasma *on page 1275*

Applies to Apo-A-I; Apo B-100; Cholesterol, Total; Chylomicronemia; High Density Lipoprotein Cholesterol (HDLC); Hypoalphalipoproteinemia; Intermediate Density Lipoprotein Cholesterol; Lipoprotein-Associated Phospholipase A_2; Low Density Lipoprotein Cholesterol (LDLC); Metabolic Syndrome; Triglycerides; Very Low Density Lipoprotein Cholesterol

Abstract Elevated serum cholesterol is one of the major risk factors for coronary heart disease (CHD). See Lipids, Overview *on page 853*. Interlaboratory variation in cholesterol measurements was a major problem in the last half of the twentieth century. To minimize the patient care errors generated by this variability, the National Cholesterol Education Program (NCEP) convened a Laboratory Standardization Panel (LSP). In 1990, the LSP published national analytical goals for clinical cholesterol measurements: (a) overall precision equivalent to a CV ≤3%, (b) total bias not to exceed 3%. To assist clinical laboratories and the manufacturers of analytical equipment in achieving these goals, the CDC created a network of laboratories (the Cholesterol Reference Method Laboratory Network - CRMLN), each using the **reference method** (see Methodology, below) and each meeting rigid external proficiency testing standards. The CRMLN provides traceability of clinical methods to the CDC reference method; there are CRMLN programs both for manufacturers of diagnostic equipment and for individual laboratories.

For information about cholesterol ratios, see Lipids, Overview *on page 853*.

Patient Preparation
- For best results: stable diet for 3 weeks, stable body weight, and fasting for at least 10 hours. (Fasting is not essential for total cholesterol but is important if triglycerides are also to be assayed.)
- Posture may be a significant factor: similar to other analytes, cholesterol values may be 10% to 15% lower after 20 minutes in a recumbent position. From standing to a sitting position values are about 6% lower after 20 minutes.
- Increases of 2% to 5% in cholesterol may be seen if tourniquet is applied for 2 minutes during sampling.

Specimen Serum or plasma

Container Red top tube or lavender top (EDTA) tube

Storage Instructions At -70°C, a decrease of 2%/year on average takes place.

Reference Interval A normative reference interval is not applicable (see Lipids, Overview *on page 853*). The Adult Treatment Panel (ATP III) of the NCEP recommends the following for adults:
- Desirable: <200 mg/dL
- Borderline high: 200-239 mg/dL
- High: ≥240 mg/dL

Actual population distributions and percentile cutoffs are available.

Use While total cholesterol has been the traditional first step in defining lipid risk factor status, current opinion favors also obtaining measurements of the LDL cholesterol and HDL cholesterol. See Lipids, Overview *on page 853*.

Methodology

Definitive method: isotope dilution gas chromatography-mass spectrometry

Reference method: chemical hydrolysis with alcoholic KOH, maximum extraction with hexane, colorimetric reaction with acetic anhydride - acetic acid - sulfuric acid (Abell-Levy-Brodie-Kendall). The reference method is considered the gold standard for clinical measurements and is the accuracy base for research trials.

Clinical methods: enzymatic-colorimetric methods from several manufacturers.

Additional Information Complete assessment of coronary heart disease (CHD) risk requires evaluation of multiple risk factors (see Lipids, Overview *on page 853*).

Plasma cholesterol values are up to 10% lower than serum values.

A point-of-care device, useful also for self-monitoring, is available which assays cholesterol in whole blood using a cholesterol oxidase colorimetric method. Investigators at the Cleveland Clinic have reported a method allowing noninvasive assessment of skin cholesterol.

Among anemic patients, cholesterol concentrations <156 mg/dL may signal celiac disease.

References

"Executive Summary of The Third Report of the National Cholesterol Education Program (NCEP) Expert Panel on Detection, Evaluation, and Treatment of High Blood Cholesterol In Adults (Adult Treatment Panel III)," *JAMA*, 2001, 285(19):2486-97.

Myers GL, Kimberly MM, Waymack PP, et al, "A Reference Method Laboratory Network for Cholesterol: A Model for Standardization and Improvement of Clinical Laboratory Measurements," *Clin Chem*, 2000, 46(11):1762-72.

Stamler J, Daviglus ML, Garside DB, et al, "Relationship of Baseline Serum Cholesterol Levels in 3 Large Cohorts of Younger Men to Long-Term Coronary, Cardiovascular, and All-Cause Mortality and to Longevity," *JAMA*, 2000, 284(3):311-8.

Internet Web Sites
www.cdc.gov/nceh/dls/crmln/crmln.htm
www.nhlbi.nih.gov

◆ **Cholinesterase, Erythrocytic** *see* Acetylcholinesterase, Red Cell and Serum *on page 93*

◆ **Cholinesterase I** *see* Acetylcholinesterase, Red Cell and Serum *on page 93*

◆ **Cholinesterase II** *see* Pseudocholinesterase, Serum *on page 1122*

◆ **Cholinesterase, Serum** *see* Pseudocholinesterase, Serum *on page 1122*

◆ **Cholinesterase, True** *see* Acetylcholinesterase, Red Cell and Serum *on page 93*

◆ **Chorionic Gonadotropin, Beta Subunit** *see* Chorionic Gonadotropin, Human, Serum and Urine *on page 397*

Chorionic Gonadotropin, Human, Serum and Urine

Related Information

Alpha₁-Fetoprotein, Serum *on page 136*
Amniotic Fluid, Chromosome and Genetic Abnormality Analysis *on page 152*
Estriol, Unconjugated, Pregnancy, Serum, Plasma, or Urine *on page 554*
Heterophilic Antibodies *on page 727*
Inhibin A, Serum *on page 799*
Pregnancy-Associated Protein A, Serum *on page 1082*
Pregnancy Test, Serum or Urine *on page 1084*
Progesterone, Serum *on page 1093*

Synonyms Beta Subunit, hCG; Chorionic Gonadotropin, Beta Subunit; hCG; Human Chorionic Gonadotropin

Applies to GnRH; Pregnancy Testing; Triple Test

Replaces Aschheim-Zondek Test

Abstract Human chorionic gonadotropin (hCG) is synthesized by syncytiotrophoblastic cells. All pregnancy tests are based on the assay of hCG.

Assay for hCG is one of the tests often employed in maternal screening programs. See table in Alpha₁-Fetoprotein, Serum *on page 136*. Other markers include hCG, uE₃, PAPP-A, inhibin A, and beta-core fragment of hCG, as well as maternal age and fetal nuchal translucency as measured by ultrasonography.

hCG is a marker for gestational trophoblastic neoplasia, for nonseminomatous germ cell tumors, and less often for seminoma.
(Continued)

Chorionic Gonadotropin, Human, Serum and Urine
(Continued)

Specimen Serum or urine; rarely cerebrospinal fluid. Blood collection on filter paper has been used for prenatal testing for PAPP-A and free β-hCG.

Container Red top tube

Sampling Time First and second trimester screening of maternal serum for Down syndrome (trisomy 21) and Edwards syndrome (trisomy 18); first trimester screening at 74-97 days gestation. Second trimester screening includes sampling at 15-21 weeks gestation when testing also includes AFP and unconjugated estriol. Optimally, second trimester collection occurs between 16 and 18 weeks gestation. Check with laboratory. (See Alpha₁-Fetoprotein, Serum *on page 136* and Estriol, Unconjugated, Pregnancy, Serum, Plasma, or Urine *on page 554*.)

Storage Instructions Serum is stable 24 hours at 25°C and 4 days at 4°C. Freeze at -20°C for longer storage. If urine is used for beta-core fragment, freezing and prolonged storage should be avoided.

Special Instructions For females, state date of last menstrual period. See special instructions in Estriol, Unconjugated, Pregnancy, Serum, Plasma, or Urine *on page 554* for maternal screening for trisomies.

Reference Interval Depends on application and methodology. <5 mIU/mL (SI: <5 IU/L) of serum usually normal (nonpregnant). Mayo Medical Laboratories recently implemented the following for serum:
- male: <0.7 IU/L
- female: premenopausal: <0.8 IU/L, postmenopausal <3.3 IU/L
- cerebrospinal fluid: <1.5 IU/L.

Concentration of hCG increases rapidly during the first 6 weeks of pregnancy, peaking at 100,000 IU/L 60-70 days following implantation. Wide individual variation is seen, partly explained by maternal plasma volume increases. The reference interval in gestation relates to maternal weight and gestational age; β-hCG >50,000 IU/L supports but does not guarantee the diagnosis of viable intrauterine pregnancy.

Critical Values In subjects with concentration <2000 IU/L, increase of serum hCG <66% in 2 days may suggest spontaneous abortion or ruptured ectopic gestation in appropriate clinical setting. Rate of increase diminishes with gestation. After the 14th week, hCG continues its rise in gestational trophoblastic disease but falls in normal pregnancy. Levels of hCG >100,000 IU/L can be found in the sera of patients with gestational trophoblastic disease.

Use The principal use of hCG is to diagnose pregnancy.

While β-hCG doubles about every 1-3 days in a normal gestation, in most ectopic pregnancies β-hCG doubles more slowly. Chorionic gonadotropin assays are sometimes used to support the diagnosis of **ectopic pregnancy**. Ectopic gestations typically secrete decreased amounts of hCG and progesterone compared to intrauterine pregnancies, but it is the slope which must be considered. Abnormally low hCG levels with abnormal rates of change coupled with transvaginal ultrasound detect many ectopic pregnancies prior to rupture. It is helpful to use hCG and progesterone sequentially every other day to detect a lack of rise in the levels. Viable and nonviable pregnancy can be distinguished by the ratio of serial hCG values separated by 48 hours. The relationship for various sampling intervals has been plotted on the accompanying chart by Dr LB Baskin, based on prior investigation by Kadar et al and Romero et al. The sampling interval is plotted on the horizontal axis and the ratio of the determined β-hCG values for that interval is plotted on the vertical axis. If the intersection of the lines extending from those two points (the sampling interval and the β-hCG ratio) is above the curve, it is likely a viable intrauterine pregnancy, and if below the line, it is likely a nonviable pregnancy. Measurement of hCG isoforms has been applied to the diagnosis of ectopic gestation.

The detection of fetal **trisomy 21 (Down syndrome)** is well established using hCG with other investigation. Detection of trisomies 21 and 18 on the basis of maternal serum concentrations of pregnancy-associated protein A and free β-hCG, maternal age, and measurement of fetal nuchal translucency as first trimester screening provides good sensitivity with some false positives. Mothers carrying fetuses with Down syndrome are more likely to have decreased serum AFP and unconjugated estriol concentrations with increased hCG at 16 weeks. Although maternal age older than 35 has been a "magic number," in fact, 70% to 80% of children with Down syndrome are born to mothers younger than age 35. Measurement of the β-subunit offers improved detection. Detection of Down syndrome is enhanced with use of inhibin A in the second trimester. Decrease in hCG is found with **trisomy 18** (Edwards syndrome).

Serum hCG levels are extremely useful in the diagnosis and management of gestational trophoblastic neoplasia (GTN), including molar gestations, placental site trophoblastic tumors, and choriocarcinomas. With α-fetoprotein (AFP) and serum lactate dehydrogenase, hCG is needed in evaluation of germ cell tumors, including embryonal carcinoma and choriocarcinoma. It is increased in about 10% of cases of seminoma.

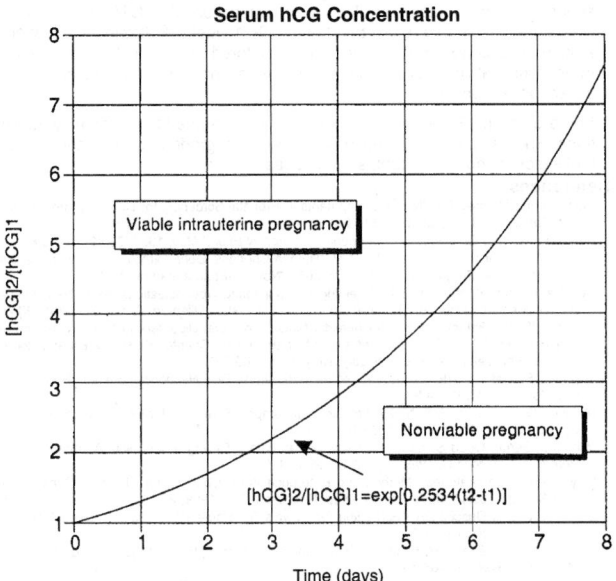

Serum hCG Concentration

Viable intrauterine pregnancy

Nonviable pregnancy

$[hCG]2/[hCG]1 = exp[0.2534(t2-t1)]$

Time (days)

Minimum increase in viable IUP during 1st trimester.
The ratio of successive serum hCG concentrations during the first trimester of a viable intrauterine pregnancy based on an exponential increase of 66% in 2 days (0.2534 days^{-1} = ln(1.66)/2 days), (sensitivity ~90%; specificity ~87%).

Courtesy of Leland B. Baskin, MD, the University of Texas Southwestern Medical Center at Dallas, with permission.

From Kadar N, Caldwell BV, and Romero R, "A Method for Screening for Ectopic Pregnancy and Its Indications," *Obstet Gynecol*, 1981, 58(2):162-5.
Romero R, Kadar N, Copel JA, et al, "The Value of Serial Human Chorionic Gonadotropin Testing as a Diagnostic Tool in Ectopic Pregnancy," *Am J Obstet Gynecol*, 1986, 155(2):392-4.

Limitations Heterophilic antibodies, also known as human antimouse antibodies (HAMA), occasionally produce a false-positive **serum** assay for hCG, and such erroneous results have led to needless therapy for gestational trophoblastic neoplasia. See discussion in Heterophilic Antibodies *on page 727*.

The same test method may not be suitable for use as a tumor marker and for pregnancy testing. Several reference preparations including World Health Organization sources may be used as standards, but different methods calibrated against different materials, and differences in circulating forms of hCG make interlaboratory comparisons difficult. At least 50 commercial kits for (Continued)

Chorionic Gonadotropin, Human, Serum and Urine
(Continued)

serum have been available in the 1990s. Although overlap exists between inevitable abortion and ectopic pregnancy, the rate of change of hCG and progesterone concentrations distinguish normal intrauterine gestation from pathologic pregnancies (ectopic gestation and inevitable abortion). A single hCG sampling falls short in pregnancy outcome determination. Increased hCG is found with multiparous gestation and with hemolytic disease of the newborn.

A positive maternal screen is not equivalent to a diagnosis. It indicates that further evaluation be considered.

Methodology Multiple immunoassays in several formats

Additional Information Ultrasonography is a sensitive additional technique for support of these diagnoses. Following evacuation of a GTN, serum β-hCG concentrations should return to normal in 9-11 weeks. Following evacuation of a mole, hCG concentrations should be monitored weekly until undetectable for three consecutive weeks, then monthly monitoring until undetectable for six consecutive months.

Some of the hCG methods available are only intended for pregnancy applications. Such assays do not necessarily detect degraded or more homogenous molecules found in trophoblastic diseases.

References
Baliff JP and Mooney RA, "New Developments in Prenatal Screening for Down Syndrome," *Am J Clin Pathol*, 2003, 120(Suppl 1):514-24.

Berkowitz RS and Goldstein DF, "Chorionic Tumors," *N Engl J Med*, 1996, 335(23):1740-8.

Borrelli PT, Butler SA, Docherty SM, et al, "Human Chorionic Gonadotropin Isoforms in the Diagnosis of Ectopic Pregnancy," *Clin Chem*, 2003, Nov 13 [Epub ahead of print].

Cole LA, Kohorn EI, and Kim GS, "Detecting and Monitoring Trophoblastic Disease. New Perspectives on Measuring Human Chorionic Gonadotropin Levels," *J Reprod Med*, 1994, 39(3):193-200.

Haddow JE and Palomaki GE, "Biochemical Markers of Fetal Disorders in Maternal Serum and Amniotic Fluid," *Medicine of the Fetus and Mother*, 2nd ed, Chapter 40, Reece EA and Hobbins JC, eds, Philadelphia, PA: Lippincott-Raven, 1999, 689-706.

Jacobs DS, DeMott WR, Oxley DK, et al, *Laboratory Test Handbook*, 5th ed, Hudson, OH: Lexi-Comp Inc, 2001, 147-9.

Malone FD and D'Alton ME, "First-Trimester Sonographic Screening for Down Syndrome," *Obstet Gynecol*, 2003, 102(5 Part 1):1066-79.

"Maternal Serum Quad Screening: Review of Laboratory Testing and Clinical Application," *Mayo Reference Services Communiqué*, 2002, 27(6):1-8.

Mayo Medical Laboratories, "Method and Reference Value Changes for Chorionic Gonadotropin, Beta-Subunit Tests," *Mayo Communiqué*, Rochester, MN, 2002, 24(11).

Mennuti MT and Driscoll DA, "Screening for Down's Syndrome - Too Many Choices?" *N Engl J Med*, 2003, 349(15):1471-3.

Wapner R, Thom E, and Simpson JL, "First-Trimester Screening for Trisomies 21 and 18," *N Engl J Med*, 2003, 349(15):1405-13.

Wenstrom K, ACOG Committee on Practice Bulletins, "Prenatal Diagnosis of Fetal Chromosomal Abnormalities," *ACOG Practice Bulletin*, 2001, 27:1-16.

Chorionic Villus Sampling, Chromosome and Genetic Abnormality Analysis

Related Information

Synonyms CVS

Test Includes Chromosomal complement of fetal trophoblast cells are examined for determination of abnormalities

Specimen Chorionic villi

Container Sterile container

Sampling Time 8-12 weeks gestation

Collection Fetal trophoblast tissue is aspirated from placental chorionic villi transcervically or transabdominally with ultrasound guidance as early as the 8th week of gestation, usually performed at 9-12 weeks of gestation.

Vascularized and budding villi of the chorion frondosum are collected transcervically or transabdominally by aseptic technique with ultrasound guidance between the 8th and 12th weeks of gestation. Maternally derived tissue such as maternal blood, decidua, and cervical mucus is carefully separated from other tissues under a dissecting microscope to prevent maternal cell contamination.

Pertinent medical findings should accompany the request, including maternal age, gestational age by sonography, reason for study, relevant history, medication history, transfusion history, note of viral infection, number of pregnancies and miscarriages, and suspected diagnosis.

Storage Instructions Specimen should be transported to the laboratory at room temperature and under sterile conditions as quickly as possible.

Causes for Rejection Specimen frozen or lacking in viable chorionic villi

Turnaround Time 1-2.5 weeks may be needed.

Reference Interval Forty-six chromosomes to include 22 sets of normal autosomal chromosomes and one set of normal sex chromosomes (XX for female; XY for male). Interpretative information is usually included.

Use Prenatal detection of chromosomal abnormalities, especially Down syndrome, in groups of pregnant women at risk. Such groups include women age 35 years or older, previous child with a chromosomal abnormality or multiple congenital abnormalities, three or more previous spontaneous abortions, familial history of a chromosomal abnormality, or known carrier of an X-linked disorder. At the same time that chorionic villi are collected for chromosomal analysis, additional sample can be obtained for testing of inherited metabolic disorders.

Limitations Failure of cells to grow in culture and/or contamination precludes complete analysis (fluorescence *in situ* hybridization may be of use if this occurs). The overall success rate of chromosome analysis is slightly lower than with amniocentesis. The rate of fetal loss due to CVS, when the fetus is viable at 8-12 weeks gestation, is approximately 1% to 3.5%. Chromosome aberrations observed in the CVS specimen but not in the fetus (confined placental mosaicism) occur in ~1% of cases. Additional invasive testing may be required in these cases.

Contraindications Environment lacking capability in ultrasonography, genetic counseling, CVS, chorionic villi culturing, and chromosomal analysis techniques

Additional Information CVS is still a limited service in many centers. A distinct advantage of CVS is the earlier gestational age in which the specimen may be collected, thus reducing the period of uncertainty and allowing termination, if elected, to be performed on an outpatient basis, in the first trimester. Additionally, there is a rapid technique to visualize spontaneous metaphases in the cytotrophoblast layer yielding results in 1-3 days if desired. A disadvantage is that amniotic fluid alpha-fetoprotein (AFAFP) measurement cannot be performed at this stage; it must be done later.

References

Goldberg JD and Golbus MS, "Chorionic Villus Sampling," *Adv Hum Genet*, 1988, 17:1-25.

Jenkins TM and Wapner RJ, "First Trimester Prenatal Diagnosis: Chorionic Villus Sampling," *Semin Perinatol*, 1999, 23(5):403-13.

Kim SK, Cho DJ, Kim JW, et al, "Adverse Pregnancy Outcome Following Postchorionic Villus Sampling Amniocentesis Compared to Chorionic Villus Sampling," *J Obstet Gynaecol Res*, 2000, 26(3):209-13.

♦ **CHr** *see* Reticulocyte Hemoglobin Content *on page 1158*

♦ **Christmas Disease Factor (Factor IX)** *see* Coagulation Factor Assays *on page 418*

Chromium, Serum

Related Information
Glucose Tolerance Test, Plasma *on page 651*

Synonyms Cr, Serum

Applies to Glucose Tolerance Factor

Abstract Chromium is used as a catalyst to harden steel alloys, to produce stainless steel, and in galvanizing, tanning, dying, and chemical manufacturing. Chromium deficiency may lead to insulin resistance and hyperlipidemia while chromium poisoning may result in vertigo, gastric irritation, conjunctivitis, vomiting, convulsions, shock, coma, and proximal tubule dysfunction. There are two forms of chromium found in biological systems: hexavalent chromium (Cr(VI)) and trivalent chromium (Cr(III)). Hexavalent chromium is more toxic and the trivalent form is poorly absorbed.

Patient Preparation Patient should be fasting for basal level.

Specimen Serum

Container Special metal-free collection tube. See the Trace Elements Introduction *on page 77*.

Sampling Time A morning fasting sample should be obtained. A glucose load, due to the insulin response it induces, drives chromium levels lower. In the nocturnal total parenteral nutrition patient, the sample should be drawn "fasting" in the afternoon, before the glucose-containing TPN solution is started for the evening.

Collection Use powder-free gloves and metal-free tubes.

Causes for Rejection Blood collection in improper tube

Reference Interval 0.05-0.15 ng/mL (SI: 1-3 nmol/L). Some laboratories report much higher "normal ranges" because the methods and/or collection techniques in use are inadequate to prevent substantial contamination. If a laboratory reports "<1 ng/mL" or some similar figure without a lower level for normal, the value can be relied upon to discover toxic states, perhaps, but not deficiency states. Almost a twofold diurnal variation is noted in serum chromium levels, with the level highest in the morning and falling after each meal as insulin levels rise. Serum chromium levels are about 60% of normal in diabetic patients, which overlaps the normal range.

Use Evaluate suspected chromium toxicity or exposure

Limitations Extreme attention to detail is needed to achieve reliable results; for many laboratories today, a high serum level more often may reflect sample contamination rather than excess chromium exposure. Any reported levels in biological materials prior to about 1979 are suspect, as the available methods did not provide sufficient sensitivity to separate normal values from the "blank." Serum chromium does not predict glucose tolerance in late pregnancy.

Methodology Stable isotope dilution, isotope ratio mass spectroscopy; graphite furnace atomic absorption spectroscopy; inductively-coupled plasma-mass spectroscopy (ICP-MS)

Additional Information Chromium is an essential element in humans, with trivalent chromium (chromium III) purported to be an integral part of "glucose tolerance factor". Deficiency of chromium can cause an acquired insulin resistance or diabetes mellitus with associated hyperlipidemia in otherwise well-nourished patients. Chromium supplementation has been shown to improve glucose tolerance and improve insulin efficiency in glucose intolerant (but not in normal or overtly diabetic) patients on diets equivalent to the lower quartile of ordinary chromium intake in the United States. Chromium deficiency with associated glucose intolerance and fasting hypoglycemia has been most often observed during refeeding of malnourished famine victims. In infants, one or more oral doses of 250 µg of chromium have been curative. Chromium deficiency associated glucose intolerance has been observed in long-term parenteral nutritional support when inadequate chromium was supplemented.

Pure metallic chromium is nontoxic. Trivalent chromium (chromium III) is poorly absorbed and much less toxic than hexavalent chromium (chromium VI). Monitoring for industry-related toxicity has relied on air samples largely for total and hexavalent chromium, the major type of concern. Workers are potentially exposed in tanneries, mines, metal plating, welding, photography, paint, dye, and explosives industries. Skin exposure may lead to dermatitis, and respiratory exposure to bronchitis, asthma, and lung cancer. Hair contains 1000-fold more chromium than serum or urine and hair chromium content correlates with industrial exposure. Intermediate levels of long-term exposure may cause tubular proteinuria in industrial workers.

The implication is that many individuals in the U.S. population have a marginal chromium intake, and that the bulk of observed cases of glucose intolerance are due to chromium deficiency. In the United States, addition of 150 µg of Cr(III) in the daily diet improved glucose tolerance in type II diabetics.

Iron competitively inhibits the binding of chromium III to transferrin. Iron overloaded patients with hemochromatosis poorly retain a radioactive tracer dose of chromium III. It has been suggested that chromium deficiency at a cellular level may play a role in the development of diabetes in hemochromatosis.

References

Chan S, Gerson B, Reitz RE, et al, "Technical and Clinical Aspects of Spectrometric Analysis of Trace Elements in Clinical Samples," *Clin Lab Med*, 1998, 18(4):615-29.

Freund H, Atamian S, and Fischer JE, "Chromium Deficiency During Total Parenteral Nutrition," *JAMA*, 1979, 241(5):496-8.

Gunton JE, Hams G, Hitchman R, et al, "Serum Chromium Does Not Predict Glucose Tolerance in Late Pregnancy," *Am J Clin Nutr*, 2001, 73(1):99-104.

Vincent JB, "The Biochemistry of Chromium," *J Nutr*, 2000, 130(4):715-8.

Chromium, Urine

Related Information

Chromium, Serum *on page 402*
Glucose Tolerance Test, Plasma *on page 651*

Synonyms Cr, Urine

Abstract Urine chromium levels are extremely low, and until recently, not reliable. Urine chromium assay is used to look for chromium toxicity in cases of potential exposure. Chromium is found in a number of oxidation states but only the trivalent state and the hexavalent state occur in biological systems. Hexavalent chromium (Cr(VI)) is more toxic than trivalent chromium (Cr(III)). See Chromium, Serum *on page 402*, for signs and symptoms of toxicity and deficiency states and for additional references.

Specimen 24-hour urine

Container Plastic metal-free container

Collection Care must be taken to avoid contact with metal. Use acid-washed plastic urinal. Stool contamination must be avoided.

Causes for Rejection Improper collection, contact with metal, dust or dirt; use of ordinary urine container without acid washing, stool contamination

Reference Interval <1 µg/24 hours. Reference ranges vary with the laboratory; ranges have declined with improved methods of avoiding contamination and assay.

Use Evaluate industrial exposure, suspected toxicity; or in conjunction with serum levels

Limitations Levels are so low in normal individuals (on the order of one part in 10 billion in urine) that many laboratories are not able to detect the lower limit of normal, and thus report "less than" some set level as being normal. Thus, for many laboratories, the test can only be used to detect toxicity but not deficiency.

Methodology Atomic absorption (AA), neutron activation, inductively-coupled plasma-mass spectroscopy (ICP-MS). ICP-MS holds promise as the most sensitive assay technique currently available and does not require predigestion.

Additional Information The main excretory pathway for chromium III is renal. Estimated safe and adequate, oral intake recommended by the U.S. National Academy of Sciences range from 50-200 µg/day, but in the U.S. 90% of people have a mean intake of 25-33 µg/day. Such intake recommendations likely derive from the 1980s when average estimated intake was determined to be 50-100 µg/day. With increasingly accurate assays, new recommended ranges may have to be set lower. Absorption of chromium III is on the order of 0.5%, by radioisotope studies. One hospital pharmacy supplies 12 µg of chromium from MTE5® trace mineral parenteral nutrition supplement per day. We are unaware of a large survey.

Diabetic patients lose threefold more chromium in the urine than nondiabetics, and despite increased intestinal absorption, diabetic patients on ordinary diets are unable to maintain the normal serum range. However, plasma chromium (Continued)

Chromium, Urine (Continued)

during pregnancy does not correlate with glucose intolerance, insulin resistance, or serum lipids.

References
Chan S, Gerson B, Reitz RE, et al, "Technical and Clinical Aspects of Spectrometric Analysis of Trace Elements in Clinical Samples," *Clin Lab Med*, 1998, 18(4):615-29.

Gunton JE, Hams G, Hitchman R, et al, "Serum Chromium Does Not Predict Glucose Tolerance in Late Pregnancy," *Am J Clin Nutr*, 2001, 73(1):99-104.

Morris BW, Blumsohn A, Mac Neil S, et al, "The Trace Element Chromium - A Role in Glucose Homeostasis," *Am J Clin Nutr*, 1992, 55(5):989-91.

♦ **Chromogranin** *see* Immunoperoxidase Procedures *on page 780*

♦ **Chromogranin A, Plasma** *see* Catecholamines, Fractionation, Plasma *on page 345*

♦ **Chromogranin A, Plasma** *see* 5-Hydroxyindoleacetic Acid, Quantitative, Urine *on page 754*

♦ **Chromogranin A, Serum** *see* Calcitonin, Serum or Plasma *on page 327*

Chromosomal Translocations, Molecular Detection

Related Information
bcl-2 Gene Rearrangement *on page 251*
Bone Marrow *on page 296*
Breakpoint Cluster Region Rearrangement in CML *on page 304*
Chromosome Analysis, Blood *on page 406*
Chromosome Analysis, Bone Marrow *on page 407*
Gene Rearrangement for Leukemia and Lymphoma *on page 633*
Polymerase Chain Reaction *on page 1069*

Synonyms Translocations, Chromosomal

Test Includes Identification of specific chromosomal translocations by polymerase chain reaction (PCR) or reverse transcriptase-polymerase chain reaction (RT-PCR) or Southern blot assay using cDNA probe

Abstract Many malignant solid tumors and leukemia/lymphomas often exhibit recurrent chromosomal translocations. These translocations may be unique for a particular type of tumor and thus provide definitive or confirmatory evidence for a specific diagnosis with implications for therapy and prognosis. Detection of specific chromosomal translocations can also play a role in post-treatment monitoring.

Specimen Peripheral whole blood, leukocytes, or bone marrow aspirate for leukemia/lymphoma; fresh and snap frozen tissue containing viable tumor cells for solid tumors/lymphoma; formalin-fixed paraffin-embedded tissue block may be used by some laboratories. The tissue must have been fixed promptly after excision and in some cases may not be subjected to prolonged formalin fixation.

Container Lavender top (EDTA) tube for liquid sample

Storage Instructions Transport to the laboratory immediately on ice or frozen; or store at -20°C (preferably -70°C) until transport.

Causes for Rejection Transport at room temperature, frozen tissue thawed during transit

Turnaround Time 3-5 days from receipt of specimen

Reference Interval Chromosomal translocation should be absent in normal tissue. Report may include an illustration and/or interpretive report.

Use Molecular detection of specific chromosomal translocation in hematopoietic malignancies and in solid tumors when the involved genes are known to provide diagnostic and prognostic information; monitor patients after therapy

Limitations The sequences at the breakpoint regions on both chromosomes involved in the translocation must be known for molecular detection. Sampling of necrotic area in a tumor is a technical limitation.

Additional Information The following table lists common chromosomal translocations in leukemia/lymphoma and solid tumors that may be detected by molecular methods (eg, PCR and RT-PCR).

Diagnosis	Chromosomal Translocation
Chronic myelogenous leukemia (CML)	t(9;22)(q34;q11)
Acute lymphoblastic leukemia (ALL)	t(9;22)(q34;q11) t(12;21)(p13;q22) t(1;19)(q23;p13)
Acute myeloid leukemia (AML)	t(8;21)(q22;q22) t(15;17)(q22;q12) inv(16)(p13;q22)
Lymphoma	t(14;18)(q32;q21) t(11;14)(q13;q32) t(2;5)(p23;q35)
Ewing sarcoma / peripheral neuroectodermal tumor	t(11;22)(q24;q12) t(21;22)(q22;q12)
Desmoplastic small round cell tumor	t(11;22)(p13;q12)
Myxoid/round cell liposarcoma	t(12;16)(q13;p11) t(12;22)(q13;q12)
Extraskeletal myxoid chondrosarcoma	t(9;22)(q22;q12) t(9;17)(q22;q11)
Malignant melanoma of soft parts / clear cell sarcoma	t(12;22)(q13;q12)
Synovial sarcoma	t(X;18)(p11;q11)
Alveolar rhabdomyosarcoma	t(2;13)(q35;q14) t(1;13)(p36;q14)
Dermatofibrosarcoma protuberans	t(17;22)(q22;q13)
Congenital fibrosarcoma	t(12;15)(p13;q25)

The following genetic abnormalities may occur in pediatric acute leukemias.

Cytogenetic Loci	Disease	Gene(s) Involved
t(9;22)(q34;q11)	B-precursor ALL	*BCR-ABL*
t(1;19)(q23;p13.3)	Pre-B-cell ALL	*E2A-PBX1*
t(12;21)(p13;q22)	P-precursor ALL	*TEL-AML1*
t(4;11)(q21;q23)	B-precursor/mixed lineage ALL	*MLL-AF4*
t(11;19)(q23;p13.3)	B-precursor/mixed lineage ALL	*MLL-ENL*
t(15;17)(q22;q12)	AML-M3 (promyelocytic)	*PML-RARα*
t(8;21)(q22;q22)	AML-M2	*AML1-ETO*
inv(16)(p13;q22); t(16;16)(p13;q22)	AML-M4Eo	*CBFβ-MYH11*
t(9;11)(p21;q23)	AML-M4 or M5, some ALL	*MLL-AF9*
t(6;9)(p23;q24)	AML-M2 or M4	*DEK-CAN*

Chromosomal translocations are of common occurrence in acute leukemias. The translocation usually results in the abnormal juxtaposition of one gene to another with creation of a unique "fusion gene". Such a chimeric gene, after transcription, results in a hybrid mRNA that produces a chimeric protein. Fusion proteins may disrupt growth or cell death in lymphoid or myeloid developing precursors with resultant leukemia. Another mechanism of leukemogenesis is post-translocation activation of a proto-oncogene with resultant overexpression of the oncogenic protein. Translocations and other gene abnormalities often correlate with clinical features and in particular with severity and prognosis. As an example, in a case of acute lymphocytic leukemia, it would be of clinical significance to identify a 12;21 translocation (which results in fusion of the *TEL* and *AML1* genes), as such cases have a relatively favorable outcome with conventional chemotherapy but are subject to late relapse. Translocations involving the *MLL* gene (chromosome 11q23) occur in 5% to 10% of human leukemias, fusion involving over 30 different partner genes. The clinical presentation and outcome of such cases is highly variable.

References

Bartolo C and Viswanatha DS, "Molecular Diagnosis in Pediatric Acute Leukemias," *Clin Lab Med*, 2000, 20(1):139-82.

Gore L, Ess J, Bitter MA, et al, "Protean Clinical Manifestations in Children With Leukemias Containing *MLL-AF10* Fusion," *Leukemia*, 2000, 14(2):2070-5.

Hokland P and Pallisgaard N, "Integration of Molecular Methods for Detection of Balanced Translocations in Diagnosis and Follow-up of Patients With Leukemia," *Semin Hematol*, 2000, 37(4):358-67.

Chromosome Analysis, Blood

Related Information

Amniotic Fluid, Chromosome and Genetic Abnormality Analysis *on page 152*

Bone Marrow *on page 296*

Breakpoint Cluster Region Rearrangement in CML *on page 304*

Chorionic Villus Sampling, Chromosome and Genetic Abnormality Analysis *on page 400*

Chromosome Analysis, High-Resolution *on page 409*

Chromosome Analysis, Lymph Node and Solid Tumor *on page 410*

Chromosome Analysis, Products of Conception *on page 412*

Fluorescence *in situ* Hybridization *on page 602*

Fragile X Syndrome DNA Test *on page 611*

Lymph Node Biopsy *on page 880*

Abstract The constitutional karyotype (chromosome complement) of each individual is determined during fertilization or during the first few cell divisions. If the karyotype is abnormal, development may be altered. In general, chromosome abnormalities with gains or losses of large amounts of chromatin will manifest early in development and often result in spontaneous abortion. Examples of such abnormalities include trisomies, monosomy X, triploids, and large unbalanced structural rearrangements. Approximately 1 in 156 live births have a major chromosome abnormality. Congenital anomalies and/or mental retardation or phenotypic abnormalities which appear later on in life are observed in about half of these cases. Some chromosome abnormalities go undetected during prenatal and perinatal periods. Physical and mental developmental delays first noted during childhood may be associated with small unbalanced rearrangements, small interstitial deletions (microdeletions), and mosaic trisomies. Sex chromosome abnormalities may not be clinically evident until puberty when inappropriate secondary sexual development occurs or when infertility is recognized later in life. Finally, normal individuals who are carriers of balanced rearrangements may remain unrecognized until adulthood, at which time they can present with multiple miscarriages or abnormal offspring.

Specimen Whole blood

Container Green top (sodium heparin) tube. **Note:** Specimens in lavender top (EDTA) tubes, blue top (sodium citrate) tubes, or green top (lithium heparin) tubes are not acceptable.

Storage Instructions Specimen should be delivered to the laboratory immediately; do not freeze.

Causes for Rejection Clotted or hemolyzed specimen, use of improper anticoagulant, improper storage

Reference Interval Forty-six chromosomes including 22 sets of normal autosomal chromosomes and one set of normal sex chromosomes (XX for female, XY for male). Interpretive information should be included.

Use Evaluate congenital anomaly (birth defect), developmental delay, ambiguous genitalia, mental retardation, cryptorchidism, hypogonadism, primary amenorrhea, infertility, multiple miscarriages, or the carrier status in relatives of patients with known chromosome abnormalities.

Limitations Failure to obtain metaphases occurs infrequently and may be due to collection in inappropriate anticoagulant or improper specimen storage (eg, frozen specimen).

Additional Information The highest proportion of chromosome abnormalities occurs in early spontaneous abortions (50%) (see Chromosome Analysis, Products of Conception *on page 412*). Approximately 7% of stillbirths and perinatal deaths are chromosomally abnormal, and 0.65% of newborns have a major chromosome abnormality. Trisomy 21 (Down syndrome) is the most frequent chromosome anomaly, with an incidence of 1 in 700-850 births. Sex chromosome aneusomies are the next most common. One XXY and one XYY is present in every 1000 male births, and one XXX is seen in every 1000 female births. Structural balanced rearrangements have a frequency of about 1 in 500 live births. Carriers of such balanced rearrangements are phenotypically normal but have an increased risk for having abnormal offspring and multiple miscarriages.

Chromosome abnormalities are present in about 10% of mentally retarded children. When mentally retarded children are examined who also have multiple birth defects or low birth weights, the incidence of chromosome abnormalities increases to 23%, half of which are Down syndrome. In addition, 3% to 6% of males and 3% to 4% of females with mental retardation will have the fragile X syndrome. (See Fragile X Syndrome DNA Test *on page 611*.)

References

Robinson A and Linden M, *Clinical Genetics Handbook*, 2nd ed, Boston, MA: Blackwell Scientific Publications, 1993.

Miller OJ and Thurman E, *Human Chromosomes*, 4th ed, New York, NY: Springer-Verlag, 2001.

Van Dyke DL and Wiktor A, "Clinical Cytogenetics," *Clinical Laboratory Medicine*, 2nd ed, McClatchey KD, ed, Baltimore, MD: Lippincott Williams & Wilkins, 2001.

Internet Web Sites
www.ncbi.nlm.nih.gov/omim/

Chromosome Analysis, Bone Marrow

Related Information

Bone Marrow *on page 296*
Breakpoint Cluster Region Rearrangement in CML *on page 304*
Chromosomal Translocations, Molecular Detection *on page 404*
Chromosome Analysis, Lymph Node and Solid Tumor *on page 410*
Fluorescence *in situ* Hybridization *on page 602*
Lymph Node Biopsy *on page 880*
White Blood Cell Count *on page 1330*

Abstract Cytogenetic analysis is often needed in the diagnostic study of patients with or suspected of having, a hematologic disorder. Numerous consistently occurring primary chromosome abnormalities have been well established in both acute and chronic hematologic diseases. Select abnormalities have specific associations with morphologic subtypes of disorders, and are therefore useful in establishing specific diagnoses. Cytogenetic findings at diagnosis have also been shown to be an independent prognostic factor associated with complete remission, duration of remission, and ultimate survival in many hematologic disorders. Chromosome analysis is also used for monitoring patients following standard treatment or bone marrow transplantation.

Specimen Bone marrow aspirate

Container 1-3 mL bone marrow should be transported in a sterile container containing preservative-free sodium heparin; specimens in EDTA, citrate, or heparin anticoagulants are not acceptable.

Storage Instructions Maintain the specimen at room temperature and transport **immediately** to the Cytogenetics Laboratory.

Causes for Rejection Clotted or hemolyzed specimen

Use Chromosome analysis of bone marrow aids in diagnosis of hematologic disorders, supplies prognostic information, and is used to monitor patients following therapy or bone marrow transplantation.

Limitations Neoplastic cells may fail to grow in culture. Since normal bone marrow stem cells can divide *in vitro*, a normal cytogenetic result may reflect either a diploid neoplastic population or the analysis of normal cells. Sensitivity may be limited in detecting minimal residual disease since only 20-30 metaphases are routinely analyzed. Fluorescence *in situ* hybridization may be used to aid in diagnosis of a genetic abnormality in all of the above stated limitations.

Additional Information Acute myelogenous leukemia: More than 30 different structural chromosome abnormalities have been implicated as primary rearrangements in AML. Several abnormalities are specifically associated with FAB (French-American-British) subclasses. For example, t(8;21) typically occurs in FAB group M2, t(15;17) in M3, inv(16)/del(16) in M4 with eosinophilia, and t(9;11) in M5. The duration of complete remission has been shown to be long for patients with t(8;21), t(15;17), and inv(16). Short durations of complete remission have been associated with abnormalities of chromosomes 5 or 7. Abnormalities of chromosomes 5 and 7, +8, and 11q23 rearrangements generally are associated with a short mean disease-free survival. In adult patients, inv(16) has been associated with relatively long survival, while in children, a more intermediate survival has been observed. Secondary AML (therapy-related or environmental mutagen-related) usually has a larger number of chromosome abnormalities than *de novo* AML. Hypodiploidy, -7, del(5q), del(11q), del(7q), and del(1;7) are among the characteristic chromosome abnormalities observed in secondary AML.
(Continued)

Chromosome Analysis, Bone Marrow (Continued)

Chronic myeloproliferative disorders: The first chromosome abnormality found to be consistently associated with a neoplastic process, was a small marker chromosome, referred to as the Philadelphia chromosome (Ph). Greater than 90% of patients with chronic myelogenous leukemia (CML) were found to be Ph-positive. It was later shown that the Ph originated from a balanced translocation, t(9;22)(q34;q11.2). Complex-variant translocations are found in 5% to 10% of CML patients and have the same molecular rearrangement as the classic t(9;22). Actual Ph-negative CML may be quite rare. Many of the cases thought to be Ph-negative in the past have been reclassified into disorders other than CML or have been shown, by molecular methods, to have a submicroscopic bcr/abl rearrangement. The detection of additional cytogenetic abnormalities (most commonly +Ph, +8, iso(17q), +19) can be helpful in predicting the accelerated phase or impending blast crisis. Secondary cytogenetic abnormalities often precede clinical blast crisis by several months. Although no karyotypic abnormality is specific for the other myeloproliferative disorders (polycythemia vera, idiopathic myelofibrosis, and essential thrombocythemia), the most frequently observed chromosome changes include rearrangements of the long arm of chromosome 1, -7, del(7q), del(5q), +8, +9, del(13q), and del(20q).

Myelodysplastic syndrome: Chromosome abnormalities are detected in approximately one-third to one-half of all de novo myelodysplastic syndromes (MDS), varying in frequency among the FAB subgroups. Among the most common recurring, nonrandom chromosome aberrations are del(5q), -7, +8, del(20q), del(7q), and del(13q). Although these abnormalities do not aid in subclassification of the FAB subgroups and can also be seen in other disorders, the detection of an acquired clonal abnormality within bone marrow cells establishes the diagnosis of a neoplastic disorder. This is especially helpful in cases of refractory anemia in which morphological changes may be subtle. Deletion of 5q is the most common chromosome abnormality in MDS (~30% of abnormal cases). 5q can be observed in any subgroup of MDS and in some other hematologic disorders. A subgroup of patients with del(5q) as the sole abnormality have specific clinicohematologic characteristics referred to as the "5q- syndrome". These patients characteristically are elderly women with refractory macrocytic anemia, elevated or normal platelet counts, and hypolobulated megakaryocytes. The clinical course is generally mild with only rare transformation to AML. A poorer prognosis is associated with del(5q) when accompanied by additional chromosome abnormalities. The frequency of chromosome abnormalities is generally higher in secondary MDS. Deletions or monosomy of chromosomes 5 and/or chromosome 7 are often observed especially following exposure to alkylating agents. Likewise, rearrangements involving 11q23 have been associated with exposure to drugs targeted against topoisomerase II.

Acute lymphocytic leukemia: Approximately two-thirds of all acute lymphocytic leukemias (ALL) have abnormal karyotypes. Loci involved in normal development of lymphocytes are often part of a chromosome rearrangement observed in both acute and chronic lymphocytic disorders. These loci include immunoglobulin genes; the heavy-chain locus (IGH) at 14q32, the kappa light-chain locus (IGK) at 2p12, the lambda light-chain locus (IGL) at 22q11, and T-cell receptor molecules; the α-chain locus (TCRA) at 14q11, the β-chain locus (TCRB) at 7q34-36, and the γ-chain locus (TCRG) at 7p13. The karyotype has been shown to be an important independent prognostic factor in ALL. Patients with a modal chromosome number >50 (hyperdiploid) have the most favorable cytogenetic prognosis. When, however, structural abnormalities coexist within the hyperdiploid karyotype, the prognosis is no longer favorable. Hypodiploidy is associated with a poor prognosis, as are specific translocations such as t(1;19), t(4;11), t(9;22) and t(8;14). A normal karyotype appears to have an intermediate prognosis. The t(12;21), observed primarily in children, has been associated with a good prognosis.

Chronic lymphoproliferative disorders: As in acute lymphocytic leukemias, immunoglobulin gene and T-cell receptor gene loci are often involved in clonal chromosome rearrangements. Other characteristic cytogenetic findings include trisomy 12 and del(13q) in chronic lymphocytic leukemia, del(6q) in numerous

B- and T-cell derived chronic disorders, and 2p rearrangements in Sézary syndrome.

References

Dewald GW, Morris MA, and Lilla VC, "Chromosome Studies in Neoplastic Hematologic Disorders," *Clinical Laboratory Medicine*, 2nd ed, McClatchey KD, ed, Baltimore, MD: Lippincott Williams & Wilkins, 2001.

Heim S and Metelman F, *Cancer Cytogenetics: Chromosomal and Molecular Genetic Aberrations of Tumor Cells*, 2nd ed, New York, NY: Wiley-Liss, 1995.

Rooney DE and Czepulkowski BH, *Human Cytogenetics: Malignancy and Acquired Abnormalities*, 3rd ed, Volume II, New York, NY: Oxford University Press, 2001.

Shah J, Theil K, and Kalaycio M, "Clinical Significance of Cytogenetics in Acute Leukemias," *Lab Med*, 2003, 34(11):796-802.

Internet Web Sites

www.infobiogen.fr/services/chromcancer/index.html

Chromosome Analysis, High-Resolution

Related Information

Chromosome Analysis, Blood *on page 406*
Fluorescence *in situ* Hybridization *on page 602*

Synonyms Microdeletion Study; Prophase Study

Abstract Standard blood culture and chromosome-staining techniques result in metaphase chromosomes with 400 to 500 total bands per haploid set of chromosomes. The recognition of various chromosome abnormalities resulting from subtle gains or losses of genetic material consisting of single bands or even smaller portions of chromosomes, has resulted in the development of techniques for detection of these minute abnormalities. High-resolution techniques arrest cells in prophase or prometaphase, resulting in elongated chromosomes with identifiable bands in the 550 to 1200 band stage. Such techniques are used when microdeletions or other subtle chromosome abnormalities are suspected.

Specimen Whole blood

Container Green top (sodium heparin) tube. **Note:** Specimens in lavender top (EDTA) tubes, blue top (sodium citrate) tubes, or green top (lithium heparin) tubes are not acceptable.

Storage Instructions Specimen should be delivered to the laboratory immediately; do not freeze.

Causes for Rejection Clotted or hemolyzed specimen, use of improper anticoagulant, improper storage (frozen specimen)

Reference Interval Forty-six chromosomes including 22 sets of normal autosomal chromosomes and one set of normal sex chromosomes (XX for female, XY for male). Interpretive information is usually included.

Use Detect small chromosomal deletions, duplications, or rearrangements

Limitations Failure to obtain metaphases occurs infrequently and may be due to collection in inappropriate anticoagulant or improper specimen storage (eg, frozen specimen). The detection limit is a function of the resolution of the microscope; therefore, submicroscopic abnormalities (deletions, duplications, or rearrangements) may remain undetected by these methods. Fluorescence *in situ* hybridization is a useful adjunct in detecting microdeletions (see Chromosome Analysis, Bone Marrow *on page 407*).

Additional Information High-resolution chromosome analysis is useful in situations in which a small chromosome deletion, duplication, or rearrangement may be clinically suspected. A number of disorders have been shown to be associated with minute deletions or duplications of chromatin. Examples of the major microdeletion/duplication syndromes and their clinical characteristics are shown in the table. These disorders are often referred to as "contiguous gene syndromes". It is thought that each of several contiguous genes involved in the abnormality is responsible for a portion of the clinical manifestations associated with the disorder. Therefore, clinical manifestations can be variable depending on the extent of material deleted or duplicated. Some individuals with these syndromes have no chromosome abnormality, and others may have abnormalities too small (submicroscopic) to be detected even by high-resolution cytogenetic methods. DNA probes for select microdeletion syndromes are now commercially available. They are used in conjunction with high-resolution chromosome analysis for confirmation of microdeletions within metaphases using fluorescence *in situ* hybridization (FISH). In some cases, FISH may confirm the
(Continued)

Chromosome Analysis, High-Resolution *(Continued)*

presence of a submicroscopic deletion that is beyond the resolution of high-resolution cytogenetic analysis.

Major Microdeletion Syndromes

Syndrome	Deletion	Features
Williams	7q11.23	Congenital heart disease, intermittent hypercalcemia, dysmorphic facial features, mental retardation
Langer-Giedion	8q24.11-q24.13	Multiple exostoses, mental retardation, sparse hair, bulbous nose
Anirida / Wilms tumor (WAGR)	11p13	Wilms tumor with aniridia, gonodoblastoma, and retardation
Retinoblastoma	13q14	Retinoblastoma, osteosarcoma, bossed head
Angelman	15q11-q13	Hypotonia, ataxia, seizures, mental retardation, excessive laughter
Prader-Willi	15q11-q13	Neonatal hypotonia, hypogonadism, obesity, small hands and feet, mental retardation
Smith-Magenis	17p11.2	Hyperactive, self-destructive behavior, facial dysmorphism, mental retardation
Miller-Dieker	17p13.3	Lissencephaly, facial dysmorphism, mental retardation
DiGeorge / Velocardiofacial	22q11.2	Hypoplasia of parathyroid and thymus, facial anomalies, congenital heart disease
Steroid sulfatase	Xp22.3	X-linked disorder, ichthyosis
Kallmann	Xp22.3	X-linked disorder, deafness, renal and cardiac anomalies

References

Gosden CM, Davidson C, and Robertson M, "Lymphocyte Culture," *Human Cytogenetics: A Practical Approach*, 3rd ed, Rooney DE and Czepulkowski BH, eds, New York, NY: Oxford University Press, 2001.

Thompson MW, McInnes RR, and Willard HF, *Genetics in Medicine*, 5th ed, Philadelphia, PA: WB Saunders Co, 1991.

Van Dyke DL and Wiktor A, "Clinical Cytogenetics," *Clinical Laboratory Medicine*, 2nd ed, McClatchey KD, ed, Baltimore, MD: Lippincott Williams & Wilkins, 2001.

Internet Web Sites

www.ncbi.nlm.nih.gov/omim/

Chromosome Analysis, Lymph Node and Solid Tumor

Related Information

Chromosome Analysis, Blood *on page 406*
Chromosome Analysis, Bone Marrow *on page 407*
Colon Cancer, Hereditary Nonpolyposis Type *on page 432*
Fluorescence *in situ* Hybridization *on page 602*
Gene Rearrangement for Leukemia and Lymphoma *on page 633*
Histopathology *on page 733*
Lymph Node Biopsy *on page 880*

Test Includes Lymph nodal or solid tumor tissue is examined cytogenetically for clonal chromosomal abnormalities

Abstract Cytogenetic analyses of both benign and malignant solid tumors and lymphoma have revealed abnormalities in the number and/or structure of chromosomes. Many such changes are specific for a particular tumor type and thus, play a direct, potentially decisive role in examination and therapy of these lesions. In addition to adding a new dimension to the formulation of diagnosis, the cytogenetic findings provide prognostic information and resolution of cellular origin. Identification of aberrant chromosomal bands has served as a basis of molecular approaches to establish the definitive genes affected and the associated consequences of these gene alterations. In some instances, recognition of the cytogenetic abnormality is the first clue that a mutated gene resides at a particular locus.

Specimen Lymph node or solid tumor

Container Sterile container

Collection A portion of lymph node or tumor specimen surgically removed for histopathologic diagnosis is submitted fresh and aseptically for cytogenetic analysis. The sample should represent the neoplastic process, and adjacent normal tissue should be discarded.

A 1-2 cm³ sample (approximately 0.5-1 g) should be provided for analysis, preferably as part of the specimen submitted for surgical pathology. Pertinent medical findings such as age and sex of the patient, location of the lesion, indication, relevant history, note of any viral infection, and suspected diagnosis should be submitted.

Common Chromosomal Abnormalities in Lymphoma and Solid Tumors

Diagnosis	Chromosomal Abnormality
B-Cell non-Hodgkin Lymphoma	
(Burkitt) small noncleaved cell	t(8;14) (q24;q32)
(Burkitt) small noncleaved cell	t(2;8) (p12;q24)
(Burkitt) small noncleaved cell	t(8;22) (q24;q11)
Small lymphocytic or diffuse large cell	+12
Diffuse large cell	t(3;14) (q27;q32)
Diffuse large cell	t(3;22) (q27;q11)
Centrocytic (variable zone) with CD5-positive cells	t(11;14) (q13;q32)
Mixed, small cleaved and large cell follicular	t(14;18) (q32;q21)
T-Cell Lymphoma	t(11;14) (p13;q11)
	inv(14) (q11;?)
	t(14;?) (q11;?)
Anaplastic large cell (Ki-1)	t(2;5) (p23;q35)
Solid Tumors	
Clear cell sarcoma / malignant melanoma of soft parts	t(12;22) (q13;q12)
Desmoplastic small round cell tumor	t(11;22) (p13;q12)
Ewing sarcoma / peripheral neuroectodermal tumor	t(11;22) (q24;q12)
	t(21;22) (q22;q12)
	t(7;22) p22;q12)
Extraskeletal myxoid chondrosarcoma	t(9;22) (q22;q12)
	t(9;17) (q22;q11)
Glioma	double minutes
	+7
	-10
Germ-cell tumors	i(12) (p10)
Leiomyoma	t(12;14) (q14;q23)
Lipoma	t(12;?) (q14;?)
Medulloblastoma	i(17) (q10)
Meningioma	-22/del(22)(q12)
Myxoid liposarcoma	t(12;16) (q13;p11)
	t(12;22) (q13;q12)
Neuroblastoma	double minutes
	homogeneously staining regions
	del(1) (p31-32)
Pleomorphic adenoma	t(3;8) (p21;q12)
Renal cell carcinoma	del(3p)
Retinoblastoma	del(13) (q14)
Rhabdomyosarcoma (alveolar)	t(2;13) (q37;q14)
	t(1;13) (p36;q14)
Small cell lung carcinoma	del(3) (p14;p23)
Synovial sarcoma	t(X;18) (p11.2;q11.2)

(Continued)

411

Chromosome Analysis, Lymph Node and Solid Tumor
(Continued)

Storage Instructions Specimen should be submitted to the laboratory as soon as possible. If overnight storage is necessary, the sample can be refrigerated in sterile isotonic saline or culture media containing serum.

Causes for Rejection Specimen frozen

Turnaround Time 2-14 days

Reference Interval Normal lymph node and solid tissue cells contain 46 chromosomes with 22 pairs of autosomal chromosomes and one set of sex chromosomes (XX for female; XY for male). Clonal numerical and/or structural chromosomal abnormalities are detected in neoplastic tissue. An abnormal clone is defined as two or more cells exhibiting a gain of the same chromosome or structural alteration, or three or more cells exhibiting loss of the same chromosome. Many of the characteristic or tumor-specific chromosomal abnormalities are listed (see table on previous page).

Use Cytogenetic analysis of lymphoma and solid tumors (in particular sarcomas), is a useful and sometimes, essential diagnostic adjunct. Cytogenetic analysis also provides prognostic information for some malignancies.

Limitations Failure of cells to grow in culture, overgrowth of normal supporting stromal cells (fibroblasts) and/or contamination precludes complete analysis (fluorescence *in situ* hybridization may be of use if this occurs).

Additional Information In some cases, molecular methods are subsequently used to further define chromosomal abnormalities.

References
Bridge JA, "Cytogenetic and Molecular Cytogenetic Techniques in Orthopedic Surgery," *J Bone Joint Surg*, 1993, 75(4):606-14.

Heim S and Metelman F, *Cancer Cytogenetics: Chromosomal and Molecular Genetic Aberration of Tumor Cells*, 2nd ed, New York, NY: Wiley-Liss, 1995.

Sandberg AA and Bridge JA, *The Cytogenetics of Bone and Soft Tissue Tumors*, Austin, TX: RG Landes Co, 1994.

Internet Web Sites
www.infobiogen.fr/services/chromcancer/index.html

Chromosome Analysis, Products of Conception

Related Information
Amniotic Fluid, Chromosome and Genetic Abnormality Analysis *on page 152*

Chorionic Villus Sampling, Chromosome and Genetic Abnormality Analysis *on page 400*

Chromosome Analysis, Blood *on page 406*

Endometrial Cytology *on page 536*

Fluorescence *in situ* Hybridization *on page 602*

Test Includes Chromosomal complement of products of conception are examined for determination of abnormalities

Abstract The overall incidence of chromosomal abnormalities in early spontaneous abortions approaches 50%. Trisomy has been identified for all the human autosomes except chromosome 1. Generally, trisomies result in spontaneous abortions. Appreciable numbers of live births occur only for trisomies 13, 18, and 21, and these children are always abnormal. Chromosomal abnormalities most commonly seen in spontaneous abortions include trisomy (62.1%), triploidy (12.4%), monosomy X (10.5%), tetraploidy (9.2%), and structural chromosome anomalies (4.7%).

Specimen Products of conception

Container Sterile container

Collection Products of conception are the evacuated contents of the uterus after termination of an early pregnancy or the spontaneously aborted material collected after early miscarriage. The latter specimen generally consists of variably recognizable fetal parts and remnants of the sac and placenta. Although some fetal specimens may be severely autolyzed (depending on the time elapsed between death and specimen collection), it is important not to be dissuaded from attempting culture because the placenta will have recently been attached to the maternal circulation and may still contain viable cells.

Causes for Rejection Specimen frozen or lacking in viable fetal cells
Turnaround Time 1-3 weeks
Reference Interval Forty-six chromosomes to include 22 sets of normal autosomal chromosomes and one set of normal sex chromosomes (XX for female; XY for male). Interpretative information is usually included.

Use Detection of chromosomal abnormality in a spontaneously aborted fetus is important for genetic counseling and to assess risks for future abnormal pregnancies. For example, young women (younger than 35 years of age) who have previously given birth to a trisomic infant or who have had a trisomic fetus detected prenatally are at high risk for having a second trisomic abortion.

Limitations Failure of cells to grow in culture and/or contamination precludes complete analysis (fluorescence *in situ* hybridization may be of use if this occurs).

Contraindications Environment lacking capability in genetic counseling, products of conception culture, and chromosomal analysis techniques

References

Griffin DK, Millie EA, Redline RW, et al, "Cytogenetic Analysis of Spontaneous Abortions: Comparison of Techniques and Assessment of the Incidence of Confined Placental Mosaicism," *Am J Med Genet*, 1997, 72(3):297-301.

Qumsiyeh MB, "Chromosome Abnormalities in the Placenta and Spontaneous Abortions," *J Matern Fetal Med*, 1998, 7(4):210-2.

Warburton D, Kline J, Stein Z, et al, "Does the Karyotype of a Spontaneous Abortion Predict the Karyotype of a Subsequent Abortion? Evidence From 273 Women With Two Karyotyped Spontaneous Abortions?" *Am J Hum Genet*, 1987, 41(3):465-83.

Citrate, Serum, Plasma, or Urine

Related Information

Abstract Citrate inhibits nephrolithiasis, partly due to binding of calcium in urine. Drug therapy with potassium citrate is useful in prevention of types of stone formation.

Specimen 24-hour urine
Container Plastic container
Collection Some laboratories require preservatives.
Storage Instructions Refrigerate
Reference Interval Adults: male: 115-921 mg (0.6-4.8 mmol)/24 hours, female: 250-1152 mg (1.3-6.0 mmol)/24 hours; 320-1240 mg (1.7-6.5 mmol)/24 hours

Hypocitraturia is defined as <300 mg/24 hours (female) and <250 mg/24 hours (male). (Other sources use other reference intervals.)

Use Hypocitraturia in patients who form kidney stones may be idiopathic, secondary to defective urine acidification, due to small intestinal malabsorption, secondary to hypokalemia or magnesemia, or due to metabolic acidosis. (Continued)

Citrate, Serum, Plasma, or Urine *(Continued)*

Methodology Gas chromatography (GC), ion chromatography, capillary electrophoresis, enzyme-mediated reactions

Additional Information See also Kidney Stone Analysis *on page 820.*

References

Goldfarb DS and Coe FL, "Prevention of Recurrent Nephrolithiasis," *Am Fam Phys*, 1999, 60(8):2269-76.

Hruska K, "Renal Calculi (Nephrolithiasis)," *Cecil Textbook of Medicine*, 21st ed, Chapter 114, Goldman L and Bennett JC, eds, Philadelphia, PA: WB Saunders Co, 2000, 622-7.

Tekin A, Tekgul S, Atsu N, et al, "A Study of the Etiology of Idiopathic Calcium Urolithiasis in Children: Hypocitruria Is the Most Important Risk Factor," *J Urol*, 2000, 164(1):162-5.

Weiss RL, *ARUP Interpretive Data Guide*, ARUP Laboratories Inc, 1999, 178-9.

♦ **Citrovorum Factor** *see* Methotrexate, Serum or Plasma *on page 905*

♦ **CK** *see* Creatine Kinase, Serum *on page 470*

♦ **CK-2** *see* Creatine Kinase MB and Other Isoenzymes, Serum *on page 469*

♦ **CK Index** *see* Creatine Kinase MB and Other Isoenzymes, Serum *on page 469*

♦ **CK Isoenzymes** *see* Creatine Kinase MB and Other Isoenzymes, Serum *on page 469*

♦ **CK Isoforms** *see* Creatine Kinase MB and Other Isoenzymes, Serum *on page 469*

♦ **CK-MB and Total CK** *see* Creatine Kinase MB and Other Isoenzymes, Serum *on page 469*

♦ **Class I and Class II Genes** *see* HLA Typing, Single Human Leukocyte Antigen *on page 739*

♦ **Clinical Decision Limit** *see page 11*

♦ **Clinitest® for Sugar, Urine** *see* Reducing Substances, Urine *on page 1148*

♦ **Clomid® Test** *see* Clomiphene Test *on page 414*

♦ **Clomiphene Citrate Challenge Test** *see* Follicle Stimulating Hormone, Serum, Plasma, or Urine *on page 609*

Clomiphene Test

Related Information

Follicle Stimulating Hormone, Serum, Plasma, or Urine *on page 609*
Luteinizing Hormone, Blood or Urine *on page 876*

Synonyms Clomid® Test

Test Includes Pre- and post-clomiphene serum FSH and LH analysis

Abstract Clomiphene is a synthetic estrogen analogue. In the presence of an intact hypothalamic-pituitary-gonadal axis, administration of clomiphene leads to increased secretion of LH and FSH, inducing ovulation in many anovulatory patients.

Patient Preparation Four weeks of basal body temperatures are recorded. Ascertain that the patient is not pregnant and that the ovaries are not enlarged. No isotopes administered 24 hours prior to venipuncture. Females initially take 50 mg clomiphene orally daily for 5 days beginning on the fifth day of the induced or spontaneous menstrual cycle.

Specimen Serum

Container Red top tube

Sampling Time Draw pre- and post-clomiphene samples. Because of the pulsatile nature of pituitary gonadotropins, it is recommended that three separate samples, 20 minutes apart, be drawn for baseline and post-clomiphene assays.

Collection Female: draw 5-9 days after last oral dose.

Storage Instructions Refrigerate serum. LH and FSH stable at least 7 days in refrigerated serum.

Special Instructions Provide information if post-clomiphene sample(s).

Reference Interval FSH and LH are expected to peak 5-9 days after completing Clomid®. FSH increase >40% above baseline; LH increase >120% above baseline. Ovulation assessed by basal body temperature or serum progesterone 2 weeks after last clomiphene dose.

Use Clomiphene may be used to evaluate the integrity of the hypothalamic-pituitary-gonadal axis and to enhance fertility in anovulatory patients with normal ovarian function.

Limitations Clomiphene has induced severe hypertriglyceridemia and pancreatitis.

Contraindications In females, observation of hyperstimulation of ovaries but unusual on doses <200 mg, and there is a small risk of multiple pregnancies, about 5%.

Methodology Radioimmunoassay (RIA) or double antibody immunoassay with chemiluminescent read out

Additional Information The structure of clomiphene is similar to that of tamoxifen.

References

Hofmann GE, Danforth DR, and Seifer DB, "Inhibin-B: The Physiologic Basis of the Clomiphene Citrate Challenge Test for Ovarian Reserve Screening," *Fertil Steril*, 1998, 69(3):474-7.

Scott RT, Leonardi MR, Hoffman GE, et al, "A Prospective Evaluation of Clomiphene Citrate Challenge Test. Screening of the General Infertility Population," *Obstet Gynecol*, 1993, 82(4 Pt 1):539-44.

Clonazepam, Serum

Related Information

Benzodiazepines, Qualitative, Urine *on page 253*
Diazepam, Serum *on page 510*

Synonyms Iktorivil®; Klonopin™; Rivatril®

Abstract The drug is in the class of benzodiazepines, which are used as tranquilizers. This drug is effective in prevention of absence seizures, myoclonic jerks, and tonic-clonic seizures. Its use is mostly for refractory myoclonic seizures. It is useful in reducing tardive dyskinesia. Its unlabeled/investigational uses include restless leg syndrome, neuralgia, multifocal tic disorder, parkinsonian dysarthria, bipolar disorder, and adjunct therapy for schizophrenia.

Specimen Serum

Container Red top tube; do not collect in gel-containing tube.

Sampling Time Serum peak levels occur ~2 hours after oral administration. The apparent half-life after a single oral dose is 20-40 hours. Use trough for monitoring. Steady-state concentration is reached in 5-10 days.

Storage Instructions Separate serum and freeze. Protect from sunlight.

Reference Interval Therapeutic: 10-50 ng/mL (SI: 32-158 nmol/L)

Critical Values >80 ng/mL

Possible Panic Range >100 ng/mL (SI: >255 nmol/L)

Use Monitor drug level and toxicity

Methodology Gas chromatography (GC) with electron capture detection, high performance liquid chromatography (HPLC) with UV detector

Additional Information

- Half-life: 20-60 hours; active metabolites have longer half-lives than the parent drug; half-lives are increased in the elderly
- Volume of distribution: 2-6 L/kg
- Protein binding: 80% to 90%.

In children, lower doses and levels have been shown to be effective. Effect of CNS depressants (eg, barbiturates, ethanol) may be augmented by concomitant use of clonazepam.

References

Dahlin MG, Amark PE, and Nergardh AR, "Reduction of Seizures With Low-Dose Clonazepam in Children With Epilepsy," *Pediatr Neurol*, 2003, 28(1):48-52.

Jacobs DS, DeMott WR, Oxley DK, et al, *Laboratory Test Handbook*, 5th ed, Hudson, OH: Lexi-Comp Inc, 2001.

Morishita S and Aoki S, "Clonazepam in the Treatment of Prolonged Depression," *J Affect Disord*, 1999, 53(3):275-8.

♦ **Clonidine Hydrochloride** *see* Clonidine, Serum or Plasma *on page 415*

Clonidine, Serum or Plasma

Related Information

Diltiazem, Serum or Plasma *on page 515*
Propranolol, Serum *on page 1096*

Synonyms Catapres®; Clonidine Hydrochloride

Abstract Clonidine is an antihypertensive. It is also used as a stimulant in a test for growth hormone release and a suppressant in a test for catecholamine release. Occasionally it is used for the treatment of attention-deficit/hyperactivity disorder, smoking cessation, and opiate withdrawal.
(Continued)

Clonidine, Serum or Plasma *(Continued)*

Specimen Plasma, serum

Container Lavender top (EDTA) tube for plasma; red top tube for serum; do not use gel separator tube for serum

Sampling Time Trough

Storage Instructions Refrigerate. Ship at room temperature.

Turnaround Time 1-4 days

Reference Interval 1-3 ng/mL

Critical Values >4 ng/mL

Use Monitor therapy or evaluate toxicity

Contraindications Unsafe in patients with porphyria

Methodology Gas chromatography (GC), high performance liquid chromatography (HPLC)

Additional Information
- Half-life: 5-20 hours
- Volume of distribution: 1.5-2.5 hours
- Protein binding: 20% to 40%

Clonidine is metabolized by the liver to inactive metabolites which are eliminated through urine and feces. Adverse reactions include bradycardia, orthostatic hypotension, drowsiness, dizziness, constipation, xerostomia, and nausea. In overdose situations, naloxone is used as an antidote.

References

Cline JC and Connelly J, "Intravenous Clonidine for Hypertensive Emergencies," *Am J Health Syst Pharm*, 1999, 56(6):572-4.

Epstein M, "Diagnosis and Management of Hypertensive Emergencies," *Clin Cornerstone*, 1999, 2(1):41-54.

Leikin JB and Paloucek FP, *Poisoning and Toxicology Compendium*, Hudson, OH: Lexi-Comp Inc, 1998, 197-8.

♦ **Clonidine Suppression Test** *see* Catecholamines, Fractionation, Plasma *on page 345*

Clostridium difficile Toxin Assay and Culture

Related Information

Bacterial Culture, Anaerobes *on page 231*
Bacterial Culture, Stool *on page 243*
Fecal Lactoferrin *on page 575*
Methylene Blue Stain, Stool *on page 906*
Ova and Parasites, Direct Exam *on page 985*
pH, Stool *on page 1034*

Test Includes Detection of toxin A and/or B and, upon special request, selective anaerobic culture for *C. difficile*

Abstract The major cause of antibiotic-associated diarrhea and pseudomembranous colitis is toxigenic *Clostridium difficile*. *C. difficile* associated diarrhea is responsible for the majority of nosocomial diarrhea and contributes to the cost of hospitalization. The *C. difficile* associated with diarrhea produce one or two toxins, toxin A and/or toxin B, that can be detected directly in stool specimens.

Specimen Diarrheal stool or proctoscopic specimen; solid stool specimens should not be tested for the presence of toxin. Material for culture may be derived from peritoneum and tissues.

Container Plastic stool container (swabs are inadequate because of small volume)

Sampling Time Repeat testing within 7 days of an initial assay provides helpful information in only ~1% of cases.

Storage Instructions Keep specimen **cold** and transport immediately. Specimens can be frozen if transportation will be delayed.

Causes for Rejection Nondiarrheal stools

Special Instructions When antibiotic-associated colitis is suspected, a toxin assay rather than a *C. difficile* stool culture should be ordered.

Reference Interval Presence of toxin is suggestive of disease. Isolation of organism (*C. difficile*) may occur in 5% to 21% of normal adults and in 50% of normal newborns. Isolation of the organism without demonstration of toxin production is a nonspecific finding and does not enhance diagnosis of diarrhea.

Use Diagnose antibiotic-related, pseudomembranous colitis caused by toxigenic *C. difficile*

Limitations The latex agglutination test is simple and rapid, but it does not detect (is not specific for) toxin A, and thus provides unreliable results. Cytotoxin assays or immunoassays for toxin A are specific but usually have sensitivities <90%. Newer immunoassays will detect both toxin A and toxin B. Toxin detection should not be used to screen hospitalized patients without diarrhea. Many hospitalized patients (44% to 63%) can have positive results and remain asymptomatic during their hospital stay.

Methodology Latex agglutination (LA) test to detect toxin, neutralization test in tissue culture (toxin), selective anaerobic culture (organism), enzyme immunoassay (EIA) (toxin A and/or B), fluorogenic immunoassay (toxin)

Additional Information Several diseases are associated with pseudomembrane formation. *C. difficile*-associated diarrhea is a common nosocomial infection. Tube-fed patients with diarrhea especially should be tested. Pseudomembranous colitis attributable to *C. difficile* is usually associated with antimicrobial therapy, anti-AIDS or antineoplastic drugs. The greatest risk of *C. difficile* diarrhea also includes the use of second- and third-generation cephalosporins, clindamycin, ampicillin, and amoxicillin.

Toxigenic strains usually produce two toxins: toxin A, an enterotoxin and toxin B, a cytotoxin that can be detected by cell culture or EIA. Not all *C. difficile* strains are toxigenic, and even toxigenic strains may not produce disease if present in insufficient numbers; consequently, culture for *C. difficile* often produces false-positive results. Strains that do not produce toxins do not cause diarrhea or colitis. The most accurate clinical laboratory test for diagnosis of pseudomembranous colitis appears to be enzyme immunoassays for toxin A/B.

References

Alcantara CS and Guerrant RL, "Update on *Clostridium difficile* Infection," *Curr Gastroenterol Rep*, 2000, 2(4):310-4.

Barr HS and Surawicz CM, "Pseudomembranous Colitis: An Update," *Can J Gastroenterol*, 2000, 14(1):51-6.

Gorbach SL, "Antibiotics and *Clostridium difficile*," *N Engl J Med*, 1999, 341(22):1690-1.

Johnson S, Kent SA, O'Leary KJ, et al, "Fatal Pseudomembranous Colitis Associated With a Variant *Clostridium difficile* Strain Not Detected by Toxin A Immunoassay," *Ann Intern Med*, 2001, 135(6):434-8.

Levy DG, Stergachis A, McFarland LV, et al, "Antibiotics and *Clostridium difficile* Diarrhea in the Ambulatory Care Setting," *Clin Ther*, 2000, 22(1):91-102.

Yassin SF, Young-Fadok TM, Zein NN, et al, "*Clostridium difficile*-Associated Diarrhea and Colitis," *Mayo Clin Proc*, 2001, 76(7):725-30.

Internet Web Sites

www.amm.co.uk/pubs/fa_cdiff.htm

Clot Retraction

Related Information

Fibrinogen *on page 583*
Platelet Aggregation *on page 1045*
Platelet Count *on page 1050*

Test Includes Test may include description of clot retraction, clot size and firmness, RBC fallout, serum "drip-out".

Abstract This test has been replaced by newer tests for platelet function and for Glanzmann thrombasthenia in most coagulation laboratories.

Specimen Whole blood

Container Red top tube

Collection Routine venipuncture; transport specimen to the laboratory immediately. (Note: Contact the laboratory prior to collecting the specimen, as the laboratory may not offer the test.)

Turnaround Time 24 hours

Reference Interval Clot retraction occurs within 4 hours.

Optional considerations: With normal clots and normal hematocrits, the clot in a red top tube occupies 40% to 60% of the original volume. The remaining 40% to 60% consists of serum as well as red cells that fall out of the clot and settle to the bottom of the tube ("red cell fall-out"). Red cell fall-out is usually <5% of the original blood sample volume (centrifuged, after removing the clot). When normal clots are removed from the tube, serum drips from the clot at a rate of two drops or less in 2 minutes.

Use Currently, it is an infrequently used clinical test. In the past, it was a test for Glanzmann thrombasthenia and platelet function.

(Continued)

Clot Retraction *(Continued)*

Limitations Platelet counts <100,000/μL, aspirin and related medications, monoclonal gammopathy (paraproteinemia), and polycythemia reduce the amount of clot retraction. Anemia increases clot retraction. With polycythemia, the increased number of red blood cells within the clot limits the extent to which the clot can retract.

Methodology The red top tube is kept at 37°C and the clot is examined at 1, 2, 4, and 24 hours for clot retraction. When the clot retracts, it pulls away from the walls of the tube. Normally, a few red blood cells fall out of the clot, and they can be seen at the bottom of the tube.

Optional approach: The initial blood specimen can be placed in a graduated tube such that volumes can be approximated. A wooden stick can be placed in the tube prior to clot formation, so that the clot can be removed from the tube for examination. A normal clot is firm and tightly attached to the stick.

Additional Information During clot formation, platelets aggregate as fibrinogen binds to platelet glycoprotein IIb/IIIa, linking platelets to each other. Normally, clot retraction occurs subsequently, as platelets within the clot contract. Glycoprotein IIb/IIIa is necessary for platelet aggregation as well as for clot retraction. In Glanzmann thrombasthenia, clot retraction and platelet aggregation are reduced because glycoprotein IIb/IIIa is deficient. With dysfibrinogenemia, hypofibrinogenemia, or disseminated intravascular coagulation (DIC), the clot can be small and an increased number of red blood cells fall out of the clot.

References

Brown BA, *Hematology: Principles and Procedures*, 6th ed, Philadelphia, PA: Lea and Febiger, 1993, 271.

Sirridge MS and Shannon R, *Laboratory Evaluation of Hemostasis and Thrombosis*, 3rd ed, Philadelphia, PA: Lea and Febiger, 1983, 83-90.

♦ **Cl, Serum** *see* Chloride, Serum, Plasma, or Blood *on page 391*

♦ **Cl, Urine** *see* Chloride, Urine *on page 394*

♦ **Clusterin (Apo J)** *see* Apolipoprotein E, Plasma *on page 204*

♦ **Clusters of Differentiation** *see* Flow Cytometry, Overview *on page 592*

♦ **CMV pp65 Detection** *see* Cytomegalovirus Antigen Detection *on page 494*

♦ **CN⁻** *see* Cyanide, Blood *on page 485*

♦ **CO** *see* Carboxyhemoglobin, Blood *on page 340*

♦ **CO₂** *see* Bicarbonate, Blood *on page 258*

♦ **CO₂** *see* Carbon Dioxide, Total, Blood *on page 339*

♦ **¹³CO₂ Breath Test** *see* Reducing Substances, Stool *on page 1147*

♦ **CO₂ Content** *see* Carbon Dioxide, Total, Blood *on page 339*

Coagulation Factor Assays

Related Information

Activated Partial Thromboplastin Time *on page 100*

Antiphospholipid Antibody (Lupus Anticoagulant and/or Anticardiolipin Antibody) *on page 193*

Factor Inhibitors *on page 566*

Factor IX Concentrate *on page 564*

Factor VIII Concentrate *on page 563*

Factor XIII *on page 565*

Fibrinogen *on page 583*

High-Molecular Weight Kininogen *on page 731*

Mixing Studies (Coagulation) *on page 918*

Prekallikrein *on page 1085*

Prothrombin Time *on page 1116*

von Willebrand Factor *on page 1321*

Warfarin, Serum or Plasma *on page 1325*

Synonyms Ac-Globulin (Factor V); Antihemophilic Factor (Factor VIII); Autoprothrombin I (Factor VII); Autoprothrombin II (Factor IX); Christmas Disease Factor (Factor IX); Hageman Factor (Factor XII); Labile Factor (Factor V); Plasma Thromboplastin Antecedent (Factor XI); Plasma Thromboplastin Component (Factor IX); Proaccelerin (Factor V); Proconvertin (Factor VII); Prothrombin (Factor II); Stable Factor (Factor VII); Stuart Factor (Factor X); Stuart-Prower Factor (Factor X)

Applies to DDAVP; Desmopressin; Factor(s) II, V, VII, VIII, IX, X, XI, XII; Factor VIII:von WIllebrand Factor Ratio; Hemophilia A (Factor VIII Deficiency); Hemophilia B (Factor IX Deficiency); INR; International Normalized Ratio; Lupus Anticoagulant; Warfarin

Abstract Isolated hereditary factor deficiencies are much less common than are multiple, acquired factor deficiencies produced by liver disease, disseminated intravascular coagulation, warfarin, or the inhibitory effects in factor assays from lupus anticoagulants, heparin, or other anticoagulants.

Specimen Plasma

Container Three blue top (sodium citrate) tubes if all factor assays are requested

Collection Routine venipuncture. If multiple tests are being drawn, draw blue top tubes after any red top tubes but before any lavender top (EDTA), green top (heparin), or gray top (oxalate/fluoride) tubes. Immediately invert tube gently at least 4 times to mix. Tubes must be appropriately filled. Deliver tubes immediately to the laboratory.

Avoid heparin contamination during specimen collection. Heparin, hirudin, or argatroban anticoagulation can interfere with factor assays by acting as an "inhibitor", resulting in falsely decreased factor levels. Heparin, if present, must be removed from specimens by the laboratory. Warfarin decreases factors II, VII, IX, and X.

Storage Instructions Separate plasma from cells as soon as possible. Store plasma at room temperature for up to 2 hours, at 2°C to 8°C for up to 4 hours, or store frozen. Factor VIII, and to a lesser extent factor V, degrade if specimens are kept unfrozen for prolonged periods.

Causes for Rejection Specimen received more than 4 hours after collection, tubes not filled, clotted specimen

Turnaround Time Less than 1 day, unless test has to be sent out to a reference laboratory

Reference Interval Factor levels are expressed as percent of normal plasma concentrations. By definition, normal plasma contains 100% (1 unit/mL) of each factor. The reference range is approximately 60% to 140%. Factor VIII levels are not decreased at birth or throughout childhood. The other factor levels are below adult reference range at birth, ranging approximately from 10% to 100%. The levels increase toward the adult reference range by age 6 months, although they may remain mildly below adult normal range throughout childhood. However, newborns and children do not normally experience bleeding, because a balance between coagulation factors and natural coagulation inhibitors is maintained throughout development. Factor XI can decrease during pregnancy, whereas fibrinogen and factor VIII increase.

Use To determine the etiology of a prolonged PT or PTT. Usually performed after a mixing study has been completed (see Mixing Studies (Coagulation) *on page 918*), to identify specific factor deficiencies or inhibitors. Assays for factors VIII, IX, XI, and XII are performed to evaluate a prolonged PTT (with normal PT). Assays for fibrinogen, factors II, V, VII, and X are performed to evaluate a prolonged PT (with normal PTT). If PT and PTT are both prolonged, all eight factors and fibrinogen may be performed to establish the cause for the prolongations. Because acquired causes of factor deficiencies are generally more common than hereditary causes, a patient found to have a factor deficiency should be evaluated for possible acquired etiologies, especially if multiple factor deficiencies are present (see Tables 1 and 2). If a hereditary etiology for the decrease appears likely, the diagnosis can be confirmed by measuring the factor in relatives.

Chromogenic factor X assays are useful for monitoring warfarin in the presence of a lupus anticoagulant, hirudin, or argatroban. Lupus anticoagulants, hirudin or argatroban can prolong the PT and therefore the international normalized ratio (INR). In these situations, the INR can overestimate the amount of warfarin anticoagulation and lead to inappropriate reductions in warfarin dose. Warfarin decreases factor X (as well as factors II, VII, and IX), and a chromogenic assay is available for factor X which has no interference from lupus anticoagulants, hirudin or argatroban. When the INR is 2-3, the chromogenic factor X level is approximately 20% to 40%. Each laboratory should determine its own chromogenic factor X therapeutic range.
(Continued)

Coagulation Factor Assays *(Continued)*

Relatives of patients with a known hereditary factor deficiency may choose to have PT, PTT, and/or factor assay(s) performed to determine if they also have the deficiency or if they are a carrier.

Factor VIII assays are part of a von Willebrand disease evaluation. Factor VIII assays, together with von Willebrand factor assays, can also assess for hemophilia A carrier status in females. In hemophilia A carriers, the factor VIII:von Willebrand factor ratio is approximately 0.5.

An increased risk for thrombosis has been reported with elevated levels of certain coagulation factors (eg, fibrinogen, factor VII, factor VIII, factor XI). In addition, factor VII levels may be affected by dietary lipids, and factor VII levels correlate with triglyceride and cholesterol levels. However, factor assays have not yet been added to most hypercoagulation panels.

Table 1. Effects of Hereditary or Acquired Factor Deficiencies on PT / PTT

PTT Prolonged, PT Normal
Deficiencies of factor(s) VIII, IX, XI, and/or XII (intrinsic pathway)
PT Prolonged, PTT Normal
Deficiency of factor VII (extrinsic pathway)
Occasionally, mild-to-moderate deficiencies of factor(s) II, V, X, and/or fibrinogen (common pathway)
Both PT and PTT Prolonged
Deficiencies of factor(s) II, V, X, and/or fibrinogen (common pathway)
Multiple factor deficiencies

Table 2. Acquired Causes of Factor Deficiencies

Acquired Conditions Affecting PT Sooner and More Significantly Than PTT	
Warfarin or vitamin K deficiency	Decreased function of factors II, VII, IX, and X
Liver dysfunction	Decreases hepatic synthesis of coagulation factors. All factors may be decreased except for factor VIII. Factor VII has the shortest half-life and therefore is often the earliest and most severely decreased factor. Factors XI and XII have the longest half-lives and therefore are often the last to be affected.
Disseminated intravascular coagulation (DIC)	All factors can be variably decreased, including factor VIII, due to factor activation and consumption.
Amyloidosis	Factor X, and occasionally other factors, can be decreased, due to binding of factor(s) to amyloid.
Acquired Conditions Affecting PTT Sooner and More Significantly Than PT	
Prolonged specimen transit to laboratory	Degradation of factors V and VIII
Proteinuria	Occasionally, decreased factors XI and XII. PT usually normal.
Acquired Conditions That Can Interfere With Factor Assays	
Heparin	Inhibits activated factors II, X, IX, XI, and XII, prolonging the PTT earlier and more than PT, and interfering in PTT-based factor assays more than PT-based factor assays, without causing a true decrease in factor levels.
Lupus anticoagulants	Inhibits the phospholipid cofactor function in coagulation, often prolonging the PTT. PT is usually normal. Can interfere in PTT-based factor assays, without causing a true decrease in factor levels. Rarely, a lupus anticoagulant can bind to factor II and cause a true decrease in factor II, prolonging the PT.
Hirudin and argatroban	Inhibit activated factor II (thrombin), prolonging both the PT and PTT, and interfering in PT- or PTT-based factor assays without causing a true decrease in factor levels.
Factor inhibitors	

Methodology Factor assays are PT- or PTT-based reactions. These assays are performed by mixing patient plasma with plasma that is deficient in the factor that is being measured. Based on the resulting PT or PTT of this mixture, the amount of factor can be determined by comparing the PT or PTT clotting time to a standard curve that plots known factor levels against clotting times. Factor VIII, IX, XI, and XII assays are PTT-based. Factor II, VII, and X assays are PT-based. PT- and PTT-based assays are both available for factor V.

The presence of an inhibitor, such as a lupus anticoagulant, can cause artifactual decreases in the *in vitro* factor level. Therefore, laboratories should perform factor assays at multiple dilutions. At higher dilutions, the inhibitor interference will decrease due to dilution of the inhibitor.

Chromogenic factor assays and immunoassays (antigen assays) are commercially available for some of the coagulation factors. For example, chromogenic assays are commercially available for factors II, VII, VIII, and X; antigen assays are commercially available for factors VII, VIII, IX, and X. Therapeutic anticoagulants, lupus anticoagulants, and other inhibitors do not interfere with these assays (except in some instances when the inhibitor or anticoagulant is directed specifically against the factor being assayed or against another factor that participates in the assay).

Additional Information Factor deficiencies can be quantitative or qualitative. In quantitative disorders, the factor level determined by routine PT- or PTT-based methods (functional activity assays) is similar to the result obtained by immunological (antigen) assays. In qualitative disorders, the PT- or PTT-based (functional) assay result is decreased, but the antigen level is normal or significantly higher than the functional level, indicating the presence of a dysfunctional protein. Antigen assays are not available for some factors.

Hereditary Deficiencies of Factor VIII (Hemophilia A) or Factor IX (Hemophilia B)

Hemophilia A (factor VIII deficiency) is the most common severe hereditary bleeding disorder, affecting 1 in 5000-10,000 males. Hemophilia B (factor IX deficiency) is also a severe hereditary bleeding disorder, affecting 1 in 25,000-30,000 males. Hemophilia A and B are X-linked recessive disorders, because the factor VIII and factor IX genes are located on the X chromosome. Therefore, typically only males are affected. Females carrying the hemophilia mutation on one of their two X chromosomes are carriers. Female carriers with factor VIII levels <50% and bleeding symptoms have been reported.

The clinical severity of hemophilia A or B depends on the factor level. With hemophilia, <1% factor VIII or IX produces severe hemophilia with spontaneous bleeding, 1% to 5% produces moderate bleeding, and >5% is considered mild hemophilia in which bleeding occurs primarily with trauma or surgery rather than spontaneously. The baseline factor level remains relatively constant within an individual and within a kindred. Bleeding manifestations include hemarthrosis, soft tissue hematomas including bleeding into muscles, easy bruising, excessive bleeding with surgery, trauma, dental extractions, and circumcision, bleeding in the gastrointestinal or genitourinary tract, epistaxis, poor wound healing, and uncommonly, umbilical stump bleeding. Intracranial hemorrhages can occur, particularly following trauma.

Hemophilia testing is suggested for male patients with unexplained bleeding, especially if the PTT is prolonged with a normal PT and platelet count. Decreases in factor VIII or IX to <20% to 30% can cause PTT prolongations, depending on the PTT reagent and instrument. Since up to 30% of hemophilia A or B cases arise from new mutations, a positive family history will not always be present. If a family history is present, the inheritance pattern is X-linked recessive. The initial tests for hemophilia are the factor VIII and factor IX assays. A von Willebrand test panel is usually also performed. In von Willebrand disease, factor VIII levels are decreased secondary to a decrease in von Willebrand factor. Thus, the tests for von Willebrand factor assess whether a decrease in factor VIII represents von Willebrand disease or hemophilia A.

(Continued)

421

Coagulation Factor Assays *(Continued)*

The von Willebrand panel is also useful in predicting hemophilia A carrier status in females. In hemophilia A carrier females, the ratio of factor VIII to von Willebrand factor is approximately 0.5:1. In normal persons, the ratio is approximately 1:1, since factor VIII circulates in the plasma with von Willebrand factor. Confirmation of carrier status often requires family or genetic studies (see below). Both factor VIII and von Willebrand factor can be elevated during acute phase reactions, including pregnancy. Therefore, if a patient is determined to have elevated acute phase reactants, testing should be repeated at a time when the acute phase reaction has subsided. Lastly, factor VIII is labile at room temperature. Consequently, a mild to moderate decrease in factor VIII may be seen in specimens that have not been processed and stored appropriately.

The factor VIII gene is quite large, and numerous mutations causing hemophilia A have been identified. Therefore, genetic testing can be difficult. An inversion mutation of intron 22 has been shown to cause up to 40% of severe hemophilia A in Caucasians, which simplifies genetic testing in these families. Restriction fragment length polymorphism (RFLP) studies or methods that directly identify the mutation may be useful in families without the intron 22 inversion.

Numerous mutations causing hemophilia B have also been identified. Like hemophilia A, genetic testing for female carrier status or prenatal detection can often be achieved with restriction fragment length polymorphism (RFLP) analysis or methods that directly demonstrate the mutation.

Many patients with hemophilia became infected with human immunodeficiency virus (HIV) as a result of treatment with factor concentrates prior to the availability of HIV testing of blood donors. Currently, factor VIII and IX concentrates are treated to destroy HIV and other viruses, and blood donors are screened for HIV. Desmopressin (DDAVP) elevates factor VIII (and von Willebrand factor) levels approximately two- to threefold over baseline for 6-12 hours. Therefore, it is often used to treat bleeding episodes in patients with mild hemophilia.

Hereditary Deficiencies of Other Coagulation Factors (Factors II, V, VII, X, XI, XII)

Unlike factor VIII and IX deficiencies, which have X-linked recessive inheritance, hereditary deficiencies of the other coagulation factors have autosomal inheritance. Hereditary deficiencies of factors II, V, VII, and X are rare. Factor XI deficiency is common among individuals of Ashkenazi Jewish descent. Factor XII deficiency is relatively common, but it is not associated with any bleeding risk. With the other factor deficiencies, bleeding symptoms may include easy bruising, epistaxis, menorrhagia, bleeding with surgery, trauma, dental extractions, postpartum, or circumcision, umbilical stump bleeding (especially with factor XIII deficiency or afibrinogenemia, described separately) and bleeding in the gastrointestinal or genitourinary tract. Intracranial hemorrhage has been reported with severe deficiencies of factor II, V, VII, or X. Hemarthrosis and bleeding into muscles, characteristic of factor VIII and IX deficiencies, are less common but can occur in other factor deficiencies. Factor deficiencies may prolong the PT and/or PTT, depending on the factor and the severity of the decrease in factor level (see Table 1).

In general, with hereditary factor deficiencies, heterozygous deficient individuals have approximately 50% (most commonly within 30% to 60%) of the normal value for the affected factor. Homozygous deficient individuals have a more severe decrease in the affected factor. As previously mentioned, heterozygous or homozygous deficiencies of factor XII do not cause bleeding. Heterozygous deficiencies of the other factors are usually either asymptomatic or have a milder bleeding tendency than homozygous deficiencies (see Table 3). Factor XI deficient heterozygotes may have bleeding symptoms. Factor II or factor X deficient heterozygotes sometimes have mild bleeding symptoms. With rare exceptions, heterozygous factor V or VII deficiencies are asymptomatic. In contrast, homozygous deficiencies of these factors (II, V, VII, X, XI) do have an increased incidence of bleeding symptoms. However, factor V, VII, or XI levels do not always correlate with severity of symptoms. In general, factor VIII and IX deficiencies tend to be the most severe, while deficiencies of factors

II, V, or XI tend to be milder than factor VIII or IX deficiencies. Severe deficiencies of factor VII or X can have a clinical presentation as severe as hemophilia A or B.

Table 3. Coagulation Factor Deficiencies and Factor Half-Lives

Factor Deficiency	Level Required for Surgical Hemostasis	Bleeding Risk in Homozygous Deficiency?	Bleeding Risk in Heterozygous Deficiency?	Biologic Half-life of Factor
Fibrinogen (factor I)	100 mg/dL	Yes	Sometimes	72-120 h
Prothrombin (factor II)	10%-40%	Yes	Sometimes	72 h (48-120 h)
Factor V	10%-30%	Yes[1]	No (rare exceptions)	12-36 h
Factor VII	10%-25%	Yes[1]	No (rare exceptions)	4-7 h
Factor VIII	Major surgery or major bleeding: 80%-100%	Yes (X-linked recessive)	No (rare exceptions; heterozygotes are carrier females)	8-12 h
	Postoperative: 30%-50%			
	Minor bleeding: 30%-50%			
Factor IX	Major surgery or major bleeding: 50%-80%	Yes (X-linked recessive)	No (rare exceptions; heterozygotes are carrier females)	18-24 h
	Postoperative: 40%			
	Minor bleeding: 30%-50%			
Factor X	10%-40%	Yes	Sometimes	24-48 h
Factor XI	15%-50%	Sometimes[1]	Sometimes	40-84 h
Factor XII	0%	No	No	48-52 h
Factor XIII	>5%-50%[2]	Yes	Sometimes	9-12 d

[1]Factor V, VII, or XI levels do not always correlate well with severity of bleeding.

[2]Factor XIII levels >1%-5% have traditionally been considered asymptomatic; however, recent evidence suggests heterozygous deficiencies with levels up to 50% can be associated with excess bleeding.

Adapted from Van Cott EM and Laposata M, "Coagulation, Fibrinolysis, and Hypercoagulation," *Clinical Diagnosis and Management by Laboratory Methods*, 20th ed, Henry JB, ed, New York, NY: WB Saunders Co, 2001, 642-59.

References:

Menitove 1995, Edmunds 1994, Laposata 1989, Roberts 1995, Roberts 1994.

Hereditary Combined Coagulation Factor Deficiencies

Combined factor deficiencies are very rare. A combined deficiency of factors V and VIII is an autosomal recessive disorder. A combined deficiency of the vitamin K-dependent factors II, VII, IX, and X has also been described.

References

Andrew M, Paes B, and Johnston M, "Development of the Hemostatic System in the Neonate and Young Infant," *Am J Pediatr Hematol Oncol*, 1990, 12(1):95-104.

Cooper DN, Millar DS, Wacey A, et al, "Inherited Factor VII Deficiency: Molecular Genetics and Pathophysiology," *Thromb Haemost*, 1997, 78(1):151-60.

Cooper DN, Millar DS, Wacey A, et al, "Inherited Factor X Deficiency: Molecular Genetics and Pathophysiology," *Thromb Haemost*, 1997, 78(1):161-72.

Edmunds LH and Salzman EW, "Hemostatic Problems, Transfusion Therapy, and Cardiopulmonary Bypass in Surgical Patients," *Hemostasis and Thrombosis: Basic Principles and Clinical Practice*, 3rd ed, Colman RW, Hirsh J, Marder VJ, et al, eds, Philadelphia, PA: Churchill Livingstone, 1994, 958.

Giangrande PLF, "Other Inherited Disorders of Blood Coagulation," Rizza C and Lowe G, eds, *Haemophilia and Other Inherited Bleeding Disorders*, London, WB Saunders Co, 1997, 291-307.

Kane WH and Davie EW, "Blood Coagulation Factors V and VIII: Structural and Functional Similarities and Their Relationship to Hemorrhagic and Thrombotic Disorders," *Blood*, 1988, 71(3):539-55.

Laposata M, Connor AM, Hicks DG, et al, *The Clinical Hemostasis Handbook*, Chicago, IL: Yearbook Medical Publishers, Inc, 1989.

Menitove JE, Gill JC, and Montgomery RR, "Preparation and Clinical Use of Plasma and Plasma Fractions," *William's Hematology*, 5th ed, Beutler E, Lichtman MA, Coller BS, et al, eds. New York, NY: McGraw-Hill, 1995, 1657.

(Continued)

Coagulation Factor Assays *(Continued)*

Modi GJ and Musclow CE, "Factor XI: A Piece of the Coagulation Puzzle," *Lab Med*, 1993, 24:353-6.

Moll S and Ortel TL, "Monitoring Warfarin Therapy in Patients With Lupus Anticoagulants," *Ann Intern Med*, 1997, 127(3):177-85.

Naylor JA, Green PM, Rizza CR, et al, "Factor VIII Gene Explains All Cases of Haemophilia A," *Lancet*, 1992, 340(8827):1066-7.

Roberts HR and Hoffman M, "Hemophilia and Related Conditions - Inherited Deficiencies of Prothrombin (Factor II), Factor V, and Factors VII to XII," *William's Hematology*, 5th ed, Beutler E, Lichtman MA, Coller BS, et al, eds, New York, NY: McGraw-Hill, 1995, 1413-39.

Roberts HR and Lefkowitz JB, "Inherited Disorders of Prothrombin Conversion," *Hemostasis and Thrombosis: Basic Principles and Clinical Practice*, 3rd ed, Colman RW, Hirsh J, Marder VJ, et al, eds, Philadelphia, PA: Churchill Livingstone, 1994, 200-18.

Cobalamin, Serum

Related Information

Synonyms B_{12}; Cyanocobalamin; Vitamin B_{12}

Applies to DIDMOAD Syndrome; Intrinsic Factor; Orotic Aciduria; Transcobalamins; Wolfram Syndrome

Abstract Radioisotopic methods have replaced microbiologic assay for detection of cobalamin deficiency, but have lower than desirable positive predictive value for clinical cobalamin deficiency. Serum methylmalonic acid and homocyst(e)ine provide a sensitive indication of early cobalamin deficiency. Additional evaluation is important in order to establish the presence of clinical cobalamin (vitamin B_{12}) deficiency. The daily requirement for cobalamin is <2 µg/day. A 4- to 5-year supply is normally stored in the liver. Thus, impaired absorption/transport may be present for several years before deficiency becomes clinically evident. There is recent interest in the ability of serum methylmalonic acid levels to identify the presence of early and/or subclinical cobalamin deficiency (see Methylmalonic Acid, Serum, Plasma, Urine, or Amniotic Fluid *on page 908*).

Patient Preparation A fasting specimen is preferred; draw before transfusions or B_{12} therapy is started.

Specimen Serum

Container Red top tube

Storage Instructions Separate serum and freeze; protect from light. Obtain hematocrit from EDTA tube before freezing the whole blood specimen.

Reference Interval The lower reference limit, which is critical to the diagnosis of cobalamin (vitamin B_{12}) deficiency/pernicious anemia, is not clearly established. It is likely in the range of 100-250 pg/mL (SI: 74-185 pmol/L). Values are method and laboratory dependent. Because of overlap in serum levels between cobalamin-deficient and normal individuals, use of an indeterminate interval is necessary. The following are interpretive intervals for serum cobalamin:

- normal: 200-900 pg/mL
- indeterminate: 160-200 pg/mL
- low: <160 pg/mL

Clinical correlation and multiple test documentation of the etiology of macrocytic anemia are advised. Occasionally, patients with significant neuropsychiatric abnormalities may have no hematologic abnormalities (absence of anemia or macrocytosis), but vitamin B_{12} level <200 pg/mL (SI: <150 pmol/L). See table for pediatric reference intervals.

Pediatric Serum B_{12} Reference Ranges

Age (y)	Male (pg/mL)		Female (pg/mL)	
	Low	High	Low	High
0-1	216	891	168	1117
2-3	195	897	307	892
4-6	181	795	231	1038
7-9	200	863	182	866
10-12	135	803	145	752
13-18	158	638	134	605

Adapted from Hicks JM, Cook J, Godwin ID, et al, "Vitamin B_{12} and Folate — Pediatric Reference Ranges," *Arch Pathol Lab Med*, 1993, 117:705.

Use Detect cobalamin (Cb1) deficiency in those patients who have hematologic (weakness, anemia, oval macrocytosis, hypersegmented neutrophils, leukopenia/thrombocytopenia) or neurologic (numbness, tingling, loss of vibratory sensation in extremities) findings suggestive of such deficiency. Because of problems associated with verifying the lower limits of "normal," this assay should not be employed as a screening test for functional cobalamin deficiency. Evaluate folic acid deficiency; evaluate hypersegmentation of granulocyte nuclei; investigate MCV >98 fL; diagnose macrocytic and megaloblastic anemia; study alcoholism, prenatal care, malabsorption, including jejunoileal bypass patients operated on for massive obesity, and certain neurological disorders.

Limitations Drugs capable of interference with absorption of Cb1 and/or folic acid include chemotherapeutic (methotrexate), antimalarial (pyrimethamine), diuretics (triamterene), protozoacides (pentamidine, isethionate), antibacterials (trimethoprim), anticonvulsants (phenytoin), sedatives (barbiturates), oral contraceptives, antituberculosis agents (cycloserine, para-aminosalicylic acid), antigout (colchicine), oral hypoglycemic, biguanide group (metformin, phenformin). Establishing functional cobalamin (B_{12}) sufficiency in any individual patient may require consideration of intra-individual variation, functional status of the gastric mucosa (in particular in elderly individuals) and transcobalamin II binding. See following discussion of application of serum methylmalonic acid.

Contraindications B_{12}/folate levels should be drawn before performance of Schilling test and before administration of any other radioactivity.

Methodology Radioimmunoassay (RIA) based on competitive protein binding: patient's unlabeled endogenous serum B_{12} competes with radiolabeled B_{12} for specific sites on a binding protein (intrinsic factor) and is compared to the behavior of a standard.

Additional Information Vitamin B_{12} (cyanocobalamin) analogues form the base compound in coenzymes having important biologic functions. The corrin system, like porphyrin (heme), is synthesized from delta aminolevulinic acid. The basic compound is named cobalamin. The term vitamin B_{12} refers to hydroxocobalamin or cyanocobalamin and in general use, "B_{12}" applies to all cobalamin forms. The form with attached cyanide group (cyanocobalamin - vitamin B_{12}) was upon original isolation from the liver, actually an artifact generated *in vitro*. Methylcobalamin predominates in the serum, deoxyadenosyl cobalamin in the cytosol. Immunoassays for "B_{12}" measure all forms after conversion to cyanocobalamin. Two metabolically important cobamides (vitamin B_{12} containing coenzymes) are deoxyadenosyl cobamide and methyl cobamide. Cobamides are required for DNA synthesis, methylation, and citric acid cycle reactions.

Cobalamin (Cb1) is not synthesized by humans. In nature, it is supplied only by Cb1-producing microorganisms. It is a requisite dietary component widely available in animal products (meat, fish, eggs, butter, milk, and cheese). The (Continued)

425

Cobalamin, Serum (Continued)

minimum daily requirement (MDR) is 1-5 µg/day, body stores are 2000-5000 µg, and obligatory daily loss is only about 0.1%. B_{12} is absorbed by mucosal epithelial cells (microvilli) of the terminal ileum, a pH and divalent cation dependent process. The vitamin is ingested in food sources nonspecifically bound to protein. Peptic digestion (at low pH in the stomach) is required for release of Cb1 from food protein. Some 25% to 50% of the elderly develop hypochlorhydria or achlorhydria with resultant decrease in proteolysis by pepsin and incomplete release of protein-bound Cb1. This situation can result in Cb1 deficiency but in normal stage I Schilling test (these individuals **can** absorb **crystalline** CN-[^{57}Co]; see Schilling Test *on page 1178*). Its absorption is dependent upon a gastric glycoprotein, intrinsic factor (IF). Gastric acid splits the B_{12} protein linkage. After the IF-B_{12} complex is absorbed by the ileal mucosa, the vitamin enters the portal circulation where it is bound by a system of carrier proteins, the transcobalamins I, II, and III. See Vitamin B_{12} Unsaturated Binding Capacity *on page 1316* and Schilling Test *on page 1178*.

Conditions associated with decreased serum cobalamin include hypochlorhydria; pernicious anemia (PA) in which cobalamin levels may vary from 0 to overlapping lower limits of patients without PA; dietary deficiency (uncommon); disorders of intestinal absorption; inflammatory bowel disease; bacterial overgrowth, small intestine; *Diphyllobothrium* fish tapeworm, small intestine; prior gastric surgery; intestinal surgery (diminished B_{12} or folate or both are found in 88% of patients with jejunoileal bypass operated for morbid obesity); resection of terminal ileum as for Crohn disease which prevents absorption of B_{12}; oral contraceptives; abnormalities of cobalamin transport or metabolism; Imerslund syndrome; and from therapeutic use of gastric H_2 blockers, chronic use of omeprazole, and possibly drug-induced lack of IF secretion. **A significant rise in red blood cell mean corpuscular volume (MCV) may be an important early indicator of B_{12} deficiency.** Conditions associated with increased serum cobalamin include chronic granulocytic leukemia (and to a lesser degree leukemoid states); chronic renal failure; severe congestive heart failure; diabetes; obesity; COPD; and cases of liver cell damage (eg, acute hepatitis). Currently, macrocytosis in a hospitalized urban population is most commonly associated with AIDS patients undergoing zidovudine therapy.

DNA Disorders With Macrocytic (Megaloblastic) Anemias Not Due to Cobalamin or Folate Deficiency

Congenital

- Orotic aciduria
- Lesch-Nyhan syndrome
- Congenital dyserythropoietic anemia

Acquired

- Wolfram (DIDMOAD) syndrome: mitochondrial DNA deletion
- Malignancy: erythroleukemia, refractory sideroblastic anemias, antineoplastic chemotherapeutics that inhibit DNA synthesis
- DNA-inhibiting zidovudine and other HIV, other viral antinucleosides
- Toxins, alcohol, others

Elevated serum or urine methylmalonic acid (MMA) concentration is probably a more definitive indication of early cobalamin (B_{12}) deficiency than the serum cobalamin level. MMA serum level, when increased, reflects decreased tissue cobalamin and is an early indicator of B_{12} deficiency. Cobalamin dependent neurologic disease with normal hematologic parameters and serum B_{12} levels may be associated with significant elevations of serum methylmalonic acid. To avoid dietary influence serum MMA levels have preference over urine studies in nonfasting patients.

Folate deficiency may be a cause of low levels of serum cobalamin. Low serum B_{12} levels may be seen in some individuals deficient only in folate. It is important to note that a normal level of cobalamin in the serum does not always exclude cobalamin deficiency. Individuals with low serum B_{12} require clinical confirmation of deficiency. In pregnancy low serum B_{12} level does not usually indicate deficiency at the biochemical level. Elevation of serum homocyst(e)ine concentrations (in the absence of folate deficiency) may be of value in establishing true B_{12} deficiency. Serum MMA level may be independent of B_{12} status in the pregnant patient.

References

Antony AC, "Megaloblastic Anemias," *Hematology: Basic Principles and Practice*, 3rd ed, Chapter 28, Hoffman R, Benz EJ Jr, Shattil SJ, et al, eds, Philadelphia, PA: Churchill Livingstone, 2000, 446-85.

Bolann BJ, Solli JD, Schneede J, et al, "Evaluation of Indicators of Cobalamin Deficiency Defined as Cobalamin-Induced Reduction in Increased Serum Methylmalonic Acid," *Clin Chem*, 2000, 46(11):1744-50.

Carmel R, "Current Concepts in Cobalamin Deficiency," *Annu Rev Med*, 2000, 51:357-75.

Fish MB, "Gastrointestinal Absorption of Cobalamin (Vitamin B_{12})," *Diagnostic Nuclear Medicine*, 3rd ed, Volume 2, Sandler MP, Coleman RE, Wackers FJ, et al, eds, Baltimore, MD: Lippincott Williams & Wilkins, 1996, 840.

Monsen ALB, Refsum H, Markestad T, et al, "Cobalamin Status and Its Biochemical Markers Methylmalonic Acid and Homocysteine in Different Age Groups From 4 Days to 19 Years," *Clin Chem*, 2003, 49(12):2067-75.

Snow CF, "Laboratory Diagnosis of Vitamin B_{12} and Folate Deficiency: A Guide for the Primary Care Physician," *Arch Intern Med*, 1999, 159(12):1289-98.

♦ **Cobalophilin** *see* Vitamin B_{12} Unsaturated Binding Capacity on page 1316

Cocaine (Cocaine Metabolite), Qualitative, Urine or Hair

Synonyms Coke; Crack; Dama Blanca; Gold Dust; Liquid Lady; Nose Candy; Rock; Snow; Toot; White Lady

Applies to Benzoylecgonine; Chewing; Mainlining; Snorting

Abstract A Schedule II controlled substance, cocaine is one of the most potent of the naturally-occurring central nervous system stimulants. It can be smoked or administered intranasally, orally, or intravenously. Other routes which have significant bioavailability include sublingual, intravaginal, or rectal administration. Prominent metabolites of cocaine are benzoylecgonine and ecgonine methyl ester. Due to its longer half-life as compared to cocaine, the former is a substance generally measured in urine to detect cocaine use.

Specimen Urine; hair can also be analyzed, reflecting long-term exposure. If blood is collected, it should be collected in tubes containing 2% sodium fluoride. Meconium is a useful sample to assess fetal exposure form maternal drug use.

Collection If forensic, observe precautions concerning surreptitious dilution or adulteration.

Storage Instructions Refrigerate

Causes for Rejection If forensic, failure to meet temperature requirements immediately after collection and/or tests for unusual dilution (specific gravity, urine creatinine) or adulteration.

Special Instructions If forensic, use chain-of-custody protocol and form. See Chain-of-Custody Protocol *on page 381* and the Chain-of-Custody form in the Introduction *on page 63*.

Reference Interval Negative (less than cutoff)

Critical Values Urine: Substance Abuse and Mental Health Services Administration (SAMHSA) screening cutoff: 300 ng/mL; confirmation cutoff: 150 ng/mL for benzoylecgonine

Use Evaluate cocaine use and toxicity; work up as part of a drug screen. Cocaine-related myocardial infarction and other cardiovascular events have increased dramatically. Blood concentrations do not correlate well with clinical effects.

Methodology Screen: immunoassay; confirmation: gas chromatography/mass spectrometry (GC/MS)

Additional Information

- Half-life: cocaine: 1 hour, benzoylecgonine: 4-9 hours; ecgonine methylester: 3-4 hours; ethylcocaine: about 2 hours
- Volume of distribution: 3-5 L/kg

Cocaine is a central nervous system stimulant. It usually appears as a fine crystal-like powder which is the hydrochloride or sulfate salt and as such is "snorted" (inhaled through the nose). When mixed with sodium bicarbonate and converted to free base, it appears as hard pieces called "crack" which can be smoked. This is currently a very prevalent form of the drug. Due to pyrolysis, only 30% to 70% of cocaine is bioavailable.

(Continued)

427

Cocaine (Cocaine Metabolite), Qualitative, Urine or Hair *(Continued)*

Benzoylecgonine is detectable in urine within 2-3 hours after cocaine intake and can be found for a period of 1-3 days.

Individuals with pseudocholinesterase deficiency may be at special risk when cocaine is used.

Alcohol inhibits cocaine degradation, enhancing its hepatotoxicity. Cocaine and alcohol are commonly used in combination, resulting in production of cocaethylene or ethylcocaine and increased euphoria, cardiotoxicity, and behavioral effects.

Complications of cocaine use may include cocaine-excited delirium, somnolence, coma, shock, disseminated intravascular coagulation, cardiomyopathy, myonecrosis, dysrhythmia, angina pectoris, myocardial ischemia, infarction, and sudden death. Noncardiogenic pulmonary edema occurs. Cocaine use may cause rhabdomyolysis, hyperthermia, acute renal failure, and adverse effects on fetal growth and development. Cocaine is hepatotoxic. A hepatotoxin, cocaine causes microvesicular steatosis and necrosis. Its abuse has been associated with arterial dissection.

Since hair grows at about 13 mm/month, results of hair analysis reflect longer-term exposure than can urine testing, indicating exposure over periods of weeks or months.

References
Boghdadi MS and Henning RJ, "Cocaine: Pathophysiology and Clinical Toxicology," *Heart Lung*, 1997, 26(6):466-83; quiz 484-5.

Hatsukami DK and Fischman MW, "Crack Cocaine and Cocaine Hydrochloride. Are the Differences Myth or Reality?" *JAMA*, 1996, 276(19):1580-8.

Isenschmid DS, "Cocaine-Effects on Human Performance and Behavior," *Forensic Sci Rev*, 2002, 14:61-100.

Lange RA and Hillis LD, "Cardiovascular Complications of Cocaine Use," *N Engl J Med*, 2001, 345(5):351-8.

Ness RB, Grisso JA, Hirschinger N, et al, "Cocaine and Tobacco Use and the Risk of Spontaneous Abortion," *N Engl J Med*, 1999, 340(5):333-9.

Coccidioidomycosis Serology

Related Information
Fine Needle Aspiration Culture *on page 589*
Fungal Culture, Biopsy or Body Fluid *on page 619*
Fungal Culture, Blood *on page 620*
Fungal Culture, Sputum *on page 624*
Fungus Smear, Stain *on page 626*
Myoglobin, Qualitative, Urine *on page 941*
Sputum Cytology *on page 1222*

Test Includes Complement fixing or precipitating antibodies

Abstract Coccidioidomycosis is an infection caused by the dimorphic fungus *Coccidioides immitis*. It is endemic in soil of certain regions of Arizona, California, Nevada, New Mexico, Texas, Utah, and Mexico, essentially in a band extending to Argentina. It is endemic in the San Joaquin Valley of California.

Diagnosis depends on identification of *C. immitis* by culture, cytologic, and/or histopathologic examination. Serology can be helpful in the diagnosis of disease when a patient is unable to produce sputum, or in the diagnosis of chronic meningitis due to *C. immitis*. Upsurges of infection follow dust storms, drought followed by heavy rain or earthquakes.

Specimen Serum, cerebrospinal fluid

Container Red top tube; clean, sterile CSF tube

Sampling Time Repeated serologic testing during the first 2 months of illness enhances likelihood of diagnosis.

Special Instructions Travel history is relevant.

Reference Interval Negative

Critical Values Increasing titers signal progressive disease.

Possible Panic Range CF antibody titer ≥1:16 may indicate disseminated disease.

Use Diagnose and evaluate the prognosis of coccidioidomycosis. Since CSF cultures are usually negative, demonstration of antibody in CSF is an important means of diagnosis. Repeat testing may be necessary.

Limitations A negative test does not exclude coccidioidomycosis. When low titers are obtained, a diagnosis of coccidioidomycosis must be based on subsequent serological tests and on clinical and mycological studies. Cross reactions may occur in sera from patients with active histoplasmosis. False-negative results often occur in patients with solitary pulmonary lesions. Subjects with meningeal coccidioidomycosis may not demonstrate elevated serum antibodies. The sensitivity, specificity, and reproducibility of enzyme immunoassay and other new tests have not been standardized.

Methodology Complement fixation (CF), tube precipitin, latex particle agglutination, double immunodiffusion (ID), enzyme-linked immunosorbent assay (ELISA).

Additional Information Transmission of *C. immitis* is by inhalation of arthroconidia (arthrospores). About 100,000 infections occur annually in the U.S., most asymptomatic, others mild, some evolving to dissemination and death. With migration to the Southwestern U.S. and increasing numbers of immunosuppressed individuals, the importance of this infection is increasing. Corticosteroids, chemotherapy, immune modulation for organ transplantation, and the AIDS epidemic all have magnified the relevance of coccidioidomycosis as an opportunistic infection. Reactivation of infection acquired years earlier is seen with suppression of cellular immunity. Symptomatic infection of coccidioidomycosis usually presents as a nonspecific illness with fever, cough, headache, and myalgia. Meningitis due to *C. immitis* can lead to permanent neurologic sequelae. Mortality is high in HIV-infected patients with diffuse lung disease. Other high risk groups include those of African-American or Filipino ancestry, diabetics, infants, pregnant women during the third trimester, and patients on immunosuppressive therapy. Seroreactivity may take place before clinically recognized coccidioidomycosis in some HIV-positive individuals.

Serology is a useful approach to the diagnosis of coccidioidal infections. Low titers are usually associated with early, residual, or meningeal coccidioidomycosis with mild and localized disease. Patients with complement fixing (CF) titers ≥1:16 should be evaluated for evidence of pulmonary or extrapulmonary dissemination. Higher CF titers are associated with poorer prognosis. CF test for IgG antibodies becomes positive later and remains positive for years. Falling CF titers indicate an improved clinical status. Specific IgM antibodies may be detected early in the course of the disease. Finding CF antibody in CSF makes the diagnosis of coccidioidal meningitis (if fungal osteomyelitis at the base of the skull can be excluded). CSF is characterized by a mononuclear pleocytosis with decreased glucose and increased protein concentration.

References

Centers for Disease Control and Prevention, "Coccidioidomycosis - Arizona, 1990-1995," *JAMA*, 1997, 277(2):104-5.

Galgiani JN, " Coccidioidomycosis," *Curr Clin Top Infect Dis*, 1997, 3(1):192-9.

Galgiani JN, "Coccidioidomycosis: A Regional Disease of National Importance. Rethinking Approaches for Control," *Ann Intern Med*, 1999, 130(4 Pt 1):293-300.

Stevens DA, "Coccidioidomycosis," *N Engl J Med*, 1995, 332(16):1077-82.

Vaz A, Pineda-Roman M, Thomas AR, et al, "Coccidioidomycosis: An Update," *Hosp Pract*, 1998, 33(9):105-15.

Internet Web Sites

www.cdc.gov/ncidod/dbmd/diseaseinfo/coccidioidomycosis_a.htm
www.acponline.org

♦ **Codate** *see* Codeine, Urine *on page 429*

♦ **Codeine** *see* Morphine, Urine *on page 921*

♦ **Codeine Phosphate** *see* Codeine, Urine *on page 429*

♦ **Codeine Sulfate** *see* Codeine, Urine *on page 429*

Codeine, Urine

Related Information

Morphine, Urine *on page 921*
Opiates, Qualitative, Urine *on page 974*

Synonyms Actacode; Codate; Codeine Phosphate; Codeine Sulfate; Codlin; Methylmorphine; Paveral; Tricodein

Applies to Morphine; Norcodeine

(Continued)

Codeine, Urine (Continued)

Test Includes Part of opiate screen

Abstract Codeine occurs naturally in opium but is produced commercially by 3-O-methylation of morphine. It is used as a narcotic analgesic and in lower doses as an antitussive. It is present in numerous proprietary preparations combined with non-narcotic analgesics (eg, aspirin and acetaminophen) and antihistamines. It is a drug of abuse.

Specimen Urine

Sampling Time Random

Storage Instructions Refrigerate specimen

Special Instructions If forensic, use Chain-of-Custody form. See Chain-of-Custody Protocol *on page 381* and the Chain-of-Custody form in the Introduction *on page 63.*

Reference Interval Negative (below cutoff)

Critical Values Substance Abuse and Mental Health Services Administration (SAMHSA) cutoffs: screen (total opiates): 2000 ng/mL; confirmation: 2000 ng/mL

Use Detect codeine abuse and toxicity

Methodology Screen: immunoassay; confirmation: gas chromatography/mass spectrophotometry (GC/MS)

Additional Information
- Half-life: 2.5-4.0 hours
- Volume of distribution: 3-4 L/kg
- Protein binding: 10% to 30%

Codeine, made by the methylation of morphine, is similar to morphine in uses, actions, contraindications, and adverse reactions. About $1/6$ to $1/10$ as potent as morphine, it is used to manage mild to moderate pain. A small amount of codeine (~10%) is converted to morphine, which accounts for analgesic properties of codeine. Thus, both codeine and morphine may be detected in the urine. Another common source of codeine and morphine is consumption of poppy seeds. To avoid false positives, SAMHSA has recently increased screen and confirmation cutoff to 2000 ng/mL. This has resulted in >300% reduction in the confirmed-positive rate for codeine and morphine. Certain poppy seeds can result in urine levels of morphine as high as 10,000 ng/mL. In low doses, it is an antitussive. After an oral dose, the onset of action is 15-30 minutes, and peak levels are reached in 1-1.5 hours. Codeine is excreted mainly in the urine as norcodeine and free and conjugated morphine.

References

Fraser AD and Worth D, "Experience With a Urine Opiate Screening and Confirmation Cutoff of 2000 ng/mL," *J Anal Toxicol*, 1999, 23(6):549-51.

Meadway C, George S, and Braithwaite R, "Opiate Concentrations Following the Ingestion of Poppy Seed Products - Evidence for the Poppy Seed Defense," *Forensic Sci Int*, 1998, 96(1):29-38.

Thevis M, Opfermann G, and Schanzer W, "Urinary Concentrations of Morphine and Codeine After Consumption of Poppy Seeds," *J Anal Toxicol*, 2003, 27(1):53-6.

♦ **Codlin** *see* Codeine, Urine *on page 429*

♦ **Coefficient of Variation (CV)** *see page 11*

♦ **Coffee Break®** *see* Caffeine, Serum *on page 326*

♦ **COHb** *see* Carboxyhemoglobin, Blood *on page 340*

♦ **Coke** *see* Cocaine (Cocaine Metabolite), Qualitative, Urine or Hair *on page 427*

Cold Agglutinin Titer

Related Information

Cold Hemolysin Test *on page 431*
Cryofibrinogen *on page 477*
Cryoglobulin, Qualitative, Serum and Plasma *on page 478*
Immunoglobulin M *on page 779*
Mycoplasma pneumoniae Culture and Serology *on page 936*
Mycoplasma pneumoniae DNA Probe Test *on page 938*
Red Blood Cells *on page 1139*
Risks of Transfusion *on page 1166*
Warming, Blood *on page 1326*

Abstract *Mycoplasma* infection activates several components of the immune system. In the majority of patients, several classes of antibody are produced.

The best recognized are the cold isohemagglutinins that are capable of clumping erythrocytes at 4°C. Cold agglutinins in *M. pneumoniae* infection are IgM antibodies directed against the I antigen on the surface of erythrocytes.

Specimen Serum

Container Red top tube

Storage Instructions After clotting at 37°C, separate serum from cells if specimen is to be stored overnight in refrigerator.

Causes for Rejection Refrigeration of the specimen before separation of serum from cells, specimen not allowed to clot at 37°C

Special Instructions Transport blood immediately to the laboratory.

Reference Interval Negative: less than a fourfold increase in titer or single titer <1:32

Critical Values Single titers ≥64 or a fourfold titer increase in specimens 5 or more days apart are relevant.

Use Detect diagnosis of primary atypical pneumonia due to *Mycoplasma pneumoniae*; investigate idiopathic cold agglutinin disease

Limitations False negatives may occur, especially if serum is refrigerated on the clot. False-positive results are associated with rubeola, adenovirus pneumonia, infectious mononucleosis, some connective tissue diseases and tropical diseases. Cold agglutinins are not to be confused with cryoglobulins or cryofibrinogens. Antibiotic therapy may interfere with antibody formation.

Methodology The highest dilution causing hemagglutination with type O blood cells at 4°C is the cold agglutinin titer.

Additional Information *M. pneumoniae* has I-like antigen specificity. Cold agglutinins are usually IgM autoantibodies directed against an altered Ii antigen on the human RBCs of *M. pneumoniae*-infected patients. The fetal i RBCs change after birth so that by 18 months red cells carry largely I. The i substance has been found in saliva, milk, amniotic fluid, ovarian cyst fluid, and serum.

Antibody to the I antigen is more specific for *Mycoplasma*, while i antigen is more commonly found in infectious mononucleosis. The most common cause of elevated cold agglutinin in high titers is an infection with *Mycoplasma pneumoniae*. Fifty-five percent of patients with disease have rising titers. In primary atypical *Mycoplasma pneumoniae*, cold agglutinins are demonstrated 1 week after onset; the titer increases in 8-10 days, peaks at 12-25 days, and rapidly falls after day 30.

The differential diagnosis of atypical pneumonias, including features of *Chlamydia*, *Mycoplasma*, and *Legionella* infections, has been reviewed by Gordon and Hindiyeh and Carroll.

References
Gordon RC, "Community-Acquired Pneumonia in Adolescents," *Adolesc Med*, 2000, 11(3):681-95.

Hindiyeh M and Carroll KC, "Laboratory Diagnosis of Atypical Pneumonia," *Semin Respir Infect*, 2000, 15(2):101-13.

Tan JS, "Role of Atypical Pneumonia Pathogens in Respiratory Tract Infections," *Can Respir J*, 1999, 6(Suppl A):15A-9A.

Cold Hemolysin Test

Related Information

Cold Agglutinin Titer *on page 430*
Cryofibrinogen *on page 477*
Cryoglobulin, Qualitative, Serum and Plasma *on page 478*
Hemosiderin Stain, Urine *on page 692*
Infectious Mononucleosis Screening Test *on page 785*
Parvovirus B19 DNA *on page 1006*
Parvovirus B19 Serology *on page 1006*
VDRL, Serum or Cerebrospinal Fluid *on page 1303*

Synonyms Donath-Landsteiner Test; Paroxysmal Cold Hemoglobinuria Test

Abstract Test for paroxysmal cold hemoglobinuria (PCH), a condition caused by sensitization of red blood cells (at temperatures less than 30°C) by a complement binding IgG biphasic hemolysin. Warming to 37°C causes hemolysis of patient's RBCs. Patients may develop fever, back/leg pain, abdominal cramps, and shaking chills with hemoglobinemia/hemoglobinuria.

Specimen Blood

Container Red top tube

(Continued)

Cold Hemolysin Test *(Continued)*

Collection Obtain 7 mL of blood by routine venipuncture from patient in previously warmed (37°C) red top tube and another 7 mL of blood in previously cooled (3°C to 4°C) red top tube in an ice water bath. Collect similar samples from a normal individual for a negative control. Deliver to hematology laboratory immediately.

Special Instructions Consult laboratory; scheduling will likely be necessary.

Reference Interval Negative

Use Diagnosis of the uncommon disorder, paroxysmal cold hemoglobinuria. PCH may occur with syphilis, infectious mononucleosis, influenza, measles, mumps, chickenpox, parvovirus B19, and other viral illnesses.

Additional Information There are two types of cold autoantibodies, the cold autoagglutinins/hemolysins (as found in cold-antibody autoimmune hemolytic anemia) and the biphasic hemolysins. These antibodies have optimal reactions at temperatures <30°C. The cold autoagglutinins are monoclonal or polyclonal IgM antibodies, usually with anti-H, anti-IH, or anti-i immunospecificity. Anti-Pr is the second most commonly associated antibody. Classic PCH is caused by an IgG complement binding biphasic hemolysin, an autoantibody that attaches to the RBC membrane at 4°C to 20°C. The antibody causes only weak agglutination of red cells in saline. When the temperature rises to 37°C, hemolysis occurs. The immunohematologic specificity of the Donath-Landsteiner type PCH antibody is usually anti-P. While PCH was the first hemolytic anemia to be recognized, it has become the least common type of autoimmune hemolytic anemia. Previously associated with syphilis, it is now more commonly seen in viral-like illness, largely in children. Parvovirus 19 infection may cause a reticulocytopenic postinfectious hemolytic anemia in children. The red cell P antigen has specificity for the D-L antibody and is also the parvovirus receptor on red cell precursors. The re-emergence of syphilis associated with AIDS may cause a resurgence of PCH.

References

Chambers LA and Rauck AM, "Acute Transient Hemolytic Anemia With a Positive Donath-Landsteiner Test Following Parvovirus B19 Infection," *J Pediatr Hematol Oncol*, 1996, 18(2):178-81.

Sivakumaran M, Murphy PT, Booker DJ, et al, "Paroxysmal Cold Haemoglobinuria Caused by non-Hodgkin's Lymphoma," *Br J Haematol*, 1999, 105(1):278-9.

- ♦ **Collagen Cross-Link-Associated C-Telopeptide** *see* Osteocalcin, Serum or Plasma *on page 983*
- ♦ **Collagen Cross-Link-Associated N-Telopeptide** *see* Osteocalcin, Serum or Plasma *on page 983*
- ♦ **Collagen Type-1** *see* N-Telopeptides, Urine *on page 967*
- ♦ **Colloidal Gold Curve** *see* Cerebrospinal Fluid Immunoglobulin G *on page 367*

Colon Cancer, Hereditary Nonpolyposis Type

Related Information

Carcinoembryonic Antigen, Serum *on page 342*
Chromosome Analysis, Lymph Node and Solid Tumor *on page 410*
Occult Blood, Stool *on page 969*

Synonyms HNPCC Gene Testing; Lynch Syndrome

Applies to Amsterdam Criteria; Muir-Torre Syndrome

Test Includes This molecular test involves end-to-end sequencing of four genes in a sequential fashion: hMLH1, hMSH2, hPMS1, and hPMS2.

Abstract Polyposis and nonpolyposis syndromes include HNPCC as well as other entities. HNPCC is an autosomal-dominant syndrome with early onset of colonic carcinoma (usually proximally situated) and with increased risk for malignancy at other (extracolonic GI and GU) sites.

Mutations in at least four mismatch repair genes (MSH2, MLH1, PMS1, PMS2) are known to cause hereditary nonpolyposis colon cancer (HNPCC). Sequencing is the most accurate method for detection of mutations. Ninety percent of HNPCC mutations are found in MSH2 and MLH1.

Specimen Whole blood

Container Blue top (sodium citrate) tube or yellow top (ACD) tube

Storage Instructions All specimens should be sent to the laboratory **immediately** after collection, preferably by overnight delivery. All specimens should be kept at room temperature or refrigerated, never frozen.

Causes for Rejection Lysed or frozen blood sample

Turnaround Time Results are available in phases: MSH2: 6 weeks; MLH1: 6 additional weeks; PMS1 and PMS2 together: 4 additional weeks

Use The laboratory generally provides an interpretive report based upon direct sequencing analysis and/or the protein truncation assay. This test is indicated for families with a history of nonpolyposis colon cancer in an autosomal dominant pattern.

Limitations Because the genes for HNPCC have only recently been identified, HNPCC testing is much newer and is generally available only in investigational settings. Such settings provide the structure needed to assure that patients are adequately informed of the risks and benefits of testing and orchestrate the provision of genetic counseling. Such testing is costly. The cost of detecting gene mutations associated with HNPCC in a family depends on the number of genes which need to be sequenced. On the basis of telephone interview, it appears that over half (57%) of subjects (members of HNPCC families) declined genetic testing, with an important barrier to test acceptance being the presence of symptoms of depression.

Additional Information Colorectal carcinoma (CRC) ranks second as a cause of cancer deaths in the United States. Some 130,200 persons in the U.S. are diagnosed with colorectal carcinoma annually, and approximately 56,300 will die from the disease. A positive family history of CRC increases risk. The empiric risk increases with the strength of the family history and with earlier ages of onset. As a result, increased surveillance in first degree relatives of an individual with CRC is currently recommended for early detection of CRC. Five-year survival rates are substantially higher for localized CRC. Survival in asymptomatic patients is greater. About 15% to 20% of CRC may be due to an autosomal dominant mutation. Two of the types of inherited CRC are hereditary nonpolyposis colon cancer (HNPCC) and familial adenomatous polyposis coli (FAP). HNPCC is characterized by CRC in the absence of large numbers of polyps, early age of onset (mean 40-45 years of age), mucinous and poorly differentiated tumors, and an excess of tumors in the proximal colon. (Carcinoma develops early in polyps associated with HNPCC.) The lifetime risk for CRC in individuals who inherit a mutation associated with HNPCC is >90%. In addition, individuals with type II HNPCC are at increased risk for extracolonic cancers including tumors of endometrium, stomach, pancreaticobiliary system, ovary, small intestine, skin, bone marrow, larynx, and upper urological tract. Because there are no distinguishing characteristics of HNPCC, historically the diagnosis has been made on the basis of the family history. The "Amsterdam Criteria" were developed to provide a uniform clinical method of diagnosis. These criteria include histologically-verified CRC in three or more relatives, one of whom is a first-degree relative of the other two; CRC in at least two generations; and at least one CRC diagnosed before the age of 50. The limitations of the Amsterdam Criteria as a means of diagnosis became apparent upon the identification of the genes responsible for HNPCC. Mutations have been found in families that do not meet the Amsterdam Criteria, in particular, with the *hMSH6* gene. Germline mutations in *hMSH6* are rare (possibly nonexistent) in HNPCC families that meet the Amsterdam Criteria. However, germline mutations in the *hMSH6* gene occur in probands with colorectal cancer and a family history of cancer that fall short of the Amsterdam Criteria. Presence of other cancers in the family (endometrial, ovarian, and ureteral tumors) is an indicator of HNPCC.

Mutations of MSH2 and MLH1 account for 90% of HNPCC. Two additional HNPCC genes (PMS1 and PMS2) were cloned in 1994. These genes encode DNA mismatch repair proteins that appear to function within a multimeric complex (to repair base pair mismatches and heteroduplex loops that develop during new DNA synthesis). The highly polymorphic *hMSH6* gene and its mutations (including missense, frameshifts, and splice-site types) appears to be associated with later-onset colon cancer with weaker familial phenotypes and not with families having clinical features of classic HNPCC. With loss of the DNA mismatch repair system, genomic instability occurs with hypermutability (microsatellite instability). The latter is linked to about 90% of the cancers in HNPCC but also occurs in some 15% of sporadic tumors. In families at risk, (Continued)

Colon Cancer, Hereditary Nonpolyposis Type
(Continued)

presymptomatic relatives should be offered testing. If a relative is found to carry the mutation, annual colonoscopy is recommended (there is a predilection for right-sided tumors in HNPCC). Due to the high incidence of multiple synchronous and metachronous cancers, subtotal colectomy is recommended in affected individuals. In an individual who has already had partial resection, aggressive surveillance of the remaining colon is warranted.

A variant of HNPCC is the Muir-Torre syndrome in which sebaceous gland lesions exhibit microsatellite instability as do the colonic and endometrial tumors from these HNPCC patients.

References
Iffit K, "Genetic Prognostic Markers for Colorectal Cancer," *N Engl J Med*, 2000, 342(2):124-5.

Lerman C, Hughes C, Trock BJ, et al, "Genetic Testing in Families With Hereditary Nonpolyposis Colon Cancer," *JAMA*, 1999, 281(17):1618-22.

O'Leary TJ, "Molecular Diagnosis of Hereditary Nonpolyposis Colorectal Cancer," *JAMA*, 1999, 282(3):281.

Swale VJ, Quinn AG, Wheeler JM, et al, "Microsatellite Instability in Benign Skin Lesions in Hereditary Nonpolyposis Colorectal Cancer Syndromes," *J Invest Dermatol*, 1999, 113(6):901-5.

♦ **Colon Washings Cytology** *see* Washing Cytology *on page 1326*

♦ **Combined Test** *see* Pregnancy-Associated Protein A, Serum *on page 1082*

♦ **Combivir®** *see* Zidovudine, Serum or Plasma *on page 1339*

♦ **Comizial®** *see* Phenobarbital, Serum or Plasma *on page 1021*

♦ **Common Pathway** *see* Activated Partial Thromboplastin Time *on page 100*

♦ **Common Pathway** *see* Prothrombin Time *on page 1116*

♦ **Comparative Genomic Hybridization** *see* Fluorescence *in situ* Hybridization *on page 602*

♦ **Compazine®** *see* Phenothiazines, Serum *on page 1021*

♦ **Complement** *see* Myeloperoxidase, Plasma *on page 939*

Complement C3, Serum

Related Information

Anti-DNA *on page 173*
Complement C4, Serum *on page 436*
Complement Components, Overview *on page 437*
Complement, Total, Serum or Body Fluid *on page 441*
Cryoglobulin, Qualitative, Serum and Plasma *on page 478*

Synonyms C3 Complement

Applies to CH_{50}

Abstract Complement levels can be a useful index for following autoimmune disease activity. Genetic deficiencies may be associated with pyogenic infections and susceptibility to autoimmune disease.

Specimen Serum

Container Red top tube or SST™ tube

Storage Instructions Allow sample to clot 15-30 minutes at room temperature (cold activation of the complement system with loss of activity may occur at 0°C), then 30-60 minutes at 4°C. Centrifuge at 4°C. Freeze serum at -70°C if assay cannot be run at once.

Reference Interval Fresh serum: 80-170 mg/dL

Use Quantitation of C3 is used to detect individuals with a congenital deficiency or those with immunologic disease in whom complement is consumed at an increased rate. These include chronic hepatitis, certain chronic infections (including hepatitis C virus associated cryoglobulinemic vasculitis), immune complex disease, poststreptococcal and membranoproliferative glomerulonephritis, and others. It is especially useful to assess disease activity in lupus erythematosus (SLE).

Limitations Detects both biologically active and inactive C3. Thus, C3 levels determined by nephelometry may be misleading as the test reagent antisera will react with inactive forms of C3 (eg, nonfunctional split products C3c). Enzyme-linked immunoassays are commercially available and can measure C3 split products iC3b and C3dg which can indicate the extent of complement activation.

Methodology Rate nephelometry

Additional Information C3 is made in the liver and comprises about 70% of the total protein in the complement system. It is central to activation of both the classical and alternate pathways. Increased levels are found in numerous inflammatory states as an acute phase response. CH_{50} (total complement hemolytic activity), C3 and/or C4 may be decreased in cases of systemic lupus erythematosus, especially in cases with lupus nephritis, acute and chronic glomerulonephritis, infective endocarditis, and with disseminated intravascular coagulation (DIC). However, C3 level is a poor indicator of diagnosis or prognosis in many entities. The "nongamma" (C3) Coombs test may detect C3 on red cell membranes in some cases of autoimmune hemolytic anemias, but C3 levels are seldom decreased. In cases of DIC, plasmin attacks C3 directly, and C3 levels have been low. Cases of hereditary C3 deficiency, while rare, have been reported and are characterized clinically by recurrent infections and by immune complex disease, in particular, membranoproliferative glomerulonephritis. Complement components are apparently involved in the efficient clearance of apoptotic cells (by macrophages) in the systemic circulation. Deficiency of early stage complement components may predispose to autoimmunity by abnormal exposure to apoptotic cells. The central role of C3 (classical and alternate pathways) place C3 deficient patients at risk for especially severe infections by the encapsulated organisms, *S. pneumoniae, H. influenzae,* and *N. meningitidis* (both gram-positive and gram-negative bacteria). Bacteremia, sinopulmonary infections, meningitis, paronychia, and impetigo may occur. C3 levels have also been found deficient in cases of uremia, chronic liver diseases, anorexia nervosa, and celiac disease.

An undetectable C3 level suggests a congenital C3 deficiency. Decreased levels of both C3 and C4 indicate classical pathway activation. Decreased C3 levels with normal C4 levels indicate alternate pathway activation. Very low or undetectable levels of C4 are often seen in type II cryoglobulinemic vasculitis, which is most commonly associated with hepatitis C infection. C3 levels fluctuate during the course of the disease. Generally, decreased levels of individual complement proteins are due to increased catabolism. Because synthesis of complement proteins increases with inflammatory disease, normal levels do not prove that the complement sequence is not involved in tissue injury. See table.

C3 and C4 Levels in the Presence of Decreased Hemolytic Complement Activity

	Normal C4	Decreased C4
Normal C3	Inborn errors (other than C4 or C3)	Inborn C4 deficiency
	Alterations *in vitro* (eg, improper specimen handling)	Immune complex disease
	Coagulation-associated complement consumption	Hypergammaglobulinemic states
		Cryoglobulinemic vasculitis
		Hereditary angioedema
Decreased C3	Inborn C3 deficiency	Active SLE
	Acute glomerulonephritis	Serum sickness
	Membranoproliferative glomerulonephritis	Immune complex disease
	Immune complex disease	Autoimmune/chronic active hepatitis
	Active SLE	Infective endocarditis

A number of genetic defects have been reported as the cause of C3 deficiency, including nonsense mutations with decrease in normal mRNA, deletions, and frameshifts (with resultant truncated C3).

Lytic activity of the complement system and of complement components is lower in neonates as compared to adults.

References

Carney DF, "Complement and Autoimmune Diseases," *Clinical and Laboratory Evaluation of Human Autoimmune Diseases,* Chapter 6, Nakamura RM, Keren DF, and Bylund DJ, eds, Chicago, IL: American Society for Clinical Pathology, ASCP Press, 2002, 71-84.

Ferriani VP, Barbosa JE, and de Carvalho IF, "Complement Haemolytic Activity (Classical and Alternative Pathways), C3, C4, and Factor B Titres in Healthy Children," *Acta Paediatr,* 1999, 88(10):1062-6.

(Continued)

Complement C3, Serum *(Continued)*

Giclas PC, "Complement Tests," Section D, "Complement, Immune Complexes, and Cryoglobulin," Giclas PC, section ed, *Manual of Clinical Laboratory Immunology*, 5th ed, Rose NR, de Macario EC, Folds JD, et al, eds, Washington, DC: ASM Press, American Society for Microbiology, 1997, 181-6.

Lamprecht P, Gause A, and Gross WL, "Cryoglobulinemic Vasculitis," *Arthritis Rheum*, 1999, 42(12):2507-16.

Nielsen CH, Fischer EM, and Leslie RG, "The Role of Complement in the Acquired Immune Response," *Immunology*, 2000, 100(1):4-12.

Complement C4, Serum

Related Information

Antinuclear Antibodies *on page 189*
Complement C3, Serum *on page 434*
Complement Components, Overview *on page 437*
Complement, Total, Serum or Body Fluid *on page 441*
HLA-B27 *on page 738*

Synonyms C4 Complement

Specimen Serum

Container Red top tube or SST™ tube

Storage Instructions Allow sample to clot 15-30 minutes at room temperature, then 30-60 minutes at 4°C. Freeze serum at -70°C if assay cannot be run at once.

Reference Interval 18-51 mg/dL

Use Quantitation of C4 is used to detect individuals with congenital deficiency or those with autoimmune diseases such as lupus erythematosus (SLE), rheumatoid arthritis, serum sickness, certain glomerulonephritides, chronic hepatitis, cryoglobulinemia, immune complex disease, and hereditary angioedema. C4 levels are sensitive indicators of SLE and proliferative glomerulonephritis disease activity. C4 may be increased with autoimmune hemolytic anemia.

Limitations Complement proteins are acute phase reactants and have short half-lives. Serum level is a balance of synthesis and catabolism. Serial measurements are more useful than single values.

Methodology Rate nephelometry, functional assays, electrophoretic allotyping and isotyping, gene cloning and sequencing

Additional Information The C4 protein is encoded by two genes (class III region of the major histocompatibility complex). There are two isotypes, C4A and C4B, that differ by four amino acids. C4 is utilized only by the classical pathway, so that it is decreased only when this arm of the complement cascade is activated. In diseases activating the alternate pathway alone, C4 levels will be normal. Total hemolytic activity (CH_{50}), C3, and C4 are frequently decreased in a variety of conditions producing immune complexes. In hereditary angioedema, the lack of C1 esterase inhibitor allows unopposed lysis of C2 and C4 by C1 esterase, so C4 levels will be low. Hereditary C4 deficiency is associated with an increased incidence of pyogenic bacterial infections, in particular those caused by the encapsulated organism *S. pneumoniae*. See table in Complement C3, Serum *on page 434*.

Neuroinflammation with complement activation is likely importantly involved in the pathogenesis of Alzheimer disease (AD). See also Complement Components, Overview *on page 437*. There is evidence that primary human astrocytes are a source of complement C4 in the human central nervous system.

Partial deletions of the C4 gene (combined with mild upper respiratory infection) may be a risk factor for sudden infant death. The complement system, including levels of C4, C3, and serum functional hemolytic activity (CH_{50}) is apparently intact in elderly individuals. A number of cutaneous diseases are characterized by association with complete or partial C4 deficiency. Systemic lupus erythematosus and related syndromes occur in patients with complete C4 deficiency.

References

Bellavia D, Frada G, DiFranco P, et al, "C4, BF, C3 Allele Distribution and Complement Activity in Healthy Aged People and Centenarians," *J Gerontol*, 1999, 54(4):B150-3.

Opdal SH, Vege A, Stave AK, et al, "The Complement Component C4 in Sudden Infant Death," *Eur J Pediatr*, 1999, 158(3):210-2.

Rogers J, "An IL-1α Susceptibility Polymorphism in Alzheimer's Disease," *Neurology*, 2000, 55(4):464-5.

Traustadottir KH, Sigfusson A, Steinsson K, et al, "C4A Deficiency and Elevated Level of Immune Complexes: The Mechanism Behind Increased Susceptibility to Systemic Lupus Erythematosus," *J Rheumatol*, 2002, 29(11):2359-60.

Complement Components, Overview

Related Information

Antinuclear Antibodies *on page 189*
C1 Esterase Inhibitor, Serum *on page 318*
Cerebrospinal Fluid and Plasma β-Amyloid$_{(1-42)}$ *on page 359*
Complement C3, Serum *on page 434*
Complement C4, Serum *on page 436*
Complement, Total, Serum or Body Fluid *on page 441*
HLA-B27 *on page 738*

Test Includes Quantitation of antigenic (immunologic) and/or functional complement components - C1, C1q, C1r, C1s, C2, C3, C4, C5, C6, C7, C8, C9; factor B; factor D

Abstract The complement system is a major participant in inflammatory reactions. It consists of cascading protein/enzymatic activities with associated receptors and inhibitors. During this process, potent low molecular weight peptide anaphylatoxins (C4a, C3a, and C5a) are generated. Complement components and their deficiency states relate importantly to pyogenic infection, *Neisseria* infection (including, in particular, meningococcal meningitis), and connective tissue disease (eg, systemic lupus erythematosus (SLE)).

Specimen Serum

Container Red top tube

Collection Complement protein components are heat labile. Samples for complement analysis should be allowed to clot 15-30 minutes at room temperature (cold activation may occur at 0°C with loss of activity) and then 30-60 minutes at 4°C. If the assay cannot be run at once, serum should be stored at -70°C. Freezing at -20°C will result in significant loss of complement activity. If sample must be transported or there is delay in processing, a normal control specimen handled in the same manner should also be analyzed.

Reference Interval The following reference values reflect the wide range of complement protein present in most "normal" populations. Null alleles (code for nonsynthesis of complement protein) may be common, as has been shown in the case of C4. Complement proteins are acute phase reactants. Increased levels may be due to recent clinical or subclinical infections and/or other

Complement Components

Component	Serum Concentrations (µg/mL)
Classical pathway	
C1q	70-300
C1r	34-100
C1s	30-80
C2	15-30
C4	350-600
Alternative pathway	
Factor B	140-240
Factor D	1-2
Factor I	35
C3	1200-1500
Terminal pathway	
C5	70-85
C6	60-70
C7	55-70
C8	55-80
C9	50-160

(Continued)

Complement Components, Overview *(Continued)*

illnesses. Levels may be elevated during pregnancy or with use of oral contraceptives. It is difficult to establish meaningful reference intervals. Vagaries in test reagent antisera, standardization, and intralaboratory/interlaboratory technical variation make it difficult to interpret an isolated value. Interpretation should be made using reference ranges from a verified "normal" population, samples properly collected and handled by the same laboratory performing the patient's test.

Components of the classical complement system may be significantly decreased (on both an immunochemical and a functional basis) in neonates and during the first few weeks of life.

Use Assess patients with hereditary or acquired deficiency of complement components; evaluation of immunodeficiency states; management of autoimmune diseases such as SLE

Limitations The presence of cryoglobulins may cause difficulty in performance of tests of complement activation and consumption.

Methodology Nephelometry, functional analysis in hemolytic system (CH_{50} or total hemolytic assay)

Additional Information The complement system is an array of over 30 proteins, mostly enzymes, which interact sequentially to produce a number of biologically active products. Most proteins of the complement system are acute phase reactants (of the inflammatory system). Serum levels of C3 (of the "classical" pathway of activation) and of factor B (of the "alternate" pathway of activation) increase by 50% and 200% respectively during an acute phase response. Most complement components are synthesized by the liver but C7 of the membrane attack complex (MAC) is synthesized by bone marrow-derived cells (monocytes/macrophages, platelets, and in particular, granulocytes) as well as fibroblasts, synovial tissue, and endothelial cells. Microglial cells of the central nervous system are rich in (synthesize) complement factors and astrocytes produce C7. Only some 10% to 60% of circulating C7 may be produced by hepatic cells. Local synthesis of C7 (in particular by granulocytes and endothelial cells in areas of inflammation) provide for an available pool of C7 and for modulation of membrane attack by the complement system's MAC. A variety of factors regulate complement system activity, including C1 inhibitor, a family of complement receptors (CR1, CR2, CR3, and CR4), C3 convertase stabilizers, and cell surface factors such as membrane cofactor protein (MCP, CD46), delay accelerating factor (DAF, CD55) and importantly, membrane inhibitor of reactive lysis (MIRL, CD59). DAF and MIRL are glycosylphosphatidyl inositol (GPI) anchored proteins, deficient in paroxysmal nocturnal hemoglobinuria (PNH). The gene PIG-A regulates GPI synthesis. PIG-A is defective in PNH (see PNH Test (GPI-Anchored Proteins) by Flow Cytometry *on page 1064*).

Complement is most often "activated" through either the "classical" pathway, beginning with antigen-antibody immune complexes (usually on some biologic surface) or the "alternate" pathway which is largely independent of antigen/antibody reaction and commonly relates to the action of bacterial products. Either mode of activation leads to formation of C3 convertase which then leads to production of the membrane attack complex (MAC). C3 is cleaved/activated by C3 convertase (C4b2a) of the classical or C3bBb of the alternative pathways. C3a is removed and the residual C3b undergoes conformational change with exposure of a highly reactive internally situated thiolester bond which is instrumental in further reaction and degradation of C3b on the cell surface, in part by serum proteases. C3b cleavage products attach to the membrane after which C5 is bound and cleaved. The anaphylatoxin C5a is released. C6 is then bound, initiating the terminal path (formation of the membrane attack complex consisting of C5b-poly C9). The MAC cylinder consists of up to 12 C9 molecules. See references by Morgan and Harris for review of considerable additional established detail concerning the process of complement activation and control. Complement proteins account for about 10% of the serum proteins; C3 is present in the highest concentration (120-150 mg/dL). See figure on following page.

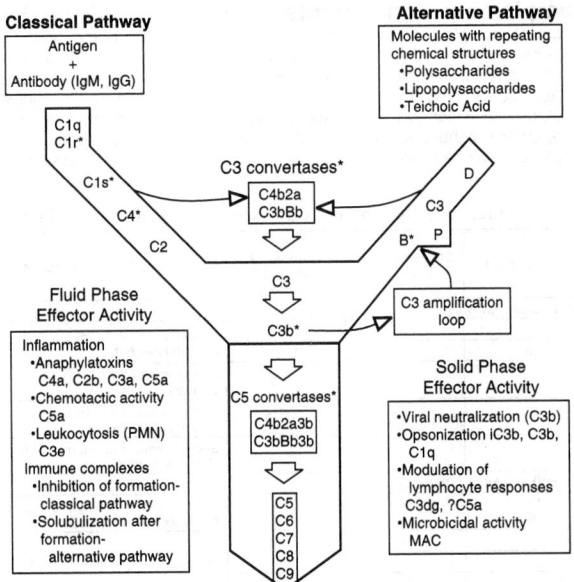

Classical Pathway

Antigen
+
Antibody (IgM, IgG)

Alternative Pathway

Molecules with repeating
chemical structures
• Polysaccharides
• Lipopolysaccharides
• Teichoic Acid

C1q
C1r*

C1s*

C4*

C2

C3 convertases*

C4b2a
C3bBb

D

C3

B*   P

Fluid Phase
Effector Activity

Inflammation
• Anaphylatoxins
  C4a, C2b, C3a, C5a
• Chemotactic activity
  C5a
• Leukocytosis (PMN)
  C3e
Immune complexes
• Inhibition of formation-
  classical pathway
• Solubulization after
  formation-
  alternative pathway

C3

C3b*

C3 amplification
loop

C5 convertases*

C4b2a3b
C3bBb3b

Solid Phase
Effector Activity

• Viral neutralization (C3b)
• Opsonization iC3b, C3b,
  C1q
• Modulation of
  lymphocyte responses
  C3dg, ?C5a
• Microbicidal activity
  MAC

C5
C6
C7
C8
C9

**Membrane attack complex (MAC) formation and insertion
→ cell death/lysis**

Asterisks (*) indicate sites of downregulation of complement activity.

From Densen P, "Complement," *Principles and Practice of Infectious Diseases*, 4th ed,
Chapter 6, Mandell GL, Bennett JE, and Dolin R, eds, New York, NY: Churchill Livingstone,
1995, 58-78.

In the course of activation several byproducts are produced which are active mediators of inflammation. C3a, C3b, C5a, and C5,6,7 are particularly important chemotactic factors and opsonins.

Measurement of total complement activity or components, particularly C3 and C4 which can reflect both complement pathways, may be useful in evaluating the activity of rheumatic disorders in which complement may be involved in pathogenesis. These include SLE, arteritis, and immune arthritis in particular.

Congenital deficiencies of complement components are associated with distinct clinical syndromes (see table on following page).

The most common infections occurring in complement deficient individuals are those due to *Neisseria meningitidis*.

Deficiency of C3 is associated with severe recurrent infections, usually with encapsulated microorganisms. Deficiencies of C1 components, C2 and C4 are associated with rheumatic diseases, including SLE, vasculitis, and dermatomyositis. Some individuals with deficiency may have no evidence of disease.

The most common complement deficiency is C2, which is a homozygous abnormality in 1 in 10,000 to 40,000 individuals, and is heterozygous in 1% to 2% of the general population. Patients with C2 deficiency and SLE often have negative or low titer ANA.

Complement components may drop in patients with active rheumatic diseases, particularly lupus nephritis, sometimes decreasing prior to the clinical attack.

The complement system (along with acute-phase proteins and cytokines) represent innate (natural) response to invading microbes, while B-cell secreted immunoglobulins represent acquired/adaptive (antigen-specific antibody based) response leading to elimination of extracellular microorganisms. Innate immunity assists in guiding the adaptive immune response with B-lymphocyte activation determined by coreceptors, including two complement receptors that are expressed on B lymphocytes, CR1 (CD35) and CR2 (CD21).

(Continued)

439

Complement Components, Overview *(Continued)*

Innate complement responses have been identified in the brain. Central nervous system inflammation relating to Alzheimer-type degenerative neuropathy involves complement components which are up-regulated in microglia. Activated microglia, failing in attempts to phagocytose senile plaques and extracellular tangles of Alzheimer disease-affected brain, nonspecifically destroy neighboring neurons and their processes (bystander lysis). This process has been seen as one of "autotoxicity", distinct from classic autoimmunity.

Genetically Determined Complement Deficiencies in Man

Component	Chromosome / Gene Location[1]	Approx Number of Patients / Kindreds	Major Clinical Associations
Activation Pathway			
C1q	1	24/14	Pyogenic infections, SLE, glomerulonephritis
C1r/C1s	12	11/7	
C4	6 (MHC class III)	21/17	Pyogenic infections, SLE, immune complex disease
C2	6 (MHC class III)	109/79	Pyogenic infections, SLE, glomerulonephritis
C3	19	19/14	Severe immune deficiency, SLE, glomerulonephritis
Membrane Attack Complex			
C5	9	27/17	Meningococcal meningitis / sepsis, gonococcal sepsis, SLE
C6	5	77/49	Meningococcal meningitis, SLE (rare)
C7	5	73/50	
C8	9	73/52	
C9	5	18/15	Usually asymptomatic, susceptible to *M. meningitis*
C1 Inhibitor	11	100s/100s	Hereditary angioedema
Factor H	1	13/8	Hemolytic uremic syndrome
Factor I	14/12		Pyogenic infections
Properdin	x-linked recessive	70/23	Meningococcal meningitis, pneumonia

SLE = systemic lupus erythematosus.

[1]Inheritance is predominantly autosomal recessive.

Adapted from:

Sims PJ and Wiedmer T, "Complement Biology," 3rd ed, Chapter 37, *Hematology: Basic Principles and Practice*, Philadelphia, PA: Churchill Livingstone, 2000, 61.

Winkelstein JA, Sullivan KE, and Colten HR, "Genetically Determined Disorders of the Complement System," *The Metabolic and Molecular Basis of Inherited Disease*, 7th ed, Chapter 130, Scriver CR, Beaudet AL, Sly WS, et al, eds, New York, NY: McGraw-Hill, Inc, 1995, 3913.

References

Delves PJ and Roitt IM, "The Immune System: First of Two Parts," Review Article-Advance in Immunology, *N Engl J Med*, 2000, 343(1):37-49.

Fearon D, "The Complement System and Adaptive Immunity," *Semin Immunol*, 1998, 10(5):355-61.

Hang L and Nakamura RM, "Inter-relationships of Immunodeficiency, Autoimmunity, and Malignancies," *Clinical and Laboratory Evaluation of Human Autoimmune Diseases*, Chapter 4, Nakamura RM, Keren DF, and Bylund DJ, eds, Chicago, IL: American Society for Clinical Pathology, ASCP Press, 2002, 47-59.

Honig LS, "Inflammation in Neurodegenerative Disease: Good, Bad, or Irrelevant?" *Arch Neurol*, 2000, 57(6):786-8.

Keren DF, "Cryoglobulins, Immune Complexes, and Autoimmune Diseases," *Clinical and Laboratory Evaluation of Human Autoimmune Diseases*, Chapter 5, Nakamura RM, Keren DF, and Bylund DJ, eds, Chicago, IL: American Society for Clinical Pathology, ASCP Press, 2002, 61-70.

Lambris JD, Reid KB, and Volanakis JE, "The Evolution, Structure, Biology and Pathophysiology of Complement," *Immunol Today*, 1999, 20(5):207-11.

Langeggen H, Pausa M, Johnson E, et al, "The Endothelium Is an Extrahepatic Site of Synthesis of the Seventh Component of the Complement System," *Clin Exp Immunol*, 2000, 121(1):69-76.

McGeer PL and McGeer EG, "Autotoxicity and Alzheimer Disease," Pleasure DE, section ed, *Arch Neurol*, 2000, 57(6):789-90.

Morgan BP and Harris CL, *Complement Regulatory Proteins*, Academic Press, Inc, 1999.

Sims PJ and Wiedmer T, "Complement Biology," *Hematology: Basic Principles and Practice*, 3rd ed, Chapter 37, Hoffman R, Benz EJ Jr, Shattil SJ, et al, eds, Philadelphia, PA: Churchill Livingstone, 2000, 651-67.

Walport MJ, "Complement: First of Two Parts," *N Engl J Med*, 2001, 344(14):1058-66.

Complement, Total, Serum or Body Fluid

Related Information

Anti-DNA *on page 173*
Antinuclear Antibodies *on page 189*
Complement C3, Serum *on page 434*
Complement C4, Serum *on page 436*
Complement Components, Overview *on page 437*

Synonyms CH_{50}

Test Includes Quantitation of total functional serum complement

Abstract The most frequently occurring alterations of complement are increased levels, since most complement proteins are acute phase reactants. The main clinical application of complement assays is the detection of decreased levels, which may indicate an on-going immunological disorder.

Specimen Serum, synovial fluid, other body fluids

Container Red top tube; for complement activation product assays, collect in EDTA, store on ice, separate, and freeze plasma.

Storage Instructions Allow sample to clot 15-30 minutes at room temperature, then 30-60 minutes at 4°C. Store serum at -70°C if assay cannot be run at once. Complement components may degrade if exposed to longer clotting times or higher temperatures.

Reference Interval 25-110 CH_{50} units with some variation between laboratories. Synovial fluid levels are 33% to 50% of serum levels in patients with nonimmune processes.

Use Total hemolytic complement (CH_{50}) is the best functional assay of the complete complement sequence. CH_{50}, assessing the integrity of the classical pathway, is often decreased in SLE, glomerulonephritis, and other immune complex diseases. C3 and C4 are important as well in autoimmune disorders. Falling complement levels are associated with increased disease activity. Decrease to undetectable activity may result from an inherited deficiency of one or more complement components.

Persistently abnormal complement concentrations in children with the hemolytic-uremic syndrome may indicate poor prognosis.

Rarely, patients with recurrent infections, especially those caused by *Neisseria*, may have genetic complement deficiencies.

Limitations A single normal result may be misleading; longitudinal studies are clinically more helpful. Decreased or absent activity may be due to a complement component deficiency but may also result from improper specimen handling (see Storage Instructions). Improper collection and handling can show *in vitro* activation, leading to misleading high values. Some fragments which are normally excreted in urine may be falsely increased in the presence of renal failure (eg, C4a and Bb).

Methodology Quantitative hemolysis (total complement hemolytic activity (CH_{50})). CH_{50} unit reflects the reciprocal of the dilution of patient's serum required to hemolyze 50% of sheep red blood cells. More recent methods include enzyme-linked immunoabsorbent (ELISA) assays and one involving lysis of liposomes used in an automated system.

Additional Information Modes of response are **innate (natural)** and **acquired (adaptive)**. Innate responses are fixed and constant while adaptive responses increase and expand with repeat activation. The soluble molecular components of the innate arm include complement, acute-phase proteins, and cytokines (including the series of interferons/interleukins). Acquired responses involve the proliferation of antigen-specific B and T cells and immunoglobulin antigen-specific antibody production. The innate immune system lacks immunologic memory and developed earlier during evolution. The innate system is ancient, appearing well before the evolution of the adaptive arm of immunity with its increasingly efficient antibody production. Innate immunity, however, guides the adaptive arm of the immune response with activation of the B lymphocyte and subsequent specific antibody production determined by coreceptors. These include the two complement receptors that are expressed on B lymphocytes, CR1 (CD35) and CR2 (CD21).

Complement activation occurs by one or more of three mechanisms.
- Classic pathway: activated by antigen/antibody complexes
- Alternative pathway: activated by microbial-cell walls

(Continued)

Complement, Total, Serum or Body Fluid (Continued)

- Lectin pathway: activated by the interaction of microbial carbohydrates with mannose-binding protein present in the plasma, mannose-binding lectin (MBL)

MBL and its two associated serine proteases (MASP-1 and MASP-2) bring about complement activation by MBL binding to carbohydrate epitopes on the surfaces of pathogenic microorganisms. The sequence of activation in the classical pathway is C1, C4, C2, C3, and C5 to C9. In the alternate pathway, C1, C4, and C2 are bypassed and C3 is activated by an initiating factor (IF), and two substances called Properdin factors D and B. Total hemolytic complement (CH_{50}) is the best functional assay of the complement sequence.

A normal CH_{50} level indicates that all the components, C1 through C9, are present. However, individual complement factors may be depleted 50% to 80% without affecting CH_{50} activity. Depletion of alternative factors is not detected. For this reason, it may be necessary to measure individual complement components. C2 deficiency is the most common genetic complement deficiency.

The **activation index** represents the ratio of the split product to the intact substance. It is used for management of SLE, cryoglobulinemia, sepsis, and glomerulonephritis type 1.

Body fluid CH_{50} activity should normally be approximately 33% to 50% of the serum value. Decreased CH_{50} titers and complement protein levels may be seen in the joint fluid of patients with rheumatoid arthritis, gout, pseudogout, Reiters syndrome, and gonococcal arthritis. Serum levels in these patients may be normal or increased.

References

Carney DF, "Complement and Autoimmune Diseases," *Clinical and Laboratory Evaluation of Human Autoimmune Diseases*, Chapter 6, Nakamura RM, Keren DF, and Bylund DJ, eds, Chicago, IL: American Society for Clinical Pathology, ASCP Press, 2002, 71-84.

"Immunobiology of Complement," *Immunological Review*, Parham P, ed, 180:April, 2001, 5-189.

Mackay I and Rosen FS, "The Immune System: First of Two Parts," *N Engl J Med*, 2000, 343(1):37-49.

Complete Blood Count

Related Information

Ammonia, Plasma *on page 150*
Anemia Flowchart *on page 35*
Bone Marrow *on page 296*
CD4/CD8 Enumeration *on page 349*
Cobalamin, Serum *on page 424*
Eosinophil Count *on page 542*
Fanconi Anemia, Chromosome Breakage Study *on page 569*
Ferritin, Serum *on page 577*
Flow Cytometry, Overview *on page 592*
Folic Acid, RBC *on page 606*
Folic Acid, Serum *on page 606*
Hematocrit *on page 674*
Hemoglobin *on page 681*
Ibuprofen, Serum *on page 764*
Iron and Total Iron Binding Capacity/Transferrin, Serum *on page 807*
Lymph Node Biopsy *on page 880*
Peripheral Blood: Differential Leukocyte Count *on page 1010*
Peripheral Blood: Red Blood Cell Morphology *on page 1016*
Platelet Count *on page 1050*
Platelet Sizing *on page 1056*
Red Blood Cell Count *on page 1133*
Red Blood Cell Indices *on page 1136*
Uric Acid, Serum *on page 1286*
White Blood Cell Count *on page 1330*

Synonyms CBC

Test Includes The components of a complete blood count (CBC) vary in different laboratories. Tests commonly included are: WBC count, differential count, Hct, Hb, RBC count, WBC and RBC morphology, RBC indices, platelet estimate, platelet count, RDW, and histograms. Variations on the theme include automated 5-part WBC differentials: granulocytes, monocytes, lympho-cytes, eosinophils, basophils, and additional RBC and platelet indices. In addi-tion, current analyzers have reticulocyte capability including determination of a set of reticulocyte indices.

Abstract The CBC is a profile of tests rather than a single test. It is the standard, broadly inclusive, usually automated test for evaluation of RBC, WBC, and platelets.

Specimen Whole blood

Container Lavender top (EDTA) tube. International Council for Standardization in Hematology recommendation is for use of K_2-EDTA, 1.5-2.2 mg/mL of blood as anticoagulant for blood cell counting and sizing.

Collection Mix specimen 10 times by gentle inversion. If specimen is not brought to the laboratory immediately refrigeration is required. If the anticipated delay in arrival is more than 4 hours, two blood smears should be prepared immediately after the venipuncture and submitted with the blood specimen.

Storage Instructions EDTA-anticoagulated sample should be analyzed within 6 hours at room temperature and within 24 hours when stored at 4°C. Blood cell parameters are stable for up to 24-48 hours (WBC differential is stable for 24 hours) at 4°C.

Causes for Rejection Improper tube, clotted specimen, hemolyzed specimen, dilution of blood with I.V. fluid

Mean (±1 SD) Reference Intervals for Hematologic Values

Age	Hb (g/dL)	Hct (%)	RBC (x 10⁶/ mm³) (x 10¹²/L)	MCV (fL)	MCH (pg)	MCHC (g/dL)	WBC (x 10³/ mm³) (x 10⁹/L)
Birth (cord blood)	17.1 ±1.8	52 ±5	4.64 ±0.5	113 ±6	37 ±2	33 ±1	
1 d	19.4 ±2.1	58 ±7	5.30 ±0.5	110 ±6	37 ±2	33 ±1	
2-6 d	19.8 ±2.4	66 ±8	5.40 ±0.7	122 ±14	37 ±4	30 ±3	
14-23 d	15.7 ±1.5	52 ±5	4.92 ±0.6	106 ±11	32 ±3	30 ±2	
24-37 d	14.1 ±1.9	45 ±7	4.35 ±0.6	104 ±11	32 ±3	31 ±3	
40-50 d	12.8 ±1.9	42 ±6	4.10 ±0.5	103 ±11	31 ±3	30 ±2	
2-3.5 mo	11.3 ±1.0	37 ±4	3.81 ±0.5	98 ±9	30 ±3	30 ±2	
5-10 mo	11.6 ±0.7	38 ±3	4.28 ±0.5	91 ±8	27 ±3	30 ±2	
1-3 y	11.9 ±0.6	39 ±2	4.45 ±0.4	87 ±7	27 ±2	30 ±2	
3-5 y	12.3 ±0.8	36 ±3	4.4 ±0.3	81 ±5	28 ±2	34 ±1	7.7 ±2.2
6-8 y	12.7 ±0.9	37 ±2	4.5 ±0.3	83 ±5	28 ±2	34 ±1	7.5 ±2.0
9-11 y	13.1 ±0.9	38 ±2	4.6 ±0.4	83 ±5	28 ±2	34 ±1	7.0 ±1.8
12-14 y							
male	13.8 ±1.0	40 ±3	4.8 ±0.4	84 ±5	29 ±2	34 ±1	7.0 ±1.8
female	13.2 ±1.0	39 ±2	4.5 ±0.3	86 ±5	29 ±2	34 ±1	7.1 ±2.1
15-17 y							
male	14.7 ±1.0	43 ±3	5.0 ±0.4	87 ±5	30 ±2	34 ±1	7.2 ±2.0
female	13.3 ±1.0	39 ±3	4.5 ±0.3	88 ±5	30 ±2	34 ±1	7.7 ±2.1
18-64 y							
male	15.2 ±1.1	44 ±3	5.0 ±0.4	89 ±5	31 ±2	34 ±1	7.4 ±2.1
female	13.5 ±1.1	40.5 ±3	4.4 ±0.4	90 ±6	30 ±2	34 ±1	7.2 ±2.1
65-74 y							
male	14.8 ±1.4	44 ±4	4.8 ±0.5	91 ±6	31 ±2	34 ±1	7.1 ±2.0
female	13.7 ±1.2	40.5 ±3	4.5 ±0.4	90 ±6	31 ±2	34 ±1	6.8 ±2.4

Adapted from Johnson TR, "How Growing Up Can Alter Lab Values in Pediatric Laboratory Medicine," *Diag Med* (special issue), 1982, 5:13-8; and Second National Health and Nutrition Examination Survey (NHANES), "Hematological and National Biochemistry Reference Data for Persons 6 Months - 74 Years of Age," *DHHS Publication No (PHS) 83-1682*, Hyattsville, MD: Public Health Service, Dec 1982.

Above values are reference intervals for the population defined in NHANES II (1976-1980, see above). These intervals are not necessarily identical to "normal intervals" for local populations, but are sufficiently broad such that mean ±2 SD should include most (at least 95%) of normal subjects. NHANES III data (from a 6-year study, 1988-1994) is available, in part, as it pertains to iron deficiency. See Looker AC, Dallman PR, Carroll MD, et al, "Prevalence of Iron Deficiency in the United States," *JAMA*, 1997, 277(12):973-6. NHANES III Hb cutoff values were calculated as the mean Hb of the reference group minus 1.645 SD (corresponds to the 5th percentile value for a variable such as Hb that has a gaussian distribution). Thus, adult males (ages 20-49) had Hb level of 15.30 ±0.97 with a cutoff value <13.7 g/100 mL. Adult females (20-49 years) had Hb level of 13.48 ±0.91 with a cutoff value <12.0 g/100 mL.

(Continued)

Complete Blood Count *(Continued)*

Turnaround Time If the analyzer is operational, a stat result may be available within 5-10 minutes.

Special Instructions Blood specimen and diluent may require prewarming to obtain meaningful results if cold agglutinins are present.

Reference Interval See White Blood Cell Count *on page 1330* for review of origin of reference intervals used by many current texts. See previous table.

Critical Values Critical values: Hematocrit: <18% or >54%; hemoglobin: <6.0 g/dL or >18.0 g/dL; WBC on admission: <2500/mm^3 or >30,000/mm^3; platelets: <20,000/mm^3 or >1,000,000/mm^3

Use Evaluate anemia, leukemia, reaction to inflammation and infections, periph-eral blood cellular characteristics, state of hydration and dehydration, polycy-themia, hemolytic disease of the newborn; manage chemotherapy decisions

Limitations Hemoglobin (and thus the derived MCH and MCHC) may be falsely high if the plasma is lipemic or if the white count is >50,000 cells/mm^3. "Spun" (manual centrifuged) microhematocrits are ~3% higher (due to plasma trap-ping) compared to automated hematocrit levels. The increase is especially pronounced in cases of polycythemia (increased Hct levels) and when the cells are hypochromic and microcytic. The spun Hct level (as compared to auto-mated instruments' calculated level) may be 12% higher at Hct levels of 70% and MCV of 48 fL with decrease in change to 3% higher at Hct levels of 70% with MCV of 100 fL. In neonates, significant differences may occur between capillary and venous complete blood counts with usually higher Hb, Hct, RBC, WBC, and lymphocyte counts in capillary samples but with platelet counts higher in venous as compared to capillary blood. Cold agglutinins (high titer) may cause spurious macrocytosis and low RBC count. This results when RBC couplets are "seen" and processed as single cells by the detection circuitry. Keeping the blood warm and warming the diluent prior to and during counting can correct this problem. See also Hematocrit *on page 674.* Cryoproteinemia (cryoglobulinemia) may cause pseudoleukocytosis or pseudothrombocytosis. Malaria may be a cause of pseudoreticulocytosis.

Methodology Varies considerably between institutions. Most laboratories have high capacity multichannel instruments in place (available from multiple commercial sources). The majority measure RBC and WBC variables on the basis of changes in electrical impedance as cells and platelets are pulled through a tiny aperture with subsequent computer processing of electrical signals. Accuracy (with proper standardization) and precision (usually in the 0.5% to 2% range) is significantly improved over older manual and semiauto-mated methods. Some instruments count impulses as cells flow across a laser beam.

Additional Information Presence of one or more of the following may be indications for further investigation: hemoglobin <10 g/dL, hemoglobin >18 g/dL, MCV >100 fL, MCV <80 fL, MCHC >37%, WBC >20,000/mm^3, WBC <2000/mm^3, presence of sickle cells, significant spherocytosis, basophilic stippling, stomatocytes, significant schistocytosis, oval macrocytes, tear drop red blood cells, eosinophilia (>10%) monocytosis (>15%), nucleated red blood cells in other than the newborn, malarial organisms or the possibility of malarial orga-nisms, hypersegmented (five or more nuclear segments) PMNs, agranular PMNs, Pelger-Huët anomaly, Auer rods, Döhle bodies, marked toxic granula-tion, mononuclears in which apparent nucleoli are prominent (blast type cells), presence of metamyelocytes, myelocytes, promyelocytes, neutropenia, pres-ence of plasma cells, peculiar atypical lymphocytes, significant increase or decrease in platelets. Some quantitative elements of the CBC are related to each other, normally, such that examination of the results of any individual analysis allow for the application of a simple but effective case individualized quality control maneuver. The RBC count, hemoglobin, and hematocrit may be interpreted by applying a "rule of three." If red cells are normochromic/normo-cytic, the RBC count times three should approximately equal the hemoglobin and the hemoglobin multiplied by three should approximate the hematocrit. If there is significant deviation from these relationships, check for supporting abnormalities in RBC indices and in the peripheral blood smear; in addition, patient identification may be a problem. If patient transfusion can be excluded, then RBC indices should vary little consecutively from day to day.

Anemias have been classified on the basis of their MCV and RDW (RBC heterogeneity). This classification has been especially helpful in the separation of iron deficiency from thalassemia. Heterozygous thalassemia (thalassemia minor) when associated with normal hemoglobin has a normal RDW (13.4 ±1.2%) while RDW is high with iron deficiency (16.3 ±1.8%). RDW will be increased slightly in cases of thalassemia with slight anemia. RDW may not always separate cases of iron deficiency from thalassemia minor unless a higher cutoff value (RDW of 17.0%) is used. A recently proposed algorithm uses ethnic background and MCV to provide a "high index of suspicion" for detection of thalassemia trait when dealing with multicultural populations. See also Ferritin, Serum *on page 577*. RDW is not clinically useful in distinguishing the anemia of chronic disease from iron deficiency. In some 30% to 50% of patients with anemia of chronic disease, red cells are hypochromic and microcytic, often with decreased serum iron, iron binding capacity and transferrin saturation even with demonstrably adequate iron stores.

Histogram of MCHC is of value in the diagnosis of hereditary spherocytosis and the differentiation of α- from β-thalassemic red cells. RDW is an insensitive parameter for the diagnosis of vitamin B_{12} deficiency, as well as for the diagnosis of folate deficiency, and the RDW has no value in separating alcohol-related macrocytosis from B_{12}/folate deficiency. In a hospitalized urban patient population, zidovudine treatment of AIDS may be the most common cause of macrocytosis.

The RDW in healthy pregnant women may rise during the last 4-6 weeks before onset of labor, possibly due to increased bone marrow activity. This change could serve to indicate impending parturition.

Most recent generation automated hematology analyzers include leukocyte differential determination and a system of "flags" to indicate presence of abnormal, atypical, and possible immature granulocytes and/or blasts. Numerous studies evaluating performance of flagging systems have been published. Evaluation of this WBC discriminate function is important in detection and characterization of leukemia.

Recently available analyzers incorporate the ability to perform reticulocyte counts, nucleated red blood cell determinations, CD4:CD8, CD64, immature granulocyte, and variant lymphocyte counts. A flow cytometric analysis of platelets, "ImmunoPlt" assay is based in part on CD61 monoclonal antibody labeling (implemented on the Cell-Dyn 4000). It is especially suited for analysis of thrombocytopenic specimens (interference by nonplatelet particles is decreased).

Detailed descriptions of current analyzers have been published and include instruments offered by the following manufacturers:
- Abbott Diagnostics (Cell-Dyn 3200, 3700, and 4000)
- ABX Diagnostics Inc (Pentra 60^{c+}, 120 Retic Hematology Analyzer)
- Bayer® Diagnostics (Advia 120 Hematology System)
- Beckman Coulter, Inc (Coulter GEN-S, HmX, STK-S with Reticulocytes, MAXM with Reticulocytes)
- Roche Diagnostics Corp (Sysmex SF-3000/SF-Alpha, 9500/SE-Alpha II, SE-9500R/SE-Alpha IIR/HST, XE 2100/XE Alpha II/HST)

For consideration of the differential leukocyte count, see Peripheral Blood: Differential Leukocyte Count *on page 1010* and also reference by Krause JR.

See White Blood Cell Count *on page 1330* for text concerning reference range and data/comments under Additional Information which apply to the white cell count component of the CBC and may also apply to other components of the CBC.

References

Aller RD and Pierre RV, "Getting Better All the Time," *CAP Today*, 2000, 14(12):27, 28, 30, 32-4.

Chapman M, "Hematology Analyzers Offer New Technology and User-Friendliness," *Lab Med*, 2000, 31(3):146-50.

Fulwood R, Johnson CL, Bryner JD, et al, "Hematological and Nutritional Biochemistry Reference Data for Persons 6 Months - 74 Years of Age: United States 1976-1980," *Vital and Health Statistics*, Series 11, No. 232, DHHS Publication No (PHS) 83-1682, 1982.

Gill JE, Davis KA, Cowart WJ, et al, "A Rapid and Accurate Closed-Tube Immunoassay for Platelets on an Automated Hematology Analyzer," *Am J Clin Pathol*, 2000, 114(1):47-56.

Hoedemakers RM, Pennings JM, and Hoffmann JJ, "Performance Characteristics of Blast Flagging on the Cell Dyn 4000 Haematology Analyzer," *Clin Lab Haematol*, 1999, 21(5):347-51.

(Continued)

Complete Blood Count *(Continued)*

Hoffman JJ and Pennings JM, "Pseudoreticulocytosis as a Result of Malaria Parasites," *Clin Lab Haematol*, 1999, 21(4):257-60.

Kiss TL, Ali MA, Levin M, et al, "An Algorithm to Aid in the Investigation of Thalassemia Trait in Multicultural Populations," *Arch Pathol Lab Med*, 2000, 124(9):1320-3.

Krantz SB, "Pathogenesis and Treatment of the Anemia of Chronic Disease," *Am J Med Sci*, 1994, 307(5):353-9.

Krause JR, "The Automated White Blood Cell Differential: A Current Perspective," *Hematol Oncol Clin North Am*, 1994, 8(4):605-16.

Lee GR, "Anemia: A Diagnostic Strategy," *Wintrobe's Clinical Hematology*, 10th ed, Volume 1, Chapter 30, Lee GR, Foerster J, Lukens J, et al, eds, Philadelphia, PA: Lea & Febiger, 1999, 908-40.

"The Blood Count," *Advanced Laboratory Methods in Haematology*, Part 1, Rowan RM, van Assendelft OW, and Preston FE, eds, London, UK: Arnold, 2002, 254.

van Duijnhoven HL and Treskes M, "Marked Interference of Hyperglycemia in Measurements of Mean (Red) Cell Volume by Technicon® H Analyzers," *Clin Chem*, 1996, 42(1):76-80.

Zuiable A and Wickramasinghe SN, "RDW in Vitamin B_{12} and Folate Deficiency and in Patients With Alcohol-Related Macrocytosis," *Clin Lab Haematol*, 1992, 14(2):164-6.

♦ **Complexed PSA** *see* Prostate Specific Antigen, Free *on page 1097*

♦ **Compound F** *see* Cortisol, Serum or Plasma *on page 460*

♦ **Computer-Assisted Semen Analysis** *see* Infertility Screen *on page 786*

♦ **Computer-Assisted Semen Analysis** *see* Sperm Penetration Assay (Zona-Free Hamster Egg Penetration Test) *on page 1220*

♦ **Computer-Assisted Sperm Analysis** *see* Semen Analysis, Basic *on page 1187*

♦ **Concentrating Ability, Urine** *see* Concentration Test, Urine *on page 446*

Concentration Test, Urine

Related Information

Antidiuretic Hormone, Plasma *on page 172*
Calcium, Serum *on page 329*
Osmolality, Serum *on page 978*
Osmolality, Urine *on page 979*
Potassium, Serum or Plasma *on page 1078*
Sodium, Serum or Plasma *on page 1210*
Specific Gravity, Urine *on page 1216*

Synonyms Concentrating Ability, Urine; Fishberg Concentration Test; Urine Concentration Test

Applies to Vasopressin Concentration Test

Test Includes Assessment of renal concentrating ability by measuring specific gravity or osmolality following water deprivation and/or after administration of vasopressin. Since the test procedure may intensify the patient's pathophysiology, this test must be closely monitored to prevent patient injury (hyperosmolar state or dehydration).

Abstract This test is done to evaluate polyuria. A random urine collected without water restriction (usually first morning urine) yielding a urine osmolality ≥850 mOsm/kg (SI: ≥850 mmol/kg) or a specific gravity ≥1.027 virtually excludes a defect in concentrating ability, and thus, may be done as a screening procedure before more elaborate deprivation studies are undertaken. **Test should be supervised by the physician**.

Patient Preparation The evening meal must be high in protein and contain not more than 200 mL of liquid. Patient is to consume no fluids after the evening meal. On awakening in the morning, the patient voids and saves the specimen in container #1. All further urine passed until 1 hour later is included in specimen #2. All further urine then until 2 hours later is collected as specimen #3. If urine specific gravity plateaus but a specific gravity of 1.027 or urine osmolality of 850 mOsm/kg is not achieved, vasopressin can be administered.

The patient must not be taking diuretics or have recently received intravenous radiographic contrast material.

Aftercare Fluid restriction may decrease plasma volume and have an adverse effect on cardiac output in patients with compromised cardiac function. If diabetes insipidus is present, urine output may remain very high despite fluid deprivation. Body weight and blood pressure should be carefully followed throughout the procedure. If body weight is decreased by 5% or orthostatic hypotension occurs, the procedure should be terminated. Patients should also be watched to prevent them from unsupervised drinking.

Specimen Urine
Container Three urine containers
Storage Instructions Refrigeration

Causes of Symptomatic (Polyuric) Deficiencies in Plasma Vasopressin

Decreased Secretion

Destruction of neurohypophysis (neurogenic diabetes insipidus)

  Sporadic

    Idiopathic

    Trauma (surgical, accidental)

    Malignancy

      Primary (craniopharyngioma, dysgerminoma, meningioma, adenoma, glioma, astrocytoma)

      Secondary (metastatic from lung or breast, lymphoma, leukemia, dysplastic pancytopenia)

    Granuloma (sarcoid, histiocytosis, xanthoma disseminatum)

    Infection (viral/bacterial meningitis, encephalitis)

    Vascular (Sheehan syndrome, carotid aneurysm, hematoma, aortocoronary bypass, ischemic brain death)

    Autoimmune disease

    Dysplasia (septo-optic, microcephaly, porencephaly, etc)

    Metabolic (anorexia nervosa)

  Familial (autosomal dominant)

Excessive water intake (primary polydipsia)

  Psychogenic (schizophrenia, ? neurosis)

  Dipsogenic (abnormal thirst)

    Idiopathic

    Trauma

    Granuloma (neurosarcoid, tuberculous meningitis)

    Autoimmune (multiple sclerosis)

    Chemical (lithium)

Increased Metabolism

Gestational

Adapted from Robertson GL and Berl T, "Pathophysiology of Water Metabolism," *The Kidney*, Brenner BM and Rector FC Jr, eds, Philadelphia, PA: WB Saunders Co, 1991, 695.

Causes of Defects in Antidiuretic Action of Vasopressin

Familial nephrogenic diabetes insipidus

  X-linked recessive

Sporadic nephrogenic diabetes insipidus

  Chemical (lithium, demeclocycline, methoxyflurane)

  Metabolic (hypokalemia, hypercalcemia)

  Mechanical (ureteral obstruction)

  Vascular (sickle cell disease or trait)

  Granulomatous (sarcoid)

  Dysplastic (polycystic disease)

  Infectious (pyelonephritis)

  Infiltrative (amyloid)

  Gestational

  Malignant (fibrosarcoma)

Solute diuresis

  Metabolic (glucosuria)

  Iatrogenic (mannitol, diuretics, radiocontrast dyes, saline loading)

  Mechanical (postureteral obstruction)

Adapted from Robertson GL and Berl T, "Pathophysiology of Water Metabolism," *The Kidney*, Brenner BM and Rector FC Jr, eds, Philadelphia, PA: WB Saunders Co, 1991, 699.

Reference Interval Normal: specific gravity of at least one specimen should be >1.026 or >850 mOsm/kg (SI: >850 mmol/kg); severe renal disease: <400 mOsm/kg (SI: <400 mmol/kg). When fluids are withheld (overnight), the ratio of urine to serum osmolality >3.0 is considered normal. Elevated urine osmolality in the face of hyponatremia may indicate the presence of the syndrome of inappropriate secretion of antidiuretic hormone (SIADH) or cerebral salt wasting.

Use Useful in the differential diagnosis of polyuric states: diabetes insipidus, compulsive water drinking, and renal disease

Limitations In polydipsic patients, the urine specific gravity may not rise to 1.026 until after many hours of deprivation. False-positive tests can be avoided by confirming that serum sodium is 141-145 mmol/L before termination of the test. The test may have to be extended or adapted to the patient, and this is best done with a clear understanding of the physiology of vasopressin and of water metabolism. For details see Robertson et al.

Contraindications Glucosuria invalidates a concentration test by virtue of its diuretic effect; hypernatremia or orthostatic hypotension are contraindications.

Methodology Modified Fishberg procedure

Additional Information Glomerular disorders with proteinuria produce an osmotic diuresis resulting in decreased urine osmolality. Even subtle renal interstitial disorders may impair concentrating ability. Hypokalemia and hypercalcemia decrease renal medullary tonicity and inhibit tubular reabsorption of water. Sickle cell disease decreases medullary blood flow and interferes with loop of Henle sodium transport. Achievement of normal maximal urinary concentration requires a normal or near normal glomerular filtration rate. In central diabetes insipidus, administration of vasopressin will raise urine osmolality. In nephrogenic diabetes insipidus, the urine osmolality will not increase with vasopressin or water deprivation.

The urine:serum osmolality ratio is addressed in Osmolality, Serum *on page 978*.

References
Bichet DG, "Nephrogenic Diabetes Insipidus," *Am J Med*, 1998, 105(5):431-2.

Demers L, "Pituitary Function," *Tietz Textbook of Clinical Chemistry*, 3rd ed, Burtis CA and Ashwood ER, eds, Philadelphia, PA: WB Saunders Co, 1999, 1488-92.

Haycock GB, "Old and New Test of Renal Function," *J Clin Pathol*, 1981, 34(11):1276-81.

Price JD and Lauener RW, "Serum and Urine Osmolalities in the Differential Diagnosis of Polyuric States," *J Clin Endocrinol Metab*, 1966, 26(2):143-8.

Robertson GL and Berl T, "Pathophysiology of Water Metabolism," *Kidney*, Brenner BM and Rector FC Jr, eds, Philadelphia, PA: WB Saunders Co, 1991, 677-736.

- ♦ **Confidence Interval** *see page 11*
- ♦ **Congenital Disorders of Glycosylation** *see* Carbohydrate-Deficient Transferrin, Serum *on page 338*
- ♦ **Congenital Myotonic Dystrophy** *see* Myotonic Dystrophy DNA Test *on page 942*
- ♦ **Conjugated Hyperbilirubinemia** *see* Bilirubin, Direct, Serum *on page 262*
- ♦ **Conjunctival Smear Cytology** *see* Ocular Cytology *on page 973*
- ♦ **Conjunctivitis, Ligneous** *see* Plasminogen *on page 1042*
- ♦ **Connecting Peptide Insulin** *see* C-Peptide, Serum *on page 465*
- ♦ **Consumptive Coagulopathy Screen** *see* Disseminated Intravascular Coagulation Screen *on page 517*
- ♦ **Contac®** *see* Methamphetamine, Qualitative, Urine *on page 902*
- ♦ **Coombs Test, Direct** *see* Antiglobulin Test, Direct *on page 176*
- ♦ **Coombs Test, Direct** *see* Antiglobulin Test, Indirect *on page 177*
- ♦ **Copper, Hepatic** *see* Liver Biopsy *on page 864*
- ♦ **Copper Reduction Tablet Test** *see* Reducing Substances, Urine *on page 1148*

Copper, Serum
Related Information
Ceruloplasmin, Serum or Plasma *on page 375*
Copper, Urine *on page 452*
Heavy Metal Screen, Blood *on page 668*
Heavy Metal Screen, Urine *on page 669*

Iron and Total Iron Binding Capacity/Transferrin, Serum *on page 807*
Liver Biopsy *on page 864*
Liver Disease: Laboratory Assessment, Overview *on page 869*
Zinc, Serum or Plasma *on page 1340*

Synonyms Cu, Serum

Applies to Beta-Monooxygenase, Dopamine; Metallothionein; Transcuprein; Zinc Administration

Abstract An essential trace element, copper (Cu) is a cofactor of several key enzyme systems and is required for hemoglobin synthesis. It circulates bound to ceruloplasmin and is excreted into the bile (see Ceruloplasmin, Serum or Plasma *on page 375*).

Specimen Serum, cerebrospinal fluid, tissue

Container Royal blue top, trace metal-free, tube which contains no anticoagulant. See the Trace Elements Introduction *on page 77*.

Collection Use powder-free gloves. Draw tube prior to any other blood samples. After centrifugation, pour serum into a metal-free vial for transport to reference laboratory. CSF can be transferred directly to a royal blue top tube.

Causes for Rejection Contamination by Cu, contact with dust, or the use of ordinary collection needles, tubes, or stoppers

Reference Interval Serum: Approximately 0.7-1.5 µg/mL (SI: 11-24 µmol/L). Mean levels are slightly higher in women and children. There is diurnal variation with peak levels in the morning. Cerebrospinal fluid: 6-35 ng/mL (SI: 94-551 nmol/L). Levels in CSF are elevated up to threefold in the neurotoxicity of Wilson disease. Liver tissue: 9-45 ng/g dry weight.

Use Serum copper is used, along with serum ceruloplasmin and urine Cu to screen for Wilson disease and to monitor adequate supplementation of parenteral or enteral nutrition, especially when Cu deficiency may be suspected because of ongoing gastrointestinal losses (see table on following page).

It is also used in the differential diagnoses of primary biliary cirrhosis and primary sclerosing cholangitis.

It is used to verify suspected Cu deficiency in premature infants when they are acutely ill and may be unable to assimilate Cu in their prescribed diets. It is used in diagnosis of Indian childhood cirrhosis (ICC) and to follow such children following penicillamine chelation therapy. It is used to verify acute Cu intoxication. Serum Cu is low in Menkes syndrome and occipital horn syndrome (OHS). Serum and urine Cu are used to follow Cu status in acrodermatitis enteropathica, in which high-dose oral zinc therapy puts the patient at risk for symptomatic Cu deficiency. Serum ceruloplasmin and serum Cu tend to parallel each other in normal individuals and therefore do not provide independent information. Copper deficiency is an important etiology of iron resistant anemia in patients receiving total parenteral nutrition.

Limitations Since ceruloplasmin is an acute-phase reactant protein which binds a large portion of serum Cu, both serum Cu and ceruloplasmin increase with inflammatory conditions and estrogen exposure. Serum Cu is therefore elevated in pregnancy, in patients on contraceptive drugs, in rheumatoid arthritis, and in a number of other inflammatory pathologic entities. Serum Cu may be increased with carbamazepine, phenobarbital, phenytoin, and valproic acid. It may be low in the presence of low serum proteins as in nephrosis, malabsorption, and malnutrition without necessarily reflecting true liver Cu stores and may be reduced under the influence of ACTH or glucocorticoid therapy. Although serum Cu levels are usually ordered to work up possible cases of Wilson disease, Menkes syndrome, and Indian childhood cirrhosis, serum Cu alone is of only limited value.

Methodology Atomic absorption (AA), inductively-coupled plasma atomic emission spectrometry

Additional Information Our understanding of Cu as an essential trace element in human nutrition and a factor in several diseases has advanced substantially over the past few years.

Copper is a component of many metalloenzymes and proteins including ceruloplasmin, which also seems to serve as a major transport protein for copper (see Ceruloplasmin, Serum or Plasma *on page 375*).

Copper is absorbed in the stomach and duodenum by a process regulated by metallothionein (MT). MT synthesis, induced by Cu, binds Cu within intestinal (Continued)

Disorders of Copper Metabolism

	Deficiency, Nutritional	Menkes Syndrome	Acute Copper Toxicity	ICC and Chronic Copper Toxicity	Wilson Disease	Smoking, Inflammatory Conditions, Pregnancy, Estrogens
Serum copper	↓	↓	↑, ↑↑	↑	N or →	↑, ↑↑
Serum ceruloplasmin	↓	↓	N (early)	↑	Usually ↓; may be N in individuals <20 y	↑, ↑↑
Urine copper	↓	↑	↑	↑	↑, ↑↑	N
CSF copper	↓				N or ↑	N
Liver copper	↓	↓	N (early)	↑, ↑↑	↑↑	N

N = normal, ↑ = increase, ↑↑ = large increase, ↓ = decrease, ICC = Indian childhood cirrhosis.

mucosal cells, effectively trapping it within mucosal cells. Such cells are sloughed into the intestinal lumen and lost into the stool unabsorbed. Thus, Cu absorption is partially self-limiting.

Both copper and zinc induce tissue levels of MT. Zinc is less avidly bound by MT than Cu, but zinc is the better inducer of MT. Small increases of zinc in the diet markedly inhibit Cu absorption by stimulating synthesis of MT. Zinc also has direct action on Cu uptake by blocking transport into intestinal mucosal cells independent of changes in intracellular MT. Vitamins containing large doses of zinc cause Cu deficiency. Cadmium and iron may also inhibit Cu absorption through similar mechanisms. Molybdenum decreases absorption by forming insoluble copper-molybdenum-sulfur compounds. This interaction has been used in the detoxification of certain patients with Wilson disease by giving oral molybdenum-sulfur compounds.

After absorption, Cu appears in the blood, loosely bound to albumin and also as copper-histidine. The liver and other organs take up Cu via membrane-bound ligands. In the liver, Cu is used to synthesize ceruloplasmin, which appears within serum a few hours later. Copper uptake by peripheral tissues is dependent upon the fraction of Cu in the blood that is incorporated into ceruloplasmin rather than that fraction bound to histidine or albumin. About 65% of Cu in peripheral blood exists in the form of ceruloplasmin.

Copper is largely eliminated from the body by secretion into bile. Renal tubular reabsorption of filtered Cu is efficient; thus, normally only a small fraction of Cu is lost in urine. Overflow losses in the urine are proportional to Cu stores, except when abnormal urine losses occur as in burns, intravenous administration of amino acids, Menkes syndrome, and with chelating drugs.

Oral zinc administration inhibits Cu absorption and may cause Cu deficiency states. Oral zinc may be used to treat Cu overloading, including Wilson disease, or may inadvertently cause Cu deficiency by stimulating excess MT synthesis.

In both Wilson disease and Indian childhood cirrhosis (ICC), there are inherited and environmental factors that lead to toxic accumulation of hepatic Cu. ICC had been considered an illness of toxic exposure of the child to milk boiled in brass vessels, when vessels were not tinned sufficiently and frequently enough to prevent exposure to brass. Indeed, the incidence of ICC drops markedly as a community discards brass utensils. There is now however, good evidence that, like Wilson disease, ICC is primarily a genetic disease. In ICC, the basal production and metal-induced synthesis of MT is defective although glucocorticoid-induced MT synthesis is normal. Although Cu accumulates in the liver, pathology of the liver in ICC is distinct from that of Wilson disease. ICC has been reported in both Europe and the United States, where excess Cu intake could not be proven.

Copper deficiency from any cause markedly reduces ceruloplasmin synthesis rates and blood ceruloplasmin levels, leading to a microcytic or normocytic anemia secondary to blocks in iron metabolism. Copper deficiency anemia fails to respond to iron, but brisk reticulocytosis follows Cu administration. Copper deficiency can cause a scurvy-like bone disease (probably due to decreased lysyl oxidase), depigmentation (probably due to decreased tyrosinase), growth failure, and neutropenia.

Menkes disease is a severe X-linked Cu deficiency syndrome usually presenting by the age of 3 months. The gene product is also a P-type Cu transporting ATPase as in Wilson disease. When defective, it results in Cu accumulation in intestinal mucosa and kidney with failure to deliver adequate Cu to liver and peripheral tissues, resulting in functional Cu deficiency. There is also increased urine Cu loss due to failure of renal tubular reabsorption. Copper deficiency affects bone formation, pigmentation, CNS development, growth, and arterial connective tissues. Several reports of intravenous copper-histidine (plus intermittent penicillamine) to prevent Cu overload are encouraging, but difficult to interpret. A recently described diagnostic test reflects deficiency of the Cu enzyme dopamine β-monooxygenase, but additional strategies still are needed.

Occipital horn syndrome (OHS) (Ehlers-Danlos syndrome type IX) has been confirmed as an inherited disorder of Cu metabolism. It may be allelic to
(Continued)

Copper, Serum *(Continued)*

Menkes disease. It is usually recognized by its characteristic physical features, but can be confirmed by serum Cu, serum ceruloplasmin, and fibroblast lysyl oxidase activity, all of which are low. Intestinal absorption of Cu is poor.

There are at least two situations in which ceruloplasmin may not parallel total serum Cu. In acute Cu toxicity, in which there may not have been time for increased ceruloplasmin synthesis, free or loosely bound Cu is elevated, total serum Cu may be elevated, and ceruloplasmin may still be normal. In Wilson disease, with chronic low levels of ceruloplasmin, more Cu in serum may be loosely bound and (total) serum Cu may be normal rather than low. In this situation, it is especially important to measure both total serum Cu and ceruloplasmin. High **urine copper** is also a feature of Wilson disease secondary to decreased excretion into bile. Because no combination of noninvasive tests has proven 100% sensitive and specific for Wilson disease, molecular genetics will likely be more frequently used in diagnosis within families.

The demand for sensitive noninvasive tests for Wilson disease, especially for children in families in which the disease is known to occur, has stimulated search for newer indices of Cu metabolism. Urine Cu after penicillamine load has been proposed.

Elevations in liver tissue Cu found in Wilson disease may occur also in other types of liver disease, especially primary biliary cirrhosis. Liver tissue Cu levels remain the gold standard for diagnosis of Wilson disease and Indian childhood cirrhosis (ICC), although it is not 100% specific. Increased liver tissue Cu levels are diagnostic in cases of Wilson disease or ICC previously diagnosed in a sibling because both diseases are inherited disorders, and other differential considerations such as primary biliary cirrhosis are rare. Wilson disease can be distinguished from chronic Cu intoxication in normal individuals by serum ceruloplasmin and the finding that values for free Cu in serum and urine of normal individuals tend to normalize in an environment without Cu in tap water. Liver Cu content is used to confirm Menkes syndrome and may be useful in evaluating liver disease of uncertain etiology. Liver Cu rises with time in biliary cirrhosis, but does not confirm the diagnosis. The liver biopsy diagnosis of Wilson disease has been reviewed. See Liver Biopsy *on page 864.*

The combination of increased hepatic Cu and low ceruloplasmin occurs only in Wilson disease and normal infants, who are born at term with increased Cu stores and develop normal ceruloplasmin levels by 3-6 months of age. Since Cu stores rise during the last trimester of pregnancy, premature infants may not have elevated liver Cu stores.

References

Barceloux DG, "Copper," *J Toxicol Clin Toxicol,* 1999, 37(2):217-30.

Eife R, Weiss M, Barros V, et al, "Chronic Poisoning by Copper in Tap Water: I. Copper Intoxications With Predominantly Gastrointestinal Symptoms," *Eur J Med Res,* 1999, 4(6):219-23.

Ludwig J, Moyer TP, and Rakela J, "The Liver Biopsy Diagnosis of Wilson's Disease: Methods in Pathology," *Am J Clin Pathol,* 1994, 102(4):443-6.

Pfeil SA and Lynn DJ, "Wilson's Disease: Copper Unfettered," *J Clin Gastroenterol,* 1999, 29(1):22-31.

Spiegel JE and Willenbucher RF, "Rapid Development of Severe Copper Deficiency in a Patient With Crohn's Disease Receiving Parenteral Nutrition," *JPEN J Parenter Enteral Nutr,* 1999, 23(3):169-72.

Copper, Urine

Related Information

Ceruloplasmin, Serum or Plasma *on page 375*
Copper, Serum *on page 448*
Heavy Metal Screen, Blood *on page 668*
Heavy Metal Screen, Urine *on page 669*
Liver Biopsy *on page 864*
Liver Disease: Laboratory Assessment, Overview *on page 869*
Urine Collection, 24-Hour *on page 1295*

Synonyms Cu, Urine

Abstract Copper (Cu) is an essential trace element in human nutrition and a component of many metalloenzymes. Although the biliary system is the major pathway of Cu excretion, some Cu is excreted in urine. Urine Cu may be used as an aid to detect Cu deficiency, Wilson disease, Menkes disease, Indian childhood cirrhosis (ICC), and chronic or acute Cu toxicity.

Patient Preparation If a bedpan or urinal is necessary for collection, it must be made of plastic. Stool contamination must be avoided.

Specimen 24-hour urine

Container Acid-washed (metal-free) urine collection containers must be used; no preservatives.

Collection Collect in acid-washed plastic container, preferably polyethylene. Acidify to pH 2 with hydrochloric or nitric acid. Avoid contamination by dust or dirt.

Causes for Rejection Specimen allowed to contact metal or feces

Reference Interval 15-60 µg/24 hours (SI: 0.22-0.9 µmol/day)

Use Increased urinary Cu is found in Wilson disease, Menkes syndrome, Indian childhood cirrhosis (ICC), and in chronic and acute Cu toxicity states. The test is also used to follow the effectiveness of chelation therapy for Wilson disease, check Cu balance in patients with Wilson disease on oral zinc therapy, or check patients receiving parenteral Cu as part of parenteral nutrition. It may also be used in the differential diagnosis of primary biliary cirrhosis and primary sclerosing cholangitis.

Limitations Increased urinary Cu may occur in ICC or with chronic active hepatitis; Wilson disease and chronic hepatitis may also resemble one another. Thus, in addition to urinary Cu excretion, other tests such as ceruloplasmin, serum Cu, and sometimes liver biopsy are needed. Copper excretion by the kidneys is abnormally increased with high-dose intravenous histidine or mixed amino acids, as in total parenteral nutrition. Captopril and other medications may chelate Cu and increase urinary excretion, which is usually only a tiny fraction of total daily balance. Urinary Cu may be increased with cisplatin, dimercaprol, and with penicillamine or trientine in patients with Wilson disease. For comparison of urine Cu in Wilson disease sibships, see figure. Over time, biliary cirrhosis leads to Cu accumulation, and since biliary excretion of Cu is blocked, urinary excretion rises, potentially leading to diagnostic confusion with Wilson disease. Serum ceruloplasmin and liver biopsy aid in differential diagnosis between Wilson disease and biliary cirrhosis.

Methodology Atomic absorption (AA), inductively-coupled plasma atomic emission spectrometry

Additional Information Chronic Cu poisoning has been described as a hypercupric state characterized by increased free Cu in the serum and urine of individuals. Elevated total serum Cu, though not specific, can suggest this diagnosis. Chronic Cu poisoning can be distinguished from Wilson disease by normal levels of ceruloplasmin and the fact that free serum and urinary Cu tend to normalize in an environment free of Cu in tap water (such as a hospital). Cases of chronic Cu poisoning masquerading as gastrointestinal illness (nausea, vomiting, and diarrhea) have been described in Germany. Changes in dietary Cu were reported to have no effects on serum Cu, ceruloplasmin, osteocalcin (marker of bone formation) nor urinary creatinine in young volunteer men. However, markers for bone resorption (urinary Pyr/Cr and Dpyr/Cr) increased when dietary Cu was decreased for several weeks and vice versa. Further information is provided in Copper, Serum *on page 448* and Ceruloplasmin, Serum or Plasma *on page 375*.

References
Eife R, Weiss M, Müller Höcker M, et al, "Chronic Poisoning by Copper in Tap Water: II. Copper Intoxications With Predominantly Systemic Symptoms," *Eur J Med Res*, 1999, 4(6)224-8.

Eife R, Weiss M, Barros V, et al, "Chronic Poisoning by Copper in Tap Water: I. Copper Intoxications With Predominantly Gastrointestinal Symptoms," *Eur J Med Res*, 1999, 4(6):219-23.

Zucker SD and Flieder A, "A 23-Year-Old Man With Fulminant Hepatorenal Failure of Uncertain Cause," Case Records of the Massachusetts General Hospital, Case 1-1997, Scully RE, Mark EJ, McNeely WF, et al, eds, *N Engl J Med*, 1997, 336(2):118-25.

♦ **Coproporphyrins** *see* Porphyrins, Quantitative, Urine *on page 1074*
♦ **Cordarone®** *see* Amiodarone, Serum *on page 148*

Cord Blood Antibody Screen

Related Information

Antibody Detection/Identification, Red Cell *on page 165*
Antiglobulin Test, Direct *on page 176*
Bilirubin, Amniotic Fluid, Delta A450 *on page 261*
Bilirubin, Neonatal, Serum *on page 263*
Hemolytic Disease of the Newborn, Antibody Identification *on page 690*
(Continued)

Cord Blood Antibody Screen *(Continued)*

Kleihauer-Betke *on page 822*
Newborn Crossmatch and Transfusion *on page 952*
Prenatal Screen, Immunohematology *on page 1086*
Rh Genotype *on page 1162*
Rh₀(D) Immune Globulin (Human) *on page 1164*
Rosette Test for Fetomaternal Hemorrhage *on page 1172*

Applies to Type and Screen, Coombs, Cord Blood

Test Includes ABO group, Rh type, direct antiglobulin test (DAT); evaluation of hemolytic disease of the newborn (HDN) in cases of positive DAT

Abstract Collection of a specimen of blood from the umbilical cord is standard practice at delivery. The test panel comprising the cord blood screen is designed to indicate the presence of HDN, to guide decisions regarding treatment options, and to assess the need for administration of postpartum Rh immune globulin. The Rh type of the mother is relevant.

Specimen Cord blood, collected if possible by cannulation of an umbilical vessel

Container One lavender top (EDTA) tube or one red top tube

Causes for Rejection Gross hemolysis, sample collected in serum separator tube, improper labeling

Use The cord blood sample is tested for ABO and Rh type and DAT. If the DAT is positive, both cord blood and maternal samples should be investigated for serological evidence of HDN. Antepartum RhIG can cause a weak-positive DAT; further testing is indicated. If mother and baby are ABO incompatible, an eluate will demonstrate anti-A, anti-B, or both from neonatal red cells.

No testing may be required except to establish the candidacy of the mother to receive RhIG or unless a question of HDN arises. Thus, storing samples is usually necessary, especially if the mother is known to be Rh(D) negative. Infants of mothers who are Rh negative should be typed for Rh₀(D), with a test for weak D, using the antiglobulin phase.

Limitations Wharton's jelly may interfere with the determination of ABO blood group as well as with the DAT. (Repeat testing from capillary blood may be done.) ABO red cell grouping of neonates cannot be confirmed by serum (reverse) grouping. It may prove difficult to establish the Rh type of cord red cells heavily sensitized with IgG maternal antibodies. The presence of large numbers of maternal red cells in a cord blood sample can confuse the interpretation of blood grouping results. Similarly, serological results can be complicated in newborns who received intrauterine blood transfusion. The administration of antenatal Rh immune globulin may complicate interpretation of cord blood DAT results. Studies on paternal blood may be indicated when no antibody can be detected in the maternal serum or eluate from infant's DAT-positive red cells. The antibody in such a case may be directed to a paternal "private" antigen of low incidence (many low-incidence blood group antigens were discovered in cases of HDN).

Additional Information The fetomaternal relationship presents special immunohematological problems for the Blood Bank Laboratory. The results of initial cord blood screening tests will indicate the need for additional studies to determine the likelihood of HDN or the need for Rh immunoprophylaxis. A positive DAT is an important indicator of HDN. A positive result will evolve into antibody identification studies of maternal serum or infant red cell eluate. Antibody specificity is an important guide to disease severity. Most severe HDN is caused by anti-D alone or in combination with anti-C or anti-E. The next grade of severity is caused by antibodies to other Rh antigens (notably anti-c) or by antibodies to antigens in other blood group systems (eg, anti-K). IgG antibodies to antigens of the ABO system cause the lowest grade of HDN severity. ABO HDN can occur in any pregnancy, even the first. The DAT is often negative and the infant is rarely symptomatic at birth.

Administration of postpartum Rh immune globulin depends on maternal Rh₀(D) status, presence or absence of maternal alloimmunization, and the results of infant Rh(D) type. See Rh₀(D) Immune Globulin (Human) *on page 1164*. Rh negative nonimmunized women who are candidates for RhIG, who deliver an Rh-positive baby, should be screened for fetomaternal hemorrhage to ascertain whether a single dose of RhIG is adequate.

See Kleihauer-Betke *on page 822* and Rosette Test for Fetomaternal Hemorrhage *on page 1172*. Flow cytometry and enzyme-linked antiglobulin tests are available.

References

Hartwell EA, "Use of Rh Immune Globulin. ASCP Practice Parameter," *Am J Clin Pathol*, 1998, 110(3):281-92.

Snyder EL and Shoos-Lipton K, "Prevention of Hemolytic Disease of the Newborn Due to Anti-D," *American Association of Blood Banks Bulletin #98-2*, February 16, 1998.

♦ **Cordilox**® *see* Verapamil, Serum or Plasma *on page 1306*

♦ **Core Window Stage of Hepatitis** *see* Hepatitis B Antigen Detection *on page 699*

♦ **Core Window Stage of Hepatitis** *see* Hepatitis B Serology *on page 702*

♦ **Corneal Cytology** *see* Ocular Cytology *on page 973*

♦ **Corramedan**® *see* Digitoxin, Serum *on page 512*

♦ **Correlation Statistics** *see page 11*

♦ **Corticotropin** *see* Adrenocorticotropic Hormone, Plasma *on page 114*

♦ **Corticotropin-Releasing Hormone** *see* Insulin Tolerance Test *on page 804*

Corticotropin-Releasing Hormone Stimulation Test

Related Information

Adrenal Cortex: Laboratory Assessment Overview *on page 110*
Adrenocorticotropic Hormone, Plasma *on page 114*
Corticotropin Stimulation Test (Rapid) *on page 456*
Cortisol, Free, Urine *on page 459*
Cortisol, Serum or Plasma *on page 460*
Insulin Tolerance Test *on page 804*
Metyrapone Stimulation Test, Serum *on page 910*
Urinary Cortisol/Creatinine Increment *on page 1292*

Synonyms CRH Stimulation Test

Applies to Dexamethasone-CRH Protocol

Abstract Corticotropin-releasing hormone (CRH) is an ACTH secretagogue which is used in the evaluation of patients with suspected Cushing syndrome (CS).

Patient Preparation

Systemic protocol: The patient fasts for 4 hours and an intravenous line is inserted. After obtaining a baseline blood specimen, ovine CRH, 1 µg/kg body weight, is given intravenously as a bolus. CRH injection may be given at 9 AM or 8 PM.

Inferior petrosal protocol: An intravenous line is placed in a peripheral vein. Catheters are inserted through the femoral veins into the right and left inferior petrosal sinuses. Precise localization is critical; the technique requires a radiologist experienced in this technique.

Dexamethasone-CRH protocol: Investigators at the NIH developed a variation of the CRH stimulation test in which the patient receives pretreatment with 0.5 mg dexamethasone given orally every 6 hours for a total of 8 doses. CRH, 0.1 µg/kg body weight, is given intravenously 2 hours after the last dose of dexamethasone.

Specimen Serum

Container Red top tube

Sampling Time

Systemic protocol: A baseline sample is always obtained 15 minutes before, and then just before, the CRH injection. Subsequent specimens are obtained according to one of these schedules:

- 5, 15, 30, 60, 120, and 180 minutes after CRH
- 5, 10, 15, 30, 45, 60, 90, and 120 minutes after CRH

Samples are assayed for cortisol and/or ACTH, depending on clinical indication.

Inferior petrosal protocol: Specimens are obtained from each inferior petrosal sinus and the peripheral site at the following times: before CRH, and then at 2 and 5 minutes after CRH. Some suggest an additional specimen at 10 minutes after CRH. Specimens may be assayed for ACTH and cortisol, but the diagnostic criteria are based primarily on the ACTH values.

(Continued)

Corticotropin-Releasing Hormone Stimulation Test
(Continued)

Dexamethasone-CRH protocol: In the original description of this test, specimens for serum cortisol and ACTH were obtained at 15, 10, 5, and 0 minutes before CRH, and then at 5, 15, 30, 45, and 60 minutes after CRH. Since experience with this protocol is limited, the original description should be consulted if its use is contemplated.

Reference Interval

Systemic protocol: The following values are associated with a pituitary source of ACTH in a patient with CS.

ACTH:
- 9:30 AM: 80 ±7 pg/mL (SI:17.6 ±1.5 pmol/L)
- 8:30 PM: 29 ±2.6 pg/mL (SI: 6.38 ±0.57 pmol/L)

Cortisol:
- 10 AM: 13.1 ±1.0 µg/dL (SI: 358.8 ±27.6 nmol/L)
- 9 PM: 17 ±0.7 µg/dL (SI: 470.2 ±19.4 nmol/L)

Inferior petrosal protocol: Following CRH, a central to peripheral serum ACTH ratio >3.0 is strong evidence for a pituitary source of ACTH in an adult patient with CS. In children and adolescents, the corresponding cutoff is >2.5.

Dexamethasone-CRH protocol: The following values are from an NIH study of 58 patients, 39 of whom proved to have CS and 19 of whom had pseudo-CS.
- Serum cortisol before CRH: >1.38 µg/dL (SI: >38 nmol/L); sensitivity 90%, specificity 100%
- Serum cortisol 15 minutes after CRH: >1.38 µg/dL (SI: >38 nmol/L); sensitivity/specificity 100%
- Serum ACTH 30 minutes after CRH: >16 pg/mL (SI: >3.5 pmol/L); sensitivity 74%, specificity 100%

Use The CRH test is used to differentiate Cushing disease (ACTH-dependent) vs other causes of Cushing syndrome (eg, adrenal adenoma, ectopic ACTH - see Adrenal Cortex: Laboratory Assessment Overview *on page 110*). The systemic protocol is somewhat less sensitive than the inferior petrosal protocol (see Limitations). The dexamethasone-CRH protocol was developed to differentiate mild cases of Cushing disease from pseudo-Cushing states.

Limitations As many as 8% of patients with Cushing disease will not have the expected increase in ACTH and/or cortisol. Most of these nonresponders will, however, have a central to peripheral ratio ≥3.0 in the inferior petrosal protocol.

References

Demers LM and Whitley RJ, "Function of the Adrenal Cortex," *Tietz Textbook of Clinical Chemistry*, 3rd ed, Burtis CA and Ashwood ER, eds, Philadelphia, PA: WB Saunders Co, 1999, 1544.

Magiakou MA, Mastorakos GM, Oldfield EH, et al, "Cushing's Syndrome in Children and Adolescents," *N Engl J Med*, 1994, 331:629-36.

Oldfield EH, Doppman JL, Nieman LK, et al, "Petrosal Sinus Sampling With and Without Corticotropin-Releasing Hormone for the Differential Diagnosis of Cushing's Syndrome," *N Engl J Med*, 1991, 325:897-905 (published erratum *N Engl J Med*, 1992, 326:1172).

Orth DN and Kovacs WJ, "The Adrenal Cortex," *Williams Textbook of Endocrinology*, 9th ed, Wilson JD, Foster DW, Kronenberg HM, et al, eds, Philadelphia, PA: WB Saunders Co, 1998, 621.

Yanovski JA, Cutler GB Jr, Chrousos GP, et al, "Corticotropin-Releasing Hormone Stimulation Following Low-Dose Dexamethasone Administration," *JAMA*, 1993, 269:2232-38.

Corticotropin Stimulation Test (Rapid)

Related Information

Adrenal Cortex: Laboratory Assessment Overview *on page 110*
Adrenocorticotropic Hormone, Plasma *on page 114*
Corticotropin-Releasing Hormone Stimulation Test *on page 455*
Cortisol, Free, Urine *on page 459*
Cortisol, Serum or Plasma *on page 460*
Insulin Tolerance Test *on page 804*
Metyrapone Stimulation Test, Serum *on page 910*
Urinary Cortisol/Creatinine Increment *on page 1292*

Synonyms Cortrosyn® Test; Cosyntropin Stimulation Test; Rapid (or Short) ACTH Test; Synthetic α 1-24-ACTH Stimulation Test

Replaces ACTH Infusion Test; Bovine ACTH

Abstract Corticotropin, in the context of adrenal testing, refers to a synthetic polypeptide, consisting of the biologically active 1-24 amino acid sequence of human ACTH. This polypeptide stimulates the secretion of cortisol, and other corticosteroids, from the adrenal cortex, and, as a testing agent, is used to evaluate possible adrenocortical insufficiency (AI), and late-onset congenital adrenal hyperplasia (CAH) due to 21-hydroxylase deficiency. It has a very short half-life. In a hospital formulary, this product is often listed under the generic term, cosyntropin; in Europe, the equivalent term is tetracosactrin (see Adrenal Cortex: Laboratory Assessment Overview *on page 110*).

The rapid corticotropin (cosyntropin) stimulation test is performed at two dose levels. The older, more conventional high-dose protocol uses a pharmacologic 250 µg dose (I.V. or I.M.), with blood samples for cortisol obtained just prior to the injection (baseline), and at 30 and 60 minutes postinjection.

The newer low-dose protocol uses a physiologic 1 µg (or 0.5 µg/m^2 body surface area). Blood samples for cortisol may be obtained at 30 minutes, or at 30 and 60 minutes postinjection. Some authors include a baseline specimen, while others do not. The low-dose test is more sensitive than the high-dose test (see Use and Limitations).

Specimen Serum

Container Red top tube

Sampling Time The test is usually performed in the morning, before 10 AM.
- High-dose protocol: Baseline and 60 minutes postinjection (in certain special situations, specimens may also be obtained at 20 and 30 minutes; consult references for guidelines).
- Low-dose protocol: The 30-minutes postinjection value is critical. Some authors also include a baseline and a 60-minute value.

Storage Instructions Separate serum and freeze.

Reference Interval

Adrenal insufficiency: Both high-dose and low-dose protocols: Normal adrenal responsiveness is suggested by either:
- a serum cortisol value of 18-20 µg/dL (SI: 500-550 nmol/L) at 30 or 60 minutes, or
- a serum cortisol increment over baseline of 7 µg/dL (SI: 193 nmol/L) at 30 or 60 minutes

However, when using the high-dose protocol, a response that meets one of these criteria of normalcy does not necessarily imply that the patient will have an adequate response to a prolonged stress (such as that associated with surgical procedures, infection, neoplastic disease or trauma).

Congenital adrenal hyperplasia (21-hydroxylase deficiency):
- High-dose protocol: The diagnosis of CAH is confirmed if the 17-hydroxy-progesterone is >1500 ng/dL (SI: >45.0 nmol/L) at 30 or 60 minutes, or if an increase over baseline >1333 ng/dL (SI: >40.0 nmol/L) is seen at 30 or 60 minutes.

Prognosis in septic shock: See Use.

Use Adrenal insufficiency is a potentially life-threatening, but subtle, disorder. The integration of clinical and laboratory information by an experienced clinician is essential in recognizing the possibility of, and diagnosing, adrenal insufficiency (AI). Total reliance on either clinical or laboratory information can have disastrous consequences for the patient. The diagnosis of AI is ruled out if, in an unstressed (not acutely ill) patient, a random serum cortisol result is >20 µg/dL (SI: 550 nmol/L). The diagnosis of AI is confirmed if the basal morning serum cortisol is <3.1 µg/dL (SI: 86 nmol/L). In between these two extremes, and in an acutely ill (stressed) patient, the corticotropin stimulation test is recommended.

Primary adrenal insufficiency (PAI): Included in this category are patients with chronic PAI (Addison disease) due to destruction of the adrenal, or metabolic adrenal dysfunction (adrenomyeloneuropathy). Classical findings are low serum cortisol, elevated serum ACTH, and hyperpigmentation of the skin.

PAI may also occur as an acute process with an abrupt onset. Causes include acute hemorrhage or necrosis of the adrenal (eg, sepsis, anticoagulant therapy, adrenal vein thrombosis, antiphospholipid antibody syndrome). These patients may have catecholamine-resistant hypotension and hyponatremia. Immediate diagnosis and treatment are essential.

(Continued)

Corticotropin Stimulation Test (Rapid) *(Continued)*

Secondary adrenal insufficiency: This category includes patients with pituitary or hypothalamic dysfunction. In the former category are patients who have received long-term corticosteroid medication (including inhaled corticosteroids). Findings may be similar to primary AI, except that serum ACTH levels are not elevated and hyperpigmentation of the skin is not observed. An abnormal result in the high-dose protocol can be interpreted as definite evidence of AI and signals the need for corticosteroid treatment. However, some patients with secondary or tertiary AI (confirmed by metyrapone and insulin-tolerance testing) have a normal response to the high-dose protocol. The recognition of these patients led to the development of the low-dose protocol. **The low-dose protocol is a more sensitive test, and therefore is preferred in the setting of secondary AI.**

When the result is borderline with the low-dose protocol (eg, 16 µg/dL [SI: 441.6 nmol/L] at 1-hour postinjection), a Metyrapone Stimulation Test, Serum *on page 910* or Insulin Tolerance Test *on page 804* may be useful; caution is warranted, however, since these tests may precipitate acute adrenal crisis in patients with severe AI.

Prognosis in septic shock: A major determinant of survival in septic shock is the functional integrity of the hypothalamic-pituitary-adrenal (HPA) axis. A French study of 189 ICU patients, investigated the Corticotropin Stimulation Test (high-dose protocol) and found that high basal cortisol and small incremental responses to corticotropin indicated a poor prognosis for survival. Further analysis of their data revealed three survival patterns based on the test results. See table.

Corticotropin Stimulation Test in Septic Shock

Basal Cortisol	Delta Max Cortisol	Likelihood of Survival	28-Day Mortality
≤34 µg/dL (SI: <938.4 nmol/L)	>9 µg/dL (SI: >248.4 nmol/L)	High	26%
≤34 µg/dL (SI: <938.4 nmol/L)	<9 µg/dL (SI: <248.4 nmol/L)	Intermediate	67%
≥34 µg/dL (SI: >938.4 nmol/L)	>9 µg/dL (SI: >248.4 nmol/L)	Intermediate	67%
≥34 µg/dL (SI: >938.4 nmol/L)	<9 µg/dL (SI: <248.4 nmol/L)	Low	82%

Another group, investigating a much smaller population of 22 patients and using a different data analysis approach, did not find this test helpful.

Congenital adrenal hyperplasia, 21-hydroxylase deficiency (CAH): CAH is usually diagnosed in infancy or early childhood on the basis of physical findings accompanied by marked elevations of corticosteroid hormones proximal to the enzyme defect. A few patients, however, have an attenuated form of late-onset (nonclassical) CAH, and in these individuals the rapid corticotropin test (high-dose protocol) is useful. Testing in this situation begins with a basal morning 17-hydroxyprogesterone (17-Pr); if the result is <230 ng/dL (SI: <6.9 nmol/L), the diagnosis of CAH is excluded. If 17-Pr is >1500 ng/dL (SI: 45.0 nmol/L), the diagnosis is confirmed. The rapid corticotropin test, **high-dose protocol**, is used for patients whose basal 17aPr results fall between 230 and 1500 ng/dL.

Limitations As indicated above (see Use), the **high-dose** protocol can produce a misleadingly normal result, especially in patients with subtle secondary or tertiary AI. A normal result in the **high-dose** protocol does not imply that the cortisol output will be sufficient during a prolonged stress (eg, surgical procedures, infections, trauma, or neoplastic disease).

Methodology Immunoassays (multiple labels)

Additional Information Estrogen medication raises serum cortisol.

References

Addison GM, *Biochemical Basis of Pediatric Disease*, Soldin SJ, Rifai N, and Hicks JMB, eds, Washington, DC: AACC Press, American Association of Clinical Chemistry, 1992, 228-9.

Amatruda TT, Hollingsworth DR, D'Esopo ND, et al, "A Study of the Mechanism of the Steroid Withdrawal Syndrome: Evidence for Integrity of the Hypothalamic-Pituitary-Adrenal System," *J Clin Endocrinol Metab*, 1960, 20:339-54.

Annane D, Sebille V, Troche G, et al, "A 3-Level Prognostic Classification in Septic Shock Based on Cortisol Levels and Cortisol Response to Corticotropin," *JAMA*, 2000, 283(8):1038-45.

Bouachour G, Roy PM, and Guiraud MP, "The Repetitive Short Corticotropin Stimulation Test in Patients With Septic Shock," *Ann Intern Med*, 1995, 123(12):962-3.

Gronowski AM and Landau-Levine M, "Reproductive Endocrine Function," *Tietz Textbook of Clinical Chemistry*, 3rd ed, Burtis CA and Ashwood ER, eds, Philadelphia, PA: Saunders Co, 1999, 1619-20.

Krasner AS, "Glucocorticoid-Induced Adrenal Insufficiency," *JAMA*, 1999, 282(7):671-6.

Rittmaster RS, "Hirsutism," *Lancet*, 1997, 349(9046):191-5.

Soferman R, Kivity S, Dickstein G, et al, "Low-Dose Adrenocorticotropin Test Reveals Impaired Adrenal Function in Patients Taking Inhaled Corticosteroids," *J Clin Endocrinol Metab*, 1995, 80(4):1243-6.

Streeten DH, Anderson GH Jr, and Bonaventura MM, "The Potential for Serious Consequences From Misinterpreting Normal Responses to the Rapid Adrenocorticotropin Test," *J Clin Endocrinol Metab*, 1996, 81(1):285-90.

Thaler LM and Blevins LS, "The Low Dose (1 μg) Adrenocorticotropin Stimulation Test in the Evaluation of Patients With Suspected Central Adrenal Insufficiency," *J Clin Endocrinol Metab*, 1998, 83(8):2726-9.

◆ **Cortisol** *see* Insulin Tolerance Test *on page 804*

◆ **Cortisol-Binding Globulin** *see* Cortisol, Serum or Plasma *on page 460*

◆ **Cortisol-Binding Globulin** *see* Urinary Cortisol/Creatinine Increment *on page 1292*

Cortisol, Free, Urine

Related Information

Adrenal Cortex: Laboratory Assessment Overview *on page 110*
Corticotropin-Releasing Hormone Stimulation Test *on page 455*
Cortisol, Serum or Plasma *on page 460*
17-Hydroxycorticosteroids, Urine *on page 753*
Insulin Tolerance Test *on page 804*
Metyrapone Stimulation Test, Serum *on page 910*
Urinary Cortisol/Creatinine Increment *on page 1292*
Urine Collection, 24-Hour *on page 1295*

Synonyms Urinary Cortisol; Urine Cortisol

Test Includes Creatinine concentration and total volume, to support adequacy of collection

Abstract The diagnosis of Cushing syndrome (CS) requires evidence of cortisol hypersecretion. While serum cortisol levels fluctuate unpredictably and are strongly dependent on concurrent cortisol-binding globulin (CBG) levels, a 24-hour urine specimen integrates the cortisol production for an entire day and is not affected by CBG. Urinary cortisol reflects the portion of serum-free cortisol filtered by the kidney, and correlates well with cortisol secretion rate.

The most common causes of CS are pituitary adenoma (65% to 70%), adrenal tumor (15% to 20%), and the ectopic CS (10% to 15%). (See Adrenal Cortex: Laboratory Assessment Overview *on page 110*.)

Patient Preparation Radioisotopes may interfere if assay method is by radioimmunoassay. Patient should avoid spironolactone or quinacrine. Avoid patient stress.

Specimen 24-hour urine

Container Either 25 mL of 50% acetic acid or 10 g boric acid can be added before collection is started. If no preservative is used, it is necessary to refrigerate during collection.

Sampling Time Two or three 24-hour collections may be needed.

Collection A normal diurnal rhythm exists with highest levels in the morning, but this circadian rhythm is lost in Cushing syndrome.

When assessing the completeness of urine collection, there should be <10% variation in the creatinine concentrations of each 24-hour specimen. A variation >10% suggests incomplete collection. Because cortisol secretion is episodic, the creatinine concentration difference cannot be used to "correct" the cortisol result from an incomplete specimen.

Reference Interval Method dependent; typical intervals (age intervals not exact) include:

HPLC:
  • 0-10 years: 2-27 μg/day (SI: 5.52-74.5 nmol/day)
(Continued)

Cortisol, Free, Urine *(Continued)*

- 11-17 years: 1-55 µg/day (SI: 2.76-151.8 mmol/day)
- adults: 5-55 µg/day (SI: 13.8-151.8 nmol/day)

RIA (extracted):

- 0-10/ years: 2-27 µg/day (SI: 5.52-74.5 nmol/day)
- 11-20 years: 5-55 µg/day (SI: 13.8-151.8 nmol/day)
- adults: 20-90 µg/24 hours (SI: 55.2-248.4 nmol/day)

Use This test is useful in the initial evaluation of patients with suspected Cushing syndrome (CS). Patients with CS usually have urine free cortisol >100 µg/24 hours (SI: 276.0 nmol/day), but there is wide variation and no single cutoff can be used safely. If the 24-hour urine free cortisol is elevated, additional testing is indicated to differentiate among pituitary-dependent CS, pituitary-independent CS, and pseudo-Cushing syndrome. (See Adrenal Cortex: Laboratory Assessment Overview *on page 110*.)

Some patients with an elevated 24-hour urine cortisol do not have Cushing syndrome, and are often classified as pseudo-Cushing syndrome. Establishing this diagnosis requires additional testing which includes the low-dose dexamethasone suppression test, the CRH stimulation test, or a protocol that combines them both. (See Adrenal Cortex: Laboratory Assessment Overview *on page 110*.)

The diagnosis of CS requires a meticulous history and physical examination, and these should precede a biochemical evaluation.

Limitations In addition to the problems noted above, misleading elevations are reported in pregnancy, infections, and with tetracycline medication. Tetracyclines may cause false elevation of results.

Murphy has challenged the specificity of contemporary immunoassay methods for cortisol. He believes that immunoassay cortisol results are falsely elevated because the antibodies cross-react with other corticosteroids.

Methodology Immunoassays (multiple labels) after extraction, high performance liquid chromatography (HPLC), gas chromatography/mass spectrometry (GC/MS)

References

Bornstein SR, Stratakis CA, and Chrousos GP, "Adrenocortical Tumors: Recent Advances in Basic Concepts and Clinical Management," *Ann Intern Med*, 1999, 130(9):759-71.

Demers LM and Whitley RJ, "Function of the Adrenal Cortex," *Tietz Textbook of Clinical Chemistry*, 3rd ed, Burtis CA and Ashwood ER, eds, Philadelphia, PA: WB Saunders Co, 1997, 1559-60.

Leavelle DE, ed, *Interpretive Handbook*, Mayo Medical Laboratories, Rochester, MN, 1997, 163-4.

Murphy BEP, "How Much UFC Is Really Cortisol?" *Clin Chem*, 2000, 46(6 Pt 1):793-4.

Cortisol, Serum or Plasma

Related Information

Adrenal Cortex: Laboratory Assessment Overview *on page 110*
Adrenocorticotropic Hormone, Plasma *on page 114*
Androstenedione, Serum *on page 158*
Corticotropin-Releasing Hormone Stimulation Test *on page 455*
Corticotropin Stimulation Test (Rapid) *on page 456*
Cortisol, Free, Urine *on page 459*
Estradiol, Serum *on page 553*
17-Hydroxycorticosteroids, Urine *on page 753*
17-Hydroxyprogesterone, Whole Blood, Serum, or Plasma *on page 755*
Metyrapone Stimulation Test, Serum *on page 910*
Testosterone, Total and Free, Serum or Plasma *on page 1238*
Urinary Cortisol/Creatinine Increment *on page 1292*

Synonyms Compound F; Hydrocortisone, Serum

Applies to Cortisol-Binding Globulin; Dexamethasone Suppression Test

Abstract Cortisol, the major adrenal glucocorticoid, is secreted in a circadian pattern, with maximum values, 5-25 µg/dL, in the early morning and minimum values, 3-16 µg/dL, in the late afternoon. Most of the cortisol in the blood is protein-bound to cortisol-binding globulin (CBG) and albumin. Under normal circumstances, cortisol has a half-life in the blood of ~80-100 minutes. The serum concentration of CBG, normally 35-40 ng/L has moderate interindividual variability, and is also influenced by a number of hormones. Of particular importance is the fact that medicinal estrogens can produce a two- to threefold increase in CBG (in both women and men), and CBG is also increased during pregnancy. After metabolic inactivation in the liver, most cortisol metabolites are excreted in the urine. Only ~1% of cortisol is excreted unchanged into urine, but this urine free cortisol is important diagnostically (see Cortisol, Free, Urine *on page 459*). Cortisol secretion fluctuates rapidly in response to a variety of stimuli, including sepsis, surgical procedures, and other stressors. It stimulates catabolism of protein and fat.

Patient Preparation

Single sleeping midnight cortisol: In this protocol, which applies to hospitalized patients in a controlled environment, the patient retires ~11 PM and is awakened, as gently as possible, 1 hour later for venipuncture.

Overnight DST: The patient, not necessarily hospitalized, is given dexamethasone 1 mg orally at bedtime (~11 PM) and a blood specimen is drawn the next morning at 8 AM.

Low-dose DST: 24-hour urine specimens are collected on four consecutive days. Beginning at 8 AM on day 2, the patient is given dexamethasone 0.5 mg orally every 6 hours for a total of 8 doses. Blood specimens are obtained at 8 AM and 8 PM on day 1, and again at 8 AM on day 5. The blood specimens are assayed for cortisol. Each urine specimen is assayed for free cortisol and creatinine. (Some clinicians also measure urinary 17-hydroxycorticosteroids.)

High-dose DST: 24-hour urine specimens are collected on four consecutive days. Beginning at 8 AM on day 2, the patient is given dexamethasone 2 mg orally every 6 hours for a total of 8 doses. Blood specimens are obtained at 8 AM and 8 PM on day 1, and again at 8 AM on day 5. The blood specimens are assayed for cortisol. Each urine specimen is assayed for free cortisol and creatinine. (Some clinicians also measure urinary 17-hydroxycorticosteroids.)

Specimen Serum or plasma
Container Red top tube or green top (heparin) tube
Sampling Time See Patient Preparation.
Storage Instructions Stable 7 days at 4°C to 25°C.
Reference Interval See table.

Children[1]		
Age	5-11 AM	5-11 PM
0-24 mo	1-34 µg/dL (SI: 28-938 nmol/L)	1-30 µg/dL (SI: 28-828 nmol/L)
2-10 y	1-33 µg/dL (SI: 28-911 nmol/L)	1-24 µg/dL (SI: 28-662 nmol/L)
11-18 y	1-28 µg/dL (SI: 28-773 nmol/L)	1-22 µg/dL (SI: 28-607 nmol/L)
Adults[2]		
8 AM	5-25 µg/dL (SI: 138-690 nmol/L)	
4 PM	3-16 µg/dL (SI: 83-442 nmol/L)	
8 PM	50% of 8 AM value	
Midnight	<5 µg/dL (SI: 138 nmol/L)	

[1]Soldin SJ, Murthy JN, Agarwalla PK, et al, "Pediatric Reference Ranges for Creatine Kinase, CK-MB, Troponin I, Iron and Cortisol," *Clin Biochem*, 1999, 32(1):77-80.

[2]Demers LM and Whitley RJ, "Function of the Adrenal Cortex," *Tietz Textbook of Clinical Chemistry*, 3rd ed, Burtis CA and Ashwood ER, eds, Philadelphia, PA: WB Saunders Co, 1999, 1530-69, 1808.

Use Cortisol excess and deficiency are at the center of several major disease processes which require laboratory testing for diagnosis (see Adrenal Cortex: Laboratory Assessment Overview *on page 110*). Random measurements of serum cortisol are often useless. Even in patients with cortisol excess or deficiency, the many factors influencing cortisol can result in cortisol values which overlap the reference interval. Diagnostically useful cortisol measurements require standardized testing protocols in which one or more preanalytic variables are held constant.

High cortisol occurs in adrenocortical hypersecretion, adrenal cortical hyperplasia, adenoma, primary pigmented nodular adrenocortical disease, or carcinoma (Cushing syndrome), and with excess pituitary ACTH (Cushing disease) or production of ACTH by a nonpituitary tumor (ectopic corticotropin syndrome, (Continued)

Cortisol, Serum or Plasma *(Continued)*

most caused by small cell carcinomas of lung), adrenal cortical adenomas, hyperplasias, and carcinomas (corticotropin-independent Cushing syndrome). When **cortisol excess** (Cushing syndrome - CS) is suspected, the primary tests include one or more of the following: a single sleeping midnight serum cortisol; one of the dexamethasone suppression tests (DST - overnight or low-dose); and measurement of urine free cortisol (see Cortisol, Free, Urine *on page 459*). A midnight sleeping cortisol <1.8 µg/dL (SI: <50 nmol/L) is strong evidence against CS; a higher value may indicate the need for additional testing.

In the **overnight DST**, a serum cortisol at 8 AM <3 µg/dL (SI: 82.8 nmol/L) is strong evidence against CS. Most patients with CS will have an 8 AM serum cortisol >10 µg/dL (SI: 276.0 nmol/L). Patients with results between these two limits may need additional testing.

In the **low-dose DST**, normal persons will have serum cortisol and urine free cortisol values on day 4 that are at least 50% below the baseline (day 1) results. (Some clinicians will also measure urinary 17-hydroxycorticosteroids.)

The **high-dose DST** is one of several procedures intended to differentiate between an adrenal cause of CS, and a cause located in the pituitary, hypothalamus, or an ectopic site. The high-dose test is therefore performed only after the diagnosis of CS has been biochemically confirmed. The high-dose protocol will produce significant suppression of cortisol output in a patient with Cushing disease (a pituitary ACTH-secreting adenoma), but not in a patient with adrenal tumor secreting cortisol or the ectopic corticotropin syndrome. If the patient has Cushing disease, the plasma cortisol will be <10 µg/dL (SI: <27.6 nmol/L) on day 5, and the urine free cortisol (and 17-hydroxycorticosteroids, if obtained) will decrease to at least 50% of baseline by day 4. If the patient has an adrenal tumor secreting cortisol, a lower degree of suppression, or no suppression, will be found. See Adrenal Cortex: Laboratory Assessment Overview *on page 110* for additional information.

Causes of **low cortisol** include pituitary destruction or failure, with resultant loss of ACTH to stimulate the adrenal, and metabolic errors or destruction of the adrenal gland itself (adrenogenital syndromes, primary adrenocortical insufficiency, Addison disease, [idiopathic, tuberculosis, histoplasmosis, other diseases]). When **cortisol deficiency** is suspected, an 8 AM serum cortisol <5 µg/dL (SI: <140 nmol/L) [some authorities prefer <3 µg/dL (SI: <86 nmol/L)] confirms the diagnosis, and a result >20 µg/dL (SI: >550 nmol/L) excludes the diagnosis **in an unstressed patient.** (Stress in this context means a stimulus which approximates the intensity of severe sepsis or a major surgical procedure). Between these extremes, additional testing is required. The next step is often the Corticotropin Stimulation Test (Rapid) *on page 456.* See Adrenal Cortex: Laboratory Assessment Overview *on page 110* for additional information.

Pediatric adrenocortical insufficiency and hypofunction include a variety of congenital and acquired entities which include disorders of the hypothalamus, pituitary, or adrenal cortex. **Congenital adrenal hyperplasia** includes a group of autosomal recessive diseases. Ninety percent of patients with congenital adrenal hyperplasia have **deficiency of 21-hydroxylase**, for which neonatal screening programs have been developed.

Limitations Random serum cortisol may be misleading because of circadian variation in secretion. Falsely abnormal results in both the overnight and low-dose DST tests are associated with a wide variety of concurrent conditions and medications (eg, obesity, depression, diphenylhydantoin, phenobarbital).

Methodology Immunoassays (multiple labels), high performance liquid chromatography (HPLC); measurement by RIA after paper chromatography may more accurately estimate serum cortisol in patients with chronic renal failure.

References

Demers LM and Whitley RJ, "Function of the Adrenal Cortex," *Tietz Textbook of Clinical Chemistry*, 3rd ed, Burtis CA and Ashwood ER, eds, Philadelphia, PA: WB Saunders Co, 1999, 1530-69, 1808.

Desai SP and Isa-Pratt S, *Clinician's Guide to Laboratory Medicine*, Cleveland, OH: Lexi-Comp, 2000, Chapters 17, 20.

Merke DP and Cutler GB, "New Approaches to the Treatment of Congenital Adrenal Hyperplasia," *JAMA*, 1997, 277(13):1073-6.

Newell-Price J, Trainer P, Perry L, et al, "A Single Sleeping Midnight Cortisol Has 100% Sensitivity for the Diagnosis of Cushing's Syndrome," *Clin Endocrinol (Oxf)*, 1995, 43(5):545-50.

Orth DN and Kovacs WJ, "The Adrenal Cortex," *Williams Textbook of Endocrinology*, 9th ed, Philadelphia, PA: WB Saunders Co, 1998, 517-664.

Soldin SJ, Murthy JN, Agarwalla PK, et al, "Pediatric Reference Ranges for Creatine Kinase, CK-MB, Troponin I, Iron and Cortisol," *Clin Biochem*, 1999, 32(1):77-80.

♦ **Cortrosyn® Test** *see* Corticotropin Stimulation Test (Rapid) *on page 456*

Corynebacterium diphtheriae Throat Culture

Related Information

Bacterial Culture, Throat, and Antigen Detection Testing for Group A Streptococci *on page 245*

Test Includes Culture specifically for the bacteria *Corynebacterium diphtheriae*

Abstract *C. diphtheriae* is a gram-positive, nonsporulating bacillus. Infection with toxigenic *C. diphtheriae* is responsible for the disease diphtheria. The organism may be found in the anterior nasal mucosa (nasal diphtheria) or larynx (resembling infectious croup), but classically it is a disease of the oropharynx, pharynx, and tonsils. *C. diphtheriae* is not an invasive pathogen and the organisms remain superficial in the respiratory tract and skin. It may spread to or begin in the larynx and can involve the tracheobronchial tree. Its potent exotoxin is responsible for the virulence of the disease.

Pharyngeal diphtheria has become a rare disease in the U.S. Found among unimmunized or poorly immunized individuals, it may be suspected on clinical grounds.

Aftercare Observe for laryngospasm following collection of specimen.

Specimen Throat swab, nasopharyngeal swab; culture nose or larynx if clinically appropriate

Container Sterile Mini-Tip Culturette® or flexible calcium alginate swab, Calgiswab®, is recommended for obtaining nasopharyngeal culture.

Collection The tongue should be depressed while both the tonsillar crypts and nasopharynx and throat lesions are swabbed. If a pseudomembrane is present, the swab should be taken from beneath the membrane, or a part of the membrane.

Turnaround Time Final reports usually take at least 4 days.

Special Instructions The laboratory should be notified before collection of specimens so that special isolation media can be made available.

Reference Interval No *C. diphtheriae* isolated

Use Isolate *C. diphtheriae* from patients suspected of having diphtheria

Limitations Stain results are presumptive and are commonly reported out as "gram-positive pleomorphic bacilli suggestive of *C. diphtheriae*". Definitive diagnosis depends on isolation of the organism. Special microbiological media that are not routinely available in the laboratory are needed for optimal culture of *C. diphtheriae*.

Contraindications Lack of clinical symptoms or signs of diphtheria, valid history of immunization

Methodology Culture on selective medium (Löeffler), cystine tellurite agar, and blood agar; direct smear stained with Löeffler methylene blue stain and/or Gram stain

Additional Information *C. diphtheriae* may occasionally cause skin infections, wound infections, pulmonary infections, and endocarditis and may be recovered from the oropharynx of healthy carriers. *C. diphtheriae* is spread through respiratory secretions by convalescent and healthy carriers. The clinical presentation of respiratory diphtheria includes a grayish-brown pseudomembrane overlying superficial ulcers in the oropharynx. It may cause local lymphadenopathy. The organism is noninvasive; however, the exotoxin elaborated in the throat affects primarily the heart, kidneys, and nervous system. Mortality is 10% to 30% in untreated individuals. Only strains of *C. diphtheriae* infected by β-phage are capable of producing toxin. Confirmation of exotoxin production requires animal testing and is rarely done for clinical isolates.

References

Bisgard KM, Hardy IR, Popovic T, et al, "Respiratory Diphtheria in the United States, 1980 through 1995," *Am J Public Health*, 1998, 88(5):787-91.

Bisno AL, "Acute Pharyngitis," *N Engl J Med*, 2001, 344(3):205-11.

Golaz A, Lance-Parker S, Welty T, et al, "Epidemiology of Diphtheria in South Dakota," *S D J Med*, 2000, 53(7):281-5.

(Continued)

Corynebacterium diphtheriae Throat Culture *(Continued)*

Vitek CR, Bogatyreva EY, and Wharton M, "Diphtheria Surveillance and Control in the Former Soviet Union and the Newly Independent States," *J Infect Dis*, 2000, 181(S1):23-6.

Internet Web Sites

www.amm.co.uk/pubs/fa_diphtheria.htm
www.cdc.gov/nip/publications/pink/dip.pdf

♦ **Cosyntropin Stimulation Test** *see* Corticotropin Stimulation Test (Rapid) *on page 456*

Cotinine, Serum, Plasma, or Urine

Related Information

Carboxyhemoglobin, Blood *on page 340*
Nicotine, Serum or Plasma *on page 961*

Applies to Nicotine

Abstract Nicotine, one of the most toxic of all poisons, is a neural stimulant found in most tobacco products including transdermal patches and Nicorette® gum. Cotinine is the proximal metabolite of nicotine.

Specimen Urine, serum, or plasma

Container Sterile urine container for urine; red top tube for serum; lavender top (EDTA) tube for plasma

Storage Instructions Maintain specimen at room temperature or refrigerate. Cotinine is stable for several weeks at room temperature.

Reference Interval

- Nonsmoker: <15 ng/mL
- Light smoker: 15-100 ng/mL
- Heavy smoker: >100 ng/mL

Use Useful in situations in which smoking status assessment is of interest: evaluation of the impact of smoking cessation programs, monitoring of pregnancy and of other groups at risk, assessment of occupational exposure to industrial pollutants, validation of phase I clinical trials, selection and control of life insurance candidates, and monitoring of environmental tobacco exposure (passive smoking)

Limitations Urinary cotinine and serum cotinine measurements only provide information on smoking and passive exposure for several days prior to specimen collection.

Methodology Gas chromatography (GC), high performance liquid chromatography (HPLC), immunoassay

Additional Information The longer half-life of cotinine, 16-20 hours versus 1 hour for nicotine, makes it a more reliable marker for active smoking and environmental tobacco exposure than nicotine.

High cotinine levels are associated with poor pregnancy outcome and increased risk of abortion.

References

Caraballo RS, Giovino GA, Pechacek TF, et al, "Racial and Ethnic Differences in Serum Cotinine Levels of Cigarette Smokers," *JAMA*, 1998, 280(2):135-9.
Haufroid V and Lison D, "Urinary Cotinine as a Tobacco-Smoke Exposure Index: A Minireview," *Int Arch Occup Environ Health*, 1998, 71(3):162-8.
Rose JE, "Nicotine Addiction and Treatment," *Annu Rev Med*, 1996, 47:493-507.
Slotkin TA, "Fetal Nicotine or Cocaine Exposure: Which One Is Worse?" *J Pharmacol Exp Ther*, 1998, 285(3):931-45.

♦ **Coumadin®** *see* Heparin-Induced Thrombocytopenia *on page 695*
♦ **Coumadin®** *see* Prothrombin Time *on page 1116*
♦ **Coumadin®** *see* Warfarin, Serum or Plasma *on page 1325*

Coxsackievirus Serology

Related Information

Enterovirus Culture *on page 539*
Viral Culture *on page 1307*

Test Includes Detection of antibody titer to coxsackie A or B virus

Abstract The majority of infections with coxsackie A and B, two groups of nonpolio enteroviruses, are asymptomatic or mildly symptomatic. Patients with coxsackievirus infections can have a variety of symptoms and presentations, including meningitis, exanthems, pericarditis, myocarditis, hand-foot and mouth syndrome and conjunctivitis. Twenty-three coxsackieviruses group A and six coxsackieviruses group B are recognized.

Specimen Serum

Container Red top tube

Sampling Time Acute and convalescent sera drawn at least 14 days apart are required.

Reference Interval Less than a fourfold increase in titer in paired sera

Use Diagnose recent infections with coxsackie A or coxsackie B virus

Limitations Documentation of infection by serology is difficult, and diagnosis may depend on culture and other methods. Neutralizing antibodies develop quickly and persist for many years after infection. Complement fixation test is not sensitive.

Methodology Viral neutralization, complement fixation (CF)

Additional Information Since culture is frequently negative, diagnosis may depend on serologic studies. It is important to collect both acute and convalescent sera in order to reach an accurate diagnosis. The coxsackie A and B viruses can also be detected directly in the specimens of infected patients (see Viral Culture *on page 1307*), especially CSF, tissue, body fluids (pericardial fluid), or eye swabs or scrapings.

References
Modlin JF and Rotbart HA, "Group B Coxsackie Disease in Children," *Curr Top Microbiol Immunol*, 1997, 223:53-80.
See DM and Tilles JG, "Viral Myocarditis," *Rev Infect Dis*, 1991, 13(5):951-6.

Internet Web Sites
www.cdc.gov/ncidod/dvrd/hfmd.htm

♦ **C-Peptide** *see* Insulin, Serum *on page 803*

C-Peptide, Serum

Related Information

Glucose, Fasting, Plasma *on page 643*
Glucose Tolerance Test, Plasma *on page 651*
Insulin Antibody, Serum *on page 800*
Insulin, Serum *on page 803*
Ketone Bodies, Blood *on page 816*

Synonyms Connecting Peptide Insulin; Insulin-Connecting Peptide

Applies to Beta-Cell Peptides; Beta-Hydroxybutyrate; C-Peptide Suppression Test; Insulin; Insulin Antibodies; Proinsulin

Abstract The serum concentration of C-peptide is proportional to endogenous insulin production. C-peptide measurements are important in the differential diagnosis of hypoglycemic states. Less frequent applications are in pancreatectomy and postpancreatic transplant patients.

Patient Preparation Patients are usually fasting, but refer to the particular testing protocol being used. No recent radioactivity.

Specimen Serum

Container Red top tube

Collection Date and time must be correct. Draw in chilled tube. Keep specimen on ice.

Storage Instructions Spin in centrifuge at 4°C. Take off serum. Freeze immediately in a plastic tube.

Reference Interval Varies among laboratories. A typical interval for fasting C-peptide in serum is 0.51-2.70 ng/mL (SI: 170-900 pmol/L). After stimulation with glucose or glucagon, values rise to 5.6 ng/mL (SI: 1870 pmol/L).

Use C-peptide measurements are used in five clinical contexts:
- differential diagnosis of hypoglycemia
- classification of diabetes mellitus
- assessment of beta-cell function in diabetes
- forecasting the survival of pancreas transplants
- evaluating the completeness of surgical pancreatectomy

Hypoglycemia: Once the criteria of Whipple triad are satisfied (low blood glucose, simultaneous hypoglycemic signs and symptoms, and relief of the signs and symptoms by correcting the low blood glucose) a standard protocol should be used. It is important to recall that if, during a symptomatic episode, a normal blood glucose value is obtained, a hypoglycemic disorder is thereby excluded and further testing is not indicated.

(Continued)

C-Peptide, Serum (Continued)

Patients with insulin-secreting neoplasms have high levels of all three beta-cell peptides (insulin, C-peptide, proinsulin). In contrast, patients with factitious hypoglycemia from surreptitious insulin administration will have low levels of C-peptide and proinsulin in the presence of elevated (exogenous) serum insulin. Ingestion of an insulin secretagogue, such as a sulfonylurea drug, closely mimics insulinoma, because these drugs stimulate the secretion of proinsulin, insulin, and C-peptide. In patients who are evaluated by a standardized, supervised 72-hour fast, those with insulinoma or sulfonylurea effect have insulin levels >6.0 µU/mL (SI: >35.7 pmol/L), C-peptide levels >0.61 ng/mL (SI: >0.2 nmol/L), and proinsulin levels >5.0 pmol/L with symptoms of hypoglycemia and blood glucose <45 mg/dL (25 mmol/L). Laboratory differentiation between these two diagnoses requires detection of the sulfonylurea (or other drug) in plasma. Patients with surreptitious insulin administration will have, after the 72-hour supervised fasting protocol, insulin >6.0 µU/mL (often as high as 100 µU/mL or 1000 µU/mL (SI: 694.5 and 6945.0 pmol/L, respectively)), C-peptide <0.61 ng/mL (SI: <0.2 nmol/L), and proinsulin <5.0 pmol/L. **Patients who have noninsulin-mediated hypoglycemia** (eg, secondary to nonislet cell neoplasm) will have, after the supervised 72-hour fast, normal islet cell peptide levels (insulin <6.0 µIU/mL (<35.7 pmol/L), C-peptide <0.61 ng/mL, and proinsulin <5.0 pmol/L).

Classification of diabetes mellitus: The classification of diabetes mellitus is usually accomplished by history, physical findings, and blood glucose values. An occasional patient will require additional testing, including serum C-peptide and, even less often, serum insulin. Patients with type 2 diabetes mellitus usually have normal or elevated levels of C-peptide and insulin, and do not have beta-cell autoantibodies. In patients with type 1 diabetes mellitus, serum C-peptide (and serum insulin) are low or, in the late stages, undetectable; 85% to 90% of patients with type 1 diabetes have detectable beta-cell autoantibodies when hyperglycemia is first detected.

Beta-cell function in diabetes: C-peptide may also be useful in evaluation of residual beta-cell function in insulin-dependent diabetics, some of whom have antibodies that interfere with insulin assays. Glucagon-stimulated C-peptide concentration has been described as a discriminator between insulin-requiring and noninsulin-requiring diabetic patients.

Pancreas transplants: A relatively new use for C-peptide levels is in the evaluation of viability and survival of pancreas transplants.

Pancreatectomy: C-peptide should be undetectable following a total pancreatectomy.

Limitations C-peptide levels are increased in renal failure. (C-peptide is normally excreted by the kidneys.) Instances of insulinoma have been described in which proinsulin was increased but insulin and C-peptide were not.

Methodology Immunoassays (multiple labels)

Additional Information C-peptide can be measured in urine, but the test has few, if any, clinical applications.

The C-peptide suppression test depends on suppression of beta cell secretion during hypoglycemia to a lesser degree in patients with insulinoma than in normal individuals (Rizza et al, 2000).

References

Le Roith D, "Tumor-Induced Hypoglycemia," *N Engl J Med*, 1999, 341(10):757-8.

Rizza RA and Service JF, "Hypoglycemic/Pancreatic Islet Cell Disorders," Goldman L and Bennett JC, eds, *Cecil Textbook of Medicine*, Philadelphia, PA: WB Saunders Co, 2000, 1285-92.

Sasaki TM, Gray RS, Ratner RE, et al, "Successful Long-Term Kidney-Pancreas Transplants in Diabetic Patients With High C-Peptide Levels," Study," *Diabetes Care*, 1997, 20(2):198-201.

Service FJ, "Hypoglycemic Disorders," *N Engl J Med*, 1995, 332(17):1144-52.

Service FJ, Rizza RA, Zimmerman BR, et al, "The Classification of Diabetes by Clinical and C-Peptide Criteria. A Prospective Population-Based *Transplantation*, 1998, 65(11):1510-12.

The Expert Committee on the Diagnosis and Classification of Diabetes Mellitus, "Report of the Expert Committee on the Diagnosis and Classification of Diabetes Mellitus," *Diabetes Care*, 1999, 22(Suppl):S5-S19.

♦ **C-Peptide Suppression Test** see C-Peptide, Serum on page 465

♦ **CPK** see Creatine Kinase, Serum on page 470

- ◆ **CPK Isoenzymes** *see* Creatine Kinase MB and Other Isoenzymes, Serum *on page 469*
- ◆ **Crack** *see* Cocaine (Cocaine Metabolite), Qualitative, Urine or Hair *on page 427*
- ◆ **Crank** *see* Methamphetamine, Qualitative, Urine *on page 902*
- ◆ **Crash Syndrome** *see* Inherited Diseases of Metabolism and Cell Structure *on page 792*
- ◆ **C-Reactive Protein** *see* Myeloperoxidase, Plasma *on page 939*

C-Reactive Protein, Serum
Related Information
Albumin, Serum *on page 120*
Alpha$_1$-Antitrypsin, Serum *on page 134*
Bacterial Culture, Lower Respiratory *on page 241*
Cholesterol, Total, Serum or Plasma *on page 396*
Haptoglobin, Serum *on page 667*
Hypercoagulation Panel *on page 758*
Lipids, Overview *on page 853*
Sedimentation Rate, Erythrocyte *on page 1181*
Transthyretin, Serum *on page 1271*

Applies to Acute Phase Reactant; Amyloid A, Serum

Abstract C-reactive protein (CRP) is an acute phase reactant (APR) which begins to increase in serum a few hours after the initiation of an inflammatory process. CRP is a sensitive but nonspecific indicator of acute injury, bacterial infection, or inflammation. Since 1996 studies have been published linking serum CRP levels to coronary heart disease risk (see Use). Some current studies support the inflammatory hypothesis of the pathogenesis of atherogenesis.

Specimen Serum

Container Red top tube

Storage Instructions Do not freeze.

Reference Interval Intervals vary, depending on the method used and the reference population. Obtain this information from the laboratory performing the test. Many laboratories offer both a **routine** CRP and a **high-sensitivity** CRP, with different reference intervals.

In a study of 143 normal blood donors, performed in a university medical center laboratory, and using an in-house enzyme-labeled immunoassay calibrated with WHO reference material, the reference interval was **0.08-3.1 mg/L;** there were no significant gender- or age-related differences.

Below are selected data sets from a study of 2291 adult men and 2203 adult women. A high-sensitivity assay (minimum detectable concentration 0.05 mg/L) was used. For additional results, consult Hutchinson et al.

Central 95% of population distribution
Male:
- 25-34 years: 0.08-7.27 mg/L
- 45-54 years: 0.19-13.95 mg/L
- 65-74 years: 0.33-18.47 mg/L

Female:
- 25-34 years: 0.07-17.18 mg/L
- 45-54 years: 0.15-12.12 mg/L
- 65-74 years: 0.30-16.58 mg/L

Other sources use: adults: 0.068-8.2 mg/L; umbilical cord: 0.001-0.035 mg/L.

Within-individual variability is reported to account for 14% of total (biological) variance; thus triplicate sampling is recommended for estimating an individual's personal baseline for risk evaluation.

A threshold <10 mg/L is often proposed as a cutoff for significant inflammatory disease.

In 2003, the American Heart Association (AHA) and the Centers for Disease Control (CDC) issued recommendations for CRP cutpoints in cardiovascular risk assessment (see below), based on the average of two CRP determinations (preferably 2 weeks apart using standardized assays) in **metabolically stable**, fasting or nonfasting patients. Reporting units for this purpose should only be
(Continued)

C-Reactive Protein, Serum *(Continued)*

mg/L (Class 1 recommendation). If the CRP level is >10 mg/L, the test should be repeated and the patient examined for sources of inflammation.

- CRP level <1 mg/L: low risk
- CRP level 1.0-3.0 mg/L: average risk
- CRP level >3.0 mg/L: high risk

Use

Postoperative monitoring: By 4-6 hours after a surgical procedure the serum CRP begins to rise and reaches a peak, in the range of 25-35 mg/L, by 2-3 days after the operation. Postoperative values exceeding this range are associated with significant complications, usually inflammatory processes. Some investigators have recommended that patients who have a **preoperative** CRP >5 mg/L not undergo **elective** surgery involving cardiopulmonary bypass because of increased risk of postoperative sepsis.

Pelvic inflammatory disease (PID): In a study of 51 women receiving antibiotic treatment for PID, the CRP was a more sensitive indicator of clinically assessed severity than was the ESR, the leukocyte count, or body temperature.

Sepsis in critically ill patients: In a study of 23 critically ill patients for a total of 306 patient days, the serum CRP was more accurate than leukocyte count or body temperature in detecting sepsis and providing an index of its severity. A serum CRP >50 mg/L had a sensitivity of 98.5% and a specificity of 75% in the detection of sepsis (defined as a systemic inflammatory response syndrome plus a positive culture from one of these sites: blood, bronchoalveolar lavage, or central venous catheter).

Diagnosis of acute appendicitis: Investigators report that there is diagnostic value in adding a serum CRP to three routinely used laboratory tests (WBC, % neutrophils, ESR) for the evaluation of patients with possible acute appendicitis.

Patients with coronary heart disease (CHD): Serum CRP is elevated in patients with acute myocardial infarction (AMI), often to the very high levels seen in severe sepsis and the postoperative state.

Risk factor for vascular disease: Serum CRP may be a risk factor for coronary heart disease. In a subset of 1086 subjects, from the Physician's Health Study, the risk of a first AMI increased with increasing serum CRP. The risk was also increased for stroke and peripheral vascular disease. A point for emphasis is that these values are well within the reference interval for the routine CRP. Similar results (but using a different method) were reported in a large group of women participating in the Women's Health Study. However, the predictive power of the correlation between CRP concentrations is decreased when adjusted for other risk factors.

Limitations Lipemia or hemolysis may give false-positive results. The value of CRP in prediction of coronary events still must await confirmation in randomized trials.

Methodology Immunoassays (multiple labels and formats)

Additional Information

Relationship to obesity. Obese individuals have higher serum CRP levels than nonobese individuals.

Effect of estrogens: CRP levels are higher in those women taking hormone replacement therapy.

References

Boeken U, Feindt P, Zimmermann N, et al, "Increased Preoperative C-Reactive Protein (CRP)-Values Without Signs of an Infection and Complicated Course After Cardiopulmonary Bypass (CPB)-Operations," *Eur J Cardiothorac Surg*, 1998, 13(5):541-5.

Hutchinson WL, Koenig W, Frohlich M, et al, "Immunoradiometric Assay of Circulating C-Reactive Protein: Age-Related Values in the Adult General Population," *Clin Chem*, 2000, 46(7):934-8.

Jacobs DS, DeMott WR, Oxley DK, et al, *Laboratory Test Handbook*, 5th ed, Hudson, OH: Lexi-Comp Inc, 2001, 523-4.

Ledue TB and Rifai N, "Preanalytic and Analytic Sources of Variations in C-Reactive Protein Measurement: Implications for Cardiovascular Disease Risk Assessment," *Clin Chem*, 2003, 49(8):1258-71.

Mosca L, "C-Reactive Protein - to Screen or Not to Screen?" *N Engl J Med*, 2002, 347(20):1615-7.

Pearson TA, Mensah GA, Alexander RW, et al, "Markers of Inflammation and Cardiovascular Disease. Application to Clinical and Public Health Practice. A Statement for Healthcare Professionals From the Centers of Disease Control and Prevention and the American Heart Association," *Circulation*, 2003, 107:499-511.

Ridker PM, Hennekens CH, Rifai N, et al, "Hormone Replacement Therapy and Increased Plasma Concentration of C-Reactive Protein," *Circulation*, 1999, 100(7):713-6.

Ridker PM, Rifai N, Clearfield M, et al, "Measurement of C-Reactive Protein for the Targeting of Statin Therapy in the Primary Prevention of Acute Coronary Events," *N Engl J Med*, 2001, 344(26):1959-65.

Ridker PM, Rifai N, Rose L, et al, "Comparison of C-Reactive Protein and Low-Density Lipoprotein Cholesterol Levels in the Prediction of First Cardiovascular Events," *N Engl J Med*, 2002, 347(20):1557-65.

Creatine Kinase MB and Other Isoenzymes, Serum

Related Information

Cardiac Markers: Laboratory Assessment, Overview *on page 343*
Creatine Kinase, Serum *on page 470*
Heterophilic Antibodies *on page 727*
Myoglobin, Blood, Serum, or Plasma *on page 940*
Troponins, Serum *on page 1278*

Synonyms CK-2; CK Isoenzymes; CK Isoforms; CK-MB and Total CK; CPK Isoenzymes; Creatine Phosphokinase-MB Isoenzyme and Total Creatine Phosphokinase, Serum

Applies to Cardiac Index; CK Index; Isoform Ratio

Test Includes Separation of enzyme CK into its isoenzymes

Abstract Creatine kinase (CK) has three isoenzymes: CK-MM (CK-3), CK-MB (CK-2), and CK-BB (CK-1). CK-MB is the isoenzyme of most interest because it is present in large quantities in myocardium. Diagnostic elevations of CK-MB isoenzyme typically are not seen earlier than 6 hours after the onset of chest pain in individuals with acute myocardial infarct. CK-MB usually peaks between 15-20 hours after the onset of a myocardial infarct.

Specimen Serum, plasma; ethylenediamine tetraacetic acid (EDTA) for isoforms

Container Red top tube, green top (lithium heparin) tube

Sampling Time Varies with local protocols. Specimens for CK-MB should be obtained at baseline, and again at 6-9 hours. If these have negative results, obtain another at 12-24 hours.

When increased CK-MB values have returned to normal, CK isoenzyme determinations are usually no longer required.

Collection Avoid hemolysis.

Storage Instructions Separate serum from red cells and refrigerate.

Turnaround Time Less than 1 hour using CK-MB mass assay

Reference Interval Mass assay of CK-MB has a reference interval of 0-6 μg/L (0-6 ng/mL). When measured by electrophoresis (*vide infra*) the reference interval is usually 0% to 6% total CK (note different units).

In some methods, a **relative index** is determined to correct for CK-MB that can be released from traumatized skeletal muscle. The reference interval for the **CK relative index** is typically 0-2. CK isoforms are rarely used clinically. Laboratory should provide reference interval information.

Use Diagnosis of AMI. Using the mass assay of CK-MB, a CK-MB ≥10 ng/mL and CK relative index >3.0 has a sensitivity = 100%; specificity = 97%; positive predictive value = 100%; diagnostic efficiency = 97% (Pearson et al, 1990). Although many cardiologists prefer to measure one of the cardiac troponins, rather than CK-MB, in evaluating AMI, the CK-MB assay has a role in the diagnosis of **reinfarction**. The cardiac troponins remain elevated for as long as 14 days after an AMI. The CK-MB, by contrast, returns to baseline by 72 hours.

Limitations MB increases occur with other entities which cause damage to the myocardium. In such cases, however, CK-MB does not show the abrupt rise and fall characteristic of AMI. Increased CK-MB may occur with hypothyroidism and increases in total CK are even more common, in about 50% of cases.

Heterophilic antibodies may cause spuriously increased levels of CK-MB when an immunoassay is used.

Many laboratorians have observed very large discrepancies in CK isoenzyme determinations by electrophoresis. A major study (Henderson, 1994) has (Continued)

Creatine Kinase MB and Other Isoenzymes, Serum
(Continued)

demonstrated that electrophoresis-based assays have substantial bias (inaccuracies).

Methodology Measurement of CK-MB by electrophoresis and immunoinhibition methods are relatively insensitive methods that have been supplanted by CK-MB mass concentration assay methods. There is no place for measurement of CK-MB by electrophoretic nor immunoinhibition methods in the 21st century laboratory. Measurement of CK-MB mass concentration by monoclonal antibodies is a very sensitive method utilizing microparticulate fluorescence, enhanced luminescence, fluorescence, or chemiluminescence.

References

Alpert JS, Thygesen K, Antman E, et al, "Myocardial Infarction Redefined - A Consensus Document of the Joint European Society of Cardiology/American College of Cardiology Committee for the Redefinition of Myocardial Infarction," *J Am Coll Cardiol*, 2000, 36:959-69.

Christenson RH, Vaidya H, Landt Y, et al, "Standardization of Creatine Kinase-MB (CK-MB) Mass Assays: The Use of Recombinant CK-MB as a Reference Material," *Clin Chem*, 1999, 45(9):1414-23.

Henderson AR, Krishnan S, Webb S, et al, "Proficiency Testing of Creatine Kinase and Creatine Kinase-2: The Experience of the Ontario Laboratory Proficiency Testing Program," *Clin Chem*, 1998, 44(1):124-33.

Henderson AR, Stark JA, McQueen MJ, et al, "Is Determination of Creatine Kinase-2 After Electrophoretic Separation Accurate?" *Clin Chem*, 1994, 40(2):177-83.

Keffer JH, "The Cardiac Profile and Proposed Practice Guidelines for Acute Ischemic Heart Disease," *Am J Clin Pathol*, 1997, 107(4):398-409.

Lee TH and Goldman L, "Evaluation of the Patient With Acute Chest Pain," *N Engl J Med*, 2000, 342(16):1187-95.

Pearson JR and Carrea F, "Evaluation of the Clinical Usefulness of a Chemiluminometric Method for Measuring Creatine Kinase MB," *Clin Chem*, 1990, 36(10):1809-11.

Thygesen K and Alpert JS, "Myocardial Infarction Redefined - A Consensus Document of The Joint European Society of Cardiology/American College of Cardiology Committee for the Redefinition of Myocardial Infarction," The Joint European Society of Cardiology/American College of Cardiology Committee,

Creatine Kinase, Serum

Related Information

Cardiac Markers: Laboratory Assessment, Overview *on page 343*
Creatine Kinase MB and Other Isoenzymes, Serum *on page 469*
Duchenne/Becker Muscular Dystrophy DNA Detection *on page 526*
Heterophilic Antibodies *on page 727*
Lactate Dehydrogenase, Serum *on page 825*
Muscle Biopsy *on page 927*
Myoglobin, Blood, Serum, or Plasma *on page 940*
Myoglobin, Qualitative, Urine *on page 941*
Myotonic Dystrophy DNA Test *on page 942*
Troponins, Serum *on page 1278*

Synonyms CK; CPK; Creatine Phosphokinase, Total, Serum

Abstract Creatine kinase (CK) is found in striated muscle, brain, heart, and kidney. CK is a sensitive though nonspecific marker for myocardial and skeletal muscle injury.

Specimen Serum, plasma

Container Red top tube, green top (lithium heparin) tube

Storage Instructions Separate serum from red cells. Store in refrigerator. Avoid hemolysis.

Reference Interval Method dependent; typical intervals:

Male:
- 0-5 years: not established
- 6-11 years: 150-499 units/L
- 12-17 years: 94-499 units/L
- ≥18 years: 52-336 units/L

Female:
- 0-5 years: not established
- 6-7 years: 134-391 units/L
- 8-14 years: 91-391 units/L
- 15-17 years: 53-269 units/L
- ≥18 years: 38-176 units/L

Extreme elevations of CK occur in the X-linked muscular dystrophies, and lesser elevations are found in female carriers. Prominent elevations also occur in myocardial infarct/injury, muscle trauma, dermatomyositis, malignant hyperthermia, cerebrovascular disease, cerebral trauma/surgery, Reye syndrome, and hypothyroidism.

Use Test for acute myocardial infarct (AMI) and for skeletal muscular disease or damage. Although CK is elevated in some individuals who have malignant hyperthermia syndrome, interval screening is most effective to detect susceptible subjects. Elevated in muscular dystrophy: CK is a marker for Duchenne muscular dystrophy, with elevations of 20-200 times normal. CK is increased in female carriers of this X-linked disease, and in muscular stress, in polymyositis, dermatomyositis, and with muscle trauma. Elevated in myocarditis. Extremely high values are seen in some instances of myositis and in the postictal state of a recent grand mal seizure. CK may be elevated in a number of entities, including the eosinophilia-myalgia syndrome. Marked increases occur with rhabdomyolysis including that with cocaine intoxication. CK is sometimes increased with cerebrovascular accident. Malignancy (advanced) may show increased CK. Cardioversion with multiple shocks may release CK-MB and may result in a false-positive diagnosis of myocardial infarction.

Limitations Total CK alone is not recommended for the evaluation of acute coronary syndromes because of nonspecificity. Better results are obtained using CK-MB (mass assays), cardiac troponins, and/or myoglobin. High CK is found after trauma, surgery, and exercise without elevation of CK-MB.

Methodology Kinetic - UV spectrophotometric

References

Lloyd-Jones DM, Camargo CA Jr, Giugliano RP, et al, "Characteristics and Prognosis of Patients With Suspected Acute Myocardial Infarction and Elevated MB Relative Index but Normal Total Creatine Kinase," *Am J Cardiol*, 1999, 84(9):957-62.

Mayo Medical Laboratories, *2001 Test Catalogue*, 2001, 168.

Polaczyk CA, Kuntz KM, Sacks DB, et al, "Emergency Department Triage Strategies for Acute Chest Pain Using Creatine Kinase-MB and Troponin I Assays: A Cost-Effectiveness Analysis," *Ann Intern Med*, 1999, 131(12):909-18.

Robinson DJ and Christenson RH, "Creatine Kinase and its CK-MB Isoenzyme: The Conventional Marker for the Diagnosis of Acute Myocardial Infarction," *J Emerg Med*, 1999, 17(1):95-104.

Rosalki SB, "Serum Enzymes in Disease of Skeletal Muscle," *Clin Lab Med*, 1989, 9(4):767-81.

Zalenski RJ, McCarren M, Roberts R, et al, "An Evaluation of a Chest Pain Diagnostic Protocol to Exclude Acute Cardiac Ischemia in the Emergency Department," *Arch Intern Med*, 1997, 157(10):1085-91.

♦ **Creatine Phosphokinase-MB Isoenzyme and Total Creatine Phosphokinase, Serum** *see* Creatine Kinase MB and Other Isoenzymes, Serum *on page 469*

♦ **Creatine Phosphokinase, Total, Serum** *see* Creatine Kinase, Serum *on page 470*

Creatinine, 12- or 24-Hour Urine

Related Information

Creatinine Clearance and Urine Creatinine *on page 473*
Creatinine, Serum or Plasma *on page 474*
Osmolality, Serum *on page 978*
Osmolality, Urine *on page 979*
Sodium, Serum or Plasma *on page 1210*
Sodium, Urine *on page 1213*
Uric Acid, Urine *on page 1287*
Urinalysis *on page 1289*
Urine Collection, 24-Hour *on page 1295*
Vanillylmandelic Acid, Urine *on page 1299*

Synonyms Urine Creatinine

Test Includes Urine creatinine in mg/dL and mg/24 hours or mg/12 hours

Specimen 12- or 24-hour urine

Container Plastic urine container

Collection If the specimen is a 24-hour collection, instruct the patient to void at 8 AM and discard the specimen. Then collect all urine including the final specimen voided at the end of the 24-hour collection period (ie, 8 AM the next morning). Keep specimen on ice during collection. Container must be labeled with patient's name, date, and time collection started and date and time collection finished.

Storage Instructions Refrigerate

(Continued)

Creatinine, 12- or 24-Hour Urine *(Continued)*

Reference Interval Children: 2-3 years: 6-22 mg/kg/24 hours (SI: 52.8-193.6 μmol/kg/day), older than 3 years: 12-30 mg/kg/24 hours (SI: 105.0-264.0 μmol/kg/day); adults: male: 1-2 g/24 hours (SI: 8.8-17.7 mmol/day), female: 0.8-1.8 g/24 hours (SI: 7.1-15.9 mmol/day). Creatinine excretion decreases with advanced age as muscle mass diminishes. Normal age-adjusted values for anticipated creatinine excretion stratified for each sex by height are published. These tables assume ideal weight.

Use The creatinine clearance test (and related clearance tests) requires measurement of serum and urine creatinine. It is used as a crude marker for completeness of 24-hour urine collections when collected for other purposes. Acute oliguria may be assessed with measurements of sodium and creatinine concentrations in plasma and urine.

Limitations Complete urine collections require vigilance on the part of patients and nursing personnel. Drug interference includes cimetidine, trimethoprim, probenecid, cefoxitin, cephalothin, and others.

Methodology Jaffé reaction (alkaline picrate); enzymatic, kinetic or endpoint

Additional Information Urine creatinine is not ordered alone. Creatinine clearance, which requires a serum creatinine, offers useful renal function data. Serum creatinine alone is not an adequate index of glomerular filtration rate. See table to help to ascertain whether acute renal failure is due to prerenal causes or acute tubular injury.

Urinary Indexes in Patients With Acute Renal Failure

Index	Prerenal Causes	Acute Tubular Injury
Urinary sodium concentration (mEq/L)	<20	>40
Fractional excretion of sodium (%)[1]	<1	>1
Ratio of urine to plasma creatinine	>40	<20
Osmolality (mOsm/kg H_2O) urine	>500	<350

[1][Urine (Na) / serum (Na)] ÷ [urine (creatinine) / serum (creatinine)] x 100.

Adapted from Klahr S and Miller SB, "Acute Oliguria," *N Engl J Med*, 1998, 338(10):671-5 and Mitch WE, "Acute Renal Failure," *Cecil Textbook of Medicine*, 21st ed, Chapter 103, Goldman L and Bennett JC, eds, Philadelphia, PA: WB Saunders Co, 2000, 567-71.

References

Klahr S and Miller SB, "Acute Oliguria," *N Engl J Med*, 1998, 338(10):671-5.

Levey AS, Perrone RD, and Madias NE, "Serum Creatinine and Renal Function," *Annu Rev Med*, 1988, 39:465-90.

Walser M, "Creatinine Excretion as a Measure of Protein Nutrition in Adults of Varying Age," *JPEN J Parenter Enteral Nutr*, 1987, 11(5 Suppl):73S-8S.

Creatinine, Amniotic Fluid

Related Information

Body Fluid Chemical Analysis *on page 291*
Lecithin:Sphingomyelin Ratio, Amniotic Fluid *on page 836*
Phosphatidylglycerol, Amniotic Fluid *on page 1030*
Pulmonary Surfactant, Amniotic Fluid *on page 1124*

Synonyms Amniotic Fluid Creatinine

Abstract Amniotic fluid creatinine directly assesses fetal kidney maturity and indirectly assesses lung maturity. Much better tests are available for assessing fetal lung maturity (see Lecithin:Sphingomyelin Ratio, Amniotic Fluid *on page 836* and Phosphatidylglycerol, Amniotic Fluid *on page 1030*).

Reference Interval See table.

Amniotic Fluid Creatinine

	mg/dL	μmol/L
Immature fetus	<1.5	<132.6
Equivocal	1.5-2.0	132.6-176.8
Mature fetus	<2.0	>176.8

Use Amniotic fluid creatinine for estimation of fetal maturity is primarily of historical interest.

Methodology Colorimetry, Jaffé reaction

References

Almeida OD and Kitay DZ, "Amniotic Fluid Urea Nitrogen in the Prediction of Respiratory Distress Syndrome," *Am J Obstet Gynecol*, 1988, 159(2):465-8.

Kjeldsberg CR and Knight JA, "Amniotic Fluid," *Body Fluids: Laboratory Examination of Amniotic, Cerebrospinal, Seminal, Serous & Synovial Fluids*, 3rd ed, Chicago, IL: ASCP Press, American Society of Clinical Pathologists, 1993, 1-63.

Pitkin RM and Zwirek SJ, "Amniotic Fluid Creatinine," *Am J Obstet Gynecol*, 1967, 98(8):1135-9.

Creatinine Clearance and Urine Creatinine

Related Information

Creatinine, 12- or 24-Hour Urine *on page 471*
Creatinine, Serum or Plasma *on page 474*
Cystatin C, Serum or Plasma *on page 489*
Kidney Biopsy *on page 818*
Kidney Stone Analysis *on page 820*
Protein, Quantitative, Urine *on page 1108*
Urea Nitrogen, Serum or Plasma *on page 1284*
Uric Acid, Serum *on page 1286*
Urine Collection, 24-Hour *on page 1295*

Applies to Body Surface Area; Glomerular Filtration Rate

Replaces Urea Clearance; Urea Nitrogen Clearance

Test Includes Serum creatinine, urine creatinine

Abstract The most common test for evaluation of renal function is serum creatinine; the next is creatinine clearance.

Patient Preparation Review all medications and discontinue those which affect creatinine measurement (eg, cefoxitin). Check with laboratory for a complete list. Have patient drink water before the clearance is begun and continue good hydration throughout the clearance. Urine flows >2 mL/minute are required for good clearance measurements. Patient's sex, age, height, and weight are needed.

Specimen 24-hour urine and serum; test can be done for shorter periods. The use of two consecutive 2-hour clearances has been advocated.

Container Urine container and red top tube

Collection Instruct the patient to void at 8 AM and discard the specimen. Then collect all urine including the final specimen voided at the end of the 24-hour collection period (8 AM the next morning). Keep specimen on ice during collection. Bottle must be labeled properly. Complete, carefully timed (usually 24-hour) collection is needed; 4- and 12-hour collections are acceptable.

Storage Instructions Refrigerate urine specimen.

Causes for Rejection Any violation of collection protocol

Reference Interval Clearance for:
- Children: 70-140 mL/minute/1.73 m^2 (SI: 1.17-2.33 mL/s/1.73 m^2)
- Adults: male: 85-125 mL/minute/1.73 m^2 (SI: 1.42-2.08 mL/s/1.73 m^2)
- Adults: female: 75-115 mL/minute/1.73 m^2 (SI: 1.25-1.92 mL/s/1.73 m^2).

For each age decade after 40, creatinine clearance decreases 6-7 mL/minute/1.73 m^2.

Critical Values Moderate renal impairment (adult): 30-40 mL/minute/1.73 m^2

Possible Panic Range Severe renal impairment (adults): <28 mL/minute/1.73 m^2

Use Renal function test to estimate glomerular filtration rate (GFR); follow possible progression of renal disease; adjust dosages of medications in which renal excretion is pivotal (eg, aminoglycosides, methotrexate, cisplatin).

When **urine creatinine** is used to evaluate the completeness of a 24-hour urine collection in connection with the measurement of other analytes (eg, urine free cortisol, other steroids, albumin, total protein, phosphate, uric acid), the following reference intervals may be applied (Painter et al):
- Infants: 8-20 mg/kg/day (SI: 71-177 µmol/kg/day)
- Children: 8-22 mg/kg/day (SI: 71-194 µmol/kg/day)
- Adolescents: 8-30 mg/kg/day (SI: 71-265 µmol/kg/day)
- Adults: male: 14-26 mg/kg/day (SI: 124-230 µmol/kg/day)
- Adults: female: 11-20 mg/kg/day (SI: 97-177 µmol/kg/day)

For each decade after 40 years of age, the urine creatinine decreases approximately 10 mg/kg/day (SI: 0.0844 µg/kg/day).

(Continued)

Creatinine Clearance and Urine Creatinine *(Continued)*

- 90 years: male: 800-2000 mg/kg/day (SI: 7.1-17.7 µg/kg/day); female: 600-1800 mg/kg/day (SI: 5.2-15.9 µg/kg/day).

Limitations Because of the exponential rise in serum creatinine concentration with decline of GFR, a 25% increase in serum creatinine actually represents a substantial diminution of GFR. When muscle mass and kidney function diminish in parallel with advancing age, an elderly woman with perceived normal creatinine concentration may have a GFR only 30% that of a young adult.

Methodology Jaffé reaction (alkaline picrate) or enzymatic. The calculation for corrected creatinine clearance in mL/minute = [(urine volume per minute x urine creatinine)/serum creatinine] x (1.73/surface area of body in square meters). Body surface area is obtained from nomograms which require height and weight.

Body surface area can be calculated from the following equation:
$$S = M^{0.425} \times H^{0.725} \times 71.84$$

where:

S = body surface in cm^2

M = mass in kg

H = height in cm

Additional Information A **prediction equation** based on three serum test results (creatinine, albumin, and urea) and three demographic characteristics (age, sex, ethnicity) has been validated in a sample of 558 patients with chronic renal disease. Using the renal clearance of I-125 iothalamate as the GFR gold standard, this equation was more accurate than the conventional creatinine clearance (24-hour urine and single serum assay) and the Cockcroft-Gault formula. The equation is:

$$GFR = 170 \ (S_{cr})^{-0.999} \times (age)^{-0.176} \times [(0.762 \text{ if female}) \text{ or } (1.180 \text{ if black})] \times (BUN)^{-0.170} \times (Alb)^{+0.318}$$

Where:

S_{cr} = serum creatinine in mg/dL

Age = age in years

BUN = serum urea nitrogen in mg/dL

Alb = serum albumin in g/dL

References

Herget-Rosenthal S, Kribben A, Pietruck F, et al, "Two by Two-Hour Creatinine Clearance - Repeatable and Valid," *Clin Nephrol*, 1999, 51(6):348-54.

Levey AS, Bosch JP, Lewis JB, et al, "A More Accurate Method to Estimate Glomerular Filtration Rate From Serum Creatinine: A New Prediction Equation," Modification of Diet in Renal Disease Study Group, *Ann Intern Med*, 1999, 130(6):461-70.

Painter PC, Cope JY, and Smith JL, "Reference Information for the Clinical Laboratory," *Tietz Textbook of Clinical Chemistry*, 3rd ed, Burtis CA and Ashwood ER, eds, Philadelphia, PA: WB Saunders Co, 1999, 1809.

Payne RB, "Biological Variation of Serum and Urine Creatinine and Creatinine Clearance," *Ann Clin Biochem*, 1989, 26(Pt 6):565-6.

Sokoll LJ, Russell RM, Sodowski JA, et al, "Establishment of Creatinine Clearance Reference Values for Older Women," *Clin Chem*, 1994, 40(12):2276-81.

Van Lente F and Suit P, "Assessment of Renal Function by Serum Creatinine and Creatinine Clearance: Glomerular Filtration Rate Estimated by Four Procedures," *Clin Chem*, 1989, 35(12):2326-30.

Young DS, *Effects of Drugs on Clinical Laboratory Tests*, 5th ed, Volume 1: Listing by Test, Washington, DC: AACC Press, American Association of Clinical Chemistry, 2000, Section 3, 258-61.

♦ **Creatinine, Ratio** *see* Body Fluid pH *on page 295*

Creatinine, Serum or Plasma

Related Information

Aminoglycosides, Serum *on page 147*

Anemia Flowchart *on page 35*

Anion Gap, Serum, Plasma, or Urine *on page 160*

Bicarbonate, Blood *on page 258*

Creatinine Clearance and Urine Creatinine *on page 473*

Cystatin C, Serum or Plasma *on page 489*

Digoxin, Serum *on page 512*

Kidney Stone Analysis *on page 820*

Lactic Acid, Whole Blood or Plasma *on page 827*

Applies to Glomerular Filtration Rate

Abstract A primary renal function test. The production of creatinine is proportional to lean body mass, and the serum level reflects the renal glomerular filtration rate. Diabetes mellitus and hypertension cause 66% of cases of end-stage kidney disease.

Specimen Serum, plasma

Container Red top tube; gray top (sodium fluoride) tube or green top (heparin) tube can be used if assay is done by Jaffé method.

Collection Pediatrics: Blood drawn from heelstick.

Causes for Rejection Hemolysis

Reference Interval Children: 1-5 years: 0.3-0.5 mg/dL (SI: 27-44 µmol/L), 5-10 years: 0.5-0.8 mg/dL (SI: 44-71 µmol/L); adults: male: up to 1.2 mg/dL (SI: 106 µmol/L), female: up to 1.1 mg/dL (SI: 97 µmol/L). The gender differences may reflect differences in muscle mass. The glomerular filtration rate increases in pregnancy; thus, serum creatinine should be slightly lower during that period.

Critical Values Chronic renal insufficiency is defined by serum creatinine concentration of 1.5-3.0 mg/dL. Chronic renal failure is defined as serum creatinine concentration >3.0 mg/dL. This value, in children, varies by age.

Use The most common clinical renal function test, providing an approximation of glomerular filtration. In children with normal muscle mass, serum creatinine can be used to estimate glomerular filtration rate (GFR) expressed in mL/minute/ 1.73 m^2 by using the formula:

$$GFR = \delta L / S_{Cr}$$

Where:

L = body length in cm

S_{Cr} = serum creatinine in mg/dL

δ = constant of proportionality (age and sex dependent)

See table.

Mean Values and Ranges of δ[1]

Age	δ (mean)	δ (range)	% Within Range
Low-birth weight infants <1 y	0.33	0.20-0.50	77
Term <1 y	0.45	0.30-0.70	79
Children 2-12 y	0.55	0.40-0.70	83
Male 13-21 y	0.70	0.50-0.90	82
Female 13-21 y	0.55	0.40-0.70	77

[1]GFR = $\delta L/S_{Cr}$

Adapted from Schwartz GJ, Brion LP, and Spitzeer A, "The Use of Plasma Creatinine Concentration for Estimating Glomerular Filtration Rate in Infants, Children, and Adolescents," *Pediatr Clin North Am*, 1987, 34:571-90.

Many drugs are potentially nephrotoxic and the serum creatinine is commonly used to monitor such situations.

Causes of high creatinine include renal diseases and insufficiency with decreased glomerular filtration (uremia or azotemia if severe); urinary tract obstruction; reduced renal blood flow including congestive heart failure, shock and dehydration; rhabdomyolysis causes high serum creatinine, which may be elevated out of proportion to BUN, or to the reduction in renal function. Creatinine >2 mg/dL is among factors for identification of patients with acute necrotizing pancreatitis at risk for adverse outcome. Creatinine with urea nitrogen and electrolytes may be indicated in patients with vomiting, diarrhea, or (Continued)

Creatinine, Serum or Plasma *(Continued)*

decreased oral intake. Increased serum creatinine concentrations are found in some patients with hypertension, diabetes mellitus, and cardiovascular disease. Increases in creatinine correlate with age older than 55 years.

Other **causes of high creatinine** include congestive heart failure, shock, dehydration, and rhabdomyolysis.

Causes of low creatinine include small stature, debilitation, decreased muscle mass, advanced liver disease, long-term corticosteroidal therapy, primary muscle diseases, dermatomyositis, and neurogenic muscle diseases.

Limitations With reduced renal blood flow, creatinine rises less quickly than urea nitrogen. Concentration of creatinine only becomes abnormal when about half or more of the nephrons have stopped functioning in chronic progressive renal disease. Thus, it is not a sensitive indicator of early renal disease. Renal failure is underestimated by serum creatinine and creatinine clearance in patients with hepatic cirrhosis.

Many drugs, and some foods and vitamins, produce interferences in the assay for creatinine. Among the more common are certain cephalosporins (especially cefoxitin), methyldopa, trimethoprim, cimetidine, guanidine, and hydantoin. Dietary causes of increased values include meat, glucose, uric acid, ascorbic acid, and fructose.

Methodology Alkaline picrate (Jaffé reaction), enzymatic, *o*-nitrobenzaldehyde (Sakaguchi reaction), imidohydrolase (Ektachem®). Many interference problems are still unresolved in the Jaffé reaction.

Additional Information Serum creatinine level is proportional to lean body muscle mass. It is unaffected by most diet or activity and is freely filtered by the glomerulus. Both BUN and creatinine are often ordered to follow renal problems. Creatinine overall is the more reliable index, but each has pitfalls. As creatinine increases in chronic renal failure, the hematocrit decreases, total carbon dioxide and bicarbonate fall, and serum phosphate and BUN increase. Uric acid increases later. When serum creatinine increases postoperatively, a group of patients may be identified who are at risk for more severe renal failure. Serum creatinine has a role in determination of dosages of many drugs.

Patients with diabetes mellitus or hypertension should be monitored with blood pressure measurement, urinalysis, serum creatinine and microalbuminuria testing.

References

Fink JC, Burdick RA, Kurth SJ, et al, "Significance of Serum Creatinine Values in New End-Stage Renal Disease," *Am J Kidney Dis*, 1999, 34(4):694-701.

Fossati P, Ponti M, Passoni G, et al, "A Step Forward in Enzymatic Measurement of Creatinine," *Clin Chem*, 1994, 40(1):130-7.

Jacobs DS, DeMott WR, Oxley DK, et al, *Laboratory Test Handbook*, 5th ed, Hudson, OH: Lexi-Comp Inc, 2001.

Luke RG, "Chronic Renal Failure - A Vasculopathic State," *N Engl J Med*, 1998, 339(12):841-3.

Savory DJ, "Reference Ranges for Serum Creatinine in Infants, Children, and Adolescents," *Ann Clin Biochem*, 1990, 27(Pt 2):99-101.

Young DS, *Effects of Drugs on Clinical Laboratory Tests*, 5th ed, Volume 1: Listing by Test, Washington, DC: AACC Press, American Association of Clinical Chemistry, 2000, Section 3, 240-58.

♦ **Creutzfeldt-Jakob Disease** *see* Cerebrospinal Fluid and Plasma β-Amyloid$_{(1-42)}$ on page 359

♦ **CRH Stimulation Test** *see* Corticotropin-Releasing Hormone Stimulation Test on page 455

♦ **CRMP** *see* Antineuronal Nuclear Antibody, Type 1 (Anti-Hu) on page 185

♦ **Crossmatch, Lymphocyte** *see* Tissue Typing on page 1259

⁵¹Cr Red Cell Survival

Related Information

Blood Volume *on page 282*
Occult Blood, Stool *on page 969*
Reticulocyte Count *on page 1156*

Synonyms Red Cell Survival

Applies to Biotin-Labeled Red Cell Survival; End-Alveolar Carbon Monoxide Assessment of Erythrocyte Survival

Patient Preparation Obtain signed procedure permit for "⁵¹Cr red cell survival." Patient's own ⁵¹Cr-labeled red cells are infused. Patient should be provided a schedule for serial blood samples to be drawn.

Specimen Whole blood is drawn, processed (tagged with ⁵¹Cr), and reinfused into the patient.

Container Lavender top (EDTA) tube

Collection Scheduled periodic blood samples are drawn for determination of residual radioactivity.

Causes for Rejection Previous isotope procedure with significant radioactivity remaining in patient's blood, significant transfused blood, intermittent bleeding episodes

Turnaround Time Time required for procedure depends on half-time of disappearance of labeled cells and averages 3 weeks.

Special Instructions At least 21 days should be allowed for this study. When selective splenic sequestration as cause of hemolysis is suspected, liver and spleen readings may be performed in conjunction with ⁵¹Cr RBC Survival Test.

Reference Interval 25-35 days

Use Support diagnosis and pace of hemolytic process

Limitations This test cannot discriminate between red cell loss due to intravascular hemolysis and red cell loss due to bleeding (blood loss from the intravascular compartment).

Methodology Patient's own RBCs incubated with ⁵¹Cr under sterile conditions are injected back into the patient's vascular system and periodic blood samples are obtained for measurement of residual radioactivity over a 2- to 3-week period.

Additional Information Because the ⁵¹Cr label is eluted, in the circulation, from labeled RBCs, the ⁵¹Cr red cell survival result is not 55-60 days (one-half the physiologic RBC lifespan) in a normal person. The RBC survival curve results from the rate of elution of the label combined with the rate of random hemolysis. As a result, the ⁵¹Cr survival time does not relate in a simple or direct manner to the red cell lifespan.

Biotinylated autologous red blood cells and endogenous carbon monoxide (CO) production (by measurement of expired CO which reflects RBC destruction) are two additional labels that have been applied to the determination of red cell survival.

References

Davey FR and Hutchison RE, "Hematopoiesis," *Clinical Diagnosis and Management by Laboratory Methods,* 20th ed, Chapter 25, Henry JB, ed, Philadelphia, PA: WB Saunders Co, 2001, 529.

Furne JK, Springfield JR, Ho SB, et al, "Simplification of the End-Alveolar Carbon Monoxide Technique to Assess Erythrocyte Survival," *J Lab Clin Med,* 2003, 142(1):52-7.

International Committee for Standardization in Haematology, "Recommended Method for Radioisotope Red-Cell Survival Studies," *Br J Haematol,* 1980, 45:659-66.

Kumpel BM, Austin EB, Lee D, et al, "Comparison of Flow Cytometric Assays With Isotopic Assays of ⁵¹Chromium-Labeled Cells for Estimation of Red Cell Clearance or Survival In Vivo," *Transfusion,* 2000, 40(2):228-39.

♦ **Cr, Serum** *see* Chromium, Serum *on page 402*

♦ **Cr, Urine** *see* Chromium, Urine *on page 403*

♦ **Crustaceans** *see* Arthropod Identification *on page 213*

♦ **Cryocrit** *see* Cryofibrinogen *on page 477*

♦ **Cryocrit** *see* Cryoglobulin, Qualitative, Serum and Plasma *on page 478*

Cryofibrinogen

Related Information

Cold Agglutinin Titer *on page 430*

Cryoglobulin, Qualitative, Serum and Plasma *on page 478*

Applies to Cryocrit

Abstract Cryofibrinogen precipitates at cold temperatures, causing predominantly cutaneous symptoms on cold-exposed areas. It is also commonly asymptomatic.

Specimen Plasma

Container Two blue top (sodium citrate) tubes or EDTA tubes; also one red top tube for cryoglobulin. Tubes may be prewarmed to 37°C if necessary.

Collection Immediately place specimens in warm water and transport to laboratory.

(Continued)

Cryofibrinogen *(Continued)*

Causes for Rejection Improper tube, specimen more than 2 hours in transit to the laboratory, specimen not warm upon arrival to laboratory

Turnaround Time 24-72 hours

Reference Interval Negative: no cryofibrinogen detected

Use Consider a cryofibrinogen assay for patients with unexplained cutaneous ulcers, ischemia or necrosis on cold-exposed areas. Occasionally, routine blood samples are noted to form a gel during or soon after blood drawing.

Contraindications Specimens containing heparin should not be used, because heparin nonspecifically precipitates fibrinogen in this assay.

Methodology Plasma is obtained by centrifuging the warm specimen at 37°C. The plasma is then refrigerated overnight, usually in a tube that can measure "cryocrit", such as a Wintrobe tube. To determine if fibrinogen precipitate has formed, the tube is centrifuged at 4°C. Each millimeter of visible precipitate represents 1% of "cryocrit" (in this case, cryofibrinogen). The cryocrit is the volume percent of the precipitate compared with the total volume of test plasma. Also, if cryofibrinogen is present, plasma fibrinogen levels are lower after refrigeration compared with fibrinogen measurements performed on the warm specimen prior to refrigeration. A cryoglobulin test is simultaneously performed, to ensure that the plasma precipitate is not cryoglobulin. Cryoglobulin precipitates in plasma or serum at cold temperatures, whereas cryofibrinogen precipitates in cold plasma but not serum (because fibrinogen is not present in serum). Cryoglobulin and cryofibrinogen disappear upon rewarming the specimen. See Cryoglobulin, Qualitative, Serum and Plasma *on page 478.*

Additional Information Cryofibrinogen consists of fibrinogen and other substances that precipitate at cold temperatures. Cryoglobulins are immunoglobulins that precipitate at cold temperatures. Cryofibrinogenemia or cryoglobulinemia both can produce cold-induced skin symptoms in the extremities, ears or nose. Such symptoms include purpura, ulceration, necrosis, gangrene, bleeding, cold urticaria, bullae, livedo reticularis, and Raynaud syndrome. In one study, 13% of cryofibrinogenemia patients had venous and/or arterial thrombosis. Cryofibrinogenemia can be a primary (essential) condition or it may arise in association with an underlying condition, such as malignancy, infection, inflammation, diabetes, pregnancy, scleroderma, or oral contraceptives. A few familial cases have been reported. Skin biopsies may show leukocytoclastic vasculitis.

References

Blain H, Cacoub P, Musset L, et al, "Cryofibrinogenaemia: A Study of 49 Patients," *Clin Exp Immunol*, 2000, 120(2):253-60.

Cryoglobulin, Qualitative, Serum and Plasma

Related Information

Cold Agglutinin Titer *on page 430*
Complement C3, Serum *on page 434*
Complement C4, Serum *on page 436*
Cryofibrinogen *on page 477*
Hepatitis C Virus RNA Detection and Quantitation *on page 705*
Hepatitis C Virus Serology *on page 706*
Immunoglobulin M *on page 779*
Rheumatoid Factor, Serum or Body Fluid *on page 1161*

Applies to Cryocrit

Abstract **Cryoglobulins** are abnormal immunoglobulins that aggregate, reversibly precipitate, or form a gel upon exposure to temperatures below 37°C *in vitro* or *in vivo*. They may fix complement and initiate an inflammatory reaction similar to antigen-antibody complexes. Cryoglobulinemia is called "essential" when not related to other recognized disease. They are most commonly associated with plasmaproliferative and lymphoproliferative diseases, and infectious and autoimmune disorders, including hepatitis C and Sjögren syndrome.

Cryofibrinogens are fibrin and fibrinogen complexes which precipitate in cold temperatures (see Cryofibrinogen *on page 477*). They can be associated with vasculitis. Other cold-related phenomena involve **cold agglutinins**.

Patient Preparation Patient should be fasting.

Specimen At least 5 mL of serum and 1 mL of plasma, the latter to identify cryofibrinogen. If sample volume is small, cryoprecipitates may not be recognized. Even trace quantities can be symptomatic.

Container Red top tube for cryoglobulins and lavender top (EDTA) tube for plasma for cryofibrinogen, each prewarmed to 37°C.

Collection Phlebotomy should be performed using prewarmed 37°C plain red top tube to collect 15 mL blood, for evaluation for cryoglobulins, and EDTA (lavender) tube prewarmed at 37°C to collect 5 mL blood for evaluation for cryofibrinogen. The tubes should be transported to the laboratory immediately in a thermos filled with 37°C water. Syringes, if used, should be prewarmed.

Storage Instructions Keep blood at 37°C for 30-60 minutes. Clot separation should be at 37°C. Centrifuge at 37°C, if possible. Separate serum from cells, recentrifuge serum if possible at 37°C and pour into clean test tube. Do not refrigerate or freeze.

Use Cryoglobulins are often associated with macroglobulinemia of Waldenström, myeloma and other lymphoproliferative/plasmaproliferative diseases with monoclonal gammopathies, autoimmune disorders including Sjögren syndrome, and chronic infections. Manifestations may include purpura, vasculitis, polyarthralgia, peripheral neuropathy, renal failure, Raynaud phenomenon, and hyperviscosity.

Limitations Failure to detect cryoglobulin may be due to:
- allowing blood to clot at temperatures below 37°C
- faulty transportation
- too short incubation time
- ionic strength, pH, temperature
- concentration of the cryoprecipitating immunoglobulin
- binding to serum lipids

Methodology Specimens are divided, one kept at 37°C, the other at 4°C. Precipitation of cryoglobulin at 1°C for 24 hours (on ice in a 4°C refrigerator). If precipitation is not seen, evaluate again in 7 days.

Type 1 monoclonal cryoglobulins generally precipitate within 24 hours, but mixed cryoglobulins, especially Type 3, are often present in low concentrations and require prolonged incubation (7 days) at low temperatures for detection. If present in sufficient quantity, cryoglobulins can be semiquantitated by performing a cryocrit. Immunoglobulin class can be determined by immunofixation. Keren provides further details.

When precipitation is found in the plasma tube but not in serum, cryofibrinogen is recognized.

Additional Information Cryoglobulins have been classified as type 1, 2, or 3 on the basis of immunoglobulin composition.

Type 1 cryoglobulinemia is often associated with a plasma cell disorder and detectable monoclonal gammopathy.

Type 2 is the most common cryoglobulinemia and is often associated with hepatitis C infection. Coinfection with GB virus C (HCV plus GBV-C) is common but GBV-C does not appear to have a primary role in the pathogenesis of type 2 mixed cryoglobulinemia. The serum concentration of the monoclonal IgM is sometimes too low to be detected by serum protein electrophoresis. The major antibody is IgM, its antigen IgG. Such activity is usually designated rheumatoid factor. Type 2 mixed cryoglobulinemia may include Raynaud phenomena, renal disease (often with nephrotic syndrome and histopathologic membranoproliferative glomerulonephritis), sensory motor neuropathy, splenomegaly, and anemia. CH_{50} is usually detected along with a significant depression of C4 and depression of C3.

Type 3 consists of mixed polyclonal cryoglobulins, most commonly IgM-IgG complexes. The concentration of the cold precipitating proteins may be too low to detect. Type 3, like type 2, is called a "mixed cryoglobulin" because the cryoglobulins contain polyclonal immunoglobulins. Unlike type 2, type 3 cryoglobulins lack a monoclonal component.

The temperature at which precipitation takes place is more clinically relevant than the quantity of precipitation (ie, a cryoglobulin which precipitates only at very low temperatures is less likely to cause clinical difficulties).

(Continued)

Cryoglobulin, Qualitative, Serum and Plasma
(Continued)

Smears made from room temperature blood of subjects with cryoglobulinemia may contain clear, light pink or basophilic substance as cytoplasmic inclusions in leukocytes. Such material is consistent with phagocytosed immunoglobulins. Identification of such bodies may provide an important clue to recognition of cryoglobulinemia.

The expression, **essential mixed cryoglobulin,** indicates cryoglobulinemia type 2 or 3 without an underlying disease (other than Sjögren syndrome). The term, **secondary mixed cryoglobulinemia,** is used to refer to cases of type 2 or 3 who have another disease process (eg, HCV, lymphoproliferative disease, connective tissue disease). (Many cases formerly called "essential..." are now known to be secondary to hepatitis C infection.)

The presence of cryoglobulins may cause difficulty in performance of serum protein electrophoresis, immunofixation, and tests of complement activation and consumption.

Other relevant initial laboratory studies include CBC, liver function tests, RF, C3, C4, ESR, serum protein electrophoresis, creatinine clearance, hepatitis C testing, testing for connective tissue diseases, urinalysis, and consideration of possible lymphoproliferative or plasmaproliferative neoplastic disease.

Type	Composition	Disease Associations	Signs / Symptoms
1 - Monoclonal IgM or IgG	May be IgA or immunoglobulin light chain; precipitates often in 24 hours; homogeneous monoclonal substances; typically, large concentrations present; usually >500 mg/dL	M components with lymphoproliferative, plasmaproliferative diseases including Waldenström macroglobulinemia, myeloma, monoclonal gammopathy of uncertain significance, B-cell lymphoma, and leukemia	Many patients asymptomatic, others have pain, Raynaud phenomenon, palpable purpura, acrocyanosis, ulceration on cold exposure
2 - Monoclonal IgM with RF activity and polyclonal IgG	Can cause immune complex disorders; precipitates 1-7 days at 4°C; usually >1 mg/mL	Hepatitis C, lymphoproliferative diseases, connective tissue diseases including Sjögren syndrome; marked depression C4 with near normal C3	Arthralgias, glomerulonephritis, vasculitis, neuropathy, weakness, and palpable purpura
3 - Polyclonal IgM with RF activity and polyclonal IgG	Can cause immune complex disorders; may require 7 days at 4°C to precipitate; usually <1 mg/mL	Hepatitis C; other chronic infections including HIV, CMV, EBV, bacterial endocarditis, leprosy, spirochetal, fungal, and parasitic diseases; autoimmune disease including SLE, RA, biliary cirrhosis; and inflammatory bowel disease	Less often symptomatic than Type 2

References

Beddhu S, Bastacky S, and Johnson JP, "The Clinical and Morphologic Spectrum of Renal Cryoglobulinemia," *Medicine*, 2002, 81(5):398-409.

"Cryoglobulinemia: Out in the Cold," *Mayo References Services Communiqué*, November 2003, 1-4.

Kallemuchikkal U and Gorevic PD, "Evaluation of Cryoglobulins," *Arch Pathol Lab Med*, 1999, 123(2):119-25.

Keren DF, "Cryoglobulins, Immune Complexes, and Autoimmune Diseases," *Clinical and Laboratory Evaluation of Human Autoimmune Diseases*, Chapter 5, Nakamura RM, Keren DF, and Bylund DJ, eds, Chicago, IL: American Society for Clinical Pathology, ASCP Press, 2002, 61-70.

Maitra A, Ward PC, Kroft SH, et al, "Cytoplasmic Inclusions in Leukocytes: An Unusual Manifestation of Cryoglobulinemia," *Am J Clin Pathol*, 2000, 113(1):107-12.

Misiani R, Mantero G, Bellavita P, et al, "GB Virus C Infection in Patients With Type II Mixed Cryoglobulinemia," *Ann Intern Med*, 1997, 127(10):891-3.

Sloan SR, Benjamin RJ, Friedman DF, et al, "Transfusion Medicine," *Hematology of Infancy and Childhood*, 6th ed, Chapter 48, Nathan DG, Orkin SH, Ginsburg G, et al, eds, Philadelphia, PA: WB Saunders Co, 2003, 1746.

♦ **Cryohemolysis Test** *see* Hypertonic Cryohemolysis Test *on page 762*

Cryoprecipitate

Related Information

Activated Partial Thromboplastin Time *on page 100*
Factor Inhibitors *on page 566*
Factor VIII Concentrate *on page 563*
Fibrinogen *on page 583*
Kidney Stone Analysis *on page 820*
Plasma, Fresh Frozen *on page 1039*
Prothrombin Time *on page 1116*
von Willebrand Factor *on page 1321*

Synonyms Cryoprecipitated Antihemophilic Factor

Applies to Cryopreservation; Fibrin Glue; Fibrinogen Therapy; Hemophilia A Therapy; von Willebrand Disease Therapy

Abstract Cryoprecipitate is a labile component containing factor VIII:C, von Willebrand factor (factor VIII:vWF), fibrinogen, factor XIII, and fibronectin. Cryoprecipitate is the only concentrated available source of fibrinogen for patients with clinical deficiencies of that factor (eg, disseminated intravascular coagulation (DIC)).

A crossmatch is not necessary. Many units are sometimes needed. One concentrate per 5 kg body weight may serve as a rough guide to initial dosage.

Patient Preparation Prothrombin time, PTT, and fibrinogen assay to document indication (eg, hemophilia). ABO group (ABO compatible cryoprecipitate is preferred). Rh type need not be considered. Cryoprecipitate should be used only if viral-inactivated factor VIII concentrates are not available for patients with hemophilia A or von Willebrand disease.

Dosage and administration: Rapid administration of about 10 mL of diluted cryoprecipitate per minute is used as a loading dose for hemophilia, followed by a smaller dose at 12-hour intervals, depending on clinical circumstances. In pooling, single containers can be rinsed with 0.9% saline, so that the volume of six units of cryoprecipitate is 100-150 mL. In the presence of circulating anticoagulants, larger doses or other special measures may be indicated. Factor VIII activity should be >80 IU/bag.

A 70 kg patient should have an increase of about 2.5% AHF for each bag of cryoprecipitate given. For minor bleeding, dosage raising the patient's level to 30% to 50% may be used. For major surgical procedures, a preoperative dose should be sufficient to raise the level to 80% to 100%, followed by postoperative maintenance calculated to keep the level constantly >50% for 10-14 days.

In treatment of **von Willebrand disease**, smaller amounts given less often will usually suffice. When using cryoprecipitate, the factor VIII levels achieved from a calculated dose will vary. Use of a factor VIII concentrate which contains vWF is preferred to cryoprecipitate for most von Willebrand patients (those requiring treatment with plasma fractions).

Cryoprecipitate must be given through a filter.

To treat **hypofibrinogenemia**, one bag can be expected to raise plasma fibrinogen level about 7-10 mg/dL. A bag of cryoprecipitate provides at least 150 mg of fibrinogen.

Cryoprecipitate may also be used as a source of topical "fibrin glue" which can stop local bleeding after surgery, including cardiothoracic surgery. While it can be derived from autologous sources, using topical thrombin and calcium chloride to convert the fibrinogen in the cryoprecipitate to fibrin, factor V inhibitors have been reported to occur if bovine thrombin is used. The volume of the individual units of cryoprecipitate used for the fibrin glue should not exceed 15 mL.

Aftercare Factor VIII assay and activated partial thromboplastin time can serve as controls in therapy of hemophilia A and von Willebrand disease, and fibrinogen levels and thrombin time when hypofibrinogenemia is being treated. Ristocetin cofactor, factor VIII antigen, and vWF multimer are useful in monitoring cases of von Willebrand disease.

Specimen Blood

Container One red top tube or one lavender top (EDTA) tube
(Continued)

Cryoprecipitate *(Continued)*

Storage Instructions (For blood component) Cryoprecipitate requires frozen storage, without thawing, for up to 1 year at -18°C or below. Before infusion, thaw for up to 15 minutes in a water bath at 37°C in a plastic overwrap, so that the precipitate is dissolved. Pool multiple units before administration. Once thawed, store at room temperature. Cryoprecipitate ideally should be transfused within 2 hours or less, but not more than 6 hours after thawing and not more than 4 hours after pooling. Once thawed it cannot be refrozen.

Causes for Rejection (Of patient sample): Gross hemolysis, sample placed in a serum separator tube, specimen tube not properly labeled

Use Treatment of deficiency of coagulation factor VIII (hemophilia A), von Willebrand disease, and hypofibrinogenemic states only if viral-inactivated concentrates are not available. Replacement of fibrinogen should be considered when levels decrease to <100 mg/dL and patient is bleeding. The physician making such decisions should be aware of the coefficient of variation for fibrinogen levels in the laboratory being used. Prolongation of the thrombin time may support indications for infusion of fibrinogen as cryoprecipitate. Fibrin surgical adhesive ("fibrin glue") derived from cryoprecipitate can stop local bleeding during surgery. Cryoprecipitate is useful as a temporary treatment of bleeding tendency in uremia. It also provides factor XIII.

Limitations Cryoprecipitate is a poor source of factors II, V, IX, X, and XI. Due to development of factor V inhibitors in patients treated with "homemade" fibrin glue using bovine thrombin, safer and more effective commercial fibrin sealants are recommended.

Contraindications Do not use, unless laboratory or clinical studies indicate a specific coagulation defect for which cryoprecipitate is appropriate.

Additional Information Hazards: The risk of hepatitis and other viral infections is less than that of nonviral-inactivated concentrate because each bag comes from a single donor. Febrile and allergic reactions may occur. Large volumes of ABO incompatible cryoprecipitate may result in a positive direct antiglobulin test with mild hemolysis. Presence of acquired inhibitors to factor VIII makes treatment with cryoprecipitate difficult or impossible. Factor VIII concentrates will be needed for such patients.

References

Martinez J, "Quantitative and Qualitative Disorders of Fibrinogen," *Hematology: Basic Principles and Practice*, 3rd ed, Chapter 112, Hoffman R, Benz EJ Jr, Shattil SJ, et al, eds, New York, NY: Churchill Livingstone, 2000, 1924-36.

Pantanowitz L, Kruskall MS, and Uhl L, "Cryoprecipitate. Patterns of Use," *Am J Clin Pathol*, 2003, 119(6):874-81.

Streiff MB and Ness PM, "Acquired FV Inhibitors: A Needless Iatrogenic Complication of Bovine Thrombin Exposure," *Transfusion*, 2002, 42(1):18-26.

♦ **Cryoprecipitated Antihemophilic Factor** *see* Cryoprecipitate *on page 481*

♦ **Cryopreservation** *see* Autologous Transfusion, Preoperative Deposit *on page 223*

♦ **Cryopreservation** *see* Cryoprecipitate *on page 481*

♦ **Cryopreservation** *see* Frozen Red Blood Cells *on page 614*

♦ **Cryopreservation** *see* Hematopoietic Progenitor Cells, Cord Blood/Placental Blood *on page 676*

♦ **Cryopreservation** *see* Hematopoietic Progenitor Cells, Marrow *on page 677*

♦ **Cryopreservation** *see* Hematopoietic Progenitor Cells, Peripheral Blood *on page 679*

♦ **Cryopreservation** *see* Plasma, Fresh Frozen *on page 1039*

♦ **Cryopreservation** *see* Plasma, Frozen, Donor Retested *on page 1041*

Cryptococcal Antigen Titer

Related Information

Bacterial Antigens, Rapid Detection Methods *on page 228*
Cerebrospinal Fluid Analysis: Overview *on page 355*
Cerebrospinal Fluid Cytology *on page 361*
Fungal Culture, Blood *on page 620*
Fungal Culture, Cerebrospinal Fluid *on page 621*
Fungal Culture, Sputum *on page 624*
Fungus Smear, Stain *on page 626*
India Ink Preparation *on page 784*

Mycobacterial Culture, Cerebrospinal Fluid *on page 931*

Test Includes Testing patient's serum, CSF, or pleural fluid for the presence of cryptococcal capsular antigen

Abstract Cryptococcosis is caused by *Cryptococcus neoformans*, a yeast. This ubiquitous yeast is associated with droppings of birds, especially those of pigeons; infection begins with inhalation of the organism. Cryptococcal antigen testing on spinal fluid is the single most useful diagnostic test for cryptococcal meningitis. (The other diagnostic fundamentals are visualization of the organism and culture.)

Specimen Serum, cerebrospinal fluid, pleural fluid

Container Red top tube, sterile CSF tube

Reference Interval Negative

Use Diagnose subacute or chronic meningitis, particularly in immunosuppressed patients; rapid diagnosis of cryptococcal meningitis; monitor response of cryptococcal meningitis to therapy. Cryptococcosis may cause pulmonary, pleural, pericardial, skin, and bone lesions, and affects other organs as well (eg, prostate).

Limitations False positives and false negatives occur. Lack of standardization between manufacturers exists; thus, titers from different kits are not comparable. False-positive results may be seen in patients with *Capnocytophaga canimorsus* and disseminated *Trichosporon beigelii* infections. False-positive reactions usually have a titer ≤1:8.

Methodology Latex agglutination (LA) with rheumatoid factor control and in some laboratories pronase pretreatment

Additional Information Serum and CSF are positive in at least 90% of patients with cryptococcal meningitis. Cryptococcal antigen assays are very sensitive and may detect cryptococcal infection in culture-negative patients. Testing for cryptococcal antigen on other body fluids (eg, bronchoalveolar lavage fluid) is not standard and has been shown to be less sensitive when compared with culture.

References

Abadi J, Nachman S, Kressel AB, et al, "Cryptococcosis in Children With AIDS," *Clin Infect Dis*, 1999, 28(2):309-13.

Hajjeh RA, Conn LA, Stephens DS, et al, "Cryptococcosis: Population-Based Multistate Active Surveillance and Risk Factors in Human Immunodeficiency Virus-Infected Persons. Cryptococcal Active Surveillance Group," *J Infect Dis*, 1999, 179(2):449-54.

Lu CH, Chang WN, Chang HW, et al, "The Prognostic Factors of Cryptococcal Meningitis in HIV-Negative Patients," *J Hosp Infect*, 1999, 42(4):313-20.

Saag MS, Graybill RJ, Larsen RA, et al, "Practice Guidelines for the Management of Cryptococcal Disease. Infectious Diseases Society of America," *Clin Infect Dis*, 2000, 30(4):710-8.

Internet Web Sites

www.cdc.gov/ncidod/dbmd/diseaseinfo/cryptococcosis_a.htm

Cryptosporidium Antigen Detection by EIA

Related Information

Acid-Fast Stain, Modified, Parasites *on page 96*
Bacterial Culture, Stool *on page 243*
Cryptosporidium Direct Staining Procedures *on page 484*
Fungus Smear, Stain *on page 626*
Giardia intestinalis Diagnostic Procedures *on page 636*
Microsporidia Diagnostic Procedures *on page 915*
Ova and Parasites, Direct Exam *on page 985*

Test Includes Direct detection of cryptosporidial antigens in stool specimens

Abstract Human infections caused by the intracellular *Coccidia* parasites, including **Cryptosporidium parvum**, are manifest as diarrhea in both normal and immunocompromised subjects. Symptoms generally begin 2-10 days after exposure and usually last 2 weeks. Disease in immunocompromised individuals is more severe and prolonged. *Cryptosporidium* infection is sometimes the cause of traveler's diarrhea.

Specimen Fresh stool; stool preserved with 10% buffered formalin

Reference Interval Negative for *Cryptosporidium* antigen

Use Establish the diagnosis of cryptosporidiosis; a part of the differential work-up of diarrhea

Limitations Test must be specifically requested. Although some assays will simultaneously detect *Giardia lamblia* antigen as well as *Cryptosporidium*, they

(Continued)

Cryptosporidium Antigen Detection by EIA (Continued)

will not detect other causes of diarrhea. Most commercial assays have high specificity (98% to 100%) and sensitivity (76% to 98%).

Methodology Enzyme immunoassay (EIA), direct immunofluorescent antibodies

Additional Information Cryptosporidiosis is a known cause of severe and chronic diarrhea in immunocompromised patients and has also been associated with food-borne and water-borne outbreaks of disease. Cryptosporidium is a coccidian parasite that can live in the intestines of many animals, including humans. The parasite is able to survive outside the body for long periods of time and is very resistant to chlorine disinfection.

References

Clark DP, "New Insights Into Human Cryptosporidiosis," Clin Microbiol Rev, 1999, 12(4):554-63.

Goldstein ST, Juranek DD, Ravenholt O, et al, "Cryptosporidiosis: An Outbreak Associated With Drinking Water Despite State-of-the-Art Water Treatment," Ann Intern Med, 1996, 124(5):459-68.

Marsh WW, "Infectious Diseases of Gastrointestinal Tract in Adolescents," Adolesc Med, 2000, 11(2):263-78.

Nichols GL, "Food-Borne Protozoa," Br Med Bull, 2000, 56(1):209-35.

Internet Web Sites

www.cdc.gov/ncidod/dpd/parasites/cryptosporidiosis/factsht_cryptosporidiosis.htm

Cryptosporidium Direct Staining Procedures

Related Information

Acid-Fast Stain, Routine or Modified on page 95
Bacterial Culture, Stool on page 243
Cryptosporidium Antigen Detection by EIA on page 483
Fecal Lactoferrin on page 575
Fungus Smear, Stain on page 626
Methylene Blue Stain, Stool on page 906
Microsporidia Diagnostic Procedures on page 915
Ova and Parasites, Direct Exam on page 985
Polymerase Chain Reaction on page 1069

Test Includes Examination of stool for the presence of Cryptosporidium by phase contrast microscopy, modified acid-fast stain, or fluorescent-labeled antibody

Abstract Human infections caused by the intracellular Coccidia parasites, including **Cryptosporidium parvum**, are manifest as diarrhea in both normal and immunocompromised subjects. In humans, Cryptosporidium can be an enteric pathogen in all age groups, but disease in immunocompromised individuals is more severe and prolonged. Methods that utilize modified acid-fast stain or fluorescent antibody are frequently employed in the clinical laboratory, because **routine ova and parasite examination will not detect these organisms**.

Specimen Fresh stool; stool preserved with 10% formalin or sodium acetate-acetic acid formalin preservative

Reference Interval Negative

Use Establish the diagnosis of cryptosporidiosis by demonstration of the oocysts; can be used as part of the differential work-up of diarrhea

Limitations Cryptosporidium is not detected by standard methods used to examine stool specimens for other ova and parasites; special stains are required for its detection, and Cryptosporidium examination must be specifically requested. The organisms are most readily demonstrated in watery diarrheal stools. A single stool specimen is not sufficient to make the diagnosis. Most recommended procedures cannot be performed on polyvinyl alcohol (PVA) preserved specimens.

Methodology Phase contrast microscopy after floatation or sedimentation concentration techniques; modified acid-fast stain on air-dried, methanol-fixed smears (decolorize with 1% H_2SO_4). Fluorescent-labeled anti-Cryptosporidium antibodies are commercially available and provide excellent sensitivity and specificity. Antigen-capture enzyme-linked immunosorbent assays are often used. The disease may also be diagnosed by colonoscopic biopsies.

Additional Information Cryptosporidium parvum is a cause of severe and chronic diarrhea in patients with hypogammaglobulinemia and the acquired immune deficiency syndrome. HIV-infected patients with CD4 counts ≤50/mm are especially at risk when exposed to *Cryptosporidium*. Although the organism is widely recognized as a disease of the immunocompromised patient, it can also cause disease in immunocompetent subjects. Recognized as risk factors for the development of cryptosporidiosis is animal contact, travel to endemic areas, living in a rural environment, day care attendance by toddlers, and exposure to contaminated public water.

References
Chen XM, Keithly JS, Paya CV, et al, "Cryptosporidiosis," *N Engl J Med*, 2002, 346(22):1723-31.
Nichols GI, "Food-Borne Protozoa," *Br Med Bull*, 2000, 56(1):209-35.
Orenstein JM and Dieterich DT, "The Histopathology of 103 Consecutive Colonoscopy Biopsies From 82 Symptomatic Patients With Acquired Immunodeficiency Syndrome. Original and Look-Back Diagnosis," *Arch Pathol Lab Med*, 2001, 125(8):1042-6.

Internet Web Sites
www.cdc.gov/ncidod/dpd/parasites/cryptosporidiosis/factsht_cryptosporidiosis. htm

Cyanide, Blood

Related Information
Thiocyanate, Serum, Plasma, or Urine *on page 1245*

Synonyms CN⁻; Hydrocyanic Acid; Potassium or Sodium Cyanide

Abstract This highly toxic substance is one of the oldest poisons known. It binds to ferric ion of cytochrome oxidase and prevents cellular respiration, causing tissue hypoxia, severe lactic acidosis, high anion gap metabolic acidosis, and death. Cyanide can also cause lipid peroxidation in the brain due to inhibition of antioxidant enzymes. Pharmacokinetic estimates vary widely, probably depending on the circumstances of poisoning.

Specimen Whole blood, since cyanide is concentrated in erythrocytes. Venous blood may appear bright red. Serum may be used in some settings.

Container Lavender top (EDTA) tube preferred; red top (for serum)

Sampling Time Stat

Storage Instructions Fill tube to capacity and keep tightly closed; analyze as soon as possible.

Reference Interval
Whole blood cyanide:
- nonsmoker: 0.016 mg/L (SI: 0.61 µmol/L)

(Continued)

Cyanide, Blood (Continued)

- smoker: 0.041 mg/L (SI: 1.57 µmol/L)

Serum cyanide:
- nonsmoker: 0.004 mg/L (SI: 0.15 µmol/L)
- smoker: 0.006 mg/L (SI: 0.23 µmol/L)
- toxic: >0.1 mg/L (SI: >3.84 µmol/L)

Use Establish the diagnosis of cyanide poisoning

Methodology Photometric analysis after conway microdiffusion, ion-specific potentiometry

Additional Information Cyanide is found in insecticides, rodenticides, vermicides, metal polishes, and electroplating baths. It constitutes 44% of sodium nitroprusside by weight. Other sources include ore refining, laetrile, synthetic rubber manufacturing, and the seeds of cherries, plums, peaches, apricots, pears, apples, crab apples, chokeberries, and lima beans. Some cyanide poisoning occurs among victims of fires, since nitrogen-containing polymers produce cyanide from combustion. In such situations, carbon monoxide poisoning may coexist with cyanide poisoning. Fires which involve urea foam insulation may produce hydrocyanic acid, which may be inhaled. Symptoms of toxicity include headache, agitation, vomiting, and confusion. A scent of bitter almonds is suggestive, but not all individuals can detect it.

The standard treatment of cyanide toxicity includes infusion of nitrites to induce methemoglobinemia and sodium thiosulfate solution to convert cyanide to less toxic thiocyanate. FDA-approved cyanide kits also include amyl nitrite inhalant for immediate use. Hydroxycobalamin is also a cyanide antidote. The serial measurement of plasma lactate concentrations is useful in assessing the severity of cyanide poisoning.

References

Baud FJ, Borron SW, Megarbane B, et al, "Value of Lactic Acidosis in the Assessment of the Severity of Acute Cyanide Poisoning," *Crit Care Med*, 2002, 30(9):2044-50.

Painter PC, Cope JY, and Smith JL, "Reference Information for the Clinical Laboratory," *Tietz Textbook of Clinical Chemistry*, 3rd ed, Burtis CA and Ashwood ER, eds, Philadelphia, PA: WB Saunders Co, 1999, 1809.

Sauer SW and Keim ME, "Hydroxycobalamin: Improved Public Health Readiness for Cyanide Disasters," *Ann Emerg Med*, 2001, 37(6):635-41.

♦ **Cyanocobalamin** see Cobalamin, Serum on page 424

♦ **Cyanogen Chloride** see Chemical Warfare Agents on page 382

♦ **3', 5'-Cyclic Adenosine Monophosphate, Plasma** see Cyclic AMP, Plasma on page 486

♦ **Cyclic Adenosine Monophosphate, Urine** see Cyclic AMP, Urine on page 486

♦ **3', 5'-Cyclic Adenosine Monophosphate, Urine** see Cyclic AMP, Urine on page 486

Cyclic AMP, Plasma

Related Information

Calcium, Ionized, Serum on page 328

Parathyroid Hormone, Serum on page 1001

Synonyms AMP, Cyclic, Plasma; cAMP, Plasma; 3', 5'-Cyclic Adenosine Monophosphate, Plasma

Abstract Cyclic adenosine monophosphate (cAMP) is an intracellular second messenger which serves as an effector in mediating the action of several peptide hormones. Although once used as an adjunct in the diagnosis of primary hyperparathyroidism, there are now better ways to approach this diagnosis. On rare occasions, urinary cAMP is measured in the evaluation of pseudohypoparathyroidism, and plasma specimens are often a required accompaniment.

References

Logue FC, Fraser WD, Gallacher SJ, et al, "The Loss of Circadian Rhythm for Intact Parathyroid Hormone and Nephrogenous Cyclic AMP in Patients With Primary Hyperparathyroidism," *Clin Endocrinol (Oxf)*, 1990, 32(4):475-83.

Cyclic AMP, Urine

Related Information

Calcium, Ionized, Serum on page 328

Calcium, Serum on page 329

Parathyroid Hormone, Serum *on page 1001*

Synonyms AMP, Cyclic, Urine; cAMP, Urine; Cyclic Adenosine Monophosphate, Urine; 3', 5'-Cyclic Adenosine Monophosphate, Urine

Abstract Cyclic adenosine monophosphate (cAMP), an intracellular second messenger, serves as an effector in mediating the action of several peptide hormones, including parathyroid hormone (PTH). In normal persons, the infusion of PTH produces a sharp increase in urinary cAMP; this spike is not seen in persons with pseudohypoparathyroidism (the syndrome of target organ resistance to PTH).

Patient Preparation Following an overnight fast, a fluid intake of 400 mL/hour should be maintained for the 2 hours preceding the test until the test is concluded.

Specimen Six consecutive 30-minute urine specimens; a plasma specimen at the midpoint of each 30-minute interval; specimens are assayed for cAMP, phosphorus, and creatinine.

Sampling Time See Cundy et al, page 114. At the beginning of the fourth urine collection interval, synthetic PTH (teriparatide - a product unavailable at the time of this writing), 3 units/kg to a maximum of 200 units, is given I.V. over a 10-minute period.

Reference Interval

Random specimen: 1.3-3.7 nmol/dL glomerular filtrate

Stimulation testing with synthetic PTH: A normal response to PTH is a 10-12-fold increase in cAMP excretion, and a 20% decrease in the ratio of the clearance of phosphate relative to the creatinine clearance (TmP/GFR). In states of resistance to PTH, the following are observed:

- Type I pseudohypoparathyroidism: Less than a fivefold increase in urinary cAMP and <10% fall in TmP/GFR.
- Type II pseudohypoparathyroidism: Normal increase in urinary cAMP and <10% fall in TmP/GFR.

Use

Pseudohypoparathyroidism: Pseudohypoparathyroidism names a group of disorders characterized by high or normal serum levels of PTH, end-organ resistance to the action of PTH, and hypocalcemia. These are among the least common causes of hypocalcemia. The biochemical confirmation of pseudohypoparathyroidism involves measuring the urinary cAMP response to intravenous synthetic PTH. The testing protocol is described in the reference, Cundy et al.

Primary hyperparathyroidism: Test protocols involving the assay of cAMP were once used to diagnose primary hyperparathyroidism in a small number of hypercalcemic patients with otherwise equivocal findings. These tests are no longer necessary since better tests are now available (see Parathyroid Hormone, Serum *on page 1001*). In addition, the cAMP-based tests are nonspecific since increased urinary cAMP is found both in primary hyperparathyroidism and in many patients with humoral hypercalcemia of malignancy.

Methodology High performance liquid chromatography (HPLC)

References

Cundy T and Reid I, "Calcium, Phosphate and Magnesium," *Clinical Biochemistry*, Marshall WJ and Bangert SK, eds, New York, NY: Churchill Livingstone, 1995, 99-115.

Spiegel AM, "The Parathyroid Glands, Hypercalcemia and Hypocalcemia," *Cecil Textbook of Medicine*, 21st ed, Goldman L and Bennett JC, eds, Philadelphia, PA: Saunders Co, 2000, 1403-4.

♦ **Cyclic Antidepressants** *see* Antidepressants, Cyclic, Serum or Plasma *on page 171*

♦ **Cyclosporine A** *see* Cyclosporine, Blood *on page 487*

Cyclosporine, Blood

Related Information

Itraconazole, Serum *on page 814*
Phenytoin, Serum or Plasma *on page 1026*
Rapamycin, Blood *on page 1132*
Tacrolimus, Whole Blood *on page 1234*
Verapamil, Serum or Plasma *on page 1306*

Synonyms Neoral®; Sandimmune®

Applies to Cyclosporine A

(Continued)

Cyclosporine, Blood (Continued)

Abstract Cyclosporine is a cyclic polypeptide widely used as an immunosuppressant, especially following organ transplants. It is generally used in combination with other immunosuppressants, such as azathioprine and corticosteroids. Other uses include treatment of methotrexate unresponsive rheumatoid arthritis, recalcitrant plaque psoriasis in nonimmunocompromised patients, and keratoconjunctivitis sicca.

Specimen Whole blood

Container Lavender top (EDTA) tube

Sampling Time Trough levels should be obtained 12-18 hours after oral dose (chronic usage), 12 hours after intravenous dose, or immediately prior to next dose. In recent years, measurement of 2-hour postdose level (called C2 level) has been suggested as a sensitive predictor of clinical outcome in organ transplantation.

Collection Draw from a different line than that through which the dose was given. If C2 levels are drawn, it is very important to draw sample exactly 2 hours after the dose.

Reference Interval 100-300 ng/mL for renal transplant; 200-350 ng/mL for cardiac, hepatic, and pancreatic transplant (12 hours after oral dose). C2 levels are 3-6 times higher than trough levels. Since cyclosporine binds to erythrocytes and lipoproteins, measurement of whole blood concentrations is preferred. Therapeutic ranges are poorly defined. They relate to the organ transplanted, time following transplantation, and organ function. They are method and specimen dependent.

Critical Values >400 ng/mL

Use Monitor blood level in management of immunosuppression for organ transplant recipients. The agent is used extensively to control rejection of organ transplants, especially of liver, heart, bone marrow, and kidney.

Monitoring of blood levels is imperative because the pharmacokinetics of cyclosporine are complex and vary over time in the same patient; thus, blood levels cannot be well predicted from dosing schedules. This drug has a narrow therapeutic window and significant toxicity at levels above that range.

Absorption of cyclosporine in the form of Sandimmune® is highly variable. The newer, microemulsion form of cyclosporine (Neoral®) provides more reproducible absorption.

Limitations Results are method dependent - some measure multiple metabolites as well as parent drug. Single assays are not as informative as a series over time.

Methodology Immunoassay, high performance liquid chromatography (HPLC), tandem mass spectrometry

Additional Information
- Half-life: 8-24 hours
- Volume of distribution: 4-6 L/kg
- Protein binding: 90%
- Time to reach steady state: 2-6 days

The exact mechanism of action of this immunosuppressive agent is not known, but its action appears to be associated with alterations in the functions of helper and effector T lymphocytes and natural killer cells. It is not myelosuppressive.

Renal toxicity with eventual renal failure is the most severe complication. Cyclosporine diminishes glomerular filtration rate. Other assays to assess renal function are recommended: BUN and creatinine clearance should be considered along with cyclosporine levels, since toxicity may begin even with "acceptable" blood levels. Other important toxic effects include hypertension (in >90% of heart transplant recipients in the first year), convulsions, tremors, pulmonary edema, and an increased risk of lymphoma. Concomitant use of gentamicin, tobramycin, vancomycin, trimethoprim, and sulfamethoxazole may potentiate renal dysfunction.

There are many drugs which affect cyclosporine pharmacokinetics, the most common being those which inhibit or induce P450 enzyme system. Drugs which enhance the potential toxicity of cyclosporine, and which are also likely to be administered to a transplant recipient, include acyclovir, aminoglycoside antibiotics, amphotericin B, cephalosporins, furosemide, ketoconazole, and trimethoprim-sulfa. **Agents which raise cyclosporine levels** by decreasing biotransformation include amphotericin B, cimetidine, erythromycin, and methylprednisolone. Other drugs which lead to increased concentrations of cyclosporine include androgens, diltiazem, ketoconazole, methotrexate, nicardipine, oral contraceptives, and verapamil. Drugs which increase hepatic metabolism and thus **lower cyclosporine levels** include carbamazepine, ethotoin, intravenous trimethoprim-sulfa, mephenytoin, phenobarbital, phenytoin, primidone, and rifampin.

Because results will vary depending on whether the assay is done on whole blood or serum/plasma, and on the method and cyclosporine antibody employed (monospecific or polyspecific), it is best for a given patient's specimens to be analyzed at a single laboratory.

References

Armstrong VW and Oellerich M, "New Developments in the Immunosuppressive Drug Monitoring of Cyclosporine, Tacrolimus, and Azathioprine," *Clin Biochem*, 2001, 34(1):9-16.

Frei U, "Overview of the Clinical Experience With Neoral® in Transplantation," *Transplant Proc*, 1999, 31(3):1669-74.

Saint-Marcoux F, Rousseau A, Le Meur Y, et al, "Influence of Sampling-Time Error on Cyclosporine Measurements Nominally at 2 Hours After Administration," *Clin Chem*, 2003, 49(5):813-5.

Streit F, Armstrong VW, and Oellerich M, "Rapid Liquid Chromatography - Tandem Mass Spectrometry Routine Method for Simultaneous Determination of Sirolimus, Everolimus, Tacrolimus, and Cyclosporin A in Whole Blood," *Clin Chem*, 2002, 48(6 Pt 1):955-8.

♦ **Cyst Aspiration Cytology** *see* Fine Needle Aspiration, Deep and Superficial Masses *on page 590*

♦ **Cystathionine** *see* Amino Acids, Urine *on page 145*

Cystatin C, Serum or Plasma

Related Information

Creatinine Clearance and Urine Creatinine *on page 473*

Creatinine, Serum or Plasma *on page 474*

Applies to Glomerular Filtration Rate

Abstract The serum level of cystatin C is proportional to glomerular filtration rate (GFR), and serum cystatin C may be more useful than creatinine and creatine clearance as a test of overall renal function.

Specimen Serum, plasma (EDTA or heparin)

Storage Instructions Stable for 7 days at 20°C, for 6 months at -80°C.

Reference Interval Children older than 1 year of age to adulthood: 0.63-1.33 mg/L

Use Cystatin C provides estimation of glomerular filtration rate; assessment of allograft function and therapeutic nephrotoxicity. It is independent of muscle mass. A cutoff concentration of 1.39 mg/L provided 90% sensitivity and 86% specificity for detection of abnormal GFR.

Limitations Greater cost than that of creatinine concentration. Assays are not widely available.

Methodology Latex particle enhanced turbidimetric or nephelometric immunoassay; other methods

Additional Information Cystatin C is freely filtered in the glomeruli and catabolized in the tubules. No extrarenal routes of elimination are known. Most of the studies of this analyte have been provided by European investigators. Its low molecular weight and high pI (9.2) enables it to be freely filtered through the glomerulus. Cystatin C is synthesized by all nucleated cells at a constant rate. Using receiver operator analysis, cystatin C is superior to creatinine in the assessment of GFR.

References

Bokenkamp A, Domanetzki M, Zinck R, et al, "Cystatin C - A New Marker of Glomerular Filtration Rate in Children Independent of Age and Height," *Pediatrics*, 1998, 101(5):875-81.

Finney H, Newman DJ, Thakkar H, et al, "Reference Ranges for Plasma Cystatin C and Creatinine Measurements in Premature Infants, Neonates, and Older Children," *Arch Dis Child*, 2000, 82(1):71-5.

Fliser D and Ritz E, "Serum Cystatin C Concentration as a Marker of Renal Dysfunction in the Elderly," *Am J Kidney Dis*, 2001, 37(1):79-83.

Helin I, Axenram M, and Grubb A, "Serum Cystatin C as a Determinant of Glomerular Filtration Rate in Children," *Clin Nephrol*, 1998, 49(4):221-5.

Keevil BG, Kilpatrick ES, Nichols SP, et al, "Biological Variation of Cystatin C: Implications for the Assessment of Glomerular Filtration Rate," *Clin Chem*, 1998, 44(7):1535-9.

Le Bricon T, Thervet E, Benlakehal M, et al, "Changes in Plasma Cystatin C After Renal Transplantation and Acute Rejection in Adults," *Clin Chem*, 1999, 45(12):2243-9.

(Continued)

Cystatin C, Serum or Plasma *(Continued)*

Norlund L, Fex G, Lanke J, et al, "Reference Intervals for the Glomerular Filtration Rate and Cell-Proliferation Markers: Serum Cystatin C and Serum Beta 2-Microglobulin/Cystatin C-Ratio," *Scand J Clin Lab Invest*, 1997, 57(6):463-70.

♦ **Cysteine, Qualitative** *see* Cystine, Urine *on page 494*

♦ **Cyst Fluid Analysis** *see* Body Fluid Analysis, Cell Count *on page 288*

♦ **Cyst Fluid Chemistry** *see* Body Fluid Chemical Analysis *on page 291*

Cyst Fluid Cytology

Related Information

Body Cavity Fluid Cytology *on page 285*
Body Fluid Amylase *on page 287*
Body Fluid Chemical Analysis *on page 291*
Breast Biopsy *on page 305*
Fine Needle Aspiration, Deep and Superficial Masses *on page 590*
Washing Cytology *on page 1326*

Applies to Brain Cyst Fluid Cytology; Breast Cyst Fluid Cytology; Hepatic Cyst fluid Cytology; Hydrocele Fluid Cytology; Ovarian Cyst Fluid Cytology; Pancreatic Cyst Fluid Cytology; Renal Cyst Fluid Cytology; Thyroid Cyst Fluid Cytology

Test Includes Cytologic evaluation of direct smears, cytocentrifuge (cytospin) preparations and/or monolayer preparations most commonly; membrane filter preparations and cell blocks are less routinely used.

Abstract Fine needle aspiration is a safe and reliable way of diagnosing non-neoplastic cysts as well as benign and malignant cystic neoplasms.

Collection Any palpable residual mass present following cyst drainage should be aspirated.

Storage Instructions If a delay in processing must occur, the cyst fluid specimen should be refrigerated.

Use The majority of cysts are non-neoplastic and aspiration may be therapeutic as well as diagnostic. However, cystic neoplasms and malignancies with a cystic component do occur and represent a common source of false-negative cytologic diagnoses. It cannot be overemphasized that any residual mass following cyst drainage must be reaspirated to ensure that the lesion has been adequately sampled.

Aspirates of hydatid hepatic cysts contain scolices. Hooklets remain in old cysts. The two major types of hepatic abscess are pyogenic and amoebic, entities which can be diagnosed by cytologic means and microbiologic procedures.

Limitations False-negative diagnoses may occur due to failure to sample malignant cells, even with appropriate aspiration technique.

Fluids may contain degenerated material. Distinction of histiocytes from epithelial cells can be difficult by light microscopy. Sometimes, in such circumstances, electron microscopy may prove useful.

References

Bardales RH, "Fine Needle Aspiration Cytology of Papillary Neoplasms," *Clinics in Laboratory Medicine: Fine Needle Aspiration*, Volume 18, No 3, Stanley MW, ed, Philadelphia, PA: WB Saunders Co, 1998, 373-99.

Tao LC, "Liver and Pancreas," *Comprehensive Cytopathology*, 2nd ed, Chapter 32, Bibbo M, ed, Philadelphia, PA: WB Saunders Co, 1997, 827-63.

♦ **Cystic Echinococcosis** *see* Echinococcosis Diagnostic Procedures *on page 530*

Cysticercosis Serology

Related Information

Ova and Parasites, Direct Exam *on page 985*

Test Includes Detection of antibodies specific for *Taenia solium*

Abstract Eggs of *Taenia solium*, the pork tapeworm acquired from contact with contaminated feces, lead to cysticercosis. *T. solium* is the only tapeworm for which humans are the intermediate host (harboring larval forms) and the definitive host (harboring the adult tapeworm). The human ingests oncospheres (embryos) which are absorbed, then embolize to striated muscle, eyes, and the central nervous system, becoming cysticerci. Cysticercosis, larval forms in tissues, is endemic in Central and portions of South America, Africa, and Asia.

Neurocysticercosis, the overwhelming clinical problem, is the expression used for involvement of *T. solium* cysts in the central nervous system.

Specimen Serum, cerebrospinal fluid

Container Red top tube

Use Support the diagnosis of cysticercosis. The most common manifestation of neurocysticercosis is epilepsy.

Limitations Cross reactions in patients with tapeworm or *Echinococcus* have been found. Sensitivity remains limited when there is low parasite burden (ie, false negatives occur). Studies have shown that only 28% of patients with a single parenchymal lesion will have a positive serologic test. The CDC immunoblot is more specific and sensitive than enzyme immunoassay (EIA). The CDC immunoblot is the immunodiagnostic test of choice for confirmation of a clinical or radiologic diagnosis of neurocysticercosis.

Methodology Immunoblotting, enzyme-linked immunosorbent assay (ELISA). The most accurate test is the enzyme-linked immunotransfer blot.

Additional Information Ingestion of eggs of the pork tapeworm (*Taenia solium*) produces cysticercosis, an infection in which larval cysts (cysticerci) are seen in various tissues. Water or food may become contaminated with eggs, especially in areas in which water purification systems are inadequate. Cysticercosis is most commonly found in brain or muscle, in which a space-occupying mass presents with local inflammatory reaction. Serious CNS involvement is often characterized as intracerebral lesions causing mass effects, seizures, or both. The CNS is involved in 90% of patients: mild CSF pleocytosis is found with increased CSF protein and decreased glucose.

References

Baily GG, "Other Cestode Infections: Intestinal Cestodes, Cysticercosis, Other Larval Cestode Infections," *Manson's Tropical Diseases*, 21st ed, Chapter 85, Cook GC and Zumla A, eds, Philadelphia, PA: WB Saunders Co, 2003, 1583-8.

Del Brutto OH, Dolezal M, Castillo PR, et al, "Neurocysticercosis and Oncogenesis," *Arch Med Res*, 2000, 31(2):151-5.

Garcia HH and Del Brutto OH, "*Taenia solium* Cysticercosis," *Infect Dis Clin North Am*, 2000, 14(1):97-119.

Sotelo J and Del Brutto OH, "Brain Cysticercosis," *Arch Med Res*, 2000, 31(3):3-14.

White AC, "Neurocysticercosis: Updates on Epidemiology, Pathogenesis, Diagnosis, and Management," *Ann Rev Med*, 2000, 51:187-206.

♦ **Cystic Fibrosis, Carrier Testing** *see* Cystic Fibrosis DNA Detection *on page 491*

Cystic Fibrosis DNA Detection

Related Information

Amniotic Fluid, Chromosome and Genetic Abnormality Analysis *on page 152*

Chloride, Sweat *on page 392*

Chorionic Villus Sampling, Chromosome and Genetic Abnormality Analysis *on page 400*

Polymerase Chain Reaction *on page 1069*

Synonyms *CFTR* Gene Mutation Analysis

Applies to Cystic Fibrosis, Carrier Testing; Cystic Fibrosis, Prenatal Diagnosis; Cystic Fibrosis Transmembrane Conductance Regulator; Delta F508

Test Includes Detection of the 10-30 most common mutations causing cystic fibrosis

Abstract Cystic fibrosis (CF) is an autosomal recessive progressive disease caused by mutations in a gene called *CFTR* (cystic fibrosis transmembrane conductance regulator). About 4% of the Caucasian population are carriers. Its prevalence is 1 in 2500-3300 live births. There are nearly 1000 known mutations in the *CFTR* gene, but the 10 most common mutations account for about 80% to 85% of all mutations in Caucasians. DNA-based testing can be used to diagnose cystic fibrosis, to detect carriers of the disease, and to perform prenatal diagnosis. It is also useful in the evaluation of fetuses with ultrasound findings suspicious for cystic fibrosis, and in the evaluation of men with infertility due to congenital absence of the vas deferens. It is occasionally used in the evaluation of patients with idiopathic chronic pancreatitis. Failure to detect gene mutations of CF does not rule out the diagnosis. CF is characterized by a broad spectrum of disease severity, characterized predominantly by frequent respiratory infections and pancreatic insufficiency.

(Continued)

Cystic Fibrosis DNA Detection *(Continued)*

Patient Preparation Informed consent is recommended for genetic testing and is required in several states.

Specimen 3-10 mL whole blood, 10-20 mL amniotic fluid, 1 T25 flask of cultured amniocytes or chorionic villi, 5-10 mg wet chorionic villi

Container Lavender top (EDTA) or yellow top (ACD) tube; avoid use of tubes containing heparin anticoagulants, which can interfere with polymerase chain reaction analysis. Amniotic fluid and chorionic villus biopsies should be collected in a sterile manner. Syringes and tubes should not contain additives that interfere with cell culture methods.

Sampling Time Blood samples can be taken at any time. Amniotic fluid should be collected between the 14th and 16th week of gestation. Chorionic villus specimens should be collected between the 8th and 12th week of gestation.

Storage Instructions Store blood samples refrigerated or at room temperature. Do not freeze. Transport amniotic fluid or chorionic villus biopsy samples to the laboratory immediately.

Turnaround Time 7-14 days

Special Instructions To provide optimal interpretation and risk calculation, the testing laboratory needs to know the patient's diagnosis, ethnic background, and family history/pedigree.

Reference Interval A normal result is the absence of detectable mutations (see Limitations). An abnormal result is the detection of either one or two mutations.

Use Indications for testing include:

- Diagnosis of suspected CF: Detection of two mutant alleles confirms the diagnosis of CF. In the Caucasian population, about 70% to 80% of patients with CF have two detectable mutations, about 20% to 30% have one detectable mutation, and about 1% to 2% have no detectable mutations.
- Carrier detection in patient with family history of CF and in his/her reproductive partner: Detection of one mutant allele indicates carrier status, and is useful for reproductive planning. In the northern European Caucasian population, about 85% to 90% of carriers have one detectable mutation. CF carrier DNA screening has been recommended.
- Carrier detection in parents of fetus with echogenic bowel: CF is one explanation for the presence of echogenic bowel detected by ultrasound during the 2nd trimester. To help rule out this explanation, both parents can be tested for *CFTR* mutations. If a fetal sample is already being obtained for other purposes (eg, chromosome analysis), the fetal sample can be tested directly.
- Prenatal diagnosis in a fetus at risk for CF: When both parents carry known mutations in the *CFTR* gene, their fetus can be tested to determine if it has inherited one or both mutant alleles.
- Investigation, baby with large foul-smelling stools.
- Investigation, baby with hepatosplenomegaly.
- Newborn screening with sweat testing.
- Meconium ileus is found in 5% to 10% of affected infants.
- Evaluation of infertile men with congenital bilateral absence of the vas deferens (CBAVD): This clinical syndrome is often caused by "mild" mutations in the *CFTR* gene that do not manifest as classical CF with pulmonary and pancreatic symptoms. CFTR gene mutations are found in 70% of individuals with CBAVD or atrophy of the vas deferens. CBAVD is the cause of 2% to 6% of male infertility.
- Evaluation of individuals with progressive/recurrent pulmonary infection, thick sputum, with other evidence of possible CF.
- Evaluation of chronic sinusitis.
- Evaluation of patients with pancreatic exocrine deficiency; evaluation of patients with idiopathic chronic pancreatitis: Some 15% to 35% of patients with idiopathic chronic pancreatitis have at least one detectable *CFTR* mutation but do not fulfill the criteria for a diagnosis of cystic fibrosis.
- Evaluation of diabetes (in appropriate clinical setting).
- DNA testing may be useful when results of pilocarpine iontophoresis sweat test are negative or equivocal, especially in subjects with clinical features of CF.

Limitations Current technology will detect 10-80 of the most common CF mutations, which account for 80% to 90% of the mutant alleles in the Caucasian population. Twenty-five to 33 mutations are recognized by commercially available systems. Thus, a negative result does not rule out the possibility that an individual is a CF carrier but can lower that probability by a factor of 5-10. Similarly, detection of zero mutations does not completely rule out CF, but makes the diagnosis unlikely. Pitfalls exist in the genetic analysis for CF.

Sweat chloride analysis can provide additional information in the diagnosis of CF (see Chloride, Sweat *on page 392*), and is particularly useful when the DNA test for CF is negative. Two sweat chloride determinations on different occasions, >60 mEq/L, have been the gold standard, but sensitivity is only 90% and carriers are not detected.

Methodology DNA is isolated from the specimen and several regions in the *CFTR* gene are amplified using polymerase chain reaction (PCR). Mutations are detected by gel electrophoresis of amplified DNA, by restriction-enzyme digestion, or by hybridization with oligonucleotide probes specific for the mutations. See 2003 reviews by Lyon and Miller and by Lewis et al for further information.

Additional Information The *CFTR* is a large gene located on chromosome 7q31.2. It encodes a transmembrane protein of 1480 amino acids called the CF transmembrane conductance regulator (CFTR). This protein is a regulated chloride channel present in the epithelia of the lung, the exocrine pancreas, and sweat glands. Mutations in the *CFTR* gene can prevent proper expression of the protein, or can impair its function as a chloride channel.

The most common mutation causing CF is called ΔF508del (deltaF508). This mutation accounts for about 70% of *CFTR* mutant alleles in Caucasians, while the 10 next most common mutations account for about 10% of the *CFTR* mutant alleles. Nearly 1000 other mutations are known, most very rare or limited to a single family. Currently available clinical test procedures focus on the 10-80 most common mutations, and thus do not detect rare mutations that may be present.

In Wang's study of patients with rhinosinusitis, the proportion of patients with a CF mutation was higher than in a control group. Nine of 10 CF carriers had the polymorphism M470V. M470V homozygotes were over-represented.

Mortality in 90% of patients relates to respiratory insufficiency.

Immunologic, microbiologic, and pathologic features of CF are described in a 2003 paper (Lewis et al).

References

Chmiel JF, Drumm ML, Konstan MW, et al, "Pitfall in the Use of Genotype Analysis as the Sole Diagnostic Criterion for Cystic Fibrosis," *Pediatrics*, 1999, 103(4 Pt 1):823-6.

Grody WW, "Cystic Fibrosis: Molecular Diagnosis, Population Screening, and Public Policy," *Arch Pathol Lab Med*, 1999, 123(11):1041-6.

Hilman BC and Constantinesco M, "Role of DNA Testing in Cystic Fibrosis," *Lab Med*, 1999, 30(1):48-55.

Lewis MJ, Lewis EH III, Amos JA, et al, "Cystic Fibrosis," *Am J Clin Pathol*, 2003, 120(Suppl 1):S3-S13.

Lyon E and Miller C, "Current Challenges in Cystic Fibrosis Screening," *Arch Pathol Lab Med*, 2003, 127(9):1133-9.

Mitchell RMS, Byrne MF, and Baillie J, "Pancreatitis," *Lancet*, 2003, 361(9367):1447-55.

Sharer N, Schwarz M, Malone G, et al, "Mutations of the Cystic Fibrosis Gene in Patients With Chronic Pancreatitis," *N Engl J Med*, 1998, 339(10):645-52.

Wang XJ, Moylan B, Leopold DA, et al, "Mutation in the Gene Responsible for Cystic Fibrosis and Predisposition to Chronic Rhinosinusitis in the General Population," *JAMA*, 2000, 284(14):1814-9.

Internet Web Sites

www.genet.sickkids.on.ca/cftr/
odp.od.nih.gov/consensus/cons/106/106_statement.htm

Cystine, Urine
Related Information
Amino Acids, Urine *on page 145*
Kidney Stone Analysis *on page 820*
Urinalysis *on page 1289*

Applies to Arginine; Cysteine, Qualitative; Homocyst(e)ine, Qualitative; Lysine; Nitroprusside Screening; Ornithine

Test Includes Homocystine, cysteine

Abstract Cystinuria is an autosomal recessive disease in which excessive dibasic amino acids are excreted in the urine. Patients with cystine urinary stones have repeated urinary tract infections.

Patient Preparation Penicillamine (a chelating agent) can cause false-negative results.

Specimen Random urine

Collection Random urine or 24-hour collection for quantitation

Storage Instructions Acidify to pH 2-3 or freeze specimen at -20°C, or 20 mL toluene can be added to the container prior to the start of a 24-hour collection.

Reference Interval Normal: 40-60 mg cystine/g creatinine; heterozygotes: <300 mg/g; homozygotes: >250 mg/g

Use Detect cystinuria, homocystinuria and other diseases related to the sulfur-containing amino acids. Work up nephrolithiasis. Cystine stones account for 1% to 3% of renal calculi.

Limitations Cystinosis, a different entity from **cystinuria,** is not detected by this test. Patients with cystinosis are diagnosed with cystine crystals in biopsies, corneal crystals on slit lamp examination or elevated leukocyte cystine levels.

Methodology Microscopic examination of the sediment of a first morning urine sample can include the hexagonal crystals in samples from homozygotes.

Nitroprusside (cyanide-nitroprusside) screening test is positive with cystine or homocystine. The urine nitroprusside test reacts positively at levels of 75-125 mg cystine/g creatinine, but false positives occur. High performance liquid chromatography (HPLC), ion exchange chromatography are used for quantitation and distinction of cystine from homocyst(e)ine, as well as for confirmation of nitroprusside results.

Additional Information Classical cystinuria is the most common inborn error of amino acid transport. A positive screening test should be followed up by a quantitative procedure for cystine. In cystinosis, plasma cystine is usually normal, but increased cystine may be found in tissues.

References
Asplin JR, Coe FL, and Favus MJ, "Nephrolithiasis," *Harrison's Principles of Internal Medicine*, 14th ed, Chapter 279, Fauci AS, Braunwald E, Isselbacher KJ, et al, eds, New York, NY: McGraw-Hill Inc, 1998, 1569-74.

Elsas LJ, Longo N, and Rosenberg LE, "Inherited Disorders of Amino Acid Metabolism and Storage," *Harrison's Principles of Internal Medicine*, 14th ed, Chapter 349, Fauci AS, Braunwald E, Isselbacher KJ, et al, eds, New York, NY: McGraw-Hill Inc, 1998, 2194-203.

Ng CS and Streem SB, "Contemporary Management of Cystinuria," *J Endourol*, 1999, 13(9):647-51.

♦ **Cystodigin**® *see* Digitoxin, Serum *on page 512*

♦ **Cytapheresis, Therapeutic** *see* Apheresis, Therapeutic *on page 201*

♦ **Cytokeratin** *see* Sentinel Lymph Node Biopsy *on page 1189*

♦ **Cytokeratins** *see* Immunoperoxidase Procedures *on page 780*

♦ **Cytology Smear Identification** *see* Specimen Identification Requirements *on page 1217*

♦ **Cytomegalic Inclusion Disease Cytology** *see* Cytomegalovirus Cytology *on page 497*

Cytomegalovirus Antigen Detection
Related Information
Cytomegalovirus Culture *on page 495*
Cytomegalovirus Cytology *on page 497*
Cytomegalovirus Nucleic Acid Detection *on page 498*
Cytomegalovirus Serology *on page 499*
Virus, Direct Detection by Fluorescent Antibody *on page 1311*

Synonyms CMV pp65 Detection

Test Includes Detection of CMV antigen in white blood cells

Abstract Active human cytomegalovirus (CMV) infection can be detected by monitoring the presence of the CMV matrix protein (pp65) in peripheral blood leukocytes. The CMV antigenemia assay has been used to quantitate the amount of active CMV in immunocompromised patients who are a high risk for severe CMV infections. This assay has been shown to be sensitive and specific for active CMV replication.

Specimen Whole blood

Container Green top (heparin) tube, blue top (sodium citrate) tube, acid citrate dextrose or lavender top (EDTA) tube

Collection Transport to laboratory at room temperature within 2 hours of collection. Process specimen within 6 hours of collection.

Storage Instructions Keep at 4°C; do not freeze.

Use Early diagnosis and monitoring of CMV infection in immunocompromised patients

Limitations Labor-intensive; time-consuming due to isolation and counting of white blood cells; microscopist must be well-trained to detect positive cells; a minimum of 50,000 cells should be available on the slide in order to determine a negative result. The assay is not standardized between laboratories. CMV may be present for a short time following acute infections. Some patients with CMV antigenemia are asymptomatic. Specimen must be processed quickly (within 6 hours) to detect viral antigen and, thus, the specimen should not be collected during laboratory off-hours.

Detection of CMV antigenemia has not been validated for congenital CMV disease.

Methodology Immunocytochemical detection of CMV antigen (lower-matrix phosphoprotein, pp65) in nuclei of peripheral blood mononuclear cells

Additional Information The detection of CMV antigen directly in the peripheral blood leukocytes has been proven to be a clinically relevant marker of CMV infection. It currently has widespread use as a clinical tool in the diagnosis and management of CMV infection in immunocompromised patients. It has been compared to viral culture methods and shown to be sensitive (76% to 88%) and specific (99% to 100%). The major disadvantage of the assay are the time-consuming methods, but a recent rapid assay has been introduced that can be completed in approximately 2 hours and requires only 2 mL of blood.

Active CMV infection can also be detected by DNA/RNA amplification methods from plasma, serum, and white blood cells, as well as other specimens (eg, CSF, for neurologic syndromes in AIDS patients and others). See Related Information at the beginning of this listing for further tests relevant to CMV.

References

Afdhal NG and Yantiss RK, "A 71-Year-Old Man With Gastric Ulcers and Ileocecal Thickening Eight Years After Renal Transplantation," Case Records of the Massachusetts General Hospital, Case 25-2001, Scully RE, Mark EJ, McNeely WF, et al, eds, *N Engl J Med*, 2001, 345(7):526.

Chiaramonte S, Pellizzer G, Rassu M, et al, "Role of Antigenemia Assay in the Early Diagnosis and Treatment of CMV Infection in Renal Transplant Patients," *Clin Nephrol*, 2000, 53(4):10-2.

Durlik M, Siennicka J, Litwinska B, et al, "Comparison of Antigenemia (pp65) Assay and Polymerase Chain Reaction in Diagnosis of Cytomegalovirus Infection in Renal Transplant Recipients Treated With ATG," *Transplant Proc*, 2000, 32(6):1350-2.

Goosen VJ, Blok MJ, Christiaans MH, et al, "Early Detection of Cytomegalovirus in Renal Transplant Recipients: Comparison of PCR, NASBA, pp65 Antigenemia, and Viral Culture," *Transplant Proc*, 2000, 32(1):155-8.

Modlin JF, Grant PE, Makar RS, et al, "A Newborn Boy With Petechiae and Thrombocytopenia," Case Records of the Massachusetts General Hospital, Case 25-2003, Scully RE, Mark EJ, McNeely WF, et al, eds, *N Engl J Med*, 2003, 349(7):691-700.

Weinberg A, Hodges TN, Li S, et al, "Comparison of PCR, Antigenemia Assay, and Rapid Blood Culture for Detection and Prevention of Cytomegalovirus Disease After Lung Transplantation," *J Clin Microbiol*, 2000, 38(2):768-72.

Internet Web Sites

www.cdc.gov/ncidod/diseases/cmv.htm

Cytomegalovirus Culture

Related Information

(Continued)

Cytomegalovirus Culture (Continued)

Sputum Cytology on page 1222
Urinary Tract Cytology on page 1293
Viral Culture on page 1307
Virus, Direct Detection by Fluorescent Antibody on page 1311

Test Includes Shell vial culture will detect CMV only using specific immunofluorescence; CMV is usually detected in a routine virus culture

Abstract Cytomegalovirus (CMV) is an ubiquitous herpes virus which infects 50% to 85% of the adult population, usually asymptomatically. CMV infection is a major problem in immunocompromised patients. CMV is the most common cause of intrauterine infections and is the virus most frequently transmitted to unborn children. Transmission of CMV to neonates is associated with serious fulminant disease consisting of jaundice, hepatosplenomegaly, and multiorgan involvement. It may be fatal.

Infectious CMV can be shed in body fluids of any previously infected person and, thus, the virus can be isolated from urine, saliva, blood, semen and breast milk. The virus can be shed asymptomatically from individuals infected with CMV, therefore, culture should be performed only from patients at high risk for severe CMV disease. Isolation of CMV in cell culture remains the diagnostic standard.

Specimen Urine, throat, bronchoalveolar lavage, bronchial washings, lung biopsy, whole blood, amniotic fluid. Urine is often the best specimen, but isolation of virus or DNA detection from amniotic fluid are the most reliable tests for congenital CMV infection. Neonatal saliva can be used for culture.

Container Cold viral transport medium for swabs

Collection Urine: A first morning clean catch urine should be submitted in a sterile screw-cap container.

Throat: Rotate swab in both tonsillar crypts and against posterior oropharynx. Place swab in tube of viral transport medium, break off end of swab and tighten cap.

Blood: Collect in a green top Vacutainer® tube containing free heparin.

Storage Instructions Do **not** freeze. Specimens should be delivered to the laboratory immediately. If freezing is absolutely necessary, most specimens can be frozen by adding an equal amount of 0.4M sucrose-phosphate to the specimen before freezing. White blood cells should be isolated from blood specimens before freezing.

Causes for Rejection Dry specimen, specimen not refrigerated during transport, specimen fixed in formalin

Turnaround Time Variable (1-14 days); negative routine viral cultures are usually not reported for 28 days; CMV rapid culture: 1-3 days

Reference Interval No virus isolated

Use Aid in the diagnosis of disease caused by CMV

Limitations Rapid culture method detect specified virus(es); negative culture does not rule out viral infection. CSF cultures are negative in encephalitis and in subjects with AIDS, and are insensitive with CMV myelitis/polyradiculopathy and with ventriculitis in AIDS. CMV culture may be positive in the absence of obvious clinical disease.

Methodology Routine culture detects CMV by cytopathic effect; rapid shell vial specifically detects CMV early viral antigen with immunofluorescence

Additional Information CMV infections are very common in normal individuals and are usually asymptomatic. However, CMV infections are frequently severe and life-threatening in immunocompromised patients, including organ recipients and AIDS patients. CMV is the major viral pathogen following renal transplantation. Blood cultures positive for CMV predict progression. Detection of CMV infection is of utmost importance so that ganciclovir can be started as soon as possible.

CMV is the most frequent cause of congenital viral infections in humans and occurs in about 1% of all newborns. CMV can be cultured from amniotic fluid or urine specimens of the newborn. Approximately 90% have no clinical symptoms at birth. Congenital infection may occur as a result of either primary or recurrent maternal infection.

Active CMV infection can also be detected by DNA/RNA amplification methods from plasma, serum, and white blood cells, as well as other specimens (eg, CSF, for neurologic syndromes in AIDS patients and others). Specific CMV antigen is also used to detect and monitor active CMV infection. See Related Information at the beginning of this listing for further tests relevant to CMV.

References

Demmler GJ, Istas A, Easley KA, et al, "Results of a Quality Assurance Program for Detection of Cytomegalovirus Infection in the Pediatric Pulmonary and Cardiovascular Complications of Vertically Transmitted Human Immunodeficiency Virus Infection Study," *J Clin Microbiol*, 2000, 38(11):3942-5.

Fowler KB, Stagno S, and Pass RF, "Maternal Immunity and Prevention of Congenital Cytomegalovirus Infection," *JAMA*, 2003, 289(8):1008-11.

Halwachs-Baumann G, Genser B, Danda M, et al, "Screening and Diagnosis of Congenital Cytomegalovirus Infection: A 5-Year Study," *Scand J Infect Dis*, 2000, 32(2):137-42.

Kovacs A, Schluchter M, Easley K, et al, "Cytomegalovirus Infection and HIV-1 Disease Progression in Infants Born to HIV-1-Infected Women," *N Engl J Med*, 1999, 341(2):77-84.

Lazzarotto T, Varani S, Guerra B, et al, "Prenatal Indicators of Congenital Cytomegalovirus Infection," *J Pediatr*, 2000, 137(1):90-5.

Modlin JF, Grant PE, Makar RS, et al, "A Newborn Boy With Petechiae and Thrombocytopenia," Case Records of the Massachusetts General Hospital, Case 25-2003, Scully RE, Mark EJ, McNeely WF, et al, eds, *N Engl J Med*, 2003, 349(7):691-700.

Internet Web Sites

www.cdc.gov/ncidod/diseases/cmv.htm

Cytomegalovirus Cytology

Related Information

Bronchoalveolar Lavage (BAL) *on page 311*
Brushings Cytology *on page 314*
Cerebrospinal Fluid Cytology *on page 361*
Cytomegalovirus Antigen Detection *on page 494*
Cytomegalovirus Culture *on page 495*
Cytomegalovirus Nucleic Acid Detection *on page 498*
Cytomegalovirus Serology *on page 499*
Urinary Tract Cytology *on page 1293*
Viral Culture *on page 1307*

Synonyms Cytomegalic Inclusion Disease Cytology

Abstract Cytomegalovirus (CMV) maybe a potentially fatal infection in infants and immunocompromised patients, but it may also be present in asymptomatic adult carriers who are otherwise healthy. CMV is a member of the herpesvirus family. CMV viral inclusions may be identified in specimens obtained from the genitourinary tract, gastrointestinal tract, and respiratory tract as well as serous fluids and cerebrospinal fluid.

Specimen Bronchoalveolar lavage fluid, fresh urine, washing fluid, alcohol-fixed brushing specimen, cervical secretions

Sampling Time Several fresh specimens are indicated when urine is to be evaluated, since shedding is intermittent but cell disintegration is rapid.

Causes for Rejection Nonfixed specimen not processed within 1 hour

Use Establish the presence of cytomegalovirus infection, especially in immunosuppressed patients, including those with bone marrow and other transplantation procedures and AIDS. A presumptive diagnosis of CMV can be provided in 25% to 50% of instances of symptomatic congenital infection by cytologic techniques.

Limitations Viral culture is often described as the method of choice for definitive diagnosis of CMV, but cytology can provide more rapid information. A negative cytologic examination for CMV does not exclude the possibility of this etiology, and culture results are usually needed.

References

Gupta PK, "Microbiology, Inflammation, and Viral Infections," *Comprehensive Cytopathology*, 2nd ed, Bibbo M, ed, Philadelphia, PA: WB Saunders Co, 1997, 125-60.

Hodinka RL, "Human Cytomegalovirus," *Manual of Clinical Microbiology*, 7th ed, Chapter 66, Murray PR, Baron EJ, Pfaller MA, et al, eds, Washington, DC: ASM Press, American Society for Microbiology, 1999, 888-99.

Rosenthal DL, "Cytologic Diagnosis of Infectious Diseases of the Lung," *Compendium on Diagnostic Cytology*, 8th ed, Wied GL, Bibbo M, Keebler CM, et al, eds, Chicago, IL: Tutorials of Cytology, 1997, 208-15.

♦ **Cytomegalovirus Low Risk Blood** *see* Newborn Crossmatch and Transfusion *on page 952*

Cytomegalovirus Nucleic Acid Detection

Related Information

Bronchoalveolar Lavage (BAL) *on page 311*
Cytomegalovirus Antigen Detection *on page 494*
Cytomegalovirus Culture *on page 495*
Cytomegalovirus Cytology *on page 497*
Cytomegalovirus Serology *on page 499*

Test Includes Direct detection of cytomegalovirus nucleic acid in white blood cells, plasma, serum, urine, tissue, CSF, and bronchial alveolar lavage fluid. CMV nucleic acid may also be quantitated to determine or evaluate risk of disease or to monitor disease progress.

Abstract Infection with cytomegalovirus (CMV) in the nonimmunosuppressed host is typically asymptomatic or limited to a mononucleosis-like syndrome. However, CMV disease is a major risk for immunosuppressed individuals. The number of patients who are immunosuppressed as result of therapeutic intervention (bone marrow transplant, solid organ transplant, and/or cancer chemotherapy) has greatly increased in addition to the number of individuals with altered immune functions due to HIV infection. Early detection of CMV infection is critical for beginning intervention. Several assays have been developed using culture or antigen detection techniques. Molecular assays for detection of CMV DNA involve the extraction of viral genome followed by nucleic acid amplification. A number of targets have been described, such as the polymerase gene or immediate early antigen gene. Detection of CMV DNA by amplification procedures has a higher sensitivity than either culture techniques or the antigenemia assay. Newer techniques have now incorporated quantitation and are useful for predicting significant risk of CMV disease.

Specimen CMV DNA may be extracted from peripheral blood lymphocytes, plasma, serum, bronchoalveolar lavage fluid, urine, cerebrospinal fluid, and various tissues including lung and brain. Most of the commercial assays use plasma (EDTA or citrate dextrose) to monitor the level of CMV nucleic acid. For congenital CMV infection, isolation of virus and DNA detection from amniotic fluid are the most reliable methods.

Container Glass or plastic containers are acceptable; anticoagulants may be used with the exception of heparin. A major laboratory requests EDTA whole blood samples.

Storage Instructions While freezing is known to reduce the viability of CMV and its isolation by culture techniques, freezing does not impair detection by amplification procedures. All specimens should be rapidly transported to the laboratory and if transportation exceeds 2 hours, the sample should be refrigerated. If transportation exceeds 8 hours, the specimen should be frozen. All tissue samples should be frozen as soon as possible after the biopsy procedure.

Causes for Rejection Collection of blood in heparinized tubes may result in inhibition of amplification.

Turnaround Time 1-3 days

Use The detection of CMV DNA is useful to identify patients in whom appropriate therapy could be instituted. Quantitative assays may be useful to establish disease risk and to monitor CMV levels after therapy.

The diagnosis of congenital CMV infection depends upon PCR assays and virus cultures. A high CMV load in amniotic fluid on quantitative PCR correlates with symptomatic infection.

Limitations A negative result does not rule out the presence of CMV in all tissues of the body. The cost of the test is high in comparison with culture, CMV antigenemia detection, or serology. Detection does not always correlate with active disease. Contamination is a risk and results in false-positive results. A variety of amplification techniques is used and results may not be interchangeable. CMV may be excreted in urine and saliva or be present in blood without clinical symptoms. Inhibitory substances may suppress viral replication.

Methodology DNA is extracted from specimens followed by amplification using either PCR, transcription-based amplification of mRNA, or signal amplification methods. Quantitation is most frequently based on simple limiting dilution quantification or coamplification of an internal reference template with the same primer binding sequences.

Additional Information Greater than 80% of adults have been infected with CMV as evidenced by the presence of antibodies. In nonimmunosuppressed individuals infection by CMV is self-limited. Infection of the fetus during the early stages of pregnancy may result in multiorgan failure, CNS abnormalities or paradoxically, continuous CMV excretion in the urine without disease. Amplification methods show good specificity and sensitivity in detecting CMV from urine of infected newborns.

The development of solid organ transplantation and bone marrow or stem cell transplantation has accelerated the need for early CMV detection. When diagnosed early, CMV infection may be effectively treated. Amplification assays for detection of CMV have been developed because of the increased sensitivity that such assays provide.

References

Caliendo AM, St. George K, Kao SY, et al, "Comparison of Quantitative Cytomegalovirus (CMV) PCR in Plasma and CMV Antigenemia Assay: Clinical Utility of the Prototype Amplicor CMV Monitor Test in Transplant Recipients," *J Clin Microbiol*, 2000, 38(6):2122-7.

Einsele H and Hebart H, "Cytomegalovirus Infection Following Stem Cell Transplantation," *Haematologica*, 1999, 84(1):46-9.

Goossens VJ, Blok MJ, Christiaans MH, et al, "Early Detection of Cytomegalovirus in Renal Transplant Recipients: Comparison of PCR, NASBA, pp65 Antigenemia, and Viral Culture," *Transplant Proc*, 2000, 32(1):155-8.

Mayo Reference Service, "Cytomegalovirus (CMV) DNA Quantitation by Polymerase Chain Reaction (PCR), Plasma," *New Test Announcement*, 82986, January 2002.

Modlin JF, Grant PE, Makar RS, et al, "A Newborn Boy With Petechiae and Thrombocytopenia," Case Records of the Massachusetts General Hospital, Case 25-2003, Scully RE, Mark EJ, McNeely WF, et al, eds, *N Engl J Med*, 2003, 349(7):691-700.

Rawlinson WD, "Broadsheet. Number 50: Diagnosis of Human Cytomegalovirus Infection and Disease," *Pathology*, 1999, 31(2):109-15.

Siennicka J, Durlik M, Litwinska B, et al, "Usefulness of Hybridization and PCR Methods in Monitoring of CMV Infection in Renal Transplant Recipients," *Ann Transplant*, 2000, 5(1):21-4.

Internet Web Sites

www.cdc.gov/ncidod/diseases/cmv.htm

Cytomegalovirus Serology

Related Information

Bronchoalveolar Lavage (BAL) *on page 311*
Cytomegalovirus Antigen Detection *on page 494*
Cytomegalovirus Culture *on page 495*
Cytomegalovirus Cytology *on page 497*
Cytomegalovirus Nucleic Acid Detection *on page 498*
Newborn Crossmatch and Transfusion *on page 952*
Red Blood Cells, Leukocytes Reduced *on page 1141*
Risks of Transfusion *on page 1166*
Sputum Cytology *on page 1222*
TORCH *on page 1262*

Test Includes Detection of cytomegalovirus-specific IgM and IgG antibodies

Abstract Human cytomegalovirus (CMV) establishes a latent infection subsequent to the primary infection. A primary CMV infection may manifest as an infectious mononucleosis-like disorder, or be asymptomatic. Primary CMV infection can also be associated with interstitial pneumonia, hepatitis, meningoencephalitis, or intrauterine infections, including congenital CMV infection. In immunocompromised patients, primary as well as reactivated CMV can cause a variety of serious diseases including retinitis, colitis, and pneumonitis.

Specimen Serum

Container Red top tube

Sampling Time Acute and convalescent sera drawn 10-14 days apart are required for IgG CMV testing. A single specimen may be sufficient for IgM testing. For determination of prior exposure to CMV (for transplantation or transfusion), a single specimen for detection of CMV IgG is satisfactory.

Special Instructions Neonatal specimens should be analyzed for specific IgM antibody only.

Reference Interval IgM: <1:8 is considered nondiagnostic. Less than a fourfold increase in CMV-IgG titer in paired sera drawn 10-14 days apart. Negative results from EIA.

Use Determine prior infection with CMV for purposes of organ transplantation, provision of blood and blood fractions to selected recipients, and screening for pregnant women

(Continued)

Cytomegalovirus Serology *(Continued)*

Limitations Heterophil antibodies and presence of rheumatoid factor may cause false-positive IgM results. Fetal IgM antibody to maternal IgG may also cause false-positive results. False negatives occur. Because of high levels of "background" antibody in adult populations, a single antibody determination is not useful. When used for diagnosis of intrauterine CMV infection, fetal IgG and IgM antibodies are difficult to interpret. IgG antibodies are passively transferred to the fetus. The sensitivity and specificity of present IgM antibody tests are not as effective as viral culture methods or PCR assays.

Methodology Enzyme immunoassay (EIA), indirect fluorescent antibody (IFA), latex agglutination (LA), hemagglutination (HA)

Additional Information The majority of CMV infections remain undetected, since most infections with CMV are asymptomatic. Individuals infected with CMV will produce antibody to the virus and antibodies can be detected throughout the lifetime of the individual. A single titer is rarely significant if past history is unknown.

CMV IgM is produced in low levels during reactivated CMV and, thus, does not always indicate primary infection. Likewise, a positive or high titer of CMV IgG should not be interpreted as representing active CMV infection. However, a fourfold or greater rise in CMV titer between acute and convalescent specimens is evidence of infection. A single IgM specific titer >1:8 is also excellent evidence of acute infection. Significant CMV titers are found almost universally in patients with AIDS.

Intrauterine transmission of CMV can occur whether or not prior maternal immunity exists. Reinfection with a different strain can lead to intrauterine transmission in seropositive women. However, the presence of maternal antibody prior to conception does provide a degree of protection against neonatal damage of congenital CMV infection.

Although serology is a useful method to detect CMV infections, the rapid CMV viral culture, CMV antigen or quantitative CMV PCR can more reliably identify symptomatic CMV infections in immunocompromised patients. See Related Information at the beginning of this listing for further tests relevant to CMV.

References

Boppana SB, Rivera LB, Fowler KB, et al, "Intrauterine Transmission of Cytomegalovirus to Infants of Women With Preconceptional Immunity," *N Engl J Med*, 2001, 344(18):1366-71.

Deorari AK, Broor S, Maitreyi RS, et al, "Incidence, Clinical Spectrum, and Outcome of Intrauterine Infections in Neonates," *J Trop Pediatr*, 2000, 46(3):155-9.

Drago F, Aragone MG, Lugani C, et al, "Cytomegalovirus Infection in Normal and Immunocompromised Humans. A Review." *Dermatology*, 2000, 200(3):189-95.

Emery VC, Cope AV, Sabin CA, et al, "Relationship Between IgM Antibody to Human Cytomegalovirus, Virus Load, Donor and Recipient Serostatus, and Administration of Methylprednisolone as Risk Factors for Cytomegalovirus Disease After Liver Transplantation," *J Infect Dis*, 2000, 182(6):1610-5.

Fowler KB, Stagno S, and Pass RF, "Maternal Immunity and Prevention of Congenital Cytomegalovirus Infection," *JAMA*, 2003, 289(8):1008-11.

Modlin JF, Grant PE, Makar RS, et al, "A Newborn Boy With Petechiae and Thrombocytopenia," Case Records of the Massachusetts General Hospital, Case 25-2003, Scully RE, Mark EJ, McNeely WF, et al, eds, *N Engl J Med*, 2003, 349(7):691-700.

Internet Web Sites

www.cdc.gov/ncidod/diseases/cmv.htm

♦ **Cytoreduction** *see* Apheresis, Therapeutic *on page 201*

♦ **Cytosol Hormone Receptors** *see* Estrogen and Progesterone Receptor Assay *on page 556*

♦ **Dalcaine®** *see* Lidocaine, Serum or Plasma *on page 850*

♦ **Dalmane®** *see* Flurazepam, Serum *on page 605*

♦ **Dama Blanca** *see* Cocaine (Cocaine Metabolite), Qualitative, Urine or Hair *on page 427*

♦ **Danaparoid** *see* Heparin Antifactor Xa Assay *on page 693*

♦ **Danaparoid** *see* Heparin-Induced Thrombocytopenia *on page 695*

♦ **Danaparoid** *see* Mixing Studies (Coagulation) *on page 918*

♦ **Danaparoid** *see* Prothrombin Time *on page 1116*

Darkfield Examination, Leptospirosis

Related Information

Leptospira Culture *on page 842*

Leptospira Serology *on page 844*

Test Includes Examination of serum, urine, or CSF for organisms

Abstract Leptospirosis is a bacterial infection that affects both humans and animals. The general term leptospirosis is preferred to the synonyms, **Weil disease** and **canicola fever**. Culture and serology are recommended for diagnosis; darkfield examination is not. Darkfield microscopic examination of specimens for leptospires can be useful to establish a rapid diagnosis, but culture and serology are recommended for confirmation of diagnosis.

Specimen Urine, serum, cerebrospinal fluid

Container Sterile plastic urine container, red top tube, or sterile CSF tube

Use Determine the presence of *Leptospira* for the diagnosis of Weil syndrome, hemorrhagic fever with renal syndrome, atypical pneumonia syndrome, aseptic meningitis, and myocarditis including cardiac arrhythmias.

Limitations The concentration of leptospires in blood and CSF is low. Therefore, concentration by centrifugation with sodium oxalate or heparin can be useful. The incidence of false positives is increased because fibrils and cellular extrusions can be mistaken for organisms.

Failure to detect leptospires does not rule out their presence. **Direct examination of blood or urine by darkfield methods frequently results in failure or misdiagnosis.** Culture has much greater value for diagnosis. Saprophytic strains as well as pathogenic ones exist.

Methodology A small drop of fluid is distributed in a thin layer between a glass coverslip and slide. The typical morphology should be observed before a presumptive diagnosis is made.

Additional Information *Leptospira* are present in blood early in the course of disease (first week only). After 10-14 days, they may be found in the urine. Urine must be neutral or alkaline. If the urine is acidic, it should be neutralized by diluting with 1% bovine serum albumin. Darkfield microscopy is best used to demonstrate leptospires in specimens in which a high concentration of organisms is present.

A DNA amplification method has recently been used to detect leptospires in clinical specimens. This assay seems to be more sensitive and specific than microscopic examination or culture. It is not currently widely available.

References

Centers for Disease Control and Prevention, "Outbreak of Acute Febrile Illness Among Participants in EcoChallenge Sabah 2000 - Malaysia," *JAMA*, 2000, 284(13):1646.

Farr RW, "Leptospirosis," *Clin Infect Dis*, 1995, 21(1):1-8.

Holk K, Nielsen SV, and Ronne T, "Human Leptospirosis in Denmark 1970-1996: An Epidemiological and Clinical Study," *Scand J Infect Dis*, 2000, 32(5):533-8.

Lomar AV, Diament D, and Torres JR, "Leptospirosis in Latin America," *Infect Dis Clin North Am*, 2000, 14(1):23-39.

Internet Web Sites

www.amm.co.uk/pubs/fa_leptospir.htm
www.cdc.gov/ncidod/dbmd/diseaseinfo/leptospirosis_g.htm

Darkfield Examination, Syphilis

Related Information

FTA-ABS, Serum *on page 618*
RPR *on page 1174*
VDRL, Serum or Cerebrospinal Fluid *on page 1303*

Abstract The sexually transmitted disease, syphilis, is caused by the spirochete *Treponema pallidum*. *T. pallidum* is a thin organism that cannot be visualized by conventional light microscopy. Darkfield examination is appropriate for the evaluation of chancre (primary lues) and condylomata lata (secondary syphilis).

Patient Preparation The surface of the chancre or condyloma is cleansed with a swab moistened with saline. This removes exudate and excess bacterial contamination. Serous fluid is collected from the surface of the chancre, using a small pipette. The fluid is placed on a slide or coverslip. Alternatively, the specimen can be collected by directly touching the slide to the lesion. Clear exudate from the subsurface of the lesion is then examined by darkfield microscopy without further preparation.

Specimen Moist serous fluid from the base of a cleansed, unhealed chancre or condyloma. The youngest lesion available is best. The chance of identification of treponemes decreases with the age of the lesion as it dries and locally heals.

Collection Collect with a pipette and put directly on a glass slide.

(Continued)

Darkfield Examination, Syphilis (Continued)

Causes for Rejection Healed chancre, previous treatment, ointment, dried-up specimen

Reference Interval Negative

Critical Values Positive darkfield examination, especially from a pregnant patient

Use Determine the presence of characteristic spirochetes to support the diagnosis of primary or secondary syphilis or of congenital syphilis.

Limitations Darkfield examination is of limited value in oral and rectal lesions, because of the presence of other, nonpathogenic spirochetes. Dry or bloody specimens render this examination worthless. The specimen should be examined within 15 minutes of collection because the organisms lose motility with decrease in temperature, exposure to oxygen, and desiccation. Experience and professional expertise are required to accurately identify *T. pallidium* by darkfield examination. Artifacts (eg, cotton fibers) can sometimes be mistaken for spirochetes. A negative result does not rule out syphilis.

Contraindications Antibiotic therapy prior to the darkfield examination. The organisms are rapidly cleared following therapy and as the lesion heals.

Methodology Darkfield microscopy

Additional Information *T. pallidum* can be found in skin lesions and lymph nodes in secondary syphilis but are more plentiful in primary chancres. They cannot be grown in culture. The Centers for Disease Control (CDC) and others use fluorescent microscopy with monoclonal or polyclonal anti-*T. pallidum* antibodies for examination of exudates.

References

Behets FM, Brathwaite AR, Hylton-Kong T, et al, "Genital Ulcers: Etiology, Clinical Diagnosis and Associated Human Immunodeficiency Virus Infection in Kingston, Jamaica," *Clin Infect Dis*, 1999, 28(5):1086-90.

Birnbaum NR, Goldschmidt RH, and Buffet WO, "Resolving the Common Clinical Dilemmas of Syphilis," *Am Fam Phys*, 1999, 59(8):2233-46.

Emmert DH and Kirchner JT, "Sexually Transmitted Diseases in Women. Gonorrhea and Syphilis," *Postgrad Med*, 2000, 107(2):181-97.

Hollier LM and Cox SM, "Syphilis," *Semin Perinatol*, 1998, 22(4):323-31.

Internet Web Sites

www.cdc.gov/std/Syphilis/STDFact-Syphilis.htm

♦ **Darvocet-N**® *see* Propoxyphene, Serum or Urine *on page 1096*

♦ **Darvon**® *see* Propoxyphene, Serum or Urine *on page 1096*

♦ **DAT** *see* Antiglobulin Test, Direct *on page 176*

♦ **Datril**® *see* Acetaminophen, Serum *on page 90*

♦ **Davidsohn Differential** *see* Infectious Mononucleosis Screening Test *on page 785*

♦ **Davidsohn Slide Test** *see* Infectious Mononucleosis Screening Test *on page 785*

♦ **DDAVP** *see* Coagulation Factor Assays *on page 418*

♦ **DDAVP** *see* von Willebrand Factor *on page 1321*

♦ **D-Dimers** *see* D-Dimers and Fibrin Degradation Products *on page 502*

♦ **D-Dimers** *see* Hypercoagulation Panel *on page 758*

D-Dimers and Fibrin Degradation Products

Related Information

Disseminated Intravascular Coagulation Screen *on page 517*
Hypercoagulation Panel *on page 758*

Synonyms D-Dimers; FBP; FDP; Fibrin Breakdown Products; Fibrin Split Products; FSP

Applies to Fibrinolysis; Plasmin; Thrombin; Thrombolysis

Abstract Fibrinolysis is mediated by plasmin, which degrades fibrin clots into D-dimers and fibrin degradation products (FDP). Plasmin can also degrade intact fibrinogen, generating fibrinogen degradation products (FDP) that are detected in FDP assays.

Specimen Plasma (some FDP assays require serum; the SimpliRed® D-dimer assay uses whole blood)

Container One blue top (citrate) tube (some FDP serum assays require special tubes that contain thrombin to clot the blood, and a fibrinolysis inhibitor to prevent FDP formation in the test tube)

Collection Routine venipuncture. If multiple tests are being drawn, draw blue top tubes after any red top tubes but before any lavender top (EDTA), green top (heparin), or gray top (oxalate/fluoride) tubes. Immediately invert tube gently at least 4 times to mix. Tubes must be appropriately filled. Deliver tubes immediately to the laboratory.

Storage Instructions Separate plasma from cells as soon as possible. Store plasma at room temperature for up to 8 hours, on ice for up to 24 hours, or store frozen. (Serum: Once serum is obtained, it may be refrigerated for up to 1 week, or it may be stored frozen.)

Causes for Rejection Clotted specimens are unsuitable for plasma-based tests.

Turnaround Time Less than 1 day (latex agglutination methods take less than 1 hour; standard ELISA methods take 5 hours; rapid ELISA methods have been developed)

Reference Interval D-dimer: approximately <0.5 µg/mL; FDP: approximately <5 µg/mL

Use D-dimers or FDP are part of a panel of tests required for diagnosing DIC. In DIC, both thrombin and plasmin are generated, causing an elevation in D-dimers and FDP. See Disseminated Intravascular Coagulation Screen *on page 517*.

D-dimers assist with the diagnosis of deep venous thrombosis (DVT) and pulmonary embolism (PE), but only if a very sensitive method is used. If the test is positive in a patient suspected to have DVT or PE, clinicians proceed with further diagnostic tests for DVT or PE. If the test is negative, depending on the clinical situation and the sensitivity of the D-dimer assay, DVT or PE is considered unlikely and further diagnostic tests for DVT or PE might not be pursued. Specificity of D-dimers is lower in inpatients as compared to outpatients. D-dimer levels have been used to assess the risk of recurrence of venous thromboembolism.

Monitoring thrombolytic therapy is not routinely required. However, if monitoring is desired, D-dimers or FDP are one of several tests that can be performed to confirm that thrombolysis (fibrinolysis) is occurring. D-dimers and FDP should become increased with thrombolytic therapy.

Limitations D-dimers and FDP can become elevated whenever the coagulation and fibrinolytic systems are activated. This occurs in a variety of conditions, and therefore the tests are not specific for any one diagnosis.

Manual latex agglutination and certain other methods are not sufficiently sensitive to exclude the diagnosis of DVT or PE when the test result is negative. Even with the most sensitive methods (eg, ELISA assays), a patient with PE or DVT may test negative for D-dimers.

High rheumatoid factor (RF) levels may cause false-positive results with some assays.

Methodology Assays are semiquantitative or quantitative immunoassays.

Latex agglutination: Patient plasma is mixed with latex particles which are coated with monoclonal anti-FDP antibodies. If FDP are present in the patient plasma, the latex particles agglutinate as FDP bind to the antibodies on the particles. These large agglutinated clumps are detected visually by the technologist. Various dilutions of patient plasma can be tested to provide an estimation of the FDP titer (semiquantitative result). Latex agglutination assays are also available for D-dimers. Automated, quantitative versions of this assay are commercially available for D-dimers (MDA® D-dimer, Organon Teknika; STA Liatest®, Diagnostica Stago), in which the agglutination is detected turbidimetrically by a coagulation analyzer rather than visually by a technologist.

Enzyme-linked immunosorbent (ELISA) assays: Quantitative ELISA assays are available for FDP, D-dimers, or fibrinogen degradation products. An automated, rapid ELISA assay for D-dimers is available (VIDAS®, bioMerieux Inc).

Other methods: SimpliRed® D-dimer (American Diagnostica) is a semiquantitative red blood cell agglutination assay that can be performed on whole blood. Other methods have been developed that are not yet available in the United States.
(Continued)

D-Dimers and Fibrin Degradation Products (Continued)

Additional Information A D-dimer is a specific FDP that is formed only by plasmin degradation of fibrin, and not by plasmin degradation of intact fibrinogen. Thus, the presence of D-dimers indicates that fibrin has been formed and degraded. In contrast, a positive FDP assay indicates that fibrin and/or fibrinogen is being degraded by plasmin, because the FDP assay detects fibrin degradation products, including D-dimers, and fibrinogen degradation products. D-dimers and FDP can be positive with DIC or thrombosis, including DVT, PE and myocardial infarction. They also may be positive in liver disease due to decreased hepatic clearance. They can become elevated postoperatively, and with significant bleeding, hemodialysis, eclampsia, sickle cell crisis, and other conditions. Cancer patients often have positive D-dimers and FDP, usually representing low-grade, chronic DIC. D-dimers and FDP mildly increase in pregnancy.

In the past, FDP had to be performed on serum samples, because the polyspecific antibodies cross-react with fibrinogen in plasma. Fibrinogen is not present in serum because it has been converted into fibrin clot, centrifuged, and discarded. Currently, monoclonal antibodies specific for FDP, without cross-reactivity against fibrinogen, allow the test to be performed in plasma. There is some evidence that the older, serum-based assays are not as reliable as the newer, plasma-based assays, because serum-based FDP may give low values due to trapping of FDP in the clot, or high values due to generation of FDP during clot formation.

References

Eichinger S, Minar E, Bialonczyk C, et al, "D-Dimer Levels and Risk of Recurrent Venous Thromboembolism," *JAMA*, 2003, 290(8):1071-4.

Gaffney PJ and Perry MJ, "Unreliability of Current Serum Fibrin Degradation Product (FDP) Assays," *Thromb Haemost*, 1985, 53(3):301-2.

Schrecengost JE, LeGallo RD, Boyd JC, et al, "Comparison of Diagnostic Accuracies in Outpatients and Hospitalized Patients of D-Dimer Testing for the Evaluation of Suspected Pulmonary Embolism," *Clin Chem*, 2003, 49(9):1483-90.

van der Graaf F, van den Borne H, van der Kolk M, et al, "Exclusion of Deep Venous Thrombosis With D-dimer Testing - Comparison of 13 D-dimer Methods in 99 Outpatients Suspected of Deep Venous Thrombosis Using Venography as Reference Standard," *Thromb Haemost*, 2000, 83(2):191-8.

Wells PS, Anderson DR, Rodger M, et al, "Evaluation of D-Dimer in the Diagnosis of Suspected Deep-Vein Thrombosis," *N Engl J Med*, 2003, 349(13):1227-35.

Wilson DB and Gard KM, "Evaluation of an Automated, Latex-Enhanced Turbidimetric D-Dimer Test (Advanced D-Dimer) and Usefulness in the Exclusion of Acute Thromboembolic Disease," *Am J Clin Pathol*, 2003, 120(6):930-7.

♦ **Decentralized Testing** *see* Point-of-Care Testing on page 1065

♦ **Decision Analysis** see page 11

Deferoxamine Infusion Test

Related Information

Aluminum, Bone and Bone Biopsy *on page 139*
Aluminum, Serum or Urine *on page 141*
Iron and Total Iron Binding Capacity/Transferrin, Serum *on page 807*

Test Includes Determination of serum aluminum (Al) prior to and 48 hours postinfusion of deferoxamine

Abstract This test is advocated in suspected Al toxicity, especially among dialysis patients with known Al exposure, who lack diagnostically elevated serum Al concentrations. The test is used to exclude Al bone disease, which a negative test will do. A positive test requires confirmation by bone histomorphometry (see Aluminum, Bone and Bone Biopsy *on page 139*). The test is felt to better reveal the total body burden of Al than does serum Al, which is more sensitive to recent exposure. Urine Al cannot be effectively used in patients with renal failure.

Specimen Serum

Container Trace metal collection tubes

Collection Draw any trace metal tubes prior to collecting other blood samples. Serum is separated and stored or submitted to reference laboratory for analysis of aluminum (see Aluminum, Serum or Urine *on page 141*).

Causes for Rejection Failure to collect in trace metal tube and to store specimen in special acid-washed plastic vials

Special Instructions Most patients for whom this test is indicated are on dialysis, and the test is done in conjunction with their dialysis procedure (hemodialysis or peritoneal dialysis). The patient is instructed to stop any aluminum-containing antacids 3 days prior to the test. The first blood sample is obtained, then 0.5 g deferoxamine is administered in 100 mL 0.9% sodium chloride intravenously during the last 2 hours of dialysis through the venous blood line. Forty-eight hours later, prior to the next hemodialysis, a second blood sample is drawn. The timing is the same for peritoneal dialysis patients, for whom peritoneal dialysis is halted during the 48-hour period.

Reference Interval The test is considered positive if the second sample more than triples the first, or exceeds it by 150 ng/mL (SI: 5.6 µmol/L)

Critical Values Serum levels >200 ng/mL (SI: >7.4 µmol/L)

Use Screen patients with known Al exposure and abnormal (but not diagnostically elevated) serum Al levels for high body burden of Al, which is known to be correlated with Al bone disease. Patients with serum Al levels ≤75 ng/mL (SI: <2.8 µmol/L) are less likely to benefit from the test, and patients with persistent serum Al levels >150 ng/mL (SI: >5.6 µmol/L) should probably go directly to bone biopsy for confirmation of bone Al burden. By restricting its use to patients who probably have a moderate but not extreme body burden of Al, the test will help select patients needing further study by bone biopsy, a more invasive procedure. The test appears to be more sensitive and specific with greater positive predictive value when serum parathormone levels are decreased.

Limitations Use of this test is controversial. Originally proposed as a 2 g deferoxamine infusion test, complications have been reported, including permanent visual disturbances after a single 2 g dose. Due to multiple reported complications (ocular, auditory, anaphylaxis, hematopoietic, infectious) of high dose deferoxamine therapy for iron overload in thalassemia and for Al overload and toxicity in chronic renal failure, therapeutic doses have generally been reduced. The 0.5 g dose infusion test has so far not been reported to cause complications. Some authors endorse the use of an infusion test, others do not. Sensitivity is suboptimal; false-negative results are seen in many patients. Serum Al values in children appear to be a better estimation of total Al body burden than does the deferoxamine challenge test.

Iron status influences response to the deferoxamine test. Patients with high iron levels will have less increase in serum Al after challenge with deferoxamine, causing false-negative results.

Contraindications When there is a heavy bone burden of Al, often reflected by high basal levels, the serum Al may acutely rise to toxic levels after deferoxamine dose is given as "challenge" test or as treatment, and encephalopathy may worsen. In cases of actual or anticipated intolerance to deferoxamine, hemoperfusion over specially treated charcoal (an Al removal device) would be an alternative treatment modality.

Methodology Atomic absorption (AA), or inductively-coupled plasma mass spectrometry (ICP-MS)

Additional Information Deferoxamine is the chelator of choice for acute and chronic iron overload from multiple transfusions, as may occur in severe chronic anemias (eg, thalassemias). It chelates iron by forming a stable complex that prevents the iron from entering into further chemical reactions. Deferoxamine has also been used successfully to promote Al excretion or removal and to provide symptomatic and objective improvement in the treatment of aluminum-associated neurotoxicity and/or bone abnormalities (osteomalacia and osteodystrophy) in patients with chronic renal failure undergoing hemodialysis.

References

Cannata JB, Fernandez-Martin JL, Diaz-Lopez B, et al, "Influence of Iron Status in the Response to the Deferoxamine Test," *J Am Soc Nephrol*, 1996, 7(1):135-9.

Menendez Fraga P, Fernandez Martin JL, Blanco Gonzalez E, et al, "Low Percentage of Aluminoxamine and Ferrioxamine in Uremic Serum After Desferrioxamine Administration," *Clin Chem*, 1998, 44(6):1262-8.

Yaqoob M, Ahmad R, Roberts N, et al, "Low-Dose Desferrioxamine Test for the Diagnosis of Aluminum-Related Bone Disease in Patients on Regular Haemodialysis," *Nephrol Dial Transplant*, 1991, 6(1):484-6.

♦ **Dehydroepiandrosterone** *see* 17-Ketosteroids Fractionation, Urine *on page 818*

Dehydroepiandrosterone and Dehydroepiandrosterone Sulfate, Serum or Plasma

Related Information

Adrenocorticotropic Hormone, Plasma *on page 114*
Androstenedione, Serum *on page 158*
Cortisol, Free, Urine *on page 459*
Cortisol, Serum or Plasma *on page 460*
Estradiol, Serum *on page 553*
17-Hydroxyprogesterone, Whole Blood, Serum, or Plasma *on page 755*
Testosterone, Total and Free, Serum or Plasma *on page 1238*

Synonyms DHEA; DHEA-S

Applies to Androstenedione; Dehydroepiandrosterone, Serum; Dehydroepiandrosterone Sulfate, Serum; Epitestosterone; Estradiol; 17-Ketosteroids; Testosterone

Abstract Dehydroepiandrosterone (DHEA) and its conjugated sulfate metabolite, DHEA-S, are synthesized in and secreted by the adrenal cortex, under the influence of adrenocorticotropin (ACTH). DHEA and DHEA-S are interconvertible into each other. In healthy women, the adrenal cortex is the exclusive site of DHEA and DHEA-S synthesis. In men, the adrenal cortex is the principal site of DHEA and DHEA-S synthesis; in addition, ~10% to 25% of DHEA and ~5% of DHEA-S originates in the testes. While both DHEA and DHEA-S are themselves weak androgens, both can be converted, in peripheral tissues, into more potent androgens (eg, androstenedione or testosterone), and into estrogens (eg, estradiol). In normal persons, the serum concentration (molar basis) of DHEA-S is approximately 300-500 times higher than the concentration of DHEA. This concentration difference is, at least in part, due to the much slower metabolic clearance of DHEA-S: the half-life of DHEA-S is 10-20 hours, while the half-life of DHEA is 1-3 hours. DHEA has a diurnal variation similar to cortisol, while DHEA-S exhibits only very minimal diurnal variation. DHEA and DHEA-S serum levels at various ages are summarized in the following table. A cross-sectional analysis of older adults (Abassi et al) has found that men with DHEA-S levels in the highest quartile are more fit, more lean, and have higher levels of total and free testosterone, than do comparable men with DHEA-S values in the lowest quartile. No comparable correlation was found in women.

Specimen Serum or plasma

Container Red top tube or lavender top (EDTA) tube

Storage Instructions Separate within 1 hour of collection. Serum or plasma stable 24 hours at 4°C. Freeze for longer storage.

Causes for Rejection Recently administered radioisotopes if RIA is used for assay

Special Instructions Often both tests are ordered at the same time. When only one is obtained, it is usually DHEA-S.

Use

Adrenal hyperfunction. DHEA and DHEA-S are elevated in congenital adrenal hyperplasias (CAH) (both the 11-beta hydroxylase and the 21-beta hydroxylase forms), and virilizing adrenal neoplasms. In the clinical context of female hirsutism, elevated basal DHEA-S and testosterone suggest an adrenal cause. When a woman with hirsutism has normal basal DHEA-S and testosterone values, an adrenal tumor is unlikely.

Virilization in girls. Signs of virilization in pubertal or prepubertal girls are most commonly due to the onset of adrenarche, a normal physiologic condition reflecting the onset of adrenal androgen secretion. Less commonly, virilization in girls is due to late-onset CAH or a virilizing adrenal neoplasm. All three conditions are associated with elevated serum DHEA-S. Biochemical differentiation among these requires additional testing.

Adrenal insufficiency. Many patients, especially women, with adrenal insufficiency have low basal DHEA and DHEA-S values.

Reversing certain effects of aging. The dramatic decrease in serum DHEA and DHEA-S values after age 30 and research results showing a positive correlation between DHEA and DHEA-S serum levels with markers of good health in older persons have led many to speculate that DHEA dietary supplements, accompanied by laboratory monitoring, might make a beneficial contribution to the physiology of older persons. Others warn, to the contrary, that there may be detrimental effects on hormone-sensitive organs (eg, breast and prostate).

Reference Interval See table.

Reference Intervals for Dehydroepiandrosterone Sulfate and Unconjugated Dehydroepiandrosterone in Serum

Age	Male	Female	Male	Female
Dehydroepiandrosterone Sulfate				
	(µg/dL)	(µg/dL)	(µmol/L)	(µmol/L)
Children				
1-5 d	12-254	10-248	0.3-6.9	0.3-6.7
1 mo to 5 y	1-41	5-55	0.03-1.1	0.1-1.5
6-9 y	2.5-145	2.5-140	0.07-3.9	0.07-3.8
10-11 y	15-115	15-260	0.4-3.1	0.4-7.0
12-17 y	20-555	20-535	0.5-15.0	0.5-14.4
Pubertal Levels (Tanner Stage)				
1	5-265	5-125	0.1-7.2	0.1-3.4
2	15-380	15-150	0.4-10.3	0.4-4.0
3	60-505	20-535	1.6-13.6	0.5-14.4
4	65-560	35-485	1.8-15.1	0.9-13.1
5	165-500	75-530	4.4-13.5	2.0-14.3
Adults				
18-30 y	125-619	45-380	3.4-16.7	1.2-10.3
31-50 y	59-452	12-379	1.6-12.2	0.8-10.2
51-60 y	20-413		0.5-11.1	
61-83 y	10-285		0.36-7.7	
Postmenopausal female		30-260		0.8-7.0
Dehydroepiandrosterone, Unconjugated				
	(ng/dL)	(ng/dL)	(nmol/L)	(nmol/L)
6-9 y	13-187	18-189	0.45-6.49	0.62-6.55
10-11 y	31-205	112-224	1.07-7.11	3.88-7.77
12-14 y	83-258	98-360	2.88-8.95	3.40-12.5
Adults	180-1250	130-980	6.25-43.4	4.51-34.0

Gronowski AM and Landau-Levine M. "Reproductive Endocrine Function," *Tietz Textbook of Clinical Chemistry*, 3rd ed, Burtis CA and Ashwood ER, eds, Philadelphia, PA: WB Saunders Co, 1999, 1632.

Methodology Immunoassay (multiple labels and formats) and gas chromatography/mass spectrometry (GC/MS)

Additional Information Extensive compilations of DHEA and DHEA-S levels in a wide range of medical conditions and DHEA and DHEA-S blood levels associated with the administration of DHEA by various dosages and routes are available (Kroboth et al).

Athletic performance enhancement. In December 1996, DHEA was added to the list of prohibited compounds by the International Olympic Commission.

Of historical interest is the fact that urinary DHEA and DHEA-S are major constituents of the urine test, "17-ketosteroids," once a mainstay of the laboratory evaluation of adrenal function.

References

Abbasi A, Duthi EH, Sheldahl L, et al, "Association of Dehydroepiandrosterone Sulfate, Body Composition, and Physical Fitness in Independent Community-Dwelling Older Men and Women," *J Am Geriatr Soc*, 1998, 46(3):263-73.

Bowers LD, "Oral Dehydroepiandrosterone Supplementation Can Increase the Testosterone/ Epitestosterone Ratio," *Clin Chem*, 1999, 45(2):295-7.

Casson PR, Andersen RN, Herrod HG, et al, "Oral Dehydroepiandrosterone in Physiologic Doses Modulates Immune Function in Postmenopausal Women," *Am J Obstet Gynecol*, 1993, 169(6):1536-9.

Kroboth PD, Salek FS, Pittenger AL, et al, "DHEA and DHEA-S: A Review," *J Clin Pharmacol*, 1999, 39(4):327-48.

(Continued)

Dehydroepiandrosterone and Dehydroepiandrosterone Sulfate, Serum or Plasma

(Continued)

Morales AJ, Haubrich RH, Hwang JY, et al, "The Effect of Six Months Treatment With a 100 mg Daily Dose of Dehydroepiandrosterone (DHEA) on Circulating Sex Steroids, Body Composition and Muscle Strength in Age-Advanced Men and Women," *Clin Endocrinol*, 1998, 49(4):421-32.

Morales AJ, Nolan JJ, Nelson JC, et al, "Effects of Replacement Dose of Dehydroepiandrosterone in Men and Women of Advancing Age," *J Clin Endocrinol Metab*, 1994, 78(6):1360-7.

Skolnick AA, "Scientific Verdict Still Out on DHEA," *JAMA*, 1996, 276(17):1365-7.

♦ **Dehydroepiandrosterone, Serum** *see* Dehydroepiandrosterone and Dehydroepiandrosterone Sulfate, Serum or Plasma *on page 506*

♦ **Dehydroepiandrosterone Sulfate, Serum** *see* Dehydroepiandrosterone and Dehydroepiandrosterone Sulfate, Serum or Plasma *on page 506*

Delta (5)-Aminolevulinic Acid, Urine

Related Information

Lead, Blood *on page 832*
Lead, Urine *on page 835*
Porphobilinogen, Urine *on page 1073*
Porphyrins, Quantitative, Urine *on page 1074*
Protoporphyrin, Free Erythrocyte *on page 1121*
Protoporphyrin, Zinc, Blood *on page 1121*

Synonyms ALA; Aminolevulinic Acid; 5-Aminolevulinic Acid; Delta-ALA

Applies to ALA Dehydratase

Abstract Delta (5)-aminolevulinic acid (ALA) is a porphyrin precursor. The acute neurological porphyrias are associated with elevated ALA and porphobilinogen (PBG).

Specimen 24-hour urine

Container Dark urine container, kept on ice or refrigerated; check with laboratory.

Collection Acidify with acetic acid to pH 3-4.5; some laboratories use 2 g barbituric acid as a preservative. Sodium carbonate is appropriate as a preservative also for porphyrins and porphobilinogen. Check with laboratory.

Storage Instructions Refrigerate or freeze according to specific laboratory instructions. Protect from light.

Reference Interval Method dependent; normal: 1.3-7.0 mg/24 hours urine (SI: 9.9-53.4 µmol/day)

Possible Panic Range >20 mg/24 hours (SI: >155.2 µmol/day)

Use Diagnose porphyrias: delta-ALA may be increased in attacks of acute intermittent porphyria, hereditary coproporphyria, porphyria variegata, and ALA dehydratase porphyria. Evaluate certain neurological problems with abdominal pain. **Porphobilinogen, delta aminolevulinic acid, and porphyrins in urine** are first order tests for acute intermittent porphyria. Erythrocyte porphobilinogen deaminase (uroporphyrinogen-1 synthase) is a second order test to confirm the diagnosis.

While urinary ALA has, in the past, been used to diagnose lead poisoning, such practice is **not recommended**. This is because lead poisoning in children does not produce increases in urinary ALA until the blood lead is 40 µg/dL, substantially above the recommended cutoff of <10 µg/dL. Urinary ALA is not a satisfactory test to screen for, or confirm, the diagnosis of lead poisoning.

Limitations ALA may be normal during latent period of the acute porphyrias. For the diagnosis of lead poisoning, measurement of blood and urine lead, free erythrocyte protoporphyrin and ALA dehydratase in red cells are available.

Methodology Ion-exchange resin columns, colorimetry. An alternative photometric method for rapid testing has been proposed.

Additional Information Toxins including alcohol, lead and other heavy metals cause increase in ALA by inhibiting porphobilinogen synthase.

References

Buttery JE, Stuart S, and Pannall PR, "An Improved Direct Method for the Measurement of Urinary Delta-Aminolevulinic Acid," *Clin Biochem*, 1995, 28(4):477-80.

Foran SE and Abel G, "Guide to Porphyrias. A Historical and Clinical Perspective," *Am J Clin Pathol*, 2003, 119(Suppl 1):S86-S93.

Sassa S, "ALAD Porphyria," *Semin Liver Dis*, 1998, 18(1):95-101.

♦ **Delta Agent Serology** *see* Hepatitis D Serology *on page 711*

- **Delta-ALA** *see* Delta (5)-Aminolevulinic Acid, Urine *on page 508*
- **Delta Aminolevulinic Acid** *see* Porphobilinogen, Urine *on page 1073*
- **Delta Aminolevulinic Acid Dehydratase** *see* Lead, Blood *on page 832*
- **Delta Bilirubin** *see* Bilirubin, Direct, Serum *on page 262*
- **Delta F508** *see* Cystic Fibrosis DNA Detection *on page 491*
- **Demerol®** *see* Meperidine, Serum or Urine *on page 895*
- **Demolox®** *see* Amoxapine, Serum or Plasma *on page 153*
- **Deoxycorticosterone** *see* 17-Hydroxyprogesterone, Whole Blood, Serum, or Plasma *on page 755*
- **11-Deoxycortisol** *see* Metyrapone Stimulation Test, Serum *on page 910*
- **Deoxypyridinoline** *see* Osteocalcin, Serum or Plasma *on page 983*
- **Deoxypyridinoline** *see* Pyridinolines (Pyridinoline and Deoxypyridinoline), Urine *on page 1126*
- **Deoxypyridoline** *see* N-Telopeptides, Urine *on page 967*
- **Depacon®** *see* Valproic Acid, Serum or Plasma *on page 1297*
- **Depakene®** *see* Valproic Acid, Serum or Plasma *on page 1297*
- **Depakote®** *see* Valproic Acid, Serum or Plasma *on page 1297*
- **Depakote® XR** *see* Valproic Acid, Serum or Plasma *on page 1297*
- **Depamide®** *see* Valproic Acid, Serum or Plasma *on page 1297*
- **Deprax®** *see* Trazodone, Serum or Plasma *on page 1272*
- **Deralin®** *see* Propranolol, Serum *on page 1096*
- **De Ritis Ratio** *see* Liver Disease: Laboratory Assessment, Overview *on page 869*
- ***Dermacentor andersoni*** *see* Arthropod Identification *on page 213*
- **Dermatan Sulfates** *see* Mucopolysaccharides, Urine *on page 922*
- **Dermatitis Herpetiformis Antibodies** *see* Skin Biopsy, Immunofluorescence *on page 1203*
- **Desipramine** *see* Antidepressants, Cyclic, Serum or Plasma *on page 171*
- **Desipramine** *see* Imipramine, Serum or Plasma *on page 767*
- **Desmin** *see* Immunoperoxidase Procedures *on page 780*
- **Desmopressin** *see* Coagulation Factor Assays *on page 418*
- **Desmopressin** *see* von Willebrand Factor *on page 1321*
- **11-Desoxycortisol** *see* 17-Hydroxyprogesterone, Whole Blood, Serum, or Plasma *on page 755*
- **Desoxyephedrine Hydrochloride** *see* Methamphetamine, Qualitative, Urine *on page 902*
- **Desoxyn®** *see* Methamphetamine, Qualitative, Urine *on page 902*
- **Desoxyphenobarbital** *see* Primidone, Serum or Plasma *on page 1091*
- **Desyrel®** *see* Trazodone, Serum or Plasma *on page 1272*
- **Dexamethasone-CRH Protocol** *see* Corticotropin-Releasing Hormone Stimulation Test *on page 455*
- **Dexamethasone Suppression** *see* Adrenocorticotropic Hormone, Plasma *on page 114*
- **Dexamethasone Suppression Test** *see* Adrenal Cortex: Laboratory Assessment Overview *on page 110*
- **Dexamethasone Suppression Test** *see* Cortisol, Serum or Plasma *on page 460*
- **Dexedrine®** *see* Amphetamine, Qualitative, Urine *on page 154*
- **Dexedrine®** *see* Methamphetamine, Qualitative, Urine *on page 902*
- **Dexies** *see* Amphetamine, Qualitative, Urine *on page 154*
- **Dexitac®** *see* Caffeine, Serum *on page 326*
- **1,4-α-D Glucanohydrolase, Serum** *see* Amylase, Serum *on page 155*
- **1,4-α-D Glucanohydrolase, Urine** *see* Amylase, Urine *on page 157*
- **DHEA** *see* Dehydroepiandrosterone and Dehydroepiandrosterone Sulfate, Serum or Plasma *on page 506*
- **DHEA-S** *see* Dehydroepiandrosterone and Dehydroepiandrosterone Sulfate, Serum or Plasma *on page 506*
- **Dialysis** *see* Urea Nitrogen, Serum or Plasma *on page 1284*
- **Diastat®** *see* Diazepam, Serum *on page 510*
- **Diazemuls®** *see* Diazepam, Serum *on page 510*

Diazepam, Serum

Related Information

Benzodiazepines, Qualitative, Urine *on page 253*
Ethanol, Blood, Urine, and Other Sources *on page 558*

Synonyms Aliseum®; Alupram®; Atensine®; Diastat®; Diazemuls®; Di-Tran®; Lamra®; Solis®; Stesolid®; Tensium®; T-Quil®; Valium®; Valrelease®; Vatran®; Vazepam®; Vivol®; Zetran®

Abstract Diazepam is a benzodiazepine used as a sedative-hypnotic (tranquilizer) to treat anxiety, panic attacks, and muscle spasms. It is also frequently used to control seizures in emergency situations and treatment of ethanol withdrawal symptoms. It is metabolized to nordiazepam, an active metabolite. Both should be measured. It is a commonly abused benzodiazepine and is detectable in urine by benzodiazepine screening.

Specimen Serum

Container Red top tube; do not use gel-containing tube.

Sampling Time For peak level, 1 hour after oral dose or 15 minutes after I.V.

Storage Instructions Do not freeze

Reference Interval Therapeutic: diazepam: 0.2-1.0 µg/mL (SI: 0.7-3.5 µmol/L), N-desmethyldiazepam (nordiazepam): 0.1-0.5 µg/mL (SI: 0.35-1.8 µmol/L)

Critical Values Toxic: sum of diazepam plus N-desmethyldiazepam >3.0 µg/mL (SI: >11 µmol/L)

Possible Panic Range Total of diazepam and nordiazepam >5 µg/mL (SI: >18 µmol/L) is toxic.

Use Therapeutic monitoring and toxicity assessment

Methodology Immunoassay, high performance liquid chromatography (HPLC), gas chromatography (GC)

Additional Information

- Half-life: 20-50 hours; bioactive half-life in CNS: 55 minutes
- Volume of distribution: 1.0-1.5 L/kg
- Protein binding: 96% to 99%

Peak blood levels are achieved within an hour after oral dose. The major metabolite (N-desmethyldiazepam) has a half-life in adults of 50-99 hours. It is also the major metabolite of clorazepate and prazepam. Diazepam exhibits synergism with other CNS depressants such as barbiturates and ethanol. Many cases of overdose are seen but few fatalities result from use of this drug alone. In fact, it is so safe that it is frequently administered by paramedics for out-of-hospital status epilepticus in adults.

A combination of this drug and ethanol is frequently found.

References

Alldredge BK, Gelb AM, Isaacs SM, et al, "A Comparison of Lorazepam, Diazepam, and Placebo for the Treatment of Out-of-Hospital Status Epilepticus," *N Engl J Med*, 2001, 345(9):631-7.

Al Tahan A, "Paradoxic Response to Diazepam in Complex Partial Status Epilepticus," *Arch Med Res*, 2000, 31(1):101-4.

Moller HJ, "Effectiveness and Safety of Benzodiazepines," *J Clin Psychopharmacol*, 1999, 19(6 Suppl 2):2S-11S.

Dibucaine Number, Serum or Plasma

Related Information

Acetylcholinesterase, Red Cell and Serum *on page 93*
Organophosphate Pesticides, Urine, Blood, or Serum *on page 975*
Pseudocholinesterase, Serum *on page 1122*

Synonyms Pseudocholinesterase Inhibition by Dibucaine

Applies to Dibucaine Toxicity; Fluoride Inhibition of Cholinesterase

Abstract Pseudocholinesterase names a serum enzyme and distinguishes it from red cell, or true, cholinesterase. Dibucaine inhibits the normal form of pseudocholinesterase. Abnormal variants are less inhibited. Percent inhibition (dibucaine number) is obtained by comparing results from the inhibited reaction with noninhibited reaction. Administration of succinylcholine may pose a risk to patients with abnormal pseudocholinesterase, because there is abnormal persistence of the succinylcholine effect.

Specimen Serum or plasma

Container Red top tube or green top (heparin) tube

Collection Do not collect within 24 hours of administration of muscle relaxant.

Storage Instructions Serum pseudocholinesterase is stable and may be stored at 0°C to 4°C or at room temperature for 1 year or more and may be frozen/thawed a number of times without changing activity of the enzyme.

Reference Interval Normal individuals have normal (high) amounts of serum pseudocholinesterase activity which can be inhibited by dibucaine. Approximately 70% to 86% inhibition is normal; atypical enzyme shows resistance to inhibition, at about the level of only 20%. See reference interval of laboratory doing the test.

Use Assess presence of homozygous or heterozygous "atypical" pseudocholinesterase variant; such individuals may experience apnea when given succinylcholine. Dibucaine inhibition provides identification of an abnormal allele. Sensitivity of pseudocholinesterase to dibucaine inhibition may distinguish congenital from acquired forms of abnormal pseudocholinesterase activity. About 4% of the population have abnormal inherited forms.

Limitations No single simple test currently exists that can detect all enzyme variants. Traditional tests including dibucaine inhibition are not adequate to identify all variants. Instances of prolonged response to succinylcholine still go without explanation.

Methodology Hydrolysis of propionylthiocholine or butyrylthiocholine with and without dibucaine at 20°C to 40°C; fluoride inhibition at 25°C.

Additional Information The degree of serum cholinesterase inhibition produced by dibucaine (and fluoride) is under genetic control. Sensitivity to succinylcholine is dependent upon at least four allelic genes (see table). The E_1^a gene is responsible for the atypical enzyme which resists inhibition by dibucaine. The E_1^f gene is responsible for the enzyme that is fluoride resistant. The E_1^s (silent) gene results in an enzyme with little or no activity. An international gene nomenclature conference has proposed a system designating the four alleles as "CHE1*U," "CHE1*A," "CHE1*F," and "CHE1*QO." Another phenotypic nomenclature designates those at risk, AF; FS and FF (moderate risk); and AA, AS, and SS (severe risk).

Genotype	Phenotype	Previous Term	Dibucaine Number
$E_1^u E_1^u$	U	Usual	84
$E_1^u E_1^a$	I	Intermediate	73
$E_1^a E_1^a$	A	Atypical	32
$E_1^u E_1^s$	U	Usual	
$E_1^s E_1^s$	S	Silent	81
$E_1^u E_1^f$	UF		
$E_1^f E_1^f$	F		

Dibucaine and fluoride numbers indicate the percent inhibition of enzyme activity by these agents when a serum sample is tested under standard conditions (inhibition expressed as a percent). This approach to screening for presence of serum cholinesterase variants does not entirely avoid the problem of variation in reactivity with some atypical enzymes. Individuals with the genotype E_1^u, E_1^f show resistance to fluoride inhibition (low fluoride number) but do not show resistance to dibucaine inhibition.

References

Evans RT and Wroe J, "Is Serum Cholinesterase Activity a Predictor of Succinyl Choline Sensitivity? An Assessment of Four Methods," *Clin Chem*, 1978, 24(10):1762-6.

Harris H and Whittaker M, "Differential Inhibition of Human Serum Cholinesterase With Fluoride: Recognition of Two New Phenotypes," *Nature*, 1961, 496-8.

Holownia P, Newman DJ, Bruno C, et al, "Automated Dibucaine Number Measurement With DuPont Dimension® ES and AR Analyzers," *Clin Chem*, 1995, 41(5):664-7.

Pantuck EJ, "Plasma Cholinesterase: Gene and Variations," *Anesth Analg*, 1993, 77(2):380-6.

♦ **Dibucaine Toxicity** *see* Dibucaine Number, Serum or Plasma *on page 510*

♦ **Dicorynan®** *see* Disopyramide, Serum or Plasma *on page 516*

♦ **DIC Screen** *see* Disseminated Intravascular Coagulation Screen *on page 517*

♦ **DIDMOAD Syndrome** *see* Cobalamin, Serum *on page 424*

♦ **Diet** *see page 11*

♦ **Differential Smear** *see* Peripheral Blood: Differential Leukocyte Count *on page 1010*

- ♦ **Digacin** *see Digoxin, Serum on page 512*
- ♦ **Digitalis** *see Digoxin, Serum on page 512*
- ♦ **Digitalis Glycosides** *see Digoxin, Serum on page 512*
- ♦ **Digitalis-Like Immunoreactive Substances** *see Digoxin, Serum on page 512*
- ♦ **Digitoxine®** *see Digoxin, Serum on page 512*

Digitoxin, Serum

Related Information

Digoxin, Serum *on page 512*

Synonyms Corramedan®; Cystodigin®; Digitalis; Digitoxine®; Digitrin®; Lanotoxin®; Nativelle®; Purodigin®; Tardigal®

Abstract A plant glycoside, digitoxin is used in the treatment of congestive heart failure, atrial fibrillation, atrial flutter, paroxysmal atrial tachycardia, and cardiogenic shock. The drug exerts positive inotropic effect. This drug is not commercially available in the United States, however, it is in use in other parts of the world.

Specimen Serum

Container Red top tube

Sampling Time 6-12 hours after dose

Storage Instructions Separate serum and store in refrigerator.

Reference Interval Therapeutic: 18-35 ng/mL (SI: 24-46 nmol/L)

Critical Values Levels >35 ng/mL (SI: >46 nmol/L) are associated with clinical toxicity in 80% of patients.

Use Therapeutic monitoring and toxicity assessment. See Table C in the Therapeutic Drug Monitoring Introduction *on page 62.*

Limitations Do not order digitoxin level on a patient receiving digoxin.

Contraindications Patient on **digoxin** or recent radioactive tracer (if RIA is used)

Methodology Immunoassay, gas chromatography (GC), high performance liquid chromatography (HPLC)

Additional Information

- Half-life: 150-250 hours
- Volume of distribution: 0.7 L/kg
- Protein binding: 90% to 95%

Optimal sampling time after dosage is 6 hours. Optimal resampling time after change in dosage is 48-96 hours. Digoxin-like immunoreactive substance (DLIS), an endogenous natriuretic substance, may cause false elevation.

References

Belz GG, Breihaupt-Grogler K, and Osowski U, "Treatment of Congestive Heart Failure," *Clin Invest,* 2001, 31:10-7.

Kulick DL and Rahimtoola SH, "Current Role of Digitalis Therapy in Patients With Congestive Heart Failure," *JAMA,* 1991, 265(22):2995-7.

Roever C, Ferrante J, Gonzalez EC, et al, "Comparing the Toxicity of Digoxin and Digitoxin in a Geriatric Population: Should an Old Drug Be Rediscovered?" *South Med J,* 2000, 93(2):199-202.

- ♦ **Digitrin®** *see Digitoxin, Serum on page 512*

Digoxin, Serum

Related Information

Amiodarone, Serum *on page 148*
B-Type Natriuretic Peptide *on page 314*
Digitoxin, Serum *on page 512*
Flecainide, Serum or Plasma *on page 592*
Magnesium, Serum *on page 885*
Potassium, Serum or Plasma *on page 1078*
Quinidine, Serum *on page 1129*
Verapamil, Serum or Plasma *on page 1306*

Synonyms Allocar®; Cardioreg®; Digacin; Lanocor®; Lanoxicaps®; Lanoxin®; Lenoxin®; Purgoxin®

Applies to Botanical Dietary Supplements; Chan Su; Digitalis Glycosides; Digitalis-Like Immunoreactive Substances; Kyushin; Liu-Shan-Wan

Abstract Digoxin is a widely-used cardiac glycoside utilized in management of atrial fibrillation with rapid ventricular response, and in treatment of heart failure. Digoxin is generally thought to ameliorate symptoms of heart failure and to improve exercise tolerance. A revised therapeutic range of 0.5-0.8 ng/mL has recently been recommended.

Specimen Serum

Container Red top tube

Sampling Time Blood specimen must be drawn **at least** 6 hours after the administration of the last dose. The average half-life is 48 hours in patients whose renal function is normal. The steady-state is usually reached in 5 days. After this time, a steady-state estimate is best obtained by a specimen drawn just before next dose.

Storage Instructions Separate serum and refrigerate.

Causes for Rejection Patient on a cardiac glycoside other than digoxin, recently administered radioisotopes if RIA is used for assay, hemolysis, sample collected in serum integrator gel tube

Reference Interval A therapeutic interval from 0.5-2.0 ng/mL (SI: 1.0-2.6 mmol/L), has been used. However, recent data suggests an optimal range of 0.5-0.8 ng/mL in patients with heart failure. A higher concentration of approximately 1.0 ng/mL may not provide any additional benefit, and concentrations ≥1.2 ng/mL are presently considered harmful.

Critical Values Toxic: >1.2 ng/mL (SI: >1.5 nmol/L). The lethal dose is about double the level causing minor toxic manifestations. In patients with hypokalemia or hypomagnesemia, toxicity may occur at a level <2 ng/mL. The most common manifestations of suspected toxicity are nausea and vomiting, ventricular fibrillation, tachycardia, supraventricular arrhythmia, and second- or third-degree atrioventricular block.

Possible Panic Range >3.0 ng/mL (SI: >3.8 nmol/L). See Table C in the Therapeutic Drug Monitoring Introduction *on page 62.*

Use Therapeutic drug monitoring and toxicity assessment

Limitations Digitoxin should not be confused with **digoxin**. Since there is cross reactivity between the two drugs, **results will not be valid if digoxin is measured when the patient is taking digitoxin.** All other digitalis derivatives will also cross react with this test and give invalid results. Toxic levels of digitoxin when assayed as digoxin give low results. Endogenous digitalis-like immunoreactive substances (DLIS) are found in digoxin-free patients with a variety of clinical states associated with salt and fluid retention, such as renal failure, hepatic failure, low renin hypertension, and pregnancy. They are also present at birth in neonates and infants. These compounds (DLIS) cross react with digoxin-specific immunoassays and give falsely elevated plasma digoxin concentrations. Current assay methods (as below) do not entirely avoid this interference. Some assay methods give a falsely lowered digoxin concentration in the presence of DLIS.

Dan Shen, a Chinese medicine, and Asian and Siberian ginsengs available in the U.S. without prescription, may interfere with serum assay of digoxin.

Methodology Immunoassays

Additional Information
- Half-life: 20-60 hours
- Volume of distribution: 7 L/kg
- Protein binding: 20% to 25%

Until recently, digoxin levels ≥1.2 ng/mL have been accepted as therapeutic. Since the DIG trial, concerns have been raised about digoxin levels >1.0 ng/mL. A recent post-hoc data analysis on 3782 men enrolled in DIG trial suggests that in men with heart failure the optimal digoxin concentration is 0.5-0.8 ng/mL. Levels of approximately 1.0 ng/mL do not provide additional benefit, and levels ≥1.2 ng/mL may be harmful. It is suggested that as the serum digoxin concentration increases, the stronger inotropic action offsets the therapeutic benefits of neurohormonal modulation.

Digoxin is indicated for long-term management of congestive heart failure and control of supraventricular tachyarrhythmias. The average volume of distribution is 7 L/kg and is decreased in patients on quinidine, with renal failure and with hypothyroidism. Plasma concentrations do not accurately reflect pharmacologic effects of the drug until it is completely distributed. If obtained before complete distribution, concentration will be misleading.

Digitalis levels must always be interpreted in light of clinical data. Older, smaller patients require less digoxin. Proportionally lower loading doses are (Continued)

Digoxin, Serum *(Continued)*

advocated in the elderly. The most common cause of digitalis intoxication is potassium depletion. The primary cause of digoxin toxicity in the aged is decreased renal function. Maintenance doses should be adjusted to the glomerular filtration rate (GFR). Renal failure, hypercalcemia, alkalosis, myxedema, hypomagnesemia, recent MI and other heart disease, hypokalemia, and hypoxia may increase sensitivity to the toxic effects of digoxin. Renal function and serum electrolytes should be frequently checked in patients receiving digoxin. Potassium-depleting diuretics are a major contributing factor to digitalis toxicity.

Quinidine may cause elevation of digoxin level by decreasing its excretion. It is recommended that serum digoxin concentration be measured before initiation of quinidine therapy and again in 4-6 days. **Verapamil** and **amiodarone** cause increased digoxin levels, also by decrease in clearance. (Clearance parallels GFR.) Other drugs which decrease its clearance include cyclosporine, spironolactone, and propafenone.

When confronted with unexpectedly low digoxin levels, consider noncompliance, thyroid disease, malabsorption, and reduced intestinal blood flow from mesenteric arteriosclerosis. **Drugs** which diminish its bioavailability include metoclopramide, cholestyramine, colestipol, kaolin-pectin, neomycin, and sulfasalazine. Other agents that decrease digoxin levels include antacids, bran and para-aminosalicylic acid (PAS). Consider, as well, congestive failure and anticholinergic drug effects when low digoxin levels are encountered. Other drug interactions include antacids, cathartics, neomycin, phenytoin, and metoclopramide which may decrease absorption of digoxin. Indomethacin, diltiazem, erythromycin, tetracycline, itraconazole, nicardipine, triamterene, and spironolactone may increase digoxin serum concentration; penicillamine may decrease pharmacologic effects of digoxin. Propantheline and atropine may increase digoxin absorption.

Fab fragments of digoxin-specific antibodies are available for the treatment of serious digoxin toxicities. Digibind® (antidigoxin Fab fragments-digoxin immune Fab) will increase total serum digoxin concentrations about 50-fold though as an inactive compound.

Patients with **digitalis resistance** may require larger doses and may have higher than usual serum levels (eg, patients with hyperthyroidism).

Extracts of Chan Su, an over-the-counter Chinese medicine, and Asian and Siberian ginsengs contain glycosides and interfere with certain immunoassays. The most effected immunoassays are FPIA and MEIA.

References

Canas F, Tanasijevic MJ, Maluf N, et al, "Evaluating the Appropriateness of Digoxin Level Monitoring," *Arch Intern Med*, 1999, 159(4):363-8.

Cohn JN, "The Management of Chronic Heart Failure," *N Engl J Med*, 1996, 335(7):490-8.

Dasgupta A, Wu S, Actor J, et al, "Effect of Asian and Siberian Ginseng on Serum Digoxin Measurement by Five Digoxin Immunoassays. Significant Variation in Digoxin-Like Immunoreactivity Among Commercial Ginsengs," *Am J Clin Pathol*, 2003, 119(2):298-303.

Eichhorn EJ and Gheorghiade M, "Digoxin - New Perspective on an Old Drug," *N Engl J Med*, 2002, 347:1394-5.

Garg R, Gorlin R, Smith T, et al, The Digitalis Investigation Group, "The Effect of Digoxin on Mortality and Morbidity in Patients With Heart Failure," *N Engl J Med*, 1997, 336(8):525-33.

Haji SA and Movahed A, "Update on Digoxin Therapy in Congestive Heart Failure," *Am Fam Phys*, 2000, 62(2):409-16.

Packer M, "End of the Oldest Controversy in Medicine. Are We Ready to Conclude the Debate on Digitalis?" *N Engl J Med*, 1997, 336(8):575-6.

Rathore SS, Curtis JP, Wang Y, et al, "Association of Serum Digoxin Concentrations and Outcomes in Patients With Heart Failure," *JAMA*, 2003, 289(7):871-8.

Williamson KM, Thrasher KA, Fulton KB, et al, "Digoxin Toxicity: An Evaluation in Current Clinical Practice," *Arch Intern Med*, 1998, 158(22):2444-9.

♦ **Dihydropteridine Reductase** *see* Newborn Screen for Phenylketonuria *on page 954*

♦ **Dihydrotestosterone** *see* Testosterone, Total and Free, Serum or Plasma *on page 1238*

Dihydrotestosterone, Serum

Related Information

Testosterone, Total and Free, Serum or Plasma *on page 1238*

Applies to 5-Alpha Reductase

Abstract Dihydrotestosterone (DHT) is produced from testosterone by the action of 5-alpha reductase, an enzyme found in many androgen-sensitive tissues (eg, skin, prostate, other internal genitalia). In the 46X,Y fetus, the development of the external genitalia depends on the action of DHT; whereas testosterone may be metabolized to DHT or to the estrogen, estradiol, DHT is not converted into estrogen and is much more potent an androgen than is testosterone. Most disorders involving excess or insufficient testicular androgen are well evaluated by measuring Testosterone, Total and Free, Serum or Plasma *on page 1238*.

Specimen Serum

Container Red top tube

Storage Instructions Separate serum and freeze.

Special Instructions No recent radioisotopes

Reference Interval The reference intervals are not well defined at present. Below are two sets of intervals from two reference laboratories.

Male:
- ≤20 years: not established
- 20-39 years: 150-1240 pg/mL
- 20-49 years: 155-553 pg/mL
- >40 years: 150-980 pg/mL
- >50 years: 36-573 pg/mL

Female:
- ≤20 years: not established
- 20-39 years: 50-250 pg/mL
- 15-49 years: 5.5-170 pg/mL
- >40 years: 50-137 pg/mL
- >50 years: 36-573 pg/mL

Use

Drugs which inhibit 5-alpha reductase: Certain drugs (eg, finasteride) used to treat benign prostatic hyperplasia exert their effect by inhibiting 5-alpha reductase, an enzyme in peripheral tissues which converts testosterone to DHT. Therefore, serum DHT may be useful in monitoring treatment with this drug class.

Congenital 5-alpha reductase deficiency: In this rare autosomal recessive disorder, a phenotypic male, 46XY, infant has hypospadias, a urogenital sinus opening on the perineum, a blind vaginal pouch, and normal appearing testes which may be cryptorchid or in labioscrotal folds. With the onset of puberty, plasma testosterone (total and free) values are normal, but dihydrotestosterone values are very low, reflecting the absence of 5-alpha reductase.

Methodology Radioimmunoassay (RIA)

Additional Information Of all the androgenic hormones, serum DHT is the only one that has been shown to correlate with male sexual functioning.

References
ARUP Laboratories, *1999-2000 User's Guide*, 262.
Hughes IA, "A Novel Explanation for Resistance to Androgens," *N Engl J Med*, 2000, 343(12):881-2
Mayo Medical Laboratories, *2001 Test Catalogue*, Rochester, MN, 196.

♦ **1,25-Dihydroxy Vitamin D₃** *see* Vitamin D, Serum *on page 1318*

♦ **Dilacor® XR** *see* Diltiazem, Serum or Plasma *on page 515*

♦ **Dilantin®** *see* Phenytoin, Serum or Plasma *on page 1026*

♦ **Dilocaine®** *see* Lidocaine, Serum or Plasma *on page 850*

♦ **Diltia XT®** *see* Diltiazem, Serum or Plasma *on page 515*

Diltiazem, Serum or Plasma
Related Information
Clonidine, Serum or Plasma *on page 415*
Propranolol, Serum *on page 1096*

Synonyms Acalix®; Britiazim®; Cardizem®; Cartia XT™; Dilacor® XR; Diltia XT®; Latiazem Hydrochloride; Taztia XT™; Tiazac®

Abstract Diltiazem, a benzodiazepine analog, is a calcium channel blocker used in the treatment of angina pectoris, hypertension, and supraventricular arrhythmias. The drug is extensively metabolized and primarily excreted in the urine.

Specimen Serum or plasma

Container Red top tube, lavender top (EDTA) tube

(Continued)

Diltiazem, Serum or Plasma *(Continued)*

Sampling Time Trough

Storage Instructions Store in refrigerator. Ship at room temperature.

Turnaround Time 2-3 days

Reference Interval 50-200 ng/mL

Use Monitor therapy

Methodology Gas chromatography (GC), high performance liquid chromatography (HPLC)

Additional Information
- Half-life: 4-6 hours
- Volume of distribution: 4-8 L/kg
- Protein binding: 70% to 90%
- Oral bioavailability: 40% to 60%

Serious toxic effects include arrhythmias, shortness of breath, and fatigue due to heart failure. The common, less severe side effects are headache, drowsiness, swelling of feet and ankles, constipation, and nausea.

References

Lacy CF, Armstrong LL, Goldman MP, et al, *Drug Information Handbook*, 12th ed, Hudson, OH: Lexi-Comp Inc, 2004.

Pool PE, "Anomalies in the Dosing of Diltiazem," *Clin Cardiol*, 2000, 23(1):18-23.

Sage PR, Kiosoglous AJ, Wuttke RD, et al, "Early Treatment With Verapamil or Diltiazem in Patients With Acute Myocardial Infarction: Safety and Possible Beneficial Effects," *Cardiovasc Drugs Ther*, 1999, 13(4):309-13.

♦ **Dima-Fen®** *see* Fenfluramine, Serum *on page 577*

♦ **Dimetapp®** *see* Methamphetamine, Qualitative, Urine *on page 902*

♦ **Dimethylarsine** *see* Arsenic, Urine *on page 211*

♦ **Dimipressin®, Iprogen®** *see* Imipramine, Serum or Plasma *on page 767*

♦ **Dintoina®** *see* Phenytoin, Serum or Plasma *on page 1026*

♦ **Diphenylan Sodium®** *see* Phenytoin, Serum or Plasma *on page 1026*

♦ **Diphenylhydantoin** *see* Phenytoin, Serum or Plasma *on page 1026*

♦ **2,3-Diphosphoglycerate (2,3-DPG)** *see* Oxygen Saturation, Blood *on page 991*

♦ **Dipropylacetic Acid** *see* Valproic Acid, Serum or Plasma *on page 1297*

♦ **Direct Antiglobulin Test** *see* Antiglobulin Test, Direct *on page 176*

♦ **Direct Coombs Test** *see* Antiglobulin Test, Direct *on page 176*

♦ **Disaturated Phosphatidylcholine** *see* Lecithin:Sphingomyelin Ratio, Amniotic Fluid *on page 836*

Disopyramide, Serum or Plasma

Related Information

Verapamil, Serum or Plasma *on page 1306*

Synonyms Dicorynan®; Napamide®; Norpace®; Rhythmodan®; Ritmilen®

Abstract Disopyramide is a class IA antiarrhythmic agent. It is used in prevention of atrial flutter and fibrillation, and to prevent ventricular tachycardia and fibrillation.

Specimen Serum or plasma

Container Red top tube, green top (heparin) tube, or lavender top (EDTA) tube

Special Instructions Other cardiac medications should be made known to the laboratory.

Reference Interval Therapeutic: trough: atrial arrhythmias: 2.8-3.2 μg/mL (SI: 8.3-9.4 μmol/L), ventricular arrhythmias: 3.3-5.0 μg/mL (SI: 9.7-15.0 μmol/L)

Critical Values Toxic: >7.0 μg/mL (SI: >20.6 μmol/L)

Possible Panic Range Fatalities are seen with concentrations >20.0 μg/mL. Bradycardia and asystole occur.

Use Therapeutic monitoring and toxicity assessment

Limitations Arrhythmias may occur at low levels. Disopyramide exhibits nonlinear binding in the therapeutic range. This, combined with the fact that disopyramide is administered as a racemic mixture, makes total concentration difficult to correlate with pharmacodynamic effects.

Methodology Immunoassay, high performance liquid chromatography (HPLC)

Additional Information
- Half-life: 4-10 hours
- Volume of distribution: 0.7-0.9 L/kg

- Protein binding: 20% to 60% (inversely proportional to concentration)

Disopyramide shares electrophysiologic properties with quinidine and procainamide. More than 80% of an oral dose is absorbed, and only a small fraction of the drug undergoes first-pass metabolism. Fifty percent of the drug is excreted unchanged in the urine. Some of the remainder is metabolized by the liver to inactive products. Dosage must be modified (dosage intervals prolonged) in patients with renal failure, and dosage interval relates to creatinine clearance. Serious toxic effects are depression of myocardial contractility and disturbances in myocardial conduction. Other effects include dry mouth, constipation, urinary hesitancy, and blurred vision. Metabolite N-desisopropyl disopyramide is also pharmacologically active, with activity approximately 25% that of disopyramide. The most common adverse effects are related to cholinergic blockade. The most serious adverse effects of disopyramide are hypotension and congestive heart failure.

Disopyramide and its metabolite may cause decrease in glucose levels and interfere with metabolism of other drugs.

References
Duff HJ, Mitchell LB, Nath CF, et al, "Concentration-Response Relationships of Disopyramide in Patients With Ventricular Tachycardia," *Clin Pharmacol Ther*, 1989, 45(5):542-7.

Hasegawa J, Mori A, Yamamoto R, et al, "Disopyramide Decreases the Fasting Serum Glucose Level in Man," *Cardiovasc Drugs Ther*, 1999, 13(4):325-7.

Trujillo TC and Nolan PE, "Antiarrhythmic Agents: Drug Interactions of Clinical Significance," *Drug Saf*, 2000, 23(6):509-32.

Disseminated Intravascular Coagulation Screen

Related Information
Activated Partial Thromboplastin Time *on page 100*
Anemia Flowchart *on page 35*
D-Dimers and Fibrin Degradation Products *on page 502*
Fibrinogen *on page 583*
Platelet Count *on page 1050*
Prothrombin Time *on page 1116*

Synonyms Consumptive Coagulopathy Screen; DIC Screen; Screen for Disseminated Intravascular Coagulation

Applies to Fibrinolysis; Schistocytes

Abstract The most useful panel of tests to screen for disseminated intravascular coagulation (DIC) includes D-dimer or fibrin degradation products (FDP), prothrombin time (PT), activated partial thromboplastin time (PTT), platelet count, examination of peripheral blood smear, and fibrinogen. **These tests are not, however, specific for DIC.**

Specimen Plasma (and whole blood for platelet count and peripheral blood smear)

Container Blue top (sodium citrate) tube; lavender top (EDTA) tube for platelet count and peripheral blood smear

Collection Routine venipuncture. Draw blue top before lavender top (EDTA) tube. If a red top is drawn, draw red top tube before blue top tube. Immediately invert tube gently at least 4 times to mix. Blue top tubes must be appropriately filled. Deliver tubes immediately to the laboratory.

Storage Instructions Blue top tube: separate plasma from cells as soon as possible; plasma may be stored on ice for up to 4 hours; otherwise store frozen

Causes for Rejection Blue top tube received more than 4 hours after collection, blue top tube not filled, clotted specimens

Turnaround Time 1-2 hours (often less than 1 hour if requested stat)

Reference Interval See individual tests.

Use Diagnose DIC in patients with an underlying disorder known to cause DIC and/or with clinical suspicion of DIC

Limitations D-dimer and FDP are positive with physiologic clot formation and lysis, and they may be positive in liver disease because they are normally cleared by the liver. Therefore, DIC can be difficult to diagnose in the presence of liver disease in some cases. Results should be reviewed in relation to the clinical situation.

Methodology See individual tests.

Additional Information DIC is a common acquired coagulation disorder resulting from excessive activation of the coagulation system, usually due to massive tissue injury, sepsis, or certain pregnancy complications. The normal (Continued)

Disseminated Intravascular Coagulation Screen
(Continued)

anticoagulant and fibrinolytic systems are overwhelmed and cannot contain the coagulation activation, which becomes systemic, resulting in disseminated microvascular thrombi. Thrombosis consumes platelets, coagulation factors, and the natural anticoagulants, which consequently become depleted. The decrease in coagulation factors causes PT and PTT prolongations, and may lead to bleeding. Depletion of platelets also contributes to the bleeding risk. The fibrinolytic system is activated to dissolve the fibrin thrombi, resulting in consumption of plasminogen as it is converted into plasmin, and the formation of fibrin degradation products (FDP) including D-dimers as plasmin degrades fibrin clots. FDP can contribute to bleeding, because they impair fibrin clot formation and interfere with platelet function. In acute DIC, the most obvious clinical symptom is bleeding, although the insidious underlying disseminated microvascular thrombosis may lead to tissue ischemia and consequently multi-organ failure. **The key laboratory findings** are elevated D-dimers or FDP, prolonged PT and/or PTT, and decreased or decreasing platelets and fibrinogen. Repeat testing may be needed to show that fibrinogen and/or platelets are decreasing over time. Fibrinogen is decreased in ~50% of acute DIC cases; the PT is prolonged in ~70%; and the PTT is prolonged in ~50% of acute DIC cases. **Thus, it is important to note that these tests can be normal in a substantial percentage of DIC cases.** D-dimer or FDP should be positive in DIC. See D-Dimers and Fibrin Degradation Products *on page 502.*

Chronic DIC may develop when the activation of the coagulation system is low-grade and prolonged, as occurs with malignancy, retained dead fetus, aneurysm, or hemangioma. The clinical features and laboratory findings in chronic DIC can be much more subtle than with acute DIC. Fibrinogen and platelet levels are commonly elevated, because they can increase during acute phase reactions in response to illness (including malignancy), injury, or other conditions. The PT and PTT may actually be short, possibly due to increased circulating activated coagulation factors. Large-vessel thrombosis can occur in chronic DIC of malignancy. The main laboratory abnormality for acute or chronic DIC is positive D-dimers or FDP, neither of which are specific for DIC.

Schistocytes are present on the peripheral blood smear in 50% or more of acute DIC cases. Schistocytes are generated by microangiopathic hemolysis of red blood cells severed by flowing through fibrin strands. A large number of other coagulation tests may be abnormal in acute or chronic DIC, but their clinical utility for DIC diagnosis remains uncertain. These include decreases in the natural anticoagulant proteins antithrombin, protein C, and protein S; prolonged thrombin time; elevated markers of coagulation activation (eg, prothrombin fragment 1.2, fibrinopeptide A, fibrinopeptide B, fibrin monomers, thrombin-antithrombin complexes), and the appearance of markers of fibrinolysis (plasmin-antiplasmin complexes, decreased plasminogen, and antiplasmin) (see Hypercoagulation Panel *on page 758*). The thrombin time is often prolonged because of decreased fibrinogen and/or elevated FDP. Elevated FDP interfere with fibrin polymerization, prolonging the thrombin time. Plasma markers of platelet activation, such as platelet factor 4 and beta-thromboglobulin, may also be detected. None of these markers are specific for DIC.

Treatment of the underlying condition is the primary treatment of DIC, along with supportive care including transfusions if needed for bleeding. Heparin use is controversial. New strategies, such as antithrombin concentrates or recombinant activated protein C, are under investigation.

References

Levi M and ten Cate H, "Disseminated Intravascular Coagulation," *N Engl J Med*, 1999, 341(8):586-92.

Spero JA, Lewis JH, and Hasiba U, "Disseminated Intravascular Coagulation: Findings in 346 Patients," *Thromb Haemost*, 1980, 43:28-33.

- **Ditan®** *see* Phenytoin, Serum or Plasma *on page 1026*
- **Dithionite Test** *see* Sickle Cell Tests *on page 1195*
- **Di-Tran®** *see* Diazepam, Serum *on page 510*
- **Divalproex Sodium** *see* Valproic Acid, Serum or Plasma *on page 1297*
- **D-Lactate** *see* Lactic Acid, Whole Blood or Plasma *on page 827*

◆ **DM1 Gene Mutation Analysis** *see* Myotonic Dystrophy DNA Test *on page 942*

◆ **DMA** *see* Arsenic, Urine *on page 211*

◆ **DNA Amplification** *see* Polymerase Chain Reaction *on page 1069*

◆ **DNA Analysis for Parentage Evaluation** *see* Identification DNA Testing *on page 765*

◆ **DNA Antibody** *see* Anti-DNA *on page 173*

DNA Banking

Related Information

Polymerase Chain Reaction *on page 1069*

Synonyms DNA Storage

Test Includes Isolation and storage of DNA specimens for future diagnostic testing

Abstract Understanding of the molecular basis of disease is proceeding at a rapid rate, and will continue to progress at an even faster rate in the future with the success of the Human Genome Project. Advances expected from this project include an increase in diagnostic tests for inherited diseases and further characterization of genetic abnormalities associated with neoplasms. Thus, the storage of DNA from individuals or tumors will be invaluable both to scientists and to individuals interested in their family history of disease. Presently, DNA applications are increasingly being used in the forensic field as well. Several states in the U.S. have policies in place for collection and storage of DNA from convicted offenders.

Specimen Whole blood, tissue, cultured cells, buccal swabs, hair follicles, blood spots

Container Blood should be collected in yellow top (ACD) or lavender top (EDTA) tubes; tissue should be frozen at -70°C; amniotic cells, fibroblasts, or lymphocytes should be grown in appropriate media. As DNA diagnostic techniques have become very sensitive, buccal swabs, hair follicles and blood spots on filter paper have become acceptable samples for DNA banking.

Collection 5-10 mL of blood should be collected. 0.1-1 g specimen should be obtained, which should then be put into a sealable plastic freezer bag and frozen at -70°C. It should be kept frozen until shipped to the laboratory. Cell cultures should be grown to confluency and tightly sealed before shipping. Buccal cells are collected on cotton swabs and smeared onto glass slides and dried. Blood can be spotted onto filter paper and dried before shipping to the laboratory.

Storage Instructions Store tissue at -70°C or on dry ice. Samples can be stored at -70°C for an unlimited amount of time. Peripheral blood should be stored and shipped at 4°C. Do **not** freeze blood.

Causes for Rejection Thawing of the tissue specimen during transport to the laboratory or before shipping

Use The storage of DNA isolated from individuals provides purified genetic material that can be used either for identification or for future diagnostic testing.

Limitations Inappropriate shipping or processing

Methodology DNA is released and isolated from white blood cells, tissue, or cultured cells by lysing the cells and extracting the cell lysate with phenol and chloroform. Purified, intact DNA is precipitated with salt in the presence of alcohol. Other methods of DNA purification are also available. The DNA is then stored indefinitely at -70°C, usually at two separate facilities. Buccal cells, hair follicles, and blood spots are stored without DNA extraction and can be directly used for polymerase chain reaction (PCR).

Additional Information Stored DNA is always the property of the person from whom it was isolated. When family testing for either a genetic disease or tumor characterization is desired, signed permission is usually required before the sample is released. If the person owning the DNA is deceased, then disposition is under the control of a legal guardian or heir. All information received from DNA tests performed on any DNA sample is completely confidential and is released only to the individual requesting the test (through an appropriate medical professional). Anonymous genetic testing at academic institutions is a common practice.

References

Butler D, "Tensions Grow Over Access to DNA Bank," *Nature*, 1998, 391(6669):727.

(Continued)

DNA Banking *(Continued)*

Deschenes M, Cardinal G, Knoppers BM, et al, "Human Genetic Research, DNA Banking and Consent: A Question of Form?" *Clin Genet*, 2001, 59(4):221-39.

Farkas DH, Kaul KL, Wiedbrauk DL, et al, "Specimen Collection and Storage for Diagnostic Molecular Pathology Investigation," *Arch Pathol Lab Med*, 1996, 120(6):591-6.

Harty LC, Garcia-Closas M, Rothman N, et al, "Collection of Buccal Cell DNA Using Treated Cards," *Cancer Epidemiol Biomarkers Prev*, 2000, 9(5):501-6.

Knoppers BM, Hirtle M, Lormeau S, et al, "Control of DNA Samples and Information," *Genomics*, 1998, 50(3):385-401.

McEwen JE, "Forensic DNA Data Banking by State Crime Laboratories," *Am J Hum Genet*, 1995, 56(6):1487-92.

McQueen MJ, "Ethical and Legal Issues in the Procurement, Storage and Use of DNA," *Clin Chem Lab Med*, 1998, 36(8):545-9.

Steinberg KK, Sanderlin KC, Ou CY, et al, "DNA Banking in Epidemiologic Studies," *Epidemiol Rev*, 1997, 19(1):156-62.

♦ **DNA Fingerprinting** *see* Identification DNA Testing *on page 765*
♦ **DNA Hybridization Test for HPV** *see* Human Papillomavirus (HPV) DNA Tests *on page 749*
♦ **DNA Ploidy** *see* Breast Biopsy *on page 305*

DNA-Probe Assay for Thalassemia (BeTha Gene Test)

Related Information

Complete Blood Count *on page 442*
Fetal Hemoglobin *on page 581*
Hemoglobin A_2 *on page 682*
Hemoglobin Electrophoresis *on page 684*
Sickle Cell Tests *on page 1195*

Test Includes Multiplex polymerase chain reaction (PCR), β-globin gene amplification, allele-specific oligonucleotide hybridization, and enzyme-linked immunosorbent assay (ELISA) visualization/detection

Abstract This assay, available commercially in kit form, provides for the qualitative detection of eight (includes the most common) Mediterranean β-thalassemia mutations. The thalassemias are a spectrum of disease states characterized by an imbalance of globin chain synthesis. There is a reduced rate of synthesis by one or more globin chain genes. Dependent upon the underlying molecular mechanism, a variety of clinical conditions may result, including some that are fatal. Thalassemia is an important cause of hypochromic microcytic anemia which must be differentiated from other causes (eg, iron deficiency and sideroblastic anemia). A variety of DNA-based analytic techniques are applicable to the diagnosis of thalassemia but are not usually required to establish the diagnosis.

Specimen Whole blood

Container Lavender top (EDTA) tube

Storage Instructions There is evidence that samples are stable at room temperature for 2 days, at 2°C to 8°C for 6 days, and at -20°C for 28 days.

Turnaround Time 2-3 hours (with use of Instagene™ chelex resin kit, Bio-Rad Laboratories, a DNA preparation procedure for provision of PCR-ready DNA)

Special Instructions This procedure may be used for fetal diagnosis using fetal DNA prepared from chorionic villi obtained during amniocentesis. Villus samples must be free of maternal tissue and transported in tissue culture medium. Amniotic fluid (15-20 mL) must be received by the laboratory within 24 hours of collection. If the specimen is to be transported, consult the laboratory performing the analysis to obtain any recommended special buffer for chorionic villus samples or tissue culture medium for transport of amniocytes.

Reference Interval Positive hybridization signal with normal allele-specific oligonucleotides (in the case of BeTha Gene 1 kit, an ELISA soluble yellow product with absorption maximum at 450 nm)

Use Screen for individuals heterozygous for mutations that may cause β-thalassemia. Molecular diagnostic techniques such as DNA probe assays can be used to confirm the diagnosis of β-thalassemia. They are especially useful for prenatal diagnosis which can identify situations in which genetic counseling may be applicable.

Limitations Availability of DNA analysis for thalassemia is likely to vary with geography. Different groups of mutations that cause β-thalassemia are associated with different geographic sites. Each area of the world in which thalassemia is prevalent has a few common mutations and many uncommon or rare varieties. The allele-specific nucleotides incorporated in a particular DNA probe assay will usually target only the common local forms of thalassemia; the test will not detect all forms (ie, the uncommon types) of the condition.

Additional Information More than 180 molecular defects in the β-globin genes are responsible for reduced synthesis of β-globin chains with resultant excess accumulation of α-globin chains. This mechanism of globin chain imbalance leads to formation of insoluble α-globin tetramers that precipitate within red blood cells. Intrasplenic sequestration and hemolysis results in red cell fragmentation, microcytosis, and increased hemoglobin A_2. There is geographic clustering of the gene abnormalities which, while numerous and diverse, include only a few common forms. Thus, detection of the majority of clinically significant cases of β-thalassemia is possible using allele-specific oligonucleotides tailored to the geographic site. In the Mediterranean area, eight mutations account for >90% of cases of β-thalassemia. These mutations are targeted/detected by the BeTha Gene 1 kit.

Homozygous β-thalassemia patients (β-thalassemia major) have a severe hemolytic disease process requiring transfusion therapy. While a few patients have been treated successfully with bone marrow transplantation techniques, and gene therapy is being developed, most require supportive care.

References
Olivieri NF, "The Beta-Thalassemias," *N Engl J Med*, 1999, 341(2):99-109.

Ugozzoli LA, Lowery JD, Reyes AA, et al, "Evaluation of the BeTha Gene 1 Kit for the Qualitative Detection of the Eight Most Common Mediterranean β-Thalassemia Mutations," *Am J Hematol*, 1998, 59(3):214-22.

Weatherall DJ and Provan AB, "Red Cells I: Inherited Anaemias," *Lancet*, 2000, 355(9210):1169-75.

♦ **DNA Storage** *see* DNA Banking *on page 519*

♦ **DNA Testing** *see* Identification DNA Testing *on page 765*

♦ **Doe** *see* Methamphetamine, Qualitative, Urine *on page 902*

♦ **Dolantin**® *see* Meperidine, Serum or Urine *on page 895*

♦ **Dolantina**® *see* Meperidine, Serum or Urine *on page 895*

♦ **Dolantine**® *see* Meperidine, Serum or Urine *on page 895*

♦ **Dolophine**® *see* Methadone, Serum or Urine *on page 900*

♦ **Dolosal**® *see* Meperidine, Serum or Urine *on page 895*

♦ **Donath-Landsteiner Test** *see* Cold Hemolysin Test *on page 431*

Donation, Blood

Related Information

Autologous Transfusion, Preoperative Deposit *on page 223*

Cytomegalovirus Culture *on page 495*

Donation, Blood, Directed *on page 523*

Hepatitis B Serology *on page 702*

Hepatitis C Virus RNA Detection and Quantitation *on page 705*

Hepatitis C Virus Serology *on page 706*

HIV-1/HIV-2 Serology *on page 736*

HTLV-I/II Antibody *on page 745*

Neutrophils, Apheresis, Donation *on page 950*

Phlebotomy, Therapeutic *on page 1029*

Platelet Transfusion *on page 1058*

RPR *on page 1174*

VDRL, Serum or Cerebrospinal Fluid *on page 1303*

Whole Blood *on page 1333*

Synonyms Phlebotomy, Blood Donor

Applies to Allogeneic Blood Transfusion

Test Includes Each donation intended for allogeneic use must be typed for ABO and Rh. Donors with a history of transfusion or pregnancy must be tested for unexpected antibodies. Each donation must be tested for syphilis; hepatitis B surface antigen; antibodies to HIV-1, HIV-2, hepatitis B core antigen (HB_c), hepatitis C virus (HCV), human T-cell lymphotrophic virus (HTLV-I/II), and nucleic acid testing for detection of HIV-1, HCV, and West Nile virus. Tests likely to be added over the next several years include nucleic acid testing for HBV, hepatitis A virus, and parvovirus B19. A variety of tests, such as a Western blot or a recombinant immunoblot assay, may be performed in an attempt to confirm any initially positive infectious disease screening test. (Continued)

Donation, Blood (Continued)

Although frequently performed, a test for antibodies to cytomegalovirus (CMV) is not required.

Abstract Accurate information volunteered by the blood donor during the health assessment, together with infectious disease testing, are necessary for the exclusion of donors whose blood may transmit infectious diseases to recipients.

Patient Preparation Donors should be at least 18 years of age. Depending on state law, donors between 17 and 18 may donate with or without parental consent. The upper age limit usually is decided by the Blood Bank physician. Donor should weigh at least 110 pounds, should have a light meal before donation, no alcoholic beverages for 12 hours, **be in generally good health**, and afebrile. Donor reactions may be precipitated by emotional factors. Personnel should be cheerful and the donor room pleasant. The donor should be asked about drugs being taken (eg, antihypertensives).

Aftercare Donors should be asked about their occupations, since fainting can be especially hazardous for some (eg, bus drivers, air crew). Activities are restricted for certain hazardous occupations for 24 hours. Donor reactions occur, but are rarely severe. Vasovagal reactions occur in 2% to 5% of blood donors, a figure unchanged in the past half century. Severe reactions occur in about 1 per 1000 donations. The vasovagal attack can be provoked in some individuals by the sight of blood, and withdrawal of sufficient blood can lead to such effects. Known risk factors of vasovagal reactions include youth, low weight, and first-time donor status. Risk relates to the blood volume withdrawn.

Immediately after donation of blood, the donor should remain lying down for several minutes. He/she should never be left alone. When moved to a chair with assistance, the donor should remain seated for at least 15 minutes and should be offered refreshments (eg, orange juice, and often cookies). The donor should be observed for at least 15 minutes and should be encouraged not to leave earlier than 20 minutes following completion of the donation.

The delayed syncopal reaction may take place up to an hour following donation. On occasion, they seem idiosyncratic, independent of any identifiable neglect.

Specimen Blood

Container Blood bag of appropriate configuration

Collection Drawn by Blood Bank personnel

Causes for Rejection History of hepatitis or yellow jaundice, history of HCV, HB$_s$Ag, or HIV positive, drug addiction involving injection, male-to-male sexual activity, sex for money or drugs, major organ disease, residence in the United Kingdom for a total of 3 months or longer between 1980-1996: permanently deferred. Temporary deferments include hypotension, hypertension, anemia (hemoglobin <12.5 g/dL), positive syphilis serology (STS), travel to malaria endemic areas, exposure to hepatitis, pregnancy, childbirth within the last week, recent surgery, recent transfusion, inmate of penal or mental institution, and certain other medical conditions. FDA advisors suggested that individuals who have received transplanted tissue from animals (xenotransplant recipients) be forbidden to donate blood or plasma fractions. Use of vitamins, thyroid preparations, or oral contraceptives does **not** disqualify donors. However, many other drugs, medications, and even immunizations can either temporarily or permanently defer a donor. Contact a Blood Bank physician for further information. Blood Banks must present would-be donors with educational materials explaining the risk of HIV/AIDS in blood transfusion. If confidential self-deferment (confidential unit exclusion) is used at donation, the donor must be informed that testing will be performed and there will be notification of any positive infectious disease test result.

Special Instructions The patient's arm should be examined for scarred veins and/or many small puncture marks, possible signs of drug addiction.

Use Obtain blood and its components for allogeneic transfusion to patients.

Limitations Once every 8 weeks for whole blood donation; every 16 weeks for two-unit red cell donation; every 4 weeks for infrequent apheresis donation; and every 2 days for plasma-, platelet-, or leukapheresis. Total donations per year may be limited depending on type of donation. Truthful information must be provided by the prospective donor.

Contraindications Donation is not advised for those who must immediately return to a hazardous activity (eg, heavy machinery operation), because delayed reaction may occur. The exercise of common sense and good judgment are advised. Adverse consequences of donation are much less among experienced donors than in those who are donating for the first time.

Methodology Donors are selected based upon a medical history and a limited physical examination. Both are performed on the date of donation. The medical history questions may be completed by the donors themselves or obtained by a qualified interviewer. Questions pertaining to HIV-associated risk behaviors should be presented to the donors by a qualified interviewer.

Additional Information The word "allogeneic" denotes identical species but with antigenic difference. Allogeneic transfusions are more often called "homologous". The expression is used in contrast to autologous transfusion, in which antigenic differences do not exist.

References
Churchill WH and Kurtz SR, *Transfusion Medicine*, Chapter 2, Boston, MA: Blackwell Scientific Publications, 1989, 26.

Huestis DW, Bove JR, and Case J, *Practical Blood Transfusion*, 4th ed, Boston, MA: Little, Brown and Co, 1988.

Klein HG, "Transfusion Safety," *Mayo Clin Proc*, 2000, 75:769-70.

Klein HG, "Will Blood Transfusion Ever Be Safe Enough?" *JAMA*, 2000, 284(2):238-40.

Kleinman S and Williams AE, "Donor Selection Procedures: Is it Possible to Improve Them?" *Transfus Med Rev*, 1998, 12(4):288-302.

Mollison PL, Engelfriet CP, Contreras M, et al, "The Withdrawal of Blood," *Blood Transfusion in Clinical Medicine*, 10th ed, Chapter 1, Oxford, UK: Blackwell Scientific Publication, 1997, 1-36.

Schmidt PJ, "Blood and Disaster - Supply and Demand," *N Engl J Med*, 2002, 346(8):617-20.

Soteriades ES, Evans JC, Larson MG, et al, "Incidence and Prognosis of Syncope," *N Engl J Med*, 2002, 347(12):878-85.

Donation, Blood, Directed

Related Information
Autologous Transfusion, Preoperative Deposit *on page 223*

Donation, Blood *on page 521*

Irradiated Blood Components *on page 811*

Test Includes Donor phlebotomy, ABO grouping, Rh typing, antibody screen, HB$_s$Ag, anti-HB$_c$, anti-HTLV-I, anti-HTLV-II, anti-HIV-1, anti-HIV-2, anti-HCV, and a serologic test for syphilis. Nucleic acid testing for hepatitis C, HIV-1, and West Nile virus is also performed.

Abstract The designation of friends or relatives to provide donations was initiated secondary to the public's concerns about the safety of the blood supply due to HIV. Currently, there are certain circumstances in which it may be important to utilize designated donor(s) for a designated recipient (eg, patients with rare blood types or difficulty finding compatible blood components due to alloimmunization).

Patient Preparation Donors must meet all the requirements of a regular blood donor.

Aftercare Activities are restricted as for a routine blood donation. Reactions occur occasionally, as with any other donations.

Collection As for regular blood donors, although in some instances, the frequency of donation may be increased if the donor is certified by a physician to be in good health.

Causes for Rejection As for regular blood donation (see Donation, Blood *on page 521*).

Special Instructions The attending physician should specify to the Blood Bank which blood component(s) he/she wishes prepared from each donation.

Use Obtain blood or components for later use by a designated patient. Some intended recipients have a highly emotional fixation and are not moved by logic. Although the procedure may be an administrative nuisance, it does make the patient feel better about transfusions.

Limitations Friends or family members recruited to be directed donors may not meet the eligibility requirements to give blood. If eligibility requirements are met and a unit is obtained, the unit may still not be made available because infectious disease testing requirements were not passed.

Other limitations include the following.
- Directed donors cannot supply blood in an emergency.
- Blood from directed donations generally cannot be available in less than 72 hours.

(Continued)

Donation, Blood, Directed *(Continued)*

- Directed donations are neither safer nor riskier than regular volunteer blood donations.
- More units are likely needed than the directed donor(s) can provide (because of differences in ABO/Rh types and donor loss due to other incompatibilities/ineligibilities).
- Graft-vs-host disease occurs occasionally in immunocompetent recipients of directed donations from blood relatives. Thus, all cellular components should be gamma irradiated with 15-25 Gy.
- Husband-to-wife transfusions incur increased likelihood of hemolytic disease of the newborn.
- Directed donors risk losing the anonymous position of the conventional (homologous) donor and may become subject to legal complications.
- Administrative costs increase when directed donors are requested, because added efforts are required.

Additional Information Some states have enacted laws which state that a directed donation service must be offered on a nonemergency basis.

References

1. *Code of Federal Regulations, 21CFR640.3(f)*, Washington, DC: U.S. Government Printing Office, 2003.

Pink J, Thomson A, and Wylie B, "Infectious Disease Markers in Autologous and Directed Donations," *Transfus Med*, 1994, 4:135-8.

Wagner FF and Flegel WA, "Transfusion-Associated Graft-Versus-Host Disease: Risk Due to Homozygous HLA Haplotypes," *Transfusion*, 1995, 35(4):284-91.

♦ **Donor Plasmapheresis** *see* Plasmapheresis, Donor *on page 1041*

♦ **Dopamine** *see* Catecholamines, Fractionation, Plasma *on page 345*

♦ **Dopamine** *see* Vanillylmandelic Acid, Urine *on page 1299*

♦ **Dopamine, Urine** *see* Catecholamines, Fractionation, Urine *on page 347*

♦ **Double Test** *see* Pregnancy-Associated Protein A, Serum *on page 1082*

♦ **Doxepin and Nordoxepine** *see* Antidepressants, Cyclic, Serum or Plasma *on page 171*

Doxepin, Serum or Plasma

Related Information

Antidepressants, Cyclic, Serum or Plasma *on page 171*

Synonyms Adapin®; Novoxapin®; Sinequan®; Triadapin®; Zonalon®

Test Includes Doxepin and desmethyldoxepin (nordoxepine)

Abstract This is a tricyclic antidepressant, a dibenzoxepin analogue of amitriptyline. It is used for various forms of depression, often with psychotherapy. It is used for anxiety disorders, and a topical preparation is used for short-term (<8 days) management of moderate pruritus in adults with atopic dermatitis or lichen simplex chronicus.

Specimen Serum or plasma

Container Red top tube or green top (heparin) tube

Sampling Time Trough levels at steady-state (50-125 hours). Time to peak serum concentration is 2-4 hours.

Causes for Rejection Specimen collected in gel tube

Reference Interval Sum of doxepin and desmethyldoxepin: 150-250 ng/mL (SI: 540-900 nmol/L)

Critical Values Toxic: >500 ng/mL (SI: >1800 nmol/L)

Use Therapeutic monitoring (compliance) and toxicity assessment

Methodology Immunoassay, high performance liquid chromatography (HPLC), gas chromatography (GC)

Additional Information

- Half-life: 10-25 hours
- Volume of distribution: 10-30 L/kg
- Protein binding: 75% to 85%

Side effects are similar to those of other tricyclic antidepressants. The most common include dry mouth, urinary retention, blurred vision, and constipation. These effects are due to anticholinergic properties of the drug. Respiratory depression, hypotension, coma, cardiac arrhythmias, and tachycardia occur in severe intoxication. Its use is contraindicated in patients with narrow-angle glaucoma, urinary retention, and patients using MAO inhibitors.

References

Geller B, Reising D, Leonard HL, et al, "Critical Review of Tricyclic Antidepressant Use in Children and Adolescents," *J Am Acad Child Adolesc Psychiatry*, 1999, 38(5):513-6.

Lacy CF, Armstrong LL, Goldman MP, et al, *Drug Information Handbook*, 12th ed, Hudson, Ohio: Lexi-Comp Inc, 2004.

Newton EH, Shih RD, and Hoffman RS, "Cyclic Antidepressant Overdose: A Review of Current Management Strategies," *Am J Emerg Med*, 1994, 12(3):376-9.

♦ **Dozic®** *see* Haloperidol, Serum or Plasma *on page 664*

♦ **Dozine®** *see* Chlorpromazine, Serum *on page 395*

♦ **DPD** *see* Osteocalcin, Serum or Plasma *on page 983*

♦ **2,3-DPG** *see* Oxygen Saturation, Blood *on page 991*

♦ **2,3-DPG** *see* P$_{50}$, Blood *on page 993*

♦ **Drug Screen, Comprehensive Drug Panel or Analysis** *see* Toxicology Screen, Serum or Plasma *on page 1263*

♦ **Drug Screen, Comprehensive Panel or Analysis, Urine** *see* Toxicology Screen, Urine *on page 1264*

Drugs of Abuse Testing, Urine

Related Information

Amphetamine, Qualitative, Urine *on page 154*
Barbiturates, Qualitative, Urine *on page 248*
Benzodiazepines, Qualitative, Urine *on page 253*
Cannabinoids (Marijuana Metabolites), Qualitative, Urine *on page 335*
Cocaine (Cocaine Metabolite), Qualitative, Urine or Hair *on page 427*
Ethanol, Blood, Urine, and Other Sources *on page 558*
Flunitrazepam, Urine *on page 601*
Methadone, Serum or Urine *on page 900*
Methamphetamine, Qualitative, Urine *on page 902*
Methaqualone, Urine *on page 903*
Morphine, Urine *on page 921*
Opiates, Qualitative, Urine *on page 974*
Phencyclidine, Qualitative, Urine *on page 1019*
Propoxyphene, Serum or Urine *on page 1096*
Toxicology Screen, Urine *on page 1264*

Test Includes Screens for commonly abused drugs and classes of abused drugs - amphetamines, barbiturates, benzodiazepines, cannabinoids, cocaine, methadone, methaqualone, opiates, phencyclidine, propoxyphene. In some laboratories, urine ethanol is included.

Abstract The usual drug-of-abuse screening panel consists of the 10 drugs listed above. Substance Abuse and Mental Health Services Administration (SAMHSA) screen includes only the following 5 drugs: marijuana metabolite, cocaine metabolite, opiates, phencyclidine, and amphetamines.

Specimen Urine

Collection If forensic, observe precautions and follow chain-of-custody protocol.

Storage Instructions Refrigerate

Causes for Rejection If forensic, failure to meet temperature requirements and tests for unusual urine dilution or adulteration

Turnaround Time Screen: 1-2 hours if done in-house; confirmation: 1-2 days

Special Instructions Specify the drug or drugs suspected in an emergency situation. If forensic, use chain-of-custody protocol and form. See Chain-of-Custody Protocol *on page 381* and the Chain-of-Custody form in the Introduction *on page 63*.

Reference Interval Negative (less than cutoff)

Critical Values See individual drug entries for cutoff values.

Use Screen for drug overdose and toxicity; screen for the presence of drugs of abuse

Methodology Screen: immunoassay; confirmation: gas chromatography/mass spectrometry (GC/MS)

Additional Information For specific drug classes see the listing by specific drug name.

Adulteration of a specimen remains a major challenge. Common tactics to beat the test include substitution of urine, dilution of specimen, or addition of some chemical to the specimen to interfere with the immunoassay or destroy the
(Continued)

Drugs of Abuse Testing, Urine *(Continued)*

drug. Direct observation of urine collection is the best safeguard against adulteration, but generally not performed due to hesitation relevant to intrusion of individual privacy and dignity. Most laboratories perform other tests (eg, temperature check, pH, specific gravity, creatinine) to rule out specimen adulteration.

References

Gerson B and Subramaniam S, "Drug Testing as Part of the War on Drugs," *Clin Lab Med*, 1998, 18(4):781-803.

O'Neal CL, Crough DJ, and Fatah AA, "Validation of Twelve Chemical Spot Tests for the Detection of Drugs of Abuse," *Forensic Sci Int*, 2000, 109(3):189-201.

Substance Abuse and Mental Health Administration. "Mandatory Guidelines for Workplace Programs: Revision to Mandatory Guidelines," *Fed Regist*, 1994, 59:29916-31.

Duchenne/Becker Muscular Dystrophy DNA Detection

Related Information

Aldolase, Plasma or Serum *on page 121*
Alpha$_1$-Fetoprotein, Amniotic Fluid *on page 135*
Amniotic Fluid, Chromosome and Genetic Abnormality Analysis *on page 152*
Chorionic Villus Sampling, Chromosome and Genetic Abnormality Analysis *on page 400*
Creatine Kinase, Serum *on page 470*
Muscle Biopsy *on page 927*
Myotonic Dystrophy DNA Test *on page 942*
Polymerase Chain Reaction *on page 1069*

Applies to Dystrophin

Abstract Duchenne and Becker progressive muscular dystrophies are X-linked recessive disorders caused by mutations in the dystrophin gene. Most of the cases are familial but sporadic cases are seen. The dystrophin gene has been located on Xp21.3-p21.2. In a majority of the cases, diagnosis can be made by molecular testing without muscle biopsy. The remaining cases are diagnosed by clinical findings, family history, serum creatine kinase concentration, and muscle biopsy. Duchenne muscular dystrophy (DMD) progresses more rapidly than Becker muscular dystrophy (BMD). Approximately 70% of males with DMD and 85% of males with BMD have deletions or duplications of one or more exons of the dystrophin gene.

Patient Preparation Consultation with a medical geneticist is desirable.

Specimen Whole blood, amniotic fluid, chorionic villus

Container Blood should be collected in yellow top (ACD) Vacutainer® tubes, blood collected in lavender top (EDTA) Vacutainer® tubes is also acceptable; amniotic fluid and chorionic villus should be collected in a sterile manner and transferred to a sterile tube for transport or to a T25 culture flask.

Sampling Time Amniotic fluid should be collected between the 14th and 17th week of pregnancy. Chorionic villus specimens should be dissected free of maternal tissue and blood clot. Transport medium is needed.

Storage Instructions All specimens should be sent to the laboratory **immediately** after collection, preferably by overnight delivery. All specimens should be kept at room temperature or refrigerated, never frozen.

Causes for Rejection Amniotic fluid specimens that are bloody may be unsuitable.

Turnaround Time Approximately 1 week

Special Instructions A complete family pedigree and clinical information are needed. Prior testing of family members is usually necessary for prenatal testing.

Reference Interval An interpretive report which includes a risk analysis is usually provided.

Use This test is indicated for patients and families with history of Duchenne or Becker muscular dystrophy. This test is also indicated for prenatal diagnosis in females known to be carriers. It is helpful for diagnosis of neonates suspected of Duchenne or Becker muscular dystrophy.

Limitations DNA analysis for Duchenne or Becker muscular dystrophy can only detect ~80% of the abnormalities responsible for the disease. Thus, a negative result does not rule out the possibility that an individual is a muscular dystrophy carrier or affected, but can lower that probability. Gonadal mosaicism occurs in about 10% of cases.

Methodology DNA is isolated from the specimen and several regions within the dystrophin gene are detected, using Southern blotting techniques. Multiplex polymerase chain reaction (PCR) is used to detect many of the most common deletions. Several of the DNA regions can be amplified in the same PCR and a change (either loss or varied mobility) in the DNA fragments indicates an abnormality. Examinations of 18 exons in this way detects up to 98% of the deletions identifiable by cDNA hybridization. In affected families in which there is no detectable deletion, linkage analysis can be done using Southern blotting to detect linkage to several known mutations.

Additional Information Duchenne muscular dystrophy, the most common of the childhood dystrophies, is the most severe type of progressive primary muscular degeneration. A crippling muscle disorder, it is associated with an abnormality in band 1 of region 2 of the short arm on the X chromosome, a locus designated Xp21.3-p21.2. In most cases, it is clinically evident by 5 years and wheelchair dependency occurs before 13 years of age. The mean age of diagnosis is 4 years 10 months. CK is very high during its early phase. Cardiac involvement leads to ECG abnormalities; heart failure and arrhythmias may occur. Cardiomyopathy may be severe. Some degree of nonprogressive cognitive impairment is common in children with DMD.

Becker muscular dystrophy is a milder form with a similar clinical course, as of Duchenne muscular dystrophy, but followed at a much slower rate. Wheelchair dependency, if present, occurs after 16 years of age. Preservation of neck flexor muscle strength in BMD differentiates from DMD. Despite milder skeletal involvement, heart failure is a common cause of morbidity and mortality. A complete cardiac evaluation is recommended at least once in all carriers. CK bears a less marked increase in BMD as compared to DMD.

The gene responsible for these disorders has been cloned and the protein product has been identified as the dystrophin protein, a muscle cytoskeletal protein. This protein was found to be 427 kilodalton (kDa). The gene spans 2-4 mb of DNA and is comprised of 79 exons. It has at least four promoters. It is the largest known human gene.

When deletion is not identified, linkage analysis is usually successful in provision of risk assessment. In cases with a positive family history but with no detectable mutation, a more intensive search using restriction fragment length polymorphism (RFLP)-linkage analysis can be done to determine the existence of known point mutations or alterations not detected by the other assay. This requires the participation of several family members, both affected and unaffected, to correlate the inheritance pattern of the RFLPs with inheritance of disease.

Muscle biopsy with dystrophin analysis may be needed in selected patients.

References

Beggs AH, Koenig M, Boyce FM, et al, "Detection of 98% of DMD/BMD Gene Deletions by Polymerase Chain Reaction," *Hum Genet*, 1990, 86(1):45-8.

Bushby KM, "The Limb-Girdle Muscular Dystrophies - Multiple Genes, Multiple Mechanisms," *Hum Mol Genet*, 1999, 8(10):1875-82.

Fassati A, Tedeschi S, Bordoni A, et al, "Rapid Direct Diagnosis of Deletions Carriers of Duchenne and Becker Muscular Dystrophies," *Lancet*, 1994, 344(8918):302-3.

Rininsland F and Reiss J, "Microlesions and Polymorphisms in the Duchenne/Becker Muscular Dystrophy Gene," *Hum Genet*, 1994, 94(2):111-6.

Zalaudek I, Bonelli RM, Koltringer P, et al, "Early Diagnosis in Duchenne Muscular Dystrophy," *Lancet*, 1999, 353(9168):1975.

Internet Web Sites

www.mdausa.org

www.muscular-dystrophy.org

♦ **Duodenal Drainage Examination** *see* Bile Fluid Examination *on page 259*

♦ **Duo-Trach®** *see* Lidocaine, Serum or Plasma *on page 850*

♦ **Duramorph®** *see* Morphine, Urine *on page 921*

♦ **Durvitan®** *see* Caffeine, Serum *on page 326*

d-Xylose Absorption Test, Serum, Urine

Related Information

Endomysial Antibodies *on page 537*

Gliadin IgG/IgA Antibodies *on page 637*

(Continued)

d-Xylose Absorption Test, Serum, Urine *(Continued)*

Synonyms Xylose Absorption Test; Xylose Tolerance Test

Applies to [^{14}C] d-Xylose Breath Test

Abstract d-Xylose, a five-carbon monosaccharide not normally present in blood, is passively absorbed unchanged by the normal duodenum and jejunum, and it is excreted unmetabolized by the kidneys. The d-xylose absorption test is used as a screening test for intestinal carbohydrate malabsorption.

Patient Preparation Urea nitrogen, creatinine, and first morning urinalysis should be normal. Patient must fast a minimum of 8 hours prior to administration of d-xylose. Pediatric patients must be fasting for at least 4 hours. Patient must remain in a supine position for duration of the test, except during urine collection. No food is permitted during the test. Since d-xylose is a pentose, the patient should refrain from eating foods containing pentoses (eg, fruits, jams, jellies, pastries). Many medications, including aspirin, indomethacin, other nonsteroidal anti-inflammatory drugs, neomycin, glipizide, or atropine interfere. These and preferably all medications should be discontinued for 24 hours prior to the test. No water restriction is required; in fact, patients should be encouraged to drink during the fasting period and during the test. Start the test in the AM. If a urine collection is to be done, instruct patient to void completely and discard this urine. If a blood determination is to be done and the testing laboratory so requires, draw a fasting blood specimen.

Administer the dose of d-xylose orally (adults, usually 25 g or 5 g, if specified by physician; children younger than 12 years, 5 g or a weight-based dosage equal to 0.5 g/kg body weight up to a maximum of 25 g). The dose should be dissolved in water, making an ~10% (w/v) solution with a maximum of 500 mL. Patient should drink the entire amount. Fill cup with 250 mL of water and have patient drink this as well. Have patient drink another cup with 250 mL of water after 1 hour. Collect urine for 5 hours after administration of d-xylose.

Specimen The specimens usually include a 5-hour, postdose urine collection, (preserved by refrigeration or freezing, according to the instructions of the testing laboratory) and a 1-hour, postdose serum or plasma specimen. (Some protocols call for a 2-hour postdose blood collection rather than a 1-hour, and others call for serial, hourly blood draws up to 5 hours postdose. Check with the testing laboratory beforehand.)

Craig and Atkinson recommended a 25 g dose with a 5-hour urine collection and a 1-hour serum specimen for adults with normal renal function, and a 1-hour serum specimen only in adult patients with renal insufficiency. However, Peled et al showed that the 5-hour urine collection much more accurately reflects abnormal intestinal absorption than the 1-hour serum specimen; therefore, many laboratories recommend that only the 5-hour urine collection be done in adults (12 years or older) with normal renal function.

In children and infants younger than 12 years of age, only a 1-hour serum specimen is recommended. Urine specimens should not be collected due to difficulties with collections in these age groups.

Container Containers include an appropriately labeled urine container, red top tube, green top (heparin) tube, or gray top (sodium fluoride/potassium) tube as required by testing needs and/or testing laboratory.

Reference Interval Reference intervals vary with the protocol used. Reference interval for a 5-hour urine collection in adults ≥12 years (25-gram dose) is **≥4 g/ 5 hours (SI: ≥26.6 mmol/5 hours)**. The table below summarizes the serum intervals recommended by Craig and Atkinson. See table.

1-Hour d-Xylose Absorption Test

Serum	mg/dL	mmol/L (SI)
Adult, 1 h (dose, 25 g)	≥25	≥1.7
Adult with intermediate renal insufficiency, 1 h (dose, 25 g)	≥20	≥1.3
Pediatric <12 y, 1 h (dose, 5 g)	≥20	≥1.3

Since a 25-gram dose can cause diarrhea, nausea, and abdominal discomfort in some adult patients, a 5-gram dose is sometimes used. For a 5-gram dose in an adult, a level of serum d-xylose between 20-40 mg/dL (SI: 1.3-2.7 mmol/L) should be reached in 30-60 minutes and maintained for a further 60 minutes; the 5-hour urine collection should yield results >1.2 grams (SI: 8.0 mmol).

Use The d-xylose test is used primarily to differentiate enterogenous malabsorption from malabsorption of chronic pancreatic disease. The differential diagnosis of an abnormal result includes celiac disease, tropical sprue, Crohn disease, surgical bowel resection, AIDS, and many other less common small bowel disorders. Craig and Atkinson report that the test (25-gram dose with a 5-hour urine collection and a 1-hour serum specimen) is 91% sensitive and 98% specific for detecting intestinal malabsorption in the absence of renal impairment or bacterial overgrowth.

Limitations False-positive results are seen with ascites, poor renal function, vomiting, decreased gastric emptying, intestinal stasis syndromes (eg, surgical blind loops), dehydration/hypovolemia, and certain drugs. The usefulness of the test has been somewhat controversial. Krawitt and Beeken concluded that no reason exists to do such d-xylose testing when jejunal biopsies are available.

Methodology Methods include enzymatic procedures with colorimetric or fluorometric measurements, gas chromatography, or gas chromatography/mass spectrometry (GC/MS).

Additional Information Normal renal function is necessary if urine values alone are determined. Pancreatic enzymes are **not** required for absorption of d-xylose. The d-xylose test is normal in the chronic nonspecific diarrhea syndrome of infancy. To differentiate patients with pancreatic insufficiency from those with small intestinal malabsorption, the d-xylose absorption test has been paired with the N-benzoyl-L-tyrosyl-p-aminobenzoic acid test (BZ-TY-PABA, PABA test, bentiromide, or BTP test) for pancreatic function.

References
Craig RM and Atkinson AJ Jr, "D-xylose Testing: A Review," *Gastroenterology*, 1988, 95(1):223-31.

Deutsch JC, Santhosh-Kumar CR, and Kolli VR, "A Noninvasive Stable-Isotope Method to Simultaneously Assess Pancreatic Exocrine Function and Small Bowel Absorption," *Am J Gastroenterol*, 1995, 90(12):2182-5.

Krawitt EL and Beeken WL, "Limitations of the Usefulness of the d-Xylose Absorption Test," *Am J Clin Pathol*, 1975, 63(2):261-3.

Peled Y, Doron O, Laufer H, et al, "D-xylose Absorption Test. Urine or Blood?" *Dig Dis Sci*, 1991, 36(2):188-92.

Eastern Equine Encephalitis Virus Serology

Related Information

Abstract Eastern equine encephalitis (EEE) virus is an alphavirus limited geographically to the Eastern and Gulf coasts of the United States. Encephalitis due to EEE is a summertime disease spread by mosquitos (*Cusliseta melanura*). EEE is a low incidence disease, with only a few cases occurring each year, but with a 50% to 70% fatality rate.

Specimen Serum or cerebrospinal fluid
Container Red top tube
(Continued)

Eastern Equine Encephalitis Virus Serology
(Continued)

Sampling Time Acute and convalescent sera drawn 10-14 days apart

Reference Interval Less than a fourfold increase in titer in paired sera

Use Support the diagnosis of Eastern equine encephalitis virus infection

Limitations Absence of IgM antibodies does not rule out the infection. Extensive cross-reactions with related arboviruses (eg, Western equine encephalitis virus and Venezuelan equine encephalitis virus) makes definitive interpretation difficult.

Methodology Complement fixation (CF), hemagglutination inhibition (HAI), virus neutralization testing, enzyme-linked immunosorbent assay (ELISA) for IgM antibodies

Additional Information EEE is carried by a mosquito vector. The mosquitos known to carry EEE breed in fresh water swamps and infect numerous bird species as well as horses. Infected birds typically have very high titer viremia and act as a reservoir for human infection.

In humans, EEE causes an acute illness that is either fatal or self-limited; chronic illness should suggest a different diagnosis. Syndromes include headache with fever, meningitis, and meningoencephalitis. The other alphavirus agents causing disease in the U.S. are Western equine encephalitis and Venezuelan equine encephalitis. These have been classified as group A arboviruses.

References

Centers for Disease Control, "Eastern Equine Encephalitis Virus - Florida," *JAMA*, 1992, 267(10):1324.

Garen PD, Tsai TF, and Powers JM, "Human Eastern Equine Encephalitis: Immunohistochemistry and Ultrastructure," *Mod Pathol*, 1999, 12(6):646-52.

Johnson AJ, Martin DA, Karabatsos N, et al, "Detection of Antiarboviral Immunoglobulin G by Using a Monoclonal Antibody-Based Capture Enzyme-Linked Immunosorbent Assay," *J Clin Microbiol*, 2000, 38(5):1827-31.

Internet Web Sites

www.cdc.gov/ncidod/dvbid/arbor/index.htm

Echinococcosis Diagnostic Procedures

Related Information

Bronchoalveolar Lavage (BAL) *on page 311*
Brushings Cytology *on page 314*
Liver Disease: Laboratory Assessment, Overview *on page 869*
Ova and Parasites, Direct Exam *on page 985*
Sputum Cytology *on page 1222*

Applies to Alveolar Echinococcosis; Cystic Echinococcosis

Abstract In humans, larval stages of *E. granulosus*, *E. multilocularis*, or *E. vogeli* cause infection. Cystic echinococcosis is a cestode parasitic disease important in livestock-raising areas in which dogs (the definitive hosts) are used. The adult *E. granulosus*, the cause of cystic echinococcosis, resides in the intestine of dogs and other canids. Sheep, caribou, deer, moose, pigs, or man are intermediate hosts. *Echinococcus granulosus* causes unilocular cysts. The cause of multilocular alveolar disease (alveolar echinococcosis), a more invasive form, is a larval form of *E. multilocularis*. Polycystic hydatid disease, caused by *E. vogeli*, only rarely occurs in humans.

Specimen Serum; aspiration specimens can be used for immunologic testing and microscopy

Container Red top tube for serology

Use Support a diagnosis of echinococcosis or exposure to the cestode. Most patients with *E. granulosus* have single organ involvement, including cysts in liver, lung, bone, brain, or other organs. Alveolar echinococcosis occurs in the liver in >98% of cases. Serological testing is more reliable for evaluation of alveolar echinococcosis. Available immunodiagnostic procedures usually permit distinction between alveolar echinococcosis and cystic echinococcosis.

Limitations Serologic methods for echinococcosis are compromised by nonspecific cross-reactivity with other helminths. Serum from 50% of patients with cysticercosis cross react. False positives are occasionally seen in patients with cirrhosis and lupus; false negatives with some large cysts, lung cysts, or dead cysts. Immunoblotting currently provides the best performance for *E. granulosus* diagnosis.

Methodology Serology: complement fixation (CF), bentonite flocculation assay (BFA), indirect hemagglutination (IHA), latex agglutination (LA), enzyme-linked immunosorbent assay (ELISA), immunofluorescence assay, immunoblot. The newer ELISA tests are more sensitive than the more traditional hemagglutination and latex assays. A number of advanced immunodiagnostic methods are addressed by Gottstein and Reichen.

Cytologic methods, including bronchoalveolar lavage, brushings, sputum, or aspiration cytology can sometimes reveal scolices or degenerated hooklets.

Additional Information Echinococcosis is usually suspected based on imaging studies and supported with serological assays. Sensitivity and specificity of serologic assays are greater for hepatic than for pulmonary infections or for those in other organs. There is generally a rapid decline in antibody after surgical removal of the cyst.

References
Gadea I, Ayala G, Diago MT, et al, "Immunological Diagnosis of Human Cystic Echinococcosis: Utility of Discriminant Analysis Applied to the Enzyme-Linked Immunoelectrotransfer Blot," *Clin Diag Lab Immunol*, 1999, 6(4):504-8.

Gottstein B and Reichen J, "Echinococcosis/Hydatidosis," *Manson's Tropical Diseases*, 21st ed, Chapter 84, Cook GC and Zumla A, eds, Philadelphia, PA: WB Saunders Co, 2003, 1561-82.

Internet Web Sites
www.cdc.gov/ncidod/dpd/parasites/alveolarhydatid/default.htm

Ehrlichiosis Serology

Related Information

Synonyms HGE Antibody; Human Granulocytotropic Ehrlichiosis Serology

Abstract Small, intracellular bacteria, the Ehrlichiae are zoonotic agents transmitted both to humans and animals via tick bites. They are rickettsia-like bacteria which localize in leukocyte phagosomes. Distinct forms of ehrlichiosis recognized in the United States include **human monocytic ehrlichiosis (HME)** and **human granulocytic ehrlichiosis (HGE)**. The disease is an acute febrile illness which resembles Rocky Mountain spotted fever. It can be mild, but about 33% of cases requires hospitalization.

Specimen Serum

Container Red top tube

Collection Collect at time of illness, then 2-4 weeks after onset (ie, acute and convalescent samples).

Storage Instructions Refrigerate or freeze serum.

Reference Interval Titer <1:80 with the source of antigen *E. chaffeensis*.

Critical Values Fourfold rise or fall in titer

Use Aid in the diagnosis of ehrlichiosis

Limitations Positive cutoff titer for disease has not been standardized and varies between laboratories performing the test. At times, serologic results are not reproducible

Methodology The most frequently used methodology is indirect fluorescent antibody (IFA) for IgG and IgM antibodies.

Ehrlichia-infected leukocytes can be detected on peripheral blood smears that are stained with Giemsa or Wright stains. Inclusions (morulae) are found in neutrophils in HGE. Isolation of *E. chaffeensis* can be accomplished with tissue culture methods and can detect bacteria even in patients with negative examination of peripheral blood.
(Continued)

531

Ehrlichiosis Serology (Continued)

Additional Information Clinical features of ehrlichiosis may include rapid onset of fever with chills (97% of patients), myalgias, headache (81% of patients), and malaise (84% of patients). A rash is seen in only 36% of patients. Abnormal laboratory findings (leukopenia, lymphopenia, neutropenia, thrombocytopenia, increased aspartate transaminase) are found in at least 86% of patients. Nearly 1200 cases have been confirmed in the U.S. with most of the cases of HME occurring in Southeastern and South Central states. Cases of HGE were found in Northeastern and upper Midwestern states. The organism responsible for HME (*E. chaffeensis*) is most commonly found in the Lone Star tick (*Amblyomma americanum*) while *E. equi*, responsible for HGE, has been found in the deer tick (*Ixodes scapularis*) and the dog tick (*Dermacentor variabilis*). *E. ewingii* was recently described in humans. HGE is found in the Northern U.S., where Lyme disease and babesiosis are endemic.

References

Belman AL, "Tick-Borne Diseases," *Semin Pediatr Neurol*, 1999, 6(4):249-66.

Dobbenburgh AV, van Dam AP, and Fikrig E, "Human Granulocytic Ehrlichiosis in Western Europe," *N Engl J Med*, 1999, 340(15):1214-5.

Fritz CL and Glaser CA, "Ehrlichiosis," *Infect Dis Clin North Am*, 1998, 12(1):123-36.

Glaser C and Johnson E, Images in Clinical Medicine, "Ehrlichiosis," *N Engl J Med*, 1995, 332(21):1417.

Goodman JL, "Ehrlichiosis - Ticks, Dogs, and Doxycycline," *N Engl J Med*, 1999, 341(3):195-6.

Jerrad D, "Ehrlichiosis," *J Emerg Med*, 1999, 17(1):27-30.

McQuiston JH, Paddock CD, Holman RC, et al, "The Human Ehrlichioses in the United States," *Emerg Infect Dis*, 1999, 5(5):635-42.

Schutze GE and Jacobs RF, "Human Monocytic Ehrlichiosis in Children," *Pediatrics*, 1997, 100(1):E10.

Internet Web Sites

www.cdc.gov/ncidod/dvrd/ehrlichia/index.htm

♦ **Elastase, Fecal Pancreatic** *see Fecal Pancreatic Elastase 1 on page 576*

♦ **Elastase, Serum** *see Lipase, Serum on page 851*

♦ **Elavil®** *see Amitriptyline, Serum or Plasma on page 149*

Electrolyte Panel, Serum

Related Information

Anion Gap, Serum, Plasma, or Urine *on page 160*
Bicarbonate, Blood *on page 258*
Blood Gases and pH, Arterial *on page 275*
Blood Gases and pH, Capillary *on page 277*
Blood Gases and pH, Venous *on page 279*
Carbon Dioxide, Total, Blood *on page 339*
Chloride, Serum, Plasma, or Blood *on page 391*
Drugs of Abuse Testing, Urine *on page 525*
Ibuprofen, Serum *on page 764*
Magnesium, Serum *on page 885*
Osmolality, Calculated, Serum or Plasma *on page 976*
Osmolality, Serum *on page 978*
pCO₂, Blood *on page 1008*
pH, Blood *on page 1018*
Potassium, Serum or Plasma *on page 1078*
Sodium, Serum or Plasma *on page 1210*

Synonyms Plasma Electrolytes; Serum Electrolytes

Test Includes The HCFA-defined electrolyte panel includes **sodium, potassium, chloride,** and **carbon dioxide**. **Anion gap** is calculated from the panel results and included on the report from many laboratories.

Abstract The electrolyte panel is used to evaluate electrolyte and acid-base balance in a wide variety of disorders.

Specimen Serum, plasma, or whole blood

Container Red top tube or green top (heparin) tube

Collection Specimen is best collected without a tourniquet if possible. Do **not** allow patient to clench-unclench his/her hand. See Venous Blood Collection *on page 1304*. Collect specimen anaerobically and avoid hemolysis during collection. Separate serum or plasma from cells as soon as possible following venipuncture.

Storage Instructions Specimen should be delivered to the laboratory and testing should occur as soon as possible. If delay is inevitable, refrigerate the specimen.

Causes for Rejection Gross hemolysis or lipemia

Use Critically important in detecting and defining metabolic abnormalities. See individual listings for tests involved.

Limitations Hemolysis and prolonged contact of serum with cells produces elevation of potassium. Exposure of specimen to air causes loss of carbon dioxide.

References
Current Procedural Terminology (CPT™) 2001, American Medical Association, 2001.

♦ **Electrolytes, Urine** *see* Chloride, Urine *on page 394*

Electron Microscopic Examination for Viruses, Stool

Related Information
Bacterial Culture, Stool *on page 243*
Electron Microscopy *on page 533*
Histopathology *on page 733*
Rotavirus, Direct Detection *on page 1173*
Skin Biopsy *on page 1200*
Viral Culture *on page 1307*

Abstract Most viruses causing gastroenteritis are readily detected by electron microscopy (EM). The viruses are present in very high concentration ($\geq 10^6$ virus particles/g) and are stable in stool.

Specimen Stool from the acute, diarrheal phase of disease; skin lesions; tissues (biopsy/autopsy)

Turnaround Time 1 day to 1 week

Reference Interval Viruses not observed

Use Demonstrate viral particles (eg, rotavirus, Norwalk virus, calcivirus, astrovirus, and coronavirus) in stool specimens; examination of tissue from biopsy or autopsy

Limitations Generally, EM visualization of viral particles is not as sensitive as is cell culture, except for detection of nonculturable viruses such as rotavirus. EM can detect viruses if they are present in concentrations of 10^6 to 10^7 particles/mL. Such sensitivity is appropriate for agents causing diarrhea, but not for detection of other potential pathogens. Very few clinical microbiology/virology laboratories have immediate access to electron microscopes.

Methodology Diluted stool is mixed with an electron-opaque heavy metal solution. The stool solution is placed on an electron-lucent grid support and examined by electron microscopy.

Additional Information The electron microscopic observation of viruses is the basis for identification. Electron microscopy is a useful procedure when viruses have not been identified by other methods. Many of the viruses detected by electron microscopy are now detected by nucleic acid methods or specific detection of viral antigen.

References
MacRae J and Srivastava M, "Detection of Viruses by Electron Microscopy: An Efficient Approach," *J Virol Methods*, 1998, 72(1):105-8.
Putzker M, Sauer H, Kirchner G, et al, "Community Acquired Diarrhea - The Incidence of Astrovirus Infections in Germany," *Clin Lab*, 2000, 46(5-6):269-73.

Internet Web Sites
www.astdhpphe.org/infect/norwalk.html
www.cdc.gov/ncidod/dvrd/revb/

Electron Microscopy

Related Information
Electron Microscopic Examination for Viruses, Stool *on page 533*
Fine Needle Aspiration, Deep and Superficial Masses *on page 590*
Kidney Biopsy *on page 818*
Microsporidia Diagnostic Procedures *on page 915*
Muscle Biopsy *on page 927*
Skin Biopsy *on page 1200*

Applies to Viral Diseases by EM

Abstract A major application of electron microscopy (EM) is to delineate histogenesis of poorly differentiated neoplasms when light microscopy is equivocal
(Continued)

Electron Microscopy (Continued)

and when proper therapy and prognosis depend on precise diagnosis. Although applications have diminished with development of immunoperoxidase methods, EM remains useful and provides information which otherwise cannot be obtained.

Specimen Fresh unfixed tissue, blood, bone marrow aspirate

Container Vial containing glutaraldehyde or other appropriate fixative, depending upon institution or reference laboratory. Blood and bone marrow aspirate may be collected in heparinized or EDTA tubes and submitted immediately.

Collection Specimen obtained by surgical biopsy should be cut within minutes of excision, minced into cubes 1 mm or less, and placed in glutaraldehyde, paraformaldehyde, or other special fixative. (See Kidney Biopsy on page 818.) Two percent to 4% phosphate or cacodylate-buffered glutaraldehyde is recommended. Formaldehyde-fixed tissue may be used if glutaraldehyde is not available. All EM fixatives, particularly glutaraldehyde, should be refrigerated until used to retard oxidative damage to fixative. Discard if a precipitate forms.

Use The distinction between poorly differentiated carcinoma, amelanotic melanoma, and lymphoma can be reliably established by electron microscopy. Neuroendocrine neoplasms contain neurosecretory granules. Small cell carcinomas contain few neurosecretory granules while they are numerous in more differentiated tumors, such as carcinoids and islet cell tumors. Mesotheliomas are differentiated from adenocarcinomas by elongated, delicate microvilli. Birbeck granules characterize the Langerhans cells of eosinophilic granuloma and other forms of Langerhans histiocytoses. Electron microscopic demonstration of platelet peroxidase in the nuclear envelope of leukemic blasts establishes a key diagnostic feature of megakaryoblastic leukemia (FAB M7). When limited material is available, such as in fine needle aspiration biopsies, electron microscopy may allow a more precise classification in selected cases.

Electron microscopy is useful in certain viral and other infectious diseases. In AIDS encephalopathy, EM of brain biopsies helps to determine the responsible agent (ie, HIV, CMV, herpes, Jakob Creutzfeldt, progressive multifocal leukoencephalopathy, and so forth). EM may be invaluable in the evaluation of muscle biopsies (eg, mitochondrial myopathies) and peripheral nerve biopsies. Storage/metabolic diseases sometimes are well evaluated with EM. Ceroid lipofuscinoses and Pompe type II glycogenosis bear characteristic inclusions in peripheral blood lymphocytes by EM. Many other lysosomal storage diseases can be identified by EM of skin, conjunctival, or gum biopsies. The immotile cilia syndrome is diagnosed by the absence of one or both dynein arms in cross sections of ciliary microtubules in nasal or bronchial biopsies. Some liver biopsies have lesions in which EM may be useful (eg, Dubin-Johnson syndrome, Rotor disease). With heart biopsy, EM can reveal Adriamycin® cardiotoxicity.

Limitations EM is labor intensive and limited by cost and sampling errors. This method is usually not useful to distinguish benign from malignant neoplasms. Utility is diminished by poor fixation, crush or drying artifact. The role of EM has recently been eroded by rapid advances in immunocytochemistry for the differential diagnosis of tumors and infectious agents. For instance immunodiagnostic methods are definitive and rapid for herpes encephalitis, but EM supports diagnosis.

Additional Information Concomitant light microscopic evaluation is usually required for ultrastructural correlation. Often light microscopic findings lead to a decision of whether or not EM is indicated (eg, tumors). EM is not a substitute for light microscopy. When renal biopsies are obtained, immunofluorescence studies are often performed, as well as light microscopy. Immunofluorescence and EM are mutually complementary. Applications in immunoelectron microscopy are invaluable tools for research scientists.

References

Herrera GA, Lowery MC, and Turbat-Herrera EA, "Immunoelectron Microscopy in the Age of Molecular Pathology," Appl Immunohistochem Molecul Morphol, 2000, 8(2):87-97.

Lieberman PH, Jones CR, Steinman RM, et al, "Langerhans Cell (Eosinophilic) Granulomatosis - A Clinicopathologic Study Encompassing 50 Years," Am J Surg Pathol, 1996, 20(5):519-552.

♦ **Electrophoresis, Protein, Urine** see Protein Electrophoresis, Urine on page 1107

♦ **Electrophoresis, Serum** see Protein Electrophoresis, Serum on page 1104

- ♦ **Elephant Tranquilizers** *see* Phencyclidine, Qualitative, Urine *on page 1019*
- ♦ **Elixophyllin®** *see* Theophylline, Serum *on page 1243*
- ♦ **E-Lor®** *see* Propoxyphene, Serum or Urine *on page 1096*
- ♦ **Emergency Blood** *see* Uncrossmatched Blood, Emergency *on page 1281*
- ♦ **Emergency Issue of Uncrossmatched Blood** *see* Uncrossmatched Blood, Emergency *on page 1281*
- ♦ **Emergency Transfusion** *see* Uncrossmatched Blood, Emergency *on page 1281*
- ♦ **Empirin®** *see* Salicylate, Serum or Plasma *on page 1176*
- ♦ **ENA** *see* Antinuclear Antibodies *on page 189*
- ♦ **ENA** *see* Sjögren Antibodies *on page 1199*
- ♦ **ENA** *see* Smith (Sm) and Ribonucleoprotein (RNP) Antibodies *on page 1206*

Encephalitis Viral Serology

Related Information
California Encephalitis Virus Serology *on page 334*
Cerebrospinal Fluid Analysis: Overview *on page 355*
Eastern Equine Encephalitis Virus Serology *on page 529*
Herpes Simplex Virus Culture and Antigen Detection *on page 721*
Herpes Simplex Virus DNA Detection *on page 723*
St Louis Encephalitis Virus Serology *on page 1224*
Viral Culture *on page 1307*
Western Equine Encephalitis Virus Serology *on page 1328*
West Nile Virus Diagnostic Procedures *on page 1329*

Applies to Encephalomyelitis Viral Serology; Meningoencephalitis Viral Serology

Abstract Encephalitogenic arboviruses commonly seen in North America are California encephalitis (that is closely related to LaCrosse), Western equine encephalitis, Eastern equine encephalitis, St Louis encephalitis, and now West Nile viruses. Arboviruses (arthropod-borne viruses) are a taxonomically heterogeneous group of viruses grouped together because they are all transmitted to humans via an arthropod vector. Around the world, a number of different arboviruses can cause encephalitis (ie, Japanese encephalitis, Dengue, Sindbis, Semliki forest, Rift Valley Fever, and numerous others). See West Nile Virus Diagnostic Procedures *on page 1329*. Herpes simplex type 1 is the most common cause of sporadic encephalitis in the U.S.

Viral encephalitis is usually accompanied by inflammation of the meninges. The expression **viral meningoencephalitis** may be used when both brain parenchyma and meninges are infected. With spinal cord involvement, the term **viral encephalomyelitis** is applied. The expression **aseptic meningitis** indicates inflammation of the meninges not caused by bacteria.

Specimen Serum, cerebrospinal fluid (CSF)

Container Red top tube

Collection Serology: acute and convalescent sera drawn 10-14 days apart

Reference Interval Serology: less than a fourfold titer increase in paired sera. CSF IgM: negative

Use Differential diagnosis of diffuse or focal CNS dysfunction including confusion or coma, fever, headache, and evidence of meningeal irritation, seizures, weakness, and other manifestations.

Limitations Cross-reacting antibodies from previous infections or from immunization for yellow fever may produce false-positive results, particularly when assays are performed on unpaired sera. Information regarding travel history of the patient will help to more accurately identify the virus responsible for infection.

Methodology Evaluation of acute arboviral infections includes serologic assays, IgM and IgG antibody-capture enzyme immunoassays or indirect immunofluorescence. Plaque-reduction neutralization assays in cell culture are used for confirmation. PCR methods are sensitive but can be falsely negative early. An HSV PCR enzyme-linked immunosorbent assay following extraction from CSF is available. Current methods for West Nile virus are provided in that monograph.
(Continued)

Encephalitis Viral Serology *(Continued)*

Additional Information Central nervous system infection by California encephalitis (LaCrosse), Western equine encephalitis, Eastern equine encephalitis, or St Louis encephalitis viruses may manifest as aseptic meningitis, encephalitis, or meningoencephalitis. There is a seasonal distribution for these infections that reflects their mode of transmission to humans by mosquitos. In the United States, the incidence of arboviral infection is low, as is the prevalence of antibodies to these agents in the general population. Consequently, a positive result in an unpaired specimen is **presumptive** evidence for a recent infection. A fourfold increase in titer or a positive CSF IgM test is confirmatory. Antibody detection is the diagnostic test of choice, as these viruses are essentially nonculturable in routine diagnostic virology laboratories. Any positive serologic results should be reported to the state health departments to monitor for a possible outbreak of viral encephalitis.

Other causes of CNS viral infection include enteroviruses (including polio, Coxsackie viruses, echoviruses, enteroviruses 70, 71), rubella, lymphocytic choriomeningitis, rabies, influenza, parainfluenza, measles, mumps, HIV-1, herpes viruses, and others. Most patients with HSV-1 DNA found in CSF have encephalitis, while most with HSV-2 DNA develop meningitis. Cytomegalovirus and varicella-zoster virus cause encephalitis in immunocompromised individuals.

Nearly 100 types of organisms cause acute meningoencephalitis.

References

Calisher CH, "Medically Important Arboviruses of the United States and Canada," *Clin Microbiol Rev*, 1994, 7(1):89-116.

Hirsch MS and Werner B, "A 38-Year-Old Woman With Fever, Headache, and Confusion," Case Records of the Massachusetts General Hospital, *N Engl J Med*, Case 17-2003, Scully RE, Mark EJ, McNeely WF, et al, 2003, 348(22):2239-47.

Nash D, Mostashari F, Fine A, et al, "The Outbreak of West Nile Virus Infection in the New York City Area," *N Engl J Med*, 2001, 344(24):1807-14.

O'Sullivan CE, Aksamit AJ, Harrington JR, et al, "Clinical Spectrum and Laboratory Characteristics Associated With Detection of Herpes Simplex Virus DNA in Cerebrospinal Fluid," *Mayo Clin Proc*, 2003, 78(11):1347-52.

Tyler KL, "West Nile Virus Encephalitis in America," *N Engl J Med*, 2001, 344(24):1859-9.

Uhlmann EJ and Storch GA, "Viral Encephalitis," *Lab Med*, 2001, 32(6):317-23.

Internet Web Sites

www.cdc.gov/ncidod/dvbid/arbor/index.htm

- **Encephalitis Virus** *see* California Encephalitis Virus Serology *on page 334*
- **Encephalomyelitis Viral Serology** *see* Encephalitis Viral Serology *on page 535*
- **End-Alveolar Carbon Monoxide Assessment of Erythrocyte Survival** *see* ^{51}Cr Red Cell Survival *on page 476*
- **Endep®** *see* Amitriptyline, Serum or Plasma *on page 149*

Endometrial Cytology

Related Information
Bacterial Culture, Genital Specimen *on page 239*
Cervical/Vaginal Cytology *on page 376*
Chromosome Analysis, Products of Conception *on page 412*

Test Includes Smears, filter, cytocentrifuge, and cell block preparations

Abstract The frequency of endometrial carcinoma has increased during the last several decades. The Pap test, proven so very successful as a screening test for cervical cancer, is not a reliable source for the diagnosis of endometrial neoplasia, although the diagnosis is rendered from time to time. Nonetheless, cytology can provide a means of diagnosis of endometrial cancer, if a direct and optimal sample is obtained. Widespread use of the office endometrial biopsy has diminished the need for endometrial cytology.

Collection Smear cellular material from collecting instrument thinly and evenly on a clean glass slide. Immediately spray or wet fix. Any remaining cellular material is deposited in a formalin bottle for cell block preparation. Slides and specimen bottle are labeled with patient's name and identification number. Aspirated fluid must be brought to the laboratory at once.

Special Instructions Include all relevant clinical data on requisition including LMP, age, prior diagnoses, history of bleeding, hypertension, diabetes, parity, as well as hormone use.

Use Evaluate possible endometrial carcinoma or hyperplasia

Contraindications Cervical or vaginal infections, cervical stenosis

Methodology Smears and cytocentrifuged specimens are examined microscopically; cell blocks and tissue fragments are processed as surgical tissue specimens for histopathologic examination.

Additional Information A series utilizing the Tao Brush has recently been reported. It has been approved by the U.S. Food and Drug Administration.

The diagnosis of extrauterine cancer in specimens from the endometrial cavity, cervical canal, or posterior vaginal fornix may include primaries from the ovary, gastrointestinal tract, fallopian tube, pancreas, urethra, breast, peritoneum, and other organs.

References

DeMay RM, "Cytology of the Glandular Epithelium," *The Art and Science of Cytopathology*, Chapter 6, Chicago, IL: ASCP Press, American Society of Clinical Pathologists, 1996, 120-35.

Ng ABP, "Endometrial Hyperplasia and Carcinoma and Extrauterine Cancer," *Comprehensive Cytopathology*, 2nd ed, Chapter 12, Bibbo M, ed, Philadelphia, PA: WB Saunders Co, 1997, 251-77.

Wu HHJ, Harshbarger KE, Berner HW, et al, "Endometrial Brush Biopsy (Tao Brush): Histologic Diagnosis of 200 Cases With Complementary Cytology - an Accurate Sampling Technique for the Detection of Endometrial Abnormalities," *Am J Clin Pathol*, 2000, 114(3):412-18.

Endomysial Antibodies

Related Information

Gliadin IgG/IgA Antibodies *on page 637*

Skin Biopsy, Immunofluorescence *on page 1203*

Applies to Gluten; Transglutinase Antibody (Anti-tTG)

Abstract The major cause of sprue in Western civilization is celiac disease, an autoimmune disorder in genetically susceptible individuals. Celiac disease (celiac sprue), or gluten-sensitive enteropathy, is a malabsorptive disease of the jejunum and proximal ileum histopathologically often, but not always, characterized by villous atrophy. High sensitivity as well as specificity of endomysial IgA antibodies for celiac disease offers a worthwhile screening test. Those whose results are positive are candidates for small bowel endoscopic biopsy.

Specimen Serum; a minimum of 0.25 mL is required for pediatric samples

Container Red top tube or SST™ tube

Special Instructions Maintain sterility.

Reference Interval Negative

Use Diagnosis of celiac disease, characterized by malabsorption, diarrhea, abdominal distension, flatus, steatorrhea, intestinal cramping, impaired growth, delayed puberty, anemia, weight loss, and osteoporosis. Dermatitis herpetiformis may coexist with celiac disease. Testing for serum endomysial antibodies (EMA), tissue transglutaminase, and for gliadin antibodies is used in evaluation of children in whom a failure to thrive problem is recognized.

Limitations The testing is subjective. A negative result does not completely rule out celiac disease or dermatitis herpetiformis, because patients with mild gluten-sensitive enteropathy may not develop detectable antibody levels. IgA deficiency is found in a significant number of celiac disease patients, who cannot develop IgA endomysial antibodies. Measurement of transglutaminase and IgG gliadin antibodies is helpful in making the diagnosis in this situation.

Methodology Indirect immunofluorescence, enzyme-linked immunosorbent assay (EIA)

Additional Information Celiac disease, or gluten enteropathy, is caused by sensitivity to gliadin, a gluten protein, the major constituent of wheat protein and to other cereal proteins in barley and rye. It is mediated by an inappropriate T-cell immune response. Celiac disease usually begins classically in infants between 4 and 24 months of age, soon after the introduction of cereal to the diet. Symptoms may disappear in later childhood or early adolescence despite continued signs of malabsorption. It sometimes appears in the third or fourth decade of life or later. Gliadin is the alcohol soluble fraction of gluten that actually causes celiac disease. Affected individuals develop villous atrophy of the small intestine and malabsorption. Symptoms resolve following removal of gluten from the diet. The definitive diagnosis of celiac disease requires jejunal biopsies before and after elimination of dietary gluten.

Ingestion of gluten triggers the production of IgG and IgA antibodies to gliadin and IgA antibodies to endomysium. (Endomysium is the sheath of reticular fibrils surrounding each smooth muscle fiber.) The IgA EMA test is both more (Continued)

Endomysial Antibodies (Continued)

specific and more sensitive than antigliadin assays. Sensitivity is 85% to 98% and specificity 97% to 100%, respectively. See Gliadin IgG/IgA Antibodies on page 637.

Tissue transglutinase is the autoantigen recognized by endomysial antibodies. **Tissue transglutinase antibodies (anti-tTG)** are highly specific for celiac disease and are useful for dermatitis herpetiformis. Results correlate with the level of severity of gluten-sensitive enteropathy, but negative results do not exclude diagnosis. The test uses IgA enzyme-linked immunosorbent assays (ELISA). Anti-tTG is more sensitive than EMA, but is less specific.

A strong association with either HLA-DQ2 or HLA-DQ8 haplotype is recognized, but most individuals expressing HLA-DQ2 do not develop celiac disease.

Dermatitis herpetiformis is a bullous skin disease closely associated with gluten-sensitive enteropathy. Strict adherence to gluten-free diet often induces remission of both the skin and small bowel abnormalities. Endomysial antibodies are also useful in diagnosis of dermatitis herpetiformis.

The titer of IgA-EMA as well as biopsy findings correlate with the severity of gluten-sensitive enteropathy. If patients strictly adhere to a gluten-free diet, the titer of IgA-EMA should begin to decrease within 6-12 months after the onset of dietary therapy.

Cholesterol concentrations which are high to normal may rule out celiac disease. Serum iron <60 µg/dL, ferritin <50 µg/dL, cholesterol <156 mg/dL, and hypochromic anemia (Hb: male <13 g/dL, female <12 g/dL) are anticipated in celiac disease.

An approach to the diagnosis of celiac sprue is provided in a 2002 review paper by Farrell and Kelly.

References

Ciacci C, Cirillo M, Giorgetti G, et al, "Low Plasma Cholesterol: A Correlate of Nondiagnosed Celiac Disease in Adults With Hypochromic Anemia," *Am J Gastroenterol*, 1999, 94(7):1888-91.

Farrell RJ and Kelly CP, "Celiac Sprue," *N Engl J Med*, 2002, 346(3):180-8.

Green PRH and Bana J, "Coeliac Disease," *Lancet*, 2003, 362(9381):383-91.

Keren DF, "Autoimmune Disease of the Gastrointestinal Tract," *Clinical and Laboratory Evaluation of Human Autoimmune Diseases*, Chapter 19, Nakamura RM, Keren DF, and Bylund DJ, eds, Chicago, IL: ASCP Press , 2002, 277-90.

Mäki M, Mustalahti K, Kokkonen J, et al, "Prevalence of Celiac Disease Among Children in Finland," *N Engl J Med*, 2003, 348(25):2517-24.

McManus R and Kelleher D, "Celiac Disease - The Villian Unmasked?" *N Engl J Med*, 2003, 348(25):2573-4.

Moskaluk CA, "Sailing Past the Horizon: The Histologic Diagnosis of Celiac Disease in "Nonflat" Intestinal Mucosa," *Am J Clin Pathol*, 2001, 116(1):7-9.

♦ **Enphenemalum** *see* Methylphenobarbital, Serum *on page 909*

Entamoeba histolytica Antigen Detection and Serology

Related Information

Bacterial Culture, Stool *on page 243*
Methylene Blue Stain, Stool *on page 906*
Ova and Parasites, Direct Exam *on page 985*

Test Includes Detection of *Entamoeba histolytica* specific antigen in stool or antibodies to *E. histolytica* in serum

Abstract *Entamoeba histolytica*, a protozoan parasite, is capable of invading the intestinal mucosa, and sometimes spreads to the liver. It causes hepatic amebic abscess. It is estimated that approximately 70 thousand deaths per year are due to amebiasis. Infection with *E. histolytica* is found worldwide, but is much more common in tropical and subtropical regions. Humans are the primary reservoir and infection occurs by ingestion of cysts from contaminated food or water.

Specimen Freshly collected stool for antigen detection; serum for antibody detection

Container Sterile, leakproof, wide-mouth container for stool; red top tube for serum

Sampling Time Repeat serologic testing in 5-7 days when early results are negative, if clinically indicated.

Reference Interval Negative for stool antigen; serology: IHA titer: <1:128; CF titer: <1:8; immunodiffusion test: negative

Critical Values IHA titer ≥1:256

Use These methods help in establishing the diagnosis of amebiasis. Serologic testing for amebiasis is the best single test to distinguish between the two major types of liver abscesses: amebic and pyogenic.

Limitations Sensitivity of the serologic test is highest in extraintestinal amebiasis, lower in amebic dysentery, and lowest in asymptomatic carriers. Some false positives occur in patients with ulcerative colitis. The utility of the serologic marker is diminished in those parts of the world in which amebiasis is highly endemic with persistence of antibodies; *vide infra*. Serologic results in acute cases may be negative early.

Extraintestinal amebiasis is frequently found without trophozoites or cysts in stool; patients then may have a negative reaction to antigen detection.

Methodology Several serologic methods are available. They include indirect hemagglutination (IHA), indirect immunofluorescent antibody (IFA), and countercurrent immunoelectrophoresis, polymerase chain reaction (PCR), riboprinting. Enzyme-linked immunosorbent assay (ELISA) is sensitive and does not yield false negatives in patients with amebic liver abscesses. ELISA provides only 3.6% false positives in endemic areas.

Additional Information The laboratory diagnosis of amebiasis can be made by examination of a permanent stained stool specimen. The detection of ingested red blood cells in the cytoplasm of the trophozoites is diagnostic for pathogenic *E. histolytica*. The antigen detection assays seem to be more sensitive and specific for amebiasis than microscopy. (See Ova and Parasites, Direct Exam on page 985.)

Positive serologic tests are seen in 75% of patients with invasive colonic amebiasis, while a positive serology is seen in >90% of patients with amebic liver abscesses. Indirect hemagglutination is positive in 87% to 100% of patients with amebic liver abscesses and >85% of patients with acute amebic dysentery. Fewer than 6% of uninfected individuals react in the test.

Molecular biology-based methods may become the "gold standard" but are costly, difficult, and time-consuming to perform.

Morphologic diagnosis in fecal specimens is still needed.

References

Colmer-Hamood JA, "Fecal Microscopy Artifacts Mimicking Ova and Parasites," *Lab Med*, 2001, 32(2):80-4.

Espinosa-Cantellano M and Martinez-Palomo A, "Pathogenesis of Intestinal Amebiasis: From Molecules to Disease," *Clin Microbiol Rev*, 2000, 13(2):318-31.

Farthing MG, Cevallos AM, and Kelly P, "Intestinal Protozoa," *Manson's Tropical Diseases*, 21st ed, Chapter 77, Cook GC and Zumla A, eds, Philadelphia, PA: WB Saunders Co, 2003, 1373-87.

Haque R, Huston CD, Hughes M, et al, "Amebiasis," Current Concepts, *N Engl J Med*, 2003, 348(16):1565-73.

Leber AL, "Intestinal Amebae," *Clin Lab Med*, 1999, 19(3):601-19.

Pillai DR, Keystone JS, Sheppard DC, et al, "*Entamoeba histolytica* and *Entamoeba dispar*: Epidemiology and Comparison of Diagnostic Methods in a Setting of Nonendemicity," *Clin Infect Dis*, 1999, 29(5):1315-8.

Tanyuksel M and Petri WA Jr, "Laboratory Diagnosis of Amebiasis," *Clin Microbiol Rev*, 2003, 16(4):713-25.

Internet Web Sites

www.cdc.gov/ncidod/dpd/parasites/amebiasis/default.htm

♦ **Enterobacteriaceae Culture** see Bacterial Culture, Aerobes on page 229

Enterovirus Culture

Related Information

Bacterial Culture, Stool on page 243
Cerebrospinal Fluid Analysis: Overview on page 355
Coxsackievirus Serology on page 464
Enterovirus Polymerase Chain Reaction on page 540
Ova and Parasites, Direct Exam on page 985
Poliomyelitis I, II, III Serology on page 1069
Viral Culture on page 1307

Test Includes Inoculation of specimen onto several appropriate cell cultures; isolated viruses are identified using specific monoclonal antibodies or neutralization.

(Continued)

Enterovirus Culture *(Continued)*

Abstract Enterovirus infections are among the most common human viral illnesses. Poliovirus, the prototypic enterovirus, was responsible for summer epidemics of paralytic poliomyelitis. Before the introduction of polio vaccines, epidemic polio occurred regularly in the U.S. with approximately 25 cases per 100,000 people annually. Currently, due to widespread vaccination, wild type poliovirus has been eliminated in the Western hemisphere and nearly eliminated in the rest of the world. Other enteroviruses cause a variety of clinical diseases, such as aseptic meningitis in young children, myocarditis, hemorrhagic conjunctivitis, and hand-foot-mouth syndrome (herpangina) (enterovirus 71).

Specimen Stool, rectal swab, cerebrospinal fluid, upper and lower respiratory tract specimens, whole blood, throat swab, various organs and tissues (heart, muscle, brain)

Container Sterile container

Sampling Time It is important to obtain specimens very early in the disease; however, virus is shed in the stool for weeks.

Storage Instructions Enteroviruses are rather hardy; however, specimens should be refrigerated or placed into cold virus transport medium and delivered immediately to the clinical laboratory since other viruses that cause similar clinical diseases are not so resilient.

Turnaround Time Negative cultures often are reported after 2 weeks for conventional culture; negative results are reported after 72 hours for rapid culture methods.

Reference Interval No virus isolated

Use Aid in the diagnosis of disease caused by enteroviruses (eg, polio, congenital viral infections, myocarditis, viral pericarditis, encephalitis, aseptic meningitis, herpangina, pharyngitis, tonsillitis, and rhinitis)

Limitations Cell culture generally does not support the growth of certain coxsackie A enteroviruses. Infrequently, aseptic meningitis is caused by coxsackie A virus types which require animal inoculation for isolation. The isolation of enterovirus in stool specimens may not indicate an active clinical disease.

Methodology Inoculation of specimens into cell culture, incubation of cultures, and observation for characteristic cytopathic effect (CPE)

Additional Information Most enteroviral infections are mild and asymptomatic. Clinical diseases in humans range from a slightly increased temperature to severe CNS disease and paralysis. Humans are the only known reservoir for enteroviruses. Thus, infection occurs by direct contact with infected individuals, the oral-fecal route, through contaminated water, or the respiratory route. Enteroviral infections occur most commonly between July and September. Children are more likely to become infected than are adults.

The detection of enteroviral nucleic acid directly from specimens is a new technique that enables the detection of enteroviruses within a few hours. This assay has been studied in the clinical setting and has been found to have clinical utility. It is currently available in many larger clinical laboratories.

References

Dolin R, "Enterovirus 71 - Emerging Infections and Emerging Questions," *N Engl J Med*, 1999, 341(13): 984-5.

Grabow WO, Botma KL, de Villiers JC, et al, "Assessment of Cell Culture and Polymerase Chain Reaction Procedures for the Detection of Polioviruses in Wastewater," *Bull World Health Organ*, 1999, 77(12):973-80.

Huang CC, Liu CC, Chang YC, et al, "Neurologic Complications in Children With Enterovirus 71 Infection," *N Engl J Med*, 1999, 341(13):936-42.

Rotbart HA, McCracken GH Jr, Whitley RJ, et al, "Clinical Significance of Enteroviruses in Serious Summer Febrile Illnesses of Children," *Pediatr Infect Dis J*, 1999, 18(10):869-74.

Sayer MH, "Enterovirus Infections: Diagnosis and Treatment," *Pediatr Infect Dis J*, 1999, 18(12):1033-9.

Internet Web Sites

www.cdc.gov/ncidod/dvrd/entrvirs.htm

Enterovirus Polymerase Chain Reaction

Related Information

Bacterial Culture, Cerebrospinal Fluid *on page 236*
Cerebrospinal Fluid Analysis: Overview *on page 355*
Enterovirus Culture *on page 539*

Polymerase Chain Reaction *on page 1069*
Viral Culture *on page 1307*

Test Includes Detection of enterovirus RNA by amplification of nucleic acid

Abstract Enteroviruses cause a wide variety of clinical disease, including upper respiratory tract infection, aseptic meningitis, myocarditis, conjunctivitis, herpangina, and pleurodynia. Enteroviruses account for >80% of U.S. cases of aseptic meningitis. The direct detection of enterovirus RNA in specimens can be accomplished with PCR amplification of the enterovirus genome. The amplification steps are followed by specific hybridization with an oligonucleotide probe. This assay is available in many larger clinical laboratories.

Specimen Blood, cerebrospinal fluid, and selected tissues including lymph nodes and myocardial tissue

Container Blood: lavender top (EDTA) tube; CSF: plastic or glass tube is acceptable

Collection Tissue samples should be snap-frozen immediately and stored at -80°C.

Storage Instructions Specimens should be transported to the laboratory as soon as they become available. If specimens cannot be processed immediately, they should be frozen and maintained at -80°C.

Causes for Rejection Specimens collected using heparin

Turnaround Time 1-3 days

Reference Interval No enterovirus genome detected

Use Identify enterovirus associated with meningitis, myocarditis, and other enterovirus-related diseases

Limitations A negative result does not rule out the presence of enterovirus genome. Inhibitors of the amplification assay may be present which lead to false-negative results. Enteroviral RNA can quickly be digested by RNases present in all specimens. Thus, the specimen should be transported in a timely fashion or handled appropriately to prevent the loss of target RNA. Limited cross-reactivity and amplification of other viruses within the Picornavirus family has been described, including amplification of some rhinovirus serotypes. It is not available in all local laboratories.

Methodology Enterovirus RNA is extracted from the specimen and a homologous DNA strand is generated using reverse transcriptase. Specific product is amplified using primers that anneal to the conserved 5' nontranslated region of most enteroviruses.

Additional Information The enterovirus genus includes 67 serotypes, including poliovirus 1-3, coxsackievirus A1-A22, A24, B1-B6, echovirus 1-9, 11-27, 29-33, and enterovirus 68-71, all of which are members of the large Picornaviridae family. To date no specific disease has been accepted to be uniquely associated with any one enterovirus serotype. Many nonpolioenteroviruses are capable of causing the paralytic disease particularly associated with polioviruses. Enteroviruses are responsible for a wide-spectrum of clinical diseases, of which the most significant include neonatal sepsis, aseptic meningitis, poliomyelitis, encephalitis, and myocarditis. Enterovirus 71 epidemics have recently been described in which 91% of deaths were in patients age 5 years or younger. Meningitis, due to enteroviral infections, has a classic pattern of appearance in the late summer and fall in the northern hemisphere.

The PCR assay can be performed quickly and thus positive results for enterovirus RNA have the potential to impact patient management.

References

Dolin R, "Enterovirus 71 - Emerging Infections and Emerging Questions," *N Engl J Med*, 1999, 341(13): 984-5.

Grabow WO, Botma KL, de Villiers JC, et al, "Assessment of Cell Culture and Polymerase Chain Reaction Procedures for the Detection of Polioviruses in Wastewater," *Bull World Health Organ*, 1999, 77(12):973-80.

Hamilton MS, Jackson MA, and Abel D, "Clinical Utility of Polymerase Chain Reaction Testing for Enteroviral Meningitis," *Pediatr Infect Dis J*, 1999, 18(6):533-7.

Ramers C, Billman G, Hartin M, et al, "Impact of a Diagnostic Cerebrospinal Fluid Enterovirus Polymerase Chain Reaction Test on Patient Management," *JAMA*, 2000, 283(20):2680-5.

Romero JR, "Reverse-Transcription Polymerase Chain Reaction Detection of the Enteroviruses," *Arch Pathol Lab Med*, 1999, 123(12):1161-9.

Rotbart HA, McCracken GH Jr, Whitley RJ, et al, "Clinical Significance of Enteroviruses in Serious Summer Febrile Illnesses of Children," *Pediatr Infect Dis J*, 1999, 18(10):869-74.

Sayer MH, "Enterovirus Infections: Diagnosis and Treatment," *Pediatr Infect Dis J*, 1999, 18(12):1033-9.

(Continued)

Enterovirus Polymerase Chain Reaction *(Continued)*

Internet Web Sites
www.cdc.gov/ncidod/dvrd/entrvirs.htm

♦ **Entomology** *see* Arthropod Identification *on page 213*

Eosinophil Count

Related Information
Complete Blood Count *on page 442*
Eosinophil Granule Proteins *on page 545*
Eosinophil Smear *on page 548*
Eotaxin, Serum, Plasma, or Urine *on page 549*
Immunoglobulin E *on page 773*
Lysozyme, Blood and Urine *on page 884*
Ova and Parasites, Direct Exam *on page 985*
Peripheral Blood: Differential Leukocyte Count *on page 1010*

Abstract Eosinophils are granulocytic type white blood cells distinguished by prominent reddish-orange cytoplasmic granules on Wright-Giemsa (Romanowsky-type) stains, frequently with bilobed nuclei, and with both circulating and tissue forms. Eosinophil count is increased in a wide variety of conditions including especially, allergy, drug reaction, parasitism, collagen vascular disease, and some malignant states. Eosinophils are decreased with hyperadrenalism.

Specimen Whole blood

Container Lavender top (EDTA) tube

Causes for Rejection Clotted specimen, specimen more than 4 hours old

Reference Interval 15-650/mm^3 (0.015 to 0.65 x 10^9/L). There is diurnal variation with lowest levels in the morning. There may be within-day physiologic variation >40%.

Use Aid in the diagnosis of allergy, drug reaction, parasitic infestations, collagen disease, Hodgkin disease, and myeloproliferative diseases. Increased also in a broad range of less common conditions including sarcoidosis, the acute hypereosinophilic syndrome, angioneurotic edema, acute renal allograft rejection, eosinophilic nonallergic rhinitis, anisakiasis, eosinophilic gastroenteritis, eosinophilia-myalgia syndrome, and others. Decrease in eosinophils (eosinopenia) occurs in Cushing disease (cortisol excess) and is seen with a variety of infections, correlating with severity of the infectious process. Absence of eosinophils has unfavorable prognostic implications in cases of infection.

Limitations Manual method is subject to an inherent error of 20% to 30%.

Additional Information The major cause of eosinophilia is allergy (atopy) which is clinically common and frequently asymptomatic or minimally symptomatic. An important cause of eosinophilia is parasitic infection. Eosinophilia is seen especially with tissue invasion by parasites as occurs with trichinosis, schistosomiasis, filariasis, echinococcal disease, spirochetal infection, liver infestation (as with *Clonorchis sinensis*, *Fasciola hepatica*, *Capillaria hepatica*) and many other parasitic organisms. Occult neoplasm is in the differential consideration but most tumor-associated eosinophilia occurs in cases with widespread metastases. Classification of eosinophilia by degree of severity may be helpful in differential diagnosis. See table.

Toxocaral disease (visceral larva migrans) is a typical parasitic disease in which eosinophil counts (eosinophils >30% on differential) are usually markedly elevated. However, ~25% of children with toxocariasis have normal eosinophil counts. Thus, normal eosinophil counts do not rule out toxocaral disease or other parasitic infestations.

There is recent interest in the clinical applicability of eosinophil granule protein levels. In some situations, eosinophil proteins may serve as markers for previous eosinophil reaction. The four principal eosinophil granule proteins can be considered as "proinflammatory" and consist of major basic protein (MBP), eosinophil cationic protein (ECP), eosinophil-derived neurotoxin (EDN), and eosinophil peroxidase (EPO). See Eosinophil Granule Proteins *on page 545*. The possibility that ECP and EDN are also synthesized and/or localized to neutrophil granules may temper enthusiasm for use of these proteins as markers of eosinophil-associated inflammation.

Likely and Less Likely Causes of Eosinophilia
on the Basis of Severity
(absolute eosinophil count)

Likely Causes	Less Likely Causes
Mild (0.7-1.5 x 10⁹/L)	
Allergic rhinitis	Neoplasm
Hay fever or atopy	Gastrointestinal disease
Extrinsic asthma	Skin disease
Drug reaction	Certain infectious diseases
Parasitic disease	Long-term dialysis
Occupational lung disease	Radiation therapy
	Immunodeficiency state
Moderate (1.5-5 x 10⁹/L)	
Parasitic disease	Polyarteritis nodosa
Intrinsic asthma	Other connective tissue disorders
Drug reaction	Neoplasm
Pulmonary eosinophilia syndrome	Hypereosinophilic syndrome
Marked (>5 x 10⁹/L)	
Parasitic diseases	Disorder usually associated with moderate eosinophilia
Visceral larva migrans associated with *Toxocara canis* or *Toxocara cati* infestation	Trichinosis, hookworm infection, ascariasis, strongyloidiasis, neoplasm
Tissue migration during larval stage (eg, *Ascaris*, *Trichinella*, hookworm, *Strongyloides* sp)	Neoplasm
Hypereosinophilic syndrome	Polyarteritis nodosa
Eosinophilic leukemia	Drug reaction

Adapted from Brigden ML, "A Practical Work-up for Eosinophilia: You Can Investigate the Most Likely Causes Right in Your Office," *Postgrad Med*, 1999, 105(3); 193-210.

An eosinophil specific chemokine (an eosinophil chemoattractant) has recently been identified, purified and named eotaxin. The gene and cDNA of human eotaxin have been isolated with gene mapping to chromosome 17. Specificity of eotaxin owes to its specific receptor, CCR-3, which is present only on eosinophils. See Eotaxin, Serum, Plasma, or Urine *on page 549*.

An important although rare cause of increased eosinophils in the peripheral blood is the acute hypereosinophilic syndrome (HES). Reported mortality ranges from 81% to 95% in 1-3 years. The HES syndrome includes high peripheral WBC count, circulating early eosinophil forms without blast cells, mental confusion, delusions, near coma, and severe cardiac symptoms. If the absolute eosinophil count is >1.5 x 10⁹/L and persists for over 6 months, HES must be considered. Recent reports indicate that some cases of HES respond to treatment by imatinib mesylate (Gleevec™, Novartis). Rather than a *bcr/abl* translocation, a new fusion gene FIP1L1-PDGFRA is formed, the product of which is a tyrosine kinase (inhibited by imatinib). There are important implications to understanding the mechanisms of malignancy.

Infiltrative lung diseases, in which peripheral blood eosinophils may be increased, include eosinophilic pneumonia, Löffler's syndrome (often related to *Ascaris* infestation), and tropical eosinophilia (usually related to filariasis). See following flowchart.

Idiopathic chronic eosinophilic pneumonia is of rare occurrence and unknown cause. There may be no blood eosinophilia but lung biopsy is characterized by interstitial tissue eosinophil infiltrate.

Eosinophilic gastroenteritis may occur with blood eosinophilia.

Eosinophilia-myalgia syndrome (EMS), characterized by an eosinophil count ≥1000 cells/mm³, severe often incapacitating myalgia, fatigue, cough, shortness of breath, rash and headache was associated with ingestion of large amounts of one form of L-tryptophan. Sarcoidosis, granulomatous myositis, (Continued)

Eosinophil Count *(Continued)*

collagen vascular diseases, neoplastic myositis, and other entities are in the differential consideration. An EMS-like syndrome has been considered to result from several factors including ingestion of tryptophan, inactivation of indoleamine-2,3-dioxygenase, and possible impairment of the hypothalamic-pituitary-adrenal axis. EMS is potentially fatal (Guillain-Barré like ascending polyneuropathy) with a clinical course resembling the toxic oil syndrome that was epidemic in Spain in 1981 and affected over 20,000 individuals resulting in 277 deaths.

The population of patients with "idiopathic" (and persistent) eosinophilia may harbor subclinical, low grade or frank evolving Sézary syndrome, or other T-cell lymphomas (including those caused by HTLV-I infection). Thus, it has been suggested that patients with "idiopathic" eosinophilia should be closely observed for lymphoproliferative disorders.

Infiltrative lung diseases, in which peripheral blood eosinophils may be increased, include eosinophilic pneumonia, Löffler's syndrome (often related to *Ascaris* infestation), and tropical eosinophilia (usually related to filariasis).

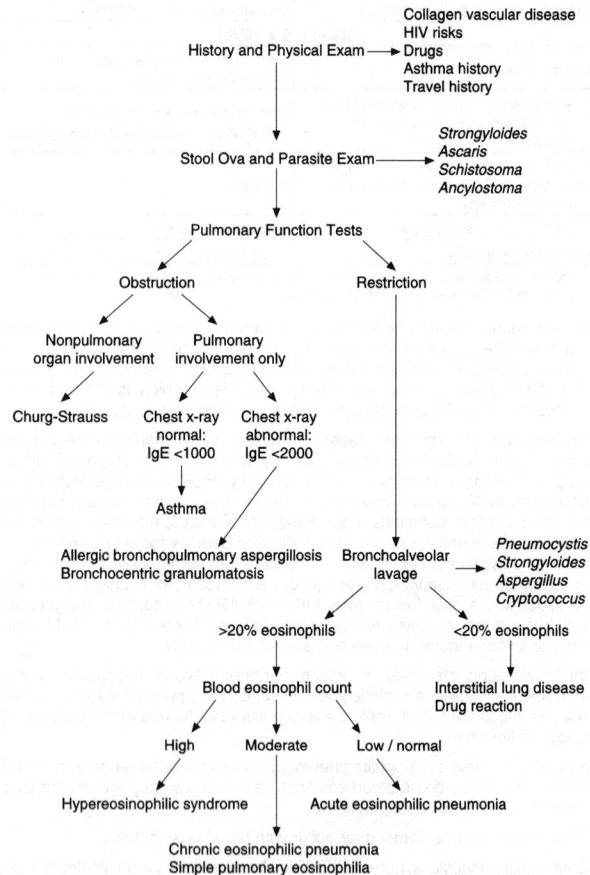

Adapted from Rochester CL, "The Eosinophilic Pneumonias," *Fishman's Pulmonary Diseases and Disorders*, 3rd ed, Chapter 74, Fishman AP, Elias JA, Fishman JA, et al, eds, New York, NY: McGraw-Hill, Health Professions Division, 1998, 1:113-50.

Idiopathic chronic eosinophilic pneumonia is of rare occurrence and unknown cause. There may be no blood eosinophilia but lung biopsy is characterized by interstitial tissue eosinophil infiltrate.

Eosinophilic gastroenteritis may occur with blood eosinophilia.

Eosinophilia-myalgia syndrome (EMS), characterized by an eosinophil count ≥1000 cells/mm^3, severe often incapacitating myalgia, fatigue, cough, shortness of breath, rash and headache may be associated with ingestion of large amounts of one form of L-tryptophan.

The population of patients with "idiopathic" (and persistent) eosinophilia may harbor subclinical, low grade or frank evolving Sézary syndrome, or other T-cell lymphomas (including those caused by HTLV-I infection). Patients with "idiopathic" eosinophilia should be observed for lymphoproliferative disorders.

References

Ackerman SJ and Butterfield JH, "Eosinophilia, Eosinophil-Associated Diseases, and the Hypereosinophilic Syndrome," *Hematology: Basic Principles and Practice*, 3rd ed, Chapter 40, Hoffman R, Benz EJ, Shattil SJ, et al, eds, Philadelphia, PA: Churchill Livingstone, 2000, 702-20.

Brigden ML, "A Practical Work-up for Eosinophilia: You Can Investigate the Most Likely Causes Right in Your Office," *Postgrad Med*, 1999, 105(3):193-210.

Cools J, DeAngelo DJ, Gotlib J, et al, "A Tyrosine Kinase Created by Fusion of the PDGFRA and FIP1L1 Genes as a Therapeutic Target of Imatinib in Idiopathic Hypereosinophilic Syndrome," *N Engl J Med*, 2003, 348(13):1201-14.

Kelly KJ, "Eosinophilic Gastroenteritis," *J Pediatr Gastroenterol Natur*, 2000, 30:Suppl:S28-35.

Marone G, "Human Eosinophils," *Chemical Immunology*, Volume 76, Adorini L, Arai K, Berek C, et al, eds, Switzerland: Karger, 2000.

Suzuki R, Seto M, and Nakaura S, "Idiopathic Eosinophilia," *N Engl J Med*, 2000, 342(9):660-1.

Terrell JC, "Laboratory Evaluation of Leukocytes: Absolute Eosinophil Counting Procedure," *Clinical Hematology: Principles, Procedures, Correlations*, 2nd ed, Chapter 24, Stiene-Martin EA, Lotspeich-Steininger CA, and Koepke JA, eds, Philadelphia, PA: JB Lippincott Co, 1998, 342-3.

Venge P, Byström J, Carlson M, et al, "Eosinophil Cationic Protein (ECP): Molecular and Biological Properties and the Use of ECP as a Marker of Eosinophil Activation in Disease," *Clin Exp Allergy*, 1999, 29(9):1172-86.

Eosinophil Granule Proteins

Related Information

Eosinophil Count *on page 542*
Eosinophil Smear *on page 548*
Eotaxin, Serum, Plasma, or Urine *on page 549*
Immunoglobulin E *on page 773*
Pregnancy-Associated Protein A, Serum *on page 1082*

Test Includes ECP and/or EDN and/or others

Abstract Eosinophils are of primary importance in the immune response to parasites and in allergic reactions, especially asthma. The intracytoplasmic eosinophil secondary granules consist largely of four basic proteins. **Major basic protein (MBP)** forms the crystalloid core while **eosinophil cationic protein (ECP)**, **eosinophil-derived neurotoxin (EDN)**, and **eosinophil peroxidase (EPO)** are present in the granule matrix In addition an array of surface molecules reflects broad participation in inflammation, including utilization of chemokines, cytokines, and adhesion molecules.

Specimen ECP and EDN, the granule proteins of clinical interest, can be measured in a variety of clinical specimens. In particular, urine and feces are appropriate specimens for the determination of EDN. Circadian variation in EDN levels introduces a variable that must be interpreted if such levels are measured in urine. The use of urine is advantageous in the pediatric population as specimen procurement is noninvasive. One is advised to consult with those performing the test as to what specimens may be acceptable and as to specimen handling. The following indicates the variety of specimens that may be of interest:

- serum/plasma (EDTA or citrate), serum is preferred
- bronchoalveolar lavage fluid, sputum
- nasal lavage fluid
- tears (from patient with vernal keratoconjunctivitis, allergic conjunctivitis, and giant papillary conjunctivitis)
- jejunal perfusion fluids (for evaluation of inflammatory bowel disease)
- fecal samples (study of infants with atopic eczema and food allergy and patients with inflammatory bowel disease)
- urine samples (especially useful in pediatric patients being monitored during respiratory therapy)

(Continued)

Eosinophil Granule Proteins *(Continued)*

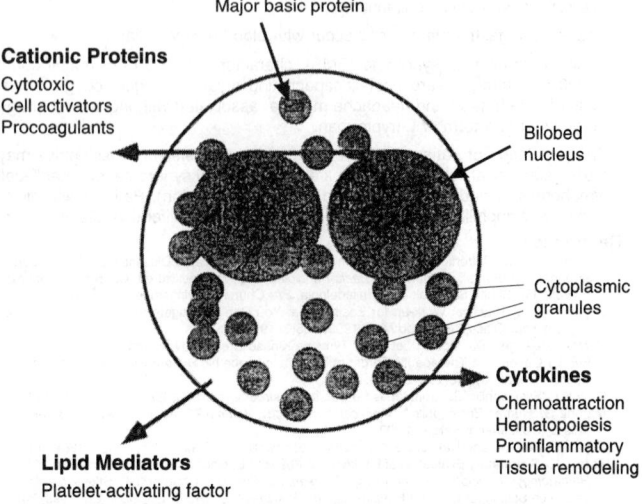

Major basic protein

Cationic Proteins
Cytotoxic
Cell activators
Procoagulants

Bilobed
nucleus

Cytoplasmic
granules

Cytokines
Chemoattraction
Hematopoiesis
Proinflammatory
Tissue remodeling

Lipid Mediators
Platelet-activating factor
Leukotriene C_4

Figure 1. Schematic diagram of the human eosinophil.

Adapted from Thomas LL and Page SM, "Inflammatory Cell Activation by Eosinophil Granule Proteins," *Human Eosinophils, Biological and Clinical Aspects*, Volume 76, Marone G, ed, Basel, Switzerland: Karger, 2000, 99-117.

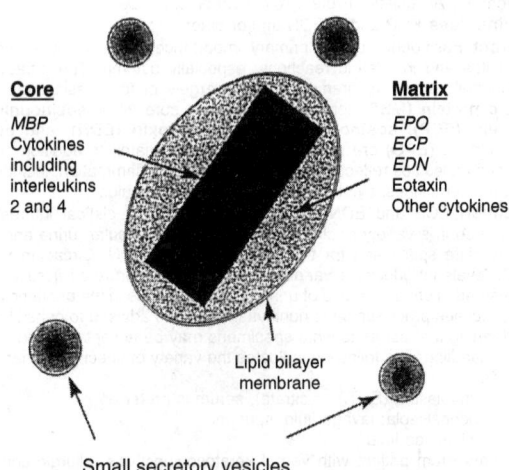

Core
MBP
Cytokines
including
interleukins
2 and 4

Matrix
EPO
ECP
EDN
Eotaxin
Other cytokines

Lipid bilayer
membrane

Small secretory vesicles

Figure 2. The crystalloid granule contains two internal compartments, the core, and the matrix.

Adapted from Lacy P and Moqbel R, "Eosinophil Cytokines," *Human Eosinophils, Biological and Clinical Aspects*, Volume 76, Marone G, ed, Basel, Switzerland: Karger, 2000, 134-55.

Storage Instructions Serum, 24-hour storage at 4°C, following a 1-hour clotting period at 20°C to 22°C (specimens from healthy controls or allergic patients). Freeze serum at -20°C if not analyzed within 1 day of collection. Freeze spot urine samples (for EPX/EDN determination) at -70°C. Simultaneous determination of urine creatinine is necessary to allow evaluation of the extent of dilution of urine.

Reference Interval Children (nonatopic): serum ECP ≤6.5 µg/L (95th percentile, 19 µg/L). Higher levels of ECP are found in serum than in EDTA plasma due to continued release of granule proteins from eosinophils in blood without additives whereas with EDTA the cells are inactivated and do not release granule proteins. There is diurnal variation in serum ECP with highest levels in the early morning.

Adults: serum ECP ≤6.0 µg/L (geometric mean), 95% interval is 2.3-15.9 µg ECP/L

Use Evaluation of respiratory conditions with opportunity to correlate serum, urine, sputum, and bronchial lavage specimens for diagnostic and/or monitoring purposes. Evaluation and monitoring of response to therapy in cases of idiopathic inflammatory bowel disease, measure clinical activity of Wells syndrome (eosinophilic cellulitis). Potential use as a maternal serum marker for Down syndrome (eosinophil MBP complexed with pregnancy-associated plasma protein-A).

Limitations Variables that may influence test results include choice of anticoagulant, time allowed for coagulation, ambient temperature and other factors relating to sample handling that may affect the extracellular release of eosinophil proteins.

Additional Information CP was first isolated from cells of chronic myelogenous leukemia. Subsequently, the protein was localized to the matrix of the larger specific eosinophil cytoplasmic granule. ECP is a zinc metalloenzyme with a molecular weight of 16-22 kDa. The gene for ECP, located on chromosome 14 (q24-31) is close to a sister protein, EPX/EDN with which it has nearly 70% amino acid sequence homology.

The release of eosinophil granule proteins reflect the activation of eosinophils. ECP has a variety of functions including cytotoxic effects (eg, destruction of parasites and tumor cells, neurotoxicity, cardiovascular and respiratory injury, and antibacterial/antiviral activity). Nontoxic effects include inhibition of T-cell proliferation, release of histamine from basophils, stimulation of secretion of airway mucus, effects on coagulation and anticoagulants, and interaction with the complement system and adhesion molecules. Cell-killing action likely relates to ECP's ability to make pores in cell membranes, with resultant cytotoxic cell death by osmotic lysis. The 6 molecular forms of ECP allow for functional heterogeneity. The half-life of ECP in circulation is about 65 minutes.

Serum ECP (and blood eosinophil counts) bears a positive correlation with the severity of asthma, but does not accurately reflect functional indices of severity in chronic stable patients. Urinary EPX/EDN levels may have value in the monitoring of bronchial inflammation in asthmatic children.

ECP and EPX/EDN radioimmunoassays have been applied to the analysis of feces in inflammatory bowel disease (IBD). Fecal ECP and EPX/EDN were increased in both active ulcerative colitis (UC) and active Crohn disease (CD) as compared to their inactive counterparts. Fecal ECP and EPX/EDN have also been suggested as a noninvasive method for monitoring allergic intestinal inflammation and disease activity in infants with atopic eczema and food allergy.

Wells syndrome (eosinophilic cellulitis with dermal edema and eosinophil infiltrate) is associated with increase in ECP and interleukin-5 (IL-5) in peripheral blood, levels of which bear a close correlation with clinical activity.

The precursor of eosinophil MBP (proMBP) is synthesized by the placenta and secreted into the maternal circulation where it forms a complex with pregnancy-associated plasma protein-A (PAPP-A). It may serve as a maternal

(Continued)

Eosinophil Granule Proteins *(Continued)*

serum marker for Down syndrome. A significant challenge has been raised as to the specificity of ECP and EPX/EDN for eosinophil granules with the finding that these apparent eosinophil markers are demonstrable also in neutrophils.

Eosinophil granule proteins and eosinophil peroxidase have procoagulant properties, inhibit heparin, and may underlie the occurrence of thrombosis in the hypereosinophilic states.

References

Christiansen M, Oxvig C, Wagner JM, et al, "The Proform of Eosinophil Major Basic Protein: A New Maternal Serum Marker for Down Syndrome," *Prenat Diagn*, 1999, 19(10):905-10.

Espana A, Sanz ML, Sola J, et al, "Wells' Syndrome (Eosinophilic Cellulitis): Correlation Between Clinical Activity, Eosinophil Levels, Eosinophil Cation Protein and Interleukin-5," *Br J Dermatol*, 1999, 140(1):127-30.

Levy AM, Gleich GJ, Sandborn WJ, et al, "Increased Eosinophil Granule Proteins in Gut Lavage Fluid From Patients With Inflammatory Bowel Disease," *Mayo Clin Proc*, 1997, 72(2):117-23.

O'Sullivan S and Kumlin M, "Eosinophil Markers in Childhood Asthma," *Clin Exp Allergy*, 1999, 29(11):1454-6.

Samoszuk M, Corwin M, and Hazen SL, "Effects of Human Mast Cell Tryptase and Eosinophil Granule Proteins on the Kinetics of Blood Clotting," *Am J Hematol*, 2003, 73(1):18-25.

Van's Gravesande KS, Mattes J, Grüntjens T, et al, "Circadian Variation of Urinary Eosinophil Protein X in Asthmatic and Healthy Children," *Clin Exp Allergy*, 1999, 29(11):1497-501.

Eosinophil Smear

Related Information

Eosinophil Count *on page 542*

Eosinophil Granule Proteins *on page 545*

Eotaxin, Serum, Plasma, or Urine *on page 549*

Abstract The eosinophil is a major effector in allergic processes. Study of appropriately stained smears of clinical material may help to define the role of allergy in any particular patient's reactive process.

Specimen Two slides of nasal secretion; smear or swab of feces; sputum or induced (with saline) sputum. No fixation is required for slides. Nasal secretions may be submitted on wax paper or plastic wrap.

Container Slides of smeared specimen or nasal secretions on wax paper

Causes for Rejection Slides received in cytology fixative, no specimen on slide, smear, or swab

Special Instructions Requisition must state site of specimen.

Reference Interval No eosinophils identified

Use Investigate allergy, asthmatic disorders, and parasitic infestations

Methodology Wright or May-Grünwald-Giemsa stain and microscopic examination of smear. Gram-stained smears of microbiology specimens will not stain eosinophils.

Additional Information Eosinophils are often increased in the blood and sputum of patients with asthma, usually in relation to the severity of the process. There is no percentage of eosinophils in sputum diagnostic of asthma, but levels >80% (related to proportion of neutrophils) are very suggestive of asthma or of chronic bronchitis with wheezing. There is evidence of an inverse correlation between the numbers of eosinophils in the circulation and/or sputum and pulmonary function (eg, airway flow rates).

Patients who are smokers and who have chronic airflow limitation (severe obstructive bronchitis) may benefit from sputum eosinophil study. Sputum eosinophilia ≥3% predicted a beneficial response to prednisone. Effort dyspnea, quality of life, and forced expiratory volume$_1$ were improved. There was accompanying decline in median sputum eosinophil percentage from 9.7% to 0.5%

Sputum eosinophil count and eosinophil cationic protein level may not reflect the severity of asthma in cases of chronic stable asthma. Change in sputum eosinophils may be useful in predicting loss of asthma control as reflected by loss of airway function.

In cases of idiopathic hypereosinophilic syndrome (sustained hypereosinophilia with organ involvement and with inapparent etiology - see Eosinophil Count *on page 542*), eosinophils, usually bilobed, may show morphologic abnormalities including hypersegmented nuclei and cytoplasmic hypogranularity.

References
Jatakanon A, Lim S, and Barnes PJ, "Changes in Sputum Eosinophils Predict Loss of Asthma Control," *Am J Respir Crit Care Med*, 2000, 161(1):64-72.

Pizzichini E, Pizzichini MM, Gibson P, et al, "Sputum Eosinophilia Predicts Benefit From Prednisone in Smokers With Chronic Obstructive Bronchitis," *Am J Respir Crit Care Med*, 1998, 158(5 Pt 1):1511-7.

Ronchi MC, Piragino C, Rosi E, et al, "Do Sputum Eosinophils and ECP Relate to the Severity of Asthma?" *Eur Respir J*, 1997, 10(8):1809-13.

Eotaxin, Serum, Plasma, or Urine

Related Information
Eosinophil Count *on page 542*
Eosinophil Smear *on page 548*
Immunoglobulin E *on page 773*

Abstract Eotaxin is an eosinophil specific chemoattractant. Eotaxin is a chemokine with a specific receptor, CCR-3, which is present only on eosinophils. The chemokine stimulates the growth and development of myeloid precursors in the bone marrow and induces rapid release of eosinophils from the marrow. As one of a number of ligands for the CCR-3 receptor, eotaxin plays a role in the pathogenesis of asthma. Plasma eotaxin levels are elevated in patients with acute asthma.

Specimen Plasma

Container Lavender top (EDTA) tube or red top tube

Storage Instructions Separate plasma from cells within 2 hours. May store at -20°C until time of assay.

Special Instructions Serum eotaxin levels increase by some 50% to 100% (as compared with plasma levels) over a 2-hour period if the serum is not immediately harvested after obtaining the specimen. The eotaxin is apparently released from red blood cells during clotting. Eotaxin values in EDTA plasma do not change even after incubation at 25°C for 6 hours.

Reference Interval 200-378 pg/mL

Use Diagnosis of asthma; evaluate lung function in patients with asthma; study the pathogenesis of asthma; diagnosis of nephritis with tissue eosinophilia (see Additional Information)

Limitations Testing available only from research and/or a few reference laboratories. Multiple cytokines stimulate eosinophils such that the role of eotaxin may be difficult to assess.

Additional Information While there are many eosinophil-active chemoattractants, eotaxin is unique in that it specifically attracts eosinophils, due to its specific receptor, CCR-3. Eotaxin stimulates myeloid development and release from the bone marrow, attracts eosinophils (eg, in inflammatory reactions, reactions to parasites), and mediates tissue eosinophilia. Eotaxin is apparently produced by epithelial cells, endothelial cells, and activated leukocytes (including eosinophils).

Demonstration of eotaxin in urine may assist in the diagnosis of diffuse interstitial nephritis (with tissue eosinophils).

References
Jundt F, Anagnostopoulos I, Bommert K, et al, "Hodgkin/Reed Sternberg Cells Induce Fibroblasts to Secrete Eotaxin, A Potent Chemoattractant for T Cells and Eosinophils," *Blood*, 1999, 94(6):2065-71.

Lilly CM, Woodruff PG, Camargo CA Jr, et al, "Elevated Plasma Eotaxin Levels in Patients With Acute Asthma," *J Allergy Clin Immunol*, 1999, 104(No 4, Pt 1):786-90.

Morita A, Shimosako K, Kikiuoka S, "Development of a Sensitive Enzyme-Linked Immunosorbent Assay for Eotaxin and Measurement of it Levels in Human Blood," *J Immunol Methods*, 1999, 226(1-2):159-67.

Wada T, Furuichi K, Sakai N, et al, "Eotaxin Contributes to Renal Interstitial Eosinophilia," *Nephrol Dial Transplant*, 1999, 14(1):76-80.

- ♦ **Epanutin®** *see* Phenytoin, Serum or Plasma *on page 1026*
- ♦ **Epilim®** *see* Valproic Acid, Serum or Plasma *on page 1297*
- ♦ **Epimorph Dolcontin®** *see* Morphine, Urine *on page 921*
- ♦ **Epinat®** *see* Phenytoin, Serum or Plasma *on page 1026*
- ♦ **Epinephrine** *see* Metanephrines, Urine or Plasma *on page 899*

- ◆ **Epinephrine, Norepinephrine, Dopamine** *see* Catecholamines, Fractionation, Plasma *on page 345*
- ◆ **Epinephrine, Urine** *see* Catecholamines, Fractionation, Urine *on page 347*
- ◆ **Epitestosterone** *see* Dehydroepiandrosterone and Dehydroepiandrosterone Sulfate, Serum or Plasma *on page 506*
- ◆ **Epithelial Membrane Antigen (EMA)** *see* Immunoperoxidase Procedures *on page 780*
- ◆ **Epitrol®** *see* Carbamazepine, Serum *on page 337*
- ◆ **Epoetin Beta** *see* Erythropoietin, Serum *on page 551*

Epstein-Barr Virus Serology

Related Information

Infectious Mononucleosis Screening Test *on page 785*

Abstract Since Epstein-Barr virus (EBV) was found in a Ugandan child with Burkitt lymphoma three decades ago, a role of the virus has been shown in several diseases in addition to infectious mononucleosis. These include hairy leukoplakia (a disorder of the tongue), carcinoma of nasopharynx, polyclonal lymphoid hyperplasias, lymphomas in patients following stem cell or solid organ transplantation and other immunomodulation, and in other neoplastic entities. Relationships to some T-cell lymphomas and to Hodgkin disease have been recognized. A 2003 study found a casual, but very infrequent, increased risk of infectious mononucleosis-related EBV infection and the EBV-positive subgroup of Hodgkin disease in young adults.

AIDS patients are at risk for EBV infection and lymphoma.

Specimen Serum; for assessment of clonality of EBV-infected lymphoid cells by Southern blot: bone marrow; cerebrospinal fluid; tissue (frozen promptly)

Container Red top tube for serum serology; lavender top (EDTA) tube for bone marrow (if indicated)

Reference Interval See table.

Epstein-Barr Virus Serology

	Uninfected	Current Infection	Previous Infection
IgG anti-VCA	<1:10 or negative	>1:10 or positive	≥1:10 or positive
IgM anti-VCA[1]	<1:10 or negative	>1:10 or positive[1]	≤1:10 or negative
Anti-EBNA	<1:5 or negative	<1:5 or negative	≥1:5 or positive

[1]An IgM anti-VCA titer >1:10 or positive by EIA is a key reaction indicative of current or recent infection.

Use Diagnose Epstein-Barr virus infection; evaluate heterophil-negative mononucleosis, other lymphoproliferative disease: lymphomas including Burkitt lymphoma and X-linked lymphoproliferative disease

Limitations Despite much publicity, these tests are neither sensitive nor specific for chronic fatigue syndrome. EBV infects >90% of humans, persisting for their lifetimes, and chronic fatigue is a commonplace complaint. It is unclear if such investigation is useful for the clinical diagnosis of nasopharyngeal carcinoma or post-transplant lymphomas. It is not needed in the diagnosis of hairy leukoplakia, which has a distinct clinical presentation.

Contraindications Epstein-Barr viral testing need not be done on patients who have heterophil antibodies with the symptoms, physical findings and lymphocyte morphology consistent with infectious mononucleosis.

Methodology Serology: indirect fluorescent antibody (IFA), enzyme-linked immunosorbent assay (ELISA).

EBV can be cultured, but such cultures are not widely available or generally clinically indicated. EBV can also be detected by immunocytochemistry and by polymerase chain reaction. EBV DNA can be detected in blood.

Additional Information EBV is a ubiquitous human lymphotropic DNA-containing herpesvirus that infects epithelial cells and B lymphocytes, stimulating proliferation of the latter. It causes classic infectious mononucleosis and is associated with >90% of the cases of Burkitt lymphoma, nearly 100% of poorly differentiated nasopharyngeal carcinomas, a subset of gastric carcinomas, lymphoproliferative disorders in immunocompromised patients, hairy leukoplakia, non-Hodgkin lymphomas in patients with congenital or acquired immunodeficiency, 40% to 60% of U.S. cases of Hodgkin disease, and smooth muscle tumors in immunosuppressed children.

Most cases of infectious mononucleosis can be diagnosed on the basis of clinical findings, blood count and morphology (lymphocytosis with virocytes), and a positive test for conventional (Davidsohn) serology (heterophil antibody) (infectious mononucleosis screening test). As many as 20% of cases may be heterophil-negative at presentation but the heterophil often becomes positive when repeated in a few days.

The serologic response to EBV includes antibody to early antigen, which is usually short lived, IgM and IgG antibodies to viral capsid antigen (VCA), and antibodies to nuclear antigen (EBNA); of these, VCA is the most useful. A high presenting VCA titer is good evidence for EBV infection. Since titers are generally high by the time a patient is symptomatic, it may not be possible to demonstrate fourfold rise in titer. Since a very high titer may be due to past infection, IgM titers should be measured to establish acute infection. Persistent absence of antibody to viral capsid is good evidence against EBV infection.

Antibody to EBV nuclear antigen (EBNA) usually develops 4-6 weeks after infection. Its presence early during an acute illness should lead one to consider a diagnosis other than EBV infectious mononucleosis.

References

Ambinder R, "Infection and Lymphoma," *N Engl J Med*, 2003, 349(14):1309-10.

Cohen JI, "Epstein-Barr Virus Infection," *N Engl J Med*, 2000, 343(7):481-92.

Godshall SE and Kirchner JT, "Infectious Mononucleosis. Complexities of a Common Syndrome," *Postgrad Med*, 2000, 107(7):175-86.

Hjalgrim H, Askling J, Rostgaard K, et al, "Characteristics of Hodgkin's Lymphoma After Infectious Mononucleosis," *N Engl J Med*, 2003, 349(14):1324-32.

Jacobs DS, DeMott WR, Oxley DK, et al, *Laboratory Test Handbook*, 5th ed, Hudson, OH: Lexi-Comp Inc, 2001.

Pagano JS, "Viruses and Lymphomas," *N Engl J Med*, 2002, 347(2):78-9.

Timms JM, Bell A, Flavell JR, et al, "Target Cells of Epstein-Barr Virus (EBV)-Positive Post-transplant Lymphoproliferative Disease: Similarities to EBV-Positive Hodgkin's Lymphoma," *Lancet*, 2003, 361:217-23.

Erythropoietin, Serum

Related Information

(Continued)

Erythropoietin, Serum *(Continued)*

Red Cell Mass *on page 1144*
Reticulocyte Count *on page 1156*
Thrombopoietin, Serum or Plasma *on page 1247*
Viscosity, Blood *on page 1312*
Vitamin B$_{12}$ Unsaturated Binding Capacity *on page 1316*

Applies to Epoetin Beta

Test Includes Serum iron

Abstract A glycoprotein formed mainly in the kidney, erythropoietin (EPO) has been purified and its gene, found on chromosome 7, has been cloned. EPO is the primary regulatory hormone for red cell production in marrow. Hypoxia increases EPO production; bilateral nephrectomy drastically reduces EPO synthesis and thereby inhibits erythropoiesis.

Polycythemia vera (PV) (primary polycythemia) is a clonal disease in which erythropoiesis is autonomous, independent of EPO. EPO in that disorder is decreased or normal in the presence of polycythemia, involving a negative feedback loop to the kidneys. Other (secondary) types of polycythemia are characterized by increased or normal EPO. Distinction of PV from relative polycythemia and secondary erythrocytosis also includes application of family history, patient history, arterial oxygen saturation, P$_{50}$, measurement of red cell mass, and plasma volume. PV-related features include the presence of persistent leukocytosis, thrombocytosis, microcytosis, unusual thromboses, splenomegaly, pruritus, and erythromelalgia.

Recombinant EPO (epoetin beta) is now available for treatment of anemia associated with chronic renal failure, HIV infection treated with zidovudine, cancer (including therapy-associated anemia), surgical procedures, and autologous blood donation.

Patient Preparation Recent exposure to radioisotopes may interfere if assay method is RIA.

Specimen Serum

Container Red top tube

Special Instructions Done only by a few laboratories

Reference Interval 5-36 mIU/mL (SI: 5-36 IU/L). EPO increases in pregnancy, in which significantly higher levels are found before the 24th week. Reference values in children from a study of 1122 subjects aged 1-18 years: male: 1.0-21.0 mIU/mL; female: 1.1-20.5 mIU/mL.

Use The test is used to investigate obscure anemias and the anemia of end-stage renal disease. Failure of renal erythropoietin production is the essential cause of the anemia of renal failure. This anemia can be successfully treated with exogenous erythropoietin. The availability of assays for EPO should provide the means to identify patients who will benefit from epoetin beta therapy.

EPO is used to differentiate secondary from primary polycythemia. Low EPO concentrations are fairly specific for PV. EPO is increased with states such as cyanotic heart disease, venous/arterial shunts, hypoxic pulmonary diseases, habitation at high altitudes, and with mutant hemoglobins with high affinity for oxygen. It may be increased in cases of Cushing syndrome, renal artery stenosis, renal cysts, and certain tumors (eg, hemangioblastoma of cerebellum, pheochromocytoma, hepatoma, nephroblastoma, rarely leiomyomas, and renal adenocarcinoma). Some overlap exists between these groups.

The physiologic effects of EPO have not escaped the attention of elite athletes seeking preternatural endurance and speed. EPO may have been related to the mysterious deaths of 19 cyclists between 1987 and 1990. The International Olympic Committee announced testing for EPO randomly selected participants in the 2000 Olympics, using both a urine test for EPO itself and a blood test which detects an inappropriate proportion of young erythrocytes.

Limitations Serum EPO levels may be increased by phlebotomy, anabolic steroids, androgens, TSH, ACTH, angiotensin, epinephrine, daunorubicin, fenoterol, growth hormone levels, and other drugs. Decreases follow acetazolamide, amphotericin B, cisplatin, enalapril, furosemide, and theophylline. Transfusions and estrogens may lower EPO levels. Normal EPO concentrations do not absolutely rule out PV.

Methodology Radioimmunoassay and immunoradiometric assay (RIA, IRMA), enzyme-linked immunoassay (ELISA), immunoprecipitin assay, and immunochemiluminometric assay

Additional Information Bone marrow examination with cytogenetic evaluation is indicated to establish the diagnosis of PV, recognize the presence of marrow fibrosis or a clonal cytogenetic disorder, or both.

A rare familial EPO receptor mutant disorder occurs with low EPO.

In chronic renal disease, the serum EPO level is generally lower than expected from the magnitude of the anemia. Serum EPO is inappropriately low in adult nephrotic syndrome mostly because of renal/urinary loss of the protein which contributes to the anemia. In chronic iron deficiency, serum EPO is increased, but the increase may not be as high as expected for the degree of anemia.

EPO response to anemia (excluding renal disease and pregnancy) in older subjects is similar to that of younger subjects.

Epoetin beta has diminished the need for transfusions in very low birthweight infants, but complications may occur including vascular thrombosis, neutropenia, thrombocytopenia, and infection.

Epoetin beta has been used to improve the yield of autologous units of blood before orthopedic surgery.

In contrast to patients with PV, most subjects with apparent polycythemia lack splenomegaly.

References
Adamson JW and Eschbach JW, "Erythropoietin for End-Stage Renal Disease," *N Engl J Med*, 1998, 339(9):625-7.

Clarey C, "EPO Tests Approved for Games in Sydney," *NY Times*, August 29, 2000.

Krafte-Jacobs B, Williams J, and Soldin S, "Plasma Erythropoietin Reference Ranges in Children," *J Pediatr*, 1995, 126(4):601-3.

Tefferi A, "Diagnosing Polycythemia Vera: A Paradigm Shift," *Mayo Clin Proc*, 1999, 74(2):159-62.

♦ **Eskalith®** *see* Lithium, Serum *on page 863*
♦ **Esophageal Washings Cytology** *see* Washing Cytology *on page 1326*
♦ **Esterase, Leukocyte, Urine** *see* Leukocyte Esterase, Urine *on page 849*
♦ **Estradiol** *see* Dehydroepiandrosterone and Dehydroepiandrosterone Sulfate, Serum or Plasma *on page 506*
♦ **Estradiol** *see* Testosterone, Total and Free, Serum or Plasma *on page 1238*
♦ **17β-Estradiol** *see* Estradiol, Serum *on page 553*

Estradiol, Serum

Related Information
Estriol, Unconjugated, Pregnancy, Serum, Plasma, or Urine *on page 554*
Estrogens, Urine *on page 557*
Follicle Stimulating Hormone, Serum, Plasma, or Urine *on page 609*
Heterophilic Antibodies *on page 727*
Luteinizing Hormone, Blood or Urine *on page 876*
Progesterone, Serum *on page 1093*

Synonyms 17β-Estradiol; Estradiol-17β

Applies to Estrone

Abstract Estradiol (E2) is the most potent endogenous estrogen. In nonpregnant females, most of the E2 originates in the ovaries, with smaller contributions from the adrenals. The placenta is an additional source in pregnancy. In males, 75% of E2 is produced from testosterone (and other androgens) by an aromatase-catalyzed reaction in the peripheral tissues; 25% of E2 is produced in the testes.

Patient Preparation Recent exposure to radioactivity (eg, scan) may interfere if assay method is RIA.

Specimen Serum

Container Red top tube

Sampling Time In females, the phase of the menstrual cycle should be recorded for interpretation. In assisted reproduction protocols, specimens for E2 must be obtained according to a specific schedule.

Storage Instructions Serum specimen is stable if refrigerated for 24 hours or frozen for up to 2 months.

(Continued)

Estradiol, Serum (Continued)

Reference Interval Children 6 months to 10 years: <15 pg/mL (SI: <55 pmol/L); adult male: 10-50 pg/mL (SI: 37-184 pmol/L); female: premenopausal: 30-400 pg/mL (SI: 110-1468 pmol/L) (depending on phase of menstrual cycle); postmenopausal: 0-30 pg/mL (SI: 0-110 pmol/L)

Use

Disorders of ovarian function: Values are decreased in ovarian failure (all causes). When ovarian failure is secondary to hypopituitary states, FSH and LH also are decreased. When ovarian failure is primary (ie, due to ovarian disease), FSH and LH are increased. In patients with infertility or irregular menses (including amenorrhea), E2 values, interpreted together with serum FSH, LH, progesterone, other hormones, and menstrual history, are useful in differential diagnosis.

Assisted reproduction: The magnitude of the E2 increment after the first dose of gonadotropins is a major determinant of the size of the second dose; in like fashion, the E2 increment following the second dose is a determinant of the next dose, etc. Very high serum E2 levels are not detrimental to clinical outcome of *in vitro* fertilization, and change in E2 level following gonadotropin-releasing hormone analogue stimulation is an effective predictor of ovarian reserve when accompanied by basal, serum FSH evaluation.

Hormonally active tumors: Rare tumors of the ovary, testis, adrenal, or nonendocrine sites may cause high E2 levels.

Limitations Significant interlaboratory variability is observed, though efforts are underway toward method standardization.

In menopausal females, estrone (E_1), rather than E2, is the predominant circulating estrogen.

E2 increases with hepatic cirrhosis. Oral contraceptives decrease serum levels. E2 levels can be normal in women who have hypogonadism.

Methodology Routine testing methods include isotopic and nonisotopic immunoassays with and without extraction. The reference/definitive method is isotope-dilution-gas chromatography/mass spectrometry (ID-GC/MS).

References

Chenette PE, Sauer MV, and Paulson RJ, "Very High Serum Estradiol Levels Are Not Detrimental to Clinical Outcome of *In Vitro* Fertilization," *Fertil Steril*, 1990, 54(5):858-63.

Gronowski AM and Landau-Levine M, "Reproductive Endocrine Function," *Tietz Textbook of Clinical Chemistry*, 3rd ed, Chapter 45, Burtis CA and Ashwood ER, eds, Philadelphia, PA: WB Saunders Co, 1999, 1601-41.

♦ **Estriol, Free** see Estriol, Unconjugated, Pregnancy, Serum, Plasma, or Urine on page 554

Estriol, Unconjugated, Pregnancy, Serum, Plasma, or Urine

Related Information

Alpha$_1$-Fetoprotein, Serum on page 136
Chorionic Gonadotropin, Human, Serum and Urine on page 397
Estradiol, Serum on page 553
Inhibin A, Serum on page 799
Pregnancy-Associated Protein A, Serum on page 1082
Urine Collection, 24-Hour on page 1295

Synonyms Estriol, Free; 16-Hydroxyestradiol; uE$_3$; Unconjugated Estriol, Pregnancy; Unconjugated Estrogen, Serum

Applies to Multiple Marker Screening Test; Triple Test

Abstract Estriol is the major estrogen of pregnancy, but its role in laboratory evaluation has become somewhat controversial. Replacement of unconjugated estriol (uE$_3$) by inhibin A for identification of Down syndrome, in the multiple marker screening test, is advocated by some.

Specimen Serum or plasma, 24-hour urine

Container Plain red top tube (not SST™ tube) or (depending on laboratory) green top (heparin) tube for blood; 24-hour urine container

Sampling Time Second trimester screening of maternal serum for Edwards syndrome (trisomy 18) and Down syndrome (trisomy 21) includes sampling at 15-21 or 15-22 weeks gestation when testing also includes AFP and hCG.

Optimally, these tests should be drawn between 16 and 18 weeks. Check with laboratory.

Collection Since circadian rhythms exist, serum estriol should be drawn at the same time of day on each visit.

Storage Instructions 4°C for up to 24 hours; for longer periods freeze at -20°C.

Special Instructions Include maternal birth date, first day of last menstrual period, gestational age by ultrasound and by physical examination, expected date of delivery, maternal weight, race, and diabetic status with request. Status of *in vitro* fertilization is relevant. Gestational age is best derived from ultrasound dating.

Reference Interval Urine concentrations of estriol increase with gestation, from 2 mg/24 hours (SI: 7 nmol/day) at 16 weeks gestation to 10-40 mg/24 hours (SI: 35-139 nmol/day) at term. A wide normal range exists. Reference intervals depend upon maternal weight as well as upon gestational age, like hCG and AFP. **Serum** levels in the table do not represent reference intervals for all laboratories. Different assays lead to wide differences of estriol in the same sample.

Normal Serum or Plasma Unconjugated Estriol Values (Fetal Well-Being)

Weeks Gestation	µg/L	SI: nmol/L
25	3.5-10.0	12-35
28	4.0-12.5	14-43
30	4.5-14.0	16-49
32	5.0-16.0	17-55
34	5.5-18.5	19-64
36	7.0-25.0	24-87
37	8.0-28.0	28-97
38	9.0-32.0	31-111
39	10.0-34.0	35-118
40	5.0-40.0	17-139

Note: This table is to be used to monitor fetal well-being.

Possible Panic Range Value of urinary estriol <4 mg/24 hours or 40% below mean of three prior values demands immediate evaluation of fetal well-being.

Use Serial estriol values, depending upon the integrity of the fetal-placental-maternal unit, had been thought to assess fetal well-being, a role no longer in wide use.

Estriol may decrease with pregnancy-induced hypertension, small for gestational age pregnancies, fetal growth restriction, molar gestation, chromosomal abnormalities, fetal demise, placental sulfatase deficiency, fetal adrenal aplasia or hypoplasia, and in anencephaly. Midtrimester uE_3 decrease with hCG increase may be associated with HELLP syndrome in patients with severe preeclampsia. See table in Alpha$_1$-Fetoprotein, Serum *on page 136*.

Limitations Single values are almost impossible to interpret; trends in a series of measurements are much more important. Other causes of decreased estriol levels include subjects living at high altitudes, on penicillin or related drugs, corticosteroids, dexamethasone, betamethasone, diuretics, Mandelamine®, probenecid, estrogens, phenazopyridine, meprobamate, phenolphthalein, cascara, senna, and glutethimide. It is decreased with anemia and severe liver disease. Estriol may be increased with multiple pregnancy and with oxytocin. It is not reliable in the presence of renal disease. Use of the test has become controversial. It is no longer done in a number of laboratories. Screening for Down syndrome without uE_3 is proposed.

Methodology Radioimmunoassay (RIA) or high performance liquid chromatography (HPLC)

Additional Information Combined evaluation of unconjugated serum estriol, maternal serum hCG, maternal serum AFP, and maternal age has been used in prediction of risk for fetal chromosomal abnormalities during the second trimester of pregnancy (the **triple screen test** for Down syndrome), but note a paper on unconjugated estriol use by Macri et al. Serum uE_3 is low in about
(Continued)

Estriol, Unconjugated, Pregnancy, Serum, Plasma, or Urine (Continued)

25% of instances of trisomy 21 and in about 55% of cases of trisomy 18. Triple screen testing identifies nearly 60% of pregnancies bearing Down syndrome conceptions with a 5% false-positive rate.

The application of likelihood ratios and adjusted risk are discussed by Baliff and Mooney.

References

Baliff JP and Mooney RA, "New Developments in Prenatal Screening for Down Syndrome," *Am J Clin Pathol*, 2003, 120(Suppl 1):514-24.

Haddow JE, "Antenatal Screening for Down's Syndrome: Where Are We and Where Next?" *Lancet*, 1998, 352(9125):336-7.

Haddow JE and Palomaki GE, "Biochemical Markers of Fetal Disorders in Maternal Serum and Amniotic Fluid," *Medicine of the Fetus and Mother*, 2nd ed, Chapter 40, Reece EA and Hobbins JC, eds, Philadelphia, PA: Lippincott-Raven, 1999, 689-706.

Haddow JE, Palomaki GE, Knight GJ, et al, "Screening of Maternal Serum for Fetal Down's Syndrome in the First Trimester," *N Engl J Med*, 1998, 338(14):955-61.

Kowalczyk TD, Cabaniss ML, and Cusmano L, "Association of Low Unconjugated Estriol in the Second Trimester and Adverse Pregnancy Outcome," *Obstet Gynecol*, 1998, 91(3):396-400.

Macri JN, Kasturi RV, Krantz DA, et al, "Maternal Serum Down Syndrome Screening: Unconjugated Estriol Is Not Useful," *Am J Obstet Gynecol*, 1990, 162(3):672-3.

Malone FD and D'Alton ME, "First-Trimester Sonographic Screening for Down Syndrome," *Obstet Gynecol*, 2003, 102(5 Part 1):1066-79.

"Maternal Serum Quad Screening: Review of Laboratory Testing and Clinical Application," *Mayo Reference Services Communiqué*, 2002, 27(6):1-8.

Shenhav S, Gemer O, Volodarsky M, et al, "Midtrimester Triple Test Levels in Women With Severe Preeclampsia and HELLP Syndrome," *Acta Obstet Gynecol Scand*, 2003, 82(10):912-5.

Wald NJ, Watt HC, and Hackshaw AK, "Integrated Screening for Down's Syndrome on the Basis of Tests Performed During the First and Second Trimesters," *N Engl J Med*, 1999, 341(7):461-7.

Estrogen and Progesterone Receptor Assay

Related Information

Breast Biopsy *on page 305*
Breast Cancer, Hereditary, BRCA1, BRCA2 *on page 307*
Histopathology *on page 733*
Immunoperoxidase Procedures *on page 780*

Applies to Breast Cancer; Cytosol Hormone Receptors; Ovarian Cancer; Tamoxifen

Abstract Estrogen receptor (ER) and progesterone receptor (PR) analysis provides invaluable predictive and prognostic information in patients with breast cancer and with certain gynecologic malignancies.

Specimen Fresh or fresh frozen tissue is required for cytosol assays for either estrogen receptor (ER) or progesterone receptor (PR). Paraffin-embedded tissue or alcohol-fixed cytology specimens are suitable substrates for immunohistochemical procedures.

Use Detection of ER and PR is used primarily to determine the potential responsiveness of breast cancer or some gynecologic tumors to estrogen receptor antagonists. Secondarily, expression of ER or PR has prognostic value in breast cancer.

Limitations Quantitation of cytosol hormone receptors may result in significant false-positive and false-negative results. In these assays, hormone receptor activity is standardized to the amount of protein in the extract. Typical scirrhous infiltrating carcinomas are protein rich but may have an exceedingly low density of receptor-positive tumor cells. **Thus, a hormone receptor-positive tumor may be reported as negative. Conversely, hormone receptor-negative tumors may be contaminated with receptor-positive normal epithelium and be falsely reported as positive. Such artifacts have profound impact on therapeutic decisions.**

Methodology Historically, estrogen and progesterone receptors have been assayed using cytosol extracts in a competitive ligand-binding assay. Fresh or frozen tissue is homogenized and the cytosol fraction is isolated on a sucrose density gradient. A competitive radioimmunoassay is used to accurately quantitate hormone receptor activity. Drawbacks to this approach include relatively large tissue requirements, extreme thermolability caused by short delays in freezing tissues, sensitivity to exposure to fixatives, low tumor cell density, contamination by normal breast tissue, and interference by endogenous hormone levels or exogenous hormone therapy.

Subsequently, enzyme-linked immunoassays supplanted the more fastidious competitive radioimmunoassays. These assays utilize monoclonal antibodies to accurately measure hormone-binding epitopes without relying on functional activity. These assays require less tissue, are not subject to aberrations conferred by exogenous hormone therapy and are not as thermosensitive as their predecessor. Intrinsic to the use of homogenates, these assays also require fresh or frozen tissue and are adversely influenced by normal tissue contamination or low tumor cell density.

More recently, immunohistochemical procedures have gained favor as the assay of choice. Currently available monoclonal antibodies and antigen retrieval techniques allow accurate detection of ER and PR activity on routinely processed histologic and cytologic materials. Immunohistochemical procedures correlate well with ligand-binding assays. Small amounts of neoplastic tissue can be assayed without interference by normal tissue. Results are reported using either qualitative assessment or quantitative measures aided by image cytometry. Immunohistochemistry is considered superior to the traditional ligand-binding assay in prediction of response to endocrine therapy. Low levels of ER expression define a population of patients with potential responsiveness that may be considered ER negative in other assay systems.

Additional Information ER expression is the most important predictive factor for a beneficial response to endocrine therapy.

The frequency of hormone receptor phenotype in pre- and postmenopausal women is depicted in the following table.

	Premenopausal Receptor Status	Postmenopausal Receptor Status
ER$^+$, PR$^+$	45%	63%
ER$^+$, PR$^-$	12%	15%
ER$^-$, PR$^-$	28%	17%
ER$^-$, PR$^+$	15%	5%

Progesterone receptor is one of many estrogen inducible proteins. Hence, PR expression is most commonly seen in parallel with ER expression.

References

Harvey JM, Clark GM, Osborne CK, et al, "Estrogen Receptor Status by Immunohistochemistry Is Superior to the Ligand-Binding Assay for Predicting Response to Adjuvant Endocrine Therapy in Breast Cancer," J Clin Oncol, 1999, 17(5):1474-81.

Osborne CK, Zhao H, and Fuqua S, "Selective Estrogen Receptor Modulators: Structure, Function and Clinical Use," J Clin Oncol, 2000, 18(17):3172-86.

♦ **Estrogen Effect** see Hormonal Evaluation, Cytologic on page 744

♦ **Estrogens** see page 11

Estrogens, Urine

Related Information

Estradiol, Serum on page 553
Estriol, Unconjugated, Pregnancy, Serum, Plasma, or Urine on page 554
Follicle Stimulating Hormone, Serum, Plasma, or Urine on page 609
Hormonal Evaluation, Cytologic on page 744
Luteinizing Hormone, Blood or Urine on page 876

Abstract Urinary total estrogens (estradiol, estrone, estriol) have been used to predict ovulation and evaluate numerous hypo- and hyperestrogenic states. Total urinary estrogen testing largely has been replaced by measurements of serum Estriol, Unconjugated, Pregnancy, Serum, Plasma or Urine on page 554, Estradiol, Serum on page 553, Lutenizing Hormone, Blood or Urine on page 876, and Follicle Stimulation Hormone, Serum, Plasma, or Urine on page 609 measurements.

Use The test is mainly of historic interest.

♦ **Estrone** see Estradiol, Serum on page 553

♦ **1,2-Ethanediol** see Ethylene Glycol, Serum or Plasma on page 561

♦ **Ethanol** see page 11

♦ **Ethanol** see Volatile Screen, Blood or Urine on page 1320

♦ **Ethanol, Blood** see Ethanol, Blood, Urine, and Other Sources on page 558

Ethanol, Blood, Urine, and Other Sources

Related Information

Acetaminophen, Serum *on page 90*
Alanine Aminotransferase, Serum *on page 116*
Alkaline Phosphatase, Serum *on page 127*
Anion Gap, Serum, Plasma, or Urine *on page 160*
Aspartate Aminotransferase, Serum *on page 216*
Cannabinoids (Marijuana Metabolites), Qualitative, Urine *on page 335*
Carbohydrate-Deficient Transferrin, Serum *on page 338*
Cocaine (Cocaine Metabolite), Qualitative, Urine or Hair *on page 427*
Drugs of Abuse Testing, Serum *on page 525*
Ethylene Glycol, Serum or Plasma *on page 561*
Gamma-Glutamyl Transferase, Serum *on page 629*
Gamma Hydroxybutyrate, Serum or Urine *on page 630*
Ketone Bodies, Blood *on page 816*
Lactic Acid, Whole Blood or Plasma *on page 827*
Liver Disease: Laboratory Assessment, Overview *on page 869*
Osmolality, Serum *on page 978*
Volatile Screen, Blood or Urine *on page 1320*

Synonyms Alcohol; Ethanol, Blood; Ethyl Alcohol, Blood; EtOH

Abstract Ethyl alcohol (EtOH) is a central nervous system depressant. This is the most commonly encountered toxic substance in forensic toxicology and is perhaps the most widely used psychoactive drug. It is estimated that it is the number one cause of fatal automobile crashes. Whole blood ethanol values are often required in law enforcement.

Light-to-moderate alcohol consumption may prevent coronary artery disease and stroke.

Patient Preparation Do not use alcohol wipe to clean venipuncture site. Hexachlorophene-based, iodine-based, or mercury-based antiseptics not containing alcohol may be used.

Specimen Whole blood, serum or plasma, urine. Saliva is used in some settings. Vitreous humor is utilized in forensic necropsy cases. Expired air is used in breath ethanol testing and is legally acceptable.

Container Red top tube for clinical testing, gray top (sodium fluoride) tube recommended for medicolegal specimens and prolonged storage; plastic urine container

Storage Instructions Refrigerate in a tightly stoppered tube.

Special Instructions Concentrations of ethanol are 12% to 18% higher in serum and plasma than in whole blood. For forensic purposes, only whole blood values are legally acceptable.

Reference Interval

Blood: negative. While mg/dL are usually used for medical purposes, percentages (g/dL) are usually used for legal needs [eg, driving under the influence (DUI)] statutes. In most laboratories, values <5-10 mg/dL (SI: <1-2 mmol/L) are considered negative. Endogenous alcohol production in the gastrointestinal tract might reach 0.005%, while cutoffs between 0.01% and 0.02% are commonplace. Signs of impairment can be observed at levels of 30-80 mg/dL (0.03% to 0.08%).

Urine: <5-10 mg/dL is considered negative. It is generally accepted that urine alcohol measurements cannot be used to determine impairment in the U.S. However, a challenge to this claim has been made. Urine ethanol levels are legally acceptable in some parts of Europe.

Critical Values Fatal blood concentration is usually considered to be >400 mg/dL (SI: >86.8 mmol/L). Lethal blood levels vary greatly and may be substantially lower when ingested with hypnotics or tranquilizers. Whole blood levels of 300 mg/dL (SI: 65.1 mmol/L) are associated with coma and can be associated with fatalities. In the U.S., levels ≥80-100 mg/dL (0.08% to 0.1%) are considered evidence of impairment for driving. The concentration which defines intoxication is lower in some other countries. Each state statute provides a specific concentration.

Individuals should not be released without accompaniment with EtOH concentrations reaching ≥0.02% (20 mg/dL) (4.34 mmol/L), by virtue of possible impairment and physical injury. Ethanol is metabolized at 10-25 mg/dL/hour by zero order kinetics.

Possible Panic Range ≥300 mg/dL (SI: ≥65.1 mmol/L)

Use Quantitation of ethanol level is important both for medical and for legal purposes. It is used to diagnose alcohol intoxication and to screen for alcoholism. Persons in rehabilitation programs may be subject to serious penalties.

EtOH concentrations are used to monitor intravenous ethanol treatment for methanol and ethylene glycol intoxication. Ethanol must be tested as a possible cause of coma of unknown etiology, since alcohol intoxication may mimic diabetic coma, cerebral trauma, and drug overdose.

Limitations Certain other alcohols (in high concentration) can interfere with enzymatic methods. Gas chromatography is the most specific methodology because it can separate, identify, and quantitate each type of alcohol present.

Methodology Enzymatic analysis (alcohol dehydrogenase) and gas chromatography (GC). Freezing point osmometry is sometimes used for qualitative estimation. GC is the method of choice for legal testing. Infrared spectrophotometry and electrochemical oxidation are used in breath analyzers. Laposata has recently published a useful tabular presentation of markers and methods. See also Soderberg et al.

Additional Information Ethanol is absorbed rapidly from the GI tract. Peak blood levels usually occur within 20-40 minutes on an empty stomach. Once peak blood ethanol levels are reached, disappearance is linear by zero-kinetics; a 70 kg man metabolizes 7-10 g of ethanol/hour (15 ±5 mg/dL/hour). The plasma:whole blood ratio varies from 1.10-1.35 with an average of about 1.20. Symptoms of intoxication in the presence of low alcohol levels could indicate a serious acute medical problem requiring immediate attention. Ethanol ingestions are discussed in Osmolality, Serum *on page 978*; Anion Gap, Serum, Plasma, or Urine *on page 160*; and Ketone Bodies, Blood *on page 816*. Breath alcohol analyzers are used by law enforcement personnel; their results are accepted as legal evidence of intoxication. They must not be used less than 15 minutes after the last ethanol ingestion.

Because saliva is easy and collection is noninvasive, increased interest in saliva ethanol testing has developed. The average saliva:blood ratio is 1:1.

Electrolyte and acid-base problems found with ethanol abuse include hypophosphatemia, hypomagnesemia, hypocalcemia, hypokalemia, hypoglycemia, metabolic acidosis, diminished urinary excretion of uric acid, secondary hyperuricemia, and compensatory respiratory alkalosis. See Liver Disease: Laboratory Assessment, Overview *on page 869*. The ethanol-acetaminophen syndrome is addressed in Acetaminophen, Serum *on page 90*.

The following table lists various agents used for the treatment of alcohol withdrawal.

Therapy for Alcohol Withdrawal

Class	Examples	Effects
Benzodiazepines (preferably long-acting)	Chlordiazepoxide, diazepam, oxazepam, lorazepam	Decreased severity of withdrawal symptoms; reduced risk of seizures and delirium tremens
Anticonvulsants	Carbamazepine	Decreased severity of withdrawal symptoms
Adjunctive agents		
Beta-blockers	Atenolol, propranolol	Improvement in vital signs; reduction in craving
Alpha-agonists	Clonidine	Decreased severity of withdrawal symptoms

Adapted from Kosten TR and O'Connor PG, "Current Concepts: Management of Drug and Alcohol Withdrawal," *N Engl J Med*, 2003, 348(18):1786-95.

Acetaminophen, EtOH, and fasting cause synergistic effects. Only 2.5-4 g/day of acetaminophen, at therapeutic dosage, can cause liver injury in alcoholics. Ethanol exerts synergistic effects with other CNS depressants, such as barbiturates and benzodiazepines.

(Continued)

Ethanol, Blood, Urine, and Other Sources (Continued)

References

Jacobs DS, DeMott WR, Oxley DK, et al, *Laboratory Test Handbook*, 5th ed, Hudson, OH: Lexi-Comp Inc, 2001.

Kosten TR and O'Connor PG, "Current Concepts: Management of Drug and Alcohol Withdrawal," *N Engl J Med*, 2003, 348(18):1786-95.

Laposata M, "Assessment of Ethanol Intake. Current Tests and New Assays on the Horizon," *Am J Clin Pathol*, 1999, 112(4):443-50.

Manno JE, "Reporting Blood Ethanol Levels," *Lab Med*, 2000, 31(8):429-30.

Schneekloth TD, Morse RM, Herrick LM, et al, "Point Prevalence of Alcoholism in Hospitalized Patients: Continuing Challenges of Detection, Assessment, and Diagnosis," *Mayo Clin Proc*, 2001, 76:460-6.

Soderberg BL, Salem RO, Best CA, et al, "Fatty Acid Ethyl Esters. Ethanol Metabolites That Reflect Ethanol Intake," *Am J Clin Pathol*, 2003, 119(Suppl):S94-S99.

Ethchlorvynol, Serum or Plasma

Synonyms Placidyl®

Abstract Ethchlorvynol is a sedative hypnotic. With the advent of better and safer sedative hypnotics, its use is limited. In overdose, it may produce deep coma, severe respiratory depression, hypotension, and bradycardia.

Specimen Serum or plasma

Container Red top tube or green top (heparin) tube

Sampling Time After 2-4 days at steady state.

Reference Interval 2-8 µg/mL (SI: 14-55 µmol/L)

Possible Panic Range Toxic: levels >20 µg/mL (SI: >138 µmol/L)

Use Monitor therapeutic drug level and toxicity assessment

Methodology High performance liquid chromatography (HPLC), gas chromatography (GC). Spot tests are available for overdose situations.

Additional Information

- Half-life: 10-20 hours
- Volume of distribution: 4 L/kg
- Protein binding: 60%

Ethchlorvynol is a nonbarbiturate sedative-hypnotic drug. The half-life increases to 100 hours when high levels of drug are present. Most of the drug is metabolized in the liver. Exaggerated hypnotic effects occur if taken with ethanol. Toxic effects include severe sedation, coma, respiratory depression, hypotension, bradycardia, and hypothermia. Toxic effects are increased by concomitant use of alcohol, CNS depressants, MAO inhibitors, and tricyclic antidepressants.

References

Baselt RC, *Disposition of Toxic Drugs and Chemicals in Man*, 6th ed, Foster City, CA: Biomedical Publications, 2002, 394-6.

Porter WH, "Clinical Toxicology," *Tietz Textbook of Clinical Chemistry*, 3rd ed, Chapter 27, Burtis CA and Ashwood ER, eds, Philadelphia, PA: WB Saunders Co, 1999, 906-81.

Yell RP, "Ethchlorvynol Overdose," *Am J Emerg Med*, 1990, 8(3):246-50.

Ethosuximide, Serum or Plasma

Related Information

Phenobarbital, Serum or Plasma *on page 1021*
Phenytoin, Serum or Plasma *on page 1026*

Synonyms Suxinutin®; Zarontin®

Abstract Ethosuximide is the drug of choice for uncomplicated absence (petit mal) seizures.

Specimen Serum or plasma

Container Red top tube or green top (heparin) tube

Sampling Time At steady state (5-14 days)

Reference Interval 40-100 µg/mL (SI: 284-710 µmol/L)

Possible Panic Range >150 µg/mL (SI: >1062 µmol/L); toxicity may manifest with lethargy or psychotic behavior; significant drug interactions are uncommon. At steady-state, each 1 mg/kg will result in a serum rise of 2 µg/mL. See Table A in the Therapeutic Drug Monitoring Introduction *on page 60*.

Use Monitor for compliance, efficacy, or possible toxicity

Methodology Immunoassay, gas chromatography (GC), high performance liquid chromatography (HPLC)

Additional Information
- Half-life: 25-70 hours
- Volume of distribution: 0.7 L/kg
- Protein binding: 0% to 5%

Acute overdose can cause CNS depression, ataxia, stupor, coma, and hypotension. Chronic overdose can cause skin rash, confusion, ataxia, proteinuria, hepatic dysfunction, and hematuria. Such effects do not regularly correlate with high drug concentrations.

See table in Antiepileptic Drugs Overview *on page 176*.

References
Brodie MJ and Dichter MA, "Antiepileptic Drugs," *N Engl J Med*, 1996, 334(3):168-75.

Capovilla G, Beccaria F, Veggioti P, et al, "Ethosuximide Is Effective in the Treatment of Epileptic Negative Myoclonus in Childhood Partial Epilepsy," *J Child Neurol*, 1999, 14(6):395-400.

Wallace SJ, "Myoclonus and Epilepsy in Childhood: A Review of Treatment With Valproate, Ethosuximide, Lamotrigine, and Zonisamide," *Epilepsy Res*, 1998, 29(2):147-54.

Ethotoin, Serum
Related Information
Mephenytoin, Serum *on page 896*
Phenytoin, Serum or Plasma *on page 1026*

Synonyms Ethylphenylhydantoin; Peganone®

Test Includes Ethotoin. Antiepileptic activity is due to parent compound.

Abstract Ethotoin is occasionally used as an adjunctive anticonvulsant. Patients suffer dose-related gingival hyperplasia and hirsutism less often than with phenytoin.

Specimen Serum

Container Red top tube

Reference Interval 14-34 µg/mL

Contraindications Abnormalities of liver, hematologic disorders

Additional Information Like phenytoin, ethotoin has zero-order kinetics and small dosage changes may produce large changes in clinical response. The half-life of ethotoin is approximately 5-10 hours (dose dependent).

Test interactions include an increase in alkaline phosphatase and a decrease in calcium.

References
Kupferberg HJ, "Other Hydantoins: Mephenytoin and Ethotoin," *Antiepileptic Drugs*, 3rd ed, Levy RH, Dreifuss FE, Mattson RH, et al, eds, New York, NY: Raven Press, 1989, 257-65.

Leikin JB and Paloucek FP, *Poisoning and Toxicology Compendium*, 3rd ed, Hudson, OH: Lexi-Comp Inc, 2002.

♦ **Ethyl Alcohol, Blood** *see* Ethanol, Blood, Urine, and Other Sources *on page 558*

♦ **Ethyl and Methyl Thiocyanate (Thanite® and Lethane®)** *see* Thiocyanate, Serum, Plasma, or Urine *on page 1245*

♦ **Ethylenediamine** *see* Theophylline, Serum *on page 1243*

Ethylene Glycol, Serum or Plasma
Related Information
Anion Gap, Serum, Plasma, or Urine *on page 160*
Osmolality, Calculated, Serum or Plasma *on page 976*
Osmolality, Serum *on page 978*
Oxalate, Urine *on page 989*
Urinalysis *on page 1289*

Synonyms 1,2-Ethanediol

Applies to Antifreeze

Abstract Ethylene glycol, a colorless, odorless, and sweet-tasting compound is used as a radiator antifreeze. Ethylene glycol by itself is only mildly toxic. Its major toxicity is due to its metabolites, glycolate and oxalate. Ethylene glycol poisoning should be suspected in an intoxicated patient with increased anion gap acidosis, hypocalcemia, and urinary crystals.

Specimen Serum or plasma

Container Red top tube or green top (heparin) tube

Reference Interval Negative

Critical Values Toxic: ≥50 mg/dL

(Continued)

Ethylene Glycol, Serum or Plasma *(Continued)*

Possible Panic Range Values between 300-400 mg/L have been observed in fatal cases.

Use Detect and quantitate ingestion of ethylene glycol

Methodology Gas chromatography (GC), enzymatic assay

Additional Information

- Half-life: 3-6 hours; 16-18 hours with ethanol therapy
- Volume of distribution: 0.8 L/kg

Children and domestic animals are attracted to ethylene glycol because of its sweetness. 1.5 mL/kg is lethal but rapid treatment may prevent serious organ damage. Toxicity is manifested by CNS depression (1-12 hours after ingestion), cardiopulmonary symptoms (12-24 hours after ingestion), and renal damage (24-72 hours after ingestion), but toxicity may be delayed by the concurrent ingestion of ethanol. Delayed sequellae may occur as late as 10 days after ingestion. In addition to elevated serum ethylene glycol, hypocalcemia, severe high anion gap metabolic acidosis (low chloride, low bicarbonate, low pH), and osmolal gap elevation are observed.

Levels at 50 mg/dL with metabolic acidosis or with renal failure indicate a need for hemodialysis, and therapeutic ethanol is recommended at concentrations >20 mg/dL. Fomepizole is a new antidote for ethylene glycol poisoning. Compared with traditional ethanol treatment, advantages of fomepizole include lack of depression of the central nervous system and hypoglycemia, and easier maintenance of effective plasma levels.

References

Brent J, McMartin K, Phillips S, et al, "Fomepizole for the Treatment of Ethylene Glycol Poisoning," *N Engl J Med*, 1999, 340(11):832-8.

Church AS and Witting MD, "Laboratory Testing in Ethanol, Methanol, Ethylene Glycol, and Isopropanol Toxicities," *J Emerg Med*, 1997, 15(5):687-92.

de Chazal I, Houghton B, and Frock J, "The Sweet Killer. Can You Recognize the Symptoms of Ethylene Glycol Poisoning?" *Postgrad Med*, 1999, 106(4):221-4, 227, 230.

Scalley RD, Ferguson DR, Piccaro JC, et al, "Treatment of Ethylene Glycol Poisoning," *Am Fam Phys*, 2002, 66(5):807-12.

- ◆ **Ethylphenylhydantoin** *see* Ethotoin, Serum *on page 561*
- ◆ **Etiocholanolone** *see* 17-Ketosteroids Fractionation, Urine *on page 818*
- ◆ **EtOH** *see* Ethanol, Blood, Urine, and Other Sources *on page 558*
- ◆ **Etrafon®** *see* Amitriptyline, Serum or Plasma *on page 149*
- ◆ **Etrafon®** *see* Antidepressants, Cyclic, Serum or Plasma *on page 171*
- ◆ **Etrafon®** *see* Phenothiazines, Serum *on page 1021*
- ◆ **Euthyroid Sick Syndrome** *see* Reverse T$_3$, Serum *on page 1160*
- ◆ **Excedrin®** *see* Ibuprofen, Serum *on page 764*
- ◆ **Exchange Transfusion** *see* Hemolytic Disease of the Newborn, Antibody Identification *on page 690*
- ◆ **Exchange Transfusion** *see* Newborn Crossmatch and Transfusion *on page 952*
- ◆ **Exercise** *see page 11*
- ◆ **Expanded Newborn Screening** *see* Newborn Screening by Tandem Mass Spectrometry (MS/MS) *on page 957*
- ◆ **Exsanguinating Emergency** *see* Uncrossmatched Blood, Emergency *on page 1281*
- ◆ **Extractable Nuclear Antigens** *see* Antinuclear Antibodies *on page 189*
- ◆ **Extractable Nuclear Antigens** *see* Sjögren Antibodies *on page 1199*
- ◆ **Extractable Nuclear Antigens (ENA)** *see* Smith (Sm) and Ribonucleoprotein (RNP) Antibodies *on page 1206*
- ◆ **Extrinsic Allergic Alveolitis Serology** *see* Hypersensitivity Pneumonitis Serology *on page 761*
- ◆ **Extrinsic Pathway** *see* Activated Partial Thromboplastin Time *on page 100*
- ◆ **Extrinsic Pathway** *see* Prothrombin Time *on page 1116*
- ◆ **Eye Smear for Cytology** *see* Ocular Cytology *on page 973*

Factor V Leiden

Related Information

Activated Protein C Resistance and the Factor V Leiden Mutation *on page 104*

Coagulation Factor Assays *on page 418*
Homocyst(e)ine, Plasma *on page 741*

Abstract See Activated Protein C Resistance and the Factor V Leiden Mutation *on page 104*. The term "factor V Leiden" is often used interchangeably with activated protein C resistance, although they are not true synonyms. The factor V Leiden mutation causes the vast majority of cases of activated protein C resistance. Whether or not additional mutations other than factor V Leiden can cause clinically-significant activated protein C resistance is under continued investigation.

Factor VIII Concentrate

Related Information

Activated Partial Thromboplastin Time *on page 100*
Coagulation Factor Assays *on page 418*
Cryoprecipitate *on page 481*
Factor Inhibitors *on page 566*
Factor IX Concentrate *on page 564*
Parvovirus B19 DNA *on page 1006*
Parvovirus B19 Serology *on page 1006*
Plasma, Fresh Frozen *on page 1039*
von Willebrand Factor *on page 1321*

Synonyms AHF, Lyophilized; Antihemophilic Factor (Human)

Abstract Factor VIII concentrates are typically prepared as derivatives from pools of plasma, which undergo further processing to inactivate contaminating viruses. Virus-inactivating procedures include the use of heat, chemical solvents and detergents, and affinity column purification. Concentrates may be produced using recombinant DNA technology. These factor VIII concentrates, which have been prepared to decrease or remove the risk of viral transmission, should be utilized in place of cryoprecipitate in patients who require factor VIII replacement (eg, patients with hemophilia A).

Patient Preparation Perform factor VIII assays to calculate dosage or, in emergencies, use the activated partial thromboplastin time (aPTT) as a therapeutic guide. See Coagulation Factor Assays *on page 418*.

Reference Interval The half-life for factor VIII concentrates is ~12 hours in the absence of inhibitors or active bleeding.

Use Treatment of acute bleeding and prevention of bleeding in patients with deficiency of clotting factor VIII (hemophilia A) and with low-titer factor VIII inhibitors. Some factor VIII concentrates (eg, Alphanate®, Humate-P®, Koāte DVI®) can be used to treat patients with von Willebrand disease because they contain von Willebrand factor.

Limitations The presence of inhibitors to factor VIII makes treatment more difficult. The most common target of monospecific-acquired anticoagulant antibodies is factor VIII. Conditions associated with such antibodies include systemic lupus erythematosus, rheumatoid arthritis, psoriasis, pemphigus vulgaris, gestation, lymphoproliferative diseases, plasma cell disorders, and use of penicillin, sulfas, chloramphenicol, and phenytoin. Idiopathic causes are recognized as well. Porcine factor VIII is available for patients with factor VIII inhibitors. In bleeding patients with inhibitors, very large doses of human factor VIII or factor IX complex (prothrombin complex) have been administered with varying degrees of success.

Contraindications Normal coagulation studies or bleeding unrelated to factor VIII deficiency. Not all factor VIII concentrates are suitable for treatment of von Willebrand disease.

Additional Information Drohan and Clark have recently tabulated the antihemophilic factor concentrates marketed in the U.S., with information relevant to purification methods and virus inactivation/removal methods. The activated partial thromboplastin time is useful for both hemophilia and von Willebrand disease and may be more readily available than are factor assays. All these tests can guide therapy, as can the clinical response of the patient.

Calculation of dosage: Each bottle is labeled with the quantity of factor VIII coagulant activity it contains in terms of International Units (IU). 1 mL of normal plasma contains 1 IU of factor VIII coagulant activity. The initial dose (30% to 100%) varies depending upon the clinical circumstance, in general, with deeper (Continued)

Factor VIII Concentrate (Continued)

hemorrhage and hemarthrosis requiring higher levels of activity. The following formula may be used:

plasma volume (PV, mL) = 40 mL/kg x body weight (kg)

desired units of factor VIII = PV x [desired level (%) - initial level (%)] divided by 100

Alternately, 1 factor VIII unit/kg body weight may raise the factor VIII level by 2%.

Administration: Factor VIII concentrates are prepared lyophilized and must be aseptically reconstituted using the manufacturer's diluent. Filter prior to administration. Administer as quickly as possible after reconstitution.

Hazards: Currently available factor VIII concentrates (excluding cryoprecipitate) utilize manufacturing procedures to reduce the risk of viral transmission and have not been associated with HIV transmission. All viruses, however, particularly the nonlipid enveloped viruses (eg, hepatitis A and parvovirus B19), may not be inactivated by these treatments. Due to the presence of anti-A or anti-B, development of a positive direct antiglobulin test with resulting hemolysis is possible in human plasma-derived preparations.

References

DiMichele D and Neufeld EJ, "Hemophilia. A New Approach to an Old Disease," *Hematol Oncol Clin North Am*, 1998, 12(6):1315-44.

Drohan WN and Clark DB, "Preparation of Plasma-Derived and Recombinant Human Plasma Proteins," *Hematology: Basic Principles and Practice*, 3rd ed, Chapter 139, Hoffman R, Benz EJ Jr, Shattil SJ, et al, eds, New York, NY: Churchill Livingstone, 2000, 2273-82.

Grosset ABM and Rodgers GM, "Acquired Coagulation Disorders," *Wintrobe's Clinical Hematology*, 10th ed, Volume 2, Chapter 69, Lee GR, Forester J, Lukens J, et al, eds, Baltimore, MD: Lippincott Williams & Wilkins, 1999, 1233-80.

Triulzi DJ, *Blood Transfusion Therapy: A Physician's Handbook*, 7th ed, Bethesda, MD: American Association of Blood Banks Press, 2002, 39-41.

Factor IX Concentrate

Related Information

Activated Partial Thromboplastin Time *on page 100*
Coagulation Factor Assays *on page 418*
Factor Inhibitors *on page 566*
Plasma, Fresh Frozen *on page 1039*

Synonyms Prothrombin Complex Concentrates

Abstract Factor IX is also called hemophilia B factor, Christmas factor, and plasma thromboplastin component.

Factor IX concentrates for the treatment of hemophilia B are available using recombinant technology or monoclonal antibody purification, thus, reducing the risk of viral transmission. Recombinant factor IX, available since 1999, contains no animal or human proteins. Factor IX complex contains factors II, VII, IX, and X.

Second generation factor IX concentrates are essentially free of the other vitamin K-dependent coagulation factors, and are designated coagulation factor IX (human).

Patient Preparation Perform factor IX assay to calculate dosage before administration. See Coagulation Factor Assays *on page 418*.

Reference Interval The half-life for factor IX concentrates is ~24 hours in the absence of inhibitors or active bleeding.

Use Treatment of acute bleeding and prevention of bleeding in patients with deficiency of clotting factor IX (hemophilia B).

Limitations Patients with factor inhibitors may be treated with factor IX complex concentrate. These should be used with caution in patients with liver disease. Disseminated intravascular coagulation and thrombosis are among the risks. Factor IX concentrates appear to be less thrombogenic than factor IX complex. Rapid infusion of factor IX complex may induce side effects, including fever, chills, headaches, nausea, and flushing. Recombinant concentrates are the purest available, but most expensive.

Contraindications Do not use in liver disease. Do not use in vitamin K deficiency, for which vitamin K preparations are appropriate, or in patients with overdose of coumarin. See Plasma, Fresh Frozen *on page 1039*.

Calculation of dosage: Each bottle of factor IX is labeled in terms of activity units, with one unit equivalent to that found in 1 mL of normal human plasma. The dose required depends upon the type and severity of bleeding. A formula used for calculating factor VIII dose can also be used to calculate factor IX dose (see Factor VIII Concentrate *on page 563*). However, double the number of units to be given since half of the infused factor IX disappears immediately after the infusion for unknown reasons.

Administration: Factor IX concentrates are prepared lyophilized and must be aseptically reconstituted using the manufacturer's diluent. Filter prior to administration. Administer as quickly as possible after reconstitution.

Additional Information Drohan and Clark have recently provided tabulation of factor IX concentrates and related fractions marketed in the U.S.

References

Bolton-Maggs PH and Pasi KJ, "Haemophilias A and B," *Lancet*, 2003, 361(9371):1801-9.

Drohan WN and Clark DB, "Preparation of Plasma-Derived and Recombinant Human Plasma Proteins," *Hematology: Basic Principles and Practice*, 3rd ed, Chapter 139, Hoffman R, Benz EJ Jr, Shattil SJ, et al, eds, New York, NY: Churchill Livingstone, 2000, 2273-82.

Greenberg CS and Orthner CL, "Blood Coagulation and Fibrinolysis," *Wintrobe's Clinical Hematology*, 10th ed, Volume 1, Chapter 24, Lee GR, Forester J, Lukens J, et al, eds, Baltimore, MD: Lippincott Williams & Wilkins, 1999, 684-764.

Linden JV, Kolakoski MH, Lima JE, et al, "Factor Concentrate Usage in Persons With Hemophilia in New York State," *Transfusion*, 2003, 43(4):470-5.

Factor XIII

Related Information

Coagulation Factor Assays *on page 418*
Cryoprecipitate *on page 481*
Plasma, Fresh Frozen *on page 1039*

Synonyms Fibrin Stabilizing Factor; Fibrinoligase; Laki-Lorand Factor

Abstract Activated factor XIII stabilizes fibrin clots by cross-linking fibrin strands. Factor XIII deficiency can cause a hereditary bleeding disorder with features including delayed bleeding, umbilical stump bleeding, and miscarriages.

Specimen Plasma

Container One blue top (sodium citrate) tube

Collection Routine venipuncture. If multiple tests are being drawn, draw blue top tubes after any red top tubes but before any lavender top (EDTA), green top (heparin), or gray top (oxalate/fluoride) tubes. Immediately invert tube gently at least 4 times to mix. Tubes must be appropriately filled. Deliver tubes immediately to the laboratory.

Storage Instructions Separate plasma from cells as soon as possible. Plasma may be stored on ice for up to 4 hours; otherwise, store frozen.

Causes for Rejection Specimen received more than 4 hours after collection, tubes not filled, clotted specimens

Turnaround Time Screening assay: 24 hours after incubation begins; quantitative assay: several days (typically it is a send-out test)

Reference Interval Screening assay: Clot stable in 5 M urea for at least 24 hours. If factor XIII deficiency is present, clot will usually dissolve in 1-2 hours. Quantitative assay: 70% to 140% of normal (some newborns have lower levels than adults).

Use Consider this test in patients with evidence for a familial bleeding disorder and a normal PT, PTT, and von Willebrand panel (because factor XIII deficiencies do not prolong the PT or PTT, and von Willebrand disease is a much more common disorder than factor XIII deficiency).

Factor XIII deficiency is rare. It causes delayed bleeding because although fibrin clots can form initially, they are weak and subsequently lyse. Factor XIII consists of two catalytic A subunits and two noncatalytic B subunits. Most mutations causing factor XIII deficiency have so far been found in the A subunit. The PT and PTT are normal in factor XIII deficiency because factor XIII stabilizes the clot after a fibrin clot has formed, whereas the PT and PTT measure the clotting time through initial fibrin formation. Inheritance is autosomal. Formerly, it was believed that heterozygotes are asymptomatic, but more recent evidence suggests they can have bleeding symptoms. Symptoms include poor wound healing, umbilical stump bleeding, miscarriage, prolonged bleeding from superficial wounds, and intracranial hemorrhage, in addition to a number of other bleeding symptoms.

(Continued)

Factor XIII (Continued)

Limitations Screening assay: will not detect heterozygotes; quantitative assay: expensive, not readily available, and high ammonia levels may falsely decrease the result

Methodology A qualitative factor XIII assay evaluates clot stability in 5 M urea. The patient sample is clotted by adding calcium, and then after 30 minutes at 37°C the clot is placed in 5 M urea for 24 hours at room temperature. Clots formed by normal individuals remain stable in 5 M urea, while clots from factor XIII deficient patients dissolve in urea. This assay detects only the most severely affected homozygous patients with 1% to 2% factor XIII activity or less. A quantitative assay can detect heterozygous deficiencies (with values of ~50%), but this test is not yet readily available in most U.S. laboratories. In the quantitative assay, factor XIII is activated by thrombin. Activated factor XIII then attaches glycine ethyl ester to a specific peptide substrate, releasing ammonia. The released ammonia generates a subsequent reaction that is detected by a photometer.

Additional Information When fibrin initially forms, fibrin monomers are held together by weak noncovalent hydrogen bonds. Factor XIII is a transglutaminase that stabilizes fibrin clot by cross-linking fibrin monomers with covalent bonds. Calcium is required for its activation by thrombin as well as its activity. It also cross-links antiplasmin to fibrin, which protects the clot from fibrinolysis by plasmin.

A factor XIII polymorphism (Val34Leu), present in nearly half of the population, is suspected to protect against deep venous thrombosis and is somewhat more frequent in patients with intracranial hemorrhage. The thrombotic and cardioprotective effect of factor XIII Leu 34 is controversial. (See Fibrinogen *on page 583*.) Factor XIII has a long half-life of 10-12 days. Therefore, treatment of factor XIII deficiency can be successful with infrequent doses of cryoprecipitate, fresh frozen plasma, or if available, factor XIII concentrates. Acquired decreases in factor XIII can arise in liver disease (decreased hepatic synthesis), disseminated intravascular coagulation (DIC), certain inflammatory diseases (Crohn, ulcerative colitis, Henoch-Schönlein purpura), leukemia, myelodysplasia, and myeloproliferative disorders. Over 25 cases of inhibitors (antibodies) against factor XIII have been reported.

References

Catto AJ, Kohler HP, Coore J, et al, "Association of a Common Polymorphism in the Factor XIII Gene With Venous Thrombosis," *Blood*, 1999; 93(3):906-8.

Lim BC, Ariëns RA, Carter AM, et al, "Genetic Regulation of Fibrin Structure and Function: Complex Gene-Environment Interactions May Modulate Vascular Risk," *Lancet*, 2003, 361(9367):1424-31.

Schroeder V and Kohler HP, "Factor XIII Activation by Thrombin Depends on FXIII Val34Leu Genotype," *Blood*, 2003, 101(1):371.

Seitz R, Duckert F, Lopaciuk S, et al, "ETRO Working Party on Factor XIII Questionnaire on Congenital Factor XIII Deficiency in Europe: Status and Perspectives," *Semin Thromb Hemost*, 1996, 22(5):415-8.

♦ **Factor Assays** *see* Mixing Studies (Coagulation) *on page 918*

♦ **Factor I** *see* Fibrinogen *on page 583*

Factor Inhibitors

Related Information

Anticardiolipin Antibody *on page 167*
Antinuclear Antibodies *on page 189*
Antiphospholipid Antibody (Lupus Anticoagulant and/or Anticardiolipin Antibody) *on page 193*
Coagulation Factor Assays *on page 418*
Cryoprecipitate *on page 481*
Factor IX Concentrate *on page 564*
Factor VIII Concentrate *on page 563*
Mixing Studies (Coagulation) *on page 918*

Synonyms Bethesda Assay; Circulating Anticoagulant; Modified Bethesda Assay

Abstract Specific factor inhibitors are antibodies that inhibit the activity of a specific coagulation factor. A severe acquired bleeding disorder may develop.

Specimen Plasma

Container Three blue top (sodium citrate) tubes

Collection Routine venipuncture. If multiple tests are being drawn, draw blue top tubes after any red top tubes but before any lavender top (EDTA), green top (heparin), or gray top (oxalate/fluoride) tubes. Immediately invert tube gently at least 4 times to mix. Tubes must be appropriately filled. Deliver tubes immediately to the laboratory. Avoid heparin contamination of specimens during specimen collection.

Storage Instructions Separate plasma from cells as soon as possible. Plasma may be stored on ice for up to 4 hours; otherwise, store frozen.

Causes for Rejection Specimen received more than 4 hours after collection, tubes not filled, clotted specimens

Turnaround Time 1 full day or longer (usually performed in a specialized laboratory)

Special Instructions Specific factor inhibitors, such as factor VIII inhibitors, can cause a severe bleeding disorder and treatment can be difficult. Therefore, the specimen should be sent to a laboratory that can perform the assay promptly.

Reference Interval The inhibitor is quantitated in Bethesda units (BU). Each Bethesda unit of inhibitor decreases the factor concentration in the assay by 50%. For example, one unit of factor VIII inhibitor decreases factor VIII from 100% (normal) to 50%, two units decrease it to 25%, three units decrease it to 12.5%, and so on.

Use Performed when mixing studies and factor assays suggest the presence of a specific factor inhibitor, and the findings are not due to a lupus anticoagulant, heparin, or other anticoagulants (see Mixing Studies (Coagulation) *on page 918*). For example, if a mixing study shows the characteristic pattern for a factor VIII inhibitor, and factor VIII is the only markedly decreased factor (usually <1% to 10%; mean normal value is 100%), a Bethesda assay should be performed to identify and titer the factor VIII inhibitor.

Methodology Bethesda assay for factor VIII inhibitor: Serial patient plasma dilutions in citrated saline are prepared, from 1:1 up to 1:160 (or higher if necessary for high-titer factor inhibitors). The purpose of these dilutions is to dilute out the inhibitor. The patient plasma dilutions are then mixed with an equal volume of normal plasma containing a normal amount of coagulation factors. The mixed dilutions are usually incubated for up to 2 hours, because certain inhibitors show an inhibitory effect only after prolonged incubation (particularly factor V and factor VIII inhibitors). Factor VIII assays are then performed on each mixed dilution. The dilution that inhibits 50% of factor VIII in the assay defines the titer of the inhibitor. For example, if the 1:40 dilution inhibits 50% of the factor VIII in the assay, the patient is reported to have a titer of 40 BU of factor VIII inhibitor.

Porcine factor VIII can be substituted for normal plasma (which contains human factor VIII) in the Bethesda assay to determine if the factor VIII inhibitor cross-reacts with porcine factor VIII. If there is little or no cross-reactivity, porcine factor VIII is often used to treat bleeding due to a factor VIII inhibitor.

The Bethesda assay can be modified to identify and titer other specific factor inhibitors. For example, if a factor V inhibitor is suspected, factor V assays are performed on the mixed dilutions instead of factor VIII assays.

Additional Information Antibodies that inhibit the activity of a specific coagulation factor can develop spontaneously or in association with certain medications, autoimmune diseases, or other conditions. These antibodies may also arise when a patient with a hereditary factor deficiency is transfused with a product containing the factor, such as a factor concentrate or fresh frozen plasma. The immune system in the patient with the deficiency views the transfused factor as foreign, and forms an antibody against the transfused factor. This complication makes treatment of bleeding episodes difficult in such patients. The most common clinically significant factor inhibitor is a factor VIII inhibitor. Factor VIII inhibitors develop in approximately 10% to 20% of patients with severe hemophilia A and less commonly with mild or moderate hemophilia A, following the infusion of factor VIII-containing products. Rarely, factor VIII inhibitors can also arise spontaneously in persons without hereditary hemophilia. Factor VIII inhibitors cause decreased factor VIII activity and consequently a prolonged PTT. Factor VIII inhibitors exhibit a characteristic pattern in the PTT mixing study in which the mixed plasma PTT is initially normal (or significantly more normal than the patient plasma's PTT) but becomes (Continued)

Factor Inhibitors *(Continued)*

prolonged (typically by increasing at least 8-10 seconds) over the course of a 1- to 2-hour incubation.

Factor IX inhibitors develop in approximately 2% to 12% of patients with severe hemophilia B, and less commonly with mild or moderate hemophilia B, following transfusion of factor IX-containing products. Very rarely, factor IX inhibitors can also arise spontaneously in persons without hereditary hemophilia B. Factor IX inhibitors cause decreased factor IX activity and consequently a prolonged PTT. The prolonged PTT caused by a factor IX inhibitor is immediately prolonged in the PTT mixing study.

Other factor inhibitors arise occasionally following exposure to "fibrin glue" preparations, which are administered topically and intraoperatively to help achieve hemostasis. Fibrin glue is prepared by adding bovine thrombin to human fibrinogen, in the form of cryoprecipitate. The affected patient's immune system recognizes the bovine thrombin as foreign, and forms an antibody against it. Frequently, traces of bovine factors V, VII, or X are also present and antibodies can be generated against these factors as well. The antibodies to bovine coagulation factors sometimes cross-react against the corresponding human coagulation factor, which can lead to bleeding. In one series, 1.7% of patients exposed to bovine thrombin preparations developed a clinically significant inhibitor with bleeding.

Other specific factor inhibitors have also been observed, but most are exceedingly rare. These include inhibitors to factors I (fibrinogen), II, V, VII, X, XI, XII, XIII, and prekallikrein.

Factor V inhibitors may behave like factor VIII inhibitors in the mixing study, with increasing PTT (or PT) prolongation over a 1- to 2-hour incubation. Other factor inhibitors most likely behave like factor IX inhibitors in mixing studies, with immediate prolongation of the PTT (or PT) in the mixed plasma.

Note: Factor inhibitors can cause artifactual decreases in the *in vitro* factor level of other coagulation factors. Therefore, laboratories should perform factor assays at multiple dilutions. At higher dilutions, the inhibitor interference will decrease due to dilution of the inhibitor. For example, a factor VIII inhibitor sometimes causes false decreases in factor IX, XI, or XII assays. Typically, the false decreases, if any, are mild to moderate, whereas the decrease in the truly inhibited factor is typically severe.

Rarely, nonspecific factor inhibition is found with monoclonal gammopathy (paraproteinemia) which can appear to nonspecifically inhibit clotting reactions in the laboratory without targeting any particular coagulation factor. The PT and PTT may be prolonged, and multiple factor assays are nonspecifically inhibited.

References

Brown BA, *Hematology: Principles and Procedures*, 6th ed, Philadelphia, PA: Lea & Febiger, 1993, 256-8.

Dorion RP, Hamati HF, Landis B, et al, "Risk and Clinical Significance of Developing Antibodies Induced by Topical Thrombin Preparations," *Arch Pathol Lab Med*, 1998, 122(10):887-94.

Shapiro SS and Hultin M, "Acquired Inhibitors to the Blood Coagulation Factors," *Semin Thromb Hemost*, 1975, 336-85.

Fanconi Anemia, Chromosome Breakage Study

Related Information
Alpha$_1$-Fetoprotein, Serum *on page 136*
Bone Marrow *on page 296*
Complete Blood Count *on page 442*
Fluorescence *in situ* Hybridization *on page 602*
Lymphocyte Transformation Test *on page 882*
Platelet Count *on page 1050*

Synonyms Chromosome Breakage Syndrome, Fanconi Anemia

Applies to Bloom Syndrome; TRACE

Test Includes Detection of increased spontaneous and chemically-induced chromosome breakage in the patient's lymphocytes compared to control lymphocytes

Abstract Fanconi anemia (FA) is an autosomal recessive disorder affecting approximately 1 in 360,000 people. In FA there is cytogenetic hypersensitivity to bifunctional alkylating agents, suggestive of a defect in the repair of DNA interstrand cross-links. Clinically, there is bone marrow failure, predisposition to the development of malignancy, and presence of multiple congenital anomalies. With marrow failure there is involvement of all marrow elements resulting in anemia, leukopenia, and thrombocytopenia. FA is associated with increased risk of malignancy (especially acute myeloid leukemia which occurs in ~9% of patients) but cancers may also develop in skin, gastrointestinal, and gynecologic systems. Frequent congenital malformations include radial ray defects, microcephaly, renal malformations, and mental retardation. Some 33% of FA patients lack congenital malformation.

Specimen Whole blood; antenatal diagnosis may utilize chorionic villus cells or fetal blood sample.

Container Green top (sodium heparin) tube. **Note:** Specimens in lavender top (EDTA) tubes, blue top (sodium citrate) tubes, or green top (lithium heparin) tubes are not acceptable.

Storage Instructions Specimen should be delivered to the laboratory immediately.

Causes for Rejection Clotted or hemolyzed specimen, specimen more than 24 hours old, use of improper anticoagulant

Special Instructions Call laboratory so that test can be arranged and scheduled.

Reference Interval An increased frequency of chromosomal damage is observed compared to control specimens for both spontaneous breakage and chemically-induced chromosome breakage following exposure to diepoxybutane (DEB) or mitomycin C. Interpretation is usually provided with report. Presence or absence of monosomy 7 is of especial interest, as it is a frequent cytogenetic finding in the bone marrow of patients with FA and acute myeloid leukemia.

Use Increased levels of chromosome instability provide an important aid to the clinical diagnosis of Fanconi anemia (FA)

Limitations Discrepancies exist between the clinical classification of some patients and the diagnosis suggested by *in vitro* cytogenetic findings. Therefore, the definitive diagnosis should not be based solely on cytogenetic results.

Methodology Peripheral T cells from PHA-stimulated blood cultures of the patient and a control are exposed to a bifunctional DNA cross-linking agent (Continued)

Fanconi Anemia, Chromosome Breakage Study
(Continued)

such as mitomycin C or DEB. In addition, spontaneous breakage is evaluated. Cells are harvested and evaluated for chromosome breakage and rearrangements.

Additional Information Seven complementation groups (different genetic defects with the ability to correct for one another) have been identified in FA on the basis of results from somatic cell fusion studies. The number of FA genes grew from four to eight over the past few years. FA-H (represented by a single patient) has been reassigned to group A (see table).

Complementation Groups of Fanconi Anemia

Group	Estimated % of FA patients	Chromosome Location
A	66	16q24.3
B	4	
C	12	9q22.3
D	4	3p25.3
E	12	
F	Rare	
G	Rare	9p13

Four of the groups (A, C, D, and G) have been mapped to discrete chromosomal loci, respectively, 16q24.3, 9q22.3, 3p22-26 and 9p13. Chromosome 3q aberrations (largely the result of unbalanced 3q translocations to other chromosomes) were shown to be common in one group of FA subjects. Gain of 3q material was associated with a poor prognosis (increased risk of myelodysplastic syndrome and acute myeloid leukemia). Thus, FA is genetically heterogeneous in contrast to ataxia telangiectasia and Bloom syndrome (which are chromosomal breakage syndromes arising from single gene mutations). FA genes A, C, and G have been cloned and encode "orphan" proteins (no similar proteins in the GenBank).

FA cells, in addition to DNA cross-linking sensitivity, also show defects in cell cycle regulation and apoptosis. The cells are hypersensitive to interferon gamma and tumor necrosis factor α likely with pathogenic significance to the development of aplastic anemia in FA patients. Defects (including mutated involved proteins) may result in chromosome instability. With the availability of the FA genes, mutations can be identified at the molecular level providing an adjunctive procedure to the diepoxybutane test.

Carriers (heterozygotes) have an increased risk of cancer, although no excess of any specific cancer type has been noted. The study of chromosome breakage by use of DEB and mitomycin C is usually a reliable technique for identification of homozygotes. However, cytogenetic analysis is not dependable for identification of FA heterozygotes.

The hematologic manifestations in FA (bone marrow dependent) undergo variable progression. Macrocytosis and pancytopenia develop during the first decade of life. Platelet and red cell deficiencies usually antedate white cell abnormalities. Patients demonstrate progressively impaired erythropoiesis with fetal-like characteristics (increased i antigen and hemoglobin F). Serum erythropoietin levels may be increased. There is variable progression to pancytopenia.

Median survival is about 25% (in untreated individuals) with FA. Some 20% of patients with FA develop malignancy. About 10% of patients develop acute myeloblastic leukemia (at an average age of 15 years). Leukemia is the presenting feature in some 25% of these affected FA patients. Cancer usually develops in older FA patients (mean age of 23 years). Squamous cell carcinoma is especially common, occurring in about 5% of patients. The oropharynx and gastrointestinal/genitourinary tracts are most commonly involved.

Elevated serum alpha-fetoprotein (AFP) has been noted in FA patients.

References

Freedman MH, "Inherited Forms of Bone Marrow Failure," 3rd ed, Chapter 18, *Hematology: Basic Principles and Practice*, Hoffman R, Benz EJ Jr, Shattil SJ, et al, eds, Philadelphia, PA: Churchill Livingstone, 2000, 260-7.

Garcia-Higuera I, Kuang Y, and D'Andrea AD, "The Molecular and Cellular Biology of Fanconi Anemia," *Curr Opin Hematol*, 1999, 6(2):83-8.

Joenje H, Levitus M, Waisfisz Q, et al, "Complementation Analysis in Fanconi Anemia: Assignment of the Reference FA-H Patient to Group A," *Am J Hum Genet*, 2000, 67(3):759-62.

Thurston VC, Ceperich TM, Vance GH, et al, "Detection of Monosomy 7 in Bone Marrow by Fluorescence *in situ* Hybridization. A Study of Fanconi Anemia Patients and Review of the Literature," *Cancer Genet Cytogenet*, 1999, 109(2):154-60.

Tönnies H, Huber S, Kühl JS, et al, "Clonal Chromosomal Aberrations in Bone Marrow Cells of Fanconi Anemia Patients: Gains of the Chromosomal Segment 3q26q29 as an Adverse Risk Factor," *Blood*, 2003, 101(10):3872-4.

♦ **Farber Disease** *see* Inherited Diseases of Metabolism and Cell Structure *on page 792*

♦ **Farmer's Lung Disease Serology** *see* Hypersensitivity Pneumonitis Serology *on page 761*

♦ **FAS Receptor** *see* Apoptosis Assays *on page 205*

♦ **Fast Hemoglobins** *see* Glycated Hemoglobin (Hemoglobin A₁c), Blood *on page 655*

♦ **Fasting Bilirubin Test** *see* Bilirubin, Total, Serum *on page 265*

♦ **Fasting Blood Sugar** *see* Glucose, Fasting, Plasma *on page 643*

♦ **Fat, Quantitative, 72-Hour Stool Collection** *see* Fecal Fat, Quantitative, 72-Hour Collection *on page 574*

Fat, Semiquantitative, Stool, Acid Steatocrit

Related Information

d-Xylose Absorption Test, Serum, Urine *on page 527*
Fat, Semiquantitative, Stool, Sudan III Stain *on page 572*
Fecal Fat, Quantitative, 72-Hour Collection *on page 574*
Fecal Pancreatic Elastase 1 *on page 576*
Meat Fibers, Stool *on page 894*
Methylene Blue Stain, Stool *on page 906*
pH, Stool *on page 1034*

Applies to Steatocrit

Abstract The steatocrit and its improved version, the acid steatocrit, are simple, low cost methods for the estimation of fecal fat. Routine glass hematocrit tubes and a standard hematocrit centrifuge (readily available laboratory equipment) are used in performance of these tests. Results are used to assess the presence of steatorrhea and to study malabsorptive/maldigestive processes, including exocrine pancreatic function in cystic fibrosis and chronic pancreatitis.

Patient Preparation To provide for adequate sensitivity in testing stool for fat, test subjects must be maintained on a high fat diet.

Specimen Random spot stool

Container Glass or plastic tube or specimen container

Storage Instructions Store at -20°C if there is a delay in analysis.

Turnaround Time 12-24 hours

Reference Interval

Steatocrit:
- small for gestational age premature infants: 29% ±1%
- appropriate for gestational age premature infants: 17% ±1%
- children 3-12 years of age (mean ±SE) 1.1% ±0.4%

Acid steatocrit:
- premature and many formula-fed term infants to 6 months of age (but clinically well, thus "physiologic steatorrhea"): >60%. **Note:** Human milk-fed infants have steatocrit and acid steatocrit values lower than those of formula-fed babies.
- children: 6 months to 3 years: <10%; 3-12 years (mean ±SEM): 3.8% ±1%
- adults (median with interquartile range): 14.5(8-23)%

Use Confirm the presence of steatorrhea; assist with the diagnostic study of malabsorption and/or maldigestion (as with exocrine pancreatic failure in some cases of cystic fibrosis, chronic pancreatitis, and pancreatectomy); monitor results of therapy (use in clinical follow-up) in patients with chronic diarrhea and in patients receiving exogenous enzyme therapy

(Continued)

Fat, Semiquantitative, Stool, Acid Steatocrit
(Continued)

Limitations There may be concern for safety of laboratory personnel (handling of stool specimens in breakable hematocrit tubes, aerosol production by hematocrit centrifuge). Plastic hematocrit tubes are available. (See Limitations in listing Fat, Semiquantitative, Stool, Sudan III Stain *on page 572*.)

Additional Information Candidates for alternatives to the only clinically proven quantitative measurement for fecal fat (the 72-hour chemical titrimetric method of van de Kamer et al) have recently been developed. The acid steatocrit appears to provide a practical, low-cost alternative that requires no special equipment. A **spot stool acid steatocrit** has been considered to estimate the quantitative fecal fat (g/24 hours) by the equation:

fecal fat = -0.43 + [0.45 x (acid steatocrit %)]

The upper limit of normal fecal fat excretion in adults is 7 g/day (on a diet of 100 g/fat daily).

The acid steatocrit (performed on random spot stools) has been reported to have a sensitivity of 100%, specificity of 95%, and a positive predictive value of 90% in the detection of steatorrhea (mixed population of subjects from normal to those with severe steatorrhea).

References

Amann S, Josephson SA, and Toskes PP, "Acid Steatocrit: A Simple, Rapid Gravimetric Method to Determine Steatorrhea," *Am J Gastroenterol*, 1997, 92(12):2280-4.

Tran M, Forget P, van den Neucker A, et al, "The Acid Steatocrit: A Much Improved Method," *J Pediatr Gastroenterol Nutr*, 1994, 19(3):299-303.

Van den Neucker A, Pestel N, Tran TD, et al, "Clinical Use of Acid Steatocrit," *Acta Paediatr*, 1997, 86(5):466-9.

Fat, Semiquantitative, Stool, Sudan III Stain
Related Information

d-Xylose Absorption Test, Serum, Urine *on page 527*
Fat, Semiquantitative, Stool, Acid Steatocrit *on page 571*
Fecal Fat, Quantitative, 72-Hour Collection *on page 574*
Fecal Lactoferrin *on page 575*
Fecal Pancreatic Elastase 1 *on page 576*
Meat Fibers, Stool *on page 894*
Methylene Blue Stain, Stool *on page 906*
pH, Stool *on page 1034*

Synonyms Fecal Fat Stain, Sudan III Stain, Stool; Stool Fat, Semiquantitative

Applies to Acid Steatocrit; Fecal Fat Analysis, 72-Hour Steatocrit

Abstract In this simple and low cost procedure, fecal fat is stained with Sudan III. The test is used as a screen for the presence of fecal neutral fat and fatty acids and may assist in the determination of the cause of steatorrhea. The diagnosis of steatorrhea should still be defined by results of a quantitative 72-hour fecal fat determination.

Patient Preparation An adult patient should be on a diet containing about 100-150 g of dietary fat (60 g/m² body surface area) per day for about 1 week before and during the test and should avoid high fiber for a few days prior to the test. The patient should not use suppositories or oily material prior to specimen collection.

Specimen Fresh random stool or, preferably, aliquot of homogenized 72-hour fecal collection. Some methods are claimed to provide useful test results without use of the ideal 3-day stool specimen.

Storage Instructions Maintain specimen at 4°C until analysis. For prolonged storage, keep at -20°C.

Causes for Rejection Contamination with water or urine; absence of patient identification or of duration of collection

Reference Interval Neutral fat: <50 fat globules/hpf, reported as normal. Fatty acids: <100 fat globules/hpf is considered normal.

Use Confirm the presence of steatorrhea; assist with the diagnostic study of malabsorption and/or maldigestion (as with exocrine pancreatic failure in some cases of cystic fibrosis); monitor results of therapy in patients with chronic diarrhea and in patients receiving exogenous enzyme therapy

Limitations Castor oil or mineral oil droplets can mimic neutral fat. Ingestion of the fat substitute olestra (a nonabsorbable substance consisting of 6-8 fatty acids esterified to a sucrose molecule) may cause false-positive results with resultant erroneous diagnosis.

This is not a definitive test; *vide infra*.

Contraindications Administration of barium, bismuth, Metamucil®, castor oil, or mineral oil within 1 week prior to collection of the specimen

Methodology A small amount of stool sample is mixed with two drops of water, two drops of 95% ethanol, and three to four drops of Sudan III stain. Increased yellow-orange refractile fat globules (direct Sudan III stain) identifies neutral fats. Fatty acids and fat soaps are detected after hydrolysis by mixing stool sample with two to three drops each of Sudan III and glacial acetic acid, followed by heating before microscopic examination.

Additional Information The test consists of determination of the presence of neutral fats and of total fats representing fatty acids. The results are reported semiquantitatively. Results of Sudan stain for fecal fat lack sensitivity and may be misleading with some dietary fatty acids or with constipation.

Presence of steatorrhea can be established by the results of a 72-hour fecal fat analysis. Maldigestion or malabsorption may cause steatorrhea. Some patients with maldigestion may excrete excess triglyceride while patients with malabsorption excrete excess fatty acid. The 72-hour fecal fat determination involves saponification and does not usually provide for selective quantitation of triglyceride and fatty acid. One may not be able to differentiate maldigestion from malabsorption (pancreatic vs intestinal steatorrhea) by comparing fecal triglyceride/fatty acid or fecal fat concentration. The influence of extrapancreatic lipase (eg, gastric lipase) must be considered. Fecal acidification enhances the sensitivity of the Sudan fecal staining method.

Recently introduced infrared reflectance spectroscopy methods are direct, easy to perform, and provide results comparable with the standard 72-hour chemical method.

References
Fine KD and Ogunji F, "A New Method of Quantitative Fecal Fat Microscopy and Its Correlation With Chemically Measured Fecal Fat Output," *Am J Clin Pathol*, 2000, 113(4):528-34.

Riley SA and Marsh MN, "Maldigestion and Malabsorption," *Sleisenger and Fordtran's Gastrointestinal and Liver Disease*, 6th ed, Chapter 88, Feldman M, Scharschmidt BF, and Sleisenger MH, eds, Philadelphia, PA: WB Saunders Co, 1998, 1505-6.

Sugai E, Srur G, Vazquez H, et al, "Steatocrit: A Reliable Semiquantitative Method for Detection of Steatorrhea," *J Clin Gastroenterol*, 1994, 19(3):206-9.

Fat, Urine

Related Information
Ethylene Glycol, Serum or Plasma *on page 561*
Glucose, Semiquantitative, Urine *on page 650*
Kidney Biopsy *on page 818*
Mercury, Blood *on page 897*
Mercury, Urine *on page 898*
Protein, Quantitative, Urine *on page 1108*
Urinalysis *on page 1289*

Synonyms Free Fat, Urine; Lipiduria

Test Includes Light and polarized microscopy of urine sediment and staining with Sudan III or IV

Abstract Lipiduria (oval fat bodies, fatty casts, and free fat) are found in nephrotic syndromes (proteinuria ≥3.5 g/24 hours). Oval fat bodies are lipid-laden, pathologic, renal tubular epithelial cells which bear a strong association with marked proteinuria.

Patient Preparation Avoid contamination of the specimen with oils and lubricants from catheters, soaps, and glove powder.

Specimen Random urine

Causes for Rejection Contamination of specimen with oils, as specified above.

Reference Interval Negative

Use Evaluate nephrotic syndrome, acute renal failure, mercury poisoning, ethylene glycol ingestion (which may produce oval fat bodies), and fatty casts in the urine; evaluate bone marrow and fat embolism, which may produce gross fat globules in urine

(Continued)

Fat, Urine *(Continued)*

Refractile lipid bodies have been found in nonglomerular renal disease at relatively low levels of proteinuria and, rarely, even in patients without renal disease. The frequency of urine lipid bodies in patients with nonglomerular renal diseases include chronic interstitial nephritis (26%), polycystic kidney disease (38%), prerenal azotemia (20%), acute tubular necrosis (15%), and acute interstitial nephritis (33%). Presence of refractile urine lipid bodies in numbers >5 per 20 high power microscopic fields may be required to differentiate glomerular from nonglomerular renal disease. See **Fatty casts** and **oval fat bodies ("lipiduria")** in Urinalysis *on page 1289.*

Limitations Urinary fat globules, like air bubbles and yeasts, can be confused with erythrocytes if Sudan stain is not used. Structures resembling oval fat bodies may be from vaginal secretions, from the seminal vesicles, or as external contaminants.

Methodology Sudan III and IV or oil red O staining of urine sediment brings out triglycerides in lipiduria. Microscopically fat globules appear as spherical or ovoid dark glistening bodies. When they include substantial quantities of cholesterol, under polarized light they appear doubly refractile and give a Maltese cross or cross pattée pattern. True urine fat globules are usually seen in urines with increased protein.

References
Braden GL, Sanchez PG, Fitzgibbons JP, et al, "Urinary Doubly Refractile Lipid Bodies in Nonglomerular Renal Diseases," *Am J Kidney Dis*, 1988, 11(4):332-7.
Larson TS, "Evaluation of Proteinuria," *Mayo Clin Proc*, 1994, 69(12):1154-8.

♦ **FBP** *see* D-Dimers and Fibrin Degradation Products *on page 502*

♦ **FBS** *see* Glucose, Fasting, Plasma *on page 643*

♦ **5-FC** *see* Flucytosine, Serum *on page 600*

♦ **F Cells** *see* Fetal Cell Detection by Flow Cytometry *on page 579*

♦ **F Distribution** *see page 11*

♦ **FDP** *see* D-Dimers and Fibrin Degradation Products *on page 502*

♦ **FDP** *see* Hypercoagulation Panel *on page 758*

♦ **Fe and TIBC** *see* Iron and Total Iron Binding Capacity/Transferrin, Serum *on page 807*

♦ **Febrile Transfusion Reaction** *see* Filters for Blood *on page 588*

♦ **Fecal Fat Analysis, 72-Hour Steatocrit** *see* Fat, Semiquantitative, Stool, Sudan III Stain *on page 572*

Fecal Fat, Quantitative, 72-Hour Collection

Related Information
Cholesterol, Total, Serum or Plasma *on page 396*
d-Xylose Absorption Test, Serum, Urine *on page 527*
Endomysial Antibodies *on page 537*
Fat, Semiquantitative, Stool, Acid Steatocrit *on page 571*
Fat, Semiquantitative, Stool, Sudan III Stain *on page 572*
Fecal Pancreatic Elastase 1 *on page 576*
Gliadin IgG/IgA Antibodies *on page 637*
Hemoglobin *on page 681*
Meat Fibers, Stool *on page 894*
Methylene Blue Stain, Stool *on page 906*
pH, Stool *on page 1034*
Vitamin D, Serum *on page 1318*

Synonyms Fat, Quantitative, 72-Hour Stool Collection; Quantitative Fecal Fat, 72-Hour Collection; Stool Fat, Quantitative

Abstract Test for the investigation of malabsorption and steatorrhea.

Patient Preparation 100-150 g/day fat diet for 3 days before and during 72-hour collection period. Barium interferes.

Specimen 72-hour stool collection, usually in the fourth, fifth, and sixth days of the 100 g/day fat diet

Container Plastic stool container, preweighed

Sampling Time 72 hours; shorter collection periods are not usually acceptable

Collection Specimen should be refrigerated during its collection.

Storage Instructions Freeze on dry ice if analysis is not to be done promptly.

Causes for Rejection Inappropriate container

Special Instructions Date and time collection started, date and time collection finished are needed.

Reference Interval 2-7 g/24 hours (SI: 2-7 g/day); <20% of total solids

Use Diagnose causes of malabsorption (eg, nontropical sprue, Crohn disease, chronic pancreatitis, cystic fibrosis, Whipple disease, or others)

Limitations Test has poor sensitivity. Increased concentrations of fecal fat do not distinguish between pancreatic (maldigestion) and intestinal (malabsorption) steatorrhea. Stool fat collection is an unpleasant experience for the patient and staff.

Contraindications Patient taking mineral oil

Methodology Extraction and titration of long chain fatty acids by sodium hydroxide. Fatty acids represent 60% to 80% of total fecal lipids.

References

Bai JC, Andrush A, Matelo G, et al, "Fecal Fat Concentration in the Differential Diagnosis of Steatorrhea," *Am J Gastroenterol*, 1989, 84(1):27-30.

Holmes GK and Hill PG, "Do We Still Need to Measure Faecal Fat?" *Br Med J (Clin Res Ed)*, 1988, 296(6636):1552-3.

♦ **Fecal Fat Stain, Sudan III Stain, Stool** *see* Fat, Semiquantitative, Stool, Sudan III Stain *on page 572*

Fecal Lactoferrin

Related Information

Bacterial Culture, Stool *on page 243*
Clostridium difficile Toxin Assay and Culture *on page 416*
Cryptosporidium Direct Staining Procedures *on page 484*
d-Xylose Absorption Test, Serum, Urine *on page 527*
Entamoeba histolytica Antigen Detection and Serology *on page 538*
Fat, Semiquantitative, Stool, Acid Steatocrit *on page 571*
Fat, Semiquantitative, Stool, Sudan III Stain *on page 572*
Meat Fibers, Stool *on page 894*
Methylene Blue Stain, Stool *on page 906*
Occult Blood, Stool *on page 969*
Ova and Parasites, Direct Exam *on page 985*

Synonyms Lactoferrin, Fecal; Leukotest®, Stool

Abstract Lactoferrin, an iron-binding glycoprotein present in the cytoplasmic granules of polymorphonuclear white blood cells (PMNs), can be used as a marker for fecal leukocytes. A commercially available (Leukotest®) latex agglutination procedure has undergone evaluation as a screening test for inflammatory bacterial diarrhea including, *Clostridium difficile* colitis.

Specimen Stool

Container Plastic specimen container or rayon tipped swab

Storage Instructions Specimen may be stored refrigerated (4°C) for up to 6 days

Reference Interval

- Pediatrics: titer up to 1:200 in children without diarrhea
- Adults: fecal lactoferrin titer ≤1:50

Titers from 1:50-1:200 may relate to mild inflammation (eg, enteric parasites) or protein malabsorption (including subclinical malabsorption of milk). Sensitivity of detection (*in vitro* suspension, not in clinical stool specimens) is <1 ng (0.31 ng/μL) of purified lactoferrin per μL (equivalent to 60 PMNs per mm^3).

Use Marker for the presence of fecal leukocytes; useful in detection of and screening for inflammatory bacterial diarrhea

Limitations Breast-fed infants (stools with lactoferrin containing human milk) may produce false positives even when high titers of fecal lactoferrin are detected.

Additional Information By chelating iron, lactoferrin inhibits the growth of microorganisms. For the diagnosis of inflammatory diarrhea, detection of a constituent of PMNs (eg, lactoferrin) can serve as an alternative to microscopic study of stool for leukocytes with some inherent advantages. Sensitivity is maintained after refrigeration of specimens for up to 6 days. Swab specimens do not show the loss of sensitivity that characterizes microscopic examination for fecal PMNs.

(Continued)

Fecal Lactoferrin *(Continued)*

While the lactoferrin stool test may be more sensitive than microscopic identification of PMNs or lactoferrin studies, neither stool PMNs or lactoferrin studies are sufficiently sensitive to serve as screening tests for *Campylobacter*, *Salmonella*, *Shigella*, or *Clostridium difficile* pathogens.

References

Dowdy LM, "Infectious Diarrhea," *Clinical Practice of Gastroenterology*, Chapter 60, Brandt LJ, ed, Philadelphia, PA: Current Medicine, Inc, 1999, 529-30.

Nachamkin I, "Fecal Lactoferrin Screening Assay for Inflammatory Bacterial Diarrhea," *J Clin Microbiol*, 1996, 34(9):2337-8

♦ **Fecal Leukocyte Stain** *see* Methylene Blue Stain, Stool *on page 906*
♦ **Fecal Occult Blood Test** *see* Occult Blood, Stool *on page 969*

Fecal Pancreatic Elastase 1

Related Information

Fat, Semiquantitative, Stool, Acid Steatocrit *on page 571*
Fat, Semiquantitative, Stool, Sudan III Stain *on page 572*
Fecal Fat, Quantitative, 72-Hour Collection *on page 574*

Synonyms Cholesterol-Binding Pancreatic Proteinase; Elastase, Fecal Pancreatic

Abstract Pancreatic elastase 1, an endoprotease and sterol-binding protein, is present in pancreatic secretions and feces and is not degraded during intestinal transit. It can serve as a noninvasive tubeless test of pancreatic function.

Specimen Small sample of stool, aliquot of an homogenized 24-hour stool collection is preferred.

Storage Instructions Stable up to at least 3 days at room temperature (24°C). May be frozen at -20°C for up to 30 days (possibly longer) prior to analysis. Enzyme is stable for months at 4°C.

Reference Interval Normal value is considered as >200 μg elastase/g stool (cutoff value). Values rise during the first month after birth to 586 μg/g ±65 μg/g and are generally >500 μg/g during childhood.

Use Noninvasive (tubeless) test for the evaluation of pancreatic function; evaluate severity of chronic pancreatitis; assist in the differentiation of maldigestion from malabsorption in cases of steatorrhea.

Limitations Reproducibility of fecal elastase 1 test (FET) results may relate to variable dilution of the enzyme in stool and variable dietary pancreatic stimulation. A significant correlation has been shown, however, between the results of FET and the secretin and pancreozymin test (SPT).

Additional Information Human pancreatic elastase 1 (E1) is a sterol-binding proteinase produced by the exocrine pancreas. It combines with bile acids and neutral sterols to transport cholesterol and derived metabolites along the intestinal lumen. E1 survives intestinal passage and its presence in stool is considered to reflect exocrine pancreatic function. ELISA assay has been developed for serum and is available commercially. The FET has been applied to the assessment of exocrine pancreatic insufficiency in children with cystic fibrosis; high levels of sensitivity and specificity (90% to 100%) have been reported. The FET, studied in relation to graded severity of exocrine pancreatic insufficiency (ie, mild, moderate, severe), may lack sensitivity for mild-to-moderate forms of the disease, which are the more frequent and difficult clinical problems.

References

Mitchell RMS, Byrne MF, and Baillie J, "Pancreatitis," *Lancet*, 2003, 361(9367):1447-55.

Soldan W, Henker J, and Sprössig C, "Sensitivity and Specificity of Quantitative Determination of Pancreatic Elastase 1 in Feces of Children," *J Pediatr Gastroenterol Nutr*, 1997, 24(1):53-5.

Stein J, Jung M, Sziegoleit A, et al, "Immunoreactive Elastase 1: Clinical Evaluation of a New Noninvasive Test of Pancreatic Function," *Clin Chem*, 1996, 42(2):222-6.

Terbrack HG, Gürtler KH, Klör HU, et al, "Human Pancreatic Elastase 1 Concentration in Faeces of Healthy Children and Children With Cystic Fibrosis," *Gut*, 1995, 37(Suppl 2):A253.

♦ **Fecal pH** *see* pH, Stool *on page 1034*
♦ **Felbamate** *see* Antiepileptic Drugs Overview *on page 176*
♦ **Felbatol®** *see* Antiepileptic Drugs Overview *on page 176*
♦ **FENA** *see* Sodium, Urine *on page 1213*

Fenfluramine, Serum

Synonyms Dima-Fen®; Pesos®; Ponderx®; Pondimin®; Ponflural®

Abstract Fenfluramine is an anorectic agent which has been withdrawn from the U.S. market; however, it is available in other countries.

Specimen Serum

Container Red top tube

Reference Interval Therapeutic: 0.05-0.15 µg/mL

Critical Values >6.5 µg/mL

Methodology Thin-layer chromatography (TLC), gas chromatography (GC); in overdose situations, immunoassay for amphetamine may give a positive result with fenfluramine.

Additional Information
- Half-life: 11-20 hours
- Volume of distribution: 15 L/kg
- Protein binding: 34%

Fenfluramine is a sympathomimetic amine, structurally related to amphetamine. However, unlike amphetamine, it lacks a stimulatory effect on the CNS.

Acute overdose causes tachycardia, asystole, ventricular fibrillation, respiratory failure, mydriasis, nystagmus, seizures, and coma. Treatment is mostly supportive.

References

Connolly HM, Crary JL, McGoon MD, et al, "Valvular Heart Disease Associated With Fenfluramine-Phenteramine," *N Engl J Med*, 1997, 337(9):581-8.

Derby LE, Myers MW, and Jick H, "Use of Dexfenfluramine, Fenfluramine, and Phenteramine and the Risk of Stroke," *Br J Clin Pharmacol*, 1999, 47(5):565-9.

Mast ST, Jollis JG, Ryan T, et al, "The Progression of Fenfluramine-Associated Valvular Heart Disease Assessed by Echocardiography," *Ann Intern Med*, 2001, 134:261-6.

◆ **Fenilcal®** *see* Phenobarbital, Serum or Plasma *on page 1021*

◆ **Fenitoina** *see* Phenytoin, Serum or Plasma *on page 1026*

◆ **Fenytoin®** *see* Phenytoin, Serum or Plasma *on page 1026*

◆ **FEP** *see* Protoporphyrin, Free Erythrocyte *on page 1121*

◆ **Ferndex®** *see* Amphetamine, Qualitative, Urine *on page 154*

◆ **Ferric Chloride Test** *see* Phenylalanine, Urine *on page 1024*

◆ **Ferric, Perchloric, Nitric (FPN) Urine Spot Test** *see* Phenothiazines, Serum *on page 1021*

Ferritin, Serum

Related Information

Anemia Flowchart *on page 35*
Bone Marrow *on page 296*
Complete Blood Count *on page 442*
Erythropoietin, Serum *on page 551*
Hereditary Hemochromatosis DNA Test *on page 718*
Iron and Total Iron Binding Capacity/Transferrin, Serum *on page 807*
Iron Stain, Bone Marrow *on page 810*
Lead, Blood *on page 832*
Lead, Urine *on page 835*
Liver Biopsy *on page 864*
Liver Disease: Laboratory Assessment, Overview *on page 869*
Occult Blood, Stool *on page 969*
Phlebotomy, Therapeutic *on page 1029*
Protoporphyrin, Free Erythrocyte *on page 1121*
Protoporphyrin, Zinc, Blood *on page 1121*
Red Blood Cell Count *on page 1133*
Transferrin Receptor, Soluble, Serum or Plasma *on page 1267*

Applies to Hepatic Iron Index; HLA-A3; Liver Iron Concentration

Abstract Used to assist in establishing the presence of iron deficiency anemia. Ferritin serum level is generally proportional to the body's iron store and reflects cellular iron stores. Serum ferritin is the best single test for the diagnosis of iron deficiency. It is also used to support diagnosis and follow therapy of patients with hemochromatosis.

Patient Preparation Ferritin determinations are not fully reliable when the patient is on iron therapy.

(Continued)

Ferritin, Serum *(Continued)*

Specimen Serum

Container Red top tube

Storage Instructions Separate serum from clot and refrigerate specimen.

Reference Interval 1 ng/mL of serum ferritin in normal subjects corresponds to ~8 mg of storage iron. Ferritin increases in adulthood in men to about the fifth decade and in women after menopause. Typical reference intervals:

- newborns: 25-200 ng/mL (SI: 25-200 µg/L)
- 1 month: 200-600 ng/mL (SI: 200-600 µg/L)
- 2-5 months: 50-200 ng/mL (SI: 50-200 µg/L)
- 6 months to 15 years: 7-140 ng/mL (SI: 7-140 µg/L)
- adult male: 20-250 ng/mL (SI: 20-250 µg/L)
- adult female: younger than 40 years: 12-122 ng/mL (SI: 12-122 µg/L); older than 40 years: 12-250 ng/mL (SI: 12-250 µg/L).

Reference intervals of ferritin vary with age and sex.

Critical Values Serious gastrointestinal disease has been reported in subjects whose serum ferritin is ≤50 ng/mL, and endoscopic examination is considered warranted in such patients.

Use Useful in the differential diagnosis of hypochromic, microcytic anemias. **Decreased** in iron deficiency anemia and **increased** in iron overload. Ferritin levels correlate with and are useful in the evaluation of total body storage iron.

In hemochromatosis, both ferritin and iron saturation are increased. Ferritin levels in hemochromatosis may be >1000 ng/mL (SI: >1000 µg/L), but a normal serum ferritin cannot rule out homozygous hemochromatosis. Screening for hemochromatosis is better done with transferrin saturation. If saturation is consistently >60% in men or >50% in women, serum ferritin should be assayed. If it is high for age and sex, liver biopsy may be considered. Genetic tests have become available in aiding the diagnosis of hemochromatosis (see Hereditary Hemochromatosis DNA Test *on page 718* and the following text). Normal ferritin does not exclude iron deficiency anemia.

Limitations Ferritin escapes from necrotic hepatocytes. It may be increased in alcoholics who are actively abusing alcohol, in individuals with other liver diseases such as autoimmune hepatitis and hepatitis C. A quarter of patients with chronic hepatitis have increased ferritin. In the presence of liver disease, with inflammation such as rheumatoid arthritis, with malignancy or with iron therapy, iron deficiency may not be reflected by low serum ferritin. Ferritin, an acute phase reactant, is increased in inflammatory and infectious disorders and with acute renal failure. It is elevated in hyperthyroidism. Bone marrow aspiration may be needed in some settings, such as low-normal ferritin and low serum iron in the presence of apparent anemia of chronic disease, low-normal ferritin in the presence of liver disease.

The differential diagnosis of transfusional siderosis versus hemochromatosis is largely dependent on clinical history; serum ferritin is less reliable than other available diagnostic measures.

Contraindications Evaluation of iron stores in alcoholics with liver disease. Ferritin is higher in abusing cirrhotics than in abstaining cirrhotics. Extremely high ferritin may be seen in cases of acute hepatitis.

Additional Information Other than a bone marrow examination, serum ferritin is the most reliable indicator of total body iron stores. When combined with serum iron and percent saturation of iron binding capacity/transferrin, it can usually differentiate the microcytic hypochromic anemias into **iron deficiency anemia** (ferritin low, iron low, saturation low, TIBC high, transferrin high), the **anemia of chronic disease** (ferritin normal or high, iron low, normal to low transferrin or TIBC), or **thalassemia** (ferritin normal or high). A low serum ferritin level and elevated serum transferrin-receptor indicate iron deficiency anemia, while the ferritin may be normal or raised with low-normal to normal transferrin receptor levels in anemia of chronic disease. (See table in Iron and Total Iron Binding Capacity/Transferrin, Serum *on page 807* and see Anemia Flowchart *on page 34*.) In iron deficiency, the **red cell distribution width** is increased, while it is usually normal with heterozygous alpha or beta thalassemia trait. The **MCV** is reduced in iron deficiency and alpha or beta thalassemia trait; each is normal with lead poisoning. Ferritin is low with combined iron deficiency and thalassemia. **In adults**, serum ferritin level ≤20 ng/mL indicates iron deficiency.

High serum ferritin may be associated with inflammation, liver disease, megaloblastic anemia, hemolytic anemia, sideroblastic anemia, thalassemia, iron overload (hemochromatosis, hemosiderosis), and malignant diseases. The latter include leukemia and malignant lymphoma. Very high levels usually indicate iron overload but may also be seen with hemophagocytosis. The reactive hemophagocytic syndrome with monocyte activation and cytokine release has been associated with ferritin concentrations of up to 400,000 µg/L. Very high levels may also occur with hemophagocytosis and/or disseminated histoplasmosis in patients who have the acquired immunodeficiency syndrome. Oral and injected iron increase ferritin levels. Increased serum ferritin is considered a risk factor in primary hepatocellular carcinoma. An hereditary hyperferritinemia-cataract syndrome has been reported and relates to mutation in the iron-responsive element of ferritin light-chain mRNA. Serum ferritin is elevated to the level of about 1000 µg/L but there is absence of iron overload.

Primary hemochromatosis is inherited as an autosomal recessive trait, but clinically it is much more common in males. Only homozygotes bear full clinical expression of hemochromatosis. Hemochromatosis can be recognized before disease develops when only homozygosity for the mutant allele is required. The gold standard for diagnosis is liver biopsy with liver iron concentration and calculation of hepatic iron index (ratio of hepatic iron to patient age). Effective therapy (phlebotomies) is available (see Phlebotomy, Therapeutic *on page 1029*). **HLA-A3** alloantigen is found in about 70% of subjects who have hemochromatosis. The gene is present on the short arm of chromosome 6. Genetic tests for detection of hemochromatosis mutations (C282Y and H63D) have become available. Homozygosity for C282Y mutation is responsible for up to 90% of hemochromatosis patients. Another mutation, H63D, has been seen in some patients with C282Y mutation, but has reduced penetrance of <2%. Inappropriate increase in iron absorption and parenchymal tissue deposition may eventuate in hepatic cirrhosis, diabetes, testicular atrophy, cardiomyopathy, arthropathy, and bronze to slate gray skin pigmentation and very high serum ferritin levels (usually >1000 ng/mL).

Serum transferrin receptor measurements distinguish iron deficiency anemia from the anemia of chronic disease, used with assays of hemoglobin, hematocrit, and ferritin. See Transferrin Receptor, Soluble, Serum or Plasma *on page 1267*.

References

Cook JD, "Iron-Deficiency Anemia," *Baillieres Clin Haematol*, 1994, 7(4):787-804.

Edwards CQ, Griffen LM, Ajioka RS, et al, "Screening for Hemochromatosis: Phenotype Versus Genotype," *Semin Hematol*, 1998, 35(1):72-6

Felitti VJ and Beutler E, "New Developments in Hereditary Hemochromatosis," *Am J Med Sci*, 1999, 318(4):257-68.

Looker AC, Dallman PR, Carroll MD, et al, "Prevalence of Iron Deficiency in the United States," *JAMA*, 1997, 277(12):973-6.

Worwood M, "The Measurement of Ferritin," *Advanced Laboratory Methods in Haematology*, Chapter 11, Rowan RM, van Assendelft OW, and Preston FE, eds, London, UK: Arnold, 2002, 254.

Fetal Cell Detection by Flow Cytometry

Related Information

Fetal Hemoglobin *on page 581*
Flow Cytometry, Overview *on page 592*
Hemoglobin Electrophoresis *on page 684*
Kleihauer-Betke *on page 822*
Prenatal Screen, Immunohematology *on page 1086*
Rh$_o$(D) Immune Globulin (Human) *on page 1164*
Rosette Test for Fetomaternal Hemorrhage *on page 1172*

Applies to F Cells

Abstract

Fluorescent-tagged antibody to an epitope of fetal red blood cells is used to detect and quantify fetomaternal hemorrhage (FMH). Estimate of FMH volume guides therapeutic dose of Rh(D) immune globulin (human). See Rh$_o$(D) Immune Globulin (Human) *on page 1164*.

This test is also relevant to certain of the hemolytic anemias.
(Continued)

Fetal Cell Detection by Flow Cytometry *(Continued)*

Specimen 5 mL maternal whole blood (red blood cells from EDTA, heparin, or citrate anticoagulated tubes or red cells freed from clotted specimens can be analyzed)

Container Lavender top (EDTA) tube

Storage Instructions Transport specimen to the laboratory promptly for testing or transfer to local/regional center for imminent batch testing. Washed, fixed erythrocytes suspended in phosphate-buffered saline/2% albumin can be stored at -70°C.

Causes for Rejection Gross hemolysis

Turnaround Time 2-3 hours

Reference Interval Normal adults (includes subjects without hemoglobinopathy): Hb F cells <0.01%; full-term newborns: Hb F cells >90%

Use Test of maternal blood for evidence of fetomaternal hemorrhage and to assess magnitude of such hemorrhage, to assist in calculation of dosage of Rh(D) immune globulin; aid in the diagnosis of some hemoglobinopathies. See Kleihauer-Betke *on page 822.*

Limitations Effect of transfusion prior to obtaining specimen for analysis must be considered in interpretation of results. Flow cytometer and associated support personnel are required for analysis.

Methodology Flow cytometry (FC); antibodies to different antigens have been used to detect fetal cells admixed within the maternal cell population. These include anti-D antigen, anti-CD71 (transferrin receptor), and antihemoglobin F (anti-Hb F). Early methods relied upon the detection of D antigen and showed greater sensitivity and precision than manual methods (Kleihauer-Betke). These tests are applicable only to cases with D incompatibility, the common form of hemolytic disease of the newborn (HDN) and cannot be used as a test for all cases of suspected FMH.

The antigen CD71 is expressed only on maternal nucleated RBCs and immature reticulocytes. Methods based on this antigen find application in genetic testing. Detection of fetal cells based on antibodies to Hb F has a number of advantages, most significant of which is allowance of a broad global approach to fetal cell detection (not restricted to Rh-negative women). In addition, the method can identify different forms of Hb F containing cells (eg, true fetal cells with Hb F as the major form of hemoglobin vs adult cells with only a small proportion of Hb F). As Hb F is intracellular, permeabilization is necessary to allow antibody access.

Additional Information Documentation of fetomaternal hemorrhage (FMH) is the most common application of methods that measure presence of fetal cells. For over 40 years, most clinical laboratories have used the Kleihauer-Betke (K-B) procedure (slide-based microscopic detection and counting of the results of acid elution treated erythrocytes) for the evaluation of FMH. While relatively rapid, simple, and logistically practical, the K-B method suffers from inaccuracy and imprecision. Proficiency testing results from K-B tests have coefficients of variation of 40% to 60%. Flow cytometric-based methods are more accurate and precise.

Flow cytometry is uniquely suited for F-cell quantitation in the study of hemoglobinopathies, in particular, sickle cell disease (SCD) and its variants. The percent of Hb F in whole blood (measured by alkali denaturation or comparable method) is one of many determinants of clinical severity in SCD. With higher levels of Hb F, clinical severity is decreased. Presence of Hb F dilutes intracellular Hb S in addition to which Hb F and its γ-globin chains impair polymerization of Hb S. At some 10% to 20% or higher level of Hb F there are fewer deleterious clinical events in cases of SCD.

Hb F is not present in all red cells. It is located in a subset of erythrocytes referred to as Hb F containing cells ("F cells"). An exception is hereditary persistence of hemoglobin F in which all red cells are F cells. In normal adults, the percentage of F cells is from 0.5% to 7%. The percentage of F cells (which survive preferentially in the blood of SCD patients) must be considered as well as the absolute amount of Hb F when considering the effect on clinical severity of sickle hemoglobinopathy. A recent study found a logarithmic correlation between the percentage of F cells and the percentage of Hb F in children with SCD with implications for pharmacologic therapy to increase Hb F. Thus, a growing application of flow cytometry may be to monitor the effect of chemical (drug) therapy of SCD on the F-cell population.

The use of hydroxyurea and butyrate as drugs to increase the level of Hb F in cases of sickle disease and β-thalassemia has led to amelioration of the disease process and interest in assessment of the efficacy of treatment programs (eg, determination of F reticulocytes).

References
Davis BH, "Is There an Alternative to the Kleihauer-Betke Method for Quantitation of Fetal-Maternal Hemorrhage?" Q and A, *Lab Med*, 1999, 30(5):307-8.

Davis BH, Olsen S, Bigelow NC, et al, "Detection of Fetal Red Cells in Fetomaternal Hemorrhage Using a Fetal Hemoglobin Monoclonal Antibody by Flow Cytometry," *Transfusion*, 1998, 38(8):749-56.

Marcus SJ, Kinney TR, Schultz WH, et al, "Quantitative Analysis of Erythrocytes Containing Fetal Hemoglobin (F Cells) in Children With Sickle Cell Disease," *Am J Hematol*, 1997, 54(1):40-6.

♦ **Fetal Fibronectin, Cervicovaginal Secretions** *see* Fibronectin, Fetal, Cervicovaginal Secretions *on page 585*

Fetal Hemoglobin
Related Information
Apt-Downey Test *on page 208*
Fetal Cell Detection by Flow Cytometry *on page 579*
Glycated Hemoglobin (Hemoglobin A$_{1c}$), Blood *on page 655*
Hemoglobin Electrophoresis *on page 684*
Kleihauer-Betke *on page 822*
Sickle Cell Tests *on page 1195*

Synonyms Hb F; Hemoglobin, Fetal

Applies to Sudden Infant Death Syndrome

Abstract Fetal hemoglobin is formed of two α-chains and two γ-chains. It is the major hemoglobin during fetal life. Hb F levels decrease after birth by about 3% to 4% per week. In 2-3 weeks fetal hemoglobin is about 65%. By 6 months of age fetal hemoglobin is <2% of the total hemoglobin.

Specimen Whole blood

Container Lavender top (EDTA) tube for venipuncture specimen; lavender top Microtainer™ tube for capillary specimen

Reference Interval 6 months to adult: 0.1% to 1.5% of the total hemoglobin; 0-6 months: up to 75% (alkali denaturation method). Term newborn with high performance liquid chromatography (HPLC) method (Diamat, Bio-Rad, Hercules, CA), 61% to 77%. About 0.5% to 7% of red cells are "F cells" (contain Hb F) in normal adults.

Use Evaluate hemoglobinopathies, hemolytic anemia; diagnose hereditary persistence of fetal hemoglobin, thalassemia; evaluate sickling hemoglobins; monitor Hb F levels in sickle cell patients receiving hydroxyurea to increase Hb F production.

Limitations Carboxyhemoglobin A is also resistant to alkali denaturation. Assay for Hb F should initially convert carboxyhemoglobin to cyanmethemoglobin or false-positive elevations of Hb F may be obtained.

Methodology Alkali denaturation, high resolution hemoglobin electrophoresis (some methods), acid elution (Kleihauer-Betke), radial immunodiffusion (RID), isoelectric focusing, high performance liquid chromatography (HPLC), enzyme immunoassay (EIA), capillary electrophoresis, capillary isoelectric focusing

Additional Information The oxygen dissociation curve of Hb F is shifted to the left as compared with normal Hb A. This facilitates placental oxygen transfer. With erythroblastosis fetalis and anoxic states of the newborn, however, Hb F is proportionally lower than in a normal newborn. Multiple inherited abnormalities of γ-chain structure have been described, but most are without clinical significance (fetal Hb normally forms <1.5% to 2.0% of total hemoglobin). An exception is Hb F Poole which has been reported as a cause of hemolytic disease of the newborn.

In the adult, hereditary persistence of fetal hemoglobin (HPFH) of multiple varieties, is associated with varying elevations of Hb F. The homozygous form of HPFH is found only in African-Americans. In the heterozygous state, the Hb F level is 15% to 35% in the black type, and 5% to 20% in the Greek type. Homozygous β-thalassemia is associated with Hb F levels >10% to <90%. About 50% of heterozygotes for β-thalassemia have elevated levels around 2%, rarely >5%. The remainder have normal Hb F. Using elements of the CBC, (Continued)

Fetal Hemoglobin *(Continued)*

discriminant analysis and use of neural networks have been applied to the differentiation of heterozygous thalassemia from iron deficiency anemia. Heterozygous S/β thalassemia may have Hb F in the 5% to 20% range. With homozygous Hb S disease the level of Hb F varies from 1% to 30%. Hb F inhibits the polymerization of sickle hemoglobin, exerting an ameliorating effect on the sickle disease process. Thus, there is interest in the development and use of drugs that stimulate the production of Hb F (enhance expression of the γ-globin gene in erythroid precursors). Hydroxyurea stimulates Hb F production and has significant clinical benefit in sickle disease. Analogues of butyrate and other short-chain fatty acid derivatives also increase Hb F and may be useful in sickle disease therapy. Gene therapy is under development.

Other conditions associated with elevated Hb F include various anemias (spherocytosis, Fanconi, acquired aplastic, hemolytic, hypoplastic, myelophthisic, megaloblastic including untreated pernicious anemia, paroxysmal nocturnal hemoglobinuria, Blackfan-Diamond anemia); all types of leukemia (especially erythroleukemia and juvenile chronic myelogenous leukemia), extramedullary hematopoiesis as with myelofibrosis and myelodysplasia, multiple myeloma and lymphomas, metastatic disease of the bone marrow; bone marrow transplantation; pregnancy; miscellaneous disorders reported include infants small for gestational age, infants with chronic intrauterine anoxia with developmental

Nonmalignant Conditions Associated With Increased Proportions of Hb F

Condition	Hb F Value (%)
Anemias	
Aplastic anemia (both congenital and acquired)	5-25
Pernicious anemia	2-6
Hereditary spherocytosis	2-5
Hereditary elliptocytosis	2-5
Congenital nonspherocytic hemolytic anemia	3-4
Anemia of chronic infection	2-3
Anemia of blood loss	2-8
Erythropoietic porphyria	2-10
Paroxysmal nocturnal hemoglobinuria	2-25
Hemoglobinopathies	
Unstable hemoglobins	<10
Homozygous Hb S disease	<20
Hb Lepore trait	<5
Hb Kenya trait	6-13
Thalassemias	
β-thalassemia minor	<5
δβ-thalassemia minor	5-20
β-thalassemia major	30-95
α-thalassemia minor	~1
Hb H disease	5-15
Hemoglobinopathy-thalassemia interactions	
S/β-thalassemia	10-30
E/β-thalassemia	10-50
C/β-thalassemia	10-30
Hereditary persistence of fetal hemoglobin (HPFH)	
African-type	
heterozygous	15-40
homozygous	100
Greek-type	
heterozygous	10-20
Swiss-type	
heterozygous	1-3

anomalies; during anticonvulsant drug therapy; diabetes; hyper- and hypothyroidism; and macroglobulinemia. Elevation of Hb F should, then, raise the question of possible underlying disease. Even after exclusion of this variety of disease states, some apparently normal individuals will have elevated Hb F in the 1.5% to 4.0% range. A majority of these result from mutations involving the gamma (γ) chain of Hb F. As of 2003, 52 variants (single amino acid substitutions) for the Gγ gene, 45 variants of the Aγ gene, and 6 variants of the AγT gene (hemoglobin F Sardinia) have been discovered.

Mutant Hb F with decreased oxygen affinity is in the differential diagnosis of newborns presenting with cyanosis of unknown cause. Hb F also causes interference with the laboratory measurement of percent saturation with O_2 and with measurement of fraction of oxygenated hemoglobin. Modern CO-Oximeter™ instruments nearly eliminate interference by Hb F through appropriate selection of light wavelengths.

Hemoglobin F levels have been studied in relation to risk factors for sudden infant death syndrome (SIDS). Higher levels of Hb F occurred in newborns with shorter gestations and lower birth weights. Newborns older than 38 weeks, whose mothers smoked, had higher Hb F levels than newborns of nonsmoking mothers. Factors associated with high risk for SIDS were found similar to factors leading to higher Hb F levels. The goal is to determine if Hb F levels can identify infants at high risk for SIDS.

References

Cochran DL, Conrad ME, and Matney J, "Hemoglobin F and Risk Factors for Sudden Infant Death Syndrome," *Lab Med*, 1997, 28(1):53-7.

Erler BS, Vitagliano P, and Lee S, "Superiority of Neural Networks Over Discriminant Functions for Thalassemia Minor Screening of Red Blood Cell Microcytosis," *Arch Pathol Lab Med*, 1995, 119(4):350-4.

Leonova JY, Kazanetz EG, Smetanina NS, et al, "Variability in the Fetal Hemoglobin Level of the Normal Adult," *Am J Hematol*, 1996, 53(2):59-65.

Nagel RL, "Hemoglobins: Normal and Abnormal," *Hematology of Infancy and Childhood*, 6th ed, Chapter 18, Nathan DG, Orkin SH, Ginsburg D, et al, eds, Philadelphia, PA: WB Saunders Co, 2003, 759-61.

Weatherall D, "Beginnings: The Molecular Pathology of Haemoglobin," *Molecular Haematology*, Chapter 1, Provan D and Gribben J, eds, Malden, MA: Blackwell Science, 2000, 1-17.

- ◆ **Fetal Hemoglobin Test in Newborn** *see* Apt-Downey Test *on page 208*
- ◆ **Fetalscreen**™ *see* Rosette Test for Fetomaternal Hemorrhage *on page 1172*
- ◆ **FFP** *see* Plasma, Fresh Frozen *on page 1039*
- ◆ **FFP** *see* Plasma, Frozen, Donor Retested *on page 1041*
- ◆ **Fibrin Breakdown Products** *see* D-Dimers and Fibrin Degradation Products *on page 502*
- ◆ **Fibrin Glue** *see* Cryoprecipitate *on page 481*
- ◆ **Fibrin Monomer** *see* Hypercoagulation Panel *on page 758*

Fibrinogen

Related Information

Activated Partial Thromboplastin Time *on page 100*
Coagulation Factor Assays *on page 418*
Cryoprecipitate *on page 481*
D-Dimers and Fibrin Degradation Products *on page 502*
Disseminated Intravascular Coagulation Screen *on page 517*
Factor XIII *on page 565*
Hypercoagulation Panel *on page 758*
Mixing Studies (Coagulation) *on page 918*
Plasma, Fresh Frozen *on page 1039*
Prothrombin Time *on page 1116*
Sedimentation Rate, Erythrocyte *on page 1181*
Thrombin Time *on page 1246*

Synonyms Factor I

Applies to Acute Phase Reactant; Afibrinogenemia; Dysfibrinogenemia; Plasmin; Sedimentation Rate; Thrombin

Abstract Fibrinogen is converted into fibrin clot by thrombin. Fibrinogen levels <100 mg/dL can be associated with bleeding. Acquired decreases in fibrinogen (eg, with liver dysfunction or DIC) are much more common than hereditary deficiencies.

Specimen Plasma

(Continued)

Fibrinogen (Continued)

Container Blue top (sodium citrate) tube

Collection Routine venipuncture. If multiple tests are being drawn, draw blue top tubes after any red top tubes but before any lavender top (EDTA), green top (heparin), or gray top (oxalate/fluoride) tubes. Immediately invert tube gently at least 4 times to mix. Tubes must be appropriately filled. Deliver tubes immediately to the laboratory.

Storage Instructions Separate plasma from cells as soon as possible. Store plasma at room temperature for up to 2 hours, at 2°C to 8°C for up to 4 hours, or store frozen.

Causes for Rejection Specimen received more than 4 hours after collection, tubes not filled, clotted specimen

Turnaround Time Less than 1 day

Reference Interval Approximately 150-400 mg/dL

Use One of several tests performed in a DIC panel, a prolonged PT or PTT evaluation, and an evaluation of a patient with an unexplained bleeding history

Limitations Heparin concentrations >0.6 units/mL can falsely decrease the result with the Clauss method (described below). Usual therapeutic doses of heparin do not significantly affect PT-based methods. The Ellis method is more sensitive to heparin than the Clauss method. Some reagents contain hexadimethrine bromide (Polybrene) to neutralize heparin, allowing fibrinogen to be measured in specimens containing heparin. Fibrin degradation products (FDP) >30-100 μg/mL may decrease fibrinogen values with the Clauss method. Hirudin or argatroban anticoagulation may falsely decrease fibrinogen levels levels with the Clauss and Ellis method, and possibly the PT-based method.

Methodology

Functional (activity) assays: The majority of clinical laboratories use the **Clauss** method, which is essentially a dilute thrombin time. A high concentration of thrombin is added to dilute patient plasma, which converts fibrinogen into fibrin clot. The clotting time is inversely proportional to the amount of fibrinogen in the sample. In the **Ellis method**, a lower amount of thrombin is added to undiluted patient plasma and change in turbidity is measured in a spectrophotometer. In the **PT-based method**, thromboplastin (tissue factor with phospholipid) is added to undiluted patient plasma to generate endogenous thrombin, and light scatter or turbidity is measured. The measured optical change (before and after fibrin clot formation) is proportional to the amount of fibrinogen in the sample.

Antigen assays (immunoassays) for fibrinogen measure the quantity of fibrinogen without assessing fibrinogen function. This method is not routinely indicated and is usually a send-out test (see Additional Information for its use in dysfibrinogenemia evaluations).

Additional Information Fibrinogen decreases with liver disease, due to decreased hepatic synthesis. However, fibrinogen may be normal or even elevated until late stages of hepatic disease. Fibrinogen decreases in DIC due to excessive thrombin generation, which converts fibrinogen into fibrin. The plasma fibrinogen level, however, is not a sensitive marker for DIC. Presence of a high fibrinogen level in patients with DIC is associated with a poor prognosis. Fibrinogen also decreases with thrombolytic therapy and fibrinolysis because plasmin breaks down fibrinogen in addition to fibrin.

Fibrinogen becomes elevated during acute phase reactions and during pregnancy, smoking, and physical inactivity. As with certain other acute phase reactants (eg, C-reactive protein), elevated fibrinogen has been associated with an increased risk of myocardial infarction. There is evident complexity involving gene to gene and gene to environment interactions in the genesis and modulation of cardiovascular risk. Thrombogenicity may relate to characteristics of the fibrin network (thin fibers and small pores with restricted entry of fibrinolytic enzymes) also affected by fibrinogen and factor XIII polymorphisms.

Hereditary deficiencies of fibrinogen are rare. The PT and PTT may be prolonged. Bleeding symptoms may include bruising, epistaxis, menorrhagia, bleeding with surgery, trauma, dental extractions, and postpartum, and bleeding in the gastrointestinal or genitourinary tract. Miscarriage and poor wound healing are also complications of fibrinogen deficiency. Umbilical stump bleeding and bleeding with circumcision may be noted in newborns with afibrinogenemia. Intracranial hemorrhage has been reported with afibrinogenemia. In general, deficiencies of fibrinogen tend to be milder than factor VIII or IX deficiencies (hemophilia).

There are three major types of fibrinogen deficiency. The homozygous quantitative form, called afibrinogenemia, results in a severe quantitative deficiency of fibrinogen and an increased risk for bleeding. The heterozygous form of this deficiency is hypofibrinogenemia, with less severe reductions in the fibrinogen level and little or no bleeding. Fibrinogen consists of two copies of each of three polypeptide chains called α, β, and γ. Among the afibrinogenemia mutations that have been characterized thus far, most have been found in the α-fibrinogen chain gene.

Dysfibrinogenemia is a qualitative fibrinogen deficiency, characterized by the production of dysfunctional fibrinogen. Many different mutations are known to cause hereditary dysfibrinogenemia. Most patients with hereditary dysfibrinogenemia are heterozygous. Rare homozygous cases have been reported. Dysfibrinogenemia patients are usually asymptomatic or have mild bleeding, but severe bleeding has been reported. Interestingly, some dysfibrinogenemia cases are associated with thrombosis, with or without bleeding. Dysfibrinogenemia has an estimated prevalence of 0.8% in patients with venous thrombosis. Arterial thrombosis is less frequent than venous thrombosis in these patients. Acquired forms of dysfibrinogenemia, of uncertain clinical significance, can be seen with liver disease or acute phase reactions with generation of high levels of fibrinogen (Galanakis D, personal communication 1999). The thrombin time and Reptilase® time, which measure the clotting time during the conversion of fibrinogen into fibrin, are often prolonged in dysfibrinogenemia. The PT and PTT may also be prolonged. In dysfibrinogenemia, assays that measure fibrinogen function show lower levels than assays that measure fibrinogen quantity (immunological or "antigen" assays), because fibrinogen function is impaired but fibrinogen quantity is not. This potentially diagnostic disparity between functional and antigen levels may be less pronounced with PT-based functional fibrinogen assays than with Clauss-based functional assays. See Table 3 in Coagulation Factor Assays *on page 418*.

See Thrombin Time *on page 1246*.

References

Haverkate F and Samama M, "Familial Dysfibrinogenemia and Thrombophilia. Report on a Study of the SSC Subcommittee on Fibrinogen," *Thromb Haemost*, 1995, 73(1):151-61.

Kamath S and Lip GYH, "Fibrinogen: Biochemistry, Epidemiology, and Determinants," *Q J Med*, 2003, 96(10):711-29.

Lim BC, Arlëns RA, Carter AM, et al, "Genetic Regulation of Fibrin Structure and Function: Complex Gene-Environment Interactions May Modulate Vascular Risk," *Lancet*, 2003, 361(9367):1424-31.

Rossi E, Mondonico P, Lombardi A, et al, "Method for the Determination of Functional (Clottable) Fibrinogen by the New Family of ACL Coagulometers," *Thromb Res*, 1988, 52(5):453-68.

Wada H, Mari Y, Okabayashi K, et al, "High Plasma Fibrinogen Level Is Associated With Poor Clinical Outcome in DIC Patients," *Am J Hematol*, 2003, 72(1):1-7.

♦ **Fibrinogenolysis** *see* Plasminogen *on page 1042*

♦ **Fibrinogen Therapy** *see* Cryoprecipitate *on page 481*

♦ **Fibrinoligase** *see* Factor XIII *on page 565*

♦ **Fibrinolysis** *see* D-Dimers and Fibrin Degradation Products *on page 502*

♦ **Fibrinolysis** *see* Disseminated Intravascular Coagulation Screen *on page 517*

♦ **Fibrinolysis** *see* Plasminogen *on page 1042*

♦ **Fibrinopeptide A** *see* Hypercoagulation Panel *on page 758*

♦ **Fibrinopeptide A** *see* Thrombin Time *on page 1246*

♦ **Fibrinopeptide B** *see* Hypercoagulation Panel *on page 758*

♦ **Fibrinopeptide B** *see* Thrombin Time *on page 1246*

♦ **Fibrin Split Products** *see* D-Dimers and Fibrin Degradation Products *on page 502*

♦ **Fibrin Stabilizing Factor** *see* Factor XIII *on page 565*

Fibronectin, Fetal, Cervicovaginal Secretions

Related Information

Bacterial Culture, Genital Specimen *on page 239*

Point-of-Care Testing *on page 1065*

Synonyms Fetal Fibronectin, Cervicovaginal Secretions

(Continued)

Fibronectin, Fetal, Cervicovaginal Secretions
(Continued)

Abstract Fetal fibronectin (fFN) is distinguishable from other fibronectins by a specific epitope (the "oncofetal domain"). FN is found in fetal connective tissue, amniotic fluid, and placenta. Placental fFN is localized to the region at which the placenta and its membranes contact the wall of the uterus. This location permits fFN leakage into the vagina prior to the onset of labor. fFN is normally present in cervicovaginal fluid before 21 weeks gestation. However, after 21 weeks, fFN ceases to be present in cervicovaginal secretions until just before delivery at about 37 weeks. **The abnormal presence of fFN in cervicovaginal secretions between 21-37 weeks gestation is associated with preterm delivery.**

Patient Preparation Patients with signs and symptoms of premature labor should be sampled between 24 weeks, 0 days and 34 weeks, and 6 days gestation. Amniotic membranes should be intact, and cervical dilation should be minimal (<3 cm). There should be no cervical cerclage; moderate or gross vaginal bleeding should also be absent.

Asymptomatic patients should be sampled between 22 weeks, 0 days and 30 weeks, 6 days in order to achieve maximum sensitivity. There should be no cervical cerclage, no placenta previa, and no history of sexual intercourse within the 24 hours prior to specimen collection.

The sampling area should be free of lubricants, soaps, disinfectants, or creams. Specimens should be obtained prior to digital cervical examination or vaginal probe ultrasound examination as manipulation of the cervix may cause the release of fFN.

Aftercare Fetal fibronectin is said to identify subjects at risk for delivery in the following 7 days.

Specimen Cervicovaginal secretions, swab of vaginal vault

Sampling Time Clinical criteria for testing:
- intact amniotic membranes
- cervical dilatation <2 cm and effacement <80%
- sampling not earlier than 24 weeks and not later than 34 weeks, 6 days gestation

Collection During speculum examination, a Dacron® swab is used to obtain specimen from the posterior vaginal fornix or cervix. The specimen is eluted from the swab into a buffer solution.

Storage Instructions Store for up to 3 days at 2°C to 8°C until analysis.

Turnaround Time ≤2 hours is recommended

Reference Interval Results are reported as negative or positive. Negative results correspond to concentrations <50 ng/mL and are associated with a low likelihood of preterm delivery. Positive results correspond to concentrations ≥50 ng/mL and are indicative of increased risk of preterm labor and delivery.

Use Fetal fibronectin is approved by the United States FDA to assess the risk of preterm delivery in symptomatic and asymptomatic pregnant women. Off-label uses include confirmation of ruptured membranes, prediction of term and post-term delivery, and prediction of successful labor induction. With identification of expectant mothers likely to go into early premature labor, intervention can support continued maturation of the fetus with diminution of risk of neonatal respiratory distress syndrome.

Limitations The sensitivity in screening for the likelihood of preterm delivery in low risk, asymptomatic patients is poor. However, sensitivity in this group improves somewhat by sampling during the early gestational stages (eg, 22-26 weeks). Specificity and negative predictive value of the fFN test in asymptomatic patients is high.

Methodology Enzyme-linked immunosorbent assay (ELISA) and lateral flow, solid-phase immunosorbent assay. Adeza® rapid assay is in use in a tertiary care center.

Additional Information The presence of fFN in the cervicovaginal secretions of symptomatic patients at risk for preterm delivery is a particularly sensitive and specific indicator of delivery within 7 days, with sensitivities of 90.5% to 93% and negative predictive values of 99% to 99.7%. Sensitivity of the test decreases in predicting longer intervals to delivery, whereas the negative predictive values remain high. A prospective study of symptomatic patients showed reduced preterm labor admissions, lengths of stay, and prescriptions for tocolytic agents as a result of testing.

See Bacterial Culture, Genital Specimen *on page 239.*

References

"Fetal Fibronectin Preterm Labor Risk Test," ACOG Committee Opinion Number 187, Committee on Obstetric Practice, Washington, DC: The American College of Obstetricians and Gynecologists, September 1997.

Giles W, Bisits A, Knox M, et al, "The Effect of Fetal Fibronectin Testing on Admissions to a Tertiary Maternal-Fetal Medicine Unit and Cost Savings," *Am J Obstet Gynecol,* 2000, 182(2):439-42.

Heise RH and Rommel S, "Fetal Fibronectin: Bedside and In-House Testing for Prediction of Preterm Labor," *Mayo Reference Services Communiqué,* Rochester, MN, 2000.

Joffe GM, Jacques D, Bemis-Heys R, et al, "Impact of the Fetal Fibronectin Assay on the Admissions for Preterm Labor," *Am J Obstet Gynecol,* 1999, 180(3 Pt 1):581-6.

Lukes AS, Thorp JM Jr, Eucker B, et al, "Predictors of Positivity for Fetal Fibronectin in Patients With Symptoms of Preterm Labor," *Am J Obstet Gynecol,* 1997, 176(3):639-41.

♦ **Filaggrin** *see* Anticyclic Citrullinated Peptide Antibody *on page 169*

♦ **Filarial Infestation** *see* Microfilariae, Peripheral Blood Preparation *on page 915*

Filariasis Diagnostic Procedures

Related Information

Microfilariae, Peripheral Blood Preparation *on page 915*
Ova and Parasites, Direct Exam *on page 985*

Abstract Filariasis (caused by nematodes *Wuchereria bancrofti, Brugia malayi, Brugia timori, Loa loa, Mansonella ozzardi, Mansonella perstans, Mansonella streptocerca,* or *Onchocerca volvulus*), a group of tropical diseases transmitted by arthropods, is diagnosed primarily by clinical presentation, Giemsa-stained blood smears, and presence or absence of nocturnal periodicity. Filarial nematodes live in body cavities, subcutaneous tissue, or as adults, in the lymphatics of the host. Infection often leads to lymphatic inflammation and/or chronic lymphatic obstruction. Hydrocoele and elephantiasis develop secondarily in *Wuchereria* infestation. Microfilariae (embryos) are ingested by bloodsucking arthropods, primarily mosquitoes, and develop to an infective phase. It is the microfilaria stage which is accessible for diagnosis.

Specimen Serum or plasma for serological testing; blood and/or hydrocoele fluid for smear diagnosis

Container Red top tube for serology

Sampling Time Smear diagnosis: periodicity is relevant. *W. bancrofti* and *B. malayi* are mostly nocturnal.

Special Instructions Travel history is needed.

Use Support a diagnosis of filariasis, microfilariasis causing elephantiasis, hydrocele, chyluria, tropical pulmonary eosinophilia

Limitations Cross-reactivity occurs between filarial antigens and antigens of other helminthes. False positives occur in subjects residing in endemic areas. False-negative blood smears occur.

Methodology Enzyme-linked immunosorbent assay (ELISA); polymerase chain reaction (PCR) for *W. bancrofti, B. malayi,* and oncocerciasis; immunochromatographic card test; counting chamber method; filtration method; snip diagnosis and biopsy for oncocerciasis

Additional Information For screening, an antigen prepared from the dog heartworm, *Dirofilaria immitis,* detects antibody responses to several clinically significant microfilariae, but sensitivity and specificity are poor. **Morphologic examination of a blood film or hydrocoele fluid remains the foundation of diagnosis.**

References

Boatin BA, Toe L, Alley ES, et al, "Diagnosis in Onchocerciasis: Future Challenges," *Ann Trop Med Parasitol,* 1998, 92(Suppl 1):S41-5.

Hall LR and Pearlman E, "Pathogenesis of Onchocercal Keratitis (River Blindness)," *Clin Microbiol Rev,* 1999, 12(3):445-53.

Harnett W, Bradley JE, and Garate T, "Molecular and Immunodiagnosis of Human Filarial Nematode Infections," *Parasitology,* 1998, 117(S):S59-71.

Kale OO, "Onchocerciasis: The Burden of Disease," *Ann Trop Med Parasitol,* 1998, 92(Suppl 1):S101-15.

Mayo Reference Services Publication, "Filarial Antigens, Serum," *New Test Announcement,* #82412, Rochester, MN: Mayo Medical Laboratories, 2001.

Simonsen PE, "Filariases," *Manson's Tropical Diseases,* 21st ed, Chapter 82, Cook GC and Zumla A, eds, Philadelphia, PA: WB Saunders Co, 2003, 1486-1526.

(Continued)

Filariasis Diagnostic Procedures (Continued)

Internet Web Sites

www.astdhpphe.org/infect/lymphfil.html

www.cdc.gov/ncidod/dpd/parasites/lymphaticfilariasis/default.htm

♦ **Filariasis, Peripheral Blood Preparation** *see* Microfilariae, Peripheral Blood Preparation *on page 915*

Filters for Blood

Related Information

Platelets, Apheresis, Donation *on page 1054*

Red Blood Cells *on page 1139*

Red Blood Cells, Leukocytes Reduced *on page 1141*

Synonyms Filters, Leukocyte Reduction; Filters, Microaggregate; Leukocyte-Depletion Filters

Applies to Alloimmunization, Leukocyte; Febrile Transfusion Reaction; Transfusion Reaction, Febrile

Special Instructions Routine blood administration (or clot filters) and special blood filters (microaggregate and leukocyte reduction) are available, as a rule, at hospital transfusion services or from pharmacy and supplies services. Leukocyte reduction filters are designed to be component specific, thus, take care to select the appropriate filter for intended use. Prestorage leukocyte-reduced blood components, which may be available from the blood supplier, negate the need for laboratory or bedside leukocyte reduction filtration.

Use All usual blood transfusions are given through an administration set which includes a filter. Clots may form in any unit of blood and are readily removed by the clot filters in all regular blood infusion sets.

Microaggregate filters remove debris composed of platelets with admixed granulocytes and fibrin in massive transfusions of older stored units of blood. Usage remains controversial, as these filters cannot prevent HLA alloimmunization nor CMV transmission.

Leukocyte filters help reduce, but do not eliminate febrile nonhemolytic reactions. Two consecutive febrile reactions may be an indication for leukocyte-poor blood. Leukocyte reduction filters also reduce the likelihood of HLA alloimmunization and CMV transmission.

The importance of leukocyte reduction in platelet fractions is discussed in Platelet Transfusion *on page 1058* and in other monographs relevant to platelets. Leukocyte reduction by filtration and ultraviolet B irradiation of platelets were equally effective in prevention of refractoriness to platelets in a 1997 study.

White particulate matter, described in 2003, is composed of aggregates of platelets and variable amounts of fibrin and trapped red and white cells. Leukocyte filtration removes the most observable particles and may be considered as a measure to reduce the frequency of particles in blood units.

Limitations Do **not** use microaggregate or leukocyte reduction filters for granulocyte transfusions. Leukocyte reduction filtration after various periods of storage of RBCs or platelets may not be as effective as at the time of collection. Leukocyte reduction filters are not adequate to prevent graft-versus-host disease in susceptible blood recipients. Both microaggregate and leukocyte reduction filters have the potential to clog and become resistant to increased blood flow. Use of blood components leukocyte reduced prior to issue from the Blood Bank may alleviate this difficulty.

Additional Information There are several different types of blood filters:

Clot filter (170 micron pore size): All blood and components must be given through this filter, intended to remove clots and fibrin shreds.

Microaggregate filters (20-40 microns pore size). These filters remove the microaggregates of leukocytes and platelets that form in stored blood, particularly for blood that is recirculated in cardiac bypass devices. The aim is to prevent microembolization and respiratory distress syndrome, although there is little support for routine use of microaggregate filters. They have not been proven to influence development of the adult respiratory distress syndrome.

Leukocyte-depletion filters (3-100 micron pore size). These remove up to three logs (99.9%) of WBCs from platelets or RBCs. The purpose of leukocyte reduction is to prevent febrile nonhemolytic transfusion reactions, alloimmunization to HLA antigens, and transmission of viruses carried by leukocytes (eg, cytomegalovirus). For such purposes, it appears that filtration must be so efficient that the platelet or RBC concentrates contain no more than 5×10^6 WBCs per transfusion. Since electronic particle counters are grossly inaccurate in those count ranges, quality control of leukocyte-reduced products requires special techniques. Cost and other practical matters are still evolving in the decision over universal leukocyte reduction and the manner (bedside filtration vs prestorage filtration) in which it will be achieved. For optimal effect, all leukocyte reduction filters must be carefully used according to the manufacturer's instructions.

References

Lipson SM, Shepp DH, Match ME, et al, "Cytomegalovirus Infectivity in Whole Blood Following Leukocyte Reduction by Filtration," Am J Clin Pathol, 2001, 116(1):52-5.

Slichter S, "Leukocyte Reduction and Ultraviolet B Irradiation of Platelets to Prevent Alloimmunization and Refractoriness to Platelet Transfusions," The Trial to Reduce Alloimmunization to Platelets Study Group, N Engl J Med, 1997, 337(26):1861-9.

Snyder EL, "Transfusion Reactions," Hematology: Basic Principles and Practice, 3rd ed, Hoffman R, Benz EJ Jr, Shattil SJ, et al, eds, New York, NY: Churchill Livingstone, 2000, 2300-10.

Triulzi DJ, Blood Transfusion Therapy: A Physician's Handbook, 7th ed, Bethesda, MD: American Association of Blood Banks Press, 2002, 81.

Update on Particulate Matter in Blood Bags, October 31, 2003. www.fda.gov/cber/infosheets/bldpartic.htm

♦ **Filters, Leukocyte Reduction** see Filters for Blood on page 588

♦ **Filters, Microaggregate** see Filters for Blood on page 588

♦ **Finasteride** see Prostate Specific Antigen, Serum on page 1098

Fine Needle Aspiration Culture

Related Information

Coccidioidomycosis Serology on page 428
Fine Needle Aspiration, Deep and Superficial Masses on page 590
Fungal Culture, Biopsy or Body Fluid on page 619
Fungal Culture, Sputum on page 624
Mycobacterial Culture, Biopsy or Body Fluid on page 929
Mycobacterial Culture, Cutaneous and Subcutaneous Tissue on page 932

Synonyms CT Guided Needle Aspiration Culture; FNA Culture

Applies to Swabs for Culture

Test Includes Culture for bacterial, mycobacterial, or fungal organisms. Gram stain is commonly indicated if sufficient material is available.

Container Clean, sterile container

Use Swabs of ulcers or fistulas may provide misleading information when deep infections are evaluated. Culture of abscesses or masses suspected of being infectious, especially tuberculous masses, requires adequate sampling. Fine needle aspiration culture or surgical sampling is needed to obtain critical material for culture (eg, in osteomyelitis). Fine needle aspiration is also used for molecular techniques of infectious disease diagnosis, and often are highly desirable for cytologic evaluation.

Limitations May not obtain sufficient material for culture

Methodology The skin is cleansed to remove any skin residual flora. A needle is guided into the mass through the skin. The specimen is aspirated into the syringe and taken directly to the laboratory for culture and examination.

Additional Information Fine needle aspiration is a simple diagnostic tool used for the evaluation of superficial and deep seated lesions or masses. Cultures from needle aspiration specimens have a high diagnostic yield and avoid invasive procedures. See specimen requirements for each specific type of culture.

Refer to Lymphadenopathy in the Disease Index.

References

Alkan S, Eltoum IA, Tabbara S, et al, "Usefulness of Molecular Detection of Human Herpesvirus-8 in the Diagnosis of Kaposi Sarcoma by Fine Needle Aspiration," Am J Clin Pathol, 1999, 111(1):91-6.

Ellison E, LaPuerta P, and Martin SE, "Fine Needle Aspiration Diagnosis of Mycobacterial Lymphadenitis. Sensitivity and Predictive Value in the United States," Acta Cytol, 1999, 43(2):153-7.

Ellison E, LaPuerta P, and Martin SE, "Supraclavicular Masses: Results of a Series of 309 Cases Biopsied by Fine Needle Aspiration," Head Neck, 1999, 21(3):239-46.

(Continued)

Fine Needle Aspiration Culture *(Continued)*

Francis IM, Das DK, Luthra UK, et al, "Value of Radiologically Guided Fine Needle Aspiration Cytology (FNAC) in the Diagnosis of Spinal Tuberculosis: A Study of 29 Cases," *Cytopathology*, 1999, 10(6):390-401.

Lew DP and Waldvogel FA, "Osteomyelitis," *N Engl J Med*, 1997, 336(14):999-1007.

Fine Needle Aspiration, Deep and Superficial Masses

Related Information

Bronchoalveolar Lavage (BAL) *on page 311*
CA 15-3, Serum *on page 320*
CA 19-9, Serum *on page 321*
CA 125, Serum *on page 323*
Carcinoembryonic Antigen, Serum *on page 342*
Cyst Fluid Cytology *on page 490*
Electron Microscopy *on page 533*
Fine Needle Aspiration Culture *on page 589*
Flow Cytometry, Overview *on page 592*
Fluorescence *in situ* Hybridization *on page 602*
Fungal Culture, Biopsy or Body Fluid *on page 619*
Immunoperoxidase Procedures *on page 780*
Mycobacterial Culture, Biopsy or Body Fluid *on page 929*
Polymerase Chain Reaction *on page 1069*
Thyroid Stimulating Hormone, Serum *on page 1250*
Transbronchial Fine Needle Aspiration *on page 1266*

Synonyms FNA

Applies to Abdominal Mass Aspiration; Adrenal Mass Aspiration; Bone Needle Aspiration Cytology; Brain Needle Aspiration; Breast Aspiration Cytology; Cyst Aspiration Cytology; Liver Needle Aspiration Cytology; Lung Needle Aspiration Cytology; Lymph Node Aspiration Cytology; Mediastinal Mass Aspiration; Neck Mass Aspiration; Needle Biopsy Cytology; Pancreas Needle Aspiration Cytology; Prostate Needle Aspiration; Renal Mass Aspiration; Retroperitoneal Mass Aspiration; Salivary Gland Needle Aspiration; Soft Tissue Mass Aspiration; Thyroid Needle Aspiration Cytology

Test Includes Examination of air-dried, Diff-Quik™ stained, and/or ethanol-fixed, Papanicolaou-stained direct smears, cytospin or cell block preparations. Useful adjuncts include microbiological cultures, special stains, flow cytometry, laser scanning cytometry, and immunohistochemistry.

Abstract Techniques for imaging deep organs have led to a reliable means of sampling, using fine needle aspiration biopsy. Successful cytologic diagnosis is dependent upon adequate sampling. As sample size is small, aspirated material can only be accurately interpreted when cytologic findings are placed in a clinical context.

Patient Preparation Signed informed consent from the patient is required. The patient is prepped surgically to produce a sterile field. One percent lidocaine is used for local anesthesia of skin and overlying subcutaneous tissue. Sedation and/or analgesics may be needed. The suite in which the procedures are done should be equipped for the handling of complications, such as pneumothorax (chest tube trays, etc).

Aftercare Superficial masses: Adequate pressure is essential to prevent significant deep hematoma, especially in hyperplastic lymph nodes, thyroid gland, and salivary gland aspirates. If the patient regularly takes aspirin, longer pressure with an ice pack may be required, as in patients on anticoagulants. Mild analgesics may be needed by the patients for 24-48 hours postbiopsy if the area is painful.

Deep masses: Patient should be advised of possible discomfort, local pain, and bleeding. A postpulmonary biopsy routine chest x-ray is obtained, as are routine cuts of liver, and/or other sites; post-CT biopsy, to exclude hematoma. Occasionally, antibiotics may be required, depending upon clinical circumstances.

Collection Most important is the accurate localization of the needle tip within the mass. Scans of deep masses should be performed to confirm appropriate needle placement prior to performing FNA. Smears and needle rinses are usually prepared by a cytotechnologist or cytopathologist. Immediate evaluation of air-dried Diff-Quik™ stained smears is possible, which rapidly determines the adequacy of the specimen, and ensures appropriate specimen triage. Material for culture, unfixed, may be obtained by this method.

Storage Instructions All material should be immediately brought to the Cytology Laboratory. The needle rinse, if immediate delivery to the laboratory is not possible, should be refrigerated at 4°C.

Use FNA is most often used to diagnose benign and malignant neoplasms, and may be particularly helpful in identifying metastasis and tumor recurrence as well as in staging. Its applications include head and neck cancer, especially when carcinoma is suspected, no primary site is identified, and a neck mass is found. FNA is useful in evaluation of selected thyroid nodules, in which adequate tissue is obtained in ~90% of patients. Differential diagnostic considerations include cystic lesions, goiter, thyroiditis, benign neoplasms, papillary/follicular carcinoma, medullary carcinoma, anaplastic carcinoma and lymphoma, as well as other entities. Fine needle aspiration of soft tissue lesions may be challenging. FNA may also assist in the diagnosis of infectious and inflammatory processes (eg, human herpesvirus 8).

Fine needle aspiration is used for culture (see Fine Needle Aspiration Culture on page 589) and for molecular detection of pathogens.

Limitations Sampling error, particularly with small (1 cm or less) pulmonary nodules; lesions completely surrounded by bone; or when the procedure must be terminated due to complications or patient discomfort.

Excisional biopsy of lymph nodes for lymphoma provides evaluation of lymph node architecture. Flow cytometry with cytomorphology enhances FNA diagnoses, but false-negative diagnosis remains possible.

The limitations of FNA are largely site dependent. For example, aspirations of thyroid follicular lesions cannot reliably indicate if the lesion is an adenoma or carcinoma; histopathologic examination is needed.

Contraindications Severe chronic obstructive pulmonary disease is a contraindication to pulmonary aspiration, which has a 15% to 30% risk of pneumothorax; coagulopathy; adrenal or extra-adrenal mass in which the diagnosis of pheochromocytoma is being considered.

Methodology Immunocytochemistry is described as the backbone of the Cytopathology Section of the Laboratory of Pathology of the National Institutes of Health. Fluorescence in situ hybridization (FISH) lends itself to cytopathology. The polymerase chain reaction is sometimes utilized.

Additional Information Needle tracking of malignancy has been reported in the literature but occurs only rarely. Although the majority of such cases used 18-gauge or greater needles, incidents of tracking with 22- and 23-gauge needles, especially of high-grade pancreatic carcinoma, have been reported. Other significant complications include pneumothorax, empyema, hemorrhage, nerve damage, and sudden death due to aspiration or to pheochromocytoma. A neck mass in an elderly patient may represent a calcified atherosclerotic carotid. Thyroid laceration with extensive bleeding may occur if the patient moves during thyroid aspiration.

Cell blocks may contribute to diagnosis.

References

Alkan S, Eltoum IA, Tabbara S, et al, "Usefulness of Molecular Detection of Human Herpesvirus 8 in the Diagnosis of Kaposi Sarcoma by Fine Needle Aspiration," Am J Clin Pathol, 1999, 11(1):91-6.

Frable WJ, "Fine Needle Aspiration Biopsy Techniques," Comprehensive Cytopathology, 2nd ed, Chapter 25, Bibbo M, ed, Philadelphia, PA: WB Saunders Co, 1997, 623-42.

Nicol TL, Silberman M, Rosenthal DL, et al, "The Accuracy of Combined Cytopathologic and Flow Cytometric Analysis of Fine-Needle Aspirates of Lymph Nodes," Am J Clin Pathol, 2000, 114(1):18-28.

Ost D, Fein AM, and Feinsilver SH, "The Solitary Pulmonary Nodule," N Engl J Med, 2003, 348(25):2535-42.

Schlinkert RT, van Heerden JA, Goellner JR, et al, "Factors That Predict Malignant Thyroid Lesions When Fine-Needle Aspiration Is Suspicious for Follicular Neoplasm," Mayo Clin Proc, 1997, 72(10):913-6.

♦ **Fingerstick Blood Collection** see Skin Puncture Blood Collection on page 1203

♦ **FiO₂** see Blood Gases and pH, Arterial on page 275

♦ **Fiorinal®** see Barbiturates, Quantitative, Serum or Plasma on page 248

+ **First Trimester Combined Test** *see* Pregnancy-Associated Protein A, Serum *on page 1082*
+ **FISH** *see* Fluorescence *in situ* Hybridization *on page 602*
+ **FISH Assays** *see* HER-2/neu *on page 716*
+ **Fishberg Concentration Test** *see* Concentration Test, Urine *on page 446*
+ **FK-506** *see* Tacrolimus, Whole Blood *on page 1234*
+ **Flea Identification** *see* Arthropod Identification *on page 213*

Flecainide, Serum or Plasma

Related Information
Amiodarone, Serum *on page 148*
Digoxin, Serum *on page 512*
Propranolol, Serum *on page 1096*

Synonyms Almartyn®; Tambocor®

Abstract Flecainide is a class IC antiarrhythmic drug. FDA recommends that the drug be reserved for life-threatening ventricular arrhythmias unresponsive to conventional therapy. Not recommended for patients with chronic atrial fibrillation. A worsening or new arrhythmia may occur (proarrhythmic effect).

Specimen Serum or plasma

Container Red top tube (preferred) or green top (heparin) tube; do not use serum separator tube.

Sampling Time Draw sample just prior to dose (trough).

Storage Instructions Separate serum or plasma within 2 hours of the time the specimen was drawn from the patient.

Special Instructions Collect sample immediately prior to next dose for trough levels.

Reference Interval Therapeutic: trough: 0.2-1.0 µg/mL (SI: 0.5-2.4 µmol/L)

Critical Values Levels >1.0 µg/mL (SI: >2.4 µmol/L) have been related to higher incidence of adverse experiences. Monitor with ECG (QRS width, QTc prolongation, presence of first-degree or greater heart block) as well as with serum concentrations.

Methodology High performance liquid chromatography (HPLC)

Additional Information
- Half-life (normal adult): 7-19 hours
- Volume of distribution: 5-13 L/kg
- Protein binding: 40% to 50% to plasma protein, primarily alpha-1-acid glycoprotein

Since 10% to 50% of the drug is eliminated in the urine as unchanged drug, impaired renal function will significantly prolong the plasma half-life. Half-life is also increased with congestive heart failure. Clearance of flecainide can be accelerated by phenobarbital and rifampin. Coadministration with digoxin increases serum digoxin concentrations by about 15% to 25%; coadministration with propranolol increases serum concentrations of both drugs (flecainide's by about 20%, propranolol's by about 30%), and the possibility of additive negative inotropic effects should be considered.

The role of oral loading single-dose flecainide for pharmacological cardioversion of recent-onset atrial fibrillation has recently been evaluated. The success rate ranged from 57% to 68% at 2-4 hours and 75% to 91% at 8 hours after drug administration.

References
Capucci A, Villani GQ, Piepoli MF, et al, "The Role of Oral 1C Antiarrhythmic Drugs in Terminating Atrial Fibrillation," *Curr Opin Cardiol*, 1999, 14(1):4-8.

Khan IA, "Oral Loading Single Dose Flecainide for Pharmacological Cardioversion of Recent-Onset Atrial Fibrillation," *Int J Cardiol*, 2003, 87(2-3):121-8.

Valdes R Jr, Jortani SA, and Gheorghiade M, "Standards of Laboratory Practice: Cardiac Drug Monitoring," *Clin Chem*, 1998, 44(5):1096-109.

+ **Fletcher Factor** *see* Prekallikrein *on page 1085*
+ **Flies** *see* Arthropod Identification *on page 213*

Flow Cytometry, Overview

Related Information
Apoptosis Assays *on page 205*
Body Cavity Fluid Cytology *on page 285*
Body Fluid Chemical Analysis *on page 291*

Applies to Ber-EP$_4$; CD; Cell Cycle Analysis by Flow Cytometry; Clusters of Differentiation; Immunophenotypic Analysis of Tissues by Flow Cytometry; Prion Protein; Suspension Array Technology

Abstract Flow cytometry (FCM) is a technique for the identification, enumeration, and/or separation of particles (including cells) on the basis of light-scattering and/or fluorescence characteristics. Operationally, a column (shaped and contained by a "sheathing" fluid) of particles flows single file past the flow cell analysis point of the cytometer. Important established and expanding uses include immunophenotyping for classification of leukemia/lymphoma, tumor ploidy and S-phase determination, diagnosis of immune deficiency disorders including acquired immunodeficiency syndrome (AIDS), stem cell enumeration, diagnosis of paroxysmal nocturnal hemoglobinuria (PNH), detection of minimal residual disease following therapy of malignancy, enumeration of reticulocyte and platelet parameters, semen analysis, and an evergrowing number of clinical as well as research applications. Multicolor flow cytometry-based suspension array technology (SAT) can perform rapid simultaneous multianalyte immunoassay.

Specimen Whole blood, heparinized. EDTA-anticoagulated blood may be used for many applications. A variety of other clinical specimens can be analyzed (eg, bone marrow, cerebrospinal fluid, serous body fluids/peritoneal washings). Cells must be viable and present as a suspension of single cellular elements. Fresh or fresh frozen tissue, needle aspirate specimen, as well as paraffin-embedded tissue may be used. Tissue must be dissociated mechanically and/or by enzyme treatment into an individual cell suspension.

Container Frozen tissue submitted for DNA ploidy studies should not be embedded in O.C.T.®

Storage Instructions Fresh tissue generally may require immediate processing. Fresh tissues submitted for immunophenotypic studies are sufficiently stable to be transported by overnight courier on ice pack to a reference laboratory. Frozen tissue submitted for DNA ploidy studies should be maintained frozen during transport to a reference laboratory. Frozen and paraffin-embedded tissues may be suitable for months to years.

Causes for Rejection Extensive necrosis, poor fixation, or other degenerative or chemical interfering substances (eg, fixatives such as B5, Zenkers)

(Continued)

Flow Cytometry, Overview *(Continued)*

Special Instructions Cell fixation and permeabilization are required for analysis of intracellular targets, see Fetal Hemoglobin *on page 581.*

Reference Interval Application dependent

Use

Hematology

- Diagnosis/classification of **leukemia** (myeloid, lymphoid, erythroid)
- Diagnosis/classification of lymphoproliferative disorders (**malignant lymphoma** and leukemia)
- Detection of clonality and minimal residual disease
- DNA content and cell cycle analysis
- Reticulocyte count, levels of maturation and indices (reticulocyte mean volume, hemoglobin concentration, and hemoglobin content), see Reticulocyte Count *on page 1156* and Hemoglobin *on page 681*
- Granulocyte function and antineutrophil antibody studies
- Assay for leukocyte alkaline phosphatase (a "PIG" protein - see PNH Test (GPI-Anchored Proteins) by Flow Cytometry *on page 1064*)
- Platelet count, maturation, and antiplatelet antibody studies
- Diagnosis of PNH by flow cytometry, see PNH Test (GPI-Anchored Proteins) by Flow Cytometry *on page 1064*
- F-cell analysis by flow cytometry, see Fetal Cell Detection by Flow Cytometry *on page 579*

Immunology

Identification and enumeration of cell populations identified by surface antigens classified as "clusters of differentiation" (CD) and including numerous members of immunoglobulin, receptor or other "superfamilies" (eg, CD34, member of a family of adhesion molecules-sialomucins and present on hemopoietic stem cells). See CD34⁺ Hematopoietic Stem Cells by Flow Cytometry *on page 351* and Cluster of Differentiation Antigens Table *on page 44* in the Introduction.

Cell Cycle Analysis

Assessment of ploidy and S phase fraction (SPF) may contribute to the estimation of prognosis and planning of therapy for patients with limited stage neoplastic disease.

Immunophenotypic Analysis of Tissues

Characterization of cells by results of immunophenotypic predominantly surface membrane studies as well as DNA cell cycle analyses for diagnostic and prognostic applications in surgical pathology, cytopathology, and hematopathology.

Limitations Eosinophils may bind unconjugated fluorescein isothiocyanate (trace amounts of which may contaminate reagent kits) and act as a potential source of analytic error. Sensitivity of immunophenotyping is dependent upon characteristics of the monoclonal antibody (eg, specificity, affinity, and properties of the fluorochrome label). Background fluorescence must be minimized. Nonspecific binding of antibody (mediated through the Fc fragment and nonspecies specific) must be decreased by pretreatment with human plasma. FCM may be less than ideally sensitive in detection of T-cell lymphoid malignancy. If malignant cells are few and present admixed with many benign/reactive hematopoietic cells (as occurs with most forms of Hodgkin disease and with T-cell rich B-cell lymphoma) malignancy may not be detected.

Use of necrotic or fibrotic tissue may result in uninterpretable results. Studies of paraffin block tissues may lack sensitivity.

Flow cytometry is of limited use when nonlymphoid or nonhematopoietic neoplasms are evaluated. The nature of the tissue may also confer severe limitations. Generally, endoscopic biopsies are too small to extract sufficient cells for a meaningful evaluation. However, blood, bone marrow, and body fluids are readily suitable for flow cytometric studies. Tissues which are fibrotic or sclerotic, such as skin, are frequently too difficult to dissociate and extract enough viable cells for evaluation. Additionally, necrotic tissues frequently produce poor results. Finally, important diagnostic morphologic and architectural features are lost when single cell suspensions are made. Therefore, the suitability of each biopsy needs to be individually considered before tissues are allocated for special studies beyond that of conventional histopathology. Interpretation of the phenotypic profile must always be correlated with pathological features of individual cases. In those cases in which flow cytometry fails to demonstrate clonality, the use of gene rearrangement studies may be helpful.

Methodology A single cell suspension is required for flow cytometric evaluation. When using tissues, cells are isolated from stromal elements by gentle mechanical dissociation. Most lymphoid tissues release lymphocytes with relative ease while other tissues, such as skin, pose significant difficulties in extracting viable lymphocytes. Isolated lymphocytes are washed and viability may be enriched by density gradient centrifugation. In some specimens, overnight culture is useful to decrease nonspecific staining caused by cytophilic antibody binding mediated by immunoglobulin Fc receptors expressed by lymphoid and other inflammatory cells. However, necrotic tissues and those with high grade neoplasm often lose viable tumor cells and become enriched by reactive T cells. Therefore, discretion is necessary to optimize the preparatory aspects of tissue processing. Ultimately, the single cell suspension is stained with fluorochrome conjugated antibodies, washed, and analyzed on the flow cytometer. Panels of antibodies are utilized to quantitate the numbers of B cells, T cells, and myelomonocytic cells. B-cell clonality is assessed with immunoglobulin light chains while T-cell clonality is inferred by abnormal expression of T-cell antigens.

Additional Information Data gathered from photomultiplier tubes monitoring the flow cell analysis point of the cytometer is digitized and can be stored in digital format (computer memory) where it is subject to manipulation by software. The data is commonly displayed as a "dot plot." On the basis of light scatter characteristics, the weak analog signals are amplified (linear or logarithmic) and converted to digital signals, a reflection of which can be represented as a dot on a display device. The example below shows white blood cells identified and displayed on the basis of size (y-axis, ordinate) representing forward light scatter (FS) and on the basis of granularity (x-axis, abscissa) representing side (90°C) light scatter (SS).

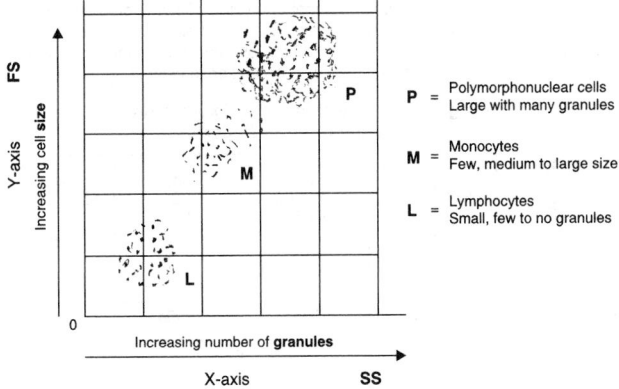

The cell populations can be further delineated on the basis of fluorescence from fluorochrome conjugated antibody to specific cell surface (membrane) markers, and in some cases, intracellular targets after cell fixation and permeabilization. FCM can detect and measure antigens which are catalogued by "CD" designation. CD refers to cluster designation, "clusters of differentiation", meaning a cluster of antibodies binding to a known antigen.

Perhaps the most valuable and widely used application of FCM is in the study of malignant (including importantly, leukemic) cells. Dispersed cells from tissue as well as circulating cells may be immunophenotyped and ploidy/S phase fraction studies performed. Immunophenotyping of leukemic cells is important in lineage determination (eg, myeloid, lymphoid), differentiation of acute B-cell from acute T-cell leukemia, establishing presence of mixed lineage leukemia, (Continued)

Flow Cytometry, Overview (Continued)

assisting in the determination of prognosis and in detection of minimal residual disease. Of 30 or so B-lymphoid cell markers, some 12 have value in analysis of hematologic malignancy. Of 26 T-lymphoid cell markers, 9 are useful in lymphoid malignancy. Of 23 myeloid/monocytic markers, some 9 are useful in hematologic malignancy. See table.

Immunophenotype Markers (Lineage Associations)
Useful in Study of Hematologic Malignancy

B-Cell Markers	T-Cell Markers	Myeloid / Monocytic Markers
CD5	CD1-CD5	CD10
CD10	CD7	CD13-CD16
CD19	CD8	CD24
CD20-CD25	CD25	CD33
CD40	CD28	CD34
CD79	CD40L	CD117 (c-kit)
CD103		

CD = Clusters of differentiation, see text and references.

Lack of CD 45 (leukocyte common antigen) expression on a cellular population should be considered as evidence against a hematopoietic neoplasm.

Expression of a rare cell marker may be absent or misleading in some cases of leukemia and lymphoma. Therefore, no single marker allows definitive diagnosis, and the composite phenotype should be utilized (see Leukocyte Cytochemistry *on page 846* for examples of immunophenotypic characteristics of some leukemias).

The utility of flow cytometric immunophenotyping of hematologic malignancy (for diagnosis, classification, prognosis, and post-therapeutic monitoring) may be under utilized in routine practice. The following table indicates the prognostic implication of antigen expression on leukemic cells in some cases.

Prognostic Value of Antigen Expression
(Composite of Eight Reports)

Prognostic Influence	ALL[1]		AML[2]	MM[3]
	Childhood	Adults		
Favorable	CD34	CD58	CD58	CD11a
	CD2		CD95	CD56
	CD9			
	CD54			
Adverse	CD45	CD34	CD34	CD20
			CD11b	sIg
			CD7	CD13/CD33
			CD9	CD28
			CD44v6	CD44
			CD56	
			MDRI[4]	
			LRP[5]	

[1]Acute lymphocytic leukemia.

[2]Acute myelogenous leukemia.

[3]Multiple myeloma.

[4]MDRI, multidrug-resistant protein.

[5]LRP, lung-resistant protein, only in childhood AML.

Adapted from Orfao A, Schmitz G, Brando B, et al, "Clinically Useful Information Provided by the Flow Cytometric Immunophenotyping of Hematologic Malignancies: Current Status and Future Directions,' *Clin Chem*, 1999, 45(10):1708-17.

Sensitive methods are required to detect minimal residual disease in cases of acute leukemia. Even though in "clinical" remission, acute leukemia may have up to 10^{10} residual malignant cells. FCM can detect one leukemic cell among 10,000 normal bone marrow cells. Because of their sensitivity, FCM and PCR find application in detection of these submicroscopic levels of leukemia.

Non-neoplastic tissues characteristically contain a dominant population of diploid cells containing a stainable amount of DNA defined as "2C" (see figure).

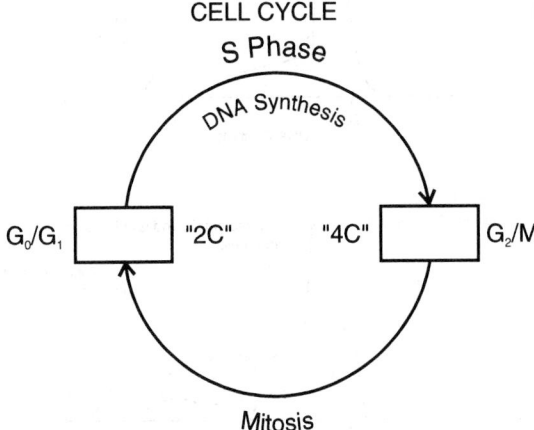

Mitosis

This tissue will also contain a smaller population of cells actively synthesizing DNA, defined as the S-phase fraction. Another small population of cells will be in either premitotic or mitotic phase with a stainable DNA content twice that of the diploid population, designated "4C". The proportions of these populations will vary with the tissue source. DNA analysis by either flow cytometry or image analysis generally shows high concordance, but disparate results may occur.

When using fresh or frozen tissue, a single cell suspension is prepared and the cells rendered permeable to a fluorescent dye that binds stoichiometrically to DNA. When paraffin-embedded tissues are used, nuclei are isolated from thick deparaffinized tissue sections and stained in a similar fashion. Suspensions are analyzed by flow cytometry and fluorescence intensity is quantitated for each nucleus passing through the laser beam. Typically, at least 10,000 nuclei are evaluated and a histogram is generated depicting cell number versus fluorescence intensity. In routine clinical applications, SPF is calculated, not directly measured.

Nuclear hyperchromasia, anisocytosis, and polylobations are examples of microscopic morphologic correlates of flow cytometry determined DNA aneuploidy. The presence of nucleoli and mitoses are counterparts of an increased S-phase fraction. However, such correlations are imperfect. Neoplasms that appear low grade or well differentiated may display aneuploid DNA content or high SPF. In general, tumor aneuploidy or increased SPF is associated with poor outcome. Notable exceptions include childhood acute lymphoblastic leukemia and neuroblastoma in which aneuploidy is associated with a more favorable outcome.

The predictive value of ploidy and SPF determinations is most comprehensively studied in breast cancer. In limited stage breast cancer, both DNA ploidy and SPF are prognostically relevant. When evaluated as independent prognostic variables, high SPF is a statistically significant poor prognostic finding. Numerous technical limitations occur however. Recent College of American Pathologists guidelines for the interpretation of prognostic findings in breast cancer consider documentation of proliferation status a desirable goal but minimize the importance of DNA ploidy.

(Continued)

Flow Cytometry, Overview *(Continued)*

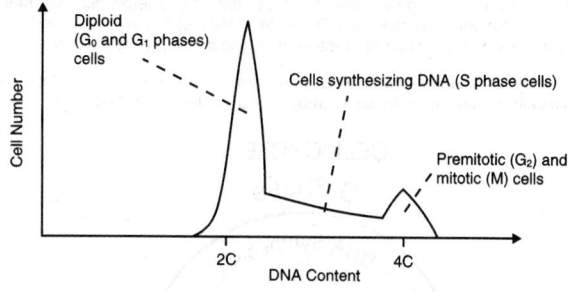

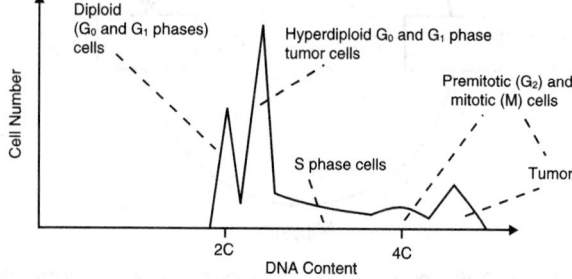

Top: A distribution typical of those obtained by flow cytometry. Bottom: A distribution such as is obtained from a tumor exhibiting DNA aneuploidy.
From Shapiro HM, "Flow Cytometry of DNA Content and Other Indicators of Proliferative Activity," *Arch Pathol Lab Med*, 1989, 113:591-7, with permission.

Cancers of the colon, prostate, urinary bladder, and lung have also been the subject of ploidy and S phase analyses.

Immunophenotypic studies are most useful in evaluation of hematologic or lymphoid tissues and their malignancies.

Approximately 90% of non-Hodgkin lymphomas are derived from monoclonal B cells. Characteristically, B-cell lymphomas will express pan B-cell antigens and express only one immunoglobulin light chain, kappa or lambda, providing clonality. B-cell lymphomas never express both kappa and lambda light chains, but approximately 5% to 10% of lymphomas are surface immunoglobulin negative. Loss of normal pan B-cell antigen expression or acquisition of T-cell antigen expression represents phenotypic aberrancy, a characteristic of malignancy. B-cell lymphomas which coexpress the T-cell associated antigen CD5 characterize lymphomas of small lymphocytic and mantle cell lymphocytic varieties. Expression of CD10 is commonly documented in follicular lymphomas.

Approximately 8% to 10% of non-Hodgkin lymphomas are derived from T cells. Clonality may be inferred by documenting abnormal pan T-cell antigen expression, abnormal T-cell subset antigen expression, or expression of thymocyte antigens. When flow cytometry cannot prove clonality, use of gene rearrangement studies may be required; see Gene Rearrangement for Leukemia and Lymphoma *on page 633*.

Flow cytometric studies in cases of Hodgkin lymphoma are nondiagnostic. Immunophenotypic verification of Hodgkin disease is best accomplished in paraffin section using an appropriate panel of antibodies correlated with morphological features of the neoplastic cells.

When considering the possibility of a neoplasm of granulocytic/monocytic precursors, expressions of CD13, CD14, or CD33 are usually documented. Additionally, expression of CD45 is characteristically weaker than that typically seen in lymphoid neoplasms. Lack of reactivity for other markers of T- or B-cell lineage is also expected and should be confirmed.

Frequently Used Antigens for the Evaluation of Lymphoma / Leukemia

Lineage Association	Antigenic Specificity / Predominate Antigen Distribution
B-cell associated markers	
CD19	Pan B cell
CD20	Pan B cell
CD21	C3d and EBV receptor, resting B cell
CD10	CALLA, follicular center cells
Kappa, lambda	Mature B cells
Ig heavy chains	Mature B cells
T-cell associated antigens	
CD2	Sheep erythrocyte receptor, pan T cell
CD3	T-cell antigen receptor complex, pan T cell
CD5	Pan T cell, B-CLL, B-cell small lymphocytic lymphoma, B-cell mantle cell lymphoma
CD7	Pan T cell
CD4	Helper / inducer subset
CD8	Cytotoxic, suppressor subset
CD1	Cortical thymocyte
Myeloid / monocytic antigens	
CD13	Predominately myeloid
CD15	Predominately myeloid, Reed-Sternberg cells
CD14	Predominately monocytic
CD33	Predominately monocytic
Miscellaneous antigens	
CD11c	Predominately granulocytic / monocytic; hairy cell leukemia, some CLL
CD25	IL-2 receptor, activated T cells, hairy cell leukemia
HLA-Dr	Immune response associated antigen, most B cells, activated T cells, early granulocytic and most monocytic cells
Glycophorin A	Erythroid precursors
CDw41	GPIIb / IIIa complex, megakaryocytes

CD indicates cluster designations.

See also the table, Cluster of Differentiation (CD) Antigens, in the introductory text.

See also a tabular presentation of CD antigens in the Introduction *on page 44*.

Disorders of granulocyte function (eg, phagocytosis, adhesion defects, production of oxygen radicals in chronic granulomatous disease - CGD) can be assessed by FCM-based methods. The Nitroblue Tetrazolium Test *on page 963*, for diagnosis of CGD, has been largely replaced by FCM-based procedures.

Blood can be analyzed by FCM to determine the percentage of fetal red cells in cases of suspected fetomaternal hemorrhage (FMH). The method uses an isothiocyanate conjugated monoclonal antibody against Hb F with access to its intracellular location gained by a process of "permeabilization" of fixed cells using Triton x-100/PBS solution. FCM analysis is then used to determine the percentage of F cells.

(Continued)

Flow Cytometry, Overview *(Continued)*

FCM may be of value in study of serous effusions harboring malignant cells. Use of Ber-EP$_4$ (monoclonal antibody directed against two glycopeptides on human epithelial cells) has, in some studies, shown specificity and sensitivity in differentiating between malignant epithelial cells and mesothelial cells.

Multicolor flow cytometer-based suspension array technology (SAT) is ideal for cytokine panel quantitation. Microparticle-based multiplexed immunoassays, affinity assays, and DNA hybridization assays utilizing low-cost miniaturized laser systems have led to commercial FCM dedicated to SAT with the potential to replace most ELISA (enzyme-linked immunoassay) based methods.

FCM's ability to detect and follow cell differentiation has been applied in an ever growing number of research investigations. An example is elucidation of prion protein's role in the transmissible spongiform encephalopathies (TSE) which includes Creutzfeldt-Jakob disease (CJD) in humans. Prion protein is etiologically related to TSE. An isoform of prion protein (PrP^{Sc}), which accumulates in CJD, is formed as the result of a post-translational process with conformational changes from the **normal** cellular isoform of the prion protein (PrP^{c}). PrP^{Sc} apparently replicates by converting PrP^{c} into PrP^{Sc} by a process of autocatalysis. FCM studies have shown that PrP^{c} plays an important role in function and development of some lymphoid hematopoietic cells. Recently, it has been shown that plasminogen selectively binds PrP^{Sc}. This finding has neuropathogenetic significance and may open avenues to prion removal from blood-derived biological products and to the development of diagnostic procedures. There is interest (especially in Europe) in the development and application of a screening test to detect individuals infected with variant CJD.

References

Bakke AC, "Clinical Applications of Flow Cytometry," *Lab Med*, 2000, 31(2):97-102.

Bakke AC, "The Principles of Flow Cytometry," *Lab Med*, 2001, 32(4):207-11.

Bedner E, Halicka HD, Cheng W, et al, "High Affinity Binding of Fluorescein Isothiocyanate to Eosinophils Detected by Laser Scanning Cytometry: A Potential Source of Error in Analysis of Blood Samples Utilizing Fluorescein-Conjugated Reagents in Flow Cytometry," *Cytometry*, 1999, 36(1):77-82.

Bostwick DG and Foster CS, "Predictive Factors in Prostate Cancer: Current Concepts From the 1999 College of American Pathologists Conference on Solid Tumor Prognostic Factors and the 1999 World Health Organization Second International Consultation on Prostate Cancer," *Semin Urol Oncol*, 1999, 17(4):222-72.

Dürig J, Giese A, Schulz-Schaeffer W, et al, "Differential Constitutive and Activation-Dependent Expression of Prion Protein in Human Peripheral Blood Leukocytes," *Br J Haematol*, 2000, 108(3):488-95.

Finn WG, Peterson LC, James C, et al, "Enhanced Detection of Malignant Lymphoma in Cerebrospinal Fluid by Multiparameter Flow Cytometry," *Am J Clin Pathol*, 1998, 110(3):341-6.

Fitzgibbons PL, Page DL, Weaver D, et al, "Prognostic Factors in Breast Cancer. College of American Pathologists Consensus Statement 1999," *Arch Pathol Lab Med*, 2000, 124(7):966-78.

Gusev Y, Sparkowski J, Raghunathan A, et al, "Rolling Circle Amplification: A New Approach to Increase Sensitivity for Immunohistochemistry and Flow Cytometry," *Am J Pathol*, 2001, 159(1):63-9.

Kussick SJ and Wood BL, "Four-Color Flow Cytometry Identifies Virtually All Cytogenetically Abnormal Bone Marrow Samples in the Workup of Non-CML Myeloproliferative Disorders," *Am J Clin Pathol*, 2003, 120(6):854-65.

Mandy F and Minkus T, "New Applications for Flow Cytometry," *ADVANCE for Administrators of the Laboratory*, 2000, 9(6):80-8.

Orfao A, Schmitz G, Brando B, et al, "Clinically Useful Information Provided by the Flow Cytometric Immunophenotyping of Hematological Malignancies: Current Status and Future Directions," *Clin Chem*, 1999, 45(10):1708-17.

Risberg B, Davidson B, Dong HP, et al, "Flow Cytometric Immunophenotyping of Serous Effusions and Peritoneal Washings: Comparison With Immunocytochemistry and Morphological Findings," *J Clin Pathol*, 2000, 53(7):513-7.

Stelzer GT, Marti G, Hurley A, et al, "U.S.-Canadian Consensus Recommendations on the Immunophenotypic Analysis of Hematologic Neoplasia by Flow Cytometry: Standardization and Validation of Laboratory Procedures," *Cytometry*, 1997, 30(5):214-30.

Shapiro HM, "How Flow Cytometers Work," *Practical Flow Cytometry*, 3rd ed, Chapter 4, New York, NY: Wiley-Liss, 1995, 75-178.

Flucytosine, Serum

Related Information

Itraconazole, Serum *on page 814*

Synonyms Ancobon®; 5-FC; 5-Fluorocytosine

Abstract Flucytosine is a synthetic antifungal agent often used in conjunction with amphotericin B, for treatment of fungal (primarily cryptococcal) meningitis, candidiasis, and chromoblastomycosis. Neutropenia and thrombocytopenia may be seen in patients treated with flucytosine. The drug should be used with caution in patients with renal insufficiency.

Specimen Serum

Container Red top tube

Sampling Time Peak concentrations: 30-60 minutes after last dose; trough: just prior to dose. Peak samples may be most informative in patients with renal failure if obtained 2 hours after a dose. (Absorption is delayed in these patients.)

Reference Interval Therapeutic: 25-100 μg/mL (SI: 194-775 μmol/L) (50-100 mcg/mL in general, 25-60 μg/mL in immunocompromised patients)

Possible Panic Range ≥100-125 μg/mL (SI: 775-970 μmol/L)

Use Monitor weekly for bone marrow toxicity, if patient has normal renal function, more often if renal function is abnormal (diminution of renal function can lead to toxic levels)

Limitations Serum levels fail to correlate well with clinical toxicity.

Methodology High performance liquid chromatography (HPLC), gas chromatography/mass spectrometry (GC/MS)

Additional Information
- Half-life: 3-8 hours (much longer in patients with renal dysfunction)
- Protein binding: 2% to 4%
- Volume of distribution: 0.68 L/kg (perhaps 50% less in patients with renal failure)

Elevated drug levels may predispose patients to abdominal pain, probably by adversely affecting rapidly reproducing cells in the gastrointestinal tract. Severe side effects include hepatotoxicity and bone marrow depression. As the drug and metabolites are excreted through the kidneys, decreased renal function will increase toxicity.

References

Patel R, "Antifungal Agents. Part I. Amphotericin B Preparations and Flucytosine," *Mayo Clin Proc*, 1998, 73(12):1205-25.

Vermes A, Guchelaar HJ, and Dankert J, "Flucytosine: A Review of its Pharmacology, Clinical Indications, Pharmacokinetics, Toxicity, and Drug Interactions," *J Antimicrob Chemother*, 2000, 46(2):171-9.

Vermes A, van Der Sijs H, and Guchelaar HJ, "Flucytosine: Correlation Between Toxicity and Pharmacokinetics Parameters," *Chemotherapy*, 2000, 46(2):86-94.

◆ **Fluid, Pericardial** *see* Body Fluid Chemical Analysis *on page 291*

◆ **Fluid, Peritoneal** *see* Body Fluid Chemical Analysis *on page 291*

◆ **Fluid, Pleural** *see* Body Fluid Chemical Analysis *on page 291*

◆ **Fluids Cytology** *see* Body Cavity Fluid Cytology *on page 285*

Flunitrazepam, Urine

Related Information
Benzodiazepines, Qualitative, Urine *on page 253*

Synonyms Rohypnol®

Abstract Flunitrazepam, a N-methyl-2-fluoro analogue of nitrazepam, is a benzodiazepine with sedative and hypnotic activities. The drug is prohibited in the United States and is now used for illicit purposes only, but it is available in other parts of the world. The drug has been identified as an agent used for date rape.

Specimen Random urine

Storage Instructions Refrigerate or freeze if not analyzing immediately. As flunitrazepam is not stable, addition of sodium fluoride is recommended.

Special Instructions If forensic, use chain-of-custody protocol and form. See Chain-of-Custody Protocol *on page 381* and the Chain-of-Custody form in the Introduction *on page 63*.

Reference Interval None present; confirmation cutoff: 200 ng/mL

Methodology Enzyme immunoassay (EIA), radioimmunoassay (RIA), gas chromatography/mass spectrometry (GC/MS)

Additional Information
- Half-life: 9-25 hours
- Protein binding: 80% to 90%
- Volume of distribution: 3.5-5.5 L/kg

(Continued)

Flunitrazepam, Urine *(Continued)*

In solution, flunitrazepam is degraded to 7-aminoflunitrazepam, which is also not very stable. Instability is accelerated by increased temperature and bacteria. Sodium fluoride inhibits this conversion.

The drug is known to impair short-term memory. In an overdose, the drug may cause myocardial depression, hypotension, lethargy, ataxia, nausea, diarrhea, tremor, and apnea. Poor correlation exists between blood levels and impairment.

The drug is not detected by commonly used benzodiazepine screening assays. Specific immunoassays and GC/MS are needed to detect very small quantities of the drug.

References

Anglin D, Spears KL, and Hutson HR, "Flunitrazepam and Its Involvement in Date or Acquaintance Rape," *Acad Emerg Med*, 1997, 4(4):323-6.

Baselt RC, *Disposition of Toxic Drugs and Chemicals in Man*, 6th ed, Foster City, CA: Biomedical Publications, 2002, 441-4.

ElSohly MA and Salamone SJ, "Prevalence of Drugs Used in Cases of Alleged Sexual Assault," *J Anal Toxicol*, 1999, 23(3):141-6.

Schwartz RH, Milteer R, and LeBeau MA, "Drug Facilitated Sexual Assault (Date Rape)," *South Med J*, 2000, 93(6):558-61.

♦ **Fluorescein-Tagged Antibodies** *see* Kidney Biopsy *on page 818*

Fluorescence *in situ* Hybridization

Related Information

Alpha$_1$-Fetoprotein, Amniotic Fluid *on page 135*
Amniotic Fluid, Chromosome and Genetic Abnormality Analysis *on page 152*
Bone Marrow *on page 296*
Chorionic Villus Sampling, Chromosome and Genetic Abnormality Analysis *on page 400*
Chromosome Analysis, Blood *on page 406*
Chromosome Analysis, Bone Marrow *on page 407*
Chromosome Analysis, High-Resolution *on page 409*
Chromosome Analysis, Lymph Node and Solid Tumor *on page 410*
Chromosome Analysis, Products of Conception *on page 412*
HER-2/*neu* *on page 716*
Histopathology *on page 733*
Immunoperoxidase Procedures *on page 780*
Lymph Node Biopsy *on page 880*

Synonyms FISH; Molecular Cytogenetics

Applies to Comparative Genomic Hybridization; Spectral Karyotyping

Abstract Fluorescence *in situ* hybridization (FISH) is a technique in which specific nucleic acid sequences can be visualized, utilizing fluorescent-labeled probes in individual metaphase or interphase cells from fresh or aged samples such as blood smears, touch and cytospin preparations, or paraffin-embedded tissue. FISH, a powerful technique which has revolutionalized the field of cytogenetics, has numerous applications. The three main areas of clinical use are diagnosis of individuals with birth defects and mental retardation, prenatal diagnosis and screening, and identification and monitoring of acquired chromosome abnormalities in leukemia/cancer. Importantly, analysis with FISH is not contingent on dividing or mitosing cells and provides cellular localization of DNA sequences in a heterogeneous cell population.

Several different types of chromosomal probes are commercially available. Those most commonly used include probes to chromosome-specific repeated sequences (such as alpha satellite and satellite III DNA, regions around the chromosomal centromere; sequence or loci-specific probes such as those which are unique for the different chromosomal regions of deletion in the microdeletion syndromes (see table in Chromosome Analysis, High-Resolution *on page 409*); or translocation breakpoint flanking or spanning probes or amplified regions of DNA; telomere-specific probes; and whole chromosome "painting" probes. In addition to chromosome-specific and region-specific probes, FISH techniques are also available for detection of alterations on a genome-wide scale. Comparative genomic hybridization (CGH) is a technique that permits the detection of chromosome gains or losses throughout the entire genome. Spectral karyotyping (SKY) and multicolor FISH (MFISH) are techniques that provide distinct identification of all 24 human chromosomes, thereby greatly facilitating the recognition of chromosome aberrations.

Specimen The specimen required will depend on the reason the test is requested; it may include blood, peripheral blood lymphocytes, bone marrow, amniotic fluid, chorionic villus sample, products of conception, lymph nodes, or solid tumors. All specimens should be sent fresh if standard cytogenetic studies will also be performed. Frozen, fixed, or paraffin-embedded samples are acceptable for detection of numerical abnormalities and some translocations and deletions, but are not suitable for all types of structural rearrangements.

Container Blood should be collected in a green top (sodium heparin) tube, 5 mL minimum; bone marrow should be collected in a heparinized syringe (20-25 units heparin), 1-2 mL minimum; amniotic fluid, chorionic villi, products of conception, lymph node and solid tumor tissue should be submitted in a sterile container.

Storage Instructions All fresh specimens must be sent to the laboratory immediately after collection. Maintain at room temperature.

Causes for Rejection Specimen more than 48 hours old, specimen clotted or hemolyzed due to the use of improper anticoagulant will yield suboptimal results.

Turnaround Time 24-72 hours

Reference Interval Normal chromosome number and structure. Interpretation is usually provided with the report.

Use Numerous applications of FISH exist. Prenatal screening for detection of aneusomy (loss or gain of one or more chromosomes) such as Turner syndrome (45,X), trisomy 21 (Down syndrome), or other autosomal or sex chromosomal disorders such as Klinefelter syndrome, trisomy 13, and trisomy 18 is becoming increasingly common. An advantage of FISH is that mitosing cells are not required and results for a suspected disorder can be obtained within 24 hours if necessary. FISH can uncover small rearrangements that are not detectable with standard karyotypic analysis. For instance, the presence of a microdeletion (see table in Chromosome Analysis, High-Resolution *on page 409*) can be detected by the absence of signal on one of a homologous chromosome pair.

FISH is also useful in the evaluation of neoplasia. Numerous leukemia/lymphoma or solid tumor associated chromosome translocations, gene deletions, or gene amplifications can be detected in both metaphase and interphase cells with high sensitivity and may be seen in the absence of the translocation visible cytogenetically (cryptic rearrangement). FISH can be used to determine bone marrow transplant engraftment in sex-mismatched donors and recipients using sex chromosome-specific probes. FISH is often used to characterize chromosomal abnormalities that are difficult to define with traditional cytogenetic analysis, such as marker and supernumerary chromosomes using specific FISH probes or the SKY or MFISH techniques. The utility of FISH continues to rapidly expand as evidenced by its recent application for detection of HER-2/*neu* amplification in breast cancer and detection of recurrent bladder cancer using multicolor FISH probe mixtures.

Limitations FISH can only provide information with respect to the specific probe being utilized. For example, using an X chromosome-specific probe to rule out Turner syndrome will not disclose chromosomal abnormalities involving other chromosomes such as trisomy 21, 18, or 13.

Methodology Cells from the specimen are first immobilized on a microscope slide and fixed with a methanol:acetic acid fixative. FISH is performed using probes selected for specific chromosomes or chromosomal regions. DNA sequences in the target and probe (which are complementary) are denatured and then mixed together so that the probe binds to the chromosomal regions in which it has high homology. The bound probe is detected using a series of fluorescent-labeled reagents and fluorescence microscopy.

References

Bubendorf L, Grilli B, Sauter G, et al, "Multiprobe FISH for Enhanced Detection of Bladder Cancer in Voided Urine Specimens and Bladder Washings," *Am J Clin Pathol*, 2001, 116(1):79-86.

Gozzetti A and Le Beau MM, "Fluorescence *in situ* Hybridization: Uses and Limitations," *Semin Hematol*, 2000, 37(4):320-33.

Haddad BR, Schrock E, Meck J, et al, "Identification of *de novo* Chromosomal Markers and Derivatives by Spectral Karyotyping," *Hum Genet*, 1998, 103(5):619-25.

(Continued)

Fluorescence *in situ* Hybridization *(Continued)*

Pergament E, Chen PX, Thangavelu M, et al, "The Clinical Application of Interphase FISH in Prenatal Diagnosis," *Prenat Diagn*, 2000, 20(3):215-20.

♦ **Fluoride Inhibition of Cholinesterase** *see* Dibucaine Number, Serum or Plasma *on page 510*

♦ **Fluorochrome Stain** *see* Acid-Fast Stain, Routine or Modified *on page 95*

♦ **5-Fluorocytosine** *see* Flucytosine, Serum *on page 600*

♦ **Fluoxetine and Norfluoxetine** *see* Antidepressants, Cyclic, Serum or Plasma *on page 171*

Fluoxetine, Serum or Plasma

Related Information

Amoxapine, Serum or Plasma *on page 153*

Antidepressants, Cyclic, Serum or Plasma *on page 171*

Imipramine, Serum or Plasma *on page 767*

Synonyms Fontex®; Prozac®

Test Includes Fluoxetine and norfluoxetine

Abstract This is a **nontricyclic** antidepressant with a long half-life and an active metabolite. It is the most frequently prescribed antidepressant in the U.S. It is also used in the treatment of binge eating and vomiting in patients with moderate-to-severe bulimia nervosa, obsessive-compulsive disorder (OCD), premenstrual dysphoric disorder (PMDD), and panic disorder with or without agoraphobia. Interactions with tricyclic antidepressants and monoamine oxidase inhibitors are well recognized. It inhibits the cytochrome P450 system.

Specimen Serum or plasma

Container Red top tube or green top (heparin) tube; do not use serum separator tube

Sampling Time Trough at steady state (10-15 days)

Reference Interval Fluoxetine: 100-800 ng/mL (SI: 289-2314 nmol/L); norfluoxetine: 100-600 ng/mL (SI: 289-1735 nmol/L); combined: 300-1200 ng/mL

Critical Values >2000 ng/mL (SI: >5784 nmol/L) (fluoxetine and norfluoxetine)

Use Therapeutic monitoring and toxicity assessment

Methodology High performance liquid chromatography (HPLC), gas chromatography (GC)

Additional Information

- Half-life: 2-3 days (norfluoxetine: 7-9 days)
- Volume of distribution: 12-42 L/kg
- Protein binding: 90% to 98%

Fluoxetine is an antidepressant that is a potent, selective inhibitor of serotonin reuptake with minimal effect on norepinephrine and dopamine. It is metabolized via demethylation to norfluoxetine, which is an active metabolite. Because of extensive tissue binding, the parent drug and the active metabolite norfluoxetine have very long half-lives. Fluoxetine may be helpful for subjects with moderate depression treated as outpatients. The overall toxicity of the drug is considerably less than that of the tricyclics.

Symptoms of overdose include ataxia, sedation, ECG abnormalities (QT prolongation, torsade de pointes), and coma. Respiratory depression may occur, especially with coingestion of alcohol or other drugs. Its use is contraindicated in patients receiving MAO inhibitors, thioridazine, or mesoridazine.

References

Cheer Sm and Goa KL, "Fluoxetine: A Review of Its Therapeutic Potential in the Treatment of Depression Associated With Physical Illness," *Drugs*, 2001, 61(1):81-110.

Gram L, "Fluoxetine," *N Engl J Med*, 1994, 331(20):1354-61.

Pearlstein T and Yonkers KA, "Review of Fluoxetine and its Clinical Applications in Premenstrual Dysphoric Disorder," *Expert Opin Pharmacother*, 2002, 3(7):979-91.

Richelson E, "Pharmacokinetic Drug Interactions of New Antidepressants: A Review of the Effects on the Metabolism of Other Drugs," *Mayo Clin Proc*, 1997, 72(9):835-47.

Fluphenazine, Serum

Related Information

Haloperidol, Serum or Plasma *on page 664*

Phenothiazines, Serum *on page 1021*

Synonyms Moditen®; Permitil®; Prolixin®

Abstract Fluphenazine is a high potency phenothiazine antipsychotic agent used in the management of manifestations of psychotic disorders and schizophrenia.

Specimen Serum

Container Red top tube

Sampling Time Plasma concentrations during oral therapy should be measured 12 hours after the evening dose and before any morning dose. Plasma concentrations during decanoate therapy should be measured immediately before the next injection.

Reference Interval 0.3-3.0 ng/mL (SI: 0.6-6.0 ng/mL)

Critical Values >50 ng/mL (SI: >98 nmol/L)

Use Therapeutic monitoring and toxicity assessment

Methodology High performance liquid chromatography (HPLC), gas chromatography (GC), immunoassay

Additional Information
- Half-life (see below)
- Protein binding: 91% to 99%

The elimination half-life varies with the salt form (HCl about 15 hours, enanthate about 3.5-4 days, decanoate about 7-10 days). Symptoms of overdose include deep sleep, hypo- or hypertension, dystonia, seizures, extrapyramidal symptoms, and respiratory failure.

Metabolism in the liver is extensive, with metabolites contributing ~50% of antipsychotic activity. There is some conjugation with glucuronide which, along with unconjugated metabolites, are excreted in the urine. Some excretion may occur via the biliary tract and feces.

References

Lacy CF, Armstrong LL, Goldman MP, et al, *Drug Information Handbook*, 12th ed, Hudson, OH: Lexi-Comp Inc, 2004.

Leikin JB and Paloucek FP, *Poisoning and Toxicology Compendium*, 3rd ed, Hudson, OH: Lexi-Comp Inc, 2002, 592-3.

Levinson DF, Simpson GM, Singh H, et al, "Fluphenazine Dose, Clinical Response, and Extrapyramidal Symptoms During Acute Treatment," *Arch Gen Psychiatry*, 1990, 47(8):761-8.

Flurazepam, Serum

Related Information

Benzodiazepines, Qualitative, Urine *on page 253*

Synonyms Benozil®; Dalmane®; Staurodorm®

Abstract This drug is a sedative-hypnotic of the benzodiazepine class with a wide therapeutic window. The drug is generally used for short-term treatment of insomnia in adults and children.

Specimen Serum

Container Red top tube

Reference Interval Therapeutic: 0-4 ng/mL (SI: 0-9 nmol/L); metabolite N-desalkylflurazepam: 20-110 ng/mL (SI: 43-240 nmol/L)

Possible Panic Range Toxic: >400 ng/mL (SI: >1000 nmol/L)

Use Monitor therapeutic drug level (rarely), toxicity assessment

Methodology High performance liquid chromatography (HPLC), immunoassay

Additional Information
- Half-life: parent drug: 3-6 hours; metabolite (N-desalkylflurazepam): 50-100 hours
- Volume of distribution: 15-30 L/kg
- Protein binding: 96% to 98%

Most common side effect is daytime drowsiness, ataxia, dizziness, and slurred speech. The major urinary metabolite is N-1-hydroxyethyl flurazepam, which is also active. Acute overdose results in apnea, respiratory depression, ataxia, and nystagmus. Its effects are additive with CNS depressants such as ethanol, barbiturates, and narcotic analgesics.

References

Fraser AD, "Use and Abuse of the Benzodiazepines," *Ther Drug Monit*, 1998, 20(5):481-9.

Schweizer E and Rickels K, "Benzodiazepine Dependence and Withdrawal: A Review of the Syndrome and its Clinical Management," *Acta Psychiatr Scand Suppl*, 1998, 393:95-101.

Younus M and Labellarte MJ, "Insomnia in Children: When Are Hypnotics Indicated?" *Paediatr Drugs*, 2002, 4(6):391-403.

♦ **FMR1 Mutation Analysis** *see* Fragile X Syndrome DNA Test *on page 611*

♦ **FMRI Protein** *see* Fragile X Syndrome DNA Test *on page 611*

♦ **FMRP** *see* Fragile X Syndrome DNA Test *on page 611*

♦ **FNA** *see* Fine Needle Aspiration, Deep and Superficial Masses *on page 590*

♦ **FNA Culture** *see* Fine Needle Aspiration Culture *on page 589*

♦ **Foam Stability Index or Shake Test** *see* Pulmonary Surfactant, Amniotic Fluid *on page 1124*

♦ **Folate Level** *see* Folic Acid, Serum *on page 606*

♦ **Folex®** *see* Methotrexate, Serum or Plasma *on page 905*

Folic Acid, RBC

Related Information

Anemia Flowchart *on page 35*
Cobalamin, Serum *on page 424*
Complete Blood Count *on page 442*
d-Xylose Absorption Test, Serum, Urine *on page 527*
Folic Acid, Serum *on page 606*
Hemoglobin *on page 681*
Homocyst(e)ine, Plasma *on page 741*
Methylmalonic Acid, Serum, Plasma, Urine, or Amniotic Fluid *on page 908*
Parietal Cell Antibody *on page 1005*
Phenobarbital, Serum or Plasma *on page 1021*
Primidone, Serum or Plasma *on page 1091*
Schilling Test *on page 1178*
Vitamin B_{12} Unsaturated Binding Capacity *on page 1316*

Synonyms Red Cell Folate

Patient Preparation Avoid radioisotope scan prior to collection of specimen if RIA is used for assay.

Specimen Erythrocytes

Container Lavender top (EDTA) tube. Green top (heparin) tube may also be used for folate assay, but heparin interferes with serum cobalamin determinations, which are often performed simultaneously.

Sampling Time Fasting specimen is preferred.

Storage Instructions Red cells (or hemolysate) can be stored at 4°C or frozen until assay.

Reference Interval 125-600 ng/mL (SI: 283-1360 nmol/L). The megaloblastic anemia of folate deficiency is usually associated with red cell folate levels <100 ng/mL (SI: <227 nmol/L) RBCs.

Use Detect folate deficiency

Limitations Red cell folate testing methods may be unreliable. In a study of 130 samples from patients with severe folate deficiency, falsely normal red cell folate levels were found in 16% to 40% of cases (different kits).

Additional Information Since serum folate values fluctuate significantly with diet, measurement of red cell folate is a better measure of tissue folate stores. Elderly patients with decreased serum cobalamin and/or folate (serum and RBC levels) but without hematologic signs of megaloblastosis likely reflect an uncertain upper limit of normal for the MCV and/or presence of subclinical megaloblastic anemia. Attention to clinical setting is important since a normal red cell folate level can be found in a rapidly developing folic acid deficiency such as the stress of pregnancy. When RBC folate and D-xylose absorption tests are both normal, the predictive accuracy for absence of celiac disease is 100%. When used together, these tests were found ideal for selecting patients for jejunal biopsy from an otherwise unmanageable number with symptoms suggestive of celiac disease.

References

Labib M, Gama R, and Marks V, "Predictive Value of D-Xylose Absorption Test and Erythrocyte Folate in Adult Coeliac Disease: A Parallel Approach," *Ann Clin Biochem*, 1990, 27(Pt 1):75-7.

Snow CF, "Laboratory Diagnosis of Vitamin B_{12} and Folate Deficiency: A Guide for the Primary Care Physician," *Arch Intern Med*, 1999, 159(12):1289-98.

Zittoun J and Zittoun R, "Modern Clinical Testing Strategies in Cobalamin and Folate Deficiency," *Semin Hematol*, 1999, 36(1):35-46.

Folic Acid, Serum

Related Information

Anemia Flowchart *on page 35*
Cobalamin, Serum *on page 424*

Synonyms Folate Level

Abstract Folic acid exists in dihydro- and tetrahydro forms. Reduction occurs by ascorbic acid or by enzyme-folate reductases. Biologic activity in tissue metabolism involves one-carbon transfer of the tetrahydro form (dihydrofolate reductase required). Folate coenzymes are involved in oxidation-reduction and single carbon transfer reactions. Folates are coenzymes involved in many metabolic pathways with one-carbon unit transfers including purine and pyrimidine synthesis and amino acid conversions. The latter is of special importance in the conversion of homocyst(e)ine to methionine (cobalamin also required as a coenzyme). A methyl group is transferred from methyl tetrahydrofolate to cobalamin (forms methyl cobalamin followed by transfer of the methyl group to homocyst(e)ine to form methionine). Elevated homocyst(e)ine levels have important implications for the development of vascular disease (see Homocyst(e)ine, Plasma *on page 741*). Folate deficiency by interfering with nucleoside biosynthesis (includes DNA synthesis) is an important cause of megaloblastic anemia. Folate deficiency/inhibition is common in pregnancy and in those who take alcohol and/or a number of drugs (including methotrexate).

Patient Preparation Patient should be fasting overnight. Collect prior to transfusion or initiation of folate therapy.

Specimen Serum

Container Red top tube

Collection Avoid hemolysis. Transport specimen to the laboratory promptly after collection. Avoid exposure to light. Significant (12% to 19%) loss of folate occurs over 24 hours in specimens kept at room temperature and exposed to light. Specimens exposed to light for more than 8 hours should be redrawn.

Storage Instructions Stable 24 hours at 4°C or store frozen early in hospital stay (feeding malnourished patient may rapidly elevate folate level to normal). Protect from light.

Causes for Rejection Hemolyzed specimen, stored specimen not frozen or protected from light, patient having had isotope scan or Schilling's test prior to collection of specimen

Reference Interval >2 ng/mL (SI: >5 nmol/L). See table for pediatric reference ranges. Hemolysis may result in markedly elevated serum folate levels.

Pediatric Serum Folate

Age (y)	Male (nmol/L)		Female (nmol/L)	
	Low	High	Low	High
0-1	16.3	50.8	14.3	51.5
2-3	5.7	34.0	3.9	35.6
4-6	1.1	29.4	6.1	31.9
7-9	5.2	27.0	5.4	30.4
10-12	3.4	24.5	2.3	23.1
13-18	2.7	19.9	2.7	16.3

Adapted from Hicks JM, Cook J, Godwin ID, et al, "Vitamin B₁₂ and Folate – Pediatric Reference Ranges," *Arch Pathol Lab Med*, 1993, 117:705.

Use Detect folate deficiency; monitor therapy with folate; evaluate megaloblastic and macrocytic anemia; evaluate alcoholic patients and those with prior jejunoileal bypass for morbid obesity or those with intestinal blind-loop syndrome; evaluate cause of increase in serum homocyst(e)ine level
(Continued)

Folic Acid, Serum *(Continued)*

Limitations May be falsely elevated with RBC hemolysis. May be decreased in patients on oral contraceptives. Folate will deteriorate on exposure to light. Significant fluctuation with diet occurs and can result in a misleadingly normal serum folate in a patient with folate deficiency. Interpretive error may be avoided by obtaining a red cell folate level. Concurrent severe iron deficiency may mask presence of folate deficiency. Serum methylmalonic acid, increased in some early cases of B_{12} deficiency, is usually normal with folate deficiency.

Additional Information Naturally occurring folates are present widely in plant and animal foods taken in the diet and absorbed in the small intestine. Folic acid (pteroylglutamic acid) has a number of biologically active forms (largely conjugates of glutamic acid, eg, N-5-methyltetrahydrofolic acid and N-5-formyltetrahydrofolic acid - folinic acid) that function as coenzymes. Lack of folic acid inhibits DNA synthesis in rapidly dividing cells, thus producing megaloblastic anemia. While a specific folate-binding protein is present in the serum, some 90% of folate is unbound. The binding protein increases with folate deficiency and returns to normal with treatment.

Serum levels are affected by dietary intake. In the pH range of physiologic significance, folate binds to aluminum hydroxide. Chronic use of antacids or H_2-receptor antagonists by patients with diets marginal in folate may be a cause of folic acid deficiency. Folate levels are commonly high in patients with B_{12} deficiency since this vitamin is needed to allow incorporation of folate into tissue cells. Increased serum homocyst(e)ine level may occur with deficiency of either vitamin B_{12} or folate. Elderly individuals have been found to require a daily folic acid supplement of 400-600 µg to reduce homocysteine levels. Folate (folic acid) deficiency is present in some 33% of pregnant women, many alcoholics, and in patients with a wide variety of malabsorption syndromes including celiac disease, sprue, Crohn disease, and jejunal/ileal bypass procedure.

Some Drugs With Potential to Cause Megaloblastosis by Effect on Folate Metabolism

- Methotrexate and other antineoplastics
- Anticonvulsants
- Oral contraceptives
- Sulfasalazine
- Trimethoprim
- Pyrimethamine
- Alcohol abuse (most common cause of folate deficiency in U.S.)

Measurement of red cell folate levels constitutes a reliable means of determining the existence of folate deficiency. Serum folate levels are less reliable. These tests should be considered in patients who have megaloblastic anemia, as well as for patients who have anemia, hypersegmentation of granulocyte nuclei, and coincident evidence of iron deficiency. The finding of a low serum folate means that the patient's recent diet has been subnormal in folate content and/or that recent absorption of folate has been subnormal, but does not prove that the patient either has or will develop tissue folate depletion requiring folate therapy. Therefore, serum folate assays have a very poor predictive value in diagnosis and should be interpreted with caution. A low red cell folate can mean either that there is tissue folate depletion due to folate deficiency requiring folate therapy, or alternatively, that the patient has primary vitamin B_{12} deficiency blocking the ability of cells to take up folate. It has been considered advisable in the past to determine red cell folate in addition to serum folate, to establish a diagnosis of folate deficiency. Currently however, analytic limitations of red cell folate assays are cautioned. Additionally, decreases in RBC folate lack specificity for folate deficiency as they also occur in cobalamin deficiency. In some geographic areas (hospitalized urban patient populations), zidovudine, used in the treatment of acquired immune deficiency syndrome, has been the most common condition associated with macrocytosis.

Levels of serum and RBC folate may be significantly increased in hyperthyroidism.

In 1997, the U.S. government mandated fortification of grain products with folic acid. The intent was to increase dietary folate to about 100 µg/person/day (diet dependent) in order to decrease the incidence of neural tube defects without masking occult cobalamin deficiency. The result was an increase in mean serum folate level from 4.6-10.0 µg/L; at the same time, the mean serum homocyst(e)ine concentration fell from 10.1 to 9.4 µmol/L. These changes will likely affect "normal" reference intervals for folate and homocyst(e)ine.

Women with low folate levels may be at increased risk of early spontaneous abortion if the fetal karyotype is abnormal.

References

Cembrowski GS, Zhang MM, Prosser CI, et al, "Folate Is Not What it Is Cracked up to Be," *Arch Intern Med*, 1999, 159(22):2747-8.

Ford HC, Carter JM, and Rendle MA, "Serum and Red Cell Folate and Serum Vitamin B$_{12}$ Levels in Hyperthyroidism," *Am J Hematol*, 1989, 31(4):233-6.

George L, Mills JL, Johansson AL, et al, "Plasma Folate Levels and Risk of Spontaneous Abortion," *JAMA*, 2002, 288(15):1867-73.

Rothenberg SP, "Increasing the Dietary Intake of Folate: Pros and Cons," *Semin Hematol*, 1999, 36(1):65-74.

Rydlewicz A, Simpson JA, Taylor RJ, et al, "The Effect of Folic Acid Supplementation on Plasma Homocysteine in an Elderly Population," *Q J Med*, 2002, 95(1):27-35.

Zittoun J and Zittoun R, "Modern Clinical Testing Strategies in Cobalamin and Folate Deficiency," *Semin Hematol*, 1999, 36(1):35-46.

Follicle Stimulating Hormone, Serum, Plasma, or Urine

Related Information

Adrenal Cortex: Laboratory Assessment Overview *on page 110*
Estradiol, Serum *on page 553*
Estrogens, Urine *on page 557*
Heterophilic Antibodies *on page 727*
Luteinizing Hormone, Blood or Urine *on page 876*
Urine Collection, 24-Hour *on page 1295*

Synonyms Follitropin; FSH

Applies to CCCT; Clomiphene Citrate Challenge Test; FSH:LH Ratio; Gonadotropic Hormones; Pituitary Gonadotropins

Abstract Follicle stimulating hormone (FSH) and luteinizing hormone (LH) are glycoprotein gonadotropic hormones, produced by the pituitary. The alpha subunits of LH, FSH, thyroid stimulating hormone (TSH), and human chorionic gonadotropin (hCG) are identical; specificity resides in the beta subunits. In females, FSH promotes the development of ovarian follicles and, together with LH, stimulates secretion of estradiol from the maturing follicles. In males, FSH stimulates spermatogenesis.

Patient Preparation If a radioimmunoassay method is used, avoid recently administered radioisotopes.

Specimen Serum or plasma, timed urine collection

Container Red top tube or green top (heparin) tube; plastic urine container with or without preservative as required by testing laboratory

Collection Refrigerate urine during collection.

Storage Instructions Separate and refrigerate or freeze serum or plasma; avoid hemolysis. Serum FSH is stable 4 hours at 4°C to 25°C, 2 weeks at -20°C, 3 months at -70°C. In urine, FSH is stable 3 months at -20°C. Avoid repeated freeze/thaw cycles.

Special Instructions For females, menstrual history is necessary for interpretation.

Reference Interval Reference intervals for serum and urine FSH vary among laboratories and are dependent upon the units used and length of urine collection.

Serum:
• prepubertal children: <10 IU/L (SI: <10 IU/L)
• adults: male: <22 IU/L (SI: <22 IU/L)
• adults: female:
  - nonmidcycle: <20 IU/L (SI: <20 IU/L)
  - midcycle surge: <40 IU/L (SI: <40 IU/L) (ovulatory midcycle peak about twice the basal level)
  - postmenopause: 40-160 IU/L (SI: 40-160 IU/L).

Urine:
• male:
  - 0-8 years of age: <5 IU/24 hours (SI: <5 IU/day)
  - older than 9 years: <22 IU/24 hours (SI: <22 IU/day)

(Continued)

Follicle Stimulating Hormone, Serum, Plasma, or Urine (Continued)

- female:
  - 0-8 years: <5 IU/24 hours (SI: <5 IU/day)
  - 9-15 years: <22 IU/24 hours (SI: <22 IU/day)
  - older than 15 years: <30 IU/24 hours (SI: <30 IU/day)
  - postmenopausal: two to three times cycling level.

Results can only interpreted with clinical information. A recent report (Backer et al), which included 3388 women 35-60 years of age, showed that 73% of women having serum FSH levels ≥20 IU/L were postmenopausal. The same study showed that serum FSH levels increase with age and are increased in smokers.

Use Elevated FSH and LH are found in primary hypogonadism, anorchia, gonadal failure, complete testicular feminization syndrome, Klinefelter syndrome, alcoholism, and castration. FSH and LH are pituitary products and are useful in distinguishing primary gonadal failure from secondary (hypothalamic/pituitary) causes of gonadal failure. They are used in investigation of impotence, gynecomastia, and menstrual disturbances including oligomenorrhea and amenorrhea. FSH and LH are useful infertility evaluations of women and men. Both FSH and LH are low in pituitary or hypothalamic (gonadotroph) failure. Timed urinary collections for FSH mitigate the problems of pulsatile, episodic secretion. They are used mainly for women undergoing *in vitro* fertilization and children being worked up for precocious puberty.

Methodology Radioimmunoassay (RIA) and two-site immunometric assay with radioisotope, fluorometric, enzyme, or chemiluminescent detection are used.

Additional Information FSH and LH are under complex regulation by hypothalamic GnRH and by gonadal sex hormones: estrogen and progesterone in females, and testosterone in males. On the simplest level, FSH and LH are high in conditions in which sex hormones cannot be elaborated, and low in conditions of primary pituitary dysfunction. FSH acts on granulosa cells of the ovary and the Sertoli cells of testis. LH acts on Leydig (interstitial) cells of the gonads. Normally FSH increase occurs at an early stage of puberty and it is 2-4 years before LH reaches similar levels.

FSH **is high** in Klinefelter syndrome and in some subjects with precocious puberty. It is decreased with precocious puberty related to adrenal tumors or congenital adrenal hyperplasia. Normal FSH, in an adult nonovulating female, indicates dysfunction at the central nervous system hypothalamic/pituitary level.

High LH:FSH ratio (>1.5) is found in the polycystic ovary syndrome.

FSH is used as a test of ovarian reserve for the purpose of predicting potential fertility in women participating in assisted reproductive technologies.

References

Backer LC, Rubin CS, Kieszak SM, et al, "Serum Follicle-Stimulating Hormone and Luteinizing Hormone Levels in Women Aged 35-60 in the U.S. Population: The Third National Health and Nutrition Examination Survey (NHANES III, 1988-1994)," *Menopause*, 1999, 6(1):29-35.

Bancsi LF, Huijs AM, Den Ouden CT, et al, "Basal Follicle-Stimulating Hormone Levels Are of Limited Value in Predicting Ongoing Pregnancy Rates After *In Vitro* Fertilization," *Fertil Steril*, 2000, 73(3):552-7.

Layman LC, Wilson JT, Huey LO, et al, "Gonadotropin-Releasing Hormone, Follicle-Stimulating Hormone Beta, Luteinizing Hormone Beta Gene Structure in Idiopathic Hypogonadotropic Hypogonadism," *Fertil Steril*, 1992, 57(1):42-9.

Vance ML, "Hypopituitarism," *N Engl J Med*, 1994, 330(23):1651-62.

- ◆ **Follitropin** see Follicle Stimulating Hormone, Serum, Plasma, or Urine on page 609
- ◆ **Follitropin** see Luteinizing Hormone, Blood or Urine on page 876
- ◆ **Fondaparinux** see Prothrombin Time on page 1116
- ◆ **Fontex®** see Fluoxetine, Serum or Plasma on page 604
- ◆ **Fortunan®** see Haloperidol, Serum or Plasma on page 664
- ◆ **Forward Grouping** see Pretransfusion Testing on page 1088
- ◆ **Fosphenytoin** see Phenytoin, Serum or Plasma on page 1026
- ◆ **Four-Marker Test** see Inhibin A, Serum on page 799
- ◆ **Fourth Generation TSH** see Thyroid Stimulating Hormone, Serum on page 1250

- ♦ **FPN Test for Phenothiazine** *see* Phenothiazines, Serum *on page 1021*
- ♦ **FPS** *see* Glucose, Fasting, Plasma *on page 643*
- ♦ **fPSA** *see* Prostate Specific Antigen, Free *on page 1097*
- ♦ **Fractional Excretion of Sodium** *see* Sodium, Urine *on page 1213*
- ♦ **Fractional Oxyhemoglobin (FO₂Hb)** *see* Oxygen Saturation, Blood *on page 991*

Fragile X Syndrome DNA Test

Related Information

Amniotic Fluid, Chromosome and Genetic Abnormality Analysis *on page 152*

Chorionic Villus Sampling, Chromosome and Genetic Abnormality Analysis *on page 400*

Chromosome Analysis, Blood *on page 406*

Polymerase Chain Reaction *on page 1069*

Synonyms FMR1 Mutation Analysis; Martin-Bell Syndrome

Applies to FMRI Protein; FMRP; FRAXE Syndrome

Test Includes Determination of the number of CGG trinucleotide repeats in the noncoding region of the FMR1 (fragile X mental retardation 1) gene and the methylation status of the FMR1 promoter region

Abstract Fragile X syndrome is the most common cause of inherited mental retardation. Recent population studies using DNA testing have provided a more accurate prevalence of 1:5000 males and about half that in females. A clinical diagnosis is often difficult. Males with fragile X syndrome have characteristics that can vary with age, including motor and speech delays, cognitive impairment, atypical craniofacial features, and certain behaviors. Affected females exhibit a similar but generally less severe phenotype.

Fragile X syndrome is inherited in an X-linked dominant fashion with reduced penetrance. The syndrome is caused by an absence or decreased amount of the protein encoded by the FMR1 (fragile X mental retardation 1) gene (chromosome locus Xq27). A diagnosis of fragile X syndrome, or of carrier status, requires a DNA test that detects two specific abnormalities of the FMR1 gene: an abnormally large number of CGG trinucleotide repeats and aberrant methylation of the promoter region. Normal FMR1 alleles have ≤45 CGG repeats with unmethylated promoter regions and produce a normal amount of fragile X protein. Normal alleles do not change when transmitted from parent to child. Abnormal full mutation FMR1 alleles have >200 CGG repeats and aberrant methylation of the promoter region, which silences the gene resulting in no protein production. All mothers of affected sons are carriers of either a full mutation or a premutation FMR1 allele. Premutation FMR1 alleles have between 45 and ~200 CGG repeats, are not methylated, and produce a sufficient amount of protein to result in a normal phenotype. Premutation FMR1 alleles do not change when transmitted by an unaffected carrier father. In contrast, premutation alleles can change to a full mutation when transmitted by an unaffected carrier mother. Therefore, premutation carrier mothers, but not fathers, are at risk of having an affected child. All sons, but only 30% to 50% of daughters, with a full mutation are affected. Unaffected daughters that are heterozygous for a full mutation presumably have skewed X inactivation ratios that permit production of sufficient FMR1 protein. Regardless of phenotype, all women carriers of full mutations are at 100% risk of having an affected son and about 50% risk of having an affected daughter.

Patient Preparation Because the genetics of fragile X syndrome are complicated, a pretest consultation with a medical geneticist is advisable, particularly for carrier and prenatal risk assessment.

Specimen Whole blood, amniotic fluid; chorionic villus samples are not optimal for this test (see Limitations)

Container Collect anticoagulated blood in either a lavender top (EDTA) tube or yellow top (ACD) Vacutainer®. Avoid use of heparin anticoagulants, which can interfere with polymerase chain reactions and endonuclease restriction enzyme activity. Amniotic fluid and chorionic villus samples should be collected in a sterile manner and transferred to a sterile tube for transport. **Note:** If a routine cytogenetic analysis is being performed simultaneously to screen for chromosomal abnormalities, an additional blood sample in sodium heparin (green top) is required (see Chromosome Analysis, Blood *on page 406*).
(Continued)

Fragile X Syndrome DNA Test *(Continued)*

Sampling Time Blood samples can be taken at any time. Amniotic fluid samples should be collected at or after 16 weeks gestation. Chorionic villus samples should be collected between 8 and 12 weeks gestation.

Storage Instructions Blood samples can be stored and shipped at room temperature or refrigerated (4°C); do not freeze. Blood samples should be received in the testing laboratory within 4 days of the draw to ensure an adequate DNA yield. Amniotic fluid and chorionic villus samples should be maintained at room temperature and sent to the testing laboratory immediately after collection, preferably by overnight delivery, to ensure successful cell culture.

Causes for Rejection Frozen samples, whole blood in heparin anticoagulant (green top) or other inappropriate collection tube, amniotic fluid sample bloody, chorionic villus sample lacking viable chorionic villi

Turnaround Time 7-14 days

Reference Interval An interpretive report is usually provided that includes a risk analysis when appropriate. There are important exceptions to the reference ranges below (see Limitations). Normal FMR1 alleles have ≤45 CGG repeats. Premutation FMR1 alleles have 46-200 CGG repeats. Premutations, which do not cause fragile X syndrome but can expand to a full mutation upon transmission through the maternal germline, are generally observed in asymptomatic carrier males and females. Full mutation FMR1 alleles have both >200 CGG repeats and abnormal methylation. Full mutations are generally observed in symptomatic males and females, and also in some asymptomatic females.

Use Testing is indicated for males or females with developmental delay, autism, or mental retardation, especially if they show other features or behaviors commonly associated with fragile X syndrome and/or have a positive family history of fragile X or undiagnosed mental retardation. In addition to fragile X DNA testing, these patients should undergo a full genetic evaluation that also includes testing for chromosomal abnormalities (see Chromosome Analysis, Blood *on page 406*). Carrier testing is indicated for individuals seeking reproductive counseling with a family history of fragile X syndrome or undiagnosed mental retardation. Testing should be considered for the fetus of a mother who is a known carrier of an FMR1 premutation or full mutation. In addition, it may be appropriate to perform the DNA test for individuals that were tested previously by the less sensitive method of cytogenetic detection of a folate-sensitive fragile site at chromosome X band q27.3, particularly if the patients' phenotype is discordant with the cytogenetic testing result.

Limitations Rare cases (<1%) of fragile X syndrome that are caused by a deletion of, or point mutation in, the FMR1 gene will not be detected by this test. The number of CGG repeats in full mutation FMR1 alleles does not predict disease severity or cognitive ability. The probability that a premutation allele will expand upon transmission and result in fragile X syndrome depends upon the number of CGG repeats, gender of parent, and gender of the child. The available probability estimates should be used with caution due to the small sample size. Premutation and full mutation alleles cannot be defined solely by the number of CGG repeats, methylation status must be determined. This is because there are individuals with alleles that do not follow the usual patterns, such as alleles with $(CGG)_{45-200}$ that are methylated or alleles with $(CGG)_{>200}$ that are not methylated. In addition, some individuals are mosaic for cells with alleles that differ in CGG length and/or methylation. Some mentally retarded males with a $(CGG)_{>200}$ allele and methylation mosaicism are "high-functioning" compared to typical fragile X patients, presumably because the unmethylated alleles can produce a small amount of fragile X protein. FMR1 premutation alleles with $(CGG)_{46-55}$ are rare and are sometimes referred to as "gray zone" alleles. Such alleles do not cause fragile X syndrome, but can have a propensity for slight expansion of a few CGG repeats upon transmission; risk of expansion to a full mutation in one transmission is considered to be very low. Chorionic villus samples are not optimal for testing because methylation of the FMR1 gene may not yet be established in the tissue.

Methodology Genomic DNA is amplified by polymerase chain reaction, using primers that flank the CGG repeat region, and sized after electrophoresis (see Polymerase Chain Reaction *on page 1069*). This method detects normal and premutation sized FMR1 alleles and determines the number of CGG repeats present. DNA fragments are electrophoresed and transferred to a Southern membrane that is hybridized to a small fragment of the FMR1 gene. Recently, new polymerase chain reaction methods have been reported that permit detection of the methylation status of an allele; in the future this method may replace Southern blot analysis.

Additional Information The normal functions of the FMR1 protein, FMRP, are under investigation. Another gene on the X chromosome, called FMR2, which upon expansion and methylation, can cause a rare form of mental retardation, the FRAXE syndrome. DNA testing for FRAXE syndrome is available and may be an appropriate follow-up for selected FMR1-negative subjects. A listing of laboratories that perform FMR1 and/or FRAXE DNA testing can be found at GeneTests® (see Websites).

Recently, a few laboratories have begun testing for protein product of FMR1, FMRP. In some patients, assessment of FMRP has been proposed as a potential prognostic indicator of disease severity.

References
Jin P and Warren ST, "Understanding the Molecular Basis of Fragile X Syndrome," *Hum Mol Genet*, 2000, 9(6):901-8.

Kooy RF, Willemsen R, and Oostra BA, "Fragile X Syndrome at the Turn of the Century," *Mol Med Today*, 2000, 6(5):193-8.

Nolin SL, Lewis FA 3rd, Ye LL, et al, "Familial Transmission of the FMR1 CGG Repeat," *Am J Hum Genet*, 1996, 59(6):1252-61.

Internet Web Sites
www.genetests.org
www.geneclinics.org/profiles/fragilex

Free Thyroxine Index
Related Information
(Continued)

Free Thyroxine Index (Continued)

Triiodothyronine, Serum on page 1276

Synonyms Free Thyroid Index; FT_4I; FT_4 Index; FTI

Applies to Free T_3; Free T_4; Free Thyroxine Assays

Test Includes T_3 uptake and T_4

Abstract In the past, the free thyroxine index (FTI) was widely used, due to unreliability of assays for free thyroxine. **In recent years, assays for TSH have improved, free thyroxine assays have become more reliable, and use of FTI has decreased.**

With the continuous improvement of the sensitivity of measurements for TSH, free T_4, and free T_3, determination of the FTI is becoming less relevant. The FTI is no longer considered an essential test.

Specimen Serum is preferred; plasma may be used.

Reference Interval 10 years to adult: normal: 5.0-13.0, low: ≤4.8, high: ≥14.0. Normal intervals will differ somewhat between laboratories. The units are similar to total T_4 units.

Use This test is only of historic interest.

Limitations See T_3 Uptake, Serum or Plasma on page 1233 for additional limitations.

Methodology Calculation from results of T_3 uptake and T_4.

$$FTI = [\%T_3U \text{ (patient)} / \%T_3U \text{ (reference serum)}] \times T_4$$

References

Dayan CM, "Interpretation of Thyroid Function Tests," *Lancet*, 2001, 357:619-24.

♦ **Free Thyroxine Index, Calculated** see Thyroxine Binding Globulin, Serum on page 1255

♦ **Free-Zone CE** see Protein Electrophoresis, Capillary Zone on page 1103

♦ **Fresh Blood** see Whole Blood on page 1333

♦ **Friedewald Equation** see Triglycerides, Serum or Plasma on page 1275

♦ **Friedewald Formula** see Low Density Lipoprotein Cholesterol on page 874

♦ **Friedreich Ataxia** see Inherited Diseases of Metabolism and Cell Structure on page 792

Frozen Red Blood Cells

Synonyms Red Blood Cells, Deglycerolized; Red Blood Cells, Frozen

Applies to Cryopreservation

Test Includes ABO, Rh, antibody screen, crossmatch, and antibody identification when screen is positive (ie, preparation as for other transfusions)

Abstract Glycerol serves as a cryoprotective agent when added to reasonably fresh or rejuvenated red blood cells, which can then be frozen at -80°C or lower. After thawing and deglycerolization by washing, some 80% to 90% of the original red cells remain, as a more or less pure suspension in isotonic saline. The hematocrit is usually about 60%. Platelets, leukocytes (except for a few lymphocytes), and plasma constituents are almost completely removed during processing. Frozen storage time can be up to 10 years, although some data support even longer periods. Post-thaw storage time is 24 hours at 1°C to 6°C.

Patient Preparation As for transfusion of whole blood or red blood cells

Aftercare One unit should raise the hematocrit of an adult about 3 percentage points (or hemoglobin 1 g/dL). Monitor hemoglobin and hematocrit.

Specimen Blood

Container One red top tube or one lavender top (EDTA) tube

Collection (Of sample from intended recipient): At the patient's bedside, ask the patient to give his or her name. Compare with the patient's wristband. Label the sample tube with two unique forms of patient identification (eg, patient's full name, hospital number); also include date and initials of the collector. Take extra care with identification of unresponsive patients.

Storage Instructions Deglycerolized red blood cells must be transfused within 24 hours after thawing or must be discarded.

Causes for Rejection If a crack is found in the frozen plastic of the container or if there is evidence of leakage, discard the unit.

Turnaround Time Long processing time is a severe disadvantage in emergency settings.

Special Instructions After issue from the transfusion service, blood must be transfused within 4 hours.

Use Restores red cell volume. Frozen red cells are essentially free of plasma proteins; about 0.025% of the original plasma is present. Such properties have more to do with the washing, than with the freezing process itself. Frozen red cells are useful particularly for patients with very rare red cell types and antibodies to high frequency antigens or combinations of antigens. They may also be transfused to patients with severe allergic or anaphylactic reactions against IgA or other plasma constituents.

Long-term storage of autologous red cells.

Rare donor red cell depot.

Limitations About 10% to 15% of the original red cells are lost in processing; expensive - about two to three times the cost of a unit of conventional red blood cells; short dating after thawing - 24-hour shelf-life; not always available even in larger cities; slow and complex freezing and deglycerolizing processes.

Contraindications Sickling hemoglobinopathies in donors are contraindications to freezing, since red cell recovery in these conditions has been poor. As for recipients, frozen red cells should generally not be used when anemia and/ or hypoxia can be corrected with specific products (eg, iron, B_{12}, folic acid). Not suitable for correction of coagulation deficiencies.

Additional Information Red blood cells, deglycerolized, must be ABO compatible. A crossmatch is necessary. Hepatitis and some other infectious diseases remain a hazard. See Risks of Transfusion *on page 1166*. Red blood cells deglycerolized must be stored at 1°C to 6°C or no longer than 24 hours. Future freezing and deglycerolization methods may permit extending storage of deglycerolized red cells to approximately 2 weeks.

References

Brecher ME, *Technical Manual*, 14th ed, Arlington, VA: American Association of Blood Banks Press, 2002, 161-86.

Circular of Information for the Use of Human Blood and Blood Components, American Association of Blood Banks, America's Blood Centers, American Red Cross, 2002, 10-5.

Lelkens CC, Noorman F, Koning JG, et al, "Stability and Thawing of RBCs Frozen With the High-and Low-Glycerol Method," *Transfusion*, 2003, 43(2):157-64.

Mollison PL, Engelfriet CP, Contreras M, et al, *Blood Transfusion in Clinical Medicine*, 10th ed, Oxford, UK: Blackwell Scientific Publication, 1997, 300-4.

Frozen Section

Related Information

Breast Biopsy *on page 305*
Histopathology *on page 733*
Lymph Node Biopsy *on page 880*
Virus, Direct Detection by Fluorescent Antibody *on page 1311*

Test Includes Gross examination by a pathologist with specimen evaluation and possible frozen section (FS) with interpretation, followed by histopathology report. Imprints and smears may be made from fresh tissue. Further studies may be initiated depending on clinical information, gross observations, and frozen section and/or cytologic findings.

Abstract Immediate intraoperative consultation to establish or confirm diagnosis, to provide support for determination of type or extent of operation (ie, provision of immediate intraoperative diagnosis when consultation is needed to enhance patient care). Intraoperative consultation may not require a frozen section at all. It is the pathologist's responsibility to do that which is in the best interest of the patient. Tissue freezing may actually be contraindicated. When patient care is not enhanced, a frozen section is often not indicated, especially if the specimen may be compromised. More sampling limitations and technical problems exist than with fixed, paraffin sections.

Specimen Fresh tissue with **no** added fixative or fluid, **rapidly** brought to the Surgical Pathology Laboratory.

Container Sterile towel, Petri dish, or sterile jar with appropriate attention to biohazard containment

Collection Container must be labeled with patient's name, date, operating room, and name of the surgeon requesting frozen section.

Causes for Rejection Specimen in fixative. See Limitations and Contraindications.

(Continued)

Frozen Section (Continued)

Turnaround Time Delays may take place when FS is not scheduled, when earlier slides must be retrieved and reviewed, when other FS specimens simultaneously arrive from the same or another surgeon, when technical problems arise, and when additional pathologists or residents participate.

Use Establish rapid histopathologic diagnosis of the presence and nature of a pathologic process; provision of rapid intraoperative diagnosis to support immediate intraoperative decisions. The frozen section diagnosis should respond to a clear, unambiguous surgical question. FS is used to ascertain whether or not the specimen is adequate for diagnosis, even if diagnosis on FS must be deferred (ie, to establish whether or not additional sampling is needed). FS may be used to ascertain if cultures are indicated and, if so, to provide indication of the type of cultures needed; procure tissue for fat stains; procure tissue for direct immunofluorescent examination (eg, products of immune activation, viral antigens); rapid evaluation for direction of fresh tissues for possible subsequent special studies such as lymphocyte markers, flow cytometry, receptor assays, and/or electron microscopy. Determination of extent of disease may be accomplished with frozen sections in selected settings; for instance, evaluation of margins of resection. Surgeons sometimes request frozen sections to evaluate unanticipated findings (eg, a nodule in the liver) which are relevant to immediate surgical decisions.

Limitations Bone or heavily calcified tissue often cannot be cut. Tissues dominated by fat are technically difficult and may not be amenable to frozen section. Fixed tissues are technically difficult to manage for frozen section. Small biopsies pose technical difficulties and may significantly compromise evaluation of corresponding paraffin sections.

Some lesions require paraffin sections for definitive diagnosis, such as many lymphoid lesions and occasional problematic breast lesions (eg, papillary lesions, instances of lobular and intraductal proliferation). Frozen sections are more useful to provide diagnosis of a visible lesion than to rule out a possibility of an entity of microscopic proportions. The problems of frozen section for thyroid surgery include the differential diagnosis between instances of follicular adenoma versus carcinoma, as well as identification of the occasional relatively small papillary carcinoma. Differential diagnosis between reactive gliosis and low grade glioma has been a problem for many experienced surgical pathologists and may continue to pose difficulties in high quality paraffin sections. In these settings, false-negative responses are more frequent than false-positive ones. Sufficient nonfrozen tissue for routine processing and ancillary studies is desirable.

Sampling errors are important pitfalls in application of frozen sections.

Frozen sections are used to assess adequacy of resection. **Margins** of specimens in resections for cancer may be a problem for which surgeons may request frozen section support. Negative margins in tumor resections may be of very limited value, especially when such margins of substantial size, by virtue of sampling problems. The presence of fat, the geometry of multiple irregular surfaces in specimens, multiplicity of specimens in some cases, and time limitation while the patient remains under anesthesia all limit the significance of a negative frozen section report of margins. Absence of positive margins does not guarantee local control of the tumor, nor is it in any way a reliable guide to tumor behavior.

False-negative frozen section diagnoses relate to the limited sampling possible within the abbreviated time available. Published results of frozen section examinations (false positives, false negatives, deferrals) often originate from institutions which have a great deal of experience with the technique.

Lack of proper clinical information (eg, history of prior irradiation) can lead to interpretive error.

Reasons to defer diagnosis at frozen section include need for more extensive sampling, lack of adequate epithelium lining cysts, twisted and infarcted lesions, and need for special stains, immunohistochemistry, and optimal sections. In some cases, diagnosis must be delayed for permanent sections. The frequency of false-positive diagnoses relates inversely to that of deferral of diagnosis.

Contraindications Tissue is consumed in the process of frozen section. Tiny critical specimens (for example, possible breast carcinomas less than 5 mm in diameter) are best not risked. Breast specimens not grossly suspicious should not be frozen. A substantial study concluded that FS examination of breast specimens should be limited to cases with distinct lesions >1.0 cm. The freezing process may distort lymphoid as well as other tissues. Therefore, for suspected lymphoma, it is advisable to await proper fixation of the lymph node and paraffin sections for definitive diagnosis, but frozen sections are commonly utilized for immunohistochemical evaluation of lymphoid lesions. See Lymph Node Biopsy *on page 880* for further details. Frozen section artifact in paraffin sections subsequently processed may preclude definitive diagnosis. Small melanocytic lesions are among contraindications to frozen section.

Additional Information Direct communication between pathologist and surgeon must take place at the time of frozen section diagnosis, according to requirements both of regulatory agencies and of good patient care.

Imprints may be stained with H&E, Diff-Quik®, Wright stain, or by other methods. They sometimes are extremely helpful in interpretation of frozen sections. Occasionally, imprints are more diagnostic than the frozen section. They are especially helpful with lymphoid specimens, occasional breast specimens, and in diagnosis of meningioma.

References

Niemann TH, Lucas JG, and Marsh WL Jr, "To Freeze or Not to Freeze: A Comparison of Methods for the Handling of Breast Biopsies With no Palpable Abnormality," *Am J Clin Pathol*, 1996, 106(2):225-8.

Page DL and Gray GF Jr, "Intraoperative Consultations by Pathologists at the Mayo Clinic: A Unique Experience," *Mayo Clin Proc*, 1995, 70(12):1222-3.

Rosai J, *Ackerman's Surgical Pathology*, 8th ed, Volume 1, Chapter 1, St Louis, MO: CV Mosby Co, 1996, 7-9.

Fructosamine, Serum

Related Information

Glucose, Fasting, Plasma *on page 643*
Glucose, Noninvasive *on page 647*
Glucose, Whole Blood (Including Point-of-Care) *on page 653*
Glycated Hemoglobin (Hemoglobin A₁c), Blood *on page 655*
Microalbuminuria *on page 913*

Synonyms Glycated Albumin; Glycated Proteins; Ketoamines, Plasma Protein

Abstract "Fructosamine" is the term used to name proteins that have been glycated (ie, are the result of a nonenzymatic reaction of a protein (usually albumin) with glucose).

Patient Preparation Patients should not take ascorbic acid for at least 24 hours prior to collection.

Specimen Serum

Container Red top tube

Storage Instructions Refrigerate. Freeze sample if assay is not done within 2 hours.

Reference Interval High intermethod variability. Nondiabetics: 1.5-2.7 mmol/L; diabetics: ≥2.0-5.0 mmol/L depending on the degree of control.

Use Monitor diabetic control, reflecting diabetic control over a 2- to 3-week period. Fructosamine is particularly useful in subjects with abnormal hemoglobins.

Limitations Very low albumin concentrations (<3.0 g/dL) may result in falsely low fructosamine values.

Methodology Colorimetry or affinity chromatography. Methods suitable for automated analyzers have been described.

Additional Information The information provided by a fructosamine assay is similar to that provided by a glycated hemoglobin (eg, Hb A₁c). However, fructosamine reflects blood glucose levels over a 2- to 3-week period, while Hb A₁c reflects blood glucose over a 2- to 4-month period.

References

Austin GE, Wheaton R, Nanes MS, et al, "Usefulness of Fructosamine for Monitoring Outpatients With Diabetes," *Am J Med Sci*, 1999, 318(5):316-23.

Hill RP, Hindle EJ, Howey JE, et al, "Recommendations for Adopting Standard Conditions and Analytical Procedures in the Measurement of Serum Fructosamine Concentration," *Ann Clin Biochem*, 1990, 27(Pt 5):413-24.

(Continued)

Fructosamine, Serum *(Continued)*

Ko GT, Chan JC, Yeung VT, et al, "Combined Use of a Fasting Plasma Glucose Concentration and HbA1c or Fructosamine Predicts the Likelihood of Having Diabetes in High-Risk Subjects," *Diabetes Care*, 1998, 21(8):1221-5.

♦ **Fructose Biphosphate Aldolase** *see* Aldolase, Plasma or Serum *on page 121*

♦ **FSH** *see* Follicle Stimulating Hormone, Serum, Plasma, or Urine *on page 609*

♦ **FSH:LH Ratio** *see* Follicle Stimulating Hormone, Serum, Plasma, or Urine *on page 609*

♦ **FSP** *see* D-Dimers and Fibrin Degradation Products *on page 502*

♦ **FT₄** *see* Thyroxine, Free, Serum *on page 1256*

♦ **FT₄I** *see* Free Thyroxine Index *on page 613*

♦ **FT₄I** *see* Thyroxine, Serum *on page 1257*

♦ **FT₄ Index** *see* Free Thyroxine Index *on page 613*

FTA-ABS, Serum

Related Information

Darkfield Examination, Syphilis *on page 501*
MHA-TP *on page 912*
RPR *on page 1174*
VDRL, Serum or Cerebrospinal Fluid *on page 1303*

Test Includes Serum specimen is absorbed and then tested with immunofluorescence for antibody to *Treponema pallidum*

Abstract The FTA-ABS, like the MHA-TP, is a specific treponemal test. Although more sensitive than the reaginic tests, it is more expensive and more technically sophisticated. Nontreponemal or reaginic tests include the VDRL and RPR. A reactive FTA-ABS in a patient also reactive to a nontreponemal test is highly specific for syphilis (lues).

Patient Preparation Patient should be fasting if possible.

Specimen Serum

Container Red top tube

Reference Interval Nonreactive. Reported as reactive, reactive minimal, equivocal, nonreactive, or atypical fluorescence observed; a titer is not determined.

Possible Panic Range Serodiagnosis of syphilis in pregnancy

Use Confirm the presence of antibodies to *Treponema pallidum* in patients who have tested positive for nontreponemal antibodies

Limitations FTA-ABS test for syphilis is often positive in the treponemal diseases pinta, yaws and endemic syphilis (bejel), and falsely positive in patients with some diseases associated with increased or abnormal globulins, antinuclear antibodies, lupus erythematosus (beaded pattern), pregnancy, and drug addiction (although drug addicts are likely to have true positives as well). Lyme disease, leprosy, malaria, infectious mononucleosis, relapsing fever, and leptospirosis are also potential causes of false-positive FTA-ABS. Fewer than 1% of healthy individuals will have a false positive. Borderline results are inconclusive and cannot be interpreted; they may indicate a very low level of treponemal antibody or may be due to nonspecific factors.

Methodology Indirect fluorescent antibody (IFA) of killed *Treponema* after serum absorption

Additional Information FTA-ABS is a sensitive test in all stages of syphilis, and is the best standard confirmatory test for a serum reactive to a screening test such as RPR or VDRL. Occasionally, patients with ocular syphilis (uveitis) or otosyphilis will have a negative VDRL while their FTA-ABS is positive. FTA-ABS cannot be used to follow disease activity or response to treatment, since it will remain high for years or for life.

When a positive serum FTA-ABS is required before performing CSF VDRL examination, the specificity of the CSF test is markedly improved. The FTA-ABS test is not recommended for testing cerebrospinal fluid; usually the VDRL is done on CSF.

References

Birnbaum NR, Goldschmidt RH, and Buffett WO, "Resolving the Common Clinical Dilemmas of Syphilis," *Am Fam Phys*, 1999, 59(8):2233-46.
Clyne B and Jerrad DA, "Syphilis Testing," *J Emerg Med*, 2000, 18(3):361-7.
Darville T, "Syphilis," *Pediatr Rev*, 1999, 20(5):160-4.

Emmert DH and Kirchner JT, "Sexually Transmitted Diseases in Women. Gonorrhea and Syphilis," *Postgrad Med*, 2000, 107(2):181-97.

Genc M and Ledger WJ, "Syphilis in Pregnancy," *Sex Transm Infect*, 2000, 76(2):73-9.

Rome ES, "Sexually Transmitted Diseases: Testing and Treating," *Adolesc Med*, 1999, 10(2):231-41.

Young H, "Syphilis. Serology," *Dermatol Clin*, 1998, 16(4):691-8.

Internet Web Sites
www.astdhpphe.org/infect/syphilis.html
www.cdc.gov/std/Syphilis/STDFact-Syphilis.htm

♦ **F Test** *see page 11*

♦ **FTI** *see* Free Thyroxine Index *on page 613*

♦ **FTI** *see* T_3 Uptake, Serum or Plasma *on page 1233*

♦ **FTI** *see* Thyroxine, Free, Serum *on page 1256*

Fungal Culture, Biopsy or Body Fluid

Related Information

Antimicrobial Susceptibility Testing, Fungi *on page 180*
Aspergillus Serology *on page 219*
Blastomycosis Serology *on page 269*
Body Cavity Fluid Cytology *on page 285*
Body Fluid Analysis, Cell Count *on page 288*
Body Fluid Chemical Analysis *on page 291*
Body Fluid pH *on page 295*
Bone Marrow *on page 296*
Candida Antigen Detection and Serology *on page 334*
Coccidioidomycosis Serology *on page 428*
Fine Needle Aspiration Culture *on page 589*
Fine Needle Aspiration, Deep and Superficial Masses *on page 590*
Fungal Culture, Blood *on page 620*
Fungal Culture, Cerebrospinal Fluid *on page 621*
Fungal Culture, Skin *on page 623*
Fungal Culture, Sputum *on page 624*
Fungus Smear, Stain *on page 626*
Histopathology *on page 733*
Histoplasmosis Antibody *on page 734*
Histoplasmosis Antigen *on page 735*
Sporotrichosis Serology *on page 1222*
Synovial Fluid Analysis *on page 1229*

Applies to Body Fluid Fungus Culture; Bone Marrow Fungus Culture

Abstract Infections due to fungi can sometimes be detected in tissue in the absence of culture. However, fungal culture provides a more accurate diagnosis. Ideally, both histopathologic examination and culture should be done together. Fungal cultures of biopsy material are more desirable than cultures from drainage.

Patient Preparation Aseptic preparation of biopsy site or site of body fluid aspiration

Specimen Surgical tissue, bone marrow, biopsy material, sterile body fluid (synovial fluid, peritoneal fluid, pleural fluid, ascites, etc)

Collection The portion of the biopsy specimen submitted for culture should be separated in a sterile environment from the portion submitted for histopathology.

Causes for Rejection Culture specimen in fixatives; specimens collected on swabs are suboptimal.

Reference Interval No growth

Use Establish the diagnosis of localized or disseminated mycosis; isolate and identify fungi

Methodology Culture under aerobic conditions on several media, usually including Sabouraud's and brain heart infusion (BHI), biphasic media with or without lysis concentration technique, frequently incubation at room temperature or at both 30°C and 37°C

Additional Information Optimal isolation of fungi from tissue is accomplished by processing as much tissue as possible. Every attempt should be made to obtain adequate tissue for culture. Depending upon the geographic area, *Histoplasma capsulatum*, *Blastomyces dermatitidis*, and *Coccidioides immitis* are the most frequently isolated deep pathogenic fungi.

(Continued)

Fungal Culture, Biopsy or Body Fluid *(Continued)*

References

Ho PL and Yuen KY, "Aspergillosis in Bone Marrow Transplant Recipients," *Crit Rev Oncol Hematol*, 2000, 34(1):55-69.

Hunt SM, Miyamoto RC, Cornelius RS, et al, "Invasive Fungal Sinusitis in the Acquired Immunodeficiency Syndrome," *Otolaryngol Clin North Am*, 2000, 33(2):335-47.

Ribes JA, Vanover-Sams CL, and Baker DJ, "Zygomycetes in Human Disease," *Clin Microbial Rev*, 2000, 13(2):236-301.

Wingard JR, "Fungal Infections After Bone Marrow Transplant," *Biol Blood Marrow Transplant*, 1999, 5(2):55-68.

Fungal Culture, Blood

Related Information

Antimicrobial Susceptibility Testing, Fungi *on page 180*
Aspergillus Serology *on page 219*
Bacterial Culture, Blood *on page 232*
Bacterial Culture, Lower Respiratory *on page 241*
Candida Antigen Detection and Serology *on page 334*
Coccidioidomycosis Serology *on page 428*
Fungal Culture, Biopsy or Body Fluid *on page 619*
Fungal Culture, Cerebrospinal Fluid *on page 621*
Fungal Culture, Urine *on page 625*
Fungus Smear, Stain *on page 626*
Histoplasmosis Antibody *on page 734*
Histoplasmosis Antigen *on page 735*
Itraconazole, Serum *on page 814*

Abstract The most sensitive method for detection of invasive fungal or yeast infections is culture from blood specimens. Many of the yeasts (*Candida* species, *Cryptococcus neoformans*) are commonly isolated from the routine, automated blood culture systems. Laboratories not using an automated blood culture system should use the lysis centrifugation method with culture onto media enriched for fungal and yeast isolation or the biphasic blood culture media. Certain dimorphic fungi (eg, *Histoplasma capsulatum*) are reliably recovered only by the lysis centrifugation method.

Patient Preparation See preparation in Bacterial Culture, Blood *on page 232.*

Specimen Blood

Container Fungal blood culture media (eg, biphasic blood culture media) or lysis centrifugation collecting tubes (Wampole Isolator™)

Collection Remove plastic cap from biphasic bottle, cleanse stoppers with acetone alcohol and 2% iodine. Collect 8 mL blood in a yellow Vacutainer® tube containing 0.35% sodium polyanethol sulfonate as an anticoagulant. Transfer appropriate volume of blood to biphasic medium to achieve an approximate 1:10 dilution of blood in the broth medium. Alternatively, 10 mL of blood is directly collected in an Isolator™ tube.

Special Instructions The laboratory should be informed of current antifungal therapy and clinical diagnosis.

Reference Interval No growth

Use Isolate and identify fungi; establish the diagnosis of fungemia, fungal endocarditis, and disseminated mycosis

Limitations Negative fungal blood culture does not exclude disseminated fungal infection (eg, blood cultures are negative in about 50% of individuals with disseminated candidiasis). If disseminated or deep fungal infection is strongly suspected, biopsy of the appropriate tissue and/or bone marrow aspiration for sections and fungus culture should be considered. Blood cultures are not helpful in some fungal diseases (eg, mucormycosis).

Methodology Biphasic (broth and agar) blood culture medium or lysis centrifugation with prompt subculture to solid media

Additional Information Fungemia can be a complication of venous or arterial catheterization, hyperalimentation, the acquired immunodeficiency syndrome (AIDS), and therapy with steroids, antineoplastic drugs, radiation, or broad spectrum antimicrobial agents. Intravenous drug abusers are prone to *Candida* endocarditis. Certain yeasts such as *Candida* species, can be isolated from conventional bacterial blood cultures; for other agents of systemic mycoses, however, it is essential to perform blood fungal cultures as outlined above.

References

Costa SF, Marinho I, Araujo EA, et al, "Nosocomial Fungaemia: A 2-Year Prospective Study," *J Hosp Infect*, 2000, 45(1):69-72.

Edelman M and McKitrick J, Images in Clinical Medicine, "*Histoplasma capsulatum* in a Peripheral Blood Smear," *N Engl J Med*, 2000, 342(1):28.

Geha DJ and Roberts GD, "Laboratory Detection of Fungemia," *Clin Lab Med*, 1994, 14(1):83-97.

Jagarlamudi R and Kumar L, "Systemic Fungal Infections in Cancer Patients," *Trop Gastroenterol*, 2000, 21(1):3-8.

Wilson ML and Weinstein MP, "General Principles in the Laboratory Detection of Bacteremia and Fungemia," *Clin Lab Med*, 1994, 14(1):69-82.

Wahyuningsih R, Freisleben HJ, Sonntag HG, et al, "Simple and Rapid Detection of *Candida albicans* DNA in Serum by PCR for Diagnosis of Invasive Candidiasis," *J Clin Microbiol*, 2000, 38(8):3016-21.

Fungal Culture, Cerebrospinal Fluid

Related Information

Bacterial Culture, Cerebrospinal Fluid *on page 236*
Cerebrospinal Fluid Analysis: Overview *on page 355*
Cerebrospinal Fluid Cytology *on page 361*
Cerebrospinal Fluid Glucose *on page 362*
Cerebrospinal Fluid Protein *on page 371*
Cryptococcal Antigen Titer *on page 482*
Flucytosine, Serum *on page 600*
Fungal Culture, Biopsy or Body Fluid *on page 619*
Fungal Culture, Blood *on page 620*
Fungus Smear, Stain *on page 626*
Histoplasmosis Antibody *on page 734*
Histoplasmosis Antigen *on page 735*
India Ink Preparation *on page 784*
Itraconazole, Serum *on page 814*
Mycobacterial Culture, Cerebrospinal Fluid *on page 931*
Viral Culture *on page 1307*

Abstract Except for *Cryptococcus neoformans*, most fungal infections that have disseminated or localized to the CNS are seen primarily in patients with underlying immunosuppression. Fungi that are yeasts at body temperatures, such as *Histoplasma capsulatum*, *Cryptococcus neoformans*, and *Blastomyces dermatitidis*, have access to the cerebral microcirculation, enabling such organisms to invade the subarachnoid space to produce acute and chronic meningitis. *Candida* species, which exists as both yeasts and pseudofungi at body temperatures, can occasionally be meningeal pathogens.

Patient Preparation Aseptic preparation of aspiration site

Specimen Cerebrospinal fluid

Collection Contamination from skin or other body surfaces must be avoided. The third tube collected during lumbar puncture is most suitable for culture, as skin contaminants from the puncture usually are washed out with fluid collected in the first two tubes.

Turnaround Time Turnaround time of culture for cryptococcosis is slower than that of rapid latex agglutination tests for antigen.

Reference Interval No growth

Use Diagnosis and etiology of fungal meningitis

Limitations Recovery of fungi from cerebrospinal fluid is directly related to the volume of cerebrospinal fluid available. A minimum of 10 mL is recommended. *Cryptococcus* can be mistaken for small lymphocytes in the counting chamber. Culture of additional specimens increases the chance for recovery.

Methodology Aerobic culture of centrifuged sediment on noninhibitory media usually including Sabouraud's, brain heart infusion agar (BHI) with blood agar at 25°C to 30°C and also frequently at 37°C.

Additional Information The diagnosis of central nervous system fungal infections is frequently complicated by the overlapping array of signs and symptoms which may accompany other clinical entities such as tuberculous meningitis, pyogenic abscess, brain tumor, hypersensitivity or allergic reactions, collagen vascular disease, leptomeningeal malignancy, chemical meningitis, meningeal inflammation secondary to contiguous suppuration, Behçet disease, Mollaret's meningitis, and the uveomeningitic syndromes.

Cryptococcosis may present as indolent to fulminant infection terminating in death within 2 weeks. A rapidly progressive disease seems to correlate with (Continued)

Fungal Culture, Cerebrospinal Fluid *(Continued)*

immunosuppression in the patient. See Cryptococcal Antigen Titer *on page 482.* *Cryptococcus* can be seen in cytologic preparations, and its mucinous capsule stains bright red with mucicarmine. The organism may infect other tissues as well.

References

Go JL, Kim PE, Ahmadi J, et al, "Fungal Infections of the Central Nervous System," *Neuroimaging Clin N Am*, 2000, 10(2):409-25.

Jagarlamudi R and Kumar L, "Systemic Fungal Infections in Cancer Patients," *Trop Gastroenterol*, 2000, 21(1):3-8.

Jaster JH and Malecha MJ, Images in Clinical Medicine, "Cryptococcal Meningitis," *N Engl J Med*, 1996, 335(26):1962.

Mylonakis E, Paliou M, Sax PE, et al, "Central Nervous System Aspergillosis in Patients With Human Immunodeficiency Virus Infection. Report of 6 Cases and Review," *Medicine (Baltimore)*, 2000, 79(4):269-80.

Walsh TJ and Chanock SJ, "Diagnosis of Invasive Fungal Infections: Advances in Nonculture Systems," *Curr Clin Top Infect Dis*, 1998, 18:101-53.

Fungal Culture, Ocular Infections

Related Information

Fungus Smear, Stain *on page 626*
Ocular Cytology *on page 973*
Viral Culture *on page 1307*

Test Includes Culture, and if specimen is adequate, KOH preparation and PAS smear

Abstract Fungal infections of the eye can manifest as either keratitis or endophthalmitis. The filamentous fungi are the most common causes of fungal keratitis. The most common risk factor for fungal keratitis is corneal injury, usually caused by vegetative material (eg, branches or straw). Fungal endophthalmitis can occur from exogenous or endogenous routes. Fungal conjunctivitis is not common. *Candida* conjunctivitis can occur in patients receiving corticosteroid eyedrops.

Patient Preparation Avoid contamination with skin flora.

Specimen Scrapings of corneal ulcer, washings of lacrimal duct, wet swabs of conjunctiva. For diagnosis of fungal endophthalmitis vitreous fluid, tissue from a wound, or tissue from the anterior chamber are appropriate.

Collection The physician should collect corneal fragments from the edge and base of the ulcer. Swabs are insufficient.

Reference Interval No growth; normal eye flora can include *Candida* species

Use Establish the presence of keratomycosis or fungal endophthalmitis

Limitations Specimens should be collected from the appropriate site of infection. Conjunctival cultures are inadequate and can be misleading for the diagnosis of fungal keratitis or fungal endophthalmitis.

Methodology Culture on appropriate media usually including Sabouraud's medium, brain heart infusion (BHI), and blood agar incubation at 37°C and 25°C to 30°C

Additional Information The more common causes of keratomycosis include *Fusarium* species, *Candida albicans*, *Aspergillus fumigatus*, *Curvularia* species, *Aspergillus flavus*, other species of *Aspergillus*, *Penicillium*, and *Paecilomyces*. Many of these fungi cause other types of infection as well (eg, the subcutaneum). A keratomycosis-like clinical presentation may also be caused by *Nocardia asteroides* and *Mycobacterium fortuitum*. Keratomycosis is a rare complication of contact lens use. Direct microscopic observation provides a higher yield than culture for the diagnosis of keratomycosis.

References

Fahey DK, Fenton S, Cahill M, et al, "*Candida* endophthalmitis: A Diagnostic Dilemma," *Eye*, 1999, 13(Pt 4):596-8.

Garg P, Gopinathan U, Choudhary K, et al, "Keratomyosis: Clinical and Microbiologic Experience With Dematiaceous Fungi," *Ophthalmology*, 2000, 107(3):574-80.

O'Day DM and Head WS, "Advances in the Management of Keratomycosis and *Acanthamoeba keratitis*," *Cornea*, 2000, 19(5):681-7.

Samson CM and Foster CS, "Chronic Postoperative Endophthalmitis," *Int Ophthalmol Clin*, 2000, 40(1):57-67.

Wang MX, Shen DJ, Liu JC, et al, "Recurrent Fungal Keratitis and Endophthalmitis," *Cornea*, 2000, 19(4):558-60.

Fungal Culture, Skin

Related Information

Antimicrobial Susceptibility Testing, Fungi *on page 180*
Fungal Culture, Biopsy or Body Fluid *on page 619*
Fungus Smear, Stain *on page 626*
HIV-1/HIV-2 Serology *on page 736*
Itraconazole, Serum *on page 814*
KOH Preparation *on page 823*
Mycobacterial Culture, Cutaneous and Subcutaneous Tissue *on page 932*
Skin Biopsy *on page 1200*
Sporotrichosis Serology *on page 1222*

Test Includes Detection of superficial fungal infections of the skin or hair

Abstract Fungal infections of the skin, nails, and hair are often considered together because these tissues all contain keratin. Dermatophytes are capable of invading the keratinous tissues and are usually unable to penetrate deeper into tissues.

Patient Preparation Select hairs that are broken off and appear diseased, and pluck them with sterile forceps. Alternatively, scrape the edges of a scalp lesion with a sterile scalpel. Cleanse skin lesions first with 70% alcohol to reduce bacteria and saprophytic fungi. Scrape from the outer edges of skin lesions. In infections of the nails, scrape out the friable material beneath the edge of the nails, or scrape or clip off portions of abnormal appearing nail and submit for examination and culture.

Specimen Skin scrapings, exudates, nail clippings, whole nail, debris under nail, hair

Collection Enclose hair specimens, skin scrapings, or nail clippings or scrapings in clean paper envelopes, sterile urine container, or Petri dish. Do not put specimens in cotton-plugged tubes, because the specimen may become trapped among the cotton fibers and lost. Do not put specimen into closed containers, such as rubber-stoppered tubes, because this keeps the specimen moist and allows overgrowth of bacteria and saprophytic fungi. The laboratory should be informed of the fungal species suspected, and/or of the clinical diagnosis.

Turnaround Time Cultures are usually reported when positive. Negative cultures are reported after 4 weeks.

Special Instructions A Wood's lamp is useful in the collection of specimens in tinea capitis infections, since hairs infected by some members of the genus *Microsporum* exhibit fluorescence under a Wood's lamp. However, in tinea capitis due to *Trichophyton* species, infected hairs usually do not fluoresce.

Reference Interval No growth

Use Isolate and identify fungi

Limitations A single negative specimen does not rule out fungal infections. *Malassezia furfur*, which causes pityriasis (formerly tinea) versicolor, frequently fails to grow in routine culture.

Methodology Aerobic culture on selective media using nonselective Sabouraud's agar incubated at 25°C to 30°C

Additional Information Tissues that contain keratin (hair, nails, skin, etc) can become infected with dermatophytes, which are a group of related keratinophilic fungi. Infections are generally mild but may be severe as a consequence of the reaction of the patient to products made by the fungus. Yeasts or certain filamentous fungi may cause cutaneous infections that resemble dermatophytoses. *Candida* species also colonize skin. Clinical diagnosis of *Candida* infection involves consideration of predisposing factors such as occlusion, maceration or altered cutaneous barrier function. Signs of *Candida* infection include bright erythema, fragile papulopustules, and satellite lesions. Patients with defects in T-lymphocyte responses, such as AIDS patients or individuals being treated with antineoplastic drugs, are especially susceptible to many fungal infections including superficial mycoses.

References

Elewski BE, "Update on Superficial Fungal Infections. Introduction," *Postgrad Med*, 1999, Spec #5.

Goldstein AO, Smith KM, Ives TJ, et al, "Mycotic Infections. Effective Management of Conditions Involving the Skin, Hair, and Nails," *Geriatrics*, 2000, 55(5):40-52.

Jacobs DS, DeMott WR, Oxley DK, et al, *Laboratory Test Handbook*, 5th ed, Hudson, OH: Lexi-Comp Inc, 2001.

(Continued)

Fungal Culture, Skin (Continued)

Johnson RA, "The Immune Compromised Host in the Twenty-First Century: Management of Muco-cutaneous Infections," *Semin Cutan Med Surg*, 2000, 19(1):19-61.

Nowak MA and Brodell RT, "Rapid Diagnosis of Superficial Fungal Infections," *Postgrad Med*, 1999, 105(2):179-80.

Rudy SJ, "Superficial Fungal Infections in Children and Adolescents," *Nurse Pract Forum*, 1999, 10(2):56-66.

Fungal Culture, Sputum

Related Information

Antimicrobial Susceptibility Testing, Fungi *on page 180*
Aspergillus Serology *on page 219*
Bacterial Culture, Lower Respiratory *on page 241*
Blastomycosis Serology *on page 269*
Bronchoalveolar Lavage (BAL) *on page 311*
Coccidioidomycosis Serology *on page 428*
Cryptococcal Antigen Titer *on page 482*
Fine Needle Aspiration Culture *on page 589*
Fungal Culture, Biopsy or Body Fluid *on page 619*
Fungal Culture, Urine *on page 625*
Fungus Smear, Stain *on page 626*
Histoplasmosis Antibody *on page 734*
Histoplasmosis Antigen *on page 735*
Itraconazole, Serum *on page 814*
KOH Preparation *on page 823*
Mycobacterial Culture, Sputum *on page 933*
Nocardia Culture *on page 964*
Sporotrichosis Serology *on page 1222*
Sputum Cytology *on page 1222*
Viral Culture *on page 1307*

Test Includes Culture for fungal organisms associated with pulmonary infections

Abstract Fungal pulmonary infections include *Histoplasma capsulatum, Coccidioides immitis, Cryptococcus neoformans, Blastomyces dermatitidis*, and Mucormycosis. The incidence of fungal infections is largely related to geographic exposure. Cases can occur in normal hosts. Opportunistic fungal pulmonary infections are due to a variety of etiologic agents, which are ubiquitous in the environment. *Candida* and *Aspergillus* species are the most frequently isolated, however, they may be present as the result of contamination from the patient's normal flora or airborne sources. Their presence may represent colonization rather than invasion.

Patient Preparation The patient should be instructed to remove dentures, rinse mouth with water, and cough deeply expectorating sputum into the sputum collection cup.

Specimen First morning sputum, induced sputum, aspirated sputum, bronchial aspirate, tracheal aspirate, transtracheal aspirate

Container Sterile sputum cup, sputum trap, sterile tracheal aspirate or bronchoscopy tube

Collection Three first morning specimens submitted on successive days. Deeply coughed sputum, transtracheal aspirate, bronchial washing or brushing, or deep tracheal aspirate should be submitted.

Turnaround Time Negative cultures are reported after 4 weeks

Reference Interval No growth; normal flora such as yeast from the oropharynx may be present.

Use Diagnose fungal respiratory disease in patients such as oncology patients, transplant patients, patients with the acquired immunodeficiency syndrome (AIDS), and diabetics

Limitations The yield may be reduced by bacterial overgrowth during storage or on standing; therefore, fresh sputum is preferred. Negative cultures do not rule out the presence of fungal infection.

Methodology Culture on selective media using supplemented Sabouraud's agar and brain heart infusion (BHI) with antibiotics to reduce bacterial overgrowth

Additional Information Definitive diagnosis of fungal pulmonary disease depends upon the presence of clinical signs of pulmonary infection, a chest x-ray revealing abnormality such as granuloma; laboratory isolation of a potentially significant organism from a suitable specimen; histologic documentation of tissue invasion by the isolated organism. Even without invasion **Aspergillus** may cause IgE mediated asthma, allergic alveolitis cell mediated hypersensitivity, mucoid impaction, and bronchocentric granulomatosis.

References

Connolly JE Jr, McAdams HP, Erasmus JJ, et al, "Opportunistic Fungal Pneumonia," *J Thorac Imaging*, 1999, 14(1):51-62.

Jacobs DS, DeMott WR, Oxley DK, et al, *Laboratory Test Handbook*, 5th ed, Hudson, OH: Lexi-Comp Inc, 2001.

Nunley DR, Ohori P, Grgurich WF, et al, "Pulmonary Aspergillosis in Cystic Fibrosis Lung Transplant Recipients," *Chest*, 1998, 114(5):1321-9.

Ribes JA, Vanover-Sams CL, and Baker DJ, "Zygomycetes in Human Disease," *Clin Microbial Rev*, 2000, 13(2):236-301.

Roebuck DJ, Fisher DA, and Currie BJ, "Cryptococcosis in HIV Negative Patients: Findings on Chest Radiography," *Thorax*, 1998, 53(7):554-7.

Sharma OP and Chwogule R, "Many Faces of Pulmonary Aspergillosis," *Eur Respir J*, 1998, 12(3):705-15.

Fungal Culture, Urine

Related Information

Bacterial Culture, Urine *on page 246*
Blastomycosis Serology *on page 269*
Fungal Culture, Blood *on page 620*
Fungal Culture, Sputum *on page 624*
Fungus Smear, Stain *on page 626*
Itraconazole, Serum *on page 814*
Mycobacterial Culture, Urine *on page 935*
Urinary Tract Cytology *on page 1293*

Abstract Use of antibacterial, antineoplastic, and immunosuppressive drugs, corticosteroids, urinary indwelling catheters, urinary tract obstruction, or the presence of diseases such as diabetes mellitus predispose to funguria. *C. albicans* is reported to cause up to 59% of positive urinary fungal cultures. *Candida glabrata* and other *Candida* species account for many of the remainder.

Patient Preparation Usual preparation for clean catch midvoid urine specimen collection. See Bacterial Culture, Urine *on page 246*.

Specimen Urine

Collection The patient must be instructed to thoroughly cleanse skin and collect midstream specimen.

Causes for Rejection Unrefrigerated specimen more than 2 hours old is subject to overgrowth of normal microorganisms.

Reference Interval <10^4 cfu/mL

Use Detect and identify yeasts and fungi in urine specimens

Limitations A single negative culture does not rule out the presence of fungal infection.

Methodology Specimen is cultured on selective media such as supplemented Sabouraud's agar and/or brain heart infusion (BHI) with antibiotics; conventional bacterial cultures can also detect candiduria.

Additional Information Candiduria associated with hematogenous infections is observed in patients with granulocytopenia, corticosteroid therapy, and with immunosuppression. The source is frequently the gastrointestinal tract or indwelling catheters, particularly with hyperalimentation. Urine is a useful specimen for culture in cryptococcosis, blastomycosis, and candidiasis. In addition to *Candida*, opportunistic pathogens in the genitourinary tract include *Aspergillus* and *Cryptococcus*. Endemic pathogens such as *Histoplasma*, *Blastomyces*, and *Coccidioides* are also encountered. Asymptomatic funguria often ultimately clears spontaneously. Patients with candiduria may or may not have candidemia; positive urine culture for fungi often may be followed by positive blood culture for fungi. Ascending infections occur in patients with diabetes, prolonged antimicrobial therapy, or following instrumentation. Urinary obstruction due to "fungus balls" may occur in diabetes and following renal transplantation.

References

Fidel PL, Vazquez JA, and Sobel JD, "*Candida glabrata*: Review of Epidemiology, Pathogenesis and Clinical Disease With Comparison to *C. albicans*," *Clin Microbiol Rev*, 1999, 12(1):80-96.

(Continued)

Fungal Culture, Urine *(Continued)*

Kauffman CA, Vazquez JA, Sobel JD, et al, "Prospective Multicenter Surveillance Study of Funguria in Hospitalized Patients. The National Institute of Allergy and Infectious Disease (NIAID) Mycoses Study Group," *Clin Infect Dis*, 2000, 30(1):14-8.

Rabkin JM, Oroloff SL, Corless CL, et al, "Association of Fungal Infection and Increased Mortality in Liver Transplant Recipients," *Am J Surg*, 2000, 179(5):426-30.

Sobel JD, Kauffman CA, McKinsey D, et al, "Candiduria: A Randomized, Double-Blind Study of Treatment With Fluconazole and Placebo. The National Institute of Allergy and Infectious Disease (NIAID) Mycoses Study Group," *Clin Infect Dis*, 2000, 30(1):19-24.

Fungus Smear, Stain

Related Information

Acid-Fast Stain, Routine or Modified *on page 95*
Antimicrobial Susceptibility Testing, Fungi *on page 180*
Blastomycosis Serology *on page 269*
Coccidioidomycosis Serology *on page 428*
Cryptococcal Antigen Titer *on page 482*
Cryptosporidium Direct Staining Procedures *on page 484*
Fungal Culture, Biopsy or Body Fluid *on page 619*
Fungal Culture, Blood *on page 620*
Fungal Culture, Cerebrospinal Fluid *on page 621*
Fungal Culture, Ocular Infections *on page 622*
Fungal Culture, Skin *on page 623*
Fungal Culture, Sputum *on page 624*
Fungal Culture, Urine *on page 625*
Gram Stain *on page 658*
Histoplasmosis Antibody *on page 734*
Histoplasmosis Antigen *on page 735*
India Ink Preparation *on page 784*
KOH Preparation *on page 823*
Skin Biopsy *on page 1200*
Sporotrichosis Serology *on page 1222*

Test Includes Detection of fungi by smear only; fungus culture and KOH preparation are ordered separately

Abstract Direct detection of fungal elements in specimens can sometimes provide a definitive diagnosis of mycotic infections. Classic examples are the detection of *Histoplasma capsulatum* in histiocytes from the blood or bone marrow and the detection of wide, ribbon-like hyphae with few separations compatible with zygomycosis.

Patient Preparation Avoid contamination with skin flora.

Specimen The same specimen as required for fungal culture of the specific site.

Reference Interval No yeast or hyphal elements seen

Use Aid in the diagnosis of fungal disease; used in combination with fungus culture

Limitations A negative smear does not rule out the presence of fungal infection. When using calcofluor white stain, background fluorescence may be prominent. Examinations should be done by trained personnel.

Methodology Periodic acid-Schiff stain, Calcofluor white, and Gomori methenamine silver stain are used to identify fungal structures.

Additional Information Calcofluor white dye binds to cellulose and chitin and fluoresces with longwave UV light and shortwave visible light. Fungal elements viewed under UV light demonstrate a brilliant fluorescence that stands out from cells, tissue debris, and background. The preparations can subsequently be overstained with PAS or GMS. Calcofluor white can be used to screen stool specimens for **Microsporidia** spores or can detect ***Acanthamoeba*** keratitis. The PAS stain with a light green counterstain demonstrates the yeast forms, spores and the hyphae of fungi as pinkish red on a green background.

References

Connolly JE Jr, McAdams HP, Erasmus JJ, et al, "Opportunistic Fungal Pneumonia," *J Thorac Imaging*, 1999, 14(1):51-62.

Ho PL and Yuen KY, "Aspergillosis in Bone Marrow Transplant Recipients," *Crit Rev Oncol Hematol*, 2000, 34(1):55-69.

Nowak MA and Brodell RT, "Rapid Diagnosis of Superficial Fungal Infections," *Postgrad Med*, 1999, 105(2):179-80.

Ribes JA, Vanover-Sams CL, and Baker DJ, "Zygomycetes in Human Disease," *Clin Microbiol Rev*, 2000, 13(2):236-301.

♦ **G20210A Mutation** *see* Hypercoagulation Panel *on page 758*

♦ **GA** *see* Chemical Warfare Agents *on page 382*

♦ **Gabapentin** *see* Antiepileptic Drugs Overview *on page 176*

♦ **Gabitril®** *see* Antiepileptic Drugs Overview *on page 176*

♦ **GAD65** *see* Glutamic Acid Decarboxylase (GAD65) Antibody *on page 654*

♦ **Galactocerebrosidase Deficiency** *see* Hematopoietic Progenitor Cells, Marrow *on page 677*

♦ **Galactocerebroside** *see* Inherited Diseases of Metabolism and Cell Structure *on page 792*

Galactokinase, Blood

Related Information
Galactose-1-Phosphate, Blood *on page 627*
Galactose-1-Phosphate Uridyl Transferase, Blood *on page 628*
Newborn Screening Tests for Galactosemia *on page 960*

Synonyms RBC Galactokinase

Abstract An overview of the galactosemias is offered in Newborn Screening Tests for Galactosemia *on page 960*.

The activity of this enzyme is decreased in galactokinase-deficient galactosemia. This form of galactosemia is milder than the galactosemia due to galactose-1-phosphate uridyl transferase deficiency. Patients with galactokinase deficiency have juvenile cataracts without mental retardation.

Patient Preparation Avoid radioisotope scans or recently administered radioisotopes prior to collection of specimen, if radioactive method for the enzyme assay is being used.

Specimen Whole blood

Container Green top (heparin) tube

Collection Send blood immediately (on ice, not frozen) to the laboratory.

Storage Instructions Red blood cells must be washed repeatedly immediately after receipt in laboratory, therefore, transportation to the laboratory is critical.

Special Instructions Communicate with the laboratory. This test is not usually routinely available and may require referral.

Reference Interval Large interlaboratory differences exist. The following are examples.
- children: 0-2 years: 11-150 mU/g Hb
- 2-18 years: 11-54 mU/g Hb
- adults: 12-40 mU/g Hb

Use Establish the diagnosis of galactokinase-deficiency galactosemia. Galactosemia may also be caused by a deficiency of galactose-1-phosphate uridyl transferase and uridine diphosphoglucose 4-epimerase.

Methodology Radioisotopic: RBCs are hemolyzed and the hemolysate is incubated with radiolabeled galactose. The $1-^{14}C$-galactose-1-phosphate is quantitated after binding to DEAE chromatography paper. High performance liquid chromatography (HPLC).

Additional Information This condition should enter into the differential diagnosis of any child with cataracts.

References
Stevens RE, Datiles MB, Srivastava SK, et al, "Idiopathic Presenile Cataract Formation and Galactosaemia," *Br J Ophthalmol*, 1989, 73(1):48-51.

♦ **Galactokinase Deficiency** *see* Galactose-1-Phosphate Uridyl Transferase, Blood *on page 628*

Galactose-1-Phosphate, Blood

Related Information
Galactokinase, Blood *on page 627*
Galactose-1-Phosphate Uridyl Transferase, Blood *on page 628*
Newborn Screening Tests for Galactosemia *on page 960*

Abstract See Newborn Screening Tests for Galactosemia *on page 960* for an overview of the galactosemias. Red cell concentration of galactose-1-phosphate is increased in patients with galactosemia due to deficiency of galactose-1-phosphate uridyl transferase or uridine diphosphate galactose-4-epimerase. The concentration of galactose-1-phosphate is the most sensitive index of dietary control in patients with galactosemia.

Specimen Whole blood

Container Green top (heparin) tube

(Continued)

Galactose-1-Phosphate, Blood *(Continued)*

Storage Instructions Store at 4°C.

Causes for Rejection Specimen more than 3 hours old

Reference Interval Normal intervals:
- nongalactosemic: 5-49 µg/g hemoglobin
- galactosemic on galactose restricted diet; 80-125 µg/g hemoglobin
- galactosemic on unrestricted diet: >125 µg/g hemoglobin

Use Monitor galactosemic patients on a galactose-free diet. A range of characteristic abnormalities result from galactose toxicity including failure to thrive, vomiting, abnormal liver function with resultant cirrhosis, and mental retardation.

Limitations Analysis is offered by only a few specialized laboratories. Monitoring of galactose-free diet may be more simply achieved with less cost by using whole blood filter paper spot tests.

Methodology Enzymatic rate reaction (absorbance of NADH). Methods to detect galactose and galactose-1-phosphate from dried blood has been described.

Additional Information Galactosemia, the result of an inherited cellular deficiency of galactose-1-phosphate uridyl transferase or uridine diphosphate galactose-4-epimerase, is characterized by galactosuria and increased red cell galactose-1-phosphate. The level of galactose in the blood is proportional to the dietary intake of lactose.

References

Diepenbrock F, Heckler R, Schickling H, et al, "Colorimetric Determination of Galactose and Galactose-1-Phosphate From Dried Blood," *Clin Biochem*, 1992, 25(1):37-9.

Rhode H, Elei E, and Taube I, "Newborn Screening for Galactosemia: Ultramicroassay for Galactose-1-Phosphate-Uridyltransferase Activity," *Clin Chim Acta*, 1998, 274(1):71-87.

Segal S and Berry GT, "Disorders of Galactose Metabolism," *The Metabolic and Molecular Bases of Inherited Disease*, 7th ed, Chapter 25, Scriver CR, Beaudet AL, Sly WS, et al, eds, New York, NY: McGraw-Hill Inc, 1995, 967-1000.

Galactose-1-Phosphate Uridyl Transferase, Blood

Related Information

Galactokinase, Blood *on page 627*
Galactose-1-Phosphate, Blood *on page 627*
Newborn Screening Tests for Galactosemia *on page 960*
Reducing Substances, Urine *on page 1148*

Applies to Galactokinase Deficiency; UDP Galactose-4-Epimerase Deficiency

Abstract An overview of the galactosemias is provided in Newborn Screening Tests for Galactosemia *on page 960*. Deficiency of this enzyme is the most common cause of galactosemia. Infants with the enzyme deficiency demonstrate failure to thrive, with onset of vomiting and diarrhea within days of milk intake. The urine from these infants will show the presence of **reducing substance** which does not react with glucose oxidase reagents. In untreated patients, the long-term effects include liver damage, cataracts, and mental deterioration.

Patient Preparation Avoid radioisotope scans or recently administered radioisotopes prior to collection of specimen, if a radionuclide label is used.

Specimen Whole blood

Container Green top (heparin) tube, lavender top (EDTA) tube

Storage Instructions Stable 14 days at room temperature, 4 weeks at 4°C; do not freeze.

Reference Interval 17-37 units (µmol/hour/g hemoglobin)

Quantitative assays can generally recognize the following genotypes.
- Normal-normal (NN)
- Duarte heterozygote (ND)
- Classical heterozygote (NG)
- Duarte homozygote (DD)
- Duarte galactosemia compound heterozygote (DG)
- Galactosemia-galactosemia (GG)

Use Diagnose galactosemia (galactose-1-phosphate uridyl transferase deficiency). Two other enzyme deficiencies that cause galactosemia are galactokinase and UDP galactose-4-epimerase.

Methodology Radioactive with ^{14}C-galactose-1-phosphate as the substrate, fluorometric, colorimetric (dried blood)

Additional Information Dietary restriction of galactose is a very effective treatment, and liver and lens changes are reversible.

Blood for galactosemia screening should be obtained as early in life as possible (less than 3-4 days) so that effective therapy can be instituted.

References
Fujimoto A, Okano Y, Miyagi T, et al, "Quantitative Beutler Test for Newborn Mass Screening of Galactosemia Using a Fluorometric Microplate Reader," *Clin Chem*, 2000, 46(6 Pt 1):806-10.

Gitzelmann R, "Disorders of Galactose Metabolism," *Inborn Metabolic Diseases: Diagnosis and Treatment*, Fernandes J, Saudubray JM, and Van den Berghe G, eds, 3rd ed, New York, NY: Springer-Verlag, 2000, 103-9.

Shield JP, Wadsworth EJ, MacDonald A, et al, "The Relationship of Genotype to Cognitive Outcome in Galactosaemia," *Arch Dis Child*, 2000, 83(3):248-50.

Tyfield L, Reichardt J, and Fridovich-Keil J, "Classical Galactosemia and Mutations at the Galactose-1-Phosphate Uridyl Transferase (GALT) Gene," *Hum Mutat*, 1999, 13(6):417-30.

♦ **Galactosemia Screening, Filter Paper** *see* Newborn Screening Tests for Galactosemia *on page 960*

♦ **β-Galactosidase** *see* Inherited Diseases of Metabolism and Cell Structure *on page 792*

♦ **Galactosyl Hydroxylysine** *see* N-Telopeptides, Urine *on page 967*

♦ **Gamma Aminobutyric Acid** *see* Gamma Hydroxybutyrate, Serum or Urine *on page 630*

Gamma-Glutamyl Transferase, Serum

Related Information

Alanine Aminotransferase, Serum *on page 116*
Alkaline Phosphatase, Heat Stable, Serum *on page 125*
Alkaline Phosphatase, Serum *on page 127*
Aspartate Aminotransferase, Serum *on page 216*
Bilirubin, Total, Serum *on page 265*
Carbohydrate-Deficient Transferrin, Serum *on page 338*
Leucine Aminopeptidase (LAP), Serum and Urine *on page 844*
Liver Biopsy *on page 864*
Liver Disease: Laboratory Assessment, Overview *on page 869*
5′ Nucleotidase, Serum *on page 968*

Synonyms Gamma-Glutamyl Transpeptidase; GGT; GGTP; γ-Glutamyl Transferase; Glutamyl Transpeptidase; GT; GTP

Applies to AST:ALT Ratio; Body Fluid GGT

Abstract A biliary excretory enzyme (a peptidase), GGT is especially responsive to obstructive hepatobiliary diseases and is sensitive to ethanol use.

Patient Preparation The patient ideally should fast for 8 hours prior to collection of the specimen. Since elevations may occur with phenytoin or phenobarbital therapy, an alternate test, including alkaline phosphatase, leucine aminopeptidase (LAP) or 5′ nucleotidase, is preferable in such patients.

Specimen Serum, ascitic fluid

Container Red top tube

Storage Instructions Hemolysis and prolonged contact with erythrocytes do not interfere. Stable 1 month at 4°C and 1 year at -20°C.

Reference Interval Varies between laboratories. Higher in newborns, in first 3-6 months; male, 6 months and older: 2-30 units/L (SI: 0.03-0.51 µkat/L), female, 6 months and older: 1-24 units/L (SI: 0.02-0.41 µkat/L). Values in adult males are 25% higher than adult females.

Use An enzyme that is especially useful in the diagnosis of obstructive jaundice, intrahepatic cholestasis, and pancreatitis; a major application is in differential diagnosis of patients with increased serum alkaline phosphatase. See diagram in Alkaline Phosphatase, Serum *on page 127*. GGT is more responsive to biliary obstruction than are aspartate aminotransferase (AST) (SGOT) and alanine aminotransferase (ALT) (SGPT). In obstructive disease, concentrations as high as 5-50 times upper limit of normal are seen. In infectious hepatitis, values seldom go above 5 times normal.

Increased in hepatoma and carcinoma of pancreas. Useful in diagnosis of metastatic carcinoma in the liver. Increasing levels in carcinoma patients relate to tumor progression and diminishing levels to response to treatment. CEA, alkaline phosphatase, and GGT used together are useful markers especially for hepatic metastasis from primaries in breast and colon. GGT is elevated in some instances of seminoma.

(Continued)

Gamma-Glutamyl Transferase, Serum *(Continued)*

Useful in diagnosis of chronic alcoholic liver disease, but some heavy drinkers do not have GGT increases. GGT > twice the upper level of reference range with **AST:ALT** ratio >2:1 is strongly suggestive of alcohol abuse. Serial determinations of serum GGT, AST, and ALT levels can distinguish recovering alcoholics who resume drinking from those who remain abstinent. Increase in body mass is positively correlated with increased GGT levels. GGT may be a marker for visceral and hepatic fat and an independent risk factor for noninsulin-dependent diabetes mellitus. GGT, postprandial glucose, and triglycerides have some correlation in certain groups of patients, including those with alcoholism and diabetes mellitus. With MCV of red cells and carbohydrate-deficient transferrin, GGT is useful as a screen for alcoholism. AST, bilirubin, alkaline phosphatase, and AST:ALT ratio are also helpful.

GGT is the test for cholestasis during or immediately following pregnancy. It is commonly elevated in cirrhosis and hepatitis. The transaminases, AST and ALT rise higher in acute viral hepatitis; these tests with GGT and other assays are best used together in work-up of liver disease.

It is increased in systemic lupus erythematosus. Very high concentrations are common in primary biliary cirrhosis. High GGT is found in infants with biliary atresia. It is increased with hyperthyroidism and decreased in those with hypothyroidism. Several tests (including alkaline phosphatase and bilirubin) are usually necessary to evaluate the biliary tract.

In **ascitic fluid**, very high GGT is increased in some, but not all cases of hepatoma, as compared to cirrhosis or liver metastases. As in serum, it is high in the ascitic fluid of those with alcoholic cirrhosis.

Limitations Acetaminophen toxicity has been reported to cause an increase. The combination of high alkaline phosphatase and normal GGT does not rule out liver disease completely. Used alone as a preoperative screening test for metastasis from colorectal carcinoma, GGT is unsatisfactory.

Drugs that may cause levels of GGT to **diminish** include azathioprine, clofibrate, conjugated estrogens, methotrexate, and ursodiol. Many agents cause **increased** results, including acetaminophen (poisoning even in some mild cases); aminoglutethimide; anticonvulsants including phenytoin, barbiturates, carbamazepine, and diphenylhydantoin; esterified estrogens; interferon alpha-n3; medroxyprogesterone; oral contraceptives; phenothiazine; streptokinase; and valproic acid.

Additional Information GGT provides greater specificity for hepatic disease than does alkaline phosphatase. It is normal in most instances of renal failure. GGT has no origin in bone or placenta, unlike alkaline phosphatase, and age beyond infancy does not influence GGT levels. It is commonly elevated in patients with infectious mononucleosis. When GGT and alkaline phosphatase are both high, but one is disproportionately elevated, suspect the possibility of drug-induced cholestasis (including alcoholism if it is GGT which is much higher).

GGT is normal in normal children, adolescents, and in pregnant women. Unlike AST, it is not elevated in skeletal muscle disease.

References

Meerkerk GJ, Njoo KH, Bongers IM, et al, "Comparing the Diagnostic Accuracy of Carbohydrate-Deficient Transferrin, Gamma-Glutamyltransferase, and Mean Cell Volume in a General Practice Population," *Alcohol Clin Exp Res*, 1999, 23(6):1052-9.

Pratt DS and Kaplan MM, "Evaluation of Abnormal Liver-Enzyme Results in Asymptomatic Patients," *N Engl J Med*, 2000, 342(17):1266-71.

Young DS, *Effects of Drugs on Clinical Laboratory Tests*, 5th ed, Volume 1: Listing by Test, Washington, DC: AACC Press, American Association of Clinical Chemistry, 2000, Section 3, 374-9.

♦ **Gamma-Glutamyl Transpeptidase** *see* Gamma-Glutamyl Transferase, Serum *on page 629*

Gamma Hydroxybutyrate, Serum or Urine

Related Information

Flunitrazepam, Urine *on page 601*

Synonyms Georgia Home Boy; GHB; Grievous Bodily Harm; Liquid Ecstasy; Liquid X; Scoop; Somatomax®

Applies to Gamma Aminobutyric Acid

Abstract Gamma hydroxybutyrate (GHB), a CNS depressant, is an endogenous metabolite of gamma aminobutyric acid (GABA). Once approved for use as a hypnotic and weight loss promoter, GHB has been withdrawn from the U.S. market and is now used only for illicit purposes. However, its precursors, gamma butyrolactone and 1,4 butanediol are available in health food stores. It is one of the commonly used date-rape drugs.

Specimen Serum or urine

Container Red top tube for serum

Collection If forensic, observe precautions.

Storage Instructions Refrigerate. Keeping samples at room temperature increases endogenous production of GHB.

Special Instructions If forensic, use chain-of-custody protocol and form. See Chain-of-Custody Protocol *on page 381* and the Chain-of-Custody form in the Introduction *on page 63*.

Reference Interval Negative (less than cutoff)

Critical Values Blood levels:
- deep sleep/coma: >260 μg/mL
- moderate sleep: 156-260 μg/mL
- light sleep: 52-156 μg/mL

Use Evaluate GHB toxicity; drug of abuse testing. In some countries, it is used for treatment of narcolepsy, alcohol dependence, and opiate dependence.

Limitations The assay of GHB is not frequently available. Relationship between blood levels and symptoms is not well established.

Methodology Gas chromatography (GC), gas chromatography/mass spectrometry (GC/MS)

Additional Information
- Half-life: 0.3-1 hour
- Volume of distribution: 0.5 L/kg
- Protein binding: 0%

In recent years, there has been an increase in the number of reports of the use of GHB, often in conjunction with alcohol, to commit sexual assault. When added to the victim's drink, the victim becomes semiconscious or unconscious. Typical dose is 1-3 grams. Onset of effect occurs within 10-30 minutes and peak plasma concentrations are achieved within 20-45 minutes. Effects generally last 2-5 hours and complete recovery occurs within 4-8 hours.

GHB toxicity causes CNS depression, amnesia, hypotonia, GI symptoms, loss of consciousness, depressed respiration, tremor, and seizures. It acts synergistically with ethanol to intensify CNS and respiratory depression. There is no specific treatment for GHB toxicity; supportive treatment is given.

A rare genetic disorder of succinic semialdehyde deficiency causes significant accumulation of GHB.

References
Chin RL, Sporer KA, Cullison B, et al, "Clinical Course of Gamma-Hydroxybutyrate Overdose," *Ann Emerg Med*, 1998, 31(6):716-22.

Couper FJ and Marinetti LJ, "Gamma-Hydroxybutyrate (GHB) - Effects on Human Performance and Behavior," *Forensic Sci Rev*, 2002, 14:101-21.

ElSohly MA and Salamone SJ, "Prevalence of Drugs Used in Cases of Alleged Sexual Assault," *J Anal Toxicol*, 1999, 23(3):141-6.

O'Connell T, Kaye L, and Plosay JJ, "Gamma-Hydroxybutyrate (GHB): A Newer Drug of Abuse," *Am Fam Phys*, 2000, 62(11):2478-83.

- **Ganja** *see* Cannabinoids (Marijuana Metabolites), Qualitative, Urine *on page 335*

- **Gardenal®** *see* Phenobarbital, Serum or Plasma *on page 1021*

- **Gases, Arterial** *see* Blood Gases and pH, Arterial *on page 275*

- **Gastric Washings Cytology** *see* Washing Cytology *on page 1326*

Gastrin, Serum

Related Information

Cobalamin, Serum *on page 424*

Helicobacter pylori Biopsy-Based Tests: The Urease Tests, Culture, Cytology, and PCR *on page 672*

Pepsinogen I and II, Serum or Plasma *on page 1009*

Schilling Test *on page 1178*

Applies to Secretin Test

(Continued)

Gastrin, Serum *(Continued)*

Abstract Gastrin, a hormone secreted by neuroendocrine G cells located in the gastric antrum (and occasionally at other gastrointestinal sites), is known primarily as a gastric acid secretagogue. Gastrin is also trophic for histamine-secreting enterochromaffin-like cells in the gastric mucosa, and, in chronic hypergastrinemia, these cells undergo hyperplasia leading, in extreme cases, to the development of gastric carcinoid tumors. Gastrin is measured clinically to diagnose gastrin-secreting carcinoid tumors (gastrinomas). Gastrinomas are most commonly located in the duodenum or pancreas and are responsible for the Zollinger-Ellison syndrome (ZES). Gastric carcinoids also occur in association with chronic atrophic gastritis and, on rare occasions, sporadically.

Recent evidence supports the concept that G cells may rarely be found in lymph nodes within the "gastrinoma triangle."

Patient Preparation For **basal** values, a 12-hour overnight fast is required. For dynamic testing (the **secretin challenge** - see below), follow a specific protocol.

Specimen Serum (plasma is not acceptable)

Container Red top tube

Sampling Time For the **secretin challenge** test, specimens are obtained at 2, 5, 10, 15, 20, and 30 minutes following the bolus injection of porcine secretin, 2 units/kg.

Collection Transport specimen immediately to the laboratory following collection. Postprandial specimens should be so indicated.

Storage Instructions Separate in a refrigerated centrifuge and freeze immediately. Stable 4 hours at 4°C and 30 days at -20°C. For long-term storage, specimens should be kept at -70°C. Serum specimens loose up to 50% of immunoreactivity during 48 hours at 4°C.

Causes for Rejection Anticoagulated specimen

Reference Interval

Fasting:
- cord blood: 20-290 ng/L
- 0-4 days: 120-183 ng/L
- childhood: <10-125 ng/L
- adults 16-60 years: 25-90 ng/L
- adults older than 60 years: <100 ng/L

Secretin challenge test: In normal persons, serum gastrin levels decrease, or increase by <200 ng/L in comparison with the preinjection value. An increase >200 ng/L is indicative of gastrinoma.

Use The key to the diagnosis of the Zollinger-Ellison syndrome (ZES) is the combination of high fasting gastrin (often >1000 ng/L) and fasting gastric hyperacidity (often with gastric pH <2.5) in a patient who does not have a retained gastric antrum. When these criteria are fulfilled, in the appropriate clinical context, the diagnosis of gastrinoma is confirmed and further biochemical testing is not needed. Patients whose values do not meet these criteria are candidates for a secretin challenge test. In normal persons, a bolus of secretin decreases serum gastrin (or produces a small increase), but in patients with a gastrinoma there is a paradoxical increase in serum gastrin, >200 ng/L, in response to secretin.

Approximately 90% of gastrinomas are located in the duodenum, pancreas, or a peripancreatic lymph node; the other 10% are widely scattered and may be found in the liver, ovary, heart, stomach, omentum, common bile duct, or small bowel.

Methodology Radioimmunoassay (RIA). Most assays are specific for C-terminal of gastrin and react equally with G-34, G17, and G-14.

Additional Information The combination of elevated gastrin plus gastric hyperacidity also occurs in primary gastrin cell (G cell) hyperplasia, a disease that can be difficult to distinguish from the Z-E syndrome. Other diseases with elevated gastrin include atrophic gastritis, pernicious anemia, postvagotomy syndrome, retained antrum, post small bowel resection, renal failure, and cirrhosis.

References

Henderson AR and Rinker AD, "Gastric, Pancreatic, and Intestinal Function," *Tietz Textbook of Clinical Chemistry*, 3rd ed, Burtis CA and Ashwood ER, eds, Philadelphia, PA: Saunders Co, 1999, 1273-8.

Jacobs DS, DeMott WR, Oxley DK, et al, *Laboratory Test Handbook*, 5th ed, Hudson, OH: Lexi-Comp Inc, 2001.

Norton JA, Fraker DL, Alexander R, et al, "Surgery to Cure the Zollinger-Ellison Syndrome," *N Engl J Med*, 1999, 341(9):635-44.

Stabile BE, Morrow DJ, and Passaro E Jr, "The Gastrinoma Triangle: Operative Implications," *Am J Surg*, 1984, 147(10):25-31.

♦ **GB** *see* Chemical Warfare Agents *on page 382*

♦ **GCDFP-15** *see* Immunoperoxidase Procedures *on page 780*

♦ **GD** *see* Chemical Warfare Agents *on page 382*

♦ **Gemonil®** *see* Barbiturates, Quantitative, Serum or Plasma *on page 248*

♦ **Gemonil®** *see* Methylphenobarbital, Serum *on page 909*

♦ **Genagesic®** *see* Propoxyphene, Serum or Urine *on page 1096*

♦ **GeneChip p53 Assay** *see* p53, Functional Assay/Sequencing *on page 995*

♦ **Gene Rearrangement, bcr** *see* Breakpoint Cluster Region Rearrangement in CML *on page 304*

Gene Rearrangement for Leukemia and Lymphoma

Related Information

bcl-2 Gene Rearrangement *on page 251*
Body Cavity Fluid Cytology *on page 285*
Bone Marrow *on page 296*
Chromosomal Translocations, Molecular Detection *on page 404*
Chromosome Analysis, Lymph Node and Solid Tumor *on page 410*
Flow Cytometry, Overview *on page 592*
Fluorescence *in situ* Hybridization *on page 602*
Histopathology *on page 733*
Immunoperoxidase Procedures *on page 780*
Leukocyte Cytochemistry *on page 846*
Lymph Node Biopsy *on page 880*
Polymerase Chain Reaction *on page 1069*
Skin Biopsy *on page 1200*

Synonyms Leukemia Gene Rearrangement; Lymphoma Gene Rearrangement

Applies to Angelman Syndrome; Catch 22 Syndrome; Newborn Aneuploidy Detection; Prader-Willi Syndrome; Velocardiofacial Syndrome; Williams Syndrome

Test Includes Detection of unique DNA rearrangements associated with T- and B-cell leukemias and lymphomas

Abstract Molecular biology techniques allow for detection of receptor gene rearrangement in germline DNA of maturing T or B cells. As lymphoid cells mature they go through rearrangements of variable, joining, and constant DNA coding regions of immunoglobulin or T-cell receptors. Such recombinations allow for almost unlimited diversity of immune responses, allowing response to literally millions of antigens. Lymphoma and lymphocytic leukemia are neoplastic disorders, diagnosis of which sometimes requires demonstration of clonal expansion of lymphoid cells. In the typical case, demonstration of clonality may be satisfied immunophenotypically by demonstration of surface T- and B-cell markers. When immunophenotypic methods fail to demonstrate clonality, the detection of T-cell and B-cell receptor gene rearrangements may be an invaluable test in cases in which morphologic diagnosis is difficult.

Specimen Peripheral whole blood for leukemia or lymphoma cells. **Unfixed, fresh or frozen lymph node biopsy** of suspected lymphoma is obtained during surgery for histopathologic diagnosis, immunophenotyping, and gene rearrangement assay. Other tissue (eg, skin biopsies, gastrointestinal tissue, and bone marrow) may also be studied for presence of variant gene rearrangement. Use of paraffin-embedded tissue is possible in some laboratories. Consultation with the laboratory (or laboratories) performing or providing referral of the specimen is critical **PRIOR** to sampling of blood, marrow, and/or tissue.

Container Lavender top (EDTA) tube for fresh whole blood sample

Storage Instructions Isolated white cells, lymph nodes, or tissue can be frozen at -70°C until DNA is extracted.

(Continued)

Gene Rearrangement for Leukemia and Lymphoma
(Continued)

Causes for Rejection Insufficient DNA isolated. Muscle tissue yields little DNA for analysis.

Turnaround Time PCR: 2-4 days; Southern blot: 10 days to 3 weeks

Reference Interval No unique rearrangement of genes for T- and B-cell receptors is found in normal white blood cells. An interpretive report is usually included with results.

Use Gene rearrangement analysis may be used to supplement and complement the results of more conventional studies that rely on histopathologic, cytogenetic, and/or immunophenotypic techniques in difficult diagnoses of leukemia/lymphoma. Gene rearrangement studies have revealed that (of cases previously considered non-B, non-T cell) most lymphoid leukemias are pre-B cell in origin rather than non-B, non-T cell. Such leukemic cells have rearrangements of genes coding for B-cell receptors, but are too immature to express cytoplasmic or surface immunoglobulins. Lymphoid leukemia and lymphoma may be of T-cell phenotype, as indicated by the use of probes designed to detect rearrangements of genes coding for T-cell receptors. Rarely, lymphoid neoplasms may have rearrangements of both T- and B-cell genes.

Table 1: Comparison of Conventional Karyotypic (Cytogenetic) Analysis With Molecular (PCR-Based) Analysis

	Karyotypic Analysis	Molecular Analysis
Requirement for fresh, viable, dividing cells	Yes	No
Average turnaround time	2-3 wk	2-4 d
Ability to detect submicroscopic abnormalities	No	Yes
Ability to detect numeric abnormalities	Yes	No
Approximate sensitivity for minimal disease detection (%)	5-10	0.001-1

Adapted from Bagg A and Kallakury B, "Molecular Pathology of Leukemia and Lymphoma," *Am J Clin Pathol*, 1999, 112(Suppl 1):S76-S92.

Table 2: Examples of Non-neoplastic Disorders Associated With Clonal Antigen Receptor Gene Rearrangements

Autoimmune diseases
   Sjögren syndrome
   Rheumatoid arthritis
Immunodeficiency states
   Congenital
   Post-transplantation immunosuppression
   Human immunodeficiency virus
Miscellaneous immunologic dysregulation
   Castleman disease
   Angioimmunoblastic lymphadenopathy
Dermatologic disorders
   Lymphomatoid papulosis
   Acute lichenoid pityriasis

Adapted from Bagg A and Kallakury B, "Molecular Pathology of Leukemia and Lymphoma," *Am J Clin Pathol*, 1999, 112(Suppl 1):S76-S92.

Limitations Some tissues yield little DNA or DNA that is degraded. Lymph nodes with <1% tumor cells cannot provide evidence of gene rearrangements when the Southern blotting method is used. Conventional cytogenetic and molecular (PCR-based) analyses are complimentary in many respects, but karyotypic analysis is unable to detect submicroscopic abnormalities (see

Table 1). Result of PCR analysis, positive for antigen receptor gene rearrangement, may not always be associated with malignancy. Disorders with immunologic dysregulation may occur with antigen receptor rearrangements (see Table 2).

Additional Information These procedures are useful to determine whether T- or B-cell gene rearrangements exist in lymphoid neoplasms. Most commonly used probes are for the joining region of B-cell receptors (J_H) and the constant region of the T-cell receptor (C_BT). B-cell maturation may be further categorized by determining if kappa and/or lambda light chain genes have undergone rearrangement, using probes directed against their constant or joining DNA regions. Gene rearrangements may be detected in minute quantities of tissue, sometimes as little as 200 mg. The assays are sensitive such that gene rearrangements may be detected in larger specimens even if the percentage of cancer cells is 1%. Gene rearrangement studies are invaluable adjunctive tests that may provide evidence of clonality in an atypical lymphoid infiltrate when other methods fail. Application of PCR to detection of minimal residual disease is of increasing import. There is need for standardization. General consensus holds that in both pediatric and adult acute lymphocytic leukemia (ALL), a level >0.1% leukemic cells at the end of induction chemotherapy is strongly predictive of subsequent marrow relapse. Continuing rearrangement of antigen receptor (AR) loci in ALL may confound interpretation. A number of variables must be considered. PCR for AR gene rearrangements may, in some conditions (eg, B-cell precursor acute lymphoblastic leukemia) avoid misleading results of flow cytometric monitoring (by immunophenotypic analysis) of post-therapy bone marrow specimens.

Guidelines for interpretation and use of molecular pathology-based procedures in the diagnosis of lymphomas and leukemias have been published (Bagg and Kallakury).

References

Bagg A and Kallakury BL, "Molecular Pathology of Leukemia and Lymphoma," *Am J Clin Pathol*, 1999, 112(Suppl 1):S76-S92.

Bartolo C and Viswanatha DS, "Molecular Diagnosis in Pediatric Acute Leukemias," *Clin Lab Med*, 2000, 20(1):139-82.

Kallakury BV, Hartman DP, Cossman J, et al, "Post-therapy Surveillance of B-Cell Precursor Acute Lymphoblastic Leukemia. Value of Polymerase Chain Reaction and Limitations of Flow Cytometry," *Am J Clin Pathol*, 1999, 111(6):759-66.

Kirsch IR and Kuehl WM, "Lymphopoiesis: Gene Rearrangements in Lymphoid Cells," *The Molecular Basis of Blood Diseases*, 3rd ed, Part III, Stamatoyannopoulos G, Majerus PW, Perlmutter RM, et al, eds, Philadelphia PA: WB Saunders Co, 2001, 389-430.

Vega F, Medeiros LJ, Jones D, et al, "A Novel Four-Color PCR Assay to Assess T-Cell Receptor Gamma Gene Rearrangements in Lymphoproliferative Lesions," *Am J Clin Pathol*, 2001, 116(1):17-24.

Wickremasinghe RG and Hoffbrand AV, "Molecular Basis of Leukaemia and Lymphoma," *Molecular Haematology*, Chapter 3, Provan D and Gribben J, eds, Malden, MA: Philadelphia, PA: Blackwell Science Ltd, 2001, 25-41.

♦ **Genetic Identification by DNA Fingerprinting** *see* Identification DNA Testing on page 765

Genital Culture for *Ureaplasma urealyticum*

Related Information
Bacterial Culture, Genital Specimen *on page 239*
Neisseria gonorrhoeae Culture and Smear *on page 945*

Test Includes Culture for *Ureaplasma urealyticum* only

Abstract *Ureaplasma urealyticum* are small bacteria that do not have a cell wall. *U. urealyticum* are commonly isolated from the lower genital tract of sexually active adults. Even though these organisms are recovered from asymptomatic patients, *U. urealyticum* does play an etiologic role in genital tract diseases of both men and women.

Specimen Culturette® swab of urethra or cervix

Container Culturette® swab

Collection Contact laboratory to obtain special collection and transport medium.

Storage Instructions Keep specimen refrigerated. **Organism is remarkably sensitive to drying**. Swab must be placed promptly into collection media and hand delivered to the Microbiology Laboratory. If stored longer than 24 hours, the specimens should be frozen at -70°C.

Turnaround Time 8 days if negative, up to 2 weeks if positive
(Continued)

Genital Culture for *Ureaplasma urealyticum*
(Continued)

Special Instructions Specimen should be collected without contact with lubricants, analgesics, or antiseptics.

Reference Interval Less than 10^4 organisms in genital tract

Use Establish the diagnosis of *Ureaplasma urealyticum* infection in suspected cases of nongonococcal urethritis and cervicitis

Limitations Culture can be negative in the presence of infection, and the presence of *Ureaplasma urealyticum* or *Mycoplasma hominis* does not always indicate infection, although there is a significant association with symptomatic disease. Culture is only offered in a few specialty laboratories.

Methodology Culture on selective media

Additional Information *Ureaplasma* and *Mycoplasma* can be isolated from urethral and genital swabs and from urine of sexually active individuals. It is also associated with chorioamnionitis and with perinatal morbidity and mortality. It can be isolated from the central nervous system and the lower respiratory tract of infected neonates. Sixty percent or more of all women asymptomatically carry *U. urealyticum* in their genital tract. Usual prevalence of these organisms in patients with urethral symptoms also is high; thus, conclusions regarding the etiologic role of an isolate in a given patient are difficult to make.

References

McKee KT Jr, Jenkins PR, Garner R, et al, "Features of Urethritis in a Cohort of Male Soldiers," *Clin Infect Dis*, 2000, 30(4):736-41.

Paul VK, Gupta U, Singh M, et al, "Association of Genital *Mycoplasma* Colonization With low Birth Weight," *Int J Gynaecol Obstet*, 1998, 63(2):109-14.

Potts JM, Ward AM, and Rackley RR, "Association of Chronic Urinary Symptoms in Women and *Ureaplasma urealyticum*," *Urology*, 2000, 55(4):486-9.

Sethi S, Sharma M, Narang A, et al, "Isolation Pattern and Clinical Outcomes of Genital *Mycoplasma* in Neonates From a Tertiary Care Neonatal Unit," *J Trop Pediatr*, 1999, 45(3):143-5.

Giardia intestinalis Diagnostic Procedures

Related Information

Cryptosporidium Antigen Detection by EIA *on page 483*

Ova and Parasites, Direct Exam *on page 985*

Synonyms *Giardia lamblia* Diagnostic Procedures

Abstract *Giardia intestinalis* is a flagellated enteric protozoan that infects both humans and animals. It is the most common cause of protozoal diarrhea throughout the world. Infection occurs through fecal-oral transmission and ingestion of contaminated food or water. Asymptomatic carriage is the most common form of giardiasis. Stool examination for the trophozoites and cysts is the traditional method used for the diagnosis. Antigen detection assays are now available and have become the method of choice for diagnosis of diarrhea due to this organism.

Specimen Stool, duodenal fluid, endoscopic brush biopsy material

Container Clean dry container with a wide mouth

Collection Collect the specimen directly in the container and submit to the laboratory quickly.

Storage Instructions Maintain specimen at room temperature.

Turnaround Time 24-48 hours

Special Instructions Collect at least three stool specimens 2-3 days apart. Travel history is important.

Use Evaluation of acute or chronic diarrhea, especially in campers, travelers, children in daycare centers, in male homosexuals, and in hypogammaglobulinemic individuals; evaluation of malabsorption, steatorrhea

Limitations Antigen detection will only detect *G. intestinalis*; other important enteric pathogens will not be detected with this method. Some commercial assays have decreased specificity if the stool is fixed with EcoFix, a formalin and mercury-free fixative.

Methodology Immunofluorescence, enzyme-linked immunoassay (EIA), trichrome or iron hematoxylin staining for light microscopy

Additional Information *G. intestinalis* antigen assays are comparable in cost to the stool ova and parasite examination. Both immunofluorescence and EIA methods provide sensitivities between 85% to 98% and specificities that exceed 90%. All of these assays detect specific *G. intestinalis* cyst cell wall proteins which are stable in human stool specimens.

References
Chan R, Chen J, York MK, et al, "Evaluation of a Combination Rapid Immunoassay for Detection of *Giardia* and *Cryptosporidium* Antigens," *J Clin Microbiol*, 2000, 38(1):393-4.

Colmer-Hamood JA, "Fecal Microscopy Artifacts Mimicking Ova and Parasites," *Lab Med*, 2001, 32(2):80-4.

Farthing MJG, Cevallos AM, and Kelly P, "Intestinal Protozoa," *Manson's Tropical Diseases*, 21st ed, Chapter 77, Cook GC and Zumla A, eds, Philadelphia, PA: WB Saunders Co, 2003, 1373-1410.

Fedorko DP, Williams EC, Nelson NA, et al, "Performance of Three Immunoassays and Two Direct Fluorescence Assays for Detection of *Giardia lamblia* in Stool Specimens Preserved in EcoFix," *J Clin Microbiol*, 2000, 38(7):2781-3.

Maraha B and Buiting AG, "Evaluation of Four Enzyme Immunoassays for the Detection of *Giardia lamblia* Antigen in Stool Specimens," *Eur J Clin Microbiol Infect Dis*, 2000, 19(6):485-7.

Internet Web Sites
vm.cfsan.fda.gov/~mow/chap22.html
www.astdhpphe.org/infect/giardiasis.html
www.cdc.gov/ncidod/dpd/parasites/giardiasis/default.htm

♦ *Giardia lamblia* Diagnostic Procedures *see Giardia intestinalis* Diagnostic Procedures *on page 636*

Gliadin IgG/IgA Antibodies

Related Information
Endomysial Antibodies *on page 537*
Immunoglobulin A *on page 770*
Skin Biopsy *on page 1200*
Skin Biopsy, Immunofluorescence *on page 1203*

Abstract Celiac disease (celiac sprue) is the major cause of sprue in Western civilization. Celiac disease, or gluten-sensitive enteropathy, is caused by ingestion of wheat, rye, and barley in patients with hypersensitivity to gliadin, a gluten protein. Ingestion of gluten triggers the production of antibodies. Measurement of IgG and IgA gliadin, IgA endomysial antibodies (EMA), and transglutinase antibodies offers a useful diagnostic approach. IgG antigliadin antibodies are especially needed in IgA-deficient patients. The gold standard for diagnosis is the small intestinal biopsy. In **early** stages there is subtotal atrophy with intraepithelial lymphocytes. In **late** stages villous atrophy is total.

Specimen Serum; a minimum of 0.25 mL is required for pediatric samples.

Container Red top tube or SST™ tube

Causes for Rejection Heat-treated specimens, icteric or lipemic specimens, specimens containing microbial contamination

Reference Interval IgA and IgG:
<2 years:
- negative: <50 units/mL
- weak positive: 50-100 units/mL
- positive: >100 units/mL

(Continued)

Gliadin IgG/IgA Antibodies *(Continued)*

≥2 years:
- negative: <25 units/mL
- weak positive: 25-50 units/mL
- positive: >50 units/mL

Use Diagnosis of celiac disease, a malabsorptive small intestinal disease characterized by diarrhea, abdominal distension, flatus, steatorrhea, anemia, intestinal cramping, delayed puberty, impaired growth, weight loss, rickets, and osteoporosis; diagnosis of dermatitis herpetiformis; monitoring adherence to a gluten-free diet

IgA gliadin antibody levels can be followed to monitor response to a gluten-free diet. Decreasing titers indicate a favorable response.

Limitations A negative IgA gliadin antibody result does not completely rule out celiac disease or dermatitis herpetiformis. Measurements of IgG gliadin antibodies are indicated when IgA gliadin antibody is not detected. Antigliadin antibody provides less sensitivity and specificity in comparison to endomysial antibody, tissue transglutinase antibody, and histopathologic evaluation.

All intestinal sprue is not caused by gluten sensitivity.

Methodology Indirect immunofluorescence, enzyme-linked immunosorbent assay (ELISA)

Additional Information IgG antibodies to gliadin are less specific than IgA antibodies. They are needed when the IgA result is negative, because selective IgA deficiency is found in ~10% of patients with celiac disease. See Endomysial Antibodies *on page 537*, in which transglutinase antibody is discussed as well.

The major complication of celiac disease is development of enteropathy-associated T-cell lymphoma.

Dermatitis herpetiformis is a bullous skin disease closely associated with gluten-sensitive enteropathy. Strict adherence to a gluten-free diet often includes remission of both the skin and small bowel abnormalities. Gliadin antibodies are also useful in diagnosis of dermatitis herpetiformis.

Recent reviews (Farrell and Kelly; Green and Jabri) provide guides to and pitfalls in the diagnosis of celiac disease.

References

Farrell RJ and Kelly CP, "Celiac Sprue," *N Engl J Med*, 2002, 346(3):180-8.

Green PHR and Jabri B, "Coeliac Disease," *Lancet*, 2003, 362(9381):383-91.

Keren DF, "Autoimmune Disease of the Gastrointestinal Tract," *Clinical Laboratory Evaluation of Human Autoimmune Diseases*, Chapter 19, Nakamura RM, Keren DF, and Bylund DJ, eds, Chicago, IL: ASCP Press , 2002, 277-90.

Moskaluk CA, "Sailing Past the Horizon - The Histologic Diagnosis of Celiac Disease in "Nonflat" Intestinal Mucosa," *Am J Clin Pathol*, 2001, 116(1):7-9.

♦ **Glial Fibrillary Acidic Protein** *see* Immunoperoxidase Procedures *on page 780*

♦ **Globulin, Serum** *see* Protein Electrophoresis, Serum *on page 1104*

♦ **Globulin, Serum** *see* Protein, Total, Serum *on page 1114*

♦ **Globulins, Urine** *see* Protein Electrophoresis, Urine *on page 1107*

Glomerular Basement Membrane Antibody

Related Information

Antineutrophil Cytoplasmic Antibody *on page 187*
Antinuclear Antibodies *on page 189*
Blood, Urine *on page 281*
Hemoglobin, Qualitative, Urine *on page 688*
Kidney Biopsy *on page 818*

Synonyms Goodpastures Antibody

Abstract Immunologic aspects of renal disease can be divided into antibody-mediated entities such as Goodpasture syndrome (antiglomerular basement membrane disease) and cell-mediated glomerulonephritis; see tables. Antibodies recognizing the α_3 domain of type IV collagen (the Goodpasture autoantigen) lead to proliferative glomerulonephritis in Goodpasture syndrome, and less commonly, cause pulmonary hemorrhage and idiopathic pulmonary hemosiderosis. Antiglomerular basement membrane antibody (anti-GBM) assays are positive in about 95% of patients with Goodpasture syndrome. Immunofluorescent studies of renal biopsy are confirmatory.

Antibody-Mediated Glomerulonephritis

Disease	Special Features
Antiglomerular Basement Membrane Diseases	
Goodpasture syndrome	Pulmonary hemorrhage and renal disease
Anti-GBM GN	No pulmonary disease
Alport disease, status following renal transplantation	Transplant contains "new" antigens
Immune Complex-Mediated Diseases	
IgA nephropathy	Male predominance; no vasculitis; may be related to cirrhosis; dermatitis herpetiformis
Henoch-Schönlein purpura	IgA in GBM and mesangium; systemic vasculitis involving skin, bowel, and kidney
Systemic lupus erythematosus	Photosensitive skin lesions, arthralgias, autoantibodies, other systemic features
Acute postinfectious GN	Recent streptococcal or staphylococcal infection; resolves with supportive care
Type 1 membranoproliferative GN (MPGN)	Mesangiocapillary changes
Type 2 MPGN	GBM dense deposits
Type 3 MPGN	Type 1 MPGN with overlapping features
Membranous GN	Subepithelial deposits; slow progression of renal dysfunction
Fibrillary GN	Membranous GN with 20-mm fibrils

Adapted from Ambrus JL and Sridhar NR, "Immunologic Aspects of Renal Disease," *JAMA*, 1997, 278(22):1938-45.

Cell-Mediated Glomerulonephritis

Disease	Special Features
Wegener granulomatosis	Granulomatous vasculitis involving upper airways, lungs, and kidneys; often rapidly progressive glomerulonephritis; often positive C-ANCA
Microscopic polyarteritis	Nongranulomatous vasculitis involving skin, lung, and kidney; often rapidly progressive glomerulonephritis; often positive P-ANCA
Churg-Strauss syndrome	Granulomatous vasculitis involving upper airway, peripheral nerves, and bowel; associated asthma and eosinophilia
ANCA-positive glomerulonephritis	No systemic vasculitis; often P-ANCA
Scleroderma	Skin tightening, gastrointestinal dysmotility, Raynaud phenomenon, autoantibodies

ANCA: antineutrophil cytoplasmic antibodies; C-ANCA: cytoplasmic ANCA; P-ANCA: perinuclear ANCA.

Adapted from Ambrus JL and Sridhar NR, "Immunologic Aspects of Renal Disease," *JAMA*, 1997, 278(22):1938-45.

Specimen Serum

Container Red top tube or SST™ tube

Special Instructions Tissue for immunofluorescence should be transported frozen in liquid nitrogen.

Reference Interval
- Negative: ≤5 EU/mL
- Borderline: 5.1-20.0 EU/mL
- Positive: >20.1 EU/mL

Use Evaluate patients with rapidly progressive microscopic hematuria and proteinuria; glomerulonephritis with or without pulmonary hemorrhage. The glomerulonephritis caused by anti-GBM is usually rapidly progressive. Elevated levels, often >250 EU/mL, are detected in patients with glomerulonephritis with pulmonary hemorrhage (Goodpasture syndrome), glomerulonephritis without pulmonary involvement, and in idiopathic pulmonary hemosiderosis.

Limitations Weakly positive results (5-30 EU/mL) may occur in some patients without antiglomerular basement membrane antibody-mediated disease. The (Continued)

Glomerular Basement Membrane Antibody *(Continued)*

use of crude basement membrane preparations in a laboratory can lead to lower sensitivity or misdiagnosis.

Methodology Direct (DFA) or indirect fluorescent antibody (IFA), enzyme-linked immunosorbent assay (ELISA)

Additional Information The two principal pathogenic mechanisms of autoimmune renal disease are immune complex mediation and specific autoantibody-mediated damage to fixed renal antigens: the renal glomerular basement membrane, Goodpasture syndrome, and antitubulointerstitial antibodies. Antitubulointerstitial antibodies mediate tubulointerstitial nephritis (eg, following methicillin).

Other entities which may present with or as Goodpasture syndrome include Wegener granulomatosis, SLE, vasculitis, and Henoch-Schönlein purpura. Low serum complement would support diagnoses of SLE or postinfectious glomerulonephritis. Normal concentrations of complement would support vasculitis or anti-GBM disease; ANCA and anti-GBM antibodies prove helpful. Wegener granulomatosis and anti-GBM antibodies may coexist.

References

Ball JA and Young KR Jr, "Pulmonary Manifestations of Goodpasture's Syndrome. Antiglomerular Basement Membrane Disease and Related Disorders," *Clin Chest Med*, 1998, 19(4):777-91.

Kalluri R, Meyers K, Mogyorosi A, et al, "Goodpasture Syndrome Involving Overlap With Wegener's Granulomatosis and Antiglomerular Basement Membrane Disease," *J Am Soc Nephrol*, 1997, 8(11):1795-800.

Lee SM and Marks EA, "The Emerging Spectrum of IgA-Mediated Renal Diseases: Is There an IgA Variant of Goodpasture's Syndrome?" *Am J Kidney Dis*, 1999, 34(3):565-8.

Levy JB, Turner AN, Rees AJ, et al, "Long-Term Outcome of Antiglomerular Basement Membrane Antibody Disease Treated With Plasma Exchange and Immunosuppression," *Ann Intern Med*, 2001, 134(11):1033-42.

♦ **Glomerular Filtration Rate** *see* Creatinine Clearance and Urine Creatinine *on page 473*

♦ **Glomerular Filtration Rate** *see* Creatinine, Serum or Plasma *on page 474*

♦ **Glomerular Filtration Rate** *see* Cystatin C, Serum or Plasma *on page 489*

Glucagon, Plasma

Related Information

Adrenocorticotropic Hormone, Plasma *on page 114*

Gastrin, Serum *on page 631*

Glucose, Fasting, Plasma *on page 643*

5-Hydroxyindoleacetic Acid, Quantitative, Urine *on page 754*

Insulin, Serum *on page 803*

Pancreatic Polypeptide, Human, Serum or Plasma *on page 998*

Vasoactive Intestinal Polypeptide, Plasma *on page 1302*

Abstract Glucagon is a peptide hormone which opposes the action of insulin and provides primary defense against hypoglycemia. Glucagonoma is an extremely rare neuroendocrine tumor usually found in the pancreas.

Patient Preparation Overnight fasting for basal levels. If diabetic, patient should be in good control before specimen is drawn. Avoid recent radioactive tracer (eg, for radioactive scan).

Specimen Plasma

Container Draw blood into a chilled lavender top (EDTA) tube. Deliver to the laboratory immediately.

Storage Instructions Freeze. Stable 2 months at -20°C.

Special Instructions Mix the blood immediately and centrifuge in a refrigerated centrifuge.

Reference Interval ≤60 pg/mL (SI: ≤60 ng/L) at one laboratory, but other intervals are in use (eg, 20-100 pg/mL)

Critical Values Most patients with glucagonoma have levels >500 pg/mL (SI: >500 ng/L); >1000 pg/mL (SI: >1000 ng/L) is diagnostic. The highest values are found with glucagonoma syndrome and with insulinomas. Fasting hyperglucagonemia has been defined as glucagon concentrations ≥120 pg/mL.

Use Diagnose glucagonoma. The most common clinical presentations of hyperglucagonemia are the Zollinger-Ellison syndrome and a "glucagonoma syndrome". Glucagonoma syndromes include a characteristic skin rash, diabetes mellitus (or impaired glucose tolerance), weight loss, abdominal pain, diarrhea, peptic ulcer disease, anemia, and venous thrombosis. This form usually is characterized by very high glucagon levels, >1000 pg/mL (SI: >1000 ng/L). Other presentations include severe diabetes and carcinoid tumors.

Contraindications Recent radioisotopes

Methodology Radioimmunoassay (RIA); ethanol extraction removes "big" glucagon, which is not considered biologically active.

Additional Information Elevated glucagon is seen in diabetic ketoacidosis, stress, uremia, Cushing syndrome, hepatic cirrhosis, hyperosmolality, acute pancreatitis, burns, trauma, surgery, and hypoglycemia. Decreased values are found in cystic fibrosis, chronic pancreatitis, and in the postpancreatectomy state.

References

Krejs GJ, "Noninsulin-Secreting Tumors of the Gastroenteropancreatic System," *Williams Textbook of Endocrinology*, 9th ed, Wilson JD, Foster DW, Kronenberg HM, et al, eds, Philadelphia, PA: WB Saunders Co, 1998, 1663-6.

Service FJ, "Hypoglycemic Disorders," *N Engl J Med*, 1995, 332(17):1144-52.

Wermers RA, Fatourechi V, and Kvols LK, "Clinical Spectrum of Hyperglucagonemia Associated With Malignant Neuroendocrine Tumors," *Mayo Clin Proc*, 1996, 71:1020-8.

♦ **Glucocerebroside** *see* Inherited Diseases of Metabolism and Cell Structure *on page 792*

♦ **Glucose, 2-Hour Postprandial, Plasma** *see* Glucose, Postglucose Load, Plasma *on page 647*

Glucose-6-Phosphate Dehydrogenase, Blood

Related Information

Autohemolysis Test *on page 221*
Heinz Body Stain *on page 669*
Hemosiderin Stain, Urine *on page 692*
Red Blood Cell Enzyme Deficiency *on page 1134*

Abstract Glucose-6-phosphate dehydrogenase (G6PD) is a red cell enzyme with an X-linked mode of inheritance that is important in maintaining RBC proteins in the reduced state. There are over 440 different mutations recorded, of which 60 by themselves or in combination with other mutations result in premature hemolysis of red cells when the mutant enzyme is stressed. G6PD quantitation may be useful if a screening test is positive.

Specimen Erythrocytes

Container Lavender top (EDTA) tube, green top (heparin) tube, or acid-citrate-dextrose (ACD) solution

Storage Instructions In above anticoagulants, RBC enzymes stable at 4°C for at least 6 days and stable at 25°C for at least 24 hours.

Special Instructions If sample must be sent to a reference laboratory, ship on wet ice, do not freeze.

Reference Interval Adults: 8.34 ±1.59 IU/g hemoglobin. Large interlaboratory differences, and the existence of several units of measurement are interpretive hazards. Normal levels in newborns are ~50% higher.

Use Evaluate G6PD deficiency; determine the cause of drug-induced hemolysis. G6PD deficient hemolysis may be secondary to acute bacterial or viral infection and metabolic disorder such as acidosis.

Limitations A blood enzyme screen, performed after a hemolytic episode may not detect G6PD deficiency even if present because the most deficient cells have been destroyed (young G6PD-deficient red cells just released from the bone marrow have relatively high enzyme activity). The screening procedure can only differentiate between normal and grossly deficient samples. **Test may need to be repeated (if initial result is normal) after the patient recovers from an undiagnosed episode of anemia.** Alternatively, family studies can be performed. Recently developed tests based on allele-specific polymerase chain reaction (AS-PCR) avoid this disadvantage.

Methodology Fluorescent NADPH spot test (screening test). Fluorescence is due to reduced nicotine adenine dinucleotide phosphate (NADPH) formed from NADP in the presence of G6PD. Recently developed polymerase chain reaction based methods can identify G6PD-deficient individuals even after blood loss or acute hemolytic activity which can produce false-negative results with the older generation of enzyme activity based tests. AS-PCR-based G6PD deficiency tests, however, will not screen for all variants of G6PD, and are usually designed to detect one or a few mutant forms common to a particular geographic area.
(Continued)

Glucose-6-Phosphate Dehydrogenase, Blood
(Continued)

Additional Information The active G6PD enzyme exists as a dimer, each monomer consisting of 515 amino acids. Aggregation of monomers into active forms (largely dimers) is NADP dependent. G6PD, a housekeeping enzyme, catalyzes the first step in the hexose monophosphate pathway which acts to remove peroxide, thus protecting the red cell from oxidative damage. The enzyme provides reducing power in the form of NADPH which with glutathione is requisite for detoxification of H_2O_2. Total deficiency is thought to be incompatible with life.

G6PD deficiency is the most common red cell defect associated with hemolysis and as such is the most common metabolic disorder of red blood cells. Some 400 million people are affected world-wide. There are many genetic variants of G6PD, some causing marked clinical manifestations, others none. Over 440 G6PD enzyme variants (many, on molecular analysis, are identical) have been subclassified by the World Health Organization, largely on the basis of clinical severity.

- Class I: Severe enzyme deficiency with chronic hemolytic anemia (<10% of normal enzyme activity)
- Class II: Severe enzyme deficiency with intermittent hemolysis
- Class III: Moderate enzyme deficiency with intermittent hemolysis usually associated with infection or drugs (10% to 60% of normal enzyme activity)
- Class IV: No enzyme deficiency or hemolysis

$G6PD^{A-}$ (Class III) and $G6PD^{A+}$ (Class IV) are high frequency G6PD-deficient variants (each found in 10% to 15% of African-Americans). The $G6PD^{A-}$ variant is associated with acute intermittent hemolysis and primaquine sensitivity. $G6PD^{A+}$, although a common and well-studied isoenzyme variant, has no evident associated hematological phenotype. The common variant $G6PD^{Mediterranean}$ (Class II) is found in Caucasians of Mediterranean/Far East ethnic origin.

A G6PD screen is recommended before quantitative G6PD is requested. G6PD hemolysis is associated with formation of Heinz bodies in peripheral red blood cells. It is the older erythrocytes which are most G6PD deficient in affected individuals. These cells are the first eliminated in a hemolytic crisis. The younger cells and reticulocytes contain more G6PD. For these reasons, following a hemolytic episode, when only younger erythrocytes and reticulocytes are present, G6PD values may be spuriously normal. Quantitative assay of G6PD may be helpful in establishing the diagnosis in female patients (who have two RBC populations) or in males with mild G6PD deficiency who have had recent hemolysis. In such cases, assay of the reticulocyte-poor bottom fraction of a centrifuged blood sample may be useful.

Drugs and Chemicals That Should Be Avoided by Persons With G6PD Deficiency

Acetanilid	Phenylhydrazine
Acetylsalicylic acid (aspirin)	Primaquine
Chloromycetin	Quinidine
Doxorubicin	Quinine
Furazolidone (Furoxone®)	Sulfacetamide
Isobutyl nitrite	Sulfamethoxazole (Gantanol®)
Methylene blue	Sulfanilamide
Nalidixic acid (NegGram®)	Sulfapyridine
Naphthalene	Sulfisoxazole
Niridazole (Ambilhar®)	Thiazolesulfone
Nitrofurantoin (Furadantin®)	Toluidine blue
Para-aminosalicylic acid	Trinitrotoluene (TNT)
Pentaquine	Urate oxidase
Phenazopyridine (Pyridium®)	

While usually mild in black children G6PD deficiency may be severe and life-threatening with oxidative stress, especially that relating to infection (viral in particular), fava bean ingestion, and less often relating to naphthalene exposure. A number of commonly used drugs/chemicals can induce hemolysis in individuals with G6PD deficiency (see table), while evidence has accumulated that some occasionally suspect drugs can be given in therapeutic doses without inducing hemolysis. If deficiency is severe, impaired granulocyte function occurs with increased susceptibility to infection (G6PD *Barcelona*). Molecular heterogeneity, as reflected by the numerous mutant isoenzymes, is accompanied by biochemical functional diversity including decreased catalytic effectiveness, impaired substrate and cofactor kinetics, variable reactivity with substrate analogues, and variations in electrophoretic migration rates and pH optima.

Mutations responsible for the G6PD deficient state have been identified by use of molecular biologic techniques, in particular, PCR based methods.

Cloning of the G6PD gene has been accomplished, and there is ongoing activity in cDNA sequencing of G6PD variants.

References

Beutler E, "G6PD Deficiency," *Blood*, 1994, 84(11):3613-36.

Liese AM, Siddiqi MQ, and Spolarics Z, "Rapid Detection of Glucose-6-Phosphate Dehydrogenase Type A-[202A/376G] Deficiency by Allele-Specific Polymerase Chain Reaction," *Am J Hematol*, 2000, 63(3):159-62.

Luzzatto L, "Glucose-6-Phosphate Dehydrogenase Deficiency and Hemolytic Anemia," *Hematology of Infancy and Childhood*, 5th ed, Chapter 18, Nathan DG and Orkin SH, eds, Philadelphia, PA: WB Saunders Co, 1998, 704-26.

Prchal JT and Gregg XT, "Red Cell Enzymopathies," *Hematology: Basic Principles and Practice*, 3rd ed, Chapter 32, Hoffman R, Benz EJ, Jr, Shattil SJ, et al, eds, Philadelphia, PA: Churchill Livingstone, 2000, 561-76.

♦ **Glucose, Body Fluid** *see* Body Fluid Glucose *on page 294*

♦ **Glucose, Cerebrospinal Fluid** *see* Cerebrospinal Fluid Glucose *on page 362*

♦ **Glucose, Dipstick, Urine** *see* Glucose, Semiquantitative, Urine *on page 650*

♦ **Glucose, Fasting, Impaired (IFG)** *see* Glucose, Fasting, Plasma *on page 643*

Glucose, Fasting, Plasma

Related Information

Body Fluid Glucose *on page 294*
C-Peptide, Serum *on page 465*
Fructosamine, Serum *on page 617*
Glucose, Postglucose Load, Plasma *on page 647*
Glucose, Random, Plasma *on page 649*
Glucose, Semiquantitative, Urine *on page 650*
Glucose Tolerance Test, Plasma *on page 651*
Glucose, Whole Blood (Including Point-of-Care) *on page 653*
Glutamic Acid Decarboxylase (GAD65) Antibody *on page 654*
Glycated Hemoglobin (Hemoglobin A$_{1c}$), Blood *on page 655*
Insulin, Serum *on page 803*
Ketone Bodies, Blood *on page 816*
Microalbuminuria *on page 913*
Myotonic Dystrophy DNA Test *on page 942*
pH, Blood *on page 1018*
Point-of-Care Testing *on page 1065*

Synonyms Blood Sugar, Fasting; Fasting Blood Sugar; FBS; FPS; Sugar, Fasting

Applies to Glucose, Fasting, Impaired (IFG); Glucose:Insulin Ratio; Tolbutamide Test

Abstract The three tests recommended by the American Diabetes Association (ADA) for the diagnosis of **diabetes mellitus** are fasting plasma glucose (FPG), random plasma glucose, and two-hour postglucose load glucose. Testing for **gestational diabetes mellitus** involves a first step screening plasma glucose **1 hour after a 50 g glucose load**, and if this result is abnormal, a **100 g glucose load test** (see Glucose, Postglucose Load, Plasma *on page 647*).

(Continued)

Glucose, Fasting, Plasma *(Continued)*

Patient Preparation Patient should be fasting (no food or beverage other than water) for at least 8 hours before testing.

Aftercare WHO has recommended an oral glucose tolerance test for individuals with impaired fasting glucose.

Specimen Plasma or serum

Container Gray top (sodium fluoride or iodacetate) tube is preferred; heparin (green top) and red top tubes are acceptable only if specimen is rapidly separated from the red cells and analyzed promptly.

Sampling Time Morning

Collection Venous specimens are recommended from all age groups, except for neonates, whose specimens are most often drawn from heelsticks.

Storage Instructions Glucose will decrease at a rate of 5-10 mg/dL (SI: 0.3-0.6 mmol/L) per hour in unseparated, room temperature blood not collected in gray top tubes.

Reference Interval Premature infants: may have glucose values as low as 30 mg/dL (SI:1.6 mmol/L); newborns: 40-60 mg/dL (SI: 2.2-3.3 mmol/L); children: 60-100 mg/dL (SI: 3.3-5.6 mmol/L); adults: 60-109 mg/dL (SI: 3.3-6.0 mmol/L)

ADA Criteria for the Diagnosis of Diabetes Mellitus

Any one of the following findings is diagnostic of diabetes, if confirmed by any one of the following findings on a subsequent day:

- Symptoms of diabetes plus a random plasma glucose level ≥200 mg/dL (11.1 mmol/L). Symptoms include polyuria, polydipsia, and unexplained weight loss.
- Fasting plasma glucose ≥126 mg/dL (7 mmol/L) after a minimum 8-hour fast.
- Two-hour postload glucose ≥200 mg/dL (11.1 mmol/L) during an oral glucose tolerance test conducted as described by the World Health Organization, using glucose load containing the equivalent of 75-g anhydrous glucose dissolved in water

In an unstressed, nonpregnant individual, fasting plasma glucose values between 100-125 mg/dL (SI: 5.6-6.9 mmol/L) are classified as **impaired fasting glucose**, a term intended to convey the presence of a metabolic abnormality between normal and diabetes.

Fasting plasma glucose results in the range of 47-60 mg/dL (SI: 2.6-3.3 mmol/L) are consistent with, but not fully diagnostic of, **hypoglycemia**, a diagnosis which has no established biochemical criterion, and which requires careful correlation with clinical features (see table, Clinical Classification of Hypoglycemic Disorders).

Possible Panic Range Infants: <40 mg/dL (SI: <2.2 mmol/L); adults: male: <50 mg/dL (SI: <2.75 mmol/L), female: <40 mg/dL (SI: <2.2 mmol/L); adults: male and female: >400 mg/dL (SI: >22 mmol/L)

Use This is the primary test to evaluate risk of diabetes, to establish the diagnosis of diabetes mellitus, and to monitor therapy and support control of diabetes. In addition, fasting glucose measurements are useful in the diagnosis and treatment of certain metabolic disorders (eg, acidosis, ketosis, dehydration, coma). Elevated fasting plasma glucose is a component of the **metabolic syndrome** (see definition given in Triglycerides, Serum or Plasma *on page 1275*).

Causes of **high plasma glucose** other than diabetes include nonfasting specimen, recent or current intravenous infusions of glucose, stress states, Cushing disease, acromegaly, pheochromocytoma, glucagonoma, severe liver disease, pancreatitis, and drugs (thiazide and other diuretics, glucocorticoids, β-blockers, nicotinic acid, estrogen-containing products, and many others).

For the evaluation of hypoglycemia, symptoms must be correlated with plasma glucose. Hypoglycemia has a lengthy differential diagnosis (see table).

Clinical Classification of Hypoglycemic Disorders

Healthy-Appearing Patient
No Coexisting Disease
Cause of predisposing condition: Drugs: ethanol, salicylates, quinine, haloperidol
Insulinoma; nonislet insulin-secreting tumors
Factitious hypoglycemia induced by insulin
Intense exercise
Ketotic hypoglycemia
Coexisting Disease Under Treatment
Diabetes mellitus
Cause of predisposing condition: Drugs: dispensing error, disopyramide, β-adrenergic-blocking agents, drugs containing sulfhydryl or thiol and autoimmune insulin syndrome; Ackee-fruit poisoning and undernutrition
Ill-Appearing Patient
Cause or Predispositing Condition
Drugs
Pentamidine for pneumocystis pneumonia
Trimethoprim-sulfamethoxazole and renal failure
Propoxyphene and renal failure
Quinine for cerebral malaria
Quinidine for malaria
Topical salicylates and renal failure
Illness or condition
Small size for gestational age in infants
Beckwith-Wiedemann syndrome
Erythroblastosis fetalis
Hyperinsulinemia in infants due to maternal diabetes
Glycogen storage disease
Defects in amino acid and fatty acid metabolism
Reye syndrome
Cyanotic congenital heart disease
Hypopituitarism
Isolated growth hormone deficiency
Isolated corticotropin deficiency
Addison disease
Galactosemia
Hereditary fructose intolerance
Carnitine deficiency
Defective type 1 glucose transporter in the brain
Acquired severe liver disease
Large nonbeta-cell tumor
Sepsis
Renal failure
Congestive heart failure
Lactic acidosis
Starvation, malnutrition
Anorexia nervosa
Surgical removal of pheochromocytoma
Insulin-antibody hypoglycemia
Hospitalized Patient
Cause or Predisposing Condition: Hospitalization for a predisposing condition; total parenteral nutrition and insulin therapy; Interference of cholestyramine with glucocorticoid absorption; shock

Adapted from Service FH, "Hypoglycemic Disorders," *N Engl J Med*, 1995, 332(17):1144-52.

The overnight fasting glucose level is the optimal test, supplemented by additional glucose specimens drawn during symptoms. Outlined below are common diagnostic problems.

- Pancreatic islet cell tumors: insulinomas cause hypoglycemia in fasting individuals or after exercise. Measurement of simultaneous glucose, C-peptide, and insulin levels at the time of spontaneous hypoglycemia help to differentiate insulinoma from other conditions. The **glucose:insulin ratio** is useful in the diagnosis of insulinoma as insulin levels are inappropriately increased for plasma glucose (see Insulin, Serum *on page 803*). An intravenous tolbutamide test with plasma glucose and serum insulin determinations may be used for evaluation of insulin-secreting islet cell

(Continued)

Glucose, Fasting, Plasma (Continued)

tumors. The test is positive in ~75% of patients with these tumors. Glucagon and leucine stimulation tests are less frequently utilized.

- Extrapancreatic tumors - very rare
- Adrenal insufficiency and congenital adrenal hyperplasia
- Hypopituitarism, isolated growth hormone or ACTH deficiency
- Hereditary fructose intolerance, galactosemia, leucine sensitivity
- Drugs including insulin (see above), oral hypoglycemic agents, salicylates, quinine, haloperidol, and many other drugs, and conditions can depress glucose levels.
- Liver disease

Infancy and childhood: The causes of neonatal hypoglycemia include delayed first feeding. Rapid glucose measurement is required for infants with tremor, convulsions, and/or respiratory distress, particularly in the presence of maternal diabetes and hemolytic disease of the newborn (erythroblastosis fetalis). Newborns too large or small for gestational age should have a glucose level measured in the first 24 hours of life. A large number of entities cause neonatal hypoglycemia, including glycogen storage diseases, galactosemia, hereditary fructose intolerance, ketotic hypoglycemia of infancy, fructose-1,6-diphosphatase deficiency, carnitine deficiency (treatable disease presenting as Reye syndrome), and nesidioblastosis (Insulin, Serum on page 803).

Control of diabetes is needed to avoid complications (see Glucose, Whole Blood (Including Point-of-Care) on page 653).

Limitations Artifactual hypoglycemia is caused by leukocytosis, hemolysis, glycolysis in specimens overheated or old, and delay in separating serum or heparinized plasma from red cells. Very prompt removal of plasma or serum, followed by prompt glucose analysis, is necessary for accurate results.

Methodology Specific enzyme-based assays using glucose oxidase or hexokinase

Additional Information Glycated hemoglobin, self-monitoring of blood glucose, and urine microalbumin determinations are recommended for monitoring diabetes control and complications. Hb A_{1c} is not presently recommended as a diagnostic test for diabetes.

Etiologic Classification of Diabetes Mellitus

I.	Type 1 diabetes (β-cell destruction, usually leading to absolute insulin deficiency)
	a. immune-mediated
	b. idiopathic
II.	Type 2 diabetes (may range from predominantly insulin resistance with relative insulin deficiency to a predominantly secretory defect with insulin resistance)
III.	Other specific types
	a. genetic defects of β-cell function
	b. genetic defects in insulin action
	c. diseases of the exocrine pancreas
	d. endocrinopathies
	e. drug- or chemical-induced
	f. infections
	g. uncommon forms of immune-mediated diabetes
	h. other genetic syndromes sometimes associated with diabetes
IV.	Gestational diabetes mellitus (GDM)

Adapted from America Diabetes Association, "Report of the Expert Committee on the Diagnosis and Classification of Diabetes Mellitus," *Diabetes Care*, 1997, 20(7):1183-97.

References

American Diabetes Association, "Report of the Expert Committee on the Diagnosis and Classification of Diabetes Mellitus," *Diabetes Care*, 1997, 20(7):1183-97.

American Diabetes Association, "Standards of Medical Care for Patients With Diabetes Mellitus," *Diabetes Care*, 2000, 23(1 Suppl):S32-S42.

Palumbo PJ, "Glycemic Control, Mealtime Glucose Excursions, and Diabetic Complications in Type 2 Diabetes Mellitus," *Mayo Clin Proc*, 2001, 76(6):609-18.

Service FJ, "Hypoglycemic Disorders," *N Engl J Med*, 1995, 332(17):1144-52.

The Expert Committee on the Diagnosis and Classification of Diabetes Mellitus, "Follow-up Report on the Diagnosis of Diabetes Mellitus," *Diabetes Care*, 2003, 26(11):3160-7.

Young DS, *Effects of Drugs on Clinical Laboratory Tests*, 5th ed, Volume 1: Listing by Test, Washington, DC: AACC Press, American Association of Clinical Chemistry, 2000, Section 3, 349-67.

♦ **Glucose:Insulin Ratio** *see* Glucose, Fasting, Plasma *on page 643*

Glucose, Noninvasive

Related Information

Glucose, Fasting, Plasma *on page 643*
Glucose, Postglucose Load, Plasma *on page 647*
Glucose, Random, Plasma *on page 649*
Glucose Tolerance Test, Plasma *on page 651*
Glucose, Whole Blood (Including Point-of-Care) *on page 653*
Point-of-Care Testing *on page 1065*

Abstract Self-monitoring of blood glucose by diabetic patients is now common-place. Traditionally, such monitoring has been accomplished by point-of-care measuring devices requiring a small quantity of whole blood obtained by skin micropuncture. Now available is a less invasive device based on iontophoresis; this device provides a continuous recording of glucose in interstitial fluid and does not require a blood sample.

Methodology Reverse iontophoresis. Another approach uses transdermal ultrasound to release interstitial fluid with subsequent assay of glucose (and other analytes).

References

Khalil OS, "Spectroscopic and Clinical Aspects of Noninvasive Glucose Measurements," *Clin Chem*, 1999, 45(2):165-77.

Rao G, Guy RH, Glikfeld P, et al, "Reverse Iontophoresis: Noninvasive Glucose Monitoring *In Vivo* in Humans," *Pharm Res*, 1995, 12(12):1869-73.

Tamada JA, Garg S, Jovanovic L, et al, "Noninvasive Glucose Monitoring: Comprehensive Clinical Results. Cygnus Research Team," *JAMA*, 1999, 282(19):1839-44.

Glucose, Postglucose Load, Plasma

Related Information

Fructosamine, Serum *on page 617*
Glucose, Quantitative, Urine *on page 649*
Glucose, Random, Plasma *on page 649*
Glucose, Semiquantitative, Urine *on page 650*
Glucose Tolerance Test, Plasma *on page 651*
Glutamic Acid Decarboxylase (GAD65) Antibody *on page 654*
Glycated Hemoglobin (Hemoglobin A$_{1c}$), Blood *on page 655*
Islet Cell Antibody *on page 813*
Ketones, Urine *on page 817*
Microalbuminuria *on page 913*
Reducing Substances, Urine *on page 1148*

Synonyms 2-Hour PP Glucose; Oral Glucose Tolerance Test; Postprandial Glucose; PP, 2-Hour

Replaces Glucose, 2-Hour Postprandial, Plasma

Test Includes Glucose level 2 hours after a meal or after a measured glucose load

Abstract The three tests recommended by the American Diabetes Association (ADA) for the diagnosis of **diabetes mellitus** are fasting plasma glucose, random plasma glucose, and the **2-hour postload** plasma glucose. Testing for **gestational diabetes mellitus** involves a first step screening plasma glucose **1 hour after a 50 g glucose load**, and if this result is abnormal, a **100 g glucose load test** (see below).

Patient Preparation Patient should be fasting (no food or beverage, except for water and prescribed medications) for 8 hours. Testing is best done in the early morning.

• **75 g load:** This test is for diabetes mellitus. Patient should drink the glucose, usually a commercially available product specific for this test, and observe labeling precautions.

• **50 g load:** This is the first stage screen for **gestational diabetes mellitus**. The patient need not be fasting, but fasting is acceptable. Patient should drink the glucose, usually a commercially available product specific for this test, and observe labeling precautions.

(Continued)

Glucose, Postglucose Load, Plasma (Continued)

- **100 g load:** This is the second stage in the diagnosis of **gestational diabetes mellitus**. The patient should be fasting and drink the glucose, usually a commercially available product specific for this test, and observe the labeling precautions.

Specimen Plasma or serum

Container Gray top (sodium fluoride or iodoacetate) tube; red top tubes are acceptable only if specimen is rapidly separated from the red cells and analyzed promptly.

Sampling Time

- **75 g load:** Draw specimen 2 hours after glucose.
- **50 g load:** Draw specimen 1 hour after glucose
- **100 g load:** Draw specimens fasting (preglucose), 1-hour, 2-hour, and 3-hour postglucose (3-hour OGTT).

Reference Interval

ADA Criteria for the Diagnosis of Diabetes Mellitus

Any one of the following findings is diagnostic of diabetes, if confirmed by any one of the following findings on a subsequent day:

- Symptoms of diabetes plus a random plasma glucose level ≥200 mg/dL (11.1 mmol/L). Symptoms include polyuria, polydipsia, and unexplained weight loss.
- Fasting plasma glucose ≥126 mg/dL (7 mmol/L) after a minimum 8-hour fast.
- Two-hour postload glucose ≥200 mg/dL (11.1 mmol/L) during a 75 g glucose load.

ADA Criteria for the Diagnosis of Gestational Diabetes Mellitus

Step 1. A plasma glucose >140 mg/dL (SI: >7.8 mmol/L) 1 hour after a 50 g load indicates the need for additional testing (Step 2).

Step 2. Gestational diabetes mellitus is diagnosed if two or more values, in the 3-hour OGTT, exceed these criteria:

- Fasting: 105 mg/dL, 5.8 mmol/L
- 1-hour: 190 mg/dL, 10.5 mmol/L
- 2-hour: 165 mg/dL, 9.2 mmol/L
- 3-hour: 145 mg/dL, 8.0 mmol/L

Use Diagnose diabetes mellitus

Methodology Specific enzyme-based assays using glucose oxidase or hexokinase

Additional Information When the fasting plasma glucose is in the range of 100-125 mg/dL (SI: 5.6-6.9 mmol/L), the impaired fasting glucose category, some authorities (eg, the World Health Organization, WHO), but not the American Diabetes Association, recommend a 2-hour oral glucose tolerance test, performed as follows: following an overnight fast, a **75 g** glucose load is taken over a 5-minute period. For children, the dose is 1.75 g glucose/kg body weight. Results are classified as follows.

- **Impaired fasting glucose (IFG):** 0-hour specimen: 100-125 mg/dL (SI: 5.6-6.9 mmol/L)
- **Impaired glucose tolerance (IGT):** 2-hour specimen: 140-199 mg/dL (SI: 7.8-11.0 mmol/L), with fasting plasma glucose **(FBG)** <126 mg/dL
- **Diabetes mellitus:** 0-hour specimen: ≥126 mg/dL (SI: ≥7.0 mmol/L); 2-hour specimen: ≥200 mg/dL (SI: ≥11.1 mmol/L)

The patients classified in the IGT group bear association with risk factors for cardiovascular disease and cardiovascular events, while the IFG subjects are much less strongly associated with those risks.

References

American Diabetes Association, "Report of the Expert Committee on the Diagnosis and Classification of Diabetes Mellitus," *Diabetes Care*, 1997, 20(7):1183-201.

McCance DR, Hanson RL, Charles MA, et al, "Comparison of Tests for Glycated Haemoglobin and Fasting and Two-Hour Plasma Glucose Concentrations as Diagnostic Methods for Diabetes," *BMJ*, 1994, 308(6940):1323-8.

The Expert Committee on the Diagnosis and Classification of Diabetes Mellitus, "Follow-up Report on the Diagnosis of Diabetes Mellitus," *Diabetes Care*, 2003, 26(11):3160-7.

♦ **Glucose, Qualitative, Urine** see Glucose, Semiquantitative, Urine on page 650

Glucose, Quantitative, Urine

Related Information

Fructosamine, Serum *on page 617*
Glucose, Fasting, Plasma *on page 643*
Glucose, Postglucose Load, Plasma *on page 647*
Glucose, Random, Plasma *on page 649*
Glucose, Semiquantitative, Urine *on page 650*
Glucose Tolerance Test, Plasma *on page 651*
Glycated Hemoglobin (Hemoglobin A$_{1c}$), Blood *on page 655*
Ketone Bodies, Blood *on page 816*
Ketones, Urine *on page 817*
Microalbuminuria *on page 913*
Osmolality, Urine *on page 979*
Point-of-Care Testing *on page 1065*
Reducing Substances, Urine *on page 1148*
Urine Collection, 24-Hour *on page 1295*

Synonyms Sugar, Quantitative, Urine; Urinary Sugar Test

Abstract Glucose can be accurately assayed in urine. There are, however, very few indications for such testing.

Specimen 24-hour urine or other specific timed collections

Container Plain urine container, sodium fluoride preservative

Storage Instructions Refrigerate

Reference Interval ≤100 mg/24 hours (SI: ≤5.6 mmol/day); normal ranges are not available on random specimens.

Use This test may be useful in the evaluation of nondiabetic patients whose urine tests positive for reducing substances (see Reducing Substances, Urine *on page 1148*).

Limitations This test has **no** place in the management of patients with diabetes mellitus.

Methodology Glucose oxidase

Glucose, Random, Plasma

Related Information

C-Peptide, Serum *on page 465*
Drugs of Abuse Testing, Urine *on page 525*
Glucose, Fasting, Plasma *on page 643*
Glucose, Noninvasive *on page 647*
Glucose, Postglucose Load, Plasma *on page 647*
Glucose Tolerance Test, Plasma *on page 651*
Glucose, Whole Blood (Including Point-of-Care) *on page 653*
Insulin, Serum *on page 803*
Ketone Bodies, Blood *on page 816*
Ketones, Urine *on page 817*
Microalbuminuria *on page 913*
Point-of-Care Testing *on page 1065*
Salicylate, Serum or Plasma *on page 1176*

Abstract The three tests recommended by the American Diabetes Association (ADA) for the diagnosis of **diabetes mellitus** are fasting plasma glucose, random plasma glucose, and the two-hour postglucose load. Testing for **gestational diabetes mellitus** involves a first step screening plasma glucose **1 hour after a 50 g glucose load**, and if this result is abnormal, a **100 g glucose load test** (see Glucose, Postglucose Load, Plasma *on page 647*).

Specimen Plasma or serum

Container Gray top (sodium fluoride or iodoacetate) tube is preferred; red top tubes are acceptable only if specimen is rapidly separated from the red cells and analyzed promptly.

Collection Venous specimens are recommended from all age groups, except for neonates, whose specimens are usually obtained from heelsticks. (In postprandial states, a capillary specimen will have a slightly higher glucose concentration than a venous specimen.)

Storage Instructions Glucose will decrease at a rate of 5-10 mg/dL (SI: 0.3-0.6 mmol/L) per hour in unseparated, room temperature blood not collected in gray top tubes.

(Continued)

Glucose, Random, Plasma (Continued)

Reference Interval Newborns: <115 mg/dL (SI: <6.4 mmol/L); adults and children: <200 mg/dL (SI: <11.1 mmol/L)

ADA Criteria for the Diagnosis of Diabetes Mellitus - see Glucose, Fasting, Plasma *on page 643*.

Random plasma glucose results 47-60 mg/dL (SI: 2.6-3.3 mmol/L) are consistent with, but not fully diagnostic of, **hypoglycemia**, a diagnosis which has no established biochemical criterion, and which requires careful correlation with clinical features (see table in Glucose, Fasting, Plasma *on page 643*).

Possible Panic Range Neonates: <40 mg/dL (SI: <2.2 mmol/L); adults: male: <50 mg/dL (SI: <2.8 mmol/L), >400 mg/dL (SI: >22.2 mmol/L); adults female: <40 mg/dL (SI: <2.2 mmol/L), >400 mg/dL (SI: >22.2 mmol/L)

Use This is one of the tests used to diagnose diabetes mellitus and to monitor therapy and support control of diabetes. In addition, random glucose measurements are useful in the diagnosis and treatment of certain metabolic disorders (eg, acidosis, ketosis, dehydration, coma). **Hypoglycemic** values approximately <45 mg/dL (SI: <2.5 mmol/L) if present, should be investigated with C-peptide, insulin and proinsulin levels as well (see C-Peptide, Serum *on page 465*). The diagnosis of **insulinoma** is suggested when random glucose <40 mg/dL (SI: <2.2 mmol/L) is found with inappropriate plasma insulin levels following prolonged fasting (see Insulin, Serum *on page 803*). See Glucose, Fasting, Plasma *on page 643* for the differential diagnosis of hypoglycemia. Determination of blood glucose on admission in patients who have had an out-of-hospital cardiac arrest can serve as a predictor of neurologic recovery. Higher levels are indicative of more severe brain ischemia and difficult resuscitation.

Limitations Glucose will decrease in samples left on the clot, and in tubes other than gray top tubes, if not processed and analyzed promptly.

Methodology Specific enzyme-based assays using glucose oxidase or hexokinase

Additional Information **Whole blood glucose** values are not equivalent to **plasma glucose**, unless the whole blood assay has been calibrated to match plasma values (see Glucose, Whole Blood (Including Point-of-Care) *on page 653* and Point-of-Care Testing *on page 1065*). Small, handheld, whole blood glucose meters are designed for use at the point-of-care and for patient self-monitoring. The wide use of such devices has underscored the need for reliable instrumentation and quality control procedures to ensure valid results from this type of testing (see Glucose, Whole Blood (Including Point-of-Care) *on page 653*). Evaluation of glycated hemoglobin, self-monitoring of blood glucose, and use of microalbuminemia testing are ongoing means for monitoring glycemic control in diabetic patients. Noninvasive devices for monitoring glycemic status are being developed (see Glucose, Noninvasive *on page 647*).

In addition to random (casual) plasma glucose levels, fasting plasma glucose levels and 2-hour postload (75-gram, glucose tolerance test in nonpregnant subjects) plasma glucose levels are used to diagnose diabetes mellitus.

If glucose is >400 mg/dL (SI: >22 mmol/L), the possibility of ketonemia should be considered. The incidence of hypoglycemia in hospitalized patients appears to be significant, but may be better controlled if frequent monitoring of glucose levels is employed.

See Diabetes Mellitus, Hypoglycemia, and Hyperglycemia in the Disease Index.

References

Jacobs DS, DeMott WR, Oxley DK, et al, *Laboratory Test Handbook*, 5th ed, Hudson, OH: Lexi-Comp Inc, 2001.

Kiechle FL, "Blood Glucose: Measurement in the Point-of-Care Setting," *Lab Med*, 2000, 31(5):276-82.

The Expert Committee on the Diagnosis and Classification of Diabetes Mellitus, "Follow-up Report on the Diagnosis of Diabetes Mellitus," *Diabetes Care*, 2003, 26(11):3160-7.

Glucose, Semiquantitative, Urine

Related Information

Fat, Urine *on page 573*

Synonyms Sugar, Qualitative, Urine; Urinary Sugar Test

Applies to Glucose, Dipstick, Urine; Glucose Tolerance Test Urines; Urines for Glucose Tolerance

Replaces Glucose, Qualitative, Urine

Test Includes Dipstick glucose is usually a part of urinalysis.

Abstract Glycosuria occurs when the renal tubular threshold for glucose is exceeded.

Specimen Random urine, double-void technique preferred

Sampling Time Random or following a glucose load

Storage Instructions If the specimen cannot be processed promptly, it should be refrigerated.

Reference Interval None detected (by reagent strips). (Quantitative limits of normal are discussed by Li and Huang).

Use The usefulness is limited to: 1) the immediate evaluation of a comatose patient - and then it is only useful until the blood glucose measurement is available, possibly 5-10 minutes later; 2) diagnostic evaluation of a newborn who has a positive reducing substance test (see Reducing Substances, Urine *on page 1148*).

This test has **no** role in the management of diabetes mellitus.

Limitations Color vision deficiency may cause erroneous reading of the strip. Each brand of commercial dipsticks lists interfering substances. Such package inserts should be reviewed.

Methodology Double sequential enzyme analysis, specific for glucose (glucose oxidase/peroxidase). Sensitivity is 50-100 mg glucose/dL (SI: 2.8-5.6 mmol/L) urine.

References
Pugia MJ, "Technology Behind Diagnostic Reagent Strips," *Lab Med*, 2000, 31(2):92-6.
Sherwin RS, "Diabetes Mellitus," *Cecil Textbook of Medicine*, 21st ed, Chapter 242, Goldman L and Bennet JC, eds, Philadelphia, PA: WB Saunders Co, 2000, 1263-85.

♦ **Glucose Suppression Test** *see* Growth Hormone, Serum *on page 662*
♦ **Glucose Tolerance Factor** *see* Chromium, Serum *on page 402*

Glucose Tolerance Test, Plasma

Related Information

Synonyms GTT; OGTT; Oral Glucose Tolerance Test

Applies to Gestational Diabetes Screening Test

Test Includes Fasting blood glucose followed by glucose levels drawn at timed intervals after administration of a glucose load. Urine is no longer collected or examined for the GTT.

Abstract The oral glucose tolerance test (OGTT) is used for the diagnosis of gestational diabetes mellitus. It has also been used, but is not optimal, for the diagnosis of types 1 and 2 diabetes mellitus.

Patient Preparation Patient should be active and have had adequate food intake with adequate carbohydrates (at least 150 g carbohydrate daily) for 3 days, and then fast 12 hours prior to test. Many drugs interfere (eg, steroids, diuretics, antihypertensives, anticonvulsants, psychoactive drugs, antituberculous agents, and anti-inflammatory drugs).

(Continued)

Glucose Tolerance Test, Plasma (Continued)

In pregnant patients, a screening test (fasting **not** required) is conducted as follows: A 50-g oral-glucose load is administered, and a blood sample is drawn after 1 hour. The gestational screening test is positive when the postload plasma glucose result is ≥140 mg/dL (SI: ≥7.8 mmol/L). Additional testing is then performed as specified in Glucose, Postglucose Load, Plasma *on page 647*.

Specimen Plasma

Container Gray top (sodium fluoride or iodoacetate) tube

Collection Nongestational OGTT: After a fasting blood specimen is obtained, administer the oral glucose solution. (The adult dose is 75 g. Children receive 1.75 g/kg body weight up to 75 g). Collect a blood specimen at 2-hours postglucose administration.

The patient should remain seated and consume nothing but water after the glucose solution is administered. Physical activity should be minimized and some recommend that the patient be kept supine throughout. Vomiting or diarrhea may alter test results.

Causes for Rejection Incorrect or no anticoagulant, time not marked on tubes, nonfasting patient. Stressed patients (eg, following surgery, with infections, on corticosteroids) should not have a OGTT.

Reference Interval

Nongestational OGTT - normal glucose tolerance:
- fasting: <100 mg/dL (SI: <5.6 mmol/L)
- 2-hour: <140 mg/dL (SI: <7.8 mmol/L)

Impaired glucose tolerance:
- fasting: 100-125 mg/dL (SI: 5.6-6.9 mmol/L)
- 2-hour: 140-199 mg/dL (SI: 7.8 -11.0 mmol/L)

Diabetes mellitus:
- fasting: ≥126 mg/dL (SI: ≥7.0 mmol/L)
- 2-hour: ≥200 mg/dL (SI: ≥11.1 mmol/L)

Criteria for interpretation for gestational diabetes mellitus: See Glucose, Postglucose Load, Plasma *on page 647*.

For further information, the reader is encouraged to review the important 2003 reference in *Diabetes Care*.

Use The OGTT is used for, but is not optimal for, the diagnosis of diabetes mellitus (type 1, type 2) and has been used in the evaluation of unexplained hypertriglyceridemia, neuropathy, impotence, diabetes-like renal diseases, retinopathy, and necrobiosis lipoidica diabeticorum. The gestational OGTT is used in pregnancy to predict perinatal morbidity, risk of fetal abnormality, and perinatal mortality.

Limitations Drawbacks of the OGTT include poor reproducibility, patient inconvenience, and strong propensity for overdiagnosis of diabetes mellitus. The American Diabetes Association (ADA) recommends the fasting plasma glucose determination rather than the OGTT for the diagnosis of type 1 and type 2 diabetes mellitus.

Contraindications The OGTT is contraindicated in the presence of obvious diabetes mellitus. Emesis is often an indication to cancel the remainder of an OGTT.

Methodology Glucose is usually determined by specific, enzyme-based assays using glucose oxidase or hexokinase.

Additional Information Fasting plasma glucose levels ≥126 mg/dL (SI: ≥7.0 mmol/L) and/or symptoms of diabetes plus random plasma glucose levels ≥200 mg/dL (SI: ≥11.1 mmol/L) on two, separate occasions are diagnostic of diabetes mellitus and obviate the need for a OGTT. However, studies of obese and postpartum, gestational diabetic subjects have shown greater diagnostic sensitivity of the OGTT compared to the fasting plasma glucose in these populations.

References

Kjos SL and Buchanan TA, "Gestational Diabetes Mellitus," *N Engl J Med*, 1999, 341(23):1749-56.

Mannucci E, Bardini G, Ognibene A, et al, "Comparison of ADA and WHO Screening Methods for Diabetes Mellitus in Obese Patients. American Diabetes Association," *Diabet Med*, 1999, 16(7):579-85.

The Expert Committee on the Diagnosis and Classification of Diabetes Mellitus, "Follow-up Report on the Diagnosis of Diabetes Mellitus," *Diabetes Care*, 2003, 26(11):3160-7.

♦ **Glucose Tolerance Test Urines** *see* Glucose, Semiquantitative, Urine *on page 650*

Glucose, Whole Blood (Including Point-of-Care)

Related Information

Glucose, Fasting, Plasma *on page 643*
Glucose, Postglucose Load, Plasma *on page 647*
Glucose, Random, Plasma *on page 649*
Glucose Tolerance Test, Plasma *on page 651*
Glycated Hemoglobin (Hemoglobin A$_{1c}$), Blood *on page 655*
Ketones, Urine *on page 817*
Microalbuminuria *on page 913*
Point-of-Care Testing *on page 1065*
Reducing Substances, Urine *on page 1148*

Abstract Situations in which whole blood glucose measurements are utilized include self-monitoring of diabetes (see also Additional Information), point-of-care (POC) testing in acute and chronic care facilities, hospital intensive care units, and hospitalized patients having arterial blood gas measurements (glucose can be assayed without drawing an additional specimen).

Most of the information in the monograph Glucose, Fasting, Plasma *on page 643* also applies to measurements in whole blood. Described below are important **differences** between whole blood and plasma measurements.

Specimen Whole blood, obtained by venipuncture or skin micropuncture

Container Green top (heparin) tube or none (drop of blood placed directly in/on assay device or on reagent container)

Causes for Rejection Failure to meet instrument criteria, specimen delayed (more than 15 minutes) in reaching testing laboratory

Turnaround Time Within 30 minutes. Most of the POC instruments give whole blood glucose values in <2 minutes.

Special Instructions Must be assayed immediately

Reference Interval Adults: 65-95 mg/dL (SI: 3.5-5.3 mmol/L)

Glucose concentrations in plasma are ~11% higher than in simultaneously obtained whole blood specimen when the hematocrit is normal. To mitigate the potential confusion which can result from the fact that whole blood and plasma glucose concentrations are different, the whole blood instrument can be calibrated using a serum-based reference material. When this is done, the whole blood result is approximately the same as the simultaneously obtained plasma sample. Most manufacturers of whole blood glucose instruments have adopted this approach.

Use Whole blood glucose assays are used for monitoring the treatment of patients with an established diagnosis of diabetes mellitus, to determine insulin dose, and for monitoring the metabolic status of seriously ill patients. **The initial diagnosis of diabetes mellitus or gestational diabetes should be made on the basis of glucose assays in fasting or randomly obtained venous plasma** (see Glucose, Fasting, Plasma *on page 643*).

Limitations Some devices used in POC testing for blood glucose have larger imprecision and bias characteristics than the analyzers used for plasma glucose measurements in medical laboratories. The National Committee for Clinical Laboratory Standards (NCCLS) has made the following recommendation: For test readings ≥75 mg/dL (SI: ≥4.2 mmol/L), the discrepancy between ancillary blood glucose testing (ABGT) concentrations and laboratory concentrations on the same specimen should be <20%; for test readings <75 mg/dL (SI: <4.2 mmol/L), the discrepancy should be no more than 15 mg/dL (SI: 0.83 mmol/L). Mannitol causes interference in trilayer electrochemical biosensor testing. Hematocrit levels are relevant. Drug interference was recently addressed. Lack of bidirectional communication between POC devices and information systems represents a major disadvantage.

The most frequent errors include inadequate instrument cleaning, incorrect quality control, improper technique, and inappropriate match with test strip calibration. Other factors include drug interference, hematocrit extremes, pO$_2$ (for glucose oxidase methods), and low total protein concentrations in extracorporeal circulation procedures.

Methodology Specific enzyme-based assays using glucose oxidase or hexokinase, trilayer biosensors
(Continued)

Glucose, Whole Blood (Including Point-of-Care)
(Continued)

Additional Information Because clinical research has demonstrated the advantages of intensive diabetes management, self-monitoring of blood glucose by diabetic patients is now commonplace. Such glucose testing is most frequently performed using POC instruments requiring a small amount of whole blood obtained by skin micropuncture. A new generation of less invasive self-monitoring devices, which are based on the technique of iontophoresis and do not require a blood sample, are becoming available. See Glucose, Noninvasive *on page 647*.

Urine ketone testing remains important for patients at risk of diabetic ketoacidosis. Glycated hemoglobin and testing for microalbuminuria are needed to maintain control in diabetic patients.

References

Kost GJ, Nguyen TH, and Tang Z, "Whole-Blood Glucose and Lactate. Trilayer Biosensors, Drug Interference, Metabolism, and Practice Guidelines," *Arch Pathol Lab Med*, 2000, 124(8):1128-34.

McCall AL, Allison N, and Stephens E, "The Monitoring of Metabolic Control for Patients With Diabetes Mellitus," *Lab Med*, 2001, 32(7):378-83.

National Committee for Clinical Laboratory Standards (NCCLS), *Point-of-Care Blood Glucose Testing in Acute and Chronic Care Facilities, Approved Guidelines*, 2nd ed, NCCLA Document C30-A2 (ISBN: 1-56238-471-6), 2002.

Tang Z, Du X, Louie RF, et al, "Effects of Drugs on Glucose Measurements With Handheld Glucose Meters and a Portable Glucose Analyzer," *Am J Clin Pathol*, 2000, 113(1):75-86.

Tang Z, Lee JH, Louie RF, et al, "Effects of Different Hematocrit Levels on Glucose Measurements With Handheld Meters for Point-of-Care Testing," *Arch Pathol Lab Med*, 2000, 124(8):1135-40.

♦ β-**Glucosidase** *see* Inherited Diseases of Metabolism and Cell Structure *on page 792*

Glutamic Acid Decarboxylase (GAD65) Antibody
Related Information

Glucose, Fasting, Plasma *on page 643*
Glucose, Postglucose Load, Plasma *on page 647*
Glucose Tolerance Test, Plasma *on page 651*
Islet Cell Antibody *on page 813*
Parietal Cell Antibody *on page 1005*
Thyroglobulin Antibody *on page 1249*
Thyroperoxidase Autoantibody *on page 1253*

Synonyms GAD65

Abstract Antibodies specific for the 65 kDa isoform of glutamic acid decarboxylase (GAD65) appear up to 10 years before the onset of clinical diabetes. Individuals with autoantibodies to GAD65 are at greater risk of developing type 1 diabetes than individuals without these antibodies. (Type 1 is that entity formerly called insulin-dependent diabetes mellitus.) Anti-GAD65 autoantibodies are detectable in the sera of about 75% of patients at the time of diagnosis of type 1 diabetes mellitus. They are described as a correlate of susceptibility to type 1 diabetes.

Circulating antibodies against GAD65 characterize the stiff-person (stiff-man) (Moersch-Woltman) syndrome, in which muscle spasms of the back and legs occur with spinal hyperlordosis. It is a marker for a cluster of autoimmune disorders including thyroiditis, Graves disease, hypothyroidism, pernicious anemia, autoimmune cerebellitis, myasthenia gravis, Lambert-Eaton syndrome, rare acquired encephalopathies, Addison disease, premature ovarian failure and vitiligo, as well as for type 1 diabetes susceptibility.

Specimen Serum

Container Red top tube or SST™ tube

Reference Interval <0.02 nmol/L (serum); CSF concentrations are less than those of serum.

Possible Panic Range Concentrations ≥0.03 nmol/L are consistent with susceptibility to autoimmune diabetes mellitus, thyroiditis, and pernicious anemia.

Use Assess susceptibility to type 1 (autoimmune) diabetes mellitus; distinguish subjects with type 2 (noninsulin-dependent) diabetes mellitus who will subsequently evolve with type 1 diabetes; confirm the diagnosis of stiff-person syndrome (Moersch-Woltman syndrome) in 98% of patients

Limitations GAD65 antibodies are detected in 20% of nondiabetic twins who remain disease free extended periods of time and in 8% of healthy individuals.

Anti-GAD65 antibody titers fail to correlate with disease severity in the stiff-person syndrome.

Clinical utility is yet to be determined.

Additional Information Antibodies specific for the 65 kDa isoform of glutamic acid decarboxylase (GAD65) comprise the majority of pancreatic islet cell auto-antibodies. Anti-GAD65 antibodies are detected in 75% of patients who have type I (insulin-dependent) diabetes mellitus and 98% of patients who have the rare disorder, stiff-man syndrome. Patients with type 1 diabetes mellitus usually have antibody levels between 0.02 and 20 nmol/L. Levels >20 nmol/L are usually found in patients with stiff-man syndrome and related autoimmune neurologic disorders such as encephalomyelopathy.

No GAD65 antibody-negative sample was positive for islet cell antibodies in a small series. A positive association exists between GAD65 antibodies and antibodies to gastric parietal cells, thyroglobulin, and thyroperoxidase antibody. These four antibodies support distinction between type 1 and type 2 diabetes. GAD65 antibody is found in 24% of patients with Lambert-Eaton myasthenic syndrome and in patients with idiopathic acquired cerebellar ataxia.

Most antibody-positive relatives of patients with type I diabetes mellitus do not develop diabetes, but one or more immune markers are usually found among those who do.

References

Barker RA, Revesz T, Thom M, et al, "Review of 23 Patients Affected by the Stiff Man Syndrome: Clinical Subdivision Into Stiff Trunk (Man) Syndrome, Stiff Limb Syndrome, and Progressive Encephalomyelitis With Rigidity," *J Neurol Neurosurg Psychiatry*, 1998, 65(5):633-40.

Hatziagelaki E, Jaeger C, Petzoldt R, et al, "The Combination of Antibodies to GAD-65 and IA-2ic Can Replace the Islet-Cell Antibody Assay to Identify Subjects at Risk of Type 1 Diabetes Mellitus," *Horm Metab Res*, 1999, 31(10):564-9.

Littorin B, Sundkvist G, Hagopian W, et al, "Islet Cell and Glutamic Acid Decarboxylase Antibodies Present at Diagnosis of Diabetes Predict the Need for Insulin Treatment. A Cohort Study in Young Adults Whose Disease Was Initially Labeled as Type 2 or Unclassifiable Diabetes," *Diabetes Care*, 1999, 22(3):409-12.

Mayo Reference Services, "Glutamic Acid Decarboxylase (GAD65) Antibody Assay, Spinal Fluid - Revised," #84221, *New Test Announcement*, December 2003.

Walikonis JE and Lennon VA, "Radioimmunoassay for Glutamic Acid Decarboxylase (GAD65) Autoantibodies as a Diagnostic Aid for Stiff-Man Syndrome and a Correlate of Susceptibility to Type 1 Diabetes Mellitus," *Mayo Clin Proc*, 1998, 73(12):1161-6.

♦ **Glutamic Oxaloacetic Transaminase, Serum** *see* Aspartate Aminotransferase, Serum *on page 216*

♦ **Glutamic Pyruvate Transaminase** *see* Alanine Aminotransferase, Serum *on page 116*

♦ **Glutamine, Spinal Fluid** *see* Cerebrospinal Fluid Glutamine *on page 363*

♦ **γ-Glutamyl Transferase** *see* Gamma-Glutamyl Transferase, Serum *on page 629*

♦ **Glutamyl Transpeptidase** *see* Gamma-Glutamyl Transferase, Serum *on page 629*

♦ **Glutathione Peroxidase 1** *see* Myeloperoxidase, Plasma *on page 939*

♦ **Gluten** *see* Endomysial Antibodies *on page 537*

♦ **Glycated Albumin** *see* Fructosamine, Serum *on page 617*

Glycated Hemoglobin (Hemoglobin A₁c), Blood

Related Information

Fetal Hemoglobin *on page 581*
Fructosamine, Serum *on page 617*
Glucose, Fasting, Plasma *on page 643*
Glucose, Semiquantitative, Urine *on page 650*
Glucose, Whole Blood (Including Point-of-Care) *on page 653*
Hemoglobin Electrophoresis *on page 684*
Microalbuminuria *on page 913*
Triglycerides, Serum or Plasma *on page 1275*

Synonyms Fast Hemoglobins; GHb; Glycohemoglobin; Hb A₁; Hemoglobin A₁ₐ, A₁ᵦ, A₁c

Applies to Protein Glycosylation

(Continued)

Glycated Hemoglobin (Hemoglobin A$_{1c}$), Blood
(Continued)

Abstract Glycated hemoglobins (GHb) comprise a heterogeneous group of substances formed by the chemical reaction between sugars and hemoglobin. The rate at which GHb is formed is proportional to the concentration of blood glucose. The GHb provides an index of the average blood glucose concentration during a 2- to 4-month period.

Some assays measure all GHb species in a sample, while other assays measure only one or two species. To standardize the clinical measurements, most assays now in use clinically measure Hb A$_{1c}$, or are calibrated to produce a result equivalent to such a measurement.

The measurement of Hb A$_{1c}$ is important because of evidence showing that tight glycemic control results in a reduced incidence of diabetic nephropathy and other long-term complications of diabetes mellitus.

Specimen Whole blood

Container Lavender top (EDTA) tube; check with the laboratory.

Sampling Time Testing at 3-month intervals is recommended for patients with type I diabetes. For patients with type II diabetes, glycated hemoglobins at diagnosis and at 6-month intervals, or as often as required for good control, are desirable.

Storage Instructions Stable 7 days at 4°C.

Reference Interval Method-dependent. For assays that measure **total GHb** reported reference intervals include 5.3% to 7.5% and 4% to 7%.

With assays that measure **Hb A$_{1c}$**, reported reference intervals include 4.5% to 5.7% (HPLC) and 4.5% to 8.5% (column chromatography).

Critical Values The risk of microalbuminuria with insulin-dependent diabetes mellitus increases when Hb A$_{1c}$ exceeds 8.1% (equivalent to Hb A$_1$ >10.1%). This value corresponds with average daily blood glucose levels ~200 mg/dL (SI: ~11.1 mmol/L). See Microalbuminuria *on page 913* with graphic.

Use Glycated hemoglobin values are used to assess long-term glucose control in diabetes. The test should be performed at the time of initial diagnosis, and then **at least quarterly** in insulin-dependent diabetes, and **as frequently as needed** in noninsulin dependent diabetes. GHb measurements reflect the level of control present over the preceding 60-120 days; more recent levels have greater influence. Continued high levels of blood glucose are reflected in high GHb concentrations.

Davidson et al advocate application of Hb A$_{1c}$ concentrations to support the diagnosis of diabetes in subjects whose fasting plasma glucose concentrations are <140 mg/dL (7.8 mmol/L) unless excessive glycosylation is demonstrated. Vinicor, however, notes that both false-positive and false-negative diagnoses would result, assuming that a fasting plasma concentration <110 mg/dL (6.1 mmol/L) or >139 mg/dL (7.7 mmol/L) is definitely normal or indicative of diabetes, respectively.

Limitations The various analytical methods for glycated hemoglobin are differently affected in the presence of hemoglobinopathies. Check with your laboratory if your patient has an abnormal hemoglobin.

Methodology Immunoassay, high performance liquid chromatography (HPLC), isoelectric focusing, electrospray ionization mass spectrometry, ion-exchange chromatography, boronate affinity with either column chromatographic or ion capture separation, electrophoresis, colorimetry, and spectrophotometry. The American Diabetes Association (ADA) recommends that laboratories use only methods certified as traceable to the Diabetes Control and Complications Trial (DCCT) reference method and participate in proficiency testing programs that use whole blood specimens with targets set by the National Glycohemoglobin Standardization Program Laboratory Network.

Additional Information This test is essential for optimal management of diabetes mellitus. If a result does not seem consistent with the clinical findings, an assay for abnormal hemoglobins and hemoglobin variants should be performed.

The ADA recommends that the goal of therapy is a GHb <7% in patients with diabetes and that treatment be re-evaluated if values consistently are >8%.

These values apply only to DCCT-traceable GHb methods. The ADA does not currently recommend GHb testing for screening or for diagnosis of diabetes.

References

American Diabetes Association, "Standards of Medical Care for Patients With Diabetes Mellitus," Position Statement, *Diabetes Care*, 2000, 23(Suppl 1):S32-S42.

American Diabetes Association, "Tests of Glycemia in Diabetes," Position Statement, *Diabetes Care*, 2000, 23(Suppl 1):S80-S82.

Davidson MB, Schriger DL, Peters AL, et al, "Relationship Between Fasting Plasma Glucose and Glycosylated Hemoglobin," *JAMA*, 1999, 281(13):1203-10.

Krishnamurti U and Steffes MW, "Glycohemoglobin: A Primary Predictor of the Development or Reversal of Complications of Diabetes Mellitus," *Clin Chem*, 2001, 47(7):1157-65.

Krolewski AS, Laffel LMB, Krolewski M, et al, "Glycosylated Hemoglobin and the Risk of Microalbuminuria in Patients With Insulin-Dependent Diabetes Mellitus," *N Engl J Med*, 1995, 332(19):1251-5.

The Expert Committee on the Diagnosis and Classification of Diabetes Mellitus, "Follow-up Report on the Diagnosis of Diabetes Mellitus," *Diabetes Care*, 2003, 26(11):3160-7.

Viberti G, "A Glycemic Threshold for Diabetic Complications?" *N Engl J Med*, 1995, 332(19):1293-4.

Vinicor F, "When Is Diabetes Diabetes?" *JAMA*, 1999, 281(13):1222-4.

♦ **Glycated Proteins** *see* Fructosamine, Serum *on page 617*

♦ **Glycine** *see* Amino Acids, Urine *on page 145*

♦ **Glycine, Cerebrospinal Fluid** *see* Cerebrospinal Fluid Glycine *on page 364*

♦ **Glycogen Storage Diseases** *see* Inherited Diseases of Metabolism and Cell Structure *on page 792*

♦ **Glycohemoglobin** *see* Glycated Hemoglobin (Hemoglobin A$_{1c}$), Blood *on page 655*

♦ **Glycolic Acid, Urine** *see* Oxalate, Urine *on page 989*

♦ **β2-Glycoprotein 1** *see* Anticardiolipin Antibody *on page 167*

♦ **Glycosaminoglycans** *see* Mucopolysaccharides, Urine *on page 922*

♦ **Glyoxylic Acid, Urine** *see* Oxalate, Urine *on page 989*

♦ **GnRH** *see* Chorionic Gonadotropin, Human, Serum and Urine *on page 397*

♦ **Go** *see* Methamphetamine, Qualitative, Urine *on page 902*

♦ **Goiter** *see* Thyroxine, Serum *on page 1257*

♦ **Gold Dust** *see* Cocaine (Cocaine Metabolite), Qualitative, Urine or Hair *on page 427*

Gold, Serum

Synonyms Auranofin; Aurothioglucose; Chrysotherapy; Gold Sodium Thiomalate; Myochrysine®; Ridaura®; Sodium Aurothiomalate; Solganal®

Abstract Gold is sometimes used as a treatment for rheumatoid arthritis, but its value is questioned because of serious side effects and availability of more efficacious therapies.

Specimen Serum

Container Red top tube

Reference Interval Therapeutic: 1.0-3.0 µg/mL (SI: 5.1-15.2 µmol/L)

Limitations Blood levels do not correlate well with therapeutic or toxic effects.

Methodology Atomic absorption spectrometry (AA)

Additional Information The following values are for gold sodium thiomalate, which is one of the most commonly used gold compounds.

- Half-life: 5-10 days
- Volume of distribution: 0.1 L/kg
- Protein binding: 95%

Symptoms of general toxicity include fever, nausea, vomiting, diarrhea, proteinuria, hematuria, and blood dyscrasias. Aplastic anemia is one of the most feared side effects.

References

Jones G and Brooks PM, "Injectable Gold Compounds: An Overview," *Br J Rheumatol*, 1996, 35(11):1154-8.

Lacaille D, Stein HB, Raboud J, et al, "Long-Term Therapy of Psoriatic Arthritis: Intramuscular Gold or Methotrexate?" *J Rheumatol*, 2000, 27(8):1922-7.

Menninger H, Herborn G, Sander O, et al "A 36 Month Comparative Trial of Methotrexate and Gold Sodium Thiomalate in the Treatment of Early Active and Erosive Rheumatoid Arthritis," *Br J Rheumatol*, 1998, 37(10):1060-8.

♦ **Gold Sodium Thiomalate** *see* Gold, Serum *on page 657*

♦ **Gonadotropic Hormones** *see* Follicle Stimulating Hormone, Serum, Plasma, or Urine *on page 609*

Gram Stain

Related Information

Abstract The first step in identification of bacterial isolates is study of their Gram stain properties. The Gram stain is a differential stain used to classify bacteria. Gram-positive bacteria retain crystal violet after decolorization and appear deep blue to purple. Gram-negative bacteria do not retain crystal violet after decolorization and are counterstained red by safranin. Gram staining characteristics may be atypical in very young, old, dead, or degenerating cultures.

Patient Preparation Same as for routine culture of specific site

Specimen Duplicate of specimen appropriate for routine culture of the specific site

Collection Collection procedure same as for routine culture of the specific site. Specimen must be collected to avoid contamination with skin, adjacent structures, and nonsterile surfaces.

Critical Values Organisms detected in aseptically obtained specimens

Possible Panic Range Organisms detected in cerebrospinal fluid, especially those within leukocyte cytoplasm; those detected in peripheral blood smears.

Use For many clinical specimens, a direct Gram stain is performed upon specimen receipt. The direct Gram stain reveals information about specimen quality (eg, by comparing numbers of squamous epithelial cells and polymorphonuclear neutrophils in sputum specimens), presence or absence of potential pathogens, and initial presumptive morphologic categorization of potential pathogens (eg, yeasts, gram-positive cocci vs gram-negative bacilli, etc).

Limitations The information from a Gram stain is presumptive. Organism detection with specimen Gram stain requires that large numbers of potential pathogens be present; consequently, sensitivity with many specimens is <100% and direct Gram stain of blood is unwarranted, because organisms are almost never present in sufficient numbers for detection. Additionally, clinical specimens that are heavily contaminated with normal flora rarely provide diagnostically useful information unless probable pathogens are morphologically distinct from normal flora. Thus, **direct Gram stain of throat swabs, stool, and rectal swabs is usually not indicated.** Certain organisms (eg, *Rickettsia* species, *Treponema pallidum*, *Legionella* species, *Mycobacterium* species, *Mycoplasma* species) stain poorly, or not at all with Gram stain.

Overdecolorization and undercolorization may cause difficulty in interpretation.

Methodology Stain of clinical material using the Gram stain technique; examination of multiple representative fields, especially those rich in neutrophils

Additional Information Gram stain results can provide a guide for initial empiric therapy, and should be correlated with subsequent culture results to help to ascertain the significance of organisms isolated from clinical specimens. Gram stains are usually scanned for the presence or absence of white blood cells (indicative of infection) and squamous epithelial cells (indicative of mucosal contamination). A **sputum specimen** showing >25 squamous epithelial cells/lpf, regardless of the number of white blood cells, indicates that the specimen is grossly contaminated with saliva and bacterial culture should not be performed. Gram stains revealing an occasional bacterium per high powered field in an uncentrifuged **urine specimen** suggest a colony count of 10,000 bacteria/mL. Bacteria in the majority of fields suggests >100,000 bacteria/mL. Gram stain is the most valuable diagnostic test in **bacterial meningitis** that is immediately available. Organisms are detectable in 60% to 80% of patients who have not been treated and in 40% to 60% of those who have been given antibiotics. Its sensitivity relates to the number of organisms present. The sensitivity of the Gram stain is greater in gram-positive infections and is only positive in half of the instances of gram-negative meningitis. Important causes of bacterial meningitis include *Haemophilus influenzae, Streptococcus pneumoniae, Neisseria meningitidis,* group B *Streptococcus, Listeria monocytogenes,* and gram-negative bacilli. Culture and Gram stain must have priority over antigen detection methods if only a small volume of CSF is available.

Barenfanger and Drake have recently provided a useful review.

References

Barenfanger J and Drake CA, "Interpretation of Gram Stains for the Nonmicrobiologist," *Lab Med*, 2001, 32(7):368-75.

Blot F, Raynard B, Chachaty E, et al, "Value of Gram Stain Examination of Lower Respiratory Tract Secretions for Early Diagnosis of Nosocomial Pneumonia," *Am J Respir Crit Care Med*, 2000, 162(5):1731-7.

Dunbar SA, Eason RA, Musher DM, et al, "Microscopic Examination and Broth Culture of Cerebrospinal Fluid in Diagnosis of Meningitis," *J Clin Microbiol*, 1998, 36(6):1617-20.

Flournoy DJ, "Interpreting the Sputum Gram Stain Report," *Lab Med*, 1998, 29:763-8.

McNair RD, MacDonald SR, Dooley SL, et al, "Evaluation of the Centrifuged and Gram-Stained Smear, Urinalysis, and Reagent Strip Testing to Detect Symptomatic Bacteriuria in Obstetric Patients," *Am J Obstet Gynecol*, 2000, 182(5):1076-9.

Schuchat A, Robinson K, Wenger JD, et al, "Bacterial Meningitis in the United States in 1995. Active Surveillance Team," *N Engl J Med*, 1997, 337(14):970-6.

Steen RM, Hanson-Steen AS, St. John C, et al, "The Use of the Gram Stain to Screen Platelet Concentrates for Bacterial Contamination," *Lab Med*, 2003, 34(8):609-11.

♦ **Granulocytapheresis** *see* Apheresis, Therapeutic *on page 201*

♦ **Granulocyte Agglutination Test** *see* Antineutrophil Alloantibody and Autoantibody *on page 186*

♦ **Granulocyte Antibody** *see* Antineutrophil Alloantibody and Autoantibody *on page 186*

♦ **Granulocyte-Colony-Stimulating Factor** *see* Hematopoietic Progenitor Cells, Peripheral Blood *on page 679*

♦ **Granulocyte-Colony-Stimulating Factor** *see* Neutrophils, Transfusion *on page 951*

♦ **Granulocyte Immunofluorescence Test** *see* Antineutrophil Alloantibody and Autoantibody *on page 186*

♦ **Granulocytes, Apheresis, Donation** *see* Neutrophils, Apheresis, Donation *on page 950*

♦ **Granulocytes, Transfusion** *see* Neutrophils, Transfusion *on page 951*

♦ **Grievous Bodily Harm** *see* Gamma Hydroxybutyrate, Serum or Urine *on page 630*

♦ **Gross Cystic Disease Fluid Protein-15** *see* Immunoperoxidase Procedures *on page 780*

♦ **Gross Cystic Disease Fluid Protein-15 (GCDFP-15)** *see* Breast Biopsy *on page 305*

Group A *Streptococcus* Screen, Rapid

Related Information

Antideoxyribonuclease-B Titer, Serum *on page 170*

Antistreptolysin O Titer, Serum *on page 197*

(Continued)

Group A *Streptococcus* Screen, Rapid *(Continued)*

Bacterial Antigens, Rapid Detection Methods *on page 228*
Bacterial Culture, Throat, and Antigen Detection Testing for Group A Streptococci *on page 245*

Test Includes Direct enzyme immunoassay for group A *Streptococcus* antigen

Abstract *Streptococcus pyogenes*, also known as group A *Streptococcus*, is a common cause of acute bacterial pharyngitis as well as other infections. Rapid antigen detection assays will detect group A *Streptococcus* antigen directly from throat swabs. This assay can be performed in a few minutes, while throat culture requires overnight incubation. The specificity of these newer tests is very high (>95%) and thus, a positive result does not require culture confirmation. In contrast, the sensitivity of these assays is lower than that of throat culture, and negative specimens should be followed with throat culture.

Specimen Throat swab; many laboratories request two swabs, one for culture if the rapid screen is negative

Container Rayon or dacron swabs rather than cotton swabs enhance the chance of detection.

Collection Rigorous swabbing of the tonsillar pillars and posterior throat increases the probability of detection of streptococcal antigen.

Special Instructions Some laboratories favor submission of dry swabs for antigen testing. Consult the laboratory for their specific recommendations.

Use Rapidly screen for the presence of group A streptococci using antigen detection methods

Limitations Many reviews have indicated a sensitivity of 75% to 80% and a specificity of 95% to 98% for the rapid methods. Specimens that yield less than 10 colonies on culture usually are negative by rapid method. Adequate specimen collection on younger patients may be difficult, and thus, contribute to the false-negative rate. A positive result can be relied upon as a rational basis to begin therapy. **A negative result is only presumptive, and a culture should be performed to reasonably exclude the diagnosis of group A streptococcal infection.** Group A streptococcal antigen disappears within 3 days following antibiotic therapy.

Contraindications This test should not be ordered unless results available within 1-2 hours of specimen collection will impact therapeutic decisions.

Methodology The streptococcal group A antigen is extracted from the swab and is detected by enzyme immunoassay (EIA) or latex agglutination (LA).

Additional Information Rheumatic fever remains a concern in the United States and serious complications including sepsis, soft tissue invasion, and toxic shock-like syndrome have been reported to be increasing in frequency. Therefore, timely diagnosis and early institution of appropriate therapy remains important. Timely therapy may reduce the acute symptoms and overall duration of streptococcal pharyngitis. The sequelae of poststreptococcal glomerulonephritis and rheumatic fever are diminished by early therapy.

See also Bacterial Culture, Throat, and Antigen Detection Testing for Group A Streptococci *on page 245*.

References

Chen FM, "Culture Confirmation of Negative Rapid Strep Test Results," *J Fam Pract*, 2000, 49(4):371-2.

Cunningham MW, "Pathogenesis of Group A Streptococcal Infections," *Clin Microbiol Rev*, 2000, 13(3):470-511.

Durbin WJ and Mark EJ, "A 19-Month-Old Boy With Fever and Soft-Tissue Masses," Case Records of the Massachusetts General Hospital. Case 18-2001, Scully RE, Mark EJ, McNeely WF, et al, eds, *N Engl J Med*, 2001, 344(24):1851-6.

Greiver M, "Practice Tips. Incorporating a Rapid Group A *Streptococcus* Assay With the Sore Throat Score," *Can Fam Physician*, 1999, 45:1181-2.

Pichero ME, Green JL, Francis AB, et al, "Recurrent Group A Streptococcal Tonsillopharyngitis," *Pediatr Infect Dis J*, 1998, 17(9):809-15.

Internet Web Sites

www.astdhpphe.org/infect/strepa.html
www.cdc.gov/ncidod/dbmd/diseaseinfo/groupastreptococcal_g.htm

Group B *Streptococcus* Screen, Rapid

Related Information

Bacterial Antigens, Rapid Detection Methods *on page 228*
Bacterial Culture, Cerebrospinal Fluid *on page 236*
Bacterial Culture, Genital Specimen *on page 239*

Test Includes Antigen detection of group B beta *Streptococcus*

Abstract Group B streptococci is found in 5% to 40% of genital and lower gastrointestinal tract specimens of women. In most cases, this constitutes asymptomatic colonization, however, in pregnant women these bacteria can be transmitted to newborns during delivery. Neonates exposed to group B streptococci may develop disseminated infection that results in a mortality rate of 5% to 10%.

Specimen Cerebrospinal, endocervical, vaginal, or rectal fluid

Container Sterile container, red top tube

Storage Instructions If a specimen for antigen detection cannot be tested immediately, it may be stored at 2°C to 8°C for 1 day or frozen at -20°C for longer storage. Storage is inconsistent with the role of the test for rapid diagnosis.

Turnaround Time About 1 hour

Critical Values Positive intrapartum test

Possible Panic Range Positive neonatal test

Use Culture-independent detection of group B *Streptococcus* is used to diagnose patients (usually in the neonatal period) suspected of group B *Streptococcus* invasive disease and as a rapid intrapartum test to identify maternal group B *Streptococcus* carriers

Limitations Rapid group B *Streptococcus* detection has a sensitivity range of 70% to 90% in neonatal meningitis. Identification of maternal group B *Streptococcus* carriers using rapid group B *Streptococcus* tests often fails to identify women colonized with low numbers of organisms. A 1996 study showed that rapid immunoassays were not sufficiently accurate for routine use in detecting vaginal colonization with group B *Streptococcus*.

Methodology Latex agglutination (LA), enzyme immunoassay (EIA), PCR assay

Additional Information Group B *Streptococcus* is one of the most common human pathogens in the neonatal period. Neonatal infection follows exposure to maternal genital flora *in utero* through ruptured membranes, or by colonization during passage through the birth canal. Neonatal infection presents as either early or late onset disease (EOD or LOD, respectively). EOD, which occurs within 5 days of birth (mean period to onset of symptoms is 20 hours), manifests as pneumonia and sepsis, occasionally accompanied by meningitis. LOD, which occurs 7 days to 3 months after birth (mean period to onset is 24 days) usually presents as meningitis.

Intrapartum antimicrobial therapy has been shown to be an effective means of reducing the incidence of EOD. It is currently not clear how best to identify individuals who should be treated. An approach to prevention of neonatal group B streptococci has been endorsed by the American College of Obstetricians and Gynecologists and by the American Academy of Pediatrics. Women selected for chemoprophylaxis are determined by the following criteria.

- Positive screening cultures of vaginal and rectal sites at 35-37 weeks gestation
- Women with group B streptococcal bacteriuria during pregnancy
- Women who have previously delivered an infant with group B streptococcal infection
- Women with significant risk factors (eg, onset of labor or membrane rupture earlier than 37 weeks of gestation, rupture of membranes for 18 hours or more before delivery, intrapartum fever)

Use of these guidelines should reduce the number of cases of neonatal group B streptococcal infection, but will not prevent all cases. The development of a group B streptococcal vaccine is being investigated which may aid in eliminating this disease.

See also Bacterial Culture, Genital Specimen *on page 239*.

References
Baker CJ, "Inadequacy of Rapid Immunoassays for Intrapartum Detection of Group B Streptococcal Carriers," *Obstet Gynecol*, 1996, 88(1):51-5.

Bergeron MG, Ke D, Ménard C, et al, "Rapid Detection of Group B Streptococci in Pregnant Women at Delivery," *N Engl J Med*, 2000, 343(3):175-9.

Centers for Disease Control, "Prevention of Perinatal Group B Streptococcal Disease: A Public Health Perspective," *MMWR Morb Mortal Wkly Report*, 1996, 45:1-24.

(Continued)

Group B *Streptococcus* Screen, Rapid *(Continued)*

Donders GG, Vereecken A, Salembier G, et al, "Accuracy of Rapid Antigen Detection Test for Group B Streptococci in the Indigenous Vaginal Bacterial Flora," *Arch Gynecol Obstet,* 1999, 263(1-2):34-6.

Reisner DP, Haas MJ, Zingheim RW, et al, "Performance of a Group B Streptococcal Prophylaxis Protocol Combining High-Risk Treatment and Low-Risk Screening," *Am J Obstet Gynecol,* 2000, 182(6):1335-43.

Internet Web Sites

www.astdhpphe.org/infect/strepb.html
www.cdc.gov/groupbstrep/

Growth Hormone, Serum

Related Information

Insulin-Like Growth Factor-1 (IGF-1), Serum or Plasma *on page 800*
Insulin-Like Growth Factor Binding Protein 3, Serum *on page 802*
Insulin Tolerance Test *on page 804*
Prolactin, Serum *on page 1094*

Synonyms GH; hGH; Somatotropin

Applies to GHRH; GHRH Plus; GHRP-6; GHRP-6 Stimulation Test; Glucose Suppression Test; Insulin-Like Growth Factors; Somatomedin-C; Somatomedins; Somatostatin

Abstract Human growth hormone (GH) is secreted from the anterior pituitary in multiple short spikes, often secondary to environmental stimuli. The half-life of GH in the blood is ~20 minutes. Maximum GH secretion occurs after the onset of deep (Stages III and IV) sleep; smaller spikes occur after exercise and after eating. Release of GH from the pituitary is influenced by three hypothalamic factors: GH releasing hormone (GHRH); GH-releasing peptide-6 (GHRP-6); and GH inhibitory hormone (GHIH), a pancreatic peptide, also known as somatostatin.

In adults, GH causes multiple physiologic effects and influences lipolysis, protein synthesis, cardiac function, muscle mass, and red cell mass. GH causes its effects both directly and indirectly, via the action of insulin-like growth factors (IGF), formerly called somatomedins, but renamed because of striking chemical similarities to insulin.

Excessive GH secretion, from a pituitary adenoma produces acromegaly in adults; in childhood, the term pituitary gigantism is used. Growth hormone deficiency is one cause of short stature in childhood. GH deficiency in adulthood causes a metabolic disorder with an increase in body fat, decrease in muscle mass, reduced muscle strength, reduced bone density, abnormalities of lipoprotein and carbohydrate dynamics, and altered renal and cardiac function.

Patient Preparation

Glucose suppression test (suspected GH excess). The patient fasts overnight and remains in bed for the test. A baseline blood specimen is obtained, after which the patient drinks a solution containing 100 g of glucose.

Suspected GH deficiency. As specified in the following table (see Use).

GHRH plus GHRP-6 stimulation test. Following an overnight fast, an indwelling catheter is placed in a forearm vein and kept open with slow infusion of 150 mmol/L sodium chloride. GHRH 1 µg/kg body weight plus GHRP-6 1 µg/kg body weight is given as a bolus at time 0.

Specimen Serum

Container Red top tube

Sampling Time

Glucose suppression test: 0, 30, 60 minutes

GHRH plus GHRP-6 stimulation test. A baseline specimen is obtained at -30 min. Poststimulation blood samples are obtained at 30 and 60 minutes.

Other stimulation protocols. As specified in the following table (see Use).

Storage Instructions Label tube with time and date of collection and identifying data. Separate serum and freeze in plastic container. Stable 4 hours at 25°C and 1 year at -20°C.

Reference Interval

Basal or random specimens:
- cord: 8-41 ng/mL
- newborns: 5-53 ng/mL

- infants: 1-12 months: 2-10 ng/mL
- adults: male: 0-4 ng/mL; female: 0-18 ng/mL
- >60 years: male: 1-9 ng/mL; female: 1-16 ng/mL

Glucose suppression testing. The 60-minute or the 120-minute postglucose specimen GH is <2 ng/mL.

Insulin tolerance test: A normal response is a poststimulation value of at least 10.0 ng/mL. (For complete protocol, see Insulin Tolerance Test *on page 804*.)

Other dynamic tests. A normal response is any poststimulation value of at least 20 ng/mL.

Use Growth hormone excess. GH secretion is so unpredictable and episodic that random serum GH values may be within the reference interval in patients with acromegaly or pituitary gigantism. The key to the diagnosis is to demonstrate that GH is not suppressed normally in response to a standard glucose load. Using the 100 g glucose suppression test (see above, Patient Preparation and Sampling Time) a normal person will have a serum GH <2 ng/mL.

Growth hormone deficiency. Basal and random specimens do not discriminate between normals and persons with GH deficiency. The traditional gold standard dynamic test is the Insulin Tolerance Test *on page 804*. Since the ITT involves some risk for the patient and requires the presence of a physician, it is rarely used in practice. All the other tests are less sensitive than the ITT; therefore, many authorities recommend using two of the stimulation tests. The stimulation testing protocols which may be used are listed below.

Dynamic Tests for Growth Hormone Insufficiency

Stimulus	Protocol	Sampling Time
Exercise	Vigorous exercise for 20 minutes	20 minutes after starting to exercise
Sleep	Patient goes to sleep at usual time	1 hour after onset of deep sleep (Stage III or IV), documented by EEG
Arginine	Arginine hydrochloride, 0.5 g/kg body weight intravenously over 30 minutes	60-120 minutes
Glucagon	0.03 mg/kg body weight (not to exceed 1.0 mg), intramuscularly or subcutaneously	120-180 minutes
L-dopa	0.5 g/1.73 m^2 body surface area, orally with lunch	30-120 minutes
Clonidine	0.15 mg/m^2 body surface, orally	90 minutes
Diazepam	0.15 mg/kg body weight, orally	60 minutes
Pentagastrin	1.5 mg/kg body weight/hour for 75 minutes	75 minutes

Adapted from Demers L, "Pituitary Function," *Tietz Textbook of Clinical Chemistry*, 3rd ed, Philadelphia, PA: WB Saunders Co, 1999, 1470-95.

GHRH plus GHRP-6 Stimulation Test. A newly described, well documented testing protocol (Popovic et al) uses two agents, growth hormone releasing hormone (GHRH), 1 μg/kg body weight, plus growth hormone-releasing peptide-6 (GHRP-6), 1 microgram/kg body weight, both given intravenously at time 0. This combination constitutes the most potent stimulation of GH available. Patients with a poststimulation value >20 ng/mL are classified as normal. Patients with both poststimulation values <10 ng/mL are classified as GH-deficient. Values in between constitute an intentional gray zone, indicating the need for additional studies and the exercise of clinical judgment. This protocol was compared with the insulin tolerance test and found to produce higher and more clear cut GH peaks than the ITT.

Limitations Patients with a GH deficiency because of hypothalamic disease may have a normal response to the GHRH plus GHRP-6 stimulation protocol. In patients with suspected hypothalamic disease, the ITT may be more useful than the GHRH plus GHRP-6 test.

Methodology Immunoassays (multiple labels and formats)

Additional Information GH is a substance which may enhance athletic performance. Some athletes surreptitiously inject themselves with recombinant human GH (rhGH) and believe that the recombinant material is chemically (Continued)

Growth Hormone, Serum *(Continued)*

undetectable. Physician-investigators in Germany (Wu et al), however, have reported a dual immunoassay system with measurements of both total human GH and a 22 kilodalton isoform of human GH. Persons who inject rhGH have relatively more of the 22kD isoform, and thus the value of their ratio is higher than in persons who do not inject.

References

Demers L, "Pituitary Function," *Tietz Textbook of Clinical Chemistry*, 3rd ed, Burtis CA and Ashwood ER, eds, Philadelphia, PA: WB Saunders Co, 1999, 1470-95.

Healy ML and Russel-Jones D, "Growth Hormone and Sport: Abuse, Potential Benefits and Difficulties in Detection," *Br J Sports Med*, 1997, 31(4):267-8.

Ho KKY, "Diagnosis of Adult GH Deficiency," *Lancet*, 2000, 356(9236):1125-6.

Popovic V, Leal A, Micic D, et al, "GH-Releasing Hormone and GH-Releasing Peptide-6 for Diagnostic Testing in GH-Deficient Adults," *Lancet*, 2000, 356(9236):1137-42.

Wu Z, Bidlingmaier M, Dall R, et al, "Detection of Doping With Human Growth Hormone," *Lancet*, 1999, 353(9156):895-6.

♦ **GT** *see* Gamma-Glutamyl Transferase, Serum *on page 629*

♦ **GTP** *see* Gamma-Glutamyl Transferase, Serum *on page 629*

♦ **GTT** *see* Glucose Tolerance Test, Plasma *on page 651*

♦ **Guaiac, Stool** *see* Occult Blood, Stool *on page 969*

♦ **Guanosine Triphosphate Cyclohydrolase 1 Deficiency** *see* Newborn Screen for Phenylketonuria *on page 954*

♦ **Guanosine Triphosphate Cyclohydrolase 1 Deficiency** *see* Phenylalanine, Blood *on page 1022*

♦ **Guthrie Test** *see* Phenylalanine, Blood *on page 1022*

♦ **Guthrie Test** *see* Phenylalanine, Urine *on page 1024*

♦ **GX** *see* Lidocaine, Serum or Plasma *on page 850*

♦ **Hageman Factor (Factor XII)** *see* Coagulation Factor Assays *on page 418*

♦ **Hair Analysis** *see* Arsenic, Blood *on page 209*

♦ **Hair Analysis** *see* Arsenic, Hair, Nails *on page 210*

♦ **Hair Analysis** *see* Heavy Metal Screen, Urine *on page 669*

♦ **Hairy Cell Leukemia** *see* Tartrate Resistant Leukocyte Acid Phosphatase *on page 1236*

♦ **Haldol®** *see* Haloperidol, Serum or Plasma *on page 664*

♦ **Haldol® Decanoate** *see* Haloperidol, Serum or Plasma *on page 664*

♦ **Haloneural®** *see* Haloperidol, Serum or Plasma *on page 664*

Haloperidol, Serum or Plasma

Related Information

Chlorpromazine, Serum *on page 395*

Lithium, Serum *on page 863*

Synonyms Dozic®; Fortunan®; Haldol®; Haldol® Decanoate; Haloneural®; Serenace®

Abstract An antipsychotic agent, this drug is used in the treatment of Tourette syndrome, severe behavioral problems in children, and for emergency sedation of severely agitated or delirious patients.

Specimen Serum or plasma

Container Red top tube or green top (heparin) tube

Sampling Time Time to peak serum concentration: Oral: 3-6 hours; I.M.: 10-20 minutes; I.M. (long-acting): 3-9 days

Reference Interval 5-20 ng/mL (SI: 10-40 nmol/L) (psychotic disorders - less for Tourette syndrome and mania)

Critical Values >50 ng/mL (SI: >100 nmol/L) (variable)

Methodology High performance liquid chromatography (HPLC), gas chromatography (GC)

Additional Information

- Half-life: 15-40 hours
- Volume of distribution: 18-30 L/kg
- Protein binding: 90%

Toxicity of haloperidol is similar to that of phenothiazines. Among the most dangerous effects of haloperidol overdose are cardiovascular alterations including myocardial depression and EKG changes such as depression of T and ST waves. Torsade de pointes associated with the administration of intravenous haloperidol has recently been reviewed (Hassaballa and Balk).

Haloperidol acute overdose may cause hyperglycemia, hypoglycemia, arrhythmias, exacerbation or precipitation of myasthenia gravis, and other signs, symptoms, and abnormalities.

References

Blin O, "A Comparative Review of New Antipsychotics," *Can J Psychiatry*, 1999, 44(3):235-44.

Hassaballa HA and Balk RA, "Torsade de Pointes Associated With the Administration of Intravenous Haloperidol," *Am J Ther*, 2003, (1):58-60.

Kudo S and Ishizaki T, "Pharmacokinetics of Haloperidol: An Update," *Clin Pharmacokinet*, 1999, 37(6):435-56.

Lawson GM, "Monitoring of Serum Haloperidol," *Mayo Clin Proc*, 1994, 69(2):189-90.

♦ **Haltran**® *see* Ibuprofen, Serum *on page 764*
♦ **Hamster Egg Penetration Test** *see* Sperm Penetration Assay (Zona-Free Hamster Egg Penetration Test) *on page 1220*

Ham Test

Related Information

Hemosiderin Stain, Urine *on page 692*
Hypertonic Cryohemolysis Test *on page 762*
Peripheral Blood: Red Blood Cell Morphology *on page 1016*
PNH Test (GPI-Anchored Proteins) by Flow Cytometry *on page 1064*
Red Blood Cells, Washed *on page 1143*
Sugar Water Test Screen *on page 1226*

Synonyms Acidified Serum Test for PNH

Abstract A positive Ham test (lysis of patient red cells in acidified serum) may be used as a screening test for paroxysmal nocturnal hemoglobinuria (PNH) and indicates unusual sensitivity of such red cells to the action of complement. PNH is characterized clinically by nocturnal hemoglobinuria, chronic hemolytic anemia, hypoplastic or aplastic hematopoiesis, and tendency to venous thrombosis. It is the result of an acquired defect of hematopoietic stem cells. Affected cells have lost the glycosyl-phosphatidylinositol (GPI) anchor to the outer cell membrane. GPI acts to anchor proteins (CD55 and CD59) which protect red blood cells from the action of complement. These changes are the result of an acquired somatic cell mutation that inactivates a gene (PIG-A) that codes for an enzyme needed for GPI biosynthesis.

Specimen Erythrocytes from EDTA anticoagulated whole blood

Container Lavender top (EDTA) tube

Causes for Rejection Specimen hemolyzed

Reference Interval A positive result shows lysis of red cells in acidified serum samples with patient's cells (not with normal cells). Positive in PNH: 10% to 50% lysis in acidified noninactivated serum. Can be as low as 5% or as much as 80%.

Use Evaluate patients with suspected PNH or suspected congenital dyserythropoietic anemia, type II (HEMPAS); evaluate hemolytic anemia, especially with hemosiderinuria, pancytopenia, decreased RBC acetylcholinesterase, decreased leukocyte alkaline phosphatase, negative direct Coombs' test, and/or apparent marrow failure.

Limitations Results may me low or negative after transfusion. False-positive results may occur in other hematologic diseases: hereditary and acquired spherocytosis, hereditary dyserythropoietic anemia (CDA type II, HEMPAS, *vide infra*), aged red cells (as with old transfused blood), aplastic anemia, leukemia, and myeloproliferative syndromes. In these conditions hemolysis will also occur in the acidified inactivated serum. The latter is negative in PNH since hemolysis is complement dependent. Flow cytometry (see PNH Test (GPI-Anchored Proteins) by Flow Cytometry *on page 1064*) is much more sensitive than the Ham test. In populations of bone marrow failure syndromes (in particular myelodysplasia), PNH cells as detected by flow cytometry are relatively common. In patients who have received massive red cell transfusion (most of the patients RBCs are of donor origin), flow cytometry allows study of and identification of PNH neutrophils.

Contraindications Transfusion

Additional Information PNH red cells are unusually susceptible to lysis by complement. The Ham (acidified serum) and sucrose hemolysis test can demonstrate this lysis *in vitro*.
(Continued)

Ham Test *(Continued)*

In some cases, three populations of cells exist in patients with PNH. One is markedly hypersensitive to complement (type III cells), one has a midlevel of sensitivity (type II cells), and the third population has normal sensitivity (type I cells). Type III cells are variably present, are the population which undergoes lysis in the Ham test and relate to the severity of illness. The young PNH cells (reticulocyte-rich) are more susceptible to lysis than the older red cells. PNH RBCs will undergo lysis in acidified normal serum and in the patient's acidified serum.

The membrane defect involves a protein, the membrane inhibitor of reactive lysis (MIRL, "protectin", CD59). It is one of some 18 proteins anchored to the external cell surface by a GPI moiety consisting of phosphatidylinositol, glucosamine, and three mannose molecules. The protein is also expressed in granulocytes, monocytes, and on platelets; its absence plays a role in the hypercoagulable state present in PNH. The underlying gene defect (in the gene PIG-A) is located on the short arm of the X chromosome. See PNH Test (GPI-Anchored Proteins) by Flow Cytometry *on page 1064*.

A positive Ham test may occur with the congenital dyserythropoietic anemias. In CDA type II (HEMPAS - hereditary erythroblastic multinuclearity with positive acidified serum test) the red cells undergo lysis in only a proportion (about 30%) of normal sera, and these RBCs do not undergo lysis in the patient's own acidified serum. The sucrose lysis test and PNH test by flow cytometry are negative in cases of HEMPAS. Heating at 56°C, which destroys the complement in serum, inactivates the lytic system so that if lysis occurs with inactivated serum this cannot be considered positive.

Another type of cell that may lyse in inactivated serum is the spherocyte. Spherocytes may lyse in acidified serum possibly due to the lowered pH.

The relationship between aplastic anemia and PNH has been considered for over two decades. A 55% to 65% incidence of PNH occurs in primary myelofibrosis and myeloid metaplasia.

A diagnostic test using flow cytometry and monoclonal antibodies was developed in the early 1990s and is now the preferred method for establishing a diagnosis of PNH. This technique detects missing proteins from granulocytes (GPI-anchored proteins) in patients with PNH. See PNH Test (GPI-Anchored Proteins) by Flow Cytometry *on page 1064*.

References

Kinoshita T, Inoue N, and Takeda J, "Role of Phosphatidylinositol-Linked Proteins in Paroxysmal Nocturnal Hemoglobinuria Pathogenesis," *Annu Rev Med*, 1996, 47:1-10.

Luzzato L and Hillmen P, "Laboratory Methods Used in the Investigation of Paroxysmal Nocturnal Hemoglobinuria (PNH)," *Practical Laboratory Haematology*, Dacie JV and Lewis SM, eds, 8th ed, Chapter 15, New York, NY: Churchill Livingstone, 1995, 287-96.

Rosse WF, "Paroxysmal Nocturnal Hemoglobinuria," *Hematology: Basic Principle and Practice*, 3rd ed, Chapter 20, Hoffman R, Benz EJ Jr, Shattil SJ, et al, eds, Philadelphia, PA: Churchill Livingstone, 2000, 331-42.

♦ **H and H** *see* Hematocrit *on page 674*

♦ **Hank's Stain for *Nocardia* Species** *see* Acid-Fast Stain, Routine or Modified *on page 95*

Hantavirus Serology

Test Includes Detection of IgM and IgG antibody specific for the Sin Nombre hantavirus

Abstract Numerous cases of hantavirus pulmonary syndrome (HPS) have been recognized in the Western U.S. and Canada. HPS begins with nonspecific symptoms (eg, fever and myalgia), which are followed in 3-6 days by progressive cough and shortness of breath. Common findings during this later stage include tachypnea, tachycardia, fever, and hypotension. Abnormalities on chest radiographs are detected bilaterally, and pleural effusions are common. Hemoconcentration, thrombocytopenia, prolonged activated partial thromboplastin time, an increased proportion of immature granulocytes on the peripheral blood smear, leukocytosis, and elevated levels of serum lactate dehydrogenase and aspartate aminotransferase are found. Serum antibodies can be detected at the time of clinical presentation.

Specimen Serum from acute phase of illness and a follow-up serum specimen 1-3 weeks later.

Container Red top tube

Storage Instructions Serum can be stored at 4°C up to 1 week; serum should be stored at -70°C after 1 week and during shipping.

Reference Interval No detectable hantavirus IgM or less than a fourfold increase in IgG

Use Confirm the diagnosis of hantavirus pulmonary syndrome

Limitations Serology for HPS are available only through reference laboratories. Results will not be available quickly enough to establish a diagnosis before treatment.

Methodology Western blot, enzyme-linked immunosorbent assay (ELISA), indirect immunofluorescence (IIF)

Additional Information Hantaviruses are single-stranded RNA viruses of the family Bunyaviridae. Four distinct hantaviruses have been described in North America with over 20 serotypes or genotypes world-wide. Rodents serve as the reservoir for hantaviruses, and infected rodents shed the virus in saliva, urine, and feces. Transmission to humans occurs most often by inhalation of infected rodent excreta. Among the greatest risks are occupational exposures to rodents in trappers, forestry workers, farmers, or the military. Recommendations for prevention include avoidance of contact with rodent urine or saliva. Hantavirus person-to-person transmission has been shown in an outbreak of 18 people in Argentina, some of whom were treating physicians.

References

Centers for Disease Control and Prevention, "Hantavirus Pulmonary Syndrome - Panama, 1999-2000," *JAMA*, 2000, 283(17):2232-3.

Colby TV, Zaki SR, Feddersen RM, et al, "Hantavirus Pulmonary Syndrome Is Distinguishable From Acute Interstitial Pneumonia," *Arch Pathol Lab Med*, 2000, 124(10)1463-6.

Goodman DA and Griego LC, "Hantavirus Pulmonary Syndrome: Implications for Critical Care Nurses," *Crit Care Nurse*, 1998, 18(1):18, 23-30.

Nichol ST, Arikawa J, and Kawaoka Y, "Emerging Viral Diseases," *Proc Natl Acad Sci U S A*, 2000, 97(23):12411-2.

Peters CJ, Simpson GL, and Levy H, "Spectrum of Hantavirus Infection: Hemorrhagic Fever With Renal Syndrome and Hantavirus Pulmonary Syndrome," *Annu Rev Med*, 1999, 50:531-45.

Internet Web Sites

www.cdc.gov/ncidod/diseases/hanta/hps/index.htm

Haptoglobin, Serum

Related Information

Alpha$_1$-Acid Glycoprotein, Serum *on page 133*
Anemia Flowchart *on page 35*
C-Reactive Protein, Serum *on page 467*
Hemoglobin, Plasma *on page 687*
Myoglobin, Blood, Serum, or Plasma *on page 940*
Transfusion Reaction Work-up *on page 1269*

Abstract Haptoglobin is synthesized in the liver (and possibly other sites), migrates electrophoretically in the alpha$_2$ position, decreases with hemolysis, and is an acute phase reactant.

Specimen Serum

Container Red top tube

Causes for Rejection Hemolysis from traumatic venipuncture

Reference Interval

- Newborns: 5-48 mg/dL
- Adults: 26-185 mg/dL
- Older than 60 years: male: 35-164 mg/dL; female: 40-175 mg/dL

Use

Hemolysis: Haptoglobin is used to detect episodes of subtle or mild hemolysis, in which situation haptoglobin levels are **decreased**. Haptoglobin is typically absent from serum when the rate of hemolysis is such that the red cell half-life, measured by radioactive chromium-51 labeled autologous red cells, is <17.5 days (normal range by this technique is a half-time of ~26 days). It is useful to recall that ineffective erythropoiesis (megaloblastic anemias, hemoglobinopathies), as well as hematomas and soft tissue hemorrhage are conditions with a hemolytic component and are associated with a decrease in haptoglobin. Haptoglobin is decreased by a number of drugs which cause hemolytic anemia. It may be decreased even with extravascular hemolysis.

Limitations Haptoglobin is an acute phase reactant and the increase which occurs during episodes of **acute inflammation** may mask the decrease (Continued)

Haptoglobin, Serum *(Continued)*

produced by concurrent hemolysis. The simultaneous measurement of another acute phase reactant (eg, C-reactive protein or alpha$_1$ acid glycoprotein) can help resolve this interpretive difficulty. **Corticosteroids, androgens,** and **protein-losing states** (eg, nephrotic syndrome, protein-losing enteropathies) produce a similar elevation in haptoglobin, and masking of hemolysis. **Liver disease** and **estrogen** medication produce decreased levels of haptoglobin which can be misinterpreted as evidence of hemolysis. **Rare examples** of genetic absence of haptoglobin and genetic hypohaptoglobinemia have been reported.

Methodology Immunologic methods including radial immunodiffusion (RID), nephelometry, automated immunoprecipitation, hemoglobin binding capacity.

References

Dobryszycka W, "Biological Functions of Haptoglobin--New Pieces to an Old Puzzle," *Eur J Clin Chem Clin Biochem*, 1997, 35(9):647-54.

Painter PC, Cope JY, and Smith JL, "Reference Information for the Clinical Laboratory," *Tietz Textbook of Clinical Chemistry*, 3rd ed, Burtis CA and Ashwood ER, eds, Philadelphia, PA: WB Saunders Co, 1999, 1816.

Patzelt D, Geserick G, and Schröder H, "The Genetic Haptoglobin Polymorphism: Relevance of Paternity Assessment," *Electrophoresis*, 1988, 9(8):393-7.

♦ **Hashish** *see* Cannabinoids (Marijuana Metabolites), Qualitative, Urine *on page 335*

♦ **Hazards of Transfusion** *see* Risks of Transfusion *on page 1166*

♦ **Hb A$_1$** *see* Glycated Hemoglobin (Hemoglobin A$_{1c}$), Blood *on page 655*

♦ **Hb F** *see* Fetal Hemoglobin *on page 581*

♦ **Hb Köln** *see* Hemoglobin, Unstable, Heat Labile Test *on page 689*

♦ **Hb Zurich** *see* Hemoglobin, Unstable, Heat Labile Test *on page 689*

♦ **Hb Zurich** *see* Hemoglobin, Unstable - Isopropanol Precipitation Test *on page 690*

♦ **hCG** *see* Chorionic Gonadotropin, Human, Serum and Urine *on page 397*

♦ **hCG** *see* Immunoperoxidase Procedures *on page 780*

♦ **β-hCG** *see* Immunoperoxidase Procedures *on page 780*

♦ **hCG, Slide Test, Stat** *see* Pregnancy Test, Serum or Urine *on page 1084*

♦ **hCG, Urine** *see* Pregnancy Test, Serum or Urine *on page 1084*

♦ **HCO$_3$$^-$** *see* Bicarbonate, Blood *on page 258*

♦ **HCO$_3$$^-$** *see* Blood Gases and pH, Arterial *on page 275*

♦ **HCV Antibody** *see* Hepatitis C Virus Serology *on page 706*

♦ **HD DNA Test** *see* Huntington Disease DNA Test *on page 752*

♦ **HDL** *see* High Density Lipoprotein Cholesterol, Serum *on page 729*

♦ **HDLC** *see* High Density Lipoprotein Cholesterol, Serum *on page 729*

♦ **HDL Cholesterol** *see* High Density Lipoprotein Cholesterol, Serum *on page 729*

♦ **HDL:LDL Cholesterol** *see* Triglycerides, Serum or Plasma *on page 1275*

♦ **Heavy Metal Screen, Arsenic** *see* Arsenic, Blood *on page 209*

Heavy Metal Screen, Blood

Related Information

Arsenic, Blood *on page 209*
Arsenic, Hair, Nails *on page 210*
Arsenic, Urine *on page 211*
Cadmium, Blood or Urine *on page 325*
Heavy Metal Screen, Urine *on page 669*
Lead, Blood *on page 832*
Mercury, Blood *on page 897*
Mercury, Urine *on page 898*
Thallium, Urine or Blood *on page 1242*

Synonyms Metals, Blood; Poisonous Metals, Blood; Toxic Metals, Blood

Test Includes Antimony, arsenic, bismuth, boron, cadmium, cobalt, copper, lead, mercury, thallium

Abstract Used principally to detect arsenic, cadmium, mercury, and lead poisoning. See individual entries for detailed information.

Specimen Whole blood (EDTA) plus serum

Container Special metal-free tube and red top tube

Storage Instructions Refrigerate: do not spin down.

Special Instructions Check with laboratory performing the assay to determine what elements will be detected and for special instructions.

Reference Interval See individual test listings.

Use Screen for heavy metal poisoning

Methodology Atomic absorption spectrometry (AA), inductively-coupled plasma (ICP)

Additional Information See individual test listings.

References

Baldwin DR and Marshall WJ, "Heavy Metal Poisoning and its Laboratory Investigation," *Ann Clin Biochem*, 1999, 36(Pt 3):267-300.

Graeme KA and Pollack CV Jr, "Heavy Metal Toxicity, Part I: Arsenic and Mercury," *J Emerg Med*, 1998, 16(1):45-56.

Graeme KA and Pollack CV Jr, "Heavy Metal Toxicity, Part II: Lead and Metal Fume Fever," *J Emerg Med*, 1998, 16(2):171-7.

Heavy Metal Screen, Urine

Related Information

Arsenic, Blood *on page 209*
Arsenic, Hair, Nails *on page 210*
Arsenic, Urine *on page 211*
Cadmium, Blood or Urine *on page 325*
Heavy Metal Screen, Blood *on page 668*
Lead, Blood *on page 832*
Mercury, Blood *on page 897*
Mercury, Urine *on page 898*
Molybdenum, Blood *on page 919*
Thallium, Urine or Blood *on page 1242*

Synonyms Metal Screen; Metals, Toxic; Poisonous Metals, Urine; Toxic Metals, Urine

Applies to Hair Analysis

Test Includes Arsenic, cadmium, mercury, lead (could also include nickel and cadmium)

Abstract Used most often to detect arsenic, mercury, lead, and cadmium poisoning. See individual entries for detailed information.

Specimen 24-hour urine

Container Plastic, acid-washed urine container (preferably polyethylene), no preservative, 20-25 mL 6 N HCl (low metal content)

Storage Instructions Refrigerate

Reference Interval Arsenic: <50 µg/L; lead: <80 µg/L; mercury: <20 µg/L; nickel: <25 µg/L; cadmium: <10 µg/L

Use Screen for heavy metal poisoning and toxic exposure

Methodology Atomic absorption spectrometry (AA), inductively-coupled plasma (ICP)

Additional Information Please see Heavy Metal Screen, Blood *on page 668* for further information.

References

Baldwin DR and Marshall WJ, "Heavy Metal Poisoning and its Laboratory Investigation," *Ann Clin Biochem*, 1999, 36(Pt 3):267-300.

Graeme KA and Pollack CV Jr, "Heavy Metal Toxicity, Part I: Arsenic and Mercury," *J Emerg Med*, 1998, 16(1):45-56.

Graeme KA and Pollack CV Jr, "Heavy Metal Toxicity, Part II: Lead and Metal Fume Fever," *J Emerg Med*, 1998, 16(2):171-7.

♦ **Heelstick Blood Collection** *see* Skin Puncture Blood Collection *on page 1203*

♦ **Heinz Bodies** *see* Hemoglobin, Unstable, Heat Labile Test *on page 689*

♦ **Heinz Bodies** *see* Hemoglobin, Unstable - Isopropanol Precipitation Test *on page 690*

♦ **Heinz Bodies** *see* n-Butanol Stability Test *on page 945*

Heinz Body Stain

Related Information

Glucose-6-Phosphate Dehydrogenase, Blood *on page 641*
Hemoglobin, Unstable, Heat Labile Test *on page 689*
Hemoglobin, Unstable - Isopropanol Precipitation Test *on page 690*
n-Butanol Stability Test *on page 945*
(Continued)

Heinz Body Stain *(Continued)*

Peripheral Blood: Red Blood Cell Morphology *on page 1016*
Red Blood Cell Enzyme Deficiency *on page 1134*
Reticulocyte Count *on page 1156*

Abstract Heinz bodies (HBs) are microscopically visible intraerythrocyte insoluble aggregates of oxidized denatured hemoglobin that attach to the internal surface of the red blood cell (RBC) membrane. They reflect the presence of a metabolic derangement of or abnormality in the secondary structure of hemoglobin. The afflicted red cell has shortened survival due to injury and/or removal from the circulation by the spleen. Demonstration of HBs in a patient's RBCs indicates the presence of a red cell biochemical defect, commonly glucose-6-phosphate dehydrogenase (G6PD) deficiency, thalassemia, or unstable hemoglobin. Drug-induced oxidative stress may lead to Heinz body (HB) formation.

Specimen Whole blood

Container Lavender top (EDTA) tube, green top (heparin) tube

Collection Obtain a tube of normal control blood at the time patient sample is drawn.

Storage Instructions Refrigerate

Causes for Rejection Clotted specimen, hemolyzed specimen

Reference Interval No Heinz bodies (or only rare HBs) identified. Using blood incubated with acetylphenylhydrazine, normal control may have one to a few (under five) HBs in about one-third of the RBCs. A positive result (indicative of a defective reducing system) will find five or more HBs in about one-third or more of the RBCs. The definition of "abnormal" may vary somewhat between different laboratories.

Use Test for hemolytic disorders associated with Heinz body formation (eg, G6PD deficiency, thalassemia, unstable hemoglobin)

Possible Causes of Heinz Body Formation

Intrinsic (Intraerythrocyte Molecular) Abnormalities
Glucose-6-phosphate dehydrogenase deficiency (G6PD)
Unstable hemoglobinopathies (eg, Hb Köln)
Intrinsic Plus Extrinsic (Drug or Chemical) Abnormalities
Unstable hemoglobinopathies (eg, Hb Zurich = sulphonamides)
G6PD deficiency and oxidant stress-producing drugs (eg, acetanilid, dapsone, methylene blue, nalidixic acid, nitrofurantoin, pamaquine, pentaquine, sulphonamides,[1] thiazolesulfone).
Chemical Exposure
Arsine (arsenic hydride)
Benzene derivatives
Heavy metals (lead, mercury, copper, arsenic)
Potassium chlorate
Naphthalene
Toluidine blue
Trinitrotoluene
Drug Exposure (see also Intrinsic Plus Extrinsic)
Daunorubicin
Diaminodiphenyl sulfone
Niradazole
Omeprazole
Pamaquine
Pentaquine
Quinidine
Sulphonamides[1]

[1]Sulphonamides including sulfacetamide, sulfamethoxazole (Gantanol®), sulfapyridine, and sulfanilamide.

Methodology Supravital stain (methyl violet, new methylene blue, crystal violet, or brilliant cresyl blue) using blood incubated (60 minutes or more) at room temperature with acetylphenylhydrazine or sterile blood incubated 24 and 48 hours at 37°C. Heinz bodies are intraerythrocytic, purple, vary in shape (round, oval, serrated), 1-3 µm across, single or multiple, and close to the cell membrane.

Additional Information Heinz bodies (HBs) are uncommon except with G6PD deficiency immediately following hemolysis, postsplenectomy, and in patients with unstable hemoglobin variants. They are present characteristically in the congenital Heinz body hemolytic anemias (CHBHA - the unstable hemoglobinopathies). There are now some 200 different identified molecular variants of hemoglobin underlying CHBHA. Less than one-half of these are of clinical significance. The three major causes for HB formation and increased hemolysis are exposure to certain chemicals and drugs, deficiency of one of the reducing systems of blood, and presence of an unstable hemoglobin.

Heinz bodies are efficiently extracted ("pitted") from red blood cells during their transit through the splenic microvasculature. Study of peripheral blood smears from susceptible individuals may show only morphologic residua or suggestion of HB formation such as presence of "bite" cells (degmacytes). These are RBCs with a semicircular bite-like defect along the edge, resultant from "pitting" or extrusion of HBs with irregular contraction of the remaining RBC.

HBs may be found after the administration of sulfonamides, nitrofurans, Dilantin®, streptomycin, fava beans, chlorates, phenylhydrazine, primaquine (in sensitive individuals), and other compounds; omeprazole (in a case of renal failure) has been incriminated. See previous table.

References
Beutler E, "Heinz Body Staining," *Williams Hematology*, 5th ed, Chapter 5, Beutler E, Lichtman MA, Coller BS, et al, eds, New York, NY: McGraw-Hill Inc, 1995, 356 and L26.

Lukens JN and Lee GR, "Unstable Hemoglobin Disease," *Wintrobes' Clinical Hematology*, 10th ed, Part IV, Disorders of Red Cells, Chapter 52, Lee GR, Foerster J, Lukens J, et al, eds, Baltimore, MD: Lippincott Williams & Wilkins, 1999, 1398-1404.

Helicobacter pylori Antigen and Serology

Related Information
Helicobacter pylori Biopsy-Based Tests: The Urease Tests, Culture, Cytology, and PCR *on page 672*
Pepsinogen I and II, Serum or Plasma *on page 1009*

Applies to Breath Testing for *H. pylori*

Test Includes Antigen detection in the stool of infected individuals; serology detects IgG, IgA, and IgM antibodies specific for *Helicobacter pylori*

Abstract *H. pylori* is very strongly associated with duodenal and gastric ulcer and chronic active gastritis. Patients with peptic ulcer disease associated with *H. pylori* have elevated levels of serum antibody against this bacterium and specific *H. pylori* antigen in the stool. The *H. pylori* fecal assay by EIA is a noninvasive qualitative test with high sensitivities (81% to 97%) and specificities (98% to 100%). Thus, it compares favorably with other techniques that require invasive procedures of endoscopy or venipuncture. The FDA has approved this test for initial diagnosis and for test of cure. Even after treatment, antibodies to *H. pylori* may persist up to a year.

Specimen Stool for antigen; serum for serology

Container Large open-mouthed stool container; red top tube for serology

Collection Acute and convalescent samples may be useful for serology.

Storage Instructions 2°C to 8°C; freeze if more than a 72-hour delay is anticipated.

Reference Interval Negative

Use Support the diagnosis of *H. pylori* infection in adults

Limitations Confirmation of eradication of *H. pylori* infection is not required. Negative results do not rule out a possibility of *H. pylori* infection. Performance characteristics are not established for patients younger than 18 years of age, or for watery, diarrheal stools.

Serologic findings only provide evidence of past or present infection. A large number of people are infected with the organism but do not have apparent disease. There is strain-to-strain antigenic variability in *H. pylori*, which makes (Continued)

Helicobacter pylori Antigen and Serology *(Continued)*

the test potentially insensitive. Sensitivities of commercially available assays range from 59% to 100%, while specificities range from 29% to 100%.

Contraindications The use of testing for *H. pylori* in asymptomatic individuals is not advised.

Methodology Enzyme immunoassay (EIA) for stool antigen; enzyme-linked immunosorbent assay (ELISA) for total antibody, IgG antibody titers

Additional Information Large numbers of small, spiral-shaped bacteria, *Helicobacter pylori*, can be cultured from, or seen microscopically in gastric biopsies from most patients with chronic active gastritis and/or peptic ulcers. They can also be found in significant numbers of asymptomatic patients who have histopathologic gastritis, and from some individuals with no abnormality. Similarly, patients with chronic gastritis usually have elevated titers of IgG antibodies to *H. pylori*. The *H. pylori* stool antigen is a rapid, easy to use, and noninvasive test. Both sensitivity and specificity were about 93%. After therapy, the antigen disappeared from feces in a few days, but the specificity was 79%. Standard diagnostic studies for *H. pylori* are compared in the following table.

Diagnostic Tests for *H. pylori*

Test	Advantages	Disadvantages
EIA on stool	Noninvasive	Less experience
Culture	Allows susceptibilities	Insensitive, slow
Serology	Noninvasive, inexpensive	Unreliable for active disease
Urease detection	Results in 2 hours	False positives, endoscopy
Urea breath tests	Rapid, evaluates response	Expensive

The association of *H. pylori* infection in carcinogenesis of gastric epithelium and primary malignant lymphoma of stomach is recognized. IgG antibody to *H. pylori* is increased in sera of patients with gastric cancer. A gold standard for the presence of *H. pylori* differs depending on the reference, but may include culture of biopsy, urea breath tests, rapid urease tests, PCR, or others.

References

Graham DY, Rakel RE, Kendrick AM, et al, "Recognizing Peptic Ulcer Disease. Keys to Clinical and Laboratory Diagnosis," *Postgrad Med*, 1999, 105(3):113-28.

Shiotani A, Nurgalieva ZZ, Yamaoka Y, et al, "*Helicobacter pylori*," *Med Clin North Am*, 2000, 84(5):1125-36.

Vaira D, Holton J, Menegatti M, et al, "Invasive and Noninvasive Tests for *Helicobacter pylori* Infection," *Aliment Pharmacol Ther*, 2000, 14(Suppl 3):13-22.

Vaira D, Holton J, Menegatti M, et al. "New Immunological Assays for the Diagnosis of *Helicobacter pylori* Infection," *Gut*, 1999, 45(Suppl I):I23-I27.

Internet Web Sites

www.cdc.gov/ulcer/md.htm
www.helico.com

Helicobacter pylori Biopsy-Based Tests: The Urease Tests, Culture, Cytology, and PCR

Related Information

Cobalamin, Serum *on page 424*
Gastrin, Serum *on page 631*
Helicobacter pylori Antigen and Serology *on page 671*
Intrinsic Factor Blocking Antibody *on page 806*
Methylmalonic Acid, Serum, Plasma, Urine, or Amniotic Fluid *on page 908*
Parietal Cell Antibody *on page 1005*
Pepsinogen I and II, Serum or Plasma *on page 1009*
Polymerase Chain Reaction *on page 1069*

Test Includes Screening for the presence of urease activity in gastric biopsies; culture of the organism from gastric biopsy specimens, and PCR for detection of *H. pylori*

Abstract *H. pylori* is a gram-negative motile spiral-shaped bacteria, frequently associated with gastritis. *H. pylori* infection is associated with gastric atrophy and intestinal metaplasia. It is acquired earlier in life in lower socioeconomic conditions, or in areas with poorer hygiene. It plays a major role in pathogenesis of peptic ulcer disease, but the majority of persons colonized remain asymptomatic. *H. pylori* infection is an independent risk factor for gastric adenocarcinoma and gastric mucosa-associated lymphoid-tissue (MALT) lymphomas.

Specimen Gastric mucosal biopsies

Container Sterile container, **no fixative** for the microbiologic tests

Storage Instructions If specimens cannot be transported immediately to the laboratory, they should be placed in 0.5 mL transport medium (normal saline).

Reference Interval Negative for urease activity, negative culture, biopsies negative for gastritis, and negative for *H. pylori*

Use Establish the presence of *Helicobacter pylori* in cases of chronic gastric ulcer, chronic active gastritis, and a relationship with duodenal ulcer

Limitations Culture and urease testing alone, without biopsies for histopathologic examination, may permit occult neoplasms to go undetected. The sensitivity of urease testing varies. Newer tests have been marketed (eg, CLO-test, HUT-test, Pyloritek) and have sensitivities up to 90% to 99%, specificities of 95% to 99%.

Methodology Urease test: Gastric biopsies are incubated on slightly buffered medium. A change of phenol red to alkaline (pink color) persisting more than 5 minutes is considered positive and indicative of the presence of *Helicobacter pylori*.

Breath isotope methods, measuring bacterial urease by detection of labeled CO_2. The breath test is preferred as a means of documenting presence of active infection and eradication of infection after therapy in the absence of endoscopy.

Culture: Culture media include enriched chocolate, Thayer-Martin with antibiotics, brain heart infusion (BHI) with 7% horse blood, and/or Mueller-Hinton with 5% sheep blood. The organism is microaerophilic and grows best in a reduced O_2 atmosphere or in a Campy-Pak™ system at 35°C. Cultures are usually observed for 7 days before being reported as negative.

Biopsies: Biopsies of the middle of the body of the stomach are reported to provide optimal specimens for the diagnosis of *H. pylori*. Biopsies of the greater curvature of the midantrum are suitable to assess colonization. Both the angulus and middle body are suitable sites for biopsies to ascertain extent of atrophic gastritis. Atrophy and intestinal metaplasia are identified most frequently in the antrum. They are regarded as preneoplastic conditions, associated with *H. pylori* infections as well as other entities. The organism may be seen in biopsies stained with Gram stain, hematoxylin-eosin (H&E), Giemsa, Dieterle, or Warthin-Starry silver stain. Newer methods include the Genta stain and immunocytochemistry with application of a specific antibody. It is most often recognized in biopsies of the antrum but may also be seen in the fundic mucosa, metaplastic gastric mucosa of esophagus (Barrett esophagus), or duodenum. Biopsies may also establish the diagnosis of carcinoma or lymphoma.

Cytology: Touch cytology preparations (ie, imprints from biopsies) provide rapid diagnosis; the biopsy specimens can be examined by histopathologic sections or sent for culture. The current gold standard for the diagnosis is arbitrarily defined as two specially stained biopsy specimens taken from the antrum.

Smear: A direct smear can be Gram stained.

Polymerase chain reaction (PCR). The most sensitive technique to detect *H. pylori* is considered to be PCR. Either gastric aspirates or biopsy specimens have shown sensitivity and specificity >95%. False positives do occur following inadequate cleaning of endoscopes.

Additional Information More than 90% of subjects who have duodenal ulcer have *H. pylori* infection. Only 50% of patients with Zollinger-Ellison syndrome with duodenal ulcer have *H. pylori*. Of patients with gastric ulcers, 75% to 80% have *H. pylori*.

References

Blaser MJ, "In a World of Black and White, *Helicobacter pylori* Is Gray," *Ann Intern Med*, 1999, 130(8):695-7.

Fennerty MB, "A Review of Tests for the Diagnosis of *Helicobacter pylori* Infection," *Lab Med*, 1998, 29(9):561-6.

(Continued)

Helicobacter pylori Biopsy-Based Tests: The Urease Tests, Culture, Cytology, and PCR *(Continued)*

Glupczynski Y, "Microbiological and Serological Diagnostic Tests for *Helicobacter pylori*: An Overview," *Brit Med Bull*, 1998, 54(1):175-86.

Jacobs DS, DeMott WR, Oxley DK, et al, *Laboratory Test Handbook*, 5th ed, Hudson, OH: Lexi-Comp Inc, 2001.

McColl K, Murray L, El-Omar E, et al, "Symptomatic Benefit From Eradicating *Helicobacter pylori* Infection in Patients With Nonulcer Dyspepsia," *N Engl J Med*, 1998, 339(26):1869-74.

Ohkusa T, Takashimizu I, Fujiki K, et al, "Disappearance of Hyperplastic Polyps in the Stomach After Eradication of *Helicobacter pylori*. A Randomized, Clinical Trial," *Ann Intern Med*, 1998, 129(9):712-5.

Parsonnet J, Shmuely H, and Haggerty T, "Fecal and Oral Shedding of *Helicobacter pylori* From Healthy Infected Adults," *JAMA*, 1999, 282(23):2240-5.

Toulaymat M, Marconi S, Garb J, et al, "Endoscopic Biopsy Pathology of *Helicobacter pylori* Gastritis. Comparison of Bacterial Detection by Immunohistochemistry and Genta Stain," *Arch Pathol Lab Med*, 1999, 123(9):778-81.

Zucca E, Bertoni F, Roggero E, et al, "Molecular Analysis of the Progression From *Helicobacter pylori*-Associated Chronic Gastritis to Mucosa-Associated Lymphoid-Tissue Lymphoma of the Stomach," *N Engl J Med*, 1998, 338(12):804-10.

Internet Web Sites
www.cdc.gov/ulcer/md.htm
www.helico.com

♦ **Hemadsorbing Virus** *see* Influenza Detection and Culture *on page 789*

Hematocrit
Related Information
Anemia Flowchart *on page 35*
Blood Volume *on page 282*
Complete Blood Count *on page 442*
Hemoglobin *on page 681*
Peripheral Blood: Red Blood Cell Morphology *on page 1016*
Red Blood Cell Count *on page 1133*
Red Blood Cell Indices *on page 1136*
Red Blood Cells *on page 1139*
Red Cell Mass *on page 1144*
Reticulocyte Count *on page 1156*
Uncrossmatched Blood, Emergency *on page 1281*

Synonyms Packed Cell Volume

Applies to H and H

Abstract Percent of whole blood that is red blood cells. A determination that is of importance in the detection and follow-up of anemia and polycythemia. The hematocrit value is used in the calculation of the MCV and MCHC.

Specimen Whole blood

Container Lavender top (EDTA) tube

Collection Routine venipuncture. Invert tube gently to mix. For capillary puncture, establish free flow of blood to minimize dilution with tissue fluid.

Hematocrit Values
First Postnatal Day

Gestational Age (wk)	Hct (%)
24-25	63
26-27	62
28-29	60
30-31	60
32-33	60
34-35	61
36-37	64
Term	61

Adapted from Zaizor R and Matoth Y, "Red Cell Values on the First Postnatal Day During the Last 16 Weeks of Gestation," *Am J Hematol*, 1976, 1(2):275-8.

Normal Hematocrit Values
Newborn

Age	Hct (%)
Birth - 2 d	54-68
2-3 d	54-66
3-4 d	52-71
4-5 d	39-55
5-6 d	50-64
6-7 d	47-61
7-8 d	47-64
1-2 wk	50-62
2-3 wk	39-53
3-4 wk	37-49

Storage Instructions If specimen is not brought to the laboratory within 4 hours, refrigeration should be provided. Perform manual Hct within 6 hours after collection of blood.

Causes for Rejection Clotted or hemolyzed specimen

Reference Interval See table in Complete Blood Count *on page 442*. In general, spun hematocrits are 2% to 3% higher than automated hematocrits, due to plasma trapping. There is evidence that Hct shows slight seasonal variation, lower in summer in nonsmokers but increased in smokers. Mean values for the two groups (summer vs winter) were, however, within the reference range of normal. See tables above and on previous page.

Use Evaluate anemia, blood loss, hemolytic anemia, polycythemia, and other conditions

Limitations During centrifugation, plasma is not completely extruded from the red cell column, plasma trapping amongst red blood cells (RBCs) which may increase the PCV of normal blood by nearly 2%. Plasma trapping is increased if abnormal RBCs (such as microcytes, macrocytes, spherocytes, thalassemic or sickled cells) are present (cells with increased rigidity). With automated instruments, falsely high results may occur when cryoproteins, significant leukocytosis, or giant platelets are present; false low results may be seen with microcytosis, *in vitro* hemolysis or in presence of autoagglutinins.

Methodology Manual microhematocrit centrifugation using disposable 75 mm long capillary tubes of 1 mm bore filled with EDTA anticoagulated whole blood and centrifuged at 10,000-12,000 g for 5 minutes (analyst must wear gloves and observe precautions). The Hct is included (by calculation) in the menu of test results provided by a variety of multiparameter hematology analyzers including different models from Coulter, Sysmex, Abbott/Cell Dyne, Bayer®/Technicon, ABX Diagnostics/Pentra, and Cobas-Helios (see Complete Blood Count *on page 442*). The Hct is derived by electronic calculation considering that Hct = RBC x MCV / 10 (latter two are directly measured).

Additional Information The degree of plasma trapping is increased in disease with less deformable RBCs (eg, sickle cell disease, hereditary spherocytosis, and iron deficiency).

The lower mean Hct and Hb value in African-American children (as compared to white children) is the result not only of α- and β-thalassemia but also of the presence of Hb AS and Hb AC.

Point-of-care devices have been developed. The Spuncrit™ is portable (battery operated), consists of an infra-red analyzer with centrifuge, and can provide Hct and hemoglobin values within 90 seconds. The Spuncrit™ estimates the hemoglobin concentration by dividing the measured Hct by a factor of 2.82. For hemoglobin concentration, 2 SD of Spuncrit™ estimates were between 1.6 g/dL below to 1.92 g/dL above comparison methods.

In a renal dialysis population, Hb vs Hct comparison studies, found Hct levels high (by 3% but possibly due to plasma trapping) with adverse effect on patient welfare as the result of inappropriate (low) erythropoietin dosage. Comparison of centrifuged microhematocrit with calculated packed cell volume using (Continued)

675

Hematocrit *(Continued)*

different automated hematology analyzers at four Swedish university hospitals found the average difference between the means amounting to 1.9% (centrifuged microhematocrit vs calculated PCV - due to trapped plasma in the former). It was concluded that the therapeutic goal should be to maintain calculated PCV <43 (rather than 45, the goal in the past based on use of centrifuged microhematocrit).

A study of 1000 Israeli airmen over a 15-year period with average number of annual Hct of 13.2 per person (Wintrobe's microhematocrit method) found that variations of up to 3% in Hct over time can be considered within normal in young males.

The red cell indices, MCV and MCHC, depend on the Hct for their derivation and are of use in the evaluation of anemia. See discussion in Complete Blood Count *on page 442.*

A six parameter (includes Hct) point-of-care handheld portable analyzer that requires just 2 minutes for analyses is available (i-STAT).

References

Andréasson B, Wahlstrom E, Jacobsson S, et al, "The Measurement of Venous Haematocrit in Patients With Polycythaemia Vera," *J Intern Med,* 1999, 246(3):293-7.

Froom P, Benbassat J, Kiwelowicz A, et al, "Significance of Low Hematocrit Levels in Asymptomatic Young Adults: Results of 15 Years Follow-up," *Aviat Space Environ Med,* 1999, 70(10):983-6.

Kristal-Boneh E, Froom P, Harari G, et al, "Seasonal Differences in Blood Cell Parameters and the Association With Cigarette Smoking," *Clin Lab Haematol,* 1997, 19(3):177-81.

Morris MW and Davey FR, "Basic Examination of Blood," *Clinical Diagnosis and Management by Laboratory Methods,* 20th ed, Chapter 24, Henry JB, ed, Philadelphia, PA: WB Saunders Co, 2001, 483.

Weatherall MS and Sherry KM, "An Evaluation of the Spuncrit™ Infra-red Analyzer for Measurement of Haematocrit," *Clin Lab Haematol,* 1997, 19(3):183-6.

♦ Hematology Introduction *see page 34*

Hematopoietic Progenitor Cells, Cord Blood/Placental Blood

Related Information

CD34⁺ Hematopoietic Stem Cells by Flow Cytometry *on page 351*

Hematopoietic Progenitor Cells, Marrow *on page 677*

Hematopoietic Progenitor Cells, Peripheral Blood *on page 679*

Synonyms Placental Blood; Transplant, Cord Blood; Umbilical Cord Blood

Applies to Cryopreservation

Test Includes For unrelated allogeneic units, the following tests must be performed: ABO and Rh type, HLA type, total nucleated cell count, CD34 count, % viability, and bacterial culture. The following tests intended to prevent infectious disease transmission must be tested using a blood sample obtained from the donor's birth mother: HBV, HTLV-I/II, HIV-1/2, HCV, CMV, and syphilis. A donor health history and a screening test for hemoglobin disorders also must be performed. Refer to appropriate Standards for additional specifications.

Abstract Hematopoietic progenitor cells (HPC) can be obtained from the umbilical cord at the time of delivery, when umbilical cord blood is immediately placed in anticoagulant and cryopreserved to serve as a source of stem cells. Patients in need of allogeneic HPC transplant have a <85% chance of finding an HLA-matched donor, and umbilical cord blood stem cells may serve as the alternate source.

Patient Preparation Informed consent from the biologic mother must be obtained according to applicable laws.

Container Appropriate anticoagulated collection bag or syringe.

Collection Collect cord blood during the third stage of labor or after delivery of the placenta if the collection interferes with care of the mother or baby. After disinfecting the cord, a large-bore needle connected to a sterile blood collection bag containing CPDA is frequently used to collect the cord blood. Approximately 65-70 mL of umbilical cord blood is typically collected in such a closed system.

Storage Instructions Since umbilical cord banks store large numbers of units, processing to reduce unit volume is usually employed prior to cryopreservation and storage in liquid nitrogen.

Causes for Rejection Too small a volume collected, unsuitable donor medical condition. Legal issues of testing and privacy exist.

Use Allogeneic cord blood HPC are used as a source of hematopoietic stem cells for both related and unrelated transplants. HPC are for pediatric and adult patients, for certain neoplastic and non-neoplastic disease, including immuno-deficiency states, Lesch-Nyhan syndrome, Hurler syndrome, and Diamond-Blackfan syndrome. Engraftment of stem cells in recipients with neoplastic diseases can be achieved, given an adequate dose of nucleated cells. A decreased risk of graft-vs-host disease permits an increase in levels of histoincompatibility.

HPC are used for β-thalassemia, sickle cell anemia, Fanconi anemia, and other lymphoid and hematopoietic genetic disorders.

Limitations Repopulation of marrow depends on the number of stem cells in the collection. The relatively small number of HPC in cord blood may be inade-quate to repopulate nonpediatric recipients. Umbilical cord blood recipients receive about 10% as many CD34⁺ cells as do those who receive allogeneic marrow.

Additional Information

- Reduced graft-vs-host disease in recipients of cord blood transplants is reported. Recipients of cord blood progenitor cells from HLA-identical siblings enjoy a lower incidence of graft-vs-host disease than recipients of bone marrow transplants from HLA-identical siblings.
- Days to engraftment in unrelated transplants are routinely longer than other sources (eg, marrow or peripheral blood).
- Standards for HPC have been published by the American Association of Blood Banks, the Foundation for the Accreditation of Cellular Therapy, and the National Marrow Donor Program. More regulations, including good tissue practices, will be forthcoming from the FDA.
- Complications include infectious diseases, hemorrhagic disorders, diffuse alveolar damage, hepatic veno-occlusive disease, and graft-vs-host disease.
- Commercialism and market-based medicine, threatening to replace ethics as a cornerstone of medical practice, affect portions of the practice of placental blood collection. Marketing practices for HPC from cord blood deserve close attention.

A *Circular of Information for the Use of Hematopoietic Progenitor Cell Products* is available from blood banks or blood centers.

References

Annas GJ, "Waste and Longing - the Legal Status of Placental-Blood Banking," *N Engl J Med*, 1999, 340(19):1521-4.

Circular of Information for the Use of Progenitor Cell Products, American Association of Blood Banks, America's Blood Centers, American Red Cross, January 2002.

Ende N, Lu S, Mack R, et al, "The Feasibility of Using Blood Bank-Stored (4°C) Cord Blood, Unmatched for HLA for Marrow Transplantation," *Am J Clin Pathol*, 1999, 111(6):773-81.

International Standards for Cord Blood Collection, Processing, Testing, Banking, Selection, and Release, 2nd ed, Omaha, NE: Foundation for the Accreditation of Cellular Therapy, 2001.

Laughlin MJ, Barker J, Bambach B, et al, "Hematopoietic Engraftment and Survival in Adult Recipients of Umbilical Cord Blood From Unrelated Donors," *N Engl J Med*, 2001, 344(24):1815-22.

Nuckols JD, "Autopsy Findings in Umbilical Cord Blood Transplant Recipients," *Am J Clin Pathol*, 1999, 112(3):335-42.

Rocha V, Wagner JE, Sobocinski KA, et al, "Graft-Versus-Host Disease in Children Who Have Received a Cord Blood or Bone Marrow Transplant From and HLA-Identical Sibling," *N Engl J Med*, 2000, 342(25):1846-54.

Rubinstein P, Carrier C, Scaradovou A, et al, "Outcomes Among 562 Recipients of Placental-Blood Transplants From Unrelated Donors," *N Engl J Med*, 1998, 339(22):1628-9.

Standards for Cord Blood Services, 1st ed, Bethesda, MD: American Association of Blood Banks Press, 2001, 32-7.

Sugarman J, Kaalund V, Kodish E, et al, "Ethical Issues in Umbilical Cord Blood Banking," *JAMA*, 1997, 278(11):938-43.

Hematopoietic Progenitor Cells, Marrow

Related Information

CD34⁺ Hematopoietic Stem Cells by Flow Cytometry *on page 351*
Hematopoietic Progenitor Cells, Cord Blood/Placental Blood *on page 676*
Hematopoietic Progenitor Cells, Peripheral Blood *on page 679*

Synonyms Bone Marrow Transplant, Allogeneic; Bone Marrow Transplant, Autologous; Transplant, Bone Marrow

Applies to CD34 Analysis; Cryopreservation; Galactocerebrosidase Deficiency

(Continued)

Hematopoietic Progenitor Cells, Marrow *(Continued)*

Test Includes For autologous donors, the following tests must be performed: ABO and Rh type, antibody screen, HLA type, relevant cell count, antigen expression analysis, cell viability, and sterility. The following tests intended to prevent infectious disease transmission must be performed: HBV, HTLV-I/II, HIV-1/2, and HCV. A donor health history screen, complete blood count, and pregnancy test (when applicable) are also performed.

For allogeneic donors, the following tests must be performed: ABO and Rh type, antibody screen, HLA type, relevant cell count, antigen expression analysis, cell viability, and sterility. The following tests intended to prevent infectious disease transmission must be performed: HBV, HTLV-I/II, HIV-1/2, HCV, CMV, and syphilis. A donor health history screen, complete blood count, and pregnancy test (when applicable) will also be performed. Refer to appropriate Standards for additional specifications.

Abstract Bone marrow was the leading source of hematopoietic progenitor cells (HPC) for patients in need of transplantation before 1990. However, other stem cell sources (eg, peripheral blood) are replacing marrow as a sole source of HPC in both autologous and allogeneic transplantation. Human marrow contains mature as well as pluripotent cells which are capable of reconstituting the hematologic and lymphoid systems.

Patient Preparation Because of blood loss in marrow collection, it is customary to collect 2 units of autologous red cells 2-3 weeks prior to the scheduled collection.

Aftercare Red cell transfusions usually take place after the collection, if possible (avoids marrow dilution). Irradiate all blood fractions given during a collection for allogeneic transplantation.

Collection The volume of marrow collected is determined by the recipient's body weight and any postcollection product manipulation. Collection targets, in general, are as follows: 1.0×10^8 nucleated cells/kg for autologous transplant and 2.0×10^8 nucleated cells/kg for allogeneic transplant. The patient/donor is placed under general or spinal anesthesia and the procedure is performed under sterile conditions in the operating room. The posterior iliac crest provides the richest source. Collecting approximately 10-15 mL of marrow per kg of recipient body weight will achieve the target dose. The aspirated marrow is anticoagulated, filtered to remove bony spicules and transported to the processing laboratory for quality assurance testing and any further manipulation or cryopreservation.

Storage Instructions Marrow may be processed prior to cryopreservation or infusion to remove plasma, red cells, or tumor. In an ABO-incompatible allogeneic transplant, processing to remove plasma or red cells prevents a hemolytic reaction. A number of manual or automated techniques are available for processing. An expiration date has not been defined, however, marrow stored for 11 years has been transplanted with resultant engraftment.

Causes for Rejection Poor anesthesia risk, obesity, malignant involvement of bone marrow, fibrosis of marrow at usual sites of collection (usually caused by prior radiation therapy of the pelvis), and patient refusal. In these cases, consider using an alternative (peripheral blood) HPC collected by apheresis.

Use Collection and storage of autologous bone marrow make it possible to treat malignant disease with heavy doses of chemotherapy and/or irradiation that would otherwise destroy the patient's marrow function. After such treatment, marrow is repopulated from the stored autologous supply. Allogeneic transplants are performed for indications such as marrow failure (eg, aplastic anemia), hemoglobinopathies, inborn errors of metabolism, and other disorders.

Central nervous system manifestations of globoid-cell leukodystrophy (deficiency of galactocerebrosidase) were reversed with allogeneic marrow and umbilical cord blood transplantation.

Marrow transplantation from an HLA-identical donor or T-cell depleted haploidentical marrow from related donors provides life-sustaining therapy for severe combined immunodeficiency.

Chemotherapy with stem cell transplantation is in use and/or in evaluation in neoplastic disease. Autologous stem cell transplantation is best with low tumor burden. A favorable impact is described with subsets of patients who have acute myelogenous leukemia, myelodysplasia, Hodgkin disease, and non-Hodgkin lymphoma.

Bone marrow contains not only HPC, but also marrow stromal cells, which can differentiate into nonhematopoietic cells such as heart myocytes. The use of stem cells in tissue regeneration is being explored.

Limitations Repopulation of marrow depends on the number of stem cells in the collection. Current enumeration is performed by CD34 analysis (1% to 3% of marrow cells). Processing may result in loss of stem cells. Inadequacy of stem cells will mean failure of repopulation. Complications include graft-vs-host disease and death. See Hematopoietic Progenitor Cells, Cord Blood/Placental Blood *on page 676.*

Additional Information Elimination of T cells from the marrow graft is reported to prevent graft-vs-host disease in management of severe combined immunodeficiency. For patients treated with HLA haploin-compatible marrow, graft-vs-host disease is a major determinant of outcome. Standards for HPC have been published by the American Association of Blood Banks, the Foundation for the Accreditation of Cellular Therapy, and the National Marrow Donor Program. Additional regulations will be forthcoming from the FDA. A Circular of Information for the Use of Hematopoietic Progenitor Cell Products is available from blood banks or blood centers.

References

Arico M, Valsecchi MG, Camitta B, et al, "Outcome of Treatment in Children With Philadelphia Chromosome-Positive Acute Lymphoblastic Leukemia," *N Engl J Med,* 2000, 342(14):998-1006.

Benesch M and Deeg HJ, "Hematopoietic Cell Transplantation for Adult Patients With Myelodysplastic Syndromes and Myeloproliferative Disorders," *Mayo Clin Proc,* 2003, 78(8):981-90.

Bensinger WI, Martin PJ, Storer B, et al, "Transplantation of Bone Marrow as Compared With Peripheral Blood Cells From HLA-Identical Relatives in Patients With Hematologic Cancers," *N Engl J Med,* 2001, 344(3):175-81.

Buckley RH, Schiff SE, Schiff RI, et al, "Hematopoietic Stem-Cell Transplantation for the Treatment of Severe Combined Immunodeficiency," *N Engl J Med,* 1999, 340(7):508-16.

Fischer A, "Thirty Years of Bone Marrow Transplantation for Severe Combined Immunodeficiency," *N Engl J Med,* 1999, 340(7):559-61.

Hansen JA, Gooley TA, Martin PJ, et al, "Bone Marrow Transplants From Unrelated Donors for Patients With Chronic Myeloid Leukemia," *N Engl J Med,* 1998, 338(14):962-8.

Herzog EL, Chai L, and Krause DS, "Plasticity of Marrow-Derived Stem Cells," *Blood,* 2003, 102(10):3483-93.

Körbling M and Estrov Z, "Adult Stem Cells for Tissue Repair - A New Therapeutic Concept?" *N Engl J Med,* 2003, 349(6):570-82.

Krivit W, Shapiro EG, Peters C, et al, "Hematopoietic Stem-Cell Transplantation in Globoid-Cell Leukodystrophy," *N Engl J Med,* 1998, 338(16):1119-26.

"Standards for Hematopoietic Progenitor Cell Services," 3rd ed, Bethesda, MD: American Association of Blood Banks Press, 2002, 23-41.

"Standards for Hematopoietic Progenitor Cell Collection, Processing and Transplantation," 2nd ed, Omaha, NE: Foundation for the Accreditation of Cellular Therapy, 2002, 22-5.

Theise ND, "Stem Cell Research: Elephants in the Room," *Mayo Clin Proc,* 2003, 78(8):1004-9.

Hematopoietic Progenitor Cells, Peripheral Blood

Related Information

Alpha$_1$-Fetoprotein, Amniotic Fluid *on page 135*

Alpha$_1$-Fetoprotein, Serum *on page 136*

Breast Biopsy *on page 305*

CD34$^+$ Hematopoietic Stem Cells by Flow Cytometry *on page 351*

Chorionic Gonadotropin, Human, Serum and Urine *on page 397*

Hematopoietic Progenitor Cells, Cord Blood/Placental Blood *on page 676*

Hematopoietic Progenitor Cells, Marrow *on page 677*

Synonyms Peripheral Blood Stem Cells, Autologous; Peripheral Stem Cells; Progenitor Cells; Stem Cell Collection

Applies to CD34; Cryopreservation; Granulocyte-Colony-Stimulating Factor

Test Includes For autologous donors, the following tests must be performed: ABO and Rh type, antibody screen, HLA type, relevant cell count, antigen expression analysis, cell viability, and sterility. The following tests intended to prevent infectious disease transmission must be performed: HBV, HTLV-I/II, HIV-1/2, and HCV. A donor health history screen, complete blood count, and pregnancy test (when applicable) are also performed.

For allogeneic donors, the following tests must be performed: ABO and Rh type, antibody screen, HLA type, relevant cell count, antigen expression analysis, cell viability, and sterility. The following tests intended to prevent infectious disease transmission must be performed: HBV, HTLV-I/II, HIV-1/2, HCV, CMV, (Continued)

Hematopoietic Progenitor Cells, Peripheral Blood
(Continued)

and syphilis. A donor health history screen, complete blood count, and pregnancy test (when applicable) will also be performed. Refer to appropriate Standards for additional specifications.

Abstract Peripheral blood progenitor cells (PBPC) are collected after mobilizing hematopoietic stem cells from the marrow using hematopoietic growth factors (eg, G-CSF) and/or treatment with chemotherapy. PBPC are then collected using a stem-cell protocol with any suitable apheresis instrument. One to five apheresis procedures may be necessary to obtain the desired number of hematopoietic progenitor cells (HPC). If the desired number of HPC are obtained in 1-3 collections, transplants using PBPC are less expensive than a marrow transplant.

Patient Preparation Excellent venous access is a daily requirement. Indwelling dual- or triple-lumen central venous catheters suitable for apheresis procedures are frequently necessary. Red cells may be required to prime the apheresis instrument for the pediatric donor.

Aftercare Complete blood counts should be performed to monitor hematocrit and platelet counts. Thrombocytopenia can be a complication. A majority of G-CSF stimulated donors experience bone pain, headaches, and fatigue associated with cytokine therapy.

Collection As described by the manufacturer of the apheresis instrument used. The optimal time to begin collection is controversial. Some institutions monitor using total white cell count, while others use CD34 count (*vide infra*). Collection target for engraftment ranges from a minimum total cell dose of 2-5 x 10^6 CD34$^+$ cells/kg.

Storage Instructions PBPC may be further processed prior to cryopreservation or infusion. A number of techniques for positive or negative selection are available to remove tumor cells or reduce the risk of graft-vs-host disease. Controlled rate (-180°C storage) and noncontrolled rate (-80°C storage) cryopreservation techniques are available. An expiration date has not been defined, however, marrow stored for 11 years has been transplanted with resultant engraftment.

Causes for Rejection Cancer or leukemia cells in peripheral blood. Sickle cell trait may be contraindicated in PBPC donors.

Use Collection and storage of autologous PBPC make it possible to treat malignant disease with heavy doses of chemotherapy and/or irradiation that would otherwise destroy the patient's marrow function. After such treatment, marrow is repopulated from the stored autologous supply. Allogeneic peripheral blood cells used for hematopoietic rescue restored blood counts faster than allogeneic bone marrow, without an increase in risk of graft-vs-host disease, in subjects treated with high-dose chemotherapy with or without radiation in treatment of hematologic malignant disease. Allogeneic transplants are also performed for indications such as marrow failure (eg, aplastic anemia), hemoglobinopathies or inborn errors of metabolism. The system is also useful for collecting stem cells for gene therapy.

Limitations Repopulation of marrow depends on the number of stem cells in the collection. Current enumeration is performed by CD34 analysis (0.01% to 0.1% of unstimulated peripheral blood cells). (CD34 is a cell surface antigen which is expressed on hematopoietic progenitor cells and on vascular endothelium.) Processing may result in loss of stem cells. Inadequacy of stem cells will mean failure of marrow repopulation. See also Hematopoietic Progenitor Cells, Cord Blood/Placental Blood *on page 676* for a discussion of different stem cell classes (clones) in relation to function.

Additional Information Compared to marrow or umbilical cord as the source of HPC, use of mobilized PBPC reduces the time to hematopoietic recovery.

Standards for HPC have been published by the American Association of Blood Banks, the Foundation for the Accreditation of Cellular Therapy, and the National Marrow Donor Program. Additional regulations will be forthcoming form the FDA.

A Circulator of Information for the Use of Hematopoietic Progenitor Cell Products is available from blood banks or blood centers.

References

Bensinger WI, Martin PJ, Storer B, et al, "Transplantation of Bone Marrow as Compared With Peripheral Blood Cells From HLA-Identical Relatives in Patients With Hematologic Cancers," *N Engl J Med*, 2001, 344(3):175-81.

Circular of Information for the Use of Progenitor Cell Products, American Association of Blood Banks, America's Blood Centers, American Red Cross, January, 2002.

Körbling M and Estrov Z, "Adult Stem Cells for Tissue Repair - A New Therapeutic Concept?" *N Engl J Med*, 2003, 349(6):570-82.

Lippman ME, "High-Dose Chemotherapy Plus Autologous Bone Marrow Transplantation for Metastatic Breast Cancer," *N Engl J Med*, 2000, 342(15):1119-20.

Nash RA, "Hematopoietic Stem Cell Transplantation," *Wintrobe's Clinical Hematology*, Volume 1, Chapter 28, Lee GR, Foerster J, Lukens J, et al, eds, Baltimore, MD: Lippincott Williams & Wilkins, 1999, 875-93.

Neito Y and Shpall EJ, "Clinical Results of Autologous Stem Cell Transplantation for Solid Tumors in Adults," *Hematology: Basic Principles and Practice*, 3rd ed, Hoffman R, Benz EJ Jr, Shattil SJ, et al, eds, New York, NY: Churchill Livingstone, 2000, 1597-609.

Stadtmauer EA, O'Neil A, Goldstein L, et al, "Conventional-Dose Chemotherapy Compared With High-Dose Chemotherapy Plus Autologous Hematopoietic Stem-Cell Transplantation for Metastatic Brease Cancer," *N Engl J Med*, 2000, 342(15):1069-76.

Standards for Hematopoietic Progenitor Cell Services, 3rd ed, Bethesda, MD: American Association of Blood Banks Press, 2002.

Standards for Hematopoietic Progenitor Cell Collection, Processing and Transplantation, 2nd ed, Omaha, NE: Foundation for the Accreditation of Cellular Therapy, 2002.

♦ **Hematopoietic Stem Cell Transplant** *see* Tissue Typing *on page 1259*

♦ **Hemipterae** *see* Arthropod Identification *on page 213*

♦ **Hemizona Binding Assay** *see* Infertility Screen *on page 786*

♦ **Hemochromatosis, DNA Testing** *see* Hereditary Hemochromatosis DNA Test *on page 718*

♦ **Hemoflagellates** *see* Microfilariae, Peripheral Blood Preparation *on page 915*

Hemoglobin

Related Information

Abstract This procedure determines the concentration of hemoglobin (Hb) in whole blood. Hb is the major component of the red cell and functions to transport oxygen. It also acts to buffer carbon dioxide formed during metabolic activity. The Hb level is important in the detection and follow up of anemia and polycythemia. The Hb value is used in the calculation of the mean corpuscular hemoglobin (MCH) and mean corpuscular hemoglobin concentration (MCHC).

Specimen Whole blood

Container Lavender top (EDTA) tube

Collection Routine venipuncture. Invert tube gently to mix.

Causes for Rejection Clotted or hemolyzed specimen

Reference Interval See tables in Complete Blood Count *on page 442*.

Use Evaluate anemia, blood loss, hemolysis, polycythemia, and other conditions (Continued)

Hemoglobin (Continued)

Limitations Hyperlipemic plasma (especially if chylomicronemia is present) or white count >50,000/mm³ may falsely elevate the Hb result with corresponding increase in the MCH and MCHC. A method correcting for lipemia has been suggested. An ABL 700 series blood gas analyzer can be applied to the measurement of Hb in cases of leukocytosis. Increased turbidity (with resultant interference in sample absorbance) may also be due to presence of a paraprotein or of an abnormal Hb (S or C). A variety of corrective procedures are available.

Methodology While oxyhemoglobin and other chemical approaches to hemoglobinometry exist, nearly all current procedures involve a one or two step procedure in which RBC lysis/dilution occurs with the formation of a cyanmethemoglobin compound. Dilutions are read by spectrophotometer at 540 nm. The majority of routine hematology laboratories obtain the Hb level as one of a number of measurements from an automated multichannel instrument. Near infrared spectroscopy has been applied to the noninvasive *in situ* measure of Hb concentration.

Additional Information The Hb determination is one of the best standardized and accurate of available clinical laboratory analyses. The results of current College of American Pathologists (CAP) surveys show good interlaboratory performance for the Hb procedure with low standard deviation and coefficient of variation values. Subject-specific reference values are of especial clinical value.

In cyanide poisoned individuals treated with methemoglobin-forming agents (to protect cytochrome oxidase) oxygen carrying capacity is decreased in direct proportion to the amount of methemoglobin (nonoxygen carrying) that is formed. A multiwavelength spectrophotometric method has been developed which allows monitoring of hemoglobin derivatives present in the blood of treated cyanide poisoned patients.

The red cell indices, MCH and MCHC, depend on the Hb for their derivation and are of use in the evaluation of anemia. A report of multiple myeloma (with an IgA-K paraprotein) found MCH and MCHC elevated values that failed Coulter STK-S machine internal limits.

Red cell measurements have been noted to undergo seasonal changes probably relating to increase in plasma volume in the summer with corresponding lower levels of Hb and Hct. In smokers, however, the increase in plasma volume did not occur and Hct increased while Hb levels did not change. Thus, smokers did not show the expected summer changes seen with nonsmokers. See also the discussion under Complete Blood Count *on page 442.*

As the red cell ages, Hb is gradually lost. The effects of glycation, carbamylation, and loss of RBC water and Hb (with age) result in changes in the RBC indices and must be taken into consideration in assessment of the expected recovery following blood transfusion.

References

"Haemoglobinometry: Screening and Routine Practice," *Advanced Laboratory Methods in Haematology*, Chapter 8, Rowan RM, van Assendelft OW, and Preston FE, eds, London, UK: Arnold, 2002, 182-90.

Morris MW and Davey FR, "Basic Examination of Blood," *Clinical Diagnosis and Management by Laboratory Methods*, 20th ed, Henry JB, ed, Philadelphia, PA: WB Saunders Co, 2001, 479-82.

Rendell M, Anderson E, Schleuter W, et al, "Determination of Hemoglobin Levels in the Finger Using Near Infrared Spectroscopy," *Clin Lab Hematol*, 2003, 25(2):93-7.

Roberts WL, Fontenot JD, and Lehman CM, "Overestimation of Hemoglobin in a Patient With an IgA-K Monoclonal Gammopathy," *Arch Pathol Lab Med*, 2000, 124(4):616-18.

Scharnhorst V, van de Laar PJ, and Vader HL, "Hemoglobin in Samples With Leukocytosis Can Be Measured on ABL 700 Series Blood Gas Analyzers," *Clin Chem*, 2003, 49(12):2107-8.

Zijlstra WG and Buursma A, "Rapid Multicomponent Analysis of Hemoglobin Derivatives for Controlled Antidotal Use of Methemoglobin-Forming Agents in Cyanide Poisoning," *Clin Chem*, 1993, 39(8):1685-9.

♦ **Hemoglobin A₁ₐ, A₁ᵦ, A₁ᵨ** see Glycated Hemoglobin (Hemoglobin A₁ᵨ), Blood *on page 655*

Hemoglobin A₂

Related Information

Anemia Flowchart *on page 35*
DNA-Probe Assay for Thalassemia (BeTha Gene Test) *on page 520*

Hemoglobin Electrophoresis *on page 684*
Peripheral Blood: Red Blood Cell Morphology *on page 1016*
Red Blood Cell Indices *on page 1136*

Abstract Hemoglobin A$_2$ (Hb A$_2$) is a tetramer of α- and δ-globulin chains (α$_2$ δ$_2$). Concentration fluctuates in the thalassemia syndromes and some acquired diseases.

Specimen Whole blood

Container Lavender top (EDTA) tube

Causes for Rejection Specimen clotted

Reference Interval The stable adult Hb A$_2$ level is 2.0% to 3.2% of total hemoglobin; ≤5 months: 0% to 2.5%; 6 months to 1 year: 1.5% to 2.9%. See table.

Alterations in Hb A$_2$ in Various Disorders

	Elevated	Reduced
Congenital	β-thalassemia trait	α-thalassemia
	Unstable hemoglobin variants	δβ-thalassemia
	Sickle trait (AS)	δ-thalassemia
	SS with α-thalassemia	HPFH
Acquired	Megaloblastic anemias	Iron deficiency
	Hyperthyroidism	Sideroblastic anemias

Adapted from Bunn HF and Forget BG, *Hemoglobin: Molecular, Genetic, and Clinical Aspects*, Philadelphia, PA: WB Saunders Co, 1986, 61-7.

Use Investigate microcytic anemia, for hemoglobinopathies, especially thalassemia, particularly beta-thalassemia trait

Limitations Blood transfusion prior to hemoglobin electrophoresis may make interpretation inconsistent. High levels of hemoglobin F usually are accompanied by lower levels of A$_2$. Sickle cell trait range is from 1.7% to 4.5% hemoglobin A$_2$. Presence of Hb S or Hb C will interfere with column chromatographic method. Presence of Hb C interferes with routine electrophoretic method. Quantitation of Hb A$_2$ by densitometric scanning of electrophoretic pattern may result in misleading (high) results.

Contraindications Recent blood transfusion

Methodology High performance liquid chromatography (HPLC), electrophoresis

Additional Information This test is done in some laboratories as part of hemoglobin electrophoresis. Hemoglobin A$_2$ levels have special application to the diagnosis of beta-thalassemia trait, which may be present even though peripheral blood smear is normal. This reflects the underlying genetic spectrum of beta-thalassemia which in reality is a complex of multiple distinct (and some less distinct) conditions, the result of over 180 different mutations. The microcytosis and other morphologic changes of beta-thalassemia trait must be differentiated from iron deficiency. Low MCV may include the majority of beta-thalassemia trait patients but does not differentiate iron deficient individuals. Low Hb A$_2$ levels occur in untreated iron deficiency. If iron deficiency occurs with beta-thalassemia, the Hb A$_2$ level may fall to within the normal range.

The most definitive evidence for presence of beta-thalassemia trait is genetic (family study). A well documented report, however, indicates the occurrence of beta-thalassemia minor as a result of a spontaneous initiation codon mutation. Offspring of a person with thalassemia major will have beta-thalassemia trait. Apart from such family/genetic studies (which are subject to practical difficulties) gene probes are the most definitive method for identifying beta-thalassemia trait. This method identifies "silent" carriers. See DNA-Probe Assay for Thalassemia (BeTha Gene Test) *on page 520*. Elevated percent Hb A$_2$ is the next best evidence for the diagnosis of beta-thalassemia trait. Sufficient criteria for the diagnosis of thalassemia trait are an elevated Hb A$_2$ level by a reliable method (Hb electrophoresis with elution and quantitation by spectrophotometry, HPLC, or column chromatography) - assuming Hb S, C, or an unstable hemoglobin are not present.

(Continued)

Hemoglobin A₂ (Continued)

Hb A₂ may be increased in megaloblastic anemia and may be decreased in sideroblastic anemia, Hb H disease, and erythroleukemia. Approximately 33% of zidovudine (AZT) treated human immunodeficiency virus-1 positive individuals have elevated Hb A₂ levels.

References

Chui DH, Fucharoen S, and Chan V, "Hemoglobin H Disease: Not Necessarily a Benign Disorder," *Blood*, 2003, 101(3):791-800.

Craver RD, Abermanis JG, Warrier RP, et al, "Hemoglobin A₂ Levels in Healthy Persons, Sickle Cell Disease, Sickle Cell Trait, and β-Thalassemia by Capillary-Isoelectric Focusing," *Am J Clin Pathol*, 1997, 107(1):88-91.

Huisman TH, "Levels of Hb A₂ in Heterozygotes and Homozygotes for Beta-Thalassemia Mutations: Influence of Mutations in the CACCC and ATAAA Motifs of the Beta-Globin Gene Promoter," *Acta Haematol*, 1997, 98(4):187-94.

Shokrani M, Terrell F, Turner EA, et al, "Chromatographic Measurements of Hemoglobin A₂ in Blood Samples That Contain Sickle Hemoglobin," *Ann Clin Lab Sci*, 2000, 30(2):191-4.

Weatherall DJ, "The Thalassemias," *The Molecular Basis of Blood Diseases*, 3rd ed, Chapter 6, Stamatoyannopoulos G, Majerus PW, Perlmutter RM, et al, eds, Philadelphia, PA: WB Saunders Co, 2001, 183-226.

Hemoglobin Electrophoresis

Related Information

DNA-Probe Assay for Thalassemia (BeTha Gene Test) *on page 520*
Fetal Hemoglobin *on page 581*
Fructosamine, Serum *on page 617*
Glycated Hemoglobin (Hemoglobin A₁c), Blood *on page 655*
Hemoglobin A₂ *on page 682*
Hemosiderin Stain, Urine *on page 692*
Methemoglobin, Whole Blood *on page 903*
P₅₀, Blood *on page 993*
Peripheral Blood: Red Blood Cell Morphology *on page 1016*
Reticulocyte Count *on page 1156*
Sickle Cell Tests *on page 1195*

Test Includes Electrophoresis for separation and distribution of hemoglobins, fetal hemoglobin (Hb F) and hemoglobin A₂ (Hb A₂) often by other methods, in particular column or high performance liquid chromatography (HPLC).

Abstract In this procedure, hemoglobins are caused to separate and migrate. A variety of techniques are utilized. Most commonly hemoglobins are separated as discrete bands as they move through a substrate in a buffer solution across an electric field with subsequent visualization by fixation and staining (hemoglobin electrophoresis). Clinical applications include detection and identification of hemoglobin variants and the investigation of some hemolytic anemias resulting from red cell intracorpuscular defects.

Specimen Whole blood

Container Lavender top (EDTA) tube for venipuncture specimen; lavender top Microtainer™ tube for capillary specimen

Causes for Rejection Specimen clotted

Reference Interval Hemoglobin A: 95% to 98%; hemoglobin A₂: 1.5% to 3.5%; hemoglobin F: 0% to 2%; hemoglobin C: absent; hemoglobin S: absent

Use Diagnose hemoglobinopathies; evaluate hemolytic anemia; diagnose thalassemia; evaluate sickling hemoglobins, hemoglobin C; with other and specialized techniques, evaluate unstable, low and high oxygen affinity hemoglobinopathies (the latter representing one cause of polycythemia)

Limitations Blood transfusion prior to hemoglobin electrophoresis may make interpretations inconsistent. Many abnormal hemoglobins do not separate from normal adult Hb A during application of routine electrophoretic techniques. Rarely, there may be lack of specificity for Hb (eg, as may occur when a monoclonal immunoglobulin is present).

Additional Information In this procedure, hemoglobin (Hb), released from lysed red blood cells, is caused to migrate through a substrate in a buffer by application of an electric current. Different types of hemoglobin are separated into specific bands that are subsequently visualized by application of one of a variety of staining procedures. Migration of hemoglobin is defined by interaction of a specific hemoglobin molecule with substrate structure, buffer pH, ionic strength, and other characteristics. By referring to the accompanying table of patterns for cellulose acetate (alkaline pH), one can determine that Hb A, in the allotted time period, migrates relatively far towards the anode (+ charged pole). It is thus "faster" in migration than most other hemoglobins. Even more anodal migrating are the fast Hbs H, I, and Bart's. Intermediate migrating Hbs are D, G, S, and Lepore while slow Hb include C, E, A_2, and O. The test is of central importance in establishing the presence of common hemoglobinopathies (Hb S, C, D, and E) and in the evaluation of some cases of hemolytic anemia.

Hemoglobin Electrophoresis Pattern

Sample Identity	Cellulose Acetate Alkaline, pH 8.4-8.6							Citrate Agar Gel Acid, pH 6.0-6.2			
	Origin	Carbonic Anhydrase	A_2, C, E, O	S, D, G	F	A_1	Ratio of S/A	C	S	A_1, A_2, D, G, E	F
Control	I	I	■	■		■	-	■	■	■	
Normal (A_1, A_2)	I	I	I							■	
Cord Blood	I	I			■					■	■*
Sickle Cell Disease (SCD)	I	I		■	I*		100/0		■		■*
Sickle Cell Trait (SA_1A_2)	I	I	■	■		■	40/60		■	■	
SC Disease	I	I	■	■			-	■	■		
SE Disease	I	I	■	■					■		
S/B⁺ Thalassemia	I	I	I	■		■	60/40		■	■	
S/B° Thalassemia	I	I	I	■	I		100/0		■		■
CC	I	I	†I■		I			■			
C Trait	I	I	†I■			■		■		■	
C_Harlem Trait	I	I	†I■		I	■			■	■	
D Trait	I	I	■		I	■				■	

*Amount varies   †A_2 may be slightly separate

Study of peripheral blood smear RBC morphology can assist in the decision to order hemoglobin electrophoresis. Hemoglobin electrophoresis (including determination of Hb F) is indicated if a positive sickle screening test has been obtained. Hb F and Hb A_2 quantitation (often included as part of hemoglobin evaluation) are important in establishing the presence of thalassemia. In some cases, additional study will be needed. Depending upon the abnormality encountered this might include reticulocyte count, haptoglobin level, citrate agar gel electrophoresis at acidic pH, globin chain electrophoresis, mRNA studies, and family studies.

Complete characterization of an abnormal hemoglobin may require sophisticated laboratory studies usually available only in a research setting (eg, as with some of the thalassemia syndromes). Amino acid chain sequencing may be used to establish the presence of a hemoglobinopathy.

Hemoglobin electrophoresis of umbilical cord blood can detect α-chain variants Hb F/G and Hb G as well as S and C gene products. Results have been reported as consistent with the predicted frequency (1 in 625) of sickle anemia at birth. Techniques of analysis, such as globin chain electrophoresis, isoelectric focusing, HPLC, amino acid sequencing, restriction endonuclease studies, and polymerase chain reaction (PCR) are additional methods, focused at the genetic level, allowing the laboratory to detect and identify hemoglobin variants. A multiplex PCR for detection of α-thalassemia (caused by α-globin gene deletion - the most common genetic abnormality in the world) is said to be applicable to routine performance by clinical laboratories. (See also DNA-Probe Assay for Thalassemia (BeTha Gene Test) on page 520.)

Some alkaline gel electrophoretic systems may offer advantages in glycohemoglobin quantitation because of their ability to discriminate Hb F and to simultaneously detect common hemoglobinopathies. Cation exchange HPLC has delineated over 10 minor hemoglobins including Hb A_{1c} and Hb A_{1d3} with evidence that the latter is useful for assessment of the uremic state.

Most states screen newborns for SCD. Problem areas include false-negative results (associated with the use of dried blood filter paper samples) and (Continued)

Hemoglobin Electrophoresis *(Continued)*

maternal contamination of cord blood. Molecular genetic analysis-based screening programs have been proposed in some cases in consideration of the lifetime cost of treatment (eg, in cases of β-thalassemia major). The California Department of Health Services, Genetic Disease Laboratory has determined the distribution of Hb F, A, S, C, E, and D quantities in 4 million dried blood spot newborn screening specimens using an automated 2-minute cation-gradient HPLC method.

A number of interactions between pathologic hemoglobin and glycated hemoglobin (Hb A_{1c}), the latter used as a measure of long-term blood glucose control in diabetic patients, have been described. A comparison of Hb A_{1c} results using cation exchange chromatography (Bio-Rad Variant) vs a method based on an immunological reaction (Bayer® DCA 2000) found close agreement of Hb A_{1c} results on specimens from individuals with Hb S, C, D, and E trait. A mathematical correction, required for correlation of Hb A_{1c} Bio-Rad Variant results in individuals with D trait has required application of a new correction calculation or use of an immune-based methodology (eg, Bayer® DCA 2000) or of a fructosamine measurement as an alternate to diabetes control (when Hb D or Hb $D_{Los Angeles}$) causes diagnostic confusion as a result of falsely low or high Hb A_{1c}.

New Hb variants continue to be described, some, such as hemoglobin Old Dominion/Burton-Upon-Trent (reported in four unrelated persons of Irish or Scots-Irish ancestry) may cause spurious increase in glycated Hb A_{1c} (with use of ion-exchange chromatography methods). The amino acid substitution is β-143 (H21) His→Tyr which affects a 2,3-diphosphoglycerate binding site with minimal increase in oxygen affinity but without clinical erythrocytosis. Hemoglobin Etobicoke, an alpha chain substitution, α-84 (F5) Ser-Arg may also compromise the measurement of Hb A_{1c}, as may beta chain substitutions at position 66 [β66(E10)].. The reference by Elder et al (as of April 1998) provides a table (with references) of 26 hemoglobin variants and Hb F that interfere with glycated Hb measurements by ion-exchange chromatography.

Hb A_{1c}, a minor Hb variant, is formed *in vivo*, a post-translational modification by glucose with glycated N-terminal β-chains. A recent sophisticated analysis using electrospray ionization mass spectrometry affirmed a curvilinear relationship of patient glucose with Hb A_{1c}. Multiple species of glycated Hb, however, cochromatographed with Hb A_{1c} on cation exchange. As glycation increased, β:α-chain ratio glycation increased, and the number of β-chain glycation sites increased. Glycated and nonglycated, both α- and β-chain varieties including some multiply glycated β-chains, were encountered, complicating the standardization of glycated hemoglobin clinical assays.

Pulse oximetry, used as a noninvasive method for monitoring arterial oxygen saturation, may produce misleading results in individuals who harbor an abnormal hemoglobin of the type that causes abnormal oxygen binding (eg, hemoglobin Rothschild).

References

Arcasoy MO and Gallagher PG, "Molecular Diagnosis of Hemoglobinopathies and Other Red Blood Cell Disorders," *Semin Hematol*, 1999, 36(4):328-39.

Bain JB, *Haemoglobinopathy Diagnosis*, Malden, MA: Blackwell Science Ltd, 2001.

Bruns CM, Thet LA, Woodson RD, et al, "Hemoglobinopathy Case Finding by Pulse Oximetry," *Am J Hematol*, 2003, 74(2):142-3.

Bunn HF, "Human Hemoglobins: Sickle Hemoglobin and Other Mutants," *The Molecular Basis of Blood Diseases*, 3rd ed, Chapter 7, Stamatoyannopoulos G, Majerus PW, Perlmutter RM, et al, eds, Philadelphia, PA: WB Saunders Co, 2001, 227-73.

Eastman JW, Lorey F, Arnopp J, et al, "Distribution of Hemoglobin F, A, S, C, E, and D Quantities in 4 Million Newborn Screening Specimens," *Clin Chem*, 1999, 45(5):683-4.

Elder GE, Lappin TR, Horne AB, et al, "Hemoglobin Old Dominion/Burton-Upon-Trent, β143 (H21) His→Tyr, Codon 143 CAC→TAC - A Variant With Altered Oxygen Affinity That Compromises Measurement of Glycated Hemoglobin in Diabetes Mellitus: Structure, Function, and DNA Sequence," *Mayo Clin Proc*, 1998, 73(4):321-8.

Friess U, Beck A, Kohne E, et al, "Novel Hemoglobin Variant [β66 (F10) Lys → Asn], With Decreased Oxygen Affinity, Causes Falsely Low Hemoglobin A_{1c} Values by HPLC," *Clin Chem*, 2003, 49(8):1412-5.

McCabe ER and McCabe LL, "State-of-the-Art for DNA Technology in Newborn Screening," *Acta Paediatr Suppl*, 1999, 88(432):58-60.

Nagel RL, "Haemoglobinopathies Due to Structural Mutations," *Molecular Haematology*, Chapter 10, Provan D and Gribben J, eds, Malden, MA: Blackwell Science Ltd, 2000, 121-33.

O'Brien DA, Flynn CM, Chuk P, et al, "Haemoglobin Etobicoke, an Incidental Finding in an Irish Diabetic," *Clin Lab Haematol*, 2003, 25:259-62.

Veillon DM, Kaltenbach JE, Hall CG, et al, "Assays for Hemoglobin S," *Lab Med*, 2000, 31(2):68-9.

Weatherall DJ and Clegg JB, "Genetic Disorders of Hemoglobin," *Semin Hematol*, 1999, 36(4 Suppl 7):24-37.

♦ **Hemoglobin, Fetal** *see* Fetal Hemoglobin *on page 581*

♦ **Hemoglobin Munchausen** *see* Sickle Cell Tests *on page 1195*

Hemoglobin, Plasma

Related Information

Haptoglobin, Serum *on page 667*

Hemoglobin *on page 681*

Hemoglobin, Qualitative, Urine *on page 688*

Transfusion Reaction Work-up *on page 1269*

Synonyms Free Hemoglobin

Abstract Test used to detect intravascular hemolysis. Increase of hemoglobin in the plasma is a reliable sign of intravascular hemolysis only if lysis of red blood cells during and after obtaining the sample (ie, traumatic puncture) can be definitely excluded. Plasma hemoglobin levels are used to assess damage to blood cells (erythrocytes) caused by a variety of medical devices.

Patient Preparation Special precautions and patient preparation are usually required to draw the specimen. Laboratory should be contacted directly.

Aftercare A pressure bandage should be applied to the site of 18-gauge needle puncture (following the puncture) to stop residual bleeding.

Specimen Plasma

Container Lavender top (EDTA) tube

Collection Recommended procedure for collecting sample without inducing hemolysis: Use 18-gauge needle with attached infusion tubing. Observing HIV precautions, place tourniquet lightly around the upper arm. Puncture antecubital vein with as little trauma as possible. Release tourniquet and clamp tubing off as soon as blood return is seen. Collect 3 mL of blood first in a red top tube with the rubber stopper off. Follow by a 5 mL collection in a green top (heparin) tube with the stopper off. Clamp tubing, withdraw needle, and apply pressure to the site until residual bleeding is stopped. Cap green top tube and gently mix three to five times. Use this specimen for the plasma hemoglobin determination.

Storage Instructions Separate and freeze plasma as soon as possible if test is not run immediately.

Causes for Rejection Traumatic venipuncture causing hemolysis

Reference Interval Under optimal conditions (with absence of hemolysis due to collection of blood) <0.6 mg/dL (<6 mg/L). Practically (allowing for minimal hemolysis during collection), a plasma (or serum) level of hemoglobin >20 mg/dL (200 mg/L) is likely indicative of a pathologic hemolytic process.

Use Evaluate hemolytic anemia, especially intravascular hemolysis; evaluate red cell lysis caused by medical devices (cardiopulmonary bypass machines, left ventricular assist devices, pumps, pulsatile assist devices); evaluate use of human hemoglobin-derived blood substitutes

Limitations Plasma hemoglobin is increased with intravascular hemolysis, in particular as occurs with paroxysmal nocturnal hemoglobinuria, paroxysmal cold hemoglobinuria, the cold-hemagglutinin syndrome, and "blackwater fever" but also as occurs with ABO incompatible transfusion, traumatic hemolysis, falciparum malaria, burns, and march hemoglobinuria. Increase may occur in some cases of extravascular hemolysis, delayed transfusion reaction, with slight increase in sickle cell anemia, and β-thalassemia. In hereditary spherocytosis (hemolysis occurs largely in the spleen), plasma hemoglobin levels are normal or only slightly increased. High bilirubin, turbidity, methemalbuminemia (method measures methemalbumin and hemoglobin together), lipemic plasma, and hemolysis during or after venipuncture may cause falsely elevated values in the plasma hemoglobin test (method based on peroxide oxidation of benzidine). Use of benzidine has been restricted because of reports that it is carcinogenic. An alternative is the use of tetramethylbenzidine which is more readily available but which should also be handled with care.

Methodology An established method utilizes fractional absorbance of oxyhemoglobin at 578 nm. A method utilizing first-derivative spectroscopy (procedure of Soloni et al) has undergone evaluation. It is rapid and not affected by (Continued)

Hemoglobin, Plasma *(Continued)*

bilirubin, myoglobin, lipemia, or turbidity. It is sensitive down to a level of 1 mg/dL.

Additional Information High bilirubin (up to 36 mg/dL), turbidity of the specimen, or a fair amount of methemalbumin will not affect the method based on fractional absorbance of oxyhemoglobin at 578 nm or the method of Soloni et al. Hemoglobinemia can be detected by gross examination of centrifuged blood. Plasma will have a pink tinge when hemoglobin is present in a concentration ≥200 mg/L.

A variety of pump-based devices are employed to support cardiovascular operative procedures. Plasma hemoglobin measurements assess trauma to the circulating cells. The U.S. Food and Drug Administration (FDA) found over 20 different assays in clinical use for measurement of plasma hemoglobin. They studied nine currently used plasma hemoglobin assays for accuracy, reproducibility, sensitivity, interference effects and ease of use. The older direct optical techniques (eg, Cripps method from 1968) had overall good performance while "added chemical" methods (eg, cyanmethemoglobin method) were less satisfactory. The Soloni et al procedure noted above is a "first-derivative absorbance" method and was found equivalent (by FDA) to the Cripps procedure.

The plasma hemoglobin level along with haptoglobin, indirect bilirubin, LD isoenzymes, reticulocyte count, and perhaps other clinical laboratory studies can be applied to evaluate parts of the spectrum of clinical hemolytic processes.

References

Dacie JV and Lewis SM, "Laboratory Methods Used in the Investigation of the Haemolytic Anaemias," *Practical Haematology*, 8th ed, Chapter 12, Dacie JV and Lewis SM, eds, New York, NY: Churchill Livingstone, 1995, 199-201.

Lee GR, "The Hemolytic Disorders: General Considerations," *Wintrobe's Clinical Hematology*, 10th ed, Volume 1, Chapter 40, Lee GR, Foerster J, Lukens J, et al, eds, Philadelphia, PA: Lippincott Williams & Wilkins, 1999, 1109-31.

Malinauskas RA, "Plasma Hemoglobin Measurement Techniques for the *In Vitro* Evaluation of Blood Damage Caused by Medical Devices," *Artif Organs*, 1997, 21(12):1255-67.

Hemoglobin, Qualitative, Urine

Related Information

Anemia Flowchart *on page 35*
Blood, Urine *on page 281*
Glomerular Basement Membrane Antibody *on page 638*
Hemoglobin, Plasma *on page 687*
Hemosiderin Stain, Urine *on page 692*
Kidney Biopsy *on page 818*
Myoglobin, Qualitative, Urine *on page 941*
Transfusion Reaction Work-up *on page 1269*
Urinalysis *on page 1289*

Test Includes Dipstick screening of urine for hemoglobin

Abstract When the urine dipstick is positive for "blood", the differential includes red cells in the urine (hematuria), hemoglobinuria, and myoglobinuria. Microscopy, preferably phase-contrast, of the urine sediment should follow such a positive dipstick test.

Specimen Random urine

Reference Interval Negative

Use A positive dipstick test for "blood" should be followed by microscopy to determine whether red cells are present. If no red cells are found, the differential includes hemoglobin vs myoglobin. Patients with hemoglobinuria will usually have pink plasma and a positive test for plasma hemoglobin; these findings are absent in myoglobinuria.

Limitations False-positive dipstick tests for "blood" are caused by iodine, bleach (hypochlorite), menstrual blood, and bacterial peroxidases. False-negative dipstick results are caused by vitamin C, proteinuria, nitrites, Pyridium® (phenazopyridine) and Serenium® (ethoxazene hydrochloride), and a prolonged delay before testing (especially if the specimen is not refrigerated).

Methodology Hemoglobin peroxidase catalyzes oxidation of a chromogen.

Additional Information Hemoglobinuria is consonant with intravascular hemolysis, such as that caused by a hemolytic transfusion reaction. Myoglobinuria often occurs in clinical settings so distinctive that further testing is not needed. Serum myoglobin is usually elevated in myoglobinuria.

See Blood, Urine *on page 281.*

References
Bartlett RC, Zern DA, Ratkiewicz I, et al, "Reagent Strip Screening for Sediment Abnormalities Identified by Automated Microscopy in Urine From Patients Suspected to Have Urinary Tract Disease," *Arch Pathol Lab Med*, 1994, 118(11):1096-101.

Hemoglobin, Unstable, Heat Labile Test

Related Information

Heinz Body Stain *on page 669*
Hemoglobin, Unstable - Isopropanol Precipitation Test *on page 690*
Hemosiderin Stain, Urine *on page 692*
n-Butanol Stability Test *on page 945*

Synonyms Unstable Hemoglobins

Applies to Hb Köln; Hb Zurich; Heinz Bodies

Abstract A simple, low cost test for detection of unstable hemoglobins. A variety of mutations (>75% involve the β-chain) are responsible for this inherited condition (some, however, are the result of spontaneous mutations). The resultant "unstable" hemoglobins impair hemoglobin solubility or increase susceptibility to oxidation of globin chains. Dependent upon the specific mutation, globin chain folding abnormality, α-β-interaction, and/or globin to heme destabilization occurs with formation of Heinz bodies (precipitates of α- and β-globin chains, globin fragments, and heme). Most of such variant hemoglobins will not separate from Hb A on routine electrophoresis.

Specimen Whole blood

Container Lavender top (EDTA) tube

Causes for Rejection Specimen more than 4 hours old

Reference Interval Less than 1% unstable hemoglobin (quantitation by hemoglobinometry of centrifuged lysate). Normally, test should result in little or no precipitate. A positive result (denatured hemoglobin present) is the presence of turbidity and/or fine flocculation, a readily visible precipitate.

Use Determine the presence of unstable hemoglobins, most of which will not be identified by routine hemoglobin electrophoresis

Limitations False-negative results may occur if only a small amount of abnormal hemoglobin is present. The visual end point in this test may be difficult to interpret. Test should be run along with a normal control. Some degree of slight precipitation may occur in an erratic manner in normals.

Methodology Washed cells lysed, acidified, lysate heated at 50°C for 1 hour and examined for turbidity/flocculation (compared to control). Percent unstable hemoglobin may be reported.

Additional Information Another approach to detection of unstable hemoglobins is to search for Heinz bodies. Heinz body test is less specific than heat instability study. Hemolysates containing unstable hemoglobin may precipitate spontaneously on standing a few days in the refrigerator. Some unstable hemoglobins are associated with hemolytic anemia and an appropriate clinical picture including intermittent jaundice and usually splenomegaly. If hypersplenism is present there may be thrombocytopenia. There have been many different unstable hemoglobins reported after the identification of Hb Köln as a cause of "congenital Heinz body anemia". Lukens and Lee in the 10th edition (1999) of *Wintrobe's Clinical Hematology* indicate that some 200 different unstable hemoglobin mutants have been identified. The most common unstable Hb variant is Hb Köln which has wide geographic distribution. Hb Köln is the result of replacement of the amino acid valine by methionine at position 98 of the β-chain (a β98 val → met substitution). A number of unstable hemoglobins have altered oxygen affinity, with Hb Köln, affinity is increased. Splenectomy (performed in some cases of unstable hemoglobin disease to ameliorate number and severity of hemolytic episodes) may (if the hemoglobin also has high oxygen affinity) result in postsplenectomy polycythemia and thrombotic vascular complications. Hb Zurich (a β63 His → Arg substitution), is an unstable hemoglobin with drug-induced (usually sulfonamide) hemolytic anemia that is ameliorated by smoking. Nearly all unstable hemoglobin (Continued)

Hemoglobin, Unstable, Heat Labile Test *(Continued)*

diseases have an autosomal dominant mode of inheritance; most patients are heterozygotes.

References

Bunn HF, "Human Hemoglobins: Sickle Hemoglobin and Other Mutants," *The Molecular Basis of Blood Diseases*, 3rd ed, Chapter 7, Stamatoyannopoulos G, Majerus PW, Perlmutter RM, et al, eds, Philadelphia, PA: WB Saunders Co, 2001, 227-73.

Lukens J and Lee GR, "Unstable Hemoglobin Disease," *Wintrobe's Clinical Hematology*, Lee GR, Foerster J, Lukens J, et al, eds, 10th ed, Volume 1, Chapter 52, Baltimore, MD: Lippincott Williams & Wilkins, 1999, 1398-404.

Thuret I, Bardakdjian J, Badens C, et al, "Priapism Following Splenectomy in an Unstable Hemoglobin: Hemoglobin Olmsted β141 (H19) Leu→Arg," *Am J Hematol*, 1996, 51(2):133-6.

Hemoglobin, Unstable - Isopropanol Precipitation Test

Related Information

Hein Body Stain *on page 669*
Hemoglobin, Unstable, Heat Labile Test *on page 689*
n-Butanol Stability Test *on page 945*

Synonyms Unstable Hemoglobins

Applies to Hb Zurich; Heinz Bodies

Abstract A simple, low cost test for unstable hemoglobins. Most of such variant hemoglobins will not be detected by routine hemoglobin electrophoresis as they migrate with Hb A.

Specimen Whole blood

Container Lavender top (EDTA) tube

Storage Instructions Fresh blood should be used. Sample must be no more than 1 week old.

Reference Interval Absence of a precipitate in buffered isopropanol. Solution should remain clear for 30-40 minutes.

Use Differential diagnosis of hemolytic anemias; detect unstable hemoglobins, many of which will not be identified (separated from Hb A) by standard electrophoretic techniques

Limitations Hemoglobin F begins to precipitate about halfway through the incubation period - about the time that one expects unstable Hb to appear. If the patient's Hb F is increased (>4%), a false-positive result for unstable Hb may result. Unstable Hb Zurich has increased affinity for carbon monoxide which stabilizes the molecule against denaturating effect of isopropanol. The rate of isopropanol-induced precipitation varies inversely with carboxyhemoglobin levels in patients with Hb Zurich.

Methodology Mix 2 mL of 17% (unit/unit) isopropanol with 0.2 mL of hemolysate. Incubate at 37°C; check for precipitate at 20 minutes.

Additional Information Heat lability, n-butanol, and Heinz body tests are also applicable to the detection and study of unstable hemoglobins. While simple to perform, interpretation may be difficult. A multitest approach has merit.

References

Brozovic M and Henthorn J (revised by), "Investigation of Abnormal Haemoglobins and Thalassemia," *Practical Haematology*, 8th ed, Chapter 14, Dacie JV and Lewis SM, eds, New York, NY: Churchill Livingstone, 1995, 268-70.

Lukens JN and Lee GR, "Unstable Hemoglobin Disease," *Wintrobe's Clinical Hematology*, 10th ed, Volume 1, Chapter 52, Lee GR, Foerster J, Lukens J, et al, Philadelphia, PA: Lippincott Williams & Wilkins, 1999, 1401.

♦ **Hemoglobin, Urine** *see Blood, Urine on page 281*

♦ **Hemolysis** *see page 11*

Hemolytic Disease of the Newborn, Antibody Identification

Related Information

Antibody Detection/Identification, Red Cell *on page 165*
Antiglobulin Test, Indirect *on page 177*
Bilirubin, Amniotic Fluid, Delta A450 *on page 261*
Bilirubin, Neonatal, Serum *on page 263*
Cord Blood Antibody Screen *on page 453*
Newborn Crossmatch and Transfusion *on page 952*
Rh Genotype *on page 1162*
Rh₀(D) Immune Globulin (Human) *on page 1164*

Synonyms Newborn/Maternal Antibody Work-up

Applies to Exchange Transfusion

Test Includes Infant ABO group, Rh type; direct antiglobulin test; antibody elution, antibody identification. Mother ABO group, Rh type; antibody detection/identification.

Abstract Hemolytic disease of the newborn (HDN) takes place when fetal red cells cross the placenta and enter the maternal circulation, with immunization of the mother to a fetal red cell antigen absent from her own erythrocytes. Production of maternal IgG antibodies causes hemolysis of fetal red cells. Fetomaternal (transplacental) hemorrhage is detectable in about 5% of pregnant women. Other events that enhance the risk of transplacental hemorrhage include spontaneous and therapeutic abortion, ectopic gestation, amniocentesis, intrauterine surgery, abdominal trauma, and peripartum hemorrhage.

Maternal antibody specificity influences therapeutic decisions, especially selection of blood for exchange transfusion and whether an Rh-negative mother is a candidate for Rh immunoprophylaxis. See Newborn Crossmatch and Transfusion *on page 952*.

Specimen Blood from infant and mother (collect maternal sample before administration of Rh immune globulin)

Container Lavender top (EDTA) tube and red top tube(s)

Causes for Rejection Gross hemolysis, sample collected in serum separator tube, improper labeling

Use Investigation of HDN, selection of blood for exchange transfusion

Limitations ABO HDN may not be detected by testing at birth. It may be indicated by subsequent clinical impression. Maternal antibody to a "private" paternal antigen may not be demonstrated with standard red cell antibody identification tests. Wharton's jelly may interfere with infant ABO tests. It may prove difficult to interpret results of Rh type of infant red cells heavily sensitized with IgG maternal antibodies. The presence of large numbers of maternal red cells in a cord blood sample can confuse interpretation of blood grouping results. Similarly, serological results can be complicated in newborns that received intrauterine blood transfusions. Administration of antenatal Rh immune globulin may complicate interpretation of cord blood DAT results.

Additional Information The fetal-maternal relationship presents special immunohematological problems for the Blood Bank Laboratory which are important for patient care. The results of initial cord blood screening tests will indicate need for additional studies to determine the likelihood of HDN or need for Rh immunoprophylaxis. Administration of postpartum Rh immune globulin will depend on maternal alloimmunization and the results of infant Rh type.

A positive DAT is an important indicator of HDN. A positive result will evolve into antibody identification studies of maternal serum or infant red cell eluate. Antibody specificity is an important guide to disease severity. Most severe HDN is caused by anti-D alone or in combination with other Rh antibodies, anti-C, or anti-E. The next grade of severity is associated with antibodies to other Rh antigens (notably anti-c) or with antibodies to antigens in other blood group systems (eg, anti-K).

IgG antibodies to antigens of the ABO system usually cause the lowest grade of HDN severity. ABO HDN can occur in any pregnancy, even the first. The DAT is often negative and the infant is rarely symptomatic at birth.

Blood group antibodies other than those in the ABO group and Rh type also cause HDN, which may be even more dangerous because they may be unsuspected. They include antibodies directed against Duffy, Kidd, and MNSs systems.

Fetal genotyping can be performed.

References

Bowman JM, Pollock JM, and Edmonds LD, "Epidemiology of Rh Disease of the Newborn in the United States," *JAMA*, 1991, 265(24):3270-4.

Brecher ME, *Technical Manual*, 14th ed, Bethesda, MD: American Association of Blood Banks Press, 2002, 497-515.

Lo YMD, Hjelm NM, Fidler C, et al, "Prenatal Diagnosis of Fetal RhD Status by Molecular Analysis of Maternal Plasma," *N Engl J Med*, 1998, 339(24):1734-8.

Luban NLC, "Hemolytic Disease of the Newborn: Progenitor Cells and Late Effects," *N Engl J Med*, 1998, 338(12):830-1.

Saade GR, "Noninvasive Testing for Fetal Anemia," *N Engl J Med*, 2000, 342(1):52-3.

Hemosiderin Stain, Urine

Related Information

Synonyms Iron Stain, Urine

Abstract Hemosiderinuria is a marker for intravascular hemolysis including importantly, chronic hemolytic anemia. With hemolysis, hemoglobin passes the glomerulus, is present in the glomerular filtrate, and is absorbed by renal tubular epithelial in which it is converted to hemosiderin. Thus, hemosiderin may not be found in the urine at the onset of acute hemolysis or a hemolytic episode, even if hemoglobinemia and hemoglobinuria are present (unless there has been a previous hemolytic episode). On the other hand, a few days after a hemolytic attack, iron-containing granules may be present in the urine (hemosiderinuria), even in the absence of hemoglobinuria.

Specimen Random urine

Container Use iron-free container and centrifuge tubes if results of initial examination are difficult to interpret.

Storage Instructions Refrigeration

Reference Interval There should be absence of stainable iron in normal urine. Normal value for urinary iron excretion is <0.1 mg/day.

Use A sign of recent intravascular hemolysis, hemosiderinuria is an important clue in the diagnosis of unexplained anemia. Urine sediment stained for hemosiderin is a screen for increased iron excretion, which can be quantitated and is increased in hemochromatosis, hemolytic anemia, and nephrotic syndrome.

Limitations Hemosiderin is first shed in the urine a few days after hemolysis begins. Hemosiderinuria declines slowly after hemolysis stops. Quantitatively, iron excretion remains slightly elevated for months after replacement of a cardiac valve which has caused hemolytic anemia, although hemosiderinuria is usually absent after a few weeks.

Methodology Microscopic examination of slide stained for hemosiderin (Prussian blue reaction). Environmental iron contaminants must be distinguished from true iron-positive granular deposits of intracellular origin. The inexperienced observer may generate a false-positive report. A true-positive test result shows blue-staining granules in an intracellular situation in intact tubular epithelial cells. Hemosiderin may appear as a brown coarsely granular pigment in tubular epithelial cells, epithelial cell casts, or free in the urine sediment.

Additional Information When haptoglobin is saturated, part of the hemoglobin in the plasma is filtered by the glomerulus and presented to the renal tubular cell. If tubular capacity is not exceeded, all hemoglobin in the urine may be absorbed by the proximal tubular cells, where hemoglobin iron is converted to hemosiderin. When these cells are shed, hemosiderin appears in urine sediment. Hemosiderinuria with or without hemoglobinuria may be seen in chronic hemolytic anemia, paroxysmal nocturnal hemoglobinuria, hemochromatosis, multiple transfusions, and other conditions which result in deposition of iron in the renal parenchyma. Hemosiderinuria may occur when the degree of hemolysis is at such a low level that hemoglobinuria is not detected.

With intravascular hemolysis, urinary iron values increase to between 3-11 mg/day. Pernicious anemia and hereditary spherocytosis have normal values.

References
CRC Desk Reference for Hematology, Shinton NK, ed, Boca Raton, FL: CRC Press, 1998, 320-1.

Dacie Sir JV and Lewis SM, "Laboratory Methods Used in the Investigation of the Haemolytic Anaemias," *Practical Haematology,* 8th ed, Chapter 12, 1995, 204-5.

Lee GR, "Hemolytic Disorders: General Considerations," *Wintrobe's Clinical Hematology,* 10th ed, Volume 1, Chapter 40, Lee GR, Foerster J, and Lukens J, eds, Baltimore, MD: Lippincott Williams & Wilkins, 1999, 1119-20.

Heparin Antifactor Xa Assay

Related Information
Activated Clotting Time *on page 98*
Activated Partial Thromboplastin Time *on page 100*
Antithrombin *on page 198*
Heparin Neutralization *on page 697*
Thrombin Time *on page 1246*

Synonyms Antifactor Xa Assay; Anti-Xa Assay; Heparin Assay

Applies to Danaparoid; Heparin; LMWH; Low-Molecular Weight Heparin; Organ®; PF4; Platelet Factor 4; PTT

Abstract Two relatively new anticoagulants, low-molecular weight heparin (LMWH) and danaparoid (Organ®), when present at therapeutic levels, usually do not significantly prolong the activated partial thromboplastin time (PTT). Therefore, when laboratory tests are used to monitor therapeutic anticoagulant levels of LMWH or danaparoid, antifactor Xa assays are necessary. In addition, in some instances the PTT cannot be used to monitor unfractionated heparin. For example, lupus anticoagulants* or certain factor deficiencies (eg, factor XII deficiencies) may prolong the baseline PTT and/or accentuate the PTT prolongation when heparin is added. In these cases, unfractionated heparin may be monitored with antifactor Xa assays (*Note: if the antifactor Xa assay demonstrates that the heparinized PTT is not affected by the lupus anticoagulant, cautious use of the PTT may be tried in that patient).

Specimen Plasma

Container One blue top (sodium citrate) tube

Sampling Time Draw specimen 4 hours after subcutaneous injection of LMWH or 6 hours after subcutaneous injection of danaparoid, otherwise, falsely low values may occur. The therapeutic antifactor Xa ranges with subcutaneous LMWH and danaparoid are defined for the peak levels.
(Continued)

Heparin Antifactor Xa Assay *(Continued)*

Collection Routine venipuncture. Deliver tube to laboratory immediately, otherwise falsely low values may occur (because platelets release platelet factor 4 (PF4) which can neutralize heparin, LMWH, or danaparoid). If multiple tests are being drawn, draw blue top tubes after any red top tubes but before any lavender top (EDTA), green top (heparin), or gray top (oxalate/fluoride) tubes. Immediately invert tube gently at least 4 times to mix. Tubes must be appropriately filled.

Storage Instructions Separate plasma from cells as soon as possible, ideally within 1 hour of specimen collection. Otherwise, falsely low values may occur (because platelets release PF4, which can neutralize heparin, LMWH, or danaparoid). Plasma can be stored for 2 hours at room temperature or on ice; otherwise, store frozen.

Turnaround Time 1 day, unless testing is batched less frequently

Special Instructions Notify the laboratory specifically as to which drug should be measured (heparin, LMWH, or danaparoid), because the laboratory must construct a standard curve using the same drug that the patient is receiving.

Reference Interval

Patients not on anticoagulants: 0 units/mL

Therapeutic range for **treatment** of existing deep venous thrombosis (DVT):
- heparin 0.3-0.7 units/mL
- LMWH: 0.4-1.1 units/mL for twice daily subcutaneous dosing. For once daily subcutaneous LMWH dosing, the therapeutic range is less certain but is approximately 1-2 units/mL.
- danaparoid: 0.5-0.8 units/mL

Target range for deep vein thrombosis (DVT) prophylaxis (prevention): There is no defined target range for **prophylaxis** of deep vein thrombosis (DVT) because such anticoagulation is not usually monitored. When anti-Xa levels have been measured, mean values have been <0.45 units/mL.

Use Determine if the patient is at the desired level of anticoagulation with therapeutic doses of heparin, LMWH, or danaparoid

Limitations More expensive and less readily available than the PTT for heparin monitoring

Methodology Chromogenic. Patient plasma is added to a known amount of excess factor Xa with excess antithrombin. If heparin (or LMWH or danaparoid) is present in the patient plasma, it will bind to antithrombin and inhibit factor Xa. The amount of residual factor Xa is inversely proportional to the amount of heparin in the plasma. The amount of residual factor Xa is detected by adding a chromogenic substrate that resembles the natural substrate of factor Xa. Factor Xa cleaves the chromogenic substrate, releasing a colored compound that can be detected by a spectrophotometer. Results are reported as anticoagulant concentration in antifactor Xa units/mL, such that high antifactor Xa values indicate high levels of anticoagulation and low antifactor Xa values indicate low levels of anticoagulation. Deficiencies of antithrombin in the patient do not affect the assay, because excess antithrombin is provided in the reaction.

Additional Information Therapeutic doses of unfractionated heparin require intense laboratory monitoring, because the amount of *in vivo* anticoagulation for a given dose is variable. That is, the dose-response for heparin is unpredictable. In contrast, LMWH and danaparoid do have a predictable dose-response, therefore, laboratory monitoring is usually not essential. In fact, if a LMWH or danaparoid antifactor Xa level is subtherapeutic, the most common causes are drawing the specimen at the wrong time (see below) or specimen transportation longer than 2 hours. Most of the time, LMWH and danaparoid antifactor Xa levels are in the appropriate range when specimens are drawn correctly. Occasions in which periodic monitoring of LMWH might be considered include renal failure, pregnancy (increased dosage requirement in the third trimester), pediatric patients (increased dosage requirement in newborns), obesity, underweight patients, prolonged use, or patients at high risk for bleeding or thrombosis. It is probably also advisable to periodically monitor danaparoid in these same conditions.

The dose-response for unfractionated heparin is unpredictable because many of the heparin chains are long. The long chains can bind nonspecifically to a variety of proteins and cells, and the amounts of these heparin-binding proteins in particular vary considerably among patients, and even vary within the same patient at different times. In contrast, LMWH and danaparoid consist of shorter chains (ie, low-molecular weight) that have much less nonspecific binding.

Causes of subtherapeutic antifactor Xa level:
- specimen drawn at incorrect time (collection times are 4 hours after injection of LMWH, 6 hours after injection of danaparoid)
- specimen transportation longer than 2 hours
- patient receiving prophylactic dose, therefore, therapeutic range is not applicable and anti-Xa level is actually appropriate for dose
- higher dose needed (uncommon with LMWH or danaparoid, more common with heparin, eg, an acute phase state often increases the heparin dose requirement)

Causes of supratherapeutic antifactor Xa level:
- renal failure (with LMWH or danaparoid) (decreased renal clearance)
- heparin contamination, if specimen was drawn from a line
- lower dose needed (uncommon with LMWH or danaparoid, more common with heparin)

References

Laposata M, Green D, Van Cott EM, et al, "College of American Pathologists Conference XXXI on Laboratory Monitoring of Anticoagulant Therapy. The Clinical Use and Laboratory Monitoring of Low-Molecular Weight Heparin, Danaparoid, Hirudin and Related Compounds, and Argatroban," *Arch Pathol Lab Med*, 1998, 122(9):799-807.

Olson JD, Arkin CF, Brandt JT, et al, "College of American Pathologists Conference XXXI on Laboratory Monitoring of Anticoagulant Therapy. Laboratory Monitoring of Unfractionated Heparin Therapy," *Arch Pathol Lab Med*, 1998, 122(9):782-98.

Teien AN and Lie M, "Evaluation of an Amidolytic Heparin Assay Method: Increased Sensitivity by Adding Purified Antithrombin III," *Thromb Res*, 1977, 10:399-410.

♦ **Heparinase** see Heparin Neutralization *on page 697*
♦ **Heparin Assay** see Heparin Antifactor Xa Assay *on page 693*
♦ **Heparin Cofactor II** see Antithrombin *on page 198*
♦ **Heparin Cofactor II** see Hypercoagulation Panel *on page 758*

Heparin-Induced Thrombocytopenia
Related Information
Platelet Count *on page 1050*
Platelet Transfusion *on page 1058*
Warfarin, Serum or Plasma *on page 1325*

Synonyms HIT

Applies to Argatroban; Coumadin®; Danaparoid; Heparin; Hirudin; PF4; Platelet Factor 4; Serotonin Release Assays

Abstract Heparin-induced thrombocytopenia (HIT) is a common, serious complication of heparin therapy, with a high risk of potentially catastrophic venous or arterial thrombosis and high mortality.

Specimen Plasma (some laboratories may use serum)

Container Blue top (sodium citrate) tube for plasma (or red top tube if serum is requested); one tube suffices for ELISA or platelet aggregation, more tubes may be required for serotonin release assay.

Collection Routine venipuncture

Storage Instructions Plasma (or serum) can be stored for 24 hours at room temperature; otherwise, store frozen.

Turnaround Time 1 day, unless testing is batched less frequently

Special Instructions For serotonin release or platelet aggregation assays, notify laboratory if patient is receiving heparin. Such specimens should ideally not contain heparin. If heparin is present, it may be removed (adsorbed) by the laboratory prior to testing.

Reference Interval Negative for HIT antibody (HIT antibody not present)

Use Determine if thrombocytopenia or thrombosis in a patient exposed to heparin is due to heparin-induced thrombocytopenia

Limitations The antibody disappears after heparin is discontinued, usually within weeks to months but occasionally longer. Therefore, testing should be performed in the acute setting when HIT is presently suspected.

Methodology Three methods are commonly in use. Enzyme-linked immunosorbent assays (ELISA) use heparin complexed to platelet factor 4 (PF4) as the antigen. In platelet aggregation assays, patient plasma (or serum) is added to normal donor platelets and heparin. If the HIT antibody is present, it stimulates (Continued)

Heparin-Induced Thrombocytopenia (Continued)

the platelets to aggregate. In serotonin release assays, patient plasma (or serum) and heparin are added to normal platelets that contain radiolabeled serotonin. If the HIT antibody is present, it activates the platelets which then release their serotonin. The released radiolabeled serotonin can then be detected.

Additional Information Thrombocytopenia and thrombosis are the predominant clinical features of HIT. Despite the thrombocytopenia, bleeding complications are uncommon. Up to 8% of heparinized patients develop the antibody that causes HIT without becoming thrombocytopenic. Another 1% to 5% of patients on heparin progress further to HIT with thrombocytopenia, and of those, a third or more develop venous and/or arterial thrombosis. Thrombosis usually occurs only in HIT patients who are thrombocytopenic. However, thrombosis has been reported in HIT patients with normal platelet counts. Thrombosis in HIT is associated with a mortality of approximately 20% to 30%, with an equal number becoming permanently disabled by amputation, stroke, or other causes. HIT can develop from even small amounts of heparin, such as line flushes or heparin-coated catheters.

In patients receiving heparin for the first time, the platelet count begins to decrease in HIT 4-20 days after initiation of heparin exposure, most commonly between days 5 and 12, with the median on day 10. In patients who were sensitized to heparin in the past, platelet counts may decrease within the first 3 days or even hours after re-exposure to heparin. A progressive decline in platelet count >50% from baseline or to <100,000/µL is typical of HIT. The median nadir is 50,000/µL (range 20,000-150,000/µL). In patients developing HIT for the first time, the nadir is reached about 5 days after the onset of the decline, although this is variable. In previously sensitized patients, the nadir can be reached as soon as the first day or two after heparin re-exposure. After discontinuing heparin, the platelet count starts to rise after 2-3 days and usually returns to normal within 4-10 days. However, occasionally the recovery requires up to 25 days.

HIT is due to an antibody that recognizes heparin bound to platelet factor 4 (PF4) on the platelet surface. The antibody binds to the heparin-PF4 complex, which then allows the antibody to bind the Fc receptor on the platelet. Interaction with the Fc receptor activates the platelet, resulting in platelet loss (thrombocytopenia) and platelet aggregation (thrombosis). A minority of cases of HIT may involve an antigen other than the PF4-heparin complex.

Among the HIT tests, the ELISA is the most sensitive and platelet aggregation the least sensitive. The sensitivity of serotonin release or ELISA is ≥90%. Thus, a negative test for HIT does not rule out the diagnosis with complete certainty if HIT is suspected clinically. All three tests have high specificity. However, the significance of a positive ELISA in the absence of thrombocytopenia or thrombosis is uncertain. There is at least one case reported in which the antibody was detected by ELISA 5 days prior to the onset of thrombocytopenia.

Heparin should be permanently discontinued in HIT patients (special arrangements are made for patients who require bypass surgery). Platelet transfusions should be avoided. Patients with HIT are often treated with danaparoid, hirudin, or argatroban. Low-molecular weight heparin (LMWH) has a lower incidence of HIT than unfractionated heparin. However, the cross-reactivity of the HIT antibody against LMWH is high enough that LMWH is also contraindicated for HIT patients, now that the newer alternatives mentioned above are available. Coumadin® should not be used alone in the setting of acute HIT, because it may precipitate venous limb gangrene. If Coumadin® is used, an immediate-acting alternative anticoagulant (ie, hirudin, danaparoid, argatroban) should be used with it until Coumadin® is therapeutic.

References

Amiral J, Bridey F, Wolf M, et al, "Antibodies to Macromolecular Platelet Factor 4-Heparin Complexes in Heparin-Induced Thrombocytopenia: A Study of 44 Cases," *Thromb Haemost*, 1995, 73(1):21-8.

Greinacher A, "Antigen Generation in Heparin-Associated Thrombocytopenia: The Nonimmunologic Type and the Immunologic Type Are Closely Linked in Their Pathogenesis," *Semin Thromb Hemost*, 1995, 21(1):106-16.

Lindhoff-Last E and Bauersachs R, "Heparin-Induced Thrombocytopenia - Alternative Anticoagulation in Pregnancy and Lactation," *Semin Thromb Hemost*, 2002, 28(5):439-46.

Lubenow N and Greinacher A, "Hirudin in Heparin-Induced Thrombocytopenia," *Semin Thromb Hemost*, 2002, 28(5):431-8.

Warkentin TE, Chong BH, and Greinacher A, "Heparin-Induced Thrombocytopenia: Toward Consensus," *Thromb Haemost*, 1998, 79(1):1-7.

Heparin Neutralization

Related Information

Activated Clotting Time *on page 98*
Activated Partial Thromboplastin Time *on page 100*
Heparin Antifactor Xa Assay *on page 693*

Synonyms Heparinase; Hepzyme®

Applies to ACT; Heparin; PTT

Abstract Heparin contamination of specimens is a common cause of an unexpected PTT prolongation. Heparinase (Hepzyme®) can be used to determine if the PTT prolongation is due to heparin. In addition, while patients are receiving heparin, it is sometimes necessary to perform coagulation tests that are affected by heparin. In such cases, heparinase can be used to remove heparin from the specimen so that coagulation tests can be performed without heparin interference.

Specimen Plasma

Container Blue top (sodium citrate) tube

Collection Routine venipuncture. If multiple tests are being drawn, draw blue top tubes after any red top tubes but before any lavender top (EDTA), green top (heparin), or gray top (oxalate/fluoride) tubes. Immediately invert tube gently at least 4 times to mix. Tubes must be appropriately filled. Deliver tubes immediately to the laboratory.

Storage Instructions Separate plasma from cells as soon as possible, ideally within 1 hour of collection. Store plasma according to the guidelines for the individual postheparinase coagulation tests that will be performed.

Causes for Rejection Specimen received more than 4 hours after collection, tube not full, specimen clotted

Turnaround Time 1 day, unless testing is batched less frequently

Reference Interval If heparin is the explanation for the prolonged PTT, the PTT will become normal after treatment of the specimen with heparinase.

Use Determine if an unexpected PTT prolongation is due to heparin contamination; remove known heparin from a specimen so that coagulation tests can be performed without heparin interference

Methodology To determine if a prolonged PTT is due to heparin, measure the PTT before and after heparinase treatment. 1 mL of patient plasma is added to one vial of heparinase and kept at room temperature for 15 minutes. Heparinase is an enzyme that degrades unfractionated heparin (and low-molecular weight heparin) by cleaving it at multiple sites including within the pentasaccharide sequence. The pentasaccharide sequence is the antithrombin binding site and therefore is required for heparin anticoagulation. Heparinase degradation yields small fragments of about 1000 daltons that lack anticoagulant activity. Heparinase is produced by the bacterium *Flavobacterium heparinum*. Up to 2 units/mL heparin can be degraded. As an alternative to heparinase, heparin-binding cellulose can be used to remove heparin from specimens. The cellulose material is added to the specimen, where it binds to heparin. The specimen is then centrifuged, bringing cellulose and heparin into the pellet. The supernatant plasma is then free of heparin.

Additional Information Some laboratories use thrombin time to detect heparin. The thrombin time is very sensitive to heparin, therefore, if the thrombin time is normal, heparin cannot account for a PTT prolongation.

In one study, heparin contamination accounted for 39% of unexpected PTT prolongations in patients who were not receiving heparin or Coumadin® treatments. Heparin contamination may account for an unexplained PTT prolongation even when a heparinized line is carefully flushed to remove heparin prior to specimen collection.

If the PTT shortens significantly but remains prolonged after heparinase, a coagulation abnormality may be present in addition to heparin contamination. If a markedly prolonged PTT (eg, >150 seconds) shortens significantly but remains slightly prolonged after heparinase, a small amount of residual heparin is a possible explanation, because the initial amount of heparin contamination (Continued)

Heparin Neutralization *(Continued)*

was very high. If a second heparinase treatment of the specimen produces a normal PTT, heparin is the confirmed explanation.

References

Lindhardt R, Grant A, Cooney CL, et al, "Differential Anticoagulant Activity of Heparin Fragments Prepared Using Microbial Heparinase," *J Biol Chem*, 1982, 257:7310-3.

Newman RS and Fagin AR, "Heparin Contamination in Coagulation Testing and a Protocol to Avoid It and the Risk of Inappropriate FFP Transfusion," *Am J Clin Pathol*, 1995, 104(4):447-9.

+ **Heparin Resistance** *see* Activated Partial Thromboplastin Time *on page 100*
+ **Heparin Resistance** *see* Antithrombin *on page 198*
+ **Hepatic Cyst fluid Cytology** *see* Cyst Fluid Cytology *on page 490*
+ **Hepatic Iron Index** *see* Ferritin, Serum *on page 577*
+ **Hepatic Iron Index** *see* Liver Biopsy *on page 864*

Hepatitis A Antibodies, IgM and IgG

Related Information

Aspartate Aminotransferase, Serum *on page 216*
Bilirubin, Total, Serum *on page 265*
Hepatitis B DNA Detection *on page 701*
Hepatitis: Laboratory Assessment, Overview *on page 713*
Liver Disease: Laboratory Assessment, Overview *on page 869*

Test Includes Detection of antibody to hepatitis A virus

Abstract Risk factors for hepatitis A include travel to an endemic area and/or ingestion of contaminated food such as shellfish.

Hepatitis A is an acute self-limited disease that manifests with fever, jaundice, anorexia, and diarrhea. The hepatitis A-specific IgM antibody appears in acute hepatitis A at the clinical onset of symptoms and usually persists 3-6 months, and in 25% of patients for 12 months. It appears at the same time that serum aminotransferase concentrations are increased. Persistence for longer than 1 year suggests a false-positive test, a rare event. Past evidence of HAV infection is found using IgG-specific HAV antibody, or total (IgG + IgM) HAV antibody. Immunization results in HAV IgG antibody. While hepatitis A and E are transmitted by the fecal-oral route, hepatitis B, C, and D are blood-borne, and types of viral hepatitis (A-E) are due to very different viral pathogens.

Specimen Serum

Container Red top tube

Storage Instructions Remove serum from clot and freeze.

Reference Interval Negative

Use Differential diagnosis of hepatitis and other causes of jaundice

Methodology Enzyme-linked immunosorbent assay (ELISA), microparticle enzyme immunoassay (MEIA)

Additional Information Hepatitis A is transmitted by contaminated food or water. Its incubation period is 2-7 weeks. Fecal excretion of HAV peaks before symptoms develop. Presence of IgM antibody to hepatitis A virus is good evidence for acute or subacute hepatitis A. IgM antibody develops within a week of symptom onset, peaks in 3 months, and is usually gone after 6 months. Hepatitis A antibody of IgG type is indicative of old infection, is found in almost half of adults, persists throughout life, and is not usually clinically relevant. Recovery from hepatitis A never results in chronic liver disease. Hepatitis A rarely causes fulminant disease, but when it does, it is sometimes fatal. Many cases of hepatitis A are subclinical, particularly in children. Presence of IgG antibody to HAV does not exclude acute hepatitis B or hepatitis C.

References

Kemmer NM and Miskovsky EP, "Hepatitis A," *Infect Dis Clin North Am*, 2000, 14(3):605-15.

Maddrey WC, "Update in Hepatology," *Ann Intern Med*, 2001, 134(3):216-23.

OiConnor JA, "Acute and Chronic Viral Hepatitis," *Adolesc Med*, 2000, 11(2):279-92.

Ryder SD and Beckingham IJ, "ABC of Diseases of Liver, Pancreas, and Biliary System: Acute Hepatitis," *BMJ*, 2001, 322(7279):151-3.

Sacher RA, Peters SM, and Bryan JA, "Testing for Viral Hepatitis. A Practice Parameter," *Am J Clin Pathol*, 2000, 113(1):12-7

Shovein JT, Damazo RJ, and Hyams I, "Hepatitis A: How Benign Is It?" *Am J Nurs*, 2000, 100(3):43-7.

Internet Web Sites

www.cdc.gov/ncidod/diseases/hepatitis/index.htm
www.cdc.gov/ncidod/diseases/hepatitis/slideset/index.htm

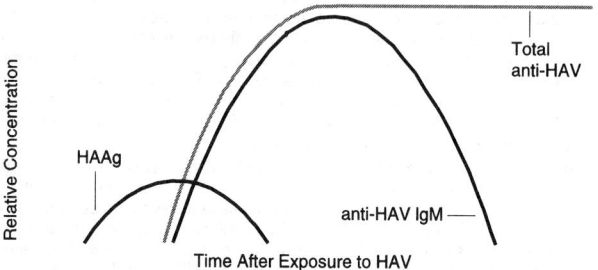

Duration	Incubation	Early Acute	Acute	Recovery
	15-45 Days	0-14 Days	3-6 Months	Years

Reprinted from Abbott Diagnostics

Hepatitis B Antigen Detection

Related Information

Aspartate Aminotransferase, Serum *on page 216*
Bilirubin, Total, Serum *on page 265*
Hepatitis B DNA Detection *on page 701*
Hepatitis B Serology *on page 702*
Hepatitis C Virus RNA Detection and Quantitation *on page 705*
Hepatitis C Virus Serology *on page 706*
Hepatitis D Serology *on page 711*
Hepatitis: Laboratory Assessment, Overview *on page 713*
Liver Biopsy *on page 864*
Liver Disease: Laboratory Assessment, Overview *on page 869*
Polymerase Chain Reaction *on page 1069*

Applies to Core Window Stage of Hepatitis; Hepatitis B$_e$ Antigen; Hepatitis B Surface Antigen (HB$_s$Ag); Hepatitis Carrier Status

Test Includes Detection of hepatitis B antigen in patient's serum

Abstract Hepatitis B virus (HBV) is a DNA virus with a protein coat, surface antigen (HB$_s$Ag), and a core consisting of nucleoprotein (HB$_c$Ag is the core antigen). Hepatitis B$_e$ antigen (HB$_e$Ag) is a circulating peptide derived from the core hepatitis B gene which is a signal for active viral replication. It is used as a marker for increased infectivity when present in serum. HB$_e$Ag appears early in hepatitis B infection. Infectivity of a patient with hepatitis B virus (HBV) can be evaluated with HB$_e$Ag and HB$_s$Ag. Measurement of serum HBV DNA also provides evidence of infectivity. Persistence of HB$_s$Ag longer than 6 months indicates chronic hepatitis B infection. For evaluation of acute hepatitis, serologic testing is recommended for HB$_s$Ag, anti-HB$_s$, anti-HB$_c$ (IgM), and HAV (IgM). HB$_s$Ag, anti-HB$_c$, HB$_e$Ag, and HBV DNA each appear even before liver enzyme elevation. Other useful tests include the transaminases; see Liver Disease: Laboratory Assessment, Overview *on page 869*.

Specimen Serum

Container Red top tube

Storage Instructions Serum must be removed from clot within 3 hours and stored frozen or refrigerated, as HB$_e$Ag is thermolabile.

Reference Interval Negative
(Continued)

699

Hepatitis B Antigen Detection *(Continued)*

Use Differential diagnosis, infectivity, and prognosis of hepatitis B infection; screen blood donors (HB$_s$Ag-positive individuals are rejected); differential diagnosis of hepatitis; evaluate risk in needlestick injuries in healthcare facilities, and guide use of hepatitis B immune globulin

Limitations Absence does not rule out infectivity or chronic hepatitis B carrier state. Patients who are negative for HB$_s$Ag may still have acute type B viral hepatitis. There is sometimes a "window" stage ("core window") when HB$_s$Ag has become negative and the patient has not yet developed antibody (anti-HB$_s$Ag). On such occasions the anti-HB$_c$Ag (IgM) is usually positive; and the patient should be treated as potentially infectious until anti-HB$_s$Ag is detected, at which time immunity is probable. In cases with strong clinical suspicion of viral hepatitis, serologic testing should not be limited to detecting HB$_s$Ag but should include a battery of tests to evaluate different stages of acute and convalescent hepatitis. These should include a test for hepatitis A antibody (IgM), and HB$_s$Ag, anti-HB$_s$, anti-HB$_c$ (IgM), and hepatitis C virus (anti-HCV).

Methodology Enzyme immunoassay (EIA)

Additional Information Transmission of hepatitis B is parenteral, sexual, or perinatal. The incubation period of hepatitis B is 2-6 months. HB$_s$Ag can be detected 1-7 weeks **before** liver enzyme elevation or the appearance of clinical symptoms. Three weeks after the onset of acute hepatitis about 50% of the patients will still be positive for HB$_s$Ag, while at 17 weeks only 10% are positive. The best available markers for infectivity are HB$_s$Ag and HB$_e$Ag. Hepatitis B$_e$ antigen is found during the most infectious period of hepatitis B. It is usually found for only 3-6 weeks. The appearance of anti-HB$_e$ and disappearance of HB$_e$Ag, indicates resolution of hepatitis B infection. Persistence beyond 10 weeks suggests development of the chronic carrier state and chronic hepatitis B. HB$_e$Ag is a proteolytic product of HB$_c$Ag and is found only in HB$_s$Ag positive sera. During the HB$_e$Ag-positive state, patients are at increased risk of transmitting the virus. Exposure to serum or body fluid positive for HB$_e$Ag and HB$_s$Ag is associated with three to five times greater risk of infectivity than with HB$_s$Ag positive serum without HB$_e$Ag. Increased transmission from HB$_e$Ag persons is probably related to increased amounts of circulating viral DNA. Persistence of HB$_e$Ag is also associated with chronic liver disease. See figures.

Hepatitis B surface antigen is the earliest indicator of acute hepatitis B infection and is persistently found in chronic carriers. Early in infection, HB$_s$Ag, HBV DNA, and DNA polymerase can all be detected in serum. The presence of HB$_s$Ag without positivity for IgM anti-HB$_c$ is suggestive of active chronic hepatitis B or carrier status. Such a condition has the potential to lead to serious liver damage and risk of hepatoma. Chronic carrier states are found in up to 10% of cases. Prevention of hepatitis B for those at risk is available via vaccination, as well as antiviral treatment for some chronic carriers.

The relative risk of hepatocellular carcinoma was 9.6 among males in Taiwan positive for HB$_s$Ag alone, and 60 among those also reactive for HB$_e$Ag.

Hepatitis B Serological Profile
Core Window Identification

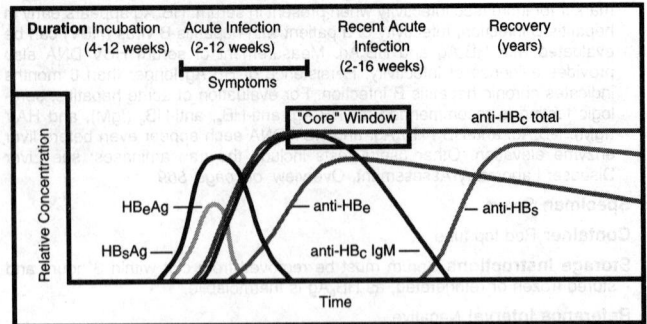

Hepatitis B Chronic Carrier
No Seroconversion

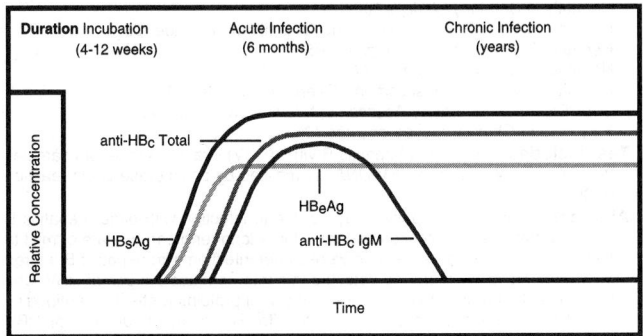

Hepatitis B Chronic Carrier
Late Seroconversion

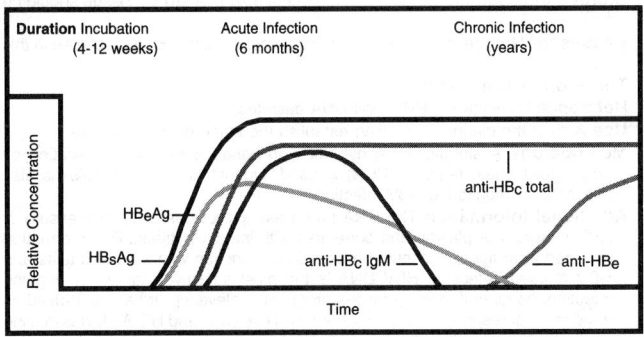

References

Befeler AS and Di Bisceglie AM, "Hepatitis B," *Infect Dis Clin North Am*, 2000, 14(3):617-32.

Grob PJ, "Hepatitis B: Virus, Pathogenesis and Treatment," *Vaccine*, 1998, 16(Suppl):S11-4.

Herrera JL, "Serologic Diagnosis of Viral Hepatitis," *South Med J*, 1994, 87(7):677-84.

Maddrey WC, "Hepatitis B: An Important Public Health Issue," *J Med Virol*, 2000, 61(3):362-6.

Mahoney FJ, "Update on Diagnosis, Management, and Prevention of Hepatitis B Virus Infection," *Clin Microbiol Rev*, 1999, 12(2):351-66.

Sjögren MH, "Serologic Diagnosis of Viral Hepatitis," *Gastroenterol Clin North Am*, 1994, 23(3):457-77.

Yang HI, Lu NS, Liaw YF, et al, "Hepatitis B$_e$ Antigen and the Risk of Hepatocellular Carcinoma," *N Engl J Med*, 2002, 347(3):168-74.

Zuckerman JN and Zuckerman AJ, "Current Topics in Hepatitis B," *J Infect*, 2000, 41(2):130-6.

Internet Web Sites

www.cdc.gov/ncidod/diseases/hepatitis/index.htm

www.cdc.gov/ncidod/diseases/hepatitis/slideset/index.htm

www.hepb.org/02-0103.hepb

♦ **Hepatitis B Core Antibody** *see* Hepatitis B Serology *on page 702*

Hepatitis B DNA Detection
Related Information

Aspartate Aminotransferase, Serum *on page 216*

Bilirubin, Total, Serum *on page 265*

(Continued)

Hepatitis B DNA Detection *(Continued)*

Hepatitis A Antibodies, IgM and IgG *on page 698*
Hepatitis B Antigen Detection *on page 699*
Hepatitis B Serology *on page 702*
Hepatitis C Virus RNA Detection and Quantitation *on page 705*
Hepatitis C Virus Serology *on page 706*
Hepatitis D Serology *on page 711*
Hepatitis: Laboratory Assessment, Overview *on page 713*
Liver Disease: Laboratory Assessment, Overview *on page 869*
Polymerase Chain Reaction *on page 1069*

Test Includes Detection of hepatitis B virus (HBV) viral DNA in serum samples or tissue. Sometimes the HBV DNA is amplified by polymerase chain reaction (PCR).

Abstract Chronic viral hepatitis may be due to infection with either hepatitis B (HBV) or hepatitis C. Infection may result in a long-term carrier state of mild to severe chronic liver disease. Two weeks after infection with hepatitis B a large excess of viral protein can be detected in serum (surface antigen of HBV) and is followed by an antibody response to the viral proteins (anti-HB$_s$). Antibodies to HBV will persist in most patients for life. However, about 10% of HBV infections result in a chronic carrier state marked by a lack of anti-HB$_s$, but persistent anti-HB$_c$ and usually HB$_s$Ag. Diagnosis in these cases should be supplemented by detection of HBV DNA in serum or liver tissue.

Specimen Serum or plasma, liver tissue

Container Red top tube, sterile container

Storage Instructions Serum should be kept frozen at -20°C. Tissue should be frozen at -70°C.

Causes for Rejection Samples containing sodium azide cannot be used in this test.

Turnaround Time 4-7 days

Reference Interval No HBV viral DNA detected

Use Aids in the diagnosis of HBV; establish the stage of HBV disease

Methodology A slot-blot DNA hybridization based assay or amplification by polymerase chain reaction (PCR) is used. Quantitation of HBV DNA is also available for monitoring HBV infection.

Additional Information The DNA probe assay provides a direct measure of HBV in serum or plasma and correlates with infectivity titers. The information provided from this test should be used in conjunction with serologic tests and HBV antigen detection. HBV DNA is the most sensitive marker for ongoing infection. Its quantification provides prognostic relevance. It has been used as a measure of response to antiviral therapy. HBV DNA and HB$_e$Ag levels decline as the body clears the infection. Viral DNA becomes negative by hybridization techniques, but is often detectable by PCR in the later integrative stage of infection when aminotransferase levels are normal. DNA detection should not replace serologic testing.

References

Grob PJ, "Hepatitis B: Virus, Pathogenesis and Treatment," *Vaccine*, 1998, 16(Suppl):S11-6.
Maddrey WC, "Hepatitis B: An Important Public Health Issue," *J Med Virol*, 2000, 61(3):362-6.
Mahoney FJ, "Update on Diagnosis, Management, and Prevention of Hepatitis B Virus Infection," *Clin Microbiol Rev*, 1999, 12(2):351-66.
Sacher RA, Peters SM, and Bryan JA, "Testing for Viral Hepatitis. A Practice Parameter," *Am J Clin Pathol*, 2000, 113(1):12-7
Zuckerman JN and Zuckerman AJ, "Current Topics in Hepatitis B," *J Infect*, 2000, 41(2):130-6.

Internet Web Sites

www.cdc.gov/ncidod/diseases/hepatitis/index.htm
www.cdc.gov/ncidod/diseases/hepatitis/slideset/index.htm
www.hepb.org/02-0103.hepb

♦ **Hepatitis B$_e$ Antibody** *see* Hepatitis B Serology *on page 702*

♦ **Hepatitis B$_e$ Antigen** *see* Hepatitis B Antigen Detection *on page 699*

Hepatitis B Serology

Related Information

Alanine Aminotransferase, Serum *on page 116*
Alkaline Phosphatase, Serum *on page 127*
Aspartate Aminotransferase, Serum *on page 216*
Bilirubin, Direct, Serum *on page 262*

Applies to Core Window Stage of Hepatitis; Hepatitis B Core Antibody; Hepatitis B$_e$ Antibody; Hepatitis B Surface Antibody

Test Includes Detection of serologic response to hepatitis B infection; specifically, the antibody specific for the core protein, e antigen, and surface proteins. Detection of serologic response to hepatitis B immunizations; specifically antibody response to surface protein

Abstract The onset of hepatitis A is abrupt, while that of hepatitis B is insidious. The onset of hepatitis C is even more insidious and is apt to be subclinical.

For diagnosis of acute hepatitis, hepatitis B core IgM antibody and HB$_s$Ag are helpful. Hepatitis B$_e$ antibody (anti-HB$_e$) appears in early convalescence in hepatitis B infection. HB$_e$ antigen is a circulating peptide derived from the core hepatitis B gene and demonstrates viral replication. It is used as a marker for increased infectivity when present in serum. Antibody to HB$_e$ demonstrates reduced infectivity.

Antibody to HB$_e$ nucleocapsid or core antigen (anti-HB$_c$) is found in early acute hepatitis B, in resolved hepatitis B, and in chronic hepatitis B. The IgG antibody specific for hepatitis B virus (HBV) core protein, may be the only serologic test that remains positive years after initial infection with hepatitis B.

Hepatitis B surface antibody (anti-HB$_s$) develops following resolved hepatitis B. It is responsible for immunity. Highly effective vaccines are used throughout the world and may eventually lead to eradication of hepatitis B. Detection of antibody to HB surface antigen in a vaccinated patient is associated with protective immunity.

Specimen Serum

Container Red top tube

Storage Instructions Freeze to ship.

Reference Interval Negative for anti-HB$_e$ and anti-HB$_c$; negative for anti-HB$_s$ in unvaccinated individuals (may be positive in vaccinated persons)

Use Differential diagnosis of hepatitis syndromes. Differential diagnosis, staging, and prognosis of hepatitis B infection. At the onset of symptoms of acute hepatitis B, anti-HB$_c$ IgM becomes detectable and may persist for up to 1 year. With HB$_s$Ag, IgM anti-HB$_c$ is used to confirm the diagnosis of hepatitis B. Anti-HB$_e$ and anti-HB$_c$ together confirm the convalescent stage of hepatitis B after the disappearance of HB surface antigen (HB$_s$Ag). Occasionally, in chronic hepatitis B, anti-HB$_c$ may persist indefinitely. Anti-HB$_c$ IgG usually persists for life. Anti-HB$_c$ may be used to screen volunteer blood donors for hepatitis B infection, detecting donors in the window period after HB$_s$Ag has disappeared before anti-HB$_s$ has appeared. Anti-HB$_c$ is used to look for past hepatitis B infections, since immunization for hepatitis B produces antibodies to hepatitis B surface antigen but not antibodies to core antigen. Up to 50% of persons immunized with hepatitis B vaccine may have serum HB$_s$Ag false positive for as long as 4 weeks. Presence of hepatitis B surface antibody indicates past infection with resolution of previous hepatitis B infection, in which case anti-HB$_c$ is also present or vaccination against hepatitis B.

Limitations Weak positives without other positive markers or abnormalities in liver-related enzymes may represent false-positive reactions. The diagnostic significance of this test is increased when they are tested together as hepatitis serologic markers. Absence of anti-HB$_e$ does not rule out chronic hepatitis B carrier state or infectivity. Presence of anti-HB$_s$ is not an absolute indicator of resolved hepatitis, nor of protection from future infection. Since there are different serologic subtypes of hepatitis B virus, it is possible (and has been (Continued)

Hepatitis B Serology *(Continued)*

reported) for a patient to have antibody to one surface antigen type and to be acutely infected with virus of a different subtype. Thus, a patient may have coexisting HB_sAg and HB_sAb. Transfused individuals or hemophiliacs receiving plasma components may give false-positive tests for antibody to hepatitis B surface antigen, and passively acquired reactivity from transfusion or globulin therapy does not indicate immunity. Most individuals vaccinated with HBV vaccine develop anti-HB_s, although some do not.

Methodology Enzyme-linked immunosorbent assay (ELISA). IgG and IgM antibodies may be differentiated.

Hepatitis B Profile

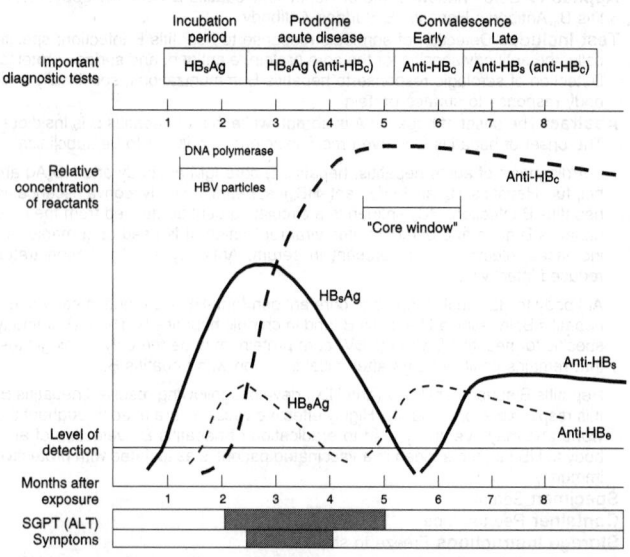

Serologic and clinical patterns observed during acute hepatitis B viral infection. From Hollinger FB and Dreesman GR, *Manual of Clinical Immunology*, 2nd ed, Rose NR and Friedman H, eds, Washington, DC: American Society for Microbiology, 1980, with permission.

Additional Information The prevalence of hepatitis B in many parts of the world can be as high as 15% to 20%. About 66% of subjects infected with HBV are asymptomatic, especially if the patient is infected early in life. Acute hepatitis develops in ~25% of patients while the rest become carriers. Chronic carrier status bears risk for ultimate development of cirrhosis and possible hepatoma.

Early in hepatitis B infection before liver enzyme elevation, HBV proliferates without hepatocyte damage. At this stage, HBV DNA, HB_sAg, HB_eAg, and anti-HB_c are detectable. Later, this immune tolerance is replaced by host response and hepatocellular damage of hepatitis. The period between the disappearance of HB_sAg and the appearance of HB_sAb is often called the "core window." Anti-HB_c persists for months to years after resolution of acute hepatitis B and persists in cases of chronic infection along with HB_eAg and HBV DNA. Hepatitis B_e antigen usually is present only when there is circulating serum HBV DNA. The appearance of anti-HB_e indicates a reduced risk of infectivity. Antibody to e antigen can persist for years, but usually disappears earlier than anti-HB_s or anti-HB_c. Anti-HB_s usually can be detected several weeks to several years after HB_sAg is no longer found, and it usually persists for life after acute infection has resolved. Presence of anti-HB_s without the presence of HB_sAg is evidence for immunity from reinfection, with virus of the same subtype (*vide supra*). Hepatitis vaccine is safe and effective in protecting most recipients from acute hepatitis B. Vaccine recipients tested for HB$_s$Ag during the 2 weeks after immunization may test positive transiently from the vaccine itself. The majority of patients with reactivation hepatitis will have detectable serum anti-HB$_c$ IgM. See figure.

References

Barash C, Conn MI, DiMarino AJ, et al, "Serologic Hepatitis B Immunity in Vaccinated Healthcare Workers," Arch Intern Med, 1999, 159(13):1481-3.

Befeler AS and Di Bisceglie AM, "Hepatitis B," Infect Dis Clin North Am, 2000, 14(3):617-26.

Maddrey WC, "Hepatitis B: An Important Public Health Issue," J Med Virol, 2000, 61(3):362-6.

Mahoney FJ, "Update on Diagnosis, Management, and Prevention of Hepatitis B Virus Infection," Clin Microbiol Rev, 1999, 12(2):351-66.

Sacher RA, Peters SM, and Bryan JA, "Testing for Viral Hepatitis. A Practice Parameter," Am J Clin Pathol, 2000, 113(1):12-7

Zuckerman JN and Zuckerman AJ, "Current Topics in Hepatitis B," J Infect, 2000, 41(2):130-6.

Internet Web Sites

www.cdc.gov/ncidod/diseases/hepatitis/index.htm
www.cdc.gov/ncidod/diseases/hepatitis/slideset/index.htm
www.hepb.org/02-0103.hepb

♦ **Hepatitis B Surface Antibody** see Hepatitis B Serology on page 702

♦ **Hepatitis B Surface Antigen (HB$_s$Ag)** see Hepatitis B Antigen Detection on page 699

♦ **Hepatitis Carrier Status** see Hepatitis B Antigen Detection on page 699

Hepatitis C Virus RNA Detection and Quantitation

Related Information

Alanine Aminotransferase, Serum on page 116
Alkaline Phosphatase, Serum on page 127
Aspartate Aminotransferase, Serum on page 216
Bilirubin, Direct, Serum on page 262
Bilirubin, Total, Serum on page 265
Donation, Blood on page 521
Hepatitis B Antigen Detection on page 699
Hepatitis B DNA Detection on page 701
Hepatitis B Serology on page 702
Hepatitis C Virus Serology on page 706
Hepatitis D Serology on page 711
Hepatitis: Laboratory Assessment, Overview on page 713
Lactate Dehydrogenase, Serum on page 825
Liver Biopsy on page 864
Liver Disease: Laboratory Assessment, Overview on page 869
Polymerase Chain Reaction on page 1069
Prothrombin Time on page 1116
Risks of Transfusion on page 1166
Smooth Muscle Antibody on page 1207

Test Includes Amplification of viral DNA (or amplification of HCV-specific probe) to detect and/or quantitate HCV RNA in plasma.

Abstract Hepatitis C now infects (chronic HCV carriers) over 4 million people in the U.S. and 170 million worldwide. Presently, 60% of HCV infections are acquired through sharing intravenous or intranasal drug paraphernalia. The initial diagnosis is by testing for antibody to HCV. If positive, HCV genotyping and quantitative viral load testing can be obtained to assist in treatment decisions. See Hepititis C Virus Serology on page 706 for integration of HCV serologic (antibody) and NAT (nucleic acid) testing and results. The onset of hepatitis C is often slow and insidious. Untreated, about 85% of people remain persistently positive. Twenty percent of subjects with chronic hepatitis C eventually develop cirrhosis. The risk of hepatocellular carcinoma is 1% to 4% annually in those with established cirrhosis.

Several commercial assays are available for detection of HCV RNA in serum.

Specimen Serum or liver tissue

Container Red top tube; tissue in a sterile container or sealed plastic bag

Sampling Time HCV RNA appears within days of exposure, before ALT concentrations become increased. Most assays are performed after detection of a positive HCV antibody.
(Continued)

Hepatitis C Virus RNA Detection and Quantitation
(Continued)

Storage Instructions Serum should be removed from clot within 2 hours of specimen collection. Store or ship serum a -20°C. Tissue should be stored or shipped at -70°C.

Causes for Rejection Serum samples not separated within 2 hours of collection; tissue specimens that have thawed during storage or shipping

Turnaround Time 4-7 days; may vary between different laboratories

Reference Interval No HCV viral RNA detected

Use Aid in the diagnosis of HCV infection, predict responsiveness to therapy, and monitor therapy

Limitations HCV RNA degrades quickly and may cause a false-negative result if specimens are not handled properly.

Methodology Viral RNA is extracted from serum or tissue and after transcription the genetic material is amplified using PCR. Quantitation of the HCV RNA is also available either by direct binding with detection by branched chain DNA (bDNA) probes or by quantitation of the RT-PCR product.

Additional Information At present, 6 HCV genotypes and 50 subtypes have been described for this RNA virus. Genotypes 1, 2, and 3 are found worldwide. Genotype 4 is principally in Egypt and Zaire, genotype 5 in South Africa, and genotype 6 in Asia. Variations in the six major types are designated with letters a, b, c, etc. Genotypes 1b and 4 cause more aggressive liver disease and respond less to therapy, compared to other types. The results of serologic and virologic assays may vary according to HCV genotype.

The HCV RNA assays use various methodologies (ie, RT-PCR, bDNA) and show similar sensitivities. Each of these detect HCV RNA of the serotypes commonly found in the U.S. In chronic HCV infections, in types 2 and 3, low levels of viremia as detected by RT-PCR or bDNA indicate a response to therapy after treatment with interferon and ribavirin.

The best means to monitor HCV activity and prognosis is liver biopsy.

References

Alter MJ, Kuhnert WL, and Finelli L, "Guidelines for Laboratory Testing and Result Reporting of Antibody to Hepatitis C Virus. Centers for Disease Control and Prevention," *MMWR Recomm Rep*, 2003, 52(RR-3):1-13.

Boyer N and Marcellin P, "Pathogenesis, Diagnosis and Management of Hepatitis C," *J Hepatol*, 2000, 32(Suppl 1):98-112.

Cheney DP, Chopra S, and Graham C, "Hepatitis C," *Infect Dis Clin North Am*, 2000, 14(3):633-42.

Germer JJ and Zein NN, "Advances in the Molecular Diagnosis of Hepatitis C and Their Clinical Implications," *Mayo Clin Proc*, 2001, 76(9):911-20.

Lauer GM and Walker BD, "Hepatitis C Virus Infection," *N Engl J Med*, 2001, 3345(1):41-52.

Liang TJ, Rehermann B, Seeff LB, et al, "Pathogenesis, Natural History, Treatment, and Prevention of Hepatitis C," *Ann Intern Med*, 2000, 132:296-305.

Thomas DL, Astemborski J, Rai RM, et al, "The Natural History of Hepatitis C Virus Infection: Host, Viral, and Environmental Factors," *JAMA*, 2000, 284(4):450-6.

Zein NN, "Clinical Significance of Hepatitis C Virus Genotypes," *Clin Microbiol Rev*, 2000, 13(2):223-35.

Internet Web Sites

www.cdc.gov/ncidod/diseases
www.hepatitis-central.com/index.html

Hepatitis C Virus Serology

Related Information

Alpha$_1$-Fetoprotein, Serum *on page 136*
Aspartate Aminotransferase, Serum *on page 216*
Bilirubin, Total, Serum *on page 265*
Cryoglobulin, Qualitative, Serum and Plasma *on page 478*
Donation, Blood *on page 521*
Hepatitis B Antigen Detection *on page 699*
Hepatitis B DNA Detection *on page 701*
Hepatitis B Serology *on page 702*
Hepatitis C Virus RNA Detection and Quantitation *on page 705*
Hepatitis: Laboratory Assessment, Overview *on page 713*
Liver Biopsy *on page 864*
Liver Disease: Laboratory Assessment, Overview *on page 869*
Risks of Transfusion *on page 1166*

Synonyms Anti-HCV; HCV Antibody

Test Includes Detection of antibody specific for hepatitis C virus in serum

Algorithm for Antibody to Hepatitis C Virus (Anti-HCV) Testing and Result Reporting

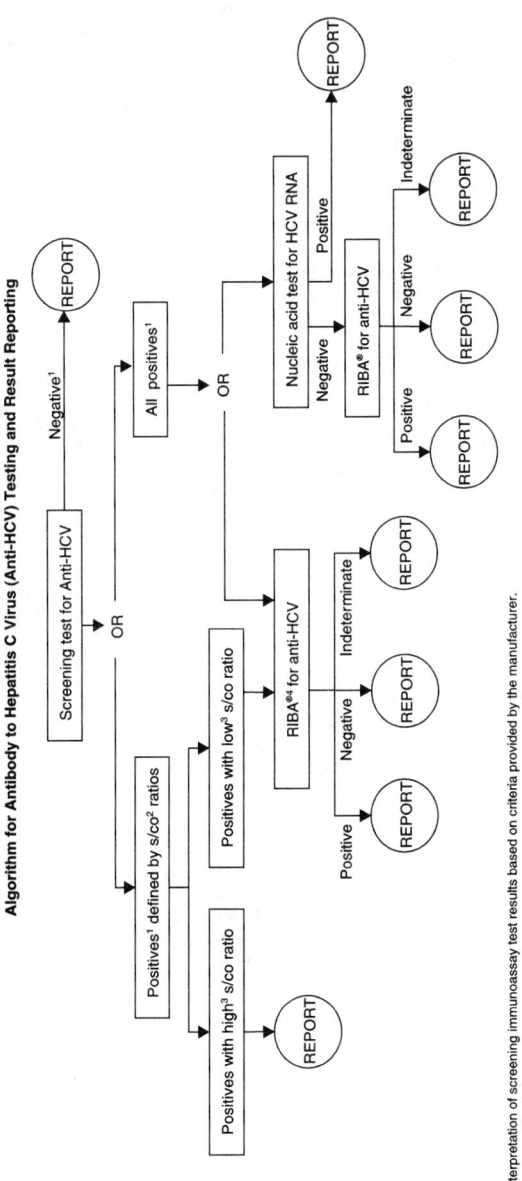

[1]Interpretation of screening immunoassay test results based on criteria provided by the manufacturer.
[2]Signal-to-cut-off.
[3]Screening-test-positive results are classified as having high s/co ratios if their ratios are at or above a predetermined value that predicts a supplemental-test-positive result ≥95% of the time among all populations tested; screening-test-positive results are classified as having low s/co ratios if their ratios are below this value.
[4]Recombinant immunoblot assay.

Abstract Hepatitis C is an enveloped single-stranded RNA virus, that causes disease which in general is slowly progressive and is often asymptomatic. Some 30% to 40% of patients recover or have a benign outcome. Although it can be self-limited, long-term chronic sequelae are well known. Chronic disease is seen in 50% to 85% of patients. Cirrhosis evolves in ~20% to 30% of patients whose chronic infection persists for longer than 20 years. In the (Continued)

Hepatitis C Virus Serology (Continued)

U.S., HCV is responsible for 20% of cases of acute hepatitis, 60% of cases of chronic hepatitis, and 30% of cases of cirrhosis. Hepatitis C infection has become the major indication for liver transplantation. Antibody tests become positive 4-10 weeks after exposure to HCV. The presence of antibody indicates infection, not necessarily immunity. Confirmation of a positive EIA may be with the RIBA (recombinant immunoblot assay which is considered obsolete by some authorities, in some circumstances) or by HCV reverse transcriptase polymerase chain reaction (RT-PCR) or HCV bDNA testing (vide infra). Nearly 50% of blood donors who test EIA positive do not have detectable HCV RNA nor a positive RIBA, and are considered false positive, with implications to "look back" programs.

Coinfection with HIV-1 accelerates the course of HCV infection and appears to increase the risk of development of cirrhosis.

Specimen Serum

Container Red top tube

Storage Instructions Freeze to ship.

Reference Interval MMWR, Vol 52/RR-3 (published by the Centers for Disease Control, information originating in the National Center for Infectious Diseases) includes recommendations for the reporting of results that emphasize the use of signal to cutoff ratios (see below) in anti-HCV test reports. See table.

Use Differential diagnosis of acute and chronic hepatitis; screen blood units for transfusion safety; evaluate patients with essential mixed cryoglobulinemia, membranoproliferative glomerulonephritis, and porphyria cutanea tarda. Individuals who received blood transfusions before 1992 should be tested.

Limitations False-positive HCV immunoassays are found in 0.2% of pregnant women while true positives are 1% to 2%. Associations with false-positive tests include recent immunization against influenza virus, hypergammaglobulinemia, rheumatoid factor, and connective tissue disorders. By virtue of problems of low specificity, ELISA-positive specimens (dependent on signal:cutoff ratio - if low, confirm) require confirmation with a different method of detecting antibody or RNA (eg, recombinant immunoblot assay - RIBA, or nucleic acid test for HCV RNA).

False-negative results occur in early acute infection due to the long incubation period and the occurrence of seroconversion (80% of patients develop detectable serum HCV antibody within 15 weeks after exposure). Antibodies may not be detectable early in the course (4 weeks after infection) due to this window period.

False negatives are common in patients with essential mixed cryoglobulinemia (type II). HCV RNA assays are recommended for such individuals. False negatives also occur in hemodialysis patients and those with immunodeficiency.

Lack of anti-HCV does not exclude the possibility of HCV hepatitis.

Methodology Enzyme immunoassay (EIA) confirmed (in low-risk populations) with RIBA for anti-HCV (RIBA has been largely replaced in high-risk populations) by HCV RNA testing. Second- and third-generation EIA assays are presently in use. A positive or indeterminate result in a setting of low prevalence is checked with HCV RT-PCR. Direct screening for transfusion blood of pooled specimens by PCR has become available. On February 7, 2003, the CDC published recommendations for reporting results of testing for anti-HCV that emphasize the use of S:CO ratios. The ratio consists of:

OD value of sample / OD value of assay cutoff (for that run)

See table on following page.

Additional Information When specific testing for HCV became available, the numbers of new cases annually dropped from 175,000 to about 30,000. Yet, additional cases keep occurring, chiefly from sharing paraphernalia in intravenous or intranasal drug use. In an individual with abnormal ALT and positive risk factors (intravenous drug use, intranasal cocaine use, transfusions before 1992, liver disease, body piercings, multiple sexual partners, sexually transmitted diseases, or long-term sex partner who is HCV positive), positive anti-HCV is highly suggestive of current infection and should be evaluated with HCV RNA.

In developed countries, most HCV infections are found in injection drug users. HCV is rarely transmitted by sexual contact, much less often than HBV.

Recommendations (CDC) for Reporting Results of Testing for Antibody to Hepatitis C Virus (Anti-HCV) by Type of Reflex Supplemental Testing Performed

Anti-HCV Screening Test Results	Supplemental Test Results	Interpretation	Comments
Negative[1]	Not applicable	Anti-HCV negative	Not infected with HCV, unless recent infection is suspected or other evidence exists to indicate HCV infection.
Positive[1] with high signal:cutoff (S:CO) ratio	Not done	Anti-HCV positive	Probably indicates past or present HCV infection; supplemental serologic testing is not performed. Samples with high S:CO ratios usually (≥95%) confirm positive, but <5 of every 100 might represent false positives; more specific testing can be requested, if indicated.
Positive	Recombinant immunoblot assay (RIBA) positive	Anti-HCV positive	Indicates past or present HCV infection.
Positive	RIBA negative	Anti-HCV negative	Not infected with HCV, unless recent infection is suspected or other evidence exists to indicate HCV infection.
Positive	RIBA indeterminate	Anti-HCV indeterminate	HCV antibody and infection status cannot be determined; another sample should be collected for repeat anti-HCV testing (>1 month) or for HCV RNA testing.
Positive	Nucleic acid test (NAT) positive	Anti-HCV positive HCV-RNA positive	Indicates active HCV infection.
Positive	NAT negative RIBA positive	Anti-HCV positive HCV-RNA negative	The presence of anti-HCV indicates past or present HCV infection; a single negative HCV RNA result does not rule out active infection.
Positive	NAT negative RIBA negative	Anti-HCV negative HCV-RNA negative	Not infected with HCV.
Positive	NAT negative RIBA indeterminate	Anti-HCV indeterminate HCV-RNA negative	Screening test anti-HCV result probably a false positive, which indicates no HCV infection.

[1]Screening immunoassay test results interpreted as negative or positive on the basis of criteria provided by the manufacturer.

Adapted from Alter MJ, Kuhnert WL, and Finelli L, "Guidelines for Laboratory Testing and Result Reporting of Antibody to Hepatitis C Virus," *MMWR Recomm Rep*, 2003, 52(RR-3):11.

However, transmission rates for multiple partners, unprotected intercourse, and coinfection with other sexually transmitted diseases may increase risk of sexual transmission as high as 20%. The risk of transmission from mother to infant is <5%. The risk of acquisition from a random needlestick in hospital workers is about 0.1%. There is no postexposure prophylaxis presently available. The risk of transfusion-associated hepatitis C has become almost negligible in developed countries. Improved blood screening measures were introduced in 1990 and 1992.

GB virus type C (GBV-C), a recently discovered distant relative of HCV, is transmitted by the parenteral route and is encountered commonly in patients infected with human immunodeficiency virus (HIV) and HCV. Some reports suggest that coinfection with GBV-C and HIV has a favorable effect on the course and survival of HIV disease. GBV-C infection, however, does not appear to influence liver disease caused by HCV.

"Look back" programs have overcome a number of hurdles with resultant Food and Drug Administration (FDA) guidance, March 1998 with revised guidelines issued September 1998. Signal:cutoff (S:CO) ratios of HCV EIA 1.0 tests were

(Continued)

Hepatitis C Virus Serology (Continued)

utilized (in cases of EIA 1.0 - repeatedly reactive donations) in place of HCV RIBA 2.0 testing. Look back results from Denmark and Canada have been published.

The apolipoprotein E gene has three common alleles, e2, e3, and e4. ApoE may determine the severity of some viral infections (eg, herpes simplex virus and human immunodeficiency virus). Carriage of an apoE - e4 allele may be protective against liver damage caused by HCV. Competition for low-density lipoprotein receptors may be involved (as a mechanism of cell entry).

Diagnostic EIA Assays for Detection of Anti-HCV

Assay	Manufacturer	Technique	Epitopes (origin of antigens)	FDA Approval at Date of Writing (June 2002)
Ortho HCV 3.0 ELISA Test System with enhanced SAVe	Ortho-Clinical Diagnostics, Raritan, NJ	EIA, microtiter plate	Core, NS3, NS4, NS5 (Chiron Corp)	FDA approved
Vitros anti-VHC	Ortho-Clinical Diagnostics, Raritan, NJ	EIA, automated on Vitros ECI device	Core, NS3, NS4, NS5 (Chiron Corp)	FDA approved
Abbott HCV EIA 2.0	Abbott Diagnostic, Chicago, IL	ELISA, microbeads	Core, NS3, NS4 (Chiron Corp)	FDA approved
Abbott HCV EIA 3.0	Abbott Diagnostic, Chicago, IL	ELISA, microbeads	Core, NS3, NS4, NS5 (Chiron Corp)	Not approved
IMx HCV 3.0	Abbott Diagnostic, Chicago, IL	EIA, automated on IMx device	Core, NS3, NS4, NS5 (Chiron Corp)	Not approved
AxSYM HCV 3.0	Abbott Diagnostic, Chicago, IL	EIA, automated on AxSYM device	Core, NS3, NS4, NS5 (Chiron Corp)	Not approved
Monolisa anti-HCV Plus version 2	Bio-Rad, Marnes-la-Coquette, France	EIA, microtiter plate	Core, NS3, NS4 (Bio-Rad)	Not approved
Access HCV AB Plus	Bio-Rad, Marnes-la-Coquette, France	EIA, automated on ACCESS device	Core, NS3, NS4 (Bio-Rad)	Not approved
Innotest HCV AB IV	Innogenetics, Ghent, Belgium	EIA, microtiter plate	Core, NS3, NS4, NS5 from genotypes 1, 2, and 3a (Innogenetics)	Not approved

Adapted from Pawlotsky JM, "Use and Interpretation of Virological Tests for Hepatitis C," *Hepatology*, 2002, 36(5 Suppl 1):S65-S73.

References

Alter MJ, Kuhnert WL, and Finelli L, "Guidelines for Laboratory Testing and Result Reporting of Antibody to Hepatitis C Virus," *MMWR Recomm Rep*, 2003, 52(RR-3):1-15.

Boyer N and Marcellin P, "Pathogenesis, Diagnosis and Management of Hepatitis C," *J Hepatol*, 2000, 32(Suppl 1):98-112.

Cheney DP, Chopra S, and Graham C, "Hepatitis C," *Infect Dis Clin North Am*, 2000, 14(3):633-42.

Christensen PB, Groenbaek K, and Krarup HB, "Transfusion-Acquired Hepatitis C: The Danish Lookback Experience. The Danish HCV [hepatitis C virus] Lookback Group," *Transfusion*, 1999, 39(2):188-93.

Epstein J, "Hepatitis C Virus Lookback: Emerging Science and Public Policy," *Transfusion*, 2000, 40(1):3-5.

Lauer GM and Walker BD, "Hepatitis C Virus Infection," *N Engl J Med*, 2001, 345(1):41-52.

Liang TJ, Rehermann B, Seeff LB, et al, "Pathogenesis, Natural History, Treatment, and Prevention of Hepatitis C," *Ann Intern Med*, 2000, 132:296-305.

Long A, Spurll G, Demers H, et al, "Targeted Hepatitis C Lookback: Quebec, Canada," *Transfusion*, 1999, 39(2):194-200.

Sacher RA, Peters SM, and Bryan JA, "Testing for Viral Hepatitis. A Practice Parameter," *Am J Clin Pathol*, 2000, 113(1):12-7

Stosor V and Wolinsky S, "GB Virus C and Mortality From HIV Infection," *N Engl J Med*, 2001, 345(10):761-2.

Thomas DL, Astemborski J, Rai RM, et al, "The Natural History of Hepatitis C Virus Infection: Host, Viral, and Environmental Factors," *JAMA*, 2000, 284(4):450-6.

Tobler LH, Tegtmeier G, Stramer SL, et al, "Lookback on Donors Who Are Repeatedly Reactive on First-Generation Hepatitis C Virus Assays:Justification and Rational Implementation," *Transfusion*, 2000, 40(1):15-24.

Wozniak MA, Itzhaki RF, Faragher EB, et al, "Apolipoprotein E-e4 Protects Against Severe Liver Disease Caused by Hepatitis C Virus," *Hepatology*, 2002, 36(2):456-63.

Internet Web Sites

www.cdc.gov/ncidod/diseases/hepatitis/index.htm

Hepatitis D Serology

Related Information

Aspartate Aminotransferase, Serum *on page 216*
Bilirubin, Total, Serum *on page 265*
Hepatitis B Antigen Detection *on page 699*
Hepatitis B DNA Detection *on page 701*
Hepatitis B Serology *on page 702*
Hepatitis C Virus RNA Detection and Quantitation *on page 705*
Hepatitis: Laboratory Assessment, Overview *on page 713*

Applies to Delta Agent Serology

Abstract Hepatitis delta agent (HDV) was first described in 1977. HDV always occurs in a person infected with hepatitis B (HBV). Patients coinfected with HDV and HBV have fulminant hepatitis more often than patients infected only with HBV, and have increased risk for chronic hepatitis, cirrhosis, and hepatocellular carcinoma.

Transmission of this defective virus is by injection drug use, and less commonly through sexual transmission. Because HDV is a satellite virus of HBV, the prevalence of HDV is low in countries with a low prevalence of HBV. In Southern Italy and parts of Russia and Romania, the prevalence of HDV is high among HBV carriers (>20%) and as high as 60% among patients with chronic liver disease due to HBV.

Hepatitis D Superinfection

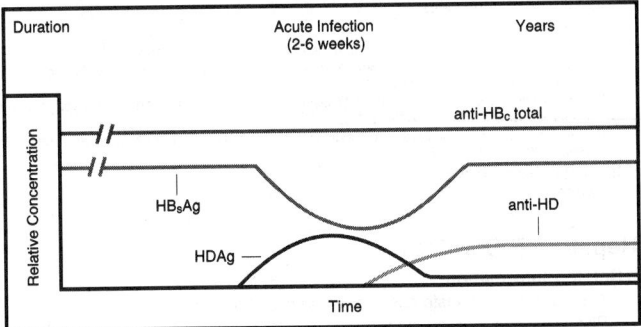

Reprinted from Abbott Diagnostics

Hepatitis D Coinfection

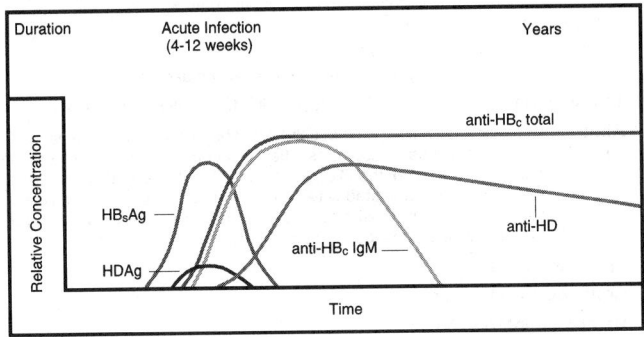

Reprinted from Abbott Diagnostics

(Continued)

Hepatitis D Serology *(Continued)*

Specimen Serum

Container Red top tube

Storage Instructions Freeze to ship.

Reference Interval Negative

Use Differential diagnosis of chronic, recurrent, and acute viral hepatitis. Testing for serological markers of HDV should be considered when a patient shows clinical signs of acute or fulminant hepatitis, or when deterioration takes place in chronic hepatitis B infection.

Limitations False-positive EIA results have been reported in patients with lipemia or high titer rheumatoid factor. Most of the immunoassays lack desirable levels of sensitivity.

Methodology Enzyme-linked immunosorbent assay (ELISA); Total (IgG and IgM) and IgM antidelta antibody assays are available.

Additional Information Hepatitis D virus ("delta" agent) is an incomplete RNA virus, that can only infect livers already infected by hepatitis B virus. HDV may occur, therefore, as a coinfection with acute HBV hepatitis, which usually resolves, or as a superinfection in chronic HBV infection, following which HDV also causes chronic hepatitis. It uses HBV surface antigen as its viral envelope. IgG and IgM antibodies to HDV develop 5-7 weeks after infection. IgM antibody is most useful in distinguishing those patients with active liver disease. IgM anti-HDV is transient and is rapidly replaced by IgG anti-HDV, which persists. Generally, laboratories use total anti-HDV ELISAs containing both IgM and IgG for diagnostic purposes.

References

Maddrey WC, "Update in Hepatology," *Ann Intern Med*, 2001, 134(3):216-23.

Modahl LE and Lai MM, "Hepatitis Delta Virus: The Molecular Basis of Laboratory Diagnosis," *Crit Rev Clin Lab Sci*, 2000, 37(1):45-92.

Ryder SD and Beckingham IJ, "ABC of Diseases of Liver, Pancreas, and Biliary System: Acute Hepatitis," *BMJ*, 2001, 322(7279):151-3.

Sacher RA, Peters SM, and Bryan JA, "Testing for Viral Hepatitis. A Practice Parameter," *Am J Clin Pathol*, 2000, 113(1):12-7.

Internet Web Sites

www.cdc.gov

Hepatitis E Serology

Related Information

Aspartate Aminotransferase, Serum *on page 216*

Bilirubin, Total, Serum *on page 265*

Hepatitis A Antibodies, IgM and IgG *on page 698*

Hepatitis B Antigen Detection *on page 699*

Hepatitis B DNA Detection *on page 701*

Hepatitis B Serology *on page 702*

Hepatitis C Virus RNA Detection and Quantitation *on page 705*

Hepatitis C Virus Serology *on page 706*

Hepatitis D Serology *on page 711*

Liver Disease: Laboratory Assessment, Overview *on page 869*

Test Includes Detection of IgM and IgG antibody specific for hepatitis E

Abstract Hepatitis E causes an acute, self-limited hepatitis similar to hepatitis A. Like hepatitis A, this virus is also transmitted fecal-orally, predominantly in the Indian subcontinent, Mexico, the Middle East, and Southeast and Central Asia. The usual vehicle is contaminated water. Food and personal contact have also been implicated. Although most hepatitis E occurs in sporadic cases, it has been known to cause epidemics in tropical and subtropical regions.

Specimen Serum

Container Red top tube

Reference Interval Negative

Use Evaluation of the etiology of hepatitis in travelers who do not have hepatitis A, B, C

Methodology ELISA or Western blot; a positive IgM titer remains present for as long as 3 months in 50% of patients. A rising titer, or a very high titer of IgG is also diagnostic.

Additional Information The incubation period following HEV exposure is 15-60 days. No evidence of chronic infection has been detected. Acute liver failure is rare except during the third trimester of pregnancy. A high incidence of spontaneous abortion and severe liver disease occurs in infected mothers. Seroprevalence in endemic regions varies between 3% and 27%, while in nonendemic regions seroprevalence has ranged between 1% and 28%. Acquisition of antibody seems to occur predominately during the teenage years.

During early hepatitis E infection, IgM antibody can be detected beginning 1-4 weeks after exposure. IgG antibodies follow soon afterwards and can persist for up to 2 years. Hepatitis E reverse transcriptase PCR (RT-PCR) technology is available, but is of limited value because no chronic illness occurs. A positive test by HEV RT-PCR is not the same as infectivity, because the PCR is more sensitive at measuring HEV than is clinical infectivity.

References

Goncales HS, Pinho JR, Moreira RC, et al, "Hepatitis E Virus Immunoglobulin G Antibodies in Different Populations in Campinas, Brazil," *Clin Diag Lab Immunol*, 2000, 7(5):813-6.

Krawczynski K, Aggarwal R, and Kamili S, "Hepatitis E," *Infect Dis Clin North Am*, 2000, 14(3):669-80.

Sacher RA, Peters SM, and Bryan JA, "Testing for Viral Hepatitis. A Practice Parameter," *Am J Clin Pathol*, 2000, 113(1):13-7.

Internet Web Sites

www.hepnet.com
www.cdc.gov

Hepatitis: Laboratory Assessment, Overview

Related Information

Synonyms Viral Hepatitis

Abstract Broadly defined, hepatitis is inflammation of liver parenchyma. A large number of infectious agents, metabolic diseases, drugs, toxic exposures, and immune derangements are included in the differential diagnosis. Some of these are addressed in the monographs, Liver Disease: Laboratory Assessment, Overview *on page 869* and in Liver Biopsy *on page 864*. Algorithms in the listings Bilirubin, Total, Serum *on page 265* and Aspartate Aminotransferase, Serum *on page 216* may be helpful. The listings Aspartate Aminotransferase, Serum *on page 216* and Alanine Aminotransferase, Serum *on page 116* include information relevant to the differential diagnosis of hepatitis. Aminotransferases (transaminases) may reach 100 times the upper poles of their reference ranges. With jaundice, total and direct bilirubin increase in proportion
(Continued)

Hepatitis: Laboratory Assessment, Overview
(Continued)

one to the other in hepatitis. The prothrombin time is close to normal in uncomplicated cases. Prothrombin time more than 4 seconds above the upper limit of reference range may indicate hepatic failure. Alkaline phosphatase is not usually greatly increased with hepatitis. Serum LDH more than slightly or moderately increased may signal an alternative diagnosis (eg, infectious mononucleosis or neoplastic disease). An abrupt onset of hepatitis suggests hepatitis A rather than hepatitis B or C. The onset of hepatitis C is classically insidious.

Chronic hepatitis by definition means increased serum transaminases (aminotransferases) for longer than 6 months, but fluctuation is seen. Fluctuation between abnormal and normal especially characterizes the course of hepatitis C. The differential diagnosis of possible chronic hepatitis includes hepatitis B, C, and D; alcoholism; autoimmune hepatitis; drug-induced chronic hepatitis; hemochromatosis; alpha$_1$ antitrypsin deficiency; and Wilson disease. See the Disease Index.

If HB$_s$Ag is positive in evaluation of chronic hepatitis B, testing for HB$_e$Ag is advocated to ascertain the presence of active viral replication. If appropriate viral hepatitis markers are not found, other evaluation is indicated, including work-up for autoimmune hepatitis. See the listing for Hepatitis (Autoimmune) in the Disease Index, in which a list of assays is included.

Algorithm 1. Serologic Testing in Suspected Acute Viral Hepatitis

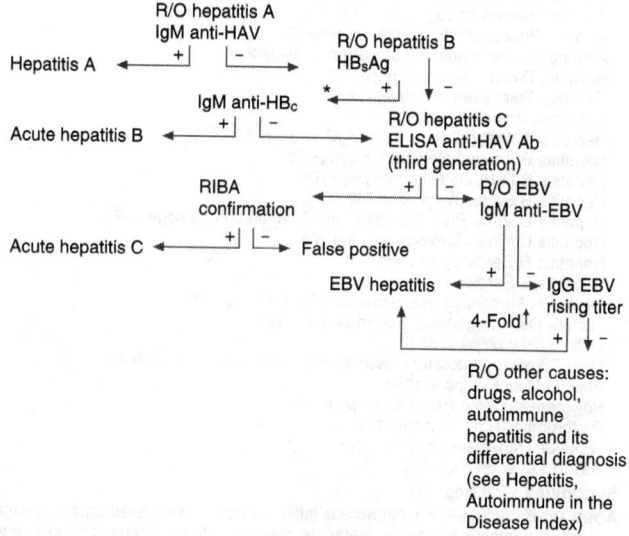

*Refer also to Algorithm 2. EBV = Epstein-Barr virus; ELISA = enzyme-linked immunosorbent assay; HAV = hepatitis A virus; anti-HB$_c$ = antibody to hepatitis B core; HB$_s$Ag = hepatitis B surface antigen; R/O = rule out; RIBA = recombinant immunoblot assay.

Adapted from Sacher RA, Peters SM, and Bryan JA, "Testing for Viral Hepatitis," *Am J Clin Pathol*, 2000, 113(1):13-7.

Algorithm 2. Serologic Testing in Suspected Chronic Viral Hepatitis

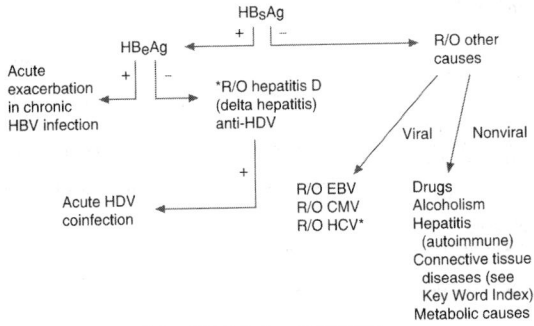

Rule out acute HAV, HBV, and HCV infection; see Algorithm 1.
*See text. CMV = cytomegalovirus; EBV = Epstein-Barr virus; HAV = hepatitis A virus; HB_sAg = hepatitis B surface antigen; HBV = hepatitis B virus; HB_eAg = hepatitis B_e antigen; HCV = hepatitis C virus; HDV = hepatitis D virus; R/O = rule

Adapted from Sacher RA, Peters SM, and Bryan JA, "Testing for Viral Hepatitis," *Am J Clin Pathol*, 2000, 113(1):13-7.

Viral Hepatitis Summary

Hepatitis Type	Clinical Features	AST, ALT	Serologic Findings	
			Acute Disease	Immunity
A (RNA)	Acute self-limiting; fecal-oral transmission; no carrier state	Markedly elevated, 3+ to 4+	IgM anti-HAV	IgG anti-HAV
B (DNA)	Acute or chronic; often asymptomatic; 5% infected blood and body fluids	1+ to 3+ (acute); variable, chronic	IgM anti-HB_c; HB_sAg; HB_e Ag; HBV DNA	IgG anti-HB_s; anti-HB_e
C (RNA)	Usually chronic	Mild to moderate elevation, even in acute cases	Total anti-HCV; HCV RNA	Unknown
D (RNA)	Requires HBV coinfection; acute or chronic; rare in U.S.	Often markedly elevated	IgM anti-HDV; total anti-HDV; may require multiple assays	Unknown
E (RNA)	Epidemic similar to HAV; acute self-limiting; waterborne and enteric; transmission in India and Southeast Asia	Variable	Testing not widely available	Unknown
G (RNA)	Close homology to HCV; high prevalence (2% of U.S. donors); usually mild disease, if at all; disease spectrum uncertain	Variable	HGV-RNA and EIA (reference laboratory only)	Unknown
Non A-E, exclude EBV, CMV	Possibly acute or chronic	Variable	None	Unknown

CMV: cytomegalovirus; EBV: Epstein-Barr virus; EIA: enzyme immunoassay; HAV: hepatitis A virus; anti-HB_c: antibody to hepatitis B core; anti-HB_e: antibody to hepatitis B_e antigen; HB_eAg: hepatitis B_e antigen; anti-HB_s: antibody to hepatitis B surface antigen; HB_sAg: hepatitis B surface antigen; HBV: hepatitis B virus; HCV: hepatitis C virus; HDV: hepatitis D virus; HGV: hepatitis G virus.
Adapted from Sacher RA, Peters SM, and Bryan JA, "Testing for Viral Hepatitis. A Practice Parameter," *Am J Clin Pathol*, 2000, 113(1):12-7.

(Continued)

Hepatitis: Laboratory Assessment, Overview
(Continued)

References

Centers for Disease Control and Prevention, "Revision of Acute Hepatitis Panel," *MMWR Morb Mortal Wkly Rep*, 2000, 49:424.

Ishak KG, "Pathologic Features of Chronic Hepatitis. A Review and Update," *Am J Clin Pathol*, 2000, 113(1):40-55.

OiConnor JA, "Acute and Chronic Viral Hepatitis," *Adolesc Med*, 2000, 11(2):279-92.

Ryder SD and Beckingham IJ, "ABC of Diseases of Liver, Pancreas, and Biliary System: Acute Hepatitis," *BMJ*, 2001, 322(7279):151-3.

Sacher RA, Peters SM, Bryan JA, "Testing for Viral Hepatitis. A Practice Parameter," *Am J Clin Pathol*, 2000, 113(1):12-7.

Internet Web Sites

www.ama-assn.org/ama/pub/category/3113.html

♦ **Hepzyme®** *see* Heparin Neutralization *on page 697*

HER-2/*neu*

Related Information

Breast Biopsy *on page 305*
Fluorescence *in situ* Hybridization *on page 602*
Immunoperoxidase Procedures *on page 780*

Synonyms c-*erb*B-2

Applies to FISH Assays

Abstract HER-2/*neu* is intimately associated with cellular proliferative capacity. Numerous studies have demonstrated an adverse effect of oncoprotein overexpression on prognosis, particularly in breast and ovarian cancer. More recently, this protein has been targeted for monoclonal antibody therapy in patients with metastatic breast carcinoma.

Specimen The HER-2/*neu* proto-oncogene is commonly evaluated by immunoperoxidase techniques using paraffin-embedded tissue. Paraffin-embedded tissue is also suitable for fluorescence *in situ* hybridization (FISH) assays.

Storage Instructions Routinely stored paraffin blocks are suitable for use over many years. Unstained slides may lose immunoreactivity for some antigens over time.

Causes for Rejection Lack of demonstrable tumor in paraffin section, extensive tissue necrosis

Limitations Immunoperoxidase methods are limited by tumor necrosis or sample size. Fluorescence *in situ* hybridization (FISH) techniques may not be performed by all laboratories and are more costly than immunoperoxidase procedures. Suboptimal tissue fixation or processing can adversely affect all immunostaining and probe techniques.

Methodology Immunoperoxidase staining procedures are most commonly used to determine HER-2/*neu* overexpression. Production of consistent staining results was initially challenging. Various staining patterns were produced by different antibodies and were influenced by different fixatives. Improved antigen retrieval methodologies, enhanced detection systems, and automated staining procedures have circumvented many of the early technical difficulties. However, interpretation of results, in some cases, may be problematic.

FISH has proven reliable for the determination of HER-2/*neu* gene amplification. Tissues are fixed in formalin, optimally for 24-48 hours, and embedded in paraffin. Sections of 4-5 microns thick are deparaffinized, treated with protease, denatured, and hybridized with fluorescent-labeled DNA probes. A color-coded fluorescent probe specific for the HER-2/*neu* gene and a second color-coded fluorescent probe specific for the centromeric region of chromosome 17 are used. The slides are then washed, counterstained, and examined using a fluorescent microscope for hybridization signals in the nuclei of neoplastic cells. The microscopist can distinguish benign epithelium, *in situ* carcinoma, and invasive carcinoma and record the ratio of HER-2/*neu* signal to chromosome 17 signal. A ratio ≥2 is interpreted as positive for HER-2/*neu* gene amplification.

An alternate (also commercially available) system provides data as an average number of HER-2/*neu* signals per cell. Recent studies using FISH have shown independent prognostic significance in multivariate analysis. Immunoassay and FISH results correlate well in most cases. Considerable controversy has arisen as to which method (immunohistochemistry vs FISH) should be employed for decisions concerning therapy. The College of American Pathologists Consensus Statement of 1999 notes that there is a current lack of standardization and comparability data. A specific method for c-*erb*B-2 (HER-2/*neu*) testing "cannot yet" be recommended. The CAP statement refers to *erb*B-2 testing as a "work in progress." Future recommendations will likely relate to the results of large multicenter outcome-based trials comparing FISH and immunohistochemical methods (currently in progress). FISH-based assays have won FDA approval for breast cancer prognosis (Inform; Ventana Medical Systems, Tucson, AZ) and for predicting response to docetaxel (Path Vysion; Vysis, Downers Grove, IL).

Quantitative PCR has been applied to the determination of HER-2/*neu* gene amplification, recently utilizing a fluorescent hybridization probe and melting curve analysis.

Chromogenic *in situ* hybridization (CISH) is a relatively new procedure that combines characteristics of immunohistochemical analysis and FISH. The method identifies gene amplification (semiquantifies gene copy number), provides a permanent slide, and uses a standard light microscope (visualization is not dependent on fluorescence microscopy). CISH assays are based on "subtractive hybridization" in which a DNA probe is visualized by a peroxidase reaction.

Additional Information HER-2/*neu* is a protein encoded by chromosome 17 and expressed as a 185 kd transmembrane receptor with tyrosine kinase activity. HER-2/*neu* shares close structural and functional similarities with the epidermal growth factor family of proteins, of which four are currently known. These membrane protein receptors transduce signals conferred by peptides in the extracellular milieu to promote cellular proliferation and differentiation by initiating a complex cascade of intracellular pathways. However, the specific ligand for HER-2/*neu* is unknown.

HER-2/*neu* protein expression is closely related to gene copy number. Normal cells have little or no demonstrable surface expression of HER-2/*neu* by immunohistochemical methods.

HER-2/*neu* is most studied in breast carcinoma, in which overexpression is detected in approximately 25% to 30% of neoplasms. Most early studies focused on the prognostic role of HER-2/*neu* overexpression, while more recent studies have emphasized its predictive attributes. Protein overexpression is highly correlated with high nuclear grade, negative hormone receptor status, high S-phase fraction, and tumor aneuploidy. Despite these associations, HER-2/*neu* expression has been shown in many studies to represent an independent adverse prognostic finding. Other studies have confirmed its negative impact on prognosis but failed to validate its value as an independent prognostic factor.

HER-2/*neu* protein overexpression has gained special significance since the 1998 FDA approval of trastuzumab (Herceptin®; Genentech, San Francisco, CA) for the therapy of metastatic breast cancer. This humanized monoclonal antibody is directed against HER-2/*neu* protein and has shown significant response rates in patients with stage IV breast cancer. The role of Herceptin® in the adjuvant setting is currently under investigation. As a prerequisite for Herceptin® therapy, the demonstration of HER-2/*neu* expression is mandatory.

An investigation of weakly positive (2+) immunohistochemical (IHC) staining reactions (based on tissue from 1556 breast biopsy specimens) was conducted by Mayo Medical Laboratories between August and December 2000. Of 216 tumors scoring 2+ (by IHC), 57% had no apparent FISH demonstrable abnormality. These results and those of numerous other comparison studies have led to the conclusion that IHC intermediate (2+) results should be regarded as borderline and should receive additional analysis (confirmatory FISH test) to be

(Continued)

HER-2/*neu* (Continued)

eligible for Herceptin therapy. Intermediate cases may result from artifactually high sensitivity and are possibly a variation of normal expression of the protein.

HER-2/*neu* protein over-expression also has predictive value for other therapeutic modalities. HER-2/*neu* overexpression is thought to predict increased resistance to cyclophosphamide therapy and enhanced sensitivity to chemotherapeutic regimens containing Adriamycin®.

Functional status of the HER-2 receptor may have more significance than gene amplification or protein overexpression. HER-2 in its phosphorylated state may be a requirement for adverse prognosis but its relation to trastuzumab therapy is unknown.

References

Arnould L, Denoux Y, MagGrogan D, et al, "Agreement Between Chromogenic *in situ* Hybridization (CISH) and FISH in the Determination of HER-2 Status in Breast Cancer," *Br J Cancer*, 2003, 88(10):1587-91.

Bilous M, Dowsett M, Hanna W, et al, "Current Perspectives on HER2 Testing: A Review of National Testing Guidelines," *Mod Pathol* 2003, 16(2):173-82.

Carney WP, Neumann R, Lipton A, et al, "Potential Clinical Utility of Serum HER-2/*neu* Oncoprotein Concentrations in Patients With Breast Cancer," *Clin Chem*, 2003, 49(10):1579-98.

Fitzgibbons PL, Page DL, Weaver D, et al, "Prognostic Factors in Breast Cancer. College of American Pathologists Consensus Statement 1999," *Arch Pathol Lab Med*, 2000, 124(7):966-78.

Hoang MP, Sahin AA, Ordonez NG, et al, "HER-2/*neu* Gene Amplification Compared With HER-2/*neu* Protein Overexpression and Interobserver Reproducibility in Invasive Breast Carcinoma," *Am J Clin Pathol*, 2000, 113(6):852-9.

Lyon E, Millson A, Lowery MC, et al, "Quantification of HER-2/*neu* Gene Amplification by Competitive PCR Using Fluorescent Melting Curve Analysis," *Clin Chem*, 2001, 47(5):844-51.

Pauletti G, Dandekar S, Rong H, et al, "Assessment of Methods for Tissue-Based Detection of the HER-2/*neu* Alteration in Human Breast Cancer: A Direct Comparison of Fluorescence *in situ* Hybridization and Immunohistochemistry," *J Clin Oncol*, 2000, 18(21):3651-64.

Perez EA, Roche PC, Jenkins RB, et al, "HER2 Testing in Patients With Breast Cancer: Poor Correlation Between Weak Positivity by Immunohistochemistry and Gene Amplification by Fluorescence *in situ* Hybridization," *Mayo Clin Proc*, 2002, 77(2):148-54.

Ross JS, Fletcher JA, Bloom KJ, et al, "HER-2/*neu* Testing in Breast Cancer," *Am J Clin Pathol*, 2003, 120(Suppl 1):S53-S71.

Varshney D, Zhou YY, Geller SA, et al, "Determination of HER-2 Status and Chromosome 17 Polysomy in Breast Carcinomas Comparing HercepTest and PathVysion FISH Assay," *Am J Clin Pathol*, 2003, 121(1):70-7.

Yeh IT, "Measuring HER-2 in Breast Cancer: Immunohistochemistry, FISH, or ELISA?" *Am J Clin Pathol*, 2002, 117(Suppl 1):S26-S35.

Hereditary Hemochromatosis DNA Test

Related Information

Ferritin, Serum *on page 577*
Iron and Total Iron Binding Capacity/Transferrin, Serum *on page 807*
Liver Biopsy *on page 864*
Liver Disease: Laboratory Assessment, Overview *on page 869*
Phlebotomy, Therapeutic *on page 1029*
Red Blood Cells *on page 1139*
Transferrin Receptor, Soluble, Serum or Plasma *on page 1267*

Synonyms Hemochromatosis, DNA Testing; HFE Genotyping

Test Includes Detection of the two most common mutations in the *HFE* gene, C282Y (nucleotide 845G→A) and H63D (nucleotide 187C→G)

Abstract Hereditary hemochromatosis (HH) is an autosomal recessive disease that is very common among people of European ethnicity. Among Caucasians, about 1 in 400 has the disease and about 1 in 10 is a carrier. It should be considered in patients with unexplained increases in AST or ALT. DNA-based testing can diagnose hemochromatosis in people with persistently elevated serum transferrin-iron saturation values. Testing is also useful in the differential diagnosis of liver diseases with increased iron loading and for evaluating at-risk relatives in families with hemochromatosis. In northern European patients with a diagnosis of hereditary hemochromatosis, about 80% to 90% have two copies of a mutation in the HLA-linked *HFE* gene referred to as C282Y. A second mutation, H63D, occurs in a smaller percentage of these patients (40% to 70% of non-C282Y cases of HH) and is associated with a much lower penetrance (likelihood of developing clinical disease). Since the C282Y mutation is very uncommon in Asian and African populations, this test is less useful in people from those ethnic backgrounds.

The classical signs of hemochromatosis include the triad of cirrhosis, diabetes mellitus and skin bronzing. It is now recognized that these late manifestations are preventable with early recognition and treatment.

Specimen 3-10 mL whole blood

Container Lavender top (EDTA) tube or yellow top (ACD) tube; avoid tubes containing heparin anticoagulants, which interfere with polymerase chain reaction analysis

Sampling Time Blood samples can be collected at any time.

Storage Instructions Store samples refrigerated or at room temperature. Do not freeze. Blood samples should normally be received in the testing laboratory within 4 days of collection to ensure adequate yield of DNA.

Turnaround Time 7-14 days

Special Instructions The following information helps the laboratory to provide an interpretive report: clinical diagnosis; ethnic background (Caucasian, Asian, African, etc); relevant laboratory values: serum iron, TIBC, and ferritin; presence or absence of a family history of hemochromatosis; presence or absence of a history of therapeutic phlebotomy.

Reference Interval

- Normal: absence of any detectable mutations
- Carrier status: detection of a single mutation
- Hereditary hemochromatosis or a genetic predisposition to develop the disease: detection of two mutations

Use Evaluation of suspected hemochromatosis: When asymptomatic patients have persistently elevated values for transferrin-iron saturation on two separate fasting blood samples, they should be evaluated further for possible hereditary hemochromatosis. Patients with clinical signs or symptoms suggesting hemochromatosis should also be evaluated. Molecular-genetic testing is being used increasingly for diagnostic purposes; detection of two mutant alleles will confirm the diagnosis in the setting of elevated biochemical iron measurements. Liver biopsy is still appropriate as a diagnostic test for patients with iron overload who lack two *HFE* mutations, and is useful for prognosis in patients who have already developed liver disease. It can provide quantitative iron determination as well as histopathologic evaluation. Because the development of cirrhosis in subjects with hemochromatosis leads to substantial risk of development of hepatocellular carcinoma, liver biopsy is essential when cirrhosis is suspected. Following studies of 32 individuals who underwent liver biopsies and testing for HFE mutations, Nash et al concluded that tests for HFE mutations and liver biopsies are complementary.

Testing at-risk relatives in families known to carry mutant alleles in the *HFE* gene: It is generally appropriate to evaluate first-degree and other relatives of an index case to determine if they are also at risk for the disease. Usually, siblings have a 1 in 4 risk of having the disease, while Caucasian parents and offspring usually have a 1 in 20 chance of being affected. Early detection and treatment of the disease by therapeutic phlebotomy can prevent development of significant organ damage.

Differential diagnosis of liver diseases with increased iron loading: Genetic testing can be useful in ruling out hereditary hemochromatosis as a causative factor in patients with cirrhosis who have other risk factors such as alcohol abuse or chronic viral hepatitis.

Limitations Expression of C282Y/C282Y homozygosity is not invariable; individuals have been reported who are homozygous for this mutation who fail to meet clinical criteria for the diagnosis of hereditary hemochromatosis. Hepatic fibrosis is not usually found before age 40 without a cofactor, eg, excessive ethanol consumption or hepatitis C infection. Age and phenotypic expression are relevant.

The penetrance of the H63D allele is much less than that of the C282Y allele. No more than 2% of people with a C282Y/H63D or H63D/H63D genotype develop clinically significant iron overload. Thus, most asymptomatic patients with these genotypes may only require periodic monitoring of serum ferritin every few years, and few will require therapeutic phlebotomy.

All homozygotes for the major hemochromatosis mutation are not detected by screening for transferrin saturation and ferritin concentrations. The absence of HFE mutations does not rule out hemochromatosis. This test is not useful for (Continued)

Hereditary Hemochromatosis DNA Test *(Continued)*

the diagnosis of neonatal or juvenile hemochromatosis, which are due to mutations in different genes.

Methodology DNA from portions of the *HFE* gene is amplified by polymerase chain reaction (PCR). Mutations are detected by restriction-enzyme digestion or by hybridization with oligonucleotide probes specific for the mutations. Analysis by automated capillary electrophoresis is described.

Additional Information The *HFE* gene is located in the HLA cluster on chromosome 6p21. It encodes a cell-surface protein of 321 amino acids that has structural similarity to HLA class I molecules. The normal protein forms a heterodimer with β-2-microglobulin; this heterodimer interacts with the transferrin receptor and may modulate its affinity for transferrin. The C282Y mutation disrupts an important disulfide bond in the *HFE* protein, and thereby prevents expression of the protein on the cell surface.

Conditions relating to hemochromatosis include increased ALT and AST without other identifiable cause, hemochromatotic arthropathy, cardiomyopathy, hyperpigmentation (bronzing), hypothyroidism, testicular atrophy and abnormalities of the anterior pituitary and pancreas. See Hemochromatosis in the Disease Index.

Non-HFE-related iron overload conditions include juvenile hemochromatosis (see above), African iron overload, iron overload in African-Americans, aceruloplasminemia, and Hallervorden-Spatz disease (a degenerative neurologic disorder).

References

Andrews NC, "Iron Metabolism: Iron Deficiency and Iron Overload," *Annual Review of Genomics and Human Genetics*, Volume 1, Palo Alto, CA: Annual Reviews, 2000, 75.

Beutler E, Felitti V, Gelbart T, et al, "The Effect of *HFE* Genotypes on Measurements of Iron Overload in Patients Attending a Health Appraisal Clinic," *Ann Intern Med*, 2000, 133(5):329-37.

Green RM and Flamm S, American Gastroenterological Association, "AGA Technical Review on the Evaluation of Liver Chemistry Tests," *Gastroenterology*, 2002, 123(4):1367-84.

Koziol JA, Ho NJ, Felitti VJ, and Beutler E, "Reference Centiles for Serum Ferritin and Percentage of Transferrin Saturation, With Application to Mutations of the *HFE* Gene," *Clin Chem*, 2001, 47(10):1804-10.

Lyon E and Frank EL, "Hereditary Hemochromatosis Since Discovery of the *HFE* Gene," *Clin Chem*, 2001, 47(7):1147-56.

Nash S, Marconi S, Sikorska K, et al, "Role of Liver Biopsy in the Diagnosis of Hepatic Iron Overload in the Era of Genetic Testing," *Am J Clin Pathol*, 2002, 118(1):73-81.

Pietrangelo A, Montosi G, Totaro A, et al, "Hereditary Hemochromatosis in Adults Without Pathogenic Mutations in the Hemochromatosis Gene," *N Engl J Med*, 1999, 341(10):725-32.

Snover DC, "Hepatitis C, Iron, and Hemochromatosis Gene Mutations," *Am J Clin Pathol*, 2000, 113(4):475-8.

Internet Web Sites

www.geneclinics.org/profiles/hemochromatosis/

♦ **Heroin** *see* Morphine, Urine *on page 921*

♦ **Heroin** *see* Opiates, Qualitative, Urine *on page 974*

♦ **Heroin Detoxification** *see* Methadone, Serum or Urine *on page 900*

♦ **Heroin Metabolite, Urine** *see* Morphine, Urine *on page 921*

Herpes Simplex Antibody

Related Information

Herpes Simplex Virus Culture and Antigen Detection *on page 721*
Herpes Simplex Virus DNA Detection *on page 723*
Herpesvirus Cytology *on page 727*
Polymerase Chain Reaction *on page 1069*
TORCH *on page 1262*
Viral Culture *on page 1307*

Test Includes Detection of IgG and IgM antibodies specific for herpes simplex 1 and/or 2

Specimen Serum, cerebrospinal fluid

Container Red top tube, sterile CSF tube

Reference Interval Interpretation depends on whether episode is initial or reinfection. Generally, IgG: <1:5; IgM: <1:10.

Critical Values Increased IgM antibodies or a fourfold rise in titer may indicate recent infection.

Use Determine a previous infection with herpes simplex virus 1 or 2; assess pregnant women for herpes simplex antibodies; identify seropositive organ transplant recipients

Limitations Extensive background antibody in the population, and cross reaction of HSV-1 and HSV-2 responses have made this test difficult to interpret. False negatives occur. HSV PCR on cerebrospinal fluid is preferred for diagnosis of herpes simplex encephalitis.

Methodology Indirect fluorescent antibody (IFA), hemagglutination, complement fixation (CF), enzyme immunoassay (EIA), Western blot assay

Additional Information A primary HSV-1 or HSV-2 infection will produce a classical rising antibody titer. Use of antiviral drugs may blunt the antibody response. Exposure to herpesvirus is very common, thus the presence of antibodies to HSV is usually not helpful. Collecting paired sera to delineate rising titers generally adds nothing to clinical management, and thus, **herpes serology cannot usually be recommended in routine clinical cases.**

Neonatal herpes is most often caused by type 2. Identification of seronegative pregnant women, near term, who have seropositive partners, has been addressed. Infants born to seronegative mothers who acquire primary genital HSV type 2 infection are at more risk than are infants born to seropositive mothers with recurrent genital herpesvirus infection. Primary cases late in gestation are regarded as especially dangerous. Herpes serology has not so far been useful in determining whether Caesarean delivery should be undertaken in pregnant patients with possible herpetic lesions. Nor is herpes serology helpful in the diagnosis of a very sick infant with possible congenital herpes. Virus culture and PCR are preferred, but therapy should not be withheld even for the results of these tests. Because of the fulminant course, even IgM antibody is not present in time to contribute to care. See Herpes Simplex Virus DNA Detection *on page 723.*

References

Ashley R and Wald A, "Genital Herpes: Review of the Epidemic and Potential Use of Type-Specific Serology," *Clin Microbiol Rev,* 1999, 12(1):1-8.

Brown ZA, Selke S, Zeh J, et al, "The Acquisition of Herpes Simplex Virus During Pregnancy," *N Engl J Med,* 1997, 337(8):509-15.

Corey L and Handsfield HH, "Genital Herpes and Public Health. Addressing a Global Problem," *JAMA,* 2000, 283(6):791-4.

Cowan FM, "Testing for Type-Specific Antibody to Herpes Simplex Virus - Implications for Clinical Practice," *J Antimicrob Chemother,* 2000, 45(Topic T3):9-13.

Internet Web Sites

www.cdc.gov/std/Herpes/STDFact-Herpes.htm

Herpes Simplex Virus Culture and Antigen Detection

Related Information

Test Includes Direct (nonculture) detection of HSV-infected cells in smears of specimens and culture for HSV only; HSV also is detected in a routine viral culture; rapid method includes specific staining with a fluorescent monoclonal antibody

Abstract Herpes simplex viruses cause a wide variety of lesions. More severe forms of infection include herpetic dendritic eye ulcers. Neonatal herpes and encephalitis are entities which may lead to serious sequelae and are often fatal. Herpes infections may be severe in subjects who are immunocompromised. Culture is the method of choice for many clinical presentations, including vesicles. Classic primary HSV-1 infection includes herpetic gingivostomatitis, which may be accompanied by fever and local lymphadenopathy. HSV-1 also causes conjunctivitis, keratitis, and sporadic encephalitis.

About 85% of instances of primary genital HSV infections are caused by HSV-2, and 99% of recurrent disease is HSV-2. Neonatal herpes is usually caused by exposure during vaginal delivery. When the mother has a primary (Continued)

Herpes Simplex Virus Culture and Antigen Detection
(Continued)

infection at delivery, the attack rate is greater than half. Untreated neonates with disseminated infection suffer mortality rates >70%. Most HSV-2 seropositive individuals intermittently shed virus from mucosal surfaces, and represent reservoirs for spread to uninfected sex partners.

Specimen Vesicle fluid, swab of base of lesion, tissue biopsy, nasopharyngeal swab, tonsillar swab, oropharyngeal swab, conjunctival swab, cervical swab, urine, cerebrospinal fluid

Container Specimen on glass slide for antigen detection; cold viral transport medium for swabs

Sampling Time Preferably within 3 days of lesion eruption. Specimens taken after 5 days are less likely to contain viral particles.

Collection

Direct antigen detection: Cells from the bottom of an ulcer or vesicle should be scraped with a swab, scalpel, or curette. For direct detection of HSV antigen, swabs should be **rolled (not smeared)** across a small area of the slide several times, and cells scraped with a scalpel should be gently dabbed onto the slide. **The smear should be air dried at room temperature.** The success of direct detection procedures depends on the careful preparation of cell smears.

HSV culture: All specimens should be kept cold and moist. Specimens should be collected in the acute stage of the disease, preferably within 3 days and no longer than 7 days after the onset of illness. All specimens should be collected on a sterile swab as described and the swab should be placed into cold viral transport medium immediately after collection.

Endocervical: Swab cervix with a rolling/scraping motion to assure obtaining epithelial cells.

Vesicular lesion: Wash vesicles with sterile saline. Carefully open several vesicles and soak up vesicular fluid with swab. If vesicles are absent, vigorously swab base of lesion (specimen should be collected during first 3 days of eruption. Specimens collected later in the course of disease rarely yield virus).

Conjunctival: Using a moistened swab, firmly rub conjunctiva using sufficient force to obtain epithelial cells.

Throat, respiratory, oral: Rotate swab in both tonsillar crypts and against posterior oropharynx.

Storage Instructions Specimens should be delivered to the laboratory and handed to a technologist within 30 minutes of collection. Outpatient specimens: If transport is to be delayed more than 30 minutes after collection, specimen **must** be refrigerated (held at 4°C to 8°C) until it can be transported to the laboratory. If inoculation onto cell cultures is not possible within 48 hours, specimens should be frozen at -70°C. Do not freeze at -20°C.

Turnaround Time Two to 4 hours for antigen detection; routine culture takes 1-5 days; rapid culture takes 16 hours to 2 days

Special Instructions Operative biopsy specimens and spinal fluid specimens for fluorescent antibody (FA) testing should be **processed immediately**. The laboratory should be **notified in advance** when either of these specimen types will be sent for FA. Special viral transport medium must be obtained from the laboratory prior to collection of specimen.

Reference Interval No virus isolated or herpes simplex virus-infected cells detected

Critical Values Detection of HSV in CSF

Use Aid in the diagnosis of disease caused by HSV

Limitations The detection of HSV directly in patient material depends in great part on the collection of a sufficiently large number of intact infected cells from the lesion. Specimen smears that are too thick can retain or trap the fluorescent reagent and make the test difficult to interpret. If at all possible, it is important to obtain cells from the base of an intact vesicle. Fluorescent antibody staining of cells from an early lesion is 80% sensitive in acute vesicles, but only 60% to 75% sensitive in resolving lesions. The presence of infected cells decreases as the lesion heals, and crusted lesions may have little or no herpes antigenic material remaining. Antigen detection methods have excellent specificity (~98%), thus, a high positive predictive value, but are less sensitive than culture.

Standard culture methods suffer 30% false negatives for identification of those neonates who subsequently develop neonatal herpes. Sensitivity of PCR assay for HSV detection is greater than that of culture methods.

Methodology Direct fluorescent antibody (DFA) or enzyme immunoassay (EIA) on patient specimens. HSV culture requires inoculation of specimen into cells, incubation of cultures, observation for characteristic cytopathic effect (CPE), and identification of HSV by fluorescein-labeled monoclonal antibodies specific for type 1 or 2. In the shell vial isolation technique, the specimen is centrifuged on a tissue culture cell layer, incubated for 12-18 hours, and stained with direct immunofluorescent staining for HSV-1 and HSV-2; characteristic fluorescent foci indicate the presence of virus.

Additional Information In general, the antigen detection test is only ~70% as sensitive as cell culture. In critical situations, use of both methods deserves consideration. Air-dried preparations on slides can also be stained with Giemsa, Pap, Wright, or Diff-Quik™ stains (Tzanck) (see Herpesvirus Cytology *on page 727*). The HSV can be cultured from most infections (eg, gingivostomatitis, herpes labialis, genital herpes, skin lesions, keratoconjunctivitis, neonatal herpes, aseptic meningitis, encephalitis). Culture can provide evidence of acyclovir resistance, and detect potential for transmission of HSV to neonates. Culture is useful for diagnosis of unusual lesions, such as herpetic glossitis in immunocompromised patients. The buccal mucosa, floor of the mouth, and soft palate may be infected as well. Otherwise, herpes simplex lesions in the mouth are not common in immunocompetent persons, except for primary herpetic gingivostomatitis.

HSV-1 can be cultured from CSF in herpes meningitis but only rarely can be cultured from the CSF in encephalitis. The HSV PCR is usually positive in encephalitis. HSV-2 may be isolated from CSF in neonatal meningitis. The detection of HSV DNA by PCR is more sensitive than detection of specific antibody for early diagnosis of herpes simplex encephalitis (97% to 99% sensitivity) and provides 95% to 100% specificity.

References
Benedetti JK, Zeh J, and Corey L, "Clinical Reactivation of Genital Herpes Simplex Virus Infection Decreases in Frequency Over Time," *Ann Intern Med*, 1999, 131(1):14-20.

Corey L and Handsfield HH, "Genital Herpes and Public Health. Addressing a Global Problem," *JAMA*, 2000, 283(6):791-4.

Hook EW 3d, Cannon RO, Nahmias AJ, et al, "Herpes Simplex Virus Infection as a Risk Factor for Human Immunodeficiency Virus Infection in Heterosexuals," *J Infect Dis*, 1992, 165(2):251-5.

Jacobs DS, DeMott WR, Oxley DK, et al, *Laboratory Test Handbook*, 5th ed, Hudson, OH: Lexi-Comp Inc, 2001.

Koelle DM and Wald A, "Herpes Simplex Virus: The Importance of Asymptomatic Shedding," *J Antimicrob Chemother*, 2000, 45(T3):1-8.

Wald A, Zeh J, Selke S, et al, "Reactivation of Genital Herpes Simplex Virus Type 2 Infection in Asymptomatic Seropositive Persons," *N Engl J Med*, 2000, 342(12):844-50.

Internet Web Sites
www.cdc.gov/std/Herpes/STDFact-Herpes.htm

Herpes Simplex Virus DNA Detection
Related Information
Encephalitis Viral Serology *on page 535*
Herpes Simplex Antibody *on page 720*
Herpes Simplex Virus Culture and Antigen Detection *on page 721*
Herpesvirus Cytology *on page 727*
Polymerase Chain Reaction *on page 1069*
Viral Culture *on page 1307*

Test Includes Direct detection of herpes simplex virus (HSV) nucleic acid in cells, blood, or CSF. HSV may also be detected in cytologic specimens or tissue samples obtained by biopsy. See Herpesvirus Cytology *on page 727*.

Abstract Herpes simplex virus (HSV) infects epithelial cells and is responsible for causing vesicular lesions, followed by small ulcers and crusts. It may involve the mouth, intestinal tract, urogenital tract, or central nervous system. HSV establishes a latent infection of nerve dorsal root ganglion cells from which it reactivates, travels retrograde through axons, and reinfects epithelial cells. Transmission is through direct contact and the virus may be spread to the newborn via the birth canal. HSV is a common sexuall-transmitted disease.
(Continued)

Herpes Simplex Virus DNA Detection *(Continued)*

Infection may cause a wide variety of clinical entities, including encephalitis, for which timely diagnosis and treatment are critical. Detection of HSV may be accomplished by a variety of molecular techniques including amplification, such as by polymerase chain reaction (PCR), or through *in situ* hybridization. HSV may be detected by culture in only 40% to 70% of genital lesions and in only 25% to 40% of neonates with encephalitis. Detection of HSV by PCR has been shown to have greater sensitivity than routine culture methods. As with culture techniques, it is possible to distinguish HSV-1 from HSV-2. Molecular assays have been used to establish the presence of latent infection, when the virus is not replicating.

Culture is the method of choice for many acute presentations.

Specimen Bronchial alveolar lavage fluid, cerebrospinal fluid, vesicle fluid, blood or serum, various tissues

Container Sterile containers are acceptable; lavender top (EDTA) tube for blood

Storage Instructions Any of the above specimens should be rapidly transported to the laboratory. However, if transportation is expected to exceed 2 hours, the sample should be refrigerated and if transportation exceeds 8 hours, a specimen should be frozen. All tissue samples should be frozen as soon as possible after the biopsy procedure.

Causes for Rejection Blood collected in heparinized tubes results in inhibition of amplification

Critical Values Detection of HSV DNA in CSF

Use Useful for rapidly identifying active infection with HSV. Untreated, herpes encephalitis and neonatal herpes are fatal, or cause severe morbidity in a majority of patients. Neurologic sequellae are common. PCR is considered the most reliable means of precision of a laboratory diagnosis of CNS HSV, but *vide infra*. Most HSV encephalitis is caused by HSV type 1. HSV meningitis is more frequent than is HSV encephalitis, and usually is caused by type 2.

Limitations A negative result does not rule out the presence of HSV, since factors such as degradation of the nucleic acid or interference with amplification by substances that may be present. False-positive results are possible.

Methodology Molecular techniques used are PCR, Hybrid capture®, and *in situ* hybridization. The most common method used is PCR.

Additional Information In most nonimmunosuppressed individuals, infection by HSV is self-limited, one important exception is found in individuals who develop encephalitis (HSE). Thirty-three percent of adults who develop herpes encephalitis have primary infection. Of those who have herpes antibodies at the onset of encephalitis, >90% have not had recurrent labial herpes. Neonates who are infected by passage through the birth canal may have skin, eye, and mucous membrane involvement leading to dissemination and encephalitis. Untreated, herpes encephalitis and neonatal herpes are fatal, or cause severe morbidity in a majority of patients. Neurologic sequellae are common. PCR is considered the most reliable method for a laboratory diagnosis of CNS HSV, but *vide infra*. Most HSV encephalitis is caused by HSV type 1. HSV meningitis is more frequent than is HSV encephalitis, and usually is caused by type 2.

Amplification assays for detection of HSV have been developed because of the increased sensitivity that these assays provide. The standard noninvasive approach (without brain biopsy) to diagnosis of HSE shows that serum antibodies, usually positive, are not helpful. HSV cultures are usually negative in HSE, but may be positive in HSV meningitis or meningoencephalitis of immunosuppressed hosts. HSV PCR, which has shown a higher sensitivity than culture, is the noninvasive diagnostic technique of choice.

Brain biopsy with culture, and histopathology are useful in diagnosis of HSE, and help to exclude clinical syndromes that mimic HSE. Studies have confirmed the ability of PCR to detect HSV DNA when culture results are negative.

References

Corey L and Handsfield HH, "Genital Herpes and Public Health: Addressing a Global Problem," *JAMA*, 2000, 283(6):791-4.

Langenberg AG, Corey L, Ashley RL, et al, "A Prospective Study of New Infections With Herpes Simplex Virus Type 1 and Type 2, Chiron HSV Vaccine Study Group," *N Engl J Med*, 1999, 341(19):1432-8.

Skoldenberg B, "Herpes Simplex Encephalitis," *Scand J Infect Dis*, 1996, 100:8-13.

Storch GA, "Identifying HSV Infections of the CNS," *Lab Med*, 2000, 31(6):316-7.

Internet Web Sites
www.cdc.gov/std/Herpes/STDFact-Herpes.htm

♦ **Herpes Smear** *see* Cervical/Vaginal Cytology *on page 376*

Herpesvirus 6 Serology

Related Information

Infectious Mononucleosis Screening Test *on page 785*

Test Includes Detection and quantitation of antibodies to human herpesvirus 6 (HHV-6)

Abstract Human herpesvirus 6 (HHV-6) is a herpes DNA virus with affinity for T lymphocytes. It usually infects infants between age 6 months and 2 years. The primary infection often manifests as roseola infantum, but HHV-6 is also the most common cause of febrile convulsions in children. Adult infections are usually recognized in immunocompromised hosts. Such infections may include several syndromes with fever, rash, pneumonitis, encephalitis, and other findings.

Specimen Serum

Container Red top tube or serum separator tube

Collection Acute and convalescent specimens are recommended.

Storage Instructions Refrigerate serum.

Use IgM HHV-6 may aid in the diagnosis of acute or recent infection with HHV-6.

Methodology Indirect fluorescent antibody (IFA)

Additional Information HHV-6 antibodies are very common. Most healthy children have antibody by age 2 or 3 years. Since the virus persists in lymphocytes indefinitely, nearly all adults have antibody, although seropositivity may decrease with age.

Primary infection with HHV-6 is associated with roseola infantum (exanthem subitum) a febrile illness with high fever which lasts for 3-5 days. Subsequently, a rash abruptly appears. It also causes a primary infection with fever without rash. HHV-6 can also causes febrile convulsions in infants, encephalitis, a heterophil-negative mononucleosis syndrome indistinguishable from that caused by EBV, or hepatitis, sometimes fulminant. HHV-6 is shed in saliva and urine.

References

Campadelli-Fiume G, Mirandola P, Menotti L, "Human Herpesvirus 6: An Emerging Pathogen," *Emerg Infect Dis*, 1999, 5(3):353-66.

Clark DA, "Human Herpesvirus 6," *Rev Med Virol*, 2000, 10(3):155-73.

Dockrell DH, Smith TF, and Paya CV, "Human Herpesvirus 6," *Mayo Clin Proc*, 1999, 74:163-170

Leach CT, "Human Herpesvirus-6 and -7 Infections in Children: Agents of Roseola and Other Syndromes," *Curr Opin Pediatr*, 2000, 12(3):269-74.

Stoeckle MY, "The Spectrum of Human Herpesvirus 6 Infection: From Roseola Infantum to Adult Disease," *Annu Rev Med*, 2000, 51:423-30.

Internet Web Sites
www.emedicine.com

Herpesvirus 8

Related Information

Body Cavity Fluid Cytology *on page 285*
Body Fluid Lactate Dehydrogenase *on page 294*
Epstein-Barr Virus Serology *on page 550*
Fine Needle Aspiration, Deep and Superficial Masses *on page 590*
Polymerase Chain Reaction *on page 1069*
Skin Biopsy *on page 1200*

Synonyms HHV-8; Human Herpesvirus Type 8; Kaposi Sarcoma-Associated Herpesvirus; KSHV

Abstract Kaposi sarcoma was originally described over a century ago, but the HHV-8 virus has only recently been identified.

Surveys show HHV-8-specific antibody in up to 10% of healthy U.S. blood donors, 2% to 4% of hemophiliacs, 20% to 30% of HIV-positive gay men without Kaposi sarcoma (KS), 70% to 90% of patients with KS, and almost 100% of immunocompetent patients with the disease.

Infection rates of HHV-8 are parallel with the incidence of KS: low in the U.S. and much of Europe and Asia; intermediate in Mediterranean countries, with the highest rates in Uganda, Zambia, and South Africa. KS is the most frequent
(Continued)

Herpesvirus 8 *(Continued)*

neoplasm in patients with AIDS. KS is much more frequent among males than females who are HIV positive. Nonhomosexual HIV-infected individuals (eg, hemophiliacs, heterosexual injection drug users) also have a lower prevalence. The genome of HHV-8 has been found in KS tumor cells from AIDS patients and as well in specimens from KS patients who are HIV negative.

Container Red top tube or SST™ tube for HHV-8 IgG antibody

Use Diagnose HHV-8 with or without KS.

KS, an angioproliferative disease, was originally described as an uncommon entity by a Hungarian physician in 1872, many decades before AIDS was recognized. In addition to its classic form, KS is prevalent among individuals with AIDS and in patients treated with immunosuppressives associated with organ transplantation. HHV-8 is transmitted through renal allografts. It is a risk factor for transplantation-associated KS. HHV-8 antibodies can be used to assess organ transplant patients. Subjects positive for HHV-8 antibodies, who are recipients of an organ transplanted from a donor positive for HHV-8, are at increased risk for post-transplantation development of KS.

The risk of KS is 73,000 times greater in homosexual patients with AIDS.

HHV-8 is found as well in **body cavity-based B-cell lymphoma (primary effusion lymphoma)**, and some plasma cell forms of multicentric Castleman disease. In primary-effusion lymphoma, malignant peritoneal, pericardial, or pleural effusions occur without lymph node involvement or identification of a distinctive, conventional neoplastic mass. It is aggressive.

Limitations Some authorities have criticized lack of sensitivity among the serologic tests. When only low concentrations of IgG are present, in the acute phase, false negatives may be seen.

Although KSHV DNA is found in only about 50% of infected patients with standard PCR assays, PCR and Southern blot hybridization assay can detect viral DNA in essentially all lesions of KS.

Methodology Second generation serological assays, indirect immunofluorescence assay, KSHV PCR, quantitative PCR

Additional Information HHV-8 is a Kaposi sarcoma tumor virus prevalent in southern Europe and in Africa. It is virologically similar to EBV, is persistent in the body after infection, reproduces in peripheral blood mononuclear cells, and may be found in saliva. This virus may be even more prevalent than suspected previously. The use of HHV-8 lytic-cycle antigens has shown nearly all KS patients are positive, as are 90% of HIV-infected gay men, 20% of HIV-positive intravenous drug addicts (IVDA), up to 25% of healthy adults, and 8% of children.

KSHV is a sexually transmitted disease, but other means of transmission predominate elsewhere (eg, in Africa, infection can occur in childhood). Maternal-infant transmission is seen in countries in which HHV-8 infection is endemic.

The risk of transmission through transfusion is unknown at this time, but is unequivocally lower than that of HIV.

References

Alkan S, Eltoum IA, Tabbara S, et al, "Usefulness of Molecular Detection of Human Herpesvirus-8 in the Diagnosis of Kaposi Sarcoma by Fine-Needle Aspiration," *Am J Clin Pathol*, 1999, 111(1):91-6.

Antman K and Chang Y, "Kaposi's Sarcoma," *N Engl J Med*, 2000, 342(14):1027-38.

Hoang MP, Rogers BB, Dawson DB, et al, "Quantitation of 8 Human Herpesviruses in Peripheral Blood of Human Immunodeficiency Virus-Infected Patients and Healthy Blood Donors by Polymerase Chain Reaction," *Am J Clin Pathol*, 1999, 111(5):655-9.

Jaffe HW and Pellett PE, "Human Herpesvirus 8 and Kaposi's Sarcoma - Some Answers, More Questions," *N Engl J Med*, 1999, 340(24):1912-3.

Jones D, Ballestas ME, Kaye KM, et al, "Primary-Effusion Lymphoma and Kaposi's Sarcoma in a Cardiac-Transplant Recipient," *N Engl J Med*, 1998, 339(7):444-9.

Mayo Reference Services, "Human Herpesvirus-8 Antibodies, IgG, Serum," *New Test Announcement*, #81971, Rochester, MN: Mayo Medical Laboratories, September 2000.

Regamey N, Tamm M, Wernli M, et al, "Transmission of Human Herpesvirus 8 Infection From Renal-Transplant Donors to Recipients," *N Engl J Med*, 1998, 339(19):1358-63.

Sitas F, Carrara H, Beral V, et al, "Antibodies Against Human Herpesvirus 8 in Black South African Patients With Cancer," *N Engl J Med*, 1999, 340(24):1863-71.

Internet Web Sites

www.emedicine.com

Herpesvirus Cytology

Related Information

Herpes Simplex Antibody *on page 720*
Herpes Simplex Virus Culture and Antigen Detection *on page 721*
Herpes Simplex Virus DNA Detection *on page 723*
Polymerase Chain Reaction *on page 1069*
Skin Biopsy *on page 1200*
Varicella-Zoster Virus Culture and Serology *on page 1300*
Viral Culture *on page 1307*

Applies to Tzanck Smears

Abstract Herpes simplex virus (HSV) infection may be asymptomatic or lead to ulcerated painful lesions of epithelial lined surfaces. Infected epithelial cells demonstrate characteristic nuclear changes including multinucleation, molding, and chromatin margination. In the Tzanck preparation, cells obtained from herpesvirus-induced vesicles are cytologically examined. This test may lead to a rapid diagnosis, but is only positive in ~50% of cases. Cytologic examination of gastrointestinal tract and genitourinary tract specimens can be used for diagnosis, but tissue biopsy is complimentary to these studies. Herpes simplex virus can also be detected in urine and CSF fluid. Genital herpes simplex virus infection can be associated with neonatal morbidity and mortality and therefore, its diagnosis in pregnant patients is important. Life-threatening herpesvirus infections include those in neonates and encephalitis.

Collection Firmly scrape the edge of the lesion, preferably a bullous lesion after removal of the bulla. The edge of normal skin and ulcer is to be scraped. In sites other than skin, a direct scrape is done. The scrape may be done with a wooden spatula or tongue blade.

Use The most common cause of viral esophagitis is herpes simplex virus, which causes multiple shallow ulcers.

Limitations Tzanck smears cannot provide distinction between HSV-1 and HSV-2. Neonatal herpes is most often due to type 2 infection. Herpes inclusions may not be seen in 50% of active lesions. Interpretation can be difficult. Viral culture has been the definitive diagnostic method. Polymerase chain reaction has been utilized successfully in detection of HSV and VZV DNA sequences and has been reported as equivalent or superior to viral culture. See Herpes Simplex Virus DNA Detection *on page 723*. Biopsy is also useful in particular cases.

Methodology Air-dried, Diff-Quik™ stained or alcohol-fixed, Pap-stained smear. Giemsa or Wright stain may also be used.

Additional Information Diagnostic yield is increased by immunoperoxidase or immunofluorescent procedures, which become positive before characteristic viral cytopathic changes develop.

References

Atkinson BF and Silverman JF, eds, *Atlas of Difficult Diagnoses in Cytopathology*, Philadelphia, PA: WB Saunders Co, 1998.

Geisinger KR, "Alimentary Tract (Esophagus, Stomach, Small Intestine, Colon, Rectum, Anus, Biliary Tract)," *Comprehensive Cytopathology*, 2nd ed, Chapter 18, Bibbo M ed, Philadelphia, PA: WB Saunders Co, 1997, 413-44.

Gupta PK, "Microbiology, Inflammation, and Viral Infections," *Comprehensive Cytopathology*, 2nd ed, Chapter 9, Bibbo M, ed, Philadelphia, PA: WB Saunders Co, 1997, 125-60.

♦ **Heterophil Antibody** *see* Infectious Mononucleosis Screening Test *on page 785*

Heterophilic Antibodies

Related Information

CA 125, Serum *on page 323*
Carcinoembryonic Antigen, Serum *on page 342*
Chorionic Gonadotropin, Human, Serum and Urine *on page 397*
Creatine Kinase, Serum *on page 470*
Erythropoietin, Serum *on page 551*
Estradiol, Serum *on page 553*
Follicle Stimulating Hormone, Serum, Plasma, or Urine *on page 609*
Free Thyroxine Index *on page 613*
Infectious Mononucleosis Screening Test *on page 785*
Luteinizing Hormone, Blood or Urine *on page 876*
Progesterone, Serum *on page 1093*
(Continued)

Heterophilic Antibodies (Continued)

Prolactin, Serum on page 1094
Thyroid Stimulating Hormone, Serum on page 1250
Thyroxine, Serum on page 1257
Triiodothyronine, Serum on page 1276
Troponins, Serum on page 1278

Synonyms Human Antianimal Antibodies; Human Antimouse Antibodies

Abstract

Terminology: Heterophilic antibodies are circulating human antibodies with specificity for some particular animal protein, usually an immunoglobulin. Such antibodies exhibit a high degree of species and class specificity. **Heterophilic antibodies** must be distinguished from heterophil antibodies which have broad species reactivity and are useful in the diagnosis of infectious mononucleosis. (For diagnosis of infectious mononucleosis, see Infectious Mononucleosis Screening Test on page 785.) For clarity in communication, **human antianimal antibodies** (HAAA) may be a better term. Because mouse monoclonal antibodies are extensively used in diagnosis and treatment, the vast majority of HAAA are murine; as a result, the term **human antimouse antibodies** (HAMA) is used often in the medical literature.

Clinical issues: HAAA can result from medications (eg, antibody-targeted drugs, antithymocyte globulin), radionuclide imaging agents, blood transfusions, and vaccination. These antibodies can cross the placenta. HAAA can cause interference in any immunoassay. Interferences have been reported in assays for all the analytes listed above (see Related Information). Depending on the assay format, the interference can result in a false increase or decrease in the results. Moreover, such interferences are largely unpredictable. The FDA now recommends that a warning be included in package inserts if a diagnostic test kit employs mouse monoclonal antibodies.

Interference from HAAA is unpredictable, varies among different assays, and usually is identified when a clinician reports that a result appears inexplicably discordant with the clinical situation. The most dramatic of the HAAA errors have been in assays for serum human chorionic gonadotropin (hCG). In this situation, HAAA circulating in the patient react with mouse monoclonal IgG used as the capture antibody in the hCG assay. Such patients have had falsely elevated serum hCG results, typically 20-500 IU/L, which have led to erroneous diagnoses of malignant trophoblastic disease, followed by chemotherapy.

Specimen The ideal specimen is the specimen which produced the result that is being questioned. Alternatively, a new specimen can be used; ordinarily it should be collected in the same way as was the original.

Critical Values Whenever an erroneous result due to HAAA is suspected, the clinician(s) should be notified promptly. Often, of course, the clinician will be the source of the information that a result is suspect.

Use

General guidelines: When a clinician reports that an immunoassay result is inconsistent with the clinical situation, or with another assay result, HAAA interference should be suspected. Appropriate preliminary evaluation might include: a) assaying multiple dilutions of the suspected sample; b) assaying for other molecular species that usually accompany the analyte of interest; or c) assaying the analyte of interest with another method, as different assays show highly variable interference effects. Assays for human antimouse antibodies (HAMA), the most common subset of HAAA, can be performed using any one of six commercially available kits, most of which are ELISA formats. These kits will not, however, detect all of the possible HAAA interferents. At the present time, however, most laboratorians will submit the specimen in question to the reagent manufacturer for definitive evaluation. Some reference laboratories offer assays for HAAA.

hCG assays: Rotmensch and Cole have recommended that protocols for the diagnosis and treatment of gestational trophoblastic disease include a test for hCG in the urine. Such a requirement would have avoided the problems caused by HAAA cited in their report. In addition, hCG specimens may be referred to the HCG Reference Service, Department of Obstetrics and Gynecology, University of New Mexico, Albuquerque, NM 87131, USA

(505-272-6137) Additional information about this reference service is available at the following website: www.hcglab.com.

Limitations Undoubtedly, there are erroneous results due to HAAA which will never be detected, a situation which will continue unless and until it becomes feasible to test all patients for HAAA.

Methodology Enzyme-linked immunosorbent assay (ELISA), radioimmuno-assay (RIA)

References

Bagshawe KD, "Limitations of Tests for Human Chorionic Gonadotropin," *Lancet*, 2000, 355(9205):671.

Dayan CM, "Interpretation of Thyroid Function Tests," *Lancet*, 2001, 357:619-24.

Kazmierczak SC, Catrou PG, and Briley KP, "Transient Nature of Interference Effects From Hetero-phil Antibodies: Examples of Interference With Cardiac Marker Measurements," *Clin Chem Lab Med*, 2000, 38(1):33-9.

Kricka L, "Human Antianimal Antibody Interferences in Immunological Assays," *Clin Chem*, 1999, 45(7):942-56.

Rotmensch S and Cole LA, "False Diagnosis and Needless Therapy of Presumed Malignant Disease in Women With False-Positive Human Chorionic Gonadotropin Concentrations," *Lancet*, 2000, 355(9205):712-5.

Internet Web Sites

www.hcglab.com

♦ **Hexamidinum**   *see* Primidone, Serum or Plasma *on page 1091*

♦ **Hexosaminidase**   *see* Beta-Hexosaminidase, Serum, White Blood Cells *on page 255*

♦ **Hexosaminidase A**   *see* Inherited Diseases of Metabolism and Cell Structure *on page 792*

♦ **Hexosaminidase B**   *see* Inherited Diseases of Metabolism and Cell Structure *on page 792*

♦ **β-Hexosaminidase, Serum**   *see* Beta-Hexosaminidase, Serum, White Blood Cells *on page 255*

♦ *HFE* **Genotyping**   *see* Hereditary Hemochromatosis DNA Test *on page 718*

♦ **Hg, Blood**   *see* Mercury, Blood *on page 897*

♦ **HGE Antibody**   *see* Ehrlichiosis Serology *on page 531*

♦ **hGH**   *see* Growth Hormone, Serum *on page 662*

♦ **Hg, Urine**   *see* Mercury, Urine *on page 898*

♦ **HHV-8**   *see* Herpesvirus 8 *on page 725*

♦ **5-HIAA, Quantitative, Urine**   *see* 5-Hydroxyindoleacetic Acid, Quantitative, Urine *on page 754*

♦ **Hibanil**®   *see* Chlorpromazine, Serum *on page 395*

♦ **High Density Lipoprotein Cholesterol (HDLC)**   *see* Cholesterol, Total, Serum or Plasma *on page 396*

♦ **High Density Lipoprotein Cholesterol (HDLC)**   *see* Lipids, Overview *on page 853*

High Density Lipoprotein Cholesterol, Serum

Related Information

Apolipoprotein A-I, Serum *on page 202*
Cholesterol, Total, Serum or Plasma *on page 396*
Lipid Panel, Serum *on page 852*
Lipids, Overview *on page 853*
Lipoprotein (a), Serum *on page 861*
Low Density Lipoprotein Cholesterol *on page 874*
Triglycerides, Serum or Plasma *on page 1275*

Synonyms Alpha₁ Lipoprotein Cholesterol; HDL; HDLC; HDL Cholesterol

Applies to Apolipoprotein A-I; Cholesterol, Total; Metabolic Syndrome

Abstract High density lipoprotein cholesterol (HDLC) refers to a class of hetero-geneous lipoprotein particles. HDLC is a major risk factor for coronary heart disease (CHD). High CHD risk is associated with low levels of HDLC, and low risk is associated with high levels of HDLC.

Patient Preparation For best results, patient should be on a stable diet for 3 weeks, stable body weight, and fasting for at least 10 hours. (Fasting is not essential for total cholesterol but is important if triglycerides are also to be assayed.)
(Continued)

High Density Lipoprotein Cholesterol, Serum
(Continued)

See also Preparation in listing Cholesterol, Total, Serum or Plasma *on page 396*. HDLC is usually done as part of a Lipid Panel, Serum *on page 852*.

Specimen Serum or plasma

Container Red top tube or lavender top (EDTA) tube

Storage Instructions Analysis promptly after sampling is best. The serum or plasma specimen can be refrigerated up to several days at 4°C or frozen for several weeks. For long-term storage use -70°C.

Reference Interval A normative reference interval is not applicable (see Lipids, Overview *on page 853*). The Adult Treatment Panel (ATP III) of the National Cholesterol Education Program (NCEP III) recommends the following interpretive criteria for adults:

- Low (high risk): <40 mg/dL
- High (favorable or protective): ≥60 mg/dL

Actual population distributions and percentile cutoffs are available in the Lipid Research Clinics Prevalence Study and were determined with a method yielding slightly higher values than would be obtained today.

Use See Lipids, Overview *on page 853*. HDLC is a major CHD risk factor and its principal use is in this context. **CHD risk increases 2% to 3% for every 1 mg/dL decrease in HDLC.** HDLC results are method-dependent. The CDC has created a reference laboratory network, the Cholesterol Reference Method Laboratory Network (CRMLN), which offers to manufacturers of diagnostic products and clinical laboratories a testing protocol comparing individual methods to the Designated Comparison Method (DCM). Participation in this program allows manufacturers and laboratory directors to adjust calibrator setpoints so that results are comparable and consistent across different methods and different laboratories, thus, minimizing the method-dependence of HDLC results. Information about participation is available at: www.aacc.org/standards/cdc/cholesterolinfo.stm

The NCEP has published these analytical goals for HDLC tests:

- Bias ≤5% of the reference value
- SD ≤17 at <42 mg/dL
- CV ≤4% at ≥42 mg/dL
- Total error ≤13%

There are five rare **genetic disorders of HDLC metabolism**.

- Familial hyperalphalipoproteinemia is a deficiency of cholesterol ester transfer protein. Affected patients have markedly elevated HDLC with low levels of LDLC and triglycerides. They appear to be protected from CHD and are typically long-lived.
- Four rare disorders are associated with low or absent HDLC: Tangier disease, apoprotein A-I deficiency, lecithin:cholesterol acyltransferase (LCAT) deficiency, and fish-eye disease. Biochemical identification requires ultracentrifugation and patients are often referred to a subspecialty lipid clinic.

Fasting HDLC <40 mg/dL is a component of the **metabolic syndrome** (see definition given in Triglycerides, Serum or Plasma *on page 1275*).

Methodology

CDC reference method (RM) comprises three separate steps:

a) ultracentrifugation to remove chylomicrons and VLDL

b) apo-B containing lipoproteins (mainly LDL) are precipitated with heparin/$MnCl_2$

c) resulting HDL-containing supernatant is analyzed for cholesterol by a modified Abell-Kendall procedure

This reference method is now used as the accuracy target for the standardization of clinical assays, but this is due to historical reasons and not because of demonstrated accuracy.

Designated comparison method (DCM) avoids the need for ultracentrifugation, and uses dextran sulfate-magnesium chloride as a precipitating agent; final step is an enzymatic color reaction.

Homogeneous assays (third generation) are those which are fully automated and do not require an off-line separation step. All of these employ some reagent which remove or inhibits the non-HDL cholesterol from the final color reaction. These are the most widely used HDLC assays.

Principally of historic interest are first- and second-generation heterogeneous assays that employed selective precipitation of LDL and VLDL, followed by cholesterol analysis of the HDL-containing supernatant by enzymatic methods. Precipitating agents include heparin/$MnCl_2$, heparin/$CaCl_2$, dextran sulfate/$MgCl_2$, sodium phosphotungstate/$MgCl_2$, and polyethylene glycol.

Electrophoretic methods, including HPLC, are rarely used.

Additional Information Much of the interindividual biologic variability is the result of genetic factors. HDLC values are higher in women than men. Modifiable factors which **increase** serum HDLC include physical exercise, weight loss, beverage alcohol (moderate quantities), and high dietary fat. Medications which **increase** HDLC as a primary effect include, in decreasing order of potency: niacin, fibrates, statins, and resins. Estrogens also **increase** HDLC. Probucol **decreases** HDLC. Modifiable factors which **decrease** HDLC include smoking, obesity, pregnancy, stress, hospitalization, and a few medications (corticosteroids, androgens, progestins, diuretics, and propranolol). Disease states associated with **decreased** HDLC include chronic renal insufficiency, diabetes (type 2), myocardial infarction, and some forms of thyroid dysfunction.

During the past 35 years, a number of ratios using HDLC and either LDLC or total cholesterol (TC) enjoyed some popularity. When HDLC is the numerator of a ratio, a high value indicates a favorable risk status. When HDLC is the denominator, the reverse is true. The use of such ratios was justified by the thought that they would "reduce the cognitive complexity" of considering the two variables separately. But this argument is untenable given the fact that different interventions are used to increase HDLC and decrease LDLC: it is necessary to consider each of these important risk factors separately.

See Apolipoprotein A-I, Serum *on page 202.*

References
"Executive Summary of the Third Report of the National Cholesterol Education Program (NCEP) Expert Panel on Detection, Evaluation, and Treatment of High Blood Cholesterol in Adults (Adult Treatment Panel III)," *JAMA*, 2001, 285(19):2486-97.

Gilbert-Barness E and Barness LA, *Metabolic Diseases*, Volume I, Natick, MA: Eaton Publishing Co, 2000, 283-322.

Kimberly MM, Leary ET, Cole TD, et al, "Selection, Validation, Standardization, and Performance of a Designated Comparison Method for HDL-Cholesterol for Use in the Cholesterol Reference Method Laboratory Network," *Clin Chem*, 1999, 45(10):1803-12.

Internet Web Sites
www.cdc.gov/nceh/dls/crmln/crmln.htm
www.nhlbi.nih.gov

High-Molecular Weight Kininogen

Related Information
Activated Partial Thromboplastin Time *on page 100*
Coagulation Factor Assays *on page 418*
Mixing Studies (Coagulation) *on page 918*
Prekallikrein *on page 1085*

Synonyms HMWK, Fitzgerald Factor; HMW Kininogen; Williams-Fitzgerald-Flaujeac Factor

Abstract High molecular weight kininogen (HMWK) is a coagulation protein involved in the early stages of intrinsic pathway activation. HMWK deficiency can cause a marked prolongation of the PTT, but it does not cause bleeding. The same is true for factor XII deficiency and prekallikrein deficiency.

Specimen Plasma

Container One blue top (sodium citrate) tube

Collection Routine venipuncture. If multiple tests are being drawn, draw blue top tubes after any red top tubes but before any lavender top (EDTA), green top (heparin), or gray top (oxalate/fluoride) tubes. Immediately invert tube gently at least 4 times to mix. Tubes must be appropriately filled. Deliver tubes immediately to the laboratory.

Storage Instructions Separate plasma from cells as soon as possible. If test is not performed within 4 hours, freeze plasma.
(Continued)

High-Molecular Weight Kininogen *(Continued)*

Causes for Rejection Specimen received more than 4 hours after collection, tubes not filled, clotted specimens

Turnaround Time Less than 1 day (longer if test is a send-out)

Special Instructions Patients cannot be on hirudin or argatroban anticoagulation, which can interfere with mixing studies and HMWK assays. Danaparoid may also interfere with these assays. If heparin is present, notify the laboratory because heparin must be removed prior to testing.

Reference Interval 60% to 140% of normal. Newborns have lower levels than adults; the values increase to near adult normal range by age 6 months.

Use May be performed when a routine prolonged PTT evaluation finds no explanation for the prolongation (see Mixing Studies (Coagulation) *on page 918*). The assay for HMWK (and prekallikrein) can be considered when the following findings are present: the PTT is normal in the mixing study; factors VIII, IX, XI, and XII are normal; the PT and fibrinogen are normal; and lupus anticoagulant assays are negative.

Methodology A factor assay for HMWK can be performed which is similar to other coagulation factor assays. Patient plasma is mixed with HMWK-deficient plasma and a PTT is performed on the mixture. The amount of HMWK in the patient plasma is determined from a standard curve that plots known amounts of HMWK against PTT values.

Additional Information HMWK is one of the contact factors that participates in the activation of the intrinsic pathway of coagulation when blood is exposed to a negatively charged foreign surface. With contact activation, activated factor XII (XIIa) converts prekallikrein into kallikrein. Kallikrein then activates more factor XII. HMWK acts as a cofactor in both of these reactions. HMWK also acts as a cofactor in the activation of factor XI by factor XIIa. Kallikrein releases bradykinin from HMWK, which has vasoactive activities. Fibrinolysis is also activated by contact activation. Recent evidence suggests that, *in vivo*, activation of prekallikrein occurs before activation of factor XII.

HMWK deficiency is rare and does not cause bleeding, despite PTT prolongations. The lack of bleeding is presumably because the extrinsic pathway of coagulation, via factor VII and tissue factor, remains intact, and factor XI can be activated by thrombin generated from the extrinsic pathway. Thus, factor XI can be activated without the need for HMWK, prekallikrein, or factor XII. This is consistent with the observation that deficiencies of the latter three factors are not associated with bleeding. Acquired, usually mild-to-moderate decreases in HMWK may be found in liver disease or disseminated intravascular coagulation (DIC).

Also see the graphic in Activated Partial Thromboplastin Time *on page 100*.

References

Schmaier AH, Rojkjaer R, and Shariat-Madar Z, "Activation of the Plasma Kallikrein/Kinin System on Cells: A Revised Hypothesis," *Thromb Haemost*, 1999, 82(2):226-33.

♦ **Hill Plots** *see P$_{50}$, Blood on page 993*

♦ **Hirudin** *see Activated Partial Thromboplastin Time on page 100*

♦ **Hirudin** *see Heparin-Induced Thrombocytopenia on page 695*

♦ **Hirudin** *see Mixing Studies (Coagulation) on page 918*

♦ **Hirudin** *see Prothrombin Time on page 1116*

Histamine, Urine, Plasma, or Whole Blood

Related Information

Immunoglobulin E *on page 773*

Abstract Histamine is pathogenetically important in anaphylaxis and other allergic states, including urticaria, flushing, asthma-like wheezing, and tachycardia. Histamine release occurs from sensitized basophil and mast cells. In sensitized individuals (ie, persons who have formed immunoglobulin E (IgE) antibodies to an antigen), much of the IgE is bound to specific receptors on basophils and mast cells. When the antigen appears again and combines with cell-bound IgE, histamine is among the vasoactive substances that are released.

Patient Preparation Patient must be on a diet of microbially processed foods, such as cheeses or sauerkraut.

Specimen Urine, plasma, whole blood

Collection Fifty percent acetic acid used as preservative should be added at beginning of 24-hour urine collection to maintain pH 2.0-4.0. Collect plasma or whole blood in EDTA tube.

Storage Instructions Store urine frozen.

Reference Interval The test is not widely available. One referral laboratory uses these reference intervals:

- plasma: 0-6 nmol/L
- whole blood: 200-2000 nmol/L
- urine: 0-321 nmol/g creatinine

Another laboratory assays urine for histamine, with the following reference intervals:

- 24-hour urine: <45 μg/g creatinine
- random urine: <100 μg/g creatinine

Use Assays for histamine are little used outside of research settings. Elevated values are found in systemic mastocytosis, urticaria, anaphylaxis, and following provocative testing.

Limitations Test is not very sensitive. There are false-positive results associated with urinary tract infections. Histamine concentration alone may not fully reflect the role of histamine in a disease process. Measurement of histamine and its metabolites may be necessary.

Additional Information Mast cells produce numerous biologically active materials, including histamine. Histamine may be elevated in myeloproliferative disorders and carcinoid tumors (particularly of gastric origin). Measurement of urinary methylated and other histamine metabolites may be more sensitive and specific.

References

ARUP Laboratories, *1999-2000 Users Guide*, 384-5.

Kaplan AP, "Anaphylaxis," *Cecil Textbook of Medicine*, 21st ed, Goldman L and Bennett JC, eds, WB Saunders Co, 2000, 1450-2.

Mayo Medical Laboratories, *2000 Test Catalog*, Rochester, MN, 292.

♦ **Histocompatibility Testing** *see* Tissue Typing *on page 1259*

♦ **Histomorphometry** *see* Aluminum, Bone and Bone Biopsy *on page 139*

Histopathology

Related Information

Aluminum, Bone and Bone Biopsy *on page 139*
Breast Biopsy *on page 305*
Chromosome Analysis, Lymph Node and Solid Tumor *on page 410*
Electron Microscopic Examination for Viruses, Stool *on page 533*
Estrogen and Progesterone Receptor Assay *on page 556*
Fine Needle Aspiration, Deep and Superficial Masses *on page 590*
Fluorescence *in situ* Hybridization *on page 602*
Frozen Section *on page 615*
Fungal Culture, Biopsy or Body Fluid *on page 619*
Gene Rearrangement for Leukemia and Lymphoma *on page 633*
Human Papillomavirus (HPV) DNA Tests *on page 749*
Liver Biopsy *on page 864*
Lymph Node Biopsy *on page 880*
Muscle Biopsy *on page 927*
Mycobacterial Culture, Biopsy or Body Fluid *on page 929*
p53, Functional Assay/Sequencing *on page 995*
Retinoblastoma Gene DNA Detection *on page 1159*
Skin Biopsy *on page 1200*
Viral Culture *on page 1307*
Virus, Direct Detection by Fluorescent Antibody *on page 1311*

Synonyms Biopsy; Surgical Pathology

Test Includes Gross and microscopic examination and diagnosis with comments, notes, prognostic and other information in selected cases. Imprints may be made if the tissue is fresh and unfixed and if indications for imprints exist.

Abstract Surgical pathology has been defined as the discipline which deals with the anatomic pathology of tissues removed from living patients. Smears, aspirates, special stains, histochemistry, immunocytochemistry, flow cytometry, electron microscopy, and/or molecular pathology may be needed.

(Continued)

Histopathology *(Continued)*

Patient Preparation Each specimen should be accompanied by an adequate description of what it is thought to represent, as well as an appropriate clinical history. The appearances of disease processes may be seriously misleading out of context.

Specimen Fresh tissue, tissue fixed in phosphate-buffered formalin or other appropriate fixative. Each specimen container must be labeled to include source as well as patient's name. Each specimen from a different anatomic site must be placed in a separate, correctly labeled container, designated "left," "right," "proximal," "distal," "ventral," "dorsal," and so forth. Crushed or damaged specimens may lead to incorrect interpretations.

Collection Use approximately 5-20 times as much fixative solution as the bulk of the tissue. Small tissues such as those from bronchoscopic biopsy, bladder biopsy, and endometrium can be ruined in a very short time by drying out.

Storage Instructions Fixation in formalin solution or other appropriate fixative

Turnaround Time Biopsy reports commonly require a day or more. Delays are caused by need for proper fixation, clinical information, deeper sections, decalcification, immunochemistry or other special stains.

Special Instructions See specific handling instructions in test listings such as Muscle Biopsy *on page 927*, Estrogen and Progesterone Receptor Assay *on page 556*, Frozen Section *on page 615*, Kidney Biopsy *on page 818*, and Liver Biopsy *on page 864*.

Use Histopathologic diagnosis; distinguish benign from malignant entities when possible; evaluate extent of lesions, adequacy of resection, provision of classification and, when appropriate, grading in the case of tumors

Limitations Tissue fixed in formalin **cannot** be used for microbial culture, certain types of histochemistry, frozen sections, gene rearrangement, or optimal electron microscopy.

The practice of surgical pathology often depends on the clinical input of other physicians, including surgeons. Miracles of extrapolation cannot consistently be provided from incomplete clinical information and/or inadequate biopsies. Medicolegal activity may follow inadequate sampling by surgeon or pathologist.

Additional Information A major advantage of conventional over frozen sections is that extensive sampling of the entire specimen can take place.

Cultures of tissue are best taken in the O.R., where a sterile field exists. A piece of tissue (eg, a curetting of a fistulous tract) should be placed in an appropriate sterile tube with requests for smear, culture, anaerobic culture, AFB, and/or fungus culture if appropriate.

Of major importance in handling bullets and other specimens of possible forensic significance, including vaginal swabs obtained in rape cases, is the scrupulous maintenance of a chain-of-custody. Specimens must be accurately labeled, and transfer and receipt must be documented. Specimens must be kept under safeguards in the laboratory until turned over to law enforcement officials. See Chain-of-Custody Protocol *on page 381*.

Bone biopsy for metabolic bone disease requires special handling.

References

Rosai K, *Ackerman's Surgical Pathology*, 8th ed, Volume 1, Chapter 1, St Louis, MO: CV Mosby Co, 1996, 1-12.

Weidner N, Cote RJ, Suster S, et al, *Modern Surgical Pathology*, Philadelphia, PA: WB Saunders Co, 2002,

Histoplasmosis Antibody

Related Information

Bacteremia Detection, Buffy Coat Micromethod *on page 227*
Bronchoalveolar Lavage (BAL) *on page 311*
Fungal Culture, Biopsy or Body Fluid *on page 619*
Fungal Culture, Blood *on page 620*
Fungal Culture, Cerebrospinal Fluid *on page 621*
Fungal Culture, Sputum *on page 624*
Fungus Smear, Stain *on page 626*
Histoplasmosis Antigen *on page 735*

Test Includes Detection of antibody specific for yeast and mycelial antigens

Abstract Antibodies to *H. capsulatum* may be found in 20% to 80% of individuals who live in endemic areas of the Ohio, Missouri, and Mississippi River valleys. Use of serology to diagnose histoplasmosis is **not reliable**. Results are positive in >90% of cases of acute pulmonary disease, 70% to 90% in cavitary lung disease, but only 30% to 50% in acute disseminated disease. Serologic testing should be interpreted in conjunction with other tests and clinical symptoms

Specimen Serum, cerebrospinal fluid

Container Red top tube, sterile CSF tube

Collection Acute and convalescent sera are recommended. Specimens taken 3-4 weeks apart are desirable.

Reference Interval Less than a fourfold change in titer between acute and convalescent samples

Use Diagnose chronic/self-limited histoplasmosis

Limitations Other diagnostic approaches are preferable. A negative result does not rule out histoplasmosis, nor do all who have encountered the fungus develop CF antibodies. False negatives occur in normals and in immunosuppressed individuals. Histoplasmin skin testing may interfere with results, giving a positive test for mycelial phase antibodies. Testing with both mycelial and yeast phase antigens must be performed. Antibodies cross react with other fungi. Anticomplementary sera cannot be tested for complement fixing antibodies. The latex agglutination test gives some false positives, and must be confirmed with another procedure.

Contraindications Previous skin testing, made from supernatant of mycelial growth

Methodology Immunodiffusion (ID), complement fixation (CF), latex agglutination (LA), enzyme immunoassay (EIA)

Additional Information Fungal stains and cultures of sputum, bone marrow examination and culture, *Histoplasma* antigen, and *Histoplasma* PCR should also be performed when histoplasmosis is suspected. Giemsa or methenamine silver preparations of bone marrow, mucosal ulcers, biopsies of liver, lung, skin, lymph nodes, or bronchoalveolar lavage may provide rapid diagnosis. Cultures are recommended, especially of blood, bone marrow, sputum, urine, and portions of biopsies. Isolates can be identified by DNA probe.

A serum titer ≥1:16 for *Histoplasma* mycelial antigen, or serum titer ≥1:32 for yeast antigen provide evidence of active disease. Increasing titers provide evidence of active infection. Titers between 1:8 to 1:32 are less diagnostic, and require clinical and other laboratory correlation.

References

Deepe GS, "*Histoplasma capsulatum*," *Principles and Practice of Infectious Diseases*, 5th ed, Mandell GL, Bennett JE, and Dolan R, eds, New York, NY: Churchill Livingstone, 2000, 2718-33.

Edelman M and McKitrick J, Images in Clinical Medicine, "*Histoplasma capsulatum* in a Peripheral Blood Smear," *N Engl J Med*, 2000, 342(1):28.

Wheat J, "Histoplasmosis. Experience During Outbreaks in Indianapolis and Review of the Literature," *Medicine*, 1997, 76(5):339-54.

Internet Web Sites

www.cdc.gov/ncidod/dbmd/diseaseinfo/histoplasmosis_g.htm
www.emedicine.com

Histoplasmosis Antigen

Related Information

Fungal Culture, Biopsy or Body Fluid *on page 619*
Fungal Culture, Blood *on page 620*
Fungal Culture, Cerebrospinal Fluid *on page 621*
Fungal Culture, Sputum *on page 624*
Fungus Smear, Stain *on page 626*
Histoplasmosis Antibody *on page 734*

Abstract *Histoplasma* antigen can be detected in the urine of 90% of disseminated infection, and 75% of persons with acute pulmonary histoplasmosis. In those with AIDS, antigenuria is detected in 95% and antigenemia in 85% of patients infected with histoplasmosis. This assay is less helpful in chronic pulmonary or self-limited forms of histoplasmosis, in which cultures and serologic tests are needed.

Specimen Serum, urine, cerebrospinal fluid

Container Red top tube, urine container, sterile CSF tube

(Continued)

Histoplasmosis Antigen *(Continued)*

Reference Interval Negative

Use Diagnose disseminated *Histoplasma capsulatum* infection

Limitations The test is now widely performed. It occasionally produces false-negative results with less severe forms of histoplasmosis, and may give false-negative results in patients with proven disseminated disease.

Cross reactions have occurred in patients who have other fungal infections due to similar antigens, including blastomycosis, paracoccidioidomycosis, and penicilliosis, but not with *Aspergillus*, *Candida*, or *Cryptococcus*. Rheumatoid factor may cause a false-positive antigen test.

Methodology Microtiter plates with anti-*H. capsulatum* antibody; enzyme immunoassay (EIA) for *Histoplasma* polysaccharide antigen

Additional Information The immunoassay for *Histoplasma* glycoprotein antigen, the precise nature of which remains unknown, has proven diagnostically useful in a variety of clinical settings (particularly in AIDS patients), and is also useful for monitoring response to therapy. Biopsies, examination of peripheral blood smears for organisms, blood cultures, and serology are also indicated for investigation of disseminated histoplasmosis.

Ninety percent of patients with progressive disseminated histoplasmosis have positive antigen, 40% of patients with cavitary lung disease have a positive antigen test, but only 20% who have acute pulmonary histoplasmosis are positive. In the clinical setting of suspected histoplasmosis, fungal cultures and antibodies are desirable.

References

Limaye AP, Connolly PA, Sagar M, et al, "Transmission of *Histoplasma capsulatum* by Organ Transplantation," *N Engl J Med*, 2000, 343(16):1163-6.

Wheat LJ, Connolly-Stringfield P, Blair R, et al, "Histoplasmosis Relapse in Patients With AIDS: Detection Using *Histoplasma capsulatum* Variety *capsulatum* Antigen Levels," *Ann Intern Med*, 1991, 115(12):936-41.

Internet Web Sites

www.aidsinfo.nih.gov/
www.apha.org/public_health/aids.htm
www.cdc.gov/ncidod/dbmd/diseaseinfo/histoplasmosis_g.htm
www.emedicine.com

♦ **HIT** *see* Heparin-Induced Thrombocytopenia *on page 695*

HIV-1/HIV-2 Serology

Related Information

Antiphospholipid Antibody (Lupus Anticoagulant and/or Anticardiolipin Antibody) *on page 193*
Beta$_2$-Microglobulin, Serum or Urine *on page 254*
Blood and Fluid Precautions, Specimen Collection *on page 271*
CD4/CD8 Enumeration *on page 349*
Donation, Blood *on page 521*
Fungal Culture, Skin *on page 623*
HIV p24 Antigen Detection *on page 738*
HTLV-I/II Antibody *on page 745*
Human Immunodeficiency Virus DNA Amplification *on page 746*
Risks of Transfusion *on page 1166*
Toxoplasmosis Diagnostic Procedures *on page 1265*
White Blood Cell Count *on page 1330*
Zidovudine, Serum or Plasma *on page 1339*

Test Includes Detection of antibody to HIV and confirmation of positives by Western blot

Abstract Present screening tests for HIV antibodies use ELISA procedures which utilize recombinant antigen products. Both the sensitivity and specificity of these tests are extremely high, but positive results on a screen should be repeated using a new specimen. If positive a second time they should be confirmed with a Western blot procedure. Because of the grave implications of a positive result, it is recommended that a second sample be assayed to eliminate false positives due to switched samples or sample contamination.

A second virus that causes AIDS, HIV-2, is very closely related to HIV-1 with 40% nucleic acid homology. HIV-2 is endemic in West Africa, however, it has spread to other countries. Occasional cases have been confirmed in the United

States. It produces the same clinical disease as HIV-1, although the incubation period before clinical AIDS develops in HIV-2 may be longer.

Patient Preparation Some states require written or informed consent of the patient or guardian.

Specimen Serum or plasma; some test systems use saliva or urine.

Container Red top tube or lavender top (EDTA) tube

Special Instructions Blood and body fluid precautions must be observed.

Reference Interval Negative

Use Document exposure to HIV-1 and HIV-2; screen blood and blood fractions for transfusion; screen organ transplant donors; test patients after documented needlestick exposure of healthcare personnel.

Limitations Positive screening tests must be confirmed by more specific follow-up procedures. If the Western blot or immunofluorescent antibody for HIV-1 is negative or indeterminate with a positive HIV-1/HIV-2 serology, further testing is required for HIV-2. This may include HIV-2 EIA and HIV-2 Western blot. There are cross reactions in some test systems due to histocompatibility antigen mismatches (in particular, antibodies to HLA-DR4). Cross reactions have been observed to other viral antigens as well. A recent influenza vaccination can result in reactivity against p24 antigen and cause a false-positive enzyme-linked immunosorbent assay.

Serology cannot be used to determine if an infant born to an HIV-positive mother is infected, because of maternal transfer of antibodies. These infants should be tested using molecular assays or culture. In early disease before patients develop antibody (window phase), HIV serology will be negative, however, nucleic acid assays can be used for diagnosis.

Methodology Enzyme-linked immunosorbent assay (ELISA), Western blot, indirect fluorescent antibody (IFA). Other test systems include HIV-1/HIV-2 combination ELISA kit, SUDS (single use diagnostic system) HIV diagnostic kit.

Additional Information Human immunodeficiency virus (HIV-1), a retrovirus, is the etiologic agent of AIDS. Acute infection is characterized by a flu-like illness, or may be asymptomatic. Generally, following infection with the virus, there is local replication of the virus in regional lymph nodes and viral nucleic acid can be detected in the plasma. By ~17 days after infection, viral replication can be detected by p24 antigen test or by PCR for genomic DNA. By ~23 days, antibody to the HIV virus can be detected in the serum or plasma.

The third generation ELISA assay, using HIV-1 env and gag proteins and HIV-2 env proteins, is now the standard for screening and detects IgM, IgA, and IgG anti-HIV. It reduces the window period in early HIV. Confirmation with Western blot (WB) is still required.

Rapid testing with results in 30 minutes can be done using the single use diagnostic system (SUDS). Results may be used in STD clinics or emergency departments, but require follow-up with standard HIV testing for positives. HIV-2 antibodies are **not** detected.

Oral mucosa transudate test (OMT) uses a cotton pad placed between cheek and gum for 5 minutes under direct supervision. The pad is tested by a special micro-ELISA system. WB is needed to confirm positives.

Urine based HIV tests to detect HIV-1 antibodies may be useful when serum specimens are nearly impossible to obtain. A positive must be confirmed by standard serum tests, ELISA and WB.

HIV home bleed test systems allow individuals to prick their own fingers, place three drops of blood on a card, and mail it to the manufacturer where an ELISA is performed anonymously. Positive results are followed by telephone counseling. Other systems have an indicator to show positives on the card 5 minutes after blood is placed on the card. False positives and negatives remain problems with these systems.

References

Helbert M and Breuer J, "Monitoring Patients With HIV Disease," *J Clin Pathol*, 2000, 53(4):266-72.

Henry K, "The Case for More Cautious, Patient-Focused Antiretroviral Therapy," *Ann Intern Med*, 2000, 132(4):306-11.

Hidalgo JA, MacArthur RD, and Crane LR, "An Overview of HIV Infection and AIDS: Etiology, Pathogenesis, Diagnosis, Epidemiology, and Occupational Exposure," *Semin Thorac Cardiovasc Surg*, 2000, 12(2):130-9.

Laufer M and Scott GB, "Medical Management of HIV Disease in Children," *Pediatr Clin North Am*, 2000, 47(1):127-53.

(Continued)

HIV-1/HIV-2 Serology *(Continued)*

Sepkowitz KA, "AIDS - The First 20 Years," *N Engl J Med*, 2001, 344(23):1764-72.

Internet Web Sites

www.aidsinfo.nih.gov/
www.apha.org/public_health/aids.htm
www.emedicine.com

HIV p24 Antigen Detection

Related Information

HIV-1/HIV-2 Serology *on page 736*
Human Immunodeficiency Virus DNA Amplification *on page 746*
Risks of Transfusion *on page 1166*
Viral Culture *on page 1307*

Test Includes Detection of HIV p24 antigen in serum, plasma, or cerebrospinal fluid

Abstract Detection of the p24 antigen of the human immunodeficiency virus (HIV) in serum of blood donors will shorten the time needed to detect HIV infection to ~2 weeks after initial infection. The plasma or serum level of HIV p24 antigen is also useful as a marker of AIDS progression.

Specimen Serum, plasma, or cerebrospinal fluid

Container Red top tube, sterile CSF tube

Special Instructions In some states written or informed patient consent is a prerequisite for the test.

Reference Interval Negative

Use Diagnose recent acute infection with HIV; may also be of prognostic significance in AIDS

Limitations Test is not as sensitive as culture or nucleic acid amplification assays for detecting HIV infection.

Methodology Enzyme immunoassay (EIA)

Additional Information The p24 antigen is a 24 kD protein product of the **gag** gene of HIV. As a viral rather than host product, it appears concomitant with initial infection, and then becomes undetectable during periods of viral latency. The reappearance of p24 antigen in serum generally heralds progression of clinical disease in AIDS. Measuring antigen may also be useful to assess therapy. The detection of p24 antigen in serum precedes seroconversion by a few days.

A new EIA has been developed recently that will simultaneously detect HIV p24 antigen and HIV antibody in a direct assay format. This assay has been evaluated for large scale blood screening.

References

Martinez-Martinez P, Martin del Barrio E, DeBenito J, et al, "New Lineal Immunoenzymatic Assay for Simultaneous Detection of p24 Antigen and HIV Antibodies," *Eur J Clin Microbiol Infect Dis*, 1999, 18(8):591-4.

Nadal D, Boni J, Kind C, et al, "Prospective Evaluation of Amplification-Boosted ELISA for Heat-Denatured p24 Antigen for Diagnosis and Monitoring of Pediatric Human Immunodeficiency Virus Type 1 Infection," *J Infect Dis*, 1999, 180(4):1089-95.

Ortigao-de-Sampaio MB, Abreu TF, Linhares-de-Carvalho MI, et al, "Surrogate Markers of Disease Progression in HIV-Infected Children in Rio de Janeiro, Brazil," *J Trop Pediatr*, 1999, 45(5):299-302.

van Binsbergen J, Siebelink A, Jacobs A, et al, "Improved Performance of Seroconversion With a 4th Generation HIV Antigen/Antibody Assay," *J Virol Methods*, 1999, 82(1):77-84.

Internet Web Sites

www.who.int/health-topics/hiv.htm

♦ **HLA-A3** *see* Ferritin, Serum *on page 577*

HLA-B27

Related Information

Complement C4, Serum *on page 436*
Complement Components, Overview *on page 437*
HLA Typing, Single Human Leukocyte Antigen *on page 739*
Rheumatoid Factor, Serum or Body Fluid *on page 1161*
Tissue Typing *on page 1259*
Yersinia enterocolitica Antibody *on page 1337*

Applies to Human Leukocyte Antigen

Test Includes Human leukocyte antigen testing for B27

Abstract HLA-B27 (HLA-antigen B27) is an allele of the human HLA-B locus that is present in a small percentage of the general population. Positive patients have a greater likelihood of developing spondyloarthritis.

Specimen Lymphocytes

Container Yellow top (ACD) tube

Storage Instructions Store at room temperature. Refrigerate for DNA typing.

Special Instructions Sample must be tested as soon as possible unless DNA typing is used.

Reference Interval Negative

Use Evaluate spondyloarthritis, juvenile rheumatoid arthritis, Reiter's syndrome, psoriatic arthritis and anterior uveitis

Methodology Flow cytometry (FC), complement dependent cytotoxicity, DNA-based methods

Additional Information HLA-B27 is present in 3% to 4% of African-Americans, 6% to 8% of Caucasians, and 1% of Asians. HLA-B27 bears a strong association with but is not found in all patients with ankylosing spondylitis (AS; Marie-Strumpell disease); over 90% of such patients are positive. Most, but not all of the currently recognized B27 subtypes are associated with AS. The disease-associated B27 alleles share a specific B pocket in the peptide binding groove with a strong specificity for an arginine side chain. This strict selection of peptides is consistent with the role of B27 in the process of binding and presentation of "arthritogenic" peptides. A B27-positive patient with consistent clinical and radiographic findings has ~100 times greater likelihood of having or developing ankylosing spondylitis than has a negative patient. However, the antigen is not causative, and 10% of normal subjects are B27 positive. **This test should not be used as a screening procedure for ankylosing spondylitis.** The antigen is less strongly associated with Reiter syndrome, psoriatic arthritis, juvenile rheumatoid arthritis, and other forms of postinfectious arthritis associated with several other gram-negative organisms. B27 has also been associated with congenital deficiency of C4 and C2, adrenal hyperplasia, and with inflammatory bowel disease.

References

Klein J and Sato A, "The HLA System. Second of Two Parts," *N Engl J Med*, 2000, 343(11):782-6.

Lamas JR, Paradela A, Roncal F, et al, "Modulation at Multiple Anchor Positions of the Peptide Specificity of HLA-B27 Subtypes Differentially Associated With Ankylosing Spondylitis," *Arthritis Rheum*, 1999, 42(9):1975-85.

Mear JP, Schreiber KL, Munz C, et al, "Misfolding of HLA-B27 as a Result of its B Pocket Suggests a Novel Mechanism for its Role in Susceptibility to Spondyloarthropathies," *J Immunol*, 1999, 163(12):6665-70.

Reveille JD, Ball EJ, and Khan MA, "HLA-B27 and Genetic Predisposing Factors in Spondyloarthropathies," *Curr Opin Rheumatol*, 2001, 13(4):265-72.

"Third International Congress on Spondyloarthropathies, Gent, Belgium, October 2-5, 2002," *Clin Exp Rheumatol*, 2002, 20(4):581-611.

♦ **HLA Class II Antigens** *see* Mixed Lymphocyte Culture *on page 917*

♦ **HLA-DP** *see* Mixed Lymphocyte Culture *on page 917*

♦ **HLA-DQ** *see* Mixed Lymphocyte Culture *on page 917*

♦ **HLA-DR** *see* Mixed Lymphocyte Culture *on page 917*

♦ **HLA Typing** *see* Tissue Typing *on page 1259*

HLA Typing, Single Human Leukocyte Antigen

Related Information

HLA-B27 *on page 738*
Identification DNA Testing *on page 765*
Mixed Lymphocyte Culture *on page 917*
Platelet Transfusion *on page 1058*
Rheumatoid Factor, Serum or Body Fluid *on page 1161*
Tissue Typing *on page 1259*

Synonyms Human Leukocyte Antigen System

Applies to Class I and Class II Genes; Leukocyte Antigens; Major Histocompatibility Complex

Test Includes Identification of human leukocyte antigens (HLA) or class I or class II alleles

Abstract The major histocompatibility complex (MHC) exists as a group of closely-linked genes that encode the HLA antigens, which are the major histocompatibility antigens. They determine compatibility between donor and recipient in transplantation, and therefore are expressed as "major." The HLA (Continued)

HLA Typing, Single Human Leukocyte Antigen
(Continued)

antigen captures foreign peptides, to present to antigen-binding T-cell receptors. Malfunction of the HLA system exerts effects on a variety of human disorders. Over 200 genes are included in the HLA complex on chromosome 6. Over 40 encode leukocyte antigens. Autoimmune phenomena are prominent features in many of the HLA and disease associations.

Specimen Lymphocytes (for serology typing), white cells (for DNA typing)

Container Lavender top (EDTA) tube for DNA testing, yellow top (ACD) tube for serology and DNA testing

Collection Deliver immediately to the laboratory.

Storage Instructions Maintain at room temperature for serology testing. Refrigerate for DNA-based testing.

Special Instructions Test requires viable lymphocytes for serologic testing, either viable or dead cells for DNA testing.

Reference Interval Identification of specific leukocyte antigens, class I and II alleles

Use The HLA system is used as an epidemiologic marker, in paternity exclusion testing, transplantation donor and recipient matching to diminish likelihood of rejection and graft-vs-host disease (GVHD), and for compatible platelet transfusions for refractory patients. Investigation of the HLA system promises applications in immunodiagnosis and in immunotherapy. Many of these antigens have been more or less closely associated statistically with a wide variety of diseases. The most striking association is that of HLA-B27 with ankylosing spondylitis (AS).

Other associations with autoimmune diseases include birdshot retinochoroidopathy, reactive arthropathy including Reiter syndrome, celiac disease, dermatitis herpetiformis, idiopathic membranous glomerulonephritis, Goodpasture syndrome, and pemphigus vulgaris. Lower levels of relative risk include rheumatoid arthritis, Behçet syndrome, systemic lupus erythematosus, insulin-dependent (type 1) diabetes mellitus, idiopathic Addison disease, Graves disease, Hashimoto disease, postpartum thyroiditis, sicca syndrome, myasthenia gravis, and multiple sclerosis.

Mutations in the HFE gene lead to hereditary hemochromatosis.

The HLA system is involved in protection against severe malaria and in other infectious disease problems.

Another significant association is DRB1*-15, DQA1*-0102, and DQB1*-0602 with narcolepsy, a wake disorder.

Limitations Except AS and narcolepsy, none is presently an aid to diagnosis.

Methodology HLA class I typing (A, B, and C loci) conventionally is done by use of a dye exclusion (eosin, trypan blue) method to assess lymphocyte viability after reaction with various HLA-directed antisera and complement (serology). Gene products defined by serological methods are termed antigens.

Loci in class II are designated by three letters. The first, D, indicates the class. The second letter, M, O, P, Q, R, indicates the family, and the third, A or B, the chain: respectively, alpha or beta. HLA class II typing (DR, DQ, and DP) by serological methods similar to those described above; B cells must be used, since quiescent T cells lack the expression of class II antigens.

Molecular genetic HLA typing is rapidly replacing serologic methods. Gene products defined by DNA-based methods are termed alleles. The advantages of DNA-based methods include: dead (nonviable) white cells and smaller samples; fingersticks; tissues such as buccal swabs, formalin-fixed tissues, or hair follicles can be used.

Additional Information The HLA system has been generally categorized into class I, class II, and class III gene regions. HLA class I genes including the A, B, and C loci and their respective antigens are expressed on cell surfaces of most nucleated cells. The HLA class III genes encode the complement proteins, C2, C4, and factor B of the alternative complement pathway. Classes I and II are involved in control and modulate immune responses. HLA class I and II molecules bind processed self and foreign peptides for presentation to

CD8- and CD4-positive T cells, respectively. Most HLA genes are highly poly-morphic, which accounts for their recognition in immune responses and matching for transplantation. Many of the alleles of these genes are in linkage disequilibrium. Many of these antigens have been more or less closely associ-ated statistically with a wide variety of diseases.

HLA typing is also performed for cadaveric renal transplantation, in conjunction with ABO typing and crossmatching. The degree of HLA mismatch between the recipient and donor is one component of the kidney allocation algorithm. These HLA molecules are the targets of rejection and of GVHD.

The system is useful in exclusion of paternity and in forensic medicine, since the HLA system includes many alleles which can be identified in addition to the many red cell antigenic systems.

References

Klein J and Sato A, "The HLA System. First of Two Parts," *N Engl J Med*, 2000, 343(10):702-9.

Klein J and Sato A, "The HLA System. Second of Two Parts," *N Engl J Med*, 2000, 343(11):782-6.

Marsh SG, Albert ED, Bodmer WF, et al, "Nomenclature for Factors of the HLA System, 2002," *Tissue Antigens*, 2002, 60(5):407-64.

Raulet DH, "Does a Low Level of Expression of HLA Molecules Engender Autoimmunity?" *N Engl J Med*, 1999, 340(4):314-5.

Rodey GE, "HLA Beyond Tears," *Introduction to Human Histocompatibility*, Durango, CO: De Novo, Inc, 2000.

Homocyst(e)ine, Plasma

Related Information

Abstract Homocyst(e)ine circulates in plasma in several different forms: 80% to 90% is bound to protein; 5% to 10% is in the oxidized form, homocyst(e)ine; 5% to 10% is combined with cysteine-forming mixed disulfides; and <2% is free. Most clinical assays measure the sum of all these entities, and some authors refer to this measurement as **total homocysteine**, although that term is techni-cally incorrect. In this book we follow the practice of using the intentionally ambiguous term homocyst(e)ine (abbreviated Hcy) to refer to the clinical measurement.

Homocyst(e)ine is formed in the metabolism of dietary methionine, an essential amino acid. Once formed, Hcy may undergo **remethylation** to form methionine (a process requiring vitamin B_{12}, folic acid, and riboflavin); **transulfuration** resulting in the synthesis of cysteine and glutathione (a process requiring vitamin B_6 and riboflavin); and **oxidation** to form homocystine and mixed disulfides.

(Continued)

Homocyst(e)ine, Plasma *(Continued)*

Specimen Fasting EDTA plasma with immediate centrifugation is optimal. Anticoagulation with citrate or heparin/sodium fluoride is also acceptable, and investigation continues into other anticoagulation procedures.

Storage Instructions Plasma Hcy is increased when the separation of cells from plasma (or serum) is delayed. Plasma Hcy increases ~10% in 1 hour at room temperature. This process is slowed by placing the specimen on ice.

Causes for Rejection Plasma not separated from cells within 1 hour

Reference Interval Current information is tentative. Values in adult men are ~25% higher than in premenopausal women. Based on 182 carefully studied individuals, with total plasma Hcy assayed by an isotope dilution method, age- and gender-specific reference intervals have been proposed as follows:

- 0-30 years: 4.6-8.1 µmol/L
- 30-59 years: male: 6.3-11.2 µmol/L; female: 4.5-7.9 µmol/L
- older than 59 years: 5.8-11.9 µmol/L

Based on results from 1437 adult white males, with total plasma Hcy assayed by HPLC methodology, a reference interval of 4.9-11.7 µmol/L has been suggested.

Similar to the situation with cholesterol measurements, the population-based 95th percentile range, approximately 5-15 µmol/L for plasma Hcy, has little relevance to the desirable, or optimal target. Unlike the situation with cholesterol measurements, there has been no large prospective study from which to derive desirable target values. There are also problems resulting from interlaboratory and intermethod precision and bias (see Limitations).

Use Raised blood Hcy is a strong, independent, dose-related risk factor for coronary, aortic, carotid, and peripheral vascular **atherosclerosis**. It is not fully established, however, if increase in Hcy level is a marker of or cause of the extent of atherothrombotic cardiovascular disease.

Raised blood Hcy is a **thrombophilic state** associated with arterial thrombosis and venous thromboembolic disease.

Raised blood Hcy is a marker for four **dietary vitamin deficiency states** (B_6, B_{12}, folic acid, and riboflavin). Persons with raised blood Hcy due to a nutritional vitamin deficiency (folic acid, B_6, or B_{12}) will have a normalization of the Hcy when adequate vitamins are supplied. Combination therapy with folic acid, vitamin B_{12} and B_6 lowers Hcy levels and improves the outcome of coronary angioplasty.

In pregnancy, raised blood Hcy, reflecting maternal folate deficiency, is associated with an increased incidence of **neural-tube defects**.

Raised blood Hcy is typically present in certain **inborn errors of cobalamin** and **folate metabolism;** it is also found in **homocystinuria**, a rare autosomal recessive metabolic disease resulting from a defect in one of the enzymes involved in homocyst(e)ine metabolism.

Long-term monitoring: Since many persons with raised Hcy will be placed on long-term interventions to reduce their Hcy levels, the long-term, within-person variability is a crucial issue. In one study of healthy volunteers, the biological CV for **fasting** Hcy was 7%, and the critical between-measurement difference (at a 90% probability) was 32% over 4 weeks. However, much greater variability is likely to be encountered when serial measurements are made in different laboratories or by different methods (see Limitations).

Limitations Clinical interpretation of Hcy values is complicated by inter- and intramethod and inter- and intralaboratory imprecision and bias. A proficiency testing program for Hcy is available from the College of American Pathologists. The 1998 results from a Scandinavian proficiency testing program (Moller et al) are available.

Methodology Enzyme immunoassay (EIA), fluorescence polarization immunoassay (FPIA), high performance liquid chromatography (HPLC), gas chromatography/mass spectrometry (GC/MS), ion-exchange chromatography, liquid chromatography electrospray tandem mass spectrometry

Additional Information Approximately 5% to 15% of the general population is homozygous for a thermolabile variant of the enzyme 5,10-methylenetetrahydrofolate reductase (MTHFR). These individuals often have raised blood Hcy.

Most importantly, they require folic acid intakes that exceed those recommended for the general population.

Elevation of Hcy is regularly present in renal insufficiency and hypothyroidism (elevation occurs with uremia). Unbalanced methylation is involved with effects on DNA metabolism and importantly, thereby, epigenetic control of gene expression, reversible by folate therapy.

Individuals with hereditary homocystinuria should probably be tested for factor V Leiden because of the greatly increased risk of thrombosis.

References

Hankey GJ and Eikelboom JW, "Homocysteine and Vascular Disease," *Lancet*, 1999, 354(9176):407-13.

Jacobs DS, DeMott WR, Oxley DK, et al, *Laboratory Test Handbook*, 5th ed, Hudson, OH: Lexi-Comp Inc, 2001.

Key NS and McGlennen RC, "Hyperhomocyst(e)inemia and Thrombophilia," *Arch Pathol Lab Med*, 2002, 126(11):1367-75.

Marcucci R, Brunelli T, Giusti B, et al, "The Role of Cysteine and Homocysteine in Venous and Arterial Thrombotic Disease," *Am J Clin Pathol*, 2001, 116(1):52-5.

Moller J, Rasmussen K, and Christensen L, "External Quality Assessment of Methylmalonic Acid and Total Homocysteine," *Clin Chem*, 1999, 45(9):1536-42.

Molloy AM, Daly S, Mills JL, et al, "Thermolabile Variant of 5,10-Methylenetetrahydrofolate Reductase Associated With Low Red-Cell Folates: Implications for Folate Intake Recommendations," *Lancet*, 1997, 349(9065):1591-3.

Mudd SH and Levy HL, "Plasma Homocyst(e)ine or Homocysteine?" *N Engl J Med*, 1995, 333(5):325 (letter).

Nygard O, Nordrehaug JE, Refsum H, et al, "Plasma Homocysteine Levels and Mortality in Patients With Coronary Artery Disease," *N Engl J Med*, 1997, 337(4):230-6.

Rasmussen K, Moller J, Lyngbak M, et al, "Age- and Gender-Specific Reference Intervals for Total Homocysteine and Methylmalonic Acid in Plasma Before and After Vitamin Supplementation," *Clin Chem* 1996, 42(2):630-6.

Schnyder G, Roffi M, Flammer Y, et al, "Effect of Homocysteine-Lowering Therapy With Folic Acid, Vitamin B_{12}, and Vitamin B_6 on Clinical Outcome After Percutaneous Coronary Intervention - The Swiss Heart Study: A Randomized Controlled Trial," *JAMA*, 2002, 288(8):973-9.

Yu HH, Joubrab R, Asmi M, et al, "Agreement Among Four Homocysteine Assays and Results in Patients With Coronary Atherosclerosis and Controls," *Clin Chem*, 2000, 46(2):258-64,

♦ **Homocyst(e)ine, Qualitative** *see* Cystine, Urine *on page 494*

Homovanillic Acid, Urine

Related Information

Catecholamines, Fractionation, Plasma *on page 345*
Catecholamines, Fractionation, Urine *on page 347*
Metanephrines, Urine or Plasma *on page 899*
Urine Collection, 24-Hour *on page 1295*
Vanillylmandelic Acid, Urine *on page 1299*

Synonyms HVA

Test Includes Measurement of creatinine excretion as well as HVA

Abstract Homovanillic acid (HVA) is a major terminal metabolite of dopamine. Patients with neuroblastoma and pheochromocytoma excrete increased HVA and/or vanillylmandelic acid (VMA).

Patient Preparation Patients should avoid aspirin, disulfiram, reserpine, and pyridoxine, if possible, at least 48 hours prior to collection of the specimen. Levodopa should be avoided for 2 weeks before collection.

Specimen 24-hour urine is preferred. Smaller collections for adults and pediatric patients are acceptable.

Container Plastic urine container

Collection The specimen should have pH 2.0-4.0. Check with your laboratory for volume and concentration.

Storage Instructions Measure 24-hour urine volume, adjust to pH 2-4 and aliquot 100 mL sample and refrigerate. Stable 7 days at 4°C.

Reference Interval Adults: <8.0 mg/24 hours (SI: <44.0 µmol/day). Pediatric values are recorded in the table.

Homovanillic Acid, Urine

Age	µg HVA/mg creatinine
<1 y	<35.0
1 y	<23.0
2-4 y	<13.5
5-9 y	<9.0
10-14 y	<12.0

(Continued)

Homovanillic Acid, Urine *(Continued)*

Special Instructions For work-up for neuroblastoma, excretion of VMA should also be measured. Patient's age is needed, as the values are age dependent.

Use Detect neuroblastoma and pheochromocytoma; follow course of tumor treatment

Limitations Almost all patients with neuroblastoma have elevations of HVA, while only about 80% have elevations of urinary catecholamines. Increased HVA levels, however, are not specific for neuroblastoma.

Methodology High performance liquid chromatography (HPLC), gas chromatography-mass spectroscopy (GC/MS), solvent extraction, colorimetry, and capillary electrophoresis.

References

Fauler G, Leis HJ, and Huber E, "Determination of Homovanillic Acid and Vanillylmandelic Acid in Neuroblastoma Screening by Stable Isotope Dilution GC/MS," *J Mass Spectrom*, 1997, 32(5):507-14.

Garcia A, Heinanen M, Jimenez LM, et al, "Direct Measurement of Homovanillic, Vanillylmandelic and 5-Hydroxyindoleacetic Acids in Urine by Capillary Electrophoresis," *J Chromatogr*, A25, 2000, 871(1-2):341-50.

Mayo Medical Laboratories, *2001 Test Catalogue*, Rochester, MN, 305.

Hormonal Evaluation, Cytologic

Related Information

Cervical/Vaginal Cytology *on page 376*
Estrogens, Urine *on page 557*
Follicle Stimulating Hormone, Serum, Plasma, or Urine *on page 609*
Luteinizing Hormone, Blood or Urine *on page 876*

Synonyms Estrogen Effect; Maturation Index

Test Includes Results are reported as a ratio or percentage of the three cell types (parabasal, intermediate, or superficial) to equal 100: P:I:S = 100. The maturation index may also be expressed as an estimated percentage (estimogram) of the overall cell pattern of the entire sample.

Abstract Due to the responsiveness of vaginal epithelium to multihormonal substances, hormonal cytology can be useful to suggest a need for hormonal therapy, monitor the effects of hormonal therapy, suggest the estimated time of ovulation, and draw attention to cases in whom the hormonal pattern is not compatible with the patient's age or clinical history.

Patient Preparation Douches should be avoided for 24 hours prior to obtaining the smear.

Specimen The preferred specimen is a scrape of the distal third of the lateral vaginal wall.

Container Bottle filled with 95% ethanol or cardboard mailer if specimen is spray-fixed

Collection Using a graphite pencil, write the patient's name and vaginal wall on the frosted end of the glass slide. Smear recently shed exfoliated cells from the lateral vaginal wall across the glass slide and fix immediately. 95% ethanol is a suitable fixative. Include patient's age and menstrual status and any indications of hormonal therapy with the patient's history.

Use Establish hormonal status; evaluate ovarian function; institute evaluation for hormone-producing tumors which can alter estrogen effect in children and postmenopausal women

Limitations The presence of infectious organisms or inflammation greatly diminishes the accuracy of the maturation index. The maturation index is of limited value as an isolated procedure, and hormonal assays prove much more sensitive and indicative of hormonal function.

Additional Information Estrogens, androgens, and progestogens influence the appearance of vaginal epithelial cells.

References

Wied GL and Bibbo M, "Hormonal Cytology," *Comprehensive Cytopathology*, 2nd ed, Chapter 8, Bibbo M, ed, Philadelphia, PA: WB Saunders Co, 1997, 101-24.

Wied GL, Bibbo M, and Keebler CM, "Evaluation of the Endocrinologic Condition of the Female Genital Tract by Exfoliative Cytology," *Compendium on Diagnostic Cytology*, 8th ed, Wied GL, Bibbo M, Keebler CM, et al, eds, Chicago, IL: Tutorials of Cytology, 1997, 55-64.

♦ **2-Hour PP Glucose** *see* Glucose, Postglucose Load, Plasma *on page 647*

- **hPP** *see* Pancreatic Polypeptide, Human, Serum or Plasma *on page 998*
- **HPV Hybrid Capture 2** *see* Human Papillomavirus (HPV) DNA Tests *on page 749*
- **HPV Screen** *see* Human Papillomavirus (HPV) DNA Tests *on page 749*
- **HPV Test** *see* Human Papillomavirus (HPV) DNA Tests *on page 749*
- **HPV Type** *see* Human Papillomavirus (HPV) DNA Tests *on page 749*
- **HSC Transplant** *see* Tissue Typing *on page 1259*
- **5-HT** *see* Serotonin, Blood, Cerebrospinal Fluid *on page 1190*

HTLV-I/II Antibody

Related Information

Donation, Blood *on page 521*
HIV-1/HIV-2 Serology *on page 736*
Human Immunodeficiency Virus DNA Amplification *on page 746*
Risks of Transfusion *on page 1166*

Abstract Human T-lymphotropic virus type I (HTLV-I) and type II (HTLV-II) are human retroviruses which are associated with adult T-cell leukemias and lymphomas as well as HTLV-associated myelopathy.

Specimen Serum

Container Red top tube

Reference Interval Negative

Use Screen blood and blood fractions for transfusion; evaluation of injection drug users; differential diagnosis of spastic paraparesis

Limitations The combined assay for anti HTLV-I/II is used mainly to screen blood donors. The assay cross reacts with only 80% of patients with antibody to HTLV-II. The 20% of blood donors who are not detected by the assay for HTLV-II are not believed to be at sufficiently high risk to transmit the disease to warrant a separate assay. Western blot is associated with a large number of indeterminate results, neither positive nor negative.

Methodology Screen: enzyme immunoassay (EIA); confirmation: Western blot (WB) or radioimmunoprecipitation (RIPA)

Additional Information HTLV-I is a pathogenic retrovirus found in Japan, the Caribbean, Eastern South America, West and Central Africa, and Papua New Guinea/Melanesia. Viral infection can be asymptomatic for prolonged periods (20 years) but is strongly associated with myelopathies and adult T-cell leukemia. Fewer than 5% of those infected with HTLV-I develop myelopathies or leukemia even after 20 years. Adult T-cell leukemia is an aggressive malignancy often associated with skin infiltrates and hypercalcemia. The viruses are tropic for T4 lymphocytes and are passed by sexual contact, blood fractions, injection drug abuse, from mother to fetus, and by breast milk. Pretransfusion testing for antibody to HTLV-I is now mandated by blood banks in order to avoid transfusion transmitted HTLV-I infection from asymptomatic infected donors. Retrospective studies have concluded that about 700 individuals per year received HTLV-I/II blood prior to 1988 when donor testing began. This risk is extremely low (0.024% per unit).

Co-infection of HIV with HTLV-II has been associated with cutaneous T-cell lymphoma. In contrast, most reports of co-infection with HTLV-I and HIV suggest a more favorable prognosis than for infection with HIV alone.

References

Blattner WA, "Human Retroviruses: Their Role in Cancer," *Proc Assoc Am Physicians*, 1999, 111(6):563-72.
Dow BC, "Noise in Microbiological Screening Assays," *Transfus Med*, 2000, 10(2):97-106.
Edlich RF, Arnette JA, and Williams FM, "Global Epidemic of Human T-Cell Lymphotropic Virus Type-I (HTLV-I)," *J Emerg Med*, 2000, 18(1):109-19.
Glynn SA, Murphy EL, Wright DJ, et al, "Laboratory Abnormalities in Former Blood Donors Seropositive for Human T-lymphotropic Virus Types 1 and 2: A Prospective Analysis," *Arch Pathol Lab Med*, 2000, 124(4):550-5.

Internet Web Sites

www.aidsinfo.nih.gov/
www.apha.org/public_health/aids.htm
www.emedicine.com

- **Human Antianimal Antibodies** *see* Heterophilic Antibodies *on page 727*
- **Human Antimouse Antibodies** *see* Heterophilic Antibodies *on page 727*
- **Human Chorionic Gonadotropin** *see* Chorionic Gonadotropin, Human, Serum and Urine *on page 397*

♦ **Human Chorionic Gonadotropin, Urine** *see* Pregnancy Test, Serum or Urine *on page 1084*

♦ **Human Granulocytotropic Ehrlichiosis Serology** *see* Ehrlichiosis Serology *on page 531*

♦ **Human Herpesvirus Type 8** *see* Herpesvirus 8 *on page 725*

Human Immunodeficiency Virus DNA Amplification

Related Information

Antiphospholipid Antibody (Lupus Anticoagulant and/or Anticardiolipin Antibody) *on page 193*

Beta$_2$-Microglobulin, Serum or Urine *on page 254*

CD4/CD8 Enumeration *on page 349*

Donation, Blood *on page 521*

Fungal Culture, Skin *on page 623*

HIV-1/HIV-2 Serology *on page 736*

HIV p24 Antigen Detection *on page 738*

HTLV-I/II Antibody *on page 745*

Risks of Transfusion *on page 1166*

Toxoplasmosis Diagnostic Procedures *on page 1265*

White Blood Cell Count *on page 1330*

Zidovudine, Serum or Plasma *on page 1339*

Applies to RNA Concentration, HIV; Viral Load Testing, HIV-1

Test Includes Detection of HIV nucleic acid from plasma and quantitation based on internal standard or HIV proviral DNA can be detected by amplifying specific DNA sequences from peripheral blood lymphocytes and subsequent hybridization with a specific HIV DNA probe.

Abstract Serial HIV viral load monitoring is the current standard of care for patients with HIV disease, both for initiation of therapy, and for revision of therapy. The HIV viral load detects cell-free plasma viral RNA by using molecular techniques such as PCR or branched chain DNA (bDNA). These quantitative techniques are very sensitive, allowing detection of virus as low as 50 Eq/mL. During therapy, viral loads are usually quite stable, and often become undetectable (lower limits of assays are usually <50 or <400 Eq/mL). A progressive increase in loads over an interval of 1-3 months, or a single high viral load (>10,000) is taken to mean that a revision in therapy may be appropriate.

Human immunodeficiency virus (HIV) contains an RNA genome that will incorporate into host DNA as proviral DNA. The target cells of this virus are the CD4 (T4) T-lymphocytes (helper T cells), monocytes/macrophage populations, and lymph node dendritic cells. To detect the incorporated viral genome, the DNA must be amplified from these cells. The amplified DNA product is detected by specific binding to an HIV probe. The DNA detection assay is used for the diagnosis of HIV in infants. It is also used as a method to recover the HIV genome when the HIV viral load is undetectable in plasma.

Specimen Blood, plasma for quantitative assay; peripheral blood lymphocytes from 10-20 mL whole blood for proviral DNA

Container Two yellow top (ACD) or lavender top (EDTA) tubes should be collected. Consult the laboratory performing the test.

Storage Instructions Tubes of blood can be sent directly to the laboratory at ambient temperature.

Causes for Rejection Specimens with inadequate volume or more than 48 hours old may be rejected. HIV-1 plasma viral load testing performed as initial screening for diagnosis of HIV may show false-positive results. Reject such specimens unless the pretest chance for HIV is high, and the viral load is also high (>100,000 Eq/mL).

Turnaround Time 2 weeks is usually required. Turnaround time may vary with individual laboratories.

Special Instructions Plasma should be separated from clot within 3-4 hours and frozen to prevent a false decrease in viral load.

Reference Interval No HIV viral DNA detected in peripheral blood lymphocytes.

Use Monitor progression of HIV disease and detect therapy failure. HIV detection in patients with unusual or indeterminant HIV serology. It may also be useful in patients with immunodeficiency syndromes characterized by a negative HIV serology and Western blot tests. It can be used in infants who have positive maternal antibody.

Contraindications Informed consent is required in many areas.

Methodology DNA amplification is used, polymerase chain reaction (PCR). DNA is extracted from peripheral blood lymphocytes. The amplified HIV DNA is confirmed by hybridization with an HIV-specific DNA probe. Hybridization with the HIV DNA probe can be detected using autoradiography or enzymatic detection procedures.

Viral load: Usually reverse-transcriptase polymerase chain reaction (RT-PCR) technology (Roche Amplicor Monitor, Basel, Switzerland) or branched chain DNA assay (Bayer Quantiplex, Emeryville, California) is used.

Additional Information The initial step in diagnosis of HIV infection is dependent on the detection of specific antibodies. The antibody screening test commonly used is the enzyme-linked immunosorbent assay (ELISA) with a confirmatory Western or immunoblot. These serologic assays will identify individuals with prior exposure to HIV or passively obtained antibody, such as babies born to HIV-positive mothers. The viral load with the CD4 lymphocyte counts are used together to determine whether therapy should be initiated in early illness, and whether therapy should be changed later. These recommendations are subject to change as information accumulates on the balance between risk of developing AIDS versus risk of problems with medications. Plasma HIV-1 RNA load correlates with rapidity of progression to AIDS after seroconversion. Very low viral loads are associated with long-term disease-free states even without antiretroviral therapy. The goal of highly active antiretroviral therapy (HAART) of HIV is to reduce the viral load to an undetectable level (>50 copies/mL). This is believed to prevent acquisition of resistance mutations.

The PCR test for HIV DNA is useful in resolution of unsatisfactory HIV antibody test results and in determination of the status of children born to mothers with positive HIV serology without the need for viral culture.

In persons who will seroconvert, the HIV quantitative load may be detected by day 4-11 after exposure. The use of screening ELISA followed by Western blot has a very low false-positive rate (approximately 0.0006%). Antibodies are not detectable for about 3-4 weeks after exposure to the virus. This has led to the use of the quantitative HIV test to detect early HIV infection. However, commercial assays now available are not approved for use in antibody-negative patients. When used in this manner, both the number of viral copies and the time after presumed exposure should be considered. Viral loads, when the person is not HIV infected, have been reported to be low (<2000) (ie, false-positive viral loads). An infected person would be expected to have a viral load in excess of 10,000 and usually much greater, if the viral load were used to screen for HIV.

References

Engels EA, Rosenberg PS, O'Brien TR, et al, "Plasma HIV Viral Load in Patients With Hemophilia and Late-Stage HIV Diseases: A Measure of Current Immune Suppression," *Ann Intern Med*, 1999, 131(4):256-64.

Ginocchio CC, "HIV-1 Viral Load Testing," *Lab Med*, 2001, 32(3):142-52.

Helbert M and Breuer J, "Monitoring Patients With HIV Disease," *J Clin Pathol*, 2000, 53(4):266-72.

Henry K, "The Case for More Cautious, Patient-Focused Antiretroviral Therapy," *Ann Intern Med*, 2000, 132(4):306-11.

International Perinatal HIV Group, "The Mode of Delivery and the Risk of Vertical Transmission of Human Immunodeficiency Virus Type 1," *N Engl J Med*, 1999, 340(13):977-87.

Kane B, "Beyond HIV Viral Load Testing," *Ann Intern Med*, 1999, 131:637-8.

Laufer M and Scott GB, "Medical Management of HIV Disease in Children," *Pediatr Clin North Am*, 2000, 47(1):127-53.

Lindegren ML, Byers RH, Thomas P, et al, "Trends in Perinatal Transmission of HIV/AIDS in the United States," *JAMA*, 1999, 282(6):531-8.

Nolte FS, "Impact of Viral Load Testing on Patient Care," *Arch Pathol Lab Med*, 1999, 123(11):1011-4.

Quinn TC, Wawer MJ, Sewankambo N, et al, "Viral Load and Heterosexual Transmission of Human Immunodeficiency Virus Type 1," Rakai Project Study Group, *N Engl J Med*, 2000, 342(13):921-9.

Sterling TR, Vlahov D, Astemborski J, et al, "Initial Plasma HIV-1 RNA Levels and Progression to Aids in Women and Men," *N Engl J Med*, 2001, 344(10):720-5.

Internet Web Sites

www.aidsinfo.nih.gov/
www.apha.org/public_health/aids.htm
www.emedicine.com

Human Immunodeficiency Virus, Resistance (Susceptibility) Testing

Related Information

Antiphospholipid Antibody (Lupus Anticoagulant and/or Anticardiolipin Antibody) *on page 193*

Beta$_2$-Microglobulin, Serum or Urine *on page 254*

CD4/CD8 Enumeration *on page 349*

Donation, Blood *on page 521*

Fungal Culture, Skin *on page 623*

HIV p24 Antigen Detection *on page 738*

HTLV-I/II Antibody *on page 745*

Human Immunodeficiency Virus DNA Amplification *on page 746*

Risks of Transfusion *on page 1166*

Toxoplasmosis Diagnostic Procedures *on page 1265*

White Blood Cell Count *on page 1330*

Zidovudine, Serum or Plasma *on page 1339*

Test Includes Genotypic or phenotypic studies on HIV for antiviral susceptibility

Abstract Resistance of HIV to antiretroviral medications is a common cause of treatment failure and results in increased viral load. Use of genotypic or phenotypic susceptibility tests may be used to help select change of therapy in failing regimens.

Specimen Plasma

Container Lavender top (EDTA) tube or SPS® tube

Turnaround Time Consult reference laboratory.

Use Detect mutations in HIV that correlate with therapy failure

Limitations The test system evaluates the predominant viruses present in the specimen. HIV exists as a mixture of virus with multiple strains being present, often with varying susceptibilities. Resistance demonstrated by testing is useful evidence to avoid that therapeutic agent. Lack of resistance on the test may not confer susceptibility. Assays for resistance may provide information only on the predominant circulating variants, and may miss minor variants. Such minor variants may predominate when antiviral therapy effectively controls the previously predominant virus. The cost of testing is high.

Methodology Genotyping requires the amplification of the reverse transcriptase gene and/or the protease gene followed by detection of mutations. Mutations can be detected by direct sequencing (costly) or by blotting techniques such as Line Probe Assay (LIPA).

Phenotyping requires growth of the HIV isolate and testing the viral replication with various drugs in tissue culture.

Additional Information There are many reasons HIV therapy fails, including missing doses of complex drug schedules, lack of adsorption of certain drugs due to interference from foods or other medications, and low serum or tissue levels of medications due to metabolism interactions, usually at the P450 level. When the HIV viral load increases to significant levels, HIV susceptibility testing has become an important adjunct.

Interpretation of genotype susceptibility testing is complicated by the need to know which codon changes are likely to convey resistance to each medication. Interpretation of phenotype susceptibility is more readily understandable by giving the susceptibility or resistance pattern of the predominantly tested virus.

References

Alcorn TM and Faruki H, "HIV Resistance Testing: Methods, Utility, and Limitations," *Mol Diagn*, 2000, 5(3):159-68.

O'Brien WA, "Resistance Against Reverse Transcriptase Inhibitors," *Clin Infect Dis*, 2000, 30(S2):S185-92.

Omrani AS and Pillay D, "Multidrug Resistant HIV-1," *J Infect*, 2000, 41(1):5-11.

Saag MS, "HIV Resistance Testing in Clinical Practice: A QALY-fied Success," *Ann Intern Med*, 2001, 134(6):475-7.

Weinstein MC, Goldie SJ, Losina E, et al, "Use of Genotypic Resistance Testing to Guide HIV Therapy: Clinical Impact and Cost-Effectiveness," *Ann Intern Med*, 2001, 134(6):440-50.

Wilson JW, Bean P, Robins T, et al, "Comparative Evaluation of Three Human Immunodeficiency Virus Genotyping Systems: The HIV-Genotyp® Method, the HIV PRT GeneChip Assay, and the HIV-1 RT Line Probe Assay," *J Clin Microbiol*, 2000, 38(8):3022-8.

Internet Web Sites

www.aidsinfo.nih.gov/

www.apha.org/public_health/aids.htm

www.emedicine.com

- **Human Leukocyte Antigen** *see* HLA-B27 *on page 738*
- **Human Leukocyte Antigen** *see* Tissue Typing *on page 1259*
- **Human Leukocyte Antigen System** *see* HLA Typing, Single Human Leukocyte Antigen *on page 739*
- **Human Leukocyte Cell Surface (Membrane) Markers** *see page 34*
- **Human Pancreatic Polypeptide** *see* Pancreatic Polypeptide, Human, Serum or Plasma *on page 998*
- **Human Papillomaviruses (HPV)** *see* Cervical/Vaginal Cytology *on page 376*

Human Papillomavirus (HPV) DNA Tests

Related Information
Cervical/Vaginal Cytology *on page 376*

Synonyms DNA Hybridization Test for HPV; HPV Hybrid Capture 2; HPV Screen; HPV Test; HPV Type; Probe Test for HPV DNA

Applies to Koilocytosis

Test Includes Screening specimens for the presence of HPV DNA for high-risk types important in cervical carcinogenesis

Abstract Rapid culture methods will only detect specified virus(es), negative cul (Stoler). More than 100 types of HPV have been defined on the basis of DNA sequence heterology (zur Housen). Approximately 30 HPV subtypes have been found in the anogenital tract.

Infection with HPV has been established as the primary cause of cervical cancer in nearly all cases. Over their lifetime, many women are infected with HPV, which is a sexually-transmitted disease. Of those infected, only a fraction develop high-grade squamous intraepithelial lesions, and even fewer progress to carcinoma (Davey). Infection with HPV is the necessary agent required for development of cervical cancer, but other cofactors are required. Other corisks suspected include smoking, oral contraceptive use, nulliparity, nutritional factors, other sexually-transmitted infections (eg, *Chlamydia trachomatis*) and, possibly, inflammation.

The cervical lesions caused by HPV range from koilocytosis, condyloma acuminatum, low-grade squamous intraepithelial lesions (LGSIL), high-grade squamous intraepithelial lesions (HGSIL), and invasive cervical squamous cell carcinoma and adenocarcinoma (Cuzick). Most women infected with HPV, especially younger women, have an effective immune response that clears the infection or lowers the viral load to undetectable levels in an average of 8-24 months (Woodman). While progressing from infection to neoplasia, changes of the cervix can be observed by cytologic screening or histopathology seen on cervical biopsies. It is known that virtually 100% of women with HGSIL will test positive for HPV. However, many women harbor HPV in their lower genital tracts without evidence of cytologic or histopathologic changes. Furthermore, the Pap test may not identity or correctly classify such changes due to errors in sampling, interpretation, or follow-up.

HPV does not lend itself to culture *in vitro*. Thus, detection of HPV relies strictly on molecular analysis of HPV. Furthermore, HPV DNA testing has emerged as another testing modality for cervical cancer and its precursor lesions.

Currently, three primary methods of testing for HPV are available. Each of these assays uses a different technique to probe for HPV. The Hybrid Capture® 2 HPV (HC2) DNA Test (Digene Corporation, Gaithersburg, MD) is the only presently (2003) FDA-approved test for detecting HPV and is currently the industry standard (Bolick). The other two methods are the Ventana Inform HPV (Ventana Medical Systems, Inc, Tucson, AZ) and polymerase chain reaction. Details for HC2 testing are described in the remainder of this monograph. More information on the other methods appears in the Additional Information section.

Patient Preparation Specimens collected for HC2 analysis are taken from patients seen during routine obstetric examinations or culposcopic exams which include Pap tests or cervical biopsies. No additional patient preparation is needed. If a cervical biopsy specimen is being sent for HC2, the cervix should not be treated with acetic acid before the biopsy is made. The cervical samplings or biopsies can be taken at any stage of the patient's menstrual cycle.
(Continued)

749

Human Papillomavirus (HPV) DNA Tests *(Continued)*

Specimen The HC2 test has been FDA approved for the use with the following types of specimens:

- Cervical specimens (liquid based) collected with a broom collection device and rinsed in the ThinPrep® System PreservCyt® solution, manufactured by Cytyc (Boxborough, MA)
- Cervical specimens (not liquid based) collected with the Digene Cervical Sampler™
- Cervical biopsies placed in Digene Specimen Transport Medium™

Container Depending on the type of specimen collected, specific containers and solutions for HC2 are used. Do not substitute. Contact receiving laboratory for collection materials and requirements before sample is taken. Some materials may have to be purchased directly from Digene (800) 344-3631 or Cytyc (800) 442-9892.

Collection A broom collection device is used at the time of Pap testing and is rinsed in PreserveCyt solution. This solution is used for preparation of a Pap-stained, monolayer-technology slide (liquid based) for cytological analysis. The remaining solution is available for the HC2 test, if indicated. These materials (broom and solution) are available from the testing laboratory or from Cytec.

If the Pap test is performed using the conventional cervical/vaginal cytology smear (nonliquid based), the Digene Cervical Sampler™ should be used for HPV sampling in addition to the standard brush and/or spatula collection tools. The clinician must purchase the Cervical Sampler directly from Digene. The Cervical Sampler includes a broom-style brush and container with transport medium.

If a cervical biopsy is obtained by standard methods (eg, culposcopic guidance), then the biopsy material is stored and transported in the Digene Specimen Transport Medium. The biopsy must be between the sizes of 0.2 cm and 0.5 cm. The specimen will be rejected if the size of the biopsy exceeds 0.5 cm. The clinician must purchase the medium directly from Digene.

Storage Instructions All specimens intended for HC2 testing are stored and transported at room temperature.

Turnaround Time Depending on the laboratory throughput, the HC2 test can be performed the same day as receipt of the specimen. The receiving laboratory can be contacted regarding turnaround time and the method of receiving the report generated from the lab.

Special Instructions The HC2 test must be performed on the specimen within 21 days of collection. The HC2 HPV test is reimbursed by most insurance providers and third-party payors using CPT code 87621. For low-risk and high-risk HPV results, the code 87621 (x2) is approved for use with the HC2 HPV test.

Use The cocktail approach of the HC2 test provides an excellent tool for triage of patients demonstrating minor cytological abnormalities by Pap test, but the test cannot determine the specific HPV type present (Hubbard). HCV genotypes detected by the HC2 assay can be grouped into the follow risk strata: 13 types are implicated in the pathogenesis of HGSIL and invasive cancer (16, 18, 31, 33, 35, 39, 45, 51, 52, 56, 58, 59, an 68). The most common high-risk types are HPV 16, 18, 31, 33, 35, and 45. The five low-risk viral types associated with LGSIL are 6, 11, 42, 43, and 44.

HPV testing can be used as an adjunct for atypical cytology results, as a primary screening tool, as a follow-up test after colposcopy or treatment, and as a quality assurance measure. HPV testing is currently recommended when a cervical screen is given the interpretation of ASCUS, equivocal for HGSIL, since it has a high positive predictive value for identifying women with an underlying high-grade cervical lesion (Schiffman). Furthermore, the American Cancer Society has recently released new guidelines for the early detection of cervical neoplasia and cancer in the United States that incorporates adjunctive HPV DNA testing in women older than 30 years of age at intervals of 3 years or longer (Saslow).

Limitations The HC2 test detects only 18 of the nearly 30 known anogenital HPV types. There is a limited test platform due to licensed and patented technologies. The cocktail approach to genotyping does not allow determination of the specific HPV type present. HC2 is a labor-intensive test, which requires skilled testing personnel. Reimbursement by third-party payers is marginal in some regions of the country (Davey).

Methodology Digene's HC2 uses a signal amplification method for detection of HPV nucleic acids. In general, the signal amplification technique depends on proprietary technologies, which are not in the public domain. In the lab, clinical specimens are combined with a base solution, which disrupts the virus and releases target DNA. No special specimen preparation is necessary prior to forwarding it to the lab. The HC2 assay uses specific RNA probes, which bind individual DNA of particular HPV genotypes. Digene owns a proprietary antibody that specifically binds the DNA-RNA complex. Ultimately, the antibody is marked with a reported molecule detected by a chemiluminescent method. A single patient specimen can also be tested for *Chlamydia trachomatis* and *Neisseria gonorrhoeae* while the specimen is being assayed by the Digene HC2 system. A similar probe for herpes simplex virus (HSV) is currently under development. A successor HC3 assay is also under development.

Additional Information Tests for HPV using nucleic acid probes have been commercially available since the late 1980s. However, the early tests were cumbersome and used nucleic acid probes labeled with radioactive phosphorus (^{32}P). Direct probing, signal amplification, and target amplification are the three principle methods used currently for molecular analysis of HPV (Hubbard). The HC2 assay, discussed above, uses a signal amplification technique.

Direct probing methods include Southern blotting and *in situ* hybridization (ISH). Southern blotting is the "gold standard" technique for analysis of HPV and was used early in studies of HPV. Ventana has developed an automated ISH method for use on both cytology and histology samples. The weaknesses of these direct probe methods are low sensitivity and time-consuming protocols (Hubbard). In addition, the Southern blot method requires large amounts of highly purified DNA and cannot be performed on fixed, histological tissue.

Target amplification is a method that holds great promise for addressing many aspects of HPV testing. Polymerase chain reaction (PCR) is the most well known assay in this testing classification. While PCR is versatile and highly specific, it is currently a "home-brew" assay, without interlaboratory standardization, and is in the realm of research use only. Real-time PCR is used to estimate viral load but currently has no known prognostic value (Davey). Licensing or patents of HPV sequences may limit the applicability of PCR as *in vitro* diagnostic tests. However, that may change in the future. In June 2002, ownership of many HPV sequences was transferred to the sister company of Roche Molecular Systems, the owner of PCR technology (Hubbard). This exchange may be the harbinger of *in vitro* diagnostic use of PCR and open a pathway for future HPV testing methods incorporating PCR methods.

References

Bolick DR, Bolick RE, Coates F, et al, "Laboratory Implementation of Human Papillomavirus Testing," *Arch Pathol Lab Med*, 2003, 127(8):984-90.

Cuzick J, "Human Papillomavirus Testing for Primary Cervical Cancer Screening," *JAMA*, 2000, 283(1):108-9.

Davey DD and Zarbo RJ, "Human Papillomavirus Testing - Are You Ready for a New Era in Cervical Cancer Screening?" *Arch Pathol Lab Med*, 2003, 127(8):927-9.

Goodman A and Wilbur DC, "A 37-Year-Old Woman With Atypical Squamous Cells on a Papanicolaou Smear," Case Record of the Massachusetts General Hospital, Case 32-2003, *N Engl J Med*, 2003, 349(16):1555-64.

Hubbard RA, "Human Papillomavirus Testing Methods," *Arch Pathol Lab Med*, 2003, 127(8):940-5.

Mork J, Lie AK, Glattre E, et al, "Human Papillomavirus Infection as a Risk Factor for Squamous-Cell Carcinoma of the Head and Neck," *N Engl J Med*, 2001, 344(15):1125-31.

Saslow D, Runowicz CD, Solomon D, et al, "American Cancer Society Guideline for the Early Detection of Cervical Neoplasia and Cancer," *CA Cancer J Clin*, 2002, 52(6):342-62.

Schiffman MH and Solomon D, "Findings to Date From the ASCUS-LSIL Triage Study (ALTS)," *Arch Pathol Lab Med*, 2003, 127(8):946-9.

Skyldberg B, Fujioka K, Hellström AC, et al, "Human Papillomavirus Infection, Centrosome Aberration, and Genetic Stability in Cervical Lesions," *Mod Pathol*, 2001, 14(4):279-84.

Stoler MH, "Human Papillomavirus Biology and Cervical Neoplasia: Implications for Diagnostic Criteria and Testing," *Arch Pathol Lab Med*, 2003, 127(8):935-9.

Woodman CB, Collins S, Winter H, et al, "Natural History of Cervical Human Papillomavirus Infection in Young Women: A Longitudinal Cohort Study," *Lancet*, 2001, 357(9271):1831-6.

zur Hausen H, "Papillomavirus and Cancer: From Basic Studies to Clinical Application," *Nat Rev Cancer*, 2002, 2(5):342-50.

(Continued)

Human Papillomavirus (HPV) DNA Tests *(Continued)*

Internet Web Sites
hpv-web.lanl.gov
www.niaid.nih.gov/factsheets/stdhpv.htm

♦ **Humster (Human + Hamster) Test** *see* Sperm Penetration Assay (Zona-Free Hamster Egg Penetration Test) *on page 1220*

♦ **Hunter Syndrome** *see* Mucopolysaccharides, Urine *on page 922*

Huntington Disease DNA Test

Related Information
Polymerase Chain Reaction *on page 1069*

Synonyms HD DNA Test

Test Includes Determination of the number of CAG trinucleotide repeats in the HD (alias IT-15) gene at chromosome 4 band p16.

Abstract Huntington disease (HD) is a progressive disorder of the central nervous system characterized primarily by motor, cognitive, and psychiatric disturbances. It is an autosomal dominant disorder. Each child of an affected, or asymptomatic individual carrying a mutant HD allele, has a 50% chance of inheriting the mutant gene. Average age of onset is 35-44 years. Severe disease, however, has been reported in juveniles who present with marked rigidity, intellectual decline, and prominent motor and cerebellar symptoms.

Affected, or genetically predisposed, individuals typically have one normal HD allele and one mutant allele. The mutant allele has an abnormally large number of a naturally-occurring CAG trinucleotide repeat motif in the HD gene. The DNA test detects 100% of cases by measuring the number of CAG repeats in each HD allele. The number of CAG repeats is polymorphic in the normal population but is always ≤26; normal alleles are stable and show no changes in CAG repeat number upon transmission to offspring. The presence of ≥40 CAG repeats indicates that an individual has, or is genetically predisposed to develop, HD. Mutant alleles with ≥40 CAG repeats are often unstable upon transmission, with offspring having an increased or decreased number of repeats; increases are more common with paternal transmission. Nearly all cases of juvenile HD are associated with paternal transmission. A greater number of CAG repeats is generally correlated with an earlier age at onset and more severe disease, thus, in some families, the disease appears to be more severe in affected individuals of the most recent generation. However, because of clinical variability, the number of CAG repeats cannot reliably predict characteristics (course) of the disease in an individual case.

Patient Preparation Testing of asymptomatic at-risk adults is predictive testing and the patient should have formal genetic counseling to explain the meaning, benefits, and risks of testing.

Aftercare For asymptomatic at-risk adults, post-test genetic counseling is appropriate to explain the test result and its significance.

Specimen Whole blood (3-10 mL)

Container Collect anticoagulated blood in either a lavender top (EDTA) tube or yellow top (ACD) Vacutainer® tube. Avoid use of heparin anticoagulants, which can interfere with polymerase chain reactions and endonuclease restriction enzyme activity.

Storage Instructions Blood samples can be stored and shipped at room temperature or refrigerated (4°C); do not freeze. Blood samples should be received in the testing laboratory within 4 days of draw to ensure an adequate DNA yield.

Causes for Rejection Frozen samples, whole blood in heparin anticoagulant (green top) or other inappropriate collection tube

Turnaround Time 7-14 days

Special Instructions To provide an optimal test interpretation, the testing laboratory needs to know the clinical diagnosis, family history, and whether DNA testing has confirmed the clinical diagnosis of an affected family member.

Reference Interval Normal HD alleles have <26 CAG repeats. Affected individuals have ≥40 CAG repeats. An asymptomatic adult with an HD allele with ≥40 CAG repeats is predicted to develop Huntington disease. Alleles with between 27-35 CAG repeats do not cause disease in the carrier individual, but these alleles can be unstable with the CAG repeats increasing in number to the abnormal pathogenic range upon transmission to offspring. Alleles with between 36 and 39 CAG repeats show reduced penetrance and may or may not cause HD. Juvenile HD cases typically have about 80-250 CAG repeats.

Use Results are useful in evaluation of individuals with progressive motor disability, cognitive decline, or personality changes, particularly if there is a positive family history. Testing of asymptomatic adults at-risk for HD should be preceded by formal genetic counseling to explain the meaning, benefits, and risks of such testing. Can be used to identify which progenitor in a family carries the HD mutation; this is important information for genetic counseling and identification of other at-risk family members. Testing of asymptomatic at-risk children under the legal age, for an adult onset disorder for which no treatment exists, may not be appropriate. Formal genetic counseling of parents requesting testing of asymptomatic at-risk children should be provided. Children who present with symptoms usually benefit from a specific diagnosis. Prenatal testing is available, but should be preceded by genetic counseling to discuss difficult ethical issues related to testing for a (typically) adult-onset disease.

Limitations The number of CAG repeats cannot reliably predict disease onset, severity, or cognitive ability in an individual patient. In cases with juvenile onset with typically larger numbers of CAG repeats, it is more difficult for the testing laboratory to detect the large mutant allele. In these cases, concerns about allele size detection should be discussed with the testing laboratory, especially when only a single sized allele is observed in an affected child (ie, an apparent HD homozygote).

Methodology Polymerase chain reaction (PCR). A segment of the HD gene is amplified from the patient's DNA, using primers that flank the CAG repeat region of the gene. After electrophoresis, the number of repeats is calculated from the product size. This reaction can detect both normal and mutant HD alleles.

Additional Information HD has a prevalence of about 5 per 100,000 in populations of Western European descent; frequencies vary by geography and ethnic background. The CAG trinucleotide repeat motif is located in the coding region of the gene and encodes a tract of polyglutamine residues, which when abnormally long, confer an unknown new function to huntingtin, protein product of the HD gene. How this "gain-of-function" causes the disease is not clear. HD testing is widely available; a list of clinical laboratories that perform this test can be found at GeneTests® (http://www.genetests.org).

References
Brandt J, Bylsma FW, Gross R, et al, "Trinucleotide Repeat Length and Clinical Progression in Huntington's Disease," *Neurology*, 1996, 46(2):527-31.

Kremer B, Goldberg P, Andrew SE, et al, "A Worldwide Study of the Huntington's Disease Mutation. The Sensitivity and Specificity of Measuring CAG Repeats," *N Engl J Med*, 1994, 330(20):1401-6.

"The American College of Medical Genetics/American Society of Human Genetics Huntington Disease Genetic Testing Working Group. ACMG/ASHG Statement. Laboratory Guidelines for Huntington Disease Genetic Testing," *Am J Hum Genet*, 1998, 62:1243-7.

Internet Web Sites
www.genetests.org

17-Hydroxycorticosteroids, Urine

Related Information

Synonyms 17-OHCS; Porter-Silber Chromogens, Urine

(Continued)

17-Hydroxycorticosteroids, Urine *(Continued)*

Abstract This presently rarely performed test measures the metabolites of cortisol and other adrenal corticosteroids.

Specimen 24-hour urine

Storage Instructions Refrigerate during collection, refrigerate or freeze after collection. Stable 45 days if refrigerated and acidified.

Reference Interval Moderate intermethod variability.

Use This test has, for the most part, been replaced by the measurement of urinary free cortisol, which is a more sensitive and specific test for hypercortisolism.

Limitations Subject to variability due to unreliable 24-hour urine collections and interferences. Serum or urine cortisol measurements are preferred.

17-OHCS (Porter-Silber) does not measure pregnanetriol.

Methodology Porter-Silber color reaction

References

Flack MR, Oldfield EH, Cutler GB Jr, et al, "Urine Free Cortisol in the High-Dose Dexamethasone Suppression Test for the Differential Diagnosis of the Cushing Syndrome," *Ann Intern Med*, 1992, 116(3):211-7.

Wilson J and Foster D, *Williams Textbook of Endocrinology*, 9th ed, Philadelphia, PA: WB Saunders Co, 1998, 615.

♦ **16-Hydroxyestradiol** *see* Estriol, Unconjugated, Pregnancy, Serum, Plasma, or Urine *on page 554*

♦ **11-Hydroxyetiocholanolone** *see* 17-Ketosteroids Fractionation, Urine *on page 818*

5-Hydroxyindoleacetic Acid, Quantitative, Urine

Related Information

Serotonin, Blood, Cerebrospinal Fluid *on page 1190*
Urine Collection, 24-Hour *on page 1295*

Synonyms 5-HIAA, Quantitative, Urine

Applies to Chromogranin A, Plasma; Serotonin, Cerebrospinal Fluid; Serotonin, Metabolite

Abstract Carcinoid tumors are neuroendocrine neoplasms which arise in the gastrointestinal and respiratory tracts and in a broad range of additional sites. Most are less aggressive than are usual adenocarcinomas. Indolent versus aggressive characteristics correlate with site of origin, histopathologic characteristics, size, and concentration of 5-HIAA, but survival in large series is difficult to project. 5-HIAA is a serotonin metabolite used as a marker for carcinoid tumors.

Patient Preparation A number of foods are high in serotonin. For a 48-hour period or more prior to and during collection, **avoid** avocados, bananas, butternuts, cantaloupe, chocolate, dates, eggplant, grapefruit, hickory, honeydew, kiwi fruit, melon, nuts (including pecans and walnuts), pineapples, plantain, plums, tomatoes, acetaminophen, salicylates, phenacetin, cough syrup containing glyceryl guaiacolate, naproxen (Naprosyn®, Anaprox®), mephenesin, methocarbamol, imipramine, isoniazid, MAO inhibitors, methenamine, methyldopa, reserpine, and phenothiazines. Drug interference relates to method; check with laboratory.

Specimen 24-hour urine

Container Check with testing laboratory. Usually, a urine container with no preservative is given to the patient.

Collection Check with testing laboratory. A 24-hour urine specimen is collected, usually without preservative. The specimen should be kept refrigerated during collection. When the specimen is received in the laboratory, its pH may be adjusted to 2-3, depending upon the protocol of the laboratory.

Storage Instructions Stable 14 days if acidified and refrigerated. Acidification is done with hydrochloric or boric acids, depending upon individual laboratory methods. Acetic acid has been shown to decrease recovery of 5-HIAA and is not generally recommended.

Reference Interval Approximately 1-9 mg/24 hours (SI: 5-48 µmol/day); some variation in reference intervals is found.

Critical Values >25 mg/24 hours, without dietary interference, is diagnostic of carcinoid.

Use Urine 5-HIAA measurement is used as an adjunct in the management of carcinoid tumors. Values >25 mg/24 hours (higher if the patient has malabsorption) are strong evidence for carcinoid. Serotonin, and consequently 5-HIAA, are produced in excess by most carcinoid tumors, especially those producing the carcinoid syndrome of flushing, hepatomegaly, diarrhea, bronchospasm, and ultimately right-sided valvular heart disease.

Limitations Urine 5-HIAA concentrations is often normal with nonmetastatic carcinoid tumor. Some patients with the carcinoid syndrome excrete nonhydroxylated indolic acids, not measured as 5-HIAA. Patients with renal disease may have falsely low urinary 5-HIAA levels. 5-HIAA is increased in untreated patients with malabsorption, who have increased urinary tryptophan metabolites (eg, patients with celiac disease, tropical sprue, Whipple disease, stasis syndrome, and cystic fibrosis), chronic intestinal obstruction, pregnancy, ovulation, and ingestion of foods rich in serotonin (see Patient Preparation).

Methodology Spectrophotometry, gas chromatography (GC), high performance liquid chromatography (HPLC), fluorescence polarization immunoassay (FPIA). A carbon fiber electrode technique allows *in vivo* monitoring of 5-HIAA in brain tissue.

References
Kulke MH and Mayer RJ, "Carcinoid Tumors," *N Engl J Med*, 1999, 340(11):858-68.

Lechago J, "Neuroendocrine Cells of the Gut and Their Disorders," *Gastrointestinal Pathology*, Goldman H, Appelman HD, and Kaufman N, eds, Baltimore, MD: Lippincott Williams & Wilkins, 1990, 181-219.

Young DS, "Effects of Drugs on Clinical Laboratory Tests," 5th ed, Volume 1: Listing by Test, Washington DC: AACC Press, American Association of Clinical Chemistry, 2000, Section 3, 450-1.

♦ **21-Hydroxylase** *see* 17-Hydroxyprogesterone, Whole Blood, Serum, or Plasma *on page 755*

21-Hydroxylase Antibodies, Serum
Related Information
Adrenal Cortex: Laboratory Assessment Overview *on page 110*
Adrenocorticotropic Hormone, Plasma *on page 114*
Corticotropin Stimulation Test (Rapid) *on page 456*
Cortisol, Serum or Plasma *on page 460*
Insulin Tolerance Test *on page 804*
Metyrapone Stimulation Test, Serum *on page 910*
Urinary Cortisol/Creatinine Increment *on page 1292*

Abstract Antibodies directed against the microsomal antigen, 21-hydroxylase, serve as a marker of autoimmune destruction of the adrenals.

Specimen Serum

Container Red top tube

Reference Interval <1 unit/mL

Use This test is useful in the evaluation of patients with adrenal insufficiency (AI-Addison disease). A positive result identifies autoimmune destruction as the etiology of the patient's AI.

References
Mayo Reference Services Publication, "21-Hydroxylase antibodies, Serum," *New Test Announcement*, Rochester, MN: Mayo Medical Laboratories, 2000.

Tanaka H, Perez MS, Powell M, et al, "Steroid 21-Hydroxylase Autoantibodies: Measurements With a New Immunoprecipitation Assay," *J Clin Endocrinol Metab*, 1997, 82(5):1440-6.

♦ **Hydroxylysylpyridinoline** *see* Pyridinolines (Pyridinoline and Deoxypyridinoline), Urine *on page 1126*

♦ **Hydroxymethylbilane Synthase** *see* Porphobilinogen Deaminase, Erythrocyte *on page 1072*

♦ **17α-Hydroxyprogesterone** *see* 17-Hydroxyprogesterone, Whole Blood, Serum, or Plasma *on page 755*

17-Hydroxyprogesterone, Whole Blood, Serum, or Plasma
Related Information
Adrenal Cortex: Laboratory Assessment Overview *on page 110*
Adrenocorticotropic Hormone, Plasma *on page 114*
Androstenedione, Serum *on page 158*
Cortisol, Free, Urine *on page 459*
Cortisol, Serum or Plasma *on page 460*
(Continued)

17-Hydroxyprogesterone, Whole Blood, Serum, or Plasma (Continued)

Pregnanetriol, Urine on page 1085
Progesterone, Serum on page 1093
Renin Activity, Plasma on page 1149

Synonyms 17α-Hydroxyprogesterone; 17-OHP

Applies to 11-Beta-Hydroxylase; Deoxycorticosterone; 11-Desoxycortisol; 21-Hydroxylase; 17-OHP, Amniotic Fluid; 17-OHP, Saliva

Replaces Urine 17-Ketogenic Steroids; Urine Pregnanetriol Assay

Abstract 17-hydroxyprogesterone (17-OHP), produced by adrenal cortices, ovaries, testes, and placenta, is a precursor of cortisol, and is markedly elevated in patients with congenital adrenal hyperplasia (CAH), a group of autosomal recessive diseases in which deficiency of cortisol leads to ACTH-induced adrenal hyperplasia. Ninety percent of CAH cases or more are caused by 21-hydroxylase deficiency.

A number of states in the U.S. screen for CAH by assaying 17-OHP in blood collected onto filter paper for newborn screening.

Specimen Serum or plasma, whole blood

Container Red top tube or lavender top (EDTA) tube; check with laboratory. Whole blood, used for screening neonatal specimen, may be collected and dried onto filter paper. Check with newborn screening laboratory for proper collection details.

Sampling Time Newborn screening is ideally performed at 2-3 days of age. Blood levels peak in the morning.

Storage Instructions Separate serum or plasma within 4 hours. Serum or plasma stable for up to 4 days at 4°C and for up to 1 month at -20°C.

Causes for Rejection Incorrect/insufficient filter paper saturation, inadequate specimen labelings

Reference Interval Intervals vary between laboratories. The following are examples.

- Cord blood: 7.40-18.70 ng/mL
- 3 days to 2 months: 0.10-9.40 ng/mL
- 3 months to 11 years: not detectable to 0.90 ng/mL
- 12-20 years: not detectable to 1.80 ng/mL
- Adult male: 0.40-3.30 ng/mL
  Adult female:
  - follicular: 0.10-1.20 ng/mL
  - luteal: 0.40-4.80 ng/mL
  - postmenopausal: 0.10-0.60 ng/mL

Allen et al recommend the following birth-weight-dependent thresholds of abnormality for 17-OHP in newborn screening using dried whole blood samples collected onto filter paper. See table.

Newborn Screening for CAH
17-OHP in Dried Whole Blood Specimens

Weight (g)	Possibly Abnormal 17-OHP ng/dL (SI: nmol/L)	Definitely Abnormal 17-OHP ng/dL (SI: nmol/L)
<1299	≥13,500 (≥405)	
1300-1699	11,500-13,400 (345-402)	≥13,500 (≥405)
1700-2199	6500-8900 (195-267)	≥9000 (≥270)
≥2200	4000-8900 (120-267)	≥9000 (≥270)

Use 17-OHP is used

- to diagnose CAH and to monitor the effectiveness of cortisol replacement therapy in patients with the disease.
- in newborn screening programs that include testing for CAH.
- to evaluate hirsutism, infertility, and/or hermaphroditism in female patients with possible 21-hydroxylase deficiency.
- to assess certain adrenal or ovarian tumors which may have endocrine activity.

Limitations 17-OHP is elevated, but less so, in the 11-beta-hydroxylase deficiency form of CAH. Measurement of serum 11-desoxycortisol (substance S) and deoxycorticosterone (DOC) differentiates these two forms of CAH.

An effective newborn screening program must minimize false-negative results. Therefore, high false-positive rates are observed from newborn screening analyses of 17-OHP in dried blood on filter paper. Unfortunately, this leads to a relatively large number of false-positive results, which are identified as such only after follow-up testing. False positives are particularly common in preterm, low-birth-weight neonates. Utilization of different decision thresholds for different birth weights improves screening performance.

False-negative results occur when the screening blood specimen is obtained before 2 days of age.

Methodology Radioimmunoassay (RIA), fluoroimmunoassay (FIA), enzyme-linked immunosorbent assay (ELISA), chemiluminescent immunoassay (CIA), high performance liquid chromatography (HPLC), gas chromatography mass spectrometry (GC/MS).

Additional Information In suspected CAH, a positive newborn screening result for 17-OHP should be confirmed with one of the following: 1) another whole blood, or serum, assay for 17-OHP; or 2) a molecular test, if it is available with reasonable turnaround time, for the most common mutation affecting the 21-hydroxylase gene. If the follow-up test is abnormal, a Corticotropin Stimulation Test (Rapid) *on page 456* is recommended. Most commonly, the high-dose protocol (250 µg dose) is used, but some authorities recommend a 125 µg dose.

CAH due to 21-hydroxylase deficiency is characterized by virilization and mineralocorticoid deficiency. It is the most common cause of female pseudohermaphroditism. Basal 17-OHP levels can be normal in late-onset 21-hydroxylase deficiency presenting as hirsutism. Such patients have a dramatically increased 17-OHP response to ACTH. Patients with 21-hydroxylase deficiency also have increased 17-ketosteroids and urine pregnanetriol.

The heterozygous state of 21-hydroxylase deficiency can be identified by measuring 17-OHP at 0, 30, and 60 minutes after cosyntropin stimulation.

Prenatal diagnosis of congenital adrenal hyperplasia is possible by HLA typing, by DNA analysis, or by hormone measurements from amniotic fluid, including 17-OHP.

References
Allen DB, Hoffman GL, Fitzpatrick R, et al, "Improved Precision of Newborn Screening for Congenital Adrenal Hyperplasia Using Weight-Adjusted Criteria for 17-Hydroxyprogesterone Levels," *J Pediatr*, 1997, 130(1):128-33.

al Saedi S, Dean H, Dent W, et al, "Screening for Congenital Adrenal Hyperplasia: The Delfia Screening Test Overestimates Serum 17-hydroxyprogesterone in Preterm Infants," *Pediatrics*, 1996, 97(1):100-2.

ARUP Laboratories, *2001-2002 User's Guide*, 405.

Speiser PW and White PC, "Congenital Adrenal Hyperplasia," *N Engl J Med*, 2003, 349(8):776-88.

Zarkovic M, Ciric J, Stojanovic M, et al, "Optimizing the Diagnostic Criteria for Standard (250-Microgram) and Low Dose (1-Microgram) Adrenocorticotropin Tests in the Assessment of Adrenal Function," *J Clin Endocrinol Metab*, 1999, 84(9):3170-3.

Internet Web Sites
www.aap.org

♦ **Hydroxyproline** *see* Amino Acids, Urine *on page 145*

Hydroxyproline, Total, Urine
Related Information
Alkaline Phosphatase, Heat Stable, Serum *on page 125*
Alkaline Phosphatase Isoenzymes, Serum *on page 125*
Alkaline Phosphatase, Serum *on page 127*
Calcium, Serum *on page 329*
Calcium, Urine *on page 332*
N-Telopeptides, Urine *on page 967*
Osteocalcin, Serum or Plasma *on page 983*
Osteocalcin (Undercarboxylated), Serum *on page 984*
Pyridinolines (Pyridinoline and Deoxypyridinoline), Urine *on page 1126*
Urine Collection, 24-Hour *on page 1295*

Abstract Urinary excretion of hydroxyproline is presumed to reflect bone matrix metabolism/catabolism, and is high in periods of rapid bone turnover (eg, Paget
(Continued)

Hydroxyproline, Total, Urine *(Continued)*

disease of bone). This test has been replaced by more specific markers (pyridi-nolines and N-telopeptides) of bone resorption.

Patient Preparation Avoid foods containing gelatin (cooked collagen) and meat at least 24 hours prior to and during urine collection. These include gelatin desserts (Jello®), ice creams, candies. Patient must avoid aspirin-containing drugs. Hormonal agents affect quantitation.

Specimen 24-hour urine

Collection Add 30 mL of 6N HCl to container before the beginning of collection. Refrigerate during collection. A 2-hour collection after overnight fast may also be used.

Special Instructions Provide patient's age and sex to the laboratory.

Reference Interval Adult male: 15-45 mg/24 hours (1-4 mg/2 hours) (7-20 μg/mg creatinine) (SI: 76-381 μmol/day). Adult female values are about one-half of the male ranges. Infants, children, and adolescents variably higher, depending on growth spurts. Normal range is higher in infancy, childhood, and adolescence, especially during growth spurts. **Care must be taken to note the type of diet employed in establishing the reference interval of a given laboratory**. Most reference intervals have been obtained on low-dose gelatin, not gelatin-free diets.

Use Evaluate collagen metabolism of bone, bone resorption, bone destruction. High in Paget disease of bone, healing fracture, primary and secondary hyperparathyroidism. The results, however, are nonspecific, due to changes induced by diet and comorbidities.

Methodology Extraction/colorimetry, high performance liquid chromatography (HPLC)

References

Beck-Jensen JE, Kollerup G, Sorensen HA, et al, "A Single Measurement of Biochemical Markers of Bone Turnover Has Limited Utility in the Individual Person," *Scand J Clin Lab Invest*, 1997, 57(4):351-9.

Eastell R and Blumsohn A, "The Value of Biochemical Markers of Bone Turnover in Osteoporosis," *J Rheumatol*, 1997, 24(6):1215-7.

Van Daele PL, Birkenhager JC, and Pols HA, "Biochemical Markers of Bone Turnover: An Update," *Neth J Med*, 1994, 44(2):65-72.

Watts NB, "Clinical Utility of Biochemical Markers of Bone Remodeling," *Clin Chem*, 1999, 45(8B):1359-68.

♦ **5-Hydroxytryptamine, Blood** *see* Serotonin, Blood, Cerebrospinal Fluid *on page 1190*

♦ **Hydroxyurea** *see* Sickle Cell Tests *on page 1195*

♦ **25-Hydroxy Vitamin D$_3$** *see* Vitamin D, Serum *on page 1318*

♦ **Hymenopterae** *see* Arthropod Identification *on page 213*

♦ **Hypercapnia** *see* pCO$_2$, Blood *on page 1008*

♦ **Hypercoagulable State, Platelet Aggregation** *see* Platelet Hyperaggregation *on page 1053*

Hypercoagulation Panel

Related Information

Activated Protein C Resistance and the Factor V Leiden Mutation *on page 104*
Anticardiolipin Antibody *on page 167*
Antiphospholipid Antibody (Lupus Anticoagulant and/or Anticardiolipin Antibody) *on page 193*
Antiplasmin *on page 196*
Antithrombin *on page 198*
C-Reactive Protein, Serum *on page 467*
Heparin Neutralization *on page 697*
Homocyst(e)ine, Plasma *on page 741*
Lipoprotein (a), Serum *on page 861*
Plasminogen *on page 1042*
Plasminogen Activator Inhibitor 1 *on page 1043*
Platelet Hyperaggregation *on page 1053*
Protein C *on page 1101*
Protein S *on page 1110*
Prothrombin G20210A Mutation *on page 1115*
Reptilase® Time *on page 1152*

Thrombin Time *on page 1246*

Synonyms Screen for Hypercoagulation; Thrombophilia Panel; Thrombotic Disease Screen

Applies to Aα Fragment; Activated Protein C Resistance; Antithrombin Deficiency; Bβ 1-42 Fragment; Beta-Thromboglobulin; D-Dimers; Dysfibrinogenemia; FDP; Fibrin Monomer; Fibrinopeptide A; Fibrinopeptide B; G20210A Mutation; Heparin Cofactor II; Hyperhomocyst(e)inemia; PAI-1; PAP; PF4; Plasmin-Antiplasmin Complexes; Platelet Factor 4; Protein C Deficiency; Protein S Deficiency; Prothrombin Fragment 1.2; Thrombin-Antithrombin Complexes; Tissue Plasminogen Activator; tPA

Abstract Testing is often performed in panels, because the presence of more than one predisposition to thrombosis further increases the risk for thrombosis.

Specimen Plasma (and serum if including anticardiolipin antibody and whole blood if including DNA tests)

Container Three blue top (sodium citrate) tubes (and one red top tube if including anticardiolipin antibody)

Collection Routine venipuncture. If a red top tube is being drawn, draw blue top tubes after red top tube. Immediately invert tubes gently at least 4 times to mix. Blue top tubes must be appropriately filled. Deliver tubes immediately to the laboratory.

Storage Instructions Separate plasma from cells as soon as possible. Plasma may be stored on ice for up to 4 hours; otherwise, store frozen.

Causes for Rejection Specimen received more than 4 hours after collection, blue top tubes not filled, blue top tubes clotted

Turnaround Time Several days

Special Instructions Notify laboratory if patient is on any anticoagulant (eg, heparin, warfarin, danaparoid, hirudin, or argatroban). Heparin should be removed from the specimen by the laboratory, and not all tests can be performed when other anticoagulants are present.

Reference Interval See individual tests.

Use Evaluate hypercoagulable states (eg, a young person with spontaneous or recurrent deep venous thrombosis, or a family with multiple members affected by deep venous thrombosis)

Methodology See individual tests.

Additional Information Venous thromboembolism affects 0.1% of the general population in the United States annually, resulting in over 50,000 deaths every year. Hereditary and acquired predisposing conditions are listed in the following tables.

Venous Thrombosis: Acquired Predisposing Conditions

Advanced age
Collagen / vascular disorders
Heparin-induced thrombocytopenia
Hyperhomocyst(e)inemia
Estrogen (oral contraceptives, pregnancy, and estrogen replacement therapy)
Hyperviscosity
Immobilization
Trauma
Inflammatory bowel disease
Antiphospholipid antibodies
Neoplastic disease and chronic disseminated intravascular coagulation (DIC)
Nephrotic syndrome
Myeloproliferative disorders
Paroxysmal nocturnal hemoglobinuria
Postoperative status
Previous episode of thromboembolism
Indwelling catheter
Obesity

(Continued)

Hypercoagulation Panel *(Continued)*

Venous Thrombosis: Hereditary Predisposing Conditions

Disorder	Prevalence in General Population (%)	Prevalence in Venous Thrombosis (%)
Antithrombin deficiency	0.17	1-5
Protein C deficiency	0.14-0.50	3-9
Protein S deficiency	0.7	2-8
Prothrombin G20210A mutation	2	6
Hyperhomocyst(e)inemia (hereditary or acquired)	5-10	10-25
Activated protein C resistance	5 (Caucasians)	20-50

Van Cott EM and Laposata M, "Laboratory Evaluation of Hypercoagulable States," *Hematol Oncol Clin North Am*, 1998, 12(6):1141-66.

A test panel to evaluate a patient with familial venous thrombosis typically includes assays for activated protein C resistance (and if the result is abnormal, factor V Leiden testing), protein C, protein S, antithrombin, homocysteine, prothrombin G20210A, and antiphospholipid antibodies (assays for both lupus anticoagulant and anticardiolipin antibodies are recommended as one may occur without the other). Activated protein C resistance, discovered in 1993, is the most common known hereditary predisposition to thrombosis. Discovered in 1996, the prothrombin G20210A mutation assay is becoming increasingly included in the test panel as this mutation is one of the most common hereditary predispositions to thrombosis. Assays for antiphospholipid antibodies (lupus anticoagulant and anticardiolipin antibodies) are also recommended, although they are not familial conditions. Homocyst(e)ine is often included, as elevated homocyst(e)ine can be a hereditary or acquired predisposition to venous thrombosis. Elevated homocyst(e)ine is unique among the hypercoagulable states in that it may be treated with vitamins B_{12}, B_6 and folate. If all these initial tests are normal and the suspicion for a hereditary hypercoagulable state remains high, assays for plasminogen, dysfibrinogenemia, heparin cofactor II, or platelet hyperaggregability may be considered. These latter four conditions are rare and/or not well characterized. Dysfibrinogenemia test results are characterized by prolonged thrombin time and/or Reptilase® time, and fibrinogen levels higher by antigen assay than by functional assay.

If a patient is undergoing an evaluation for arterial thrombosis, the panel of tests may be different. Antiphospholipid antibodies should be included, as these are associated with arterial and/or venous thrombosis. Homocyst(e)ine levels can also be considered, as the evidence linking hyperhomocyst(e)inemia with arterial thrombosis (particularly coronary artery disease) is even more extensive than it is for venous thrombosis. When arterial thrombosis occurs in the setting of atherosclerosis (eg, coronary artery disease/myocardial infarction, stroke), lipoprotein (a) may be considered in addition to the conventional lipid panel and clinical cardiovascular risk factors (family history, diabetes, hypertension, smoking). Other cardiovascular risk markers are under investigation, including C-reactive protein (or other markers of inflammation) and LDL subclasses (small, dense LDL). The other tests described above for evaluation of venous thrombosis (eg, activated protein C resistance) have an uncertain association with arterial thrombosis. It is possible that the markers of venous thrombosis increase the risk for arterial thrombosis only when a second risk factor for arterial thrombosis is present, such as smoking, hypertension, or hypercholesterolemia.

Just as deficiencies of certain coagulation factors may cause bleeding, elevated levels of certain coagulation factors have been implicated in thrombotic risk. For example, high levels of fibrinogen and factor VII have been associated with an increased risk of myocardial infarction, and high levels of factor VIII or XI have been implicated in venous thrombosis. Coagulation factor levels have not yet been added to many hypercoagulation panels, at least partly because the levels are difficult to interpret in individual patient cases.

Markers of coagulation activation are also commercially available, mostly on a research basis. These tests, when elevated, indicate on-going coagulation activation, as may occur in the setting of thrombosis or disseminated intravascular coagulation (DIC). Such tests, not routinely used clinically, include prothrombin fragment 1.2, fibrinopeptide A, fibrinopeptide B, fibrin monomers, thrombin-antithrombin complexes (TAT), platelet factor 4 (PF4) and beta-thromboglobulin. As prothrombin is converted into thrombin, a peptide is released from prothrombin, called prothrombin fragment 1.2. As fibrinogen is converted into fibrin, two peptides called fibrinopeptide A and fibrinopeptide B are released from fibrinogen. The remaining portion of fibrinogen is called a fibrin monomer. Fibrin monomers then polymerize to form fibrin clot. As thrombin is formed, antithrombin binds to thrombin, forming a thrombin-antithrombin complex (TAT), thereby inhibiting thrombin to prevent excessive clotting. Platelet consumption (thrombocytopenia) and platelet activation markers (eg, platelet factor 4 and beta-thromboglobulin) may also be present. Fibrinogen and antithrombin may be consumed, as well as protein C and protein S.

Markers of fibrinolysis are also present in patients with thrombosis or DIC. Tests for these markers are commercially available but, except for the D-dimer and FDP, they are not commonly used clinically. Such tests include: plasminogen, antiplasmin, plasmin-antiplasmin complexes (PAP), tissue plasminogen activator (tPA), plasminogen activator inhibitor (PAI-1), Bβ 1-42 fragment and Aα fragment. When fibrinolysis is activated, plasminogen levels may decrease as plasminogen is converted into plasmin. As plasmin degrades fibrin, two peptide fragments, called the Bβ 1-42 fragment and Aα fragment, are released, and FDP and D-dimers are formed. As plasmin is formed, antiplasmin binds to plasmin, forming a plasmin-antiplasmin complex (PAP), thereby inhibiting plasmin to prevent excessive fibrinolysis. Antiplasmin and tPA activity can become decreased, and PAI-1 can increase.

References

Brandt JT, "Plasminogen and Tissue-Type Plasminogen Activator Deficiency as Risk Factors for Thromboembolic Disease," *Arch Pathol Lab Med*, 2002, 126(11):1376-81.

Francis CW, "Plasminogen Activator Inhibitor-1 Levels and Polymorphisms. Association With Venous Thromboembolism," *Arch Pathol Lab Med*, 2002, 126(11):1401-4.

Inbal A, Freimark D, Modan B, et al, "Synergistic Effects of Prothrombotic Polymorphisms and Atherogenic Factors on the Risk of Myocardial Infarction in Young Males," *Blood*, 1999, 93(7):2186-90.

Simioni P, Sanson BJ, Prandoni P, et al, "Incidence of Venous Thromboembolism in Families With Inherited Thrombophilia," *Thromb Haemost*, 1999, 81(2):198-202.

Thomas RH, "Update on Thrombophilic Disorders," *Lab Med*, 2003, 34(9):672-9.

Van Cott EM and Laposata M, "Algorithms for Hypercoagulation Testing," *Lab Med*, 2003, 34(3):216-22.

Van Cott EM and Laposata M, "Laboratory Evaluation of Hypercoagulable States," *Hematol Oncol Clin North Am*, 1998, 12(6):1141-66.

♦ **Hyperhomocyst(e)inemia** *see* Hypercoagulation Panel *on page 758*

♦ **Hyperimmunoglobulin M Syndrome** *see* Immunoglobulin M *on page 779*

♦ **Hyperphenylalaninemia, Maternal** *see* Newborn Screen for Phenylketonuria *on page 954*

♦ **Hyperphenylalaninemia Screen** *see* Phenylalanine, Blood *on page 1022*

Hypersensitivity Pneumonitis Serology

Related Information

Aspergillus Serology *on page 219*
Bacterial Culture, Lower Respiratory *on page 241*
Lymphocyte Transformation Test *on page 882*

Synonyms Extrinsic Allergic Alveolitis Serology; Farmer's Lung Disease Serology

Applies to *Aspergillus fumigatus* Precipitating Antibodies; *Aspergillus niger* Precipitating Antibodies; *Micropolyspora faeni* Precipitating Antibodies; *Thermoactinomyces vulgaris* Precipitating Antibodies; *Thermolospora viridis* Precipitating Antibodies

Replaces *Thermoactinomyces* Precipitating Antibodies

Test Includes *Micropolyspora faeni, Aspergillus fumigatus, Alternaria* species, *Aspergillus niger, Thermoactinomyces vulgaris, Thermolospora viridis* antigen testing by immunodiffusion of combined antigenic extract.
(Continued)

Hypersensitivity Pneumonitis Serology *(Continued)*

Abstract The diagnostic term, **hypersensitivity pneumonitis**, applies to patients with interstitial lung disease caused by inhaled organic dusts derived from living sources (*vide supra*) as well as chemical agents (eg, isocyanates and trimellitic anhydride).

Specimen Serum

Container Red top tube

Reference Interval Negative

Use Support the clinical diagnosis of hypersensitivity pneumonitis in those who, following repeated exposure to moist hay or grains, develop cough, chills, dyspnea, and sometimes fever, without bronchospasm.

Limitations A positive test does not establish the diagnosis of hypersensitivity pneumonitis, nor does the absence of precipitins eliminate it. Open lung biopsy may be needed to establish the diagnosis.

Methodology Immunodiffusion (ID)

Additional Information Some individuals become sensitized to inhaled antigens and develop acute bronchospastic symptoms 4-6 hours following exposure. Many of these have been diagnosed with disease names indicating the nature of the exposure. They include bird-fancier's disease, farmer's lung, mushroom-picker's disease, silo-filler's disease, maple bark-stripper's disease, paprika-slicer's lung, sauna-taker's lung, as well as bagassosis (particles of sugar cane fiber). Other sources include redwood tree bark, cheese, and dust from air conditioners. The antigenic material is usually an *Aspergillus* species or one of the thermophilic actinomycetes. Individuals with precipitating antibodies may have no symptoms, and patients with severe symptoms may not show antibody while their disease is inactive. Thus, there must be careful correlation of clinical and laboratory results.

References
Colby TV, Lombard C, Yousen SA, et al, *Atlas of Pulmonary Surgical Pathology*, Philadelphia, PA: WB Saunders Co, 1991, 260-3.

Hypertonic Cryohemolysis Test

Related Information

Autohemolysis Test *on page 221*
Glucose-6-Phosphate Dehydrogenase, Blood *on page 641*
Osmotic Fragility *on page 980*
Osmotic Fragility, Incubated *on page 982*
Peripheral Blood: Red Blood Cell Morphology *on page 1016*
Pyruvate Kinase Assay, Erythrocytes *on page 1127*
Red Blood Cell Enzyme Deficiency *on page 1134*

Synonyms Cryohemolysis Test

Abstract Red blood cells from patients with hereditary spherocytosis (HS) are susceptible to hemolysis when suspended in hypertonic solution, incubated at 37°C, and then kept at 0°C. This hypertonic cryohemolysis test is said to be sensitive to essentially all cases of HS including asymptomatic carriers. Specificity of the test is <100% and is not fully established.

Specimen Fresh whole blood, at least 3 mL

Container Lavender top (EDTA) tube

Storage Instructions Do not store sample over one day.

Causes for Rejection Anticoagulated blood, hemolysis

Reference Interval 3% to 5% (hemolysis of red cells)

Use Evaluation for hereditary spherocytosis

Limitations Specificity of this test for HS is not yet fully defined.

Methodology Red cells washed in cold saline are placed in warm (37°C) buffered 0.7 M sucrose, incubated for 10 minutes, placed in an ice cold bath for 10 minutes, vortexed, and centrifuged. Absorbance of the supernatant is read at 540 nm and % cryohemolysis is calculated as a ratio of 100% hemolysis (determined as a result of red cells placed in deionized water).

Additional Information Red blood cells in a hypertonic environment undergo significant hemolysis when cooled rapidly from 37°C to 0°C. This may relate to the lipid bilayer of the cell membrane changing from a fluid to a gel. Erythrocytes of hereditary spherocytosis are especially susceptible to hypertonic cryohemolysis. This sensitivity of HS red cells has been confirmed and forms the basis of a new test (subject of this listing) for hereditary spherocytosis.

Increased cryohemolysis is also seen in some forms of hereditary elliptocytosis and in some cases of HEMPAS (hereditary erythroblast multinuclearity with positive acidified serum test) - a form of congenital dyserythropoietic anemia (CDA Type II). See Ham Test *on page 665*. The heterogeneous nature of HS relates to a variety of underlying inherited molecular defects in red cell structural proteins including spectrin, ankyrin, band 3, and protein 4.2.

References

Gallagher PG, Forget BC, and Lux SE, "Disorders of the Erythrocyte Membrane," *Hematology of Infancy and Childhood*, 5th ed, Chapter 16, Nathan DG and Oski SH, eds, Philadelphia, PA: WB Saunders Co, 1998, 594-6.

Hassoun H, Vassiliadis JN, Murray J, et al, "Characterization of the Underlying Molecular Defect in Hereditary Spherocytosis Associated With Spectrin Deficiency," *Blood*, 1997, 90(1):398-406.

Streichman S and Gesheidt Y, "Cryohemolysis for the Detection of Hereditary Spherocytosis: Correlation Studies With Osmotic Fragility and Autohemolysis," *Am J Hematol*, 1998, 58(3):206-12.

♦ **Hypertonicity** *see* Sodium, Serum or Plasma *on page 1210*

♦ **Hyperventilation** *see* pCO$_2$, Blood *on page 1008*

♦ **Hypoalphalipoproteinemia** *see* Cholesterol, Total, Serum or Plasma *on page 396*

♦ **Hypoalphalipoproteinemia** *see* Lipids, Overview *on page 853*

♦ **Hypoglycorrhachia** *see* Cerebrospinal Fluid Glucose *on page 362*

Hypo-osmotic Swelling Test (Spermatozoa)

Related Information

Infertility Screen *on page 786*

Semen Analysis, Basic *on page 1187*

Sperm Mucus Penetration Test (Human or Bovine Cervical Mucus) *on page 1219*

Sperm Penetration Assay (Zona-Free Hamster Egg Penetration Test) *on page 1220*

Applies to Sperm Penetration Assay

Abstract The hypo-osmotic swelling test (HOS test) evaluates the functional integrity of the sperm tail membrane. Results are correlated with the *in vitro* fertilizing ability of spermatozoa (ie, hamster egg sperm penetration assay, sperm motility and morphology).

Specimen Semen

Container Clean, dry, wide-mouth glass or plastic bottle maintained warm and free of detergent or other toxic compounds

Collection Physician usually provides instruction for collection. Specimen quality is enhanced when collected in physician's office or laboratory, obviating delay in testing and exposure to extremes of temperature occasioned by transportation. See Semen Analysis, Basic *on page 1187*.

Storage Instructions Patient should be instructed to bring specimen to the laboratory within 30-60 minutes after collection maintaining warmth (37°C) during transport. Patient should be instructed to transport specimen in a pocket close to the skin. Low temperature during transport may decrease motility of sperm.

Causes for Rejection Specimen older than 60 minutes; semen specimen should be tested within 60 minutes after complete liquefaction

Special Instructions Semen, as with all blood, urine, and body fluid specimens, because of the risk of AIDS, should be received and handled with attention to universal precautions.

Reference Interval The test result is normal if >60% of spermatozoa undergo tail swelling. The semen specimen is abnormal if <50% of spermatozoa show tail swelling.

Use This test measures the functional integrity of the sperm membrane. Change in the properties of the sperm membrane is a requirement of successful spermatocyte/oocyte interaction (ie, sperm capacitation, acrosome reaction, and binding of spermatozoa to the surface of the ovum).

Limitations Rarely, a semen specimen may contain >5% swollen spermatozoa or coiled tails (swollen spermatozoa may be seen occasionally in untreated semen, usually in the range of 3% to 5%), or require more than 30-60 minutes for complete liquefaction. Test results in these circumstances may lack significance.

Methodology See graphic.

(Continued)

Hypo-osmotic Swelling Test (Spermatozoa)
(Continued)

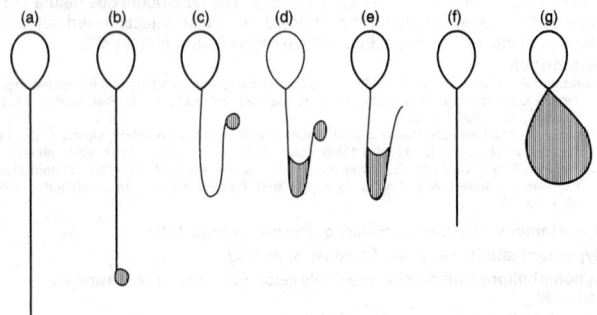

Schematic representation of typical morphological changes of human spermatozoa subjected to hypo-osmotic stress: a = no change; b-g = various types of tail changes. Tail region showing swelling is indicated by the hatched area.

Adapted from Jeyendran RS, Van der Ven HH, Perez-Pelaez M, et al, "Development of an Assay to Assess the Functional Integrity of the Human Sperm Membrane and Its Relationship to Other Semen Characteristics," *J Reprod Fertil*, 1984, 70(1):219-28.

Additional Information Successful fertilization of a human ovum involves penetration of that ovum and subsequent fusion activities by a normally functioning spermatozoon. The HOS test was developed to provide a simple assessment of the overall functional characteristics of the sperm membrane. The test involves microscopic detection and percent quantification of morphologic swollen spermatozoa resulting from exposure to a hypotonic sugar/salt solution (half and half each 150 mOsm fructose and sodium citrate). If there is swelling under hypo-osmotic conditions intact membrane function is implied. Sperm swelling relates poorly to sperm morphology (r=0.30) but somewhat better to sperm motility (r=0.61).

References

Jeyendran RS, Van der Ven HH, Perez-Pelaez M, et al, "Development of an Assay to Assess the Functional Integrity of the Human Sperm Membrane and Its Relationship to Other Semen Characteristics," *J Reprod Fertil*, 1984, 70(1):219-28.

Sarkar S and Henry JB, "Andrology Laboratory and Fertility Assessment," *Clinical Diagnosis and Management by Laboratory Methods*, 20th ed, Part 3, Chapter 20, Henry JB, ed, Philadelphia, PA: WB Saunders Co, 2001, 430.

World Health Organization, *WHO Laboratory Manual for the Examination of Human Semen and Sperm-Cervical Mucus Interaction*, 4th ed, Cambridge, UK: Cambridge University Press, 1999, 4-11, 29, 60-1, 69-70.

♦ **Hypothalamic-Pituitary-Adrenal Axis** *see* Insulin Tolerance Test *on page 804*

♦ **Hypoventilation** *see* pCO$_2$, Blood *on page 1008*

♦ **Hypoxanthine, Urine** *see* Molybdenum, Blood *on page 919*

♦ **Ibuprin®** *see* Ibuprofen, Serum *on page 764*

Ibuprofen, Serum

Related Information
Lactic Acid, Whole Blood or Plasma *on page 827*

Synonyms Advil®; Excedrin®; Genpril®; Haltran®; Ibuprin®; Ibuprohm®; Medipren®; Menadol®; Midol®; Motrin®; Nuprin®; Pamprin®; Rufen®; Trendar®; Unipro®

Abstract Ibuprofen is a leading non-narcotic, nonsteroidal analgesic and anti-inflammatory agent. As compared to acetylsalicylic acid, it has less intense effects of gastrointestinal irritation and bleeding. In neonates, it is also used in the treatment of patent ductus arteriosus and is as efficacious as indomethacin.

Specimen Serum

Container Red top tube

Critical Values See nomogram.

Ibuprofen Toxicity Nomogram

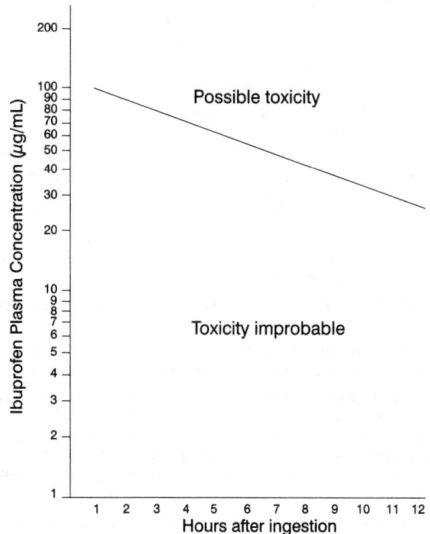

Adapted from Hall AH, Smolinske SC, Stover B, et al, "Ibuprofen Overdose in Adults," *J Toxicol Clin Toxicol*, 1992, 30:34.

Use Evaluate possible toxicity

Limitations Assays for ibuprofen are not readily available.

Methodology Thin layer chromatography (TLC), high performance liquid chromatography (HPLC), gas chromatography (GC)

Additional Information

- Half-life: 0.9-2.5 hours
- Volume of distribution: 0.14 L/kg
- Protein binding: 99%

Side effects of ibuprofen are predominantly gastritis; they include nausea, epigastric pain, diarrhea, vomiting, dizziness, blurred vision, and edema. Single doses have caused severe anaphylactic reactions in persons allergic to ibuprofen. Some patients develop coma, metabolic acidosis, and renal failure after ibuprofen overdose. Overdose treatment includes gut decontamination and supportive therapy.

References

Casalaz D, "Ibuprofen Versus Indomethacin for Closure of Patent Ductus Arteriosus," *N Engl J Med*, 2001, 344(6):457-8.

Fenton J, "Ibuprofen," *The Laboratory and the Poisoned Patient*, Washington, DC: AACC Press, American Association of Clinical Chemistry, 1998, 188-92.

Hall AH, Smolinske SC, Stover B, et al, "Ibuprofen Overdose in Adults," *J Toxicol Clin Toxicol*, 1992, 30(1):23-37.

♦ **Ibuprohm®, Medipren®** *see* Ibuprofen, Serum *on page 764*

♦ **Ice** *see* Methamphetamine, Qualitative, Urine *on page 902*

♦ **ICSH** *see* Luteinizing Hormone, Blood or Urine *on page 876*

♦ **Ictotest®** *see* Bilirubin, Urine *on page 268*

♦ **ICTP** *see* Osteocalcin, Serum or Plasma *on page 983*

Identification DNA Testing

Related Information

HLA Typing, Single Human Leukocyte Antigen *on page 739*

Polymerase Chain Reaction *on page 1069*

Tissue Typing *on page 1259*

(Continued)

Identification DNA Testing *(Continued)*

Synonyms DNA Analysis for Parentage Evaluation; DNA Fingerprinting; DNA Testing; Genetic Identification by DNA Fingerprinting; Paternity Testing by DNA Testing; RFLP Analysis for Parentage Evaluation

Test Includes Identification of individuals by using DNA polymorphic regions

Abstract Progress in the field of DNA technology and Human Genome Project has resulted in tremendous knowledge about the genetic material that makes each individual unique. The DNA from both maternal and paternal sources may be normal, but will have slight variations in character. These variations can be detected and used to map heredity much like the variations in blood group antigens and the human leukocyte antigen (HLA) system. By using between 20-30 different polymorphic sites on different chromosomes, identity or parentage can be established with up to 99.99% exclusion probability. Other DNA identification applications include identification of suspects in forensic cases, and origin and migration history of modern humans.

Patient Preparation Patient should receive no transfusions 90 days prior to testing.

Specimen Peripheral whole blood, tissue, amniotic fluid, semen, or cultured cells; dried blood, hair and skin scrapings are frequently used in forensic cases.

Collection Blood should be collected in a yellow top (ACD) tube or lavender top (EDTA) tube; tissue should be frozen at -70°C; amniotic cells, fibroblasts, or lymphocytes should be grown in appropriate media in T25 tissue culture flasks. Collection in heparin tubes should be avoided as heparin interferes in polymerase chain reaction. Dried blood, hair and skin scrapings are collected in a plastic sealable bag.

Storage Instructions Store tissue at -70°C or on dry ice. Peripheral blood should be stored and shipped at 4°C. Do **not** freeze blood. Dried blood, hair and skin scraping can be stored at room temperature.

Causes for Rejection If the tissue specimen thaws during transport to the laboratory or before shipping, DNA may not be obtained from the specimen. Blood samples that have been frozen and thawed will yield low quality DNA. Specimens inadequately identified will be rejected.

Turnaround Time 2-4 weeks. Samples of DNA can be stored for an unlimited amount of time.

Reference Interval The test provides a 99.99% exclusion probability.

Use The analysis of highly polymorphic regions of human DNA can clarify the relationships between individuals and verify the identify of unknown individuals (such as suspects in criminal investigations or unidentified victims of murder).

Methodology DNA is released and isolated from the white blood cells, tissue, or cultured cells by lysing. The DNA is then digested with various restriction enzymes and electrophoresed through an agarose gel. DNA is then transferred to a solid support such as a nylon membrane and hybridized with a radioactive or fluorescent DNA probe. After washing the unhybridized DNA probe off the membrane, the target DNA is exposed to x-ray or fluorescence sensitive film to detect the polymorphic regions of DNA. Certain regions of the human genome show a high degree of polymorphism in that >85% of the population show heterogeneity. These regions are highly informative in determining DNA identification. Upon digestion with different restriction enzymes, the size and pattern of the DNA fragments vary with each individual, an inherited trait. If the appropriate family members are tested, the inheritance pattern can be established, and applied to determination of the paternity of a child or the zygosity of twins. There is application to the identification of an unknown criminal or victim.

Sufficient material to provide a source of DNA may not be available at the scene of the crime. When the quantity of DNA (eg, from dried blood, hair and skin scraping) is very small, the polymerase chain reaction is used to amplify DNA before identification by digestion with restriction enzymes. See Polymerase Chain Reaction *on page 1069.*

Additional Information The genetic material of humans is highly polymorphic, and an individual's genotype represents a unique pattern which determines that person's identity and heredity. The only exception to this rule is identical twins, since they are derived from a single fertilized egg and hence have the same DNA profile.

Healthcare professionals should exercise care in the collection and storage of specimens for DNA testing to prevent contamination and to preserve evidence which may have crucial legal significance.

References

Baird ML, "Use of DNA Identification for Forensic and Paternity Analysis," *J Clin Lab Anal*, 1996, 10(6), 350-8.

Reeder DJ, "Impact of DNA Typing on Standards and Practice in the Forensic Community," *Arch Pathol Lab Med*, 1999, 123(11):1063-5.

Schneider PM, "Basic Issues in Forensic DNA Typing," *Forensic Sci Int*, 1997, 88(1):17-22.

Taroni F and Aitken CG, "DNA Evidence, Probabilistic Evaluation and Collaborative Tests," *Forensic Sci Int*, 2000, 108(2):121-43.

- ◆ **Identification Requirements, Specimen** *see* Specimen Identification Requirements *on page 1217*
- ◆ **Idiopathic Thrombocytopenic Purpura** *see* Platelet Antibodies *on page 1047*
- ◆ **IgA Antibodies** *see* Antibodies to IgA *on page 164*
- ◆ **IgE Allergen Specific** *see* Allergen Specific IgE Antibody *on page 131*
- ◆ **IGF-1** *see* Insulin-Like Growth Factor-1 (IGF-1), Serum or Plasma *on page 800*
- ◆ **IGFBP-3** *see* Insulin-Like Growth Factor Binding Protein 3, Serum *on page 802*
- ◆ **IGF-I** *see* Insulin-Like Growth Factor-1 (IGF-1), Serum or Plasma *on page 800*
- ◆ **IgG₁** *see* Immunoglobulin G Subclasses *on page 778*
- ◆ **IgG₂** *see* Immunoglobulin G Subclasses *on page 778*
- ◆ **IgG₃** *see* Immunoglobulin G Subclasses *on page 778*
- ◆ **IgG₄** *see* Immunoglobulin G Subclasses *on page 778*
- ◆ **IgG Ratios and IgG Index, Cerebrospinal Fluid** *see* Cerebrospinal Fluid IgG:Albumin Ratio, IgG Index, and IgG Synthesis Rate *on page 365*
- ◆ **IgG Subclasses** *see* Immunoglobulin G Subclasses *on page 778*
- ◆ **IgG Synthesis Rate** *see* Cerebrospinal Fluid IgG:Albumin Ratio, IgG Index, and IgG Synthesis Rate *on page 365*
- ◆ **Ikacor®** *see* Verapamil, Serum or Plasma *on page 1306*
- ◆ **Iktorivil®** *see* Clonazepam, Serum *on page 415*
- ◆ **iMg²⁺ₛ (Ionized Serum Magnesium)** *see* Magnesium, Serum *on page 885*
- ◆ **Imipramine and Desipramine** *see* Antidepressants, Cyclic, Serum or Plasma *on page 171*

Imipramine, Serum or Plasma

Related Information

Antidepressants, Cyclic, Serum or Plasma *on page 171*
Doxepin, Serum or Plasma *on page 524*
Fluoxetine, Serum or Plasma *on page 604*
Warfarin, Serum or Plasma *on page 1325*

Synonyms Berkomine®; Dimipressin®; Iprogen®; Janimine®; Pertofrane®; Presamine®; SK-Pramine®; Tofranil®; Tofranil-PM®

Applies to Desipramine; Norpramin®

Test Includes Desipramine levels

Abstract Imipramine is a tricyclic antidepressant used in the treatment of endogenous depression. Its investigational uses include enuresis in children, analgesic for certain chronic and neuropathic pain, panic disorder, and attention-deficit/hyperactivity disorder (ADHD). It is metabolized to desipramine, an active metabolite. Both parent drug and the metabolite should be monitored.

Specimen Serum or plasma

Container Red top tube or green top (heparin) tube

Sampling Time Trough levels at steady-state (30-90 hours)

Causes for Rejection Sample collected in gel tube

Reference Interval Imipramine and desipramine: 100-300 ng/mL (SI: 350-1070 nmol/L). Metabolism may be impaired in geriatric patients.

Critical Values >500 ng/mL (SI: 1780 nmol/L)

Possible Panic Range ≥1000 ng/mL (SI: 3570 nmol/L)

Use Therapeutic monitoring and toxicity assessment

Contraindications Concomitant use of MAO inhibitors is contraindicated.

(Continued)

Imipramine, Serum or Plasma *(Continued)*

Methodology Immunoassay, high performance liquid chromatography (HPLC), gas chromatography (GC)

Additional Information
- Half-life: 6-18 hours
- Volume of distribution: 9-23 L/kg
- Protein binding: 60% to 95%

Anticholinergic side effects such as blurred vision, dry mouth, constipation, and urinary retention are common with this drug. They are not severe. They may diminish with continued therapy or can be treated with other pharmacologic and nonpharmacologic therapies. Anticholinergic side effects are more troublesome in the elderly. Sedation may also decrease with continued use. Imipramine can also lower the seizure threshold and cause orthostasis and arrhythmias. The cardiovascular effects are more common in patients with underlying cardiovascular disorders. Recently, hyperpigmentation has been associated with long-term drug use of imipramine.

Drug interactions are common with the tricyclic antidepressants. Concomitant treatment with cimetidine, fluoxetine, and antipsychotics produce unexpectedly elevated concentrations of imipramine. Enzyme inducers (eg, phenytoin, chloral hydrate, smoking, and the barbiturates) decrease imipramine concentrations. Additive anticholinergic side effects occur when cyclics are combined with antihistamines, anti-Parkinson drugs, and antipsychotics.

References
Geller B, Reising D, Leonard HL, et al, "Critical Review of Tricyclic Antidepressant Use in Children and Adolescents," *J Am Acad Child Adolesc Psychiatry*, 1999, 38(5):513-6.

Kerr GW, McGuffie AC, and Wilkie S, "Tricyclic Antidepressant Overdose: A Review," *Emerg Med J*, 2001, 18(4):236-41.

Linder MW and Keck PE Jr, "Standards of Laboratory Practice: Antidepressant Drug Monitoring. National Academy of Clinical Biochemistry," *Clin Chem*, 1998, 44(5):1073-84.

- ♦ **Immunocapture** *see Allergen Specific IgE Antibody on page 131*
- ♦ **Immunocytochemistry** *see Immunoperoxidase Procedures on page 780*
- ♦ **Immunocytochemistry** *see Sentinel Lymph Node Biopsy on page 1189*
- ♦ **Immunoelectrophoresis** *see Immunofixation Electrophoresis, Serum or Urine on page 768*
- ♦ **Immunoelectrophoresis** *see Protein Electrophoresis, Urine on page 1107*

Immunofixation Electrophoresis, Serum or Urine

Related Information

Bone Marrow *on page 296*
Cerebrospinal Fluid IgG:Albumin Ratio, IgG Index, and IgG Synthesis Rate *on page 365*
Cerebrospinal Fluid Protein Electrophoresis *on page 374*
Immunoglobulin A *on page 770*
Immunoglobulin D *on page 772*
Immunoglobulin G *on page 775*
Immunoglobulin G Subclasses *on page 778*
Immunoglobulin M *on page 779*
Lymph Node Biopsy *on page 880*
Protein Electrophoresis, Serum *on page 1104*
Protein Electrophoresis, Urine *on page 1107*
Protein, Quantitative, Urine *on page 1108*
Protein, Total, Serum *on page 1114*
Syndecan-1, Serum *on page 1228*
Viscosity, Serum or Plasma *on page 1313*

Applies to Monoclonal Gammopathy; M Protein; Paraprotein Evaluation

Replaces Bence Jones Protein Test; Immunoelectrophoresis

Test Includes Identification of monoclonal gammopathies, pathologist interpretation

Abstract A monoclonal protein (M-protein) population is the product of an expanded single clone of B lymphocytes and/or plasma cells that can be detected in serum or urine by protein electrophoresis. Serum protein electrophoresis is the simplest and lowest cost means of excluding the presence of a monoclonal protein but is not particularly sensitive (can detect bands ≤0.5 g/dL). An M-protein detected by Protein Electrophoresis, Serum *on page 1104* and/or Protein Electrophoresis, Urine *on page 1107* can be further characterized by immunofixation electrophoresis (IFE) as to its immunoglobulin class (G, A, M, D, or E) and/or the light chain type (kappa or lambda). Biclonal, or even triclonal, patterns may occur, reflecting the presence of two or more M-proteins or formation of paraprotein fragments or complexes.

Specimen Serum, urine, or other body fluid, including cerebrospinal fluid

Container Red top or SST™ tube, urine container

Collection Urine: No preservative required.

Storage Instructions Separate serum from cells, centrifuge and/or filter urine.

Reference Interval Not applicable; each study should be accompanied by an interpretive report.

Use Evaluate and characterize clinical conditions that may be associated with monoclonal gammopathy (M-protein) detected initially on protein electrophoretic study of serum, urine, cerebrospinal fluid, or other body fluids. The differential diagnosis of a monoclonal serum or urine protein includes solitary plasmacytoma, plasma cell myeloma, B-cell lymphoma/leukemia, amyloidosis, Waldenström macroglobulinemia, heavy-chain diseases, the POEMS syndrome (**P**eripheral neuropathy, **O**rganomegaly, **E**ndocrine deficiency, **M**onoclonal gammopathy, **S**kin pigmentation and sclerotic bone lesions (sclerosing myeloma)), and a heterogeneous default group called "monoclonal gammopathy of undetermined significance" (MGUS). Serum protein immunofixation is also recommended for patients with a normal serum protein electrophoresis, but whose clinical symptoms suggest the possibility of amyloidosis, myeloma, or POEMS.

Detect, quantitate, and analyze abnormal specific (but non-M-protein) serum protein populations. Can be used for detection of leakage of CSF into nose or ear, in which a beta-2 transferrin band may be detected by use of an antiserum against human transferrin. There may be problems with sensitivity (a negative result cannot exclude CSF leakage). Congenital atransferrinemia is of rare occurrence; a serum control should be used.

Methodology The immunofixation electrophoresis (IFE) process is, essentially, one of protein electrophoretic separation with detection of the specific protein(s) of interest (the antigen) by appropriate monoclonal or polyclonal antibody (antigen-antibody reaction) followed by a staining procedure for visualization.

Additional Information Guidelines for evaluation of M-proteins (presented at the College of American Pathologists (CAP) Conference XXXII, May 1998) recommended use of IFE to "define the abnormal protein type" and, in some cases, to detect small M-proteins. A flowchart in Immunoglobulin G *on page 775* may be helpful. Immunoelectrophoresis is "discouraged" and is no longer a listing in this book.

An article of the CAP consensus conference indicates that IFE of urine (for monoclonal light chain - Bence Jones protein) should be performed on all patients with a serum M-protein >1.5 g/dL (irrespective of urinary total protein) and/or if a plasma cell proliferative process is suspect. It is advised that kappa and lambda antisera that are "monospecific and potent" for both free and combined (intact immunoglobulin molecule) light chains be used. If urine protein electrophoresis shows a localized globulin spike but IFE is negative for monoclonal light chain, the possibility of gamma heavy-chain disease should be considered and IFE using antisera to IgG (includes γ heavy chains) should be performed. The demonstration of low levels of Bence Jones protein may be difficult with both sensitivity and specificity problems relating to mechanical concentration and ladder pattern backgrounds (polyclonal free light chains).

There is evidence that urine levels of free light chain can be used to track disease-related B-cell activity in subacute lupus erythematosus. Normally occurring proteins may mimic serum M-proteins in high resolution protein electrophoresis requiring IFE for differentiation. Such "pseudoparaproteins" may have characteristic features (migration positions and appearances) in many cases such that further laboratory or clinical evaluation may not be required.

In a Mayo Clinic series, patients with monoclonal proteins had the following (see table on following page) diagnoses at the time of detection.

The most common M-protein in myeloma is IgG. The most common M-protein in B-cell lymphoma/leukemia is IgM but cases with IgG or IgA are frequent. The (Continued)

Immunofixation Electrophoresis, Serum or Urine
(Continued)

Disease	% of cases
MGUS	56
Myeloma	22
Amyloid	10
Lymphoma	5
Plasmacytoma	3
CLL	2
Macroglobulinemia	2

most common M-protein in amyloidosis (unaccompanied by plasma cell myeloma) is either IgG or IgA typically in low concentration (<10 g/L).

Monoclonal gammopathy of unknown significance (MGUS) is common, occurring in 1% of patients >50 years and 3% of patients >70 years. The risk of malignant transformation (eg, development of or transformation into myeloma, lymphoma, or amyloidosis) is 17% at 10 years after detection and 33% at 20 years. Patients with MGUS typically have a serum IgG M-protein <20 g/L or an IgA M-protein <10 g/L. The associated normal polyclonal immunoglobulins are not suppressed, Bence Jones proteinuria is absent, and urine beta-2 microglobulin is <3 mg/dL.

References

Alexanian R, Weber D, and Liu F, "Differential Diagnosis of Monoclonal Gammopathies," *Arch Pathol Lab Med*, 1999, 123(2):108-13.

Attaelmannan M and Levinson SS, "Understanding and Identifying Monoclonal Gammopathies," *Clin Chem*, 2000, 46(8 Pt 2):1230-8.

Gertz MA, Lacy MQ, and Dispenzieri A, "Amyloidosis: Recognition, Confirmation, Prognosis, and Therapy," *Mayo Clin Proc*, 1999, 74(5):490-4.

Keren DF, Alexanian R, Goeken JA, et al, "Guidelines for Clinical and Laboratory Evaluation of Patients With Monoclonal Gammopathies," *Arch Pathol Lab Med*, 1999, 123(2):106-7.

Kyle RA, "Plasma Cell Disorders," *Cecil Textbook of Medicine*, 21st ed, Goldman L and Bennett JC, eds, WB Saunders Co, 2000, 977-87.

Strobel SL, "The Incidence and Significance of Pseudoparaproteins in a Community Hospital," *Ann Clin Lab Sci*, 2000, 30(3):289-94.

♦ **Immunofluorescence** *see* Immunoperoxidase Procedures *on page 780*

♦ **Immunofluorescence Skin Biopsy** *see* Skin Biopsy, Immunofluorescence *on page 1203*

Immunoglobulin A

Related Information

Antibodies to IgA *on page 164*
Endomysial Antibodies *on page 537*
Gliadin IgG/IgA Antibodies *on page 637*
Immunofixation Electrophoresis, Serum or Urine *on page 768*
Immunoglobulin D *on page 772*
Immunoglobulin G Subclasses *on page 778*
Protein Electrophoresis, Capillary Zone *on page 1103*
Protein Electrophoresis, Serum *on page 1104*
Protein, Total, Serum *on page 1114*
Risks of Transfusion *on page 1166*

Abstract Immunoglobulin A (IgA) comprises about 13% of serum gamma globulin. It is the major secretory immunoglobulin, with an important role in mucosal immunity. There are two subclasses, IgA_1 and IgA_2. IgA molecules have a half-life of some 6 days. IgA does not fix complement and does not cross the placenta.

Specimen Serum

Container Red top tube

Storage Instructions If there is clinical suspicion of cryoglobulinemia, or of presence of macroglobulins (polymeric IgA M-protein), the sample should be drawn and held at 37°C. See Cryoglobulin, Qualitative, Serum and Plasma *on page 478.* Such samples should not be refrigerated prior to serum separation from clot.

Reference Interval Ranges may vary among laboratories.

Pediatrics: See table.

Adults: 85-385 mg/dL

Age	Male (mg/dL)	Female (mg/dL)
1-30 d	1-20	1-19
31-182 d	7-56	1-59
183-365 d	9-107	15-90
1-3 y	18-171	25-141
4-6 y	60-231	47-206
7-9 y	77-252	41-218
10-12 y	61-269	73-239
13-15 y	42-304	82-296
16-18 y	89-314	90-322

Use Evaluate humoral immunity; diagnose and monitor therapy in IgA myeloma; evaluate anaphylaxis associated with transfusion of blood and blood components (see Antibodies to IgA *on page 164*). Association of IgA deficiency with celiac disease is recognized.

Limitations If samples containing cryoglobulins or cold agglutinins are handled at incorrect temperatures, false low values may result. Of individuals subject to anaphylaxis on exposure to IgA, some have deficiency only to certain subclasses of IgA.

Additional Information IgA, the secretory immunoglobulin, is the primary antibody in saliva, tears, colostrum, and gastrointestinal/respiratory/urinary tract mucosal membrane related fluids. IgA_1 and IgA_2 are the two subclasses of IgA. IgA_1 is 85% of total plasma IgA. IgA usually circulates in the serum as a single 7S unit. In secretions, however, it is a dimer or tetramer linked by a J (for joining) peptide chain.

Increased monoclonal IgA may be produced in lymphoproliferative disorders, especially multiple myeloma and "Mediterranean" lymphoma involving bowel. An IgA monoclonal peak >2 g/dL is a major criterion for IgA myeloma. Of subjects with myeloma, about 25% have monoclonal IgA. Hypercalcemia is a prominent feature of IgA myeloma. The hyperviscosity syndrome is prominent as IgA M-components tend to form polymers. IgA myeloma is the second most common cause of hyperviscosity syndrome.

IgA may be elevated in a wide range of conditions affecting mucosal surfaces, where IgA is largely produced. Some clinically significant IgA deficiencies have concomitant deficiencies of IgG_2 and IgG_4. IgA may be decreased in patients with chronic sinopulmonary disease, in ataxia-telangiectasia, or congenitally. Patients with congenital IgA deficiency are prone to autoimmune diseases, and may develop antibody to IgA and anaphylaxis if transfused; see Antibodies to IgA *on page 164* and Risks of Transfusion *on page 1166*. IgA levels may rise with exercise and fall during pregnancy.

IgA deficiency bears association with respiratory and other bacterial infection, especially with IgG_2 deficiency. IgA deficiency relates also to chronic diarrheal disease including giardiasis (hypogammaglobulinemia in general as with common variable immunodeficiency) and autoimmune diseases such as rheumatoid arthritis. In genetically susceptible individuals IgA deficiency may occur following therapy with phenytoin or penicillamine.

Idiopathic IgA nephropathy (Berger disease) is the most common primary glomerulonephritis. About 50% of patients have elevated serum IgA. There are clinical and glomerular histopathologic similarities to the nephritis of Henoch-Schönlein purpura. Frequently presenting as recurrent painless gross hematuria, often following an acute upper respiratory infection or flu-like episode, the subsequent course is quite variable with some 20% to 50% of cases developing renal failure over the course of many (20-30) years. At the glomerular level, IgA nephropathy is characterized by mesangial deposition of IgA_1 (with possible subsequent deposition of IgG) followed by variable mesangial expansion and proliferation.

(Continued)

Immunoglobulin A *(Continued)*

References

D'Amico G, "Natural History of Idiopathic IgA Nephropathy: Role of Clinical and Histological Prognostic Factors," *Am J Kidney Dis*, 2000, 36(2):227-37.

Soldin SJ, Bailey J, Beatey J, et al, "Pediatric Reference Ranges for Immunoglobulins G, A, and M on the Behring Nephelometer," *Clin Chem*, 1996, 42(S1):308.

Tricot G, "Multiple Myeloma and Other Plasma Cell Disorders," *Hematology: Basic Principles and Practice*, 3rd ed, Chapter 76, Hoffman R, Benz EJ Jr, Shattil SJ, eds, Philadelphia, PA: Churchill Livingstone, 2000, 1398-1416.

Immunoglobulin D

Related Information

Cerebrospinal Fluid Oligoclonal Bands *on page 369*
Immunofixation Electrophoresis, Serum or Urine *on page 768*
Immunoglobulin A *on page 770*
Immunoglobulin G *on page 775*
Mevalonic Acid, Urine or Amniotic Fluid *on page 911*

Abstract Immunoglobulin (Ig) D has a very low serum concentration but is an important lymphocyte membrane receptor, functioning as an antigen receptor and leading to B-lymphocyte activation and antibody production. IgD myeloma is rare but should be a diagnostic consideration when protein studies do not identify one of the more common monoclonal gammopathies. An IgD immune response (with increase in serum IgD or with cerebrospinal fluid IgD oligoclonal bands) may occur in at least some patients with central nervous system (CNS) tumors.

Specimen Serum

Container Red top tube or two microbilirubin tubes

Storage Instructions Store serum at 4°C. May be stored at -20°C for 10-20 years without significant degradation.

Reference Interval 0-14 mg/dL (SI: 0-140 mg/L); IgD is <1% of plasma immunoglobulin.

Use Investigation of monoclonal gammopathy for diagnosis of IgD myeloma

Methodology Radial immunodiffusion (RID), rate nephelometry, radioimmunoassay (RIA), enzyme immunoassay (EIA). Using the latter two sensitive assays, IgD can be detected in the serum of nearly all individuals, whereas with the less sensitive RID procedure, IgD is not detected in some sera.

Additional Information IgD, formed of two heavy chains and two light chains, is a 7S immunoglobulin with a molecular weight of 180 kDa. It has a short half-life of 2.8 days and does not cross the placental barrier. While IgD has a very low serum concentration, it is a major surface Ig of lymphocytes, functioning as an antigen receptor in B-cell activation, and may lead to antibody production.

When looking for myeloma, initial investigation usually includes serum protein electrophoresis and quantification of IgG, IgA, IgM, kappa, and lambda. In cases in which an unexplained restriction is seen (not explained by heavy chain studies), an IgD assay should be performed. IgD-producing myelomas are rare, <2% of reported cases. They are usually characterized by a serum M-spike that is small or absent and by heavy light-chain proteinuria. With the IgD M-protein and the prominent urine lambda light chain as the only distinctive features, IgD myeloma has been considered as a variant of Bence Jones myeloma. The median survival is nearly 2 years, but one-fifth of patients survive for more than 5 years.

In a study by Mavra et al of unconcentrated cerebrospinal fluid from a variety of central nervous system tumors, oligoclonal IgD bands were identified upon (isoelectric focusing, 7 of 25 cases). Two patients with CNS malignancy had demonstrable intrathecal synthesis of both IgD and IgG oligoclonal bands.

A syndrome, hyperimmunoglobulinemia D and periodic fever (HIDS) has been characterized at the biomolecular level. Most patients develop recurrent high fever within the first year of life. The fever lasts from 1-7 days with variable frequency of episodes between patients. There may be associated rash, diarrhea, abdominal pain, arthralgia, vomiting, and headache. Mevalonic acid is increased in the urine, during but not between, febrile crises. Investigation has shown that mutations in MVK, encoding mevalonate kinase, cause HIDS. Fibroblasts from HIDS patients have decreased mevalonate kinase activity.

The International Hyper IgD Study Group has considered an IgD value >6 mg/dL as elevated. A syndrome of episodic fever, malaise, aphthous stomatitis, tonsillitis, pharyngitis, and cervical adenopathy (PFAPA syndrome) is also associated with elevated levels of IgD. IgD-related syndromes are of rare occurrence. Some are likely overlooked.

References

Blade J and Kyle RA, "Nonsecretory Myeloma, Immunoglobulin D Myeloma, and Plasma Cell Leukemia," *Hematol Oncol Clin North Am*, 1999, 13(6):1259-72.

Drenth JP, Cuisset L, Grateau G, et al, "Mutations in the Gene Encoding Mevalonate Kinase Cause Hyper-IgD and Periodic Fever Syndrome," *Nat Genet*, 1999, 22(2):178-81.

Mavra M, Drulovic J, Levic Z, et al, "CNS Tumours: Oligoclonal Immunoglobulin D in Cerebrospinal Fluid and Serum," *Acta Neurol Scand*, 1999, 100(2):117-8.

Immunoglobulin E

Related Information

Allergen Specific IgE Antibody *on page 131*
Aspergillus Serology *on page 219*
Eosinophil Count *on page 542*
Histamine, Urine, Plasma, or Whole Blood *on page 732*

Applies to Kimura Disease

Abstract Significant elevations of serum IgE concentrations are seen in patients with allergic disease, parasitism, bronchopulmonary aspergillosis, extrinsic asthma, urticaria, atopic eczema, IgE myeloma, and hyperimmunoglobulinemia E *Staphylococcus* abscess syndrome (Buckley syndrome, Job syndrome).

Patient Preparation Avoid exposure to radioisotopes when assay method is RIA.

Specimen Serum

Container Red top tube or SST™ tube

Reference Interval <100 IU/mL (100-700 µg/L with mean of 300 µg/L); age-specific ranges vary by laboratory and method; cord serum, usually <2 IU/mL. Total serum IgE values may be expressed in international units/mL, in weight/volume measures, in terms of molar concentration, or less commonly in SI units. One IU = 2.4 ng of protein, one SI = 1 µg/L.

Use Very high IgE levels may be helpful diagnostically. Atopic disease is likely if the total IgE level ≥375 IU/mL. Extreme elevations of 800-25,000 IU/mL are found in asthma associated with severe atopic dermatitis, allergic bronchopulmonary aspergillosis, parasitic infections, IgE myeloma, and Buckley syndrome.

For allergic patients, serial testing may be useful to confirm expected seasonal variations that occur with allergen contact. There is some evidence that elevated serum IgE levels in cord blood are predictive of subsequent development of atopy later in life.

Limitations Total IgE antibody levels can be correlated with clinical allergy only in a general way. Patients with high concentrations of total IgE usually demonstrate sensitivity to many specific allergens. However, many patients with clinical evidence of allergy and detectable allergen-specific IgE have total IgE levels within the reference range. If the total IgE level is very low, atopic disease can usually be excluded. Patients with total IgE level ≤20 IU/mL rarely have detectable specific IgE levels.

Additional Information IgE is a monomeric immunoglobulin formed of two epsilon heavy chains and two kappa or lambda light chains (epsilon$_2$kappa$_2$ or epsilon$_2$lambda$_2$). IgE does not fix complement, does not cross the placenta, and has a short serum half-life of 2.4 days. Like other immunoglobulins, it is formed in response to antigenic stimulation. Unlike other immunoglobulins, IgE routinely binds via its Fc region, at least partly through the high-affinity IgE receptor, Fc epsilon RI (FceRI), to the surface membrane of basophils. The FceRI complex provides a receptor for the Fc region of antigen-specific IgE, participates in IgE-mediated antigen presentation, and controls activation of basophils, mast cells, and T cells. Subsequent trapping of antigen with cross-linking by cell-bound IgE molecules triggers basophilic cells to release a variety of bioactive molecules, including histamine, prostaglandin D2, kallikrein, and leukotrienes C, D, and E. These compounds produce the classic allergic reaction. Signs and symptoms of this immediate (type I hypersensitivity) reaction may include hives or urticaria, sneezing, rhinorrhea, conjunctival edema with itching and irritation, attacks of shortness of breath/wheezing (asthma), (Continued)

Immunoglobulin E *(Continued)*

and/or anaphylaxis (difficulty in breathing with potentially catastrophic fall in blood pressure/shock).

Plasma cells that produce IgE are present especially along respiratory and gastrointestinal membranes. IgE is a form of secretory immunoglobulin. IgE-dependent reactions likely represent protective mechanisms against parasitic invaders. Biologically powerful affector molecules require tight control to avoid serious pathologic consequences of IgE-dependent inflammation. Recent investigations have partially defined interactions between protein chemical mediators (cytokines) and specific inflammatory cells and their membrane receptors. Definition of this immunoregulatory process has (and will) allow identification of dysregulation on a genetic basis (mutant regulatory molecules and cell membrane receptors). Interleukin-4 (IL-4) and Interleukin-13 (IL-13) are important cytokines that mediate IgE synthesis. IL-4 effects B-lymphocyte switching to IgE antibody production. IL-13 also acts on B cells to produce IgE. Variants of the interleukin-4 receptor (IL-4R) are under study for their involvement in the hyper-IgE syndrome, severe eczema, and other conditions. In addition to IL-4, IgE synthesis by B cells requires interaction of the B cell surface antigen CD40 with its ligand (CD40-L) which is expressed on activated T cells. There is interest at the pharmacologic level in the mechanisms by which IL-4 and CD40 signaling switch on IgE production. With intervention there is the possibility that the IgE antibody response might be suppressed. See graphic.

The IgE System as a Model of Allergic Inflammation

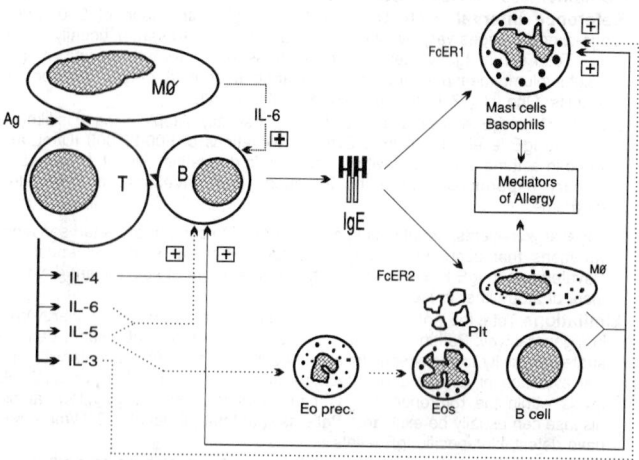

Adapted from Vercelli D and Geha RS, *J Clin Immunol*, 1989, 9:75-83.

Further definition of the cellular/molecular mechanisms referred to but briefly above may assist in understanding the involvement of IgE in:

- The hyper-IgE syndrome (#s 147060 and 2437000 in McKusick's *Mendelian Inheritance in Man (MIM)*, 12th ed, 1998), a rare multisystem disorder characterized by marked increase in serum levels of IgE (>2000 IU/mL), recurrent staphylococcal skin abscesses, pneumonia often with residual pneumatocoele formation, dental and skeletal abnormalities including facial asymmetry, prominent forehead, broad nasal bridge, mild prognathism, tendency to fracture with minor trauma, hyperextensible joints, and other changes. There is usually blood eosinophilia but without correlation to serum IgE levels. Occasionally, an adult with hyper-IgE syndrome has a normal serum IgE result. Referred to in the past as Job syndrome, there is some question as to the existence of variant or incomplete forms of the hyper-IgE syndrome, rendering reports of incidence (generally considered as over 200 case reports) uncertain. There has been recent interest in a mutant form of the IL-4 α-subunit and its possible role in the genesis of this syndrome (IL-4 receptor variant Q576R).

- Kimura disease, a rare condition characterized in most cases by eosinophilia and elevated IgE levels with enlargement of salivary glands, lymph nodes, and subcutaneous soft tissue. The condition can be confused with lymphoma or infection. Atypical follicular hyperplasia with intense eosinophil infiltrate often with foci of necrosis and/or hyaline deposits within follicles is seen histopathologically. Th2 cytokines may be involved in the development of Kimura disease.

- Allelic variants of the IL-4 receptor α-chain (IL-4Rα), notably the R576 IL-4α allele have been reported. R576 IL-4Rα is an evident risk factor for atopy.

- Some patients with low levels of serum IgE (<2.5 IU/mL), considered to have IgE deficiency (IgE hypogammaglobulinemia) were found to have an increased prevalence of multiple immunoglobulin deficits, autoimmune disease, and nonallergic reactive airway disease.

- Study of a 12-year old long-term (since birth) survivor of vertically acquired HIV found presence of anti-HIV-specific IgE interpreted as possibly representing "a protective mechanism against HIV replication..."

- IgE myeloma is rare, some 40 cases having been reported. "Benign" IgE monoclonal gammapathies and biclonal/triclonal gammapathies (with IgE as one component) have been described. A recent survey found that IgE myeloma presents symptoms similar to those of the more common forms of myeloma. Specific or typical symptoms/findings associated with IgE myeloma were not noted. IgE myeloma generally has a more malignant clinical course than other forms of multiple myeloma.

References

Kairemo KJ, Lindberg M and Prytz M, "IgE Myeloma: A Case Presentation and a Review of the Literature," *Scand J Clin Lab Invest*, 1999, 59(6):451-6.

Karavattathayyil SJ and Krause JR, "Kimura's Disease: A Case Report," *Ear Nose Throat J*, 2000, 79(3):195-6, 199.

Presotto F, Trentin L and Agostini C, "Hyper-IgE Syndrome," *N Engl J Med*, 1999, 341(5):375-7.

Seroogy CM, Wara DW, Bluth MH, et al, "Cytokine Profile of a Long-Term Pediatric HIV Survivor With Hyper-IgE Syndrome and a Normal CD4 T-Cell Count," *J Allergy Clin Immunol*, 1999, 104(5):1045-51.

Immunoglobulin G

Related Information

Immunofixation Electrophoresis, Serum or Urine *on page 768*
Immunoglobulin D *on page 772*
Immunoglobulin G Subclasses *on page 778*
Protein Electrophoresis, Capillary Zone *on page 1103*
Protein Electrophoresis, Serum *on page 1104*
Protein Electrophoresis, Urine *on page 1107*
Protein, Total, Serum *on page 1114*

Abstract There are five major types of immunoglobulins classified on the basis of their heavy chain: IgG, IgM, IgA, IgE, and IgD. IgG is present in plasma in the highest concentration. Elevation of all immunoglobulin classes (polyclonal gammopathy) is seen in chronic inflammatory and autoimmune diseases. An increase of a single immunoglobulin (monoclonal gammopathy) may be associated with a benign condition or a malignancy such as plasma cell myeloma or lymphoma. In these latter conditions, other immunoglobulins may be suppressed. Reduced immunoglobulins are seen in various immune deficiency states.

Specimen Serum

Container Red top tube

Storage Instructions Serum specimens may be kept up to 5 days at 2°C to 8°C. For longer period of time prior to analysis freeze samples at temperature of -20°C or less. Do not refreeze thawed specimens.

Samples suspected of having macroglobulins or cryoglobulins should be drawn and held at 37°C. Samples suspected of containing cold agglutinins should not be refrigerated prior to separation of serum from the clot.
(Continued)

Immunoglobulin G *(Continued)*

Reference Interval Ranges are individual method, age, and laboratory dependent.

Pediatrics: See table.

Adults: 564-1765 mg/dL

Age	Male (mg/dL)	Female mg/dL)
1-30 d	260-986	221-1031
31-182 d	195-643	390-794
183-365 d	184-974	407-774
1-3 y	507-1305	550-1407
4-6 y	571-1550	675-1540
7-9 y	700-1680	589-1717
10-12 y	818-1885	705-1871
13-15 y	709-1861	891-1907

Possible Panic Range Very high levels of immunoglobulin (as occur in some cases of multiple myeloma) may be associated with hyperviscosity and/or hypercalcemia and represent a medical emergency.

Use Quantitate IgG in patient's serum to evaluate humoral immunity; establish diagnosis and monitor therapy in IgG myeloma; evaluate patients, including especially children and those with lymphoma, with propensity to infections. Detection, evaluation, and follow-up of patients with various immunodeficiency states and for hyper-IgM syndrome (in which IgG is usually decreased). In asymptomatic plasma cell myeloma, monoclonal IgG is usually >3 g/dL.

Limitations If samples containing macroglobulins, cryoglobulins, or cold agglutinins are inappropriately handled (prematurely exposed to cold), false low values may result.

Methodology Rate nephelometry, enzyme immunoassay (EIA), capillary electrophoresis with or without immunosubtraction. Capillary zone electrophoresis measures protein by absorbance (proteins are not stained). The electrophoretograms, while similar to those of high resolution gel electrophoresis, are slightly more sensitive. If an M-protein is identified, immunotyping can be performed by immunosubtraction. In this relatively new procedure, the serum is incubated with anti-γ, α, μ, kappa, or lambda conjugated sepharose beads after which the supernatants are studied (again by capillary electrophoresis) to determine which reagent(s) removed the abnormal protein.

Additional Information Immunoglobulin G is the major antibody containing protein fraction of blood. There are four subtypes, of which IgG$_1$ and IgG$_2$ comprise 85% of the total.

The four subclasses of IgG differ in the constant regions of their heavy chains. A patient may have a normal total IgG yet still have a significant decrease in one subclass. IgG$_1$ deficiencies are associated with EBV infections, IgG$_2$ with sinorespiratory infections and infections with encapsulated bacteria, IgG$_3$ with sinusitis and otitis media, and IgG$_4$ with allergies, ataxia telangiectasia, and sinorespiratory infections.

With significant decreases in IgG level, on either a congenital or acquired basis, there is an increased susceptibility to infectious processes ordinarily dealt with by humoral antibody. Thus, patients with repeated infection should have their immunoglobulins, and specifically IgG, measured. Therapy with exogenous gamma globulins may be efficacious in such patients.

Conversely, IgG levels will be increased in immunocompetent individuals responding to a wide variety of infections or inflammatory insults (indeed, this represents the basis of the serologic diagnosis of infectious diseases). Today, a major cause for a polyclonal increase in IgG is the acquired immunodeficiency syndrome.

Oligoclonal IgG can occur in multiple sclerosis and in some cases of chronic hepatitis, in particular those with features of autoimmune hepatitis in which hypergammaglobulinemia (increased IgG) occurs.

Monoclonal IgG can be demonstrated in ~60% of cases of multiple myeloma. 3 g/dL of monoclonal IgG is a major diagnostic criterion for myeloma. A monoclonal gammopathy may be present when the total IgG value is in the normal range. While many of these patients do not have multiple myeloma, evaluation for such gammopathy and the presence of Bence Jones protein in urine is important. The differential diagnosis of myeloma includes essential monoclonal gammopathy (monoclonal gammopathy of undetermined significance, "MGUS"), which is characterized by lack of symptoms, lack of anemia, lack of bone lesions, monoclonal gammopathy with M component <3.5 g/dL IgG or <2.5 g/dL IgA and marrow plasma cells <10%.

Guidelines for the evaluation of monoclonal gammopathy have been developed by the College of American Pathologists (conference XXXII). The initial study, obtained on all patients with clinical suspicion of a plasma cell dyscrasia, is high resolution electrophoresis of serum/urine. **Use of low resolution electrophoretic procedures is discouraged.** Immunofixation is recommended to define the abnormal protein, use of the less sensitive immunoelectrophoresis is discouraged. With suspicion of a plasma cell dyscrasia, immunofixation with kappa and lambda light chain antisera is recommended for detection of small M-proteins. Nephelometry is recommended to establish the levels (on initially substantiating the presence of a plasma cell dyscrasia) of immunoglobulins. **Use of radial immunodiffusion is discouraged.** Additional guidance is provided including evaluation/therapy of the hyperviscosity syndrome and cryoglobulinemia.

A recommended sequence for utilization of serum and urine assays has also been published by the College of American Pathologists as a part of Conference XXXII, see flowchart.

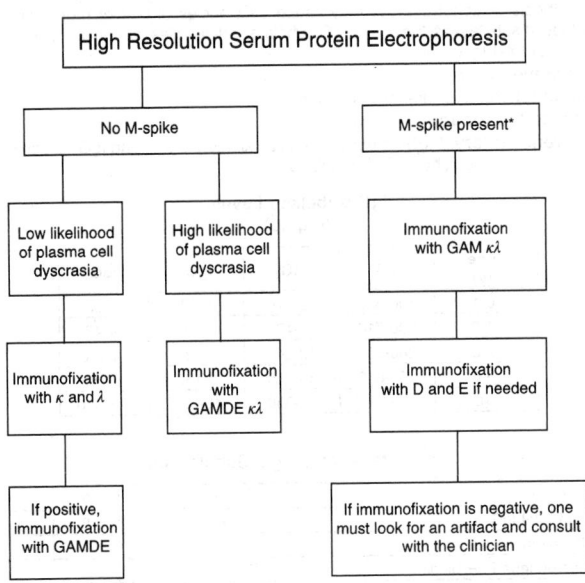

If the M-spike is >1.5 g/dL (15 g/L), a 24-hour urine specimen should be collected for electrophoresis and immunofixation. If the M-spike is >1.5 g/dL (15 g/L), nephelometric measurement of immunoglobulin (Ig) G, IgA, and IgM is indicated.

Adapted from Kyle RA, "Sequence of Testing for Monoclonal Gammopathies: Serum and Urine Assays," Arch Pathol Lab Med, 1999,123(2):114-8

(Continued)

Immunoglobulin G (Continued)

Reduction of IgG, usually <300 mg/dL, with normal or increased IgM and IgD levels (the hyper-IgM syndrome), leads to susceptibility to pyogenic and opportunistic (including *P. carinii*, candidal, and mycobacterial) infections. This condition is inherited as X-linked (X-HIM) or autosomal recessive and is due to failure of activated T lymphocytes to express CD40 ligand that binds CD40 on B cells signaling them to proliferate, form lymphoid follicle germinal centers, and develop into memory B cells (see Immunoglobulin M *on page 779*).

References

Keren D, Alexanian R, Goeken JA, et al, "Guidelines for Clinical and Laboratory Evaluation of Patients With Monoclonal Gammopathies," *Arch Pathol Lab Med*, 1999, 123(2):106-7.

Kyle RA, "Sequence of Testing for Monoclonal Gammopathies: Serum and Urine Assays," *Arch Pathol Lab Med*, 1999, 123(2):114-8.

Soldin SJ, Bailey J, Beatey J, et al, "Pediatric Reference Ranges for Immunoglobulins G, A, and M on the Behring Nephelometer," *Clin Chem*, 1996, 42(S1):308.

♦ **Immunoglobulin G, Cerebrospinal Fluid** *see* Cerebrospinal Fluid Immunoglobulin G *on page 367*

Immunoglobulin G Subclasses

Related Information

Immunofixation Electrophoresis, Serum or Urine *on page 768*
Immunoglobulin A *on page 770*
Immunoglobulin G *on page 775*
Protein, Total, Serum *on page 1114*

Synonyms IgG Subclasses

Applies to IgG_1; IgG_2; IgG_3; IgG_4

Abstract Approximately 80% of total serum immunoglobulin (Ig) in adults is IgG, which is divided into four subclasses on the basis of structural differences in the hinge region. IgG_3 and IgG_1 are closely related, as are IgG_2 and IgG_4. Impaired synthesis or abnormal loss may cause deficiencies of one or more immunoglobulin subclasses. Selective IgG subclass deficiencies can occur in spite of a normal total IgG level.

Specimen Serum

Container Red top tube or SST™ tube

Storage Instructions Store at 4°C.

Reference Interval IgG_1 is 65%, IgG_2 is 25%, IgG_3 is 6%, and IgG_4 is some 4% of total immunoglobulin G. See tables.

IgG Subclass Levels (mg/dL)

Age (y)	IgG_1	IgG_2	IgG_3	IgG_4
0-1	190-620	30-140	9-62	6-63
1-2	230-710	30-170	11-98	4-43
2-3	280-830	40-240	6-130	3-120
3-6	350-810	50-310	9-160	5-180
>6	270-1740	30-630	13-320	11-620

Characteristics of IgG Subclasses

Subtype	1	2	3	4
% of total IgG	66	22	8	4
Complement fixation	+	±	++	−
Macrophage FcR binding	++	±	++	+
Crosses placenta	+	±	+	+
Anti-Ig abs (rheumatoid factors)	+	+	−	+
Aggregation with hyperviscosity	−	−	++	−
Retention in sera of patients with generalized hypogammaglobulinemia	−	−	+	−
Average 1/2 life of circulating subtype	21 d	21 d	7-8 d	21 d

Use Study of patients with recurrent bacterial infections. Selective deficiencies may be found among the subclasses in some individuals who suffer repeated infections, especially IgG$_1$.

Additional Information IgG antibody responses to certain antigens occur to a greater extent in one type of IgG subclass than another. Therefore, some patients with normal total IgG levels may have problems with pyogenic infections because they do not produce IgG$_2$ or combinations of IgG$_2$, IgG$_3$, and/or IgG$_4$. IgG$_2$-deficient patients usually have recurrent respiratory infections. Rarely, severe neutropenia with marrow T-lymphocyte infiltrate (apparently benign) has been noted. Some 120 cases of selective IgG$_1$-deficient patients have been reported from France. Infections, largely mild to moderate, were predominantly sinorespiratory and caused by the encapsulated bacteria *Pneumococcus* and *Haemophilus*. Some clinically significant IgG subclass deficiencies occur in patients who have IgA deficiency. IgA and IgG$_2$/IgG$_4$ subclass deficiency have been reported with phenytoin therapy. Some 5% of epileptic patients maintained on phenytoin develop an IgA deficiency. Combined IgA and IgG$_2$ deficiency in association with zonisamide therapy has been noted.

IgG$_4$ level is increase in patients with sclerosing pancreatitis, which finding may serve to separate this condition from other diseases of the pancreas or biliary tract, in particular pancreatic cancer.

References

Hamano H, Kawa S, Horiuchi A, et al, "High Serum IgG4 Concentrations in Patients With Sclerosing Pancreatitis," *N Engl J Med*, 2001, 344(10):732-8.

Lacombe C, Aucouturier P, Preudhomme JL, "Selective IgG$_1$ Deficiency," *Clin Immunol Immunopathol*, 1997, 84(2):194-201.

Lassoued K, Oksenhendler E, Lambin JP, et al, "Severe Neutropenia Associated With IgG$_2$ Subclass Deficiency and Bone Marrow T-Lymphocyte Infiltration," *Am J Hematol*, 1998, 57(3):241-4.

Paraskevas F, "Cell Interactions in the Immune Response," *Wintrobe's Clinical Hematology*, 10th ed, Chapter 21, Lee RG, Foerster J, Lukes J, et al, eds, Baltimore MD: Williams and Wilkins, 1999, 544-614.

Immunoglobulin M

Related Information

Antimitochondrial Antibody *on page 183*
Cold Agglutinin Titer *on page 430*
Cryoglobulin, Qualitative, Serum and Plasma *on page 478*
Immunofixation Electrophoresis, Serum or Urine *on page 768*
Protein Electrophoresis, Capillary Zone *on page 1103*
Protein Electrophoresis, Serum *on page 1104*
Protein, Total, Serum *on page 1114*
Rheumatoid Factor, Serum or Body Fluid *on page 1161*
Viscosity, Blood *on page 1312*

Applies to Hyperimmunoglobulin M Syndrome

Replaces Macroglobulins, Ultracentrifuge Determination

Abstract With a large molecular mass, about 900 daltons, the distribution of IgM is essentially limited to the vascular compartment. IgM composes 5% to 10% of immunoglobulins. IgM molecules are clinically important as rheumatoid factors, cold agglutinins and as isoagglutinins.

IgM synthesis against intravascular antigens includes that against parasites and red cell surface antigens. It is a complement activator.

Reference Interval
Pediatrics: See table; Adults: 53-375 mg/dL

Age	Male (mg/dL)	Female (mg/dL)
1-30 d	12-117	19-104
31-182 d	27-147	9-212
183-365 d	41-197	4-216
1-3 y	63-240	70-298
4-6 y	64-248	81-298
7-9 y	49-231	62-270
10-12 y	58-249	81-340
13-15 y	57-298	69-361
16-18 y	59-291	86-360

(Continued)

Immunoglobulin M *(Continued)*

Specimen Serum, cerebrospinal fluid

Container Red top tube or SST™ tube, sterile CSF tube

Storage Instructions Samples suspected of having macroglobulins or cryoglobulins should be drawn and held at 37°C. Samples suspected of containing cold agglutinins should not be refrigerated prior to serum separation from clot.

Use Evaluate humoral immunity; establish the diagnosis and monitor therapy in macroglobulinemia of Waldenström and other lymphoid and lymphoplasmacytic neoplasms. Differential diagnosis includes plasma cell myeloma (the rare IgM myeloma), essential macroglobulinemia, and other entities. Like IgG and IgA, IgM increase can present as a monoclonal gammopathy.

IgM is increased in primary biliary cirrhosis with high concentrations of serum alkaline phosphatase and with antimitochondrial antibodies.

IgM levels are used to evaluate likelihood of *in utero* infections or acuteness of infection. IgM deficiency is associated especially with gram-negative infections.

Limitations If samples containing macroglobulins, cryoglobulins, or cold agglutinins are handled at incorrect temperatures, false low values may result.

Additional Information Immunoglobulin M is a pentamer of 7S gamma globulin, and is an efficient complement binder. It is the antibody type produced initially in the immune response and the first immunoglobulin class to be synthesized by a fetus or newborn. IgM antibodies do not cross the placenta. For these reasons the demonstration of IgM-specific antibody is useful in assessing whether a particular infection is acute (in which case IgM antibodies will be present) or chronic (IgG antibodies will predominate) and whether a newborn has a congenital infection (a newborn with IgM antibody is infected; a newborn with IgG antibody has passively acquired maternal antibody, which crossed the placenta).

In the hyper-IgM combined primary immunodeficiency syndrome, there is an absence or low levels of IgG, IgA, and IgE in serum and a marked increase in IgM with neutropenia. Patients with the syndrome develop serious pyogenic infections in which the pathogens are encapsulated bacteria, and intracellular organisms including *Pneumocystis carinii*, *Cryptosporidium parvum*, and *Leishmania*. Macroglobulins produced in Waldenström's disease are IgM, and may produce hyperviscosity syndrome. In contrast, monoclonal IgG or IgA are found in most cases of myeloma. Increased IgM (with other immunoglobulins) may develop in inflammatory/infectious conditions. The majority of rheumatoid factors are IgM. IgM is decreased in congenital or acquired hypogammaglobulinemia, associated with increased, recurrent infection. In patients with bacterial meningitis, CSF IgM is usually elevated along with C-reactive protein.

In macroglobulinemia of Waldenström, weight loss and epistaxis are found, in contrast to essential macroglobulinemia. In the former, hepatosplenomegaly, purpura, and lymphadenopathy occur; IgM is >3 g/dL; and anemia and hyperviscosity are seen.

References

Foerster J, "Waldenström Macroglobulinemia," *Wintrobe's Clinical Hematology*, 10th ed, Chapter 100, Lee GR, Foerster J, Lukens J, et al, eds, Baltimore, MD: Lippincott Williams & Wilkins, 1999, 2681-92.

Hadzic N, Pagliuca A, Rela M, et al, "Correction of the Hyper-IgM Syndrome After Liver and Bone Marrow Transplantation," *N Engl J Med*, 2000, 342(5):320-4.

Ropper AH and Gorson KC, "Neuropathies Associated With Paraproteinemia," *N Engl J Med*, 1998, 338(22):1601-7.

Soldin SJ, Bailey J, Beatey J, et al, "Pediatric Reference Ranges for Immunoglobulins G, A, and M on the Behring Nephelometer," *Clin Chem*, 1996, 42(S1):308.

♦ **Immunoglobulins** *see* Protein Electrophoresis, Capillary Zone *on page 1103*
♦ **Immunoglobulins** *see* Protein Electrophoresis, Serum *on page 1104*
♦ **Immunohistochemistry** *see* Immunoperoxidase Procedures *on page 780*
♦ **Immunomicroscopy** *see* Immunoperoxidase Procedures *on page 780*

Immunoperoxidase Procedures

Related Information

Body Cavity Fluid Cytology *on page 285*

Synonyms Immunocytochemistry; Immunohistochemistry; Immunomicroscopy; Immunostains

Applies to Alpha Fetoprotein; CA-125; Chromogranin; Cytokeratins; Desmin; Epithelial Membrane Antigen (EMA); Factor VIII Related Antigen; GCDFP-15; Glial Fibrillary Acidic Protein; Gross Cystic Disease Fluid Protein-15; hCG; β-hCG; HMB-45; Immunofluorescence; *in situ* Hybridization; Intermediate Filaments; Kappa Light Chains; Ki-67; Lambda Light Chains; Lectins; Leukocyte Common Antigen; Leu M1; Light Chains; Lysozyme; MIB-1; Monoclonal Immunoglobulins; Myoglobin; Myosin; PCNA; Peptide Hormones; Prostate Specific Acid Phosphatase; S-100; Synaptophysin; Thyroglobulin; Tissue Antigens; T Lymphocytes; TTF-1; Vimentin

Test Includes Immunostains are often used in conjunction with flow cytometry in the characterization of hematopoietic and lymphoid neoplasms.

Abstract Immunohistochemical procedures provide important diagnostic tools in all disciplines of surgical pathology. Results of such tests often document lineage of poorly differentiated neoplasms, offer insights to the origin of carcinomas of unknown primary site, document expression of biomarkers of prognostic and predictive importance, and document phenotypic evidence of clonal lymphoid proliferation. In addition to classification of undifferentiated tumors and lymphomas, immunohistochemistry supports recognition of some neuroendocrine and soft tissue neoplasms. Additionally, it is possible to identify and characterize infectious agents using these methods.

Container Sterile Petri dish is ideal for submitting fresh tissue for lymphoma evaluation.

Storage Instructions Neutral-buffered formalin for most specimens; avoid overfixation.

Special Instructions See Lymph Node Biopsy *on page 880*.

Use A panel of immunostains is selected to provide complimentary positive and negative results. Use of overly restricted panels may lead to false interpretive results (eg, S100 positive carcinoma, as in some primary tumors of breast). A commonly used panel for a poorly differentiated neoplasm generally includes pan-cytokeratin, CD45 (leukocyte common antigen), S100 protein, and vimentin. See Table 1 *on page 782*.

Carcinoma of unknown primary site is a common diagnostic challenge, particularly in today's era of image guided needle biopsy. Immunohistochemical stains often provide significant clues as to the site of origin. Although such an approach may not firmly establish the site of origin, correlation with clinical and radiographic findings will often allow a working diagnosis with reasonable medical certainty.

Cytokeratins 7 and 20 have proven extremely useful in this regard. The majority of adenocarcinomas show a CK7+/CK20- phenotype. Neoplasms which are CK7+/CK20+, CK7-/CK20+, or CK7-/CK20- form a select group of tumors whose inclusion or exclusion may have a significant impact on the final diagnosis. See Table 2 *on page 783*.

Thyroid transcript factor 1 (TTF-1) is a recently characterized marker with extreme utility in diagnostic surgical pathology. Distribution of the marker is limited and is among the first clinically useful markers for adenocarcinoma of (Continued)

Immunoperoxidase Procedures *(Continued)*

lung origin. TTF-1 is also useful in differentiating small cell carcinoma of lung origin from Merkel cell carcinoma, but not in discriminating small cell carcinoma of lung from extrapulmonary small cell carcinoma. TTF-1 is also useful in distinguishing pulmonary adenocarcinoma from malignant mesothelioma.

Gross cystic disease fluid protein-15 (GCDFP-15) is useful in identifying carcinoma of breast origin. Apart from breast cancer, salivary gland tumors and sweat gland tumors are commonly immunoreactive. Many sweat gland carcinomas also express estrogen receptors. Appropriate caution is necessary when considering this differential diagnosis.

Other tissue selective antigens useful for addressing primary site of origin include **prostate specific antigen**, prostatic acid phosphatase, **thyroglobulin**,

Table 1. Characteristic Phenotypic Profile of Neoplasms Based on Histogenesis[1]

	CK	Intermediate Filaments				CD45 (LCA)	S100	Synapto	Chromo A
		Vim	Des	NF	GFAP				
Carcinoma[2] NOS	+	-/+	-	-	-	-	-/+	-	-
Neuroendocrine[3]	+/-	-/+	-	-/+	-	-	-	+	+
Lymphoma[4]	-	-/+	-	-	-	+	-	-	-
Melanoma[5]	-	+	-	-	-	-	+	-	-
Soft tissue tumors									
Fibrous histiocytoma	-/+	+	-	-	-	-	-	-	-
Nerve sheath	-	+	-	-	-	-	+	-	-
Muscle	-	+	+	-	-	-	-	-	-
Vascular[6]	-	+	-	-	-	-	-	-	-
Glioma	-	+	-	-	+	-	+	-	-

Abbreviations: CK = cytokeratin, Vim = vimentin, Des = desmin, NF = neurofilament, GFAP = glial acid fibrillary protein, LCA = leukocyte common antigen, Synapto = synaptophysin, Chromo A = chromogranin A.

Reactions: (+) characteristically positive; (+/-) characteristically positive but sometimes negative; (-) characteristically negative; (-/+) characteristically negative but sometimes positive.

[1] There are many exceptions to the indicated reactions. The phenotypic findings must always be put in perspective with light microscopic and other clinical findings.

[2] Cytokeratin profile and expression of other tissue-associated antigens often help identify origin of metastatic neoplasms.

[3] Expression of various peptide hormones may further aid in the classification of the neoplasm.

[4] For additional information, see listings Lymph Node Biopsy and Immunophenotypic Analysis of Tissue by Flow Cytometry.

[5] Expression of melanoma-associated antigen HMB-45 may help distinguish from nerve sheath tumors.

[6] Many vascular tumors express factor VIII-related antigen, CD31 and CD34.

Table 2. Distinguishing Phenotypic Features of Selected Carcinoma[1]

	Pan-CK	CK7	CK20	TTF-1	GCDFP-15	PSA
Lung						
Adenoca	+	+	−	+	−	−
Small cell	+	−/+	−	+	−	−
Squamous	+	−	−	−	−	−
Breast	+	+	−	−	+	−
Colon	+	−	+	−	−	−
Pancreas	+	+	+/−	−	−	−
Liver	+	−	−	−	−	−
Kidney	+	−	−	−	−	−
Prostate	+	−	−	−	+	+
Bladder	+	+	+/−	−	−	−
Ovary	+	+	−	−	−	−

Abbreviations: Pan-CK = pan cytokeratin, CK7 = cytokeratin 7, CK20 = cytokeratin 20, TTF-1 = thyroid transcription factor 1, GCDFP-15 = gross cystic disease fluid protein 15, PSA = prostate specific antigen.

Reactions: (+) characteristically positive; (+/−) characteristically positive but sometimes negative; (−) characteristically negative; (−/+) characteristically negative but sometimes positive.

[1]There are many exceptions to the indicated reactions. The phenotypic findings must always be put in perspective with light microscopic and other clinical findings.

CA-125 (ovary), and **placental alkaline phosphatase** (gynecologic and germ cell tumors).

Markers of potential prognostic and predictive value have been extensively studied in a variety of neoplasms. Suffice it to say that only a few markers have achieved recognized utility in clinical practice. These include **HER-2/neu** (see HER-2/neu on page 716) and **estrogen and progesterone receptors** (see Estrogen and Progesterone Receptor Assay on page 556). Proliferative markers such as **MIB-1** are under intense investigation and are considered useful in certain malignancies.

Immunohistochemical stains are also extremely useful in the diagnosis and characterization of hematopoietic and lymphoid neoplasms. In recent years, a number of useful commercially available antibodies with activity in paraffin embedded tissue have been characterized. The availability of these markers often circumvents the need for fresh frozen tissue. Nonetheless, documentation of light chain restriction is still best achieved by frozen section immunohistochemistry. Furthermore, frozen tissue blocks can be utilized for molecular genetic studies if needed. See Lymph Node Biopsy on page 880.

Limitations Inadequate tissue sampling and deficient basic histology are among the most commonly encountered problems limiting interpretation of immunostains. Variation in tissue fixation and processing, selection of most appropriate antigen retrieval techniques, and variable antibody reactivities supplied by different vendors are a few of the confounding factors that highlight the need for vigilant quality control and skilled technical personnel. Relative to routine histochemical procedures, immunostains are costly. Finally, intrinsic variability of antigen expression owing to the neoplastic state creates the potential for significant interpretive errors.

Additional Information **Immunofluorescent procedures** address similar issues, but require a relatively expensive fluorescence microscope, are less sensitive, lack the resolution afforded by light microscopy, and do not produce an archivable slide. However, immunofluorescence is the method of choice for the localization of immunoglobulins, complement, and fibrin in the evaluation of renal biopsies and inflammatory dermatoses. **Lectins**, plant proteins with specificity for given carbohydrate moieties, are useful for antigen localization and can be used much like a primary antibody. More recently, *in situ* hybridization has been employed to identify nucleic acid sequences in cells. Following procedures generally similar to immunoperoxidase stains, biotin-labeled, genetically-engineered sequences of nucleic acids localize complementary gene

(Continued)

Immunoperoxidase Procedures *(Continued)*

sequences to detect viral genes and oncogenes of potential diagnostic significance.

References

Chu P, Wu E, and Weiss LM, "Cytokeratin 7 and Cytokeratin 20 Expression in Epithelial Neoplasms: A Survey of 435 Cases," *Mod Pathol*, 2000, 13(9):962-72.

Dabbs DJ, *Diagnostic Immunohistochemistry*, New York, NY: Churchill Livingstone, 2001.

DeYoung BR and Wick MR, "Immunohistologic Evaluation of Metastatic Carcinomas of Unknown Origin: An Algorithmic Approach," *Semin Diagn Pathol*, 2000, 17(3):184-93.

Ordonez NG, "Thyroid Transcription Factor-1 Is a Marker of Lung and Thyroid Carcinomas," *Adv Anat Pathol*, 2000, 7(2):123-7.

Werner M, Chott A, Fabiano A, et al, "Effect of Formalin Tissue Fixation and Processing on Immunohistochemistry," *Am J Surg Pathol*, 2000, 24(7):1016-9.

- ♦ **Immunophenotypic Analysis of Tissues by Flow Cytometry** *see* Flow Cytometry, Overview *on page 592*
- ♦ **Immunoreactive Insulin** *see* Insulin, Serum *on page 803*
- ♦ **Immunoreactive PTH** *see* Parathyroid Hormone, Serum *on page 1001*
- ♦ **Immunostains** *see* Immunoperoxidase Procedures *on page 780*
- ♦ **Immunosubtraction Electrophoresis** *see* Protein Electrophoresis, Capillary Zone *on page 1103*
- ♦ **Imprecision** *see page 11*
- ♦ **IM Serology** *see* Infectious Mononucleosis Screening Test *on page 785*
- ♦ **Inborn Errors of Metabolism** *see* Inherited Diseases of Metabolism and Cell Structure *on page 792*
- ♦ **Inborn Errors of Metabolism Screen** *see* Amino Acids, Plasma *on page 143*
- ♦ **Inclusion Conjunctivitis** *see* Ocular Cytology *on page 973*
- ♦ **Inderal®** *see* Propranolol, Serum *on page 1096*

India Ink Preparation

Related Information

Bacterial Culture, Cerebrospinal Fluid *on page 236*
Cerebrospinal Fluid Analysis: Overview *on page 355*
Cerebrospinal Fluid Cytology *on page 361*
Cryptococcal Antigen Titer *on page 482*
Fungal Culture, Blood *on page 620*
Fungal Culture, Cerebrospinal Fluid *on page 621*
Fungus Smear, Stain *on page 626*

Test Includes Staining of CSF sediment with India ink to detect the polysaccharide capsule surrounding the yeast

Abstract *Cryptococcus neoformans* is the most common central nervous system fungal disease in both immunocompetent hosts and patients with the acquired immunodeficiency syndrome (AIDS). Skin, lungs, spleen, kidneys, liver, and/or bone may also be infected. Pigeon droppings act as a year-round vector for dispersion of encapsulated yeast cells.

Patient Preparation Same as for culture of specific site

Specimen Cerebrospinal fluid

Storage Instructions Do **not** refrigerate.

Reference Interval No *Cryptococcus* identified

Critical Values Presence of encapsulated yeast

Use Establish a diagnosis of cryptococcal meningitis; organisms may be identified in cytologic preparations with Papanicolaou stain as well as with India ink.

Limitations This technique is only 30% to 75% sensitive in cases of cryptococcal meningitis. Cultures and rapid latex agglutination (LA) methods are more sensitive than direct preparations; therefore, the India ink preparation may be negative when the culture or LA test is positive. Many laboratories have abandoned the use of the India ink preparation in favor of LA. In facilities in which the LA test is not available, microscopic examination of centrifuged CSF stained with Giemsa, periodic acid - Schiff and/or Gram stain are more sensitive than India ink stain for the rapid detection of cryptococcal meningitis. The mucinous capsules of *Cryptococcus* appear bright red on mucicarmine staining. (Mucicarmine is available in almost all histopathology laboratories.)

Leukocytes in the specimen may be mistaken for yeast cells leading to a false-positive result, and overall there is a substantial incidence of false positives.

Methodology Wet mount with India ink (nigrosin) for contrast. Centrifugation will concentrate organisms and improve sensitivity (10-20 minutes at 1500 g). Some sources of India ink are better than others for this purpose. Finer particles are desirable. Addition of 2% chromium mercury to the India ink preparation allows for the identification of characteristic structures of the organism.

References

Jaster JH and Malecha MJ, Images in Clinical Medicine, "Cryptococcal Meningitis," *N Engl J Med*, 1996, 335(26):1962.

Perfect JR, Wong B, Chang YC, et al, "*Cryptococcus neoformans*: Virulence and Host Defenses," *Med Mycol*, 1998, 36(S1):79-86.

Roebuck DJ, Fisher DA, and Currie BJ, "Cryptococcosis in HIV-Negative Patients: Findings on Chest Radiography," *Thorax*, 1998, 53(7):554-7.

Sato Y, Osabe S, Kuno H, et al, "Rapid Diagnosis of Cryptococcal Meningitis by Microscopic Examination of Centrifuged Cerebrospinal Fluid Sediment," *J Neurol Sci*, 1999, 164(1):72-5.

Internet Web Sites

www.cdc.gov/ncidod/dbmd/diseaseinfo/cryptococcosis_a.htm

♦ **Indirect Coombs Test** *see* Antiglobulin Test, Indirect *on page 177*

Infectious Mononucleosis Screening Test

Related Information

Complete Blood Count *on page 442*
Cytomegalovirus Serology *on page 499*
Epstein-Barr Virus Serology *on page 550*
Herpesvirus 6 Serology *on page 725*
Heterophilic Antibodies *on page 727*
Lactate Dehydrogenase Isoenzymes, Serum *on page 824*
Lactate Dehydrogenase, Serum *on page 825*
Lymph Node Biopsy *on page 880*
Peripheral Blood: Differential Leukocyte Count *on page 1010*

Synonyms Davidsohn Slide Test; Heterophil Antibody; IM Serology; Monospot™ Test; Monosticon® Dri-Dot® Test

Replaces Davidsohn Differential; Paul-Bunnell Davidsohn Test

Test Includes Screening for the presence of heterophil antibodies

Abstract Patients with infectious mononucleosis classically have fever, sore throat, and lymphadenopathy. Abnormal lymphocytes are found. Patient recovery is spontaneous.

Heterophil antibodies agglutinate sheep erythrocytes. They were first described in association with infectious mononucleosis (IM) by Paul and Bunnell in 1932. These antibodies are usually detected at the onset of illness. A delayed appearance of heterophil antibodies may be associated with a prolonged illness.

Most cases involve patients 15-24 years of age.

Specimen Serum

Use Diagnose infectious mononucleosis (IM) which is a self-limiting disease caused by the Epstein-Barr virus (EBV)

Limitations Correlation with clinical findings is imperative since false-positive and negative results have been reported. About 10% of the adult population with IM will not develop heterophil antibodies. Failure to develop heterophil antibodies occurs even more frequently in younger children. In such instances, the presence of EBV antibodies is relevant. Less than 2% false positives have been reported with Hodgkin disease, lymphoma, acute lymphocytic leukemia, infectious hepatitis, pancreatic carcinoma, cytomegalovirus, Burkitt lymphoma, rheumatoid arthritis, malaria, and rubella. Rare unexplained positive horse cell screening tests have been reported with negative differential absorptions. Overall, the heterophil test has 95% specificity.

Methodology Heterophil antibodies are IgM. They can agglutinate sheep or horse red blood cells. In IM, heterophil antibodies can be absorbed by beef erythrocytes but not by guinea pig kidney. The classical present tests were introduced by Dr Israel Davidsohn. The rapid slide test is widely used. It has essentially replaced the older Paul-Bunnell-Davidsohn technique, which is more sensitive but time consuming.

(Continued)

Infectious Mononucleosis Screening Test *(Continued)*

Additional Information The IM heterophil antibody appears in the serum of patients by the sixth to tenth day of illness. Highest titers are usually found in the second to third week. Antibody levels may remain detectable as briefly as 1 week, or persist for as long as a year; usual persistence is 4-8 weeks. The level of antibody activity is not correlated with the severity of disease or the degree of lymphocytosis. A positive test with the appropriate clinical and hematologic setting is sufficient to make the diagnosis of IM. If there is a clinical mononucleosis syndrome, but the screening test is negative, consider tests for EBV specific antibodies, CMV, HHV-6, and toxoplasmosis antibodies. Consider especially repeating this test after a short delay. Laboratory criteria for IM include peripheral blood lymphocytosis, >50% of WBCs with at least 10% of lymphocytes appearing reactive ("reactive lymphs") ie, virocytes. Serologic, hematologic, and clinical criteria all ideally should be met for diagnosis. The peripheral blood smear should be carefully examined.

Other tests often elevated in IM include aminotransferase with disproportionately greater elevations of LDH and alkaline phosphatase. The isomorphic pattern is found with LDH isoenzyme separation. Hemoglobin is usually normal in IM, while platelets are rarely worse than slightly diminished, findings in contrast to those of leukemia.

Cold agglutinins, cryoglobulins, ANA, or rheumatoid factor may be increased in IM.

References

Bisno AL, "Pharyngitis," *N Engl J Med*, 2001, 344(3):205-11.

Cohen JI, "Epstein-Barr Virus Infection," *N Engl J Med*, 2000, 343(7):481-92.

Lee CL, Davidsohn I, and Slaby R, "Horse Agglutinins in Infectious Mononucleosis," *Am J Clin Pathol*, 1968, 49:3-11.

Infertility Screen

Related Information

Chloride, Sweat *on page 392*

Cystic Fibrosis DNA Detection *on page 491*

Hypo-osmotic Swelling Test (Spermatozoa) *on page 763*

Semen Analysis, Advanced *on page 1186*

Semen Analysis, Basic *on page 1187*

Sperm Mucus Penetration Test (Human or Bovine Cervical Mucus) *on page 1219*

Sperm Penetration Assay (Zona-Free Hamster Egg Penetration Test) *on page 1220*

Swim Up/Swim Down Procedures (Spermatozoa) *on page 1227*

Testosterone, Total and Free, Serum or Plasma *on page 1238*

Applies to Antispermatozoal Antibody Test; Computer-Assisted Semen Analysis; Franklin-Dukes Test; Hemizona Binding Assay; Leukocytospermia; Prosaposin; Sperm Agglutination and Inhibition; Sperm Antibodies; Sperm-Oolemma Binding Test

Test Includes Semen analysis. Specialized infertility/andrology laboratories perform infertility testing, but there is not a standardized test menu.

Abstract Extensive and sophisticated methodology for the evaluation of spermatozoa has been developed during the last decade. Expanding capabilities related to continuing growth in *in vitro* fertilization/artificial insemination/intracytoplasmic sperm-spermatid injection has provided impetus for the study of sperm function, in particular by specialized andrology laboratories. The following is an overview of infertility evaluation, emphasis largely on the "male factor".

Patient Preparation Follow physician's instructions. Ejaculation should be avoided for 2-3 days prior to collection of the specimen. The entire ejaculate must be collected (initial portion may contain the majority of spermatozoa).

Specimen Serum (both partners) and semen

Container Blood: red top tube; semen: clean, dry, wide-mouth glass or plastic container known to be free of detergent or other toxic agents. Storage in a plastic container may reduce motility of the spermatozoa. Examination should be conducted immediately upon liquefaction (or specimen placed in a glass container) if collection has been made initially in plastic.

Collection Semen: Postcoital or masturbation using condom-like Silastic seminal fluid collection device (see Semen Analysis, Basic *on page 1187*).

Storage Instructions Tests should be started as soon as possible after collection of semen at least within 2-3 hours, preferably within 1 hour or immediately after liquefaction.

Causes for Rejection Semen specimen more than 2 hours old, specimen subjected to extremes of temperature or toxic material

Turnaround Time Test dependent, usually 1-2 days

Special Instructions Semen, as with all blood, urine, and body fluid specimens, because of the risk of AIDS, should be received and handled with attention to universal precautions. Specialized infertility/andrology laboratories perform infertility testing.

Use Evaluate infertility; guide application of assisted reproductive techniques

Limitations While the existence of sperm antibodies relating to infertility seems firmly established, methodology of some early testing procedures has not proven to be highly technically reliable. Tests utilizing methanol-fixed spermatozoa may give unpredictable and nonreproducible results.

Computer-assisted semen analysis results have been found to be unreliable if sperm count is <20 million/mL. Sperm motility may be underestimated by computer-assisted analysis on post-thaw (after freezing) specimens.

Methodology Routine semen analysis protocols, semen function tests, computer-assisted sperm morphology/motility studies, Kibrick agglutination assay with or without the use of gelatin, method of Franklin and Dukes, microagglutination tests, sperm immobilization/cytotoxicity tests, (eg, Isojima assay), mixed antiglobulin reaction assay (MAR), enzyme-linked immunosorbent assay (ELISA) using as antigen glutaraldehyde-fixed spermatozoa, Immunobead™ binding assay, antisperm antibodies by indirect fluorescence using flow cytometry, fluorescence *in situ* hybridization, human sperm activation assay

Additional Information Approximately 95% of normal couples should conceive within 13-15 months. Many factors may be responsible for infertility. These may be divided into **"female factors"** and **"male factors"**. Common causes of female infertility include endometriosis, tubal factors (often pelvic inflammatory disease), ovulation and cervical/uterine factors. A "male factor" is responsible in some 20% of cases. Basic evaluation (American Fertility Society) includes history and examination of the female, postcoital test, evaluation of tubal patency (hysterosalpingogram), evaluation of hormonal factors, history and examination of the male, and semen analyses. To overcome the effect of transient variables, multiple and/or periodic study of spermatozoa may be necessary. Most cases of male infertility are classified as **idiopathic oligospermia** and **asthenospermia**, defined as <50 million/ejaculate and <50% motility respectively. Immunologic factors are a potential important cause of infertility.

Tests listed below deal largely with "male factor" causes of infertility, some in support of assisted reproduction programs, including *in vitro* fertilization and embryonic transfer and intracytoplasmic sperm injection. Tests of human sperm function (most not commonly available) include:

- resistance of sperm to decondensation in sodium dodecyl sulfate
- acidic aniline blue stain for immature sperm nuclei
- acridine orange stain for abnormal nuclear chromatin
- trypan blue or eosin Y membrane dye exclusion test
- follicular fluid induction of acrosome reaction
- calcium ionophore A23187 induction of acrosome reaction
- acrosome assessment by *Pisum sativum* agglutinin fluorescein stain
- measurement of acrosomal proteinase activity
- assessment of hyaluronidase activity
- sperm-oolemma binding test
- hypo-osmotic swelling test
- hemizona binding assay
- variety of tests for determination of antisperm antibodies

Presence of **antisperm antibodies** has not been clearly associated with disease states but bears association with diminished fertility. When clumping of sperm or sperm with poor motility is found, antibodies should be considered. Such antibodies may be found in the circulation, free or as immune complexes, (Continued)

787

Infertility Screen *(Continued)*

in seminal plasma, and/or attached to the sperm surface. Tests that measure immunoglobulin on the sperm surface have greater sensitivity than assays of serum antisperm antibodies. Some males may have blood negative for sperm antibodies but have demonstrable antibody on the surface of spermatozoa. Antibodies are found in some males with testicular disease. They are also seen in some cases of autoimmune aspermatogenesis experimentally induced by immunization with semen, spermatozoa, or testicular homogenates. There may be a cause and effect relationship between spermatozoal antibodies in serum of females and unexplained infertility. Both members of the couple should have their serum tested. Use of a single type of test may not be adequate.

Because of shortcomings in currently available antisperm antibody (ASA) tests, Helmerhorst et al conclude "...it is difficult to consider the routinely used ASA tests as an essential procedure in the fertility work-up. It is even more difficult to justify a treatment on the basis of such tests." Future generations of more specific tests may detect ASA directed against defined fertilization-related antigens such as rSMP-B, a sperm protein localized to the surface of the tail and midpiece region of human sperm. Immunologic mechanisms may, theoretically, interfere with the sperm-egg binding protein, prosaposin, and/or its sperm receptors.

Electro-optical and computer-assisted devices and systems are available for the semiautomated/automated study of sperm morphology and motility, that is, computer-assisted semen analysis (CASA). These are utilized by a growing number of andrology laboratories. Sperm concentration, motility and concentration of progressively motile cells, along with CASA determined simple sperm morphometry characteristics appear to relate to time to conception. Sperm concentration determination by CASA is improved by use of fluorescent DNA stains.

The finding of apparent decline of sperm counts in fertile men over the past 50 years is disputed by recent studies.

Exposure to aromatic solvents appears to be associated with reduced sperm quality (low sperm count, low % motility, and low % normal forms).

Infertility may be the first clinical sign of male genitourinary tuberculosis. Semen/sperm parameters are usually normal with the exception of leukocytospermia (pyospermia), which occasionally may be massive. Results of a morning urine culture can establish the diagnosis.

The brown color of semen in men with spinal cord injury (27% incidence on at least one ejaculation) may relate to seminal-vesicle dysfunction involving phagocytosis, degradation, and processing of sperm by macrophages with generation of lipofusion-like pigment.

Evaluation of spermatozoal function has been approached by measuring surrogate cervical mucus penetration (see also Sperm Mucus Penetration Test (Human or Bovine Cervical Mucus) *on page 1219*) and the determination of sperm capacitation index, based on the degree of polyspermy in penetration of zona-free, pellucida-free hamster ova. See also Sperm Penetration Assay (Zona-Free Hamster Egg Penetration Test) *on page 1220*.

Heavy smoking appears to be associated with detrimental effects on sperm viability and morphology. This effect may be mediated by smoker's seminal plasma, replacement of which, by physiological media, may possibly enhance the fertilizing capability of involved spermatozoa in assisted reproductive procedures. The use of cocaine has been associated with low sperm concentration, low sperm motility, and presence of abnormal morphology.

Increased frequency of chromosomal as well as numerical abnormalities has been found in sperm from infertile males.

Hypogonadotropic hypogonadism is an uncommon explanation for male infertility, for which therapy is available.

Clinical features of cystic fibrosis (CF) include male infertility. Over 95% of men with CF are azoospermic due to a variety of regressive abnormalities of the mesonephric duct (Wolffian duct-derived anomalies) of which congenital bilateral absence of the vas deferens (CBAVD) is most common. Over 60% of patients with CBAVD have one or more mutations of the cystic fibrosis transmembrane conductance regulator (CFTR) gene. CBAVD patients may carry the most common (ΔF508) mutation but may also harbor rare mutations (ie, R117H). The CFTR gene abnormalities associated with obstructive azoospermia anomalies are found uncommonly in the classic CF population. As routine testing for CFTR mutations may miss the unusual abnormalities and as epididymal/testicular aspiration and intracytoplasmic sperm injection are now available, it has been recommended that genetic analyses and counselling be pursued due to the possibility of iatrogenic transmission of CFTR mutations.

References

Barroso G, Mercan R, Ozgur K, et al, "Intra- and Interlaboratory Variability in the Assessment of Sperm Morphology by Strict Criteria: Impact of Semen Preparation, Staining Techniques, and Manual Versus Computerized Analysis," Hum Reprod, 1999, 14(8):2036-40.

Helmerhorst FM, Finken MJ, and Erwich JJ, "Detection Assays for Antisperm Antibodies: What Do They Test?" Bronson R, "Detection of Antisperm Antibodies: An Argument Against Therapeutic Nihilism," Debate: Antisperm Antibodies, Hum Reprod, 1999, 14(7):1669-73.

Mak V, Zielenski J, Tsui LC, et al, "Proportion of Cystic Fibrosis Gene Mutations Not Detected by Routine Testing in Men With Obstructive Azoospermia," JAMA, 1999, 281(23):2217-24.

World Health Organization Laboratory Manual for the Examination of Human Semen and Sperm-Cervical Mucus Interaction, 4th ed, Cambridge, UK: Cambridge University Press, 1999.

♦ **Influenza A and B Antigen Testing** see Influenza Detection and Culture on page 789

Influenza Detection and Culture

Related Information

Bronchoalveolar Lavage (BAL) on page 311
Influenza Virus Detection, Culture and Serology on page 791
Parainfluenza Viral Culture and Serology on page 999
Viral Culture on page 1307
Virus, Direct Detection by Fluorescent Antibody on page 1311
Washing Cytology on page 1326

Applies to Hemadsorbing Virus; Influenza A and B Antigen Testing

Test Includes Rapid (<30 minutes) antibody-based tests; conventional (5-10 days) and "rapid" (2 days) culture; immunoassays (2 hours); immunofluorescence (direct fluorescent antibody staining, 2-4 hours); reverse transcriptase polymerase chain reaction (RT-PCR, 2 days)

Abstract Influenza is a highly contagious acute respiratory disease. Epidemics occur almost every year, typically from December to March in the Northern Hemisphere and from May to August in the Southern Hemisphere. Annual vaccination with trivalent vaccine reduces the severity of illness. Several antiviral agents are available for the treatment and prevention of influenza viral infections.

Specimen Nasopharyngeal swab, throat swab, bronchial washing, or nasal aspirate; sputum is not a good specimen.

Container Viral transport medium for culture. For rapid antibody-based tests at point-of-care, follow manufacturer's instructions.

Sampling Time Specimens should be collected within 3 days of the onset of illness.

Storage Instructions Specimens for culture should be placed into viral transport medium and kept cold at all times, and delivered immediately to the laboratory. Do not freeze specimens.

Causes for Rejection Dry specimen; specimen submitted on bacterial Culturettes®; sputum, drainages of bile, pus, or exudate specimens

Reference Interval No virus isolated

Use Isolate and identify influenza virus as an etiologic agent in croup, bronchitis, primary viral pneumonia, and combined influenza and bacterial pneumonia. Isolation permits surveillance of epidemic viral strains and aids in selection of influenza strains used in the production of the annual vaccine.

Influenza may cause complications in patients with other disorders, including bronchopulmonary disease, heart disease, and those with compromised immune status.

Limitations Negative culture does not rule out a viral etiology.

Methodology

Rapid antibody-based tests: Enzyme-linked immunosorbent assay
(Continued)

Influenza Detection and Culture *(Continued)*

Conventional culture: Inoculation of specimens into cell cultures, incubation of cultures, observation for characteristic cytopathic effect, and identification and/or speciation by methods such as hemadsorption and fluorescent monoclonal antibodies

Rapid culture: Inoculation of cells in a shell vial, centrifugation, culture, and staining with a fluorescent monoclonal antibody

Immunoassay: Enzyme-labeled antibody

RT-PCR: Reverse transcriptase polymerase chain reaction

Additional Information The influenza pandemic of 1918 killed 21-40 million people, including 675,000 in the U.S. A book review providing a medical perspective is recommended.

Influenza is transmitted from person to person by inhalation of aerosols, especially in crowded conditions. In the U.S., the Centers for Disease Control (CDC), in collaboration with the World Health Organization (WHO), conducts ongoing surveillance to monitor influenza activity. This surveillance assists the WHO in detecting antigenic changes in the influenza strains. Information is available from the CDC regarding influenza surveillance and vaccination composition through the toll-free CDC phone number (888) 232-3228, fax (888) 232-3299 (request document number 361100), or through the CDC web site, http://www.cdc.gov/ncidod/diseases/flu/weekly.htm.

The shell vial technique for rapid detection of viruses has been adopted to detect influenza A and B viruses. Serology for detection of influenza antibodies is available. See Influenza A and B Serology *on page 791.*

In a year 2000 report, 99.8% were type A and 0.2% were type B.

Some laboratories provide a rapid culture that will detect seven of the most common respiratory viruses (parainfluenza type 1, 2, and 3; influenza Type A and type B; adenovirus; and respiratory syncytial virus) in a single culture. Conventional viral culture is useful in that it may detect other viruses not detected by the rapid culture, and provides for influenza surveillance that is useful in typing for vaccine development.

References

Bachman CA, Doyle WJ, and Skoner DP, "Influenza A Virus-Induced Acute Otitis Media," *J Infect Dis*, 1995, 172(5):1348-51.

CDC, "Prevention and Control of Influenza: Recommendations of the Advisory Committee on Immunization Practices (ACIP)," *MMWR Morb Mortal Wkly Rep*, 1999, 48(RR-4):1-22.

Centers for Disease Control and Prevention, "Update: Influenza Activity - United States, 1999-2000 Season," *JAMA*, 2000, 283(13):1681-2.

Guarner J, Shieh WJ, Dawson J, et al, "Immunohistochemical and *In Situ* Hybridization Studies of Influenza A Virus Infection in Human Lungs," *Am J Clin Pathol*, 2000, 114(2):227-33.

Hostetter MK, "Influenza 1918: The Worst Epidemic in American History," *N Engl J Med*, 1999, 341(9):703 (book review - see Footnote 6).

Iezzoni L, *Influenza 1918: The Worst Epidemic in American History*, New York, NY: TV Books, 1999.

Izurieta HS, Thompson WW, Kramarz P, et al, "Influenza and the Rates of Hospitalization for Respiratory Disease Among Infants and Young Children," *N Engl J Med*, 2000, 342(4):232-9.

Kohn MA, Farley TA, Sundin D, et al, "Three Summertime Outbreaks of Influenza Type A," *J Infect Dis*, 1995, 172(1):246-9.

Magnard C, Valette M, Aymard M, et al, "Comparison of Two Nested PCR, Cell Culture, and Antigen Detection for the Diagnosis of Upper Respiratory Tract Infections Due to Influenza Viruses," *J Med Virol*, 1999, 59(2):215-20.

Mayo Reference Services, "Influenza A and B Antigen, Nasopharyngeal Aspirate," *New Test Announcement*, #81856, Rochester, MN: Mayo Medical Laboratories, December 1998.

Neuzil KM, Mellen BG, Wright PF, et al, "The Effect of Influenza on Hospitalizations, Outpatient Visits, and Courses of Antibiotics in Children," *N Engl J Med*, 2000, 342(4):225-31.

Ruiz M, Ewig S, Marcos MA, et al, "Etiology of Community-Acquired Pneumonia: Impact of Age, Comorbidity, and Severity," *Am J Respir Crit Care Med*, 1999, 160(2):397-405.

Shih SR, Tsao KC, Ning HC, et al, "Diagnosis of Respiratory Tract Viruses in 24 Hours by Immunofluorescent Staining of Shell Vial Cultures Containing Madin-Darby Canine Kidney (MDCK) Cells," *J Virol Methods*, 1999, 81(1-2):77-81.

Sintchenko V and Dwyer DE, "The Diagnosis and Management of Influenza. An Update," *Aust Fam Physician*, 1999, 28(4):313-7.

Yungbluth M, "The Laboratory Diagnosis of Pneumonia. The Role of the Community Hospital Pathologist," *Clin Lab Med*, 1995, 15(2):209-34.

Zambon M, "Cell Culture for Surveillance on Influenza," *Dev Biol Stand*, 1999, 98:65-71.

Ziegler T and Cox NJ, "Influenza Viruses," *Manual of Clinical Microbiology*, 7th ed, Murray PR, Baron EJ, Pfaller MA, et al, eds, Washington, DC: AMS Press, American Society for Microbiology, 1999, 928-35.

Internet Web Sites

www.amm.co.uk/pubs/fa_influenza.htm
www.astdhpphe.org/infect/flu.html
www.cdc.gov/ncidod/diseases/flu/fluvirus.htm
www.who.int/health-topics/influenza.htm

Influenza Virus Detection, Culture and Serology

Related Information

Viral Culture *on page 1307*
Virus, Direct Detection by Fluorescent Antibody *on page 1311*

Test Includes Rapid (<30 minutes) antibody-based tests; conventional (5-10 days) and "rapid" (2 days) culture; immunoassays (2 hours); immunofluorescence (direct fluorescent antibody staining, 2-4 hours); reverse transcriptase polymerase chain reaction (RT-PCR, 2 days). Serology detects both IgG and IgM antibody to influenza A and/or B.

Abstract Influenza is a highly contagious acute respiratory disease. Epidemics occur almost every year, typically from December to March in the Northern Hemisphere and from May to August in the Southern Hemisphere. Annual vaccination with trivalent vaccine reduces the severity of illness. Several antiviral agents are available for the treatment and prevention of influenza viral infections.

Specimen Nasopharyngeal swab, throat swab, bronchial washing, or nasal aspirate; sputum is not a good specimen.

Container Viral transport medium for culture. For rapid antibody-based tests at point-of-care, follow manufacturer's instructions.

Collection Specimens for culture should be collected within 3 days of the onset of illness. Acute and convalescent sera drawn 10-14 days apart are required.

Storage Instructions Specimens for culture should be placed into viral transport medium and kept cold at all times. Specimens should be delivered immediately to the laboratory. Do not freeze.

Causes for Rejection Dry specimen; specimen submitted on bacterial Culturette®; sputum, drainages of bile, pus, or exudate specimens

Turnaround Time See Test Includes.

Reference Interval No virus isolated in culture; less than a fourfold increase in antibody titer in paired sera; IgG <1:10, IgM <1:10

Use Aid in the diagnosis of influenza virus as an etiologic agent in croup, bronchitis, primary viral pneumonia. Isolation of virus permits surveillance of epidemic viral strains and aids in selection of influenza strains used in the production of the annual vaccine.

Limitations Negative culture does not rule out a viral etiology

Methodology

Rapid antibody-based tests: Enzyme-linked immunosorbent assay

Conventional culture: Inoculation of specimens into cell cultures, incubation of cultures, observation for characteristic cytopathic effect, and identification and/or speciation by methods such as hemadsorption and fluorescent monoclonal antibodies

Rapid culture: Inoculation of cells in a shell vial, centrifugation, culture, and staining with a fluorescent monoclonal antibody

Immunoassay: Enzyme-labeled antibody

RT-PCR: Reverse transcriptase polymerase chain reaction

Additional Information Although an annual vaccine is available, epidemics still occur each year. Influenza is transmitted from person to person by inhalation of aerosols, especially in crowded conditions. In the U.S., the Centers for Disease Control (CDC), in collaboration with the World Health Organization (WHO), conducts ongoing surveillance to monitor influenza activity. This surveillance assists the WHO in detecting antigenic changes in the influenza strains. In the year 2000 report, 99.8% were type A and 0.2% were type B.

Information is available from the CDC regarding influenza surveillance and vaccination composition through the toll-free CDC phone number (888) 232-3228, fax (888) 232-3299 (request document number 361100), or through the CDC web site, http://www.cdc.gov/ncidod/diseases/flu/weekly.htm. (Continued)

Influenza Virus Detection, Culture and Serology
(Continued)

References

Belshe RB, "Influenza Prevention and Treatment: Current Practices and New Horizons," *Ann Intern Med*, 1999, 131(8):621-4.

Centers for Disease Control and Prevention, "Prevention and Control of Influenza: Recommendations of the Advisory Committee on Immunization Practices (ACIP)," *MMWR Morb Mortal Wkly Report*, 1999, 48(RR-4):1-22.

Centers for Disease Control and Prevention, "Update: Influenza Activity - United States, 1999-2000 Season," *JAMA*, 2000, 283(13):1681-2.

Nelson JK, Shields MD, Stewart MC, et al, "Investigation of Seroprevalence of Respiratory Virus Infections in an Infant Population With a Multiantigen Fluorescence Immunoassay Using Heel-prick Blood Samples Collected on Filter Paper," *Pediatr Res*, 1999, 45(6):799-802.

Pertmer TM and Robinson HL, "Studies on Antibody Responses Following Neonatal Immunization With Influenza Hemagglutinin DNA or Protein," *Virology*, 1999, 257(2):406-14.

Ruiz M, Ewig S, Marcos MA, et al, "Etiology of Community-Acquired Pneumonia: Impact of Age, Comorbidity, and Severity," *Am J Respir Crit Care Med*, 1999, 160(2):397-405.

Internet Web Sites

www.amm.co.uk/pubs/fa_influenza.htm
www.astdhpphe.org/infect/flu.html
www.cdc.gov/flu/
www.who.int/health-topics/influenza.htm

◆ **INH** *see* Isoniazid, Serum or Plasma *on page 813*

Inherited Diseases of Metabolism and Cell Structure

Related Information

Amniotic Fluid, Chromosome and Genetic Abnormality Analysis *on page 152*
Beta-Hexosaminidase, Serum, White Blood Cells *on page 255*
Chorionic Villus Sampling, Chromosome and Genetic Abnormality Analysis *on page 400*
Methylmalonic Acid, Serum, Plasma, Urine, or Amniotic Fluid *on page 908*
Mevalonic Acid, Urine or Amniotic Fluid *on page 911*
Mucopolysaccharides, Urine *on page 922*
Muscle Biopsy *on page 927*
Newborn Screening by Tandem Mass Spectrometry (MS/MS) *on page 957*
Urinalysis *on page 1289*

Synonyms Inborn Errors of Metabolism; Large Molecule Diseases; Lipidoses; Lysosomal Storage Diseases; Mucolipidoses; Small Molecule Diseases

Applies to ABC Proteins; Adrenoleukodystrophy; Berry Spot Test; Crash Syndrome; Farber Disease; Friedreich Ataxia; Galactocerebroside; β-Galactosidase; Glucocerebroside; β-Glucosidase; Glycogen Storage Diseases; Hexosaminidase A; Hexosaminidase B; Krabbe Disease; Leigh Syndrome; Merosin; Metachromatic Leukodystrophy; Mitochondrial Disorders; Mucopolysaccharidoses; Organic Acidurias; Peroxisomes; Sandhoff Disease; Shindler Disease; Sphingolipidoses; Sphingomyelinase; Steroid Sulfatase Deficiency; Zellweger Syndrome

Test Includes Under this heading recognition is given to an ever growing number of genetically determined disorders caused by a metabolic defect (usually an enzyme deficiency or abnormality). The limitations of space allow only brief mention. Most of these disorders can be categorized as to biochemical type (eg, sphingolipidoses, mucopolysaccharidoses, lysosomal storage diseases, glycogen storage diseases, mitochondrial disorders) with overlap between these concepts and with some independent entities.

Of some 500 diseases with a defined biochemical basis, a majority involve abnormalities in enzymes, receptors, and/or structural proteins. The 7th (1995) edition of the monumental text, *The Metabolic and Molecular Bases of Inherited Disease* (with 302 authors) has descriptions of inborn errors of metabolism that include some 900 entities tabulated over 4605 pages. They are grouped by biochemical type (eg, carbohydrate, amino acid, lipoprotein/lipid, purine/pyrimidine, acid lipase), tissue/function type (eg, blood and blood forming organs, transport, peroxisome, immune, etc) and include some 70 **disorders of lysosomal enzymes**, with which this listing will be largely concerned. Clinical features (age, age at onset, severity, signs and symptoms) may show considerable variation within diagnostic groups. The 12th edition of McKusick's *Mendelian Inheritance in Man (MIM)* lists 8587 (including 2082 new entries) human genes and their disorders as of early June 1997 with 56,163 journal reference citations and including 1644 clinical disorders that have been mapped to specific chromosomal sites. This synopsis of the map of the human genome is updated nearly daily in a computer accessible on line database (catalog) available from the William H Welch Medical Library (Johns Hopkins University School of Medicine). OMIM™ (Online Mendelian Inheritance in Man), an online version of MIM (which is continuously updated), is available on the World Wide Web (Internet) from the National Center for Biotechnology Information (NCBI) at the National Library of Medicine, Bethesda, Maryland. Access is provided by NCBI at http://www.nlm.nih.gov/omim/. Questions about OMIM™ access may be directed to NCBI at (301) 496-2475. The OMIM™ database can be searched by an author's name, a word, or combination of words. A CD-Rom version of OMIM™ (MIM-CD™) is available from the Johns Hopkins University Press, Hampden Station, Baltimore, MD, #21211, or http://www.press.jhu.edu/home.html.

Abstract Inherited diseases of metabolism and cell structure include a very large array of clinical disorders with a variety of underlying genetic bases. Many represent a specific enzyme deficiency with resultant excessive accumulation of substances (substrates) usually present in only small amounts. The disorders range from acute life-threatening crises to episodic conditions with prolonged asymptomatic periods or with developmental delay. Presentation in the pediatric age group is common and as indicated in the reference by Lindor and Karnes, rapid diagnosis and institution of therapy may be lifesaving or may be important to optimizing long-term outcome.

Nonorganic Acid Origins of Organic Acidurias

Diagnostic Organic Acid	Associated Organic Acidurias[1]	Nonorganic Acid Origins
2-oxoglutaric acid	Dihydrolipolyl dehydrogenase (E3) deficiency	Krebs cycle and acid-base alterations
3-OH isovaleric acid	3-methylcrotonyl CoA carboxylase deficiency, 3-methylglutaconyl-CoA hydratase deficiency, 3-OH 3-methylglutaric aciduria, biotinidase deficiency, holocarboxylase synthetase deficiency	Severe ketosis
3-OH propionic acid	Biotinidase deficiency, holocarboxylase synthetase deficiency, methylmalonicaciduria, propionic acidemia	Bacterial origin
4-OH phenylacetic acid	Tyrosinemia	Bacterial origin
5-oxoproline	5-oxoprolinuria, hawkinsinuria	Drug depletion of glutathione
Acetyl-L-tyrosine	Tyrosinemia	Amino acid solution therapy
Adipic acid	Medium-chain acyl-CoA dehydrogenase deficiency, glutaric aciduria type II	Dietary: gelatin and MCT feedings
Azelaic and pimelic acids	Short-chain acyl-CoA dehydrogenase deficiency	Plastic container storage
Dicarboxylic aciduria	Short-chain acyl-CoA dehydrogenase deficiency	MCT feedings; container storage
Lactic acid	Biotinidase deficiency, holocarboxylase synthetase deficiency, lactic acidosis	Bacterial origin and artifactual isomers
Fumaric acid	Fumarase deficiency	Krebs cycle and acid-base alterations
Glycerol	Glyceroluria	Pharmaceutical preparations and ointments
Glycolic acid	Hyperoxaluria type I	Ethylene glycol poisoning
Long-chain organic acids	Lysosomal storage disorders	Equipment conditions

[1]Sweetman L, "Organic Acid Analysis," *Techniques in Diagnostic Human Biochemical Genetics: A Laboratory Manual,* Hommes FA, ed, New York, NY: Wiley-Liss, 1991, 165-71.

MCT = medium-chain triglyceride containing diet.

Adapted from Joseph F and Russo TM, "Origins of Spurious Organic Acidurias," *Clin Chem,* 2000, 31(11):622.

Patient Preparation Appropriate preliminary studies may be critically important to narrow the range of diagnostic possibilities. Such investigation might include eye examination (cherry red macula occurs in some gangliosidoses; corneal opacities in Fabry disease; optic atrophy in metachromatic leukodystrophy and

(Continued)

Inherited Diseases of Metabolism and Cell Structure
(Continued)

Krabbe disease); blood/urine screening tests (Berry spot test positive in G_{M1} gangliosidosis; anemia; vacuolated lymphocytes in fucosidosis and other lysosomal storage diseases - see Mucopolysaccharides, Urine *on page 922*); x-ray studies (for developmental changes in bone as with mucopolysacchari-doses; EEG, nerve conduction time, and bone marrow in search of inclusion bearing or foamy histiocytes, eg, as with Gaucher cells). A variety of dietary, pharmacologic, microbiologic, and environmental, as well as analytic factors, may complicate urine testing for organic acidurias (see table on previous page).

Urine specimens submitted for organic acid testing should be accompanied by a record of patient's diet, medication and specimen handling information (eg, when was specimen frozen).

Sphingolipid Storage Diseases (Sphingolipidoses)

Disease	Signs and Symptoms	Enzyme Defect
Anderson-Fabry disease	Reddish-purple skin rash, kidney failure, pain in lower extremities	Ceramidetrihexo-side-α-galactosidase
Farber disease	Hoarseness, dermatitis, skeletal deformation, mental retardation	Ceramidase
Fucosidosis	Cerebral degeneration, muscle spasticity, thick skin	α-fucosidase
Gaucher disease	Spleen and liver enlargement, erosion of long bones and pelvis, mental retardation only in infantile form	Glucocerebro-side-β-glucosidase
Generalized gangliosi-dosis	Mental retardation, liver enlargement, skeletal deformities, about 50% with red spot in retina	β-galactosidase
Krabbe disease (globoid leukodystrophy)	Mental retardation, almost total absence of myelin, globoid bodies in white matter of brain	Galactocerebro-side-β-galactosidase
Niemann-Pick disease type I, II, and subtypes	Liver and spleen enlargement, mental retardation, about 30% with red spot in retina	Sphingomyelinase
Metachromatic leuko-dystrophy	Mental retardation, psychological disturbances in adult form, nerves stain yellow-brown with cresyl violet dye	Sulfatidase
Sandhoff disease	Same as Tay-Sachs disease but progressing more rapidly	Hexosaminidase A and B
Shindler disease	Neurodegeneration, psychomotor retardation, cortical blindness, myoclonic seizures	α-N-acetyl-galactosaminidase
Tay-Sachs disease	Mental retardation, red spot in retina, blindness, muscular weakness	Hexosaminidase A

Adapted from Brady RO and Kolodny EH, "The Sphingolipid Storage Disorders: Diagnosis and Detection," *Lab Management*, 1982, 20:28.

Specimen The majority of tests in this area involve lysosomal enzymes, present in body tissues and fluids. Blood (serum, plasma, or white cells), urine, and tears are the most easily obtained samples for analysis. Solid tissue may be biopsied (eg, skin, liver, muscle). Most commonly used are serum, leukocytes, and cultured fibroblasts (from tissue biopsy). Heparin anticoagulated whole blood, usually at least 5 mL is needed, along with the serum. Specimen preference may be disease dependent (eg, the α-glucosidase deficiency of Pompe disease is best detected by using cultured skin fibroblasts or skeletal muscle). Leukocytes provide a favorable substrate for sphingolipidosis testing but for detection of heterozygotes of Niemann-Pick or Krabbe disease DNA mutation analysis may be required. Use of cultured fibroblasts (with analysis of fatty acid oxidation intermediates) may be preferable for the diagnosis of mitochondrial oxidation disorders. Referral of a leukocyte pellet or biopsy in culture media or even a growing culture of fibroblasts may be required. Molecular analysis for identification of a specific gene defect may be applicable usually with more readily obtained specimen (eg, whole blood).

Turnaround Time Approximately 1 week

Special Instructions For a number of reasons (a sampling follows), the reference laboratory should be contacted before specimens are sent. Details of the preliminary findings can be reviewed, appropriate tests recommended, the preferred samples obtained, need for a clinical photograph established, mode of transport decided, and any other special requirement arranged.

Use Assist in the diagnosis of inherited diseases of metabolism, in particular, sphingolipid and mucopolysaccharide lysosomal storage diseases, by demonstrating presence of a partial or complete enzyme deficiency. Major symptoms, lipids accumulating, and enzymes involved in the sphingolipidoses are given in the table *on page 794*.

Methodology Tests for enzyme deficiency utilizing synthetic substrates (eg, monosaccharide derivatives of 4-methylumbelliferone) have found growing application in this group of diseases. In addition to serum or plasma assays, tests utilizing pelleted leukocytes and fibroblast cultures are used. Molecular analyses for identification of a specific gene defect utilizing specific restriction fragment length polymorphism (RFLP), polymerase chain reaction (PCR), synthetic oligonucleotide probes, DNA sequencing, and/or study of linked chromosomal anomalous DNA sequences may be used. They may be particularly helpful for identification of heterozygosity. In the past few years, electrospray tandem mass spectrometry (MS/MS) has been applied to the quantification of amino acids and acylcarnitines in blood spots to screen for metabolic disorders (in particular aminoacidopathies and organic acidurias) of the neonate. Postmortem dried blood studied by MS/MS has shown that some unexplained infant deaths can be attributed to inborn errors of metabolism. MS/MS provides an automated, high throughout and broad spectrum screening method (that can replace bacterial inhibition assays) and requires just 2 minutes of analytic time per specimen. See Newborn Screening by Tandem Mass Spectrometry (MS/MS) *on page 957*.

Additional Information The phrase "inborn errors of metabolism" is relatively nonspecific. With continually evolving insight, as provided at the level of molecular biology, the terms "inborn" and "errors of metabolism" are likely overly inclusive. In general usage, "inborn errors of metabolism" refers to numerous disorders that have a genetic basis and are characterized by accumulation of a metabolite that cannot be degraded, usually the result of an enzyme deficiency (absence, partial absence, or functional absence of specific enzyme molecules). The lysosomal storage diseases are the primary examples of such "inborn errors," resulting from an intralysosomal accumulation of metabolites that would normally be degraded by a lysosomal enzyme.

One group of lysosomal storage diseases, the mucopolysaccharidoses are due to genetic defects in enzymes that degrade connective tissue glycosaminoglycan. The table relates the type of mucopolysaccharidosis to the enzyme defect. Urine screening tests are available for initial diagnosis of these diseases (see Mucopolysaccharides, Urine *on page 922*). See table on following page.

Wolman disease and cholesteryl ester storage disease are autosomal recessive disorders with defective lysosomal acid lipase, the result of several mutations involving chromosome 10q23.2-q23. Cholesteryl ester and triglycerides accumulate in many cells and tissues including fibroblasts. Electrophoresis of cultured cell extracts reveal acid lipase components A, B, and C, isoenzymes that have an abnormal pattern in Wolman and cholesteryl ester storage disease. Prenatal diagnosis is based on the study of amniotic and chorionic villus cells. Bilateral adrenal calcification, hepatosplenomegaly, and gastrointestinal involvement (foam cells in the intestinal mucosa) are suggestive of Wolman disease. Diet low in cholesterol and triglycerides, and the use of cholestyramine and simvastatin have been beneficial. Bone marrow transplant may be helpful in Wolman disease while lovastatin and, in a few cases, liver transplant have been of benefit in the treatment of cholesteryl ester storage disease.

(Continued)

Mucopolysaccharidoses

Type	Eponymic Designation	Lysosomal Enzyme Defect
I	Three allelic disorders Hurler-Scheie	α-L-iduronidase
II	Hunter severe, mild	Iduronate sulfatase
III (A-D)	Four nonallelic disorders Sanfilippo syndromes A-D	IIIA Heparan N-sulfatase
		IIIB N-acetyl-α-D-glucosaminidase
		IIIC Acetyl-CoA: α-glucosaminide N-acetyl transferase
		IIID N-acetyl-α-D-glucosaminide 6-sulfate sulfatase
IV (A,B)	Two nonallelic disorders Morquio syndromes A,B	IVA Galactosamine 6-sulfate sulfatase
		IVB β-galactosidase
VI	Several allelic types Maroteaux-Lamy syndrome	Arylsulfatase B
VII	Sly syndrome	β-glucuronidase

A table of glycogen storage diseases follows. Most of these disorders of carbohydrate metabolism are not lysosomal storage diseases. They are all autosomal recessive except one form of liver phosphorylase kinase deficiency in which only males are affected and the inheritance is X-linked.

Glycogen Storage Diseases

I	von Gierke Ia, Ib	Glucose-6-phosphatase
II	Pompe infantile, adult form	Lysosomal α-1,4-glucosidase
III	Cori Forbe	Amylo-1,6-glucosidase (debrancher enzyme)
IV	Andersen	Amylo-(1,4:1,6)-transglucosidase (brancher enzyme)
V	McArdle	Muscle phosphorylase
VI	Hers, glycogenoses	Hepatic phosphorylase X-linked phosphorylase-β-kinase Autosomal phosphorylase-β-kinase
VII	Tarui	Muscle phosphofructokinase

Disease severity and symptoms vary widely with type and organ site of the defective enzyme activity.

The spectrum of inherited abnormalities of metabolism and structure is continually enlarging. The majority are uncommon to rare, so that resources of equipment and experienced personnel for testing are justifiably limited to the specialized laboratory. Even so there are enough laboratories performing these assays so as to raise the question if any **one** can develop necessary case experience. Dialogue between the referring physician and the specialty laboratory is essential (discussed above in relation to technical considerations) in particular because of the biochemical heterogeneity frequently seen with these conditions. Clinical expression may be variable and unpredictable.

Each of the sphingolipidoses represents not one but several diseases differing in clinical signs and/or enzyme activity. They are characterized by differing age of onset, site of pathology, and amount of residual enzyme activity (total or partial deficiency). As in Tay-Sachs disease more than one form of the involved enzyme may be present. Hexosaminidase A is formed of subunits α and β, each under different chromosome control. Hexosaminidase, the enzyme involved exists as two isoenzymes, A and B. In Tay-Sachs disease, hexosaminidase A is decreased or absent while hexosaminidase B is increased. In Sandhoff's disease, a variant form of Tay-Sachs, there is deficiency of both hexosaminidase A and B due to hexosaminidase β-subunit defect (encoded on chromosome 5). In addition, the usual sources of variance may affect enzyme deficiency testing. The activity of β-N-acetyl hexosaminidases in Tay-Sachs is affected by pregnancy, chronic diseases of liver, heart, joints, endocrine system, skin and medications including oral contraceptives, some steroids, thyroid, and Butazolidin®. These factors do not affect the result of hexosaminidase assays performed on leukocytes, fibroblasts, or tears.

Sphingolipid storage diseases involve most cells of the body and thus are expressed as multisystem diseases. Gaucher disease is the most common and may show hepatosplenomegaly, thrombocytopenia, erosion of bone with tendency to pathologic fracture and in a few cases, CNS involvement. All of the lipid storage diseases show autosomal recessive inheritance with the exception of Anderson-Fabry disease (AFD) which is transmitted as X-linked.

AFD is the second most common lysosomal storage disease. It results from an inborn deficiency of α-galactosidase A (α-GalA) with progressive accumulation of globotriasylceramide (Gb$_3$). The condition becomes symptomatic in childhood/adolescence followed by premature death. There are proteon clinical manifestations, diagnosis is difficult, often missed, and in males, is confirmed by serum/tissue or DNA analysis. Heterozygous females may have normal α-GalA levels and require measurement of Gb$_3$ in blood or urine and genetic analysis. With enzyme replacement therapy (eg, agalsidase alfa (Replagal®) and agalsidase beta (Fabrazyme®)) available for treatment, early diagnosis is important.

The advent of treatment strategies (eg, enzyme infusion, use of recombinant enzymes, and gene therapy) may add further impetus to establishing the diagnosis, detection of carriers, and monitoring of pregnancies at risk (through amniocentesis and culturing of epithelial cells in the amniotic fluid).

Peroxisomes, cellular organelles involved in oxidative functions are deficient in cases of Zellweger syndrome. There is accumulation of long chain fatty acids, phytanic acid, pipecolic acid, bile acid intermediates, and lack of plasmalogen biosynthesis. Other diseases in this group include adrenoleukodystrophy and a form of chondrodysplasia punctata.

Increased understanding of the molecular pathophysiology responsible for conditions previously catalogued as peroxisomal, mitochondrial, or other (site) disorders has led to the development of classifications (of inborn errors), and in some cases, cross-classifications that are focused upon basic structural biochemical mechanisms. Such clinically diverse conditions as adrenoleukodystrophy (ALD), the cerebro-hepato-renal syndrome of Zellweger, persistent hyperinsulinemic hypoglycemia of infancy (PHHI), a rare disease due to a mutation of the gene for the sulfonylurea receptor (SUR1), and cystic fibrosis (CF) can be considered as the result of mutated human ABC proteins.

ABC (ATP binding cassette) proteins are one of the largest protein families. Over half of human ABC proteins have been discovered in the past few years. Some 30 ABC proteins have been fully sequenced. ABC proteins are formed of transmembrane domains (TMDs) and nucleotide binding domains (NBDs or ATP-binding cassettes) and are defined by presence of the "ABC unit." ABC proteins serve largely as membrane transporters (translocate a variety of substrates to various compartments) accounting for a wide spectrum of functions.

See the references by Applegarth et al, Burlina et al, and the text by Scriver et al for investigation of small molecule diseases; organic acid, urea cycle, and peroxisomal disorders.

HUMAN MITOCHONDRIAL DISORDERS

Disorders considered to be mitochondrial based include defects in fatty acid oxidation, pyruvate metabolism, and the respiratory chain (due to abnormal enzyme structure/function) and are of especial importance in neonates relating in part to their high energy requirements. Mitochondrial-based disorders may (Continued)

Inherited Diseases of Metabolism and Cell Structure
(Continued)

result from defects in mitochondrial DNA (mtDNA) or in nuclear DNA (nDNA). Mitochondria contain their own genetic material, 16,569 base pairs long and encoding 13 respiratory chain proteins, 2 ribosomal proteins, and tRNAs required for assembly of these proteins. Mitochondrial function is thus under dual genetic control (from both mtDNA and nDNA). Over 100 different rearrangements and 50 different point mutations involving mtDNA have been associated with human disease.

Mutated and wild type (normal) mtDNA may exist in cells together (heteroplasmy) complicating the analysis and clinical expression of mtDNA and its genetic variants. The mutated mtDNA must reach threshold levels before cellular expression can occur. It may be difficult to document the presence of mtDNA disease. Careful consideration and assimilation of clinical and laboratory data is necessary. In some cases, analysis of skeletal muscle may be required (eg, histochemical study that may reveal subsarcolemmal accumulation of mitochondria ("ragged-red fibers") or mosaic deficiency of cytochrome c oxidase - absence of these findings does not exclude the diagnosis). Analysis of DNA extracted from muscle may be necessary for diagnosis.

The neonate with a mitochondrial-based metabolic disorder may present clinically with hypotonia, lethargy, feeding and respiratory difficulties, failure to thrive, psychomotor delay, seizures, and/or vomiting. Laboratory results that may lead to a diagnosis include abnormal levels of lactate, pyruvate, lactate:pyruvate ratio, glucose, and ketone bodies.

Inborn errors of metabolism may have broad and devastating effects on the development and function of the nervous system. A recent compilation of inherited metabolic disorders having neurologic involvement and with adult onset includes nearly 40 entities (exclusive of subtypes). Included are lysosomal storage diseases, amino acid disorders, organic acid disorders, peroxisomal disorders, lactic acidaemias, disorders of the glycogenolytic and glycolytic pathway and a number of miscellaneous conditions. The adhesion molecules merosin (laminin-2), Po, and L_1, and their associated clinical conditions, congenital muscular dystrophy, peripheral neuropathies, and crash syndrome, respectively, are detailed in a recent review by Kamiguchi et al. Inherited diseases involving glycan chemistry and associated enzymes include I-cell disease, congenital disorders of glycosylation, leukocyte adhesion deficiency type II, hereditary erythroblastic multinuclearity with a positive acidified serum test, and Wiskott-Aldrich syndrome.

References

Applegarth DA, Dimmick JE, and Toone JR, "Laboratory Detection of Metabolic Disease," *Pediatr Clin North Am*, 1989, 36(1):49-65.

Burlina AB, Bonafé L, and Zacchello F, "Clinical and Biochemical Approach to the Neonate With a Suspected Inborn Error of Amino Acid and Organic Metabolism," *Semin Perinatol*, 1999, 23(2):162-73.

Chace DH, DiPerna JC, Mitchell BL, et al, "Electrospray Tandem Mass Spectrometry for Analysis of Acylcarnitines in Dried Postmortem Blood Specimens Collected at Autopsy From Infants With Unexplained Cause of Death," *Clin Chem*, 2001, 47(7):1166-82.

Charrow J, Andersson HC, Kaplan P, et al, "The Gaucher Registry: Demographics and Disease Characteristics of 1698 Patients With Gaucher Disease," *Arch Intern Med*, 2000, 160(18):2835-43.

Chinnery PF and Turnbull DM, "Mitochondrial DNA and Disease," *Lancet*, 1999, 354(Suppl 1):SI17-SI21.

Gilbert-Barness E and Barness LA, *Metabolic Diseases: Foundations of Clinical Management, Genetics, and Pathology*, Natick, MA: Eaton Publishing, 2000.

Gravel RA, Clarke JT, and Kabuck MM, "The GM2 Gangliosidoses," *The Metabolic Basis of Inherited Disease*, 7th ed, Chapter 92, Scriver CR, Beaudet AL, Sly WS, et al, eds, New York, NY: McGraw-Hill Inc, 1995, 2839-79.

Gray RGF, Preece MA, Green SH, et al, "Inborn Errors of Metabolism as a Cause of Neurological Disease in Adults: An Approach to Investigation," *J Neurol Neurosurg Psychiatry*, 2000 69(1):5-12.

Johnson DW, "A Rapid Screening Procedure for the Diagnosis of Peroxisomal Disorders: Quantification of Very-Long-Chain Fatty Acids, as Dimethylaminoethyl Esters, in Plasma and Blood Spots, by Electrospray Tandem Mass Spectrometry," *J Inherit Metab Dis*, 2000, 23(5):475-86.

Joseph F and Russo TM, "Origins of Spurious Organic Acidurias," *Lab Med*, 2000, 31(11):622.

Kamiguchi H, Hlavin ML, Yamasaki M, et al, "Adhesion Molecules and Inherited Diseases of the Human Nervous System," *Annu Rev Neurosci*, 1998, 21:97-125.

Klein I, Sarkadi B, and Váradi A, "An Inventory of the Human ABC Proteins," *Biochim Biophys Acta*, 1999, 1461(2):237-62.

Kolodny EH and Charria-Ortiz G, "Storage Diseases of the Reticuloendothelial System," *Hematology of Infancy and Childhood*, 6th ed, Volume 2, Chapter 35, Nathan DG and Oski FA, eds, Philadelphia, PA: WB Saunders Co, 2003, 1439-40.

Lindor NM and Karnes PS, "Initial Assessment of Infants and Children With Suspected Inborn Errors of Metabolism," *Mayo Clin Proc*, 1995, 70(10):987-8.

Mehta A, "New Developments in the Management of Anderson-Fabry Disease," *Q J Med*, 2002, 95(10):647-53.

O'Brien JF, "Lysosomal Storage Disease: Method for the Preparation of Leukocytes," *Tietz Textbook of Clinical Chemistry*, 3rd ed, Chapter 49, Burtis CA and Ashwood ER, eds, Philadelphia, PA: WB Saunders Co, 1999, 1776-84.

Rashed MS, Rahbeeni Z, and Ozand PT, "Application of Electrospray Tandem Mass Spectrometry to Neonatal Screening," *Semin Perinatol*, 1999, 23(2):183-93.

Rinaldo P, "Inherited Metabolic Disorders in the Neonate," *Semin Perinatol*, 1999, 23(2):99-204.

Sue CM, Hirano M, DiMauro S, et al, "Neonatal Presentations of Mitochondrial Metabolic Disorders," *Semin Perinatol*, 1999, 23(2):113-24.

Surtees R, "Inborn Errors of Neurotransmitter Receptors," *J Inherit Metab Dis*, 1999, 22(4):374-80.

Unger ER and Piper MA, "Nucleic Acid Biochemistry and Diagnostic Applications," *Tietz Textbook of Clinical Chemistry*, 3rd ed, Chapter 18, Burtis CA and Ashwood ER, eds, Philadelphia, PA: WB Saunders Co, 1999, 421-43.

Internet Web Sites
www.nlm.nih.gov/omim/
www.press.jhu.edu/

Inhibin A, Serum

Related Information

Alpha$_1$-Fetoprotein, Serum *on page 136*
Amniotic Fluid, Chromosome and Genetic Abnormality Analysis *on page 152*
Chorionic Gonadotropin, Human, Serum and Urine *on page 397*
Chorionic Villus Sampling, Chromosome and Genetic Abnormality Analysis *on page 400*
Estriol, Unconjugated, Pregnancy, Serum, Plasma, or Urine *on page 554*
Pregnancy-Associated Protein A, Serum *on page 1082*
Testosterone, Total and Free, Serum or Plasma *on page 1238*

Applies to Four-Marker Test; Integrated Test; Multiple Marker Screening Test; Quadruple Test; Triple Test

Abstract Inhibin is the major gonadal inhibitory peptide regulating secretion of FSH in both sexes. It is synthesized in the placenta and plateaus during weeks 14-30. The α subunit combines with a β subunit to form either inhibin A (Aβ) or inhibin B (Bβ). The measurement of inhibin A in serum is used with other studies for prediction of Down syndrome. Studies discussed as second trimester tests include AFP, hCG, and inhibin A with or without unconjugated estriol (uE$_3$). Replacement of estriol with inhibin A in the multiple marker screening test has been advocated by some in screening for Down syndrome.

Specimen Serum

Sampling Time Measurements of inhibin A are used only after 14 weeks. Check with laboratory.

Special Instructions Maternal age is needed.

Use In conjunction with other tests, to screen for Down syndrome after 14 weeks gestation. Dimeric inhibin A is elevated about twofold in the second trimester of Down syndrome gestations. Down syndrome is also associated with low maternal AFP and uE$_3$ and high hCG levels. Maternal AFP, dimeric inhibin A with maternal age, as a combination, also detect autosomal trisomies other than Down syndrome. An **integrated test** includes nuchal translucency studies with pregnancy-associated plasma protein A in the first trimester, with serum AFP, hCG, unconjugated estriol, and inhibin A in the second trimester. Dimeric inhibin A is reported to detect 92% of trisomy 18 and 71% of trisomy 13 afflicted gestations. It is also used for Turner syndrome with AFP, at a 53% detection rate, with a false-positive rate of ~20%.

Methodology Radioimmunoassay, enzyme-linked immunoassay (ELISA). A two-site enzyme immunoassay, selective for inhibin A ($\alpha\beta$ A dimer) is described.

Additional Information The **triple test** includes maternal AFP, uE$_3$, hCG, and age. The **four-marker test** also includes inhibin A. The second trimester **quadruple test**, including these four tests, is the superior screening battery in the second trimester. Addition of dimeric inhibin A enhances the rate of detection of Down syndrome gestations to nearly 75% at a 5% false-positive rate. (Continued)

Inhibin A, Serum *(Continued)*

Threefold increased concentrations of maternal inhibin A in the second trimester predict pre-eclampsia.

References

Baliff JP and Mooney RA, "New Developments in Prenatal Screening for Down Syndrome," *Am J Clin Pathol*, 2003, 120(Suppl 1):514-24.

Benn PA, Fang M, Egan JFX, et al, "Incorporation of Inhibin A in Second-Trimester Screening for Down Syndrome," *Obstet Gynecol*, 2003, 101(3):451-4.

Demers LM, "Dimeric Inhibin A, A Fourth Maternal Serum Screening Marker for Down's Syndrome," *Diagnostic Endocrinology, Immunology, and Metabolism*, 2000, 18(5):131-3.

Malone FD and D'Alton ME, "First-Trimester Sonographic Screening for Down Syndrome," *Obstet Gynecol*, 2003, 102(5 Part 1):1066-79.

Wald NJ, Watt HC, and Hackshaw AK, "Integrated Screening for Down Syndrome Based on Tests Performed During the First and Second Trimesters," *N Engl J Med*, 1999, 341(7):461-7.

Wenstrom KD, Chu DC, Owen J, et al, "Maternal Serum Alpha-Fetoprotein and Dimeric Inhibin A Detect Aneuploidies Other Than Down Syndrome," *Am J Obstet Gynecol*, 1998, 179(4):966-70.

♦ **Inhibitor Screen** *see* Mixing Studies (Coagulation) *on page 918*

♦ **INR** *see* Coagulation Factor Assays *on page 418*

♦ **INR** *see* Prothrombin Time *on page 1116*

♦ **Insecticides** *see* Organophosphate Pesticides, Urine, Blood, or Serum *on page 975*

♦ **Insect Identification** *see* Arthropod Identification *on page 213*

♦ *in situ* **Hybridization** *see* Immunoperoxidase Procedures *on page 780*

♦ **Insulin** *see* C-Peptide, Serum *on page 465*

♦ **Insulin Antibodies** *see* C-Peptide, Serum *on page 465*

Insulin Antibody, Serum

Related Information

Insulin, Serum *on page 803*

Test Includes Antibodies to both beef and pork insulin

Abstract Nearly all patients treated with beef or pork insulin develop anti-insulin antibodies. Clinically apparent antibody-mediated insulin resistance, however, develops in only 0.01% of treated persons. Patients with insulin antibodies require large doses of insulin. Most anti-insulin antibodies are IgG, but a few are IgE. Assays are performed only in reference laboratories.

Patient Preparation Avoid radioisotopes prior to collection.

Specimen Serum

Container Red top tube

Storage Instructions Serum stable for 7 days at 4°C.

Reference Interval Method dependent. Results may be reported as percent binding of patient's serum to labeled insulin, in which case the reference interval is <3%.

Use Determine the presence of antibodies against heterologous insulins. Insulin antibodies may cause misleading results of assays for insulin.

Limitations Anti-insulin antibodies have little clinical significance. Clinical and other biochemical findings are now used to evaluate the syndrome of insulin resistance.

Methodology Radioimmunoassay (RIA) or enzyme-linked immunosorbent assay (ELISA)

Additional Information Human insulin is less antigenic than are those derived from other species. Insulin antibodies may interfere with insulin assay.

References

Unger RH and Foster DW, "Diabetes Mellitus," *Williams Textbook of Endocrinology*, 9th ed, Wilson JD, Foster DW, Kronenberg HM, et al, eds, Philadelphia, PA: WB Saunders Co, 1998, 1034.

Williams KV, Erbey JR, Becker D, et al, "Can Clinical Factors Estimate Insulin Resistance in Type 1 Diabetes?" *Diabetes*, 2000, 49(4):626-32.

♦ **Insulin-Connecting Peptide** *see* C-Peptide, Serum *on page 465*

♦ **Insulin-Induced Hypoglycemia Test** *see* Insulin Tolerance Test *on page 804*

Insulin-Like Growth Factor-1 (IGF-1), Serum or Plasma

Related Information

Growth Hormone, Serum *on page 662*

Insulin-Like Growth Factor Binding Protein 3, Serum *on page 802*

Insulin Tolerance Test *on page 804*

Synonyms IGF-1; IGF-I; Sm-C; Somatomedin-C; Sulfation Factor

Abstract Two IGFs are of medical interest: IGF-1 (somatomedin C) and IGF-2. The former is used in evaluation of growth disorders, and is generally a better test than is assay for basal- and glucose-suppressed growth hormone. **The serum level of IGF-1 reflects the effects of growth hormone.**

Patient Preparation Overnight fast is preferable, no recent administration of radioactivity if assay is to be done by RIA.

Specimen Plasma or serum, depending upon method of assay

Container Lavender top (EDTA) tube (check with the laboratory), red top tube

Storage Instructions Separate plasma immediately by centrifuging at 4°C. Freeze plasma in a plastic tube at -20°C.

Reference Interval See table.

Insulin-Like Growth Factor 1

	Male (ng/mL)	Female (ng/mL)
2 mo - 5 y	17-248	17-248
6-8 y	88-474	88-474
9-11 y	110-565	117-771
12-15 y	202-957	261-1096
16-24 y	182-780	
25-39 y	114-492	
40-54 y	90-360	
≥55 y	71-290	
Tanner Stages		
I	109-485	128-470
II	174-512	186-695
III	230-818	292-883
IV	396-776	394-920
V	402-839	308-1138

Adapted from Mayo Medical Laboratories, *2001 Test Catalogue*, Rochester, MN.

Use Diagnose acromegaly (IGF-1 and growth hormone elevated), and monitor the response to growth hormone (GH) treatment. Low values of IGF-1, in combination with elevated GH are characteristic of **Laron dwarfism**, the prototype for GH insensitivity.

Limitations Low values of IGF-1 are occasionally encountered in hypopituitarism, malnutrition, delayed puberty, diabetes mellitus, old age, cirrhosis, and some cases of idiopathic short stature; elevated values. Elevated values (dependent, unfortunately on the reference intervals employed) are occasionally associated with adolescent growth spurt, precocious puberty, pregnancy, obesity, and diabetes mellitus.

Methodology Radioimmunoassay (RIA) following dissociation from binding protein and chromatography, immunoradiometric assay (IRMA), chemiluminescence immunoassay

Additional Information

Low values are described with the extremes of age (first 5-6 years and advanced age), hypopituitarism, malnutrition, diabetes mellitus, hypothyroidism, maternal deprivation syndrome, pubertal delay, cirrhosis, hepatoma, Laron dwarfism, and some cases of short stature and normal GH response to pharmacologic tests. Low values may be found with nonfunctioning pituitary tumors, with constitutional delay of growth and development, and with anorexia nervosa.

High values occur with adolescence, true precocious puberty, pregnancy, obesity, pituitary gigantism, **acromegaly**, and diabetic retinopathy.

Since IGF-1 is decreased with malnutrition, its concentration provides a useful index with which to monitor therapy for food deprivation.

Provocative testing is done for assays of growth hormone; IGF-1 provides another approach for evaluation of pituitary GH secretion. Treatment of patients (Continued)

Insulin-Like Growth Factor-1 (IGF-1), Serum or Plasma *(Continued)*

who have acromegaly, with growth-hormone receptor antagonists, results in decrease in IGF-1 and clinical improvement.

References
Jones JI and Clemmons DR, "Insulin-Like Growth Factors and Their Binding Proteins: Biological Actions," *Endocr Rev*, 1995, 16(1):3-34.

Mayo Medical Laboratories, *2001 Test Catalogue*, Rochester, MN, 329.

Shalet SM, Toogood A, Rahim A, et al, "The Diagnosis of Growth Hormone Deficiency in Children and Adults," *Endocr Rev*, 1998, 19(2):203-23.

Insulin-Like Growth Factor Binding Protein 3, Serum

Related Information
Growth Hormone, Serum *on page 662*
Insulin-Like Growth Factor-1 (IGF-1), Serum or Plasma *on page 800*
Insulin, Serum *on page 803*

Synonyms IGFBP-3; Somatomedins

Abstract Insulin-like growth factors (IGFs), formerly known as somatomedins, circulate in plasma bound to proteins, the insulin-like growth factor binding proteins (IGFBPs), a category which now has 10 members. The serum levels of the IGFs and IGFBPs are, in general, proportional to output of growth hormone (GH). Unlike GH, however, the IGFs and IGFBPs have long serum half-lives and have been investigated as markers of GH secretion. Of all the IGFBPs, IGFBP-3 is the most intensely studied. IGFBP-3 is the most abundant IGFBP in the circulation and binds ~95% of the IGFs in the blood. Originally, it was believed that the major, or only, function of the IGFBPs was to transport the IGFs and modulate their availability to IGF receptors. Recently, however, IGF-independent activities of IGFBP-3 have been identified. In particular, IGFBP-3 is now known to be a potent apoptotic agent, thus inhibiting cell proliferation.

Specimen Serum

Container Red top tube

Storage Instructions Refrigerate up to 2 days; freeze thereafter.

Causes for Rejection Recently administered isotopes, if testing is to be done by radioimmunoassay methods; EDTA or heparinized plasma specimen

Reference Interval See table (ARUP Laboratories).

Insulin-Like Growth Factor Binding Protein 3, Serum

Age (y)	Male (mg/L)	Female (mg/L)	Age (y)	Male (mg/L)	Female (mg/L)
0-1	0.94-1.79	0.66-2.51	22-23	1.45-4.75	1.45-5.69
2-3	1.12-2.33	0.84-3.77	24-25	1.15-4.27	1.51-4.47
4-5	1.16-3.13	1.32-3.60	26-27	1.24-5.18	1.38-4.70
6-7	1.32-3.38	1.21-4.66	28-29	1.23-4.27	1.19-5.43
8-9	1.35-3.94	1.58-3.99	30-34	1.29-4.06	1.29-4.06
10-11	1.53-5.02	1.93-5.46	35-39	1.50-3.44	1.50-3.44
12-13	1.73-5.11	1.78-6.08	40-44	1.33-3.58	1.33-3.58
14-15	1.90-6.40	2.02-5.44	45-49	1.44-2.75	1.44-2.75
16-17	1.70-6.04	1.88-5.29	50-54	1.31-2.52	1.31-2.52
18-19	1.52-6.01	1.63-6.02	55-59	1.53-2.43	1.53-2.43
20-21	1.79-5.41	1.82-5.35	60-70	1.40-3.22	1.40-3.22

Use Growth hormone deficiency: Originally proposed as a convenient (because of long half-life) marker of GH activity in the diagnosis of GH deficiency, the clinical accuracy is less than ideal. For example, Juul et al report the following results, comparing IGFBP-3 against dynamic testing with arginine provocation and clonidine provocation:
- <10 years: 60% sensitivity, 97.9% specificity
- 10-20 years: 56.5% sensitivity, 78.7% specificity

Limitations Poor predictive values (see Use)

Methodology Immunoassays (multiple labels)

References

ARUP Laboratories, *1999-2000 User's Guide*, 415.

Diamandi A, Mistry J, Krishna RG, et al, "Immunoassay of Insulin-Like Growth Factor-Binding Protein-3 (IGFBP-3): New Means to Quantifying IGFBP-3 Proteolysis," *J Clin Endocrinol Metab*, 2000, 85(6):2327-33.

Juul A and Skakkebaek NE, "Prediction of the Outcome of Growth Hormone Provocative Testing in Short Children by Measurement of Serum Levels of Insulin-Like Growth Factor 1 and Insulin-Like Growth Factor Binding Protein 3," *J Pediatr*, 1997, 130(2):197-204.

Rechler MM, "Growth Inhibition by Insulin-Like Growth Factor (IGF) Binding Protein 3 - What's IGF Got to Do With It?" *Endocrinology*, 1997, 138(7):2645-7.

♦ **Insulin-Like Growth Factor-II (IGF-II)** *see* Insulin, Serum *on page 803*

♦ **Insulin-Like Growth Factors** *see* Growth Hormone, Serum *on page 662*

Insulin, Serum

Related Information

Ammonia, Plasma *on page 150*
C-Peptide, Serum *on page 465*
Glucose, Fasting, Plasma *on page 643*
Glucose, Random, Plasma *on page 649*
Glucose Tolerance Test, Plasma *on page 651*
Insulin Antibody, Serum *on page 800*
Insulin-Like Growth Factor Binding Protein 3, Serum *on page 802*
Microalbuminuria *on page 913*
Pancreatic Polypeptide, Human, Serum or Plasma *on page 998*

Synonyms Immunoreactive Insulin

Applies to C-Peptide; Insulin-Like Growth Factor-II (IGF-II); Pancreatic Polypeptides; Proinsulin; SUR 1 Mutation

Test Includes Glucose must be drawn simultaneously.

Abstract Within the pancreatic beta cell, the polypeptide, **preproinsulin**, is metabolized via a complex mechanism to three substances which are measured clinically: **proinsulin**, **insulin**, and **C-peptide**, often collectively referred to as **pancreatic polypeptides**.

Causes of fasting hypoglycemia in both adults and children include islet cell tumor, exogenous insulin or oral hypoglycemic drugs, alcohol use, pituitary or adrenal insufficiency, bulky extrapancreatic tumor, and instances of very severe hepatic disease. In infancy and childhood, there are also familial and sporadic diseases known collectively as persistent hyperinsulinemic hypoglycemia of infancy (PHHI).

Patient Preparation In most instances, patients should be fasting.

Specimen Serum or plasma

Container Red top tube

Storage Instructions Separate serum and freeze until time for assay.

Reference Interval Fasting:
- infants: 0-13 µIU/mL (SI: 0-90 pmol/L)
- adults: 0-17 µIU/mL (SI: 0-118 pmol/L)

There is moderate interlaboratory variability. Consult with the selected reference laboratory.

Use

Hypoglycemia: Diagnostic criteria for the hypoglycemic disorders (islet cell tumor, surreptitious insulin administration, and hypoglycemic drug effect) are presented in the discussion of C-peptide. (See C-Peptide, Serum *on page 465*.)

Hypoglycemic disorders of infancy and childhood, in patients who do not have ketosis or acidosis, are usually due to hyperinsulinemia and are collectively referred to as **persistent hyperinsulinemic hypoglycemia of infancy** (PHHI). Clinically subdivided into three forms of varying severity, these disorders share the features of serum glucose <54 mg/dL (SI:<2.9 mmol/L), serum insulin >10 µU/mL (SI: >69.45 pmol/L), and increases in serum C-peptide and proinsulin. The autosomal recessive, clinically most severe form, is usually detected within the first few hours of life. An autosomal dominant, less severe form, may be detected anytime in infancy or childhood. A third form, also less severe, is accompanied by hyperammonemia (100-200 µmol N/L). In some children, an additional study involves measuring the glucose response to intravenous glucagon (30 µg/kg) at the time of hypoglycemia; an increase in serum glucose
(Continued)

Insulin, Serum (Continued)

of 40 mg/dL (SI: 2.2 mmol/L) at 10 minutes postinjection is further evidence for PHHI, and also implies that the patient's glycogen stores and glycogenolytic enzymes are intact. Infants with severe PHHI are candidates for pancreatic resection; the decision between partial and near-total resection is guided by intraoperative frozen section of biopsies from the head, body and tail.

Tumor-induced hypoglycemia: Insulinoma (islet cell tumor) is the most common type of islet cell tumor and is the most common tumor causing hypoglycemia. About 90% are benign. Diagnostic criteria are provided in the listing C-Peptide, Serum *on page 465*. In >85% of cases of insulinoma, proinsulin is >25%; normally it is <25% of total immunoreactive insulin.

Other tumors occasionally cause hypoglycemia by mechanisms other than insulin hypersecretion.

Syndrome of insulin resistance: Fasting hyperinsulinemia may be an important attribute in a clinical syndrome which includes hypertension, an unfavorable lipid profile, type 2 diabetes, obesity, and coronary heart disease. (See the Disease Index for Diabetes Mellitus, Hypertension, and Obesity.)

Limitations See Insulin Antibody, Serum *on page 800*.

Methodology Immunoassays (multiple labels)

Additional Information See graphic in listing Microalbuminuria *on page 913* relevant to years of insulin-dependent diabetes and glomerular filtration rate.

References

Daneman MB, "Disorders of Carbohydrate Metabolism in Infants and Children," *Biochemical Basis of Pediatric Disease*, Soldin SJ, Rifai N, and Hicks JMB, eds, Washington, DC: AACC Press, American Association of Clinical Chemistry, 1992, 261-91.

de Lonlay-Debeney P, Poggi-Travert F, Fournet JC, et al, "Clinical Features of 52 Neonates With Hyperinsulinism," *N Engl J Med*, 1999, 340(15):1169-75.

Isomaa B, Almgren P, Tumoi T, et al, "Cardiovascular Morbidity and Mortality Associated With the Metabolic Syndrome," *Diabetes Care*, 2001, 24(4):683-9.

Painter PC, Cope JY, and Smith JL, "Reference Information for the Clinical Laboratory," *Tietz Textbook of Clinical Chemistry*, 3rd ed, Philadelphia, PA: WB Saunders Co, 1999.

Shulman GI, "Cellular Mechanisms in Insulin Resistance," *J Clin Invest*, 2000, 106(2):171-6.

Insulin Tolerance Test

Related Information

Adrenal Cortex: Laboratory Assessment Overview *on page 110*
Adrenocorticotropic Hormone, Plasma *on page 114*
Corticotropin-Releasing Hormone Stimulation Test *on page 455*
Corticotropin Stimulation Test (Rapid) *on page 456*
Cortisol, Free, Urine *on page 459*
Cortisol, Serum or Plasma *on page 460*
Growth Hormone, Serum *on page 662*
Insulin-Like Growth Factor-1 (IGF-1), Serum or Plasma *on page 800*
Insulin-Like Growth Factor Binding Protein 3, Serum *on page 802*
Insulin, Serum *on page 803*
Metyrapone Stimulation Test, Serum *on page 910*
Urinary Cortisol/Creatinine Increment *on page 1292*

Synonyms Insulin-Induced Hypoglycemia Test; ITT

Applies to Corticotropin-Releasing Hormone; Cortisol; Hypothalamic-Pituitary-Adrenal Axis

Abstract The insulin tolerance test (ITT) was introduced in 1969 (Plumpton et al) to provide an objective basis for the diagnosis of relative adrenal insufficiency (AI) caused by long-term exogenous corticosteroid medication. The original study included both clinical and biochemical endpoints, and the ITT has become the gold standard for investigation of the hypothalamic-pituitary-adrenal (HPA) axis. If the HPA axis is intact, hypoglycemia is followed by increases in serum ACTH and cortisol (as well as growth hormone (GH) and prolactin). The ITT is used clinically as a test to evaluate either the functional integrity of the HPA axis or the capacity of the pituitary to release GH.

Patient Preparation The patient must be attended by a physician during this procedure. The patient should fast for at least 8 hours before the test, and remain supine during the procedure. As a precaution, an intravenous preparation of 50% glucose (at least 30 mL) must be available at the bedside, ready for immediate infusion. An intravenous line should be used to inject the insulin (and glucose, if needed) and to obtain blood specimens.

After a baseline blood sample has been obtained, the patient is given regular insulin, 0.1-0.15 units/kg body weight, intravenously. The 0.1 unit/kg dose is preferred for patients with suspected pituitary or adrenal insufficiency. Some authorities recommend increasing the dose to 0.25 units/kg for patients with obesity, "insulin resistance" (see discussion for Insulin, Serum *on page 803*), suspected Cushing syndrome, or suspected acromegaly.

Blood glucose values must be measured stat, ideally at the bedside (see Sampling Time). An adequate hypoglycemic response is variously defined as 40 mg/dL (SI: 2.2 mmol/L) or 35 mg/dL (SI: 1.9 mmol/L). If this degree of hypoglycemia is not achieved, a second dose of insulin is given.

Specimen Serum

Container Red top tube

Sampling Time A baseline sample is always obtained, just before insulin injection. Subsequent specimens are obtained according to one of these schedules: 30 and 45 minutes after insulin; at 30, 60, and 90 minutes after insulin; or 30, 45, 60, and 90 minutes after insulin. Specimens are assayed for glucose (stat), and sent to the laboratory for either cortisol or growth hormone, depending on the purpose of the test.

Reference Interval If the HPA axis is functioning normally, serum cortisol should reach, or exceed, 20 µg/dL (SI: 550 nmol/L). (The original procedure had the additional criterion of a peak cortisol value which was 5 µg/dL (SI: 138 nmol/L) over baseline.) In normal persons, serum growth hormone should rise to 10 ng/mL.

Use

Adrenal insufficiency (AI): Although the ITT is the gold standard for HPA axis evaluation, in practice the ITT is a tertiary test and, therefore, used only occasionally. For example, if the morning serum cortisol is <5 µg/dL (SI: 140 nmol/L) or if the urine free cortisol (in an adult) is <20 µg/day (SI: 55 nmol/day), the diagnosis of AI is biochemically confirmed and the ITT is not needed. Most patients with AI will be correctly diagnosed with some combination of **primary tests** (see Cortisol, Free, Urine *on page 459*, and Corticotropin Stimulation Test (Rapid) *on page 456*) and **secondary tests** (see Metyrapone Stimulation Test, Serum *on page 910*, Corticotropin-Releasing Hormone Stimulation Test *on page 455*, and Urinary Cortisol/Creatinine Increment *on page 1292*).

AI is confirmed if, during the ITT, the cortisol fails to reach 20 µg/dL (SI: 550 nmol/L).

Growth hormone (GH) deficiency: Most patients with GH or insulin-like growth factor (IGF) deficiency are diagnosed using **primary tests** (see Growth Hormone, Serum *on page 662*, Insulin-Like Growth Factor-1 (IGF-1), Serum or Plasma *on page 800* and Insulin-Like Growth Factor Binding Protein 3, Serum *on page 802*) and **secondary tests** (ie, GH response to various stimuli). Only an occasional patient will require the ITT. **GH deficiency** is confirmed if, during the ITT, serum GH does not reach 10 ng/mL.

Cushing syndrome (CS): The ITT is very rarely indicated in the evaluation of patients with suspected CS. The ITT may be used in a patient to resolve the differential diagnosis of CS vs pseudo-CS. A normal cortisol response to the ITT favors the diagnosis of pseudo-CS (see Adrenal Cortex: Laboratory Assessment Overview *on page 110*).

Contraindications The ITT is contraindicated in persons with seizure disorders, coronary heart disease, and cardiac failure.

Additional Information Burke has published his view that the ITT has no usefulness except in suspected acromegaly.

References

Burke CW, "The Pituitary Megatest: Outdated?" *Clin Endocrinol (Oxf)*, 1992, 36(2):133-4.

Orth DN and Kovacs WJ, "The Adrenal Cortex," *Williams Textbook of Endocrinology*, 9th ed, Wilson JD, Foster DW, Kronenberg HM, et al, eds, Philadelphia, PA: WB Saunders Co, 1998, 517-664.

Plumpton FS and Besser GM, "The Adrenocortical Response to Surgery and Insulin-Induced Hypoglycemia in Corticosteroid-Treated and Normal Subjects," *Br J Surg*, 1969, 56(3):216-9.

Thorner M0, Vance ML, Laws ER Jr, et al, "The Anterior Pituitary," *Williams Textbook of Endocrinology*, 9th ed, Wilson JD, Foster DW, Kronenberg HM, et al, eds, Philadelphia, PA: WB Saunders Co, 1998, 249-340.

(Continued)

Insulin Tolerance Test *(Continued)*

Vance ML and Mauras N, "Growth Hormone Therapy in Adults and Children," *N Engl J Med*, 1999, 341(16):1206-16.

Intrinsic Factor Blocking Antibody

Related Information

Abstract A glycoprotein, intrinsic factor (IF) is generated in gastric parietal cells. It tightly binds cyanocobalamin (vitamin B_{12}) and facilitates its absorption. About 50% of patients with pernicious anemia (PA) develop antibodies to intrinsic factor. Secretion of IF parallels that of gastric HCl.

Patient Preparation Avoid recent radioactive scan, B_{12} injection within the past week.

Specimen Serum

Container Red top tube or SST™ tube

Reference Interval None detected

Use Differentiate pernicious anemia (PA) from other megaloblastic anemias; investigate patients with low vitamin B_{12} levels

Limitations Negative results do not rule out PA.

Additional Information Pernicious anemia is the most common cause of cobalamin (vitamin B_{12}) deficiency and is associated with antibodies to gastric parietal cells and with anti-intrinsic factor autoantibodies. Two types of antibodies to intrinsic factor have been described. Type 1 antibodies block the binding of cobalamin to intrinsic factor. Type 2 (precipitating antibody) react with the intrinsic factor binding site to ileal receptors. Type 2 antibodies are rarely found in the absence of type 1 antibodies. Fifty percent of patients with pernicious anemia have the type 1 intrinsic factor blocking antibody. About 35% (probably a much greater number) of patients with PA have type II antibodies. False-positive results are rare; some patients with Graves disease and atrophic gastritis have detectable antibodies.

Although parietal cell antibodies are present in 90% of patients with pernicious anemia, they are less specific than intrinsic factor antibodies.

References

Antony AC, "Megaloblastic Anemias," *Hematology: Basic Principles and Practice*, 3rd ed, Chapter 28, Hoffman R, Benz EJ Jr, Shattil SJ, eds, Philadelphia, PA: Churchill Livingstone, 2000, 462-3.

Galperin C and Gershwin E, "Immunopathogenesis of Gastrointestinal and Hepatobiliary Diseases," *JAMA*, 1997, 278(22):1946-55.

Waters HM, Dawson DW, Howarth JE, et al, "High Incidence of Type II Autoantibodies in Pernicious Anaemia," *J Clin Pathol*, 1993, 46(1):45.

♦ **Intrinsic Pathway** *see* Activated Partial Thromboplastin Time *on page 100*

♦ **Intrinsic Pathway** *see* Prothrombin Time *on page 1116*

♦ **Ionized Calcium** *see* Calcium, Ionized, Serum *on page 328*

♦ **Iontophoresis** *see* Chloride, Sweat *on page 392*

♦ **Iproveratril Hydrochloride** *see* Verapamil, Serum or Plasma *on page 1306*

Iron and Total Iron Binding Capacity/Transferrin, Serum

Related Information

Anemia Flowchart *on page 35*
Complete Blood Count *on page 442*
Copper, Serum *on page 448*
Deferoxamine Infusion Test *on page 504*
Erythropoietin, Serum *on page 551*
Ferritin, Serum *on page 577*
Hemoglobin *on page 681*
Hemosiderin Stain, Urine *on page 692*
Hereditary Hemochromatosis DNA Test *on page 718*
Iron Stain, Bone Marrow *on page 810*
Lead, Blood *on page 832*
Lead, Urine *on page 835*
Liver Biopsy *on page 864*
Liver Disease: Laboratory Assessment, Overview *on page 869*
Occult Blood, Stool *on page 969*
Phlebotomy, Therapeutic *on page 1029*
Porphyrins, Quantitative, Urine *on page 1074*
Protoporphyrin, Free Erythrocyte *on page 1121*
Protoporphyrin, Zinc, Blood *on page 1121*
Schilling Test *on page 1178*
Transferrin Receptor, Soluble, Serum or Plasma *on page 1267*

Synonyms Fe and TIBC; TIBC

Applies to Iron Poisoning; Transferrin; Transferrin Receptor, Soluble

Test Includes Serum iron, total iron binding capacity and/or transferrin, percent transferrin saturation

Abstract Diseases of iron homeostasis rank among the most common of human diseases. These include states of **iron deficiency**, including those characterized by **inadequate absorption** (eg, celiac disease, inflammatory bowel disease, bowel resection, dietary causes, bioavailability disorders, and intrinsic red cell defects), and those related to **increased loss**. The latter includes patients with tumors, inflammatory bowel disease, varices, gastritis, ulcer, parasitic infestations, and other gastrointestinal disorders. Genitourinary and other routes of loss are relevant as well.

Diseases of **iron overload** include two major settings. When erythropoiesis is normal but iron exceeds iron binding capacity of transferrin (eg, hemochromatosis), iron is deposited in parenchymal cells of liver and other organs. When catabolism of red cells causes iron overload (eg, transfusional iron overload), Fe is deposited in macrophages of the reticuloendothelial system.

Tests of iron status also include erythrocyte protoporphyrin, serum ferritin, and transferrin saturation, with hemoglobin and other hematologic parameters.

Patient Preparation Specimen should be drawn fasting in the morning (circadian rhythm affects iron; levels are lower in the evening). Sample should be drawn before patient is given therapeutic iron or blood transfusion. Iron determinations on patients who have had blood transfusions should be delayed several days.

Specimen Serum

Container Red top tube

Sampling Time Morning; marked daily variation occurs. Serum iron levels are 30% higher in the morning and blood levels should be determined on fasting AM samples.

(Continued)

Iron and Total Iron Binding Capacity/Transferrin, Serum *(Continued)*

Collection Blood should be drawn before other specimens which require anticoagulated tubes. Separate serum from cells as soon as possible.

Storage Instructions Stable 1 week at 4°C

Causes for Rejection Hemolysis

Reference Interval Normal intervals vary between laboratories (method dependent). Iron: 50-160 µg/dL (SI: 9.0-28.8 µmol/L) for adult males; slightly lower (5% to 10%) values for adult females. **Iron binding capacity:** 250-350 µg/dL (SI: 45-63 µmol/L). **Percent saturation (transferrin saturation):** 20% to 50%, lower in children. **Transferrin:** 200-380 mg/dL (SI: 2.0-3.8 g/L). **TIBC is an approximation of transferrin.** When transferrin value is known, TIBC can be calculated. TIBC (µg/dL) = transferrin (mg/dL) x 1.25.

Critical Values Transferrin saturation >62% predicts homozygous genotype for hemochromatosis in 92% of cases, but in women, >50% is recommended; *vide infra.*

- Mild iron toxicity: ≥350 µg/dL (SI: ≥63 µmol/L)
- Serious iron toxicity: 500 µg/dL (SI: 89.5 µmol/L)
- Death from iron toxicity: 1000 µg/dL (SI: 179 µmol/L)

Use Differential diagnosis of anemia, especially with hypochromia and/or low MCV. The **percent saturation** is more helpful than the serum iron to estimate iron stores and iron deficiency anemia. Evaluate thalassemia and possible sideroblastic anemia; work up hemochromatosis, in which iron is increased and iron saturation is high. Decrease in iron level after performance of a Schilling test supports the diagnosis of vitamin B_{12} deficiency, *vide infra.* Evaluate iron poisoning (toxicity) and overload in renal dialysis patients or patients with transfusion dependent anemias. Use of TIBC in iron toxicity may be less useful than previously believed. TIBC or transferrin is a useful index of nutritional status. See table for iron status indicators in various disease states.

Iron Status Indicators in Various Disease States

Disease	Ferritin	Transferrin / TIBC	Serum Iron	Iron Saturation
Uncomplicated iron deficiency	↓	↑	↓	N/↓
Anemia of chronic disease	N/↑	N/↓	↓	N/↓
Sideroblastic anemias	↑	N/↓	N/↑	↑
Hemolytic anemias	↑	N/↓	↑	↑
Hemochromatosis	↑	Slight ↓	↑	↑↑
Protein depletion		N/↓	N/↓	N/↓
Acute liver disease	↑	Var	↑	↑

↑ = increase
↓ = decrease
N = normal
Var = variable

In inflammatory, infectious, or malignant disease, measurement of usual parameters of iron status may not be sufficient for differentiating iron deficiency anemia from anemia of chronic disease. While bone marrow iron stain is the most definitive assessment of iron status, measurement of serum soluble **transferrin receptor (sTfR)** may avoid marrow aspiration and its potential complications. Serum soluble transferrin receptor increases in iron deficiency anemia and is usually unaffected in chronic disease; see Transferrin Receptor, Soluble, Serum or Plasma *on page 1267.*

Hemochromatosis: Transferrin saturation >55% with serum ferritin >400 µg/L establishes the diagnosis in the appropriate clinical setting. Liver biopsy can confirm iron overload and represents the gold standard when hepatic iron concentration is measured. **Increased saturation** occurs with HLA-related (classical) hemochromatosis before ferritin is greatly increased, and also with iron overload (eg, cirrhosis and portacaval shunt), in hemolytic anemias, and with iron therapy. Sample contamination and fluctuation in serum iron levels can make percent saturation misleading on occasion. Genetic tests for detecting hemochromatosis mutations (C282Y and H63D) are available. Homozygosity for C282Y mutation is responsible for up to 90% of hemochromatosis patients; see Hereditary Hemochromatosis DNA Test *on page 718.*

Chronic dialysis for renal failure: Monitor iron levels in patients undergoing dialysis. To follow treatment for iron overload with deferoxamine or with regimen of recombinant human erythropoietin and phlebotomy.

Limitations Except for iron poisoning, a serum iron without TIBC or transferrin is of limited value. Ferritin levels are also useful for iron deficiency. Low iron level may not indicate iron deficiency in acute infection with leukocytosis. Low iron concentrations may be misleading in chronic infection, inflammation, and malignancy; high ferritin levels occur in many such states. TIBC and transferrin are increased in patients on oral contraceptives, with normal saturation. Transferrin saturation is depressed with acute and chronic inflammation. Gross hemolysis may interfere with serum iron.

A group of patients exists who have iron-deficiency anemia with normal plasma transferrin concentration.

Deferoxamine, used in therapy of iron toxicity, interferes with TIBC. TIBC may be overestimated in the presence of excessive free iron (ie, iron toxicity). It falsely increases TIBC, and interferes with colorimetric iron methods, causing spuriously low results.

Contraindications Parenteral iron before sample is drawn will cause misleading high iron results. Recent blood transfusion may have only a small positive effect on the serum iron level.

Methodology Ferrozine, bathophenanthroline (iron), nephelometry, turbidimetry (transferrin), $MgCO_3$ column, other methods (TIBC), atomic absorption, anodal stripping, inductively-coupled plasma atomic emission spectroscopy (iron)

Additional Information Serum iron is **increased** in hemosiderosis, hemolytic anemias (especially thalassemia), sideroblastic anemias, hepatitis, acute hepatic necrosis, hemochromatosis, and with inappropriate iron therapy.

Iron may reach high levels with **iron poisoning**, which presents with emesis and severe abdominal pain. Metabolic acidosis with increased anion gap, leukocytosis and hyperglycemia may be found with increased bilirubin, AST, ALT and LD. Pill fragments may be found on abdominal x-ray examination.

Some patients who receive multiple transfusions (eg, some hemolytic anemias, thalassemia, renal dialysis patients) will have increased serum iron levels.

Serum iron is **decreased** with insufficient dietary iron, chronic blood loss (including the hemolytic anemias, paroxysmal nocturnal hemoglobinuria), inadequate absorption of iron, and impaired release of iron stores as in inflammation, infection, and chronic diseases. The combination of low iron, high TIBC and/or transferrin, and low saturation indicates iron deficiency. Without all of these findings together, iron deficiency is unproven. Low ferritin confirms the diagnosis of iron deficiency. **Detection of iron deficiency may lead to discovery of malignancy involving the gastrointestinal tract, a point which cannot be overemphasized.**

In recovery from pernicious anemia, especially just after B_{12} dose, iron levels are low. The iron level falls just after B_{12} is administered to a patient with pernicious anemia. **The fall in serum iron after the Schilling test flushing dose** of vitamin B_{12} may be more useful in diagnosis of B_{12} deficiency than the radioactivity of the 24-hour urine collection.

TIBC is increased in iron-deficiency, use of oral contraceptives, and in pregnancy.

TIBC decreased in hypoproteinemia from many causes, including kwashiorkor, and in a number of inflammatory states.

The serum **ferritin** is usually a more sensitive test than the serum iron or TIBC for iron deficiency and for iron overload. When all these tests are used together, as is often necessary, they usually can distinguish between iron deficiency anemia and the anemia of chronic disease. (See the Hematology introduction for the Anemia Flowchart *on page 34.*) The best and most reliable evaluation of total body iron stores is by **bone marrow aspiration and** (Continued)

Iron and Total Iron Binding Capacity/Transferrin, Serum *(Continued)*

biopsy. The best evaluation of iron deficiency in childhood (unless lead toxicity is suspected) is **free erythrocyte porphyrins**.

With recombinant erythropoietin therapy serum iron, transferrin saturation, and ferritin levels decline due to rapid utilization by stimulated erythropoiesis with resultant decrease in storage iron.

"Iron chips" (cDNA-based microarrays representing genes involved in iron metabolism and pathways) have been used in gene-expression profiling of iron regulatory patterns including those related to hemochromatosis.

References
Andrews NC, "Disorders of Iron Metabolism," *N Engl J Med*, 1999, 341(26):1986-94.

Brandhagen DJ, Fairbanks VF, Batts KP, et al, "Update on Hereditary Hemochromatosis and the HFE Gene," *Mayo Clin Proc*, 1999, 74(9):917-21.

Looker AC, Dallman PR, Carroll MD, et al, "Prevalence of Iron Deficiency in the United States," *JAMA*, 1997, 277(12):973-6.

Muckenthaler M, Richter A, Gunkel N, et al, "Relationships and Distinctions in Iron-Regulatory Networks Responding to Interrelated Signals," *Blood*, 2003, 101(9):3690-8.

Press RD, "Hereditary Hemochromatosis: Impact of Molecular and Iron-Based Testing on the Diagnosis, Treatment, and Prevention of a Common, Chronic Disease," *Arch Pathol Lab Med*, 1999, 123(11):1053-9.

Wharton BA, "Iron Deficiency in Children: Detection and Prevention," *Br J Haematol*, 1999, 106(2):270-80.

♦ **Iron, Hepatic** *see* Liver Biopsy *on page 864*

♦ **Iron Poisoning** *see* Iron and Total Iron Binding Capacity/Transferrin, Serum *on page 807*

♦ **Iron Stain** *see* Siderocyte Stain *on page 1198*

Iron Stain, Bone Marrow

Related Information
Bone Marrow *on page 296*
Ferritin, Serum *on page 577*
Hereditary Hemochromatosis DNA Test *on page 718*
Iron and Total Iron Binding Capacity/Transferrin, Serum *on page 807*
Liver Biopsy *on page 864*
Reticulocyte Hemoglobin Content *on page 1158*
Siderocyte Stain *on page 1198*
Transferrin Receptor, Soluble, Serum or Plasma *on page 1267*

Synonyms Hemosiderin Stain; Prussian Blue Stain; Sideroblast Stain

Applies to Perls' Reaction; Ringed Sideroblasts; Zinc Protoporphyrin:Heme Ratio

Test Includes Iron stain on sections of marrow aspirate clot and/or bone marrow biopsy and iron stain marrow cover slip smears

Abstract A stain (the Prussian blue reaction) applied to smears of bone marrow particles demonstrate storage iron as blue to blue-green granular precipitates in macrophages/histiocytes. Such iron is seen also in a proportion of red cell precursors in the bone marrow (sideroblasts), in some peripheral red blood cells (siderocytes), and in some marrow immature red cells with impaired iron utilization ("ringed sideroblasts").

Specimen Bone marrow glass coverslip or slide smears, marrow aspirate, or biopsy

Container Coverslips or glass microslides are prepared at the bedside. Biopsy and clot fixed in formalin or other fixative (eg, B5 or Zenker's solution).

Collection Physician obtains bone marrow aspirate specimen by aseptic aspiration technique. Phlebotomist simultaneously obtains blood specimen for the preparation of peripheral blood smears.

Causes for Rejection No marrow obtained ("dry tap") or no bone marrow particles on smears

Special Instructions Requisition should include a brief clinical history.

Reference Interval Results should be interpreted in light of clinical background. Peripheral blood: no stainable iron is usually present (ie, no siderocytes are normally found). Bone marrow: stainable iron present as extracellular granules/

globules and/or intracellular in cytoplasm of histiocytes - cells of the reticuloendothelial (RE) system. About one-third of the rubricytes in the marrow may be iron-positive sideroblasts (but not "ringed sideroblasts").

Use Semiquantitation of bone marrow iron stores; sensitive test for the evaluation of iron reserve; aid in the diagnosis of iron deficiency and its differentiation from another hypochromic/microcytic condition, thalassemia, in which iron stores are often increased; aid in the diagnosis of hemosiderosis/hemochromatosis; aid in the diagnosis of sideroblastic anemia (including refractory anemia with ringed sideroblasts) and in the detection of hemophagocytosis

Limitations Specimen should include sufficiently large spicules of marrow. True stainable marrow iron must be differentiated from iron positive artifacts.

Methodology Ferrocyanide ion reacts in acid with ferric ion to form a dark blue-green precipitate called Prussian blue. Presence of this pigment is reported semiquantitatively (and subjectively) as absent, decreased, normal, or increased (or on a scale of 0 to 4+). Some (20% to 40%) of red cell precursors will also have some iron-positive granules. These cells are called sideroblasts. In the common condition, "anemia of chronic disease," iron stores are normal or increased, but sideroblasts are absent.

Additional Information A bone marrow biopsy is more accurate than an aspirate for detecting storage iron: 65% of iron-positive biopsies are associated with iron-negative aspirates. Reliance on aspirates alone may result in an erroneous diagnosis of iron deficiency. An important caveat is that if a biopsy is left in decalcifying solution more than 2 hours, the stainable iron may be removed, resulting in an erroneous diagnosis of iron deficiency. Without evaluation of both types of specimens a significant overdiagnosis of iron deficiency may occur. Hemochromatosis, hemolytic anemias, and those with ineffective erythropoiesis (eg, thalassemia, megaloblastic and sideroblastic anemias) and anemias of chronic disease (especially inflammation) are characterized by increase in iron stores. The usual sideroblast has small iron positive granules without pattern in the cytoplasm. Ringed sideroblasts are rubricytes with tiny particles of iron located in mitochondria forming a ring around at least two-thirds of the nucleus. These pathologic sideroblasts occur in cases of normoblastic refractory anemia, B_6 responsive anemia including inherited (X-linked) pyridoxine-responsive sideroblastic anemia due to mutant erythroid 5-aminolevulinate synthase, thalassemia, a variety of sideroblastic anemias, in some cases of B_{12}/folic acid deficiency, and in chloramphenicol toxicity.

There has been little study of marrow iron stores in healthy children. It appears that stainable marrow iron is quite limited in apparently normal children during the first 5 years of life. Thus, iron stains of marrow in early childhood may not be helpful in establishing diagnosis of iron deficiency anemia while presence of classical stainable iron may assist in excluding such diagnosis.

A cost-efficient and noninvasive alternative to bone marrow iron study is the proposed combined determination of zinc protoporphyrin:heme ratio and serum ferritin.

References

Geaghan SM, "Hematologic Values and Appearances in the Healthy Fetus, Neonate, and Child," *Clinics in Lab Medicine: Diagnostic Pediatric Hematology*, Geagham SM, ed, 1999, 19(1):1-37.

Koduri PR, "Prussian Blue Reaction and Hemophagocytosis: A New Use for an Old Test," *Am J Hematol*, 1995, 49(2):167.

Perkins S, "Hypochromic Microcytic Anemias," *Practical Diagnosis of Hematologic Disorders*, 3rd ed, Chapter 2, Kjeldsberg C, ed, Chicago, IL: ASCP Press, American Society of Clinical Pathologists, 2000, 29-32.

♦ **Iron Stain, Urine** *see* Hemosiderin Stain, Urine *on page 692*

Irradiated Blood Components

Related Information

Donation, Blood, Directed *on page 523*
Neutrophils, Transfusion *on page 951*
Newborn Crossmatch and Transfusion *on page 952*
Platelet Transfusion *on page 1058*
Red Blood Cells *on page 1139*
Whole Blood *on page 1333*

Test Includes Irradiation of cellular blood components with a gamma radiation source, usually cesium-137 or cobalt-60

(Continued)

Irradiated Blood Components *(Continued)*

Abstract Transfusion-associated graft-versus-host disease (TA-GVHD) may occur when donor T lymphocytes from transfused blood attack recipient tissues, beginning 2-30 days after transfusion. It is characterized by fever, maculopapular rash, diarrhea, hepatitis, and pancytopenia. Mortality is ~90%.

Patient Preparation Same as for other cellular component pretransfusion testing.

Aftercare Same as for transfused component.

Use Avoid TA-GVHD in blood recipients at risk for development of this condition. Indications for using irradiated blood components include immunocompromised transplant (stem cell or organ) recipients, allogeneic stem cell transplant candidates, intrauterine transfusions, neonatal exchange transfusion or extracorporeal membrane oxygenation, recipients with Hodgkin disease, recipients with congenital cell-mediated immunodeficiencies, recipients of directed donations from biologic relatives or HLA-matched donors, and recipients who are heterozygous at an HLA locus for which the donor is homozygous and shares an allele. Other possible indications include individuals receiving immunosuppressive therapy or who are immunosuppressed, patients undergoing high-dose chemotherapy, low birthweight infants/premature infants, and patients with AIDS who have opportunistic infections. See table.

Indications for Gamma Irradiation of Cellular Blood Components

Well-Defined

- Bone marrow or peripheral blood stem cell transplant
- Current or anticipated congenital cell-mediated immunodeficiencies (eg, severe combined immunodeficiency disease, Wiscott-Aldrich, DiGeorge)
- Intrauterine or postintrauterine transfusions
- Directed donations from blood relative or HLA-matched donors
- Hodgkin disease
- Adult/childhood acute lymphocytic leukemia
- Immunocompromised organ transplant recipients

Relative (possible)

- Malignancies and organ transplants treated with immunosuppressive chemotherapy or radiotherapy
- Exchange transfusion or use of extracorporeal membrane oxygenation in neonates
- Low-birth-weight neonates (<1200 g)
- Neonates with possible immunodeficiency
- Neonates receiving intrauterine transfusions
- Human immunodeficiency virus-infected patients with opportunistic infections

Probably Not Indicated

- Full-term neonates (exceptions noted above)
- Human immunodeficiency virus-infected patients

Modified from "Practice Parameter for the Use of Red Blood Cell Transfusions. Developed by the Red Blood Cell Administration Practice Guideline Development Task Force of the College of American Pathologists," *Arch Pathol Lab Med*, 1998, 122(2):130-8.

Limitations Irradiation induces RBC membrane damage and causes higher potassium levels in the supernatant. Such increase in K^+ is not clinically relevant except for exchange transfusions or when infants receive massive transfusions. After irradiation, shelf-life is reduced to 28 days.

Methodology Irradiation of cellular blood components prevents proliferation of donor lymphocytes. The central portion of the canister should receive 2500 cGy, while no less than 1500 cGy should be delivered to the periphery of the canister.

Additional Information TA-GVHD occurs when viable lymphocytes are transfused into severely immunosuppressed patients. The patient is unable to destroy these incoming lymphocytes, which attack the host cells, recognizing them as foreign. TA-GVHD may occur in immunocompetent patients if they receive blood from a blood relative who is homozygous for an HLA haplotype for which the patient is heterozygous. Preventive irradiation is a wise resort in the case of directed donations from blood relatives, even if the HLA types are unknown. Available filtration methods of leukocyte removal are not adequate to prevent TA-GVHD. GVHD also occurs after allogeneic bone marrow transplantation, however, it is a distinct disease entity from TA-GVHD. TA-GVHD results in pancytopenia, is resistant to therapy, and usually results in rapid death.

References
Mollison PL, Engelfriet CP, Contreras M, et al, "Some Unfavorable Effects of Transfusion," *Blood Transfusion in Clinical Medicine*, 10th ed, Chapter 15, Oxford, UK: Blackwell Scientific Publications, 1997, 487-508.

Przepiorka D, LeParc GF, Stovall MA, et al, "Use of Irradiated Blood Components. Practice Parameter," *Am J Clin Pathol*, 1996, 106(1):6-11.

Simon TL, Alverson DC, AuBuchon J, et al, "Practice Parameter for the Use of Red Blood Cell Transfusions. Developed by the Red Blood Cell Administration Practice Guideline Development Task Force of the College of American Pathologists," *Arch Pathol Lab Med*, 1998, 122(2):130-8.

♦ **Irregular Antibody Detection/Identification** *see* Antibody Detection/Identification, Red Cell *on page 165*

Islet Cell Antibody
Related Information
Glucose, Postglucose Load, Plasma *on page 647*
Glucose Tolerance Test, Plasma *on page 651*
Glutamic Acid Decarboxylase (GAD65) Antibody *on page 654*
Liver/Kidney Microsomal Type 1 Antibodies *on page 873*

Abstract Islet cell autoantibodies (ICA) were the first markers of islet cell-specific autoimmunity applied to diabetes research. They are detectable in about 80% of patients with new onset type 1 diabetes mellitus. When they occur in high titer in unaffected individuals, they indicate a 40% to 50% risk of developing type 1 diabetes within 5 years.

Specimen Serum

Container Red top tube or SST™ tube of blood

Use Assess risk of developing type 1 diabetes mellitus

Limitations Not all patients with islet cell antibodies will develop hyperglycemia. The test is often negative at the time of diagnosis in children who develop diabetes before the age of 2 years. The test frequently becomes negative within 2-10 years after the onset of overt disease. This assay is much less sensitive than the assay for glutamic acid decarboxylase antibodies (GAD65).

Methodology Indirect immunofluorescence using fresh frozen human pancreatic tissue

Additional Information ICA can occur in type 2 diabetes. Variability of assays between laboratories has led to confusion in the medical literature about the significance of islet antibodies. Its usefulness has become debatable following the introduction of the assay for GAD65 antibodies. No GAD65 antibody-negative serum was islet cell antibody positive in a 1998 paper.

Eighty-nine percent of subjects with stiff-man syndrome were detected by ICA, compared to 98% by GAD65.

References
Decochez K, Tits J, Coolens JL, et al, "High Frequency of Persisting or Increasing Islet-Specific Autoantibody Levels After Diagnosis of Type 1 Diabetes Presenting Before 40 Years of Age," *Diabetes Care*, 2000, 23:838-44.

Turner R, Stratton I, Horton V, et al, "UKPDS 25: Autoantibodies to Islet-Cell Cytoplasm and Glutamic Acid Decarboxylase for Prediction of Insulin Requirement in Type 2 Diabetes," UK Prospective Diabetes Study Group, *Lancet*, 1997, 350(9087):1288-93.

Walikonis JE and Lennon VA, "Radioimmunoassay for Glutamic Acid Decarboxylase (GAD65) Autoantibodies as a Diagnostic Aid for Stiff-Man Syndrome and a Correlate of Susceptibility to Type 1 Diabetes Mellitus," *Mayo Clin Proc*, 1998, 73(12):1161-6.

♦ **Isoform Ratio** *see* Creatine Kinase MB and Other Isoenzymes, Serum *on page 469*

♦ **Isolation Patients, Precautions for Specimen Collection** *see* Blood and Fluid Precautions, Specimen Collection *on page 271*

♦ **Isoleucine** *see* Amino Acids, Urine *on page 145*

Isoniazid, Serum or Plasma
Related Information
Liver Disease: Laboratory Assessment, Overview *on page 869*
Synonyms INH; Isonicotinic Acid Hydrazide; Laniazid®; Nydrazid®
(Continued)

Isoniazid, Serum or Plasma *(Continued)*

Abstract Isoniazid is one of the drugs of choice in the therapy and prophylaxis of tuberculosis. Use of this drug has increased in recent years in the United States due to resurgence of tuberculosis. Contraindications include acute liver disease or previous history of hepatic damage during isoniazid therapy.

Specimen Serum or plasma

Container Serum: red top tube; plasma: green top (heparin) tube

Sampling Time Trough

Storage Instructions Refrigerate. Ship in plastic container in dry ice.

Turnaround Time 2-5 days

Reference Interval Therapeutic: 2-5 µg/mL; toxic: >20 µg/mL

Use Evaluate therapy or possible toxicity

Methodology Gas chromatography (GC), high performance liquid chromatography (HPLC)

Additional Information

- Half-life: 1-1.5 hours (fast acetylators); 2-4 hours (slow acetylators)
- Volume of distribution: 0.6 L/kg
- Protein binding: zero

Mild overdose causes nausea and vomiting. Peripheral neuropathy is seen in moderate overdose. Severe overdose causes refractory seizures, metabolic acidosis (from lactate and beta-hydroxybutyric acid), and deep coma. Severe and sometimes fatal hepatitis associated with isoniazid therapy may occur and may develop even after many months of treatment. Pyridoxine has been shown to be effective in the treatment of intoxication, especially when seizures occur.

References

Aguado JM, Pulido F, Moreno S, et al, "Isoniazid Prophylaxis for High Risk Patients With Anergy and HIV Infection," *N Engl J Med*, 1997, 337(23):1696-7.

Amsterdam D, "The Laboratory Diagnosis of Tuberculosis in a Period of Resurgence: Challenge for the Laboratory," *Clin Lab Sci*, 1996, 9(4):207-12.

Lacy CF, Armstrong LL, Goldman MP, et al, *Drug Information Handbook*, 12th ed, Hudson, OH: Lexi-Comp Inc, 2004.

Tulsky JP, Pilote L, Hahn JA, et al, "Adherence to Isoniazid Prophylaxis in the Homeless: A Randomized Controlled Trial," *Arch Intern Med*, 2000, 160(5):697-702.

♦ **Isonicotinic Acid Hydrazide** *see* Isoniazid, Serum or Plasma *on page 813*

♦ **Isonipecaine Hydrochloride** *see* Meperidine, Serum or Urine *on page 895*

♦ **Isopropanol** *see* Volatile Screen, Blood or Urine *on page 1320*

♦ **Isopropyl Alcohol Intoxication** *see* Osmolality, Calculated, Serum or Plasma *on page 976*

♦ **Isoptin®** *see* Verapamil, Serum or Plasma *on page 1306*

♦ **Isosulfan Blue** *see* Sentinel Lymph Node Biopsy *on page 1189*

♦ **Isovaleric Aciduria** *see* Cerebrospinal Fluid Glycine *on page 364*

♦ **Itano Solubility Test** *see* Sickle Cell Tests *on page 1195*

♦ **ITP** *see* Platelet Antibodies *on page 1047*

Itraconazole, Serum

Related Information

Flucytosine, Serum *on page 600*

Synonyms Sporanox®

Abstract Itraconazole is an orally administered triazole antifungal agent with a broad spectrum of activity, including most pathologic fungi. Efficacy in candidiasis, blastomycosis, *Blastomyces brasiliensis* (paracoccidioidomycosis), chromoblastomycosis, coccidioidomycosis, cryptococcosis, histoplasmosis, sporotrichosis, maduramycotic mycetomas, and many cases of *Aspergillus* infections is described. It has also been effective as adjunctive therapy in patients with corticosteroid-dependent allergic bronchopulmonary aspergillosis. Its role in clinical medicine is still evolving.

Specimen Serum

Container Red top tube

Sampling Time 4-5 hours after an oral dose. Steady-state concentrations are achieved after approximately 5-10 days.

Reference Interval Therapeutic: varies with methodology; see Additional Information. Tissue levels are 3- to 20-fold higher than plasma concentrations. Only negligible concentrations were reported in CSF and urine.

Use Serum therapeutic levels may be useful if poor absorption is suspected, or in cases of therapeutic failure or relapse.

Methodology Bioassay, high performance liquid chromatography (HPLC)

Additional Information Half-life is dose-dependent (longer half-life with higher serum concentrations) and ranges from 24-42 hours. More than 99% binds to plasma proteins. Itraconazole is metabolized mainly in the liver to ~30 metabolites. One of these, hydroxy-itraconazole, has antifungal activity. Serum levels as determined by bioassay are ~10 times (0.3-7 µg/mL at peak draw) the levels determined by HPLC, presumably because bioassay also detects active metabolites.

References
Negroni R and Arechavala AI, "Itraconazole: Pharmacokinetics and Indications," *Arch Med Res*, 1993, 24(4):387-93.

Stevens DA, Schwartz HJ, Lee JY, et al, "A Randomized Trial of Itraconazole in Allergic Broncho-pulmonary Aspergillosis," *N Engl J Med*, 2000, 342(11):756-62.

Warnock DW, "Itraconazole Pulse: An Overview of Current Use," *Hosp Med*, 1998, 59(4):309-11.

♦ **ITT** *see* Insulin Tolerance Test *on page 804*

♦ *Ixodes dammini* **Identification** *see* Arthropod Identification *on page 213*

♦ **Jamestown Canyon Virus** *see* California Encephalitis Virus Serology *on page 334*

♦ **Janimine**® *see* Imipramine, Serum or Plasma *on page 767*

Jo-1 Antibody

Related Information
Antinuclear Antibodies *on page 189*
Muscle Biopsy *on page 927*
Sjögren Antibodies *on page 1199*
Topoisomerase I Antibody *on page 1261*

Synonyms Antihistidyl Transfer tRNA Synthetase

Applies to Antisynthetases

Abstract An autoantibody identifiable in ANA-positive sera, Jo-1 is more common than the other antisynthetases.

Specimen Serum

Container Red top tube or SST™ tube

Storage Instructions Refrigerate

Reference Interval Negative

Use Jo-1 antibody is the most common marker for idiopathic inflammatory myopathies, including polymyositis and dermatomyositis. It is found especially in those patients in whom disease advances to pulmonary fibrosis.

Methodology Double immunodiffusion (DID), enzyme immunoassay (EIA)

Additional Information Patients with dermatomyositis and polymyositis have inflammatory infiltration and destruction of muscle and other organ systems. Autoantibodies to the antigen, Jo-1, have been associated with pulmonary involvement and arthropathy. The Jo-1 antigen resides on the enzyme, histidyl-tRNA synthetase, which is usually located in the cytoplasm of cells, rather than in the nucleus. Antibodies to the Jo-1 antigen are detected in ~25% of adult patients with myositis including polymyositis, dermatomyositis, and overlap syndromes. Jo-1 is detected in polymyositis more frequently than in dermatomyositis.

The enzymes CPK (CK), AST, ALT, and LDH (LD) with isoenzyme LD-5 are useful in evaluation of muscle disease. CK and LD-5 are especially helpful. Some metabolic myopathies can resemble myositis. A broad differential diagnosis exists, which includes dermatomyositis, polymyositis, overlap syndromes, inclusion body myositis, and cancer-associated myositis.

References
Homburger HA, "Advances in the Diagnosis and Laboratory Evaluation of Systemic Rheumatic Diseases Other Than Rheumatoid Arthritis," *Clinical and Laboratory Evaluation of Human Auto-immune Diseases*, Chapter 11, Nakamura RA, Keren DF, and Bylund DJ, eds, Chicago, IL: American Society for Clinical Pathology, 2002, 153-64.

Moder KG, "Use and Interpretation of Rheumatologic Tests: A Guide for Clinicians," *Mayo Clin Proc*, 1996, 71(4):391-6.

von Mühlen CA and Nakamura RM, "Guidelines for Selecting and Using Laboratory Tests for Autoantibodies to Nuclear, Nucleolar, and Other Related Cytoplasmic Antigens," *Clinical and Laboratory Evaluation of Human Autoimmune Diseases*, Chapter 13, Nakamura RA, Keren DF, and Bylund DJ, eds, Chicago, IL: American Society for Clinical Pathology, 2002, 183-98.

♦ **Joint Fluid Analysis** *see* Body Fluid Analysis, Cell Count *on page 288*

- ◆ **Joint Fluid Analysis** *see* Synovial Fluid Analysis *on page 1229*
- ◆ **Jones Stain** *see* Kidney Biopsy *on page 818*
- ◆ **Kallikrein** *see* Prekallikrein *on page 1085*
- ◆ **Kaposi Sarcoma-Associated Herpesvirus** *see* Herpesvirus 8 *on page 725*
- ◆ **Kappa Light Chains** *see* Immunoperoxidase Procedures *on page 780*
- ◆ **Kay Jay** *see* Phencyclidine, Qualitative, Urine *on page 1019*
- ◆ **Keppra®** *see* Antiepileptic Drugs Overview *on page 176*
- ◆ **Keratan Sulfate** *see* Mucopolysaccharides, Urine *on page 922*
- ◆ **Ketoamines, Plasma Protein** *see* Fructosamine, Serum *on page 617*
- ◆ **11-Ketoandrosterone** *see* 17-Ketosteroids Fractionation, Urine *on page 818*
- ◆ **11-Ketoetiocholanolone** *see* 17-Ketosteroids Fractionation, Urine *on page 818*

17-Ketogenic Steroids, Urine

Synonyms 17-KGS

Abstract 17-KGS, once a mainstay in the work-up of suspected disorders involving the hypothalamic-pituitary-adrenal axis, is now rarely, if ever, needed for this purpose. The test measures the metabolic products of cortisol plus other 21-hydroxysteroids, and this is probably why the results are nonspecific. The test is still available at some reference laboratories; they can be consulted regarding appropriate specimen collection and result interpretation.

Additional Information See Adrenal Cortex: Laboratory Assessment Overview *on page 110.*

Ketone Bodies, Blood

Related Information

Anion Gap, Serum, Plasma, or Urine *on page 160*
Bicarbonate, Blood *on page 258*
Blood Gases and pH, Arterial *on page 275*
Blood Gases and pH, Capillary *on page 277*
Blood Gases and pH, Venous *on page 279*
Glucose, Fasting, Plasma *on page 643*
Glucose, Random, Plasma *on page 649*
Ketones, Urine *on page 817*
Osmolality, Calculated, Serum or Plasma *on page 976*
Osmolality, Serum *on page 978*
pH, Blood *on page 1018*
Urea Nitrogen, Serum or Plasma *on page 1284*

Synonyms Ketones, Blood; Nitroprusside Reaction, Blood

Applies to Acetoacetate; Acetone; Beta-Hydroxybutyrate; β-Hydroxybutyrate

Abstract Carbohydrate deprivation and increased catabolism of fatty acids leads to increases in the ketone bodies (acetoacetate and acetone). Beta-hydroxybutyrate is also increased and is usually included, although it is not a ketone. Blood beta-hydroxybutyrate and acetoacetate are among tests indicated to assess an ill infant or child in whom an inborn error of metabolism is suspected.

Specimen Serum, plasma, or whole blood are acceptable, depending upon assay methodology.

Container Red top tube or green top (heparin) tube

Collection Capillary tubes should be filled as much as possible using technique to avoid air bubbles. Heelsticks should be free flowing. Avoid hemolysis.

Causes for Rejection Hemolysis

Reference Interval Negative in normal nutritional states by semiquantitative/qualitative nitroprusside screening tests (eg, Bayer Acetest® tablets and Bayer Ketostix® reagent strips).

Random quantitative beta-hydroxybutyrate levels in healthy individuals are <0.4 mmol/L.

Possible Panic Range Positivity of Acetest® in 1:32 dilution indicates severe ketosis; β-hydroxybutyrate level >5.0 mmol/L is consistent with ketoacidosis.

Use Ketones are elevated in lipolytic metabolic states such as chronic starvation and diabetes mellitus. Lactic acid, glucose, electrolytes, urea nitrogen, venous or arterial pH should also be measured in possible ketoacidosis, with alcohol level, CBC, and urinalysis if clinically indicated. Up to 33% of patients with diabetic ketoacidosis also have lactic acidosis. Serum osmolality is often needed.

Limitations False negatives or falsely weak reactions may occur. Often these are due to an equilibrium shift away from acetone and acetoacetate, toward beta-hydroxybutyrane, a ketone not measured by the nitroprusside reaction. **Thus, as the ketoacidosis is treated, a paradoxically more positive positive Acetest® is observed while there is an actual reduction of total plasma ketone body concentration.** Quantitative beta-hydroxybutyrate measurement, therefore, is preferable to the nitroprusside test. Ketostix® false positives occur with large amounts of levodopa. Drugs containing free-sulfhydryl groups can give false-positive results in the Acetest®.

Methodology The nitroprusside reaction with colorimetric endpoint provides the basis of the qualitative/semiquantitative testing that is done using Bayer Acetest® tablets or the qualitative testing done with the Bayer Ketostix® reagent strips. Gas chromatography (GC) and enzymatic methods are used. A rapid, bedside, quantitative beta-hydroxybutyrate test system (GDS STAT-Site®) is available that uses a dry-reagent, enzymatic method. Very recently, a hand-held device for home monitoring has become available that uses an electrochemical sensor for measuring β-hydroxybutyrate. A multipoint kinetic method allows determination of acetoacetate, beta-hydroxybutyrate, lactate and pyruvate in a single cuvette.

Additional Information Strongly positive serum acetone without severe acidosis, with normal anion gap, bicarbonate, and plasma glucose suggests the possibility of isopropanol (rubbing alcohol) intoxication.

References

Byrne HA, Tieszen KL, Hollis S, et al, "Evaluation of an Electrochemical Sensor for Measuring Blood Ketones," *Diabetes Care*, 2000, 23(4):500-3.

Foreback CC, "β-Hydroxybutyrate and Acetoacetate Levels," *Am J Clin Pathol*, 1997, 108(5):602-4.

Mayo Medical Laboratories, *2000 Test Catalogue*, Rochester, MN.

Nuwayhid NF, Johnson GF, and Feld RD, "Multipoint Kinetic Method for Simultaneously Measuring the Combined Concentrations of Acetoacetate-Beta-Hydroxybutyrate and Lactate-Pyruvate," *Clin Chem*, 1989, 35(7):1526-31.

Porter WH, Yao HH, and Karounos DG, "Laboratory and Clinical Evaluation of Assays for β-Hydroxybutyrate," *Am J Clin Pathol*, 1997, 107(3):353-8.

♦ **Ketones, Blood** *see* Ketone Bodies, Blood *on page 816*

Ketones, Urine

Related Information

Synonyms Acetest®; Nitroprusside Reaction for Ketones, Urine; Urine Ketones

Applies to Acetoacetic Acid, Urine; Acetone, Semiquantitative, Urine; Beta-Hydroxybutyric Acid, Urine

Abstract Ketones are intermediates in the metabolism of fats. Ketonuria may warn of impending diabetic coma, or signal the presence, in an infant, of an inborn metabolic error.

Specimen Random urine

Container Plastic urine container

Storage Instructions Test immediately.

Reference Interval Negative; ketonuria accompanies very low carbohydrate diets.

(Continued)

Ketones, Urine *(Continued)*

Use Detect ketoacidosis (alcoholism and diabetes mellitus, fasting, starvation, high protein diets, and isopropanol ingestion).

In infants and children, ketonuria can occur with febrile illnesses and toxic states accompanied by vomiting or diarrhea. Genetic disorders resulting in ketonuria include propionyl CoA carboxylase deficiency, glycogen storage disease, branched-chain ketonuria, and methylmalonic aciduria.

Limitations False-positive results are associated with ascorbic acid, levodopa metabolites, 2-mercaptoethane sulfonic acid, valproic acid, N-acetylcysteine, phenazopyridine (Pyridium®), PSP dye, phenylketones, or phthalein compounds. Beta-hydroxybutyric acid (the third of the three ketone bodies) is not detected.

Methodology Acetoacetic acid and acetone react with nitroprusside to create a color change

Additional Information In adult healthy men, a fast of 18 hours or greater produces ketonemia at a level that would result in detectable ketonuria. Aging is associated with increased susceptibility to fasting-induced hyperketonemia. Ketonuria may be noted in normal pregnancy.

References
Lindor NM and Karnes PS, "Initial Assessment of Infants and Children With Suspected Inborn Errors of Metabolism," *Mayo Clin Proc*, 1995, 70:987-8.

♦ **17-Ketosteroids** *see* Dehydroepiandrosterone and Dehydroepiandrosterone Sulfate, Serum or Plasma *on page 506*

17-Ketosteroids Fractionation, Urine

Related Information
Adrenal Cortex: Laboratory Assessment Overview *on page 110*

Synonyms 17-KS Fractionation

Applies to Androsterone; Dehydroepiandrosterone; Etiocholanolone; 11-Hydroxyandrosterone; 11-Hydroxyetiocholanolone; 11-Ketoandrosterone; 11-Ketoetiocholanolone

Test Includes Quantitation of some or all of the following: androsterone, etiocholanolone, and dehydroepiandrosterone (DHEA); these are the three major metabolites of androgens in the urine. Such fractionation may also include 11-ketoandrosterone, 11-ketoetiocholanolone, 11-hydroxyandrosterone, 11-hydroxyetiocholanolone, pregnanediol, pregnanetriol, delta-5-pregnanetriol, and 11-ketopregnanetriol.

Abstract This test is rarely, if ever, needed in the evaluation of endocrine abnormalities. Assays can be obtained from some reference laboratories, which can be consulted regarding appropriate specimen collection and result interpretation.

17-Ketosteroids, Total, Urine

Related Information
Adrenal Cortex: Laboratory Assessment Overview *on page 110*
Dehydroepiandrosterone and Dehydroepiandrosterone Sulfate, Serum or Plasma *on page 506*
Testosterone, Total and Free, Serum or Plasma *on page 1238*

Synonyms 17-KS

Abstract Once widely used to investigate endocrine disorders, the test is now rarely, if ever, needed. The assay is available from some reference laboratories, which can be consulted for information relevant to appropriate specimen collection and result interpretation.

♦ **17-KGS** *see* 17-Ketogenic Steroids, Urine *on page 816*

♦ **Ki-67** *see* Immunoperoxidase Procedures *on page 780*

♦ **Kiditard®** *see* Quinidine, Serum *on page 1129*

Kidney Biopsy

Related Information
Antineutrophil Cytoplasmic Antibody *on page 187*
Antinuclear Antibodies *on page 189*
Creatinine Clearance and Urine Creatinine *on page 473*
Electron Microscopy *on page 533*

Fat, Urine *on page 573*
Glomerular Basement Membrane Antibody *on page 638*
Hemoglobin, Qualitative, Urine *on page 688*
Immunoperoxidase Procedures *on page 780*
Protein, Quantitative, Urine *on page 1108*
Protein, Semiquantitative, Urine *on page 1113*
Topoisomerase I Antibody *on page 1261*
Urea Nitrogen, Serum or Plasma *on page 1284*
Urinalysis *on page 1289*

Synonyms Renal Biopsy

Applies to Fluorescein-Tagged Antibodies; Jones Stain; Michael's Solution; PAS Stain; Zeus Fixative

Test Includes Light microscopy: H&E, PAS, methenamine silver, trichrome, congo red, and other stains; immunofluorescent studies; electron microscopy

Abstract Renal biopsy provides evaluation of type and extent of renal disease.

Patient Preparation CBC, prothrombin time, activated thromboplastin time, and urine Gram stain are prerequisite, with appropriate imaging and sometimes with a template bleeding time.

Specimen Fresh kidney tissue obtained by percutaneous needle biopsy or open surgery.

Specimen handling: The core(s) of renal tissue or a wedge obtained by open biopsy is(are) immediately placed in a Petri dish containing physiologic saline solution or sterile culture media to prevent drying. An alternative is wrapping the specimen(s) in saline-moistened gauze. The specimen(s) should be sent to the laboratory within 5-10 minutes. If the specimen(s) cannot be sent to the laboratory within this time frame, it(they) should be divided into three parts and prepared in the appropriate fixative for light microscopy, electron microscopy, and immunofluorescence. Avoid compression of the specimen.

Specimen preparation: Several means of collection may be used. For immunofluorescence studies, one core or portion of a wedge (open biopsy) is placed in a foil or plastic bag, snap-frozen in liquid nitrogen or in a cryostat, shipped on dry ice to the laboratory, and stored at -76°C until processed. The frozen state must be maintained. An alternative method is to immerse the biopsy in a half-saturated ammonium sulfate buffer at room temperature. Michael's solution is used to transport the specimen to another institution. Zeus fixative is sometimes used. The tissue should not be held in this fixative for more than 5 days, preferably less. For **light microscopy**, the second core or fragment is commonly fixed in 4% formaldehyde that is 10 times the volume of the tissue, but a variety of fixatives are in use. For **electron microscopy**, the third core or fragment is fixed in 2.5% glutaraldehyde fixative.

Use There are no absolute indications for renal biopsy. Clinical judgment is required to determine necessity of biopsy. A single disease can lead to different patterns of abnormality, and several disease entities can cause similar clinical presentations. In general, renal biopsy is useful to establish diagnosis in subjects with renal dysfunction, ascertain prognosis, evaluate disease severity and extent, and guide therapy in conditions which include the following:

- acute renal failure in cases in which clinical diagnosis cannot be established and/or which are unresponsive
- asymptomatic non-nephrotic progressive proteinuria
- nephrotic syndrome in adults, especially including SLE; selected diabetic patients
- nephrotic syndrome in children older than 6 years of age, and children of any age who do not respond to therapy as anticipated
- acute nephritic syndrome; characteristics of acute nephritic syndrome include acute onset of hematuria with red cell casts, hypertension, edema, and proteinuria often with deteriorating renal function. Causes include post-infectious glomerulonephritis, antiglomerular basement membrane disease, membranoproliferative glomerulonephritis, IgA nephropathy, hemolytic uremic syndrome, Henoch-Schönlein purpura, SLE, and vasculitis.
- hematuria of uncertain etiology, selected cases
- systemic diseases with renal involvement
- drug toxicity
- candidacy for renal transplantation in patients in chronic renal failure

(Continued)

Kidney Biopsy *(Continued)*

- evaluation of dysfunction in recipients of renal allografts; transplantation reactions, rejection, or failure

Contraindications
- Bleeding diathesis
- Neoplasm
- Cystic disease, large cysts
- Obstructive uropathy
- Acute pyelonephritis
- Abscess
- Uncontrolled hypertension
- Anatomic abnormalities
- Pregnancy
- Chronic renal disease with very small kidneys
- Renal arterial aneurysm

This list of contraindications is more relative than absolute. In several of these situations, the patient may be considered for biopsy after receiving appropriate therapy. In some cases, an open biopsy may be performed.

Methodology Light microscopy, immunofluorescence, and electron microscopy are complementary and necessary in most cases. EM is essential in about 25% of cases and helpful in half. In patients with suspected kidney transplant rejection, light microscopy is usually sufficient.

Additional Information Complications of renal biopsy:
- Hematuria: Microscopic hematuria is a common complication seen in most patients. It resolves spontaneously. Gross hematuria is seen in 5% to 9% of the cases and is more common in patients with uncontrolled hypertension or uremia. It usually resolves spontaneously in 2-3 days. In 0.5% of the patients, hematuria will persist for 2-3 weeks, occasionally occurring a few days after the biopsy. Blood transfusions are only necessary in about 1% to 3% of the cases, and medical or surgical nephrectomy for massive or persistent bleeding is necessary in only 1 of 2000-5000 cases.
- Perinephric hematoma is not uncommon, however, only 1% to 2% of patients develop a local mass, hypotension, or diminution in hematocrit. The hematoma usually resolves within a few months.
- Arteriovenous fistula is considered frequent in arteriographic studies. Most cases are clinically silent and resolve spontaneously within 2 years.
- Flank pain may last 2-3 days and responds to pain medications.
- A large blood clot may form in the bladder, causing obstruction and requiring an indwelling catheter for a few days.
- Other complications: Postbiopsy aneurysm appears in <1% of patients. Other rare complications that have been described include infection; ileus; lacerations of the liver, spleen, pancreas, gallbladder, intestine, visceral and subcostal arteries; pancreatitis; pneumothorax; and dissemination of carcinoma. Death has occurred in 0.12% of patients.

References
Feneberg R, Schaeffer F, Zieger B, et al, "Percutaneous Renal Biopsy in Children: A 27-Year Experience," *Nephron*, 1998, 79(4):438-46.

Jennette JC, Olson JL, Schwartz MM, et al, *Heptinstall's Pathology of the Kidney*, Baltimore, MD: Lippincott Williams & Wilkins, 1998.

Kern W, Silva F, Laszik Z, et al, *Atlas of Renal Pathology*, Philadelphia, PA: WB Saunders Company, 1999.

Khajehdehi P, Junaid SM, Salinas-Madrigal L, et al, "Percutaneous Renal Biopsy in the 1990s: Safety, Value, and Implications for Early Hospital Discharge," *Am J Kidney Dis*, 1999, 34(1):92-7.

Striker G, Striker LJ, and D'Agati V, "Major Problems in Pathology," *The Renal Biopsy*, Philadelphia, PA: WB Saunders Company, 1997.

Internet Web Sites
www.ajkd.org
www.kidneyatlas.org

Kidney Stone Analysis

Related Information

Synonyms Calculus Analysis; Nephrolithiasis Analysis

Applies to Sulfate, Urine

Test Includes Analysis for calcium, carbonate, citrate, cystine, magnesium, oxalate, phosphates, and urates

Abstract As many as 75% of renal stones are calcium oxalate, of which about 30% include calcium phosphate (apatite); 3% to 5% are uric acid stones; ≤2% are cystine. Up to 20% are magnesium ammonium phosphate (struvite) (infection stones). Uric acid can lead to formation of calcium oxalate stones without incorporation into the crystals. Five percent to 20% of Americans will form kidney stones. Risk factors for stone formation include hypercalciuria (including primary hyperparathyroidism). diets rich in protein and salt, low urine volume, positive family history, osteoporosis, pathologic fracture, urinary tract infection, gout, magnesium deficiency, and Crohn disease with prior small bowel resection. Rarely, genetic disorders (eg, cystinuria, xanthinuria, or oxaluria) can cause nephrolithiasis.

Since the majority of individuals with nephrolithiasis suffer recurrence (95% at 25 years), prevention holds high priority, and stone evaluation is needed for management.

Specimen Kidney/ureteral stones

Collection Specimen should be washed free of tissue and blood, and submitted in a clean, dry container. If necessary, urine should be filtered to recover gravel or stone.

Storage Instructions Do **not** apply any tape to stones. Adhesives interfere with infrared spectroscopy.

Special Instructions Urinalysis can provide useful information including detection of crystalluria and presence of red blood cells. Urine culture is often indicated.

Use Evaluation of stone composition is indicated to decrease morbidity, prevent development of new stones, and support recognition of underlying abnormalities.

Methodology Crystallographic analysis, chemical analysis, infrared spectroscopy, polarization microscopy, x-ray diffraction analysis

Additional Information Cystinuria and xanthinuria are rare causes of renal calculi. See Cystine, Urine *on page 494*. Other uncommon causes of renal stones include sarcoidosis, Cushing syndrome, excessive calcium or vitamin D ingestion, steroids, immobilization, bone disease, including Paget disease of bone, hyperthyroidism, and glycogen storage disease. An increasing but still low rate of triamterene stones has been noted. Most calcium stones relate to idiopathic hypercalciuria and hyperuricosuria, which may coexist. Hyperoxaluria is a factor requiring evaluation in patients with oxalate nephrolithiasis.
(Continued)

Kidney Stone Analysis *(Continued)*

Higher intake of calcium is associated with a reduced risk of oxalate nephrolithiasis. Such inverse relationship between dietary calcium and kidney stone is probably caused by increased binding of oxalate by ingested calcium.

Control of urine pH is an important facet of management of nephrolithiasis (eg, acid urine supports development of uric acid stones, while calcium carbonate stones develop in alkaline urine).

Other recommended studies include serum calcium and phosphorus, alkaline phosphatase, uric acid, albumin, magnesium, BUN, creatinine, sodium, potassium, chloride, CO_2, and sometimes parathormone concentrations, vitamin D_3 levels and CBC. Urine studies which may be indicated include culture, urine nitroprusside (for cystinuria), volume, fasting AM pH, urine electrolytes, creatinine clearance, uric acid, oxalate, calcium, ammonium, magnesium, and phosphate.

See Citrate, Serum, Plasma, or Urine *on page 413.*

References

Baggio B, Plebani M, and Gambaro G, "Pathogenesis of Idiopathic Calcium Nephrolithiasis: Update 1997," *Crit Rev Clin Lab Sci*, 1998, 35(2):153-87.

Bihl G and Meyers A, "Recurrent Renal Stone Disease - Advances in Pathogenesis and Clinical Management," *Lancet*, 2001, 358:651-6.

Bushinsky DA, "Nephrolithiasis," *J Am Soc Nephrol*, 1998, 9(5):917-24.

Heller HJ, "The Role of Calcium in the Prevention of Kidney Stones," *J Am Coll Nutr*, 1999, 18(5 Suppl):373S-378S.

+ **Killer Weed** *see* Phencyclidine, Qualitative, Urine *on page 1019*
+ **Kimura Disease** *see* Immunoglobulin E *on page 773*
+ **Kinidin®** *see* Quinidine, Serum *on page 1129*
+ **Kinyoun Stain** *see* Acid-Fast Stain, Routine or Modified *on page 95*

Kleihauer-Betke

Related Information

Alpha₁-Fetoprotein, Amniotic Fluid *on page 135*
Fetal Cell Detection by Flow Cytometry *on page 579*
Fetal Hemoglobin *on page 581*
Prenatal Screen, Immunohematology *on page 1086*
Rhₒ(D) Immune Globulin (Human) *on page 1164*
Rosette Test for Fetomaternal Hemorrhage *on page 1172*

Synonyms Acid Elution for Fetal Hemoglobin

Abstract This test identifies fetal red blood cells in the maternal blood, thus allowing the diagnosis and quantitation of fetomaternal hemorrhage (FMH). Quantitation of FMH is required to determine the proper dose of Rh immune globulin.

Specimen Whole blood

Container Lavender top (EDTA) tube

Storage Instructions Blood must be less than 6 hours old. Smears must be fixed within 1 hour after preparation.

Causes for Rejection Clotted specimen, gross hemolysis

Special Instructions A cord blood specimen should also be obtained and submitted for use as a positive control (contains fetal cells).

Reference Interval Normal adults: Hb F cells are <0.01% (blood of newborns is largely Hb F cells)

Use Diagnose fetomaternal hemorrhage (FMH); assess the magnitude of fetal-maternal hemorrhage; calculate dosage of Rh immune globulin; aid in diagnosis of certain types of anemia in adults. A study by Emery concludes on the basis of 523 tests that the Kleihauer-Betke (KB) should be performed on all screening test (rosetting of D-positive fetal cells was utilized) positive Rh-negative mothers of Rh-positive infants and in cases of maternal trauma, unexplained increased maternal alpha-fetoprotein levels, fetal distress with abnormal cardiac tracings, intrauterine fetal death, and in cases of unexplained neonatal anemia. Testing should also be considered in cases of cordocentesis (in which FMH may occur).

Limitations Possibility of the presence of a hemoglobinopathy with increase in Hb F must be considered when this test is used to assess FMH. Specimens must be obtained prior to transfusion. The KB assay lacks precision, coefficient of variation (CV) has been found to range from 40% to 60% on samples with fetal cells at the 0.2% to 1% level.

Contraindications Known pre-existing elevation of maternal Hb F (eg, mothers with hereditary persistence of fetal hemoglobin)

Methodology Acid elution. After fixation with alcohol Hb F remains as a precipitate within the cell while Hb A is soluble in citric acid phosphate buffer. The adult RBCs containing little or no Hb F appear as ghosts under the microscope. Automated (flow cytometric) assays are becoming commercially available. See Fetal Cell Detection by Flow Cytometry *on page 579.*

Additional Information The KB test is helpful in distinguishing some forms of thalassemia from hereditary persistence of fetal hemoglobin (HPFH). The hereditary persistence of fetal hemoglobin reveals a uniform distribution of fetal hemoglobin in each red cell. deltaβ-thalassemia, in contrast, demonstrates a heterogeneous distribution of fetal hemoglobin (ie, some cells are stained and others are ghost RBCs).

Some Rh immunoglobulin failures are due to a failure to suspect and diagnose FMH that may require more than one dose of Rh immunoglobulin. The ultimate purpose is to prevent generation of anti-D antibodies in the postpartum woman and subsequent evolution of hemolytic disease of the newborn (erythroblastosis fetalis). The amount of fetal blood contamination can be calculated. Each vial of Rh immunoglobulin contains 300 μg of anti-D, enough to prevent maternal immunization when the fetal bleed is as large as 30 mL of whole blood (15 mL packed cells). One vial of Rh immunoglobulin is given to the Rh-negative mother for every 30 mL of FMH from an Rh-positive fetus.

In cases of maternal hereditary persistence of fetal hemoglobin, a flow cytometric-based method for detection and quantification of fetal Rh(D) positive cells can be used. An indirect immunofluorescent reaction is involved with IgG anti-D as the primary antibody.

FMH of more than 30 mL (of fetal blood lost into the maternal circulation) occurs in 2.5% of patients following Cesarean section (on the basis of KB testing). Rh-negative women, after having a C-section, should be tested for fetomaternal hemorrhage. Testing should also be considered when cordocentesis may have led to alloimmunization to D antigen.

References

Duckett JR and Constantine G, "The Kleihauer Technique: An Accurate Method of Quantifying Fetomaternal Haemorrhage?" *Br J Obstet Gynaecol,* 1997, 104(7):845-6.

Emery CL, Morway CF, Chung-Park M, et al, "The Kleihauer-Betke Test. Clinical Utility, Indication, and Correlation in Patients With Placental Abruption and Cocaine Use," *Arch Pathol Lab Med,* 1995, 119(11):1032-7.

Nelson M, Zarkos K, Popp H, et al, "A Flow-Cytometric Equivalent of the Kleihauer Test," *Vox Sang,* 1998, 75(3):234-41.

Samadi R, Greenspoon JS, Gviazda I, et al, "Massive Fetomaternal Hemorrhage and Fetal Death: Are They Predictable?" *J Perinatol,* 1999, 19(3):227-9.

♦ **Klonopin**™ *see* Clonazepam, Serum *on page 415*

KOH Preparation

Related Information

Fungal Culture, Skin *on page 623*
Fungal Culture, Sputum *on page 624*
Fungus Smear, Stain *on page 626*
Gram Stain *on page 658*
Skin Biopsy *on page 1200*

Test Includes Detection of fungal structures after potassium hydroxide (KOH) hydrolysis of proteinaceous debris and cells; microscopic examination under 10x and 40x objectives

Specimen The appropriate specimen for KOH preparation is the same as for fungal culture. See specific site fungus culture listing for details.

Special Instructions The laboratory should be informed of the specific source of the specimen and the clinical diagnosis. Aseptic technique should be used to collect the specimen. Use sterile scissors, forceps, and nail clippers and disinfect the specimen site with 70% alcohol before collection.

Reference Interval No fungus elements identified

Use Determine the presence of fungi in skin, nails, or hair. Exudates from abscesses, sinus tracts, aspirates, etc, can be examined by KOH preparation and also smeared for Gram stain.

(Continued)

KOH Preparation *(Continued)*

Limitations Cultures are usually more sensitive than KOH preparations. The test may require overnight incubation for complete disintegration of hair, nail, or skin debris.

Methodology 10% KOH with gentle heat, alternately 10% to 20% KOH and 40% dimethyl sulfoxide (DMSO)

Additional Information Fungal infections of the keratinized tissues such as skin, nails, and hair are called dermatophytoses. Different fungus species may cause similar diseases.

Diagnostic specimens should be collected before antifungal therapy is instituted. Topical steroids should not be prescribed until fungal infection is excluded.

References

Chapman SW and Daniel CR 3rd, "Cutaneous Manifestations of Fungal Infections," 1994, 8(4):879-910.

Kemna ME and Elewski BE, "A U.S. Epidemiologic Survey of Superficial Fungal Diseases," *J Am Acad Dermatol*, 1996, 35(4):539-42.

Schwinn A, Ebert J, and Brocker EB, "Frequency of *Trichophyton rubrum* in Tinea Capitis," *Mycoses*, 1995, 38(1-2):1-7.

Lactate Dehydrogenase Isoenzymes, Serum

Related Information

Synonyms LDH Isoenzymes; LD Isoenzymes

Test Includes Total serum LD (LDH) and electrophoretic quantitation of isoenzymes

Abstract LD is found in all body cells and exists in five molecular forms (isoenzymes). Changes of LD isoenzymes have historically been serially measured following onset of chest pain, to study the relationships of the anodic fractions and to provide information for the differential diagnosis of acute infarct of myocardium (AMI) ("LD_1/LD_2 flip" indicating AMI, peaking after CK and CK-MB). **The troponins are elevated for as long as 4-6 days after myocardial infarct and have essentially eliminated LD isoenzymes for detection of myocardial infarct.** The differential diagnosis of certain other diseases is enhanced with use of LD isoenzymes. LD_5 provides a degree of specificity when liver problems are investigated, and LD_1 is useful in work-up of hemolytic and megaloblastic anemias.

Specimen Serum

Container Red top tube

Collection Avoid hemolysis

Reference Interval LD_1 and LD_2 (anodal fractions) are associated with cardiac and RBC origin. LD_5 and LD_4 are associated with hepatic and skeletal muscle origin. LD_2 is normally greater than LD_1. Thus, the $LD_1:LD_2$ ratio is normally 0.50-0.80. In myocardial damage, such as AMI, and in hemolytic and megaloblastic anemias, there is flip or inversion of $LD_1:LD_2$ (LD_1 becoming greater than LD_2).

Use Useful in the differential diagnosis of megaloblastic anemia (folate deficiency, pernicious anemia), hemolytic anemia, and very occasionally renal infarct. These entities are characterized by LD_1 increases, often with $LD_1:LD_2$ inversion.

The **isomorphic pattern** (total LD significantly high with no significant increase in percentage, of any fraction) is seen with neoplasia, cardiorespiratory diseases, hypothyroidism, infectious mononucleosis, and other inflammatory states, uremia, and necrosis.

LD_5 increases are seen with striated muscle lesions (eg, trauma) and with liver diseases (eg, hepatic congestion, congestive heart failure, hepatitis, cirrhosis, alcoholism). LD_5 increase is probably more significant when the $LD_5:LD_4$ ratio is increased. LD_5 is considerable more specific for liver disease or injury than are the aminotransferases (AST, ALT), especially when CK is normal. See Liver Disease: Laboratory Assessment, Overview *on page 869.*

Additional Information Patterns of LD isoenzymes in acute pulmonary edema include the isomorphic pattern and LD_5 increases.

There is evidence that when LD_5 sufficiently exceeds LD_4, liver disease may exist. Such liver disease might be primary or secondary (eg, congestive heart failure). Additional tests which may be useful, if clinically indicated, to work up such possible liver disease or injury might include ALT (SGPT), GGT, serum protein electrophoresis, and prothrombin time. LD_5 is the striated muscle fraction as well as the liver fraction. Although striated muscle problems are usually clinically obvious, occasionally the physician does not get a clinical history of the postictal state or of various withdrawal syndromes. In such situations, serum CK may be helpful.

The association between LD_1 and testicular seminoma has been widely recognized. Its relationship to nonseminomatous testicular tumors as well are described. The ovarian equivalent of seminoma is dysgerminoma, which also may relate to LD_1 increases. A variety of malignant tumors are characterized by total LD increases, sometimes with isomorphic patterns or with LD_5 increases.

LD with LD isoenzymes is useful as a tumor marker. Applications in adenocarcinoma of lung, colorectal carcinoma, malignant germ cell tumors, and in lymph nodes are recognized. LD_3 may be useful in chronic granulocytic leukemia.

References

Jacobs DS, Robinson RA, Clark GM, et al, "Clinical Significance of the Isomorphic Pattern of the Isoenzymes of Serum Lactate Dehydrogenase," *Ann Clin Lab Sci*, 1977, 7(5):411-21.

Young DS, *Effects of Drugs on Clinical Laboratory Tests*, 5th ed, Volume 1: Listing by Test, Washington, DC: AACC Press, American Association for Clinical Chemistry, 2000, Section 3, 491.

Zimmerman HJ, *Hepatotoxicity: The Adverse Effects of Drugs and Other Chemicals on the Liver*, 2nd ed, Baltimore, MD: Lippincott Williams & Wilkins, 1999, 208-13.

Lactate Dehydrogenase, Serum

Related Information

(Continued)

Lactate Dehydrogenase, Serum *(Continued)*

Synonyms LD; LDH

Applies to LDH:AST Ratio

Abstract Released with cell damage (eg, hypoxia, necrosis), LD is increased in a wide variety of neoplastic states and in other disease entities. Some correlation with tumor bulk exists. High serum LDH before treatment is an adverse risk factor for non-Hodgkin lymphoma.

Specimen Serum, body fluid

Container Red top tube, serum separator tube, or green top (heparin) tube

Storage Instructions Stable 2-3 days at room temperature.

Causes for Rejection Hemolysis in collection of sample

Reference Interval Normal intervals for serum LD vary among methods. They are higher in childhood. For adults, in most laboratories, the range is up to ~200 units/L for lactate to pyruvate assays which are by far the most common.

Lactate Dehydrogenase

Age (y)	Units/L
0-2	125-275
2-3	166-232
3-4	112-221
4-5	108-206
5-6	104-205
6-7	100-204
7-8	95-203
8-12	90-201
12-14	90-199
14-16	Up to 168
16-17	Up to 161
17-43	90-156
≥43	90-176

Use Causes of **high LD: Neoplastic states** (especially with high alkaline phosphatase, very high total LD, and isomorphic pattern of LD isoenzymes) or with LD_5 increase. High LDH may be found with lymphomas, leukemias, tumors of testis, neuroblastoma, and with a variety of carcinomas, including primaries of lung, breast, pancreas, and gastrointestinal tract. High LDH may be found with **hypoxia; cardiorespiratory diseases,** including cardiac failure and myocarditis (LDH is especially elevated in patients with HIV and *Pneumocystis* pneumonia); **hemolytic anemia,** including that secondary to prosthetic cardiac valves; **megaloblastic anemias,** including pernicious anemia (levels may be >2000 units/L and LD isoenzymes reveal LD_1:LD_2 flip); **infectious mononucleosis; inflammation; hypothyroidism** (some cases); **myocardial infarct:** LD begins to rise about 12 hours after infarct and remains elevated for up to 1-2 weeks after CK and AST have returned to normal; isoenzymes usually most useful 48 hours from onset of infarct to reveal LD_1:LD_2 ≥1.0 ("flip"); **pulmonary infarct** (LD increase is typically present, sometimes with the triad of LD, bilirubin, AST increases); other **lung diseases.**

Diseases of **liver,** including cirrhosis. Total LD in cirrhosis is usually not greatly increased. In acute viral hepatitis, LD is not notably elevated and AST is usually three or more times higher (in relation to the upper limit of normal) than LD; **chronic alcoholism** is usually associated with some combination of elevated MCV (mean corpuscular volume), triglyceride, alkaline phosphatase, AST (SGOT), ALT (SGPT), GGT, and bilirubin with low folate.

Renal infarct - high LD, out of proportion to AST and alkaline phosphatase; **seizures, other CNS diseases;** acute **pancreatitis; connective tissue (collagen) diseases;** excessive **destruction of cells; fracture,** other **trauma,** including head trauma, **muscle damage; muscular dystrophy; focal necrosis; shock, hypotension; intestinal obstruction.**

LD isoenzymes may be useful in the diagnosis of a number of the disease states mentioned above including neoplastic states, hemolytic anemia, megaloblastic anemias including pernicious anemia, infectious mononucleosis, some cases of hypothyroidism, diseases of the liver, renal infarct, myocardial infarction, and excessive destruction of cells, especially in hematopoietic neoplasms. See Lactate Dehydrogenase Isoenzymes, Serum *on page 824*.

Other causes of increased LD include specimen tube artifact, such as serum contact with clot or exposure to heat. Test profile with very high LD and no glucose may relate to unseparated serum and cells in a tube at room temperature or higher.

Useful with protein and cytologic examination in initial assessment of pleural effusion.

Limitations Artifactual as well as clinical hemolysis elevates LD results. Oxalate inhibits LD. Physiological serum decrease may be seen with anticonvulsants. Analytical serum increases include method-dependent observations relevant to acetaminophen, fluosol-DA, and phenobarbital. Physiological serum increases may be seen with amiodarone, anabolic steroids, dicumarol, gentamicin, isotretinoin, methotrexate, nitrofurantoin, penicillamine, sulfisoxazole, valproic acid, and other drugs.

Additional Information In **infectious mononucleosis**, LD is usually more elevated than AST, and there is commonly an isomorphic pattern of LD isoenzymes. In **viral hepatitis**, by contrast, AST and ALT (the aminotransferases or transaminases) are much more increased than is LD, about three or more times higher than total LD, and LD_5 is high. The differential diagnosis of acute infarct of myocardium includes pericarditis and angina, entities in which enzymes are usually not substantially increased. LD is useful in selected settings as a tumor marker, but LD is not helpful as a screening test for cancer. Tumor burden in Hodgkin disease and non-Hodgkin lymphoma may be estimated by serum LDH concentration and disease stage. High pretreatment concentrations of serum LDH are an important adverse risk factor in subjects with non-Hodgkin lymphoma and with small cell carcinoma of lung. Increases may be found with dysgerminoma of ovary, seminoma of testis and other germ cell tumors, neuroblastoma, and a wide variety of other neoplastic states. High serum LDH has been described as a marker for drug resistance with high tumor volume in multiple myeloma. (Other applications as a tumor marker are included in the listing Lactate Dehydrogenase Isoenzymes, Serum *on page 824*.)

LDH:AST ratio >18, in patients with biliary pancreatitis, has been proposed as an indicator of pancreatic necrosis.

References

Isogai M, Yamaguchi A, Hori A, et al, "LDH:AST Ratio in Biliary Pancreatitis - a Possible Indicator of Pancreatic Necrosis: Preliminary Results," *Am J Gastroenterol*, 1998, 93(3):363-7.

Sandlund JT, Downing JR, and Crist WM, "Non-Hodgkin Lymphoma in Childhood," *N Engl J Med*, 1996, 334(19):1238-48.

Stokkel MP, van Eck-Smit BL, Zwinderman AH, et al, "Pretreatment Serum LDH as Additional Staging Parameter in Small-Cell Lung Carcinoma," *Neth J Med*, 1998, 52(2):65-70.

♦ **Lactic Acid, Body Fluid** *see* Body Fluid Chemical Analysis *on page 291*

♦ **Lactic Acid, Cerebrospinal Fluid** *see* Cerebrospinal Fluid Lactate *on page 369*

Lactic Acid, Whole Blood or Plasma

Related Information

Ammonia, Plasma *on page 150*
Anion Gap, Serum, Plasma, or Urine *on page 160*
Cyanide, Blood *on page 485*
Ethanol, Blood, Urine, and Other Sources *on page 558*
Ibuprofen, Serum *on page 764*
pH, Blood *on page 1018*
Salicylate, Serum or Plasma *on page 1176*

Synonyms Blood Lactate; Lactate, Blood

Applies to Biotin; D-Lactate; Metformin; Oxygen Transport; Phenformin

Abstract Lactate (L-lactic acid) is formed from pyruvate in glycolysis. Strenuous exercise can produce a 10- or 15-fold increase in venous plasma lactate within several seconds. Blood lactate is lowest during fasting and reaches the upper (Continued)

Lactic Acid, Whole Blood or Plasma (Continued)

end of the reference interval in the postprandial state. The D-isomer of lactic acid does not occur in human metabolism, but may be present in rare instances (see Additional Information). **Lactic acidosis**, with elevated blood lactate, occurs in two clinical contexts. **Type A** lactic acidosis is due to hypoxia and is the more common form; **type B** lactic acidosis may be due to drugs, inborn errors of metabolism, severe liver disease, or a metabolic myopathy.

Patient Preparation Mannitol interference is described for a whole-blood analysis method.

Specimen Whole blood, arterial or venous, or plasma, *vide infra*. Arterial blood is preferred, since contraction of muscles can cause increase in lactate in venous blood.

Container Gray top (sodium fluoride) tube; heparinized syringe, heparin-containing tube, anaerobic draw, depending upon available instrumentation

Collection Avoid hand-clenching, and if possible use of a tourniquet. A tourniquet or a patient clenching and unclenching his/her hand will lead to build-up of potassium and lactate from the hand muscles.

Lactate is commonly needed with or as stat follow-up to venous or arterial pH. Serial determinations are often valuable. **Send specimen on wet ice.**

Storage Instructions Centrifuge immediately and take off plasma (unless laboratory uses a whole blood method). Keep plasma on ice or at 2°C to 8°C, analyze promptly.

Causes for Rejection Specimen not received on ice

Turnaround Time Use of whole blood improves turnaround time.

Special Instructions Keep tube on ice until delivered. Tube must be processed within 15 minutes of being drawn.

Reference Interval See table.

Whole Blood Lactate		
	mmol/L[1]	mg/dL[1]
Venous		
at rest	0.5-1.3	5-11
in hospital	0.9-1.7	8-15
Arterial		
at rest	0.36-0.75	3-7
in hospital	0.36-1.25	3-11
Plasma Lactate		
	mmol/L[2]	mg/dL[2]
Venous	0.5-2.2	4.5-19.8
Arterial	0.5-1.6	4.5-14.4

[1]Sacks DB, "Carbohydrates," *Tietz Textbook of Clinical Chemistry*, 3rd ed, Burtis CA and Ashwood ER, eds, Philadelphia, PA: WB Saunders Co, 1999, 789.

[2]Painter PC, Cope JY, and Smith JL, "Reference Information for the Clinical Laboratory," *Tietz Textbook of Clinical Chemistry*, 3rd ed, Burtis CA and Ashwood ER, eds, Philadelphia, PA: WB Saunders Co, 1999, 1822.

Critical Values There is an inverse relationship between hyperlactatemia and survival. Lactate >36 mg/dL (4 mmol/L) is a strong predictor of need for hospital admission from Emergency Department (ED) as well as predictor of mortality.

Possible Panic Range ≥45.0 mg/dL

Use The differential diagnosis of type A lactic acidosis includes hypoxemia (eg, carbon monoxide, anemia, methemoglobinemia, respiratory failure), hypotension, shock, decreased perfusion, and strenuous exercise. Type B may be caused by ethanol, methanol, ethylene glycol, phenformin, cyanide, nitroprusside, salicylate, nalidixic acid, streptozocin, diabetes, liver failure, renal failure, infection, systemic malignancy, and inborn errors of metabolism.

Lactate determination is generally indicated if anion gap is >20 mmol/L and if pH is <7.25 and the pCO_2 is not elevated.

When lactate is <45 mg/dL (SI: <5.0 mmol/L), suspect carbohydrate infusions, exercise, diabetic ketoacidosis, or ethanol. When lactate is >45 mg/dL, suspect shock, severe anemia, severe congestive failure, or systemic malignancy.

Limitations Assays for L-lactate provide no information about the isomer, D-lactate (see Additional Information).

Contraindications Absence of acidosis is **not** a contraindication for this test.

Methodology Enzymatic; other methods include gas chromatography (GC), amperometric, enzymatic, substrate-specific electrode. Whole blood analysis by trilayer-biosensors is available.

Additional Information Most assays for lactic acid measure only the L-isomer of lactic acid, and all the discussion above relates to the L-isomer. The **D-isomer of lactic acid**, while not produced in human metabolism, can, on rare occasions, be absorbed from gastrointestinal microorganisms. Serious disease from D-lactic acidosis has been reported after intestinal bypass operations. D-lactic acidosis is rare, and difficult to diagnose. D-lactic acid can be measured by gas-liquid chromatography and by microbiologic assay.

Normal L-lactate occurs with high D-lactate in **D-lactic acidosis**. Metabolic acidosis following jejunoileal bypass for obesity, related to altered gastrointestinal flora, may develop in subjects who develop dysarthria, cerebellar ataxia, and confusion as well, in whom D-lactate is the causative anion.

References

Adrogué HJ and Madias NE, "Medical Progress: Management of Life-Threatening Acid-Base Disorders (First of Two Parts)," *N Engl J Med*, 1998, 338(1):26-34.

Kost GJ, Nguyen TH, and Tang Z, "Whole-Blood Glucose and Lactate. Trilayer Biosensors, Drug Interference, Metabolism, and Practice Guidelines," *Arch Pathol Lab Med*, 2000, 124(8):1128-34.

Marshall WJ, "Hydrogen Ion Homeostasis, Tissue Oxygenation and Their Disorders," *Clinical Biochemistry*, Marshall WJ and Bangert SK, eds, New York, NY: Churchill Livingstone, 1995, 61-86.

Uribarri J, Oh MS, and Carroll HJ, "D-Lactic Acidosis. A Review of Clinical Presentation, Biochemical Features, and Pathophysiologic Mechanisms," *Medicine*, 1998, 77(2):73-82.

Vella A and Farrugia G, "D-Lactic Acidosis: Pathologic Consequences of Saprophytism," *Mayo Clin Proc*, 1998, 73(5):451-6.

♦ **Lactoferrin, Fecal** *see* Fecal Lactoferrin *on page 575*

Lactose Tolerance Test

Related Information

d-Xylose Absorption Test, Serum, Urine *on page 527*
Fat, Semiquantitative, Stool, Sudan III Stain *on page 572*
Fecal Fat, Quantitative, 72-Hour Collection *on page 574*
pH, Stool *on page 1034*
Reducing Substances, Stool *on page 1147*

Synonyms Tolerance Test, Lactose

Applies to Breath Hydrogen Analysis

Abstract Lactose intolerance, a deficiency of lactase in the gut epithelium, results in an osmotic diarrhea when lactose-containing foods are ingested. The disease may be primary, due to decreased synthesis of lactase, or secondary to any disease characterized by diffuse damage to the intestinal epithelium. The diagnosis of lactose intolerance is usually made from the clinical history. Laboratory testing, which is rarely indicated, involves measuring the patient's response to an oral dose of lactose. Two protocols are available: 1) a noninvasive hydrogen breath test (preferred), and 2) measuring the glucose in sequential blood specimens.

Patient Preparation Patient should fast for 8 hours before testing, usually overnight. For adults, 50 g of lactose in 200 mL of lemon-flavored water; children, 2 g/kg up to 50 g total. Patient is encouraged to drink a moderate amount of water during the test, one to two glasses. Patient should remain seated or in bed.

Aftercare Test may produce diarrhea and cramps.

Specimen Plasma or 24-hour urine

Container Gray top (sodium fluoride) tube, plastic 24-hour urine container

Collection

Sequential blood glucose protocol: Draw specimens in gray top tubes: fasting, 15 minutes, 30 minutes, 45 minutes, 60 minutes, and 90 minutes after the lactose dose. Record patient symptoms (especially cramps, nausea, watery diarrhea).

(Continued)

Lactose Tolerance Test *(Continued)*

Hydrogen breath protocol: Breath is monitored for 5 hours.
Reference Interval

Hydrogen breath protocol: Breath samples are monitored over a 5-hour period. Normal response is a peak rise in hydrogen <10 (some references use <12) parts/million.

Sequential blood glucose protocol: Normal response is a plasma glucose increase >30 mg/dL (SI: >1.7 mmol/L). An abnormal result must be followed by a similar protocol in which the absorption of each of the constituent monosaccharides (glucose and galactose) is verified.

Use Lactose intolerance is usually diagnosed by clinical history. The hydrogen breath protocol is strongly recommended for the rare patient who requires biochemical testing. When such testing is unavailable locally, referral to another facility should be considered. The rigors of the sequential blood glucose protocol militate against the use of such testing and suggest that it should be relegated to the annals of medical history.

Methodology Hydrogen breath analysis is done by gas chromatography (GC). Plasma glucose by specific enzyme-based assay using glucose oxidase or hexokinase.

References

Arola H, "Diagnosis of Hypolactasia and Lactose Malabsorption," *Scand J Gastroenterol Suppl*, 1994, 202:26-35.

Henderson AR and Rinker AD, "Gastric, Pancreatic, and Intestinal Function," *Tietz Textbook of Clinical Chemistry*, 3rd ed, Chapter 41, Burtis CA and Ashwood ER, eds, Philadelphia, PA: WB Saunders Co, 1999, 1319.

Malagelada JR, "Lactose Intolerance," *N Engl J Med*, 1995, 333(1):53-4.

Mascolo R and Saltzman JR, "Lactose Intolerance and Irritable Bowel Syndrome," *Nutr Rev*, 1998, 56(10):306-8.

Shaw AD and Davies GJ, "Lactose Intolerance: Problems in Diagnosis and Treatment," *J Clin Gastroenterol*, 1999, 28(3):208-16.

Suarez FL, Savaiano DA, and Levitt MD, "A Comparison of Symptoms After the Consumption of Milk or Lactose-Hydrolyzed Milk by People With Self-Reported Severe Lactose Intolerance," *N Engl J Med*, 1995, 333(1):1-4.

♦ **Laki-Lorand Factor** *see* Factor XIII *on page 565*

♦ **Lambda Light Chains** *see* Immunoperoxidase Procedures *on page 780*

Lamellar Bodies, Amniotic Fluid

Related Information

Lecithin:Sphingomyelin Ratio, Amniotic Fluid *on page 836*
Phosphatidylglycerol, Amniotic Fluid *on page 1030*
Pulmonary Surfactant, Amniotic Fluid *on page 1124*

Synonyms Amniotic Fluid Lamellar Bodies

Abstract Amniotic fluid lamellar body density is a test for fetal lung maturity. The procedure is rapid, inexpensive, and uses cell counting equipment that is usually readily available.

Specimen Fresh amniotic fluid

Container Specimen from amniocentesis must be submitted in a sterile container.

Collection Specimen is obtained at amniocentesis by an obstetrician.

Storage Instructions Keep at room temperature. Specimen will usually be analyzed within 8 hours of sample collection but can be kept refrigerated at 2°C to 8°C for at least 2 weeks or frozen at -20°C for some 4-10 weeks. After 30 days, the concentric layers appear loose by electron microscopy with a trend to decrease in number of lamellar bodies.

Causes for Rejection Presence of visible blood, meconium or mucous specimen, specimen not in sterile container

Turnaround Time 1 hour (procedure itself can be performed in 15-20 minutes)

Reference Interval Lamellar body count >30,000/μL (decision threshold) is predictive of pulmonary maturity. Count <10,000/μL has a 67% positive predictive value for respiratory distress syndrome.

Use Predict fetal lung maturity and the risk of developing respiratory distress syndrome. The test is most useful between 34 and 36 weeks gestation.

Limitations Presence of mucous (as may be present in vaginal pool samples) can produce artifactual increase in lamellar body (LB) counts. Contamination with meconium or lysed blood do not appear to affect LB count. Varying decision thresholds have been recommended as the result of different studies and may relate in part to use of variable centrifugation protocols.

Methodology Lamellar bodies range in size from 1.7-7.3 fL. They are conveniently and accurately counted utilizing the platelet apertures of multichannel hematology analyzers (flow/impedance method of particle counting). Dependent upon instrument manufacturer, counting aperture size used for platelets is 50-60 μm, and upper limit of volume range used for platelet classification is 20-35 fL. Within-run coefficients of variation have ranged from 1.3% to 3.5% for different count levels.

Additional Information Lamellar bodies are concentrically layered protein, cholesterol, and phospholipid structures, 1-5 μm in diameter, largely some 2 μm across. They are produced in the lung by type II pneumocytes and are the storage form of pulmonary surfactant. As the fetal lung matures, lamellar bodies increase in number with, as expected, increased amniotic fluid phospholipids and lecithin:sphingomyelin ratio. Lamellar bodies have size and volume characteristics similar to platelets which allow the use of readily available multichannel cell counters in their enumeration. A number of studies have shown a high correlation between lamellar body count and biochemical methods (eg, phosphatidyl glycerol and lecithin:sphingomyelin ratio). When a threshold value of 30,000/μL (or greater) was used to indicate fetal lung maturity, no false-mature results were noted. The significance of numerous false-immature results could be decreased by use of a separate cutoff value (<10,000/μL) as indicative of risk for respiratory distress syndrome.

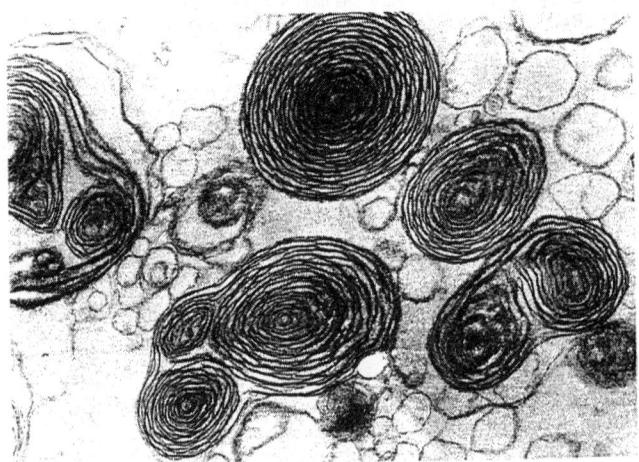

Transmission electron micrograph of lamellar bodies from fresh human amniotic fluid. Uranyl acetate, x 21,000.

From Lafler D, Mendoza A, Cousins L, et al, "Refrigerated and Frozen Amniotic Fluid for Fetal Lung Maturity Testing and Lamellar Body Density Counts," *Lab Med*, 1996, 27(11):770-4.

References

Bowie LJ, Shammo J, Dohnal JC, et al, "Lamellar Body Number Density and the Prediction of Respiratory Distress," *Am J Clin Pathol*, 1991, 95(6):781-6.

Lafler D, Mendoza A, Poeltler D, et al, "Coulter STK-S vs Abbott Cell-Dyn 3500 for Counting Lamellar Bodies in Amniotic Fluid," *Lab Med*, 1998, 29(5):298-301.

Pearlman ES, Baiocchi JM, Lease JA, et al, "Utility of a Rapid Lamellar Body Count in the Assessment of Fetal Maturity," *Am J Clin Pathol*, 1991, 95(6):778-80.

Richardson DK and Heffner LJ, "Fetal-Lung Maturity Tests: Tests Mature, Interpretation Not," *Lancet*, 2001, 358:684-6.

♦ **Lamellar Body Count** *see* Pulmonary Surfactant, Amniotic Fluid *on page 1124*

- **Lamictal**® *see* Antiepileptic Drugs Overview *on page 176*
- **Lamotrigine** *see* Antiepileptic Drugs Overview *on page 176*
- **Lamra**® *see* Diazepam, Serum *on page 510*
- **Laniazid**® *see* Isoniazid, Serum or Plasma *on page 813*
- **Lanocor**® *see* Digoxin, Serum *on page 512*
- **Lanotoxin**® *see* Digitoxin, Serum *on page 512*
- **Lanoxicaps**® *see* Digoxin, Serum *on page 512*
- **Lanoxin**® *see* Digoxin, Serum *on page 512*
- **LAP** *see* Leucine Aminopeptidase (LAP), Serum and Urine *on page 844*
- **LAP Score** *see* Leukocyte Alkaline Phosphatase *on page 845*
- **Largactil**® *see* Chlorpromazine, Serum *on page 395*
- **Large Molecule Diseases** *see* Inherited Diseases of Metabolism and Cell Structure *on page 792*
- **L-Aspartate-2-Oxoglutarate Aminotransferase** *see* Aspartate Aminotransferase, Serum *on page 216*
- **Latex Sensitization** *see* Allergen Specific IgE Antibody *on page 131*
- **Latiazem Hydrochloride** *see* Diltiazem, Serum or Plasma *on page 515*
- *Latrodectus hasselti* *see* Arthropod Identification *on page 213*
- **LATS** *see* Thyrotropin Receptor Antibody, Serum *on page 1254*
- **Lavage Cytology** *see* Washing Cytology *on page 1326*
- **LD** *see* Lactate Dehydrogenase, Serum *on page 825*
- **LD, Body Fluid** *see* Body Fluid Chemical Analysis *on page 291*
- **LD, Body Fluid** *see* Body Fluid Lactate Dehydrogenase *on page 294*
- **LDH** *see* Lactate Dehydrogenase, Serum *on page 825*
- **LDH** *see* Troponins, Serum *on page 1278*
- **LDH:AST Ratio** *see* Lactate Dehydrogenase, Serum *on page 825*
- **LDH Isoenzymes** *see* Lactate Dehydrogenase Isoenzymes, Serum *on page 824*
- **LDH Isoenzymes** *see* Troponins, Serum *on page 1278*
- **LD Isoenzymes** *see* Cardiac Markers: Laboratory Assessment, Overview *on page 343*
- **LD Isoenzymes** *see* Lactate Dehydrogenase Isoenzymes, Serum *on page 824*
- **LDLC** *see* Low Density Lipoprotein Cholesterol *on page 874*
- **LDLC:HDLC Ratio** *see* Low Density Lipoprotein Cholesterol *on page 874*
- **LDL Cholesterol:HDL Cholesterol** *see* Triglycerides, Serum or Plasma *on page 1275*

Lead, Blood

Related Information
Anemia Flowchart *on page 35*
Delta (5)-Aminolevulinic Acid, Urine *on page 508*
Lead, Urine *on page 835*
Porphobilinogen, Urine *on page 1073*
Protoporphyrin, Free Erythrocyte *on page 1121*
Protoporphyrin, Zinc, Blood *on page 1121*

Synonyms Pb, Blood

Applies to Delta Aminolevulinic Acid Dehydratase; Lead, Hair

Abstract Blood lead level is measured to detect recent lead exposure. It does not necessarily provide lead body burden. Children are at higher risk of lead toxicity as they may absorb up to 50% of dietary lead intake and have greater likelihood of lead exposure. It has been increasingly recognized that lead has harmful effects at concentrations previously considered safe. The current CDC cutoff for lead in children is 10 µg/mL, but recent studies show that lead has detrimental effects when <10 µg/mL. The CDC estimates that there are about 900,000 children in the U.S. with lead levels >10 µg/mL.

Specimen Whole blood; venous blood is recommended. Capillary blood is also used.

Container Special lead-free tube with heparin; trace metal Vacutainer® tubes containing lithium heparin can be used. EDTA is satisfactory.

Storage Instructions Do not separate red cells.

Causes for Rejection Improper draw

Special Instructions Avoid contact with leaded glass during collection

Reference Interval Children: <10 µg/dL (whole blood) (SI: <0.5 µmol/L). Recent data indicates that even blood lead concentrations <10 µg/dL are inversely associated with children's IQ (Canfield 2003). **Adults:** WHO has defined whole blood levels >30 µg/dL (1.5 µmol/L) as indicative of significant exposure. Lead levels >60 µg/dL (3 µmol/L) require chelation therapy. The American Academy of Pediatrics provides the recommendations in the following table for venous blood lead management.

Recommended Follow-up Services in Children, According to Diagnostic Blood Lead Level (BLL)

BLL (µg/dL)	Action
<10	No action required, likely to change in the future
10-14	Obtain a confirmatory venous BLL within 1 month. If still within this range, provide education to decrease blood lead exposure, and repeat BLL test within 3 months.
15-19	Obtain a confirmatory venous BLL within 1 month. If still within this range, take a careful environmental history, provide education to decrease blood lead exposure and to decrease lead absorption, and repeat BLL test within 2 months.
20-44	Obtain a confirmatory venous BLL within 1 week. If still within this range, conduct a complete medical history (including an environmental evaluation and nutritional assessment) and physical examination, provide education to decrease blood lead exposure and to decrease lead absorption, and either refer the patient to the local health department or provide case management that should include a detailed environmental investigation with lead hazard reduction and appropriate referrals for support services. If BLL is >25 µg/dL, consider chelation (not currently recommended for BLLs <45 µg/dL) after consultation with clinicians experienced in lead toxicity treatment.
45-69	Obtain a confirmatory venous BLL within 2 days. If still within this range, conduct a complete medical history (including an environmental evaluation and nutritional assessment) and physical examination, provide education to decrease blood lead exposure and to decrease lead absorption, and either refer the patient to the local health department or provide case management that should include a detailed environmental investigation with lead hazard reduction and appropriate referrals for support services. Begin chelation therapy in consultation with clinicians experienced in lead toxicity treatment.
≥70	Hospitalize the patient and begin medical treatment immediately in consultation with clinicians experienced in lead toxicity therapy. Obtain a confirmatory BLL immediately. The rest of the management should be as noted for management of children with BLLs between 45 and 69 µg/dL

OSHA uses a lead level of 40 µg/dL for occupational exposure. Lead level >40 µg/dL requires the employer to notify the worker in writing within 5 days. The employee should be removed from work and enter a chelation program if lead level is >60 µg/dL, or with certain other circumstances. In an occupational setting, OSHA requires both whole-blood lead and erythrocyte zinc protoporphyrin (ZPP) testing. ZPP is not thought to be an adequate indicator for lead toxicity at lead blood levels <25 µg/dL, and therefore is not currently recommended by CDC for lead screening in children 6 years of age and younger. Inhibition of erythrocyte **delta aminolevulinic acid dehydratase** is also a measure of lead toxicity, but is not inhibited until blood lead levels are >15-20 µg/dL. ZPP and delta aminolevulinic acid dehydratase are affected by other factors, including iron deficiency and metal toxicity. Therefore, direct measurement of whole blood lead is a preferred method for evaluation of lead toxicity.

Possible Panic Range >70 µg/dL (SI: >3.34 µmol/L) in acute lead poisoning; toxicity at lower levels in chronic poisoning

Use Evaluate recent lead exposure
(Continued)

Lead, Blood (Continued)

Limitations In chronic lead exposure, blood levels do not correlate with the severity of toxicity. The EDTA lead mobilization test or x-ray fluorescence may be needed for diagnosis of lead nephropathy.

Methodology Electrothermal atomic absorption spectrometry (AA), anodic stripping voltametry, inductively coupled plasma-mass spectrometry (ICP-MS), x-ray fluorescence spectroscopy, neutron activation

Additional Information In 1997, CDC recommended a basic three-question questionnaire for parents as a starting point to evaluate risk for lead exposure.

1. Does your child live in or regularly visit a house that was built before 1950? This question can apply to a facility such as home day-care or the home of a babysitter or relative.
2. Does your child live in or regularly visit a house built before 1978 with recent or ongoing renovations or remodeling (within the last 6 months)?
3. Does your child have a sibling or playmate who has or did have lead poisoning?

Screen all children whose parent/guardian responds "yes" or "don't know" to any question.

Lead expresses toxicity by several mechanisms and the toxicity follows a progressive pattern, as shown in the figure on the following page.

Of sources of lead poisoning in children, paint remains the most important. A single paint chip can contain as much as 10,000 µg of lead. Although the lead content of paints for household application has been limited to <0.5% since 1972, older paint is still in place on wood and even in adjacent soil and dust (pica).

Other sources of lead include air, soil, dust, drinking water, food (especially in lead-soldered cans), solder, storage batteries, ammunition (including retained bullets) and other metal objects, welding, printing, pottery making, gasoline additives, candles with lead-containing wicks, third world cosmetics, herbal remedies, moonshine, and other chemicals.

Lead absorption is influenced by iron deficiency. Greater than 90% of absorbed lead is deposited in the skeleton and in teeth. Bone lead concentration can be estimated with K x-ray fluorescence or L x-ray fluorescence.

Acute exposures are commonly associated with symptoms of anorexia, malaise, nausea, vomiting, and abdominal pain (lead colic). Fecal discoloration (black or red) may occur. Constipation, anemia, tremor, headache, coma, hearing loss, tinnitus, alopecia, and bradycardia may be seen. Severe exposures can result in encephalopathy and death; chronic exposures manifest with hypertension, arthralgias, teratogenesis, and impotence. Microcytic hypochromic anemia and basophilic stippling are characteristic, but basophilic stippling of red blood cells occurs in lead toxicity as well as in lead poisoning, and is not consistently found.

Saturnine gout is a hyperuricemic form of interstitial nephritis of chronic lead exposure.

Lead crosses the placenta; cord blood concentrations are 85% to 90% those of the mother.

Some nutritional factors affect lead absorption and excretion. High iron and calcium and low fat in diet reduce lead absorption. Vitamin C increases lead excretion.

References

Campbell C and Osteolindt KC, "Prevention of Childhood Lead Poisoning," *Curr Opin Pediatr*, 2000, 12:428-37.

Canfield RL, Henderson CR Jr, Cory-Slechta DA, et al, "Intellectual Impairment in Children With Blood Lead Concentrations <10 µg per Deciliter," *N Engl J Med*, 2003, 348(16):1517-26.

Rosen JF and Mushak P, "Primary Prevention of Childhood Lead Poisoning - The Only Solution," *N Engl J Med*, 2001, 344(19):1470-1.

Tong S, Baghurst PA, Sawyer MG, et al, "Declining Blood Lead Levels and Changes in Cognitive Function During Childhood," *JAMA*, 1998, 280(22):1915-9.

Williams RH and Erickson T, "Evaluating Lead and Iron Intoxication in an Emergency Setting," *Lab Med*, 1998, 29(4):224-31.

Internet Web Sites

www.cdc.gov/nceh/lead/lead.htm

Effects of Inorganic Lead on Children and Adults-- Lowest Observable Adverse Effect Levels

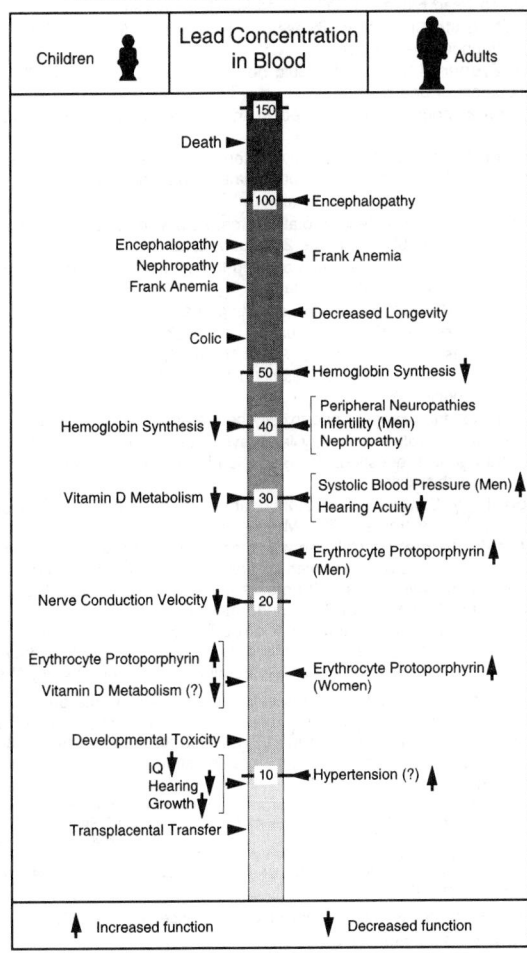

| Children | Lead Concentration in Blood | Adults |

Children — Death ► — 150

100 — ◄ Encephalopathy

Encephalopathy ►
Nephropathy ►
Frank Anemia ► — ◄ Frank Anemia

◄ Decreased Longevity

Colic ►

50 — ◄ Hemoglobin Synthesis ▼

Hemoglobin Synthesis ▼ ► — 40 — ⎡ Peripheral Neuropathies
Infertility (Men)
Nephropathy ⎣

Vitamin D Metabolism ▼ — 30 — ⎡ Systolic Blood Pressure (Men) ▲
Hearing Acuity ▼ ⎣

◄ Erythrocyte Protoporphyrin ▲ (Men)

Nerve Conduction Velocity ▼ ► — 20

Erythrocyte Protoporphyrin ▲
Vitamin D Metabolism (?) ▼ ► — ◄ Erythrocyte Protoporphyrin ▲ (Women)

Developmental Toxicity ►

IQ ▼
Hearing ▼
Growth ▼ — 10 — ◄ Hypertension (?) ▲

Transplacental Transfer ►

▲ Increased function   ▼ Decreased function

Note that blood levels are usually >40 µg/dL when anemia is recognized (ie, an apparently normal CBC does not rule out plumbism).

Adapted from U.S. Department of Health and Human Services, Royce SE and Needleman HL, eds, "Case Studies in Environmental Medicine: Lead Toxicity," Agency for Toxic Substances and Disease

- ♦ **Lead Excretion Ratio** *see* Lead, Urine *on page 835*
- ♦ **Lead, Hair** *see* Lead, Blood *on page 832*
- ♦ **Lead Mobilization Test** *see* Lead, Urine *on page 835*
- ♦ **Lead Toxicity** *see* Porphyrins, Quantitative, Urine *on page 1074*

Lead, Urine

Related Information

Anemia Flowchart *on page 35*
Delta (5)-Aminolevulinic Acid, Urine *on page 508*
Lead, Blood *on page 832*
Protoporphyrin, Free Erythrocyte *on page 1121*
(Continued)

Lead, Urine *(Continued)*

Synonyms Pb, Urine

Applies to Lead Excretion Ratio; Lead Mobilization Test

Abstract This test is used to assess lead body burden (lead mobilization test), not to diagnose lead poisoning.

Patient Preparation Patient should be instructed to use a specially cleaned plastic urinal or bedpan

Specimen 24-hour urine is preferred; in an emergency, random specimens may be acceptable

Container Plastic (preferably polyethylene) acid-washed (nitric acid) urine container copiously rinsed with appropriate (nontap deionized) water

Collection Avoid a catheter if possible.

Storage Instructions Record total volume. Acidify to pH 2 with concentrated HCl or add 20 mL 6N HCl to the 24-hour volume.

Causes for Rejection Specimen allowed to contact glass or metal, specimen not collected in acid-washed containers

Special Instructions Indicate if a chelating agent has been administered

Reference Interval ≤80 µg/24 hours (SI: ≤0.39 µmol/day)

Critical Values >125 µg/24 hours (SI: >0.60 µmol/day) is considered excessive and associated with toxicity; values of 80-125 µg/24 hours (SI: 0.39-0.60 µmol/day) are inconclusive

Use Evaluate lead exposure and toxicity before and following chelation therapy

Limitations The value of urinary lead levels without prior administration of a chelating agent is questionable. Blood lead concentrations provide the best correlation with toxicity.

Methodology Electrothermal atomic absorption (AA), inductively-coupled plasma-mass spectrometry (ICP-MS)

Additional Information Children with blood lead levels between 25-40 µg/dL should be evaluated by a lead mobilization test to determine need for chelation therapy. Those with levels >40 µg/dL should receive chelation therapy. The **lead mobilization test (LMT)** is performed by administering 500 mg/m^2 of CaNa$_2$-ethylenediaminetetraacetic acid (EDTA) and then collecting urine for 8 or 24 hours; a positive result is defined as a ratio of urinary lead/dose EDTA >0.6 (µg lead/mg EDTA given) or a total urinary excretion >200 µg of lead.

The **lead excretion ratio (LER)** is calculated by dividing the amount of lead excreted (in µg/24 hours) by the amount of Ca EDTA given (in mg). A ratio >0.60 is considered positive for the LER. Blood for lead and free erythrocyte protoporphyrin should be obtained before chelation.

References

Berger OG, Gregg DJ, and Succop PA, "Using Unstimulated Lead Excretion to Assess the Need for Chelation in the Treatment of Lead Poisoning," *J Pediatr*, 1990, 116(1):46-51.

Gulson BL, Jameson CW, Mahaffey KR, et al, "Pregnancy Increases Mobilization of Lead From Maternal Skeleton," *J Lab Clin Med*, 1997, 130(1):51-62.

Markowitz ME and Rosen JF, "Need for the Lead Mobilization Test in Children With Lead Poisoning," *J Pediatr*, 1991, 119(2):305-10.

♦ **LE Antibodies** *see* Skin Biopsy, Immunofluorescence *on page 1203*

♦ **Least Square Analysis** *see page 11*

♦ **LE Cell Preparation** *see* Anti-DNA *on page 173*

♦ **LE Cell Preparation** *see* Antinuclear Antibodies *on page 189*

♦ **Lecithin:Sphingomyelin Ratio** *see* Lecithin:Sphingomyelin Ratio, Amniotic Fluid *on page 836*

Lecithin:Sphingomyelin Ratio, Amniotic Fluid

Related Information

Creatinine, Amniotic Fluid *on page 472*

Lamellar Bodies, Amniotic Fluid *on page 830*

Phosphatidylglycerol, Amniotic Fluid *on page 1030*

Pulmonary Surfactant, Amniotic Fluid *on page 1124*

Synonyms Amniotic Fluid Lecithin:Sphingomyelin Ratio; Lecithin:Sphingomyelin Ratio; L:S Ratio; Lung Profile, Amniotic Fluid; Phospholipid Profile, Amniotic Fluid

Applies to Disaturated Phosphatidylcholine; Phosphatidylglycerol; Phosphatidylinositol

Test Includes L:S ratio; may include qualitative determination of phosphatidyl-glycerol (PG) and phosphatidylinositol (PI).

Abstract Amniotic fluid (AF) phospholipid testing assesses fetal lung maturity from which physicians infer the probability of neonatal respiratory distress syndrome (RDS) (hyaline membrane disease). Lecithin (L) concentrations in AF increase with advancing gestational age and lung maturation, whereas sphingomyelin (S) concentrations remain relatively stable. The L:S ratio increases with increasing fetal lung maturity, rising sharply during the final weeks of gestation.

Aftercare All nonsensitized Rh-negative patients should receive anti-D immuno-globulin after amniocentesis.

Specimen Amniotic fluid

Sampling Time The incidence of RDS at ≥37 weeks gestation is extremely low in most pregnancies (with the exception of poorly controlled maternal diabetic pregnancies), whereas the incidence of RDS risk becomes significantly high at ≤34 weeks. Consequently, fetal lung maturity testing is most useful in reliably dated, high-risk pregnancies that would benefit from early delivery during the 34-37 week gestational period.

Collection Ultrasound-guided, transabdominal amniocentesis is performed by the physician. Vaginal pool specimens are discouraged.

Storage Instructions The specimen should be centrifuged at low speed in a refrigerated centrifuge. The supernatant may be stored at 4°C for up to 10 days or it may be frozen indefinitely.

Causes for Rejection Blood contamination

Special Instructions Fetal sacs of multiple pregnancies should be sampled and analyzed individually. Send sample(s) to the laboratory **immediately** after collection.

Reference Interval Moderate intermethod variability to the analytical methods used. The following are intervals reported by Kulovich et al.

- Mature: ratio ≥2.0
- Transitional: ratio 1.5-1.9
- Immature, premature: ratio: <1.5

Possible Panic Range An L:S ratio <1.5 predicts RDS on delivery and is associated with a gestational age of 34 weeks or less.

Use Assess the likelihood of neonatal RDS. Results are also used to determine the optimal time for obstetrical intervention in cases of possible fetal distress due to maternal diabetes, toxemia, hemolytic disease of the newborn, or post-maturity.

The predictive value of a mature result is 95% to 100% and the predictive value of an immature result is 33% to 50%.

Limitations Thin-layer chromatographic patterns are subjective and require experienced interpretation.

False predictions of maturity using the L:S ratio traditionally occur in uncontrolled diabetic gestations. Measurement of PG (see Phosphatidylglycerol, Amniotic Fluid *on page 1030*) may be used to improve diagnostic accuracy.

Blood contamination of the AF specimen invalidates the L:S result due to the presence of lecithin and sphingomyelin in plasma. Though meconium recently was shown to contain neither lecithin nor sphingomyelin, its presence in AF will cause misleading results. The measurement of PG is unaffected by blood or meconium contamination, and its **accurate** detection in AF is highly predictive of lung maturity.

Methodology One- or two-dimensional, thin-layer chromatography (TLC) is used for the determination of L/S and other AF phospholipids. High performance liquid chromatography (HPLC), also, has been used, albeit less commonly.

Additional Information With the availability of rapid tests for PG (see Phosphatidylglycerol, Amniotic Fluid *on page 1030*) and other rapid tests for assessing fetal lung maturity (see Pulmonary Surfactant, Amniotic Fluid *on page 1124*), cascade-testing approaches have been suggested to minimize the cost, time, and effort of performing L:S ratios. It has been recommended that the L:S ratio and PG by TLC be done only by laboratories that have at least fifteen requests per week.

(Continued)

Lecithin:Sphingomyelin Ratio, Amniotic Fluid
(Continued)

References

"ACOG Educational Bulletin, Assessment of Fetal Lung Maturity, Number 230, November 1996," *Int J Gynecol Obstet*, 1997, 56(2):191-8.

Berkowitz K, Reyes C, Saadat P, et al, "Fetal Lung Maturation. Comparison of Biochemical Indices in Gestational Diabetic and Nondiabetic Pregnancies," *J Reprod Med*, 1997, 42(12):793-800.

Delgado JC, Greene MF, Winkelman JW, et al, "Comparison of Disaturated Phosphatidylcholine and Fetal Lung Maturity Surfactant/Albumin Ratio in Diabetic and Nondiabetic Pregnancies," *Am J Clin Pathol*, 2000, 113(2):233-9.

Dubin SB, "Assessment of Fetal Lung Maturity Practice Parameter," *Am J Clin Pathol*, 1998, 110(6):723-32.

Field NT and Gilbert WM, "Current Status of Amniotic Fluid Tests of Fetal Maturity," *Clin Obstet Gynecol*, 1997, 40(2):366-86.

Longo SA, Towers CV, Strauss A, et al, "Meconium Has No Lecithin or Sphingomyelin But Affects the Lecithin/Sphingomyelin Ratio," *Am J Obstet Gynecol*, 1998, 179(6 Pt 1):1640-42.

♦ **Lectins** *see* Immunoperoxidase Procedures *on page 780*

Legionella Culture and Serology

Related Information

Bacterial Culture, Lower Respiratory *on page 241*
Bronchoalveolar Lavage (BAL) *on page 311*
Fine Needle Aspiration Culture *on page 589*
Legionella Direct Fluorescent Antigen Smear *on page 839*
Legionella Urine Antigen *on page 840*
Psittacosis Serology *on page 1123*
Q Fever Serology *on page 1128*

Test Includes Culture and direct fluorescent antigen (DFA) detection for *Legionella* spp; serology detects antibody (IgG or IgM) to *Legionella pneumophila*

Abstract Legionnaires' disease is a systemic disease with pneumonia as the prominent clinical finding. Currently, it is estimated that each year 8000-18,000 people in the U.S. are infected with *Legionella* sp and approximately 5% to 30% of these patients die from the disease (see CDC website below). Sputum characterized by acute inflammatory features, without a classical pattern of bacteria, may represent *Legionella*.

Detection of antibodies to the *Legionella* species is useful for epidemiologic purposes, but is rarely employed for the diagnosis of *Legionella pneumoniae*. The differential diagnosis of the atypical pneumonias includes *Chlamydia pneumoniae*, Legionnaires' disease, *Mycoplasma pneumoniae*, and rickettsial pneumonia. *Legionella*, *Mycoplasma pneumoniae*, and *Chlamydia pneumoniae* collectively cause 10% to 20% of cases of pneumonia. Hypersensitivity pneumonitis and metal fume fever deserve consideration.

Specimen Culture: Lung tissue, other body tissue, pleural fluid, other body fluid, transtracheal aspiration, bronchoalveolar lavage, bronchial brushing, sputum, content of abscess. **Serology:** Serum, acute and convalescent

Container Serology requires a red top tube

Collection Two specimens for serology are recommended, a convalescent sample 10-14 days after the acute sample is drawn.

Turnaround Time Positive results require 2-5 days. Primary plates are often held for 7-14 days before a final negative report is issued.

Special Instructions Clinical suspicion of *Legionella* infection should be reported to the laboratory.

Reference Interval No *Legionella* recovered in culture; less than a fourfold change in serology titer between acute and convalescent samples; <1:256 in a single sample

Use Aid in the diagnosis of *Legionella pneumophila* infection; determine the prevalence of disease

Limitations Sputum (expectorated), bronchial aspirates, and other specimens having normal flora are subject to bacterial overgrowth and are not as desirable as transtracheal aspirates, pleural fluid, and biopsy material for culture. Sensitivity of cultures is generally ~80% and specificity approaches 100%.

When using serology it is important to test for both IgG and IgM; with both, overall sensitivity is ~80%. Specificity is fairly high (approaching 95%), but predictive values may be low in low prevalence populations. Serologic diagnosis is often retrospective and should be considered presumptive. Antibody may persist for years. False positives are found due to serologic cross-reactions with numerous bacteria, such as *Pseudomonas* species, *Burkholderia* species, and other gram-negative bacteria. Seroconversion may be delayed.

Methodology Culture on selective and nonselective media (buffered charcoal yeast extract). Growth cannot be anticipated on conventional media. Serology uses indirect fluorescent antibody (IFA) assay or latex agglutination (LA).

Additional Information Laboratory diagnosis of *Legionella* infection can be attempted with three distinct methodological approaches. First, a rapid diagnosis can be attained by methods that detect bacterial products in clinical specimens (eg, direct fluorescent antibody tests on respiratory specimens, DNA amplification, radioimmunoassay, or enzyme immunoassays on urine specimens). Second, indirect fluorescence antibody methods can be used to detect the patient's antibody response to *Legionella* spp. Finally, *Legionella* cultures can be performed; *Legionella* cultures have sensitivities and specificities of approximately 80% and 100%, respectively. Culture for *Legionella* may require 2 weeks of incubation before a result is available, thus it is not helpful in making clinical decisions. Other nonculture tests are more useful for the rapid diagnosis of Legionnaires' disease. See table below and items listed under Related Information on previous page.

Clinical Clues to the Diagnosis of Legionnaires' Disease

- Gram stain of respiratory secretions reveals numerous neutrophils, but few organisms
- Presence of hyponatremia (serum sodium ≤130 mmol/L)
- Failure to respond to β-lactam and aminoglycoside antibiotics
- Occurrence in hospital where potable water system is known to be contaminated with *Legionella*
- History of smoking and alcohol use
- Pleuritic chest pain
- Fever malaise, myalgia, headache

Adapted from Harrison TG and Taylor AG, "Timing of Seroconversion in Legionnaires' Disease," *Lancet*, 1988, 2(8614):795.

References
Bernstein JM, "Treatment of Community-Acquired Pneumonia-IDSA Guidelines. Infectious Diseases Society of America," *Chest*, 1999, 115(3S):9S-13S.
Breiman RF and Butler JC, "Legionnaires' Disease: Clinical, Epidemiological, and Public Health Perspectives," *Semin Respir Infect*, 1998, 13(2):84-9.
Castellani Pastoris M, Lo Monaco R, et al, "Legionnaires' Disease on a Cruise Ship Linked to the Water Supply System: Clinical and Public Health Implications," *Clin Infect Dis*, 1999, 28(1):33-8.
Centers for Disease Control, "Legionnaires' Disease Associated With Potting Soil - California, Oregon, and Washington, May-June 2000," *MMWR Morb Mortal Wkly Report*, 2000, 49:7778.
Centers for Disease Control, "Outbreak of Legionnaires' Disease Among Automotive Plant Workers - Ohio, 2001," *MMWR Morb Mortal Wkly Report*, 2001, 285(22):2848-9.
Cunha BA, "Clinical Features of Legionnaires' Disease," *Semin Respir Infect*, 1998, 13(2):116-27.

Internet Web Sites
www.amm.co.uk/pubs/fa_legionnaires.htm
www.astdhpphe.org/infect/legion.html
www.cdc.gov/ncidod/dbmd/diseaseinfo/legionellosis_g.htm

Legionella Direct Fluorescent Antigen Smear

Related Information
Bronchoalveolar Lavage (BAL) *on page 311*
Legionella Culture and Serology *on page 838*
Legionella Urine Antigen *on page 840*

Abstract Direct fluorescent antigen (DFA) used to detect organisms directly in specimens for rapid diagnosis of *Legionella* pneumonia.

Specimen Lung tissue, pleural fluid, other body fluid, transtracheal aspirate, sputum, bronchial washing, bronchoalveolar lavage, abscess content

Container Sterile container

Reference Interval No *Legionella* species detected

Use Rapid detection of the presence of *Legionella* organisms in direct smear of specimen using fluorescent antibody specific for *Legionella* bacteria

(Continued)

Legionella Direct Fluorescent Antigen Smear
(Continued)

Limitations DFA is not as sensitive as culture and false-positive results occur due to cross-reactions with other bacterial infections and with environmental *Legionella* species. False-positive reactions can occur with other bacterial species, especially gram-negative bacteria.

Since culture has a higher sensitivity, DFA should never replace culture for the diagnosis of *Legionella* infections. It is not helpful for all species of *Legionella*.

Methodology Direct fluorescent antibody (DFA); detection of *Legionella* by PCR in respiratory specimens is available.

Additional Information Community-acquired and nosocomial infections caused by multiple serogroups of *Legionella* are increasingly recognized. Culture is now possible on buffered charcoal yeast extract agar. However, the demonstration of organisms in sputum, tissue, or brushings is a rapid means to make a diagnosis. It also has the advantage of applicability to specimens contaminated with other bacteria. Development of monoclonal antibodies has increased sensitivity and specificity. A combination of both culture and antigen detection is recommended. A direct FA should not be ordered alone; it should always be accompanied by a request for *Legionella* culture.

References
Chow JW and Yu VL, "*Legionella*: A Major Opportunistic Pathogen in Transplant Recipients," *Semin Respir Infect*, 1998, 13(2):132-9.

Plouffe JF, McNally C, and File TM Jr, "Value of Noninvasive Studies in Community-Acquired Pneumonia," *Infect Dis Clin North Am*, 1998, 12(3):689-99.

Shelton BG, Kerbel W, Witherell L, et al, "Review of Legionnaires' Disease," *AIHAJ*, 2000, 61(5):738-42.

Waterer GW, Baselski VS, and Wunderink RG, "*Legionella* and Community-Acquired Pneumonia: A Review of Current Diagnostic Tests From a Clinician's Viewpoint," *Am J Med*, 2001, 110(1):41-8.

Internet Web Sites
www.amm.co.uk/pubs/fa_legionnaires.htm
www.astdhpphe.org/infect/legion.html
www.cdc.gov/ncidod/dbmd/diseaseinfo/legionellosis_g.htm

Legionella Urine Antigen

Related Information
Legionella Culture and Serology *on page 838*
Legionella Direct Fluorescent Antigen Smear *on page 839*

Test Includes Detection of excreted *Legionella pneumophila* serogroup 1 antigen in urine

Abstract Antigen detection in urine provides an additional rapid approach to diagnosis of infections to *Legionella pneumophila* serogroup 1.

Specimen Urine

Container Sterile urine container

Reference Interval Negative

Use Aid in the diagnosis of *Legionella pneumophila* serogroup 1 infection; other *Legionella* species are not detected with this assay.

Limitations Commercially available enzyme immunoassays have a specificity >99% and sensitivity of 70% to 90%. Concentration of the urine increases the sensitivity. Latex agglutination tests have a specificity of 85% to 99% and sensitivity of 55% to 90%. Antigen excretion in urine may persist for months after recovery from infection. This test should be obtained in conjunction with other, more proven, laboratory tests such as *Legionella* culture and/or antibody assays. It is not useful for detection of all species of *Legionella*.

Methodology Enzyme immunoassay (EIA) or latex agglutination (LA)

Additional Information The rapid detection of *Legionella* antigen in urine is currently recommended to assist in establishing a diagnosis of *Legionella* infection. The commercially available tests have an acceptable sensitivity and specificity. However, as with many diagnostic tests, a negative test does not definitively rule out *Legionella*. *Legionella* antigen may be detected in urine for months after the acute infection; thus a positive result with these assays does not always indicate a current infection.

References
Bernstein JM, "Treatment of Community-Acquired Pneumonia-IDSA Guidelines. Infectious Diseases Society of America," *Chest*, 1999, 115(3S):9S-13S.

Plouffe JF, McNally C, and File TM Jr, "Value of Noninvasive Studies in Community-Acquired Pneumonia," *Infect Dis Clin North Am*, 1998, 12(3):689-99.

Shelton BG, Kerbel W, Witherell L, et al, "Review of Legionnaires' Disease," *AIHAJ*, 2000, 61(5):738-42.

Waterer GW, Baselski VS, and Wunderink RG, "*Legionella* and Community-Acquired Pneumonia: A Review of Current Diagnostic Tests From a Clinician's Viewpoint," *Am J Med*, 2001, 110(1):41-8.

Internet Web Sites
www.amm.co.uk/pubs/fa_legionnaires.htm
www.astdhpphe.org/infect/legion.html
www.cdc.gov/ncidod/dbmd/diseaseinfo/legionellosis_g.htm

♦ **Leigh Syndrome** *see* Inherited Diseases of Metabolism and Cell Structure *on page 792*

Leishmaniasis Diagnostic Procedures

Related Information

Bone Marrow *on page 296*
Chagas Disease Diagnostic Procedures *on page 381*
Lymph Node Biopsy *on page 880*
Protein Electrophoresis, Serum *on page 1104*
Skin Biopsy *on page 1200*

Abstract Infections with *Leishmania* species are commonly zoonotic. Leishmaniasis is endemic in certain parts of the world. Visceral leishmaniasis (kala-azar) is typically caused by *Leishmania donovani* which parasitizes macrophages. Infections are found in spleen, bone marrow, liver, and lymph nodes. Other infections include cutaneous leishmaniasis (*L. tropica, L. major, L. aethiopica*) and new world cutaneous leishmaniasis (*L. mexicana, L. braziliensis, L. venezuelensis, L. peruviana*), and mucocutaneous leishmaniasis (espundia) caused by *L. braziliensis*.

Specimen Serum

Container Red top tube

Turnaround Time Cultures are incubated as long as 28 days.

Reference Interval Negative

Use Evaluation of hepatosplenomegaly, fever, lymphoadenopathy, diarrhea, ascites, normocytic or normochromic anemia. Cutaneous leishmaniasis is found on exposed surfaces accessible to sandflies, commonly presenting as ulcerative lesions. Lymphoangitic dissemination may develop.

Limitations Serology is helpful in epidemiology studies but does not support the clinical diagnosis of visceral, cutaneous, or mucocutaneous leishmaniasis. Serologic cross-reactivity with Chagas disease; false positives in malaria and other diseases. Sensitivity of serologic tests for cutaneous leishmaniasis is poor.

Methodology Microscopy, utilizing Giemsa stains of bone marrow aspiration, skin aspiration, or other sources; smears and tissue sections are used. Cultivation methods are available. PCR methods and Western blot are used. Species identification is done with isoenzymes. Many serologic methods are available, including indirect hemagglutination (IHA), indirect fluorescent antibody (IFA), and immunoblot assay. Of these, the enzyme-linked immunosorbent assay (ELISA) presently is considered the most sensitive.

Additional Information *Leishmania* species are intracellular parasites that are transmitted to humans by bites from phlebotomine sand flies. Occasionally, nonvector transmissions have been reported through blood transfusions, intravenous drug use with syringe sharing, sexual intercourse, organ transplantation, and congenital transmission.

Other laboratory abnormalities in visceral leishmaniasis may include polyclonal gammopathy, normocytic normochromic anemia, leukopenia, and thrombocytopenia.

References

Belli A, Garcia D, Palacios X, et al, "Widespread Atypical Cutaneous Leishmaniasis Caused by *Leishmania (L.) chagasi* in Nicaragua," *Am J Trop Med Hyg*, 1999, 61(3):380-5.

Chatterjee M, Jaffe CL, Sundar S, et al, "Diagnostic and Prognostic Potential of a Competitive Enzyme-Linked Immunosorbent Assay for Leishmaniasis in India," *Clin Diagn Lab Immunol*, 1999, 6(4):550-4.

Dedet JP and Pratlong F, "Leishmaniasis," *Manson's Tropical Diseases*, 21st ed, Chapter 75, Cook GC and Zumla A, eds, Philadelphia, PA: WB Saunders Co, 2003, 1339-64.

Kenner JR, Aronson NE, Bratthauer GL, et al, "Immunohistochemistry to Identify *Leishmania* Parasites in Fixed Tissues," *J Cutan Pathol*, 1999, 26(3):130-6

Meinecke CK, Schottelius J, Oskam L, et al, "Congenital Transmission of Visceral Leishmaniasis (Kala Azar) From an Asymptomatic Mother to Her Child," *Pediatrics*, 1999, 104(5):e65.

(Continued)

Leishmaniasis Diagnostic Procedures *(Continued)*
Internet Web Sites
www.cdc.gov/ncidod/dpd/parasites/leishmania/default.htm

♦ **Lenoxin®** *see* Digoxin, Serum *on page 512*
♦ **Lepidopterae** *see* Arthropod Identification *on page 213*
♦ **Leptilan®** *see* Valproic Acid, Serum or Plasma *on page 1297*

Leptin, Serum or Plasma
Applies to Body Mass Index (BMI)

Abstract Leptin, an obesity-related cytokine protein was discovered in 1994 and has been the subject of hundreds of research papers since then. In humans, leptin is produced only in white adipose tissue. Increased levels suppress appetite and increase thermogenesis. Leptin gene mutations producing leptin deficiency lead to massive obesity.

Specimen Serum; plasma may be used.

Container Red top tube, green top (heparin) tube

Sampling Time 12-hour fast

Collection If heparin is used, add no more than 10 IU/mL of blood.

Storage Instructions Store at -20°C or lower.

Reference Interval Varies with the laboratory performing the test. One RIA procedure uses 7.5 ±9.3 ng/mL in lean subjects, and 31.3 ±24.1 ng/mL in obese subjects. Values in women are higher than in men. Serum leptin is directly proportional to body-mass index (BMI) (see Additional Information). Determining a well-defined reference interval is made more difficult by relatively large short-term biologic coefficients of variation (10.9% in lean subjects and 22.5% in obese subjects).

Use Assays of serum leptin are available from a few reference laboratories, but the significance of the results is not understood. Leptin is still both a tool for research into obesity and an object of obesity research. Leptin may not be involved in human obesity since lean individuals have lower levels than obese individuals.

Methodology Radioimmunoassay (RIA) (Linco Research, Inc), limit of quantification is 0.2 ng/mL; enzyme-linked immunosorbent assay (ELISA)

Additional Information Many studies of obesity use the body-mass index (BMI) as a measurement of obesity. The BMI is calculated as follows:

BMI (kg) / (m²) = body weight (kg) / height (m²)

The normal reference interval for the BMI is usually stated as 18.5-24.9 kg/m². The interval of 25.0-29.9 kg/m² is often called "overweight", while the term "obesity" is reserved for those with values >30 kg/m². Another approach to obesity classification measures the distribution of body fat - the ratio of waist circumference to hip circumference. A waist:hip ratio reflecting a "central" distribution of fat (eg, >0.90 in women and >1.0 in men) is correlated with a higher risk for morbidity than a ratio reflecting a "peripheral" distribution of fat (eg, <0.75 in women and <0.85 in men.) Body fat can also be measured by magnetic resonance imaging, but such measurements are not widely available. The ideal index of obesity remains undefined.

References
Auwerx J and Staels B, "Leptin," *Lancet*, 1998, 351(9104):737-42.

Considine RV, Sinha MK, Heiman ML, et al, "Serum Immunoreactive-Leptin Concentrations in Normal-Weight and Obese Humans," *N Engl J Med*, 1996, 334(5):292-5.

Heymsfield SB, Greenberg AS, Fujioka K, et al, "Recombinant Leptin for Weight Loss in Obese and Lean Adults," *JAMA*, 1999, 282(16):1568-75.

Rosenbaum M and Leibel RL, "The Role of Leptin in Human Physiology," *N Engl J Med*, 1999, 341(12):913-5.

Leptospira Culture
Related Information
Bacterial Culture, Blood *on page 232*
Darkfield Examination, Leptospirosis *on page 500*
Leptospira Serology *on page 844*

Abstract Leptospires are spiral-shaped bacteria that can be found as free-living organisms in freshwater, soil, or mud, or they live in association with animal hosts. The bacteria can infect a variety of wild and domestic animals which excrete the organisms in urine. Leptospirosis results from the direct or indirect exposure to the urine of infected animals. The bacteria enter the body through breaks in the skin or through the mucous membranes and conjunctivae. Clinical manifestations can vary from a mild self-limiting illness to a fulminating fatal illness including hepatorenal failure (Weil syndrome).

Patient Preparation Thoroughly instruct the patient in the proper collection technique for a midvoid urine specimen. Avoid contamination with skin flora, since normal urine contaminants will overgrow leptospires in the specimen. See also Bacterial Culture, Urine *on page 246* and Bacterial Culture, Blood *on page 232* for detailed instructions.

Specimen Urine, indicate midvoid, catheter or suprapubic puncture specimen; **blood** and **cerebrospinal fluid** may also be cultured

Container Sterile, urine container; for blood, lavender top (EDTA) Vacutainer® or green top (heparin) Vacutainer®. Avoid collecting blood in citrate solutions, which may inhibit the growth of leptospires.

Collection Specimen should be transported to the laboratory within 1 hour of collection. For midvoid urine culture, patient should be instructed to clean skin thoroughly, do not collect first portion of stream, **collect midportion of stream**, and do not collect final portion of stream. Catheter or suprapubic puncture specimen may also be used. If possible, specimens should be collected prior to antibiotic treatment and while the patient is febrile.

Storage Instructions Specimens should not be refrigerated. They should be left at room temperature. Urine with an acid pH should be alkalinized if it cannot be set up immediately.

Turnaround Time 4-8 weeks

Special Instructions The laboratory should be informed of the specific request for *Leptospira* culture, collection time, current antibiotic therapy, and date of onset of illness. Urine must be alkaline; *Leptospira* do not survive for more than a few hours in acid urine. Repeated cultures may be required.

Reference Interval No *Leptospira* isolated

Use Diagnose leptospirosis (Weil disease)

Limitations Other organisms are not isolated or identified. The organism is slow growing; results of cultures may be prolonged.

Contraindications Leptospiremia occurs during the septicemic acute phase of infection. This phase last 4-7 days after which organisms are not recoverable from blood.

Methodology Specimen is inoculated onto specially prepared media containing rabbit serum or albumin and fatty acids. Incubation is for 4-6 weeks in the dark at 28°C to 29°C. Cultures are examined with darkfield or phase microscopy for motile leptospires at weekly intervals.

Additional Information Leptospirosis in humans is usually associated with exposure to water or soil contaminated by infected animal urine. Outbreaks have been associated with canoeing, rafting, and swimming in contaminated lakes and rivers. In addition to recreational and accidental exposure, occupational exposure of veterinarians, dairymen, swineherds, abattoir workers, miners, fish and poultry processors, and those who work in rat-infested environments are at increased risk.

The septicemic phase of the disease lasts from 4-7 days. During this first week of disease the most reliable means of detecting leptospires is by direct culture of blood or spinal fluid on appropriate media. The initial phase is followed by a quiescent period of 1-3 days. The last phase of the disease is the immune phase in which as many as 40% of the patients will have meningitis and the majority will have *Leptospira* in the urine. Urine can remain positive for several months. Concentration of *Leptospira* in human urine is low and shedding may be intermittent. Therefore, repeated isolation attempts should be made. Serology (acute and early convalescent) is recommended. Direct darkfield examination is no longer recommended.

References

Guidugli F, Castro AA, and Atallah AN, "Systemic Reviews on Leptospirosis," *Rev Inst Med Trop Sao Paulo*, 2000, 42(1):47-9.

Holk K, Nielsen SV, and Ronne T, "Human Leptospirosis in Denmark 1970-1996: An Epidemiological and Clinical Study," *Scand J Infect Dis*, 2000, 32(5):533-8.

Schwartz DA, "Emerging and Reemerging Infections. Progress and Challenges in the Subspecialty of Infectious Disease Pathology," *Arch Pathol Lab Med*, 1997, 121(8):776-84.

Internet Web Sites

www.amm.co.uk/pubs/fa_leptospir.htm
www.cdc.gov/ncidod/dbmd/diseaseinfo/leptospirosis_g.htm

Leptospira Serology

Related Information

Darkfield Examination, Leptospirosis *on page 500*
Leptospira Culture *on page 842*

Abstract Leptospirosis is a disease that occurs in all parts of the world. The disease can present as a mild flu-like form or as a severe fatal disease characterized by renal failure, liver dysfunction, and hemorrhages (Weil syndrome).

Specimen Serum

Container Red top tube

Sampling Time Acute and convalescent sera drawn 10-14 days apart are suggested. Agglutinins peak at 3-4 weeks of illness.

Reference Interval Negative. A fourfold increase in titer in paired sera is positive.

Critical Values IgM titer ≥1:100 provides evidence of recent or active disease

Use Diagnose leptospirosis (Weil disease)

Limitations The antigens used in the test are from the serovars most commonly causing disease, but there are many other serovars that might not be detected. A battery of antigens should be used.

Methodology Microscopic agglutination test (MAT), macroagglutination, complement fixation (CF), hemagglutination, enzyme-linked immunosorbent assay (ELISA), LEPTO dipstick

Additional Information Leptospirosis is an acute febrile zoonotic illness caused primarily by *Leptospira interrogans*, a large spirochete with more than 180 serologic variants. Patients with extensive animal contact or contact with contaminated water are particularly at risk. Although leptospires can be cultured from blood or urine during the first week of illness, this interval is often missed. Antibody appears at the end of the first week of illness and peaks at 3-4 weeks, after which it slowly disappears.

References

Centers for Disease Control and Prevention, "Outbreak of Acute Febrile Illness Among Participants in EcoChallenge Sahah 2000 - Malaysia," *JAMA*, 2000, 284(13):1646.

Farr RW, "Leptospirosis," *Clin Infect Dis*, 1995, 21(1):1-8.

Smits HL, Ananyina YV, Chereshsky A, et al, "International Multicenter Evaluation of the Clinical Utility of a Dipstick Assay for Detection of *Leptospira*-Specific Immunoglobulin M Antibodies in Human Serum Specimens," *J Clin Microbiol*, 1999, 37(9):2904-9.

Internet Web Sites

www.amm.co.uk/pubs/fa_leptospir.htm
www.cdc.gov/ncidod/dbmd/diseaseinfo/leptospirosis_g.htm

♦ **Leucine** *see Amino Acids, Urine on page 145*

Leucine Aminopeptidase (LAP), Serum and Urine

Related Information

Alkaline Phosphatase, Serum *on page 127*
Bilirubin, Total, Serum *on page 265*
Gamma-Glutamyl Transferase, Serum *on page 629*
Liver Disease: Laboratory Assessment, Overview *on page 869*

Synonyms Arylamidase; Arylamidase Naphthylamidase; LAP

Abstract LAP has been used occasionally to determine whether an elevated serum alkaline phosphatase value was due to a process in the liver/biliary tract vs bone or other site. A potential, but as yet unproven, use is the assay for LAP in urine to detect early renal tubular injury in diabetes. Serum LAP levels are elevated in patients with systemic lupus erythematosus (SLE). It has been proposed as a potential activity indicator for SLE.

Specimen Serum, ascitic fluid

Container Red top tube

Use LAP increases in cholestasis; it is a biliary excretory enzyme which is not increased with bone disease.

Limitations LAP increases in late pregnancy.

References

Bedir A, Ozener IC, and Emerk K, "Urinary Leucine Aminopeptidase Is a More Sensitive Marker of Early Renal Damage in Noninsulin-Dependent Diabetics Than Is Microalbuminuria," *Nephron*, 1996, 74(1):110-3.

Inokuma S, Setoguchi K, Ohta T, et al, "Serum Leucine Aminopeptidase as an Activity Indicator in Systemic Lupus Erythematosus: A Study of 46 Consecutive Cases," *Rheumatology (Oxford)* 1999, 38(8):705-8.

♦ **Leukapheresis, Therapeutic** *see* Apheresis, Therapeutic *on page 201*
♦ **Leukemia Gene Rearrangement** *see* Gene Rearrangement for Leukemia and Lymphoma *on page 633*

Leukocyte Alkaline Phosphatase

Related Information
Bone Marrow *on page 296*
Breakpoint Cluster Region Rearrangement in CML *on page 304*
Leukocyte Cytochemistry *on page 846*
Phlebotomy, Therapeutic *on page 1029*
White Blood Cell Count *on page 1330*

Synonyms LAP Score

Abstract A cytochemical reaction (LAP score) useful in differential diagnosis of myeloproliferative diseases, in particular, distinguishing leukemoid reaction from chronic myelogenous leukemia (CML). The LAP score reflects the intensity of alkaline phosphatase (AP) staining (of a neutrophil apparently nongranular but intracytoplasmic component) scored on a scale of 0 to 4+.

Specimen Whole blood

Container Slides with smears of blood

Collection Make six smears on long slides from fingerstick blood. Air dry the slides. Transport to Hematology immediately (ie, within 30 minutes).

Storage Instructions Slides must be fixed with cold 10% formalin methanol or citrated buffered acetone, rinsed, air dried, and frozen within 8 hours (preferably within 30 minutes) after obtaining the blood. After fixation, smears can be stored for up to 8 weeks before staining. The enzyme activity may, in some cases, be stable for up to 1 year when stored at -20°C.

Causes for Rejection Blood collected in EDTA anticoagulant, transit time to the laboratory in excess of 30 minutes, neutrophil count <1000/mm³ in peripheral blood

Reference Interval 15-130 (variable between laboratories)

Use Aid in the differential diagnosis of chronic myelogenous leukemia (CML) versus leukemoid reaction; aid in the evaluation of polycythemia and myelofibrosis

Limitations Pregnancy, increased number of immature forms of neutrophils, and postoperative or "stressful" states are associated with increased scores. The differential must have adequate numbers of mature neutrophilic granulocytes to perform the LAP.

Methodology Enzyme reaction with leukocyte alkaline phosphatase liberating naphthol or a substituted naphthol compound which then couples with fast blue RR or other chromogen to form an insoluble precipitate. Color of the precipitate relates to the type of substituted naphthol substrate and diazonium dye used (color is reagent dependent). Cells are scored as to the degree of phosphatase activity present, 0 to 4+. One hundred cells are counted and the score totaled.

Additional Information Low LAP scores have been associated with CML, myelodysplasia, PNH, thrombocytopenic purpura, and hereditary hypophosphatasia. In chronic phase CML, regardless of the total white count, the score remains low but with therapy or progression of disease, the LAP score may increase. In chronic neutrophilic leukemia (a rare myeloproliferative disorder with neutrophil granulocytosis and an indolent course) LAP score is normal or increased.

In nonleukemic neutrophilia, the LAP rises as the WBC rises. High LAP scores have been seen in leukemoid reactions, polycythemia vera, myelofibrosis, aplastic anemia, mongolism, hairy cell leukemia, and neutrophilia either physiological or secondary to infection. LAP is also increased in Hodgkin disease.

References
Catovsky D, "Leukocyte Cytochemical and Immunological Techniques," *Practical Haematology*, 8th ed, Chapter 9, Dacie JV and Lewis SM, eds, New York, NY: Churchill Livingstone, 1995, 143-74.
Enright H and McGlave P, "Chronic Myelogenous Leukemia," *Hematology: Basic Principles and Practice*, 3rd ed, Chapter 62, Hoffman R, Benz EJ Jr, Shattil SJ, et al, eds, Philadelphia, PA: Churchill Livingstone, 2000, 1155-71.
Grozdea J, Vergnes H, Cambus JP, et al, "Neutrophil Alkaline Phosphatase Activity in Turner Syndrome," *Acta Haematol*, 1999, 102(4):201-2.

♦ **Leukocyte Antigens** *see* HLA Typing, Single Human Leukocyte Antigen *on page 739*

♦ **Leukocyte Common Antigen** *see* Immunoperoxidase Procedures *on page 780*

♦ **Leukocyte Concentrate** *see* Neutrophils, Transfusion *on page 951*

♦ **Leukocyte Count** *see* White Blood Cell Count *on page 1330*

Leukocyte Cytochemistry

Related Information

Bone Marrow *on page 296*
Breakpoint Cluster Region Rearrangement in CML *on page 304*
Chromosomal Translocations, Molecular Detection *on page 404*
Flow Cytometry, Overview *on page 592*
Gene Rearrangement for Leukemia and Lymphoma *on page 633*
Leukocyte Alkaline Phosphatase *on page 845*
Lysozyme, Blood and Urine *on page 884*
Peripheral Blood: Differential Leukocyte Count *on page 1010*
Tartrate Resistant Leukocyte Acid Phosphatase *on page 1236*
White Blood Cell Count *on page 1330*

Test Includes Stain reactions, potentially including: Stain for Intracellular Pigment; Acid Phosphatase With and Without Tartrate; Alpha-Naphthyl Esterase With and Without Fluoride; Amyloid; ASD Chloroacetate Esterase; Beta-Glucuronidase; Methenamine Silver; Methyl Green-Pyronine; Myeloperoxidase (MPO); Nonspecific Esterase; Oil Red O; PAS; Sudan Black B (SBB); and as indicated by examination of routinely stained preparations or specific request

Abstract Cytochemical reactions are useful in differential diagnosis of conditions in which leukocytes are involved, in particular, in the study and characterization of acute leukemia. Immunophenotypic and cytochemical studies of the leukemias are generally complementary. Identification of cell antigens with flow cytometry techniques usually provides a more detailed and definitive analysis of cell populations. Cytogenetic analysis may be required to confirm or establish a diagnosis.

Specimen Blood or bone marrow smears, imprints or smears of cell suspensions

Container Green top (heparin) tube

Collection Smears are prepared at the patient's bedside.

Storage Instructions Transport specimen to the laboratory immediately.

Turnaround Time Variable

Reference Interval Interpretation and significance usually requires correlation of cytomorphologic, cytochemical, immunophenotypic, cytogenetic, and clinical features. The French-American-British (FAB) cooperative group classification of acute leukemias requires that SBB or MPO staining be positive in at least 3% of blasts to establish myeloid lineage. If staining is below the 3% level (FAB M0, see below), immunophenotyping must be performed to provide myeloid identification. Generally, intensity of granule staining increases with cell maturation. See Table 1.

Table 1. Cytochemistry – Normal White Blood Cells

	MPO/ SBB	Nonspecific Esterase		Chloroacetate Esterase	PAS	Acid Phos
		α-N Acetate	α-N Butyrate			
Promyelocyte	+/++	-/±	-	+/++	±	+/++
Myelocyte	++	±		++		
Neutrophil	+++	±		+++	+++	+
Monocyte	-/±	+++[1]	++/+++[1]	±	±	++
Lymphocyte	-	-/±	±	-	-/+	-/++
Erythroblast	-	-/±	-	-	-	±
Megakaryocyte	-	+++	±	-	++	++

[1]Strongly inhibited by sodium fluoride.

Use Cytochemically evaluate neoplasms and abnormal cells in bone marrow, peripheral blood, or other specimens such as imprints; detect amyloidosis; classification of leukemias and plasma cell dyscrasias, in particular the acute myeloid leukemias; evaluate myeloproliferative/lymphoproliferative disorders

Limitations Enzyme cytochemical positive reaction may not have diagnostic specificity.

Methodology Generally, based on cell cytoplasmic granule enzymatic activity. Test conditions include provision of substrate, which when acted upon by the appropriate enzyme, generates reaction product(s) that react with color reagent(s) allowing visualization (see Leukocyte Alkaline Phosphatase *on page 845*). Sudan black B (SBB) is a lipophilic dye reacting with phospholipids, neutral fats, and steroids. It allows visualization of the phospholipid membranes of myeloid granules. SBB is more sensitive but less specific than MPO in demonstrating granulocytic lineage.

Additional Information See also Tartrate Resistant Leukocyte Acid Phosphatase *on page 1236*, for hairy cell leukemia. In the technique of Yam et al for esterase reactions, a single slide preparation is consecutively stained for two different enzyme activities. The nonspecific esterase (α-naphthol acetate substrate - black granulation) is monocyte specific, while the chloroacetate esterase (naphthol ASD chloroacetate - red granulation) is granulocyte specific. These reactions should be helpful in some cases in distinguishing acute myelogenous from acute monocytic leukemia and are of value in distinguishing acute myelomonocytic leukemia from acute myelogenous leukemia and acute monocytic leukemia. In acute myelomonocytic leukemia, both granulocytic and monocytic markers are present simultaneously in the leukemic cells. Nonspecific esterase activity inhibited by fluoride has been described in red cell precursors - including megaloblasts in cases of untreated pernicious anemia, megaloblastoid rubricytes in cases of DiGuglielmo syndrome, and rubricytes in cases of severe untreated iron deficiency. The rubricyte series of cells from normal marrow lack nonspecific esterase activity. The fluoride inhibited nonspecific esterase reaction indicates monocytic origin and is unusual in T-lymphocyte cell malignancy.

Cytochemistry (CC) is especially employed in identification and classification of acute myeloid leukemia (AML). This role of CC is increasingly shared and diluted by the application of immunophenotypic and cytogenetic findings which

Table 2. Proposed WHO Classification of Acute Myeloid Leukemias (AML)

I.	AML with recurrent cytogenetic translocations
	AML with t(8;21) (q22; q22),
	Acute promyelocytic leukemia (AML with t(15;17) (q22; q11-12) and variants)
	AML with abnormal bone marrow eosinophils (inv (16) (p13; q22) or t(16;16) (p13; q11))
	AML with 11q23 abnormalities
II.	Acute myeloid leukemia with multilineage dysplasia with or without prior myelodysplastic syndrome
III.	AML, therapy related
	Alkylating agent related
	Epipodophyllotoxin related (some lymphoid)
	Other types
IV.	AML not otherwise categorized (extension of FAB groups)
	M0-M7 (see Table 3)
	Acute basophilic leukemia
	Acute panmyelosis with myelofibrosis
	Acute leukemia and transient myeloproliferative disorder followed by acute megakaryoblastic leukemia in Down syndrome
	Myeloid sarcoma

Adapted from Willman CL and Harris NL et al (see References).

(Continued)

Leukocyte Cytochemistry *(Continued)*

have led to accelerated activity in reclassification of the hematopoietic malignancies. The accompanying tables are derived from current and proposed classifications of the acute myeloid leukemias, notably those of the World Health Organization (WHO) and the French-American-British Cooperative Group (FAB).

Cytochemical stains are generally negative or do not contribute significantly to the diagnosis of acute lymphoblastic leukemia (ALL). A positive PAS reaction combined with negative myeloperoxidase, Sudan black B, and alpha-naphthyl butyrate esterase results continues to have a diagnostic role in differentiating lymphoblastic and myeloblastic leukemia.

Table 3. FAB / WHO Morphologic Classification With Cytochemistry of Acute Myeloid Leukemias

Subtype of AML		Morphology[1]	Cytochemistry[2]		
			MPO or SBB	CAE	NSE
M0	Myeloblastic minimally differentiated	≥20% myeloblasts without granules	-	-	-
M1	Myeloblastic without maturation	≥20% myeloblasts, some with scant granules <10% with maturation beyond blast stage	+	±	-
M2	Myeloblastic with maturation	≥20% myeloblasts with granules; ≥10% more mature cells; <20% monocytic cells	+	±	-
M3	Acute promyelocytic hypergranular	≥20% myeloblasts and promyelocytes with prominent granules	+++	+++	-
M3v	Acute promyelocytic microgranular (hypogranular) variant	Granules with procoagulant substances → DIC[3]			+ in 20% to 25% of cases
M4	Myelomonocytic (MM)	>20% myeloblasts, monoblasts, and promyelocytes	+	+	+ NaF inhibits
M4eo	MM with marrow eosinophilia	>20% monocytes; blood monocyte count >5 x 10³/mm³ if <20% monocytes			
M5a	Monocytic without differentiation	>80% monocytic cells of which >80% are large monoblasts	-	-	+ NaF inhibits
M5b	Monocytic with differentiation	>80% monocyte cells with monoblasts, promonocytes, monocytes			
M6	Erythroleukemia[4]	Erythroblasts (megaloblastic) >50% of nucleated marrow cells; >20% of remaining cells - myeloblasts	+	-	±
M7	Megakaryoblastic[5]	>20% blasts, megakaryoblasts; cytoplasmic budding may be present	-	±	±

[1]FAB criteria require 30% blasts. WHO proposed classification of 20% to 30% blasts as sufficient for lower limit in recognition that some cases with a 20% blast level have similar clinical outcome to those satisfying the higher (30%) FAB criteria.

[2]MPO = myeloperoxidase; SBB = Sudan black B; CAE = chloroacetate esterase; NSE = nonspecific esterase.

[3]Disseminated intravascular coagulation.

[4]Cytoplasmic vacuoles, strongly PAS positive may be present.

[5]On tissue sections, megakaryocytes may be PAS positive. Cytochemistry is nonspecific. Immunologic and/or ultrastructural (electron microscopic) studies may be needed to show platelet peroxidase in the leukemic blasts.

Table 4. Acute Myeloid Leukemias, Immunophenotypic Characteristics, Cytogenetic Abnormalities

FAB Category	Immunophenotype[1]	Chromosomal Abnormality
M0	CD13, CD33, CD11b, CD15	None unique
M1	CD13, CD33, HLA-DR, CD19	None unique
M2	CD13, CD15, CD33, HLA-DR, CD19	t(8;21) (q22; q22) in 20% to 25% of cases relatively good prognosis
M3	CD13, CD15, CD33	t(15;17), nearly diagnostic retinoic acid-α fusion retinoic acid may induce maturation
M4 M4eo	CD11, CD13, CD14, CD15, CD33, HLA-DR	inv(16); t(16;16) (p13;q22) (seen with m4eo) relatively good prognosis
M5 M5a	CD11b, CD11c, CD13, CD14, CD15, CD33, HLA-DR; may show CD34 expression	t(9,11) (p22; q23)
M6	CD71, glycophorin A mixed populations	Clonal abnormalities present in most cases, none specific -5/5q-, -7/7q-, and +8 are common and are also seen in therapy-induced leukemia
M7	CD33, CD41, CD42, CD61	inv(3), t(3;-) in older infants and children; t(1;22) (p13, q13)

[1]The CD antigens may be expressed either singly or in combination by the leukemic cells.

References

Harris NL, Jaffé ES, Diebold J, et al, "The World Health Organization Classification of Hematological Malignancies Report of the Clinical Advisory Committee Meeting, Arlie House, Virginia, November 1997," *Mod Pathol*, 2000, 13(2):193-207.

Manaloor EJ, Neiman RS, Heilman DK, et al, "Immunohistochemistry Can Be Used to Subtype Acute Myeloid Leukemia in Routinely Processed Bone Marrow Biopsy Specimens. Comparison With Flow Cytometry," *Am J Clin Pathol*, 2000, 113(6):814-22.

McKenna R, "Acute Myeloid Leukemia," *Practical Diagnosis of Hematologic Disorders*, 3rd ed, Chapter 27, Kjeldsberg CR, ed, Chicago, IL: ASCP Press, 2000, 399-430.

Nguyen AND, Milam JD, Johnson KA, et al, "A Relational Database for Diagnosis of Hematopoietic Neoplasms Using Immunophenotyping by Flow Cytometry," *Am J Clin Pathol*, 2000, 113(1):95-106.

Rosenthal N and Farhi DC, "Special Stains in the Diagnosis of Acute Leukemia," *Clin Lab Med*, 2000, 20(1):29-38.

Willman CL, "Acute Leukemias: A Paradigm for the Integration of New Technologies in Diagnosis and Classification," *Mod Pathol*, 1999, 12(2):218-28.

♦ **Leukocyte-Depletion Filters** see Filters for Blood on page 588

Leukocyte Esterase, Urine

Related Information

Bacterial Culture, Urine on page 246

Gram Stain on page 658

Nitrite, Urine on page 962

Urinalysis on page 1289

Synonyms Bacteria Screen, Urine; Esterase, Leukocyte, Urine

Test Includes Screening of urine for leukocyte esterase activity by dipstick is usually a part of urinalysis

Abstract A rapid indirect test for detection of bacteriuria. A positive test reflects the presence of neutrophils. Evaluation of urinary tract infection includes nitrite (also on reagent strips) microscopy, Gram stain, urine culture with colony count, and other methods. **The best test for urinary tract infection is culture.**

Specimen Random clean catch urine; preferably midstream, clean catch collection, catheterized specimen, bladder aspiration

Storage Instructions If the specimen cannot be processed within 2 hours, it should be refrigerated for other portions of urine evaluation.

Reference Interval Negative

(Continued)

Leukocyte Esterase, Urine (Continued)

Use The leukocyte esterase test is positive with intact or lysed white blood cells. The lysis of leukocytes that occurs when urine is allowed to stand intensifies the color reaction from release of esterase. The test performs best for specimens in which colony counts are >10^5 cfu/mL and when combined with nitrite: together the most enthusiastic reports find a specificity of 98.3%, sensitivity of 84%, positive predictive value of 84%, and negative predictive value of 98.3%. Sensitivity, however, was much worse in a Belgian study (Zaman, 1998).

Limitations There are numerous false positives and false negatives. **Leukocyte esterase** is unreliable as a screen. The urinalysis, including leukocyte esterase and nitrite, should not replace culture in symptomatic patients. For example, in women with symptomatic cystitis, both leukocyte esterase and nitrite miss approximately 1 in 5 patients who have positive cultures.

When laboratory testing is needed for a patient with symptoms of urinary sepsis, urine culture should be performed.

Methodology The substrate on the strip is indoxyl carbonic acid ester. Indoxyl is oxidized by atmospheric oxygen to indigo blue. The reaction time is 1 minute, but high sensitivity requires interpretation 5 minutes after immersion in the sample.

References

Bachman JW, Heise RH, Naessens JM, et al, "A Study of Various Tests to Detect Asymptomatic Urinary Tract Infections in an Obstetric Population," *JAMA*, 1993, 270(16):1971-4.

Semeniuk H and Church D, "Evaluation of the Leukocyte Esterase and Nitrite Urine Dipstick Screening Tests for Detection of Bacteriuria in Women With Suspected Uncomplicated Urinary Tract Infections," *J Clin Microbiol*, 1999, 37(9):3051-2.

Shaw ST Jr, Poon SY, and Wong ET, "Routine Urinalysis, Is the Dipstick Enough?" *JAMA*, 1985, 253(11):1596-600.

Van Nostrand JD, Junkins AD, and Bartholdi RK, "Poor Predictive Ability of Urinalysis and Microscopic Examination to Detect Urinary Tract Infection," *Am J Clin Pathol*, 2000, 113(5):709-13.

Zaman Z, Borremans A, Verhaegen J, et al, "Disappointing Dipstick Screening for Urinary Tract Infection in Hospital Inpatients," *J Clin Pathol*, 1998, 51(6):471-2.

Lidocaine, Serum or Plasma

Related Information

Synonyms Anestacon®; Baylocaine®; Dalcaine®; Dilocaine®; Duo-Trach®; LidoPen®; Lignocaine; Nervocaine®; Norocaine®; Octocaine®; Xylocaine®

Applies to GX

Abstract Lidocaine is a class IB antiarrhythmic used in therapy of ventricular arrhythmias. It is extensively metabolized to two active metabolites, monoethylglycinexylidide (MEGX) and glycinexylidide (GX). It is also used as a local anesthetic agent (5% lidocaine patch, Lidoderm®) for relief of pain associated with postherpetic neuralgia.

Specimen Serum or plasma

Container Red top tube, green top (heparin) tube, or lavender top (EDTA) tube

Sampling Time Draw specimens 12 hours after initiating therapy for arrhythmia prophylaxis, then every 24 hours thereafter. Obtain specimens every 12 hours when cardiac or hepatic insufficiency exists.

Collection Do not collect in gel-containing tubes.

Reference Interval Therapeutic: 1.5-5.0 µg/mL (SI: 6.4-21.4 µmol/L), up to 6.0 µg/mL (SI: 25.6 µmol/L) if necessary.

Critical Values At levels >6.0 µg/mL (SI: >25.6 µmol/L), there may be seizure activity. See Table C in the Therapeutic Drug Monitoring Introduction *on page 62*.

Possible Panic Range Toxic: >8.0 µg/mL (SI: >34.2 µmol/L); levels >15.0 µg/mL (SI: 64.5 µmol/L) are associated with **fatalities**.

Use Monitor therapeutic drug level and toxicity assessment

Methodology Immunoassay, gas chromatography (GC), high performance liquid chromatography (HPLC)

Additional Information

- Half-life: 1.5-2 hours
- Volume of distribution: 1.0-1.5 L/kg
- Protein binding: 60% to 80%

Oral absorption is rapid, but it is a high extraction or "first-pass" drug; therefore lidocaine is not bioavailable when given orally. It is 60% to 80% serum protein bound and lidocaine becomes more highly bound after myocardial infarction.

Adverse effects can be concentration related in many cases. Central nervous system effects such as confusion, dizziness, or blurred vision can be seen at the high end of the therapeutic range. Seizures, cardiovascular depression, tremors, and coma are seen usually at levels >8 µg/mL. Such effects may be due to an accumulation of metabolites, particularly MEGX. Elderly patients with CHF or acute myocardial infarct are at highest risk for these toxicities.

Interactions occur with drugs that change liver blood flow or plasma protein binding. Beta-blockers and cimetidine decrease liver blood flow, causing increased lidocaine levels. Anticonvulsants, quinidine, and oral contraceptives can change the plasma protein binding of lidocaine, causing toxicity. Other cardiovascular drugs can potentiate the cardiovascular effects of lidocaine when given concomitantly.

References

Alexander JH, Granger CB, Sadowski Z, et al, "Prophylactic Lidocaine Use in Acute Myocardial Infarction: Incidence and Outcomes From Two International Trials. The Gusto-I and Gusto-IIb Investigators," *Am Heart J*, 1999, 137(5):799-805.

Argoff CE, "New Analgesics for Neuropathic Pain: The Lidocaine Patch," *Clin J Pain*, 2000, 16(2 Suppl):S62-S66.

De Toledo JC, "Lidocaine and Seizures," *Ther Drug Monit*, 2000, 22(3):320-2.

♦ **LidoPen®** *see* Lidocaine, Serum or Plasma *on page 850*

♦ **Li-Fraumeni Syndrome** *see* p53, Functional Assay/Sequencing *on page 995*

♦ **Light Chains** *see* Immunoperoxidase Procedures *on page 780*

♦ **Light Chains** *see* Protein Electrophoresis, Serum *on page 1104*

♦ **Light Chains, Urine** *see* Protein Electrophoresis, Capillary Zone *on page 1103*

♦ **Light Chains, Urine** *see* Protein Electrophoresis, Urine *on page 1107*

♦ **Lignocaine** *see* Lidocaine, Serum or Plasma *on page 850*

♦ **Liley Test** *see* Bilirubin, Amniotic Fluid, Delta A450 *on page 261*

♦ **Limbitrol®** *see* Amitriptyline, Serum or Plasma *on page 149*

♦ **Linear Regression** *see page 11*

Lipase, Serum

Related Information

Amylase, Serum *on page 155*
Amylase, Urine *on page 157*
Bilirubin, Total, Serum *on page 265*
Body Fluid Amylase *on page 287*
Fat, Semiquantitative, Stool, Acid Steatocrit *on page 571*
Fat, Semiquantitative, Stool, Sudan III Stain *on page 572*

Synonyms Triacylglycerol Acylhydrolase

Applies to Body Fluid Lipase; Elastase, Serum

(Continued)

Lipase, Serum *(Continued)*

Abstract Lipase is synthesized in pancreatic acinar cells. Simultaneous determination of both lipase and amylase is widely recommended for patients with abdominal pain.

Specimen Serum; lipase (unlike amylase) is not applicable to urine. Lipase may be measured in pleural or peritoneal fluid.

Container Red top tube

Storage Instructions Stable 1 week at 25°C, 3 weeks at 4°C.

Reference Interval Substantial intermethod variability. An enzymatic colorimetric assay has an interval of 3-73 units/L.

Use Diagnose acute and chronic pancreatitis. Since amylase levels return to normal range first, assay of serum lipase is especially helpful in subjects who appear several days after onset.

Limitations Elevated in many diseases (cirrhosis, cholecystitis) and conditions (hemodialysis, elevated lipids). Elevated in patients taking many drugs (eg, acetaminophen, valproic acid, oral contraceptives, bethanechol, cholinergics, codeine, meperidine, methacholine, narcotics including morphine, pentazocine, secretin, calcitriol, cerivastatin, chlorothiazide, clozapine, diazoxide, didanosine, dideoxyinosine, estropipate, felbamate, hydrocortisone, mercaptopurine, metolazone, metronidazole, minocycline, nitrofurantoin, pegaspargase, prednisolone, and sulfamethoxazole).

Methodology Turbidimetric (using triolein), spectrophotometric, fluorometric, titrimetric, immunoassay

Additional Information Serum lipase is usually normal in those patients with elevated serum amylase, but without pancreatitis.

Laboratory and other factors useful to project severity of pancreatitis include C-reactive protein. Important concepts in recognition and management of clinically **severe acute pancreatitis** were recently tabulated. They include age older than 55 years, WBC >16,000/mm^3, glucose >200 mg/dL, LDH >350 IU/L, AST >250 IU/L, decreased Hct >10%, increased BUN >5 mg/dL, serum calcium <8 mg/dL, arterial PaO$_2$ <60 mm Hg, base deficit >4 mmol/L, and fluid sequestration >6 L. Amylase and lipase may be normal or minimally increased in **chronic pancreatitis**. Bilirubin and alkaline phosphatase may be abnormal. Fecal fat excretion may be increased if malabsorption develops.

References

Baron TH and Morgan DE, "Acute Necrotizing Pancreatitis," *N Engl J Med*, 1999, 340(18):1412-17 (review paper).

Mayo Medical Laboratories, "Method and Reference Value Change for Lipase, Serum," *Mayo Reference Services Communiqué*, Rochester, MN, 1999, 24(11).

Steer ML, Waxman I, and Freedman S, "Chronic Pancreatitis," *N Engl J Med*, 1995, 332(22):1482-90 (review paper).

Steinberg W and Tenner S, "Acute Pancreatitis," *N Engl J Med*, 1994, 330(17):1198-210.

Young DS, *Effects of Drugs on Clinical Laboratory Tests*, 5th ed, Volume 1: Listing by Test, Washington, DC: AACC Press, American Association for Clinical Chemistry, 2000, Section 3, 526-8.

♦ **Lipidoses** *see* Inherited Diseases of Metabolism and Cell Structure *on page 792*

Lipid Panel, Serum

Related Information

Apolipoprotein A-I, Serum *on page 202*
Apolipoprotein B-100, Serum *on page 203*
Cholesterol, Total, Serum or Plasma *on page 396*
High Density Lipoprotein Cholesterol, Serum *on page 729*
Lipids, Overview *on page 853*
Low Density Lipoprotein Cholesterol *on page 874*
Triglycerides, Serum or Plasma *on page 1275*

Test Includes This is an HCFA-defined panel which must include these three measurements:

- Cholesterol (total), serum
- High density lipoprotein cholesterol, serum
- Triglycerides, serum

In addition, many laboratories will calculate a low density lipoprotein cholesterol value based on these measurements, using the Friedewald formula:

LDLC (mg/dL) = cholesterol, total (mg/dL) - HDLC (mg/dL) - [triglycerides (mg/dL) / 5]

However, the Friedewald formula must **not** be used when any one of three conditions is present:
1) fasting TG ≥400 mg/dL
2) chylomicrons are present (usually a visible layer of "cream" following overnight refrigerated storage)
3) the patient has familial dysbetalipoproteinemia (see Lipids, Overview *on page 853*).

Patient Preparation An early morning specimen after a 14-hour fast is ideal.

Specimen Serum or plasma

Container Red top tube or lavender top (EDTA) tube

Storage Instructions If testing is not performed the same day the specimen is drawn, store at 4°C.

Reference Interval See tables in Lipids, Overview *on page 853*.

Use See Lipids, Overview *on page 853*.

Methodology See individual analytes.

References
Current Procedural Terminology (CPT™) 2002, American Medical Association, 2002, 71.

Internet Web Sites
www.cdc.gov/nceh/dls/crmln/crmln.htm
www.nhlbi.nih.gov

Lipids, Overview

Related Information
Apolipoprotein A-I, Serum *on page 202*
Apolipoprotein B-100, Serum *on page 203*
Cholesterol, Total, Serum or Plasma *on page 396*
C-Reactive Protein, Serum *on page 467*
High Density Lipoprotein Cholesterol, Serum *on page 729*
Homocyst(e)ine, Plasma *on page 741*
Lipid Panel, Serum *on page 852*
Lipoprotein (a), Serum *on page 861*
Low Density Lipoprotein Cholesterol *on page 874*
Mevalonic Acid, Urine or Amniotic Fluid *on page 911*
Triglycerides, Serum or Plasma *on page 1275*

Applies to Apo-A-I; Apo B-100; Cholesterol, Total; Chylomicronemia; High Density Lipoprotein Cholesterol (HDLC); Hypoalphalipoproteinemia; Intermediate Density Lipoprotein Cholesterol; Lipoprotein-Associated Phospholipase A_2; Low Density Lipoprotein Cholesterol (LDLC); Metabolic Syndrome; Triglycerides; Very Low Density Lipoprotein Cholesterol

Abstract The Adult Treatment Panel (ATP III) of the National Cholesterol Education Program (NCEP III) recommends that all adults have a fasting blood lipid profile in order to assess their risk for coronary heart disease (CHD). The results of this testing are interpreted as follows:

LDL cholesterol (LDLC):
- Optimal: <100 mg/dL
- Above optimal: 100-129 mg/dL
- Borderline high: 130-159 mg/dL
- High: 160-189 mg/dL
- Very high: ≥190 mg/dL

Total cholesterol (TC):
- Desirable: <200 mg/dL
- Borderline high: 200-239 mg/dL
- High: ≥240 mg/dL

HDL cholesterol (HDLC):
- Low (undesirable): <40 mg/dL
- High (desirable): ≥60 mg/dL

Complete assessment of coronary heart disease (CHD) risk requires the simultaneous consideration of six major risk factors:
- Age (women 55 years and older, men 45 years and older)
- LDLC >100 mg/dL

(Continued)

Lipids, Overview (Continued)

- HDLC <40 mg/dL
- Cigarette smoking
- Hypertension
- Family history of premature CHD (male first degree relative younger than 55 years, female first degree relative younger than 65 years)

The **highest risk status** in the ATP III framework is the presence of clinical CHD or other major clinical manifestations of atherosclerosis, such as abdominal aortic aneurysm, peripheral vascular disease, or carotid arterial disease. Such individuals have a >20% risk of a major coronary event within 10 years. Prior to 2001, diabetes mellitus was classified as a CHD risk factor, but in NCEP III, consonant with the findings in a 1998 study (Rader 2000), the presence of diabetes mellitus has been upgraded to a CHD equivalent. The wisdom of this has been challenged by the results of a study with both cross-sectional and cohort components (Law et al 2002).

Actual population distributions of TC, LDLC, HDLC, and triglycerides are available.

Patient Preparation Patient should be on stable diet for 3 weeks and should fast for at least 10 hours before collection of the specimen. However, when LDLC and HDLC are assayed by **direct** methods, fasting may not be required. See Preparation in the listing Cholesterol, Total, Serum or Plasma *on page 396.*

Specimen Serum or plasma

Container Red top tube or lavender top (EDTA) tube

Storage Instructions Lipoproteins are labile. Even stored at 4°C, analysis should not be delayed more than a few days.

Reference Interval A normative reference interval, traditionally the central 95% of a population distribution, is not applicable to cholesterol measurements. This is because, in the U.S., CHD is so prevalent, and has such a long symptom-free latent period, that it is not possible to define a reference population free of the disease. In addition, the relationship between LDLC and CHD, and between HDLC and CHD, is graded and continuous, with no apparent cutoffs or inflection points, separating normal persons from those with clinical CHD. Therefore, we recommend the interpretive guidelines indicated below. See Low Density Lipoprotein Cholesterol *on page 874,* High Density Lipoprotein Cholesterol, Serum *on page 729,* and Triglycerides, Serum or Plasma *on page 1275.*

Use

CORONARY HEART DISEASE (CHD) RISK

In the context of CHD, TC, and more importantly LDLC and HDLC are risk factors, **not** diagnostic tests. This important distinction is illustrated in the results of the original Framingham Study cohort (Castelli 1984).

A complete CHD risk assessment, based on Framingham Study data, can be obtained using the following tables. The term "smoker" refers to any cigarette usage in the past month.

Estimate of 10-Year Risk for Men
(Framingham Point Scores)

Total Cholesterol (mg/dL)	Points				
	20-39 (y)	40-49 (y)	50-59 (y)	60-69 (y)	70-79 (y)
<160	0	0	0	0	0
160-199	4	3	2	1	0
200-239	7	5	3	1	0
240-279	9	6	4	2	1
≥280	11	8	5	3	1

Adapted from "Executive Summary of the Third Report of the National Cholesterol Education Program (NCEP) Expert Panel on Detection, Evaluation, and Treatment of High Blood Cholesterol in Adults (Adult Treatment Panel III)," *JAMA,* 2001, 285(19):2486-97.

Estimate of 10-Year Risk for Men
(Framingham Point Scores)

Age (y)	Points
20-34	-9
35-39	-4
40-44	0
45-49	3
50-54	6
55-59	8
60-64	10
65-69	11
70-74	12
75-79	13

Adapted from "Executive Summary of the Third Report of the National Cholesterol Education Program (NCEP) Expert Panel on Detection, Evaluation, and Treatment of High Blood Cholesterol in Adults (Adult Treatment Panel III)," *JAMA*, 2001, 285(19):2486-97.

Estimate of 10-Year Risk for Men
(Framingham Point Scores)

	Points				
	20-39 (y)	40-49 (y)	50-59 (y)	60-69 (y)	70-79 (y)
Nonsmoker	0	0	0	0	0
Smoker	8	5	3	1	1

Adapted from "Executive Summary of the Third Report of the National Cholesterol Education Program (NCEP) Expert Panel on Detection, Evaluation, and Treatment of High Blood Cholesterol in Adults (Adult Treatment Panel III)," *JAMA*, 2001, 285(19):2486-97.

Estimate of 10-Year Risk for Women
(Framingham Point Scores)

	Points				
	20-39 (y)	40-49 (y)	50-59 (y)	60-69 (y)	70-79 (y)
Nonsmoker	0	0	0	0	0
Smoker	9	7	4	2	1

Adapted from "Executive Summary of the Third Report of the National Cholesterol Education Program (NCEP) Expert Panel on Detection, Evaluation, and Treatment of High Blood Cholesterol in Adults (Adult Treatment Panel III)," *JAMA*, 2001, 285(19):2486-97.

Estimate of 10-Year Risk for Men
(Framingham Point Scores)

HDL (mg/dL)	Points
≥60	-1
50-59	0
40-49	1
<40	2

Adapted from "Executive Summary of the Third Report of the National Cholesterol Education Program (NCEP) Expert Panel on Detection, Evaluation, and Treatment of High Blood Cholesterol in Adults (Adult Treatment Panel III)," *JAMA*, 2001, 285(19):2486-97.

(Continued)

Estimate of 10-Year Risk for Men
(Framingham Point Scores)

Point Total	10-Year Risk (%)
<0	<1
0	1
1	1
2	1
3	1
4	1
5	2
6	2
7	3
8	4
9	5
10	6
11	8
12	10
13	12
14	16
15	20
16	25
≥17	≥30

Adapted from "Executive Summary of the Third Report of the National Cholesterol Education Program (NCEP) Expert Panel on Detection, Evaluation, and Treatment of High Blood Cholesterol in Adults (Adult Treatment Panel III)," *JAMA*, 2001, 285(19):2486-97.

Estimate of 10-Year Risk for Men
(Framingham Point Scores)

Systolic BP (mm Hg)	Untreated	Treated
<120	0	0
120-129	0	1
130-139	1	2
140-159	1	2
≥160	2	3

Adapted from "Executive Summary of the Third Report of the National Cholesterol Education Program (NCEP) Expert Panel on Detection, Evaluation, and Treatment of High Blood Cholesterol in Adults (Adult Treatment Panel III)," *JAMA*, 2001, 285(19):2486-97.

Estimate of 10-Year Risk for Women
(Framingham Point Scores)

Age (y)	Points
20-34	-7
35-39	-3
40-44	0
45-49	3
50-54	6
55-59	8
60-64	10
65-69	12
70-74	14
75-79	16

Adapted from "Executive Summary of the Third Report of the National Cholesterol Education Program (NCEP) Expert Panel on Detection, Evaluation, and Treatment of High Blood Cholesterol in Adults (Adult Treatment Panel III)," *JAMA*, 2001, 285(19):2486-97.

Estimate of 10-Year Risk for Women
(Framingham Point Scores)

Total Cholesterol (mg/dL)	Points				
	20-39 (y)	40-49 (y)	50-59 (y)	60-69 (y)	70-79 (y)
<160	0	0	0	0	0
160-199	4	3	2	1	1
200-239	8	6	4	2	1
240-279	11	8	5	3	2
≥280	13	10	7	4	2

Adapted from "Executive Summary of the Third Report of the National Cholesterol Education Program (NCEP) Expert Panel on Detection, Evaluation, and Treatment of High Blood Cholesterol in Adults (Adult Treatment Panel III)," *JAMA*, 2001, 285(19):2486-97.

Estimate of 10-Year Risk for Women
(Framingham Point Scores)

HDL (mg/dL)	Points
≥60	-1
50-59	0
40-49	1
<40	2

Adapted from "Executive Summary of the Third Report of the National Cholesterol Education Program (NCEP) Expert Panel on Detection, Evaluation, and Treatment of High Blood Cholesterol in Adults (Adult Treatment Panel III)," *JAMA*, 2001, 285(19):2486-97.

Estimate of 10-Year Risk for Women
(Framingham Point Scores)

Systolic BP (mm Hg)	Untreated	Treated
<120	0	0
120-129	1	3
130-139	2	4
140-159	3	5
≥160	4	6

Adapted from "Executive Summary of the Third Report of the National Cholesterol Education Program (NCEP) Expert Panel on Detection, Evaluation, and Treatment of High Blood Cholesterol in Adults (Adult Treatment Panel III)," *JAMA*, 2001, 285(19):2486-97.

(Continued)

Lipids, Overview (Continued)

Estimate of 10-Year Risk for Women
(Framingham Point Scores)

Point Total	10-Year Risk (%)
<9	<1
9	1
10	1
11	1
12	1
13	2
14	2
15	3
16	4
17	5
18	6
19	8
20	11
21	14
22	17
23	22
24	27
≥25	≥30

Adapted from "Executive Summary of the Third Report of the National Cholesterol Education Program (NCEP) Expert Panel on Detection, Evaluation, and Treatment of High Blood Cholesterol in Adults (Adult Treatment Panel III)," *JAMA*, 2001, 285(19):2486-97.

After a risk assessment, many individuals are placed on lipid-lowering therapy which may include diet, exercise and, and drugs. For most patients, LDLC is the primary target of therapy. ATP III defined treatment goals depend on the total risk factor profile and these are indicated in the table *on page 859*.

Interlaboratory variation in cholesterol measurements was a major problem in the last half of the twentieth century. To minimize the patient care errors generated by this variability, the National Cholesterol Education Program (NCEP) convened a Laboratory Standardization Panel (LSP). In 1990, the LSP published national analytical goals for total cholesterol (TC): a) precision equivalent to a CV ≤3%, and b) total bias not to exceed 3%. Analytical goals have also been developed by the NCEP for LDLC and HDLC. For LDLC, the total analytic error should not exceed 12%; this criterion can be met when the bias does not exceed ±4%, and the analytical imprecision (CVa) does not exceed 4%. For HDLC, the total analytical error should not exceed 13%; this can be achieved if the bias is ≤5%, and the analytical imprecision (CVa) is <4%.

To assist clinical laboratories and the manufacturers of analytical equipment in achieving these goals, the CDC created a network of laboratories (the Cholesterol Reference Method Laboratory Network - CRMLN), each using the **reference method** and each meeting very stringent external proficiency testing standards. CRMLN provides traceability of clinical methods to the CDC reference method; there are CRMLN programs both for manufacturers of diagnostic equipment and for individual laboratories.

PRIMARY LIPID DISORDERS

Patients with **familial hypercholesterolemia** (an autosomal dominant disease) have elevated TC as a result of deficient or defective LDL receptors. Heterozygotes (1 in 500 persons) have TC values ~400-500 mg/dL, and, if untreated, often experience their first coronary event and signs of peripheral

LDL Cholesterol Goals and Cutpoints
for Therapeutic Lifestyle Changes (TLC)
and Drug Therapy in Different Risk Categories[1]

Risk Category	Goal (mg/dL)	Initiate Therapeutic Lifestyle Changes (mg/dL)	Consider Drug Therapy (mg/dL)
CHD or CHD risk equivalents (10-year risk >20%)	<100	≥100	≥130 (100-129: drug optional)[2]
2+ risk factors (10-year risk ≤20%)	<130	≥130	10-year risk 10%-20%: ≥130 10-year risk <10%: ≥160
0-1 risk factors[3]	<160	≥160	≥190 (160-189: LDL-lowering drug optional)

[1]LDL: low-density lipoprotein; CHD: coronary heart disease.

[2]Some authorities recommend use of LDL-lowering drugs in this category if an LDL cholesterol level <100 mg/dL cannot be achieved by therapeutic lifestyle changes. Others prefer use of drugs that primarily modify triglycerides and HDL (eg, nicotinic acid or fibrate). Clinical judgment also may call for deferring drug therapy in this subcategory.

[3]Almost all people with a 0-1 risk factor have a 10-year risk <10%; thus, 10-year risk assessment in people with a 0-1 risk factor is not necessary.

Adapted from "Executive Summary of the Third Report of the National Cholesterol Education Program (NCEP) Expert Panel on Detection, Evaluation, and Treatment of High Blood Cholesterol in Adults (Adult Treatment Panel III)," *JAMA*, 2001, 285(19):2486-97.

vascular disease in their fourth or fifth decade. Homozygotes (much rarer) have TC ~1000 mg/dL, premature vascular disease, and often have tendon xanthomas. **Familial defective apolipoprotein B** (also autosomal dominant) has a very similar biochemical phenotype, but is due to defective ligand binding with markedly delayed clearance of LDL cholesterol. Patients with **familial hyperalphalipoproteinemia** have elevated TC, but fractionation reveals that the major increase is in HDL cholesterol, not LDL cholesterol. Such individuals have a decreased incidence of CAD.

Lipoprotein lipase deficiency and **apoprotein C II deficiency** present with markedly elevated triglycerides (see Triglycerides, Serum or Plasma *on page 1275*); TC may be normal or slightly elevated.

Familial chylomicronemia is a rare, autosomal recessive disorder characterized by very high serum triglycerides (TG). Either lipoprotein lipase or apo-C II is defective or deficient.

In **familial dysbetalipoproteinemia,** an autosomal recessive condition also known as type III hyperlipoproteinemia, serum TG is very high and may be numerically equal to serum TC which is **not** elevated. Xanthomas are often present, together with premature coronary and peripheral atherosclerosis. Electrophoresis reveals an abnormal "broad beta" band. Definitive diagnosis, which includes ultracentrifugation and identification of an aberrant apoE genotype requires referral to a specialized lipid laboratory.

Familial hypoalphalipoproteinemia (low HDLC) is probably a heterogeneous group of individuals having the common denominator of low serum HDLC. Within this group are marked differences in premature CAD, xanthomas, and many other features.

Polygenic hypercholesterolemia is the name given to individuals with high LDLC, but whose findings do not support a diagnosis of FH or any other of the foregoing diagnostic categories. Since this is a default category, it is almost certain that the individuals assigned thereto are a heterogeneous grouping, and one that will be altered on the basis of additional research.

SECONDARY HYPERLIPIDEMIAS

Numerous drugs and certain diseases produce changes in blood lipids. Of most concern are those which produce elevations of TC (eg, high fat diet, hypothyroidism, Cushing syndrome, oral contraceptives, chronic liver disease) and TG (high carbohydrate diet, excessive ethanol, Cushing syndrome, hypothyroidism, pregnancy, obesity, pancreatitis, diabetes mellitus, diuretics, estrogen use, beta blocker therapy).

(Continued)

Lipids, Overview *(Continued)*

Methodology See individual test listings.

Additional Information The gold standard for the assay of cholesterol subsets (HDL, IDL, LDL, VLDL) is ultracentrifugation. The cholesterol methods used clinically are close approximations to the gold standard, but the results are not precisely the same. For example, clinical laboratory methods for LDLC and TG both include variable proportions of intermediate-density lipoprotein cholesterol (IDLC). Some patients with rare dyslipidemias (*vide supra*) require ultracentrifugal analysis for diagnosis and appropriate management.

Since an individual will have many cholesterol measurements over a lifetime, **biologic variability** is important and should be kept in mind by clinicians advising patients. The mean intraindividual biologic coefficient of variation (CVb) for TC ranges from 5% to 10%. The CVb is higher in individuals with very high cholesterol values. Cooper and colleagues (CDC) have recommended that intraindividual CVb can be minimized by measuring two or more TC values and calculating a relative range (RR); the RR is "...the difference between the lowest and highest concentration values observed for a person, divided by the mean of all the observed values." Equally important for multiple measurements over the long-term is **intermethod variability**. Historically, this has been a major obstacle to useful cholesterol results. The Laboratory Standardization Panel (LSP), created by the National Cholesterol Education Program (NCEP), recommended in 1988 that laboratories achieve a bias not exceeding ±3%, and in addition, precision sufficient that the analytic coefficient of variation (CVa) also not exceed ±3%. An additional recommendation was that all clinical cholesterol methods be traceable to the CDC reference method (a modified Abell-Kendall technique) or to the National Institute for Standards and Technology (NIST) definitive method (isotope-dilution mass spectrometry).

Some experts believe that the most accurate risk information is provided by measuring the apoproteins themselves (eg, apo-A-I instead of HDLC, and apo B-100 instead of LDLC).

Quite a different approach (Ramsay et al 1996) to cholesterol testing, recommended by some British scientists, involves assessing individuals for blood pressure, smoking, diabetes, and left ventricular enlargement risk factors. Cholesterol testing is provided only for those individuals who have, based on these other risk factors, a >3% estimated likelihood of a coronary event in the present year. While we do not endorse this practice, we present it for completeness.

A different view is that that physicians have focused excessively on "normalizing" the individual components of risk, such as cholesterol. Instead, they assert, attention should focus on **all** modifiable risk factors so that overall risk is reduced to the greatest degree possible. For example, a 30-year old woman with LDLC of 160 mg/dL and no other sources of increased risk may receive no treatment, or may receive a diet or medication which targets a mild reduction in LDLC. On the other hand, a 50-year old man with the same LDLC, but who also has hypertension and diabetes, will receive multiple modalities of therapy (statin drugs, diet, exercise) aimed at a maximum reduction in total risk.

Increased risk for CHD is also conferred by elevated serum Homocyst(e)ine, Plasma *on page 741* and elevations in markers of inflammation and coagulation, such as C-Reactive Protein *on page 467*, Fibrinogen *on page 583*, and Plasminogen *on page 1042*. The following graphic compares the efficiency of these markers in the prediction of CHD.

Cholesterol ratios. Between 1970 and 1990, certain cholesterol **ratios** enjoyed some popularity as ways to report cholesterol fractions. Most commonly used was the **ratio of TC:HDLC**; less often the **ratio of LDLC:HDLC**. In both cases, a low value was desirable, since this implied a relatively large concentration of "good cholesterol", HDLC, and/or a small concentration of TC or "bad cholesterol", LDLC. These ratios are no longer used for the obvious reason that appropriate clinical management requires knowing, and acting on, each value separately. For example, an "elevated" **TC:HDLC ratio** of 5 can be due to 1) a very low HDLC, 2) a very high LDLC, or 3) both; and each of these possibilities has a different clinical management. But using the ratio, instead of the individual values, obliterates the clinically crucial information about which component (or both) is at an undesirable level. Moreover, cholesterol ratios are not used in contemporary management protocols, such as that described in **Coronary Heart Disease (CHD) Risk.**

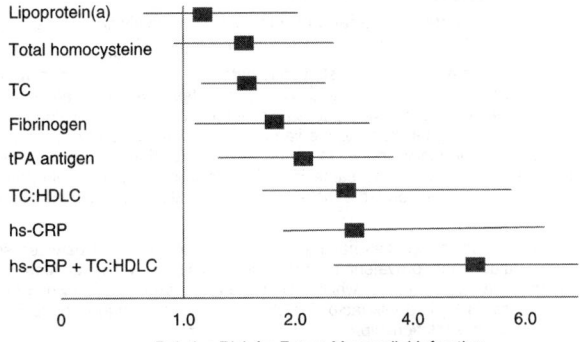

Relative Risk for Future Myocardial Infarction

Relative risk for future myocardial infarction among apparently healthy middle-aged men in the *Physicians' Health Study*, according to baseline levels of lipoprotein(a), total plasma homocysteine, total cholesterol (TC), fibrinogen, tissue-type plasminogen activator (tPA) antigen, the ratio of total cholesterol to high-density lipoprotein cholesterol (HDLC), and high-sensitivity C-reactive protein (hs-CRP). For consistency, risks are computed for men in the top compared with the bottom quartile for each marker.

From Ridker PM, "Evaluating Novel Cardiovascular Risk Factors: Can We Better Predict Heart Attacks?" *Ann Intern Med*, 1999, 130(11):933-7.

References
Castelli WP, "Epidemiology of Coronary Heart Disease: The Framingham Study," *Am J Med*, 1984, 76(2A):4-12.

"Executive Summary of the Third Report of the National Cholesterol Education Program (NCEP) Expert Panel on Detection, Evaluation, and Treatment of High Blood Cholesterol in Adults (Adult Treatment Panel III)," *JAMA*, 2001, 285(19):2486-97.

Myers GL, Kimberly MM, Waymack PP, et al, "A Reference Method Laboratory Network for Cholesterol: A Model for Standardization and Improvement of Clinical Laboratory Measurements," *Clin Chem*, 2000, 46(11):1762-72.

Rader DJ, "Inflammatory Markers of Coronary Risk," *N Engl J Med*, 2000, 343(16):1179-82.

Ramsay LE, Haq IU, Jackson PR, et al, "Targeting Lipid-Lowering Drug Therapy for Primary Prevention of Coronary Disease: An Updated Sheffield Table," *Lancet*, 1996, 348(9024):387-8.

Internet Web Sites
www.cdc.gov/nceh/dls/crmln/crmln.htm
www.nhlbi.nih.gov

♦ **Lipiduria** *see* Fat, Urine *on page 573*

Lipoprotein (a), Serum
Related Information
Apolipoprotein A-I, Serum *on page 202*
Apolipoprotein B-100, Serum *on page 203*
Apolipoprotein E, Plasma *on page 204*
Lipid Panel, Serum *on page 852*
Lipids, Overview *on page 853*
Low Density Lipoprotein Cholesterol *on page 874*

Synonyms Lp(a)

Applies to Lp(a)-C

Abstract Lp(a) is thought to be a risk factor for coronary heart disease (CHD) and cerebrovascular disease. Lp(a) can be measured using a variety of methods, but there is a high degree of intermethod variability in results. At present, two approaches are widely used: immunochemical mass assays [reported as Lp(a)] and cholesterol-based mass assays [reported as Lp(a)-C]. Presently, three assay formats are used:

   1) an ELISA format with monoclonal antibodies against unique apo(a) epitopes

(Continued)

Lipoprotein (a), Serum *(Continued)*

2) a sandwich ELISA with an antiapo(a) capture antibody and an antiapoB signal antibody

3) a cholesterol-based assay which requires a relatively complex separation of lipoproteins prior to the cholesterol measurement

Across all human populations studied to date, the naturally occurring values of Lp(a) vary by more than 1000-fold. Values in whites have a skewed distribution while values in blacks have a gaussian distribution. The median value in blacks is three times higher than the median in whites. The fact that CHD is equally prevalent in the black and white populations implies that the significance of a particular value is not the same in whites as it is in blacks. Therefore, the reference intervals, and the risk-factor cutoff values, should be ethnically based.

Many case-control studies have shown a positive correlation between serum Lp(a) values and prevalent coronary heart disease. While at least seven prospective studies (one of which included only postmenopausal women) have shown that Lp(a) is a risk factor for CHD; at least three studies have failed to confirm such a relationship.

Specimen Serum

Container Red top tube

Storage Instructions Serum is stable at 4°C for 1 week and frozen for months.

Reference Interval There are no well documented reference values based on large populations, and there is no large database establishing risk percentiles. Traditionally, Lp(a) mass values >30 mg/dL (SI: 1.05 mmol/L) have been classified as reflecting a high risk. Other authorities prefer a cutoff >25 mg/dL (SI: 0.88 mmol/L).

A report from the Framingham Offspring Study, including 3332 white subjects, and using a cholesterol-based assay, found the following reference intervals, based on the central 90% of values: Lp(a)-cholesterol:

- male: 1.35-19.62 mg/dL (SI: 0.035-0.508 mmol/L)
- female: 1.24-20.06 mg/dL (SI: 0.032-0.519 mmol/L)

Lp(a) mass assays performed on a subset of 1000 of these subjects showed that, although the Lp(a) mass result was always higher than the corresponding Lp(a)-C result, the results were strongly correlated.

Use Lp(a) measurements are not recommended for general screening. Some investigators do, however, recommend Lp(a) measurements in persons who have a family history of CHD, stroke or "elevated lipid values". **To be useful in practice, Lp(a) results should be reported with ethnically-specific, and gender-specific, population based reference values; moreover, the result should be accompanied by an interpretive statement indicating which reference population percentile corresponds to the patient result.** Results that exceed the 80th percentile are often interpreted as indicative of high risk.

Methodology

- Lp(a) mass: Enzyme-linked immunosorbent assay (ELISA), radial immunodiffusion, radioimmunoassay, latex immunoassay, immunonephelometry, electroimmunodiffusion, immunoturbidimetric, fluorescence assay
- Lp(a) cholesterol: Ultracentrifugation followed by spectrophotometry
- Apo(a) phenotypes: Immunoblotting

Additional Information

Apo(a) phenotypes. A case-controlled study demonstrates a strong correlation between apo(a) phenotype and the age at onset of CHD.

Correlates of high Lp(a) values include a wide array of pathologic conditions, including preeclampsia, recurrent fetal loss, early renal insufficiency, left atrial thrombus, childhood thromboembolism, and children without growth hormone deficiency who are treated with growth hormone.

Risk factor modification. Niacin reduces Lp(a) levels. In a large study (n=2759) of postmenopausal women, treatment with estrogen and progesterone was associated with a lowering of Lp(a) values; this effect was most pronounced in women with the highest baseline Lp(a) values. Other drugs which may lower Lp(a) levels include tamoxifen and anabolic steroids.

References

Bostom AG, Gagnon DR, Cupples LA, et al, "A Prospective Investigation of Elevated Lipoprotein(a) Detected by Electrophoresis and Cardiovascular Disease in Women: The Framingham Heart Study," *Circulation*, 1994, 90(45):1688-95.

Bostom AG, Cupples LA, Jenner JL, et al, "Elevated Plasma Lipoprotein(a) and Premature Coronary Heart Disease in Framingham Men: A Prospective Study," *JAMA*, 1996, 276(7):544-8.

Gazzaruso C, Garzaniti A, Buscaglia P, et al, "Association Between Apolipoprotein(a) Phenotypes and Coronary Heart Disease at Young Age," *J Am Coll Cardiol*, 1999, 33(1):157-63.

Marcovina SM and Koschinsky ML, "Lipoprotein(a) as a Risk Factor for Coronary Artery Disease," *Am J Cardiol*, 1998, 82(12A):57U-66U.

Schaefer EJ, Lamon-Fava S, Jenner JL, et al, "Lipoprotein(a) Levels and Risk of Coronary Artery Disease in Men: The Lipid Research Clinics Coronary Primary Prevention Trial," *JAMA*, 1994, 271:999-1003.

Seman LJ, DeLuca C, Jenner JL, et al, "Lipoprotein(a) Cholesterol and Coronary Heart Disease in the Framingham Heart Study," *Clin Chem*, 1999, 45(7):1039-46.

Shlipak MG, Simon JA, Vittinghoff E, et al, "Estrogen and Progestin, Lipoprotein(a), and the Risk of Recurrent Coronary Heart Disease Events After Menopause," *JAMA*, 2000, 283(14):1845-52.

Wald NJ, Law M, Watt HC, et al, "Apolipoproteins and Ischaemic Heart Disease: Implications for Screening," *Lancet*, 1994, 343:75-9.

♦ **Lipoprotein-Associated Phospholipase A$_2$** *see* Cholesterol, Total, Serum or Plasma *on page 396*

♦ **Lipoprotein-Associated Phospholipase A$_2$** *see* Lipids, Overview *on page 853*

♦ **Liquid Ecstasy** *see* Gamma Hydroxybutyrate, Serum or Urine *on page 630*

♦ **Liquid Lady** *see* Cocaine (Cocaine Metabolite), Qualitative, Urine or Hair *on page 427*

♦ **Liquid X** *see* Gamma Hydroxybutyrate, Serum or Urine *on page 630*

♦ **Lithane®** *see* Lithium, Serum *on page 863*

Lithium, Serum

Related Information

Antidepressants, Cyclic, Serum or Plasma *on page 171*

Haloperidol, Serum or Plasma *on page 664*

Verapamil, Serum or Plasma *on page 1306*

Synonyms Carbolith®; Cibalith-S®; Eskalith®; Lithane®; Lithobid®; Lithonate®; Lithotabs®; Phasal®

Abstract Lithium is used in the treatment of the manic phase of affective disorders, mania and particularly for manic-depressive illness. It is also used for depression, aggression, post-traumatic stress disorder, and conduct disorder in children. It affects cation transport across cell membrane in nerve and muscle cells and influences reuptake of serotonin and/or norepinephrine.

Aftercare Follow urine osmolality, EKGs, thyroid profile, BUN, creatinine, and sodium. Avoid sodium depletion.

Specimen Serum

Container Red top tube

Sampling Time Draw sample 12 hours after the last dose. Steady-state occurs at 90-120 hours.

Storage Instructions Refrigerate.

Causes for Rejection Specimen collected in tube containing lithium heparin, hemolysis

Reference Interval Therapeutic: 0.6-1.2 mEq/L (SI: 0.6-1.2 mmol/L), for acute mania; 0.8-1.0 mEq/L (SI: 0.8-1.0 mmol/L) for protection against future episodes in most patients with bipolar disorder. Elderly patients are maintained on the lower side of the therapeutic range.

Possible Panic Range See table.

Lithium (Acute Ingestion)

Serum Level	Symptoms
1.5-2.5 mEq/L	Polyuria, blurred vision, weakness, lethargy, dizziness, increased reflexes, fasiculations
2.5-3.0 mEq/L	Myoclonic twitching, incontinence, stupor, restlessness, coma
>3.0 mEq/L	Seizures, hypotension, cardiac arrhythmias
>4.0 mEq/L	Coma, death

(Continued)

Lithium, Serum *(Continued)*

Use Monitor therapeutic drug level, support compliance, avoid intoxicating levels; evaluate coma

Limitations Lithium toxicity, including severe neurotoxic effects, can occur with normal serum lithium levels. Instances of acute intoxication may be accompanied by high serum lithium levels without clinical evidence of neurotoxic effects. Lithium penetrates neurons slowly. Thiazides can cause significant rise in serum lithium.

Methodology Flame photometry, atomic absorption spectrophotometry (AA), ion-selective electrode (ISE)

Additional Information
- Half-life: 18-24 hours
- Volume of distribution: 0.7-1.0 L/kg
- Protein binding: 0%
- Bioavailability: 100%
- Elimination: unchanged, >98% renal
- Peak concentration: 2-5 hours

Peak concentrations occur 1-2 hours after a dose of regular lithium and 4-5 hours after a dose of the sustained-release form. Lithium is cleared by the kidney. It is filtered at the glomerulus and actively reabsorbed at the proximal tubule, much like sodium. The clearance of lithium is increased in pregnancy and when sodium supplements are given. The clearance is decreased in renal impairment, dehydration, and when patients are hyponatremic. Patients should try to maintain a consistent intake of sodium while on this drug.

Acute lithium toxicity is neuro- and nephrotoxic. Concentration-related side effects include weakness, muscle weakness, tremor, and confusion. Gastrointestinal effects such as nausea and vomiting can be lessened if the patient is given an extended release product. A fully developed case of intoxication is characterized by coma to semicoma, rigidity, hyperactive reflexes, and seizures at times. A high incidence of pulmonary complications is recognized.

Varying degrees of nephrogenic diabetes insipidus have been reported to occur in 33% of lithium treated patients due to inhibition of antidiuretic-hormone-induced water transport in the kidney. Chronic lithium administration has a goitrogenic effect in 4% of lithium-treated patients. Lithium administration results in slightly decreased serum T_4 levels and transiently elevated levels of TSH in nearly 33% of these subjects. Thyroxine treatment before or during lithium treatment has been proposed. Lithium affects the cardiac conduction system by incomplete substitution for other cations, especially sodium and potassium. These electrolyte changes account for the usually unimportant and reversible T-wave depressions observed in 10% to 20% of patients on lithium therapy.

Creatinine for monitoring glomerular filtration rate, beta-2-microglobulin for tubular dysfunction, and thyroid function testing should be done on a regular basis in patients on long-term lithium treatment.

References

Dunner DL, "Optimizing Lithium Treatment," *J Clin Psychiatry*, 2000, 61(Suppl 9):76-81.

Fawcett JA, "Lithium Combinations in Acute and Maintenance Treatment of Unipolar and Bipolar Depression," *J Clin Psychiatry*, 2003, 64(Suppl 5):32-7.

Groleau G, "Lithium Toxicity," *Emerg Med Clin North Am*, 1994, 12(2):511-31.

Sadosty AT, Groleau GA, and Atcherson MM, "The Use of Lithium Levels in the Emergency Department," *J Emerg Med*, 1999, 17(5):887-91.

♦ **Lithobid**® *see* Lithium, Serum *on page 863*

♦ **Lithonate**® *see* Lithium, Serum *on page 863*

♦ **Lithotabs**® *see* Lithium, Serum *on page 863*

♦ **Liu-Shan-Wan** *see* Digoxin, Serum *on page 512*

Liver Biopsy

Related Information

Acetaminophen, Serum *on page 90*

Alanine Aminotransferase, Serum *on page 116*

Albumin, Serum *on page 120*

Alkaline Phosphatase, Serum *on page 127*

Synonyms Needle Biopsy of Liver

Applies to Copper, Hepatic; Hepatic Iron Index; Iron, Hepatic; Toxic Reactions, Liver; Transjugular Needle Biopsy of Liver

Test Includes Needle biopsies of the liver may be acquired by percutaneous or transjugular routes. Investigation includes light microscopy, commonly with a number of special stains. Immunohistochemistry is often helpful. Electron microscopy is occasionally needed. *In situ* hybridization techniques may be helpful. Liver concentrations of copper or iron can be provided for evaluation of Wilson disease or hemochromatosis, respectively. Microbiologic culture may be required.

Abstract Liver biopsy is often needed in patients who have persistent/unexplained abnormalities of liver function tests. Results of history, physical examination, clinical evaluation, and testing must be provided to the histopathologist who works up and holds responsibility for interpretation of the liver biopsy.

Patient Preparation Procedures and risks of the procedure are explained and consent is required. All aspirin products and nonsteroidal agents must be discontinued at least 7 days beforehand. If taking oral anticoagulants (Coumadin®), hospitalization is required to convert to heparin therapy before biopsy. Screening laboratory studies ordered 24-48 hours in advance commonly include CBC (with platelet count), PT/PTT, BUN, bleeding time, type and screen or type and crossmatch for possible transfusion, additional to careful history and physical examination. Patients with diffuse liver disease and at significant risk for bleeding complications may be candidates for transjugular needle biopsy. See table in listing Bilirubin, Total, Serum *on page 265* and diagrams in listings Alkaline Phosphatase, Serum *on page 127* and Aspartate Aminotransferase, Serum *on page 216*. Electrolytes are usually optional. If pneumonia or pleural effusion is suspected, PA and lateral chest x-ray is obtained.

Aftercare In general, patient is monitored in a recovery area with frequent vital signs postbiopsy. If hypotension, tachycardia, fever, rigidity of abdomen, or uncontrolled pain occurs, physician should be notified immediately and an intravenous line placed. Some physicians recheck hematocrit 24 hours after the procedure.

Specimen At least two to three liver cores, each >2 cm in length are desirable.

Collection Tissue fixation for light microscopy: specimen is usually fixed in 10% buffered formalin within 1 minute; alternatives include Zenker's fluid or a (Continued)

Liver Biopsy (Continued)

Zenker modification. A specimen from subjects with cystinosis should be separately alcohol-fixed and so labeled. For transmission electron microscopy, 1 mm cubes of specimen are fixed immediately in glutaraldehyde, but EM is not often needed. Copper concentrations can be assayed; see below.

Use This procedure, by nature, is invasive. In most cases, noninvasive imaging studies such as CT scan or ultrasound are obtained first. **Indications for liver biopsy include:**

- persistent elevations of **AST and ALT** concentrations to > twice the upper limits of reference interval, abnormalities of other liver-related chemistry assays (eg, in selected patients with steatosis/steatohepatitis following other measures). Definitive diagnosis of nonalcoholic fatty liver disease requires a liver biopsy, which also can provide evidence of advanced disease.
- chronic hepatitis, with or without cirrhosis, to identify cases of autoimmune hepatitis, and the entities included in its differential diagnosis; evaluation of nonalcoholic steatohepatitis (NASH); grading and staging of disease severity (eg, chronic hepatitis B,C). Patients with positive results for hepatitis B virus DNA and hepatitis B$_e$ antigen should be considered for liver biopsy. Chronic hepatitis is itself defined as a necroinflammatory disease lasting longer than 6 months.
- suspected cases of hepatic cirrhosis, to confirm and grade the diagnosis and, if possible, establish etiology (eg, chronic hepatitis C infection, alcohol, alpha$_1$-antitrypsin deficiency, primary biliary cirrhosis vs primary sclerosing cholangitis); assess and stage level of activity; assess complications
- liver biopsy is used in diagnosis of hemochromatosis (when the HFE analysis (genetic test) is negative in the presence of iron overload, and for detection of cirrhosis) and is used in Wilson disease; quantitative estimation of iron and copper respectively. See Hereditary Hemochromatosis DNA Test on page 718 and Wilson Disease in the Disease Index.
- portal hypertension
- intrahepatic cholestasis of unknown etiology in which other studies of biliary obstruction are negative; staging of primary biliary cirrhosis (see Antimitochondrial Antibody on page 183)
- instances of disorders of bilirubin metabolism (eg, Dubin-Johnson syndrome)
- selected cases of fever of unknown origin (eg, tuberculosis, brucellosis); a portion of biopsy can be cultured for appropriate organisms
- infectious disease (eg, Q fever); extrapulmonary P. carinii infection is reported in liver and spleen in immunocompromised individuals
- suspected liver disease in the known alcoholic patient, to confirm alcoholic liver disease, exclude alternative causes of liver disease, stage and assess disease activity
- diagnosis of benign and malignant tumors (eg, hepatoma, metastatic neoplasms)
- recognition and staging of lymphoma
- suspected multisystem and/or infiltrative disease with liver involvement in which other diagnostic techniques have not been fruitful (eg, sarcoidosis, amyloidosis, tuberculosis, parasitic infestation, glycogen storage disease)
- unexplained hepatomegaly
- selected cases of hepatitis of unknown etiology, in order to try to differentiate viral from drug-induced etiologies (not always possible) or to assess complications, such as cholestasis; differential diagnosis of drug-induced hepatic disease; see tables on following page.
- acute hepatitis without explained etiology; in protracted cases
- evaluation of efficacy of or adverse response to treatment (eg, patients on methotrexate)
- following bone marrow or liver transplantation

- investigate inborn errors of metabolism (eg, Wilson disease, α_1-antitrypsin deficiency, glycogen storage disease, Gaucher disease, other storage diseases)
- instances of certain vascular abnormalities (eg, veno-occlusive disease)

Liver biopsy is less useful in:

- acute hepatitis A or B infection, unless the diagnosis is in question
- extrahepatic biliary obstruction in which percutaneous transhepatic cholangiography and endoscopic retrograde cholangiopancreatography (ERCP) are considered first-line procedures
- fluid-filled liver cysts detected on ultrasound or CT scan, probably more amenable to guided thin needle aspiration first

Table 1. Types of Toxic Reactions Occurring in the Liver

Type of Reaction	Examples of Agents
Direct reaction	Acetaminophen, carbon tetrachloride, mushrooms, phosphorus
Idiosyncratic reaction	Isoniazid, disulfiram, propylthiouracil[1]
Toxic-allergic reaction	Halothane, isoflurane, ticrynafen
Allergic hepatitis	Phenytoin, amoxicillin-clavulanic acid, sulfonamides
Cholestatic reaction	Chlorpromazine, erythromycin estolate, estradiol, captopril, sulfonamides
Granulomatous reaction	Diltiazem, quinidine, phenytoin, procainamide
Chronic hepatitis	Nitrofurantoin, methyldopa, isoniazid, trazodone
Alcoholic hepatitis-like reaction	Amiodarone, perhexiline maleate, valproic acid
Microvesicular steatosis	Tetracyclines, aspirin, zidovudine, didanosine, fialuridine
Fibrosis or cirrhosis alone	Methotrexate, vitamin A, methyldopa
Veno-occlusive disease	Cyclophosphamide, other chemotherapeutic agents, herbal teas
Ischemic damage	Cocaine, sustained-release nicotinic acid, methylenedioxyamphetamine

[1]There are hundreds of other agents that can cause idiosyncratic reactions.

Adapted from Lee WM, "Drug-Induced Hepatotoxicity," *N Engl J Med*, 1995, 333(17):1121.

Table 2. Relative Degree of Increase of Aminotransferases in Toxic Hepatic Injury

Toxicant	Lesion		Degree of Increase in Serum Enzyme Levels	
	Zonal Necrosis	Steatosis	AST	ALT
CCl_4	+	+	4+	3+
Thioacetamide	+	−	4+	3+
Tetracycline	−	+	2	+
Ethionine	−	+	+	−
Phosphorous	±	+	1-2+	1-2+

ALT = alanine aminotransferase; AST = aspartate aminotransferase; CCl_4 = carbon tetrachloride; 4+ refers to relative degree of increase.

Modified from Zimmerman HJ, *Hepatotoxicity: The Adverse Effects of Drugs and Other Chemicals on the Liver*, 2nd ed, Philadelphia, PA: Lippincott Williams and Wilkins, 1999.

Limitations Failure to heed accepted contraindications may lead to bleeding episodes.

Contraindications to percutaneous liver biopsy include:

- if AST and ALT are not > twice the upper limit of reference interval and a chronic liver disease has not been found, observation may be the best course.

(Continued)

Liver Biopsy *(Continued)*

- impaired hemostasis, accepted as prothrombin time more than 3 seconds over control, PTT more than 20 seconds over control, thrombocytopenia <50,000/mm^3, or markedly prolonged bleeding time (≥10 minutes)
- history of bleeding which has gone without diagnosis
- severe anemia (Hb <9.5 g/dL)
- local infection near needle entry site, such as right-sided empyema, right lower lobe pneumonia, right subphrenic abscess, local cellulitis, infected ascites or peritonitis
- tense ascites (low yield technically, risk of leakage)
- high-grade extrahepatic biliary obstruction with jaundice (increased risk of bile peritonitis)
- septic cholangitis
- possible hemangioma, other vascular tumors
- possible echinococcal (hydatid) cyst
- lack of adequate facilities for blood transfusion, lack of appropriate blood
- uncooperative patient
- liver biopsy is more hazardous when performed on outpatients who have cirrhosis or neoplasms, since these are the categories associated with mortality (from hemoperitoneum)
- relative contraindications include morbid obesity, hemophilia

Complications: Significant morbidity has been estimated at 1%. Fatality rate of up to 0.1% is recognized with thick needle biopsy. More commonly seen complications include:

- pain
- hemorrhage - minor episodes are common. Significant hemorrhage is infrequent but is the most common cause of death from liver biopsy. Several series have estimated an incidence of ~0.2%, but Sherlock (1984) reported 40 patients out of 6379 who required transfusion for intraperitoneal bleeding. Specific sites include the abdominal cavity (hemoperitoneum), liver capsule (capsular hematoma), liver parenchyma (intrahepatic hematoma), biliary tree (hemobilia), or into the pleural space. Postulated risk factors include cirrhosis, coagulopathy, amyloid liver, hepatocellular injury, hemangioma and vascularized tumor. However, bleeding may be massive even when no risk factors are present.
- bile leakage with peritonitis - associated with severe obstruction of the larger bile ducts.
- laceration of internal organs and viscera - right kidney, gallbladder, colon, pancreas, and others
- others: right-sided pneumothorax, bacteremia, sepsis, arteriovenous fistula, drug toxicity

Contraindications Frozen section is usually contraindicated for needle biopsy material except for recognition of tumor in the course of open surgical procedures.

Methodology Several biopsy needles are available: Menghini needle, "Trucut" needle, Vim-Silverman needle.

- Tissue stains including: **H&E**; **reticulin** preparation; **Masson's trichrome**; **pentachrome**; **iron stain** (eg, Perls' stain) - useful for hemosiderosis and hemochromatosis; **PAS stain** with and without diastase - useful for alpha$_1$-antitrypsin globules, bile ducts; **orcein**; stains for amyloid and various stains for organisms; other stains occasionally needed as well. Immunohistochemistry is sometimes needed (eg, CEA).
- Cytologic preparation - fluid from aspirating syringe may be smeared on clean microscope slide, fixed, and sent to Cytology Laboratory
- Microbiological culture - send specimen without fixative in sterile container. Special stains (AFB, KOH, etc) and cultures (tuberculosis, viral, *Brucella*, parasites, fungi) as needed. See Fungal Culture, Biopsy or Body Fluid *on page 619* and Mycobacterial Culture, Biopsy or Body Fluid *on page 929*. Liver biopsies are most likely to contain acid-fast organisms in HIV-positive subjects when the patient is febrile, AIDS is longstanding and serum alkaline phosphatase is very high.
- Hepatic copper is assayed by graphite furnace atomic absorption spectrometry or neutron activation.

Additional Information Although **fine needle aspiration** is useful for the diagnosis of carcinoma, it is not usually adequate for evaluation of other hepatic disease entities.

In Wilson disease, **tissue copper** values are >250 µg/g (4 µmol/g) dry weight, while in other entities characterized by increased copper (chronic cholestatic liver diseases including primary biliary cirrhosis and primary sclerosing cholangitis) liver copper concentration is <80 µg/g (1.26 µmol/g) dry weight. Normal interval is reported as 20-50 µg/g. The only other entity with very high copper levels, Indian childhood cirrhosis, is extremely rare in the U.S. If copper levels are needed, specific paraffins are required to embed the needle biopsy. See Copper, Serum *on page 448*; Copper, Urine *on page 452*; and Ceruloplasmin, Serum or Plasma *on page 375*.

The hepatic **iron index** is the ratio of liver concentration of iron to patient age:

hepatic iron (µg/g dry weight) / 56 x age of patient

A result >2.0 is diagnostic of genetic **hemochromatosis**.

References

Bravo AA, Sheth SG, and Chopra S, "Liver Biopsy," *N Engl J Med*, 2001, 344(7):495-500.

Clark JM and Diehl AM, "Nonalcoholic Fatty Liver Disease. An Underrecognized Cause of Cryptogenic Cirrhosis," *JAMA*, 2003, 289(22):3000-4.

El-Youssef M, "Wilson Disease," *Mayo Clin Proc*, 2003, 78(9):1126-36.

Ferrell L, "Liver Pathology: Cirrhosis, Hepatitis, and Primary Liver Tumors. Update and Diagnostic Problems," *Mod Pathol*, 2000, 13(6):679-704.

Green RM and Flamm S, American Gastroenterological Association, "AGA Technical Review on the Evaluation of Liver Chemistry Tests," *Gastroenterology*, 2002, 123(4):1367-84.

Ishak KG, "Pathologic Features of Chronic Hepatitis: A Review and Update," *Am J Clin Pathol*, 2000, 113(1):40-55.

Kamath PS, "Clinical Approach to the Patient With Abnormal Liver Test Results," *Mayo Clin Proc*, 1996, 71(11):1089-95.

Ludwig J, Moyer TP, and Rakela J, "The Liver Biopsy Diagnosis of Wilson's Disease. Methods in Pathology," *Am J Clin Pathol*, 1994, 102(4):443-6.

Pratt DS, and Kaplan MM, "Evaluation of Abnormal Liver-Enzyme Results in Asymptomatic Patients," *N Engl J Med*, 2000, 342(17):1266-71.

Zimmerman HJ, *Hepatotoxicity: The Adverse Effects of Drugs and Other Chemicals on the Liver*, 2nd ed, Baltimore, MD: Lippincott Williams & Wilkins, 1999.

Liver Disease: Laboratory Assessment, Overview

Related Information

Acetaminophen, Serum *on page 90*
Alanine Aminotransferase, Serum *on page 116*
Albumin, Serum *on page 120*
Alkaline Phosphatase, Serum *on page 127*
Alpha$_1$-Antitrypsin Phenotyping *on page 133*
Alpha$_1$-Antitrypsin, Serum *on page 134*
Amiodarone, Serum *on page 148*
Antimitochondrial Antibody *on page 183*
Aspartate Aminotransferase, Serum *on page 216*
Bilirubin, Total, Serum *on page 265*
Bilirubin, Urine *on page 268*
Body Fluid Chemical Analysis *on page 291*
Carbohydrate-Deficient Transferrin, Serum *on page 338*
Carcinoembryonic Antigen, Serum *on page 342*
Ceruloplasmin, Serum or Plasma *on page 375*
Copper, Serum *on page 448*
Copper, Urine *on page 452*
Echinococcosis Diagnostic Procedures *on page 530*
Ethanol, Blood, Urine, and Other Sources *on page 558*
Ferritin, Serum *on page 577*
Gamma-Glutamyl Transferase, Serum *on page 629*
Hepatitis B Antigen Detection *on page 699*
Hepatitis B DNA Detection *on page 701*
Hepatitis B Serology *on page 702*
Hepatitis C Virus RNA Detection and Quantitation *on page 705*
Hepatitis C Virus Serology *on page 706*
Hepatitis D Serology *on page 711*
Hepatitis E Serology *on page 712*
Hepatitis: Laboratory Assessment, Overview *on page 713*
Hereditary Hemochromatosis DNA Test *on page 718*
(Continued)

Liver Disease: Laboratory Assessment, Overview
(Continued)

Applies to Acetaminophen; ALT:AST Ratio; AST:ALT Ratio; De Ritis Ratio; Ecstasy; LFTs; Liver Function Tests, So-Called; Liver Profile

Test Includes The differential diagnosis of liver disease often requires albumin; bilirubin, total; bilirubin, direct; alkaline phosphatase; aspartate aminotransferase (AST/SGOT); alanine aminotransferase (ALT/SGPT); protein, total; and often gamma-glutamyl transferase (GGT/GGTP). It may also include serum protein electrophoresis, prothrombin time, and hepatitis serological tests when indicated. Alpha$_1$-antitrypsin phenotype and quantitation on occasion explains cases otherwise difficult to classify but is rarely, if ever, included in liver profiles. Other tests which may be helpful in evaluation of liver disease include immunoglobulins IgG, IgA, IgM; ANA; antimitochondrial antibody; smooth muscle antibody; liver/kidney microsomal type 1 antibodies and alpha-fetoprotein. Ammonia is useful in selected cases, **(eg, Reye syndrome, urea cycle disorders**, and certain **organic acidurias)**. Lactate dehydrogenase and LD isoenzymes may be useful in the differential diagnosis of icterus.

Abstract Characterization of liver disease requires correlation of the medical history; inventory of medications including over-the-counter drugs, vitamins, supplements, and herbs; alcohol consumption; comorbid conditions; physical examination; laboratory test results; imaging if needed; and, when indicated, liver biopsy. Classification of liver disease includes hepatocellular, cholestatic (intra- or extrahepatic) and infiltrative states. Clinical assessment includes patient history and physical examination. Family history is especially relevant in cases of hemolytic anemia, Gilbert syndrome, Dubin-Johnson syndrome, Wilson disease, hemochromatosis, and α_1-antitrypsin deficiency.

Storage Instructions Protect specimens for bilirubin from light.

Causes for Rejection Hemolysis interferes with certain tests.

Special Instructions The specimens should be handled with extra precaution, especially if there is a greater than usual possibility of viral hepatitis.

Use Evaluate hepatobiliary disease, hepatoma, autoimmune hepatitis and cirrhosis, including biliary cirrhosis; investigate otherwise unexplained increases in such tests as AST, ALT, alkaline phosphatase, hyperbilirubinemia, or prolongation of prothrombin time; work up possible alcoholism. Clinical evidence of chronic liver disease and/or hepatic decompensation may include ascites, encephalopathy, coagulopathy, and/or portal hypertension. A number of useful tests include those to work up **pancreatitis**, such as serum and urine amylase and serum lipase.

In **hepatocellular injury**, abnormalities of the aminotransferases (ALT, AST) are conspicuous. The most common causes of persistent abnormalities of liver function in Western countries are nonalcoholic fatty liver disease, hepatitis C, and heavy alcohol consumption. Use of blood alcohol determinations for investigation of liver disease has been advocated. Other tests useful in alcoholism include carbohydrate deficient transferrin, MCV, albumin, and folate levels, triglycerides, GGT, and bilirubin. AST increment is greater than that of ALT in alcoholic hepatitis; the **AST:ALT ratio** is >2.0 and AST is not increased more than 250 units/L. Prothrombin time is often helpful. Alcoholism causes malnutrition, including deficiencies of vitamins, including thiamine and vitamin A as well as folate. It may lead to hyperlacticacidemia, hyperuricemia, ketosis, and hyperlipidemia. Alcoholics are vulnerable to a wide variety of substances, solvents, and medications including acetaminophen. Alcoholism may lead to a variety of complications including cirrhosis, gastritis, malnutrition, pancreatitis, and cardiomyopathy.

In **viral hepatitis**, ALT is greater than AST, but in chronic liver diseases the **AST:ALT ratio** is less useful. In cirrhosis caused by hepatitis B and other agents, the AST:ALT ratio may be >1.0. In acute viral hepatitis, AST is usually three to five times or more higher (as multiples of the upper limit of normal) than LD; in cases which clinically resemble hepatitis, but in which LD equals or exceeds AST, LD isoenzymes may be useful. LD_4 and LD_5 are the hepatic fractions. An isomorphic pattern, if detected, may suggest infectious mononucleosis, CMV infection, neoplasm, or cirrhosis/alcoholism, depending on clinical setting. Transaminases increase in viral and drug-induced hepatitis and peak within 7-14 days, usually in the low thousands range, returning to normal in about 6 weeks in uncomplicated viral hepatitis. At onset they are generally more than 10 times the upper limit of the reference interval. Alkaline phosphatase is only moderately elevated. Acute viral hepatitis includes at least five separate disease entities. See Hepatitis: Laboratory Assessment, Overview *on page 713.*

Appropriate positive serological tests support a diagnosis of viral hepatitis, while negative ones provide support for other disorders including drug-induced hepatitis. Resolution of liver disease with removal of the offending agent enhances the latter diagnosis; see Related Information. Ecstasy, a synthetic amphetamine, may cause hepatic damage resembling acute viral hepatitis.

Table 2 in the listing Liver Biopsy *on page 864,* summarizes histopathologic findings with AST and ALT in toxic hepatic injury.

Chronic hepatitis includes chronic viral hepatitis and autoimmune hepatitis. It clinically may resemble alcoholic and nonalcoholic steatohepatitis. A review of its classification, criteria, and grading is available (Batts and Ludwig). See Hepatitis in the Disease Index.

Nonalcoholic steatohepatitis (NASH) (nonalcoholic fatty liver disease) a common cause of liver disease in Western countries, is associated with type 2 diabetes, obesity, hyperlipidemia, and jejunoileal bypass. The only clinical manifestations of hepatic steatosis and NASH may be moderate increases of aminotransferase concentrations. The ratio of **ALT to AST (ALT:AST ratio** or De Ritis ratio) is >1, but <1.0 in alcoholic liver disease (it is normally <1). Liver biopsy is recommended. See Nonalcoholic Steatohepatitis and Steatosis (Liver) in the Disease Index. Fatty liver disease is caused by a diverse variety of nutritional, metabolic, and genetic entities and drugs including amiodarone and perhexilene.

Cholestasis and biliary tract: alkaline phosphatase, total bilirubin, conjugated bilirubin, GGT, eosinophil count, urine bile (as part of urinalysis) are used. Alkaline phosphatase is increased in **infiltrative diseases** of the liver, as well as in states of cholestasis, impaired conjugation, and biliary obstruction (with bilirubin and GGT). In **extrahepatic biliary obstruction** the serum alkaline phosphatase is increased two to three times or more while AST remains <300 units/L. Very high alkaline phosphatase levels may be found with intrahepatic cholestasis, such that alkaline phosphatase which is high out of proportion to the severity of jaundice, may indicate an intrahepatic disease. Viral, alcoholic, or drug-related cholestatic hepatitis may give rise to chemistry tests indistinguishable from those of extrahepatic obstruction. Primary biliary cirrhosis, primary sclerosing cholangitis, overlap syndrome, and autoimmune cholangiopathy sometimes represent difficult differential diagnosis. See Biliary Function Tests, Cholangitis (Primary Sclerosing), and Cirrhosis (Primary Biliary) in the Disease Index.

Liver excretory function: urine urobilinogen, total bilirubin, conjugated bilirubin.

Immunologic stimulation: Protein electrophoresis: features suggestive of cirrhosis but not always present in that disease include low albumin, low $alpha_2$, polyclonal or oligoclonal gammopathy, and beta/gamma bridging. Oligoclonal gammopathy is found in <50% of cases of autoimmune hepatitis. HLA phenotypes may be relevant.

Autoimmune hepatitis, found predominantly in young to middle-aged women, is recognized by the presence of increased immunoglobulins and the presence (Continued)

Liver Disease: Laboratory Assessment, Overview
(Continued)

of autoantibodies. More than 80% have hypergammaglobulinemia. Two other major groups of autoimmune liver diseases are recognized, primary biliary cirrhosis and primary sclerosing cholangitis. The distinction between autoimmune hepatitis and chronic viral hepatitis with hyperglobulinemia and/or autoantibodies is sometimes blurred. Wilson disease also enters this differential diagnosis. In autoimmune hepatitis, classical laboratory features include elevation of serum aminotransferases, bilirubin and alkaline phosphatase, and increased gamma globulins, especially IgG with autoantibodies. (ANA and antismooth muscle antibodies in autoimmune hepatitis type 1; in type 2 autoimmune hepatitis, anti-liver-kidney-microsomal 1 antibodies are detected. Soluble liver antigen and smooth muscle antibody are anticipated in type 3.) See discussion and table in Smooth Muscle Antibody *on page 1207*. Liver biopsy is recommended for disease confirmation. Testing is summarized under Hepatitis (Autoimmune) in the Disease Index.

In **ischemic hepatitis** , as develops in instances of cardiac failure and hypotension, the transaminases may be >10,000 IU/L, and vast increases of LDH may be found. Such increased transaminase concentrations may be seen as well with **acetaminophen overdose** and with **herpes simplex hepatitis**.

In the presence of liver disease with Coombs-negative hemolysis, **Wilson disease** (WD) must be considered. Differential diagnosis of WD with fulminant hepatic failure is supported by a high concentration of bilirubin out of proportion to modest increase in the aminotransferases (transaminases), normal to low alkaline phosphatase, AST:ALT ratio >4, as well as high urine copper level. With hypoceruloplasminemia, hypocupremia, and hypercupruria, a diagnosis of Wilson disease is expedited, but liver biopsy for microscopy including rhodamine and orcein stains and tissue copper analysis are worthwhile. See Copper, Urine *on page 452*, Copper, Serum *on page 448*, and Wilson Disease in the Disease Index.

Hemochromatosis: Mild abnormalities in liver profile tests (AST, ALT, ALP) may occur in hemochromatosis. Iron, IBC, transferrin, ferritin, and molecular testing are available. Liver biopsy is considered essential for confirmation of the diagnoses of Wilson disease and of many cases of hemochromatosis, save that biopsy may be unnecessary in patients younger than age 40 with normal results. See Hereditary Hemochromatosis DNA Test *on page 718* and Hemochromatosis in the Disease Index.

Liver disease related to **alpha$_1$-antitrypsin deficiency** is associated with periodic acid-Schiff positive diastase-resistant globules and an abnormal α_1-antitrypsin phenotype, PiZZ homozygotes and several other alleles. (The MM phenotype (PiMM) is normal.) See Alpha$_1$-Antitrypsin Deficiency in the Disease Index.

Hepatic functional reserve (hepatic synthetic function): Both albumin and prothrombin time are useful in evaluation of the liver, but they are nonspecific. Albumin reflects hepatic synthesis and nutritional status but is lost in a variety of gastrointestinal and renal diseases.

Liver metabolic function: Serum ammonia may increase in liver necrosis and cirrhosis as well as in Reye syndrome.

Hepatoma (carcinoma, liver) and other tumors: See Carcinoma (Liver) in the Disease Index. Alpha$_1$-fetoprotein may increase moderately in nonmalignant liver diseases; rising or high levels may indicate hepatoma. Such clinical laboratory tests do not prove tumor without imaging and biopsy. Other primary tumors of liver, benign and malignant, are found.

Metastatic tumors are the most common malignant neoplasms of the liver. The most frequent primaries include carcinoma of the stomach, colon, pancreas, esophagus, lung, and breast. The liver is a site of spread of malignant lymphomas.

Limitations Some types of hepatic injury may not be accompanied by very much increase in enzymes (eg, injury from ethionine or phosphorus).

The enzymes which are sometimes called "LFTs" ("liver function tests") often lack specificity for the liver when used alone, and reflect injury or disease of other organs as well. Use of LDH isoenzymes supports specificity: LDH_5 reflects disease of liver or of striated muscle, and CK arises from injury to the latter. See table.

Groups of Serum Enzymes According to Their Sensitivities

Enzyme	Cholestasis[1]	Hepatocellular Necrosis	Chronic Injury	Injury of Other Organs[2]
Group I Cholestasis > Hepatic Injury				
ALP, GGT, 5′N, LAP	↑↑↑	↑	↑	±
Group II Hepatic Injury > Cholestasis				
A - Extrahepatic and Hepatic Disease				
AST, LDH	↑	↑↑↑	↑	↑
B - More Specificity for the Liver				
ALT	↑	↑↑↑	↑	↑
C - Still Greater Specificity for Hepatic Injury/Disease				
LDH_5	↑	↑↑↑	↑	Muscle
Group III Insensitivity to Liver Injury				
CK	Normal	Normal	Normal	↑

5′N = 5′-nucleotidase; ALP = alkaline phosphatase; ALT = alanine aminotransferase; AST = aspartate aminotransferase; CK = creatine kinase; GGT = gamma-glutamyl transferase; LAP = leucine aminopeptidase; LDH_5 = least anodic isoenzyme of lactic dehydrogenase; LDH = lactase dehydrogenase; ↑ = increased; ↑↑↑ = markedly increased; ± = little change.

[1]Obstructive jaundice or intrahepatic cholestasis.

[2]Cardiac or skeletal muscle, brain, or kidney.

Modified from Zimmerman HJ, *Hepatotoxicity: The Adverse Effects of Drugs and Other Chemicals on the Liver*, 2nd ed, Philadelphia, PA: Lippincott Williams and Wilkins, 1999.

More specialized tests are often necessary to establish an etiologic diagnosis. Liver biopsy is often needed to provide precise diagnosis in subjects with subacute to chronic abnormalities.

Additional Information Relevant topics in the Disease Index include Acetaminophen Toxicity, Carcinoma (Liver), Cirrhosis, Cirrhosis (Primary Biliary), Hemochromatosis, Hepatic Necrosis/Failure, Hepatitis, Hepatitis (Autoimmune), Jaundice, Liver, and Wilson Disease.

References

American Gastroenterological Association, "American Gastroenterological Association Medical Position Statement: Evaluation of Liver Chemistry Tests," *Gastroenterology*, 2002, 123(4):1364-6.

Andrew V, Mao A, Bruguera M, et al, "Ecstasy: A Common Cause of Severe Acute Hepatotoxicity," *J Hepatol*, 1998, 29(3):394-7.

Angulo P, "Nonalcoholic Fatty Liver Disease," *N Engl J Med*, 2002, 346(16):1221-31.

Batts KP and Ludwig J, "Chronic Hepatitis: An Update on Terminology and Reporting," *Am J Surg Pathol*, 1995, 19(12):1409-17.

El-Youssef M, "Wilson Disease," *Mayo Clin Proc*, 2003, 78(9):1126-36.

Green RM and Flamm S, American Gastroenterological Association, "AGA Technical Review on the Evaluation of Liver Chemistry Tests," *Gastroenterology*, 2002, 123(4):1367-84.

James O and Day C, "Nonalcoholic Steatohepatitis; Another Disease of Affluence," *Lancet*, 1999, 353(9165):1634-6.

Kamath PS, "Clinical Approach to the Patient With Abnormal Liver Test Results," *Mayo Clin Proc*, 1996, 71:1089-95.

Ludwig J, Moyer TP, and Rakela J, "The Liver Biopsy Diagnosis of Wilson's Disease. Methods in Pathology," *Am J Clin Pathol*, 1994, 102(4):443-6.

Pratt DS and Kaplan MM, "Evaluation of Abnormal Liver-Enzyme Results in Asymptomatic Patients," *N Engl J Med*, 2000, 342(17):1266-71.

Zimmerman HJ, *Hepatotoxicity: The Adverse Effects of Drugs and Other Chemicals on the Liver*, 2nd ed, Baltimore, MD: Lippincott Williams & Wilkins, 1999.

♦ **Liver Function Tests, So-Called** see Liver Disease: Laboratory Assessment, Overview on page 869

♦ **Liver Iron Concentration** see Ferritin, Serum on page 577

Liver/Kidney Microsomal Type 1 Antibodies

Related Information

Antimitochondrial Antibody on page 183
Bilirubin, Total, Serum on page 265
Islet Cell Antibody on page 813
Liver Biopsy on page 864
(Continued)

Liver/Kidney Microsomal Type 1 Antibodies
(Continued)

Liver Disease: Laboratory Assessment, Overview *on page 869*
Parietal Cell Antibody *on page 1005*
Smooth Muscle Antibody *on page 1207*
Thyroglobulin Antibody *on page 1249*
Thyroperoxidase Autoantibody *on page 1253*

Synonyms Antiliver/Kidney Microsomal Antibodies; Anti-LKM1

Applies to Antiliver Cytosol 1 Antibodies; P45011D6; Soluble Liver Antigen (Anti-SLA)

Abstract Chronic liver diseases include **autoimmune hepatitis** characterized by chronic inflammatory changes and autoantibodies. Three diagnostic categories are recognized. **Type 1** is the classical type, corresponding to the entity lupoid hepatitis described in 1950-1951. Type 1 patients have smooth muscle antibodies and/or ANA reactivity. **Type 2** patients lack those antibodies but are characterized by anti-LKM1 antibodies. **Type 3** have antibodies to soluble liver antigen (SLA) and SMA, but lack anti-LKM1. See the table in Smooth Muscle Antibody *on page 1207* for clinical and immunologic features.

Specimen Serum

Container Red top tube or SST™ tube

Reference Interval Negative

Use One of the major immunologic tests for autoimmune liver disease, especially in children. Others include smooth muscle antibody, antimitochondrial antibody, and antinuclear antibody.

Additional Information A fraction of cases of autoimmune hepatitis, autoimmune hepatitis Types 2a and 2b, are positive for these antibodies. These reactive subsets include a number with related immunologic disorders. Type 2a patients are characterized as well by the presence of antiliver cytosol 1 antibodies. Other immunologic disorders in LKM1-positive patients include thyroiditis, diabetes mellitus, hemolytic anemia, arthritis, and ulcerative colitis. Antithyroid, antiparietal cell, and anti-islet antibodies are common. Low IgA levels in serum are often seen.

Patients with type 2a are predominantly children, female, lack HCV, and have higher titers of anti-LKM1. Their serum is often reactive for P45011D6, which is recognized by the antibody LKM1. Patients classified as type 2b are characterized by seroreactivity to HCV, features of chronic viral disease, and infrequent positivity to P45011D6.

References
Burgart LJ, Batts KP, Ludwig J, et al, "Recent-Onset Autoimmune Hepatitis. Biopsy Findings and Clinical Correlations," *Am J Surg Pathol*, 1995, 19(6):699-708.
Krawitt EL, "Autoimmune Hepatitis," *N Engl J Med*, 1996, 334(14):897-903.

♦ **Liver/Kidney Microsomes Antibody** *see* Smooth Muscle Antibody *on page 1207*

♦ **Liver Needle Aspiration Cytology** *see* Fine Needle Aspiration, Deep and Superficial Masses *on page 590*

♦ **Liver Profile** *see* Liver Disease: Laboratory Assessment, Overview *on page 869*

♦ **LKM Antibody** *see* Smooth Muscle Antibody *on page 1207*

♦ **LMWH** *see* Heparin Antifactor Xa Assay *on page 693*

♦ **Long-Acting Thyroid Stimulator** *see* Thyrotropin Receptor Antibody, Serum *on page 1254*

♦ **Loperamide Inhibition Test** *see* Adrenal Cortex: Laboratory Assessment Overview *on page 110*

♦ **Lotusate®** *see* Barbiturates, Quantitative, Serum or Plasma *on page 248*

♦ **Lovastatin** *see* Mevalonic Acid, Urine or Amniotic Fluid *on page 911*

Low Density Lipoprotein Cholesterol
Related Information

Apolipoprotein A-I, Serum *on page 202*
Apolipoprotein B-100, Serum *on page 203*
Cholesterol, Total, Serum or Plasma *on page 396*
High Density Lipoprotein Cholesterol, Serum *on page 729*
Lipid Panel, Serum *on page 852*

Lipids, Overview *on page 853*
Triglycerides, Serum or Plasma *on page 1275*

Synonyms Beta Lipoproteins; LDLC

Applies to Familial Hypercholesterolemia; Friedewald Formula; LDLC:HDLC Ratio

Abstract The concentration of low density lipoprotein cholesterol (LDLC) is a major risk factor for coronary artery disease (CAD). Lowering LDLC is the primary target of treatment in individuals with elevated (>160 mg/dL) or borderline high (130-159 mg/dl) LDLC. For additional information, see Lipids, Overview *on page 853*.

Patient Preparation For best results, Patient should be on a stable diet for 3 weeks, stable body weight, and fasting for at least 10 hours. (Fasting is not essential for LDLC, but is important if triglycerides are also to be assayed.) See also Preparation in the listing Cholesterol, Total, Serum or Plasma *on page 396*. HDLC is usually done as part of Lipid Panel, Serum *on page 852*.

Specimen Serum or plasma

Container Red top tube or lavender top (EDTA) tube

Reference Interval A normative reference interval is not applicable (see Lipids, Overview *on page 853*). The Adult Treatment Panel (ATP III) of the National Cholesterol Education Program (NCEP III) recommends the following for adults:

- Optimal: <100 mg/dL
- Near or above optimal: 100-129 mg/dL
- Borderline high: 130-159 mg/dL
- High: 160-189 mg/dL
- Very high: ≥190 mg/dL

Actual population distributions and percentile cutoffs are shown in the Lipid Research Clinics Prevalence Study.

Use The principal use for LDLC measurements is in the assessment of CHD risk and LDLC is the major target for lipid lowering treatments by diet and/or drugs. **Indirect** estimation of LDLC is often made by the Friedewald formula:

LDLC (mg/dL) = cholesterol, total (mg/dL) - HDLC (mg/dL) - [triglycerides (mg/dL) / 5]

The Friedewald formula must **not** be used when any one of three conditions is present:

1) fasting TG ≥400 mg/dL
2) chylomicrons are present (usually a visible layer of "cream" following overnight refrigerated storage)
3) patient has familial dysbetalipoproteinemia (see Lipids, Overview *on page 853*).

In these situations, the Friedewald formula underestimates the LDLC. When the Friedewald formula cannot be used, and a **direct** method is unavailable, the specimen should be referred to a specialized lipid laboratory.

Limitations It should be emphasized that routine clinical assays for LDLC also include intermediate-density lipoprotein cholesterol (IDLC) and lipoprotein (a). Ordinarily these substances make a very small contribution to the LDLC result. Problem cases should be referred to a specialized lipid laboratory.

LDLC may be increased by thiazides, beta blockers, estrogens, and other drugs. LDLC may be decreased by fish oils, statins, niacin, fibrates, estrogens, and other drugs.

Methodology

Direct: Enzymatic colorimetry following removal of HDLC and VLDLC by (a) precipitating antibodies, (b) detergent inactivation, or (c) selective ultracentrifugation.

Indirect: The Friedewald formula (presented above in Use) produces a calculated value for LDLC.

References

"Executive Summary of the Third Report of the National Cholesterol Education Program (NCEP) Expert Panel on Detection, Evaluation, and Treatment of High Blood Cholesterol in Adults (Adult Treatment Panel III)," *JAMA*, 2001, 285(19):2486-97.

Lee TH, Cleeman JI, Grundy SM, et al, "Clinical Goals and Performance Measures for Cholesterol Management in Secondary Prevention of Coronary Heart Disease," *JAMA*, 2000, 283(1):94-8.

(Continued)

Low Density Lipoprotein Cholesterol (Continued)

Nauck M, Graziani MS, Bruton D, et al, "Analytical and Clinical Performance of a Detergent-Based Homogeneous LDL-Cholesterol Assay: A Multicenter Evaluation," *Clin Chem*, 2000, 46(4):506-14

Internet Web Sites
www.cdc.gov/nceh/dls/crmln/crmln.htm
www.nhlbi.nih.gov

- **Low Density Lipoprotein Cholesterol (LDLC)** *see* Cholesterol, Total, Serum or Plasma *on page 396*
- **Low Density Lipoprotein Cholesterol (LDLC)** *see* Lipids, Overview *on page 853*
- **Low-Molecular Weight Heparin** *see* Heparin Antifactor Xa Assay *on page 693*
- **Loxapine** *see* Amoxapine, Serum or Plasma *on page 153*
- **Lp(a)** *see* Lipoprotein (a), Serum *on page 861*
- **Lp(a)-C** *see* Lipoprotein (a), Serum *on page 861*
- **LSD** *see* Lysergic Acid Diethylamide, Urine *on page 883*
- **LSD-2** *see* Lysergic Acid Diethylamide, Urine *on page 883*
- **L:S Ratio** *see* Lecithin:Sphingomyelin Ratio, Amniotic Fluid *on page 836*
- **Lude®** *see* Methaqualone, Urine *on page 903*
- **Ludes** *see* Methaqualone, Urine *on page 903*
- **Ludiomil®** *see* Maprotiline, Serum *on page 892*
- **Lumbar Puncture** *see* Cerebrospinal Fluid Analysis: Overview *on page 355*
- **Luminal®** *see* Barbiturates, Quantitative, Serum or Plasma *on page 248*
- **Luminal®** *see* Phenobarbital, Serum or Plasma *on page 1021*
- **Lumpectomy** *see* Breast Biopsy *on page 305*
- **Lung Needle Aspiration Cytology** *see* Fine Needle Aspiration, Deep and Superficial Masses *on page 590*
- **Lung Profile, Amniotic Fluid** *see* Lecithin:Sphingomyelin Ratio, Amniotic Fluid *on page 836*
- **Lupus Anticoagulant** *see* Anticardiolipin Antibody *on page 167*
- **Lupus Anticoagulant** *see* Antiphospholipid Antibody (Lupus Anticoagulant and/or Anticardiolipin Antibody) *on page 193*
- **Lupus Anticoagulant** *see* Coagulation Factor Assays *on page 418*
- **Lupus Anticoagulant** *see* Prothrombin Time *on page 1116*
- **Lupus Band Test** *see* Skin Biopsy, Immunofluorescence *on page 1203*
- **Lupus Inhibitor** *see* Antiphospholipid Antibody (Lupus Anticoagulant and/or Anticardiolipin Antibody) *on page 193*
- **Luteinizing Hormone** *see* Testosterone, Total and Free, Serum or Plasma *on page 1238*

Luteinizing Hormone, Blood or Urine

Related Information

Adrenal Cortex: Laboratory Assessment Overview *on page 110*
Clomiphene Test *on page 414*
Estradiol, Serum *on page 553*
Estrogens, Urine *on page 557*
Follicle Stimulating Hormone, Serum, Plasma, or Urine *on page 609*
Growth Hormone, Serum *on page 662*
Heterophilic Antibodies *on page 727*
Progesterone, Serum *on page 1093*
Prolactin, Serum *on page 1094*
Testosterone, Total and Free, Serum or Plasma *on page 1238*
Urine Collection, 24-Hour *on page 1295*

Synonyms Follitropin; ICSH; Interstitial Cell Stimulating Hormone; LH

Applies to Gonadotropic Hormones; Leydig Cells; Pituitary Gonadotropins

Abstract Luteinizing hormone (LH) and follicle stimulating hormone (FSH) are under complex regulation by hypothalamic gonadotropin releasing hormone (GnRH) and by the gonadal sex hormones: estrogen and progesterone in females, and testosterone in males. FSH and LH, respectively, stimulate spermatogenesis and production of testosterone by the Leydig cells of the testes.

Levels of LH and FSH are used to assess anterior pituitary gonadotropic function, sexual differentiation, fertility, and pseudohermaphroditism. In general, FSH and LH are elevated in conditions in which sex hormones cannot be elaborated and are decreased in conditions of primary pituitary dysfunction.

Patient Preparation Avoid radioisotope administration to patient prior to collection of specimen if RIA is used for assay.

Specimen Serum or 24-hour urine

Container Red top tube; plastic urine container containing boric acid

Storage Instructions Separate serum from cells. Stable 14 days at 4°C to 25°C.

Special Instructions In females, date of last menstrual period should be supplied. It is important to measure both FSH and LH.

Reference Interval Check with the testing laboratory as there is considerable variation among laboratories. Results can only be interpreted with clinical information. The following intervals are from Mayo Medical Laboratories.

Serum:
- prepubertal children: <1.0 IU/L
- adults, male: 1.0-9.0 IU/L
- adults, female:
  - follicular: 1.0-18.0 IU/L
  - midcycle: 20.0-80.0 IU/L
  - luteal: 0.5-18.0 IU/L
  - postmenopausal: 12.0-55.0 IU/L

Urine:
- prepubertal children: <0.2 IU/24 hours
- adults, male: 0.2-5.0 IU/24 hours
- adults, female:
  - nonmidcycle: <5.0 IU/24 hours
  - postmenopausal: >5.0 IU/24 hours

Use Excessive FSH and LH production occurs in anorchia, gonadal failure, complete testicular feminization syndrome, and menopause. Both FSH and LH are low with primary pituitary or hypothalamic failure. When one is high and the other low, a gonadotropin-producing pituitary tumor is possible.

FSH and LH are useful in infertility evaluation of women and testicular dysfunction in men. Elevated basal LH with high LH:FSH ratio (>2), accompanied by an increase of ovarian androgens in a nonovulatory adult female is presumptive evidence of Stein-Leventhal syndrome (sensitivity 75%) in the appropriate clinical setting. High concentrations of LH (during the follicular phase) in patients with polycystic ovary syndrome interfere with conception and may contribute to early pregnancy loss in these patients.

Urinary LH and FSH are used in children who have precocious puberty, in many of whom serum FSH and LH levels overlap the normal range. An advantage of 24-hour urine collections is that they overcome problems of pulsatile secretion spikes.

FSH, LH and testosterone are low in Kallmann syndrome. Isolated LH deficiency is a variant of Kallmann syndrome. Patients who have growth hormone deficiency have FSH and/or LH deficiency as well.

Limitations Secretion of both LH and FSH are pulsatile, in response to the normal intermittent release of gonadotropin releasing hormone (GnRH). Problems of secretion spikes are minimized with 24-hour urine collections. While both are pulsatile, LH exhibits a circadian rhythm while FSH does not. In addition, in females both FSH and LH vary over the course of the menstrual cycle, with peaks at time of ovulation. Thus, interpretation of a single determination may be difficult.

Methodology Radioimmunoassay (RIA), immunoradiometric assay (IRMA), radioreceptor assay, dissociation enhanced lanthanide fluoroimmunoassay (DELFIA), chemiluminescence immunoassay (CIA), immunofluorometric assay

Additional Information LH increases in urine after the pituitary LH surge that precedes ovulation by 24-36 hours. Home use test kits for qualitative urine LH are available for predicting when ovulation is likely to occur and have been reported to be effective for this use.

References

Mayo Medical Laboratories, *2000 Test Catalogue*, Rochester, MN.

Miller PB and Soules MR, "The Usefulness of a Urinary LH Kit for Ovulation Prediction During Menstrual Cycles of Normal Women," *Obstet Gynecol*, 1996, 87(1):13-7.

(Continued)

Luteinizing Hormone, Blood or Urine (Continued)

Young WF, Scheithauer BW, Kovacs KT, et al, "Gonadotroph Adenoma of the Pituitary Gland: A Clinicopathologic Analysis of 100 Cases," Mayo Clin Proc, 1996, 71(7):649-56.

Lyme Disease DNA Detection

Related Information

Borrelia burgdorferi Culture *on page 301*

Lyme Disease Serology *on page 879*

Test Includes DNA from *Borrelia burgdorferi* is amplified from a patient specimen or tick and identified.

Abstract Lyme disease is a tick-borne zoonosis caused by infection with the spirochete *Borrelia burgdorferi* in the U.S. Epidemics tend to occur in the spring and fall when a tick vector, *Ixodes ricinus*, is proliferating. Ticks transmit the spirochete to humans through bites. The diagnosis of Lyme disease is difficult due to the insensitivity and unreliability of serologic diagnostic tests. The detection of *Borrelia burgdorferi* genetic material by amplifying DNA directly specimens can often confirm the diagnosis of Lyme disease when serologic tests are equivocal.

Specimen Serum or plasma, cerebrospinal fluid, skin biopsy, synovial fluid, urine

Container Red top tube or lavender top (EDTA) tube for blood samples. Spinal fluid, urine, and synovial fluid should be collected in a sterile container. Skin biopsy should be frozen quickly and should not be fixed with formalin.

Sampling Time Specimens should be collected before antibiotic therapy is initiated.

Storage Instructions All specimens should be sent to the laboratory immediately or kept at 4°C until shipped to the laboratory.

Causes for Rejection Samples containing sodium azide cannot be used in this test

Reference Interval No *Borrelia burgdorferi* DNA detection

Use Detect the presence of *Borrelia burgdorferi* genomic material in patients with signs and symptoms of Lyme disease

Limitations *B. burgdorferi* DNA has been found only in a small number of U.S. patients in cerebrospinal fluid.

Methodology Isolation of DNA amplification using specific primers for *Borrelia burgdorferi* sequences. The amplified DNA is then confirmed to be *B. burgdorferi* by hybridization with a DNA probe.

Additional Information Transmission of *Borrelia burgdorferi*, a pathogenic spirochete, to humans occurs primarily by way of infected *Ixodid* ticks, including *I. dammini*, *I. pacificus*, *Ixodes ricinus* complex, and others. The signs and symptoms of Lyme disease vary, but the most common clinical manifestation following the bite of an infected tick is a distinctive skin lesion, erythema chronicum migrans. This initial stage of Lyme disease is benign and is usually successfully treated with oral antibiotics. Symptoms sometimes persist or reappear after antibiotic treatment. The later stage disease may include chronic progressive encephalomyelitis, chronic severe arthritis, as well as various cardiac manifestations. Direct microscopic detection of *Borrelia burgdorferi* is difficult; therefore the most widely used indicator of infection is the detection of *Borrelia burgdorferi*-specific antibody. Lyme disease serologic testing suffers from lack of specificity and reproducibility. Amplification of *Borrelia burgdorferi* DNA detects the organism in a specific and sensitive assay from patient specimens.

Culture in Barbour-Stoenner-Kelly medium may be positive early, mostly from biopsies of erythema migrans. PCR detection is superior later in joint fluid.

References

Brettschneider S, Bruckbauer H, Klugbauer N, et al, "Diagnostic Value of PCR for Detection of *Borrelia burgdorferi* in Skin Biopsy and Urine Samples From Patients With Skin Borreliosis," *J Clin Microbiol*, 1998, 36(9):2658-65.

Mouritsen CL, Wittwer CT, Litwin CM, et al, "Polymerase Chain Reaction Detection of Lyme Disease. Correlation With Clinical Manifestations and Serologic Responses," *Am J Clin Pathol*, 1996, 105(5):647-54.

Oksi J, Marjamaki M, Nikoskelainen J, et al, "*Borrelia burgdorferi* Detected by Culture and PCR in Clinical Relapse of Disseminated Lyme Borreliosis," *Ann Med*, 1999, 31(3):225-32.

Tugwell P, Dennis DT, Weinstein A, et al, "Laboratory Evaluation in the Diagnosis of Lyme Disease," *Ann Intern Med*, 1997, 127(12):1109-23.

Internet Web Sites
www.amm.co.uk/pubs/fa_lyme.htm
www.astdhpphe.org/infect/lyme.html
www.cdc.gov/ncidod/dvbid/lyme/index.htm

Lyme Disease Serology

Related Information

Test Includes Detection of serological response to *Borrelia burgdorferi*

Abstract Lyme disease is a zoonosis due to *Borrelia burgdorferi*, a spirochete transmitted to humans in the U.S. by tick bites. The other agents are *B. afzelii* and *B. garinii*. *B. burgdorferi* infection is characterized by a rash (erythema migrans) which occurs in up to 80% of patients. Weeks to months later some patients develop arthralgia, neuroborreliosis, myocarditis, or encephalopathy. Early, clinical identification of erythema migrans is helpful. Therapy may be appropriate in seronegative individuals with clinical evidence of Lyme disease.

Specimen Serum or cerebrospinal fluid

Container Red top tube, sterile CSF tube

Use Serologic evaluation of possible Lyme disease; investigate rash, arthritis, carditis, polyneuropathy, and encephalopathy.

Limitations False positives occur in patients infected with tick-borne relapsing fever, human granulocytic ehrlichiosis, syphilis, leptospirosis, periodontal disease, and possibly multiple sclerosis. Early in Lyme disease, test sensitivities are low and serologic testing is not considered useful. Positives remain so for years after successful treatment. **Serologic evidence should not be the sole criterion for a diagnosis of Lyme disease; rather, it deserves consideration with clinical evaluation and risk of exposure. Negative serologic results should never be used as a reason to withhold antibiotic treatment when Lyme disease is suspected.** Positive serologic results in apparently healthy subjects are likely to be false positives.

Contraindications Lyme disease serology does not become positive until weeks or more following a tick bite; *vide supra*. Patients who have only the nonspecific symptoms of myalgia, fatigue, and arthralgia need not be tested for Lyme disease.

Methodology Indirect immunofluorescent antibody (IFA), enzyme-linked immunosorbent assay (ELISA), Western blot assay. A two-test approach for detection of active Lyme disease and of previous infection is recommended by the Centers for Disease Control. The specimen should first be tested using a sensitive enzyme immunoassay (EIA) or immunofluorescent antibody (IFA) followed by Western blot of positive or equivocal specimens. See Lyme Disease DNA Detection *on page 878*.

Additional Information Lyme disease is a multisystem disorder, with rash and arthritis as conspicuous symptoms.

Response may be mitigated by antibiotics. Most patients with chronic disease usually have strong serologic responses to *B. burgdorferi* antigens. Immunoassays using recombinant outer surface protein C (22 kDa), flagellin protein (14 kDa), as well as a high molecular weight protein, p83, seem to have increased sensitivity when compared to use of an EIA with sonicated antigen from *B. burgdorferi*. All positive EIA or IFA results should be confirmed using a Western blot assay and detection of antibodies specific for *B. burgdorferi* proteins.

Two Lyme disease vaccines have been developed, LYMErix™ (SmithKline Beecham Pharmaceuticals) and ImuLyme™ (Pasteur Merieux Connaught). Only LYMErix™ has been licensed by the U.S. Food and Drug Administration. Both vaccines use the *B. burgdorferi* outer-surface protein A (rOspA) as the immunogen. The Advisory Committee on Immunization Practices for the CDC has indicated that the Lyme disease vaccine does not protect all individuals against infection with the *B. burgdorferi* bacteria and that vaccinated persons should continue to practice personal protective measures when participating in outdoor activities that put them at risk for tick bites. Unfortunately, the vaccines have not met an enthusiastic public response.

(Continued)

Lyme Disease Serology (Continued)

Relevant tick species are illustrated in a 2003 paper by Hayes.

Prevention of Lyme disease is addressed by Poland and by Hayes and Piesman.

Steere has published a very helpful review paper.

References

Brown SL, Hansen SL, and Langone JJ, "Role of Serology in the Diagnosis of Lyme Disease," *JAMA*, 1999, 282(1):62-6

Fix AD, Strickland GT, and Grant J, "Tick Bites and Lyme Disease in an Endemic Setting: Problematic Use of Serologic Testing and Prophylactic Antibiotic Therapy," *JAMA*, 1998, 279(3):206-10.

Gardner P, "Long-Term Outcomes and Management of Patients With Lyme Disease," *JAMA*, 2000, 283(5):658-9.

Hayes EB and Piesman J, "How Can We Prevent Lyme Disease?" *N Engl J Med*, 2003, 348(29):2424-30.

Huppertz HI, "Lyme Disease in Children," *Curr Opin Rheumatol*, 2001, 13(5):434-39.

Kaiser R and Rauer S, "Serodiagnosis of Neuroborreliosis: Comparison of Reliability of Three Confirmatory Assays," *Infection*, 1999, 27 (3):177-82.

Poland GA, "Prevention of Lyme Disease: A Review of the Evidence," *Mayo Clin Proc*, 2001, 76(7):713-24.

Steere AC, "Lyme Disease," *N Engl J Med*, 2001, 345(2):115-24.

Van Solingen RM and Evans J, "Lyme Disease," *Curr Opin Rheumatol*, 2001, 13(4):293-9.

Willis MS and Bai Y, "Confounding Issues in the Diagnosis of Multiple Sclerosis: Lyme Disease Testing," *Lab Med*, 2003, 34(6):467-75.

Wormser GP, Aguero-Rosenfeld ME, and Nadelman RB, "Lyme Disease Serology: Problems and Opportunities," *JAMA*, 1999, 282(1):79-80.

Internet Web Sites

www.amm.co.uk/pubs/fa_lyme.htm
www.astdhpphe.org/infect/lyme.html
www.cdc.gov/ncidod/dvbid/lyme/index.htm

♦ **Lymph Node Aspiration Cytology** see Fine Needle Aspiration, Deep and Superficial Masses *on page 590*

Lymph Node Biopsy

Related Information

Test Includes Histopathology, frozen section, imprint cytology, immunoperoxidase procedures, flow cytometry, cytogenetics, molecular genetics may be indicated

Abstract Lymph nodes are commonly involved in a variety of infectious, inflammatory, neoplastic, and other infiltrative disorders, and are commonly biopsied to establish a diagnosis. The diagnosis of lymphoma and related disorders necessitates special processing to optimize diagnostic studies.

Specimen Lymph node or other tissues suspected of harboring lymphoma, ideally submitted fresh within minutes of the biopsy

Container Petri dish with sterile saline-moistened gauze

Collection Important clinical considerations include age, location, duration, and associated manifestations. Larger nodes are more likely diagnostic than smaller ones, especially if smaller ones are superficial to deeper large nodes. Needle aspiration and biopsy can yield diagnostic information but is suboptimal for initial lymphoma classification. Excisional biopsy is preferred.

Initial triage can usually be confidently directed by imprint cytology. If a frozen section evaluation is necessary to initiate a "lymphoma protocol", the tissues used for this rapid diagnosis are often unsuitable for immunophenotypic analysis. If the size of biopsy is limiting, a routine frozen section evaluation should be discouraged, as freezing distorts lymphoid tissue and may result in errors in final interpretation.

Ideally, sufficient tissue must be available for both permanent sections and snap frozen for possible immunophenotypic analysis. Flow cytometry is ideally suited for characterization of most lymphomas, particularly small cell variants. Frozen tissues are also suitable for genotypic studies if necessary. If tissues are to be sent to a reference laboratory for immunotyping, three basic options are available. First, the tissues may be snap frozen and stored at -70°C or colder until such time as immunotyping is considered necessary. Second, the tissues may be delivered in carrier media or saline-soaked gauze on ice immediately by courier to the reference laboratory, where experienced personnel will process the tissue. Third, tissues may be placed in carrier media that may circumvent the need for immediate action for 24 hours without significantly compromising the immunologic studies.

Conventional histopathologic study remains the gold standard in diagnostic hematopathology and optimal histology begins with prompt and proper fixation. Fine nuclear detail is best achieved using B5, zinc formalin, or a Zenker-like fixative. These fixatives are also best for cell marker analysis in paraffin section. Deficient basic histology is a common cause of interpretive error in hematopathology.

Storage Instructions Snap frozen tissues should be maintained at -70°C or colder until immunophenotypic analysis can be performed. If frozen tissues are to be transported to a reference laboratory, they should be shipped on dry ice, using an overnight courier if necessary. Tissues placed in carrier media should be maintained on wet ice or at room temperature and packaged in insulated containers to avoid large fluctuations in temperature, if sent to a reference laboratory.

Special Instructions The specimen should not be placed in fixative if it can be delivered immediately to the laboratory. All such specimens should be brought to the immediate attention of a pathologist.

Use Diagnose various lymphadenopathies, including malignant lymphoma and metastatic neoplasia

Limitations Formalin-fixed tissue cannot be used for culture or imprints and is suboptimal for electron microscopy. Classification of lymphoma is complicated by small sample size. Drug reactions may cause confusion with other entities.

Additional Information A multiparameter approach is essential in the accurate diagnosis and classification of lymphoid and hematopoietic neoplasms, including the need for immunophenotypic and genetic studies.

At times, sufficient fresh tissue may not be available for flow cytometric studies or frozen section immunohistochemistry. However, an ever-expanding selection of antibodies is useful for establishing lineage of hematopoietic cells in paraffin section. Recent studies have documented the utility of these markers in the paraffin section evaluation of small B-cell lymphoid malignancies, T-cell malignancies, Hodgkin disease, and blastic hematopoietic neoplasms. Nonetheless, phenotypic indicators of clonal proliferation are most reliably established by flow cytometry or frozen section immunohistochemistry.

References
Abbondanzo SL, "Paraffin Immunohistochemistry as an Adjunct to Hematopathology," *Ann Diagn Pathol*, 1999, 3(5):318-27.
(Continued)

Lymph Node Biopsy *(Continued)*

Association of Directors of Anatomic and Surgical Pathology, "ADASP Recommendations for Processing and Reporting Lymph Node Specimens Submitted for Evaluation of Metastatic Disease," *Am J Surg Pathol*, 2001, 25(7):961-3.

Chen CC, Raikow RB, Sonmez-Alpan E, et al, "Classification of Small B-Cell Lymphoid Neoplasms Using a Paraffin Section Immunohistochemical Panel," *Appl Immunohistochem Molecul Morphol*, 2000, 8(1):1-11.

de Leon ED, Alkan S, Huang JC, et al, "Usefulness of an Immunohistochemical Panel in Paraffin-Embedded Tissues for the Differentiation of B-Cell non-Hodgkin's Lymphomas of Small Lymphocytes," *Mod Pathol*, 1998, 11(11):1046-51.

El-Zimaity HM, El-Zaatari FA, Dore MP, et al, "The Differential Diagnosis of Early Gastric Mucosa-Associated Lymphoma: Polymerase Chain Reaction and Paraffin Section Immunophenotyping," *Mod Pathol*, 1999, 12(9):885-93.

Habermann TM and Steensma DP, "Lymphadenopathy," *Mayo Clin Proc*, 2000, 75(7):723-32.

Harris NL, Jaffe ES, Diebold J, et al, "The World Health Organization Classification of Hematological Malignancies Report of the Clinical Advisory Committee Meeting, Airlie House, Virginia, November 1997," *Mod Pathol*, 2000, 13(2):193-207.

Ioachim HL and Ratech H, *Ioachim's Lymph Node Pathology*, 3rd ed, Baltimore, MD: Lippincott Williams & Wilkins, 2002.

Jaffe ES, "Hematopathology: Integration of Morphologic Features and Biologic Markers for Diagnosis," *Mod Pathol*, 1999, 12(2):109-15.

Rudiger T, Ott G, Ott MM, et al, "Differential Diagnosis Between Classic Hodgkin's Lymphoma, T-Cell-Rich B-Cell Lymphoma, and Paragranuloma by Paraffin Immunohistochemistry," *Am J Surg Pathol*, 1998, 22(10):1184-91.

♦ **Lymph Node Mapping** *see* Sentinel Lymph Node Biopsy *on page 1189*

♦ **Lymphocyte CD4 Counts** *see* CD4/CD8 Enumeration *on page 349*

♦ **Lymphocyte Crossmatch** *see* Tissue Typing *on page 1259*

Lymphocyte Transformation Test

Related Information

Aspergillus Serology *on page 219*
Fanconi Anemia, Chromosome Breakage Study *on page 569*
Hypersensitivity Pneumonitis Serology *on page 761*
Mixed Lymphocyte Culture *on page 917*

Abstract Lymphocyte proliferation normally occurs early in an immune response. Lymphocyte transformation assays test the integrity of the early proliferative response using either nonspecific mitogens or specific antigens to induce blastogenesis. Antigen induced lymphocyte proliferation also correlates with previous exposure and acquisition of cellular immunity.

Specimen Whole blood

Container Yellow top (ACD) tube or green top (heparin) tube. Check with the laboratory performing the assay for special instructions.

Storage Instructions Do not refrigerate, freeze, or expose to extreme temperatures.

Causes for Rejection Old specimen, specimen without viable lymphocytes, specimen refrigerated or frozen

Special Instructions Schedule procedure in advance with laboratory. Specimens to evaluate therapy should include three baseline samples.

Reference Interval Mitogen: phytohemagglutinin (PHA), stimulation index >130; mitogen: pokeweed mitogen (PWM), stimulation index >20; mitogen: concanavalin A (con A), stimulation index >40

Use Detect and classify congenital or acquired immunodeficiency disorders, study the integrity of lymphokine production, monitor immunosuppressive or immunoenhancing therapy. Document cellular hypersensitivity reactions to environmental allergens or antigens. Similar methods are used to predict allograft compatibility in the transplantation setting. See Mixed Lymphocyte Culture *on page 917*.

Methodology Most commonly, lymphocyte transformation is measured by the incorporation of tritiated thymidine into lymphocytes. Lymphocytes are isolated from peripheral blood and set up in microtiter plate cultures for a period of 3-7 days with and without mitogen or antigen. Lymphocytes are pulsed with tritiated thymidine. Incorporated thymidine is measured and a stimulation index calculated based on control values.

Flow cytometric assays have also been developed. In these procedures, measurement of S-phase fraction or incorporation of bromodeoxyuridine correlates with blastogenesis.

A number of other immunodiagnostic methods (eg, agarose gel double diffusion, immunoenzymetric (IEMA/ELISA) assays, and Western blot analysis) have been developed to demonstrate serum antibodies to respiratory allergens, notably antibody to *Aspergillus* antigens and to organic dust (see Hypersensitivity Pneumonitis Serology *on page 761*).

Additional Information PHA and con A are potent T-cell mitogens while PWM, lipopolysaccharide and staphylococcal protein A are selective B-cell mitogens. Patients with DiGeorge anomaly, Nezelof syndrome, and severe combined immunodeficiency syndrome typically show selectively impaired blastogenic responses to T-cell mitogens, while patients with pure humoral immunodeficiencies generally are characterized by normal responses. Variable blastogenic responses to B-cell selective mitogens are found in humoral immunodeficiency disorders.

Patients with chronic mucocutaneous candidiasis may show relatively normal mitogenic responses with T- and B-cell selective mitogens but have impaired blastogenesis to *Candida* antigen.

Patients with pulmonary berylliosis show beryllium induced lymphocyte blastogenesis.

References
Kurup VP and Fink JN, "Immunological Tests for Evaluation of Hypersensitivity Pneumonitis and Allergic Bronchopulmonary Aspergillosis," Allergic Diseases, Hamilton RG, section ed, de Macario EC, volume ed, *Manual of Clinical Laboratory Immunology*, 5th ed, Chapter 112, Section N, Rose NR, de Macario E, Folds JD, et al, eds, Washington, DC: ASM Press, American Society for Microbiology, 1997, 908-15.

Newman LS, Bobka C, Schumacher B, et al, "Compartmentalized Immune Response Reflects Clinical Severity of Beryllium Disease," *Am J Respir Crit Care Med*, 1994, 150(1):135-42.

Sarma PU, Banerjee B, Bir N, et al, "Immunodiagnosis of Allergic Bronchopulmonary Aspergillosis," *Immunol Allergy Clin North Am*, 1998, 18(3):525-47.

Lysergic Acid Diethylamide, Urine

Related Information
  Phencyclidine, Qualitative, Urine *on page 1019*

Synonyms LSD; LSD-2

Abstract LSD is one of the most potent hallucinogenic agents and is currently a DEA schedule I drug.

Specimen Urine; levels can be done on blood specimens as well.

Storage Instructions Refrigerate

Special Instructions If forensic, observe precautions and use chain-of-custody protocol.

Reference Interval None present in normal urine; confirmation cutoff: 0.5 ng/mL

Possible Panic Range ≥4 ng/mL

Methodology Screen: immunoassay; confirmation: gas chromatography/mass spectrophotometry (GC/MS)

Additional Information
  • Half-life: 3-4 hours
  • Volume of distribution: 0.28 L/kg
  • Protein binding: 90%
  • Onset of action: oral: 20-90 minutes, I.V.: 10 minutes
  • Duration of action: 6-8 hours

Fatalities due to LSD are very rare, as the doses required for hallucination are much lower than the amounts that cause fatalities. Signs and symptoms of LSD overdose include hallucinations, delusions, rhabdomyolysis, fear, tremors, (Continued)

Lysergic Acid Diethylamide, Urine *(Continued)*

delirium, and psychosis. There is no specific antidote for LSD; overdose treatment is supportive.

The detection of LSD in the laboratory is challenging due to very low concentrations of LSD in body fluids. Several manufacturers now make ultrasensitive immunoassay kits, which can detect LSD concentrations as low as 0.1 ng/mL.

References
Batzer W, Ditzler T, and Brown C, "LSD Use and Flashbacks in Alcoholic Patients," *J Addict Dis*, 1999, 18(2):57-63.

Ritter D, Cortese CM, Edwards LC, et al, "Interference With Testing for Lysergic Acid Diethylamide," *Clin Chem*, 1997, 43(4):635-7.

♦ **Lysine** *see* Cystine, Urine *on page 494*

♦ **Lysophatidic Acid** *see* CA 125, Serum *on page 323*

♦ **Lysosomal Storage Diseases** *see* Inherited Diseases of Metabolism and Cell Structure *on page 792*

♦ **Lysozyme** *see* Immunoperoxidase Procedures *on page 780*

Lysozyme, Blood and Urine
Related Information
Bone Marrow *on page 296*
Eosinophil Count *on page 542*
Leukocyte Cytochemistry *on page 846*
Lymph Node Biopsy *on page 880*
Urine Collection, 24-Hour *on page 1295*

Synonyms Muramidase, Blood and Urine

Specimen Serum, 24-hour urine

Container Red top tube or lavender top (EDTA plasma) tube; 24-hour urine container

Collection Collect 24-hour urine on ice.

Storage Instructions Separate serum or plasma and freeze **immediately** in plastic vial on dry ice. Upon receipt of urine specimen, freeze on dry ice **immediately** in plastic vial.

Reference Interval Serum: 4.0-15.6 µg/mL (0.28-1.10 µmol/L); urine: 0-1.4 µg/mL (0-0.097 µmol/L)

Use Differential diagnosis of leukemia; present in M4 and M5 types of acute myeloid (AML) leukemia. Serum lysozyme has been proposed as a marker for monitoring disease progression/regression in cases of sarcoidosis.

Revised FAB (French, American, British) criteria indicate that serum or urine lysozyme levels three times normal fulfill one of three criteria for presence of M4/M5 (acute myeloid leukemia with monocytic differentiation) vs M2 (acute myeloblastic leukemia with maturation) (Sexton et al).

Limitations Test may lack specificity when applied to classification of acute leukemia (occasional false positive in cases of M1, M2, and M6).

Additional Information Lysozyme, an hydrolytic enzyme - a bacteriolytic glycosidase, when present in large amounts may appear as a far cathodal migrating ("cationic") band on serum or urine protein electrophoresis. Lysozyme has been found in all three human neutrophil granules (azurophil, specific, and gelatinase types). It is elevated in some cases of AML, and most cases of the M4 and M5 types of AML. The elevation is proportional to the degree of monocytic differentiation and to tumor cell burden and if marked, can result in potassium wasting and hypokalemia. Lysozyme has been found within the granules of normal and leukemic eosinophils by immunoelectron microscopic study. Elevated serum lysozyme may not establish presence of monocytic differentiation in cases of acute myeloid leukemia with eosinophilia. Serum lysozyme has been shown to be elevated in a number of conditions, including tuberculosis and sarcoidosis as well as leukemia. Serum lysozyme, however, is less specific for sarcoidosis than serum angiotensin-converting enzyme. Utilizing a turbidimetric method for measurement of serum lysozyme activity, there was evidence that such an assay was useful in differentiating infection from rejection in transplant recipients.

References
Jones JW, Su S, Jones MB, et al, "Serum Lysozyme Activity Can Differentiate Infection From Rejection in Organ Transplant Recipients," *J Surg Res*, 1999, 84(2):134-7.

Sexton C, Buss D, Powell B, et al, "Usefulness and Limitations of Serum and Urine Lysozyme Levels in the Classification of Acute Myeloid Leukemia: An Analysis of 208 Cases," *Leuk Res*, 1996, 20(6):467-72.

Tomita H, Sato S, Matsuda R, et al, "Serum Lysozyme Levels and Clinical Features of Sarcoidosis," *Lung*, 1999, 177(3):161-7.

♦ **Lysylpyridinoline** *see* Pyridinolines (Pyridinoline and Deoxypyridinoline), Urine *on page 1126*

♦ **Macroglobulins, Ultracentrifuge Determination** *see* Immunoglobulin M *on page 779*

♦ **Magnesium:Creatinine Ratio, Urine** *see* Magnesium, Urine *on page 886*

♦ **Magnesium-Loading Test** *see* Magnesium, Urine *on page 886*

♦ **Magnesium Retention Test** *see* Magnesium, Urine *on page 886*

Magnesium, Serum

Related Information

Aminoglycosides, Serum *on page 147*
Calcium, Serum *on page 329*
Digoxin, Serum *on page 512*
Magnesium, Urine *on page 886*
Potassium, Urine *on page 1080*

Synonyms Mg, Serum

Applies to iMg^{2+}_s (Ionized Serum Magnesium); tMg_s (Total Magnesium, Serum); tMg (Total Magnesium)

Abstract Magnesium (Mg) is a major inorganic cation. It is infrequently measured, but may be important in critically ill patients. Up to 40% of hypokalemic patients are also hypomagnesemic. Magnesium assays are available for both **total** and **ionized** components, tMg and iMg, respectively.

Specimen Serum

Container Red top tube

Collection Draw without venous stasis. Separate serum from red cells as soon as possible.

Storage Instructions Refrigerate. Serum separated from cells is stable at 2°C to 6°C for several days.

Causes for Rejection Hemolysis

Reference Interval

tMg: 1.5-2.3 mg/dL (SI: 0.62-0.95 mmol/L or 1.2-1.9 mEq/L), although slightly different intervals are reported by different laboratories. **Four sets of units are in use to express concentration of Mg: 1.0 mEq/L = 1.22 mg/dL = 0.5 mmol/L = 12.2 mg/L.** Mg is partly protein bound. Slightly low values in the presence of hypoalbuminemia or hypoproteinemia should not, therefore, be of major concern.

iMg: 1.1-1.6 mg/dL (SI: 0.47-0.65 mmol/L)

Possible Panic Range Symptoms appear at <1.2 mg/dL (SI: <0.5 mmol/L); serum concentrations at <1.2 mg/dL are regarded as severe depletion. Toxic symptoms appear >4.9 mg/dL (SI: >2.0 mmol/L). Possible death from respiratory failure >14.6 mg/dL (SI: >6.0 mmol/L).

Hypomagnesemia iMg^{2+}_s <1.1 mg/dL (<0.46 mmol/L).

Use Summary of Indications for the measurement of serum magnesium concentration:

- myocardial infarct
- refractory cardiac arrhythmia
- alcoholism, other types of malnutrition
- refractory hypokalemia, hypocalcemia, hyponatremia
- diuretic therapy, particularly with multiple agents
- digoxin toxicity
- aminoglycoside, cyclosporine, cisplatin, amphotericin B therapy
- parenteral nutrition
- severe or chronic diarrhea
- unexplained electrolyte disturbance
- unexplained neuromuscular irritability, particularly in the absence of hypocalcemia
- treatment with nephrotoxic or cytotoxic agents
- neurological abnormalities in the neonate
- preeclampsia or eclampsia

(Continued)

Magnesium, Serum *(Continued)*

Magnesium deficiency produces neuromuscular spasm, fasciculations, hyperactivity and may cause weakness, dizziness, tremors, tetany, and convulsions.

Limitations Hemolysis will yield elevated results as levels in erythrocytes are two to three times higher than serum. Serum magnesium constitutes only a small fraction of total body stores and may not predict magnesium status correctly. **Magnesium deficiency sufficiently severe to lead to hypocalcemia and cardiac arrhythmias may exist in the presence of normal levels of serum magnesium.** Declining before serum Mg, urinary Mg is reported as an earlier and more reliable signal of evolving Mg deficiency. Huijgen et al found that about 70% of patients hypomagnesemic by tMg_s are no longer so when iMg^{2+}_s is measured.

Methodology Atomic absorption spectrophotometry (preferred), fluorometry, enzymatic methods also available, inductively-coupled plasma emission spectroscopy

Additional Information Approximately 50% of the total body magnesium is present in soft tissues with the other half stored in bone. Less than 1% of the total body magnesium is present in serum.

References

Foley C and Zaritsky A, "Should We Measure Ionized Magnesium?" *Crit Care Med*, 1998, 26(12):1949-50.

Huijgen HJ, Soesan M, Sanders R, et al, "Magnesium Levels in Critically Ill Patients. What Should We Measure?" *Am J Clin Pathol*, 2000, 114(5):688-95.

Ryan MF and Barbour H, "Magnesium Measurement in Routine Clinical Practice," *Ann Clin Biochem*, 1998, 35(Pt 4):449-59.

Siddiqui MN, Zafar H, Alvi R, et al, "Hypomagnesemia in Postoperative Patients: An Important Contributing Factor in Postoperative Mortality," *Int J Clin Pract*, 1998, 52(4):265-7.

Magnesium, Urine

Related Information

Aminoglycosides, Serum *on page 147*
Calcium, Serum *on page 329*
Calcium, Urine *on page 332*
Citrate, Serum, Plasma, or Urine *on page 413*
Cyclosporine, Blood *on page 487*
Digoxin, Serum *on page 512*
Ketone Bodies, Blood *on page 816*
Kidney Stone Analysis *on page 820*
Magnesium, Serum *on page 885*
Oxalate, Urine *on page 989*
Potassium, Serum or Plasma *on page 1078*
Urine Collection, 24-Hour *on page 1295*
Vitamin D, Serum *on page 1318*

Synonyms Mg, Urine

Applies to Magnesium:Creatinine Ratio, Urine; Magnesium-Loading Test; Magnesium Retention Test

Abstract Urinary magnesium may provide earlier indication of evolving deficiency than serum levels. Magnesium deficiency sufficiently severe to lead to hypocalcemia and cardiac arrhythmias can exist in the presence of normal serum magnesium concentrations.

Patient Preparation Patient should be instructed to use a plastic bedpan; *vide infra*.

Container Plastic acid-washed urine container; addition of an acidifying agent such as hydrochloric acid as preservative is desirable; check with laboratory. Acidification to pH 1.0 is recommended.

Collection Since a circadian rhythm exists for magnesium excretion, 24-hour collections are indicated. In children, in whom a timed collection may not be feasible, **urine magnesium:creatinine ratio** can be used. See graphic.

Storage Instructions Refrigerate

Causes for Rejection Specimen allowed to contact metal.

Reference Interval 6.0-10.0 mEq/day (SI: 3-5 mmol/day). Some laboratories report 120-160 mg/day. Values are slightly higher in males.

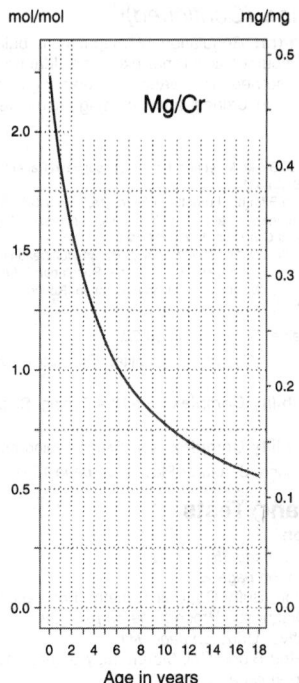

mol/mol   mg/mg

Mg/Cr

Age in years

Estimated 95th percentiles for urinary
Mg/Cr ratios in relation to age.

Use Urinary collection is valuable to detect magnesium deficiency due to gastro-intestinal causes (eg, gut failure or Crohn disease). Replacement of Mg in patients with gut failure should be directed at normalization of urinary Mg. The propensity to form calcium oxalate kidney stones is related to urine concentrations of oxalate and calcium, and inversely to those of citrate and Mg.

Urinary excretion is not diminished if depletion is caused by a renal leak. Magnesium urinary excretion is enhanced by increasing blood alcohol levels, diuretics, Bartter syndrome, Gitelman syndrome, corticosteroids, cis-platinum therapy, and aldosterone. Renal magnesium wasting occurs in renal transplant recipients who are on cyclosporine and prednisone. Other drugs which may lead to magnesium depletion include aminoglycosides, cyclosporine, pentamidine, and foscarnet. Renal conservation of magnesium is diminished by hypercalciuria, salt-losing conditions, and the syndrome of inappropriate secretion of antidiuretic hormone. Urinary magnesium analyses have been advocated before and after therapeutic magnesium administration to further investigate the significance of an apparent low serum concentration. Significantly diminished urinary magnesium concentration is found in some subjects whose serum magnesium is within normal limits.

Limitations A lack of uniformity exists among references regarding boundaries of reference interval for urine magnesium. Lack of consistent normal intervals for serum magnesium between published papers, laboratories, and geographic areas is also evident. High urinary concentrations of Mg can lead to errors by calmagite photometric assays if such specimens are not appropriately diluted.

Methodology Atomic absorption (AA), inductively-coupled plasma emission spectroscopy, enzymatic methods, photometric assays. Methods have been compared and tabulated (Ryan and Barbour, 1998).
(Continued)

Magnesium, Urine *(Continued)*

Additional Information Regulation of magnesium balance is via intestinal absorption (dietary source) and renal excretion. Magnesium is filtered at the glomerulus and reabsorbed along various tubular segments. The loop of Henle plays the major role in orchestration of magnesium resorption and urinary magnesium excretion.

References

Chernow B, Bamberger S, Stoiko M, et al, "Hypomagnesemia in Patients in Postoperative Intensive Care," *Chest*, 1989, 95(2):391-7.

Elin RJ, "Magnesium: The Fifth but Forgotten Electrolyte," *Am J Clin Pathol*, 1994, 102(5):616-22.

Fleming CR, George L, Stoner GL, et al, "The Importance of Urinary Magnesium Values in Patients With Gut Failure," *Mayo Clin Proc*, 1996, 71(1):21-4.

Matos V, Van Melle G, Boulat O, et al, "Urinary Phosphate/Creatinine, Calcium/Creatinine and Magnesium/Creatinine Ratios in a Healthy Pediatric Population," *J Pediatr*, 1997, 131(2):252-7.

Ryan MF and Barbour H, "Magnesium Measurement in Routine Clinical Practice," *Ann Clin Biochem*, 1998, 35(Pt 4):449-59.

♦ **Magnum**® *see Caffeine, Serum on page 326*

♦ **Mainlining** *see Cocaine (Cocaine Metabolite), Qualitative, Urine or Hair on page 427*

♦ **Major Histocompatibility Complex** *see HLA Typing, Single Human Leukocyte Antigen on page 739*

♦ **Major Histocompatibility Complex** *see Tissue Typing on page 1259*

♦ **Majsolin**® *see Primidone, Serum or Plasma on page 1091*

Malaria Smear and Tests

Related Information

Anemia Flowchart *on page 35*
Babesiosis Serology *on page 225*
Bacteremia Detection, Buffy Coat Micromethod *on page 227*
Chagas Disease Diagnostic Procedures *on page 381*
Myoglobin, Qualitative, Urine *on page 941*
Peripheral Blood: Red Blood Cell Morphology *on page 1016*
Reticulocyte Count *on page 1156*

Applies to Hemozoin; OptiMAL®; *Plasmodium* Species

Test Includes Examination of thick and thin smears

Abstract Malaria is still the most common infectious disease in the world. Its rapid diagnosis in the laboratory is extremely important. With continuing increase in international travel and liberalization of world trade (social and economic homogeneity) the incidence of malaria in the United States and Europe may be expected to increase.

Specimen Fresh blood (no anticoagulant) - fresh fingerstick smears (two or three of each thick and thin film type) preferably made at bedside using clean oil-/grease-free slides. Air dry and fix by brief dip in methyl alcohol. EDTA anticoagulated whole blood should be obtained for saponin lysis.

To make a thick film smear, spread a drop of blood centrally placed on a slide into a square about four times the area of original drop. Spreading is conveniently achieved with the corner of another slide. Ideally, blood should be thinned until small newsprint is just visible.

Whole blood is generally the specimen of choice for a variety of recently developed methods (see below).

Container Slides and lavender top (EDTA) tube

Sampling Time Specimen should be drawn immediately before a fever spike is anticipated.

Causes for Rejection Specimen clotted

Special Instructions If the patient has traveled to a malaria endemic area the date and area traveled should be specified on the requisition. Most cases of malaria seen in the U.S. are found in foreign nationals traveling in the United States.

Reference Interval No organisms identified

Use Diagnose malaria, parasitic infestation of blood; evaluate febrile disease of unknown origin

Limitations One negative result does not rule out the possibility of parasitic infestation. If protozoal, filarial, or trypanosomal infection is strongly suspected, test should be performed at least three times with samples obtained at different times in the fever cycle.

Methodology Microscopic examination of thick and thin peripheral blood Romanowsky dye (in particular Giemsa) stained smears. Thick films are more difficult to interpret but greatly increase sensitivity (by concentrating cells and organisms). The thin/thick film study is a sensitive technique which under ideal circumstances can detect as few as 10 parasites/μL of blood.

Screening for malaria can also be accomplished by fluorescent microscopy using acridine orange or benzothiocarboxypurine. These methods are sensitive and can provide consistent results without the need for highly experienced observers. DNA hybridization probes for detection of malaria have been described, but sensitivity may not be comparable to that obtained with use of thick films. Thin smears, although not as sensitive, are far superior for determining the species of *Plasmodium* on morphological grounds. A rapid and sensitive "magnet test" has been developed. The Abbott Cell-Dyn 4000 can detect intraleukocytic malaria-associated hemozoin pigment by depolarization analysis (recent study of 831 blood samples indicating a specificity of 97.4% and sensitivity of 80%).

Malaria Species Infecting Human Red Cells

Plasmodium Species	Type of Malaria	Time, Bite to Symptoms	Length of Cycle (h)
P. vivax	Tertian	10-14 d	45
P. falciparum	Malignant tertian	10-14 d	48
P. ovale	Ovale		48
P. malariae	Quartan	18-42 d	72

Additional Information Malaria is a protozoan parasitic infection caused in humans by four members of the species *Plasmodium*: *P. vivax*, *P. falciparum*, *P. malariae*, and *P. ovale*. The disease is endemic in subtropical/tropical areas of the world, corresponding to the distribution of the mosquito insect vector, members of the genus *Anopheles*, in which the sexual phase of the life cycle transpires. There has been a resurgence of cases in the United States and Europe beginning in the late 1960s and relating importantly to the problems in Vietnam and political/economic unrest on the African continent.

Proper therapy depends upon identification of the specific variety of malaria parasite. Release of trophozoites and RBC debris results in a febrile response. Periodicity of fever correlates with type of malaria (see table). Organisms are most likely to be detected just before onset of fever which is predictable in many cases. Sampling immediately upon onset of fever is the most desirable time to obtain blood. Alternatively in cases negative by these means but with a strong clinical history, multiple sampling at different times in the fever cycle may prove successful.

Malarial parasites are destroyed in AS and SS patients. Resistance to malaria is also seen with glucose-6-phosphate dehydrogenase deficiency, with increase in Hb E and/or F, and with Southeast Asian ovalocytosis. Malarial resistance to *Plasmodium vivax* occurs in Duffy ($Fy^a Fy^b$) negative individuals.

While stained thin and thick blood smears are not difficult to prepare and have low space and cost requirements, technical expertise in identification of malarial organisms is difficult to assure, especially in those geographic areas of the world where it is most needed. Tests that would not require human detection/identification of malarial organisms by microscopy have been developed and assessed but cost, unfortunately, may limit their broad application. During the past decade, the following have undergone field evaluation:

- QBC tube Buffy coat with acridine orange, this test could not identify species in 40% of cases.
- Para-Sight™-F test (a laminated nitrocellulose stick using a single drop of blood, with results in about 10 minutes and using antigen capture technique with mouse monoclonal antibody to a portion of histidine-rich protein-2 (HRP-2) of *P. falciparum*. Rheumatoid factor or post-therapy persistent antigenemia may give false-positive results).
- ICT Malarial Pf™ (for *Plasmodium falciparum*) or Pv™ (for *Plasmodium vivax*) rapid methods for detection and diagnosis using 10 μL whole blood,

(Continued)

Malaria Smear and Tests *(Continued)*

placed on a sample pad with colloidal gold-labeled antibody to histidine-rich protein (HRP)-2. Rheumatoid factor or post-therapy persistent antigenemia may give false-positive results.

- OptiMAL® (based on detection of parasite lactate dehydrogenase), a 15 minute test that is claimed to detect and differentiate *falciparum* from non-*falciparum* malaria.

Changes in Infected RBCs
Useful in Identification of Malaria Species

Plasmodium Species	Infected RBC Enlarged	Presence of Schüffner Dots	Presence of Maurer Dots	Multiple Parasites per RBC	Parasite With Double Chromatin Dots	Gametocytes
P. vivax	+	+	—	Rare	Rare	Spherical, compact
P. falciparum	—	—	+	+	+	Crescentic
P. ovale	±	+	—	—	—	Oval, fills 3/4 cell
P. malariae	—	—	+	—	—	Round, fills 2/3 cell

The HRP-2 dependent tests may be misleading in post-treatment follow-up studies (to confirm cure) in that many patients have persistent HRP-2 antigenemia for up to 7-14 days post-therapy even though microscope study and clinical evidence indicates cure. A number of disadvantages of the new generation of alternative tests (as compared to conventional microscopy) are reviewed by Hänscheid.

Reticulocyte methods used by automated routine hematology analyzers that stain intraerythrocytic nucleic acid may give falsely increased reticulocyte counts (pseudoreticulocytosis) in patients with severe malaria infection.

Malaria during pregnancy, while potentially the most controllable aspect of global malaria, is not commonly the target of programs to prevent the disease. The need to work with traditional birth attendants, in attempts to provide control in those geographic regions in which the problem exists, has been emphasized.

References

Araz E, Tanyuksel M, Ardic N, et al, "Performances of a Commercial Immunochromatographic Test for the Diagnosis of Vivax Malaria in Turkey," *Trans R Soc Trop Med Hyg*, 2000, 94(1):55-6.

British Committee for Standards in Haematology, Malaria Working Party of the General Haematology Task Force, "The Laboratory Diagnosis of Malaria," *Clin Lab Haematol*, 1997, 19(3):165-70.

Dacie JV and Lewis SM, "Preparation and Staining Methods for Blood and Bone-Marrow Films," *Practical Haematology*, 8th ed, Chapter 6, New York, NY: Churchill Livingstone, 1995, 89-96.

Hänscheid T, "Diagnosis of Malaria: A Review of Alternatives to Conventional Microscopy," *Clin Lab Haematol*, 1999, 21(4):235-45.

Hoffman JJML and Pennings JMA, "Pseudoreticulocytosis as a Result of Malaria Parasites," *Clin Lab Haematol*, 1999, 21(4):257-60.

Kakkilaya BS, "Rapid Diagnosis of Malaria," *Lab Med*, 2003, 34(8):602-8.

Nahlen BL, "Rolling Back Malaria in Pregnancy," *N Engl J Med*, 2000, 343(9):651-2.

Lanar DE, McLaughlin GL, Wirth DF, et al, "Comparison of Thick Films, *In Vitro* Culture and DNA Hybridization Probes for Detecting *Plasmodium falciparum* Malaria," *Am J Trop Med Hyg*, 1989, 40(1):3-6.

Palmer CJ, Validum L, Lindo J, et al, "Field Evaluation of the OptiMAL® Rapid Malaria Diagnostic Test During Antimalarial Therapy in Guyana," *Trans R Soc Trop Med Hyg*, 1999, 93(5):517-8.

Scott CS, van Zyl DV, Ho E, et al, "Automated Detection of Malaria-Associated Intraleucocytic Haemozoin by Cell-Dyn CD4000 Depolarization Analysis," *Clin Lab Haematol*, 2003, 25(2):77-86, Erratum: 2003, 25(2):205-7.

Tarimo DS, Moshiro C, Mpembeni R, et al, "Field Trial of the Direct Acridine Orange Method and Para-Sight™-F Test for the Rapid Diagnosis of Malaria at District Hospitals in Dar es Salaam, Tanzania," *Trans R Soc Trop Med Hyg*, 1999, 93(5):521-2.

White NJ, "The Treatment of Malaria: Current Concepts, Review Article," *N Engl J Med*, 1996, 335(11):800-6.

Manganese, Serum or Blood

Related Information

Manganese, Urine *on page 892*

Synonyms Mn, Serum

Abstract Manganese, essential to life in many species, is part of many human enzyme systems. It is considered essential for human nutrition, even though no well-documented example of a manganese deficiency state has been found in humans. Manganese (Mn) is routinely included in enteral and parenteral formula.

Specimen Serum, whole blood

Container Special metal-free blood collection tube. See the Trace Elements Introduction *on page 77.*

Collection Use powder-free gloves. Draw this and any other trace metal blood sample first before using needle to perforate the rubber stopper of an ordinary blood tube. For isolation of serum, allow time for clotting and centrifuge. Carefully pour serum into special plastic metal-free vial or transfer with acid-washed all-plastic pipet, being careful not to disturb the clot, or buffy coat. Store sample frozen if it is to be transported to reference laboratory.

Causes for Rejection Failure to collect specimen as stated above.

Reference Interval Serum: 0.43-0.76 ng/mL (SI: 7.8-13.8 nmol/L); whole blood: 10-11 ng/mL (SI: 190-200 nmol/L). Levels are about threefold higher in very low birth weight babies at birth and fall slightly over the first 3 months of life. Other reference ranges have been recommended for ages 1 month to 18 years and 22-75 years (Rukgauer et al).

Critical Values Exposed manganese workers who developed signs of Mn toxicity had whole blood Mn levels of 20-400 ng/mL when measured during ongoing exposure. A protective limit of 100 $\mu g/m^3$ average annual exposure has been recommended in long-term ferroalloy workers. In this cohort of exposed workers, the mean blood Mn concentration was 9.18 $\mu g/L$ compared to 5.74 $\mu g/L$ in nonexposed individuals.

Use Evaluate manganese deficiency and toxicity

Limitations In evaluating toxicity, serum concentrations may return to normal while elevated cerebral Mn concentrations and neurologic damage persist. Mn levels are 60% lower than normal in hemodialysis patients. Plasma uptake of Mn is reduced by concomitant ingestion of calcium but increased by concomitant ingestion of zinc.

Methodology Neutron activation, atomic absorption (AA) spectrophotometry with Zeeman background correction, and inductively-coupled plasma mass spectrometry (ICP-MS)

Additional Information In the serum, essentially all manganese is transported bound to transferrin. Transferrin receptors on the cell internalize the Mn as efficiently as iron transferrin complexes. Once in the cell, Mn is 80% bound to ferritin.

In humans, low blood concentrations are associated with epilepsy (regardless of type of anticonvulsant) and reported with the skeletal deformities of Perthes disease. Some have argued that the late hip dislocations in infants may be related to the low Mn content in cow's milk.

Manganese toxicity causes nausea, vomiting, headache, and psychiatric disturbances with central nervous system damage manifested by disorientation, memory loss, anxiety, compulsive laughing or crying, dementia, and psychosis. In the more chronic form, Mn toxicity resembles Parkinson disease with akinesia, rigidity, tremors, and mask-like faces. Normalization of serum levels may or may not reverse neurological damage.

Several medications are thought to chelate Mn, including valproate, and hydralazine. Dialysis lowers Mn levels and patients on hemodialysis usually have lower basal serum concentrations. Ninety-nine percent of Mn excretion occurs with bile in the feces.

References

Jacobs DS, DeMott WR, Oxley DK, et al, *Laboratory Test Handbook*, 5th ed, Hudson, OH: Lexi-Comp Inc, 2001.

Lucchini R, Apostoli P, Perrone C, et al, "Long-Term Exposure to 'Low Levels' of Manganese Oxides and Neurofunctional Changes in Ferroalloy Workers," *Neurotoxicity*, 1999, 20(2-3):287-97.

Mason JB, "Consequences of Altered Micronutrient Status," *Cecil Textbook of Medicine*, 21st ed, Goldman L and Bonnett JC, eds, Philadelphia, PA: WB Saunders Co, 2000, 1170-8.

Neve J and Leclercq N, "Factors Affecting Determinations of Manganese in Serum by Atomic Absorption Spectrometry," *Clin Chem*, 1991, 37(5):723-8.

Reynolds AP, Kiely E, and Meadows N, "Manganese in Long-Term Paediatric Nutrition," *Arch Dis Child*, 1994, 71(6):527-8.

(Continued)

Manganese, Serum or Blood *(Continued)*

Rukgauer M, Klein J, and Kruse-Jarres JD, "Reference Values for the Trace Elements Copper, Manganese, Selenium, and Zinc in the Serum/Plasma of Children, Adolescents, and Adults," *J Trace Elem Med Biol*, 1997, 11(2):92-8.

Manganese, Urine

Related Information

Manganese, Serum or Blood *on page 890*

Synonyms Mn, Urine

Abstract Urine manganese is used in conjunction with serum manganese (Mn) to evaluate possible toxicity or deficiency. An essential mineral which, in high concentration, can cause neurological damage similar to that of Parkinson disease. See Manganese, Serum or Blood *on page 890* for signs and symptoms of deficiency and toxicity.

Specimen 24-hour or random ("spot") urine. Consider simultaneous determination of urine creatinine, especially on "spot" samples.

Container Acid-washed plastic urine container, avoid contamination by stool, dust, and metal.

Causes for Rejection Contamination by metal, stool, or dust

Reference Interval <2.0 µg/L (SI: <36 nmol/L) (97.5% confidence). Varies with laboratory. Level may fall to 0.2 µg/L (tenfold) in experimental deficiency. Some of the best data on industrial exposure have been normalized to the creatinine content of urine or to µg/hour excretion due to the impracticality of 24-hour urine samples in the industrial setting. Ten free-living, nonexposed young men in Wisconsin had urinary excretion varying from approximately 0.17-0.66 µg/g creatinine. Levels higher than this but <9.0 µg/g creatinine may reflect increased exposure, not necessarily at a toxic level. Another group of factory workers exposed to manganese, some of whom were symptomatic with parkinsonian signs, had urine levels of Mn varying from 11.2-216.0 µg/L. Bile and feces are the main routes of excretion, accounting for 99% of excretion when intake is low, and excretion rather than absorption appears to be regulated. Urine losses appear to be overflow losses, representing a higher fraction of total loss when intake is high.

Use Confirm manganese exposure, toxicity, or poisoning by documentation of excessive urine excretion. Also used to individualize Mn dosing in long-term parenteral nutrition, especially in liver disease. Used to follow the success of chelation therapy with para-aminosalicylate sodium in manganism.

Limitations Low urine levels are not good indicators of Mn deficiency. Manganese toxicity may leave residual neurologic damage after serum and urine levels have returned to normal.

Methodology Atomic absorption (AA) with Zeeman background correction

Additional Information As much as twofold diurnal variation is present in urine Mn, especially in workers occupationally exposed to Mn. Low urinary Mn excretion may play a role in liver Mn overload in alcoholics. Hair Mn content has been utilized as a biological index for exposure in Mn refinery employees.

References

Foo SC, Khoo NY, Heng A, et al, "Metals in Hair as Biological Indices for Exposure," *Int Arch Occup Environ Health*, 1993, 65(1 Suppl):S83-6.

Lucchini R, Apostoli P, Perrone C, et al, "Long-Term Exposure to "Low Levels" of Manganese Oxides and Neurofunctional Changes in Ferroalloy Workers," *Neurotoxicity*, 1999, 20(2,3):287-97.

Rodriguez-Moreno F, Gonzalez-Reimers E, and Santolaria-Fernandez F, "Zinc, Copper, Manganese and Iron in Alcoholic Liver Disease," *Alcohol*, 1997, 14(1):39-44.

♦ **Maprotiline** *see* Antidepressants, Cyclic, Serum or Plasma *on page 171*

Maprotiline, Serum

Related Information

Antidepressants, Cyclic, Serum or Plasma *on page 171*

Synonyms Ludiomil®

Abstract Maprotiline is a tetracyclic antidepressant with a long half-life. Like the tricyclics, maprotiline is an inhibitor of the reuptake of norepinephrine. It is metabolized to an active metabolite, desmethyl maprotiline. Its investigational uses include treatment of bulimia, duodenal ulcers, enuresis, urinary symptoms of multiple sclerosis, panic attacks, and cocaine withdrawal.

Specimen Serum

Container Red top tube

Sampling Time Peak serum values are reached in 12 hours and steady-state is achieved in 5-10 days.

Storage Instructions Separate serum from clot and refrigerate.

Reference Interval 200-400 ng/mL (SI: 721-1442 nmol/L)

Critical Values >1000 ng/mL (SI: >3605 nmol/L)

Use Evaluate toxicity and therapeutic drug monitoring

Contraindications It should not be given in combination with MAO inhibitors.

Methodology High performance liquid chromatography (HPLC), gas chromatography (GC)

Additional Information

- Half-life: Maprotiline: 27-58 hours; active metabolite desmethyl maprotiline: 60-90 hours
- Volume of distribution: 15-35 L/kg
- Protein binding: 80% to 90%

Maprotiline is often used in patients who do not respond to tricyclics. The drug is taken orally at an average adult dose of 75-300 mg/day (50-75 mg/day in elderly patients). The drug possesses moderate anticholinergic activity and cardiovascular toxicity. It also may lower seizure threshold. It has a higher incidence of seizures than the tricyclic antidepressants, trazodone, fluoxetine, and fluvoxamine. Adverse effects include vertigo, blurred vision, seizures, drowsiness, and urinary retention. Concomitant use of risperidone increases plasma concentration of maprotiline.

References

Normann C, Lieb K, and Walden J, "Increased Plasma Concentration of Maprotiline by Coadministration of Risperidone," *J Clin Psychopharmacol*, 2002, 22(1):92-4.

Pisani F, Spina E, and Oteri G, "Antidepressant Drugs and Seizure Susceptibility: From *In Vitro* Data to Clinical Practice," *Epilepsia*, 1999, 40(Suppl 10):S48-56.

Rotzinger S, Bourin M, Akimoto Y, et al, "Metabolism of Some "Second"- and "Fourth"-Generation Antidepressants: Iprindole, Viloxazine, Bupropion, Mianserin, Maprotiline, Trazodone, Nefazodone, and Venlafaxine," *Cell Mol Neurobiol*, 1999, 19(4):427-42.

Measles Serology

Related Information

Viral Culture on page 1307

Test Includes Antibodies specific for rubeola, either IgG and IgM, in patient's serum or cerebrospinal fluid

Abstract Measles is an acute viral disease which causes generalized infection involving the respiratory tract and lymphoreticular tissues. It includes a papular eruption, lymphadenopathy, cough, and fever. Due to the widespread use of childhood immunization, measles has been nearly eliminated in most developed countries, including the U.S.

Specimen Serum or cerebrospinal fluid

Container Red top tube, sterile CSF tube

Reference Interval Less than fourfold rise in IgG titer, absent IgM titer

Use Diagnose infection with measles in a nonvaccinated individual; establish immunity subsequent to vaccination

Limitations The presence of measles-specific IgG in a single serum specimen indicates past or present infection or past vaccination.

Methodology Enzyme-linked immunosorbent assay (ELISA). Less frequently used methods are hemagglutination inhibition (HAI) and viral neutralization (NT).

(Continued)

Measles Serology (Continued)

Additional Information Measles (rubeola) is caused by a paramyxovirus, and is rapidly spread from person-to-person via aerosol. Occasional small epidemics have occurred in vaccinated populations and revaccination appears to be of value. In many individuals, **detectable** immunity does not persist. In acute illness, hemagglutinating and neutralizing antibody peak 2 weeks after the rash appears. It is necessary to demonstrate rising titers over 2 weeks, or identify IgM antibody to establish diagnosis. Very high serum titers in the absence of acute illness, or high CSF titer, are seen in subacute sclerosing panencephalitis. Due to massive immunization of children, measles is no longer an indigenous disease in the U.S. Information on childhood immunizations including the combination measles, mumps, and rubella vaccine can be found in the CDC publication "Combination Vaccines for Childhood Immunizations, Recommendations of the Advisory Committee on Immunization Practices (ACIP), the American Academy of Pediatrics (AAP), and the American Academy of Family Physicians (AAFP)".

References

Centers for Disease Control and Prevention, "Combination Vaccines for Childhood Immunizations, Recommendations of the Advisory Committee on Immunization Practices (ACIP), the American Academy of Pediatrics (AAP), and the American Academy of Family Physicians (AAFP)," *MMWR Morb Mortal Wkly Rep*, 1999, 48(RR-5):1-15.

Centers for Disease Control and Prevention, "Epidemiology of Measles - United States, 1998," *MMWR Morb Mortal Wkly Rep*, 1999, 48(34):749-53.

Helfand RF, Kebede S, Gary HE Jr, et al, "Timing of Development of Measles-Specific Immunoglobulin M and G After Primary Measles Vaccination," *Clin Diagn Lab Immunol*, 1999, 6(2):178-80.

Whittle H, Aaby P, Samb B, et al, "Poor Serologic Responses Five to Seven Years After Immunization With High and Standard Titer Measles Vaccines," *Pediatr Infect Dis*, 1999, 18(1):53-7.

Internet Web Sites

www.astdhpphe.org/infect/measles.html
www.cdc.gov/nip/publications/pink/meas.pdf
www.who.int/vaccines-diseases/research/virus1.shtml

♦ **Measurin®** *see* Salicylate, Serum or Plasma *on page 1176*

Meat Fibers, Stool

Related Information

d-Xylose Absorption Test, Serum, Urine *on page 527*
Fat, Semiquantitative, Stool, Sudan III Stain *on page 572*
Fecal Fat, Quantitative, 72-Hour Collection *on page 574*
Fecal Lactoferrin *on page 575*
Fecal Pancreatic Elastase 1 *on page 576*
Methylene Blue Stain, Stool *on page 906*
pH, Stool *on page 1034*
Reducing Substances, Stool *on page 1147*

Synonyms Muscle Fiber, Stool

Abstract Stool is examined microscopically for the presence of striated muscle fibers as an indicator of malabsorption/pancreatic insufficiency. While simple and inexpensive, this test lacks sensitivity and specificity.

Patient Preparation Patient is required to eat adequate amounts of red meat for 24-72 hours before testing. Specimens may be obtained with warm saline or Fleet Phospho®-Soda enema. Specimens obtained with mineral oil, bismuth, or magnesium compounds are unsatisfactory. Barium procedures or laxatives should be avoided for 1 week prior to collection of the specimen.

Specimen Stool

Causes for Rejection Purgatives other than saline or Fleet® Phospho®-Soda

Reference Interval Negative for muscle fibers

Use Initial evaluation of malabsorption syndromes, pancreatic exocrine dysfunction, or gastrocolic (fecal) fistula

Additional Information The presence of fecal undigested muscle fibers implies impaired intraluminal digestion. There is good correlation with stool fat determinations. Study of stool for muscle fibers when a high meat intake has been maintained is an inexpensive but necessarily nonspecific test for malabsorption. The presence of fecal muscle fibers cannot differentiate pancreatic insufficiency from other causes of malabsorption. Muscle fibers reported as present in urine are suggestive of fecal contamination of the specimen (eg, fecal fistula).

References

Birch DF, Fairley KF, Becker GJ, et al, *A Color Atlas of Urine Microscopy*, London: Chapman and Hall Medical, 1994, 8, 11.

Heisig DG, Threatte GA, and Henry JB, "Laboratory Diagnosis of Gastrointestinal Tract and Pancreatic Disorders," *Clinical Diagnosis and Management by Laboratory Methods*, 20th ed, Part 3, Chapter 23, Henry JB, ed, Philadelphia, PA: WB Saunders Co, 2001, 475.

Kalser MH, "Malabsorption Syndromes," *Bockus Gastroenterology*, 5th ed, Chapter 58, Haubrich WS, Schaffner F, and Berk JE, eds, Philadelphia, PA: WB Saunders Co, 1995, 996-8.

♦ **Mebaral**® *see* Barbiturates, Quantitative, Serum or Plasma *on page 248*

♦ **Mebaral**® *see* Methylphenobarbital, Serum *on page 909*

♦ **Median** *see page 11*

♦ **Mediastinal Mass Aspiration** *see* Fine Needle Aspiration, Deep and Superficial Masses *on page 590*

♦ **Medical Legal Specimens** *see* Chain-of-Custody Protocol *on page 381*

♦ **Mee's Lines** *see* Arsenic, Hair, Nails *on page 210*

♦ **Mee's Lines** *see* Thallium, Urine or Blood *on page 1242*

♦ **Megathrombocytes** *see* Platelet Histogram Maximum *on page 1053*

♦ **Melagatran** *see* Prothrombin Time *on page 1116*

♦ **Mellaril**® *see* Phenothiazines, Serum *on page 1021*

♦ **MEN 2A** *see* Multiple Endocrine Neoplasia/Familial Medullary Thyroid Carcinoma *on page 924*

♦ **MEN 2B** *see* Multiple Endocrine Neoplasia/Familial Medullary Thyroid Carcinoma *on page 924*

♦ **MEN2/FMTC** *see* Multiple Endocrine Neoplasia/Familial Medullary Thyroid Carcinoma *on page 924*

♦ **MEN A** *see* Calcitonin, Serum or Plasma *on page 327*

♦ **Menadol**® *see* Ibuprofen, Serum *on page 764*

♦ **MEN B** *see* Calcitonin, Serum or Plasma *on page 327*

♦ **Meningoencephalitis Viral Serology** *see* Encephalitis Viral Serology *on page 535*

Meperidine, Serum or Urine

Related Information

Codeine, Urine *on page 429*

Morphine, Urine *on page 921*

Propoxyphene, Serum or Urine *on page 1096*

Synonyms Centralgine®; Demerol®; Dolantin®; Dolantina®; Dolantine®; Dolosal®; Isonipecaine Hydrochloride; Pethidine Hydrochloride

Applies to Normeperidine

Test Includes This test is included in the comprehensive Urine Drug Screen.

Abstract Meperidine is a synthetic narcotic analgesic with about one-tenth the potency of morphine. It produces less smooth muscle spasm, constipation, and depression of the cough reflex than morphine.

Specimen Serum or urine

Container Red top tube, plastic urine container

Sampling Time Trough for serum

Storage Instructions Refrigerate

Special Instructions If forensic, use chain-of-custody protocol and form. See Chain-of-Custody Protocol *on page 381* and the Chain-of-Custody form in the Introduction *on page 63*.

Reference Interval Serum: 70-500 ng/mL (SI: 0.28-2.0 μmol/L); urine: negative when not on therapy

Critical Values Serum toxic levels: >1000 ng/mL (SI: >4.0 nmol/L)

Possible Panic Range Serum levels: >5000 ng/mL (SI: >20 nmol/L)

Use Evaluate toxicity; detect an abused drug. Meperidine is a shorter-acting synthetic morphine-like compound used in the management of moderate to severe pain and as an adjunct to anesthesia and preoperative sedation.

Contraindications This drug is contraindicated in individuals with renal dysfunction or those on monoamine oxidase inhibitors.

Methodology Gas chromatography (GC) or high performance liquid chromatography (HPLC). High concentrations in overdose can be detected by thin-layer chromatography (TLC).

Additional Information

- Half-life: 2-4 hours

(Continued)

Meperidine, Serum or Urine *(Continued)*

- Volume of distribution: 3-4 L/kg
- Protein binding: 40% to 60%

Analgesic effects after oral ingestion or I.M. injection peak in 1 hour. It is biotransformed to normeperidine, which is a toxic metabolite. Adverse effects include tachycardia, CNS and respiratory depression, nausea and vomiting, hypotension, bradycardia, miosis, increased intracranial pressure, and physical and psychological dependence. When evaluating therapeutic levels, order Meperidine, Serum.

References

Belgrade M, "Postherpetic Pain Control With Concomitant Illness?" *Postgrad Med*, 2001, 109(3):31-2.

Clark RF, Wei EM, and Anderson PO, "Meperidine: Therapeutic Use and Toxicity," *J Emerg Med*, 1995, 13(6):797-802.

Mephenytoin, Serum

Related Information

Phenytoin, Serum or Plasma *on page 1026*

Synonyms Mesantoin®; Methoin; Methylphenylethylhydantoin; Phenantoin; Sedantoinal®

Applies to 5-Phenyl-5-Ethylhydantoin (Nirvanol®)

Abstract Mephenytoin has a pharmacologic effect similar to that of phenytoin. Mephenytoin has serious toxicity but less dose-related effects compared to phenytoin. This drug is generally used only for patients refractory to or unable to tolerate other anticonvulsants.

Specimen Serum

Container Red top tube

Sampling Time Consistent sampling time. Steady-state is reached in ~40 hours.

Reference Interval 25-40 µg/mL for the sum of the parent drug and 5-phenyl-5-ethylhydantoin metabolite, which is usually present at a higher level than the parent compound

Possible Panic Range Toxic level about 50 µg/mL (SI: 230 µmol/L). See Table A in the Therapeutic Drug Monitoring Introduction *on page 60*.

Use Monitor for compliance, efficacy, and possible toxicity. Mephenytoin may be suited for patients who respond to phenytoin but who cannot tolerate dose-related side effects. It is used for treatment of tonic-clonic and partial seizures.

Contraindications Indications for discontinuation of drug include WBC <1600.

Methodology Gas chromatography (GC)

Additional Information

- Half-life: parent compound: ~8 hours, metabolite: >100 hours
- Protein binding: 20% to 50%

Acute overdosage results in sedation and eventually coma. Most adverse effects are not dose-related, but are due to the accumulation of arene-oxide intermediates. They are reported to include rash, hepatotoxicity, blood dyscrasias, systemic lupus erythematosus, periarteritis nodosa, and fever.

References

Ko JW, Desta Z, and Flockhart DA, "Human N-Demethylation of (S)-Mephenytoin by Cytochrome P450s 2C9 and 2B6," *Drug Metab Dispos*, 1998, 26(8):775-8.

Leikin JB and Paloucek FP, *Poisoning and Toxicology Compendium*, Hudson, OH: Lexi-Comp Inc, 1998.

Product Information: Mesantoin®, Mephenytoin, East Hanover, NJ: Sandoz Pharmaceuticals Corporation, 1995.

- ♦ **Mephobarb** *see* Barbiturates, Qualitative, Urine *on page 248*
- ♦ **Mephobarbital** *see* Barbiturates, Quantitative, Serum or Plasma *on page 248*
- ♦ **Mephobarbital** *see* Methylphenobarbital, Serum *on page 909*
- ♦ **Mephobarbitone** *see* Methylphenobarbital, Serum *on page 909*
- ♦ **Meprobam®** *see* Meprobamate, Serum *on page 896*

Meprobamate, Serum

Synonyms Equagesic®; Equanil®; Meprobam®; Meprospan®; Miltown®; Neuramate®; Tenavoid®

Abstract Meprobamate is a sedative-anxiolytic, producing effects similar to those of the benzodiazepines and barbiturates. It is a schedule IV controlled substance.

Specimen Serum

Container Red top tube

Reference Interval Sedative dose: 6-12 µg/mL (SI: 28-55 µmol/L)

Critical Values Toxic: >60 µg/mL (SI: >275 µmol/L)

Possible Panic Range Coma is associated with levels >70 µg/mL (SI: >321 µmol/L); **fatalities** can occur at >142 µg/mL (SI: >650 µmol/L); lethal: 200 µg/mL (SI: 916 µmol/L)

Use Therapeutic monitoring and toxicity assessment

Methodology Gas chromatography (GC), high performance liquid chromatography (HPLC). Thin-layer chromatography (TLC) is used for qualitative analysis in overdose situations.

Additional Information

- Half-life: 6-15 hours
- Volume of distribution: 0.5-1.0 L/kg
- Protein binding: 20%

Meprobamate is well absorbed from the gastrointestinal tract and reaches peak concentration in 2-3 hours. It may also be detected in urine or gastric juice. Carisoprodol is a noncontrolled muscle relaxant that is metabolized to meprobamate.

Adverse effects in overdose include hypotension, drowsiness, dizziness, hangover, and seizures. A benzodiazepine antagonist, flumazenil, has been used as an antidote in overdose cases.

References

Gaillard Y, Billault F, and Pepin G, "Meprobamate Overdosage: A Continuing Problem. Sensitive GC/MS Quantitation After Solid Phase Extraction in 19 Fatal Cases," *Forensic Sci Int*, 1997, 86(3):173-80.

Logan BK, Case GA, and Gordon AM, "Carisoprodol, Meprobamate, and Driving Impairment," *J Forensic Sci*, 2000, 45(3):619-23.

Reeves RR, Carter OS, Pinkofsky HB, et al, "Carisoprodol (Soma): Abuse Potential and Physician Unawareness," *J Addict Dis*, 1999, 18(2):51-6.

♦ **Meprospan®** *see* Meprobamate, Serum *on page 896*

Mercury, Blood

Related Information

Mercury, Urine *on page 898*

Synonyms Cinnabar; Hg, Blood; Organic Mercury; Quicksilver

Applies to Methylmercury

Abstract Elemental mercury is toxic if inhaled or injected, but may safely pass through the digestive tract if small amounts are swallowed. Blood analysis for mercury is used principally for evaluation of toxicity from organic mercury, found in wood preservatives, paints, fungicides, cosmetics, foods, seeds, and in contaminated fish. The mercury catastrophe in Minamata Bay, Japan, and Iraq involved ingestion of methylmercury-contaminated fish and grain respectively. Accidental and occupational exposures occur. Sources include production of chlorine and sodium hydroxide and use in batteries, switches, and fluorescent lights. Other sources include control devices for domestic gas supplies.

The fetal brain is more susceptible to damage induced by mercury.

Relative order of toxicity:

$Hg^0 < Hg^{++} <$ methylmercury $<$ dimethylmercury

Specimen Whole blood, hair, nails

Container Special metal-free EDTA tube

Collection See Blood Collection Methods for Trace Elements in the Trace Elements Introduction *on page 77.*

Causes for Rejection Failure to collect blood in a special metal-free container or exposure to metal containing dusts

Special Instructions Whole blood is analyzed.

Reference Interval Normally whole blood mercury is <10 µg/L (SI: <50 mmol/L), but this number varies widely in a large population of healthy unexposed individuals.

(Continued)

Mercury, Blood (Continued)

Critical Values >50 µg/L (SI: 250 nmol/L) if exposure is to alkyl mercury compounds and >200 µg/L (SI: 1000 nmol/L) if exposure is to Hg^{+2} compounds. The lethal dose of dimethylmercury is ~400 mg.

Possible Panic Range >100.0 µg/L (SI: 500 nmol/L)

Use Evaluate for mercury toxicity

Limitations Once exposure ceases, blood levels may not provide a good indicator of remaining body burden. There are marked variations in levels considered toxic by different investigators.

Methodology Electrothermal atomic absorption (AA); hair analysis may be used for poisoning or chronic exposure.

Additional Information Organic methyl mercury is an important environmental contaminant. It was discovered that elemental mercury could be oxidized into inorganic mercury (Hg^{+2}) and then into organic mercury (methylmercury) from industrial wastes. Organic mercury then accumulates in large amounts in predator fish, and thus, into the human food chain. The highest concentrations are found in long-lived predatory fish such as shark, tile fish, marlin, king mackerel, swordfish, pike, and bass. Canned tuna is the subject of active debate (Foran).

The major physical forms of mercury to which humans are exposed are (elemental) mercury vapor and methylmercury compounds. Severe acute poisoning leads to pulmonary distress with acute pneumonitis, hemoptysis, and cyanosis. Mechanical ventilation is often required. Permanent sequelae may include irreversible lung impairment and pulmonary fibrosis. Paresthesias, visual-field constriction, and ataxia are found.

The half-life of inorganic mercury is 24 days and of methylmercury (organic mercury) is 54 days. Ethyl mercury causes renal damage. Half-life of ethylmercury (used as thimerosal in vaccines) is 8 days. In the U.S., thimerosal has been removed from vaccines based on the claims that it may be a cause of autism and related disorders in children. However, the WHO advisory committee in 2002 concluded that thimerosal use in vaccines is safe (Clarkson).

A likely victim of occupational mercury intoxication is found in *Alice's Adventures in Wonderland: The Mad Hatter*.

References

Carroll L, *Alice's Adventures in Wonderland: A Mad Tea-Party*, Great Britain: Hedder and Staughton, 1922, 89-103.

Clarkson TW, Magos L, and Myers GJ, "Current Concepts: The Toxicology of Mercury - Current Exposures and Clinical Manifestations," *N Engl J Med*, 2003, 349(18):1731-7.

Eto K, Takizawa Y, Akagi H, et al, "Differential Diagnosis Between Organic and Inorganic Mercury Poisoning in Human Cases - the Pathological Point of View," *Toxicol Path*, 1999, 27(6):664-71.

Foran SE, Flood JG, and Lewandrowski KB, "Measurement of Mercury Levels in Concentrated Over-the-Counter Fish Oil Preparations," *Arch Pathol Lab Med*, 2003, 127(12):1603-5.

Graeme KA and Pollack CV, "Heavy Metal Toxicity, Part I: Arsenic and Mercury," *J Emerg Med*, 1998, 16(1):45-56.

Ozuah PO, "Mercury Poisoning," *Curr Probl Pediatr*, 2000, 30(3):91-9.

Mercury, Urine

Related Information

Mercury, Blood *on page 897*

Synonyms Hg, Urine

Abstract See Mercury, Blood *on page 897*.

Specimen 24-hour urine

Container Plastic (preferably polyethylene) acid-washed container, no preservative

Storage Instructions Store in special metal-free container.

Causes for Rejection Failure to collect sample in a special metal-free container or exposure to metal-containing dusts

Reference Interval 24-hour urine: <20 µg/L (SI: <100 nmol/L)

Critical Values Significant exposure is indicated when daily urine mercury reaches 50 µg/day.

Possible Panic Range >150 µg/L (SI: >748 nmol/L); symptoms are found with levels >600 µg/L (>3000 nmol/L); lethal urine levels: >800 µg/L (SI: >3992 nmol/L).

Use Urine mercury is best used for evaluation of inorganic mercury exposure.

Limitations Organic mercury is found mostly in red cells and is lipid soluble. Urine mercury may not be useful for evaluating pure organic mercury poisoning. Once exposure ceases, urine levels may not represent a good indicator of remaining body burden.

Methodology Electrothermal atomic absorption (AA)

Additional Information See Mercury, Blood *on page 897*

References

Clarkson TW, Magos L, and Myers GJ, "Current Concepts: The Toxicology of Mercury - Current Exposures and Clinical Manifestations," *N Engl J Med*, 2003, 349(18):1731-7.

Graeme KA and Pollack CV, "Heavy Metal Toxicity, Part I: Arsenic and Mercury," 1998, 16(1):45-56.

Metanephrines, Urine or Plasma

Related Information

Synonyms Metanephrine; Normetanephrine; Total Metanephrines

Applies to Catecholamines; Epinephrine; Metanephrine:Creatinine Ratio, Urine; Norepinephrine

Test Includes Metanephrine and normetanephrine

Abstract The collective term "metanephrines" includes metanephrine and normetanephrine, the *O*-methylated metabolites of the catecholamines, epinephrine, and norepinephrine, respectively. Clinical interest centers on cate-cholamine-secreting neoplasms predominantly arising from the adrenal medulla: pheochromocytomas, paragangliomas and neuroblastomas. Measurement of metanephrines is sensitive for the diagnosis of pheochromo-cytoma, and the urine **metanephrine:creatinine ratio** may be even better.

Patient Preparation All methylxanthine-containing foods should be avoided for 24 hours. Many drugs interfere. Check with testing laboratory for specific instructions (see Limitations).

Specimen 24-hour urine, plasma

Container Plastic urine container, lavender top (EDTA) tube; check with the laboratory.

(Continued)

Metanephrines, Urine or Plasma (Continued)

Collection

Urine: 50 mL from a 24-hour collection. Add 25 mL of 50% acetic acid at start of collection. Use 15 mL for children younger than 5 years of age. Check with laboratory for instructions.

Blood: Draw specimen after the patient has rested supine for at least 20 minutes; arterial blood can be used. Collect blood into precooled tubes containing EDTA; centrifuge within 30 minutes.

Storage Instructions Keep urine collection cold. Freeze plasma until assay.

Reference Interval Reference intervals vary among laboratories.

Urine: metanephrines: <1.3 mg/24 hours

Urinary metanephrine:creatinine ratios:
- men: 0.152 ±0.074
- women: 0.181 ±0.090
- upper limit of normal: 0.354

Plasma:
- normetanephrine: <0.90 nmol/L
- metanephrine: <0.5 nmol/L

Use Measurement of metanephrines is usually performed to diagnose pheochromocytoma. Normal plasma concentration by contemporary methods is strong evidence against pheochromocytoma.

Limitations Urine: Overestimation or underestimation of time of collection of a urine specimen remain problematic. False-negative **urinary** results occur (eg, interference by methylglucamine in x-ray contrast medium). False-positive **urinary** results occur; results should be confirmed by repeat testing, imaging or other means, such as urine catecholamines and VMA.

Methodology Liquid chromatography with electrochemical detection (preferred), gas chromatography (GC), high performance liquid chromatography (HPLC), radioenzymatic methods, and immunoassay

Additional Information Because the secretion from a pheochromocytoma is usually episodic, plasma catecholamine levels drawn during asymptomatic periods may be normal, while total urinary metanephrines collected over 24 hours are likely to be abnormal. On the other hand, a tumor could be intermittently active, but secrete over a full 24-hour period total metanephrines that are within reference ranges. Analyses of 24-hour urine specimens are preferable to spot (random) urine samples for catecholamines, VMA, and metanephrines.

References

Bravo EL, "Plasma or Urinary Metanephrines for the Diagnosis of Pheochromocytoma? That Is the Question," *Ann Intern Med*, 1996, 125(4):331-2.

Eisenhofer G, Lenders JWM, Linehan WM, et al, "Plasma Normetanephrine and Metanephrine for Detecting Pheochromocytoma in Von Hippel-Lindau Disease and Multiple Endocrine Neoplasia Type 2," *N Engl J Med*, 1999, 340(24):1872-9.

Héron E, Chatellier G, Billaud E, et al, "The Urinary Metanephrine-to-Creatinine Ratio for the Diagnosis of Pheochromocytoma," *Ann Intern Med*, 1996, 125(4):300-3.

Krakoff LR, "Searching for Pheochromocytoma: A New and Better Test?" *Ann Intern Med*, 1995, 123(2):150-1.

Lenders JW, Eisenhofer G, Armando I, et al, "Determination of Metanephrines in Plasma by Liquid Chromatography With Electrochemical Detection," *Clin Chem*, 1993, 39(1):97-103.

Lenders JW, Keiser HR, Goldstein DS, et al, "Plasma Metanephrines in the Diagnosis of Pheochromocytoma," *Ann Intern Med*, 1995, 123(2):101-9.

Mayo Medical Laboratories, *2001 Test Catalogue*, Rochester, MN, 372.

Pacak K, Linehan WM, Eisenhofer G, et al, "Recent Advances in Genetics, Diagnosis, Localization, and Treatment of Pheochromocytoma," *Ann Intern Med*, 2001, 134(4):315-29.

Peaston RT, Lennard TW, and Lai LC, "Overnight Excretion of Urinary Catecholamines and Metabolites in the Detection of Pheochromocytoma," *J Clin Endocrinol Metab*, 1996, 81(4):1378-84.

Wassell J, Reed P, Kane J, et al, "Freedom From Drug Interference in New Immunoassays for Urinary Catecholamines and Metanephrines," *Clin Chem*, 1999, 45(12):2216-23.

Young DS, *Effects of Drugs on Clinical Laboratory Tests*, 5th ed, Volume 1: Listing by Test, Washington, DC: AACC Press, 2000, 3:551-2.

♦ **Metasedin®** *see* Methadone, Serum or Urine *on page 900*
♦ **Metformin** *see* Lactic Acid, Whole Blood or Plasma *on page 827*
♦ **Meth** *see* Methamphetamine, Qualitative, Urine *on page 902*

Methadone, Serum or Urine

Related Information

Antidiuretic Hormone, Plasma *on page 172*
Opiates, Qualitative, Urine *on page 974*

Propoxyphene, Serum or Urine *on page 1096*

Synonyms Dolophine®; Eptadone®; Metasedin®; Methadose®; Physeptone®; Symoron®

Applies to Heroin Detoxification; Naloxone

Abstract This drug is a synthetic opiate agonist used during World War II as a morphine substitute. Structurally related to propoxyphene, it is used for detoxification of opiate addicts. There are more than 100,000 patients enrolled in methadone maintenance programs. Overdose is a common problem. Methadone is also a frequently abused drug.

Specimen Random urine collection; serum quantitation has become available.

Container Red top tube or urine container

Storage Instructions Refrigerate

Special Instructions If forensic, use chain-of-custody protocol and form. See Chain-of-Custody Protocol *on page 381* and the Chain-of-Custody form in the Introduction *on page 63.*

Reference Interval Negative (less than cutoff); when used therapeutically for pain; serum/plasma levels are in the range of 0.10-0.40 µg/mL (SI: 0.32-1.29 µmol/L).

Critical Values Cutoff for urine screening: 300 ng/mL; confirmation: typically 200 ng/mL, variable between laboratories

Use Evaluate toxicity and detection as drug of abuse. Methadone is a drug of abuse and is included in most drug-of-abuse screening panels. It is measured to monitor compliance.

Methodology Screen: immunoassay; confirmation: gas chromatography/mass spectrometry (GC/MS)

Additional Information

- Half-life: 15-25 hours
- Volume of distribution: 4-6 L/kg
- Protein binding: 85% to 95%
- Oral bioavailability: 40% to 95%

Onset of action is 30-60 minutes after an oral dose and 10-20 minutes following parenteral administration. Adverse effects include marked sedation after repeated administration, CNS and respiratory depression, nausea and vomiting, bradycardia, hypotension, increased intracranial pressure, miosis, antidiuretic hormone release, and physical and psychological dependence. Treatment for overdose is supportive and administration of the opioid antagonist naloxone.

The analgesic and pharmacologic properties of methadone are similar to those of morphine.

This drug is used in the management of narcotic (particularly heroin) detoxification programs. Methadone maintenance programs have effectively reduced heroin dependency and transmission of HIV and are available in most countries. Once tolerance is developed, it may be used continually with less harmful side effects.

The primary metabolic pathway for methadone is by mono- and di-N-demethylation, followed by spontaneous cyclization to form 2-ethylidene-1,5-dimethyl-3,3-diphenylpyrrolidine (EDDP) and 2-ethyl-5-methyl-3,3-diphenylpyrroline (EMDP). It can be detected for 2-4 days in urine.

References

Garrido MJ and Troconiz IF, "Methadone: A Review of its Pharmacokinetic/Pharmacodynamic Properties," *J Pharmacol Toxicol Methods*, 1999, 42(2):61-6.

Stout PR and Farrell LJ, "Opioids: Effects on Human Performance and Behavior," *Forensic Sci Rev*, 2002, 15:29-59.

Strain EC, Bigelow GE, Liebson IA, et al, "Moderate- vs High-Dose Methadone in the Treatment of Opioid Dependence: A Randomized Trial," *JAMA*, 1999, 281(11):1000-5.

Wolff K, "Characterization of Methadone Overdose: Clinical Considerations and the Scientific Evidence," *Ther Drug Monit*, 2002, 24(4):457-70.

♦ **Methadose®** *see* Methadone, Serum or Urine *on page 900*

♦ **Methaminodiazepoxide Hydrochloride** *see* Chlordiazepoxide, Serum *on page 391*

♦ **Methampex®** *see* Methamphetamine, Qualitative, Urine *on page 902*

Methamphetamine, Qualitative, Urine

Related Information

Amphetamine, Qualitative, Urine *on page 154*

Synonyms Crank; Crystal; Desoxyephedrine Hydrochloride; Desoxyn®; Doe; Ecstasy; Go; Ice; Meth; Methampex®; Methedrine®; Speed; Zip

Applies to Benzedrine®; Contac®; Dexedrine®; Dimetapp®; Vicks Inhaler®

Test Includes Amphetamine, methamphetamine

Abstract Methamphetamine, a schedule II drug, is a CNS stimulant with some legitimate therapeutic uses. The d-form of the drug is used as a weight-reducing agent and for the treatment of hyperactivity and attention-deficit disorder in children, and the l-form is used as a nasal decongestant. Methamphetamine is highly addictive and is widely abused.

Specimen Random urine

Collection If forensic, observe precautions.

Storage Instructions Refrigerate

Special Instructions If forensic, use chain-of-custody protocol and form. See Chain-of-Custody Protocol *on page 381* and the Chain-of-Custody form in the Introduction *on page 63.*

Reference Interval Negative (less than cutoff); therapeutic, serum: 20-30 ng/mL

Critical Values Substance Abuse and Mental Health Services Administration (SAMHSA) cutoff: screen: 1000 ng/mL; confirmation: 500 ng/mL. For methamphetamine, a positive report requires methamphetamine ≥500 ng/mL and amphetamine (a methamphetamine metabolite) ≥200 ng/mL in the same sample.

Use Evaluate for drug abuse; assess toxicity

Limitations Screening test may give false positives with common cold and antiallergy medications.

Methodology Screening: immunoassay; confirmation: gas chromatography/mass spectrometry (GC/MS)

Additional Information

- Half-life: 10-30 hours, dependent on urinary pH (urine alkalinization increases half-life of sympathomimetic amines, acidification decreases it)
- Volume of distribution: 3-4 L/kg
- Protein binding: 10% to 40%

In this class, the most abused drug is d-methamphetamine. The optical isomer, l-methamphetamine (sold under the label l-desoxyephedrine), has more peripheral effects and less pronounced central effects and is used as a nasal decongestant in Vicks Inhaler® (legal, over-the-counter). Amphetamine isomers are present in Dexedrine® and Benzedrine®. In order to rule out the false-positive given by l-methamphetamine (legal nasal decongestant), a chiral column or procedure, which separates the "l" and "d" isomers, must be used in the GC/MS confirmation. Current immunoassays are geared towards detection of d-methamphetamine. Their cross reactivity with l-form is very low.

In interpretation of methamphetamine and amphetamine, positive test knowledge of legitimate and illicit sources is important. Amphetamines are sometimes prescribed as weight-reducing agents. Many substances (amphetaminil, benzphetamine, clobenzorex, deprenyl, dimethylamphetamine, ethylamphetamine, famprofazone, fencamine, fenethylline, fenproporex, furfenorex, mefenorex, mesocarb, and prenylamine) which are available as prescription drugs contain or are metabolized in the body to methamphetamine and/or amphetamine.

References

Jirovsky D, Lemr K, Sevcik J, et al, "Methamphetamine - Properties and Analytical Methods of Enantiomer Determination," *Forensic Sci Int*, 1998, 96(1):61-70.

Logan BK, "Methamphetamines - Effects on Human Performance and Behavior," *Forensic Sci Rev*, 2002, 14(2):133-51.

Williams RH, Erickson T, and Broussard LA, "Evaluating Sympathomimetic Intoxication in an Emergency Setting," *Lab Med*, 2000, 31(9):497-507.

♦ **Methanol** *see* Osmolality, Calculated, Serum or Plasma *on page 976*

♦ **Methanol** *see* Volatile Screen, Blood or Urine *on page 1320*

Methaqualone, Urine

Synonyms Lude®; Ludes; Quaalude®; Sopor™

Abstract This drug is a sedative-hypnotic and is currently a DEA schedule I drug. Once approved as a sedative and hypnotic, it was taken off the U.S. market in 1984. Its use has dramatically decreased and it is presently a rarely encountered drug.

Specimen Serum, urine

Container Red top tube, plastic urine container

Special Instructions If forensic, use chain-of-custody protocol and form. See Chain-of-Custody Protocol *on page 381* and the Chain-of-Custody form in the Introduction *on page 63.*

Reference Interval Urine: negative (less than cutoff)

Critical Values Cutoff for urine: screen: 300 ng/mL; confirmation: typically 200 ng/mL, variable between laboratories

Possible Panic Range Serum values >8 µg/mL (SI: >32 nmol/L) associated with unconsciousness

Use Evaluate for toxicity, evaluate for drug abuse

Methodology Screen: immunoassay; confirmation: gas chromatography/mass spectrophotometry (GC/MS)

Additional Information

- Half-life: 10-40 hours
- Volume of distribution: 5-7 L/kg
- Protein binding: 70% to 90%

It is rapidly absorbed from the GI tract. Hyperexcitability, coma, and cardiovascular and respiratory depression characterize overdosage. Once a common drug of abuse, in recent years the use of methaqualone has become rare.

References

Baselt RC, *Disposition of Toxic Drugs and Chemicals in Man*, 6th ed, Foster City, CA: Biomedical Publications, 2002, 657-9.

Brenner C, Hui R, Passarelli J, et al, "Comparison of Methaqualone Excretion Patterns Using Abuscreen ONLINE and EMIT II Immunoassays and GC/MS," *Forensic Sci Int*, 1996, 79(1):31-41.

♦ **Metharbital** *see* Barbiturates, Quantitative, Serum or Plasma *on page 248*

♦ **MetHb** *see* Methemoglobin, Whole Blood *on page 903*

♦ **Methedrine®** *see* Methamphetamine, Qualitative, Urine *on page 902*

Methemoglobin, Whole Blood

Related Information

Carboxyhemoglobin, Blood *on page 340*
Hemoglobin *on page 681*
Phlebotomy, Therapeutic *on page 1029*

Synonyms MetHb

Applies to Carboxyhemoglobin; Reduced Hemoglobin; Sulfhemoglobin

Abstract Methemoglobin (metHb) is a form of hemoglobin in which the iron has been oxidized from the normal ferrous (Fe^{++}) to the ferric (Fe^{+++}) state. MetHb cannot bind oxygen to act as an oxygen carrier.

Specimen Whole blood

Container Green top (heparin) tube

Storage Instructions Keep tube on ice. Run as promptly as possible after draw because there is significant decrease with time. . Studies have shown up to 10% drop in 4 hours, up to 16% drop in 8 hours, in samples kept on ice.

Reference Interval Up to 1% of total hemoglobin. Smokers have slightly higher values than nonsmokers.

Possible Panic Range Headache and other symptoms occur at levels >30%. Methemoglobinemia levels >70% may be lethal.

Use Evaluate causes of cyanosis. Examination for metHb is used to monitor patients on high dose nitrate therapy. Its measurement in CSF may detect small cerebral and subdural hematomas.

Methodology Co-oximetry, spectrophotometry. Hb M variants are best detected by electrophoresis.

Additional Information MetHb is an inactive, oxidized form of hemoglobin (Hb) which does not contribute to the oxygen-carrying capacity of blood. Therefore, arterial %O_2 saturation will be inappropriately low for a given inhaled air (Continued)

Methemoglobin, Whole Blood (Continued)

oxygen concentration and p_aO_2, if the calculation for $\%O_2$ saturation is based on total Hb. Concentrations of metHb of over 10% to 25% of Hb cause cyanosis. The most common cause of cyanosis is the presence of excessive reduced Hb which becomes clinically apparent (as a bluish discoloration of skin and mucous membranes) when the capillary level is >5 g/dL. Cyanosis appears with metHb concentrations of 10% to 25%, but symptoms are minimal until metHb rises to 35% to 40%, at which patients may experience fatigue, dizziness, dyspnea, headache, and tachycardia. At the 60% level, lethargy and stupor may occur; levels >70% in adults may be fatal.

Methemoglobinemia may be hereditary or acquired. Polycythemia is occasionally present as a compensatory mechanism. Most instances of methemoglobinemia are acquired from drugs and chemicals. Nitro and amino groups are especially involved, eg, aniline and derivatives, nitrites, nitroglycerin, nitrate salts in burn patients, flutamide, metoclopramide, phenazopyridine, dapsone (perhaps the most common cause of drug-induced methemoglobinemia), phenacetin, acetophenetidin, prilocaine, some sulfonamides, sulfones, chlorates, primaquine, quinones, large doses of ferrous sulfate, and many other drugs and some intestinal bacteria.

Hereditary methemoglobinemia is uncommon. The most common phenotype is due to a deficiency of red cell NADH-methemoglobin reductase (diaphorase, also termed cytochrome b_5 reductase), which is inherited as an autosomal recessive trait. Homozygotes have metHb levels of 15% to 20%. Heterozygotes are apt to develop toxic methemoglobinemia when exposed to substances which can oxidize hemoglobin iron. Certain hemoglobin variants (particularly M variant) with abnormal tendency to stabilize Fe^{+++} cause methemoglobinemia.

References

Jacobs DS, DeMott WR, Oxley DK, et al, *Laboratory Test Handbook*, 5th ed, Hudson, OH: Lexi-Comp Inc, 2001.

Lukens JN and Lee GR, "Unstable Hemoglobin Disease," *Wintrobe's Clinical Hematology*, 10th ed, Chapter 52, Lee GR, Foerster J, Lukens J, et al, eds, Baltimore, MD: Lippincott Williams & Wilkins, 1999, 1398-404.

Wentworth P, Madan R, Wilson B, et al, "Toxic Methemoglobinemia in a 2-Year-Old Child," *Lab Med*, 1999, 30(5):311-15.

Wright RO, Lewander WJ, and Woolf AD, "Methemoglobinemia: Etiology, Pharmacology and Clinical Management," *Ann Emerg Med*, 1999, 34(5):645-56.

♦ **Methionine** see Amino Acids, Urine on page 145

Methionine Loading Test

Related Information

Cobalamin, Serum on page 424
Folic Acid, RBC on page 606
Folic Acid, Serum on page 606
Homocyst(e)ine, Plasma on page 741
Vitamin B$_6$, Plasma or Serum on page 1315

Abstract This test includes the measurement of plasma homocyst(e)ine (Hcy - see Homocyst(e)ine, Plasma on page 741) at certain intervals (eg, 2, 4, or most commonly, 6 hours) following an oral dose of methionine (100 mg L-methionine/kg body weight).

Reference Interval Not well defined. Increased risk of atherogenesis is correlated with 6-hour postmethionine plasma homocyst(e)ine values >40 µmol/L.

Use This test is used in three different clinical situations:

1. Detection of mild (heterozygous) cases of homocystinuria, by stressing the transulfuration pathway of methionine metabolism.
2. Assessment of vitamin (B$_6$, B$_{12}$, and folate) status
3. More recently, to identify individuals at increased risk for thrombotic events despite normal fasting Hcy levels (see Homocyst(e)ine, Plasma on page 741)

Limitations For all applications, the appropriate reference intervals are uncertain. Testing results appear to be much less specific than was anticipated by theoretical consideration.

Additional Information For examples of how this test is used in clinical research studies, see the references by Chambers et al and Vermeulen et al.

References
Chambers JC, Obeid OA, Refsum H, et al, "Plasma Homocysteine Concentrations and Risk of Coronary Heart Disease in UK Asian and European Men," *Lancet*, 2000, 355(9203):523-7.

Mayo Medical Laboratories, "New Test Available - Methionine Load, Plasma," Rochester, MN, October 1998.

Still RA and McDowell IFW, "Clinical Implications of Plasma Homocysteine Measurement in Cardiovascular Disease," *J Clin Pathol*, 1998, 51:183-8.

Vermeulen EG, Stehouwer CD, Twisk JW, et al, "Effect of Homocysteine-Lowering Treatment With Folic Acid Plus Vitamin B$_6$ on Progression of Subclinical Atherosclerosis: A Randomized, Placebo-Controlled Trial," *Lancet*, 2000, 355(9203):517-22.

♦ **Methionine, Serum** *see Molybdenum, Blood on page 919*

♦ **Methoin** *see Mephenytoin, Serum on page 896*

Methotrexate, Serum or Plasma
Related Information
C-Reactive Protein, Serum *on page 467*

Synonyms Amethopterin; Folex®; Mexate®; MTX; Rheumatrex®

Applies to Cerebrospinal Fluid Methotrexate; Citrovorum Factor

Abstract Methotrexate (MTX) is an anticancer drug acting through competitively inhibiting folic acid reductase, an enzyme necessary for cellular replication. A role for MTX is recognized as a part of therapy of acute lymphoblastic leukemia; small cell carcinoma of lung; and carcinomas of urinary bladder, breast, and head and neck. In high dosage with leucovorin rescue, it is among drugs used for osteogenic sarcoma. Low-dose methotrexate is also used in a number of other diseases, such as rheumatoid arthritis, psoriatic arthritis, polymyositis, and Reiter syndrome. It has a limited role in management of inflammatory bowel disease. Methotrexate should be monitored in patients receiving high-dose therapy (>100 mg/m^2). Since methotrexate is extremely toxic, leucovorin, a folate analogue, is generally used to rescue host cells. Such rescue makes possible administration of much higher doses of methotrexate.

Specimen Serum or plasma

Container Red top tube, green top (heparin) tube, or lavender top (EDTA) tube

Sampling Time Varies according to dosing protocol. Time to peak serum concentration: oral: within 1-2 hours; parenteral: within 30-60 minutes.

Storage Instructions Separate serum or plasma and freeze.

Reference Interval Therapeutic range is dependent upon therapeutic approach. "**High dose**" regimens produce drug levels between 0.1-1 µmol/L, 24-72 hours after drug infusion.

Critical Values Blood concentrations are generally monitored at 24, 48, and 72 hours after the single dose. Leucovorin is administered if the concentrations are higher than:
- 10 µmol/L 24 hours after dose
- 1 µmol/L 48 hours after dose
- 0.1 µmol/L 72 hours after dose

Leucovorin rescue is not complete until plasma concentrations fall below the levels described above.

Use Monitor therapeutic drug level of methotrexate; evaluate potential toxicity

Methodology Immunoassay, high performance liquid chromatography (HPLC)

Additional Information
- Half-life: distribution: 0.5-1 hour; elimination: 5-9 hours
- Volume of distribution: 0.4-1.0 L/kg
- Protein binding: 50% to 70%

Methotrexate is excreted unchanged in the urine in 12 hours when kidney function is adequate and the patient is hydrated. Urinary alkalinization is recommended.

Signs of fatal MTX toxicity include extensive erosions in the oral cavity and other mucous membranes, gastrointestinal hemorrhage, interstitial pneumonia, and progressive renal insufficiency. Accompanying laboratory findings may include agranulocytosis, thrombocytopenia, anemia, and hyperbilirubinemia. MTX use is contraindicated in patients with severe renal or hepatic impairment, pre-existing myelosuppression, and pleural/peritoneal effusions.

References
Barnhart K, Coutifaris C, and Esposito M, "The Pharmacology of Methotrexate," *Expert Opin Pharmacother*, 2001, 2(3):409-17.

(Continued)

Methotrexate, Serum or Plasma (Continued)

Bathon JM, Martin RW, Fleischmann RM, et al, "A Comparison of Etanercept and Methotrexate in Patients With Early Rheumatoid Arthritis," *N Engl J Med*, 2000, 343(22):1586-93.

Lipsky PE, van der Heijde DM, St Clair EW, et al, "Infliximab and Methotrexate in the Treatment of Rheumatoid Arthritis," *N Engl J Med*, 2000, 343(22):1594-602.

Moe PJ and Holen A, "High-Dose Methotrexate in Childhood ALL," *Pediatr Hematol Oncol*, 2000, 17(8):615-22.

♦ 3-Methoxy-4-Hydroxymandelic Acid *see* Vanillylmandelic Acid, Urine *on page 1299*

Methylene Blue Stain, Stool

Related Information

Bacterial Culture, Stool *on page 243*
Clostridium difficile Toxin Assay and Culture *on page 416*
Cryptosporidium Direct Staining Procedures *on page 484*
d-Xylose Absorption Test, Serum, Urine *on page 527*
Entamoeba histolytica Antigen Detection and Serology *on page 538*
Fat, Semiquantitative, Stool, Sudan III Stain *on page 572*
Fecal Fat, Quantitative, 72-Hour Collection *on page 574*
Fecal Lactoferrin *on page 575*
Meat Fibers, Stool *on page 894*
Occult Blood, Stool *on page 969*
Ova and Parasites, Direct Exam *on page 985*
pH, Stool *on page 1034*

Synonyms Fecal Leukocyte Stain

Applies to Leukotest® (Fecal Lactoferrin)

Test Includes Methylene blue, Gram, or Wright stain of stool smear

Abstract Evaluation for fecal leukocytes is a part of the initial evaluation of chronic diarrhea. In general, the presence of fecal leukocytes provides indication for stool culture. Absence of fecal leukocytes, however, does not exclude bacterial diarrhea or the need for stool culture.

Patient Preparation Collect specimen prior to barium procedures.

Specimen Fresh random stool, rectal swab

Container Cup specimen is more sensitive than swab specimen in detection of fecal leukocytes.

Storage Instructions Refrigerate

Reference Interval No predominance of yeast, cocci in clusters, or leukocytes

Use Examine fecal specimens for the presence of leukocytes as an indicator of inflammatory diarrhea (eg, invasive enteric infection). Diarrhea for more than 4 weeks is an indication for evaluation.

Limitations Ten percent to 15% of stools which yield an invasive bacterial pathogen on culture have an absence of fecal leukocytes. Fecal leukocytes are present in idiopathic inflammatory bowel disease.

Methodology Smear of stool (preferably mucus) with one drop methylene blue, coverslip, and observe for the presence of leukocytes.

Additional Information Conditions associated with marked fecal leukocytes, blood, and mucus include predominantly bacterial infections including invasive *E. coli*, shigellosis, salmonellosis, *Helicobacter*, *Yersinia* infection, ulcerative colitis, and cases of antibiotic-associated colitis and pseudomembranous colitis. *Salmonella typhi* may evoke a monocyte response. Conditions associated with modest numbers of fecal leukocytes include early shigellosis involving small bowel, and cases of antibiotic-associated colitis. In amebiasis, stool leukocytes are variable. Diarrhea can be watery or bloody. Conditions associated with an absence of fecal leukocytes include toxigenic bacterial infection including *Vibrio cholerae*, giardiasis, and viral infections. The methylene blue stain for polymorphonuclear leukocytes has a high sensitivity (85%) and specificity (88%) for bacterial diarrhea (*Shigella, Salmonella, Helicobacter*). Positive predictive value is 59%. Negative predictive value is 97%. Combined with a history of abrupt onset, more than four stools per day, and no vomiting before the onset of diarrhea, the stool methylene blue stain for fecal polymorphonuclear leukocytes is a very effective presumptive diagnostic test for bacterial diarrhea. A positive occult blood test may also be suggestive of acute bacterial diarrhea being more sensitive (79% vs 42%) than the fecal leukocyte test (in detection of invasive bacteria in pediatric patients). The occult blood test in this setting, however, lacks specificity and has been found to have a positive predictive value of only 24%.

Neither absence of fecal occult blood and/or leukocytes should pre-empt the use of culture. When both tests were positive, there was a sensitivity of 81% and specificity of 74% for bacterial diarrhea.

Slightly over 50% of patients with collagenous colitis have demonstrable fecal leukocytes.

A commercially available latex agglutination screening test (Leukotest®) for the leukocyte marker lactoferrin has several advantages over the microscopic-based detection of fecal leukocytes. Both tests, however, are insufficiently sensitive to be used as screening tests for such pathogens as *Campylobacter*, *Salmonella*, or *Shigella* spp. See Fecal Lactoferrin *on page 575*.

References
Dowdy LM, "Infectious Diarrhea," *Clinical Practice of Gastroenterology*, Chapter 60, Brandt LJ, ed, Philadelphia, PA: Current Medicine Inc, 1999, 529-30.
McNeely WS, Dupont HL, Mathewson JJ, et al, "Occult Blood Versus Fecal Leukocytes in the Diagnosis of Bacterial Diarrhea: A Study of U.S. Travelers to Mexico and Mexican Children," *Am J Trop Med Hyg*, 1996, 55(4):430-3.
Zins BJ, Tremaine WJ, and Carpenter HA, "Collagenous Colitis: Mucosal Biopsies and Association With Fecal Leukocytes," *Mayo Clin Proc*, 1995, 70(5):430-3.

♦ **Methylene Chloride** *see* Carboxyhemoglobin, Blood *on page 340*

♦ **Methylenedioxyamphetamine** *see* 3,4 Methylenedioxymethamphetamine, Urine *on page 907*

♦ **Methylenedioxyethamphetamine** *see* 3,4 Methylenedioxymethamphetamine, Urine *on page 907*

3,4 Methylenedioxymethamphetamine, Urine

Related Information

Amphetamine, Qualitative, Urine *on page 154*
Methamphetamine, Qualitative, Urine *on page 902*

Synonyms Adam; E; Ecstasy; MDMA; X; XTC

Applies to Methylenedioxyamphetamine; Methylenedioxyethamphetamine

Abstract 3,4 methylenedioxymethamphetamine (MDMA) is a derivative of methamphetamine. It produces a relaxed, euphoric state with heightened feelings within 20-40 minutes of oral intake. However, memory and working skills are impaired. Recently, due to its widespread use as a recreational drug in rave parties and dance clubs, the U.S. Drug Enforcement Administration placed the drug in Schedule I. In 1999, U.S. Customs seized 5.4 million hits of MDMA, 720% increase from 1998.

Specimen Random urine

Collection If forensic, observe precautions. See the Introduction *on page 63*.

Storage Instructions Refrigerate

Reference Interval Negative (less than cutoff)

Possible Panic Range Blood levels >1000 ng/mL

Use Evaluate drug abuse; assess toxicity

Limitations When screened by immunoassays for amphetamines, MDMA or MDA concentrations must be 5-20 times higher than cutoff values for amphetamine assay to cause positive results.

Methodology Screen: immunoassays, thin-layer chromatography (TLC), high performance liquid chromatography (HPLC); confirmation: gas chromatography/mass spectrometry (GC/MS)

Additional Information
- Half-life: 6-10 hours, longer in alkaline urine
- % excreted unchanged in urine: 65%

MDMA is metabolized to an active metabolite 3,4 methylenedioxyamphetamine (MDA). Both MDMA and MDA have marked sympathomimetic activities similar to amphetamines. They cause CNS stimulation, peripheral vasoconstriction, tachycardia, and pupillary dilation. Overdoses result in convulsions, hyperthermia, and behavioral changes. Severe complications include rhabdomyolysis, intravascular coagulation, arrhythmias, seizures, acute renal failure, and hepatonecrosis.

References
Doyon S, "The Many Faces of Ecstasy," *Curr Opin Pediatr*, 2001, 13(2):170-6.
(Continued)

3,4 Methylenedioxymethamphetamine, Urine
(Continued)

Gulledge C, Phillips J, and Hammett-Stabler C, "New Kids on the Block: An Update on Selected Club Drugs," *Therapeutic Drug Monitoring and Clinical Toxicology Division Newsletter*, 2000, 15(4):1-6.

Pedersen W and Skrondal A, "Ecstasy and New Patterns of Drug Use: A Normal Population Study," *Addiction*, 1999, 94(11):1695-706.

Methylmalonic Acid, Serum, Plasma, Urine, or Amniotic Fluid

Related Information

Abstract The predominant cause of cobalamin deficiency is pernicious anemia. Recommended initial testing in macrocytic/megaloblastic anemia includes serum levels of cobalamin with serum and red cell folate, utilizing methylmalonic acid (MMA) and homocyst(e)ine concentrations as follow-up evaluation when necessary. Recently, a more important role for the latter two tests (as initial test modalities, that can detect early and subclinical functional cobalamin deficiency) has been pursued for these metabolites. Since cobalamin is required for conversion of methylmalonic acid to succinic acid, subjects with cobalamin deficiency have increased serum and urine concentrations of methylmalonic acid which decrease following cobalamin administration.

MMA is also a diagnostic marker for methylmalonic acidemias and acidurias, a group of some eight inherited disorders, inborn errors of organic acid metabolism.

Patient Preparation Broad spectrum antibiotics can cause decreased to normal levels in patients who have cobalamin deficiency by inhibition of gut flora, but such drugs do not change increased homocyst(e)ine concentrations.

Specimen Serum, plasma (EDTA), urine; amniocentesis obtained between 16-19 weeks gestational age.

Reference Interval 70-270 nmol/L; the median concentration in cases of cobalamin deficiency is about 3500 nmol/L, but values as great as 2,000,000 nmol/L are described. In cobalamin responsive methylmalonic aciduria, following initiation of treatment, concentrations return to normal in 5-10 days.

Use Confirm cobalamin deficiency; detect early/subclinical cobalamin deficiency; distinction between cobalamin and folate deficiency; methylmalonic acid is increased in >95% of subjects who have cobalamin deficiency. Methylmalonic acid concentrations are not increased with folate deficiency. (Serum homocyst(e)ine concentrations are increased in both). Folate therapy in cobalamin-deficient individuals does not cause a decrease of methylmalonic acid concentrations. Cobalamin deficiency, prevalent following gastric surgery, occurs following resection of distal small intestine (eg, as with Crohn disease) as well as in Addisonian pernicious anemia.

Limitations Renal failure and volume depletion may increase levels of both methylmalonic acid and homocyst(e)ine. Patients with cobalamin deficiency following gastric surgery have elevation of serum MMA and homocyst(e)ine but some 20% have normal MMA concentrations.

Methodology Gas chromatography-mass spectrometry (GC/MS). Recently, liquid chromatography-electrospray tandem mass spectrometry (LC-MS/MS) methods have been developed for the routine determination of total homocyst(e)ine and of methylmalonic acid.

Additional Information Homocyst(e)ine accumulates with deficiency of either vitamin B_{12} or folate. Conversion of methylmalonyl coenzyme A to succinyl coenzyme A is impaired by deficiency of vitamin B_{12}, leading to accumulation of methylmalonic acid. Methylmalonic acid and homocyst(e)ine concentrations provide strong evidence of deficiency. They have a high level of sensitivity to early and subclinical cobalamin deficiency, greater than that of the traditional cobalamin level and Schilling test studies.

Isolated methylmalonyl CoA mutase deficiency is associated with clinical neonatal or infantile metabolic ketoacidosis. There may be absence of mutase (mut^0) or structurally altered mutase (mut$^-$) in affected cells. Affected children have methylmalonic acidemia and methylmalonic aciduria that do not respond to cobalamin treatment, but may be treated by protein restriction. In infants with methylmalonic acidurias, cobalamin deficiency must be excluded, particularly in infants who are breast-fed by a mother who is either a strict vegetarian or who has subclinical pernicious anemia.

References

Allen RH, "Megaloblastic Anemias," *Cecil Textbook of Medicine*, 21st ed, Goldman L and Bennett JC, eds, Philadelphia, PA: WB Saunders Co, 2000, 859-67.

Carmel R, "Current Concepts in Cobalamin Deficiency," *Annu Rev Med*, 2000, 51:357-75.

Chanarin I and Metz J, "Diagnosis of Cobalamin Deficiency: The Old and the New," *Br J Haematol*, 1997, 97(4):695-700.

deBaulny HO and Saudubray JM, "Branched-Chain Organic Acidurias," *Inborn Metabolic Diseases: Diagnosis and Treatment*, Fernandes J, Saudubray JM, and Van den Berghe G, eds, New York, NY: Springer-Verlag, 2000, 195-212.

Elin RJ and Winter WE, "Methylmalonic Acid. A Test Whose Time Has Come?" *Arch Pathol Lab Med*, 2001, 125(6):824-7.

Hvas AM, Vestergaard H, Gerdes LU, et al, "Physicians' Use of Plasma Methylmalonic Acid as a Diagnostic Tool," *J Intern Med*, 2000, 247(3):311-7.

Klee GG, "Cobalamin and Folate Evaluation: Measurement of Methylmalonic Acid and Homocysteine vs Vitamin B$_{12}$ and Folate," *Clin Chem*, 2000, 46(8 Pt 2):1277-83.

♦ **Methylmalonic Aciduria** see Cerebrospinal Fluid Glycine on page 364

♦ **Methylmercury** see Mercury, Blood on page 897

♦ **Methylmorphine** see Codeine, Urine on page 429

Methylphenobarbital, Serum

Related Information

Barbiturates, Quantitative, Serum or Plasma on page 248
Phenobarbital, Serum or Plasma on page 1021

Synonyms Enphenemalum; Gemonil®; Mebaral®; Mephobarbital; Mephobarbitone

Abstract Mephobarbital is metabolized to phenobarbital, which accounts for most pharmacologic effects. However, mephobarbital has slightly different pharmacokinetics.

Specimen Serum

Container Red top tube

Sampling Time Consistent sampling time. Steady-state is reached between 8-10 days.

Reference Interval Monitored through phenobarbital levels. Phenobarbital level should be in the range of 15-40 µg/mL.

Critical Values See Phenobarbital, Serum or Plasma on page 1021. Levels >80 µg/mL correlate with decreased mental status.

Use Mephobarbital is used for prophylactic management of tonic-clonic (grand mal) seizures and for absence (petit mal) seizures.

Methodology Immunoassay, gas chromatography (GC)

Additional Information

- Half-life: 45-55 hours
- Volume of distribution: 2-3 L/kg
- Protein binding: 40% to 60%

Like other barbiturates, methylphenobarbital is extensively metabolized in the liver, by demethylation, to phenobarbital. About 75% of a single dose is converted to phenobarbital in 24 hours.

Methylphenobarbital has a more linear response between dosage and blood phenobarbital level than does phenobarbital.

References

Ceccato A, Boulanger B, Chiap P, et al, "Simultaneous Determination of Methylphenobarbital Enantiomers and Phenobarbital in Human Plasma by On-Line Coupling of an Achiral Precolumn to a Chiral Liquid Chromatographic Column," *J Chromatogr A*, 1998, 819(1-2):143-53.

Leikin JB and Paloucek FP, *Poisoning and Toxicology Compendium*, Hudson, OH: Lexi-Comp Inc, 1998.

Willis J, Nelson A, Black FW, et al, "Barbiturate Anticonvulsants: A Neuropsychological and Quantitative Electroencephalographic Study," *J Child Neurol*, 1997, 12(3):169-71.

♦ **Methylphenylethylhydantoin** *see* Mephenytoin, Serum *on page 896*

Metyrapone Stimulation Test, Serum

Related Information

Adrenal Cortex: Laboratory Assessment Overview *on page 110*
Adrenocorticotropic Hormone, Plasma *on page 114*
Cortisol, Free, Urine *on page 459*
Cortisol, Serum or Plasma *on page 460*
17-Hydroxycorticosteroids, Urine *on page 753*

Applies to 11-Deoxycortisol

Test Includes Oral metyrapone at midnight, followed at 8 AM by serum levels of cortisol, 11-deoxycortisol, and ACTH

Abstract The metyrapone stimulation test (MST) is a secondary test of adrenal function which is used in the setting of suspected adrenal insufficiency (AI) or Cushing syndrome (CS) (see Adrenal Cortex: Laboratory Assessment Overview *on page 110*). There are three MST protocols: overnight, 2-day, and 3-day. In this book, we deal only with the overnight protocol, the one most commonly used. MST tests the functional integrity of the hypothalamic-pituitary-adrenal (HPA) axis. Metyrapone inhibits the conversion of 11-deoxycortisol to cortisol and thereby produces a marked decrease in serum cortisol which, when sensed by the pituitary, results in an increase in pituitary output, and serum level, of ACTH. The endpoints of the overnight MST are the serum levels of cortisol, 11-deoxycortisol, and ACTH, drawn at 8 AM the morning after a midnight dose of metyrapone.

Patient Preparation At midnight, the patient is given metyrapone by mouth (with milk or a snack) on a weight-adjusted scale: <70 kg receives 2 grams; 70-90 kg receives 2.5 grams; >90 kg receives 3 grams. Since metyrapone is metabolized by enzymes that are induced by drugs that enhance steroid metabolism (eg, phenytoin, rifampin, phenobarbital, mitotane, and corticosteroids), these drugs should be discontinued before conducting the MST.

Specimen Serum

Container Red top tube

Sampling Time 8 AM on the morning after oral metyrapone

Storage Instructions Freeze serum if assay is not performed within 24 hours.

Reference Interval The 8 AM serum cortisol should be <3 μg/dL (SI: <83 nmol/L). Failure to achieve this level makes the other values uninterpretable. Possible causes of a serum cortisol above this level include failure to take the metyrapone and rapid metyrapone clearance. Several drugs cause rapid clearance (see Patient Preparation) and 4% of otherwise normal persons also exhibit this phenomenon. When the HPA is intact, the following thresholds are met or exceeded in the 8 AM specimen:

- serum 11-deoxycortisol: >7 μg/dL (SI: >210 nmol/L)
- serum ACTH: >75 pg/mL (SI: >17 pmol/L)

Use

Adrenal insufficiency (AI): The MST is useful in patients being evaluated for secondary AI (pituitary- or hypothalamic-based AI). Patients with secondary AI will have serum 11-deoxycortisol and ACTH values less than the criteria indicated above, indicating pituitary or hypothalamic dysfunction. The MST is **not** used in the diagnosis of primary AI (see Adrenal Cortex: Laboratory Assessment Overview *on page 110*).

Cushing syndrome: The MST is useful in distinguishing pituitary-based CS from adrenal-based CS (see Adrenal Cortex: Laboratory Assessment Overview). When a patient with CS has MST results which meet or exceed the criteria indicated above, the diagnosis is usually pituitary-based CS (Cushing disease). In CS patients whose MST results do not meet these criteria, the differential diagnosis includes adrenal tumor vs the ectopic corticotropin syndrome (ECS). Patients with adrenal tumor typically have very low, or undetectable, serum ACTH, while those with ECS have normal or elevated (sometimes markedly elevated) serum ACTH level

Methodology See individual analytes.

References

Avgerinos PC, Yanovski JA, Oldfield EH, et al, "The Metyrapone and Dexamethasone Suppression Tests for the Differential Diagnosis of the Adrenocorticotropin-Dependent Cushing Syndrome: A Comparison," *Ann Intern Med*, 1994, 121(5):318-27.
Orth DN, "Cushing's Syndrome," *N Engl J Med*, 1995, 332(12):791-803.

Orth DN and Kovacs WJ, "The Adrenal Cortex," *Williams Textbook of Endocrinology*, 9th ed, Wilson JD, Foster DW, Kronenberg HM, et al, eds, Philadelphia, PA: WB Saunders Co, 1998, 620-1.

Mevalonic Acid, Urine or Amniotic Fluid

Related Information

Amino Acids, Plasma *on page 143*
Amino Acids, Urine *on page 145*
Immunoglobulin D *on page 772*
Inherited Diseases of Metabolism and Cell Structure *on page 792*
Lipids, Overview *on page 853*
Urine Collection, 24-Hour *on page 1295*

Synonyms Urinary MVA

Applies to HMG-CoA Reductase Inhibitor; Lovastatin; Pravastatin

Abstract Urinary mevalonic acid (MVA) is a marker of cholesterol biosynthesis *in vivo*. Mevalonic aciduria (a rare form of organic aciduria) is the result of an inherited deficiency of mevalonate kinase.

Specimen Urine, amniotic fluid

Reference Interval See table.

Subject	Age	MVA μmol/day mean ±SD (range)
Male	21-58	2.32 ±0.82 (0.70-4.76)
Female	19-58	1.85 ±0.47 (0.73-3.19)
Male (before pravastatin)	25-50	2.32 ±0.65 (1.50-3.58)
Male (after prevastatin)	25-50	1.47 ±0.49 (1.03-2.38)

Adapted from Hiramatus M, Hayashi A, Hidaka H, et al, "Enzyme Immunoassay of Urinary Mevalonic Acid and Its Clinical Application," *Clin Chem*, 1998, 44(10):2152-7.

Use Urinary MVA is an indicator of *in vivo* cholesterol biosynthesis. Such tests may be useful to monitor LDL cholesterol-lowering drugs.

Limitations The presence of time-related variation in cholesterol synthesis and simultaneous increase in plasma and urinary MVA (with peak at midnight) has been noted with earlier methods, but was not observed with enzyme immunoassay.

Methodology Enzyme immunoassay (with intra-assay coefficient of variation, 3.4% and interassay valuation (VC) of 6.9%); radioimmunoenzyme assay; liquid partition chromatography with subsequent gas chromatography and gas chromatography-mass spectrometry of the trimethylsilyl and methylester trimethylsilyl ether derivatives

Additional Information Mevalonic aciduria is a rare autosomal recessive disorder. The gene is present on chromosome 12. Prenatal diagnosis is supported by the finding of increased MVA in amniotic fluid and confirmed by the assay of cultured amniocytes.

References

Hiramatsu M, Hayashi A, Hidaka H, et al, "Enzyme Immunoassay of Urinary Mevalonic Acid and its Clinical Application," *Clin Chem*, 1998, 44(10):2152-7.

Naoumova RP, Marais AD, Mountney J, et al, "Plasma Mevalonic Acid, an Index of Cholesterol Synthesis In Vivo, and Responsiveness to HMG-CoA Reductase Inhibitors in Familial Hypercholesterolaemia," *Atherosclerosis*, 1996, 119(2):203-13.

Pappu AS, Illingworth DR, and Bacon S, "Reduction in Plasma Low-Density Lipoprotein Cholesterol and Urinary Mevalonic Acid by Lovastatin in Patients With Heterozygous Familial Hypercholesterolemia," *Metabolism*, 1989, 38(6):542-9.

Sweetman L and Williams JC, "Branched Chain Organic Acidurias," *The Metabolic and Molecular Bases of Inherited Disease*, 7th ed, Volume 1, Chapter 40, Scriver CR, Beaudet AL, Sly WS, et al, eds, New York, NY: McGraw-Hill Inc, 1995, 1387-8, 1402-5.

♦ **Mexate®** *see* Methotrexate, Serum or Plasma *on page 905*

Mexiletine, Serum

Related Information

Lidocaine, Serum or Plasma *on page 850*

Synonyms Mexitil®

Abstract Mexiletine is a class 1B antiarrhythmic agent used to treat ventricular arrhythmia. It is structurally similar to lidocaine, but is orally active. It has also been used in treatment-resistant bipolar disorder and as an analgesic used in diabetic neuropathy.

Specimen Serum
(Continued)

Mexiletine, Serum *(Continued)*

Container Red top tube

Sampling Time Draw 2-4 hours after last dose for peak level. Draw immediately prior to next dose for trough levels.

Reference Interval Therapeutic: 0.75-2.0 µg/mL (SI: 4-11 µmol/L)

Possible Panic Range >2.0 µg/mL (SI: >9 µmol/L). Levels >20 µg/mL (SI: >90 µmol/L) are associated with seizures.

Use Therapeutic monitoring and toxicity assessment

Methodology Fluorometry, high performance liquid chromatography (HPLC), gas chromatography (GC)

Additional Information
- Half-life: 7-15 hours
- Volume of distribution: 5-7 L/kg
- Protein binding: 60% to 70%

Half-life is urine pH dependent. Acidic urine accelerates elimination. Hepatic impairment prolongs the elimination half-life of mexiletine.

Mexiletine has no active metabolites. Toxic effects include dizziness, vomiting, confusion, tremor, bradycardia, and hypotension. At very high concentrations, it causes seizures.

References

Jarvis B and Coukell AJ, "Mexiletine. A Review of Its Therapeutic Use in Painful Diabetic Neuropathy," *Drugs*, 1998, 56(4):691-707.

Labbe L and Turgeon J, "Clinical Pharmacokinetics of Mexiletine," *Clin Pharmacokinet*, 1999, 37(5):361-84.

Manolis AS, Deering TF, Cameron J, et al, "Mexiletine: Pharmacology and Therapeutic Use," *Clin Cardiol*, 1990, 13(5):349-59.

♦ **Mexitil®** *see* Mexiletine, Serum *on page 911*

♦ **Mg, Serum** *see* Magnesium, Serum *on page 885*

♦ **Mg, Urine** *see* Magnesium, Urine *on page 886*

MHA-TP

Related Information

Darkfield Examination, Syphilis *on page 501*

FTA-ABS, Serum *on page 618*

RPR *on page 1174*

VDRL, Serum or Cerebrospinal Fluid *on page 1303*

Test Includes Detection of serologic response specific for *Treponema pallidum*

Abstract In the diagnosis and management of patients with syphilis, the MHA-TP is used as a treponemal-specific (confirmatory) test for syphilis. The FTA-ABS and the MHA-TP are the standard treponemal tests. A reactive treponemal test in a subject also reactive in a nontreponemal test is highly specific.

Patient Preparation Patient should be fasting if possible.

Specimen Serum

Container Red top tube

Storage Instructions Remove serum from clot and freeze to ship.

Causes for Rejection Plasma collected

Reference Interval Nonreactive

Critical Values Reactive results in a pregnant patient

Use Confirmatory serologic test for syphilis

Limitations The test cannot be used to test CSF specimens. The test has decreased sensitivity in early (primary) stages of syphilis, but in the secondary and latent stages of syphilis, the sensitivity is equal to that of the FTA-ABS tests. The MHA-TP assay gives fewer false-positive results than the other treponemal test methods. False positives may occur in patients with systemic lupus, infectious mononucleosis, drug addiction, collagen disease, or lepromatous leprosy. Potential causes of false-positive serologic tests for syphilis are tabulated in VDRL, Serum or Cerebrospinal Fluid *on page 1303*.

Methodology Hemagglutination - sensitized sheep cells are coated with *T. pallidum* antigen

Additional Information This is a *Treponema*-specific test and should not be used as a screening test except in late stages of syphilis. It is as sensitive and specific as FTA-ABS in all stages of syphilis except primary, in which it is less sensitive. It provides fewer false positives than the FTA-ABS. It is positive with treponemal infections other than syphilis (bejel, pinta, yaws). In general, treponemal-specific serologic assays will not determine if a patient has been adequately treated for syphilis.

References

Augenbraun MH, Rolfs R, Johnson R, et al, "Treponemal Specific Tests for the Serodiagnosis of Syphilis. Syphilis and HIV Study Group," *Sex Transm Dis*, 1998, 25(10):549-52.

Singh AE and Romanowski B, "Syphilis: Review and Emphasis on Clinical, Epidemiologic, and Some Biologic Features," *Clin Microbiol Rev*, 1999, 12(2):187-209.

Young H, "Syphilis Serology," *Dermatol Clinic*, 1998, 16(4):691-8.

Internet Web Sites

www.amm.co.uk/pubs/fa_leptospir.htm

www.cdc.gov/ncidod/dbmd/diseaseinfo/leptospirosis_g.htm

♦ **MHC** *see* Tissue Typing *on page 1259*

♦ **MIB-1** *see* Immunoperoxidase Procedures *on page 780*

♦ **MIC** *see* Antimicrobial Susceptibility Testing, Aerobic and Facultatively Anaerobic Bacteria *on page 178*

♦ **MIC** *see* Antimicrobial Susceptibility Testing, Unusual Isolates/Fastidious Organisms *on page 182*

♦ **Michael's Solution** *see* Kidney Biopsy *on page 818*

Microalbuminuria

Related Information

Glucose, Fasting, Plasma *on page 643*

Glucose, Postglucose Load, Plasma *on page 647*

Glycated Hemoglobin (Hemoglobin A$_{1c}$), Blood *on page 655*

Insulin, Serum *on page 803*

Protein Electrophoresis, Urine *on page 1107*

Protein, Quantitative, Urine *on page 1108*

Protein, Semiquantitative, Urine *on page 1113*

Urinalysis *on page 1289*

Urine Collection, 24-Hour *on page 1295*

Synonyms Oligoalbuminuria

Applies to Albumin:Creatinine Ratio; Albumin Excretion

Test Includes Creatinine clearance may be advised from the same urine collection.

Abstract "Microalbumin" is not the name of a substance, but refers to the urinary excretion of a small, but diagnostically and prognostically important, quantity of protein. Specifically, the term, "microalbuminuria", refers to the urinary excretion of 30-300 mg/24 hours (or 20-200 µg/minute). This concentration is too low to be detected accurately by dipstick testing which has a threshold of ~150-300 mg/L. It is very important for patients with diabetes mellitus to undergo microalbuminuria testing, because a positive result indicates early, but potentially reversible, diabetic nephropathy.

Specimen A 24-hour urine specimen is optimal. A timed overnight 10-hour collection, or spot AM urine after initial voiding may be used, but they are much less desirable. Albumin:creatinine ratio in a random specimen is best done on a first-morning voiding.

Storage Instructions Refrigeration and freezing are usually acceptable; review instructions from the laboratory which will perform the test.

Turnaround Time Often sent to a reference laboratory

Reference Interval ≤30 mg/24 hours (SI: ≤0.03 g/day). For spot AM samples <0.03 mg albumin/mg creatinine.

Use "Microalbuminuria" is defined as albuminuria of 20-200 µg/minute, or 30-300 mg/24 hours. Higher levels than these are diagnostic of diabetes nephropathy. **Albumin:creatinine ratio** >30 mg/g predicts an overnight excretion rate >30 µg/minute.

Microalbuminuria predicts the development of proteinuria, diabetic nephropathy, serious microvascular disease, and early mortality in type I and/or II diabetes. Time between appearance of microalbuminuria and full-blown proteinuria is 1-5 years. This testing is useful in patients with early diabetes mellitus, when it is possible to avoid or delay the onset of diabetic renal disease. Intensive treatment with tight control of blood glucose has delayed onset and reduced progression of microalbuminuria and albuminuria in (Continued)

Microalbuminuria *(Continued)*

subjects with IDDM. The risk of microalbuminuria increases when hemoglobin A_{1c} exceeds of 8.1%.

Microalbumin testing is also used in selected patients with hypertension, preeclampsia, and systemic lupus erythematosus.

Limitations The relationship between microalbuminuria and glycemia may not be linear. The Clinitek® Microalbumin System reacts with Tamm-Horsfall mucoprotein. Diurnal variation in urine albumin excretion exists; it is 30% to 50% less at night.

Increased excretion may be secondary to exercise, pregnancy, febrile/inflammatory disorders, urinary tract infection, bleeding in the urinary tract, or benign postural proteinuria.

Methodology Radial immunodiffusion (RID), enzyme-linked immunosorbent assay (ELISA), immunoturbidimetric methods, radioimmunoassay (RIA)

Additional Information In insulin-dependent diabetes mellitus, detectable diabetic nephropathy begins with urine albumin ~40-300 mg/24 hours.

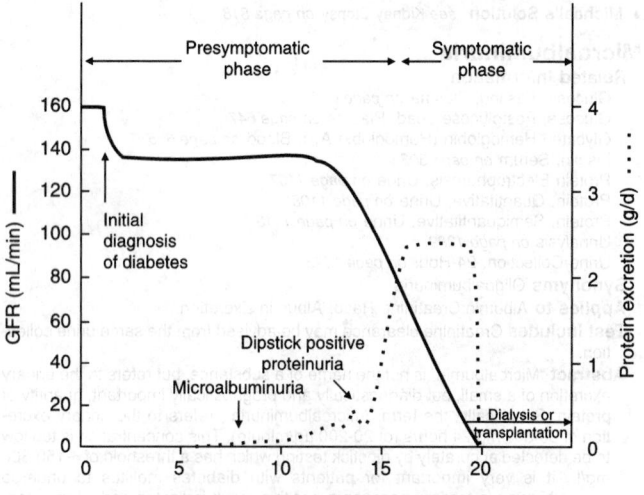

Adapted from Hostetter TH, "Diabetes and the Kidney," *Cecil Textbok of Medicine*, 21st ed, Chapter 110, Goldman L and Bennett JC, eds, Philadelphia, PA: WB Saunders Co, 2000, 610-3.

See Glycated Hemoglobin (Hemoglobin A_{1c}), Blood *on page 655.*

References

Ellis D, Coonrod BA, Dorman JS, et al, "Choice of Urine Sample Predictive of Microalbuminuria in Patients With Insulin-Dependent Diabetes Mellitus," *Am J Kidney Dis*, 1989, 13(4):321-8.

Emancipator K, "Laboratory Diagnosis and Monitoring of Diabetes Mellitus," *Am J Clin Pathol*, 1999, 112(5):665-74.

Hostetter TH, "Diabetes and the Kidney," *Cecil Textbook of Medicine*, 21st ed, Chapter 110, Goldman L and Bennett JC, eds, Philadelphia, PA: WB Saunders Co, 2000, 610-3.

Mathiesen ER, Ronn B, Storm B, et al, "The Natural Course of Microalbuminuria in Insulin-Dependent Diabetes: A 10-Year Prospective Study," *Diabet Med*, 1995, 12(6):482-7.

Mogensen CE, "Microalbuminuria Predicts Clinical Proteinuria and Early Mortality in Maturity-Onset Diabetes," *N Engl J Med*, 1984, 310(6):356-60.

♦ **Microbilirubin** *see* Bilirubin, Neonatal, Serum *on page 263*

♦ **Microdeletion Study** *see* Chromosome Analysis, High-Resolution *on page 409*

Microfilariae, Peripheral Blood Preparation

Related Information

Bacteremia Detection, Buffy Coat Micromethod *on page 227*
Filariasis Diagnostic Procedures *on page 587*
Peripheral Blood: Red Blood Cell Morphology *on page 1016*

Synonyms Filariasis, Peripheral Blood Preparation; Trypanosomiasis, Peripheral Blood Preparation

Applies to Filarial Infestation; Hemoflagellates

Test Includes Examination of both thick and thin smears, wet preparation

Specimen Fresh blood from fingerstick

Container Slides

Collection Recommended procedure is for specimen to be obtained when patient spikes a fever. Optimal yield results from examination of a daytime specimen (ie, noon), and a night time specimen (ie, midnight). Timing of sampling relates to geographic place of exposure.

Causes for Rejection Specimen clotted

Special Instructions If patient has traveled to an endemic area, the date of travel, the area, and the suspect parasite should be specified.

Reference Interval No parasites identified

Use Diagnose trypanosomiasis or microfilariasis; work up of elephantiasis, parasitic infestation of blood

Limitations One negative result does not rule out the possibility of parasitic infestation. Since some species of blood parasites can be found during the day and others are nocturnal, both day and night specimens enhance identification. Most filariae generate microfilariae which can be found in peripheral blood, but *Onchocerca volvulus* and *Dipetalonema streptocerca* give rise to microfilariae which do not circulate.

Methodology Fresh wet blood film, with a coverslip, in which motile microfilariae cause agitation of adjacent red cells. Stained films are used as well. Buffy coat methods may be applicable. In some cases, diagnosis may be made by aspiration or brush cytology.

Additional Information Anemia, thrombocytopenia and disseminated intravascular coagulation may characterize the clinical course after some forms of therapy. Biopsy of skin and subcutaneous mass is used in diagnosis of *D. streptocerca* and *O. volvulus*. Differential diagnosis of species of circulating microfilariae requires distinction between the presence or absence of a sheath, the pattern of nuclei in the tail and sometimes the history of geographic exposure and time of sampling. The QBC tube (quantitative buffy coat, utilizing tubes precoated with acridine orange) may provide rapid detection of microfilarial organisms but has low sensitivity.

References

Dacie JV and Lewis SM, "Blood Parasites in Preparation and Staining Methods for Blood and Bone-Marrow Films," *Practical Haematology*, 8th ed, Chapter 6, New York, NY: Churchill Livingstone, 1995, 93-6.

Freedman DO and Berry RS, "Rapid Diagnosis of Bancroftian Filariasis by Acridine Orange Staining of Centrifuged Parasites," *Am J Trop Med Hyg*, 1992, 47(6):787-93.

Maddocks S and O'Brien R, Images in Clinical Medicine, "African Trypanosomiasis in Australia," *N Engl J Med*, 2000, 342(17):1254.

♦ *Micropolyspora faeni* **Precipitating Antibodies** *see* Hypersensitivity Pneumonitis Serology *on page 761*

♦ **Microsomal Antibody, Thyroid** *see* Thyroperoxidase Autoantibody *on page 1253*

Microsporidia Diagnostic Procedures

Related Information

Bacterial Culture, Stool *on page 243*
Cryptosporidium Direct Staining Procedures *on page 484*
Electron Microscopy *on page 533*
Ova and Parasites, Direct Exam *on page 985*

Test Includes Examination of stool, body fluids, or biopsy for Microsporidia using special stains, immunofluorescence, or electron microscopy

Abstract Microsporidiosis refers to diseases produced by Microsporidia, a group of primitive, obligate intracellular protozoan parasites belonging to the phylum Microspora. Five different genera (*Enterocytozoon*, *Encephalitozoon*, *Nosema*, *Septata*, *Pleistophora*) have been implicated in human infections. They are (Continued)

Microsporidia Diagnostic Procedures *(Continued)*

best known as causes of diarrhea in patients with AIDS and also cause acute bilateral keratoconjunctivitis, sinonasal disease, bronchiolitis, pneumonia, infection of biliary and pancreatic ducts, acalculus cholecystitis, and other disease states.

Specimen Feces, urine, sputum, corneal or conjunctival scrapings, biopsy tissue, cerebrospinal fluid

Collection Biopsies should be fixed in formalin as soon as possible.

Special Instructions If microsporidiosis is suspected, consult with the laboratory. These organisms may not be detected with usual ova and parasite stains.

Use Aid in the diagnosis Microsporidia-associated diseases in immunocompromised patients

Limitations Detection of Microsporidia is entirely dependent on the adequacy of the specimen, staining and preparation of the specimen, and experience of the microscopist. They are small organisms and easily missed in biopsies and cytology specimens, as well as in stool specimens by light microscopy. Conventional H&E and Papanicolaou stains do not lend themselves to recognition of these organisms.

Methodology Organisms can be detected in stained cytologic smears, tissue biopsies, and stool.

There are several good stains for Microsporidia. Microsporidia in paraffin-embedded tissues are seen with a tissue Gram stain (Brown and Hopps stain, Brown and Brenn stain, Steiner stains, chromotrope methods). Microsporidia in plastic-embedded tissues stain well with toluidine blue and with methylene blue-azure II-basic fuchsin.

Microsporidia in cytologic centrifugation, smears and scraping preparations usually stain well with Gram stain for specimens with little or no bacterial contamination. In these preparations, most Microsporidia and bacteria are dark purple; some Microsporidia are gram-negative or gram-variable. Weber's modified trichrome (chromotrope-based) stain works well with specimens with bacterial contamination; Microsporidia are magenta-pink and the background (including bacteria) is blue-green.

Some laboratories use Giemsa stain for stool smears and body fluids. Calcofluor white is useful for screening stools for Microsporidia spores. Gram stain is useful for demonstration of spores in sputum. Most stains cause Microsporidia to appear as extremely fat bacteria which have a uniform oval shape, do not show budding, contain polar densities, and have a central clear band or area. Immunofluorescence staining techniques appear to work well in detecting Microsporidia in clinical specimens.

Additional Information Microsporidiosis has been recognized as one of the major causes of AIDS enteropathy, especially in homosexual patients. The diarrhea in these patients is of gradual onset and can persist for months. The stools are usually watery without blood or mucus, and undigested food may be seen in feces. The diarrhea may lead to dehydration, hypokalemia, and hypomagnesemia.

References

Chioralia G, Trammer T, Kampen H, et al, "Relevant Criteria for Detecting Microsporidia in Stool Specimens," *J Clin Microbiol*, 1998, 36(8):2279-83.

Conteas CN, Sowerby T, Berlin GW, et al, "Fluorescence Techniques for Diagnosing Intestinal Microsporidiosis in Stool, Enteric Fluid, and Biopsy Specimens From Acquired Immunodeficiency Syndrome Patients With Chronic Diarrhea," *Arch Pathol Lab Med*, 1996, 120(9):847-53.

Croppo GP, Visvesvara GS, Leitch GJ, et al, "Identification of the Microsporidian *Encephalitozoon hellem* Using Immunoglobulin G Monoclonal Antibodies," *Arch Pathol Lab Med*, 1998, 122(2):182-6.

Goodgame R, Stager C, Marcantel B, et al, "Intensity of Infection in AIDS-Related Intestinal Microsporidiosis," *J Infect Dis*, 1999, 180(3):929-32.

Lamps LW, Bronner MP, Vnencak-Jones CL, et al, "Optimal Screening and Diagnosis of Microsporidia in Tissue Sections: A Comparison of Polarization, Special Stains, and Molecular Techniques," *Am J Clin Pathol*, 1998, 109(4):404-10.

Moura H, Schwartz DA, Bornay-Llinares F, et al, "A New and Improved Quick-Hot Gram-Chromotrope Technique That Differentially Stains Microsporidian Spores in Clinical Samples, Including Paraffin-Embedded Tissue Sections," *Arch Pathol Lab Med*, 1997, 121(8):888-93.

Internet Web Sites

www.cdc.gov/ncidod/dpd/parasites/microsporidia/default.htm

- **Midnight to Morning Urinary Cortisol Increment** *see* Urinary Cortisol/Creatinine Increment *on page 1292*

- **Midol®** *see* Ibuprofen, Serum *on page 764*

- **Millipedes** *see* Arthropod Identification *on page 213*

- **Miltown®** *see* Meprobamate, Serum *on page 896*

- **Minimal Inhibitory Concentration** *see* Antimicrobial Susceptibility Testing, Aerobic and Facultatively Anaerobic Bacteria *on page 178*

- **Minimal Inhibitory Concentration** *see* Antimicrobial Susceptibility Testing, Unusual Isolates/Fastidious Organisms *on page 182*

- **Mist** *see* Phencyclidine, Qualitative, Urine *on page 1019*

- **Mite Identification** *see* Arthropod Identification *on page 213*

- **Mitochondrial Antibody** *see* Antimitochondrial Antibody *on page 183*

- **Mitochondrial Disorders** *see* Inherited Diseases of Metabolism and Cell Structure *on page 792*

- **Mitran®** *see* Chlordiazepoxide, Serum *on page 391*

Mixed Lymphocyte Culture

Related Information

Flow Cytometry, Overview *on page 592*
HLA Typing, Single Human Leukocyte Antigen *on page 739*
Lymphocyte Transformation Test *on page 882*
Tissue Typing *on page 1259*

Applies to HLA Class II Antigens; HLA-DP; HLA-DQ; HLA-DR

Test Includes Blood lymphocytes (and some monocytes - dendritic cells) from potential donors and the recipient are cultured together and tested for reactivity against each other.

Abstract Mixed lymphocyte culture (MLC) is used primarily to predict histocompatibility in the transplantation setting. Results of the MLC reaction are determined by the D locus of the human leukocyte antigen (HLA) system. If cells share common D loci, they do not activate one another. If the D loci are different, lymphocyte cell activation occurs.

Specimen Leukocytes

Container Green top (heparin) tubes, usually 20-30 mL of **sterile** heparinized blood

Collection Blood must be delivered to the laboratory immediately.

Storage Instructions Do not refrigerate, freeze, or expose to extreme heat or cold. Specimen should be processed within 24 hours of collection.

Turnaround Time The classical MLC reaction includes an incubation period of 4-6 days.

Special Instructions Blood specimen must be collected fresh on day of test. Check with the laboratory performing the assay for special instructions.

Reference Interval Response compared with that of simultaneously evaluated normal control; requires interpretation

Use Tissue matching for transplantation, largely bone marrow transplantation; evaluate cellular immunocompetence

Limitations Test will be negative if donor or responder cells have a severe cellular immunodeficiency. Test depends on viability of lymphocytes. With the advent and use of DNA-based HLA typing methods, MCL is used less for donor selection but can be used to follow the transplant recipients post-transplant donor antigen-specific immune status.

Methodology Lymphocytes from two individuals are typically cocultured to measure unidirectional lymphocyte blastogenesis (lymphocyte activation). Lymphocytes from the allograft donor are typically rendered nonresponsive by treating with mitomycin-C or radiation. Responder lymphocytes are derived from the prospective allograft recipient. Proliferating cells are pulsed with tritiated thymidine, incorporated thymidine is a function of responder (allograft recipient) lymphocyte blastogenesis. Multiparameter flow cytometry and other methods have been developed as variations of MLC. The procedures are less time consuming and avoid use of radioisotopes.

Additional Information MLC measures the ability of CD4 positive lymphocytes to recognize HLA class II antigens encoded by the HLA-DR, DQ, and DP loci. Incompatibility of these antigens are among the most potent of blastogenic stimuli. Unidirectional or bidirectional histocompatibility is reflected by little or
(Continued)

Mixed Lymphocyte Culture *(Continued)*

no blastogenesis. Prominent blastogenic responses predict tissue incompatibility and poor graft survival. Recent advances in molecular genetic typing has also shown high predictive value and assays for the identification of HLA proteins are being replaced by DNA-based typing methods.

The role of a number of cytokines, costimulatory and modulating molecules upon the MLC are under study.

References

James SP, "Measurement of Proliferative Responses of Cultured Lymphocytes," *Current Protocols in Immunology*, Volume 2, Section II, Coico R, series ed, John Wiley & Sons, Inc, 1994, National Institutes of Health (1991-1999) Supplement 11, 7.10.1-7.10.10.

Margulies DH, "The Major Histocompatibility Complex," *Fundamental Immunology*, 4th ed, Chapter 8, Paul WE, ed, Philadelphia, PA: Lippincott-Raven, 1999, 263-85.

Umemoto M, Azuma E, Hirayama M, et al, "Cytokine-Enhanced Mixed Lymphocyte Reaction (MLR) in Cord Blood," *Clin Exp Immunol*, 1998, 112(3):459-63.

Mixing Studies (Coagulation)

Related Information

Activated Partial Thromboplastin Time *on page 100*
Antiphospholipid Antibody (Lupus Anticoagulant and/or Anticardiolipin Antibody) *on page 193*
Coagulation Factor Assays *on page 418*
Factor Inhibitors *on page 566*
Heparin Neutralization *on page 697*
High-Molecular Weight Kininogen *on page 731*
Prekallikrein *on page 1085*
Prothrombin Time *on page 1116*
Thrombin Time *on page 1246*
Warfarin, Serum or Plasma *on page 1325*

Synonyms Circulating Anticoagulant Screen; Inhibitor Screen

Applies to Argatroban; Danaparoid; Factor Assays; Heparin; Hirudin

Abstract Mixing studies can be performed when the PT or PTT is prolonged, to determine if the etiology of the prolongation is a factor deficiency or an inhibitor.

Specimen Plasma

Container Blue top (sodium citrate) tubes

Collection Routine venipuncture. If multiple tests are being drawn, draw blue top tubes after any red top tubes but before any lavender top (EDTA), green top (heparin), or gray top (oxalate/fluoride) tubes. Immediately invert tube gently at least 4 times to mix. Tubes must be appropriately filled. Deliver tubes immediately to the laboratory.

Storage Instructions Separate plasma from cells as soon as possible. Plasma may be stored on ice for up to 4 hours; otherwise, store frozen.

Causes for Rejection Specimen received more than 4 hours after collection, tubes not filled, clotted specimen

Turnaround Time Several hours; longer if additional follow-up tests are indicated

Special Instructions Notify the laboratory if patient is on heparin (including low-molecular-weight heparin), hirudin, danaparoid, or argatroban anticoagulation, any of which can prolong PTT and/or PT.

Reference Interval There are three types of results in the PTT mixing study:

1. If the PTT of the mixture is normal, and remains normal after prolonged (2-hour) incubation, the results indicate the presence of factor deficiency(ies). The PTT is normal in such mixtures because the normal plasma supplies the factor that is deficient in the patient plasma. There may be one or more deficient factors. Assays for factors VIII, IX, XI, and XII should then be performed to identify the specific factor deficiency(ies). If the PT is also prolonged, common pathway factor assays can also be considered.

2. If the PTT of the mixture remains prolonged, the results suggest the presence of an inhibitor, most commonly, a lupus anticoagulant. Therefore, lupus anticoagulant assays should then be performed. Heparin, hirudin, argatroban, or high-dose danaparoid, if present, will also show this type of result in a mixing study. Specific factor inhibitors against a particular coagulation factor (eg, factor IX, XI, or XII), are very rare possibilities.

3. If the PTT of the mixture is initially normal (or significantly shorter than the patient plasma's PTT) but becomes prolonged after a 1- or 2-hour incubation, the results are characteristic of a factor VIII inhibitor (factor VIII inhibitors show an inhibitory effect only after prolonged incubation). A factor VIII assay should then be performed and, if decreased (usually to <10%), a factor VIII inhibitor assay (Bethesda assay) should be performed.

When the PT is prolonged and the PTT is normal, a PT mixing study may also be useful in determining if the etiology is a factor deficiency or a factor inhibitor, similar to that described for the PTT. However, factor inhibitors that affect only the PT and not the PTT are rare. The results of PT mixing studies in patients on warfarin are consistent with factor deficiencies, because warfarin acts as an anticoagulant by decreasing the activity of factors II, VII, IX, and X.

Methodology When the PTT is prolonged, the laboratory should first determine if the prolongation is due to heparin by treating the specimen to remove heparin (see Heparin Neutralization *on page 697*). Alternatively, some laboratories perform the thrombin time, which is prolonged when even a small amount of heparin is in the sample. If a prolonged PTT is not due to heparin, patient plasma is then mixed with an equal volume of normal plasma, and the PTT is repeated. The resulting PTT of this mixture indicates whether the prolongation is due to a factor deficiency or an inhibitor. Inhibitors are substances that inhibit clotting reactions. They are usually antibodies (eg, lupus anticoagulants or specific factor inhibitors) or anticoagulants such as heparin, hirudin, or argatroban. Based on the mixing study results, factor assays, lupus anticoagulant tests, or tests for factor inhibitors may be indicated. PT mixing studies can be similarly performed to evaluate PT prolongations.

References

Van Cott EM and Laposata M, "Coagulation, Fibrinolysis and Hypercoagulation," *Clinical Diagnosis and Management by Laboratory Methods*, 20th ed, Henry JB, ed, New York, NY: WB Saunders Co, 2001, 644-6.

♦ **MMA** *see* Arsenic, Urine *on page 211*
♦ **MMF** *see* Mycophenolic Acid, Serum *on page 936*
♦ **Mn, Serum** *see* Manganese, Serum or Blood *on page 890*
♦ **Mn, Urine** *see* Manganese, Urine *on page 892*
♦ **Mo, Blood** *see* Molybdenum, Blood *on page 919*
♦ **Mode** *see page 11*
♦ **Modified Acid-Fast Stain** *see* Acid-Fast Stain, Routine or Modified *on page 95*
♦ **Modified Bethesda Assay** *see* Factor Inhibitors *on page 566*
♦ **Moditen®** *see* Fluphenazine, Serum *on page 604*
♦ **Mole®** *see* Caffeine, Serum *on page 326*
♦ **Molecular Cytogenetics** *see* Fluorescence *in situ* Hybridization *on page 602*
♦ **Molecular Genetic HLA Typing** *see* Tissue Typing *on page 1259*
♦ **Molipaxin®** *see* Trazodone, Serum or Plasma *on page 1272*

Molybdenum, Blood

Related Information

Uric Acid, Serum *on page 1286*

Synonyms Mo, Blood

Applies to Hypoxanthine, Urine; Methionine, Serum; Molybdopterin; S-Sulfocysteine, Urine; Sulfite, Test; Xanthine, Urine

Abstract Molybdenum (Mo) is vital to human health through its inclusion in at least three human enzymes: xanthine oxidase, aldehyde oxidase, and sulfite oxidase. Although relatively nontoxic, heavy exposures have been associated with increased serum uric acid.

Specimen Whole blood, serum, plasma

Container Trace metal collection tube. If other samples are to be collected at the same blood draw, draw the trace metal tube first so as not to contaminate the needle by puncture through ordinary rubber stoppers. Use powder-free gloves. See the Trace Metals Introduction *on page 77*.

Reference Interval Whole blood Mo: <60 ng/mL (SI: 625 nmol/L); lower limit not established. Blood level parallels Mo intake. Apparently normal individuals vary over a 100-fold from 0.5-60 ng/mL (SI: 5-625 nmol/L), depending on Mo intake. 170 ng/mL (SI: 1771 nmol/L) appears to border on the toxic level based (Continued)

Molybdenum, Blood (Continued)

on the report below. Seventy-five percent of people in the United States have levels ≤5 ng/mL (SI: 52 nmol/L), but some geographical areas show 70% of the population >5 ng/mL. In children 2-12 years of age, the plasma Mo concentrations have been reported to vary from <1 to 3 ng/mL (SI: 10-31 nmol/L) with plasma concentration of 1.75 ng/mL ±0.8 ng/mL (mean ±SD) (SI: 18 nmol/L ±8.3 nmol/L). There are no age nor sex differences.

Limitations Data are new and sketchy for this trace metal, and clinical syndromes poorly defined. Serum or plasma Mo norms are still being developed. Levels in apparently healthy people vary enormously based on intake, and blood levels are only significant when extremely high or extremely low levels (yet undefined) are encountered.

Methodology Neutron activation; graphite furnace atomic absorption (AA) spectrophotometry after extraction into 8-hydroxyquinoline

Additional Information Molybdenum interferes with copper metabolism, especially in the presence of dietary sulfides by the formation of insoluble copper thiomolybdenates in the gut lumen; absorbed thiomolybdenates may also interfere with copper metabolism.

Molybdenum cofactor deficiency: This is a recessively inherited error of metabolism involving failure to synthesize molybdopterin. It is diagnosed by noting combined xanthine oxidase deficiency (low serum uric acid <1 mg/dL, increased urine hypoxanthine and xanthine) and sulfite oxidase deficiency (marked by increased urinary sulfite, decreased to absent urine inorganic sulfate, increased urinary S-sulfocysteine). These patients have severe neurologic abnormalities from infancy on the basis of the sulfite oxidase deficiency including seizures, anterior lens dislocations, opisthotonos, decreased brain weight, decreased brain myelin, and usually death prior to 1 year of age. Molybdenum is virtually absent from the liver of such patients, suggesting that molybdopterin is an important storage form of Mo in soft tissues. Molybdenum may not be retained at all in soft tissue without Mo cofactor. Serum Mo in this disease is reported to be normal. Screening test to detect this among neonates with intractable seizures is the urine sulfite test.

Molybdenum toxicity: This has only rarely been reported. Two situations are known to expose human populations chronically to excess Mo: industrial exposure and through the food chain in areas of the world with high local soil Mo. Serum Mo concentrations are increased in patients with chronic renal failure. Dialysis reduces serum Mo concentrations but serum Mo still remains high when compared to patients with good renal function. High serum Mo may contribute to dialysis related arthritis. Serum Mo concentrations are elevated in liver disease including but not limited to chronic active hepatitis, cirrhosis, alcoholic liver disease, liver metastases, and gallstones.

Recent data shows that Mo in different premature infant formulas vary over 20 times. It is not clear if some infants on these formulas are not getting enough Mo or if others are being overexposed.

References

Barceloux DG, "Molybdenum," *J Toxicol Clin Toxicol*, 1999, 37(2):231-7.

Sardesai VM, "Molybdenum: An Essential Trace Element," *Nutr Clin Pract*, 1993, 8(6):277-81.

Sievers E, Oldigs HD, Dorner K, et al, "Molybdenum Balance Studies in Premature Male Infants," *Eur J Pediatr*, 2001, 160(2):109-13.

Vyskocil A and Viau C, "Assessment of Molybdenum Toxicity in Humans," *J Appl Toxicol*, 1999, 19(3):185-92.

♦ **Monomethylarsine** *see* Arsenic, Urine *on page 211*

♦ **Monospot™ Test** *see* Infectious Mononucleosis Screening Test *on page 785*

♦ **Monosticon® Dri-Dot® Test** *see* Infectious Mononucleosis Screening Test *on page 785*

♦ **Morphine** *see* Codeine, Urine *on page 429*

Morphine, Urine

Related Information

Codeine, Urine *on page 429*

Opiates, Qualitative, Urine *on page 974*

Synonyms Astramorph™ PF; Duramorph®; Epimorph Dolcontin®; Heroin Metabolite, Urine; MS Contin®; MSIR®; MST®; OMS®; Oramorph SR®; RMS®; Roxanol®; Roxanol SR™; Sevredol®; Statex®

Applies to 6-0-Acetyl Morphine; Codeine; Heroin; Naloxone; Opium

Test Includes Codeine, heroin metabolite (6-O-acetyl morphine), hydromorphone (Dilaudid®), oxycodone

Abstract This drug is widely used therapeutically as an analgesic for the treatment of moderate to severe pain. Morphine itself is not an extensively used drug of abuse but two derivatives, heroin and codeine, are. It is a metabolite of ethylmorphine, heroin, and codeine.

Storage Instructions Refrigerate sample

Special Instructions If forensic, use chain-of-custody protocol and form. See Chain-of-Custody Protocol *on page 381* and the Chain-of-Custody form in the Introduction *on page 63*.

Reference Interval Negative (less than cutoff)

Critical Values Substance Abuse and Mental Health Services Administration (SAMHSA) cutoff: screen (total opiates): 2000 ng/mL; confirmatory: morphine: 2000 ng/mL. For clinical purposes, many laboratories use a cutoff of 300 ng/mL.

Use Concentrations are measured to evaluate toxicity or detect drug of abuse. Heroin is metabolized to morphine, but morphine detection may only suggest heroin use. To **prove** heroin use, 6-O-acetyl morphine must be identified in the urine.

Methodology Screen: immunoassay; confirmation: gas chromatography/mass spectroscopy (GC/MS)

Additional Information

- Half-life: 2-4 hours (adults), 5-13 hours (neonates)
- Volume of distribution: 2-4 L/kg
- Protein binding: 30% to 40%

Adverse effects include CNS depression, nausea and vomiting, hypotension, bradycardia, histamine release, increased intracranial pressure, miosis, respiratory depression, antidiuretic hormone release, and physical and psychological dependence. Naloxone is a specific antidote. Morphine is generally detectable for 1-2 days after use.

Ingestion of poppy seeds (bagels, pastries including Danish) can cause positive opiate screens at ≥300 ng/mL. Certain poppy seeds can result in urine levels of morphine as high as 10,000 ng/mL. In an attempt to address this problem, SAMHSA has recently increased the cutoff of morphine and codeine to 2000 ng/mL.

References

Joranson DE, Ryan KM, Gilson AM, et al, "Trends in Medical Use and Abuse of Opioid Analgesics," *JAMA*, 2000, 283(13):1710-4.

Stout PR and Farrell LJ, "Opioids - Effects on Human Performance and Behavior," *Forensic Sci Rev*, 2002, 15:29-59.

Thevis M, Opfermann G, and Schanzer W, "Urinary Concentrations of Morphine and Codeine After Consumption of Poppy Seeds," *J Anal Toxicol*, 2003, 27(1):53-6.

♦ **Motrin®** *see* Ibuprofen, Serum *on page 764*

♦ **Moxadil®** *see* Amoxapine, Serum or Plasma *on page 153*

♦ **MPA** *see* Mycophenolic Acid, Serum *on page 936*

♦ **M Protein** *see* Immunofixation Electrophoresis, Serum or Urine *on page 768*

♦ **M-Proteins** *see* Protein Electrophoresis, Capillary Zone *on page 1103*

♦ **M-Proteins** *see* Protein Electrophoresis, Serum *on page 1104*

♦ **MPV** *see* Platelet Sizing *on page 1056*

- ♦ **MS Contin®** *see* Morphine, Urine *on page 921*
- ♦ **MSIR®** *see* Morphine, Urine *on page 921*
- ♦ **MST®** *see* Morphine, Urine *on page 921*
- ♦ **MTX** *see* Methotrexate, Serum or Plasma *on page 905*
- ♦ *MUC1* *see* CA 15-3, Serum *on page 320*
- ♦ *MUC1* Gene *see* CA 27.29, Serum *on page 322*

Mucin Clot Test

Related Information

Synovial Fluid Analysis *on page 1229*

Synonyms Synovial Fluid Viscosity

Abstract Qualitative test for the nature of hyaluronic acid in synovial fluid. This test lacks specificity, but bears correlation with the presence of inflammation. It is considered by some to be only of historical interest, but is still in use as a simple, rapid, and practical measure of inflammatory change in joint fluid. The test estimates the degree of polymerization of hyaluronate such that clots can be categorized as to their quality (eg, good, fair, or poor).

Specimen Synovial fluid

Container Sterile tube or lavender top (EDTA) tube; avoid oxalate anticoagulants.

Storage Instructions Refrigerate if there is delay in analysis.

Reference Interval Mucin clot - positive (firm clot)

Use An adjunct in differential diagnosis of joint disease

Limitations The mucin clot test lacks specificity (ie, it is not indicative of a single entity). It has low specificity (49%) and positive predictive value (52%).

Methodology Evaluate clot formed on reaction of synovial fluid with acetic acid. Equal amounts of fluid and 5% acetic acid are mixed on a glass slide. Quality of the clot may be graded as "good", "fair", or "poor". This test reflects the physical chemical status of hyaluronic acid in a qualitative manner. Inflammation degrades the quality of the mucin clot.

Additional Information In osteoarthritis the clot is firm and the surrounding fluid remains clear even after agitation. In rheumatoid arthritis the clot is friable and the surrounding fluid is turbid. In acute rheumatic fever the clot is firm, and in lupus erythematosus the clot is firm (ie, normal).

References

McCarty DJ, "Synovial Fluid," *Arthritis and Allied Conditions: A Textbook of Rheumatology,* Koopman WJ, ed, 13th ed, Chapter 4, Philadelphia, PA: Lea & Febiger, 1997, 84-5.

Smith GP and Kjeldsberg CR, "Cerebrospinal, Synovial, and Serous Body Fluids," *Clinical Diagnosis and Management by Laboratory Methods,* 20th ed, Part 3, Chapter 19, Henry JB, ed, Philadelphia, PA: WB Saunders Co, 2001, 414.

- ♦ **Mucolipidoses** *see* Inherited Diseases of Metabolism and Cell Structure *on page 792*

Mucopolysaccharides, Urine

Related Information

Amniotic Fluid, Chromosome and Genetic Abnormality Analysis *on page 152*

Chorionic Villus Sampling, Chromosome and Genetic Abnormality Analysis *on page 400*

Inherited Diseases of Metabolism and Cell Structure *on page 792*

Muscle Biopsy *on page 927*

Peripheral Blood: Differential Leukocyte Count *on page 1010*

Urinalysis *on page 1289*

Urine Collection, 24-Hour *on page 1295*

Applies to Dermatan Sulfates; Glycosaminoglycans; Heparan; Hunter Syndrome; Hurler Syndrome; Keratan Sulfate; Lymphocyte, Vacuolated; Sulfatase

Abstract Increased excretion of urinary mucopolysaccharides (dermatan and/or heparan sulfates) occurs in lysosomal storage diseases. The mucopolysaccharides are currently referred to as glycosaminoglycans (GAGs). "Mucopolysaccharidosis" is still generally used in the medical literature. Inherited deficiency or absence of lysosomal enzymes results in a number of clinical pathologic entities, the mucopolysaccharidoses.

Specimen 24-hour urine collection is preferred.

Container Plastic urine container

Disorder	Clinical Feature	Enzyme Defect	Assay[1]
MPS I-H Hurler syndrome	Mental retardation, progressive, beginning at age 1; corneal opacities; coarse facies; stiff joints; dwarfing; organomegaly; death, usually by age 14	α-L-iduronidase	L, F, Ac, CV
MPS I-S Scheie syndrome	Mild form of I-H; stiff joints; corneal opacity; mild-to-absent mental retardation; aortic stenosis; survive to adult	α-L-iduronidase	L, F, Ac, CV
MPS I-H/S Hurler/Scheie syndrome	Intermediate phenotype between I-H and I-S; some are Hurler-Scheie double heterozygotes	α-L-iduronidase	L, F, Ac, CV
MPS II-XR Hunter syndrome	Similar to MPS I-H but with clear cornea; may be deafness; later onset and longer survival (to adulthood)	Iduronate sulfate sulfatase	S, F, Af, Ac, CV
MPS III-A MPS III-B	Most common of the MPS with behaviorial problems, progressive dementia, seizures, intrafamilial variability, hirsute with coarse hair; survival to 2nd or 3rd decade of life	Heparan N-sulfatase α-N-acetylglucosaminidase	L, F, Ac, CV
MPS III-C MPS III-D Sanfilippo A-D syndromes		α-glucosaminide-N-acetyltransferase N-acetylglucosamine-6-sulfate sulfatase	F, Ac
MPS IV-A Morquio A syndrome	Short trunk dwarfism; fine corneal opacities; short neck; odontoid anomalies; normal intellect	Galactosamine-6-sulfate sulfatase	L, F, Ac
MPS IV-B Morquio B syndrome	Mild form of IV-A	β-galactosidase	L, F, Ac, CV
MPS V	Formerly Scheie disease		
MPS VI Maroteaux-Lamy syndrome	Hurler phenotype with severe-to-mild dysostosis multiplex; gross corneal opacity; growth retardation; normal intellect	Arylsulfatase B	L, F, Ac
MPS VII Glucuronidase deficiency Sly syndrome	Highly variable with mild mental retardation; tendency to coarse facies; gingivitis; organomegaly; inclusions in granulocytes	β-glucuronidase	S, F, Ac

L = leukocytes, S = serum, F = cultured fibroblasts, Ac = cultured amniotic cells, Af = amniotic fluid, CV = chorionic villi.

[1]With the exception of MPS IV-B and MPS VII, the above mucopolysaccharidoses are characterized by the presence of excess urinary mucopolysaccharides, dermatan and/or heparan sulfates. Keratan sulfate is present in urine from cases of MPS IV-A. Quantitative enzyme assay is applicable to each of the above conditions.

Adapted from O'Brien J, "Lysosomal Storage Disease," *Tietz Textbook of Clinical Chemistry,* 3rd ed, Chapter 49, Burtis CA and Ashwood ER, eds, Philadelphia, PA: WB Saunders Co, 1999, 1777-80 and from Spranger J, "Mucopolysaccharidoses," *Emery and Rimoin's Principles and Practice of Medical Genetics,* 3rd ed, Chapter 96, Rimoin DL, Connor JM, and Pyeritz RE, eds, New York, NY: Churchill Livingstone, 1997, 2073.

(Continued)

Storage Instructions Freeze

Special Instructions Do not collect specimen while patient is receiving intravenous heparin.

Reference Interval
- Younger than 1 year: 20-40 mg/mmol creatinine
- 1-5 years: 10-15 mg/mmol creatinine
- 5 years and older: 3-8 mg/mmol creatinine

Consult laboratory performing the test.

Use Screen for the presence of mucopolysaccharidose

Limitations "Spot" tests may be associated with a 20% to 30% incidence of false-positive or false-negative results.

Methodology Spectrophotometry (quantitative) based on a colorimetric reaction with 1,9-dimethylmethylene blue; electrophoretic separation of glycosaminoglycans (GAG)

Additional Information If enzyme activity is missing in the stepwise degradation of a GAG, undegraded molecules gradually accumulate in the lysosome, the underlying basis for the multiple forms of mucopolysaccharidosis, MPS I-VII. See table on previous page.

These disorders, often clinically apparent, can be detected by the identification of specific GAG segments in the urine and verified by assay of the specific hydrolases in leukocytes or fibroblasts. To assist with the diagnosis, assays for defective enzyme activity can be applied to serum, cultured fibroblasts, amniotic fluid, cultured amniotic cells and/or chorionic villi. Inheritance is autosomal recessive with the exception of MPS II (Hunter) which is X-linked.

References

Kakkis ED, Muenzer J, Tiller GE, et al, "Enzyme-Replacement Therapy in Mucopolysaccharidosis I," *N Engl J Med*, 2001, 344(3):182-8.

O'Brien JF, "Lysosomal Storage Disease," *Tietz Textbook of Clinical Chemistry*, 3rd ed, Chapter 49, Burtis CA and Ashwood ER, eds, Philadelphia, PA: WB Saunders Co, 1999, 1776-9.

Spranger J, "Mucopolysaccharidoses," *Emery and Rimoin's Principles and Practice of Medical Genetics*, 3rd ed, Volume 11, Rimoin DL, Connor MJ, and Pyeritz RE, eds, New York, NY: Churchill Livingstone, 1997, 2071-9.

Ullrich K, "Screening for Lysosomal Disorders," *Eur J Pediatr*, 1994, 153(Suppl 1):S38-S43.

Whitfield PD, Sharp P, Meikle PJ, et al, "Characterization of Ganglioside Storage in Mucopolysaccharidosis IIIA by Tandem Mass Spectrometry," *J Inherit Metab Dis*, 2000, 23(Suppl 1):237.

♦ **Mucopolysaccharidoses** *see* Inherited Diseases of Metabolism and Cell Structure *on page 792*

♦ **Muir-Torre Syndrome** *see* Colon Cancer, Hereditary Nonpolyposis Type *on page 432*

♦ **Multimer Assay** *see* von Willebrand Factor *on page 1321*

Multiple Endocrine Neoplasia/Familial Medullary Thyroid Carcinoma

Related Information

Calcitonin, Serum or Plasma *on page 327*
Calcium, Serum *on page 329*
Catecholamines, Fractionation, Urine *on page 347*
Metanephrines, Urine or Plasma *on page 899*
Pancreatic Polypeptide, Human, Serum or Plasma *on page 998*
Parathyroid Hormone, Serum *on page 1001*
Polymerase Chain Reaction *on page 1069*

Synonyms Familial Medullary Thyroid Carcinoma/Multiple Endocrine Neoplasia; MEN 2A; MEN 2B; MEN2/FMTC; RET Gene Testing

Applies to Multiple Endocrine Neoplasia (MEN), Types 2A, 2B; RET Proto-oncogene

Test Includes This molecular test can detect point mutations in the RET gene.

Abstract Medullary thyroid carcinoma (MTC) is an entity distinct from papillary or follicular carcinoma of thyroid. It may be sporadic as well as familial. See Calcitonin, Serum or Plasma *on page 327*. MTC is the major cause of death in individuals with MEN 2A and with familial medullary thyroid carcinoma (FMTC).

Mutations in the RET proto-oncogene associated with FMTC and multiple endocrine neoplasia (type 2A and 2B) (MEN 2A and 2B) are detected by polymerase chain reaction (PCR) followed by restriction digestion or enzyme sequencing the DNA of the RET gene. Testing for RET mutations in all individuals with MTC is recommended and has become the clinically accepted approach to the diagnosis of MEN 2 and FMTC. RET analysis followed by linkage analysis in the absence of RET mutations identify subjects with MEN 2A and FMTC.

Specimen Whole blood

Container Blue top (sodium citrate) tube or yellow top (ACD) tube or EDTA tube. Avoid collection in heparin tube; heparin inhibits PCR.

Storage Instructions All specimens should be sent to the laboratory **immediately** after collection, preferably by overnight delivery. All specimens should be kept at room temperature or refrigerated, never frozen.

Use The laboratory generally provides an interpretive report based upon direct sequencing analysis. This test is indicated for families with a history of MEN 2A or MEN 2B or FMTC (autosomal dominant). Hyperparathyroidism also occurs in MEN type 2A (see Parathyroid Hormone, Serum *on page 1001*). Mutation analysis may be indicated in cases of apparently sporadic MTC and pheochromocytoma.

Limitations If a high-risk affected individual tests negative for RET mutations, an inherited form of cancer may still exist and appropriate genetic counseling should be provided. Mutation analysis is positive about 95% of the time in families with clinically apparent MEN 2A. This evaluation is costly.

Methodology DNA is isolated from peripheral blood leukocytes. Regions within the RET gene (exons 10, 11, 13, 14, and 16) are amplified by PCR and sequenced in entirety.

Additional Information Medullary thyroid carcinoma (MTC) is a malignancy of the calcitonin-secreting cells (C cells) of the thyroid and accounts for about 10% of thyroid cancer. One in 5000 individuals is affected with MTC each year; about 50% of patients have metastases at the time of diagnosis. The historical 10-year survival is about 50%. About 20% of MTC occurs as part of one of three familial syndromes: **multiple endocrine neoplasia type 2A (MEN 2A)** characterized by MTC, pheochromocytoma ($\sim$50%), and hyperparathyroidism ($\sim$10%); **multiple endocrine neoplasia type 2B (MEN 2B)** consisting of MTC, pheochromocytoma ($\sim$50%), ganglioneuromatosis, and marfanoid habitus; and **familial medullary thyroid carcinoma syndrome (FMTC)** characterized by MTC alone. All are autosomal dominantly inherited. They are caused by germline mutations in the RET proto-oncogene, which is located in the pericentromeric region of chromosome 10, band q11.2. Virtually everyone who inherits a mutation in RET will develop MTC. Prior to the identification of the RET gene in 1993, it was the standard of practice to perform annual biochemical screening on all individuals in definite or suspected FMTC, MEN 2A, and MEN 2B families. Because it is not possible to distinguish sporadic from familial tumors histopathologically and because the family history is often unreliable, biochemical screening has been performed on first-degree relatives of many individuals affected with apparently sporadic MTC. Such screening has been performed annually from 5 years of age to about 40 years to detect C-cell hyperplasia (the precursor of medullary thyroid carcinoma) as early as possible, followed by total thyroidectomy. Pentagastrin stimulation testing yields false positives, false negatives, and equivocal results requiring test repetition. See Calcitonin, Serum or Plasma *on page 327*. The majority of cases of MTC are **not familial**, and in those families in which it is inherited, 50% of first-degree relatives would be expected not to have inherited the gene. DNA testing for mutations in the RET gene in all individuals with MTC accurately diagnoses a heritable form of MTC in >90% of cases and identifies family members who have inherited the mutation. Family members who have not inherited the mutation require no further screening. It is recommended that those family members who do inherit the mutation undergo thyroidectomy. Additionally, with the knowledge that such individuals are at risk for other MEN 2-associated tumors, biochemical screening for premorbid detection of pheochromocytoma and hyperparathyroidism can be initiated.

The risk of C-cell disease, pheochromocytoma, and hyperparathyroidism was higher in subjects with codon 634 mutations than in patients with other mutations in a series of 348 patients and family members at risk. Codon 634 is often involved in MEN 2A.

(Continued)

Multiple Endocrine Neoplasia/Familial Medullary Thyroid Carcinoma *(Continued)*

References

Heshmati HM, Gharib H, Khosla S, et al, "Genetic Testing in Medullary Thyroid Carcinoma Syndromes: Mutation Types and Clinical Significance," *Mayo Clin Proc*, 1997, 72(5):430-6.

Noll WW, "Utility of RET Mutation Analysis in Multiple Endocrine Neoplasia Type 2," *Arch Pathol Lab Med*, 1999, 123(11):1047-9.

Pacak K, Linehan M, Eisenhofer G, et al, "Recent Advances in Genetics, Diagnosis, Localization, and Treatment of Pheochromocytoma," *Ann Intern Med*, 2001, 134(4):315-29.

Internet Web Sites

www.uwcm.ac.uk/uwcm/mg/search/120346.html

♦ **Multiple Endocrine Neoplasia (MEN) Type 1** *see* Parathyroid Hormone, Serum *on page 1001*

♦ **Multiple Endocrine Neoplasia (MEN), Types 2A, 2B** *see* Multiple Endocrine Neoplasia/Familial Medullary Thyroid Carcinoma *on page 924*

♦ **Multiple Marker Screening Test** *see* Estriol, Unconjugated, Pregnancy, Serum, Plasma, or Urine *on page 554*

♦ **Multiple Marker Screening Test** *see* Inhibin A, Serum *on page 799*

Mumps Culture and Serology

Related Information

Viral Culture *on page 1307*
Virus, Direct Detection by Fluorescent Antibody *on page 1311*

Test Includes Viral culture for mumps virus; detection of serologic response to mumps infection or vaccination

Abstract Mumps is caused by a paramyxovirus and man is the only known reservoir. Mumps is a self-limited illness characterized by parotitis, high fever, and fatigue. In the U.S. and many other developed countries, mumps has decreased dramatically due to the mumps vaccine. It is recommended that children 12-15 months of age receive the trivalent vaccine that contains live, attenuated measles, mumps, and rubella. A second dose of vaccine should be administered at 4-6 or 11-12 years of age.

Specimen Culture: Saliva, urine, CSF, viral swab of Stensen's duct; Serology: Serum

Container Culture: Sterile container for urine and CSF; tube with cold viral transport medium for swabs. Serology: Red top tube.

Sampling Time Culture: At or within 5 days of the onset of illness. Serology: Acute and convalescent sera drawn 10-14 days apart.

Collection It is desirable to collect culture specimens as early in the disease as possible. Collect saliva 9 days before onset to 8 days after onset. Swabs must immediately be placed into cold viral transport medium. CSF from patients with meningoencephalitis should be collected 6 days after onset. Virus is also excreted in urine for as long as 14 days after the onset of illness. All specimens must immediately be placed on ice and sent to the laboratory.

Storage Instructions Do **not** freeze at -20°C. Storage at -20°C rapidly inactivates the mumps virus. If inoculation is delayed by more than 48 hours, specimens should be frozen at -70°C.

Turnaround Time Culture: 3-14 days; negative results are reported at 14 days. Serology: Within a week to days.

Reference Interval Culture: No virus isolated. Serology: Positive IgG indicates immunity; a fourfold or greater increase in titer is indicative of recent mumps infection.

Use Support for the diagnosis of mumps virus infection; serology can document previous exposure to mumps virus or mumps vaccination

Limitations Negative viral culture does not rule out the involvement of mumps virus in disease process. Several test systems that detect IgG are not specific for mumps and may cross-react with other paramyxovirus antibodies.

Methodology Culture includes inoculation of cell cultures, incubation and observation of cytopathic effect (CPE), and identification by hemadsorption and fluorescent monoclonal antibodies. Serology may be performed by complement fixation (CF), enzyme-linked immunosorbent assay (ELISA), indirect fluorescent antibody (IFA), hemagglutination inhibition (HAI), hemolysis-in-gel, or virus neutralization.

Additional Information Although virus isolation is the most certain means for establishing the laboratory diagnosis, serologic methods are also useful and technically easier. Demonstration of IgM antibodies in acute serum is diagnostic of primary infection. In most instances, the diagnosis of mumps is made on the basis of exposure, history of immunization (or lack of immunization), and the presence of parotid swelling.

Complications of mumps include aseptic meningitis, encephalitis, orchitis, oophoritis, and pancreatitis. Serologic study may be undertaken to confirm a diagnosis in acute disease or to demonstrate established immunity. Since the introduction of the trivalent vaccine (MMR) in 1967, the number of mumps infections has decreased in the U.S. to only 308 in 1999. However there are still some populations that do not accept immunization. The number of vaccines and the complexity of vaccination schedules may make provision of vaccinations difficult. The immunization provider may change during the course of a vaccination series. In order to improve vaccination coverage, the CDC has recently published a report on vaccine-preventable diseases.

References

Centers for Disease Control and Prevention, "Combination Vaccines for Childhood Immunizations, Recommendations of the Advisory Committee on Immunization Practices (ACIP), the American Academy of Pediatrics (AAP), and the American Academy of Family Physicians (AAFP)," *MMWR Morb Mortal Wkly Rep*, 1999, 48:RR-5.

Centers for Disease Control and Prevention, "Vaccine-Preventable Diseases: Improving Vaccination Coverage in Children, Adolescents, and Adults," *MMWR Morb Mortal Wkly Rep*, 1999, 48:RR-8.

Narita M, Matsuzono Y, Takekoshi Y, et al, "Analysis of Mumps Vaccine Failure by Means of Avidity Testing for Mumps Virus-Specific Immunoglobulin G," *Clin Diagn Lab Immunol*, 1998, 5(6):799-803.

Pipkin PA, Afzal MA, Heath AB, et al, "Assay of Humoral Immunity to Mumps Virus," *J Virol Methods*, 1999, 79(2):219-25.

Internet Web Sites

www.astdhpphe.org/infect/mumps.html
www.cdc.gov/nip/publications/pink/mumps.pdf
www.thecommunityguide.org/

♦ **Muramidase, Blood and Urine** *see* Lysozyme, Blood and Urine *on page 884*

♦ **Murphy-Pattee** *see* Thyroxine, Serum *on page 1257*

Muscle Biopsy

Related Information

Aldolase, Plasma or Serum *on page 121*
Cerebrospinal Fluid Analysis: Overview *on page 355*
Creatine Kinase, Serum *on page 470*
Duchenne/Becker Muscular Dystrophy DNA Detection *on page 526*
Electron Microscopy *on page 533*
Histopathology *on page 733*
Inherited Diseases of Metabolism and Cell Structure *on page 792*
Jo-1 Antibody *on page 815*
Mucopolysaccharides, Urine *on page 922*
Myoglobin, Blood, Serum, or Plasma *on page 940*
Myoglobin, Qualitative, Urine *on page 941*
Myotonic Dystrophy DNA Test *on page 942*
Potassium, Serum or Plasma *on page 1078*
Thyroid Stimulating Hormone, Serum *on page 1250*
Trichinosis Diagnostic Procedures *on page 1272*

Applies to Dystrophin

Abstract Diagnosis and classification of muscle disease includes medical history, examination, laboratory assessment, electromyogram with nerve conduction studies, and muscle biopsy.

Patient Preparation Clinical data is required and should include the patient's age and sex; the pattern, severity, duration, and tempo of the muscle involvement; relevant laboratory results (eg, aldolase, creatine kinase, lactate dehydrogenase, LDH isoenzymes, aspartate aminotransferase, erythrocyte sedimentation rate, antinuclear antibodies, and rheumatoid factor); nerve conduction and electromyographic (EMG) findings; and the presence of significant related conditions (ie, dermatitis, neoplasm, AIDS); all current medications, and particularly any exposure to corticosteroids during the previous 3 months; and the site of muscle biopsy. Family history may be essential.
(Continued)

Muscle Biopsy *(Continued)*

Sampling Time The biopsy should be performed early in the day, as the specimen will immediately require special handling and should arrive when histotechnical personnel are available.

Collection Selection of muscle biopsy site: The site for muscle biopsy should be one that bears well characterized features (ie, quadriceps femoris or biceps brachii (preferred) or gastrocnemius). Unusual muscle groups such as oculo-motor or pharyngeal muscles should be avoided. Biopsy should be from an accessible muscle that is involved by the disease but has not reached "end-stage" atrophy. If more distal muscles are involved, a more distal biopsy site may be required. EMG or injection sites and sites near the myotendinous junction should be avoided, as these biopsies commonly exhibit artifactual changes. It is the muscle belly that should be sampled, not the tendon insertion. Needle biopsies of muscle provide inferior specimens.

Deliver on a saline-moistened gauze pad immediately to the Pathology Department. Moistened gauze is used to prevent drying. The specimen must not become saturated as this will cause severe ice crystal artifact during snap freezing. **The tissue should not be placed in fixative and should ideally reach the Pathology Laboratory as quickly as possible.**

Use Evaluate neurogenic atrophy, muscular dystrophies, myositis (infectious and "idiopathic," or autoimmune), hereditary and acquired metabolic and endocrine myopathies, ischemic, traumatic, and drug-induced problems, acquired diseases of the neuromuscular junction, and congenital/hereditary myopathies and enzyme deficiencies. Very rarely, muscle biopsy may shed light on a systemic condition such as systemic vasculitis in the absence of overt clinical muscle disease.

Methodology A portion of the clamped muscle is oriented, frozen in isopentane/liquid nitrogen, and transverse sections are obtained for H&E, trichrome, and various histochemical preparations. Electron microscopy may be needed.

Dystrophin is usually prominent in sarcolemmal membranes. It can be demonstrated by immunostains and is markedly reduced or absent in patients with Duchenne muscular dystrophy.

Additional Information See the Disease Index for Dermatomyositis, Duchenne Muscular Dystrophy, Guillain-Barré Syndrome, Muscle Disease, Muscular Dystrophy, Myasthenia Gravis, and Myositis.

References
Buchbinder R, Forbes A, Hall S, et al, "Incidence of Malignant Disease in Biopsy-Proven Inflammatory Myopathy. A Population-Based Cohort Study," *Ann Intern Med*, 2001, 134(12):1087-95.

Heffner RR Jr, "Muscle Biopsy in Neuromuscular Diseases," *Diagnostic Surgical Pathology*, Sternberg SS, ed, 3rd ed, Philadelphia, PA: Lippincott and Williams, 1999, 109-29.

Pearl GS and Ghatak NR, "Muscle Biopsy," *Arch Pathol Lab Med*, 1995, 119(4):303-6.

Rollins S, Prayson RA, and McMahon JT, "Diagnostic Yield of Muscle Biopsy in Patients With Clinical Evidence of Mitochondrial Cytopathy," *Am J Clin Pathol*, 2001, 116(3):326-30.

Vladutiu GD and Heffner RR, "Succinate Dehydrogenase Deficiency," *Arch Pathol Lab Med*, 2000, 124(12):1755-8.

♦ **Muscle Fiber, Stool** *see* Meat Fibers, Stool *on page 894*
♦ **Mychel-S®** *see* Chloramphenicol, Serum *on page 390*

Mycobacteria by DNA Probe

Related Information
Acid-Fast Stain, Routine or Modified *on page 95*
Adenosine Deaminase, CSF, Pleural Fluid, Pericardial Fluid, Peritoneal Fluid *on page 107*
Mycobacterial Culture, Biopsy or Body Fluid *on page 929*
Mycobacterial Culture, Cerebrospinal Fluid *on page 931*
Mycobacterial Culture, Cutaneous and Subcutaneous Tissue *on page 932*
Mycobacterial Culture, Sputum *on page 933*
Mycobacterial Culture, Urine *on page 935*

Test Includes Direct detection of mycobacterial DNA in specimens or from isolated colony

Abstract It has been estimated that world-wide the number of people infected with *Mycobacterium tuberculosis* (TB) each year is ~8 million. In the U.S., the number of tuberculosis cases dramatically increased between 1985 and 1992. Since 1992, the number of cases has consistently declined, primarily due to the prevention and control measures instituted by the CDC. One of these measures was the use of new and improved laboratory methods to both culture and detect TB. Culture and identification of this organism can take up to 8 weeks and direct staining procedures are insensitive. Thus, the ability to detect specific mycobacterial DNA directly from the patient specimen with early diagnosis has great advantages in diagnosis, treatment, and prevention of spread to others.

Specimen Sputum, pleural fluid, cerebrospinal fluid, bronchial aspirates, urine, and tissue biopsy or isolate from growth of acid-fast bacteria

Container Sputum, pleural fluid, and cerebrospinal fluid should be collected and transported in a tightly sealed plastic container. This container should be transferred into a secondary sealed container for transport.

Collection Samples should be collected as for mycobacteria culture. Sputum should be collected as early in the morning as possible. First morning urine should be collected.

Storage Instructions The specimens should be kept refrigerated if not immediately processed. Do not freeze.

Causes for Rejection Containers with external contamination are rejected. Samples that are left at room temperature for more than 12 hours are rejected due to overgrowth of other bacteria.

Turnaround Time Approximately 24-72 hours

Reference Interval No *M. tuberculosis* nucleic acid detected

Use Rapid detection and identification of *M. tuberculosis* in clinical specimens or culture.

Limitations A potential for false-positive results exists with molecular assays due to cross contamination of specimens. False negatives may occur as well due to inhibitors in the specimen. **Nonviable mycobacterial DNA can lead to positive PCR. Thus, this assay cannot be used to monitor the efficacy of therapy.** The currently available nucleic acid assays do not differentiate among members of the *M. tuberculosis* complex (*M. tuberculosis*, *M. bovis*, *M. bovis* BCG, *M. africanum*, *M. microti*, and *M. canetti*).

Methodology Culture confirmation: DNA probe is used to detect species specific rRNA from cultured acid-fast bacteria.

Direct detection: Specific amplification of *Mycobacterium tuberculosis* nucleic acid by polymerase chain reaction (PCR) or transcription-based amplification is currently FDA approved for direct detection in respiratory specimens from untreated patients.

Additional Information Mycobacteria are aerobic rod-shaped bacteria noted for their very slow growth. The laboratory diagnosis of mycobacterial disease is currently based on a positive acid-fast stain and culture of the mycobacterial organism. Because of the long culture periods required, therapeutic decisions are often made before a laboratory diagnosis is available. To improve upon the detection of mycobacteria, DNA detection assays have been developed that have increased sensitivity and specificity and decreased turnaround time when compared with culture.

References

Della-Latta P and Whittier S, "Comprehensive Evaluation of Performance, Laboratory Application, and Clinical Usefulness of Two Direct Amplification Technologies for the Detection of *Mycobacterium tuberculosis* Complex," *Am J Clin Pathol*, 1998, 110(3):301-10.

Kiechle FL, "DNA Technology, The Clinical Laboratory and the Future," *Arch Pathol Lab Med*, 2001, 125(1):78-6.

Salian NV, Rish JA, Eisenach KD, et al, "Polymerase Chain Reaction to Detect *Mycobacterium tuberculosis* in Histologic Specimens," *Am J Respir Crit Care Med*, 1998, 158(4):1150-5.

Smith MB, Bergmann JS, Onoroto M, et al, "Evaluation of the Enhanced Amplified *Mycobacterium tuberculosis* Direct Test for Direct Detection of *Mycobacterium tuberculosis* Complex in Respiratory Specimens," *Arch Pathol Lab Med*, 1999, 123(11):1101-3.

Soini H and Musser JM, "Molecular Diagnosis of Mycobacteria," *Clin Chem*, 2001, 47(5):809-14.

Troesch A, Nguyen H, Miyada CG, et al, "*Mycobacterium* Species Identification and Rifampin Resistance Testing With High-Density DNA Probe Arrays," *J Clin Microbiol*, 1999, 37(1):49-55.

"Update: Nucleic Acid Amplification Tests for Tuberculosis," *JAMA*, 2000, 284(7):826.

Internet Web Sites

www.astdhpphe.org/infect/tb.html
www.cdc.gov/nchstp/tb/faqs/qa.htm
www.who.int/health-topics/tb.htm

Mycobacterial Culture, Biopsy or Body Fluid

Related Information

Acid-Fast Stain, Routine or Modified *on page 95*

(Continued)

Mycobacterial Culture, Biopsy or Body Fluid
(Continued)

Test Includes Preparation of the tissue specimen by grinding and/or mincing and inoculation of appropriate media and incubation at temperatures that support the growth of most known pathogenic mycobacterial species.

Abstract In addition to *Mycobacterium tuberculosis*, a number of nontuberculous mycobacteria can cause granulomatous inflammatory diseases. In general, mycobacteria are cultured from biopsies that contained necrotizing or non-necrotizing granulomas. Tissue with fibrotic or hyalinized granulomas, nonspecific chronic inflammation, nonspecific reactive or reparative changes, or malignancy failed usually are not positive for culture.

Specimen Surgical tissue, bone marrow, biopsy material, endometrial curettings, aspirated fluid; **swab specimens are not acceptable specimens.**

Container Sterile container

Collection The surgical specimen submitted for culture should be separated from the portion submitted for histopathology. Up to a liter of ascitic fluid is needed to achieve 80% sensitivity for mycobacterial culture.

Turnaround Time Negative cultures are reported after 8 weeks.

Reference Interval No growth

Use Isolate and identify mycobacteria; establish the etiology of granulomatous disease and fever of unknown origin (FUO)

Limitations Transbronchial biopsy cultures may be of assistance in diagnosing tuberculosis in sputum smear negative cases; however, sputum and bronchial washing cultures have a higher yield.

Mycobacterium marinum may cause a localized cutaneous lesion that may be nodular, verrucous, ulcerative, or sporotrichoid, and which may rarely involve deeper structures. If it is suspected, the laboratory must be notified so that the culture may be incubated at an appropriate temperature (30°C); *vide infra*. *M. haemophilum*, which has been recovered in a variety of biopsy specimens from immunosuppressed patients, also requires special media and conditions to grow; inform the laboratory if this organism is suspected.

Methodology Culture on specialized selective media, incubated at 35°C with 5% to 10% CO_2. Cutaneous and subcutaneous tissues are incubated at room temperature to enhance recovery of *M. marinum* and *M. ulcerans*.

Additional Information Occult infections with **atypical mycobacteria**, particularly *Mycobacterium avium* and *Mycobacterium intracellulare*, occur in immunosuppressed patients. Mycobacteria have been recovered from several types of tissue in which the characteristic granulomatous reaction has been absent. Optimal isolation of mycobacteria from tissue is accomplished by processing as much tissue as possible for culture.

Rapidly growing mycobacteria have been implicated in cases of sternal wound infection, early prosthetic valve endocarditis, infections complicating mammary augmentation surgery, and other cutaneous/subcutaneous infections. *M. fortuitum* is most commonly implicated in these infections. Rapidly growing mycobacteria grow on routine bacterial culture media within incubation periods used in routine bacterial cultures. Such organisms can be misidentified as "diphtheroids" and disregarded as contaminants.

References

Das DK, "Fine-Needle Aspiration Cytology in the Diagnosis of Tuberculous Lesions," *Lab Med*, 2000, 31(11):425-32.

"Diagnosis and Treatment of Disease Caused by Nontuberculous Mycobacteria. The Official Statement of the American Thoracic Society was Approved by the Board of Directors, March 1997. Medical Section of the American Lung Association," *Am J Respir Crit Care Med*, 1997, 156(2 Pt 2):S1-25.

Hussong J, Peterson LR, Warren JR, et al, "Detecting Disseminated *Mycobacterium avium* Complex Infections in HIV-Positive Patients: The Usefulness of Bone Marrow Trephine Biopsy Specimens, Aspirate Cultures, and Blood Cultures," *Am J Clin Pathol*, 1998, 110(6):806-9.

Levendoglu-Tugal O, Munoz J, Brudnicki A, et al, "Infections Due to Nontuberculous Mycobacteria in Children With Leukemia," *Clin Infect Dis*, 1998, 27(5):1227-30.

Olivier KN, "Nontuberculous Mycobacterial Pulmonary Disease," *Curr Opin Pulm Med*, 1998, 4(3):148-53.

Weitzul S, Eichhorn PJ, and Pandya AG, "Nontuberculous Mycobacterial Infections of the Skin," *Dermatol Clin*, 2000, 18(2):359-77.

Internet Web Sites

www.astdhpphe.org/infect/tb.html
www.cdc.gov/nchstp/tb/faqs/qa.htm
www.who.int/health-topics/tb.htm

Mycobacterial Culture, Cerebrospinal Fluid

Related Information

Acid-Fast Stain, Routine or Modified *on page 95*
Adenosine Deaminase, CSF, Pleural Fluid, Pericardial Fluid, Peritoneal Fluid *on page 107*
Bacterial Antigens, Rapid Detection Methods *on page 228*
Bacterial Culture, Cerebrospinal Fluid *on page 236*
Cerebrospinal Fluid Analysis: Overview *on page 355*
Cerebrospinal Fluid Cytology *on page 361*
Cerebrospinal Fluid Glucose *on page 362*
Cerebrospinal Fluid Protein *on page 371*
Cryptococcal Antigen Titer *on page 482*
Fungal Culture, Cerebrospinal Fluid *on page 621*
Mycobacteria by DNA Probe *on page 928*
Viral Culture *on page 1307*

Applies to Nucleic Acid Amplification

Test Includes Culture for mycobacteria

Abstract Conventional cultures and direct acid-fast stain for mycobacteria rarely detect the bacteria in cerebrospinal fluid and, thus, are of limited value in the diagnosis of tuberculosis meningitis. The direct acid-fast stain is usually negative due to the low number of organisms in meningitis. Currently, the best laboratory assay for detection of *M. tuberculosis* in cerebrospinal fluid is nucleic acid amplification.

Patient Preparation Usual sterile preparation

Specimen Cerebrospinal fluid; 10 mL is optimum, 5 mL the minimum volume

Reference Interval No growth

Use Investigate cases of meningitis with subacute or chronic onset, cases in which a history of contact with tuberculosis

Limitations Recovery of mycobacteria is directly related to the volume of specimen available for culture; 5-10 mL is recommended for optimal yield. Recovery of organisms can require 4-6 weeks. Usually CSF glucose ≤60 mg/dL (3.3 mmol/L) and CSF WBC is increased; if not, culture is unlikely to be useful.

Methodology Culture in broth media is recommended for primary mycobacterial isolation. Additionally, solid media that supports the growth of most mycobacteria may also be used.

Additional Information Factors associated with tuberculosis meningitis include subacute or chronic onset, positive tuberculin skin test, previous active tuberculosis, recent exposure to tuberculosis, and suspicion of tuberculosis on imaging procedures. Early in the course, neutrophils predominate in the CSF. Lymphocytes, mononuclear cells, and granulocytes are found later. Rarely does the cell count exceed 1000 cells/mm³. The CSF is clear and colorless early; later, a pellicle forms on standing. Low CSF glucose, <40 mg/dL, can be observed. Increased protein is often >300 mg/dL. Acid-fast organisms can be identified only rarely on centrifuged sediments.

(Continued)

Mycobacterial Culture, Cerebrospinal Fluid *(Continued)*

Untreated tuberculous meningitis is fatal, usually within 3 weeks of presentation.

References

Christie JD and Callihan DR, "The Laboratory Diagnosis of Mycobacterial Diseases. Challenges and Common Sense," *Clin Lab Med*, 1995, 15(2):279-306.

Kelly JJ, Horowitz EA, Destache CJ, et al, "Diagnosis and Treatment of Complicated Tubercular Meningitis," *Pharmacotherapy*, 1999, 19(10):1167-72.

Smith MB, Boyars MC, Veasey S, et al, "Generalized Tuberculosis in the Acquired Immune Deficiency Syndrome. A Clinicopathologic Analysis Based on Autopsy Findings," *Arch Pathol Lab Med*, 2000, 124(9):1267-74.

Wolinsky E, "Mycobacterial Diseases Other Than Tuberculosis," *Clin Infect Dis*, 1992, 15(1):1-10.

Internet Web Sites

www.astdhpphe.org/infect/tb.html
www.cdc.gov/nchstp/tb/faqs/qa.htm
www.who.int/health-topics/tb.htm

Mycobacterial Culture, Cutaneous and Subcutaneous Tissue

Related Information

Acid-Fast Stain, Routine or Modified *on page 95*
Antimicrobial Susceptibility Testing, Mycobacteria *on page 181*
Fine Needle Aspiration Culture *on page 589*
Fungal Culture, Skin *on page 623*
Mycobacteria by DNA Probe *on page 928*
Mycobacterial Culture, Biopsy or Body Fluid *on page 929*
Mycobacterial Culture, Sputum *on page 933*
Skin Biopsy *on page 1200*
Viral Culture *on page 1307*

Test Includes Culture and identification of acid-fast bacteria including a direct acid-fast smear

Abstract Cutaneous or subcutaneous mycobacterial infections may result from either direct inoculation or hematogenous dissemination of infecting organisms. A variety of *Mycobacterium* spp may cause infections of this type; they are most likely to occur in patients at risk due to immunosuppression (eg, *M. tuberculosis*, *M. haemophilum*, *M. kansasii*), poverty and geographic exposure (eg, *M. leprae*, *M. ulcerans*), prior surgical treatment (eg, *M. fortuitum*), or hobbies or occupational exposure (eg, *M. marinum*).

Container Sterile tube containing 0.5 mL sterile saline

Turnaround Time As early as 2-3 days for a rapid growing mycobacteria and as long as 8 weeks in specimens.

Reference Interval No growth

Use Isolate and identify mycobacteria

Limitations Scrapings, curettings, or biopsy tissue rather than swabs of lesions should be submitted to the laboratory. Mycobacteria adhere tightly to material on a swab and thus are difficult to remove from the swab. The yield on cultures is proportional to the volume of specimen submitted. *M. leprae* does not grow in culture.

Methodology Acid-fast stain of smear prepared from clinical specimen, culture of specimen on solid media and liquid media. Incubation at both 35°C and 25°C to 30°C. Hemin-containing media is required to recover *M. haemophilum*. Identification of bacterial isolates based on routine methods.

Additional Information *Mycobacterium marinum* causes granulomatous cutaneous lesions. Its lesions are similar to those seen with sporotrichosis. A careful clinical history addressing occupational or recreational activities usually yields important clues to the diagnosis (eg, swimming pool or seawall abrasions, barnacle scrapes, fish fin punctures, exposure to tropical salt water fish tanks).

Members of the *Mycobacterium fortuitum* complex (*M. fortuitum* and *M. chelonae*) can cause surgical wound infections and cutaneous abscesses and osteomyelitis in trauma victims and debilitated hosts. The key clinical feature is that symptoms of infection, localized cellulitis, or abscess formation appear 4-6 weeks after traumatic injury or surgery.

Mycobacterium ulcerans causes a chronic granulomatous skin lesion called Buruli ulcer. *M. ulcerans* is uncommon in North America. It is most frequently isolated in Australia and Africa.

Isolates of *Mycobacterium avium-intracellulare* (MAI) and *Mycobacterium kansasii* cause skin lesions in patients with the acquired immunodeficiency syndrome (AIDS).

Cutaneous tuberculosis has been found in patients with neoplastic or immunosuppressive diseases. The most common clinical presentations are lupus vulgaris, scrofuloderma, tuberculids, tuberculous verrucosa cutis, and tuberculous gumma. *M. haemophilum* produces disseminated cutaneous disease and infections of the bones, joints, and lymphatics in immunocompromised individuals.

References

Bedlow AJ, Vittay GI, Stephenson J, et al, "Deep Cutaneous Infection With *Mycobacterium avium-intracellulare* Complex in an Immunosuppressed Patient With Dermatomyositis," *Br J Dermatol*, 1998, 139(5):920-2.

Escalonilla P, Esteban J, Soriano ML, et al, "Cutaneous Manifestations of Infection by Nontuberculous Mycobacteria," *Clin Exp Dermatol*, 1998, 23(5):214-21.

Kumar B and Muralidhar S, "Cutaneous Tuberculosis: A Twenty-Year Prospective Study," *Int J Tuberc Lung Dis*, 1999, 3(6):494-500.

Ramesh V, Misra RS, Beena KR, et al, "A Study of Cutaneous Tuberculosis in Children," *Pediatr Dermatol*, 1999, 16(4):264-9.

Weitzul S, Eichhorn PJ, and Pandya AG, "Nontuberculous Mycobacterial Infections of the Skin," *Dermatol Clin*, 2000, 18(2):359-77.

Internet Web Sites

www.astdhpphe.org/infect/tb.html
www.cdc.gov/nchstp/tb/faqs/qa.htm
www.who.int/health-topics/tb.htm

Mycobacterial Culture, Sputum

Related Information

Acid-Fast Stain, Routine or Modified *on page 95*
Antimicrobial Susceptibility Testing, Mycobacteria *on page 181*
Bacterial Culture, Lower Respiratory *on page 241*
Bronchoalveolar Lavage (BAL) *on page 311*
Fungal Culture, Sputum *on page 624*
Mycobacteria by DNA Probe *on page 928*
Mycobacterial Culture, Biopsy or Body Fluid *on page 929*
Mycobacterial Culture, Cutaneous and Subcutaneous Tissue *on page 932*
Mycobacterial Culture, Urine *on page 935*
Nocardia Culture *on page 964*
Sputum Cytology *on page 1222*

Test Includes Mycobacteria (AFB) stain, culture, and identification

Abstract Tuberculosis is a pulmonary disease caused by *Mycobacterium tuberculosis*. It is spread from person to person by airborne droplets. Patients can be latently infected with *M. tuberculosis* for many years, have a positive skin test but remain asymptomatic. Such patients are not infectious and have only a 10% risk for development of active tuberculosis.

The World Health Organization (WHO) estimates there are over 8 million new cases of tuberculosis in the world each year. Causes of the enormous global burden of tuberculosis include poor control in Southeast Asia, Sub-Saharan Africa, and Eastern Europe, and include the high incidence of *M. tuberculosis* and HIV coinfection in parts of Africa. In the U.S., cases of tuberculosis are found more commonly in populations that are medically underserved, homeless persons, elderly, prison inmates, alcoholics, people who inject illegal drugs, and foreign-born people from areas of high tuberculosis prevalence.

Patient Preparation Instruct patient to remove dentures, rinse mouth with water, and cough deeply.

Specimen First morning sputum or induced sputum, fasting gastric aspirate, bronchial aspirate, tracheal aspirate, transtracheal aspirate. In neonates, gastric and endotracheal aspirates may be used but are not optimal.

Container Sputum cup, sputum trap, sterile tracheal aspirate or bronchoscopy tube

Sampling Time In children, gastric aspiration should be done early in the morning as the child awakens before the stomach empties. Collect samples on three separate mornings.

(Continued)

Mycobacterial Culture, Sputum *(Continued)*

Collection A newly recommended screening procedure includes two first morning specimens submitted on successive days. The patient should be instructed to brush his/her teeth and/or rinse mouth well with water before attempting to collect the specimen to reduce the possibility of contamination of the specimen. After the specimen has been collected, it should be examined to make sure it contains a sufficient quantity (at least 5 mL) of thick mucus (**not saliva**). If a two-part collection system has been used, only the screw-cap tube should be submitted to the laboratory. (The outer container is considered contaminated and its transport through the hospital or by courier constitutes a health hazard!) The specimen should be properly labeled and accompanied by properly completed requisition. The specimen can be divided in the laboratory for fungal, mycobacterial, and routine cultures.

Storage Instructions The specimen should be refrigerated if it cannot be promptly processed. If a gastric aspirate cannot be processed immediately, its pH should be neutralized for storage.

Turnaround Time Negative cultures are reported after 6-8 weeks; *vide infra*.

Reference Interval No growth

Use Diagnose pulmonary tuberculosis or other *Mycobacterium* species

Limitations Postbronchoscopy expectorated specimens may provide a better yield of organisms than those obtained during the procedure. Acid-fast smear of gastric aspirates provides a useful clinical diagnosis only if positive. The relative yield of mycobacteria from clinical specimens is prebronchoscopy sputum > bronchial washings > postbronchoscopy sputum > bronchial biopsy.

Poor sensitivity of smears (45%) and prolonged times required for culture are limitations recognized with traditional diagnostic approaches for diagnosis of mycobacteria. A positive culture currently requires 7-20 days, even with the newer culture systems. It is recommended that a direct molecular assay be used on acid-fast smear positive specimens to detect *M. tuberculosis* complex. The sensitivity of this assay is ~95% and the specificity is 100%.

Methodology Concentration and decontamination, then inoculation of broth media is required for the initial isolation of mycobacteria. It is useful to also include a solid culture medium.

Identification of culture isolates may be accomplished by traditional biochemical methods, gas liquid chromatography, or by DNA probe detection.

Additional Information Tuberculosis must be considered in patients who have chronic cough and fever, regardless of results of tuberculin testing, especially in the presence of HIV infection. When tuberculosis occurs as a first or case-defining opportunistic infection, 75% to 100% of HIV-positive patients have pulmonary disease. After the diagnosis of AIDS has been made, 25% to 70% of HIV-associated tuberculosis patients have an extrapulmonary site of infection. In addition, the emergence of *M. avium-intracellulare* infections in patients with acquired immunodeficiency syndrome has contributed to the increasing need for rapid, accurate diagnosis of mycobacterial infections.

Two specimens processed for acid-fast stain and culture identifies all cases of active tuberculosis within the time required for culture. The most infective cases are identified immediately by the acid-fast stain. Tuberculin tests and chest x-rays should also be performed.

Nosocomial transmission of multidrug-resistant *Mycobacterium tuberculosis* has been noted to occur from patient to patient and from patient to healthcare worker. Acid-fast bacilli isolation precautions and adherence to appropriate infection control procedures are recommended.

While *M. tuberculosis* is contagious and is usually transmitted from person to person, most of the other disease-causing mycobacteria are not characterized by person-to-person spread. Bacteria found in the environment are considered opportunistic pathogens, including *M. avium*, *M. intracellulare*, *M. asiaticum*, *M. flavescens*, *M. fortuitum* complex, *M. gordonae*, *M. haemophilum*, *M. kansasii*, *M. malmoense*, *M. marinum*, *M. scrofulaceum*, *M. simiae*, *M. smegmatis*, and *M. xenopi*.

See Mycobacteria by DNA Probe *on page 928*.

References

American Thoracic Society, "Diagnosis and Treatment of Disease Caused by Nontuberculous Mycobacteria," *Am J Respir Crit Care Med*, 1997, 156(2 Pt 2):S1-25.

Barnes PF, "Reducing Ongoing Transmission of Tuberculosis," *JAMA*, 1998, 280(19):1702-3.

Dye C, Scheele S, Dolin P, et al, "Global Burden of Tuberculosis. Estimated Incidence, Prevalence, and Mortality by Country," *JAMA*, 1999, 282(7):677-86.

Ginsberg AM, "The Tuberculosis Epidemic. Scientific Challenges and Opportunities," *Public Health Rep*, 1998, 113(2):128-36.

Griffith DE, "Mycobacteria as Pathogens of Respiratory Infection," *Infect Dis Clin North Am*, 1998, 12(3):593-611.

Olivier KN, "Nontuberculous Mycobacterial Pulmonary Disease," *Curr Opin Pulm Med*, 1998, 4(3):148-53.

Raviglione MC, Snider DE Jr, and Kochi A, "Global Epidemiology of Tuberculosis. Prevalence and Mortality of a Worldwide Epidemic," *JAMA*, 1995, 273(3):220-6.

Reichman LB, "On Target: A Tuberculosis Control Strategy Whose Time Has Come," *Ann Intern Med*, 1999, 131(8):617-8.

Internet Web Sites

www.astdhpphe.org/infect/tb.html
www.cdc.gov/nchstp/tb/faqs/qa.htm
www.cdc.gov/ncidod/dbmd/diseaseinfo/mycobacteriumavium_t.htm
www.who.int/health-topics/tb.htm

Mycobacterial Culture, Urine

Related Information

Acid-Fast Stain, Routine or Modified *on page 95*
Antimicrobial Susceptibility Testing, Mycobacteria *on page 181*
Bacterial Culture, Urine *on page 246*
Fungal Culture, Urine *on page 625*
Mycobacteria by DNA Probe *on page 928*
Mycobacterial Culture, Biopsy or Body Fluid *on page 929*
Mycobacterial Culture, Sputum *on page 933*
Viral Culture *on page 1307*

Patient Preparation Usual preparation for clean catch midvoid urine specimen collection. See Bacterial Culture, Urine *on page 246* for detailed information.

Specimen At least 40 mL of first morning voided urine

Collection Three first morning voided urine specimens should be submitted.

Turnaround Time Negatives are reported after 6-8 weeks.

Reference Interval No growth

Use Isolate and identify mycobacteria from the urinary tract

Limitations Positive acid-fast stained smears are not diagnostic, because of the presence of *Mycobacterium smegmatis* in genital secretions of normal patients.

Contraindications A 24-hour pooled urine and catheter bag specimens are unacceptable because of increased chance of bacterial contamination.

Methodology Culture in broth media and on solid media that supports the growth of mycobacteria

Additional Information Mycobacterial genitourinary tract infections represented about 20% of extrapulmonary tuberculosis cases. Most patients with genitourinary tuberculosis have symptoms of urinary tract disease, but some are asymptomatic. If mycobacteria are cultured, isolates are definitively identified, and susceptibility testing performed on request. Although it has been thought that tuberculosis of the urinary tract should be suspected when hematuria and pyuria occur without recovery by routine culture of usual urinary tract pathogens (sterile pyuria), concomitant infections with ordinary pathogens are not rare. Cultures of urine for mycobacteria are ~90% sensitive. The kidney is the most frequent site of such genitourinary infection; prostate, salpinx, and endometrial involvement also occurs. Genitourinary infections with atypical mycobacteria, particularly *M. kansasii* and *M. avium-intracellulare*, occur.

References

Dye C, Scheele S, Dolin P, et al, "Global Burden of Tuberculosis. Estimated Incidence, Prevalence, and Mortality by Country," *JAMA*, 1999, 282(7):677-86.

Ginsberg AM, "The Tuberculosis Epidemic. Scientific Challenges and Opportunities," *Public Health Rep*, 1998, 113(2):128-36.

Raviglione MC, Snider DE Jr, and Kochi A, "Global Epidemiology of Tuberculosis. Prevalence and Mortality of a Worldwide Epidemic," *JAMA*, 1995, 273(3):220-6.

Reichman LB, "On Target: A Tuberculosis Control Strategy Whose Time Has Come," *Ann Intern Med*, 1999, 131(8):617-8.

Internet Web Sites

www.astdhpphe.org/infect/tb.html
www.cdc.gov/nchstp/tb/faqs/qa.htm
www.who.int/health-topics/tb.htm

♦ *Mycobacterium* Smear *see* Acid-Fast Stain, Routine or Modified *on page 95*

Mycophenolic Acid, Serum

Related Information

Cyclosporine, Blood *on page 487*

Tacrolimus, Whole Blood *on page 1234*

Synonyms CellCept™; MMF; MPA

Abstract Mycophenolate mofetil (MMF), 2-morpholinoethyl ester of mycophenolic acid (MPA), is an immunosuppressant used for heart, kidney, and liver transplants. It is generally used concomitantly with cyclosporine, tacrolimus, and/or corticosteroids. After gastrointestinal absorption, it is rapidly metabolized to MPA, an active metabolite. MPA is measured for therapeutic drug monitoring. Some laboratories also measure MPA metabolite, MPA-glucuronide. It has also been used as an antirheumatic agent and in several other inflammatory conditions, including systemic lupus erythematosus.

Specimen Serum

Container Plain red top tube; do not use gel-containing tube.

Sampling Time Trough level drawn just before next dose.

Storage Instructions Store and ship refrigerated sample.

Causes for Rejection Sample drawn at time other than trough

Reference Interval Renal transplant MPA: 1.0-3.5 µg/mL; MPA-glucuronide (patients with normal phase II metabolic activity): 35-100 µg/mL

MPA-glucuronide levels may be higher or lower in patients with altered phase II metabolic activity. Therapeutic levels of MPA in patients with liver or heart transplant are higher.

Use Therapeutic drug monitoring and toxicity evaluation

Methodology High performance liquid chromatography (HPLC) with UV detector or mass spectrometry, immunoassay; HPLC is a preferred method.

Additional Information

- Half-life: 18±6 hours
- Volume of distribution: 4±1.2 L/kg
- Protein binding: 97% to plasma albumin
- Time to reach steady state: ~2 weeks

After oral administration, MMF is rapidly and completely absorbed. Very rapidly (in less than 10 minutes) the prodrug is converted to MPA, an active metabolite, by the hepatic phase II metabolizing system. MPA interferes with DNA synthesis and proliferation of T- and B-lymphocytes by inhibition of inosine monophosphate dehydrogenase, a key enzyme in the *de novo* synthesis of guanosine nucleotide. It is recommended that MMF be administered along with cyclosporine or tacrolimus or a corticosteroid such as prednisone or prednisolone. MMF is also useful in patients who do not tolerate cyclosporine or tacrolimus as the main immunosuppressant drug.

MPA is primarily metabolized to the glucuronide (MPAG), which does not have immunosuppressant properties. More than 90% of the dose is eliminated in the urine as MPAG. Failure of MPAG excretion in renal impairment results in increased concentrations of MPA.

The primary side effects of MMF include diarrhea, leukopenia, sepsis, vomiting, and higher frequency of infections. Intravenous administration causes the same effects as oral dose.

References

Holt DW, "Monitoring Mycophenolic Acid," *Ann Clin Biochem*, 2002, 39(3 Pt 3):173-83.

Moder KG, "Mycophenolate Mofetil: New Applications for This Immunosuppressant," *Ann Allergy Asthma Immunol*, 2003, 90(1):15-9.

Schutz E, Shipkova M, Armstrong VW, et al, "Identification of a Pharmacologically Active Metabolite of Mycophenolic Acid in Plasma of Transplant Recipients Treated With Mycophenolate Mofetil," *Clin Chem*, 1999, 45(3):419-22.

Shaw LM, Pawinski T, Korecka M, et al, "Monitoring of Mycophenolic Acid in Clinical Transplantation," *Ther Drug Monit*, 2002, 24(1):68-73.

Shipkova M, Schutz E, Armstrong VW, et al, "Stability of Mycophenolic Acid and Mycophenolic Acid Glucuronide in Human Plasma," *Clin Chem*, 1999, 45(1):127-9.

Mycoplasma pneumoniae Culture and Serology

Related Information

Bacterial Culture, Lower Respiratory *on page 241*

Cold Agglutinin Titer *on page 430*
Mycoplasma pneumoniae DNA Probe Test *on page 938*

Test Includes Culture for *Mycoplasma* organisms only and detection of serologic response to *Mycoplasma pneumoniae* infection

Abstract *Mycoplasma* lacks the rigid cell walls of other bacteria, thus, they do not react with Gram stain. Due to the lack of a cell wall, the organisms are not susceptible to killing by beta-lactam antibiotics. These fastidious bacteria are the smallest free-living organisms known and are smaller than some viruses. *Mycoplasma pneumoniae* accounts for up to 20% of hospitalized adults with community-acquired pneumonia, and a larger percentage of those treated as outpatients.

Specimen Culture requires throat or nasopharyngeal swabs, sputum, broncho-alveolar lavage. Serology requires serum.

Container Culture specimens are collected in a sterile container. Serum is collected in a red top tube.

Sampling Time Culture should be collected during acute infection. Acute and convalescent sera drawn 10-14 days apart are desirable.

Collection Throat or nasopharyngeal swabs should be collected vigorously to obtain as many cells as possible. Throat or nasopharyngeal swabs should be placed **immediately** in special transport medium (often obtained from the laboratory) and sent immediately to the laboratory. All specimens should be kept at 4°C during transport.

Storage Instructions If storage longer than 24 hours is needed, the specimen should be frozen at -70°C.

Turnaround Time 2-3 weeks for culture; 1 week for serology

Reference Interval No *Mycoplasma pneumoniae* cultured; negative, IgG and IgM. A fourfold increase in titer in paired sera, drawn 2-4 weeks apart, provides a definitive diagnosis.

Use Aid in the diagnosis of pneumonia or other respiratory diseases caused by *Mycoplasma pneumoniae*

Limitations The culture procedure is not often used because it is slow and somewhat insensitive; 2-3 weeks or more are often required for isolation and definitive identification of positive cultures.

False-positive serology occurs when antibody from prior infection is detected, which is especially a problem in patients older than 40 years of age. The complement fixation procedures are based on lipid extracts of the organism, which cross react with antigens in other bacteria, human tissues, and some plants. False positives can occur in patients with cross-reactive autoantibodies. False negatives are found when testing is done too early and may be found in immunocompromised individuals. The commercial serology assays are not equivalent, due to variations in the antigen preparation. Newer EIAs use recombinant antigens that have less cross reactions, which increases the specificity of the assay.

Elevated cold agglutinins have been used to indicate acute *M. pneumoniae* disease, but this test lacks sensitivity. Titers ≥1:64 are considered significant. All patients with *Mycoplasma* infections do not have cold agglutinins. Cold agglutinins are IgM antibodies against the I antigen of red cells.

Methodology Isolates are cultured in special broth and on special agar media and are identified by biochemical tests and ability to hemolyze erythrocytes. Serology is tested using complement fixation (CF), indirect fluorescent antibody (IFA), enzyme immunoassay (EIA), specific IgM antibody by agglutination, IgM anti-P1 immunoblotting, microtiter procedure utilizing anti-*M. pneumoniae* IgM

Additional Information *Mycoplasma pneumoniae* is a cause of "primary atypical pneumonia". *Mycoplasma pneumoniae* infection is acquired via the respiratory route from small-particle aerosols or large droplets of secretions. The organism can penetrate the mucociliary barrier of respiratory epithelium and produce cellular injury and ciliostasis that may account for the prolonged cough observed clinically. Most infections are observed in older children and young adults.

The *Mycoplasma* organisms are more difficult to culture than ordinary bacteria and thus serologic confirmation of the diagnosis is often desirable. A DNA amplification method has recently become available. A number of reference

(Continued)

Mycoplasma pneumoniae Culture and Serology
(Continued)

laboratories offer this test as a rapid alternative to culture. See *Mycoplasma pneumoniae* DNA Probe Test *on page 938*.

References

Bisno AL, "Acute Pharyngitis," *N Engl J Med*, 2001, 344(3):205-11.

Ewing S and Torres A, "Severe Community-Acquired Pneumonia," *Clin Chest Med*, 1999, 20(3):575-87.

File TM, Tan JS, and Plouffe JF, "The Role of Atypical Pathogens: *Mycoplasma pneumoniae*, *Chlamydia pneumoniae*, and *Legionella pneumophila* in Respiratory Infection," *Infect Dis Clin N Amer*, 1998, 12(3):569-592.

Foy HM, "*Mycoplasma pneumoniae* Pneumonia: Current Perspectives," *Clin Infect Dis*, 1999, 28(2):237.

Ruuskanen O and Mertsola J, "Childhood Community-Acquired Pneumonia," *Semin Respir Infect*, 1999,14(2):163-72.

Taylor-Robinson D, "Infections Due to Species of *Mycoplasma* and *Ureaplasma*: An Update," *Clin Infect Dis*, 1996, 23(4):671-84.

Mycoplasma pneumoniae DNA Probe Test

Related Information

Cold Agglutinin Titer *on page 430*

Mycoplasma pneumoniae Culture and Serology *on page 936*

Test Includes Direct detection of *Mycoplasma pneumoniae* nucleic acids in clinical specimens

Abstract Respiratory infections due to *Mycoplasma pneumoniae* are difficult to assess because current laboratory techniques lack sensitivity or require long periods of time (3 weeks) for results. Serological procedures are currently the most widely used diagnostic tests. Culture of this microorganism is rarely done due to the difficulty of recovery and the long incubation time required for growth. A rapid and sensitive test for the diagnosis of *M. pneumoniae* infection is important, since effective antibiotic therapy is available. The nucleic acid based test is a sensitive method for rapid diagnosis of respiratory infections of *M. pneumoniae*.

Specimen Sputum, throat swab, bronchial wash, lung biopsy

Container Special DNA transport medium may be provided by the laboratory. For some specimens (eg, sputum, bronchial wash or biopsy), a sterile container is acceptable. Sterile viral swabs placed in specific *Mycoplasma* media (eg, SP4, 10 B, or 2SP) can be used to collect throat specimens.

Collection Sputum specimens should be collected early in the day so they can be sent directly to the laboratory.

Storage Instructions The specimens should be maintained at room temperature or refrigerated. Do not freeze.

Turnaround Time 1-2 days

Reference Interval Negative for *Mycoplasma pneumoniae*

Use Rapid detection of *Mycoplasma pneumoniae* in clinical specimens from respiratory sites.

Limitations This assay cannot determine whether or not the microorganism is viable. The organism may persist for varying lengths of time following acute infection. As with other molecular tests, costs and the potential for contamination must be controlled.

Methodology This test detects *Mycoplasma pneumoniae* nucleic acid directly from respiratory specimens using amplification methods.

Additional Information A wide range of clinical manifestations of *Mycoplasma pneumoniae* respiratory infections is recognized, from mild infection to severe pneumonia. This microorganism causes approximately 20% of community-acquired lower respiratory infection. Mortality is rare and infection is usually self-limited. Laboratory diagnosis of *Mycoplasma pneumoniae* is usually based on measurement of specific antibodies in paired sera and/or isolation of the organism by culture. Culture is tedious, labor intensive, and usually requires several weeks. The DNA detection assay for *Mycoplasma pneumoniae* is rapid and facilitates treatment with appropriate antibiotics.

References

Dorigo-Zetsma JW, Zaat SA, Wertheim-van Dillen PM, et al, "Comparison of PCR, Culture, and Serological Tests for Diagnosis of *Mycoplasma pneumoniae* Respiratory Tract Infection in Children," *J Clin Microbiol*, 1999, 37(1):14-7.

Foy HM, "*Mycoplasma pneumoniae* Pneumonia: Current Perspectives," *Clin Infect Dis*, 1999, 28(2):237.

Tong CY, Donnelly C, Harvey G, et al, "Multiplex Polymerase Chain Reaction for the Simultaneous Detection of *Mycoplasma pneumoniae*, *Chlamydia pneumoniae*, and *Chlamydia psittaci* in Respiratory Samples," *J Clin Pathol*, 1999, 52(4):257-63.

♦ **Myelin Basic Protein** *see* Cerebrospinal Fluid Oligoclonal Bands *on page 369*

♦ **Myelokathexis** *see* White Blood Cell Count *on page 1330*

♦ **Myeloperoxidase** *see* Myeloperoxidase, Plasma *on page 939*

♦ **Myeloperoxidase Antibody (MPO-ANCA)** *see* Antineutrophil Cytoplasmic Antibody *on page 187*

Myeloperoxidase, Plasma

Applies to Complement; C-Reactive Protein; Glutathione Peroxidase 1; Myeloperoxidase

Abstract Acute coronary artery syndrome relating to underlying arteriosclerotic plaques (usually at least partially calcified) may have an associated inflammatory component. Increased numbers of myeloperoxidase-containing macrophages are present in eroded, fissured, and/or ruptured plaques. A recent report concludes that a "single initial measurement of plasma myeloperoxidase independently predicts the early risk of myocardial infarction...". In addition, plasma myeloperoxidase predicted the risk of an adverse cardiac event over the next 30-day and 6-month periods. Red-cell glutathione peroxidase 1 activity has also been noted to be independently associated with increase in risk for adverse cardiovascular events.

Specimen Plasma, EDTA anticoagulant; assay of neutrophil myeloperoxidase and/or red cell glutathione peroxidase (GP) requires immediate separation of cells with gradient isolation of neutrophils, lysis, and snap freezing in liquid nitrogen. For GP analysis, red cells must be hemolyzed and stored frozen.

Storage Instructions Store specimens, plasma or cells, after processing (lysis) at -80°C.

Reference Interval 120 pM (median of 115 control subjects) with interquartile range of 97-146 pM, men 213, women 184 pM (median levels)

Critical Values Patients with chest pain, 198 pM (median level); 119-394 pM (interquartile range)

Use Prediction (short-term risk) of pathologic (arteriosclerotic) coronary artery-based major adverse cardiac events, in particular, early risk of myocardial infarction

Limitations Increased with history of hyperlipidemia; contamination of plasma by leukocyte myeloperoxidase

Methodology Enzyme-linked immunosorbent assay (ELISA)

Additional Information Evidence has been presented that an initial plasma level of myeloperoxidase (MPO) predicts risk of adverse cardiovascular events at 30 days and at 6 months. Leukocytes, which are rich in MPO release this enzyme in relation to injury (erosion, fissure formation, hemorrhage, and rupture) occurring during the evolution of arteriosclerotic plaques. Of especial significance is the apparent ability of MPO level to predict risk of myocardial infarction in patients who do not have significant elevation of troponin T (patients without evidence of myocardial necrosis based on troponin T level). Median myeloperoxidase levels did not relate to smoking history, presence of diabetes mellitus, hypertension, or past history of myocardial infarction or coronary artery disease. A number of previous studies have delineated the role of MPO in the evolution of leukocyte macrophages to activated phagocytes and to foam cell mononuclears that populate arteriosclerotic/atherosclerotic plaques. A broad family of mediators of inflammation (eg, complement components, including the membrane attack complex, and C-reactive protein) are active in smooth muscle-like cells and macrophages in the irregularly thickened intima of arteriosclerotic plaques.

An inverse relationship between erythrocyte glutathione peroxidase 1 activity and risk of cardiovascular events (over a period of 4.7 years) has been noted. Patients with coronary artery disease and a low level of red-cell glutathione peroxidase-1 have an independent increased risk of adverse cardiac events.

References

Blankenberg S, Rupprecht HJ, Bickel C, et al, "Glutathione Peroxidase 1 Activity and Cardiovascular Events in Patients With Coronary Artery Disease," *N Engl J Med*, 2003, 349(17):1605-13.

(Continued)

Myeloperoxidase, Plasma (Continued)

Brennan ML, Penn MS, Van Lente F, et al, "Prognostic Value of Myeloperoxidase in Patients With Chest Pain," N Engl J Med, 2003, 349(17):1595-604.

Hazen SL, Hsu FF, and Heinecke JW, "p-Hydroxyphenylacetaldehyde Is the Major Porduct of L-Tyrosine Oxidation by Activated Human Phagocytes," J Biol Chem, 1996, 271(4):1861-7.

Podrez EA, Febbraio M, Sheibani N, et al, "Macrophage Scavenger Receptor CD36 Is the Major Receptor for LDL Modified by Monocyte-Generated Reactive Nitrogen Species," J Clin Invest, 2000, 105(8):1095-108.

Sugiyama S, Okada Y, Sukhova GK, et al, "Marcophage Myeloperoxidase Regulation by Granulo-cyte Macrophage Colony-Stimulating Factor in Human Atherosclerosis and Implications in Acute Coronary Syndromes," Am J Pathol, 2001, 158(3):879-91.

Yasojima K, Schwab C, McGeer EG, et al, "Generation of C-Reactive Protein and Complement Components in Atherosclerotic Plaques," Am J Pathol, 2001, 158(3):1039-51.

Zhang R, Brennan ML, Fu X, et al, "Association Between Myeloperoxidase Levels and Risk of Coronary Artery Disease," JAMA, 2001, 286(17):2136-42.

♦ **Mylepsin®** see Primidone, Serum or Plasma on page 1091

♦ **Myocardial Infarct Panel** see Cardiac Markers: Laboratory Assessment, Overview on page 343

♦ **Myochrysine®** see Gold, Serum on page 657

♦ **Myoglobin** see Immunoperoxidase Procedures on page 780

Myoglobin, Blood, Serum, or Plasma

Related Information

Carboxyhemoglobin, Blood on page 340
Cardiac Markers: Laboratory Assessment, Overview on page 343
Creatine Kinase MB and Other Isoenzymes, Serum on page 469
Creatine Kinase, Serum on page 470
Haptoglobin, Serum on page 667
Lactate Dehydrogenase, Serum on page 825
Muscle Biopsy on page 927
Myoglobin, Qualitative, Urine on page 941
Troponins, Serum on page 1278

Abstract Myoglobin is an oxygen-binding protein found in cardiac and striated muscle. A rapid increase in its concentration is a useful early marker for acute myocardial infarct (AMI).

Other applications for myoglobin are applied to urinary myoglobins (see Myoglobin, Qualitative, Urine on page 941).

Specimen Serum, plasma

Container Red top tube, green top (lithium heparin) tube

Sampling Time Myoglobin is best measured in patients whose chest pain began less than 6 hours before evaluation.

Storage Instructions Serum may be stored at 4°C for 2 years or frozen at -20°C.

Turnaround Time Usually 1 hour

Reference Interval 0-0.09 µg/mL; varies with method.

Use Serum myoglobin levels have their greatest utility as an early indicator of AMI, especially when used with cardiac troponins or CK-MB. After an AMI, the myoglobin returns rapidly to normal, while the troponins remain elevated (see Troponins, Serum on page 1278).

Serum myoglobin is rapidly cleared by the kidneys. It is increased in cocaine use, high voltage electrical accident, intramuscular injection, shock, cardiac surgery, and thrombolytic therapy, as well as myocardial infarction. Its **main advantage** is as a sensitive (99% to 100%) marker for **early myocardial injury** because it is released earlier from necrotic cells than the cardiac troponins and CK-MB, allowing for earlier detection of myocardial infarction. Levels rise as early as 1 hour after infarct and peak within 4-12 hours. Repeat myoglobin that has doubled within 1-2 hours after presentation, even if still within the normal range, may signify an acute myocardial infarct.

Urinary as well as serum myoglobin are used in the diagnosis of rhabdomyolysis. It is found with trauma, malignant hypothermia, ischemia, dermatomyositis, polymyositis, and muscular dystrophy.

Methodology Immunoassay, multiple formats; nephelometry

References

Alpert JS, Thygesen K, Antman E, et al, "Myocardial Infarction Redefined. A Consensus Document of the Joint European Society of Cardiology/American College of Cardiology Committee for the Redefinition of Myocardial Infarction," *J Am Coll Cardiol*, 2000, 36:959-69.

Kost GJ, Kirk JD, and Omand K, "A Strategy for the Use of Cardiac Injury Markers (Troponin I and T, Creatine Kinase-MB Mass and Isoforms, and Myoglobin) in the Diagnosis of Acute Myocardial Infarction," *Arch Pathol Lab Med*, 1998, 122(3):245-51.

Plebani M and Zaninotto M, "Diagnostic Strategies Using Myoglobin Measurement in Myocardial Infarction," *Clin Chim Acta*, 1998, 272(1):69-77.

Myoglobin, Qualitative, Urine

Related Information

Blood, Urine *on page 281*
Carboxyhemoglobin, Blood *on page 340*
Cocaine (Cocaine Metabolite), Qualitative, Urine or Hair *on page 427*
Coccidioidomycosis Serology *on page 428*
Creatine Kinase, Serum *on page 470*
Hemoglobin, Qualitative, Urine *on page 688*
Lactate Dehydrogenase Isoenzymes, Serum *on page 824*
Malaria Smear and Tests *on page 888*
Muscle Biopsy *on page 927*
Myoglobin, Blood, Serum, or Plasma *on page 940*
Troponins, Serum *on page 1278*

Synonyms Myoglobin Screen, Urine

Abstract Myoglobin appears in plasma following damage to cardiac or striated muscle. Myoglobin in plasma is rapidly cleared into urine, and has a plasma half-time of 1-3 hours. Release of large quantities of myoglobin into the circulation, especially in the presence of shock, can lead to acute renal failure (eg, in massive crush injury).

The finding of dark red to dark brown to tea-colored urine positive on reagent strip for blood, without red cells on microscopy, suggests the presence of myoglobin or hemoglobin.

Specimen Random urine

Container Clean, chemical-free, plastic (preferable) urine container

Storage Instructions Stable for 12 days in urine when the pH is adjusted to between 8.0 and 9.5; stable for 1 month in serum.

Reference Interval Negative (<5 ng/mL)

Use Assay for the presence of myoglobinuria is used to investigate myositis and other entities which damage muscle. Extensive injury to striated muscle is accompanied by high serum creatine kinase (CK), myoglobin, and may be accompanied by myoglobinuria. **Serum testing is recommended.** See table.

Causes of Myoglobinuria: Summary

Metabolic – impaired substrate utilization for energy metabolism	Enzyme deficiencies (LD and others), substrate deficiency, hypokalemia, hypophosphatemia, hypomagnesemia
Excessive muscle use	Severe / unaccustomed exercise, seizures, march hemoglobinuria with myoglobinuria
Hyperpyrexia	Heat stroke, exertional hyperthermia, hyperthermia associated with drug use (eg, cocaine), heat injury
Postinfections viral	Influenza A, herpes simplex, Epstein-Barr, Coxsackie, AIDS
bacterial	Fever and sepsis, clostridial with gangrene; *Legionella, Streptococcus sp, Francisella tularemia, Salmonella sp*
Primary muscle disease	Muscular dystrophy, McArdle disease, polymyositis, dermatomyositis, familial paroxysmal myoglobinuria, steroids and other drugs

(Continued)

Causes of Myoglobinuria: Summary *(Continued)*

Poisoning	
drug	Carbon monoxide, alcohol, barbiturate, cocaine, amphetamine, phencyclidine, neuroleptic malignant syndrome
animal	Hoff disease (fish poisoning), sea snake bite (*Enhydrina schistosa*), trichinosis
Ischemia	Myocardial infarction, vascular occlusion (thromboembolism or external vascular compression); infarction of large muscle, anterior tibial syndrome
Traumatic	Crush injury, wounds, surgical muscle trauma, beatings, high voltage or lightning electrical injury, electrocution, limb compression with prolonged immobilization due to sleep, anesthesia, or coma

Limitations Presence of hypochlorite or microbial peroxidase or other oxidizing contaminants may cause false-positive reactions. Presence of ascorbic acid (high concentrations) may decrease sensitivity.

Methodology Nephelometry, radial immunodiffusion, differential ultrafiltration, differential solubility

Additional Information Myoglobin, especially in very high concentrations, can cause acute renal failure. Contrariwise, renal failure from any cause can be associated with elevated plasma myoglobin.

The most common cause, however, of rhabdomyolysis related to use of abused drugs is limb compression during sleep or coma. Serum CK is usually normal with hemolysis, in which serum LD is generally increased with high LD_1. With circulating myoglobin, serum CK is usually very high, serum LD may be moderately increased, but it is LD_5 that is usually elevated.

Rhabdomyolysis with myoglobinuria as well as hemoglobinuria may complicate *Plasmodium falciparum* malaria.

References

Loun B, Astles R, Copeland KR, et al, "Adaptation of a Quantitative Immunoassay for Urine Myoglobin. Predictor in Detecting Renal Dysfunction," *Am J Clin Pathol*, 1996, 105(4):479-86.

Slater MS and Mullins RJ, "Rhabdomyolysis and Myoglobinuric Renal Failure in Trauma and Surgical Patients: A Review," *J Am Coll Surg*, 1998, 186(6):693-716.

♦ **Myoglobin Screen, Urine** *see* Myoglobin, Qualitative, Urine *on page 941*

♦ **Myosin** *see* Immunoperoxidase Procedures *on page 780*

Myotonic Dystrophy DNA Test

Related Information

Amniotic Fluid, Chromosome and Genetic Abnormality Analysis *on page 152*

Chorionic Villus Sampling, Chromosome and Genetic Abnormality Analysis *on page 400*

Creatine Kinase, Serum *on page 470*

Duchenne/Becker Muscular Dystrophy DNA Detection *on page 526*

Glucose, Fasting, Plasma *on page 643*

Muscle Biopsy *on page 927*

Polymerase Chain Reaction *on page 1069*

Synonyms DM1 Gene Mutation Analysis

Applies to Congenital Myotonic Dystrophy

Test Includes Determination of the number of CTG trinucleotide repeats in the DM1 (dystrophia myotonica 1) gene at chromosome 19 band 13.3

Abstract Myotonic dystrophy is the most common form of muscular dystrophy in adults, with a prevalence of 1 in 10-20,000. It is an autosomal dominant myotonic myopathy that can also affect the eyes, heart, gastrointestinal, endocrine, and central nervous systems. Weakness in extremities usually begins distally, progressing slowly, ultimately affecting proximal limb-girdle muscles. Normal sensory examination is characteristic. Affected, or genetically predisposed, individuals have one normal DM1 allele and one mutated allele. In virtually all cases, the mutation is an abnormal number of repeats of a naturally-occurring CTG trinucleotide motif in the DM1 gene. The DNA test provides a definitive diagnosis by measuring the number of CTG repeats in each DM1 allele. The number of CTG repeats is polymorphic in the normal population and varies from 5-37; normal alleles in this size range do not change upon transmission to offspring. The presence of 50 or more CTG repeats indicates that an individual has, or is genetically predisposed to develop, myotonic dystrophy. Alleles with ≥50 repeats can be unstable upon transmission; the number of repeats can increase or decrease, with increases being much more common. A greater number of CTG repeats correlates with an earlier age at onset and more severe disease, thus, in some families, the disease appears to be more severe in the most recent generation. Neonates with severe congenital myotonic dystrophy, characterized by hypotonia, muscle weakness, respiratory insufficiency, and developmental/mental delay, typically have DM1 alleles with about 1000 or more CTG repeats. Congenital myotonic dystrophy is inherited almost exclusively from an affected, or presymptomatic carrier, mother. The probability that a woman's abnormal DM1 allele will expand sufficiently to cause the congenital form of the disease in her child cannot be predicted, but the probability does appear to increase if the mother's DM1 allele has >300 CTG repeats and/or she has a previous child with congenital disease.

Patient Preparation For asymptomatic and prenatal testing, pretest counseling with a medical geneticist or genetic counselor is recommended to explain the meaning, benefits, and risks of testing.

Aftercare For asymptomatic and prenatal testing, post-test genetic counseling is recommended to explain the test result and its significance.

Specimen Whole blood (3-10 mL), amniotic fluid (10-20 mL), chorionic villus (3-5 mg wet weight), or one T25 flask of cultured amniocytes or chorionic villus

Container Collect anticoagulated blood in either a lavender top (EDTA) tube or yellow top (ACD) Vacutainer®. Avoid use of heparin anticoagulants, which can interfere with polymerase chain reactions and endonuclease restriction enzyme activity. Amniotic fluid and chorionic villus samples should be collected in a sterile manner and transferred to a sterile tube for transport.

Storage Instructions Blood samples can be stored and shipped at room temperature or refrigerated (4°C); do not freeze. Blood samples should be received in the testing laboratory within 4 days of the draw to ensure an adequate DNA yield. Amniotic fluid and chorionic villus samples should be maintained at room temperature and sent to the testing laboratory immediately after collection, preferably by overnight delivery, to ensure successful cell culture.

Causes for Rejection Frozen samples, whole blood in heparin anticoagulant (green top) or other inappropriate collection tube, amniotic fluid sample bloody, chorionic villus sample lacking viable chorionic villi

Turnaround Time 7-14 days

Special Instructions To provide an optimal test interpretation, the testing laboratory needs to know the clinical diagnosis, the family history or pedigree, and whether DNA testing has confirmed the clinical diagnosis of an affected family member.

Reference Interval Normal DM1 alleles have ≤37 CTG repeats. Abnormal mutant DM1 alleles have from 50 to >2000 CTG repeats. Rarely, DM1 alleles with 38-49 CTG repeats are detected; these are considered intermediate or premutation alleles. Individuals with premutation alleles are not known to develop myotonic dystrophy. Upon transmission, however, a premutation allele can expand to >50 CTG repeats and cause myotonic dystrophy in offspring.

Use To confirm or clarify a clinical diagnosis, to perform asymptomatic testing for adults at risk for myotonic dystrophy, and to perform prenatal testing for fetuses at risk for myotonic dystrophy. Asymptomatic children under the legal age should not be tested. Testing can be used to identify which progenitor in a family carries the DM1 mutation; this is important information for genetic counseling and the identification of other at-risk family members. DNA testing is part of the differential diagnosis of neonates with unexplained hypotonia, poor feeding, and/or respiratory difficulties, and of children with developmental delay and myopathic facies. This test is used occasionally to evaluate unexplained perioperative pulmonary complications, which occur at increased frequency in affected individuals, and idiopathic polyhydramnios, which can be associated with an affected fetus.

Limitations The number of CTG repeats cannot reliably predict disease onset, severity, or cognitive ability in an individual patient. This is because there are (Continued)

Myotonic Dystrophy DNA Test *(Continued)*

significant overlaps in the number of CTG repeats in congenital, childhood, adult and late onset cases. Prenatal testing cannot reliably predict whether a fetus will have the severe congenital form of the disease based on the number of CTG repeats in the DM1 gene of the fetus. This test will not detect rare cases of myotonic dystrophy-like disease (probably <1% of clinical cases) that are caused by mutations at the DM2 locus on chromosome 3.

Methodology Polymerase chain reaction (PCR). A segment of the DM1 gene is amplified from genomic DNA, using primers that flank the CTG region of the gene. After electrophoresis, the number of repeats is calculated from the product size. This reaction can detect both normal and mutated DM1 alleles with <200 CTG repeats. Alleles with a greater number of CTG repeats are generally detected by Southern blot and hybridization to a fragment of the DM1 gene. The fragment size is indicative of the number of CTG repeats.

Additional Information Every affected individual has inherited an abnormally expanded DM1 gene from one of their parents; new mutations are not known to occur in this gene. Myotonic dystrophy occurs worldwide, with a prevalence of about 1 in 10-20,000. The DM1 protein is a serine-threonine protein kinase whose normal functions are not known. The CTG trinucleotide repeat motif is located in the 3′ untranslated part of the DM1 gene; how CTG expansion affects DM1 gene expression and causes the disease is not clear. There is evidence that long CTG repeat tracts in DM1 affect the expression of several nearby genes; whether these genes contribute to the myotonic dystrophy phenotype remains to be determined. A listing of laboratories that perform myotonic dystrophy (DM1) DNA testing can be found at GeneTests® (see Websites).

Other characteristics of the disease include cataracts, testicular atrophy, intellectual impairment, difficulties with ventilation and hypoxia. Dysphagia, cardiac conduction defects, and endocrine problems occur. Serum CK is normal to moderately high.

References

Cobo AM, Poza JJ, and Martorell L, "Contribution of Molecular Analyses to the Estimation of the Risk of Congenital Myotonic Dystrophy," *J Med Genet*, 1995, 32(2):105-9.

Groenen P and Wieringa B, "Expanding Complexity in Myotonic Dystrophy," *Bioessays*, 1998, 20(11):901-12.

The International Myotonic Dystrophy Consortium (IDMC), "New Nomenclature and DNA Testing Guidelines for Myotonic Dystrophy Type 1 (DM1)," *Neurology*, 2000, 54(6):1218-21.

Internet Web Sites

www.genetests.org
www.geneclinics.org/profiles/myotonic-d

- ◆ **Mysoline®**   *see* Primidone, Serum or Plasma *on page 1091*
- ◆ **Na⁺**   *see* Sodium, Serum or Plasma *on page 1210*
- ◆ **N-Acetyl-β-Hexosaminidase A**   *see* Beta-Hexosaminidase, Serum, White Blood Cells *on page 255*
- ◆ **N-Acetyl Procainamide**   *see* Procainamide, Serum *on page 1092*
- ◆ **NAG**   *see* Beta-Hexosaminidase, Serum, White Blood Cells *on page 255*
- ◆ **β-NAG A**   *see* Beta-Hexosaminidase, Serum, White Blood Cells *on page 255*
- ◆ **Nagao Isoenzyme**   *see* Alkaline Phosphatase, Heat Stable, Serum *on page 125*
- ◆ **Nagao Isoenzyme**   *see* Alkaline Phosphatase Isoenzymes, Serum *on page 125*
- ◆ **NAIT**   *see* Platelet Antibodies *on page 1047*
- ◆ **Naloxone**   *see* Methadone, Serum or Urine *on page 900*
- ◆ **Naloxone**   *see* Morphine, Urine *on page 921*
- ◆ **Naloxone Stimulation Test**   *see* Adrenal Cortex: Laboratory Assessment Overview *on page 110*
- ◆ **Nantucket Fever Serological Test**   *see* Babesiosis Serology *on page 225*
- ◆ **NAPA**   *see* Procainamide, Serum *on page 1092*
- ◆ **Napamide®**   *see* Disopyramide, Serum or Plasma *on page 516*
- ◆ **Naproxen**   *see* Porphyrins, Quantitative, Urine *on page 1074*
- ◆ **Narcotics**   *see* Opiates, Qualitative, Urine *on page 974*
- ◆ **National Institute for Drug Abuse** *see page 63*
- ◆ **Nativelle®**   *see* Digitoxin, Serum *on page 512*

♦ **Natriuretic Peptide, Brain** *see* B-Type Natriuretic Peptide *on page 314*

♦ **Na, Urine** *see* Sodium, Urine *on page 1213*

♦ **NBT Dye Test** *see* Nitroblue Tetrazolium Test *on page 963*

n-Butanol Stability Test

Related Information
Heinz Body Stain *on page 669*
Hemoglobin, Unstable, Heat Labile Test *on page 689*
Hemoglobin, Unstable - Isopropanol Precipitation Test *on page 690*

Synonyms Unstable Hemoglobins

Applies to Heinz Bodies

Abstract A simple, low-cost test for unstable hemoglobins. The van der Waals bonds of the molecular structure of hemoglobin in n-butanol solvent (a relatively nonpolar environment as compared with water) are weakened and molecular stability is decreased. Under the test conditions, unstable hemoglobins precipitate.

Specimen Whole blood

Container Lavender top (EDTA) tube

Storage Instructions Fresh blood should be used.

Reference Interval Absence of formation of a precipitate at 120 minutes is a negative test for unstable hemoglobin.

Use Screening test for unstable hemoglobins

Limitations False-positive results may occur if the sample contains ≥10% of Hb F. Increased methemoglobin (which may occur with storage) may also cause a false-positive result.

Additional Information Hemoglobin that is significantly unstable develops marked precipitation at 90 minutes with flocculation at 2 hours. Slightly unstable hemoglobins (Hb E) develop diffuse precipitation, in some cases minimal, at 2 hours. See Hemoglobin, Unstable, Heat Labile Test *on page 689* and Hemoglobin, Unstable - Isopropanol Precipitation Test *on page 690* for additional comments about unstable hemoglobins.

References
Brozovic M and Henthorn J, "Investigation of Abnormal Haemoglobins and Thalassemia," *Practical Haematology*, 8th ed, Chapter 14, Dacie JV and Lewis SM, eds, New York, NY: Churchill Livingstone, 1995, 269-70.

Molchanova TP, "A New Screening Test for Unstable Hemoglobins Using n-Butanol and Red Blood Cells," *Hemoglobin*, 1993, 17(1):81-4.

♦ **n-DNA** *see* Anti-DNA *on page 173*

♦ **Near Patient Testing** *see* Point-of-Care Testing *on page 1065*

♦ **Neck Mass Aspiration** *see* Fine Needle Aspiration, Deep and Superficial Masses *on page 590*

♦ **Needle Biopsy Cytology** *see* Fine Needle Aspiration, Deep and Superficial Masses *on page 590*

♦ **Needle Biopsy of Liver** *see* Liver Biopsy *on page 864*

Neisseria gonorrhoeae Culture and Smear

Related Information
Bacterial Culture, Genital Specimen *on page 239*
Beta-Lactamase Test *on page 256*
Cervical/Vaginal Cytology *on page 376*
Chlamydia trachomatis Culture *on page 385*
Chlamydia trachomatis Direct Antigen Test *on page 387*
Chlamydia trachomatis Nucleic Acid Detection *on page 388*
Genital Culture for *Ureaplasma urealyticum* *on page 635*
Gram Stain *on page 658*
Herpes Simplex Virus Culture and Antigen Detection *on page 721*
HIV-1/HIV-2 Serology *on page 736*
Neisseria gonorrhoeae Nucleic Acid Detection *on page 947*
RPR *on page 1174*
Synovial Fluid Analysis *on page 1229*
VDRL, Serum or Cerebrospinal Fluid *on page 1303*

Test Includes Selective culture for *Neisseria gonorrhoeae*; Gram stain of specimen from normally sterile site and male urethral swabs

(Continued)

Neisseria gonorrhoeae Culture and Smear (Continued)

Abstract Gonorrhea is a sexually-transmitted disease which can be transmitted during birth. Most gonococcal infections are uncomplicated genital tract infections that can be treated with antimicrobial agents. Of concern is the large number of asymptomatic infections in both men and women. If untreated, gonococcal infection may develop into epididymitis, prostatitis, or urethral stricture. In women, untreated gonorrhea may cause pelvic inflammatory disease manifested as endometritis, salpingitis, pelvic peritonitis, or tubo-ovarian abscesses. A small percentage of patients (1% to 3%) will develop disseminated gonococcal infection.

Patient Preparation Preparation same as for clean catch urine. See Bacterial Culture, Urine *on page 246* for detailed information. *Neisseria gonorrhoeae* is very sensitive to lubricants and disinfectants.

Specimen Body fluid, discharge, pus, swab of genital lesions, urethral discharge (best when available for men); endocervix (best when available for female); throat swab, rectal swab; sediment of first 10 mL of centrifuged urine collected at least 2 hours after last micturition, or first few drops of urine voided into a sterile cup for "first voided urine specimen" for asymptomatic males, or first void overnight urine, centrifuged.

Container Swab with transport medium, a dacron or rayon swab should be used, cotton swabs may be used only if the the cotton is treated to neutralize toxicity, sterile container for tissue or pus. The best method for growth of *N. gonorrhoeae* is to directly plate the specimen on Transgrow, Jembec™, or Thayer-Martin medium immediately after collection.

Collection

Urethral discharge: Collect male urethral discharge by endourethral swab after stripping toward the orifice to express exudate.

Rectal swab: Collect anorectal specimens from the crypts just inside the anal ring. Direct visualization with anoscopy is useful. Insert the swab past the anal sphincter. Move the swab circumferentially around the anal crypts. Allow 15-30 seconds for organisms to adsorb onto the swab.

Prostatic fluid yields fewer positives than culture of urethral discharge.

Urethral or vaginal cultures are indicated from females when endocervical culture is not possible.

Urethra in women: Massage the urethra against the pubic symphysis to express discharge or use endourethral swab.

Vagina: Obtain the specimen from the vaginal vault. Allow 15-30 seconds for organisms to adsorb onto the swab.

Endocervical/cervical: Gently compress cervix between speculum blades to express any endocervical exudate. Swab in a circular pattern.

Bartholin gland: Express exudate from duct. Abscesses should be aspirated with needle and syringe.

Oropharyngeal and tonsillar specimens are obtained via swab, preferably under direct vision.

Specimens should be transported to the laboratory within 1 hour of collection or plated on appropriate media in the examination room.

Storage Instructions Specimen should not be refrigerated or exposed to a cold environment. Growth of the organism is less likely following refrigeration. If the specimen is directly inoculated on Thayer-Martin medium, it should be transported to the laboratory as soon as possible and placed directly in CO_2 incubator or candle jar.

Turnaround Time Gram stain results are usually available in less than 1 hour. Cultures usually require 48 hours for completion.

Reference Interval Gram stain: no intracellular gram-negative diplococci seen; no *Neisseria gonorrhoeae* isolated

Critical Values Positive culture for *Neisseria gonorrhoeae* during pregnancy, or in young children or neonates

Use Isolate and identify *Neisseria gonorrhoeae*; establish the diagnosis of gonorrhea

Limitations Cultures are usually screened only for *Neisseria gonorrhoeae*. Other organisms are usually not identified. Overgrowth by *Proteus* and yeast may make it impossible to rule out presence of *N. gonorrhoeae*. The vancomycin in Thayer-Martin media may inhibit some strains of *N. gonorrhoeae* and the trimethoprim in New York City media may inhibit the growth of other strains of *N. gonorrhoeae*.

Methodology Culture on selective medium, Thayer-Martin, or New York City (NYC). DNA probes, monoclonal antibodies, enzyme immunoassays (EIA), and chromogenic substrate assays are used as alternatives or adjuncts to culture in some laboratories.

Additional Information Gram stain smear has a high sensitivity in a symptomatic male with urethral discharge (95% to 99%). Endocervical Gram stain is of little value in the female as the sensitivity is lower (50%), and endemic normal flora have a similar morphologic appearance, causing false positives. The Gram stain smear will detect 75% of gonococcal conjunctivitis and 10% to 20% of gonococcal skin lesions. It is of no value in pharyngitis. Anal and throat cultures are recommended for individuals engaging in anal or oral sex. Cervical cultures have a sensitivity of 80% to 90%. Although demonstration of gram-negative diplococci in leukocytes in a urethral smear from a symptomatic male is presumptive evidence of gonorrhea and is sufficiently diagnostic to initiate therapy, culture confirmation should be considered if available.

Serologic tests for syphilis (VDRL, RPR), HIV, Cervical/Vaginal Cytology, and a diagnostic test for *Chlamydia* should be performed in patients suspected of having gonorrhea. As many as 45% of women with gonorrheal infection have chlamydial infection as well.

References
Angulo JM and Espinoza LR, "Gonococcal Arthritis," *Compr Ther*, 1999, 25(3):155-62.

Centers for Disease Control and Prevention, "1998 Guidelines for the Treatment of Sexually Transmitted Diseases," *MMWR Morb Mortal Wkly Report*, 1998, 47(RR1):59-63.

Centers for Disease Control and Prevention, "Gonorrhea - United States, 1998," *JAMA*, 2000, 284(2):173-4.

Gunn RA, Rolfs RT, Greenspan JR, et al, "The Changing Paradigm of Sexually Transmitted Disease Control in the Era of Managed Healthcare," *JAMA*, 1998, 279(9):680-4.

Ingram DL, Everett VD, Flick LAR, et al, "Vaginal Gonococcal Cultures in Sexual Abuse Evaluations: Evaluation of Selective Criteria for Preteenaged Girls," *Pediatrics*, 1997, 99(6):E8.

Olsen CC, Schwebke JR, Benjamin WH Jr, et al, "Comparison of Direct Inoculation and Copan Transport Systems for Isolation of *Neisseria gonorrhoeae* From Endocervical Specimens," *J Clin Microbiol*, 1999, 37(11):3583-5.

Internet Web Sites
www.cdc.gov/ncidod/dastlr/gcdir/gono.html

www.medinfo.ufl.edu/year2/mmid/bms5300/bugs/neigonor.html

Neisseria gonorrhoeae Nucleic Acid Detection

Related Information
Bacterial Culture, Genital Specimen *on page 239*
Chlamydia trachomatis Culture *on page 385*
Chlamydia trachomatis Direct Antigen Test *on page 387*
Chlamydia trachomatis Nucleic Acid Detection *on page 388*
Neisseria gonorrhoeae Culture and Smear *on page 945*

Test Includes Detection of *Neisseria gonorrhoeae* nucleic acid in clinical specimens

Abstract *Neisseria gonorrhoeae* is one of the most common sexually transmitted infections and commonly may be asymptomatic. Disease in women is associated with a high rate of tubal pregnancies, pelvic inflammatory disease and infertility. Molecular assays, either direct detection or after amplification, have greatly decreased the time for identification of *N. gonorrhoeae*. *N. gonorrhoeae* is fastidious, and optimal growth conditions must be maintained for its recovery. In recognition of common dual infections by *Chlamydia* and gonorrhea, commercial assays have been developed which permit detection of these organisms with a single swab or specimen.

Patient Preparation For urethral specimens, the patient should not have urinated for 1 hour prior to collection.

Specimen Either direct swabs of a potentially infected site or urine. A swab specimen may be collected from the genital/urinary tract of males or females. Commercial assays have also been approved for detection of gonorrhea from urine.
(Continued)

Neisseria gonorrhoeae Nucleic Acid Detection
(Continued)

Container Special collection and transport kits must be used with the appropriate commercial detection assay. A commercial kit typically contains a swab and transport media or device.

Collection Two swabs are provided in the typical commercial kit for use in females. The cervix or endocervix is first cleaned with one swab and the second swab is used to collect the specimen. The swab is inserted into the endocervical canal and rotated to collect epithelial cells from the infected site. The swab is then placed into transport media and sent to the laboratory. In males, the swab is inserted into the urethral meatus and rotated to collect epithelial cells. The swab is then placed into transport media and sent to the laboratory.

Use of urine in some of the amplification assays requires first voided urine. The patient should not have urinated for 2 hours prior to collection of urine for testing.

Storage Instructions Some specimens may be stored at room temperature or refrigerated and sent without freezing to the laboratory. Specimens collected for commercial molecular assays are typically stable for up to 1 week after collection. Urine specimens should be refrigerated immediately after collection.

Causes for Rejection Collection of a specimen from a nonapproved site, excessively bloody specimens (for some assays)

Reference Interval Negative for *Neisseria gonorrhoeae*

Use This test provides for the rapid detection of *N. gonorrhoeae.*

Limitations Gonococcal nucleic acid can be detected for up to 3 weeks after successful treatment; thus, do not use to determine cure. Results by molecular assays have not been accepted as evidence by the courts and legal system. For these purposes, a culture is required. Antimicrobial resistance in *N. gonorrhoeae* cannot be detected by the commercial probe assays. Although the rectum and the nasopharynx are also sites that may be infected by gonorrhea, swabs of these sites have not been approved for use with molecular assays.

Some of the amplification assays produce false-positive results due to the detection of nonpathogenic strains of *Neisseria* which are known to be normal human flora.

Methodology Several molecular assays have been approved or are under review by the Food and Drug Administration. Most of the assays utilize a nucleic amplification or probe amplification step, and the specific products are then detected using various immunochemical methods or chemiluminescence.

Additional Information Gonorrhea is one of the most important sexually transmitted diseases in the United States. Infection is characterized by acute urethritis in males and as cervicitis in females. The molecular probe assays provide several advantages over traditional culture assays or antibody-based tests. One of the most significant advantages is the ability to detect nonviable organisms, since only the presence of nucleic acids is necessary. This latter property also minimizes the need to rapidly transfer the specimen to culture media. The sensitivity of the test exceeds that for culture and other nonmolecular assays. Detection of organisms in urine provides an additional advantage over alternative methods, and may have a significant impact on public health attempts at limiting the spread of disease.

References
Brown TJ, Yen-Moore A, and Tyring SK, "An Overview of Sexually Transmitted Diseases. Part I," *J Am Acad Dermatol*, 1999, 41(4):511-32.

Farrell DJ, "Evaluation of AMPLICOR *Neisseria gonorrhoeae* PCR Using cppB Nested PCR and 16S rRNA PCR," *J Clin Microbiol*, 1999, 37(2):386-90.

Koumans EH, Johnson RE, Knapp JS, et al, "Laboratory Testing for *Neisseria gonorrhoeae* by Recently Introduced Nonculture Tests: A Performance Review With Clinical and Public Health Considerations," *Clin Infect Dis*, 1998, 27(5):1171-80.

Molodysky E, "Urethritis and Cervicitis," *Aust Fam Physician*, 1999, 28(4):333-8.

Internet Web Sites
www.cdc.gov/ncidod/dastlr/gcdir/gono.html
www.medinfo.ufl.edu/year2/mmid/bms5300/bugs/neigonor.html

♦ **Neisseria sp** *see* Bacterial Culture, Aerobes *on page 229*
♦ **Nembutal®** *see* Barbiturates, Quantitative, Serum or Plasma *on page 248*

- ◆ **Neonatal Alloimmune Thrombocytopenia** *see* Platelet Antibodies *on page 1047*
- ◆ **Neonatal Reference Ranges** *see page 11*
- ◆ **Neoral**® *see* Cyclosporine, Blood *on page 487*
- ◆ **Nephrolithiasis Analysis** *see* Kidney Stone Analysis *on page 820*
- ◆ **Nervocaine**® *see* Lidocaine, Serum or Plasma *on page 850*
- ◆ **Neuramate**® *see* Meprobamate, Serum *on page 896*
- ◆ **Neurofibrillary Tangles** *see* Apolipoprotein E, Plasma *on page 204*
- ◆ **Neuroglycopenia** *see* Cerebrospinal Fluid Glucose *on page 362*
- ◆ **Neuronal Thread Proteins** *see* AD7c Neural Thread Protein, CSF or Urine *on page 106*

Neuron-Specific Enolase, Serum

Synonyms NSE; Phosphopyruvate Hydratase

Applies to S-100, Serum

Abstract Neuron-specific enolase (NSE) is undergoing evaluation as a serum marker for neuroendocrine neoplasms and for cerebral injury.

Specimen Serum

Container Red top tube

Collection Must avoid hemolysis. Place blood on ice immediately after collecting.

Storage Instructions Centrifuge within 30-45 minutes. Maintain at 4°C and analyze same day or freeze at -70°C until assayed.

Causes for Rejection Specimen with hemolysis (RBCs contain γ-enolase). Hemolysis will cause false-positive results.

Turnaround Time This analysis is usually performed by a reference laboratory.

Reference Interval Not well established. Check with reference laboratory.

Serum:
- 0-30 ng/mL
- 0-12.5 ng/mL
- 7.1 ±3.6 ng/mL

CSF: 0-13 years: 0-4.8 ng/mL; values may be age- and sex-dependent

Use

Tumor marker: NSE has no established role in screening for, or early detection of, neoplastic disease. In a study of 770 patients with small cell lung carcinoma treated in 9 centers, NSE has been used, together with disease stage and performance status, in a formula to calculate a prognostic index. Investigators from the National Cancer Institute in Milan have found serum NSE useful as in treatment monitoring and prognosis assessment in patients with neuroblastoma.

Brain injury and dysfunction: Serum NSE is elevated in patients with traumatic brain injury and can be used to predict outcome. Likewise, serum NSE (S-NSE) is elevated in patients with stroke, but S-100, serum (a related protein) appears to correlate better with infarct size and outcome. The NSE level in CSF correlates with the duration and outcome of status epilepticus. The NSE level in amniotic fluid is reported to be a potentially useful marker of brain injury in neonates.

Methodology Immunoassays (multiple formats)

References

Ashwood ER, personal communication. July 2000.

Elimian A, Figueroa R, Verma U, et al, "Amniotic Fluid Neuron-Specific Enolase: A Role in Predicting Neonatal Neuronal Injury?" *Obstet Gynecol*, 1998, 92(4 Pt 1):546-50.

Jorgensen LG, Osterlind K, Gomm J, et al, "Serum Neuron Specific Enolase (S-NSE) and the Prognosis in Small Cell Lung Cancer (SCLS): A Combined Multivariable Analysis on Data From Nine Centers," *Br J Cancer*, 1996, 74:(4)463-7 (published erratum *Br J Cancer*, 1996, 74:2043).

Massaron S, Seregni E, Luksch R, et al, "Neuron Specific Enolase Evaluation in Patients With Neuroblastoma," *Tumor Biology*, 1998, 19:261-8.

Mayo Medical Laboratories, *Test Catalogue*, Rochester, MN, 1999, 394.

McKeating EG, Andrews PJ, and Mascia L, "Relationship of Neuron Specific Enolase and Protein S-100 Concentrations in Systemic and Jugular Venous Serum to Injury Severity and Outcome After Traumatic Brain Injury," *Acta Neurochirurgica*, 1998, 71(Suppl):117-9.

Nygaard O, Langbakk B, and Romner B, "Neuron-Specific Enolase Concentrations in Serum and Cerebrospinal Fluid in Patients With no Previous History of Neurological Disease," *Scand J Clin Lab Invest*, 1998, 58(3):183-6.

Rodriguez-Nunez A, Cid E, Eiris J, et al, "Neuron-Specific Enolase Levels in the Cerebrospinal Fluid of Neurologically Healthy Children," *Brain and Development*, 1999, 21(1):16-9.

♦ **Neurontin®** *see* Antiepileptic Drugs Overview *on page 176*

♦ **Neutropenia** *see* Antineutrophil Alloantibody and Autoantibody *on page 186*

♦ **Neutrophil Antibody** *see* Antineutrophil Alloantibody and Autoantibody *on page 186*

Neutrophils, Apheresis, Donation

Related Information

Donation, Blood *on page 521*

Neutrophils, Transfusion *on page 951*

Synonyms Granulocytes, Apheresis, Donation; Leukocytes, Apheresis

Test Includes As for regular blood donation. Many of the infectious disease tests, however, may not be completed prior to transfusion.

Abstract Granulocytes may be useful in septic patients with severe neutropenia ($<0.5 \times 10^9$/L) who have not responded to appropriate antibiotic therapy and who have a reasonable chance of marrow recovery.

Patient Preparation The more granulocytes collected, the more effective the granulocyte transfusions (GTX). To increase granulocyte yields, donors may be stimulated with corticosteroids and granulocyte colony-stimulating factor (G-CSF). Donors should be questioned regarding history of hypertension, diabetes, or peptic ulcer disease before beginning corticosteroid stimulation. Use of growth factors for allogeneic granulocyte donation is not yet approved by the FDA. Thus, corticosteroids alone may be given to a donor to increase granulocyte yields.

Aftercare Donors receiving G-CSF experience bone pain, myalgia, headaches, and nausea/vomiting, which usually respond to acetaminophen. Headaches or peripheral edema from an increased circulatory volume may also occur secondary to the sedimenting agent used during the apheresis procedure.

Specimen Donor granulocytes including therapeutic doses of platelets

Collection A granulocyte unit should contain a minimum of 1×10^{10} granulocytes per transfusion. Thus, most collections are usually prepared from a single donor using an apheresis instrument. They may also be prepared as a "buffy coat" from a single unit of fresh whole blood for a neonatal transfusion.

Selection of donors who are both red cell and leukocyte compatible is important.

Storage Instructions Store granulocytes at 20°C to 24°C for up to 24 hours without agitation. Granulocyte concentrates should be transfused as soon as possible after collection.

Causes for Rejection As for regular blood donation.

Special Instructions Donors selected for this procedure are often family members. ABO and Rh compatibility are desirable; HLA compatibility is desirable in the case of alloimmunized recipients but is seldom practical.

Use Severely neutropenic subjects with life-threatening infections should be considered for GTX provided in adequate doses. Septic neonatal patients may also benefit from granulocyte transfusions. The importance of high patient doses of PMNs ($\geq 1.7 \times 10^{10}$/day) is stressed.

Contraindications Donors with intolerance to stimulating or sedimenting agents or donor reactions during apheresis procedure (see Platelets, Apheresis, Donation *on page 1054* for a table listing donor reactions specific to leukapheresis).

Methodology Centrifugation leukapheresis

Additional Information ABO and Rh compatibility are desirable; if more than 2 mL of red cells are present in the product, the component should be crossmatched. Many platelets are present in a granulocyte concentrate, which is beneficial because most neutropenic patients are also thrombocytopenic. Irradiate prior to infusing into immunoincompetent recipients. Do **not** administer through a leukocyte-reduction filter.

References

Strauss RG, "Principles of Neutrophil (Granulocyte) Transfusions," *Hematology: Basic Principles and Practice*, 3rd ed, Chapter 137, Hoffman R, Benz EJ Jr, Shattil SJ, et al, eds, New York, NY: Churchill Livingstone, 2000, 2257-63.

Triulzi DJ, *Blood Transfusion Therapy: A Physician's Handbook*, 7th ed, Bethesda, MD: American Association of Blood Banks Press, 2002, 22-4.

Vamvakas EC and Pineda AA, "Meta-analysis of Clinical Studies of the Efficacy of Granulocyte Transfusions in the Treatment of Bacterial Sepsis," *J Clin Apheresis*, 1996, 11(1):1-9.

Neutrophils, Transfusion

Related Information

Irradiated Blood Components *on page 811*
Neutrophils, Apheresis, Donation *on page 950*

Synonyms Granulocytes, Transfusion; Leukocyte Concentrate

Applies to Granulocyte-Colony-Stimulating Factor

Test Includes ABO and Rh type, antibody screen, and crossmatch; antibody identification if indicated

Abstract Granulocyte transfusions (GTX) may be useful in septic patients with severe neutropenia (<0.5 x 10^9/L) as an adjunct to antimicrobial therapy, in treatment and/or possible prevention of infections. Severe neutropenia and disordered PMN function may lead to infections with bacteria, yeasts, and fungi. In neonates, WBC as high as 3.0 x 10^9/L may prompt candidacy for GTX, but GTX application in neonates is controversial.

Patient Preparation Give daily for a minimum of 4 days to demonstrate clinical benefit. Granulocytes obtained from granulocyte-colony-stimulating-factor (G-CSF) stimulated donors may allow for every-other-day transfusion. However, use of growth factors for allogeneic granulocyte donation is not yet approved by the FDA.

Dosage and administration: Administer through a standard blood filter. Daily infusion of 4-8 x 10^{10} PMNs is advocated for patients with severe persistent neutropenia and infections which have not responded to reasonable courses of antibiotic therapy. Slow the infusion rate or give an antihistamine, antipyretic, or steroid for the fever, chills, or allergic reactions that may occur with granulocyte infusions. Gamma irradiate the granulocyte component to prevent graft-vs-host disease if administering to an immunoincompetent recipient. Do not administer in conjunction with amphotericin, as severe pulmonary reactions may occur.

Specimen Blood from recipient and donor

Container One red top tube or one lavender top (EDTA) tube

Collection (Of sample from intended recipient): As for other red-cell-containing blood components.

Storage Instructions Storage should be at 20°C to 24°C, without agitation, for a maximum of 24 hours.

Causes for Rejection (Of patient sample): Gross hemolysis, sample placed in a serum separator tube, specimen tube not properly labeled

Special Instructions Expiration date is 24 hours. Transfuse granulocytes as soon as possible after collection.

Use Granulocytes in adequate doses may be useful in therapy of serious infections with bacteria, yeast, or fungi in patients with severe neutropenia (<0.5 x 10^9/L) as an adjunct to antimicrobial drugs, in subjects who have a reasonable chance of marrow recovery. Indications have included bacterial septicemia, pneumonia, serious localized infections, fever of unknown origin, and possibly invasive fungal and yeast infections. Especially, septic patients with persistent neutropenia caused by continuing marrow failure may be helped by GTX in adequate doses with antibiotic therapy. Septic neonatal patients may also benefit from granulocyte transfusions.

Limitations There is a risk of viral transmission, particularly cytomegalovirus (CMV). Select appropriate CMV seronegative donors, if needed, since leukocyte reduction filters may not be used with these components to reduce the risk of CMV transmissions (see Filters for Blood *on page 588*). Granulocyte transfusions will not be of benefit to the patient whose bone marrow is unlikely to recover. They are expensive. GTX are recommended only for progressive infections uncontrolled with antimicrobials.

Risk of alloimmunization to HLA and red cell antigens exists.

Contraindications Not indicated for infections that can be managed successfully with antibiotics.

Methodology Leukocyte crossmatching. See Neutrophils, Apheresis, Donation *on page 950*.

Additional Information Normal production of granulocytes is about 1 x 10^{11}/day in an adult. Only 5% to 10% of the PMN pool is in circulation.

ABO and Rh compatibility are desirable; if more than 2 mL of red cells are present in the fraction, the component should be crossmatched. Many platelets
(Continued)

Neutrophils, Transfusion (Continued)

are present in a granulocyte concentrate, which is beneficial because most neutropenic patients are also thrombocytopenic.

References

Adkins D, Spitzer G, Johnston M, et al, "Transfusions of Granulocyte-Colony-Stimulating-Factor Mobilized Granulocyte Components to Allogeneic Transplant Recipients: Analysis of Kinetics and Factors Determining Post-transfusion Neutrophil and Platelet Counts," *Transfusion*, 1997, 37(7):737-48.

Strauss RG, "Principles of Neutrophil (Granulocyte) Transfusions," *Hematology: Basic Principles and Practice*, 3rd ed, Hoffman R, Benz EJ Jr, Shattil SJ, et al, eds, New York, NY: Churchill Livingstone, 2000, 2257-63.

♦ **Newborn Aneuploidy Detection** see Gene Rearrangement for Leukemia and Lymphoma on page 633

Newborn Crossmatch and Transfusion

Related Information

Bilirubin, Amniotic Fluid, Delta A450 on page 261
Bilirubin, Neonatal, Serum on page 263
Cord Blood Antibody Screen on page 453
Cytomegalovirus Serology on page 499
Hemolytic Disease of the Newborn, Antibody Identification on page 690
Irradiated Blood Components on page 811
Rh$_o$(D) Immune Globulin (Human) on page 1164
Rosette Test for Fetomaternal Hemorrhage on page 1172
Warming, Blood on page 1326

Synonyms Exchange Transfusion; Type and Crossmatch for Exchange Transfusion of Newborn

Applies to Cytomegalovirus Low Risk Blood; Hemolytic Disease of the Newborn, Crossmatch; Satellite Bags

Test Includes For exchange transfusion or routine transfusion, the unit should be compatible with the mother's ABO group and Rh type, and with any unexpected blood group antibodies present in mother's serum. Anti-A and anti-B may not correspond to the newborn's ABO blood group. For routine transfusions of nongroup O red cells, include an antiglobulin test when testing for the presence of anti-A and/or anti-B. If the initial red cell antibody screen is negative, it is unnecessary to crossmatch donor red cells for the initial or subsequent transfusion and repeat testing may be omitted for an infant younger than 4 months of age.

Abstract See Hemolytic Disease of the Newborn, Antibody Identification on page 690, Bilirubin, Neonatal, Serum on page 263, and Bilirubin, Amniotic Fluid, Delta A450 on page 261.

Patient Preparation For exchange transfusion, blood should be passed through a warming device to raise the temperature of the blood to about 37°C during administration. Transfusion-associated graft-vs-host disease (TA-GVHD) has been reported in neonates who have received intrauterine transfusions (with or without subsequent exchange transfusions); neonates with known or suspected T-cell immune deficiencies; and, rarely, in term infants. Irradiation of blood components prevents TA-GVHD. Directed donor blood from blood relatives and HLA-matched platelets must be irradiated. Neonates weighing less than 1200 grams should receive CMV-reduced-risk units; vide infra. Consider selection of blood known to lack hemoglobin S for exchange transfusion. For smaller, nonexchange transfusions, typically administer group O, Rh-specific blood, 10-15 mL/kg, over 2-3 hours. The type of red cell storage medium is believed not to pose a risk to the neonate.

Aftercare Fatal cardiac arrhythmias have been reported during exchange transfusion secondary to hyperkalemia. Irradiation increases red cell potassium leakage. Irradiate blood as close to the time of transfusion as possible if large quantities of irradiated blood will be transfused. The quantity of potassium administered in routine (10-15 mL/kg) transfusions is comparatively insignificant. Hypocalcemia may be avoided by measurement of postexchange ionized calcium. It may be necessary to determine drug levels or repeat doses after exchange transfusion.

Specimen Blood from mother and infant

Container One red top tube or one lavender top (EDTA) tube from mother; appropriate pediatric tubes from newborn

Causes for Rejection (Of patient sample): Gross hemolysis, sample placed in a serum separator tube, specimen tube not properly labeled

Special Instructions Advance notice permits collection of appropriate donor blood into a bag with multiple attached satellite bags. This permits multiple small transfusions to be given to the infant from the same donor (ie, reduces donor exposure and reduces the risk of transfusion-transmitted infectious disease). Additional satellite bags may also be attached by means of a sterile-connecting device. Although many feel that washed red cells should be limited for low-volume transfusions in neonates to those, for example, with T-activation or in renal failure who would be harmed by the potassium load (see Red Blood Cells, Washed *on page 1143*), others prefer washed red cells or fresh (<5 days old) red cells relevant to potassium load. When needed, ABO-compatible plasma-containing fractions (eg, platelets, fresh-frozen plasma) are desirable for the neonate. If unavailable, platelets may be centrifuged and plasma removed.

Use Hemolytic disease of the newborn is due to transplacental passage of maternal antibodies - ABO, Rh (D, C, c), Kell, Duffy, Kidd, or other blood group system antibodies which are directed against antigens expressed on the neonate's red cells.

Immediate exchange laboratory criteria for term infants include significant anemia and rapidly increasing hyperbilirubinemia; see Bilirubin, Neonatal, Serum *on page 263*. The second indication for exchange is the presence of congestive heart failure secondary to severe anemia. A classic indication for exchange transfusion in full-term infants is an indirect bilirubin level ≥20 mg/dL. At this level, brain damage may occur. In premature babies or those with other complications, brain damage may occur at lower levels of bilirubin. An exchange transfusion may then be appropriate at levels <20 mg/dL. Severe bilirubinemia may also be seen in multiorgan failure. Consult pediatric literature for more detailed information.

Neonatal Red Blood Cell Transfusion Guidelines[1]

Transfuse with ≤20 mL/kg (not to exceed hematocrit of 0.45 or hemoglobin of 15 g/dL)
1. Hct ≤0.20 or Hb ≤7 g/dL and reticulocyte count <4% (or absolute <1,000,000/mL)
2. Hct ≤0.25 or Hb ≤8 g/dL and any of the following conditions:
a. Episodes of apnea/bradycardia ≥10 episodes/24 hours or ≥2 episodes requiring bag-mask ventilation
b. Sustained tachycardia (>180 beats/minute) or sustained tachypnea (>80 breaths/minute) (x 24 hours by averaging every 3-hour measurements)
c. Cessation of adequate weight gain x 4 days (≤10 g/24 hours despite ≥420 kJ/kg/24 hours)
d. Mild RDS + FiO$_2$ 0.25-0.35 or nasal cannula 1/8-1/4 L/minute or IMV or NCPAP with Paw <6 cm H$_2$O.
3. Hct ≤0.30 or Hb ≤10 g/dL with moderate RDS + FiO$_2$ >35% or nasal cannula O$_2$ or intermittent mandatory ventilation with Paw 6-8 cm + H$_2$O.
4. Hct ≤0.35 or Hb ≤12 g/dL with severe RDS requiring mechanical ventilation and Paw >8 cm H$_2$O and FiO$_2$ >50%, or severe congenital heart disease associated with cyanosis or heart failure.
5. Acute blood loss with shock: blood replacement to reestablish adequate blood volume and Hct of 0.40.
6. CMV-seronegative or leukocyte-reduced blood advocated in preterm infants and immunodeficient babies.
7. Irradiated blood components recommended, especially in preterm infants, those with possible immunodeficiency and possible transplant candidates, and those who are receiving intrauterine transfusions.

[1]RDS = respiratory distress syndrome; FiO$_2$ = inspired oxygen content; IMV = intermittent mandatory ventilation; NCPAP = nasal continuous positive airway pressure; Paw = mean airway pressure

Adapted from Simon TL, Alverson DC, AuBuchon J, et al, "Practice Parameter for the Use of Red Blood Cell Transfusions," *Arch Pathol Lab Med*, 1998, 122(2):130-8.

(Continued)

Newborn Crossmatch and Transfusion *(Continued)*

The blood volume of neonates is 85 mL/kg for full-term babies and 100-105 mL/kg for preterm infants. The anemia of prematurity responds to recombinant human erythropoietin.

Cytomegalovirus (CMV)-seronegative or leukocyte-reduced blood is indicated, and irradiated blood components are recommended.

Limitations Relatively mild jaundice beginning 1-2 weeks after delivery with a weakly positive direct antiglobulin test in a baby of group A or B and a mother of group O usually indicates ABO hemolytic disease of the newborn. An eluate in these instances (group A or B infant born to a group O mother) is of no assistance in prediction of cases of hemolytic disease of the newborn. Exchange transfusion is seldom necessary in ABO hemolytic disease of the newborn.

Methodology Blood less than 7 days old is usually provided for an exchange transfusion. This component will likely be reconstituted whole blood (hemoglobin S-negative red cells reconstituted to a hematocrit of 50% to 60% with group AB FFP), which is CMV-safe (either leukocyte-reduced or from a CMV-seronegative donor) and irradiated.

Additional Information Red cell transfusions are not infrequently given to premature infants weighing less than 1300 g for anemia of prematurity and for loss from repeated blood sampling.

Use of blood components with low risk for CMV infection is advocated for intrauterine transfusion and transfusion in neonates with birth weight <1200 g, unless the mother is CMV antibody positive.

Blood components at low risk for CMV include those from CMV-negative donors. See Red Blood Cells, Leukocytes Reduced *on page 1141.*

References
Chambers LA and Luban NC, "Neonatal and Intrauterine Transfusion," *Transfusion Therapy: Clinical Principles and Practice*, Mintz PD, ed, Bethesda, MD: American Association of Blood Banks Press, 1999, 299-311.

Simon TL, Alverson DC, AuBuchon J, et al, "Practice Parameter for the Use of Red Blood Cell Transfusions: Developed by the Red Blood Cell Administration Practice Guideline Development Task Force of the College of American Pathologists," *Arch Pathol Lab Med*, 1998, 122(2):130-8.

Smith DM and Shoos Lipton K, "Leukocyte Reduction for the Prevention of Transfusion-Transmitted Cytomegalovirus (TT-CMV)," Association Bulletin #97-2, Bethesda, MD: American Association of Blood Banks Press, 1997, 1-15.

♦ **Newborn/Maternal Antibody Work-up** *see* Hemolytic Disease of the Newborn, Antibody Identification *on page 690*

Newborn Screen for Phenylketonuria

Related Information
Amino Acids, Plasma *on page 143*
Newborn Screen for T$_4$, Filter Paper *on page 956*
Newborn Screen for TSH, Filter Paper *on page 957*
Phenylalanine, Blood *on page 1022*

Synonyms Phenylketonuria, Newborn Screen; PKU, Neonatal

Applies to BH$_4$ Cofactor Deficiencies; Dihydropteridine Reductase; Guanosine Triphosphate Cyclohydrolase 1 Deficiency; Hyperphenylalaninemia, Maternal; Phenylalanine Hydroxylase Activity; Pyruvoyl Tetrahydropterin Synthase Deficiency; Tetrahydrobiopterin Pathway

Test Includes Phenylalanine screen using filter paper collection. Samples for newborn screens for congenital hypothyroidism, galactosemia, and other diseases (varies among different states and jurisdictions) are usually obtained at the same time.

Abstract Autosomal recessive aminoacidopathy due to deficiency of phenylalanine hydroxylase (≥97% cases) or biopterin (folic acid constituent). Detection by dried blood spot screening results in early treatment, intended to prevent mental retardation.

Patient Preparation Newborns should have milk (protein) feeding ideally for 48 hours before testing; sample should be taken as late as possible prior to discharge from hospital (see Use).

Specimen Whole blood, serum or plasma

Container State-approved papercard

Collection NCCLS guidelines (NCCLS, 1992) and instructions on blood collection form

Storage Instructions Follow instructions provided; avoid environmental extremes.

Causes for Rejection Instructions on blood collection form not followed

Reference Interval

Screening cutoffs:
- >4 mg/dL in 40 states and jurisdictions
- >3 mg/dL in 6 states
- >2 mg/dL in 7 states

Confirmation: Phenylalanine:
- >10 mg/dL suggests classic variant PKU
- ≤10 mg/dL suggests milder variant
- ≥6 mg/dL - consider for dietary restriction of phenylalanine

Also test for: serum tyrosine, urinary pteridines, and blood dihydropteridine;
To exclude cofactor variants that require additional individualized therapy.

Use There is a tradeoff between screening time and appropriate cutoff: the earlier the screening, the lower the cutoff should be. Thus, if screening is done during the first 24 hours of life, the cutoff of 2 mg/dL is appropriate, although a significant number of false positives will occur (American Academy of Pediatrics 1996; updated information at www.aap.org). The false negatives at various cutoffs are included in the following table.

Collection Time	Cutoff	False Negatives
During first 12 hours	4 mg/dL	33%
	2 mg/dL	2.6%
At 24 hours	4 mg/dL	10%
	2 mg/dL	<1%
24-48 hours	4 mg/dL	2.4%
>48 hours	4 mg/dL	0.15%

In actual practice, infants screened before 24 hours of age are rescreened at some time during the first two weeks of life to identify early screening failures.

Not all babies with elevated blood phenylalanine have classical PKU. There are several PKU subtypes and optimal treatment depends on accurate subclassification, typically involving referral to a physician with expertise in genetics.

Screening must be integrated with follow-up, confirmation of diagnosis, and treatment. Problems have included failure to screen all neonates and imperfect compliance with follow-up screening.

Methodology Guthrie bacterial inhibition assay (BIA); fluorometry

Additional Information Successful detection of phenylketonuria prevents central nervous system damage due to excessive levels of phenylalanine. Cofactor variants must be correctly diagnosed to provide appropriate therapy. Once identified, harmful CNS effects can be largely avoided by dietary measures, notably a semisynthetic diet low in phenylalanine in the usual cases. In young PKU patients, the tolerance for dietary phenylalanine (to maintain nontoxic plasma levels) is about 250-550 mg/day. Widespread institution of PKU screening programs, worldwide, is an outstanding public health triumph of the 20th century. Incidence is 1:10,000 to 1:15,000 in the United States. For African-Americans in Maryland, the reported incidence is 1:50,000.

Intrauterine fetal injury results from exposure of the developing fetus to increased intrapartum maternal plasma phenylalanine levels. There is a high incidence of resultant fetal damage including microcephaly, intrauterine growth retardation, mental retardation, and congenital heart disease, as a result of **maternal hyperphenylalaninemia**. Such individuals should receive genetic counselling before becoming pregnant. Dietary management of mothers identified by newborn screening programs has as its goal the maintenance of near normal maternal phenylalanine levels throughout pregnancy.

References

American Academy of Pediatrics, Committee on Genetics, "Newborn Screening Fact Sheets," *Pediatrics*, 1996, 98:473-501.

Gilbert-Barness E and Barness LA, *Metabolic Diseases: Foundations of Clinical Management, Genetics, and Pathology*, Natick, MA: Easton Publishing, 2000, 27-31.

(Continued)

Newborn Screen for Phenylketonuria (Continued)

Levy HL, "Phenylketonuria: Old Disease, New Approach to Treatment," *Proc Natl Acad Sci U S A*, 1999, 96(5):1811-3.

National Committee for Clinical Laboratory Standards, "Blood Collection on Filter Paper for Neonatal Screening Programs: Approved Standard," Villanova, PA: National Committee for Clinical Laboratory Standards, 1992, NCCLS Publication:LA4-A2.

Rezvani I, "Defects in Metabolism of Amino Acids," Behrman RE, Kliegman RM, and Jenson HB, eds, *Nelson Textbook of Pediatrics*, 16th ed, Philadelphia, PA: WB Saunders Co, 2000, 344-7.

Scriver CR and Waters PJ, "Monogenic Traits Are Not Simple: Lessons From Phenylketonuria," *Trends Genet*, 1999, 15(7):267-72.

Waisbren SE, Hanley W, and Levy HL, "Outcome at Age 4 Years in Offspring of Women With Maternal Phenylketonuria: The Maternal PKU Collaborative Study," *JAMA*, 2000, 283(6):756-62.

Internet Web Sites
www.aap.org

Newborn Screen for T₄, Filter Paper

Related Information

Newborn Screen for Phenylketonuria *on page 954*
Newborn Screen for TSH, Filter Paper *on page 957*
Newborn Screening Tests for Galactosemia *on page 960*
Phenylalanine, Blood *on page 1022*
Thyroxine, Free, Serum *on page 1256*
Thyroxine, Serum *on page 1257*

Synonyms T₄ Neonatal; Thyroid Screen for Newborns

Abstract Untreated newborns with congenital hypothyroidism suffer from irreversible mental retardation and physical deformities. Successful outcome from early treatment makes this disorder an excellent candidate for newborn screening. Screening for congenital hypothyroidism now occurs in all 50 states of the U.S., in most of Europe, and in many other parts of the world.

In the U.S. the primary screen is with blood thyroxine (T₄); specimens with results in the lowest 5% to 10% are then assayed for thyrotropin (TSH). In Europe and Japan, the primary screen is often with TSH. Assay of both hormones is optimal.

Specimen Whole blood collected on newborn screening filter paper

Container Special state-(or jurisdiction-) approved filter paper collection card

Sampling Time Just before infant is discharged from the nursery, and at least 12 hours old

Collection Follow instructions on blood collection form (NCCLS, 1992).

Causes for Rejection Failure to follow instructions on blood collection form

Special Instructions The T₄ specimen is usually collected at the same time the PKU specimen is obtained. Optimal collection time is 3-7 days after birth, 4-10 days after birth is recommended for low birth weight infants.

Reference Interval

Newborn screen:
- Thyroxine: 6-22 µg/dL (SI: 80-283 nmol/L)
- TSH:
    <1 week: 0.7-27.0 mIU/L
    2-20 weeks: 1.7-9.1 mIU/L

Many jurisdictions will use local population-based decision levels.

Use Screen for congenital hypothyroidism. Updated information available at: www.aap.org

Limitations Congenital thyroxine binding globulin deficiency, an X-linked trait, will result in low total T₄ values even though the patient is euthyroid. These patients have normal free thyroxine and TSH. Severe acute illnesses (respiratory distress syndrome, sepsis) cause false-positive results.

Methodology Radioimmunoassay (RIA), enzyme immunoassay (EIA), fluorescent immunoassay

Additional Information Congenital hypothyroidism is a common preventable cause of mental retardation. The overall incidence is approximately 1 in 4000 in the Caucasian population; females are affected about twice as often as males. Persons from African descent have a lower incidence (1 in 32,000). Patients who do not receive early therapy will develop mental retardation, variable growth failure, metabolic changes of hypothyroidism, deafness and neurologic abnormalities.

Hypothyroidism in pregnant women adversely affects the fetus, therefore, screening for thyroid deficiency during pregnancy may be warranted.

References

American Academy of Pediatrics. Committee on Genetics, "Newborn Screening Fact Sheets," *Pediatrics*, 1996, 98(3 Pt 1):473-501.

American Academy of Pediatrics. Section on Endocrinology and Committee on Genetics, and American Thyroid Association Committee on Public Health, "Newborn Screening for Congenital Hypothyroidism: Recommended Guidelines," *Pediatrics*, 1993, 91:1203-9.

Dugbartey AT, "Neurocognitive Aspects of Hypothyroidism," *Arch Intern Med*, 1998, 158(13):1413-8.

Haddow JE, Palomaki GE, Allan WC, et al, "Maternal Thyroid Deficiency During Pregnancy and Subsequent Neuropsychological Development of the Child," *N Engl J Med*, 1999, 341(8):549-55.

National Committee for Clinical Laboratory Standards, "Blood Collection on Filter Paper for Neonatal Screening Programs: Approved Standard," Villanova, PA: National Committee for Clinical Laboratory Standards, 1992, NCCLS Publication:LA4-A2.

Nicholson JF, et al, "Reference Ranges for Laboratory Tests and Procedures," Behrman RE, Kliegman RM, and Jenson HB, eds, *Nelson Textbook of Pediatrics*, 16th ed, Philadelphia, PA: WB Saunders Co, 2000, 2209-10.

Stoddard JJ and Farrell PM, "State-to-State Variations in Newborn Screening Policies," *Arch Pediatr Adolesc Med*, 1997, 151(6):561-4.

Internet Web Sites

www.aap.org

Newborn Screen for TSH, Filter Paper

Related Information

Newborn Screen for Phenylketonuria *on page 954*
Newborn Screen for T$_4$, Filter Paper *on page 956*
Newborn Screening Tests for Galactosemia *on page 960*
Phenylalanine, Blood *on page 1022*
Thyroxine, Free, Serum *on page 1256*
Thyroxine, Serum *on page 1257*

Synonyms Thyroid Stimulating Hormone Screen, Filter Paper; TSH, Filter Paper

Applies to Thyroid Screen for Newborns

Abstract See Newborn Screen for T$_4$, Filter Paper *on page 956*.

Limitations Although an elevated TSH level is a better indicator of hypothyroidism than a decreased level of T$_4$, the TSH surge at birth can cause false-positive results if the sample is collected too early. When TSH is used as the primary screen, cases of secondary or tertiary hypothyroidism are not detected.

References

Hannon WH and Therrell BL Jr, "Laboratory Methods for Detecting Congenital Hypothyroidism," *Laboratory Methods for Neonatal Screening*, Therrell BL Jr, ed, Washington, DC: American Public Health Association, 1993, 139-54.

Kopp P, van Sande J, Parma J, et al, "Brief Report: Congenital Hyperthyroidism Caused by a Mutation in the Thyrotropin-Receptor Gene," *N Engl J Med*, 1995, 332(3):150-4.

Morreale de Escobar G and Escobar del Rey F, "Maternal Thyroid Deficiency During Pregnancy and Subsequent Neuropsychological Development of the Child," *N Engl J Med*, 1999, 341(26):2015-6.

Rovet JF, "Congenital Hypothyroidism: Long-Term Outcome," *Thyroid*, 1999, 9(7):741-8.

Wang ST, Pizzolato S, and Demshar HP, "Diagnostic Effectiveness of TSH Screening and of T$_4$ With Secondary TSH Screening for Newborn Congenital Hypothyroidism," *Clin Chim Acta*, 1998, 274(2):151-8.

♦ **Newborn Screening by MS/MS** *see* Newborn Screening by Tandem Mass Spectrometry (MS/MS) *on page 957*

Newborn Screening by Tandem Mass Spectrometry (MS/MS)

Related Information

Adrenal Cortex: Laboratory Assessment Overview *on page 110*
Amino Acids, Plasma *on page 143*
Ammonia, Plasma *on page 150*
Anion Gap, Serum, Plasma, or Urine *on page 160*
Blood Gases and pH, Arterial *on page 275*
Homocyst(e)ine, Plasma *on page 741*
17-Hydroxyprogesterone, Whole Blood, Serum, or Plasma *on page 755*
Inherited Diseases of Metabolism and Cell Structure *on page 792*
Methylmalonic Acid, Serum, Plasma, Urine, or Amniotic Fluid *on page 908*
Newborn Screen for Phenylketonuria *on page 954*
Newborn Screening Tests for Galactosemia *on page 960*
(Continued)

Newborn Screening by Tandem Mass Spectrometry (MS/MS) *(Continued)*

Synonyms Expanded Newborn Screening; Metabolic Screen by Tandem Mass Spectrometry; Newborn Screening by MS/MS; Supplemental Newborn Screening

Applies to Acylcarnitines; Organic Acids

Abstract Presymptomatic diagnosis by newborn screening for certain diseases has been shown to significantly decrease morbidity and mortality. Although the number of diseases screened varies from state to state, phenylketonuria (PKU) and hypothyroidism are screened in all 50 states. Other commonly screened diseases include galactosemia, congenital adrenal hyperplasia (CAH), and hemoglobinopathies. Less commonly screened diseases include maple syrup urine disease (MSUD), homocystinuria, biotinidase deficiency, cystic fibrosis, and tyrosinemia.

In recent years, a few states have started expanded newborn screening by tandem mass spectrometry (MS/MS). With this powerful technique, more than 20 metabolic diseases and disorders can be screened from a drop of blood spotted to a newborn screening filter paper.

Specimen Blood spotted on to newborn screening filter paper

Sampling Time Before the newborn is discharged from the hospital

Storage Instructions Air dry the filter paper. Once air dried, the sample can be shipped by regular mail.

Causes for Rejection Sample not uniformly spotted on to the filter paper

Turnaround Time 2-4 days

Reference Interval Results are generally given as qualitative; negative or positive

Critical Values Positive results for any of the screened diseases

Use Screening for metabolic diseases, particularly aminoacidopathies, disorders of fatty acid oxidation, organic acidurias, and urea cycle defects

Limitations This is a screening assay only. The positives identified by MS/MS must be confirmed by more definitive methods, such as plasma amino acids, urine organic acids, and plasma acylcarnitines. This generally requires consultation with a specialist and additional specimen collection. The equipment is expensive and expertise is limited. Sensitivity (false-negative) and specificity (false-positive) data on most diseases screened by MS/MS are not available. Standardization between laboratories performing newborn screening by tandem mass spectrometry is lacking. The addition of 20-30 new disorders to the newborn screening menu will increase the number of patients identified, requiring more clinical biochemical geneticists, genetic counselors, and nutritionists. Many diseases identified by MS/MS do not respond consistently to treatment.

Methodology Tandem mass spectrometry (MS/MS); see also Inherited Diseases of Metabolism and Cell Structure *on page 792.*

Additional Information The following table shows the disorders detectable by MS/MS.

Although the incidence of any individual disorder is low, the combined incidence of these diseases is approximately 1:4000 (excluding phenylketonuria). This is equivalent to the incidence of hypothyroidism in newborn screening. Most of the MS/MS identifiable disorders are treatable. Presymptomatic diagnosis and treatment result in improved prognosis in affected patients. Descriptions of some of these disorders follow.

Argininemia, argininosuccinic aciduria, and citrullinemia: These urea cycle disorders are characterized by deficiency of arginase, argininosuccinate lyase, and argininosuccinate synthase respectively. Clinical features include confusion, slurred speech, Reye's syndrome, mental retardation, ataxia, hyperammonemic coma, and stupor. Laboratory findings common to all of these disorders include increased ammonia, glutamine, AST and ALT, and decreased urea. In addition in argininemia, there is increased urine orotic acid.

Metabolic Disorders Detectable by MS/MS

Amino Acid Metabolism	Fatty Acid Metabolism	Organic Acid Metabolism	Other
Argininemia	Short-chain acyl-CoA dehydrogenase (SCAD) deficiency	Glutaric acidemia Type I (GA-1)	Galactosemia
Argininosuccinic aciduria	Medium-chain acyl-CoA dehydrogenase (MCAD) deficiency	Isovaleric acidemia	Congenital adrenal hyperplasia
Citrullinemia	Long-chain 3-hydroxy acyl-CoA dehydrogenase (LCHAD) deficiency	3-ketothiolase deficiency	Cholestatic hepatobiliary disease
Homocystinuria	Very long-chain acyl-CoA dehydrogenase (VLCAD) deficiency	3-methylcrotonyl-CoA dydratase deficiency	
Maple syrup urine disease (MSUD)	Multiple acyl-CoA dehydrogenase deficiency (previously called glutaric aciduria type II)	3-methylglutaconyl-CoA hydratase deficiency	
Phenylketonuria	Carnitine palmitoyl-transferase type II (CPT-2) deficiency	3-hydroxy-3-methylglutaryl-CoA (HMG-CoA) lyase deficiency	
Tyrosinemia type I and II	Carnitine-acylcarnitine translocase deficiency	Multiple CoA carboxylase deficiency	
	Mitochondrial Trifunctional protein (TFP) deficiency	Methylmalonic acidemias	
	2,4 dienoyl-CoA reductase deficiency	Propionic acidemia	

Homocystinuria: Generally characterized by cystathionine-beta-synthase (CBS) deficiency, but may also be due to deficiency of 5,10-methylenetetrahydrofolate reductase or cobalamin or defects in cobalamin metabolism. Clinical findings may include mental retardation, developmental delay, and thromboembolic episodes. CBS deficiency is also characterized by ectopia lentis. Cobalamin deficiency or metabolism defects also result in macrocytic anemia and methylmalonic aciduria (see Methylmalonic Acid, Serum, Plasma, Urine, or Amniotic Fluid *on page 908*).

Maple syrup urine disease (also called branched-chain aminoacidemia), **phenylketonuria, and tyrosinemia:** See Amino Acids, Plasma *on page 143*.

Fatty acids oxidation disorders: Fatty acid oxidation is activated after fasting. Over a dozen fatty acids oxidation defects are known. Nearly all of these defects, particularly in early infancy, present as acute, life-threatening episodes of hypoketotic, hypoglycemic coma induced by fasting. Medium-chain acyl-CoA dehydrogenase (MCAD) deficiency is one of the most common inborn errors of metabolism in Caucasians. It is characterized by frequent episodes of Reye-like syndrome with high incidence of sudden and unexpected death. Attacks become less frequent with age as tolerance to fasting improves. In general, avoidance of prolonged fasting helps these patients.

Glutaric aciduria type I (GA I): Due to a deficiency of glutaryl-CoA dehydrogenase, clinical features include dystonia, macrocephaly, and episodic encephalopathy. Metabolic acidosis, hypoglycemia, ketosis, and hepatopathy may also be present. Protein restriction and treatment with carnitine and riboflavin seem to be effective in some patients but not in all.

Isovaleric acidemia, 3-ketothiolase deficiency, 3-methylcrotonyl-CoA hydratase deficiency, 3-methylglutaconyl-CoA hydratase deficiency, 3-hydroxy-3-methylglutaryl-CoA (HMG-CoA) lyase deficiency, multiple CoA carboxylase deficiency, methylmalonic acidemias, and propionic acidemia: These are disorders of branched-chain amino acids (leucine, isoleucine, and valine). Patients with these disorders generally present with acidosis, (Continued)

Newborn Screening by Tandem Mass Spectrometry (MS/MS) *(Continued)*

increased anion gap, hypoglycemia, and hyperammonemia. Treatment includes low protein diet and carnitine supplementation.

Galactosemia: See Galactokinase, Blood *on page 627*, Galactose-1-Phosphate, Blood *on page 627*, and Galactose-1-Phosphate Uridyl Transferase, Blood *on page 628*.

Congenital adrenal hyperplasia (CAH): See Adrenal Cortex: Laboratory Assessment Overview *on page 110* and 17-Hydroxyprogesterone, Whole Blood, Serum, or Plasma *on page 755*.

References

Chace DH, Kalas TA, and Naylor EW, "Use of Tandem Mass Spectrometry for Multianalyte Screening of Dried Blood Specimens From Newborns," *Clin Chem*, 2003, 49(11):1797-817.

Charrow J, Goodman SI, McCabe ERG, et al, "Tandem Mass Spectrometry in Newborn Screening," *Genet Med*, 2000, 2:267-9.

Clague A and Thomas A, "Neonatal Biochemical Screening for Disease," *Clin Chim Acta*, 2002, 315(1-2):99-110.

Clarke S, "Tandem Mass Spectrometry: The Tool of Choice for Diagnosing Inborn Errors of Metabolism?" *Br J Biomed Sci*, 2002, 59(1):42-6.

Hoffman G, "Tandem Mass Spectrometry in the Newborn Screening Laboratory," *Lab Med*, 2003, 34(7):505-7.

Leonard JV and Dezateux C, "Screening for Inherited Metabolic Disease in Newborn Infants Using Tandem Mass Spectrometry," *BMJ*, 2002, 324(7328):4-5.

Levy HL and Albers S, "Genetic Screening of Newborns," *Annu Rev Genomics Hum Genet*, 2000, 1:139-77.

Seashore MR, "Tandem Spectrometry in Newborn Screening," *Curr Opin Pediatr*, 1998, 10(6):609-14.

Waisbren SE, Albers S, Amato S, et al, "Effect of Expanded Newborn Screening for Biochemical Genetic Disorders on Child Outcomes and Parental Stress," *JAMA*, 2003, 290(19):2564-72.

Internet Web Sites

www.genes-r-us.uthscsa.edu
www.genetests.org
www.marchofdimes.com
www.mayoclinic.com
www.savebabies.org

Newborn Screening Tests for Galactosemia

Related Information

Amino Acids, Urine *on page 145*
Galactokinase, Blood *on page 627*
Galactose-1-Phosphate, Blood *on page 627*
Galactose-1-Phosphate Uridyl Transferase, Blood *on page 628*
Newborn Screen for T$_4$, Filter Paper *on page 956*
Reducing Substances, Urine *on page 1148*

Synonyms Beutler-Baluda Test; Beutler Test; Blood Spot Screen for Galactose/Galactose-1-Phosphate; Galactosemia Screening, Filter Paper; Paigen Test (*E. coli* Bacteriophage Resistance to Lysis Assay)

Abstract Galactosemia is an autosomal recessive disease which, if untreated, causes failure to thrive, vomiting, cataracts, mental retardation, liver disease, aminoaciduria, hypoglycemia, and death. Incidence is about 1:60,000 but estimates worldwide range from 1:18,000 to 1:180,000. Newborn screening is now available in 44 states, the District of Columbia and the Virgin Islands (for updated information see www.aap.org).

Useful tests include those that assay blood galactose, and others that assay galactose-1-phosphate uridyl transferase (GALT). In some screening programs both tests are performed.

Specimen Whole blood, dried as a spot on filter paper

Sampling Time Screening tests for GALT can be done at any time. Tests for galactose should be performed within the first 3 days of life after milk feeding has begun

Collection Follow instructions on blood collection form.

Causes for Rejection Failure to follow instructions on blood collection form

Reference Interval Newborns:
- Galactose, serum: 0-2 mg/dL (SI: 0-1.11 mmol/L)
- GALT, whole blood: 18-26 units/g Hb

Use Detect galactosemia; monitor dietary therapy of galactosemia

Limitations False-negative results are rare in patients with classic homozygous deficiency, but are more frequent in patients with partial deficiencies. **Transfusion may result in a false-negative test for GALT up to 3 months post-tranfusion.**

Methodology For galactose: microbiologic assay (Piagen test), fluorometry (Hill test); for GALT: fluorometry (Beutler test)

Additional Information Urine from normal newborns may have levels of galactose as high as 60 mg/dL (SI: 3.33 mmol/L) (physiologic melituria); in premature infants this may occur over the first 2 weeks of life. High level of milk intake may also produce galactosuria. In cases of galactosemia, galactosuria may be intermittent (partly relating to the intake of milk) and may be missed if urine is very dilute. A screening program based on a copper reduction test, then, may result in false negatives.

Urine can be screened for galactose by a cupric ion reduction method (eg, Benedict's Test, Clinitest® tablets). If positive, a glucose oxidase specific method is applied. If the specific test for glucose is negative, but a reducing substance is present, there is presumptive evidence for one of the three forms of galactosemia.

References

"American Academy of Pediatrics, Committee on Genetics, Newborn Screening Fact Sheets," *Pediatrics*, 1996, 98:473-501.

Beutler E, "Galactosemia: Screening and Diagnosis," *Clin Biochem*, 1991, 24(4):293-300.

Gilbert-Barness E and Barness LA, *Metabolic Diseases: Foundations of Clinical Management, Genetics and Pathology*, Natick, MA: Easton Publishing, 2000, 185-92.

Gitzelmann R, "Disorders of Galactose Metabolism," *Inborn Metabolic Diseases: Diagnosis and Treatment*, Fernandes J, Saudubray JM, and Van den Berghe G, eds, 3rd ed, New York, NY: Springer-Verlag, 2000, 103-9.

Jacobs DS, DeMott WR, Oxley DK, et al, *Laboratory Test Handbook*, 5th ed, Hudson, OH: Lexi-Comp Inc, 2001.

Jensen UG, Brandt NJ, Christensen E, et al, "Neonatal Screening for Galactosemia by Quantitative Analysis of Hexose Monophosphates Using Tandem Mass Spectrometry: A Retrospective Study," *Clin Chem*, 2001, 47(8):1364-72.

Paigen K, Pacholec F, and Levy HL, "A New Method of Screening for Inherited Disorders of Galactose Metabolism," *J Lab Clin Med*, 1982, 88:895-907.

Internet Web Sites

www.aap.org

♦ **NH₃, Blood** *see* Ammonia, Plasma *on page 150*

♦ **Nicotine** *see* Cotinine, Serum, Plasma, or Urine *on page 464*

Nicotine, Serum or Plasma

Related Information

Carboxyhemoglobin, Blood *on page 340*

Cotinine, Serum, Plasma, or Urine *on page 464*

Abstract Nicotine, one of the most toxic of all poisons, is a neural stimulant found in most tobacco products including transdermal patches and Nicorette® gum. Its role in atherosclerosis and pulmonary disease is well established.

Specimen Serum or plasma

Container Red top tube

Storage Instructions Store at -20°C for 1 week.

Reference Interval Mean plasma concentration after smoking one cigarette: 5-30 ng/mL. In smokers, peak level is 30-50 ng/mL.

Possible Panic Range Serum concentrations >50 ng/mL may be associated with toxicity. Plasma concentrations >13,000 ng/mL have been associated with **fatality.**

Use Work-up of acute poisoning in children ingesting cigarettes or cigarette butts

Methodology Gas chromatography (GC), high performance liquid chromatography (HPLC), immunoassay

Additional Information

- Half-life: 24-84 minutes
- Volume of distribution: 1.0 L/kg

Nicotine is metabolized in the liver to cotinine and trans-3-hydroxy cotinine. The metabolites have a half-life >15 hours and are frequently used to assess the effectiveness of nicotine addiction treatment. Acute toxicity may include, but is not limited to: nausea, cyanosis, insomnia, hyponatremia, blurred vision, hyperventilation, nystagmus, dementia, abdominal pain, apnea, and respiratory (Continued)

Nicotine, Serum or Plasma (Continued)

depression. Chronic exposure to nicotine in the form of cigarette smoke also has other deleterious effects including exacerbation of atherogenesis.

References

Kilaru S, Frangos SG, Chen AH, et al, "Nicotine: A Review of Its Role in Atherosclerosis," *J Am Coll Surg*, 2001, 193(5):538-46.

Molander L, Hansson A, and Lunell E, "Pharmacokinetics of Nicotine in Healthy Elderly People," *Clin Pharmacol Ther*, 2001, 69(1):57-65.

Pérez-Stable EJ, Herrera B, Jacob P, et al, "Nicotine Metabolism and Intake in Black and White Smokers," *JAMA*, 1998, 280(2):152-6.

♦ NIDA *see page 63*

♦ Nipride® *see* Thiocyanate, Serum, Plasma, or Urine *on page 1245*

Nitrite, Urine

Related Information

Bacterial Culture, Urine *on page 246*
Leukocyte Esterase, Urine *on page 849*
Urinalysis *on page 1289*

Synonyms Bacteria Screen, Urine

Applies to Urinary Tract Infection Screen; UTI Screen

Test Includes This test is usually part of a routine urinalysis.

Abstract A rapid screening method for detection of bacteriuria, reacting with those bacteria which reduce urinary nitrate to nitrite. The test has poor sensitivity, which limits its effectiveness. **The best test for urinary tract infection is culture, which remains the definitive test to ascertain whether or not urinary tract infection is present.**

Specimen Urine, first morning specimen is preferred; random urine is acceptable; preferably midstream, clean catch collection; catheterized specimen; bladder aspiration

Storage Instructions Test immediately.

Causes for Rejection A random urine is less likely than an overnight specimen to produce a positive reaction.

Reference Interval Negative

Use A reagent strip method is intended to detect the presence of potentially significant bacteriuria. Dipstick screen for detection of bacteriuria performs best when there are >10^5 cfu/mL, when leukocyte esterase and urinary nitrite are used together; however, sensitivity when so used has been disappointing.

Limitations There are numerous causes of false-negative and false-positive results. Some urinary tract infections are caused by organisms which do not contain reductase to convert nitrate to nitrite. (Negative results are found when infecting organisms do not convert nitrate to nitrite). These include infections caused by *Streptococcus faecalis* and other gram-positive cocci, *N. gonorrhoeae*, and *M. tuberculosis*.

In addition, the urine may not have been retained in the bladder for 4 hours or more to allow adequate reduction of nitrate to occur. In detection of asymptomatic urinary tract infections in obstetric patients, dipstick nitrite testing identified only 50% of the patients with such infections and was less sensitive than Gram staining.

Methodology Nitrite from the urine reacts with *p*-arsanilic acid forming a diazonium compound.

Other Screening Methods
for Detection of Urinary Tract Infection

	Sensitivity (%)	Specificity (%)
Microscopic analysis on spun urine	79	93
Methylene blue stain for pyuria	60	99
Gram stain for pyuria	45	93
Gram stain for bacteriuria	65	75

Adapted from Carroll KC, Hale DC, Von Boerum DH, et al, "Laboratory Evaluation of Urinary Tract Infections in an Ambulatory Clinic," *Am J Clin Pathol*, 1994, 101(1):100-3.

Additional Information The use of nitrate and leukocyte esterase together is more extensively discussed in Leukocyte Esterase, Urine *on page 849*. See also Related Information.

For >10^5 colony counts/mL, the sensitivity of the nitrite test was recently reported as 27% and specificity 94%. The authors concluded that leukocyte esterase and nitrite dipstick tests are unsuitable for screening for urinary tract infections.

References

Semeniuk H and Church D, "Evaluation of the Leukocyte Esterase and Nitrite Dipstick Screening Tests for Detection of Bacteriuria in Women With Suspected Uncomplicated Urinary Tract Infections," *J Clin Microbiol*, 1999, 37(9):3051-2.

Zaman Z, Borremans A, Verhaegen J, et al, "Disappointing Dipstick Screening for Urinary Tract Infection in Hospital Inpatients," *J Clin Pathol*, 1998, 51(6):471-2.

Nitroblue Tetrazolium Test

Related Information

Flow Cytometry, Overview *on page 592*

Synonyms NBT Dye Test

Abstract NBT test is used mainly for the diagnosis of chronic granulomatous disease (CGD). In CGD there is failure of the respiratory burst with decreased or absent superoxide production. Reactive oxygen elements (eg, H_2O_2, hypohalous acids, and ·OH) are produced from superoxide and have potent microbial activity. The condition has a heterogeneous genetic basis, the result of mutations in any of the components of the NADPH oxidase complex (including gp 91-phox, p22-phox, p47-phox, p67-phox, and p40-phox). CGD is characterized by disabled phagocyte NADPH oxidase with inability to efficiently kill phagocytized bacteria.

Specimen Whole blood

Container Green top (heparin) tube

Storage Instructions Specimen cannot be stored (test utilizes live granulocytes).

Causes for Rejection Transit to the laboratory of more than 1 hour

Special Instructions Advance scheduling with the laboratory may be required, as the specimen must be tested while the neutrophils are still viable. Transport to the laboratory **immediately** following collection.

Reference Interval Without *Staphylococcus* activation, 2% to 8% segmented neutrophils reduce dye. In patients with bacterial infection, NBT-positive cells may be increased (>10%).

Use Diagnose chronic granulomatous disease (CGD) of childhood

Limitations The NBT test is unreliable in the differentiation of bacterial from viral and other infections, producing unacceptable false-negative and false-positive results.

Methodology Assessment of reduction of a tetrazolium dye by stimulated and unstimulated neutrophils. Neutrophils reduce the dye to a dark blue-black formazan pigment upon phagocytosis. Flow cytometry and chemiluminescence-based methods may also be applicable. Chronic granulomatous disease can be diagnosed using restriction fragment length polymorphism with labeled gene probes.

Additional Information The NBT test is usually a reliable aid in the diagnosis of CGD, in which neutrophils are unable to reduce the NBT dye (which correlates with their inability to kill bacteria). In patients with CGD, the NADPH oxidase system fails to generate superoxide and derived oxygen intermediates, with resultant susceptibility to recurrent bacterial (in particular, *Staphylococcus aureus*) and fungal (in particular, *Aspergillus* sp) infections. Catalase-positive microorganisms are especially aggressive pathogens in CGD patients because catalase degrades hydrogen peroxide (they neutralize their own H_2O_2).

CGD is rare, with an incidence of some 1:500,000 individuals. It manifests shortly after birth as recurrent suppurative infections with skin involvement, infected eczematoid rash, lymphadenitis, multiple abscesses (lung, liver, other viscera, epidural space) and osteomyelitis. Common pathogens are *S. aureus*, *S. serratia*, and *Salmonella* species. Alternate inflammatory response produces granulomas, occasionally bulky, which can lead to gastrointestinal obstruction. Severity of clinical features is variable. Some mild forms of CGD are not evident until adolescence or adulthood.

(Continued)

Nitroblue Tetrazolium Test *(Continued)*

The heterogeneous nature of CGD derives from multiple different mutations involving different components of the NADPH oxidase enzyme complex. In the inactive state of the oxidase, two membrane-bound components (gp91-phox-phagocyte oxidase and p22-phox) form a unique cytochrome (cytochrome b_{558}), the central redox component of phagocyte oxidase. It is unique in that it has a very low redox potential. With neutrophil phagocytic activation, three cytosolic factors (p47-phox, p67-phox, and p40-phox) translocate from the cytosol and associate with membrane cytochrome b_{558}.

In this manner the superoxide-generating enzyme complex, respiratory burst oxidase, is formed. Mutations in the gp91-phox gene (x chromosome at p21.1), account for 60% to 65% of cases of CGD. The other 35% to 40% of patients have autosomal recessive mode of inheritance. See table.

Genetic Forms of Chronic Granulomatous Disease

NADPH Oxidase Subunit Involved	Site of Gene Mutation	% of Cases of CGD	Cell Locale
gp91^{phox}	X at p21.1	60% to 65%	Membrane
p47^{phox}	7q 11.23	25%	Cytosol
p67^{phox}	1q 25	5%	Cytosol
p22^{phox}	16q 24	5%	Membrane
p40^{phox}	None known	Unknown	Cytosol

Neutrophil function is altered (chemotaxis and chemiluminescence decreased) in diabetic patients, especially those with vascular complications and hyperglycemia.

Treatment of CGD patients with recombinant interferon-γ has been shown to result in a near-normal level of superoxide production and return of granulocyte bactericidal capacity to normal control levels. Interferon-γ (rIFN-γ) stimulates progenitor cells and their mature progeny. Colonies of such cells regain the ability to generate superoxide. Therapy with rIFN-γ is considered safe and effective with decrease in the number and severity of infections, but is expensive.

Patients with advanced CGD, who have failed treatment with sulphamethazole trimethoprim, multiple courses of antibiotic/antifungal agents, and interferon-γ therapy may be treated successfully by bone marrow transplant, reports of morbidity/mortality not withstanding. Gene transfer studies using viral vectors in murine models, and cultured CD34$^+$ hematopoietic progenitor cells have been performed.

References

Leung TF, Chik KW, Li CK, et al, "Bone Marrow Transplantation for Chronic Granulomatous Disease: Long-term Follow-up and Review of Literature," *Bone Marrow Transplant*, 1999, 24(5):567-70.

Paraskevas F, "Phagocytosis," *Wintrobe Clinical Haematology*, Chapter 17, Lee GR, Foerster J, Lukens J, et al, ed, Baltimore, MD: Lippincott Williams & Wilkins, 1999, 426.

Segal BH, Leto TL, Gallin JI, et al, "Genetic, Biochemical, and Clinical Features of Chronic Granulomatous Disease," *Reviews in Molecular Medicine, Medicine*, 2000, 79(3):170-200.

♦ **Nitrogen Mustard** *see* Chemical Warfare Agents *on page 382*

♦ **Nitroprusside** *see* Thiocyanate, Serum, Plasma, or Urine *on page 1245*

♦ **Nitroprusside Reaction, Blood** *see* Ketone Bodies, Blood *on page 816*

♦ **Nitroprusside Reaction for Ketones, Urine** *see* Ketones, Urine *on page 817*

♦ **Nitroprusside Screening** *see* Cystine, Urine *on page 494*

♦ **Nits Identification** *see* Arthropod Identification *on page 213*

Nocardia Culture

Related Information

Acid-Fast Stain, Routine or Modified *on page 95*
Actinomyces Culture *on page 98*
Cerebrospinal Fluid Analysis: Overview *on page 355*
Fine Needle Aspiration Culture *on page 589*
Fungal Culture, Sputum *on page 624*

Mycobacterial Culture, Sputum *on page 933*

Test Includes Culture for *Nocardia* spp and direct microscopic examination of clinical specimens by Gram stain and/or modified acid-fast stains

Abstract *Nocardia* infections in humans are rare and usually occur in patients who are severely immunocompromised or in patients with a traumatic injury. Human nocardial infections may be divided into six categories based on body site involved, and clinical and pathological characteristics:

- pulmonary nocardiosis
- systemic nocardiosis
- central nervous system nocardiosis
- extrapulmonary, localized nocardiosis
- cutaneous, subcutaneous, and lymphocutaneous nocardioses
- mycetoma

The usual portal of entry is the lung. Skin manifestations may be complications of disseminated disease and should not always be attributed to local inoculation.

Specimen Pus, tissue, cerebrospinal fluid or other body fluid, aspirate, sputum

Special Instructions Consultation with laboratory prior to collection of the specimen is recommended when nocardiosis is suspected clinically. **Culture should be specifically ordered as "Culture for *Nocardia*."**

Reference Interval No *Nocardia* species isolated

Critical Values *Nocardia* species recovered from a central nervous system specimen

Use Establish the diagnosis of nocardiosis or mycetoma; identify its etiologic agent, especially in patients following transplantation surgery, those on steroids, and with other states of immunosuppression

Limitations *Nocardia* species will not be recovered by routine bacterial culture techniques because of its relatively slow growth. Growth of *Nocardia* may be obscured by overgrowth of other organisms in mixed culture (ie, sputum). The diagnosis may not be made unless the laboratory is advised of the clinical suspicion of nocardiosis. *Nocardia* species are not strongly gram-positive, but their branching pattern when visible is helpful. A modified acid-fast stain is needed, since *Nocardia* are weakly acid-fast and may not be found with conventional acid-fast staining. Staining may be positive when cultures fail. *Nocardia* species are variably acid-fast and may be frequently confused with *Actinomyces* species or saprophytic fungi in Gram stains of clinical specimens. Silver stains, tissue Gram stains, and Fite-Ferraco acid-fast stain are among techniques used in tissue sections. Routine blood cultures do not detect *Nocardia* species, but *Nocardia* septicemia has been reported.

Methodology Aerobic culture on various bacterial media. Cultures are usually held for 10-30 days, when the laboratory is aware of clinical suspicion of nocardiosis.

Additional Information *Nocardia* species are aerobic, filamentous gram-positive bacteria which are relatively slow growing. *N. asteroides*, *N. brasiliensis*, *N. farcinica*, *N. nova*, *N. otitidiscaviarum* (formerly *N. caviae*), and *N. transvalensis* are human pathogens. Many laboratories do not distinguish organisms in the *N. asteroides* complex (ie, *N. asteroides*, *N. nova*, *N. farcinica*) from one another, and identify all these organisms as *N. asteroides*. Human infection is seen most frequently in patients whose immune systems are suppressed by HIV infection, lymphoreticular malignancy, or chemotherapy. However, pulmonary nocardiosis may occur in immunocompetent patients. The clinical picture may be similar to that observed with systemic mycobacterial or fungal infections. Infections may be acute, subacute, or chronic; and they may be disseminated or localized to cutaneous sites or the respiratory tract. Metastatic infection in brain, bone, joints, soft tissue, heart, kidneys, skin, or subcutaneous infection in the presence of pulmonary involvement is suggestive of nocardiosis.

References

Cremades MJ, Menendez R, Santos M, et al, "Repeated Pulmonary Infection by *Nocardia asteroides* Complex in a Patient With Bronchiectasis," *Respiration*, 1998, 65(3):211-3.

Filipiuk D, Weisenberg E, and Malecki Z, "Pulmonary Nocardiosis in a Patient With Human Immunodeficiency Virus," *Arch Pathol Lab Med*, 2001, 125(7):979-80.

Lerner PI, "Nocardiosis," *Clin Infect Dis*, 1996, 22(6):891-903.

Sharma S and Sridhar MS, "Diagnosis and Management of *Nocardia* Keratitis," *J Clin Microbiol*, 1999, 37(7):2389.

(Continued)

Nocardia Culture *(Continued)*

Internet Web Sites
www.cdc.gov/ncidod/dbmd/diseaseinfo/nocardiosis_t.htm

Nortriptyline, Serum

Related Information
Amitriptyline, Serum or Plasma *on page 149*
Antidepressants, Cyclic, Serum or Plasma *on page 171*

Synonyms Allegron®; Aventyl® Hydrochloride; Nortrilen®; Norval®; Pamelor®

Abstract Nortriptyline is a tricyclic antidepressant and is the active metabolite of amitriptyline. It has analgesic properties, and is included in the group of adjuvant analgesic drugs. It is used as a prophylactic agent for migraine, chronic pain, anxiety disorders, enuresis, and attention-deficit/hyperactivity disorder (ADHD).

Specimen Serum

Container Red top tube

Sampling Time Trough levels at steady state (100-300 hours)

Storage Instructions Separate serum and refrigerate.

Causes for Rejection Sample collected in gel tube

Reference Interval Therapeutic: 50-150 ng/mL (SI: 190-570 nmol/L)

Critical Values Toxic: >500 ng/mL (SI: >1900 nmol/L)

Possible Panic Range ≥1000 ng/mL

Use Monitor therapeutic drug level; evaluate toxicity, overdoses. Nortriptyline is a derivative and metabolite of amitriptyline and is used to treat endogenous depression.

Contraindications Coadministration of monoamine oxidase inhibitors

Methodology High performance liquid chromatography (HPLC), immunoassay

Additional Information
- Half-life: 20-60 hours (pharmacokinetic parameters may have wide ranges for the tricyclic antidepressants)
- Volume of distribution: 15-23 L/kg
- Protein binding: 90% to 95%

Nortriptyline, primarily detoxified in the liver, may be associated with cholestasis and cholestatic jaundice. Hematologic consequences include agranulocytosis, purpura, and thrombocytopenia.

Nortriptyline is completely absorbed but undergoes some first-pass elimination. The bioavailability of nortriptyline is 45% to 70% with a peak concentration 2-6 hours after an oral dose.

As with other tricyclic antidepressants, anticholinergic side effects are common with this drug. Due to side effects of tricyclic antidepressants, low doses of the drugs have been tried. A meta-analysis on effects and side effects of low-dose tricyclic antidepressants has been published (Furukawa et al, 2002).

Drug interactions are common with the tricyclic antidepressants. Concomitant treatment with cimetidine, fluoxetine, and antipsychotics produce unexpectedly increased concentrations of nortriptyline. Enzyme inducers (eg, phenytoin, chloral hydrate, smoking, and the barbiturates) will decrease nortriptyline concentrations.

References

Feldman MD, "Therapeutic Blood Monitoring of Tricyclic Antidepressants," *South Med J,* 1994, 87(1):101.

Furukawa TA, McGuire H, and Barbui C, "Meta-analysis of Effects Side Effects of Low Dosage Tricyclic Antidepressants in Depression: Systematic Review," *BMJ,* 2002, 325(7371):991.

Hirschfeld RM, "Antidepressants in Long-Term Therapy: A Review of Tricyclic Antidepressants and Selective Serotonin Reuptake Inhibitors," *Acta Psychiatr Scand Suppl,* 2000, 403:35-8.

- ◆ **Norval®** *see* Nortriptyline, Serum *on page 966*
- ◆ **Norverapamil** *see* Verapamil, Serum or Plasma *on page 1306*
- ◆ **Nose Candy** *see* Cocaine (Cocaine Metabolite), Qualitative, Urine or Hair *on page 427*
- ◆ **Novocainamidum** *see* Procainamide, Serum *on page 1092*
- ◆ **Novocamid®** *see* Procainamide, Serum *on page 1092*
- ◆ **Novoxapin®** *see* Doxepin, Serum or Plasma *on page 524*
- ◆ **Novrad®** *see* Propoxyphene, Serum or Urine *on page 1096*
- ◆ **NPT** *see* Point-of-Care Testing *on page 1065*
- ◆ **nRNP** *see* Smith (Sm) and Ribonucleoprotein (RNP) Antibodies *on page 1206*
- ◆ **NSE** *see* Neuron-Specific Enolase, Serum *on page 949*
- ◆ **N-Telopeptides of Type-1 Collagen** *see* N-Telopeptides, Urine *on page 967*

N-Telopeptides, Urine

Related Information

Alkaline Phosphatase, Heat Stable, Serum *on page 125*
Alkaline Phosphatase Isoenzymes, Serum *on page 125*
Alkaline Phosphatase, Serum *on page 127*
Calcium, Serum *on page 329*
Calcium, Urine *on page 332*
Hydroxyproline, Total, Urine *on page 757*
Osteocalcin, Serum or Plasma *on page 983*
Parathyroid Hormone, Serum *on page 1001*
Phosphorus, Serum *on page 1031*
Phosphorus, Urine *on page 1032*
Pyridinolines (Pyridinoline and Deoxypyridinoline), Urine *on page 1126*
Urine Collection, 24-Hour *on page 1295*

Synonyms N-Telopeptides of Type-1 Collagen; NTx

Applies to Collagen Type-1; C-Telopeptide; Deoxypyridinoline; Galactosyl Hydroxylysine; Pyridinoline; Telopeptides

Abstract N-telopeptide (NTx) is a marker of bone resorption.

Sampling Time Second morning void or 24-hour urine. Due to diurnal variation, 24-hour urine sample is preferred. NTx excretion is higher at night. Following baseline assay(s), repeat assay about 3 months following beginning of antiresorptive treatment, repeat at 12 months, then annually if medically necessary.

Collection Refrigerate during collection

Storage Instructions Refrigerate. Freeze for longer storage (>2 days).

Reference Interval NTx levels are higher in children. In the literature, there is a wide variation in reference intervals, even with the same methodology, particularly in children. Approximate adult reference intervals for 24-hour urine are:
- male: <65 pmol/μmol creatinine
- female: premenopausal: <65 pmol/μmol creatinine; postmenopausal: <131 pmol/μmol creatinine

(Continued)

N-Telopeptides, Urine *(Continued)*

Use The rate of NTx excretion in urine is used as an index of bone resorption, most commonly a monitor of response to therapy. However, clinical usefulness is **not** effective in establishing the diagnosis of osteoporosis.

The Negotiated Rulemaking Committee of HCFA has listed the following indications for collagen cross-link assays (any method):

- Identifying individuals with elevated bone resorption, who have osteoporosis, in whom response to treatment is being monitored
- Predict response (as assessed by bone mass measurements) to FDA-approved antiresorptive therapy in postmenopausal women
- Assess effectiveness of osteoporosis treatment including FDA-approved antiresorptive therapies in postmenopausal women, individuals with osteoporosis, Paget disease of bone and antiestrogen or selective estrogen therapies.

Limitations Substantial variability between methods and specimens is recognized. Therefore, a second baseline assay is not inappropriate. NTx is affected by renal clearance.

Methodology Enzyme-linked immunosorbent assay (ELISA), competitive immunoassay, radioimmunoassay (RIA)

Additional Information Urinary deoxypyridinoline, pyridinoline, galactosyl hydroxylysine, and N- and C-telopeptides, all released by osteoclast activity, are used as markers of bone resorption.

NTx levels are increased in osteoporosis, Paget disease, metastatic bone disease, primary and secondary hyperparathyroidism, hyperthyroidism, and other diseases with increased bone resorption. NTx levels are increased in aging individuals, predominantly postmenopausal women, as compared to premenopausal controls.

Treatment of osteoporotics and postmenopausal women with biphosphonates and/or estrogen replacement therapy and other means decreases bone resorption markers. As compared to pyridinolines, NTx responds better to antiresorptive therapy. A typical response to antiresorptive therapy may be seen as a decrease of 30% to 40% from baseline after 3 months of treatment. An adequate therapeutic response is signaled by a ≥50% reduction in the resorption marker. Clinical utility has not been established.

References

Hurley DL and Khosla S, "Update on Primary Osteoporosis," *Mayo Clin Proc*, 1997, 72(10):943-9.

Mayo Reference Services Publication, "NTx Telopeptide, Urine," *New Test Announcement*, Rochester, MN: Mayo Medical Laboratories, 2000.

Miller PD, Baran DT, Bilezikian JP, et al, "Practical Clinical Application of Biochemical Markers of Bone Turnover: Consensus of an Expert Panel," *J Clin Densitom*, 1999, 2(3):323-42.

Watts NB, "Clinical Utility of Biochemical Markers of Bone Remodeling," *Clin Chem*, 1999, 45(8 Pt 2):1359-68.

Zanze M, Souberbielle JC, Kindermans C, et al, "Procollagen Propeptide and Pyridinium Cross-Links as Markers of Type I Collagen Turnover: Sex- and Age-Related Changes in Healthy Children," *J Clin Endocrinol Metab*, 1997, 82(9):2971-7.

♦ **NTx** *see* N-Telopeptides, Urine *on page 967*

♦ **NTx** *see* Osteocalcin, Serum or Plasma *on page 983*

♦ **Nuclear Ribonucleoprotein** *see* Smith (Sm) and Ribonucleoprotein (RNP) Antibodies *on page 1206*

♦ **Nucleic Acid Amplification** *see* Mycobacterial Culture, Cerebrospinal Fluid *on page 931*

♦ **Nucleolar Antibody** *see* Antinuclear Antibodies *on page 189*

5′ Nucleotidase, Serum

Related Information

Alkaline Phosphatase, Serum *on page 127*

Gamma-Glutamyl Transferase, Serum *on page 629*

Liver Disease: Laboratory Assessment, Overview *on page 869*

Abstract A plasma membrane enzyme relevant to the biliary tract.

Patient Preparation A fasting specimen is preferred.

Specimen Serum

Container Red top tube

Reference Interval Varies with laboratory. Adults: 2-17 units/L.

Use 5′NT is used in the differential diagnosis of diseases involving the liver and the porta hepatic area. Marked increases (4-6 times the upper limit of normal) are found in hepatobiliary diseases (eg, obstruction of common duct, biliary cirrhosis, intrahepatic cholestasis). Normal values, or very small increases, are found in parenchymal liver disease (eg, hepatitis) and skeletal diseases. In the past, 5′NT was among the tests used to evaluate persons with an elevated alkaline phosphatase result on a biochemical screen, but this application is decreasing.

Limitations The test is not widely available.

Additional Information The diseases identified by elevations of 5′NT parallel closely those identified by Gamma-Glutamyl Transferase, Serum *on page 629*, and there does not appear to be a need for two tests providing similar information. In patients receiving antiepileptic drugs, the frequency of enzyme elevation was similar to that of alkaline phosphatase but lower than that of GGT.

It has recently been reported that a ratio of GGT:5′NT is significantly lower in patients with intrahepatic cholestasis as compared to extrahepatic cholestasis. A threshold of GGT:5′NT <1.9 had a sensitivity of 40% and specificity of 100% for diagnosis of intrahepatic cholestasis.

References
Sapey T, Mendler MH, Guyader D, et al, "Respective Value of Alkaline Phosphatase, Gamma-Glutamyl Transpeptidase and 5′-Nucleotidase Serum Activity in the Diagnosis of Cholestasis: A Prospective Study of 80 Patients," *J Clin Gastroenterol*, 2000, 30(3):259-63.

♦ **Nuprin®** *see* Ibuprofen, Serum *on page 764*

♦ **Nydrazid®** *see* Isoniazid, Serum or Plasma *on page 813*

♦ **O₂ Binding Capacity (BO₂)** *see* Oxygen Saturation, Blood *on page 991*

♦ **O₂ Content (ctO₂)** *see* Oxygen Saturation, Blood *on page 991*

♦ **6-O-Acetyl Morphine** *see* Opiates, Qualitative, Urine *on page 974*

♦ **Occult Blood, Semiquantitative, Urine** *see* Urinalysis *on page 1289*

Occult Blood, Stool

Related Information

Carcinoembryonic Antigen, Serum *on page 342*
Colon Cancer, Hereditary Nonpolyposis Type *on page 432*
⁵¹Cr Red Cell Survival *on page 476*
Fecal Lactoferrin *on page 575*
Ferritin, Serum *on page 577*
Iron and Total Iron Binding Capacity/Transferrin, Serum *on page 807*
Methylene Blue Stain, Stool *on page 906*

Synonyms Blood, Occult, Stool; Fecal Occult Blood Test; Stool, Occult Blood

Applies to Guaiac, Stool

Abstract Occult bleeding can be defined as blood in the stool, not apparent to the patient, for which detection requires chemical testing. It may present as iron deficiency anemia. The most common global cause is hookworm infestation. In industrialized nations, the most important cause is colorectal adenocarcinoma. False positives as well as false negatives occur. In spite of its shortcomings, annual fecal occult blood testing (FOBT) does diminish deaths from colorectal carcinomas, especially when used in concert with episodic colonoscopy and sigmoidoscopy. Positive FOBT may derive from the upper gastrointestinal tract or small intestine as well as from colorectal disease.

Patient Preparation Patient should not receive vitamin C (ascorbic acid) for 5 days prior to FOBT by guaiac. 250 mg vitamin C intake leads to false-negative FOBT based on guaiac methods. Vitamin C does not affect the HemoQuant® test. A high bulk, red meat free diet with restriction of peroxidase-rich vegetables (turnips, horseradish, artichokes, mushrooms, radishes, broccoli, bean sprouts, cauliflower, apples, oranges, bananas, cantaloupes and other melons, grapes) has been recommended for 3-5 days prior to guaiac testing and during testing, to decrease the incidence of false positives. Ingestion of red meat or use of aspirin or other nonsteroidal anti-inflammatory drugs for 3 days before or during the collection may affect results for guaiac tests or HemoQuant®, but low-dose aspirin, in doses used for cardiovascular prophylaxis, probably should not be stopped. Alcohol and aspirin, especially together, and other gastric irritants should also be avoided. Social use of ethanol alone is unlikely to cause false-positive HemoQuant® results but therapeutic doses of aspirin may do so.
(Continued)

Occult Blood, Stool *(Continued)*

Halogens and cimetidine can cause reactions with guaiac tests. Oral iron may not cause positive guaiac or HemoQuant® stool assay results. Positive stool reactions from subjects on therapeutic doses of oral iron should not be dismissed as false positives.

Specimen Guaiac tests: Follow manufacturer's instructions. HemoQuant®: 1 g feces from a single defecation, collected with a sampler provided with the kit.

Container For HemoQuant®, the sampler is inserted in a screw-capped tube.

Collection Stool collection is awkward. Some patients are unwilling or unable to comply. Up to 75% of blood leaches from the fecal surface into toilet water in 4-12 minutes. Many toilet sanitizers generate chlorine and may cause false-positive reactions to guaiac tests. False positives can be traced to toilet bowl water containing blood from menstruation or urine. Toilet sanitizers and detergents may reduce hemoglobin concentration or antigenicity of immuno-logic-based tests.

Storage Instructions Delay in examination can adversely affect Hemoccult® results and also affects HemoQuant® results. Refrigerated HemoQuant® spec-imen must arrive at reference laboratory within 24 hours of collection, or be sent frozen on dry ice.

Special Instructions Tests for stool occult blood are not appropriately applied to detection of blood in gastric juice by virtue of possible ingestion of drugs and the pH of gastric juice.

Reference Interval Normal blood loss into the gastrointestinal tract is 0.5-1.5 mL/day, an amount below detection by FOBT.

Guaiac: negative. (A positive report has much more significance than does a negative). No consistent fecal hemoglobin level exists above which guaiac tests are reliably positive or below which they are not. HemoQuant®: <2 mg total hemoglobin/g of feces is considered normal, 2-3 mg total hemoglobin/g of feces is marginal, >3 mg total hemoglobin/g of feces is elevated.

Use Detection of gastrointestinal bleeding, especially that from large bowel adenocarcinoma. FOBT is applied to case finding as well as to population screening. Colorectal carcinoma is the second leading cause of death from cancer in the U.S. Fecal occult blood testing (FOBT) is applied both to screening and to case detection. It is relatively insensitive as a screen for colorectal neoplasia and is even less sensitive for detection of polyps. Fewer than 30% of carcinomas and larger polyps bleed sufficiently to be detected by occult blood screening. Successful detection and therapy of even a fraction of the ~150,000 new U.S. cases annually is meaningful. However, technical and other serious flaws work against implementation of broad-based population surveillance at this time. The importance of fecal blood and leukocytes as markers for the inflammatory diarrheas is discussed in Methylene Blue Stain, Stool *on page 906*.

Disease of the **upper** gastrointestinal tract may first be recognized by FOBT. Even gastritis associated with *H. pylori*, as well as gastric cancer, can cause positive FOBT. Bleeding from the upper gastrointestinal tract is best detected by hemeporphyrin FOBT; guaiac-based FOBT are not as sensitive, and immu-nochemical tests are insensitive to this source. Celiac disease is also a cause of occult bleeding. Other causes of positive FOBT in the small intestine include tumors, Meckel's diverticulum, and vascular ectasias. However, both upper gastrointestinal diseases and colorectal carcinoma are common; identification of the former should not necessarily exclude colonoscopy.

Limitations Most methods lack sensitivity to small amounts of blood and might fail to detect slow rates of blood loss. Many adenomas and carcinomas do not bleed. When occult GI bleeding is suspected, at least three samples, preferably of separate bowel movements, should be submitted. Many substances and conditions interfere with guaiac tests. Vitamin C (ascorbic acid) and antacids may cause false negatives to guaiac tests. False-positive results may be caused by excessive dietary intake of vegetable peroxidases, especially horse-radish. Drugs shown to be associated with gastrointestinal blood loss in normal subjects include salicylates (aspirin), steroids, rauwolfia derivatives, all nonste-roidal anti-inflammatory drugs, and colchicine. Sensitivity of the slides is increased by rehydration prior to development. Other means to increase sensi-tivity include testing on three consecutive days or use of more sensitive tests.

Guaiac tests present a number of problems (see table). Acid pH, heat, and dry stools lead to some false negatives, while watery stools are more apt to test positive. **Intestinal converted fraction** is an expression which describes the fraction of heme converted to porphyrins and iron during fecal transit, a phenomenon which is not detected by guaiac-based tests, but is detected by heme-porphyrin assay.

Colorectal adenomas and carcinomas cause a minority of positive guaiac tests; 70% of carcinomas and 80% of adenomas larger than 2 cm are missed. Hemoccult® II with rehydration yields false-positive rates ≥10%.

Patients with unexplained iron deficiency anemia or with evidence or symptoms of gastrointestinal disease must be investigated, whether or not their FOBTs are positive.

FOBT are not sufficiently sensitive to replace endoscopy in high-risk situations such as familial colorectal carcinoma.

Methodology Guaiac is a leuko-dye. Commercial tests include Colo-Screen® (Helena Laboratories), Colo-Rect® (Roche Diagnostics), Hema-Chek® (Miles Laboratories), Quick-Cult® (Laboratory Diagnostics), Hemoccult® II (SmithKline) and Hemoccult® II Sensa. The peroxidase-like activity of hemoglobin or nonspecific oxidants catalyze the reaction of peroxide and the chromogen ortho-toluidine to form blue oxidized orthotolidine.

New methodology, based on heme-derived porphyrin fluorescence, is more expensive (**HemoQuant®**) and somewhat controversial. It detects peroxidase-negative heme-derived porphyrins, peroxidase-positive hemoglobin, and free heme spectrofluorometrically.

Immunologic tests have been developed, but enteric degradation of hemoglobin interferes with immunologic as well as guaiac tests. Immunologic tests do not react with drugs, red meats, or nonhemoglobin peroxidase compounds. They utilize antibodies against human globin epitopes. Sensitivity compares favorably with that of a guaiac test. See table on following page.

When paper slides impregnated with guaiac are not promptly processed, drying occurs, which may diminish sensitivity. Such slides may be rehydrated with a drop of deionized water. Rehydration increases the number of positives but decreases specificity.

Additional Information Methods for guaiac tests of stool for occult blood utilize peroxidation of a chromogen by stool peroxidases. Hemoglobin acts as a peroxidase, but stool may also contain meat, bacterial and plant peroxidases. Hemoccult® begins to turn positive at about 5 mg hemoglobin/g feces, which is considered to be the upper limit of normal stool peroxidase activity. This method is capable of detection of 6 mg of added hemoglobin/g feces in 90% of observations, but will fail 80% of the time to detect up to 1.5 mg/g feces. The heme-derived porphyrin based method (HemoQuant®) has been shown to correlate closely with ^{51}Cr-labeled RBC radioisotope measurements of short-term (12 day) quantitation of fecal blood loss.

Stool obtained by digital rectal examination did not increase the number of false positives in asymptomatic patients at average risk. New lesions have been identified by this means. Alternatives to FOBT include colonoscopy, which is an expensive gold standard, and serum ferritin.

Approximately 56,000 deaths occur in the U.S. annually from colorectal cancer. Family history may be relevant.

References

Allison JE, Tekawa IS, Ransom LJ, et al, "A Comparison of Fecal Occult-Blood Tests for Colorectal-Cancer Screening," *N Engl J Med*, 1996, 334(3):155-9.

Bini EJ, Rajapaksa RC, and Weinshel EH, "The Findings and Impact of Nonrehydrated Guaiac Examination of the Rectum (FINGER) Study: A Comparison of 2 Methods of Screening for Colorectal Cancer in Asymptomatic Average-Risk Patients," *Arch Intern Med*, 1999, 159(17):2022-6.

Bond JH, "Fecal Occult Blood Tests in Occult Gastrointestinal Bleeding," *Semin Gastrointest Dis*, 1999, 10(2):48-52.

Mandel JS, Church TR, Bond JH, et al, "The Effect of Fecal Occult-Blood Screening on the Incidence of Colorectal Cancer," *N Engl J Med*, 2000, 343(22):1603-7.

Rockey DC, "Occult Gastrointestinal Bleeding," *N Engl J Med*, 1999, 341(1):38-46.

Rockey DC, Auslander A, and Greenberg PD, "Detection of Upper Gastrointestinal Blood With Fecal Occult Blood Tests," *Am J Gastroenterol*, 1999, 94(2):344-50.

(Continued)

Occult Blood, Stool

	Guaiac	Heme-Derived Porphyrin Fluorescence	Immunoassays
Cost	Low	High	Intermediate
Reliability	False-positives, false-negatives; best at detection of large, distal lesions; questions about accuracy	Little better	Intermediate; loss of globin antigenicity at room temperature
Specificity	Poor	Measures both heme and porphyrins, but a high false-positive rate limits application of this test	Improved over guaiac
Recognition of porphyrin derived from heme	No	Yes	N/A
Recognition of bleeding from right colon (intestinal converted fraction)	Like immunologic tests, especially insensitive-enteric heme degradation	Sensitive	Similar to guaiac
Sensitivity deteriorates with time, fecal storage (eg, mailed in specimens)[1]	Yes (Hemoccult II Sensa® more sensitive than Hemoccult II®)	No	Similar to guaiac
Such reducing substances as ascorbic acid cause false-negatives	Yes	No	No
Antacids cause false-negatives[1]	Yes	No	No
Dietary (red meats) hemoglobin[2]	Yes	Yes	No
Vegetable peroxidases[2]	False-positives	Not affected	Not affected
Rehydration[2]	Yes Raises sensitivity but reduces specificity	No	No
Iron, dietary	No	No	No
Ease of test performance	Not difficult	Complex	Complex - an enzyme immunoassay is available and counterimmunoelectrophoresis is described.
Commonly used tests	Hemoccult II® Hemoccult II Sensa®	HemoQuant®	Heme-Select® FlexSure OBT®

[1]Causes of false-negative results.

[2]Causes of false-positive results.

Rozen P, Knaani J, and Samuel Z, "Performance Characteristics and Comparison of Two Immuno-chemical and Two Guaiac Fecal Occult Blood Screening Tests for Colorectal Neoplasia," *Dig Dis Sci*, 1997, 42(10):2064-71.

Sinatra MA, St John DJ, and Young GP, "Interference of Plant Peroxidases With Guaiac-Based Fecal Occult Blood Tests Is Avoidable," *Clin Chem*, 1999, 45(1):123-6.

Woolf SH, "The Best Screening Test for Colorectal Cancer - A Personal Choice," *N Engl J Med*, 2000, 343(22):1641-3.

♦ **Occult Blood, Urine** *see* Blood, Urine *on page 281*

♦ **Octocaine®** *see* Lidocaine, Serum or Plasma *on page 850*

Ocular Cytology

Related Information

Chlamydia trachomatis Direct Antigen Test *on page 387*
Herpes Simplex Virus Culture and Antigen Detection *on page 721*
Viral Culture *on page 1307*

Synonyms *Chlamydia* Smears Cytology; Conjunctival Smear Cytology; Corneal Cytology; Eye Smear for Cytology; Vitreous Fluid

Applies to Inclusion Conjunctivitis

Test Includes Cytologic evaluation of direct smears, cytocentrifuge (cytospin) preparations, and/or liquid-based monolayer preparations most commonly; membrane filter preparations are used less routinely. Air-dried smears stained with Giemsa or Diff-Quik stain are commonly used.

Abstract Useful for diagnosis of infections, inflammatory conditions, and primary as well as metastatic malignances

Specimen Direct smear (scraping, washing) of ocular lesion or fine needle aspiration specimen

Storage Instructions Keep specimen refrigerated until it can be delivered to the laboratory.

Use Diagnose trachoma-inclusion conjunctivitis and evaluate possible dysplastic or malignant conjunctival lesions; evaluation of certain intraocular and/or orbital tumors. Cytologic examination of material aspirated from the aqueous or vitreous has proven useful for large cell lymphoma, retinoblastoma, and phago-lytic glaucoma.

Additional Information Diagnosis of viral and chlamydial infections is consid-erably improved by immunofluorescent and immunoperoxidase stains for orga-nisms. Leibowitz has provided a beautifully illustrated, recent (published in 2000) clinical review paper.

References
Leibowitz HM, "The Red Eye," *N Engl J Med*, 2000, 343(5):345-51.

♦ **OD 450 Method** *see* Bilirubin, Amniotic Fluid, Delta A450 *on page 261*

♦ **Off-Site Testing** *see* Point-of-Care Testing *on page 1065*

♦ **OGTT** *see* Glucose Tolerance Test, Plasma *on page 651*

♦ **1,25-(OH)$_2$ D$_3$** *see* Vitamin D, Serum *on page 1318*

♦ **17-OHCS** *see* 17-Hydroxycorticosteroids, Urine *on page 753*

♦ **25-(OH) D$_3$** *see* Vitamin D, Serum *on page 1318*

♦ **17-OHP** *see* 17-Hydroxyprogesterone, Whole Blood, Serum, or Plasma *on page 755*

♦ **17-OHP, Amniotic Fluid** *see* 17-Hydroxyprogesterone, Whole Blood, Serum, or Plasma *on page 755*

♦ **17-OHP, Saliva** *see* 17-Hydroxyprogesterone, Whole Blood, Serum, or Plasma *on page 755*

♦ **Oligoalbuminuria** *see* Microalbuminuria *on page 913*

♦ **Oligoclonal Bands, Cerebrospinal Fluid** *see* Cerebrospinal Fluid Oligoclonal Bands *on page 369*

♦ **Oligoclonal Bands, CSF** *see* Cerebrospinal Fluid IgG:Albumin Ratio, IgG Index, and IgG Synthesis Rate *on page 365*

♦ **Omnipres®** *see* Amoxapine, Serum or Plasma *on page 153*

♦ **OMS®** *see* Morphine, Urine *on page 921*

♦ **Oncogenes** *see* bcl-2 Gene Rearrangement *on page 251*

♦ **o-Phosphoric-Monoester Phosphohydrolase** *see* Acid Phosphatase, Plasma *on page 97*

Opiates, Qualitative, Urine

Related Information

Codeine, Urine *on page 429*
Morphine, Urine *on page 921*

Applies to Heroin; Narcotics; 6-O-Acetyl Morphine; Poppy Seeds

Test Includes Morphine, codeine, hydrocodone (Hycodan®), hydromorphone (Dilaudid®), oxycodone, oxymorphone

Abstract Opioids are among the most common and effective analgesics for the treatment of mild to severe pain. They have a high potential for addiction, leading to physical and psychological dependence. Opioids include derivatives of the poppy plant, semisynthetic substances, and synthetic compounds. The qualitative detection of urine opiates is used almost exclusively to demonstrate presence of drugs of abuse in this class. Morphine and codeine are commonly used therapeutically for pain.

Specimen Random urine

Collection If forensic, observe precautions (see the Introduction *on page 63*).

Storage Instructions Refrigerate

Special Instructions If forensic, use chain-of-custody protocol and form. See Chain-of-Custody Protocol *on page 381* and the Chain-of-Custody form in the Introduction *on page 63*.

Reference Interval Negative (less than cutoff)

Critical Values Substance Abuse and Mental Health Services Administration (SAMHSA) cutoff: screen: 2000 ng/mL; confirmation: 2000 ng/mL for codeine and morphine. For clinical purposes, many laboratories use a cutoff of 300 ng/mL.

Use Evaluate opiate use

Methodology Screening: immunoassay; confirmation: gas chromatography/mass spectrometry (GC/MS)

Additional Information For morphine:

- Half-life: 2-4 hours (adults), 5-13 hours (neonates)
- Volume of distribution: 2-4 L/kg
- Protein binding: 30% to 40%

Its presence in urine, even after confirmation, must be interpreted very carefully. Ingestion of poppy seeds (bagels, pastries including Danish) can often cause positive opiate screens at a 300 ng/mL cutoff and sometimes at a cutoff of 2000 ng/mL. Due to this, SAMHSA has recently increased the cutoff of morphine and codeine to 2000 ng/mL. The intake of heroin by the user can only be proved by the detection of 6-O-acetyl morphine by the urine confirmatory test.

References

Joranson DE, Ryan KM, Gilson AM, et al, "Trends in Medical Use and Abuse of Opioid Analgesics," *JAMA*, 2000, 283(13):1710-4.

Samet JH, "Drug Abuse and Dependence," *Cecil Textbook of Medicine*, 21st ed, Chapter 17, Goldman L and Bennett JC, eds, Philadelphia, PA: WB Saunders Co, 2000, 54-9.

"Substance Abuse and Mental Health Administration. Mandatory Guidelines for Workplace Programs: Revision to Mandatory Guidelines," *Fed Regist*, 1998, 63:483.

♦ **Organic Mercury** *see* Mercury, Blood *on page 897*

Organophosphate Pesticides, Urine, Blood, or Serum

Related Information

Acetylcholinesterase, Red Cell and Serum *on page 93*

Pseudocholinesterase, Serum *on page 1122*

Applies to Carbamate Toxicity; Insecticides; Pesticides; Pralidoxime

Test Includes Azinphos-methyl, carbophenthion, chlorpyrifos, coumaphos, diazinon, dichlorvos, dimethoate, ethion, fenchlorphos, fenthion, fonofos, malathion, metasystox, methyl parathion, mevinphos, *p*-nitrophenol, paraoxon, parathion, phorate, terbufos

Abstract The organophosphates and the carbamates are the insecticide groups. Organophosphorus insecticide poisoning is a major global health problem with ~3 million poisonings or more and 220,000 deaths annually. These agents cause poisoning by irreversibly inhibiting acetylcholinesterase. Organophosphate poisoning causes CNS intoxication and polyneuropathy.

There are two cholinesterase enzymes in blood: **acetylcholinesterase** (also called true cholinesterase) in red cells; and "**pseudocholinesterase**" (acylcholine acylhydrolase) in serum. Organophosphates inhibit both enzymes, but the toxic effect on the serum enzyme, pseudocholinesterase, is more rapid and intense, so that it is somewhat more useful in the initial diagnosis of organophosphate toxicity. Red cell acetylcholinesterase is often preferred for evaluating chronic organophosphate exposure; red cell acetylcholinesterase levels normalize more slowly than do serum pseudocholinesterase values.

Specimen Urine and blood or serum

Container Lavender top (EDTA) tube or red top tube

Storage Instructions Freeze sample if analysis cannot be performed immediately.

Use Determine occupational, accidental, and intentional poisoning. Insecticides are among the most toxic pesticides and cause most pesticide intoxication.

Limitations Correlation between severity of toxicity and the amount of pesticide metabolite in urine is poor.

Methodology High performance thin layer chromatography (HPTLC). Urinary metabolites, the di-alkyl-phosphates, can be measured in urine by gas chromatography.

Additional Information Treatment of organophosphate poisoning includes atropine treatment and suctioning of oral secretions as required until atropinization is achieved. (Atropine is a physiologic antidote, used to treat muscarinic effects.) Pralidoxime (Protopam®, 2-PAM), a specific antidote, should be administered to seriously ill organophosphate-poisoned patients. If induction of paralysis with muscle-relaxing agents is required for intubation, succinylcholine should be avoided because of potential prolonged duration of paralysis secondary to pseudocholinesterase inhibition by the organophosphate.

The chemical terrorist agents Sarin, Soman, Tabun, and VX are organophosphate nerve gases.

References

Keifer MC, "The Clinical Laboratory in the Diagnosis of Overexposure to Agrochemicals," *Lab Med*, 1998, 29(11):689-95.

Khurana D and Prabhakar S, "Organophosphorus Intoxication," *Arch Neurol*, 2000, 57(4):600-2.

Senanayake N and Karalliedde L, "Neurotoxic Effects of Organophosphorus Insecticides. An Intermediate Syndrome," *N Engl J Med*, 1987, 316(13):761-3.

♦ **Organophosphates** *see* Chemical Warfare Agents *on page 382*

♦ **Organophosphate Toxicity** *see* Acetylcholinesterase, Red Cell and Serum *on page 93*

♦ **Organan®** *see* Heparin Antifactor Xa Assay *on page 693*

♦ **Ormazine®** *see* Chlorpromazine, Serum *on page 395*

♦ **Ormazine®** *see* Phenothiazines, Serum *on page 1021*

♦ **Ornithine** *see* Amino Acids, Urine *on page 145*

♦ **Ornithine** *see* Cystine, Urine *on page 494*

♦ **Orosomucoid** *see* Alpha₁-Acid Glycoprotein, Serum *on page 133*

♦ **Orotic Aciduria** *see* Cobalamin, Serum *on page 424*

♦ **Osmolal Gap** *see* Osmolality, Calculated, Serum or Plasma *on page 976*

♦ **Osmolal Gap** *see* Osmolality, Serum *on page 978*

♦ **Osmolal Gap** *see Sodium, Serum or Plasma on page 1210*

♦ **Osmolal Gap, Urine** *see Osmolality, Urine on page 979*

Osmolality, Calculated, Serum or Plasma

Related Information

Anion Gap, Serum, Plasma, or Urine *on page 160*
Antidiuretic Hormone, Plasma *on page 172*
Bicarbonate, Blood *on page 258*
Electrolyte Panel, Serum *on page 532*
Ethanol, Blood, Urine, and Other Sources *on page 558*
Ethylene Glycol, Serum or Plasma *on page 561*
Ketone Bodies, Blood *on page 816*
Osmolality, Serum *on page 978*
Osmolality, Urine *on page 979*
pH, Blood *on page 1018*
Sodium, Serum or Plasma *on page 1210*
Sodium, Urine *on page 1213*
Volatile Screen, Blood or Urine *on page 1320*

Applies to Isopropyl Alcohol Intoxication; Methanol; Osmolal Gap; Pseudohyponatremia

Test Includes Sodium, urea nitrogen (BUN), glucose

Abstract Serum osmolality may be directly measured, or predicted, using one of several equations based on the concentrations of serum sodium, glucose, and urea. The purpose of each equation is the necessity, now unique to the U.S., of converting the conventionally reported glucose and urea results into molar (or SI) units. A formula widely used to calculate osmolality in mOsm/kg H_2O, is:

Calculated osmolality = 2[Na] + BUN / 2.8 + GLU / 18

in which BUN is the serum urea nitrogen in mg/dL; and GLU is the plasma or serum glucose in mg/dL. Dividing the BUN by 2.8 converts the measurement from mg/dL to mmol/L. The same is true for dividing the glucose by 18. Multiplying the sodium, already in mmol/L, by 2 is done to account for the osmotically active anions (mostly chloride and bicarbonate) associated with the sodium. The **osmolal gap** is the arithmetic difference between the measured osmolality and the calculated osmolality.

The **sodium** value used in the equation should be one measured without dilution by a sodium-selective ion electrode. This avoids using the erroneous value (euphemistically called pseudohyponatremia) which results from a sodium measurement made after dilution by flame photometry on a specimen with an increased concentration of lipid or protein.

Patient Preparation Patient ideally should be fasting for 8 hours, a setting not usually possible when the need for investigation of osmolality arises.

Specimen Serum or plasma

Container Red top tube, green top (heparin) tube

Collection Keep one green top tube on ice, should a pH be needed.

Causes for Rejection Gross hemolysis

Reference Interval

- Calculated osmolality: 290 mOsm/kg H_2O
- Measured osmolality: up to age 60 years: 275-295 mOsm/kg H_2O, older than 60 years: 280-301 mOsm/kg H_2O
- Osmolal gap: 9.0 (±6.4) mOsm/kg H_2O

Possible Panic Range A gap >20 mOsm/kg H_2O (SI: >20 mmol/kg H_2O)

Use Hyponatremia is a laboratory finding which raises the possibility of an osmolal disturbance and prompts additional laboratory studies. Since some, but by no means all, patients with hyponatremia are at risk for acute cerebral edema, prompt diagnosis is essential. The following table details typical findings in various clinical conditions having hyponatremia as a common denominator. See Sodium, Serum or Plasma *on page 1210*.

Limitations Considerable variability occurs in the normal osmolal gap.

Methodology The tests above can be done on a variety of instruments, that have in common the items necessary to calculate osmolality and to follow many of the patients seen in hospital practice, with other tests ordered, as clinically indicated. See individual listings for sodium, BUN, and glucose.

Typical Serum Sodium, Osmolality, and Effective Osmolality (Tonicity) in Different Clinical States[1]

Condition	Serum Sodium mmol/L	Blood Glucose mmol/L (mg/dL)	Serum Urea Nitrogen mmol/L (mg/dL)	Mannitol or Ethanol mmol/L	Osm_c mmol/kg H_2O	Osm_m mmol/kg H_2O	Osmolal Gap mmol/kg H_2O	Effective Osmolality Osm_m mmol/kg H_2O[2]	Risk of Cerebral Edema[3]
Normal	140	5 (90)	5 (14)	0	290	290	0	285 (normal)	None
Hyponatremia (without abnormal amounts of other solutes)	120	5 (90)	5 (14)	0	250	250	0	245 (low)	Increased
Pseudohyponatremia (eg, from extreme hypertriglyceridemia)	120	5 (90)	5 (14)	0	250	290	40	285 (normal)	Unchanged
Hyponatremia caused by severe hyperglycemia	120	75 (1350)	5 (14)	0	320	320	0	315 (high)	Variable[4]
Hyponatremia caused by retention of mannitol	120	5 (90)	5 (14)	75 See Footnote 5.	250	325	75	320 (high)	Decreased
Hyponatremia together with high serum urea nitrogen[6]	120	5 (90)	45 (126)	0	290	290	0	245 (low)	Increased
Hyponatremia together with high blood ethanol level[6]	120	5 (90)	5 (14)	40 See Footnote 7.	250	290	40	245 (low)	Increased
Hypernatremia	160	5 (90)	5 (14)	0	330	330	0	325 (high)	Decreased

[1] Osm_c indicates calculated osmolality (calculated as $2[Na] + SUN / 2.8 + GLU / 18$, where SUN indicates serum urea nitrogen, and GLU glucose); Osm_m indicates measured osmolality (assume that normal osmolal gap is zero). Osmolal gap is $Osm_m - Osm_c$.

[2] Effective osmolality (tonicity) is that portion of osmolality inducing transmembrane water movement; the cited values for effective osmolality were calculated by subtracting the contributions of urea and ethanol (if present) from the measured osmolality.

[3] Immediate risk of the osmotic type of cerebral edema before treatment (the more acute the hyponatremia, the greater the risk, since osmotic adaptation is less advanced).

[4] Effect on intracellular fluid volume depends on clinical circumstances.

[5] Neglecting the correction factor for serum water content. 75 mmol of mannitol per liter corresponds to approximately 1365 mg/dL.

[6] Substances (eg, urea and ethanol) that easily cross cell membranes contribute to measured osmolality, but not to tonicity.

[7] Neglecting the correction factor for serum water content. 40 mmol of ethanol per liter corresponds to approximately 184 mg/dL.

Adapted from Oster JR and Singer I, "Hyponatremia, Hypo-osmolality, and Hypotonicity," Arch Intern Med, 1999, 159(4):333-6.

(Continued)

Osmolality, Calculated, Serum or Plasma (Continued)

Causes of Increased Osmolal Gap, Overview

Exogenous Causes

    Acetone

    Methanol

    Ethanol

    Isopropanol

    Ethylene glycol

    Propylene glycol

    Ethyl ether

    Dimethyl sulfoxide[1]

    Glycine[1]

    Mannitol[1]

    Osmotic contrast dyes

Diseases

    Chronic renal failure

    Lactate acidosis?

    Alcoholic ketoacidosis

    Diabetic ketoacidosis

    Hyperlipidemia,[2] hyperglobulinemia[2]

Artifactual

    Lavender top (EDTA[3]) tube 15 mOsm/L[4]

    Gray top (sodium fluoride - potassium oxalate) tube 150 mOsm/L[4]

    Blue top (citrate) tube 10 mOsm/L[4]

    Green top (lithium heparin) tube 6 mOsm/L[4]

[1]These agents are given by intravenous infusion.

[2]Sodium was measured by flame emission spectrophotometry.

[3]EDTA indicates ethylenediaminetetraacetic acid.

[4]Values are reported as calculated contributions.

Adapted from Osterloh JD, Kelly TJ, Khayam-Bashi H, et al, "Discrepancies in Osmolal Gaps and Calculated Alcohol Concentrations.," *Arch Pathol Lab Med*, 1996, 120(7):637-41.

References

Desai SP and Isa-Pratt S, *Clinician's Guide to Laboratory Medicine*, Cleveland, OH: Lexi-Comp Inc, 2000, 228-35.

Oster JR and Singer I, "Hyponatremia, Hypo-osmolality, and Hypotonicity: Tables and Fables," *Arch Intern Med*, 1999, 159(4):333-6.

Osterloh JD, Kelly TJ, Khayam-Bashi H, et al, "Discrepancies in Osmolal Gaps and Calculated Alcohol Concentrations," *Arch Pathol Lab Med*, 1996, 120(7):637-41.

Osmolality, Serum

Related Information

Anion Gap, Serum, Plasma, or Urine *on page 160*

Antidiuretic Hormone, Plasma *on page 172*

Carbohydrate-Deficient Transferrin, Serum *on page 338*

Drugs of Abuse Testing, Urine *on page 525*

Electrolyte Panel, Serum *on page 532*

Ethanol, Blood, Urine, and Other Sources *on page 558*

Ethylene Glycol, Serum or Plasma *on page 561*

Ketone Bodies, Blood *on page 816*

Osmolality, Calculated, Serum or Plasma *on page 976*

Osmolality, Urine *on page 979*

pH, Blood *on page 1018*

Sodium, Serum or Plasma *on page 1210*

Specific Gravity, Urine *on page 1216*

Urea Nitrogen:Creatinine Ratio *on page 1283*

Synonyms Serum Osmolality

Applies to Osmolal Gap; Osmolality:Serum Ratio; Sodium, Serum:Osmolality, Serum Ratio; Urine:Serum Osmolality Ratio

Replaces Osmolarity

Abstract The osmolality of a solution is the number of particles in a liter of solution. Osmolality is independent of particle size or charge.

Ethanol ingestion is a common cause of increased osmolality and is responsible for most osmolal gap testing.

Specimen Serum

Container Red top tube

Collection Pediatrics: Blood drawn from heelstick

Storage Instructions Refrigerate or freeze serum if not run within 4 hours.

Reference Interval 275-295 mOsm/kg H_2O (SI: 275-295 mmol/kg H_2O)
- Urine:serum osmolality ratio: 1.0-3.0 with fluid restriction: 3.0-4.7
- Sodium, serum:osmolality:serum ratio: 0.43-0.50

Possible Panic Range <265 mOsm/kg H_2O (SI: <265 mmol/kg H_2O), >320 mOsm/kg H_2O (SI: >320 mmol/kg H_2O). Result of 385 mOsm/kg H_2O (SI: 385 mmol/kg H_2O) may reflect stupor in hyperglycemia. Values 400-420 mOsm/kg H_2O (SI: 400-420 mmol/kg H_2O) may reflect grand mal seizures. Values >420 mOsm/kg H_2O (SI: >420 mmol/kg H_2O) may be lethal.

Use Serum osmolality is used to evaluate electrolyte and water balance, hyperosmolar status, hydration/dehydration status, acid-base balance, seizures, antidiuretic hormone function, liver disease, and hyperosmolar coma. Osmolality is proportional to the concentration of particles in solution. Freezing point depression serum osmolality with calculated osmolal gap is useful in screening for and approximating the serum concentrations of certain low molecular weight toxins, such as ethanol, ethylene glycol, isopropanol, and methanol, especially as a rapid approximation for emergent situations. See Limitations.

High serum osmolality may result from hypernatremia, dehydration, hypovolemia, hyperglycemia, mannitol therapy, azotemia, and ingestion of ethanol, methanol, or ethylene glycol. Thus, osmolality has a role in toxicology and in coma evaluation. Very low birth weight infants may have elevated serum osmolality for the first week of life.

Low serum osmolality may be secondary to overhydration, hyponatremia, and the syndrome of inappropriate antidiuretic hormone secretion (SIADH).

Serum osmolality measurements do not measure the fraction of serum that is water. Osmolality measurement by freezing point depression is also indifferent to permeability of solutes to cell membranes.

Limitations When vapor pressure osmometry is used, volatile solutes (eg, alcohols and glycols) may remain in the vapor phase and not be detected.

Methodology Freezing point depression (more often used) or vapor pressure elevation

Additional Information Measured osmolality usually exceeds the calculated osmolality. If measured osmolality is >15 mOsm/kg H_2O (SI: >15 mmol/kg H_2O) greater than calculated, the differential diagnosis includes: methanol, ethylene glycol, ethanol, or other toxicity; shock and trauma. Elevated serum osmolality with normal sodium suggests hyperglycemia, uremia, or alcoholism. Both serum and urine values and calculated osmolality (see Osmolality, Calculated, Serum or Plasma *on page 976*) are sometimes required for accurate diagnosis.

After overnight dehydration, **urine:serum osmolality ratio** is usually ≥3. Even with fluid restriction, the ratio is 0.2-0.7 in diabetes insipidus. It is usually normal in neurogenic polyuria without fluid restriction and is increased with fluid restriction.

The expression "urinary:plasma ratio" is also used. See Osmolality, Urine *on page 979*. Laboratory criteria for hypovolemia include elevation of urea nitrogen:creatinine ratio >25 and/or increased serum osmolality and sodium concentration. Serum sodium and osmolality increase in dehydration; the **serum sodium:serum osmolality ratio** remains within normal limits.

References

DuFour DR, "Laboratory Recognition and Testing in Acid-Base Disorders," *Lab Med*, 1999, 30(12):776-82.

Osterloh JD, Kelly TJ, Khayam-Bashi H, et al, "Discrepancies in Osmolal Gaps and Calculated Alcohol Concentration," *Arch Pathol Lab Med*, 1996, 120(7):637-41.

♦ **Osmolality:Serum Ratio** *see* Osmolality, Serum *on page 978*

Osmolality, Urine

Related Information

Antidiuretic Hormone, Plasma *on page 172*

(Continued)

Osmolality, Urine *(Continued)*

Osmolality, Calculated, Serum or Plasma *on page 976*
Osmolality, Serum *on page 978*
Sodium, Serum or Plasma *on page 1210*
Specific Gravity, Urine *on page 1216*

Synonyms Urine Osmolality

Applies to Osmolal Gap, Urine; U:P Ratio; Urine Osmolar Gap; Urine:Serum Osmolality Ratio

Abstract Osmolality is the number of particles in a liter of solution.

Specimen Random or timed urine (1 mL minimum)

Storage Instructions Refrigerate

Reference Interval Random urine: neonates: 75-300 mOsm/kg H_2O (SI: 75-300 mmol/kg H_2O); children and adults: 250-900 mOsm/kg H_2O (SI: 250-900 mmol/kg H_2O). Normal range of serum sodium (mmol/L) to osmolality (mOsm/kg) ratio is 0.43-0.50. Patients with normal renal function after 14-hour restriction of fluids should be able to concentrate to >800 mOsm/kg H_2O (SI: >800 mmol/kg H_2O).

Critical Values <400 mOsm/kg H_2O (SI: <400 mmol/kg H_2O) is interpreted by Weisberg as severe renal impairment. Prolonged dehydration may be dangerous for some patients.

Possible Panic Range <100 mOsm/kg H_2O (SI: <100 mmol/kg H_2O) in overhydration, >800 mOsm/kg H_2O (SI: >800 mmol/kg H_2O) in dehydration

Use Urine osmolality is use to evaluate the concentrating ability of the kidneys (eg, in acute and chronic renal failure), electrolyte and water balance, renal disease, syndrome of inappropriate antidiuretic hormone secretion (SIADH), diabetes insipidus, dehydration, and amyloidosis. Estimation of urinary ammonium concentration and detection of increased osmolarity due to unusual molecules are possible using the urine osmolal gap. Osmolality is desirable in examination of neonatal urine when protein or glucose are present.

Limitations Serum osmolality is often needed to interpret urine osmolality.

Methodology Freezing point depression

Additional Information Osmolality is a better measurement of urine concentration than specific gravity and is a measure of renal tubular concentration, depending on the state of hydration. Simultaneous determination of urine and serum osmolalities facilitates interpretation of results. High **urinary:plasma ratio** (U:P) is seen in concentrated urine. Normal ranges for the U:P ratio are given by Dr Weisberg as approximately 0.2-4.7, and >3.0 with overnight dehydration. With poor concentrating ability the ratio is low but still ≥1.0. In SIADH urine sodium and urine osmolality are high for plasma osmolality. (The expression urine:serum osmolality ratio is also used, see Osmolality, Serum *on page 978*.)

Specifically derived regression equations are advocated in neonates to predict urine osmolality from specific gravity measurements.

The urine osmolal gap is the sum of urinary concentrations of sodium, potassium, bicarbonate, chloride, glucose, and urea compared to measured urine osmolality. The gap is normally 80-100 mOsm/kg H_2O (SI: 80-100 mmol/kg H_2O) greater for measured than for calculated. Determination of the urine osmolal gap is used to characterize metabolic acidosis. High urine osmolal gap can be used semiquantitatively.

References

Halperin ML, Margolis BL, Robinson LA, et al, "The Urine Osmolal Gap: A Clue to Estimate Urine Ammonium in "Hybrid" Types of Metabolic Acidosis," *Clin Invest Med*, 1988, 11(3):198-202.

Leech S and Penney MD, "Correlation of Specific Gravity and Osmolality of Urine in Neonates and Adults," *Arch Dis Child*, 1987, 62(7):671-3.

Weisberg HF, "Unraveling the Laboratory Model of a Syndrome: The Osmolality Model," *Clinician and Chemist. The Relationship of the Laboratory to the Physician*, Young DS, Hicks J, Nipper H, et al, eds, Washington, DC: AACC Press, American Association of Clinical Chemistry, 1979, 200-43.

♦ **Osmolarity** *see* Osmolality, Serum *on page 978*

Osmotic Fragility

Related Information

Autohemolysis Test *on page 221*
Osmotic Fragility, Incubated *on page 982*

Peripheral Blood: Red Blood Cell Morphology *on page 1016*
Red Blood Cell Indices *on page 1136*
Reticulocyte Count *on page 1156*

Synonyms Fragility, Red Blood Cells

Applies to Pyropoikilocytosis, Hereditary

Abstract Osmotic fragility, when increased, is consistent with the presence of spherocytes in the peripheral blood. Spherocytes (spherical red blood cells) form as a result of loss of red cell membrane surface area relative to intracellular volume. Osmotic fragility is increased when spherocytic red cells are present as occurs with hereditary spherocytosis (HS) heterogeneous clinically and pathologically, as the result of diversity at the molecular level. Abnormal osmotic fragility occurs with, but is not exclusively diagnostic of, hereditary spherocytosis.

Specimen Whole blood

Container Lavender top (EDTA) tube or green top (heparin) tube

Storage Instructions Store refrigerated (4°C) if test performance must be delayed.

Causes for Rejection Hemolysis, clotted specimen, blood more than 6 hours old, oxalate or citrate anticoagulated blood collected

Special Instructions May need to schedule this test in advance.

Reference Interval Hemolysis begins 0.45%, hemolysis complete 0.35%. In some 10% to 20% of individuals with HS, the osmotic fragility curve (unincubated red cells) will be in the normal range. Osmotic fragility test after incubation (see Osmotic Fragility, Incubated *on page 982*) will identify the great majority of these "false-negative" cases.

Use Evaluate hemolytic anemia, especially hereditary spherocytosis; evaluate immune hemolytic states

Limitations Any severe anemia including iron deficiency will yield an abnormal curve. Test measures presence of spherocytes and "spheroidal" cells. Test is **not** specific for hereditary spherocytosis. Trauma-free venipuncture is needed.

Additional Information HS is a common hemolytic anemia with a broad spectrum of severity. Patients have chronic hemolysis but may be clinically asymptomatic or may suffer severe uncompensated hemolysis that responds to splenectomy. HS is the result of inherited defects of the erythrocyte membrane skeleton. A series of proteins interact with each other and form the red cell membrane and interact with lipid of the outer lipid bilayer. Vertical interactions (perpendicular to the membrane) stabilize the lipid bilayer and include spectrin-ankryin-band 3 interactions and protein 4.1-glycophorin C linkage. Horizontal interactions (parallel to the plane of the membrane) are responsible for structural integrity of the red cell membrane. Horizontal interactions affect the assembly of tetramers from spectrin heterodimers at the spectrin heterodimer site and contact with actin and protein 4.1. HS is considered a disorder of vertical interactions. The lipid bilayer is destabilized, lipid is lost, and the membrane surface area becomes deficient with resultant spherocytosis. Studies have identified structural membrane abnormalities corresponding to four different groups of patients. These are combined spectrin and ankyrin deficiency, the most common defect, followed by band 3 deficiency, isolated spectrin deficiency, and protein 4.2 deficiency.

HS with uncompensated hemolytic anemia may be associated with reticulocytosis (usually mild to moderate) and mild elevation of serum indirect bilirubin. Patients are susceptible to pigment stones of the gallbladder and to aplastic marrow crises. Spherocytes are more susceptible than are normal red cells to hemolysis in dilute (hypotonic) saline. They show increased osmotic fragility. Spherocytes of any origin (including conditions other than hereditary spherocytosis, eg, autoimmune hemolytic anemia) will cause increased osmotic fragility. Generally, fully expanded cells (ie, spheroidal cells or spherocytes) have increased osmotic fragility while cells with higher surface area to volume ratios (eg, thin cells, hypochromic, target) have decreased osmotic fragility, including some cases of stomatocytosis. See following table for some hematologic findings that may assist in the characterization of some red cell membrane disorders.

The molecular pathology of spherocytosis and other red cell membrane structural protein defects has been partially established and described. Most hereditary hemolytic anemias (including spherocytosis and elliptocytosis) involve
(Continued)

Osmotic Fragility (Continued)

Laboratory Differentiation of RBC Membrane Disorders

Test	Result in Patients With:				
	HS	HE	HPP	HST	HX
RBC morphology	Spherocyte	Elliptocyte	Spherocyte, elliptocyte, poikilocyte	Stomatocyte	Target cell
MCV	↑, ↓, or normal	Slightly ↑ or normal	↓	↑	↑ or normal
MCHC	↑	Normal	↑	↓	↑
Osmotic fragility	↑	Normal	↑	↑	↓
Thermal sensitivity	NA	↑, fragments at 47°C-48°C	↑, fragments at 46°C	NA	NA

HS = hereditary spherocytosis; HE = hereditary elliptocytosis; HPP = hereditary pyropoikilocytosis; HST = hereditary stomatocytosis; HX = hereditary xerocytosis; MCV = mean corpuscular volume; MCHC = mean corpuscular hemoglobin concentration; NA = not applicable.

Adapted from Hassoun H, Vassiliadis JN, Murray J, et al, "Hereditary Spherocytosis With Spectrin Deficiency Due to an Unstable Truncated β Spectrin," *Blood*, 1996, 87(6):2538-45.

mutations of membrane structural proteins, the majority code for abnormal spectrin molecules. In hereditary elliptocytosis, the defect involves horizontal membrane protein interactions which may relate to defective spectrin chains, defective or deficient-based 4.1, glycophorin C deficiency, or presence of abnormal band 3 (anion transport protein). In hereditary pyropoikilocytosis there are two inherited abnormalities, an α-spectrin deficiency (causing defective vertical interactions) and a mutant spectrin that causes atypical horizontal interactions. Red cell protein 4.2 deficiency has been described. It has been reported in Japanese individuals who have related anemia and whose red cells show osmotic fragility.

Osmotic fragility is increased in cases of malaria infestation. Both infected and uninfected cells show the increased osmotic fragility. Both osmotic and mechanical fragility of RBCs in patients with multiple sclerosis has been reported as increased.

Decreased osmotic fragility (resistance to lysis) may be seen with iron deficiency (hypochromic cell population), other hemoglobinopathies (especially hemoglobin C disease) likely due to the target cell population, and is characteristic of thalassemia.

See Red Blood Cell Indices *on page 1136* for value of histogram of MCHC in detection of hereditary spherocytosis.

References
Delhommeau F, Cynober T, Schischmanoff PO, et al, "Natural History of Hereditary Spherocytosis During the First Year of Life," *Blood*, 2000, 95(2):393-7.

Elghetany MT and Davey FR, "Erythrocytic Disorders," *Clinical Diagnosis and Management by Laboratory Methods*, 20th ed, Henry JB, ed, Philadelphia, PA: WB Saunders Co, 2001, 557-9.

Gallagher PG and Jarolim P, "Red Cell Membrane Disorders," *Hematology: Basic Principles and Practice*, 3rd ed, Chapter 33, Hoffman R, Benz EJ Jr, Shattil SJ, et al, eds, Philadelphia, PA: Churchill Livingstone, 2000, 576-610.

Osmotic Fragility, Incubated

Related Information
Autohemolysis Test *on page 221*
Osmotic Fragility *on page 980*

Abstract The same as the previous test (osmotic fragility) except blood is incubated at 37°C for 24 hours. The test is mainly for diagnosis of hereditary spherocytosis.

Specimen Whole blood

Container Lavender top (EDTA) tube or green top (heparin) tube

Collection Sterile technique must be used.

Causes for Rejection Specimen hemolyzed, specimen clotted, improper anticoagulant (oxalate or citrate), improper venipuncture technique

Reference Interval See Additional Information.

Use Evaluate hemolytic anemia, particularly hereditary spherocytosis, congenital nonspherocytic hemolytic anemia, thalassemia

Additional Information Incubation accentuates increased osmotic fragility. In cases of nonspherocytic hemolytic anemia, fragility may be normal in the unincubated osmotic fragility test but increased after incubation. See graph.

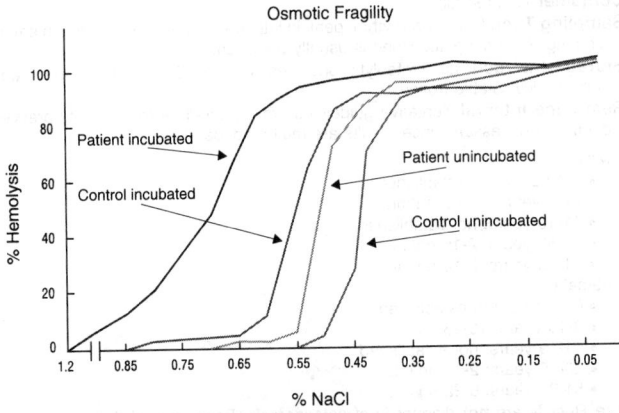

Osmotic fragility of unincubated and incubated RBCs from a normal individual and from a patient with hereditary spherocytosis. Note the increase in fragility produced by incubation of hereditary spherocytosis RBCs.

From Rapaport SI, *Introduction to Hematology*, 2nd ed, Philadelphia, PA: J.B. Lippincott Co., 1987, with permission.

References

Luzzato L and Roper D, "Osmotic Fragility as Measured by Lysis in Hypotonic Saline," *Practical Haematology*, 8th ed, Chapter 13, Dacie JV and Lewis SM, eds, New York, NY: Churchill Livingstone, 1995, 216-20.

Osteocalcin, Serum or Plasma

Related Information

Alkaline Phosphatase Isoenzymes, Serum *on page 125*
Alkaline Phosphatase, Serum *on page 127*
Aluminum, Bone and Bone Biopsy *on page 139*
Calcium, Serum *on page 329*
Calcium, Urine *on page 332*
Hydroxyproline, Total, Urine *on page 757*
N-Telopeptides, Urine *on page 967*
Osteocalcin (Undercarboxylated), Serum *on page 984*
Parathyroid Hormone, Serum *on page 1001*
Pyridinolines (Pyridinoline and Deoxypyridinoline), Urine *on page 1126*
Vitamin D, Serum *on page 1318*

Synonyms BGP; Bone GLA Protein

Applies to Bone Formation; Bone Resorption; Collagen Cross-Link-Associated C-Telopeptide; Collagen Cross-Link-Associated N-Telopeptide; Deoxypyridinoline; DPD; ICTP; NTx; PYD; Pyridinoline

Abstract Osteocalcin (OC), the major noncollagen calcium-binding protein of bone matrix, is synthesized by osteoblasts and has a high affinity for hydroxyapatite. The serum concentration of OC is believed to reflect concurrent bone formation. OC synthesis is vitamin D dependent. The osteogenic activity of OC requires post-translational carboxylation, a process which is vitamin K dependent, and is biochemically analogous to the carboxylation of the vitamin K-dependent coagulation factors. OC serum levels are high during childhood, and peak levels occur in early puberty. A second peak occurs in females at menopause.
(Continued)

Osteocalcin, Serum or Plasma *(Continued)*

Specimen Serum or plasma; because there are method-dependent decreases in immunoreactivity with some antibodies, collect on ice, and separate serum immediately. Keep frozen until ready to assay.

Container Red top tube

Sampling Time Circadian rhythm: peak in late afternoon-evening; nadir in early morning. A morning specimen is usually preferred.

Storage Instructions Proteolytic enzymes degrade OC. Separate from red cells quickly. Stable frozen.

Reference Interval Tentative guidelines are indicated as follows. Interpretive caution is necessary since results are method-dependent.

Male:
- 0-1 years: not established
- 2-10 years: 10-43 ng/mL
- 11-19 years: not established
- 20-50 years: 2-15 ng/mL
- 51-70 years: 2-10 ng/mL

Female:
- 0-1 years: not established
- 2-10 years: 10-43 ng/mL
- 11-19 years: not established
- 20-50 years: 2-15 ng/mL
- 51-80 years: 6-22 ng/mL

Use Results are **not** diagnostic of osteoporosis. Bone mineral densitometry is the gold standard for the diagnosis of osteoporosis. Results may be helpful in:
- **monitoring drug, exercise and/or hormone treatment of osteoporosis**; decreasing serum levels of osteocalcin are interpreted as evidence of a favorable response to treatment
- monitoring bone metabolic changes secondary to Cushing syndrome, primary hyperparathyroidism, malabsorption syndromes, including inflammatory bowel disease and plasma cell myeloma
- monitoring bone metabolic changes occurring as a result of resistance exercise training
- selecting the most appropriate treatment for patients with hypercalciuria

Limitations Most assays measure total OC. Interpretation of total OC results is confounded when patients are on treatment with a vitamin K antagonist such as warfarin, or are consuming a diet deficient in vitamin K. In such situations, a significant fraction of the osteocalcin is undercarboxylated and, therefore lacks osteogenic functionality. These patients may have a normal or high total OC serum level which does not reflect osteogenic activity. One approach to resolve such interpretive ambiguity is to use an ELISA test with specificity for undercarboxylated osteocalcin; an assay is not widely available at this time. (See Osteocalcin (Undercarboxylated), Serum *on page 984*.)

Methodology Immunoassays (multiple labels), hydroxyapatite binding

Additional Information Results are higher in summer than winter.

References

Diaz Diego ED, Guerrero R, and de la Piedra C, "Six Osteocalcin Assays Compared," *Clin Chem*, 1994, 40(11):2071-7.

Hurley DL and Khosla S, "Update on Primary Osteoporosis," *Mayo Clin Proc*, 1997, 72(10):943-9.

Kakonen SM, Hellman J, Karp M, et al, "Development and Evaluation of Three Immunofluorometric Assays That Measure Different Forms of Osteocalcin in Serum," *Clin Chem*, 2000, 46(3):332-7 (biochemical heterogeneity).

Leavelle DE, *Mayo Medical Laboratories Interpretive Handbook*, Rochester, MN: Mayo Medical Laboratories, 1997, 406-7.

Mora S, Pitukcheewanont P, Kaufman FR, et al, "Biochemical Markers of Bone Turnover and the Volume and the Density of Bone in Children at Different Stages of Sexual Development," *J Bone Miner Res*, 1999, 14(10):1664-71.

Watts NB, "Clinical Utility of Biochemical Markers of Bone Remodeling," *Clin Chem*, 1999, 45(8B):1359-68.

Osteocalcin (Undercarboxylated), Serum

Related Information

Alkaline Phosphatase Isoenzymes, Serum *on page 125*
Alkaline Phosphatase, Serum *on page 127*
Aluminum, Bone and Bone Biopsy *on page 139*
Calcium, Serum *on page 329*

Calcium, Urine *on page 332*
Hydroxyproline, Total, Urine *on page 757*
N-Telopeptides, Urine *on page 967*
Osteocalcin, Serum or Plasma *on page 983*
Parathyroid Hormone, Serum *on page 1001*
Prothrombin Time *on page 1116*
Pyridinolines (Pyridinoline and Deoxypyridinoline), Urine *on page 1126*
Vitamin D, Serum *on page 1318*

Abstract Osteocalcin (OC) is a bone-specific protein, which is dependent on both vitamin D and vitamin K (see Osteocalcin, Serum or Plasma *on page 983*). Under ordinary circumstances, total serum OC is assumed to be a reflection of concurrent bone formation. However, this assumption does not hold for persons taking a vitamin K antagonist (warfarin) or persons consuming a diet deficient in vitamin K. Total serum OC in such individuals will contain a substantial proportion of undercarboxylated OC (uOC), a peptide which lacks osteoblastic functionality.

Specimen Serum; because there are method-dependent decreases in immunoreactivity with some antibodies, collect on ice, and separate serum immediately. Keep frozen until ready to assay.

Container Red top tube

Sampling Time Circadian rhythm: peak in late afternoon-evening; nadir in early morning. A morning specimen is usually preferred.

Reference Interval Interpretive caution is essential, since methods are only now being validated. One source, using an in-house enzyme-linked immunoassay (ELISA) method and a hydroxyapatite-binding (HAP) method, report following from a study of elderly French women:

ELISA: uOC: 5.8 (±4.1) ng/mL (±1 SD)
HAP: uOC: 4.4 (±3.3) ng/mL (±1 SD)

Use Results are **not** diagnostic of osteoporosis.

Clinical investigators have studied the role of uOC in prediction of the risk of femoral fracture in elderly persons. In these evaluations, the serum uOC in considered together with bone mineral density in arriving at a predictive metric.

Another potential application for serum uOC is in the evaluation of vitamin K nutritional status.

Limitations Bone marker assays must be used in concert with bone mineral densitometry, other biochemical markers, and various imaging studies. Bone mineral densitometry is the gold standard for the diagnosis of osteoporosis.

Methodology immunoassays (various labels), hydroxyapatite binding

References
Sokoll LJ, Booth SL, O'Brien ME, et al, "Changes in Serum Osteocalcin, Plasma Phylloquinone, and Urinary Gamma-Carboxyglutamic Acid in Response to Altered Intakes of Dietary Phylloquinone in Human Subjects," *Am J Clin Nutr*, 1997, 65(3):779-84.

Vergnaud P, Garnero PJ, Meunier PJ, et al, "Undercarboxylated Osteocalcin Measured With a Specific Immunoassay Predicts Hip Fracture in Elderly Women: The EPIDOS Study," *J Clin Endocrinol Metab*, 1997, 82(3):719-24.

Ova and Parasites, Direct Exam

Related Information
Acid-Fast Stain, Modified, Parasites *on page 96*
Bacterial Culture, Stool *on page 243*
Clostridium difficile Toxin Assay and Culture *on page 416*
Cryptosporidium Antigen Detection by EIA *on page 483*
Cryptosporidium Direct Staining Procedures *on page 484*
Cysticercosis Serology *on page 490*
Echinococcosis Diagnostic Procedures *on page 530*
Entamoeba histolytica Antigen Detection and Serology *on page 538*
Enterovirus Culture *on page 539*
Eosinophil Count *on page 542*
Fecal Lactoferrin *on page 575*
Giardia intestinalis Diagnostic Procedures *on page 636*
HIV-1/HIV-2 Serology *on page 736*
Methylene Blue Stain, Stool *on page 906*
Microsporidia Diagnostic Procedures *on page 915*
Pinworm Preparation *on page 1036*
Rotavirus, Direct Detection *on page 1173*
(Continued)

Ova and Parasites, Direct Exam *(Continued)*

Schistosomiasis Diagnostic Procedures *on page 1180*
Toxoplasmosis Diagnostic Procedures *on page 1265*
Trichinosis Diagnostic Procedures *on page 1272*
Viral Culture *on page 1307*

Test Includes Gross appearance, direct wet mounts, saline and iodine, concentration procedure, hematoxylin smear, or trichrome smear. Many laboratories do not do all tests on all specimens (eg, most laboratories do not routinely examine for *Cryptosporidium* or *Cyclospora* without direct orders by a physician). Evaluation for intestinal parasites would include detection of cysts and trophozoites of protozoa and larvae, ova and adults (including proglottids). Urine specimens can be examined for the identification of certain parasitic infections (eg, *Schistosoma haematobium*, filarial infections, and *Trichomonas vaginalis*).

Abstract The two most common protozoal infections seen worldwide are *Giardia lamblia* and *Entamoeba histolytica*. The symptoms produced by pathogenic intestinal protozoa are similar (eg, diarrhea, cramping, abdominal pain). These symptoms are neither specific nor diagnostic. Additionally, clinical symptoms can vary depending on the type of protozoal infection and the immune status of the patient. The definitive diagnosis of intestinal protozoal infections has depended on the microscopic examination of stool specimens.

Patient Preparation Specimens obtained with a warm saline enema or Fleet® Phospho®-Soda are acceptable. Specimens obtained with mineral oil, bismuth, or magnesium compounds are unsatisfactory. Wait 1 week or more after barium procedures or laxative administration before collecting specimen.

Aftercare Warning: Any stool collected by or from the patient may harbor pathogens which are **immediately infective.** Use gloves and extreme caution when collecting stool specimens and hands should be washed after collecting specimens.

Specimen Fresh or preserved random stool or duodenal aspirate. If pinworm is suspected, a Scotch® Tape preparation should be submitted to the laboratory instead of stool (see Pinworm Preparation *on page 1036*). Early voided midday urine for detection of certain parasites.

Container Use a clean, dry plastic stool container with a wide mouth. Patients are provided with collection devices that contain polyvinyl alcohol (PVA) and formalin for fresh stool. Zinc sulfate can replace mercuric chloride in PVA. Sodium acetate-acetic acid-formalin (SAF) fixative also avoids use of mercury. Preservation of the feces specimen assures that ova and trophozoites will be well preserved for examination. Sterile, plastic container for urine.

Collection Unpreserved specimen need to be delivered within 30 minutes to 1 hour of collection. Stools that are processed by the laboratory in less than 1 hour need not be preserved. Direct exams for motile trophozoites can only be performed on fresh stools.

Mushy, loose, or watery stools that cannot reach the laboratory within 1 hour should be preserved. Polyvinyl alcohol (PVA) will preserve the trophozoite stage of protozoa. The Merthiolate® iodine formalin (MIF) kit will preserve protozoan cysts, helminth eggs, and larvae. It is intended to be sent home with the patient and mailed back to the laboratory.

Formed stools may be preserved in formalin or refrigerated in a secure container until they can be transported to the laboratory.

Examination of a single specimen per day has been recommended with examination of additional specimens if indicated. The shedding of parasites in the stool varies and may be intermittent and in cases of high suspicion (travel in endemic countries) several specimens should be collected over 7-10 days (see table *on page 988*).

For *Trichomonas vaginalis*, collect freshly passed urine and immediately transport to the laboratory within 1 hour without refrigeration. For *Schistosoma haematobium*, the terminal portion of the urine specimen may contain numerous eggs trapped in mucus. Peak egg excretion occurs between noon and 3 PM.

OVA AND PARASITES, AMEBAE

	Entamoeba histolytica	Entamoeba hartmanni	Entamoeba coli	Entamoeba polecki[1]	Endolimax nana	Iodamoeba bütschlii	Dientamoeba fragilis[2]
Trophozoite							
Cyst							No cyst

[1] Rare, probably of animal origin   Scale: 0   5   10 µm
[2] Flagellate

Amebae found in human stool specimens.

OVA AND PARASITES, COCCIDIA

CILIATE	COCCIDIA			BLASTOCYSTIS
Balantidium coli	Isospora belli	Sarocystis spp	Cryptosporidium spp	Blastocystis hominis
Trophozoite	immature oocyst	mature oocyst	mature oocyst	
Cyst	mature oocyst	single sporocyst		

Ciliate, coccidia, and *B. hominis* found in human stool specimens.

OVA AND PARASITES, FLAGELLATES

FLAGELLATES				
Trichomonas hominis	Chilomastix mesnili	Giardia lamblia	Enteromonas hominis	Retortamonas intestinalis
Trophozoite				
Cyst — No cyst				

Flagellates found in human stool specimens.

Parasite	Cyclical Peak
Ascaris lumbricoides	Constant
Dientamoeba fragilis	Irregular
Diphyllobothrium latum	Irregular
Entamoeba histolytica	7-10 days
Giardia lamblia	3-7 days
Hookworm	Constant
Trichuris trichiura	Constant
Schistosoma species	Irregular

Adapted from Miller JM and Holmes HT, "Specimen Collection, Transport and Storage," *Manual of Clinical Microbiology*, 7th ed, Murray PR, Baron EJ, Pfaller MA, et al, eds, Washington, DC: American Society for Microbiology, 1999, 33-63.

Storage Instructions Liquid specimens should be brought directly to the laboratory. Wet mounts are performed immediately, and the specimen placed in PVA and/or MIF preservatives to maintain ova and trophozoite states. Urine specimens should not be refrigerated.

Causes for Rejection Patients who develop diarrhea after 3 days of hospitalization rarely have bacterial or parasitic diseases identified by routine examination. The most frequent etiologic agent implicated in this setting is *Clostridium difficile*.

Because of risk to laboratory personnel, specimens sent on diaper or tissue paper, specimen contaminating outside of transport container may not be acceptable. Specimen containing interfering substances (eg, castor oil, bismuth, Metamucil®, barium, specimens delayed in transit, and those contaminated with urine) lack optimal yield. Stool specimens that are extremely hard (nonpuncturable) will not yield a useful parasitic examination.

Special Instructions Geographic and travel history is helpful to the laboratory in order to examine the specimen for likely parasites.

Reference Interval No parasites seen

Use Establish the diagnosis of intestinal parasitic infestation or infection

Limitations A negative result does not rule out the possibility of parasitic infestation. *Entamoeba dispar* and *E. moshkovskii* are morphologically identical to *E. histolytica* but are nonpathogenic and noninvasive. Molecular biology-based diagnostic procedures are preferred methods but they are costly and time-consuming to perform. Stool examination for *Giardia* may be negative in early stages of infection, in patients who shed organisms cyclically, and in chronic infections. The sensitivity of microscopic methods for detection of *Giardia* range from 46% to 95%. *Giardia* are found predominantly in the upper small intestine. Tests for *Giardia* antigen may have a much higher yield (see *Giardia intestinalis* Diagnostic Procedures *on page 636*). Differential diagnosis of pathogens from artifacts, accidental parasites and spurious infection may be challenging. Stool must be collected directly into a dry container or into fixative. Urine and water destroy amebas.

Contraindications Parasite exams on stool from patients hospitalized more than 3 days are not productive and should not be ordered unless special circumstances exist. Administration of barium, bismuth, Metamucil®, castor oil, mineral oil, tetracycline therapy, administration of antiamebic drugs within 1 week prior to test. Purgation contraindicated for pregnancy, ulcerative colitis, cardiovascular disease, child younger than 5 years of age, appendicitis or possible appendicitis.

Methodology Wet mounts and trichrome stains after concentration are routine. An acid-fast-trichrome stain for *Microsporidium*, *Cryptosporidium parvum*, *Cyclospora cayetanensis*, and *Isospora belli* is performed if requested (see *Cryptosporidium* Direct Staining Procedures *on page 484* and Microsporidia Diagnostic Procedures *on page 915*).

Additional Information Of the pathogenic protozoa, *E. histolytica* and *G. lamblia* are the two most common infections identified worldwide. An accurate diagnosis of infestation is critical for the management of these diseases as well as the prevention of new protozoal infections. The major causes of diarrhea in the U.S. are due to *Cryptosporidium* and *Giardia*, spread in daycare centers and through municipal water supplies. Parasites identified in the stool of **immunocompromised subjects** (eg, AIDS patients) include *Cryptosporidium*, *Microsporidia*, *Entamoeba histolytica*, *Giardia lamblia*, *Isospora belli*, and *Strongyloides stercoralis*.

Certain parasites cannot be seen in stools containing barium, and urine or water in the specimen can destroy amebic structures. Optimal diagnostic yield is obtained by the examination of fresh, warm stool by an experienced technologist. Amebic cysts, *Giardia* cysts, and helminth eggs can be recovered from formed stools. Mushy or liquid stools (either normally passed or obtained by purgation) often yield trophozoites. The observation of erythrophagocytic trophozoites in bloody, mucoid stools provides optimal evidence of the presence of invasive **amebiasis**. A smear stained by trichrome or iron hematoxylin is confirmatory. *E. histolytica* is recognized in endoscopic biopsies in only 50% of the cases. Amebae are not always present in the stools of patients who have amebic abscess of liver. In such patients, serologic tests are more reliable (see *Entamoeba histolytica* Antigen Detection and Serology *on page 538*).

References

Ash L and Orihel T, "Wherever Parasites Are Found, Diagnose Them Here," *Atlas of Human Parasitology*, 4th ed, 1997.

Cartwright CP, "Utility of Multiple-Stool-Specimen Ova and Parasite Examinations in a High-Prevalence Setting," *J Clin Microbiol*, 1999, 37(8):2408-11.

Colmer-Hamood JA, "Fecal Microscopy Artifacts Mimicking Ova and Parasites," *Lab Med*, 2001, 32(2):80-4.

Cooke RA and Stewart B, *A Colour Atlas of Medical Parasitology: Human Disease Series*, Queensland, Australia: Knowledge Books & Software, 1999.

Haque R, Huston CD, Hughes M, et al, "Amebiasis," Current Concepts, *N Engl J Med*, 2003, 348(16):1565-73.

Long EG and Christie JD, "The Diagnosis of Old and New Gastrointestinal Parasites," *Clin Lab Med*, 1995, 15(2):307-31.

Reisner BS and Spring J, "Evaluation of a Combined Acid-Fast-Trichrome Stain for Detection of Microsporidia and *Cryptosporidium parvum*," *Arch Pathol Lab Med*, 2000, 124(5):777-9.

Tanyuksel M and Petri WA Jr, "Laboratory Diagnosis of Amebiasis," *Clin Microbiol Rev*, 2003, 16(4):713-25.

Internet Web Sites

www.cdc.gov/ncidod/diseases/list_parasites.htm

◆ **Ovarian Cancer** *see* Estrogen and Progesterone Receptor Assay *on page 556*

◆ **Ovarian Cyst Fluid Cytology** *see* Cyst Fluid Cytology *on page 490*

Oxalate, Urine

Related Information

Ascorbic Acid, Serum or Plasma *on page 215*
Citrate, Serum, Plasma, or Urine *on page 413*
Ethylene Glycol, Serum or Plasma *on page 561*
Kidney Stone Analysis *on page 820*
Magnesium, Serum *on page 885*
Magnesium, Urine *on page 886*
Urinalysis *on page 1289*
Urine Collection, 24-Hour *on page 1295*

Synonyms Calcium Oxalate, Urine

Applies to Glycolic Acid, Urine; Glyoxylic Acid, Urine; L-Glyceric Acid, Urine

Abstract Calcium oxalate stones are common in the urinary tract. Oxalate excretion is a predictor of oxalate nephrolithiasis; hyperoxaluria can be detected in up to 33% of patients with calcium oxalate stones.

Patient Preparation Avoid vitamin C for 24 hours before collection. Pyridoxine is said to diminish oxaluria. The patient should be ambulatory, preferably at home, on usual fluid and food intake, to best interpret risk factors for nephrolithiasis.

Specimen 24-hour urine; first morning urine may give oxalate concentrations similar to 24-hour collections. Total amount of oxalate excreted might be estimated using first morning urinary oxalate concentration and an estimate of daily urine output. Oxalate can also be measured in plasma, but such testing is not widely available at the present time.

Container Acid-washed plastic container with 20 mL of 6N HCl added prior to collection (depending upon laboratory). Acid prevents oxalate crystallization and conversion of ascorbate to oxalate. No metal cap.

(Continued)

Oxalate, Urine *(Continued)*

Collection Urine creatinine is often determined. In children, second morning urine specimen can be used to determine oxalate/creatinine ratio. Between ages 0-6 years, the ratio is inversely related to age. The ratio plateaus at age 6 years.

Reference Interval 20-60 mg/24 hours (SI: 0.23-0.68 mmol/day) in those whose diets include oxalate-rich foods; *vide infra*. Without excess of such foods, normal men excrete up to 45 mg (500 µmol) and women up to 40 mg (444 µmol). Greater excretion of oxalic acid in men is recognized in healthy subjects as well as in stone formers. The differences were unexplained on the basis of body surface. A relationship with age was not found.

Use Calcium oxalate renal stones are common. Patients who form calcium oxalate kidney stones appear to absorb and excrete a higher fraction of dietary oxalate in urine than do normals. Hyperoxaluria is not uncommon in subjects with malabsorption, including fat malabsorption. Twenty-four hour urine collections for oxalate are indicated in patients with surgical loss of distal small intestine, especially those with Crohn disease. The incidence of nephrolithiasis in patients who have inflammatory bowel disease is 2.6% to 10%.

Oxaluria is characteristic of ethylene glycol intoxication.

Hyperoxaluria is regularly present after jejunoileal bypass for morbid obesity; such patients may develop nephrolithiasis and oxalate nephrosis.

Limitations Interference by ascorbate is a major impediment in developing a simple assay. Urine specimens containing significant amounts of ascorbic acid (10-325 µg/dL) experience interference.

Additional Information Oxalic acid excretion is increased with methoxyflurane.

Hyperoxaluria may occur with high intake of gelatin, strawberries, pepper, rhubarb, beans, beets, spinach, tomatoes, chocolate, cocoa, tea, pecans, peanuts, okra, and lime peel. Hyperoxaluria is described with pyridoxine deficiency. Urinary oxalate derives from the metabolism of glycine and ascorbic acid more than from dietary ingestion. Oxalate excretion is increased in vegetarians, despite low animal protein ingestion. Calcium taken orally with oxalate loads decreases urinary oxalate excretion in patients with ileal disease. Calcium supplements taken with meals are less likely to lead to nephrolithiasis.

Vitamin C increases oxalate excretion and may be a risk factor for calcium oxalate nephrolithiasis in individuals consuming "megadose" vitamin C.

Rare genetic disorders which increase endogenous oxalate production; there are two types of primary hyperoxaluria. They are characterized by elevated urinary oxalate excretion and recurrent oxalate nephrocalcinosis. In type I, an autosomal recessive, a defect in glyoxalate metabolism is found, leading to increased oxalate synthesis. Excessive quantities of urinary glyoxylic and glycolic acid excretion occur. Type II is rare; it is characterized by excessive urinary excretion of oxalic and L-glyceric acids with normal excretion of glycolic acid. type I causes renal failure and systemic oxalosis, but type II rarely does so. Urinary oxalate excretion in both forms is >135-270 mg (SI: >1523-3000 µmol) daily.

References

Hesse A, Schneeberger W, Engfeld S, et al, "Intestinal Hyperabsorption of Oxalate in Calcium Oxalate Stone Formers: Application of a New Test With $^{(13C2)}$Oxlate," *J Am Soc Nephrol*, 1999, 10(Suppl 14):S329-33.

Levine M, Rumsey SC, Daruwala R, et al, "Criteria and Recommendations for Vitamin C Intake," *JAMA*, 1999, 281(15):1415-23.

Matos V, Van Melle G, Werner D, et al, "Urinary Oxalate and Urate to Creatinine Ratios in Healthy Pediatric Population," *Am J Kidney Dis*, 1999, 34(2):6E.

Oxazepam, Serum

Related Information

Benzodiazepines, Qualitative, Urine *on page 253*

Synonyms Adumbran®; Serax®; Serenid® Forte

Abstract Oxazepam is a benzodiazepine used as an antianxiety agent and as an anticonvulsant for management of simple partial seizures. Benzodiazepines are also used in the treatment of insomnia and alcohol withdrawal. It is an active metabolite of several other benzodiazepines.

Specimen Serum

Container Red top tube; do not collect in gel-containing tubes.

Sampling Time Collect specimen immediately prior to next dose unless specified otherwise.

Storage Instructions Separate serum and refrigerate.

Reference Interval 0.2-1.4 µg/mL (SI: 0.7-4.9 µmol/L)

Use Monitor therapeutic drug level; evaluate toxicity

Contraindications Hypersensitivity to benzodiazepines; CNS depression; narrow angle glaucoma

Methodology Gas chromatography (GC), high performance liquid chromatography (HPLC), immunoassay for semiquantitation

Additional Information
- Half-life: ~4-12 hours
- Volume of distribution: 0.5-2.0 L/kg
- Protein binding: 95% to 98%

Adverse effects are mild and infrequent. They include drowsiness, vertigo, ataxia, headache, tremor, slurred speech, nausea, hypotension, and leukopenia. Ethanol and other CNS depressants increase the CNS effects of oxazepam.

References
Garretty DJ, Wolff K, Hay AW, et al, "Benzodiazepine Misuse by Drug Addicts," *Ann Clin Biochem*, 1997, 34(Pt 1):68-73.

Malcolm R, Myrick H, Brady KT, et al, "Update on Anticonvulsants for Treatment of Alcohol Withdrawal," *Am J Addict*, 2001, 10(Suppl):16-23.

Woods JH and Winger G, "Current Benzodiazepine Issues," *Psychopharmacology*, 1995, 118(3):107-15; discussion 118, 120-1.

Oxygen Saturation, Blood

Related Information

Synonyms SaO$_2$; SO$_2$; sO$_2$

Applies to 2,3-Diphosphoglycerate (2,3-DPG); 2,3-DPG; Fractional Oxyhemoglobin (FO$_2$Hb); O$_2$ Binding Capacity (BO$_2$); O$_2$ Content (ctO$_2$); Oximetry, Pulse; Oxygen-Hemoglobin Dissociation Curve (ODC); Pulse Oximetry, Transcutaneous

Test Includes A complete O$_2$ status evaluation includes concentration of total hemoglobin (ctHb), hematocrit (Hct), O$_2$ content (ctO$_2$), O$_2$ binding capacity (BO$_2$), partial pressure of O$_2$ or O$_2$ tension (pO$_2$), partial pressure of O$_2$ at which SO$_2$ is 50% (p50), SO$_2$, fractional oxyhemoglobin (FO$_2$Hb), fractional carboxyhemoglobin (FCOHb), fractional methemoglobin (FMetHb), and fractional deoxyhemoglobin (FHHb). pH and pCO$_2$ are often included. Electrolytes and other analytes are relevant.

Abstract Hemoglobin oxygen saturation (SO$_2$) is the percentage or fraction of functional hemoglobin (ie, hemoglobin able to bind oxygen) that is oxygenated. It is determined by the following equation:

$$SO_2\ (\%) = [cO_2Hb\ /\ (cO_2Hb + cHHb)] \times 100$$

(Continued)

991

Oxygen Saturation, Blood *(Continued)*

where cO_2Hb is the concentration of oxyhemoglobin and $cHHb$ is the concentration of deoxyhemoglobin.

Specimen Whole blood, arterial

Container Heparinized syringe, capillary tubes, or green top (heparin) tube

Collection Draw specimen anaerobically into heparinized syringe, avoid clots and air bubbles, and stopper tightly. Place heparinized specimen on ice. Take to the laboratory immediately.

Special Instructions If capillary tubes are used to collect the specimen, warm skin 10-15 minutes prior to puncture to obtain free flow of blood with a sufficiently deep puncture ("arterialized capillary blood"). Fill heparinized capillary tubes completely full, cap the tubes, and mix well.

Reference Interval Newborns: 85% to 90% (SI: 0.85-0.90); thereafter: 95% to 99% (SI: 0.95-0.99)

Possible Panic Range Arterial pO_2 of 20 mm Hg and SO_2 of 35% (SI: 0.35) are critically low, life-endangering levels; the same values from mixed venous blood indicate tissue hypoxia. Arterial pO_2 of 40 mm Hg and SO_2 of 75% (SI: 0.75) are panic values that correlate with cyanosis, but these values are normal for mixed venous blood.

Use SO_2 (together with pO_2, $FHbO_2$, ctO_2, $FMetHb$, and $FCOHb$) is used to assess the extent of oxygenation of hemoglobin and adequacy of tissue oxygenation in the evaluation of hypoxia (due to lung and/or cardiac disease or dysfunction, cyanosis, or toxic exposure) and monitor respiratory function during mechanical ventilation. It allows for the evaluation of oxygenation and oxyhemoglobin dissociation of blood with use of the oxygen dissociation curve (ODC).

Limitations Accuracy and precision may be affected by the presence of other pigments in the blood, low hemoglobin concentrations, plasma turbidity, and presence of cell fragments. Interpretation is more meaningful if the nature of patient's inspired gas is recorded (ie, room air or mixture of gases with a controlled oxygen content).

Though reference intervals are virtually identical, SO_2 **should not** be used interchangeably with fractional oxyhemoglobin (FO_2Hb) or estimated oxygen saturation (abbreviated as "O_2Sat") as these can have substantially different values in critically ill patients or patients with abnormal hemoglobins. SO_2, by definition, is not affected by dyshemoglobinemia or hemoglobin variants, whereas FO_2Hb, which is the concentration of oxyhemoglobin (cO_2Hb) divided by the concentration of total hemoglobin ($ctHb$), is decreased under these circumstances. Consequently, the NCCLS (1997) recommends that SO_2 results be reported together with FO_2Hb results or only after ruling out dyshemoglobinemia. Simple oximeters (eg, pulse oximeters, indwelling fiber optic oximeters) that use only two wavelengths of light measure only O_2Hb and HHb and report only a result for SO_2; these instruments are incapable of determining FO_2Hb or detecting the presence of dyshemoglobins and should be used with caution and preferably only after dyshemoglobinemia is ruled out.

Estimated oxygen saturation (O_2Sat) is calculated empirically from the results of pH, pO_2, and hemoglobin measurements. Errors result from the fact that the equations assume normal O_2 affinity for hemoglobin, normal erythrocyte 2,3-diphosphoglycerate (2,3-DPG) concentration, and absence of dyshemoglobinemia. Values obtained by these equations may vary significantly from SO_2 and should not be used in further calculations (eg, shunt fraction) or assumed to be equivalent to FO_2Hb.

Contraindications Contraindications of arterial puncture

Methodology At minimum, blood is analyzed spectrophotometrically at two wavelengths, one with a large difference in absorbance between oxygenated and deoxygenated hemoglobin, the other at the isobestic point (molar absorbance identical for the oxygenated and deoxygenated forms). Modern oximetry employs multiple wavelengths for detection of not only these functional hemoglobins, but for detection of common, interfering dyshemoglobins (eg, carboxyhemoglobin, methemoglobin, sulfhemoglobin) as well. Transcutaneous pulse oximetry measures the absorption of different wavelengths of light passed through living tissue.

Additional Information The terms, "O$_2$ content (ctO$_2$)," "O$_2$ binding capacity (BO$_2$)," and "SO$_2$", refer to various definitions of the amount of O$_2$ carried in the blood. ctO$_2$ is the **total** amount of O$_2$ present (bound to hemoglobin and dissolved in plasma) in the blood. BO$_2$ is the maximum amount of O$_2$ that can be carried by hemoglobin if **all** the hemoglobin capable of binding O$_2$ were oxygenated. SO$_2$ is defined in the Abstract (see above).

The binding of O$_2$ by hemoglobin is dependent upon (primarily) pH (and thereby CO$_2$ parameters that contribute to the control of pH), temperature, the concentration of 2,3-DPG, and the molecular species of hemoglobin. When graphed, the relationship results in a sigmoid (S-shaped) O$_2$-hemoglobin disso-ciation curve (ODC). The SO$_2$ (in %) is represented on the Y axis with the pO$_2$ (in mm Hg) on the X axis with the curve shifted to the right or left (isobars) by changes in pH or other parameters. This is a biochemically fixed relation such that if any two of the three determinants, pH, pO$_2$, or SO$_2$ are known, the other may roughly be predicted. With the ODC in hand one can assess whether the three reported values are consistent with each other as defined by the ODC (assuming normal molecular species of hemoglobin, concentration of 2,3-DPG, and constant temperature).

Red cell 2,3-DPG concentration plays an important role in regulation of hemo-globin's affinity for O$_2$. 2,3-DPG binds to the beta chains of O$_2$Hb and results in displacement of O$_2$ by the following equation:

$$Hb\ O_2 + 2,3\text{-}DPG = Hb\text{-}2,3\text{-}DPG + O_2$$

An increase in 2,3-DPG will cause a shift of the reaction to the right. Greater affinity of fetal hemoglobin (Hb F) for O$_2$ has been ascribed to the poor binding of 2,3-DPG by the gamma chains of Hb F. Increased erythrocyte 2,3-DPG concentrations decrease intracellular pH resulting in a further reduction in O$_2$ affinity. 2,3-DPG is increased with hypoxia.

Red cells of newborns contain ~80% Hb F. Hb F has a slightly higher O$_2$ affinity compared to hemoglobin A (normal adult hemoglobin) and binds 2,3-DPG less strongly than hemoglobin A. Following birth, O$_2$ affinity decreases as red cell 2,3-DPG concentrations rise (20% during the first week of life). At 1-4 weeks, healthy prematures have p50 values approaching those of normal adults.

There is widespread use of **transcutaneous pulse oximetry** to determine SO$_2$, particularly in premature and in critically ill newborns and children. This noninvasive technique avoids the rigor and hazard of arterial puncture and necessity of subsequent proper sample handling, and provides continual moni-toring. It is reliable and useful in monitoring adequacy of oxygenation, effective-ness of resuscitative efforts, detection of development of prolonged periods of decreased SO$_2$ in neonates, and monitoring preterm infant's response to phys-ical therapy. Pulse oximetry has also been applied to detection of hyperoxemia in newborns but has low specificity. Limitations of pulse oximetry have included overestimation of SO$_2$ at values 65% and less, and variation from *in vitro* determined SO$_2$ in samples with >50% Hb F as compared with samples having <25% Hb F. A study of pulse oximeter determined SO$_2$ in pregnant patients and their newborns has found that SO$_2$ in neonates is commonly ≤90% within 10 minutes after birth and may not always be indicative of pathologic hypoxia. Specialized devices (eg, balloon-tipped, thermodilution, fiberoptic, pulmonary arterial catheter) have been developed for the intraoperative monitoring of mixed venous oxygen saturation.

See P$_{50}$, Blood *on page 993*.

References

National Committee for Clinical Laboratory Standards: Fractional Oxyhemoglobin, Oxygen Content and Saturation, and Related Quantities in Blood: Terminology, Measurement, and Reporting, Approved Guideline, C25-A, Wayne, PA: National Committee for Clinical Laboratory Standards, 1997.

Shannon DC, "Rational Monitoring of Respiratory Function During Mechanical Ventilation of Infants and Children," *Intensive Care Med*, 1989, 15(Suppl 1):S13-6.

♦ **Oxygen Transport** *see* Lactic Acid, Whole Blood or Plasma *on page 827*

♦ **P-5'-P** *see* Vitamin B$_6$, Plasma or Serum *on page 1315*

P$_{50}$, Blood
Related Information
Blood Volume *on page 282*
(Continued)

P$_{50}$, Blood *(Continued)*

Carboxyhemoglobin, Blood *on page 340*
Erythropoietin, Serum *on page 551*
Fetal Hemoglobin *on page 581*
Methemoglobin, Whole Blood *on page 903*
Oxygen Saturation, Blood *on page 991*

Synonyms Blood Gas P-50; pO$_2$ (0.5); pO$_2$ at Half Saturation

Applies to 2,3-DPG; Hill Plots; Oxygen-Hemoglobin Dissociation Curve (ODC)

Abstract P$_{50}$ is that pO$_2$ (partial pressure of oxygen) at which hemoglobin is 50% saturated with oxygen (O$_2$). It is affected by pH, temperature, pCO$_2$ (partial pressure of carbon dioxide), concentration of 2,3-diphosphoglycerate (2,3-DPG), and the presence of fetal hemoglobin, variant hemoglobins, or dyshemoglobins (eg, carboxyhemoglobin and methemoglobin). P$_{50}$ is inversely related to the binding affinity of hemoglobin for oxygen. The affinity of hemoglobin for O$_2$ is graphically illustrated with a sigmoid O$_2$ dissociation curve (ODC), which is a plot of O$_2$ saturation (SO$_2$) versus pO$_2$. See figure.

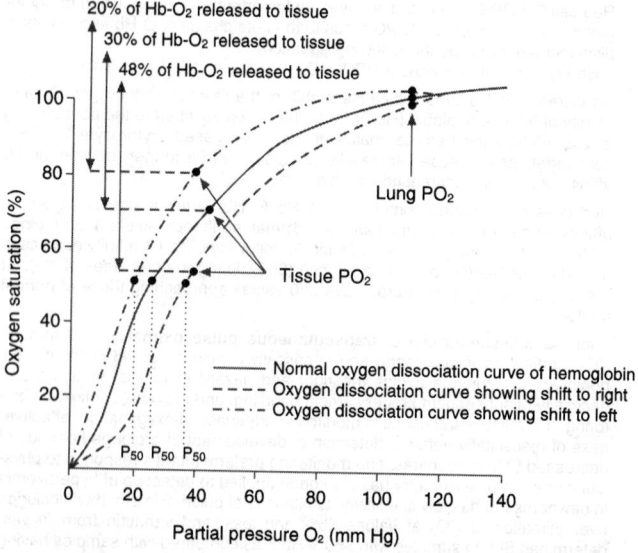

Oxygen dissociation curves under different conditions.

Adapted from *Technical Manual*, 13th ed, Chapter 8, Bethesda, MD: American Association of Blood Banks, 1999, 165.

Shifts in the curves to the left and right reflect, respectively, increased and decreased affinity of hemoglobin for O$_2$. Determination of P$_{50}$ value assumes that a tightly maintained physiologic relation between pO$_2$ and oxygen saturation exists and by implication that hemoglobin function is normal. This assumption does not necessarily apply to all hemoglobin types.

Specimen Arterial or venous whole blood

Container Heparinized syringe

Collection Avoid contact with air, insert needle into cork or hard rubber block, place on ice, and deliver to the laboratory immediately.

Causes for Rejection Specimen clotted, air bubbles in syringe, needle not tightly capped

Reference Interval Newborns: 18-24 mm Hg (because Hb F is present); adults: 25-29 mm Hg (corrected to pH 7.4, measured at 37°C); 26.3 mm Hg at sea level

Use The P_{50} is used to detect abnormalities in the affinity of hemoglobin for oxygen, secondary to disease or from hemoglobin variants. Decreased affinities (ie, **increased P_{50}** values with curve shifts to the right) result from acidemia, hypercapnia, hyperthermia, conditions associated with increased 2,3-DPG, and some hemoglobinopathies (eg, hemoglobin Seattle). See figure. Increased affinities (ie, **decreased P_{50}** values with curve shifts to the left) are observed in alkalemia, hypocapnia, conditions associated with decreased 2,3-DPG, dyshemoglobinemias, and some hemoglobinopathies. Thus, P_{50} is an indirect measure of 2,3-DPG. Change in 2,3-DPG concentration is the most common cause of change in P_{50}. Other causes of altered P_{50} include carboxyhemoglobin, methemoglobin, or increased hemoglobin F.

Limitations *In vivo* P_{50} changes with body temperature, CO_2T, and pH.

Methodology The P_{50} is most accurately determined by construction of the O_2 dissociation curve, but these methods are cumbersome and not routinely available. With the availability of blood gas analyzer/multi-wavelength oximeter combinations, single point calculation methods for determining P_{50} have been described. These methods are suitable for detection of clinically significant abnormalities in O_2 affinity such as those that occur in patients with abnormal hemoglobins and carbon monoxide poisoning, but do not provide truly accurate P_{50} and Hill values and are unable to discriminate between high and low values for P_{50} within the reference interval.

References

Hsia CCW, "Respiratory Function of Hemoglobin," *N Engl J Med*, 1998, 338(4):239-46.

Kwant G, Oeseburg B, and Zijistra WG, "Reliability of the Determination of Whole-Blood Oxygen Affinity by Means of Blood-Gas Analyzers and Multiwavelength Oximeters," *Clin Chem*, 1989, 35(5):773-7.

p53, Functional Assay/Sequencing

Related Information

Apoptosis Assays *on page 205*
Breast Biopsy *on page 305*
Breast Cancer, Hereditary, BRCA1, BRCA2 *on page 307*
Histopathology *on page 733*
Immunoperoxidase Procedures *on page 780*
Polymerase Chain Reaction *on page 1069*
Retinoblastoma Gene DNA Detection *on page 1159*

Applies to GeneChip *p53* Assay; Li-Fraumeni Syndrome

Test Includes Functional gene assay of *p53* and sequencing of the gene for ascertainment of mutations.

Abstract The *p53* gene is located on the short arm of chromosome 17 (17p13.1) and codes for a nuclear protein that plays a role in the regulation of cell growth and division. *p53* is a tumor suppressor gene and a sequence-specific transactivator (a transcriptional regulator). Mutations of *p53* have been implicated in many inherited and sporadic forms of cancer, including premalignant conditions, and are particularly common in bladder, breast, colorectal, lung cancer, brain tumors, and adrenocorticocarcinoma in children. Li-Fraumeni syndrome, a rare autosomal dominant disorder, is characterized by a germline mutation of *p53* and high incidence of malignancies of the breast, soft tissue, and brain. Functional and sequencing assays are available to assess *p53* mutations providing information which can be used to monitor and manage disease.

Specimen 30 mL whole blood; 100 mg solid tumor, frozen or paraffin-embedded; body fluids including bladder washings may be appropriate as specimens; other specimens including feces

Container Yellow top (ACD) tube for blood

Storage Instructions Transport whole blood at ambient temperature to the laboratory immediately. Fresh solid tumor should be frozen and transported on dry ice.

Turnaround Time 3-4 weeks

Reference Interval The *p53* functional assay will detect mutations located within codons 67 and 347 (the DNA-binding domain) of the *p53* gene, where >95% of *p53* mutations have been found. Sequence analysis will provide a (Continued)

p53, Functional Assay/Sequencing *(Continued)*

complete analysis of the *p53* sequence. Interpretative reports are usually provided by laboratories.

Use Detection of *p53* mutations in families at high risk of developing cancer (Li-Fraumeni syndrome) and as a prognostic parameter in patients with cancer (particularly gastrointestinal, including esophagus, stomach, gallbladder, colon, and rectum; lung, urinary bladder, ovary, breast, and prostate). *p53* has been detected in a variety of gynecologic tumors. Some studies have shown that *p53* mutations are associated with short survival and resistance to chemo- or radiation-induced DNA damage (colorectal carcinoma).

Association with tumor progression is reported with immunocytochemistry in mucinous borderline tumors of ovary. *p53* accumulation was found in some but not most mucinous, serous, and endometrioid carcinomas.

Limitations The functional assay cannot detect mutations outside of codons 67 and 347 including mutations in the regulatory domains. Association of *p53* mutations with *in situ* bladder tumors which bear a propensity for progression and with high grade or advanced bladder neoplasms is recognized, but *p53* mutations may occur late in the natural history of some tumors. About 40% of carcinomas of bladder lack *p53* mutations.

Additional Information Most human tumors have defects in the *p53* or retinoblastoma (RB) pathways. In human malignancy, *p53* is the most commonly mutated gene. *p53* is a tumor suppressor gene, at least in part due to its role in the induction of apoptosis in response to DNA damage. The *p53* gene is a key regulator in many cellular processes, including cell cycle control, DNA repair, genome stability, apoptosis (programmed cell death), differentiation, cell senescence, and angiogenesis. Conformation-specific monoclonal antibodies have shown that single point mutations in *p53* can change the conformation of the entire resultant protein molecule. p53 is a sequence-specific transactivator, containing an acidic domain near its N-terminus similar to those in other transcription factors, and an activation sequence that functions in both yeast and mammalian cells and is situated within codons 1 and 42.

Monoclonal antibodies PAb1801, PAb421, and DO-7 (DAKO, Carpinteria, California) may be used for immunohistochemistry. They react with the *p53* gene product, a p53 phosphoprotein barely detectable in the nucleus of normal cells. With damage to DNA, *p53* can arrest the cell cycle, allowing repair of DNA or progression to apoptosis. Inactive mutant p53 protein is more stable than wild type p53 and accumulates in the nucleus of neoplastic cells. Positive immunostaining indicates presence of an abnormal *p53* gene and gene product. Frozen sections or paraffin-embedded tissues may be used.

A host of interactions between p53 and other cell-cycle regulating and apoptotic-inducing molecules are under investigation. Rb exerts control of the cell cycle from G1 to S phase. Rb is regulated via phosphorylation (mediated by cell cycle-dependent kinases and cyclins with inhibition by the cell cycle inhibitor, p21$^{cip/waf}$). The latter gene (*p21$^{cip/waf}$*) is a *p53* target gene and thus, *p53* is implicated in the upstream control and regulation of Rb. The *mdm2* gene is also a *p53* target. The oncoprotein mdm2, once induced by *p53* can then interact with and inhibit p53 as well as target it for degradation, thus establishing a negative feedback loop involving *p53* activity. Thus, the concept of a Rb-mdm2-*p53* trimeric complex that governs the apoptotic function of *p53* (see diagram).

The identification and study of two homologues of *p53*, *p63*, and *p73*, indicate that they are members of a family of transcription factors. Members of this family have amino acid sequence identity up to 63% in the DNA-binding domain. They (*p53* and *p73*) are both induced by DNA damage and *p73* can activate *p53*-regulated genes as well as suppress growth or induce apoptosis. The genes *p63* and *p73* are rarely mutated in human cancer. They should probably not be considered as tumor suppressors.

Chemotherapy, stimulating hair follicle epithelial apoptosis, results in hair loss. The process is *p53* dependent.

The GeneChip *p53* assay has been applied to the analysis of *p53* mutation in a cancer of the colon. An accumulation of *p53* mutations was noted, a sample from the primary site having a mutation at codon 273 in exon 8 (of chromosome

17) while a sample of the tumor from a hepatic metastasis had two point mutations, one at codon 217 in exon 6 as well as the exon 8 abnormality.

Pathways to p53-Induced Apoptosis

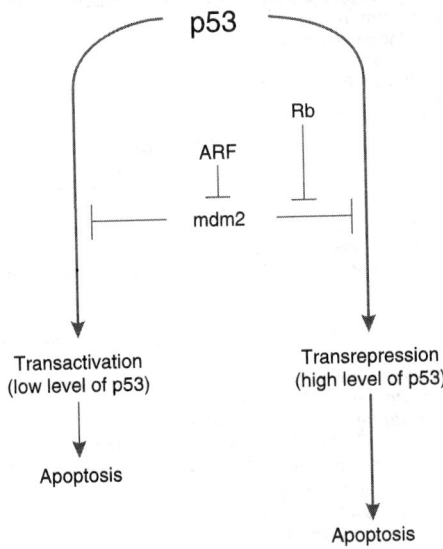

Adapted from Yap DB, Hsieh JK, Chan FS, et al, "mdm2: A Bridge Over the Two Tumor Suppressors, p53 and Rb," *Oncogene*, 1999, 18(53):7681-9.

References

Botchkarev VA, Komarova EA, Siebenhaar F, et al, "*p53* Is Essential for Chemotherapy-Induced Hair Loss," *Cancer Res*, 2000, 60(18):5002-6.

Fitzgerald PJ, "The Oncogene," Chapter 20, and "Suppressor Genes," Chapter 21, *From Demons and Evil Spirits to Cancer Genes*, Washington, DC: American Registry of Pathology, Armed Forces Institute of Pathology, 2000, 207-24, 225-40.

Kadakia M, Slader C, and Berberich SJ, "Regulation of p63 Function by Mdm2 and MdmX," *DNA Cell Biol*, 2001, 20(6):321-30.

Shen Y and White E, "p53-Dependent Apoptosis Pathways," *Advances in Cancer Research*, 2001, 82:55-84.

Soussi T, "The *p53* Tumor Suppressor Gene: From Molecular Biology to Clinical Investigation," *Ann N Y Acad Sci*, 2000, 910:121-37

Takahasi Y, Nagata T, Asai S, et al, "Detection of Aberrations of 17p and *p53* Gene in Gastrointestinal Cancers by Dual (Two-Color) Fluorescence *in situ* Hybridization and GeneChip *p53* Assay," *Cancer Genet Cytogenet*, 2000, 121(1):38-43.

+ **Pancreas Needle Aspiration Cytology** *see* Fine Needle Aspiration, Deep and Superficial Masses *on page 590*
+ **Pancreatic Cyst Fluid Cytology** *see* Cyst Fluid Cytology *on page 490*

Pancreatic Polypeptide, Human, Serum or Plasma

Related Information
Glucagon, Plasma *on page 640*
Insulin, Serum *on page 803*
Multiple Endocrine Neoplasia/Familial Medullary Thyroid Carcinoma *on page 924*
Vasoactive Intestinal Polypeptide, Plasma *on page 1302*

Synonyms hPP; Human Pancreatic Polypeptide

Abstract Human pancreatic polypeptide (hPP) is a hormone produced primarily by the endocrine type F cells of the pancreatic islets. Plasma levels rise in a biphasic manner after a meal, regulated initially by vagal cholinergic stimulation and later by cholecystokinin release. The exact physiologic role of hPP is unknown and the assay is of very limited clinical importance.

Patient Preparation Patient must fast overnight for basal evaluations. Avoid recent radioisotope scan.

Specimen Serum or plasma

Container Red top tube or lavender top (EDTA) tube; check with reference laboratory for specific instructions.

Storage Instructions Process blood promptly in a refrigerated centrifuge. Serum or plasma should be frozen until analysis.

Reference Interval There is considerable variation among laboratories and literature reports. A study of basal serum levels from 623 fasting, adult subjects gave the following age- and gender-specific reference intervals.

Male (pg/mL = pmol/L x 4.1817):
- 20-29 years: 3-47 pmol/L; 13-197 pg/mL
- 30-39 years: 4-74 pmol/L; 17-309 pg/mL
- 40-49 years: 5-102 pmol/L; 21-427 pg/mL
- 50-59 years: 7-156 pmol/L; 29-652 pg/mL
- 60-69 years: 5-146 pmol/L; 21-611 pg/mL

Female (pg/mL = pmol/L x 4.1817):
- 20-29 years: 2-48 pmol/L; 8-201 pg/mL
- 30-39 years: 2-71 pmol/L; 8-297 pg/mL
- 40-49 years: 3-84 pmol/L; 13-351 pg/mL
- 50-59 years: 5-121 pmol/L; 21-506 pg/mL
- 60-69 years: 10-76 pmol/L; 42-318 pg/mL

Basal hPP levels of 233 ±147 pg/mL were reported from a study of 45 healthy, pediatric patients ranging in age from 0-15 years.

Use Human pancreatic polypeptide is used, but very rarely, as a pancreatic endocrine tumor marker and as an indicator of vagal function after a meal or sham feeding.

Limitations Low diagnostic sensitivity is reported for pancreatic endocrine tumors.

Methodology Radioimmunoassay (RIA)

Additional Information An atropine suppression test is used to distinguish tumor-related secretion of hPP from normal. Release of hPP from normal cells is under cholinergic control, inhibited by atropine, while autonomous secretion from a tumor is anticipated. Pancreatic endocrine tumors often secrete more than a single peptide. Peptides associated with such tumors include hPP, vasoactive intestinal polypeptide, gastrin, glucagon, somatostatin, and neurotensin. Though a distinct clinical syndrome due to excessive secretion of hPP has not been established, a recent case report suggests that hPP hyperplasia may cause a watery diarrhea syndrome. Most hPP-secreting neuroendocrine tumors originate from the pancreas; however, there was a recent case reported of a pure, hPP-secreting neuroendocrine carcinoma of the gallbladder in a patient whose high levels of hPP may have contributed to presenting symptoms of cholestasis and cholelithiasis. Insulin-dependent diabetic patients predisposed to development of autonomic neuropathy may demonstrate decreased hPP (and epinephrine) levels in response to insulin-induced hypoglycemia. Blunted hPP responses after meal ingestion in type 1 diabetics may be due to vagal neuropathy or islet cell dysfunction. Human pancreatic poly-peptide may be a glucoregulatory hormone. Its release appears to be dependent upon intraluminal starch digestion. Measurement of hPP has been used as an indicator or pancreatic endocrine function post-transplantation.

References
Brimnes Damholt M, Rasmussen BK, Hilsted L, et al, "Basal Serum Pancreatic Polypeptide Is Dependent on Age and Gender in an Adult Population," *Scand J Clin Lab Invest*, 1997, 57(8):695-702.

Gehlert DR, "Multiple Receptors for the Pancreatic Polypeptide (PP-Fold) Family: Physiological Implications," *Proc Soc Exp Biol Med*, 1998, 218(1):7-22.

Hazelwood RL, "The Pancreatic Polypeptide (PP-Fold) Family: Gastrointestinal, Vascular, and Feeding Behavioral Implications," *Proc Soc Exp Biol Med*, 1993, 202(1):44-63.

- ♦ **Pancreatic Polypeptides** *see* Insulin, Serum *on page 803*
- ♦ **Panwarfin®** *see* Warfarin, Serum or Plasma *on page 1325*
- ♦ **PaO₂** *see* Blood Gases and pH, Arterial *on page 275*
- ♦ **PAP** *see* Acid Phosphatase, Plasma *on page 97*
- ♦ **PAP** *see* Hypercoagulation Panel *on page 758*
- ♦ **PAPP-A** *see* Pregnancy-Associated Protein A, Serum *on page 1082*
- ♦ **Pappenheimer Body Stain** *see* Siderocyte Stain *on page 1198*
- ♦ **Pap Smear** *see* Cervical/Vaginal Cytology *on page 376*
- ♦ **Pap Test** *see* Cervical/Vaginal Cytology *on page 376*
- ♦ **Paracentesis Fluid Analysis** *see* Body Fluid Analysis, Cell Count *on page 288*
- ♦ **Paracentesis Fluid Analysis** *see* Body Fluid Chemical Analysis *on page 291*
- ♦ **Paracentesis Fluid Cytology** *see* Body Cavity Fluid Cytology *on page 285*
- ♦ **Paracetamol** *see* Acetaminophen, Serum *on page 90*

Parainfluenza Viral Culture and Serology

Related Information
Viral Culture *on page 1307*
Virus, Direct Detection by Fluorescent Antibody *on page 1311*

Test Includes Culture and identification of parainfluenza viruses. Often includes the concurrent culture for other respiratory viruses (influenza, adenovirus, and respiratory syncytial viruses). Serology includes antibody titers to parainfluenza virus types 1, 2, 3, and 4.

Abstract Parainfluenza viruses types 1, 2, and 3 are the most common cause of croup (laryngotracheitis) in infants and children. They are known to cause a wide variety of both lower and upper respiratory infections. Parainfluenza infection is associated with an acute self-limited upper airway disease of children. Croup presents with barking cough, inspiratory stridor, and hoarseness. Parainfluenza viruses are second only to respiratory syncytial virus as a cause of serious infantile respiratory diseases. This group of viruses also routinely causes otitis media, pharyngitis, conjunctivitis, and the common cold. All four types of human parainfluenza viruses can reinfect individuals throughout their lives. Thus, immunity does not confer lifetime protection.

Specimen Culture: Throat or nasopharyngeal swab, nasopharyngeal washes, and secretions. Serology: Serum.

Container Culture: Viral transport medium; cold viral transport medium for swabs. Serology: Red top tube.

Sampling Time Culture specimens should be collected during acute infection. Antibody detection should be performed on acute and convalescent sera drawn 10-14 days apart.

Collection For culture, place swabs into cold viral transport medium and keep cold. With infants and small children, a soft catheter or suction device (syringes and suction bulbs) can be used to collect nasal secretions from far back in the nose. Another excellent method is to introduce 3-7 mL of sterile saline into the child's posterior nasal cavity and immediately aspirate the fluid. **Do not use cold sterile saline when aspirating samples. Warm to room temperature.**

Storage Instructions Keep culture specimens cold after collection, but do not freeze specimens at -20°C. Specimens that cannot be inoculated into cell culture within 48 hours should be frozen at -70°C.

Turnaround Time Conventional culture: 5-14 days; rapid culture: 2-4 days
(Continued)

Parainfluenza Viral Culture and Serology *(Continued)*

Reference Interval A single low titer or a less than fourfold change in titer in paired sera is considered normal. Culture: no virus isolated.

Critical Values Positive results in a child younger than 5 years of age

Use Support the diagnosis of parainfluenza virus infection. Serologic studies are of value in epidemiology.

Limitations Negative viral culture does not rule out viral etiology; rapid methods will only detect specified virus(es). Serology requires convalescent specimen which delays diagnosis. Heterotypic rises in parainfluenza titers may occur in infections with other viruses such as mumps. Infant antibody response may be undetectable. Parainfluenza viral serology is not recommended for clinical diagnostic purposes.

Methodology Viral culture: Inoculation of specimens into cell cultures, incubation of cultures, observation for hemadsorption or characteristic cytopathic effect (CPE), and identification/speciation by fluorescent monoclonal antibodies specific for types 1, 2, or 3 or by virus neutralization.

Rapid culture: Inoculation of cells in shell vial and detection of specific virus with an immunofluorescent monoclonal antibody.

Serology: Complement fixation (CF), hemagglutination inhibition (HAI), enzyme-linked immunosorbent assay (ELISA).

Additional Information Parainfluenza viruses are rarely isolated from healthy individuals; thus their detection is usually diagnostic. Conventional and rapid cultures generally detect other respiratory viruses in addition to the parainfluenza viruses. Currently, the standard for many laboratories is to do a rapid shell vial assay for detection of several respiratory viruses. Serology for detection of parainfluenza antibodies is often difficult to interpret and results are not available in a timely fashion.

References

Chan PW, Goh AY, Chua KB, et al, "Viral Aetiology of Lower Respiratory Tract Infection in Young Malaysian Children," *J Paediatr Child Health*, 1999, 35(3):287-90.

Hall CB, "Respiratory Syncytial Virus and Parainfluenza Virus," *N Engl J Med*, 344(25):1917-28.

Marx A, Gary HE Jr, Marston BJ, et al, "Parainfluenza Virus Infection Among Hospitalized for Lower Respiratory Tract Infection," *Clin Infect Dis*, 1999, 29(1):134-40.

Reed G, Jewett PH, Thompson J, et al, "Epidemiology and Clinical Impact of Parainfluenza Virus Infections in Otherwise Healthy Infants and Young Children <5 Years Old," *J Infect Dis*, 1997, 175(4):807-13.

Rosencrans JA, "Viral Croup: Current Diagnosis and Treatment," *Mayo Clin Proc*, 1998, 73(11):1102-7.

Shih SR, Tsao KC, Ning HC, et al, "Diagnosis of Respiratory Tract Viruses in 24 h by Immunofluorescent Staining of Shell Vial Cultures Containing Madin-Darby Canine Kidney (MDCK) Cells," *J Virol Methods*, 1999, 81(1-2):77-81.

♦ **Paraprotein Evaluation** *see* Immunofixation Electrophoresis, Serum or Urine *on page 768*

♦ **Parathormone** *see* Parathyroid Hormone, Serum *on page 1001*

♦ **Parathormone** *see* Vitamin D, Serum *on page 1318*

♦ **Parathyroid Hormone, C-Terminal** *see* Parathyroid Hormone, Serum *on page 1001*

♦ **Parathyroid Hormone, Intact** *see* Parathyroid Hormone, Serum *on page 1001*

♦ **Parathyroid Hormone, N-Terminal** *see* Parathyroid Hormone, Serum *on page 1001*

♦ **Parathyroid Hormone (PTH), Whole Molecule** *see* Parathyroid Hormone, Serum *on page 1001*

♦ **Parathyroid Hormone-Related Protein** *see* Calcium, Serum *on page 329*

Parathyroid Hormone-Related Protein, Serum

Related Information

Calcium, Ionized, Serum *on page 328*

Calcium, Serum *on page 329*

Calcium, Urine *on page 332*

Parathyroid Hormone, Serum *on page 1001*

Synonyms PTHrP

Abstract Parathyroid hormone-related protein (PTHrP) is the substance which is elaborated by many diverse neoplasms and which causes the syndrome of the humoral hypercalcemia of malignancy (HHM).

Specimen EDTA plasma (check with reference laboratory for additional preservatives)

Container Lavender top (EDTA) tube

Collection Centrifuge promptly and freeze immediately (check with reference laboratory)

Reference Interval 0-1.5 pmol/L

Use PTHrP assists in the differential diagnosis of hypercalcemia, and is most useful for those patients in whom primary hyperparathyroidism (PHPT) has been ruled out, or appears unlikely, on the basis of assay for PTH. Approximately one-third to one-half of patients presenting with hypercalcemia have HHM.

Limitations In addition to patients with MAH, PTHrP is also elevated in women who are lactating and correlates with the associated decrement in bone density. PTHrP production by benign tumors occurs rarely.

Methodology Immunoradiometric assay (IRMA)

References
Knecht TP, Behling CA, Burton DW, et al, "The Humoral Hypercalcemia of Benignancy. A Newly Appreciated Syndrome," *Am J Clin Pathol*, 1996, 105(4):487-92.

Strewler GJ, "Mechanisms of Disease. The Physiology of Parathyroid Hormone-Related Protein," *N Engl J Med*, 2000, 342(3):177-85.

Internet Web Sites
www.arup-lab.com

Parathyroid Hormone, Serum

Related Information

Synonyms Immunoreactive PTH; Parathormone; PTH

Applies to Calcium:Creatinine Ratio, Urinary; Multiple Endocrine Neoplasia (MEN) Type 1; Parathyroid Hormone, C-Terminal; Parathyroid Hormone, Intact; Parathyroid Hormone, N-Terminal; Parathyroid Hormone (PTH), Whole Molecule

Test Includes Concomitant serum calcium is required for interpretation. Age, sex, serum phosphorus and creatinine are also important (*vide infra*, see Use).

Abstract Parathyroid hormone (PTH) is synthesized and stored in the parathyroid glands, from which it is secreted at a rate inversely proportional to ambient concentration of ionized serum calcium. The **half-life** of PTH in the plasma is 2-5 minutes. The overall biochemical effects of PTH, increasing the serum concentrations of both ionized and total calcium and decreasing the serum concentration of phosphorus, are mediated by actions in bone, kidney, and gut. PTH mobilizes calcium from bone, increases reabsorption of calcium from tubular urine, decreases reabsorption of phosphorus from tubular urine, and increases gastrointestinal absorption of calcium (by stimulating the formation of $1,25(OH)_2$ vitamin D).

Patient Preparation Patient should be fasting.

Specimen Serum; plasma is acceptable in some protocols.

Container Red top tube for serum; check with reference laboratory for preferred anticoagulant if plasma is to be used.

Sampling Time Because a circadian rhythm is present, specimen collection times should be standardized to reduce biologic variation.

Collection Centrifuge promptly. Freeze immediately.

(Continued)

Parathyroid Hormone, Serum *(Continued)*

Reference Interval The reference intervals vary with the method, patient age, and gender (in children).

Children: See table.

Serum Intact PTH in Children

Age (y)	n	Median (ng/L)	2.5th-97.5th Percentiles (ng/L)
2.1-4	M: 48 F: 42	M: 12.46 F: 12.80	M: 5.7-34.2 F: 3.6-32.0
4.1-6	M: 46 F: 39	M: 9.65 F: 8.71	M: 4.4-15.6 F: 1.0-13.0
6.1-8	M: 48 F: 54	M: 9.93 F: 10.91	M: 2.5-27.3 F: 2.7-24.6
8.1-10	M: 60 F: 66	M: 13.11 F: 12.88	M: 4.6-33.8 F: 2.0-30.2
10.1-12	M: 57 F: 67	M: 13.02 F: 16.53	M: 2.5-25.4 F: 4.3-33.9
12.1-14	M: 54 F: 65	M: 12.84 F: 15.62	M: 1.4-25.5 F: 1.6-36.5
14.1-16	M: 72 F: 76	M: 13.57 F: 13.98	M: 4.5-35.8 F: 1.2-39.0

Adapted from Cioffi M, Corradino M, Gazzerro P, et al, "Serum Concentrations of Intact Parathyroid Hormone in Healthy Children," *Clin Chem*, 2000, 46(6):863.

Adults:

- whole molecule, immunochemiluminometric assay (ICMA): 1.0-5.2 pmol/L
- whole molecule, radioimmunoassay (RIA): 10.0-65.0 pg/mL
- whole molecule, immunoradiometric, double antibody (IRMA): 1.0-6.0 pmol/L

Use

Differential diagnosis of hypercalcemia: The most common causes of hypercalcemia are primary hyperparathyroidism (**PHP**) and humoral hypercalcemia of malignancy (HHM). These two diagnoses account for >90% of patients with hypercalcemia. Less common causes include familial hypocalcuric hypercalcemia (**FHH**), granulomatous diseases, certain lymphomas, thyrotoxicosis, and vitamin D intoxication. Rare causes of hypercalcemia include lithium medication, thiazide therapy, Addison disease, hypothyroidism, milk-alkali syndrome, immobilization, PTH receptor defects, recovery from renal failure and following renal transplant.

PHP and FHH: If the serum PTH is elevated, PHP is the most likely diagnosis. A family history will help identify the small number of patients with FHH, an autosomal dominant genetic disease caused by a mutation in the calcium-sensing receptor. Patients with FHH have low urine calcium, and the **urinary calcium:creatinine ratio** is usually <0.01. Interpretation of the PTH result is not optimal without simultaneous measurements of serum calcium, creatinine and phosphorus.

HHM: If the serum PTH is appropriate for the concurrent calcium, then HHM is the most likely diagnosis. HHM occurs by two mechanisms, and both mechanisms can be operative in the same patient.

- Many tumors secrete Parathyroid Hormone-Related Protein, Serum *on page 1000* (**PTHrP**) which shares many biologic properties with PTH. PTHrP mobilized calcium from bone (by osteoclast activation), produces renal conservation of calcium, and stimulates the formation of $1,25(OH)_2$ vitamin D (calcitriol).
- Tumors in bone release cytokines which activate osteoclasts and produce hypercalcemia by this mechanism, sometimes called **local osteolytic hypercalcemia** (LOH).

The LOH mechanism is particularly prominent in myeloma. For most patients with MAH the diagnosis can be made with a PTH assay (including serum calcium, phosphorus, and creatinine) **in addition to** a thorough work-up for malignancy. Only rarely is it necessary to obtain a PTHrP assay.

Endogenous vitamin D intoxication: In normal individuals, 25 hydroxy vitamin D (25OH vitamin D) is converted in the kidney to 1,25 dihydroxy vitamin D [1,25(OH)$_2$ vitamin D]. In some patients with granulomatous diseases (and a few with lymphomas not involving bone), lesional, non-neoplastic, macro-phages also have the ability to effect this conversion, which then leads to hypercalcemia.

When PHP, HHM, FHH, and endogenous vitamin D intoxication have been ruled out, the search for a diagnosis focuses on the rare conditions listed above (Differential Diagnosis).

Intraoperative Monitoring of PTH: The surgical treatment of primary hyper-parathyroidism requires an intraoperative decision about whether the disease involves just one gland (adenoma) or multiple glands (hyperplasia or multiple adenomas). Pathologic evaluation by frozen section sometimes is an imperfect tool for this purpose. Investigators at the Mayo Clinic originally described a modification of their immunochemiluminometric (ICMA) whole molecule PTH assay, which required only 15 minutes of assay time and therefore allowed the assay results to be reported quickly enough for intraoperative monitoring. These investigators also found that when a parathyroid gland(s) is hyperfunc-tioning, the other (normal) glands are suppressed and do not regain function for up to 130 minutes after resection of the hyperfunctioning gland; therefore, intraoperative PTH levels are not affected by the function of previously suppressed parathyroid tissue. Improvements in assay technique now provide a 10 minute assay, and actual experience with that assay indicates that a decrease of 50% or more from the baseline PTH predicts operative success with an overall accuracy of 97%. When the PTH decrement is <50% of base-line, further exploration to find additional hyperfunctioning tissue is indicated. Intraoperative PTH assays, combined with differential venous sampling assists in the localization of difficult-to-find parathyroid glands.

Differential diagnosis of hypocalcemia: Causes include hypoparathyroidism (including familial hypocalcemia due to a mutation in the calcium-sensing receptor), rickets, low serum albumin, acute pancreatitis, sepsis, medication effect, hungry bone syndrome, tumor lysis syndrome, rhabdomyolysis, chronic renal insufficiency, and magnesium deficiency.

Hypocalcemia from hypoparathyroidism is a recognized risk following thyroid surgery. Hypoparathyroidism may be due to organ destruction (iron overload or autoimmunity, granulomatous disease, metastatic neoplasm), and failure to develop (DiGeorge syndrome). These patients have low serum calcium with inappropriately low serum PTH. Rickets is most often due to dietary Vitamin D deficiency (but may also be caused by an inborn error of vitamin D metabo-lism). Dietary vitamin D deficiency is usually confirmed by the serum 25OH vitamin D assay (see Vitamin D, Serum *on page 1318*); such patients are expected to have elevated serum PTH due to the negative feedback loop described above (see Abstract). An inborn error of metabolism responsive to high doses of Vitamin D may be due to a mutation in the 25OH vitamin D 1-alpha-hydroxylase gene. Dietary calcium deficiency is less common than vitamin D deficiency. Some hypocalcemic patients have malabsorption of vitamin D, calcium, or both; gluten-sensitive enteropathy and postsurgical bowel are leading causes.

When serum albumin is low, the total serum calcium can be corrected using the following formula:

Corrected calcium = measured calcium [mg/dL] + 0.8 (4 - patient albumin [g/dL])

Drugs causing hypocalcemia include calcitonin, mithramycin, biphosphonates, phosphates, phenytoin plus phenobarbital, and foscarnet.

Pseudohypoparathyroidism refers to conditions in which there is end organ unresponsiveness to PTH. Typical biochemical findings are hypocalcemia, accompanied by elevated serum phosphate and PTH.

Limitations Not all patients with PHP have elevated serum PTH at the time of testing. Both serum calcium and PTH may be intermittently elevated.
(Continued)

Parathyroid Hormone, Serum *(Continued)*

Hypercalcemia may also be intermittent. In an evaluation of 60 patients with surgically proven primary hyperparathyroidism (PHP), 59 (98.3%) patients had elevated ionized calcium and 49 (82%) had elevated serum PTH.

Methodology Immunochemiluminometric assay (ICMA), two-site immunoradiometric assay. Now that reliable assays are available for **whole molecule PTH** (also called intact PTH), there are few, if any, indications for N-terminal and C-terminal assays.

Additional Information About 85% of individuals with **Multiple endocrine neoplasia (MEN) type 1** by age 40 have hyperparathyroidism. Primary hyperparathyroidism occurs in **MEN type 2A**.

Parathyroid carcinoma should be considered when a patient presents with features of hyperparathyroidism, a palpable neck mass, both bone disease and nephrolithiasis, and marked increases of both serum calcium and PTH levels.

Neonatal severe primary hyperparathyroidism is rare. There are very high PTH and calcium concentrations.

The interrelationships between serum PTH and calcium in three common clinical situations and in normals are illustrated in the figure.

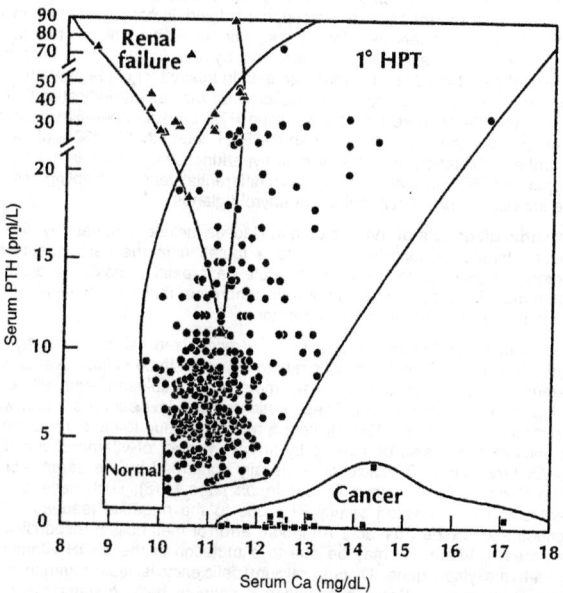

Overlapping domains are seen, especially early in the course of the disorders shown.

SUMMARY: The two most important tests for primary hyperparathyroidism are PTH and serum ionized calcium.

References

Cioffi M, Corradino M, Gazzerro P, et al, "Serum Concentrations of Intact Parathyroid Hormone in Healthy Children," *Clin Chem*, 2000, 46(6 Pt 1):863-4.

Glendenning P, Gutteridge DH, Retallack RW, et al, "High Prevalence of Normal Total Calcium and Intact PTH in 60 Patients With Proven Primary Hyperparathyroidism: A Challenge to Current Diagnostic Criteria," *Aust N Z J Med*, 1998, 28(2):173-8.

Irvin GL and Carneiro DM, "Management Changes in Primary Hyperparathyroidism," *JAMA*, 2000, 284(8):934-6.

Irvin GL and Carneiro DM, "Rapid Parathyroid Hormone Assay Guided Exploration," *Operative Tech Gen Surg*, 1999, 1:18-27.

Mayo Medical Laboratories, *2000 Test Catalogue*, Rochester, MN, 403.

Painter PC, Cope JY, and Smith JL, "Reference Information for the Clinical Laboratory," *Tietz Textbook of Clinical Chemistry*, Burtis CA and Ashwood ER, eds, Philadelphia, PA: WB Saunders Co, 1999, 1788-1846.

Selby PL and Adams PH, "The Investigation of Hypercalcemia," *J Clin Pathol*, 1994, 47:579-84.

Parietal Cell Antibody

Related Information

Anemia Flowchart *on page 35*
Cobalamin, Serum *on page 424*
Folic Acid, RBC *on page 606*
Folic Acid, Serum *on page 606*
Gastrin, Serum *on page 631*
Glutamic Acid Decarboxylase (GAD65) Antibody *on page 654*
Helicobacter pylori Biopsy-Based Tests: The Urease Tests, Culture, Cytology, and PCR *on page 672*
Intrinsic Factor Blocking Antibody *on page 806*
Liver/Kidney Microsomal Type 1 Antibodies *on page 873*
Methylmalonic Acid, Serum, Plasma, Urine, or Amniotic Fluid *on page 908*
Phosphorus, Urine *on page 1032*
Schilling Test *on page 1178*
Vitamin B_{12} Unsaturated Binding Capacity *on page 1316*

Applies to Intrinsic Factor Autoantibodies

Test Includes Titers are performed if the test is positive.

Abstract The gastric parietal cell secretes intrinsic factor which combines with ingested vitamin B_{12} to facilitate absorption in the ileum. Parietal cells also secrete hydrochloric acid, blood group substances, transforming growth factors (TGFα), and cathepsin. **Autoimmune gastritis** is characterized by disturbance of parietal cell function, resulting in reduced production of gastric acid. Progressive destruction of fundic glands ultimately leads to atrophic gastritis with achlorhydria and vitamin B_{12} (cobalamin) deficiency (pernicious anemia).

Specimen Serum

Container Red top tube or SST™ tube

Storage Instructions Keep specimen cool.

Reference Interval Negative

Use In classical pernicious anemia with complete gastric atrophy, autoantibodies to parietal cells and to intrinsic factor may be identified. Testing for parietal cell antibody is occasionally used in the differential diagnosis of pernicious anemia, atrophic gastritis, and autoimmune gastritis.

Limitations Parietal cell antibodies are not disease specific, because they are also detectable in 30% to 60% of patients with chronic atrophic gastritis, 20% with gastric ulcers, 30% with Sjögren syndrome, 30% of first-degree relatives of patients with pernicious anemia, and 7% of healthy adults.

Patients with common variable immunodeficiency may develop a pernicious anemia-like disorder in which antibodies to intrinsic factor and to parietal cells are not found.

Methodology Indirect immunofluorescence

Additional Information Parietal cells are histologically distinctive. They are pyramidal with eosinophilic to clear cytoplasm.

Antibodies to parietal cells are present in about 90% of adults with pernicious anemia and in up to 60% of subjects with atrophic gastritis; they do not correlate with malabsorption of vitamin B_{12}. They are also present in occasional patients with gastric ulcer or gastric cancer. There is cross positivity of parietal cell and thyroid antibodies in patients with thyroiditis and pernicious anemia. With time, the titer of parietal cell antibodies will decline in some patients with pernicious anemia (possibly related to loss of parietal cells) whereas intrinsic factor antibodies persist.

Anti-intrinsic factor, as well as antiparietal cell autoantibodies, can be detected with substantial frequency in unaffected family members and in subjects who have other autoimmune disorders.

References

Davidson RJ, Atrah HI, and Sewell HF, "Longitudinal Study of Circulating Gastric Antibodies in Pernicious Anemia," *J Clin Pathol*, 1989, 42(10):1092-5.

Galperin C and Gershwin E, "Immunopathogenesis of Gastrointestinal and Hepatobiliary Diseases," *JAMA*, 1997, 278(22):1946-55.

♦ **Paroxysmal Cold Hemoglobinuria Test** *see* Cold Hemolysin Test *on page 431*

♦ **Partial Thromboplastin Time** *see* Activated Partial Thromboplastin Time *on page 100*

Parvovirus B19 DNA

Related Information
Parvovirus B19 Serology *on page 1006*

Test Includes Detection of parvovirus B19 by nucleic acid amplification and detection of products

Abstract Infections with parvovirus B19 can be asymptomatic or cause mild nonspecific symptoms. The most common manifestation caused by parvovirus B19 is erythema infectiosum, an illness of children, also known as fifth disease. The more serious clinical manifestations of parvovirus B19 include arthralgias, severe anemia, fetal hydrops, severe pancytopenia, and thrombocytopenic purpura. These serious manifestations are generally seen in patients with underlying conditions such as sickle cell anemia, organ or bone marrow transplantation, or those who are immunocompromised. Replication of parvovirus B19 is in erythroid progenitor cells in marrow.

Specimen Serum, amniotic fluid, tissue, bone marrow aspirate

Container Red top tube for serum and amniotic fluid, EDTA tube for bone marrow, fresh frozen tissue

Use Detection of parvovirus B19 infection is associated with stillbirth, arthritis, aplastic crisis and fifth disease

Methodology Amplification of virus specific nucleic acid and detection of specific products of amplification

Additional Information Parvovirus B19 is a DNA virus that can cause a wide spectrum of disease including self-limited erythema infectiosum, bone marrow failure, and fetal death. In most people, the infection is associated with mild symptoms and a red cell aplasia which may be subclinical. The clinical presentation may include facial rash in ≥50% of patients. Because the virus destroys red blood cell precursors, there is a reduction in erythrocyte production. This transient aplastic crisis may be particularly severe in patients with hemoglobinopathies associated with decreased erythrocyte lifespan (sickle cell disease, spherocytosis, and β-thalassemia).

Parvovirus B19 DNA can be detected in serum of infected patients 6 days after infection. Viremia usually peaks 2-3 days later and may last for 1 week. After that time, the viral titer declines and low levels of parvovirus B19 DNA can be detected for several months only with nucleic acid amplification techniques. The presence of parvovirus DNA provides definite evidence of recent infection.

References

Boggino H and Payne DA, "Quantitative Direct Probe Method for the Detection of Parvovirus B19," *J Clin Lab Anal*, 2000, 14(1):38-41.

Dieck D, Schild RL, Hansmann M, et al, "Prenatal Diagnosis of Congenital Parvovirus B19 Infection: Value of Serological and PCR Techniques in Maternal and Fetal Serum," *Prenat Diagn*, 1999, 19(12):1119-23.

Lundqvist A, Tolfvenstam T, Bostic J, et al, "Clinical and Laboratory Findings in Immunocompetent Patients With Persistent Parvovirus B19 DNA in Bone Marrow," *Scand J Infect Dis*, 1999, 31(1):11-6.

Marchand S, Tchernia G, Hiesse C, et al, "Human Parvovirus B19 Infection in Organ Transplant Recipients," *Clin Transplant*, 1999, 13(1 Pt 1):17-24.

Vadlamudi G, Rezuke WN, Ross JW, et al, "The Use of Monoclonal Antibody R92F6 and Polymerase Chain Reaction to Confirm the Presence of Parvovirus B19 in Bone Marrow Specimens of Patients With Acquired Immunodeficiency Syndrome," *Arch Pathol Lab Med*, 1999, 123(9):768-73.

Internet Web Sites
www.amm.co.uk/pubs/fa_parvovirus.htm
www.cdc.gov/ncidod/diseases/b19&preg.htm
www.cdc.gov/ncidod/diseases/b19.htm

Parvovirus B19 Serology

Related Information
Bone Marrow *on page 296*
Parvovirus B19 DNA *on page 1006*

Test Includes Assays for parvovirus B19 may include detection of IgM and IgG antibodies

Abstract Detection of antibodies specific for parvovirus B19 are used to diagnose recent infections and to determine immune status. In normal hosts, specific IgM antibodies can be detected during the second week of infection and can be detected for 4-6 months after an active infection. Specific IgG antibodies appear shortly after IgM and can persist for years. In immunocompromised patients, the immune response to parvovirus B19 may be normal, altered, or even absent.

Specimen Serum, amniotic fluid for IgM antibody

Container Red top tube

Sampling Time B19-specific IgM antibodies appear 10-14 days after onset and persist for only 4-6 months. Collect acute and convalescent sera for IgG antibodies.

Storage Instructions Separate serum and freeze.

Reference Interval Negative for IgG and IgM.

Use Investigation of maculopapular rash with fever, arthralgias, red cell aplastic crisis in subjects with hemolytic anemias; work-up of fetal hydrops and spontaneous abortion

Limitations The absence of antiparvovirus B19 antibodies cannot exclude infection. Immunocompromised patients may show poor antibody response. Unpaired sera tested for IgG are of little use for diagnosis of infection. Fetal blood sampling prior to 22 weeks of gestation often produces false-negative IgM results. Some cross-reactions with patient sera containing rubella virus-specific IgM.

Methodology Radioimmunoassay (RIA), indirect enzyme immunoassay (EIA) for IgG, or immunoblot assay (Western blot) for detection of IgM antibodies to parvovirus B19

Additional Information Parvovirus B19 is a DNA virus and can cause a wide spectrum of disease, ranging from erythema infectiosum (fifth disease) to bone marrow failure and to fetal death. Intrauterine transfusion has been suggested when there is evidence of B19 parvovirus-associated anemia and hydrops. Because the virus destroys erythroid precursors, which leads to a reduction in normal red blood cell production, infection with parvovirus B19 can cause aplastic crisis in patients already at maximum red cell production and in those with increased red cell destruction (sickle cell disease, β-thalassemia, and spherocytosis). In immunocompromised patients, parvovirus B19 infection can cause life-threatening anemia or thrombocytopenic purpura. The presence of IgM antibodies to parvovirus B19 provide definite evidence of recent infection.

References

Dieck D, Schild RL, Hansmann M, et al, "Prenatal Diagnosis of Congenital Parvovirus B19 Infection: Value of Serological and PCR Techniques in Maternal and Fetal Serum," *Prenat Diagn*, 1999, 19(12):1119-23.

Hemauer A, Gigler A, Searle K, et al, "Seroprevalence of Parvovirus B19 NS1-Specific IgG in B19-Infected and Uninfected Individuals and in Infected Pregnant Women," *J Med Virol*, 2000, 60(1):48-55.

Kido S, Ito Y, Nishimura N, et al, "Human Parvovirus B19-Associated Thrombocytopenic Purpura," *Acta Paediatr Jpn*, 1998, 40(5):486-8.

Lin KH, You SL, Chen CJ, et al, "Seroepidemiology of Human Parvovirus B19 in Taiwan," *J Med Virol*, 1999, 57(2):169-73.

Valeur-Jensen AK, Pedersen CB, Westergaard T, et al, "Risk Factors for Parvovirus B19 Infection in Pregnancy," *JAMA*, 1999, 281(12):1099-105.

Internet Web Sites

www.amm.co.uk/pubs/fa_parvovirus.htm
www.cdc.gov/ncidod/diseases/parvovirus/b19&preg.htm
www.cdc.gov/ncidod/diseases/parvovirus/b19.htm

♦ **PAS Stain** *see* Kidney Biopsy *on page 818*

♦ **Paternity Testing by DNA Testing** *see* Identification DNA Testing *on page 765*

♦ **Pathology Specimen Identification** *see* Specimen Identification Requirements *on page 1217*

♦ **Paul-Bunnell Davidsohn Test** *see* Infectious Mononucleosis Screening Test *on page 785*

♦ **Paveral** *see* Codeine, Urine *on page 429*

♦ **Pb, Blood** *see* Lead, Blood *on page 832*

♦ **Pbg-D** *see* Porphobilinogen Deaminase, Erythrocyte *on page 1072*

♦ **PBI** *see* Thyroxine, Serum *on page 1257*

♦ **Pb, Urine** *see* Lead, Urine *on page 835*

- **PCA-1** *see* Antineuronal Nuclear Antibody, Type 1 (Anti-Hu) *on page 185*
- **PCA-2** *see* Antineuronal Nuclear Antibody, Type 1 (Anti-Hu) *on page 185*
- **PCA-Tr** *see* Antineuronal Nuclear Antibody, Type 1 (Anti-Hu) *on page 185*
- **PChE** *see* Pseudocholinesterase, Serum *on page 1122*
- **PCNA** *see* Immunoperoxidase Procedures *on page 780*
- **pCO₂** *see* Blood Gases and pH, Arterial *on page 275*

pCO₂, Blood

Related Information
Arterial Blood Collection *on page 211*
Bicarbonate, Blood *on page 258*
Blood Gases and pH, Arterial *on page 275*
Blood Gases and pH, Capillary *on page 277*
Blood Gases and pH, Venous *on page 279*
Carbon Dioxide, Total, Blood *on page 339*
pH, Blood *on page 1018*
Point-of-Care Testing *on page 1065*

Applies to Acid-Base Status; Hypercapnia; Hyperventilation; Hypoventilation; Respiratory Acidosis; Respiratory Alkalosis

Test Includes Test is part of blood gas panels and often, electrolyte panels

Abstract pCO_2 is the partial pressure of dissolved carbon dioxide in blood and is useful in evaluating the respiratory component acid-base status.

Specimen Whole blood (arterial, venous, or capillary)

Collection Specimen should be collected anaerobically. See Arterial Blood Collection *on page 211*, Venous Blood Collection *on page 1304*, and Skin Puncture Blood Collection *on page 1203*. Deliver specimen immediately to the laboratory. When pCO_2 is drawn as a venous specimen, avoid a tourniquet if possible. Especially, **avoid fist clenching**.

Storage Instructions Specimen should be analyzed immediately. However, if delay in analysis is unavoidable, the specimen should be placed in a slurry of ice chips and water. Analysis should not be delayed beyond 1 hour. *In vitro* pCO_2 will increase by about 0.5 mm Hg/hour at 4°C.

Causes for Rejection Specimen with clots, air bubbles, or not received on ice; needle not tightly stoppered

Turnaround Time Usually 30 minutes or less

Reference Interval
- Newborns, infants, and children up to ~2 years of age, with **arterialized capillary blood** (heel, fingertip, big toe) or arterial blood: 27-40 mm Hg
- Children older than 2 years of age (arterial) 27-41 mm Hg
- Adults, arterial: male: 35-48 mm Hg, female: 32-45 mm Hg
- Adults, venous: 40-45 mm Hg

Critical Values Partial pressure of arterial carbon dioxide ($PaCO_2$) >45 mm Hg may be a risk factor for pulmonary complications. In limited numbers of reviewed cases, such patients had significant airway obstruction on spirometry.

Possible Panic Range <20 mm Hg, >70 mm Hg

Use Alveolar ventilation varies inversely as arterial pCO_2; therefore, pCO_2 is an indication of adequacy of CO_2 elimination by the lungs. **Respiratory alkalosis** (hyperventilation) results in low pCO_2 with elevated pH. **Respiratory acidosis** (hypoventilation) results in high pCO_2 with lowered pH.

Limitations Venous and arterial values are sensitive to sampling technique.

Methodology pCO_2 electrode

Additional Information Respiratory compensation for metabolic acidosis and alkalosis involves adjustment of the pCO_2 level by hypoventilation (as in cases of metabolic alkalosis with resultant rise in pCO_2) or by hyperventilation (as in cases of metabolic acidosis with decreasing pCO_2). As CO_2 is highly soluble, there is a quickly attained equilibrium between arterial carbon dioxide tension **($PaCO_2$)** and alveolar carbon dioxide tension. Therefore, the arterial CO_2 measurement also determines the status of ventilation. $PaCO_2$ has a reverse relationship to alveolar ventilation.

The expression **"hypercapnia"** indicates the presence of excessive carbon dioxide in the blood. Disorders associated with hypercapnia include central respiratory depression, abnormal neuromuscular function, chest wall abnormality, upper or lower respiratory tract disease, or hypercapnia secondary to cardiac disease. Arterial pCO_2 may be an indicator of systemic perfusion during cardiopulmonary resuscitation. Increased venous-arterial pCO_2 gradients do not appear to be a reliable indicator of inadequate tissue perfusion during cardiopulmonary bypass. Transcutaneous measurements of pCO_2 may be a convenient means of monitoring the neonate in intensive care units. Acidosis does affect the ability to correlate transcutaneous and arterial pCO_2 values.

References

Narins RG and Emmett M, "Simple and Mixed Acid-Base Disorders: A Practical Approach," *Medicine (Baltimore)*, 1980, 59:161-87.

Preuss HG, "Fundamentals of Clinical Acid-Base Evaluation," *Clin Lab Med*, 1993, 13(1):103-16.

Smetana GW, "Preoperative Pulmonary Evaluation," *N Engl J Med*, 1999, 340(12):937-53.

Pepsinogen I and II, Serum or Plasma

Related Information

Cobalamin, Serum *on page 424*
Gastrin, Serum *on page 631*
Helicobacter pylori Antigen and Serology *on page 671*
Helicobacter pylori Biopsy-Based Tests: The Urease Tests, Culture, Cytology, and PCR *on page 672*

Synonyms Pepsinogen A; Pepsinogen C; PG I; PG II

Applies to Pepsinogen A:C Ratio; Pepsinogen I:II Ratio; PG I:PG II Ratio

Abstract Pepsinogen I (PG I) and pepsinogen II (PG II) are groups of gastric mucosal proenzymes involved in protein digestion. Both PG I and PG II are secreted by chief cells and mucous neck cells of the gastric fundus and corpus; PG-II is also secreted by the pyloric glands of the antrum and Brunner's glands of the proximal duodenum. In general, acute or mild chronic gastric inflammation causes increased secretion of both PG I and PG II; atrophy causes decreased secretion. However, corpus atrophy results in chief cells being replaced by pyloric glands, resulting in decreased serum levels of PG I with no change or increase in PG II and decreased PG I:PG II ratios.

Patient Preparation Patient should fast for 10-12 hours and discontinue antacids or other medications affecting stomach acidity or gastrointestinal mobility for at least 48 hours prior to collection.

Specimen Serum, plasma

Container Red top tube, lavender top (EDTA) tube

Storage Instructions Freeze specimen immediately after centrifugation and transport to referral laboratory on dry ice.

Reference Interval PG I: 20-107 µg/L; PG II: 3-19 µg/L; PG I:PG II ratio: 5-6

Use These tests are very rarely used. Ratios are decreased in patients with atrophic gastritis, gastric cancer, and gastric body ulcer. Ratios are increased in duodenal ulcer patients and in patients with achlorhydria (eg, pernicious anemia).

(Continued)

Pepsinogen I and II, Serum or Plasma *(Continued)*

Limitations The test is not widely available, even among the large referral laboratories. Considerable overlap in values is found between healthy individuals and those with abnormality.

Methodology Radioimmunoassay (RIA), fluorometric immunoassay, enzyme-linked immunosorbent assay (ELISA)

References

Kitahara F, Kobayashi K, Sato T, et al, "Accuracy of Screening for Gastric Cancer Using Serum Pepsinogen Concentrations," *Gut*, 1999, 44(5):693-7.

Konishi N, Matsumoto K, Hiasa Y, et al, "Tissue and Serum Pepsinogen I and II in Gastric Cancer Identified Using Immunohistochemistry and Rapid ELISA," *J Clin Pathol*, 1995, 48(4):364-7.

Matsumoto K, Konishi N, Ohshima M, et al, "Association Between *Helicobacter pylori* Infection and Serum Pepsinogen Concentrations in Gastroduodenal Disease," *J Clin Pathol*, 1996, 49:1005-8.

♦ **Pepsinogen I:II Ratio** *see* Pepsinogen I and II, Serum or Plasma *on page 1009*

♦ **Peptide Hormones** *see* Immunoperoxidase Procedures *on page 780*

♦ **Percoffedrinol N®** *see* Caffeine, Serum *on page 326*

♦ **Percutafeine®** *see* Caffeine, Serum *on page 326*

♦ **Pericardial Fluid Analysis** *see* Body Fluid Analysis, Cell Count *on page 288*

♦ **Pericardial Fluid Analysis** *see* Body Fluid Chemical Analysis *on page 291*

♦ **Pericardial Fluid Cytology** *see* Body Cavity Fluid Cytology *on page 285*

♦ **Pericardial Fluid pH** *see* Body Fluid pH *on page 295*

Peripheral Blood: Differential Leukocyte Count

Related Information

Apoptosis Assays *on page 205*
Bacteremia Detection, Buffy Coat Micromethod *on page 227*
Bone Marrow *on page 296*
Buffy Coat Smear Study of Peripheral Blood *on page 316*
CD4/CD8 Enumeration *on page 349*
Complete Blood Count *on page 442*
Cryoglobulin, Qualitative, Serum and Plasma *on page 478*
Eosinophil Count *on page 542*
Infectious Mononucleosis Screening Test *on page 785*
Leukocyte Cytochemistry *on page 846*
Lymph Node Biopsy *on page 880*
Peripheral Blood: Red Blood Cell Morphology *on page 1016*
Platelet Count *on page 1050*
Platelet Sizing *on page 1056*
White Blood Cell Count *on page 1330*

Synonyms Differential Smear; White Blood Cell Morphology

Applies to *Capnocytophaga canimorsus*

Test Includes Relative frequency (%) of and also (in some laboratories) the absolute number of the different types of white blood cells in the peripheral blood, RBC morphology, platelet evaluation

Abstract This procedure, usually a part of the complete blood count, determines the relative and/or absolute number of the different types of leukocytes circulating in the peripheral blood. There are significant differences between pediatric and adult reference ranges.

Specimen Whole blood, fresh, anticoagulated (EDTA preferred)

Container International Council for Standardization in Hematology recommendation is for use of K_2-EDTA, 1.5-2.2 mg/mL of blood as anticoagulant for blood cell counting and sizing. Lavender top (EDTA) tube or smears prepared directly from fingerstick or heelstick blood. Heparin or oxalate may produce artifactual distortion of morphology, especially of white blood cells.

Reference Interval See the tables that follow throughout this monograph.

Use Determine qualitative and quantitative variations in white cell numbers and morphology, morphology of red cells and platelet evaluation; evaluate anemia, leukemia, infections, inflammatory states, and inherited disorders of red cells, white cells, and platelets; with automated instruments, generation of cell histograms

Leukocyte Values From Birth to Maturity

Age	Leukocyte count ($\times 10^3/mm^3$)	Neutrophils			Eosinophils	Basophils	Lymphocytes	Monocytes
		Total	Band	Segmented				
Birth	18.1 (9.0–30.0)	11.0 (6.0–26.0) 61%	1.65 9.1%	9.4 52%	0.40 (0.02–0.85) 2.2%	0.10 (0–0.64) 0.6%	5.5 (2.0–11.0) 31%	1.05 (0.40–3.1) 5.8%
12 h	22.8 (13.0–38.0)	15.5 (6.0–28.0) 68%	2.33 10.2%	13.2 58%	0.45 (0.02–0.95) 2.0%	0.10 (0–0.50) 0.4%	5.5 (2.0–11.0) 24%	1.20 (0.40–3.6) 5.3%
24 h	18.9 (9.4–34.0)	11.5 (5.0–21.0) 61%	1.75 9.2%	9.8 52%	0.45 (0.05–1.00) 2.4%	0.10 (0–0.30) 0.5%	5.8 (2.0–11.5) 31%	1.10 (0.20–3.1) 5.8%
1 wk	12.2 (5.0–21.0)	5.5 (1.5–10.0) 45%	0.83 6.8%	4.7 39%	0.50 (0.07–1.10) 4.1%	0.05 (0–0.25) 0.4%	5.0 (2.0–17.0) 41%	1.10 (0.30–2.7) 9.1%
1 mo	10.8 (5.0–19.5)	3.8 (1.0–9.0) 35%	0.49 4.5%	3.3 30%	0.30 (0.07–0.90) 2.8%	0.05 (0–0.20) 0.5%	6.0 (2.5–16.5) 56%	0.70 (0.15–2.0) 6.5%
6 mo	11.9 (6.0–17.5)	3.8 (1.0–8.5) 32%	0.45 3.8%	3.3 28%	0.30 (0.07–0.75) 2.5%	0.05 (0–0.20) 0.4%	7.3 (4.0–13.5) 61%	0.58 (0.10–1.3) 4.8%
1 y	11.4 (6.0–17.5)	3.5 (1.5–8.5) 31%	0.35 3.1%	3.2 28%	0.30 (0.05–0.70) 2.6%	0.05 (0–0.20) 0.4%	7.0 (4.0–10.5) 61%	0.55 (0.05–1.1) 4.8%
6 y	8.5 (5.0–14.5)	4.3 (1.5–8.0) 51%	0.25 3.0%	4.0 48%	0.23 (0–0.65) 2.7%	0.05 (0–0.20) 0.6%	3.5 (1.5–7.0) 42%	0.40 (0–0.8) 4.7%
12 y	8.0 (4.5–13.5)	4.4 (1.8–8.0) 55%	0.25 3.0%	4.2 52%	0.20 (0–0.55) 2.5%	0.04 (0–0.20) 0.5%	3.0 (1.2–6.0) 38%	0.35 (0–0.8) 4.4%
18 y	7.7 (4.5–12.5)	4.4 (1.8–7.7) 57%	0.23 3.0%	4.2 54%	0.20 (0–0.45) 2.6%	0.04 (0–0.20) 0.5%	2.7 (1.0–5.0) 35%	0.40 (0–0.8) 5.2%
21 y	7.4 (4.5–11.0)	4.4 (1.8–7.7) 59%	0.22 3.0%	4.2 56%	0.20 (0–0.45) 2.7%	0.04 (0–0.20) 0.5%	2.5 (1.0–4.8) 34%	0.30 (0–0.8) 4.0%

Adapted from Altman PL and Dittmer DS, eds, *Biology Data Book*, 2nd ed, Volume 3, Bethesda, MD: Federation of American Societies for Experimental Biology. 1974.

Peripheral Blood: Differential Leukocyte Count
(Continued)

Review of Peripheral Blood Smear

Red Cell Variant*†	Clinical Associations
Crenated cell (Echinocyte)	Variant form of normal RBC
Burr cell	DIC, I.V. fibrin deposition
Schizocyte	Microangiopathic hemolytic anemia
Helmet cell	Hypertension
(Schizocyte)	Cardiac valve disease
	Uremia, burns
	Metastatic malignancy
	Severe iron deficiency / bleeding lesion
	Normal newborn
Elliptocyte	Few seen normally
Ovalocyte	Many may mean primary elliptocytosis
	Iron deficiency
	Thalassemia
	Hb S or C
	Other hemolytic anemias
Target cell (Codocyte)	Hemoglobinopathies (S, C, D, thalassemia, esp)
	Iron deficiency
	Liver disease
	LCAT deficiency
Oval macrocyte (Megalocyte)	Megaloblastic anemia
	B_{12} / folate deficiency
	Myeloproliferative disease
	Chemotherapy patients
Spherocyte	Hereditary spherocytosis
	Immune and other hemolytic states
Tear drop cell (Dacryocyte)	Myeloproliferative diseases
	Myelophthisic processes
	Pernicious anemia
	Thalassemia
Sickle cell (Drepanocyte)	Sickle cell disease and variants (ie, sickle / thalassemia, SD disease, SC disease)
Acanthocyte	Abetalipoproteinemia
	Alcoholic cirrhosis with hemolysis
	Pyruvate kinase deficiency
	Postheparin in some individuals[1]

[1]Silber R, "Of Acanthocytes, Spurs, Burrs, and Membranes," Blood, 1969, 34:111.

*Established terminology is followed (parentheses) by that of Bessis M et al, introduced on the basis of ultrastructural analyses.

†Some abnormal red cell forms may represent artifact introduced during preparation of the blood smear (especially stomatocytes and elliptocytes, occasionally target-like cells).

Limitations Because of sampling, large statistical variation exists, particularly with 100 cell count manual method and with low incidence cells. Day-to-day changes should be interpreted in relation to known method-related variation.

Methodology One or a combination of methods: manual enumeration of white cells on Wright-stained peripheral blood smear; automated WBC computer image analysis (such instruments are no longer being manufactured); continuous flow system (automated) using cytochemical/light scattering measurements; cell volume (impedance related/conductivity/light scattering) measurements, resultant electronic signals of combined methods are further manipulated with computer-assisted synthesis and derivations.

Additional Information Significantly abnormal findings (automated or manual method) should be the subject of further study and review. Changes in leukocyte fractions are a window to a spectrum of minor to serious physiologic and pathologic changes.

The past decade has seen significant contributions to the definition of WBC differential reference values. Variations, often not clinically significant, relate to differences in sex, race, physiologic state, and method of analysis.

Review of Peripheral Blood Smear (Continued)

Red Cell Variant*†	Clinical Associations
Stomatocyte	Hereditary stomatocytosis
	Alcoholism
	Rh null disease
Schistocyte	Microangiopathic hemolysis
Helmet cell	Cardiac valve disease
Spurr cell	DIC
(Schizocyte) (Keratocyte)	Severe burns
	Uremia
Triangulocytes[2]	Alcoholism
	Rarely, Hb C disease
	Thalassemia
	Nonalcohol liver disease
	TTP
	Antimitotic chemotherapy
Eccentrocytes[3] (Asymmetric distribution of Hb)	G6PD deficiency
Bite cells[4,5]	Heinz body hemolytic anemia
(Degmacyte)	Oxidative hemolysis
	Methemoglobinemia due to phenazopyridine sulfanilamide
	Unstable hemoglobin (eg, Hb Köln)
	Thalassemia
Hemighosts[6]	Severe oxidative injury
	Heinz body hemolytic anemia
	Oxidative hemolysis
Polychromatophil	Increased erythropoiesis
Reticulocyte	Myelophthisic states
Nucleated RBC	Hemolytic states
	Postsplenectomy
Basophilic stippling	Lead poisoning
(Punctate basophilis)	Hemolytic states, other anemias
	Thalassemia
	Pyrimidine-5'-nucleotidase deficiency
Pappenheimer bodies	Some hemolytic anemias
(Siderocytes)	Postsplenectomy
	Some megaloblastic anemias
	Some sideroblastic states
Parasites	Plasmodium (malaria)
	Bartonella
	Microfilaria (not intracellular)
	Whipple's disease bacillus (Tropheryma whippelii)
	Babesia and Babesia-like organisms
	Spirochetes of Borrelia species (relapsing fever)

[2]Schumacher HR, Khanna S, and Moyer B, "Letter: Triangulocytes in Alcoholism," JAMA, 1976, 235:2285-6.

[3]Ham TH, Grauel JA, Dunn RF, et al, "Physical Properties of Red Cells as Related to Effects In Vivo. IV. Oxidant Drugs Producing Abnormal Intracellular Concentration of Hemoglobin (Eccentrocytes) With a Rigid Red-Cell Hemolytic Syndrome," J Lab Clin Med, 1973, 82:898-910.

[4]Ward PC, Schwartz BS, and White JG, "Heinz Body Anemia: Bite Cell Variant — A Light and Electron Microscopic Study," Am J Hematol, 1983, 15:135-46.

[5]Greenberg MS, "Heinz Body Hemolytic Anemia: "Bite Cells — A Clue to Diagnosis," Arch Intern Med, 1976, 136:153-5.

[6]Chan TK, Chan WC, and Weed RI, "Erythrocyte Hemighosts: A Hallmark of Severe Oxidative Injury In Vivo," Br J Haematol, 1982, 50:575.

*Established terminology is followed (parentheses) by that of Bessis M et al, introduced on the basis of ultrastructural analyses.

†Some abnormal red cell forms may represent artifact introduced during preparation of the blood smear (especially stomatocytes and elliptocytes, occasionally target-like cells).

Enumeration of band population is definition dependent. Automated differential determinations may vary between instruments and manual techniques but clinical significance of such differences may be minimal. Black individuals have lower neutrophil values than whites. The greatest variation, both in relative and (Continued)

Review of Peripheral Blood Smear

Red Cell Variant*†	Clinical Associations
Rouleaux of RBCs	Reflects increased protein concentration
	May be associated with multiple myeloma
	Waldenström's macroglobulinemia blue staining background
Howell-Jolly bodies	Hemolytic anemia
(Nuclear fragments)	Hyposplenism / asplenism (splenectomy)
	Megaloblastic anemia
Cabot Ring (Nuclear remnants)	Megaloblastic anemia
Heinz bodies	Some drug sensitive oxidative hemolytic anemias
(Denatured Hb)	Unstable hemoglobinopathies
Hb H inclusions	Unstable Hb β_4 tetramers in rare RBCs; stain with cresyl blue[7]

WBC Abnormalities	Clinical Associations
Leukocytosis	Acute reactive state metabolic basis infections (esp bacterial) basis
Increase in % bands (left shift)	
Toxic granulation	
Toxic vacuolation	
Döhle bodies	Acute infection, esp pneumonia
	Scarlet fever
	Measles,
	Septicemia
	May-Hegglin anomaly
Chediak-Higashi anomaly	Inherited, giant lysosomal granules
Pseudo-Chediak-Higashi anomaly	Few cases of acute myeloid leukemia
Hypersegmented neutrophils	Megaloblastic states as pernicious anemia
Hypogranular neutrophils	Some cases of chronic myelogenous leukemia
Intraleukocyte microorganisms	Bacteria
	Ehrlichia sennetsu (Sennetsu fever)
Intragranulocytic	Ehrlichia species
Intramonocytic and small lymphocytic	Ehrlichia chaffeensis (human ehrlichiosis)
Auer rods (present in blast cells)	Acute myelogenous leukemia
Chediak-Higashi inclusions	Congenital deficiency
	Lysosomal membrane
	Phospholipid
Alder-Rielly anomaly	Mucopolysaccharidosis
Pelger-Huët anomaly (mono and bilobed neutrophils with clumped nuclear chromatin)	Congenital form
	Acquired form - associated with myelogenous leukemia
Howell-Jolly body-like basophilic neutrophil inclusions	AIDS
Monocytes with disorganized cytoplasmic fibrils[8]	AIDS
Neutrophil cytoplasmic inclusion bodies	Human ehrlichiosis

Platelet Abnormalities	Clinical Associations
Platelet satellitosis	No definite causal clinical association
Platelet clumping	May cause spurious leukocytosis and thrombocytopenia[9]

[7]Sabath DE, Cross ST, and Mamiya LY, "An Improved Method for Detecting Red Cells With Hemoglobin H Inclusions That Does Not Require Glass Capillary Tubes," Clin Lab Hematol, 2003, 25(1):87-91.

[8]Kass L, "Cytoplasmic Abnormalities in Monocytes From Patients With AIDS," Lab Med, 2001, 32(3):139-42.

[9]Solanki DL and Blackburn BC, "Spurious Leukocytosis and Thrombocytopenia. A Dual Phenomenon Caused by Clumping of Platelets In Vitro," JAMA, 1983, 250:2514-5.

*Established terminology is followed (parentheses) by that of Bessis M et al, introduced on the basis of ultrastructural analyses.

†Some abnormal red cell forms may represent artifact introduced during preparation of the blood smear (especially stomatocytes and elliptocytes, occasionally target-like cells).

Adapted from Bessis M, Blood Smears Reinterpreted, New York, NY: Springer International, 1977.

Adapted from Bessis M, Weed RI, and Leblond PF, Red Cell Shape: Physiology, Pathology, Ultrastructure, New York, NY: Springer Verlag, 1973.

absolute terms occurs as the result of age. High WBC levels are present in the newborn and lymphocytes are increased in childhood (as compared to adult values). Correlation of the manually performed vs automated differential is somewhat poorer in pediatric as compared to adult populations. Both relative (%) and absolute (actual number of cells/mm^3) values need to be considered in relation to the clinical situation.

Current hematology instruments can produce automated differentials by cytochemistry/light scattering or cell volume/conductivity/light scattering techniques. Conductivity measurements utilize a high frequency electromagnetic probe. Leukocytes are separated into granulocytes, lymphocytes, monocytes, eosinophils, and basophil categories and an immaturity index (not exactly bands) of the granulocytes. With modern hematology instruments, manual differential determinations are largely redundant. High white counts and fever are better indicators of clinical infection than the percentage of bands. This is because of the high variability of manual differential band counts from technologist to technologist. The band count was found superior to the immature to total neutrophil count ratio, the total WBC count, and the neutrophil count in the diagnosis of inflammatory and infective disease when compared to the C-reactive protein level (see reference by Seebach et al).

Any particular patient (however uncommonly encountered) may present with very few but very significantly abnormal peripheral red or white cells allowing timely diagnosis not possible by other initial methods of evaluation. Some morphologic abnormalities of peripheral blood neutrophils (degree of hypogranulation and of Pelger-Huët-like changes) reflect the degree of marrow dysplasia in myelodysplastic syndromes. Detection of circulating myeloma cells is clinically important as they correlate with disease activity. The cells of multiple myeloma are best detected by the use of sensitive immunofluorescent, flow cytometric, or molecular genetic techniques. Cryoglobulins, while of rare occurrence, may be due to multiple myeloma in some 50% of cases and are also seen as a complication of hepatitis C. Cryoglobulins are circulating immunoglobulins that are reversibly cold precipitable. Clinically, they may be associated with arthralgia, skin lesions, the Raynaud phenomena and liver/kidney changes. Associated laboratory abnormalities include elevated erythrocyte sedimentation rate, positive rheumatoid factor test, and hypocomplementemia. There may be pseudoleukocytosis and pseudothrombocytosis. The routine Wright-stained peripheral blood smear (prepared from blood collected at room temperature) shows neutrophil and monocyte intracytoplasmic vacuole-like inclusions and extracellular globules of lightly pink to slightly basophilic amorphous material in the background. These morphologic changes, present on the peripheral blood smear, may be the initial finding in some cases of myeloma and should prompt evaluation for cryoglobulinemia.

The cytoplasmic vacuoles seen in cases of cryoglobulinemia must be distinguished from artifact due to EDTA, and most importantly, from those that occur with overwhelming sepsis. With septicemia, the neutrophil intracytoplasmic vacuoles are clear while those seen with cryoglobulinemia are usually pink to lightly basophilic. Clinical correlation with other findings is essential (eg, presence of "left shift", toxic granulation, Döhle bodies). A number of intracellular bodies may occur in cells involved by lymphoproliferative disorders. There is a recent report of filamentous inclusions in a case of chronic lymphocytic leukemia.

A number of other uncommon but potentially critically important conditions may leave footprints on the peripheral blood smear (see table and Buffy Coat Smear Study of Peripheral Blood *on page 316*). A recent example is the report of potentially rapidly fatal septicemia with Waterhouse-Friderichsen syndrome due to *Capnocytophaga canimorsus*. The PBS may show neutrophil toxic granulation with intracytoplasmic and extracellular bacterial rods. This gram-negative microaerophil has been associated with infection and with hemolytic uremic syndrome following animal bites, particularly dogs.

Cytoplasmic abnormalities ("disorganized" cytoplasmic fibrils) have been observed in monocytes and found to be associated with clinical AIDS.

See also comments under Additional Information of Peripheral Blood: Red Blood Cell Morphology *on page 1016*. See Platelet Count *on page 1050* and (Continued)

Peripheral Blood: Differential Leukocyte Count
(Continued)

Platelet Sizing *on page 1056* for discussion of platelet morphology and abnormalities.

References

Dorian RP and Shaw JH, "Intracytoplasmic Filamentous Inclusions in the Peripheral Blood of a Patient With Chronic Lymphocytic Leukemia - A Bright-Field, Electron Microscopic, Immunofluorescent, and Flow Cytometric Study," *Arch Pathol Lab Med*, 2003, 127(5):618-20.

Howard MR and Smith RA, "Early Diagnosis of Septicaemia in Preterm Infants From Examination of Peripheral Blood Films," *Clin Lab Haematol*, 1999, 21(5):365-8.

Kass L, "Cytoplasmic Abnormalities in Monocytes From Patients With AIDS," *Lab Med*, 2001, 32(3):139-142.

Lesesve JF and Goasguen J, "Cryoglobulin Detection From a Blood Smear Leading to the Diagnosis of Multiple Myeloma," *Eur J Haematol*, 2000, 65(1):77.

Maitra A, Ward PC, Kroft SH, et al, "Cytoplasmic Inclusions in Leukocytes: An Unusual Manifestation of Cryoglobulinemia," *Am J Clin Pathol*, 2000, 113(1):107-12.

Mirza I, Wolk J, Toth L, et al, "Waterhouse-Friderichsen Syndrome Secondary to *Capnocytophaga canimorsus* Septicemia and Demonstration of Bacteremia by Peripheral Blood Smear," *Arch Pathol Lab Med*, 2000, 124(6):859-63.

Seebach JD, Morant R, Rüegg R, et al, "The Diagnostic Value of the Neutrophil Left Shift in Predicting Inflammatory and Infectious Disease," *Am J Clin Pathol*, 1997, 107(5):582-91.

Peripheral Blood: Red Blood Cell Morphology
Related Information

Anemia Flowchart *on page 35*
Autohemolysis Test *on page 221*
Blood Volume *on page 282*
Complete Blood Count *on page 442*
Ham Test *on page 665*
Heinz Body Stain *on page 669*
Hematocrit *on page 674*
Hemoglobin A$_2$ *on page 682*
Hemoglobin Electrophoresis *on page 684*
Malaria Smear and Tests *on page 888*
Microfilariae, Peripheral Blood Preparation *on page 915*
Osmotic Fragility *on page 980*
Peripheral Blood: Differential Leukocyte Count *on page 1010*
Platelet Sizing *on page 1056*
Red Blood Cell Count *on page 1133*
Red Blood Cell Indices *on page 1136*
Schilling Test *on page 1178*

Synonyms Red Blood Cell Morphology

Applies to Pyropoikilocytosis

Specimen Whole blood

Container Lavender top (EDTA) tube or smears prepared directly from fingerstick or heelstick blood. Heparin or oxalate may produce artifactual distortion especially of white blood cells.

Collection Routine venipuncture. Invert tube gently to mix.

Storage Instructions Refrigerate

Causes for Rejection Clotted or hemolyzed specimen

Reference Interval Normal morphology. It may not be possible to correlate minor changes in RBC morphology (eg, 5% to 10% elliptocytosis) with identifiable disease.

Use Evaluate red cell disorders, white cell disorders, platelet disorders, and correlation of findings with CBC parameters as quality control function

Methodology Study of red blood cell morphology as it presents on Wright stained peripheral blood. Red blood cell indices as determined by automated cell counters (eg, MCV, MCH, MCHC, RDW) also give insight into morphologic red cell abnormalities.

Additional Information Diverse trends at all levels of the medical care system have combined to focus attention on the utility/cost-effectiveness of manual review of the peripheral blood smear (PBS), in particular, as a routine incorporated in the CBC. The CBC has evolved from highly labor intensive to highly capital intensive, while the PBS has remained highly labor intensive. Automated devices continue to expand their repertoires, adding new measurements that digitize and in some cases improve upon information formerly gleaned from study of the PBS (eg, red cell distribution width which quantitates the morphologic observation anisocytosis). Study of PBS is a part of the quality control system built into each CBC and available to the medical technologist performing the test and to physicians and others who may subsequently question the accuracy of the results. Each CBC (when truly complete with study of PBS, WBC differential count, and RBC indices) is a self-contained case individualized quality control unit. Automated counters when confronted with uncertainty (capable of generating spurious results) respond with a set of "flags", some parameters of which can be set or adjusted by laboratory staff professionals. A flagged result indicates the need for manual review of that patient's peripheral blood smear. A large study (467 samples) from various inpatient/outpatient environments found that at a detection limit ≥1% abnormal WBCs, >20% of samples were not correctly flagged (1997 study).

Efforts expended on technical and economic aspects of delivering a CBC may cloud the fact that some information may be hidden (eg, presence of target, sickle, or tear drop cells, Howell-Jolly or Pappenheimer bodies, intracellular parasites, and other abnormalities). This information, beneficial to the patient, may be lost without manual study of the PBS. Examples of blood cell morphologic abnormalities (not comprehensive) are given in tabular form in Peripheral Blood: Differential Leukocyte Count *on page 1010*. The possibility of confusing ring forms of *Babesia* (an intraerythrocytic protozoan which is also characterized by presence of red cell tetrad forms) with ring forms of the malarial organism *Plasmodium falciparum* is notable. While babesiosis is usually transmitted by tick bite, infection can also be caused by transfusion of blood (or blood components) from donors (who themselves are without evident illness).

The numerous abnormal morphologic red cell forms present potentially on a patient's PBS must be identified and correlated with clinical features to determine the need for additional laboratory investigation. An example is the finding of elliptocytes. Occasional elliptocytes (ovalocytes), present on the peripheral smear, may not have meaningful clinical association. Hereditary elliptocytosis (HE) may account for greater numbers of such cells. A study based on morphology of 1000 RBCs per iron deficient patient found that as the percentage of elliptocytes increased, Hb, Hct, RBC count, and MCH levels decreased. As the percentage of tailed poikilocytes increased, Hb, Hct, and RBC count decreased while the RBC distribution width increased. Serum ferritin levels did not correlate with the morphologic abnormalities, severity of anemia, or RBC indices. It was concluded that "...microscopic evaluation of RBC morphology remains an important tool... to evaluate the severity of anemia in patients with iron deficiency." Concerning elliptocytes, their presence on peripheral smear along with spherocytes and schistocytes (micropoikilospherocytes) and with family history of hereditary elliptocytosis may indicate the presence of hereditary pyropoikilocytosis (HPP). A positive thermal sensitivity test (TST) will support diagnosis of this rare condition. The TST is performed by incubating patient's blood for 10 minutes at 46°C. The test is positive if there is marked increase in elongation and fragmentation of the patient's RBCs (see table in Osmotic Fragility *on page 980*).

References

Cochran DL and Burnside LK, "Detecting and Identifying Hereditary Pyropoikilocytosis," Clinical Pathology Rounds, *Lab Med*, 1999, 30(1):26-9.

Linden JV, Wong SJ, Chu FK, et al, "Transfusion-Associated Transmission of Babesiosis in New York State," *Transfusion*, 2000, 40(3):285-9.

Rodgers MS, Chang C-C, and Kass L, "Elliptocytes and Tailed Poikilocytes Correlate With Severity of Iron-Deficiency Anemia," *Am J Clin Pathol*, 1999, 111(5):672-5.

Thalhammer-Scherrer R, Knöbl P, Korninger L, et al, "Automated Five-Part White Blood Cell Differential Counts. Efficiency of Software-Generated White Blood Cell Suspect Flags of the Hematology Analyzers Sysmex SE-9000, Sysmex NE-8000, and Coulter STK-S," *Arch Pathol Lab Med*, 1997, 121(6):573-7.

pH, Blood

Related Information

Abstract Blood pH is used to indicate the presence of acidemia or alkalemia.

Specimen Whole blood (arterial, venous, or capillary)

Container Green top (heparin) tube or heparinized syringe

Collection See Arterial Blood Collection *on page 211*, Venous Blood Collection *on page 1304*, and Skin Puncture Blood Collection *on page 1203*. For a venous sample, it is best to collect without a tourniquet if possible. **Do not allow the patient to clench/unclench his/her hand**, an activity which builds up lactate. Draw the specimen into an air-free heparinized syringe with needle quickly stoppered, making sure that no air bubbles remain. Capillary tubes should be filled as much as possible, metal flea inserted, capped, and mixed well with a magnet. All specimens should be on ice and brought to the laboratory immediately. For capillary collection, the skin area to be punctured should be warmed 10-15 minutes. The puncture should be deep enough to allow a free flow of blood.

Storage Instructions Specimen should be analyzed immediately. However, if delay in analysis is unavoidable, the specimen should be placed in a slurry of ice chips and water. Analysis should not be delayed beyond 1 hour. *In vitro* pH will decrease by <0.01 pH unit/hour at 4°C.

Causes for Rejection Specimen with clots, air bubbles, and not properly iced; needle not tightly stoppered

Turnaround Time Usually reported in less than 30 minutes.

Reference Interval Newborns (<2 months), with **arterialized capillary blood** (heel, fingertip, big toe) or arterial blood: 7.32-7.49; 2 months to 2 years, arterialized capillary or arterial blood: 7.34-7.46; children (>2 years) and adults: **arterial: 7.35-7.45, venous: 7.32-7.43.** Blood pH can be measured from either arterial or venous blood samples, usually with only very small differences.

Possible Panic Range <7.20, >7.60. Occasionally, patients with acidosis as severe as pH 6.80 survive. Many of these present with diabetic ketoacidosis, for which effective therapy is available.

Use The measurement of pH is used to diagnose acidosis (eg, ketoacidosis) and alkalosis (eg, emesis with loss of gastric juice), and to evaluate acid-base balance; to assess significance of serum or plasma potassium levels (eg, in hypokalemia); and in interpretation of oxyhemoglobin dissociation curves. It may be useful for assessment of birth asphyxia in the depressed newborn.

Blood pH is **increased** with uncompensated metabolic and respiratory alkalosis and **decreased** with uncompensated metabolic and respiratory acidosis.

Limitations pH values are sensitive to sampling technique.

Methodology Glass pH electrode

Additional Information pH should be judged in relation to other measurements such as pCO_2, HCO_3, Na^+, K^+, Cl^-, glucose, ketone bodies, phosphorus, lactate, BUN, creatinine, and osmolality of serum and urine. The osmolal gap is addressed in the listing Osmolality, Calculated, Serum or Plasma *on page 976.*

Causes of metabolic acidosis with normal and with increased anion gap are addressed in Anion Gap, Serum, Plasma, or Urine *on page 160.* See Blood Gases and pH, Arterial *on page 275.*

Methanol, ethylene glycol, paraldehyde, salicylate toxicity, diabetic ketoacidosis, alcoholic ketoacidosis, lactic acidosis, renal failure, and starvation are among causes of **high anion gap metabolic acidosis**.

Additional to electrolytes and other tests listed above, relevant laboratory findings in acidemia may also include ketones, ethanol concentration, uric acid, albumin, CBC, urinalysis with examination for oxalate crystals, and salicylate concentration.

Hypoproteinemia causes a metabolic alkalosis.

Umbilical arterial pH may be useful in the assessment of birth asphyxia of the depressed neonate. Others contend that infants must be severely depressed with Apgar scores ≤3 at 1 and 5 minutes to be reflected in a decreased serum pH. Blood may be obtained from a clamped umbilical segment up to 1 hour after delivery. See Blood Gases and pH, Umbilical Cord *on page 278.*

In hypotensive patients, tissue hypoxia may be assessed by measurements of arterial pH, mixed venous pH, and bicarbonate concentrations. Such measurements do not appear to reliably assess tissue hypoxia in patients with fulminant hepatic failure. For such patients, oxygen flux (the difference between arterial and venous oxygen) remains the best way to detect the presence of covert tissue hypoxia.

References

Adrogué HJ and Madias NE, "Management of Life-Threatening Acid-Base Disorders," First of Two Parts, *N Engl J Med*, 1998, 338(1):26-34, Second of Two Parts, *N Engl J Med*, 1998, 338(2):107-11.

Adrogué HJ, Wilson H, Boyd AE III, et al, "Plasma Acid-Base Patterns in Diabetic Ketoacidosis," *N Engl J Med*, 1982, 307:1603-10.

Brandenburg MA and Dire DJ, "Comparison of Arterial and Venous Blood Gas Values in the Initial Emergency Department Evaluation of Patients With Diabetic Ketoacidosis," *Ann Emerg Med*, 1998, 31(4):459-65.

Goldaber KG and Gilstrap LC 3d, "Correlations Between Obstetric Clinical Events and Umbilical Cord Blood Acid-Base and Blood Gas Values," *Clin Obstet Gynecol*, 1993, 36(1):47-59.

♦ **pH, Body Fluid** *see* Body Fluid pH *on page 295*

♦ **Phenacetin** *see* Acetaminophen, Serum *on page 90*

♦ **Phenantoin** *see* Mephenytoin, Serum *on page 896*

Phencyclidine, Qualitative, Urine

Related Information

Lysergic Acid Diethylamide, Urine *on page 883*

(Continued)

Phencyclidine, Qualitative, Urine *(Continued)*

Synonyms Angel Dust; Crystal Joint; Elephant Tranquilizers; Goon; Hog; Kay Jay; Killer Weed; Mist; PCP; Peace Pills; Peace Weed; Rocket Fuel; Sheets; Sherm; Snorts; Supergrass; Wickistick

Abstract This is a widely used drug of abuse which was formerly sold as a veterinary tranquilizer. Since 1978, legal manufacture and sale has been stopped. It is a dissociative anesthetic mimicking the symptoms of schizophrenia. It is classified by DEA as a Schedule I controlled substance with no clinical role.

Specimen Random urine

Collection If forensic, observe precautions (see Introduction *on page 63*).

Storage Instructions Refrigerate

Special Instructions If forensic, use chain-of-custody protocol and form. See Chain-of-Custody Protocol *on page 381* and the Chain-of-Custody form in the Introduction *on page 63*.

Reference Interval Negative (less than cutoff)

Critical Values Cutoff: screen: 25 ng/mL; confirmation: 25 ng/mL

Possible Panic Range Blood: excitation: 20-30 ng/mL; coma: 30-100 ng/mL; seizures, **fatalities:** >500 ng/mL

Use Evaluate phencyclidine abuse and toxicity

Methodology Screen: immunoassay; confirmation: gas chromatography/mass spectrometry (GC/MS)

Additional Information

- Half-life: 10-50 hours, depending on urine pH (it is water soluble and lipophilic)
- Volume of distribution: 5-7 L/kg
- Protein binding: 65% to 80%

PCP is available in a number of forms. It can be a pure white crystal-like powder, a tablet, liquid, spray, or capsule, and it can be swallowed, smoked (alone or with marijuana), sniffed, or injected. Although PCP is illegal, it is easily manufactured. A number of PCP analogues including cyclohexamine, phenylcyclohexylpyrrolidine, phenylcyclopentylpiperidine, and thienylcyclohexylpiperidine are manufactured as street drugs. It is metabolized mostly by hydroxylation and excreted through the kidneys. Urinary acidification enhances PCP renal clearance.

PCP combinations usually involve other drugs, most frequently cocaine, ethanol, and opiates.

Speech, muscle coordination, and vision are affected; sense of touch and pain are dulled; and body movements are slowed. Time seems to "space out." Effects include increased heart rate and blood pressure, flushing, sweating, dizziness, and numbness. When large doses are taken, effects include drowsiness, convulsions, and coma. PCP can be detected for up to 7 days after administration; 2-4 weeks in chronic users.

References

Mozayani A, "Phencyclidine - Effects on Human Performance and Behavior," *Forensic Sci Rev*, 2002, 15:61-74.

Rogowski R and Krenzelok E, "Averting The Medical, Social, and Legal Implications of a False Positive Phencyclidine Determination," *J Toxicol Clin Toxicol*, 1997, 35:551.

Schneider S, Kuffer P, and Wennig R, "Determination of Lysergide (LSD) and Phencyclidine in Biosamples," *J Chromatogr B Biomed Sci Appl*, 1998, 713(1):189-200.

- ◆ **Phenemal** *see* Phenobarbital, Serum or Plasma *on page 1021*
- ◆ **Phenemalum** *see* Phenobarbital, Serum or Plasma *on page 1021*
- ◆ **Phenformin** *see* Lactic Acid, Whole Blood or Plasma *on page 827*
- ◆ **Phenistix®** *see* Phenylalanine, Urine *on page 1024*
- ◆ **Phenobarb** *see* Barbiturates, Qualitative, Urine *on page 248*
- ◆ **Phenobarb** *see* Phenobarbital, Serum or Plasma *on page 1021*
- ◆ **Phenobarbital** *see* Barbiturates, Quantitative, Serum or Plasma *on page 248*
- ◆ **Phenobarbital** *see* Phenytoin, Serum or Plasma *on page 1026*
- ◆ **Phenobarbital** *see* Valproic Acid, Serum or Plasma *on page 1297*
- ◆ **Phenobarbital:Primidone Ratio** *see* Primidone, Serum or Plasma *on page 1091*

Phenobarbital, Serum or Plasma

Related Information

Barbiturates, Quantitative, Serum or Plasma *on page 248*
Benzodiazepines, Qualitative, Urine *on page 253*
Ethosuximide, Serum or Plasma *on page 560*
Methylphenobarbital, Serum *on page 909*
Phenytoin, Serum or Plasma *on page 1026*
Primidone, Serum or Plasma *on page 1091*
Valproic Acid, Serum or Plasma *on page 1297*

Synonyms Barbita®; Comizial®; Fenilcal®; Gardenal®; Luminal®; Phenemal; Phenemalum; Phenobarb; Phenobarbitone; Phenylethylmalonylurea; Solfoton®

Abstract Phenobarbital is indicated for generalized tonic-clonic and partial seizures. It is also used for febrile seizures in children and for treatment of neonatal hyperbilirubinemia. It is a major metabolite of primidone.

Specimen Serum or plasma

Container Red top tube, green top (heparin) tube, or lavender top (EDTA) tube

Sampling Time Consistent sampling time is desirable but less important than for other anticonvulsants, due to its long half-life. The time to reach steady-state is 10-30 days in adults and 8-15 days in children.

Reference Interval Infants and children: 15-30 µg/mL (SI: 65-129 µmol/L); **adults:** 20-40 µg/mL (SI: 86-172 µmol/L)

Critical Values Toxic: >40 µg/mL (SI: >172 µmol/L); levels >80 µg/mL (SI: >344 µmol/L) are associated with coma. **Fatal:** 50-130 µg/mL (SI: 215-559 µmol/L).

Use Monitor patients for compliance, efficacy, and possible toxicity

Methodology Immunoassay, gas chromatography (GC), high performance liquid chromatography (HPLC)

Additional Information

- Half-life: children: 40-70 hours; adults: 50-140 hours
- Volume of distribution: 0.5-1.0 L/kg
- Protein binding: 40% to 50%

Phenobarbital induces metabolism of a number of drugs including phenytoin, ethosuximide, carbamazepine, valproic acid, clonazepam, vitamin D and K, cimetidine, dicumarol, chloramphenicol, theophylline, oral anticoagulants (warfarin), cyclosporine, and oral contraceptives; where appropriate, the use of these drugs in patients on phenobarbital should be monitored clinically and through the laboratory.

When combined with other CNS depressants, ethanol, narcotic analgesics, antidepressants, or benzodiazepines, additive respiratory and CNS depression may occur.

Drugs that antagonize folic acid include antiepileptic drugs (AEDs): carbamazepine, phenytoin, phenobarbital, primidone, and non-AEDs: trimethoprim and triamterene. These drugs may increase the baseline rate of neural tube defects in mothers when used in the first month of pregnancy, before the neural tube closes. Use of these drugs during the second or third month following the last menstrual period can increase the relative risk of cardiovascular defects, oral clefts, and urinary tract defects. The use of folic acid supplements from the time of conception can reduce the incidence of such birth defects. Consequently, all women of childbearing age prescribed the above medicines should also take supplemental folic acid.

References

Brodie MJ and Dichter MA, "Antiepileptic Drugs," *N Engl J Med*, 1996, 334(3):168-75.
Hernández-Diaz S, Werler MM, Walker AM, et al, "Folic Acid Antagonists During Pregnancy and the Risk of Birth Defects," *N Engl J Med*, 2000, 343(22):1608-14.
Lerman-Sagie T and Lerman P, "Phenobarbital Still Has a Role in Epilepsy Treatment," *J Child Neurol*, 1999, 14(12):820-1.
Painter MJ, Scher MS, Stein AD, et al, "Phenobarbital Compared With Phenytoin for the Treatment of Neonatal Seizures," *N Engl J Med*, 1999, 341(7):485-9.

♦ **Phenobarbitone** *see* Phenobarbital, Serum or Plasma *on page 1021*

♦ **Phenothiazines** *see* Chlorpromazine, Serum *on page 395*

Phenothiazines, Serum

Related Information

Antidepressants, Cyclic, Serum or Plasma *on page 171*
(Continued)

Phenothiazines, Serum *(Continued)*

Chlorpromazine, Serum *on page 395*
Fluphenazine, Serum *on page 604*

Applies to Chlorpromazine; Compazine®; Etrafon®; Ferric, Perchloric, Nitric (FPN) Urine Spot Test; FPN Test for Phenothiazine; Mellaril®; Mesoridazine; Ormazine®; Prochlorperazine; Prolixin®; Serentil®; Stelazine®; Thioridazine; Thorazine®; Trifluoperazine

Abstract The drugs in this class are used as antipsychotic agents and tranquilizers. These are tricyclic compounds with some properties in common with tricyclic antidepressants. In addition, they are used to control nausea or vomiting.

Specimen Serum

Container Red top tube

Sampling Time Obtain serum at least 3 hours after last dose.

Special Instructions In overdose situations, phenothiazines can also be measured in urine and gastric contents. Qualitative detection of drug or metabolites is usually sufficient.

Reference Interval Therapeutic: chlorpromazine: 50-300 ng/mL (SI: 157-942 nmol/L), thioridazine: 1.0-1.5 µg/mL (SI: 2.7-4.1 µmol/L), fluphenazine: 2-4 ng/mL

Possible Panic Range Toxic: chlorpromazine: >750 ng/mL (SI: >2350 nmol/L), thioridazine: >10 ng/mL (SI: >27 nmol/L), fluphenazine: >20 ng/mL

Use Evaluate possibility of phenothiazine toxicity

Limitations These drugs are not usually monitored, because poor correlation exists between serum level and pharmacologic effect.

Methodology Immunoassay, high performance liquid chromatography (HPLC); The ferric, perchloric, nitric (FPN) urine spot test is frequently used for phenothiazine screening.

Additional Information

- Half-life: 20-50 hours
- Volume of distribution: 15-25 L/kg
- Protein binding: 70% to 90%

There are three different classes of phenothiazines: aliphatic (chlorpromazine), piperidine (thioridazine), and piperazine (fluphenazine). All are effective in therapy in appropriate doses, but differ in frequency, type, and severity of side effects. Phenothiazine toxicity principally involves the central nervous and cardiovascular systems. Side effects include drowsiness, ataxia, respiratory depression, hypotension, tachycardia, cardiac arrest, and bone marrow depression. There is also a "dysphoric" response by some patients. Phenothiazines have anticholinergic symptoms similar to those of the tricyclic antidepressants.

References

Dyer KS and Woolf AD, "Use of Phenothiazines as Sedatives in Children: What Are the Risks?" *Drug Saf*, 1999, 21(2):81-90.

Krishel S and Jackimczyk K, "Cyclic Antidepressants, Lithium, and Neuroleptic Agents. Pharmacology and Toxicology," *Emerg Med Clin North Am*, 1991, 9(1):53-86.

Porter WH, "Clinical Toxicology," *Tietz Textbook of Clinical Chemistry*, 3rd ed, Chapter 27, Burtis CA and Ashwood ER, eds, Philadelphia, PA: WB Saunders Co, 1999, 906-81.

♦ **5-Phenyl-5-Ethylhydantoin (Nirvanol®)** *see* Mephenytoin, Serum *on page 896*

♦ **Phenylalanine** *see* Amino Acids, Urine *on page 145*

Phenylalanine, Blood

Related Information

Amino Acids, Plasma *on page 143*
Amino Acids, Urine *on page 145*
Newborn Screen for Phenylketonuria *on page 954*
Newborn Screen for T₄, Filter Paper *on page 956*
Newborn Screen for TSH, Filter Paper *on page 957*
Phenylalanine, Urine *on page 1024*

Synonyms Guthrie Test; Hyperphenylalaninemia Screen; Phenylalanine Screening Test, Blood; Phenylketonuria Test; PKU Test

Applies to Biopterin Cofactor Deficiency; Guanosine Triphosphate Cyclohydrolase 1 Deficiency; Pyruvoyl Tetrahydropterin Synthase Deficiency; Tetrahydrobiopterin Cofactor Deficiency

Abstract Phenylketonuria (PKU) is an autosomal recessive aminoacidopathy due to phenylalanine hydroxylase or a cofactor deficiency. Detection by dried blood spot screening results in early treatment, intended to prevent mental retardation. Clinical phenotypes reflect genetic heterogeneity.

Patient Preparation Newborn should have milk (protein) feeding ideally for 48 hours before testing; sample as late as possible prior to discharge from hospital. Collection is recommended at 4-10 days for low birth weight infants. Antibiotics interfere with the bacterial inhibition assay. See Newborn Screen for Phenylketonuria *on page 954*.

Specimen Whole blood, serum or plasma

Container Green top (heparin) tube; check with laboratory

Sampling Time Fasting; or, in infants, at least 4 hours after a meal

Storage Instructions Centrifuge quickly; plasma is stable for 5 days at 4°C.

Reference Interval
- Premature: 98-213 µmol/L
- 0-1 month: 38-137 µmol/L
- 1-24 months: 31-75 µmol/L
- 2-18 years: 26-91 µmol/L
- Adults: 35-85 µmol/L

Use Evaluate patients for phenylketonuria, monitor therapy with phenylalanine restricted diet

Limitations Cases have been missed because blood phenylalanine was not increased, even after the third day of life. Identification of non-PKU forms of hyperphenylalaninemia (see following information) requires additional testing for tetrahydrobiopterin pathway enzyme defects. **Not all individuals with increased blood phenylalanine have phenylketonuria.** When the infant is tested for PKU before 24 hours of age, there is a 16% chance of missing a positive case. When screened between 24 and 48 hours of birth, there is a 2.2% chance of missing a positive, between 48 and 72 hours, 0.3% chance.

Methodology Ion exchange, gas chromatography-mass spectrometry (GC/MS), high performance liquid chromatography (HPLC), fluorometry, enzymatic; bacterial inhibition (Guthrie test)

Additional Information Presence of hyperphenylalaninemia implies a disorder of phenylalanine hydroxylation. PKU due to phenylalanine hydroxylase (PAH) deficiency is the common example. Patients with classic PKU have plasma phenylalanine >1000 µmol/L; in mild forms, the plasma phenylalanine is 500-1000 µmol/L. However, in addition to PAH, phenylalanine hydroxylation requires oxygen and tetrahydrobiopterin (BH_4) as a cofactor. A defect in the metabolism of BH_4 that results in BH_4 deficiency will impair hydroxylation and result in increased plasma phenylalanine concentrations (800-1000 µmol/L). Three types of inborn errors of BH_4 metabolism ("atypical PKU") have been identified. Defects in BH_4 synthesis are guanosine triphosphate cyclohydrolase I deficiency and pyruvoyl tetrahydropterin synthase deficiency. The third type of defect involves the regeneration of BH_4 catalyzed by the enzyme dihydropteridine reductase. Experience with treatment of PKU has shown that some 3% (variable between different populations) fail to respond. These are largely cases of BH_4 cofactor deficiency. There are important differences in therapy between classical PKU and the various BH_4 cofactor deficiencies. "PKU positive" cases (identified as the result of phenylalanine screening tests) should be additionally tested for BH_4 deficiency. Clinical features; urine, blood, and enzyme analyses; prenatal diagnosis; and therapy of BH_4 deficiencies have been reviewed.

Consult references (below) by Scriver et al, Güttler et al, and Gilbert-Barness et al for review of the genetic/molecular biology of the hyperphenylalaninemias. Clinical heterogeneity of PKU relates to the existence of multiple different mutations in the PAH gene and to combinations of mutations. Such genetic considerations have important implications for therapy and its outcome.

Women with PKU should receive genetic counseling before becoming pregnant. Maternal hyperphenylalaninemia produces a serious embryopathy. In order to retain the achievements of early detection and treatment of PKU, (Continued)

Phenylalanine, Blood *(Continued)*

increasing attention is being turned to control of maternal hyperphenylalaninemia.

References

"American Academy of Pediatrics, Committee on Genetics, Newborn Screening Fact Sheets: Phenylketonuria," *Pediatrics*, 1996, 98:473-501.

Gilbert-Barness E and Barness LA, *Metabolic Diseases: Foundations of Clinical Management,
Genetics, and Pathology*, Natick, MA: Eaton Publishing, 2000, 27-31.

Güttler F and Guldberg P, "Mutations in the Phenylalanine Hydroxylase Gene: Genetic Determinants for the Phenotypic Variability of Hyperphenylalaninemia," *Acta Paediatr Suppl*, 1994,
407:49-56.

Jacobs DS, DeMott WR, Oxley DK, et al, *Laboratory Test Handbook*, 5th ed, Hudson, OH:
Lexi-Comp Inc, 2001.

Matalon R, Michals K, Blau N, et al, "Hyperphenylalaninemia Due to Inherited Deficiencies of
Tetrahydrobiopterin," *Advanced Pediatrics*, eds, Barness LA, DeViro DC, Morrow G, et al,
Chicago, IL: Year Book Medical Publishers, 1989, 36:67-89.

Mayo Medical Laboratories, *2001 Test Catalogue*, Rochester, MN, 424.

Scriver CR, Kaufman S, Eisensmith RC, et al, "The Hyperphenylalaninemias," *The Metabolic and
Molecular Bases of Inherited Disease*, 7th ed, Volume 1, Chapter 27, Scriver CR, Beaudet AL,
Sly WS, et al, eds, New York, NY: McGraw-Hill Inc, 1995, 1015-75.

Internet Web Sites

www.aap.org

♦ **Phenylalanine Hydroxylase Activity** *see* Newborn Screen for Phenylketonuria
on page 954

♦ **Phenylalanine Screening Test, Blood** *see* Phenylalanine, Blood *on*
page 1022

Phenylalanine, Urine

Related Information

Newborn Screen for Phenylketonuria *on page 954*
Phenylalanine, Blood *on page 1022*

Synonyms Phenylpyruvic Acid, Urine; PKU, Urine Test

Applies to Ferric Chloride Test; Guthrie Test; Phenistix®; Tyrosyluria

Abstract Phenylketonuria (PKU) is a disorder of phenylalanine catabolism,
primarily due to deficiency of phenylalanine hydroxylase. Phenylalanine inhibits
synthesis of catecholamines and serotonin, resulting in neurological damage.

Specimen Freshly voided random urine

Collection Transport specimen to the laboratory within 1 hour of collection.
Container must provide date and time of collection.

Reference Interval Negative (level of detection with Phenistix® is 5-10 mg/100
mL)

Use In a patient with an established diagnosis of PKU, a urine test may be
employed to follow adequacy of dietary control, including monitoring of dietary
intake of pregnant women who lack phenylalanine hydroxylase. In most
instances, however, monitoring is accomplished by blood levels.

Limitations Many cases of false-positive and false-negative results.

Methodology Screening tests:

- Ferric chloride method - green color
- Phenistix® urine reagent strips (a ferric chloride method) - a persistent
  blue-gray to gray-green color is produced with phenylpyruvic acid. (Salicylates or phenothiazine derivatives give pink to purple colors.) Tyrosyluria in
  the premature or newborn due to liver disease will produce a fading green
  color with ferric chloride screening tests.

Confirmatory tests may be done by one of several methods:

- High voltage electrophoresis followed by chromatography
- Cation exchange column chromatography for plasma and urine phenylalanine quantitation
- More sensitive and specific methods, detecting phenylpyruvic acid in the
  urine by chemiluminescent or fluorometric techniques. High performance
  liquid chromatography with fluorescent detection is also used, and can
  detect the low levels of phenylpyruvic acid in normal adults or newborns.
- Phenylpyruvic acid can be detected by gas chromatography-mass spectrometry during organic acids analysis.

Additional Information Monitoring of urinary excretion of phenylalanine
metabolites (by gas chromatography/mass spectrometry) to avoid neurotoxic
effects in children with PKU has been undertaken with the goal of identifying a

range of blood phenylalanine that is associated with normal levels of excretion of such products. The importance of controlling blood phenylalanine levels by diet during pregnancy in women with hyperphenylalaninemia (including those with diagnosed and undiagnosed phenylketonuria) has been emphasized in order to decrease the risk of maternal phenylketonuria syndrome. See Phenylalanine, Blood *on page 1022.*

References
"American Academy of Pediatrics, Committee on Genetics, Newborn Screening Fact Sheets: Phenylketonuria," *Pediatrics,* 1996, 98:473-501.

Erlandsen H and Stevens RC, "The Structural Basis of Phenylketonuria," *Mol Genet Metab,* 1999, 68(2):103-25.

Waisbren SE, Hanley W, Levy HL, et al, "Outcome at Age 4 Years in Offspring of Women With Maternal Phenylketonuria: The Maternal PKU Collaborative Study," *JAMA,* 2000, 283(6):756-62.

Internet Web Sites
www.aap.org

♦ **Phenylethylmalonamide** *see* Primidone, Serum or Plasma *on page 1091*

♦ **Phenylethylmalonylurea** *see* Phenobarbital, Serum or Plasma *on page 1021*

♦ **Phenylketonuria, Newborn Screen** *see* Newborn Screen for Phenylketonuria *on page 954*

♦ **Phenylketonuria Test** *see* Phenylalanine, Blood *on page 1022*

♦ **Phenylpyruvic Acid, Urine** *see* Phenylalanine, Urine *on page 1024*

Phenytoin, Free, Serum or Plasma

Related Information
Phenytoin, Serum or Plasma *on page 1026*

Synonyms Free Dilantin®; Free Phenytoin

Abstract Measurement of free phenytoin may be clinically important in situations associated with altered binding of phenytoin or decrease in albumin.

Specimen Serum or plasma

Container Red top tube or lavender top (EDTA) tube

Reference Interval 1-2 µg/mL (SI: 4-8 µmol/L)

Critical Values >5 µg/mL

Possible Panic Range Toxicity may be progressive at levels >2 µg/mL.

Limitations Free drug levels are not useful when total phenytoin level is <3 µg/mL. Even in severe hepatorenal syndrome, free fraction is seldom >50%.

Additional Information
- Half-life (adults): 20-40 hours
- Volume of distribution: 0.6-0.7 L/kg
- Protein binding: 85% to 95% (significant interindividual variation exists)

Phenytoin is 90% bound to serum proteins, but only the free fraction circulates through plasma membranes and is biologically active. Because of rapid equilibration between free and bound portions of drugs, free levels are potentially important only in antiepileptic drugs (AEDs) that are highly bound. There is significant interindividual variation in free phenytoin. In one study, the free phenytoin fraction ranged from 7% to 35%. Binding kinetics may be altered in uremia, hepatic disease, late pregnancy or postpartum, cases of head injury associated with a hypermetabolic state, and certain instances of polypharmacy, described below. Free phenytoin concentration is increased in patients with HIV. Determination of free levels may also be helpful in overdosages, since only the free portion can be cleared by dialysis.

Drugs which compete for binding sites on albumin and which may displace phenytoin include valproic acid, acetazolamide, high doses of salicylic acid, phenylbutazone, ceftriaxone, nafcillin, and sulfamethoxazole. In a clinical setting in which one of these drugs is used with phenytoin and toxicity is suspected despite normal phenytoin levels, a free level may be useful.

Due to significant interpatient and intrapatient variability in phenytoin protein binding, monitoring of total serum concentrations is unreliable and free phenytoin serum concentrations should be considered for monitoring in hospitalized patients (Banh et al).

References
Banh HL, Burton ME, and Sperling MR, "Interpatient and Intrapatient Variability in Phenytoin Protein Binding," *Ther Drug Monit,* 2002, 24(3):379-85.

Burt M, Anderson DC, Kloss J, et al, "Evidence-Based Implementation of Phenytoin Therapeutic Drug Monitoring," *Clin Chem,* 2000, 46(8 Pt 1):1132-5.

(Continued)

Phenytoin, Free, Serum or Plasma *(Continued)*

Dasgupta A and McLemore JL, "Elevated Free Phenytoin and Free Valproic Acid Concentrations in Serum of Patients Infected With Human Immunodeficiency Virus," *Ther Drug Monit*, 1998, 20(1):63-7.

Soldin SJ, Wang E, Verjee Z, et al, "Phenytoin Overview - Metabolite Interference in Some Immunoassays Could Be Clinically Important. Results of a College of American Pathologists Study," *Arch Pathol Lab Med*, 2003, 127(12):1623-5.

Phenytoin, Serum or Plasma

Related Information

Carbamazepine, Serum *on page 337*
Ethosuximide, Serum or Plasma *on page 560*
Mephenytoin, Serum *on page 896*
Phenobarbital, Serum or Plasma *on page 1021*
Phenytoin, Free, Serum or Plasma *on page 1025*
Primidone, Serum or Plasma *on page 1091*
Valproic Acid, Serum or Plasma *on page 1297*

Synonyms Antisacer®; Cerebyx®; Dilantin®; Dintoina®; Diphenylan Sodium®; Diphenylhydantoin; Ditan®; Epanutin®; Epinat®; Fenitoina; Fenytoin®

Applies to Carbamazepine; Fosphenytoin; Phenobarbital; Primidone

Abstract Phenytoin is effective for general motor (tonic-clonic or grand mal) and focal seizures and status epilepticus. It is less effective for complex partial seizures. Its investigational uses include ventricular arrhythmias, including those associated with digitalis intoxication, prolonged QT interval, and surgical repair of congenital heart diseases in children. Recently, a new phosphate ester prodrug of phenytoin, called **fosphenytoin** (Cerebyx®) was introduced as a therapeutic form of phenytoin. Advantages of fosphenytoin include more convenient and rapid intravenous administration, availability for intramuscular injection, and low potential for adverse local reactions at injection sites.

Specimen Serum or plasma

Container Red top tube or lavender top (EDTA) tube

Sampling Time In monitoring patients maintained on chronic therapy, a trough level or consistent sampling time should be used. Peak concentrations are reached in 2-8 hours.

Special Instructions Designate the drug in use in a particular patient: fosphenytoin concentrations cannot be accurately assayed by immunoassays for phenytoin.

Reference Interval Although 10-20 µg/mL (SI: 40-79 µmol/L) is commonly utilized, toxicity is measured clinically and some patients require levels outside the suggested therapeutic range.

Possible Panic Range Toxic: 25-50 µg/mL (SI: 100-200 µmol/L); lethal: >100 µg/mL (SI: >400 µmol/L). Phenytoin can precipitate seizures at levels >40 µg/mL (>158 µmol/L).

Use Monitor for compliance, efficacy, and possible toxicity. There is a good relationship between serum/plasma levels, efficacy, and toxicity.

Methodology Immunoassay, gas or liquid chromatography, high performance liquid chromatograpy; most laboratories use immunoassay.

Additional Information

- Half-life: adults: 20-40 hours, children: ~10 hours
- Volume of distribution: 0.6-0.7 L/kg
- Protein binding: 85% to 95%

As the drug is highly protein bound, disease states or conditions which alter albumin concentration or affect phenytoin binding to albumin will alter total phenytoin concentrations. In such patients, measurement of unbound (active) phenytoin is recommended.

Most of a dose of phenytoin is excreted into the bile as inactive metabolites which are then reabsorbed from the intestines and excreted into the urine. Despite normal levels, phenytoin may interfere with the actions of other drugs, including cyclosporine, oral anticoagulants, oral contraceptives, and theophylline; appropriate laboratory monitoring of some of these agents is advised.

Phenytoin, phenobarbital, primidone, and carbamazepine, enzyme-inducing antiepileptic drugs, can cause transient deficiency of vitamin K-dependent clotting factors in the neonate. It has been recommended that expectant mothers on these drugs should be given vitamin K in the last several weeks of gestation, and it is recommended that babies be given vitamin K immediately after birth, to avoid neonatal central nervous system hemorrhage. However, use of phenytoin during pregnancy has been regarded as contraindicated; it can lead to the fetal hydantoin syndrome (Soldin, 2003).

Women of childbearing age prescribed this medication should take supplemental folic acid, as phenytoin increases folic acid clearance.

Phenytoin may decrease the serum concentration or effectiveness of other drugs including valproic acid, carbamazepine, ethosuximide, primidone, corticosteroids, chloramphenicol, rifampin, doxycycline, quinidine, mexiletine, disopyramide, dopamine, or nondepolarizing skeletal muscle relaxants. Protein binding of phenytoin can be affected by valproic acid or salicylates. Serum phenytoin concentrations may be increased by cimetidine, disulfiram, trazodone, ethanol, halothane, phenylbutazone, azapropazone, ibuprofen, amiodarone, imipramine, miconazole, metronidazole, nifedipine, chloramphenicol, INH, trimethoprim, or sulfonamides and decreased by rifampin, cisplatin, vinblastine, bleomycin, folic acid, continuous NG feeds, oxacillin, or nitrofurantoin.

References

Bachmann KA and Belloto RJ Jr, "Differential Kinetics of Phenytoin in Elderly Patients," *Drugs Aging*, 1999, 15(3):235-50.

Brodie MJ and Dichter MA, "Drug Therapy: Antiepileptic Drugs," *N Engl J Med*, 1996, 334(3):168-75.

Dasgupta A, Handy BC, and Datta P, "Mathematical Models to Calculate Fosphenytoin Concentrations in the Presence of Phenytoin Using Phenytoin Immunoassays and Alkaline Phosphatase," *Am J Clin Pathol*, 2000, 113(1):87-92.

Fischer JH, Patel TV, and Fischer PA, "Fosphenytoin: Clinical Pharmacokinetics and Comparative Advantages in the Acute Treatment of Seizures," *Clin Pharmacokinet*, 2003, 42(1):33-58.

Hernández-Diaz S, Werler MM, Walker AM, et al, "Folic Acid Antagonists During Pregnancy and the Risk of Birth Defects," *N Engl J Med*, 2000, 343(22):1608-14.

Painter MJ, Scher MS, Stein AD, et al, "Phenobarbital Compared With Phenytoin for the Treatment of Neonatal Seizures," *N Engl J Med*, 1999, 341(7):485-9.

Roberts WL, De BK, Coleman JP, et al, "Falsely Increased Immunoassay Measurements of Total and Unbound Phenytoin in Critically Ill Uremic Patients Receiving Fosphenytoin," *Clin Chem*, 1999, 45(6 Pt 1):829-37.

Soldin SJ, Wang E, Verjee Z, et al, "Phenytoin Overview - Metabolite Interference in Some Immunoassays Could Be Clinically Important. Results of a College of American Pathologists Study," *Arch Pathol Lab Med*, 2003, 127(12):1623-5.

pH, Gastric Intramucosal

Related Information

Bicarbonate, Blood *on page 258*
Blood Gases and pH, Arterial *on page 275*
Lactic Acid, Whole Blood or Plasma *on page 827*
pCO₂, Blood *on page 1008*
pH, Blood *on page 1018*

Synonyms Intramural pH; pH$_i$

Abstract Gastric intramucosal pH (pH$_i$) is an indirect indicator of splanchnic hypoxia and is used as an index of therapeutic tissue oxygenation and a predictor of multiorgan dysfunction syndrome.

Patient Preparation A nasogastric tonometer with a balloon-tip catheter is inserted into the stomach and its position is confirmed radiographically. A histamine-2-receptor antagonist may be administered and enteral feedings should be stopped for the duration of the testing. Once inserted the tonometer is filled with 2.5 mL of normal saline solution. The silicone balloon is permeable only to gases and, after proper placement and filling, is allowed to equilibrate for 60 to ≥90 minutes, permitting carbon dioxide (CO_2) produced by the gastric mucosal cells to diffuse across the balloon wall and dissolve in the saline solution.

Specimen After the equilibration period the saline is aspirated anaerobically and submitted to the laboratory for pCO₂ measurement. A concurrent arterial blood specimen is drawn for bicarbonate (HCO_3^-) determination (see Arterial Blood Collection *on page 211*, Blood Gases and pH, Arterial *on page 275*, and Bicarbonate, Blood *on page 258*).

Container Glass or plastic syringe with no anticoagulant for the saline specimen; glass or plastic heparinized blood gas syringe for the arterial blood specimen

Sampling Time The equilibration time should be recorded.

(Continued)

pH, Gastric Intramucosal *(Continued)*

Collection Collection of saline and arterial blood specimens should be done anaerobically. Expel air from syringe tip and seal securely.

Storage Instructions Analysis should be done immediately. However, if delay in analysis of more than 10-15 minutes is unavoidable or otherwise anticipated, place specimens in a slurry of ice chips and water. Delay of analysis should not exceed 1 hour.

Causes for Rejection Air bubbles in syringes, unsealed/open syringes, specimen incorrectly iced

Special Instructions The equilibration time should be noted on the requisition as it affects the calculation of pH_i.

Reference Interval pH_i ≥7.32

Critical Values A pH_i result <7.32 has been shown to have a sensitivity of 89% and a specificity of 77% for predicting early mortality in critically ill adult patients. This threshold also predicts mortality in children ages 1 month to 16 years of age.

Use The test is used as an assessment of the adequacy of systemic tissue oxygenation, a prognostic indicator in critically ill patients. Monitoring pH_i guides treatment to increase systemic O_2 transport or to reduce O_2 demand when the pH_i drops into the critical range.

Limitations Since standard blood gas instrumentation is designed for analyses of whole blood specimens, pCO_2 measurements in a matrix of saline may be unreliable. Takala et al showed that saline pCO_2 measurement can be an important source of error in pH_i measurement, depending upon the blood gas instrument in use and the pCO_2 level, and that only changes in pH_i ≥0.06 are clinically meaningful given the accuracy and precision observed from their study. Some blood gas instruments significantly underestimate the pCO_2 in the saline specimen and should not be used.

The pH_i reflects not only mucosal perfusion but also metabolic abnormalities and is, therefore, affected unpredictably by therapy with vasoactive drugs and should not be used to guide resuscitation efforts.

The assumption that the arterial HCO_3^- accurately reflects the mucosal HCO_3^- has recently been brought into question, and its use has been shown to cause underestimations of pH_i.

Methodology Gas tonometry

Additional Information Other methods for assessing adequate tissue perfusion include hemodynamic monitors; systemic indicators of oxygen transport; capillary filling time; central-peripheral temperature difference; urine output; arterial, venous, and capillary pH; and lactate measurements.

References

Casado-Flores J, Mora E, Perez-Corral F, et al, "Prognostic Value of Gastric Intramucosal pH in Critically Ill Children," *Crit Care Med*, 1998, 26(6):1123-7.

Gutierrez G, Bismar H, Dantzker DR, et al, "Comparison of Gastric Intramucosal pH With Measures of Oxygen Transport and Consumption in Critically Ill Patients," *Crit Care Med*, 1992, 20(4):451-7.

Morgan TJ, Venkatesh B, and Endre ZH, "Accuracy of Intramucosal pH Calculated From Arterial Bicarbonate and the Henderson-Hasselbalch Equation: Assessment Using Simulated Ischemia," *Crit Care Med*, 1999, 27(11):2495-9.

♦ **pH_i** *see* pH, Gastric Intramucosal *on page 1027*

♦ **Philadelphia Chromosome** *see* Breakpoint Cluster Region Rearrangement in CML *on page 304*

Phlebotomist Procedures

Related Information

Blood and Fluid Precautions, Specimen Collection *on page 271*
Blood Collection Tube Information *on page 275*
Skin Puncture Blood Collection *on page 1203*
Specimen Identification Requirements *on page 1217*
Venous Blood Collection *on page 1304*

Synonyms Specimen Collection Policy, Phlebotomist

Collection Phlebotomists are generally required to adhere to the following procedures. See Specimen Identification Requirements *on page 1217*. Phlebotomists are limited to attempting venipunctures in upper extremities unless so ordered by the physician. **Phlebotomists may not perform a venipuncture above an I.V. site or in an arm with a heparin lock or shunt.** Only by a physician's order may a trained phlebotomist collect a sample from a fistula or shunt. Phlebotomists should not collect a sample from an arm which is on the same side as a recent mastectomy. Phlebotomists are generally limited to two attempts to obtain a blood sample. After two unsuccessful tries, the phlebotomist must call another phlebotomist or supervisor. If the second phlebotomist is unsuccessful, the physician or responsible nurse is usually notified. The phlebotomist will perform skin punctures when ordered by the physician (with some exceptions) or if venipuncture is unsuccessful or prohibited (due to I.V., etc), provided the procedure requested can be performed on a skin puncture specimen. (Dependent on the policy and equipment available in the laboratory receiving the specimen.) See Skin Puncture Blood Collection *on page 1203*.

Special Instructions The perception of pain associated with venipuncture in children increases with anxiety and is inversely correlated with the patient's age. Strategies to reduce the child's and parents' distress during venipuncture are important considerations. Approximately 30% of adult patients report needle discomfort greater than expected, which is one factor associated with patient dissatisfaction. Every effort should be made to improve patient satisfaction by reducing discomfort from phlebotomy procedures.

Additional Information See Sampling Problems in Maximizing the Information From Laboratory Tests - The Ulysses Syndrome *on page 11.*

References

Love G, "Easing the Discomfort of Venipuncture," *Nursing*, 1998, 28(3):30.

Roth D, "Venipuncture Tips for Geriatric Patients," *Nursing*, 1997, 27(10):69.

Scholz MJ, "Minimizing the Pain of Venipuncture," *RN*, 1996, 59(5):78.

Wray D and Wells G, "Procedures in Phlebotomy," *Arch Pathol Lab Med*, 2000, 124(4):641.

♦ **Phlebotomy, Blood Donor** *see* Donation, Blood *on page 521*

Phlebotomy, Therapeutic

Related Information

Abstract Blood may be electively removed via therapeutic phlebotomy in patients with erythrocytosis (eg, polycythemia vera or cyanotic heart disease) or to remove excess iron in patients with hemochromatosis. Phlebotomies are done for porphyria cutanea tarda. Phlebotomy may be necessary in acute pulmonary edema. Currently, such blood is not utilized for allogeneic transfusion, except for those units collected from individuals with hemochromatosis who have met the following criteria.

- There is no charge to the individual for the therapeutic phlebotomy.
- The indication for the therapeutic phlebotomy is hemochromatosis.
- The Food and Drug Administration has approved the program.

Patient Preparation Before the elective removal of blood, the attending physician must document the request and specify the amount of blood to be withdrawn. Record prephlebotomy and postphlebotomy vital signs. The patient and physician should both understand that a therapeutic phlebotomy is an operative procedure and not a blood donation.

Aftercare Observe patient for at least 20 minutes for vasovagal and other adverse reactions. Local discomfort and, occasionally, hematoma occur at phlebotomy site. Avoid strenuous exercise for 24 hours. Tell patient whom to contact if adverse effects develop after discharge. Multiple phlebotomies may be necessary over a long time period.
(Continued)

Phlebotomy, Therapeutic (Continued)

Collection Phlebotomies for patients with hemochromatosis should remove 500 mL of blood weekly until the patient is iron deficient, or until mild hypoferritinemia is found; then phlebotomy is done as required to maintain serum ferritin <50 µg/L.

Only five to six phlebotomies are usually needed at 1- to 2-week intervals for porphyria cutanea tarda.

Causes for Rejection Anemia, hematocrit at designated target, abnormal vital signs. Phlebotomy is contraindicated in stress polycythemia, in which plasma volume is contracted. Certain conditions may require the presence of the attending physician during phlebotomy (eg, hypertension, cardiac symptoms). The blood bank physician may decline to do the procedure on high-risk patients.

Special Instructions For polycythemia, evaluate the patient's hemoglobin, hematocrit, red cell count, platelet count, WBC count, leukocyte alkaline phosphatase, serum vitamin B_{12}, carboxyhemoglobin, and blood volume. In primary polycythemia, erythropoietin levels in blood and urine are decreased. Erythropoietin is increased in secondary polycythemia. Arterial blood gases may be helpful, with significantly decreased pO_2 and oxygen saturation pointing to secondary polycythemia. Arterial blood oxygen saturation ≥92% is expected in polycythemia vera, and 75% have splenomegaly.

The diagnosis of **hemochromatosis** requires transferrin saturation >55% with serum ferritin >400 µg/L. The diagnosis of hereditary hemochromatosis can be confirmed by direct mutation analysis of the "HFE gene." However, not all patients with hemochromatosis have mutations in the HFE gene. Each unit of 450-500 mL blood contains 200-250 mg iron.

Use Removal of blood in polycythemia to maintain the hematocrit <45%; hypoxic lung disease or cyanotic heart disease; hemochromatosis; therapy for porphyria cutanea tarda

Limitations Complications and deaths have been reported. Very thin, elderly patients with active cardiopulmonary disease may not tolerate the 500 mL phlebotomy. Consider removing only 250 mL of blood in such patients. A crystalloid or colloid solution can be simultaneously infused as well.

Contraindications Lack of documented increase of red cell mass when polycythemia is considered. Hemoglobinopathies exist in which polycythemia occurs, the abnormal hemoglobin having increased oxygen affinity. Methemoglobinemias may relate to secondary polycythemias (see Methemoglobin, Whole Blood *on page 903*). Uncommonly, certain tumors induce erythrocytosis. Renal tumors are the most widely known cause of tumor erythrocytosis.

Additional Information Transfusional siderosis is a complication of long term transfusion therapy (eg, in management of thalassemia). The diagnosis is made with clinical history, liver biopsy, or magnetic-susceptibility evaluation; serum ferritin is less reliable in this setting. Phlebotomy is usually contraindicated.

References

Andrews NC, "Disorders of Iron Metabolism," *N Engl J Med*, 1999, 341(26):1986-95.

Brittenham GM, "Disorders of Iron Metabolism: Iron Deficiency and Overload," *Hematology: Basic Principles and Practice*, 3rd ed, Hoffman R, Benz EJ Jr, Shattil SJ, et al, eds, Philadelphia, PA: Churchill Livingstone, 2000, 397-428.

Fridey J, *Standards for Blood Banks and Transfusion Services*, 22nd ed, Bethesda, MD: American Association of Blood Banks Press, 2003, 28.

Grima KM, "Therapeutic Apheresis in Hematological and Oncological Diseases," *J Clin Apheresis*, 2000, 15(1-2):28-52.

♦ **Phlebotomy, Venous** *see Venous Blood Collection on page 1304*

♦ **Phosgene** *see Chemical Warfare Agents on page 382*

♦ **Phosphatase, Acid** *see Acid Phosphatase, Plasma on page 97*

♦ **Phosphate, Blood** *see Phosphorus, Serum on page 1031*

♦ **Phosphatidylglycerol** *see Lecithin:Sphingomyelin Ratio, Amniotic Fluid on page 836*

Phosphatidylglycerol, Amniotic Fluid

Related Information

Creatinine, Amniotic Fluid *on page 472*

Lamellar Bodies, Amniotic Fluid *on page 830*
Lecithin:Sphingomyelin Ratio, Amniotic Fluid *on page 836*
Pulmonary Surfactant, Amniotic Fluid *on page 1124*

Synonyms Amniotic Fluid Phosphatidylglycerol

Abstract Phosphatidylglycerol (PG), a component of pulmonary surfactant, first appears in the amniotic fluid (AF) at 35-36 weeks gestation. Its presence is highly predictive of fetal lung maturity.

Aftercare All nonsensitized Rh-negative patients should receive anti-D immunoglobulin after amniocentesis.

Specimen Amniotic fluid; the detection of PG is unaffected by blood or meconium contamination.

Collection Ultrasound-guided, transabdominal amniocentesis is performed by the physician. Vaginal pool specimens are discouraged.

Storage Instructions The specimen should be centrifuged at low speed in a refrigerated centrifuge. The supernatant may be stored at 4°C for up to 10 days or it may be frozen indefinitely.

Special Instructions Fetal sacs of multiple pregnancies should be individually sampled and analyzed. Send sample(s) to the laboratory **immediately** after collection.

Reference Interval
- Mature: PG present
- Immature: PG absent

The predictive value of a mature result is virtually 100%, whereas the predictive value of an immature result ranges from 30% to 50%.

Use Results are used to determine the optimal time for obstetrical intervention in cases of possible fetal distress.

Limitations Normal vaginal flora can produce PG; thus causing a false-positive result.

Methodology Immunologic agglutination, enzymatic assays, thin layer chromatography (TLC)

Additional Information The incidence of the respiratory distress syndrome (RDS) at ≥37 weeks gestation is extremely low, except in maternal diabetes. Fetal lung maturity testing is most useful in reliably dated, high-risk pregnancies that would benefit from early delivery during the 34-37 week gestational period.

With the availability of rapid tests for PG and other rapid tests for assessing fetal lung maturity (see Pulmonary Surfactant, Amniotic Fluid *on page 1124*), cascade-testing approaches have been suggested to minimize the cost, time, and effort of performing lecithin:sphingomyelin ratios.

References
"ACOG Educational Bulletin, Assessment of Fetal Lung Maturity, Number 230, November 1996," *Int J Gynecol Obstet*, 1997, 56(2):191-8.
Ashwood ER, "Standards of Laboratory Practice: Evaluation of Fetal Lung Maturity," *Clin Chem*, 1997, 43(1):211-4.
Dubin SB, "Assessment of Fetal Lung Maturity Practice Parameter," *Am J Clin Pathol*, 1998, 110(6):723-32.

♦ **Phosphatidylinositol** *see* Lecithin:Sphingomyelin Ratio, Amniotic Fluid *on page 836*

♦ **Phospholipid Profile, Amniotic Fluid** *see* Lecithin:Sphingomyelin Ratio, Amniotic Fluid *on page 836*

♦ **Phosphopyruvate Hydratase** *see* Neuron-Specific Enolase, Serum *on page 949*

Phosphorus, Serum
Related Information
Amino Acids, Urine *on page 145*
Anion Gap, Serum, Plasma, or Urine *on page 160*
Calcium, Serum *on page 329*
Ethanol, Blood, Urine, and Other Sources *on page 558*
Ketone Bodies, Blood *on page 816*
Kidney Stone Analysis *on page 820*
Lactic Acid, Whole Blood or Plasma *on page 827*
Parathyroid Hormone, Serum *on page 1001*
Vitamin D, Serum *on page 1318*
(Continued)

Phosphorus, Serum (Continued)

Synonyms Phosphate, Blood

Abstract Phosphates are abundant in bone and striated muscle. Less than 1% of body phosphate is in the plasma.

Patient Preparation A fasting specimen is preferred.

Specimen Serum

Container Red top tube

Sampling Time A mean diurnal variation of 0.6 ±0.1 mg/dL with nadir about 11 AM, plateau about 4 PM, and peak at midnight.

Storage Instructions Serum should be promptly separated from the clot to avoid false elevations.

Causes for Rejection Observable hemolysis

Reference Interval Infants: 4.5-7.5 mg/dL (SI: 1.45-2.42 mmol/L). Children: ~4.0-6.0 mg/dL (SI: 1.29-1.94 mmol/L). Adults: 2.5-4.5 mg/dL (SI: 0.81-1.45 mmol/L). Some variation in reference interval exists among authorities.

Possible Panic Range <1.0 mg/dL is critical; <1.5 mg/dL indicates severe hypophosphatemia.

Use

Causes of **high phosphorus:** Exercise; dehydration and hypovolemia; high phosphorus content enema; acromegaly; hypoparathyroidism; pseudohypoparathyroidism; bone metastases; hypervitaminosis D; sarcoidosis; milk-alkali syndrome; liver disease, such as portal cirrhosis; catastrophic events such as cardiac resuscitation, pulmonary embolism, renal failure; diabetes mellitus with ketosis; serum artifact - sample not refrigerated; overheated, hemolyzed sample, or serum allowed to remain too long on the clot. Thrombocytosis causes elevated serum concentrations, but plasma phosphate remains unaltered.

Although phosphate accumulation occurs as renal disease progresses, hyperphosphatemia is not a feature of early renal failure; it does not usually develop before renal function has diminished to about 25% of normal.

Causes of **low phosphorus:** Hyperparathyroidism, humoral hypercalcemia of malignancy, antacids (especially those containing aluminum), diuretics, steroids, medicinal corticosteroids, vitamin D deficiency, carbohydrate administration, ethanol intoxication, sepsis, renal tubular disorders, dialysis, emesis, diarrhea, nasogastric suction, intravenous fluid administration, and other diseases in which phosphate is depleted.

Manifestations of Hypophosphatemia:
- Musculoskeletal: chronic myopathy, rhabdomyolysis, osteopenia, osteomalacia
- Cardiovascular: cardiomyopathy, arrhythmias
- Pulmonary: respiratory failure, failure of weaning ventilator support
- Neurologic: delirium, seizures, encephalopathy, hallucinations, peripheral neuropathy
- Hematologic: impaired oxygen release, hemolysis, leukocyte dysfunction
- Metabolic: metabolic acidosis, glucose intolerance

Limitations Ninety-seven percent of hyperparathyroid subjects with normal renal function have serum phosphate <3.3 mg/dL, 80% <3.0 mg/dL, and 40% <2.5 mg/dL. Collection of multiple data points throughout the day may help to establish the diagnosis of primary hyperparathyroidism in patients with borderline values.

Methodology Photometric

References

Jacobs DS, DeMott WR, Oxley DK, et al, *Laboratory Test Handbook*, 5th ed, Hudson, OH: Lexi-Comp Inc, 2001.

Subramanian R and Khardori R, "Severe Hypophosphatemia: Pathophysiologic Implications, Clinical Presentations, and Treatment," *Medicine*, 2000, 79(1):1-8.

Young DS, *Effects of Drugs on Clinical Laboratory Tests*, 5th ed, Volume 1: Listing by Test, Washington, DC: AACC Press, American Association of Clinical Chemistry, 2000, Section 3, 616-21.

Phosphorus, Urine

Related Information

Calcitonin, Serum or Plasma *on page 327*
Calcium, Serum *on page 329*

Synonyms Urine Phosphorus

Test Includes Phosphorus on random or timed urine specimen

Abstract Urinary phosphorus supports evaluation of calcium/phosphorus balance.

Specimen Timed or random urine. Diurnal variation exists.

Storage Instructions Refrigerate. Laboratory adjusts final pH of urine aliquot to 6.

Reference Interval Adults on unrestricted diet: 0.9-1.3 g/24 hours (SI: 29-42 mmol/day). In children, between the ages of 0-6 years, an inverse correlation exists between urine phosphate:creatinine ratio and age. The upper limit of normal then stabilizes around the value of 1.0. See graphic.

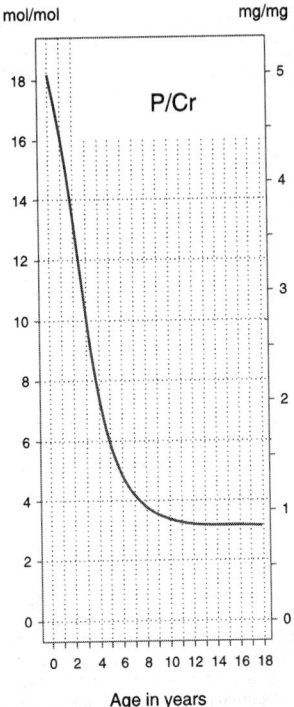

Estimated 95th percentiles for urinary
P/Cr ratios in relation to age.

Use Test is rarely used. Evaluates calcium/phosphorus balance. **High** urinary phosphorus (ie, increased renal losses) occurs in primary hyperparathyroidism, vitamin D deficiency, renal tubular acidosis, diuretic use. **Low** urinary phosphorus is seen in malnutrition, hypoparathyroidism, pseudohypoparathyroidism, and vitamin D intoxication. See table on following page.

Limitations A number of drugs and chemicals alter phosphate excretion, including aluminum salts, diltiazem, phloridzin, acetazolamide, asparaginase, (Continued)

Phosphorus, Urine *(Continued)*

aspirin, bicarbonate, calcitonin, corticosteroids, dihydrotachysterol, hydrochlorothiazide, mercurial diuretics, metolazone, and orthophosphate.

Effects on Phosphorus Transport

Atrial natriuretic peptide	↓
Calcitonin	↓
Glucocorticoid	↑
Growth hormone	↑
Insulin-like growth factor-I	↑
Metabolic acidosis (chronic)	↓
Metabolic alkalosis	↑
Parathyroid hormone	↓
Parathyroid hormone-related peptide (HHM factor)	↓
Phosphorus supply	↑ or ↓
Vasopressin	↓
Vitamin D	↑
Volume expansion	↓

Adapted from Hruska KA, "Phosphate Balance and Metabolism," *The Principles and Practice of Nephrology,* Chapter 19, Jacobson HR, Striker GE, and Klahr S, eds, Philadelphia, PA: Mosby-Year Book Inc, 1991, 122.

Methodology Photometric

References

Matos V, Van Melle G, Boulat O, et al, "Urinary Phosphate/Creatinine, Calcium/Creatinine, and Magnesium/Creatinine Ratios in a Healthy Pediatric Population," *J Pediatr,* 1997, 131(2):252-7.

Subramanian R and Khardori R, "Severe Hypophosphatemia: Pathophysiologic Implications, Clinical Presentations, and Treatment," *Medicine,* 2000, 79(1):1-8.

Young DS, *Effects of Drugs on Clinical Laboratory Tests,* 5th ed, Volume 1: Listing by Test, Washington, DC: AACC Press, American Association for Clinical Chemistry, 2000, Section 3, 621-2.

♦ **Photopheresis** *see* Apheresis, Therapeutic *on page 201*

pH, Stool

Related Information

Clostridium difficile Toxin Assay and Culture *on page 416*
Fat, Semiquantitative, Stool, Sudan III Stain *on page 572*
Fecal Fat, Quantitative, 72-Hour Collection *on page 574*
Meat Fibers, Stool *on page 894*
Methylene Blue Stain, Stool *on page 906*
Reducing Substances, Stool *on page 1147*

Synonyms Fecal pH; Stool pH

Applies to Chloridorrhea, Congenital

Abstract Evaluation for diarrhea depends first on history and physical examination. Nocturnal diarrhea, weight loss >5 kg, increased ESR, and/or decreased Hb/Hct favor presence of organic disease. Investigation of chronic diarrhea may include examination for fecal leukocytes, ova and parasites, *C. difficile* toxin assay, stool pH, stool for reducing substances; CBC and differential, ESR, electrolytes, BUN/creatinine, TSH, T₄, gastrin; vasoactive intestinal polypeptide if hypokalemia is present and diarrhea volume >1 liter/day; assay for *Giardia* antigen and other studies, including imaging and endoscopy.

Patient Preparation Barium procedures and laxatives should be avoided for 1 week prior to collection of the specimen.

Specimen Fresh random stool

Storage Instructions Refrigerate

Causes for Rejection Specimen contaminated with urine

Reference Interval Diet dependent; normal: neutral to slightly alkaline or acid. Stool pH is usually slightly acidic at a pH of ~6. Breast-fed infants have slightly acid stools; bottle-fed infants, neutral or slightly alkaline. Acid stool is formed with fat malabsorption.

Critical Values Stool pH <5.3 is diagnostic of carbohydrate intolerance; stool pH >6.8 is evidence of cholerheic enteropathy.

Use Screen for carbohydrate and fat malabsorption; evaluate small intestinal disaccharidase deficiencies

Limitations Limited value due to dependence on stool volume and transit time. The diagnosis of steatorrhea requires 72-hour specimen with diet of 75-100 g fat/24 hours; see Fecal Fat, Quantitative, 72-Hour Collection *on page 574*.

Additional Information Stool pH is dependent in part on fermentation of sugars. Colonic fermentation of normal amounts of carbohydrate sugars and production of fatty acids accounts for the normally slightly acidic pH. If disaccharide intolerance is suspect, simple screening tests may be performed. Slightly alkaline pH may occur in cases of secretory diarrhea without food intake, colitis, villous tumor, and possibly with antibiotic usage (with resultant impaired colonic fermentation). A stool pH <6 (measured by pH paper) is suggestive evidence of sugar malabsorption and is more likely to be associated with osmotic than with secretory diarrhea.

Osmotic diarrhea is due to an excess of nonabsorbable solutes, while secretory diarrhea is the result of intestinal mucosal secretory activity exceeding absorptive capacity.

Determination of fecal reducing substances is a more reliable screening test for disaccharidase deficiency. See Reducing Substances, Stool *on page 1147* and Fat, Semiquantitative, Stool, Sudan III Stain *on page 572*.

References
Castro-Rodríguez JA, Salazar-Lindo E, and León-Barúa R, "Differentiation of Osmotic and Secretory Diarrhoea by Stool Carbohydrate and Osmolar Gap Measurements," *Arch Dis Child*, 1997, 77(3):201-5.

Donowitz M, Kokke FT, and Saidi R, "Evaluation of Patients With Chronic Diarrhea," *N Engl J Med*, 1995, 332(11):725-9.

♦ **Phthirus pubis Identification** *see Arthropod Identification on page 213*
♦ **pH, Umbilical Venous Blood Gases (pCO$_2$ and pO$_2$)** *see Blood Gases and pH, Umbilical Cord on page 278*

pH, Urine

Related Information
Kidney Stone Analysis *on page 820*
Mexiletine, Serum *on page 911*
pH, Blood *on page 1018*
Salicylate, Serum or Plasma *on page 1176*
Urinalysis *on page 1289*

Test Includes pH is part of a routine urinalysis.

Abstract Acids excreted by the renal glomeruli include sulfuric, phosphoric and hydrochloric with small quantities of pyruvic, citric, and lactates, and when present, ketone bodies. Bicarbonate is reabsorbed. Sodium of the glomerular filtrate is exchanged for hydrogen ions. NH$_4$ is excreted.

Specimen Random urine

Storage Instructions Test immediately.

Reference Interval Normal kidneys can produce urine with pH from 4.6-8.0, but with ordinary diet, urine pH is about 6.0. Urine becomes more alkaline after meals. Urine is most acidic fasting in the morning. Proteins cause lower pH, and citrus fruits cause higher pH.

Use Urine pH is a crude indicator of the acid-base balance of the body. It may be helpful in determination of renal tubular disease or pyelonephritis. Urine pH is useful for identification of crystals in urine and determination of predisposition to formation of a given type of stone. Control of urinary pH is important in management of nephrolithiasis. See table. When an accurate pH assessment of acid-base status and renal response is desired, the urine should be collected under circumstances more controlled than is usual. Attention is given to the time of day, the fasting status of the patient, and transfer of sample so as to prevent degassing of sample or growth of bacteria; rapid analysis by pH meter rather than dipstick is indicated. Simultaneous serum pH may also be ordered.

Methodology Dipstick double indicator principle (methyl red and bromthymol blue) which gives a broad range of colors covering the urinary pH range 5-9 ±0.5 pH units. A pH meter is the back-up and the more accurate method. (Continued)

pH, Urine (Continued)

Conditions Associated With Acid Urine
Metabolic acidosis
Diabetes mellitus
Diarrhea
Starvation
Respiratory acidosis
Emphysema
Sleep
Renal failure with lack of NH_3 buffer
Conditions Associated With Alkaline Urine
Respiratory alkalosis
Metabolic alkalosis
Urea-splitting bacteria (*Proteus* sp)
Vegetable diet
Gastric suction and vomiting
Diuretic therapy
Urine allowed to stand
Postprandial alkaline tide (1 hour after meal)
Alkali therapy (citrate, bicarbonate)
Renal tubular acidosis

Additional Information The capacity to exchange H^+ for cation is decreased with impaired renal tubular function. In renal tubular acidosis, the distal tubules cannot effectively exchange H^+ for cations. pH of the urine may reflect attempts at correction of metabolic acid-base disturbances. In chronic acidosis such as diabetic ketoacidosis, large amounts of H^+ are excreted. In metabolic alkalosis, high levels of bicarbonate are produced. Compensation of respiratory acidosis and respiratory alkalosis is also associated with increased excretion of H^+ and bicarbonate, respectively.

Urine pH is a part of a supersaturation profile intended to enhance management of patients with nephrolithiasis. Other relevant urinary analytes included by a reference laboratory for this profile include potassium, calcium, phosphorus, oxalate, uric acid, citrate, magnesium, sodium, chloride, and sulfate. Urine volume is also important, but is not an analyte.

References

Alon US, "Renal Tubular Acidosis: Clinical Assessment," *Pediatric Nephrology in Perspective*, Strauss J, ed, Coral Gables, FL: University of Miami Press, 1995, 33-44.

Cogan MG and Rector FC Jr, "Acid-Base Disorders," *Kidney*, Chapter 18, Brenner BM and Rector FC Jr, eds, Philadelphia, PA: WB Saunders Co, 1991, 737-804.

Mayo Reference Services Publication, "Supersaturation Profile, Urine, Test," *New Test Announcement*, #32029, Rochester, MN: Mayo Medical Laboratories, 2000.

♦ **Phyllocontin**® see Theophylline, Serum *on page 1243*

♦ **Physeptone**® see Methadone, Serum or Urine *on page 900*

♦ **Pick-me-up**® see Caffeine, Serum *on page 326*

♦ **Pilocarpine Iontophoresis** see Chloride, Sweat *on page 392*

Pinworm Preparation

Related Information

Ova and Parasites, Direct Exam *on page 985*

Test Includes Detection of pinworm eggs from the perianal region

Abstract The human pinworm, *Enterobius vermicularis*, has a worldwide distribution. The parasite primarily infects young children. The eggs of the parasite rapidly develop to the infective stage and they can persist for long periods in the environment. The adult female worm migrates out of the anal orifice at night and lays its eggs in the perianal area. The eggs adhere to the skin of the perianal area and can be detected by the use of a cellulose tape pressed to the perianal area. This scotch tape is then examined under the microscope for the presence of the *E. vermicularis* eggs.

Specimen Scotch® Tape slide preparation of perianal region

Container Scotch® Tape slide must be submitted in a covered container. Commercial kit products are also available for collection of pinworm specimens. **Caution**: Pinworm eggs are very infectious.

Collection The specimen is best obtained early in the morning before a bowel movement or bath. Clear Scotch® Tape should be used; the nontransparent type is unsatisfactory. An 8 cm (3 in) piece of cellophane tape is placed over the end of a glass slide sticky side out. The anal folds are spread apart and the mucocutaneous junction is firmly pressed in all four quadrants. The tape is then pressed over the slide and the specimen is transported to the laboratory in a carefully sealed container. Refer to diagram. It is important to provide clear instructions, because these specimens are often collected at home.

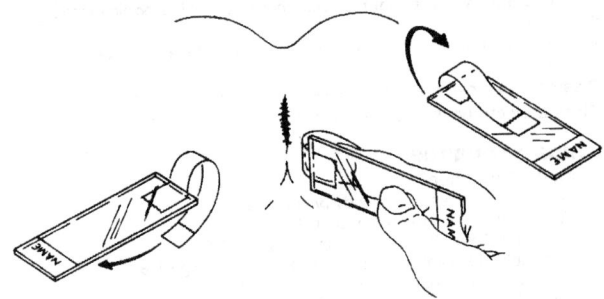

Cellophane tape slide preparation. Attach 3" piece of cellophane tape to undersurface of clear end of microscope slide, which has previously been identified (ground-glass end). Press sticky surface of tape against perianal skin. Then roll back tape onto slide, sticky surface down. Wash hands and nails well. From Bauer JD, *Clinical Laboratory Methods*, 9th ed, St Louis, MO: Mosby-Year Book Inc, 1982, 989.

Causes for Rejection Use of nontransparent Scotch® Tape, Scotch® Tape on both sides of the slide, specimen which is not inside a covered container, use of frosted slide, tape sent sticky side up. Specimens which are not properly contained pose excessive risk to laboratory personnel since the eggs of *E. vermicularis* are extremely contagious.

Reference Interval No pinworm eggs (*Enterobius vermicularis*) identified. Positives reported as few, moderate, or many eggs identified.

Use Detect cases of pinworm infestation (enterobiasis), *Enterobius vermicularis* parasitic infestation

Limitations Examination for pinworm only. One negative result does not rule out possibility of parasitic infestation. Examinations on multiple days may be required to diagnose infection. Stool specimens are not usually satisfactory for pinworm studies.

Contraindications Specimen collection at improper time

Methodology Microscopy

Additional Information The diagnosis of pinworm infection is by the recovery of eggs or female worms from the perianal region. Enterobiasis often is present in multiple family members. Therefore, it is recommended that all members of the family be tested. The responsible parent should be instructed how to collect samples, using one kit per individual. Female worms or parts of them may be demonstrated on the tape by microscopic examination. The number of positive specimens on different days correlates with the severity of disease. Eggs, if present, may be immature, embryonated (with viable or dead larvae), or (if the specimen is several days or more old) empty egg shells will be present.

References

al-Rufaie HK, Rix GH, Perez Clemente MP, et al, "Pinworms and Postmenopausal Bleeding," *J Clin Pathol*, 1998, 51(5):401-2.

Herrstrom P, Fristrom A, Karlsson A, et al, "*Enterobius vermicularis* and Finger Sucking in Young Swedish Children," *Scand J Prim Healthcare*, 1997, 15(3):146-8.

Nabulsi M, Shararah N, and Khalil A, "Perinatal *Enterobius vermicularis* Infection," *Int J Gynaecol Obstet*, 1998, 60(3):285-6.

Internet Web Sites

www.cdc.gov/ncidod/dpd/parasites/pinworm/default.htm

♦ **Pituitary Gonadotropins** *see* Follicle Stimulating Hormone, Serum, Plasma, or Urine *on page 609*

♦ **Pituitary Gonadotropins** *see* Luteinizing Hormone, Blood or Urine *on page 876*

♦ **PKU, Neonatal** *see* Newborn Screen for Phenylketonuria *on page 954*

♦ **PKU Test** *see* Phenylalanine, Blood *on page 1022*

♦ **PKU, Urine Test** *see* Phenylalanine, Urine *on page 1024*

♦ **Placental Blood** *see* Hematopoietic Progenitor Cells, Cord Blood/Placental Blood *on page 676*

♦ **Placidyl®** *see* Ethchlorvynol, Serum or Plasma *on page 560*

♦ **Plague Diagnostic Procedures** *see* Yersinia pestis Diagnostic Procedures *on page 1337*

♦ **Plasma β-Amyloid$_{(1-42)}$** *see* Cerebrospinal Fluid and Plasma β-Amyloid$_{(1-42)}$ *on page 359*

♦ **Plasma Cholinesterase** *see* Pseudocholinesterase, Serum *on page 1122*

♦ **Plasma Electrolytes** *see* Electrolyte Panel, Serum *on page 532*

Plasma Exchange

Related Information
Acetylcholine Receptor Antibody *on page 92*
Calcium, Ionized, Serum *on page 328*
Cholesterol, Total, Serum or Plasma *on page 396*
Cryoglobulin, Qualitative, Serum and Plasma *on page 478*
Low Density Lipoprotein Cholesterol *on page 874*
Plasma, Fresh Frozen *on page 1039*
Plasmapheresis, Donor *on page 1041*
Platelet Count *on page 1050*

Synonyms Plasmapheresis

Test Includes ABO group is indicated if fresh frozen plasma must be selected and thawed for use as a replacement solution.

Abstract Selective removal of a pathologic component of a patient's plasma to assist in the treatment of a disease. Separation techniques are varied and include manual (whole blood bags) and automated instruments (eg, centrifugation and adsorption). Manual techniques are no longer often used.

Patient Preparation Excellent venous access is a daily requirement. Indwelling dual- or triple-lumen central venous catheters suitable for apheresis procedures are frequently necessary. Hypotension secondary to hypovolemia will become a concern if the extracorporeal volume exceeds 15% of the patient's blood volume. Patients taking angiotensin-converting-enzyme (ACE) inhibitors may be prone to marked hypotension prior to routine or adsorption treatment. Such therapy should be removed from their medication if possible. Understanding which medications being given to the patient will be altered by the procedure is relevant so that dosage can be adjusted accordingly (withhold administration for 1 hour before and after procedure if possible). Obtain and document informed consent.

Aftercare Monitor patient for signs and symptoms of reduced plasma levels of ionized calcium (tingling, perioral paresthesias, cardiac arrhythmias) and hypotension secondary to hypovolemia.

Special Instructions Usually on a scheduled basis, but can be emergent. Typical plasma exchanges are 1-1.5 times plasma volume daily (40 mL/kg patient's body weight). Whenever possible, monitor efficacy of exchanges by monitoring levels of some marker that indicates progress or regression of the disease treated (eg, specific antibody, immunoglobulin, abnormal protein).

Use During therapeutic plasmapheresis, plasma containing the pathologic substance is removed and replaced with crystalloids, albumin, or fresh frozen plasma. Usually 1-1.5 plasma volumes are exchanged. Selective removal of pathologic substances in plasma via adsorption columns (eg, dextran sulfate removal of LDL cholesterol or staphylococcal protein A removal of IgG antibodies) eliminates the need for replacement fluid. See the table for some established clinical indications for plasma exchange. Presently, they include hyperviscosity/Waldenström macroglobulinemia, thrombotic thrombocytopenic purpura, myasthenia gravis, and chronic inflammatory demyelinating polyneuropathy.

Indications for Therapeutic Plasmapheresis

Generally Seems to Be Effective[1]
Hyperviscosity syndrome
Myasthenia gravis[1]
Goodpasture syndrome
Thrombotic thrombocytopenic purpura[1]
Cryoglobulinemia
Familial hypercholesterolemia (via selective adsorption)
Guillain-Barré syndrome[1]
Post-transfusion purpura
Efficacy Debatable or Controlled Studies Lacking
Systemic lupus erythematosus
Schizophrenia
Warm autoimmune hemolytic anemia
Amyotrophic lateral sclerosis
Rh hemolytic disease of the newborn
Multiple sclerosis
Renal transplant rejection
Thyroid storm

[1]Effective = producing significant clinical improvement that is better than transitory. Entities marked with superscript 1 are among the most frequent indications for plasma exchange in the Canadian apheresis group study. Included among these are also Waldenström macroglobulinemia and chronic inflammatory demyelinating polyneuropathy.

Adapted from Brecher ME, *Technical Manual*, 14th ed, Bethesda, MD: American Association of Blood Banks Press, 2002, 127-47.

Contraindications Consultation is especially important if the patient is pediatric, elderly, clinically unstable, has poor vascular access, or has a condition for which apheresis is of unknown benefit.

Additional Information The most common adverse effects are vasovagal reactions, hypovolemia, hypocalcemia, allergic reactions, and citrate toxicity. Allergic reactions (hives, dyspnea, wheezing) and symptoms of hypocalcemia (tingling, perioral paresthesias, cardiac arrhythmias) may occur more frequently when fresh frozen plasma is used as a replacement fluid. Symptoms of hypovolemia include hypotension, diaphoresis, and tachycardia.

References

Brecher ME, *Technical Manual*, 14th ed, Bethesda, MD: American Association of Blood Banks Press, 2002, 134-6.

Clark WF, Rock GA, Buskard N, et al, "Therapeutic Plasma Exchange: An Update From the Canadian Apheresis Group," *Ann Intern Med*, 1999, 131(6):453-62.

"Clinical Applications of Therapeutical Apheresis," *J Clin Apheresis*, 2000, 15:1-159 (special issue).

McLeod BC, Price TH, and Weinstein R, *Apheresis Principles and Practice*, 2nd ed, Bethesda, MD: American Association of Blood Banks Press, 2003, 321-410.

Plasma, Fresh Frozen

Related Information

Activated Partial Thromboplastin Time *on page 100*
Cryoprecipitate *on page 481*
Factor IX Concentrate *on page 564*
Factor VIII Concentrate *on page 563*
Factor XIII *on page 565*
Fibrinogen *on page 583*
Plasma, Frozen, Donor Retested *on page 1041*
Prothrombin Time *on page 1116*
Warfarin, Serum or Plasma *on page 1325*

Synonyms FFP

Applies to Cryopreservation; Factor IX Replacement; Factor VII Replacement; Factor V Replacement; Factor XI Replacement; Factor X Replacement; Thawed Plasma

Test Includes ABO type

Abstract Plasma from a unit of whole blood separated from the red blood cells and frozen at -18°C or lower within 8 hours of collection. The unit has a volume (Continued)

Plasma, Fresh Frozen *(Continued)*

of 150-275 mL. FFP collected via apheresis has a volume of 500 mL. Both contain all labile and stable coagulation factors (each at 1 IU/mL), but are not concentrates. Does not contain platelets. A severe deficiency of coagulation factors cannot be corrected by giving FFP. Fluid overload would likely result.

Patient Preparation Use coagulation studies as a guide to transfusion of FFP.

Dosage and administration: FFP should be ABO compatible, especially when it is to be given to infants. Depending on the ABO group, it may contain anti-A or anti-B. Rh need not be considered (but see Additional Information). Cross-match is not necessary. Administer through a filter. The usual unit (prepared from one unit of whole blood) contains about 200 units of factor VIII, and 250-400 mg of fibrinogen. It contains factor IX as well as other stable and labile coagulation factors. One unit of fresh frozen plasma will raise patient's plasma level of fibrinogen only about 10-13 mg/dL; cryoprecipitate is a better source of fibrinogen. Give FFP at about 10 mL/minute, to a total dose of about 10 mL/kg.

Specimen Blood

Container One red top tube or one lavender top (EDTA) tube

Storage Instructions Frozen at -18°C or lower, FFP has a shelf-life of 1 year. Examine the frozen plastic bag for cracks, especially the seams. Thaw at 37°C with agitation in a waterbath, using a plastic overwrap. Thawing requires 15-30 minutes depending on the number of units being thawed. Once thawed, store in Blood Bank refrigerator and transfuse within 24 hours. Plasma ideally should be transfused within 2 hours after thawing when the patient requires labile coagulation factors. After 24 hours, a thawed unit of FFP becomes a unit of "thawed plasma".

Turnaround Time Usually available on request. Requires 15-30 minutes to thaw and issue.

Use Treatment of bleeding caused by multiple labile and stable coagulation factor deficiencies (eg, liver disease, DIC); with massive transfusion and abnormal coagulation assays, in severe warfarin overdosage before vitamin K can reverse the warfarin effect, with cardiac bypass surgery and as replacement medium in plasma exchange for thrombotic thrombocytopenic purpura (TTP).

Limitations Transmission of infectious disease, including HIV, hepatitis A, B, and C, and parvovirus B19 (CMV and HTLV do not appear to be transmitted by plasma); circulatory overload; TRALI (transfusion-related acute lung injury); leukocytes 10^6 to 10^7 per bag with potential for reactions and allergic reactions are hazards of use of FFP. Patients with TTP may receive large volumes of plasma, which multiply their risks.

Considerable variation in coagulation factors per bag exists. **Fresh frozen plasma is overused.**

Contraindications Do not use FFP prophylactically to prevent dilutional coagulopathy in large transfusions. Do not use it as a plasma expander; albumin is better and safer. Do not use FFP if prothrombin time and activated partial thromboplastin time are less than 1.5 times normal and in the absence of abnormal bleeding. Coagulopathies in patients with hemophilia and von Willebrand disease are better treated with specific factor concentrates.

Additional Information Hazards: Risk of disease transmission (that of any single unit exposure), plasma volume overload, anaphylaxis in IgA deficient recipient is a remote hazard. One unit of FFP collected via apheresis contains the equivalent coagulation factor of ~2 units FFP while one exposing the patient to one donor. FFP contains anti-A or anti-B. Although FFP is basically cell-free, it is not without antigens. Recipients can have mild or severe allergic reactions and sometimes fever. Immunization can take place to soluble constituents as well as to Rh and other red cell antigens, the latter presumably from cell fragments in the plasma. See also Cryoprecipitate *on page 481.*

An alternative to increase the safety of plasma involves storage of the unit, while the donor is retested after a period longer than the window periods of known viruses. Negative results of such retesting would support lack of infectivity of the unit. This approach was approved by the FDA, September 1998, for units in which retesting takes place over a minimum of 112 days.

References

Consensus Conference, "Fresh Frozen Plasma. Indications and Risks," *JAMA*, 1985, 253(4):551-3.

Goodnough LT, Brecher ME, Kanter MH, et al, "Transfusion Medicine. Second of Two Parts: Blood Conservation," *N Engl J Med*, 1999, 340(7):525-33.

Pehta JC, "Advances in Plasma Products Use," *Lab Med*, 2001, 32(1):26-31.

Plasma, Frozen, Donor Retested

Related Information

Cryoprecipitate *on page 481*

Donation, Blood *on page 521*

Factor IX Concentrate *on page 564*

Factor VIII Concentrate *on page 563*

Plasma, Fresh Frozen *on page 1039*

Synonyms FFP

Applies to Cryopreservation; Factor IX Replacement; Factor VII Replacement; Factor V Replacement; Factor XI Replacement; Factor X Replacement

Test Includes ABO type

Abstract Donor retested (DR) plasma is collected and prepared in a similar manner to that as fresh frozen plasma (see Plasma, Fresh Frozen *on page 1039*). However, DR plasma may reduce the risk of viral transmission for hepatitis C, hepatitis B, HIV, and HTLV because it is not released for transfusion until the donor returns for repeat testing. The plasma is then released if the donor tests negative for transfusion-transmitted infectious diseases.

Patient Preparation See Plasma, Fresh Frozen *on page 1039*.

Container One red top tube or one lavender top (EDTA) tube

Storage Instructions See Plasma, Fresh Frozen *on page 1039*.

Use See Plasma, Fresh Frozen *on page 1039*.

Limitations As with fresh frozen plasma, circulatory overload, allergic reactions, and risk of infectious disease transmission continue to occur. **Fresh frozen plasma is overused.** Due to complex inventory management problems, DR plasma may not be available at every blood bank and at a constant supply.

Contraindications See Plasma, Fresh Frozen *on page 1039*.

Methodology DR plasma may be prepared from a unit of whole blood by separating and freezing the plasma within 8 hours of phlebotomy or by apheresis (similar to fresh frozen plasma). The plasma is tested for routine infectious disease testing, but held for 90 or more days (usually at least 112 days) until the donor comes back a second time. The first unit can be released once the donor tests negative the second time.

Additional Information Interest in the use of this component has declined with the use of nucleic acid testing.

Hazards: See Plasma, Fresh Frozen *on page 1039*.

References

Brecher ME, *Technical Manual*, 14th ed, Bethesda, MD: American Association of Blood Banks Press, 2002, 167.

"Fresh Frozen Plasma, Cryoprecipitate, and Platelets Administration Practice Guidelines Development Task Force of the College of American Pathologists. Practice Parameter for the Use of Fresh-Frozen Plasma, Cryoprecipitate, and Platelets," *JAMA*, 1994, 271:777-81.

Triulzi DJ, *Blood Transfusion Therapy: A Physician's Handbook*, 7th ed, Bethesda, MD: American Association of Blood Banks Press, 2002, 24-8.

♦ **Plasmapheresis** *see* Apheresis, Therapeutic *on page 201*

♦ **Plasmapheresis** *see* Plasma Exchange *on page 1038*

Plasmapheresis, Donor

Related Information

Donation, Blood *on page 521*

Plasma Exchange *on page 1038*

Synonyms Donor Plasmapheresis

Test Includes As for regular whole blood donation, but may include total protein and serum protein electrophoresis, depending upon the donation interval.

Abstract The expression, "**plasmapheresis**", means "taking away plasma" and indicates selective removal of the donor's plasma with return of all other components. Separation techniques are varied and include manual (whole blood bags) and automated instruments (centrifugation or membrane filtration). Manual techniques are no longer often used.

When sufficient plasma has been removed, it becomes advisable to infuse albumin or plasma as replacement; this is **plasma exchange**. Plasma exchange is used to remove a particular constituent.

(Continued)

Plasmapheresis, Donor *(Continued)*

Patient Preparation Obtain and document informed consent. Emergency medical care must be available.

For donors in the UK, prior blood donation is required with age older than 50 years, and normal CBC and total serum protein. Periodically CBC, total protein, and serum albumin are repeated and immunoglobulins, urine protein, and glucose are measured.

Aftercare Monitor total red cell losses during procedures. If red cells cannot be returned during a procedure, defer donor for 8 weeks.

Causes for Rejection (Of donor): As for regular whole blood donation, except for the donation interval.

Use Plasma is collected as fresh frozen plasma or for subsequent application into derivatives such as plasma protein fraction, albumin, immune globulins, and clotting factor concentrates. Plasma from particular donors hyperimmunized to particular antigens such as $Rh_o(D)$ is to provide certain specific immunoglobulins.

Limitations Must meet donor criteria specified by AABB and FDA.

Methodology May be separated manually from whole blood collections with return of RBCs to donors, or may be prepared by use of automated apheresis instruments.

Additional Information A minimum of 48 hours should elapse between successive procedures. A donor should undergo no more than two procedures in 1 week.

References

Brecher ME, *Technical Manual*, 14th ed, Bethesda, MD: American Association of Blood Banks Press, 2002, 130-1.

Code of Federal Regulations. Title 21 CFR Part 640, Washington, DC: U.S. Government Printing Office, 1999.

Mollison PL, Engelfriet CP, Contreras M, et al, "The Withdrawal of Blood," *Blood Transfusion in Clinical Medicine*, 10th ed, Chapter 1, Oxford, UK: Blackwell Scientific Publications, 1997, 1-36.

♦ **Plasma Protein Fraction (Human) (PPF)** see Albumin and Plasma Protein Fraction for Infusion *on page 118*

♦ **Plasma Thromboplastin Antecedent (Factor XI)** see Coagulation Factor Assays *on page 418*

♦ **Plasma Thromboplastin Component (Factor IX)** see Coagulation Factor Assays *on page 418*

♦ **Plasma Volume Measurement** see Blood Volume *on page 282*

♦ **Plasmin** see D-Dimers and Fibrin Degradation Products *on page 502*

♦ **Plasmin** see Fibrinogen *on page 583*

♦ **Plasmin-Antiplasmin Complexes** see Hypercoagulation Panel *on page 758*

♦ **Plasmin Inhibitor** see Antiplasmin *on page 196*

Plasminogen

Related Information

Disseminated Intravascular Coagulation Screen *on page 517*
Hypercoagulation Panel *on page 758*
Plasminogen Activator Inhibitor 1 *on page 1043*

Applies to Acute Phase Reactant; Conjunctivitis, Ligneous; Fibrinogenolysis; Fibrinolysis; Tissue Plasminogen Activator; tPA; uPA; Urokinase-Type Plasminogen Activator

Abstract Plasminogen is the precursor of plasmin, which lyses fibrin clots. Hereditary plasminogen deficiency is rare, and it may predispose to venous thrombosis.

Specimen Plasma

Container Blue top (sodium citrate) tube

Collection Routine venipuncture. If multiple tests are being drawn, draw blue top tubes after any red top tubes but before any lavender top (EDTA), green top (heparin), or gray top (oxalate/fluoride) tubes. Immediately invert tube gently at least 4 times to mix. Tubes must be appropriately filled. Deliver tubes immediately to the laboratory.

Storage Instructions Separate plasma from cells as soon as possible. Store plasma on ice for up to 4 hours, or store frozen.

Causes for Rejection Specimen received more than 4 hours after collection, tubes not filled, clotted specimens

Turnaround Time 1 day or longer, depending on how often testing is batched

Reference Interval Functional results are reported as a percent of the amount expected in normal plasma. By definition, the mean value in normal plasma is 100%. The reference range is approximately 75% to 130%. Antigen results may be reported in mg/dL, with a reference range of approximately 6-14 mg/dL. Plasminogen levels can increase during pregnancy. Newborn levels are about 60% of adult values. Newborn levels increase to near adult values by age 6 months.

Use May be considered in patients with familial venous thrombosis and no evidence for more common hypercoagulable states. Occasionally, if monitoring of thrombolytic therapy is desired, plasminogen levels are followed. Plasminogen decreases during thrombolytic therapy. Consider testing plasminogen in patients with ligneous conjunctivitis, a condition that is associated with severe plasminogen deficiency.

Limitations Plasminogen may become elevated during pregnancy and during acute phase reactions. Antigen assays will not detect qualitative (dysfunctional) deficiencies.

Methodology

Functional (activity) assays: Chromogenic assays for plasminogen are available. Streptokinase is added to patient plasma, which binds to plasminogen. The streptokinase-plasminogen complex has plasmin-like activity which cleaves a chromogenic substrate, releasing a colored-compound. The amount of color detected spectrophotometrically is proportional to the amount of plasminogen in the sample.

Antigen (immunologic) assays: Radial immunodiffusion methods are commercially available.

Additional Information Plasminogen is converted into plasmin by tissue plasminogen activator (tPA) or urokinase-type plasminogen activator (uPA). Plasmin degrades fibrin clots (fibrinolysis) as well as intact fibrinogen (fibrinogenolysis). Plasmin also inactivates factors Va and VIIIa. Plasminogen can be decreased during thrombolytic therapy, liver disease, disseminated intravascular coagulation (DIC), and rarely, with a hereditary plasminogen deficiency. The incidence of plasminogen deficiency is 0.29% to 0.73% in healthy individuals, up to 1.4% to 2.2% among patients with venous thrombosis, and 1.4% among patients with arterial thrombosis. In one study, 2.5% of a general population with plasminogen deficiency had a history of thrombosis. Hereditary deficiencies of plasminogen could result in decreased fibrinolysis. However, in some studies, plasminogen-deficient relatives of affected individuals have similar rates of thrombosis as nondeficient relatives, whereas in other studies they do have a higher rate of thrombosis. Severe hereditary plasminogen deficiency is associated with ligneous conjunctivitis, a rare chronic pseudomembranous conjunctivitis characterized histologically by massive deposits of fibrin in the affected tissues. Apparently, the fibrin depositions result from decreased or absent clearance of fibrin by plasminogen.

References

Biasiutti FD, Sulzer I, and Stucki B, "Is Plasminogen Deficiency a Thrombotic Risk Factor? A Study on 23 Thrombophilic Patients and Their Family Members," *Thromb Haemost*, 1998, 80:167-70.

Girolami A, Sartori MT, Saggiorato G, et al, "Symptomatic Versus Asymptomatic Patients in Congenital Hypoplasminogenemia: A Statistical Analysis," *Haematologia (Budap)*, 1994, 26(2):59-65.

Heijboer H, Brandjes DP, Buller HR, et al, "Deficiencies of Coagulation-Inhibiting and Fibrinolytic Proteins in Outpatients With Deep-Vein Thrombosis," *N Engl J Med*, 1990, 323(22):1512-6.

Schuster V, Zeitler P, Seregard S, et al, "Homozygous and Compound-Heterozygous Type I Plasminogen Deficiency Is a Common Cause of Ligneous Conjunctivitis," *Thromb Haemost*, 2001, 85(6):1004-10.

Tait RC, Walker ID, Conkie JA, et al, "Isolated Familial Plasminogen Deficiency May Not Be a Risk Factor for Thrombosis," *Thromb Haemost*, 1996, 76(6):1004-8.

Plasminogen Activator Inhibitor 1

Related Information

Antiplasmin *on page 196*
Plasminogen *on page 1042*

Synonyms PAI-1

Applies to Acute Phase Reactant; Tissue Plasminogen Activator; tPA

(Continued)

Plasminogen Activator Inhibitor 1 *(Continued)*

Abstract PAI-1 inhibits tissue plasminogen activator (tPA). High levels of PAI-1 may be associated with an increased risk of arterial thrombosis due to inhibition of fibrinolysis, and low levels of PAI-1 characterize a rare familial bleeding disorder due to excessive fibrinolysis. A causal effect of high levels of PAI-1 on arterial thrombosis has not yet been established.

Specimen Plasma

Container Blue top (sodium citrate) tube. Specialized tubes containing platelet inhibitors (to prevent platelet release of PAI-1) or acid (to prevent PAI-1 from forming a complex with tPA) have been recommended, but are not necessary if specimens are handled appropriately.

Sampling Time PAI-1 has a circadian rhythm: its plasma concentration is highest in the morning, and lowest in the afternoon and evening. In one study, the mean level was 23 ng/mL at 9 AM and 10 ng/mL at 4 PM.

Collection Collect blood from a steadily flowing venipuncture. Discard the first 3-5 mL if PAI-1 is the only test being drawn. If multiple tests are being drawn, draw blue top tubes after any red top tubes but before any lavender top (EDTA), green top (heparin), or gray top (oxalate/fluoride) tubes. Immediately invert tube gently at least 4 times to mix. Tubes must be appropriately filled. Deliver tubes immediately to the laboratory.

Storage Instructions Separate plasma from cells as soon as possible. Laboratories should avoid platelet contamination of plasma because platelets contain PAI-1. Centrifugation at 2000-3000 g for 15 minutes helps ensure platelet-free plasma. Store plasma on ice for up to 2 hours, or store frozen.

Causes for Rejection Specimen received more than 2 hours after collection; tubes not filled; clotted specimens; antifibrinolytic agent present in specimen, such as aprotinin or epsilon-aminocaproic acid, which interfere with functional PAI-1 assays

Turnaround Time Usually at least several days, as testing is often batched

Reference Interval Approximately 4-40 ng/mL in antigen assay and 0-12 units/mL in functional assay (see Sampling Time for note regarding circadian rhythm)

Use Not a commonly performed clinical assay. May be considered in patients with strong evidence for a familial bleeding disorder and normal test results for more common bleeding disorders (eg, von Willebrand disease). May be considered in patients with unexplained premature myocardial infarction.

Limitations PAI-1 is an acute phase reactant. Therefore, it becomes elevated following a thrombotic event and it should not be measured in the acute setting following thrombosis. A related inhibitor, PAI-2, is normally not present in plasma. However, it becomes elevated in pregnant women and can cause overestimations of PAI-1 during pregnancy. PAI-1 also becomes elevated during pregnancy. Antigen assays will not detect qualitative deficiencies.

Methodology

Functional (activity) assays: Patient plasma is added to a known amount of urokinase; PAI-1 in the patient plasma binds and inhibits the urokinase. The amount of residual urokinase is detected by adding plasminogen, which is converted to plasmin by urokinase. Plasmin cleaves a chromogenic synthetic substrate, releasing a colored compound which can be detected spectrophotometrically. The amount of released color is inversely proportional to the amount of PAI-1 in the sample. This assay contains an inhibitor of antiplasmin and other plasmin inhibitors to prevent these other inhibitors from interfering with the assay. Another version of this assay uses tPA instead of urokinase, and an acidification step to destroy antiplasmin and other plasmin inhibitors.

Antigen (enzyme-linked immunosorbent) assays are also available.

Additional Information PAI-1 is produced by the endothelium and liver and is also present in platelets. PAI-1 inhibits both tPA and urokinase-type plasminogen activator (uPA). PAI-1 may be active, inactive, or complexed with tPA.

The relationship between elevated PAI-1 and coronary artery disease may be at least partly due to its association with established cardiovascular risk factors, namely, the syndrome of insulin resistance. Elevated PAI-1 is associated with an increased incidence of myocardial infarction in prospective studies, but the association has not always remained significant after adjusting for other factors such as insulin resistance. The synthesis of PAI-1 is increased by high glucose or insulin levels. PAI-1 levels are elevated in insulin resistance, which is associated with a constellation of lipid and other abnormalities and an increased risk of coronary artery disease. Weight loss, which may reduce insulin resistance, also reduces PAI-1. Studies are conflicting regarding an association between a PAI-1 polymorphism, higher PAI-1 levels, and myocardial infarction.

References
Brandt JT, "Plasminogen and Tissue-Type Plasminogen Activator Deficiency as Risk Factors for Thromboembolic Disease," *Arch Pathol Lab Med*, 2002, 126(11):1376-81.

Contant G, Nicham F, and Martinoli JL, "Determination of Plasminogen Activator Inhibitor (PAI) by a New Venom-Based Assay," *Fibrinolysis*, 1992, 6(Suppl 3):85-6.

Francis CW, "Plasminogen Activator Inhibitor-1 Levels and Polymorphisms. Association With Venous Thromboembolism," *Arch Pathol Lab Med*, 2002, 126(11):1401-4.

Kohler HP and Grant PJ, "Plasminogen-Activator Inhibitor Type I and Coronary Artery Disease," *N Engl J Med*, 2000, 342(24):1792-801.

Macy EM, Meilahn EN, Declerck PJ, et al, "Sample Preparation for Plasma Measurement of Plasminogen Activator Inhibitor-1 Antigen in Large Population Studies," *Arch Pathol Lab Med*, 1993, 117(1):67-70.

♦ *Plasmodium* Species *see* Malaria Smear and Tests *on page 888*

Platelet Aggregation
Related Information
Platelet Count *on page 1050*
Platelet Hyperaggregation *on page 1053*
von Willebrand Factor *on page 1321*

Synonyms Aggregometer Test; Platelet Function Studies

Applies to ATP:ADP Ratio; Beta-Thromboglobulin

Test Includes Response to adenosine diphosphate (ADP), epinephrine, collagen, ristocetin, and arachidonic acid

Abstract Platelet aggregation tests are used to assess platelet function.

Patient Preparation Patients should not have aspirin (or any medication containing aspirin) for at least 7 days prior to testing. Nonsteroidal anti-inflammatory drugs or other platelet-inhibiting agents should also be avoided.

Specimen Platelet-rich plasma

Container Three blue top or plastic (sodium citrate) tubes

Collection Routine venipuncture. Immediately invert tubes gently at least 4 times to mix. Deliver tubes immediately to the laboratory at room temperature (platelets are activated at cold temperatures).

Storage Instructions Keep specimen at room temperature and perform test immediately (or within 2 hours, if transportation to a reference laboratory is required). Do not refrigerate or freeze specimen.

Causes for Rejection Specimen received more than 2 hours after collection, specimen clotted, specimen received on ice.

Turnaround Time Less than 1 day

Special Instructions Usually must be scheduled in advance with the laboratory.

Reference Interval >60% of platelets aggregate with each agonist tested. Normally no significant spontaneous aggregation. Normal newborns can have decreased aggregation compared to adults.

Use Assess platelet function. When a familial bleeding disorder is suspected, this test is usually not performed unless routine tests are normal (PT, PTT, and platelet count) and von Willebrand tests are normal, because von Willebrand disease is much more common than hereditary platelet dysfunction.

Methodology Citrated plasma is centrifuged at a gentle speed, to draw red and white blood cells into a pellet, leaving platelets suspended in the plasma. Various platelet aggregating agents (agonists) are added to aliquots of the platelet-rich plasma, and the resulting platelet aggregation is measured in an aggregometer. The aggregometer measures platelet aggregation by monitoring optical density. As platelets aggregate, more light can pass through the specimen. The platelet agonists commonly include arachidonate, ADP, collagen, epinephrine, and ristocetin. One aliquot usually has no platelet agonist added, to assess for spontaneous platelet aggregation.

A rapid, whole blood point-of-care device has been compared to platelet aggregation in monitoring platelet function during antiplatelet therapy. Another rapid whole blood platelet function analyzer has been studied in small numbers of patients with various platelet function abnormalities.
(Continued)

Platelet Aggregation (Continued)

Additional Information The most common cause of platelet dysfunction detected in this assay is medications. With aspirin and related compounds, arachidonate aggregation is markedly decreased or absent, and other aggregation tracings may be variably impaired. A variety of other platelet-inhibiting agents, such as ticlopidine, clopidogrel, and abciximab, are known to impair platelet aggregation. A vast number of other medications have been implicated in impaired *in vitro* platelet aggregation, and the clinical significance, if any, is usually uncertain. If a patient is found to have impaired platelet aggregation in this assay, a careful review of prescribed, as well as over-the-counter medications, is indicated. An on-line literature search for each medication is often informative. Other acquired causes of impaired platelet aggregation include uremia and paraproteinemia (monoclonal gammopathy). Myeloproliferative disorders can impair platelet aggregation, by epinephrine in particular. Hyperaggregation has also been reported with myeloproliferative disorders.

Hereditary platelet dysfunction is far less common than acquired dysfunction. A hereditary disorder may be considered in patients with bleeding histories and no obvious acquired etiology to account for an abnormal platelet aggregation study. Ideally, the aggregation study is repeated on a fresh specimen to determine if the abnormality is reproducible. The presence of the same abnormality in family members supports the diagnosis of a hereditary defect. Platelet storage pool disorders may variably decrease responses to epinephrine, ADP, and occasionally other agonists. Platelet storage pool disorders are characterized by deficiencies in alpha or dense platelet granules. Alpha granules normally store platelet factor 4 (PF4), beta-thromboglobulin, and other substances. Dense granules normally contain ADP, serotonin, and other compounds. Alpha granule deficiency is a rare disorder called "gray platelet syndrome", because platelets appear gray with light microscopy due to a lack of alpha granules. Alpha granules give normal platelets their purple granular appearance. In gray platelet syndrome, platelets are large; thrombocytopenia may be present; and beta-thromboglobulin (a research test) is decreased in platelets but may be elevated in plasma. A research test for dense granule deficiency is the platelet ATP:ADP ratio, which is increased with dense granule deficiency. Uncommonly, patients are deficient in both alpha and dense granules. Rare genetic disorders may underlie some cases of storage pool deficiency, including Hermansky-Pudlak syndrome (dense granule granule deficiency with pulmonary fibrosis and albinism), Chédiak-Higashi syndrome, Wiskott-Aldrich syndrome, or thrombocytopenia with absent radius syndrome.

Glanzmann thrombasthenia is a rare inherited condition in which platelet glycoprotein IIb/IIIa (GPIIb/IIIa) is deficient. GPIIb/IIIa mediates platelet aggregation using fibrinogen to link platelets to each other. Therefore, in Glanzmann thrombasthenia, aggregation is decreased with all agonists (ADP, collagen, epinephrine, arachidonate) except ristocetin. Ristocetin agglutinates platelets using von Willebrand factor and platelet glycoprotein Ib (GPIb). Bernard-Soulier disease is a rare inherited disorder characterized by GPIb deficiency and therefore decreased ristocetin-induced aggregation. Giant platelets and often thrombocytopenia are also present. With severe von Willebrand disease, ristocetin aggregation can be decreased, but most cases of von Willebrand disease are mild and ristocetin aggregation is most often normal. For that reason, platelet aggregation is not used to screen for von Willebrand disease.

Note: The term "agglutination" is often used to describe ristocetin-induced platelet aggregation, because true platelet aggregation links platelets through fibrinogen and GPIIb/IIIa, whereas ristocetin links platelets through von Willebrand factor and GPIb.

References

Brown BA, *Hematology: Principles and Procedures*, 6th ed, Philadelphia, PA: Lea & Febiger, 1993, 271-4.

Fressinaud E, Veyradier A, Truchaud F, et al, "Screening for von Willebrand Disease With a New Analyzer Using High Shear Stress: A Study of 60 Cases," *Blood*, 1998, 91(4):1325-31.

Kereiakes DJ, Broderick TM, Roth EM, et al, "Time Course, Magnitude, and Consistency of Platelet Inhibition by Abciximab, Tirofiban, or Eptifibatide in Patients With Unstable Angina Pectoris Undergoing Percutaneous Coronary Intervention," *Am J Cardiol*, 1999, 84(4):391-5.

♦ **Platelet Aggregation, Hypercoagulable State** *see* Platelet Hyperaggregation on page 1053

Platelet Antibodies

Related Information

Heparin-Induced Thrombocytopenia *on page 695*
Platelet Antibody, Immunohematologic *on page 1049*
Platelet Count *on page 1050*
Platelets, Apheresis, Donation *on page 1054*
Platelet Transfusion *on page 1058*
Quinidine, Serum *on page 1129*
Transfusion Reaction Work-up *on page 1269*

Applies to Idiopathic Thrombocytopenic Purpura; ITP; Lymphocytotoxicity Assay; NAIT; Neonatal Alloimmune Thrombocytopenia; Platelet Transfusion Refractoriness; Post-transfusion Purpura; PTP

Abstract Platelet antibodies can be autoimmune (as found in idiopathic thrombocytopenic purpura (ITP)), drug-induced, or alloimmune (as found in neonatal alloimmune thrombocytopenia (NAIT), post-transfusion purpura (PTP), platelet transfusion refractoriness). Heparin-induced thrombocytopenia (HIT), the most common drug-induced immune thrombocytopenia, occurs by a unique mechanism. Therefore, HIT is discussed separately.

Specimen Varies depending on method. Whole blood for direct antibody tests (measuring antibody attached to platelets) or for identifying platelet antigens; serum or plasma for indirect antibody tests (measuring antiplatelet antibody not bound to platelets). DNA testing requires whole blood or other source of DNA.

Container Varies depending on method. Commonly requires lavender top (EDTA - whole blood or plasma) tube and/or red top (serum) tube. Some methods require 6-8 tubes of blood; other methods need only one tube. Specimens should be transported to the laboratory immediately.

Collection Routine venipuncture

Storage Instructions Varies depending on method. Some methods require whole blood at room temperature; others recommend refrigerated whole blood, or refrigerated or frozen serum or plasma.

Turnaround Time Usually several days, because these assays are often send-out tests.

Use Confirm drug-induced thrombocytopenia, NAIT, PTP, or platelet transfusion refractoriness. The tests are not considered necessary for diagnosing ITP, according to an expert panel. For ITP, the tests may lack adequate sensitivity (particularly certain newer methods) or specificity (particularly certain older methods).

Methodology A variety of methods exist. Older methods that measure antibody associated with platelets are generally sensitive but not specific. For example, the use of **radiolabeled antibodies** that bind to other antibodies are used in some specialized coagulation laboratories as either a direct or indirect antiplatelet antibody assay. More recently, a number of enzyme-linked immunosorbent (**ELISA**) assays are available to test for specific antiplatelet antibodies in serum or plasma. The platelet antigen of interest, such as an HLA antigen or glycoprotein Ib/IX, is bound to the surface of a microtiter plate. The patient sample is added and if antibody is present, it will bind to the antigen. In **antigen capture immunoassays**, monoclonal antibodies directed against platelet antigens are used to individually capture various known platelet antigens onto a solid phase. Patient serum is added. If the corresponding antibody is present in the patient serum, it will bind. For example, if an antibody in the patient serum binds in the assay containing the Pl^{A1} antigen, then the patient is found to have an anti-Pl^{A1} antibody. **Flow cytometry** is also used in some laboratories to detect platelet-associated antibodies.

NAIT: The diagnosis of NAIT often involves typing (identifying) platelet antigens in the mother and father (and newborn), to demonstrate that the mother lacks a platelet antigen that is present on the platelets of the father (and newborn). It can also be demonstrated that there is an antibody in the mother's serum that is directed against a platelet antigen in the father (and newborn). Testing the newborn directly is typically not necessary, if the father can be tested.

PTP: The diagnosis of PTP often involves typing platelet antigens in the patient, and demonstrating that a platelet antibody in the patient's serum is directed against an antigen that is absent on the patient's platelets. Methods for (Continued)

Platelet Antibodies *(Continued)*

detecting platelet antibodies in serum have been described above. Some of the current methods used for typing platelet antigens are described below.

Platelet antigen typing by antigen-capture immunoassays: Monoclonal antibodies are used to immobilize the patient's platelet antigens onto a solid phase. Various antibodies of known antigen specificity are added. If an antibody binds, the patient's platelets have that particular antigen. For example, if an anti-Pl^{A1} antibody binds to the patient's platelet antigens in this assay, then the patient is found to carry the Pl^{A1} antigen. If the Pl^{A1} antibody does not bind, then the patient's platelets lack the Pl^{A1} antigen. Alternatively, polymerase chain reaction (**PCR**) assays can be used to identify the patient's platelet antigens. The platelet-specific antigens that cause platelet antibody formation are polymorphisms of platelet glycoproteins. Many of the alterations in DNA sequence that account for these polymorphisms are known and can be identified by PCR.

Drug-induced thrombocytopenia: The serotonin release assay, flow cytometry or other methods can be used to diagnose drug-induced thrombocytopenia. These tests are not routinely available. In serotonin release assays, patient plasma (or serum) and the suspected drug are added to normal platelets that contain radiolabeled serotonin. If antibodies against the drug are present, they stimulate the platelets to release their serotonin. The released radiolabeled serotonin can then be detected.

A **lymphocytotoxicity assay** (percent reactive antibody, PRA) can be used to detect HLA antibodies in patients who are refractory to platelet transfusions.

Additional Information

ITP: ITP is an isolated thrombocytopenia due to an autoantibody against platelets. The platelet antibodies are most commonly directed against components of platelet glycoprotein IIb/IIIa or to a lesser extent glycoprotein Ib/IX. In children, it is most often an acute disorder that resolves spontaneously. In adults, it is most often a chronic condition. Typically, the only abnormality on a peripheral blood smear is thrombocytopenia with normal to large platelets. Because ITP is a diagnosis of exclusion, laboratory tests recommended by a consensus panel to exclude other disorders include a peripheral blood smear, complete blood count (CBC), HIV testing in individuals with HIV risk factors, thyroid function tests in adults considering splenectomy, liver function tests in pregnant women to exclude HELLP syndrome, and bone marrow biopsy in persons older than age 60, adults considering splenectomy, or chronic cases in children that do not respond to IVIg.

Neonatal alloimmune thrombocytopenia (NAIT) occurs when fetal platelets have an antigen from the father that is absent in the mother, and the mother forms antibodies that cross the placenta and destroy fetal platelets. Newborn platelet counts are often <100,000/μL at birth, returning to normal within 2 weeks after birth. The antigens are usually components of platelet glycoprotein IIb/IIIa, most commonly, an antigen called Pl^{A1}. The incidence of NAIT is approximately one case per 1000-5000 live births. See Platelet Transfusion *on page 1058*.

Post-transfusion purpura (PTP) is a rare condition that occurs when a patient is transfused with platelets that express an antigen that is absent in the patient. The patient forms antibodies against the donor platelets. For unclear reasons in PTP, these antibodies also destroy the patient's own platelets, even though they lack the offending antigen. As with NAIT, the antigens are usually components of platelet glycoprotein IIb/IIIa, most commonly Pl^{A1}. PTP is characterized by the sudden onset of thrombocytopenia 5-12 days after transfusion of a platelet-containing product. The thrombocytopenia is typically severe (<10,000/μL), and it usually begins to resolve within 14 days after the transfusion.

Drug-induced immune thrombocytopenia: A vast number of drugs have been implicated in drug-induced thrombocytopenia, but a cause-effect relationship has not been proven for most drugs. Some of the drugs that cause immune thrombocytopenia include quinidine, quinine, sulfonamides, sulfonylureas, gold salts, and salicylates. Some drugs cause thrombocytopenia through nonimmune mechanisms, including marrow suppression (eg, ethanol, thiazide, procarbazine) or nonimmune destruction (eg, ristocetin, bleomycin, protamine).

In the nonimmune cases, there is no antibody and therefore no need for platelet antibody tests. With immune drug-induced thrombocytopenia, platelet counts are often severely decreased (<10,000/μL). Platelet counts typically return to normal within 7 days after discontinuing the drug.

Platelet refractoriness is a condition that occurs in thrombocytopenic patients who have received multiple platelet transfusions. The transfusions expose the patient to a variety of foreign HLA and other platelet antigens, against which the patient forms antibodies. These antibodies destroy subsequently transfused platelets, and, the patient is said to be refractory to platelet transfusion. Platelet refractoriness is most often due to antibodies against HLA-A or HLA-B antigens; less common causes include antibodies against ABO blood group antigens or platelet glycoproteins.

References
George JN, Woolf SH, Raskob GE, et al, "Idiopathic Thrombocytopenic Purpura: A Practice Guideline Developed by Explicit Methods for the American Society of Hematology," *Blood*, 1996, 88(1):3-40.

Warner M and Kelton JG, "Laboratory Investigation of Immune Thrombocytopenia," *J Clin Pathol*, 1997, 50(1):5-12.

Platelet Antibody, Immunohematologic

Related Information
Platelet Antibodies *on page 1047*
Platelet Count *on page 1050*
Platelet Transfusion *on page 1058*
Transfusion Reaction Work-up *on page 1269*

Synonyms Antiplatelet Antibody; Platelet-Associated IgG; Platelet-Bound IgG; Platelet Serology

Abstract Like red cells, antigens on platelets may be the targets of antibody assault. The consequences of antiplatelet antibody action include refractoriness to platelet transfusion, neonatal alloimmune thrombocytopenia, autoimmune thrombocytopenic purpura, post-transfusion purpura, and drug-induced immune thrombocytopenia. Antibodies may be directed to antigens shared with red cells (eg, ABO), to antigens shared with white cells (eg, HLA), or to platelet glycoprotein antigens not found on other blood cells (eg, HPA).

Specimen Blood

Collection Methods for the collection of blood samples for platelet serology are very test specific - consult with the laboratory.

Storage Instructions Storage of blood for platelet serology is very test specific - consult with the laboratory.

Use Platelet serology is indicated in the investigation of thrombocytopenia and refractoriness to platelet transfusion. Platelet crossmatching may be indicated for refractory patients. Solid-phase antibody detection systems have proven to be the most practical for crossmatching. Many blood transfusion centers utilize both crossmatch and HLA-compatible platelets for transfusion to refractory patients. (See Platelet Transfusion *on page 1058*.)

Limitations Platelet serology is an evolving discipline. A particular obstacle to the development of assays based on the detection of platelet-bound antibody is the inherent binding of inert IgG to the platelet through membrane Fc receptors. Methods that measure antibody binding to isolated platelet glycoproteins (radioimmunoprecipitation, immunoblotting, and antigen capture by monoclonal antibodies) represent attempts to overcome this.

The collection of sufficient platelets for testing from thrombocytopenic patients may be problematic.

Methodology Current test systems are based on the detection of platelet-bound IgG. Test systems that utilize fresh whole platelets can be employed to detect platelet-associated IgG. This is equivalent to the direct antiglobulin test on red cells. The tests used for platelet crossmatching are ELISA, platelet immunofluorescence test, and solid-phase assays.

Platelet serology is in the development stage and new test modalities can be anticipated.

Additional Information See Platelet Antibodies *on page 1047*.

Platelets bear only class I, and not class II, HLA proteins. Simultaneously transfused WBCs in platelet and/or red cell fractions of fetal blood entering the
(Continued)

Platelet Antibody, Immunohematologic *(Continued)*

maternal circulation in pregnancy cause primary immunization. Alloimmunization is almost eliminated by use of leukocyte-depleted fractions among recipients without prior WBC exposure.

References

Brecher ME, *Technical Manual*, 14th ed, Bethesda, MD: American Association of Blood Banks Press, 2002, 341-59.

Kruskall MS, "The Perils of Platelet Transfusions," *N Engl J Med*, 1997, 337(26):1914-5.

The Trial to Reduce Alloimmunization to Platelets Study Group (TRAP Trial Study Group), "Leukocyte Reduction and Ultraviolet B Irradiation of Platelets to Prevent Alloimmunization and Refractoriness to Platelet Transfusions," *N Engl J Med*, 1997, 337(26):1861-9.

♦ **Platelet-Associated IgG** *see* Platelet Antibody, Immunohematologic *on page 1049*

♦ **Platelet Autoaggregation** *see* Platelet Hyperaggregation *on page 1053*

♦ **Platelet-Bound IgG** *see* Platelet Antibody, Immunohematologic *on page 1049*

Platelet Count

Related Information

Activated Partial Thromboplastin Time *on page 100*
Antiphospholipid Antibody (Lupus Anticoagulant and/or Anticardiolipin Antibody) *on page 193*
Blood, Urine *on page 281*
Complete Blood Count *on page 442*
Fanconi Anemia, Chromosome Breakage Study *on page 569*
Peripheral Blood: Differential Leukocyte Count *on page 1010*
Platelet Aggregation *on page 1045*
Platelet Antibodies *on page 1047*
Platelet Antibody, Immunohematologic *on page 1049*
Platelets, Apheresis, Donation *on page 1054*
Platelet Sizing *on page 1056*
Platelet Transfusion *on page 1058*
Quinidine, Serum *on page 1129*
Reticulated Platelet Count *on page 1155*
Thrombopoietin, Serum or Plasma *on page 1247*

Abstract Enumeration of platelets in the circulating peripheral blood. Platelet count is important to assess bleeding, thrombotic, and malignant neoplastic processes, in evaluation of marrow function, and in study of effects of some diseases involving autoimmune mechanisms.

Specimen Whole blood

Container Lavender top (EDTA) tube

Causes for Rejection Clotted specimen, platelet clumping

Reference Interval 150,000-450,000/mm^3 (150-450 x 10^9/L or 150,000-450,000/µL). Platelet count in healthy term infants and in preterm infants weighing <1500 g is comparable to that in adults with use of either venous or capillary blood and by either phase microscopy or automated impedance counting methods. The count gradually rises during the first few months of life.

Count is method dependent; results of manual count have high coefficient of variation as compared to automated methods. Occasional, apparently normal children, in particular those under 24 months of age, may have platelet counts in the 500,000-750,000 range.

Possible Panic Range <50,000/mm^3 or >1,000,000/mm^3

Use Evaluate, diagnose, and follow up bleeding disorders, purpura/petechiae, drug-induced thrombocytopenia, idiopathic thrombocytopenic purpura, disseminated intravascular coagulation, leukemia, hypercoagulable states, and chemotherapeutic management of malignant disease

Limitations Clumping may cause false low count. Platelet satellitism around neutrophils may cause pseudothrombocytopenia. An IgG autoantibody is apparently involved with specificity against the platelet membrane glycoprotein IIb/IIIa complex and also against the neutrophil Fcγ receptor III (CD16). Formation of circulating platelet-neutrophil complexes, however, may have more ominous implications and may also be associated with a decline in platelet count (see Additional Information). Pseudoincrease in automated platelet counts may occur due to hypertriglyceridemia. The increase is in the range of

2-40 x 10^9/L, a relatively small change if the platelet count is normal but of potential significance with low platelet numbers. The combination of thrombocytopenia and hypertriglyceridemia (as may occur with L-asparaginase treated childhood acute leukemia) may necessitate manual platelet count. RBC (eg, microspherocytes) or WBC fragments including fragmented fragile leukemic cells and neutrophil pseudoplatelets may cause falsely elevated counts. EDTA-induced platelet clumping is a frequent cause of spuriously low platelet counts resulting from various platelet antigen/antibody reactions including antiplatelet with antiphospholipid antibodies and occurring also *in vitro* (eg, IgM autoantibody against 78 kD platelet glycoprotein). A small whole blood clot or very small fibrin clots in the EDTA anticoagulated specimen will usually be associated with clumping of platelets on the slide and with a false low platelet count.

Methodology A variety of automated/semiautomated devices are in use. Counts are performed on platelet-rich plasma or whole blood by optical or impedance matching counting techniques. Carefully controlled phase microscopy manual count has been considered the reference method but has a wide coefficient of variation (CV) of 7% to 17%. When the platelet count is low, CV may be 30%. Automated platelet counts using immunologic markers with detection of fluorescence by flow cytometry have recently been developed (see Additional Information).

Additional Information The platelet, of growing practical clinical importance in hemostatic considerations and a variety of medical/surgical processes, is also fundamental to etiologic considerations of arteriosclerotic and malignant disease. Platelets are generally 2-3 microns in diameter but large forms (megathrombocytes) appear when production is increased. The production of platelets is controlled by thrombopoietin (see Thrombopoietin, Serum or Plasma *on page 1247*). Platelets survive for 8-10 days and are subject to circadian periodicity, highest platelet counts occurring during midday. Some drugs may increase the platelet count by stimulating thrombopoietin production. Deaths from cardiovascular disease may relate temporally to the circadian rhythm of platelet production.

Quantitative platelet disorders have varied etiology. Thrombocytopenia may have an immunologic basis, may be the result of production deficiency due to the effect of drugs or physical agents, abnormal platelet pooling or increased destruction (eg, sequestration by large vascular tumor), or result from a variety of probably nonimmunologic mechanisms (eg, hypersplenism). Decreases may occur after severe bleeding, transfusion, infections (thrombocytopenia may be the presenting finding of typhoid fever), or relating to defective production of or regulation by thrombopoietin. Serum lactate dehydrogenase and platelet count may predict survival in thrombotic thrombocytopenic purpura.

Drugs and chemicals associated with thrombocytopenia, often on an immune mediated basis or as the result of marrow suppression, include quinidine, quinine, heparin, gold salts, sulfas, rifampicin, ASA (which acts by acetylating cyclo-oxygenase), digitoxin, apronal, chlorothiazides, chlorpropamide, meprobamate, antihistamines, chloramphenicol, penicillin, DDT, benzol, a variety of other industrial organic chemicals, diphenylhydantoin, PAS, hydrochlorothiazide, phenylbutazone, and a variety of antineoplastic chemotherapeutic agents. Abciximab (a platelet receptor antagonist) may cause severe thrombocytopenia 2-4 hours after administration. A recent review concluded that most reports of drug-induced thrombocytopenia do not present evidence that the drug has a **definite** or probable role as a cause of thrombocytopenia.

Thrombocytosis is less common, but likewise varied in etiology: physiologic (eg, postpartum or after exercise); myeloproliferative syndromes (eg, thrombocythemia, some cases of chronic myelogenous leukemia, myelofibrosis with myeloid metaplasia); rebound following thrombocytopenia, marrow regenerative activity after bleeding episode, hemophilia, iron deficiency; asplenism, infections, inflammatory or malignant disease, in particular, carcinomatosis but also in early (stage IB) cervical carcinoma.

Thrombocytosis may be primary or secondary. Nearly half of primary cases are due to essential thrombocythemia. Other cases of primary thrombocytosis are the result of myeloproliferative disorders such as chronic myeloid leukemia, polycythemia vera, and osteomyelofibrosis. Secondary thrombocytosis relates to tissue damage, infection, malignancy, and chronic inflammation. Cases of (Continued)

Platelet Count (Continued)

primary thrombocytosis are associated with higher platelet counts and both arterial and venous thromboembolic complications. In secondary thrombocytosis, thromboembolic complications involve only the venous system and occur only in the presence of other risk factors. Artifactual serum hyperkalemia and hypercalcemia may occur with essential thrombocythemia, apparently the result of in vitro secretion of calcium from the large number of abnormally activated platelets.

Oral contraceptives may cause slight increase in platelet count. Slight to moderate decrease in platelet count has been noted during pregnancy in most women who have essential thrombocythemia.

Congenital causes of thrombocytopenia include Wiskott-Aldrich syndrome, May-Hegglin anomaly, thrombocytopenia with absent radius, Bernard-Soulier syndrome, and Paris-Trousseau syndrome. Thrombocytopenia, while variable, usually accompanies the inherited giant platelet disorders.

See Platelet Sizing on page 1056 for discussion of changes in platelet count and size in pre-eclampsia and the HELLP syndrome (hemolysis, elevated liver tests, and low platelet count).

Automated platelet counting methods in current use may not be able to discriminate true platelets from nonplatelet particles as may be generated by a variety of pathologic processes including the tumor lysis syndrome. Two-dimensional optical platelet analysis and immunoplatelet counting have recently become available. In the former, platelets are "sphered" and volume (platelet size) and refractive index (platelet density) are simultaneously determined, cell by cell using a flow system with determination of two angles of light scatter. Optical two-dimensional platelet analysis shows a high degree of correlation with the phase microscopy current reference method.

The Advia 120 Hematology System (Bayer®, Tarrytown, NY), in addition to determining the platelet count, mean platelet volume, platelet-crit, and platelet distribution width also provides the additional following parameters:
- MPC: mean platelet component concentration
- PCDW: platelet component distribution width
- MPM: mean platelet mass
- PMDW: platelet mass distribution width

The clinical significance of these new measurements must be determined and is complicated by the status of platelet activation. Studies using measurement of platelet CD62P expression by flow cytometry (to establish the absence of sample activation) at time intervals of 20 minutes and 3.5 hours (to simulate "routine" and "urgent" sample test requests) confirm that some platelet characteristics change with sample age (in vitro, EDTA anticoagulated). As the MPV increases, MPC decreases, stabilizing at 15-20 minutes. This finding is consistent with previous observations that platelets swell and lose their discoid shape (becoming more spherical) during the first 2 hours of EDTA anticoagulation. Platelet parameters may not be effected similarly by all anticoagulants. Platelet discoid shape tends to persist in sodium citrate resulting in a lower MPC and wider PCDW.

ImmunoPlt (CD61) assay is based on an antiglycoprotein IIIa monoclonal antibody, a flow cytometric assay that has been automated and implemented on the Cell-Dyn 4000 analyzer. The method can provide a rapid result (<5 minutes from closed-tube aspiration to report). For low platelet counts, the CD61-based assay is more accurate and precise than optical scatter or impedance count methods. The RBC:platelet ratio method for platelet counting is now recommended by the International Council for Standardization in Hematology and the International Society of Laboratory Hematology (Task Force on Platelet Counting) as a reference method for the enumeration of platelets.

References

Cines DB and Blanchette VS, "Immune Thrombocytopenic Purpura," N Engl J Med, 2002, 346(13):995-1008.

Geaghan SM, "Hematologic Values and Appearances in the Healthy Fetus, Neonate, and Child," Clin Lab Med, 1999, 19(1):1-37.

George JN, Raskob GE, Shah SR, et al, "Drug-Induced Thrombocytopenia: A Systematic Review of Published Case Reports," Ann Intern Med, 1998, 129(11):886-90.

Greisshammer M, Bangerter M, Sauer T, et al, "Aetiology and Clinical Significance of Thrombocytosis: Analysis of 732 Patients With an Elevated Platelet Count," *J Intern Med*, 1999, 245(3):295-300.

Griffiths C and Fisher M, "Abciximab-Induced Thrombocytopenia," *Q J Med*, 2002, 95(9):635-8.

International Council for Standardization in Haematology Expert Panel on Cytometry and International Society of Laboratory Hematology Task Force on Platelet Counting, "Platelet Counting by the RBC/Platelet Ratio Method. A Reference Method," *Am J Clin Pathol*, 2001, 115(3):460-4.

Kabutomori O, Iwatani Y, and Kabutomori M, "Effects of Hypertriglyceridemia on Platelet Counts in Automated Hematologic Analysis," *Ann Intern Med*, 1999, 130(5):452.

Li S and Salhany KE, "Spurious Elevation of Automated Platelet Counts in Secondary Acute Monocytic Leukemia Associated With Tumor Lysis Syndrome," *Arch Pathol Lab Med*, 1999, 123(11):1111-4.

Mhawech P and Saleem A, "Inherited Giant Platelet Disorders. Classification and Literature Review," *Am J Clin Pathol*, 2000, 113(2):176-90.

Serefhanoglu K, Kaya E, Sevinc A, et al, "Isolated Thrombocytopenia: The Presenting Finding of Typhoid Fever," *Clin Lab Haematol*, 2003, 25(1):63-5.

Shehata N, Burrows R, and Kelton JG, "Gestational Thrombocytopenia," *Clin Obstet Gynecol*, 1999, 42(2):327-34.

♦ **Platelet Factor 4** see Heparin Antifactor Xa Assay *on page 693*

♦ **Platelet Factor 4** see Heparin-Induced Thrombocytopenia *on page 695*

♦ **Platelet Factor 4** see Hypercoagulation Panel *on page 758*

♦ **Platelet Function Studies** see Platelet Aggregation *on page 1045*

Platelet Histogram Maximum

Related Information

Platelet Count *on page 1050*
Platelet Sizing *on page 1056*
Platelet Transfusion *on page 1058*
Reticulated Platelet Count *on page 1155*
Reticulocyte Count *on page 1156*

Applies to Megathrombocytes

Abstract The highest peak of the platelet volume distribution curve applied to the study of thrombocytopenia, is less effected by particulate artifact than the mean platelet volume (MPV). The test is of value in discriminating between decreased platelet production and increased platelet destruction and thus in establishing the presence of idiopathic thrombocytopenic purpura (ITP).

Specimen Whole blood

Container Lavender top (EDTA) tube

Turnaround Time 1-2 hours

Reference Interval Maxima ±SD: ITP: 7.9 ±0.93; hypoproduction: 5.12 ±0.71

Use Differential diagnosis of thrombocytopenia, in particular, differentiation of decreased platelet production from increased platelet destruction

Limitations Appropriate instrumentation for producing a platelet histogram must be available.

Methodology Platelet volume distribution measured by automated platelet counting equipment

Additional Information An increased percentage of large platelets in the peripheral blood reflects increased numbers of marrow megakaryocytes and a state of platelet hyperproduction, as occurs with idiopathic thrombocytopenic purpura. Few or absent megathrombocytes (platelets larger than 3.0 microns across) are seen with hypoproductive thrombocytopenia (eg, aplastic anemia, chemotherapy). Sensitivity/specificity studies using receiver operating characteristic curves show that the maximum of the platelet histogram is superior to the MPV in the evaluation of thrombocytopenia. Use of the "maximum" avoids the effect of artifacts (debris and/or cell fragments from red cells or leukemic blasts) that artificially change the platelet count and MPV (especially in cases of very low platelet count).

References

Niethammer AG and Forman EN, "Use of the Platelet Histogram Maximum in Evaluating Thrombocytopenia," *Am J Hematol*, 1999, 60(1):19-23.

Platelet Hyperaggregation

Related Information

Antithrombin *on page 198*
Hypercoagulation Panel *on page 758*
Platelet Aggregation *on page 1045*
(Continued)

Platelet Hyperaggregation *(Continued)*

Synonyms Hypercoagulable State, Platelet Aggregation; Platelet Aggregation, Hypercoagulable State; Platelet Autoaggregation

Test Includes Evaluation for spontaneous aggregation, and aggregation in response to low concentrations of adenosine diphosphate (ADP), epinephrine, arachidonate, and collagen

Abstract Platelet hyperaggregation in response to platelet agonists, and/or spontaneous platelet aggregation (aggregation without a platelet agonist) has been described in association with hypercoagulability, including strokes, myocardial infarction, and less commonly venous thrombosis.

Patient Preparation Patients should not have aspirin (or any medication containing aspirin) for at least 7 days prior to testing. Nonsteroidal anti-inflammatory drugs or other platelet-inhibiting agents should also be avoided.

Specimen Platelet-rich plasma

Container Four to six blue top or plastic (sodium citrate) tubes

Collection Routine venipuncture. Immediately invert tubes gently at least 4 times to mix. Deliver tubes immediately to the laboratory at room temperature (platelets are activated at cold temperatures).

Storage Instructions Keep specimen at room temperature and perform test within 2 hours of collection. Do not refrigerate or freeze specimen.

Causes for Rejection Specimen received more than 2 hours after collection, specimen clotted, specimen received on ice

Turnaround Time Less than 1 day

Special Instructions Test usually must be scheduled in advance with the laboratory.

Reference Interval No spontaneous platelet aggregation and no hyperaggregation compared to a normal control. Normal newborns can have decreased aggregation compared to adults.

Use Evaluation for excessive platelet aggregation may be useful in patients with evidence of unexplained hypercoagulability and normal values in the routine hypercoagulation test panel.

Limitations Subjective. Results are compared to a normal control. Variable results among patients and controls. Platelet-inhibiting medications interfere. Labor-intensive for the laboratory, therefore, not suitable for high volume clinical testing.

Methodology As described for platelet aggregation, except each platelet agonist is tested at multiple lower-than-usual concentrations. For example, instead of testing epinephrine only at 10 µM, it is tested at 10 µM, 5 µM, 1 µM, and 0.5 µM. No agonist is added to one aliquot to assess for spontaneous aggregation.

Additional Information Various medications can cause increased *in vitro* platelet aggregation. If a patient is found to have increased platelet aggregation in this assay, a careful review of prescribed, as well as over-the-counter medications, is indicated. An on-line literature search for each medication is often informative. Hyperaggregation has also been reported with myeloproliferative disorders. Hereditary hyperaggregation as a cause of hypercoagulability is not well characterized.

References

Landolfi R, Marchioli R, and Patrono C, "Mechanisms of Bleeding and Thrombosis in Myeloproliferative Disorders," *Thromb Haemost*, 1997, 78(1):617-21.

Mammen EF, "Sticky Platelet Syndrome," *Semin Thromb Hemost*, 1999, 25(4):361-5.

♦ **Platelet Indices** *see* Platelet Sizing *on page 1056*

♦ **Plateletpheresis** *see* Platelets, Apheresis, Donation *on page 1054*

♦ **Platelet-Rich Plasma** *see* Platelet Transfusion *on page 1058*

♦ **Platelets** *see* CD40 Ligand (Soluble), Serum or Plasma *on page 353*

♦ **Platelets** *see* Platelet Transfusion *on page 1058*

Platelets, Apheresis, Donation

Related Information

Donation, Blood *on page 521*

HLA Typing, Single Human Leukocyte Antigen *on page 739*

Neutrophils, Apheresis, Donation *on page 950*

Synonyms Plateletpheresis; Platelets, Pheresis (FDA); Single-Donor Platelets

Applies to Platelets, Single-Donor

Test Includes As for regular whole blood donation, but donations may take place more frequently. The donor's platelet count should be >150,000/μL if the donation interval is less than 4 weeks. In addition, the AABB requires Blood Banks or Transfusion Services to limit and detect bacterial contamination in all platelet components, beginning March 1, 2004.

Abstract Selective removal of the donor's platelets with return of all other components using automated apheresis instruments. Donors include random volunteer donors, recipient's family members, and HLA-matched donors.

Patient Preparation Obtain and document informed consent. Emergency medical care must be available.

Aftercare Monitor total red cell losses during procedures. If red cells cannot be returned during a procedure, defer donor for 8 weeks.

Storage Instructions Store in the Blood Bank at 20°C to 24°C with continuous agitation for 5 days. Do not refrigerate.

Causes for Rejection (Of donor): As for regular blood donation. In addition, donors taking aspirin-containing medication within 36 hours of donation are usually deferred.

(Of units): Do not transfuse units containing excessive platelet aggregates. "Swirling" (observed while holding bag against a light source and gently squeezing) correlates well with adequate platelet *in vivo* viability.

Special Instructions Single-donor platelets (platelets, pheresis) are available from most blood centers and larger hospital blood banks on a more or less regular basis. In some institutions they are reserved for patients who regularly receive platelet transfusions, or who may be refractory to regular platelet transfusions.

Reactions Encountered in Apheresis Donors

Adverse Reactions	Prevention/Initial Treatment
General	
Vasovagal	Donor reassurance; Trendelenburg position; temporarily stop procedure
Hypovolemia	Avoid excessive extracorporeal volume; Trendelenburg position; saline administration
Hyperventilation	Donor reassurance; breathe into paper bag
Citrate	Decrease whole blood flow rate and/or citrate infusion; temporarily stop procedure
Hemolysis	Proper instrument set-up; monitor plasma color
Chills	Keep donor warm (use light blanket), use blood warmer
Air embolism	Proper instrument set-up; left-side in Trendelenburg position
Specific to Granulocyte Donation	
Headache, edema, hypervolemia	Avoid too frequent use of sedimenting agent
Hydroxyethylstarch accumulation	Consider use of Pentastarch as sedimenting agent
Steroid effect	Avoid excessive dosage
Growth factor effect	Usually relieved by acetaminophen

Adapted from Randels MJ, "Selection of Care of Apheresis Donors," *Apheresis: Principles and Practice*, 2nd ed, McLeod RC, Price TH, and Weinstein R, eds, Bethesda, MD: American Association of Blood Banks Press, 2003.

Use Single-donor platelets are suitable for any patient needing platelet transfusions. Single-donor platelets have the advantage of providing an entire platelet transfusion for an adult from one donor, hence with a single antigenic combination and a single donor exposure, decreasing the risk of alloimmunization or to

(Continued)

Platelets, Apheresis, Donation *(Continued)*

infectious agents. Furthermore, the donor can be selected (eg, from a patient's relatives or by HLA type), providing easier selection of CMV-negative units when indicated.

Limitations A minimum of 48 hours should elapse between successive donation procedures and a maximum of 24 donations should take place in 1 year. Platelets, pheresis are more expensive than platelet concentrates, but were comparable in the TRAP study.

Methodology Exact procedure depends on apheresis instrument used. Use of a leukocyte reduction filter frequently is not necessary as the platelets produced by the instruments are leukocyte-reduced products. Platelet apheresis donation differs from leukocyte donation in that a sedimenting agent is not used and there is no need to stimulate the donor with steroids.

Additional Information One apheresis platelet unit may contain the equivalent in platelet dosage to four or five or more random donor platelet concentrates, containing at least 3.0×10^{11} platelets in ~300 mL of plasma. See Platelet Transfusion *on page 1058* for transfusion information.

References

Kruskall MS, "The Perils of Platelet Transfusions," *N Engl J Med*, 1997, 337(26):1914-5.

Silberman S, "Platelets: Preparations, Transfusion, Modifications, and Substitutes," *Arch Pathol Lab Med*, 1999, 123(10):889-94.

The Trial to Reduce Alloimmunization to Platelets Study Group (TRAP Trial Study Group), "Leukocyte Reduction and Ultraviolet B Irradiation of Platelets to Prevent Alloimmunization and Refractoriness to Platelet Transfusions," *N Engl J Med*, 1997, 337(26):1861-9.

♦ **Platelet Serology** *see* Platelet Antibody, Immunohematologic *on page 1049*

Platelet Sizing

Related Information

Complete Blood Count *on page 442*
Peripheral Blood: Differential Leukocyte Count *on page 1010*
Peripheral Blood: Red Blood Cell Morphology *on page 1016*
Platelet Count *on page 1050*
Platelet Transfusion *on page 1058*
Reticulated Platelet Count *on page 1155*
Thrombopoietin, Serum or Plasma *on page 1247*

Synonyms MPV; PDW; Platelet Indices

Applies to May-Hegglin Anomaly

Test Includes MPV (mean platelet volume), platelet count, PDW (platelet distribution width)

Abstract Modern automated cell counters may generate platelet sizing results (eg, MPV and PDW) which may be abnormal in some clinical situations. Platelet indices are analogous to red blood cell indices (eg, MCV and RDW) but have only modest clinical application. MPV and PDW are increased in patients with idiopathic thrombocytopenic purpura (ITP).

Specimen Whole blood

Container Lavender top (EDTA) tube; green top (heparin) tube may cause platelet clumping

Storage Instructions In EDTA anticoagulated whole blood, platelets undergo a change in shape from discoid to spherical (discocyte-to-echinocyte transformation) with increase in apparent volume such that the MPV increases about 20% with time in the first 1-2 hours (*in vitro*), is then stable for 1-3 hours, and then undergoes further increase with time. Thus, MPV should be determined between 1 and 3 hours after the sample is obtained.

Use Differential diagnosis of hematologic disease; assess platelet function; evaluation of thrombocytopenia; guide the need for platelet transfusion in thrombocytopenic patients

Limitations May be unreliable if platelet count is <10,000/mm³. MPV may be unstable in EDTA. With impedance counting methods, increase in platelet volume may occur with increasing time from venous sampling while MPV may decrease with storage when a light scattering method is used.

Reference Interval See tables for mean platelet volume and platelet "crit" reference values. Overall, MPV reference value for adults is 6.5 to 12 fL. Platelet size normally varies inversely with platelet count.

Platelet Parameters in Females (mean ±SD)[1]

Age (y)	n	Platelet Count (x 10³/mm³) (x 10⁹/L)	MPV (fL)	PCT (%)
1-5	25	381 ±76	8.9 ±0.8	0.337 ±0.069
6-10	18	336 ±76	9.7 ±1.1	0.326 ±0.080
11-15	31	298 ±72	9.8 ±1.2	0.288 ±0.058
16-20	22	270 ±58	9.7 ±0.7	0.262 ±0.058
21-30	43	270 ±58	9.8 ±1.0	0.261 ±0.046
31-40	30	282 ±56	9.8 ±1.2	0.271 ±0.046
41-50	26	279 ±65	9.8 ±0.9	0.274 ±0.072
51-60	21	285 ±54	9.7 ±0.7	0.276 ±0.045
61-70	30	274 ±61	9.6 ±0.9	0.262 ±0.052
71-83	24	279 ±65	9.5 ±1.0	0.261 ±0.054

[1]SI conversion units for platelet count x 10³/mm³ is platelet count x 10⁹/L.
Adapted from Graham SS, Traub B, and Minic IB, "Automated Platelet-Sizing Parameters on a Normal Population," *Am J Clin Pathol*, 1987, 87:365-9.

Platelet Parameters in Males (mean ±SD)[1]

Age (y)	n	Platelet Count (x 10³/mm³) (x 10⁹/L)	MPV (fL)	PCT (%)
1-5	24	357 ±70	8.6 ±0.7	0.304 ±0.059
6-10	24	351 ±85	8.6 ±0.8	0.300 ±0.058
11-15	16	282 ±63	9.8 ±1.0	0.274 ±0.053
16-20	16	266 ±63	10.2 ±1.1	0.266 ±0.049
21-30	24	238 ±49	9.6 ±0.6	0.277 ±0.045
31-40	12	244 ±56	9.8 ±1.2	0.237 ±0.044
41-50	17	271 ±66	9.4 ±1.0	0.250 ±0.045
51-60	22	258 ±61	9.8 ±1.2	0.248 ±0.045
61-70	29	256 ±53	9.4 ±1.1	0.238 ±0.047
71-86	23	237 ±49	9.6 ±1.0	0.226 ±0.048

[1]SI conversion units for platelet count x 10³/mm³ is platelet count x 10⁹/L.
Adapted from Graham SS, Traub B, and Minic IB, "Automated Platelet-Sizing Parameters on a Normal Population," *Am J Clin Pathol*, 1987, 87:365-9.

Methodology Flow cytometry allows measurement of platelet volume and size by changes in electrical impedance and microprocessor assisted mathematical analysis.

Additional Information Large platelets are generally young platelets and have better hemostatic function than average age or old platelets. MPV (in normal subjects) bears an inverse relation to platelet count. MPV rises with increased platelet turnover, due to production of megathrombocytes. MPV and PDW are increased with ITP. MPV tends to be low in reactive thrombocytosis. Use of increased platelet volume may assist in the differential diagnosis of ITP versus acute leukemia. Large platelets are present in the recovery stage of alcohol-induced thrombocytopenia. Small platelets are seen in the Wiskott-Aldrich syndrome. Autoimmune thrombocytopenia and leukemia may be associated with presence of platelet fragments and decreased platelet volume. Hypersplenetic patients have smaller platelet size. In the Bernard-Soulier syndrome, there is thrombocytopenia with large platelets. Large platelets, occasionally with abnormal morphology, occur in the myeloproliferative syndromes. Increased MPV may also be seen in hyperthyroidism but only with low platelet count. Generally, the platelet count is normal in hyperthyroidism, MPV is normal, but reticulated platelets are increased, suggestive of increase in thrombopoiesis. Mean platelet volume correlates with bleeding tendency in thrombocytopenic patients. A significantly lower frequency of
(Continued)

Platelet Sizing *(Continued)*

bleeding occurs with mean platelet volumes >6.4 fL. This measure, then, may be of use in assessing the need for platelet transfusion.

Platelet volume has been noted to decrease during cardiopulmonary bypass and to increase in atherosclerotic smokers. Increase in MPV due to smoking has been proposed as a risk factor for atherosclerotic disease.

References

Eldor A, Avitzour M, Or R, et al, "Prediction of Haemorrhagic Diathesis in Thrombocytopenia by Mean Platelet Volume," *Br Med J (Clin Res Ed)*, 1982, 285(6339):397-400.

Osselaer JC, Jamart J, and Scheiff JM, "Platelet Distribution Width for Differential Diagnosis of Thrombocytosis," *Clin Chem*, 1997, 43(6):1072-6.

Stiegler G, Stohlawetz P, Brugger S, et al, "Elevated Numbers of Reticulated Platelets in Hyperthyroidism: Direct Evidence for an Increase of Thrombopoiesis," *Br J Haematol*, 1998, 101(4):656-8.

♦ **Platelets, Pheresis (FDA)** *see* Platelets, Apheresis, Donation *on page 1054*
♦ **Platelets, Single-Donor** *see* Platelets, Apheresis, Donation *on page 1054*

Platelet Transfusion

Related Information

Bleeding Time *on page 270*
HLA Typing, Single Human Leukocyte Antigen *on page 739*
Irradiated Blood Components *on page 811*
Platelet Aggregation *on page 1045*
Platelet Antibodies *on page 1047*
Platelet Antibody, Immunohematologic *on page 1049*
Platelet Count *on page 1050*
Platelet Hyperaggregation *on page 1053*
Platelets, Apheresis, Donation *on page 1054*
Platelet Sizing *on page 1056*
Rh$_o$(D) Immune Globulin (Human) *on page 1164*
Risks of Transfusion *on page 1166*
Transfusion Reaction Work-up *on page 1269*

Synonyms Platelet-Rich Plasma; Platelets; Pooled Platelets; Random Platelets

Applies to Antiglobulin-Augmented Lymphotoxicity Assay; Solid Phase Red Cell Adherence Assay

Test Includes ABO and Rh type

Abstract Two major types of platelet fractions have been available: **random donor platelets** or **platelets** and **platelets, pheresis** or **single-donor platelets**. Other types include **filtered, pooled platelet concentrates from random donors**, and **filtered platelets from apheresis from single random donors**. Platelet concentrates (**platelets**) consist of platelets, suspended in about 50 mL of plasma, separated from whole blood collected from a single whole blood donation, known as **random donor platelets**. It contains at least 5.5×10^{10} platelets per unit. The platelets in a single bag of random donor platelets are insufficient to provide an adequate therapeutic dose in a thrombocytopenic adult. Storage life is 5 days. It contains stable coagulation factors, the presence of which may be significant, since a common dose is about 1 concentrate per 10 kg body weight. Thus, platelet concentrates from 4-10 donors are combined as **pooled concentrates**.

Platelets, pheresis contain at least 3×10^{11} platelets per unit, The dose is usually one unit per adult. The donor plasma should be ABO compatible with erythrocytes of the intended recipient, especially when the recipient is an infant. When platelets of a specific antigenic type are needed, apheresis is especially useful. Use of leukocyte-depleted fractions should diminish the incidence of platelet transfusion reactions.

Patient Preparation Follow standard transfusion patient identification procedures. It is not unusual for a platelet concentrate to have a pink tinge, indicating the presence of some RBCs. Despite this, a red cell crossmatch is not indicated. If the patient is likely to receive many platelet transfusions, it is wise to have the patient HLA-typed early, in case platelet refractoriness occurs.
Dosage and administration: In a bleeding adult with a platelet count <20,000/mm^3, 1 concentrate per 10 kg body weight or 1 unit of platelets, pheresis is a good dose. Use a 19-gauge needle or larger and administer through a routine administration filter designed for platelets. Give platelets rapidly, with an average of 10 minutes per platelet concentrate. Isotonic saline may be used to

flush the container and filter. Do not warm platelets. Do not add any medications to platelet packs.

Causes of Refractoriness to Platelet Transfusions

Nonimmune	Immune
Infection, sepsis, fever	ABO
Hemorrhage, purpura, DIC	Prior transfusions
Splenomegaly	Pregnancy
Antibiotic therapy	HLA alloantibodies
Amphotericin B	Platelet-specific alloantibodies (uncommon)
Bone marrow transplant	Autoimmunity

Aftercare Close clinical/nursing observation for bleeding, petechiae is indicated. **To evaluate efficacy of platelet transfusions, get platelet counts prior to and within 1 hour after the transfusions are completed**. They are useful to evaluate the response to platelet transfusions and to calculate the corrected count increment (CCI). The latter expresses the platelet increment per 10^{11} platelets transfused per meter of body surface area (BSA). Where post = post-transfusion platelet count x $10^3/mm^3$; pre = pretransfusion platelet count x $10^3/mm^3$; and PTx = number of platelets transfused (x 10^{11}).

$$CCI \times 10^3 = (post-pre) \times BSA/PTx$$

A value above 7.5 is usually considered satisfactory. Thus, a patient with a BSA of 1.5 m^2 receives 6 platelet concentrates with a total of 4.2 x 10^{11} platelets. The pre- and postcounts are 10 x 10^3 and 40 x 10^3 respectively.

$$CCI = (40 - 10) \times 1.5/4.2 = 10.7 \times 10^3$$

That would be considered a good response. This formula is not needed if the raw increment is zero or close to it, or if the response is obviously satisfactory. But it is helpful when the patient is small (eg, a child), or when the postcounts appear to show small or moderate increments. Poor response to platelet transfusions (refractoriness) is common, usually from nonimmune causes (see table). When refractoriness seems to be due to alloimmunization, then it will be necessary to demonstrate the presence (and specificity, if possible) of those antibodies by antiglobulin-augmented lymphocytotoxicity assay (LCT) or solid phase red cell adherence assay (SPRCA). If the percentage of reactive antibody (PRA) is <70%, then the patient can receive either crossmatch compatible units or units which lack the antigen to which the patient has antibody. Patients with PRA >70% should be placed on a transfusion schedule because they need to receive "A" and "BU" crossmatch compatible units from donors who are selectively recruited from an HLA-typed donor pool. Alloimmunization to platelets seems to be largely caused by contaminating leukocytes in the concentrates and may be prevented by the removal of leukocytes by special filtration (see Filters for Blood *on page 588*), or reduced experimentally by ultraviolet-B irradiation of platelets.

Selection of Platelets for Immune-Refractory Patients

1. Recruit blood relatives of patient as donors, if available.

2. HLA type patient, if possible.

3. Test patient for cytotoxic antibodies.

4. Test patient for platelet-specific antibodies.

5. Select crossmatch compatible platelet units (check postplatelet counts).

6. Select donors by HLA type. The HLA does not have to be identical; a reasonable match by cross-reacting groups is usually as effective (check postplatelet counts).

7. Select patient donors lacking antigens corresponding to any antibodies detected in steps 3 and 4, if feasible (check postplatelet counts).

8. Check whether ABO compatibility affects outcome. Sometimes it does, especially type A platelets to an O patient and especially if no antibody is detected in steps 5 and 6.

9. If PRA is >70% or CCIs poor: Crossmatch antigen-negative units and selectively recruit 'A' and/or 'BU' donors.

10. Recognize that this is a thorny problem. Any or all of the above may work, not necessarily in the order given; or none of them may be effective.

(Continued)

Platelet Transfusion *(Continued)*

Fever within 6 hours of platelet transfusion may signal bacterial contamination. The organisms which contaminate platelets are often the same as those involved in catheter-related sepsis.

Specimen Blood

Container One red top tube or one lavender top (EDTA) tube

Collection (Of sample from intended recipient): At the patient's bedside, ask the patient to give his or her name. Compare with the hospital wristband. Label the Blood Bank wristband (if there is one) with the patient's full name, hospital number, date, and initials of the collector. Label sample tube with the same information, including identification number from the wristband. Label requisition form with identification number. The collector signs the requisition, verifying the patient's identity with hospital wristband and Blood Bank wristband. Some hospitals require additional information. It is always best to stamp the requisition with the patient's identification plate to avoid transcription errors. Take extra care with identification of unresponsive patients.

Storage Instructions Platelets lose ability to aggregate at refrigerator temperature; such platelets become suboptimal for transfusion. Store in the Blood Bank at 20°C to 24°C (room temperature) with continuous agitation for 5 days (agitation facilitates gas exchange). Pooled platelets must be transfused within 4 hours after pooling.

Special Instructions Do not refrigerate platelets.

Use Platelets may be used therapeutically to stop bleeding or prophylactically to prevent it. Treatment of bleeding, petechiae, and ecchymoses when platelet count is <20,000/mm³ or when platelets are functionally abnormal; prophylaxis against bleeding due to thrombocytopenia when platelet count is 5000-10,000/mm³, associated with chemotherapy or other marrow hypoplasias; in splenectomy for ITP when abnormal bleeding occurs; in acute blood loss or prior to invasive surgical procedures with platelet count <50,000/mm³. Platelets, pheresis can be used for any patient needing platelet transfusions. They have the advantage of providing an entire platelet transfusion for an adult from one donor, hence, a single donor exposure. Furthermore, the donor can be selected (eg, from a patient's relatives or by HLA type). Platelets, pheresis are available from most blood centers and larger hospital blood banks on a more or less regular basis. Some institutions reserve them for patients who regularly receive platelet transfusions or who may be refractory to regular platelet transfusions.

Limitations Hazards: As for transfusion of other blood components. Platelets may contain both red and white blood cells, and immunization to any of these antigens may occur. If feasible, Rh-negative women of childbearing age should receive Rh-negative platelets to avoid Rh immunization. Otherwise, consider giving them Rh immune globulin. The dosage can be calculated by knowing that platelets seldom contain more than 0.5 mL RBC each and platelets, pheresis rarely more than 5 mL each. Although ABO antigens are poorly developed on platelets, ABO-incompatible platelets sometimes have decreased post-transfusion survival. ABO-incompatible platelet units have caused intravascular hemolysis by passive infusion of anti-A₁ or anti-B.

Immunosuppressed patients, and rarely immunocompetent patients, receiving platelets from closely related donors can suffer graft-versus-host disease; this is prevented by irradiation of platelets (see Irradiated Blood Components *on page 811*). Bacterial contamination and growth during storage are a real risk, as platelets are stored at room temperature. Such contamination may derive from skin organisms of the donor or from occult bacteremia. Platelet-containing fractions are most often implicated in transmission of bacteria and caused 21 of 29 deaths reported to the FDA between 1986 and 1991. Bacterial contamination is found in as many as 1 in 2000 units.

White blood cells present in platelet units can cause febrile reactions, HLA alloimmunization, adverse immunomodulatory effects, and transmission of CMV and human T-lymphotrophic virus 1.

Contraindications Not usually useful in idiopathic or immune thrombocytopenia, thrombotic thrombocytopenic purpura or certain stages of disseminated intravascular coagulation. Not to be used if bleeding is not caused by thrombocytopenia or abnormal platelet function. Patients with hypersplenism or septicemia may also fail to respond.

Methodology Random donor platelets are obtained by centrifugation of whole blood within 8 hours of collection, at low centrifugal force, at room temperature, to provide platelet-rich plasma with or without contaminating white blood cells. Separation of such plasma from erythrocytes in a satellite bag and recentrifugation for platelet concentration follow. After most of the plasma is separated for use as fresh frozen plasma, the residual 50 mL of plasma is used to resuspend the platelets. Platelets, pheresis are discussed in Platelets, Apheresis, Donation *on page 1054.*

Leukocyte reduction by filtration and ultraviolet B irradiation are equally effective in prevention of alloantibody-mediated platelet refractoriness, respectively to remove or inactivate cells bearing alloantigens. Ultraviolet B irradiation is not, however, available on a routine basis. See Platelet Antibodies *on page 1047.*

The AABB requires Blood Bank or Transfusion Services to limit and detect bacterial contamination in all platelet components beginning March 1, 2004. Methods which limit bacterial contamination include the following:
- Careful phlebotomy technique - use of an iodine-based scrub and avoiding areas of scarring or dimpling
- Phlebotomy diversion - diverting the initial 20-30 mL of blood collection
- Use of apheresis platelets

Methods which detect bacterial contamination include the following:
- Culture - detects one bacterial organism per mL (1 cfu/mL). Two methods are currently FDA approved, the BacT/ALERT (bioMerieux, Inc, Durham, NC) and Pall BDS (Medsep Corp, Covina, CA).
- Staining - detects 10^4 cfu/mL (for acridine orange) to 10^5 to 10^6 cfu/mL (for Gram stain and Wright stain).
- Dipstick - glucose and pH multireagent strips. Detects >10^7 cfu/mL; in some cases may detect 10^3 cfu/mL.

Additional Information One unit of platelets usually increases the platelet count of an adult with a blood volume of 5000 mL by about 5000/mm³. One unit of platelets, pheresis usually increases the platelet count of an adult by about 30,000-60,000/mm³.

Neonatal and fetal alloimmune thrombocytopenia are usually caused by PI[A1] (HPA-1a) antigen inherited from the father by the baby, who then immunizes the mother so that she makes antibody to the baby's platelets (ie, anti-PI[A1]). The mother's platelet count is normal, but the baby's platelet count is low. It may not, however, be sufficiently low to cause symptoms, and so the diagnosis may be missed. The first born child may be affected. The most serious complication is intracranial hemorrhage, which may occur during pregnancy, but the risk is probably greatest during delivery. If the diagnosis is made after delivery and the baby is unharmed, remember that later babies are also at risk.

For confirmatory diagnosis, send mother's serum and father's platelets to a reference laboratory for antibody identification. See Platelet Antibodies *on page 1047.*

Treatment: The most convenient and readily available source of compatible platelets is the mother. A large number of platelets, relative to the baby's size, can be obtained by maternal platelet apheresis. If maternal platelets are used, the antibody-containing plasma should be removed and the unit irradiated and leukocyte reduced.

Neonatal **autoimmune thrombocytopenia**, in which maternal thrombocytopenia also occurs, is treated by exchange transfusion or intravenous immune globulin.

Eight to 10 platelets may contain the equivalent amount of stable clotting factors found in two plasma units. This volume may need to be reduced for children.

Blumberg and Heal recently addressed the costs and benefits of apheresis versus random donor platelets.

(Continued)

Platelet Transfusion (Continued)

References

Blumberg N and Heal JM, "Mortality Risks, Costs, and Decision Making in Transfusion Medicine," *Am J Clin Pathol*, 2000, 114(6):934-7.

Bussel JB, Zabusky MR, Berkowitz RL, et al, "Fetal Alloimmune Thrombocytopenia," *N Engl J Med*, 1997, 337(1):22-6.

Dodd Ry and Lipton KS, "Guidance on Implementation of New Bacteria Reduction and Detection Standard," *AABB Association Bulletin 03-07*, May 16, 2003.

Goodnough LT, Brecher ME, Kanter MH, et al, "Transfusion Medicine. First of Two Parts: Blood Transfusion," *N Engl J Med*, 1999, 340(6):438-48.

Kickler TS and Herman JH, *Current Issues in Platelet Transfusion Therapy and Platelet Alloimmunity*, Bethesda, MD: American Association of Blood Banks Press, 1999.

McManigal S and Sims KL, "Intravascular Hemolysis Secondary to ABO Incompatible Platelet Products: An Underrecognized Transfusion Reaction," *Am J Clin Pathol*, 1999, 111(2):202-6.

Murphy S, "Platelet Transfusion Therapy," *Thrombosis and Hemorrhage*, 2nd ed, Loscalzo J and Schafer J, eds, Baltimore, MD: Lippincott Williams & Wilkins, 1998, 1119-34.

Murphy S and Varma M, "Selecting Platelets for Transfusion of the Alloimmunized Patient: A Review," *Immunohematology*, 1998, 14:117-23.

Rothenberger SS and McCarthy LJ, "Neonatal Alloimmune Thrombocytopenia: From Prediction to Prevention," *Lab Med*, 1997, 28(9):592-6.

Silberman S, "Platelets: Preparations, Transfusion, Modifications, and Substitutes," *Arch Pathol Lab Med*, 1999, 123(10):889-94.

Slichter S, "Leukocyte Reduction and Ultraviolet B Irradiation of Platelets to Prevent Alloimmunization and Refractoriness to Platelet Transfusions," The Trial to Reduce Alloimmunization to Platelets Study Group, *N Engl J Med*, 1997, 337(26):1861-9.

Snyder EL and Rinder HM, "Platelet Storage - Time to Come in From the Cold?" *N Engl J Med*, 2003, 348(20):2032-3.

♦ **Platelet Transfusion Refractoriness** *see* Platelet Antibodies *on page 1047*
♦ **Pleural Fluid Analysis** *see* Body Fluid Analysis, Cell Count *on page 288*
♦ **Pleural Fluid Analysis** *see* Body Fluid Chemical Analysis *on page 291*
♦ **Pleural Fluid Cytology** *see* Body Cavity Fluid Cytology *on page 285*
♦ **Pleural Fluid pH** *see* Body Fluid pH *on page 295*
♦ **Ploidy** *see* Breast Biopsy *on page 305*

Pneumocystis carinii Preparation

Related Information

Bronchoalveolar Lavage (BAL) *on page 311*
Pneumocystis Immunofluorescence *on page 1063*
Polymerase Chain Reaction *on page 1069*
Sputum Cytology *on page 1222*
Viral Culture *on page 1307*

Test Includes Papanicolaou, Diff-Quik™, Giemsa, methenamine silver stains, or monoclonal antibodies to *Pneumocystis*

Abstract *Pneumocystis carinii* pneumonia is essentially an alveolar disease. The cornerstone of diagnosis is bronchoalveolar lavage (BAL). Of value in the diagnosis of *P. pneumonia* in any immunocompromised subject, including individuals with AIDS, postorgan transplant patients, those with hematologic malignant diseases, inflammatory disorders, and those receiving chemotherapeutic regimens.

Patient Preparation Additional to BAL, specimens may be obtained by induced sputum, bronchoscopic lung biopsy, or other means. These procedures have generally replaced open lung biopsy for this disease.

Specimen Lung biopsy, transthoracic needle aspirate, bronchoalveolar lavage fluid, or induced sputum. **Routine sputum is not acceptable** for identification of *Pneumocystis*.

Collection Saccomanno fixative has been used. The addition of bronchoscopic lung biopsy to BAL leads to diagnosis in a few subjects in whom *Pneumocystis* was not identified by BAL. For **tissue lung biopsy**, touch preparations made from the fresh surgical specimen are made by lightly touching the fresh tissue in rapid succession along the length of three to four slides. Induced sputum is sent to the laboratory fresh, and prepared in the laboratory.

Use Identify *Pneumocystis carinii* organisms, predominantly in immunocompromised patients

Methodology In the hands of an experienced cytopathologist, *P. carinii* can be identified with Pap, Giemsa, Diff-Quik™, and Grocott methenamine silver stains. Direct immunofluorescence monoclonal antibody-stained bronchoalveolar lavage specimens provide higher yields. See *Pneumocystis* Immunofluorescence *on page 1063*. PCR detection of *P. carinii* is more sensitive than cytology or immunofluorescence.

Additional Information Impaired cellular immunity is the major predisposing background for *Pneumocystis carinii* infection, including AIDS, malnutrition, prematurity, immunodeficiency disease entities, and use of immunosuppressive drugs and/or corticosteroids.

P. carinii is the most common infection and most common cause of death in patients with AIDS. If a patient has been receiving prophylactic therapy for *P. carinii* and presents with pneumonitis, the cytopathologist must be notified, because if organisms are present, they will have destroyed or partially destroyed cyst walls. Examination must be extremely meticulous to identify the residual organisms.

P. carinii occasionally causes extrapulmonary infections in patients with or without AIDS, usually in patients with advanced HIV infection.

References
Rotaeche JG and Costabel U, "Bronchoalveolar Lavage in Diagnostic Cytology," *Compendium on Diagnostic Cytology*, 8th ed, Wied GL, Bibbo M, Keebler CM, et al, eds, Chicago, IL: Tutorials of Cytology, 1997.
Zimmerman RL, "Testing for *Pneumocystis carinii* Pneumonia," *Lab Med*, 2000, 31(9):477-8.

Pneumocystis Immunofluorescence

Related Information
Bacterial Culture, Lower Respiratory *on page 241*
Blood Gases and pH, Arterial *on page 275*
Bronchoalveolar Lavage (BAL) *on page 311*
Pneumocystis carinii Preparation *on page 1062*
Polymerase Chain Reaction *on page 1069*
Sputum Cytology *on page 1222*

Test Includes Direct and indirect immunofluorescence for detection of *Pneumocystis carinii* in clinical specimens

Abstract The most common lung complication in untreated AIDS patients is pneumonia caused by *Pneumocystis carinii*. Entities associated with *P. carinii* pneumonia in patients without HIV infection include systemic corticosteroid therapy, malignant hematologic diseases, transplantation, inflammatory disorders, and solid tumors. The extrapulmonary organs most often affected are lymph nodes, liver, spleen, and bone marrow.

Specimen Tissue, sputum, bronchoalveolar lavage

Container Sterile container

Special Instructions Tissue specimens should be received fresh or snap-frozen.

Reference Interval No organisms observed

Use Diagnose *Pneumocystis carinii* pneumonia

Limitations A negative result does not exclude the diagnosis. Patients may have other concurrent pulmonary infections not detected by this assay.

Methodology Indirect (IFA) and direct fluorescent antibody (DFA)

Additional Information Pneumonia due to *Pneumocystis carinii* (PCP) usually only occurs in immunosuppressed patients, primarily patients with advanced HIV infection. The prevalence of PCP has declined due to improved treatment of patients infected with HIV; however, detection of PCP in a patient may warrant HIV testing, since it is often the first sign of immunosuppression.

Diagnosis depends primarily on seeing either cysts in tissue or cytology preparations. The *Pneumocystis* cell wall contains chitin and glucan polymers, and thus stains that are used to detect fungi (eg, methenamine silver, toluidine blue, calcofluor) can be used to detect *P. carinii*. Direct immunofluorescence is more sensitive than silver stains. PCR detection of *P. carinii* is more sensitive than conventional cytologic stains or immunofluorescence.

References
Kiska DL, Bartholoma NY, and Forbes BA, "Acceptability of Low-Volume, Induced Sputum Specimens for Diagnosis of *Pneumocystis carinii* Pneumonia," *Am J Clin Pathol*, 1998, 109(3):335-7.
Metersky ML, Aslenzadeh J, and Stelmach P, "A Comparison of Induced and Expectorated Sputum for the Diagnosis of *Pneumocystis carinii* Pneumonia," *Chest*, 1998, 113(6):1555-9.
Salzman SH, "Bronchoscopic Techniques for the Diagnosis of Pulmonary Complications of HIV Infection," *Semin Respir Infect*, 1999, 14(4):318-26.

(Continued)

Pneumocystis Immunofluorescence *(Continued)*

Torres J, Goldman M, Wheat LJ, et al, "Diagnosis of *Pneumocystis carinii* Pneumonia in Human Immunodeficiency Virus-Infected Patients With Polymerase Chain Reaction: A Blinded Comparison to Standard Methods," *Clin Infect Dis*, 2000, 30(1):141-5.

Wilkin A and Feinberg J, "*Pneumocystis carinii* Pneumonia: A Clinical Review," *Am Fam Phys*, 1999, 60(6):1699-708, 1713-4.

Internet Web Sites
www.cdc.gov/ncidod/dpd/parasites/pneumocystis/default.htm

PNH Test (GPI-Anchored Proteins) by Flow Cytometry

Related Information

Flow Cytometry, Overview *on page 592*
Ham Test *on page 665*
Hemosiderin Stain, Urine *on page 692*
Peripheral Blood: Red Blood Cell Morphology *on page 1016*
Red Blood Cells, Washed *on page 1143*
Sugar Water Test Screen *on page 1226*

Applies to GPI-Anchored Proteins

Test Includes Flow cytometric analysis of peripheral red and/or white blood cells using monoclonal antibodies against glycosyl-phosphatidylinositol anchored proteins, in particular the CD55 (decay accelerating factor - DAF) and/or CD59 (membrane inhibitor of reactive lysis - MIRL, protectin) antigens

Abstract This test determines the presence or absence of the glycosyl-phosphatidylinositol (GPI) anchored protein decay-accelerating factor (DAF, CD55), membrane inhibitor of reactive lysis (MIRL, CD59), and others (individual laboratory dependent). DAF and MIRL are complement regulatory proteins. Their absence accounts for the complement sensitivity of PNH red cells. PNH is characterized by chronic intravascular hemolysis which is the result of abnormal sensitivity of PNH red cells to complement-mediated lysis. Failure of regulation of the alternative pathway of complement activation results in hemolysis (in patients with PNH). The test involves flow cytometric analysis (immunophenotyping) of red cells or white cells (in particular, neutrophils) of peripheral blood utilizing monoclonal antibodies (conjugated with fluorescent dye) to tag the appropriate GPI-anchored proteins. Results of such analyses are considered to define the presence of PNH. Flow cytometry for PNH is an important advance over the Ham (acidified serum) test which in comparison lacks sensitivity as well as specificity and is subject to operational inconsistency.

Specimen 5 mL whole blood

Container Lavender top (EDTA) tube or green top (heparin) tube; consult laboratory as to their specimen container preference.

Storage Instructions Analysis should be performed within 24 hours of specimen collection for optimal results. Test may be performed on samples obtained 48-72 hours prior to analysis. Keep specimen at room temperature. Refrigeration may cause loss of cell surface antigen.

Causes for Rejection Aged and/or cold stored specimens may cause false-positive results.

Turnaround Time 1 day

Special Instructions This relatively recently developed assay is likely to be performed by a research or reference laboratory. Consult your clinical laboratory for availability and requirements.

Reference Interval Normal individuals (those without PNH) have presence of flow cytometric detected GPI-anchored proteins while patients with PNH have absence of such proteins (and thus are positive for PNH).

Use To establish the presence of cells diagnostic of PNH. Patients with congenital dyserythropoietic anemia (CDA) give a false-positive result (for PNH) with the Ham test but test (appropriately) negative for PNH with flow cytometry analysis (CDA patients show presence of GPI-anchored proteins).

Limitations Requires availability of flow cytometer, appropriate reagents, and support personnel.

Methodology Flow cytometry (FC) using a variety of fluorescent-tagged monoclonal antibodies to one or more GPI-anchored proteins with accompanying controls. A well-controlled three-color flow cytometric protocol developed and used at the National Heart, Lung, and Blood Institute including delineation of (and commercial sources for) antibody reagents has been published (see Dunn et al). CD59, CD55, C24, CD16, CD14, and/or CD66b may be targeted for analysis in flow cytometry tests for PNH.

Monoclonal antibody CD59 (membrane inhibitor of reactive lysis - MIRL - or "protectin") is the preferred single monoclonal antibody to use for GPI-anchored protein deficiency. CD59 is present on granulocytes and is the most deficient GPI protein in cases of PNH in which cells are only partially deficient.

Demonstration of the absence of GPI-anchored protein (one or more) by flow cytometry currently provides the most reliable method for diagnosis of PNH. The method is sensitive and specific and should be used in preference to the Ham test. The Ham and sugar water hemolysis tests detect only PNH erythrocytes which may be variably decreased due to shortened life-span with resultant false-negative test results.

Additional Information The conundrum of apparently disparate clinical and laboratory features that characterize paroxysmal nocturnal hemoglobinuria (PNH) have been largely (albeit still incompletely) resolved during the past 1-3 decades by significant (and brilliant) advances in delineation of the underlying molecular pathology. The PNH condition has been established as a loss of multiple membrane proteins, all of which share the common feature of being anchored to the cell surface by glycosyl-phosphatidylinositol.

Some 18 proteins are currently identified as deficient in PNH. These include leukocyte alkaline phosphatase, acetylcholinesterase, decay accelerating factor (DAF, CD55), membrane inhibitor of reactive lysis (MIRL, CD59) and FcγRIIIB (CD16b). The absence of CD55 and CD59 account for the susceptibility of erythrocytes to lysis by complement in patients with PNH. PNH is caused by an acquired somatic mutation. The defective gene, identified in 1993, is present on the short arm of the X chromosome (Xp 22.1). The gene, PIG-A (phosphatidylinositol glycan-class A) encodes a protein that is needed for the normal synthesis of N-acetylglucosaminylphosphatidlinositol (G1cNAc-PI), an early intermediate in the synthesis of the GPI anchor. The somatic mutation occurs at an early stage in the evolution of a pluripotent hematopoietic stem cell, accounting for involvement of multiple cell lines (ie, erythrocytes, neutrophils, monocytes, lymphocytes, and platelet precursors).

The clinical features of PNH relate (albeit at times somewhat tortuously) to missing GPI-anchored proteins and include intravascular hemolysis, aplastic anemia/pancytopenia, and venous thrombosis. The relationship of PNH to apparently associated bone marrow failure syndromes (in particular myelodysplasia), aplastic anemia, autoimmunity, and apoptosis is under investigation.

References
Benz EJ Jr, "Clonal Variation, Autoimmunity, and Neoplasia: An Ecology Lesson From Paroxysmal Nocturnal Hemoglobinuria," *Ann Intern Med*, 1999, 131(6):467-8.
Dunn DE, Tanawattanacharoen P, Boccuni P, et al, "Paroxysmal Nocturnal Hemoglobinuria Cells in Patients With Bone Marrow Failure Syndromes," *Ann Intern Med*, 1999, 131(6):401-8.
Rosse WF, "Paroxysmal Nocturnal Hemoglobinuria," *Hematology: Basic Principles and Practice*, 3rd ed, Chapter 20, Hoffman R, Benz EJ Jr, Shattil SJ, et al, eds, Philadelphia, PA: Churchill Livingstone, 2000, 331-42.

♦ **PNH Test Screen** *see* Sugar Water Test Screen *on page 1226*

♦ **pO₂** *see* Blood Gases and pH, Arterial *on page 275*

♦ **pO₂ (0.5)** *see* P₅₀, Blood *on page 993*

♦ **pO₂ at Half Saturation** *see* P₅₀, Blood *on page 993*

♦ **POCT** *see* Point-of-Care Testing *on page 1065*

Point-of-Care Testing

Related Information
B-Type Natriuretic Peptide *on page 314*
Carboxyhemoglobin, Blood *on page 340*
Fibronectin, Fetal, Cervicovaginal Secretions *on page 585*
Glucose, Fasting, Plasma *on page 643*
Glucose, Noninvasive *on page 647*
Glucose, Random, Plasma *on page 649*
Glucose, Whole Blood (Including Point-of-Care) *on page 653*
Troponins, Serum *on page 1278*

Synonyms Alternate Site Testing; Ancillary Testing; AST; Bedside Testing; Decentralized Testing; Near Patient Testing; NPT; Off-Site Testing; POCT

Applies to Pulse Oximetry
(Continued)

Point-of-Care Testing (Continued)

Test Includes Point-of-care assays may include hematocrit, hemoglobin, hemoglobin A$_{1c}$, glucose, beta-hCG, CK-MB, blood gases (pH, pCO$_2$, pO$_2$, SO$_2$%), electrolytes (Na$^+$, K$^+$, Cl$^-$, ionized calcium, ionized magnesium), lactate, blood urea nitrogen, creatinine, ethanol, drugs of abuse, activated clotting time, troponin T, myoglobin, PTT, aPTT, bilirubin, and other assays.

Abstract Development of diagnosis-related groups (DRGs), manufacturer's incentives to provide instrumentation to less regulated sites (eg, physicians' office laboratories prior to CLIA '88), advancing technology including development of biosensors with improved electronics, and evolving requirements for abbreviated turnaround times have driven interest in near patient testing. Point-of-care testing (POCT) is finding a niche in terms of clinical utility. It has been proven practicable and some aspects desirable. Growth in POCT has been climbing at a rate of about 18% compared to the overall growth of diagnostic testing of 4% to 5% per year. Its shortcomings include use by individuals with limited training in medical technology and quality control needs.

Specimen Whole blood, urine, saliva

Turnaround Time Minutes

Reference Interval See reference range for each particular analyte. Ranges may differ from those of the laboratory hospital or reference laboratory depending on the specimen utilized, (ie, whole blood versus plasma or serum).

Use Bedside monitoring of tests listed above in areas such as emergency departments, and surgical theaters and other settings where need for rapid turnaround times is great or desirable. Physician offices with on-site availability of results can avoid patient revisits and provide early treatment. The major use of POCT glucose testing at bedside or home is to monitor individuals with diabetes mellitus, not to establish that diagnosis.

Limitations Point-of-care testing generally is more expensive than analysis of specimens in a centralized laboratory. Expense is related in part to volume of testing and the number of personnel involved in such testing; handheld bedside-type devices tend to be less precise than larger instruments in centralized laboratories. Assay range is generally narrower than that of central laboratory instruments. Information handling, storage, and billing may be problematical if POCT instruments are not linked to laboratory or hospital information systems. Such challenges may be overcome by incorporation of radiofrequency and infrared links to the central laboratory as part of the point-of-care testing devices.

Preanalytical problems may exist when POCT devices are used by nonlaboratorian personnel with little or no formal training in laboratory technology and no exposure to the concepts of quality control. Inspection and accreditation requirements, quality assurance and proficiency testing, supervision and management for POCT continue to evolve. Difficulties include as well needs to monitor such programs and requirements to maintain standards with training and competency requirements. Only 34.2% of hospitals in one series had a multidisciplinary POCT team, yet 88.7% of glucose testing, 41% of electrolyte testing, 58.3% of coagulation testing, 45.5% of hematocrit, and 47.7% of blood gas testing was performed at the bedside.

Maintaining adequate quality control of POCT instruments, in a hospital setting, is a major concern. The main laboratory may be held responsible for quality of results, without direct laboratory staff involvement or authority in testing.

Pulse oximetry, a practical example of POCT, has been unable to distinguish carboxyhemoglobin from oxyhemoglobin at the wavelengths used by most oximeters. See Carboxyhemoglobin, Blood *on page 340.*

Most POCT devices calculate hematocrit by conductivity. Measurement of hematocrit by conductivity in seriously sick patients with severe electrolyte imbalance can lead to incorrect results.

Methodology Methodology ranges from reflectance measurements of glucose oxidase strips for some of the simplest bedside glucometers to microelectrode technology for some of the most sophisticated devices measuring electrolytes and blood gases. The technology is ever-evolving and newer devices are being developed with technology that parallels central laboratory methodology but which may be more robust. Some bedside devices for example measure cardiac troponin T by a biotinylated antitroponin T antibody/streptavidin methodology much as for some central laboratory methods. Immunoassays for proteins, such as beta-hCG, and for drugs of abuse are becoming available. Methods include ion-selective electrodes, substrate-specific electrodes, electrical conductance sensors, and analyte-specific optical sensors.

Additional Information In the past, laboratory tests have mainly been performed on large automated instruments with the capacity of interfaces to laboratory and hospital information systems. Recently, the advent of microminiature electronics and very powerful microprocessors has led to a new generation of compact portable instruments that are relatively easy to use and reported to be capable of reasonable precision even in the hands of relatively inexperienced users.

Need for decreased turnaround times in areas such as anticoagulation during balloon angioplasty, blood gases and electrolyte measurement in emergency department, bedside coagulation monitoring, and bedside glucose testing to guide insulin therapy has driven the site of testing from the central laboratory to the bedside. Tests felt to be the most critical in managing unstable and critically ill patients include blood gas measurements, electrolytes, and coagulation and hematology tests. Use of POCT for cardiac markers and drugs of abuse is increasing in emergency departments.

Other issues that likely will remain controversial include selection of performance standards for POCT, definition of acceptable reproducibility (precision), information handling, reporting and billing issues, responsibility for POCT programs in hospitals, and definition of realistic turnaround times. A report on timeliness of clinical laboratory tests makes it clear that the perception of reasonable turnaround time is much different for clinical care providers than for laboratory staff.

In the past, glucose values obtained by POCT analyzers were lower due to whole blood glucose values. Currently, most of the glucose meters are adjusted to give plasma equivalent values. Thus, glucose values obtained by glucose meters match well with main laboratory chemistry analyzers. However, the upper limit of result reporting is still limited. Difficulties include variation in hematocrit and interference from drugs. See Glucose, Whole Blood (Including Point-of-Care) *on page 653*.

POCT devices for coagulation and infectious disease testing are becoming available. Due to the number of different reagents and technologies in use, comparisons with main lab results are critical.

Use of anticoagulated samples versus capillary specimens, and use of different anticoagulants between central laboratory and point-of-care procedures may create confusion, possibly harming patient care.

Vital to the effective administration of any point-of-care testing program is the realization that someone must be designated as responsible for point-of-care testing service. Responsibilities would include, but not be limited to, the following:

- knowing which point-of-care tests are being performed, by whom and in which locations
- appropriate administrative responsibilities for delegation of the program
- monitoring quality control documentation
- evaluation and selection of appropriate equipment
- troubleshooting all aspects of the point-of-care testing program
- coordination of training
- serving as a liaison between the nursing service, other units such as emergency department, operating rooms, and the laboratory

It is recommended that any institution with point-of-care testing have a standing committee to ensure appropriate oversight. Such a committee would be composed of representatives from the laboratory, nursing service, medical staff, and the quality assurance officer. The committee should have the authority to approve the methodology to be used and control the purchasing of point-of-care devices.

(Continued)

Point-of-Care Testing *(Continued)*

COST BENEFIT ISSUES

The literature is filled with opinions of the costs and effectiveness of point-of-care testing. Deficiencies in a number of published cost analyses include the following.

- Only incremental costs of labor and consumables are included with many assumptions about the costs of nonlaboratory personnel.
- Studies do not differentiate the differences between fixed and incremental costs as they impact the central laboratory and point-of-care testing.
- They do not consider the cost of the episode of care or the patient's long-term follow-up costs.

It is important to recognize that assessment of the costs of point-of-care testing should include potential savings to the institution as decreased inpatient days, more efficient care leading to decreased utilization of services, and reduced numbers of follow-up visits to the clinic. There are other costs which may be hard to quantify such as the savings of a diabetic patient remaining in good control and thus avoiding or delaying the onset of microangiopathy. Other benefits from point-of-care testing include:

- improved turnaround time
- improved patient management
- improved patient satisfaction and compliance
- improved throughput of the clinic/institution
- improved clinic/laboratory relations
- improved job satisfaction of physician and nurse empowerment
- improved clinic reputation
- decreased transfusions in surgery and decreased re-explorations
- decreased operating room time
- improved morbidity and mortality
- decreased dietetic services awaiting early morning glucose testing
- small sample volume
- reduced pre- and postanalytical errors

It is clear that decisions must be made daily, even in the absence of hard data. In most cases, cost must be weighed against quality impact of point-of-care methods.

A current editorial by Kost notes that a major disadvantage of POCT is lack of bidirectional communication between information systems and POCT devices.

References

Bennett J, Cervantes C, and Pacheco S, "Point-of-Care Testing: Inspection Preparedness," *Perfusion*, 2000, 15(2):137-42.

Chance, JJ, Li DJ, Jones AA, et al, "Technical Evaluation of Five Glucose Meters With Data," *Am J Clin Pathol*, 1999, 111(4):547-56.

"Coagulation at the Point of Care," *CAP Today*, October 2000, 80-2, 86-7, 90.

Gallichan M, "Self Monitoring of Glucose by People With Diabetes: Evidence Based Practice," *BMJ*, 1997, 314(7085):964-7.

Gonzales Y and Kampa IS, "The Effect of Various Storage Environments on Reagent Strips," *Lab Med*, 1999, 28(2):136-7.

Hicks JM, Haeckel R, Price CP, et al, "Recommendations and Opinions for the Use of Point-of-Care Testing for Hospitals and Primary Care: Summary of a 1999 Symposium," *Clin Chim Acta*, 2001, 303(1-2):1-17.

Jacobs DS, DeMott WR, Oxley DK, et al, *Laboratory Test Handbook*, 5th ed, Hudson, OH: Lexi-Comp Inc, 2001.

Koepke JA, "Point-of-Care Coagulation Testing," *Lab Med*, 2000, 31:343-6.

Kost GJ, "Connectivity - The Millenium Challenge for Point-of-Care Testing," *Arch Pathol Lab Med*, 2000, 124(8):1108-10.

Louie RF, Tang Z, Shelby DG, et al, "Point-of-Care Glucose Testing: Millenium Technology for Critical Care," *Lab Med*, 2000, 31:402-8.

Main RI and Kiechle FL, "Point-of-Care Testing: Administration Within a Health System," *Lab Med*, 2000, 31(8):453-9.

Quality Point-of-Care Testing: A Joint Commission Handbook, Oakbrook Terrace, IL: Joint Commission on Accreditation of Healthcare Organizations, 1999, 140.

Wilson F, "Minimally Invasive Testing Expands Beyond Glucose," *Lab Med*, 2000, 31(8):436-41.

Poliomyelitis I, II, III Serology

Related Information

Cerebrospinal Fluid Analysis: Overview *on page 355*
Enterovirus Culture *on page 539*
Viral Culture *on page 1307*

Test Includes Detection of antibodies to poliovirus in patient's serum

Abstract The eradication of poliomyelitis by vaccination is the aim of the World Health Organization (WHO) Poliomyelitis Eradication Initiative (PEI). Due to ongoing efforts to provide vaccination throughout the world, a dramatic reduction of cases of poliomyelitis has been noted in regions of the world which were once endemic for polio. Currently, wild type poliovirus is rarely found in the Western hemisphere and nearly eliminated in the rest of the world. Although poliovirus infection is rare, it has not been completely eradicated. Thus, the differential diagnosis of individuals with acute aseptic meningitis, with or without paralysis, should consider poliomyelitis if the patient has not been immunized.

Specimen Serum

Container Red top tube

Sampling Time Acute and convalescent sera drawn 10-14 days apart

Reference Interval Presence of neutralizing antibody indicates adequate immunization; normal <1:8

Possible Panic Range A fourfold increase in antibody titer in paired sera is diagnostic of poliomyelitis. Specific IgM antibody is diagnostic as well. Any suspected case of poliomyelitis should be reported to the local health department as well as to the CDC.

Use Support for the diagnosis of acute poliovirus infection, documentation of previous exposure to poliovirus (complement fixing antibodies); documentation of immunization (neutralizing antibodies)

Limitations Complement-fixing antibodies can be broadly cross-reactive with other enteroviruses.

Methodology Viral neutralization, complement fixation (CF)

Additional Information Poliovirus, an enterovirus, may also be cultured, producing a characteristic cytopathic effect in tissue culture. Culture is more suitable than serology for diagnosis of acute infection.

References

American Academy of Pediatrics, "Prevention of Poliomyelitis: Recommendations for Use of Only Inactivated Poliovirus Vaccine for Routine Immunization," *Pediatrics*, 1999, 104(6):1404-6.

Centers for Disease Control and Prevention, "Developing and Expanding Contributions of the Global Laboratory Network for Poliomyelitis Eradication, 1997-1999," *JAMA*, 2000, 283(13):1683-4.

John TJ, "The Final Stages of the Global Eradication of Polio," *N Engl J Med*, 2000, 343(11):806-7.

Soto NE and Lutwick LI, "Poliovirus Immunizations. What Goes Around, Comes Around," *Infect Dis Clin North Am*, 1999, 13(1):265-78.

Internet Web Sites

www.astdhpphe.org/infect/polio.html
www.cdc.gov/nip/publications/vis/vis-IPV.pdf
www.who.int/health-topics/poliomyelitis.htm

♦ **Polycythemia of the Newborn** *see* Viscosity, Blood *on page 1312*

Polymerase Chain Reaction

Related Information

Activated Protein C Resistance and the Factor V Leiden Mutation *on page 104*
Amniotic Fluid, Chromosome and Genetic Abnormality Analysis *on page 152*
Bartonella Diagnostic Procedures *on page 250*
bcl-2 Gene Rearrangement *on page 251*
Body Cavity Fluid Cytology *on page 285*
Bone Marrow *on page 296*
Breakpoint Cluster Region Rearrangement in CML *on page 304*
Cerebrospinal Fluid Cytology *on page 361*
Chlamydia trachomatis Nucleic Acid Detection *on page 388*
Chorionic Villus Sampling, Chromosome and Genetic Abnormality Analysis *on page 400*
Cystic Fibrosis DNA Detection *on page 491*
Cytomegalovirus Nucleic Acid Detection *on page 498*
Duchenne/Becker Muscular Dystrophy DNA Detection *on page 526*
(Continued)

Polymerase Chain Reaction *(Continued)*

Enterovirus Polymerase Chain Reaction *on page 540*
Fragile X Syndrome DNA Test *on page 611*
Helicobacter pylori Biopsy-Based Tests: The Urease Tests, Culture, Cytology, and PCR *on page 672*
Hepatitis B Antigen Detection *on page 699*
Hepatitis B Serology *on page 702*
Hepatitis C Virus RNA Detection and Quantitation *on page 705*
Hepatitis C Virus Serology *on page 706*
Hepatitis D Serology *on page 711*
Herpes Simplex Virus DNA Detection *on page 723*
Herpesvirus 6 Serology *on page 725*
HIV-1/HIV-2 Serology *on page 736*
HTLV-I/II Antibody *on page 745*
Human Immunodeficiency Virus DNA Amplification *on page 746*
Huntington Disease DNA Test *on page 752*
Identification DNA Testing *on page 765*
Lyme Disease DNA Detection *on page 878*
Mixed Lymphocyte Culture *on page 917*
Myotonic Dystrophy DNA Test *on page 942*
p53, Functional Assay/Sequencing *on page 995*
Pneumocystis Immunofluorescence *on page 1063*
Retinoblastoma Gene DNA Detection *on page 1159*

Synonyms DNA Amplification; PCR

Test Includes Amplification of target DNA sequences as much as a millionfold

Abstract The polymerase chain reaction (PCR) is a technique with unlimited potential use in the medical laboratory. The major diagnostic applications of PCR include disease detection by mutational analysis and detection and identification of bacteria and viruses. The PCR technique permits over a millionfold amplification of target DNA in several hours. The amplification of target DNA is achieved by repetition of three steps: denaturation of target DNA, annealing (binding) of primers to specific sequences, and elongation of primers.

Specimen The specimen for the PCR assay will depend on the type of analysis. For example, prenatal diagnosis will require amniotic fluid or chorionic villus biopsy (see Amniotic Fluid, Chromosome and Genetic Abnormality Analysis *on page 152*), whole blood will be required for human immunodeficiency virus (HIV) detection, other specimens such as cerebrospinal fluid, sputum, serum, biopsies, or discharge from wounds for other infectious agents, or solid tissue by biopsy for cancer diagnosis. With the improved PCR assays, DNA purification is generally not required.

Collection Varies with type of specimen

Use Use of the technique to amplify short fragments of DNA is nearly unlimited. See individual entries in Related Information for some applications of PCR.

Limitations Because PCR is very sensitive, the potential for contamination is great. The tests must be carefully monitored with appropriate controls (especially negative controls). The most common cause of contamination is from the products previously amplified in the laboratory. **Never open the tubes containing amplified product in the same room being used for PCR reactions.** The PCR technique requires knowledge of at least partial base sequence (where primers bind) of the DNA of interest to be amplified.

Methodology From the sequence data of target DNA, oligonucleotide primers approximately 18-25 nucleotides in length can be constructed using oligonucleotide synthesizers. These primers generally flank a 100-2000 base sequence of the DNA sequence of interest. The primers are constructed so that the primers bind (anneal) to opposite strands of the target double helix. A special DNA polymerase, purified from *Thermus aquaticus* (*Taq*), is used because it can withstand the many denaturing, reannealing, and elongation cycles without the need for replenishment. The reaction requires the target DNA, the primers, *Taq* polymerase, and the four deoxynucleotide triphosphates. The mixture is heated several minutes to 95°C to separate the target DNA double strands. The primers are then allowed to bind to the target DNA at 50°C to 60°C and the polymerase reaction allowed to proceed for several minutes at 72°C. This cycle of denaturation, annealing, and elongation is repeated over and over, as many as 25-35 times, amplifying the sequence between the primers hundreds of thousands to millions of times (see figure). The amplified DNA can then be detected by agarose electrophoresis followed by ethidium bromide staining. The amplified bands can be seen with a UV light and photographed for analysis. The amplified DNA can also be detected by other techniques, such as hybridization or direct measurement of fluorescence, if fluorescent primers and deoxynucleotides are used.

DNA can be extracted from paraffin-embedded tissue for PCR analysis. Such tissue is best fixed in 10% formalin. Genotype can be ascertained by selective ultraviolet radiation fractionation.

Blood spotted on to filter paper, saliva, and tissue extract can be directly used as a source of DNA.

Polymerase Chain Reaction Cycles

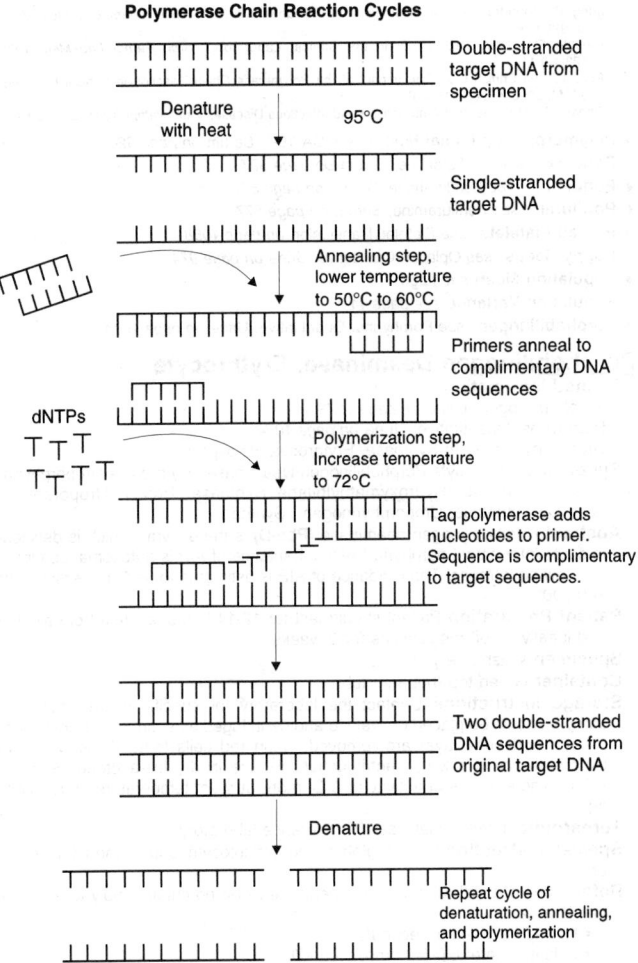

Double-stranded target DNA from specimen

Denature with heat — 95°C

Single-stranded target DNA

Annealing step, lower temperature to 50°C to 60°C

Primers anneal to complimentary DNA sequences

dNTPs

Polymerization step, Increase temperature to 72°C

Taq polymerase adds nucleotides to primer. Sequence is complimentary to target sequences.

Two double-stranded DNA sequences from original target DNA

Denature

Repeat cycle of denaturation, annealing, and polymerization

Additional Information The technique is continuously being expanded and refined. The current PCR techniques can amplify DNA sequences over 10 kilobases. Reverse transcription PCR has been useful in detection of a number (Continued)

Polymerase Chain Reaction *(Continued)*

of viral diseases. The procedure is automated with programmable heating blocks to cycle the reaction automatically. The technique has unprecedented sensitivity, being able to use nanogram quantities of target DNA, and could theoretically be used to amplify the DNA from a single cell. Other amplification reactions (eg, ligase chain reaction) are also being developed for use in the diagnostic laboratory. In the past, PCR was used only for qualitative analysis, but in recent years, PCR is increasingly used for quantitative analyses, such as the measurement of viral load in a sample.

References

Baumforth KR, Nelson PN, Digby JE, et al, "Demystified...the Polymerase Chain Reaction," *Mol Pathol*, 1999, 52(1):1-10.

Hahn S, Zhong XY, Troeger C, et al, "Current Applications of Single-Cell PCR," *Cell Mol Life Sci*, 2000, 57(1):96-105.

Jung R, Soondrum K, and Neumaier M, "Quantitative PCR," *Clin Chem Lab Med*, 2000, 38(9):833-6.

Kiechle FL, "DNA Technology in the Clinical Laboratory," *Arch Pathol Lab Med*, 1999, 123(12):1151-3.

Post JC and Ehrlich GD, "The Impact of the Polymerase Chain Reaction in Clinical Medicine," *JAMA*, 2000, 283(12):1544-6.

Rogers BB, "Nucleic Acid Amplification and Infectious Disease," *Hum Pathol*, 1994, 25(6):590-3.

- ♦ **Polymorphic Epithelial Mucin** *see* CA 15-3, Serum *on page 320*
- ♦ **Ponderx®** *see* Fenfluramine, Serum *on page 577*
- ♦ **Pondimin®** *see* Fenfluramine, Serum *on page 577*
- ♦ **Ponflural®** *see* Fenfluramine, Serum *on page 577*
- ♦ **Pooled Platelets** *see* Platelet Transfusion *on page 1058*
- ♦ **Poppy Seeds** *see* Opiates, Qualitative, Urine *on page 974*
- ♦ **Population Mean** *see page 11*
- ♦ **Population Variance** *see page 11*
- ♦ **Porphobilinogen** *see* Porphyrins, Quantitative, Urine *on page 1074*

Porphobilinogen Deaminase, Erythrocyte

Related Information

Porphobilinogen, Urine *on page 1073*
Porphyrins, Quantitative, Urine *on page 1074*
Uroporphyrinogen III Synthase, Erythrocyte *on page 1297*

Synonyms Erythrocyte Porphobilinogen Deaminase; Erythrocyte Uroporphyrinogen I Synthase; Hydroxymethylbilane Synthase; Pbg-D; Uroporphyrinogen-Cosynthetase; Uroporphyrinogen I Synthase

Abstract Porphobilinogen deaminase (Pbg-D) is the enzyme which is deficient in acute intermittent porphyria (AIP). Inheritance of AIP is autosomal dominant with low penetrance. The incidence of AIP is estimated to be 5-10 persons per 100,000.

Patient Preparation Patient should fast for 12-14 hours, abstain from alcohol and ideally be off medications for 2 weeks.

Specimen Washed erythrocytes

Container Green top (heparin) tube

Storage Instructions Contact the laboratory for specific instructions. The sample should be placed on wet ice and centrifuged as soon as possible. After plasma and buffy layer are removed, wash red cells three times with cold isotonic saline. Pack by centrifugation. Freeze in dry ice-acetone. Store at -20°C. There is loss of activity at 4°C, more at room temperature as red cells age.

Turnaround Time Usually sent to reference laboratory

Special Instructions Hemoglobin and reticulocyte count should also be ordered.

Reference Interval Method dependent; the following cutoffs apply to a fluorometric assay.
- Normal: ≥7.0 nmol/second/L
- Indeterminate: 6.0-6.9 nmol/second/L
- Diminished: <6.0 nmol/second/L

Use Assay of Pbg-D is the definitive test for AIP, and will detect carriers as well as persons with clinical AIP. Suspected clinical AIP is usually first evaluated with urinary porphobilinogen (see Porphyrins, Quantitative, Urine *on page 1074*).

Limitations Abnormal results may be encountered in other diseases. Only a minority of carriers of the defective enzyme experience clinical AIP.

Methodology Fluorometric assay, ELISA assay, electrophoresis for molecular analysis

Additional Information See tables in Porphyrins, Quantitative, Urine *on page 1074*. More than 130 mutations causing AIP are described. Most individuals with a defective enzyme do **not** develop clinical ALP. Pbg-D is the enzyme which converts porphobilinogen to uroporphyrinogen I.

References
Foran SE and Abel G, "Guide to Porphyrias. A Historical and Clinical Perspective," *Am J Clin Pathol*, 2003, 119(Suppl 1):S86-S93.

Hindmarsh JT, Oliveras L, and Greenway DC, "Biochemical Differentiation of the Porphyrias," *Clin Biochem*, 1999, 32(8):609-19.

Kauppinen R and von und zu Fraunberg M, "Molecular and Biochemical Studies of Acute Intermittent Porphyria in 196 Patients and Their Families," *Clin Chem*, 2002, 48(11):1891-900.

Pierach CA, Weimer MK, Cardinal RA, et al, "Red Blood Cell Porphobilinogen Deaminase in the Evaluation of Acute Intermittent Porphyria," *JAMA*, 1987, 257(1):60-1.

Porphobilinogen, Urine

Related Information
Delta (5)-Aminolevulinic Acid, Urine *on page 508*
Lead, Blood *on page 832*
Porphobilinogen Deaminase, Erythrocyte *on page 1072*
Porphyrins, Quantitative, Urine *on page 1074*
Protoporphyrin, Free Erythrocyte *on page 1121*
Uroporphyrinogen III Synthase, Erythrocyte *on page 1297*

Applies to Delta Aminolevulinic Acid; Hoesch Test; Porphyrins, Fecal; Uroporphyrinogen Synthase; Watson-Schwartz Test

Test Includes Qualitative screen for porphobilinogen and/or urobilinogen; semiquantitative or quantitative porphobilinogen

Abstract Porphobilinogen (PBG) is a heme precursor which is excreted in the urine in the acute porphyrias. It is the single most useful test to screen for acute porphyria in the symptomatic patient.

Specimen Random or 24-hour urine

Container Clean dark container; check with laboratory regarding preservative.

Sampling Time PBG levels in the urine should be measured during acute attacks of abdominal pain, extremity pain or paresthesias, tachycardia, hypertension, nausea and vomiting, neurologic abnormalities, and in the investigation of dark urine.

Collection Keep refrigerated during collection. Prevent exposure to light. Some laboratories require a preservative. Check with the laboratory.

Storage Instructions Specimen may be stored for a brief period in refrigerator, but must be analyzed promptly. If analysis must be delayed or sent to a referral laboratory, freezing is often recommended. Check with the laboratory.

Causes for Rejection Specimen exposed to light, specimen not received frozen, improper preservative

Reference Interval
- Qualitative, random: Negative
- Quantitative, random: <2.0 mg/L (SI: ≤8.8 µmol/L); ≤0.5 mg/g creatinine (SI: ≤0.025 µmol/mmol creatinine)
- Quantitative, 24-hour: ≤1.5 mg/24 hours (SI: ≤6.6 µmol/24 hours)

Use PBG is the best screen for the acute porphyrias: acute intermittent porphyria (AIP), hereditary coproporphyria, and variegate porphyria. **Positive results from qualitative tests must be confirmed by a quantitative method, ideally on the same urine specimen.**

Limitations False negatives and false positives occur. The major drawback of the qualitative Watson-Schwartz and Hoesch tests is the need for subjective interpretation of the visual endpoint.

Sometimes negative or within reference interval in the patient with asymptomatic (latent) phase of AIP, in whom porphobilinogen deaminase may detect the presence of AIP (see Porphobilinogen Deaminase, Erythrocyte *on page 1072* for this assay).

The intoxication porphyrinurias (including lead poisoning) are better detected by delta aminolevulinic acid and other tests.
(Continued)

Porphobilinogen, Urine *(Continued)*

Methodology PBG reacts with 4-dimethylaminobenzaldehyde (Ehrlich's reagent) to form a red product. The Watson-Schwartz test relied on visual recognition of the colored product, but that test was insensitive and nonspecific. The procedure is enhanced by anion exchange chromatography which removes non-PBG reactive substances. Eluates may be quantified.

A commercial semiquantitative kit (Trace, Miami, FL) for urinary porphobilinogen (PBG), in which urine is pretreated with ion-exchange resin and the color of the Ehrlich-PBG adduct matched against a set of surrogate standards, has been described. The method was compared with qualitative screening methods (Watson-Schwartz) in common use. For 129 urine samples with raised PBG, 95% samples were positive with Trace kit as compared to only 38% positive by qualitative tests. Sixteen out of 91 results for pigmented urine samples with normal PBG were reported as positive using qualitative screening tests, but only one using the Trace kit. Therefore, the Trace method seems far more sensitive and specific than the qualitative screening tests. The study recommended that Watson-Schwartz-type screening tests should be abandoned and, ideally, all urine samples analyzed by quantitative methods. However, the Trace method is a convenient alternative which is adequate for the initial screening of symptomatic patients.

References

Deacon AC and Peters TJ, "Identification of Acute Porphyria: Evaluation of a Commercial Screening Test for Urinary Porphobilinogen," *Ann Clin Biochem*, 1998, 35(Pt 6):726-32

Deacon AC and Elder GH, "ACP Best Practice No 165: Front Line Tests for the Investigation of Suspected Porphyria," *J Clin Pathol*, 2001, 54(7):500-7.

Hindmarsh JT, Oliveras L, and Greenway DC, "Biochemical Differentiation of the Porphyrias," *Clin Biochem*, 1999, 32(8):609-19

Kauppinen R and von und zu Fraunberg, "Molecular and Biochemical Studies of Acute Intermittent Porphyria in 196 Patients and Their Families," *Clin Chem*, 2002, 48(11):1891-900.

♦ **Porphyrins, Erythrocytes** *see* Porphyrins, Quantitative, Urine *on page 1074*

♦ **Porphyrins, Fecal** *see* Porphobilinogen, Urine *on page 1073*

♦ **Porphyrins, Fecal** *see* Porphyrins, Quantitative, Urine *on page 1074*

♦ **Porphyrins, Plasma** *see* Porphyrins, Quantitative, Urine *on page 1074*

Porphyrins, Quantitative, Urine

Related Information

Delta (5)-Aminolevulinic Acid, Urine *on page 508*
Iron and Total Iron Binding Capacity/Transferrin, Serum *on page 807*
Lead, Blood *on page 832*
Lead, Urine *on page 835*
Liver Biopsy *on page 864*
Phlebotomy, Therapeutic *on page 1029*
Porphobilinogen Deaminase, Erythrocyte *on page 1072*
Porphobilinogen, Urine *on page 1073*
Protoporphyrin, Free Erythrocyte *on page 1121*
Urine Collection, 24-Hour *on page 1295*
Uroporphyrinogen III Synthase, Erythrocyte *on page 1297*

Synonyms Coproporphyrins; Porphobilinogen; Uroporphyrins

Applies to Lead Toxicity; Naproxen; Porphyrins, Erythrocytes; Porphyrins, Fecal; Porphyrins, Plasma; Uroporphyrinogen Decarboxylase

Test Includes Uroporphyrins (octacarboxylporphyrins), heptacarboxylporphyrins, hexacarboxylporphyrins, pentacarboxylporphyrins, coproporphyrins (tetracarboxylporphyrins)

Abstract Porphyrins are heme precursors which accumulate in excess when there is a defective enzyme in the heme biosynthetic pathway. Depending on which enzyme is defective, the accumulation of excessive substrate produces a characteristic pattern of porphyrins in the urine.

The hereditary tyrosinemias represent another group of diseases with a similar excretion pattern, but the cause is inhibition of PBG synthetase by succinylacetone.

Secondary porphyrias: Increased porphyrin excretion may also be encountered in patients with lead poisoning, chronic ethanol intoxication, multiple liver diseases, drug treatment, and other conditions. In most of these situations, there is a toxic inhibition of porphobilinogen (PBG) synthetase, with the excretion of excess PBG and delta-aminolevulinic acid (ALA).

Patient Preparation Abstain from alcohol. Avoid medications 1-2 weeks prior to testing.

Specimen 24-hour urine

Container Clean, dark container. Must be kept covered.

Collection Check with laboratory; 5 g sodium carbonate is usually added to container before collection. Specimen should be kept cool and protected from light during collection. Transport specimen immediately to the laboratory upon completion of collection.

Storage Instructions Refrigerate (some laboratories require specimen be sent frozen). **Protect specimen from light.**

Reference Interval See table. See also literature from individual laboratory.

	μg/24 hours	SI: nmol/d
Uroporphyrins (octacarboxyl)	3-25	3.6-30.0
Heptacarboxylporphyrins	≤7	≤8.9
Hexacarboxylporphyrins	≤6	≤8.0
Pentacarboxylporphyrins	≤7	≤10.0
Coproporphyrins (tetracarboxyl)		
male	25-150	38-230
female	8-110	12-168

Use The determination of porphyrins in a 24-hour urine specimen accompanied by a urine PBG determination is the most effective, first order testing approach when porphyria is suspected but the type of porphyria is unknown. The only porphyria for which both of these tests would be expected to be negative is erythropoietic protoporphyria (EPP), a disorder that accumulates protoporphyrin, a water insoluble substance that is not excreted in urine, and ALA dehydratase deficiency porphyria (ADP), a porphyria in which only ALA accumulates and is excreted in the urine (see Porphyrias: Overview table on *page 1077*).

In **congenital erythropoietic porphyria** (CEP), severe hemolysis and photosensitivity are found; see Uroporphyrinogen III Synthase, Erythrocyte *on page 1297*.

In **acute intermittent porphyria** (AIP), PBG and ALA are elevated in acute attacks, and mild increases of urinary uroporphyrin and coproporphyrin may be found. Urine PBG is a better test than urine ALA overall for AIP, but both are used (see Porphobilinogen Deaminase, Erythrocyte *on page 1072*).

Coproporphyrin and PBG excretion in urine are markedly increased during acute attacks of **hereditary coproporphyria** (HCP) and increased fecal coproporphyrin III is described. Urinary ALA is increased with acute attacks as well.

In **variegate porphyria (VP)**, results are similar to those of AIP. Urine coproporphyrin exceeds uroporphyrin excretion during acute attacks. Plasma can be used to identify **porphyria cutanea tarda** (PCT) and distinguish it from VP. Fecal porphyrins can be used to differentiate AIP from VP or HCP.

In **lead poisoning** elevation of ALA greater than that of PBG occurs, and PBG may be normal. Urinary coproporphyrin characteristically is increased. Free erythrocyte protoporphyrin and zinc protoporphyrin (ZPP) are increased. (See Lead, Blood *on page 832*.) Toxins such as lead interfere with heme synthesis and cause porphyrinuria.

Increased urine excretion of uroporphyrinogen, uroporphyrin, and coproporphyrin occurs in **porphyria cutanea tarda (PCT)**. These patients do not excrete increased PBG. Iron overload is found with the acquired form, for which treatment includes repeated phlebotomies.

Limitations Porphyrin excretion is increased in many diseases (see Abstract).

Methodology High performance liquid chromatography (HPLC) is the method of choice; spectrophotometry and fluorometry are also used.

Additional Information Porphyrias are classified according to whether neurovisceral and/or cutaneous symptomatology are present and whether the presentation is acute or chronic. Neurovisceral symptomatology includes abdominal pain, dark urine, vomiting, constipation, muscle weakness, mental status changes, limb, head, neck, or chest pain, hypertension, tachycardia, (Continued)

Porphyrins, Quantitative, Urine *(Continued)*

seizures, sensory loss, fever, respiratory paralysis, and diarrhea. Cutaneous symptomatology includes severe dermatologic photosensitivity, skin fragility, blistering lesions, cutaneous thickening, scarring hypertrichosis, hyperpigmentation, erythrodontia, and edema. See table on following page.

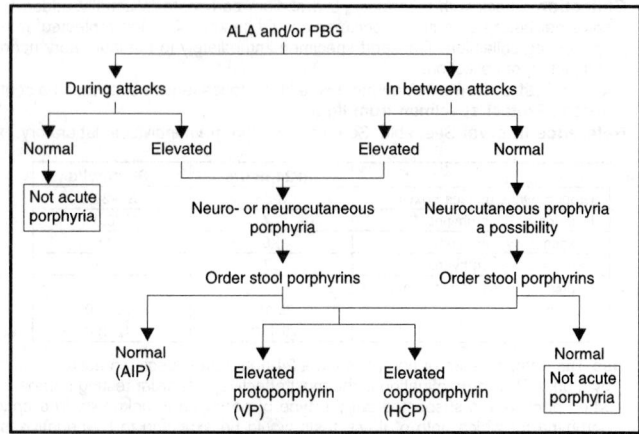

Interpretation of urine studies, as the cause of symptoms, during evaluation of acute porphyria. AIP = acute intermittent porphyria; ALA = Δ- aminolevulinic acid; HCP = hereditary coproporphyria; PBG = porphobilinogen; VP = variegate porphyria.

From Tefferi A, Solberg LA, and Ellefson RD, "Porphyrias: Clinical Evaluation and Interpretation of Laboratory Tests," *Mayo Clin Proc*, 1994, 69:289-90 with permission.

Useful Tests for Porphyrias

Porphyria	First Order Tests	Second Order Tests
Suspected porphyria; type unknown	PBG, random or 24-hour urine Porphyrins, 24-hour urine	
Acute intermittent porphyria (AIP) Variegate porphyria (VP) Hereditary coproporphyria (HCP) 5-Aminolevulinic acid dehydratase porphyria (ADP)	PBG, random or 24-hour urine ALA, 24-hour urine Porphyrins, 24-hour urine	Porphyrins, feces (differentiates AIP, VP, and HCP) Porphobilinogen deaminase (uroporphyrinogen I synthase), erythrocytes (deficient in AIP) ALA dehydratase, erythrocytes (deficient in ADP)
Porphyria cutanea tarda (PCT) Hepatoerythropoietic porphyria (HEP)	Porphyrins, 24-hour urine Porphyrins, plasma	Porphyrins, feces Uroporphyrinogen decarboxylase, erythrocytes (deficient in HEP)
Erythropoietic protoporphyria (EPP)	Porphyrins, erythrocytes	Protoporphyrins, erythrocytes Porphyrins, feces
Congenital erythropoietic porphyria (CEP)	Porphyrins, 24-hour urine Prophyrins, erythrocytes	Porphyrins, feces Porphyrins, plasma Uroporphyrinogen III cosynthase, erythrocytes

References

Ahmed I, Cockerill F, Krause P, et al, "The Challenges of Testing for and Diagnosing Porphyrias," *Mayo Reference Services Communiqué*, 2002, 27(11):1-10.

Bonkovsky HL and Barnard GF, "Diagnosis of Porphyric Syndromes: A Practical Approach in the Era of Molecular Biology," *Semin Liver Dis*, 1998, 18(1):57-66.

Deacon AC and Elder GH, "ACP Best Practice No 165: Front Line Tests for the Investigation of Suspected Porphyria," *J Clin Pathol*, 2001, 54(7):500-7.

Foran SE and Abel G, "Guide to Porphyrias. A Historical and Clinical Perspective," *Am J Clin Pathol*, 2003, 119(Suppl 1):S86-S93.

Gilbert-Barness E and Barness LA, *Metabolic Diseases: Foundations of Clinical Management, Genetics, and Pathology*, Natick, MA: Eaton Publishing, 2000, 443-55.

Porphyrias: Overview

Disorder	Inheritance	Deficient Enzyme	Usual Age of Clinical Onset	Porphyria Classification	Increased Levels Seen in Affected Patients			
					Urine	Feces	Erythrocytes	Plasma
Porphyria cutanea tarda (PCT) (most common porphyria)	Type I, acquired; types II and III, autosomal dominant	Uroporphyrinogen dexcarboxylase	Adulthood	Nonacute Cutaneous	Uroporphyrin Heptacarboxylporphyrin	Isocoproporphyrin Heptacarboxylporphyrin III		Uroporphyrin
Hepatoerythropoietic porphyria (HEP) (rare)	Autosomal recessive	Uroporphyrinogen decarboxylase (markedly deficient)	Infancy; early childhood	Nonacute Cutaneous	Uroporphyrin Heptacarboxylporphyrin	Isocoproporphyrin Heptacarboxylporphyrin III	Zinc protoporphyrin	Uroporphyrin
Erythropoietic protoporphyria (EPP)	Autosomal dominant	Ferrochelatase	Childhood	Nonacute Cutaneous	—	Protoporphyrin	Free erythrocyte protoporphyrin	Protoporphyrin
Congenital erythrocytic porphyria (CEP) Günther disease (rare)	Autosomal recessive	Uroporphyrinogen III cosynthase (markedly deficient)	Early infancy	Nonacute Cutaneous	Uroporphyrin Coproporphyrin (pink to red urine)	Uroporphyrin I Coproporphyrin I	Uroporphyrin Coproporphyrin Zinc protoporphyrin	Uroporphyrin Coproporphyrin I
Acute intermittent porphyria (AIP) (second most common porphyria)	Autosomal dominant	Porphobilinogen deaminase	Adulthood	Acute Neurovisceral	ALA PBG Uroporphyrin Coproporphyrin (±)	—		Increased porphyrins during acute episodes
Variegate porphyria (VP)	Autosomal dominant	Protoporphyrinogen oxidase	Adulthood	Acute Neurovisceral Cutaneous	Coproporphyrin PBG ALA	Protoporphyrin > coproporphyrin Coproporphyrin III to coproporphyrin I ratio	—	—
Hereditary coproporphyria (HCP) (rare)	Autosomal dominant	Coproporphyrinogen oxidase	Adulthood	Acute Neurovisceral Cutaneous	Coproporphyrin ALA PBG	Coproporphyrin Coproporphyrin III to coproporpyrin I ratio	—	Coproporphyrin
5-Aminolevulinic acid dehydratase porphyria (ADP) (rare)	Autosomal recessive	ALA dehydratase	Variable; few reported cases	Acute Neurovisceral	ALA	—	—	—

Hindmarsh JT, Oliveras L, and Greenway DC, "Plasma Porphyrins in the Porphyrias," *Clin Chem*, 1999, 45(7):932-3.

Sassa S and Kappas A, "Molecular Aspects of the Inherited Porphyrias," *J Intern Med*, 2000, 247(2):169-78.

Schanbacher CF, Vanness ER, Daoud MS, et al, "Pseudoporphyria: A Clinical and Biochemical Study of 20 Patients," *Mayo Clin Proc*, 2001, 76(5):488-92.

Young DS, "Effects of Drugs on Clinical Laboratory Tests," 5th ed, Volume 1: Listing by Test, Washington, DC: AACC Press, American Association of Clinical Chemistry, 2000, Section 3, 474-5.

Internet Web Sites

www.porphyriafoundation.com/

♦ **Porter-Silber Chromogens, Urine** *see* 17-Hydroxycorticosteroids, Urine *on page 753*

♦ **Postoperative Cell Salvage** *see* Autologous Transfusion, Intraoperative Blood Salvage *on page 222*

♦ **Postprandial Glucose** *see* Glucose, Postglucose Load, Plasma *on page 647*

♦ **Post-transfusion Purpura** *see* Platelet Antibodies *on page 1047*

♦ **Posture** *see page 11*

♦ **Pot** *see* Cannabinoids (Marijuana Metabolites), Qualitative, Urine *on page 335*

♦ **Potassium, Arterial** *see* Potassium, Serum or Plasma *on page 1078*

♦ **Potassium or Sodium Cyanide** *see* Cyanide, Blood *on page 485*

Potassium, Serum or Plasma

Related Information

Adrenal Cortex: Laboratory Assessment Overview *on page 110*
Aldosterone, Serum or Plasma *on page 122*
Aldosterone, Urine *on page 124*
Amino Acids, Urine *on page 145*
Bicarbonate, Blood *on page 258*
Calcium, Serum *on page 329*
Chloride, Serum, Plasma, or Blood *on page 391*
Chloride, Urine *on page 394*
Concentration Test, Urine *on page 446*
Digoxin, Serum *on page 512*
Electrolyte Panel, Serum *on page 532*
Magnesium, Serum *on page 885*
Magnesium, Urine *on page 886*
Potassium, Urine *on page 1080*
Renin Activity, Plasma *on page 1149*
Salicylate, Serum or Plasma *on page 1176*
Sodium, Serum or Plasma *on page 1210*

Synonyms K^+, Serum or Plasma

Applies to Potassium, Arterial

Abstract The major intracellular cation, K^+ is very commonly measured as one of the serum/plasma or urine electrolytes.

Specimen Serum; plasma; capillary, venous, or arterial blood

Container Red top tube or green top (heparin) tube

Collection Avoid very small needles if possible. Avoid stasis, use of tourniquet if possible and **avoid hand clenching**. Fist clenching increases K^+. If a tourniquet must be used, sample blood 1-2 minutes after the hand is relaxed and the tourniquet removed. Avoid potassium-containing tubes such as potassium oxalate. If arterial puncture is done for pO_2, plasma can be tested for Na^+, K^+, and Cl^- so long as lithium and not potassium heparinate anticoagulant is used, sparing the patient a venipuncture.

Storage Instructions Storage of unspun blood at 4°C causes serum and plasma K^+ to increase.

Causes for Rejection Hemolyzed specimen, serum specimen not removed from clot in patient with high platelet count

Reference Interval Plasma: 3.5-5.0 mmol/L. Add **approximately** 0.1 to normal ranges if serum is sampled rather than plasma. Pediatric ranges are sometimes reported as slightly higher than adult levels. Even slight hemolysis can dramatically increase K^+ results; red cells have an intracellular K^+ concentration of 100-120 mmol/L or more.

Possible Panic Range Newborns: <2.5 mmol/L (SI: <2.5 mmol/L), >7.0 mmol/L (SI: >7.0 mmol/L); adults: <2.5 mmol/L (SI: <2.5 mmol/L), >6.5 mmol/L (SI: >6.5 mmol/L). If potassium is high and serum was used, examine peripheral blood smear for thrombocytosis and/or leukocytosis.

Use Evaluate electrolyte balance; K^+ level should be followed especially in elderly patients, those on intravenous hyperalimentation, in patients on diuretic therapy and in cases of renal disease, particularly patients with acute renal failure and those on hemodialysis. K^+ concentrations are part of regular assessment of acid-base balance, management of intravenous therapy, and evaluation of patients with hypertension.

Evaluate muscular weakness and irritability, mental confusion, weakness; manage leukemia, diseases of gastrointestinal tract including laxative abuse, hepatic encephalopathy, fistulas and tube drainage; evaluate and prevent cardiac arrhythmias; evaluate alcoholism with delirium tremens; detect, diagnose, and manage mineralocorticoid excess or deficiency, renal tubular abnormalities, heat stroke, licorice ingestion mineralocorticoid effect.

Hypokalemia (low potassium) is found in 80% to 90% of hypertensive patients with primary aldosteronism. Hypokalemia occurs with secondary hyperaldosteronism (eg, renal artery stenosis). Low K^+ occurs with endogenous or exogenous increase in other corticosteroids, including that in Cushing syndrome as well as with dietary or parenteral deprivation of K^+ (eg, parenteral therapy without adequate K^+ replacement). Hypokalemia occurs with vomiting, diuretics, β_2-adrenergic agonist medications, burns, excessive perspiration, alkalosis, Bartter syndrome, Gitelman syndrome, malnutrition, ureterosigmoidostomy, alcoholism, theophylline overdose, toluene inhalation, anabolic states, insulin overdose, folic acid deficiency, and renal tubular disorders.

Other drugs causing hypokalemia include bronchodilators; tocolytic agents; caffeine; verapamil intoxication; chloroquine intoxication; diuretics; mineralocorticoids and substances such as licorice; high-dose glucocorticoids; high-dose penicillin; nafcillin, ampicillin, and carbenicillin; aminoglycosides; cisplatin; foscarnet; amphotericin B; phenolphthalein; and sodium polystyrene sulfonate.

Causes of K^+ loss in stools include laxatives, diarrhea, tumors (eg, VIPoma, colorectal villous tumor, and Zollinger-Ellison syndrome), jejunoileal bypass, enteric fistulas, malabsorption, chemotherapy, radiation enteropathy, and other entities (see Potassium, Urine *on page 1080*).

Hyperkalemia (high potassium) reflects generally inadequate renal excretion, mobilization of potassium from the tissues, or excessive intake or administration. Hypertensive patients who suddenly change to the use of KCl-containing "low salt" substitutes may develop symptomatic severe hyperkalemia (with cardiac rhythm and ECG abnormalities). Hyperkalemia occurs with trauma, with administration of K^+ salts of some drugs, ACE inhibitors, Addison disease, acidosis including ketoacidosis as in diabetes mellitus, insulin lack, with increased osmolality (eg, glucose, mannitol), and in other entities as well as in renal diseases with azotemia, with malignant hyperthermia, and with renal tubular acidosis. Increased K^+ can occur with potassium-sparing diuretics, nonsteroidal anti-inflammatory drugs, especially in the presence of renal disease. Systemic heparin therapy can suppress aldosterone release and increase K^+, especially in the presence of other factors. Hyperkalemia is reported with high-dose trimethoprim-sulfamethoxazole therapy. Other factors which may cause hyperkalemia include dehydration, exercise, pregnancy, standing posture, and hyperventilation. Hyperkalemia can be caused by ingestion of large quantities of water from a potassium-based water softener in subjects in renal failure. **Artifact causing hyperkalemia includes hemolysis from collection, storage, or processing.**

Limitations Inadequate sodium intake may mask the hypokalemia of aldosteronism; sodium loading in that setting may make hypokalemia recognizable. Heparinized plasma is probably the specimen of choice for K^+, because clotting causes cytolysis and may elevate serum values, usually but not always only slightly. Artifactual hyperkalemia may be found in serum (but not plasma) in patients with essential thrombocythemia. Plasma K^+ is subject to spurious increase if hemolysis has occurred.

(Continued)

Potassium, Serum or Plasma (Continued)

Samples left unattended for several hours may have falsely elevated K+ in the absence of visible hemolysis.

While acute hypokalemia is reflected largely by serum/plasma K+ concentration, chronic hypokalemia is more apt to be accompanied by reduction of total body stores, as well as serum/plasma concentrations.

Methodology Ion-selective electrode (ISE) in most laboratories

Additional Information Preoperative patients may benefit from serum/plasma potassium measurement. Low K+ is much more significant with a low pH than with a high pH. When pH increases by 0.1, K+ decreases ~0.6 mmol/L. With low pH, as in ketoacidosis, as therapeutic adjustment towards normal pH is made, plasma/serum K+ concentrations will decrease. Phosphorus concentrations tend to follow potassium downwards during therapy of diabetic ketoacidosis; both are largely intracellular. With insulin therapy (and increased utilization of carbohydrate), potassium moves into cells and serum/plasma concentrations fall. Hyperalimentation may have a similar effect. Hypokalemia has been reported in slightly >50% of a series of 32 patients with acute myelogenous leukemia, but thrombocytosis can increase serum K+ levels.

Consider magnesium status in patients who have hypokalemia.

Since platelets release K+ during coagulation, samples from patients who have thrombocytosis (eg, some cases of polycythemia vera and other myeloproliferative diseases) yield spuriously elevated K+ concentrations. This "pseudohyperkalemia" may also occur in cases of leukemia with high WBC count (notably chronic myelogenous leukemia) as potassium is released from WBCs and platelets during clot formation. For such patients it is best to assay K+ on a heparinized sample. Pseudohyperkalemia in serum specimens may be due to increased platelets. Serum potassium increases with the platelet count in normal subjects and in those with thrombocytosis and spherocytosis, and such pseudohyperkalemia increment is an artifact. Plasma concentrations are not affected.

References

Gennari FJ, "Hypokalemia," *N Engl J Med*, 1998, 339:451-58.

Halperin ML and Kamel KS, "Potassium," *Lancet*, 1998, 352(9122):135-40.

Hawkins R, "Measurement of Whole-Blood Potassium - Is It Clinically Safe?" *Clin Chem*, 2003, 49(12):2105-6.

Howard MR, Ashwell S, Bond LR, et al, "Artifactual Serum Hyperkalaemia and Hypercalcaemia in Essential Thrombocythaemia," *J Clin Pathol*, 2000, 53(2):105-9.

Hoye A and Clark A, "Iatrogenic Hyperkalaemia," *Lancet*, 2003, 361(9375):2124-5.

Young DS, "Effects of Drugs on Clinical Laboratory Tests," 5th ed, Volume 1: Listing by Test, Washington, DC: AACC Press, American Association of Clinical Chemistry, 2000, Section 3, 644-54.

♦ **Potassium Thiocyanate (KSCN)** see Thiocyanate, Serum, Plasma, or Urine *on page 1245*

Potassium, Urine

Related Information

Synonyms K+, Urine; Urine K+

Abstract Urine potassium studies provide explanation for disturbances of serum or plasma values. Urinary potassium losses may reflect primary hyperaldosteronism; congenital adrenal hyperplasia (22-β-hydroxylase deficiency; 17-α-hydroxylase deficiency); renin-secreting tumors, ectopic corticotropin syndrome, Cushing syndrome (pituitary or adrenocortical disease); glucocorticoid-responsive aldosteronism, renovascular hypertension, malignant hypertension, vasculitis; Liddle syndrome; 11-β-hydroxysteroid dehydrogenase deficiency; Bartter syndrome and Gitelman syndrome.

Specimen Random or timed urine (ie, 8-, 12-, or 24-hour)

Container No preservative

Storage Instructions Refrigerate

Reference Interval 26-123 mmol/24 hours, markedly intake dependent. If significantly decreased serum or plasma K⁺ has existed for days or more, urine K⁺ excretion should be low: ≤15 mmol/L (SI: ≤15 mmol/L), or ≤30 mmol/24 hours (SI: ≤30 mmol/day). There is significant diurnal variation, output greater at night.

Differential Diagnosis of Hypokalemia

Diuretic therapy is the most common cause of hypokalemia
(see Gennari reference).

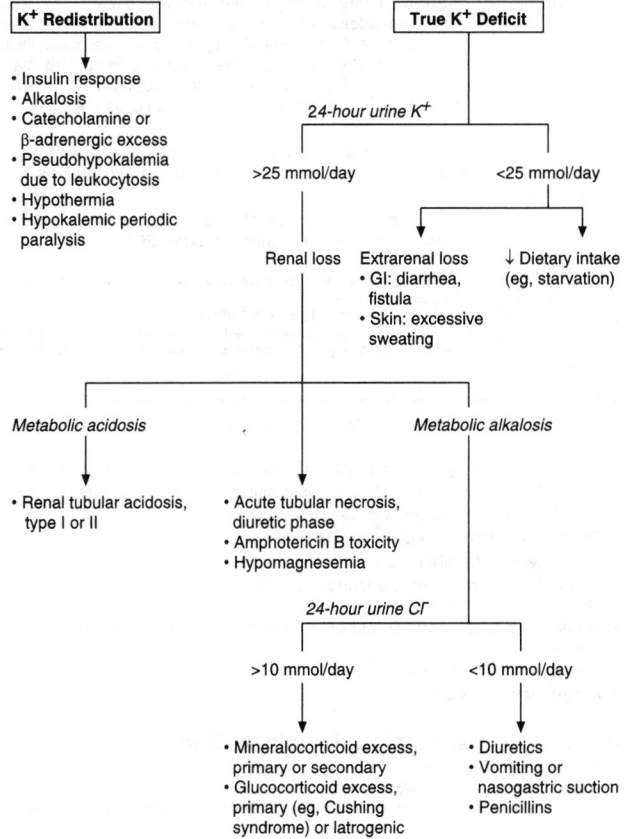

Adapted from Heusel JW, Siggard-Anderson O, and Scott MG, "Physiology and Disorders of Water, Electrolyte and Acid-Base Metabolism," *Tietz Textbook of Clinical Chemistry*, 3rd ed, Burtis CA and Ashwood ER, eds, Philadelphia, PA: WB Saunders Co, 1999, 1095-124.

Use Evaluate electrolyte balance, acid-base balance; evaluate hypokalemia; urinary loss of 40 mmol/24 hours (SI: 40 mmol/day) in the presence of hypokalemia <3 mmol/L is excessive. In the presence of such hypokalemia, urinary K⁺ determination is helpful to separate renal from nonrenal losses. Excretion <20 mmol/24 hours (SI: <20 mmol/day) is evidence that hypokalemia is not from renal loss. Causes include vomiting, nasogastric tube suctioning, villous tumor, (Continued)

Potassium, Urine (Continued)

VIPoma, laxative abuse, and hyperhidrosis. Renal loss >50 mmol/L in a hypokalemic, hypertensive patient not on a diuretic may indicate primary or secondary hyperaldosteronism/hyper-reninemia. The kidneys do not respond quickly to potassium deprivation. There is renal wastage of K⁺ in secondary aldosteronism. Glucocorticoids, including endogenous steroids in Cushing syndrome, are among the causes of kaliuresis. A 24-hour urine collection for potassium represents an appropriate screening test for hyperaldosteronism, preferably following K⁺ repletion to achieve serum level >3 mmol/L. Urine sodium:potassium ratio helps in the evaluation of children with hypercalciuria.

Limitations A number of agents cause alterations in urinary K⁺ excretion.

Methodology Ion-selective electrode (ISE) in most laboratories

Additional Information Urinary K⁺ may be elevated with dietary (food and/or medicinal) increase, hyperaldosteronism, renal tubular acidosis, onset of alkalosis, and with other disorders. Characteristics of primary aldosteronism include hypertension, hypokalemia, decreased plasma renin and increased plasma aldosterone concentrations. Such patients have inappropriate kaliuresis, 24-hour urinary K⁺ >30 mmol, in the presence of hypokalemia (serum K⁺ <3.0 mmol/L).

Time relationships are important in interpretation. K⁺ will decrease in Addison disease and in renal disease with decreased urine flow (nephrosclerosis, pyelonephritis, glomerulonephritis).

An algorithm for the differential diagnosis of hypokalemia is given in the flowchart on previous page. See Chloride, Urine *on page 394.*

References

Carroll HJ and Oh MS, *Water Electrolyte and Acid-Base Metabolism: Diagnosis and Management*, Philadelphia, PA: JB Lippincott Co, 1978.

Gennari FJ, "Hypokalemia," *N Engl J Med*, 1998, 339(7):451-8.

Young DS, "Effects of Drugs on Clinical Laboratory Tests," 5th ed, Volume 1: Listing by Test, Washington, DC: AACC Press, American Association of Clinical Chemistry, 2000, Section 3, 654-5.

- ◆ **PP, 2-Hour** *see* Glucose, Postglucose Load, Plasma *on page 647*
- ◆ **Prader-Willi Syndrome** *see* Gene Rearrangement for Leukemia and Lymphoma *on page 633*
- ◆ **Pralidoxime** *see* Organophosphate Pesticides, Urine, Blood, or Serum *on page 975*
- ◆ **Pravastatin** *see* Mevalonic Acid, Urine or Amniotic Fluid *on page 911*
- ◆ **Prealbumin** *see* Transthyretin, Serum *on page 1271*
- ◆ **Prealbumin, CSF** *see* Cerebrospinal Fluid Protein *on page 371*
- ◆ **Preanalytical Variability** *see page 11*
- ◆ **Precautions, Specimen Collection** *see* Blood and Fluid Precautions, Specimen Collection *on page 271*
- ◆ **Predeposit Autologous Donation** *see* Autologous Transfusion, Preoperative Deposit *on page 223*
- ◆ **Predictive Value (PV)** *see page 11*
- ◆ **Pregnancy** *see page 11*

Pregnancy-Associated Protein A, Serum

Related Information

Alpha₁-Fetoprotein, Serum *on page 136*

Amniotic Fluid, Chromosome and Genetic Abnormality Analysis *on page 152*

Chorionic Gonadotropin, Human, Serum and Urine *on page 397*

Chorionic Villus Sampling, Chromosome and Genetic Abnormality Analysis *on page 400*

Estriol, Unconjugated, Pregnancy, Serum, Plasma, or Urine *on page 554*

Inhibin A, Serum *on page 799*

Synonyms PAPP-A

Applies to Combined Test; Double Test; First Trimester Combined Test; Integrated Test; Triple Test

Abstract In the first trimester of pregnancy, Down syndrome is associated with high concentrations of maternal free β-hCG or hCG, low concentrations of pregnancy-associated plasma protein A (PAPP-A), and high values for fetal nuchal translucency by ultrasound. Application of such multiple testing has increased sensitivity of first trimester screening to as much as 80%. The reader is urged to review the monographs listed above under Related Information.

Specimen Serum

Sampling Time Measure only at 10-14 weeks gestation. The power of discrimination of PAPP-A seems to be lost after that period.

Storage Instructions Separate serum and refrigerate.

Special Instructions Maternal age is needed.

Reference Interval PAPP-A values are low in affected pregnancies throughout the 8- to 14-week period, more at 8-11 weeks than at 12-14 weeks.

Use The most common reason for provision of prenatal genetic diagnosis is risk of trisomy 21 (Down syndrome). Screening for Down syndrome in the first trimester with PAPP-A and either hCG or its free β-subunit in maternal serum is feasible. Such screening, with fetal nuchal translucency by ultrasonography, detects most cases of trisomy 18 as well.

Limitations At 8-14 weeks gestation, the detection rate for trisomy 21 is 62% to 85% with a false-positive rate of 5% to 9%. For trisomy 18, 90% of cases were identified with a 2% false-positive rate. Down syndrome screening in the first trimester is currently not widely accepted as standard of care.

Methodology Two-site immunometric assay, enzyme-linked immunosorbent assay (ELISA)

Additional Information The PAPP-A and β-hCG values at 10 weeks are combined with maternal age to estimate risk of Down syndrome. The results using these two tests are similar to the **triple test** (α-fetoprotein, unconjugated estriol and hCG with maternal age) at 15-22 weeks.

The combined test proposes application of PAPP-A, free β subunit of hCG, and nuchal translucency at 10-13 weeks. Without AFP, screening for open neural tube or abdominal wall defects is lost.

Rates of detection of Down syndrome were 42% for PAPP-A, 29% for free hCG, 25% for free β-subunit of hCG, 17% for α-fetoprotein, and 4% for unconjugated estriol at 5% false-positive rates. In combination with maternal age and PAPP-A, the detection rate of hCG was 63% while 60% was the rate for free β-subunit hCG.

Patients with positive screens are often referred for chorionic villus sampling or amniocentesis.

The **integrated test**, combining results of first and second trimester markers, achieves low false-positive rates. Baliff and Mooney provide very helpful discussion.

A 2003 multicenter study of 8216 patients with singleton, first-trimester pregnancies used maternal age, maternal serum levels of free βhCG, PAPP-A, and ultrasonographic fetal nuchal translucency measurements as screening criteria. This combination of tests identified 85.2% of 61 Down syndrome (trisomy 21) cases with a false-positive rate of 9.4% and 90.9% of 11 trisomy 18 cases with a 2% false-positive rate (Wapner et al). Down syndrome, the most common cause of mental retardation, is the most prevalent chromosomal autosomal abnormality in live births.

References

Baliff JP and Mooney RA, "New Developments in Prenatal Screening for Down Syndrome," *Am J Clin Pathol*, 2003, 120(Suppl 1):S14-24.

Copel JA and Bahado-Singh RO, "Prenatal Screening for Down's Syndrome - A Search for the Family's Values," *N Engl J Med*, 1999, 341(7):521-2.

Haddow JE, Palomaki GE, Knight GJ, et al, "Screening of Maternal Serum for Fetal Down's Syndrome in the First Trimester," *N Engl J Med*, 1998, 338(14):955-61.

Malone FD and D'Alton ME, "First-Trimester Sonographic Screening for Down Syndrome," *Obstet Gynecol*, 2003, 102(5 Part 1):1066-79.

"Maternal Serum Quad Screening: Review of Laboratory Testing and Clinical Application," *Mayo Reference Services Communiqué*, 2002, 27(6):1-8.

Wald NJ, George L, Smith D, et al, "Serum Screening for Down's Syndrome Between 8 and 14 Weeks of Pregnancy," *Br J Obstet Gynaecol*, 1996, 103(5):407-12.

Wald NJ, Watt HC, and Hackshaw AK, "Integrated Screening for Down's Syndrome Based on Tests Performed During the First and Second Trimesters," *N Engl J Med*, 1999, 341(7):461-7.

Wapner R, Thom E, and Simpson JL, "First-Trimester Screening for Trisomies 21 and 18," *N Engl J Med*, 2003, 349(15):1405-13.

Wenstrom K, ACOG Committee on Practice Bulletins, "Prenatal Diagnosis of Fetal Chromosomal Abnormalities," *ACOG Practice Bulletin*, 2001, 27:1-16.

♦ **Pregnancy Testing** *see* Chorionic Gonadotropin, Human, Serum and Urine *on page 397*

Pregnancy Test, Serum or Urine
Related Information
Chorionic Gonadotropin, Human, Serum and Urine *on page 397*
Progesterone, Serum *on page 1093*

Synonyms Beta Subunit Human Chorionic Gonadotropin, Urine or Serum; hCG, Slide Test, Stat; hCG, Urine; Human Chorionic Gonadotropin, Urine

Abstract The gold standard for the ascertainment of pregnancy is a urine or serum hCG test.

Specimen Serum or urine; first voided morning specimen is preferred if urine is tested (to obtain most concentrated specimen).

Container Red top tube or urine container

Storage Instructions Serum should be frozen at -20°C if not run within 48 hours. Urine is stable 4 hours at 25°C and 3 days at 4°C.

Causes for Rejection Urine specimen grossly contaminated, low urinary specific gravity, proteinuria (applicable to certain tests), gross lipemia or turbidity

Turnaround Time 1 hour

Special Instructions Centrifuge turbid urine specimens prior to testing.

Reference Interval Normal male/nonpregnant female: negative; normal pregnant female: positive.

Sensitivity and specificity of β-subunit two point RIA or EIA tests may allow detection of pregnancy 6 days after conception. Qualitative tests are positive when intact hCG exceeds 25 IU/L (International Reference Preparation, IRP). Concentration of hCG in urine is parallel to that of serum, with about equal quantities in 1 L of serum and in a 24-hour urine collection.

Use Detect pregnancy. A quantitative test for the presence of hCG is preferable to a pregnancy test for diagnosing or following a patient with gestational trophoblastic neoplasia.

Limitations Results may be negative in early pregnancy or with low specific gravity urine. In early pregnancy, incomplete abortion, recent complete abortion, ectopic pregnancy (in which hCG level is low), slide test endpoints may be difficult to interpret. Methods using covalent bonded latex particles and tests producing macroagglutination are more reliably interpreted. Reliability of serum and urine tests are comparable.

Methodology Immunoassay, multiple formats, largely replaced by enzyme immunoassay (EIA); β-subunit hCG by radioimmunoassay (RIA) (serum or urine); immunoradiometric (IRMA)

Additional Information In normal pregnancy, hCG levels rise at implantation and peak at 8-12 weeks. Although newer urine pregnancy tests are quite sensitive, false negatives can occur early in gestation. In such cases, if ectopic pregnancy is suspected, serial serum hCG and progesterone assays may be of value (see Chorionic Gonadotropin, Human, Serum and Urine *on page 397*).

Early in the first trimester of pregnancy (1-2 weeks) serum hCG concentrations are from 50-500 mIU/mL (SI: 50-500 IU/L). Current generation sensitive tests can detect pregnancy shortly (2-3 days) after implantation of the ovum. By 3-4 weeks of gestation, hCG is at the 500-10,000 mIU/mL level (SI: 500-10,000 IU/L). Serum hCG level peaks during the second to third month of gestation (30,000-100,000 mIU/mL) (SI: 30,000-100,000 IU/L). Use of serum for pregnancy testing may provide greater sensitivity, of special value in cases of early pregnancy, and is of greater value in serial testing for follow-up of an abnormal gestation or gestational trophoblastic neoplasia. If hCG concentrations do not correlate with the clinical situation, periodic repeat hCG and progesterone determinations may be helpful. If there is demise of the developing embryo/fetus, hCG and progesterone concentrations will fall. Because of slow clearance, hCG may be detected in serum/urine for as long as 4 weeks following abortion. When a gestation is large enough to produce in excess of 1600 IU/L hCG (IRP) (at about 5 weeks gestation) one may anticipate that in normal gestation the sac should be visible on transvaginal ultrasonography.

Serum progesterone levels used with β-hCG levels may assist in differentiating normal intrauterine from abnormal intrauterine or **ectopic pregnancy**

(cutoff: 15 ng/mL) (SI: 48 nmol/L). β-hCG and progesterone levels are lower in abnormal pregnancies. Less overlap occurs using progesterone values. When a positive pregnancy test is obtained, differential considerations should include the possibility of simultaneous intrauterine and extrauterine gestations and the vanishingly rare possibility of passively acquired hCG (an individual recently transfused with fresh frozen plasma prepared from pregnant donors).

A number of commercially successful home pregnancy tests are available. They vary widely in performance.

References

Baskin LB and Charles RA, "Detecting Ectopic Pregnancy," *Lab Med*, 1997, 28(2):103-5.

Bastian LA and Piscitelli JT, "Is This Patient Pregnant? Can You Reliably Rule In or Rule Out Early Pregnancy by Clinical Examination?" *JAMA*, 1997, 278(7):586-91.

Bastian LA, Nanda K, Hasselblad V, et al, "Diagnostic Efficiency of Home Pregnancy Test Kits: A Meta-analysis," *Arch Fam Med*, 1998, 7(5):465-9.

Smikle CB, Sorem KA, Wians FH, et al, "Measuring Quantitative Serum Human Chorionic Gonadotropin: Variations in Levels Between Kits," *J Repro Med*, 1995, 40(6):439-42

Pregnanetriol, Urine

Related Information

17-Hydroxyprogesterone, Whole Blood, Serum, or Plasma *on page 755*

Abstract Pregnanetriol is a metabolite of 17-hydroxyprogesterone. Increased urinary excretion is observed when the serum level of 17-hydroxyprogesterone is elevated.

Patient Preparation Avoid exercise before and during collection.

Specimen 24-hour urine

Collection Preserve with boric acid.

Reference Interval Varies with laboratory.
- 0-5 years: <0.1 mg/day (SI: <0.3 µmol/day)
- 6-9 years: <0.3 mg/day (SI: <0.9 µmol/day)
- adult male: 0.4-2.5 mg/day (SI: 1.2-7.5 µmol/day)
- adult female: follicular: 0.1-0.8 mg/day (SI: 0.3-5.3 µmol/day); luteal: 0.9-2.2 mg/day (SI: 2.7-6.5 µmol/day)

Use Urinary pregnanetriol is increased in congenital adrenal hyperplasia (21-hydroxylase type), but this test is not usually required in the diagnostic evaluation for this disorder.

Limitations The expected elevation in urinary pregnanetriol is not observed in infants. Muscular exercise may increase urinary pregnanetriol.

Methodology Gas-liquid chromatography (GLC)

References

Grumbach MM and Conte FA, "Disorders of Sex Differentiation," *Williams Textbook of Endocrinology*, 9th ed, Wilson JD, Foster DW, Kronenberg HM, et al, eds, Philadelphia, PA: WB Saunders Co, 1998, 1303-425.

Painter PC, Cope JY, and Smith JL, "Reference Information for the Clinical Laboratory," *Tietz Textbook of Clinical Chemistry*, 3rd ed, Burtis CA and Ashwood ER, eds, Philadelphia, PA: WB Saunders Co, 1999,1831.

Shackleton CH, Irias J, McDonald C, et al, "Late-Onset 21-Hydroxylase Deficiency: Reliable Diagnosis by Steroid Analysis of Random Urine Collections," *Steroids*, 1986, 48(3-4):239-50.

Prekallikrein

Related Information

Activated Partial Thromboplastin Time *on page 100*

Coagulation Factor Assays *on page 418*

High-Molecular Weight Kininogen *on page 731*

Synonyms Fletcher Factor

Applies to Kallikrein

Abstract Prekallikrein is a coagulation protein involved in the early stages of intrinsic pathway activation. Prekallikrein deficiency can cause a marked prolongation of the PTT, but it does not cause bleeding. The same is true for factor XII deficiency and high-molecular weight kininogen (HMWK) deficiency.

Patient Preparation Patients cannot be on hirudin or argatroban anticoagulation, which can interfere with mixing studies and PTT-based prekallikrein assays. Danaparoid may also interfere with these assays. If heparin is present, notify the laboratory because heparin must be removed prior to testing.

Specimen Plasma

Container One blue top (sodium citrate) tube

Collection Routine venipuncture. If multiple tests are being drawn, draw blue top tubes after any red top tubes but before any lavender top (EDTA), green top
(Continued)

Prekallikrein *(Continued)*

(heparin), or gray top (oxalate/fluoride) tubes. Immediately invert tube gently at least 4 times to mix. Tubes must be appropriately filled. Deliver tubes immediately to the laboratory.

Storage Instructions Separate plasma from cells as soon as possible. If assay is not performed within 4 hours, freeze plasma specimen.

Causes for Rejection Specimen received more than 4 hours after collection, tubes not filled, clotted specimens

Turnaround Time Less than 1 day (longer if test is a send-out)

Reference Interval 60% to 140% of normal. Newborns have lower levels than adults; the values increase to near adult normal range by age 6 months.

Use May be performed when a routine prolonged PTT evaluation finds no explanation for the prolongation (see Mixing Studies (Coagulation) *on page 918*). The PTT is normal in the mixing study; factors VIII, IX, XI, and XII are normal; the PT and fibrinogen are normal; and lupus anticoagulant assays are negative. If prekallikrein assays are normal, HMWK assays may be considered.

Methodology

Screening assay: Preincubate a PTT test sample for 10 minutes prior to adding calcium. A prolonged PTT that shortens after the 10-minute incubation is suspicious for a prekallikrein deficiency. In addition, if a PTT is performed with ellagic acid as the intrinsic pathway activator, the PTT will usually be normal in prekallikrein deficiencies.

Specific factor assay for prekallikrein: Can be performed similar to other coagulation factor assays. Patient plasma is mixed with prekallikrein-deficient plasma and a PTT is performed on the mixture. The amount of prekallikrein in the patient plasma is determined from a standard curve that plots known amounts of prekallikrein against PTT values. **Note:** PTT tests are normally incubated for 3 minutes prior to adding calcium. When using the PTT in a prekallikrein factor assay, the incubation period is shortened to 1 minute. A chromogenic prekallikrein assay is also available.

Additional Information Prekallikrein is one of the contact factors that participates in the activation of the intrinsic pathway of coagulation when blood is exposed to a negatively charged foreign surface. With contact activation, activated factor XII (XIIa) converts prekallikrein into kallikrein. Kallikrein then activates more factor XII. HMWK acts as a cofactor in both of these reactions. HMWK also acts as a cofactor in the activation of factor XI by factor XIIa. Kallikrein releases bradykinin from HMWK, which has vasoactive activities. Fibrinolysis is also activated by contact activation. Recent evidence suggests that, *in vivo*, activation of prekallikrein occurs before activation of factor XII.

Prekallikrein deficiency is rare and does not cause bleeding, despite prolongation of the PTT. The lack of bleeding is presumably because the extrinsic pathway of coagulation, via factor VII and tissue factor, remains intact, and factor XI can be activated by thrombin generated from the extrinsic pathway. Thus, factor XI can be activated without the need for prekallikrein, HMWK, or factor XII. This is consistent with the observation that deficiencies of the latter three factors are not associated with bleeding.

Acquired, usually mild-to-moderate decreases in prekallikrein may be found in liver disease or disseminated intravascular coagulation (DIC). Rarely, a prekallikrein inhibitor (antibody) has been reported.

References

Andrew M, Paes B, and Johnston M, "Development of the Hemostatic System in the Neonate and Young Infant," *Am J Pediatr Hematol Oncol*, 1990,12(1):95-104.

Sanfelippo MJ, Carafo AJ, and Hollister WN, "APTT Prolonged by Prekallikrein Deficiency," *Lab Med*, 1998, 29(5):274-6.

Prenatal Screen, Immunohematology

Related Information

Antibody Titer *on page 167*
Antiglobulin Test, Indirect *on page 177*
Bilirubin, Amniotic Fluid, Delta A450 *on page 261*
Cord Blood Antibody Screen *on page 453*
Fetal Cell Detection by Flow Cytometry *on page 579*

Hemolytic Disease of the Newborn, Antibody Identification *on page 690*
Kleihauer-Betke *on page 822*
Polymerase Chain Reaction *on page 1069*

Synonyms Prenatal Serology

Test Includes ABO group, Rh type, antibody screen, and antibody identification if antibody screen is positive

Abstract The purpose of prenatal serological screening is to identify mothers at risk for delivering an infant with hemolytic disease of the newborn (HDN). HDN occurs when the fetus possesses a paternal blood group antigen to which the mother is sensitized. The fetus may be affected by pre-existing maternal alloantibody as well as antibody provoked by the current pregnancy. Maternal alloantibodies, capable of causing HDN, can be detected. All positive antibody screens require identification of the antibody. Antibody specificity can be correlated with clinical significance. Most severe HDN is caused by anti-Rh(D) alone or in combination with other Rh antibodies (eg, anti-C or anti-E). The next grade of severity is associated with antibodies to other Rh antigens (notably anti-c) or with antibodies to other blood group antigens (eg, anti-K). IgG antibodies to antigens of the ABO system cause the lowest grade of HDN severity.

Patient Preparation All pregnant women should be screened at the initial visit. Previous pregnancy or transfusion is an indication for follow-up antibody screens. Rh(D)-negative patients who are not immunized to Rh(D) should have antibody screening repeated at 28-30 weeks gestation before Rh immunoprophylaxis (Rh_o(D) immune globulin) is given. It is not necessary to repeat the antibody screen on Rh(D)-positive women unless antibodies associated with HDN other than anti-D are present. The presence of a clinically significant antibody in any patient is cause for evaluation of fetal risk (eg, coordination of antibody serial titration values with possible amniocentesis).

Specimen Blood

Container Red top tube

Causes for Rejection Gross hemolysis, improper labeling

Reference Interval Antibody screen negative (indirect antiglobulin test)

Use Identify women at risk for delivery of an infant with HDN: identify candidates for antenatal Rh immunoprophylaxis

Limitations Antibody detection tests will not expose antibodies to "private" paternal antigens or predict HDN due to ABO fetal/maternal incompatibility. ABO HDN can occur in pregnancy, even the first. The direct antiglobulin test is often negative and the infant is rarely symptomatic at birth in ABO HDN.

Methodology Methods to detect IgG alloantibodies, especially the indirect antiglobulin test

Additional Information Antibody screening tests are necessary to identify obstetric patients who may possess alloantibodies capable of causing HDN. Antibody screening should also be carried out on all Rh(D)-negative cases of ectopic pregnancy, incomplete abortion, or any situation that might immunize the mother to fetal Rh(D) antigen, in order to ascertain the need for Rh immunoprophylaxis.

Antepartum administration of Rh_o(D) immune globulin causes positive antibody screens in most women, due to passively acquired anti-Rh_o(D). This passively acquired antibody may remain detectable at the time of delivery. Its presence is not a contraindication for further Rh_o(D) immune globulin administration.

Molecular biological techniques (eg, allele-specific polymerase chain reaction (PCR) or PCR-restriction fragment length polymorphisms) are beginning to be utilized in the clinical laboratory to predict the likelihood of HDN. Fetal DNA can be prepared from cells obtained by conventional invasive techniques (eg, amniocentesis or chorionic villus sampling) by noninvasive procedures from trophoblasts collected by transcervical sampling, or from fetal erythroblasts derived from the maternal circulation. These procedures will prove valuable in cases in which the maternal serum contains an antibody associated with HDN and the father's antigenic status is heterozygous, indeterminable, or unknown. Molecular biological techniques hold great promise to radically change approaches to prenatal laboratory testing. Caution must be used in some cases of molecular typing, as an Rh(D) pseudogene has been identified in Rh-negative Africans and African-Americans.

(Continued)

Prenatal Screen, Immunohematology *(Continued)*

References

Hartwell EA, "Use of Rh Immune Globulin. ASCP Practice Parameter," *Am J Clin Pathol*, 1998, 110(3):281-92.

Lo YMD, Hjelm NM, Fidler C, et al, "Prenatal Diagnosis of Fetal RhD Status by Molecular Analysis of Maternal Plasma," *N Engl J Med*, 1998, 339(24):1734-8.

Singleton BK, Green CA, Avent ND, et al, "The Presence of an Rh(D) Pseudogene Containing a 37 Base Pair Duplication and a Nonsense Mutation in Africans With the Rh D-Negative Blood Group Phenotype," *Blood*, 2000, 95(1):12-8.

♦ **Prenatal Serology** *see* Prenatal Screen, Immunohematology *on page 1086*

♦ **Prenatal Testing** *see* Rh Genotype *on page 1162*

♦ **Presamine®** *see* Antidepressants, Cyclic, Serum or Plasma *on page 171*

♦ **Presamine®** *see* Imipramine, Serum or Plasma *on page 767*

♦ **Presenilin** *see* Cerebrospinal Fluid and Plasma β-Amyloid$_{(1-42)}$ *on page 359*

♦ **Pressor Amines** *see* Catecholamines, Fractionation, Plasma *on page 345*

Pretransfusion Testing

Related Information

Antibody Detection/Identification, Red Cell *on page 165*
Antiglobulin Test, Direct *on page 176*
Antiglobulin Test, Indirect *on page 177*
Red Blood Cells *on page 1139*
Rh Genotype *on page 1162*
Risks of Transfusion *on page 1166*
Uncrossmatched Blood, Emergency *on page 1281*
Whole Blood *on page 1333*

Applies to ABO Group and Rh Type; Blood Grouping and Rh Typing; Forward Grouping; Reverse Grouping; Rh(D) Typing; Type and Crossmatch; Type and Screen

Test Includes ABO group, Rh type, antibody screen (indirect antiglobulin test), crossmatch, and "type and screen"

Abstract Pretransfusion testing permits the appropriate selection of blood fractions which will have the intended therapeutic effect and cause no harm to the recipient. The *Standards* of the American Association of Blood Banks requires that the recipient be ABO and Rh grouped and screened for unexpected antibodies before transfusion. Pretransfusion testing procedures may be modified in life-threatening situations (see Uncrossmatched Blood, Emergency *on page 1281*).

Patient Preparation Procedures must be in place to positively identify the patient. Identification may include a wristband containing two unique identifiers, physically attached to the patient. Such identifiers should not be on the wall or clipped to the chart.

Specimen Blood

Container Red top tube, lavender top (EDTA) tube

Collection The vast majority of hemolytic transfusion reactions are the consequences of erroneous patient identification or sample labeling. The collection of a properly labeled blood sample from the intended recipient is the foundation of transfusion safety. At the patient's bedside, ask the patient to give his/her name and compare with the hospital wristband. Label the Blood Bank wristband (if there is one) with patient's full name, hospital number, date, and initials of the phlebotomist. Label sample tubes with the same information, including identification number from the wristband. Label requisition form with identification number. The sample collector must sign the requisition, verifying the patient's identity with hospital wristband and Blood Bank wristband. Some hospitals require additional information. It is always best to stamp the requisition with the patient's identification plate to avoid transcription errors.

Causes for Rejection Hemolysis, improper labeling, sample collected in serum separator tube

Turnaround Time In life-threatening emergencies, group O RBCs can be issued immediately to patients whose blood group is unknown (see Uncrossmatched Blood, Emergency *on page 1281*), but the more complete the pretransfusion testing, the safer the transfusion. In emergencies, the Blood Bank can issue blood at any stage of testing. The time requirements are shown in the table.

Timetable for Obtaining Emergency Blood From Blood Bank

Time Available (approx minutes)	Blood Bank Can Issue	Extent of Testing Done
<5	Type O RBCs, Rh-neg if possible	None
5-10	RBCs of patient's own ABO and Rh types	Patient's ABO and Rh typing, no crossmatch
45-60 (assuming no unexpected antibody detected)	Serologically compatible RBCs	Full pretransfusion testing, including antibody screen

Use Determine patient's blood group to enable blood selection before transfusion; detect and, if present, identify any unexpected blood group antibodies; detect any incompatibility with donor units before transfusion. ABO, Rh, and antibody screen are used in prenatal testing to detect fetal-maternal incompatibility that might cause hemolytic disease of the newborn.

Limitations Abnormal plasma proteins, cold autoagglutinins, positive direct antiglobulin test, and in some cases, bacteremia may interfere. These tests do not assure normal red cell survival, will not detect all red cell antibodies or incompatibilities, do not prevent all transfusion reactions, and do not prevent reactions to blood components other than red cells. Clerical and technical competence is requisite. A great many pitfalls exist which may cause false-positive or false-negative reactions.

Contraindications See Risks of Transfusion *on page 1166.*

Methodology Serologic pretransfusion testing attempts to reproduce *in vitro* biologic manifestations of incompatibility. The test systems involve reacting the patient's separated serum and red cells with known standardized antisera and phenotyped red cells. These procedures and reagents are common to ABO grouping and Rh typing, to testing of the serum for the presence of unexpected blood group antibodies, and to compatibility testing (crossmatching). All testing reagents are FDA licensed and procedures are subject to review and inspection by accrediting agencies, including any computerized programs for recording test results or for tracking of patients and donor blood components. Methods vary according to reagents, procedures, and equipment in use; testing may be in test tubes, microplates, or gel columns depending on testing techniques in use.

Selection of Donor Red Cells (RBCs) for Transfusion to Recipients of Various ABO and Rh Types

Patient Blood Type	First Choice	Second Choice	Third Choice
O pos	O pos	O neg	
O neg	O neg	None	O pos
A pos	A pos	A neg, O pos, O neg	
A neg	A neg	O neg	O pos, A pos
B pos	B pos	B neg, O pos, O neg	
B neg	B neg	O neg	O pos, B pos
AB pos	AB pos	AB neg, A pos, B pos, A neg, B neg, O pos, O neg	
AB neg	AB neg	A neg, B neg, O neg	AB pos, A pos, B pos, O pos

Note: The technologist may always substitute Rh-negative donor RBC for Rh-positive patients, if supplies permit. **Physician approval is required for third choice donor blood selection in the event that first and second choice are unavailable.** In this table, "pos" and "neg" refer to Rh-positive and Rh-negative respectively.

(Continued)

Pretransfusion Testing *(Continued)*

ABO grouping consists of testing the patient's red cells with reagent anti-A, anti-B, and anti-AB (forward grouping). The patient's serum (other than infants) is also tested with reagent group A and B red cells (reverse grouping).

Rh typing is done by testing red cells with anti-D. Red cells positive for the Rh(D) antigen are referred to as Rh positive; those lacking Rh(D) are termed Rh negative. Because Rh(D) is by far the most antigenic of all the Rh antigens, this is the only distinction routinely necessary for transfusion. Do not be concerned about the numerous other Rh antigens unless the patient is immunized to one of them (which would usually become apparent from the results of antibody screening or crossmatching tests).

As a rule, the Blood Bank issues blood of the patient's own ABO and Rh types, but some leeway is permissible; see table. When difficulties occur in blood selection, consultation is necessary between the Blood Bank physician and the patient's physician.

Antibody screen: (See Antiglobulin Test, Indirect *on page 177.*) About 1.5% of patients possess unexpected blood group antibodies with the potential for causing hemolytic transfusion reactions. The antibody screening test is the procedure used to detect these antibodies. Antibody screening may be performed in advance of or simultaneous with crossmatching tests. A positive antibody screen indicates the need for further tests to determine the specificity of the antibody. Determination of antibody specificity permits the appropriate selection of donor blood, lacking the offending antigen, for compatibility tests.

Basically, the antibody screen can be thought of as a crossmatch against a small panel of red cells, selected to contain the clinically important antigens, instead of donor red cells. AABB *Standards* require that screening tests employ methods that detect clinically significant antibodies and include an indirect antiglobulin test preceded by incubation at 37°C. Antibody screening tests are not perfect and can miss antibodies to low-incidence antigens as well as weakly reacting antibodies. Transfusion of antigen-incompatible red cells to a patient with a weakly reactive, undetected, antibody can result in rapid immune recall of antibody leading to a hemolytic transfusion reaction.

Crossmatch: (See Antiglobulin Test, Indirect *on page 177* and Uncrossmatched Blood, Emergency *on page 1281.*) The crossmatch is a direct test of compatibility between the patient's serum and donor red cells. Except for the source of the red cells, it is like the antibody screen. The crossmatch may utilize various techniques to detect IgG clinically significant antibodies, but its prime purpose is to detect ABO incompatibility. When antibody screening tests (both current and prior) have demonstrated no evidence of unexpected antibodies, the antiglobulin test may be omitted and the crossmatch restricted to a simple test for ABO incompatibility (eg, immediate spin test or computer-assisted crossmatch). The rarity of the exposure of a clinically significant antibody by the antiglobulin phase of the crossmatch, when the antibody screening test is negative, provides the rationale for the abbreviated crossmatch. The benefit of abbreviation of the crossmatch includes reduced turn-around time. It greatly lessens the amount of laboratory work per unit of blood, at little or no increase in risk.

Computer-assisted crossmatching: Under specified circumstances, accrediting agencies have sanctioned the use of the so-called "electronic or computer crossmatch". This is not a serologic test at all, it is rather a computer check of donor and recipient ABO and Rh compatibility. The computer-assisted crossmatch may be employed when antibody screening and history review have detected no clinically significant antibodies. The computer program must be validated on site to ensure that only ABO compatible whole blood or red cells have been selected for transfusion. The program must verify correct entry of data prior to release of blood and contain logic to signal discrepancies. FDA approval of a variance to regulations on compatibility testing is required.

Type and screen: When a patient is undergoing a procedure or treatment in which transfusion is very unlikely, it is wasteful to crossmatch blood and put it aside for that patient. Instead, it is more appropriate to order a "type and screen". The type and screen includes ABO and Rh typing and an antibody screen. If the antibody screen is negative and hemorrhage occurs, the Blood

Bank may issue blood of the patient's type immediately, without awaiting the crossmatch. Of course, if the antibody screen detects an unexpected antibody, crossmatch becomes necessary, and the patient's physician is alerted to the situation beforehand.

Additional Information Allow additional time for patients known to be immunized to red cell antigens. Unanticipated problems can occur with antibody reidentification and selection of blood of appropriate phenotype. Patients receiving a series of transfusions are at risk of forming red cell antibodies. For this reason, *Standards* require a new blood sample and repeat antibody screening every 3 days.

References

Fridey J, *Standards for Blood Banks and Transfusion Services*, 22nd ed, Bethesda, MD: American Association of Blood Banks Press, 2003, 44-8.

Linden JV, "Errors in Transfusion Medicine: Scope of the Problem," *Arch Pathol Lab Med*, 1999, 123(7):563-5.

♦ **Primaclone** *see Primidone, Serum or Plasma on page 1091*
♦ **Primidone** *see Phenytoin, Serum or Plasma on page 1026*

Primidone, Serum or Plasma

Related Information

Antiepileptic Drugs Overview *on page 176*
Carbamazepine, Serum *on page 337*
Folic Acid, RBC *on page 606*
Folic Acid, Serum *on page 606*
Phenobarbital, Serum or Plasma *on page 1021*
Phenytoin, Serum or Plasma *on page 1026*
Valproic Acid, Serum or Plasma *on page 1297*

Synonyms Desoxyphenobarbital; Hexamidinum; Majsolin®; Mylepsin®; Mysoline®; Primaclone; Prysolin®

Applies to Metabolites of Primidone; PEMA; Phenobarbital:Primidone Ratio; Phenylethylmalonamide

Test Includes Phenobarbital, PEMA

Abstract Primidone is indicated for generalized tonic-clonic and partial seizures. Concurrent monitoring of its major active metabolite, phenobarbital, is necessary. A second metabolite, phenylethylmalonamide (PEMA), also exerts activity.

Specimen Serum or plasma

Container Red top tube, green top (heparin) tube, or lavender top (EDTA) tube

Sampling Time Trough sampling is preferable. Consistent sampling time is desirable. Levels of phenobarbital and PEMA can be measured simultaneously.

Reference Interval Children younger than 5 years of age: 7-10 µg/mL (SI: 32-46 µmol/L); adults: 5-12 µg/mL (SI: 23-55 µmol/L). Phenobarbital concentration should also be used to guide dosing. Phenobarbital, serum: 15-40 µg/mL.

Critical Values At levels >12 µg/mL (SI: >55 µmol/L) primidone produces CNS depression, vertigo, visual disturbances, areflexia, somnolence, and lethargy. Clinical toxicity correlates with primidone rather than metabolite concentrations. In overdosage, a biphasic peak may be seen with highest toxicity a few hours after ingestion and again 48 hours afterwards. Crystalluria is a feature of overdosage.

Possible Panic Range >15 µg/mL (SI: >69 µmol/L). See Table A in the Therapeutic Drug Monitoring Introduction *on page 60.*

Use Monitor efficacy, compliance, and possible toxicity

A table comparing antiepileptic drugs is included in Antiepileptic Drugs Overview *on page 176.*

Methodology Immunoassay, gas chromatography (GC), high performance liquid chromatography (HPLC)

Additional Information

- Half-life: adults: 4-12 hours; children: 4-6 hours
- Volume of distribution: 0.5-1.0 L/kg
- Protein binding: 20%

Since phenobarbital requires a longer interval (48 hours) to achieve therapeutic blood levels, checking its concentrations can be used to determine chronic compliance. The phenobarbital:primidone ratio normally is 2.5, can be higher *(Continued)*

1091

Primidone, Serum or Plasma *(Continued)*

(4.3 mean) in patients on other anticonvulsants (phenytoin, carbamazepine) and lower than normal among patients discontinued from those medicines or who are chronically noncompliant. Primidone decreases the effects of oral anticoagulants.

Both primidone and phenobarbital undergo renal biotransformation and excretion. Patients with renal impairment have higher levels and need lower dosage.

All women of childbearing age prescribed this medication should also take supplemental folic acid. See Phenobarbital, Serum or Plasma *on page 1021* for more information.

References

Brodie MJ and Dichter MA, "Antiepileptic Drugs," *N Engl J Med*, 1996, 334(3):168-75.

Hernández-Diaz S, Werler MM, Walker AM, et al, "Folic Acid Antagonists During Pregnancy and the Risk of Birth Defects," *N Engl J Med*, 2000, 343(22):1608-14.

Shafer LC, Schaffer CB, and Caretto J, "The Use of Primidone in the Treatment of Refractory Bipolar Disorder," *Ann Clin Psychiatry*, 1999, 11(2):61-6.

♦ **Prion Protein** *see* Flow Cytometry, Overview *on page 592*

♦ **Proaccelerin (Factor V)** *see* Coagulation Factor Assays *on page 418*

♦ **Probability** *see page 11*

♦ **Probe Test for HPV DNA** *see* Human Papillomavirus (HPV) DNA Tests *on page 749*

Procainamide, Serum

Related Information

Amiodarone, Serum *on page 148*
Antinuclear Antibodies *on page 189*
Nortriptyline, Serum *on page 966*
Quinidine, Serum *on page 1129*

Synonyms Biocoryl®; Novocainamidum; Novocamid®; Procaine Amide Hydrochloride; Procanbid®; Procan® SR; Pronestyl®; Pronestyl-SR®; Retard®; Rhythmin®

Applies to N-Acetyl Procainamide; NAPA

Test Includes Procainamide and its metabolite, N-acetyl procainamide (NAPA)

Abstract This drug is class 1A antiarrhythmic used in the treatment of ventricular tachycardia (VT), premature ventricular contractions, paroxysmal atrial tachycardia (PSVT), and atrial fibrillation (AF). It is metabolized to N-acetyl procainamide (NAPA), an active metabolite. Both should be measured for therapeutic drug monitoring.

Specimen Serum

Container Red top tube

Sampling Time Oral treatment: peak: 75 minutes after dose; trough: immediately before next dose. I.V. treatment: immediately after loading dose; 0.5, 2, 6, 12, and 24 hours after starting I.V. maintenance.

Reference Interval Therapeutic: procainamide: 4.0-10.0 µg/mL (SI: 17-42 µmol/L), NAPA: 10-30 µg/mL, sum of procainamide and N-acetyl procainamide: <30 µg/mL (SI: <127 µmol/L). Optimal ranges must be ascertained for individual patients with ECG monitoring.

Possible Panic Range Toxic: procainamide: >12 µg/mL (SI: >59.5 µmol/L); sum of procainamide and N-acetyl procainamide: >30 µg/mL (SI: >127 µmol/L). See Table C in the Therapeutic Drug Monitoring Introduction *on page 62*.

Use Monitor therapeutic drug level. Procainamide may be used in instances instead of quinidine. The reverse is also true.

Limitations Long-term administration leads to the development of a positive antinuclear antibody test in 50% of patients. A lupus erythematosus-like syndrome may evolve in 20% to 30% of patients.

Methodology Immunoassay, high performance liquid chromatography (HPLC), gas chromatography (GC)

Additional Information

- Half-life: procainamide: 2-6 hours, NAPA: 8 hours
- Volume of distribution: 2-4 L/kg
- Protein binding: 10% to 20%
- Bioavailability: close to 100% in 85% of patients

The cardiac actions of this drug are similar to those of quinidine. It is used in a variety of arrhythmias. Optimal plasma sampling time after oral dosage is 1-2 hours. Optimal sampling time after I.V. administration of dose is 30 minutes. The drug is converted by the liver to its active metabolite, NAPA. Rate of metabolism is genetically determined (slow and fast acetylator types) contributing to significant interindividual variability. Fast metabolizers have a NAPA concentration equal to or greater than procainamide 3 hours after dosing. Impairment of renal function has pronounced effect on drug disposition, especially for NAPA. Patients with severe renal dysfunction generally have prolonged and highly variable half-life characteristics.

The prolonged administration of procainamide often leads to the development of a positive antinuclear antibody (ANA) test, with or without symptoms of a lupus erythematosus-like syndrome.

References

Gold MR, O'Gara PT, Buckley MJ, et al, "Efficacy and Safety of Procainamide in Preventing Arrhythmias After Coronary Artery Bypass Surgery," *Am J Cardiol*, 1996, 78(9):975-9.

Jacobs LO, Andrews TC, Pederson DN, et al, "Effect of Intravenous Procainamide on Direct-Current Cardioversion of Atrial Fibrillation," *Am J Cardiol*, 1998, 82(2):241-2.

Valdes R Jr, Jortani SA, Gheorghiade M, et al, "Standards of Laboratory Practice: Cardiac Drug Monitoring. National Academy of Clinical Biochemistry," *Clin Chem*, 1998, 44(5):1096-109.

♦ **Procaine Amide Hydrochloride** *see* Procainamide, Serum *on page 1092*

♦ **Procanbid®** *see* Procainamide, Serum *on page 1092*

♦ **Procan® SR** *see* Procainamide, Serum *on page 1092*

♦ **Prochlorperazine** *see* Phenothiazines, Serum *on page 1021*

♦ **Proconvertin (Factor VII)** *see* Coagulation Factor Assays *on page 418*

♦ **Progenitor Cells** *see* Hematopoietic Progenitor Cells, Peripheral Blood *on page 679*

Progesterone, Serum

Related Information

Chorionic Gonadotropin, Human, Serum and Urine *on page 397*
Estradiol, Serum *on page 553*
Estrogens, Urine *on page 557*
Heterophilic Antibodies *on page 727*
17-Hydroxyprogesterone, Whole Blood, Serum, or Plasma *on page 755*
Pregnancy Test, Serum or Urine *on page 1084*

Abstract Progesterone is synthesized by the corpus luteum. The serum level of progesterone rises during the luteal (secretory) phase of the menstrual cycle (indicating ovulation). During pregnancy, it increases dramatically from the end of the first trimester to term. Lower values occur with ectopic gestation or miscarriage.

Specimen Serum, saliva

Container Red top tube

Storage Instructions Serum stable 4 days at 4°C and 3 months at -20°C.

Special Instructions Patient's LMP (last menstrual period) and trimester of pregnancy are relevant.

Reference Interval

Serum:
- prepubertal: 7-52 ng/dL (SI: 0.2-1.7 nmol/L)
- adult male: 13-97 ng/dL (SI: 0.4-3.1 nmol/L)
- female: follicular: 15-70 ng/dL (SI: 0.5-2.2 nmol/L); luteal 200-2500 ng/dL (SI: 6.4-79.5 nmol/L)
- pregnancy: 1st trimester: 725-4400 ng/dL (SI: 23.0-140.0 nmol/L); 2nd trimester: 1950-8250 ng/dL (SI: 62.0-262.4 nmol/L); 3rd trimester: 6500-22,900 ng/dL (SI: 206.7-728.2 nmol/L)

Saliva: Assays not widely available. Check with your reference laboratory.
- follicular: 0.5-4.0 ng/dL (SI: 0.016-0.127 nmol/L)
- luteal: 6.0-12.0 ng/dL (SI: 0.19-0.38 nmol/L)

Use In a pregnant patient with first trimester bleeding, a serum progesterone >250 ng/dL (SI: >7.95 nmol/L) predicts a normal intrauterine pregnancy, while a value <150 ng/dL (SI: <4.77 nmol/L) predicts an abnormal gestation. The serum progesterone cannot, however, distinguish between an ectopic pregnancy and an abnormal intrauterine pregnancy.
(Continued)

Progesterone, Serum *(Continued)*

Serum progesterone levels used with β-hCG levels may assist in differentiating normal intrauterine from abnormal intrauterine or **ectopic pregnancy** (cutoff: 15 ng/mL) (SI: 48 nmol/L). β-hCG and progesterone levels are lower in abnormal pregnancies. Less overlap occurs using progesterone values.

Evaluation of infertility: Midluteal luteinizing hormone and progesterone on day 21: progesterone concentration ≥10 μg/mL usually indicates adequate luteinization. Endometrial biopsy between days 24 and 26 provides assessment of secretory phase endometrium.

Assisted reproduction: The test is useful for monitoring patients having ovulation during induction with hCG, hMG, FSH/LHRH, or clomiphene.

Limitations High concentrations of 17-hydroxyprogesterone falsely increase progesterone levels in a number of assays.

Methodology Immunoassays (multiple labels)

References

Baskin LB, "Pregnancy and Prenatal Testing," *The Handbook of Clinical Pathology*, 2nd ed, McKenna RW and Keffer JH, eds, Chicago, IL: ASCP Press, American Society of Clinical Pathologists, 2000, 281-92.

Gronowski AM and Landau-Levine M, "Reproductive Endocrine Function," *Tietz Textbook of Clinical Chemistry*, 3rd ed, Burtis CA and Ashwood ER, eds, Philadelphia, PA: WB Saunders Co, 1999, 1601-41.

Perkins SL, Al-Ramahi M, and Claman P, "Comparison of Serum Progesterone as an Indicator of Pregnancy Nonviability in Spontaneously Pregnant Emergency Room and Infertility Clinic Patient Populations," *Fertil Steril*, 2000, 73(3):499-504.

Rosevear S, "Bleeding in Early Pregnancy," *High Risk Pregnancy*, James DK, Steer PJ, Weiner CP, et al, eds, Philadelphia, PA: WB Saunders Co, 1999, 61-89.

♦ **Prograf®** *see* Tacrolimus, Whole Blood *on page 1234*

♦ **Progressive Systemic Sclerosis Antibody** *see* Topoisomerase I Antibody *on page 1261*

♦ **Proinsulin** *see* C-Peptide, Serum *on page 465*

♦ **Proinsulin** *see* Insulin, Serum *on page 803*

Prolactin, Serum

Related Information

Amoxapine, Serum or Plasma *on page 153*

Dehydroepiandrosterone and Dehydroepiandrosterone Sulfate, Serum or Plasma *on page 506*

Estradiol, Serum *on page 553*

Follicle Stimulating Hormone, Serum, Plasma, or Urine *on page 609*

Heterophilic Antibodies *on page 727*

Abstract Prolactin (PRL), so named because it initiates and maintains lactation, is one of the anterior pituitary hormones. Hypothalamic dopamine inhibits the release of PRL. PRL is secreted in pulses which are superimposed on a circadian rhythm. A number of physiologic stimuli (eg, sleep, exercise, and hyperglycemia) cause short-term increases in serum PRL. The most common pituitary tumor secretes PRL.

Amenorrhea, irregular menses, and galactorrhea provide sensitive signals for possible presence of prolactinoma in reproductive age women. Such tumors in males and older females lack such indicators and appear later, often with invasion and mass effect.

Presence of macroprolactin may complicate the interpretation of prolactinemia.

Specimen Serum

Container Red top tube

Collection Draw in chilled tube between 8 AM and 10 AM. Keep specimen on ice.

Storage Instructions Separate serum in refrigerated centrifuge and freeze. Stable 3 months at -20°C.

Reference Interval Reference intervals are method-dependent. The normal interval has little medical importance.

Male:

- 0-1 month: 0-90 ng/mL
- 2-11 months: 0-30 ng/mL
- 1-18 years: 1-15 ng/mL
- >19 years: 4-23 ng/mL

Female:
- 0-1 month: 0-90 ng/mL
- 2-11 months: 0-30 ng/mL
- 1-18 years: 1-15 ng/mL
- >19 years: 4-30 ng/mL

Use Pituitary tumors: The principal clinical use for serum PRL assays is in the evaluation of PRL-secreting pituitary tumors. Interpretive caution is warranted because any nonsecretory pituitary tumor large enough to compress the pituitary stalk will, via stalk compression, cause increased PRL. Such tumors are commonly called "pseudoprolactinomas" since the increased PRL comes not from the tumor but from suppression of trophic hormones which normally inhibit the secretion of PRL. In this setting, the PRL values **must** be interpreted with the imaging findings in order to avoid misinterpretation. Thus, in a patient with a pituitary adenoma >10 mm in diameter (macroadenoma), a PRL >1000 ng/mL confirms the diagnosis of a PRL-secreting tumor. In this setting, a PRL value <200 ng/mL (some recommend <500 ng/mL) suggests pseudoprolactinoma. PRL values between 500 and 1000 ng/mL require careful evaluation. Medical therapy usually produces some measurable decrease in the size of a PRL-secreting tumor, but not in a pseudoprolactinoma.

Male Infertility: Some physicians include PRL measurements in their evaluation of subfertile males. When elevated PRL levels are found in this setting, treatment with bromocriptine may be effective in carefully selected patients. Men with **gynecomastia** usually have normal prolactin.

Limitations Many drugs produce increased PRL as a side effect.

Methodology Immunoassay (multiple labels)

Additional Information The differential diagnosis of **hyperprolactinemia** includes drug side effect, PRL-secreting pituitary adenoma, presence of macroprolactin, and other diseases of the hypothalamic-pituitary stalk region. Rare causes include renal disease, primary hypothyroidism, and endogenous estrogen overproduction. Prolactin circulates in the serum in different forms. Monomeric prolactin is the usual product released from the anterior pituitary. Macroprolactin (big, big prolactin, a prolactin-IgG complex - autoantibody) may account for a small but variable percentage of total prolactin. Macroprolactin, while biologically inactive, retains immunoreactivity. Prolactin immunoassays may vary in their ability to detect macroprolactin. Up to about 25% of cases of hyperprolactinemia may be due to the presence of macroprolactin. In order to avoid misdiagnosis of and unnecessary treatment for a suspect prolactinoma, it has been recommended that additional laboratory study include repeat prolactin assay after polyethylene glycol immunoprecipitation of the serum specimen and use of an appropriate reference interval (see reference by Suliman et al).

References

Calle-Rodrigue RD, Giannini C, Scheithauer BW, et al, "Prolactinomas in Male and Female Patients: A Comparative Clinicopathologic Study," *Mayo Clin Proc*, 1998, 73:1046-52.

Freeman ME, Kanyicska B, Lerant A, et al, "Prolactin: Structure, Function, and Regulation of Secretion," *Physiol Rev*, 2000, 80(4):1523-631.

Mayo Medical Laboratories, *2000 Test Catalogue*, Rochester, MN, 423.

Schlechte JA, "Prolactinoma," *N Engl J Med*, 2003, 349(21):2035-41.

Suliman AM, Smith TP, Gibney J, et al, "Frequent Misdiagnosis and Mismanagement of Hyperprolactinemic Patients Before the Introduction of Macroprolactin Screening: Application of a New Strict Laboratory Definition of Macroprolactinemia," *Clin Chem*, 2003, 49(9):1504-9.

Young DS, "Effects of Drugs on Clinical Laboratory Tests," 5th ed, Volume 1: Listing by Test, Washington, DC: AACC Press, American Association of Clinical Chemistry, 2000, Section 3, 662.

Propoxyphene, Serum or Urine

Related Information

Methadone, Serum or Urine *on page 900*

Synonyms Darvocet-N®; Darvon®; E-Lor®; Genagesic®; Novrad®; Propacet®; Wygesic®

Applies to Norpropoxyphene

Test Includes Quantitation of propoxyphene and metabolite norpropoxyphene

Abstract Propoxyphene is a narcotic analgesic that is somewhat less potent than codeine. It is structurally and pharmacologically close to methadone. It is used clinically with nonopioid analgesics. It is biotransformed to norproxyphene, a potentially toxic metabolite. Overdoses may lead to seizures.

Specimen Serum (TDM), urine (drugs of abuse)

Container Red top tube, urine container

Sampling Time Urine: random; serum: trough

Special Instructions If forensic, use chain-of-custody protocol and form. See Chain-of-Custody Protocol *on page 381* and the Chain-of-Custody form in the Introduction *on page 63*.

Reference Interval Therapeutic: serum: 0.1-0.4 µg/mL (SI: 0.3-1.2 µmol/L) (therapeutic ranges published vary between laboratories and may not correlate with clinical effect); urine (for drugs of abuse): negative (less than cutoff)

Critical Values Cutoff for urine: screen: 300 ng/mL; confirmation: 200 ng/mL

Possible Panic Range Toxic: serum: >0.5 µg/mL (SI: >1.5 µmol/L); **minimal fatal:** 1.0 µg/mL (SI: 2.9 µmol/L)

Use Therapeutic monitoring, toxicity assessment, and drug-of-abuse testing

Methodology Immunoassay, gas chromatography (GC), high performance liquid chromatography (HPLC), gas chromatography/mass spectrometry (GC/MS)

Additional Information

- Half-life: 8-24 hours
- Volume of distribution: 10-25 L/kg
- Protein binding: 70% to 80%

Toxic effects include nausea, vomiting, and progressive central nervous system depression. Toxicity is additive with ethanol and other CNS depressants such as barbiturates and benzodiazepines. Toxicity can be neutralized by narcotic antagonists such as naloxone. Peak serum level occurs 2 hours postoral dose. Propoxyphene can also be measured in urine as part of a drug of abuse screen.

Recent data show that propoxyphene is one of the most common inappropriately prescribed drugs for the elderly.

References

Aparasu RR and Sitzman SJ, "Inappropriate Prescribing for Elderly Outpatients," *Am J Health Syst Pharm*, 1999, 56(5):433-9.

Perin ML, "Problems With Propoxyphene," *Am J Nurs*, 2000, 100(6):22.

Schnitzer TJ, "Non-NSAID Pharmacologic Treatment Options for the Management of Chronic Pain," *Am J Med*, 1998, 105(1B):45S-52S.

Propranolol, Serum

Related Information

Flecainide, Serum or Plasma *on page 592*

Synonyms Angilol®; Apsolol®; Beaden®; Bedranol®; Berkolol®; Deralin®; Inderal®

Abstract Propranolol is a relatively short-acting beta-blocker. It is used in the management of angina pectoris, arrhythmias, and hypertension. It is also used for pheochromocytoma and essential tremors. African-Americans are more sensitive to propranolol than are Caucasians.

Specimen Serum

Container Red top tube

Sampling Time Trough: immediately prior to next dose

Reference Interval Therapeutic: 50-100 ng/mL (SI: 190-390 nmol/L) at end of dose interval

Possible Panic Range >1000 ng/mL (SI: >3860 nmol/L); **fatal:** >2000 ng/mL (SI: >7720 nmol/L). See Table C in the Therapeutic Drug Monitoring Introduction *on page 62*.

Use Monitor therapeutic drug level in patients with cardiac arrhythmias, angina pectoris, and hypertension; evaluate for potential toxicity

Contraindications Uncompensated congestive heart failure, cardiogenic shock, bradycardia, and asthma

Methodology Gas chromatography (GC), high performance liquid chromatography (HPLC)

Additional Information
- Half-life: 4-6 hours
- Volume of distribution: 3-4 L/kg
- Protein binding: 90% to 95%

Adverse effects of this drug include precipitation of heart failure, bronchospasm, bradycardia, and hypoglycemia. Hyperthyroidism exerts an age-dependent inducing effect on the metabolism of propranolol.

Phenobarbital and rifampin may increase propranolol clearance and may decrease its activity. Cimetidine may reduce propranolol clearance and may increase its effects.

References
Lacy CF, Armstrong LL, Goldman MP, et al, *Drug Information Handbook*, 12th ed, Hudson, OH: Lexi-Comp Inc, 2004.

Levy S, "Pharmacologic Management of Atrial Fibrillation: Current Therapeutic Strategies," *Am Heart J*, 2001, 141(2 Suppl):S15-21.

Reith DM, Dawson AH, Epid D, et al, "Relative Toxicity of Beta Blockers in Overdose," *J Toxicol Clin Toxicol*, 1996, 34(3):273-8.

♦ **2-Propylpentanoic Acid** *see* Valproic Acid, Serum or Plasma *on page 1297*

♦ **2-Propylvaleric Acid** *see* Valproic Acid, Serum or Plasma *on page 1297*

♦ **Prosaposin** *see* Infertility Screen *on page 786*

♦ **Prostate Needle Aspiration** *see* Fine Needle Aspiration, Deep and Superficial Masses *on page 590*

♦ **Prostate Specific Acid Phosphatase** *see* Immunoperoxidase Procedures *on page 780*

Prostate Specific Antigen, Free

Related Information

Prostate Specific Antigen, Serum *on page 1098*

Synonyms fPSA; Free PSA

Applies to Complexed PSA; Free PSA:Total PSA Ratio

Abstract In patients with adenocarcinoma of the prostate, a higher proportion of PSA is bound (>90%). The unbound fraction is called **"free PSA"** while the bound fraction is called **"complexed PSA"**. The measurement of both free and complexed fractions may be useful in screening men with equivocal findings. Goals include reduction of the number of unnecessary biopsies in men with total PSA concentrations between 4 and 10 ng/mL, the gray zone. It is not helpful when total PSA is <2.5 or 3.0 or 4.0. Free PSA increases specificity with only slight loss of sensitivity, providing stratification of cancer risk in the gray zone. Although low levels of fPSA have been reported to be associated with adverse characteristics, they may be seen in benign states such as prostatitis.

Patient Preparation Fasting specimen is preferred. Avoid recent exposure to radioactivity if RIA is used.

Specimen Serum

Container Red top tube

Sampling Time Recent prostatic digital rectal examination and/or needle biopsy should be avoided. PSA has little diurnal variation. Although it has often been sampled together with prostatic acid phosphatase (PAP), the latter has fallen into disrepute.

Collection Rectal examination within 48 hours of specimen collection may cause elevation of results.

Storage Instructions PSA is stable in serum for 48 hours if refrigerated. For longer periods, store at -20°C or colder. No special treatment of serum is required. Ship to a reference laboratory in a plastic vial on dry ice.

Reference Interval Method dependent. Median free PSA % is lower in subjects with prostatic carcinoma. Free PSA/total PSA is typically >25% in normal men and <25% in men with prostatic adenocarcinoma. It has been recommended that men with mildly elevated total PSA (4-10 ng/mL) and a free PSA/total PSA <20% to 30% be evaluated further. Use of percentage free PSA cutoff of 25% (Continued)

Prostate Specific Antigen, Free *(Continued)*

detected 98% of carcinomas at ages 50-59 years, 94% at 60-69 years, 90% at 70-75 years, and in men 50-75 years of age with a palpably benign prostate and total PSA of 4-10 ng/mL.

Use The free PSA:total PSA ratio is a screening test for classifying men with equivocal laboratory or physical findings who may benefit from prostate biopsy. This group includes those with mildly elevated total PSA (4-10 ng/mL) or those with enlarged prostate volumes detected by ultrasound.

An increased risk of carcinoma is described when the percentage of free PSA is very low. Use of the ratio in monitoring and determining prognosis of men with prostate carcinoma has not been established. A role in prostatic intraepithelial neoplasia is described.

Limitations As with other immunoassays, it is important to use the same method when following a patient or at least to be aware of the relationship between the operational characteristics of different immunoassay methods. This is even more important when using a ratio such as the free PSA:total PSA.

Low free PSA and low ratio occur also in benign situations (eg, prostatitis).

Addition of free PSA in a 6-year screening study provided enhancement of sensitivity over total PSA from 52% to only 56%.

Additional Information Assays for complexed PSA are also available. These may recognize either PSA complexed to alpha$_1$ antichymotrypsin or to alpha$_2$ macroglobulin.

References

Bostwick DG, Grignon DJ, Hammond MEH, et al, "Prognostic Factors in Prostate Cancer: College of American Pathologists Consensus Statement 1999," *Arch Pathol Lab Med*, 2000, 124(7):995-1000.

Carlson GD, Calvanese CB, and Childs SJ, "The Appropriate Lower Limit for the Percent Free Prostate-Specific Antigen Reflex Range," *Urology*, 1998, 52(3):450-4.

Catalona WJ, Partin AW, Slawin KM, et al, "Use of the Percentage of Free Prostate-Specific Antigen to Enhance Differentiation of Prostate Cancer From Benign Prostatic Disease: A Prospective Multicenter Clinical Trial," *JAMA*, 1998, 279(19):1542-7.

Kilic S, Kukul E, Danisman A, et al, "Ratio of Free to Total Prostate-Specific Antigen in Patients With Prostatic Intraepithelial Neoplasia," *Eur Urol*, 1998, 34(3):176-80.

Lechevallier E, Echazarian C, Ortega JC, et al, "Kinetics of Postbiopsy Levels of Serum Free Prostate-Specific Antigen and Percent Free Prostate-Specific Antigen," *Urology*, 1999, 53(4):731-5.

Wald NJ, Watt HC, George L, et al, "Adding Free to Total Prostate-Specific Antigen Levels in Trials of Prostate Cancer Screening," *Br J Cancer*, 2000, 82:731-6.

Prostate Specific Antigen, Serum

Related Information

Acid Phosphatase, Plasma *on page 97*
Immunoperoxidase Procedures *on page 780*
Prostate Specific Antigen, Free *on page 1097*

Synonyms PSA

Applies to Finasteride; PSA Density; PSA Velocity

Abstract The best, most accurate and the most useful marker for adenocarcinoma of the prostate, PSA is increased in most men with clinically significant prostate cancer, but it is sometimes increased in benign entities and may fall within normal range even with advanced prostatic adenocarcinoma. With digital rectal examination (DRE) it increases rates of **detection** of prostatic adenocarcinoma, but some of the carcinomas it detects are biologically insignificant. It is best when the blood sample is collected prior to digital rectal examination, to which it is complementary. Serially measured, PSA is extremely useful in monitoring presurgical, as well as postsurgical patients. PSA level, PSA density, ratio of free to total PSA, and needle biopsy histopathologic observations represent predictors of extent of tumor. PSA density, PSA velocity and fractions of bound or free PSA are called "PSA derivatives" and a consensus does not yet exist about their precise utility. Ultimately, it is clinical outcome which is most important. Category I of the College of American Pathologists rankings, factors proven to be of prognostic importance and useful in clinical patient management, include preoperative serum PSA, TNM stage, Gleason score, and surgical margin status. (Other important factors associated with progression of prostatic carcinoma include positive lymph nodes, intraprostatic vascular invasion, and seminal vesicle invasion). PSA has become an indispensable predictor for recurrent adenocarcinoma in postsurgical patients and those treated with radiation therapy.

Patient Preparation Fasting specimen is preferred. The practice of sampling at least 4 weeks following DRE and needle biopsy is recommended. Ejaculation may cause transient, minor increase.

Specimen Serum

Container Red top tube

Special Instructions Individual patients should be followed with the same assay consistently. When a decision is made to follow patients without immediate therapy, serial assays are recommended to ascertain rate of change of PSA. This is called **PSA velocity**, defined as change in PSA at 1-year intervals.

Reference Interval Male: <4 ng/mL; female <0.5 ng/mL (immunoassay method); *vide infra*. Age adjusted reference intervals have been advocated; PSA results increase moderately with age; *vide infra*. Use of upper limit of 4.5 ng/mL for men 60-69 years of age and 6.5 ng/mL for those 70 years and older would decrease sensitivity from 86% to 77% but would enhance specificity from 56% to 67%. It would decrease the numbers of biopsies.

Other reference ranges:
- <50 years: <2.6 ng/mL
- 50-59 years: <3.6 ng/mL
- 60-69 years: <4.6 ng/mL
- >70 years: <6.6 ng/mL
- PSA velocity: *vide infra*

For most PSA assays, sensitivity = ~73% to 84%; specificity = ~59% to 93%.

Use Prostate specific antigen, human kallikrein-3 (hK3), is a serine protease that is produced almost exclusively by epithelial cells of prostatic tissue. It is present at high concentrations in seminal fluid and functions in the liquefaction of seminal coagulum. Preoperative PSA serum levels correlate (but imperfectly) with extent of disease in patients with prostate cancer. **PSA density** is obtained by division of serum PSA result by prostatic weight utilizing transrectal ultrasound **(TRUS)** volumetric measurements:

PSA density = [PSA] / total prostate volume; units = ng/mL/cc

Results of PSA density are not definitive; it is increased with prostatitis but may be useful in patient selection for radical prostatectomy. With PSA assay and histopathologic interpretation from prostate needle biopsy, correlation of **extent of tumor** can be improved. (Of these, TRUS has not consistently been found to predict tumor extent.) PSA is useful in detection of residual tumor and disease progression in postoperative stage of prostate cancer. When PSA rises in patients who have previously had prostatectomy and who have no detectable disease, metastases are likely, in spite of inability to detect them.

PSA has several advantages over prostatic acid phosphatase (PAP). It is more stable, lacks significant diurnal variation, and provides better correlation with carcinoma of prostate.

PSA relates to increasing Gleason score and outcome. With DRE, PSA is successfully widely used in screening selected populations of patients with or without symptoms indicative of prostate cancer and has led to increased detection of prostate cancer. It detects incidental, as well as, aggressive neoplasms. PSA lacks sufficient sensitivity and specificity to be used alone as a screening test for prostatic carcinoma in the present screening format. In conjunction with digital rectal examination and TRUS, PSA greatly increases prostatic carcinoma detection rates.

A significant shift to lower stage carcinomas, fewer nondiploid, more frequently organ-confined tumors at initial assessment has been recognized since the advent of PSA testing at a large referral center. A new clinical stage category, stage T1c, has evolved following experience with PSA; it is nonpalpable carcinoma detected by increased serum concentration of PSA and diagnosed by examination of needle biopsies.

A PSA assay is indicated prior to initiation of finasteride therapy for benign hyperplasia of prostate, since this drug causes an ~50% decrease in serum PSA concentration.

(Continued)

Prostate Specific Antigen, Serum (Continued)

PSA is used in forensic medicine as a marker for semen (eg, in cases of alleged rape).

Limitations Approximately 25% to 46% of men with benign prostatic hyperplasia have an elevated serum PSA concentration. One-third of men with PSA >4 ng/mL will have carcinoma on prostate biopsy but two-thirds will not. Between 20% and 40% of patients with organ-confined prostatic carcinoma have serum PSA concentrations within the reference interval. PSA is not acceptable used alone for staging and alone should not be used to select candidates for radical prostatectomy. Physiologic fluctuation ≤30% is described. PSA is not specific for prostatic adenocarcinoma, but serum levels are specific for prostatic tissue. Elevations may also be associated with DRE, prostatic massage, urethral instrumentation, cystoscopy, prostatitis, TUR, prostatic needle biopsy, urinary retention, or prostatic ischemia or infarct.

PSA values in the gray zone, between 4 and 10 ng/mL, include some carcinomas. Of men whose PSA values fall in this range and whose DRE is negative, 75% do not have carcinoma in the initial biopsy.

It is now known that 50% to 90% of serum PSA is bound to an inhibitor called α_1-antichymotrypsin. (See Prostate Specific Antigen, Free on page 1097.) The antibodies used in assays have different affinities for bound and free PSA. When following patient values, it is most important to continue use of the same assay.

Confirmation several weeks later, of a newly recognized PSA increase is suggested before invasive tests, such as biopsies, are considered.

Methodology Radioimmunoassay (RIA), enzymetric immunoassay, immunofluorometric assay, monoclonal two-site immunoradiometric assay (IRMA). An ultrasensitive RIA has been developed. Over 30 different assays are available.

Additional Information Following prostatectomy for carcinoma, persistent increase in PSA signals residual disease, but undetectable results do not assure cure. Increasing PSA concentrations can be seen months or even years before other evidence of progression. Clinical objectives in management of rising postoperative PSA include prevention of possible symptoms, metastases, or death from prostate carcinoma. "Advanced" prostatic carcinoma indicates PSA increase only; "lethal" indicates detection of metastases on imaging.

Use of age-specific reference intervals may make PSA density superfluous. Use of age-specific reference intervals for PSA has been claimed to provide greater sensitivity in patients younger than the age of 60 and greater specificity in older patients.

PSA velocity is an expression used to indicate rate of change of PSA. It may provide an index capable of earlier detection of adenocarcinoma of prostate with distinction from benign hyperplasia and normal and may support clinical decisions following radical prostatectomy. PSA increase of 0.75-0.8 ng/mL/year in a man whose PSA is within reference interval, with a minimum of three assays at least a year apart, is considered an indication for further investigation. **PSA doubling time** correlates with risk of progression in patients whose tumors recur after primary radiation or surgical therapy.

Prostatic specific antigen is a reliable immunocytochemical marker for primary and metastatic adenocarcinoma of prostate, reacting with at least some cells in almost all adequate biopsies.

References

Amling CL, Blute ML, Lerner SE, et al, "Influence of Prostate-Specific Antigen Testing on the Spectrum of Patients With Prostate Cancer Undergoing Radical Prostatectomy at a Large Referral Practice," Mayo Clin Proc, 1998, 73(5):401-6.

Barry MJ, "Prostate-Specific-Antigen Testing for Early Diagnosis of Prostate Cancer," N Engl J Med, 2001, 344(18):1373-7.

Bostwick DG, Grignon DJ, Hammond MEH, et al, "Prognostic Factors in Prostate Cancer: College of American Pathologists Consensus Statement 1999," Arch Pathol Lab Med, 2000, 124(7):995-1000.

Catalona WJ, Partin AW, Slawin KM, et al, "Use of the Percentage of Free Prostate-Specific Antigen to Enhance Differentiation of Prostate Cancer From Benign Prostatic Disease: A Prospective Multicenter Clinical Trial," JAMA, 1998, 279(19):1542-7.

Eastham JA, Riedel E, Scardino PT, et al, "Variation of Serum Prostate-Specific Antigen Levels - An Evaluation of Year-to-Year Fluctuations," JAMA, 2003, 289(20):2695-700.

Epstein JI, *Prostate Biopsy Interpretation*, 3rd ed, Baltimore, MD: Lippincott Williams & Wilkins, 2002.

Nelson WG, De Marzo AM, and Isaacs WB, "Mechanisms of Disease - Prostate Cancer," *N Engl J Med*, 2003, 349(4):366-81.

Pound CR, Partin AW, Eisenberger MA, et al, "Natural History of Progression After PSA Elevation Following Radical Prostatectomy," *JAMA*, 1999, 281(17):1591-7.

"Prostate Disease and PSA Testing," *Mayo Communiqué*, 2001, 26(1):1-4.

Schröder FH and Kranse R, "Verification Bias and the Prostate-Specific Antigen Test - Is There a Case for a Lower Threshold for Biopsy?," *N Engl J Med*, 2003, 349(4):393-5.

Slovacek KJ, Riggs MW, Spiekerman AM, et al, "Use of Age-Specific Normal Ranges for Serum Prostate-Specific Antigen," *Arch Pathol Lab Med*, 1998, 122(4):330-2.

Weider JA and Belldegrun AS, "The Utility of PSA Doubling Time to Monitor Prostate Cancer Recurrence," *Mayo Clin Proc*, 2001, 76(6):571-2.

Internet Web Sites

http://kidney.niddk.nih.gov/kudiseases/pubs/prostateenlargement/index.htm

♦ **Prostatic Acid Phosphatase** *see* Acid Phosphatase, Plasma *on page 97*

♦ **Protease Inhibitors** *see* Alpha₁-Antitrypsin Phenotyping *on page 133*

♦ **Protease Inhibitors** *see* Alpha₁-Antitrypsin, Serum *on page 134*

♦ **Proteinase 3 (PR3)** *see* Antineutrophil Cytoplasmic Antibody *on page 187*

♦ **Protein, Body Fluids** *see* Body Fluid Chemical Analysis *on page 291*

Protein C

Related Information

Activated Protein C Resistance and the Factor V Leiden Mutation *on page 104*

Antithrombin *on page 198*

Hypercoagulation Panel *on page 758*

Protein S *on page 1110*

Warfarin, Serum or Plasma *on page 1325*

Abstract Protein C, with protein S as a cofactor, is a natural anticoagulant protein. A hereditary deficiency of protein C leads to a hypercoagulable state with an increased risk for venous thrombosis. Type I deficiency is a quantitative deficiency of protein C. Type II deficiencies result from qualitatively abnormal (but often quantitatively normal) protein C.

Patient Preparation Determine if patient is on oral anticoagulants. Protein C levels are decreased by warfarin (Coumadin®).

Specimen Plasma

Container One blue top (sodium citrate) tube

Sampling Time Testing should be deferred until patients have not received Coumadin® for at least 10 days, because Coumadin® decreases protein C levels.

Collection Routine venipuncture. If multiple tests are being drawn, draw blue top tubes after any red top tubes but before any lavender top (EDTA), green top (heparin), or gray top (oxalate/fluoride) tubes. Immediately invert tube gently at least 4 times to mix. Tubes must be appropriately filled. Deliver tubes immediately to the laboratory

Storage Instructions Separate plasma from cells as soon as possible. Plasma may be stored on ice for up to 4 hours; otherwise, store frozen.

Causes for Rejection Specimen received more than 4 hours after collection, tubes not filled, clotted specimens

Turnaround Time Several hours to several days, depending on how often test batches are performed

Reference Interval Results are reported as a percent of the amount expected in normal plasma. By definition, the mean value in normal plasma is 100%. The reference range is approximately 70% to 140%. At birth, protein C levels are only 35% (range 17% to 53%) of adult normal values. Mean protein C levels rise to above 50% of adult normal values by age 6 months, but may remain below adult normal range until the age of 10-16 years. Women may have slightly decreased protein C levels in comparison to men, and premenopausal women may have slightly lower levels than postmenopausal women.

Use A functional assay should be performed first, because both type I and type II protein C deficiencies will be detected. The antigen assay is needed only if the functional assay is decreased, in order to determine if the deficiency is type I or type II. If the antigen assay is performed without the functional assay, type II deficiencies will not be detected (see Additional Information).

(Continued)

Protein C *(Continued)*

Limitations Acquired protein C deficiencies are more common than hereditary deficiencies (see Additional Information).

Chromogenic (functional) assays: Certain type II protein C deficiencies may not be detected in the chromogenic assay but will be detected by clot-based assays. Assays are usually designed to tolerate up to 1 unit/mL heparin. The advantage of this assay is that it is not affected by lupus anticoagulants, factor VIII levels, factor V Leiden, or other coagulation abnormalities that can interfere with clot-based protein C assays.

Clot-based (functional) assays: Commonly encountered coagulation conditions can interfere. For example, lupus anticoagulants can artifactually increase the protein C test result. Elevations in factor VIII (>200%) can artifactually decrease the result; factor VIII elevations occur in patients with an acute phase reaction. Falsely low values have been reported in patients with the factor V Leiden mutation. The advantage of this assay is that all known type I and type II variants should be detected. Assays that tolerate up to 1 unit/mL heparin are available. Cannot be performed in patients on hirudin or argatroban anticoagulation.

Antigen (immunologic) assays: If not used in conjunction with a functional assay, type II deficiencies will not be detected (see Additional Information).

Methodology Assays are functional (chromogenic or clot-based) or antigenic.

Chromogenic assays: Protein C in the patient plasma sample is activated, usually by a specific snake venom. The activated protein C cleaves a synthetic substrate that resembles the natural substrate of protein C, liberating a chromogenic substance that can be measured spectrophotometrically.

Clot-based assays: Protein C in the patient plasma sample is activated, usually with a specific snake venom. The activated protein C then degrades factors Va and VIIIa, thereby prolonging a PTT-based clotting time.

Antigenic (immunoassay): Enzyme-linked immunosorbent assay (ELISA)

Additional Information Protein C, a vitamin K dependent zymogen of a serine protease (activated protein C), has a molecular weight of 62,000 daltons. Protein C functions as an anticoagulant by using protein S as a cofactor to degrade activated factors V and VIII. Protein C must first be converted into activated protein C by interacting with a thrombin-thrombomodulin complex on the surface of endothelial cells. Protein C also indirectly promotes fibrinolysis.

Hereditary protein C deficiency is present in 0.14% to 0.50% of the general population. It accounts for 3% of unselected patients with venous thrombosis and up to 9% of patients younger than 70 years of age with thrombosis. Protein C deficiency does not appear to cause arterial thrombosis. Many different protein C gene mutations causing hereditary protein C deficiency have been identified. Individuals heterozygous for protein C deficiency have a sevenfold increased risk for venous thrombosis. Heterozygotes generally have protein C levels between 35% to 65%, although higher levels have been reported. The risk for thrombosis is further increased in the presence of a second risk factor with the possible exception of prothrombin G20210A mutation. The age at onset of thrombosis is usually between 10-50 years in heterozygous individuals. **Coumadin®-induced skin necrosis** may occur if protein C deficient patients are treated with Coumadin® without the addition of an immediate-acting anticoagulant (eg, heparin) until the Coumadin® levels are therapeutic. Homozygous deficiencies are rare (incidence of 1:500,000 to 1:750,000) and are fatal if untreated. They present in the newborn period with severely decreased protein C, **purpura fulminans**, and disseminated intravascular coagulation (DIC).

Decreased protein C can also arise from acquired conditions, such as:
- decreased hepatic synthesis from liver disease or L-asparaginase treatment
- synthesis of a dysfunctional protein due to vitamin K deficiency or warfarin (Coumadin®) use
- consumption from thrombosis, DIC, or surgery

Rarely, an acquired inhibitor (autoantibody) to protein C has been reported. If a patient with low protein C has any of the conditions listed above, the test should be repeated once the condition is no longer present. Confirmation of a hereditary protein C deficiency may require documenting protein C deficiency in a relative. In nephrotic syndrome, protein C may increase, decrease, or remain unchanged. Malm et al. reported that protein C can increase with oral contraceptives and pregnancy, whereas Kjellberg et al reported no significant increase in protein C during pregnancy.

Protein C has a relatively short half-life of 6-8 hours; therefore, it is one of the first hepatic coagulation proteins to decrease with liver dysfunction as well as with Coumadin® initiation.

Protein C deficiencies are quantitative (type I) or qualitative (type II). In type I deficiencies, normal protein C molecules are made, but in reduced quantity. In type II deficiencies, normal amounts of protein C are made, but the protein C is defective. Functional assays measure protein C function (activity). Antigenic assays are immunoassays that measure the quantity of protein C, regardless of the quality of its function. Accordingly, type I deficiencies have decreased protein C in both functional and antigenic assays. Type II deficiencies have normal antigenic protein C levels, with decreased functional protein C. Thus, if only antigenic assays are performed, type II deficiencies will not be detected. Therefore, a functional assay should be used as the initial screening assay. If the result is decreased, an antigenic assay may be performed to determine if the deficiency is type I or type II.

In addition to its role as an anticoagulant, protein C has anti-inflammatory properties. Activated protein C (APC) plays a central role in defense against sepsis. The use of recombinant APC improves survival in cases of adult sepsis.

References
Andrew M, Paes B, Milner R, et al, "Development of the Human Coagulation System in the Full-Term Infant," *Blood*, 1987, 70(1):165-72.

Bernard GR, Vincent JL, Laterre PF, et al, "Efficacy and Safety of Recombinant Human Activated Protein C for Severe Sepsis," *N Engl J Med*, 2001, 344(10):699-709.

Melissari E, Monte G, Lindo VS et al, "Congenital Thrombophilia Among Patients With Venous Thromboembolism," *Blood Coagul Fibrinolysis*, 1992, 3(6):749-58.

Miletich J, Sherman L, and Broze G Jr, "Absence of Thrombosis in Subjects With Heterozygous Protein C Deficiency," *N Engl J Med*, 1987, 317:991-6.

Roberts D and Schwartz RS, "Clotting and Hemorrhage in the Placenta - A Delicate Balance," *N Engl J Med*, 2002, 347(1):57-9.

Tait RC, Walker ID, Reitsma PH, et al, "Prevalence of Protein C Deficiency in the Healthy Population," *Thromb Haemost*, 1995, 73(1):87-93.

van der Meer FJ, Koster T, Vandenbroucke JP, et al, "The Leiden Thrombophilia Study (LETS)," *Thromb Haemost*, 1997, 78(1):631-5.

♦ **Protein C Deficiency** *see* Hypercoagulation Panel *on page 758*
♦ **Protein, Cerebrospinal Fluid** *see* Cerebrospinal Fluid Protein *on page 371*
♦ **Protein:Creatinine Ratio** *see* Protein, Semiquantitative, Urine *on page 1113*
♦ **Protein C Resistance, Activated** *see* Activated Protein C Resistance and the Factor V Leiden Mutation *on page 104*

Protein Electrophoresis, Capillary Zone

Related Information
Immunofixation Electrophoresis, Serum or Urine *on page 768*
Immunoglobulin G *on page 775*
Immunoglobulin M *on page 779*
Protein Electrophoresis, Serum *on page 1104*
Protein Electrophoresis, Urine *on page 1107*

Synonyms Capillary Electrophoresis; Free-Solution CE; Free-Zone CE

Applies to Bisalbuminemia; Immunoglobulins; Immunosubtraction Electrophoresis; Light Chains, Urine; Monoclonal Gammopathy; M-Proteins; Serum and Urine Protein Electrophoresis

Abstract Advances in capillary technology have permitted application in electrophoresis systems. Capillary zone electrophoresis (CZE) may be used to analyze proteins for reasons analogous to other electrophoretic methods. Monoclonal gammopathy includes monoclonal immunoglobulin heavy chain (gamma, alpha, mu, delta, epsilon) and/or light chain (kappa or lambda). During long-term follow-up of asymptomatic subjects with a small monoclonal gammopathy (monoclonal gammopathy of undetermined significance), up to 25% develop multiple myeloma, macroglobulinemia of Waldenström, or amyloidosis. Monoclonal proteins are a tumor marker in subjects with myeloma and Waldenström macroglobulinemia.

(Continued)

Protein Electrophoresis, Capillary Zone (Continued)

Specimen Serum, urine

Container Red top tube

Storage Instructions Refrigerate

Reference Interval

- Albumin: 3.7-5.2 g/dL
- Alpha₁: 0.2-0.57 g/dL
- Alpha₂: 0.3-0.76 g/dL
- Beta: 0.5-1.1 g/dL
- Gamma: 0.6-1.48 g/dL
- A:G ratio: 1.62 (1.09-2.16)

Use As with protein electrophoresis, the primary applications are to detect and monitor monoclonal gammopathies (paraproteins). It is more sensitive for small monoclonal protein abnormalities, especially IgA paraproteins, which can hide in the beta area in agarose gel separations. It has also been reported to be useful for therapeutic drug monitoring, hemoglobin, and amino acid quantitation. It provides better resolution with separation of the beta region into transferrin and C3 components, and offers advantages in cost and sensitivity (eg, in detection of bisalbuminemia).

Limitations Monoclonal IgA may not be detected in all cases by agarose gel or capillary electrophoresis.

Methodology Capillary electrophoresis (CE) uses high voltage applied across a buffer-filled fused-silica small bore capillary (20-100 microns diameter x ~100 cm long) to cause differential migration of the sample components. Typically, analytes are detected by passage of a light beam near one end of the capillary. Using CE, one can rapidly achieve high-resolution separation of molecules ranging in size from small ions to nucleotides and even to whole cells using small samples.

A method called **"immunosubtraction"** has been introduced in which the serum or urine is incubated with specific antibodies that are immobilized. This removes the specific protein, and the resultant electropherogram can then be compared to one performed without incubation. Comparison of the two electropherograms allows quantitation of the specific protein. Immunosubtraction electrophoresis is reported to be less accurate than immunofixation electrophoresis in determination of the immunotype of a monoclonal protein.

References

Bossuyt X, Bogaerts A, Schiettekatte G, et al, "Detection and Classification of Paraproteins by Capillary Immunofixation/Subtraction," *Clin Chem*, 1998, 44(4):760-4.

Jaeggi-Groisman SE, Byland C, and Gerber H, "Improved Sensitivity of Capillary Electrophoresis for Detection of Bisalbuminemia," *Clin Chem*, 2000, 46(6):880-1.

Katzmann JA, Clark R, Sanders E, et al, "Prospective Study of Serum Protein Capillary Zone Electrophoresis and Immunotyping of Monoclonal Proteins by Immunosubtraction," *Am J Clin Pathol*, 1998, 110(4):503-9.

Keren DF, "Capillary Zone Electrophoresis in the Evaluation of Serum Protein Abnormalities," *Am J Clin Pathol*, 1998, 110(2):248-52.

Litwin CM, Anderson SK, Philipps G, et al, "Comparison of Capillary Zone and Immunosubtraction With Agarose Gel and Immunofixation Electrophoresis for Detecting and Identifying Monoclonal Gammopathies," *Am J Clin Pathol*, 1999, 112(3):411-7.

Internet Web Sites

www.ceandcec.com

Protein Electrophoresis, Serum

Related Information

Viscosity, Serum or Plasma *on page 1313*

Synonyms Electrophoresis, Serum; Serum Protein Electrophoresis

Applies to Beta-Gamma Bridging; Globulin, Serum; Immunoglobulins; Light Chains; Monoclonal Gammopathy; M-Proteins

Test Includes Serum electrophoresis; a total protein value is needed, since the fractions are otherwise available only as percentages.

Abstract Serum proteins have different net charges and can be separated by electrophoresis into several distinct bands: albumin, alpha 1-globulin (α_1G), alpha 2-globulin (α_2G), beta globulin (βG), and gamma globulins (γG). Protein concentrations are altered as results of different disease states.

Serum and urine protein electrophoresis are among pivotal studies in the diagnosis and differential diagnosis of multiple myeloma, Waldenström macroglobulinemia, amyloidosis, and monoclonal gammopathy of undetermined significance (MGUS). The latter is the most common form of monoclonal gammopathy.

Specimen Serum

Container Red top tube or SST™ tube

Storage Instructions Refrigerate separated serum.

Reference Interval Values in the table are representative, but variation between methods and laboratories exists. The figures in the table are based on an agarose system. Values in infancy and early childhood are not identical to adult reference intervals.

Table 1. Protein Electrophoresis, Serum

Component	Relative (%) Normal Range	Absolute (g/dL) Normal Range
Total protein		5.90-8.00
Albumin	58.0-74.0	4.00-5.50
Alpha$_1$	2.0-3.5	0.15-0.25
Alpha$_2$	5.4-10.6	0.43-0.75
Beta	7.0-14.0	0.50-1.00
Gamma	8.0-18.0	0.60-1.30
A:G ratio		1.4-2.6

Critical Values Monoclonal globulin levels >3.0 g/dL indicate a malignant disease, IgG or IgA myeloma, or IgM of Waldenström macroglobulinemia.

Use Serum protein electrophoresis is a part of evaluation of patients when multiple myeloma, macroglobulinemia of Waldenström or primary amyloidosis is considered. Interpretation of serum protein electrophoresis patterns is helpful in screening for or evaluation of the diagnosis of disease states such as acute phase reaction, chronic inflammation, autoimmune hepatitis, cirrhosis, humoral immunodeficiency, α_1-antitrypsin abnormalities, and monoclonal, oligoclonal, and polyclonal gammopathies.

Narrow, intensely stained bands appearing as tall, narrow peaks in densitometric tracings are **monoclonal gammopathies** (paraproteins, monoclonal immunoglobulins, M-proteins), found in multiple myeloma, macroglobulinemia of Waldenström, amyloidosis, lymphoproliferative diseases, and in heavy-chain diseases. They occur also in 11% of patients with hepatitis C. **Monoclonal gammopathy of undetermined significance** (MGUS) is characterized by the presence of monoclonal IgG or IgA without evidence of multiple myeloma. Hematologic malignancy sometimes develops. Special efforts to establish diagnosis are indicated in the presence of bone lesions, hypercalcemia, or anemia, which have prognostic relevance in cases of myeloma. Serum protein electrophoresis should be repeated in 1 year for asymptomatic patients with a monoclonal protein <1.5 g/dL who have normal values of hemoglobin, calcium, and creatinine. Electrophoresis should be repeated in 2-3 months if the monoclonal protein is between 1.5 and 2.5 g/dL. Patients being treated for multiple myeloma, Waldenström macroglobulinemia or amyloidosis should be monitored at 1- to 2-month intervals.

Smoldering myeloma is a designation applied to cases with monoclonal gammopathy, with atypical marrow plasma cells, without anemia, bone lesions, or renal failure.

(Continued)

Protein Electrophoresis, Serum *(Continued)*

Indications for Protein Electrophoresis

- Consideration of multiple myeloma, macroglobulinemia of Waldenström, or primary amyloidosis
- Back pain, especially in subjects >50 years old
- Osteoporosis
- Osteolytic lesions by imaging
- Hypercalcemia
- Presence of Bence Jones protein (monoclonal light chains) in urine
- Increasing serum creatinine
- Recurrent infections
- Unexplained peripheral neuropathy
- Congestive heart failure refractory to usual therapy
- Nephrotic syndrome
- Malabsorption in subjects >50 years old
- Unexplained hepatomegaly and/or splenomegaly and/or anemia
- Initial screening for α_1-antitrypsin deficiency (most of α_1 globulin is α_1-antitrypsin)
- Evaluation of chronic liver disease: decreased albumin, polyclonal gammopathy, β- and γ-bridging are prototypical findings of hepatic cirrhosis; the γ increase is IgM in primary biliary cirrhosis
- Increased ESR

Limitations Serum protein electrophoresis may be normal in some patients with plasma cell dyscrasia. Such individuals may have monoclonal light chains (Bence Jones protein) in urine. See Protein Electrophoresis, Urine *on page 1107* and Immunofixation Electrophoresis, Serum or Urine *on page 768*. Capillary zone electrophoresis is more sensitive than agarose gel in detection of low levels of monoclonal proteins.

Methodology Cellulose acetate and agarose electrophoresis are widely used methods, quantitated by densitometry. Stains include Ponceau S and Coomassie brilliant blue. Capillary zone and immunosubtraction electrophoretic methods are gaining acceptance (see Protein Electrophoresis, Capillary Zone *on page 1103*).

Table 2.

Pattern	Protein Changes	Frequently Associated Diseases
Acute inflammation	Normal or ↓ albumin ↑ α_1G and/or α_2G	Acute infection and inflammatory disorders (acute phase reaction)
Chronic inflammation	Normal or ↓ albumin ↑ α_1G and/or α_2G ↑ γG	Autoimmune diseases, chronic liver disease including chronic autoimmune hepatitis, primary biliary cirrhosis, chronic infection, cancer
Hypoalbuminemia	↓ albumin	Metastatic cancer, CHF, malnutrition, protein-losing disorders
Hypogammaglobulinemia	Normal or ↓ albumin ↓ γG	Lymphoproliferative disorders, inflammatory bowel disease, congenital immunodeficiencies
Polyclonal gammopathy	↑ γG	Autoimmune disease, chronic infections, infestations such as visceral leishmaniasis, liver disease including autoimmune hepatitis and cirrhosis
Cirrhosis	Often ↓ albumin ↑ γG, Beta-gamma bridging	Cirrhosis Autoimmune hepatitis
Protein-losing disorder	↓ albumin ↑ α_2G ↓ γG	Nephrotic syndrome, exudative skin disorders, gastroenteropathies
Monoclonal gammopathy	Normal or ↓ albumin ↑ γG	Myeloma, macroglobulinemia, MGUS, CLL, lymphoma, amyloidosis, Gaucher disease (25% of patients), AIDS (15% of patients)
Antitrypsin deficiency	Absent or low α_1G	Alpha-1 antitrypsin deficiency
Hyperbetaglobulinemia	Normal or ↓ albumin ↑ βG	Hyperlipidemia, diabetes mellitus, iron deficiency anemia
Immune deficiency	↓ or absent γG	Congenital or acquired immunodeficiency states

α_1G = alpha-1 globulin; α_2G = alpha-2 globulin; βG = beta globulin; γG = gamma globulin.

Additional Information The most commonly recognized electrophoresis patterns are summarized in Table 2.

The most common presentations of myeloma include osteolytic lesions, anemia, renal insufficiency, and recurrent bacterial infections. The diagnosis is supported by the presence of a monoclonal immunoglobulin in serum or urinary light chains. Immunofixation is needed to characterize monoclonal gammopathies.

Most patients with immunoglobulin amyloidosis have monoclonal immunoglobulin in serum, urine, or both.

References

Alexanian R, Weber D, and Liu F, "Differential Diagnosis of Monoclonal Gammopathies," *Arch Pathol Lab Med*, 1999, 123(2):108-13.

Bataille R and Harousseau JL, "Multiple Myeloma," *N Engl J Med*, 1997, 336(23):1657-64.

Goeken JA and Keren DF, "Introduction to the Report of the Consensus Conference on Monoclonal Gammopathies," *Arch Pathol Lab Med*, 1999, 123(2):104-5.

Katzmann JA, Clark R, Sanders E, et al, "Prospective Study of Serum Protein Capillary Zone Electrophoresis and Immunotyping of Monoclonal Proteins by Immunosubtraction," *Am J Clin Pathol*, 1998, 110(4):503-9.

Keren DF, "Procedures for the Evaluation of Monoclonal Immunoglobulins," *Arch Pathol Lab Med*, 1999, 123(2):126-32.

Keren DF, Alexanian R, Goeken JA, et al, "Guidelines for Clinical and Laboratory Evaluation of Patients With Monoclonal Gammopathies," *Arch Pathol Lab Med*, 1999, 123:106-7.

Kyle RA, "Sequence of Testing for Monoclonal Gammopathies," *Arch Pathol Lab Med*, 1999, 123(2):114-8.

Litwin CM, Anderson SK, Philipps G, et al, "Comparison of Capillary Zone and Immunosubtraction With Agarose Gel and Immunofixation Electrophoresis for Detecting and Identifying Monoclonal Gammopathies," *Am J Clin Pathol*, 1999, 112(3):411-7.

♦ **Protein Electrophoresis, Spinal Fluid** *see* Cerebrospinal Fluid Protein Electrophoresis *on page 374*

Protein Electrophoresis, Urine

Related Information

Immunofixation Electrophoresis, Serum or Urine *on page 768*
Immunoglobulin G *on page 775*
Microalbuminuria *on page 913*
Protein Electrophoresis, Serum *on page 1104*
Protein, Quantitative, Urine *on page 1108*
Protein, Semiquantitative, Urine *on page 1113*
Urine Collection, 24-Hour *on page 1295*
Viscosity, Serum or Plasma *on page 1313*

Synonyms Electrophoresis, Protein, Urine; Globulins, Urine; Urine Electrophoresis; Urine Protein Electrophoresis

Applies to Immunoelectrophoresis; Light Chains, Urine; Monoclonal Gammopathy; Monoclonal Light Chains

Replaces Bence Jones Protein

Test Includes Quantitative total urine protein, urine albumin, urine alpha$_1$, urine alpha$_2$, urine beta, and urine gamma globulin fractions

Abstract Normal glomerular and tubular function results in excretion <150 mg of protein/day. Two thirds of the filtered protein is comprised of albumin, transferrin, low molecular weight proteins, and some immunoglobulins. Renal injury may result in proteinuria. Urine protein electrophoresis separates proteins according to charge and allows classification of the type of renal injury. Protein patterns are interpreted by a clinical pathologist and may be reported as glomerular, tubular, or mixed patterns.

Excretion of monoclonal light chains >50 mg/day is evidence against the benign character of monoclonal gammopathy.

Container No preservative

Storage Instructions Refrigerate during and after the collection

Causes for Rejection Total protein too low to measure or to yield usable electrophoretic pattern

Reference Interval A normal urine protein pattern consists of albumin and occasionally faint alpha-1 and beta bands.

- Total protein: <150 mg/24-hour specimen
- Albumin: <50% of total

(Continued)

Protein Electrophoresis, Urine (Continued)

• Total globulin: 60% to 67% of total

Use Work-up for monoclonal gammopathy

Limitations May not detect pathologic light chains due to insufficient sensitivity of this method; immunofixation electrophoresis is the next step.

Contraindications Reagent strips, sulfosalicylic acid, and acidified heat precipitation are unsuitable for characterization of monoclonal free light chains.

Methodology Electrophoresis, cellulose acetate, and agarose gel are most commonly used. Evaluation of monoclonal free light chains is best accomplished with quantitation of 24-hour urinary protein excretion, densimetry, and immunofixation.

Additional Information Glomerular filtration produces fluid containing very little protein with molecular weights >40,000 daltons. Proteins with molecular weights <15,000 daltons pass freely through the glomerulus but are then almost completely reabsorbed in the proximal tubules. Two thirds of the filtered protein is comprised of albumin, transferrin, low molecular weight proteins, and some immunoglobulins. The remainder, such as Tamm-Horsfall glycoprotein, is derived from the urinary tract itself.

Monoclonal gammopathy (M-protein) is found with myeloma, Waldenström macroglobulinemia, lymphoproliferative diseases including chronic lymphocytic leukemia, and primary systemic amyloidosis. A more common entity, monoclonal gammopathy of undetermined significance (MGUS), is an important consideration in differential diagnosis. Bence Jones protein (monoclonal light chains) in MGUS is absent or <50 mg/day.

References

Alexanian R, Weber D, and Liu F, "Differential Diagnosis of Monoclonal Gammopathies," *Arch Pathol Lab Med*, 1999, 123(2):108-13.

Baskin LB and Hsu RM, "Laboratory Evaluation of Proteinuria," *MLO*, 1999, 30-6.

Keren DF, Alexanian R, Goeken JA, et al, "Guidelines for Clinical and Laboratory Evaluation of Patients With Monoclonal Gammopathies," *Arch Pathol Lab Med*, 1999, 123(2):106-7.

Roach BM, Meinke JS, Sridhar N, et al, "Multiple Narrow Bands in Urine Protein Electrophoresis," *Clin Chem*, 1999, 45(5):716-8.

♦ **Protein Glycosylation** see Glycated Hemoglobin (Hemoglobin A_{1c}), Blood on page 655

Protein, Quantitative, Urine

Related Information

Albumin, Serum on page 120

Blood, Urine on page 281

Cadmium, Blood or Urine on page 325

Chromium, Serum on page 402

Creatinine Clearance and Urine Creatinine on page 473

Fat, Urine on page 573

Glycated Hemoglobin (Hemoglobin A_{1c}), Blood on page 655

Immunofixation Electrophoresis, Serum or Urine on page 768

Kidney Biopsy on page 818

Microalbuminuria on page 913

Osmolality, Urine on page 979

Protein Electrophoresis, Urine on page 1107

Protein, Semiquantitative, Urine on page 1113

Protein, Total, Serum on page 1114

Urinalysis on page 1289

Urine Collection, 24-Hour on page 1295

Applies to Tamm-Horsfall Protein

Test Includes Concomitant creatinine clearance is often indicated. Total urine creatinine should be **routinely** included to help assure that a complete 24-hour collection was tested.

Abstract Quantitation of urinary protein loss provides evaluation of renal diseases, including nephrotic syndromes.

Specimen 24-hour urine

Container Plain urine container; consult the laboratory performing the assay about whether or not there is a need for preservative.

Storage Instructions Refrigeration and freezing may be desirable; check with the laboratory performing the assay.

Reference Interval 1-14 mg/dL; 50-80 mg/day (at rest); <250 mg/day (after intense exercise). Urinary protein normally tends to increase with age, exercise, and standing posture.

Critical Values Nephrotic syndromes: children: >1.0 g/m^2/day; adults: ≥3.5 g/24 hours

Use Evaluate proteinuria; evaluate renal diseases, including proteinuria complicating diabetes mellitus, the nephrotic syndromes, metal poisoning, renal vein thrombosis, systemic lupus erythematosus (SLE), constrictive pericarditis, and amyloidosis; work up other renal diseases including hypertension, glomerulonephritis, Goodpasture syndrome, Henoch-Schönlein purpura, thrombotic thrombocytopenic purpura, collagen diseases, cryoglobulinemia, preeclampsia, drug nephrotoxicity, hypersensitivity reactions, allergic reactions, and renal tubular lesions; management of myeloma and macroglobulinemia of Waldenström (Bence Jones proteinuria); evaluate hypoproteinemia; tubular proteinurias include Wilson disease and Fanconi syndrome. In the table shown,

Some Causes of Proteinuria

Normal proteinuria	Albumin ≤30 mg/24 h	
	Other (Tamm-Horsfall) ≤50 mg/24 h	
Prerenal proteinuria	Congestive heart failure	
	Orthostatic proteinuria	
	Transient, associated with febrile illness, surgery, anemia, hyperthyroidism, stroke, exercise, seizures	
	Bence Jones proteinuria associated with myeloma, Waldenström macroglobulinemia, amyloidosis (light chain proteinuria)	
	Lysozyme associated with myelocytic leukemia	
Renal proteinuria	Renovascular hypertension	
	Malignant hypertension of any cause	
	Glomerular Proteinuria >3.5 g/24 h usually reflects a glomerular lesion (in children >1.0 g/m^2/day)	Membranous nephropathy and proliferative glomerulonephritis
		Chronic pyelonephritis
		Polycystic disease
		Diabetic nephropathy
		Amyloidosis
		Lupus erythematosus (SLE)
		Goodpasture syndrome
		Renal vein thrombosis
		Minimal change nephropathy
		Focal segmental glomerulosclerosis
		HIV nephropathy
		Alport syndrome
		Preeclampsia
		High molecular weight proteinuria
	Tubular usually <1 g/24 h	Fanconi syndrome
		Wilson disease
		Renal tubular acidosis
		Heavy metal poisoning: lead, mercury, cadmium
		Galactosemia
		Low molecular weight (<60,000) proteinuria
		Beta$_2$-microglobulinemia (molecular weight 11,800)
	Interstitial	Bacterial pyelonephritis
		Uric acid, urate or calcium deposition
		Idiosyncratic drug reaction: methicillin, phenindione, sulfonamides, phenytoin, others
		Interstitial diseases generally reflected as tubular defects or mixed tubular interstitial
Postrenal proteinuria	Tumors of the bladder or renal pelvis	
	<1 g/24 h, IgM excretion significant marker, amount of proteinuria related to size and spread of tumor	
	Cystitis, severe	

Note that this brief outline is intended only to provide an overview, and is incomplete.

(Continued)

Protein, Quantitative, Urine (Continued)

some renal lesions are not easily categorized (eg, the glomerular lesions of chrysotherapy) and of toxemia of pregnancy. All important entities are not shown in the table, including for instance absence of one kidney and vasculitis.

Limitations Although evaluation for proteinuria may be useful in chronic renal disease, proteinuria may wax and wane. Preeclampsia, characterized by hypertension, proteinuria, and edema in a woman more than 20 weeks pregnant, is a state in which urine protein excretion is commonly measured.

Twenty-four hour urine collections are subject to collection errors. The laboratory method, depending on an aliquot and varying dilutions, is subject to calculation errors. When protein is determined by precipitation methods, x-ray contrast media, tolbutamide, penicillin or cephalosporin analogs and sulfonamides may cause false positives. Pyridium® interferes with the reaction by causing color interference. Functional and postural proteinuria occur.

Methodology A number of methods are in use including trichloroacetic acid, sulfosalicylic acid precipitation, biuret method with phosphotungstic acid, and Coomassie blue dye binding. The standard for most methodologies is albumin. Different methods are more or less sensitive to globulin than to albumin. Thus, for nonselective proteinurias, in which a variety of proteins are present, different methodologies yield different results.

Additional Information Some patients exhibit orthostatic proteinuria (ie, recumbent urine protein 100-180 mg in a 12-hour overnight urine collection and up to 1 g in the subsequent 12 hours while ambulatory). The presence of urinary protein >200 mg in the overnight specimen or equally increased amounts of urine protein in both specimens indicates need for further work-up.

In a study of blood pressure and the kidney, proteinuria was identified as an independent risk factor for progression of renal disease. Proteinuria was greater in subjects with glomerular disease entities, diabetes, and hereditary nephritis. Proteinuria is a predictor of decline of glomerular filtration rate.

References

Larson TS, "Evaluation of Proteinuria," Mayo Clin Proc, 1994, 69(12):1154-8.

Painter PC, Cope JY and Smith JL, "Reference Information for the Clinical Laboratory," Tietz Textbook of Clinical Chemistry, 3rd ed, Burtis CA and Ashwood ER, eds, Philadelphia, PA: WB Saunders Co, 1999, 1832.

Peterson JC, Adler S, Burkart JM, et al, "Blood Pressure Control, Proteinuria, and the Progression of Renal Disease. The Modification of Diet in Renal Disease Study," Ann Intern Med, 1995, 123(10):754-62.

Protein S

Related Information

Activated Protein C Resistance and the Factor V Leiden Mutation on page 104

Antithrombin on page 198

Hypercoagulation Panel on page 758

Protein C on page 1101

Warfarin, Serum or Plasma on page 1325

Abstract Protein S is a required cofactor for the anticoagulant activity of protein C. A hereditary deficiency of protein S leads to a hypercoagulable state with an increased risk for venous thrombosis. Protein S deficiencies are quantitative (type I) or qualitative (type II).

Patient Preparation Determine if patient is on oral anticoagulants or estrogen (eg, oral contraceptives, estrogen replacement) or if the patient is pregnant. Protein S levels are decreased by estrogen, pregnancy, and warfarin (Coumadin®).

Specimen Plasma

Container Blue top (sodium citrate) tube

Sampling Time Testing should be deferred until patients have not received Coumadin® for at least 10 days, because Coumadin® decreases protein S levels.

Collection Routine venipuncture. If multiple tests are being drawn, draw blue top tubes after any red top tubes but before any lavender top (EDTA), green top (heparin), or gray top (oxalate/fluoride) tubes. Immediately invert tube gently at least 4 times to mix. Tubes must be appropriately filled. Deliver tubes immediately to the laboratory.

Storage Instructions Separate plasma from cells as soon as possible. Plasma may be stored on ice for up to 4 hours; otherwise, store frozen.

Causes for Rejection Specimen received more than 4 hours after collection, tubes not filled, clotted specimens

Turnaround Time Several days (because testing is usually batched)

Special Instructions Elevated factor VIII (>200%) is a common cause of artifactually decreased protein S in PTT-based functional assays. It is recommended to measure factor VIII on the same specimen when the functional protein S is decreased by PTT-based methods, to determine if the decrease is due to elevated factor VIII.

Reference Interval Results are reported as a percent of the amount expected in normal plasma. By definition, the mean value in normal plasma is 100%. The reference range is approximately 70% to 140%; lower for women than for men. At birth, protein S (total antigen) levels are only 36% (range 12% to 60%) of adult normal values. Protein S rises into the adult reference range by age 6 months.

Use A functional assay should be performed first, because all subtypes of protein S deficiencies will be detected. The free antigen assay is needed only if the functional assay is decreased, and the total antigen assay is needed only if the free antigen is decreased, in order to determine the deficiency subtype. If the antigen assays are performed without the functional assay, patients with certain subtypes will not be detected (see Additional Information and the table).

Limitations Acquired protein S deficiencies are more common than hereditary deficiencies (see Additional Information).

Functional assays: Commonly encountered coagulation conditions interfere. For example, lupus anticoagulants can falsely increase the protein S test result. Elevations in factor VIII (>200%) can artifactually decrease PTT-based results; factor VIII elevations occur in patients with an acute phase reaction. In some assays, falsely low values have been reported in patients with the factor V Leiden. Assays that tolerate up to 1-2 units/mL heparin are available. The functional assay cannot be performed in patients on hirudin or argatroban anticoagulation.

Antigen assays: If not used in conjunction with a functional assay, patients with some subtypes will not be detected (see Additional Information and table).

Methodology

Functional (activity) assays: Protein S is measured by its ability to serve as a cofactor required for activated protein C-mediated degradation of activated factors V and VIII, thereby prolonging a PTT- or PT-based clotting time.

Free antigen (immunoassay): Monoclonal antibodies specific for free (unbound) protein S are used in an enzyme-linked immunosorbent assay (ELISA). In an older assay, free protein S was determined by first treating specimens with polyethylene glycol (PEG), which precipitates bound protein S and leaves free protein S in the supernatant. In the new ELISA using monoclonal antibodies specific for free protein S, the PEG step is no longer necessary. Elimination of the PEG-precipitation step has significantly improved the accuracy of the test result.

Total antigen (immunoassay): Measures total (free and bound) protein S by ELISA. An alternative method uses latex particles coated with antibodies directed against protein S. In the presence of protein S, the latex particles form aggregates that absorb light passing through the specimen. The amount of light absorbance is directly related to the amount of protein S in the specimen. A third method, rocket immunoelectrophoresis, is an older method that is still in use in some laboratories.

Additional Information Protein S is a vitamin K dependent protein that is a required cofactor for activated protein C. Activated protein C, with protein S as a cofactor, acts as an anticoagulant by degrading activated factors V and VIII. Sixty percent of total protein S is bound to C4b-binding protein and is inactive. The remainder, called free protein S, is the functionally active form.

Hereditary protein S deficiency is present in 0.7% of the general population. It accounts for 2% of unselected patients with venous thrombosis and up to 7.6% of patients younger than 70 years of age with thrombosis. Many different mutations in the protein S gene are known to cause hereditary protein S (Continued)

Protein S *(Continued)*

deficiency. Individuals heterozygous for protein S deficiency have an increased risk for venous thrombosis, and the risk is further increased in the presence of a second risk factor. Heterozygotes generally have protein S levels between 20% to 65%. The age at onset of thrombosis is usually between 10-50 years in heterozygous individuals. **Coumadin®-induced skin necrosis** has been reported in protein S deficient patients who are started on Coumadin® without the addition of an immediate-acting anticoagulant (eg, heparin) until the Coumadin® levels are therapeutic. Homozygous deficiencies are rare, and are fatal if untreated. They present in the newborn period with severely decreased protein S, purpura fulminans, and disseminated intravascular coagulation (DIC).

Decreased protein S can also arise from acquired conditions, such as:

- decreased hepatic synthesis from liver disease or L-asparaginase treatment
- synthesis of a dysfunctional protein due to vitamin K deficiency or warfarin (Coumadin®) use
- consumption from thrombosis, DIC or invasive procedures
- estrogen, including oral contraceptives, estrogen replacement therapy, or pregnancy (decreased protein S may persist for up to 2 months after delivery or estrogen discontinuation)
- acute phase reactions (due to elevated C4b-binding protein, which decreases free and consequently functional protein S)

May also become decreased in nephrotic syndrome, varicella infection or HIV infection. Acquired inhibitors (autoantibodies) to protein S have been reported, some of which arose in association with varicella infections. If an acquired cause is present, the test should be repeated once the condition is no longer present, if possible. Confirmation of a hereditary protein S deficiency may require documenting protein S deficiency in a relative.

In liver disease, protein S is occasionally normal despite decreased protein C and antithrombin (all three proteins are synthesized in the liver). It is speculated that this is because protein S is synthesized in endothelial cells and megakaryocytes in addition to the liver, whereas protein C and antithrombin are synthesized predominantly or exclusively in the liver.

Protein S deficiencies are quantitative (type I) or qualitative (type II). In type I deficiencies, normal protein S molecules are made, but in reduced quantity. In type II deficiencies, normal amounts of protein S are made, but the protein S is defective. Functional assays measure protein S function. The total antigen assay is an immunoassay that measures the total quantity of protein S, regardless of the quality of its function. Free antigen assays are immunoassays that measure only unbound (free) protein S, regardless of the quality of its function. Only free (unbound) protein S is active; protein S that is bound to its binding protein (C4b-binding protein) is inactive. Accordingly, type I deficiencies have decreased protein S in both functional and antigenic assays. Type II deficiencies have normal total antigen levels, with decreased functional protein S. A further type II subtype (known as type IIa or type III) is characterized by decreased functional and free antigen levels with normal total antigen levels (see table). This subtype may be due to mutations causing increased binding of protein S to C4b-binding protein. In summary, if only antigenic assays are performed, type II deficiencies will not be detected. Therefore, a functional assay should be used as the initial screening assay. If the result is decreased, a free antigen assay should be performed to determine the deficiency subtype.

If the free antigen is decreased, a total antigen assay may be performed to further determine the deficiency subtype (see table).

Classification of Hereditary Protein S Deficiencies

Type	Functional Protein S	Free Protein S (Free Antigen Assay)	Total Protein S (Total Antigen Assay)
I	Low	Low	Low
II (also called IIb)	Low	Normal	Normal
III (also called IIa)	Low	Low	Normal

References

Aillaud MF, Pouymayou K, Brunet D, et al, "New Direct Assay of Free Protein S Antigen Applied to Diagnosis of Protein S Deficiency," *Thromb Haemost*, 1996, 75(2):283-5.

Andrew M, Paes B, Milner R, et al, "Development of the Human Coagulation System in the Full-Term Infant," *Blood*, 1987, 70(1):165-72.

Heijboer H, Brandjes DPM, Buller HR, et al, "Deficiencies of Coagulation-Inhibiting and Fibrinolytic Proteins in Outpatients With Deep-Vein Thrombosis," *N Engl J Med*, 1990, 323:1512-6.

Melissari E, Monte G, Lindo VS, et al, "Congenital Thrombophilia Among Patients With Venous Thromboembolism," *Blood Coagul Fibrinolysis*, 1992, 3(6):749-58.

Rodeghiero F and Tosetto A, "The Epidemiology of Inherited Thrombophilia: The VITA Project," *Thromb Haemost*, 1997, 78(1):636-40.

Van Cott EM and Laposata M, "Laboratory Evaluation of Hypercoagulable States," *Hematol Oncol Clin North Am*, 1998, 12(6):1141-66.

♦ **Protein, Screen, Urine** *see* Protein, Semiquantitative, Urine *on page 1113*

♦ **Protein S Deficiency** *see* Hypercoagulation Panel *on page 758*

Protein, Semiquantitative, Urine

Related Information

Chromium, Serum *on page 402*
Kidney Biopsy *on page 818*
Microalbuminuria *on page 913*
Osmolality, Urine *on page 979*
Protein Electrophoresis, Urine *on page 1107*
Protein, Quantitative, Urine *on page 1108*
Urinalysis *on page 1289*

Synonyms Albumin, Urine; Protein, Screen, Urine; Protein, Urine, Sulfosalicylic Acid; Urine Screen for Albumin; Urine Screen for Protein

Applies to Protein:Creatinine Ratio

Test Includes Screening for urine protein by dipstick and, in some laboratories, sulfosalicylic acid method for confirmation is part of routine urinalysis.

Abstract A screening test for renal disease. Microscopic examination of urine sediment is essential to evaluate patients with proteinuria.

Specimen Random urine

Collection Early morning specimen is recommended to provide maximally concentrated urine, when immunoglobulin light chain (Bence Jones protein) detection is important and when orthostatic proteinuria must be ruled out. For other renal disease, daytime urine is satisfactory or even preferred. Transport specimen to the laboratory within 2 hours of collection. Container should state date and time of collection.

Storage Instructions If not run promptly, specimen should be refrigerated.

Reference Interval Dipstick results include grades negative and trace (10-20 mg/dL) (SI: 0.1-0.2 g/L). The sensitivity of the dipstick is in the range of 150-300 mg/L.

Critical Values Dipstick 1+ is about 30 mg/dL, 2+ 100 mg/dL, 3+ (300 mg/dL), or 4+ (1000 mg/dL)

Use Reagent strips for detection of urinary protein are used to screen for pree-clampsia and other disorders, including nephrotic syndromes, complications of diabetes mellitus, glomerulonephritis, amyloidosis, and other entities. Some causes of proteinuria are found in the table in Protein, Quantitative, Urine *on page 1108*. Proteinuria is probably the single most important indicator of renal disease and its severity.

Limitations The reagent strip method is very sensitive to albumin, but much less so to positively charged proteins; thus, dipsticks commonly will not detect immunoglobulin light chains (Bence Jones protein) or myeloma protein, to which sulfosalicylic acid procedures are usually sensitive. False negatives may be found with highly dilute urines. False-negative rates for dipstick protein can be substantial. Both false-negative and false-positive reagent strip results are commonplace in testing pregnant patients.

False-positive results may be obtained with highly alkaline (pH ≥7) urines on dipsticks, with hematuria, and in highly concentrated urine. Contaminating quaternary ammonium groups or chlorhexidine present in disinfectants may also give false-positive dipstick results. A negative result does not rule out the presence of nonalbumin proteins. Pyridium® metabolites may mask the reaction. X-ray contrast media, tolbutamide, nafcillin, massive doses of penicillin, (Continued)

Protein, Semiquantitative, Urine *(Continued)*

sulfisoxazole (Gantrisin®), para-aminosalicylic acid, and high levels of cephalosporins may cause false-positive reactions with the sulfosalicylic acid method. Dipstick methods may be unreliable in unusually colored urines or when a great deal of sediment is present. The detection limit of Albustix® (Ames Division, Miles Laboratories) is reported as 300 mg/L or 500 mg/day protein. Since normal albuminuria is <20 mg/L, the screening dipstick lacks sensitivity for early detection of protein loss in diabetic nephropathy. See Microalbuminuria *on page 913.* Use of reagent strips is sometimes delegated to personnel whose level of training and whose supervision are limited.

Methodology Dipstick, and in some laboratories sulfosalicylic acid, are run on all urinalyses. The dipstick test is based on the color development of indicators, usually bromphenol blue in citric acid buffer. The sulfosalicylic acid test is based on the acid precipitation of protein. Immunofixation or immunoelectrophoresis is indicated when Bence Jones protein is suspected.

Additional Information The protein:creatinine ratio corrects protein concentration for urine creatinine, correcting for dilution effects in a random specimen. It provides a modicum of correlation with 24-hour collections for protein. A ratio <0.20 is regarded as normal.

If the dipstick for protein is negative and the sulfosalicylic acid test is positive, immunoglobulin light chains may be present. If clinically indicated in this situation, a urine examination for electrophoresis and immunoelectrophoresis for light chains or immunofixation should be considered.

Normal newborns may have proteinuria during first 3 days of life.

Transient proteinuria may be secondary to fever, congestive heart failure, and following exercise or cold exposure. Repeat collection of first morning urine with sediment microscopy can direct further investigation.

References
Larson TS, "Evaluation of Proteinuria," *Mayo Clin Proc,* 1994, 69(12):1154-8.

Misdraji J and Nguyen PL, "Urinalysis. When - and When Not - To Order," *Postgrad Med,* 1996, 100(1):173-6, 181-2, 185-8 passim.

Protein, Total, Serum

Related Information
Albumin:Globulin Ratio, Serum *on page 119*
Albumin, Serum *on page 120*
Cerebrospinal Fluid Protein Electrophoresis *on page 374*
Immunofixation Electrophoresis, Serum or Urine *on page 768*
Immunoglobulin A *on page 770*
Immunoglobulin G *on page 775*
Immunoglobulin G Subclasses *on page 778*
Immunoglobulin M *on page 779*
Protein Electrophoresis, Capillary Zone *on page 1103*
Protein Electrophoresis, Serum *on page 1104*
Protein, Quantitative, Urine *on page 1108*

Synonyms Total Protein, Serum

Applies to Globulin, Serum

Abstract Used to evaluate protein nutritional status and protein altering diseases

Specimen Serum or plasma

Container Red top tube, green top (heparin) tube

Collection Pediatrics: Blood drawn from heelstick for capillary.

Storage Instructions Separate serum from cells. Refrigerate at 2°C to 8°C.

Reference Interval 6.0-8.0 g/dL (SI: 60-80 g/L) in later childhood and adults. Lower intervals occur in early childhood. Ambulatory values are slightly higher than are those found in recumbency. If normal intervals are set for inpatients, then many outpatients appear to be a little above the upper limit. Because plasma contains fibrinogen, the plasma protein concentration may be up to 0.4 g/dL higher than serum protein concentration.

Use Evaluate nutritional status; investigate edema.

In the entities which follow, the diseases listed are sometimes increased or decreased as indicated, but are not always so.

Causes of **high total protein:** dehydration; some cases of chronic liver disease, including autoimmune hepatitis and cirrhosis; neoplasms, especially myeloma; macroglobulinemia of Waldenström; tropical diseases (eg, kala-azar, leprosy, and others); granulomatous diseases, such as sarcoidosis; diseases in which total protein is sometimes high include collagen disease (eg, lupus erythematosus (SLE), and other instances of acute or chronic infection/inflammation).

Causes of **low total protein:** pregnancy; intravenous fluids; cirrhosis or other liver disease, including chronic alcoholism; prolonged immobilization; heart failure; nephrotic syndromes; glomerulonephritis; neoplasia; protein losing enteropathies; Crohn disease and chronic ulcerative colitis; starvation, malabsorption, or malnutrition; hyperthyroidism; burns; severe skin disease; and other chronic diseases.

Very low total protein (<4.0 g/dL (SI: <40 g/L)) and low albumin cause edema (eg, nephrotic syndromes).

Limitations Venous stasis during venipuncture can lead to increased values. Hemolysis can falsely elevate total protein. Clinical interpretation is greatly enhanced by examination of the fractions composing total protein, when such separation is clinically indicated (ie, serum protein electrophoresis, methods for IgG, IgA, IgM, immunofixation).

Additional Information Total protein and albumin normally decrease by 5% to 10% upon recumbency, as in hospitalization. "Globulin" may be provided as a calculation, total protein - albumin = globulin.

Following an acute phase stimulus such as infection or trauma, many liver derived plasma proteins increase in concentration ("acute phase reactants"), while albumin decreases. Thus, while the total serum protein concentration may remain the same in acute phase reactions, the composition of the serum proteins is altered.

Drug effects are summarized. **Increases** related to analytical methods include fluosol-DA, phenazopyradine, radiographic agents and sulfasalazine. Pharmacological increases include effects of anabolic steroids, angiotensin, bumetanide, corticosteroids, digitalis, furosemide, insulin isoretinoin, oral contraceptives, progesterone, and other drugs. **Decreases** are related to carvedilol, hetastarch, laxatives (with continued use), tacrolimus, and other drugs.

References
Young DS, *Effects of Drugs on Clinical Laboratory Tests*, 5th ed, Volume 1: Listing by Test, Washington, DC: AACC Press, American Association of Clinical Chemistry, 2000, Section 3, 672, 677.

♦ **Protein, Urine, Sulfosalicylic Acid** *see* Protein, Semiquantitative, Urine *on page 1113*

♦ **Prothrombin Complex Concentrates** *see* Factor IX Concentrate *on page 564*

♦ **Prothrombin (Factor II)** *see* Coagulation Factor Assays *on page 418*

♦ **Prothrombin Fragment 1.2** *see* Hypercoagulation Panel *on page 758*

Prothrombin G20210A Mutation

Related Information
Hypercoagulation Panel *on page 758*
Polymerase Chain Reaction *on page 1069*

Abstract This mutation is the cause of a common hereditary predisposition to venous thrombosis. DNA-based methods, such as the polymerase chain reaction (PCR)-based assay, are used to determine the presence of a specific mutation at nucleotide position 20210 in the prothrombin gene.

Specimen Whole blood

Container Varies with laboratory

Collection Routine venipuncture

Storage Instructions Do not centrifuge or freeze specimen. Store at 4°C or room temperature.

Turnaround Time Several days or longer (depending on how often test batches are performed)

Reference Interval Normal: prothrombin G20210A mutation not present
(Continued)

Prothrombin G20210A Mutation *(Continued)*

Use The test identifies individuals who have the prothrombin G20210A mutation. The results indicate whether an affected individual is heterozygous or homozygous for the mutation. The heterozygous form of the mutation is present in 2.3% of the general population and 6.2% of patients with venous thrombosis. It is present in 18% of cases of familial venous thrombosis.

Methodology Commonly, polymerase chain reaction (PCR). The prothrombin G20210A mutation is a point mutation in which the guanine at nucleotide position 20210 is replaced by an adenine. The nucleotide change also allows the introduction of a new Hind III restriction site during PCR. To perform the test, DNA is isolated from whole blood and the mutation site is amplified by PCR. The PCR product is digested with Hind III and then subjected to agarose gel electrophoresis to separate the DNA bands based on size. The presence of a Hind III site at position 20210 can be determined by the pattern of DNA bands detected on the gel. The presence of a Hind III site at position 20210 indicates the presence of the prothrombin G20210A mutation. Heterozygotes and homozygotes can be specifically identified.

Additional Information The mutation involves a guanine to adenine transition at nucleotide position 20210 in an untranslated region of the gene. It is associated with elevated prothrombin levels and an increased risk for **venous** thrombosis. Individuals heterozygous for the prothrombin G20210A mutation have a two- to threefold increased risk for venous thrombosis while homozygous individuals likely have an even higher risk. The liklihood for **venous** thrombosis is further increased in the presence of a second risk factor. Some studies have shown an increased risk for **arterial** thrombosis while other studies have not. It is possible that an increased risk for **arterial** thrombosis exists only when additional risk factors are present.

References

Inbal A, Freimark D, Modan B, et al, "Synergistic Effects of Prothrombotic Polymorphisms and Atherogenic Factors on the Risk of Myocardial Infarction in Young Males," *Blood*, 1999, 93(7):2186-90.

McGlennen RC and Key NS, "Clinical and Laboratory Management of the Prothrombin G20210A Mutation," *Arch Pathol Lab Med*, 2002, 126(11):1319-25.

Poort SR, Rosendaal FR, Reitsma PH, et al, "A Common Genetic Variation in the 3'-Untranslated Region of the Prothrombin Gene Is Associated With Elevated Plasma Prothrombin Levels and an Increase in Venous Thrombosis," *Blood*, 1996, 88(10):3698-703.

van der Meer FJ, Koster T, Vandenbroucke JP, et al, "The Leiden Thrombophilia Study (LETS)," *Thromb Haemost*, 1997, 78(5):631-5.

Prothrombin Time

Related Information

Activated Partial Thromboplastin Time *on page 100*
Blood, Urine *on page 281*
Coagulation Factor Assays *on page 418*
Disseminated Intravascular Coagulation Screen *on page 517*
Factor Inhibitors *on page 566*
Fibrinogen *on page 583*
Liver Disease: Laboratory Assessment, Overview *on page 869*
Mixing Studies (Coagulation) *on page 918*
Plasma, Fresh Frozen *on page 1039*
Point-of-Care Testing *on page 1065*
Protein C *on page 1101*
Protein S *on page 1110*
Reptilase® Time *on page 1152*
Thrombin Time *on page 1246*
Warfarin, Serum or Plasma *on page 1325*

Synonyms Protime; PT

Applies to Argatroban; Common Pathway; Coumadin®; Danaparoid; Extrinsic Pathway; Fondaparinux; Heparin; Hirudin; INR; International Normalized Ratio; Intrinsic Pathway; Lupus Anticoagulant; Melagatran; Thromboplastin; Vitamin K; Ximelagatran

Abstract The prothrombin time (PT) measures the clotting time from the activation of factor VII, through the formation of fibrin clot (see figure). This test measures the integrity of the extrinsic and common pathways of coagulation, whereas the activated partial thromboplastin time (PTT) measures the integrity of the intrinsic and common pathways of coagulation. PT prolongations are most commonly caused by factor deficiencies involving fibrinogen or factors II, V, VII, or X. Less commonly, PT prolongations are due to an inhibitor, such as therapeutic anticoagulants including heparin, hirudin, or argatroban. Rarely, PT prolongations are caused by lupus anticoagulants or by specific factor inhibitors against fibrinogen or factor II, V, VII, or X.

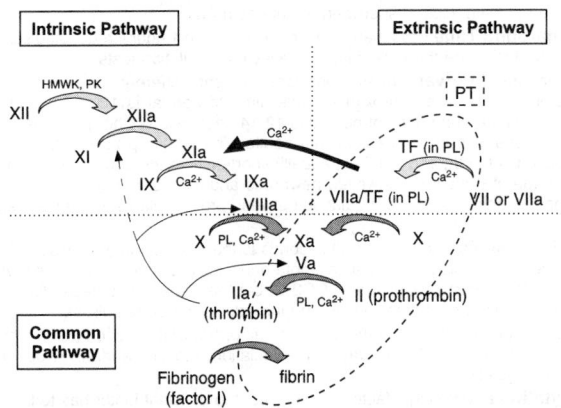

The PTT measures the clotting time from factor XII through fibrin formation (intrinsic and common pathways of coagulation). The PT measures the clotting time from factor VII through fibrin formation (extrinsic and common pathways of coagulation). The intrinsic pathway is activated when factor XII binds to a negatively charged "foreign" surface exposed to the blood, with sequential activation of factor XI, then IX, then X, followed by II, and finally fibrinogen is converted to fibrin. Factors V and VIII and phospholipid serve as cofactors. Many steps also require calcium. It is now believed that *in vivo*, coagulation is primarily initiated through the extrinsic pathway, upon exposure of blood to tissue factor (TF) at sites of tissue injury. In this model of coagulation, the ability to activate factor IX (by TF/VIIa) and factor XI (by thrombin) without factor XII indicates that factor XII, prekallikrein, and HMWK of the intrinsic pathway are not needed in normal procoagulant pathways. This is consistent with the observation that deficiencies of the latter three factors are not associated with bleeding symptoms, whereas deficiencies of the other factors may cause a bleeding tendency.

Key:
TF = tissue factor (a transmembrane protein; thus, it is associated with phospholipid *in vivo*).
PK = prekallikrein.
HMWK = high molecular weight kininogen.
PL = phospholipid.
Ca^{2+} = calcium.

Adapted from Van Cott EM and Laposata M, "Coagulation, Fibrinolysis and Hypercoagulation," *Clinical Diagnosis and Management by Laboratory Methods*, 20th ed, Henry JG, ed, Philadelphia, PA: WB Saunders Co, in press.

Specimen Plasma

Container Blue top (sodium citrate) tube; 3.2% citrate tubes are now recommended instead of 3.8% citrate tubes.

Collection Routine venipuncture. If multiple tests are being drawn, draw blue top tubes after any red top tubes but before any lavender top (EDTA), green top (heparin), or gray top (oxalate/fluoride) tubes. Recent data suggest that an initial discard tube is not necessary. Immediately invert tube gently at least 4 times to mix. Tubes must be appropriately filled. Deliver tubes immediately to the laboratory.

Specimens drawn from a heparinized line are easily contaminated with heparin, even when the initial volume drawn is discarded. Although heparin prolongs the PTT, it can also prolong the PT to a lesser extent. Hirudin and argatroban prolong the PT and PTT. Therefore, coagulation tests are best drawn directly from a peripheral vein, avoiding the arm in which heparin, hirudin or argatroban is being infused (if relevant).
(Continued)

Prothrombin Time (Continued)

Storage Instructions Separate plasma from cells as soon as possible. Plasma (or uncentrifuged specimen) may be stored at room temperature or on ice for up to 24 hours, otherwise store frozen.

Causes for Rejection Specimen received more than 24 hours after collection, tube not filled, clotted specimen, visible hemolysis

Turnaround Time Less than 1 day; often less than 1 hour if requested stat. The PT and PTT are the most readily available coagulation tests.

Reference Interval Varies significantly among different reagent-instrument combinations. The approximate lower limit of normal is 10-12 seconds; the approximate upper limit of normal is 12-14 seconds. Newborns normally have prolonged PTs in comparison with adults. The PT is up to approximately 16 seconds at birth, and the PT gradually shortens into the adult normal range by the age of 6 months. However, newborns and infants do not normally experience bleeding, because a balance between procoagulants and natural anticoagulants is maintained.

Critical Values Longer than 30 seconds is the most commonly used PT panic value in specialized coagulation laboratories, according to the College of American Pathologists 1999 Survey CG2-C, but the value varies depending on the reagent-instrument combination and individual laboratory policies.

Use Screen the integrity of the extrinsic (factor VII) and common (fibrinogen and factors II, V, and X) pathways of coagulation; monitor warfarin (Coumadin®) anticoagulation

Limitations With single factor deficiencies, the deficient factor has to be below 15% to 45% before the PT becomes prolonged, depending on the reagent. With multiple factor deficiencies, the PT becomes prolonged with less severe decreases in factor levels.

Deficiencies of factors VIII, IX XI, XII, prekallikrein, or high-molecular weight kininogen do not affect the PT, but do affect the PTT. Factor XIII does not affect the PT nor PTT. A specific factor XIII assay can screen for factor XIII deficiencies.

Heparin can prolong the PT, depending on the reagent. Some reagents contain a heparin neutralizer to reduce or eliminate heparin interference.

Lupus anticoagulants uncommonly prolong the baseline PT. Most PT reagents contain excess phospholipid such that lupus anticoagulants (which are antiphospholipid antibodies) do not prolong the PT. However, with some PT reagents, lupus anticoagulants can accentuate the prolongation of the PT when patients are on warfarin. In these situations, an alternative assay such as a chromogenic factor X assay can be used rather than (or in addition to) the PT to monitor warfarin (see Additional Information).

Methodology PT reagent is called thromboplastin (phospholipid with tissue factor and calcium). It is added to patient plasma, and the time until clot formation is measured in seconds. Tissue factor activates the extrinsic pathway of coagulation. Phospholipid and calcium are required cofactors in the coagulation cascade. Citrate in the blue top tube prevents clotting by chelating calcium. PT reagents contain excess calcium to overcome the citrate. More recently, point-of-care PT test methods have become available which use a single drop of whole blood, and these methods are undergoing evaluation.

Additional Information If indicated, a vitamin K trial may be performed in a patient with an unexplained PT prolongation. If the PT prolongation is due to vitamin K deficiency, the PT becomes normal or significantly shorter within 12-24 hours after vitamin K administration.

To determine the etiology of an unexplained PT prolongation, a mixing study is usually the first step (if the PTT is also prolonged, the presence of heparin or related anticoagulants must first be excluded - see Mixing Studies (Coagulation) *on page 918*). Mixing studies can predict whether the cause of the PT prolongation is a factor deficiency or an inhibitor. The majority of PT prolongations are due to factor deficiencies. If the PT mixing study suggests a factor deficiency, assays for fibrinogen and factors II, V, VII, and X can be performed to identify the deficient factor(s). Inhibitors that prolong the PT are rare. Factor VII inhibitors prolong the PT but not the PTT. Factor II, V, or X inhibitors typically prolong the PTT as well as the PT (see Coagulation Factor Assays *on page 418* for more information). As mentioned above, lupus anticoagulants are inhibitors that commonly prolong the PTT, but uncommonly prolong the PT.

In patients with both a lupus anticoagulant and a prolonged PT, a factor II assay could be considered, because occasionally lupus anticoagulants cause decreased factor II due to increased clearance.

Acquired causes of PT prolongations are much more common than hereditary causes, especially among inpatients (see list below). The liver synthesizes all of the coagulation factors. Therefore, with liver disease, multiple factor deficiencies can develop which prolong the PT earlier and more than the PTT. Coumadin® or vitamin K deficiency impair the function of factors II, VII, IX, and X, leading to PT and eventually PTT prolongations. In disseminated intravascular coagulation (DIC), multiple factor deficiencies may arise due to activation and consumption of factors, prolonging the PT more often than the PTT. Heparin inhibits activated factors II, X, IX, XI, XII, and kallikrein by enhancing antithrombin activity, prolonging the PTT more than the PT. Hirudin and argatroban inhibit only activated factor II (thrombin), prolonging the PT and PTT.

CAUSES OF PT PROLONGATIONS:

Hereditary:

- Deficiency of factor VII *(PTT is normal)*
- Deficiency of fibrinogen or factors II, V, or X *(PTT may also be prolonged)*

Acquired:

- Liver dysfunction *(PT affected earlier and more than PTT)*
- Vitamin K deficiency *(PT affected earlier and more than PTT)*
- Warfarin *(PT affected earlier and more than PTT)*
- Disseminated intravascular coagulation (DIC) *(PT affected earlier and more than PTT)*
- Lupus anticoagulants *(may or may not prolong the PTT; PT is rarely prolonged)*
- Heparin *(PT less affected than PTT, PT may be normal)*
- Hirudin or argatroban *(PTT also prolonged)*
- Specific factor inhibitors *(PTT also prolonged except in the rare case of an inhibitor against factor VII)*

The effects of hereditary or acquired factor deficiencies on PT and PTT are shown in Tables 1 and 2 in Coagulation Factor Assays *on page 418*. Factor half-lives are summarized in Table 3 in that listing.

Monitoring warfarin: Warfarin is monitored by the international normalized ratio (INR). The usual therapeutic goal is an INR of 2-3. The INR is calculated from the PT and is intended to allow valid comparisons of results regardless of the type of PT reagent used among different laboratories:

INR = [patient PT / mean normal PT]^{ISI}

The international sensitivity index (ISI) is a measure of the sensitivity of a particular PT reagent. Different PT reagents have different sensitivities to factor deficiencies. For example, with an insensitive reagent, the PT will not become prolonged until the factor levels are very decreased, whereas with a sensitive reagent, the PT will become prolonged with milder factor deficiencies. Insensitive reagents have higher ISI values, up to about 3.0. Sensitive reagents have lower ISI values, down to about 1.0. The ISI for each reagent is determined by the manufacturer. The INR standard of reporting has been widely accepted but several problems have been recognized. The significance of, and approach to, correction of these problems has been reviewed by Hirsh, et al (see References).

During warfarin initiation, the PT/INR is typically checked daily or at least 4-5 times per week until the dose and INR are therapeutic and stable. The interval between PT/INR tests can then be gradually decreased to as infrequently as every 4 weeks, depending on the stability of the dose and the PT/INR result. It takes 4-5 days for warfarin's antithrombotic action to take effect, because the half-lives of factors II and X are relatively long. For this reason, patients who need immediate anticoagulation are treated with an immediate-acting anticoagulant (eg, heparin) while waiting for warfarin to become therapeutic. Heparin is typically continued until the INR is in the desired range for two consecutive days.

(Continued)

Prothrombin Time *(Continued)*

To treat warfarin overdose (bleeding), vitamin K or fresh frozen plasma can be administered. If the INR is >5 without bleeding, vitamin K administration can be considered. If large doses of vitamin K are administered, patients can become temporarily warfarin resistant.

Warfarin (Coumadin®) and vitamin K deficiency share the same molecular basis for their effects. Warfarin is used as a therapeutic anticoagulant because it impairs the regeneration of active vitamin K, thereby decreasing the amount of active vitamin K. Vitamin K in its active form is a cofactor in a reaction which carboxylates glutamic acid residues to form gamma carboxyglutamic acid residues on factors II, VII, IX, and X as well as protein C and protein S. This carboxylation step is necessary for normal activity of these proteins. As a result, vitamin K deficiency or warfarin therapy decreases the activity of these proteins and prolongs the PT. Patients with mild vitamin K deficiency or low levels of warfarin anticoagulation can have a normal PTT.

In certain situations (eg, lupus anticoagulants or the concomitant use of hirudin or argatroban with warfarin) alternative assays may be used to monitor warfarin because the PT/INR will be elevated by hirudin, argatroban, and occasionally by lupus anticoagulants. Alternative assays (eg, chromogenic factor X assays) are not affected by these interferences. However, alternative assays have not yet been well studied in these settings. An INR of 2-3 corresponds approximately to a chromogenic factor X of 20% to 40%.

Changes in dietary vitamin K (see website reference below) and many medications (many of which are listed in reference from Hirsh 1998) can alter the warfarin dose requirement. Hyperthyroidism, liver failure, cancer, fever, or vitamin K deficiency (from malabsorption, steatorrhea, poor nutrition, certain antibiotics etc) tend to decrease the dose required to increase the PT. Hypothyroidism or certain genetic polymorphisms tend to increase the dose requirement. Some patients have hereditary warfarin resistance, an uncommon condition in which very high doses of warfarin are needed to maintain a therapeutic INR.

Warfarin should not be used alone in the acute setting of heparin-induced thrombocytopenia, because paradoxical thrombosis can occur. If warfarin is used in this setting, a rapid-acting anticoagulant (eg, hirudin, danaparoid, or argatroban) must also be used until the INR is therapeutic. A similar approach is used for patients with hereditary protein C or protein S deficiency, to prevent Coumadin®-induced skin necrosis.

Some potentially important new anticoagulants that may not require laboratory monitoring are undergoing evaluation. While low-molecular-weight heparins are replacing unfractionated heparin in many clinical situations (in part because they are less likely to cause immunologic thrombocytopenia) a "next-generation heparin-derived anticoagulant, fondaparinux has been synthesized and undergone initial clinical evaluation. Fondaparinux is a synthetic pentasaccharide (the minimal antithrombin-binding unit of heparin), binds only to antithrombin, and is a specific inhibitor of factor Xa. Anticoagulants that are direct thrombin inhibitors (and act independently of other plasma proteins) include hirudin, argatroban, melagatran, and ximelagatran (a chemically modified form of melagatran that is more readily absorbed resulting in a new oral anticoagulant).

References

Andrew M, Paes B, and Johnston M, "Development of the Hemostatic System in the Neonate and Young Infant," *Am J Pediatr Hematol Oncol*, 1990, 12(1):95-104.

Bajaj SP and Joist JH, "New Insights Into How Blood Clots: Implications for the Use of APTT and PT as Coagulation Screening Tests and in Monitoring of Anticoagulant Therapy," *Semin Thromb Hemost*, 1999, 25(4):407-18.

Fairweather RB, Ansell J, van den Besselaar AM, et al, "College of American Pathologists Conference XXXI on Laboratory Monitoring of Anticoagulant Therapy. Laboratory Monitoring of Oral Anticoagulant Therapy," *Arch Pathol Lab Med*, 1998, 122(9):768-81.

Fischer KG and Sutor AH, "Hirudin," *Semin Thromb Hemost*, 2002, 28(5):403-89.

Gottfried EL and Adachi MM, "Prothrombin Time and Activated Partial Thromboplastin Time Can Be Performed on the First Tube," *Am J Clin Pathol*, 1997, 107(6):681-3.

Hirsh J, Dalen JE, Anderson DR, et al, "Oral Anticoagulants. Mechanism of Action, Clinical Effectiveness, and Optimal Therapeutic Range," *Chest*, 1998, 114(5):445-69S.

Hirsh J, Fuster V, Ansell J, et al, "American Heart Association/American College of Cardiology Foundation Guide to Warfarin Therapy," *Circulation*, 2003, 107(12):1692-711.

Moll S and Ortel TL, "Monitoring Warfarin Therapy in Patients With Lupus Anticoagulants," *Ann Intern Med*, 1997, 127(3):177-85.

National Committee for Clinical Laboratory Standards, "Collection, Transport, and Processing of Blood Specimens for Coagulation Testing and General Performance of Coagulation Assays: Approved Guideline 3rd edition," NCCLS Document H21-A3, NCCLS, 940 West Valley Road, Wayne, Pennsylvania 19087, USA 1998.

van der Besselaar, AMHP, "Standardization and Control of Oral Anticoagulant Therapy," *Advanced Laboratory Methods in Haematology*, Part 6, Chapter 8, Rowan RM, van Assendelft OW, and Preston FE, eds, London, UK: Arnold, 2002, 386-417.

Internet Web Sites
www.nal.usda.gov (foods with vitamin K)

♦ **Protime** *see* Prothrombin Time *on page 1116*

Protoporphyrin, Free Erythrocyte
Related Information
Delta (5)-Aminolevulinic Acid, Urine *on page 508*
Ferritin, Serum *on page 577*
Iron and Total Iron Binding Capacity/Transferrin, Serum *on page 807*
Lead, Blood *on page 832*
Lead, Urine *on page 835*
Porphobilinogen, Urine *on page 1073*
Porphyrins, Quantitative, Urine *on page 1074*
Protoporphyrin, Zinc, Blood *on page 1121*
Transferrin Receptor, Soluble, Serum or Plasma *on page 1267*

Synonyms FEP; Free Erythrocyte Protoporphyrin; Protoporphyrins, Fractionation, Erythrocytes; RBC Protoporphyrin

Abstract Free erythrocyte protoporphyrin is a heme precursor which is increased in iron deficiency, lead toxicity, and erythropoietic porphyria.

Patient Preparation Patient should fast for 12-14 hours, abstain from alcohol for 24-hours, and ideally be off all medications for at least 1 week.

Specimen Whole blood or washed erythrocytes; check with laboratory.

Container Lavender top (EDTA) tube or green top (heparin) tube; check with laboratory.

Storage Instructions Check with laboratory.

Special Instructions Current hematocrit may be needed.

Reference Interval 1-10 µg/dL packed cells

Use FEP is used in the evaluation of erythropoietic protoporphyria (see Porphyrins, Quantitative, Urine *on page 1074*).

Methodology Hematofluorometer, extraction method, and high performance liquid chromatography (HPLC). The hematofluorometer measures porphyrins unbound in erythrocytes. A definitive fluorometric method has been published by the National Committee for Clinical Laboratory Standards (NCCLS).

Additional Information Prior to 1991, FEP was used to detect lead poisoning, but contemporary practice (and CDC guidelines) now require that lead screening should be done with blood lead assays. Anemias due to iron deficiency or lead toxicity are better evaluated with zinc protoporphyrin (see Protoporphyrin, Zinc, Blood *on page 1121*).

References
Ahmed I, Cockerill F, Krause P, et al, "Childhood Porphyrias," *Mayo Clin Proc*, 2002, 77:825-36.
Downey DC, "Porphyria and Chemicals," *Med Hypotheses*, 1999, 53(2):166-71.
National Committee for Clinical Laboratory Standards, "Erythrocyte Protoporphyrin Testing: Approved Guidelines," *NCCLS Document*, C42-A, Wayne, PA: NCCLS, 1996.
Thadani H, Deacon A, and Peters T, "Diagnosis and Management of Porphyria," *BMJ*, 2000, 320(7250):1647-51.

♦ **Protoporphyrins, Fractionation, Erythrocytes** *see* Protoporphyrin, Free Erythrocyte *on page 1121*

Protoporphyrin, Zinc, Blood
Related Information
Delta (5)-Aminolevulinic Acid, Urine *on page 508*
Ferritin, Serum *on page 577*
Iron and Total Iron Binding Capacity/Transferrin, Serum *on page 807*
Lead, Blood *on page 832*
Protoporphyrin, Free Erythrocyte *on page 1121*

Synonyms Protoporphyrin, Zinc, Erythrocyte; Zinc Protoporphyrin; ZPP

Abstract Zinc protoporphyrin (ZPP) measurement is occasionally useful in the differentiation of iron deficiency from thalassemia. ZPP is not useful in
(Continued)

Protoporphyrin, Zinc, Blood *(Continued)*

screening programs for lead intoxication. ZPP may also be increased in erythropoietic protoporphyria, however free erythrocyte protoporphyrin concentrations are much higher than ZPP.

Specimen Whole blood

Container Lavender top (EDTA) tube, green top (heparin) tube

Collection Routine venipuncture

Storage Instructions Do not centrifuge. Refrigerate and protect from light. Stable 1 week at 4°C.

Causes for Rejection Hemolysis, icterus

Reference Interval 10-38 µg/dL packed cells

Critical Values >100 µg/dL (SI: >1.6 µmol/L)

Use ZPP is elevated in iron deficiency, but not in thalassemia.

Limitations Zinc protoporphyrin is increased in lead toxicity. High concentrations of aluminum in subjects on hemodialysis can cause increased levels. It may be increased with the anemia of chronic disease, including cases of lymphoma, sideroblastic and hemolytic anemia, and secondary polycythemia.

Methodology Hematofluorometry (front-face); if washed erythrocytes are used, the assay becomes more specific and sensitive.

Additional Information The Centers for Disease Control has lowered the cutoff level for lead intoxication in children younger than 6 years of age to **10** µg/dL (SI: 0.48 µmol/L), and this level is so low that **ZPP is not useful**. It is mandatory to measure lead levels in any screening program, rather than ZPP.

References

Centers for Disease Control, "Preventing Lead Poisoning in Young Children: A Statement by the Center for Disease Control," *U.S. Government Printing Office Department 1996-533-410*, Atlanta, GA: U.S. Department of Healthcare and Human Service, 1991.

Labbe RF, Vreman HJ, and Stevenson DK, "Zinc Protoporphyrin: A Metabolite With a Mission," *Clin Chem*, 1999, 45(12):2060-72.

Mayo Medical Laboratories, *2001 Test Catalogue*, Rochester, NM, 448.

♦ **Protoporphyrin, Zinc, Erythrocyte** *see* Protoporphyrin, Zinc, Blood *on page 1121*

♦ **Protriptyline** *see* Antidepressants, Cyclic, Serum or Plasma *on page 171*

♦ **Prozac®** *see* Fluoxetine, Serum or Plasma *on page 604*

♦ **Prozil®** *see* Chlorpromazine, Serum *on page 395*

♦ **Prussian Blue Stain** *see* Iron Stain, Bone Marrow *on page 810*

♦ **Prysolin®** *see* Primidone, Serum or Plasma *on page 1091*

♦ **PSA** *see* Prostate Specific Antigen, Serum *on page 1098*

♦ **PSA Density** *see* Prostate Specific Antigen, Serum *on page 1098*

♦ **PSA Velocity** *see* Prostate Specific Antigen, Serum *on page 1098*

♦ **Pseudocholinesterase Inhibition by Dibucaine** *see* Dibucaine Number, Serum or Plasma *on page 510*

Pseudocholinesterase, Serum

Related Information

Acetylcholinesterase, Red Cell and Serum *on page 93*
Dibucaine Number, Serum or Plasma *on page 510*
Organophosphate Pesticides, Urine, Blood, or Serum *on page 975*

Synonyms Acylcholine Acylhydrolase; Benzoyl Cholinesterase; Cholinesterase II; Cholinesterase, Serum; PChE; Plasma Cholinesterase

Abstract Two types of cholinesterase are found in blood: **"true"** cholinesterase in red cells, lung, and brain, while **"pseudocholinesterase" (PChE)** is found in serum. Organophosphorus-containing insecticides decrease both RBC cholinesterase and serum pseudocholinesterase. Serum pseudocholinesterase is more sensitive than the red cell enzyme to organophosphate compounds. See Acetylcholinesterase, Red Cell and Serum *on page 93*.

Specimen Serum

Container Red top tube

Storage Instructions Stable at room temperature for 6 hours, 1 week at 4°C, and 6 months at -70°C. Avoid repeat freezing and thawing.

Causes for Rejection Gross hemolysis

Reference Interval Low in infancy, then increasing to adult levels by the second month. Intervals vary between methods and laboratories. Typical values are 5-12 units/mL (with propionyl thiocholine); 7-19 units/mL (with butylthiocholine). Intermediate levels are found in heterozygotes.

Possible Panic Range Serious neuromuscular symptoms occur at decreases of ~80%.

Use Screen preoperative patients for inherited succinylcholine anesthetic sensitivity, genetic or secondary to insecticide exposure, in appropriate circumstances. The diagnosis of the cholinergic syndrome is based on clinical findings as well as PChE activity. Prevent or evaluate prolonged anesthetic effect, prolonged apnea, after surgery. Very small amounts (0.04-0.06 mg/kg) of succinylcholine are needed to obtain 90% of neuromuscular blockade in patients with an abnormal allele who have low levels of PChE activity.

Monitor and diagnose organophosphorous exposure and poisoning, in which PChE level is decreased; establish patient's baseline value before exposure. See Organophosphate Pesticides, Urine, Blood, or Serum *on page 975.*

Family studies may be done when an individual with a genetically abnormal type is documented by PChE deficiency and, ideally, confirmed by phenotyping. See Dibucaine Number, Serum or Plasma *on page 510.*

Limitations Serum PChE may be decreased in patients on estrogens and oral contraceptives. Fluoride interferes. PChE is also low in some instances of liver disease and in malnutrition. Red cell cholinesterase is more useful for chronic insecticide exposure. Carbamate-poisoned persons can appear to have near normal or normal levels of PChE.

Contraindications Not useful to screen for toxicity from chlorinated insecticides.

Methodology Colorimetry, kinetic enzyme utilizing different substrates, fluorometry

Additional Information One patient in 1500 is susceptible to succinylcholine anesthetic mishap, but not all of these patients have a low serum PChE value.

References
Bardin PG, van Eeden SF, Moolman JA, et al, "Organophosphate and Carbamate Poisoning," *Arch Intern Med*, 1994, 154(13):1433-41.
Lessenger JE and Reese BE, "Rational Use of Cholinesterase Activity Testing in Pesticide Poisoning," *J Am Board Fam Pract*, 1999, 12(4):307-14.

♦ **Pseudohyponatremia** *see* Osmolality, Calculated, Serum or Plasma *on page 976*

♦ **Pseudomonas aeruginosa, Mucoid** *see* Chloride, Sweat *on page 392*

♦ **Pseudomonas spp** *see* Bacterial Culture, Aerobes *on page 229*

Psittacosis Serology

Related Information
Chlamydia Group Serology *on page 385*

Test Includes Detection of antibody specific for *Chlamydia psittaci*

Abstract Human psittacosis is usually contracted from exposure to birds, including domestic fowl. Infection in birds usually involves the intestinal tract and the organism is shed in the feces, which can contaminate the environment. *C. psittaci* is a zoonotic infection which can cause diseases in many domestic mammals.

Specimen Serum

Container Red top tube

Collection Acute and convalescent samples are needed.

Reference Interval Less than a fourfold increase in titer in paired sera

Use Diagnose psittacosis; serology is the best available laboratory diagnostic test for psittacosis.

Limitations The microimmunofluorescence test is technically difficult. Complement fixation tests identify antibody to an antigen common to all members of the genus and a fourfold rise in CF antibody titer cannot distinguish psittacosis from other chlamydial infections. Interpretation of complement fixation test results performed on unpaired sera is complicated by high "background" levels of antibody in the general population. The presence of IgG titer may indicate prior exposure. Assays are only performed in large reference laboratories or research facilities.

Methodology Complement fixation (CF), enzyme immunoassay (EIA), microimmunofluorescence (MIF)
(Continued)

Psittacosis Serology (Continued)

Additional Information Most patients with psittacosis develop high titers of complement fixing antibody. Specific IgM antibody can sometimes be demonstrated.

C. psittaci is generally very stable and can persist in the environment for months without losing viability. Culture for *C. psittaci* is available from some reference laboratories, however, care should be taken when collecting the specimen. The organism, difficult to isolate, is contagious and dangerous.

References

Hedberg K, White KE, Hedberg CW, et al, "Persistence of *Chlamydia* Complement Fixation Antibody After an Outbreak of Psittacosis," *J Infect Dis*, 1993, 167(2):502-3.

Peeling RW and Brunham RC, "Chlamydiae as Pathogens: New Species and New Issues," *Emerg Infect Dis*, 1996, 2(4):307-19.

Russell EG, "Evaluation of Two Serologic Tests for the Diagnosis of Chlamydial Respiratory Disease," *Pathology*, 1999, 31(4):403-5.

Internet Web Sites

www.cdc.gov/ncidod/dbmd/diseaseinfo/psittacosis_t.htm

- **PT** see Prothrombin Time *on page 1116*
- **PTH** see Parathyroid Hormone, Serum *on page 1001*
- **PTHrP** see Parathyroid Hormone-Related Protein, Serum *on page 1000*
- **PTP** see Platelet Antibodies *on page 1047*
- **PTT** see Activated Partial Thromboplastin Time *on page 100*
- **PTT** see Antiphospholipid Antibody (Lupus Anticoagulant and/or Anticardiolipin Antibody) *on page 193*
- **PTT** see Heparin Antifactor Xa Assay *on page 693*
- **PTT** see Heparin Neutralization *on page 697*

Pulmonary Surfactant, Amniotic Fluid

Related Information

Lamellar Bodies, Amniotic Fluid *on page 830*
Lecithin:Sphingomyelin Ratio, Amniotic Fluid *on page 836*
Phosphatidylglycerol, Amniotic Fluid *on page 1030*

Synonyms Amniotic Fluid Pulmonary Surfactant

Applies to Foam Stability Index or Shake Test; Lamellar Body Count; Optical Density at 650 nm; Surfactant:Albumin Ratio

Abstract As the fetal lungs mature, fetal pulmonary surfactant is secreted into the amniotic fluid. Analyses of amniotic fluid pulmonary surfactant assess fetal lung maturity and the probability of respiratory distress syndrome (RDS) (hyaline membrane disease). This section addresses commonly used amniotic fluid pulmonary surfactant tests other than the L:S ratio and PG.

Aftercare All nonsensitized Rh-negative patients should receive anti-D immunoglobulin after amniocentesis.

Specimen Amniotic fluid; specimen volume required varies with analytical method (1 mL minimum).

Container The use of siliconized collection tubes will interfere with the shake test and should not be used.

Collection Ultrasound-guided amniocentesis is performed by the physician. Vaginal pool specimens are discouraged.

Reference Interval The following table summarizes the reference intervals for the various pulmonary surfactant tests.

Amniotic Fluid Pulmonary Surfactant

Pulmonary Surfactant Test	Maturity Threshold
Surfactant:albumin ratio (S/A)	>55 mg/g
Lamellar body count (LB)	>50,000/μL
Foam stability index (FSI)	$\geq$47
Optical density at 650 nm (OD$_{650}$)	$\geq$0.15

Adapted from "ACOG Educational Bulletin. Assessment of Fetal Lung Maturity, Number 230, November 1996," *Int J Gynaecol Obstet*, 1997, 56(2):191-8; and

Dubin SB, "Assessment of Fetal Lung Maturity Practice Parameter," *Am J Clin Pathol*, 1998, 110(6):723-32.

Use The analysis of amniotic fluid pulmonary surfactant is done in high-risk pregnancies to evaluate fetal lung maturity and the newborn's risk for developing RDS.

Limitations Contamination of specimen with blood, meconium, or vaginal mucus and debris may falsely affect results by all methods. One study showed that blood contamination tends to yield falsely immature or borderline values in truly mature specimens by the automated S/A assay, but mature results from blood contaminated specimens are valid. Another study concluded that all bloody specimens with S/A results <90 mg/g be tested for PG by the rapid immunoagglutination method for confirmation.

There is considerable variation in diagnostic thresholds for lamellar body (LB) counts due to variation in centrifugation speeds and times and commercial hematology counters. Neither bilirubin nor meconium interferes with this test, but red cells decrease results. Consistent centrifugation conditions are necessary for both LB counts and OD_{650} determinations.

Though the predictive value of mature results are close to 100% for any of the pulmonary surfactant tests, the predictive value of immature results range from 30% to 50%. Sequential (cascade) testing procedures are sometimes used, employing combinations of these procedures and/or L/S or PG determinations to further characterize initial, "immature" results.

Methodology Fluorescence polarization (S/A), platelet channel counting (LB), light scattering (LB, OD_{650}), ethanol dilution (FSI, Shake Test)

Additional Information Fetal lung maturity testing by fluorescence polarization in preterm labor has been shown to be most cost-effective between 34 and 36 weeks gestation and to have diagnostic power equivalent to or better than the L:S ratio.

Lamellar bodies are surfactant-containing structures that are secreted from type II pneumocytes. Because of their size (1-5 µm), estimations of lamellar body counts in amniotic fluid may be made using the platelet counting chamber of standard cell counters. See Lamellar Bodies, Amniotic Fluid *on page 830.*

References
"ACOG Educational Bulletin. Assessment of Fetal Lung Maturity. Number 230, November 1996," *Int J Gynaecol Obstet*, 1997, 56(2):191-8.

Dubin SB, "Assessment of Fetal Lung Maturity Practice Parameter," *Am J Clin Pathol*, 1998, 110(6):723-32.

Lee IS, Cho YK, Ahm K, et al, "Lamellar Body Count in Amniotic Fluid as a Rapid Screening Test for Fetal Lung Maturity," *J Perinatol*, 1996, 16(3 Pt 1):176-80.

Lewis PS, Lauria MR, Dzieczkowski J, et al, "Amniotic Fluid Lamellar Body Count: Cost-Effective Screening for Fetal Lung Maturity," *Obstet Gynecol*, 1999, 93(3):387-91.

Pulsed-Field Gel Electrophoresis Genotyping

Test Includes Analysis of individual bacterial isolates by careful isolation of high molecular weight DNA, digestion with a restriction enzyme, and electrophoresis in a special pulsed-field gel electrophoresis (PFGE) apparatus that separates high molecular weight DNA fragments

Abstract Genetic analysis of bacteria is usually performed to determine if bacterial isolates associated with an outbreak are genetically related. The technique has been used to distinguish contamination of food products to detect a common source of infection, dissemination of antibiotic resistant bacterial strains between different hospitals, to study bacterial isolates associated with a specific clinical syndrome, and, most commonly, to detect nosocomial spread of common bacterial infections within a hospital or healthcare system.

Specimen Bacterial isolates from an outbreak

Storage Instructions Most bacterial isolates can be stored frozen at -20°C for several months or at -70°C for years before testing

Turnaround Time Usually 1-2 weeks

Reference Interval Results are expressed as percent of relatedness between at least two different bacterial isolates.

Use Identification of an outbreak of closely related bacterial infections, usually for epidemiologic purposes

Limitations There is currently no gold standard for clonal typing methods. Clonal typing of bacterial genetic material by PFGE takes from 4-7 days to complete analysis. This technique requires specialized electrophoresis equipment. Certain common strains of bacteria, such as methicillin-resistant *Staphylococcus aureus* and *E. coli* O157:H7, represent small genetic subsets of

(Continued)

Pulsed-Field Gel Electrophoresis Genotyping
(Continued)

strains within the species and isolates unrelated epidemiologically may show very similar genetic patterns.

Methodology High molecular bacterial DNA is obtained and then cut at very specific sites with restriction enzymes. The fragments resulting from restriction enzyme digestion will range from 10-800 kilobases. These fragments can be separated in a specialized agarose gel electrophoresis in which the orientation of the electrical field is changed periodically, or pulsed, rather than kept constant. After staining the agarose gel for DNA (usually with ethidium bromide) a DNA "banding" pattern can be distinquished. Bacterial strains may contain from 5-20 distinct, well resolved "bands". Each band represents a specific product of DNA digestion and the pattern of all bands is used to compare two or more isolates for genetic relatedness or clonality. Criteria have been established for interpreting strain patterns and can be used to evaluate a possible outbreak.

Additional Information Generally, bacterial species show a substantial amount of genetic diversity, and clinical isolates are usually quite divergent genetically. By using PFGE, an investigator can determine the likely relatedness between bacterial isolates of the same species. PFGE has been used with a wide variety of microorganisms and seems to be highly discriminatory and reproducible. Currently, PFGE is the typing method of choice for most of the commonly encountered bacterial pathogens such as staphylococci, enterococci, and common gram-negative rods such as bacteria from the family *Enterobacteriaceae* and *Pseudomonas aeruginosa*.

References

Arakawa E, Murase T, Matsushita S, et al, "Pulsed-Field Gel Electrophoresis-Based Molecular Comparison of *Vibrio Cholerae* O1 Isolates From Domestic and Imported Cases of Cholera in Japan," *J Clin Microbiol*, 2000, 38(1):424-6.

Guyot A, Barrett SP, Threlfall EJ, et al, "Molecular Epidemiology of Multi-Resistant *Escherichia coli*," *J Hosp Infect*, 1999, 43(1):39-48.

Horvat RT, Potter LM, and Bartholomew WR, "Clonal Dissemination of Vancomycin-Resistant Enterococci and Comparison of Susceptibility Testing Methods," *Diagn Microbiol Infect Dis*, 1998, 30(4):235-41.

Saiman L, Jakob K, Holmes KW, et al, "Molecular Epidemiology of Staphylococcal Scalded Skin Syndrome in Premature Infants," *Pediatr Infect Dis*, 1998, 17(4):329-34.

♦ **Pulse Oximetry** *see* Point-of-Care Testing *on page 1065*

♦ **Pulse Oximetry, Transcutaneous** *see* Oxygen Saturation, Blood *on page 991*

♦ **Purgoxin®** *see* Digoxin, Serum *on page 512*

♦ **Purodigin®** *see* Digitoxin, Serum *on page 512*

♦ **PYD** *see* Osteocalcin, Serum or Plasma *on page 983*

♦ **Pyridinium Collagen Cross-Links** *see* Pyridinolines (Pyridinoline and Deoxypyridinoline), Urine *on page 1126*

♦ **Pyridinoline** *see* N-Telopeptides, Urine *on page 967*

♦ **Pyridinoline** *see* Osteocalcin, Serum or Plasma *on page 983*

Pyridinolines (Pyridinoline and Deoxypyridinoline), Urine

Related Information

Alkaline Phosphatase Isoenzymes, Serum *on page 125*
Alkaline Phosphatase, Serum *on page 127*
Calcium, Serum *on page 329*
Calcium, Urine *on page 332*
Hydroxyproline, Total, Urine *on page 757*
N-Telopeptides, Urine *on page 967*
Osteocalcin, Serum or Plasma *on page 983*
Parathyroid Hormone, Serum *on page 1001*
Phosphorus, Serum *on page 1031*
Phosphorus, Urine *on page 1032*
Urine Collection, 24-Hour *on page 1295*
Vitamin D, Serum *on page 1318*

Synonyms Deoxypyridinoline; Hydroxylysylpyridinoline; Lysylpyridinoline; Pyridinium Collagen Cross-Links

Test Includes Pyridinoline and deoxypyridinoline

Abstract Pyridinolines are markers of bone resorption. The rate of excretion of pyridinolines in urine is believed to be an index of bone matrix degradation and resorption.

Specimen Urine

Sampling Time Due to diurnal variation, 24-hour urine collection is preferred. However, for deoxypyridinoline the variation is less as compared to N-telopeptide. Excretion of deoxypyridinoline is greater at night.

Resorption markers fall faster than formation markers when a change in rate of remodeling occurs: 2-12 weeks for resorption markers, 3-6 months for formation markers.

Storage Instructions Refrigerate. Freeze for longer storage (>2 days).

Special Instructions Avoid exposure to light.

Reference Interval

Male:
- 0-21 years: not established
- 22-80 years: 18-40 nmol/mmol creatinine
- older than 80 years: not established

Female:
- 0-19 years: not established
- 20-50 years: 20-62 nmol/mmol creatinine
- older than 50 years: not established

Use The major roles of biochemical markers of bone remodeling are as monitors of response to therapy. These markers can also be used to evaluate bone metabolism, bone resorption (eg, in osteopenia, osteoporosis); identify patients at high risk of fracture; assess patients with carcinoma as markers of bone metastasis as well as indicators of response to therapy.

Limitations Day-to-day variation is ~20% for bone resorption markers. Pyridinium cross-links are affected by renal clearance.

Methodology High performance liquid chromatography (HPLC), enzyme-linked immunosorbent assay (ELISA), chemiluminescence immunoassay, radioimmunoassay

Additional Information The measurement of bone mineral density is the gold standard for the diagnosis of osteoporosis.

Pyridinolines are increased in osteoporosis, Paget disease, metastatic bone disease, primary and secondary hyperparathyroidism, hyperthyroidism, and other diseases with increased bone resorption. Urinary excretion of pyridinium cross-links is diminished in hypothyroidism.

References

Ju HS, Leung S, Brown B, et al, "Comparison of Analytical Performance and Biological Variability of Three Bone Resorption Assays," *Clin Chem*, 1997, 43(9):1570-6.

Mayo Medical Laboratories, *2001 Test Catalogue*, Rochester, MN, 452.

Mora S, Prinster C, Proverbio MC, et al, "Urinary Markers of Bone Turnover in Healthy Children and Adolescents: Age-Related Changes and Effect of Puberty," *Calcif Tissue Int*, 1998, 63(5):369-74.

Watts NB, "Clinical Utility of Biochemical Markers of Bone Remodeling," *Clin Chem*, 1999, 45(8 Pt 2):1359-68.

♦ **Pyridoxal-5-Phosphate** *see* Vitamin B₆, Plasma or Serum *on page 1315*

♦ **Pyridoxine** *see* Vitamin B₆, Plasma or Serum *on page 1315*

♦ **Pyropoikilocytosis** *see* Peripheral Blood: Red Blood Cell Morphology *on page 1016*

♦ **Pyropoikilocytosis, Hereditary** *see* Osmotic Fragility *on page 980*

♦ **Pyruvate Kinase Assay** *see* Pyruvate Kinase Assay, Erythrocytes *on page 1127*

Pyruvate Kinase Assay, Erythrocytes

Related Information

Autohemolysis Test *on page 221*
Red Blood Cell Enzyme Deficiency *on page 1134*

Synonyms Pyruvate Kinase Assay

Abstract A deficiency of this glycolytic enzyme causes a decrease in RBC ATP level and resultant nonspherocytic chronic hemolytic anemia. Signs and symptoms usually develop in early childhood and consist of anemia, icterus and splenomegaly. An autosomal recessive condition, parents will be heterozygotes. Hemolytic disease of the newborn may occur.

(Continued)

Pyruvate Kinase Assay, Erythrocytes *(Continued)*

Specimen Erythrocytes (washed)

Container Yellow top (ACD) tube, green top (heparin) tube, or lavender top (EDTA) tube

Collection Mix tube three times by gentle inversion, place on ice.

Causes for Rejection Specimen not fresh

Special Instructions Notify laboratory before specimen collection. Deliver specimen on ice. Specimen **must** be received in the laboratory within 30 minutes of collection.

Reference Interval Adults: 6-19 μmol NAD(H)$_2$/min/g Hb (37°C)

Use Evaluate chronic hemolytic anemia

Limitations Some patients with pyruvate kinase-deficient variant hemolytic disease may not be identified if the enzyme is assayed only under conditions of substrate saturation.

Methodology Spectrophotometric kinetic assay

Additional Information Pyruvate kinase (PK) deficiency is the most common enzyme defect in anaerobic red cell glycolysis and the most common cause of congenital nonspherocytic hemolytic anemia. Glycolysis is a major energy source for red cells, thus, enzyme deficiency leads to hemolysis. The substrate for anaerobic glycolysis is glucose-6-phosphate (phosphorylated by hexokinase to glucose) which is processed through the Embden-Meyerhof pathway to pyruvate or may undergo oxidative (aerobic) glycolysis in the pentose phosphate pathway. Some 20 enzymes are involved in the anaerobic and oxidative glycolytic pathways. PK is a glycolytic enzyme that converts phosphoenolpyruvate to pyruvate. There are four isoenzymes, M$_1$, M$_2$, L, and R, the latter expressed only in red blood cells. The deficiency is inherited as an autosomal recessive trait, about 400 cases have been reported and about 100 gene mutations have been identified.

PK deficiency causes hemolytic anemia of varying severity. About 33% of patients are symptomatic in the neonatal period with hemolysis and jaundice. Splenomegaly may be present. Exchange transfusion may be required. Subsequently, hemolysis may require transfusion. Hemolysis may be mild and go unnoticed for years. Echinocytes and (if severe hemolysis is present) spherocytes and nucleated red cells may be seen on peripheral blood smear. Anemia is normochromic/normocytic. Reticulocyte count may be very high (40% to 70%), especially, paradoxically, after splenectomy.

Application of molecular biologic techniques (polymerase chain reaction and restriction endonuclease analysis) may allow prenatal diagnosis (using amniotic fluid cells and/or cord blood). Slight to moderate decrease of red cell PK activity can be seen in some patients with leukemia and in bone marrow aplasia.

References
McMullin MF, "The Molecular Basis of Disorders of Red Cell Enzymes," *J Clin Pathol*, 1999, 52(4):241-4.

Mentzer WC, "Pyruvate Kinase Deficiency and Disorders of Glycolysis," *Nathan and Oski's Hematology of Infancy and Childhood*, 5th ed, Chapter 17, Nathan DG and Orkin SH, eds, Philadelphia, PA: WB Saunders Co, 1998, 665-703.

♦ **Pyruvoyl Tetrahydropterin Synthase Deficiency** *see* Newborn Screen for Phenylketonuria *on page 954*

♦ **Pyruvoyl Tetrahydropterin Synthase Deficiency** *see* Phenylalanine, Blood *on page 1022*

Q Fever Serology

Related Information
Arthropod Identification *on page 213*
Bone Marrow *on page 296*
Liver Biopsy *on page 864*

Test Includes Detection of antibody specific for *Coxiella burnetii*, the organism responsible for Q fever

Abstract *Coxiella burnetii*, originally called *Rickettsia burneti*, is a pleomorphic coccobacillus with a gram-negative cell wall, that does not stain with the Gram stain. It is an obligate intracellular pathogen. The primary reservoirs for Q fever are cattle, sheep, and goats. The organism withstands heat and drying, survives on inanimate surfaces, and can persist in the environment long after an infected animal has vacated an area. It is highly infectious and its distribution is worldwide.

A broad clinical spectrum exists, from subclinical to fatal. Transmission is by inhalation of the organism. Person to person transmission has been documented but it is rare.

This is a potential agent of bioterrorism.

Specimen Serum

Container Red top tube

Sampling Time Acute and convalescent samples are recommended, the latter 2-4 weeks from onset.

Reference Interval Titer: <1:10; comparison of acute and convalescent titers is of greatest diagnostic value

Critical Values Titer ≥1:10; fourfold or greater increase in titer provides evidence of recent infection

Use Support the diagnosis of Q fever; the most common symptoms include fever, chills, and headache.

Limitations Reagents prepared from fresh isolates (phase I organisms) react differently from those from multiply-passaged organism, a laboratory artifact (phase II). Enzyme-linked immunosorbent assay (ELISA) is more sensitive than indirect immunofluorescent antibody (IFA) or complement fixation (CF) assay. Serology tests for the diagnosis of Q fever have not been standardized. Cross reactions with *Legionella* have been described.

Methodology Complement fixation (CF), indirect immunofluorescent antibody (IFA), enzyme-linked immunosorbent assays (ELISA)

Additional Information Q fever should be considered in the presence of fever with negative blood cultures. Serology is preferable to culture for detection of infection with *C. burnetii*. Nucleic acid amplification has also been used to detect *C. burnetii* in blood specimens of infected patients. Sera from patients with Q fever do not react in the Weil-Felix test with *Proteus* antigen. Convalescent sera react best with phase II organism (see above), but sera from chronic persistent infection react best with phase I organisms. Chronic Q fever is uncommon and tends to affect patients with valvular heart disease.

References

Franz DR, Jahrling PB, Friedlander AM, et al, "Clinical Recognition and Management of Patients Exposed to Biological Warfare Agents," *JAMA*, 1997, 278(5):399-411.

Levy PY, Carrieri P, and Raoult D, "*Coxiella burnetii* Pericarditis: Report of 15 Cases and Review," *Clin Infect Dis*, 1999, 29(2):393-7.

Marrie TJ and Raoult D, "Q Fever - A Review and Issues for the Next Century," *Int J Antimicrob Agents*, 1997, 8:145-61.

Maurin M and Raoult D, "Q Fever," *Clin Microbiol Rev*, 1999, 12(4):518-53.

Tissot-Dupont H, Torres S, Nezri M, et al, "A Hyperendemic Focus of Q Fever Related to Sheep and Wind," *Am J Epidemiol*, 1999, 150(1):67-74.

Internet Web Sites

www.cdc.gov/ncidod/dvrd/qfever/index.htm

♦ **Quaalude®** *see* Methaqualone, Urine *on page 903*

♦ **Quadruple Test** *see* Inhibin A, Serum *on page 799*

♦ **Quantitative Fecal Fat, 72-Hour Collection** *see* Fecal Fat, Quantitative, 72-Hour Collection *on page 574*

♦ **Quicksilver** *see* Mercury, Blood *on page 897*

♦ **Quinaglute® Dura-Tabs®** *see* Quinidine, Serum *on page 1129*

♦ **Quinalan®** *see* Quinidine, Serum *on page 1129*

♦ **Quinidex® Extentabs®** *see* Quinidine, Serum *on page 1129*

Quinidine, Serum

Related Information

Amiodarone, Serum *on page 148*
Digoxin, Serum *on page 512*
Procainamide, Serum *on page 1092*
Verapamil, Serum or Plasma *on page 1306*

Synonyms Biquin®; Cardioquin®; Cin-Quin®; Kiditard®; Kinidin®; Quinaglute® Dura-Tabs®; Quinalan®; Quinidex® Extentabs®; Quini® Durules®; Quinora®; Systodin®

(Continued)

Quinidine, Serum (Continued)

Abstract Quinidine is a class 1 antiarrhythmic agent that is frequently used for management of life-threatening ventricular and supraventricular arrhythmias. It is active against *Plasmodium falciparum* malaria.

Specimen Serum

Container Red top tube. The stoppers on some tubes contain plasticizers which affect measured drug levels.

Sampling Time Trough: collect just before next dose. Steady state is reached between 30 and 40 hours.

Causes for Rejection Sample collected in gel barrier tube

Reference Interval Therapeutic: 2-5 µg/mL (SI: 6.2-15.4 µmol/L). Patient dependent therapeutic response occurs at levels of 3-6 µg/mL (SI: 9.2-18.5 µmol/L). Optimal therapeutic level is method dependent.

Possible Panic Range Toxic: >8 µg/mL (SI: >24.7 µmol/L). Some patients may have toxic effects at levels near of just above the upper limit of therapeutic range. Levels >14 µg/L are associated with cardiac toxicity. See Table C in the Therapeutic Drug Monitoring Introduction *on page 62.*

Use Therapeutic monitoring and toxicity assessment

Methodology Immunoassay, high performance liquid chromatography (HPLC), gas chromatography (GC)

Additional Information
- Half-life: 6-8 hours
- Volume of distribution: 2-3 L/kg
- Protein binding: 70% to 90%
- Bioavailability: sulfate: 80%; gluconate: 70%

Absorption of quinidine is complete and rapid. Peak concentrations are reached in 1.5-2 hours after oral intake. With slow preparations (quinidine gluconate) peak levels are reached 4-5 hours after intake. Optimal resampling time after change in dosage is 1-2 days. **Doses >250 mg/day of quinidine result in increased serum digoxin concentrations about 2.5 times the digoxin concentration before quinidine is added.** The new steady-state digoxin concentration occurs in 7-14 days, with signs of toxicity beginning to appear in 3-7 days after initiation of quinidine therapy. Therefore, **serum digoxin concentrations should be measured before initiation of quinidine therapy and again in 4-6 days.** Measure trough because of variability of peak interval. Enzyme inducers, such as aminoglutethimide, carbamazepine, phenobarbital, phenytoin, primidone, rifabutin, and rifampin, decrease quinidine blood levels. Verapamil, amiodarone, alkalinizing agents, and cimetidine may increase quinidine serum concentrations. **Warfarin effects may be increased by quinidine; monitor INR closely during addition or withdrawal of quinidine.** Quinidine is among the most common causes of thrombocytopenia (~1 patient/1000 at risk). Quinidine clearance depends on both liver and renal function. Severe heart failure prolongs half-life, as does decreased liver or renal function.

References

Albrecht H, Lennox J, and del Rio C, "Quinidine and Malaria," *Arch Intern Med*, 2001, 161(8):1118-9.

Capucci A, Aschieri D, Villani GO, "Clinical Pharmacology of Antiarrhythmic Drugs," *Drugs Aging*, 1998, 13(1):51-70.

Cines DB and Laposata M, "A 70-Year-Old Woman With Atrial Fibrillation and the Rapid Onset of Hemorrhagic Manifestations," Case Records of the Massachusetts General Hospital, Case 15-1995, Scully RE, Mark EJ, McNeely WF, et al, eds, *N Engl J Med*, 1995, 332(20):1363-70.

Grace AA and Camm AJ, "Quinidine," *N Engl J Med*, 1998, 338(1):35-45.

♦ **Quini® Durules®** *see* Quinidine, Serum *on page 1129*

♦ **Quinora®** *see* Quinidine, Serum *on page 1129*

♦ **Rabbit Fever Diagnostic Procedures** *see* Tularemia Diagnostic Procedures *on page 1279*

Rabies

Related Information

Virus, Direct Detection by Fluorescent Antibody *on page 1311*

Test Includes Microscopy for Negri bodies, human skin biopsy using direct immunofluorescent antibody

Abstract Rabies has been recognized as a fatal infection in humans and animals for more than 25 centuries. It is usually transmitted through the bite of a rabid animal or exposure to rabid bats. Individuals with a high risk of contact with rabid animals (eg, veterinarians, animal control officers) should consider vaccination against rabies.

Specimen Head of large animal or entire small animal suspected of rabies. Use gloves and mask when handling an animal carcass suspected of rabies.

Human diagnosis is possible through laboratory investigation of cerebro-spinal fluid, serum, saliva, biopsy of brain, corneal impressions or nuchal skin.

Container Double-sealed container

Collection A 6-8 mm full thickness wedge or punch biopsy specimen from the neck containing as many hair follicles as possible should be sampled, snap frozen, and shipped frozen at -70°C to a reference laboratory. Consult with reference laboratory for shipping instructions.

Storage Instructions Ideally, animal brain should be examined in the fresh state. Transport using wet ice or place in absorbent material, then in two plastic bags, or, place half the brain in 50% glycerol, half in 10% formalin, depending on instructions from state laboratory. The local state laboratory must be consulted. Rabies virus may also be demonstrated by immunofluorescence in skin biopsies of patients suspected of having rabies (*vide infra*).

Use Diagnose rabies; evaluate animal bites and exposure to possibly rabid animals for candidacy for rabies immune globulin and/or rabies vaccine.

Limitations Negri bodies (viral inclusions in neurons) are found in about 90% of rabid animals.

Contraindications Formalin fixation precludes fluorescent antibody application

Methodology Direct immunofluorescent antibody (DFA) for detection of rabies virus antigen in brain or skin tissue; Negri bodies can be seen in H&E; nucleo-tide sequence analysis, PCR testing of saliva and skin. Inoculation of mice with suspension of brain tissue is not available in conventional clinical laboratories.

Additional Information Human exposure to bats, followed by skunks, foxes, and dogs has been responsible for most of the cases of rabies in the U.S. Twenty-one of 24 cases of human rabies diagnosed in the U.S. since 1981 are associated with variants of rabies virus related to bats. A clear history of bite was only documented in six. Bites of rabbits, squirrels, hamsters, guinea pigs, gerbils, chipmunks, rats, mice, and other rodents have seldom if ever resulted in human rabies in the United States and are regarded as low risk.

Domestic animals suspected of rabies should be kept alive in quarantine if possible. Unprovoked animal bites are more likely to transmit rabies. Survival of an animal for 10 days makes rabies unlikely. Signs of rabies among wild carnivorous animals cannot be reliably interpreted; any such animal that bites or scratches a person should be killed at once and the head submitted for rabies testing.

Around the world, almost all rabies follows a rabid animal bite exposure. However, rabies virus can enter through nonbite exposure, such as an open wound, or by inhalation of aerosolized bat urine (eg, cave explorers) or by corneal transplantation.

References

Centers for Disease Control and Prevention, "Human Rabies - California, Georgia, Minnesota, New York, and Wisconsin, 2000," *JAMA*, 2001, 285(2):158-60.

Fleisher GR, "The Management of Bite Wounds," *N Engl J Med*, 1999, 340(2):138-40.

Hanlon VA, Olson JG, and Clark CJ, "Article I: Prevention and Education Regarding Rabies in Human Beings. National Working Group on Rabies Prevention and Control," *J Am Vet Med Assoc*, 1999, 215(9):1276-80.

Haupt W, "Rabies - Risk of Exposure and Current Trends in Prevention of Human Cases," *Vaccine*, 1999, 17(13-14):1742-9.

Moran GJ, Talan DA, Mower W, et al, "Appropriateness of Rabies Postexposure Prophylaxis Treatment for Animal Exposures," *JAMA*, 2000, 284(8):1001-7.

Mrak RE and Young L, "Rabies Encephalitis in Humans: Pathology, Pathogenesis and Pathophysiology," *J Neuropathol Exp Neurol*, 1994, 53(1):1-10.

Smith JS, "Rabies," *Clin Microbiol Newslet*, 1999, 21(3):17-23.

Internet Web Sites

www.amm.co.uk/pubs/fa_rabies.htm
www.astdhpphe.org/infect/rabies.html
www.cdc.gov/ncidod/dvrd/rabies

Rapamycin, Blood

Related Information

Cyclosporine, Blood *on page 487*
Mycophenolic Acid, Serum *on page 936*
Tacrolimus, Whole Blood *on page 1234*

Synonyms Rapamune®; Sirolimus

Abstract Rapamycin is a hydrophobic macrolide antibiotic with structural similarities to other macrocyclic lactones such as tacrolimus and erythromycin. It is generally used in combination with cyclosporine and corticosteroids, as an immunosuppressant. Although mostly used in renal transplants, it has also been used in liver and lung transplants.

Specimen Whole blood

Container Lavender top (EDTA) tube

Sampling Time Immediately before next dose.

Storage Instructions Refrigerate

Reference Interval 3-10 ng/mL for renal transplant and 12-20 ng/mL for liver transplant

Use Therapeutic drug monitoring

Contraindications Rapamycin causes hyperlipidemia and decreases glomerular filtration rate. It interacts with CYP3A4 inhibitors. Cases of bronchial anastomotic dehiscence and deaths have been reported in lung transplant patients when sirolimus was used as part of an immunosuppressive regimen.

Methodology Immunoassay, liquid chromatography with ultraviolet detector, liquid chromatography with tandem mass-spectrometry (LC-MS-MS) detector. Chromatographic methods are preferred. As compared to chromatographic methods, immunoassays overestimate drug concentrations (approximately by 30%) due to cross-reactivity with rapamycin metabolites.

Additional Information

- Half-life: 60 hours
- Volume of distribution: 12 ± 7 L/kg
- Protein binding: >95%
- Time to reach steady state: ~2 weeks

The drug is metabolized in the liver by cytochrome CYP3A4 and P-glycoprotein and has an elimination half-life of about 60 hours. Drugs inhibiting P4503A enzymes increase rapamycin levels. Usual blood:plasma ratios are about 30, making whole blood the specimen of choice for therapeutic drug monitoring. Following 2-4 months of combined therapy, withdrawal of cyclosporine may be considered in low-to-moderate risk patients. Cyclosporine withdrawal is not recommended in high immunological risk patients. Cyclosporine should be discontinued over 4-8 weeks, and a necessary increase in the dosage of sirolimus (up to fourfold) should be anticipated due to removal of metabolic inhibition by cyclosporine and to maintain adequate immunosuppressive effects.

The major side effects of rapamycin are hyperlipidemia and thrombocytopenia. Hypertension, anemia, diarrhea, and rash have also been reported.

Recently, uses of rapamycin other than transplantation have been described. As rapamycin inhibits cell proliferation, its role in prevention of neointimal growth has been studied. Compared to bare-metal stents, the sirolimus-coated stents resulted in less smooth muscle cell colonization and minimal intimal hyperplasia. Patients with angina who received sitolimus-eluting stents for the treatment of single, primary lesions in native coronary arteries had no angiographic evidence of late luminal loss or in-stent restenosis at 6 months, no episodes of thrombosis, and a very low rate of cardiac events at 1 year.

References

Gallant-Haidner HL, Trepanier DJ, Freitag DG, et al, "Pharmacokinetics and Metabolism of Sirolimus," *Ther Drug Monit*, 2000, 22(1):31-5.

Johnson RW, "Sirolimus (Rapamune) in Renal Transplantation," *Curr Opin Nephrol Hypertens*, 2002, 11(6):603-7.

Morice MC, Serruys PW, Sousa JE, et al, "A Randomized Comparison of a Sirolimus-Eluting Stent With a Standard Stent for Coronary Revascularization," *N Engl J Med*, 2002, 346(23):1773-80.

Ruygrok PN, Muller DW, and Serruys PW, "Rapamycin in Cardiovascular Medicine," *Intern Med J*, 2003, 33(3):103-9.

Watson CJ, Friend PJ, Jamieson NV, et al, "Sirolimus: A Potential New Immunosuppressant for Liver Transplantation," *Transplantation*, 1999, 67(4):505-9.

- **Rapid (or Short) ACTH Test** *see* Corticotropin Stimulation Test (Rapid) *on page 456*
- **RAST®** *see* Allergen Specific IgE Antibody *on page 131*
- **RBC Galactokinase** *see* Galactokinase, Blood *on page 627*
- **RBC Protoporphyrin** *see* Protoporphyrin, Free Erythrocyte *on page 1121*
- **RB Tumor Suppressor Gene** *see* Retinoblastoma Gene DNA Detection *on page 1159*
- **[^{14}C] d-Xylose Breath Test** *see* d-Xylose Absorption Test, Serum, Urine *on page 527*
- **Receiver Operating Characteristic Curves** *see page 11*
- **Receptor Blocking Antibody** *see* Acetylcholine Receptor Antibody *on page 92*
- **Receptor Modulating Antibody** *see* Acetylcholine Receptor Antibody *on page 92*

Red Blood Cell Count

Related Information

Cobalamin, Serum *on page 424*
Complete Blood Count *on page 442*
Erythropoietin, Serum *on page 551*
Ferritin, Serum *on page 577*
Hematocrit *on page 674*
Hemoglobin *on page 681*
Peripheral Blood: Red Blood Cell Morphology *on page 1016*
Red Blood Cell Indices *on page 1136*
Red Cell Mass *on page 1144*
White Blood Cell Count *on page 1330*

Abstract The red blood cell (RBC) count is the number of RBCs per unit volume of whole blood. The RBC count is a fundamental measurement which, by itself, has importance in medical diagnosis and is also used to calculate red cell indices.

Specimen Whole blood

Container Lavender top (EDTA) tube for venipuncture specimen; properly filled lavender top Microtainer™ tubes for capillary specimen

Causes for Rejection Hemolyzed or clotted specimen

Reference Interval Male: 4.6-6.0 x 10^6/mm^3; female: 3.9-5.5 x 10^6/mm^3. See Complete Blood Count *on page 442* for age-related reference intervals.

Use Evaluate anemia, polycythemia

Limitations Presence of cold agglutinins may result in falsely low RBC counts. Modern electronic cell counters have a level of precision and accuracy greatly improved over manual counting techniques. With correct threshold setting, properly controlled and functioning instrumentation, reliability, and reproducibility of the RBC count is equivalent to or better than most laboratory tests. Surveys (College of American Pathologists) indicate a coefficient of variation (CV) of 1% to 2%. Some instruments commonly perform with CVs <1%. Between laboratory precision has been reported as 2.8%.

Methodology Manual hemocytometer chamber count of diluted blood sample or electronic counting and sizing of red cells made to flow through a fine aperture or capillary, microcomputer control and analysis of data developed from changes in impedance or flow past a laser beam

Additional Information Decrease in RBC count may be the result of red cell loss by bleeding or hemolysis (intravascular or extravascular), failure of marrow production (due to a broad variety of causes), or may be secondary to dilutional factors (eg, intravenous fluids). Increase in RBC count may be the result of primary polycythemia (polycythemia vera) or secondary polycythemia (hypoxemia of lung or cardiovascular disease, increased erythropoietin production associated with renal cyst, renal cell carcinoma, cerebellar hemangioblastoma, (Continued)

Red Blood Cell Count *(Continued)*

or high O_2 affinity hemoglobinopathy) including stress polycythemia (hemoconcentration associated with exercise, exertion, fright, etc). The uncommon condition, "inapparent polycythemia vera" is characterized by normal Hb, Hct, and RBC count levels, but increased red cell mass and clinical features of polycythemia vera. RBC count is normally higher in individuals residing at high altitudes.

References

Lamy T, Devillers A, Bernard M, et al, "Inapparent Polycythemia Vera: An Unrecognized Diagnosis," *Am J Med*, 1997, 102(1):14-20.

Morris MW and Davey FR, "Basic Examination of Blood," *Clinical Diagnosis and Management by Laboratory Methods*, 20th ed, Chapter 24, Henry JB, ed, Philadelphia, PA: WB Saunders Co, 2001, 484-5, 489-99.

"Quality Assurance," *Practical Hematology*, 8th ed, Chapter 4, Dacie Sir JV and Lewis SM, eds, New York, NY: Churchill Livingstone, 1995, 35-47.

Red Blood Cell Enzyme Deficiency

Related Information

Anemia Flowchart *on page 35*
Autohemolysis Test *on page 221*
Glucose-6-Phosphate Dehydrogenase, Blood *on page 641*
Heinz Body Stain *on page 669*
Hemosiderin Stain, Urine *on page 692*
Pyruvate Kinase Assay, Erythrocytes *on page 1127*
Reticulocyte Count *on page 1156*

Test Includes Glucose-6-phosphate dehydrogenase (G6PD), pyruvate kinase (PK), and any of over 20 RBC enzymes included in the following reference by E. Beutler

Abstract Clinical suspicion of red cell enzyme deficiency as a cause of hemolytic anemia but with normal G6PD and PK levels indicates the need to consider use of assays (screening or quantitative) for the more uncommon red cell enzyme deficiencies.

Specimen Erythrocytes; about 1 mL of blood is required for each enzyme assay.

Container Lavender top (EDTA) tube; heparin or acid-citrate-dextrose samples may also be used.

Storage Instructions Normal enzymes are stable for 24 hours at 25°C and for 6-20 days at 4°C.

Causes for Rejection Clotted or hemolyzed specimen

Special Instructions Most of these assays will need to be sent to a reference laboratory. Ship on wet ice, do not freeze.

Reference Interval Screening tests: enzyme present (reported as normal); quantitative G6PD: 8.6-18.6 IU/g hemoglobin; phosphohexoisomerase: 14.7-42.2 IU/g hemoglobin; pyruvate kinase: 2.0-8.8 IU/g hemoglobin

Use Investigation of hemolytic state. Quantitative assays confirm results of screening tests and generally allow for identification of heterozygotes (which may be of value in genetic analyses). Phenotyping of RBC enzymes has been applied to paternity testing.

Limitations False-normal results may occur if testing is performed on a sample obtained just after a hemolytic episode (deficiency may be obscured by the presence of a young, enzyme-rich population of RBCs, reticulocytes, the older enzyme-deficient red cells having been destroyed).

In vitro assay (with optimal pH, substrate concentration, and other conditions) may not reflect *in vivo* enzyme performance. Contamination by leukocytes (that may have enzymes with high specific activity) can cause spurious normal RBC values. Transfused RBCs may mask a deficient state. Mixed reticulocytes and senescent red cells (with reduced enzyme activity) may obscure the level of enzyme deficiency.

Contraindications Sample obtained just after an acute hemolytic episode has occurred

Methodology Standardized methods have been recommended by the International Committee for Standardization in Haematology.

Additional Information See graphic.

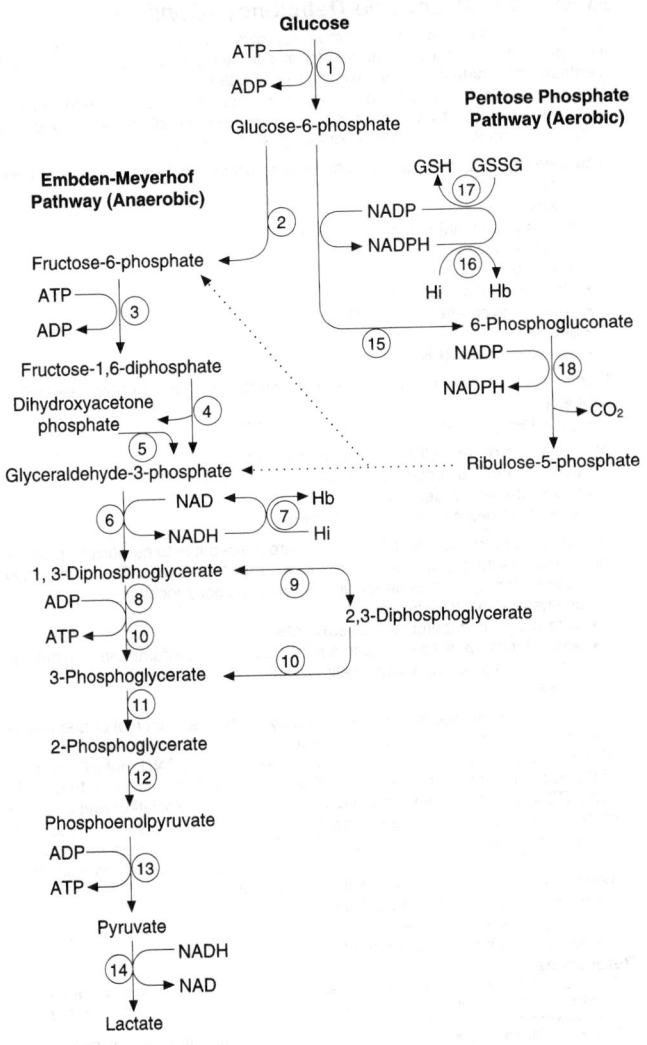

Schematic representation of red cell glycolytic pathways. The enzymes are indicated as follows: (1) hexokinase, (2) glucosephosphate isomerase, (3) phosphofructokinase, (4) aldolase, (5) triose phosphate isomerase, (6) glyceraldehyde-3-phosphate dehydrogenase, (7) NADH-methaemoglobin reductase, (8) phosphoglycerate kinase, (9) diphosphoglyceromutase, (10) diphosphoglycerate phosphatase, (11) phosphoglyceromutase, (12) enolase, (13) pyruvate kinase, (14) lactate dehydrogenase, (15) glucose-6-phosphate dehydrogenase, (16) NADPH-methaemoglobin reductase, (17) glutathione reductase, (18) 6-phosphogluconate dehydrogenase.

Adapted from Luzzatto L, Roper D, Dacie Sir JV, et al, *Practical Haematology*, 8th ed, Chapter 13, New York, NY: Churchill Livingstone, 1995, 226.

The mature red blood cell cannot synthesize protein or lipids and is dependent upon glycolysis to obtain energy from glucose. Glucose is initially phosphorylated by hexokinase with the resultant glucose-6-phosphate (G6P) further
(Continued)

Red Blood Cell Enzyme Deficiency *(Continued)*

processed as substrate for anaerobic glycolysis (Embden-Meyerhof pathway with production of pyruvate and ATP) or alternately G6P is processed in the pentose phosphate pathway (oxidative glycolysis). Of the some 20 enzymes that function in these two major pathways, the two most common enzyme disorders associated with hemolysis are glucose-6-phosphate dehydrogenase (G6PD) deficiency and pyruvate kinase (PK) deficiency.

The following enzymes of the anaerobic glycolytic pathway cause hemolysis when deficient:

- hexokinase (HK)
- glucose phosphate isomerase (GPI)
- phosphofructokinase (PFK)
- aldolase
- triose phosphate isomerase (TPI)
- phosphoglycerate kinase (PGK)
- enolase
- pyruvate kinase (PK)

Enzymes of the oxidative glycolytic pathway that may cause hemolysis when deficient:

- glucose-6-phosphate dehydrogenase (G6PD)

Abnormalities of red cell nucleotide metabolism:

- pyrimidine-5'-nucleotidase deficiency
- adenylate kinase deficiency
- adenosine deaminase overproduction

Individuals with low levels of RBC G6PD are susceptible to hemolytic episodes after exposure to certain chemicals, drugs, and fava beans. Drugs that may precipitate hemolysis in patients with G6PD deficiency include:

- analgesics/antipyretics: aspirin
- sulfa drugs: sulfapyridine, sulfisoxazole
- antimalarias: primaquine, pentaquine, quinine, nitrofurantoin, Chloromy-cetin®, quinidine, para-aminosalicylic acid
- others

Deficient or absent RBC enzyme activity may relate to absence of or decreased level of the enzyme, presence of an inactive molecular form or of an isoenzyme with altered activity. In some hematologic diseases (eg, PNH, aplastic anemia and acute leukemia), acquired red cell enzyme deficiency is not infrequently seen. RBC glutathione reductase deficiency bears an association with a variety of chemotherapeutically treated malignant states (up to 43% in a study of hospitalized patients). Glutathione reductase deficiency also occurs in some cases of malnutrition, liver disease, and sepsis. Disorders of red cell nucleotide metabolism possibly associated with hemolytic anemia:

- pyrimidine-5-nucleotidase deficiency
- adenylate kinase deficiency
- adenosine deaminase overproduction

References

Beutler E, "Glucose-6-Phosphate Deficiency and Other Enzyme Abnormalities," *Williams Hematology*, 5th ed, Chapter 54, Beutler E, Lichtman MA, Coller BS, et al, eds, New York, NY: McGraw-Hill Inc, 1995, 564-81.

Beutler E, Blume KG, Kaplan JC, et al, "International Committee for Standardization in Haematology: Recommended Methods for Red-Cell Enzyme Analysis," *Br J Haematol*, 1977, 35(2):331-40.

Layton DM and Bellingham AJ, "Disorders of Erythrocyte Metabolism," *Pediatric Hematology*, 2nd ed, Lilleyman JS, Hann IM, and Blanchette VS, eds, New York, NY: Churchill Livingstone, 1999, 285.

Luzzatto L and Karadimitris A, "The Molecular Basis of Anaemia," *Molecular Haematology*, Chapter 11, Provan D and Gribben J, eds, Malden, MA: Blackwell Science Ltd, 2000, 150-7.

McMullin MF, "The Molecular Basis of Disorders of Red Cell Enzymes," *J Clin Pathol*, 1999, 52(4):241-4.

Red Blood Cell Indices

Related Information

Applies to Red Cell Distribution Width (RDW)

Test Includes MCV, MCH, MCHC, RDW, RBC, Hct, Hb

Abstract The RBC indices are measured or mathematically derived from hemoglobin (Hb), hematocrit (Hct), and red blood cell (RBC) count. The values can be used for quick assessment of anemia and have application in the provision of internal quality control.

Specimen Whole blood

Container Lavender top (EDTA) tube for venipuncture specimen; lavender top Microtainer™ tube for capillary specimen

Collection Routine venipuncture. Invert the tube 5-10 times gently to mix. There must be no clots.

Storage Instructions Specimen cannot be used if stored over 10 hours at room temperature or 18 hours at 4°C refrigerated temperature. Specimen must not be frozen.

Causes for Rejection Hemolyzed or clotted specimen

Reference Interval Because of slight physiologic and machine analytic variation, the following reference values, derived from review of multiple sources, are given as general ranges. For extensive tables of reference ranges delineating variation with age, race, altitude, and analyzer consult texts by Kjeldsberg and Shinton. The upper limit of MCV is important in the evaluation of cobalamin (and possible early or subclinical cobalamin) deficiency. Upper limit for MCV might better be considered 96 (see reference by Chanarin). See Cobalamin, Serum *on page 424*.

Red Blood Cell Indices

	MCV (fL)	MCH (pg)	MCHC (g/dL)	RDW (%)
Neonates	110-130	31-37	29-37	
1 y	70-86	23-31	30-36	
2-12 y	75-95	24-33	31-37	
12-18 y	78-100	25-35	31-37	
Adult male	80-100	26-34	31-37	11.7-14.2
Adult female	79-98	26-34	30-36	11.7-14.2

Use Differential diagnosis of anemia, iron deficiency, hereditary spherocytosis (about 50% of such individuals have MCHC >36%), immune spherocytosis, thalassemia, chronic lead poisoning, folate deficiency, vitamin B$_{12}$ deficiency, vitamin B$_6$ deficiency, pernicious anemia, anemia of chronic disease, and anemia of pregnancy

Limitations Patients with macrocytosis and microcytosis (of RBCs) may have indices within the reference range due to the averaging method used by machine determination. Patients with autoagglutination will show spurious results. Although MCV is commonly elevated early in pernicious anemia, it may be normal, especially later when micropoikilocytosis develops. In a recent review of the value of MCV in detection of cobalamin (vitamin B$_{12}$) deficiency (uncertainty over the upper limit of normal for MCV as an acknowledged limiting factor), sensitivity of MCV was less than desirable. At best (patients with established cobalamin deficiency and anemia), sensitivity was 75%, positive predictive value for B$_{12}$ deficiency ranged (5 different studies) from 0% to 55%.

Methodology RBC indices are usually obtained by microcomputerized, highly automated electronic and pneumatic multichannel analyzers based on aperture-impedance cell sizing and counting.

(Continued)

Red Blood Cell Indices *(Continued)*

Additional Information The group of three red cell indices are a productive and economically efficient approach to screening for hematologic abnormality, in particular the compensated and uncompensated anemias. The MCV (mean corpuscular volume) is the size (volume) of the average red cell. The MCH (mean corpuscular hemoglobin) is the weight of hemoglobin in the average red cell. The MCHC (mean corpuscular hemoglobin concentration) is the amount of hemoglobin present in the average red cell as compared to its size. The RBC indices are a valuable guide to the choice of more specific measurements such as serum iron, ferritin, folic acid, and/or vitamin B_{12} levels. The MCV decreases before MCHC in evolving iron deficiency anemia while the MCHC decreases before the MCV in evolving anemia of chronic disease. In hemolytic anemias, particularly the hemoglobinopathies, the RBC indices are less helpful. Decreased MCV levels may be due to thalassemia minor although they are not specific for this condition; A_2 hemoglobin levels should be studied to follow-up low MCV levels in appropriate clinical settings. Screening for high A_2 hemoglobin levels will discriminate a population of beta-thalassemics with normal MCV levels. Technicon® (Bayer® Diagnostics) series H analyzers provide a measure of heterogeneity in hemoglobin concentration of individual red cells (histogram of MCHCs). Decreased cell Hb concentration (increased cell hydration) has been noted to characterize α-thalassemia while both decreased and increased cell Hb concentration (cell hydration and dehydration) characterize β-thalassemia red cells. Cell dehydration as reflected by histogram of MCHCs can also be used to identify most patients with hereditary spherocytosis.

Changes in RBC Indices With Disease

Condition	MCV	MCH	MCHC	RDW
Iron deficiency anemia	↓	↓	↓	↑
Chronic inflammation	↓	N±	N±	N±
Pernicious anemia	↑	N or high N	high N	↑
B_{12} / folate deficiency	↑	N or high N	high N	↑
Hereditary spherocytosis	N or ↓	↑	↑	N±
Hemolytic / aplastic anemia	N±	N±	N±	N±
Anemia secondary to acute blood loss	N±	N±	N±	N±
Polycythemia	N±	N±	N±	N±
Hemochromatosis	↑	↑	↑	N±

A number of conditions (usually characterized by the generation of numerous RBC fragments) may show "relative microcytosis" as reflected by slightly decreased to low normal MCV. This has been described with sickle cell anemia. In children, low MCV for age suggests iron deficiency, lead poisoning, thalassemia syndrome, or very rarely, a pyridoxine-responsive anemia. Examination of the peripheral smear, family history, dietary history, and stool guaiac are helpful in this setting. Megaloblastic hematopoiesis is <10% of cases of macrocytosis (increase in MCV). Mean values of MCV, MCH, and MCHC (as well as Hb and Hct) are increased in hemochromatosis. MCV may be significantly increased in diabetic ketoacidosis (this is due to plasma hyperosmolarity). The high molar concentration of glucose is the main contributing factor to the increased plasma osmolarity, producing a hypertonic intracellular state of RBCs. When such cells are put into a relatively hypotonic diluent (eg, Coulter cell counting), water enters the cell, it swells and may produce erroneously high MCV. Recognition of increase in MCV may alert the clinician to presence of the hyperosmolar state. Runner's anemia is characterized by macrocytosis (elevated MCV), with plasma volume expansion and mild anemia.

Aluminum toxicity, occurring in uremic patients on chronic hemodialysis, is associated in some cases with decreased MCV, microcytic anemia.

RDW is an electronic measurement of anisocytosis (red cell size variability). RDW is typically elevated in iron deficiency anemia while usually normal in beta thalassemia minor (heterozygous thalassemia). RDW is elevated in beta thalassemia major. In establishing nomograms of hematologic parameters in pregnancy, study of the RDW found that it was significantly increased between 34 weeks of gestation and the onset of labor.

Drugs, in particular DNA inhibiting anti-AIDs therapeutics and other viral antinucleoside therapeutics are the most common cause of macrocytosis (increase in MCV) in the hospitalized urban patient population.

Statistical analysis of RBC indices, notably a set of linear discriminant functions, are said to effectively differentiate between α-thalassemia, β-thalassemia, and iron deficiency anemia with a high degree of accuracy.

References

Chanarin I and Metz J, "Diagnosis of Cobalamin Deficiency: The Old and the New," *Br J Haematol*, 1997, 97(4):695-700.

Dang CV, "Runner's Anemia," *JAMA*, 2001, 286(6):714-6.

Eldibany MM, Totonchi KF, Joseph NJ, et al, "Usefulness of Certain Red Blood Cell Indices in Diagnosing and Differentiating Thalassemia Trait From Iron-Deficiency Anemia," *Am J Clin Pathol*, 1999, 111(5):676-82.

Kjeldsberg CR, Elenitoba-Johnson K, Foucar K, et al, *Practical Diagnosis of Hematologic Disorders*, 3rd ed, Kjeldsberg CR, ed, Chicago, IL: ASCP Press, 2000.

Oosterhuis WP, Niessen RW, Bossuyt PM, et al, "Diagnostic Value of the Mean Corpuscular Volume in the Detection of Vitamin B_{12} Deficiency," *Scand J Clin Lab Invest*, 2000, 60(1):9-18.

Savage DG, Ogundipe A, Allen RH, et al, "Etiology and Diagnostic Evaluation of Macrocytosis," *Am J Med Sci*, 2000, 319(6):343-52.

Shehata HA, Ali MM, Evans-Jones JC, et al, "Red Cell Distribution Width (RDW) Changes in Pregnancy," *Int J Gynaecol Obstet*, 1998, 62(1):43-6.

♦ **Red Blood Cell Morphology** *see* Peripheral Blood: Red Blood Cell Morphology on page 1016

Red Blood Cells

Related Information

Activated Partial Thromboplastin Time *on page 100*
Blood Gases and pH, Arterial *on page 275*
Coagulation Factor Assays *on page 418*
Cold Agglutinin Titer *on page 430*
Filters for Blood *on page 588*
Hematocrit *on page 674*
Hemoglobin *on page 681*
Hereditary Hemochromatosis DNA Test *on page 718*
Irradiated Blood Components *on page 811*
Oxygen Saturation, Blood *on page 991*
Pretransfusion Testing *on page 1088*
Prothrombin Time *on page 1116*
Red Blood Cells, Leukocytes Reduced *on page 1141*
Red Blood Cells, Washed *on page 1143*
Risks of Transfusion *on page 1166*
Transfusion Reaction Work-up *on page 1269*
Uncrossmatched Blood, Emergency *on page 1281*
Warming, Blood *on page 1326*
Whole Blood *on page 1333*

Synonyms Packed Red Cells, Transfusion

Test Includes ABO and Rh type, antibody screen, crossmatch, antibody identification when screen is positive (ie, preparation as for other transfusions)

Abstract A unit of red blood cells has a volume of 230-350 mL. RBCs may be prepared from whole blood by the removal of 200-250 mL of plasma or obtained by apheresis collection. Red cells with CPDA-1 anticoagulant have a hematocrit of ~70% and expire 35 days after the date of collection, when stored continuously between 1°C to 6°C. With additional adenine supplementation after removal of plasma, AS-1 red blood cells have a hematocrit of 55% to 60% and a storage period of 42 days at 1°C to 6°C. If the hermetic seal is broken during preparation, the red blood cells must be infused within 24 hours.

Patient Preparation The patient should have an identification wristband. Emergency Departments (ERs) may use special or temporary identification. **Dosage and administration:** Give red cells through a standard 170 micron filter. Most transfusions should not exceed 4 hours duration; 2 hours or less per unit is preferable. Units collected with CPDA-1 anticoagulant can speed up the infusion by adding 50-100 mL of sterile isotonic sodium chloride solution, USP, just before administration. **Do not add or transfuse with lactated Ringer's** (Continued)

Red Blood Cells *(Continued)*

solution, 5% aqueous dextrose, 5% dextrose in 0.225% saline, or other calcium-containing, hypotonic, or glucose-containing fluids through the same tubing because clumping, hemolysis, or clotting may occur. Drugs or medications may **not** be added to blood or blood components.

Aftercare One unit should raise the hematocrit of a 70 kg adult about 3 percentage points (or hemoglobin 1 g/dL). Monitor hemoglobin and hematocrit.

Specimen Blood

Container One red top tube or one lavender top (EDTA) tube

Collection (Of sample from intended recipient): At the patient's bedside, ask the patient to give his or her name. Compare with the patient's wristband. Label the sample tube with two unique forms of patient identification (eg, patient's full name, hospital number), also include date and initials of the collector. Further information may be required on a requisition form. Take extra care with identification of unresponsive patients.

Storage Instructions Store in Blood Bank monitored refrigerator only until issue. When it is not possible to transfuse immediately after issue, return blood to Blood Bank within ~30 minutes. Otherwise, blood that has been out of monitored refrigeration must be discarded. Appropriate designated refrigeration includes specified storage conditions, temperature recorders, and alarm signals.

Beware of blood which appears darker than usual; it may be contaminated.

If stored as packed red cells in CPDA-1 anticoagulant, red cells can be stored for up to 5 weeks. If stored as an additive system unit, unit can be stored for up to 6 weeks.

Causes for Rejection (Of patient sample): Gross hemolysis, sample placed in a serum separator tube, specimen tube not properly labeled

Turnaround Time For routine situations, red blood cells can be ready for transfusion within 30-45 minutes from the time the Blood Bank gets the type and crossmatch sample, if blood of the appropriate type is on hand. Presence of unexpected antibodies may require hours to a day or two for identification.

Special Instructions Blood banks and hospital transfusion services hold crossmatched blood only for 24 hours, after which they make it available for other patients. There will be some exceptions. Notify the Blood Bank as soon as possible if the patient will not need transfusion so the blood can be available for some other patient.

Critical Values Perioperative patients almost always require transfusion at Hb <6 g/dL, but no magic number exists to trigger indication for transfusion for all patients. Thirty percent to 40% loss of blood volume is associated with increased signs of shock and >40% relates to severe shock.

Possible Panic Range In otherwise stable patients, the risk of mortality without transfusion rises very significantly at hemoglobin levels of 3.5-4.0 g/dL.

Use Replace red cell volume; rapid loss of >30% to 40% of blood volume or fall in Hb to <6 g/dL requires RBC transfusions in most subjects. Hemoglobin >10 g/dL rarely indicates transfusion, whereas hemoglobin <6 g/dL almost always does. Transfusion of patients with heart, liver, or renal disease in whom restriction of plasma volume or of sodium may be desirable, eg, to decrease the likelihood of volume overload (as compared with effect of whole blood); transfusion of patients with chronic anemias; replace blood lost in surgical operations. Type O RBCs may be given in emergencies to recipients of unknown ABO type (RBC units have less anti-A and anti-B than whole blood). Indications for transfusion include maintenance of perfusion, arterial oxygenation with augmentation of oxygen delivery to the tissues and maintenance of cardiac output and blood volume.

Some recommend red cell transfusions for critically ill patients when Hb falls to <7 g/dL. Others suggest a threshold for transfusion of Hb 7-8 g/dL in patients who are not critically ill and have no risk of ischemia, and 10 g for those at risk. Intraoperative or postoperative ischemia of myocardium is more likely to take place in patients whose hematocrits fall to <28%, especially in the presence of tachycardia.

RBC transfusions are used in those patients with chronic anemias unresponsive to pharmacologic agents (eg, iron, cobalamin, folate, recombinant human erythropoietin). Neonates require a hematocrit >0.30-0.35 in the presence of respiratory distress.

See Red Blood Cells, Leukocytes Reduced *on page 1141*.

A summary of transfusion guidelines in included in Whole Blood *on page 1333*.

Limitations Red cells prepared in an "open" system expire in 24 hours. Most RBCs, however, are prepared in closed systems with full dating.

The potential problems of provision of blood include alloimmunization, autoimmunization, and requirements of competent professional credentialing and staffing at all levels, from phlebotomist through and inclusive of directorship.

Contraindications With the AIDS epidemic and increasing knowledge of the infectious disease risks of transfusion, as well as Transfusion Committee surveillance, documentation of the necessity for transfusion is required. Do not give blood transfusion when anemia and/or hypoxia can be corrected with specific and safer therapy such as iron, B_{12}, or folic acid. For correction of coagulation deficiencies, specific viral-inactivated concentrates are appropriate.

Methodology Attention to electrolytes, blood gases and blood warmers is indicated with rapid and/or massive transfusion; see Warming, Blood *on page 1326*.

Red cells should not be infused rapidly in elective situations; rather, they should be infused over a period not less than 2 hours or more than 4 hours/unit.

Additional Information Red blood cells must be ABO and Rh compatible. A crossmatch is necessary unless life-threatening urgency exists. See Risks of Transfusion *on page 1166*. An advantage of RBCs as compared to whole blood is greater safety in treatment of patients likely to suffer complications of volume excess.

Problems of iron overload are partly addressed in Hereditary Hemochromatosis DNA Test *on page 718* and related monographs.

Anemia in pregnancy is defined as Hb <10 g/dL. The mean loss after vaginal delivery is 500 mL and after caesarean section, 1000 mL. CMV-seronegative or leukocyte-reduced blood is indicated when a CMV-negative or CMV status unknown pregnant patient requires transfusion.

References

Ely EW and Bernard GR, "Transfusions in Critically Ill Patients," *N Engl J Med*, 1999, 340(6):467-8.

Goodnough LT, Brecher ME, Kanter MH, et al, "Transfusion Medicine. First of Two Parts: Blood Transfusion," *N Engl J Med*, 1999, 340(6):438-48.

Goodnough LT, Monk TG, and Andriole GL, "Erythropoietin Therapy," *N Engl J Med*, 1997, 336(13):933-8.

Hebert PC, Wells G, Blajchman MA, et al, "A Multicenter, Randomized, Controlled Clinical Trial of Transfusion Requirements in Critical Care," *N Engl J Med*, 1999, 340(6):409-17.

Simon TL, Alverson DC, AuBuchon J, et al, "Practice Parameter for the Use of Red Blood Cell Transfusions: Developed by the Red Blood Cell Administration Practice Guideline Development Task Force of the College of American Pathologists," *Arch Pathol Lab Med*, 1998, 122(2):130-8.

♦ **Red Blood Cells, Deglycerolized** *see* Frozen Red Blood Cells *on page 614*

♦ **Red Blood Cells, Frozen** *see* Frozen Red Blood Cells *on page 614*

Red Blood Cells, Leukocytes Reduced

Related Information

Filters for Blood *on page 588*
Frozen Red Blood Cells *on page 614*
Red Blood Cells *on page 1139*
Red Blood Cells, Washed *on page 1143*
Risks of Transfusion *on page 1166*
Whole Blood *on page 1333*

Synonyms Leukocyte-Reduced Red Blood Cells

Test Includes See Red Blood Cells *on page 1139*.

Abstract The most common adverse effect of transfusion is the nonhemolytic febrile reaction.

AABB Standards specify that leukocyte-reduced red cells must contain <5 x 10^6 leukocytes per unit while retaining 85% of the original red cells. This is usually accomplished by the use of a special, leukocyte-reduction filter (see Filters for Blood *on page 588*). Indications for use of leukocyte-reduced red cells include patients with repeated febrile transfusion reactions, to decrease

(Continued)

Red Blood Cells, Leukocytes Reduced *(Continued)*

the incidence of HLA alloimmunization, prevention of alloimmunization to leukocyte antigens and hence, refractoriness to platelet transfusions, and prevention of cytomegalovirus transmission.

Patient Preparation See Red Blood Cells *on page 1139*.

Aftercare See Red Blood Cells.

Container One red top tube or one lavender top (EDTA) tube

Collection See Red Blood Cells.

Storage Instructions See Red Blood Cells.

Causes for Rejection (Of patient sample:) Gross hemolysis, sample placed in a serum separator tube, specimen tube not properly labeled

Special Instructions See Red Blood Cells.

Use See Red Blood Cells. Leukocyte-reduced red cells are specifically indicated for use in multitransfused patients and multiparous females with repeated febrile nonhemolytic transfusion reactions, to prevent alloimmunization to leukocyte antigens (HLA) (and, hence, refractoriness to platelet transfusions), and to prevent cytomegalovirus transmission. Leukocyte-reduced red cells may provide benefits to patients with paroxysmal nocturnal hemoglobinuria. Leukocyte-reduced blood or CMV-seronegative blood is indicated, when pregnant patients known to be CMV negative or when CMV status is unknown, are to be transfused. See tables.

Leukocyte-reduced RBCs reduce postoperative complications, morbidity, and mortality. Use of these fractions is highly cost effective in cardiac surgery.

Table 1. Indications for Leukocyte-Reduced Red Blood Cells

Prevention of Alloimmunization
Congenital hemolytic anemias
Hypoproliferative anemias likely to need multiple transfusions
Aplastic anemia
Myelodysplasias
Myeloproliferative syndrome
Plasma cell dyscrasias
Bone marrow / peripheral blood stem cell transplants
Hematopoietic malignancies
Therapy in Pre-existing Conditions
Recurrent severe febrile hemolytic transfusion reaction
Known HLA alloimmunization
Possible Use
Alternative to cytomegalovirus-seronegative components;
Human immunodeficiency virus-infected patients

Adapted from Simon TL, Alverson DC, AuBuchon J, et al, "Practice Parameter for the Use of Red Blood Cell Transfusions," *Arch Pathol Lab Med*, 1998, 122(2):130-8. *vide infra*

Table 2. Indications for Prevention of Cytomegalovirus Transmission by Seronegative or Leukocyte-Reduced ($<5 \times 10^6$) Red Blood Cells

Well-Defined (patient seronegative for cytomegalovirus or status unknown)
Low-birth-weight neonates (<1200 g)
Human immunodeficiency virus-infected patients
Recipients of seronegative allogeneic, organ, marrow, or stem cell transplants or likely candidates for such transplants
Pregnant women
Intrauterine transfusion
Relative (patient seronegative for cytomegalovirus or status unknown)
Hodgkin disease and non-Hodgkin lymphoma
Recipients of immunosuppressive therapy
Candidates for autologous bone marrow/stem cell transplants
Hereditary or acquired cellular immunodeficiencies
Not Indicated
Seronegative full-term infants, birth weight >1200 g
Seropositive pregnant women

Adapted from Simon TL, Alverson DC, AuBuchon J, et al, "Practice Parameter for the Use of Red Blood Cell Transfusions," *Arch Pathol Lab Med*, 1998, 122(2):130-8.

Limitations Leukocyte reduction does **not** prevent graft-versus-host disease. It is an added expense.

Methodology Filtration is presently the most widely used method of leukoreduction, performed either in the Blood Bank or at the bedside. Personnel administering red cells through a bedside leukocyte reduction filter must be familiar with manufacturer's requirements for use in order to maximize leukocyte reduction and ensure against inordinate loss of red cells. Reproducibility and effectiveness is best sustained when filtration is performed in the laboratory.

Additional Information The use of universal leukocyte reduction in preventing changes in host immune functions is controversial. Animal models have supported the concept that early leukodepletion may prevent alloimmunization to HLA, a practice which may be especially relevant for candidates for bone marrow transplantation, and for other patients likely to need multiple transfusions (eg, those with certain hemoglobinopathies, lymphoplasmacytic and myeloproliferative diseases, and patients being treated for solid tumors with myelosuppressive adjuvant chemotherapy). Early leukodepletion may diminish bacterial contamination, but further investigation is needed. The roles of transfused WBCs in other clinical settings require further studies. Certain of the proposed roles for leukoreduced blood are controversial.

References
Blumberg N and Heal JM, "Mortality Risks, Costs, and Decision Making in Transfusion Medicine," *Am J Clin Pathol*, 2000, 114(6):934-7.
Goodnough LT, "Universal Leukoreduction of Cellular Blood Components in 2001? No," *Am J Clin Pathol*, 2001, 115(5):674-7.
Lipson SM, Shepp DH, Match ME, et al, "Cytomegalovirus Infectivity in Whole Blood Following Leukocyte Reduction by Filtration," *Am J Clin Pathol*, 2001, 116(1):52-5.
Ness PM and Rothko K, "Principles of Red Blood Cell Transfusion," *Hematology, Basic Principles and Practice*, 3rd ed, Hoffman R, Benz EJ Jr, Shattil SJ, et al, eds, New York, NY: Churchill Livingstone, 2000, 2241-8.
Snyder EL, "Transfusion Reactions," *Hematology: Basic Principles and Practice*, 3rd ed, Hoffman R, Benz EJ Jr, Shattil SJ, et al, New York, NY: Churchill Livingstone, 2000, 2300-10.
Sweeney JD, "Universal Leukoreduction of Cellular Blood Components in 2001? Yes," *Am J Clin Pathol*, 2001, 115(5):666-73.

Red Blood Cells, Washed

Related Information

Antibodies to IgA *on page 164*
Ham Test *on page 665*
Red Blood Cells *on page 1139*
Red Blood Cells, Leukocytes Reduced *on page 1141*

Applies to Transfusion Reaction, Allergic

Test Includes ABO and Rh type, antibody screen, crossmatch, and antibody identification when screen is positive, as for other transfusions.

Abstract Washed red cells are prepared using sterile saline to a final volume of 180 mL and a hematocrit of 70% to 80%. The primary indications for use include 1) prevention of allergic or anaphylactic transfusion reactions; and 2) removal of antibodies in plasma which are harmful to the recipient. Do not use when leukocyte reduction is indicated, as leukocyte reduction filters are more efficacious.

Patient Preparation Same as for other RBC transfusions.

Aftercare Same as for other RBC transfusions.

Specimen Blood

Container One red top tube or one lavender top (EDTA) tube

Collection (Of sample from intended recipient): Same as for other RBC transfusions.

Storage Instructions Once prepared, store in a monitored Blood Bank refrigerator only until issue. Same as for other RBC transfusions.

Causes for Rejection (Of patient sample): Gross hemolysis, sample placed in a serum separator tube, specimen tube not properly labeled
(Continued)

Red Blood Cells, Washed (Continued)

Turnaround Time Allow about 1 hour after selection of compatible red cells.

Special Instructions Must be used within 24 hours after preparation.

Use Useful for patients who have severe allergic reaction to conventional transfusion even with antihistamines. Prevention of transfusion reaction to plasma proteins, especially IgA, in patients with IgA immunoglobulin deficiency. In such subjects, with preformed anti-IgA, infusion of IgA-containing plasma can cause anaphylaxis. Washed red cells can be used for patients with paroxysmal nocturnal hemoglobinuria (PNH), although for transfusion of patients with PNH, leukocyte-reduced red cells may be preferable.

Limitations High outdating rate (24 hours). Cannot be ordered "on hold." Time consuming and expensive. There is up to 20% loss of red cells.

Methodology Various methods are available. The best is probably a system using a continuous-flow centrifuge and 2-3 L of isotonic saline per unit.

Additional Information This component is comparable to deglycerolized red blood cells, from which 99% of WBCs are removed. The washing process removes most of the plasma proteins, platelets, and leukocytes, including lymphocytes. The effectiveness of washing depends on the method and on the volume of isotonic saline used. Red cells have been shown to have normal survival after washing. See also Filters for Blood *on page 588*. Use of washed RBCs has given way to leukocyte filters for removal of WBCs, but washed cells are appropriate for patients who have repeated allergic reactions or who have anti-IgA. Washing RBCs is not satisfactory for the prevention of graft-versus-host disease.

References

Circular of Information for the Use of Human Blood and Blood Components, American Association of Blood Banks, America's Blood Centers, American Red Cross, 2002, 10-5.

Ness PM and Rothko K, "Principles of Red Blood Cell Transfusion," *Hematology: Basic Principles and Practice*, 3rd ed, Chapter 135, Hoffman R, Benz EJ Jr, Shattil SJ, et al, eds, New York, NY: Churchill Livingstone, 2000, 2241-48.

Triulzi DJ, *Blood Transfusion Therapy: A Physician's Handbook*, 7th ed, Bethesda, MD: American Association of Blood Banks Press, 2002, 13-4.

♦ **Red Cell Distribution Width (RDW)** *see* Red Blood Cell Indices *on page 1136*

♦ **Red Cell Exchange** *see* Apheresis, Therapeutic *on page 201*

♦ **Red Cell Folate** *see* Folic Acid, RBC *on page 606*

Red Cell Mass

Related Information

Blood Volume *on page 282*
Erythropoietin, Serum *on page 551*
Hematocrit *on page 674*
Hemoglobin *on page 681*
Red Blood Cell Count *on page 1133*

Synonyms Red Cell Volume

Test Includes Red cell mass, plasma volume, and total blood volume

Abstract In the evaluation of patients with reproducibly high hemoglobin (Hb) or hematocrit (Hct), it must be determined if there is an absolute increase in the red cell mass. Direct measurement of RBC mass with an isotope dilution method will obtain definitive data.

Specimen Laboratory will usually manage sampling and reinjecting.

Special Instructions May require scheduling with laboratory in order to procure radioisotope. Patient's weight and height must be made available to the laboratory.

Reference Interval Male: 28.2 ±4 mL/kg; female: 24.2 ±2.6 mL/kg. See table for reference intervals relating to body surface area. At higher elevations (2000-3000 meters above sea level), RBC mass is higher in otherwise normal individuals (ie, in the 30-32 ±6-7 mL/kg range).

An elevated RCM can be defined as being 25% or more over the normal value predicted for any individual. Blood volume increases gradually over the course of pregnancy.

Formulas for Calculating RBC Mass Reference Values From Body Surface Area(s)

Male	Mean normal RCM (mL) = (1486 x S) - 825
	98% limits = ±25%
	Mean normal PV (mL) = 1578 x S
	99% limits = ±25%
Female	Mean normal RCM (mL) = (1.06 x age) + (822 x S)
	99% limits = ±25%
	Mean normal PV (mL) = 1395 x S
	99% limits = ±25%

$S = W^{0.425} \times h^{0.725} \times 0.007184$

S = surface area (m^2)

age = age (years)

h = height (cm)

w = weight (kg)

From International Committee for Standardization in Hematology, "Interpretation of Measured Red Cell Mass and Plasma Volume in Adults," *Br J Haematol*, 1995, 89(4):748-56.

Determination of elevated RBC volume in mL/kg body weight is confounded by obesity as RBC volume per kg fat is only 1/10 that of total RBC volume of the lean body mass (LBM). Application of biologic impedance measuring devices to directly measure a person's body composition and percentage of fat has allowed normalization of total RBC volume (TRCV) for men and women to mL/kg of LBM. TRCV is 36 mL/kg; when >43 mL/kg LBM, the presence of polycythemia can be considered as established.

Schematic for Normalizing Total RBC Volume From Body Fat Content

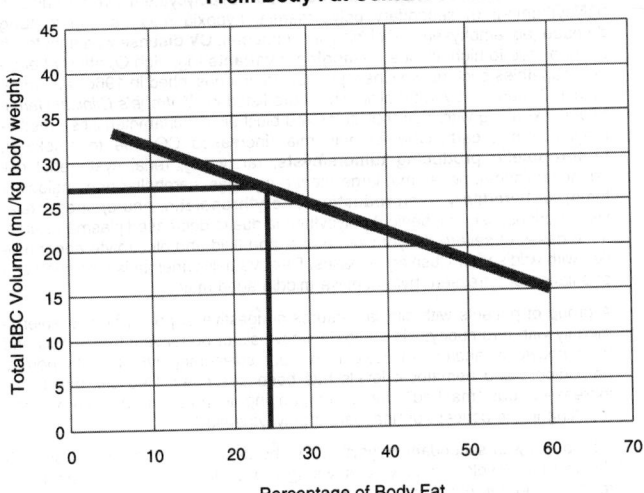

After measuring the percentage of fat, a line is drawn vertically from that number along the x-axis until it intersects the slope. At the intersect with the slope, a horizontal line is drawn to the y-axis, which gives a reading of the normalized TRCV.

Adapted from Berlin NI and Lewis SM, "Measurement of Total RBC Volume Relative to Lean Body Mass for Diagnosis of Polycythemia," *Am J Clin Pathol*, 2000, 114:922-6.

Use Determine red cell mass; support the diagnosis of polycythemia vera; monitor therapy with antineoplastic drugs
(Continued)

Red Cell Mass *(Continued)*

Limitations Radiation from isotopes used in bone scans, liver scans, brain scans, and other isotope procedures may interfere with red cell mass determination. Application of "normal range" tables based on weight and height must be made cautiously. An individual patient may not be comparable to the normal (eg, severely edematous individuals). Dilutional relationship will be unrepresentative and RBC mass overestimated. Severe edema might also be associated with isotope loss but is especially problematic due to unrepresentative normal range comparisons. Test is expensive and time consuming.

Contraindications Severe active bleeding

Additional Information The assessment of anemia and polycythemia (assessment of whether or not one of these conditions truly exists) depends foremost upon a reliable and direct determination of red cell volume. RBC count, Hb level, and Hct provide only concentration parameters, the measured number or amount relative to the solution in which it exists. In a number of clinical situations (eg, acute blood loss) the RBC count, Hb, and Hct will not reflect the actual decrease or increase in circulating red cell mass. Components (RBC count, etc) do correlate with RBC volume. Due to technical complexity (resulting in high cost and prolonged turnaround time) CBC is usually used, especially for follow-up or monitoring situations, even though RBC mass study would provide a more meaningful result. Nevertheless, some clinical situations (eg, polycythemia, complicated fluid and electrolyte management problems) will benefit from at least initial red cell volume determination.

Causes of decreased red cell volume include anemia, nutritional (iron, B_{12}, folate deficit, etc), hemolytic (intravascular or extravascular hemolysis); production deficit (marrow failure, drug or chemical related); acute and/or chronic blood loss; acute blood loss (decreased RBC volume may be present with normal or increased Hb/Hct/RBC count); chronic disease (inflammation/infection); radiation, starvation, or severe edema.

Causes of increased red cell volume include: a) polycythemia vera ("primary" polycythemia); b) secondary polycythemia, hypoxia may be due to **lung disease** (eg, emphysema, Pickwickian syndrome), **CV disease** with right to left shunt or due to high altitude, **hemoglobin variants** (eg, high O_2 affinity hemoglobinopathies such as Hb Chesapeake, the first described in 1966, with more than 110 reported by 1996 of which 38 are listed in *Wintrobe's Clinical Hematology* text along with their molecular and biochemical characteristics), methemoglobinemia, carboxyhemoglobinemia (increased CO due to smoking), **erythropoietin producing tumors/cysts**, rarely (eg, renal cyst, renal cell adenocarcinoma, hepatoma, large uterine myomas, cerebellar hemangioblastoma), or **hereditary overproduction of erythropoietin**; and c) stress (relative, spurious, pseudo, benign) polycythemia due to decreased plasma volume, as in cases of severe dehydration, burns, and fluid and electrolyte abnormalities with Addison or Cushing diseases. Relative polycythemia is also seen in a population of "stressed" hypertensive middle aged males.

A group of patients with clinical features suggestive of polycythemia (splenomegaly with thrombocytosis and/or leukocytosis or portal vein thrombosis and thrombocytosis and/or leukocytosis without splenomegaly) but with normal hemoglobin and hematocrit levels has been described. Red cell mass was increased, but "masked", by accompanying increase in plasma volume resulting in the concept of "inapparent polycythemia."

Individuals with secondary polycythemia who descend from high altitude residence may develop hemolysis of young red cells (neocytolysis) apparently relating to erythropoietin suppression.

See also Blood Volume *on page 282.*

References

Lamy T, Devillers A, Bernard M, et al, "Inapparent Polycythemia Vera: An Unrecognized Diagnosis," *Am J Med*, 1997, 102(1):14-20.

Leslie WD, Dupont JO, and Peterdy AE, "Effect of Obesity on Red Cell Mass Results," *J Nucl Med*, 1999, 40(3):422-8.

Pearson TC, Guthrie DL, Simpson J, et al, "Interpretation of Measured Red Cell Mass and Plasma Volume in Adults: Expert Panel on Radionuclides of the International Council for Standardization in Haematology," *Br J Haematol*, 1995, 89(4):748-56.

Pollycove M and Tono M, "Blood Volume," *Diagnostic Nuclear Medicine*, 3rd ed, Volume 2, Chapter 42, Sandler MP, Coleman RE, Wackers FJT, et al, eds, Baltimore, MD: Lippincott Williams & Wilkins, 1996, 827-34.

Rice L, Ruiz W, Driscoll T, et al, "Neocytosis on Descent From Altitude: A Newly Recognized Mechanism for the Control of Red Cell Mass," *Ann Intern Med*, 2001, 134(8):652-6.

Spivak JL, "Polycythemia Vera: Myths, Mechanisms, and Management," *Blood*, 2002, 100(13):4272-90.

Tefferi A, "Polycythemia Vera: A Comprehensive Review and Clinical Recommendations," *Mayo Clin Proc*, 2003, 78(2):174-94.

- ◆ **Red Cell Survival** *see* [51]Cr Red Cell Survival *on page 476*
- ◆ **Red Cell Volume** *see* Blood Volume *on page 282*
- ◆ **Red Cell Volume** *see* Red Cell Mass *on page 1144*
- ◆ **Red Devils** *see* Barbiturates, Quantitative, Serum or Plasma *on page 248*
- ◆ **Red Man Syndrome** *see* Vancomycin, Serum *on page 1298*
- ◆ **Reduced Hemoglobin** *see* Methemoglobin, Whole Blood *on page 903*

Reducing Substances, Stool

Related Information

d-Xylose Absorption Test, Serum, Urine *on page 527*
Lactose Tolerance Test *on page 829*
Meat Fibers, Stool *on page 894*
pH, Stool *on page 1034*

Synonyms Stool Reducing Substances

Applies to Breath Hydrogen Analysis; $^{13}CO_2$ Breath Test

Test Includes Stool weight, stool pH, and total reducing substances

Abstract Screening test for fecal reducing substance (eg, sugars) as an indication of disaccharidase (sucrase, lactase) deficiency and to assist in the differentiation of osmotic from secretory diarrhea

Specimen Fresh random stool

Collection Transport the specimen to the laboratory as soon as possible; delay may cause falsely low results.

Storage Instructions Freeze specimen if testing is delayed.

Causes for Rejection Specimen collected in a diaper or other absorbent surface

Reference Interval Normal: <2 mg/g stool; borderline: between 2-5 mg/g stool; abnormal: >5 mg/g stool. Even though premature infants have relative lactase deficiency and pancreatic insufficiency, there is evidence that older (32 weeks gestation) prematures have insignificant fecal loss of intact carbohydrate.

Use Detect deficiency of intestinal border enzymes, primarily sucrase and lactase (disaccharidases) due to congenital deficiency or nonspecific mucosal injury

Limitations Bacterial fermentation may give falsely low results if specimen is not analyzed within 1 hour. In the neonatal period, high Clinitest® results may be observed.

Methodology Clinitest® performed on a 1:3 dilution of supernatant from a diarrheal stool. A result >200-250 mg/dL of stool may be considered abnormal (normal is <2 mg/g of stool).

Additional Information Sugars should be rapidly absorbed in the upper small intestine. If not, however, they remain in the intestine and cause osmotic diarrhea by the osmotic pressure of the unabsorbed sugar in the intestine, drawing fluid and electrolytes into the gut. Carbohydrate malabsorption is a major cause of the watery diarrhea and electrolyte imbalance seen in patients with the short bowel syndrome. As a result of bacterial fermentation, the stools become acid with a high concentration of lactate. The pH measurement reflects this process. The unabsorbed sugars are measured as reducing substances. Although sucrose is not a reducing sugar, it is subjected to acid hydrolysis in the gut, and thus, is also measured as a reducing substance. The presence of excess fecal reducing substances in infants and children admitted with watery diarrhea is reported as a sensitive and specific indicator favoring osmotic over secretory diarrhea.

Idiopathic lactase deficiency is common. Lactase activity declines with age in humans and is controlled genetically.

The **breath hydrogen analysis** test may provide more definitive information. A positive breath hydrogen result (as with lactose intolerance) will exceed 20

(Continued)

Reducing Substances, Stool *(Continued)*

parts (hydrogen) per million. The $^{13}CO_2$ **breath test** is reportedly more sensitive and specific than the H_2 test in the detection of low jejunal lactase activity. A ^{13}C-lactose digestion test has shown (on the basis of increase in the plasma ^{13}C-glucose concentration after consumption of ^{13}C-lactose) a higher prevalence of lactose maldigestion (Caucasian population) than indicated by the H_2 breath test. There is considerable variation in lactose tolerance between lactase deficient subjects. Classically, stools from patients with disaccharidase deficiency are liquid, acid, and frothy in appearance.

References

Castro-Rodríguez JA, Salazar-Lindo E, and León-Barúa R, "Differentiation of Osmotic and Secretory Diarrhoea by Stool Carbohydrate and Osmolar gap Measurements," *Arch Dis Child*, 1997, 77(3):201-5.

Koetse HA, Stellaard F, Bijleveld CM, et al, "Noninvasive Detection of Low-Intestinal Lactase Activity in Children by Use of a Combined 13CO2/H2 Breath Test," *Scand J Gastroenterol*, 1999, 34(1):35-40.

Matthews SB and Campbell AK, "When Sugar Is Not So Sweet," *Lancet*, 2000, 355(9212):1330.

Reducing Substances, Urine

Related Information

Ammonia, Plasma *on page 150*
Galactose-1-Phosphate Uridyl Transferase, Blood *on page 628*
Glucose, Fasting, Plasma *on page 643*
Glucose, Postglucose Load, Plasma *on page 647*
Glucose, Quantitative, Urine *on page 649*
Glucose, Random, Plasma *on page 649*
Glucose, Semiquantitative, Urine *on page 650*
Glucose Tolerance Test, Plasma *on page 651*
Ketone Bodies, Blood *on page 816*
Ketones, Urine *on page 817*
Newborn Screening Tests for Galactosemia *on page 960*

Synonyms Clinitest® for Sugar, Urine; Copper Reduction Tablet Test

Applies to Benedict Test

Test Includes Clinitest® testing of urine

Abstract Many carbohydrates (including galactose, lactose, and fructose), some amino acids, and a few other abnormal urinary substances reduce copper and are referred to as reducing substances.

Specimen Random urine

Container Plastic urine container

Storage Instructions If the specimen cannot be processed promptly by the laboratory, it should be refrigerated.

Reference Interval None detected

Use This test is no longer used to diagnose or monitor patients with diabetes mellitus. The differential diagnosis of a positive test for reducing substances in urine includes fructose, lactose, galactose, maltose, arabinose, xylose, ribose, uric acid, ascorbic acid, creatinine, cysteine, ketones, sulfanilamide, oxalic acid, hippuric acid, homogentisic acid, glucuronic acid, formaldehyde, isoniazid, salicylates, cinchophen, and salicyluric acid, as well as glucose. In practice, the major use of this test is to screen infants for hereditary galactosemia.

Methodology Copper sulfate reacts with reducing substances in urine, converting cupric sulfate to cuprous oxide. The lower limit of glucose detection by Clinitest® is 200 mg glucose/dL (SI: 11.1 mmol/L). Semiquantitation is accomplished by comparison of the color generated with a reference chart. Semiquantitation of urine glucose is also readily accomplished by urine dipstick. This reaction is also utilized in detection of reducing substances in stool.

Additional Information Positive urine samples from sick newborns and children that are negative for glucose can be further confirmed by carbohydrate chromatography. In this setting, even trace positive Clinitest® results should be recognized as abnormal.

References

Gilbert-Barness E and Barness LA, *Metabolic Diseases: Foundations of Clinical Management, Genetics, and Pathology*, Natick MA, ed, Eaton Publishing, 2000, 185-92.

Lindor NM and Karnes PS, "Initial Assessment of Infants and Children With Suspected Inborn Errors of Metabolism," *Mayo Clin Proc*, 1995, 70:987-8.

Sacks DB, "Carbohydrates," *Tietz Textbook of Clinical Chemistry*, 3rd ed, Burtis CA and Ashwood ER, eds, Philadelphia, PA: WB Saunders Co, 1999, 750-808.

- ◆ **Reference Interval** *see page 11*
- ◆ **Reference Range** *see page 11*
- ◆ **Refractive Index, Urine** *see* Specific Gravity, Urine *on page 1216*
- ◆ **Regan Isoenzyme** *see* Alkaline Phosphatase, Heat Stable, Serum *on page 125*
- ◆ **Regan Isoenzyme** *see* Alkaline Phosphatase Isoenzymes, Serum *on page 125*
- ◆ **Regional Lymph Node Dissection** *see* Sentinel Lymph Node Biopsy *on page 1189*
- ◆ **Rejection Criteria, Specimen** *see* Specimen Rejection Criteria *on page 1219*
- ◆ **Relative Frequency** *see page 11*
- ◆ **Renal Biopsy** *see* Kidney Biopsy *on page 818*
- ◆ **Renal Cyst Fluid Cytology** *see* Cyst Fluid Cytology *on page 490*
- ◆ **Renal Mass Aspiration** *see* Fine Needle Aspiration, Deep and Superficial Masses *on page 590*
- ◆ **Renin** *see* Aldosterone, Serum or Plasma *on page 122*
- ◆ **Renin** *see* Aldosterone, Urine *on page 124*

Renin Activity, Plasma

Related Information

Adrenal Cortex: Laboratory Assessment Overview *on page 110*
Aldosterone, Serum or Plasma *on page 122*
Aldosterone, Urine *on page 124*
B-Type Natriuretic Peptide *on page 314*
Electrolyte Panel, Serum *on page 532*
Potassium, Serum or Plasma *on page 1078*
Potassium, Urine *on page 1080*
Sodium, Serum or Plasma *on page 1210*

Applies to Angiotensin; Captopril Test

Test Includes Fasting supine or upright specimens, catheterization studies

Abstract Renin catalyzes the conversion of angiotensinogen to angiotensin I, which is, in turn, converted to angiotensin II, a biologically active peptide, which both 1) stimulates adrenocortical secretion of aldosterone, and 2) has direct vasopressor activity. Clinical interest in measuring plasma renin (PR) centers on patients who have aldosterone excess. There are two types of aldosterone excess: 1) **primary hyperaldosteronism** (Conn syndrome) in which the aldosterone excess is autonomously produced by an adrenal adenoma or hyperplasia, and 2) **secondary hyperaldosteronism** in which the increased aldosterone is a physiological response to a disease process such as cardiac failure, cirrhosis, renovascular hypertension, a renin-secreting tumor, diuretic medication, or protracted vomiting. In primary hyperaldosteronism PR is characteristically low, while in secondary hyperaldosteronism PR is characteristically high. Interpretation of a PR result is difficult because 1) some assays are indirect and therefore nonspecific, 2) many preanalytic variables affect renin production (sodium balance, posture, medications), and 3) the circadian variation in renin production (maximum in early morning, minimum in late afternoon). Renin secretion is stimulated by upright posture, low sodium intake, and diuretic medication. (See also Aldosterone, Serum or Plasma *on page 122*.) Renin and aldosterone concentrations with other studies, including especially serum/plasma potassium, are needed to evaluate the renin-angiotensin-aldosterone system; *vide infra*.

Patient Preparation Preanalytic variables which must be controlled are sodium balance, posture, blood pressure medications, and time of day. Samples for PR are commonly drawn at the end of a 24-hour collection of urine for sodium and creatinine and after several days of stable sodium intake controlled by the physician. Check with the laboratory for particular patient preparation instructions.

Specimen Plasma, peripheral venous blood, bilateral renal vein samples

Container Lavender top (EDTA) tube

(Continued)

Renin Activity, Plasma *(Continued)*

Sampling Time Timing of sampling (morning) and the posture of the patient before sampling (upright) require standardization. Maximum activity is found early in the morning, during sleep. Minimum renin activity occurs in the late afternoon.

Collection Draw specimen into a prechilled syringe. Place in chilled lavender top tubes (with the rubber stopper off). Recap, mix, and immediately place on ice and deliver to the laboratory. **Posture of the patient must be recorded.**

Storage Instructions Place in an ice-water bath. After the specimen is well cooled, centrifuge at 4°C. Separate plasma immediately and freeze in a plastic container. Avoid freeze-thaw cycles.

Causes for Rejection Clotted sample, patient preparation incorrect for the analysis needed; hemolysis

Reference Interval Method dependent, with large interlaboratory variation.
Indirect assay of angiotensin I:
Sodium replete:
- 18-39 years: <0.6-4.3 ng/mL/hour
- >40 years: <0.6-3.0 ng/mL/hour

Sodium depleted:
- 18-39 years: 2.9-24.0 ng/mL/hour
- >40 years: 2.9-10.8 ng/mL/hour

Direct immunoassay of active renin: see table.

Critical Values Ratio of plasma aldosterone to renin >30-50. A ratio >30 deserves further evaluation.

Use

Primary hyperaldosteronism: Basal PR is low and does not increase in response to normal physiologic stimuli (ie, volume depletion, hyponatremia, and upright posture). This can be tested for by:

1. Placing the patient on a low sodium diet for 5-7 days and measuring basal (8 AM) PR 2 hours after the patient has been upright and walking.
2. Same as above plus a diuretic (such as furosemide). This constitutes a maximum stimulation. Some patients with Conn syndrome may show a misleading increase in PR.
3. Failure of an elevated aldosterone value to be suppressed by a saline infusion.

When primary hyperaldosteronism is suspected, PR **must** be interpreted with concurrent serum aldosterone values and information about preanalytic variables. Serum or plasma electrolytes are needed.

Secondary hyperaldosteronism: Basal PR is high, reflecting a normal physiologic response to volume depletion or decreased effective renal blood flow.

Other hypertensive patients: Although it was once popular to classify all hypertensive patients into high-renin, normal-renin, or low-renin categories, this is rarely done today **except** when there is clinical suspicion of primary hyperaldosteronism as the cause.

Limitations Nonspecificity is a problem since many reference laboratories use an indirect assay, measuring the generation of angiotensin I, rather than renin itself. Since preanalytic variables are difficult to control, reference intervals are wide.

Methodology Immunoassay (various labels); an immunoradiometric assay specific for active renin utilizes two monoclonal antibodies.

References
Ganguly A, "Primary Aldosteronism," *N Engl J Med*, 1998, 339(25):1828-34.

Mayo Medical Laboratories, *2000 Test Catalogue*, Rochester, MN, 436.

Schrier RW and Abraham WT, "Hormones and Hemodynamics in Heart Failure," *N Engl J Med*, 1999, 341(8):577-84.

♦ **Repazine®** *see* Chlorpromazine, Serum *on page 395*

Renin: Direct Immunoassay
Reference Values for Healthy Adults

Range of Active Renin (ng/L)	Upright			Supine		
	Number of Persons	Frequency	Cumulative Frequency	Number of Persons	Frequency	Cumulative Frequency
<10	20	0.20	0.20	48	0.48	0.48
10-14	22	0.42	0.42	30	0.30	0.78
15-19	16	0.58	0.58	14	0.14	0.92
20-24	14	0.72	0.72	8	0.08	1.00
25-29	16	0.88	0.88			
30-34	4	0.92	0.92			
35-39	4	0.96	0.96			
>40	4	1.00	1.00			

Adapted from Simon D, Hartman DJ, Badouaille G, et al, "Two-Site Direct Immunoassay Specific for Active Renin," *Clin Chem*, 1992, 38:1959-62.

Reptilase® Time

Related Information

D-Dimers and Fibrin Degradation Products *on page 502*
Fibrinogen *on page 583*
Hypercoagulation Panel *on page 758*
Thrombin Time *on page 1246*

Abstract Clotting time similar to thrombin time except that a snake venom (Reptilase®) is used instead of thrombin. The Reptilase® time is prolonged by decreased or dysfunctional fibrinogen, or high levels of fibrin degradation products (FDP). Dysfibrinogenemia is an uncommon hereditary or acquired condition characterized by dysfunctional fibrinogen.

Specimen Plasma

Container One blue top (sodium citrate) tube

Collection Routine venipuncture. If multiple tests are being drawn, draw blue top tubes after any red top tubes but before any lavender top (EDTA), green top (heparin), or gray top (oxalate/fluoride) tubes. Immediately invert tube gently at least 4 times to mix. Tubes must be appropriately filled. Deliver tubes immediately to the laboratory.

Storage Instructions Separate plasma from cells as soon as possible. Plasma may be stored at room temperature or on ice for up to 8 hours; otherwise, store frozen.

Causes for Rejection Specimen received more than 4 hours after collection, tube not filled, clotted specimens

Turnaround Time Less than 1 day

Reference Interval 16-24 seconds

Use Performed with thrombin time to diagnose dysfibrinogenemia in patients undergoing evaluation for hypercoagulability and/or a bleeding tendency. Often performed only if an initial panel of tests excludes more common disorders, as dysfibrinogenemia is uncommon. Unlike the thrombin time, Reptilase® time is not prolonged by heparin or hirudin.

Methodology Reptilase® is added to patient plasma and the clotting time is measured in seconds. Reptilase® cleaves fibrinogen, releasing fibrinopeptide A from fibrinogen and converting fibrinogen into fibrin clot. In contrast, when thrombin cleaves fibrinogen, fibrinopeptide A and fibrinopeptide B are both released from fibrinogen.

Additional Information Many different mutations are known to cause hereditary dysfibrinogenemia. Dysfibrinogenemia mutations can cause bleeding, thrombosis, or both, or they may be clinically asymptomatic. If bleeding is present, it is usually mild, but severe bleeding has been reported. Dysfibrinogenemia has an estimated prevalence of 0.8% in patients with venous thrombosis. Arterial thrombosis is less frequent than venous thrombosis in these patients. Most patients with hereditary dysfibrinogenemia are heterozygous. Rare homozygous cases have been reported. The Reptilase® time and thrombin time, which measure the clotting time during the conversion of fibrinogen into fibrin, are often prolonged in dysfibrinogenemia because fibrinogen is dysfunctional. Assays that measure fibrinogen function show lower levels than assays that measure fibrinogen quantity (immunological or "antigen" assays), because fibrinogen function is impaired but fibrinogen quantity is not. The PT and PTT may be prolonged in dysfibrinogenemia. Causes of acquired dysfibrinogenemia include liver disease, hepatoma, or acute phase reactions with generation of high levels of fibrinogen. The bleeding and thrombosis risk with acquired dysfibrinogenemia is uncertain. Prolongation of the thrombin time and Reptilase® time has been commonly observed with amyloidosis due to inhibition of fibrinogen conversion to fibrin.

References

Galanakis DK, "Fibrinogen Anomalies and Disease. A Clinical Update," *Hematol Oncol Clin North Am*, 1992, 6(5):1171-87.

Gastineau DA, Gertz MA, Daniels TM, et al, "Inhibitor of the Thrombin Time in Systemic Amyloidosis: A Common Coagulation Abnormality," *Blood*, 1991, 77(12):2637-40.

Haverkate F and Samama M, "Familial Dysfibrinogenemia and Thrombophilia. Report on a Study of the SSC Subcommittee on Fibrinogen," *Thromb Haemost*, 1995, 73(1):151-61.

♦ **Resin Triiodothyronine Uptake** *see* T₃ Uptake, Serum or Plasma *on page 1233*

♦ **Resin Uptake Ratio** *see* T₃ Uptake, Serum or Plasma *on page 1233*

♦ **Respiratory Acidosis / Alkalosis** *see* pCO₂, Blood *on page 1008*

Respiratory Syncytial Virus Antigen Detection

Related Information

Respiratory Syncytial Virus Culture and Serology *on page 1153*
Viral Culture *on page 1307*
Virus, Direct Detection by Fluorescent Antibody *on page 1311*

Test Includes Detection of RSV antigen in specimens using enzyme immunoassay (EIA) or immunofluorescence (IFA)

Abstract Respiratory syncytial virus (RSV), followed by parainfluenza viruses, is the leading cause of hospitalization for recognizing illness in small children. Clinical presentations due to RSV include bronchiolitis, pneumonia, and tracheobronchitis. Hypoxemia is common.

Specimen Nasopharyngeal secretions (nasal washings or aspirates) or nasopharyngeal swab

Collection Swabs must be placed in cold viral transport medium. Soft catheters and suction devices can be used to collect nasal secretions. Nasal washings are done by carefully introducing sterile saline (3-7 mL) into the nasal cavity and aspirating the fluid. **Do not use cold sterile saline when aspirating samples. Warm to room temperature.**

Storage Instructions Can be transported at room temperature without loss of viral antigens.

Turnaround Time A few hours to 1 day

Critical Values Positive results

Use Rapid diagnosis of RSV, especially in children and in immunocompromised patients

Limitations Specimens with <100 cells on the slide will yield an insensitive result when detecting RSV by immunofluorescence; immunofluorescence requires experienced personnel to read results.

Methodology Enzyme-linked immunosorbent assay (ELISA), direct fluorescent antibody (DFA). Positive results are enhanced by analysis of specimens from bronchopulmonary lavage.

Additional Information RSV direct EIAs have become the most frequently used method of detecting RSV infections. RSV is very labile and cell culture infectivity is lost rapidly when the virus is kept at 37°C. The detection of RSV antigen allows for detection of virus in specimens in which virus is not culturable. EIA and IFA are considered more sensitive than culture. Specimen handling requirements are not as stringent as those required for culture. The EIAs used to detect RSV antigen are simple, objective, and quick.

References

Baker KA and Ryan ME, "RSV Infection in Infants and Young Children. What's New in Diagnosis, Treatment, and Prevention?" *Postgrad Med*, 1999, 106(7):97-9, 103-4, 107-8 passim.

Hall CB, "Respiratory Syncytial Virus: A Continuing Culprit and Conundrum," *J Pediatr*, 1999, 135(2 Pt 2):2-7.

Hall CB, "Respiratory Syncytial Virus and Parainfluenza Virus," *N Engl J Med*, 2001, 344(25):1917-28.

Internet Web Sites

www.astdhpphe.org/infect/rsv.html
www.cdc.gov/ncidod/dvrd/revb/

Respiratory Syncytial Virus Culture and Serology

Related Information

Respiratory Syncytial Virus Antigen Detection *on page 1153*
Viral Culture *on page 1307*
Virus, Direct Detection by Fluorescent Antibody *on page 1311*

Test Includes Conventional culture or rapid culture using specific monoclonal antibodies for respiratory syncytial virus (RSV). Serology detects antibodies specific for RSV.

Abstract Respiratory syncytial virus (RSV) is the most common viral agent causing infant lower respiratory illnesses. By the first year of life, 50% of infants have experienced an RSV infection. Common symptoms are fever, wheezing, lower respiratory tract congestion, cough, and rhinorrhea. In healthy adult individuals, mortality rates due to RSV are relatively low.

In a primary infection, RSV-specific IgM appears after 5-8 days and persists for several weeks. Detectable increases in RSV-specific IgG antibody titers occur 2-4 weeks after symptomatic infections.

(Continued)

Respiratory Syncytial Virus Culture and Serology
(Continued)

Specimen Throat or nasopharyngeal swab, nasopharyngeal washes and secretions are needed for viral culture. Sputum is not an appropriate specimen for culture of RSV. Serology requires serum.

Container Culture requires cold viral transport medium; serology requires red top tube

Sampling Time Culture specimens should be collected during acute infection. Serology requires acute and convalescent specimens.

Collection Place swabs into cold viral transport medium and keep cold. Infants and small children: soft catheters and suction devices (syringes and suction bulbs) can be used to collect nasal secretions from far back in the nose (best specimens). Another excellent method is to introduce 3-7 mL of sterile saline into the child's posterior nasal cavity and immediately aspirate the fluid. **Do not use cold sterile saline when aspirating samples. Warm to room temperature.**

Storage Instructions Respiratory syncytial virus is extremely labile and will lose as much as 90% of cell culture infectivity when left at room temperature for 24 hours. **Do not freeze specimens at -20°C.** Send specimens to the laboratory **as soon as possible.** Although less than optimal conditions, specimens can be stored up to 48 hours at 4°C. If necessary specimen can be quickly frozen at -70°C, but freezing will cause loss of infectivity.

Turnaround Time Conventional culture: 1-14 days; rapid culture: 1-2 days. Serology may take up to 1 week after receipt of both acute and convalescent serum.

Reference Interval No virus isolated in culture. IgG: less than fourfold rise in titer; IgM: negative

Critical Values Isolation of RSV

Use Culture supports diagnosis of respiratory disease caused by respiratory syncytial virus. Serology detects specific RSV antibody in patients suspected of having RSV.

Limitations Rapid culture methods will only detect specified virus(es), negative culture does not rule out RSV infection. RSV is a very thermolabile virus and may not survive transport to the laboratory or extreme conditions. Thus, false-negative cultures occur. Serologic tests have limited use since young infants have maternal IgG antibody and may have a false positive. A fourfold rise in IgG titers cannot be detected in half of the children younger than 6 months of age. Enzyme immunoassays may use different antigen sources. No serologic assay has been standardized for detection of RSV infection.

Methodology Culture uses the inoculation of specimen into cell cultures, incubation and observation for characteristic cytopathic effect. The use of a rapid shell vial culture technique is reported to yield positive culture results overnight. Serology uses complement fixation (CF), enzyme immunoassay (EIA), virus-neutralization assay, or immunofluorescence assay.

Additional Information RSV is a common cause of acute respiratory disease, including serious disease among older persons. In healthy individuals, the mortality rate due to RSV infection is low; however, in patients with respiratory or cardiac compromise, immune dysfunction, and the elderly, the mortality rate is high (~37%).

Many laboratories offer enzyme immunoassay (EIA) tests for the direct detection of RSV in patient nasopharyngeal swab specimens. In general, these tests are very rapid, sensitive, and specific. Diagnosis of RSV by RSV-specific IgG depends on demonstration of a rise in antibody titer over a 2- to 3-week period. As such, the test is seldom useful in planning clinical care in an acute illness or in control of nosocomial infections.

References
Choy G, "A Review of Respiratory Syncytial Virus Infection in Infants and Children," *Home Care Provid*, 1998, 3(6):306-11.

Jones BL, Clark S, Curran ET, et al, "Control of an Outbreak of Respiratory Syncytial Virus Infection in Immunocompromised Adults," *J Hosp Infect*, 2000, 44(1):53-7.

Simoes EA, "Respiratory Syncytial Virus Infection," *Lancet*, 1999, 354(9181):847-52.

Internet Web Sites
www.astdhpphe.org/infect/rsv.html
www.cdc.gov/ncidod/dvrd/revb/

◆ **Resposan-10®** *see* Chlordiazepoxide, Serum *on page 391*

◆ **Retard®** *see* Procainamide, Serum *on page 1092*

◆ **RET Gene Testing** *see* Multiple Endocrine Neoplasia/Familial Medullary Thyroid Carcinoma *on page 924*

Reticulated Platelet Count

Related Information

Flow Cytometry, Overview *on page 592*
Platelet Aggregation *on page 1045*
Platelet Count *on page 1050*
Platelet Sizing *on page 1056*

Abstract Results of this test (which is performed on peripheral blood) reflect bone marrow thrombopoietic activity. While requiring use of a flow cytometer, the test has been proposed as a simple and rapid means to distinguish between thrombocytopenia due to excess platelet consumption vs inadequate platelet production. Currently, lack of discriminate ability in patients with platelet counts <50,000/μL and lack of RNA staining specificity may limit clinical application of this test.

Specimen 2 mL EDTA or sodium citrate anticoagulated whole blood; 1 mL anticoagulated blood is sufficient for most flow cytometer based methods (only about 2 μL of blood is required for each pass through the cytometer).

Storage Instructions If test is not performed on the sample (kept at room temperature) the same day of collection, use of 1% formaldehyde fixed platelet-rich plasma sample is recommended (stable for up to 7 days in buffer at 4°C).

Turnaround Time 1-2 hours

Reference Interval Method dependent; varies in adults from 1% to 15%

Gestational age (in %, mean ±SD):
- >36 weeks: 4.0 ±2.4
- 30-36 weeks: 4.6 ±1.7
- <30 weeks: 8.8 ±5.1

Use The percentage of reticulated platelets (% RPLT) in peripheral blood, the platelet count, and absolute reticulated platelet count reflect bone marrow megakaryocyte activity (to produce platelets). These measurements differentiate thrombocytopenia due to bone marrow failure (decreased platelet production) from increased platelet consumption (destruction). The % of RPLT has been used in the study of platelet engraftment times following peripheral blood progenitor cells compared with bone marrow transplant, in the evaluation of post-transplant thrombocytopenia, (potentially) in the timing of thrombopoietin therapy, and in the evaluation of the effects of drugs and growth factors on thrombopoiesis.

Limitations Requires use of a flow cytometer and cytometer knowledgeable personnel. May require use of formalin-fixed platelets if analysis cannot be performed within 24 hours. With platelet counts <50,000 μL, normal or decreased reticulated platelet level may not reflect decreased thrombopoiesis.

Methodology Platelets, resuspended in buffer, are stained with the RNA fluorochrome, thiazole orange (TO), and enumerated by a fluorescence-assisted cell analyzer (flow cytometer). The "platelet cloud" includes platelets, red blood cells, and cellular fragments that can have overlapping light scatter profiles. A dual-staining (two color) method has been described that circumvents problems occurring in samples that have a complex "platelet cloud".

Additional Information Reticulated platelets are analogous to reticulated red blood cells (reticulocytes). As such, they are young, ribonucleic acid (RNA) rich and larger than normal adult platelets. The RNA fluorochrome thiazole orange (TO) used to tag these young reticulated platelets renders them highly fluorescent, detected and counted by a flow cytometer, the original method utilizing laser emission at 488 nm (for excitation) and a 530/30 nm band pass filter for fluorescence detection. TO is the same dye used for erythrocyte reticulocyte determination.

Clinical utility of the reticulated platelet level is limited by the observation that at platelet counts <50,000 μL, normal or decreased absolute reticulated platelet levels may not be indicative of decreased thrombopoiesis. As the platelet count falls ≤50,000, platelet survival may be 3 days or less, reticulated platelets account for most of the circulating platelet mass. Reticulated platelets would (Continued)

Reticulated Platelet Count (Continued)

then be the major targets of destruction, and with marrow thrombopoiesis at maximum levels, a decrease in the absolute level of reticulated platelets would result (at platelet count levels <20,000 μL). Under these circumstances, a decrease in the absolute level of reticulated platelets might not be indicative of decreased thrombopoiesis.

Reticulated platelet count methodology is also complicated by recent studies showing that TO labeling is not entirely mRNA specific. TO-positive platelets are lower than normal in dense granule-deficient platelets (Hermansky-Pudlak syndrome and storage pool disease) and are at about the same level as TRAP degranulated platelets (TRAP is a nonenzymatic platelet agonist thrombin receptor activating peptide). The reticulated platelet count may not provide significant clinical advantage over the platelet parameter MPV (mean platelet volume), which is provided readily and at low cost by many automated cell counters.

Reticulated platelets are increased in hyperthyroidism, likely reflecting increase in thrombopoiesis (see Platelet Sizing on page 1056).

References

Peterec SM, Brennan SA, Rinder HM, et al, "Reticulated Platelet Values in Normal and Thrombocytopenic Neonates," J Pediatr, 1996, 129(2):269-74.

Robinson MS, Mackie IJ, Khair K, et al, "Flow Cytometric Analysis of Reticulated Platelets: Evidence for a Large Proportion of Nonspecific Labeling of Dense Granules by Fluorescent Dyes," Br J Haematol, 1998, 100(2):351-7.

Robinson MS, Mackie IJ, Machin SJ, et al, "Two Colour Analysis of Reticulated Platelets," Clin Lab Haematol, 2000, 22(4):211-3.

Stiegler G, Stohlawetz P, Brugger S, et al, "Elevated Numbers of Reticulated Platelets in Hyperthyroidism: Direct Evidence for an Increase of Thrombopoiesis," Br J Haematol, 1998, 101(4):656-8.

Reticulocyte Count

Related Information

Anemia Flowchart on page 35
⁵¹Cr Red Cell Survival on page 476
Erythropoietin, Serum on page 551
Flow Cytometry, Overview on page 592
Heinz Body Stain on page 669
Hematocrit on page 674
Hemoglobin on page 681
Hemoglobin Electrophoresis on page 684
Osmotic Fragility on page 980
Red Blood Cell Enzyme Deficiency on page 1134
Reticulocyte Hemoglobin Content on page 1158
Sickle Cell Tests on page 1195

Applies to Reticulocyte Maturity Index

Aftercare Reticulocytes are young red blood cells from which the nucleus has been extruded but which still contain some remnants of ribosomal ribonucleic acid. Ribosomes will react with some basic dyes - vital stains (azure B, brilliant cresyl blue, methylene blue) with the formation of a granular and/or filamentous blue precipitate. The number of reticulocytes present in the peripheral blood relates to erythropoietic activity occurring normally largely in the bone marrow. Recently available automated analyzers determine reticulocyte maturity by measuring fluorescence intensity. **Highly fluorescent reticulocytes** (HFR) percentage allow derivation of the **reticulocyte maturity index** (RMI), an early predictor of bone marrow regenerative activity, of use in following the results of bone marrow transplantation.

Specimen Whole blood

Container Lavender top (EDTA) tube or green top (heparin) tube for venipuncture specimen; heparinized capillary tube for capillary specimen

Storage Instructions Store EDTA anticoagulated blood at room temperature for up to 48 hours or up to 72 hours at 4°C. Immature reticulocyte fraction (IRF) parameters using the ABX-Pentra 120 Retic are reported as stable for only 8 hours at 4°C and 6 hours at room temperature.

Causes for Rejection Clotted or hemolyzed specimen

Reference Interval Adults: 0.5% to 1.5%; newborns: ≤7%, expressed as a percentage of 1000 RBCs. Normal values at birth (by flow cytometry): 1.6% to 8.3%, mean of 5.3, falling to normal adult level by the end of the second week.

The elderly (older than 70 years of age) have a slightly higher percent of reticulocytes than young individuals but still fall within the normal range.

Absolute reticulocyte count: 10-80 x 10^9/L; reticulocytes: >100 x 10^9/L indicates increased erythropoiesis.

Use Evaluate erythropoietic activity. Increased in acute and chronic hemorrhage, hemolytic anemias. Evaluate erythropoietic response to therapy of various anemias. **The test is underutilized, especially when one considers it is at a pivotal decision-making juncture.** The reticulocyte production index will decide if one is working with a hyperproliferative or nonproliferative anemia, and thus, which tests should be subsequently ordered.

The RMI has application as a criterion for the success of marrow engraftment following bone marrow transplantation. The HFR fraction appears to be the earliest and most sensitive index of engraftment. HFR might also be used as a marker of early response to immunosuppression in the therapy of severe aplastic anemia.

Limitations In recently transfused patients, reticulocytes may decrease on a dilutional basis due to transfusion. Automated flow cytometry methods using thiazole orange may give spuriously high counts in patients with chronic lymphocytic leukemia, Howell-Jolly bodies, intracellular parasites (including malarial parasites as a cause of pseudoreticulocytosis), large platelets, some drug therapies, erythropoietic protoporphyria, and with cold agglutinins (see NCCLS Document H44-P). Hb H may cause high or low interference (time dependent) in automated analyzers using new methylene blue. Increase in incubation time from 60-180 minutes results, in some cases of Hb H disease, in falsely elevated reticulocyte counts. This problem is most likely to present with patients of Southeast Asian origin.

Contraindications Patients receiving a large number of blood transfusions

Methodology Manual methods (new methylene blue is commonly used) lack precision with coefficient of variation averaging 30%. Flow cytometric and other methods using new methylene blue (Coulter STK-S), thioflavin T, thiazole orange, auramine O, acridine orange, ethidium bromide, pyronine Y, or Oxazine 750 (Technicon® H3 - Bayer® Corporation) have the advantage of reproducibility. The measured fluorescence is proportional to the amount of RNA in the cell allowing derivation of the reticulocyte maturation index. The assay is a standard part of the repetoire of some multiparameter hematology instruments. Flow cytometry allows rapid analysis of 20,000-50,000 red cells per sample with resultant markedly improved reproducibility/precision as compared to manual methods. The Technicon® H3 combines routine CBC with 5-part differential analysis and reticulocyte (r) parameters including r count, and r cellular indices (r mean cell volume, r mean cell hemoglobin concentration, r cell hemoglobin content, and their distribution widths).

Additional Information Demonstration of an increase in the number of circulating reticulocytes provides reliable and inexpensive evidence of increased red cell production. Care should be exercised during interpretation of results that an apparent increase in reticulocytes is not the result of decrease in the number of nonreticulated RBCs (ie, anemia with fewer mature red cells). A variety of corrections have been proposed and are in use. **Absolute reticulocyte count** = reticulocytes (%) x RBC count. This gives the number of reticulocytes per mm³ of blood. **Reticulocyte index** (RI) = reticulocytes (%) x patient Hct/normal Hct or patient RBC/normal RBC or patient Hb/normal Hb. This corrects the reticulocyte count for anemia. **Reticulocyte production index** (RPI) = RI x (1/maturation time), or RPI = patient's absolute reticulocyte count/ normal absolute reticulocyte count x (1/maturation time). Maturation time is usually taken as 2. RPI corrects for the premature release of reticulocytes from the marrow as might occur in cases of brisk hemolysis or significant bleeding. RPI gives a reticulocyte percent value that reliably estimates RBC production.

Current generation automated methods derive a reticulocyte maturity index (RMI) from study of reticulocyte RNA levels. Absolute reticulocyte count with RMI can be applied to classification and evaluation of anemia. Failure of marrow production results in anemia with absence of the expected increase in RPI.

See reference by d'Onofrio, et al, for an excellent review of the numerous aspects of reticulocyte counting (including manual and automated/flow

(Continued)

Reticulocyte Count (Continued)

cytometric methods), and immature reticulocyte fractions (including premature "shift" or "stress macroreticulocytes," with relation to F reticulocytes). Clinical applications are discussed, including monitoring use of human recombinant erythropoietin (rhu-EPO) and evaluation of the microhemolytic anemia occurring with intense exercise/sport activity. Measurement of reticulocyte parameters may assist in detecting illegal use of the rhu-EPO in athletes.

References

Buttarello M, Bulian P, Farina G, et al, "Flow Cytometric Reticulocyte Counting: Parallel Evaluation of Five Fully Automated Analyzers: An NCCLS-ICSH Approach," *Am J Clin Pathol*, 2001, 115(1):100-11.

d'Onofrio G, Zini G, and Rowan RM, "Reticulocyte Counting: Methods and Clinical Applications," *Advanced Laboratory Methods in Haematology*, Chapter 5, Rowan RM, van Assendelft OW, and Preston FE, eds, London, UK: Arnold, 2002, 78-126.

Grotto HZ, Vigoritto AC, Noronha JF, et al, "Immature Reticulocyte Fraction as a Criterion for Marrow Engraftment. Evaluation of a Semiautomated Reticulocyte Counting Method," *Clin Lab Haematol*, 1999, 21(4):285-7.

Lacombe F, Lacoste L, Vial JP, et al, "Automated Reticulocyte Counting and Immature Reticulocyte Fraction Measurement. Comparison of ABX PENTRA 120 Retic, Sysmex R-2000, Flow Cytometry, and Manual Counts," *Am J Clin Pathol*, 1999, 112(5):677-86.

Lai SK, Yow CMN, and Benzie IFF, "Interference of Hb H Disease in Automated Reticulocyte Counting," *Clin Lab Haematol*, 1999, 21(4):261-4.

Luzzatto L and Gordon-Smith EC, "Inherited Hemolytic Anemias," *Postgraduate Hematology*, 4th ed, Hoffbrand AV, et al, eds, Oxford, England: Butterworth-Hermenian, 1999, 120-43.

Reticulocyte Hemoglobin Content

Related Information

Flow Cytometry, Overview *on page 592*
Hematocrit *on page 674*
Hemoglobin *on page 681*
Iron and Total Iron Binding Capacity/Transferrin, Serum *on page 807*
Iron Stain, Bone Marrow *on page 810*
Red Blood Cell Indices *on page 1136*
Reticulocyte Count *on page 1156*
Transferrin Receptor, Soluble, Serum or Plasma *on page 1267*

Synonyms CHr

Abstract Reticulocyte hemoglobin content (CHr), one of a number of reticulocyte measurements available from automated flow cytometry based hematology analysis, may have application in the early detection and evaluation of iron deficient erythropoiesis and in monitoring the effectiveness of iron replacement therapy.

Specimen Whole blood

Container Lavender top (EDTA) or green top (heparin) tube

Storage Instructions Refrigerate specimen immediately after collection.

Causes for Rejection Clotted or hemolyzed specimen

Reference Interval CHr of 26 pg or more. CHr of under 26 pg is associated with other demonstrable features of iron deficiency.

Use Early detection and monitoring of treatment of iron deficiency. Detection of functional iron deficiency in patients treated with recombinant human erythropoietin (rHuEPO).

Limitations The necessary analyzer may not be in immediate close proximity to the patient candidate.

Methodology Flow cytometric technique (eg, Bayer H•3 hematology analyzer) utilizing three parameters:

- low-angle light scatter at 2°C to 3°C (correlates to cell volume)
- high-angle scatter at 5°C to 15°C (correlates to hemoglobin concentration)
- absorbance (correlates to intensity of staining and reflects RNA content)

The EDTA-anticoagulated whole blood specimen is prepared manually by dilution in a reagent that contains a surfactant (spheres the red cells) and a nucleic acid dye (oxazine 750) that stains the reticulocyte on the basis of its RNA content. Fluorescence measurement is not required.

Reticulocytes are divided into three fractions according to the level of maturity. Reticulocyte indices are determined, the mean cell volume (MCVr), mean cell hemoglobin concentration (MCHCr) and the reticulocyte hemoglobin content (CHr). The latter is the most stable. If there is a delay in testing, change in reticulocyte water content may affect the MCVr and MCHCr but not hemoglobin content.

Additional Information Reticulocyte hemoglobin content (CHr), if decreased, can provide an early indication of iron deficiency. CHr can identify iron deficient erythropoiesis in some apparently iron sufficient individuals. CHr has been studied in nondialysis patients treated with rHuEPO and found to be a sensitive and specific indicator of functional iron deficiency. With rHuEPO stimulation of erythropoiesis there is high erythroid iron use with shift in the balance between storage and erythroid iron reservoirs. With the intense erythropoietic stimulation the amount of iron immediately available for erythropoiesis may be insufficient, even though whole body iron stores are not depleted. This is a result of the massive transfer of storage iron to erythroid precursors, with such brisk mobilization of iron that storage sites cannot keep up with the demand. Thus, the concept of "functional iron deficiency" occurring despite normal appearing levels of serum ferritin and transferrin saturation. There may be associated poor response to treatment of renal anemia (with rHuEPO) which can be corrected by intravenous administration of iron. Determination that "functional" iron deficiency is present in rHuEPO treated dialysis patients may allow use of a lower erythropoietin dose.

The CHr may prove to have value as a screening test for early iron deficiency (before the onset of anemia), particularly if studies show that adverse effects of iron deficiency (eg, those on the brain such as irreversible mental/motor impairment) begin before anemia develops.

References
Brugnara C, Zurakowski D, DiCanzio J, et al, "Reticulocyte Hemoglobin Content to Diagnose Iron Deficiency in Children," *JAMA*, 1999, 281(23):2225-30.

Fishbane S, Galgano C, Langley RC, et al, "Reticulocyte Hemoglobin Content in the Evaluation of Iron Status of Hemodialysis Patients," *Kidney Int*, 1997, 52(1):217-22.

Goodnough LT, Skikne B, and Brugnara C, "Erythropoietin, Iron, and Erythropoiesis," *Blood*, 2000, 96(3):822-33.

♦ **Reticulocyte Maturity Index** *see* Reticulocyte Count *on page 1156*

Retinoblastoma Gene DNA Detection

Related Information
Apoptosis Assays *on page 205*
Histopathology *on page 733*
p53, Functional Assay/Sequencing *on page 995*
Polymerase Chain Reaction *on page 1069*

Synonyms RB Tumor Suppressor Gene

Test Includes Restriction fragment length polymorphism (RFLP) analysis and/ or polymerase chain reaction (PCR) amplification followed by sequencing of the gene to ascertain mutations.

Abstract The retinoblastoma (*Rb*1) gene is the prototype tumor suppressor gene. Located on chromosome 13q14, the gene encodes a nuclear protein that participates in the control of cell proliferation and progression through the cell cycle. Deletion or inactivation of both *Rb* alleles plays an essential role in the development of retinoblastoma, a tumor of retinoblasts affecting newborns and young children. Somatic inactivation of *Rb* is also found in other tumors not associated with retinoblastoma, including astrocytomas, several sarcomas, small cell and squamous cell carcinoma of lung, and carcinomas of breast, bladder, prostate, and parathyroid.

Specimen 30 mL whole blood; 100 mg solid tumor, frozen; amniotic cells grown in appropriate media

Container Blood: Yellow top (ACD) tube; amniotic cells: T25 tissue culture flask

Storage Instructions Transport whole blood at ambient temperature to the laboratory immediately. Store solid tumor at -70°C.

Turnaround Time 3-4 weeks

Reference Interval The laboratory usually provides an interpretive report.

Use DNA tests make it possible to predict the occurrence of tumors in offspring and siblings of patients with retinoblastoma. DNA testing is useful in several settings. Identification of unaffected relatives of patients with retinoblastoma who are carriers of the germline defect aids in accurate risk assessment. DNA testing can be performed in newborns of affected or carrier parents to determine if the newborn carries the mutation. If a mutation is present, more frequent examination for detection of the tumor is warranted. Although
(Continued)

Retinoblastoma Gene DNA Detection *(Continued)*

successful treatment is available for early diagnosed retinoblastomas, patients remain at risk for nonocular tumors. Because of these risks, prenatal diagnosis can be offered to families with germline mutations.

Limitations Failure to obtain DNA from the blood, tissue, or cultured cells due to inappropriate shipping or processing.

Additional Information Retinoblastoma occurs in either a hereditary (40% of cases) or a nonhereditary (sporadic) form (60% of cases). Hereditary predisposition to retinoblastoma is caused by a germline mutation at the *Rb*1 locus. The germline mutation is transmitted in an autosomal dominant fashion with 90% penetrance. Retinoblastoma develops following a somatic mutation or deletion affecting the remaining *Rb* allele. Bilateral disease occurs in the majority of hereditary cases and patients have an increased risk for developing nonocular tumors (mainly osteosarcoma) in later life. In nonhereditary retinoblastoma, the tumor is usually unilateral and arises following successive somatic mutations affecting the two *Rb* alleles.

The role of the *Rb* tumor suppressor gene in tumor initiation and progression is currently being studied in numerous tumors unrelated to retinoblastoma (see above). Such studies may produce prognostic information useful in tumors additional to retinoblastoma.

The *Rb* gene product, pRB, plays an important role in regulating the cell cycle (G₁ → S checkpoint), in control of cell cycle progression. Cyclin-dependent protein kinases (Cdks) and cyclins are involved. Regulation involves activating phosphorylations and dephosphorylations. pRB and its family members are targets of the Cdks. The *Rb* family (including p107 and p130) controls gene expression as mediated by the E2F family of transcription factors. The p107 and p130 proteins function as relays between cellular signals that control proliferation and nuclear transcription. They are involved in the cell differentiation process, including protection from apoptosis (E2F-1 has the ability to induce apoptosis). The possible function of p107 and p130 in tumor suppression is under investigation.

The combined loss of *p53* protein and pRB expression have been reported to indicate relatively favorable prognosis in squamous cell carcinoma of the esophagus.

References

Fitzgerald PJ, "The Oncogene," Chapter 20, and "Suppressor Genes," Chapter 21, *From Demons and Evil Spirits to Cancer Genes*, Washington, DC: American Registry of Pathology, Armed Forces Institute of Pathology, 2000, 207-24, 225-40.

Ikeguchi M, Oka S, Gomyo Y, et al, "Combined Analysis of p53 and Retinoblastoma Protein Expressions in Esophageal Cancer," *Ann Thorac Surg*, 2000, 70(3):913-7.

Lowy DR and Wolff L, "Molecular Oncology: Molecular Aspects of Oncogenesis," *The Molecular Basis of Blood Diseases*, 3rd ed, Part VI, Chapter 25, Stamatoyannopoulos G, Majerus PW, Perlmutter RM, et al, eds, Philadelphia, PA: WB Saunders Company, 2001, 792-3, 821.

♦ **Retinoids** *see* Vitamin A, Serum or Plasma *on page 1314*

♦ **Retinol, Serum** *see* Vitamin A, Serum or Plasma *on page 1314*

♦ **RET Proto-oncogene** *see* Multiple Endocrine Neoplasia/Familial Medullary Thyroid Carcinoma *on page 924*

♦ **Retroperitoneal Mass Aspiration** *see* Fine Needle Aspiration, Deep and Superficial Masses *on page 590*

♦ **Retrovir®** *see* Zidovudine, Serum or Plasma *on page 1339*

♦ **Reverse Grouping** *see* Pretransfusion Testing *on page 1088*

Reverse T₃, Serum

Related Information

Free Thyroxine Index *on page 613*
Thyroid Stimulating Hormone, Serum *on page 1250*
Thyroxine, Free, Serum *on page 1256*
Thyroxine, Serum *on page 1257*

Synonyms rT₃; Triiodothyronine Reverse

Applies to Euthyroid Sick Syndrome; Nonthyroidal Illness

Abstract Reverse T₃ (rT₃) is an inactive isomer of triiodothyronine (T₃).

Patient Preparation Avoid radioisotope administration prior to collection of specimen if testing is by RIA.

Specimen Serum

Container Red top tube

Storage Instructions Separate serum and store at 4°C if the test is done within 24 hours. Freeze for longer periods of storage. Frozen samples are stable for at least 1 month.

Reference Interval Values are higher in cord blood and newborns. Intervals are as follows:

- cord blood: 130-300 ng/dL (SI: 2.00-4.62 nmol/L)
- 1 day: 83-194 ng/dL (SI: 1.28-2.99 nmol/L)
- 2 days: 107-209 ng/dL (SI: 1.65-3.22 nmol/L)
- 3 days: 102-166 ng/dL (SI: 1.57-2.56 nmol/L)
- 1 month to 20 years: 10-35 ng/dL (SI: 0.15-0.54 nmol/L)
- adults: 10-28 ng/dL (SI: 0.17-0.51 nmol/L)

Use The rT_3 test has a very limited role in the diagnosis of the euthyroid sick syndrome. In patients with low serum thyroxine (T_4) accompanied by a low or normal serum thyrotropin (TSH), the differential diagnosis includes the euthyroid sick syndrome (in which rT_3 is elevated) and central hypothyroidism (in which rT_3 is not elevated).

Methodology Radioimmunoassay (RIA)

References

Camacho PM and Dwarkanathan AA, "Sick Euthyroid Syndrome. What to Do When Thyroid Function Tests Are Abnormal in Critically Ill Patients," *Postgrad Med*, 1999, 105(4):215-9.

Dayan CM, "Interpretation of Thyroid Function Tests," *Lancet*, 2001, 357:619-24.

Faase EM, Meacham LR, Novack CM, et al, "Decreased Reverse T_3 Levels in Neonates With Central Hypothyroidism," *J Perinatol*, 1997, 17(1):15-7.

Whitley RJ, "Thyroid Function," *Tietz Textbook of Clinical Chemistry*, 3rd ed, Burtis CA and Ashwood ER, eds, Philadelphia, PA: WB Saunders Co, 1999, 1496-529.

♦ **RF** *see* Rheumatoid Factor, Serum or Body Fluid *on page 1161*

♦ **RFLP Analysis for Parentage Evaluation** *see* Identification DNA Testing *on page 765*

♦ **Rh(D) Typing** *see* Pretransfusion Testing *on page 1088*

♦ **Rheumatoid Effusion** *see* Body Fluid pH *on page 295*

♦ **Rheumatoid Factor, Body Fluid** *see* Body Fluid Chemical Analysis *on page 291*

Rheumatoid Factor, Serum or Body Fluid

Related Information

Anticyclic Citrullinated Peptide Antibody *on page 169*
Antinuclear Antibodies *on page 189*
Body Fluid Chemical Analysis *on page 291*
Body Fluid Glucose *on page 294*
Body Fluid pH *on page 295*
C-Reactive Protein, Serum *on page 467*
Cryoglobulin, Qualitative, Serum and Plasma *on page 478*
HLA-B27 *on page 738*
HLA Typing, Single Human Leukocyte Antigen *on page 739*
Immunoglobulin M *on page 779*
Sedimentation Rate, Erythrocyte *on page 1181*
Smith (Sm) and Ribonucleoprotein (RNP) Antibodies *on page 1206*
Synovial Fluid Analysis *on page 1229*

Synonyms RF

Applies to Rheumatoid Factor, Synovial Fluid

Abstract Rheumatoid arthritis (RA) and juvenile RA are clinical syndromes diagnosed by specific criteria. They are systemic autoimmune disorders. Rheumatoid factor (RF) is an antibody (usually of the IgM class) which reacts with the Fc region of other immunoglobulins (often, but not always, of the IgG class). No single laboratory test defines RA, and no etiologic agent is established.

Specimen Serum, body fluid

Container Red top or SST™ tube

Reference Interval There are wide interlaboratory differences, some of which are method dependent.

Use Diagnosis: Approximately 65% to 85% of adult patients with the clinical diagnosis of RA have serologic evidence of RF. Unfortunately, positive serologic results are found in a number of other conditions, including clinically normal persons.

(Continued)

Rheumatoid Factor, Serum or Body Fluid *(Continued)*

Limitations Disease activity monitoring: Serial RF measurements have **not** been useful in monitoring the course of RA. Better tests for this purpose include: C-Reactive Protein, Serum *on page 467* and Sedimentation Rate, Erythrocyte *on page 1181*.

Approximately 3% of the general population has low level RF. The prevalence increases with age, up to 20% in persons older than 65 years old. Only 5% of healthy individuals with a positive RF test will eventually develop rheumatoid arthritis. The higher the RF levels, the greater the risk of disease.

A positive RF test is not specific for rheumatoid arthritis. High levels of RF are present in the majority of patients with Sjögren syndrome and essential mixed cryoglobulinemia. RF is present in low titers in other connective tissue diseases, and in a variety of chronic infections and inflammatory disorders. They include infectious endocarditis, tuberculosis, liver disease, sarcoidosis, idiopathic pulmonary fibrosis, and several hematologic diseases. Many of these conditions are associated with hypergammaglobulinemia and intense stimulation of the immune system. The specificity of RF for rheumatoid arthritis increases when the RF test is repeatedly positive and present in high titer.

Up to 35% of patients with rheumatoid arthritis have negative rheumatoid factor tests. Therefore, a negative test does not rule out the diagnosis of rheumatoid arthritis in a patient who otherwise meets clinical criteria. Female patients and patients with elderly onset rheumatoid arthritis are more likely to be seronegative.

Methodology Latex-human IgG agglutination, sheep RBC-rabbit IgG agglutination, rate nephelometry in which international units are reported

Additional Information RF is typically an immunoglobulin of the IgM or IgG class, but occasionally may be IgA or IgE. RF forms immune complexes with target immunoglobulin within the circulation or joint fluid. These complexes may reach high concentrations and mediate tissue injury.

Seropositive rheumatoid arthritis patients tend to have more severe disease than seronegative patients and high RF titers are often associated with multiple subcutaneous nodules, necrotizing vasculitis, and poorer long-term prognosis. Elevated RF titers may not be seen in the serum for the first several months in early rheumatoid arthritis and may be found in the joint fluid before it is seen in the serum.

RF is present in only 30% of children with juvenile rheumatoid arthritis. Other joint diseases, such as ankylosing spondylitis, Reiters syndrome, and psoriatic arthritis do not have elevated RF titers.

RF in pleural fluid, with titer ≥320, provides evidence of rheumatoid pleuritis in patients who have established rheumatoid arthritis.

The American Rheumatism Association 1987 Revised Criteria for the Classification of Rheumatoid Arthritis remain current and are widely available.

Although the differential diagnosis of the RA patient may include SLE, almost 50% of RA patients have antinuclear antibodies.

References

Fox RI, "Clinical and Laboratory Evaluation of Rheumatoid Arthritis," *Clinical and Laboratory Evaluation of Human Autoimmune Diseases*, Chapter 10, Nakamura RM, Keren DF, and Bylund DJ, eds, Chicago, IL: American Society for Clinical Pathology, 2002, 141-52.

Moder KG, "Use and Interpretation of Rheumatologic Tests: A Guide for Clinicians," *Mayo Clin Proc*, 1996, 71(4):391-6.

♦ **Rheumatoid Factor, Synovial Fluid** *see* Rheumatoid Factor, Serum or Body Fluid *on page 1161*

♦ **Rheumatrex®** *see* Methotrexate, Serum or Plasma *on page 905*

Rh Genotype

Related Information

Bilirubin, Amniotic Fluid, Delta A450 *on page 261*
Bilirubin, Neonatal, Serum *on page 263*
Cord Blood Antibody Screen *on page 453*
Hemolytic Disease of the Newborn, Antibody Identification *on page 690*
Newborn Crossmatch and Transfusion *on page 952*
Polymerase Chain Reaction *on page 1069*

Pretransfusion Testing *on page 1088*
Rh₀(D) Immune Globulin (Human) *on page 1164*

Synonyms Rh Zygosity

Applies to Hemolytic Disease of the Newborn (HDN) Prognosis; Prenatal Testing; Weak D

Test Includes Rh genotyping of male partners of pregnant, Rh-immunized women; testing of red cells with Rh antisera anti-D, C, E, c, e

Abstract When a woman of childbearing age is found to have anti-D, it is important for prognostic purposes to know the husband's zygosity for the gene producing the Rh antigen D. A father homozygous for Rh(D) must produce Rh(D)-positive children, whereas a father heterozygous for Rh(D) has a 50% chance of producing Rh(D)-negative children.

Specimen Blood, amniotic fluid

Container Red top tube or lavender top (EDTA) tube

Causes for Rejection Gross hemolysis

Special Instructions It is important that ethnic group of subject is recorded.

Reference Interval Rh gene frequencies vary considerably with different ethnic groups. Most published data are for European whites. It is important that the appropriate frequency table be utilized.

Incidence of the More Common Genotypes in D-Positive Persons

Phenotype	Genotype		Incidence (%)	
	DCE	Rh-hr	Whites	Blacks
DCce	DCe/ce	R^1r	31.1	8.8
	DCe/Dce	R^1R^0	3.4	15.0
	Ce/Dce	$r'R^0$	0.2	1.8
DCe	DCe/DCe	R^1R^1	17.6	2.9
	DCe/Ce	R^1r'	1.7	0.7
DcEe	DcE/ce	R^2r	10.4	5.7
	DcE/Dce	R^2R^0	1.1	9.7
DcE	DcE/DcE	R^2R^2	2.0	1.3
	DcE/cE	R^2r''	0.3	<0.1
DCcEe	DCe/DcE	R^1R^2	11.8	3.7
	DCe/cE	R^1r''	0.8	<0.1
	Ce/DcE	$r'R^2$	0.6	0.4
Dce	Dce/ce	R^0r	3.0	22.9
	Dce/Dce	R^0R^0	0.2	19.4

Adapted from Brecher ME, *Technical Manual*, 14th edition, Bethesda, MD: American Association Blood Banks Press, 2002.

Limitations Genotype frequencies are given for random populations. The partners of Rh-immunized women are a weighted population with a higher incidence of homozygosity. On the other hand, if any or all of the subject's children or if either of his parents is Rh(D) negative, then the partner must be heterozygous.

Methodology The subject's red cells are tested with the range of Rh antisera indicated earlier. From the results, his Rh phenotype is determined. The most likely genotype can be determined by reference to a frequency table for the appropriate racial group. Fetal Rh(D) status can be determined from polymerase chain reaction (PCR) analysis of amniotic fluid, an invasive procedure. Fetal Rh(D) status can be determined by analysis of maternal plasma, using a PCR assay.

Additional Information Serologic results often can determine Rh phenotype only and genotype is assigned according to frequency tables. Clearly, a significant number of "most probable genotypes" are incorrect. However, many clinically important blood groups have now been characterized at the gene level. This knowledge has permitted the development of noninvasive prenatal tests to identify the presence of fetal DNA in maternal plasma. These procedures will prove valuable when the father's antigenic status is heterozygous, indeterminable, or unknown. Furthermore, it is likely that in the near future, it will be

(Continued)

Rh Genotype *(Continued)*

feasible to screen Rh(D)-negative pregnancies by molecular biological methods, to select those requiring antenatal Rh immunoprophylaxis.

The expression "Dᵘ" is presently termed weak D. It indicates a weak expression of the D antigen in which red cells are not directly agglutinated by all anti-D sera. The frequency of weak D is about 0.2% in Caucasians.

References

Bowman JM, "RhD Hemolytic Disease of the Newborn," *N Engl J Med*, 1998, 339(24):1775-6.

Domen RE, "Policies and Procedures Related to Weak D Phenotype Testing and Rh Immune Globulin Administration: Results From Supplementary Questions to the Comprehensive Transfusion Medicine Survey of the College of American Pathologists," *Arch Pathol Lab Med*, 2000, 124(8):1118-21.

Hartwell EA, "Use of Rh Immune Globulin: ASCP Practice Parameter," *Am J Clin Pathol*, 1998, 110(3):281-92.

Lo YM, Hjelm NM, Fidler C, et al, "Prenatal Diagnosis of Fetal RhD Status by Molecular Analysis of Maternal Plasma," *N Engl J Med*, 1998, 339(24):1734-8.

Reid ME, Rios M, and Yazdanbakhsh K, "Applications of Molecular Biology Techniques to Transfusion Medicine," *Semin Hematol*, 2000, 37(2):166-76.

Saade GR, "Noninvasive Testing for Fetal Anemia," *N Engl J Med*, 2000, 342(1):52-3.

♦ **RhIG** *see* Rhₒ(D) Immune Globulin (Human) *on page 1164*
♦ **Rh Immune Globulin** *see* Rhₒ(D) Immune Globulin (Human) *on page 1164*

Rhₒ(D) Immune Globulin (Human)

Related Information

Antibody Detection/Identification, Red Cell *on page 165*
Antiglobulin Test, Direct *on page 176*
Cord Blood Antibody Screen *on page 453*
Fetal Cell Detection by Flow Cytometry *on page 579*
Hemolytic Disease of the Newborn, Antibody Identification *on page 690*
Kleihauer-Betke *on page 822*
Newborn Crossmatch and Transfusion *on page 952*
Rh Genotype *on page 1162*
Rosette Test for Fetomaternal Hemorrhage *on page 1172*

Synonyms RhIG; Rh Immune Globulin

Test Includes D/weak D type of mother and baby, test to detect excessive fetomaternal hemorrhage (rosette test followed by Kleihauer-Betke or flow cytometry), and, antibody screen on mother

Abstract Hemolytic disease of the newborn (HDN) is caused by an IgG maternal antibody which destroys the antigen-positive erythrocytes of the fetus and newborn. Anti-Rhₒ(D) is the most important such alloantibody, produced by Rhₒ(D) negative women exposed to Rhₒ(D)-positive fetal/neonatal or transfused red cells. It is given to Rhₒ(D) negative pregnant/postpartum women who have not developed anti-Rhₒ(D); *vide infra*. Rh-immune globulin (RhIG) is an immune globulin, predominantly IgG anti-D prepared from pooled human plasma. Intramuscular (I.M.) or intravenous (I.V.) preparations are available for administration. I.M. RhIG is available in 300 μg and 50 μg doses. A 300 μg dose is sufficient to prevent immunization by 30 mL of D-positive whole blood or 15 mL of D-positive red cells. The 50 μg dose protects against fetal bleed during the first trimester only. Intravenous RhIG is available in 300 μg and 120 μg doses. The 300 μg dose can suppress the immunizing potential of ~17 mL of D-positive red cells. It is administered at 28-30 weeks gestation or after invasive procedures before 34 weeks gestation, unless the father is known to be Rh(D)-negative. The 120 μg dose can be administered to the mother within 72 hours of delivery or after invasive procedures associated with increased risk of Rh(D) isoimmunization after 34 weeks gestation.

Patient Preparation Each Transfusion Service is required to have a policy for RhIG prophylaxis when Rh-negative patients are exposed to Rh-positive red cells. This must include a process to ensure that an adequate dose is administered. The rosette test is an effective screening test (see Rosette Test for Fetomaternal Hemorrhage *on page 1172*), which, if positive, must be followed by a quantitative test such as the Kleihauer-Betke test. The weak D test (formerly known as the Dᵘ test) is not recommended to identify large fetomaternal hemorrhage.

Specimen Blood from both mother and infant

Container One red top tube or one lavender top (EDTA) tube

Collection Collected postpartum. Blood required from both mother and newborn. Label each specimen.

Storage Instructions RhIG must be stored at 2°C to 8°C.

Causes for Rejection (Of patient sample): Gross hemolysis, sample placed in a serum separator tube, specimen tube not properly labeled.

RhIG is not indicated when the newborn is Rh-negative, mother is Rh-positive or weak D-positive, or when anti-D is present in mother's serum when mother has not had prenatal Rh immune globulin.

Use Given to D-negative women postpartum, or after termination of pregnancy, ectopic pregnancy, abortion, threatened abortion, obstetric complications, tubal ligation, immune thrombocytopenic purpura, or any event associated with increased risk of fetomaternal hemorrhage to prevent development of anti-D. Anti-D antibody may cause erythroblastosis fetalis (hemolytic disease of the newborn) or years later lead to transfusion reaction if Rh-positive RBCs are transfused. RhIG is given to Rh-negative women after amniocentesis, percutaneous umbilical cord sampling, or chronic villus sampling. Give RhIG antepartum at 28-32 weeks, as well as within 3 days of delivery. When it is given antepartum, then after delivery there will still be anti-D in the maternal serum. Give a postpartum dose to the appropriate mother whether or not Rh immune globulin was given antepartum. Occasionally, RhIG is given to an Rh-negative person who received Rh-positive red blood cells or an Rh-positive component (eg, platelets, granulocytes). The I.V. RhIG is also approved for treatment of nonsplenectomized D-positive patients with immune thrombocytopenic purpura (ITP). Cases of HDN have been reported in infants whose mothers had a weak D phenotype. A recent survey indicated 71% of Transfusion Services recommended RhIG be administered to pregnant weak D-positive women with a possible Rh-D-positive fetus.

Limitations In instances of large fetomaternal hemorrhage, one dose is not sufficient. When a transplacental hemorrhage is >30 mL fetal blood (by Kleihauer-Betke) the dose of $Rh_o(D)$ immune globulin must be at least 10 µg/mL of fetal blood in the maternal circulation. The weak D test is not recommended to identify large fetomaternal hemorrhage. The rosette test is more sensitive (see Rosette Test for Fetomaternal Hemorrhage *on page 1172*). A smaller Rh immune globulin dose may be used after abortion or miscarriage up to 12 weeks gestation, but not beyond; after 12 weeks of gestation a conventional dose is indicated.

Failures occur. The most common cause of Rh immunization is failure to give RhIG when it is indicated. RhIG is sometimes forgotten in ectopic gestation and in abortion in Rh-negative women. However, 1% to 2% of term mothers develop anti-D in spite of postpartum RhIG properly administered. Postpartum failure may be secondary to fetomaternal hemorrhage in the third trimester (hence, antenatal RhIG) and because of large fetomaternal hemorrhages at delivery.

Contraindications Do not give RhIG to an Rh-positive person (unless treating for ITP), or a person already immunized to the $Rh_o(D)$ blood factor whose serum contains anti-D. (However, if $Rh_o(D)$ immune globulin was given as an antenatal dose to mother, then anti-D detectable in her serum is not a contraindication to postnatal administration of RhIG.) (The usual antenatal dose of RhIG does not cause titers >4.) Women who deliver $Rh_o(D)$-negative infants are not candidates for RhIG. If there is certain documentation that the biologic father is Rh-negative, the RhIG is not needed. **Do not give RhIG to an infant.**

Methodology Read appropriate manufacturer's package inserts for dosage and administration instructions.

The product contains small amounts of IgA and other globulins.

Additional Information Give a full dose, 300 µg RhIG (I.M.) to an Rh-negative mother within 72 hours of delivery, miscarriage, or any event associated with increased risk of fetomaternal hemorrhage. If a fetomaternal hemorrhage >15 mL RBCs has taken place, then additional RhIG is indicated.

The **Kleihauer-Betke** test done on maternal blood after delivery estimates the volume of fetal-maternal hemorrhage. Calculate the dose of Rh_o immune globulin as follows.

- Percent of fetal red cells x 50 = mL fetal whole blood in maternal circulation.

(Continued)

Rh₀(D) Immune Globulin (Human) *(Continued)*

- Although the usually recommended dose is 300 µg of RhIG per 30 mL of fetal blood, always give one more dose (300 µg) than that calculated because of the poor precision of the Kleihauer-Betke test. For example:

  1.8% fetal RBCs

  1.8 x 50 = 90 mL fetal whole blood

  90 mL/30 = 3 doses

  3 + 1 = 4

  four, 300 µg doses of RhIG administered

Give RhIG to Rh-negative mothers with negative screens when cord blood is not available (ectopic pregnancies, abortions, etc). If the patient refuses, she should sign an appropriate statement to that effect. Although **antenatal doses** may have been given at 28-32 weeks, give a postpartum dose anyway. Transmission of viral infections does not occur with this preparation.

References

Bowman JM, "RhD Hemolytic Disease of the Newborn," *N Engl J Med*, 1998, 339(24):1775-6.

Bowman JM, "Antenatal Suppression of Rh Alloimmunization," *Clin Obstet Gynecol*, 1991, 34(2):296-303.

Domen RE, "Policies and Procedures Related to Weak D Phenotype Testing and Rh Immune Globulin Administration," Results From Supplementary Questions to the Comprehensive Transfusion Medicine Survey of the College of American Pathologists, *Arch Pathol Lab Med*, 2000, 124(8):118-21.

Hartwell EA, "Use of Rh Immune Globulin," ASCP Practice Parameter, *Am J Clin Pathol*, 1998, 110(3):281-92.

Rushin J, Rumsey DH, Ewing CA, et al, "Detection of Multiple Passively Acquired Alloantibodies Following Infusions of IV Rh Immune Globulin," *Transfusion*, 2000, 40(5):551-4.

Snyder EL and Shoos Lipton K, "Prevention of Hemolytic Disease of the Newborn Due to Anti-D," *AABB Association Bulletin*, 98-2. AABB News Briefs, March 1998, 16-7.

Risks of Transfusion

Related Information

Synonyms Hazards of Transfusion; Transfusion Complications

Test Includes See Transfusion Reaction Work-up *on page 1269* for a listing of tests performed for noninfectious complications of transfusion. Blood is tested for five major transfusion-transmitted viral diseases, hepatitis B, HIV, HTLV, hepatitis C, and West Nile virus. Transmission of these diseases can occur through transfusion. Any cases of suspected transfusion-transmitted disease should be reported to the Blood Bank. Other agents known to be transmitted by transfusion include cytomegalovirus, malaria, babesiosis, and Chagas disease.

Patient Preparation Obtain and document informed consent prior to transfusion.

Special Instructions Report all adverse effects of transfusion at once to the Blood Bank for follow-up and investigation (see Transfusion Reaction Work-up *on page 1269*).

Some Risks of Allogeneic Transfusion

Reactions
Hemolytic, immediate, delayed
Febrile
Allergic, anaphylactic
Sepsis
Overload
Hypothermia, cold
Air embolism
Post-transfusion purpura
Disease Transmission
Hepatitis B, C, etc
Cytomegalovirus
Parvovirus B19
HTLV-I/II
Syphilis
Malaria
Babesiosis
Yersinia
Chagas disease
HIV-1/2
West Nile virus
Other
Alloimmunization RBC, WBC, etc
Marrow suppression
Immunosuppression
Storage changes
Graft-vs-host disease
Dilutional coagulopathy
Nonimmune hemolysis
Siderosis (transfusional)

Use A physician's understanding of the estimated risks of transfusions should be explained to the patient (in nonemergency situations) as part of the informed consent process. All transfusions carry risk, including autologous transfusions.

Methodology Nucleic acid technology, introduced in the U.S. in 1998, has provided earlier detection of HCV and HIV type 1.

(Continued)

Risks of Transfusion *(Continued)*
Additional Information

NONINFECTIOUS COMPLICATIONS:

Hemolytic transfusion reactions usually result from clerical and other identification errors and frequently result in ABO incompatibility. This is why unlabeled or improperly labeled sample tubes are unacceptable to the Transfusion Service. Most blood errors in administration are caused by failure to correctly identify the recipient and blood unit, but phlebotomy errors and Blood Bank errors occur as well. Chills, fever, dyspnea, chest or back pain, headache, abnormal bleeding, or shock can all characterize acute hemolytic reactions. Hemoglobinemia heralds **intravascular** hemolysis, followed by hemoglobinuria, then jaundice. This is usually mediated by anti-A or anti-B or both. **Extravascular** hemolysis takes place mostly in the spleen as a result of the action of IgG antibodies, such as those of the Rh system. With these, hemoglobinemia and hemoglobinuria seldom occur. Renal shutdown, shock, or hemorrhage may be fatal. When this type of reaction occurs, stop the transfusion at once (see Transfusion Reaction Work-up *on page 1269*). Treat shock. Give appropriate fluids and diuretics to maintain urinary output. Treat for incipient renal failure, if indicated. Rarely, passive transfer of alloantibodies can cause unanticipated hemolytic anemia.

Delayed hemolytic reactions can occur in some patients with other serologically undetectable antibodies (frequently Kidd blood group system antibodies). Such reactions may come to the attention of the Blood Bank Staff when more blood is ordered a few days after an earlier transfusion of apparently compatible blood, usually with a poor clinical response to the prior transfusion. The Blood Bank finds a positive direct antiglobulin test and antibody that is now incompatible with the recently transfused RBCs. The antibody may be either in the patient's serum or in an eluate from the red cells. The diagnosis is easily missed. Delayed hemolytic reactions are not uncommon; they are usually mild.

Nonimmune hemolysis may occur secondary to inappropriate solutions running in the same tubing with blood components. With the exception of 0.9% sodium chloride, USP, drugs, or medications must not be added to blood unless they have been approved for this use by the FDA or unless records are available to show that such addition is safe and does not adversely affect the blood component.

Other immune reactions include **febrile nonhemolytic** reactions. Leukocyte-derived cytokines are a major cause of febrile reactions related to platelet transfusions. Leukocyte reduction by filtration is effective, but soluble mediators and cytokines are released in storage. They may be prevented by transfusion of leukocyte-reduced blood components. **Allergic transfusion reactions** usually appear in the form of hives (urticaria) without fever. Treatment and prevention is administration of antihistamine to the patient. Anaphylaxis is rare.

Bone marrow suppression of RBC production will occur after the transfusion of RBCs, another reason to avoid transfusions to patients whose anemia might respond to conventional medication.

Immunosuppression of varying degree follows allogeneic transfusion. Although it has been observed to improve renal allografts, survival concerns have been raised regarding the adverse effects of transfusion in other clinical settings, including increased rates of postoperative infections and tumor recurrence. The usefulness of leukocyte-reduction in such clinical settings remains controversial.

Transfusion-related acute lung injury ("TRALI") occurs within 6 hours following transfusion. Clinically similar to adult respiratory distress syndrome, TRALI is caused by noncardiogenic pulmonary edema. TRALI seems to be related to HLA or leukocyte antibodies in donor plasma, or to biologically active mediators.

Simple **volume overload** of the recipient's circulation may cause pulmonary edema without leukocyte antibodies.

Air embolism can result from any admission of air into intravenous tubing and can have serious consequences.

Anaphylactic reaction: See Immunoglobulin A *on page 770.* Nausea, chills, severe abdominal cramps, emesis, diarrhea, dyspnea, and flushing with hypotension may take place due to a generalized reaction associated with an IgA antibody. Washed cellular components or IgA-deficient plasma is available for these recipients.

Graft-versus-host disease (GVHD) can result from transfusion of blood components containing living donor lymphocytes that engraft and clonally expand in a susceptible host. GVHD usually occurs in immunocompromised individuals, but is occasionally seen in immunocompetent recipients. It has been seen when related directed donors are utilized, due to greater genetic homogeneity. While transfusion-associated GVHD is usually fatal, GVHD is preventable by irradiation of any blood component to be transfused to a patient at risk (see Irradiated Blood Components *on page 811).*

Immunosuppressed patients include recipients of hematopoietic stem-cell donations, recipients of other organ donations, others on immunosuppressive therapy, those with lymphomas and Hodgkin disease and leukemias undergoing chemotherapy, and those with AIDS. Such patients and other with immunodeficiency states are benefited by screening of blood fractions for CMV. See Red Blood Cells, Leukocytes Reduced *on page 1141.* Immunosuppressed individuals are sensitive to bacterial contamination, an especially relevant problem when platelets are transfused. Graft-vs-host disease is addressed above.

Complications of massive transfusion: Hemorrhagic diathesis may follow dilution and washout (dilutional coagulopathy) of coagulation factors and platelets. DIC occurs in settings in which massive transfusions are given. Rapid laboratory evaluation of hemostasis can be vital. Treatment of abnormal bleeding in this situation is primarily with platelet concentrates, sometimes also with FFP, and less often with cryoprecipitate. If fluid balance is not carefully observed, fluid overload or adult respiratory distress syndrome may occur. 2,3-DPG depletion of stored RBCs is a theoretic problem, rarely of any clinical significance. Hypothermia, caused by rapid massive transfusion of cold blood, can be prevented with blood warmers (see Warming, Blood *on page 1326).* Other possible problems faced with massive transfusions may include citrate toxicity and hyperkalemia. With massive transfusions, particularly in trauma, there is often tumult and confusion, creating a setting which may promote likelihood of clerical error and increased possibility of incompatible blood transfusion. Avoiding errors in such settings is vital.

TRANSFUSION-ASSOCIATED INFECTIOUS DISEASES:

Viral hepatitis , the incidence of which is changing. Type A is very rare, B uncommon, and C decreasing significantly. A causal relationship between hepatitis G virus and hepatitis has not been established.

Cytomegalovirus (CMV) infection can be significant in premature newborns born to CMV-seronegative mothers and immunosuppressed CMV-seronegative adults, including transplant recipients. Transfuse CMV-seronegative or leukocyte-reduced blood components.

Bacterial contamination of blood components can cause septic shock and death and must be vigorously treated if observed. Gram-positive organisms are more frequently seen in components stored at room temperature; gram-negative organisms (eg, *Yersinia enterocolitica* can grow in refrigerated blood). In 2004, Blood Banks and Transfusion Services must limit and detect bacterial contamination in platelet components.

HIV/AIDS, as a transfusion hazard, is statistically rare, but regarded by the public as a terrifying risk of transfusion. The onset of testing of the blood supply in 1985, beginning with a test for antibody to HIV and now including tests for HIV antigen and nucleic acid testing, has led to an extremely small estimated risk.

West Nile virus (WNV). Blood collection facilities implemented nucleic acid testing for WNV in 2003 after 23 individuals were reported to have acquired transfusion transmitted WNV in 2002. WNV transmission has been documented from infected RBCs, platelets, and fresh frozen plasma. Transmission through derivatives (eg, albumin, factor concentrates) is unlikely as processing destroys this lipid-coated virus. Risk of transmission is temporally and geographically dependent. Up-to-date information is available on the following
(Continued)

Risks of Transfusion (Continued)

website: www.cdc.gov/ncidod/dvbid/westnile/index.htm. See also West Nile Virus Diagnostic Procedures on page 1329.

The following transfusion-associated infections should be considered rare in the U.S.: syphilis, malaria, babesiosis (endemic in some areas of the East Coast), Trypanosoma cruzi (Chagas disease), and leishmaniasis. See table.

Type of Outcome or Infectious Agent	Estimated Risk per Unit Transfused[1]
Acute hemolytic	1:38,000-1:77,000
Allergic	1:33-1:100
Alloimmunization, RBC antigens	1:100
Anaphylactic	1:18,000-1:170,000
Bacteria, platelets	1:10,000-1:20,000
Bacteria, red cells	1:500,000
Circulatory overload	1:100-1:10,000
Delayed hemolytic	1:5000-1:12,000
Febrile, nonhemolytic	1:3-1:100 (platelets) 1:17-1:200 (RBCs)
Hepatitis A virus	1:1,000,000
Hepatitis B virus	1:220,000[2]
Hepatitis C virus	1:1,600,000 (with nucleic acid testing)[2]
Human immunodeficiency virus	1:1,800,000 (with nucleic acid testing)[2]
Human T-cell lymphotropic virus	1:650,000
Malaria, Babesia	<1:1,000,000
Transfusion-related acute lung injuy	1:5000-1:190,000
Trypanosoma cruzi	1:42,000

Modified from:

[1]Brecher ME, Technical Manual, 14th ed, Bethesda, MD: American Association of Blood Banks Press, 2002, 586-9.

[2]Busch MP, Kleinman SH, and Nemo GJ, "Current and Emerging Infectious Risks of Blood Transfusions,"JAMA, 2003, 289(8):959-62.

References

Biggerstaff BJ and Petersen LR, "Estimated Risk of Transmission of the West Nile Virus Through Blood Transfusion in the US, 2002," Transfusion, 2003, 43(8):1007-17.

Blumberg N, "The Cost and Consequences of Management Fads and Politically Driven Regulatory Oversight. The Case of Blood Transfusion," Arch Pathol Lab Med, 1999, 123(7):580-4.

Busch MP, Kelinman SH, and Nemo GJ, "Current Emerging Infectious Risks of Blood Transfusions," JAMA, 2003, 289(8):959-62.

Center for Disease Control and Prevention, "West Nile Virus Activity - United States, July 17-23, 2003," JAMA, 2003, 290(7):882.

Christensen PB, Groenbaek K, Krarup HB, et al, "Transfusion-Acquired Hepatitis C: The Danish Lookback Experience," Transfusion, 1999, 39(2):188-93.

Dobroszycki J, Herwaldt BL, Boctor F, et al, "A Cluster of Transfusion-Associated Babesiosis Cases Traced to a Single Asymptomatic Donor," JAMA, 1999, 281(10):297-30.

Dry SM, Bechard KM, Milford EL, et al, "The Pathology of Transfusion-Related Acute Lung Injury," Am J Clin Pathol, 1999, 112(2):216-21.

Ely EW and Bernard GR, "Transfusions in Critically Ill Patients," N Engl J Med, 1999, 340(6):467-8.

Garratty G, "Problems Associated With Passively Transfused Blood Group Alloantibodies," Am J Clin Pathol, 1998, 109(6):769-77.

Glynn SA, Kleinman SH, Schreiber GB, et al, "Trends in Incidence and Prevalence of Major Transfusion-Transmissible Viral Infections in US Blood Donors, 1991 to 1996," JAMA, 2000, 284(2):229-35.

Goodnough LT, Brecher ME, Kanter MH, et al, "Transfusion Medicine. First of Two Parts: Blood Transfusion," N Engl J Med, 1999, 340(6):438-48.

Goodnough LT, Brecher ME, Kanter MH, et al, "Transfusion Medicine. Second of Two Parts: Blood Conservation," N Engl J Med, 1999, 340(7):525-33.

Hirsch MS and Werner B, "17-2003: A 38-Year-Old Woman With Fever, Headache, and Confusion," Case Records of the Massachusetts General Hospital, Case 17-2003, Scully RE, Mark EJ, McNeely WF, et al, eds, N Engl J Med, 348(22):2239-47.

Leiby DA, Lenes BA, Tibbals MA, et al, "Prospective Evaluation of a Patient With Trypanosoma cruzi Infection Transmitted by Transfusion," N Engl J Med, 1999, 341(16):1237-9.

Lenahan SE, Domen RE, Silliman CC, et al, "Transfusion-Related Acute Lung Injury Secondary to Biologically Active Mediators," Arch Pathol Lab Med, 2001, 125(4):523-6.

Linden JV, "Errors in Transfusion Medicine: Scope of the Problem," College of American Pathologists Conference XXXIII, August 20-22, 1998, *Arch Pathol Lab Med*, 1999, 123(7):563-5.

Myhre BA and McRuer D, "Human Error - A Significant Cause of Transfusion Mortality," *Transfusion*, 2000, 40(7):879-85.

Ogedegbe HO and St Hill H, "West Nile Virus: Laboratory Diagnosis and FDA Guidance," *Lab Med*, 2003, 34(6):445-8, 465-7.

Petersen LR, Marfin AA, and Gubler DJ, "West Nile Virus," *JAMA*, 2003, 290(4):524-8.

Sampathkumar P, "West Nile Virus: Epidemiology, Clinical Presentation, Diagnosis, and Prevention," *Mayo Clin Proc*, 2003, 78(9):1137-44.

Sejvar JJ, Haddad MB, Tierney BC, et al, "Neurologic Manifestations and Outcome of West Nile Virus Infection," *JAMA*, 2003, 290(4):511-5.

Sazama K, DeChristopher PJ, Dodd R, et al, "Practice Parameter for the Recognition, Management and Prevention of Adverse Consequences of Blood Transfusion," *Arch Pathol Lab Med*, 2000, 124(1):61-70.

Shulman IA, "Assessing Blood Administering Practices," *Arch Pathol Lab Med*, 1999, 123(7):595-8.

Simon TL, Alverson DC, AuBuchon J, "Practice Parameter for the Use of Red Blood Cell Transfusions," Developed by the Red Blood Cell Administration Practice Guideline Development Task Force of the College of American Pathologists, *Arch Pathol Lab Med*, 1998, 122(2):130-8.

Spence RK, Jeter EK, and Mintz PD, "Transfusion in Surgery and Trauma," *Transfusion Therapy: Clinical Principles and Practice*, Mintz PD, ed, Bethesda, MD: American Association of Blood Banks Press, 1999, 171-97.

Vamvakas EC and Blajchman MA, *Immunomodulatory Effects of Blood Transfusion*, Bethesda, MD: American Association of Blood Banks Press, 1999.

- ◆ **Ristocetin Cofactor** *see* von Willebrand Factor *on page 1321*
- ◆ **Ristocetin-Induced Platelet Aggregation Assay** *see* von Willebrand Factor *on page 1321*
- ◆ **Ritmilen®** *see* Disopyramide, Serum or Plasma *on page 516*
- ◆ **Rivatril®** *see* Clonazepam, Serum *on page 415*
- ◆ **RMS®** *see* Morphine, Urine *on page 921*
- ◆ **RNA Concentration, HIV** *see* Human Immunodeficiency Virus DNA Amplification *on page 746*
- ◆ **RNP** *see* Smith (Sm) and Ribonucleoprotein (RNP) Antibodies *on page 1206*
- ◆ **RNP Antibodies** *see* Smith (Sm) and Ribonucleoprotein (RNP) Antibodies *on page 1206*
- ◆ **Ro Antibodies** *see* Sjögren Antibodies *on page 1199*
- ◆ **ROC Curves** *see page 11*
- ◆ **Rock** *see* Cocaine (Cocaine Metabolite), Qualitative, Urine or Hair *on page 427*
- ◆ **Rocket Fuel** *see* Phencyclidine, Qualitative, Urine *on page 1019*

Rocky Mountain Spotted Fever Serology

Related Information

Anemia Flowchart *on page 35*
Arthropod Identification *on page 213*
Ehrlichiosis Serology *on page 531*

Abstract Rocky Mountain spotted fever (RMSF) is an acute, febrile disease characterized by headache, fever, weakness, and (in ~90% of cases) a centipetal macular eruption. It is a life-threatening illness with a case fatality rate of 20% to 25%. The disease is caused by *Rickettsia rickettsii*, obligate intracellular, gram-negative, pleomorphic bacteria that are transmitted to humans by tics. In the United States, RMSF is seen most frequently in a narrow geographic band from North Carolina to Oklahoma. Serologic diagnosis is generally the only test performed for detection of RMSF.

Specimen Serum

Container Red top tube

Sampling Time Paired sera 7-10 days apart is recommended. A titer >1:128 can often be detected during the second week of illness.

Special Instructions Acute and convalescent specimens

Reference Interval Less than a fourfold increase in titer in paired sera; IgG <1:32, IgM <1:8 indicates that RMSF is unlikely

Critical Values Fourfold increase to specific rickettsial antigen is diagnostic

Possible Panic Range IgG: ≥1:128

Use Diagnose Rocky Mountain spotted fever

Limitations Diagnostic IgG titers may persist for years and IgM titers as high as 1:64 have occasionally been demonstrated a year after infection. IgG titer may be negative early in disease. Consequently, convincing serologic diagnosis can only be demonstrated with increasing titers or an IgM titer ≥1:128. False-positive reactions occur during pregnancy, particularly in the last two (Continued)

Rocky Mountain Spotted Fever Serology *(Continued)*

trimesters. False positives are also occasionally seen with other *Rickettsia* in the spotted fever group.

Methodology Indirect immunofluorescent antibody (IFA); enzyme-linked immunosorbent assay (EIA); latex agglutination (LA). Solid phase dot immunoassays, such as serum dipsticks, have become available. Immunohistologic methods will demonstrate the *Rickettsia* in tissue or in circulating endothelial cells. More recently, PCR techniques have been used to detect *R. rickettsii* DNA in specimens.

Additional Information RMSF occurs primarily from April through October, when ticks are active. It is a disease of variable clinical manifestation and some cases present with few or no "spots". All laboratory tests used to aid in the diagnosis are important, since there is good specific therapy available, and if left untreated the disease has serious outcomes. Serologic diagnosis may be made promptly enough to direct therapy. Tests for IgM specific antibody are helpful in early disease, since they appear 3-8 days after infection.

References

Abramson JS and Givner LB, "Rocky Mountain Spotted Fever," *Pediatr Infect Dis J*, 1999, 18(6):539-40.

Akinbami L, "Rocky Mountain Spotted Fever," *Pediatr Rev*, 1998, 19(5):171-2.

Belman AL, "Tick-Borne Diseases," *Semin Pediatr Neurol*, 1999, 6(4):249-66.

Drage LA, "Life-Threatening Rashes: Dermatologic Signs of Four Infectious Diseases," *Mayo Clin Proc*, 1999, 74(1):68-72.

Sexton DJ and Kirkland KB, "Rickettsial Infections and the Central Nervous System," *Clin Infect Dis*, 1998, 26(1):247-8.

Thorner AR, Walker DH, and Petri WA Jr, "Rocky Mountain Spotted Fever," *Clin Infect Dis*, 1998, 27(6):1353-9.

Internet Web Sites

www.astdhpphe.org/infect/rms.html
www.cdc.gov/ncidod/dvrd/rmsf/index.htm

♦ **Rohypnol®** *see* Flunitrazepam, Urine *on page 601*

Rosette Test for Fetomaternal Hemorrhage

Related Information

Fetal Cell Detection by Flow Cytometry *on page 579*
Flow Cytometry, Overview *on page 592*
Kleihauer-Betke *on page 822*
Newborn Crossmatch and Transfusion *on page 952*
Rhₒ(D) Immune Globulin (Human) *on page 1164*

Synonyms Fetalscreen™

Applies to Acid Elution Test

Abstract A postdelivery qualitative test for fetomaternal hemorrhage, the rosette test detects small numbers of Rh(D)-positive fetal red cells in Rh(D)-negative mothers. A positive result must be followed by a quantitative procedure to identify mothers needing a greater than standard postpartum dose of Rh immune globulin.

Specimen Blood

Container One red top tube and one lavender top (EDTA) tube

Sampling Time Postdelivery - preferably within 1 hour of delivery

Causes for Rejection Specimen grossly hemolyzed, improper labeling

Reference Interval Specimens in which <2.5 mL of Rhₒ(D)-positive fetal red cells are present yield negative results.

Use Determine if a fetomaternal hemorrhage of more than 15 mL has occurred

Limitations This is a screening test for detection of fetal Rh(D)-positive red cells in the circulation of Rh(D)-negative mothers. A positive result must be followed by a quantitative procedure such as an acid elution test, an enzyme-linked antiglobulin test, or flow cytometry to quantitate the number of fetal cells present. Weak D-positive (Dᵘ) red cells do not react as strongly in the rosette test as normal D-positive genotypes.

The Kleihauer-Betke test is better when fetal red cells are the weak D phenotype; rosette test results are weak to negative in that circumstance. Strongly positive results are found with the rosette test when the maternal red cells are weak D phenotype, creating confusion with massive fetomaternal hemorrhage. Specific testing for fetal RBCs is recommended.

Methodology A suspension of maternal red cells is incubated with chemically-modified or high protein reagent anti-D serum. Any fetal Rh(D)-positive cells present will become sensitized with the anti-D. Coating of fetal red cells with anti-D is recognized by adding Rh(D)-positive indicator cells, which form rosettes around the fetal cells.

Additional Information The rosette test detects about 5 mL of Rh_o(D)-positive fetal red cells (about 10 mL of Rh_o(D)-positive whole blood). Positive results are only found in about 1% to 3% of women who are candidates for RhIG.

The enzyme-linked antiglobulin test and flow cytometry are other methods to detect Rh_o(D)-positive erythrocytes.

References
Hartwell EA, "Use of Rh Immune Globulin," *Am J Clin Pathol*, 1998, 110(3):281-92.

Issitt PD and Anstee DJ, *Applied Blood Group Serology*, 4th ed, Durham, NC: Montgomery Scientific Publications, 1998, 115-63, 1049.

Rotavirus, Direct Detection

Related Information
Bacterial Culture, Stool *on page 243*
Electron Microscopic Examination for Viruses, Stool *on page 533*
Ova and Parasites, Direct Exam *on page 985*
Polymerase Chain Reaction *on page 1069*
Viral Culture *on page 1307*

Test Includes Direct detection of rotavirus antigen in stool specimens

Abstract Human rotavirus is a major cause of pediatric diarrhea. Infection is acquired by the fecal-oral route. Infections due to group A rotaviruses occur all over the world. Group B rotaviruses have been detected primarily in China. Group C rotaviruses also occur worldwide but have been detected only sporadically. Generally the incubation period is 1-2 days and the onset is abrupt. Symptoms include vomiting, diarrhea, fever, and abdominal pain. Loss of fluids is the most severe result of rotavirus infection and can lead to severe dehydration.

Specimen Stool from the diarrheal phase of disease; rectal swab

Sampling Time 3-5 days after onset

Collection Several specimens during the course of illness should be submitted in an attempt to eliminate false-negative results.

Reference Interval No virus detected

Use Differential diagnosis of acute onset gastroenteritis, diarrhea, emesis.

Limitations Commercially available EIA kit for detection of rotavirus in stool leads to false positives when used in healthy neonates, but results in symptomatic subjects are reliable. These assays detect the highly conserved internal capsid protein of the group A rotavirus.

Methodology Latex agglutination (LA), enzyme immunoassay (EIA), dot blot technology.

Additional Information Among infants and small children worldwide, rotavirus infections are the most common cause of severe gastroenteritis. Young children between 6 months-3 years of age exhibit the most severe effects of the disease. The illness is most likely to occur in winter, is highly contagious, involves 5-8 days of diarrhea, and is rarely fatal. Other viral agents causing gastroenteritis are enteric adenoviruses, caliciviruses, astroviruses, coronaviruses, and Norwalk and Norwalk-like viruses.

References
Belhorn T, "Rotavirus Diarrhea," *Curr Probl Pediatr*, 1999, 29(7):198-207.

Centers for Disease Control and Prevention, "Laboratory-Based Surveillance for Rotavirus - United States, July 1996-June 1997," *JAMA*, 1998, 279(3):192.

Desselberger U, "Rotavirus Infections: Guidelines for Treatment and Prevention," *Drugs*, 1999, 58(3):447-52.

Steele JC Jr, "Rotavirus," *Clin Lab Med*, 1999, 19(3):691-703.

Internet Web Sites
vm.cfsan.fda.gov/~mow/chap33.html
www.astdhpphe.org/infect/rot.html
www.cdc.gov/ncidod/dvrd/revb/

♦ **Roxanol**® *see* Morphine, Urine *on page 921*
♦ **Roxanol SR**™ *see* Morphine, Urine *on page 921*

RPR

Related Information

Anticardiolipin Antibody *on page 167*
Bacterial Culture, Genital Specimen *on page 239*
Darkfield Examination, Syphilis *on page 501*
FTA-ABS, Serum *on page 618*
MHA-TP *on page 912*
Neisseria gonorrhoeae Culture and Smear *on page 945*
VDRL, Serum or Cerebrospinal Fluid *on page 1303*

Test Includes Detection and titer of nontreponemal antibodies for detection of syphilis infections.

Abstract Antibodies that develop in response to *T. pallidum* infection are classified as nontreponemal antibodies and treponemal-specific antibodies. The nontreponemal antibodies react with lipoidal material such as cardiolipin. Such antibodies can be produced in other conditions (eg, autoimmune diseases and pregnancy). Any specimen that shows a positive nontreponemal antibody reaction should then be tested with a treponemal-specific test such as FTA-ABS.

Specimen Serum

Container Red top tube

Reference Interval Negative

Use Screening test for syphilis

Limitations This is a nontreponemal test and is associated with false-positive reactions due to intercurrent infections, pregnancy, drug addiction, autoimmune disease, increased age, Gaucher disease, malignancy, and a number of viral, protozoal, and *Mycoplasma* infections. This test cannot be performed on cerebrospinal fluid.

The VDRL and RPR are insensitive early (in the primary phase) and late (latent and tertiary phases). See VDRL, Serum or Cerebrospinal Fluid *on page 1303*.

Methodology Flocculation test to detect reagin antibody

Additional Information RPR is a screening test for syphilis and detects antibodies to cardiolipin, cholesterol, and lecithin, also called reagin. Such antibodies usually develop after 4-6 weeks of initial infection, peak during the secondary phase of disease, and then decrease with time. They also decrease and usually disappear with treatment. The RPR is more sensitive than the VDRL and the ART for determining efficacy of treatment.

Because of the many causes of false-positive tests, any reactive serum should be tested by a treponemal-specific test, preferably MHA-TP or FTA-ABS for confirmation. Serial serologic testing during pregnancy with maternal and neonatal serologic studies at delivery is desirable for detection of neonates at risk. In instances of negative tests even following dilutions, serology should be repeated within several weeks when suspicion of congenital syphilis exists.

References

Augenbraun MH and Rolfs R, "Treatment of Syphilis, 1998: Nonpregnant Adults," *Clin Infect Dis*, 1999, 28(S1):S21-8.

Darville T, "Syphilis," *Pediatr Rev*, 1999, 20(5):160-4.

Miller KE and Graves JC, "Update on the Prevention and Treatment of Sexually Transmitted Diseases," *Am Fam Phys*, 2000, 61(2):379-86.

Sheffield JS and Wendel GD Jr, "Syphilis in Pregnancy," *Clin Obstet Gynecol*, 1999, 42(1):97-106.

Singh AE and Romanowski B, "Syphilis: Review With Emphasis on Clinical Epidemiologic, and Some Biologic Features," *Clin Microbiol Rev*, 1999, 12(2):187-209.

Internet Web Sites

www.cdc.gov/std/Syphilis/STDFact-Syphilis.htm

♦ **rT₃** *see* Reverse T$_3$, Serum *on page 1160*

Rubella Culture and Serology

Related Information

TORCH *on page 1262*
Viral Culture *on page 1307*

Test Includes Isolation and identification of rubella virus in cell culture; detection of serologic response to rubella infection or vaccination

Abstract Rubella infection was one of the viral diseases targeted by the CDC for eradication by the end of the year 2000, because it is specifically a human disease with no animal reservoir. In the United States, the incidence of rubella infections has been very low since 1994. This has been primarily due to the wide use of mumps-measles-rubella (MMR) immunization in children. However, several recent outbreaks of rubella have occurred in populations which were not vaccinated, such as people who have immigrated to the United States from countries in which vaccination for rubella is uncommon. Infection with rubella is usually characterized by a macular exanthem, lymphadenopathy, pharyngitis, and conjunctivitis. The incubation period is 14-21 days but subclinical or asymptomatic infections are common. Severe transplacental infections can occur in the first trimester. Culture for rubella is used to identify an outbreak in conjunction with specific rubella serology.

Specimen Culture: two throat swabs, 10 mL urine, cerebrospinal fluid, tissues, amniotic fluid. Serology: serum, fetal blood

Container Culture: sterile container or viral transport medium; Serology: red top tube

Sampling Time Virus is more likely to be isolated if specimen is collected within 5 days after onset of illness. Serologic diagnosis requires an acute and convalescent serum specimen.

Collection Acute and convalescent samples for IgG.

Storage Instructions Culture specimens should not be stored. Specimens should be delivered immediately to the laboratory. If unavoidable delays occur, the specimen can be stored at 4°C for up to 3 days, but there is a loss of infectivity when culture is delayed.

Turnaround Time Positive cultures are usually detected in 3-7 days. Serology may take up to 1 week.

Reference Interval No virus isolated from culture. Absence of antibody indicates susceptibility to rubella. IgG, IgM: negative. Postvaccination: positive.

Critical Values Any positive rubella culture, especially from amniotic fluid. Presence of IgM antibody indicates acute infection or vaccination. Fourfold increase in IgG titer may indicate acute infection. **All positive rubella results should be immediately reported to the state health department.**

Possible Panic Range Evidence of susceptibility in a pregnant woman recently exposed to rubella

Use Aid in the diagnosis of rubella virus infections especially diagnosis of congenital rubella infection (IgM); Prenatal evaluation of immune status

Limitations Because of the length of time for positive detection, the isolation of rubella virus is usually of little help in the diagnosis of rubella. Exceptions may be in the cases of severe rubella complications, epidemiological purposes, and fatality. Most laboratories perform rubella serology as a qualitative test to determine susceptibility to infection and do not determine titers. This presents a problem when trying to diagnose infection by increasing IgG titers; special arrangements need to be made with the laboratory if titers are required for diagnosis. Rubella-specific antibodies are transferred from the mother to the child in utero and thus the detection of rubella-specific antibody in a child younger than 6 months of age may give a false-positive result. Rubella-specific IgM may persist for some months after vaccination.

Methodology Culture: cell culture, isolation, and confirmation/identification by antibody-specific neutralization. Serology: indirect fluorescent antibody (IFA), hemagglutination, radioimmunoassay (RIA), complement fixation (CF), latex agglutination (LA), enzyme immunoassay (EIA)

Additional Information Pregnant women who become infected with rubella have a very high risk of the virus crossing the placenta and infecting the fetus. Congenital rubella infections have disastrous effects, causing fetal death, premature delivery, and severe congenital defects including deafness and congenital heart disease. Neonates with congenital rubella excrete rubella virus in nasopharyngeal secretions and urine for many months after birth. These children pose a risk to susceptible pregnant women. Serology is available for diagnostic purposes. Usually immune status can be determined by examining a single serum sample.

The role of serologic testing for antibodies to rubella is different in different clinical settings. The simplest and most straight forward application is an assessment of immunity. If a woman has antibodies against rubella, there should be no concern about infection during subsequent pregnancy. If she is not immune, and is not pregnant, she should receive the rubella vaccine. (Continued)

Rubella Culture and Serology (Continued)

An infant born with an illness that may be congenital rubella should also be evaluated for rubella titers. Problems here include evaluation of whether antibody represents passively acquired immunity by transplacental passage or is indicative of true neonatal infection. In this setting, determination of the immunoglobulin class is particularly important, since IgM antibodies do not pass the placental barrier. The presence of rubella-specific IgM antibody strongly supports congenital infection.

References

Bar-Oz B, Ford-Jones L, and Koren G, "Congenital Rubella Syndrome. How Can We Do Better?" Can Fam Physician, 1999, 45:1865-9.

Bullens D, Smets K, and Vanhaesebrouck P, "Congenital Rubella Syndrome After Maternal Reinfection," Clin Pediatr (Phila), 2000, 39(2):113-6.

Craig SC, Broughton G 2nd, Bean J, et al, "Rubella Outbreak, Fort Bragg, North Carolina, 1995: A Clash of Preventive Strategies," Milit Med, 1999, 164(9):616-8.

Fenner F, "Candidate Viral Diseases for Elimination or Eradication," Bull World Health Organ, 1998, 76(Suppl 2):68-70.

Jacobs DS, DeMott WR, Oxley DK, et al, Laboratory Test Handbook, 5th ed, Hudson, OH: Lexi-Comp Inc, 2001.

Linder N, Sirota L, Aboudy Y, et al, "Placental Transfer of Maternal Rubella Antibodies to Full-Term and Preterm Infants," Infection, 1999, 27(3):203-7.

Internet Web Sites

www.astdhpphe.org/infect/rubella.html

♦ **Rufen®** see Ibuprofen, Serum on page 764

♦ **S-100** see Immunoperoxidase Procedures on page 780

♦ **S-100, Serum** see Neuron-Specific Enolase, Serum on page 949

♦ **Sabril®** see Antiepileptic Drugs Overview on page 176

♦ **Sabrilex®** see Antiepileptic Drugs Overview on page 176

Salicylate, Serum or Plasma

Related Information

Anion Gap, Serum, Plasma, or Urine on page 160
Lactic Acid, Whole Blood or Plasma on page 827
pH, Blood on page 1018

Synonyms Anacin®; ASA; Ascriptin®; Aspergum®; Aspirin; Bufferin®; Easprin®; Ecotrin®; Empirin®; Measurin®; Synalgos®; ZORprin®

Abstract This is the active product produced from aspirin (acetylsalicylic acid) in the body. It is an analgesic, antipyretic and anti-inflammatory drug. Salicylate was one of the top 10 drugs associated with fatalities in 1998.

Acute poisoning may include hypokalemia, respiratory alkalosis, dehydration, metabolic acidosis, and hepatotoxicity. Pulmonary and cerebral edema may develop.

Specimen Serum or plasma

Container Red top tube or lavender top (EDTA) tube

Sampling Time Time to peak serum concentration is about 1-2 hours. Optimal sampling time after dosage is 4-6 hours.

Reference Interval Therapeutic: ~10 mg/dL (SI: ~0.72 mmol/L) for analgesic; 15-20 mg/dL (SI: 1.09-1.45 mmol/L) for anti-inflammatory properties

Possible Panic Range Mild toxicity: ~30 mg/dL (SI: 2.17 mmol/L) (tinnitus, dizziness); severe toxicity: >80 mg/dL (SI: >3.62 mmol/L) 6 hours postingestion (CNS effects)

Use Monitor therapeutic drug level, evaluate aspirin toxicity. Most organ systems are affected by salicylate toxicity.

Methodology Immunoassay, high performance liquid chromatography (HPLC), gas chromatography (GC)

Additional Information
- Half-life: 2-3 hours
- Volume of distribution: 0.1-0.3 L/kg
- Protein binding: 90% to 95%

In salicylate poisoning the following findings may occur: initial respiratory alkalosis, mixed respiratory alkalosis and increased anion gap metabolic acidosis, sometimes with ketosis and possible elevated blood glucose (see table).
In children, the alkalosis phase is very short. The Done nomogram is used to estimate severity of toxicity based on blood level 6 hours or more after a single-dose ingestion, but cannot be used for chronic intoxication. See nomogram.

Serum Salicylate: Clinical Correlations

Serum Salicylate Concentration (mg/dL)	Desired Effects	Adverse Effects / Intoxication
~10	Antiplatelet Antipyresis Analgesia	GI intolerance and bleeding, hypersensitivity, hemostatic effects
15-30	Anti-inflammatory	Mild salicylism
25-40	Treatment of rheumatic fever	Nausea/vomiting, hyperventilation, salicylism, flushing, sweating, thirst, headache, diarrhea, and tachycardia
>40		Respiratory alkalosis, hemorrhage, excitement, confusion, asterixis, pulmonary edema, convulsions, tetany, metabolic acidosis, fever, coma, cardiovascular collapse, renal and respiratory failure

Patients in chronic salicylate toxicity may not exceed levels in the therapeutic range; their levels may be 10-20 mg/dL.

Serum Salicylate Level and Severity of Intoxication Single Dose Acute Ingestion Nomogram

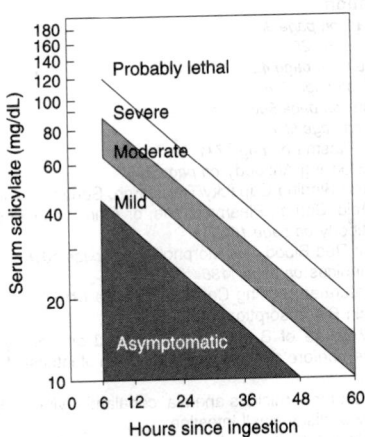

Nomogram relating serum salicylate concentration and expected severity of intoxication at varying intervals following the ingestion of a single dose of salicylate.
From Done AK, "Aspirin Overdosage: Incidence, Diagnosis, and Management," *Pediatrics*, 1978, 62:890-7 with permission.

Signs and symptoms of **acute overdose** may include nausea, vomiting, dehydration, hyperpnea, oliguria, and tinnitus. **Severe poisoning** can include coma, convulsions, severe hyperpnea, and metabolic acidosis.

Symptoms of **chronic salicylism** include fever, vomiting, and tachypnea. It is likely following doses in excess of 100 mg/kg/day for 2 days of more.

References

Done AK, "Aspirin Overdosage: Incidence, Diagnosis, and Management," *Pediatrics*, 1978, 62(5 Pt 2 Suppl):890-7.
(Continued)

Salicylate, Serum or Plasma *(Continued)*

Fenton J, *The Laboratory and the Poisoned Patient: A Guidebook for Interpreting Laboratory Data*, Washington, DC: AACC Press, 1998, 304-9.

Rexrode KM, Buring JE, Glynn RJ, et al, "Analgesic Use and Renal Function in Men," *JAMA*, 2001, 286(3):315-21.

Vane JR and Botting RM, "Mechanism of Action of Aspirin-Like Drugs," *Semin Arthritis Rheum*, 1997, 26(6 Suppl 1):2-10.

♦ **Salivary Gland Needle Aspiration** *see* Fine Needle Aspiration, Deep and Superficial Masses *on page 590*

♦ **SAMHSA** *see page 63*

♦ **Sample Space** *see page 11*

♦ **Sampling Problems** *see page 11*

♦ **Sandhoff Disease** *see* Inherited Diseases of Metabolism and Cell Structure *on page 792*

♦ **Sandimmune®** *see* Cyclosporine, Blood *on page 487*

♦ **SaO₂** *see* Oxygen Saturation, Blood *on page 991*

♦ ***Sarcoptes scabiei* Skin Scrapings Identification** *see* Arthropod Identification *on page 213*

♦ **Sarin** *see* Chemical Warfare Agents *on page 382*

♦ **Sarin Exposure** *see* Acetylcholinesterase, Red Cell and Serum *on page 93*

♦ **SARS** *see* Severe Acute Respiratory Syndrome Diagnostic Procedures *on page 1193*

♦ **SARS-CoV** *see* Severe Acute Respiratory Syndrome Diagnostic Procedures *on page 1193*

♦ **Satellite Bags** *see* Newborn Crossmatch and Transfusion *on page 952*

Schilling Test

Related Information

Anemia Flowchart *on page 35*
Bone Marrow *on page 296*
Cobalamin, Serum *on page 424*
Folic Acid, RBC *on page 606*
Folic Acid, Serum *on page 606*
Gastrin, Serum *on page 631*
Homocyst(e)ine, Plasma *on page 741*
Intrinsic Factor Blocking Antibody *on page 806*
Iron and Total Iron Binding Capacity/Transferrin, Serum *on page 807*
Methylmalonic Acid, Serum, Plasma, Urine, or Amniotic Fluid *on page 908*
Parietal Cell Antibody *on page 1005*
Peripheral Blood: Red Blood Cell Morphology *on page 1016*
Red Blood Cell Indices *on page 1136*
Vitamin B₁₂ Unsaturated Binding Capacity *on page 1316*

Synonyms Vitamin B₁₂ Absorption Test

Test Includes Measure of B₁₂ absorption (based on urinary excretion of a radiolabeled dose), before and after administration of intrinsic factor (two stage procedure).

Abstract *In vivo* test for pernicious anemia, cobalamin (vitamin B₁₂) malabsorption, and integrity of distal small intestine.

Patient Preparation Patient must be fasting from midnight the day of the test. Patient should have received no B vitamins for a period of 3 days before the test.

Aftercare The radiolabeled crystalline cobalamin (⁵⁷Co) used in the Schilling test generates such a low level of radiation as to be noninjurious to patient, attendants, or the environment.

Container Large gallon (plastic preferred) container that has been counted and shown to be free of contaminating radiation

Sampling Time 24 hours (urine)

Collection 24-hour urine

Storage Instructions Keep specimen on ice or refrigerate.

Causes for Rejection Patient having radioisotope scan prior to test, patient receiving cobalamin (Cbl) prior to the test, patient not fasting, incomplete collection of urine, failure to administer the parenteral B₁₂, contamination of urine with stool (which will contain some unabsorbed radiolabeled B₁₂). The waiting time between prior radiopharmaceutical administration and the Schilling examination varies considerably, dependent on the isotope previously administered. Only a few days may be involved or over 100 days may be required.

Special Instructions This radiopharmaceutical should not be administered to patients who are pregnant or during lactation unless the information to be gained outweighs the potential hazards. The test should not be started within 2-3 days of a therapeutic dose (1000 μg) of B_{12} or previous Schilling test. Bone marrow examinations and B_{12} and folate levels must be obtained before the Schilling test is performed.

Reference Interval Normal values vary with the laboratory and procedure used. Generally, >10% excretion in the urine of radioactive B_{12} indicates intact intrinsic factor (IF) function. When only a few percent of B_{12} absorption occurs without IF, with improvement into normal range with exogenous IF, the diagnosis of pernicious anemia is established. Poor absorption (<6% or 7%) with, as well as without, IF suggests intestinal malabsorption.

Use Assess cobalamin absorption in the diagnosis of malabsorption due to the lack of intrinsic factor (ie, Addisonian pernicious anemia [APA]), a diagnostic adjunct in other defects of small intestinal absorption; evaluate extent of Crohn disease in terminal ileum. Part I of the Schilling test is ^{57}Co by itself. If abnormal, 5 days later it is repeated with intrinsic factor administered orally at the same time (Part II). If this part of the test is still abnormal, it suggests that the cause is not gastric in origin (PA) but lower in the GI tract. A well-conducted Schilling test may establish an important diagnosis, even when the disease (APA) is in full remission.

Limitations If other isotope tests are to be performed, Schilling test should be completed first. The presence of renal dysfunction, pancreatic insufficiency, bacterial overgrowth of intestinal content, antibodies against IF in gastric secretions, myxedema, liver disease, or any other condition resulting in decreased absorption of B_{12} from the GI tract, its concentration in the liver, or its excretion in the urine may result in abnormal values. Incomplete urine collection or fecal contamination may invalidate results.

Need for the Schilling test has been called into question on theoretic grounds as well as on practical logistic and operational bases. The most common cause of an abnormal Schilling test is an incomplete urine collection. Economic necessity dictates that most Schilling tests will be performed on an outpatient basis with attendant operational shortcomings (eg, incomplete urine collection, fecal contamination of urine). Acceptance of a urine immunoassay test for intrinsic factor may also impact use of the Schilling test.

In elderly individuals with hypochlorhydria, the proteolytic activity of pepsin may be negated by absence of requisite low pH with failure of release of Cbl from food protein and a resultant false-negative stage I Schilling test (normal Stage I results in a Cbl-deficient individual). On the other hand, a modified "food-Cbl" absorption test found that in some healthy elderly individuals there was reduced absorption but normal serum levels. Renal insufficiency, with delay in excretion of labeled Cbl and potential false-positive result may be circumvented by measuring radiolabeled Cbl in plasma samples.

Methodology Patient swallows radiolabeled B_{12} dose and receives a B_{12} intramuscular injection. The patient's urine is collected for 24 hours. The total activity from the B_{12} label is measured and calculated. The percent excretion of vitamin B_{12} is then determined. This is Part I of the Schilling test. If abnormal (<7% to 10% excretion indicating impaired cobalamin absorption), Part II of the test is performed. This is a repeat of Part I, but with oral administration of intrinsic factor.

Additional Information After release from food proteins, cobalamin binds to glycoproteins (R-binders) in saliva and gastric juice. In the upper small intestine, pancreatic proteases release B_{12} from the R-proteins. B_{12} then binds to intrinsic factor (IF) secreted by gastric parietal cells. The IF-B_{12} complex resists proteolysis, and in the distal ileum it binds to receptors unique to cells of the ileal mucosal membrane microvilli. A few hours after IF-B_{12} binds to ileal absorptive cells, B_{12} appears in the circulation complexed to transcobalamin II.

Some patients with pernicious anemia may not absorb B_{12} adequately, even when given with IF, reversible after B_{12} therapy, thus complicating the diagnosis of pernicious anemia using the Schilling test. The importance of establishing the diagnosis and commencing life-long B_{12} maintenance therapy (so as to reverse the severe effects of the megaloblastic anemia and nervous system (Continued)

Schilling Test *(Continued)*

degenerative process - combined systems disease) dictates that a combination of clinical and laboratory findings be considered. The laboratory picture should include appropriate abnormalities of RBC indices (in particular high MCV), presence of anemia, peripheral blood smear findings of oval macrocytes, tear drop shaped RBCs, leukopenia, thrombocytopenia, hypersegmented PMNs, megaloblastic marrow picture, decreased serum B_{12} level, usually increased serum LD, all to be correlated with Schilling test result and the clinical presentation.

The Schilling test may be "overused" in the evaluation of Cbl deficiency and is often normal when more sensitive indicators of early or subclinical functional deficiency (eg, methylmalonic acid (MMA), "metabolic levels" are abnormal (elevated)). Homocyst(e)ine levels are increased in both Cbl and folate deficiency. See Methylmalonic Acid, Serum, Plasma, Urine, or Amniotic Fluid *on page 908*.

A peroral nonradioactive vitamin B_{12} absorption test has been utilized.

References

Brigden ML, "Schilling Test Still Useful in Pernicious Anemia?" *Postgrad Med*, 1999, 106(5):37-8.

Magnus E and Müller C, "A New, Peroral Nonradioactive Vitamin B_{12} Absorption Test Compared With the Schilling Test," *Eur J Haematol*, 1995, 54:117-9.

Snow CF, "Laboratory Diagnosis of Vitamin B_{12} and Folate Deficiency," *Arch Intern Med*, 1999, 159(12):1289-98.

Waters HM and Dawson DW, "An Enzyme Immunoassay for Intrinsic Factor in Urine," *Clin Lab Haematol*, 1999, 21(3):169-72.

♦ **Schistocytes** *see* Disseminated Intravascular Coagulation Screen *on page 517*

Schistosomiasis Diagnostic Procedures

Related Information

Ova and Parasites, Direct Exam *on page 985*
Urinalysis *on page 1289*
Urinary Tract Cytology *on page 1293*

Abstract In excess of 250 million people have schistosomiasis (bilharziasis) and it is associated with death in 200,000 cases per year. The three major species of these blood flukes are *Schistosoma haematobium*, *S. mansoni*, and *S. japonicum*. **Identification of ova in rectal or bladder biopsy, feces, or urine is definitive.** *S. mansoni* and *S. japonicum* ova are found in colorectal biopsies. *S. mansoni* and *S. japonicum* cause hepatic pipestem fibrosis, and marked splenomegaly may evolve secondary to increased portal pressure. Lesions of *S. haematobium* are found on cystoscopy and ova confirmed on biopsy. Its relationship to squamous cell carcinoma of bladder is widely recognized.

Specimen Serum for serology; urine and stool for ova; rectal, bladder, and liver biopsies for tissue diagnosis

Container Red top tube for serology

Special Instructions Geographic information is pivotal.

Use Serologic methods may support the clinical diagnosis of schistosomiasis in selected cases. The demonstration of ova in urine or stool, bladder or bowel biopsy is definitive. Evaluation of transverse myelitis, fever, hepatomegaly, splenomegaly, diarrhea, bloody feces, variceal bleeding, anemia.

Limitations Antibodies persist following cure. Serological testing does not differentiate between recently acquired infection and chronic multiple exposures and so is simply reported as positive or negative. All tests do not differentiate between intestinal and vesical schistosomiasis. Serologic diagnosis appears limited to instances in which ova cannot be found in fecal or urine specimens, or rectal or vesical biopsy. Cross-reactions with other helminths such as *Ancylostoma* species and *Ascaris lumbricoides* can occur. Multiple fecal examinations provide better sensitivity and specificity than immunodiagnostic kits.

Methodology Serology: indirect fluorescent antibody (IFA) using sections of adult worms, enzyme-linked immunosorbent assay (ELISA) with egg antigen, circulating anodic antigen and circulating cathodic antigen are antigens which are genus specific. Indirect hemagglutination (IHA). An immunoblot assay for adult-worm antigen has reported 95% sensitivity and 100% specificity.

Hatching procedures are done in which miracidic evolve from excreted ova.

Multiple fecal examinations are more sensitive and more specific than are immunologic tests.

Additional Information Schistosomiasis worldwide is one of the most common parasitic diseases (the most common cause of hematuria, for example).

References

Davis A, "Schistosomiasis," *Manson's Tropical Diseases*, 21st ed, Chapter 74, Cook GC and Zumla A, eds, Philadelphia, PA: WB Saunders Co, 2003, 1431-69.

Hamilton JV, Klinkert M, and Doenhoff MJ, "Diagnosis of Schistosomiasis: Antibody Detection, With Notes on Parasitological and Antigen Detection Methods," *Parasitology*, 1998, 117(S1):41-57.

Kaplan BS and Meyers K, Images in Clinical Medicine, "*Schistosoma haematobium*," *N Engl J Med*, 2000, 3443(15):1085.

Ropper AH and Stemmer-Rachamimov A, "A 31-Year-Old Man With an Apparent Seizure and a Mass in the Right Parietal Lobe," Case Records of the Massachusetts General Hospital. Case 21-2001, Scully RE, Mark EJ, McNeely WF, et al, *N Engl J Med*, 2001, 345(2):126-31.

Ross AG, Bartley PB, Sleigh AC, et al, "Schistosomiasis," *N Engl J Med*, 2002, 346(16):1212-20.

Salah F, Demerdash Z, Shaker Z, et al, "A Monoclonal Antibody Against *Schistosoma haematobium* Soluble Egg Antigen: Efficacy for Diagnosis and Monitoring of Cure of *S. haematobium* Infection," *Parasitol Res*, 2000, 86(1):74-80.

Southgate VR and Bray RA, "Medical Helminthology," *Manson's Tropical Diseases*, 21st ed, Appendix III, Cook GC and Zumla A, eds, Philadelphia, PA: WB Saunders Co, 2003, 1669-75.

Woolhouse ME and Hagan P, "Seeking the Ghost of Worms Past," *Nat Med*, 1999, 5(11):1225-7.

Internet Web Sites

www.cdc.gov/ncidod/dpd/parasites/schistosomiasis/default.htm

www.who.int/health-topics/schisto.htm

- ◆ **Schwachman-Diamond Syndrome** *see* White Blood Cell Count *on page 1330*
- ◆ **Scl-70 Antibody** *see* Topoisomerase I Antibody *on page 1261*
- ◆ **Scleroderma Antibody** *see* Topoisomerase I Antibody *on page 1261*
- ◆ **Scoop** *see* Gamma Hydroxybutyrate, Serum or Urine *on page 630*
- ◆ **Scorpions** *see* Arthropod Identification *on page 213*
- ◆ **Screen for Disseminated Intravascular Coagulation** *see* Disseminated Intravascular Coagulation Screen *on page 517*
- ◆ **Screen for Hypercoagulation** *see* Hypercoagulation Panel *on page 758*
- ◆ **Secobarb** *see* Barbiturates, Qualitative, Urine *on page 248*
- ◆ **Secobarbital** *see* Barbiturates, Quantitative, Serum or Plasma *on page 248*
- ◆ **Seconal**™ *see* Barbiturates, Quantitative, Serum or Plasma *on page 248*
- ◆ **Secretin Test** *see* Gastrin, Serum *on page 631*
- ◆ **Securon**® *see* Verapamil, Serum or Plasma *on page 1306*
- ◆ **Sedantoinal**® *see* Mephenytoin, Serum *on page 896*
- ◆ **Sedimentation Rate** *see* Fibrinogen *on page 583*

Sedimentation Rate, Erythrocyte

Related Information

C-Reactive Protein, Serum *on page 467*

Fibrinogen *on page 583*

Synonyms Westergren Sed Rate

Abstract The erythrocyte sedimentation rate (ESR) is a nonspecific measure of inflammation which is applied to the detection/evaluation of infectious and immune based (in particular rheumatic type) inflammatory disease. It is commonly used to follow management of rheumatologic patients. The test is a measure of the acute phase inflammatory response. Test result is derived from the sedimentation of patient's red blood cells through his/her plasma. As such, ESR is accelerated by increase in the plasma level of acute phase proteins of large molecular size, (fibrinogen in particular) and by anemia. Thus, the ESR may reflect both the hyperproteinemia and anemia of inflammatory disease. ESR methodology has undergone modification in order to increase precision and decrease biohazard exposure.

Specimen Whole blood

Container Lavender top (EDTA) tube or citrated plasma in 4:1 dilution (4 volumes of blood to 1 volume of 109 mmol/L of trisodium citrate)

Collection Specimen must be received and test carried out within 4-6 hours of collection.

Storage Instructions Keep specimen at 4°C prior to testing

Causes for Rejection Insufficient blood, clotted, hemolyzed specimen

Turnaround Time 1-2 hours, some recent automated methods can determine an ESR endpoint in as little as 20 minutes

(Continued)

Sedimentation Rate, Erythrocyte *(Continued)*

Special Instructions If EDTA anticoagulated blood is used for the standard Westergren method, 1 volume of 109 mmol/L of trisodium citrate is added to 4 volumes of blood just before performing the test.

Reference Interval Male: younger than 50 years of age: 0-15 mm/hour, older than 50 years of age: 0-20 mm/hour; female: younger than 50 years of age: 0-25 mm/hour, older than 50 years of age: 0-30 mm/hour by Westergren method. See reference by Wolfe and Michaud for their extensive study of and application of the ESR in a number of different settings. Their study indicates that in a rheumatology clinic women younger than 60 years of age who have noninflammatory disorders have upper limit of normal for ESR of 38 mm/hour.

Use Evaluate the activity of infections, inflammatory states, autoimmune disorders, and plasma cell dyscrasias; in particular to screen for, to assess severity of and follow clinical activity of rheumatic diseases, especially rheumatoid arthritis; important in clinical evaluation of temporal arteritis and polymyalgia rheumatica

Limitations Anemia and paraproteinemia invalidate results; High-thermal-amplitude cold agglutinins cause less rapid sedimentation (lower ESR) as the temperature rises towards 37°C. The ESR is usually reduced in stored blood. Some procedural methods may be associated with hazardous exposure of medical technologists to fresh whole blood.

Methodology Westergren: A 30 cm long glass tube, 2.55 mm in diameter, is filled with mixed sample of blood up to the 200 mm mark of a Westergren tube. After 1 hour the column of clear plasma above the upper limit of sedimenting red cells is measured. The result is given as number of mm in 1 hour.

In the past decade a number of semiautomated/automated procedures have been developed, many in Europe, to replace the classical ESR procedure and its variants. Most of these are closed systems that puncture the stopper of a blood collection tube to obtain a capillary sample of whole blood upon which red cell sedimentation is measured by infrared microphotometry or photoelectronically utilizing photodiodes/phototransistors. The appropriate anticoagulant must be utilized, specialized blood collection tubes may be necessary. The measurements are converted into comparable Westergren values. Results of comparison studies have been reported as generally favorable although a few have not been acceptable due to lack of sufficient correlation with Westergren type ESR. Instrumentation for an earlier alternative sedimentation rate method (the ZSR) was discontinued by its manufacturer but has recently been resurrected in China. Studies in China determined the normal ZSR (326 normal subjects) to be 52.5% ±5.05% while 106 inpatients had ZSR of 68.52% ±5.54%. Comparison with the Westergren method gave agreement of 86.6%. Advantages of the ZSR (as noted previously in Western countries) were rapid result, low sample volume, not affected by anemia, and lack of need for age/sex correction.

Additional Information Elevations in fibrinogen, alpha- and beta-globulins (acute phase reactants), and immunoglobulins increase the sedimentation rate of red cells through plasma. The test is important in the diagnosis and management of temporal arthritis and of polymyalgia rheumatica (PMR), which are nearly always but not uniformly, characterized by a significantly elevated ESR. Presence of a normal ESR in a PMR-like clinical setting could lead to delay in diagnosis and therapy.

The ESR, while lacking in specificity, is elevated in a very broad spectrum of conditions characterized by inflammation. For decades the ESR has played a prominent role in the evaluation and monitoring of rheumatoid inflammatory processes, notably rheumatoid arthritis and its variants including ankylosing spondylitis. ESR continues to be actively utilized in this manner being further defined and itself further defining rheumatoid inflammatory conditions and their therapy.

In a study by Wolfe comparing IgM rheumatoid factor (RF) methods, serial measures of RF did correlate with changes in clinical activity but repeat testing to determine clinical status was found to be only a "fair predictor" and to have little utility in clinical practice. ESR and C-reactive protein (CRP) were found to be more effective in following clinical activity.

In a study comparing the value of ESR vs CRP in measuring disease activity in ankylosing spondylitis (AS) the positive predictive values of both CRP and ESR were low, neither CRP nor ESR was considered superior in assessment of disease activity. On the basis of literature review (Ruof and Stucki, 1967 through April 1998), it was concluded that neither ESR or CRP is superior in terms of validity and it was considered that acute phase reactants "do not comprehensively represent the disease process" in AS.

Lurie et al found that an ESR >100 mm/hour, in a patient with back pain, has a likelihood ratio of 55 for presence of a serious underlying cause. See tables.

Table 1. Factors That May Affect Erythrocyte Sedimentation Rate

Factor	Effect on rate
Red blood cell aggregation (plasma proteins raised in infection, inflammation, and malignant conditions)	Increased[1]
Pregnancy	Increased[1]
Anemia (decreased hematocrit)	Increased[1]
Obesity	Possibly increased[1]
Drug therapy (eg, steroids, anti-inflammatories)	Decreased[1]
Female sex	Slightly increased[1]
Old age	Increased,[1] possibly due to higher prevalence of disease
Red blood cell abnormalities	
Macrocytosis	Increased[1]
Microcytosis	Decreased
Sickle cell disease	Decreased
Polycythemia	Decreased
Protein abnormalities (hypofibrinogenemia, hypogammaglobulinemia, dysproteinemia with hyperviscosity state)	Decreased

[1]Upper limit of normal: Age ≤50 years: men: 15 mm/hour; women: 25 mm/hour. Age >50 years: men: 20 mm/hour; women: 30 mm/hour.

Adapted from Sox and Liang, *Postgrad Med*, 1998, 103(5):258.

Table 2. Comparison of Erythrocyte Sedimentation Rate, Plasma Viscosity, and C-Reactive Protein Tests

Test	Advantages	Disadvantages
Erythrocyte sedimentation rate (ESR)	Inexpensive, quick, simple to perform	Affected by anemia and red blood cell size, not sensitive enough for screening
Plasma viscosity	Not affected by anemia or red blood cell size	Cumbersome apparatus, expensive, not widely available
C-reactive protein	Rapid response to inflammation, complementary to ESR	Wide reference range necessitates sequential recording of values, expensive, batch processing may delay individual results

References

Koepke JA, "Measuring the Erythrocyte Sedimentation Rate," *Advanced Laboratory Methods in Haematology*, Chapter 10, Rowan RM, van Assendelft OW, and Preston FE, eds, London, UK: Arnold, 2002, 225-37.

Lurie JD, Gerber PD, and Sox HC, "A Pain in the Back," *N Engl J Med*, 2000, 343(10):723-6.

Paulus HE, Ramos B, Wong WK, et al, "Equivalence of the Acute Phase Reactants C-Reactive Protein, Plasma Viscosity, and Westergren Erythrocyte Sedimentation Rate When Used to

(Continued)

Sedimentation Rate, Erythrocyte *(Continued)*

Calculate American College of Rheumatology 20% Improvement Criteria or the Disease Activity Score in Patients With Early Rheumatoid Arthritis," *J Rheumatol*, 1999, 26(11):2324-31.

Ruof J and Stucki G, "Validity Aspects of Erythrocyte Sedimentation Rate and C-reactive Protein in Ankylosing Spondylitis: A Literature Review," *J Rheumatol*, 1999, 26(4):966-70.

Su J, Peng Z, Li P, et al, "Investigation of the Zeta Reference Values by Chinese-built Zetafuge® and Preliminary Application," *Hua Hsi I Ko Ta Hsueh Hsueh Pao*, 1998, 29(4):431-4.

Wolfe F, "A Comparison of IgM Rheumatoid Factor by Nephelometry and Latex Methods: Clinical and Laboratory Significance," *Arthritis Care Res*, 1998, 11(2):89-93.

Selenium, Serum

Related Information

Selenium, Urine *on page 1185*

Synonyms Se, Serum

Applies to Selenocysteine; Selenoprotein P

Abstract Selenium (Se) is an important essential trace element in human nutrition. It is a constituent of glutathione peroxidase and iodothyronine deiodinases. It is thought to play a protective role in human carcinogenesis and cardiovascular disease.

Specimen Serum, plasma, whole blood. Hair may be analyzed: hair concentrations of selenium correlate with those of blood.

Container Trace metal-free blood collection tube. See the Trace Elements Introduction *on page 77*.

Collection Draw blood into special trace metal vacuum tube. Centrifuge and pour serum into special plastic metal-free vial for transport. Use powder-free gloves.

Causes for Rejection Failure to obtain specimen using special trace element blood collection kit or store specimen in special metal-free plastic vials

Reference Interval Serum: 95-165 ng/mL (SI: 1203-2090 nmol/L). Approximately 40% higher for whole blood. Serum reflects recent intake; red cells reflect more remote intake. Whole blood therefore reflects an average of recent and remote intake of Se. (Se-dependent glutathione peroxidase activity reflects Se available for enzyme synthesis - see below.)

Critical Values Levels >500 ng/mL (SI: >6332 nmol/L) are associated with toxicity

Use Monitor Se nutritional status, **especially in long-term parenteral nutrition**. Studies have indicated no factor or factors that can accurately predict serum levels to preclude need for measurement. May be used diagnostically in cardiomyopathy of unknown cause, especially where nutritional factors are suspected. Monitor Se status in children with propionic acidemia who are at risk of deficiency on their special diets. Monitor for acute toxicity states, characterized clinically by nausea, diarrhea, mental alterations, peripheral neuropathy, and loss of hair and nails.

Limitations Some controversy exists regarding the "best" marker for Se status. Since Se as selenomethionine is incorporated nonspecifically into protein, serum and whole blood Se concentration increases with increasing Se intake to different degrees, depending on inorganic or organic sources of Se. Glutathione peroxidase activity is more sensitive to deficiency but the test is not well standardized and therefore not reproducible from laboratory to laboratory. Hair Se may be contaminated by Se-containing shampoo. Serum Se concentration correlates best with intake and therefore with both deficiency and toxicity states, but a wide range of serum levels is compatible with apparent good health.

Methodology Graphite furnace atomic absorption spectrophotometry (GFAAS), inductively-coupled plasma-mass spectrophotometry (ICP-MS)

Additional Information Multiple cases of Se deficiency have been reported, mostly among patients given parenteral nutrition without Se supplementation. Deficiency also occurs endemically in places where soil Se is low, and low levels are thus present throughout the food chain. Endemic cretinism, Balkan nephropathy, Keschan disease (endemic dilated cardiomyopathy), and Kashin-Back disease (endemic deforming osteoarthritis) are probably all caused by endemic Se deficiency causing the host to poorly tolerate an additional environmental stress (cretinism: iodine deficiency; the others: unknown local toxins). Low Se blood levels have been shown to be a risk factor for peripartum cardiomyopathy in sub-Sahara Africa. Simple deficiency is marked by whitening of the nailbeds, erythrocyte macrocytosis, cardiomyopathy, painful weak muscles, skin and hair depigmentation, and elevations of transaminase and creatinine kinase.

Selenium is depressed in HIV infection, critical illness, kwashiorkor, inflammatory bowel disease, renal failure, hemodialysis status, low protein diet, phenylketonuria, maple syrup urine disease (possibly all in part related to poor protein intake), low birth weight, and premature infants with inadequate Se intake. Levels are increased mostly with the use of glucocorticoids.

Selenium toxicity can occur endemically, again due to high soil levels, or through accidental or industrial exposure. Symptoms include garlic breath, odor, thick brittle fingernails, dry brittle hair, red swollen skin of the hands and feet, and nervous system abnormalities of numbness, convulsions, or paralysis.

Therapeutic trials for Se indicate preventive effects for colorectal, lung, and prostate carcinomas.

References

Clark LC, Combs GF Jr, Turnbull BW, et al, "Effects of Selenium Supplementation for Cancer Prevention in Patients With Carcinoma of the Skin: A Randomized Controlled Trial," *JAMA*, 1996, 276(24):1957-63.

Colditz GA, "Selenium and Cancer Prevention: Promising Results Indicate Further Trials Required," *JAMA*, 1996, 276(4):1984-85.

Harrison I, Littlejohn D and Fell GS, "Distribution of Selenium in Human Blood Plasma and Serum," *Analyst*, 1996, 121(2):189-94.

Lin TH, Tseng WC, and Cheng SY, "Direct Determination of Selenium in Human Plasma and Seminal Plasma by Graphite Furnace Atomic Absorption Spectrophotometry and Clinical Application," *Biol Trace Elem Res*, 1998, 64(1-3):133-49.

Mason JB, "Consequences of Altered Micronutrient Status," *Cecil Textbook of Medicine*, 21st ed, Chapter 231, Goldman L and Bennett JC, eds, Philadelphia, PA: WB Saunders Co, 2000, 1176.

Suzuki KT and Itoh M, "Metabolism of Selenite Labeled With Enriched Stable Isotope in the Bloodstream," *J Chromatogr B Biomed Sci Appl*, 1997, 692(1):15-22.

Thomson CD, "Selenium Speciation in Human Body Fluids," *Analyst*, 1998, 123(5):827-31.

Selenium, Urine

Related Information

Selenium, Serum *on page 1184*

Synonyms Se, Urine

Abstract Urine selenium (Se) is used in conjunction with serum Se to assess Se nutrition or potential toxic exposure. Like other 24-hour urine collections of an essential element, this reflects recent intake, assuming the patient is in Se balance.

Specimen 24-hour urine

Container Acid-washed plastic urine container

Collection Avoid contamination by hair, since some patients use Se-containing shampoos.

Causes for Rejection Failure to store urine in acid-washed plastic containers

Reference Interval Levels <15 µg/L or >150 µg/L (SI: <190 µmol/L or >1900 µmol/L) probably represent unusually low or high intake without necessarily representing illness. Values vary widely and in apparently healthy U.S. citizens they have been reported to vary from 7 µg/L (SI: 89 µmol/L) (24-hour sample) to 231 µg/L (SI: 2925 µmol/L). Intake is partly determined by local soil content of Se and use of local vegetables as food. Healthy persons in New Guinea have been reported with levels as low as 0.9 µg/L (SI: 11.4 µmol/L) but similar patients have rapidly developed symptomatic Se deficiency when placed on total parenteral nutrition lacking Se. Urine levels of 7 µg/L have been reported from China in areas where Se deficiency is symptomatic. Levels >880 µg/L (SI: >11,144 µmol/L) have been seen in chronic selenosis and >600 µg/L (SI: 7599 µmol/L) during the first 24 hours after acute Se intoxication. Levels >500 µg/L (SI: 6332 µmol/L) probably represent toxicity. Some authors or laboratories report as µg/day.

Use Monitor nutritional therapy, especially parenteral nutrition; monitor potential toxic exposure

Limitations Spot urine Se is of little value, as urine Se goes up after each meal depending on Se intake. Fasting urine samples may give a reasonable estimate of urinary output of Se on a population basis. Twenty-four hour urines are necessary for diagnosis of deficiency. Urinary excretion may be lower in children, elderly people and pregnant women.

(Continued)

Selenium, Urine *(Continued)*

Methodology Atomic absorption (AA), inductively-coupled plasma/mass spectrometry (ICP/MS)

Additional Information In the case of Se, skin and stool losses are significant and amount to 30% to 50% of total losses; nevertheless, urine losses often represent overflow losses and can help indicate whether recent intake has been adequate or possibly toxic. At higher levels of intake, the 24-hour urine is well correlated with intake and can be used as evidence of excess intake, adequate intake, or prior toxic exposure. When Se supplementation normalizes serum Se and whole blood Se, and then Se supplementation is stopped, urine Se falls back toward baseline much faster than blood or serum levels. Impaired hepatic production of selenoproteins and other Se-containing compounds is the most likely explanation for reduced serum Se concentrations in patients with chronic liver disease.

Red cell glutathione peroxidase may be monitored as an index of Se status. Levels will be depressed in deficiency but will not be elevated in toxicity.

References

Horng CJ, Tsai JL, and Lin SR, "Determination of Urinary Arsenic, Mercury, and Selenium in Steel Production Workers," *Biol Trace Elem Res*, 1999, 70(1):29-40.

Thomson CD, "Selenium Speciation in Human Body Fluids," *Analyst*, 1998, 123(5):827-31.

Thomson CD, Smith TE, Butler KA, et al, "An Evaluation of Urinary Measures of Iodine and Selenium Status," *J Trace Elem Med Biol*, 1996, 10(4):214-22.

♦ **Selenocysteine** *see Selenium, Serum on page 1184*

♦ **Selenoprotein P** *see Selenium, Serum on page 1184*

Semen Analysis, Advanced

Related Information

Infertility Screen *on page 786*

Semen Analysis, Basic *on page 1187*

Sperm Mucus Penetration Test (Human or Bovine Cervical Mucus) *on page 1219*

Sperm Penetration Assay (Zona-Free Hamster Egg Penetration Test) *on page 1220*

Testosterone, Total and Free, Serum or Plasma *on page 1238*

Test Includes Semen Analysis, Basic plus sperm processing through a density gradient preparation plus sperm survival after 24-hour incubation

Abstract The recently described "Advanced Semen Analysis" is a set of simple procedures, results of which are useful in predicting the success of intrauterine insemination (IUI). The analysis includes basic semen analysis, assessment of total motile sperm available for insemination, and 24-hour survival of spermatozoa.

Specimen Semen specimen from a subject who has abstained from ejaculation for 2-5 days; semen from ejaculate specimen

Container Sterile specimen container; clean, dry, wide mouth glass or plastic bottle maintained warm and known to be free of detergent or other toxic compounds

Collection See Semen Analysis, Basic *on page 1187*.

Causes for Rejection Specimen more than 1 hour old, exposed to cold, contaminated, or not liquefied

Special Instructions Semen, as with all blood, urine, and body fluid specimens, because of the risk of AIDS, should be received and handled with attention to universal precautions. Gloves must be worn during the handling and manipulation of sperm/semen containing fluids. Persons with sores/open wounds of the skin must avoid contact with semen.

Reference Interval

- Processed (through Percoll or other gradient) total motile sperm count: lower limit of 5×10^6, optimum level of $\geq 10 \times 10^6$
- Sperm survival test: >70% correlates with high pregnancy rate as a result of IUI

Use Prediction of success of intrauterine insemination

Methodology Analysis includes basic studies (ie, semen volume, sperm concentration, motility, sperm morphology, and seminal leukocytes). Total mobile sperm count is determined after processing through a density gradient preparation. Subsequently, specimen is adjusted to a maximum concentration of 10 x 10^6 motile sperm/mL of medium and incubated at 37°C in 5% CO_2. After 24 hours, the percentage of sperm motility is again determined.

Additional Information The parameters of basic semen analysis (ie, semen volume, sperm concentration, motility, sperm morphology, and seminal leukocytes) correlate poorly with the success of intrauterine insemination. A recently developed and described "Advanced Semen Analysis" includes basic studies, a processed motile sperm fraction determination, and a 24-hour incubation to assess sperm survival. Sperm survival test is apparently highly predictive of successful IUI. Advanced Semen Analysis may predict results of intrauterine insemination (IUI) independent of the basic semen analysis. The procedure may also be used to screen for couples that may require intracytoplasmic sperm injection.

References
Branigan EF, Estes MA, and Muller CH, "Advanced Semen Analysis: A Simple Screening Test to Predict Intrauterine Insemination Success," *Fertil Steril*, 1999, 71(3):547-51.

Correa-Pérez JR, "Advanced Sperm Analysis - A Step in the Right Direction?" *Fertil Steril*, 1999, 72(6):1150-1.

Semen Analysis, Basic

Related Information

Synonyms Sperm Count; Sperm Morphology Study

Applies to Computer-Assisted Sperm Analysis; Spermatids, Round

Test Includes A variety of parameters may be included in a "standard" or "basic" semen analysis, generally an assessment of number, motility, and morphology of the spermatozoa. More specifically, volume of the semen specimen, concentration of sperm, total count, liquefaction status, viscosity, color, odor, assessment of motility, determination of viability, and detection of abnormal morphologic forms. Direct microscopy of wet preparations and/or a modified Papanicolaou method may be utilized.

Abstract Analysis usually consists of a number of measurements including physical characteristics (eg, volume, color, odor, consistency, etc) of the semen and the morphologic characteristics/functional ability of its constituent spermatozoa. Computer-assisted assessment of sperm morphology and motility can enrich the basic study by providing analysis of individual spermatozoon motility characteristics (sperm kinematics).

Patient Preparation Ejaculation should be avoided for 2-3 days (but not more than 7 days) prior to collection.

Specimen Semen from ejaculate specimen

Container Clean, dry, wide mouth glass or plastic bottle maintained warm (20°C to 40°C) and known to be free of detergent or other toxic compounds

Sampling Time Because of apparent diurnal variation in semen quality (see below), late afternoon seminal fluid collection is preferable.

Collection Physician usually provides instruction for collection. Specimen quality is enhanced when collected in physician's office or laboratory. Alternately, specimen may be obtained at patient's home by masturbation and delivered to the laboratory within 1 hour. Collection of semen during intercourse using a seminal collection device may yield a specimen of higher quality. Use of such a Silastic condom-type seminal pouch (with a small pencil lead-size perforation) to obtain a specimen for analysis in overcoming human infertility is (Continued)

Semen Analysis, Basic *(Continued)*

acceptable to the Catholic Church. Ordinary latex condoms may interfere with the viability of spermatozoa and should not be used. Coitus interruptus should not be used. The first portion of the ejaculate usually contains the most spermatozoa. The semen specimen should be complete.

Storage Instructions Patient should be instructed to bring specimen to the laboratory within 30-60 minutes after collection maintaining warmth (37°C) during transport. Patient should be instructed to transport specimen in a pocket close to the skin. Low temperature during transport may decrease motility of sperm.

Causes for Rejection Specimen more than 2 hours old

Turnaround Time Dependent upon component tests of analysis, minimum of 2-3 days

Special Instructions Requisition should specify infertility study or postvasectomy study. Semen, as with all blood, urine, and body fluid specimens, because of the risk of AIDS, should be received and handled with attention to universal precautions. Gloves must be worn during the handling and manipulation of sperm/semen containing fluids. Persons with sores/open wounds of the skin must avoid contact with semen.

Because semen/sperm parameters may vary significantly from day to day in any one individual, two samples should be collected and studied at initial evaluation. The sample interval should be from 7-21 days apart.

Reference Interval

- Volume: 2-5 mL
- Appearance: white to gray, opalescent, viscid, opaque
- Clotting and liquefaction: complete in 20-30 minutes, rarely over 60 minutes
- pH: 7.2-8.0
- Sperm count: ≥20 million/mL
- Total sperm count: ≥40 million/ejaculate
- Motility: at least 50% motile
- Morphology: about 70% normal oval-headed forms (<15% normal forms is usually associated with decreased *in vitro* fertilization rate)
- Leukocytes (largely neutrophils): <1 million/mL

A diurnal rhythm in sperm quality has been reported. Afternoon specimens show higher number and concentration of spermatozoa as compared to morning specimens.

Use Infertility studies, postvasectomy studies; diagnose azoospermia, oligospermia

Limitations Lack of standardization of many test parameters

Methodology Macroscopic and microscopic analysis; direct observation, enumeration, and description, preferentially using the "stricter criteria" of Menkveld et al.

Systems for computer-assisted study of morphology and motility (CASA) are increasingly employed. The improved Neubauer hemocytometer has been considered as the standard for sperm counting.

Additional Information The semen sample may be red-brown if red blood cells or heme pigments are present or in cases of spinal cord injury (see Infertility Screen *on page 786*). Increase in leukocytes seen in semen of men with spinal cord injury appear to be the result of urinary tract infection. Semen may be yellow in some cases of pyospermia, in jaundiced patients, and in some individuals taking certain vitamins. If pH is <7 and there is azoospermia, obstruction of the ejaculatory ducts or bilateral congenital absence of the vas deferens may be present. Sperm counts may be reduced in patients taking cimetidine and possibly other histamine-receptor blockers. Other drugs that may be responsible for decrease in sperm count include sulfasalazine, nitrofurantoin, cyclophosphamide, nitrogen mustard, procarbazine, vincristine, methotrexate, and possibly other chemotherapeutics. Estrogens and methyltestosterone may suppress spermatogenesis. Orchitis, testicular atrophy (as after mumps), varicocele, testicular failure, obstruction of vas deferens (as after vasectomy), and hyperpyrexia may be associated with hypospermia or azoospermia and/or morphologically aberrant forms of spermatozoa. The motility of spermatozoa is dependent upon the level of ionized calcium, in particular the intracellular translocation of calcium ions. Cigarette smoking is associated with decrease in volume of semen; coffee drinking with increase in sperm density and percentage of abnormal forms. Alcohol consumption may not affect sperm function at least as measured at semen analysis. Use of cocaine may be associated with low sperm concentration, low sperm motility, and presence of abnormal morphology. Other illicit drugs which may interfere with spermatogenesis include marijuana; anabolic steroids, lead, and arsenic.

In some cases of infertility, cytologically abnormal spermatocytes or spermatogonia may be seen. Fertility correlates most closely with the parameters of sperm motility and morphology.

Testicular biopsy and fine needle biopsy are in use, and other types of investigation are available as well.

Recent studies appear to refute previous findings indicating a decline in the quality of sperm from men living in industrialized countries.

References

Cagnacci A, Maxia N, and Volpe A, "Diurnal Variation of Semen Quality in Human Males," *Hum Reprod*, 1999, 14(1):106-9.

Sarkar S and Henry JB, "Andrology Laboratory and Fertility Assessment," *Clinical Diagnosis and Management by Laboratory Methods*, 20th ed, Part 3, Chapter 20, Henry JB, ed, Philadelphia, PA: WB Saunders Co, 2001, 428-9.

World Health Organization, *WHO Laboratory Manual for the Examination of Human Semen and Sperm-Cervical Mucus Interaction*, 4th ed, Cambridge, UK: Cambridge University Press, 1999, 4-11, 60-1.

♦ **Sensitivity** *see page 11*

♦ **Sensitization** *see* Antiglobulin Test, Direct *on page 176*

Sentinel Lymph Node Biopsy

Related Information

Breast Biopsy *on page 305*
Lymph Node Biopsy *on page 880*
Polymerase Chain Reaction *on page 1069*
Skin Biopsy *on page 1200*

Synonyms Lymph Node Mapping

Applies to Cytokeratin; HMB-45; Immunocytochemistry; Isosulfan Blue; Lymphoscintigraphy; MART-1; Regional Lymph Node Dissection; Tyrosinase

Test Includes Lymph node biopsy, immunohistology

Abstract Sentinel lymph node (SLN) mapping is a technique designed to identify the first lymph node or nodes in the lymphatic drainage of a tumor bed, which represents highest risk for metastatic disease. The procedure is considered a standard of care for patients at risk for metastatic melanoma and is gaining wide acceptance for selected patients with breast cancer. Early studies suggest that this procedure may spare a significant population from postoperative morbidity related to lymphedema. Findings may impact therapeutic decisions for lymph node dissection and/or chemotherapy.

Specimen Lymph nodes may be submitted in neutral buffered formalin or other suitable fixative or submitted fresh. Appropriate labeling for radioactive materials is necessary. Delays in processing may be required to allow radioactivity to decay to permissible levels. The half-life of ^{99}Tc is approximately 6 hours.

Limitations Even though pathologic examination is more intense than for most cases, a portion of submitted tissue remains unexamined.

Contraindications Patients undergoing mastectomy or those with palpable adenopathy are generally not suitable for SLN biopsy procedures. SLN identification is sometimes more difficult following excisional biopsy.

Methodology Pathologic examination is designed to optimize the chance of finding micrometastatic disease. Commonly each lymph node is thinly sectioned and separately submitted *in toto*. Eight to 12 cut surfaces are microscopically examined from at least three different levels. If such sections are morphologically negative, additional sections are subjected to immunohistochemical stains. In patients with breast cancer, stains for cytokeratin provide a high degree of sensitivity and specificity for malignant cells. In patients with melanoma, a cocktail of HMB-45 and MART-1 may provide the highest degree of sensitivity and specificity. The role of melanoma markers (eg, tyrosinase) detected by reverse transcriptase polymerase chain reaction, poses intriguing questions currently under investigation.

(Continued)

Sentinel Lymph Node Biopsy (Continued)

Additional Information Approximately 25% of node-negative patients with either breast cancer or melanoma eventually develop metastatic disease. While some recurrences no doubt reflect initial hematologic dissemination, many cases probably reflect intrinsic limitations of regional lymph node examination. Historically, lymph node staging is performed by pathologic examination of multiple lymph nodes isolated from a regional dissection. Microscopic examination is commonly limited to 1 or 2 cut surfaces of a lymph node at one level and typically does not include special studies to detect micrometastatic disease. This traditional approach limits microscopic examination to approximately 1% to 5% of available tissue and is thought to miss metastatic disease 25% to 50% of the time. This limitation is a practical and economic reality of examining multiple lymph nodes in a regional dissection.

SLNs are thought to represent the first lymph nodes encountered within the lymphatic drainage bed of a neoplasm and, thus, represent the highest risk for early metastatic involvement. Examination is commonly performed on multiple cut surfaces at multiple levels and may employ immunohistochemical or molecular techniques to identify micrometastatic disease not detected with usual stains.

If SLNs are negative for metastatic disease, then the patient may be spared from regional lymph node dissection and subsequent morbidity related to lymphedema. In contrast, SLN involvement justifies full regional node dissection.

The technique to identify the SLN is generally similar in patients with either melanoma or breast cancer. The tumor bed is carefully infiltrated with radiolabeled ^{99}Tc filtered colloid 2-6 hours prior to surgery. Lymphoscintigraphy is helpful for locating SLNs and may reveal unexpected lymphatic drainage patterns. Approximately 5-10 minutes prior to lymph node biopsy, the tumor bed is infiltrated with isosulfan blue (Lymphazurin Blue™) which rapidly gains access to the lymphatic channels surrounding the neoplasm. At the time of surgery, a gamma probe aids in the percutaneous localization of "hot spots". A small incision is made over a hot spot and soft tissues are dissected until the lymph node is located. Optimally, the blue dye will often highlight lymphatic channels leading into a blue stained lymph node. Confirmation is established with the gamma probe and all lymph nodes are separately excised and submitted for pathologic examination. Lymphatic bed counts confirm that all SLNs are sampled.

Most commonly, patients with melanoma undergo SLN mapping at the time of wide local excision. Patients with melanoma >1.0 mm in thickness are regarded as candidates for SLN mapping and biopsy. Recent studies have suggested that some patients with melanoma 0.67 mm in thickness may also benefit from this approach. This may be especially true with ulcerated melanomas.

References

Farshid G, Pradhan M, Kollias J, et al, "Computer Stimulations of Lymph Node Metastasis for Optimizing the Pathologic Examination of Sentinel Lymph Nodes in Patients With Breast Carcinoma," *Cancer*, 2000, 89(12):2527-37.

Messina J, Glass LF, Cruse CW, et al, "Pathologic Examination of the Sentinel Lymph Node in Malignant Melanoma," *Am J Surg Pathol*, 1999, 23(6):686-90.

Meyer JS, "Sentinel Lymph Node Biopsy: Strategies for Pathologic Examination of the Specimen," *J Surg Oncol*, 1998, 69(4):212-8.

Pendas S, Dauway E, Cox CE, et al, "Sentinel Node Biopsy and Cytokeratin Staining for the Accurate Staging of 478 Breast Cancer Patients," *Am J Surg*, 1999, 65(6):500-5.

♦ **Serax®** *see* Oxazepam, Serum *on page 990*

♦ **Serenace®** *see* Haloperidol, Serum or Plasma *on page 664*

♦ **Serenid® Forte** *see* Oxazepam, Serum *on page 990*

♦ **Serentil®** *see* Phenothiazines, Serum *on page 1021*

Serotonin, Blood, Cerebrospinal Fluid

Related Information
5-Hydroxyindoleacetic Acid, Quantitative, Urine *on page 754*

Synonyms 5-HT; 5-Hydroxytryptamine, Blood

Applies to Serotonin, Cerebrospinal Fluid

Abstract Serotonin (5-hydroxytryptamine [5-HT]) is synthesized from tryptophan in the intestinal chromaffin cells or in central and peripheral neurons. The major metabolite of serotonin, 5-HIAA, is measured more commonly than serotonin.

Patient Preparation Monoamine oxidase inhibitor drugs should be discontinued for at least 1 week. Avoid application of radioisotopes (eg, scans) before collection of specimen if RIA is used for assay. Some methods require a low indole diet for several days. Avoid eggplant, avocado, bananas, tomatoes, pineapple, walnuts, and red plums.

Specimen Whole blood, cerebrospinal fluid

Container Tube with EDTA, sometimes with ascorbic acid. Check with laboratory if the assay is available at all.

Collection Draw in chilled tubes. Keep on ice.

Storage Instructions Place whole blood in plastic bottle containing 10 mg EDTA and 75 mg ascorbic acid. Freeze within 4 hours of collection. Stable 7 days at -20°C.

Reference Interval 10-30 µg/dL (SI: 570-1700 nmol/L). Values vary among laboratories and are method dependent. In serum, serotonin (5-hydroxytryptamine) levels in females are about 1.3-fold that of males. By RIA, a study provided intervals: male: 7-12 µg/dL (SI: 380-680 nmol/L), female: 9-16 µg/dL (SI: 520-900 nmol/L).

Use This assay is rarely used in the diagnosis of carcinoid syndrome. **Urinary 5-HIAA is more sensitive and specific for diagnosis of carcinoid tumors.**

References

Engbaek F and Voldby B, "Radioimmunoassay of Serotonin (5-Hydroxytryptamine) in Cerebrospinal Fluid, Plasma, and Serum," *Clin Chem*, 1982, 28(4 Pt 1):624-8.

Meltzer HY, "The Role of Serotonin in Schizophrenia and the Place of Serotonin-Dopamine Antagonist Antipsychotics," *J Clin Psychopharmacol*, 1995, 15(1 Suppl 1):2S-3S.

♦ **Serotonin, Cerebrospinal Fluid** *see* 5-Hydroxyindoleacetic Acid, Quantitative, Urine *on page 754*

♦ **Serotonin, Cerebrospinal Fluid** *see* Serotonin, Blood, Cerebrospinal Fluid *on page 1190*

♦ **Serotonin, Metabolite** *see* 5-Hydroxyindoleacetic Acid, Quantitative, Urine *on page 754*

♦ **Serotonin Release Assays** *see* Heparin-Induced Thrombocytopenia *on page 695*

♦ **Serous Fluid Analysis** *see* Body Fluid Analysis, Cell Count *on page 288*

♦ **Sertraline Hydrochloride** *see* Sertraline, Serum *on page 1191*

Sertraline, Serum

Related Information

Antidepressants, Cyclic, Serum or Plasma *on page 171*
Trazodone, Serum or Plasma *on page 1272*

Synonyms Sertraline Hydrochloride; Zoloft®

Abstract Sertraline, an antidepressant, is a serotonin reuptake inhibitor. It is used for the treatment of major depression, obsessive-compulsive disorder (OCD), panic disorder, post-traumatic stress disorder (PTSD), premenstrual dysphoric disorder (PMDD), and social anxiety disorder.

Specimen Serum

Container Red top tube

Reference Interval Peak (4 hours after dose): 100-200 ng/mL; trough: 20-50 ng/mL; Desmethylsertraline (active metabolite): peak (4 hours after dose): 200-300 ng/mL, trough: 30-100 ng/mL

Critical Values >500 ng/mL

Use Treatment of major depression

Contraindications Do not use in combination with monoamine oxidase inhibitors.

Methodology High performance liquid chromatography (HPLC), gas chromatography (GC)

Additional Information

- Half-life: 24 hours; metabolite: 80 hours
- Volume of distribution: 20 L/kg
- Protein binding: high: 98% to 99%
- Bioavailability: oral: 90%
- Time to peak plasma concentration: 8-12 hours

(Continued)

Sertraline, Serum (Continued)

Sertraline acts by inhibiting serotonin reuptake. The drug has less sedating and anticholinergic effects as compared to tricyclic antidepressants. Sertraline is a relatively safe drug due to its lack of interaction with norepinephrine, dopamine, monoamine oxidase, and cholinergic receptors, as well as its almost complete lack of cardiovascular activity and effect on the P450 system. The signs and symptoms which develop in sertraline overdose are minor and of short duration. They include tremor, lethargy, nausea, and vomiting.

Two randomized controlled studies have demonstrated that sertraline is an effective and well-tolerated short-term treatment for children and adolescents with major depressive disorder.

References

Edwards JG and Anderson I, "Systematic Review and Guide to Selection of Selective Serotonin Reuptake Inhibitors," Drugs, 1999, 57(4):507-33.

MacQueen G, Born L, and Steiner M, "The Selective Serotonin Reuptake Inhibitor Sertraline: Its Profile and Use in Psychiatric Disorders," CNS Drug Rev, 2001, 7(1):1-24.

Richelson E, "Pharmacokinetic Drug Interactions of New Antidepressants: A Review of the Effects on the Metabolism of Other Drugs," Mayo Clin Proc, 1997, 72(9):835-47.

Wagner KD, Ambrosini P, Rynn M, et al, Sertraline Pediatric Depression Study Group, "Efficacy of Sertraline in the Treatment of Children and Adolescents With Major Depressive Disorder: Two Randomized Controlled Trials," JAMA, 2003, 290(8):1033-41.

♦ **Serum and Urine Protein Electrophoresis** see Protein Electrophoresis, Capillary Zone on page 1103

♦ **Serum-Ascites Albumin Gradient (Alb$_{s-a}$)** see Body Fluid Chemical Analysis on page 291

Serum Bactericidal Test

Related Information
Antimicrobial Susceptibility Testing, Aerobic and Facultatively Anaerobic Bacteria on page 178

Test Includes Assay of serum or body fluid for antimicrobial activity

Abstract This method is used to assess bactericidal activity of serum from patients treated for critical infections, such as endocarditis or osteomyelitis.

Patient Preparation Sterile aspiration of body fluid

Specimen Peak and trough serum from patient and bacterial isolate causing infection

Container Red top tube; sterile tube for body fluid

Sampling Time Both peak and trough levels should be obtained. The peak level is the level 60 minutes after completing an intravenous or intramuscular dose and 90 minutes after an oral dose. Trough levels are obtained immediately before the next dose. If more than one antibiotic is administered, an attempt should be made to draw the peak and trough specimens around a dose when antibiotics are administered simultaneously. If this is not feasible, draw the specimens around the dose of the more frequently administered agent.

Collection Specimen should be transported to the laboratory within 1 hour of collection. Label specimens with time and date collected, time of last antimicrobial infusion (start and completion of infusion), and whether the specimen is a peak or trough.

Storage Instructions Separate serum from cells and freeze at -70°C if test will not be performed within 2 hours.

Causes for Rejection Bacterial isolate discarded before request for bactericidal testing, serum specimen allowed to sit at room temperature for more than 2 hours

Turnaround Time 2-3 days

Special Instructions If a serum bactericidal test is desired, the physician should request that the laboratory save the patient's isolate within 48 hours of submission of the specimen for initial culture. **If the isolate has not been saved, the test cannot be performed.**

Reference Interval Peak titers ≥1:32 and trough titers ≥1:8, respectively, are considered adequate. Peak and trough titers ≤1:2 are considered inadequate. Titers between these ranges are considered intermediate.

Use Determine the maximum bactericidal dilution (MBD) and/or serum bactericidal dilution (SBD) of serum or body fluid after administration of antibiotic(s)

Limitations Technical and biological variables affecting test performance make interpretation of test results difficult. Some antimicrobial agents lose activity quickly at room temperature and if not stored appropriately, the assay results are not accurate.

Methodology Serial dilution of patient's serum with Mueller-Hinton broth, 1:1 final ratio recommended, supplemented if necessary. Each dilution is incubated with a standard inoculum of the patient's bacterial isolate and assessed for inhibitory and bactericidal activity.

Additional Information The serum bactericidal test has been applied experimentally to detect antimicrobial activity in cerebrospinal fluid, joint fluid, and amniotic fluid. It is also useful in determining whether serum antimicrobial activity remains adequate after a shift from parenteral to oral therapy. Serum bactericidal titers are useful in evaluation of synergy between antibiotics after administration of the drugs. Results reflect the combined effect of all antimicrobial agents present in the patient's serum (or body fluid) on the infecting organism(s). Results are accurate to ±1 dilution.

References
Hacek DM, Dressel DC, and Peterson LR, "Highly Reproducible Bactericidal Activity Test Results by Using a Modified National Committee for Clinical Laboratory Standards for Broth Macrodilution Technique," *J Clin Microbiol,* 1999, 37(6):1881-4.

National Committee for Clinical Laboratory Standards, *Methodology for the Serum Bactericidal Test, Approved Guideline M21-A, in press,* Villanova, PA: National Committee for Clinical Laboratory Standards, 2000.

Schentag JJ, "Antimicrobial Action and Pharmacokinetics/Pharmacodynamics: The Use of AUIC to Improve Efficacy and Avoid Resistance," *J Chemother,* 1999, 11(6):426-39.

♦ **Serum Electrolytes** *see* Electrolyte Panel, Serum *on page 532*
♦ **Serum Osmolality** *see* Osmolality, Serum *on page 978*
♦ **Serum Protein Electrophoresis** *see* Protein Electrophoresis, Serum *on page 1104*
♦ **Se, Serum** *see* Selenium, Serum *on page 1184*
♦ **Se, Urine** *see* Selenium, Urine *on page 1185*

Severe Acute Respiratory Syndrome Diagnostic Procedures

Synonyms SARS; SARS-CoV

Test Includes Molecular testing employs reverse transcriptase polymerase chain reaction (RT-PCR); serologic testing employs enzyme-linked immunosorbent assay (ELISA); viral cell culture

Abstract SARS is the cause of the first pandemic of the 21st century. It first appeared in November 2002 in mainland China, and subsequently, spread to Hong Kong, Vietnam, Singapore, Canada, and Taiwan. Within only months of its appearance, it had affected more than 8000 individuals and caused 774 deaths in 26 countries on five continents (Peiris). It appeared as an unusual pneumonia. An international crash program identified the causative agent as a novel coronavirus, SARS-CoV. During the first year, the case fatality rate was approximately 10%. At present, SARS is diagnosed when certain clinical and epidemiologic criteria are satisfied (see Web Sites).

Specimen The usual specimens for RT-PCR include nasopharyngeal aspirate, nose and throat swabs. On occasion, other respiratory secretions such as bronchial washings, and tissue from lung biopsy or autopsy may be used (contact laboratory). Urine and stool may be used. Serology uses acute and/or convalescent serum.

Sampling Time Virus isolation should take place before the third week of illness, but it can be detected by reverse-transcriptase polymerase chain reaction (RT-PCR) for >30 days after onset.

Collect a serum sample at least 21 days, preferably 28 days, after onset of symptoms to rule out SARS.

Special Instructions The timing of specimen collection relevant to disease onset is needed.

Use SARS is diagnosed on clinical, radiographic, epidemiologic, and, in selected situations, autopsy findings. Affected patients have pneumonia or acute respiratory distress syndrome (ARDS). Laboratory testing, when available, may be used to place patients in one of three categories listed below and defined by the Centers for Disease Control and Prevention (CDC).

Confirmed: One of the following is obtained.
(Continued)

1193

Severe Acute Respiratory Syndrome Diagnostic Procedures *(Continued)*

- Detection of SARS coronavirus RNA by a RT-PCR test validated by CDC, from:

  one specimen on two occasions using the original clinical specimen on each occasion, **or**

  two clinical specimens from different sources (nasopharyngeal and stool), **or**

  two clinical specimens collected from the same source on two different days

- Isolation of SARS coronavirus in cell culture, and PCR confirmation using a test validated by the CDC.

- Detection of antibody to SARS-associated coronavirus in a serum specimen.

Negative: Absence of SARS-associated coronavirus in a convalescent serum specimen (>28 days after symptom onset).

Undetermined: Testing not performed or incomplete.

Limitations Testing is not widely available. RT-PCR cannot rule out infection. Antibody can be detected only after the first week of illness.

Methodology Viral cell culture; reverse transcriptase polymerase chain reaction (RT-PCR); immunofluorescence assay or enzyme-linked immunosorbent assay (ELISA) for serology; immunohistochemistry (tissue); and *in situ* hybridization (tissue)

Additional Information A family of enveloped, single-stranded RNA viruses, coronaviruses cause disease in humans and in animals. An animal reservoir is considered likely. Live game animals, Himalayan palm civets and ferrets may well bear epidemiologic relevance.

The incubation period is 2-10 days, with a probable maximum of 14 days. The peak viral load occurs at about the 10th day of illness. Transmission has occasionally taken place in the workplace, taxis, and in airplanes.

Transmission appears to be by respiratory droplets or fomites. Aerosolization (eg, endotracheal intubation, bronchoscopy) may amplify transmission. When dried and in feces, the virus survives for many days.

The initial clinical presentation includes fever, myalgia, malaise, chills, rigor, and often, cough. Other signs and symptoms (eg, watery diarrhea) may appear subsequently.

Laboratory findings may include lymphocytopenia, thrombocytopenia, increased ALT, CK, and LDH.

A human disease reservoir is not considered likely.

References

Fowler RA, Lapinsky SE, Hallett D, et al, "Critically Ill Patients With Severe Acute Respiratory Syndrome," *JAMA*, 2003, 290(3):367-73.

Franks TJ, Chong PY, Chui P, et al, "Lung Pathology of Severe Acute Respiratory Syndrome (SARS): A Study of 8 Autopsy Cases From Singapore," *Hum Pathol*, 2003, 34(8):743-8.

Grant PR, Garson JA, Tedder RS, et al, "Detection of SARS Coronavirus in Plasma by Real-Time RT-PCR," *N Engl J Med*, 2003, 349(25):2468-9.

Ksiazek TG, Erdman D, Goldsmith CS, et al, "A Novel Coronavirus Associated With Severe Acute Respiratory Syndrome," *N Engl J Med*, 2003, 348(20):1953-66.

Kuiken T, Fouchier AM, Schutten M, et al, "Newly Discovered Coronavirus as the Primary Cause of Severe Acute Respiratory Syndrome," *Lancet*, 2003, 362(9380):263-70.

Low DE and McGeer A, "SARS - One Year Later," *N Engl J Med*, 2003, 349(25):2381-2.

Olsen SJ, Chang HL, Cheung TY, et al, "Transmission of the Severe Acute Respiratory Syndrome on Aircraft," *N Engl J Med*, 2003, 349(25):2416-22.

Peiris JM, Yuen KY, Osterhaus AE, et al, "Current Concepts: The Severe Acute Respiratory Syndrome," *N Engl J Med*, 2003, 349(25):2431-41.

Internet Web Sites

www.afip.org/Departments/Pulmonary/SARS/
www.bcgsc.ca/bioinfo/SARS/
www.cdc.gov/ncidod/sars/casedefinition.htm
www.cdc.gov/ncidod/sars/sequence.htm

♦ **Sevredol®** *see* Morphine, Urine *on page 921*

♦ **Sex-Hormone Binding Globulin** *see* Testosterone, Total and Free, Serum or Plasma *on page 1238*

♦ **SGOT** *see* Aspartate Aminotransferase, Serum *on page 216*

- ◆ **SGPT** *see* Alanine Aminotransferase, Serum *on page 116*
- ◆ **SG, Urine** *see* Specific Gravity, Urine *on page 1216*
- ◆ **Sheets** *see* Phencyclidine, Qualitative, Urine *on page 1019*
- ◆ **Sherm** *see* Phencyclidine, Qualitative, Urine *on page 1019*

Shiga Toxin Test, Direct

Related Information

Bacterial Culture, Stool *on page 243*
Clostridium difficile Toxin Assay and Culture *on page 416*

Applies to *E. coli* O157:H7; Verocytotoxins

Test Includes Detection of *E. coli* O157:H7 by assay for the Shiga-like toxin directly in stool or after isolation of a bacterial isolate

Abstract In the late 1970s, the *E. coli* serotype O157:H7 was recognized as a human pathogen and was associated with severe bloody diarrhea and the hemolytic-uremic syndrome. This serotype of *E. coli* bacteria produces a Shiga-like toxin that is similar to the Shiga toxin expressed by *Shigella dysenteriae*. The production of the toxin, or the genes encoding the toxin, can be detected by a variety of methods, and some are commercially available.

Specimen Stool, food, or bacterial isolate

Container Sterile container

Reference Interval No toxin detected

Critical Values Positive detection of Shiga-like toxin

Use Detect toxin-producing strains after they have been isolated, directly from patient specimens such as stool, directly from food (especially meat), or from fecal specimens of food animals such as cattle.

Limitations Results from such assays should be confirmed by the isolation of a Shiga-toxin producing *E. coli*. Some of the commercial assays may have false-positive results with *Pseudomonas aeruginosa*.

Methodology Enzyme immunoassays (EIA), latex agglutination, genetic detection methods

Additional Information Certain strains of *E. coli* have the ability to produce toxins (verotoxins or Shiga-like toxins) that closely resemble the Shiga toxin of *Shigella dysenteriae*. Since *E. coli* isolates that produce the Shiga-like toxin are associated with severe hemorrhagic colitis and/or the life-threatening hemolytic uremic syndrome (HUS), any laboratory detection of this toxin should be considered clinically important and should be immediately reported to the attending physician and the state health department. The detection of this toxin or the genes required for toxin production can be used to predict the presence of this bacteria.

References

Acheson DWK and Jaeger JL, "Shiga Toxin-Producing *Escherichia coli*," *Clin Microbiol*, 1999, 21(23):183-8.

Bolton FJ and Aird H, "Verocytotoxin-Producing *Escherichia coli* O157: Public Health and Microbiological Significance," *Br J Biomed Sci*, 1998, 55(2):127-35.

Kawamura N, Yamazaki T, and Tamai H, "Risk Factors for the Development of *Escherichia coli* O157:H7 Associated With Hemolytic Uremic Syndrome," *Pediatr Intern*, 1999, 41(2):218-22.

Nataro JP and Kaper JB, "Diarrheagenic *Escherichia coli*," *Clin Microbiol Rev*, 1998, 11(1):142-201.

Paton JC and Paton AW, "Pathogenesis and Diagnosis of Shiga Toxin-Producing *Escherichia coli* Infections," *Clin Microbiol Rev*, 1998, 11(3):450-79.

Internet Web Sites

vm.cfsan.fda.gov/~mow/chap15.html
www.amm.co.uk/pubs/fa_vtec.htm

- ◆ **Shindler Disease** *see* Inherited Diseases of Metabolism and Cell Structure *on page 792*
- ◆ **Sickle Cell Solubility Test** *see* Sickle Cell Tests *on page 1195*

Sickle Cell Tests

Related Information

Anemia Flowchart *on page 35*
Fetal Hemoglobin *on page 581*
Hemoglobin *on page 681*
Hemoglobin Electrophoresis *on page 684*
Hemosiderin Stain, Urine *on page 692*
Peripheral Blood: Differential Leukocyte Count *on page 1010*
Peripheral Blood: Red Blood Cell Morphology *on page 1016*
Reticulocyte Count *on page 1156*
(Continued)

Sickle Cell Tests *(Continued)*

Synonyms Dithionite Test; Itano Solubility Test; Metabisulfite Test; Sickle Cell Solubility Test

Applies to Hemoglobin Munchausen; Hydroxyurea

Test Includes A variety of similar but usually slightly modified tests have been developed, described, and achieved varying degrees of acceptance. In some institutions a positive is confirmed by performing an alternate confirmatory sickle cell test and/or hemoglobin electrophoresis (the preferred procedure).

Abstract Screening tests for sickle cell anemia and related entities which include hemoglobin S (eg, SC, SD, sickle thalassemia, others). The sickle solubility test, available in slightly different form from a number of commercial sources, is based on the decreased solubility of reduced hemoglobin S in a concentrated phosphate buffer with standardized detection of resultant turbidity. If a sickle screen test is positive, confirmatory studies (eg, Hb electrophoresis) should be performed (see following). There are continuing advances in understanding the mechanisms of pathologic damage in sickle cell disease (SCD) with resultant improvements in management and therapy.

Specimen Whole blood

Container Lavender top (EDTA) tube for venipuncture specimen; lavender top Microtainer™ for capillary specimen

Collection Routine venipuncture. Invert tube gently to mix.

Causes for Rejection Clotted specimen, hemolyzed specimen

Reference Interval Negative

Use Detect sickling hemoglobins; evaluate hemolytic anemia; evaluate undiagnosed hereditary anemia with morphologic (sickle-like) abnormalities on peripheral blood smear

Limitations **False-positive** solubility test for sickling may be due to polycythemic blood; excess blood in relation to the quantity of reagent; interference by some forms of hyperglobulinemia such as may occur with myeloma, Waldenström macroglobulinemia or cryoglobulinemia (if suspect, test should be repeated using patient's washed red cells). Abnormal hemoglobins other than "classical S" may give a positive sickle solubility test result. Included are hemoglobins I, Bart's, Alexandra, Porto Alegre, Memphis/S, and six variants which have a second mutation in the β^s gene (resulting in an abnormal hemoglobin with two amino acid substitutions). The latter include C_{Harlem} ($C_{Georgetown}$), $C_{Ziguinchor}$, S_{Travis}, $S_{Antilles}$, $S_{Providence}$, and S_{Oman}. **False-negative** solubility test reaction may occur with inadequate quantities of blood from anemic patients (hemoglobin levels <8.0 g/dL); deterioration of reducing agents (detected by negative result on positive control); deterioration of the lytic agent (detected by negativity of positive control); improper illumination and visualization of the line-reader scale and high concentration of Hb F or of phenothiazines that may inhibit the sickle reaction; quantities of hemoglobin S too small to detect, as at birth or with transfusions of nonhemoglobin S units of blood into patients who have sickle blood (or multiunit transfusion including some donor units from sickle cell trait individuals). The appearance of hemoglobin S is genetically delayed and is not usually present in sufficient quantity (above ~25%) for a positive screening test result until after 3 months of age. Maximum levels are not reached until about 6 months of age. Solubility tests and sodium metabisulfite test are unlikely to be reliably positive until after 6 months of age. Babies with initial evidence of SS, SC, SD, Sβ^+, or Sβ^0 thalassemia should be retested within 6-8 weeks of birth.

Methodology Hb high salt solubility - Sickledex™ and a number of other commercially available products; alternate: slide test with 2% sodium metabisulfite. The sodium metabisulfite sickling test involves microscopic identification of sickled red blood cells which have formed due to deoxygenation of hemoglobin (in intact red cells) after exposure to the reducing agent, sodium metabisulfite. Use of excessive metabisulfite will cause a false-positive reaction (changes resembling sickling in normal red cells). Poikilocytes may be difficult to differentiate from true sickle cells. Outdated metabisulfite or sickle solubility test reagents may be the cause of false-negative results. Sickle solubility screening procedures have evolved from the Itano solubility test. These tests are based on visual detection of turbidity resulting from decreased solubility of reduced hemoglobin S in a concentrated phosphate buffer solution.

Additional Information Hb S is the result of a single amino acid substitution (valine for the normally present glutamic acid) at the 6th position of the β-globin chain of the hemoglobin molecule. The homozygous state results in sickle cell disease, a condition with significant morbidity and mortality. The heterozygous state causes sickle trait, a condition ordinarily characterized by little or no morbidity. Sickle disease presents a spectrum of clinical severity relating in part to activation of coagulation through loss of normal membrane phospholipid asymmetry and the appearance of phosphatidylserine (PS) on cell surfaces. Surface PS, a "docking site" for hemostatic proteins, is involved in "flip-flop" activity (leading to increased procoagulant activity) as the red cell membrane skeleton is uncoupled from the lipid bilayer. High levels of fetal hemoglobin (F-cell level of some 70%) favor decreased red cell membrane microvesicle formation, PS exposure, and thrombin generation (with normalized levels of prothrombin fragment 1.2) (see Fetal Hemoglobin *on page 581* and Hypercoagulation Panel *on page 758*). Amelioration results from coincidental occurrence of other hemoglobinopathies, notably those with increased levels of Hb F (eg, thalassemia and different types of hereditary persistence of fetal hemoglobin). Variation in clinical expression may be the result of DNA polymorphism in the β-globin gene cluster.

Clinical improvement may occur in some cases of sickle cell anemia after induction of iron-deficient erythropoiesis (iron deficiency lowers the mean corpuscular deoxyhemoglobin-S concentration leading to decrease in sickling tendency and severity of hemolysis).

The incidence of sickle cell trait amongst African Americans in the U.S. is 8.5%. Because of possible anterior segment ischemia (a significant complication of retinal detachment surgery) that may occur in otherwise asymptomatic sickle trait individuals, it has been recommended that African-Americans undergo preoperative sickle tests prior to such procedures.

Distinction between Hb S beta-thalassemia and sickle cell anemia is not always possible on clinical, hematologic, or electrophoretic grounds. Thalassemia heterozygotes have hypochromia and microcytosis, but overlap values exist. Differentiation can best be made by family or molecular pathology methods. Regional prevalence in the midwest area of Hb S beta-thalassemia is estimated to be 1:23,000 of the black population. It is recommended that positive sickle cell screen patients be further evaluated with cellulose acetate or agarose gel hemoglobin electrophoresis at pH 8.6, citrate agar gel electrophoresis at pH 6.0, Hb F studies and family studies. Complete characterization may require specialized laboratory studies.

Study of peripheral blood smear has utility in detection of sickled red blood cells (RBCs) with classic findings of numerous sickled cells, schistocytes, nucleated red cells, leukocytosis, target cells, and other abnormalities dependent on the degree of hypoxia and stage of the disease (eg, Howell-Jolly bodies if splenic atrophy has occurred). In quiescent periods of sickle disease (Hb SS), however, sickle cells may present irregularly and uncommonly on peripheral smears. Review of routinely stained blood smears has been considered to have no role in the confident identification of abnormal cells in subjects with sickle cell trait (Hb AS). A recent report by Wilson et al, however, indicates that on the basis of light microscopy, abnormal cells (defined as elongated RBCs with tapering of opposite ends culminating in a point) can be identified in a majority of patients with sickle cell trait. Ninety-six percent of trait subjects had such abnormal red cells as compared to 4% of controls (96% of the control group had no abnormal RBCs in their peripheral smears).

A survey of laboratory results, methods and problems in screening for sickle cell disease with emphasis on laboratory responsibilities has been published by the US Department of Health and Human Services.

The rare condition, "hemoglobin Munchausen" has been reviewed.

Significant advances in the therapy of SCD have accompanied increased understanding of the mechanisms responsible for morbidity and mortality. Use of hydroxyurea, a ribonucleotide reductase inhibitor (RRI), for induction of Hb F production, has been followed by the search for more effective RRIs and the use of combination therapy.

(Continued)

Sickle Cell Tests *(Continued)*

Nitrous oxide may cause a neuropathy similar to that seen with pernicious anemia and has been reported in SCD patients. The neuropathy responds to intramuscular B_{12} and avoidance of nitrous oxide. Presence of SCD may be a contraindication for use of nitrous oxide.

References

Bain BJ, *Haemoglobinopathy Diagnosis*, Malden, MA: Blackwell Science, Inc, 2001.

Ballas SK, "Munchausen Sickle Cell Painful Crisis," *Ann Clin Lab Sci*, 1992, 22(4):226-8.

Bunn HF, "Human Hemoglobins: Sickle Hemoglobin and Other Mutants," *The Molecular Basis of Blood Diseases*, 3rd ed, chapter 7, Stamatoyannopoulos G, Majerus PW, Perlmutter RM, et al, eds, Philadelphia, PA: WB Saunders Co, 2001, 227-73.

Guideline: Laboratory Screening for Sickle Cell Disease, U.S. Department of Health and Human Services, *Lab Med*, 1993, 24(8):515-22.

Iyamu WE, Adunyah SE, Fasold H, et al, "Enhancement of Hemoglobin and F-Cell Production by Targeting Growth Inhibition and Differentiation of K562 Cells With Ribonucleotide Reductase Inhibitors (Didox and Trimidox) in Combination With Streptozotocin," *Am J Hematol*, 2000, 63(4):176-83.

Koduri PR, "Iron in Sickle Cell Disease: A Review Why Less Is Better," *Am J Hematol*, 2003, 73(1):59-63.

Ogundipe O, Walker M, Pearson TC, et al, "Sickle Cell Disease and Nitrous Oxide-Induced Neuropathy," *Clin Lab Haematol*, 1999, 21(6):409-12.

Setty BN, Kulkarni S, Rao AK, et al, "Fetal Hemoglobin in Sickle Cell Disease: Relationship to Erythrocyte Phosphatidylserine Exposure and Coagulation Activation," *Blood*, 2000, 96(3):1119-24.

Steinberg MH, Voskaridou E, Kutlar A, et al, "Concordant Fetal Hemoglobin Response to Hydroxyurea in Siblings With Sickle Cell Disease," *Am J Haematol*, 2003, 72(2):121-6.

Veillon DM, Kaltenbach JE, Hall CG, et al, "Assays for Hemoglobin S," *Lab Med*, 2000, 31(2):68.

Vichinsky EP, Neumayr LD, Earles AN, et al, "Causes and Outcomes of the Acute Chest Syndrome in Sickle Cell Disease," *N Engl J Med*, 2000, 342(25):1855-65.

Wilson CI, Hopkins PL, Cabello-Inchausti B, et al, "The Peripheral Blood Smear in Patients With Sickle Cell Trait: A Morphologic Observation," *Lab Med*, 2000, 31(8):445.

♦ **Sideroblast Stain** *see* Iron Stain, Bone Marrow *on page 810*

Siderocyte Stain

Related Information

Bone Marrow *on page 296*

Iron Stain, Bone Marrow *on page 810*

Synonyms Hemosiderin Stain; Iron Stain; Pappenheimer Body Stain; Perls' Test (Prussian Blue Reaction)

Abstract Siderocyte stain (Prussian blue reaction) demonstrates non-haem iron containing granules in non-nucleated red blood cells on smear of peripheral blood. Iron containing granules are called Pappenheimer bodies as seen on Wright stain of peripheral blood smear. Finding of siderocytes has application to the study of sideroblastic anemia-myelodysplastic syndrome and hemolytic anemia.

Specimen Blood: coverslip or slide smears; bone marrow: coverslip smears preferred

Reference Interval Peripheral blood: no siderocytes identified; bone marrow: stainable iron present

Use Detect sideroblastic anemias and hemolytic anemia; semiquantitation of marrow iron stores, evaluation of iron reserve; assist in the diagnosis of iron deficiency and hemosiderosis/hemochromatosis

Limitations Siderocytes may be present in asplenic patients. They occur in large numbers following splenectomy.

Methodology Prussian blue (potassium ferrocyanide) reaction in which a blue colored compound, ferriferocyanide, is formed with the siderotic material (hemosiderin).

Additional Information Siderotic granules represent iron not yet incorporated into hemoglobin and are a water insoluble complex of ferric iron, lipid, protein, and carbohydrate. They occur primarily when there is impaired hemoglobin synthesis (eg, sideroblastic anemia, lead poisoning). Numerous siderocytes are noted postsplenectomy and in some hemoglobinopathies. They are absent with iron deficiency.

In a normal individual, iron positive granules in circulating (peripheral) red blood cells (siderocytes) are not identified. After splenectomy they may be seen in up to 14% to 15% of red cells (they are normally removed by the spleen). Other abnormalities seen on study of Romanowsky type dye-stained peripheral blood smears from splenectomized patients are target cells, Howell-Jolly bodies, acanthocytes, and Pappenheimer bodies (equivalent to siderotic granules as seen with an iron stain). and may allow identification of a clinically unsuspected "medical" splenectomy. Siderocytes may be found in the peripheral blood in a variety of hemolytic anemias but not in thalassemia. Splenic activity in removing circulating siderocytes makes correlation of quantitative siderocyte counts with type of hemolysis unreliable.

References
Beutler E, "Blood, Marrow, and Urine Iron Stains," *Williams Hematology*, 5th ed, Chapter L6, Beutler E, Lichtman MA, Coller BS, et al, eds, New York, NY: McGraw-Hill Inc, 1995, L27.

Douglas AS and Dacie JV, "The Incidence and Significance of Iron Containing Granules in Human Erythrocytes and Their Precursors," *J Clin Pathol*, 1953, 6:307.

♦ **Sinequan®** *see* Antidepressants, Cyclic, Serum or Plasma *on page 171*

♦ **Sinequan®** *see* Doxepin, Serum or Plasma *on page 524*

♦ **Single-Donor Platelets** *see* Platelets, Apheresis, Donation *on page 1054*

♦ **Sipple Syndrome** *see* Calcitonin, Serum or Plasma *on page 327*

♦ **Sirolimus** *see* Rapamycin, Blood *on page 1132*

Sjögren Antibodies

Related Information
Anticardiolipin Antibody *on page 167*
Anti-DNA *on page 173*
Antinuclear Antibodies *on page 189*
Chlamydia trachomatis Culture *on page 385*
Fungal Culture, Ocular Infections *on page 622*
Jo-1 Antibody *on page 815*
Ocular Cytology *on page 973*
Topoisomerase I Antibody *on page 1261*
Viral Culture *on page 1307*

Synonyms La Antibodies; Ro Antibodies; SS-A Antibodies; SS-B Antibodies

Applies to ENA; Extractable Nuclear Antigens; Sm; SS-A/Ro; SS-B/La; U₁RNP Antibody

Abstract Sjögren syndrome (SS) is a complex immunologic, autoimmune entity. It includes keratoconjunctivitis sicca, xerostomia, parotid enlargement, and arthritis. A secondary form is found in patients with other diseases regarded as autoimmune. Lymphoproliferative disorders occur.

Extractable nuclear antigens (ENA) include nuclear ribonuclear protein (RNP, U₁ RNP), Smith (**Sm**), Sjögren syndrome A (**SS-A/Ro**), and Sjögren syndrome B (**SS-B/La**). These nique ribonucleoprotein autoantigens can be eluted by saline washing.

Specimen Serum

Container Red top or SST™ tube

Storage Instructions Store at 4°C for up to 72 hours **or** -20°C or colder, without freezing and thawing indefinitely.

Reference Interval Negative

Use SS-A (Ro) and SS-B (La) are useful in diagnosis of Sjögren syndrome (especially with vasculitis) and some forms of lupus; may be present in anti-phospholipid antibody syndrome

Limitations Some autoantibodies, especially anti-SS-A (Ro), can be found in sera of apparently normal individuals.

Methodology Double immunodiffusion (DID). Best detection for anti-SS-B (La) antibodies is by immunoblotting while anti-SS-A (Ro) antibodies are more often detected by enzyme immunoassay (EIA).

Additional Information Antibodies to the ribonucleoprotein SS-A (Ro) are detected in 35% to 60% of SLE patients and antibodies to SS-B/La in about 15% of subjects with SLE. Anti-SS-A autoantibodies have been associated with photosensitivity, sicca symptoms, thrombocytopenia, and subacute cutaneous LE rash. Subacute cutaneous lupus erythematosus is a widespread, nonscarring, often photosensitive, form of cutaneous lupus with mild systemic manifestations and a low frequency of CNS and renal involvement. Anti-SS-A (Ro) antibodies are strongly associated with neonatal lupus in babies born to mothers with SLE. Maternal IgG antibodies cross the placenta, causing disease in the neonate. The two major manifestations of neonatal lupus erythematosus are transient dermatitis and usually irreversible heart block. Photosensitive dermatitis develops after the first few weeks of life and resolves within 6

(Continued)

Sjögren Antibodies (Continued)

months, coincident with the clearing of maternal antibodies from the infant's circulation. Heart block is due to binding of SS-A antibodies in fetal conducting system tissue. The antibodies do not affect the adult heart.

Lymphocytic infiltration of exocrine glands, particularly the salivary and lacrimal glands, and other organs characterize Sjögren syndrome. The most common clinical presentation is with sicca symptoms, xerophthalmia, and xerostomia.

The autoantibodies strongly associated with Sjögren syndrome are directed against the ribonucleoproteins SS-A (Ro) and SS-B (La). SS-A antibodies are detected in 40% to 60% of patients with Sjögren syndrome when the assays are done by immunodiffusion. The frequency approaches 90% or more when EIA is used. SS-B (La) antibodies occur slightly less frequently and never occur in the absence of SS-A antibodies. The presence of SS-A and SS-B antibodies are used to support the diagnosis of Sjögren syndrome. However, their presence must be interpreted in the proper clinical context because they are also found in patients with SLE and other diseases. These antibodies also provide prognostic information. Patients with these antibodies more often have extraglandular disease including vasculitis, purpura, lymphadenopathy, leukopenia, thrombocytopenia, hypergammaglobulinemia, and the presence of rheumatoid factor.

Other laboratory abnormalities in Sjögren syndrome may include elevation of the ESR, presence of antinuclear antibodies, presence of rheumatoid factor, polyclonal gammopathy, and the presence of cryoglobulins which may include IgM kappa proteins. Hypothyroidism secondary to Hashimoto thyroiditis occurs.

Diagnosis of Sjögren syndrome may be supported by minor salivary gland biopsy (lip biopsy).

References

Bridges AJ, Lorden TE, and Havighurst TC, "Autoantibody Testing for Connective Tissue Diseases. Comparison of Immunodiffusion, Immunoblot, and Enzyme Immunoassay," Am J Clin Pathol, 1997, 108(4):406-10.

Homburger HA, "Advances in the Diagnosis and Laboratory Evaluation of Systemic Rheumatic Diseases Other Than Rheumatoid Arthritis," Clinical and Laboratory Evaluation of Human Autoimmune Diseases, Chapter 11, Nakamura RA, Keren DF, and Bylund DJ, eds, Chicago, IL: American Society for Clinical Pathology, 2002, 153-64.

Kavanaugh A, Tomar R, Reveille J, et al, "Guidelines for Clinical Use of the Antinuclear Antibody Test and Tests for Specific Autoantibodies to Nuclear Antigens," Arch Pathol Lab Med, 2000, 124(1):71-81.

Moder KG, Miller TD, and Tazelaar HD, "Cardiac Involvement in Systemic Lupus Erythematosus," Mayo Clin Proc, 1999, 74(3):275-84.

Skin Biopsy

Related Information

Antinuclear Antibodies on page 189
Electron Microscopic Examination for Viruses, Stool on page 533
Electron Microscopy on page 533
Fungal Culture, Skin on page 623
Fungus Smear, Stain on page 626
Gene Rearrangement for Leukemia and Lymphoma on page 633
Gram Stain on page 658
Herpesvirus Cytology on page 727
Histopathology on page 733
Immunoperoxidase Procedures on page 780
KOH Preparation on page 823
Leishmaniasis Diagnostic Procedures on page 841
Lymph Node Biopsy on page 880
Mycobacterial Culture, Cutaneous and Subcutaneous Tissue on page 932
Skin Biopsy, Immunofluorescence on page 1203
Topoisomerase I Antibody on page 1261
Varicella-Zoster Virus Culture and Serology on page 1300
Viral Culture on page 1307

Abstract Inflammatory and neoplastic disorders of the skin are among the most commonly encountered disorders in clinical medicine. Biopsy of the skin is often central to the diagnosis and subsequent management of the disorder.

Container 10% neutral formalin is usually satisfactory. Special requirements exist for immunofluorescence and electron microscopy.

Collection Shave biopsy: A technique for obtaining superficial samples of predominantly epidermal or projecting lesions (eg, seborrheic keratoses, verrucae) by cutting them flush with adjacent skin. **Since shave biopsy provides the most limited specimen, a serious potential for histopathologic misdiagnosis exists, especially in regard to melanocytic lesions.**

Punch biopsy: Very popular with dermatologists, biopsy punches range from 3-6 mm in size. Punch biopsy may limit one's ability to evaluate adjacent skin and subcutaneous tissue to fully characterize some disorders. Very early lesions should be sampled in cases of ulcers, pustular lesions, and vesiculobullous disorders. Avoid crushing.

Excisional biopsy: This usually implies total removal of a skin lesion, most commonly a tumor. It is the preferred technique for removal of pigmented lesions and tumors.

Incisional biopsy: Removal of a portion of a lesion by scalpel, performed when a non-neoplastic lesion (eg, necrobiosis lipoidica) is too large to be totally excised but definitive diagnosis mandates a large sample to evaluate overall architectural detail. It may be used selectively in the case of tumors for which complete excision would require extensive surgery and/or would produce cosmetic deformity that would not be warranted, until accurate histopathologic diagnosis is established.

Smears and/or aspirates: Wright or Giemsa type stains may suffice to demonstrate polymorphonuclear leukocytes or eosinophils (as in toxic erythema or pustular melanosis of newborns). Gram, acid-fast, PAS, and Grocott methenamine-silver (GMS) stains, and cultures are used to study bacterial or fungal organisms. Finding appropriate multinucleated giant cells in smears in a proper clinical context suggests herpes or related viral infection: see Herpesvirus Cytology *on page 727.* Smears of *Molluscum contagiosum* may be diagnostic. Aspirates may be adequate for cultures for bacteria, fungi, and viruses. Scrapings are often utilized to evaluate dermatophytoses. Negative KOH examination does not exclude the diagnosis of fungal infection. Cytologic techniques (Tzanck smears) are rarely used in practice to evaluate acantholytic processes or tumors.

Curettings: Most frequently used when nodular basal cell carcinoma is suspected, and used as well for actinic and seborrheic keratoses. **Contraindicated for suspicious melanotic lesions.**

Use Diagnosis of dermatologic disease

Limitations Shave biopsies are contraindicated for lesions which may be melanoma. Differential diagnosis cannot be established between squamous cell carcinoma and keratoacanthoma in shave biopsy specimens. Shave biopsies of acral skin may not even reach the basal layer.

Currettage specimens sometimes cannot be evaluated. They are irregular, scanty, superficial, without architectural relationships, and often crushed.

Contraindications Anti-inflammatory agents and other therapy may alter histopathologic appearances. It is best to biopsy lesions prior to such treatment.

Avoid old, entirely scarred areas in scarring alopecia: select an area of erythema in which hair shafts are visible.

When the low dermis and subcutaneum are not included in the biopsy, characteristic features of some disease entities will be lost.

Additional Information

Selected Problems in Dermatopathology:

A. Tumors and Pigmented Lesions of the Skin

1. Tumors and pigmented lesions should be excised with careful attention to margins. The pathologists will commonly identify surgical margins with ink for proper microscopic evaluation. Pigmented lesions should not be aspirated, curetted, shaved, or punched. In a general practice, benign nevi, seborrheic keratoses, actinic keratoses, cysts, and dermatofibromas are most commonly encountered. In the diagnosis of malignant melanoma, important prognostic information includes tumor size, Clark level, Breslow depth of invasion, adequacy of margins, presence of satellite lesions, and hemolymphatic invasion.

(Continued)

Skin Biopsy *(Continued)*

B. Specimens of Vesiculobullous Lesions

1. If the diagnostic impression is pemphigus or pemphigoid, fresh lesions are preferred. Figure 1 illustrates appropriate biopsy technique.

Figure 1. Punch Biopsy Pemphigus or Pemphigoid

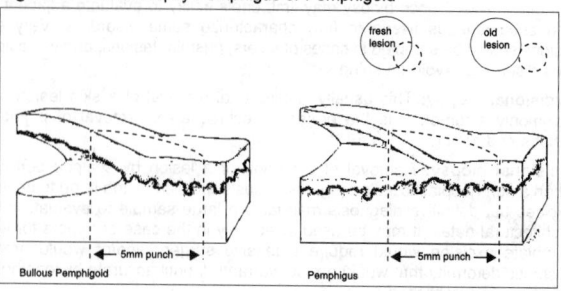

Bullous Pemphigoid — 5mm punch

Pemphigus — 5mm punch

fresh lesion / old lesion

2. If the diagnostic impression is dermatitis herpetiformis, take the biopsy at the edge of the lesion (to study the change in dermal papillae) as shown in Figure 2, rather than the lesion itself.

Figure 2. Punch Biopsy: Dermatitis Herpetiformis

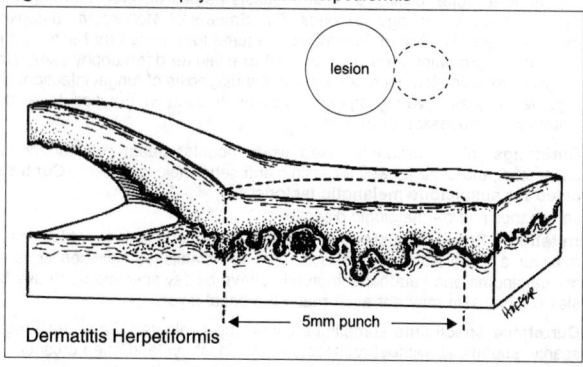

lesion

Dermatitis Herpetiformis — 5mm punch

3. Immunofluorescent studies of vesiculobullous lesions: Vesiculobullous lesions which require biopsy should be considered for immunofluorescent studies (IF). In many laboratories, skin samples for immunofluorescent studies are separately submitted in vials of isopentane prior to snap freezing in liquid nitrogen by the laboratory. Some reference laboratories provide special solutions such as Michael's solution or Zeus fixative to store specimens for their analysis. See Skin Biopsy, Immunofluorescence on page 1203.

References

Elder DE and Murphy GF, "Melanocytic Tumors of the Skin," *Atlas of Tumor Pathology*, Washington, DC: Armed Forces Institute of Pathology, 1990.

Elenitsas R and Halpern AC, "Biopsy Techniques," *Lever's Histopathology of the Skin*, 8th ed, Chapter 2, Elder D, Elenitsas R, Jaworsky C, et al, eds, Philadelphia, PA: JB Lippincott Raven, 1997, 3-4.

Murphy GF and Elder DE, "Nonmelanocytic Tumors of the Skin," *Atlas of Tumor Pathology*, Washington, DC: Armed Forces Institute of Pathology, 1991.

Wick MR and Compton C, "Protocol for the Examination of Specimens From Patients With Carcinomas of the Skin, Excluding Eyelid, Vulva, and Penis. A Basis for Checklists," *Arch Pathol Lab Med*, 2001, 125(9):1169-73.

Internet Web Sites

www.med.unc.edu/derm/nebr_site/index.htm

♦ **Skin Biopsy Antibodies** *see* Skin Biopsy, Immunofluorescence *on page 1203*

♦ **Skin Biopsy For Bullous or Collagen Disease** *see* Skin Biopsy, Immunofluorescence *on page 1203*

♦ **Skin Biopsy For Pemphigus/Pemphigoid** *see* Skin Biopsy, Immunofluorescence *on page 1203*

Skin Biopsy, Immunofluorescence

Related Information

Endomysial Antibodies *on page 537*
Skin Biopsy *on page 1200*
Viral Culture *on page 1307*

Synonyms Immunofluorescence Skin Biopsy

Applies to Basement Membrane Zone Antibodies; Dermatitis Herpetiformis Antibodies; Intercellular Substance Antibodies; LE Antibodies; Lupus Band Test; Skin Biopsy Antibodies; Skin Biopsy For Bullous or Collagen Disease; Skin Biopsy For Pemphigus/Pemphigoid

Test Includes Anti-IgG, anti-IgA, anti-IgM, anti-C3, antifibrin immunofluorescence

Abstract Two methods of immunofluorescence are commonly used in the evaluation of skin biopsies. The most common is direct immunofluorescence (DFA) when the patient's skin is tested for autoantibodies or indicators of inflammation. The second method is indirect immunofluorescence (IFA) when the patient's serum is tested for the presence of antibodies to epidermal antigens on a test substrate such as monkey or guinea pig esophagus. Both methods provide valuable information in the differential diagnosis of certain inflammatory diseases of the skin.

Specimen 3 mm^3 skin punch biopsy and patient serum

Use Useful in differential diagnosis of bullous skin diseases including epidermolysis bullosa acquisita, SLE, DLE, pemphigus, bullous pemphigoid, herpes gestationis and dermatitis herpetiformis, other skin entities, and for small vessel vasculitis

Limitations Steroid therapy can convert findings to negative in previously positive patients. Electron microscopy may be needed for diagnosis of **epidermolysis bullosa**. In the blistering disease **porphyria cutanea tarda**, uroporphyrinogen, uroporphyrin, and coproporphyrin urinary excretion is increased.

Additional Information Distinctive patterns of IgG, IgA, IgM, fibrin, and complement components in epidermis, basement membrane, and dermal vessels may contribute to the differential diagnosis of bullous skin diseases, discoid and systemic lupus erythematosus, and other entities.

Serum antibodies to skin components can be demonstrated, using a tissue substrate (usually monkey or guinea pig esophagus). Such observations must be correlated with clinical and histopathologic facets as well as findings on direct immunofluorescence. In bullous disease, antibody levels often reflect disease activity and rising titers may foretell clinical relapse, but such correlation is imperfect.

See Endomysial Antibodies *on page 537*, relevant to dermatitis herpetiformis and celiac disease. See Antinuclear Antibodies *on page 189*, relevant to systemic lupus erythematosus and related diseases.

Immunoglobulins or complement deposition in vessels may be found in skin biopsies in instances of vasculitis.

References

Cardinali C, Caproni M, and Fabbri P, "The Utility of the Lupus Band Test on Sun-Protected Nonlesional Skin for the Diagnosis of Systemic Lupus Erythematosus," *Clin Exp Rheumatol*, 1999, 17(4):427-32.

Elenitsas R, Van Belle P, and Elder D, "Laboratory Methods," *Lever's Histopathology of the Skin*, 8th ed, Chapter 4, Elder D, Elenitsas R, Jaworsky C, et al, eds, Philadelphia, PA: Lippincott-Raven, 1997, 51-60.

Wick MR, Ritter JH, Humphrey PA, et al, "Immunopathology of Non-neoplastic Skin Disease. A Brief Review," *Am J Clin Pathol*, 1996, 105(4):417-29.

Skin Puncture Blood Collection

Related Information

Blood and Fluid Precautions, Specimen Collection *on page 271*
Phlebotomist Procedures *on page 1028*

(Continued)

Skin Puncture Blood Collection *(Continued)*

Specimen Identification Requirements *on page 1217*

Synonyms Capillary Blood Collection; Fingerstick Blood Collection; Heelstick Blood Collection

Applies to Arterialized Capillary Blood; Peripheral Blood Smear Preparation

Test Includes Obtaining capillary blood from finger tip of an adult or heel in infants

Abstract If only a small volume of blood is required, skin puncture may be the optimal method of blood collection, particularly in infants and children. It is also an ideal way of collecting blood for home testing (eg, glucose testing).

Patient Preparation

Heel puncture: Select a site on the medial or lateral portion of the plantar surface of the foot. Do not puncture greater than 2.4 mm. Do not puncture the posterior curvature of the heel. Do not repuncture previous puncture sites because of the possibility of infection.

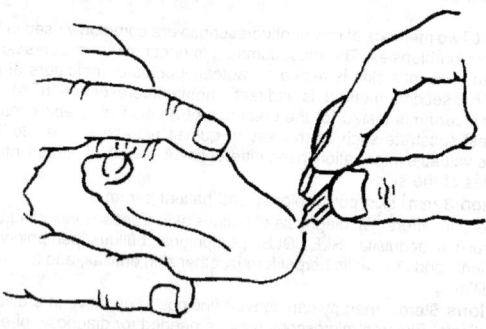

Gloves should be worn when collecting capillary blood specimens.
From JD Bauer, *Clinical Laboratory Methods*, 9th ed, St. Louis, MO: Mosby-Year Book Inc, 1982, with permission.

Finger puncture: Select a site on the palmar aspect on the center of distal phalanx. Do not puncture the side or tip of the phalanx because the skin is much thinner.

Skin preparation: The skin site selected should be cleaned with 70% isopropyl alcohol (isopropanol) and dried with sterile gauze. Infection is a frequent complication of fingersticks. Prepare the skin site carefully. Do not use iodine, which interferes with many assays.

Arterialized blood: Warming the site with a moist towel at temperatures not to exceed 42°C produces an increase in blood flow and **arterializes the capillary blood**. pH and blood gas determinations are usually performed on arterialized capillary blood in infants and children.

Aftercare Elevate the site above the body and apply direct pressure to the puncture site with sterile gauze until bleeding stops. Bandaids or bandages are generally not applied because of the risk of skin sensitization to tape and the risk of aspiration, should the bandage come loose.

Specimen Capillary blood or arterialized capillary blood

Container Capillary tube or microtube

Collection After preparation of the selected site, the skin should be punctured at a slight angle. A disposable skin puncture lancet should be used rather than a surgical blade because a surgical blade may make too deep an incision and damage underlying tissues. The first drop should be wiped away as it may be diluted by tissue fluid. Blood flow will be increased by holding the site downward. Slight pressure may be applied to the surrounding tissue. **Squeezing or milking the puncture site should not be done.** Tubes should be sealed quickly to avoid exposure to atmospheric oxygen.

Use Obtain capillary or arterialized capillary blood for analysis. Collection of blood from infants and children, patients who have had repeated venipunctures or whose veins are damaged or inadequate. Procedure of choice for preparing peripheral blood smears for morphologic examinations.

Limitations Technically, a specimen obtained by skin puncture consists of a mixture of arterial, capillary and venous blood, and tissue fluid. Specimen volume is limited. Repeat determinations often require repeat blood collection. Finger tips are sensitive; the procedure may be painful. Infection, particularly in debilitated hosts, may occur. Cell counts (ie, RBC, WBC, and platelets) are not accurate on capillary specimens. There is greater risk of infection than venipuncture.

Contraindications Use of surgical blades may create a wound deeper and larger than necessary and are contraindicated. **Finger punctures should not be performed on infants because the distance from the skin to the bone is less than 1.5 mm.**

Additional Information In recent years, blood collection by skin puncture has become more convenient and standardized due to availability of spring-loaded lancets. A cut of standard width and depth can be made with these lancets.

References

Blumenfeld TA, Turi GK, and Blanc WA, "Recommended Site and Depth of Newborn Heel Skin Punctures Based on Anatomical Measurements and Histopathology," *Lancet*, 1979, 1(8110):230-3.

National Committee for Clinical Laboratory Standards, *Procedures for the Collection of Diagnostic Blood Specimens by Skin Puncture*, 3rd ed, Approved Standard, NCCLS Publication H4-A3, Wayne, PA: National Committee for Clinical Laboratory Standards, 1991.

♦ **Skin Scrapings for *Sarcoptes scabiei* Identification** *see* Arthropod Identification *on page 213*

♦ **SK-Lygen®** *see* Chlordiazepoxide, Serum *on page 391*

♦ **SK-Pramine®** *see* Imipramine, Serum or Plasma *on page 767*

♦ **SLA Antibody** *see* Smooth Muscle Antibody *on page 1207*

♦ **Slo-Phyllin®** *see* Theophylline, Serum *on page 1243*

♦ **Sm** *see* Sjögren Antibodies *on page 1199*

♦ **Smail®** *see* Chlordiazepoxide, Serum *on page 391*

♦ **Small Molecule Diseases** *see* Inherited Diseases of Metabolism and Cell Structure *on page 792*

Smallpox Diagnostic Procedures

Related Information

Herpes Simplex Virus Culture and Antigen Detection *on page 721*
Varicella-Zoster Virus Culture and Serology *on page 1300*

Synonyms Variola Major/Variola Minor Diagnostic Procedures

Abstract The last natural case of smallpox occurred in Somalia in 1977. Two known stocks of variola virus are known to be retained, one by the Centers for Disease Control and Prevention (CDC) in Atlanta and one by the Vector Institute in Novosibirsk, Russia. Routine smallpox vaccination was stopped in the U.S. in 1972, and elsewhere since about 1982. Thus, a large pool of susceptible individuals now exists.

Oral and airway lesions occur first. Macules become papules, then vesicles, then pustules. The disease appears first on the face and extremities, and lesions follow on the trunk and extremities.

Since smallpox bears a potential for human-to-human transmission, it has a terrifying potential as a bioterrorist weapon.

Specimen Skin lesion scrapings, vesicular or pustular fluid, crusts, scabs, blood, tonsillar swabbings, or saliva sent to CDC. The CDC should be contacted anytime, day or night, at 770-488-7100. Public health officials should be contacted. Agents such as variola can be handled only in biosafety level 4 laboratories.

Container Vacutainer® tube

Collection Specimen collection should be done by a previously vaccinated individual, or one who is vaccinated that day. Mask and gloves should be worn.

Paired sera samples are needed for serologic testing to recognize recent infection.

(Continued)

Smallpox Diagnostic Procedures *(Continued)*

Storage Instructions Storage of biopsies, scabs, vesicular fluid for virology: -20°C to -70°C. Transport conditions: 4°C in 6 hours or less.

Special Instructions Guidelines for the collection and shipping of specimens are available at http://www.cdc.giv/smallpox.

Use Diagnose papulovesicular eruptions in a setting of possible bioterrorism

Methodology Immunohistochemistry for viral antigen; polymerase chain reaction (PCR) for Orthopoxvirus genetic material; live-cell cultures with nucleic acid identification; growth on chorioallantoic membranes; serologic testing detecting IgM responses. Complement fixation (CF), immunofluorescence, Ouchterlony techniques, and hemadsorption with chicken erythrocytes can be used.

Electron microscopy for recognition of virions can be used.

The inclusion bodies of smallpox, designated Guarnieri bodies, are found in intraepidermal lesions characterized by multilocular vesicles with ballooning and reticular degeneration. Guarnieri bodies or the virions on electron microscopy do not discriminate between Orthopoxvirus species.

Additional Information The differential diagnosis of papulovesicular and maculopapular eruptions includes smallpox and chickenpox. See the April 25, 2002 and January 30, 2003 issues of the *New England Journal of Medicine*. Lesions of chickenpox begin on the trunk. Herpes simplex virus as well may cause a false alarm.

References

Breman JG and Henderson DA, "Current Concepts: Diagnosis and Management of Smallpox," *N Engl J Med*, 2002, 346(17):1300-8.

Branda JA and Ruoff K, "Bioterrorism - Clinical Recognition and Primary Management," *Am J Clin Pathol*, 2002, 117(Suppl 1):S116-S123.

Carroll K, Held M, Stombler RE, et al, "Laboratory Preparedness for Bioterrorism: From the Phlebotomist to the Pathologist," *Lab Med*, 2003, 34(3):169-82.

Drazen JM, "Perspective: Smallpox and Bioterrorism," *N Engl J Med*, 2002, 346(17):1262-3.

Henderson DA, Inglesby TV, and O'Toole T, *Bioterrorism: Guidelines for Medical and Public Health Management*, Chicago, IL: AMA Press, 2002.

Mack T, "A Different View of Smallpox and Vaccination," *N Engl J Med*, 2003, 460-3.

Meyer RF and Morse SA, "Bioterrorism Preparedness for the Public Health and Medical Communities," *Mayo Clin Proc*, 2002, 77(7):619-21.

Moses AE and Cohen-Poradosu R, "Eczema Vaccinatum - A Timely Reminder," *N Engl J Med*, 2002, 346(17):1287.

Tucker JB, "Scourge: The Once and Future Threat of Smallpox," *Atlantic Monthly Press*, 2001.

Varkey P, Poland GA, Cockerill FR III, et al, "Confronting Bioterrorism: Physicians on the Front Line," *Mayo Clin Proc*, 2002, 17(7):661-72.

Internet Web Sites

www.asm.org
www.bt.cdc.gov
www.cdc.gov/smallpox
www.idsociety.org/bt/toc.htm

♦ **Sm-C** *see* Insulin-Like Growth Factor-1 (IGF-1), Serum or Plasma *on page 800*

Smith (Sm) and Ribonucleoprotein (RNP) Antibodies

Related Information

Anti-DNA *on page 173*
Antinuclear Antibodies *on page 189*

Synonyms Anti-sn RNP; Anti-U$_1$ RNP; Anti-U1 snRNP; nRNP; Nuclear Ribonucleoprotein; Ribonucleoprotein Antibodies; RNP; RNP Antibodies; U1 snRNP Antibody

Applies to ENA; Extractable Nuclear Antigens (ENA)

Abstract The Smith (**Sm**) and nuclear ribonucleoprotein (**RNP**) antigens are a particulate complex composed of small nuclear RNAs (U-RNAs, U$_1$RNPs) and proteins. This complex has also been referred to as **extractable nuclear antigens (ENA)**, since it is soluble in saline. Autoantibodies to Sm and RNP occur respectively in systemic lupus erythematosus and mixed connective tissue disease. **ENAs** include **SS-A (Ro)** and **SS-B (La)**, nuclear ribonuclear protein (**RNP**), and Smith (**Sm**).

Specimen Serum

Container Red top or SST™ tube

Causes for Rejection To avoid false positives, some laboratories will perform ENA testing only on ANA-positive cases.

Use Anti-Sm is used to confirm the diagnosis of **systemic lupus erythematosus** (SLE) or **mixed connective tissue disease** (MCTD). In the presence of possible diseases similar to SLE, with ANA positivity, anti-ds-DNA and anti-Sm are helpful. Antibodies to Sm are rare in drug-induced LE.

Anti-RNP is the defining antibody used in MCTD. It correlates with myositis, esophageal dysmotility, Raynaud phenomenon, sclerodactyly and interstitial lung disease. It is useful to establish the diagnosis of SLE and of Sjögren syndrome.

Limitations With rare exceptions, these tests should not be ordered if the ANA was negative or weakly positive, because <5% of patients with ANA titers <1:160 will have positive follow-up tests. Sm titers should not be measured as a marker of disease activity or to establish prognosis.

Methodology Immunodiffusion (ID), enzyme-linked immunosorbent assay (ELISA), immunoblotting provided best detection for RNP in a comparison study.

Additional Information The Sm (Smith) and related nuclear ribonucleoproteins (nRNPs) are targets for autoantibodies in SLE. **Anti-Sm antibodies are only present in 15% to 30% of patients with SLE, but they are highly specific for SLE.** They occur more frequently (60%) in young black females with SLE. They almost never occur in healthy individuals or patients with other diseases. (Anti-Sm antibodies should not be confused with antismooth muscle antibodies detected in autoimmune liver disease.)

Anti-RNP antibodies, which are commonly tested in conjunction with anti-Sm, are found in 30% to 40% of SLE patients but anti-RNP antibodies lack specificity for SLE.

Many patients present with clinical signs and symptoms that are compatible with more than one systemic rheumatic disease. One such overlap syndrome is **mixed connective tissue disease** (MCTD). Patients with MCTD have overlapping features of SLE, scleroderma, polymyositis, and rheumatoid arthritis. Arthritis, arthralgia, dyspnea, Raynaud phenomenon, esophageal dysfunction, and myositis are common, but renal involvement is rare. Half or more are RF positive. **Detection of high levels** of U_1RNP antibody, in the absence of other antibodies, strongly suggests the diagnosis of MCTD in the appropriate clinical setting. (Low titers of nRNP are found in SLE, scleroderma, and other diseases.) Isolated pulmonary hypertension and/or severe interstitial lung disease may evolve in patients with MCTD. Two laboratory criteria are necessary to diagnose MCTD: 1) the presence of high titer RNP antibodies and 2) the absence of anti-DNA, anti-Sm, and histone antibodies.

References

Homburger HA, "Advances in the Diagnosis and Laboratory Evaluation of Systemic Rheumatic Diseases Other Than Rheumatoid Arthritis," *Clinical and Laboratory Evaluation of Human Autoimmune Diseases*, Chapter 11, Nakamura RA, Keren DF, and Bylund DJ, eds, Chicago, IL: American Society for Clinical Pathology, 2002, 153-64.

Kavanaugh A, Tomar R, Reveille J, et al, "Guidelines for Clinical Use of the Antinuclear Antibody Test and Tests for Specific Autoantibodies to Nuclear Antigens," *Arch Pathol Lab Med*, 2000, 124(1):71-81.

Korn JH and Mauiyyedi S, "Hypertensive Encephalopathy With Impaired Renal Function in a 67-Year-Old Woman With Polymyositis," Case Records of the Massachusetts General Hospital. Case 26-2001, Scully RE, Mark EJ, McNeely WF, et al, eds, *N Engl J Med*, 2001, 345(8):596-605.

Moder K, "Use and Interpretation of Rheumatologic Tests: A Guide for Clinicians," *Mayo Clin Proc*, 1996, 71(4):391-6.

Smooth Muscle Antibody

Related Information

(Continued)

Smooth Muscle Antibody (Continued)

Synonyms Antismooth Muscle Antibody

Applies to Antiactin Antibodies; Anti-ASGPR; Anti-F Actin; Asialoglycoprotein Receptor Antibodies; Liver/Kidney Microsomes Antibody; LKM Antibody; SLA Antibody; Soluble Liver Antigen Antibody

Abstract Major types of autoimmune liver disease include primary biliary cirrhosis, autoimmune hepatitis, and primary sclerosing cholangitis. The differential diagnosis of these disorders includes chronic viral hepatitis B, D, and C, drug-induced chronic hepatitis and Wilson disease. Overlap and outlier syndromes are also recognized. Detection of antimitochondrial and antismooth muscle antibodies (SMA) has become an important facet in the differential diagnosis of chronic liver disease: SMA positivity supports classification as type 1 or 3 autoimmune hepatitis, while antimitochondrial antibodies are found more often and in higher titer in primary biliary cirrhosis.

Specimen Serum

Container Red top or SST™ tube

Storage Instructions Keep specimen cool.

Reference Interval Titers <1:20 are considered negative.

Use Smooth muscle antibody is useful in the differential diagnosis of chronic liver disease. Autoimmune type 1 hepatitis is characterized by autoantibodies (ANA, SMA, and marked polyclonal increases of serum immunoglobulins, high IgG, and often, relationship with other autoimmune entities). Antiliver/kidney microsomal type 1 (anti-LKM-1) and SMA are mutually exclusive. Anti-LKM-1 is found in type 2 autoimmune hepatitis. See table.

Limitations Antismooth muscle antibody is present, usually at titers <1:80, in 35% of patients with primary biliary cirrhosis, and in occasional cases of cryptogenic cirrhosis, infectious mononucleosis, asthma, and neoplasm. Detection of antimitochondrial or antismooth muscle antibodies does not rule out bile duct obstruction.

Methodology Indirect immunofluorescent antibody (IFA)

Additional Information Smooth muscle antibodies are IgG or IgM antibodies that are primarily directed against F-actin. Since F-actin is present in all smooth muscle fibers, these antibodies are not organ specific. Smooth muscle antibodies are found in 50% to 80% of patients with autoimmune hepatitis, with the exception of type 2. It is also associated with positive ANA. Elevated titers are also found in a minority of patients with primary biliary cirrhosis, and patients with cryptogenic cirrhosis. Elevated smooth muscle antibody titers have also been reported in a small number of patients with viral hepatitis, infectious mononucleosis, malignant tumors, alcoholic cirrhosis, and 5% of normal patients.

Antibodies to soluble liver antigens (cytokeratins 8 and 18) can be detected in about 10% of those with type 1 autoimmune hepatitis. Most subjects with autoimmune hepatitis are positive for HLA-B8, DR3, or DR4.

The differential diagnosis of autoimmune liver disease includes hemochromatosis, alpha$_1$-antitrypsin deficiency, and liver disease secondary to drugs such as alpha methyldopa, nitrofurantoin, and propylthiouracil as well as viral infection and the entities mentioned above.

See Liver/Kidney Microsomal Type 1 Antibodies *on page 873* for a discussion of autoimmune hepatitis subtypes 2a and 2b. See table.

References

Buschenefelde KH, Lohse AW, Meyer zum, "Autoimmune Hepatitis," *N Engl J Med*, 1995, 333(15):1004-5.

Galperin G and Gershwin ME, "Immunopathogenesis of Gastrointestinal and Hepatobiliary Diseases," *JAMA*, 1997, 278(22):1946-55.

♦ **Snorting** *see* Cocaine (Cocaine Metabolite), Qualitative, Urine or Hair *on page 427*

♦ **Snorts** *see* Phencyclidine, Qualitative, Urine *on page 1019*

♦ **Snow** *see* Cocaine (Cocaine Metabolite), Qualitative, Urine or Hair *on page 427*

♦ **Snowshoe Hare Virus** *see* California Encephalitis Virus Serology *on page 334*

♦ **SO$_2$** *see* Oxygen Saturation, Blood *on page 991*

♦ **Sodium, Arterial Blood** *see* Sodium, Serum or Plasma *on page 1210*

Autoimmune Hepatitis Subtypes

Characteristics	Type 1	Type 2a	Type 2b	Type 3
Female preponderance	Yes	Uncertain	No	Yes
Age, years	10-20, 45-70	2-14 (adults: 4%)	>40	30-50
Immunologic features Serologic profile Autoantibody (target antigen)	SMA (actin F and G, tubulin); ANA (centromere, 52-kd, SS-A/ Ro, histones, RNPs)	Anti-LKM-1 (cytochrome P-450, IID6 protein); antiliver cytosol 1	Anti-LKM-1; anti-HCV	Anti-SLA (hepatocyte cytokeratins 8 and 18); anti-AMA (PDC-E2)
Remarks	Anti-F-actin antibodies: increase specificity for Type 1 AIH	Anti-LKM-1 presence is mutually exclusive of SMA	Significant number of patients with anti-HV have confirmed HCV RNA in serum	Anti-SLA: increases specificity for Type 3 AIH
Serum IgA levels	Normal	Low	Normal	Normal
Extrahepatic autoimmune syndromes	+	++	−	+++
Progression to cirrhosis	+/++	+++	+++	+++
Remarks	Classic or lupoid hepatitis type	Predominantly children	Predominantly children	Antisoluble liver antigen (anti-SLA)

SMA = antismooth muscle antibodies; **anti-LKM-1** = antiliver-kidney microsomal antibodies; **anti-SLA** = antisoluble liver antigen antibodies; **ANA** = antinuclear antibodies; **AMA** = antimitochondrial antibodies; **HCV** = hepatitis C virus; **RNPs** = ribonucleoproteins; **PDC-E2** = E2 component of pyruvate dehydrogenase; + = present; − = absent. Extrahepatic syndromes include arthritis, thyroid disease, and ulcerative colitis. Type 2 patients suffer more frequent nonhepatic disorders.

Adapted from Galperin C and Gershwin ME, "Immunopathogenesis of Gastrointestinal and Hepatobiliary Diseases," *JAMA,* 1997, 278(22):1946-55.

♦ **Sodium Aurothiomalate** *see* Gold, Serum *on page 657*

♦ **Sodium, Corrected** *see* Sodium, Serum or Plasma *on page 1210*

Sodium, Serum or Plasma

Related Information

Anion Gap, Serum, Plasma, or Urine *on page 160*
Antidiuretic Hormone, Plasma *on page 172*
Blood Volume *on page 282*
Chloride, Serum, Plasma, or Blood *on page 391*
Chloride, Urine *on page 394*
Concentration Test, Urine *on page 446*
Electrolyte Panel, Serum *on page 532*
Lithium, Serum *on page 863*
Osmolality, Calculated, Serum or Plasma *on page 976*
Osmolality, Serum *on page 978*
Osmolality, Urine *on page 979*
Potassium, Serum or Plasma *on page 1078*
Potassium, Urine *on page 1080*
Renin Activity, Plasma *on page 1149*
Sodium, Urine *on page 1213*
Urea Nitrogen, Serum or Plasma *on page 1284*
Uric Acid, Serum *on page 1286*

Synonyms Na^+

Applies to Hypertonicity; Osmolal Gap; Sodium, Arterial Blood; Sodium, Corrected

Abstract Sodium with its accompanying anions is the most important extracellular osmotically active solute. The major cation of extracellular fluid, Na^+ is extremely important in maintenance of water and osmotic pressure equilibrium.

Specimen Serum or plasma

Container Red top tube or green top (lithium heparin, not sodium heparin) tube

Collection Pediatrics: Blood drawn from heelstick for capillary sample. Na^+, with K^+ and Cl^-, can be reported from arterial or venous blood. If an arterial puncture is done for pO_2, lithium heparin anticoagulant must be used.

Reference Interval

Infants:
- 0-7 days: 133-146 mmol/L
- 7-31 days: 134-144 mmol/L
- 1-6 months: 134-142 mmol/L
- 6 months to 1 year: 133-142 mmol/L
- older than 1 year: 134-143 mmol/L

Adults: 136-145 mmol/L

Possible Panic Range <125 mmol/L, >150 mmol/L

Use Evaluation of electrolytes, acid-base balance, water balance, water intoxication, dehydration

Hypernatremia occurs from loss of water or from Na^+ retention. It is found in dehydration, with diuretic use, fever, burns, hyperpnea, sweating, and high ambient temperatures. Nasogastric protein feeding with insufficient fluids may cause hypernatremia, as can vomiting and diarrhea. Hypernatremia without obvious cause may relate to Cushing syndrome, central or nephrogenic diabetes insipidus with insufficient fluids, adipsia, primary aldosteronism, and other diseases. Often, patients who have primary aldosteronism have mild hypernatremia. Severe hypernatremia may be associated with volume contraction, lactic acidosis, and azotemia. Increased hematocrit may provide evidence of dehydration. The corrected serum Na^+ is often high in nonketotic hyperosmolar coma. (A corrected Na^+ is calculated by increasing Na^+ by 1.3-1.6 mmol/L for each 100 mg/dL increment in serum or plasma glucose). The corrected serum Na^+ level should be calculated in nonketotic hyperosmolar coma. **Apparent mild hyponatremia with very high glucose may actually mean hypernatremia.** Infusion of hypertonic saline or sodium bicarbonate or ingestion of Na^+ may cause sodium retention. The pathophysiology of hypernatremia is reviewed by Adrogué et al; its differential diagnosis is presented in Table 1 from that source. Hypernatremia denotes hypertonicity.

Table 1. Causes of Hypernatremia

Net Water Loss

Pure water
  Unreplaced insensible losses (dermal and respiratory)
  Hypodipsia
  Neurogenic diabetes insipidus
    Post-traumatic
    Tumors, cysts, histiocytosis, tuberculosis, sarcoidosis
    Idiopathic
    Aneurysms, meningitis, encephalitis, Guillain-Barré syndrome
    Ethanol ingestion (transient)
  Congenital nephrogenic diabetes insipidus
  Acquired nephrogenic diabetes insipidus
    Renal disease (eg, medullary cystic disease)
    Hypercalcemia or hypokalemia
    Drugs (lithium, demeclocycline, foscarnet, methoxyflurane, amphotericin B,
      vasopressin V_2-receptor antagonists)
Hypotonic fluid
  Renal
    Loop diuretics
    Osmotic diuresis (glucose, urea, mannitol)
    Postobstructive diuresis
    Polyuric phase of acute tubular necrosis
    Intrinsic renal disease
  Gastrointestinal
    Vomiting
    Nasogastric drainage
    Enterocutaneous fistula
    Diarrhea (most common cause in infancy)
    Use of osmotic cathartic agents (eg, lactulose)
  Cutaneous
    Burns
    Excessive sweating

Hypertonic Sodium Gain

Hypertonic sodium bicarbonate infusion (eg, resuscitation)
Hypertonic feeding preparation
Ingestion of sodium chloride
Ingestion of sea water
Sodium chloride-rich emetics
Hypertonic saline enemas
Intrauterine injection of hypertonic saline
Hypertonic sodium chloride infusion
Hypertonic dialysis
Primary hyperaldosteronism
Cushing syndrome

Adapted from Adrogué HJ and Madias NE, "Hypernatremia," *N Engl J Med*, 2000, 342(20):1493-9.

Hyponatremia (serum Na^+ <136 mmol/L) can be found with low, normal, or high tonicity. The most common type is dilutional, from retention of water. Hyponatremia occurs with nephrotic syndrome, cachexia, hypoproteinemia, intravenous glucose (salt-free) infusion, congestive heart failure, mineralocorticoid deficiency, and cystic fibrosis. Mineralocorticoid deficiency leads to hyponatremia, hypovolemia, and hyperkalemia through inadequate Na^+ and water resorption and diminished potassium excretion. Serum sodium is a factor predictive of cardiovascular mortality in patients with severe congestive heart failure. Hyponatremia without congestive heart failure or dehydration may occur with hypothyroidism, the syndrome of inappropriate secretion of antidiuretic hormone (SIADH), renal failure, or renal sodium loss. See Table 2.

(Continued)

Table 2. Causes of Hypotonic Hyponatremia

Impaired Capacity of Renal Water Excretion

Decreased Volume of Extracellular Fluid

Renal sodium loss

  Diuretic agents

  Osmotic diuresis (glucose, urea, mannitol)

  Adrenocortical insufficiency

  Salt-wasting nephropathy

  Bicarbonaturia (renal tubular acidosis, disequilibrium stage of vomiting)

  Ketonuria

Extrarenal sodium loss

  Diarrhea

  Vomiting

  Blood loss

  Excessive sweating (eg, in marathon runners)

  Fluid sequestration in "third space": Bowel obstruction, peritonitis, pancreatitis, muscle trauma, burns

Increased Volume of Extracellular Fluid

Congestive heart failure

Cirrhosis

Nephrotic syndrome

Renal failure (acute or chronic)

Pregnancy

Essentially Normal Volume of Extracellular Fluid

Thiazide Diuretics[1]

Hypothyroidism

Adrenocortical insufficiency

Syndrome of Inappropriate Secretion of Antidiuretic Hormone

  Cancer: Pulmonary neoplasms, mediastinal neoplasms, extrathoracic neoplasms

  Central nervous system disorders: Acute psychosis, mass lesions, inflammatory and demyelinating diseases, stroke, hemorrhage, trauma

  Drugs: Carbamazepine, chlorpropamide, clofibrate, cyclophosphamide, desmopressin, nicotine, opiate derivatives, oxytocin, phenothiazines, prostaglandin-synthesis inhibitors, serotonin-reuptake inhibitors, tricyclics, vincristine

  Pulmonary conditions: Infections, acute respiratory failure, positive-pressure ventilation

  Miscellaneous: **Postoperative state**, pain, severe nausea, infection with the human immunodeficiency virus

Decreased intake of solutes: Beer potomania, tea-and-toast diet

Excessive Water Intake

Primary polydipsia[2]

Dilute infant formula

Sodium-free irrigant solutions (used in hysteroscopy, laparoscopy, transurethral resection of prostate)[3]

Accidental intake of large amounts of water (eg, during swimming lessons)

Multiple tap-water enemas

---

[1]Sodium or potassium depletion, stimulation of thirst, and impaired urinary dilution are implicated

[2]Often a mild reduction in the capacity for water excretion is also present.

[3]Hyponatremia is not always present.

Adapted from Androgueé HJ and Madias NE, "Hyponatremia," *N Engl J Med*, 2000, 342(21):1581-9.

The evaluation of hyponatremia includes Addison disease, hypopituitarism, liver disease including cirrhosis, hypertriglyceridemia, and psychogenic polydipsia. Diuretics and other drugs may cause hyponatremia. Sodium decreasing to levels <120 mmol/L can lead to significant neurological dysfunction with cerebral edema, increased intracranial pressure and uncal herniation, a life-threatening complication. See Osmolality, Calculated, Serum or Plasma *on page 976.*

The differential diagnosis of hyponatremia includes determination of urine sodium and osmolality and serum urea nitrogen (BUN). BUN is often normal or decreased in SIADH, but is increased in states in which hyponatremia is related to volume depletion. (Extracellular volume is normal or increased in SIADH.)

Hyperlipidemia, hyperproteinemia, and hyperglycemia must be considered; *vide infra*.

Limitations Care should be taken that one is not dealing with "pseudohyponatremia;" *vide infra*.

Blood collection in inappropriate tubes can lead to a preanalytic error, (eg, sodium fluoride - gray, sodium citrate - light blue or black, EDTA - lavender, sodium heparin - green).

Sodium contamination of specimens in the laboratory or office can derive from bleach (sodium hypochlorite), baking soda cleanser (sodium bicarbonate), and liquid soap.

Methodology Ion-selective electrode (ISE) in most laboratories

Additional Information The ratio of serum sodium to osmolality is normally 0.43-0.50; a decreased ratio is found in uremia and other states in which there are increased substances with osmotic activity (see Osmolality, Calculated, Serum or Plasma *on page 976* and Osmolality, Serum *on page 978*).

With sodium <128 mmol/L, hypo-osmolality, and low BUN, consider the syndrome of inappropriate secretion of antidiuretic hormone.

In "pseudohyponatremia" treatment may be undesirable. With pseudohyponatremia serum sodium is decreased but the serum is not hypotonic (serum osmolality is normal or even increased). This may occur as the result of other molecules replacing water in relation to sodium. The water content is effectively lowered - sodium is "diluted." In severe hypertriglyceridemia or paraprotein-related marked increase in protein, the concentration of sodium in relation to water is normal but the analytic result is determined as mmol/L of serum. Osmolality in this situation is determined as the amount of particles per kg of water and will be normal. Analysis by ion-selective electrode of the direct potentiometric type (requires no dilution) is not artifactually low in patients with hyperlipidemia. If large amounts of solute, such as glucose or mannitol, are present, movement of intracellular water into the extracellular space may produce dilutional hyponatremia. In this case, sodium concentration in relation to water is actually low. "Osmolal gap" however exists between measured and calculated serum osmolality. Other substances capable of increasing serum osmolality (eg, ethanol) may also cause increase in the osmolal gap.

Another cause of pseudohyponatremia is increased serum viscosity due to increased globulin proteins, occurring particularly in Waldenström's macroglobulinemia. Analyzers may aspirate too little sample when viscosity is so high, leading to a factitious low sodium concentration.

In the geriatric population, hyponatremia on admission to the hospital is association with poor prognosis.

References

Adrogué HJ and Madias NE, "Hypernatremia," *N Engl J Med*, 2000, 342(20):1493-9.

Adrogué HJ and Madias NE, "Hyponatremia," *N Engl J Med*, 2000, 342(21):1581-9.

Oh MS and Carroll HJ, "Disorders of Sodium Metabolism: Hypernatremia and Hyponatremia," *Crit Care Med*, 1992, 20(1):94-103.

Young DS, "Effects of Drugs on Clinical Laboratory Tests," 5th ed, Volume 1: Listing by Test, Washington, DC: AACC Press, American Association of Clinical Chemistry, 2000, Section 3, 714-20.

◆ **Sodium, Serum:Osmolality, Serum Ratio** *see* Osmolality, Serum *on page 978*

◆ **Sodium, Sweat** *see* Chloride, Sweat *on page 392*

Sodium, Urine

Related Information

(Continued)

Sodium, Urine *(Continued)*

Urine Collection, 24-Hour *on page 1295*

Synonyms Na, Urine

Applies to FENA; Fractional Excretion of Sodium

Abstract Urinary sodium excretion normally relates to intake. Body sodium stores are based upon intake and renal excretion.

Specimen Timed or random urine

Container Plain urine container

Reference Interval 24-hour urine: 27-287 mmol/day, varies markedly with dietary intake of sodium. There is diurnal variation (output is lower at night). A European study provides average sodium excretion: male: 162 mmol/day, range: 143-208 mmol/day; female: 134 mmol/day, range: 119-165 mmol/day; within person CV: male: 30%, female: 34%.

Use Work up volume depletion, acute renal failure, acute oliguria, and differential diagnosis of hyponatremia. See tables.

Limitations High urine sodium does not necessarily indicate that total body sodium is increased (eg, salt-losing nephropathy).

Methodology Ion-selective electrode (ISE), flame emission photometry.

Fractional excretion of sodium (FENa) is calculated as follows:

$$FENa = (urine\ Na^+ / urine\ Cr) \times (serum\ Cr / serum\ Na^+) \times 100$$

Additional Information In cases of hyponatremia, random urine Na^+ <10 mmol/L or FENa <1% commonly indicates extrarenal depletion: dehydration (gastrointestinal or sweat loss), congestive heart failure, liver disease or nephrotic syndrome. With renal or adrenal diseases, urinary Na^+ concentration is usually >20 mmol/L.

Random urine Na^+ >10 mmol/L may indicate diuretics, emesis, intrinsic renal diseases, Addison disease, hypothyroidism, or syndrome of inappropriate antidiuretic hormone (SIADH). In hypothyroidism and in SIADH, Na^+ and Cl^- may be >40 mmol/L. (Depending on intake, such results also can be found in normal individuals.)

In SIADH, random urinary sodium usually is >20 mmol/L. SIADH has been found in 7% of patients with small cell lung cancer. Medications, CNS disease, and other pulmonary diseases are additional causes of SIADH. Such patients have hyponatremia, often severe, with hypo-osmolar serum, absence of clinical evidence of volume depletion, high urinary sodium excretion with urine osmolality greater than that of serum. Urine may not be maximally concentrated in some patients but it should not have osmolarity less than that of serum.

The classification as presented in the tables is abbreviated. Pitfalls exist (eg, increase of Na^+ necessary to balance excretion of penicillin).

Urine Na^+ >40 mmol/L in oliguria is found in acute tubular necrosis. However, a better indicator is the fractional excretion of Na^+ based on simultaneously obtained random urine and blood Na^+ and creatinine. (See Methodology above.)

Low Na^+ excretion may be found with early obstructive uropathy and with the oliguria of acute glomerulonephritis and in some patients with x-ray contrast acute renal failure.

It is important to know the urinary sodium concentration in patients with unexplained hyperchloremic metabolic acidosis when the diagnosis of distal renal tubular acidosis is being considered.

An autosomal recessive disorder, pseudohypoaldosteronism type I, is characterized by renal salt wasting and high concentrations of sodium in stool, sweat, and saliva; hyperkalemia; increased renin activity; and high circulating aldosterone levels.

References

Harrington JT and Cohen JJ, "Measurement of Urinary Electrolytes - Indications and Limitations," *N Engl J Med*, 1975, 293:1241-3.

Scheinman SJ, Guay-Woodford LM, Thakker RV, et al, "Genetic Disorders of Renal Electrolyte Transport," *N Engl J Med*, 1999, 340(15):1177-87.

Sherman RA and Eisinger RP, "The Use (and Misuse) of Urinary Sodium and Chloride Measurements," *JAMA*, 1982, 247:3121-4.

Evaluation and Treatment of the HYPONATREMIC Patient

Condition	Clinical Presentation	Urinary Electrolytes	Etiology
Hypovolemic	Orthostatic hypotension; tachycardia; azotemia	Urinary sodium >30 mmol/L	Diuretics, RTA, mineralocorticoid deficiency, salt-wasting nephritis
		Urinary sodium <30 mmol/L	Extrarenal losses: vomiting, diarrhea, burns, sequestration
Euvolemic	No evidence of volume depletion or overload. Subclinical increase in total body water may be present.	Urinary sodium >20 mmol/L	Hypothyroidism, glucocorticoid deficiency, SIADH, drugs, acute water intoxication
Hypervolemic	Volume excess; edema	Urinary sodium >30 mmol/L	Acute and chronic renal failure
		Urinary sodium <10 mmol/L	Cirrhosis, cardiac failure, nephrotic syndrome

Adapted from Devita MV and Michelis MF, "Perturbations in Sodium Balance: Hyponatremia and Hypernatremia," *Clinics in Laboratory Medicine*, Vol 13, Preuss HG, ed, Philadelphia, PA: WB Saunders Co, 1993, 135-48, with permission.

Evaluation and Treatment of the HYPERNATREMIC Patient

Condition		Urinary Electrolytes	Etiology
Hypovolemic	Renal losses	Urinary sodium >30 mmol/L	Osmotic diuresis
	Extrarenal losses	Urinary sodium <30 mmol/L	Sweating; diarrhea in children
Euvolemic	Renal losses	Variable urinary sodium	Central diabetes insipidus; nephrogenic diabetes insipidus; partial diabetes insipidus; hypodipsia
	Extrarenal losses		Respiratory or skin losses
Hypervolemic	Increased total body sodium	Urinary sodium >30 mmol/L	Primary or secondary aldosteronism; Cushing's syndrome; hypertonic I.V. infusion; I.V. sodium bicarbonate administration; sodium chloride tablets

Adapted from Devita MV and Michelis MF, "Perturbations in Sodium Balance: Hyponatremia and Hypernatremia," *Clinics in Laboratory Medicine*, Vol 13, Preuss HG, ed, Philadelphia, PA: WB Saunders Co, 1993, 135-48, with permission.

- ◆ **Soft Tissue Mass Aspiration** *see* Fine Needle Aspiration, Deep and Superficial Masses *on page 590*
- ◆ **Solfoton**® *see* Phenobarbital, Serum or Plasma *on page 1021*
- ◆ **Solganal**® *see* Gold, Serum *on page 657*
- ◆ **Solid Phase Red Cell Adherence Assay** *see* Platelet Transfusion *on page 1058*
- ◆ **Solis**® *see* Diazepam, Serum *on page 510*
- ◆ **Solium**® *see* Chlordiazepoxide, Serum *on page 391*

- **Soluble Liver Antigen Antibody** *see* Smooth Muscle Antibody *on page 1207*
- **Soluble Liver Antigen (Anti-SLA)** *see* Liver/Kidney Microsomal Type 1 Antibodies *on page 873*
- **Soluble Serum Transferrin Receptor** *see* Transferrin Receptor, Soluble, Serum or Plasma *on page 1267*
- **Soman** *see* Chemical Warfare Agents *on page 382*
- **Somatomax®** *see* Gamma Hydroxybutyrate, Serum or Urine *on page 630*
- **Somatomedin-C** *see* Growth Hormone, Serum *on page 662*
- **Somatomedin-C** *see* Insulin-Like Growth Factor-1 (IGF-1), Serum or Plasma *on page 800*
- **Somatomedins** *see* Growth Hormone, Serum *on page 662*
- **Somatomedins** *see* Insulin-Like Growth Factor Binding Protein 3, Serum *on page 802*
- **Somatostatin** *see* Growth Hormone, Serum *on page 662*
- **Somatotropin** *see* Growth Hormone, Serum *on page 662*
- **Sopor™** *see* Methaqualone, Urine *on page 903*

Specific Gravity, Urine

Related Information

Antidiuretic Hormone, Plasma *on page 172*
Concentration Test, Urine *on page 446*
Osmolality, Serum *on page 978*
Osmolality, Urine *on page 979*
Urinalysis *on page 1289*

Synonyms Refractive Index, Urine; SG, Urine

Applies to Body Fluid, Specific Gravity

Test Includes Specific gravity is usually part of Urinalysis.

Abstract Specific gravity (SG) is a measure of the density of dissolved solids in a fluid. **Hyposthenuric urine** is urine with SG <1.007; **isosthenuric urine**, SG about 1.010, is that of fixed SG. (SG of the protein-free glomerular filtrate is about 1.010.) Loss of concentrating ability takes place in end-stage renal disease.

Specific Gravity, Urine

Increased >1.020
Water restriction
Dehydration
Fever
Sweating
Vomiting
Diarrhea
Diabetes mellitus (glycosuria)
Proteinuria
Congestive heart failure
X-ray contrast media
Adrenal insufficiency
Inappropriate antidiuretic hormone secretion syndrome
Tumors-secreting antidiuretic hormone
Decreased <1.009
Excess water ingestion
Excess I.V. fluids
Diuresis
Hypothermia
Impaired renal concentrating ability
pyelonephritis
glomerulonephritis
diabetes insipidus (central, renal)
Fixed 1.010
Severe renal damage, urine concentration fixed at 1.010, the value of glomerular filtrate

Specimen Voided urine or body fluid. Refractometer requires only a few drops of urine. Other methods require considerably more.

Collection First morning specimen is recommended, unless part of complete urinalysis.

Storage Instructions Refrigeration

Reference Interval 1.005-1.035; adult on normal fluid intake: 1.016-1.022. Following overnight fluid deprivation for 12 hours, urine SG should be ~1.022. SG decreases with increasing age.

Critical Values Although low values are not necessarily abnormal, they may reflect advanced kidney disease or diabetes insipidus.

Use SG provides evaluation of renal concentrating power and hydration status. Urine SG test strips are effective in home use to help stone formers drink sufficient water to reduce risk of stone formation. The SG of urine indicates the relative proportions of dissolved solid components to the total volume of the specimen. It reflects the relative degree of concentration or dilution of the specimen. Knowledge of the SG is needed in interpretation of results of most tests in urinalysis. SG must be interpreted in light of presence or absence of glycosuria and/or proteinuria. Patients with diabetes insipidus have marked decrease in SG and osmolality of urine. See table.

Limitations Radiographic dyes in urine increase the SG by hydrometer or refractometer. Glucose or protein also increase SG out of proportion to osmolality, as measured by hydrometer or refractometer. **Strip method urine SG** was reported as having a significant positive bias at urine pH ≤6 and negative bias at pH >7 compared to SG by refractometer. Urine osmolality is preferable in some settings. Benitez et al suggest that osmolality is the only accurate measure of urine concentration in newborn infants.

Methodology Refractometer and colorimetric (reagent strip) methods.

Dipstick method responds to the ionic strength of urine (linear relation to osmolality due to electrolytes). The strip provides a polyionic polymer with binding sites saturated with hydrogen ions that with urine testing are replaced by sodium or potassium cations, consequent release of hydrogen ions (change of pH) affecting an indicator color change; apparent pKa change. Albumin, glucose, and osmotic effects are not measured, and as such a true specific gravity measurement may not be obtained. Critical clinical decisions should be based on the more definitive methods.

Additional Information Reagent strip methods for the determination of urine SG employ a colorimetric method and are sensitive to ions but not undissociated solutes such as urea. The strip method requires a corrected reading for pH ≥6.5 and protein increases the reading. The strip methods are reported to be suitable for urine screening purposes, but not uniformly so.

References

Adam P, "Evaluation and Management of Diabetes Insipidus," *Am Fam Phys*, 1997, 55(6):2146-53.

Assadi FK and Fornell L, "Estimation of Urine Specific Gravity in Neonates With a Reagent Strip," *J Pediatr*, 1986, 108(6):995-6.

Benitez OA, Benitez M, Stijnen T, et al, "Inaccuracy in Neonatal Measurement of Urine Concentration With a Refractometer," *J Pediatr*, 1986, 108(4):613-6.

♦ **Specificity** *see page 11*

♦ **Specimen Chain-of-Custody Protocol** *see* Chain-of-Custody Protocol *on page 381*

♦ **Specimen Collection Introduction** *see page 23*

♦ **Specimen Collection Policy, Phlebotomist** *see* Phlebotomist Procedures *on page 1028*

Specimen Identification Requirements

Related Information

Arterial Blood Collection *on page 211*
Phlebotomist Procedures *on page 1028*
Skin Puncture Blood Collection *on page 1203*
Venous Blood Collection *on page 1304*

Synonyms Identification Requirements, Specimen

Applies to Bacteriology Specimen Identification; Blood Specimen Identification; Body Fluid Identification; Cytology Smear Identification; Pathology Specimen (Continued)

Specimen Identification Requirements *(Continued)*

Identification; Requisition Information; Spinal Fluid Identification; Urine Specimen Identification

Abstract Regulatory agencies require that each laboratory provides a written policy for specimen identification.

Causes for Rejection Laboratories must reserve the right to refuse improperly labeled specimens.

Special Instructions Patient identification: Inpatient: Compare the information on the order form with the patient's identification band and room and bed number. Confirm identification by asking the patient to state his/her full name. If the patient cannot state his/her name, ask a nurse or patient's relative to confirm the patient's identity. Confirm that the specimen label information is identical to the wristband and order form information. Label specimens as indicated below. Outpatient: Confirm identification by asking the patient to state his/her full name. If the patient cannot state his/her name, ask a nurse or patient's relative to confirm the patient's identity. Confirm the specimen label information is identical to the wristband and order form information. The requirements for labeling specimens and requisitions are as follows.

Blood specimens: All blood specimens received by the laboratory must have a permanently attached label with the following information written in black indelible ink: patient's name, hospital number, date and time of collection, initials of person drawing the specimen. Person obtaining blood sample must perform the proper identification check. Consult the individual test listings for specific information and also the Transfusion Service (Blood Bank) Introduction *on page 78* for specifics on Blood Bank patient identification.

Urine specimens: All urine specimens received by the laboratory must have the following information fixed to the container (not cover): patient's name, hospital number, date and time of collection. Urine specimens delivered to the laboratory usually must be placed in the specimen refrigerator. Certain urine tests require special or immediate handling after collection; eg, crystalluria is best evaluated on a fresh, warm specimen. Consult the individual test listings for specific information.

Cerebrospinal fluids: Each tube submitted must be labeled with the patient's name, hospital number, source of specimen, date and time of collection, tube identification number (#1, #2, #3 according to the order of collection). Spinal fluid tests are usually considered to be stat procedures because spinal fluid constituents are unstable and cells degrade quickly. Consequently, spinal fluid must be taken to the laboratory immediately after collection and handed to a technologist or receptionist. Consult the individual test listings for specific information; most such listings in this book bear designations beginning "Cerebrospinal Fluid...".

Body fluids: All body fluids must be labeled with the patient's name, hospital number, date and time of collection, source of fluid. Body fluid constituents are unstable, and thus, expeditious handling is required. Some may derive from patients with medical emergencies or catastrophies. Consult individual test listings for specific information; many in this book bear designations beginning "Body Fluid...".

Cytology smears: All slides for cytologic examination should be appropriately and immediately fixed, labeled with the patient's name, and placed in a cardboard folder with the cytology requisition slip containing patient's name, patient information, and physician's name wrapped around it. Further information is provided in the Cytopathology Introduction *on page 30*.

Bacteriology specimens: All specimens for bacteriology testing must be labeled with the patient's name, hospital number, date and time of collection, source of material. Bacteriology specimens should be delivered to the laboratory as soon after collection as possible to preserve the viability of bacteria or viruses and to provide optimal patient care.

Pathology (surgical) specimens: All specimens for pathology must be labeled with the patient's name, hospital number, name of physician, name of surgeon, and source of specimen. Further information is available in the

Anatomic Pathology Introduction *on page 26*. For many but not all specimens, the listing Histopathology provides detailed information. Consult individual test listings for specific information pertinent to specimens requiring special handling.

Chain-of-custody specimens: Specimens for drugs of abuse screening and specimens which may be used as legal evidence (eg, bullets removed surgically) must be collected according to a chain-of-custody protocol. For Chain-of-Custody form see Chain-of-Custody Protocol *on page 381*.

Specimen Rejection Criteria

Synonyms Rejection Criteria, Specimen; Unsatisfactory Specimens Criteria

Causes for Rejection Criteria for specimen rejection are dependent on individual tests. Generally, specimens received by a laboratory are not discarded until the physician ordering the test or responsible nursing unit is notified. Events which may lead to the rejection of a specimen include specimen improperly labeled or unlabeled, specimen improperly collected and/or preserved, specimen sample volume not sufficient for requirement of test protocol, outside of container contaminated by specimen (ie, infectious hazard), or patient not properly prepared for test requirements. An example of the latter includes a nonfasting state for assays in which fasting is needed.

Some information is provided under Causes for Rejection in individual test listings, but most requirements are provided under appropriate fields such as Preparation.

It is ultimately the responsibility of the ordering physician to make certain that the laboratory is provided with a properly collected and identified specimen for analysis. Communications regarding suboptimal specimens generally should be oriented toward concern for patient welfare rather than nonavailability or unwillingness to provide laboratory service.

- ◆ **Spectral Karyotyping** *see* Fluorescence *in situ* Hybridization *on page 602*
- ◆ **Speed** *see* Amphetamine, Qualitative, Urine *on page 154*
- ◆ **Speed** *see* Methamphetamine, Qualitative, Urine *on page 902*
- ◆ **Sperm Agglutination and Inhibition** *see* Infertility Screen *on page 786*
- ◆ **Sperm Antibodies** *see* Infertility Screen *on page 786*
- ◆ **Spermatids, Round** *see* Semen Analysis, Basic *on page 1187*
- ◆ **Sperm Count** *see* Semen Analysis, Basic *on page 1187*
- ◆ **Sperm Morphology Study** *see* Semen Analysis, Basic *on page 1187*

Sperm Mucus Penetration Test (Human or Bovine Cervical Mucus)

Related Information

Hypo-osmotic Swelling Test (Spermatozoa) *on page 763*
Infertility Screen *on page 786*
Semen Analysis, Advanced *on page 1186*
Semen Analysis, Basic *on page 1187*
Sperm Penetration Assay (Zona-Free Hamster Egg Penetration Test) *on page 1220*
Testosterone, Total and Free, Serum or Plasma *on page 1238*

Synonyms Bovine Cervical Mucus Penetration Test

Test Includes Migration distance, penetration density, migration reduction (decrease in penetration density at 4.5 cm cf at lcm), and duration of progressive movements in cervical mucus in hours

Patient Preparation Follow physician's instructions. Ejaculation should be avoided for 2-3 days prior to collection.

Specimen Liquefied semen. Normal seminal fluid should be liquefied by about 30 minutes after collection (at 37°C). Test should be started within 1 hour after collection of the sample.

Container Clean, dry, wide mouth glass or plastic bottle known to be free of detergent or substances toxic to spermatozoa

Collection Postcoital or masturbation using condom-like Silastic seminal fluid collection device (see Semen Analysis, Basic *on page 1187*) or directly into sterile glass jar.

(Continued)

Sperm Mucus Penetration Test (Human or Bovine Cervical Mucus) *(Continued)*

Storage Instructions Samples should be tested as soon as possible after collection. Time between ejaculation and start of test should not be over 2-3 hours, preferably 1 hour. Human cervical mucus (obtained during time of ovulation) may be stored for some hours at 4°C in the refrigerator.

Special Instructions See Semen Analysis, Basic *on page 1187.*

Reference Interval ≥30 mm penetration by the "vanguard sperm" (Penetrak™ test)

Use Evaluate interaction between spermatozoa and cervical mucus

Limitations Test provides information additional to other semen tests (sperm count, motility, morphology; see Semen Analysis, Basic *on page 1187*). It is not intended for use without other evaluation.

Additional Information Results of *in vitro* tests of sperm penetration have been found to correlate with fertility (pregnancy). Cervical mucus penetration testing has been found of value in assessment of fertility prognosis, in particular when modified to produce a crossmatching penetrability test. The latter study compares results of penetration using cervical mucus of the patient's wife by subject's spermatozoa and additionally with use of semen from fertile donors. The use of hormonally standardized human cervical mucus from female partners has been considered superior to bovine cervical mucus as a penetration medium and as to ability to provide information about sperm function. There are data to suggest that the ability of cervical mucus to accept spermatozoa is dependent upon the carbohydrate composition of mucus glycoproteins.

References

Eggert-Kruse W, Gerhard I, Tilgen W, et al, "Clinical Significance of Crossed *In Vitro* Sperm-Cervical Mucus Penetration Test in Infertility Investigation," *Fertil Steril,* 1989, 52(6):1032-40.

Morales P, Roco M, and Vigil P, "Human Cervical Mucus: Relationship Between Biochemical Characteristics and Ability to Allow Migration of Spermatozoa," *Hum Reprod,* 1993, 8(1):78-83.

World Health Organization Laboratory Manual for the Examination of Human Semen and Sperm-Cervical Mucus Interaction, 4th ed, Cambridge, UK: Cambridge University Press, 1999, 51-8 and 110-13.

♦ **Sperm-Oolemma Binding Test** *see* Infertility Screen *on page 786*

♦ **Sperm Penetration Assay** *see* Hypo-osmotic Swelling Test (Spermatozoa) *on page 763*

Sperm Penetration Assay (Zona-Free Hamster Egg Penetration Test)

Related Information

Hypo-osmotic Swelling Test (Spermatozoa) *on page 763*
Infertility Screen *on page 786*
Semen Analysis, Advanced *on page 1186*
Semen Analysis, Basic *on page 1187*
Sperm Mucus Penetration Test (Human or Bovine Cervical Mucus) *on page 1219*
Testosterone, Total and Free, Serum or Plasma *on page 1238*

Synonyms Hamster Egg Penetration Test; Humster (Human + Hamster) Test

Applies to Computer-Assisted Semen Analysis

Abstract The sperm penetration assay (SPA) is an *in vitro* test that can provide a measure of the fertilizing capacity of human spermatozoa. The test measures the ability of human sperm to penetrate zona-free hamster oocytes. The test thus reflects the ability of sperm to capacitate, acrosome react, fuse with the oolemma, and to undergo nuclear decondensation (within zona-free hamster oocytes). It is not a global test for male infertility.

Patient Preparation Abstinence for a period of at least 48 hours prior to test; less than 48 hours (12 or 24 hours) is associated with reduced penetration potential even though sperm count or motility may not be decreased.

Specimen Semen

Container Clean, wide-mouth glass or plastic container; condom-like device (Silastic seminal fluid collection device - available commercially) placed in a clean jar. Containers should be known free of detergent or other toxic compounds.

SPERM PENETRATION ASSAY (ZONA-FREE HAMSTER EGG PENETRATION TEST)

Collection See Semen Analysis, Basic *on page 1187.*

Storage Instructions Specimen is maintained in the laboratory at room temperature, allowed to liquefy (usually occurs within 30 minutes, abnormal if not liquefied by 60 minutes), standard semen analysis is commonly initiated, and specimen is buffered (see below) and incubated at 4°C for 18 hours or longer.

Causes for Rejection Question concerning authenticity of specimen identification (label, content), exposure to extreme of temperature, specimen more than 1 hour old (sperm should not stay in contact with seminal fluid for more than 1-2 hours prior to processing)

Special Instructions See Semen Analysis, Basic *on page 1187.*

Reference Interval Penetration of 21% to 100% - ("good category"); penetration index of over 0.2. The terms "penetration index," "fertilization index," and "sperm capacitation index," are a measure of polyspermy (the average number of penetrations per ovum).

Use Evaluate fertilizing capability of human spermatozoa; application to the study of effect that environmental factors have on male fertility; a clinical test of sperm function in conjunction with semen analysis and other studies as a screen for *in vitro* fertilization. The SPA measures components of sperm penetration including capacitation (see below), acrosome reaction, chromatin decondensation, and ability to fuse with oolemma. SPA-failed spermatozoa may still be functional in intracytoplasmic sperm injection (ICSI), as the injection procedure bypasses the need for independent sperm oocyte penetration.

Limitations SPA does not assess all functions of human spermatozoa. The test does not measure the ability of sperm to penetrate human ova with granulosa cells and zona pellucida intact. Patients with low scores on SPA may subsequently initiate pregnancy (false-negative result). The test suffers from lack of standardization and is relatively expensive.

Additional Information There is a lack of consensus and a surprisingly wide variety of opinion concerning the value of the sperm penetration assay in the study of fertility. The 1992 review by Liu and Baker noted that "the clinical significance of the SPA in predicting male fertility is still disputed" but that "the SPA has been widely used as a clinical test of sperm function." Correlation with the results of *in vitro* fertilization has been variable. Lack of standardized test parameters, small sample size, and/or variable, often poorly defined parameters of the patient test population characterize many studies. Comparisons and conclusions are problematic.

There is unanimity of opinion that the SPA does assess sperm capacitation, acrosome reaction, ability to fuse with oolemma, and chromatin decondensation with head swelling in cytoplasm of the ovum. These are major physiologic events necessary to fertilization. With the acrosome reaction there is fusion and vesiculation of membranes, formation of pores, and eventual loss of sperm, and outer acrosomal membranous envelope anterior to the equator of the sperm head. The acrosome reaction must occur before the sperm can penetrate the zona pellucida. Round headed spermatozoa (without acrosomes) are thus not capable of fertilization.

Findings from computer-assisted semen analysis have been compared with SPA penetration rates. Total motile oval count (TMO), defined as the product of total count, percent motility, and percent normal (oval) forms in the semen specimen, was a greater risk factor than percent sperm with oval morphology in relation to the outcome of SPA. Below 20% penetration in the SPA assay, both TMO and percent oval sperm were comparable predictive factors.

Acceptance of SPA for clinical assessment of fertility is hampered by interlaboratory variation, cost, and presence of false-negative results. A study of herbal effects on the penetration of zona-free hamster oocytes suggested that high concentrations of St John's wort, ginkgo, and Echinacea damage reproductive cells and that St John's wort may be mutagenic to sperm cells (results of denaturing gradient gel electrophoresis).

References

Brandeis VT, "Importance of Total Motile Oval Count in Interpreting the Hamster Ovum Sperm Penetration Assay," *J Androl*, 1993, 14(1):53-9.

World Health Organization Laboratory Manual for the Examination of Human Semen and Sperm-Cervical Mucus Interaction, 4th ed, Cambridge, UK: Cambridge University Press, 1999, 31, 94-8.

- **S Phase** *see* Breast Biopsy *on page 305*
- **Sphingolipidoses** *see* Inherited Diseases of Metabolism and Cell Structure *on page 792*
- **Sphingomyelinase** *see* Inherited Diseases of Metabolism and Cell Structure *on page 792*
- **Spiders** *see* Arthropod Identification *on page 213*
- **Spinal Fluid Analysis** *see* Cerebrospinal Fluid Analysis: Overview *on page 355*
- **Spinal Fluid Glucose** *see* Cerebrospinal Fluid Glucose *on page 362*
- **Spinal Fluid Identification** *see* Specimen Identification Requirements *on page 1217*
- **Spinal Fluid Lactate** *see* Cerebrospinal Fluid Lactate *on page 369*
- **Sporanox®** *see* Itraconazole, Serum *on page 814*

Sporotrichosis Serology

Related Information

Fungal Culture, Biopsy or Body Fluid *on page 619*
Fungal Culture, Sputum *on page 624*
Fungus Smear, Stain *on page 626*
Skin Biopsy *on page 1200*

Abstract Sporotrichosis is a fungal disease that begins in the distal extremity, often at a site of inoculation as a painless papule. It spreads proximally, involving lymphatics. Extracutaneous disease includes arthritis. Pulmonary sporotrichosis is much less frequently found than osteoarticular infection. The organisms in tissue and in 37°C culture exist as small oval to round yeast with single or multiple buds.

Specimen Serum or cerebrospinal fluid

Container Red top tube; sterile CSF tube

Reference Interval Latex agglutinating titer: <1:4; ELISA: <1:16 in serum, <1:8 in CSF. Antibody titers up to 1:40 may be found in normal subjects.

Critical Values Titers >1:128, rising titers, and persistent elevation are common with pulmonary or systemic disease. Positive reaction in CSF is diagnostic, and is useful in chronic meningitis caused by this organism, which is difficult to culture from spinal fluid.

Use Support the diagnosis of sporotrichosis

Limitations A negative test result does not rule out infection. Serial titers are not prognostically useful. There are occasional low titer false positives from nonfungal disease. This test is not widely available and is not standardized.

Methodology Tube agglutination, latex agglutination (LA), enzyme-linked immunosorbent assay (ELISA), immunoblot analysis

Additional Information Sporotrichosis is usually acquired by traumatic implantation of the dimorphic fungus *Sporothrix schenckii*, a plant saprophyte. Rose gardening, sphagnum moss, and hay bales have been implicated in some cases. The first signs of infection appear after an average incubation period of 3 weeks; the initial lesion may appear as a small ulcer or a small, hard movable, nontender and nonattached subcutaneous nodule. Disease progresses along lymphatic channels that drain the area of the initial lesion. Less frequent forms of sporotrichosis include pulmonary, osteoarticular, central nervous system, and disseminated disease.

References

Davis BA, "Sporotrichosis," *Dermatol Clin*, 1996, 14(1):69-76.

Smego RA Jr, Castiglia M, and Asperilla MO, "Lymphocutaneous Syndrome. A Review of Non-Sporothrix Causes," *Medicine (Baltimore)*, 1999, 78(1):38-63.

Werner AH and Werner BE, "Sporotrichosis in Man and Animal," *Int J Dermatol*, 1994, 33(10):692-700.

Wescott BL, Nasser A, and Jarolim DR, "*Sporothrix* Meningitis," *Nurse Pract*, 1999, 24(2):90-8

Internet Web Sites

www.cdc.gov/ncidod/dbmd/diseaseinfo/sporotrichosis_g.htm

- **Sputum Culture** *see* Bacterial Culture, Lower Respiratory *on page 241*

Sputum Cytology

Related Information

Bacterial Culture, Lower Respiratory *on page 241*
Bronchial Brushings/Washings Cytology *on page 310*
Bronchoalveolar Lavage (BAL) *on page 311*

Test Includes Three to five consecutive first morning deep cough specimens

Abstract Five major techniques are available for the cytologic diagnosis of carcinoma of lung. Sputum cytology is the most fundamental. The three bronchoscopic methods are bronchial washing, bronchial brushing, and bronchoalveolar lavage. FNA techniques are performed with imaging guidance or transbronchoscopically. Cytopathological examination of sputum aids in the evaluation of respiratory infections or neoplasms. Cancer of lung is the most common fatal malignant disease in both sexes in the U.S.

The least invasive means to establish pathologic diagnosis, sputum cytology is up to 70% sensitive for central tumors, but provides much lower sensitivity for peripheral neoplasms.

Specimen Expectorated sputum, **not saliva or nasal aspirates**

Container 50 mL screw-top plastic container; for sputum series it will contain cytologic fixative (ie, Saccomanno fixative, Carbowax®).

Collection Upon arising the patient rinses his mouth with water and expectorates a deep cough into the container. The **first** cough specimen is the most rewarding.

Storage Instructions Refrigeration; if it is not possible to bring unfixed material to the laboratory, a prefixed specimen (using Carbowax® or 70% ethanol) may be substituted.

Use Establish the presence of neoplasm; aid in the diagnosis of respiratory infections with herpesvirus, cytomegalovirus, fungal diseases, *Strongyloides*, *Echinococcus*, and *Paragonimus*; aid in the diagnosis of lipoid pneumonitis, allergic processes, hemosiderosis, Goodpasture syndrome, asbestosis, and alveolar proteinosis

Limitations If no carbon bearing histiocytes are identified in the specimen, it is considered to be an unsatisfactory specimen (not a deep cough specimen). A 2.1% false-positive rate (97.9% specificity) is recognized.

Mechanical blending decreases sensitivity for small cell undifferentiated carcinoma by disruption, and that for adenocarcinoma by shearing mucin-containing vacuoles.

Methodology Sputa may be collected fresh, without fixative; make direct smears from white flecks and blood-tinged areas; fix smears immediately in 95% ethanol. Sputa may be collected in Carbowax® if the Saccomanno technique is used. Saccomanno technique involves the collection of sputum material in a mixture of 50% ethanol and 2% polyethylene glycol. If the patient cannot produce sputum spontaneously by deep coughing, it should be induced.

Additional Information Special stains are sometimes needed. When a pulmonary lesion is suspected, a complete sputum series should be examined. The complete sputum series consists of a fresh, early morning, deep cough specimen each day for 3-5 days. A postbronchoscopy sputum should be included in the series. The complete sputum series increases the detection of primary bronchogenic carcinoma from 45% (one specimen) to 86% (three specimens). A 12- to 24-hour specimen is collected in Carbowax® in patients with scanty sputum, when a previous single sputum contains rare malignant cells, or cells highly suspicious for malignancy are present. In cases in which infectious agents are identified by cytology, confirmation with culture is advised.

References

Linder J, "Lung Cancer Cytology: Something Old, Something New," *Am J Clin Pathol*, 2000, 114(2):169-71.

Rosenthal DL, "Cytologic Diagnosis of Infectious Diseases of the Lung," *Compendium on Diagnostic Cytology*, 8th ed, Wied GL, Bibbo M, Keebler CM, et al, eds, Chicago, IL: Tutorials of Cytology, 1997, 208-15.

(Continued)

Sputum Cytology *(Continued)*

Saccomanno G, Saunders RP, Ellis H, et al, "Concentration of Carcinoma or Atypical Cells in Sputum," *Acta Cytol*, 1963, 7:305-10.

- ♦ **SS-A Antibodies** *see* Sjögren Antibodies *on page 1199*
- ♦ **SS-A/Ro** *see* Sjögren Antibodies *on page 1199*
- ♦ **SS-B Antibodies** *see* Sjögren Antibodies *on page 1199*
- ♦ **SS-B/La** *see* Sjögren Antibodies *on page 1199*
- ♦ **ss-DNA** *see* Anti-DNA *on page 173*
- ♦ **S-Sulfocysteine, Urine** *see* Molybdenum, Blood *on page 919*
- ♦ **Stable Factor (Factor VII)** *see* Coagulation Factor Assays *on page 418*
- ♦ **Standard Deviation** *see page 11*
- ♦ **Standard Deviation (S)** *see page 11*
- ♦ **Standard Error (SE)** *see page 11*
- ♦ **Staphylococcus aureus** *see* Bacterial Culture, Aerobes *on page 229*
- ♦ **Statex®** *see* Morphine, Urine *on page 921*
- ♦ **Statistics** *see page 11*
- ♦ **Staurodorm®** *see* Flurazepam, Serum *on page 605*
- ♦ **Stay Awake®** *see* Caffeine, Serum *on page 326*
- ♦ **Steatocrit** *see* Fat, Semiquantitative, Stool, Acid Steatocrit *on page 571*
- ♦ **Stelazine®** *see* Phenothiazines, Serum *on page 1021*
- ♦ **Stem Cell Collection** *see* Apheresis, Therapeutic *on page 201*
- ♦ **Stem Cell Collection** *see* Hematopoietic Progenitor Cells, Peripheral Blood *on page 679*
- ♦ **Stem Cells** *see* CD34⁺ Hematopoietic Stem Cells by Flow Cytometry *on page 351*
- ♦ **Steroid Sulfatase Deficiency** *see* Inherited Diseases of Metabolism and Cell Structure *on page 792*
- ♦ **Stesolid®** *see* Diazepam, Serum *on page 510*
- ♦ **sTfR** *see* Transferrin Receptor, Soluble, Serum or Plasma *on page 1267*

St Louis Encephalitis Virus Serology

Related Information

Arthropod Identification *on page 213*
Bacterial Culture, Cerebrospinal Fluid *on page 236*
California Encephalitis Virus Serology *on page 334*
Cerebrospinal Fluid Analysis: Overview *on page 355*
Eastern Equine Encephalitis Virus Serology *on page 529*
Encephalitis Viral Serology *on page 535*
Viral Culture *on page 1307*
Western Equine Encephalitis Virus Serology *on page 1328*

Abstract St Louis encephalitis virus causes fever with headache, aseptic meningitis, and encephalitis. Viral transmission includes birds and mosquitoes. Severity of illness increases with age. Patients older than 60 years of age bear the highest frequency of encephalitis.

Specimen Serum or cerebrospinal fluid

Container Red top tube, sterile CSF tube

Sampling Time Acute and convalescent sera drawn 10-14 days apart are recommended.

Reference Interval A less than fourfold increase in titer in paired sera; HI titer: <1:10; CF titer: <1:8; plaque reduction: <70%; no IgM antibody in CSF

Use Support the diagnosis of St Louis encephalitis

Limitations Cross reactivity between alphavirus group and flavivirus group; false reactions from yellow fever vaccination. (The yellow fever virus is found in the family Flaviviridae, as is the St Louis encephalitis virus.)

Methodology Complement fixation (CF), hemagglutination inhibition (HAI), indirect immunofluorescent antibody, enzyme-linked immunosorbent assay (ELISA) for IgM or IgG in CSF or serum antibodies

Additional Information Arboviruses, such as St Louis encephalitis, are the most frequent cause of epidemic encephalitides in the United States. In most cases, demonstration of IgM antibody in CSF rapidly establishes a diagnosis of arboviral encephalitis.

References

Day JF and Curtis GA, "Blood Feeding and Oviposition by Culex Nigripalpus (Diptera: Culicidae) Before, During, and After a Widespread St. Louis Encephalitis Virus Epidemic in Florida," *J Med Entomol*, 1999, 36(2):176-81.

Monath TP, "Flavivirus (Yellow Fever, Dengue, and St Louis Encephalitis)," *Principles and Practice of Infectious Diseases*, 3rd ed, Mandell GL, Douglas RG Jr, and Bennett JE, eds, New York, NY: Churchill Livingstone, 1990, 1248-51.

Tsai TF and Kuno G, "Arboviruses," *Manual of Clinical Laboratory Immunology*, 5th ed, Rose NR, Conway de Macario E, Folds JD, et al, eds, Washington, DC: ASM Press, American Society for Microbiology, 1997, 729-36.

Internet Web Sites

www.astdhpphe.org/infect/sle.html
www.cdc.gov/ncidod/dvbid/arbor/sle_qa.htm

♦ **Stool Fat, Quantitative** *see* Fecal Fat, Quantitative, 72-Hour Collection *on page 574*

♦ **Stool Fat, Semiquantitative** *see* Fat, Semiquantitative, Stool, Sudan III Stain *on page 572*

♦ **Stool, Occult Blood** *see* Occult Blood, Stool *on page 969*

♦ **Stool pH** *see* pH, Stool *on page 1034*

♦ **Stool Reducing Substances** *see* Reducing Substances, Stool *on page 1147*

♦ **Streptodornase** *see* Antideoxyribonuclease-B Titer, Serum *on page 170*

Streptozyme

Related Information

Antideoxyribonuclease-B Titer, Serum *on page 170*
Antistreptolysin O Titer, Serum *on page 197*
Bacterial Culture, Throat, and Antigen Detection Testing for Group A Streptococci *on page 245*
Group A *Streptococcus* Screen, Rapid *on page 659*

Abstract Streptozyme is among the serologic tests which detect immune response to extracellular products of *S. pyogenes*.

Specimen Serum

Container Red top tube

Reference Interval <100 streptozyme units

Use Screening for antibodies to streptococcal antigens NADase, DNase, streptokinase, streptolysin O, and hyaluronidase

Limitations A single determination is less useful than a series. May not be as sensitive in children as in adults. A disadvantage of the test is that borderline antibody elevations, which could be clinically significant particularly in children, may not be detected.

Methodology Hemagglutination

Additional Information With throat culture, this test is useful for evaluation of patients suspected of having poststreptococcal sequelae. They include rheumatic fever following *Streptococcus pyogenes* infection. Streptozyme is a screening test for antibodies to several streptococcal antigens. It has the advantages of detecting several antibodies in a single assay (although which one has been detected cannot be ascertained), of being technically quick and easy, and of being unaffected by several factors producing false positives in the ASO test. A serially rising titer is more significant than a single determination.

References

Ayoub EM and Harden E, "Immune Response to Streptococcal Antigens: Diagnostic Methods," *Manual of Clinical Laboratory Immunology*, 5th ed, Rose NR, Conway de Macario E, Folds JD, et al, eds, Washington, DC: ASM Press, American Society for Microbiology, 1997, 450-7.

Bisno AL, "Group A Streptococcal Infections and Acute Rheumatic Fever," *N Engl J Med*, 1991, 325(11):783-93.

Ruoff K, Whiley RA, and Beighton D, "*Streptococcus*," *Manual of Clinical Microbiology*, 7th ed, Chapter 17, Murray PR, Baron EJ, Pfaller MA, et al, eds, Washington, DC: ASM Press, American Society for Microbiology, 1999, 283-96.

Todd JK, "Group A *Streptococcus*," *Nelson Textbook of Pediatrics*, 16th ed, Chapter 184, Behrman RE, Kliegman RM, and Jenson HB, eds, Philadelphia, PA: WB Saunders Co, 2000, 802-10.

Internet Web Sites

www.astdhpphe.org/infect/strepa.html
www.cdc.gov/ncidod/dbmd/diseaseinfo/groupastreptococcal_g.htm

♦ **sTSH** *see* Thyroid Stimulating Hormone, Serum *on page 1250*

♦ **Stuart Factor (Factor X)** *see* Coagulation Factor Assays *on page 418*

♦ **Stuart-Prower Factor (Factor X)** *see* Coagulation Factor Assays *on page 418*

- ♦ **Student's t Distribution** *see page 11*
- ♦ **Substance Abuse and Mental Health Services Administration** *see page 63*
- ♦ **Succinylcholine Sensitivity** *see Acetylcholinesterase, Red Cell and Serum on page 93*
- ♦ **Sudden Infant Death Syndrome** *see Fetal Hemoglobin on page 581*
- ♦ **Sugar, Fasting** *see Glucose, Fasting, Plasma on page 643*
- ♦ **Sugar, Qualitative, Urine** *see Glucose, Semiquantitative, Urine on page 650*
- ♦ **Sugar, Quantitative, Urine** *see Glucose, Quantitative, Urine on page 649*

Sugar Water Test Screen

Related Information
Ham Test *on page 665*
Hemosiderin Stain, Urine *on page 692*
PNH Test (GPI-Anchored Proteins) by Flow Cytometry *on page 1064*

Synonyms PNH Test Screen

Test Includes Sucrose hemolysis if hemolysis is found in the sugar water screen test

Abstract Screening test for suspected paroxysmal nocturnal hemoglobinuria (PNH). Confirm with PNH Test (GPI-Anchored Proteins) by Flow Cytometry *on page 1064*.

Specimen Blood

Container Blue top (sodium citrate) tube

Causes for Rejection Hemolyzed specimens

Reference Interval Absence of hemolysis is the normal condition. If no hemolysis is present, the patient probably does not have PNH provided multiple recent transfusions have not reduced the proportion of abnormal cells. When positive, the test shows lysis of some 10% to 80% of red cells. The screening test is not definitive.

Use Screen for PNH

Limitations False positives may be seen in cases of megaloblastic anemias and autoimmune hemolytic anemias (usually <5% lysis). Some patients with leukemia or myelosclerosis may also have false-positive results (usually mild lysis, <10% of red cells). False-negative results may occur if heparin or EDTA is used as an anticoagulant.

Additional Information If the sugar water test is positive, a subsequent PNH test by flow cytometry is strongly recommended. A negative sugar water test rules out PNH in most instances, provided the proportion of patient cells has not been reduced by previous transfusion. The most complement sensitive cells in PNH are younger RBCs and reticulocytes. Sensitivity of the assay can be increased by separating young RBCs from old by centrifugation.

Patients with aplastic anemia may develop PNH. Patients with PNH may present as pancytopenia and have a dysplastic or aplastic appearing marrow. Patients may develop thrombosis with hypercoagulable state, and may bleed with thrombocytopenia and a disseminated intravascular coagulation-like picture. Renal pathology, clinically including acute renal failure, results at least in part from repeated renal microvascular thrombosis.

The molecular pathology of PNH is partially, if not largely, established and involves a number of PIG-A gene mutations. These affect a glycolipid (glycosyl-phosphatidylinositol anchor for decay accelerating factor (DAF)) or CD59. DAF and CD59 are membrane-regulatory proteins which inhibit complement activation. See Ham Test *on page 665* and PNH Test (GPI-Anchored Proteins) by Flow Cytometry *on page 1064*. Deficient or absent CD59 (which regulates assembly of the C9 complement attack complex) results in red cell hemolysis and the hypercoagulable state. Demonstration of decreased red cell (and/or granulocyte) levels of CD59 by flow cytometry is the preferred test for supporting a diagnosis of PNH.

References
Azenishi Y, Ueda E, Machii T, et al, "CD59-Deficient Blood Cells and PIG-A Gene Abnormalities in Japanese Patients With Aplastic Anemia," *Br J Haematol*, 1999, 104(3):523-9.
Hillman RS and Ault KA, "The Dysplastic and Sideroblastic Anemias," *Hematology in Clinical Practice: A Guide to Diagnosis and Management*, 2nd ed, Chapter 9, New York, NY: McGraw-Hill, 1998, 151-2.

Rosse WF, "Paroxysmal Nocturnal Hemoglobinuria," *Hematology: Basic Principles and Practice*, 3rd ed, Chapter 20, Hoffman R, Benz EJ Jr, Shattil SJ, et al, eds, Philadelphia, PA: Churchill Livingstone, 2000, 331-42.

Swim Up/Swim Down Procedures (Spermatozoa)

Related Information

Infertility Screen *on page 786*
Semen Analysis, Basic *on page 1187*
Testosterone, Total and Free, Serum or Plasma *on page 1238*

Abstract Preparatory techniques used to obtain spermatozoa from seminal plasma for use in a variety of diagnostic and therapeutic procedures. The harvested sperm are largely morphologically normal and highly motile. Nonmotile/dead spermatozoa, white blood cells, and debris are eliminated. The resultant preparations can be used for intrauterine insemination (IUI) and for attempted gender preselection utilizing IUI.

Specimen Semen

Container Sterile specimen container; clean, dry, wide mouth glass or plastic bottle maintained warm and known to be free of detergent or other toxic compounds

Collection See Semen Analysis, Basic *on page 1187*.

Storage Instructions Patient should be instructed to bring specimen to the laboratory within 30-60 minutes after collection, maintaining warmth (37°C) during transport. Patient should be instructed to transport specimen in a pocket close to the skin.

Turnaround Time 1-2 hours

Special Instructions Specimens for use in assisted reproduction procedures such as *in vitro* fertilization/intrauterine insemination (IVF)/IUI, if obtained with diluents having human serum albumin as a constituent, must use highly purified albumin that is free from viral, bacterial, and prior contamination. Requisition should specify infertility study or postvasectomy study. Semen, as with all blood, urine, and body fluid specimens, because of the risk of HIV, should be received and handled with attention to universal precautions.

(Continued)

Swim Up/Swim Down Procedures (Spermatozoa)
(Continued)

Use Separation of spermatozoa from semen for use in some therapeutic/diagnostic procedures in clinical andrology. The resultant preparations are enriched by morphologically normal and motile spermatozoa.

Methodology

Swim up: A preovulatory human tubal fluid-like diluent is placed over the liquefied semen specimen and spermatozoa are allowed to migrate into the diluent. The upper 1 mL of diluent (contains the motile spermatozoa) is removed and used to determine sperm concentration, morphology, function, and/or for insemination.

Swim down: An aliquot of a centrifuged, resuspended semen sample is layered on top of 10% bovine serum albumin and incubated at 37°C for 45-60 minutes. Sperm in the bottom two-thirds of the albumin layer are analyzed for concentration, morphology, and motility.

Gradient centrifugation: Preparation of spermatozoa by passage through Percoll gradients, or Sephadex and Ficoll columns has also been used.

Stimulation techniques: Sperm can be prepared by incubation with stimulants in order to increase motility and/or capacitation. These include pentoxifylline (phosphodiesterase inhibitor) and calcium ionophore A23187 (calcium transport modulator).

Additional Information Sperm preparation by washing (and diluting) with culture media and with multiple cycles of centrifugation serves to remove inactive, fragmented and dead spermatozoa, white blood cells, and seminal debris as well as capacitation inhibitors and prostaglandins. Cyclic centrifugation and compaction of the sample, however, is traumatic and usually results in significant loss (or absence) of sperm motility. Functional spermatozoa which are more motile and have higher density can be harvested by the relatively more gentle swim-up, swim-down, or discontinuous density gradient procedures.

Direct swim-up procedures are preferred for the separation of motile spermatozoa. Centrifugation of sperm prior to swim-up may produce reactive oxygen species/cell membrane damage and should be avoided.

Swim-up and Percoll gradient preparations result in higher pregnancy rates than the wash, swim-down and refrigeration/heparin methods.

References

Carrell DT, Kuneck PH, Peterson CM, et al, "A Randomized, Prospective Analysis of Five Sperm Preparation Techniques Before Intrauterine Insemination of Husband Sperm," *Fertil Steril*, 1998, 69(1):122-6.

World Health Organization Laboratory Manual for the Examination of Human Semen and Sperm-Cervical Mucus Interaction, 4th ed, Cambridge, UK: Cambridge University Press, 1999, 34-5, 104-6.

♦ **Symoron**® *see* Methadone, Serum or Urine *on page 900*

♦ **Synalgos**® *see* Salicylate, Serum or Plasma *on page 1176*

♦ **Synaptophysin** *see* Immunoperoxidase Procedures *on page 780*

Syndecan-1, Serum

Related Information

Beta₂-Microglobulin, Serum or Urine *on page 254*
Bone Marrow *on page 296*
Immunofixation Electrophoresis, Serum or Urine *on page 768*

Abstract Syndecan is a membrane heparin sulfate proteoglycan. It is involved in cell-matrix adhesion processes, is a low-affinity receptor for heparin-binding growth factors, and in the bone marrow, is a cell surface antigen in early and late stages of B-lymphocyte differentiation. While absent on mature B cells, it is re-expressed on plasma cells. In patients with multiple myeloma, syndecan-1 has been found to be expressed only on myeloma cells. It is also seen on malignant plasma cells circulating in the peripheral blood. The level of syndecan-1, shed from the surface of myeloma cells, has been found to correlate with tumor burden and survival. Serum syndecan-1 level may be of use in prognostic classification of myeloma.

Specimen Serum, 1-2 mL

Container Red top tube

Turnaround Time 4 hours

Reference Interval Median syndecan-1 levels in normal controls; 128 units/mL, 79% of myeloma patients had syndecan-1 levels above the mean level, ±2 SD (>370 units/mL)

Use Serum syndecan-1 levels correlate with survival in multiple myeloma. Syndecan-1 is considered to represent a new independent prognostic parameter in myeloma.

Methodology Enzyme-linked immunosorbent assay (ELISA), commercially available (Diaclone Research, Besancon, France)

Additional Information Syndecans are a family of transmembrane heparan sulfated proteoglycans. Syndecan-1 is formed of a core protein with side chains of heparan sulfate and chondroitin sulfate. Syndecan-1 is involved in the adhesion of mature plasma cells to stomal cells of bone marrow. Monoclonal antibodies against syndecan-1 are markers for malignant plasma cells and are of use in the purification of myeloma cells. The Nordic Myeloma Study Group has concluded that serum syndecan-1 levels represent a new independent prognostic parameter in multiple myeloma. Their study of 174 myeloma patients found that "low" syndecan-1 levels had a median survival of 44 months, while "high" levels had a median survival of only 20 months.

References

Jourdan M, Ferlin M, Legouffe E, et al, "The Myeloma Cell Antigen Syndecan-1 Is Lost by Apoptotic Myeloma Cells," *Br J Haematol*, 1998, 100(4):637-46.

Seidel C, Sundan A, Hjorth M, et al, "Serum Syndecan-1: A New Independent Prognostic Marker in Multiple Myeloma," *Blood*, 2000, 95(2):388-92.

Witzig TE, Kimlinger T, Stenson M, et al, "Syndecan-1 Expression on Malignant Cells From the Blood and Marrow of Patients With Plasma Cell Proliferative Disorders and B-Cell Chronic Lymphocytic Leukemia," *Leukemia and Lymphoma*, 1998, 31(1-2):167-75.

Synovial Fluid Analysis

Related Information

Antinuclear Antibodies *on page 189*
Body Cavity Fluid Cytology *on page 285*
Body Fluid Analysis, Cell Count *on page 288*
Body Fluid Glucose *on page 294*
Body Fluid Lactate Dehydrogenase *on page 294*
Chlamydia trachomatis Culture *on page 385*
Chlamydia trachomatis Direct Antigen Test *on page 387*
Chlamydia trachomatis Nucleic Acid Detection *on page 388*
Fungal Culture, Biopsy or Body Fluid *on page 619*
Gram Stain *on page 658*
Lyme Disease DNA Detection *on page 878*
Lyme Disease Serology *on page 879*
Mucin Clot Test *on page 922*
Mycobacterial Culture, Biopsy or Body Fluid *on page 929*
Neisseria gonorrhoeae Culture and Smear *on page 945*
Rheumatoid Factor, Serum or Body Fluid *on page 1161*
Uric Acid, Serum *on page 1286*

Synonyms Joint Fluid Analysis

Applies to Calcium Pyrophosphate Dihydrate Crystals; Urate Crystals

Test Includes May vary between laboratories but should include cell count and differential, cultures and Gram stain for pathogens and polarizing microscopic examination for crystals. Other tests, depending on the clinical situation, may be of value including viscosity, clot lysis, uric acid, rheumatoid factor, cytokines, culture, and cytology.

Abstract Analysis of joint fluid usually includes multiple studies to determine if joint disease is present and to assess its nature and severity. Analyses are particularly helpful in differentiating traumatic arthritis from the immune-based and crystal-induced arthritides.

Patient Preparation As per physician's usual aseptic aspiration technique

Specimen Joint fluid; simultaneously drawn venous blood in red top tube often helpful, especially with order for selected serum chemistry assays.

Container Capped syringe or three sterile tubes, one with sodium heparin (or liquid EDTA) and two red top tubes for joint fluid; one to two red top tubes of venous blood desirable; Thayer-Martin agar is best inoculated with joint fluid at bedside if gonococcal (GC) infection is suspected. Avoid anticoagulants composed of crystalline EDTA, or oxalate or lithium heparin.

(Continued)

Synovial Fluid Analysis (Continued)

Collection An experienced physician, using sterile technique, obtains the specimen (arthrocentesis). Media appropriate for culture of *N. gonorrhoeae*, *M. tuberculosis*, and other organisms should be available if indicated.

Storage Instructions Testing should be initiated, in most cases, shortly after receipt of the specimen. It may be prudent to save a portion of the specimen for possible additional analyses (eg, bacterial and crystal-induced diseases) if such testing is not performed initially. An average 42% decrease in white blood cell count has been noted during the first 6 hours after obtaining the specimen. Calcium pyrophosphate crystals decrease in a few days, while monosodium urate (MSU) crystals maintain their number, size, and birefringence over the first days but decrease over a few weeks.

Causes for Rejection Some constituent tests of the analysis may not be performed if gross contamination has occurred.

Special Instructions Specimens should be delivered immediately to the laboratory and placed in the hands of a medical technologist. **Physician should indicate clinical impression and indicate tests he/she feels are necessary.** If presence of monosodium urate (MSU) crystals (as are present with gout) is a consideration, alcohol-based fixatives and stains must be used. MSU crystals are water soluble.

Reference Interval When synovial fluid can be aspirated, abnormality is probably evident. Normal synovial fluid does not clot spontaneously, because it lacks fibrinogen. Clotting bears an implication of inflammation. There should be <200 WBCs/mm³, 0% to 25% neutrophils, protein should be ≤3.0 g/dL (SI: ≤30 g/L), uric acid should be <8.0 mg/dL (SI: <476 µmol/L) and fluid LD (LDH) should be the same or less than that of the patient's serum drawn at the same time. Glucose is significantly abnormal in the nonfasting subject when it is <40 mg/dL (SI: <2.2 mmol/L). There should be a long string produced normally when synovial fluid is poured from a container and the mucin clot test is normally positive (ie, a firm mucin clot is formed in the presence of acetic acid).

Comparison of Gout and CPPD Crystal Deposition Disease[1]

Feature	Gout	CPPD Crystal Deposition Disease
Male-female ratio	7:1	1.5:1
Age-group affected	Middle-aged men, postmenopausal women	Elderly
Hereditary forms?	Yes	Yes
Hyperuricemia present?	Yes	No
Clinical picture	Asymptomatic phase, acute and chronic arthritis, tophi, renal disease	Asymptomatic phase, acute and chronic arthritis, tophus-like collections only rarely
Typical joint localization	First MTP joints, midfoot, ankles, knees, wrists	Knees, wrists, MCP joints, elbows, shoulders
Radiologic findings	Erosions with overlying edge of displaced bone	Chondrocalcinosis
Findings on synovial fluid analysis	Evidence of inflammation, MSU crystals	Evidence of inflammation, CPPD crystals
Birefringence of synovial fluid crystals	Strong with negative elongation	Weak with positive elongation
Possible underlying disease	Diabetes, obesity, hypertension, hyperlipidemia	Hyperparathyroidism, hemochromatosis, hypomagnesemia, hypophosphatasia, severe hypothyroidism

[1]Formerly called pseudogout.

CPPD = calcium pyrophosphate dihydrate; MCP = metacarpophalangeal; MSU = monosodium urate; MTP = metatarsophalangeal; NSAID = nonsteroidal anti-inflammatory drugs.

Adapted from Beutler A and Schumacher HR Jr, "Gout and Pseudogout: When Are Arthritic Symptoms Caused by Crystal Deposition?" *Postgrad Med*, 1994, 95(2):103-16.

Use Aid the diagnosis of rheumatic disease and diseases which cause joint symptoms, pain, increase in joint fluid or destruction of joint space, including rheumatoid arthritis, joint infection, gout, and pseudogout. In cases with appropriate clinical presentation (see table), synovial fluid analysis using polarized light should be done to confirm presence of a crystal-induced arthropathy. When infection is **not** in question, routine synovial fluid analysis has a low diagnostic yield and does not contribute significantly to the management of cases with previously established rheumatic disease.

Limitations Appropriate work-up may not be accomplished unless there is discussion between physician and analyst. Oxalate anticoagulants may lead to problems in crystal identification. The anticoagulants sodium EDTA, ammonium oxalate, and lithium heparin can form crystals in joint fluid. If GC infection is in the differential consideration, the physician, ideally, should plate out the Thayer-Martin medium at the bedside at time of aspiration. Limited sensitivity and specificity exist overall for synovial fluid analysis.

Methodology Includes polarizing and phase as well as conventional microscopy. Laboratory notes volume, clarity, color, and presence of clot in centrifuged specimen. Cultures are made from centrifuged sediment and media inoculated for acid-fast bacteria, fungus, routine culture, and Gram stain smear. Rheumatoid factor titer is desirable in suspected cases of rheumatoid arthritis. Other testing methodologies may be applied as indicated or ordered. The physician should verify that laboratory personnel are experienced in the use of birefringence in crystal analysis. A polarizing microscope and first-order red compensator must be available to determine the sign of birefringence of the crystal.

Additional Information The contribution of the laboratory to diagnosis of joint disease depends importantly on its ability to identify crystals present in the patient's synovial fluid. The two most common crystals are monosodium urate (needle-shaped and found in an intracellular situation in neutrophils/monocytes in 90% of cases of acute gout) and calcium pyrophosphate dihydrate (often rhomboid and the cause of "pseudogout," for which the preferred term is now, calcium pyrophosphate deposition disease). Both MSU and CPPD crystals are birefringent.

Normal fluid is rarely obtained because of its small volume; therefore, any fluid which is aspirated is potentially a diagnostic specimen. Detection of synovial fluid monosodium urate crystals is important in establishing a diagnosis of gout. A heparinized tube is needed for cell count and differential. A red cell count is not necessary, but if grossly bloody, a hematocrit should be ordered. Joint fluid uric acid significantly higher than serum uric acid may be diagnostic of gout. Other chemistries are usually of little value with the occasional exception of protein, LD (LDH), and glucose. Decreased fluid glucose indicates inflammation, but the result should be compared to that of serum or plasma. High synovial LD but normal serum LD suggests RA, infectious arthritis, or gout. Synovial LD is normal in degenerative joint disease.

Presence of cartilage cells supports a diagnosis of traumatic arthritis or osteoarthritis. While involvement of joints by malignancy is uncommon, both primary and metastatic tumors should be included in differential consideration. Cytologic examination may support diagnosis of pigmented villonodular synovitis or Reiter's syndrome. While rhomboid cholesterol crystals have been reported in chronic joint effusions of rheumatoid arthritis and osteoarthritis, birefringent lipid bodies (lipid microspherules, liposomes, smectic mesophases, lipid liquid crystals), intra- and extracellular, at least partly formed of cholesterol ester, have been found in fluids from acute and chronic arthritis, traumatic arthritis, and pigmented villonodular synovitis.

Phase microscopy is used to look for intracellular inclusions in pus cells. These are "RA cells" only if the rheumatoid titer is positive. Synovial fluid in pseudogout, traumatic arthritis and osteoarthritis rarely also may contain intraleukocytic cytoplasmic inclusions. Polarizing microscopy is done to identify crystals of urate and pyrophosphate, the causes, respectively, of gout and pseudogout. Calcium pyrophosphate dihydrate (CPPD) crystal deposition disease does not equate exclusively with a diagnosis of pseudogout (chondrocalcinosis). It is important to recognize that CPPD disease encompasses an array of disorders (Continued)

Synovial Fluid Analysis *(Continued)*

occurring largely in the aged. These include asymptomatic chondrocalcinosis, acute pseudogout, and chronic pyrophosphate arthropathy. As an example, asymptomatic chondrocalcinosis occurs in 10% to 15% in those 65-75 years of age and in 30% to 60% of those older than 80 years of age. Cases of CPPD disease may be sporadic, familial, or secondary to degenerative or metabolic disease. Hyperparathyroidism and hemochromatosis may be associated with CPPD deposition. Cholesterol crystals occur occasionally, indicating RA. Steroid crystals after therapeutic injection may be found.

Test for viscosity and mucin test for hyaluronic acid measure the physical character of the synovial fluid. Abnormality in either of these tests indicates dilution or inflammation. Viscosity can be evaluated by the quality of stringing; inflammatory fluids of low viscosity produce very short strings, but normal or noninflammatory fluids produce long strings. Viscosity is generally equivalent to mucin hyaluronate content.

In some large urban areas, gonococcal infection is the most common cause of infectious arthritis. While peripheral blood white cell counts may not be helpful, synovial fluid white cell levels average some 50,000 cells/μL with Gram stains positive in about 25% of cases. In cases with tenosynovitis following migratory arthralgia, some 40% of patients have positive blood cultures. Nongonococcal acute bacterial arthritis, usually occurring in immune impaired or traumatized patients, is most commonly the result of *Staphylococcus aureus* infection. Group A/B streptococci, *E. coli*, and *Pseudomonas aeruginosa* should also be considered. When the white cell count in the synovial fluid is >100,000/μL, largely neutrophils, blood culture is positive in some 50% of patients.

More than 2% eosinophils in synovial fluid may be a clue to presence of Lyme disease. In rheumatoid factor-negative patients with oligoarthritis of unknown etiology, antigen-specific lymphocyte proliferation may be found in the synovial fluid of 34% of patients. *C. trachomatis* is the most frequent single agent detected.

Individual laboratories may not be proficient in the identification of crystals present in synovial fluid. Artifacts and unexplained birefringent materials may be present. Patients with hemarthrosis may develop solid and angular birefringent crystals of two different types. Rectangular hemoglobin- like crystals within red cells are weakly birefringent. Golden brown rhomboid crystals (likely hematoidin) are intensely birefringent with positive or negative elongation. These red cell-derived crystals may be confused with pathogenic crystals.

Arthritis is one of the characteristic features of Whipple disease (WD). The arthritis, episodic and migratory, is the presenting symptom (oligoarthritis or polyarthritis) in 60% of patients with WD. Synovial fluid from a patient with a clinical course consistent with WD will contain many neutrophils and bacterial DNA of *Tropheryma whippelii* (most likely causative agent of WD as identified by polymerase chain reaction).

Various aspects of psoriatic arthritis have been studied, including measurement of synovial fluid levels of IL-1β and IL-6.

References

Debets R, Hegmans JP, Deleuran M, et al, "Expression of Cytokines and Their Receptors by Psoriatic Fibroblasts. I. Altered IL-6 Synthesis," *Cytokine*, 1996, 8(1):70-9.

Judkins SW and Cornbleet PJ, "Synovial Fluid Crystal Analysis," *Lab Med*, 1997, 28(12):774-9.

Lange U and Teichmann J, "Diagnosis of Whipple's Disease by Molecular Analysis of Synovial Fluid," concise communications, *Arthritis Rheum*, 1999, 42(8):1777-8.

O'Connell JX, "Pathology of the Synovium," *Am J Clin Pathol*, 2000, 114(5):773-84.

Pal B, Nash J, Oppenheim B, et al, "Is Routine Synovial Fluid Analysis Necessary? Lessons and Recommendations From an Audit," *Rheumatol Int*, 1999, 18(5-6):181-2.

Sieper J, Braun J, Brandt J, et al, "Pathogenetic Role of *Chlamydia, Yersinia,* and *Borrelia* in Undifferentiated Oligoarthritis," *J Rheumatol*, 1992, 19(8):1236-42.

♦ **T₃U** *see* T₃ Uptake, Serum or Plasma *on page 1233*

T₃ Uptake, Serum or Plasma

Related Information

Free Thyroxine Index *on page 613*
Thyroid Stimulating Hormone, Serum *on page 1250*
Thyroxine Binding Globulin, Serum *on page 1255*
Thyroxine, Serum *on page 1257*

Synonyms Resin Triiodothyronine Uptake; Resin Uptake Ratio; T₃ Resin Uptake; T₃RU; T₃U

Applies to Free T_4 Index (FT₄I); FTI; T_7; T_{12}; Thyroid Hormone Binding Ratio; Thyroxine Ratio

Abstract T₃U is an indirect measure of unsaturated binding sites on thyroid binding globulin (TBG). T₃ uptake does **not** measure serum T₃ levels. The first T₃ uptake assay used red blood cells and was introduced as a test for pregnancy. Red cells were subsequently replaced by resins. **In recent years, better tests have become available and use of T₃ uptake has decreased.**

Patient Preparation Avoid recent isotope scan before collection of specimen if a radioactive method is used.

Specimen Serum is preferred, plasma may also be acceptable.

Container Red top tube; lavender top (EDTA) tube or green top (heparin) tube is also acceptable.

Storage Instructions Separate within 48 hours. Store at 2°C to 8°C.

Reference Interval Varies with different laboratories. An interval of 24% to 34% is frequently used. The T₃ uptake can be expressed in several ways. The Committee on Nomenclature of the American Thyroid Association recommends that raw uptake results be normalized by dividing the raw T₃ uptake by the T₃ uptake of normal pooled serum, to form the **thyroid hormone binding ratio (THBR)**. The THBR in all laboratories will have a reference interval centered on 1.00 and is usually given as 0.90-1.10. See also table in Thyroxine Binding Globulin, Serum *on page 1255*.

Use A thyroid function test for the diagnosis of hypothyroidism or hyperthyroidism, T₃U is used with T_4 or equivalent to provide free thyroxine index (free T_4 index, FT₄I). An indirect measure of binding protein, the T₃ uptake reflects available binding sites (ie, reflects TBG) and estimation of free thyroxine concentration. T₃U is **not** a measurement of serum T₃. It should never be used alone; rather, its usual application is in conjunction with total T_4 measurement. THBR is preferred and should be used instead of T₃U. Better assays for sTSH, selectively with tests for free T₃ and/or free T_4, are presently recommended for initial thyroid evaluation. See table for typical examples of use.

Diagnostic Utility of T₃U and FT₄I

Clinical Condition	T₄	T₃U	FT₄I
Normal	Normal	Normal	Normal
Hyperthyroid	Increased	Increased	Increased
Hypothyroid	Decreased	Decreased	Decreased
Increased TBG (eg, pregnancy)	Increased	Decreased	Normal
Decreased TBG (eg, nephrotic syndrome)	Decreased	Increased	Normal

Limitations An **increase** in T₃U occurs in hyperthyroidism; in situations in which drugs displace T_4 from TBG (eg, high doses of salicylates, phenytoin, phenylbutazone, etc); and in cases in which the TBG concentration decreases (eg, nephrotic syndrome, malnutrition, active acromegaly). Nicotinic acid increases T₃ resin uptake ratios. A **decrease** in T₃U occurs in hypothyroidism and in cases in which an increase in TBG occurs, such as estrogen administration (as contraceptive, during menopause, or treatment of osteoporosis), during pregnancy, and in conjunction with perphenazine.

Alterations in binding capacity of TBG are described with major illness and with high doses of salicylates and corticosteroids, and with use of heroin, methadone, phenytoin, and perphenazine. Alterations occur with malnutrition, such as in metastatic malignancy, and are found in patients with abnormal serum *(Continued)*

T_3 Uptake, Serum or Plasma *(Continued)*

protein patterns (eg, nephrotic syndromes, cirrhosis). Other states in which changes in TBG occur include infancy, acromegaly, molar and ordinary pregnancy, oral contraceptives, and with exogenous hormones including androgens, anabolic steroids, and estrogens. Hereditary increase and decrease of TBG occurs. **Most authorities have abandoned this test in favor of more specific, sensitive tests such as FT$_4$, sTSH, and FT$_3$.**

Contraindications This test should not be ordered alone; it is only useful with T$_4$ type tests.

Methodology Resin sponge uptake, charcoal bead uptake, related methods, based on *in vitro* competition for thyroid hormone between thyroid binding globulin and the added inert receptor. For THBR, divide patient's T$_3$U by T$_3$U for normal or reference serum.

Additional Information Free thyroxine index is calculated as:

$$FT_4I = [\% \ T_3U \ (patient) \ / \ \% \ T_3U \ (reference \ serum)] \ x \ T_4$$

The FT$_4$I range usually approximates the range for total T$_4$. In the presence of thyroid binding globulin abnormalities, the free thyroxine index is a useful laboratory parameter regarding clinical thyroid status. A number of pseudonyms, including thyroxine ratio, T$_7$, and T$_{12}$, for FT$_4$I are used in the literature. Use of these terms should be discouraged.

References

Attia J, Margetts P, and Guyatt G, "Diagnosis of Thyroid Disease in Hospitalized Patients: A Systematic Review," *Arch Intern Med*, 1999, 159(7):658-65.

Christofides ND, Wilkinson E, Stoddart M, et al, "Assessment of Serum Thyroxine Binding Capacity-Dependent Biases in Free Thyroxine Assays," *Clin Chem*, 1999, 45(4):520-5.

Dayan CM, "Interpretation of Thyroid Function Tests," *Lancet*, 2001, 357:619-24.

Faix JD, Rosen HN, and Velazquez FR, "Indirect Estimation of Thyroid Hormone-Binding Proteins to Calculate Free Thyroxine Index: Comparison of Nonisotopic Methods That Use Labeled Thyroxine ("T-Uptake")," *Clin Chem*, 1995, 41(1):41-7.

Feldkamp CS and Carey JL, "An Algorithmic Approach to Thyroid Function Testing in a Managed Care Setting. 3-Year Experience," *Am J Clin Pathol*, 1996, 105(1):11-6.

Tacrolimus, Whole Blood

Related Information

Cyclosporine, Blood *on page 487*
Itraconazole, Serum *on page 814*

Synonyms FK-506; Prograf®

Applies to Interleukin Production

Abstract Tacrolimus is a potent immunosuppressant used in renal, liver, heart, lung, bone marrow, and small bowel transplants. It is also used to treat atopic dermatitis in patients not responsive to conventional therapy.

Specimen Whole blood

Container Lavender top (EDTA) tube

Sampling Time Trough levels should be obtained 12-18 hours after oral dose or immediately prior to next dose.

Collection Do not draw the sample from the line used to administer the drug.

Reference Interval Trough (in whole blood): 3-20 ng/mL. Trough blood concentrations of 4-10 ng/mL for liver transplantation, 6-12 ng/mL for renal transplantation, 10-18 ng/mL for pancreas transplantation, and 10-20 ng/mL for bone marrow transplantation are recommended. Since tacrolimus binds to erythrocytes and lipoproteins, measurement of whole blood concentrations is preferred. Plasma levels are 2% to 20% of whole blood level.

Use Monitor the adequacy of drug dosage levels in management of immunosuppression for organ transplant recipients. The agent is used extensively to suppress rejection of autologous organ grafts.

Methodology Enzyme-linked immunosorbent assay (ELISA), microparticle enhanced immunoassay (MEIA), high performance liquid chromatography (HPLC), liquid chromatography with tandem mass spectrometry (LC/MS/MS)

Additional Information

- Half-life: 10-14 hours
- Volume of distribution: 1.5 L/kg
- Protein binding: 99%
- Time to reach steady state: 2-3 days
- Bioavailability: 5% to 67%

On a molar basis, tacrolimus is 10- to 100-fold more potent than cyclosporine. The immunosuppressant properties of the drug appear to interfere with T-helper cell function and secretion of lymphokines to decrease interleukin production.

Tacrolimus may be administered I.V. or orally. The absorption of tacrolimus is highly variable with bioavailability varying from 5% to 67%. The plasma protein binding of tacrolimus is ~99% and is independent of concentration over a range of 5-50 ng/mL. Tacrolimus is bound to albumin and alpha-1-acid glycoprotein, and has a high level of association with erythrocytes. The distribution of tacrolimus between whole blood and plasma depends on several factors, such as hematocrit, temperature at the time of plasma separation, drug concentration, interaction with other drugs, and plasma protein concentration.

As with cyclosporine, renal toxicity with eventual renal failure is the most severe complication. Other assays to assess renal function (ie, BUN, creatinine clearance) should be ordered along with tacrolimus levels, since toxicity may begin even with "acceptable" blood levels. In particular, to avoid excess nephrotoxicity, tacrolimus should not be used simultaneously with cyclosporine. Tacrolimus or cyclosporine should be discontinued at least 24 hours prior to initiating the other. Other common toxicities include neurotoxicity, insomnia, nausea, and hypertension.

Tacrolimus is extensively metabolized by the mixed-function oxidase system, primarily the cytochrome P450 system (CYP3A). At least nine different metabolites have been identified. There are many drugs which affect tacrolimus pharmacokinetics, the most common being those which inhibit or induce the P450 enzyme system. Drugs which increase tacrolimus concentrations include bromocriptine, cimetidine, cisapride, clarithromycin, clotrimazole, cyclosporine, danazol, diltiazem, erythromycin, fluconazole, itraconazole, ketoconazole, methylprednisolone protease inhibitors, metoclopramide, nicardipine, nifedipine, troleandomycin, and verapamil. Drugs which increase hepatic metabolism and thus lower tacrolimus levels include carbamazepine, ethotoin, mephenytoin, phenobarbital, phenytoin, primidone, rifampin, and intravenous trimethoprim-sulfa.

References

Armstrong VW and Oellerich M, "New Developments in the Immunosuppressive Drug Monitoring of Cyclosporine, Tacrolimus, and Azathioprine," *Clin Biochem*, 2001, 34(1):9-16.

de Mattos AM, Olyaei AJ, and Bennett WM, "Nephrotoxicity of Immunosuppressive Drugs: Long-Term Consequences and Challenges for the Future," *Am J Kidney Dis*, 2000, 35(2):333-46.

Gummert JF, Ikonen T, and Morris RE, "Newer Immunosuppressive Drugs: A Review," *J Am Soc Nephrol*, 1999, 10(6):1366-80.

Oellerich M, Armstrong VW, Schutz E, et al, "Therapeutic Drug Monitoring of Cyclosporine and Tacrolimus. Update on Lake Louise Consensus Conference on Cyclosporine and Tacrolimus," *Clin Biochem*, 1998, 31(5):309-16.

Rustin M, "Tacrolimus Ointment for the Management of Atopic Dermatitis," *Hosp Med*, 2003, 64(4):214-7.

Undre NA, Stevenson P, and Schafer A, "Pharmacokinetics of Tacrolimus: Clinically Relevant Aspects," *Transplant*, 1999, 31(7A):21S-24S.

Tartrate Resistant Leukocyte Acid Phosphatase

Related Information

Leukocyte Cytochemistry *on page 846*

Lymph Node Biopsy *on page 880*

Synonyms Acid Phosphatase, Tartrate Resistant, Leukocytes

Applies to Hairy Cell Leukemia

Test Includes Leukocyte acid phosphatase reaction with and without tartrate inhibition

Abstract Tartrate resistant acid phosphatase (TRAP) positive white cells identified on peripheral blood and/or bone marrow smears along with characteristic findings on study of sections from bone marrow biopsy are largely diagnostic of hairy cell leukemia (HCL).

Specimen Glass microscope slide smears prepared from fresh capillary or heparinized whole blood, fixed immediately (glutaraldehyde-acetone) after preparation

Container Green top (heparin) tube

Storage Instructions Smeared glass slides may be stored at least 1 week prior to assay if fixed immediately after preparation.

Causes for Rejection Smears unfixed, blood not fresh

Reference Interval Most white cells of peripheral blood as well as platelets are acid phosphatase positive, the reaction is inhibited by L(+) tartrate (tartrate sensitive). Tartrate resistant cells are not present in the blood of normal individuals.

Use Diagnose "hairy cell leukemia" ("leukemic reticuloendotheliosis"); differential diagnosis of small cell lymphocytic neoplasms

Limitations Patients with HCL are often leukopenic with relatively few hairy cells in the peripheral blood.

Additional Information There are some six isoenzymes of leukocyte acid phosphatase. Of the six, isoenzyme V is not inhibited by (is resistant to) L(+) tartaric acid. It has been found that the malignant mononuclear cells of leukemic reticuloendotheliosis ("hairy cell leukemia") contain isoenzyme V and are resistant to inhibition by L(+) tartaric acid. There is evidence the reaction is not entirely specific as tartrate resistant acid phosphatase reactions have been reported in cases of prolymphocytic leukemia and malignant lymphoma as well as some cases of infectious mononucleosis. There have also been reports of rare false-negative results (patients with leukemic reticuloendotheliosis having negative tartrate resistant acid phosphatase reactions). Most T-cell lymphoproliferative processes are strongly acid phosphatase positive, tartrate sensitive. In T-cell chronic leukemias the acid phosphatase reaction is variable. A few tartrate resistant acid phosphatase positive cells may be found in normal bone marrow. Osteoclasts are strongly TRAP positive. An assay based on antibody against band V acid phosphatase has been described. The immunoassay has greater specificity than the standard cytochemical procedure. A monoclonal antibody against TRAP (anti-TRAP) has been developed and can be used with paraffin-imbedded tissue. The application of immunophenotypic studies to the diagnosis of HCL has decreased the need for TRAP staining. Hairy cells are B lymphocytes and express the pan-B-cell antigens CD19, CD20, and CD22, they lack CD5 and CD10, but express CD103, considered a sensitive marker for HCL.

References

"Chronic Lymphocytic Leukemias and Other Lymphoid Leukemias: Hairy Cell Leukemia," *Practical Diagnosis of Hematologic Disorders*, 3rd ed, Chapter 33, Kjeldsberg CR, ed, Chicago, IL: ASCP Press, 2000, 543-4.

Janckila AJ, Cardwell EM, Yam LT, et al, "Hairy Cell Identification by Immunohistochemistry of Tartrate-Resistant Acid Phosphatase," *Blood*, 1995, 85(10):2839-44.

Tallman MS, and Polliack A, *Hairy Cell Leukemia*, The Netherlands: Harwood Academic Publishers, 2000.

♦ **Tau Fraction** *see* Cerebrospinal Fluid Protein *on page 371*

♦ **Tau Fraction** *see* Cerebrospinal Fluid Protein Electrophoresis *on page 374*

♦ **Tau Protein** *see* Apolipoprotein E, Plasma *on page 204*

♦ **Tau Protein** *see* Cerebrospinal Fluid and Plasma β-Amyloid$_{(1-42)}$ *on page 359*

♦ **Taztia XT™** *see* Diltiazem, Serum or Plasma *on page 515*

♦ **TBG** *see* Thyroxine Binding Globulin, Serum *on page 1255*

♦ **TBG** *see* Thyroxine, Serum *on page 1257*

Terminal Deoxynucleotidyl Transferase

Related Information

Bone Marrow *on page 296*
Lymph Node Biopsy *on page 880*

Abstract Terminal deoxynucleotidyl transferase (TdT) is an enzyme found in lymphoid thymic cells and a minor (1% to 2%) population of thymic related T-cell lymphoid precursors in the bone marrow. Some leukemic cell populations are TdT positive, notably T-cell acute lymphoblastic leukemia (TALL), CALL, and some blast cells of chronic myelogenous leukemia in crisis. TdT positivity, however, is not exclusively diagnostic of leukemia.

Specimen Blood or bone marrow (avoid heparin) for leukocyte separation, pelletization, and storage at -20°C. Glass slide smears, air dried, of blood or marrow.

Collection Store dried smears at room temperature for up to 5 days.

Storage Instructions Store leukocyte pellets (long-term) at -20°C.

Reference Interval Peripheral blood: negative for TdT-positive cells; bone marrow: <1.8% TdT-positive cells

Use Classify certain leukemias and lymphomas, normally used to distinguish acute lymphoblastic from nonlymphoblastic leukemia (ALL from AML and from other lymphoproliferative disorders such as adult T cell leukemia); diagnose acute lymphoblastic leukemia, lymphoid blast crisis of chronic myelogenous leukemia, and lymphoblastic lymphomas. A TdT positive result does not exclude acute myeloid leukemia (AML) as the latter is characterized in some 5% to 10% of cases by TdT positive myeloblasts.

Terminal Deoxynucleotidyl Transferase (TdT) in Hematologic Disease

Disease	Percent Positive
Acute lymphoblastic leukemia	>90
Lymphoblastic lymphoma	90
Chronic granulocytic leukemia in blast crisis	30
Acute undifferentiated leukemia	60
Acute nonlymphocytic leukemia	2-5

Adapted from Rubin E, MD and Farber JL, MD, *Pathology*, 2nd ed, Philadelphia, PA: JB Lippincott Co, 1994, 1077.

Limitations TdT-positive blasts are seen commonly in the bone marrow recovery phase postchemotherapy for ALL (a small number of TdT-positive cells cannot be considered as indicative of residual disease or relapse).

Additional Information TdT acts to catalyze the polymerization of deoxynucleoside triphosphates (by addition to the 3′ hydroxyl ends of oligodeoxynucleotides or polydeoxynucleotides without DNA template instructions). Thymus is the primary site of TdT-positive cells and TdT is found in the nucleus of the more primitive T cells. A thymus related population of TdT-positive cells resides in the bone marrow (normally a minor population - 1% to 2%). TdT is increased in more than 90% of the cases of ALL of childhood. This is true for even pre-B cell as well as B-cell ALL. A minor (5% to 10%) population of patients with acute nonlymphoblastic leukemia have TdT-positive blasts. **TdT-positive blasts are prominent in some cases of chronic myeloid leukemia (CML),** (Continued)

Terminal Deoxynucleotidyl Transferase *(Continued)*

in which they signal the development of an acute blast phase. TdT-positive cases of blast phase CML correlate with a positive response to chemotherapy (vincristine and prednisone).

TdT assay has been applied to the detection of DNA strand breaks, of importance in the study of apoptosis. See Apoptosis Assays *on page 205*. DNA strand breakage can be assessed by flow cytometry. The assay can be used to evaluate DNA damage occurring in chronic lymphocytic leukemia cells exposed to UV irradiation and the apoptosis-inducing chemotherapeutic agents fludarabine and 2-chloro-2'-deoxyadenosine. Thus, by determining the sensitivity of patient's leukemic lymphocytes to induction of apoptosis, the assay may have value in predicting patient response to therapy.

Tonsils, as sites of lymphopoiesis, may harbor TdT-positive precursor cells which should not necessarily be interpreted as evidence of lymphoblastic lymphoma or leukemia.

References

"Acute Lymphoblastic Leukemia," *Practical Diagnosis of Hematologic Disorders*, 3rd ed, Chapter 28, Kjeldsberg C, ed, Chicago, IL: ASCP Press, 2000, 451-2.

Bromidge TJ, Howe DJ, Johnson SA, et al, "Adaptation of the TdT Assay for Semiquantitative Flow Cytometric Detection of DNA Strand Breaks," *Cytometry*, 1995, 20(3):257-60.

Chapman RS, Chresta CM, Herberg AA, et al, "Further Characterization of the *in situ* Terminal Deoxynucleotidyl Transferase (TdT) Assay for the Flow Cytometric Analysis of Apoptosis in Drug Resistant and Drug Sensitive Leukaemic Cells," *Cytometry*, 1995, 20(3):245-56.

Strauchen JA and Miller LK, "Terminal Deoxynucleotidyl Transferase-Positive Cells in Human Tonsils," *Am J Clin Pathol*, 2001, 116(1):12-6.

♦ **Testosterone** *see* Dehydroepiandrosterone and Dehydroepiandrosterone Sulfate, Serum or Plasma *on page 506*

Testosterone, Total and Free, Serum or Plasma

Related Information

Adrenal Cortex: Laboratory Assessment Overview *on page 110*
Adrenocorticotropic Hormone, Plasma *on page 114*
Androstenedione, Serum *on page 158*
Cortisol, Serum or Plasma *on page 460*
Dehydroepiandrosterone and Dehydroepiandrosterone Sulfate, Serum or Plasma *on page 506*
Dihydrotestosterone, Serum *on page 514*
Follicle Stimulating Hormone, Serum, Plasma, or Urine *on page 609*
Infertility Screen *on page 786*
Inhibin A, Serum *on page 799*
17-Ketosteroids, Total, Urine *on page 818*
Luteinizing Hormone, Blood or Urine *on page 876*
Semen Analysis, Advanced *on page 1186*
Semen Analysis, Basic *on page 1187*
Sperm Mucus Penetration Test (Human or Bovine Cervical Mucus) *on page 1219*
Sperm Penetration Assay (Zona-Free Hamster Egg Penetration Test) *on page 1220*
Swim Up/Swim Down Procedures (Spermatozoa) *on page 1227*

Applies to Dihydrotestosterone; Estradiol; Free Testosterone; Gonadotropin-Releasing Hormone; Luteinizing Hormone; Sex-Hormone Binding Globulin

Abstract Testosterone (T), secreted by the testicular Leydig cells, is both an androgen itself, and a prohormone which can be converted into an even more potent androgen, dihydrotestosterone (DHT), and an estrogenic hormone, estradiol (E2). The conversion into DHT occurs in tissues containing the enzyme, 5-alpha reductase (skin, prostate, other internal genitalia), while the conversion into E2 occurs in tissues containing the enzyme, aromatase (fat tissue, breast). Secretion of T is primarily dependent on stimulation of the testicular Leydig cells by pituitary luteinizing hormone (LH), which is, in turn, dependent on stimulation of pituitary gonadotrophic cells by hypothalamic gonadotropin-releasing hormone (GnRH). T is part of a classical negative feedback loop: increased serum levels of T suppress serum levels of LH, while decreased T leads to marked increases in serum LH. T has a diurnal variation, with peak serum levels at 0400-0800 and minimum levels at 1600-2000. T circulates in the plasma ~65% bound to sex-hormone binding globulin (SHBG) and ~30% to 32% bound to albumin; ~1% to 4% of T in plasma is free (not bound to protein).

Patient Preparation Avoid recent administration of radioactive isotopes if assay is to be done by RIA.

Specimen Serum or plasma

Container Red top tube, green top (heparin) tube, or lavender top (EDTA) tube

Storage Instructions Separate serum or plasma and freeze.

Reference Interval See table.

Testosterone, Total and Free

Total Testosterone			
Male			
Age (y)	ng/dL	nmol/L	% Free
<1	Not established		
1-9	<40	<1.39	
10-11	<200	<6.94	
12-13	<800	<27.76	
14	<1200	<41.64	
15-16	100-1200	3.47-41.64	
17-18	300-1200	10.41-41.64	
19-40	300-950	10.41-32.97	
>40	240-950	8.32-32.97	
Female			
<1	Not established		
1-9	<40	<1.39	
10-11	<75	<2.60	
12-16	<120	<4.16	
17-18	20-120	0.69-4.16	
>18	20-80	0.69-2.77	
Free Testosterone			
Male	9-30	0.31-1.04	2.0-4.8
Female	0.3-1.9	0.01-0.07	0.9-3.8

Adapted from Mayo Medical Laboratories, *2000 Test Catalog*, Rochester, MN, 2000, 470.

Use

Total vs free T. For most clinical purposes, when an assessment of T is needed, measurement of total T provides all of the essential diagnostic information. Measurement of free (unbound) T is rarely indicated but may be useful in obese patients who sometimes have low SHBG concentrations, which make total T levels seem inappropriately low.

Polycystic ovary syndrome (PCOS). PCOS is a highly prevalent disorder characterized by signs of androgen excess (anovulation accompanied by hirsutism, acne, and male-pattern baldness). Most patients with PCOS have elevated serum T.

Female hirsutism. The differential diagnosis of hirsutism in an adult female includes;
   a) PCOS (see discussion above and Disease Index)
   b) late-onset congenital adrenal hyperplasia (CAH)
   c) Cushing syndrome
   d) hormone-secreting tumors of the adrenal or ovary
   e) idiopathic hirsutism

Idiopathic hirsutism is usually a diagnosis of exclusion; such patients typically have normal serum T. Fiet et al recommend using a coordinated set of immunoassays for eight hormones in the differential diagnosis of hirsutism and acne in women.
(Continued)

Testosterone, Total and Free, Serum or Plasma
(Continued)

Male hypogonadism. Primary testicular insufficiency or failure is characterized by low serum T and elevated serum LH and FSH.

Androgen insensitivity syndrome. In the fetus with a 46X,Y genotype, T and DHT combine with the nuclear androgen receptor to mediate the development of male internal and external genitalia. When, because of a genetic mutation, the nuclear androgen receptor is absent or inactive, the individual with a 46X,Y (male) genotype will have a female phenotype (ie, female external genitalia, breast development, and a female habitus) - a disorder historically referred to as the testicular feminization syndrome, but now renamed the androgen-insensitivity syndrome. Such patients have a vagina, no uterus, and bilaterally cryptorchid testes. Androgen insensitivity is also a possible diagnosis when a phenotypically female infant has inguinal hernias and a mass in the inguinal area or labia.

In **Klinefelter syndrome**, testosterone can be at the low end of the reference interval or lower. Even when it is almost normal, LH levels are increased.

Anabolic steroid abuse. Some athletes take androgenic hormones in an attempt to increase muscle mass; the effects, if any, on serum T and DHT are unpredictable since the exogenous hormone suppresses endogenous hormone production.

Limitations Plasma testosterone level may be elevated in patients using cimetidine.

Methodology Radioimmunoassay (RIA), chemiluminescent immunoassay (CIA). Free testosterone is generally done by RIA or CIA after ultrafiltration or equilibrium dialysis.

Additional Information Testosterone concentrations may be relevant in male infertility, erectile or sexual dysfunction, and in osteoporosis.

References

Adachi M, Takayanagi R, Tomura A, et al, "Androgen Insensitivity Syndrome as a Possible Coactivator Disease," *N Engl J Med*, 2000, 343(12):856-62.

Fiet J, Gosling JP, Soliman H, et al, "Hirsutism and Acne in Women: Coordinated Radioimmunoassays for Eight Relevant Plasma Steroids," *Clin Chem*, 1994, 40(12):2296-305.

King DS, Sharp RL, Vukovich MD, et al, "Effect of Oral Androstenedione on Serum Testosterone and Adaptations to Resistance Training in Young Men: A Randomized Controlled Trial," *JAMA*, 1999, 281(21):2020-8.

Street ME, Weber A, Camacho-Hubner C, "Girls With Virilization in Childhood: A Diagnostic Protocol for Investigation," *J Clin Pathol*, 1997, 50(5):379-83.

Tetanus Antibody

Related Information

Bacterial Culture, Anaerobes *on page 231*
Botulism Diagnostic Procedures *on page 302*

Test Includes Detection of tetanus toxoid-specific antibody in serum

Abstract In tetanus, *Clostridium tetani* proliferates at the site of a deep or penetrating injury. *C. tetani* is a gram-positive, spore-forming obligate anaerobe found in soil and in the gastrointestinal tracts of animals. In developed countries, tetanus is rare with only 33 cases reported in the United States in 1999. However, over one million cases still occur annually worldwide with a 20% to 50% mortality. Immunity to tetanus is seen in response to tetanus toxoid immunization but this immunity wanes with age. Booster immunizations are recommended to maintain adequate immunity to the toxin.

Specimen Serum

Container Red top tube

Reference Interval Hemagglutinating antibody present; hemagglutination concentrations >0.01 units/mL and EIA results >0.15 IU/mL are considered protective.

Use Assess immunocompetence and patients at risk for tetanus and epidemiology studies

Contraindications Not valid in an unimmunized individual or in patients who have recently received tetanus antitoxin serum.

Methodology Hemagglutination, enzyme immunoassay (EIA)

Additional Information Acute spastic paralysis is caused by tetanospasmin, a neurotoxin, better known as tetanus toxin. In developed countries, most individuals have been immunized against tetanus and assessing whether they have antibody to tetanus antigen is a means to document intact humoral immunity. Serologic studies have no place in management of clinical tetanus. The tests are too slow, too insensitive, and do not correlate with the course of disease or response to treatment.

References

Centers for Disease Control, "Notifiable Diseases/Deaths in Selected Cities Weekly Information," *MMWR Morb Mortal Wkly Rep*, 2000, 48(51):1183-90.

Gergen PJ, McQuillan GM, Kiely M, et al, "A Population-Based Serologic Survey of Immunity to Tetanus in the United States," *N Engl J Med*, 1995, 332(12):761-6.

O'Malley CD, Smith N, Braun R, et al, "Tetanus Associated With Body Piercing," *Clin Infect Dis*, 1998, 27(5):1343-4.

Saikh KU, Sesno J, Brandler P, et al, "Are DNA-Based Vaccines Useful for Protection Against Secreted Bacterial Toxins? Tetanus Toxin Test Case," *Vaccine*, 1998, 16(9-10):1029-38.

Sanford JP, "Tetanus - Forgotten but Not Gone," *N Engl J Med*, 1995, 332(12):812-3.

Internet Web Sites

www.astdhpphe.org/infect/tetanus.html

www.cdc.gov/nip/publications/pink/tetanus.pdf

♦ **Tetracyclic Antidepressants** *see* Antidepressants, Cyclic, Serum or Plasma *on page 171*

♦ **Tetrahydrobiopterin Cofactor Deficiency** *see* Phenylalanine, Blood *on page 1022*

♦ **Tetrahydrobiopterin Pathway** *see* Newborn Screen for Phenylketonuria *on page 954*

♦ **Tetraiodothyronine** *see* Thyroxine, Serum *on page 1257*

♦ **Tg Ab** *see* Thyroglobulin Antibody *on page 1249*

TGF-β, Serum

Related Information

Apoptosis Assays *on page 205*

Immunoglobulin G Subclasses *on page 778*

Synonyms Transforming Growth Factor β

Abstract Transforming growth factor β (TGF-β) is an immunoregulatory cytokine, one of a family of polypeptide growth factors. Most cells produce TGF-β and most have TGF-β cell membrane receptors. TGF-β is a multifunctional cytokine with both stimulatory and inhibitory effects on cells, including stimulation of fibroblast proliferation and formation of the extracellular matrix. This cytokine may have an important role in the development of bone marrow fibrosis.

Specimen Serum

Storage Instructions Store at -70°C if assay is to be delayed.

Turnaround Time Days, possibly weeks

Special Instructions Assay is likely to be available only from a research laboratory.

Reference Interval 1.4 ±0.1 ng/mL. Patients with idiopathic myelofibrosis have levels of 53 ±4 ng/mL while those with secondary fibrosis have levels of 37 ±5 ng/mL (results given as mean ±SE).

Use Evaluation of myelofibrosis (idiopathic and secondary marrow fibrosis); of potential use in evaluation of patients with fibrotic conditions of lung, liver, and/or kidney (eg, idiopathic pulmonary fibrosis and generalized scleroderma)

Methodology Bioassay (growth inhibition of mink lung epithelial cells, CCL 64), appears to measure TGF-β1 isoform; bioassay (suppression of lymphocyte proliferation by thymocytes); radioreceptor assay (inhibition of binding of radiolabeled TGF-β to its receptor); enzyme-linked immunosorbent assay (ELISA), commercially available (Quantikine kit, R & D Systems)

Additional Information TGF-βs are pleiotropic factors that have a regulatory role in somatic tissue development and renewal. They are multifunctional signaling molecules. They can have opposite (positive or negative) effects dependent upon the target cell's stage of development or upon its environment. There are three isoforms of TGF-β: TGF-β1, TGF-β2, and TGF-β3, each encoded by a specific gene. TGF-β exerts regulatory influence by binding to cell surface receptors, types I, II, and III. Types I and II receptors include serine-threonine kinases in their intracellular domains that phosphorylate transcription factors known as "SMAD" proteins (1 through 10). TGF-β has an (Continued)

TGF-β, Serum *(Continued)*

important role in cell-cycle regulation and is a potent inhibitor of cell proliferation. TGF-β has a number of effects on tumor cell growth, on fibroblasts, on the induction of fibrotic disease, and on atherosclerosis (acting as an inhibitor of this process). Mutations in the type II receptor gene render smooth muscle/endothelial cells resistant to the antiproliferative and apoptotic effects of TGF-β. The surprisingly broad involvement of TGF-β in these and other human developmental and disease processes are detailed in a recent review by Blobe et al.

TGF-β is one of a number of cytokines involved in the regulation of hematopoiesis. TGF-β is elevated in the sera of patients with marrow fibrosis including both idiopathic myelofibrosis and secondary marrow fibrosis.

Serum TGF-$β_1$ is increased in patients with chronic idiopathic neutropenia along with increase in serum IgA but decreased IgG_3.

TGF-β is involved in the regulation of inflammation. Opiate addicts are prone to infections, partly due to opiate-induced macrophage apoptosis which is enhanced by TGF-β (and inhibited by anti-TGF-β antibody). TGF-β inhibits lipopolysaccharide-stimulated expression of inflammatory cytokines.

References

Blobe GC, Schiemann WP, and Lodish HF, "Role of Transforming Growth Factor β in Human Disease," *N Engl J Med*, 2000, 342(18):1350-8.

Fortunel NO, Hatzfeld A, and Hatzfeld JA, "Transforming Growth Factor-β: Pleiotropic Role in the Regulation of Hematopoiesis," *Blood*, 2000, 96(6):2022-36.

Papadaki HA, Palmblad J, Kapsimali V, et al, "Increased Serum IgA and Decreased IgG_3 Strongly Correlate With Increased Serum TGF-$β_1$ Levels in Patients With Nonimmune Chronic Idiopathic Neutropenia of Adults," *Eur J Haematol*, 2000, 65(3):237-44.

Singhal PC, Kapasi AA, Franki N, et al, "Morphine-Induced Macrophage Apoptosis: The Role of Transforming Growth Factor-β," *Immunology*, 2000, 100(1):57-62.

Zhou S, Kinzler KW, and Vogelstein B, "Going Mad With *SMADS*," *N Engl J Med*, 1999, 341(15):1144-5.

Thallium, Urine or Blood

Applies to Mee's Lines

Abstract A byproduct of lead smelting, thallium salts are used in photomultiplier tubes, infrared detectors and transmitters, lens and glass making, and in rockets and flares. It has been, and in some areas remains, a component of insecticides, pesticides, and rodenticides. Thallium has been excluded as a rodenticide in the U.S. since 1972. It may be found in grain.

Patient Preparation The patient should be instructed to use a plastic bedpan or urinal if necessary, not metal.

Specimen 24-hour urine, serum

Container Special metal-free EDTA tube or metal-free plastic urine container. See Trace Metal Introduction *on page 77.*

Causes for Rejection Specimen allowed to contact metal or dusts with metal, use of nonmetal-free containers

Reference Interval Urine: <2 μg/L (SI: <9.8 nmol/L); serum: <0.5 μg/L (SI: <24.5 nmol/L)

Critical Values Serum value in most normal individuals is <10 μg/L or 10 ng/mL (SI: 49 nmol/L); daily urine excretion in most normal individuals is <10 μg/day (SI: 49 nmol/day). Spot urinary thallium concentration in normal unexposed individuals is <1.5 μg/L.

Possible Panic Range Blood levels >100 μg/L (SI: 490 nmol/L) or urine values >200 μg/L (SI: 978 nmol/L)

Use Diagnose thallium exposure and toxicity, including alopecia with neuropathy which may resemble that of Guillain-Barré syndrome, and abdominal colic.

Limitations There are marked variations in heavy metal levels considered toxic by different investigators.

Methodology Graphite furnace atomic absorption spectrometry (GFAAS), inductively-coupled plasma-mass spectrometry (ICP-MS)

Additional Information Thallium is almost 100% absorbed from the GI tract and, like potassium, distributes throughout the body. Symptoms of thallium poisoning initially begin with generalized nausea, vomiting, abdominal pain, diarrhea, and gastrointestinal bleeding. Other nonspecific clinical findings include polyneuritis, encephalitis, delirium, ophthalmologic symptoms, convulsions, shock, and coma. Alopecia and painful ascending peripheral neuropathy are the most characteristic components of a thallium "toxidrome." Because of the delayed development of alopecia (several weeks after poisoning), the diagnosis of thallotoxicosis is often initially overlooked until alopecia appears. Mee's lines (transverse white lines on the nails) appear on the hands and feet 1 month after exposure, but are not specific for thallium poisoning. Other dermatological findings include crusted eczematous lesions, hypohidrosis, anhidrosis, palmar erythema, painful glossitis, stomatitis, and hair discoloration.

References
Galvan-Arzate S and Santamaria A, "Thallium Toxicity," *Toxicol Lett*, 1998, 99(1):1-13.

Malbrain ML, Lambrecht GL, and Zandijk E, "Treatment of Severe Thallium Intoxication," *J Toxicol Clin Toxicol*, 1997, 35(1):97-100.

Mercurio M and Hoffman RS, "Thallium." *Goldfrank's Toxicologic Emergencies*, 6th ed, Goldfrank LR, Flomenbaum NE, Lewin NA, et al, eds, Stamford, CT: Appleton and Lange, 1998, 1349-57.

♦ **Thawed Plasma** *see* Plasma, Fresh Frozen *on page 1039*

♦ **Theo-Dur®** *see* Theophylline, Serum *on page 1243*

♦ **Theolair™** *see* Theophylline, Serum *on page 1243*

Theophylline, Serum
Related Information
Amiodarone, Serum *on page 148*
Caffeine, Serum *on page 326*
Carbamazepine, Serum *on page 337*
Phenobarbital, Serum or Plasma *on page 1021*
Phenytoin, Serum or Plasma *on page 1026*
Verapamil, Serum or Plasma *on page 1306*

Synonyms Aminophylline; Elixophyllin®; Ethylenediamine; Phyllocontin®; Slo-Phyllin®; Sustaire®; Theo-Dur®; Theolair™; Theospan®; Truphylline®

Abstract Used for over 80 years, theophylline is a bronchodilator useful in asthma and chronic obstructive pulmonary disease (COPD). Its characteristics include immunomodulatory and anti-inflammatory properties. For asthma, it is used in conjunction with inhaled beta-2 agonists. The drug is tolerated well when serum concentrations are kept within therapeutic range. It is used in neonates for idiopathic apnea/bradycardia.

Specimen Serum

Container Red top tube

Sampling Time Measure **trough** and **peak**. Draw blood at 2 hours after most recent dose for rapid dissolution preparations; 4-6 hours after sustained release preparations. See table. If toxicity is suspected, draw a level any time during a continuous I.V. infusion, or 2 hours after an oral dose.

Guidelines for Drawing Theophylline Serum Levels

Dosage Form	Time to Draw Level
P.O. liquid, fast-release tab	Peak: 1 h post 4th dose
	Trough: just before 4th dose
P.O. slow-release product	Peak: 4 h post 3rd dose
	Trough: just before 3rd dose

Time to peak serum concentration:
- oral: 1 hour
- uncoated tablets: 2 hours
- chewable tablets: 1-1.5 hours
- enteric-coated tablets: 5 hours
- extended-release capsules and tablets: 4-7 hours, in overdoses up to 27 hours
- retention enema: 1-2 hours

Theophylline levels should be initially drawn after 2 days of therapy; repeat levels are indicated 2 days after each increase in dosage or weekly if on a stabilized dosage.

Reference Interval
Therapeutic:
- asthma: 10-20 µg/mL (SI: 56-111 µmol/L); about 65% of available maximal bronchodilatory effect occurs with levels of 10 µg/mL. The role for low-dose theophylline in subjects with chronic asthma (levels 5-10 µg/mL) requires

(Continued)

Theophylline, Serum *(Continued)*

further study. Levels of 10-15 µg/mL are appropriate for management of COPD, and 8-12 µg/mL may prove to be adequate.

- neonatal apnea: 6-13 µg/mL (SI: 33-72 µmol/L)

Possible Panic Range

- 15-25 µg/mL: GI upset, diarrhea, nausea/vomiting, abdominal pain, nervousness, headache, insomnia, agitation, dizziness, muscle cramp, tremor
- 25-35 µg/mL: tachycardia, occasional premature ventricular contractions
- >35 µg/mL: ventricular tachycardia, frequent premature ventricular contractions, seizure

Use Monitor therapeutic drug level; detect noncompliance and subtherapeutic levels; attempt to predict theophylline toxicity if possible

Limitations Elderly, acutely ill subjects, and patients with severe respiratory problems, pulmonary edema, or liver dysfunction are at greater risk of toxicity because of reduced drug clearance.

Contraindications Uncontrolled arrhythmias, hypersensitivity to ethylenediamine are contraindications to use of this drug.

Methodology Immunoassay, high performance liquid chromatography (HPLC), gas chromatography (GC)

Additional Information For adult nonsmoker:

- Half-life: 6-10 hours in normal adults
- Volume of distribution: 0.4-0.6 L/kg
- Protein binding: 50% to 60%

In order to improve aqueous solubility, formulation as a complex or salt is common. Aminophylline, the ethylenediamine salt, is widely used. Aminophylline is 85% anhydrous theophylline by weight.

A higher incidence of toxicity is recognized in patients with hepatic cirrhosis and hepatitis. Prolonged half-life occurs in premature infants. **Dosage should be reduced in these situations.**

By contrast, half-life is shortened in smokers and is variable with phenobarbital administration; higher doses are tolerated also in acidemia. Smokers on the average are reported to need 1.5-2 times as much of the drugs as nonsmokers to achieve the same effects. Marijuana smoking, rifampin, carbamazepine, phenobarbital, phenytoin, and aminoglutethimide may **decrease** theophylline concentrations. Optimal resampling time after change in dosage is 48 hours for adults, 1-2 days for children. The half-life varies between individuals. See table.

Theophylline Half-Life

Half-life (h)	Patient Population
6-10	Normal healthy adults
2-9	Children
15-58	Premature infants
18-24	Severe congestive heart failure
29	Cirrhosis

Toxic effects and signs and symptoms of acute overdose include nausea, vomiting, abdominal pain, tremors, esophageal ulceration, palpitation, anorexia, diuresis, skin rash, insomnia, irritability, atrial fibrillation, tachycardia, paroxysmal supraventricular tachycardia, convulsions, hypotension, visual hallucinations, hypokalemia, hypercalcemia, lactic acidosis, fecal discoloration (black), and death.

References

Barnes PJ, "Theophylline: New Perspectives for an Old Drug," *Am J Respir Crit Care Med,* 2003, 167(6):813-8.

Kips JC, Peleman RA, and Pauwels RA, "The Role of Theophylline in Asthma Management," *Curr Opin Pulm Med,* 1999, 5(2):88-92.

Vassallo R and Lipsky JJ, "Theophylline: Recent Advances in the Understanding of its Mode of Action and Uses in Clinical Practice," *Mayo Clin Proc,* 1998, 73(4):346-54.

Weinberger M and Hendeles L, "Theophylline in Asthma," *N Engl J Med,* 1996, 334(21):1380-8.

- **Theospan®** *see* Theophylline, Serum *on page 1243*
- **Therapeutic Cytapheresis** *see* Apheresis, Therapeutic *on page 201*

- ◆ **Thermoactinomyces Precipitating Antibodies** *see* Hypersensitivity Pneumonitis Serology *on page 761*
- ◆ **Thermoactinomyces vulgaris Precipitating Antibodies** *see* Hypersensitivity Pneumonitis Serology *on page 761*
- ◆ **Thermolospora viridis Precipitating Antibodies** *see* Hypersensitivity Pneumonitis Serology *on page 761*

Thiocyanate, Serum, Plasma, or Urine

Related Information
Carboxyhemoglobin, Blood *on page 340*
Cyanide, Blood *on page 485*

Synonyms Ethyl and Methyl Thiocyanate (Thanite® and Lethane®); Potassium Thiocyanate (KSCN)

Applies to Cyanide; Nipride®; Nitroprusside

Abstract Thiocyanate is a relatively inert metabolite of the antihypertensive drug, nitroprusside. Nitroprusside-produced cyanide is converted to thiocyanate by hepatic rhodanase. It is also a product of cyanide metabolism, a byproduct of cigarette smoking, and is found in the serum and urine of individuals consuming cassava beans.

Specimen Serum or plasma, urine

Container Red top tube, lavender top (EDTA) tube, plastic urine container

Reference Interval Serum, nonsmoker: 1-4 µg/mL (SI: 0.02-0.07 mmol/L), smoker: up to 10 µg/mL (SI: up to 0.17 mmol/L); urine: nonsmoker: 1-4 mg/24 hours (SI: 0.02-0.07 mmol/day), smoker: 7-17 mg/24 hours (SI: 0.12-0.30 mmol/day)

Possible Panic Range Serum: >35 µg/mL (SI: >0.60 mmol/L); 200 µg/mL (SI: 3.44 mmol/L) is lethal

Use Follow exposure to cyanide; evaluation of clearance of thiocyanate; monitor nitroprusside toxicity and smoking or nonsmoking compliance

Limitations Because of rapid metabolism of the drug, results are usually clinically meaningless in the setting of acute CN⁻ exposure by the time they are reported.

Methodology Photometry, chromatography, ion chromatography with fluorescence and ultraviolet detection

Additional Information Thiocyanate toxicity may occur with long-term nitroprusside use (7-10 days with normal renal function and 3-6 days with renal impairment). When thiosulfate is given to treat cyanide toxicity, thiocyanate toxicity may occur. Toxic manifestations may include psychotic behavior, agitation, and convulsions.

Home fires in which urea foam insulation burns produce hydrocyanic acid and formaldehyde.

References
Apple FS, Lowe MC, Googins MK, et al, "Serum Thiocyanate Concentrations in Patients With Normal or Impaired Renal Function Receiving Nitroprusside," *Clin Chem*, 1996, 42(11):1878-9.

Baselt RC, *Disposition of Toxic Drugs and Chemicals in Man*, 6th ed, Foster City, CA: Biomedical Publications, 2002, 762-3.

Haque MR and Brandbury JH, "Simple Method for Determination of Thiocyanate in Urine," *Clin Chem*, 1999, 45(9):1459-64.

- ◆ **Thioridazine** *see* Phenothiazines, Serum *on page 1021*
- ◆ **Third Generation TSH** *see* Thyroid Stimulating Hormone, Serum *on page 1250*
- ◆ **Thoracentesis Fluid Analysis** *see* Body Fluid Analysis, Cell Count *on page 288*
- ◆ **Thoracentesis Fluid Analysis** *see* Body Fluid Chemical Analysis *on page 291*
- ◆ **Thoracentesis Fluid Cytology** *see* Body Cavity Fluid Cytology *on page 285*
- ◆ **Thoracentesis Fluid pH** *see* Body Fluid pH *on page 295*
- ◆ **Thorazine®** *see* Chlorpromazine, Serum *on page 395*
- ◆ **Thorazine®** *see* Phenothiazines, Serum *on page 1021*
- ◆ **Thrombin** *see* D-Dimers and Fibrin Degradation Products *on page 502*
- ◆ **Thrombin** *see* Fibrinogen *on page 583*
- ◆ **Thrombin-Antithrombin Complexes** *see* Hypercoagulation Panel *on page 758*

Thrombin Time

Related Information

Activated Partial Thromboplastin Time *on page 100*
D-Dimers and Fibrin Degradation Products *on page 502*
Factor Inhibitors *on page 566*
Fibrinogen *on page 583*
Heparin Antifactor Xa Assay *on page 693*
Heparin Neutralization *on page 697*
Hypercoagulation Panel *on page 758*
Reptilase® Time *on page 1152*

Applies to Fibrinopeptide A; Fibrinopeptide B

Abstract Measures clotting time of the last step in the coagulation cascade, which is the conversion of fibrinogen into fibrin by thrombin. Useful for diagnosis of dysfibrinogenemia. Very sensitive to low amounts of heparin, hirudin, or argatroban anticoagulants.

Specimen Plasma

Container One blue top (sodium citrate) tube; plastic tubes may result in higher thrombin time values, occasionally with potential clinical significance.

Collection Routine venipuncture. If multiple tests are being drawn, draw blue top tubes after any red top tubes but before any lavender top (EDTA), green top (heparin), or gray top (oxalate/fluoride) tubes. Immediately invert tube gently at least 4 times to mix. Tubes must be appropriately filled. Deliver tubes immediately to the laboratory, on ice.

Storage Instructions Separate plasma from cells as soon as possible. Plasma may be stored on ice for up to 4 hours; otherwise, store frozen.

Causes for Rejection Specimen received more than 4 hours after collection, tube not filled, clotted specimens

Turnaround Time Less than 1 day

Reference Interval Approximately 10-13 seconds or 16-24 seconds, depending on thrombin concentration and ionic strength of the reaction conditions

Use Performed together with Reptilase® time to diagnose dysfibrinogenemia in patients undergoing evaluation for hypercoagulability and/or a bleeding tendency. Often performed only if an initial panel of tests excludes more common disorders, because dysfibrinogenemia is uncommon. The thrombin time is an older method for detecting heparin contamination in specimens; direct heparin neutralizing methods are now available for this purpose. The thrombin time has occasionally been used to monitor heparin therapy in patients for whom the PTT could not be used, but now antifactor Xa assays are available for such situations. The thrombin time is often too sensitive to monitor heparin anticoagulation, and the assay is not standardized for this purpose.

Methodology Thrombin is added to patient plasma and the clotting time is measured in seconds. Thrombin cleaves fibrinogen, releasing fibrinopeptide A and fibrinopeptide B from fibrinogen and converting fibrinogen into fibrin clot. Assays use bovine or human thrombin.

Some laboratories use the thrombin time to detect unexpected heparin contamination in specimens. Even small amounts of heparin prolong the thrombin time, because heparin inhibits thrombin. If the thrombin time is prolonged, patient plasma can be mixed with an equal volume of normal plasma. The thrombin time of the mixture remains prolonged if the prolongation is due to heparin, fibrin degradation products (FDP), hirudin, argatroban, or other thrombin inhibitors. If the thrombin time of the mixture is normal, then the etiology of the prolongation is decreased fibrinogen or dysfibrinogenemia. The Reptilase® time is not prolonged by heparin. Many laboratories now use direct heparin neutralizing methods to detect heparin contamination rather than the thrombin time (see Heparin Neutralization *on page 697*).

Additional Information The thrombin time is prolonged when fibrinogen is decreased or dysfunctional, or when a thrombin inhibitor is present. **Dysfibrinogenemia** is an uncommon hereditary or acquired condition characterized by dysfunctional fibrinogen. Many different mutations are known to cause hereditary dysfibrinogenemia. Dysfibrinogenemia mutations can cause bleeding, thrombosis, or both, or they may be clinically asymptomatic. If bleeding is present it is usually mild, but severe bleeding has been reported. Dysfibrinogenemia has an estimated prevalence of 0.8% in patients with venous thrombosis. Arterial thrombosis is less frequent than venous thrombosis in these patients. Most patients with hereditary dysfibrinogenemia are heterozygous. Rare homozygous cases have been reported. The thrombin time and Reptilase® time, which measure the clotting time during the conversion of fibrinogen into fibrin, are often prolonged in dysfibrinogenemia because fibrinogen is dysfunctional. Assays that measure fibrinogen function show lower levels than assays that measure fibrinogen quantity (immunological or "antigen" assays), because fibrinogen function is impaired but fibrinogen quantity is not. The PT and PTT may be prolonged in dysfibrinogenemia. Causes of acquired dysfibrinogenemia include liver disease, hepatoma, or or acute phase reactions with generation of high levels of fibrinogen. The bleeding and thrombosis risk with acquired dysfibrinogenemia is uncertain. See Fibrinogen on page 583.

The thrombin time can be prolonged in disseminated intravascular coagulation (DIC) or thrombolytic therapy due to high levels of FDP and decreased fibrinogen. However, the thrombin time is not a necessary test for DIC diagnosis because fibrinogen and FDP can be measured directly. Prolongation of the thrombin time and Reptilase® time has been commonly observed with amyloidosis due to inhibition of fibrinogen conversion to fibrin. Patients exposed to bovine thrombin may develop thrombin inhibitors that prolong bovine-based thrombin times, and if the antibody cross-reacts against human thrombin, human-based thrombin times can also be prolonged (see Factor Inhibitors on page 566). The Reptilase® time is normal with these inhibitors. Rarely, heparin-like anticoagulants have been reported in patients with malignancies or other disorders, with prolonged thrombin times and normal Reptilase® times.

References

Flanders MM, Crist R, and Rodgers GM, "A Comparison of Blood Collection in Glass Versus Plastic Vacutainers® on Results of Esoteric Coagulation Assays," Lab Med, 2003, 34(10):732-5.

Galanakis DK, "Fibrinogen Anomalies and Disease. A Clinical Update," Hematol Oncol Clin North Am, 1992; 6(5):1171-87.

Haverkate F and Samama M, "Familial Dysfibrinogenemia and Thrombophilia. Report on a Study of the SSC Subcommittee on Fibrinogen," Thromb Haemost, 1995, 73(1):151-61.

Olson JD, Arkin CF, Brandt JT, et al, "College of American Pathologists Conference XXXI on Laboratory Monitoring of Anticoagulant Therapy. Laboratory Monitoring of Unfractionated Heparin Therapy," Arch Pathol Lab Med, 1998, 122:782-98.

♦ **Thrombolysis** see D-Dimers and Fibrin Degradation Products on page 502
♦ **Thrombophilia Panel** see Hypercoagulation Panel on page 758
♦ **Thromboplastin** see Prothrombin Time on page 1116

Thrombopoietin, Serum or Plasma

Related Information

Erythropoietin, Serum on page 551
Platelet Count on page 1050
Platelet Sizing on page 1056
Reticulated Platelet Count on page 1155

Abstract Thrombopoietin, a cytokine glycoprotein, is the primary regulator of marrow megakaryocyte and platelet development. Thrombopoietin levels may indicate whether a patient has thrombocytopenia on the basis of failure of marrow production (decrease in thrombopoietin effect) or increase in peripheral platelet destruction.

Specimen Serum or EDTA plasma; sample requirements may be method dependent. A correlation has been reported between TPO values in serum and plasma (serum TPO = -0.257 + 4.039 x plasma TPO values).

Storage Instructions Stable for up to 6 days in platelet poor and in platelet-rich plasma. For prolonged storage, remove serum from cells and store at -30°C. Free TPO is evidently quite stable. Repeated freeze/thaw (up to nine times) of EDTA plasma did not effect TPO levels. The interval between specimen collection and plasma separation (up to 25 hours) did not effect plasma TPO levels.

Reference Interval
- Male: 0.79 ±0.35 fmol/mL; female: 0.70 ±0.26 fmol/mL
- Cord blood: 3.73 ±1.48 fmol/mL
- 5 days: 4.32 ±0.94 fmol/mL
- 1 month: 3.77 ±1.45 fmol/mL
- 2-11 months: 2.10 ±0.69 fmol/mL
- 1-2 years: 2.23 ±0.89 fmol/mL

(Continued)

Thrombopoietin, Serum or Plasma *(Continued)*

- 3-15 years: 1.97 ±0.67 to 1.24 ±0.40 fmol/mL

Mean adult normal value (using plasma) has been noted as 133 pg/mL (range of 57-377).

Use May be helpful in determining the etiology of thrombocytopenia.

Methodology A number of TPO ELISA assays using serum and/or plasma have been developed. Results/normal ranges may not be directly comparable. An ELISA Kit is available commercially (R & D Systems, Abington, UK).

Additional Information Humoral control of the growth and development of megakaryocytes (and their platelet progeny) has undergone partial definition only gradually over the past 30 some years. Multiple growth factors (protein regulators - cytokines) are involved. The identification and cloning of many human hematopoietic growth factors (and the availability of their recombinant forms) in the 1980s/1990s has allowed new insight into cytokine control of megakaryocytopoiesis. In particular, interleukin-3 (IL-3), granulocyte/monocyte colony stimulating factor (GM-CSF), interleukin-6 (IL-6), interleukin-11 (IL-11), and others contribute to multifactorial synergistic control of megakaryocytopoiesis.

These growth factors, however, are not megakaryocyte/platelet lineage specific and, as in the case of IL-3, individually may not affect *in vivo* platelet production. Approval has been given for the use of IL-11 in treatment of chemotherapy-induced thrombocytopenia.

Early thrombopoietin (TPO) assays lacked specificity due to the interplay of pleiotropic cytokines (IL-3, IL-11, etc). Doubt about the existence of a lineage-specific megakaryocyte platelet humoral regulator was allayed when a transforming oncogene of the virus that induces myeloproliferative leukemia (MPL), in mice, was found also in humans (over time) to have characteristics consistent with the TPO receptor. In particular, antisense oligonucleotides against *c-mpl* RNA were shown to inhibit *in vitro* megakaryocytopoiesis. Thus, TPO (the ligand) was identified (cloned in 1994) only after its receptor (*c-mpl*) was discovered and cloned.

TPO is now recognized as the lineage specific and most important megakaryocyte/platelet regulatory protein. TPO is a 90-kDa glycoprotein encoded by a gene located on chromosome 3 (3q 26-27). Two forms of TPO have been produced for clinical study. One form, recombinant human TPO consists of the full-length polypeptide. The other form consists of the receptor-binding region only, chemically modified by the addition of polyethylene glycol (PEG) and referred to as PEG-conjugated recombinant human megakaryocyte growth and development factor, a "pegylated" molecule.

Thrombopoietin acts to increase the number and size of megakaryocytes, stimulates polyploidy, stimulates megakaryocyte colony formation, and acts in synergy with other cytokines (including IL-3, IL-11, and erythropoietin) to stimulate megakaryocyte and erythroid progenitor cell growth as well as hematopoietic stem-cell proliferation and survival. With a lag period of some 6-16 days, TPO produces an increase in platelet count and causes increase in platelet activation.

Platelets remove and degrade TPO by receptor binding and internalization. TPO is degraded in a time-dependent manner and is not recycled to the surface. The resultant plasma survival half-life is some 20-30 hours. Data indicates that the level of TPO is regulated by its binding to platelets and/or megakaryocytes (the platelet/megakaryocyte mass itself is the key regulator of TPO serum/plasma levels). When the platelet mass is normal (remains constant), circulating TPO is at a basal equilibrium level. With decrease in platelet mass (thrombocytopenia uncompensated by enlarged size of platelets) there is decreased binding of TPO (and subsequent degradation by *c-mpl*-positive cells) with resultant increase in free TPO. Reactive thrombocytosis, conversely, is occasioned by decreased levels of TPO. In chronic idiopathic thrombocytopenic purpura (in which megakaryocytes are increased), TPO levels are normal or decreased.

Hereditary thrombocythemia (HT), not the same disorder as essential thrombocythemia (ET), is a rare condition, has a constant association with increased serum/plasma TPO, and in some families, is transmitted as an autosomal dominant trait. Sporadic cases of ET with mutation of the thrombopoietin gene are reported.

Elevation of serum TPO has been noted to precede thrombocytosis in Kawasaki disease and to occur in thrombocytosis associated with inflammatory bowel disease. The latter finding may be of pathogenetic significance to the risk of thromboembolic complications in Crohn disease and ulcerative colitis.

The therapeutic use of thrombopoietin may be limited by concern over inducing thrombocytosis with attendant risk for thrombosis (which may occur in patients with myeloproliferative disorders).

References

Folman CC, von dem Borne AE, Rensink IH, et al, "Sensitive Measurement of Thrombopoietin by a Monoclonal Antibody Based Sandwich Enzyme-Linked Immunosorbent Assay," *Thromb Haemost*, 1997, 78(4):1262-7.

Ishiguro A, Nakahata T, Matsubara K, et al, "Age-Related Changes in Thrombopoietin in Children: Reference Interval for Serum Thrombopoietin Levels," *Br J Haematol*, 1999, 106(4):884-8.

Kaushansky K, "Thrombopoietin," *N Engl J Med*, 1998, 339(11):746-54.

Tahara T, Usuki K, Sato H, et al, "A Sensitive Sandwich ELISA for Measuring Thrombopoietin in Human Serum: Serum Thrombopoietin Levels in Healthy Volunteers and in Patients With Haemopoietic Disorders," *Br J Haematol*, 1996, 93(4):783-8.

♦ **Thrombotic Disease Screen** see Hypercoagulation Panel *on page 758*

♦ **Thrombran®** see Trazodone, Serum or Plasma *on page 1272*

♦ **Thymulin Assay** see Zinc, Serum or Plasma *on page 1340*

♦ **Thyrocalcitonin** see Calcitonin, Serum or Plasma *on page 327*

♦ **Thyroglobulin** see Immunoperoxidase Procedures *on page 780*

Thyroglobulin Antibody

Related Information

Acetylcholine Receptor Antibody *on page 92*
Glutamic Acid Decarboxylase (GAD65) Antibody *on page 654*
Parietal Cell Antibody *on page 1005*
Thyroperoxidase Autoantibody *on page 1253*
Thyrotropin Receptor Antibody, Serum *on page 1254*

Synonyms Antithyroglobulin Antibody; Tg Ab; Thyroid Antithyroglobulin Antibody

Applies to Thyroid Autoantibodies

Test Includes Titers on positive specimens

Abstract The improved sensitivity and specificity of current antithyroperoxidase assays obviates the need to test for Tg Ab.

Specimen Serum

Use No established clinical usefulness

References

Dayan CM and Daniels GH, "Chronic Autoimmune Thyroiditis," *N Engl J Med*, 1996, 335(2):99-107.

Kaplan LA, ed, The National Academy of Clinical Biochemistry, *Standards of Laboratory Practice: Laboratory Support for the Diagnosis & Monitoring of Thyroid Disease*, 1996.

Thyroglobulin, Serum

Abstract Thyroglobulin (Tg) is a secretory product of thyroid follicular epithelium. It is the storage form of the thyroid hormones, T_4, and T_3. Its major clinical use is as a tumor marker, in the management of well differentiated thyroid carcinomas.

Patient Preparation Avoid scans and other recent prior administration of radioisotopes before collection of specimen if RIA is used for assay. Do not draw a specimen for this test soon after needle biopsy, thyroid surgery, or radioiodine therapy. The test is most sensitive for detection of thyroid cancer when the patient is off thyroxine replacement long enough to have increased TSH. This may take as long as 6 weeks.

Specimen Serum

Container Red top tube

Storage Instructions Freeze serum.

Reference Interval Normal: <59.4 ng/mL; athyroidic PT: <5.0 ng/mL

Use The principal use for Tg is in the monitoring of patients who have had well differentiated thyroid carcinomas resected. In such patients, values after successful surgery should be <5 ng/mL. Increasing values signal recurrence. Tg is also used in the detection of thyrotoxicosis caused by intentional self-medication (thyrotoxicosis factitia), in which low values are obtained.
(Continued)

Thyroglobulin, Serum (Continued)

Limitations Thyroglobulin is not useful as a tumor marker for anaplastic or medullary carcinoma of thyroid or for thyroid lymphoma.

Methodology Radioimmunoassay (RIA), immunoradiometric assay (IRMA), sandwich enzyme immunoassay (EIA), immunochemiluminometric assays (ICMA)

References

Jacobs DS, DeMott WR, Oxley DK, et al, *Laboratory Test Handbook*, 5th ed, Hudson, OH: Lexi-Comp Inc, 2001.

Mayo Medical Laboratories, *2001 Test Catalogue*, Rochester, MN, 497.

Schlumberger MJ, "Papillary and Follicular Thyroid Carcinoma," *N Engl J Med*, 1998, 338(5):297-306.

Utiger RD, "Follow-up of Patients With Thyroid Carcinoma," *N Engl J Med* 1997, 337(13):928-30.

♦ **Thyroid Antimicrosomal Antibody** *see* Thyroperoxidase Autoantibody *on page 1253*

♦ **Thyroid Antithyroglobulin Antibody** *see* Thyroglobulin Antibody *on page 1249*

♦ **Thyroid Autoantibodies** *see* Thyroglobulin Antibody *on page 1249*

♦ **Thyroid Autoantibodies** *see* Thyroperoxidase Autoantibody *on page 1253*

♦ **Thyroid Cyst Fluid Cytology** *see* Cyst Fluid Cytology *on page 490*

♦ **Thyroid Hormone Binding Ratio** *see* T_3 Uptake, Serum or Plasma *on page 1233*

♦ **Thyroid Needle Aspiration Cytology** *see* Fine Needle Aspiration, Deep and Superficial Masses *on page 590*

♦ **Thyroid Screen for Newborns** *see* Newborn Screen for T_4, Filter Paper *on page 956*

♦ **Thyroid Screen for Newborns** *see* Newborn Screen for TSH, Filter Paper *on page 957*

♦ **Thyroid Stimulating Autoantibody** *see* Thyrotropin Receptor Antibody, Serum *on page 1254*

♦ **Thyroid Stimulating Hormone Screen, Filter Paper** *see* Newborn Screen for TSH, Filter Paper *on page 957*

Thyroid Stimulating Hormone, Serum

Related Information

Cholesterol, Total, Serum or Plasma *on page 396*
Fine Needle Aspiration, Deep and Superficial Masses *on page 590*
Free Thyroxine Index *on page 613*
Heterophilic Antibodies *on page 727*
Lithium, Serum *on page 863*
T_3 Uptake, Serum or Plasma *on page 1233*
Thyroperoxidase Autoantibody *on page 1253*
Thyrotropin Receptor Antibody, Serum *on page 1254*
Thyroxine, Free, Serum *on page 1256*
Thyroxine, Serum *on page 1257*
Triglycerides, Serum or Plasma *on page 1275*
Triiodothyronine, Serum *on page 1276*

Synonyms Fourth Generation TSH; sTSH; Third Generation TSH; Thyrotropin; TSH; Ultrasensitive TSH

Applies to TRH Stimulation Test

Abstract Produced by the anterior pituitary gland, thyroid stimulating hormone (TSH), also called thyrotropin, stimulates secretion of T_4 (thyroxine) and T_3 (triiodothyronine). TSH secretion is physiologically regulated by T_4 and T_3 (feedback inhibition) and is stimulated by TRH (thyrotropin releasing hormone) from the hypothalamus.

Patient Preparation Avoid radioisotope administration before collection of specimen if RIA is used for assay.

Specimen Serum

Container Red top tube

Sampling Time A diurnal rhythm exists. Peak levels occur at about 11 PM. TSH release is pulsatile.

Storage Instructions Separate serum within 4 hours and refrigerate. Stable 4 days at 4°C.

Reference Interval Reference intervals are age dependent and somewhat method dependent. For prematures and during the first week of life, TSH values are significantly higher. Approximate values using third generation TSH assay are as follows:

- prematures, 28-36 weeks: 0.7-27.0 mIU/L
- birth to 4 days: 1.0-39.0 mIU/L
- 2-20 weeks: 1.7-9.1 mIU/L
- 21 weeks to 20 years: 0.7-6.4 mIU/L
- 21-54 years: 0.4-4.2 mIU/L
- 55-87 years: 0.5-8.9 mIU/L

Use Primary hypothyroidism (ie, due to disease in the thyroid itself) is usually identified by finding an elevated serum TSH. The dynamics of the pituitary-thyroid feedback loop are such that an elevation of TSH precedes a recognizably decreased T_4 (total or free). Patients with **central hypothyroidism** (ie, due to pituitary failure, and also called **secondary hypothyroidism**), by contrast, usually have a TSH value that is low in relation to the simultaneously obtained T_4, but TSH may be within, or above, the adult reference interval. In such patients, the pituitary cannot fully respond to the low T_4 in the periphery. When a patient has a low T_4 (total or free) and a normal TSH, the clinician will probably investigate further; however, when only the TSH has been obtained (the practice in many clinics) and is normal, there is a danger of overlooking the diagnosis of central hypothyroidism. Such a scenario can be further complicated if the patient has a surgical procedure or intercurrent illness and thereby develops, in addition to the unrecognized central hypothyroidism, the **euthyroid sick syndrome** in which there is a transient, additional decrement in TSH; Waise et al have described six such examples. An additional problem is recognizing patients who have **resistance to thyroid hormone** (RTH), usually a familial disease. In such patients, the pituitary and most other organs and tissues are relatively insensitive to thyroid hormone, and the characteristic biochemical profile includes elevated T_4 (free and total), elevated T_3 (free and total), and an inappropriately "normal" (or elevated) TSH. An important hazard for persons with RTH is receiving the erroneous diagnosis, and treatment, for hyperthyroidism. Such a scenario is most likely to arise when a physician looks only at the laboratory results and fails to correlate them with the clinical findings.

An increasingly common practice, but still controversial, involves diagnosing and treating **subclinical hypothyroidism**, understood as individuals who appear eumetabolic by clinical criteria, have normal serum T_4 (total and free) and elevated TSH. Its usual causes include thyroiditis and inadequate treatment of hypothyroidism.

A novel, and presumably rare, form of hypothyroidism in children appears to be due to **pituitary-thyroid feedback hypersensitivity**. The patient reported by Frankton et al had low values for free T_4, total T_4, and free T_3; the total T_3 was at the lower limit of the reference interval; the TSH was inappropriately normal and the TSH response to thyrotropin-release-hormone stimulation was low for the concurrent T_4 and T_3 levels.

TSH has become the primary thyroid function test for stable ambulatory subjects who lack pituitary or neuropsychiatric illness. A TSH result within the accepted reference interval provides strong evidence for euthyroidism. A simple algorithm for thyroid testing can be found in the following chart, modified from that at the Mayo Clinic Laboratory.

Subclinical hyperthyroidism is defined as suppression of sTSH concentrations with normal levels of serum thyroxine and triiodothyronine. This disorder may lead to diminution of bone mineral density, and increases risk of atrial fibrillation in patients older than 60 years of age. The most common cause of subclinical hyperthyroidism is excessive thyroid hormone treatment.

Hyperthyroidism is characterized by low serum TSH values by third generation assays.

Limitations Many drugs and intercurrent illnesses are associated with alterations in serum TSH. Values should be obtained when patients are clinically stable and not acutely, or critically, ill.

Heterophilic antibodies are a problem (see Heterophilic Antibodies *on page 727*).
(Continued)

Thyroid Stimulating Hormone, Serum *(Continued)*

**The Mayo Medical Laboratories
Thyroid Function Cascade
(modified)**

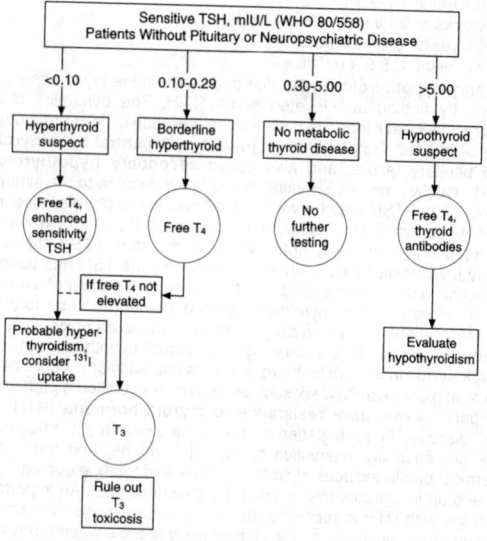

Common errors in interpretation by thyroid function tests are described in the following table.

Common Errors in Interpretation of Thyroid Function Tests

Thyroid Function Test Results	Circumstance	Wrong Interpretation	Correct Interpretation
Low TSH, low FT$_3$ or FT$_4$	Post-treatment for thyrotoxicosis	Persistent hyperthyroidism	Profoundly hypothyroid
Low TSH, high FT$_3$ or FT$_4$	Short history of neck pain	Established thyrotoxicosis	Possible self-resolving thyroiditis
Low or normal TSH, low or normal FT$_3$ or FT$_4$	Patient systemically ill	Hypothyroid	Nonthyroidal illness
Normal TSH (FT$_3$ or FT$_4$ not tested)	Pituitary disease	Euthyroid	Possibly hypothyroid - check FT$_3$ or FT$_4$
High TSH, normal FT$_3$ or FT$_4$	TSH fails to fall with T$_4$	Noncompliance with T$_4$	Possible interfering heterophil or an antibody

FT$_3$ = free T$_3$; FT$_4$ = free T$_4$.

Adapted from Dayan CM, "Interpretation of Thyroid Function Tests," *Lancet*, 2001, 357:619-24.

Methodology Immunoassays including radioimmunoassay (RIA), immunochemiluminometric assays (ICMA), sandwich immunoradiometric assays (IRMA), fluorometric enzyme immunoassay with use of monoclonal antibodies, microparticle enzyme immunoassay (MEIA) on IMx and AxSym (Abbott Laboratories). Most assays presently in use are beyond the first generation of tests.

Additional Information When TSH is used to monitor thyroxine therapy, generally 6-8 weeks are required to achieve normal TSH levels. Likewise, TSH levels lag behind the return of thyroxine concentration to normal after thyroid replacement therapy, thyroid ablation, or antithyroid medication for hyperthyroidism. The most common cause of low TSH among outpatients is excessive replacement of thyroid hormone.

As the prevalence of hypothyroidism is very high (up to 14%) in the elderly population, screening for hypothyroidism in this population is justified, and TSH is used as a primary screening test. Subclinical hypothyroidism, as defined by elevated TSH and normal FT_4, is an independent risk factor for atherosclerosis and myocardial infarction in the elderly population.

Among those age 60 and older, low TSH (<0.1 mIU/L) is a risk factor for atrial fibrillation.

Different strategies have been proposed for the laboratory investigation of thyroid function. The American Association of Clinical Endocrinologists, the American Thyroid Association, and many others endorse sensitive TSH measurements as "the single best screening test" for thyroid disease. The Royal College of Physicians of London recommends use of TSH and T_4 or FT_4.

A very rare cause of elevated TSH is a TSH-secreting tumor, usually of the pituitary.

See table in Thyroxine Binding Globulin, Serum *on page 1255* for effects of some drugs on thyroid function tests.

Subclinical hypothyroidism and subclinical hyperthyroidism are addressed in the listing, Thyroxine, Free, Serum *on page 1256*.

References

Brucker-Davis F, Skarulis MC, Grace MB, et al, "Genetic and Clinical Features of 42 Kindreds With Resistance to Thyroid Hormone," The National Institutes of Health Prospective Study, *Ann Intern Med*, 1995, 123(8):572-83.

Dayan CM, "Interpretation of Thyroid Function Tests," *Lancet*, 2001, 357:619-24.

Frankton S, Karmali R, Mirkine N, et al, "Pituitary-Thyroid Feedback Hypersensitivity as a Novel Cause of Hypothyroidism in Children," *Lancet*, 2000, 356(9237):1238-40.

Gladwin MT and Duell PB, "Inappropriate Thyroid Gland Ablation in Patients With Generalized Resistance to Thyroid Hormone. A Common Sequela of a Rare Disorder," *Arch Intern Med*, 1996, 156(1):106-9.

Ladenson PW, Singer PA, Ain K, et al, "American Thyroid Association Guidelines for Detection of Thyroid Dysfunction," *Arch Intern Med*, 2000, 160(11):1573-5.

Leavelle DE, *Mayo Medical Laboratories Interpretive Handbook*, Rochester, MN: Mayo Medical Laboratories, 1997, 502-3.

Toft AD, "Subclinical Hyperthyroidism," *N Engl J Med*, 2001, 345(7):512-6.

Vanderpump MP, Ahlquist JA, Franklyn JA, et al, "Consensus Statement for Good Practice and Audit Measures in the Management of Hypothyroidism and Hyperthyroidism," *BMJ*, 1996, 313(7056): 539-44.

Whitley RJ, "Thyroid Function," *Tietz Textbook of Clinical Chemistry*, 3rd ed, Burtis CA and Ashwood ER, eds, Philadelphia: WB Saunders Co, 1999, 1496-529.

♦ **Thyroid Stimulating Immunoglobulins** *see* Thyrotropin Receptor Antibody, Serum *on page 1254*

Thyroperoxidase Autoantibody

Related Information

Acetylcholine Receptor Antibody *on page 92*
Anemia Flowchart *on page 35*
Free Thyroxine Index *on page 613*
Glutamic Acid Decarboxylase (GAD65) Antibody *on page 654*
Parietal Cell Antibody *on page 1005*
T_3 Uptake, Serum or Plasma *on page 1233*
Thyroglobulin Antibody *on page 1249*
Thyroid Stimulating Hormone, Serum *on page 1250*
Thyroxine, Free, Serum *on page 1256*
Thyroxine, Serum *on page 1257*

Synonyms Antithyroid Peroxidase Antibody; Microsomal Antibody, Thyroid; Thyroid Antimicrosomal Antibody

Applies to Thyroid Autoantibodies

Abstract Thyroperoxidase (TPO) is the major antigen in antibody-dependent cell-mediated cytotoxicity in thyroid disease.

Specimen Serum

Container Red top or SST™ tube

Storage Instructions Refrigeration is acceptable for up to 72 hours.

(Continued)

Thyroperoxidase Autoantibody (Continued)

Reference Interval

Anti-TPO concentration:

- normal: <20 IU/mL
- positive (adults): ≥20 IU/mL
- associated with autoimmune thyroiditis, other autoimmune diseases: ≥50 IU/mL

Use In a patient with hypothyroidism, an elevated TPO-Ab supports the diagnosis of autoimmune thyroiditis (sensitivity 95%).

Limitations Approximately 10% of adults have low titers without other evidence of thyroid disease.

Methodology Enzyme-linked immunosorbent assay (ELISA), thyroperoxidase antibodies by chemiluminometric assay

References

Dayan CM and Daniels GH, "Chronic Autoimmune Thyroiditis," *N Engl J Med*, 1996, 335(2):99-107.

Mayo Reference Services Publication, "Thyroperoxidase (TPO) Antibodies, Serum," *New Test Announcement*, Rochester, MN: Mayo Medical Laboratories, November, 1998.

♦ **Thyrotropin** see Thyroid Stimulating Hormone, Serum on page 1250

Thyrotropin Receptor Antibody, Serum

Related Information

Free Thyroxine Index on page 613
T₃ Uptake, Serum or Plasma on page 1233
Thyroglobulin Antibody on page 1249
Thyroid Stimulating Hormone, Serum on page 1250
Thyroperoxidase Autoantibody on page 1253
Thyroxine, Free, Serum on page 1256
Thyroxine, Serum on page 1257

Synonyms LATS; Long-Acting Thyroid Stimulator; Thyroid Stimulating Autoantibody; Thyroid Stimulating Immunoglobulins; TRAb; Ts Antibodies; TSH-Receptor Antibodies; TSIG

Abstract This autoantibody, is important in the pathogenesis of Graves disease. It was previously known as the long-acting thyroid stimulator (LATS).

Patient Preparation Recent radioisotope administration must be avoided if test is done by competitive binding radioimmunoassay.

Specimen Serum

Container Red top tube

Reference Interval

- Negative: <10%
- Indeterminate: 10% to 14%
- Positive: ≥15%

Use This assay is occasionally used to clarify the differential diagnosis between Graves disease and toxic nodular goiter. Sensitivity in Graves disease ranges from 80% to 99%. Despite the importance of the thyrotropin receptor antibody (TRAb) in the pathogenesis of Graves disease, the test is not often used in the clinical management of patients. TRAb is probably most helpful in patients with infiltrative ophthalmopathy or dermopathy without overt hyperthyroidism. In such situations, correct diagnosis requires correlation of physical findings, imaging studies, and laboratory tests.

Limitations Negative values can be secondary to low concentrations of serum proteins or immunoglobulins. Subjects with subclinical Graves disease may have indeterminate values.

Methodology Indirect assay of thyroid cell activation, competitive assay using solubilized porcine TSH receptor

References

Larsen PR, Davies TF, and Hay ID, "The Thyroid Gland," *Williams Textbook of Endocrinology*, 9th ed, Wilson JD, Foster DW, Kronenberg HM, et al, eds, Philadelphia, PA: WB Saunders Co, 1998,389-515.

Mayo Medical Laboratories, *2001 Test Catalogue*, 499.

Weetman AP, "Graves' Disease," *N Engl J Med*, 2000, 343(17):1236-48.

♦ **Thyroxine** see Thyroxine, Serum on page 1257

Thyroxine Binding Globulin, Serum

Related Information

Free Thyroxine Index *on page 613*
T$_3$ Uptake, Serum or Plasma *on page 1233*
Thyroxine, Free, Serum *on page 1256*
Thyroxine, Serum *on page 1257*

Synonyms T$_4$-Binding Globulin; TBG

Applies to Familial Dysalbuminemic Hyperthyroxinemia; Free Thyroxine Index, Calculated; Transthyretin

Abstract Serum thyroid binding proteins include albumin, transthyretin (thyroid binding prealbumin), and, most important, thyroxine binding globulin (TBG). As >75% of T$_3$ and T$_4$ are bound to TBG, any change, either genetic or acquired, in TBG concentration affects total T$_3$ and T$_4$ values. Abnormal total, but normal free T$_3$ and T$_4$, in an euthyroid individual, suggests TBG deficiency or excess. In such circumstances, measurement of TBG may help to explain confusing results. **This test should not be confused with thyroglobulin.**

Patient Preparation No recent administration of radioactive isotopes or *in vivo* uptakes if method is by RIA.

Specimen Serum

Container Red top tube

Effects of Some Drugs on Thyroid Function Tests

Cause	Drug	Effect
Inhibit TSH secretion	Dopamine L-dopa Glucocorticoids Somatostatin	↓T$_4$; ↓T$_3$; ↓TSH
Inhibit thyroid hormone synthesis or release	Iodine Lithium	↓T$_4$; ↓T$_3$; ↑TSH
Inhibit conversion of T$_4$ to T$_3$	Amiodarone Glucocorticoids Propranolol Propylthiouracil Radiographic contrast agents	↓T$_3$; ↑rT$_3$; ↓, ←→, ↑T$_4$ and FT$_4$; ←→, ↑TSH
Inhibit binding of T$_4$/T$_3$ to serum proteins	Salicylates Phenytoin Carbamazepine Furosemide Nonsteroidal anti-inflammatory agents Heparin (*in vitro* effect)	↓T$_4$; ↓T$_3$; ↓FT$_4$E; ←→, ↑FT$_4$; ←→TSH
Stimulate metabolism of iodothyronines	Phenobarbital Phenytoin Carbamazepine Rifampicin	↓T$_4$; ↓FT$_4$; ←→TSH
Inhibit absorption of ingested T$_4$	Aluminum hydroxide Ferrous sulfate Cholesytramine Colestipol Iron sucralfate Soybean preparations Kayexalate	↓T$_4$; ↓FT$_4$; ↑TSH
Increase in concentration of T$_4$ binding proteins	Estrogen Clofibrate Opiates (heroin, methadone) 5-Fluouracil Perphenazine	↑T$_4$; ↑T$_3$; ←→FT$_4$; ←→TSH
Decrease in concentration of T$_4$ binding proteins	Androgens Glucocorticoids	↓T$_4$; ↓T$_3$; ←→FT$_4$; ←→TSH

↓, reduced serum level; ↑, increased serum level; ←→, no change.

Adapted from Whitley RJ, "Thyroid Function," *Teitz Textbook of Clinical Chemistry*, 3rd ed, Chapter 42, Burtis C and Ashwood E, eds, Philadelphia, PA: WB Saunders Co, 1999.

(Continued)

Thyroxine Binding Globulin, Serum *(Continued)*

Reference Interval The reference intervals are higher in children as compared to adults.

- 0-1 week: 3-8 mg/dL
- 1-12 months: 3-6 mg/dL
- 2-10 years: 2-5 mg/dL
- Adults: 1.2-2.5 mg/dL; adult levels are reached about age 14.

Use The major indication for TBG testing is in diagnosis of **hereditary deficiency of TBG**. Most of these subjects with TBG deficiency can be expected to have low T_4 but normal TSH. However, TBG deficiency can cause hypothyroidism.

TBG is occasionally used to clarify thyroid function status in euthyroid individuals who have increased T_4 and normal levels of free hormones. It provides documentation of cases of hereditary deficiency or increase of TBG. The most common explanation for increased TBG is increased endogenous or exogenous estrogens, many other drugs are also possibilities (see table). Increased TBG is found in certain genetically determined states, with a prevalence of 1 in 25,000.

Decreased TBG is found with some chronic diseases and certain medications (see table on previous page). The reported prevalence of familial TBG deficiency is 1 in 5000.

Limitations TBG is normal in **familial dysalbuminemic hyperthyroxinemia**, an entity which can be incorrectly identified as thyrotoxicosis. Low T_3 uptake and normal free T_4 or free T_4 index or thyroxine binding ratio often make measurement of TBG unnecessary.

Methodology Double antibody precipitation, radioimmunoassay (RIA)

References

Komatsu M, Hanamura N, Seki T, et al, "A Family With Hereditary High Serum Thyroxine-Binding Globulin," *Endocr J*, 1994, 41(4):467-70.

Tojo K, Miura Y, Mori Y, et al, "Familial Thyroxine-Binding Globulin Deficiency Associated With Hyperthyroidism," *Intern Med*, 1995, 34(5):413-7.

- ◆ **Thyroxine-Binding Prealbumin** *see* Transthyretin, Serum *on page 1271*
- ◆ **Thyroxine-Binding Protein Electrophoresis** *see* Thyroxine, Serum *on page 1257*

Thyroxine, Free, Serum

Related Information

Free Thyroxine Index *on page 613*
Newborn Screen for TSH, Filter Paper *on page 957*
Thyroid Stimulating Hormone, Serum *on page 1250*
Thyrotropin Receptor Antibody, Serum *on page 1254*
Thyroxine Binding Globulin, Serum *on page 1255*
Thyroxine, Serum *on page 1257*
Triiodothyronine, Serum *on page 1276*

Synonyms Free T_4; Free Thyroxine; FT_4; T_4, Free; Unbound T_4

Applies to FTI

Abstract Free T_4 (FT_4) is the metabolically active fraction and is the precursor of T_3. Measurement of FT_4 is very helpful in patients who are suspected of having hyperthyroidism (*vide infra*).

Patient Preparation Recent injection of radioisotope may interfere, depending on assay system in use.

Specimen Serum

Container Red top tube

Storage Instructions Separate serum within 48 hours. Stable 2 weeks at 4°C.

Reference Interval Newborns: 2.6-6.3 ng/dL (SI: 33.5-81.3 pmol/L); adults: 0.8-2.7 ng/dL (10.3-35 pmol/L)

Use A sensitive test for thyroid function, free T_4 is increased with hyperthyroidism and decreased in hypothyroidism. It is a better indicator of thyroid function than total T_4 because it is not affected by changes in thryoxine-binding proteins. Free thyroxine is normal in euthyroid subjects with elevated thyroxine binding globulin. FT_4 clarifies patient status in situations such as secondary hypothyroidism related to pituitary disease. When TSH, used as the primary screening test is low and FT_4 is normal, measurement of serum T_3 is indicated. The combination of low TSH with normal free T_3 or free T_4 occurs with ingestion of thyroxine as well as with subclinical hyperthyroidism. A diagnostic strategy for diagnostic work up of thyroid disease using TSH and FT_4 can be found in Thyroid Stimulating Hormone, Serum *on page 1250.*

Limitations FT_4 results are misleading in the presence of antithyroxine autoantibodies and rheumatoid factor, and with low molecular weight heparin treatment. FT_4 is increased in familial dysalbuminemic hyperthyroxinemia, amiodarone treatment, states of resistance to thyroid hormone, and acute psychiatric illness. Free T_4 will not detect T_3 thyrotoxicosis. Transiently increased free T_4 levels may occur in subjects with nonthyroidal diseases. Low values are reported in many patients with nonthyroidal illness. Concentrations of sTSH become abnormal before FT_4 levels do so, early in primary hypo- and hyperthyroidism.

Methodology Equilibrium dialysis is the reference method. Routinely used methods are radioimmunoassay (RIA), fluorescent immunoassay, and chemiluminescent immunoassay.

Additional Information Subclinical hypothyroidism (normal FT_4 and elevated TSH or low FT_4 and normal TSH) is an independent risk factor for atherosclerosis and myocardial infarction, and progression to overt hypothyroidism.

In **subclinical hyperthyroidism** (TSH <0.3-0.4 µU/L with normal concentrations of free thyroxine and free triiodothyronine), there is increased risk for atrial fibrillation, osteoporosis, and progression to overt hyperthyroidism.

References
Attia J, Margetts P, and Guyatt G, "Diagnosis of Thyroid Disease in Hospitalized Patients: A Systematic Review," *Arch Intern Med*, 1999, 159(7):658-65.
Dayan CM, "Interpretation of Thyroid Function Tests," *Lancet*, 2001, 357:619-24.
Hak AE, Pols HA, Visser TJ, et al, "Subclinical Hypothyroidism Is an Independent Risk Factor for Atherosclerosis and Myocardial Infarction in Elderly Women: The Rotterdam Study," *Ann Intern Med*, 2000, 132(4):270-8.
Toft AD, "Subclinical Hyperthyroidism," *N Engl J Med*, 2001, 345(7):512-6.

Internet Web Sites
www.thyroid.org/

♦ **Thyroxine Ratio** *see* T_3 Uptake, Serum or Plasma *on page 1233*

Thyroxine, Serum

Related Information
Free Thyroxine Index *on page 613*
Heterophilic Antibodies *on page 727*
Lithium, Serum *on page 863*
Newborn Screen for TSH, Filter Paper *on page 957*
T_3 Uptake, Serum or Plasma *on page 1233*
Thyroid Stimulating Hormone, Serum *on page 1250*
Thyroperoxidase Autoantibody *on page 1253*
Thyroxine Binding Globulin, Serum *on page 1255*
Thyroxine, Free, Serum *on page 1256*
Triiodothyronine, Serum *on page 1276*

Synonyms T_4; Tetraiodothyronine; Thyroxine

Applies to Free Thyroid Index; FT_4I; Goiter; TBG; Thyroxine-Binding Protein Electrophoresis

Replaces T_4 CPB; Murphy-Pattee; PBI

Abstract Thyroxine (T_4) and triiodothyronine (T_3) are the major secretory products of the thyroid gland. T_4 is carried through the blood bound (in equilibrium) to thyroxine binding globulin (TBG), prealbumin, and albumin (>99.9%). In the peripheral tissues, T_4 is converted to T_3, the active hormone.

It is important to recall that abnormal T_4 results occur in euthyroid individuals having altered thyroxine binding protein. The combination of free T_4 and TSH is better than total T_4.

Patient Preparation Avoid radioisotope administration prior to collection of specimen if testing is by RIA.

Specimen Serum

Container Red top tube

Storage Instructions Separate serum within 48 hours and refrigerate. Separated serum stable 1 week at 25°C.

Reference Interval Values are much higher in the first few weeks of life and fall with age. Approximate intervals are found in the following table.

(Continued)

Thyroxine, Serum (Continued)

Age	μg/dL	SI: nmol/L
1-3 d	11.8-22.6	152-292
1-2 wk	9.8-16.6	126-214
1-4 mo	7.2-14.4	93-186
4-12 mo	7.8-16.5	101-213
1-5 y	7.3-15.0	94-194
5-10 y	6.4-13.3	83-172
10-15 y	5.6-11.7	72-151
≥15 y	5.0-11.0	65-138

Interval is increased in women on birth control pills, owing to increased TBG. Free thyroxine index and free T_4 will still be within the reference interval. Reference interval in pregnancy: approximately 5.5-16.0 mcg/dL (SI: 71-206 nmol/L).

Use Those most likely to benefit from thyroid testing include the elderly, women in the postpartum state 4-8 weeks following delivery, patients with autoimmune disorders, and those with family history of thyroid disease. T_4 is a secondary test, most commonly ordered when an abnormal TSH and/or free T_4 (FT_4) result is found. T_4 values are **decreased** in hypothyroidism, in genetically or acquired disorders accompanied by decreased thyroxine binding globulin (TBG), decreased thyroxine-binding prealbumin, drugs (phenytoin, carbamazepine), and in the third stage of (painful) subacute thyroiditis. T_4 values are **increased** with hyperthyroidism, peripheral resistance to thyroid hormone, drugs (amiodarone, amphetamines), with subacute thyroiditis in its first stage, with thyrotoxicosis due to Graves disease, with increased TBG (pregnancy, genetically increased TBG, acute intermittent porphyria, primary biliary cirrhosis), thyrotoxicosis factitia, and occasionally in euthyroid patients with familial dysalbuminemic hyperthyroxinemia (FDH). Serum T_3 is usually normal in FDH. It is a useful assay in evaluation of apparent irritable bowel syndrome.

Euthyroid hyperthyroxinemia is an expression used as a collective term for all of these nonthyroidal diseases and states which increase thyroxine levels with normal thyroid tissue and metabolism.

Limitations T_4 levels may be abnormal in the euthyroid sick syndrome.

T_4 is less sensitive than TSH in the diagnosis of early primary hypothyroidism or hyperthyroidism.

Anti-T_4 antibodies may exist, interfering with T_4 and free T_4 determinations.

Methodology Radioimmunoassay (RIA), enzyme-linked immunosorbent assay (ELISA), fluorescence polarization immunoassay (FPIA), chemiluminescence immunoassay (CIA)

Additional Information The combination of the serum T_4 and free T_4 or T_3 uptake (an estimate of FT_4I) as an assessment of TBG, helps to determine whether an abnormal T_4 value is due to alterations in serum thyroxine binding globulin or to changes of thyroid hormone levels. The **free thyroid index** is $T_4 \times T_3$ uptake. It is used to evaluate selected patients from sTSH screening. Deviations of both tests in the same direction usually indicate that an abnormal T_4 is due to abnormalities in thyroid hormone. Deviations of the two tests in opposite directions provide evidence that an abnormal T_4 may relate to alterations in TBG. Free T_4 assays are preferred over free thyroid index. See Free Thyroxine Index *on page 613*.

Causes of increased TBG binding include neonatal state, molar and conventional pregnancy, estrogens, oral contraceptives, heroin, methadone, 5-fluorouracil, clofibrate, infectious hepatitis, autoimmune hepatitis and primary biliary cirrhosis, acute intermittent porphyria, lymphoma, and hereditary TBG increase.

Causes of decreased TBG binding include abnormal protein states. These include nephrotic syndrome, androgens, anabolic steroids, prednisone, acromegaly, liver or other systemic illness, severe stress, and hereditary TBG deficiency. Salicylates, T_3, and diphenylhydantoin may lower T_4 significantly, and nicotinic acid appears to do so. Amiodarone may cause increased thyroxine levels and can cause hypothyroidism or hyperthyroidism.

Lithium carbonate may cause goiter with or without hypothyroidism.

Carbamazepine (Tegretol®) and phenytoin are reported to cause decreased values of total thyroxine by displacing thyroxine from its binding proteins.

Relationships of drugs to thyroid testing and physiology have recently been updated. See table in Thyroxine Binding Globulin, Serum *on page 1255*.

Thyroid Tests With Disease and Varying TBG

Diagnosis	T_4	FT_4 (or FT_4I)	TSH
Normal	Normal	Normal	Normal
Hyperthyroidism	Increased	Increased	Decreased
Hypothyroidism	Decreased	Decreased	Increased
Increased TBG	Increased	Normal	Normal
Decreased TBG	Decreased	Normal	Normal

References
Dayan CM, "Interpretation of Thyroid Function Tests," *Lancet*, 2001, 357:619-24.
Klee GG and Hay ID, "Biochemical Testing of Thyroid Function," *Endocrinol Metab Clin North Am*, 1997, 26(4):763-75.

Tissue Typing

Related Information

Synonyms Crossmatch, Lymphocyte; Hematopoietic Stem Cell Transplant; Histocompatibility Testing; HLA Typing; HSC Transplant; Human Leukocyte Antigen; Lymphocyte Crossmatch; Lymphocytotoxic Antibody Screening; Molecular Genetic HLA Typing; Organ Donor Tissue Typing; Tissue Typing, Donor; Transplant Tissue Typing; White Cell Crossmatch

Applies to Lymphocytotoxicity Assay; Major Histocompatibility Complex; MHC

Test Includes Determination of compatibility between recipient and donors for organ or hematopoietic stem cell transplant

Abstract Multiple alloantigenic determinants (epitopes) are found on each HLA molecule, which can be classified as public or private. Donor-specific IgG antibodies detected in the recipient are contraindications for organ transplantation.

Specimen Blood from donor, serum from recipient for kidney or hematopoietic stem cell transplant

Container Donor: green top (heparin) tube or preferably yellow top (ACD) tube; recipient: red top tube; DNA testing: lavender top (EDTA) tube. Consult the reference laboratory which will be used.

Storage Instructions Should be tested immediately. Do **not** refrigerate or freeze unless testing by DNA.

(Continued)

Tissue Typing *(Continued)*

Use Tissue typing aids in determination of compatibility of solid organs (transplanted kidneys, hearts, and lungs) and hematopoietic stem cell transplants. HLA-B27 is addressed in a separate listing (HLA-B27 *on page 738*). DR antigens are found in diseases with immune associations. Addison disease is strongly associated with DR3. Juvenile diabetes mellitus (DR3,4 and DQ2,8), myasthenia gravis, and Graves disease also show association with DR3, as does gluten-sensitive enteropathy. Multiple sclerosis is associated with DR2.

Methodology Lymphocytotoxicity assay, mixed lymphocyte culture (MLC), polymerase chain reaction (PCR), sequence-specific primers (SSP), sequence-specific oligonucleotide probe (SSOP), sequence-based typing (SBT), enzyme-linked immunosorbent assay (ELISA), flow cytometry (FC)

Additional Information HLA antigens are glycoproteins, the product of six closely linked genes on the short arm of chromosome 6, usually inherited as an intact unit (haplotype). HLA antigens are the primary determinants of tissue graft acceptance, and thus, the major histocompatibility complex. This same HLA region contains genes of importance to complement and immune responses. The HLA loci are HLA-A, B, C (class I) and DR, DQ, and DP (class II or HLA-D region). A, B, and C antigens are expressed on nearly all nucleated human cells, D region antigens are restricted to B lymphocytes, monocytes, and endothelial cells. HLA antigens are inherited as two sets of six antigens, one set from each parent and are codominantly expressed. These antigens show linkage disequilibrium, that is certain pairs or triplets occur more frequently than expected by chance.

Tissue typing is usually undertaken to assess the "match" between donor and recipient of an organ or hematopoietic stem cell transplantation. To search for unrelated donors for hematopoietic stem cell transplant (HSCT) through the National Marrow Donor Program (NMDP) registry, typing for class II alleles by DNA-based methods is required. Since these antigens are widely expressed in tissue, mismatches result in graft rejection, or graft-vs-host disease. Estimated renal graft survival at 10 years was 52% for HLA-matched transplants, but 37% for HLA-mismatched organs. The estimated half-lives of the transplants were 12.5 and 8.6 years, respectively, in a recently published study. Other factors are relevant as well. DNA-based methods are gradually replacing serologic typing. The HLA lymphocytotoxic antibody screening is also performed by ELISA or flow methods. Many laboratories also perform crossmatching by flow cytometry.

References

Duquesnoy RJ, "Histocompatibility Testing in Organ Transplantation," *Lab Med*, 1999, 30(12):796-802.

Hurley CK, Baxter-Lowe LA, Logan B, et al, "National Marrow Donor Program HLA-Matching Guidelines for Unrelated Marrow Transplants," *Blood Marrow Transplant*, 2003, 9(10):610-5.

Klein J and Sato A, "The HLA System. First of Two Parts," *N Engl J Med*, 2000, 343(10):702-9.

Klein J and Sato A, "The HLA System. Second of Two Parts," *N Engl J Med*, 2000, 343(11):782-6.

Takemoto SK, Terasaki PI, Gjertson DW, et al, "Twelve Years' Experience With National Sharing of HLA-Matched Cadaveric Kidneys for Transplantation," *N Engl J Med*, 2000, 343(15):1078-84.

♦ **Tissue Typing, Donor** *see* Tissue Typing *on page 1259*

♦ **T Lymphocytes** *see* Immunoperoxidase Procedures *on page 780*

♦ **tMg$_s$ (Total Magnesium, Serum)** *see* Magnesium, Serum *on page 885*

♦ **tMg (Total Magnesium)** *see* Magnesium, Serum *on page 885*

♦ **TnI (Troponin I) (cTnI)** *see* Troponins, Serum *on page 1278*

♦ **TnT (Troponin T) (cTnT)** *see* Troponins, Serum *on page 1278*

♦ **Tobramycin, Serum** *see* Aminoglycosides, Serum *on page 147*

Tocainide, Serum or Plasma

Related Information

Lidocaine, Serum or Plasma *on page 850*

Synonyms Tonocard®; Xylotocan®

Abstract Tocainide is a class 1B antiarrhythmic agent closely related to lidocaine, but dissimilar from quinidine, procainamide, and disopyramide. In contrast to lidocaine, tocainide undergoes negligible first pass metabolism.

Specimen Serum or plasma

Container Red top tube or green top (heparin) tube; avoid serum separator tube

Sampling Time Peak: 1-1.5 hours after administration; trough: just before next dose

Reference Interval Therapeutic: 6-15 µg/mL (SI: 32-78 µmol/L)

Possible Panic Range >15 µg/mL (SI: >78 µmol/L)

Use Therapeutic monitoring and toxicity assessment

Methodology High performance liquid chromatography (HPLC), gas chromatography (GC)

Additional Information
- Half-life: 10-15 hours
- Volume of distribution: 2-4 L/kg
- Protein binding: 10% to 20%
- Bioavailability: ~100%

Adverse effects of tocainide are mainly neurological (faintness, tremor) and following overdose coma, seizures, edema and respiratory arrest can occur. Patients receiving recommended dosage can develop agranulocytosis, bone marrow depression, leukopenia, and aplastic anemia. Metabolites are inactive.

Tocainide is eliminated through the kidneys. Patients with renal insufficiency have prolonged half-life of tocainide.

References

Moyer TP, "Therapeutic Drug Monitoring," *Tietz Textbook of Clinical Chemistry*, 3rd ed, Burtis CA and Ashwood ER, eds, Philadelphia, PA: WB Saunders Co, 1999, 885.

Roden DM and Woosley RL, "Drug Therapy. Tocainide," *N Engl J Med*, 1986, 315(1):41-4.

- ◆ **Tocopherol** *see* Vitamin E, Serum or Plasma *on page 1319*
- ◆ **α-Tocopherol** *see* Vitamin E, Serum or Plasma *on page 1319*
- ◆ **Tofranil**® *see* Antidepressants, Cyclic, Serum or Plasma *on page 171*
- ◆ **Tofranil**® *see* Imipramine, Serum or Plasma *on page 767*
- ◆ **Tofranil-PM**® *see* Imipramine, Serum or Plasma *on page 767*
- ◆ **Tolbutamide Test** *see* Glucose, Fasting, Plasma *on page 643*
- ◆ **Tolerance Test, Lactose** *see* Lactose Tolerance Test *on page 829*
- ◆ **Tonocard**® *see* Tocainide, Serum or Plasma *on page 1260*
- ◆ **Toot** *see* Cocaine (Cocaine Metabolite), Qualitative, Urine or Hair *on page 427*
- ◆ **Topamax**® *see* Antiepileptic Drugs Overview *on page 176*
- ◆ **Topiramate** *see* Antiepileptic Drugs Overview *on page 176*

Topoisomerase I Antibody

Related Information

Anti-DNA *on page 173*
Antinuclear Antibodies *on page 189*
Centromere/Kinetochore Antibody *on page 354*
Jo-1 Antibody *on page 815*
Kidney Biopsy *on page 818*
Sjögren Antibodies *on page 1199*
Skin Biopsy *on page 1200*
Smith (Sm) and Ribonucleoprotein (RNP) Antibodies *on page 1206*

Synonyms Progressive Systemic Sclerosis Antibody; Scl-70 Antibody; Scleroderma Antibody

Applies to Anti-Ku; Anti-PM-1; Anti-PM-Scl

Abstract Scl-70 is an antigen present on DNA topoisomerase I, which is the nuclear enzyme responsible for twisting and untwisting the DNA helix during gene transcription. Antibodies directed against topoisomerase I usually give a nucleolar or speckled, positive ANA pattern or HEp-2 cells.

Progressive systemic sclerosis (diffuse scleroderma) (PSS) is a multisystem disease which includes sclerosis (fibrosis) of skin, gastrointestinal tract, lungs, vessels, heart, and renal parenchyma. Scleroderma may be localized. Topoisomerase I antibody (Scl-70 antibody) is often related to more widespread skin and internal organ disease than is anticentromere antibody. Both antibodies are useful to investigate the differential diagnosis of scleroderma. See Centromere/Kinetochore Antibody *on page 354*. The clinical differential diagnosis of scleroderma includes scleredema, eosinophilic fasciitis, scleromyxedema, as well as CREST syndrome, mixed connective tissue disease, and related entities. Scleroderma overlap syndromes are recognized disorders. Scleroderma is

(Continued)

Topoisomerase I Antibody (Continued)

among the disorders called connective tissue diseases, or systemic rheumatic diseases.

Specimen Serum

Container Red top tube

Storage Instructions Refrigerate separated serum.

Reference Interval Negative; titers are not usually performed.

Use Autoantibodies to saline ENA type I topoisomerase are specific for scleroderma (progressive systemic sclerosis). The antibody is found in up to 25% of individuals who have idiopathic Raynaud phenomenon.

Limitations Absence of scleroderma antibody does not exclude diagnosis either of PSS or of CREST syndrome. Sensitivity for the diagnosis of PSS is 34%, specificity >98% to 99%.

Methodology Indirect fluorescent antibody (IFA), immunodiffusion (ID), immunoblotting, enzyme immunoassay (EIA)

Additional Information Approximately 25% to 40% of patients with diffuse scleroderma have autoantibodies to Scl-70, mainly IgG. Patients with scleroderma whose anti-Scl-70 autoantibody result is positive are more likely to have more severe visceral disease, particularly pulmonary fibrosis. Scleroderma patients with anti-Scl-70 antibodies tend to have more widespread skin disease as well. Anti-Scl-70 and anticentromere antibodies seldom coexist in the same individual. Of patients with scleroderma, 60% to 80% have positive ANA.

Other autoantibodies associated with scleroderma include anti-PM-1, anti-Ku, and anti-PM-Scl.

References

Homburger HA, "Advances in the Diagnosis and Laboratory Evaluation of Systemic Rheumatic Diseases Other Than Rheumatoid Arthritis," *Clinical and Laboratory Evaluation of Human Autoimmune Diseases*, Chapter 11, Nakamura RA, Keren DF, and Bylund DJ, eds, Chicago, IL: American Society for Clinical Pathology, 2002, 153-64.

Kavanaugh A, Tomar R, Reveille J, et al, "Guidelines for Clinical Use of the Antinuclear Antibody Test and Tests for Specific Autoantibodies to Nuclear Antigens," *Arch Pathol Lab Med*, 2000, 124(1):71-81.

Landzberg MJ, Roberts DJ, and Mark EJ, "A 38-Year-Old Woman With Increasing Pulmonary Hypertension After Delivery," Case Records of the Massachusetts General Hospital, Case 4-1999, Scully RE, Mark EJ, McNeely WF, et al, eds, *N Engl J Med*, 1999, 340(6):455-64.

Moder KG, "Use and Interpretation of Rheumatologic Tests: A Guide for Clinicians," *Mayo Clin Proc*, 1996, 71(4):391-6.

von Mühlen CA and Nakamura RM, "Guidelines for Selecting and Using Laboratory Tests for Autoantibodies to Nuclear, Nucleolar, and Other Related Cytoplasmic Antigens," *Clinical and Laboratory Evaluation of Human Autoimmune Diseases*, Chapter 13, Nakamura RA, Keren DF, and Bylund DJ, eds, Chicago, IL: American Society for Clinical Pathology, 2002, 183-98.

TORCH

Related Information

Cytomegalovirus Serology *on page 499*
Herpes Simplex Antibody *on page 720*
Rubella Culture and Serology *on page 1174*
Toxoplasmosis Diagnostic Procedures *on page 1265*

Test Includes Detection of antibody specific for toxoplasmosis, rubella, cytomegalovirus, and herpes simplex virus

Abstract Although the acronym TORCH serves to enhance awareness of congenital infections, the disease entities must be considered separately rather than collectively. The entities are **TO**xoplasmosis, **R**ubella, **C**ytomegalovirus, **H**erpes.

Specimen Serum

Container Red top tube

Collection Paired specimens drawn 2-4 weeks apart or a single sample for monitoring immunity in pregnant females

Reference Interval IgG: less than a fourfold increase in titer; IgM: negative

Use Screening evaluation for possible congenital infection due to toxoplasmosis, rubella, cytomegalovirus, and herpesvirus infection

Limitations The presence of IgG antibodies in maternal blood indicates prior exposure to that specific agent through natural infection or immunization (rubella). Since the prevalence of antibodies to HSV and CMV is very high in the general population, a single positive IgG antibody to these agents is of little diagnostic value regardless of titer. Fetal or neonatal IgG antibodies merely indicate transplacental transfer of maternal antibodies. Negative serologic results do not exclude infection.

Methodology Indirect fluorescent antibody (IFA), enzyme-linked immunosorbent assay (ELISA), IgG and IgM specificity

Additional Information *Toxoplasma*, rubella, cytomegalovirus, and herpes simplex are all causes of congenital infections, which can be quickly fatal or lead to chronic sequelae.

In the fulminant case serologic diagnosis is of little use. The disease outstrips the immune response and even IgM antibody cannot be demonstrated in time to be clinically useful.

However, in the disease that becomes manifest weeks to months after birth, demonstration of IgM antibody or rising titers of IgG antibody can confirm a diagnosis of specific infection. The presence of IgM-specific antibody in cord, fetal, or neonatal blood indicates congenital infection. **It should be emphasized that TORCH testing is of very limited usefulness. Results must be interpreted in conjunction with complete clinical information, and such testing in no way substitutes for careful clinical examination and judgment.**

References
Crino JP, "Ultrasound and Fetal Diagnosis of Perinatal Infection," *Clin Obstet Gynecol*, 1999, 42(1):71-80.

Cullen A, Brown S, Cafferkey M, et al, "Current Use of the TORCH Screen in the Diagnosis of Congenital Infection," *J Infect*, 1998, 36(2):185-8.

Greenough A, "The TORCH Screen and Intrauterine Infections," *Arch Dis Child*, 1994, 70(3 Spec No):163-5

Helfgott A, "TORCH Testing in HIV-infected Women," *Clin Obstet Gynecol*, 1999, 42(1):149-62.

Newton ER, "Diagnosis of Perinatal TORCH Infections," *Clin Obstet Gynecol*, 1999, 42(1):59-70.

- ◆ **Total Antioxidant Capacity** *see* Antioxidant Concentrations, Plasma *on page 192*
- ◆ **Total Bilirubin, Neonatal** *see* Bilirubin, Neonatal, Serum *on page 263*
- ◆ **Total Calcium, Serum** *see* Calcium, Serum *on page 329*
- ◆ **Total Metanephrines** *see* Metanephrines, Urine or Plasma *on page 899*
- ◆ **Total Protein, Serum** *see* Protein, Total, Serum *on page 1114*
- ◆ **Total Radical Absorbin Parameter** *see* Antioxidant Concentrations, Plasma *on page 192*
- ◆ **Total T₃** *see* Triiodothyronine, Serum *on page 1276*
- ◆ **Total Urinary Catecholamines** *see* Catecholamines, Fractionation, Urine *on page 347*
- ◆ **Tourniquets** *see page 11*
- ◆ **Toxic Metals, Blood** *see* Heavy Metal Screen, Blood *on page 668*
- ◆ **Toxic Metals, Urine** *see* Heavy Metal Screen, Urine *on page 669*
- ◆ **Toxicology/Drugs of Abuse Introduction** *see page 63*

Toxicology Screen, Serum or Plasma

Related Information

Toxicology Screen, Urine *on page 1264*

Synonyms Drug Screen, Comprehensive Drug Panel or Analysis

Test Includes Acetaminophen (Tylenol®); acetone; alcohol (ethyl, methyl, isopropyl); amitriptyline (Elavil®); amobarbital (Amytal®); butabarbital (Butisol®); butalbital (Fiorinal®); caffeine; carbamazepine (Tegretol®); carisoprodol (Soma®); desipramine (Norpramin®); diltiazem (Cardizem®); diphenhydramine (Benadryl®); doxepin (Sinequan®); ephedrine; fluoxetine (Prozac®); glutethimide (Doriden®); ibuprofen; imipramine (Tofranil®); mephobarbital (Mebaral®); meprobamate (Equanil®) methaqualone (Quaalude®); nicotine; nortriptyline (Aventyl®); pentobarbital (Luminal®); phenobarbital; phenytoin (Dilantin®); salicylate (Aspirin®); secobarbital (Seconal™); theophylline (Theo-Dur®); thiopental (Pentithal®); tricyclic antidepressants; varies with the laboratory

Abstract This toxicology screen is carried out by performing individual quantitative tests for each drug or by thin layer chromatography or semiquantitative automated high performance liquid chromatography (Remedi®) or gas chromatography. Many times urine qualitative screening is faster and more useful in toxicologic emergencies, but both may be needed. Immunoassays are also frequently employed due to their fast turnaround time.

Specimen Serum or plasma

(Continued)

Toxicology Screen, Serum or Plasma *(Continued)*

Container Red top tube or lavender top (EDTA) tube

Special Instructions Do **not** collect blood in heparinized tubes.

Reference Interval See individual drug listing for therapeutic and toxic ranges.

Use Monitor toxic/overdose situations; most desirable to analyze in conjunction with urine toxicology testing

Methodology Spot tests, immunoassay, thin-layer chromatography (TLC), gas chromatography (GC), high performance liquid chromatography (HPLC)

Additional Information If only documentation of exposure to toxic drugs or drugs of abuse is desired, a urine drug screen is the most economical approach. See Toxicology Screen, Urine *on page 1264.* When Toxicology Screen, Serum or Plasma is ordered, the individual drugs are quantitated in serum. When Toxicology Screen, Urine is ordered, qualitative identification is carried out.

References

Giorgi DF and Jagoda A, "Poisoning and Overdose," *Mt Sinai J Med,* 1997, 64(4-5):283-91.

Litovitz TL, Klein-Schwartz W, White S, et al, "1999 Annual Report of the American Association of Poison Control Centers Toxic Exposure Surveillance System," *Am J Emerg Med,* 2000, 18(5):517-74.

Vernon DD and Gleich MC, "Poisoning and Drug Overdose," *Crit Care Clin,* 1997, 13(3):647-67.

Toxicology Screen, Urine

Related Information

Toxicology Screen, Serum or Plasma *on page 1263*

Synonyms Drug Screen, Comprehensive Panel or Analysis, Urine

Test Includes A variety of qualitative screens are in use. Sensitivity and specificity vary and are method dependent.

Acetaminophen (Tylenol®); acetone; alcohol (ethyl, methyl, isopropyl); amitriptyline (Elavil®); amobarbital (Amytal®); amphetamines; benzodiazepines; benztropine (Cogentin®); butabarbital (Butisol®); butalbital (Fiorinal®); caffeine; carbamazepine (Tegretol®); carisoprodol (Soma®); cocaine (as metabolite); codeine; cyclobenzaprine (Flexeril®); desipramine (Norpramin®); dextromethorphan (DN®); diltiazem (Cardizem®); diphenhydramine (Benadryl®); doxepin (Sinequan®); ephedrine; fluoxetine (Prozac®); glutethimide (Doriden®); heroin (as monoacetylmorphine); ibuprofen; imipramine (Tofranil®); lidocaine (Xylocaine®); loxapine (Loxitane®); maprotiline (Ludiomil®); meperidine (Demerol®); mephobarbital (Mebaral®); meprobamate (Equanil®); methadone; methamphetamine; methaqualone (Quaalude®); morphine; naproxen (Naprosyn®); nicotine; nortriptyline (Aventyl®); oxycodone (Percodan®); pentazocine (Talwin®); pentobarbital (Nembutal®); phencyclidine (PCP®); phenobarbital (Luminal®); phenylpropanolamine, phenytoin (Dilantin®); propoxyphene (Darvon®); propranolol; protriptyline (Vivactyl®); pseudoephedrine; salicylate (Aspirin®); secobarbital (Seconal™); sertraline (Zoloft®); THC metabolite (marijuana); theophylline (Theo-Dur®); thiopental (Pentothal®); trazodone (Desyrel®); trimipramine (Surmontil®) are generally included in the urine drug screen

Abstract This is a qualitative screen which with thin-layer chromatography, gas chromatography, or automated high performance liquid chromatography can detect any of several hundred drugs (Remedi®). Immunoassays are generally also used in conjunction with other methods.

Specimen Random urine

Storage Instructions Keep refrigerated.

Special Instructions Specify the drug or drugs suspected.

Reference Interval None detected or negative (less than cutoff for drugs of abuse)

Use Screen for drug abuse, drug toxicity alone or in conjunction with serum/plasma testing. For typical drugs-of-abuse screening, see Drugs of Abuse Testing, Urine *on page 525.*

Limitations Test provides **only** qualitative detection of drugs. Some drugs and/or metabolites are not detected or optimally detected in urine, again relating to method. Sensitivity is of the order of 0.5-1.0 µg/mL for TLC of urine. Newer methods such as GC and HPLC are more sensitive. Some substances should be quantitated in blood or serum (eg, iron overdose, methanol, acetaminophen, salicylate, carbon monoxide, ethanol, digoxin, lithium, theophylline, and methemoglobin).

Methodology Spot tests, immunoassay, thin-layer chromatography (TLC), gas chromatography (GC), high performance liquid chromatography (HPLC)

Additional Information For specific drug blood levels see the listing by specific generic drug name for drug desired. Some toxins (eg, metals, volatiles, gaseous compounds) may require specific methodology (eg, atomic absorption spectrophotometry, gas chromatography). Also see Toxicology Screen, Serum or Plasma *on page 1263*.

References

Giorgi DF and Jagoda A, "Poisoning and Overdose," *Mt Sinai J Med*, 1997, 64(4-5):283-91.

Litovitz TL, Klein-Schwartz W, White S, et al, "1999 Annual Report of the American Association of Poison Control Centers Toxic Exposure Surveillance System," *Am J Emerg Med*, 2000, 18(5):517-74.

Vernon DD and Gleich MC, "Poisoning and Drug Overdose," *Crit Care Clin*, 1997, 13(3):647-67.

♦ **Toxicology, Volatiles** *see* Volatile Screen, Blood or Urine *on page 1320*

♦ **Toxic Reactions, Liver** *see* Liver Biopsy *on page 864*

Toxoplasmosis Diagnostic Procedures

Related Information

Buffy Coat Smear Study of Peripheral Blood *on page 316*

HIV-1/HIV-2 Serology *on page 736*

TORCH *on page 1262*

Test Includes IgG and IgM antibody specific for *Toxoplasma gondii*

Abstract *Toxoplasma gondii* is an obligate intracellular protozoan parasite that infects humans and other mammals. A worldwide distribution exists and human infections are very common. Serologic studies indicate that approximately a third to a half of the U.S. population has been infected with *T. gondii*. Most human infections follow an asymptomatic chronic course. However, severe disseminated infections occur, particularly in immunocompromised hosts.

Toxoplasmosis is especially important in two clinical situations: infections transmitted to the fetus, and as the most common opportunistic infection of the central nervous system in patients with acquired immunodeficiency syndrome (AIDS). In congenital infections, it can cause hydrocephalus, mental retardation, and chorioretinitis. It may be a life-threatening complication of in organ transplantation.

Specimen Serum, cerebrospinal fluid, amniotic fluid, other body fluids, biopsy

Container Red top tube

Collection Acute and convalescent serum specimens are recommended at 3-week interval for serology. Blood contamination of amniotic fluid specimens may lead to false-negative PCR results.

Reference Interval Serology: IgG: less than a fourfold increase in titer; IgM: negative

Use Support the diagnosis of toxoplasmosis; document past exposure and/or immunity to *Toxoplasma gondii*. The absence of IgG is useful: it provides evidence of lack of infection. Painless cervical lymphadenopathy is the most common presentation of symptomatic disease. It may be accompanied by fever. The differential diagnosis includes infectious mononucleosis and lymphoma. Pericarditis or acute chorioretinitis may occur.

Limitations A high prevalence of IgG antibody to *Toxoplasma* is found in most populations. Although a single high titer may suggest active infection, titers may persist at high levels for years in healthy people. Distinction between chronic active, inactive, and past infection may be obscure. Many different tests are available, but false-positive and false-negative results are seen. Serologic responses in the presence of immunodeficiency lack reliability. Polymerase chain reaction (PCR), when positive, provides indication of the presence of *T. gondii* DNA, but negative PCR does not rule out the presence of active disease.

Methodology Serology: indirect fluorescent antibody (IFA), enzyme-linked immunosorbent assay (ELISA). PCR used for prenatal diagnosis of congenital *T. gondii* infection is rapid and accurate with a sensitivity of 81% and a specificity of 96%.

Other methods include biopsy, animal inoculation, and tissue culture. Giemsa staining permits light microscopic identification of organisms.

Additional Information *Toxoplasma gondii* is endemic in cats. Oocysts can be shed in large numbers in cat feces (100,000 per gram of feces) and can survive (Continued)

Toxoplasmosis Diagnostic Procedures *(Continued)*

in the environment for several months. Humans are easily exposed to oocysts, either in caring for pets or in casual environmental contact. The majority of individuals develop antibody without any clinical disease, and a self-limited lymphadenitis is the most common clinical presentation in symptomatic infection.

Congenital toxoplasmosis and infection in immunocompromised hosts (especially AIDS or transplant patients) are more serious, and can produce a fatal cerebritis or disseminated illness.

Blood transfusion is a rare cause of toxoplasmosis. A somewhat greater risk exists with granulocyte transfusion.

References

Cohen BA, "Neurologic Manifestations of Toxoplasmosis in AIDS," *Semin Neurol*, 1999, 19(2):201-11.

Foulon W, Pinon JM, Stray-Pedersen B, et al, "Prenatal Diagnosis of Congenital Toxoplasmosis: A Multicenter Evaluation of Different Diagnostic Parameters," *Am J Obstet Gynecol*, 1999, 181(4):843-7.

Holliman RE, "Toxoplasmosis," *Manson's Tropical Diseases*, 21st ed, Chapter 76, Cook GC and Zumla A, eds, Philadelphia, PA: WB Saunders Co, 2003, 1365-71.

Naessens A, Jenum PA, Pollak A, et al, "Diagnosis of Congenital Toxoplasmosis in the Neonatal Period: A Multicenter Evaluation," *J Pediatr*, 1999, 135(6):714-9.

Robert-Gangneux F, Commerce V, Tourte-Schaefer C, et al, "Performance of a Western Blot Assay to Compare Mother and Newborn Anti-*Toxoplasma* Antibodies for the Early Neonatal Diagnosis of Congenital Toxoplasmosis," *Eur J Clin Microbiol Infect Dis*, 1999, 18(9):648-54.

Zufferey J, Hohlfeld P, Bille J, et al, "Value of the Comparative Enzyme-Linked Immunofiltration Assay for Early Neonatal Diagnosis of Congenital *Toxoplasma* Infection," *Pediatr Infect Dis J*, 1999, 18(11):971-5.

Internet Web Sites

www.amm.co.uk/pubs/fa_toxoplasma.htm
www.astdhpphe.org/infect/toxo.html
www.cdc.gov/ncidod/dpd/parasites/toxoplasmosis/default.htm

♦ **TPA** *see* CA 19-9, Serum *on page 321*

♦ **tPA** *see* Hypercoagulation Panel *on page 758*

♦ **tPA** *see* Plasminogen *on page 1042*

♦ **tPA** *see* Plasminogen Activator Inhibitor 1 *on page 1043*

♦ **T-Quil®** *see* Diazepam, Serum *on page 510*

♦ **TRAb** *see* Thyrotropin Receptor Antibody, Serum *on page 1254*

♦ **TRACE** *see* Fanconi Anemia, Chromosome Breakage Study *on page 569*

♦ **Tranquilizers (Valium®, Librium®, Dalmane®, Tranxene®, Klonopin™, Ativan®, Serax®, Centrax®, Restoril®, Xanax®, Halcion®, Versed®, Doral®, etc)** *see* Benzodiazepines, Qualitative, Urine *on page 253*

♦ **Transaminase** *see* Alanine Aminotransferase, Serum *on page 116*

♦ **Transaminase** *see* Aspartate Aminotransferase, Serum *on page 216*

Transbronchial Fine Needle Aspiration

Related Information

Bronchial Brushings/Washings Cytology *on page 310*
Bronchoalveolar Lavage (BAL) *on page 311*
Fine Needle Aspiration Culture *on page 589*
Fine Needle Aspiration, Deep and Superficial Masses *on page 590*

Abstract Transbronchial fine needle aspiration (FNA) is useful in the evaluation of patients with pulmonary nodules that involve or are adjacent to major bronchi, as well as in sampling hilar and mediastinal lymph nodes in staging of bronchogenic carcinoma.

Specimen Prepared smears may be air-dried or immediately fixed in 95% ethanol. The needle should be rinsed in a balanced salt solution or formalin for processing by cytocentrifugation or cell block preparation.

Collection The specimen is obtained through a fiberoptic bronchoscope using a long flexible tube with an attached needle and following the techniques for fine needle aspiration. Transbronchial FNA should be performed prior to bronchial brushings, washings, or grasp biopsy to minimize contamination by blood and tracheobronchial secretions.

Storage Instructions If a delay in processing must occur, the needle rinse material should be refrigerated.

Use Transbronchial fine needle aspiration is a relatively low-risk procedure used as an adjunct to bronchoscopy, in the evaluation of patients with localized pulmonary disease. Though most useful in establishing a diagnosis of lung carcinoma, this technique may also facilitate the identification of infection (eg, TB, histoplasmosis) and inflammatory processes (eg, sarcoidosis). Transbronchial FNA provides the greatest diagnostic yield when a submucosal mass is present or when extrinsic compression of the bronchi is identified. Lesions which ulcerate the bronchial mucosa may be more accessible to direct forceps tissue biopsy or bronchial brushing. This technique may also be used to sample mediastinal and hilar lymph nodes in the staging of bronchogenic carcinoma.

Limitations A negative result does not exclude malignancy. False-negative results are usually secondary to sampling problems. False-positive results may be due to contaminating bronchial secretions from distal airways or sampling a parenchymal tumor instead of a lymph node. Specimens consisting only of respiratory cells and mucus as well as purported lymph node samples lacking lymphocytes should be considered nondiagnostic.

Contraindications Abnormal coagulation profile, severe hypoxemia, pulmonary hypertension, intrapulmonary vascular lesions, echinococcal cysts.

References

Das DK, "Fine-Needle Aspiration Cytology in the Diagnosis of Tuberculous Lesions," *Lab Med*, 2000, 31(11):625-32.

Johnston WW, "Cytopathology of the Lung: Diagnostic Applications of Sputum, Bronchial Brushings, and Fine Needle Aspiration Biopsy Specimens," *Compendium on Diagnostic Cytology*, 8th ed, Wied GL, Bibbo M, Keebler CM, et al, eds, Chicago, IL: Tutorials of Cytology, 1997, 216-29.

◆ **Transcobalamins** *see* Cobalamin, Serum *on page 424*

◆ **Transcobalamins** *see* Vitamin B$_{12}$ Unsaturated Binding Capacity *on page 1316*

◆ **Transcuprein** *see* Ceruloplasmin, Serum or Plasma *on page 375*

◆ **Transcuprein** *see* Copper, Serum *on page 448*

◆ **Transcutaneous Bilirubinometry** *see* Bilirubin, Neonatal, Serum *on page 263*

◆ **Transferrin** *see* Iron and Total Iron Binding Capacity/Transferrin, Serum *on page 807*

◆ **Transferrin Receptor, Soluble** *see* Iron and Total Iron Binding Capacity/Transferrin, Serum *on page 807*

Transferrin Receptor, Soluble, Serum or Plasma

Related Information

Anemia Flowchart *on page 35*
Ferritin, Serum *on page 577*
Hereditary Hemochromatosis DNA Test *on page 718*
Iron and Total Iron Binding Capacity/Transferrin, Serum *on page 807*
Protoporphyrin, Free Erythrocyte *on page 1121*
Reticulocyte Hemoglobin Content *on page 1158*

Synonyms Soluble Serum Transferrin Receptor; sTfR

Abstract Soluble transferrin receptor (sTfR) has origin from red blood cell precursors (normoblasts) and is found in the circulation. It is a truncated form of the cellular transferrin receptor. As iron deficiency develops, the number of cellular transferrin receptors increases, and this is reflected by an increased plasma concentration of sTfR. The test has application in differentiating iron deficiency anemia (sTfR usually increased) from the anemia of chronic disease (sTfR usually normal). Serum sTfR is also increased in conditions of high turnover erythropoiesis (eg, hemolytic anemia). A ratio of the serum transferrin receptor to serum ferritin provides an estimate of body iron in mg per kg of body weight. The role of sTfR in routine clinical practice is not fully defined.

Specimen Serum or plasma

Container Red top tube, lavender top (EDTA) tube, or green top (heparin) tube

Collection Separate from cells within 30 minutes and freeze.

Storage Instructions Specimen is stable for 30 minutes at ambient temperature and indefinitely when frozen. Avoid repeated freeze/thaw cycles.

Causes for Rejection Severe hemolysis, icterus, or lipemia

Reference Interval Check with the testing laboratory as reference intervals and units vary greatly due to intermethod standardization differences (eg, purified tissue receptor vs transferrin receptor complex vs purified soluble serum receptor).

(Continued)

Transferrin Receptor, Soluble, Serum or Plasma
(Continued)

An overall reference interval of 9.6-29.6 nmol/L is reported from a study of 225 healthy, adult subjects without hematological abnormalities (Allen et al). There were no differences relating to age or sex. However, black subjects and individuals living at high altitude had higher sTfR levels than comparable groups of nonblack subjects and sea level inhabitants.

Reference interval as provided with a commercially available method "Quantikine sTfR," R&D systems, Minneapolis, MN is 8.8-28.1 nmol/L. Reference interval provided with another commercially available method "The Ramco TfR," Ramco Laboratories, Houston, TX is 2.9-8.3 µg/mL.

Use This relatively new test is proposed as a sensitive, early indicator of iron deficiency.

Evaluation of iron restricted erythropoiesis. Guide to need for intravenous administration of iron to ensure adequate iron replacement. Of use in prediction of the erythropoietic response to increased erythropoietin dosage (in conjunction with use of serum ferritin and reticulocyte hemoglobin content). Assist in monitoring response to erythropoietin therapy.

The ultimate usefulness of this test may be limited by the large number of conditions, other than iron deficiency, in which elevated values are obtained (see Limitations).

Limitations The differential diagnosis of increased sTfR includes autoimmune hemolytic anemia, recent blood donation, recent blood loss, sickle cell anemia, hereditary spherocytosis, β-thalassemia, α-thalassemia, polycythemia vera, and other hematological malignancies, vitamin B_{12} deficiency, and folic acid deficiency. This broad differential indicates lack of specificity if the test is utilized for general hematologic diagnosis.

Methodology Enzyme-linked immunosorbent assay (ELISA)

Additional Information Transferrin receptor was isolated from the serum in 1990 using monoclonal antibodies and immunoaffinity chromatography after report of its initial detection in 1986. The serum receptor, smaller than its cell membrane counterpart, is apparently a soluble fragment of the complete molecule, missing the first 100 amino acid residues, and resulting from proteolytic cleavage by a serine protease.

In iron overload states, sTfR values are decreased. The clinical utility of this observation is not known.

The measurement of sTfR to assess anemia in rheumatoid arthritis and other inflammatory conditions is superior to traditional tests (eg, ferritin, transferrin, TIBC), because sTfR is unaffected by the acute-phase response. TfR's ability to discriminate between iron deficiency anemia (IDA) and anemia of chronic disease (ACD) (comparing serum iron concentration, total iron binding capacity, and % transferrin saturation). Statistical study, including receiver operating characteristic curve analysis, showed that ability to discriminate decreased in the order: TIBC > TfR > MCV > %TS = RDW > SIC. Area under the curve values for TIBC and TfR were not significantly different. TIBC and TfR measures provide the highest (and similar) ability to discriminate between IDA and ACD.

Iron stores (mg/kg) are predicted by the ratio of serum transferrin receptor to serum ferritin.

References

Ahluwalia N, Skikne BS, Savin V, et al, "Markers of Masked Iron Deficiency and Effectiveness of EPO Therapy in Chronic Renal Failure," *Am J Kidney Dis*, 1997, 30(4):532-41.

Cook JD, "The Measurement of Serum Transferrin Receptor," *Am J Med Sci*, 1999, 318(4):269-76.

Cook JD, Flowers CH, and Skikne BS, "The Quantitative Assessment of Body Iron," *Blood*, 2003, 101(9):3359-64.

Goodnough LT, Skikne B, and Brugnara C, "Erythropoietin, Iron, and Erythropoiesis," *Blood*, 2000, 96(3):823-33.

Skikne BS, "Circulating Transferrin Receptor Assay-Coming of Age," *Clin Chem*, 1998, 44(1):7-9.

♦ **Transforming Growth Factor** β see TGF-β, Serum on page 1241

♦ **Transfusion, Autologous** see Autologous Transfusion, Preoperative Deposit on page 223

♦ **Transfusion Complications** see Risks of Transfusion on page 1166

- **Transfusion Reaction, Allergic** *see* Red Blood Cells, Washed *on page 1143*
- **Transfusion Reaction, Febrile** *see* Filters for Blood *on page 588*

Transfusion Reaction Work-up

Related Information

Synonyms Adverse Reactions to Transfusion

Test Includes

Suspected acute hemolytic transfusion reaction work-up: Clerical check (label on blood containers and all records examined for error in identification); postreaction patient specimen - perform direct antiglobulin test and examine for presence of hemolysis in serum/plasma. Compare to pretransfusion sample if present. Repeat serologic testing (ABO group and/or Rh type on patient and donor blood; crossmatch) as needed. Analyze urine for hemoglobinuria (intact red cells will pellet out of a centrifuged urine specimen - free hemoglobin will not; a dipstick does not differentiate between intact red cells and hemoglobin).

Delayed hemolytic transfusion reaction work-up: Direct antiglobulin test, eluate and antibody identification. Bilirubin may be elevated and hematocrit may be declining.

Transfusion-related acute lung injury (TRALI): Anti-HLA (Class I and Class II) and/or antineutrophil antibody identification (from donor and recipient samples).

Anaphylactic/severe allergic: Evaluate for the presence of anti-IgA in an IgA deficient patient.

Platelet transfusion refractoriness, post-transfusion purpura, and problems of **alloimmunization** to platelets are addressed in the platelet monographs listed above under Related Information.

Use of leukocyte-depleted fractions almost eliminates alloimmunization among subjects without prior WBC exposure.

Abstract Any adverse reaction event experienced by a patient in association with a transfusion is regarded as a suspected transfusion complication until proven otherwise. Each transfusion service must have a system in place for detecting, reporting, and evaluating these complications.

Aftercare

If a suspected acute transfusion reaction occurs:

1. Stop transfusion immediately.
2. Verify that correct unit was given to the correct patient.
3. Maintain I.V. access. Ensure adequate urine output with an appropriate crystalloid or colloid solution.
4. Maintain blood pressure, pulse, adequate ventilation.
5. Notify Blood Bank and attending physician.

(Continued)

Transfusion Reaction Work-up *(Continued)*

6. If sepsis is suspected, obtain a blood culture from the patient, and treat patient as needed.

7. Send the following to the Blood Bank for transfusion reaction work-up: blood bag and administration set and blood/urine samples.

8. Blood Bank performs work-up as follows:

   a. Clerical check of paperwork to ensure correct unit transfused to right patient.

   b. Perform direct antiglobulin test.

   c. Evaluate serum/plasma for presence of hemoglobinemia.

   d. Repeat other serologic testing as needed (ABO, Rh, antibody screen, crossmatch).

   e. Examine urine for hemoglobinuria.

   f. If sepsis is suspected, culture the unit and examine a Gram-stained smear from the unit.

If intravascular hemolytic transfusion reaction is confirmed:

9. Monitor renal status (BUN, creatinine).

10. Initiate a diuresis.

11. Monitor coagulation status (prothrombin time, activated partial thromboplastin time, platelet count).

12. Monitor for signs of hemolysis (lactate dehydrogenase, bilirubin, haptoglobin).

Specimen Blood, urine

Container Red top tube, lavender top (EDTA) tube, and plastic urine container

Causes for Rejection (Of patient sample): Specimen tube not properly labeled, sample placed in serum separator tube

Turnaround Time Examination of pretransfusion and current serum or plasma for hemolysis can be done very rapidly. A repeat ABO and Rh, direct antiglobulin test, and clerical check can be done in minutes.

Special Instructions If the only adverse event noted is mild urticaria, the transfusion may be temporarily interrupted and antihistamine administered (eg, diphenhydramine 25-50 mg). If symptoms promptly subside, the transfusion may be resumed. This **does not** apply to the anaphylactic-type reactions presenting with vasomotor instability or to a suspected acute hemolytic transfusion reaction. Mild allergic reactions, as well as circulatory overload, need not be evaluated as possible hemolytic transfusion reactions.

Use Investigate cause of possible transfusion reactions. Some signs and symptoms are noted as follows:

- Acute hemolytic transfusion reaction: fever (1°C or 2°F) and/or chills, hemoglobinuria, renal failure, hypotension, DIC, back pain
- Febrile nonhemolytic transfusion reaction: fever (1°C or 2°F) and/or chills, headache, malaise
- Allergic: pruritus, urticaria, flushing
- Anaphylactic: urticaria, respiratory distress, hypotension, laryngeal edema.
- TRALI: marked respiratory distress, fever (1°C or 2°F), hypoxia, hypotension, bilateral pulmonary edema
- Bacterial contamination: high fever, rigors, shock, hypotension

Limitations Two of the most common complications of transfusion are urticaria and fever. Fever (with or without chills) most frequently indicates a febrile nonhemolytic transfusion reaction (FNHTR). FNHTR is caused by either recipient antibody to donor leukocytes or to accumulated cytokines in the transfused unit. Unfortunately, signs and symptoms of this reaction mimic an acute hemolytic transfusion reaction, and a hemolytic transfusion reaction work-up will help differentiate the two.

Faulty blood warming blood apparatus, inappropriate storage of blood components before infusion or inappropriate medication added to blood are other causes of adverse effects of transfusion. These are not always detected by the transfusion reaction work-up.

Additional Information Most fatal transfusion reactions involve clerical (labeling) error with resultant subsequent incompatibility in the ABO system, and intravascular hemolysis. Fatal transfusion reactions must be reported to the Food and Drug Administration (FDA).

Delayed hemolytic transfusion reactions may not be considered in a differential diagnosis. They occur in patients with unexplained anemia and fever and may be detected by a positive direct antiglobulin test or by the detection of an unexplained blood group antibody which was absent when the type and screen was previously performed.

Transfusion-transmitted diseases: Although blood is tested for hepatitis B, HIV, HTLV, hepatitis C, and West Nile virus, transmission of these diseases can still occur through transfusion. Any cases of suspected transfusion-transmitted disease should be reported to the Blood Bank to identify infectious donors. Other agents known to be transmitted include cytomegalovirus, malaria, babesiosis, Lyme disease, and Chagas disease. See Risks of Transfusion *on page 1166.*

References

Mollison PL, Engelfriet CP, and Contreras M, *Blood Transfusion in Clinical Medicine*, 10th ed, Chapters 10 and 11, Oxford, UK: Blackwell Scientific Publications, 1997.

Triulzi DJ, *Blood Transfusion Therapy: A Physician's Handbook*, 7th ed, Bethesda, MD: American Association of Blood Banks Press, 2002, 109-36.

♦ **Transglutinase Antibody (Anti-tTG)** *see* Endomysial Antibodies *on page 537*

♦ **Transjugular Needle Biopsy of Liver** *see* Liver Biopsy *on page 864*

♦ **Translocations, Chromosomal** *see* Chromosomal Translocations, Molecular Detection *on page 404*

♦ **Transplant, Bone Marrow** *see* Hematopoietic Progenitor Cells, Marrow *on page 677*

♦ **Transplant, Cord Blood** *see* Hematopoietic Progenitor Cells, Cord Blood/Placental Blood *on page 676*

♦ **Transplant Tissue Typing** *see* Tissue Typing *on page 1259*

♦ **Transthyretin** *see* Thyroxine Binding Globulin, Serum *on page 1255*

Transthyretin, Serum

Related Information

C-Reactive Protein, Serum *on page 467*

Synonyms Prealbumin; Thyroxine-Binding Prealbumin

Abstract Transthyretin is a serum protein commonly called **prealbumin**. It functions as a transport protein for thyroxine, triiodothyronine, and the vitamin A/retinol binding complex, and it is a negative acute phase reactant. Decreased serum levels also occur as a result of diminished protein intake, inflammation, malignancy, and other chronic diseases. Increased levels may occur as a result of decreased renal function, transthyretin-producing tumor, or Hodgkin disease.

Patient Preparation Adults should fast overnight prior to specimen collection.

Specimen Serum or plasma may be used depending upon assay methodology.

Storage Instructions Specimen may be stored at 4°C for up to 72 hours, at -20°C for up to 6 months, and indefinitely at -70°C.

Causes for Rejection Lipemia or hemolysis depending upon assay methodology

Reference Interval

- Cord blood: 8.1-18.7 mg/dL
- Newborns: 4-19 mg/dL
- Adults: 12-50 mg/dL

Use Transthyretin is used primarily as an indicator of nutritional status. Of the several circulating proteins sometimes used for this purpose (eg, albumin, transferrin, fibronectin, retinol-binding protein), transthyretin appears to be one of the best in terms of diagnostic sensitivity and specificity, largely due to its intermediate half-life (~48 hours) and relative rapid response to changes in protein/energy homeostasis.

Concentrations in adults <11 mg/dL (SI: 110 mg/L) are considered high risk, requiring major nutritional intervention. Moderate risk includes concentrations in the range of 11-17 mg/dL (SI: 110-170 mg/L), and little to no risk is associated with values >17 mg/dL (SI: >170 mg/L).

Limitations Transthyretin values are affected by conditions that alter its rate of synthesis, utilization, and excretion. Thus, decreased serum levels occur as a result of inflammation, malignancy, liver disease, and kidney disease, and increased levels may occur as a result of compromised renal function. The (Continued)

Transthyretin, Serum (Continued)

measurement of C-reactive protein together with transthyretin is sometimes used to rule in or out the presence of concomitant inflammation.

Methodology Rate immunonephelometry, particle-enhanced turbidimetric immunoassay, turbidimetric immunoassay, enzyme immunoassay (EIA), electroimmunoassay, and radial immunodiffusion (RID)

References

Bernstein LH, Leukhardt-Fairfield CJ, Pleban W, et al, "Usefulness of Data on Albumin and Prealbumin Concentrations in Determining Effectiveness of Nutritional Support," *Clin Chem*, 1989, 35(2):271-4.

Lee C, Berrett PG, and Richmond SA, "Determination of Serum Prealbumin (Transthyretin) by Competitive Enzyme Immunoassay," *Lab Med*, 1992, 23(7):473-6.

Raubenstine DA, Ballantine TVN, Greecher CP, et al, "Neonatal Serum Protein Levels as Indicators of Nutritional Status: Normal Values and Correlation With Anthropometric Data," *J Pediatr Gastroenterol Nutr*, 1990, 10(1):53-61.

◆ **Trazodone** see Antidepressants, Cyclic, Serum or Plasma on page 171

Trazodone, Serum or Plasma

Related Information

Antidepressants, Cyclic, Serum or Plasma on page 171

Synonyms Deprax®; Desyrel®; Molipaxin®; Thrombran®; Trittico®

Abstract This drug is an antidepressant chemically unrelated to the tricyclic or tetracyclic antidepressants. The mechanism of action involves inhibition of reuptake of serotonin and norepinephrine.

Specimen Serum or plasma

Container Red top tube or green top (heparin) tube

Sampling Time Trough: at steady state (20-40 hours)

Reference Interval Therapeutic: 800-1600 ng/mL

Critical Values >5000 ng/mL

Use Therapeutic monitoring and toxicity assessment

Methodology High performance liquid chromatography (HPLC), gas chromatography (GC)

Additional Information

- Half-life: 4-8 hours
- Volume of distribution: 0.9-1.5 L/kg
- Protein binding: 85% to 95%

Trazodone is a structurally unique antidepressant that is pharmacologically different from other antidepressants. The toxicities observed in tricyclic overdose (neuro- and cardiotoxicity and respiratory depression) are not commonly seen with trazodone. Chronic toxicity is very low with trazodone although it does have unique side effects including akathisia, allergic reactions, chest pain, delayed urine flow, early and delayed menses, hypersalivation, hypomania, and priapism, among others. Acute toxicity causes seizures and hyponatremia. Concurrent use of monoamine oxidase inhibitors and selective serotonin reuptake inhibitors may lead to serotonin excess, causing drowsiness, vomiting, hypotension, tachycardia, incontinence, and coma.

References

Haria M, Fitton A, and McTavish D, "Trazodone. A Review of its Pharmacology, Therapeutic Use in Depression and Therapeutic Potential in Other Disorders," *Drugs Aging*, 1994, 4(4):331-55.

Lacy CF, Armstrong LL, Goldman MP, et al, *Drug Information Handbook*, 12th ed, Hudson, OH: Lexi-Comp Inc, 2004.

Rotzinger S, Bourin M, Akimoto Y, et al, "Metabolism of Some "Second"- and "Fourth"-Generation Antidepressants: Iprindole, Viloxazine, Bupropion, Mianserin, Maprotiline, Trazodone, Nefazodone, and Venlafaxine," *Cell Mol Neurobiol*, 1999, 19(4):427-42.

◆ **Trendar®** see Ibuprofen, Serum on page 764

◆ **TRH Stimulation Test** see Thyroid Stimulating Hormone, Serum on page 1250

◆ **Triacylglycerol Acylhydrolase** see Lipase, Serum on page 851

◆ **Triacylglycerols** see Triglycerides, Serum or Plasma on page 1275

◆ **Triadapin®** see Doxepin, Serum or Plasma on page 524

◆ **Triavil®** see Amitriptyline, Serum or Plasma on page 149

Trichinosis Diagnostic Procedures

Related Information

Ova and Parasites, Direct Exam on page 985

Abstract The nematode, *Trichinella spiralis*, the agent of human trichinosis, is a parasite of carnivores. There is little host specificity and cysts have been detected in such diverse products as pork and meat of bear, walrus, badger, and other animals. Ingestion of undercooked meat products containing the encysted larvae causes human infections.

Specimen Serum

Container Red top tube

Sampling Time Serology: Measurable antibody titers are not usually reached until 2-3 weeks after infestation. Paired specimens have been recommended.

Reference Interval Serology: negative; <1:16 by ELISA; less than a fourfold increase in paired titer; dependent upon laboratory and method

Use Ingestion of undercooked pork, or other animal meat containing larvae, may lead to myalgias, fever, myocardial infestation, neurological symptoms, and peripheral blood eosinophilia.

Limitations Low serologic titers may represent antibody from previous rather than current infection. Antibody remains detectable for 2-3 years. The test may have a high false-negative rate of 15% to 22% during the first period of the infection. Sensitivity and specificity of testing depends on the type of antigen used (excretory/secretory or crude somatic antigens).

Methodology A variety of techniques is available to test for antibodies. Muscle biopsy establishes the diagnosis when positive.

Additional Information Antibody becomes detectable 3 weeks after infection, reaches peak concentrations at ~3 months, and then declines slowly so that most individuals will test negative 2-3 years after the initial infection.

ESR may be increased with infestation.

References

Gilles HM, "Soil-Transmitted Helminth's (Geohelminths)," *Manson's Tropical Diseases*, 21st ed, Chapter 83, Cook GC and Zumla A, eds, Philadelphia, PA: WB Saunders Co, 2003, 1553-60.

Moorhead A, Grunenwald PE, Dietz VJ, et al, "Trichinellosis in the United States, 1991-1996: Declining but Not Gone," *Am J Trop Med Hyg*, 1999, 60(1):66-9.

Southgate VR and Bray RA, "Medical Helminthology," *Manson's Tropical Diseases*, 21st ed, Appendix III, Cook GC and Zumla A, eds, Philadelphia, PA: WB Saunders Co, 2003, 1699-1701.

Wakelin D, "Immune Responses to Intestinal Parasites: Protection, Pathology and Prophylaxis," *Parasitologia*, 1997, 39(4):269-74.

Internet Web Sites

www.cdc.gov/ncidod/dpd/parasites/trichinosis/default.htm

Trichomonas Preparation

Related Information

Bacterial Culture, Genital Specimen *on page 239*

Cervical/Vaginal Cytology *on page 376*

Test Includes Wet mount and microscopic examination. Pap smear and/or culture may also be performed.

Abstract The pathogenic flagellate, *Trichomonas vaginalis*, is primarily a sexually transmitted protozoan disease that infects the urogenital tracts of both men and women. The incidence of *Trichomonas* varies depending on the population studied, but estimates range from 0.25% to 5.1%.

Specimen Vaginal, cervical, or urethral swabs, prostatic fluid, urine sediment

Container Sterile tube containing 1 mL of sterile nonbacteriostatic saline or specific media for *Trichomonas*

Collection The specimen should be collected using a speculum without lubricant. The mucosa of the posterior vagina may be swabbed, or the secretions may be collected with a pipette. The swab should be expressed into saline for transport. The specimen should be examined as soon as possible.

Storage Instructions Do not refrigerate. Transport immediately to the laboratory so that viable motile organisms may be examined.

Turnaround Time Same day; 48 hours if culture is performed

Reference Interval Negative: no trichomonads identified; positive: demonstration of actively motile flagellates and/or positive culture

Use Establish the presence of *Trichomonas vaginalis*

Limitations The specimen is examined for *Trichomonas vaginalis* only. A separate swab (Culturette®) must be collected for culture or DNA detection of *Neisseria gonorrhoeae* or *Chlamydia trachomatis*, if required. One negative result does not rule out the possibility of *Trichomonas vaginalis* infection. The most important factor affecting sensitivity is the time between collection of the

(Continued)

Trichomonas **Preparation** *(Continued)*

specimen and examination in the laboratory. Culture is not available in many laboratories.

Contraindications Douching within 3 days prior to specimen collection

Methodology Wet mount microscopic examination within 1 hour of collection; Pap smear; culture in Diamond's, Trichosol, or Hollanders liquid medium; use of a commercial culture system such as InPouch TV; direct immunofluorescent technique with monoclonal antibody

Additional Information The absence of the classical yellow, frothy discharge does not exclude trichomoniasis. Up to 50% of women are asymptomatic carriers and the majority of men are asymptomatic. In males, a milky white fluid discharge and urethral irritation present for more than 4 weeks is frequently associated with urethritis caused by *T. vaginalis*. Neonates usually acquire *T. vaginalis* during vaginal births.

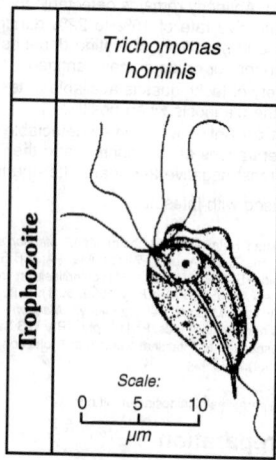

From Brooks MM and Melvin DM, *Morphology of Diagnostic Stages of Intestinal Parasites of Humans*, 2nd ed, Atlanta, GA: U.S. Department of Health and Human Services, Publication No. 84-8116, Centers for Disease Control, 1984, with permission.

Culture may yield positive results when wet preparations are negative. The rate of false negatives, ~40%, and false positives observed with stained preparations (Pap smears) requires that confirmation by wet mount or culture be considered when the reported results are inconsistent with the clinical findings.

References

Beverly AL, Venglarik M, Cotton B, et al, "Viability of *Trichomonas vaginalis* in Transport Medium," *J Clin Microbiol*, 1999, 37(11):3749-50.

Blake DR, Duggan A, and Joffe A, "Use of Spun Urine to Enhance Detection of *Trichomonas vaginalis* in Adolescent Women," *Arch Pediatr Adolesc Med*, 1999, 153(12):1222-5.

Joyner JL, Douglas JM Jr, Ragsdale S, et al, "Comparative Prevalence of Infection With *Trichomonas vaginalis* Among Men Attending a Sexually Transmitted Diseases Clinic," *Sex Transm Dis*, 2000, 27(4):236-40.

Ohlemeyer CL, Hornberger LL, Lynch DA, et al, "Diagnosis of *Trichomonas vaginalis* in Adolescent Females: InPouch TV Cultures Versus Wet-Mount Microscopy," *J Adolesc Health*, 1998, 22(3):205-8.

Internet Web Sites

www.cdc.gov/ncidod/dpd/parasites/trichomonas/default.htm

♦ **Tricodein** *see* Codeine, Urine *on page 429*

- ◆ **Tricyclic Antidepressants** see Antidepressants, Cyclic, Serum or Plasma on page 171
- ◆ **Trifluoperazine** see Phenothiazines, Serum on page 1021
- ◆ **Triglycerides** see Cholesterol, Total, Serum or Plasma on page 396
- ◆ **Triglycerides** see Lipids, Overview on page 853

Triglycerides, Serum or Plasma

Related Information

Apolipoprotein A-I, Serum on page 202
Apolipoprotein B-100, Serum on page 203
Cholesterol, Total, Serum or Plasma on page 396
Glycated Hemoglobin (Hemoglobin A$_{1c}$), Blood on page 655
High Density Lipoprotein Cholesterol, Serum on page 729
Lipid Panel, Serum on page 852
Lipids, Overview on page 853
Low Density Lipoprotein Cholesterol on page 874

Synonyms Triacylglycerols

Applies to Chylomicrons; Friedewald Equation; HDL:LDL Cholesterol; LDL Cholesterol:HDL Cholesterol; Metabolic Syndrome; VLDL

Abstract Triglycerides (TG) are a family of complex lipids composed of glycerol esterified with three fatty acids. Triglycerides are not soluble in blood and are transported as chylomicrons (TG from exogenous sources) or as very-low-density lipoprotein (VLDL-TG from endogenous sources). TG constitute 95% of tissue storage fat. See Lipids, Overview on page 853.

Patient Preparation The patient should be fasting for 10-14 hours and should be on a stable diet 3 weeks prior to collection of blood. Alcohol should be avoided for 3 days and strenuous exercise for at least 24 hours prior to collection of blood. See Preparation in Cholesterol, Total, Serum or Plasma on page 396.

Specimen Serum or plasma

Container Red top tube or lavender top (EDTA) tube

Storage Instructions 4°C; long-term at -70°C, a 2.8% decrease/year takes place.

Causes for Rejection Specimen collected in a glycerinated tube, nonfasting specimen

Reference Interval The third report of the National Cholesterol Education Program's Adult Treatment Panel (ATPIII) uses the following guidelines:
- Normal TG: <150 mg/dL
- Borderline TG: 150-199 mg/dL
- High TG: 200-499 mg/dL
- Very high TG: ≥500 mg/dL

Use

Coronary heart disease (CHD) risk: Elevated TG is an undesirable CHD risk factor, but appears somewhat less potent than LDLC and HDLC. The most common reason for TG assays is that the result is needed in the **indirect** estimate of LDLC using the Friedewald formula:

LDLC (mg/dL) = cholesterol, total (mg/dL) - HDLC (mg/dL) - [triglycerides (mg/dL) / 5]

The Friedewald equation must **not** be used when:
1) fasting TG ≥400 mg/dL
2) chylomicrons are present (usually a visible layer of "cream" following overnight storage) or
3) patient has familial dysbetalipoproteinemia (see Lipids, Overview on page 853)

The current (2002) version of the CPT-defined Lipid Panel, Serum on page 852 assumes the use of the Friedewald formula.

Other situations:
- Elevated TG (>150 mg/dL) is a defining feature of the highly prevalent **metabolic syndrome**, a disease process in which insulin resistance is a major pathogenetic factor. The metabolic syndrome is diagnosed when three or more of the following are present: waist circumference >102 cm (40.2 inches) in men and 88 cm (34.6 inches) in women; serum TG ≥150

(Continued)

Triglycerides, Serum or Plasma *(Continued)*

mg/dL; HDLC <40 mg/dL; blood pressure ≥130/85 mm Hg; or fasting serum glucose ≥110 mg/dL.

- The differential diagnosis of elevated TG includes: nonfasting specimen, dyslipidemia (see Lipids, Overview *on page 853*) hypothyroidism, nephrotic syndrome, diabetes mellitus, ethanol ingestion, pancreatitis, and some forms of glycogen storage disease.

Limitations Correction for free serum glycerol in critically ill patients or patients on hyperalimentation with glycerol-based solutions may be necessary in some enzymatic methods.

Methodology Enzymatic, colorimetric, following removal of LDL and HDL

Additional Information Theoretical mechanisms and empirical data on the interrelationships between insulin resistance, obesity, hypertension, vascular thrombosis and elevated triglycerides have been reviewed.

Drug effects have been extensively reviewed. Some women on estrogens have increased triglycerides. Increases occur with pregnancy, similar to those with oral contraceptives. Hypertriglyceridemia is associated with use of steroids, thiazide diuretics, and beta-adrenergic blocking agents. Rarely, tamoxifen has been associated with severe hypertriglyceridemia. Clomiphene has recently been reported to induce hypertriglyceridemia and pancreatitis. Fish oils and statin drugs reduce serum triglycerides.

References

"Executive Summary of the Third Report of the National Cholesterol Education Program (NCEP) Expert Panel on Detection, Evaluation, and Treatment of High Blood Cholesterol in Adults (Adult Treatment Panel III)," *JAMA*, 2001, 285(19):2486-97.

Ford ES, Giles WH, and Dietz WH, "Prevalence of the Metabolic Syndrome Among U.S. Adults," *JAMA*, 2002, 287(3):356-59.

Shuldiner AR, Yang R, and Gong D, "Resistin, Obesity, and Insulin Resistance - The Emerging Role of the Adipocyte and an Endocrine Organ," *N Engl J Med*, 2001, 345(18):1345-6.

Internet Web Sites

www.cdc.gov/nceh/dls/crmln/crmln.htm
www.nhlbi.nih.gov

♦ **Triiodothyronine Reverse** *see* Reverse T$_3$, Serum *on page 1160*

Triiodothyronine, Serum

Related Information

Free Thyroxine Index *on page 613*
Heterophilic Antibodies *on page 727*
Thyroglobulin, Serum *on page 1249*
Thyroid Stimulating Hormone, Serum *on page 1250*
Thyrotropin Receptor Antibody, Serum *on page 1254*
Thyroxine, Free, Serum *on page 1256*
Thyroxine, Serum *on page 1257*

Synonyms T$_3$, Total; Total T$_3$; Triiodothyronine, Total

Applies to Free T$_3$; Free T$_4$

Abstract T$_3$ (triiodothyronine) is a thyroid hormone produced mainly (80%) from peripheral conversion of T$_4$ (a prohormone). T$_3$ has greater biological activity than T$_4$ and binds to thyroid binding globulin (TBG) less tightly than T$_4$. Only 0.3% exists in free form. When total or free thyroxine (T$_4$ or FT$_4$) are normal and TSH is low, high T$_3$ confirms triiodothyronine toxicosis.

Reference Interval See table.

Age	ng/dL	SI: nmol/L
1-3 d	100-740	1.54-11.40
1-11 mo	105-245	1.62-3.77
1-5 y	105-269	1.62-4.14
6-10 y	94-241	1.45-3.71
11-15 y	82-213	1.26-3.28
16-20 y	80-210	1.23-3.23
20-50 y	70-204	1.08-3.14
50-90 y	40-181	0.62-2.79

Adapted from Whitley RJ, "Thyroid Function," *Tietz Textbook of Clinical Chemistry*, 3rd ed, Burtis CA and Ashwood ER, eds, Philadelphia, PA: WB Saunders Co, 1999, 1496-529.

Patient Preparation Avoid radioisotope administration prior to collection of specimen if RIA is used for assay.

Specimen Serum

Container Red top tube

Storage Instructions Separate serum within 48 hours. Stable up to 7 days at 25°C. Refrigeration is preferred. Stable for at least 30 days at -20°C.

Use This thyroid function test is indicated in patients with decreased TSH and normal free thyroxine and/or total thyroxine levels. Useful in evaluation of hyperthyroid states, particularly in the diagnosis of T_3 thyrotoxicosis, an uncommon variant of hyperthyroidism in which T_3 is increased and T_4 is within normal limits. (See a diagnostic diagram in Thyroid Stimulating Hormone, Serum *on page 1250*.) Serum T_3 is increased in, and helpful for, confirmation of the diagnosis of conventional hyperthyroidism, in which both serum T_3 and T_4 concentrations are increased. Serum T_3 is normal to slightly increased with familial dysalbuminemic hyperthyroxinemia. Recommended for patients with supraventricular tachycardia, for patients with fatigue and weight loss not otherwise explained, or for those with proximal myopathy, in whom T_4 concentrations are not elevated. It is also helpful in monitoring T_4 replacement therapy.

Limitations T_3 is decreased with nonthyroidal chronic diseases and influenced by the state of nutrition. It may be normal with thyrotoxicosis (thyroxine thyrotoxicosis) and is normal in subclinical hyperthyroidism. Variations in TBG and other binding proteins can affect T_3. In these situations, free T_3 should be measured, as free T_3 is not altered with changes in TBG. It is decreased with nicotinic acid. Increases may be found with use of oral contraceptives, pregnancy, and other binding protein abnormalities outlined in Thyroxine, Serum *on page 1257*.

T_3 is not reliable for evaluation of hypothyroidism, since T_3 typically remains normal in mild and moderate thyroid gland failure.

Methodology Radioimmunoassay (RIA), immunochemiluminometric assay, fluorescence polarization immunoassay (FPIA), fluorometric immunoassay

Additional Information Serum concentrations of the free forms of T_3 and T_4 are regulated by feedback systems and appear to parallel rates of cellular uptake. Thus, the free hormone fraction determines the thyroid status of the individual. Proteins that bind T_3 include thyroxine binding globulin, transthyretin, and albumin. As the serum concentrations of the binding proteins rise so does the total T_3, while the free T_3 fraction may be unchanged; examples include pregnancy and estrogen therapy. Reduced total T_3 levels occur in drug therapy with androgens, prednisone, dexamethasone, and glucocorticoids. T_3 levels are also reduced in iodine deficiency, nonthyroidal illness, and anorexia nervosa. T_3 has a higher metabolic potency relative to T_4. As ~33% of T_4 is converted to T_3, T_4 appears to have little intrinsic metabolic activity in humans. See table for comparison of T_3 with T_4.

Comparison of T_3 and T_4 in Humans

	T_3	T_4
Serum concentration		
total (µg/dL)	0.14	8.0
free (ng/dL)	0.4	1.6
Fraction of total serum hormone in free form (%)	0.3	0.02
Distribution volume (L)	35	10
Fraction intracellular (%)	64	10-20
Half-life (days)	1	7
Production rate (µg/day)	33	80
Fraction directly from thyroid (%)	20	100
Relative metabolic potency	1	0.3

From Larsen PR, "The Thyroid," *Cecil Textbook of Medicine*, Vol 2, Wyngarden JB, Smith LH, and Bennett JC, eds, Philadelphia, PA: WB Saunders Co, 1992, 1250, with permission.

Increased T_3 often occurs in hyperthyroidism, but in ~5% of cases only T_3 is elevated, "T_3 toxicosis."

(Continued)

Triiodothyronine, Serum (Continued)

References

Dayan CM, "Interpretation of Thyroid Function Tests," *Lancet*, 2001, 357:619-24.

Whitley RJ, "Thyroid Function," *Tietz Textbook of Clinical Chemistry*, 3rd ed, Burtis CA and Ashwood ER, eds, Philadelphia, PA: WB Saunders Co, 1999, 1496-529.

♦ **Triiodothyronine, Total** *see* Triiodothyronine, Serum *on page 1276*

♦ **Trileptal**® *see* Antiepileptic Drugs Overview *on page 176*

♦ **Triple Test** *see* Alpha₁-Fetoprotein, Serum *on page 136*

♦ **Triple Test** *see* Chorionic Gonadotropin, Human, Serum and Urine *on page 397*

♦ **Triple Test** *see* Estriol, Unconjugated, Pregnancy, Serum, Plasma, or Urine *on page 554*

♦ **Triple Test** *see* Inhibin A, Serum *on page 799*

♦ **Triple Test** *see* Pregnancy-Associated Protein A, Serum *on page 1082*

♦ **Trittico**® *see* Trazodone, Serum or Plasma *on page 1272*

♦ **Tropium**® *see* Chlordiazepoxide, Serum *on page 391*

♦ **Troponin C** *see* Troponins, Serum *on page 1278*

♦ **Troponin I** *see* Troponins, Serum *on page 1278*

Troponins, Serum

Related Information

Cardiac Markers: Laboratory Assessment, Overview *on page 343*
Creatine Kinase MB and Other Isoenzymes, Serum *on page 469*
Creatine Kinase, Serum *on page 470*
Heterophilic Antibodies *on page 727*
Myoglobin, Blood, Serum, or Plasma *on page 940*
Point-of-Care Testing *on page 1065*

Synonyms Cardiac Troponins; TnI (Troponin I) (cTnI); TnT (Troponin T) (cTnT)

Applies to Troponin C; Troponin I; Troponin T

Replaces LDH; LDH Isoenzymes

Abstract Cardiac **troponin I (cTnI)** and **troponin T (cTnT)** are primary tests in the diagnosis of acute myocardial injury because of high specificity and sensitivity. A recent consensus document should be consulted for important background information (Alpert et al).

Specimen Serum, plasma

Container Red top tube, green top (lithium heparin) tube

Sampling Time According to the consensus document, blood should be obtained for testing on admission, at 6-9 hours, and, if the first two samples have negative results, again at 12-24 hours. Markers may be negative very early in the course of acute coronary syndromes.

Storage Instructions Serum stable 4 days at 4°C.

Causes for Rejection Specimen hemolyzed

Turnaround Time Use of plasma decreases turnaround time. Point-of-care testing methods are available.

Reference Interval Undetectable

Critical Values TnI: >1.5 ng/mL. Timing and results of the rise and fall in serial sampling are more important than cutoff values.

Possible Panic Range Any elevation of troponin concentration is considered abnormal.

Use The major criteria for the diagnosis of acute myocardial infarction (AMI) have been redefined to include the typical rise and fall of biochemical markers of cardiac muscle necrosis, according to a joint consensus statement published by the American College of Cardiology and the European Society of Cardiology. These markers are: **cardiac troponin I** (cTI), **cardiac troponin T** (cTT), and the **MB fraction of creatine kinase** (CK-MB). In addition to satisfying the biochemical criteria, at least **one** of the following is required for the diagnosis of AMI:

- ischemic symptoms
- development of pathologic Q waves in the EKG
- EKG changes of ischemia (ST segment changes)
- a coronary artery intervention (eg, angioplasty)

Both of the cardiac troponins are more sensitive than CK-MB in the detection of cardiac muscle necrosis. Myoglobin, the marker elevated soon after AMI, is acceptable as an early marker, but is less specific than the troponins; if the myoglobin result is positive, confirmatory values should be obtained using one of the cardiac troponins or CK-MB. Both of the cardiac troponins remain elevated for 5-7 days after an episode of cardiac injury. The troponins provide prognostic information. They are useful in therapeutic decision making.

Limitations cTnT has been increased in patients with renal failure, and, in some assay systems, muscle injury disease. However, third generation tests predict short-term prognosis in subjects with acute coronary syndromes regardless of results of creatinine clearance.

Methodology Enzyme immunoassay (EIA), one-step; double monoclonal sandwich enzyme immunoassay (EIA); fluorescent immunoassay

Additional Information Because the cardiac troponins remain elevated for 5-7 days, a suspected reinfarction during that interval should be evaluated with creatine kinase MB and/or myoglobin.

Antman et al evaluated cardiac troponin T as a bedside assay for AMI and found it rapid and sensitive. Kratz et al used troponin I with CK-MB and myoglobin as point-of-care markers followed by central laboratory assays for TnI and CK-MB.

References

Alpert JS, Thygesen K, Antman E, et al, "Myocardial Infarction Redefined - A Consensus Document of the Joint European Society of Cardiology/American College of Cardiology Committee for the Redefinition of Myocardial Infarction," *J Am Coll Cardiol*, 2000, 36:959-69.

Antman EM, "Decision Making With Cardiac Troponin Tests," *N Engl J Med*, 2002, 346(26):2079-82.

Antman EM, Grudzien C, and Sacks DB, "Evaluation of a Rapid Bedside Assay for Detection of Serum Cardiac Troponin T," *JAMA*, 1995, 273(16):1279-82.

Aviles RJ, Askari AT, Lindahl B, et al, "Troponin T Levels in Patients With Acute Coronary Syndromes, With or Without Renal Dysfunction," *N Engl J Med*, 2002, 346(26):2047-52.

Caragher TE, Fernandez BB, and Barr LA, "Long-Term Experience With an Accelerated Protocol for Diagnosis of Chest Pain," *Arch Pathol Lab Med*, 2000, 124(10):1434-9.

Kratz A, Januzzi JL, Lewandrowski KB, et al, "Positive Predictive Value of a Point-of-Care Testing Strategy on First-Draw Specimens for the Emergency Department-Based Detection of Acute Coronary Syndromes," *Arch Pathol Lab Med*, 2002, 126(12):1487-93.

Lewandrowski K, Chen A, and Januzzi J, "Cardiac Markers for Myocardial Infarction," *Am J Clin Pathol*, 2002, 118(Suppl 1):S93-S99.

Tularemia Diagnostic Procedures

Related Information

Synonyms *Francisella tularensis* Diagnostic Procedures; Rabbit Fever Diagnostic Procedures

Abstract *Francisella tularensis* is a zoonotic bacteria and is found in a number of small mammals, including rabbits. It is transmitted to man by contact with (Continued)

Tularemia Diagnostic Procedures *(Continued)*

such animals, by inhalation, or by insect bites. The diagnosis of tularemia is often made in the absence of a positive culture for *F. tularensis*. Detection of increased titers of specific antibodies and culture remains the most definitive procedure.

If *F. tularensis* is used as a biological weapon, it will cause febrile illness 3-5 days later. Disease presents as pneumonia, followed by respiratory failure and shock. The abrupt onset of illness among large numbers of individuals with rapid disease progression would lead to consideration of possible bioterrorism.

Specimen Serum for serology; culture and Gram stain from blood, sputum, pleural fluid, lymph node or lymph node aspiration material, pharyngeal washings, skin or mucosal lesions.

Container Red top tube for serology

Sampling Time Paired sera collected 2-3 weeks apart are recommended for serology.

Storage Instructions Refrigeration is recommended if transportation will require longer than 1 hour.

Special Instructions *F. tularensis* is a risk to laboratory personnel and extra precautions are indicated.

Reference Interval Agglutination titer: <1:40; ELISA: <1:500; less than a four-fold increase in paired titer

Critical Values A single serologic result with a titer ≥1:160 in a patient having clinical tularemia or a fourfold rise in titer is diagnostic.

Use Investigation of illness characterized by an ulcerative lesion at a site of inoculation, with regional lymphadenopathy, fever, and pneumonia. Primary pneumonic tularemia follows inhalation. Ulceroglandular, glandular, oculoglandular, oropharyngeal, intestinal, pneumonic, and typhoidal disease types are recognized. Serologic testing is useful for retrospective diagnosis.

Limitations There is serologic cross reactivity with *Brucella* species, *Proteus* OX-19, and *Yersinia* species. IgM and IgG titers may remain elevated (1:20-1:80) for over a decade after infection, limiting the value of unpaired specimens; single titers may be misleading.

Methodology Agglutination, hemagglutination, enzyme-linked immunosorbent assay (ELISA), antigen detection assays, immunoblotting, pulsed-field gel electrophoresis, direct fluorescent antibody, immunohistochemistry, PCR, culture, Gram stain. It is an intracellular, nonmotile, aerobic, faintly staining, tiny gram-negative coccobacillus, or pleomorphic rod. The fastidious organism requires cysteine for growth.

Additional Information In the United States, there are approximately 100-200 tularemia cases reported annually with 1-4 deaths. Antibodies to *F. tularensis* develop 2-3 weeks after infection and peak in 4-5 weeks. Rising titers over a 2-week interval are the best indicator of recent infection.

Although laboratory diagnosis of tularemia is often established by serologic methods, *F. tularensis* may be recovered in culture from a variety of clinical specimens. Isolation is hazardous and sometimes difficult. Blood cultures are often negative. **If tularemia is clinically suspected, contact the laboratory so that appropriate precautions can be taken.**

References

Branda JA and Ruoff K, "Bioterrorism - Clinical Recognition and Primary Management," *Am J Clin Pathol*, 2001, 117(Suppl 1):S116-23.

Dennis DT, Inglesby TV, Henderson DA, et al, "Tularemia as a Biological Weapon. Medical and Public Health Management," *JAMA*, 2001, 285:2763-73.

Franz DR, Jahrling PB, Friedlander AM, et al, "Clinical Recognition and Management of Patients Exposed to Biological Warfare Agents," *JAMA*, 1997 278(5):399-411.

Henderson DA, Inglesby TV, and O'Toole T, *Bioterrorism: Guidelines for Medical and Public Health Management*, Chicago, IL: AMA Press, 2002.

McGovern TW, Christopher GW, and Eitzen EM, "Cutaneous Manifestations of Biological Warfare and Related Threat Agents," *Arch Dermatol*, 1999, 135(3):311-22.

Meyer RF and Morse SA, "Bioterrorism Preparedness for the Public Health and Medical Communities," *Mayo Clin Proc*, 2002, 77(7):619-21.

Shapiro DS and Mark EJ, "A 60-Year-Old Farm Worker With Bilateral Pneumonia," Case Records of the Massachusetts General Hospital. Case 14-2000, Scully RE, Mark EJ, McNeely WF, et al, eds, *N Engl J Med*, 2000, 342(19):1430-8.

Shapiro ED, "Tick-Borne Diseases," *Adv Pediatr Infect Dis*, 1997, 13:187-218.

Steinemann TL, Sheikholeslami MR, Brown HH, et al, "Oculoglandular Tularemia," *Arch Ophthalmol*, 1999, 117(1):132-3.

Varkey P, Poland GA, Cockerill FR III, et al, "Confronting Bioterrorism: Physicians on the Front Line," *Mayo Clin Proc*, 2002, 77(7):661-72.

Internet Web Sites
www.cdc.gov/ncidod/dvbid/tularemia.htm

Uncrossmatched Blood, Emergency

Related Information

Synonyms Emergency Blood; Emergency Transfusion; Universal Donor Blood; Urgent Transfusion

Applies to Emergency Issue of Uncrossmatched Blood; Exsanguinating Emergency; Massive Acute Blood Loss

Test Includes No testing is needed in the Transfusion Service if group O, Rh negative blood is issued. ABO group and Rh type can be completed in 5-10 minutes for issue of group specific blood. As soon as possible, complete antibody screen and crossmatch and other serologic tests (eg, antibody identification) if indicated.

(Continued)

Uncrossmatched Blood, Emergency *(Continued)*

Abstract The administration of a blood component, usually red cells, before the completion of routine pretransfusion testing, in an emergency situation when a delay in transfusion would harm the patient. Risks include those a physician must evaluate (eg, degree of atherosclerosis, nature of the disease, level of oxygenation, heart rate, blood pressure, and control or lack of control of bleeding).

Patient Preparation Proper patient identification is necessary. Emergency Department (ER) may use special or temporary identification. Care must be taken to follow transfusion protocols, especially when multiple trauma victims are being treated simultaneously, as errors in sample/patient identification may lead to fatal ABO hemolytic transfusion reactions.

Specimen Venous or arterial blood

Container One red top tube or one lavender top (EDTA) tube

Collection (Of sample from intended recipient): Identify patient by wristband(s) or other system specially set up for identification of unconscious or noncommunicating patients in emergencies. Label tube specimen with the same information, including identification number from the wristband; label requisition form with identification number. Requisition should be signed by collector, indicating that patient's identity has been verified. Positive identification of patient sample is important, even in an emergency. If it is impossible to get a blood sample, record this fact.

Turnaround Time Although uncrossmatched O Rh-negative red blood cells can be issued immediately if available, ABO and Rh type can be done in only 5-10 minutes. Antibody screen and crossmatch require as much as 1 hour, longer if antibodies are detected. The process is much quicker if patient has already had a type and screen. See Pretransfusion Testing *on page 1088* for a table of times needed for blood issuance in emergencies.

Special Instructions An emergency request for uncrossmatched blood should include name of physician requesting blood and signature of person authorized, name and location of patient, and nature of emergency. There should also be a statement that the situation was sufficiently urgent to require release of blood before completion of testing.

Critical Values 30% to 40% loss of blood volume is associated with increased signs of shock, and >40% relates to severe shock.

Use Blood replacement in exsanguinating emergency, massive acute blood loss. See discussion of use in Red Blood Cells *on page 1139*.

Limitations All parties involved need to understand that blood issued in life-threatening emergencies is clearly more dangerous than that in controlled circumstances.

Contraindications Do not use group O whole blood even in emergencies, for patients of other types, because the anti-A and anti-B can cause hemolysis of the recipient's RBCs. In life-threatening trauma or bleeding when the patient's type is unknown, group O RBCs should be used. Whenever possible, Rh-negative RBCs should be used in females of childbearing age.

Methodology When issuing uncrossmatched blood, apply a label indicating uncrossmatched status.

EMERGENCY RELEASE COMPATIBILITY TESTING INCOMPLETE

Additional Information There is no such thing as a "universal donor." Group O blood lacks A and B but has antigens of other blood group systems, any of which may be a problem for a given patient. Group O RBCs can be transfused when the blood type is unknown. A blood sample can be ABO and Rh typed in 5-10 minutes. Thus, the patient may then receive type-specific RBCs when their blood type is determined. If the patient's indirect antiglobulin test (screen for unexpected antibodies) is negative, transfusion of uncrossmatched but type-specific/compatible blood carries a very low risk of being incompatible. Although volume can be made up temporarily with plasma expanders, possible adverse effects of albumin solutions have been noted in critically ill patients. See also Transfusions in Trauma and Other Emergencies in the Transfusion Service (Blood Bank) Introduction *on page 78.*

References

Cohn SM, "Blood Substitutes in Surgery," *Surgery,* 2000, 127(6):599-602.

Goodnough LT, Brecher ME, Kanter MH, et al, "Transfusion Medicine. First of Two Parts: Blood Transfusion," *N Engl J Med,* 1999, 340(6):438-48.

Hendrix NW, Chauhan SP, Mobley J, et al, "Risk Factors Associated With Blood Transfusion in Ectopic Pregnancy," *J Reprod Med,* 1999, 44(5):433-40.

Spence RK, Jeter EK, and Mintz PD, "Transfusion in Surgery and Trauma," *Transfusion Therapy: Clinical Principles and Practice,* Bethesda, MD: American Association of Blood Banks Press, 1999, 171-97.

Simon TL, Alverson DC, AuBuchon J, et al, "Practice Parameter for the Use of Red Blood Cell Transfusions: Developed by the Red Blood Cell Administration Practice Guideline Development Task Force of the College of American Pathologists," *Arch Pathol Lab Med,* 1998, 122(2):130-8.

Urea Nitrogen:Creatinine Ratio

Related Information

Creatinine, Serum or Plasma *on page 474*

Osmolality, Serum *on page 978*

Urea Nitrogen, Serum or Plasma *on page 1284*

Synonyms BUN:Creatinine Ratio

Test Includes Serum creatinine and urea nitrogen

Abstract The urea nitrogen:creatinine ratio is used with other investigation to attempt to distinguish between prerenal failure, intrinsic failure, and obstruction.

Increased ratio occurs with prerenal and postrenal disorders, while the ratio is normal in renal disease states. It is also relevant to transit of blood through the gastrointestinal tract.

Specimen Serum

Reference Interval 10-20 (up to 30 in infants); about 14 for a person on a normal diet. The ratio is usually normal with intrinsic renal failure.

Critical Values Ratio is generally >20:1 in prerenal failure (decreased renal blood flow, decreased perfusion), gastrointestinal bleeding, or in urinary tract obstruction

Use High BUN:creatinine ratio is found in overproduction or lowered excretion of urea nitrogen. High ratios occur with prerenal azotemia (eg, congestive heart failure), decreased renal perfusion, shock, volume depletion/hypovolemia, hypotension, and dehydration. High ratio with normal creatinine concentration may be found in catabolic states. Often the BUN:creatinine ratio is greatly elevated in gastrointestinal bleeding and with swallowed blood from the upper (Continued)

Urea Nitrogen:Creatinine Ratio *(Continued)*

airway. A BUN:creatinine ratio >36 suggests upper gastrointestinal bleeding, whereas a ratio <36 is not helpful in locating the source of bleeding.

The ratio may be increased with high protein diet, with ileal conduit, and with urinary tract obstruction. It may also be increased with tetracyclines or steroids. These conditions frequently occur with normal serum creatinine levels. In post-renal obstruction and in prerenal azotemia superimposed on renal disease a high BUN:creatinine ratio is present with elevated serum creatinine.

Low BUN:creatinine ratio may be found in low protein diet, malnutrition, pregnancy, severe liver disease, rhabdomyolysis, prolonged I.V. fluid therapy, ketosis (acetoacetic acid interferes with and falsely elevates creatinine), repeated hemodialysis, inappropriate secretion of antidiuretic hormone, with drugs which increase creatinine but not urea nitrogen (eg, cimetidine, trimethoprim), and with tetracycline use (antianabolic effect).

Limitations Patients' variability in protein intake and mass of voluntary muscle can cause this ratio to be misleading. It is only a preliminary guide.

Additional Information In hypovolemia, increased serum osmolality and sodium concentrations may be anticipated, as well as increased urea nitrogen:creatinine ratio. See Osmolality, Serum *on page 978*, in which the urine:serum osmolality ratio and the sodium, serum:osmolality ratio are addressed.

References

Ernst AA, Haynes ML, Nick TG, et al, "Usefulness of the Blood Urea Nitrogen/Creatinine Ratio in Gastrointestinal Bleeding," *Am J Emerg Med*, 1999, 17(1):70-2.

Jurado R and Mattix H, "The Decreased Serum Urea Nitrogen-Creatinine Ratio," *Arch Intern Med*, 1998, 158(22):2509-11.

McGee S, Abernethy WB 3rd, and Simel DL, et al, "Is This Patient Hypovolemic?" *JAMA*, 1999, 281(11):1022-9.

Urea Nitrogen, Serum or Plasma

Related Information

Bicarbonate, Blood *on page 258*
Creatinine Clearance and Urine Creatinine *on page 473*
Creatinine, Serum or Plasma *on page 474*
Cystatin C, Serum or Plasma *on page 489*
Kidney Biopsy *on page 818*
Kidney Stone Analysis *on page 820*
Osmolality, Calculated, Serum or Plasma *on page 976*
Sodium, Serum or Plasma *on page 1210*
Urea Nitrogen:Creatinine Ratio *on page 1283*

Synonyms Blood Urea Nitrogen; BUN

Applies to Dialysis; Urea Reduction Ratio

Abstract The end product of protein metabolism, urea is synthesized by the liver. Easily filtered by renal glomeruli and highly diffusible, urea nitrogen reflects the ratio between urea **production** and **clearance**. Increased urea nitrogen may be due to increased production or decreased excretion. Although many use the expression "BUN," (for blood urea nitrogen), most laboratories use serum, occasionally plasma, but never whole blood.

Specimen Serum, plasma

Container Red top tube; EDTA is suitable as well as lithium heparin for young children.

Collection Pediatrics: Blood drawn from heelstick for capillary (lithium heparin tube)

Storage Instructions Stable 1 day at room temperature, 3 days at 4°C to 8°C, and 3 months at -20°C.

Reference Interval 1 week: 3-25 mg/dL (SI: 1.07-8.9 mmol/L); 1 year: 4-16 mg/dL (SI: 1.4-5.7 mmol/L); 1-40 years: 5-20 mg/dL (SI: 1.8-7.1 mmol/L); gradual slight increase subsequently occurs over 40 years of age. It is 8-23 mg/dL in subjects older than 60 years of age. It is decreased in pregnancy.

Possible Panic Range BUN >100 mg/dL (SI: >35.7 mmol/L) has been used in the definition of uremia.

Use Useful to assess renal function, especially with serum creatinine. **High BUN** occurs in acute and chronic renal diseases. BUN is useful to follow hemodialysis and other therapy. (A highly diffusible molecule, urea falls rapidly with dialysis.) "Uremia" is an expression of a constellation of signs and symptoms in patients with severe azotemia secondary to acute or chronic renal failure. Other causes of increased BUN include severe congestive heart failure, increased protein catabolism, tetracyclines, diuretic use, hyperalimentation, ketoacidosis, shock, and dehydration. Even moderate dehydration can cause BUN to increase up to about 24 mg/dL, usually with normal creatinine concentrations. It is dependent on renal blood flow and urine flow rates. Corticosteroids tend to increase BUN by causing increased protein catabolism. Bleeding from the gastrointestinal tract is an important cause of high urea nitrogen, commonly accompanied by disproportionate increase in BUN relative to creatinine (see Urea Nitrogen:Creatinine Ratio *on page 1283*). Nephrotoxic drugs must be considered.

Borderline high values may occur after recent ingestion of high protein meal and with muscle wasting.

Low BUN occurs in late normal pregnancy, decreased protein intake, with intravenous fluids, with some antibiotics, and in severe liver damage. BUN has a role in assessment of nutritional support.

The BUN is especially important when creatinine values are misleading (eg, with certain cephalosporins). The BUN is superior to creatinine in assessing the function of the filter in hyperbilirubinemic patients undergoing hemodiafiltration for renal failure.

In the syndrome of inappropriate secretion of antidiuretic hormone (SIADH), findings include hyponatremia with serum or plasma Na^+ $\leq$128 mmol/L, serum hypo-osmolality, <260 mOsm/kg, with urine osmolality >300 mOsm/kg (SI: >300 mmol/kg) with low BUN. Such observations occur in situations in which patients are overhydrated. Clinical findings included absence of edema or evidence of heart, liver, thyroid, renal or adrenal disease. Hypouricemia, with uric acid levels in 16 of 17 patients <4 mg/dL (SI: <238 μmol/L), is reported with the syndrome of inappropriate secretion of antidiuretic hormone. (SIADH can be seen with higher serum sodiums and higher osmolalities. Urine osmolality is greater than serum osmolality in SIADH.)

BUN is needed to assess calculated osmolality. Osmolality (mOsm/kg H_2O) may be calculated as follows:

Osmolality = $[Na^+$ (mmol/L) x 2] + urea N (mg/dL)/2.8 + glucose (mg/dL)/18

Limitations Creatinine is usually more specific for glomerular function but may be less sensitive to some types of early renal disease. Uremia and other types of renal dysfunction are best evaluated with creatinine as well as urea nitrogen. In both prerenal and postrenal azotemia, BUN is apt to be increased somewhat more than is creatinine. However, in a series of dehydrated children with gastroenteritis who had metabolic acidosis and increased anion gap, 88% had BUN concentration $\leq$18 mg/dL (SI: $\leq$6.4 mmol/L). The authors found bicarbonate and anion gap more sensitive indices in this setting. In chronic progressive renal disease, about 75% of renal parenchyma must be damaged or destroyed before azotemia develops. BUN lacks sensitivity and specificity, but still remains a useful test. It is insensitive to early diminution of glomerular filtration rate.

Additional Information Although creatinine is generally considered a more specific test to evaluate renal function, BUN and creatinine are commonly used together.

BUN before and following dialysis and between dialysis treatments is among the determinants of patients so treated. The **urea reduction ratio** (percent reduction in concentration of BUN during a dialysis treatment) is not as powerful as serum albumin as a predictor of death. Its calculation: 100 x (1-[C_t/C_o]); C_t is the BUN 5 minutes following the end of dialysis, C_o is predialysis BUN. This ratio bears relationship to blood clearance by dialysis, length of dialysis, and volume of distribution of urea in the patient. The ratio represents quantitation of an individual's urea clearance during a single dialysis.

A large number of substances and drugs affect the concentration of urea nitrogen by pharmacologic/physiologic or by analytic means, including those causing nephrotoxicity, dehydration, and volume depletion.

Other tests are relevant to the differential diagnosis of renal failure. Hypercalcemia with hyperuricemia may indicate a neoplastic state. High CK may point (Continued)

Urea Nitrogen, Serum or Plasma *(Continued)*

to rhabdomyolysis. Abnormalities in serum and/or urine proteins may indicate myeloma. Allergic interstitial nephritis may be signaled by eosinophilia. An osmolal gap and oxalate crystalluria may point to ethylene glycol toxicity. Glomerular basement membrane antibody, ANA, ANCA, and other immunologic studies may be helpful.

References

Bonadio WA, Hennes HH, Machi J, et al, "Efficacy of Measuring BUN in Assessing Children With Dehydration Due to Gastroenteritis," *Ann Emerg Med*, 1989, 18(7):755-7.

Thadhani R, Pascual M, and Bonventre JV, "Acute Renal Failure," *N Engl J Med*, 1996, 334(22):1448-60.

Young DS, *Effects of Drugs on Clinical Laboratory Tests*, 5th ed, Volume 1: Listing by Test, Washington, DC: AACC Press, American Association of Clinical Chemistry, 2000, Section 3, 806-17.

♦ **Urea Reduction Ratio** *see* Urea Nitrogen, Serum or Plasma *on page 1284*

♦ **Urgent Transfusion** *see* Uncrossmatched Blood, Emergency *on page 1281*

♦ **Uric Acid/Creatinine Ratio, Urine** *see* Uric Acid, Urine *on page 1287*

Uric Acid, Serum

Related Information

Ammonia, Plasma *on page 150*
Complete Blood Count *on page 442*
Creatinine Clearance and Urine Creatinine *on page 473*
Creatinine, Serum or Plasma *on page 474*
Ethanol, Blood, Urine, and Other Sources *on page 558*
Kidney Stone Analysis *on page 820*
Lead, Blood *on page 832*
Molybdenum, Blood *on page 919*
Sodium, Serum or Plasma *on page 1210*
Sputum Cytology *on page 1222*
Synovial Fluid Analysis *on page 1229*
Uric Acid, Urine *on page 1287*

Synonyms Urate

Abstract Uric acid is the end product of purine metabolism and is associated clinically with gout. Antoni van Leeuwenhoek described crystals from a gouty tophus in the mid-17th century. Such sodium urate crystals in synovial fluid or from a tophus are diagnostic of gout. In most cases, gout causes a monoarticular arthritis in a peripheral lower extremity joint. Factors which contribute to the risk of gout include obesity, high purine diet, high ethanol intake, thiazide and loop diuretic therapy.

Specimen Serum

Container Red top tube

Collection Separate serum. Do not collect in lavender top (EDTA) tube or gray top (sodium fluoride) tube for urease method.

Storage Instructions Urate is stable in serum for 3 days at 25°C, 3-7 days at 4°C, and 6-12 months at -20°C.

Special Instructions Alcoholic beverages are best avoided before collection.

Reference Interval An increase occurs during childhood. Adults: male: 3.4-7.0 mg/dL (SI: 202-416 µmol/L), female: 2.4-6.0 mg/dL (SI: 143-357 µmol/L), if specific laboratory methods are used.

Possible Panic Range "Severe hyperuricemia" has been classified as uric acid >12.0 mg/dL (SI: >714 µmol/L).

Use An increased uric acid level does not establish a diagnosis of gout; serum uric acid is not always elevated during an episode of clinical gout. The extensive overlap in serum uric acid levels between those with and without gout is shown in a study in which the lowest level in a gouty subject was 6 mg/dL (SI: 357 µmol/L), while the highest uric acid in a nongouty person was 9.5 mg/dL (SI: 565 µmol/L). Among subjects with serum urate concentrations of 9.0 mg/dL (SI:535.5 µmol/L), the incidence of acute gout is about 5% per year. Gouty tophi form in cooler portions of the body because uric acid solubility varies directly with the temperature.

Elevations of uric acid occur in renal diseases with renal failure, prerenal azotemia, preeclampsia, certain diets rich in protein, and starvation. Three types of kidney disease are caused by precipitation: acute uric acid nephropathy, nephrolithiasis, and chronic urate nephropathy. Lead poisoning (saturnine gout) from paint, batteries, and moonshine is a secondary type of decreased secretion. Other secondary causes include acidosis and excessive cell destruction (eg, leukemia, lymphoma, hemolytic anemia).

Overproduction of urate increases the risk of uric acid and also of calcium oxalate urolithiasis.

Asymptomatic hyperuricemia is found in the population at large, especially in family members of subjects with gout. It is much more common than is gout.

The appearance of gout in the second or third decades of life is an indication for evaluation of hereditary disorders of purine metabolism.

Drugs: A large number of drugs cause analytical and pharmacologic **elevations** in serum uric acid concentrations. Prominent among these are cancer chemotherapy agents which cause extensive tumor cell destruction. Others include cyclosporine, didanosine, some diuretics, ethambutol, niacin, pyrazinamide, and low dose salicylates.

Low levels of uric acid accompany certain renal tubular defects, xanthinuria, severe liver disease, hyponatremia and hypo-osmolar conditions, and low protein diets. A number of drugs can cause decrease of uric acid concentrations.

Limitations An elevated uric acid result does not establish the diagnosis of gout, and a normal result may be obtained even during an acute attack of gout. The best test for gout is the demonstration of uric acid crystals in a joint aspirate.

Methodology Spectrophotometry, high performance liquid chromatography (HPLC)

Additional Information The **differential diagnosis between gout and calcium pyrophosphate deposition disease** is tabulated and discussed in Synovial Fluid Analysis *on page 1229*.

Increased concentrations of uric acid have been reported to bear independent and significant association with cardiovascular mortality risk.

When gouty tophi are biopsied, their histopathologic appearance is often diagnostic in H&E. Alcohol fixation is highly desirable for such biopsies. Staining methods are available.

References

Fang J and Alderman MH, "Serum Uric Acid and Cardiovascular Mortality - The NHANES I Epidemiologic Follow-up Study, 1971-1992," *JAMA*, 2000, 283(18):2404-10.

Rott KT and Agudelo CA, "Gout," *JAMA*, 2003, 289(21):2857-60.

Terkeltaub RA, "Gout," *N Engl J Med*, 2003, 349(17):1647-55.

Waring WS, "Uric Acid: An Important Antioxidant in Acute Ischaemic Stroke," *Q J Med*, 2002, 95(10):691-3.

Young DS, *Effects of Drugs on Clinical Laboratory Tests*, 5th ed, Volume 1: Listing by Test, Washington, DC: AACC Press, American Association of Clinical Chemistry, 2000, Section 3, 817-30.

Uric Acid, Urine

Related Information

Calcium, Urine *on page 332*
Creatinine, 12- or 24-Hour Urine *on page 471*
Kidney Stone Analysis *on page 820*
Lead, Urine *on page 835*
pH, Urine *on page 1035*
Synovial Fluid Analysis *on page 1229*
Uric Acid, Serum *on page 1286*
Urine Collection, 24-Hour *on page 1295*

Synonyms Urate, Urine

Applies to Uric Acid/Creatinine Ratio, Urine

Abstract Uric acid results from *de novo* synthesis and dietary sources. Seventy-five percent of urate is eliminated through the kidney and 25% through the intestine. Renal excretion of urate involves reabsorption by the proximal tubules, secretion by the distal portion of the proximal tubules, and further reabsorption by the distal tubules.

Patient Preparation Patient on normal diet; no ethanol during collection.

Specimen 24-hour urine, random urine with creatinine determination

(Continued)

Uric Acid, Urine *(Continued)*

Container To prevent precipitation in acid urine, add 10 mL of sodium hydroxide solution (12.5 M) to specimen container prior to collection.

Storage Instructions Do not refrigerate. Stable about 3 days.

Reference Interval Approximately 250-750 mg/24 hours (SI: 1.5-4.5 mmol/day) for women. Interval for men may extend to 800 mg/24 hours (SI: 4.8 mmol/day) on ordinary diet. Increases on purine-rich diet. Of those with idiopathic gout, about 10% have overexcretion of uric acid.

In children, the upper limit of normal of second morning urine specimen uric acid to creatinine ratio has been established. See graphic.

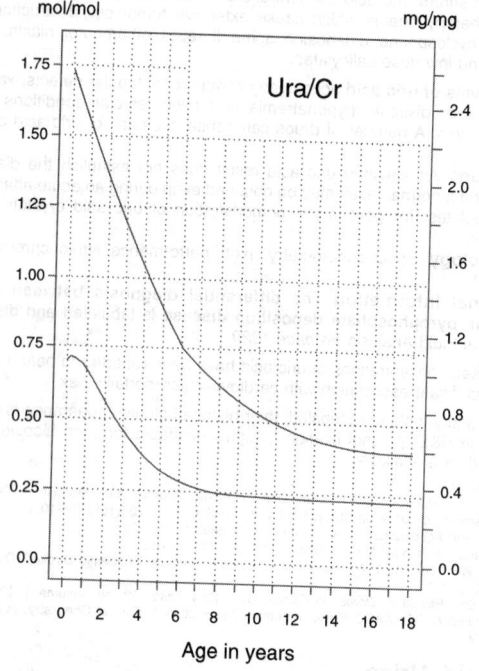

Estimated 95th and 5th percentiles for urate (Ura) to creatinine (Cr) ratios related to age.

A number of drugs affect uric acid excretion including aspirin, other anti-inflammatory preparations, x-ray contrast agents, vitamin C, and warfarin. Diuretics decrease uric acid excretion.

Use Hyperuricosuria, as well as hyperuricemia, is associated with renal calculus formation. Overproduction of urate increases risk of uric acid as well as calcium oxalate urolithiasis. Identify overexcretors to determine risk of stone formation; identify genetic defects, influence of overexcretion on therapy of gout. Uric acid nephrolithiasis occurs in primary gout or in secondary hyperuricemia and may complicate inflammatory bowel disease and jejunoileal bypass. Most subjects with uric acid stones do not have gout. Evaluate uric acid metabolism in gout.

A number of drugs affect uric acid excretion including aspirin, other anti-inflammatory preparations, x-ray contrast agents, vitamin C, and warfarin. Diuretics decrease uric acid excretion.

Methodology Phosphotungstate, uricase, high performance liquid chromatography (HPLC). The uricase method is more specific and is in wide use in modern chemistry analyzers.

Additional Information Even mild renal failure decreases uric acid excretion. Uric acid excretion is decreased with hypertension.

A young patient with acute gouty arthritis, uric acid stones, and any patient who excretes >1000 mg uric acid/24 hours (SI: >5.9 mmol/day), should be screened for hypoxanthine-guanine phosphoribosyltransferase (HPRT) deficiency. The uric acid:creatinine ratio has been used as a screen for Lesch-Nyhan syndrome (HPRTase deficiency). Normal control patients 0.21-0.59; partial enzyme deficient group 0.62-2.00; complete enzyme deficiency 1.98-5.35. Determinations of serum uric acid, as well as of urine uric acid:creatinine ratio, are of value in initiation of investigation of purine-related metabolic diseases.

The ratio of uric acid to creatinine concentration on a random urine specimen has also been shown to be >1.0 in patients with acute renal failure secondary to acute uric acid nephropathy, but <1.0 in patients with acute renal failure resulting from other causes.

References

Matos V, Van Melle G, Werner D, et al, "Urinary Oxalate and Urate to Creatinine Ratios in a Healthy Pediatric Population," *Am J Kidney Dis*, 1999, 34(2):6E.

Mayo Reference Services Publication, "Supersaturation Profile, Urine," *New Test Announcement*, Rochester, MN: Mayo Medical Laboratories, 2000.

Valik D and Jones JD, "Hereditary Disorders of Purine and Pyrimidine Metabolism: Identification of Their Biochemical Phenotypes in the Clinical Laboratory," *Mayo Clin Proc*, 1997, 72(8):719-25.

Urinalysis

Related Information

Anion Gap, Serum, Plasma, or Urine *on page 160*
Anti-DNA *on page 173*
Ascorbic Acid, Serum or Plasma *on page 215*
Bacterial Culture, Urine *on page 246*
Blood, Urine *on page 281*
Chromium, Serum *on page 402*
Cystine, Urine *on page 494*
Ethylene Glycol, Serum or Plasma *on page 561*
Fat, Urine *on page 573*
Glucose, Semiquantitative, Urine *on page 650*
Hemoglobin, Qualitative, Urine *on page 688*
Ibuprofen, Serum *on page 764*
Inherited Diseases of Metabolism and Cell Structure *on page 792*
Ketones, Urine *on page 817*
Kidney Biopsy *on page 818*
Kidney Stone Analysis *on page 820*
Leukocyte Esterase, Urine *on page 849*
Microalbuminuria *on page 913*
Mucopolysaccharides, Urine *on page 922*
Nitrite, Urine *on page 962*
Osmolality, Calculated, Serum or Plasma *on page 976*
Osmolality, Urine *on page 979*
Oxalate, Urine *on page 989*
pH, Urine *on page 1035*
Primidone, Serum or Plasma *on page 1091*
Protein, Quantitative, Urine *on page 1108*
Protein, Semiquantitative, Urine *on page 1113*
Schistosomiasis Diagnostic Procedures *on page 1180*
Specific Gravity, Urine *on page 1216*
Urinary Tract Cytology *on page 1293*

Synonyms UA

Applies to Casts, Urine; Crystals, Urine; Dysmorphic Red Cells; Occult Blood, Semiquantitative, Urine; Urine Crystals

Test Includes Opacity, color, appearance, specific gravity, pH, protein, glucose, occult blood, ketones, bilirubin, and in some laboratories, urobilinogen and microscopic examination of urine sediment. Some laboratories include screening for leukocyte esterase and nitrite and do not perform a microscopic *(Continued)*

Urinalysis *(Continued)*

examination unless one of the chemical screening (macroscopic) tests is abnormal or unless a specific request for microscopic examination is made.

Abstract The examination of urine is one of the oldest practices in medicine. A carefully performed urinalysis still provides a wealth of information about the patient, both in terms of differential diagnosis, and by exclusion of many conditions when the urinalysis is "normal."

Patient Preparation Instructions should be given in method of collection. Both males and females need instruction in cleansing the urethral meatus. "Midstream collections" are performed by initiating urination into the toilet, then bringing the collection device into the urine stream to catch the midportion of the void. In infants and young children, urine specimens can be obtained by urine bag, bladder catheterization, or bladder aspiration.

Specimen Dipstick: random urine; microscopy: centrifuged sediment from random urine

Container Plastic urine container

Collection A voided specimen is usually suitable. If the specimen is likely to be contaminated by vaginal discharge or hemorrhage, a clean catch specimen is desirable. If the specimen is collected by catheter or bladder aspiration, it should be so labeled. The timing of urine collection will vary with the purpose of the test. To check for casts or renal concentration ability, a first voided morning specimen may be preferred. For screening purposes, this is also the best time, as a later and more dilute specimen may make small increases in protein, red blood cells (RBC), or leukocytes (WBC) excretion harder to detect.

Storage Instructions Transport specimen to the laboratory as soon as possible after collection. If the specimen cannot be processed immediately by the laboratory it should be refrigerated. Refrigeration preserves formed elements in the urine, but may precipitate crystals not originally present. **Examination is best done on freshly voided, warm urine.**

Causes for Rejection Specimen delayed in transport, fecal contamination, decomposition, or bacterial overgrowth

Reference Interval See table. **Crystals** are part of the morphologic evaluation and are interpreted by an experienced microscopist.

Urinalysis

Test	Reference Range
Specific gravity	1.003-1.029
pH	4.5-7.8
Protein	Negative / trace[1]
Glucose	Negative
Ketones	Negative
Bilirubin	Negative
Occult blood	Negative
Leukocyte esterase	Negative
Nitrite	Negative
Urobilinogen	0.1-1.0 EU/dL
WBCs	0-4/hpf
RBCs	0-2 RBCs/hpf
Casts	0-4/lpf hyaline
Bacteria	Negative

[1]In concentrated urine.
hpf = high power field.
lpf = low power field.
EU = Ehrlich units.

Possible Panic Range The presence of massive amounts of oxalate crystals in fresh urine should be reported promptly to the physician, as this finding may reflect ethylene glycol intoxication.

Use Careful microscopy, preferably by phase-contrast technique, of urine sediment on freshly voided, warm urine is important whenever abnormalities have been identified by the dipstick.

Limitations Urine dipsticks are an imperfect tool. The urinalysis, with or without microscopy, is not an effective screen for urine culture. Most patients with urinary infections are treated without laboratory tests. For those who require testing, immediate culture is indicated. Please see Blood, Urine *on page 281*, Glucose, Quantitative, Urine *on page 649*, Glucose, Semiquantitative, Urine *on page 650*, Ketones, Urine *on page 817*, Nitrite, Urine *on page 962*, Leukocyte Esterase, Urine *on page 849*, pH, Urine *on page 1035*, Protein, Quantitative, Urine *on page 1108*, and Protein, Semiquantitative, Urine *on page 1113*, for information about sensitivity and specificity.

Methodology The chemical portion of the urinalysis is done by reagent strip, with confirming chemical method for protein (sulfosalicylic acid precipitation). Such dipsticks were introduced for glucose (1956), protein (1957), ketone (1957), pH (1959), occult blood (1961), bilirubin (1969), urobilinogen (1969), nitrite (1972), specific gravity (1981), and for the presence of white blood cells. Most are colorimetric, read visually, but instruments are available. Avoid contamination of specimens with any additives. See individual test entries for further information.

Additional Information

MICROSCOPY:

Crystalluria: Crystals are most diagnostically useful when observed in warm, fresh urine (*in vivo* crystal formation) in evaluation of hematuria, nephrolithiasis, or toxin ingestion. Polarizing microscopy and pH are useful in crystal identification.

Calcium oxalate crystals are classically fairly uniform small double pyramids, base to base, which under the microscope look like little crosses on a square, the octahedral shape of the dihydrate form. Ovoid and dumbbell shapes of oxalate crystals are more easily missed. Polarization helps: oxalate crystals are birefringent, but the red cells and yeasts with which the ovoid forms can be confused are not anisotropic. Acetic acid (3%) will lyse red cells but not oxalate crystals or yeasts. See Oxalate, Urine *on page 989* and Kidney Stone Analysis *on page 820*. In abundance, **calcium oxalate** and/or **hippurate crystals** may suggest ethylene glycol ingestion (especially if known to be accompanied by neurological abnormalities, appearance of drunkenness, hypertension, and a high anion gap acidosis.)

Uric acid crystals are reddish brown, rectangular, rhomboidal, or flower-like structures of narrow rectangular petals. **Ammonium urates**, in alkaline urine, are irregular blobs and crescents, sometimes resembling fragmented red cell shapes.

Calcium phosphate crystallizes in urine as flowers of narrow rectangular needles.

Cystine crystals form large irregular hexagonal plates, which may dissolve if alkalinized. They occur only in the urine of subjects with cystinuria. (See Cystine, Urine *on page 494*).

Calcium magnesium ammonium phosphate, or "triple phosphate," forms unique "coffin lid" angularly domed rectangles which may be present in massive quantities in alkaline urine. They usually are associated with urine infected by urea splitting bacteria which cause "infection," or "triple phosphate" stones.

Indinavir crystals are found in >30% of patients treated with this protease inhibitor. Manual microscopy detects such radial clusters, forming starburst shapes.

Leukocyturia may indicate inflammatory disease in the genitourinary tract.

White cell casts indicate the renal origin of leukocytes, and are most frequently found in acute pyelonephritis.

Red cell casts indicate renal origin of hematuria and suggest glomerulonephritis. Degenerated red cell casts may be called "**hemoglobin casts**". Orange to red casts may be found with myoglobinuria as well.

(Continued)

Urinalysis *(Continued)*

Dysmorphic red cells are observed in glomerulonephritis. "Dysmorphic" red cells refer to heterogeneous sizes, hypochromia, distorted irregular outlines and frequently small blebs extruding from the cell membrane. Phase contrast microscopy best demonstrates RBC and WBC morphology. Nonglomerular urinary red blood cells resemble peripheral circulating red blood cells.

Crenated RBCs provide no information regarding RBC source.

Dark brown or smoky urine suggests a renal source of hematuria.

A **pink or red urine** suggests an extrarenal source.

Hyaline casts occur in physiologic states (eg, after exercise) and many types of renal diseases.

Renal tubular (epithelial) casts are most suggestive of tubular injury, as in acute tubular necrosis. They are also found in other disorders, including eclampsia, heavy metal poisoning, ethylene glycol intoxication, and acute allograft rejection.

Granular casts: Very finely granulated casts may be found after exercise and in a variety of glomerular and tubulointerstitial diseases.; coarse granular casts are abnormal and are present in a wide variety of renal diseases.

"Dirty brown" granular casts are typical of acute tubular necrosis.

Waxy casts are found especially in chronic renal diseases, and are associated with chronic renal failure.

Fatty casts and **oval fat bodies ("lipiduria")** are generally found in the nephrotic syndromes, usually glomerular diseases including minimal change disease, focal segmental glomerulosclerosis, membranous glomerulopathy, and membranoproliferative glomerulonephritis.

Broad casts originate from dilated, chronically damaged tubules or the collecting ducts. They can be granular or waxy. **Broad waxy casts** are called "renal failure casts."

Spermatozoa may be seen in male urine related to recent or retrograde ejaculation. In female urine, the presence of spermatozoa may provide evidence of vaginal contamination following recent intercourse.

Automation of the urinalysis is routine in many laboratories. Some authors wish to abandon microscopic evaluation of the urine, which is not easily automated, on urine samples testing "normal" by dipstick screening. A urine sample that is normal to inspection and dipstick will be normal to microscopic exam 95% of the time.

Tests for **inherited diseases of metabolism** involve blood as well as urine. These subjects are summarized in Inherited Diseases of Metabolism and Cell Structure *on page 792.*

References

Bartlett RC, Zern DA, Ratkiewicz I, et al, "Reagent Strip Screening for Sediment Abnormalities Identified by Automated Microscopy in Urine From Patients Suspected to Have Urinary Tract Disease," *Arch Pathol Lab Med*, 1994, 118(11):1096-101.

Misdraji J and Nguyen PL, "Urinalysis. When - and When Not - To Order," *Postgrad Med*, 1996, 100(1):173-6, 181-2, 185-8 passim.

Threatti GA and Henry JB, "Urine and Other Body Fluids," *Clinical Diagnosis and Management by Laboratory Methods*, Henry JB, ed, Philadelphia, PA: WB Saunders Co, 2001, 367-401.

Wenz B and Lampasso JA, "Eliminating Unnecessary Urine Microscopy - Results and Performance Characteristics of an Algorithm Based on Chemical Reagent Strip Testing," *Am J Clin Pathol*, 1989, 92(1):78-81.

♦ **Urinary Cortisol** *see* Cortisol, Free, Urine *on page 459*

Urinary Cortisol/Creatinine Increment

Related Information

Insulin Tolerance Test *on page 804*
Metyrapone Stimulation Test, Serum *on page 910*

Synonyms Midnight to Morning Urinary Cortisol Increment; Urinary Cortisol Increment

Applies to Cortisol-Binding Globulin

Abstract This test for adrenal insufficiency (AI) uses **sleep** as a stimulus for ACTH release. The endpoints of the test are cortisol measurements in two urine specimens collected under standardized conditions. A double voiding technique is used to assure that the cortisol measured corresponds to the cortisol secretion at the time of urine collection, and the cortisol measurement is normalized to the concurrent creatinine to allow for interindividual differences in urinary flow rate. The urine free cortisol (see Cortisol, Free, Urine *on page 459*) in a 1-hour collection, beginning immediately after a postawakening void, is compared with the cortisol in a baseline 1-hour urine specimen collected from 11 PM to midnight the night before. The cortisol results are divided by the concurrent urinary creatinine, resulting in a ratio. The midnight ratio is subtracted from the morning ratio to obtain the urinary cortisol/creatinine increment (UCCI).

Sampling Time The specimens are usually collected at home. The patient is instructed to collect urine samples (5-20 mL) at bedtime and on awakening, and to place them in prelabeled containers. A double-voiding technique is used: the patient empties the bladder at 11 PM and collects a urine sample for testing 1 hour later; the patient next empties the bladder on awakening and collects the second sample for testing 1 hour later.

Storage Instructions Check with reference laboratory or use a preservative appropriate for free cortisol.

Reference Interval In persons with a normal hypothalamic-pituitary axis, the UCCI is >9.

Use This is a noninvasive test for **adrenal insufficiency** (AI), and is useful when equivocal results are obtained from primary testing (see Cortisol, Serum or Plasma *on page 460*, Cortisol, Free, Urine *on page 459*, and Adrenocorticotropic Hormone, Plasma *on page 114*). The use of urine cortisol overcomes some of the limitations of serum cortisol measurements, such as unpredictable short-term fluctuations and dependence on normal levels of cortisol-binding globulin (CBG). The positive predictive value of the test (in a population of 40 patients with AI and 40 controls), using the Insulin Tolerance Test *on page 804* as the gold standard, was 95%; in this study, patients who had a normal UCCI and an abnormal ITT also had low serum CBG, which may have masked a truly normal adrenal response.

Limitations This is a relatively new test. While it seems very promising, it is not possible to predict what the results of larger studies will be.

References
Kong WM, Alaghband-Zadeh J, Jones J, et al, "The Midnight to Morning Urinary Cortisol Increment Is an Accurate, Noninvasive Method for Assessment of the Hypothalamic-Pituitary-Adrenal Axis," *J Clin Endocrinol Metab*, 1999, 84(9):3093-8.

♦ **Urinary Cortisol Increment** *see* Urinary Cortisol/Creatinine Increment *on page 1292*

♦ **Urinary MVA** *see* Mevalonic Acid, Urine or Amniotic Fluid *on page 911*

♦ **Urinary Sugar Test** *see* Glucose, Quantitative, Urine *on page 649*

♦ **Urinary Sugar Test** *see* Glucose, Semiquantitative, Urine *on page 650*

Urinary Tract Cytology

Related Information
Bacterial Culture, Genital Specimen *on page 239*
Blood, Urine *on page 281*
Cytomegalovirus Culture *on page 495*
Cytomegalovirus Cytology *on page 497*
Flow Cytometry, Overview *on page 592*
Fungal Culture, Urine *on page 625*
Schistosomiasis Diagnostic Procedures *on page 1180*
Urinalysis *on page 1289*
Viral Culture *on page 1307*

Test Includes Smears, cytocentrifuge preparations, millipore filter preparations, flow cytometry, immunocytochemistry
(Continued)

Urinary Tract Cytology *(Continued)*

Patient Preparation Hydrate patient (give several glasses of water) 30 minutes to 1 hour prior to collection. Patient should **not** have had mineral oil cathartics. Inform patient to discard first early morning voided urine.

Specimen Voided or catheterized urine; intraoperative washings of urinary bladder, ureters, or renal pelvis; ileal conduit urine

Sampling Time Urine which has been in the bladder for prolonged periods shows extensive cellular degeneration. Fresh specimens left at room temperature demonstrate cellular degeneration within 1 hour.

Storage Instructions If the specimen cannot be brought immediately to the Cytology Laboratory, it must be refrigerated at 4°C and fixed with a suitable fixative.

Causes for Rejection 24-hour collection, prolonged period at room temperature with extensive degeneration of cellular detail

Special Instructions Bladder washings should not be collected in a hypotonic solution. Voided urine is the specimen of choice for male patients, and catheterized urine is the specimen of choice for female patients (to avoid vaginal-vulvar squamous contamination). If cytomegalovirus infection is suspected, this concern should be noted.

Use Recognition of primary benign or malignant as well as metastatic disease; routine surveillance for recurrent transitional cell carcinoma; follow-up patients receiving intravesicular therapy for transitional cell carcinoma; aid in diagnoses of infections with herpesvirus, polymovirus, cytomegalovirus, fungal diseases, and *Schistosoma*; detect renal hemosiderosis, cerebral metachromatic leukodystrophy, and endometriosis of the urinary tract

Limitations Low grade (grade 1) papillary transitional cell carcinoma cannot be diagnosed reliably by cytology alone. Polyoma virus infections may sometimes be confused with high grade transitional cell carcinoma. Recent instrumentation and calculi may produce atypical changes in urothelial cells simulating malignancy. History of instrumentation of the bladder must be provided. Numerous chemotherapeutic agents including cyclophosphamide (Cytoxan®), busulfan (Myleran®), thiotepa, and BCG may produce cell changes almost indistinguishable from true dysplasia or neoplasia. For these reasons, complete clinical history is of utmost importance. Urine cytology has a very low sensitivity for detection of primary renal and prostate neoplasms. Diagnostic accuracy of urine cytology appears closely related to the grade of bladder tumors, pretreatment and post-treatment status, and minimally to the type of therapy (radiation, chemotherapy, surgery).

Contraindications 24-hour urine collections are **not** recommended.

Additional Information Voided urine is much preferred over a catheterized sample due to atypical cell changes induced by trauma of the catheter itself. Barbotage (instilled saline insufflated with air in tiny bubbles to gently exfoliate the urothelium) cytology has the highest sensitivity for detection of transitional cell carcinoma. Although poorly-differentiated carcinomas are diagnosed with relative ease, well-differentiated (low grade) carcinomas may not be diagnosed by usual cytologic methods. DNA flow cytometry (DNA analysis) may detect the presence of an aneuploid population of cells with or without an increased S phase. It is a sensitive indicator for recurrent transitional neoplasia but is of no use in cases in which the primary transitional cell carcinoma is diploid. Maximum sensitivity is obtained using both cell morphology and DNA analysis.

References

Farrow GM, "Urine Cytology of Transitional Cell Carcinoma: Diagnostic Efficacy," *Compendium on Diagnostic Cytology*, 8th ed, Wied GL, Bibbo M, Keebler CM, et al, eds, Chicago, IL: Tutorials of Cytology, 1997, 280-4.

Koss LG, "Urinary Cytology," *Compendium on Diagnostic Cytology*, 8th ed, Wied GL, Bibbo M, Keebler CM, et al, eds, Chicago, IL: Tutorials of Cytology, 1997, 276-9.

Rosenthal DL and Mandell DB, "Cytologic Detection of Urothelial Lesions," *Compendium on Diagnostic Cytology*, 8th ed, Wied GL, Bibbo M, Keebler CM, et al, eds, Chicago, IL: Tutorials of Cytology, 1997, 268-75.

♦ **Urinary Tract Infection Screen** *see* Nitrite, Urine *on page 962*

♦ **Urine 17-Ketogenic Steroids** *see* 17-Hydroxyprogesterone, Whole Blood, Serum, or Plasma *on page 755*

♦ **Urine Calcium:Creatinine Ratio** *see* Blood, Urine *on page 281*

♦ **Urine Cl** *see* Chloride, Urine *on page 394*

♦ **Urine Collection, 12-Hour, 2-Hour, and Timed** see Urine Collection, 24-Hour on page 1295

Urine Collection, 24-Hour

Synonyms Twenty-Four Hour Urine Collection

Applies to Timed Urine Collection; Urine Collection, 12-Hour, 2-Hour, and Timed

Abstract Proper instruction of the patient is critically important. The individuals who provide collection materials are situated ideally to review appropriate procedures. If several tests are needed, using different preservatives, explanation is desirable.

Patient Preparation See individual entries.

Specimen The patient should understand whether or not refrigeration of the specimen during collection is needed or contraindicated.

Collection Twenty-four hour urine collections are often a problem for both patient and laboratory. A good collection regimen is as follows: Discard first morning specimen on day one. Collect all specimens during the remainder of the day and evening. Collect the first morning specimen on day two. Stop collection. Label specimen with patient's name, hospital number, room number, and date and time of collection. This presumes that time of arising is the same on day one and day two. Alternate regimen: Patient is to empty his/her bladder completely at a designated time (eg, 8 AM). This specimen is discarded. All urine is saved throughout the day and evening. Patient is to empty his/her bladder at the same time on day two as in step 1 above (eg, 8 AM). This specimen is combined with the rest of the collection for the previous 24 hours. Stop collection. Label specimen with patient's name, hospital number, room number, date and time of collection. Urines must be kept chilled at 5°C if bacteriologic activity will adversely affect results (eg, Schilling's test).

Normal fluid intake is allowed during 24-hour urine collections. Dietary restrictions are required for some procedures and are specified in the individual test listing. Since results are based on total volume, it is critical that the volume be measured accurately and the information included with the paper or electronic test requisition. Clearance tests require an estimate of body surface area; patient's height and weight must be available to the laboratory or provided to the laboratory when a clearance is requested.

For other timed collection (eg, 12-hour, etc), similar directions, as for 24-hour urine collection, should be followed.

Causes for Rejection A number of laboratories reject the sample, if >10% is lost in collection.

Special Instructions See individual entries.

Limitations Some tests require a critical minimal volume for accuracy. For example, some Schilling test protocols require a minimum of 500 mL/24 hours, and are unreliable for reduced urine volumes. Consult individual test listings for information on critical urine volumes.

Additional Information The procedure may be followed for other timed collections (ie, 12-hour, etc) as follows: Discard the initial specimen. Record time. Collect all specimens voided within the requested time frame. Label the specimen with patient's name, hospital number, room number, and date and time of collection.

Due to difficulties in 24-hour urine collection, particularly in children, an increasing number of test results (eg, organic acids, amino acids, HVA, VMA, etc) are normalized to creatinine on randomly collected urine.

♦ **Urine Concentration Test** see Concentration Test, Urine on page 446
♦ **Urine Cortisol** see Cortisol, Free, Urine on page 459
♦ **Urine Creatinine** see Creatinine, 12- or 24-Hour Urine on page 471
♦ **Urine Crystals** see Urinalysis on page 1289
♦ **Urine Electrophoresis** see Protein Electrophoresis, Urine on page 1107
♦ **Urine K⁺** see Potassium, Urine on page 1080
♦ **Urine Ketones** see Ketones, Urine on page 817
♦ **Urine Osmolality** see Osmolality, Urine on page 979
♦ **Urine Osmolar Gap** see Osmolality, Urine on page 979
♦ **Urine Phosphorus** see Phosphorus, Urine on page 1032

Urobilinogen, 2-Hour Urine

Related Information

Bilirubin, Direct, Serum *on page 262*
Bilirubin, Total, Serum *on page 265*
Bilirubin, Urine *on page 268*
Liver Disease: Laboratory Assessment, Overview *on page 869*
Phenothiazines, Serum *on page 1021*
Porphobilinogen, Urine *on page 1073*

Synonyms Urobilinogen, Quantitative, Urine

Replaces Urobilinogen, 24-Hour Urine

Abstract Urobilinogen is formed in the intestine by the action of bacteria on excreted conjugated (direct) bilirubin. A portion of the urobilinogen is absorbed from the gastrointestinal tract into the bloodstream. It returns to the liver where some is re-excreted in bile (enterohepatic circulation), and the rest (via the general circulation) is excreted into the urine.

Patient Preparation A marked diurnal peak in excretion occurs; therefore an afternoon collection ideally should be scheduled. Patient should be hydrated.

Specimen 2-hour urine

Container Dark urine container or foil wrapped container

Collection Have patient void at 2 PM and discard urine. Give patient 500 mL of water to be ingested at once. Collect all urine from 2 PM - 4 PM. Transport promptly to the laboratory. Urobilinogen is sensitive to room temperature and light.

Storage Instructions Refrigerate specimen, protect from light.

Reference Interval Male: 0.3-2.1 mg/2 hours (SI: 0.5-3.6 µmol/2 hours); female: 0.1-1.1 mg/2 hours (SI: 0.2-1.9 µmol/2 hours). Results are often expressed in Ehrlich units, 1 mg urobilinogen = 1 EU, because what is measured is a mixture of compounds.

Use Urine urobilinogen can be increased in hemolytic anemia, hepatitis, liver damage, cirrhosis, and congestive heart failure. See table in Bilirubin, Urine *on page 268*.

Limitations The test has poor sensitivity.

Methodology Urobilistix®, Watson's method, Ehrlich's aldehyde reagent. Para-dimethylaminobenzaldehyde reacts with urobilinogen in Multistix®; this reaction is not specific, reacting with substances which are detected by Ehrlich's reagent, including porphobilinogen, procaine, 5-HIAA, *p*-aminosalicylic acid metabolites, sulfonamides, indole, and methyldopa. Chemstrip® uses 4-methoxybenzene-diazonium-tetra fluoroborate. Specific for urobilinogen; nitrite or formalin can reduce the reaction.

Additional Information See Bilirubin, Urine *on page 268*.

References

Binder L, Smith D, Kupka T, et al, "Failure of Prediction of Liver Function Test Abnormalities With the Urine Urobilinogen and Urine Bilirubin Assays," *Arch Pathol Lab Med*, 1989, 113(1):73-6.

- ♦ **Uroporphyrinogen Decarboxylase** *see* Porphyrins, Quantitative, Urine *on page 1074*
- ♦ **Uroporphyrinogen III Co-synthase** *see* Uroporphyrinogen III Synthase, Erythrocyte *on page 1297*

Uroporphyrinogen III Synthase, Erythrocyte

Related Information
Porphyrins, Quantitative, Urine *on page 1074*

Synonyms U-III-S; Uroporphyrinogen III Co-synthase

Abstract Congenital erythropoietic porphyria (CEP) results from decreased uroporphyrinogen III synthetase in red blood cells. This disorder is rarely a diagnostic problem because the presentation is very specific: severe photosensitivity and red-to-dark urine.

Patient Preparation Avoid medications for 1 week before test; fasting state for 12 hours before test; abstinence from ethanol for 24 hours before test.

Specimen Blood

Container Green top (heparin) tube

Collection Place on wet ice immediately. See instructions of reference laboratory.

Storage Instructions Maintain low temperatures; follow recommendations of laboratory which will perform the test.

Reference Interval Normal: ≥75 relative units; CEP: ≤10 relative units

Use Diagnose congenital erythropoietic porphyria (erythropoietic uroporphyria); investigate dark red urine, red to brown stained diapers, hemolytic anemia, splenomegaly, photosensitivity, mostly in neonates; investigate stained teeth, hirsutism, scarring of sun-exposed skin

Methodology Incubation of washed red cells with substrate and HPLC measurement of product

Additional Information The high porphyrin content of urine has been observed to cause characteristic reddish fluorescence on a diaper examined under Wood's light (320-420 nm). Long-term biochemical and clinical effectiveness of allogenic bone marrow transplantation seems to be promising.

References
Ahmed I, Cockerill F, Krause P, et al, "Childhood Porphyrias," *Mayo Clin Proc*, 2002, 77:825-36.

Fritsch C, Bolsen K, Ruzicka T, et al, "Congenital Erythropoietic Porphyria," *J Am Acad Dermatol*, 1997, 36(4):594-610.

Gilbert-Barness E and Barness LA, *Metabolic Diseases: Foundations of Clinical Management, Genetics, and Pathology*, Natick, MA: Eaton Publishing, 2000, 509-10.

Pollock SS and Rosenthal MS, Images in Clinical Medicine, "Diaper Diagnosis of Porphyria," *N Engl J Med*, 1994, 330:114.

- ♦ **Uroporphyrinogen I Synthase** *see* Porphobilinogen Deaminase, Erythrocyte *on page 1072*
- ♦ **Uroporphyrinogen Synthase** *see* Porphobilinogen, Urine *on page 1073*
- ♦ **Uroporphyrins** *see* Porphyrins, Quantitative, Urine *on page 1074*
- ♦ **UTI Screen** *see* Nitrite, Urine *on page 962*
- ♦ **Vacutainer® Tube Description** *see* Blood Collection Tube Information *on page 275*
- ♦ **Valine** *see* Amino Acids, Urine *on page 145*
- ♦ **Valium®** *see* Diazepam, Serum *on page 510*
- ♦ **Valkote®** *see* Valproic Acid, Serum or Plasma *on page 1297*
- ♦ **Valproate Semisodium** *see* Valproic Acid, Serum or Plasma *on page 1297*
- ♦ **Valproate Sodium** *see* Valproic Acid, Serum or Plasma *on page 1297*

Valproic Acid, Serum or Plasma

Related Information
Carbamazepine, Serum *on page 337*
Phenobarbital, Serum or Plasma *on page 1021*
Phenytoin, Serum or Plasma *on page 1026*
Primidone, Serum or Plasma *on page 1091*

Synonyms Depacon®; Depakene®; Depakote®; Depakote® XR; Depamide®; Dipropylacetic Acid; Divalproex Sodium; Epilim®; Ergenyl®; Leptilan®; 2-Propylpentanoic Acid; 2-Propylvaleric Acid; Valkote®; Valproate Semisodium; Valproate Sodium

Applies to Phenobarbital
(Continued)

Valproic Acid, Serum or Plasma (Continued)

Abstract Valproic acid is useful for many seizure types, including primary generalized tonic-clonic, partial, complex partial, myoclonic, atonic, and mixed seizures. It also is used for some psychiatric conditions including bipolar affective disorder and for prophylaxis of migraine headaches.

Specimen Serum or plasma

Container Red top tube or green top (heparin) tube

Sampling Time Trough values drawn just before next dose or consistent sampling time in chronic monitoring. Steady-state levels are reached in 2-3 days.

Reference Interval Total: 50-100 µg/mL (SI: 350-690 µmol/L); free: 6-20 µg/mL (SI: 42-140 µmol/L). Some patients require higher levels for seizure control.

Critical Values Toxic concentration >200 µg/mL (SI: >1390 µmol/L). Toxicity may occur at lower levels. Patients with hypoalbuminemia may become toxic within the normal range due to the nonlinear protein saturation pharmacokinetic characteristics of valproic acid. In these patients, measurement of free valproic acid is desirable. See Table A in the Therapeutic Drug Monitoring Introduction on page 60.

Use Monitor for compliance, efficacy, and possible toxicity

Contraindications Pregnant women on valproic acid are at risk for a higher incidence of neural tube defects and transient vitamin K-dependent clotting factors in the neonate.

Methodology Immunoassay, gas chromatography (GC), high performance liquid chromatography (HPLC)

Additional Information
- Half-life (adult): 8-15 hours
- Volume of distribution: 0.1-0.5 L/kg
- Protein binding: 85% to 95%

Valproic acid (VPA) is a broad spectrum antiepileptic that may work by increasing or enhancing the inhibitory neurotransmitter γ-aminobutyric acid (GABA) in the brain. VPA is almost entirely cleared hepatically and is a low extraction drug, so its clearance is dependent upon enzyme activity and plasma protein binding. The clearance of VPA is also very susceptible to alterations in enzyme activity for its clearance. Valproic acid inhibits the hepatic P450 enzymes that are responsible for the clearance of other drugs. Hepatic failure has occurred during the first 6 months of therapy. Hepatotoxicity may be preceded by nonspecific symptoms such as malaise, weakness, lethargy, anorexia, and vomiting. Hepatotoxicity occurs in very young children, most often those on multiple anticonvulsants. Children younger than age 6 are at increased risk of fatal hepatotoxicity. Encephalopathy with hyperammonemia without liver function test abnormalities may occur. About 20% of patients on valproic acid have hyperammonemia without liver damage. Liver function tests are recommended. Valproic acid may also cause embryopathy. Neural tube, cardiac, facial (characteristic pattern of dysmorphic facial features), skeletal, and multiple other defects have been reported.

References

Acharya S and Bussel JB, "Hematologic Toxicity of Sodium Valproate," *J Pediatr Hematol Oncol*, 2000, 22(1):62-5.

Brodie MJ and Dichter MA, "Antiepileptic Drugs," *N Engl J Med*, 1996, 334(3):168-75.

Kozma C, "Valproic Acid Embryopathy: Report of Two Siblings With Further Expansion of the Phenotypic Abnormalities and a Review of the Literature," *Am J Med Genet*, 2001, 98(2):168-75.

Sztajnkrycer MD, "Valproic Acid Toxicity: Overview and Management," *J Toxicol Clin Toxicol*, 2002, 40(6):789-801.

- **Valrelease®** see Diazepam, Serum on page 510
- **Vancocin®** see Vancomycin, Serum on page 1298
- **Vancoled®** see Vancomycin, Serum on page 1298

Vancomycin, Serum

Related Information

Aminoglycosides, Serum on page 147

Synonyms Lyphocin®; Vancocin®; Vancoled®

Applies to Red Man Syndrome

Abstract Vancomycin is a glycopeptide antimicrobial agent with potent activity against most gram-positive bacteria, including some gram-positive rods. It is often bactericidal (not against enterococcal species) and interferes with cell wall synthesis. Its blockade of pepidoglycan synthesis facilitates uptake of aminoglycosides, explaining the synergism of gentamicin with vancomycin. Due to emergence of vancomycin-resistant bacteria, its use is often reserved for gravely ill patients. Hospital Infection Control Practices Advisory Committee (HICPAC) recommends that each hospital should prepare guidelines for controlled use of vancomycin.

Specimen Serum, body fluid

Container Red top tube, sterile fluid container

Sampling Time Both peak and trough are measured. **Peak:** 15 minutes to 1 hour after completion of infusion; **trough:** immediately prior to next dose.

Storage Instructions Separate serum using aseptic technique and place in freezer.

Reference Interval Therapeutic: peak: 20-40 µg/mL (SI: 14-27 µmol/L); trough: 5-10 µg/mL (SI: 3.4-6.8 µmol/L)

Critical Values Trough concentrations ≥10 µg/mL may be associated with nephrotoxicity, but a relationship of nephrotoxicity to vancomycin use is controversial.

Possible Panic Range Toxic: >80 µg/mL (SI: >54 µmol/L). See Table B in the Therapeutic Drug Monitoring Introduction *on page 61*.

Limitations Monitoring serum vancomycin concentration has not been correlated with improved efficacy or decreased toxicity.

Methodology Immunoassay, high performance liquid chromatography (HPLC), gas chromatography (GC)

Additional Information
- Half-life: 4-8 hours (much longer in patients with renal insufficiency); 90% of an intravenous dose is eliminated by glomerular filtration over 24 hours. Concentrations in bile can reach half those in plasma. Cerebrospinal fluid levels reach 1% to 30% of those in plasma, respectively, in the absence or presence of inflammation in the meninges.
- Volume of distribution: 0.62-0.80 L/kg, increased by female sex, obesity, and increasing age
- Protein binding: 55% (range: 44% to 82%)

Oral doses are poorly bioavailable. Vancomycin is currently used in its intravenous form to treat a variety of gram-positive bacterial infections. Additionally, vancomycin is often used in its oral form to treat pseudomembranous colitis due to *Clostridium difficile*. The emergence of vancomycin-resistant enterococcal species (especially *Enterococcus faecium*) in many hospitals has created substantial oversight motivation to minimize this practice (as well as inappropriate intravenous usage of vancomycin, a fairly widespread practice). Toxic effects include nephrotoxicity and ototoxicity.

References
Li JTC, Markus PJ, Osmon DR, et al, "Reduction of Vancomycin Use in Orthopedic Patients With a History of Antibiotic Allergy," *Mayo Clin Proc*, 2000, 75(9):902-6.

Moran GJ and Mount J, "Update on Emerging Infections: News From the Centers for Disease Control and Prevention," *Ann Emerg Med*, 2003, 41(1):148-51.

Murray BE, "Drug Therapy: Vancomycin-Resistant Enterococcal Infection," *N Engl J Med*, 2000, 342(10):710-21.

Palmer-Toy DE, "Therapeutic Monitoring of Vancomycin," *Arch Pathol Lab Med*, 2000, 124(2):322-3.

Perl TM, "The Threat of Vancomycin Resistance," *Am J Med*, 1999, 106(5A):26S-37S.

Wilhelm MP and Estes L, "Symposium on Antimicrobial Agents - Part XII. Vancomycin," *Mayo Clin Proc*, 1999, 74(9):928-35.

Vanillylmandelic Acid, Urine

Related Information
Catecholamines, Fractionation, Plasma *on page 345*
Catecholamines, Fractionation, Urine *on page 347*
Creatinine, 12- or 24-Hour Urine *on page 471*
Homovanillic Acid, Urine *on page 743*
Metanephrines, Urine or Plasma *on page 899*
Urine Collection, 24-Hour *on page 1295*

Synonyms 3-Methoxy-4-Hydroxymandelic Acid; VMA

Applies to Dopamine
(Continued)

Vanillylmandelic Acid, Urine *(Continued)*

Abstract Vanillylmandelic acid (VMA) is a major metabolite of the catecholamines, epinephrine and norepinephrine. VMA is less sensitive than metanephrines in evaluation for pheochromocytoma.

Patient Preparation Many foods and drugs interfere. Request a list from the laboratory.

Specimen 24-hour urine

Container Plastic container with hydrochloric or acetic acid preservative added before collection, according to the protocol of the laboratory which will perform the test. Adjust to pH 2-4 after collection, according to procedures of the laboratory doing the analysis.

Collection Creatinine is measured concomitantly to ensure adequate collection and to calculate the excretion ratio of VMA/creatinine and metanephrine/creatinine.

Storage Instructions Refrigerate. Stable up to 2 weeks when the sample is acidified.

Reference Interval
- 0-11 months: <27 µg/mg creatinine
- 12-23 months: <18 µg/mg creatinine
- 2-4 years: <13 µg/mg creatinine
- 5-9 years: <8.5 µg/mg creatinine
- 10-14 years: <7.0 µg/mg creatinine
- Adults: <8.0 mg/24 hours

Use Evaluating a number of urinary tests for pheochromocytoma, Lucon et al found sensitivity of VMA second to that of metanephrines and comparable to that of norepinephrine. Its sensitivity exceeded that of epinephrine and dopamine. Witteles et al found VMA slightly less sensitive than plasma total catecholamines and urine total metanephrines.

VMA is used for diagnosis and follow-up of neuroblastoma, ganglioneuroma, and ganglioneuroblastoma. Most neuroblastoma patients excrete excess homovanillic acid (HVA) in 24-hour collections. If VMA and HVA are both measured, up to 80% of cases will be detected. Mass screening for neuroblastoma is reported to be effective.

Methodology High performance liquid chromatography (HPLC), spectrophotometric following extraction, gas chromatography/mass spectrometry (GC/MS)

References

Mayo Medical Laboratories, *2001 Test Catalogue*, Rochester, MN, 515.

Lucon AM, Pereira MA, Mendonca BB, et al, "Pheochromocytoma: Study of 50 Cases," *J Urol*, 1997, 157:1208-12.

Sawada T, Sugimoto T, Kawakatsu H, et al, "Mass Screening for Neuroblastoma in Japan," *Pediatr Hematol Oncol*, 1991, 8(2):93-109.

Young DS, *Effects of Drugs on Clinical Laboratory Tests*, 5th ed, Volume One: Listing by Test, Washington, DC: AACC Press, 2000, 3:835-6.

♦ **Variance of a Sample** *see page 11*

Varicella-Zoster Virus Culture and Serology

Related Information

Herpesvirus Cytology *on page 727*
Viral Culture *on page 1307*
Virus, Direct Detection by Fluorescent Antibody *on page 1311*

Test Includes Rapid culture by detection of virus using immunofluorescence staining with specific monoclonal antibody. Varicella-zoster virus (VZV) can be detected in a routine viral culture. Serology includes detection of IgG and IgM antibodies specific for VZV.

Abstract VZV produces two clinical syndromes, varicella or chickenpox in children and herpes zoster or shingles (reactivated latent infection) in adults. Severe sharp persistent pain (postherpetic neuralgia) following zoster can be debilitating. Deaths have occurred from primary infection with varicella, even though this disease is preventable with the availability of an effective vaccine. Diagnosis of chickenpox and shingles is usually based upon history and physical examination. In those rare cases in which VZV infection cannot be clinically distinguished from similar infections (eg, disseminated herpes simplex infection vs disseminated zoster), the diagnostic test of choice is viral culture. Serologic tests are used primarily to confirm recent or past infections.

Specimen Culture requires swab specimens from the base of fresh, unroofed lesions, vesicle fluid, vesicle scrapings; blood or bronchial washings (immuno-compromised patients). Although VZV cannot be cultured from cerebrospinal fluid (CSF) antibody in the CSF has been detected. In addition, acute and convalescent sera should be collected at appropriate times to document a clinically significant rise in antibody titer.

Container Culture specimens should be collected in cold viral transport medium for swab specimens; green top (heparin) tube for blood; sterile CSF tube; syringe or sterile capillary pipet can be used to collect vesicular fluid. Sera should be collected in a red top tube.

Sampling Time Culture specimens should be collected during the acute phase of the disease (within 3 days of lesion eruption). Acute and convalescent sera drawn 10-14 days apart are recommended. A single specimen is satisfactory to establish immune status. Antibodies to VZV are not usually detected until 1-3 days after the exanthem appears.

Collection Unroofed lesions should be cleaned before culture specimens are taken. Vesicular fluid from several vesicles can be pooled in a single syringe and sent to the laboratory undiluted. Alternatively, the bases of several freshly unroofed lesions can be vigorously sampled with a sterile swab and then placed into cold viral transport medium.

Storage Instructions Keep specimens cold and moist. **Do not freeze specimens at -20°C.** VZV is extremely labile. If immediate inoculation onto cell cultures is not possible, specimens should be frozen quickly at -70°C. Freezing at -70°C will reduce infectivity 10% to 30%.

Turnaround Time Conventional culture: 1-14 days; rapid culture: 2-5 days; serology: 2-5 days

Special Instructions Both PCR and antibody analysis are recommended for CSF evaluation.

Reference Interval No virus isolated in culture. A single low titer or less than a fourfold increase in IgG titer in paired sera. **Immune status:** IgG alone is sufficient. Reactive result indicates immunity (either from infection or vaccination) but does not assure protection from shingles.

Critical Values Positive culture or serology from CSF or blood. Negative serology in an adult or pregnant female exposed to varicella-zoster virus.

Use Aid in the diagnosis of disease caused by varicella-zoster virus (ie, chickenpox and shingles). Serology is used to determine adult susceptibility or immunity to infection.

Limitations Rapid culture method will only detect VZV or HSV, negative culture does not rule out other viral infections. VZV is extremely labile and many times cannot be isolated from specimens that have been transported in adverse conditions.

Complement fixation test is insensitive and has heterologous reactions with herpesvirus. Primary or secondary infections with other herpesviruses may produce significant VZV titer increases in individuals who have previously had chickenpox. A positive low titer may not correlate with protection. Zoster can develop despite substantial antibody titers. Antibodies to VZV persist in serum in most adults.

Methodology Routine culture: Culture requires inoculation of specimens into cell cultures, incubation of cultures, observation for characteristic cytopathic effect (CPE), and identification by fluorescent monoclonal antibody.

Rapid culture: Specimens are centrifuged onto cell cultures to accelerate virus attachment and penetration. After incubation, fluorescein-labeled monoclonal antibodies are applied to the infected cells to detect viral antigens. Characteristic fluorescent foci indicate the presence of virus.

Serology: Fluorescent antibody to membrane antigen (FAMA), hemagglutination, anticomplement immunofluorescence, enzyme-linked immunosorbent assay (ELISA). Molecular techniques (eg, PCR) will become methods of choice for VZV when methods become widely available. Detection of viral DNA in amniotic fluid by PCR may prove to be useful.

Additional Information Varicella-zoster virus is a common virus found around the world. A primary VZV infection can be life-threatening in pregnant persons, immunocompromised persons, children who receive cancer therapy, following organ transplantation, and in fetuses exposed during pregnancy. Complications (Continued)

Varicella-Zoster Virus Culture and Serology
(Continued)

include dissemination, pneumonitis, myocarditis, cardiomyopathy, hepatitis, Guillain-Barré syndrome, myelitis, ventriculitis, granulomatous arteritis, vasculopathy, and meningoencephalitis. Congenital chickenpox can result in neonatal systemic disease and/or congenital malformations. People suffering from AIDS can have prolonged reactivated VZV infections. Zoster occurs mostly in elderly and immunocompromised individuals.

Serology may be useful in the differential diagnosis of blistering illnesses or when an unusual complication, such as hepatitis, develops. IgG antibodies can be detected 9 days after the onset of rash in primary varicella and 10 days after reactivation of zoster. IgM antibodies can be detected after 6-7 days and peak at ~14 days. It may also be important to establish whether an individual is susceptible when clinical history is unclear, or when varicella immune globulin may be needed, as in the immunocompromised host or cancer patient on toxic chemotherapy. Zoster is more common with aging and may occur in spite of significant antibody titers.

A vaccine consisting of live-attenuated varicella is now available and is recommended by the CDC for children between the ages of 1-6 years of age, persons who work in an environment where transmission of VZV is likely such as day care center, teachers, residents and staff in institutional settings, college students, prison inmates, and military personnel. Additionally, the vaccine is recommended for nonpregnant women of childbearing years and international travelers who are unsure of past exposure. Patients who are immunocompromised (eg, childhood leukemics) and other susceptible individuals should also be immunized. Positive serology should be indicative of protection.

References

Chartrand SA, "Varicella Vaccine," *Pediatr Clin North Am*, 2000, 47(2):373-94.

Centers for Disease Control and Prevention, "Varicella - Related Deaths - Florida, 1998," *MMWR Morb Mortal Wkly Report*, 1999, 48(18):379-81.

Cohen JI, Brunell PA, Straus SE, et al, "Recent Advances in Varicella-Zoster Virus Infection," *Ann Intern Med*, 1999, 130(11):922-32.

Gilden DH, Kleinschmidt-DeMasters BK, LaGuardia JJ, et al, "Neurologic Complications of the Reactivation of Varicella-Zoster Virus," *N Engl J Med*, 2000, 342(9):635-45.

Isaacs D, "Neonatal Chickenpox," *J Paediatr Child Health*, 2000, 36(1):76-7.

Kleinschmidt-DeMasters BK and Gilden DH, "Varicella-Zoster Virus Infections of the Nervous System. Clinical and Pathologic Correlates," *Arch Pathol Lab Med*, 2001, 125(6):770-80.

Leisegang TJ, "Varicella Zoster Viral Disease," *Mayo Clin Proc*, 1999, 74(10):983-8.

Niederhauser VP, "Varicella: The Vaccine and the Public Health Debate," *Nurse Pract*, 1999, 24(3):74-84.

Sauerbrei A, Eichhorn U, Schalke M, et al, "Laboratory Diagnosis of Herpes Zoster," *J Clin Virol*, 1999, 14(1):31-6.

Internet Web Sites
www.cdc.gov/ncidod/diseases/list_varicl.htm
www.who.int/vaccines-diseases/diseases/PP_Varicella.shtml

♦ **Variola Major/Variola Minor Diagnostic Procedures** *see* Smallpox Diagnostic Procedures *on page 1205*

Vasoactive Intestinal Polypeptide, Plasma
Related Information

Glucagon, Plasma *on page 640*
Pancreatic Polypeptide, Human, Serum or Plasma *on page 998*

Synonyms VIP

Abstract Vasoactive intestinal polypeptide (VIP), a nonadrenergic noncholinergic neurotransmitter, is found widely distributed in cells of the central and peripheral nervous systems. Clinical interest in VIP centers on tumors which secrete VIP (so-called **VIPomas**) and thereby cause a high-volume secretory diarrhea much like cholera. The term **pancreatic cholera** has been applied since, in adults, most VIPomas are in the tail of the pancreas. A clinically identical syndrome of secretory diarrhea, also due to VIP, occurs in children, but the tumors are most commonly in the adrenal (see Use).

Patient Preparation Avoid recent administration of radioactive isotopes. Patient must be completely fasting for 10-12 hours. Not even water may be taken. No antacids for 24 hours prior to collection. All medications should be discontinued for 24-48 hours prior to collection.

Specimen Plasma

Sampling Time 6 AM to 8 AM

Collection Collect specimen in lavender top (EDTA) tube and centrifuge immediately. Freeze plasma in plastic vial.

Storage Instructions Transport frozen on dry ice.

Reference Interval Method-dependent, with wide interlaboratory variation. One publication states <50 pg/mL, and another <75 pg/mL.

Use In the appropriate clinical context, increased VIP levels support a diagnosis of VIPoma, but provide no information about the location of the tumor(s). The most common location in adults is the tail of the pancreas. In a review of 18 adult patients, the preoperative VIP levels ranged from 15.0-5975.0 pg/mL; only one patient had a value below their laboratory cutoff of 75 pg/mL.

In children, most VIPomas are ganglioneuromas or ganglioneuroblastomas. Reported locations include the adrenal, gastrointestinal tract, lung, and paravertebral soft tissue.

Limitations Not all patients with a syndrome of pancreatic cholera have increased VIP. Increased VIP can be found in healthy controls and in laxative abusers.

Methodology Radioimmunoassay (RIA)

References

Krejs GJ, "Noninsulin-Secreting Tumors of the Pancreatic Islets," *Williams Textbook of Endocrinology*, 9th ed, Wilson JD, Foster DW, Kronenberg HM, et al, eds, Philadelphia, PA: WB Saunders, 1998, 1667.

Murphy MS, Sibal A, and Mann JR, "Persistent Diarrhea and Occult VIPomas in Children," *BMJ*, 2000, 320(7248):1524-6.

Smith SL, Branton SA, Avino AJ, et al, "Vasoactive Intestinal Polypeptide Secreting Islet Cell Tumors: A 15-Year Experience and Review of the Literature," *Surgery*, 1998, 124(6):1050-5.

♦ **Vasopressin** *see* Antidiuretic Hormone, Plasma *on page 172*

♦ **Vasopressin Concentration Test** *see* Concentration Test, Urine *on page 446*

♦ **Vatran®** *see* Diazepam, Serum *on page 510*

♦ **Vazepam®** *see* Diazepam, Serum *on page 510*

VDRL, Serum or Cerebrospinal Fluid

Related Information

Bacterial Culture, Cerebrospinal Fluid *on page 236*
Cerebrospinal Fluid Analysis: Overview *on page 355*
Cerebrospinal Fluid Protein *on page 371*
Darkfield Examination, Syphilis *on page 501*
FTA-ABS, Serum *on page 618*
MHA-TP *on page 912*
RPR *on page 1174*

Test Includes Detection of serologic response to *Treponema* infection and titer of reactive specimens

Abstract Throughout the world, syphilis is a common sexually transmitted disease despite effective therapy. The diagnosis of syphilis can present a difficult dilemma. Nontreponemal serologic tests are used for screening but can be negative if performed in the primary stage. The VDRL serum test may also be falsely negative in late syphilis. Most of the syphilis cases reported in the U.S. are diagnosed at the latent stage and many cases are not diagnosed until the patient is suffering from neurosyphilis. A reactive CSF VDRL test result is acceptable to establish the diagnosis of neurosyphilis.

Specimen Serum, cerebrospinal fluid

Container Red top tube for serum; clean, sterile CSF container

Collection The CDC recommends that when screening for congenital syphilis, the mother's serum should be tested rather than cord blood.

Causes for Rejection Plasma specimen

Special Instructions Do not heat inactivate the CSF.

Reference Interval Nonreactive

Possible Panic Range Positive results in pregnancy should be confirmed by a treponemal test.

Use VDRL is a reaginic (nontreponemal) test for syphilis. VDRL test on CSF is used for diagnosis of neurosyphilis. VDRL, CSF is the only laboratory test for neurosyphilis approved by the Centers for Disease Control.

(Continued)

VDRL, Serum or Cerebrospinal Fluid (Continued)

Limitations False-positive results may be found in advancing age, during pregnancy, malaria, infectious mononucleosis, with anticardiolipin antibodies, infectious hepatitis, leprosy, brucellosis, connective tissue diseases, atypical pneumonia, typhus, and other entities. Reactive tests due to related treponemal infections also occur: *T. pallidum* subspecies pertenue, which causes yaws; *T. carateum* which causes pinta, and *T. pallidum* subspecies endemicum, the cause of nonvenereal or endemic syphilis. False-negative results occur early in primary syphilis when the chancre is still present, and also occur in late syphilis. False-negative tests due to prozone phenomenon are identified by testing with diluted serum. This should be performed in individuals who test negative despite high clinical suspicion. A negative VDRL on CSF does not rule out the diagnosis of neurosyphilis; further investigation when clinically appropriate may include a serum FTA-ABS.

Methodology Flocculation test detects reagin, antibody to nontreponemal antigen

Additional Information Despite false-positive results, the VDRL remains an extremely useful screening test for syphilis. The VDRL test is very specific. The sensitivity of the CSF VDRL for the diagnosis of neurosyphilis is ~90%. The VDRL becomes positive 2 weeks after the chancre appears, and by 6 weeks, 90% of cases are positive. By 9-12 weeks, in the secondary stage, 100% of patients should be reactive. (The secondary stage includes rash, mucocutaneous lesions, and lymphadenopathy.) The latent stage follows. With therapy the VDRL reverts to negative. Even without treatment the VDRL may become negative years after infection. Thus, the VDRL may be negative in tertiary syphilis. Positive VDRL results should be confirmed with a *Treponema*-specific test. The diagnosis of syphilis in persons infected with HIV may be difficult: lack of serologic response in a patient with confirmed syphilis is seen and rapid progression in spite of treatment occurs.

A negative test may occur in 30% of patients with tabes dorsalis, and will be positive in only 69% to 92% of patients who had active neurosyphilis. The CSF VDRL may take years to become nonreactive after adequate therapy.

Neurosyphilis includes **meningeal syphilis**, which is usually found within a year of infection. Its characteristics include headache, stiff neck, nausea and vomiting, sometimes with cranial nerve involvement. **Syphilitic meningitis** can be localized (**gumma**). **Meningovascular syphilis** is found 4-7 years after infection. **General paresis** or **tabes dorsalis** occur late, frequently decades after infection. Uveitis, retinitis, luetic optic neuritis may occur with or without syphilitic meningitis. Central nervous system syphilis may be asymptomatic. The diagnosis of neurosyphilis is excluded if the serum FTA-ABS result is nonreactive.

References

Birnbaum NR, Goldschmidt RH, and Buffet WO, "Resolving the Common Clinical Dilemmas of Syphilis," *Am Fam Phys*, 1999, 59(8):2233-46.

Flood JM, Weinstock HS, Guroy ME, et al, "Neurosyphilis During the AIDS Epidemic, San Francisco, 1985-1992," *J Infect Dis*, 1998, 177(4):931-40.

Golden MR, Marra CM, and Holmes KK, "Update on Syphilis - Resurgence of an Old Problem," *JAMA*, 2003, 290(11):1510-4.

Jacobs DS, DeMott WR, Oxley DK, et al, *Laboratory Test Handbook*, 5th ed, Hudson, OH: Lexi-Comp Inc, 2001.

Larsen SA, Steiner BM, and Rudolph AH, "Laboratory Diagnosis and Interpretation of Tests for Syphilis," *Clin Microbiol Rev*, 1995, 8(1):1-21.

Luger AF, Schmidt BL, and Kaulich M, "Significance of Laboratory Findings for the Diagnosis of Neurosyphilis," *Int J STD AIDS*, 2000, 11(4):224-34.

Woods GL, "Update on Laboratory Diagnosis of Sexually Transmitted Diseases," *Clin Lab Med*, 1995, 15(3):665-84.

Internet Web Sites

www.cdc.gov/std/Syphilis/STDFact-Syphilis.htm

♦ **Velocardiofacial Syndrome** *see* Gene Rearrangement for Leukemia and Lymphoma *on page 633*

♦ **Venipuncture, Venous** *see* Venous Blood Collection *on page 1304*

Venous Blood Collection

Related Information

Blood and Fluid Precautions, Specimen Collection *on page 271*
Blood Collection Tube Information *on page 275*

Electrolyte Panel, Serum *on page 532*
Ethanol, Blood, Urine, and Other Sources *on page 558*
Lactic Acid, Whole Blood or Plasma *on page 827*
pCO₂, Blood *on page 1008*
pH, Blood *on page 1018*
Phlebotomist Procedures *on page 1028*
Potassium, Serum or Plasma *on page 1078*
Specimen Identification Requirements *on page 1217*

Synonyms Blood Collection, Venous; Phlebotomy, Venous; Venipuncture, Venous

Test Includes Routine method for obtaining blood when anticoagulants or larger volumes than can be obtained by capillary blood collection are required

Patient Preparation Select a suitable site for venipuncture. Prepare the site by scrubbing with 70% isopropyl alcohol (isopropanol) using a circular motion working out from the site. Dry with gauze or let air dry. Do not touch site after cleansing until blood drawing is complete. Special requirements exist for alcohol levels; see Ethanol, Blood, Urine, and Other Sources *on page 558*. Avoid hand pumping.

Aftercare Apply pressure to the venipuncture site and elevate the arm until bleeding stops. If bleeding persists, apply a pressure dressing to the site.

Specimen Venous blood

Container Syringe with a 20- or 21-gauge needle for volumes up to 10 mL, 18-gauge for larger volumes to assure adequate blood flow. A 20- or 21-gauge Butterfly® infusion set may be used for difficult draws or blood cultures with multiple tubes.

Sampling Time Some samples for analytes that exhibit diurnal variation should be drawn at specific times of the day. See individual test listings.

Collection Before collecting blood, the phlebotomist should verify that the patient is fasting, if fasting is indicated. The patient should be comfortably seated or supine for a few minutes before blood collection. If one must be used at all, apply tourniquet 4-6 inches above drawing site. Do not leave tourniquet applied for more than 1 minute. Cleanly puncture the vein, loosen the tourniquet, and apply gentle suction or insert Vacutainer® tube into holder to fill tubes. Remove the needle and fill the tubes without delay.

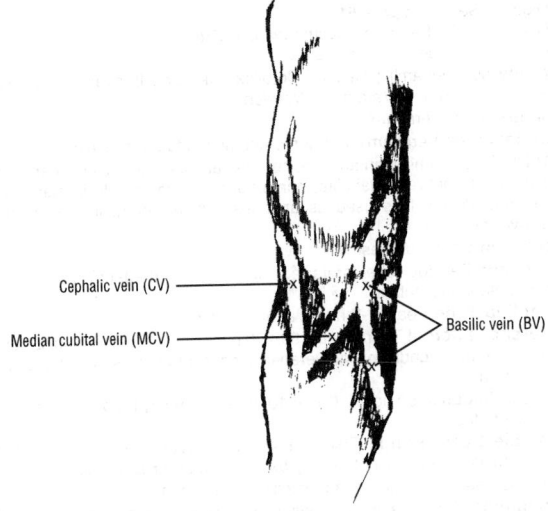

Cephalic vein (CV)
Median cubital vein (MCV)
Basilic vein (BV)

From JD Bauer, *Clinical Laboratory Methods*, 9th ed, St. Louis, MO: Mosby-Year Book Inc, 1982, with permission.

If blood is being drawn for venous pH, pCO_2, lactate, or electrolytes, it is best to draw blood without a tourniquet. If one is needed, leave it in place while the sample is drawn. Alternatively, if a tourniquet must be used, sample blood 1-2 minutes after the hand is relaxed and the tourniquet removed. **Avoid hand clenching for these tests, especially for potassium and lactate determinations.** The first aliquot is best for pH.

Gently invert tubes 10 times to assure mixing of anticoagulants.

Causes for Rejection Samples collected for coagulation studies which have <90% of the expected fill should be rejected. Samples collected for coagulation studies may require additional anticoagulant if the hematocrit is low. Grossly hemolyzed specimens may be rejected depending on tests requested. Specimens that are not properly identified cannot be accepted.

Use Obtain venous blood for analysis

Limitations Venipuncture is technically difficult in obese patients, infants, children, patients with collapsed veins such as those in shock, and occasionally, other subjects as well. Hemolysis may occur as a result of excessive suction during collection, violent mixing of the specimen, or vigorous transfer of the specimen from syringe to tube.

Contraindications Fist clenching will alter potassium, pH and pCO_2 studies. If a tourniquet is applied but released before blood is aspirated, K^+ and lactate are washed out resulting in erroneous and misleading results.

Additional Information Draw EDTA tube last. Hematology tubes contain K-EDTA. Any contamination can raise potassium and decrease pH. **See individual test listings for specific specimen collection requirements.**

References

Farber V, "Venipunctures on the Elderly: Handle With Care," *Adv Med Lab Professionals*, 2001, 13:9-11.

Flynn JC Jr, *Procedures in Phlebotomy*, 2nd ed, Philadelphia, PA: WB Saunders Co, 1999.

Gambino R, "The Correct Way to Draw Venous Blood - Does Stasis Matter?" *Lab Rep*, 1996, 18(1):1-3.

♦ **Venous Blood Gases** *see* Blood Gases and pH, Venous *on page 279*

♦ **Ventricular Hypertrophy** *see* B-Type Natriuretic Peptide *on page 314*

♦ **Veramex**® *see* Verapamil, Serum or Plasma *on page 1306*

Verapamil, Serum or Plasma

Related Information

Digoxin, Serum *on page 512*
Disopyramide, Serum or Plasma *on page 516*
Quinidine, Serum *on page 1129*

Synonyms Azupamil®; Calan®; Cordilox®; Ikacor®; Iproveratril Hydrochloride; Isoptin®; Securon®; Veramex®; Verelan®

Applies to Norverapamil

Test Includes Verapamil and norverapamil (metabolite levels)

Abstract Verapamil is an antihypertensive, antianginal calcium channel antagonist with an active metabolite, norverapamil. It has a role in management of cardiomyopathy. It is used as an antiarrhythmic drug, including therapy of tachyarrhythmias.

Specimen Serum or plasma

Container Red top tube (preferred) or green top (heparin) tube. Do not use serum separator tubes.

Sampling Time Peak: 1-2 hours after last dose

Reference Interval Therapeutic: 50-250 ng/mL (SI: 100-510 nmol/L) for parent; under normal conditions norverapamil concentration is about the same as parent drug.

Critical Values >250 ng/mL (SI: >510 nmol/L); toxicity proportional to verapamil concentration

Possible Panic Range Toxic: >845 ng/mL; **fatal:** >2000 ng/mL. A ratio of verapamil/norverapamil >2.3 may be a predictor for fatal outcome.

Use Therapeutic monitoring and toxicity assessment

Limitations Clinical signs and symptoms do not strongly correlate with drug concentrations. Verapamil is commercially available as a racemic mixture. The (S-) enantiomer is the isomer with the greatest pharmacological activity. These enantiomers exhibit stereoselective absorption, binding, and clearance. This makes for a very complex pharmacokinetic picture. It also makes it very difficult to associate measured verapamil concentrations with effect when it cannot differentiate the active (S) enantiomer from the relatively inactive (R) enantiomer.

Methodology High performance liquid chromatography (HPLC), gas chromatography (GC)

Additional Information
- Half-life: 3-5 hours
- Volume of distribution: 4-6 L/kg
- Protein binding: 85% to 95%

Verapamil is an antiarrhythmic, antihypertensive drug. Its main metabolite is norverapamil, which has about 20% of the activity of the parent drug. Verapamil is a calcium channel blocker. Coadministration of verapamil and beta-blockers should be approached with caution. Verapamil increases serum digoxin concentrations 50% to 70%. Toxicity may result when verapamil is used with carbamazepine or lithium.

Increased cardiovascular adverse effects occur with beta-adrenergic blocking agents (especially when administered intravenously), digoxin, quinidine, and disopyramide. Verapamil may increase serum concentrations of digoxin, quinidine, carbamazepine, theophylline, and cyclosporine necessitating a decrease in dosage. Phenobarbital and rifampin may decrease verapamil serum concentrations by increasing its clearance.

References
Brogden RN and Benfield P, "Verapamil: A Review of its Pharmacological Properties and Therapeutic Use in Coronary Artery Disease," *Drugs*, 1996, 51(5):792-819.
Hansen JF and Mellemgaard K, "Angina Pectoris, Myocardial Infarction, and Verapamil," *J Am Coll Cardiol*, 1999, 34(3):957-8.
Jespersen C, "Verapamil in Acute Myocardial Infarction. The Rationales of the VAMI and DAVIT III Trials," *Cardiovasc Drugs Ther*, 1999, 13(4):301-7.
Pritza DR, Bierman MH, and Hammeke MD, "Acute Toxic Effects of Sustained-Release Verapamil in Chronic Renal Failure," *Arch Intern Med*, 1991, 151(10):2081-4.

♦ **Verelan®** *see* Verapamil, Serum or Plasma *on page 1306*
♦ **Verocytotoxins** *see* Shiga Toxin Test, Direct *on page 1195*
♦ **Very Low Density Lipoprotein Cholesterol** *see* Cholesterol, Total, Serum or Plasma *on page 396*
♦ **Very Low Density Lipoprotein Cholesterol** *see* Lipids, Overview *on page 853*
♦ **Vicks Inhaler®** *see* Methamphetamine, Qualitative, Urine *on page 902*
♦ **Vigabatrin** *see* Antiepileptic Drugs Overview *on page 176*
♦ **Vimentin** *see* Immunoperoxidase Procedures *on page 780*
♦ **VIP** *see* Vasoactive Intestinal Polypeptide, Plasma *on page 1302*

Viral Culture

Related Information
Adenovirus Culture and Serology *on page 109*
Bacterial Culture, Blood *on page 232*
Bacterial Culture, Genital Specimen *on page 239*
Bacterial Culture, Urine *on page 246*
Chlamydia trachomatis Nucleic Acid Detection *on page 388*
Cytomegalovirus Culture *on page 495*
Enterovirus Culture *on page 539*
Herpes Simplex Antibody *on page 720*
Herpes Simplex Virus Culture and Antigen Detection *on page 721*
Herpes Simplex Virus DNA Detection *on page 723*
Influenza Virus Detection, Culture and Serology *on page 791*
Mumps Culture and Serology *on page 926*
Parainfluenza Viral Culture and Serology *on page 999*
Poliomyelitis I, II, III Serology *on page 1069*
Rubella Culture and Serology *on page 1174*
Varicella-Zoster Virus Culture and Serology *on page 1300*

Test Includes Inoculation of specimen onto appropriate cell cultures; isolated viruses are identified using specific monoclonal antibodies, neutralization, or hemadsorption

Abstract The primary technique for many viral infections is the isolation of virus in tissue culture cells. For other viral infections serologic diagnosis, direct antigen detection in body fluids or tissues, or nucleic acid detection is the method of choice for diagnosis. For viral culture, specimens must be collected (Continued)

Viral Culture *(Continued)*

within the first few days of an illness and specimens brought to the laboratory in viral holding medium as quickly as possible.

Specimen Whole blood, cerebrospinal fluid, dermal, ocular, genital, mucosal, respiratory, oral, stool, rectal, urine, tissue, biopsy. See table.

Viruses Typically Isolated From Clinical Specimens

Specimen	Virus[1]
Blood	CMV, enteroviruses[2,3], HSV,[3] VZV[3]
CSF and CNS tissues	Enteroviruses, mumps virus, HSV, CMV
Dermal lesions	HSV, VZV, adenovirus, enteroviruses
Eye	HSV, VZV, adenovirus, enteroviruses, CMV, *Chlamydia*
Genital	HSV, CMV, *Chlamydia*
Mucosal	HSV, VZV
Oral	HSV, VZV
Rectal	HSV, VZV, enterovirus
Respiratory tract	
upper	Adenovirus, rhinovirus, influenza, parainfluenza, enteroviruses, RSV, reovirus, HSV
lower	Adenovirus, influenza, parainfluenza, RSV, CMV[4]
Sputum not acceptable	
Stool	Enteroviruses, adenoviruses
Tissues	CMV, HSV, enteroviruses
Urine	CMV, adenovirus, enteroviruses, mumps

[1]Abbreviations:

HSV — herpes simplex virus
CMV — cytomegalovirus
VZV — varicella-zoster virus
RSV — respiratory syncytial virus

[2]Enteroviruses: Coxsackievirus, poliovirus, echovirus, and enterovirus.

[3]Rarely isolated.

[4]Usually in immunocompromised hosts.

Container Viral transport medium for swabs; sterile screw-cap tube or container for fluids, feces, nasal washings, urine, or biopsy (without preservative); green top (heparin) tube for blood, bone marrow, and buffy coat. **Keep all specimens at 4°C and moist. Do not freeze specimen at -20°C especially in a frost-free freezer.**

Collection Specimen should be collected during the acute phase of the disease, as follows.

Blood: 5 mL whole blood in heparinized tube

Cerebrospinal fluid: Collect 1 mL CSF aseptically in a sterile dry screw-cap vial. Spinal fluid and throat washings must be kept cold and must not be frozen. **Keep cold and bring to the laboratory immediately.**

Skin lesions: Open the vesicle and absorb exudate into a dry swab, and/or vigorously scrape base of freshly exposed lesion with a swab. If enough vesicle fluid is available, aspirate the fluid with a fine-gauge needle and tuberculin syringe, and place fluid into cold viral transport medium. Use Virocult® or Culturette® swabs for specimen collection. **Keep cold and bring to the laboratory immediately.** Calcium alginate swabs are toxic to *Chlamydia* and many enveloped viruses. **Do not use calcium alginate swabs for viral or chlamydial isolation.**

Eye swab or scraping: Use a Virocult® or Culturette® swab to collect conjunctival material or take conjunctival scrapings with a fine sterile spatula and transfer the scraping to a viral transport medium. **Keep cold and bring to the laboratory immediately.** Calcium alginate swabs are toxic to *Chlamydia* and many enveloped viruses. **Do not use calcium alginate swabs for viral or chlamydial isolation.**

Genital swab: See skin. **Keep cold and bring to the laboratory immediately.**

Throat swab: Carefully rub the posterior wall of the nasopharynx with a dry, sterile swab. Avoid touching the tongue or buccal mucosa. Use Virocult® or Culturette® swabs for specimen collection. **Keep cold and bring to the laboratory immediately.**

Feces: Collect 4-8 g of feces (about the size of a thumbnail), and place in a clean, leakproof container. Stool specimens should not be placed into viral transport medium or frozen. **Do not** dilute the specimen (into virus transport medium) or use preservatives. **Keep cold and bring to the laboratory immediately.**

Rectal swab: Insert a sterile swab 2-4 inches into the rectum and rub the mucosa. Use Virocult® or Culturette® swabs for specimen collection. **Keep cold and bring to the laboratory immediately.** Swab may be placed into cold virus transport medium.

Urine: Collect clean-catch, midstream urine in a sterile container. Urine specimens for CMV culture **must not** be frozen; they should be packed with an ice pack or snow gel, but not with dry ice. **Keep cold and bring to the laboratory immediately.**

Tissue: Use a fresh set of sterile instruments to collect each tissue. Place each specimen in a sterile nontoxic container. Identify each tissue with patient's name, type of tissue, and date collected. **Keep cold and bring tissue to the laboratory immediately.**

Storage Instructions Specimen must be kept cold and moist, and must be delivered to the laboratory as soon as possible. If a longer period is required, check with the laboratory for specific instruction. **Do not freeze specimens at -20°C.** For longer storage, freeze the specimen quickly at -70°C. Specimens to be cultured for influenza virus and cytomegalovirus should be sent on wet ice or with an ice pack.

Causes for Rejection Dry specimen, specimen not refrigerated during transport, specimen fixed in formalin, unlabeled specimen, use of calcium alginate swab

Turnaround Time The characteristic cytopathic effect (CPE) can be observed as soon as 1 day and as late as 28 days postinoculation of the cell culture. Virus grown in rapid cultures from shell-vials can be detected as early as 24 hours after inoculation.

Special Instructions Culture, serological tests, antigen detection, and nucleic acid detection for certain specific viruses are available. Contact the laboratory to determine availability of tests. When possible, the serological tests should be requested at same time as culture. Acute and convalescent blood samples (5 mL) are required for serologic studies. Requisition **must** state specific virus(s) suspected, source of specimen, age of patient, current antibiotic therapy, relevant vaccinations, and pertinent clinical history.

Reference Interval No virus isolated

Use Aid in the diagnosis of viral diseases

Limitations Many common viruses are not culturable. For many such viral diseases, serologic testing is required. Recovery of virus from sites where it may be found in the absence of disease may not be the causative agent of disease. Negative viral culture does not rule out a viral etiology. Some positive cultures are sent to State Health Laboratory for specific virus identification.

Viruses That Are Nonculturable or Require Animal Inoculation or Special Culture Technique

Arenaviruses	Hepatitis D
Astrovirus	Hepatitis E
Calicivirus	Lassa
California encephalitis	Marburg
Coronaviruses	*Molluscum contagiosum*
Coxsackievirus type A	Norwalk agent
Dengue	Papillomaviruses
Eastern equine encephalitis	Parvoviruses
Ebola	Polyomavirus
Filoviruses	Rabies
Hantavirus (Muerto Canyon)	Rotaviruses
Hepatitis A	St Louis encephalitis
Hepatitis B	Western equine encephalitis
Hepatitis C	Yellow fever

(Continued)

Viral Culture (Continued)

Methodology Inoculation of specimen into cell cultures, incubation of cultures, observation for characteristic cytopathic effect (CPE), and identification by methods such as hemadsorption and fluorescent monoclonal antibodies. If specific viruses such as HSV, CMV, VZV, influenza, RSV, parainfluenza, or adenovirus are suspected, the laboratory might be able to use rapid (1-2 days) culture (shell vial) methods to detect these viruses.

Additional Information

Viral cultures: Viruses are obligate intracellular organisms that require cell culture techniques for growth. The detection and identification of viruses associated with infection is important for patient management, such as choice of antiviral therapy, and for epidemiology. Some viral infections can be detected by identification of viral antigens without culture. Other diagnostic methods are the detection of specific viral nucleic acid and the detection of a specific antibody response in the patient. The laboratory diagnosis of viral diseases is changing rapidly; consultation with the laboratory for currently used methods is recommended.

Virus from blood: It is rare to detect virus from blood specimens, but when a virus such as varicella-zoster or herpes simplex is isolated from a blood specimen it usually indicates a serious condition. The most commonly isolated virus from blood is cytomegalovirus (CMV) in immunocompromised patients.

Virus from body fluids and tissue: Viruses are seldom isolated from synovial or pericardial fluid. Often when a virus causes pericarditis or arthritis, it is no longer present when symptoms are diagnosed. Viruses are most commonly isolated from fluids such as cerebrospinal fluid, amniotic fluid, and pleural fluid and occasionally lymph nodes.

Virus from cerebrospinal fluid: Viral meningitis is the most important cause of aseptic meningitis, a condition which has several clinical presentations and which has both infectious and noninfectious etiologies. Enteroviruses most often cause meningitis in the late summer and are the most common cause of aseptic meningitis in the United States. Patients are usually children and young adults. Enteroviruses include polioviruses, coxsackieviruses, and echoviruses.

Virus from skin lesions: Viral agents most commonly associated with skin lesions are herpes simplex virus (HSV), type 1 and 2, and varicella-zoster virus (VZV). These viruses are prevalent worldwide. Both viruses establish latency after a primary infection and can react to cause lesions when a host becomes immunosuppressed.

Virus from the eye: The leading cause of a red eye is viral conjunctivitis, which usually affects one eye before the other.1 Conjunctivitis is the most common eye infection and can be found in patients of all ages. Several viruses can infect the eye. Among the most common viral infections of the eye are adenovirus, influenza viruses, herpes simplex viruses (HSV), varicella-zoster virus (VZV), and cytomegalovirus (CMV) infections. Infection of the cornea, keratitis, is a much more serious infection and if left untreated can result in loss of eyesight.

Virus from respiratory tract: Viral agents cause 20% to 62% of community-acquired pneumonias. The most common viral agents are influenza viruses, parainfluenza virus, rhinoviruses, and respiratory syncytial virus (RSV). Pneumonia due to adenovirus is less common.

Virus from urogenital specimens: One of the most common viruses isolated from urogenital specimens is herpes simplex viruses (HSV). Many infected women are of childbearing age can transmit the virus to a fetus or newborn. The classic presentation of primary herpes simplex infection is the appearance of several painful genital vesicles. Primary infections are also associated with fever and inguinal lymphadenopathy.

In general, specimens for viral culture should be collected in the acute stage of the illness, kept moist, and refrigerated immediately. Always check with the laboratory to coordinate shipping/transfer times and expected arrival times. Viruses that are enveloped (herpesviruses, myxoviruses, and paramycoviruses such as RSV) are not stable outside the body. However, these viruses can survive transit for at least 24-48 hours if they are kept at 4°C. Many viruses do not survive freezing at -20°C. For example, herpes simplex virus titer can decrease by 100-fold with a single freeze-thaw cycle. Every measure should be taken to ensure the proper collection and transport of specimens for viral cultures.

References
Casteels A, Naessens A, Gordts F, et al, "Neonatal Screening for Congenital Cytomegalovirus Infections," *J Perinat Med*, 1999, 27(2):116-21.

Chien JW and Johnson JL, "Viral Pneumonias: Epidemic Respiratory Viruses," *Postgrad Med*, 2000, 107(3):41-52.

Dedicoat M and Muir D, "Viral Meningitis - or Encephalitis?" *Practitioner*, 1998, 242(1587):489-92.

Leibowitz HM, "The Red Eye," *N Engl J Med*, 2000, 343(5):345-51.

McNamee K, "Patient Education. What You Should Know About Genital Herpes," *Aust Fam Physician*, 1999, 28(11):1168.

Negrini B, Kelleher KJ, and Wald ER, "Cerebrospinal Fluid Findings in Aseptic Versus Bacterial Meningitis," *Pediatrics*, 2000, 105(2):316-9.

Norris CM, Danis PG, and Gardner TD, "Aseptic Meningitis in the Newborn and Young Infant," *Am Fam Phys*, 1999, 59(10):2761-70.

Preboth M, "ACOG Practice Bulletin on Management of Herpes in Pregnancy. American College of Obstetricians and Gynecologists," *Am Fam Phys*, 2000, 61(2):556-61.

Riley LE, "Herpes Simplex Virus," *Semin Perinatol*, 1998, 22(4):284-92.

Sawyer MH, "Enterovirus Infections: Diagnosis and Treatment," *Pediatr Infect Dis J*, 1999, 18(12):1033-9.

Sintchenko V and Dwyer DE, "The Diagnosis and Management of Influenza: An Update," *Aust Fam Physician*, 1999, 28(4):313-7.

Waisman Y, Lotem Y, Hemmo M, et al, "Management of Children With Aseptic Meningitis in the Emergency Department," *Pediatr Emerg Care*, 1999, 15(5):314-7.

Zambon M, "Cell Culture for Surveillance on Influenza," *Dev Biol Stand*, 1999, 98:65-74.

♦ **Viral Diseases by EM** *see* Electron Microscopy *on page 533*

♦ **Viral Hepatitis** *see* Hepatitis: Laboratory Assessment, Overview *on page 713*

♦ **Viral Load Testing, HIV-1** *see* Human Immunodeficiency Virus DNA Amplification *on page 746*

Virus, Direct Detection by Fluorescent Antibody

Related Information
Adenovirus Culture and Serology *on page 109*
Cytomegalovirus Antigen Detection *on page 494*
Cytomegalovirus Culture *on page 495*
Herpes Simplex Virus Culture and Antigen Detection *on page 721*
Influenza Virus Detection, Culture and Serology *on page 791*
Mumps Culture and Serology *on page 926*
Parainfluenza Viral Culture and Serology *on page 999*
Rabies *on page 1130*
Respiratory Syncytial Virus Antigen Detection *on page 1153*
Respiratory Syncytial Virus Culture and Serology *on page 1153*
Varicella-Zoster Virus Culture and Serology *on page 1300*
Viral Culture *on page 1307*

Test Includes Direct (nonculture) detection of virus-infected cells using monoclonal antibodies and immunofluorescence

Abstract Direct detection methods can provide a rapid diagnosis of viral infections (eg, herpes simplex (HSV), influenza A and B, varicella-zoster virus (VZV), respiratory viruses).

Specimen Impression smears of tissues, lesion scrapings and swabs, frozen sections, cell suspensions, upper respiratory tract swabs; nasopharyngeal aspirate (>1 mL) for influenza A and B antigen.

Storage Instructions Refrigerate nasopharyngeal aspirations.

Turnaround Time Less than 1 day

Special Instructions Make at least four impression smears or place four frozen sections on four separate slides. Cell suspensions should be centrifuged, resuspended to slight turbidity, and applied to prewelled slides.

Reference Interval No virus detected

(Continued)

Virus, Direct Detection by Fluorescent Antibody
(Continued)

Possible Panic Range Positive detection of rabies

Use Useful in the rapid diagnosis of HSV, VZV, respiratory syncytial virus (RSV), parainfluenza, influenza, adenovirus, cytomegalovirus (CMV), measles, mumps, and rabies infections

Limitations It is possible for the test to be negative in the presence of viral infection. Expertly trained and experienced personnel, excellent quality reagents, and adequate numbers of cells are required for accurate detection of virus. **Contact laboratory prior to requesting test to determine if laboratory offers a direct test for the viruses suspected.** Generally, this test is not as sensitive as cell culture.

Methodology Monoclonal antibody reagents and immunofluorescence microscopy are used to detect viruses/viral antigens in specimen cells.

Additional Information Direct detection of viruses in respiratory secretions can be diagnostically helpful because cell culture results often take several days to weeks.

Polymerase chain reaction (PCR) is available and has acceptable sensitivity for a variety of viral infections.

References

Baker KA and Ryan ME, "RSV Infection in Infants and Young Children. What's New in Diagnosis, Treatment and Prevention?" *Postgrad Med*, 1999, 106(7):97-108.

Scicchitano LM, Shetterly B, and Bourbeau PP, "Evaluation of Light Diagnostics SimulFluor HSV/VZV Immunofluorescence Assay," *Diagn Microbiol Infect Dis*, 1999, 35(3):205-8.

Viscosity, Blood

Related Information

Erythropoietin, Serum *on page 551*
Immunoglobulin M *on page 779*
Viscosity, Serum or Plasma *on page 1313*

Applies to Polycythemia of the Newborn

Abstract Viscosity refers to the internal friction of a fluid which makes it resistant to flow (as in relation to a solid surface or additional layer of fluid). Whole blood viscosity is the resistance to shearing motion of blood flow and is due, primarily, to circulating red blood cells. Whole blood viscosity is determined by the packed cell volume of erythrocytes (logarithmic relationship, plasma viscosity) aggregation of red blood cells, and deformability of red cells. Hyperviscosity accounts for the symptoms that occur in polycythemia of the newborn.

Specimen Whole blood

Container Green top (heparin) tube

Causes for Rejection Specimen clotted or hemolyzed

Special Instructions While this is an infrequently performed test and may not be routinely available, there is evidence that neonatal hyperviscosity is common, suggesting that the test should be more frequently utilized. Consult the laboratory to determine if the requisite microviscometer can be obtained.

Reference Interval See references for normal range data. Viscosity normally rises with increase in hematocrit and is lower with lower shear rates. Study has shown that umbilical cord and venous hematocrits (not capillary) correlate with microviscometer readings in newborns.

Use Detect hyperviscosity states including especially hyperviscosity in the neonatal period.

Additional Information The relatively new fields of haemorheology and clinical haemorheology focus upon the characteristics and resultant clinical effects of the flow behavior of blood. Somer and Meiselman have classified the hematological hyperviscosity syndromes as of polycythemic, sclerocythemic, or plasma type. The polycythemic category includes syndromes the result of erythrocytosis (primary or secondary) or of hyperleukocytic leukemia. Sclerocythemic cases are the result of decreased deformability of red cells (as occur with sickle hemoglobinopathies, other hemolytic anemias, some forms of malaria, and with rigid leukemic cells). In the plasma category are paraproteinemias (eg, multiple myeloma and Waldenström's macroglobulinemia) and reactive polyclonal dysproteinemias.

Neonatal hyperviscosity, usually but not always associated with polycythemia (central hematocrit ≥65%), may be accompanied by a fairly typical clinical picture while many are asymptomatic. Plethora, anorexia, feeding disturbances, hypoglycemia, lethargy, and jitteriness/seizures (CNS symptoms) occur. There may be symptoms and findings suggesting congenital heart disease (CHD) (ie, respiratory distress, cardiac enlargement, and cyanosis). False diagnoses of CHD have been made in such cases. About 50% of such infants have modest hyperbilirubinemia (bilirubin >12 mg/dL). Severe complications include pulmonary hypertension, necrotizing enterocolitis, and renal failure. Blood viscosity of small or large-for-gestational age infants does not differ from average-for-gestational age infants. About 50% of the cases have schistocytes and increased nucleated RBC on peripheral blood smear. There may be thrombocytopenia. The whole blood viscosity test can be used to follow the result of exchange transfusion therapy of neonatal hyperviscosity syndrome. Whole blood viscosity is increased after splenectomy (adults).

The incidence of neonatal polycythemia is increased with:

- high altitude
- small for gestational age infants
- twin-to-twin transfusion
- delayed clamping of the umbilical cord
- infants of diabetic mothers
- 13-, 18-, or 21 trisomy, adrenogenital syndrome, and in Beckwith-Wiedemann syndrome
- exposure to chronic fetal hypoxia

The treatment of symptomatic neonatal hyperviscosity is phlebotomy and replacement with saline/albumin or partial exchange transfusion (to decrease the hematocrit to 50%).

Hemorheological cardiovascular risk factors have been shown to improve in as short a period as 2 days following cessation of smoking. Changes in viscosity were the result in part of decreases in packed cell volume, total plasma protein, and fibrinogen concentration.

References
Ramamurthy RS and Berlanga M, "Postnatal Alteration in Hematocrit and Viscosity in Normal and Polycythemic Infants," *J Pediatr*, 1987, 110(6):929-34.

Rothwell M, Rampling MW, Cholerton S, et al, "Haemorheological Changes in the Very Short Term After Abstention From Tobacco by Cigarette Smokers," *Br J Haematol*, 1991 79(3):500-3.

Somer T and Meiselman HJ, "Disorders of Blood Viscosity," *Ann Med*, 1993, 25(1):31-9.

Stoll BJ and Kliegman RM, "The Fetus and the Neonatal Infant," *Nelson Textbook of Pediatrics*, 16th ed, Chapter 99, Behrman RE, Kliegman RM, and Jenson HB, eds, Philadelphia, PA: WB Saunders Co, 2000, 525-6.

Viscosity, Serum or Plasma

Related Information
Immunofixation Electrophoresis, Serum or Urine *on page 768*
Protein Electrophoresis, Serum *on page 1104*
Protein Electrophoresis, Urine *on page 1107*
Viral Culture *on page 1307*
Viscosity, Blood *on page 1312*

Abstract Viscosity of a fluid is its intrinsic resistance to flow due to internal friction between molecular/particles flowing through a tube or vessel. Plasma viscosity is the resistance to shearing motion of plasma mainly due to large proteins, fibrinogen, and some immunoglobulins, in particular, those with proximal atrial asymmetry.

Specimen Serum or plasma

Container Red top tube or lavender top (EDTA) tube

Reference Interval See table on following page.

Use Evaluate hyperviscosity syndromes associated with monoclonal gammopathy states (myeloma, macroglobulinemia of Waldenström and other dysproteinemias), including occasional cases of rheumatoid arthritis, systemic lupus erythematosus, hyperfibrinogenemia

Limitations Does not measure whole blood viscosity, which increases with high hemoglobin/hematocrit. Subjective endpoint, temperature dependent, large technical error, less than ideal correlation exists between measured viscosity levels and clinical symptoms.

Viscosity, Serum or Plasma *(Continued)*

Normal and Abnormal Levels of Viscosity

	Absolute Viscosity		Relative Viscosity
	37°C	25°C	
Distilled water	0.69	0.89	1.00
Plasma viscosity			
Reference range	1.16-1.35	1.50-1.72	1.67-1.94
Population range	1.14-1.50	1.46-1.94	1.65-2.17
Acute reactions	1.35-1.95	1.72-2.51	1.95-2.83
Chronic reactions	1.35-1.55	1.72-2.00	1.95-2.25
?Paraproteinemia	>2.0	>2.5	>2.9
Serum viscosity			
Reference range	1.09-1.23	1.40-1.60	1.57-1.80
Population range	1.08-1.36	1.39-1.60	1.57-1.97
?Paraproteinemia	>1.85	>2.4	>2.7

Reference and population ranges (mean ±2 SD) for plasma and serum viscosity at 37°C and 25°C, and relative to water. The usual zones for plasma viscosity in acute and chronic phase protein reactions are indicated, as are cutoff points above which paraproteins are very likely.

Adapted from *CRC Desk Reference for Hematology*, 3rd ed, Shinton NK, ed, Boca Raton, FL: CRC Press,1998, 682.

Additional Information Hyperviscosity is most frequent (33% of cases) with IgM monoclonal gammopathy (Waldenström's macroglobulinemia); next with IgA myeloma. When IgG myeloma leads to hyperviscosity IgG levels are usually very significantly elevated. Kappa light chain myeloma may (rarely) be responsible for hyperviscosity syndrome apparently as a result of true polymer formation. A relative viscosity of 6-7 usually results in symptoms of the hyperviscosity syndrome, they have however been described with lower levels of relative viscosity (ie, 4). Results of plasma viscosity obtained with an automated capillary viscometer show good precision and close correlation with the Harkness manual method (standard method selected by the International Committee for Standardization in Hematology).

There is ongoing interest in association of changes in blood/plasma viscosity with inflammatory/rheumatic diseases, heart/coronary artery/vascular diseases, and contributors (possibly through effects on viscosity) to the genesis of these disease processes.

References

Danesh J, Collins R, Peto R, et al, "Haematocrit, Viscosity, Erythrocyte Sedimentation Rate: Meta-analyses of Prospective Studies of Coronary Heart Disease," *Eur Heart J*, 2000, 21(7):515-20.

Paulus HE, Ramos B, Wong WK, et al, "Equivalence of the Acute Phase Reactants C-reactive Protein, Plasma Viscosity, and Westergren Erythrocyte Sedimentation Rate When Used to Calculate American College of Rheumatology 20% Improvement Criteria or the Disease Activity Score in Patients With Early Rheumatoid Arthritis," *J Rheumatol*, 1999, 26(11):2324-31.

Shinton NK, ed, *CRC Desk Reference for Hematology*, Boca Raton: FL, CRC Press, 1998, 681-2.

Winters JL and Pineda AA, "Hemapheresis," *Clinical Diagnosis and Management by Laboratory Methods*, Henry JB, ed, 20th ed, Chapter 32, Philadelphia, PA: WB Saunders Co, 2001, 795.

Vitamin A, Serum or Plasma

Related Information

Antioxidant Concentrations, Plasma *on page 192*
Vitamin E, Serum or Plasma *on page 1319*
Zinc, Serum or Plasma *on page 1340*
Zinc, Urine *on page 1341*

Applies to Carotene; Retinoids; Retinol, Serum

Test Includes Vitamin A and beta-carotene determination

Abstract The expression, vitamin A, refers to retinoids which have the biologic activity of retinol. Vitamin A is a fat-soluble essential vitamin which is necessary for the integrity of epithelial cells and also plays an important role in the visual cycle. Its two natural forms are **retinol** (vitamin A_1) and **3-dehydro-retinol** (vitamin A_2). β-**carotene** is the most common provitamin A of the 50-60 known compounds and represents ~25% of the total of serum carotenoids.

Patient Preparation Patient must fast a minimum of 8 hours.

Specimen Serum or plasma

Container Red top tube, green top (heparin) tube

Collection Draw in chilled tube. Protect from light. Keep specimen on ice.

Storage Instructions Separate serum or plasma in a 4°C centrifuge and freeze in a plastic vial immediately. Stable 2 years frozen. Stable 4 weeks at 4°C, although freezing is preferred. Protect from light.

Reference Interval
- Vitamin A: adults: 30-120 µg/dL (SI: 1.05-4.20 µmol/L). Normal range is less in childhood. Concentrations in males are about 20% greater than those in females.
- Beta-carotene: 10-85 µg/dL (SI: 0.19-1.58 µmol/L)

Use Differential diagnosis of hypervitaminosis A, which may include toxicity; *vide infra*. Vitamin A levels <10 µg/dL (SI: <0.35 µmol/L) may provide evidence of marked deficiency; >140 µg/dL (SI: >4.89 µmol/L) signals toxicity.

A combination of a low serum carotene level and a low vitamin A suggests inadequate vitamin A nutrition. Vitamin A deficiency is found most commonly in children younger than 5 years of age and is usually due to insufficient dietary intake. Individuals with fat malabsorption are especially likely to develop vitamin A deficiency. Vitamin A deficiency is also found with prolonged ethanol consumption.

Limitations Serum levels do not correlate with liver stores because of homeostatic control exerted by the liver. Increased in patients on oral contraceptives. β-carotene may be increased with hypothyroidism and with hyperlipidemia of diabetes mellitus.

Methodology Fluorescence or UV/VIS spectroscopy and other methods; presently, the method of choice is high performance liquid chromatography (HPLC), which permits simultaneous determination of vitamins A and E.

Additional Information Hypervitaminosis A may be due to increased intake, or conditions causing impaired disposal such as myxedema, diabetes mellitus, or renal disease. The toxic effects of increased vitamin A include elevation of intracranial pressure, skin desquamation, hair loss, joint pain, headache, nausea, fever, vertigo, and visual disorientation. Toxicity is best assessed by measuring retinyl esters, which normally comprise 5% of total vitamin A, but may comprise >30% of total vitamin A in toxicity (taking >50,000 IU/day). Toxicity appears when vitamin A levels exceed the capacity of retinol-binding protein (RBP) to bind to it.

Retinoids can be teratogenic.

References

Kritchevsky SK, "Beta-Carotene, Carotenoids, and Prevention of Coronary Heart Disease," *J Nutr*, 1999, 129:5-8.

Rothman KJ, Moore LL, Singer MR, et al, "Teratogenicity of High Vitamin A Intake," *N Engl J Med*, 1995, 333(21):1369-73.

Usha N, Sankaranarayanan A, Walia BN, et al, "Early Detection of Vitamin A Deficiency in Children With Persistent Diarrhoea," *Lancet*, 1990, 335(8686):422.

Vitamin B₆, Plasma or Serum

Related Information

Homocyst(e)ine, Plasma *on page 741*

Synonyms P-5'-P; Pyridoxal-5-Phosphate; Pyridoxine

Abstract Vitamin B₆ acts as a coenzyme in heme synthesis and intermediary metabolism.

Patient Preparation Avoid radioisotope scan prior to collection of specimen if RIA is used for assay.

Specimen Serum or plasma

Container Red top tube, lavender top (EDTA) tube

Collection Transport specimen **immediately** to the laboratory following collection. Avoid exposing specimen to light.

Storage Instructions Separate plasma or serum and freeze **immediately**. Avoid exposure to light. Stable 10 days at -80°C; 50% loss in 7 days at -20°C.

Causes for Rejection Specimen more than 30 minutes in transit to the laboratory.

Special Instructions Communicate with laboratory before this test is ordered; it is not always routinely available. Scheduling and/or use of a reference laboratory may be required.

(Continued)

Vitamin B₆, Plasma or Serum *(Continued)*

Reference Interval 5-30 ng/mL (SI: 20-121 nmol/L) (varies considerably with method).

Critical Values <5 ng/mL (SI: <20 nmol/L) indicates deficiency

Use Detect vitamin B_6 deficiency (eg, subjects with xanthurenic aciduria, primary cystathioninuria, homocystinuria, and hyperhomocystinemia). B_6 deficiency may lead to dermatitis with cheilitis and glossitis. Some drugs may lead to B_6 deficiency (eg, penicillamine, levodopa, disulfiram, oral contraceptives, theophylline, phenelzine, isoniazid, cycloserine, and pyrazinoic acid).

Neurotoxicity may be associated with pyridoxine megavitaminosis, including tingling, numbness, clumsiness, gait disturbances, and pseudoathetosis.

Methodology Microbiologic assay, fluorometry, high performance liquid chromatography (HPLC)

Additional Information Pyridoxine is required for heme synthesis. With deficiency, a hypochromic form of sideroblastic anemia may occur, characterized by the presence of ring sideroblasts (iron positive granules deposited about the nuclei of red cell precursors). Occasionally, the anemia may have megaloblastic characteristics.

In adults, elevated serum homocyst(e)ine levels due to vitamin B_6 deficiency may promote atherogenesis with resultant arteriosclerotic and thromboembolic cerebral, coronary, and/or peripheral vascular events.

References

Jacobs DS, DeMott WR, Oxley DK, et al, *Laboratory Test Handbook*, 5th ed, Hudson, OH: Lexi-Comp Inc, 2001.

Robinson K, Mayer E, and Jacobsen DW, "Homocysteine and Coronary Artery Disease," *Cleve Clin J Med*, 1994, 61(6):438-50.

van den Berg M, Franken DG, Boers GH, et al, "Combined Vitamin B₆ Plus Folic Acid Therapy in Young Patients With Arteriosclerosis and Hyperhomocysteinemia," *J Vasc Surg*, 1994, 20(6):933-40.

♦ **Vitamin B₁₂** *see* Cobalamin, Serum *on page 424*
♦ **Vitamin B₁₂ Absorption Test** *see* Schilling Test *on page 1178*

Vitamin B₁₂ Unsaturated Binding Capacity

Related Information

Cobalamin, Serum *on page 424*
Erythropoietin, Serum *on page 551*
Folic Acid, RBC *on page 606*
Folic Acid, Serum *on page 606*
Homocyst(e)ine, Plasma *on page 741*
Intrinsic Factor Blocking Antibody *on page 806*
Methylmalonic Acid, Serum, Plasma, Urine, or Amniotic Fluid *on page 908*
Parietal Cell Antibody *on page 1005*
Schilling Test *on page 1178*

Synonyms UBBC

Applies to Cobalophilin; Transcobalamins

Specimen Serum is most commonly used by most reference laboratories but use of EDTA plasma avoids increase in binding protein released from granulocytes (see following information).

Container Red top tube

Storage Instructions Refrigerate serum if not delivered to the laboratory immediately. Stable for days. Very stable when stored at -20°C.

Reference Interval 1000-2000 pg/mL binding capacity

Use Differential diagnosis of polycythemia vera from secondary and relative polycythemias; evaluate macrocytic/megaloblastic anemia; diagnose congenital absence of transcobalamin II or cobalophilin (transcobalamin I and III)

Limitations Increased with pregnancy and use of contraceptive hormones. Unrepresentative increase may occur during clotting of blood samples by release of unsaturated binding protein (cobalophilin) from granulocytes. May give low values in samples with low protein content. Usually available only at reference or research laboratories.

Additional Information Serum transport of vitamin B_{12} is accomplished by normally occurring proteins termed transcobalamins including I (an α-globulin), II (a β-globulin), and III (a group of transport factors - "R-type" binders or binder III - found also in some tissues, saliva, milk, and tears). Transcobalamin cell membrane receptors regulate cellular uptake of the cobalamins. The term "R-type" refers to binding protein with "rapid" mobility on electrophoresis. The term haptocorrin (TCO, I, II, R binder, cobalophyllin) refers to a family of immunologically similar proteins which are variably glycosylated and not all of which have rapid electrophoretic mobility. Transcobalamin I is the major B_{12} transport protein binding 80% to 90% of endogenous cobalamin which is delivered from peripheral tissues to the liver. It bears immunologic identity to granulocyte cobalophilin. Transcobalamin II binds only 10% to 25% of total plasma cobalamin but provides most of the total UBBC of plasma. Thus, transcobalamin II is the carrier protein that provides most of the cobalamin transport within the intravascular (and extracellular) space. Intrinsic factor is the carrier protein within the gastrointestinal tract. Less than 2% of plasma TC II is saturated at any one point in time. The TC II bound B_{12} is decreased in untreated B_{12} deficiency. As the B_{12} deficient state develops, TC II UBBC increases. See table for other clinical associations. Isoelectric focusing has shown that the cobalophilins (R-binders) are a microheterogenous group of plasma binding proteins. Cobalophilin is increased in diseases characterized by excess granulocyte production, reactive leukocytosis, chronic myelogenous leukemia, and other myeloproliferative states, in particular polycythemia vera. UBBC levels are increased in >66% of the cases of polycythemia vera. Most cases of secondary/relative polycythemia patients have normal levels of UBBC.

Levels and Binding Capacity of Cobalamin-Binding Proteins in Disease

Binder	Disease
Increased TC I (R protein)	Myeloproliferative disorders
	Polycythemia vera
	Myelofibrosis
	Benign neutrophilia
	Chronic myelocytic leukemia
	Hepatoma (occasionally)
	Metastatic cancer
Increased TC II	Myeloproliferative disorders
	Liver disease
	Inflammatory disorders
	Gaucher disease
	Anti-TC II antibodies
Unsaturated cobalamin binders[1] Increased	Transient neutropenia
	Elevated TC I
Decreased	Liver disease
	Elevated serum cobalamin

[1]UBBC = unsaturated B_{12} binding capacity.

Adapted from Babior BM, "Metabolic Aspects of Folic Acid and Cobalamin," *Williams Hematology*, 5th ed, Chapter 35, Beutler E, Lichtman MA, Coller BS, et al, eds, New York, NY: McGraw-Hill Inc, 1995, 390.

TC II deficiency (presents shortly after birth as failure to thrive) is an uncommon condition characterized by vomiting and weakness, pancytopenia, megaloblastic anemia, and serum cobalamin levels that are normal or nearly normal. Even more uncommon is R-binder deficiency (absent or deficient TC I) in which serum cobalamin level is decreased but there are no clinical signs of B_{12} deficiency. In such cases TC II cobalamin levels are normal.

References

Babior BM, "Metabolic Aspects of Folic Acid and Cobalamin," *Williams Hematology*, 5th ed, Chapter 35, Beutler E, Lichtman MA, Coller BS, et al, eds, New York, NY: McGraw-Hill Inc, 1995, 388-90.

Fenton WA and Rosenberg LE, "Inherited Disorders of Cobalamin Transport and Metabolism," *The Metabolic and Molecular Bases of Inherited Disease*, 7th ed, Chapter 102, McGraw-Hill Inc, 1995, 3129-49.

Zittoun J, Farcet JP, Marquet J, et al, "Cobalamin (Vitamin B₁₂) and B₁₂ Binding Proteins in Hypereosinophilic Syndromes and Secondary Eosinophilia," *Blood*, 1984, 63(4):779-83.

♦ **Vitamin C** *see* Ascorbic Acid, Serum or Plasma *on page 215*

♦ Vitamin D₁ see Vitamin D, Serum on page 1318

Vitamin D, Serum

Related Information

Aluminum, Bone and Bone Biopsy on page 139
Amino Acids, Urine on page 145
Calcium, Serum on page 329
Calcium, Urine on page 332
Fecal Fat, Quantitative, 72-Hour Collection on page 574
Kidney Stone Analysis on page 820
Magnesium, Serum on page 885
Magnesium, Urine on page 886
Osteocalcin, Serum or Plasma on page 983
Parathyroid Hormone, Serum on page 1001
Phosphorus, Serum on page 1031

Synonyms Cholecalciferol

Applies to Calcitonin; Calcitriol; 1,25-Dihydroxy Vitamin D_3; Ergocalciferol (Vitamin D_2); 25-Hydroxy Vitamin D_3; 1,25-$(OH)_2$ D_3; 25-(OH) D_3; Parathormone; Vitamin D₁

Abstract The generic term, vitamin D, includes several substances. Entities important for laboratory evaluation include the biologically inactive precursors, 25-hydroxy vitamin D (25OH vitamin D), and the biologically active 1,25-dihydroxy vitamin D (1,25OH vitamin D). 1,25OH vitamin D has a plasma half-life of 4-6 hours. 25OH vitamin D, with a plasma half-life of 2-3 weeks, is the substance usually measured to assess vitamin D deficiency or excess.
In the medical literature, **calcitriol**, is a synonym for 1,25OH vitamin D.

Specimen Serum, plasma

Container Red top tube, green top (heparin) tube

Storage Instructions Stable 3 days at 4°C to 25°C. Processed serum stable for months at -20°C, tolerates freeze-thaw cycles.

Causes for Rejection Administration of radioisotopes if RIA is used for assay

Reference Interval Varies with method and environmental conditions (eg, recent sunlight exposure and diet). Following are selected reference intervals:
25OH vitamin D: 10-50 ng/mL (SI: 25-125 nmol/L)
  • winter: 14-42 ng/mL (high performance liquid chromatography)
  • summer: 15-80 ng/mL (high performance liquid chromatography)
1,25OH vitamin D: 15-60 pg/mL (SI: 36-144 nmol/L)
  • 15-60 pg/mL (high performance liquid chromatography)

Use 25OH vitamin D should be used to assess nutritional adequacy.

1,25OH vitamin D is useful in evaluating patients with hypercalcemia due to extrarenal conversion of 25OH vitamin D to 1,25OH vitamin D (calcitriol). Such a situation occurs in sarcoidosis, cat-scratch disease, and certain lymphomas. These patients have **increased** levels of 1,25OH vitamin D. This endogenous form of vitamin D intoxication is, in many series, the third most common cause of hypercalcemia, following primary hyperparathyroidism and malignancy.

1,25OH vitamin D is also **increased** in primary hyperparathyroidism, dietary vitamin D intoxication, and type II vitamin D-dependent rickets, a disease characterized by end-organ unresponsiveness to 1,25OH vitamin D.

1,25OH vitamin D is **decreased** in hypoparathyroidism, pseudohypoparathyroidism, hypercalcemia of malignancy, renal failure, hyperphosphatemia, hypomagnesemia, and vitamin D-dependent rickets (type 1).

Methodology Competitive binding assay, radioimmunoassay (RIA), radioreceptor assay (RRA), high performance liquid chromatography (HPLC)

Additional Information The table on the following page lists vitamin D levels in hypocalcemic and hypercalcemic disorders.

References

Endres DB and Rude RK, "Mineral and Bone Metabolism," *Tietz Textbook of Clinical Chemistry*, 3rd ed, Burtis CA and Ashwood ER, eds, Philadelphia, PA: WB Saunders Co, 1999, 1395-457.
Fraser DR, "Vitamin D," *Lancet*, 1995, 345(8942):104-7.
Mayo Medical Laboratories, *2000 Test Catalogue*, Rochester, MN, 500-1.
Nellen JF, Smulders YM, Frissen PH, et al, "Hypovitaminosis D in Immigrant Women: Slow to Be Diagnosed," *BMJ*, 1996, 312(7030):570-2.
Seymour JF, Gagel RF, Hagemeister FB, et al, "Calcitriol Production in Hypercalcemic and Normocalcemic Patients With non-Hodgkin Lymphoma," *Ann Intern Med*, 1994, 121(9):633-40.
Weiss RL, "ARUP Interpretive Data Guide," ARUP Labs, 1999, 509-10.

Anticipated Vitamin D Levels in Hypocalcemic and Hypercalcemic Disorders

Clinical Disorder	Hypocalcemia	
	25(OH)D[1]	1,25(OH)$_2$D[1]
Vitamin D deficiency	D	D,N,I
Severe hepatocellular disease	D	D,N
Nephrotic syndrome	D	D,N
Renal failure	N	D
Hyperphosphatemia	N	D
Hypoparathyroidism	N	D,N
Pseudohypoparathyroidism	N	D,N
Hypomagnesemia	N	D,N
Vitamin D-dependent rickets, type I (pseudovitamin D deficiency rickets)	N,I	D
Vitamin D-dependent rickets, type II (pseudovitamin D resistant rickets)	N,I	D
	Hypercalcemia	
Vitamin D intoxication	I	D,N
1,25(OH)$_2$D intoxication	N	I
Granulomatous diseases	N	N,I
Lymphoma	N	N,I
Hypercalcemia of malignancy	N	D,N
Hyperparathyroidism	N	N,I
Idiopathic hypercalciuria	N	N,I

[1]D = decreased, N = normal, I = increased.

Adapted from Weiss RL, "ARUP Interpretive Data Guide," ARUP Labs, 1999, 509-10.

Vitamin E, Serum or Plasma
Related Information
Antioxidant Concentrations, Plasma *on page 192*
Cholesterol, Total, Serum or Plasma *on page 396*
Vitamin A, Serum or Plasma *on page 1314*

Synonyms Alpha Tocopherol; Tocopherol; α-Tocopherol

Abstract Vitamin E is an antioxidant. Naturally occurring structures with vitamin E activity include four tocopherols and four tocotrienols (alpha, beta, gamma, delta).

Specimen Serum, plasma

Container Red top tube, green top (heparin) tube

Sampling Time Oral vitamin E may circulate for 1-2 days.

Storage Instructions Separate serum or plasma within 2 hours. Protect from light. Serum or plasma is stable 2 weeks at 25°C, 14 days at 4°C, and 1 year at -20°C.

Reference Interval Adults: 0.5-1.8 mg/dL (SI: 12-42 µmol/L). Reference intervals vary with method and age.

Critical Values Values <0.2 mg/dL indicate need for supplementation.

Possible Panic Range Although vitamin E toxicity has not been clearly established, values >4 mg/dL indicate significant excess.

Use Evaluate vitamin E deficiency when hemolytic disease occurs in premature infants; evaluate neuromuscular disease in infants and adults with chronic biliary disease/chronic cholestasis; evaluate patients on long-term parenteral nutrition and those on long-term dialysis; evaluate patients with malignancy or malabsorption (eg, patients with cystic fibrosis, cases of intestinal bypass surgery); investigate brown-bowel syndrome; investigate weakness in subjects who may be vitamin E deficient. Patients with malabsorption can develop myopathy and neuropathy.

Vitamin E deficiency may occur with acanthocytosis (abetalipoproteinemia). (Continued)

Vitamin E, Serum or Plasma *(Continued)*

Methodology High performance liquid chromatography (HPLC), fluorometry after solvent extraction, colorimetry, thin layer chromatography (TLC)

Additional Information Vitamin E is an antioxidant so widely distributed in foodstuffs that deficiency rarely occurs in normal adults from diet. However, vitamin E is fat soluble, and malabsorption and deficiency may develop in cases of chronic intraluminal intestinal bile deficiency. This has been particularly noted in premature infants and children with biliary atresia or cystic fibrosis (chronic intrahepatic cholestasis). Clinically, this may lead to a hemolytic anemia, due to increased erythrocyte fragility, or to a slowly progressive neurologic disorder characterized by ataxia, areflexia, gaze disturbances, and loss of proprioception and vibratory sensation. A similar syndrome has been reported in adults with malabsorption. The syndrome may respond to treatment with parenteral vitamin E.

Early treatment with vitamin E may delay or prevent the neuropathy with ataxia that develops during the course of abetalipoproteinuria. Vitamin E therapy, in some cases, may have a favorable effect on moderate and severe cases of the retinopathy of prematurity and the retinopathy of abetalipoproteinemia. While lipid malabsorption syndrome (with steatorrhea and malabsorption of vitamin E) may be etiologically related to ataxic spinocerebellar neurologic degenerative disease, there are reports of familial and sporadic vitamin E deficiency with neurologic impairment in the absence of fat malabsorption or demonstrated plasma lipoprotein level abnormality.

See Antioxidant Concentrations, Plasma *on page 192*; some desirable effects of antioxidants are discussed.

References

Greenberg ER and Sporn MB, "Antioxidant Vitamins, Cancer, and Cardiovascular Disease," *N Engl J Med*, 1996, 334(18):1189-90.

Kushi LH, Folsom AR, Prineas RJ, et al, "Dietary Antioxidant Vitamins and Death From Coronary Heart Disease in Postmenopausal Women," *N Engl J Med*, 1996, 334(18):1156-62.

McCormick DB and Greene HL, "Vitamins," *Tietz Textbook of Clinical Chemistry*, Burtis CA and Ashwood ER, 3rd ed, Chapter 29, Philadelphia, PA: WB Saunders Co, 1999, 999-1028.

Meagher EA, Barry OP, Lawson JA, et al, "Effects of Vitamin E on Lipid Peroxidation in Healthy Persons," *JAMA*, 2001, 285(9):1178-82.

Meydani SN, Meydani M, Blumberg JB, et al, "Vitamin E Supplementation and *In Vivo* Immune Response in Healthy Elderly Subjects," A Randomized Controlled Trial, *JAMA*, 1997, 277(17):1380-6.

Yusuf S, Dagenais G, Pogue J, et al, "Vitamin E Supplementation and Cardiovascular Events in High-Risk Patients. The Heart Outcomes Prevention Evaluation Study Investigators," *N Engl J Med*, 2000, 342(3):154-60.

♦ **Vitamin K** *see* Prothrombin Time *on page 1116*

♦ **Vitamin K** *see* Warfarin, Serum or Plasma *on page 1325*

♦ **Vitreous Fluid** *see* Ocular Cytology *on page 973*

♦ **Vivactil®** *see* Antidepressants, Cyclic, Serum or Plasma *on page 171*

♦ **Vivarin®** *see* Caffeine, Serum *on page 326*

♦ **Vivol®** *see* Diazepam, Serum *on page 510*

♦ **VLDL** *see* Triglycerides, Serum or Plasma *on page 1275*

♦ **VMA** *see* Vanillylmandelic Acid, Urine *on page 1299*

Volatile Screen, Blood or Urine

Related Information

Anion Gap, Serum, Plasma, or Urine *on page 160*
Ethanol, Blood, Urine, and Other Sources *on page 558*
Ketone Bodies, Blood *on page 816*
Ketones, Urine *on page 817*
Osmolality, Calculated, Serum or Plasma *on page 976*
Osmolality, Serum *on page 978*

Synonyms Toxicology, Volatiles

Applies to Acetone; Ethanol; Isopropanol; Methanol

Test Includes Determination of volatiles by gas chromatography including acetone, ethanol, isopropanol, and methanol.

Abstract This screening profile measures acetone, ethanol, isopropanol, and methanol. Other volatiles, up to 40, may also be included. **It is important to keep in mind that ethylene glycol is not part of the volatile screen.**

Specimen Serum or plasma, urine, gastric fluid

Container Red top tube, gray top (sodium fluoride) tube; tightly stoppered container for urine and gastric fluid

Collection All containers should be tightly stoppered. The gray (oxalate/fluoride) tube top is recommended for medicolegal collections and if storage is prolonged. Sodium fluoride (50 mg) can be added as a preservative to urine and gastric samples. Other anticoagulants (eg, heparin EDTA) are acceptable.

Reference Interval None detected

Possible Panic Range Blood: acetone, methanol, isopropanol >50 mg/dL (SI: acetone: >8.6 mmol/L, methanol: >15.6 mmol/L, isopropanol: >8.32 mmol/L); ethanol: >200 mg/dL (SI: >43.4 mmol/L); urine: acetone, methanol, isopropanol >50 mg/dL (SI: acetone: >8.6 mmol/L, methanol: >15.6 mmol/L, isopropanol: 8.32 mmol/L); ethanol: >160 mg/dL (SI: >34.7 mmol/L)

Use Evaluate methanol and isopropanol toxicity, and alcohol drug abuse

Limitations Urine levels do not correlate well with blood levels.

Methodology Gas chromatography (GC). Head-space GC is a preferred method for volatile analysis.

Additional Information Both methanol and isopropanol are more intoxicating than ethanol. Like ethanol, methanol exhibits zero order elimination kinetics. Methanol is converted to formaldehyde and formic acid, which causes metabolic acidosis and retinal damage leading to blindness. Treatment may include infusion of ethanol to inhibit methanol metabolism or treatment with alcohol dehydrogenase inhibitors such as fomepizole.

References
Church AS and Witting MD, "Laboratory Testing in Ethanol, Methanol, Ethylene Glycol, and Isopropanol Toxicities," *J Emerg Med*, 1997, 15(5):687-92.

Hammett-Stabler CA, "Fomepizole, a New Treatment for Methanol and Ethylene Glycol Poisoning," *Ther Drug Monit Toxicol*, 1999, 20:31-2.

Sharp ME, "A Comprehensive Screen for Volatile Organic Compounds in Biological Fluids," *J Anal Toxicol*, 2001, 25(7):631-6.

♦ **von Willebrand Disease Therapy** see Cryoprecipitate on page 481

von Willebrand Factor

Related Information

Coagulation Factor Assays on page 418
Factor VIII Concentrate on page 563

Synonyms Multimer Assay; Ristocetin Cofactor; Ristocetin-Induced Platelet Aggregation Assay; von Willebrand Factor Antigen; von Willebrand Factor Assay; von Willebrand Factor Collagen-Binding Assay; von Willebrand Factor Multimer Assay

Applies to Acute Phase Reactant; DDAVP; Desmopressin; Factor VIII:von Willebrand Factor Ratio

Test Includes Assays for von Willebrand factor (vWF) activity (ristocetin cofactor), vWF antigen, and factor VIII should be ordered. If indicated by these results, a vWF multimer analysis and/or low-dose ristocetin aggregation assay can be ordered.

Abstract von Willebrand factor (vWF) mediates platelet adhesion to injured endothelium, the first step in hemostasis. It also helps maintain factor VIII levels. When vWF is deficient, patients have a bleeding disorder called von Willebrand disease (vWD). vWD is the most common hereditary bleeding disorder, of which several subtypes are recognized (see below).

Specimen Plasma

Container Three blue top (sodium citrate) tubes

Collection Routine venipuncture. Deliver tubes to laboratory immediately; otherwise, falsely low factor VIII values may occur (factor VIII is labile). If multiple tests are being drawn, draw blue top tubes after any red top tubes but before any lavender top (EDTA), green top (heparin), or gray top (oxalate/fluoride) tubes. Immediately invert tubes gently at least 4 times to mix. Tubes must be appropriately filled.

Storage Instructions Separate plasma from cells as soon as possible. Plasma can be stored for 2 hours on ice, otherwise store frozen. For vWF antigen only, plasma can be stored for 8 hours at room temperature or 24 hours on ice, otherwise store frozen.

Causes for Rejection Specimen received more than 4 hours after collection, tubes not filled, clotted specimens
(Continued)

von Willebrand Factor (Continued)

Turnaround Time Several days (longer if follow-up testing is needed, such as multimer analysis)

Reference Interval Varies with blood type through an unknown mechanism. African Americans have higher levels of vWF antigen than do Caucasians. Results are reported as a percent of the amount expected in normal plasma. By definition, the mean value in pooled normal plasma is 100%. In a large study of normal persons, the mean vWF level was 75% in blood type O, 106% in type A, 117% in type B, and 123% in type AB individuals. The overall mean vWF level was 100%. Newborns have higher vWF levels than do adults. Values for vWF gradually decrease into the adult normal range by age 6 months.

Use Determine if a patient with a personal or family history of bleeding has von Willebrand disease (vWD); assist in determining hemophilia A carrier status in females

Methodology

Initial tests:

The **ristocetin cofactor assay** assesses vWF function by measuring ristocetin-mediated binding of vWF to platelet GPIb, which leads to platelet agglutination. The ristocetin cofactor assay is performed by mixing patient plasma with ristocetin and formalin-fixed normal platelets and measuring the amount of platelet agglutination in an aggregometer. **Note:** The term "agglutination" is often used to describe ristocetin-induced platelet aggregation, because true platelet aggregation links platelets through fibrinogen and GPIIb/IIIa, whereas ristocetin links platelets through von Willebrand factor and GPIb. The **von Willebrand factor antigen** assay is an enzyme-linked immunosorbent assay (ELISA), which measures the quantity of vWF, independent of vWF function. An alternative automated assay involves latex particles coated with antibodies directed against vWF. In the presence of vWF, the latex particles form aggregates that absorb light passing through the specimen. The amount of light absorbance is directly related to the amount of vWF in the specimen. Rocket immunoelectrophoresis is an older antigen assay that is still in use in some laboratories. The **factor VIII assay** is a PTT-based clotting assay which measures factor VIII activity.

More recently, alternative immunoassays have been designed to assess vWF function. The collagen-binding assay is an ELISA in which collagen is the antigen. If vWF is functional, it binds to collagen and is subsequently detected. Another vWF "functional" ELISA uses monoclonal antibodies that recognize a functional epitope on vWF, but there is conflicting evidence whether this test correlates better with ristocetin cofactor or with vWF antigen. These newer functional assays are not yet as established as the ristocetin cofactor assay.

Follow-up tests (performed if indicated, see table):

Multimer analysis is performed when type 2 vWD is suspected. A plasma sample is electrophoresed on a gel to separate the multimers by size. The multimers are then visualized using ^{125}I-labeled anti-vWF antibody or other techniques. **Low-dose ristocetin platelet aggregation assay** is performed when type 2B vWD is suspected. This test is similar to the ristocetin cofactor assay, except that the patient's own platelets are used instead of normal platelets, and lower doses of ristocetin are used. The patient's own platelets and plasma are mixed with ristocetin, and platelet aggregation is measured in an aggregometer. This assay is less sensitive than the ristocetin cofactor assay for diagnosing vWD, but it is useful for confirming a diagnosis of type 2B vWD. Type 2B patients' platelets become abnormally coated with vWF *in vivo*, due to increased affinity of the mutant vWF for platelet GPIb. As a result, the patient's platelets show increased aggregation in this assay. Platelet-type vWD also shows increased aggregation in this assay, due to a mutation on platelet GPIb which increases its affinity for vWF. In contrast, other types of vWD may show decreased ristocetin-induced platelet aggregation due to decreased vWF quantity and/or function. Further specialized coagulation testing can be performed to distinguish type 2B from platelet-type vWD.

Additional Information von Willebrand disease (vWD) is the most common hereditary bleeding disorder, occurring in up to 1% of the general population. Many cases remain undiagnosed because of the mild nature of bleeding in many patients and the fact that acute phase reactions can mask the diagnosis. vWF is a polypeptide synthesized in endothelial cells and megakaryocytes, which polymerizes to form multimers containing up to 100 subunits. Bleeding symptoms resemble those of a platelet function defect, since platelet adhesion is impaired. Thus, the most common symptoms are epistaxis, easy bruising, bleeding with dental extractions, and menorrhagia.

Laboratory testing for vWD is summarized in the table. Repeat testing is often required, because both vWF and factor VIII become elevated above baseline during acute phase reactions (including even minor illnesses, injury, or stress), pregnancy, estrogen use, or in newborns. **An elevation of a low or borderline value for vWF into the normal range during any of these conditions often masks the diagnosis of vWD.** Measurement of an acute phase reactant such as fibrinogen is helpful in assessing the likelihood that a patient is in an acute phase reaction at the time of testing.

vWF serves as the carrier protein for factor VIII, and levels of factor VIII are often decreased when vWF is decreased. When vWF is markedly decreased, the factor VIII level can also become very low, which prolongs the PTT. In most vWD patients, the disease is mild or moderate and the PTT is therefore normal.

Many variants of vWD have been described, but the classification scheme has recently been simplified into three types (see table). Type 1 is by far the most common form, accounting for the majority of cases. Type 1 vWD is characterized by a partial quantitative deficiency of vWF. Although the quantity of vWF is reduced, the function of the individual vWF molecules which are synthesized is normal.

Interpretation of von Willebrand Factor Assays

RCoF + vWF Ag + FVIII + Fibrinogen (or other acute phase reaction marker)
• Normal:[1] vWD unlikely if no acute phase reaction, pregnancy, estrogen use, newborn
• Normal[1] but fibrinogen (or factor VIII) elevated: repeat vWF assays when fibrinogen and factor VIII are normal
• RCoF, WF Ag, FVIII reduced to a similar extent: **type 1** vWD likely
• RCoF, WF Ag, FVIII severely reduced (<10%) or undetectable: **type 3** vWD likely
• RCoF reduced more severely than vWF Ag and FVIII:[2] consider **type 2** vWD (2A, 2B, or 2M); perform multimer analysis and low-dose ristocetin cofactor to determine subtype:
– Multimer analysis normal: type 2M likely (subtle abnormalities in some variants)
– Multimer analysis missing high molecular weight multimers: type 2A likely
– Multimer analysis missing high and intermediate molecular weight multimers: type 2B or platelet type likely
– Increased low-dose ristocetin aggregation: type 2B or platelet type[3]
– Normal or decreased low-dose ristocetin aggregation: not type 2B or platelet type
• FVIII reduced (5% to 40%), RCoF and vWF Ag normal:[2] consider **type 2N** vWD; or in males, mild hemophilia A. In female hemophilia A carriers, factor VIII is approximately 50% with large variability. Consider also factor VIII degradation if prolonged specimen transportation.

RCoF = ristocetin cofactor assay; vWF Ag = von Willebrand factor antigen assay; FVIII = factor VIII assay.

[1]Consider blood type when determining if values are normal.

[2]Mean RCoF:vWF Ag ratio is 0.3 for type 2A, 0.6 for type 2B, and uncertain (<1) for type 2M. Mean FVIII:vWF Ag ratio is 0.28 for type 2N (see reference Meyer 1997).

[3]Thrombocytopenia may occur with type 2B or platelet-type (and rare type 2A variants).

Type 2 vWD is characterized by qualitative (functional) deficiencies of vWF. Often the quantity of vWF is also reduced. Type 2 vWD is further subdivided into four categories (see table). Type 2A and type 2B are characterized by a loss of high molecular weight multimers of vWF. The highest molecular weight multimers have more hemostatic function than the lower molecular weight multimers. Therefore, in these disorders, the overall function relative to the quantity of vWF molecules is reduced. Thus, the functional assay (ristocetin cofactor) result is reduced more than the quantitative assay (von Willebrand factor antigen). In type 2A, the loss of high molecular weight multimers is due to defective multimer assembly and secretion or increased proteolysis of multimers.

(Continued)

von Willebrand Factor *(Continued)*

Type 2B vWD mutations lead to increased binding of vWF to GPIb, the platelet vWF receptor. Platelets coated with vWF are cleared from the bloodstream at an increased rate, leading to loss of high molecular weight multimers as well as thrombocytopenia.

Platelet-type or pseudo-vWD is a rare disorder in which a mutation in the platelet GPIb gene (not the vWF gene) leads to increased binding of vWF to GPIb, resulting in the same findings described above for type 2B vWD.

Types 2M and 2N vWD are rare subtypes of type 2 vWD. Type 2M vWF mutations cause decreased function despite the presence of normal-sized multimers, often because the mutation impairs the ability of vWF to bind to platelet GPIb. In type 2N (Normandy) vWD, the factor VIII-binding ability of vWF is impaired, and the half-life of factor VIII is consequently shortened. Thus, vWF is normal in quantity (normal antigen assay) and has normal platelet-adhesion function (normal ristocetin cofactor assay), but factor VIII levels are decreased. **As a result, type 2N patients are frequently misdiagnosed as having hemophilia A.** The family history may be useful in distinguishing type 2N vWD from hemophilia A. Type 2N vWD is inherited autosomally (males and females are affected), whereas hemophilia A is an X-linked recessive disorder (males are affected and females are carriers). An assay which measures the ability of vWF to bind factor VIII is available in a limited number of specialized laboratories. Type 2N patients show decreased binding of factor VIII in this assay.

Type 3 vWD is a rare, severe bleeding disorder characterized by a severe quantitative deficiency of vWF such that vWF is typically undetectable.

The bleeding time is often prolonged in vWD. However, it is neither a necessary nor a reliable test for diagnosis.

In hemophilia A carriers (who are females only), the factor VIII:vWF ratio is ~0.5, instead of the normal ratio of 1. Definitive determination of carrier status may require DNA-based testing for mutations that cause hemophilia A.

Bleeding episodes, in most patients, can be treated with DDAVP (desmopressin) if needed, as DDAVP temporarily increases the levels of vWF and factor VIII two- to threefold. As a small percentage of patients do not respond to DDAVP, patients are usually given a trial dose of DDAVP while asymptomatic, with measurement of their vWF level before and after DDAVP, to ensure that their vWF levels do increase with DDAVP. Bleeding patients who do not respond to DDAVP or patients with severe vWD can be treated with vWF-containing factor VIII concentrates (eg, Humate-P®, Alphanate®, Koāte®). Some consider DDAVP contraindicated in type 2B because it can cause thrombocytopenia. However, others report DDAVP is a beneficial treatment for type 2B patients.

Acquired vWD is a rare condition that can occur spontaneously or in association with a variety of underlying disorders, such as hematologic neoplasms or autoimmune diseases. Severe aortic valve stenosis may be a cause of acquired type 2AvW syndrome and is improved by aortic valve replacement. Thrombotic thrombocytopenic purpura (TTP) is due to a deficiency of a vWF-cleaving protease, usually due to an autoantibody against the protease. This could account for the microvascular platelet-rich thrombi and thrombocytopenia that are characteristic of TTP. Unusually large vWF multimers may also be seen in TTP.

References

Budde U, Drewke E, Mainusch K, et al, "Laboratory Diagnosis of Congenital von Willebrand Disease," *Semin Thromb Hemost*, 2002, 28(2):173-90.

Ewenstein BM, "von Willebrand's Disease," *Annu Rev Med*, 1997, 48:525-42.

Favaloro EJ, "Collagen Binding Assay for von Willebrand Factor (VWF:CBA): Detection of von Willebrand's Disease (VWD), and Discrimination of VWD Subtypes, Depends on Collagen Source," *Thromb Haemost*, 2000, 83(1):127-35.

Furlan M, Robles R, Galbusera M, et al, "von Willebrand Factor-Cleaving Protease in Thrombotic Thrombocytopenic Purpura and the Hemolytic-Uremia Syndrome," *N Engl J Med*, 1998, 339(22):1578-84.

Gill JC, Endres-Brooks J, Bauer PJ, et al, "The Effect of ABO Blood Group on the Diagnosis of von Willebrand's Disease," *Blood*, 1987, 69(6):1691-5.

Meyer D, Fressinaud E, Gaucher C, et al, "Gene Defects in 150 Unrelated French Cases With Type 2 von Willebrand's Disease: From the Patient to the Gene," *Thromb Haemost*, 1997, 78(1):451-6.

Michiels JJ, "Diagnosis and Management of Congenital von Willebrand Disease," *Semin Thromb Hemost*, 2002, 28(2):109-69.

Ruggeri ZM and Zimmerman TS, "Variant von Willebrand's Disease: Characterization of Two Subtypes by Analysis of Multimeric Composition of Factor VIII/von Willebrand Factor in Plasma and Platelets," *Blood*, 1980, 65(6):1318-25.

Sadler JE, "A Revised Classification of von Willebrand Diseases," *Thromb Haemost*, 1994, 71:520-5.

Tsai HM and Lian EC, "Antibodies to von Willebrand Factor-Cleaving Protease in Acute Thrombotic Thrombocytopenic Purpura," *N Engl J Med*, 1998, 339(22):1585-94.

Warfarin, Serum or Plasma

Related Information

Synonyms Athrombin-K®; Coumadin®; Panwarfin®

Applies to Anticoagulants, Oral; Vitamin K

Abstract Warfarin is an oral anticoagulant used for prophylaxis and treatment of venous thrombosis, pulmonary embolism, and thromboembolic disorders. It is also used in atrial fibrillation with risk of embolism and as an adjunct in the prophylaxis of systemic embolism after myocardial infarction. Serum warfarin concentrations are seldom useful in managing therapy. International normalized ratios (INR), a process which normalizes PT results for variations in reagent activity, and less frequently now, prothrombin times (PTs) are far more helpful than serum warfarin concentrations in assessing clinical efficacy/toxicity.

Specimen Serum or plasma

Container Red top tube, lavender top (EDTA) tube

Reference Interval Therapeutic: 2-5 µg/mL (SI: 6.5-16 µmol/L)

Possible Panic Range Toxic: >10 µg/mL (SI: >32 µmol/L)

Use Selective therapeutic monitoring and toxicity assessment

Limitations This test **does not** measure bishydroxycoumarin and should not be used to monitor this drug.

Methodology High pressure liquid chromatography (HPLC), gas chromatography (GC)

Additional Information
- Half-life: 36-42 hours
- Volume of distribution: 0.14 L/kg
- Protein binding: >97%

Warfarin is highly bioavailable and is very highly bound (≥97%) to albumin. It crosses the placenta and is a known teratogen, but is not found in breast milk.

Side effects associated with warfarin include bleeding, skin necrosis and purple toe syndrome (cholesterol emboli). Patients receiving warfarin should not be given I.M. drug injections due to the risk of bleeding at the injection site.

Warfarin alone or in combination with aspirin is superior to aspirin alone in reducing the incidence of composite events after an acute myocardial infarction, but has higher risk of bleeding.

Additional relevant information is included in Prothrombin Time *on page 1116*.
(Continued)

Warfarin, Serum or Plasma *(Continued)*

References

Hirsh J, Dalen JE, Anderson DR, et al, "Oral Anticoagulants: Mechanism of Action, Clinical Effectiveness, and Optimal Therapeutic Range," *Chest*, 1998, 114(5 Suppl):445S-69S.

Hurlen M, Abdelnoor M, Smith P, et al, "Warfarin, Aspirin, or Both After Myocardial Infarction," *N Engl J Med*, 2002, 347(13):969-74.

Hylek EM, "Oral Anticoagulants. Pharmacologic Issues for Use in the Elderly," *Clin Geriatr Med*, 2001, 17(1):1-13.

Warming, Blood

Related Information

Cold Agglutinin Titer *on page 430*
Newborn Crossmatch and Transfusion *on page 952*
Red Blood Cells *on page 1139*
Uncrossmatched Blood, Emergency *on page 1281*
Whole Blood *on page 1333*

Abstract Warming should take place, when necessary, using an FDA-approved device during passage through the transfusion set. A visible thermometer and a warning system are required.

Use For very rapid, massive transfusion (>50 mL/kg/hour in adults or 15 mL/kg/hour in children), patients with severe cold agglutinin disease and infants undergoing exchange transfusion.

Limitations Uncontrolled warming of donor blood can severely damage RBCs. Although red cells must be heated to 44°C or higher to be damaged, it is probably best not to allow warming above 40°C. Warming must be done so as not to cause hemolysis.

Contraindications When moderate volumes of blood are given at ordinary rates, warming is unnecessary. It is probably unnecessary also in most patients with cold agglutinin disease or paroxysmal cold hemoglobinuria who are not seriously ill.

Additional Information Relatively large volumes of blood at refrigerator temperature, infused rapidly, can cause hypothermia and cardiac arrest. Blood subjected to excessive heat (ie, above 44°C) may be lethal. A quality assurance protocol is essential for all blood warmers and is required by accrediting agencies.

References

Calhoun L, "Blood Product Preparation and Administration," *Clinical Practice of Transfusion Medicine*, 3rd ed, Petz LD, Swisher SN, Kleinman S, et al, eds, New York, NY: Churchill Livingstone, 1996, 305-33.

Fridey J, *Standards for Blood Banks and Transfusion Services*, 22nd ed, Bethesda, MD: American Association of Blood Banks Press, 2003, 8.

Washing Cytology

Related Information

Asbestos, Lung or Sputum *on page 215*
Body Cavity Fluid Cytology *on page 285*
Body Fluid Amylase *on page 287*
Body Fluid Chemical Analysis *on page 291*
Body Fluid Glucose *on page 294*
Body Fluid pH *on page 295*
Breast Cancer, Hereditary, BRCA1, BRCA2 *on page 307*
Fine Needle Aspiration, Deep and Superficial Masses *on page 590*
Flow Cytometry, Overview *on page 592*
Fungal Culture, Biopsy or Body Fluid *on page 619*
Mycobacterial Culture, Biopsy or Body Fluid *on page 929*
Transbronchial Fine Needle Aspiration *on page 1266*

Synonyms Lavage Cytology

Applies to Breast Ductal Lavage; Colon Washings Cytology; Esophageal Washings Cytology; Gastric Washings Cytology; Peritoneal Washings Cytology

Test Includes Cytologic evaluation of cytocentrifuge (cytospin) preparations is most common; direct smears, membrane filters, and cell blocks are less often used.

Abstract Used to establish the presence of inflammatory or neoplastic lesions in various body sites.

Patient Preparation For gastric or esophageal washings, patient must be fasting at least 12 hours prior to procedure. Soft supper the night before, water *ad lib* 1 hour before. For intubation patient should be sitting upright. Dentures, if worn, should be removed. Colon washings specimens should be collected prior to barium examination.

Specimen Gastric washings, colon washings, ductal lavages, esophageal washings, peritoneal washings in a fresh unfixed state

Container Plastic, screw-top container, 50 mL; may contain 50% ethanol; **packed in ice**

Collection Gastric washing: Evaluation for neoplasm: Collect resting gastric contents and discard. Then instill 300 mL of a balanced salt solution through the gastric tube. Have patient then sit, lie on back, lie on stomach, lie on right side, and lie on left side. Aspirate as much of injected saline as possible and place in container packed in ice. Label with patient name, identification number, and date. Deliver immediately on ice to the Cytology Laboratory.

Peritoneal washings: Wash peritoneal site vigorously with several hundred mL of a balanced salt solution. Retrieve as much as possible and submit as above, labeled by anatomic site (eg, "subdiaphragm", "cul-de-sac", "left gutter wash", "right gutter wash").

Storage Instructions Due to rapid degeneration of cellular material, storage, even at 4°C for any extended length of time, is not recommended.

Use Establish the presence of primary or metastatic neoplasms. Washing cytology in **laparotomies** provides detection of occult neoplasms, recognition of tumor persistence and likelihood of recurrence as well as staging. Postive cytologic examinations in **peritoneal washings**, especially for patients with ovarian and endometrial primaries, are apt to lead to alterations in patient management. Washing cytology also supports recognition of reactive processes and infectious diseases and aids especially in staging of gyneco-logic and gastrointestinal neoplasms. Peritoneal washings are commonly done with gynecologic surgical procedures. In some cases, recognition of malignant cells in washings changes postoperative staging, greatly influencing therapy and prognosis.

Ductal lavage is a minimally invasive method, whereby breast ductal cells are aspirated using saline and are analyzed by a cytopathologist. The procedure is performed at large medical institutions under the hands of an oncological breast specialist. Ductal lavage is indicated for the following patients at high risk for breast cancer:

- previous personal history
- family history
- 5-year Gail risk ≥1.7%
- BRCA1 or BRCA2 gene mutation

Atypical findings confer a 4.3-5.3 times greater relative risk of developing breast cancer. Practitioners using ductal lavage should be prepared to counsel patients about the results, offer psychological support, and be prepared to further evaluate the patient when atypia is identified. Ductal lavage is not a cancer detection technique and should not replace standard cancer screening methods.

Limitations Nondiagnostic if epithelium is not present or poorly preserved; if specimen is grossly contaminated with food or barium sulfate; if no mesothelial cells are identified in peritoneal washings, the specimen is unsatisfactory; may be of limited value in intestinal cases in which the lesion is submucosal; a Wang transmucosal needle aspirate may be helpful. Ductal lavage requires 10 well preserved epithelial cells for adequacy. Difficulties in cytologic interpreta-tion, including false positives, may derive from the presence of reactive meso-thelial cells, endosalpingiosis, endometriosis, hemorrhage, and pelvic inflammatory disease, and from effects of chemotherapy and irradiation therapy.

False negatives remain a serious problem.

Additional Information Lavage is not as sensitive or specific as endoscopi-cally directed brushings or biopsy (aspiration biopsy or tissue forceps biopsy). However, a complete set of peritoneal, pelvic, and diaphragmatic washings are an essential part of the staging of gynecologic, particularly ovarian carcinomas.

References

Board of Directors, "The American Society of Breast Surgeons. Official Statement Ductal Lavage and Cell-Based Risk Assessment," Approved April 30,2003 (www.breastsurgeons.org).

(Continued)

Washing Cytology *(Continued)*

DeMay RM, "Fluids," *The Art and Science of Cytopathology*, Chapter 8, Chicago, IL: ASCP Press, American Society of Clinical Pathologists, 1996, 257-325.

Ducatman BS and Soisson AP, "Peritoneal Washing Cytology. How Significant?" *Arch Pathol Lab Med*, 1997, 121(9):923-4.

Dupont WD and Page DL, "Risk Factors for Breast Cancer in Women With Proliferative Breast Disease," *N Engl J Med*, 1985, 312(3):146-51.

Dupont WD, Parl FF, Hartmann WH, et al, "Breast Cancer Risk Associated With Proliferative Breast Disease and Atypical Hyperplasia," *Cancer*, 1993, 71(4):1258-65.

Fabian CJ, Kimler BF, Zalles CM, et al, "Short-Term Breast Cancer Prediction by Random Periareolar Fine Needle Aspiration Cytology and the Gail Risk Model," *J Natl Cancer Inst*, 2000, 92(15):1217-27.

Mathew S and Erozan YS, "Significance of Peritoneal Washings in Gynecologic Oncology: The Experience With 901 Intraoperative Washings at an Academic Medical Center," *Arch Pathol Lab Med*, 1997, 121(6):604-6.

Wrensch MR, Petrakis NL, King EB, et al, "Breast Cancer Incidence in Women With Abnormal Cytology in Nipple Aspirates of Breast Fluid," *Am J Epidemiol*, 1992, 135(2):130-41.

Internet Web Sites

www.breastsurgeons.org

♦ **Watson-Schwartz Test** *see* Porphobilinogen, Urine *on page 1073*

♦ **Weak D** *see* Rh Genotype *on page 1162*

♦ **Wellbutrin®** *see* Bupropion, Serum or Plasma *on page 317*

♦ **Wellbutrin SR®** *see* Bupropion, Serum or Plasma *on page 317*

♦ **Wellbutrin XL™** *see* Bupropion, Serum or Plasma *on page 317*

♦ **Westergren Sed Rate** *see* Sedimentation Rate, Erythrocyte *on page 1181*

Western Equine Encephalitis Virus Serology

Related Information

Arthropod Identification *on page 213*
Bacterial Culture, Cerebrospinal Fluid *on page 236*
California Encephalitis Virus Serology *on page 334*
Cerebrospinal Fluid Analysis: Overview *on page 355*
Eastern Equine Encephalitis Virus Serology *on page 529*
Encephalitis Viral Serology *on page 535*
St Louis Encephalitis Virus Serology *on page 1224*
Viral Culture *on page 1307*

Abstract In North America, Western equine encephalitis (WEE) is a disease of summer and it usually causes disease in states west of the Mississippi. Infants are most susceptible and are at risk for sequelae, such as infantile convulsions. The case fatality rate is 3% to 5%. Transmission is via the mosquito vector *Culex tarsalis*. The pathogenesis of WEE resembles that of Eastern equine encephalitis (EEE). Clinical features are also similar to those of St Louis encephalitis.

The encephalitic alphaviruses, naturally transmitted by mosquitos, are infectious by aerosol, thus, they lend themselves to application as bioterrorist weapons. These include Eastern, Western, and Venezuelan encephalitis viruses. They should be considered in epidemic febrile disease situations, especially when some patients progress to neurological signs and symptoms.

Specimen Serum or cerebrospinal fluid

Container Red top tube, sterile CSF tube

Collection Acute and convalescent specimens are recommended.

Reference Interval Less than a fourfold increase in titer in paired sera; HAI titer: <1:10; CF titer: <1:8; **no IgM antibody in serum or cerebrospinal fluid (CSF)**

Use Establish the diagnosis of Western equine encephalitis virus infection; generally, an encephalitis antibody panel is tested that includes WEE, EEE, SLE, and California encephalitis virus.

Limitations Cross reactions can occur to Eastern equine encephalitis (EEE) virus.

Methodology Diagnosis is best established by detection of specific IgM antibody. Complement fixation, hemagglutination inhibition (HAI), neutralization, indirect fluorescent antibody (IFA), enzyme-linked immunosorbent assay.

Additional Information The initial signs and symptoms of WEE infection resemble those of enterovirus infections. WEE infection includes fever, aseptic meningitis, and meningoencephalitis. The CSF leukocyte count is 10-300 /

mm³. Virus has been recovered from CSF, blood, and brain and recent developments show that when the patient is viremic, specific nucleic acid can be detected by polymerase chain reaction (PCR) from serum, tissue, or CSF.

References

Franz DR, Jahrling PB, Friedlander AM, et al, "Clinical Recognition and Management of Patients Exposed to Biological Warfare Agents," *JAMA*, 1997, 278(5):399-411.

Kramer LD and Fallah HM, "Genetic Variation Among Isolates of Western Equine Encephalomyelitis Virus From California," *Am J Trop Med Hyg*, 1999, 60(4):708-13.

Nasci RS and Moore CG, "Vector-Borne Disease Surveillance and Natural Disasters," *Emerg Infect Dis*, 1998, 4(2):333-4.

Internet Web Sites

www.astdhpphe.org/infect/wee.html
www.cdc.gov/ncidod/dvbid/arbor/index.htm

West Nile Virus Diagnostic Procedures

Related Information

Encephalitis Viral Serology *on page 535*
St Louis Encephalitis Virus Serology *on page 1224*

Abstract Although severe neurologic disease develops in 1 of 150 infections, most West Nile virus (WNV) infections are subclinical. Advanced age is a risk factor. Immunocompromised individuals may be at high risk for death. Encephalitis is more common than meningitis/meningoencephalitis. A poliomyelitis-like syndrome occurs as well. The agent is an arbovirus principally spread by *Culex* mosquito vectors and by birds, especially crows and jays.

Specimen Serum, plasma, cerebrospinal fluid, brain tissue (humans), mosquitoes, avian samples

Sampling Time Specimens drawn before 8 days after onset of symptoms may be serologically negative. When WNV-specific IgM is detected in serum but not cerebrospinal fluid, serum from acute and convalescent stages is needed. In that setting, increase in titer by a factor ≥4 is required for confirmation of diagnosis.

Turnaround Time The plaque reduction neutralization test usually requires up to 8 days to test for both WNV and St Louis encephalitis virus.

Special Instructions WNV is a biosafety level 3 pathogen.

The laboratory should know the date of onset and dates of sample collection to assess whether to anticipate IgM or IgG reactions. Travel history is relevant, and vaccination history is important.

Use Evidence of encephalitis, myelitis, meningoencephalitis, fever, muscle weakness, fever, and acute often asymmetrical flaccid paralysis may indicate WNV infection. Other clinical observations include roseolar or maculopapular rash, lymphadenopathy, and polyarthropathy.

Limitations Cross-reactions may occur in subjects who have recently been vaccinated against related flaviviruses (eg, yellow fever, Japanese encephalitis) or recently infected with one of these agents (eg, dengue fever); but cross-reactivity was not reported to represent a problem in acute-phase samples tested for IgM antibody. Antibody tests fail to identify recently infected individuals who are yet to become seropositive, but who are potential blood donors.

Methodology IgM capture and IgG enzyme-linked immunosorbent assays (ELISA). IgM antibody may be found without IgG virus-specific IgG in the first week of illness. Switch to IgG antibody is reported following 4-5 days of illness and is found earlier in CSF. A positive serum result can be followed by WNV IgM in CSF by ELISA, by WNV RNA in CSF, by fourfold increase in IgG, or by virus isolation. ELISA serum titers can be confirmed with plaque-reduction neutralization.

The presence of WNV RNA is confirmatory by reverse transcriptase-PCR testing, but its sensitivity in CSF is only 57%, in serum only 14%.

Screening of donors for blood donations with nucleic acid-based assays is advocated.

WNV may be identified by immunocytochemistry and immunofluorescent assays.

Additional Information Monitoring for West Nile virus activity includes mosquitoes, chickens, wild birds, and susceptible mammals, including horses as well as humans. The virus can survive in donated blood and can be transmitted by red cell, platelet and fresh frozen plasma transfusions. Transmission has been (Continued)

West Nile Virus Diagnostic Procedures *(Continued)*

conclusively linked to organ transplantation. Other means of transmission include breast-feeding, transplacental transmission, and laboratory acquisition.

The antibody response to WNV infection by IgM antibody capture enzyme-linked immunosorbent assay is positive in most subjects 7-8 days after onset. Such IgM response in serum or cerebrospinal fluid is detectable within 4 days of onset in about 75%. Such IgM antibody persists for a year or more. Detection of WNV-specific IgM in cerebrospinal fluid is confirmatory of current infection, while detection in serum is thought to provide only a diagnosis of probable infection.

References

Glass JD, Samuels O, and Rich MM, "Poliomyelitis Due to West Nile Virus," *N Engl J Med*, 2002, 347(16):1280-1.

Iwamoto M, Jernigan DB, Guasch A, et al, "Transmission of West Nile Virus From an Organ Donor to Four Transplant Recipients," *N Engl J Med*, 2003, 348(22):2196-203.

Kulas KE, "Use of Arboviral Immunofluorescent Assay in Screening for West Nile Virus," *Ann N Y Acad Sci*, 2001, 951:357-60.

Lanciotti RS, Kerst AJ, Nasci RS, et al, "Rapid Detection of West Nile Virus From Human Clinical Specimens, Field-Collected Mosquitoes, and Avian Samples by a TaqMan Reverse Transcriptase-PCR Assay," *J Clin Microbiol*, 2000, 38(11):4066-71.

Leis AA, Stokic DS, Polk JL, et al, "A Poliomyelitis-Like Syndrome From West Nile Virus Infection," *N Engl J Med*, 2002, 347(16):1279-80.

Morse DL, "West Nile Virus - Not a Passing Phenomenon," *N Engl J Med*, 2003, 348(22):2173-4.

Ogedegbe HO and St Hill H, "West Nile Virus: Laboratory Diagnosis and FDA Guidance," *Lab Med*, 2003, 34(6):445-8, 465-7.

Pealer LN, Marfin AA, Petersen LR, et al, "Transmission of West Nile Virus Through Blood Transfusion in the United States in 2002," *N Engl J Med*, 2003, 349(13):1236-45.

Petersen LR, "West Nile Virus: A Primer for the Clinician," *Ann Intern Med*, 2002, 137(3):173-9.

Petersen LR, Roehrig JT, and Hughes JM, "West Nile Virus Encephalitis," *N Engl J Med*, 2002, 347(16):1225-6.

Roos KL, "Fever and Asymmetrical Weakness in Summer: Evidence of a West Nile Virus-Associated Poliomyelitis-Like Illness," *Mayo Clin Proc*, 2003, 78(10):1205-6.

Sampathkumar P, "West Nile Virus: Epidemiology, Clinical Presentation, Diagnosis, and Prevention," *Mayo Clin Proc*, 2003, 78(9):1137-44.

Sampson BA, Ambrosi C, Charlot A, et al, "The Pathology of Human West Nile Virus Infection," *Hum Pathol*, 2003, 31(5):527-31.

Sejvar JJ, Haddad MB, Tierney BC, et al, "Neurologic Manifestations and Outcome of West Nile Virus Infection," *JAMA*, 2003, 290(4):511-5.

Tardei G, Ruta S, Chitu V, et al, "Evaluation of Immunoglobulin M (IgM) and IgG Enzyme Immunoassays in Serologic Diagnosis of West Nile Virus Infection," *J Clin Microbiol*, 2000, 38(6):2232-9.

Internet Web Sites www.phppo.cdc.gov/PHTN/webcast/westnile/

White Blood Cell Count

Related Information

Antineutrophil Alloantibody and Autoantibody *on page 186*

Apoptosis Assays *on page 205*

Bone Marrow *on page 296*

CD4/CD8 Enumeration *on page 349*

Chromosome Analysis, Bone Marrow *on page 407*

Complete Blood Count *on page 442*

HIV-1/HIV-2 Serology *on page 736*

Leukocyte Alkaline Phosphatase *on page 845*

Leukocyte Cytochemistry *on page 846*

Lymph Node Biopsy *on page 880*

Peripheral Blood: Differential Leukocyte Count *on page 1010*

Synonyms Leukocyte Count

Applies to Myelokathexis; Schwachman-Diamond Syndrome

Abstract This procedure determines the white blood cell (WBC) concentration in a body fluid, usually peripheral blood. The count is most commonly generated by an automated analyzer using aperture-impedance and/or laser beam technology. Different types of white blood cells (eg, granulocytes, monocytes, lymphocytes, etc) are included in the total count. The results have widespread application to the diagnosis and monitoring of infectious, neoplastic, and immunologic disease states.

Specimen Whole blood or other body fluid

Container Lavender top (EDTA) tube

Causes for Rejection Clotted specimen, hemolyzed specimen

Reference Interval Peripheral blood (adult): 4500-11,000/mm³ (SI: 4.5-11.0 x 10⁹/L). There is diurnal variation with lowest level of WBC count in the morning (subject at rest) and with maximum level in the afternoon. With high levels of activity, stress, exercise (associated with release of Adrenalin®), increase in WBC count of 2000-5000/mm³ is common; rises up to 30,000/mm³ may occur.

See the listing, White Blood Cell Count in Jacobs et al, *Laboratory Test Handbook*, 5th ed, page 496-500, for a discussion of origin and significance of reference intervals, WBC, and other components of the complete blood count.

It would appear that allowing for some minor exceptions, **past reference intervals for the WBC count**, and for that matter, for most elements of the "complete blood count" **are clinically useful**. Studies from the last decade, utilizing multiparameter automated analyzers, deal with necessarily smaller sample sizes (as compared to the NHANES II study) but result in similar reference intervals. Modern investigations generally focus on population subsets such as very-low-birth-weight neonates, newborns at term, infants, school-age children, and ethnic/sex differences.

A 10-year old study of the use of total and differential leukocyte counts in clinically well children found no unsuspected illness as a result of an abnormal total leukocyte count (778 CBC results during a 1-year period) and including 9.8% of 387 clinically well subjects with neutropenia. The high frequency of results outside published normal intervals led the authors of this study (Moyer and Grimes) to call for a re-evaluation of "normal ranges" for leukocyte counts in pediatric patients. Logistic and efficiency considerations with use of multichannel analyzers (in spite of low yield in case finding situations) are likely to perpetuate use of total leukocyte and differential counts in such surveys.

WBC normal ranges (SI units: x 10⁹ cells/L):

- birth: 9.0-30.0 cells x 1000/mm³
- 24 hours: 9.4-34.0 cells x 1000/mm³
- 1 month: 5.0-19.5 cells x 1000/mm³
- 1-3 years: 6.0-17.5 cells x 1000/mm³
- 4-7 years: 5.5-15.5 cells x 1000/mm³
- 8-13 years: 4.5-13.5 cells x 1000/mm³
- adults: 4.5-11.0 cells x 1000/mm³

Note: Data largely from Albritton EC, 1952

See the tables and charts reproduced in *Laboratory Test Handbook* 5th edition, for a compilation of data from recent reference range studies.

Possible Panic Range On admission <2500/mm³ (SI: 2.5 x 10⁹/L) or >30,000/mm³ (SI: >30.0 x 10⁹/L).

Use White cell enumeration; evaluate myelopoiesis, bacterial and viral infections, toxic metabolic processes; diagnose/evaluate leukemic states

Limitations If nucleated RBCs are found in differential count, the white blood cell count should be corrected if there is clinical relevance but in particular if, generally, there are significant numbers of nucleated RBCs (over 10 per 100 WBCs). The use of capillary blood samples may give WBC count some 3% to 12% higher values than those obtained with use of venous blood. Electronic machine counters are subject to spurious high WBC counts in a variety of situations including presence of cryoglobulins/cryofibrinogen, clumped platelets (which may cause false elevation of WBC count and thrombocytopenia), fibrin strands, nucleated red blood cells, nonlysed red cells (as in some cases of hemoglobinopathy), EDTA-induced platelet aggregation, monoclonal proteins, and cold agglutinins. Falsely low WBC counts may be seen with microclots or partial clotting of the specimen, loss of cell integrity due to prolonged storage, or if fragile (apoptotic) white cells are present (as may occur in specimens from patients with myelodysplastic syndromes). Fat cells (as may accompany connective tissue stores) may cause artifactual elevation of machine white cell counts on specimens from traumatic venous puncture and from bone marrow aspirate material.

Methodology Manual - hemocytometer counting chambers. Most WBC count determinations are obtained from one channel of a highly automated multichannel electronic and pneumatic analyzer using aperture-impedance and/or aperture conductance and/or laser light scattering technologies. Prior to manual or automated counting, blood is diluted with a solution that lyses red (Continued)

White Blood Cell Count *(Continued)*

cells. Visual-based manual methods are time-consuming and are less reproducible. Coefficient of variation (CV) is about 10%, compared to machine methods that have CVs in the 1% to 3% range. WBC differential determination is provided by recent generations of analyzers. Excellent performance (precision, linearity, and lack of carryover) has been found by Picard et al on field evaluation of a recent multichannel instrument, the Coulter GEN-S.

Additional Information The white blood cell count (leukocyte count) in the postmillenium is commonly determined by an automated hematology analyzer. A variety of circulating white cells (neutrophils, lymphocytes, monocytes, eosinophils, basophils) and ordinarily much less commonly encountered elements (eg, plasma cells and CD34+ blastic stem cells) form the normal leukocyte composite. Reference intervals for the total white count are age- and ethnic-dependent. Individual cell components are particularly age-dependent, knowledge of which may have significant clinical relevance. In adults, polymorphonuclear leukocytes (PMNs) are the predominant cell type (normally slightly over $1/2$ to $3/4$ of circulating white blood cells) while lymphocytes are the most common type of circulating white cell in most infants and children.

In **newborn infants** WBC counts from different vascular sources (ie, capillary vs venous vs arterial blood) are not equivalent. WBC counts from actively crying babies may show leukocytosis with left shift, **possibly erroneously suggesting bacterial infections**. Any stressful situation in newborns, children, or adults which leads to increase in endogenous epinephrine production may cause a rapid (15-30 minutes) increase in WBC count. In the evaluation of infection in newborns and young children, it is recommended that several counts be obtained from a consistent vascular source in resting individuals. Hematologic measurements have significant within subject and between subject individual variation. Screening using conventional reference limits may be misleading. Subject specific reference intervals are likely to have greater clinical utility. There is modest progressive leukocytosis (due to neutrophils) throughout **pregnancy** into the third trimester with subsequent decline in white count after about 34 weeks gestation. The count returns to normal about 1 week after delivery.

Elevated WBC count has a broad clinical differential diagnosis. Infections and/or leukemic processes are of special importance. **Acute infections** are usually associated with an increase in neutrophil type WBCs. If the white cells are **lymphocytes**, however, viral illness, leukemic process, and **pertussis** (whooping cough) are candidates for consideration. While pertussis is usually associated with lymphocytosis at the 25,000/mm^3 or so level, some cases (especially in the very young) may be seen, temporarily with 100,000/mm^3 level WBC counts. Paroxysmal cough in a nonimmunized child are important clinical findings.

Included in the broad differential consideration for the cause of neutropenia is collagen-vascular disease, notably lupus erythematosus and other autoimmune neutropenias. Many drugs result in leukopenia including bezafibrate, an antihyperlipidemic fibric acid. Thrombocytopenia, myelodysplasia, or acute myelogenous leukemia may develop in patients with congenital neutropenia on long-term therapy with recombinant granulocyte colony-stimulating factor.

Absolute total neutrophil (ATN) counts differ in infants of appropriate weight for gestational age vs those of very low birth-weight (VLBW neonates). The latter may show a fall in WBC count for the first 3 days (vs the expected postnatal rise in infants of appropriate weight). Neutropenia may be the result of or predispose to neonatal sepsis. A near doubling of VLBW neonate survival since the 1974-76 period (from 42% to over 79%) has resulted in an uncertain lower boundary of the ATN reference interval (for VLBW neonates) due to a greater number of <1500 g subjects. With new intervals in use, fewer VLBW neonates would be considered "neutropenic," with implications to the recognition and treatment of neonatal sepsis.

Several studies indicate that the baseline WBC count is a predictor of the relative risk of coronary heart disease (CHD) morbidity and mortality independent of cigarette smoking, while the latter increases the white cell count significantly within the normal range. The white count is a direct function of the amount of inhaled smoke.

Congenital disorders with severe neutropenia include cyclic neutropenia, myelokathexis, the Schwachman-Diamond syndrome, and severe congenital neutropenia. In these conditions, marrow neutrophil precursors fail to complete differentiation and fail to enter into the peripheral blood (apoptosis implied). In myelokathexis, a rare congenital cause of chronic leukopenia/neutropenia, accelerated apoptosis of marrow myeloid precursors may account for under-production of myeloid cells and resultant neutropenia.

Neutrophils have been considered to play a role in the link between pathways of coagulation and inflammation. The neutrophil membrane has binding sites for proteins of the contact system, provides a platform for assembly of the prothrombinase complex, with affects on neutrophil functions (eg, chemotaxis and degranulation). Neutrophil elastase, in turn, can degrade some coagulation proteins contributing to modulation of thrombotic and fibrinolytic systems.

References

Aprikyan AA, Liles WC, Park JR, et al, "Myelokathexis, a Congenital Disorder of Severe Neutro-penia Characterized by Accelerated Apoptosis and Defective Expression of bcl-x in Neutrophil Precursors," Blood, 2000, 95(1):320-7.

Dale DC, Cottle TE, Fier CJ, et al, "Severe Chronic Neutropenia: Treatment and Follow-up of Patients in the Severe Chronic Neutropenia International Registry," Am J Hematol, 2003, 72(2):82-93.

Houwen B, "The Blood Cell Count," Advanced Laboratory Methods in Haematology, Part 1, Chapter 2, Rowan RM, van Assendelft OW, and Preston FE, eds, London, UK: Arnold, 2002, 19-44.

Mouzinho A, Rosenfeld CR, Sanchez PJ, et al, "Revised Reference Ranges for Circulating Neutro-phils in Very-Low-Birth-Weight Neonates," Pediatrics, 1994, 94(1):76-82.

Moyer VA and Grimes RM, "Total and Differential Leukocyte Counts in Clinically Well Children. Information or Misinformation?" Am J Dis Child, 1990, 144(11):1200-3.

Picard F, Gicquel C, Marnet L, et al, "Preliminary Evaluation of the new Hematology Analyzer Coulter GEN-S in a University Hospital," Clin Chem Lab Med, 1999, 37(6):681-6.

Witko-Sarsat V, Rieu P, Descamps-Latscha B, et al, "Neutrophils: Molecules, Functions, and Pathophysiological Aspects," Lab Invest, 2000, 80(5):617-53.

♦ **White Blood Cell Morphology** see Peripheral Blood: Differential Leukocyte Count on page 1010

♦ **White Cell Crossmatch** see Tissue Typing on page 1259

♦ **White Lady** see Cocaine (Cocaine Metabolite), Qualitative, Urine or Hair on page 427

Whole Blood

Related Information

Blood Gases and pH, Arterial on page 275
Donation, Blood on page 521
Irradiated Blood Components on page 811
Oxygen Saturation, Blood on page 991
Pretransfusion Testing on page 1088
Red Blood Cells on page 1139
Red Blood Cells, Leukocytes Reduced on page 1141
Risks of Transfusion on page 1166
Transfusion Reaction Work-up on page 1269
Uncrossmatched Blood, Emergency on page 1281
Warming, Blood on page 1326

Applies to Fresh Blood; Massive Transfusions

Test Includes ABO and Rh type, antibody screen, crossmatch, and antibody identification when screen is positive, as for other transfusions

Abstract The primary indication for transfusion of whole blood is to provide both blood volume expansion and oxygen-carrying capacity. A unit of whole blood consists of about 450 mL (±10%) blood including plasma and about 63 mL of anticoagulant preservative such as citrate phosphate dextrose adenine solution (CPDA-1). A typical donor unit has a hematocrit of about 35% to 40%. The expiration date for CPDA-1 blood is 35 days after the date of collection if stored continuously at 1°C to 6°C.

Patient Preparation The patient should have an identification wristband. Emergency Departments (ERs) may use special or temporary identification. Measure blood loss if possible, as well as fluid intake and output. **Dosage and administration:** Give whole blood through a standard 170 micron filter. It can be warmed, if warming is clinically indicated. The blood should not be warmed above 38°C. The rate of infusion depends on clinical conditions but should not be slower than 4 hours per unit. **No medications or solutions should be added to blood. Never give Ringer's lactate, hypotonic, or** (Continued)

Whole Blood *(Continued)*

dextrose-containing solutions through the same tubing as blood. These solutions are incompatible with stored blood, causing clots, aggregates, and shortened cell survival.

Aftercare Same as for other RBC transfusions.

Specimen Blood

Container One red top tube or one lavender top (EDTA) tube

Collection (Of sample from intended recipient): At the patient's bedside, ask the patient to give his or her name. Compare with the patient's wristband. Label the sample tube with two unique forms or patient identification (eg, patient's full name, hospital number), also include date and initials of the collector. Further information may be required on a requisition form. Take extra care with identification of unresponsive patients.

Storage Instructions Store in Blood Bank monitored refrigerator only until issue. When it is not possible to transfuse immediately after issue, return blood to Blood Bank within 30 minutes. Otherwise, blood that has been out of monitored refrigeration must be discarded. Appropriate designated refrigeration includes specified storage conditions, temperature recorders, and alarm signals.

Causes for Rejection (Of patient sample): Gross hemolysis, sample placed in a serum separator tube, specimen tube not properly labeled

Turnaround Time Because of required testing, the time from blood donation until the blood is available for transfusion varies from about 3-5 days. For routine situations, a unit can be ready for transfusion within 30-45 minutes from the time the Blood Bank gets the type and crossmatch sample, if blood of the appropriate type is on hand. Presence of unexpected antibodies may require hours to a day or two for identification.

Special Instructions Blood Banks and hospital transfusion services hold crossmatched blood only for 24 hours, after which they usually must make it available from other patients. There may be some exceptions. Notify the Blood Bank as soon as possible if the patient will not need transfusion so the blood can be used for some other patient.

Summary of Transfusion Guidelines (Excluding Neonates)[1]

Acute Blood Loss

1. Evaluate for risk of ischemia and other concomitant disease
2. Estimate and/or anticipate degree of blood loss

    >30% to 40% rapid blood volume loss: transfuse RBCs, whole blood as available

    <30% to 40% rapid blood volume loss: RBCs usually not needed in previously healthy person

3. Measure hemoglobin

    >10 g/dL: RBCs rarely needed

    <6 g/dL: RBCs usually needed

    6-10 g/dL: RBC need depends on other factors

4. Measure vital signs and tissue oxygenation (most useful in 6-10 g/dL hemoglobin range when extent of blood loss is unknown)

    Tachycardia, hypotension not corrected by volume replacement alone: RBCs needed

    $P\bar{v} O_2$ <25 torr, extraction ratio >50%

    VO_2 <50% of baseline: RBCs often needed

Chronic Anemia

1. Treat with specific pharmacologic agents (eg, cobalamin, folic acid, recombinant human erythropoietin, iron) when diagnosis permits
2. Special strategies for sickle cell disease, thalassemia are needed
3. Transfuse to minimize symptoms and risk of anemia (usually at hemoglobin levels of 5-8 g/dL)

[1]RBCs = red blood cells; $P\bar{v} O_2$ = oxygen tension of pulmonary arterial blood at the completion of oxygen unloading; VO_2 = oxygen consumption.

Adapted from Simon TL, Alverson DC, AuBuchon J, et al, "Practice Parameter for the Use of Red Blood Cell Transfusions," *Arch Pathol Lab Med*, 1998, 122(2):130-8.

Use Replace red cell mass and plasma volume in patients in whom there is significant loss or depletion of both, improve oxygen transport (ie, treatment of acute blood loss). Therapy for acute bleeding, including massive transfusion in exsanguinating emergencies, and some surgical cases. Exchange transfusion. See table and see Red Blood Cells *on page 1139* for further discussion of medical indications for transfusion.

Limitations Some components, eg, platelets, factor V, factor VIII (AHF), are labile and not present in sufficient quantity in stored whole blood to provide adequate replacement therapy. Adenine, citrate, sodium, and antibodies are less in red blood cells, which are preferable to whole blood for patients with chronic renal or liver disease. The plasma in whole blood is unneeded in many situation.

Contraindications Do not use whole blood for anemia that can be corrected with specific, safer products (eg, iron, B_{12}, folic acid). Whole blood is contraindicated in patients with congestive heart failure, uremia or hepatic failure, or with other chronic decrease of red cell mass. Such patients should receive red cells rather than whole blood if they require transfusion. For exchange transfusion, whole blood should preferably not be more than 5 days old. Replace blood volume deficits more safely and adequately with other volume expanders (saline, Ringer's lactate, albumin, plasma protein fraction). Treat coagulation factor deficiencies with appropriate factor-specific concentrates. The infusion of large volumes of blood may cause additional bleeding due to dilution of clotting factors or platelets. Donor and recipient must be ABO compatible.

Methodology Attention to electrolytes, blood gases and blood volumes is indicated with rapid and/or massive transfusion; see Warming, Blood *on page 1326.*

Additional Information When whole blood is not available, red blood cells are substituted. **Fresh blood** is impossible to define and obtain except by reference to whatever labile component is needed. Requests for "fresh" blood necessitate consultation and are usually filled by provision of the appropriate components or fractions.

References

Ely EW and Bernard GR, "Transfusions in Critically Ill Patients," *N Engl J Med*, 1999, 340(6):467-8.

Goodnough LT, Brecher ME, Kanter MH, et al, "Transfusion Medicine. First of Two Parts: Blood Transfusion," *N Engl J Med*, 1999, 340(6):438-48.

Hebert PC, Wells G, Blajchman MA, et al, "A Multicenter, Randomized, Controlled Clinical Trial of Transfusion Requirements in Critical Care," *N Engl J Med*, 1999, 340(6):409-17.

- ♦ **Wickistick** *see* Phencyclidine, Qualitative, Urine *on page 1019*
- ♦ **Williams-Fitzgerald-Flaujeac Factor** *see* High-Molecular Weight Kininogen *on page 731*
- ♦ **Williams Syndrome** *see* Gene Rearrangement for Leukemia and Lymphoma *on page 633*
- ♦ **Wolfram Syndrome** *see* Cobalamin, Serum *on page 424*
- ♦ **Wygesic®** *see* Propoxyphene, Serum or Urine *on page 1096*
- ♦ **X** *see* 3,4 Methylenedioxymethamphetamine, Urine *on page 907*
- ♦ **Xanthine, Urine** *see* Molybdenum, Blood *on page 919*
- ♦ **Ximelagatran** *see* Prothrombin Time *on page 1116*
- ♦ **XTC** *see* 3,4 Methylenedioxymethamphetamine, Urine *on page 907*
- ♦ **Xylocaine®** *see* Lidocaine, Serum or Plasma *on page 850*
- ♦ **Xylose Absorption Test** *see* d-Xylose Absorption Test, Serum, Urine *on page 527*
- ♦ **Xylose Tolerance Test** *see* d-Xylose Absorption Test, Serum, Urine *on page 527*
- ♦ **Xylotocan®** *see* Tocainide, Serum or Plasma *on page 1260*

Yellow Fever

Related Information

Arthropod Identification *on page 213*
Bilirubin, Total, Serum *on page 265*
Hepatitis D Serology *on page 711*
Hepatitis: Laboratory Assessment, Overview *on page 713*
Prothrombin Time *on page 1116*
(Continued)

Yellow Fever (Continued)

Abstract Yellow fever is classified among the viral hemorrhagic fevers, a group which also includes hantavirus pulmonary syndrome and others.

The first clinical description of the disease caused by yellow fever (YF) was in 1648. The virus was probably spread to the Western hemisphere by trading vessels from West Africa that were infested with YF infected *Aedes aegypti*. Currently, YF is transmitted in parts of sub-Saharan Africa and South America. The disease has never been documented in Asia. *A. aegypti* mosquitoes that are infected after feeding on viremic humans spread the infection in subsequent feeding attempts. This method of transmission is responsible for epidemic (urban) YF. The illness due to YF ranges in severity from a mild self-limited disease to hemorrhagic fever which is fatal in 50% of cases.

Specimen Serum

Container Red top tube

Sampling Time Acute and convalescent phases; *vide infra*.

Storage Instructions Refrigerate at 4°C.

Reference Interval Negative

Critical Values Fourfold or greater change in titer; *vide infra*.

Use Diagnose yellow fever; document vaccination

Limitations Patients from areas of the world where there are numerous flavivirus infections (eg, Africa) can have cross-reactive antibodies that give low positive values.

Methodology Enzyme immunoassay (EIA), indirect immunofluorescent antibody (IFA), complement fixation (CF), reverse transcriptase PCR

Additional Information Yellow fever has been a disease of public health importance since the 15th century. At various times in human history it has caused epidemics of disease in the Americas, Africa, and Europe. Yellow fever is an arboviral infection with epidemiologic transmission cycles between monkeys, mosquitoes, and humans. The disease, characterized by fever, headache, and myalgias, follows an incubation period of 3-6 days. The symptoms begin abruptly, accompanied by facial flushing and a relative bradycardia (Faget's sign). Laboratory abnormalities include leukopenia, thrombocytopenia, prolongation of prothrombin time and PTT, proteinuria, increased serum total and conjugated bilirubin, AST, ALT, and creatinine. Resolution of this period of infection is the end of the illness in most patients. However, in others, the fever will fade for a few hours to several days to be followed by symptoms that include high fever, headache, back pain, nausea, vomiting, abdominal pain, and somnolence. This phase of the illness is dominated by icteric hepatitis and a hemorrhagic diathesis.

The liver is characterized by classical midzone necrosis with Councilman bodies and vacuolar fatty changes. Liver biopsy is usually considered contraindicated by virtue of possible uncontrolled bleeding.

Effective vaccines are available for the prevention of yellow fever. Travelers (including children) to endemic areas are encouraged to receive these vaccinations before visiting countries with high prevalence of YF.

Patients tested for serology will usually have positive results from 7-10 days after the onset of illness. IgM ELISA is detected in high titer only for a short interval following infection; thus, a single convalescent serum specimen is reliable for this method. In secondary infections, the IgM and IgG ELISA is usually detected as early as 4-5 days after the onset of illness.

References

Dick L, "Travel Medicine: Helping Patients Prepare for Trips Abroad," *Am Fam Phys*, 1998, 58(2):383-402.

Jong EC, "Travel Immunizations," *Med Clin North Am*, 1999, 83(4):903-22.

Robertson SE, Hull BP, Tomori O, et al, "Yellow Fever: A Decade of Reemergence," *JAMA*, 1996, 276(14):1157-62.

Sood SK, "Immunization for Children Traveling Abroad," *Pediatr Clin North Am*, 2000, 47(2):435-48.

Tomori O, "Impact of Yellow Fever on the Developing World," *Adv Virus Res*, 1999, 53:5-34.

Internet Web Sites

www.astdhpphe.org/infect/yellow.html

www.cdc.gov/ncidod/dvbid/yellowfever/index.htm

♦ **Yellow Jackets** see Barbiturates, Quantitative, Serum or Plasma on page 248

Yersinia enterocolitica Antibody

Related Information

Bacterial Culture, Blood *on page 232*
Bacterial Culture, Stool *on page 243*
HLA-B27 *on page 738*
Risks of Transfusion *on page 1166*
Viral Culture *on page 1307*

Test Includes Detection of antibody (IgG, IgM, and IgA) specific for *Yersinia enterocolitica*

Abstract Reservoirs of *Y. enterocolitica* are a variety of animals including pigs, goats, sheep, dogs, and cats. Transmission may include milk, pork, and water. Usually the organisms are ingested, but the infection can also be acquired by transfusion (red cells, platelets). Diarrhea due to *Y. enterocolitica* is infrequent in the United States but is more common in Europe. Often the differential diagnosis includes appendicitis. *Y. enterocolitica* and *Y. pseudotuberculosis* may cause similar clinical presentations.

Specimen Serum

Container Red top tube

Collection Acute and convalescent specimens are recommended.

Reference Interval Titer <1:160

Critical Values Antibodies may not be detectable for the first week of symptoms, but then rise rapidly to high titers (1:1280 is diagnostic).

Use Useful in diagnosis of *Yersinia enterocolitica* infection

Limitations Present serodiagnostic techniques are described as having only limited value due to high seroprevalence in certain healthy populations.

Methodology Enzyme immunoassay (EIA); agglutination with serotypes 03, 08, and 09

Additional Information *Yersinia enterocolitica* infection is characterized by mesenteric lymphadenitis and/or terminal ileitis with abdominal pain, gastroenteritis, and diarrhea. Bloody diarrhea may occur. It manifests often as enterocolitis in children younger than 5 years of age. It is also a cause of liver abscesses; association with hemochromatosis is recognized. Most infections are self-limited. Detection of *Yersinia*-specific antibody should be used in conjunction with culture to confirm a diagnosis of yersiniosis. After recovery of infection, low titers (1:40 or 1:80) may persist for years. **When stool is sent to the laboratory for culture, request for culture of this organism is usually needed so that an enrichment technique can be utilized.**

References

Henderson DA, Inglesby TV, and O'Toole T, *Bioterrorism: Guidelines for Medical and Public Health Management*, Chicago, IL: AMA Press, 2002.

Katz JA, "At the Focal Point. *Yersinia enterocolitica*," *Gastrointest Endosc*, 1998, 48(1):61.

Koornhof HJ, Smego RA Jr, and Nicol M, "Yersiniosis. II: The Pathogenesis of *Yersinia* Infections," *Eur J Clin Microbiol Infect Dis*, 1999, 18(2):87-112.

Smego RA, Frean J, and Koornhof HJ, "Yersiniosis I: Microbiological and Clinicoepidemiological Aspects of Plague and Nonplague *Yersinia* Infections," *Eur J Clin Microbiol Infect Dis*, 1999, 18(1):1-15.

Tuohy AM, O'Gorman M, Byington C, et al, "*Yersinia enterocolitica* Mimicking Crohn's Disease in a Toddler," *Pediatrics*, 1999, 104(3):e36.

Internet Web Sites

vm.cfsan.fda.gov/~mow/chap5.html
www.cdc.gov/ncidod/dbmd/diseaseinfo/yersinia_g.htm

♦ **Yersinia pestis Culture** *see Yersinia pestis Diagnostic Procedures on page 1337*

Yersinia pestis Diagnostic Procedures

Related Information

Bacterial Culture, Blood *on page 232*
Bacterial Culture, Lower Respiratory *on page 241*
Fine Needle Aspiration Culture *on page 589*
Viral Culture *on page 1307*

Synonyms Plague Diagnostic Procedures

Applies to *Yersinia pestis* Culture

Abstract *Yersinia pestis* (*Pasteurella pestis* until 1970) is the etiologic agent of plague. Epidemic bubonic plague has been described historically. It was responsible for the deaths of 25% or more of Europe's population in the Middle Ages, the Black Death of the 14th century and subsequent epidemics. (Continued)

Yersinia pestis Diagnostic Procedures *(Continued)*

Currently, epidemics occur throughout the world with at least 2000 cases reported annually. The cycle can be stable (enzootic) or epidemic (epizootic) in rodents, squirrels, and prairie dogs. Fleas become infected and carry the organisms. In the U.S., plague occurs west of the 100th meridian (North Dakota to Texas).

The major forms of plague include lymphadenitis (bubonic plague), and secondary and primary pneumonic disease. A present concern is the possibility that *Y. pestis* may be used as a biological weapon. Used by bioterrorists, *Y. pestis* would probably be aerosolized, causing pneumonic plague. A terrorist attack might be recognized by a sudden outbreak of severe pneumonia, possibly with sepsis.

Specimen Serum for serology; sputum, blood, or aspirated material from lymph nodes for culture and immunofluorescence; sputum, serum, urine, and bubo aspirates for rapid diagnostic test

Container Red top tube for serology; specimens for microbiology can be transported in Cary-Blair agar.

Storage Instructions Acidified serum for serology may be stored in a refrigerator.

Turnaround Time A rapid diagnostic test has recently been described by Chanteau et al. It is a hand-held immunochromatographic dipstick assay. Cultures may grow slowly, and should be held for 7 days before discard. Hemagglutinating antibodies generally appear toward the end of the first week of disease.

Special Instructions In cases of suspected plague, the Centers for Disease Control and Prevention, Vector-borne Infectious Diseases, Fort Collins, CO (970) 221-6400 should be contacted at once. Sera must be inactivated and absorbed with sheep erythrocytes prior to testing. Sputum, aspirates, and tissue are hazardous.

Reference Interval Serology: titer ≤1:10

Critical Values Serology: A fourfold or greater increase in titer or a single titer >1:16 indicates exposure to *Y. pestis* and should be immediately reported to the physician and the CDC.

Use Although serological studies can confirm the diagnosis of plague, bacteriologic methods, ELISA, and the new rapid diagnostic test are preferable; *vide infra*.

Limitations F1 antigen may diffuse into the Cary-Blair transport medium, leading to false negatives. Isolation and ELISA are specific but their sensitivity is impaired by deterioration and/or contamination during transportation and antibiotic therapy prior to sampling.

Methodology F1 antigen is found in blood, sputum, and bubo specimens from infected subjects. It is specific to *Y. pestis*. Methods include immunocapture ELISA for F1 antigen, passive hemagglutination on acute and convalescent serum for fraction 1 (F1) antigen; direct antigen staining in urine, testing for IgM and IgG antibodies by PCR, direct immunofluorescence, and tissue immunostaining. Culture: *vide infra*. The new rapid diagnostic test uses monoclonal antibodies to the F1 antigen. Gram, Wright, Giemsa, or Wayson stains, directly or after mouse inoculation.

Additional Information A hemagglutination titer ≥1:16 is presumptive evidence of an immunologic response to *Yersinia pestis*. Seeing the stained organism in clinical material (aspiration of a bubo, sputum) can also make diagnosis, using conventional stains or fluorescent antibody. Appearing as gram-negative bacilli or as coccobacilli, Wright, Giemsa, or Wayson preparations may demonstrate bipolar staining. Blood, bubo (lymph node) aspiration, sputum, bronchial washings, and other materials can be stained and cultured. The organism grows at 25°C to 28°C on blood or MacConkey agar. It grows in brain-heart broth at 37°C. Identification of the F1 capsular antigen from organisms grown at 37°C is diagnostic.

References

Branda JA and Ruoff K, "Bioterrorism - Clinical Recognition and Primary Management," *Am J Clin Pathol*, 2001, 117(Suppl 1):S116-23.

Carroll K, Held M, Stombler RE, et al, "Laboratory Preparedness for Bioterrorism: From the Phlebotomist to the Pathologist," *Lab Med*, 2003, 34(3):169-82.

Chanteau S, Rahalison L, Ralafiarisoa L, et al, "Development and Testing of a Rapid Diagnostic Test for Bubonic and Pneumonic Plague," *Lancet*, 2003, 361:211-16.

Dennis DT and Chu MC, "A Major New Test for Plague," *Lancet*, 2003, 361:191.

Franz DR, Jahrling PB, Friedlander AM, et al, "Clinical Recognition and Management of Patients Exposed to Biological Warfare Agents," *JAMA*, 1997, 278(5):399-411.

Gage KL, Dennis DT, Orloski KA, et al, "Cases of Cat-Associated Human Plague in the Western US, 1977-1998," *Clin Infect Dis*, 2000, 30(6).

Meyer RF and Morse SA, "Bioterrorism Preparedness for the Public Health and Medical Communities," *Mayo Clin Proc*, 2002, 77(7):619-21.

Ratsitorahina M, Chanteau S, Rahalison L, et al, "Epidemiological and Diagnostic Aspects of the Outbreak of Pneumonic Plague in Madagascar," *Lancet*, 2000, 355(9198):111-3.

Rollins SE, Rollins SM, and Ryan ET, "*Yersinia pestis* and the Plague," *Am J Clin Pathol*, 2003, 119(Suppl 1):S78-S85.

Varkey P, Poland GA, Cockerill FR III, et al, "Confronting Bioterrorism: Physicians on the Front Line," *Mayo Clin Proc*, 2002, 77(7):661-72.

Internet Web Sites
www.astdhpphe.org/infect/plague.html
www.cdc.gov/ncidod/dvbid/plague/index.htm

♦ **Zarontin**® *see* Ethosuximide, Serum or Plasma *on page 560*

♦ **Zellweger Syndrome** *see* Inherited Diseases of Metabolism and Cell Structure *on page 792*

♦ **Zetran**® *see* Diazepam, Serum *on page 510*

♦ **Zeus Fixative** *see* Kidney Biopsy *on page 818*

Zidovudine, Serum or Plasma

Related Information
Beta$_2$-Microglobulin, Serum or Urine *on page 254*
CD4/CD8 Enumeration *on page 349*
HIV-1/HIV-2 Serology *on page 736*

Synonyms Azidothymidine; AZT; Combivir®; Retrovir®

Abstract Azidothymidine (AZT) or zidovudine was the first FDA-approved drug for the treatment of human immunodeficiency virus (HIV) infection, the cause of acquired immunodeficiency syndrome (AIDS). The drug is frequently used in combination with other antiretroviral agents. It is also used for the prevention of mother-to-child HIV-1 transmission. The drug crosses the placenta and is also present in breast milk. It is a competitive inhibitor of HIV reverse transcriptase.

Patient Preparation Coadministration of hydrochlorothiazide and sulfapyridine may cause misleading AZT results.

Specimen Serum or plasma

Container Red top tube preferred, green top (heparin) tube acceptable

Sampling Time Trough level, just before next dose

Storage Instructions Refrigerate; label as infectious material.

Causes for Rejection Incorrect specimen sampling time; use of SST™ tube

Reference Interval Peak serum level at 60-90 minutes after 200 mg dose: ~1.5 µg/mL. Nadir serum concentration (just before the next dose) is <0.02 µg/mL. Measurement of nadir concentration is of limited clinical use.

Critical Values Not clear; acute overdoses of up to 50 grams in children and adults have been reported without fatalities.

Possible Panic Range Peak concentrations >1.8 µg/mL may provide evidence of increased risk of toxicity.

Methodology Immunoassay, high performance liquid chromatography (HPLC)

Additional Information
- Half-life: 1 hour
- Volume of distribution: 1.4 L/kg
- Protein binding: 25%

Drug is activated in the lymphocytes by phosphorylation to AZT-triphosphate, which is not found in serum. The major toxicity associated with zidovudine use is hematologic suppression which may manifest as anemia, leukopenia, and/or granulocytopenia. Macrocytic anemia is very common. Monitoring appropriate hematologic parameters is the most reasonable approach toward evaluation of toxicity; serum levels currently contribute little to evaluating toxic effects of zidovudine. Other fairly common side effects include nausea, headache, malaise, asthenia, and insomnia. Hepatic steatosis and lactic acidosis are among risks of toxicity. Hepatic and renal failure increase serum levels.

Zidovudine, as monotherapy or in combination with other antiretroviral agents, remains a first-choice therapy for the prophylaxis of mother-to-child HIV transmission as shown by substantial reductions in transmission rates. In pediatric (Continued)

Zidovudine, Serum or Plasma (Continued)

populations, zidovudine in combination with another nucleoside analogue and a protease inhibitor is a first or second choice.

References

Acosta EP, Page LM, and Fletcher CV, "Clinical Pharmacokinetics of Zidovudine. An Update," *Clin Pharmacokinet*, 1996, 30(4):251-62.

Bhana N, Ormrod D, Perry CM, et al, "Zidovudine: A Review of Its Use in the Management of Vertically-Acquired Pediatric HIV Infection," *Paediatr Drugs*, 2002, 4(8):515-53.

Peckham C and Newell ML, "Preventing Vertical Transmission of HIV Infection," *N Engl J Med*, 2000, 343(14):1036-7.

Simpson DM, "Human Immunodeficiency Virus-Associated Dementia: Review of Pathogenesis, Prophylaxis, and Treatment Studies of Zidovudine Therapy," *Clin Infect Dis*, 1999, 29(1):19-34.

♦ **Ziehl-Neelsen Stain** see Acid-Fast Stain, Routine or Modified on page 95
♦ **Zinc Administration** see Copper, Serum on page 448
♦ **Zinc Protoporphyrin** see Protoporphyrin, Zinc, Blood on page 1121
♦ **Zinc Protoporphyrin:Heme Ratio** see Iron Stain, Bone Marrow on page 810

Zinc, Serum or Plasma

Related Information

Albumin, Serum on page 120
Copper, Serum on page 448
Zinc, Urine on page 1341

Synonyms Zn, Serum

Applies to Metallothionein; Penicillamine; Thymulin Assay

Abstract Zinc (Zn) is an essential trace element in human nutrition, found in over 100 enzymes. It has effects on growth, development, weight loss, immune function, and central nervous system function. Serum Zn concentrations are not a reliable indicator of Zn status, especially in the elderly. Zn and copper are competitive for intestinal absorption.

Specimen Serum or plasma

Container Use powder-free gloves. Collect blood in metal-free tubes. Avoid contact with rubber. Separate serum or plasma and store in metal-free plastic vial. See the Trace Elements Introduction on page 77.

Collection Avoid hemolysis or stasis: red cell Zn concentrations are about 10 times those of serum.

Causes for Rejection Failure to collect specimen with special collection kit or store serum in special metal-free plastic vials

Reference Interval 66-110 µg/dL (SI: 10.0-16.8 µmol/L). Serum Zn, when low in an apparently healthy (nonstressed, nonseptic) patient who has normal serum albumin, is evidence for low Zn stores, especially if urine Zn is also low, or there is known excessive unregulated Zn loss (diarrhea, nephrotic syndrome, etc).

Use Evaluate Zn deficiency or toxicity

Limitations One should not interpret a normal serum Zn as evidence for adequate Zn stores, as serum Zn is insensitive and may be normal even after symptoms of Zn deficiency have surfaced. Levels may be low in fever, sepsis, inflammation, exogenous corticosteroids, oral contraceptives, pregnancy, stress, or myocardial infarct, reflecting mobilization from serum to the liver by interleukin. Serum concentrations are usually low in uremia with normal tissue levels. Levels may be high in familial hyperzincemia without toxicity or high Zn stores. Albumin is a binding protein for Zn and concentrations are usually low in cases of hypoalbuminemia of all causes.

Drugs which may decrease serum Zn concentrations include carbamazepine, phenytoin, prednisone, and valproic acid. Those which may cause increased serum Zn concentrations include zurzoufin, chlorthalidone, oral contraceptives, and penicillamine. Dietary supplements of Zn may cause copper deficiency.

Methodology Atomic absorption spectrometry (AA). Makino described a simple and sensitive colorimetric method utilizing a cationic porphyrin.

Additional Information Many potential markers have been examined in the search for the ideal index of Zn status. The simplest of these is serum Zn concentration, which is reduced in moderate to severe Zn deficiency. It is not sensitive to mild deficiency states and is depressed in situations often parallel with reductions of serum albumin, its major binding protein.

Both primary and conditioned nutritional deficiency is fairly common worldwide, as well as in the United States, and has been described in a large variety of clinical situations: premature infants born with low hepatic stores; in breast-fed premature and full-term infants whose mother's milk is lower than normal in Zn; in growing children, especially boys, whose height velocity has increased while under Zn therapy; in prepubertal boys who display delayed sexual maturity, especially in association with diets low in animal protein and high in phytates from grains (which reduce gastrointestinal Zn absorption); in malabsorption and diarrheal states; in diabetes, nephrotic syndrome, cirrhosis (in each of these hyperzincuria occurs), in AIDS and ARC; in burn patients, and those receiving high doses of oral histidine or intravenous amino acids (as in TPN, especially cysteine or histidine) associated with hyperzincuria; and in geophagia (in which Zn absorption is reduced). Pharmacological doses of folate increase stool losses of Zn. The most severe cases of Zn deficiency have occurred in acrodermatitis enteropathica (see below), and in the early days of total parenteral nutrition, when Zn was not specifically included in the formulation.

Pregnant women are at higher risk of developing acquired Zn deficiency due to high uptake of Zn by the fetus. Also, excessive iron and folic acid, generally prescribed in pregnancy, interfere in Zn absorption.

Babies with the disease of Zn malabsorption known as acrodermatitis enteropathica usually first develop their characteristic facial and diaper rash when weaned. Untreated, symptoms progress and include growth retardation, diarrhea, impaired T-cell immunity, poor wound healing, infections, delayed testicular development in adolescence, and early death. Parenteral or enteral Zn corrects the condition. The classic disease is associated with low serum and urine Zn, but some cases have a normal serum concentration. These cases nevertheless respond to Zn supplementation.

Serum metallothioneine can help to distinguish reductions in serum Zn due to redistribution (sepsis, stress) from that due to nutritional inadequacy. Both serum Zn and metallothioneine concentrations are reduced in nutritional inadequacy (reduced pool size) while low serum Zn concentration with high serum metallothioneine concentration reflects a redistribution of Zn stores.

Oral Zn supplementation interferes with copper absorption and chronic oral Zn supplementation may precipitate copper deficiency. Copper status should be monitored for patients on long-term Zn therapy. The anemia of zinc toxicity may be microcytic or sideroblastic and is seen with neutropenia.

Zinc deficiency in adolescents and adults is marked by slow growth or weight loss, altered taste, delayed puberty, dwarfism, impaired dark adaptation, central scotomata, alopecia, emotional instability, tremors, cerebellar ataxia, and a bullous-pustular rash over sacral areas. Candidiasis reflects impaired T-cell function. Thymulin apo-hormone is present but inactive due to lack of Zn. Lymphopenia may occur in severe deficiency and death may follow an overwhelming infection.

References

Igic PG, Lee E, Harper W, et al, "Toxic Effects Associated With Consumption of Zinc," *Mayo Clin Proc*, 2002, 77:713-6.

Makino T, "A Simple and Sensitive Colorimetric Assay of Zinc in Serum Using Cationic Porphyrin," *Clin Chim Acta*, 1999, 282(1-2):65-76.

McCall KA, Huang C, and Fierke CA, "Function and Mechanism of Zinc Metalloenzymes," *J Nutr*, 2000, 130(5S Suppl):1437S-46S.

Prasad AS, Meftah S, Abdallah J, et al, "Serum Thymulin in Human Zinc Deficiency," *J Clin Invest*, 1988, 82(4):1202-10.

Zima T, Mestek O, Nemecek K, et al, "Trace Elements in Hemodialysis and Continuous Ambulatory Peritoneal Dialysis Patients," *Blood Purification*, 1998, 16(5):253-60.

Zinc, Urine

Related Information

Zinc, Serum or Plasma *on page 1340*

Synonyms Zn, Urine

Abstract Although zinc (Zn) is mainly eliminated from the body by fecal excretion, minor quantities are excreted in urine. High urine, but low serum Zn, may be associated with hepatic cirrhosis, neoplastic disease, or increased catabolism. Low urine and serum Zn concentrations may be caused by Zn deficiency.

Specimen 24-hour urine

Container Acid-washed plastic urine container

(Continued)

Zinc, Urine *(Continued)*

Collection Use 10 mL concentrated hydrochloric acid as a preservative for 24-hour collection.

Storage Instructions Keep on ice or refrigerated. Laboratory will measure and record volume and remove aliquot for analysis.

Causes for Rejection Failure to collect specimen in metal-free (acid-washed) container, specimen allowed to contact rubber

Special Instructions Avoid contact with rubber during collection such as through rubber catheter. If a urinary catheter is absolutely essential, as in burn patients, consider the use of a silicone catheter, which has been shown to release less Zn than other types. Most catheters contribute Zn to the collection, some to a substantial degree.

Reference Interval Normal subjects: 0.14-0.80 mg/24 hours (SI: 2.1-12.2 µmol/24 hours). Compliant patients on oral Zn therapy for Wilson disease: >2.00 mg/24 hours (SI: 30.6 µmol/24 hours).

Use Evaluate Zn toxicity; evaluate low serum Zn levels; evaluate compliance in oral Zn therapy of Wilson disease

Limitations Zinc deficiency is usually accompanied by decreased urine Zn excretion. Zn deficiency, however, may be in part due to excess urine losses, especially in cirrhosis, viral hepatitis, hemolytic anemias, sickle cell disease, alcoholism, diabetes, chronic renal diseases, or parenteral nutrition. For effect of certain drugs on Zn, see Zinc, Serum or Plasma *on page 1340.*

Methodology Atomic absorption spectrometry (AA), inductively-coupled plasma-mass spectrometry (ICP-MS)

Additional Information See Zinc, Serum or Plasma *on page 1340.*

References

Henderson LM, Brewer GJ, Dressman JB, et al, "Use of Zinc Tolerance Test and 24-Hour Urinary Zinc Content to Assess Oral Zinc Absorption," *J Am Coll Nutr*, 1996, 15(1):79-83.

Jackson GE, Blewet R, Rodgers AL, et al, "Trace Metal Excretion in Patients With Homozygous Hypercholesterolemia," *J Trace Elem Med Biol*, 1999, 13(1-2):62-7.

Sallsten G and Barregard L, "Urinary Excretion of Mercury, Copper, and Zinc in Subjects Exposed to Mercury Vapour," *Biometals*, 1997, 10(4):357-61.

ACRONYMS AND ABBREVIATIONS
GLOSSARY

The following truncated glossary is intended for those interested in the clinical medical laboratory. Although the editors recognize that acronyms and abbreviations must exist, we feel that there are altogether too many of them. Some may lead to confusion, awkward misunderstanding, and even error. We think it best to communicate clearly, in English, with use of as few acronyms and abbreviations as possible.

The order of monographs in this volume may help the reader reach the information he/she may seek. The Disease Index may also serve to resolve questions.

a	atto (10^{-18})
A	a blood group antigen
AA	atomic absorption
A_1AT	alpha$_1$ antitrypsin
AABB	American Association of Blood Banks
AACC	American Association of Clinical Chemistry
[A-a]Do$_2$	alveolar-arterial oxygen gradient
AaG	alveolar arterial gradient
(A-a)O$_2$	alveolar-arterial oxygen gradient
AAP	American Academy of Pediatrics
AAPCC	American Association of Poison Control Centers
A-aP$_{co2}$	alveolar-arterial carbon dioxide difference
AAS	atomic absorption spectrometry
AAT	alpha antitrypsin
Ab	antibody
AB	a blood group antigen
ABG	arterial blood gas
ABO	ABO blood group
ACA	anticentromere antibodies
ACD	acid-citrate-dextrose; anemia of chronic disease
ACE	angiotensin converting enzyme
AChE	acetylcholinesteraseE
AChR	acetylcholine receptor antibody
aCl	anticardiolipin (antibody)
ACI	amylase creatinine clearance
ACOG	American College of Obstetrics and Gynecology
ACP	American College of Physicians
ACPA	anticytoplasmic antibodies
ACT	activated clotting time
ACTH	adrenocorticotropic hormone (corticotropin)
ADA	American Diabetes Association
ADH	antidiuretic hormone
AF	amniotic fluid
AF-AFP	amniotic fluid AFP
AFB	acid-fast bacillus
aFP, AFP	alpha-fetoprotein
Ag	antigen
AG	anion gap
A:G	albumin:globulin ratio
AGC	atypical glandular cells
AGUS	atypical glandular cells of indetermined significance
AHA	acquired hemolytic anemia; acute hemolytic anemia; American Hospital Association; autoimmune hemolytic anemia
AHF	antihemophilic factor (factor VIII)
AIDS	acquired immunodeficiency syndrome
AIDS-KS	acquired immunodeficiency syndrome with Kaposi's sarcoma
AIHA	autoimmune hemolytic anemia
AILD	angioimmunoblastic lymphadenopathy with dysproteinemia
AIP	acute intermittent porphyria
AIS	endocervical adenocarcinoma *in situ*
AITP	autoimmune thrombocytopenic purpura
Al	aluminum
ALA	aminolevulinic acid
ALCL	anaplastic large cell lymphoma
Alk phos	alkaline phosphatase
alk p'tase	alkaline phosphatase
ALL	acute lymphocytic leukemia
ALP	alkaline phosphatase
ALT	alanine aminotransferase
ALT:AST	ratio of serum alanine aminotransferase to serum aspartate aminotransferase
AMA	American Medical Association; antimitochondrial antibody
Amf	amniotic fluid
AMI	acute myocardial infarction
AML	acute myelocytic leukemia
AMML	acute myelomonocytic leukemia
ANA	antinuclear antibody

ANCA	antineutrophil cytoplasmic antibodies
anti-LKM1	liver/kidney microsomal type 1 antibodies
AOCD	anemia of chronic disease
AP	acid phosphatase; alkaline phosphatase; antepartum
APC	activated protein C
APCA	antiparietal cell antibody
APGAR	appearance, pulse, grimace, activity, and respiration (score of newborn physical status)
APhA	American Pharmaceutical Association
APR	acute phase reactant
APS	antiphospholipid syndrome
aPTT	(activated) partial thromboplastin time
ARC	acquired immunodeficiency syndrome-related complex; American Red Cross
ARD	antimicrobial removal device; acute respiratory distress
ARDS	adult/acute respiratory distress syndrome
ARF	acute renal failure
ASA	acetylsalicylic acid
ASAP	as soon as possible
ASC	atypical squamous cells
ASCF	ascitic fluid
ASC-H	atypical squamous cells cannot exclude HSIL
ASCP	American Society for Clinical Pathology
ASC-US	atypical squamous cells of undetermined significance
ASHP	American Society of Health-System Pharmacists
ASM	American Society of Microbiology
ASMA	antismooth muscle antibody
ASO	antistreptolysin O
AST	aspartate aminotransferase
AT	antithrombin
ATP	adenosine triphosphate
ATPase	adenosine triphosphatase
A-V	arteriovenous; atrioventricular
A-VO$_2$	ateriovenous oxygen difference
B	a blood group antigen
BAL	bronchoalveolar lavage
BAO	basal (gastric) acid output
BCG	bacillus Calmette-Guérin; bromcresol green
BCP	bromcresol purple
bcr	breakpoint cluster region
βhCG	beta human chorionic gonadotropin
BHI	brain heart infusion
BIA	bacterial inhibition assay
BIH	bacterial inhibition assay
BMI	body mass index
BNP	beta natriuretic peptide
BOOP	bronchiolitis obliterans organizing pneumonia
BP	blood pressure
BPH	benign prostatic hyperplasia; benign prostatic hypertrophy
Br	breath; bromine; bromide
BSA	body surface area
BUN	blood urea nitrogen
c	centi (10^{-2})
C3	complement component
C4	complement component
C5	complement component
C6	complement component
C7	complement component
CA 15-3	a tumor marker antigen
CA 19-9	a tumor marker antigen
CA 50	a tumor marker antigen
CA 125	a tumor marker antigen
CABG	coronary artery bypass graft
CAD	coronary artery disease
CAH	congenital adrenal hyperplasia
CALLA	common acute lymphoblastic leukemia antigen
cAMP	cyclic adenosine monophosphate
C-ANCA	cytoplasmic antineutrophil cytoplasmic antibody
CAP	College of American Pathologists
CBC	complete blood count
CBIL	conjugated bilirubin
C$_{Cr}$	creatinine clearance
CCU	cardiac care unit; coronary care unit
CDC	Centers for Disease Control and Prevention
CEA	carcinoembryonic antigen
CEP	congenital erythropoietic porphyria

CES	conjugated estrogenic substances
CF	complement fixation; cystic fibrosis
cfu	colony forming units
cfu-GM	colony forming units - granulocytic/monocytic
CHD	coronary heart disease
CHF	congestive heart failure
CHOP	cyclophosphamide, hydroxydaunorubicin, Oncovin® (vincristine), and prednisone
CHOR	cyclophosphamide, hydroxydaunorubicin, Oncovin® (vincristine), and radiation
CIN	cervical intraepithelial neoplasia
CISH	chromogenic *in situ* hybridization
Cit	citrate
CJD	Creutzfeldt-Jakob disease
CK	creatine kinase; creatinine phosphokinase
Cl	chlorine
CLA	certified laboratory assistant
CLL	chronic lymphocytic leukemia
CLS	clinical laboratory scientist
CLSL	chronic lymphosarcoma (cell) leukemia
cm	centimeter
cm^2	square centimeter
cm^3	cubic centimeter
CMA	Canadian Medical Association
C_{max}	maximum concentration of drug
C_{min}	minimum concentration of drug
CML	chronic myelocytic leukemia
CMML	chronic myelomonocytic leukemia
CMS	Centers for Medicare and Medicaid Services
CMV	cytomegalovirus
CN$^-$	cyanate ion
CNS	central nervous system
CO	carbon monoxide
CO_2	carbon dioxide
CO_3	carbonate
COHb	carboxyhemoglobin
COPD	chronic obstructive pulmonary disease
CPDA-1	citrate phosphate dextrose adenine solution
CPE	cytopathic effect
CPK (CK preferred)	creatinine phosphokinase
CPmax	peak (maximum) serum concentration
CPmin	trough (minimum) serum concentration
CPPB	continuous positive pressure breathing
CPR	cardiopulmonary resuscitation
CRH	corticotropin-releasing hormone
CRP	C-reactive protein
CRYO	cryoprecipitate
C/S	culture and sensitivity
CSF	cerebrospinal fluid
cTI	cardiac troponin T
cTnI	cardiac troponin I
CTnT	cardiac troponin T
cTT	cardiac troponin T
Cu	copper
CV	coefficient of variation
CVA	cerebrovascular accident
CVO_2	O_2 content in mixed venous blood
CVS	clean voided specimen
Cx	cervical, cervix
CZE	capillary zone electrophoresis
d	deci (10^{-1})
DAT	direct antiglobulin test
DCIS	ductal carcinoma *in situ*
DEA	Drug Enforcement Administration
DFA	direct fluorescent antibody
DHEA	dehydroepiandrosterone
DHEA-S	dehydroepiandrosterone sulfate
DIC	disseminated intravascular coagulation
DIP	desquamative interstitial pneumonia/pneumonitis
dL	deciliter
dm	decimeter
DM	diabetes mellitus
DMD	Duchenne muscular dystrophy
DMSO	dimethyl sulfoxide
DNA	deoxyribonucleic acid
DNase	deoxyribonuclease

DNP	deoxyribonucleoprotein
DRE	digital rectal examination
DRG	diagnostic-related group(s)
ds-DNA	double-stranded DNA
EBEA	Epstein-Barr early antigen
EBNA	Epstein-Barr nuclear antigen
EBV	Epstein-Barr virus
EBVCA	Epstein-Barr virus, capsid antigen
EBVEA	Epstein-Barr virus, early antigen
EBVNA	Epstein-Barr virus, nuclear antigen
ECG	electrocardiogram
ECV	extracellular volume
ECW	extracellular water
ED	emergency department
EDTA	ethylenediaminetetraacetic acid
EEE	Eastern equine encephalitis
EEG	electroencephalogram
EIA	enzyme immunoassay
EKG	electrocardiogram
ELISA	enzyme-linked immunosorbent assay
EM	electron microscopy
EMA	endomysial antibody; epithelial membrane antigen
EMG	electromyelogram; electromyogram
EMIT	enzyme-multiplied immunoassay technique
ENA	extractable nuclear antigen
EPA	Environmental Protection Agency
Eq	equivalent
ER	estrogen receptor
ER-	estrogen receptor-negative
ER+	estrogen receptor-positive
ERA	estrogen receptor assay
ERCP	endoscopic retrograde cholangiopancreatography
ESBL	extended-spectrum beta lactamases
ESR	erythrocyte sedimentation rate
EtOH	ethyl alcohol
EU	Ehrlich unit
f	femto (10^{-15})
F⁻	fluoride
F⁻/Ox.	fluoride and oxalate
F VIII	factor 8 (antihemophilic factor)
FA	fluorescent antibody
FAB	French-American-British
FACP	Fellow of the American College of Physicians
FACS	Fellow of the American College of Surgeons
FBG	fasting blood glucose
FBS	fasting blood sugar
Fc	portion of antibody molecule bound by membrane receptors
FCM	flow cytometry
FDA	Food and Drug Administration
FDP	fibrin degradation product; fructose diphosphate
FeCl₃	ferric chloride
FENa	fractional excretion of sodium (Na)
FEP	free erythrocyte protoporphyrin
FFP	fresh frozen plasma
fg	femtogram
FIA	fluoroimmunoassay
FiO₂	fraction of inspired O_2
FISH	fluorescence in situ hybridization
fL	femtoliter; fluid
FMH	fetomaternal hemorrhage
fmol	femtomole
FNA	fine needle aspiration
FNAB	fine needle aspiration biopsy
FOBT	fecal occult blood testing
FPG	fasting plasma glucose
FPIA	fluorescence polarization immunoassay
FSH	follicle stimulating hormone
FSI	foam stability index
FTA	fluorescent treponemal antibody
FTA-ABS	fluorescent treponemal antibody absorption
FTI	free thyroxine index
5-FU	5-fluorouracil
FUO	fever of unknown/undetermined origin
g	gram
G6PD	glucose 6-phosphate dehydrogenase

GABA	gamma aminobutyric acid
GAD	glutamic acid decarboxylase
Gast Cont	gastric contents
GastF	gastric fluid
Gast Res	gastric residue
GBM	glomerular basement membrane
GC	gonococcus; gonorrhea culture; gas chromatography
GC/MS	gas chromatography/mass spectrometry
G-CSF	granulocyte colony-stimulating factors
GCSF	granulocyte cell-stimulating factor
g/dL	grams per deciliter
GERD	gastroesophageal reflux disease
GFR	glomerular filtration rate
GGT	gamma-glutamyltransferase
GGTP	gamma-glutamyltransferase
GH	growth hormone
GHb	glycated hemoglobin
GHB	gamma hydroxybutyrate; glycohemoglobin
GHRH	growth hormone-releasing hormone
GHRIF	growth hormone release-inhibiting factor
GHRIH	growth hormone release-inhibiting hormone
GHRP-6	growth hormone-releasing peptide-6
GIFT	gamete interfallopian transfer
GIP	gastric inhibitory polypeptide
GLC	gas-liquid chromatography
GM-cfu	granulocyte-macrophage colony-forming unit
GMS	Gomori methenamine silver
GnRH	gonadotropin releasing hormone
GOT (SGOT)	glutamic-oxaloacetic transaminase
GPT (SGPT)	glutamic-pyruvic transaminase
GTT	glucose tolerance test
GVHD	graft-versus-host disease
h	hour (hora)
HAAA	human antianimal antibodies
HAART	highly active antiretroviral therapy
HAI	hemagglutination inhibition
HAMA	human antimouse antibodies
HAV	hepatitis A virus
HAV Ab	hepatitis A virus antibody
HB$_c$	hepatitis B core antibody
HB$_e$Ag	hepatitis B e antigen
HB$_s$Ab	hepatitis B surface antibody
HB$_s$Ag	hepatitis B surface antigen
HBV	hepatitis B virus
HCFA	Health Care Financing Administration
hCG	human chorionic gonadotropin
HCO$_3$	bicarbonate
HCO$_3^-$	bicarbonate concentration
HCPCS	Heathcare Common Procedure Coding System
Hct, HCT	hematocrit
HCV	hepatitis C virus
HCV DNA	hepatitis C virus RNA detection
HD	Hodgkin's disease
HDAg	hepatitis D antigen
HDL	high density lipoprotein
HDLC	high density lipoprotein cholesterol
HDN	hemolytic disease of the newborn
HDPAA	heparin-dependent platelet-associated antibody
HDV	hepatitis D virus
H&E	hematoxylin-eosin
HELLP	microangiopathic hemolytic anemia, elevated liver tests, low platelet count (see Disease Index)
Hep	heparin
Hgb	hemoglobin
HGE	human granulocytic ehrlichiosis
hGH	human growth hormone
HGSIL	high grade squamous epithelial lesion
HGV	hepatitis G virus
HHM	humoral hypercalcemia of malignancy
HHS	(Department of) Health and Human Services
HHV-6	human herpesvirus 6
HHV-7	human herpesvirus 7
HHV-8	human herpesvirus 8
HIAA	hydroxyindoleacetic acid
HIT	heparin-induced thrombocytopenia

HIV	human immunodeficiency virus
HK	hexokinase
HLA	histocompatibility leukocyte antigen; histocompatibility locus antigen; human leukocyte antigen
hLH	human luteinizing hormone
HME	human monocytic erhlichiosis
HMG-CoA	hepatic hydroxymethylglutaryl coenzyme A
HMM	hexamethylmelamine
HMO	Health Maintenance Organization
HMW	high molecular weight
HNPCC	hereditary nonpolyposis colon cancer
HPA	hypothalamic-pituitary axis
HPC	hematopoietic progenitor cells
hpf	high power field
HPFH	hereditary persistence of fetal hemoglobin
hPL	human placental lactogen
HPLC	high-performance liquid chromatography
hPP	human pancreatic polypeptide
HPT	hyperparathyroidism
hPTH	human parathyroid hormone
HPTLC	high performance thin layer chromatography
HPV	human papillomavirus
HSIL	high-grade squamous intraepithelial lesions
HSV	herpes simplex virus
HTLV	human T-lymphotropic virus
HTN	hypertension
HVA	homovanillic acid
IAT	indirect antiglobulin test
ICG	indocyanine green
ICMA	immunochemiluminometric assay
ICP-MS	inductively-coupled plasma-mass spectrometry
ICSI	intracytoplasmic sperm injection
ICT	indirect Coombs test
ICU	intensive care unit
IDDM	insulin-dependent diabetes mellitus
ID-GC/MS	isotope-dilution gas chromatography/mass spectrometry
IDL	intermediate-density lipoprotein
IEP	immunoelectrophoresis
IF	immunofluorescence; interstitial fluid; intrinsic factor
IFA	indirect fluorescent antibody
IFG	impaired glucose fasting
Ig	immunoglobulin
IgA	immunoglobulin A
IgD	immunoglobulin D
IgE	immunoglobulin E
IgG	immunoglobulin G
IgM	immunoglobulin M
IGT	impaired glucose tolerance
IHA	indirect hemagglutination
IHC	immunohistochemical
IIb-IIIa	glycoproteins found on platelet membranes
IIF	indirect immunofluorescence
IM	infectious mononucleosis
I.M.	intramuscular
IND	investigational new drug
INH	isonicotinic acid hydrazide; isoniazid
INR	international normalized ratio
IRB	institutional review board
IRMA	immunoradiometric assay
ISE	ion-selective electrode
ISI	International Sensitivity Index
ITP	idiopathic thrombocytopenic purpura; immunogenic thrombocytopenic purpura
IU	International unit
IUD	intrauterine death; intrauterine device
IUGR	intrauterine growth retardation
I.V.	intravenous
IVC	inferior vena cava; intravenous cholangiography
IVF	in vitro fertilization
IVIg	intravenous immune globulin; intravenous immunoglobulin
JAMA	The Journal of the American Medical Association
JCAHO	Joint Commission on Accreditation of Heathcare Organizations
k	kilo (10^3)
K	potassium
kcal	kilocalorie
kg	kilogram

ACRONYMS AND ABBREVIATIONS GLOSSARY

KGS . . . . ketogenic steroids
km . . . . kilometer
KOH . . . . potassium hydroxide
LA . . . . latex agglutination
LATS . . . . long-acting thyroid stimulator
LBM . . . . lean body mass
LBW . . . . lean body weight; low birth weight
LC-MS/MS . . . . liquid chromatography-electrospray tandem mass spectrometry
LCIS . . . . lobular carcinoma *in situ* (breast)
LD . . . . lactate dehydrogenase
LD_1 . . . . lactate dehydrogenase fraction 1
LDH . . . . lactate dehydrogenase
LDL . . . . low density lipoprotein
LDLC . . . . low density lipoprotein cholesterol
LDLs . . . . low density lipoproteins
Le . . . . Lewis antigen (blood group)
LE . . . . lupus erythematosus
LGSIL . . . . low grade squamous epithelial lesion
LH . . . . luteinizing hormone
LHRH . . . . luteinizing hormone releasing hormone
Li . . . . lithium
LISS . . . . low ionic strength saline
L-J . . . . Löwenstein-Jensen (microbiologic culture medium)
LKM . . . . liver-kidney microsomal antibodies
LMP . . . . last menstrual period
LMWH . . . . low molecular weight heparin
LP . . . . lumbar puncture
Lp(a) . . . . lipoprotein little a
LPA . . . . lysophosphatidic acid
lpf . . . . low power field
LRC . . . . Lipid Research Clinic
L:S . . . . lecithin:sphingomyelin ratio
LSD . . . . lysergic acid diethylamide
LSIL . . . . low-grade squamous intraepithelial lesions
LVEF . . . . left ventricular ejection fraction
m . . . . meter; milli (10^{-3})
m^2 . . . . square meter
m^3 . . . . cubic meter
M . . . . mega (10^6); mix (misce)
mA . . . . milliampere
MAI . . . . *Mycobacterium avium-intracellulare*
MALT . . . . mucosa-associated lymphoid tissue
MBP . . . . major basic protein
mc . . . . millicurie
mcg . . . . microgram
MCH . . . . mean corpuscular hemoglobin
MCHC . . . . mean corpuscular hemoglobin concentration
mCi . . . . millicurie
MCT . . . . medullary carcinoma of thyroid
MCTD . . . . mixed connective tissue disease
MCV . . . . mean corpuscular volume
MEIA . . . . microparticle enzyme immunoassay
MEN . . . . multiple endocrine neoplasia
mEq . . . . milliequivalent
mg . . . . milligram
mg% . . . . . milligrams per deciliter; milligrams per 100 milliliters; milligrams percent
Mg . . . . magnesium
MGUS . . . . monoclonal gammopathy of undetermined significance
MHb . . . . methemoglobin
MHC . . . . major histocompatibility complex
MHz . . . . megahertz
MIC . . . . minimum inhibitory concentration
μ . . . . micro (10^{-6})
μg . . . . microgram
μL . . . . microliter
μm . . . . micrometer
$μm^3$ . . . . cubic micrometer
μmol . . . . micromole
μmol/L . . . . micromolar
μOsm . . . . micro-osmolar
μU . . . . microunit
MIF . . . . microimmunofluorescence
mIU . . . . milli International unit
mL . . . . milliliter
mL/L . . . . milliliters per liter

MLC	mixed leukocyte culture; mixed lymphocyte culture
MLT	medical laboratory technician
mm	millimeter
mm^2	square millimeter
mm^3	cubic millimeter
mM	millimolar; millimole
mmol	millimole
mmol/L	millimolar
MMR	maternal mortality rate; measles, mumps, rubella
Mn	manganese
mNAP	membrane alkaline phosphatase
MNS	MNS blood group antigens
Mo	molybdenum
mol	mole
mol/L	molar
mol/m^3	mole per cubic meter
mol/s	mole per second
mOsm	milliosmole
mOsm/kg	milliosmoles per kilogram
MPH	Master of Public Health
MPO	myeloperoxidase
MPV	mean platelet volume
MRI	magnetic resonance imaging
MRSA	methicillin-resistant *S. aureus*
MS	mass spectroscopy
MSAFP	maternal serum AFP
MS/MS	tandem mass spectroscopy
MT	medical technologist
MT (ASCP)	medical technologist, certified by American Society of Clinical Pathologists
MTC	medullary thyroid carcinoma
mU	milliunit
n	nano (10^{-9})
Na	sodium
NAD	nicotinamide adenine dinucleotide
NADH	reduced form of NAD
NADP	nicotinamide adenine dinucleotide phosphate
NADPH	reduced form of NADP
NAIT	neonatal alloimmune thrombocytopenia
NAPA	*N*-acetyl procainamide
NASH	nonalcoholic steatohepatitis
NATP	neonatal autoimmune thrombocytopenic purpura
NBT	nitro blue tetrazolium
NCCLS	National Committee for Clinical Laboratory Standards
NCEP	National Cholesterol Education Program
NCI	National Cancer Institute
ng	nanogram
NHANES	National Health and Nutrition Examination Survey
NICU	neonatal intensive care unit
NIDA	National Institute on Drug Abuse
NIDDM	noninsulin-dependent diabetes mellitus
NIH	National Institutes of Health
NIOSH	National Institute for Occupational Safety and Health
nL	nanoliter
NLM	National Library of Medicine
nm	nanometer
nmol	nanomole
NMR	nuclear magnetic resonance
NMRI	nuclear magnetic resonance imaging
NRBC	nucleated red blood cell
NSAID	nonsteroidal anti-inflammatory drug
NSE	neuron-specific enolase
O	a blood group antigen
O$_2$	oxygen
ΔOD450	change of optical density at 450 nm
OGTT	oral glucose tolerance test
25(OH)D$_3$	25-hydroxy vitamin D
17-OHP	17-hydroxyprogesterone
Osm/kg	osmole per kilogram (osmolality)
Osm/L, Osm/l	osmole per liter (osmolality)
OSM S	osmolarity serum
OSM U	osmolarity urine
O/T	oral temperature
Ox	oxalate
P	phosphorus

P̄	L. *post* after; mean pressure
p24	antigen in HIV infection
p50	half saturation (oxygen)
Pa	arterial pressure; pulmonary arterial (pressure); pulmonary artery (line); Pascal (unit of pressure)
PA	pernicious anemia
$PaCO_2$	systemic arterial CO_2 tension
P-ANCA	perinuclear antineutrophil cytoplasmic antibodies
PaO_2	partial pressure of arterial oxygen; arterial O_2 tension
PAO	peak acid output; peripheral airway obstruction
PAO_2	partial pressure of oxygen in alveoli
PAO_2-PaO_2	alveolar-arterial difference in partial pressure of oxygen
Pap	Papanicolaou stain or smear
PAPP	pregnancy-associated plasma protein
PAS	periodic acid Schiff
PAWP	pulmonary arterial wedge pressure
PBC	primary biliary cirrhosis
PBG	porphobilinogen
Pco	carbon monoxide pressure or tension
pCO_2	carbon dioxide partial pressure (tension)
PChE	pseudocholinesterase
PCR	polymerase chain reaction
PCR-SBT	polymerase chain reaction sequence-based typing
PCR-SSCP	polymerase chain reaction single-strand conformation polymorphism
PCR-SSOP	polymerase chain reaction sequence-specific oligonucleotide blot hybridization
PCR-SSP	polymerase chain reaction sequence-specific primers
PE	pulmonary embolism
PEG	polyethylene glycol
PericF	pericardial fluid
PF_3	platelet factor 3
pg	picogram
PgR	progesterone receptor
pH	measurement of acidity or alkalinity
pHa	arterial blood pH
PhD	Doctor of Philosophy
PID	pelvic inflammatory disease
PIH	pregnancy-induced hypertension
PIO_2	inspired oxygen tension; partial pressure of inspiratory oxygen
PK	pyruvate kinase
PKU	phenylketonuria
Placent	placental
Plf	pleural fluid
PMNs	polymorphonuclear white blood cells
pmol	picomole
P_{N2}	partial pressure of nitrogen
PNH	paroxysmal nocturnal hemoglobinuria
p.o.	by mouth (per os)
POC	point-of-care; products of conception
POCT	point-of-care testing
Poststim	poststimulation
pp	postprandial
PPF	plasma protein fraction
ppm	parts per million
ppt	precipitate
PRBC	packed red cells
PRSP	penicillinase-resistant synthetic penicillins
PSA	prostate specific antigen
PT	prothrombin time
PTH	parathyroid hormone
PTP	post-transfusion purpura
PTT	partial thromboplastin time
PV	polycythemia vera
RA	rheumatoid arthritis
RBC	red blood cell
RCT	randomized clinical trial
RDS	respiratory distress syndrome
RDW	red cell distribution width
RF	rheumatoid factor
Rh	rhesus antigen, blood group
RhIG	$Rh_0(D)$ immune globulin
$Rh_0(D)$	major red cell antigen in rhesus system
rHuEPO	recombinant human erythropoietin
RIA	radioimmunoassay
RIBA	recombinant immunoblot assay

RIPA	radioimmunoprecipitation
RMSF	Rocky Mountain spotted fever
RN	registered nurse
RNA	ribonucleic acid
RNP	ribonucleoprotein
RPR	rapid plasma reagin
RSV	respiratory syncytial virus
RT	reverse transcriptase
rT_3	reverse T_3
RTA	renal tubular acidosis
RTH	resistance to thyroid hormone
RT-PCR	reverse transcriptase polymerase chain reaction
RVVT	Russell viper venom test
s	serum; sulfur
sal	saliva
SAMHSA	Substance Abuse and Mental Health Services Administration
SARS	severe acute respiratory syndrome
SBB	Specialist in Blood Bank technology
S-C	sickle cell-(hemoglobin) C (disease)
SCID	severe combined immunodeficiency
SCN^-	thiocyanate ion
S-D	sickle cell-(hemoglobin) D (disease)
Se	selenium
SemF	seminal fluid
SerF	serous fluid
SG	specific gravity
SIADH	syndrome of inappropriate antidiuretic hormone secretion
SIDS	sudden infant death syndrome
SIL	squamous intraepithelial lesions
SLE	systemic lupus erythematosus
Sm	Smith antigen
SMA	smooth muscle antibody
SPF	S-phase fraction
SS	Salmonella-Shigella; saturated solution; subaortic stenosis
SS-A	Sjögren syndrome A antibody
SS-B	Sjögren syndrome B antibody
ss-DNA	single-stranded DNA
stat	at once (statim); immediately
statins	HMG-CoA reductase inhibitors
STD	sexually transmitted disease; skin test dose
STS	serologic test for syphilis
swt	sweat
SynF	synovial fluid
T_3	triiodothyronine
T_3RU	triiodothyronine resin uptake
T_3UP	triiodothyronine uptake
T_3UR	triiodothyronine uptake ratio
T_4	levothyroxine; thyroxine
TBG	thyroxine binding globulin
TBIL	total bilirubin
TdT	terminal deoxynucleotidyl transferase
Tg	thyroglobulin
TG	triglycerides
TIA	transient ischemic attack
TIBC	total iron binding capacity
TLC	thin-layer chromatography
t_{max}	time of maximal concentration
TPCV	total packed cell volume
TPN	total parenteral nutrition
TPO	thyroperoxidase
TRALI	transfusion-related acute lung injury
TRAP	tartrate resistance leukocyte acid phosphatase
TRCV	total red cell volume
TRH	thyrotropin releasing hormone
TSH	thyroid stimulating hormone (thyrotropin)
TSI	thyroid stimulating immunoglobulin; total serum iron
TSS	toxic shock syndrome
TT	thrombin time
TTP	thrombotic thrombocytopenic purpura
TUNEL	terminal dUTP nick end labeling
TURP	transurethral prostatectomy; transurethral resection of prostate
Umax	maximal urinary osmolality
UNa	urinary sodium
UOP	urinary output
UOsm	urinary osmolality

URI	upper respiratory illness; upper respiratory infection
UTI	urinary tract infection
VDG, VD-G	venereal disease; gonorrhea
VDRL	test for syphilis; Venereal Disease Research Laboratory
VIP	vasoactive intestinal polypeptide
VLDL	very low density lipoprotein
VLDLC	very low density lipoprotein cholesterol
VLDL-TG	very low density lipoprotein triglyceride
VMA	vanillylmandelic acid
VTE	venous thromboembolic disease
vWf, vWF	von Willebrand factor
VZIG	varicella-zoster immune globulin
VZV	varicella-zoster virus
WBC	white blood cell
WEE	Western equine encephalitis
WHO	World Health Organization
WNV	West Nile virus
Zn	zinc
ZPP	zinc protoporphyrin
ZSR	zeta sedimentation rate

DISEASE INDEX

ABETALIPOPROTEINEMIA [272.5]

Patients have mutations in one or both apo B alleles, with LDLC <50 mg/dL. Apo B proteins are truncated. With RBC acanthosis (burr cells) and reduced serum cholesterol and triglyceride concentrations, intestinal biopsy is diagnostic.

ABORTION [637.9] see also HEMOLYTIC DISEASE OF THE NEWBORN, FETAL DEATH

ABSCESS [682.9] see also ABSCESS (AMEBIC), ABSCESS (BRAIN), ABSCESS (EPIDURAL), ABSCESS (LIVER), ABSCESS (LUNG)

ABSCESS (AMEBIC) [006.3]

ABSCESS (BRAIN) [324.0] see also ACTINOMYCOSIS, INFECTIVE ENDOCARDITIS, MENINGITIS, NOCARDIOSIS, SEPTICEMIA, SUBDURAL EMPYEMA

May present as space-occupying mass such as brain tumor, but brain abscess progresses more rapidly.

ABSCESS (EPIDURAL) [324.9] *see also* ACTINOMYCOSIS, INFECTIVE ENDOCARDITIS, MENINGITIS, OSTEOMYELITIS, TUBERCULOSIS

ABSCESS (LIVER) [572.0] *see also* ACTINOMYCOSIS, AMEBIASIS, BLASTOMYCOSIS, CANDIDIASIS, COCCIDIOIDOMYCOSIS, CROHN DISEASE, CRYPTOCOCCOSIS, FEVER UNDETERMINED ORIGIN (FUO), HISTOPLASMOSIS, INFECTIVE ENDOCARDITIS, SEPTICEMIA, STAPHYLOCOCCAL INFECTION, STREPTOCOCCAL INFECTION

Consideration for this diagnosis includes subjects with biliary tract disease with stasis, Crohn disease, intra-abdominal malignancy, diverticular disease, or postoperative complications, who present with fever, leukocytosis, and right upper quadrant pain.

ABSCESS (LUNG) [513.0] *see also* ALCOHOLISM, BRONCHIECTASIS, EMPYEMA, FUNGI, INFECTIVE ENDOCARDITIS, PNEUMONIA, SEIZURES, SEPTICEMIA, TUBERCULOSIS

An abscess within lung parenchyma, the expression usually is not intended to include cavitary lesions caused by mycobacterial, fungal, or parasitic diseases. Causes include tricuspid valve infective endocarditis in intravenous drug users, in whom septic pulmonary embolism may cause multiple lung abscesses. Lung abscess secondary to aspiration is usually solitary. Culture is best obtained by bronchoalveolar lavage or by protected specimen brush.

(Continued)

ABSCESS (LUNG) [513.0] *(Continued)*

ACANTHOCYTOSIS *see* ABETALIPOPROTEINEMIA

ACETAMINOPHEN TOXICITY [965.4] *see also* ALCOHOLISM, POISONING
In the alcohol-acetaminophen syndrome, coagulopathy and extremely abnormal transaminases are found. ALT >3500 units/L may be seen.

ACHLORHYDRIA *see* ATROPHIC GASTRITIS

ACIDOSIS/ALKALOSIS (ACID-BASE BALANCE) [276.2; 276.3; 775.7] *see also* ALCOHOL INTOXICATION, ALDOSTERONISM, CUSHING SYNDROME, DEHYDRATION/HYPOVOLEMIA, DIABETES MELLITUS, DIARRHEA, ETHYLENE GLYCOL POISONING, FANCONI SYNDROME, HEMOPTYSIS, METHANOL POISONING, PARALDEHYDE POISONING, RENAL FAILURE, RESPIRATORY FAILURE, SALICYLISM/SALICYLATE TOXICITY, VOMITING, WILSON DISEASE

(Continued)

ACQUIRED IMMUNODEFICIENCY SYNDROME (AIDS) [042] *(Continued)*

ACRODERMATITIS ENTEROPATHICA [686.8]

ACROMEGALY [253.0] *see also* PITUITARY/PITUITARY ADENOMA

ACTINOMYCOSIS [039.9] *see also* NOCARDIOSIS

ACTIVATED PROTEIN C RESISTANCE

ACUTE PHASE REACTION

ADDISON DISEASE [255.4] *see also* AMYLOIDOSIS, AUTOIMMUNE DISEASES, CONGENITAL ADRENAL HYPERPLASIA, HISTOPLASMOSIS, PITUITARY/PITUITARY ADENOMA, TUBERCULOSIS

ALCOHOL INTOXICATION [980.9] *see also* ALCOHOLISM, COMA, ISOPROPANOL POISONING, NEUROGLYCOPENIA, POISONING

ALCOHOLISM [303.9] *see also* ABSCESS (LUNG), CARCINOMA (LIVER), CARDIOMYOPATHY, CIRRHOSIS, FOLATE DEFICIENCY, HEPATITIS, KETOSIS/KETONURIA/KETOACIDOSIS, LIVER, MALNUTRITION, NEUROGLYCOPENIA, NONALCOHOLIC STEATOHEPATITIS, PANCREATITIS, STEATOSIS (LIVER)

AST, GGT, MCV, bilirubin, alkaline phosphatase may be increased in variable combinations; look for high AST:ALT ratio.

(Continued)

ALZHEIMER DISEASE [331.0] *(Continued)*

AMEBIASIS [006.9] *see also* ABSCESS (LIVER), ACQUIRED IMMUNODEFICIENCY SYNDROME (AIDS), COLITIS, COLITIS (ULCERATIVE), DIARRHEA, PARASITIC INFESTATIONS

AMEBIC ABSCESS *see* ABSCESS (AMEBIC)

AMENORRHEA [626.0] *see also* ADENOMA (ADRENAL), CONGENITAL ADRENAL HYPERPLASIA, HYPOGONADISM, HYPOPITUITARISM, OVARIAN FUNCTION TESTS, PITUITARY/PITUITARY ADENOMA, POLYCYSTIC OVARY SYNDROME, PROLACTINOMA, TURNER SYNDROME, UREMIA, VIRILIZATION, WILSON DISEASE

AMINOACIDURIAS [270.9] *see also* ALKAPTONURIA, ARGININOSUCCINIC ACIDURIA, INBORN ERRORS OF METABOLISM, WILSON DISEASE

AMINOGLYCOSIDE TOXICITY [960.8] *see also* RENAL FAILURE

AMYLOIDOSIS [277.3] *see also* MACROGLOBULINEMIA OF WALDENSTRÖM, MONOCLONAL GAMMOPATHY, MYELOMA, NEPHROSIS/NEPHROTIC SYNDROME, POEMS SYNDROME

Amyloidosis should be considered in patients older than 40 years of age with nephrotic syndrome, congestive heart failure not caused by ischemia, neuropathy, or unexplained hepatomegaly. Look for proteinuria without cellular elements in the urinalysis.

ANDERSON-FABRY DISEASE [272.7A]

ANEMIA [285.9] see also ANEMIA (DYSERYTHROPOIETIC), ANEMIA (ENZYME DEFICIENCY), ANEMIA (HEINZ BODY), ANEMIA (HEMOLYTIC), ANEMIA (HYPOPLASTIC), ANEMIA (IRON DEFICIENCY), ANEMIA (MACROCYTIC/MEGALOBLASTIC), ANEMIA (MICROANGIOPATHIC HEMOLYTIC), ANEMIA (MICROCYTIC), ANEMIA (SICKLE CELL AND VARIANT SICKLING HEMOGLOBINS), ANEMIA (SIDEROBLASTIC), ANEMIA (SPHEROCYTOSIS), ARTHRITIS (RHEUMATOID), GASTROINTESTINAL HEMORRHAGE (NEONATAL), GAUCHER DISEASE, HEMOGLOBINOPATHY, HEMOGLOBINURIA, HEMOGLOBINURIA (PAROXYSMAL COLD), HEMOGLOBINURIA (PAROXYSMAL NOCTURNAL) (PNH), HEMOLYSIS, RENAL FAILURE, THALASSEMIA, UREMIA

See the Anemia Flowchart in the introduction to the Hematology chapter.

(Continued)

ANEMIA [285.9] *(Continued)*

ANEMIA (DYSERYTHROPOIETIC) [285.8]

ANEMIA (ENZYME DEFICIENCY) [282.2]

ANEMIA (FANCONI) [284.0] *see also* FANCONI SYNDROME

ANEMIA (HEINZ BODY) [282.7]

ANEMIA (HEMOLYTIC) [282.9] *see also* AUTOIMMUNE DISEASES, DISSEMINATED INTRAVASCULAR COAGULATION (DIC), HELLP SYNDROME, HEMOGLOBINOPATHY, HEMOGLOBINURIA (PAROXYSMAL NOCTURNAL) (PNH), HEMOLYSIS, HEMOLYTIC-UREMIC SYNDROME, JAUNDICE, LEAD TOXICITY, LYMPHOCYTES/LYMPHOPROLIFERATIVE DISORDERS, MACROGLOBULINEMIA OF WALDENSTRÖM, MALARIA, SYSTEMIC LUPUS ERYTHEMATOSUS (SLE), THALASSEMIA, WILSON DISEASE

See the Anemia Flowchart in the introduction to the Hematology chapter.

(Continued)

ANEMIA (HEMOLYTIC) [282.9] *(Continued)*

ANEMIA (HYPOCHROMIC) *see* ANEMIA (MICROCYTIC)

ANEMIA (HYPOPLASTIC) [284.9]

ANEMIA (IRON DEFICIENCY) [280.9] *see also* BLEEDING (GASTROINTESTINAL), CARCINOMA (COLORECTAL), CARCINOMA (GASTROINTESTINAL TRACT), CARCINOMA (STOMACH), CYSTIC FIBROSIS, HOOKWORM INFESTATION, MENORRHAGIA, PEPTIC ULCER

See the Anemia Flowchart in the introduction to the Hematology chapter.

ANEMIA (MACROCYTIC/MEGALOBLASTIC) [281.9] *see also* ALCOHOLISM, ATROPHIC GASTRITIS

See the Anemia Flowchart in the introduction to the Hematology chapter.

ANEMIA (MICROANGIOPATHIC HEMOLYTIC) [283.19] *see also* RENAL VEIN THROMBOSIS

See the Anemia Flowchart in the introduction to the Hematology chapter.

ANEMIA (MICROCYTIC) [280.9]

See the Anemia Flowchart in the introduction to the Hematology chapter.

ANEMIA (SICKLE CELL AND VARIANT SICKLING HEMOGLOBINS) [282.60] *see also* ASPLENIA, HEMOGLOBINOPATHY

See the Anemia Flowchart in the introduction to the Hematology chapter.

(Continued)

ANEMIA (SICKLE CELL AND VARIANT SICKLING HEMOGLOBINS) [282.60] *(Continued)*

ANEMIA (SIDEROBLASTIC) [285.0]

See the Anemia Flowchart in the introduction to the Hematology chapter.

ANEMIA (SPHEROCYTOSIS) [282.0] *see also* ANEMIA (HEMOLYTIC), JAUNDICE

See the Anemia Flowchart in the introduction to the Hematology chapter.

ANENCEPHALY [740.0] *see also* FETAL DEATH

ANESTHETIC EFFECT (PROLONGED) *see* CHOLINESTERASE DEFICIENCY

ANEURYSM [442.9] *see also* ATHEROSCLEROSIS, SYPHILIS

ANGINA [413.9] *see also* ATHEROSCLEROSIS, CHEST PAIN, DIABETES MELLITUS, DYSPNEA, EMBOLISM, HYPERLIPIDEMIA, HYPERTENSION, MYOCARDIAL INFARCT, PLEURAL EFFUSION/EXUDATE/PLEURISY/ PLEURITIS, SYPHILIS, VASCULITIS

ANTISTREPTOCOCCAL ANTIBODIES *(Continued)*

ANTISYNTHETASE SYNDROME *see also* RAYNAUD PHENOMENON

ANTITHROMBIN DEFICIENCY

AORTIC VALVULAR INSUFFICIENCY/STENOSIS [424.1] *see also* SYPHILIS

AORTITIS *see* ANGINA, AORTIC VALVULAR INSUFFICIENCY/STENOSIS, MYOCARDIAL INFARCT, SYPHILIS, VASCULITIS

APPENDICITIS [541] *see also* COLITIS (PSEUDOMEMBRANOUS), CROHN DISEASE, ENTEROCOLITIS (NECROTIZING), PELVIC INFECTION/PELVIC INFLAMMATORY DISEASE (PID), PERFORATED VISCUS, PERITONITIS, *YERSINIA* INFECTION

WBC >10,000/mm³, elevated CRP, and neutrophils >75% improve clinical variables in differential diagnosis.

ARGININEMIA

ARGININOSUCCINIC ACIDURIA [270.6]

ARRHYTHMIAS (DYSRHYTHMIAS) [427.9] *see also* CHAGAS DISEASE, MYOCARDIAL INFARCT, PHEOCHROMOCYTOMA, SCLERODERMA, UREMIA

ARSENIC POISONING [985.1] *see also* ANEMIA (HEMOLYTIC), POISONING

ARTERIOSCLEROSIS *see* ATHEROSCLEROSIS

ARTERITIS (TEMPORAL) [446.5] *see also* VASCULITIS

ARTHRITIS [716.9] *see also* ALKAPTONURIA, ARTHRITIS (INFECTIOUS/ INFLAMMATORY), ARTHRITIS (RHEUMATOID), BEHÇET SYNDROME, CIRRHOSIS (PRIMARY BILIARY), GOUT, LEAD TOXICITY, LYME DISEASE, PURPURA (HENOCH-SCHÖNLEIN), SPONDYLITIS (ANKYLOSING), SYSTEMIC LUPUS ERYTHEMATOSUS (SLE)

(Continued)

ARTHRITIS [716.9] *(Continued)*

ARTHRITIS (INFECTIOUS/INFLAMMATORY) [711.9; 714.9] *see also*
COLITIS (ULCERATIVE), CROHN DISEASE, GONORRHEA, *HAEMOPHILUS INFLUENZAE* INFECTION, LYME DISEASE, STREPTOCOCCAL INFECTION, TICK BITE, TUBERCULOSIS

Campylobacter, *Salmonella*, and *Shigella* infections are also associated with arthritis.

ARTHRITIS (RHEUMATOID) [720.0] *see also* ANEMIA, ARTHRITIS

ASBESTOSIS [501] *see also* CARCINOMA (LUNG)

ASCITES [789.5] *see also* CIRRHOSIS, CONGESTIVE HEART FAILURE, EFFUSIONS/TRANSUDATES/EXUDATES, ENCEPHALOPATHY (HEPATIC), ESOPHAGEAL VARICES, HEPATITIS, HEPATITIS (AUTOIMMUNE), HYPOTHYROIDISM, LIVER, MALNUTRITION, NEPHROSIS/NEPHROTIC SYNDROME, PERICARDIAL EFFUSION/PERICARDITIS, PERITONITIS, PLEURAL EFFUSION/EXUDATE/PLEURISY/PLEURITIS, PROTEIN-LOSING ENTEROPATHY, UREMIA, WILSON DISEASE

An ascitic fluid, serum:total bilirubin ratio >6 supports distinction of exudate (eg, malignancy) from transudate.

(Continued)

AUTOIMMUNE DISEASES [279.4] *(Continued)*
ERYTHEMATOSUS (SLE), THYROIDITIS, VASCULITIS, WEGENER GRANULOMATOSIS

The expression "autoimmune diseases" indicates entities in which an autoimmune derangement, reaction, or injury contributes to pathogenesis. Look for increased gamma globulin in serum protein electrophoresis and serum autoantibodies.

AUTOIMMUNE HEMOLYTIC ANEMIA *see* ANEMIA (HEMOLYTIC)

AUTOIMMUNE LYMPHOPROLIFERATIVE SYNDROME

AZOOSPERMIA/OLIGOSPERMIA [606.0] *see also* CRYPTORCHIDISM, HYPOGONADISM

BABESIOSIS [088.82] *see also* TICK BITE

(Continued)

BENZODIAZEPINE OVERDOSE [969.4] *(Continued)*

BERNARD-SOULIER SYNDROME [287.1]

BILIARY FUNCTION NEONATAL *see* NEONATAL CHOLESTASIS WORK-UP

BILIARY OBSTRUCTION [576.2] *see also* CHOLANGITIS (PRIMARY SCLEROSING), CIRRHOSIS (PRIMARY BILIARY), JAUNDICE, LIVER

BIOLOGICAL WARFARE *see also* CHEMICAL WARFARE

Potential bioterrorism weapons may include anthrax, botulism, brucellosis, plague, Q fever, tularemia, smallpox, viral encephalitis, viral hemorrhagic fevers, and staphylococcal enterotoxin B. This book is intended to support usual clinical practice. However, a modicum of organization of topics relevant to bioterrorism presently seems appropriate. The *Handbook* does not address all of the entities on the biological warfare lists. The monographs listed below include useful citations to the literature which will provide at least introductory information.

BIOTINIDASE DEFICIENCY

BITES

BITES (INSECT) [989.5] *see also* TICK BITE

(Continued)

(Continued)

CARCINOMA (ADRENAL) [194.0] *(Continued)*

CARCINOMA (BLADDER) [188.9] *see also* HEMATURIA

CARCINOMA (BREAST) [174.9] *see also* LI-FRAUMENI CANCER SYNDROME

CARCINOMA (BREAST) RECURRENT, MARKERS FOR [V10.3; V103] *see also* CARCINOMA (BREAST)

Clinical examination and chest imaging are also pivotal for detection of the metastatic state in breast cancer.

CARCINOMA (CERVIX) [180.9] *see also* CERVICITIS

CARCINOMA (CHOLANGIOCARCINOMA) [156.1]

Complications of primary sclerosing cholangitis includes cholangiocarcinoma. Infestation by *Opisthorchis viverrini* and *Clonorchis sinensis* are very strong risk factors.

(Continued)

CARCINOMA (STOMACH) [151.9] (Continued)

CARDIAC ARRHYTHMIAS see ARRHYTHMIAS (DYSRHYTHMIAS)

CARDIAC TAMPONADE see PERICARDIAL EFFUSION/PERICARDITIS

CARDIOMYOPATHY [425.4] see also ACQUIRED IMMUNODEFICIENCY SYNDROME (AIDS), ALCOHOLISM, AMYLOIDOSIS, CONGESTIVE HEART FAILURE, DERMATOMYOSITIS, DIPHTHERIA, DUCHENNE MUSCULAR DYSTROPHY, GAUCHER DISEASE, GLYCOGEN STORAGE DISEASE, HEART, HEMOCHROMATOSIS, HEMOLYTIC-UREMIC SYNDROME, HYPERTENSION, HYPERTHYROIDISM, HYPOTHYROIDISM, INBORN ERRORS OF METABOLISM, LYME DISEASE, MYOCARDITIS, NIEMANN-PICK DISEASE, PHEOCHROMOCYTOMA, REYE SYNDROME, RICKETTSIAL INFECTION/ROCKY MOUNTAIN SPOTTED FEVER, SARCOIDOSIS, SCLERODERMA, SYSTEMIC LUPUS ERYTHEMATOSUS (SLE), TAY-SACHS DISEASE, UREMIA, VASCULITIS, WILSON DISEASE

CAT BITE *see* BITES

CAT SCRATCH DISEASE [078.3] *see also* MYCOBACTERIAL INFECTION (ATYPICAL), TUBERCULOSIS, TULAREMIA

CELIAC DISEASE [579.0] *see also* CYSTIC FIBROSIS, DIARRHEA, LACTOSE INTOLERANCE, MALABSORPTION/MALDIGESTIVE DISEASES, MALNUTRITION, PROTEIN-LOSING ENTEROPATHY

After antiendomysial antibodies and endoscopic biopsies, other relevant laboratory studies include the presence of hypochromic anemia, iron <60 µg/dL, ferritin <50 ng/dL, and cholesterol <156 mg/dL.

CELLULITIS [682.9]

CENTRAL NERVOUS SYSTEM *see* ALZHEIMER DISEASE, CEREBRAL INFARCTION/CEREBRAL THROMBOSIS/CEREBROVASCULAR ACCIDENT (CVA), COMA, CYTOMEGALOVIRUS, DEMENTIA, ENCEPHALITIS, HEAVY METAL/METAL POISONING, HERPESVIRUS INFECTION, INBORN ERRORS OF METABOLISM, LEAD TOXICITY, MENINGITIS, MENINGITIS (ASEPTIC), MULTIPLE SCLEROSIS, NEUROPATHY, NEUROSYPHILIS, NOCARDIOSIS, SYPHILIS

CEREBRAL INFARCTION/CEREBRAL THROMBOSIS/ CEREBROVASCULAR ACCIDENT (CVA) [434.0; 434.91; 436] *see also* BLEEDING, COMA, EMBOLISM, ENCEPHALITIS, THROMBOSIS, THROMBOSIS (VENOUS)

(Continued)

CEREBRAL INFARCTION/CEREBRAL THROMBOSIS/ CEREBROVASCULAR ACCIDENT (CVA) [434.0; 434.91; 436] (Continued)

CERVICITIS [616.0] see also CARCINOMA (CERVIX), CARCINOMA (ENDOCERVIX), GONORRHEA, HERPESVIRUS INFECTION (GENITAL), SEXUALLY TRANSMITTED DISEASES, UROGENITAL INFECTIONS

CHAGAS DISEASE [086.2]

CHARCOT JOINT see DIABETES MELLITUS, SYPHILIS

CHÉDIAK-HIGASHI SYNDROME [288.2]

CHEMICAL WARFARE see also BIOLOGICAL WARFARE

CHEST PAIN [786.50] see also ANGINA, CORONARY ARTERIAL DISEASE, DYSPNEA, ESOPHAGITIS, HEART, HEMOPTYSIS, MYOCARDIAL INFARCT, PANCREATITIS, PLEURAL EFFUSION/EXUDATE/PLEURISY/ PLEURITIS, PULMONARY EMBOLISM

CHIARI-FROMMEL SYNDROME see HYPERPROLACTINEMIA/ CHIARI-FROMMEL SYNDROME

CHICKENPOX see VARICELLA-ZOSTER

CHLAMYDIA INFECTION [079.98] see also CONJUNCTIVITIS, PNEUMONIA, PNEUMONITIS

CIRRHOSIS (JUVENILE) [571.5]

CIRRHOSIS (PRIMARY BILIARY) [571.6] see also ARTHRITIS, CHOLANGITIS (PRIMARY SCLEROSING), CIRRHOSIS, CIRRHOSIS (PRIMARY BILIARY), CREST SYNDROME ("LIMITED SCLERODERMA"), HEPATITIS (AUTOIMMUNE), HYPOTHYROIDISM, JAUNDICE, OSTEOPOROSIS, RAYNAUD PHENOMENON, SCLERODERMA, SJÖGREN SYNDROME

Antimitochondrial antibody is found in 95%. Some drugs cause cholestasis.

(Continued)

COLITIS (PSEUDOMEMBRANOUS) [008.45] see also COLITIS, COLITIS (ULCERATIVE), CROHN DISEASE, ENTEROCOLITIS (NECROTIZING), PERFORATED VISCUS, UREMIA

COLITIS (ULCERATIVE) [556.9] see also AMEBIASIS, COLITIS (PSEUDOMEMBRANOUS), DIARRHEA, ENTERIC FEVER, PROTEIN-LOSING ENTEROPATHY

COLLAGEN DISEASES see AUTOIMMUNE DISEASES

COLON CARCINOMA see CARCINOMA (COLORECTAL)

COMA [780.01] see also ALCOHOL INTOXICATION, ARRHYTHMIAS (DYSRHYTHMIAS), BARBITURATE POISONING, BROMIDE INTOXICATION, CEREBRAL INFARCTION/CEREBRAL THROMBOSIS/CEREBROVASCULAR ACCIDENT (CVA), ETHYLENE GLYCOL POISONING, HYPERGLYCEMIA, HYPOGLYCEMIA, HYPOGLYCEMIA (KETOTIC), INBORN ERRORS OF METABOLISM, MENINGITIS, METHANOL POISONING, OLIGURIA, POISONING, RENAL FAILURE, REYE SYNDROME, UREMIA

(Continued)

COMA [780.01] *(Continued)*

CONDYLOMA ACUMINATUM [078.11]

CONGENITAL ADRENAL HYPERPLASIA [255.2] *see also* ADENOMA (ADRENAL), CARCINOMA (ADRENAL), HIRSUTISM, INTERSEXUALITY, POLYCYSTIC OVARY SYNDROME, VIRILIZATION

See text in listing, Testosterone, Total and Free, Serum or Plasma.

CONGENITAL RUBELLA SYNDROME [771.0]

CONGESTIVE HEART FAILURE [428.0] *see also* ALCOHOLISM, AMYLOIDOSIS, ARSENIC POISONING, ASCITES, DIABETES MELLITUS, EDEMA, EFFUSIONS/TRANSUDATES/EXUDATES, HEART, HEMOCHROMATOSIS, HYPERTENSION, LEAD TOXICITY, LIVER, LYME DISEASE, MYOCARDIAL INFARCT, MYOCARDITIS, PLEURAL EFFUSION/ EXUDATE/PLEURISY/PLEURITIS, RENAL FAILURE, RHEUMATIC FEVER, SARCOIDOSIS, SCLERODERMA, UREMIA

Chest x-ray, EKG, echocardiography, or radionucleotide ventriculography are among studies recommended. Myocarditis should be ruled out in selected patients.

(Continued)

CORONARY ARTERIAL DISEASE (RISK) *see also* ATHEROSCLEROSIS, DIABETES MELLITUS, HYPERTENSION, HYPOTHYROIDISM

Important risk factors include cigarette smoking, hypertension, male gender, unfavorable cholesterol pattern, and family history. Diabetes mellitus is now a coronary disease risk equivalent, and impaired glucose tolerance is associated with cardiovascular disease risk factors.

COUMARIN THERAPY

COXSACKIE VIRUS INFECTION [079.2] see also BRONCHITIS, CONJUNCTIVITIS, ENCEPHALITIS, ENTEROVIRUS INFECTION, GASTROENTERITIS, MYOCARDITIS, PERICARDIAL EFFUSION/ PERICARDITIS

CREST SYNDROME ("LIMITED SCLERODERMA") [710.1] see also AUTOIMMUNE DISEASES, CIRRHOSIS (PRIMARY BILIARY), ESOPHAGEAL DYSFUNCTION, RAYNAUD PHENOMENON, SCLERODERMA

The CREST syndrome is Calcinosis, Raynaud phenomenon, Esophageal dysmotility, Sclerodactyly, and Telangiectasia.

CREUTZFELDT-JAKOB DISEASE [046.1]

CROHN DISEASE [555.9] see also COLITIS, COLITIS (ULCERATIVE), DIARRHEA, KIDNEY STONE, MALABSORPTION/MALDIGESTIVE DISEASES, OSTEOPOROSIS, OXALURIA, PROTEIN-LOSING ENTEROPATHY, TUBERCULOSIS

(Continued)

CROHN DISEASE [555.9] *(Continued)*

CROUP [464.4] *see also* ASTHMA, COXSACKIE VIRUS INFECTION, DIPHTHERIA, ENTEROVIRUS INFECTION, EPIGLOTTITIS, *HAEMOPHILUS INFLUENZAE* INFECTION, MEASLES (RUBEOLA), SUPRAGLOTTITIS

CRUSH INJURY [929.9]

CRYOGLOBULINEMIA [273.2] *see also* AUTOIMMUNE DISEASES, COLD AGGLUTININ DISEASE, COLD SENSITIVITY, HEPATITIS C, LYMPHOMA, MACROGLOBULINEMIA OF WALDENSTRÖM, MONOCLONAL GAMMOPATHY, RAYNAUD PHENOMENON, SJÖGREN SYNDROME, VASCULITIS

Cryoglobulinemias bear association with other diseases, including hepatitis C, Sjögren syndrome, and entities related to production of abnormal proteins such as myeloma.

CRYPTOCOCCAL MENINGITIS *see* MENINGITIS (CRYPTOCOCCAL)

(Continued)

(Continued)

DEEP VEIN THROMBOSIS (DVT) [453.9] *(Continued)*

DEHYDRATION/HYPOVOLEMIA [276.5; 775.5] *see also* ACIDOSIS/ ALKALOSIS (ACID-BASE BALANCE), ADDISON DISEASE, BARTTER SYNDROME, CONGENITAL ADRENAL HYPERPLASIA, CYSTIC FIBROSIS, DIABETES MELLITUS, DIARRHEA, INBORN ERRORS OF METABOLISM, PYLORIC STENOSIS, RENAL VEIN THROMBOSIS, SHOCK, UREMIA, VOMITING

DELIRIUM TREMENS [291.0] *see also* ALCOHOLISM, ALCOHOL INTOXICATION, NEUROGLYCOPENIA, NEUROSYPHILIS

DEMENTIA [294.8] *see also* ALZHEIMER DISEASE, HYPOGLYCEMIA, NEUROSYPHILIS, UREMIA

(Continued)

DERMATOMYOSITIS [710.3] *(Continued)*

DIABETES INSIPIDUS [253.5] *see also* ACIDOSIS/ALKALOSIS (ACID-BASE BALANCE), BARTTER SYNDROME, DEHYDRATION/HYPOVOLEMIA, LANGERHANS' CELL HISTIOCYTOSIS, POLYDIPSIA, POLYURIA, SYNDROME OF INAPPROPRIATE SECRETION OF ANTIDIURETIC HORMONE (SIADH)

DIABETES INSIPIDUS (NEPHROGENIC) [588.1] *see also* DEHYDRATION/ HYPOVOLEMIA, DIABETES INSIPIDUS, DIABETES MELLITUS, HYPOPITUITARISM, POLYDIPSIA, POLYURIA

DIABETES MELLITUS [250.0; 775.1] *see also* ACIDOSIS/ALKALOSIS (ACID-BASE BALANCE), CYSTIC FIBROSIS, DIABETES MELLITUS (GESTATIONAL), HYPERLIPIDEMIA, KETOSIS/KETONURIA/ KETOACIDOSIS, NONALCOHOLIC STEATOHEPATITIS, PRENATAL DIAGNOSIS

Diagnostic criteria depend on plasma glucose. Management requires periodic testing for glycated hemoglobin (or equivalent) and microalbuminuria.

DIABETES MELLITUS (GESTATIONAL) [648.8] see also DIABETES MELLITUS

DIABETIC NEPHROPATHY [250.4] see also DIABETES MELLITUS, UREMIA

DIARRHEA [787.91] see also ABETALIPOPROTEINEMIA, ACQUIRED IMMUNODEFICIENCY SYNDROME (AIDS), ADDISON DISEASE, ADENOVIRUS INFECTION, AMEBIASIS, CELIAC DISEASE, COLITIS, COLITIS (PSEUDOMEMBRANOUS), COLITIS (ULCERATIVE), CONGENITAL ADRENAL HYPERPLASIA, CROHN DISEASE, CRYPTOSPORIDIOSIS, CYSTIC FIBROSIS, CYTOMEGALOVIRUS, ENTERIC FEVER, ENTEROCOLITIS (NECROTIZING), GASTROENTERITIS, GIARDIASIS, GLUCAGONOMA (ALPHA-CELL TUMOR), HYPERTHYROIDISM, IGA DEFICIENCY, MALABSORPTION/MALDIGESTIVE DISEASES, MICROSPORIDIOSIS, NEUROBLASTOMA, PHEOCHROMOCYTOMA, TYPHOID FEVER, ZOLLINGER-ELLISON SYNDROME

Acute diarrhea includes infectious and noninfectious entities. Evaluation of the chronic diarrheal diseases begins with basic testing involving blood and stool.

(Continued)

DIARRHEA [787.91] *(Continued)*

DIC *see* DISSEMINATED INTRAVASCULAR COAGULATION (DIC)

(Continued)

DISSEMINATED INTRAVASCULAR COAGULATION (DIC) [776.2; 286.6]
(Continued)

DOG BITE *see* BITES

DOWN SYNDROME [758.0] *see also* CHROMOSOMAL DISORDERS, LEUKEMIA, MENTAL RETARDATION, TRISOMY

DRUG-INDUCED IMMUNE THROMBOCYTOPENIA [287.4]

DUBIN-JOHNSON SYNDROME [782.4] *see also* JAUNDICE

DUCHENNE MUSCULAR DYSTROPHY [359.1] *see also* MUSCULAR DYSTROPHY, MYASTHENIA GRAVIS, MYOSITIS

DWARFISM [259.4] *see also* GROWTH HORMONE DEFICIENCY

DYSARTHRIA [784.5] *see also* ALCOHOLISM

DYSBETALIPOPROTEINEMIA (FAMILIAL) [272.2]

DYSENTERY [009.0]

DYSENTERY (BACILLARY) [004.9; 711.3] *see also* DIARRHEA

DYSERYTHROPOIETIC ANEMIA *see* ANEMIA (DYSERYTHROPOIETIC)

DYSFIBRINOGENEMIA [286.3] *see also* BRUISING, DISSEMINATED INTRAVASCULAR COAGULATION (DIC), ECCHYMOSIS (SPONTANEOUS), FACTOR IX DEFICIENCY, HEMOPHILIA, VITAMIN K DEFICIENCY, VON WILLEBRAND DISEASE

DYSGERMINOMA (OVARY) [183.0] *see also* CARCINOMA (OVARY), SEMINOMA

This is the ovarian equivalent of seminoma of testis.
- ••Histopathology 733
- ••Lactate Dehydrogenase Isoenzymes, Serum 824
- ••Lactate Dehydrogenase, Serum 825

DYSKINESIA [781.3]
- Haloperidol, Serum or Plasma. 664
- Phenothiazines, Serum 1021

DYSPNEA [786.09] *see also* ANGINA, CONGESTIVE HEART FAILURE, HYPERSENSITIVITY PNEUMONITIS, MYASTHENIA GRAVIS, MYOCARDIAL INFARCT, PLEURAL EFFUSION/EXUDATE/PLEURISY/PLEURITIS, PNEUMONIA, PULMONARY EMBOLISM
- •B-Type Natriuretic Peptide 314
- Digoxin, Serum 512

DYSPROTEINEMIA *see* MACROGLOBULINEMIA OF WALDENSTRÖM, MYELOMA

DYSRHYTHMIAS *see* ARRHYTHMIAS (DYSRHYTHMIAS)

DYSURIA [788.1] *see also* CYSTITIS/PYELONEPHRITIS, GONORRHEA, KIDNEY STONE, URETHRITIS
- Arterial Blood Collection 211
- Bacterial Culture, Genital Specimen 239
- •Bacterial Culture, Urine 246
- *Chlamydia trachomatis* Culture 385
- *Chlamydia trachomatis* Nucleic Acid Detection 388
- Creatinine, Serum or Plasma 474
- Genital Culture for *Ureaplasma urealyticum* 635
- Leukocyte Esterase, Urine 849
- *Neisseria gonorrhoeae* Culture and Smear 945
- *Neisseria gonorrhoeae* Nucleic Acid Detection 947
- Nitrite, Urine 962
- *Trichomonas* Preparation 1273
- Urea Nitrogen, Serum or Plasma 1284
- ••Urinalysis 1289

ECCHYMOSIS (SPONTANEOUS) [782.7] *see also* BRUISING, DISSEMINATED INTRAVASCULAR COAGULATION (DIC), DYSFIBRINOGENEMIA, FACTOR IX DEFICIENCY, HEMOPHILIA, LEUKEMIA, LYMPHOMA, MACROGLOBULINEMIA OF WALDENSTRÖM, PLATELET FUNCTION TESTS, PURPURA, THROMBOCYTOPENIA, UREMIA
- •Activated Partial Thromboplastin Time. 100
- Complete Blood Count 442
- D-Dimers and Fibrin Degradation Products 502
- Disseminated Intravascular Coagulation Screen 517
- •Fibrinogen 583
- Platelet Aggregation 1045
- ••Platelet Count. 1050
- •Prothrombin Time 1116
- Salicylate, Serum or Plasma 1176

ECHINOCOCCOSIS [122.9] *see also* LIVER, PARASITIC INFESTATIONS
- •Cerebrospinal Fluid Cytology 361
- ••Echinococcosis Diagnostic Procedures 530
- ••Histopathology 733

ECHOVIRUSES *see* ENTEROVIRUS INFECTION

ECTOPIC ACTH SYNDROME [255.0] *see also* ACIDOSIS/ALKALOSIS (ACID-BASE BALANCE), CUSHING SYNDROME
- Adrenal Cortex: Laboratory Assessment Overview 110
- •Adrenocorticotropic Hormone, Plasma. 114
- Calcium, Serum 329
- Calcium, Urine 332
- Corticotropin-Releasing Hormone Stimulation Test. 455
- Cortisol, Free, Urine 459
- •Cortisol, Serum or Plasma 460
- ••Histopathology 733
- 17-Hydroxycorticosteroids, Urine 753
- •Metyrapone Stimulation Test, Serum. 910
- •Potassium, Serum or Plasma 1078
- Sputum Cytology 1222

ECTOPIC PREGNANCY *see* PREGNANCY (ECTOPIC)

ELEPHANTIASIS [457.1] *see also* PARASITIC INFESTATIONS

EMBOLISM [444.9] *see also* ANTICOAGULANT THERAPY, DEEP VEIN THROMBOSIS (DVT), HEMOPTYSIS, RENAL VEIN THROMBOSIS, THROMBOSIS

EMBOLISM (AIR) [958.0]

EMBRYONAL CARCINOMA *see* CARCINOMA (EMBRYONAL)

EMPYEMA [510.9] *see also* ABSCESS (LUNG), CARCINOMA (LUNG), NOCARDIOSIS, PLEURAL EFFUSION/EXUDATE/PLEURISY/PLEURITIS, PNEUMONIA, STAPHYLOCOCCAL INFECTION, TUBERCULOSIS

A collection of exudate within a body space, including the pleural space. (In lung abscess, the collection is in pulmonary parenchyma.)

ENCEPHALITIS [323.9] *see also* BROMIDE INTOXICATION, COMA, COXSACKIE VIRUS INFECTION, ENCEPHALITIS (HERPESVIRUS), MENINGITIS, MENINGOENCEPHALITIS, SUBACUTE SCLEROSING LEUKOENCEPHALITIS, WEST NILE VIRUS INFECTION

(Continued)

ENCEPHALITIS [323.9] *(Continued)*

ENCEPHALITIS (HERPESVIRUS) [054.3] *see also* ENCEPHALITIS

ENCEPHALOPATHY (HEPATIC) [572.2] *see also* ALCOHOLISM, ASCITES, CIRRHOSIS, COMA, DELIRIUM TREMENS, ESOPHAGEAL VARICES, HEPATITIS, HYPOKALEMIA, INBORN ERRORS OF METABOLISM, LEAD TOXICITY, LIVER, POISONING, REYE SYNDROME, SYSTEMIC LUPUS ERYTHEMATOSUS (SLE), TYPHOID FEVER, WILSON DISEASE

ENCEPHALOPATHY (TOXIC) [349.82] *see also* ACIDOSIS/ALKALOSIS (ACID-BASE BALANCE), ASCITES, CIRRHOSIS, COMA, HYPERNATREMIA, HYPEROSMOLALITY, LIVER, POISONING, UREMIA

ENDOCARDITIS *see* INFECTIVE ENDOCARDITIS

ENDOCERVICAL CARCINOMA *see* CARCINOMA (ENDOCERVIX)

ENDODERMAL SINUS TUMOR [186.9; 183.0] *see also* CARCINOMA (EMBRYONAL)

ETHYLENE GLYCOL POISONING [982.8]
Severe, high anion gap metabolic acidosis is seen.

EXANTHEM SUBITUM see HUMAN HERPESVIRUS-6 INFECTION, ROSEOLA INFANTUM

EYE see CONJUNCTIVITIS, CORNEAL ULCER, RETINITIS, RETINOBLASTOMA

FACTOR DEFICIENCY [286.9] see also CIRCULATING ANTICOAGULANTS, DYSFIBRINOGENEMIA, HEMOPHILIA

FACTOR II DEFICIENCY [286.3]

FACTOR V DEFICIENCY [286.9]

FACTOR V LEIDEN MUTATION

FACTOR VII DEFICIENCY [286.3]

(Continued)

FARMER'S LUNG DISEASE [495.0]

FASCIITIS (EOSINOPHILIC) [729.1]

FAT EMBOLISM [958.1]

FATTY LIVER OF PREGNANCY, ACUTE see also HELLP SYNDROME
In differential diagnosis from HELLP syndrome: prothrombin time, aPTT, and fibrinogen are usually within normal ranges in HELLP syndrome.

FELTY SYNDROME [714.1]

FETAL ABNORMALITY see CHROMOSOMAL DISORDERS, PRENATAL DIAGNOSIS, TRISOMY

FETAL DEATH [779.9] see also ABORTION, CHROMOSOMAL DISORDERS, FETAL DISTRESS, FETAL MATURITY, HEMOLYTIC DISEASE OF THE NEWBORN, PREMATURITY, PRENATAL DIAGNOSIS, RESPIRATORY DISTRESS SYNDROME, SYSTEMIC LUPUS ERYTHEMATOSUS (SLE)

(Continued)

FIRE EXPOSURE *(Continued)*

HEAVY METAL/METAL POISONING [985.9] *see also* LEAD TOXICITY, MERCURY POISONING, TICK PARALYSIS

HEINZ BODY ANEMIA *see* ANEMIA (HEINZ BODY)

HELICOBACTER PYLORI [041.86]

HELLP SYNDROME [642.5] *see also* ANEMIA (HEMOLYTIC), DISSEMINATED INTRAVASCULAR COAGULATION (DIC), FATTY LIVER OF PREGNANCY, ACUTE, HEMOLYTIC-UREMIC SYNDROME, IDIOPATHIC THROMBOCYTOPENIC PURPURA (ITP), LIVER, PREECLAMPSIA, THROMBOTIC THROMBOCYTOPENIC PURPURA (TTP)

Characterized by right upper quadrant pain, nausea and vomiting, microangiopathic hemolytic anemia, elevated concentrations of liver-related enzymes, and low platelets (thrombocytopenia), the HELLP syndrome is a cause of maternal and fetal morbidity and mortality. It is found with severe pre-eclampsia or eclampsia. Other types of liver disease unique to pregnancy are hyperemesis gravidarum, cholestasis, and acute fatty liver.

HEMATURIA [599.7] *see also* BLEEDING, CARCINOMA (BLADDER), CYSTITIS/PYELONEPHRITIS, GLOMERULONEPHRITIS, GOODPASTURE SYNDROME, HEMOGLOBINOPATHY, HEMOGLOBINURIA, HEMOGLOBINURIA (PAROXYSMAL NOCTURNAL) (PNH), KIDNEY STONE,

(Continued)

HEMOLYSIS [283.2] *see also* ANEMIA (HEMOLYTIC), HEMATURIA, HEMOGLOBINURIA

HEMOLYTIC ANEMIA *see* ANEMIA (HEMOLYTIC)

HEMOLYTIC DISEASE OF THE NEWBORN [773.2] *see also* HYPOGLYCEMIA, JAUNDICE

HEMOLYTIC-UREMIC SYNDROME [283.11] *see also* ANEMIA (MICROANGIOPATHIC HEMOLYTIC), GLOMERULONEPHRITIS, HELLP SYNDROME, HYPERTENSION, PREGNANCY, SCLERODERMA, SYSTEMIC LUPUS ERYTHEMATOSUS (SLE), THROMBOCYTOPENIA, THROMBOTIC THROMBOCYTOPENIC PURPURA (TTP)

Typically follows infections with *S. dysenteriae* and *Shiga*-toxin-producing *E. coli*, especially 0157:H7. The cardinal features include hemolytic anemia, thrombocytopenia, and acute renal failure, most often in children soon after the onset of diarrhea. See the Anemia Flowchart in the introduction to the Hematology chapter.

(Continued)

HEMOLYTIC-UREMIC SYNDROME [283.11] *(Continued)*

HEMOPHILIA [286.0] *see also* ACQUIRED IMMUNODEFICIENCY SYNDROME (AIDS), BRUISING, FACTOR IX DEFICIENCY, FACTOR VIII DEFICIENCY, VON WILLEBRAND DISEASE

HEMOPHILIA A *see* FACTOR VIII DEFICIENCY

HEMOPHILIA B *see* FACTOR IX DEFICIENCY

HEMOPTYSIS [786.3] *see also* ABSCESS (LUNG), ASPERGILLOSIS, BLASTOMYCOSIS, BRONCHIECTASIS, CARCINOMA (LUNG), CHEST PAIN, COCCIDIOIDOMYCOSIS, CYSTIC FIBROSIS, DYSPNEA, EMBOLISM, FUNGI, GOODPASTURE SYNDROME, PNEUMONIA, PULMONARY EMBOLISM, SYSTEMIC LUPUS ERYTHEMATOSUS (SLE), TUBERCULOSIS, WEGENER GRANULOMATOSIS

HEMORRHAGE *see* ANEMIA, BLEEDING, BLEEDING (GASTROINTESTINAL), BRUISING, DISSEMINATED INTRAVASCULAR COAGULATION (DIC), ECCHYMOSIS (SPONTANEOUS), ESOPHAGEAL VARICES, FACTOR DEFICIENCY, FIBRINOLYSIS, HEMOGLOBINOPATHY, HEMOLYSIS, IDIOPATHIC THROMBOCYTOPENIC PURPURA (ITP), PEPTIC ULCER, PETECHIA, THROMBOTIC THROMBOCYTOPENIC PURPURA (TTP)

(Continued)

HEPATIC NECROSIS/FAILURE [570] *(Continued)*

HEPATITIS [573.3] *see also* ACQUIRED IMMUNODEFICIENCY SYNDROME (AIDS), CIRRHOSIS, CYTOMEGALOVIRUS, HEPATITIS (AUTOIMMUNE), HERPESVIRUS INFECTION, JAUNDICE, LYME DISEASE, MONONUCLEOSIS (INFECTIOUS), REYE SYNDROME, SEXUALLY TRANSMITTED DISEASES, SYPHILIS, TYPHOID FEVER, WILSON DISEASE

HEPATITIS (AUTOIMMUNE) [571.49] *see also* ALCOHOLISM, ALPHA₁-ANTITRYPSIN DEFICIENCY, AUTOIMMUNE DISEASES, CHOLANGITIS (PRIMARY SCLEROSING), CIRRHOSIS, CIRRHOSIS (PRIMARY BILIARY), HEMOCHROMATOSIS, HEPATITIS, HYPOTHYROIDISM, LIVER, NONALCOHOLIC STEATOHEPATITIS, STEATOSIS (LIVER), WILSON DISEASE

Antinuclear antibody >1:160, especially with the homogeneous pattern, with positive smooth muscle antibody provides evidence of autoimmune liver disease, especially when disease has existed for more than 6 months. Distinction among autoimmune hepatitis, alcohol-induced liver disease, nonalcoholic steatohepatitis (nonalcoholic fatty liver disease), chronic viral hepatitis, primary biliary cirrhosis, sclerosing cholangitis, hepatic involvement in systemic disease, and instances of drug-induced hepatitis requires study of clinical, histopathologic, and laboratory characteristics.

HEPATITIS C [070.51]

HEPATOLENTICULAR DEGENERATION *see* WILSON DISEASE

HEPATOMA *see* CARCINOMA (LIVER)

HEPATORENAL SYNDROME [572.4]
Functional renal failure in subjects with end-stage hepatic disease. Creatinine is increased; look for hyperosmolar urine and Na⁺ excretion <10 mEq/L.

(Continued)

1437

HEPATORENAL SYNDROME [572.4] *(Continued)*

HEREDITARY PERSISTENCE OF HEMOGLOBIN F [282.7]

HEREDITARY SPHEROCYTOSIS *see* ANEMIA (SPHEROCYTOSIS)

HERMANSKY-PUDLAK SYNDROME

HERPANGINA [074.0] *see also* ADENOVIRUS INFECTION, COXSACKIE VIRUS INFECTION, ENTEROVIRUS INFECTION, HERPESVIRUS INFECTION, STREPTOCOCCAL INFECTION

HERPESVIRUS-6 *see* HUMAN HERPESVIRUS-6 INFECTION

HERPESVIRUS 8 *see* KAPOSI SARCOMA, PRIMARY EFFUSION LYMPHOMA

HERPESVIRUS INFECTION [054.9] *see also* ACQUIRED IMMUNODEFICIENCY SYNDROME (AIDS), BULLOUS SKIN DISEASE/PEMPHIGOID/PEMPHIGUS, HEPATITIS, HERPESVIRUS INFECTION (GENITAL), PNEUMONIA (VIRAL), VESICLE

HERPESVIRUS INFECTION (GENITAL) [054.10] *see also* ACQUIRED IMMUNODEFICIENCY SYNDROME (AIDS), GONORRHEA, SEXUALLY TRANSMITTED DISEASES, SYPHILIS

HERPES ZOSTER [053.9] *see also* VARICELLA-ZOSTER

HG-PRT DEFICIENCY [277.2]

HHH SYNDROME

HUMAN HERPESVIRUS-6 INFECTION (Continued)
• Herpesvirus 6 Serology 725
HUMAN IMMUNODEFICIENCY SYNDROME (HIV) see ACQUIRED
IMMUNODEFICIENCY SYNDROME (AIDS)
HUMAN T-CELL LYMPHOTROPIC VIRUS-I/II (HTLV-I/II) [079.51; 079.52]
• HTLV-I/II Antibody 745
HUMORAL HYPERCALCEMIA OF MALIGNANCY see also CARCINOMA
(BREAST), CARCINOMA (LUNG), CARCINOMA (RENAL),
HYPERPARATHYROIDISM, MYELOMA, SARCOIDOSIS
Squamous cell carcinomas of head and neck are among causes of this entity.
•• Calcium, Serum 329
•• Parathyroid Hormone-Related Protein, Serum 1000
• Parathyroid Hormone, Serum 1001
• Phosphorus, Serum 1031
• Vitamin D, Serum 1318
HUMORAL IMMUNE DEFICIENCY [279.00] see also ACQUIRED
IMMUNODEFICIENCY SYNDROME (AIDS), GRANULOMATOUS DISEASE
OF CHILDHOOD
• Immunoglobulin A 770
• Immunoglobulin G 775
• Immunoglobulin G Subclasses 778
• Immunoglobulin M 779
• Protein Electrophoresis, Capillary Zone 1103
• Protein Electrophoresis, Serum 1104
Protein Electrophoresis, Urine 1107
HUNTINGTON DISEASE [333.4]
Huntington Disease DNA Test... 752
HURLER SYNDROME [272.7]
•• Mucopolysaccharides, Urine 922
HYALINE MEMBRANE DISEASE see RESPIRATORY DISTRESS
SYNDROME
HYDATID CYST see CYST (HYDATID)
HYDATID DISEASE see ECHINOCOCCOSIS
HYDATIDIFORM MOLE [630] see also CHORIOCARCINOMA, GESTATIONAL
TROPHOBLASTIC DISEASE
Trophoblastic cells express the RhD factor.
•• Chorionic Gonadotropin, Human, Serum and Urine 397
•• Flow Cytometry, Overview 592
•• Histopathology 733
Methotrexate, Serum or Plasma 905
• Pregnancy Test, Serum or Urine 1084
HYDRATION/OVERHYDRATION [999.9] see also DEHYDRATION/
HYPOVOLEMIA
Albumin and Plasma Protein Fraction for Infusion 118
Albumin, Serum 120
Chloride, Serum, Plasma, or Blood 391
Complete Blood Count 442
Concentration Test, Urine 446
Electrolyte Panel, Serum 532
Hematocrit 674
Hemoglobin 681
• Osmolality, Calculated, Serum or Plasma 976
• Osmolality, Serum 978
• Osmolality, Urine 979
pH, Blood 1018
Plasma, Fresh Frozen 1039
Plasma, Frozen, Donor Retested 1041
Protein, Total, Serum 1114
Sodium, Serum or Plasma 1210
Specific Gravity, Urine 1216
• Urea Nitrogen:Creatinine Ratio 1283
Urea Nitrogen, Serum or Plasma 1284
Urinalysis 1289
HYDRONEPHROSIS [591]
Bacterial Culture, Urine 246
Creatinine Clearance and Urine Creatinine 473
Creatinine, Serum or Plasma 474
Electrolyte Panel, Serum 532
Prostate Specific Antigen, Serum 1098
Urea Nitrogen, Serum or Plasma 1284
Urinalysis 1289
HYPERAGGREGABILITY see PLATELET FUNCTION TESTS

HYPERCOAGULABLE STATES *see also* ANTICOAGULANT THERAPY, CARCINOMA (PANCREAS), EMBOLISM, NEPHROSIS/NEPHROTIC SYNDROME, THROMBOSIS

HYPEREMESIS GRAVIDARUM [643.0]

Occurring in the first trimester of pregnancy, symptoms are nausea and vomiting.

HYPERGAMMAGLOBULINEMIA [289.8] *see also* CIRRHOSIS, HEPATITIS (AUTOIMMUNE), LEISHMANIASIS, LEPROSY, MACROGLOBULINEMIA OF WALDENSTRÖM, MALARIA, MYELOMA, SARCOIDOSIS, SYSTEMIC LUPUS ERYTHEMATOSUS (SLE)

HYPERGLYCEMIA [790.6] *see also* COMA, DIABETES MELLITUS, PANCREATITIS, PHEOCHROMOCYTOMA

HYPERGLYCINEMIAS [270.7]

(Continued)

HYPERNATREMIA [276.0] *(Continued)*

HYPEROSMOLALITY [276.0] *see also* ACIDOSIS/ALKALOSIS (ACID-BASE BALANCE), HYPERGAMMAGLOBULINEMIA, HYPERKALEMIA, HYPERLIPIDEMIA, HYPERNATREMIA

The nonketotic hyperosmolar syndrome includes hyperglycemia, plasma osmolality >320 mOsm/L without ketonemia. See individual monographs.

HYPERPARATHYROIDISM [252.0] *see also* CARCINOMA (LUNG), HYPERCALCEMIA, MEDULLARY CARCINOMA OF THYROID, MULTIPLE ENDOCRINE NEOPLASIA (MEN 1, MEN 2), PANCREATITIS, PHEOCHROMOCYTOMA, RENAL FAILURE

HYPERPHENYLALANINEMIA [270.1] 143
  Amino Acids, Plasma 954
  ••Newborn Screen for Phenylketonuria 1022
  ••Phenylalanine, Blood 1024
  Phenylalanine, Urine 1318
  Vitamin D, Serum

HYPERPHOSPHATASIA (HEREDITARY) *see also* PAGET DISEASE OF
  BONE 127
  ••Alkaline Phosphatase, Serum

HYPERPIGMENTATION (GENERALIZED) *see also* ADDISON DISEASE, ARSENIC
  POISONING, CARCINOMA (LUNG), CARCINOMA (PANCREAS),
  CIRRHOSIS (PRIMARY BILIARY), CUSHING SYNDROME, ECTOPIC ACTH
  SYNDROME, HEMOCHROMATOSIS, HYPERTHYROIDISM, NELSON
  SYNDROME, SCLERODERMA, SYSTEMIC LUPUS ERYTHEMATOSUS
  (SLE)

HYPERPLASIA (LYMPH NODE) [785.6] *see also* LYMPHADENOPATHY,
  LYMPHOCYTES/LYMPHOPROLIFERATIVE DISORDERS, LYMPHOMA 249
  Bartonella Culture 250
  Bartonella Diagnostic Procedures 442
  ••Complete Blood Count 499
  Cytomegalovirus Serology 736
  HIV-1/HIV-2 Serology 785
  •Infectious Mononucleosis Screening Test 880
  ••Lymph Node Biopsy 1265
  Toxoplasmosis Diagnostic Procedures

HYPERPROLACTINEMIA/CHIARI-FROMMEL SYNDROME [253.1; 453.0;
  676.6] *see also* AMENORRHEA, GALACTORRHEA, HYPOGONADISM,
  HYPOTHYROIDISM, PITUITARY/PITUITARY ADENOMA, POLYCYSTIC
  OVARY SYNDROME, PROLACTINOMA 114
  Adrenocorticotropic Hormone, Plasma 153
  Amoxapine, Serum or Plasma 171
  Antidepressants, Cyclic, Serum or Plasma 474
  Creatinine, Serum or Plasma 609
  Follicle Stimulating Hormone, Serum, Plasma, or Urine 664
  Haloperidol, Serum or Plasma 890
  Manganese, Serum or Blood 892
  Manganese, Urine 974
  Opiates, Qualitative, Urine 1021
  Phenothiazines, Serum 1084
  Pregnancy Test, Serum or Urine 1094
  ••Prolactin, Serum 1257
  Thyroxine, Serum

HYPERSEGMENTATION OF GRANULOCYTE NUCLEI [288.2] *see also*
  ANEMIA (MACROCYTIC/MEGALOBLASTIC) 296
  Bone Marrow 424
  •Cobalamin, Serum 442
  ••Complete Blood Count 908
  Methylmalonic Acid, Serum, Plasma, Urine, or Amniotic Fluid ... 1010
  ••Peripheral Blood: Differential Leukocyte Count 1016
  •Peripheral Blood: Red Blood Cell Morphology 1178
  Schilling Test

HYPERSENSITIVITY PNEUMONITIS [495.9] *see also* ACQUIRED
  IMMUNODEFICIENCY SYNDROME (AIDS), ASPERGILLOSIS, ASTHMA,
  DYSPNEA, PNEUMONIA, PNEUMONITIS, TUBERCULOSIS 219
  Aspergillus Serology 275
  Blood Gases and pH, Arterial 311
  Bronchoalveolar Lavage (BAL) 442
  Complete Blood Count 542
  Eosinophil Count 624
  Fungal Culture, Sputum 733
  •Histopathology 761
  •Hypersensitivity Pneumonitis Serology 1010
  Peripheral Blood: Differential Leukocyte Count 1103
  Protein Electrophoresis, Capillary Zone 1104
  Protein Electrophoresis, Serum

HYPERSPLENISM [289.4] *see also* ALCOHOLISM, ALDOSTERONISM,
  AMYLOIDOSIS, ANEMIA (SPHEROCYTOSIS), CIRRHOSIS, GAUCHER
  DISEASE, LEUKEMIA, LYMPHOMA, SPLENOMEGALY, THALASSEMIA,
  WILSON DISEASE 442
  •Complete Blood Count 674
  •Hematocrit 681
  •Hemoglobin 1010
  Peripheral Blood: Differential Leukocyte Count

(Continued)

HYPERSPLENISM [289.4] *(Continued)*

HYPERTENSION [401.9] *see also* ACROMEGALY, ALDOSTERONISM, ATHEROSCLEROSIS, CARDIOMYOPATHY, CONGENITAL ADRENAL HYPERPLASIA, CUSHING SYNDROME, DIABETIC NEPHROPATHY, GLOMERULONEPHRITIS, HEMOLYTIC-UREMIC SYNDROME, HYDRONEPHROSIS, HYPERTHYROIDISM, HYPOTHYROIDISM, LEAD TOXICITY, NEPHRITIS, NEPHROSIS/NEPHROTIC SYNDROME, PHEOCHROMOCYTOMA, POLYCYSTIC DISEASE (KIDNEY), PORPHYRIA, PREGNANCY, PURPURA (HENOCH-SCHÖNLEIN), RENAL FAILURE, RENAL VEIN THROMBOSIS, SYSTEMIC LUPUS ERYTHEMATOSUS (SLE), TURNER SYNDROME, UREMIA, URINARY TRACT INFECTION, VASCULITIS, WILMS TUMOR

HYPERTENSION (PAROXYSMAL) *see* PHEOCHROMOCYTOMA

HYPERTENSION (PORTAL) [572.3] *see also* LIVER, SPLENOMEGALY

HYPERTHERMIA (MALIGNANT) [778.4]

HYPERTHYROIDISM [242.9; 775.3] *see also* CHORIOCARCINOMA, GOITER, HYDATIDIFORM MOLE, PITUITARY/PITUITARY ADENOMA, PROPTOSIS, THYROIDITIS

(Continued)

(Continued)

(Continued)

IMMUNODEFICIENCY (SEVERE COMBINED) [279.2] (Continued)

IMMUNODEFICIENCY (T-CELL) [279.10] see also LYMPHOCYTES/
LYMPHOPROLIFERATIVE DISORDERS

IMMUNOGLOBULIN

IMPOTENCE [607.84] see also DIABETES MELLITUS, HYPOGONADISM

INAPPROPRIATE ANTIDIURETIC HORMONE see SYNDROME OF
INAPPROPRIATE SECRETION OF ANTIDIURETIC HORMONE (SIADH)

INBORN ERRORS OF METABOLISM see also AMINOACIDURIAS,
CYSTINOSIS/CYSTINURIA, GALACTOSEMIA/GALACTOSURIA, GAUCHER
DISEASE, GLYCOGEN STORAGE DISEASE, HEMOCHROMATOSIS,
HYPERLIPIDEMIA, MENKE SYNDROME, PHENYLKETONURIA,
PORPHYRIA, TANGIER DISEASE, TAY-SACHS DISEASE, VITAMINS,
WILSON DISEASE

INDIAN CHILDHOOD CIRRHOSIS see CIRRHOSIS (JUVENILE)

INFANTILE HYPOTONIA see FLOPPY-INFANT SYNDROME

INFECTION (INTRAUTERINE) see also ACQUIRED IMMUNODEFICIENCY
SYNDROME (AIDS), HERPESVIRUS INFECTION, RUBELLA, SYPHILIS,
TOXOPLASMOSIS, VARICELLA-ZOSTER

Cytomegalovirus is the most common cause of intrauterine infections.

(Continued)

INFERTILITY (MALE) [606.9] *(Continued)*

INFLAMMATORY DISEASES AND STATES *see also* ACUTE PHASE REACTION

INFLAMMATORY MYOPATHIES *see* DERMATOMYOSITIS, MYOSITIS, POLYMYOSITIS

INFLUENZA [487.1; 760.2] *see also* BRONCHITIS, PNEUMONIA (VIRAL)

INSECTICIDE POISONING [989.4] *see also* TICK PARALYSIS

INSECTS *see* BITES (INSECT)

INSULINOMA (BETA-CELL TUMOR) [157.4; 211.7] *see also* GLUCAGONOMA (ALPHA-CELL TUMOR), HYPOGLYCEMIA, ISLET-CELL TUMORS

INSULIN RESISTANCE *see also* DIABETES MELLITUS, HYPERTENSION, OBESITY

INTERSEXUALITY [752.7] *see also* CONGENITAL ADRENAL HYPERPLASIA

(Continued)

JAUNDICE [782.4] *(Continued)*

JAUNDICE (HEMOLYTIC) *see* ANEMIA (HEMOLYTIC)

KALA-AZAR *see* LEISHMANIASIS

KALLMANN SYNDROME [253.4] *see also* HYPOGONADISM

KAPOSI SARCOMA [176.9] *see also* ACQUIRED IMMUNODEFICIENCY SYNDROME (AIDS), TRANSPLANTATION

KARYOTYPING *see* CYTOGENETICS

KAWASAKI DISEASE [446.1]

KERATITIS *see* CONJUNCTIVITIS

KERATOCONJUNCTIVITIS SICCA *see* CIRRHOSIS (PRIMARY BILIARY), SJÖGREN SYNDROME

KETOSIS/KETONURIA/KETOACIDOSIS [276.2] *see also* ACIDOSIS/ ALKALOSIS (ACID-BASE BALANCE), ALCOHOLISM, DIABETES MELLITUS, HYPOGLYCEMIA, STARVATION

Diabetic ketoacidosis includes hyperglycemia, metabolic acidosis, increased ketones, and increased anion gap. Other types of ketoacidosis include starvation and alcoholism/alcohol withdrawal.

LEAD TOXICITY [984.9] *see also* ANEMIA (HEMOLYTIC), NEUROPATHY, PORPHYRIA

LEGIONNAIRES' DISEASE [482.83] *see also* HYPONATREMIA, *MYCOPLASMA* INFECTION, PNEUMONIA, PSITTACOSIS, Q FEVER, SYNDROME OF INAPPROPRIATE SECRETION OF ANTIDIURETIC HORMONE (SIADH)

LEISHMANIASIS [085.9] *see also* PARASITIC INFESTATIONS

LEPROSY [030.9] *see also* AMYLOIDOSIS, MYCOBACTERIAL INFECTION (ATYPICAL), TUBERCULOSIS

LEPTOSPIROSIS (WEIL SYNDROME) [100.9] *see also* FEVER UNDETERMINED ORIGIN (FUO), JAUNDICE

LESCH-NYHAN SYNDROME [277.2]

(Continued)

LEUKEMIA (CHRONIC MYELOGENOUS/CHRONIC MYELOCYTIC) [205.10]
(Continued)

LEUKEMIA (CHRONIC NEUTROPHILIC) *see* LEUKEMIA (CHRONIC MYELOGENOUS/CHRONIC MYELOCYTIC)

LEUKEMOID REACTIONS [288.8] *see also* MONONUCLEOSIS (INFECTIOUS)

LEUKOPENIA [776.7; 288.0] *see also* AUTOIMMUNE DISEASES, GAUCHER DISEASE, SYSTEMIC LUPUS ERYTHEMATOSUS (SLE), SHWACHMAN-DIAMOND SYNDROME, THROMBOCYTOPENIA, UREMIA
Autoimmune diseases and a variety of drugs may lead to neutropenia.

LI-FRAUMENI CANCER SYNDROME [758.3]

LIPIDS *see* ATHEROSCLEROSIS, HYPERLIPIDEMIA, HYPERTRIGLYCERIDEMIA

LIPOSARCOMA (MYXOID)

LITHIUM POISONING [985.8]

LIVER *see also* ACETAMINOPHEN TOXICITY, ALCOHOLISM, AMYLOIDOSIS, ANEMIA (HEMOLYTIC), BILIARY OBSTRUCTION, CIRRHOSIS, CONGESTIVE HEART FAILURE, DIABETES MELLITUS,

ECHINOCOCCOSIS, ENCEPHALOPATHY (HEPATIC), GALACTOSEMIA/
GALACTOSURIA, GILBERT DISEASE, HELLP SYNDROME,
HEMOCHROMATOSIS, HEPATIC NECROSIS/FAILURE, HEPATITIS,
HEPATITIS (AUTOIMMUNE), JAUNDICE, LEPTOSPIROSIS (WEIL
SYNDROME), NONALCOHOLIC STEATOHEPATITIS, REYE SYNDROME,
STEATOSIS (LIVER), TUBERCULOSIS, WILSON DISEASE

See table in listing, Bilirubin, Total, Serum.

LIVER CARCINOMA see CARCINOMA (LIVER)

LOUIS-BAR SYNDROME see ATAXIA-TELANGIECTASIA

LOW MOLECULAR WEIGHT HEPARIN THERAPY

LUNG see also ALPHA$_1$-ANTITRYPSIN DEFICIENCY, BRONCHIECTASIS,
BRONCHITIS, CARCINOMA (LUNG), CYSTIC FIBROSIS, HYPERCAPNIA,
LANGERHANS' CELL HISTIOCYTOSIS, LEGIONNAIRES' DISEASE,
OBSTRUCTIVE LUNG DISEASE, PNEUMONIA, RESPIRATORY DISTRESS
SYNDROME (ADULT), RESPIRATORY FAILURE, TUBERCULOSIS

LUNG ABSCESS see ABSCESS (LUNG)

LUPUS ANTICOAGULANT [710.0] see also ANTICOAGULANT THERAPY,
SYSTEMIC LUPUS ERYTHEMATOSUS (SLE)

LUTEAL CELL DYSFUNCTION see also HYPOGONADISM,
HYPOPITUITARISM

LYME DISEASE [088.81] see also ARTHRITIS (INFECTIOUS/
INFLAMMATORY), EHRLICHIOSIS, MYOCARDITIS, RHEUMATIC FEVER,
TICK BITE

LYMPHADENOPATHY [785.6] see also ACQUIRED IMMUNODEFICIENCY
SYNDROME (AIDS), ACTINOMYCOSIS, AMYLOIDOSIS, ARTHRITIS
(RHEUMATOID), ASPERGILLOSIS, CANDIDIASIS, CAT SCRATCH
DISEASE, CHLAMYDIA INFECTION, COCCIDIOIDOMYCOSIS,
CRYPTOCOCCOSIS, CYTOMEGALOVIRUS, DIPHTHERIA, EPSTEIN-BARR
VIRUS, GAUCHER DISEASE, HEPATITIS, HERPESVIRUS INFECTION,
HISTOPLASMOSIS, HODGKIN DISEASE, HUMAN HERPESVIRUS-6
INFECTION, HYPERPLASIA (LYMPH NODE), HYPERTHYROIDISM,
LANGERHANS' CELL HISTIOCYTOSIS, LEPROSY, LEUKEMIA, LYME
DISEASE, LYMPHOGRANULOMA VENEREUM, LYMPHOMA,
MACROGLOBULINEMIA OF WALDENSTRÖM, MEASLES (RUBEOLA),
MONONUCLEOSIS (INFECTIOUS), MUMPS, MYCOBACTERIAL INFECTION
(ATYPICAL), NEUROBLASTOMA, NIEMANN-PICK DISEASE,
NOCARDIOSIS, PARASITIC INFESTATIONS, PERINAUD
OCULOGLANDULAR SYNDROME, RUBELLA, SARCOIDOSIS, SCROFULA,

(Continued)

LYMPHADENOPATHY [785.6] *(Continued)*

SERUM SICKNESS, SPOROTRICHOSIS, STREPTOCOCCAL INFECTION, SYPHILIS, SYSTEMIC LUPUS ERYTHEMATOSUS (SLE), TOXOPLASMOSIS, TUBERCULOSIS, TULAREMIA

Look first for local infection in the drainage area of an enlarged lymph node. Without local infection, a single prominent lymph node in a febrile subject is suggestive of *Francisella tularensis*, *Bartonella henselae*, and mycobacterial infection.

LYMPHANGITIS (NODULAR)

LYMPHOCYTES/LYMPHOPROLIFERATIVE DISORDERS [238.7] *see also*

IMMUNE STATUS, IMMUNODEFICIENCY (T-CELL), LEUKEMIA, LEUKEMIA (CHRONIC LYMPHOCYTIC), LYMPHADENOPATHY, LYMPHOMA, MACROGLOBULINEMIA OF WALDENSTRÖM, MYELOMA

(Continued)

LYMPHOMA [202.8] *(Continued)*

LYMPHOMA (PRIMARY EFFUSION) *see* PRIMARY EFFUSION LYMPHOMA

MACROCYTIC ANEMIA *see* ANEMIA (MACROCYTIC/MEGALOBLASTIC)

MACROGLOBULINEMIA (ESSENTIAL) [273.3] *see also* MACROGLOBULINEMIA OF WALDENSTRÖM

MACROGLOBULINEMIA OF WALDENSTRÖM [273.3] *see also* HYPERGAMMAGLOBULINEMIA, HYPERVISCOSITY, LYMPHADENOPATHY, LYMPHOMA, MONOCLONAL GAMMOPATHY, MYELOMA, SPLENOMEGALY

MACROGLOSSIA *see* AMYLOIDOSIS

MALABSORPTION/MALDIGESTIVE DISEASES [579.9] *see also* ALCOHOLISM, AMYLOIDOSIS, CELIAC DISEASE, CROHN DISEASE, CYSTIC FIBROSIS, DIARRHEA, MALNUTRITION, PARASITIC INFESTATIONS, VITAMINS, VOMITING

(Continued)

MALNUTRITION [263.9] *(Continued)*

MALTASE DEFICIENCY *see* POMPE DISEASE

MANIC-DEPRESSIVE PSYCHOSIS [296.80]

MAPLE SYRUP URINE DISEASE [270.3] *see also* ACIDOSIS/ALKALOSIS (ACID-BASE BALANCE), KETOSIS/KETONURIA/KETOACIDOSIS

MASTOCYTOSIS [757.33] *see also* ANGIOEDEMA, PEPTIC ULCER

MATERNAL ANTIBODIES [656.20] *see also* PREGNANCY, PRENATAL DIAGNOSIS

MEASLES (RUBEOLA) [055.9]

MECONIUM ILEUS [777.1] *see also* CYSTIC FIBROSIS

MEDULLARY CARCINOMA OF THYROID [193] *see also* CARCINOMA (THYROID), HYPERPARATHYROIDISM, MULTIPLE ENDOCRINE NEOPLASIA (MEN 1, MEN 2), PHEOCHROMOCYTOMA

(Continued)

MENTAL RETARDATION [319] *(Continued)*

MERCURY POISONING [985.0] *see also* HEAVY METAL/METAL POISONING, POISONING

MESENTERIC LYMPHADENITIS [289.2]

MESOTHELIOMA (PLEURA) [163.9] *see also* ASBESTOSIS, CARCINOMA (LUNG)

METABOLIC BONE DISEASE [277.8] *see also* BONE DISEASE, HYPERPARATHYROIDISM, OSTEOMALACIA, OSTEOPOROSIS, PAGET DISEASE OF BONE

METABOLIC SYNDROME *see also* NONALCOHOLIC STEATOHEPATITIS

METAL POISONING *see* HEAVY METAL/METAL POISONING

METASTASIS [199.1]

METASTASIS (LIVER) [197.7] see also METASTASIS

METHANOL POISONING [980.1] see also ALCOHOLISM, ALCOHOL INTOXICATION, ISOPROPANOL POISONING, POISONING

Severe, high anion gap metabolic acidosis is seen.

METHYLENE TETRAHYDROFOLATE (MTHFR) DEFICIENCY

MGUS see MONOCLONAL GAMMOPATHY OF UNKNOWN SIGNIFICANCE (MGUS)

MICROANGIOPATHIC HEMOLYTIC ANEMIA see ANEMIA (MICROANGIOPATHIC HEMOLYTIC)

MICROCYTIC ANEMIA see ANEMIA (MICROCYTIC)

MICROFILARIASIS [125.9] see also PARASITIC INFESTATIONS

MICROSPORIDIOSIS [110.9]

MILK-ALKALI SYNDROME [275.4]

MISCARRIAGE see ABORTION

MITES [133.8]

(Continued)

MITES [133.8] *(Continued)*
••Arthropod Identification 213

MITRAL INSUFFICIENCY/MITRAL STENOSIS [394.0; 424.0] *see also*
INFECTIVE ENDOCARDITIS
Antistreptolysin O Titer, Serum 197
Bacterial Culture, Blood 232

MIXED CONNECTIVE TISSUE DISEASE (MCTD) [710.9] *see also*
ARTHRITIS, ARTHRITIS (RHEUMATOID), AUTOIMMUNE DISEASES,
CREST SYNDROME ("LIMITED SCLERODERMA"), DERMATOMYOSITIS,
ESOPHAGEAL DYSFUNCTION, FASCIITIS (EOSINOPHILIC), MUSCLE
DISEASE, MYOSITIS, RAYNAUD PHENOMENON, SCLERODERMA,
SYSTEMIC LUPUS ERYTHEMATOSUS (SLE)
MCTD includes predominantly features of SLE, scleroderma, and myositis.
Arthritis, Raynaud phenomenon, esophageal dysfunction, as well as myositis
occur but kidney disease is infrequent. High concentrations of anti-nRNP
(U_1RNP) autoantibodies are characteristic.
Anti-DNA 173
Antinuclear Antibodies 189
Jo-1 Antibody 815
Protein Electrophoresis, Serum 1104
Rheumatoid Factor, Serum or Body Fluid 1161
Sedimentation Rate, Erythrocyte 1181
••Smith (Sm) and Ribonucleoprotein (RNP) Antibodies 1206

MOERSCH-WOLTMAN SYNDROME *see* STIFF-PERSON SYNDROME

MOLAR PREGNANCY [631]
•Chorionic Gonadotropin, Human, Serum and Urine 397
•Flow Cytometry, Overview 592
••Histopathology 733

MONILIASIS *see* CANDIDIASIS

MONOCLONAL GAMMOPATHY
A flowchart in the Immunoglobulin G monograph may be helpful.
Apheresis, Therapeutic 201
Complete Blood Count 442
Cryoglobulin, Qualitative, Serum and Plasma 478
Hemoglobin 681
••Immunofixation Electrophoresis, Serum or Urine 768
••Immunoglobulin A 770
•Immunoglobulin D 772
••Immunoglobulin G 775
•Immunoglobulin G Subclasses 778
••Immunoglobulin M 779
Lymph Node Biopsy 880
•Peripheral Blood: Differential Leukocyte Count 1010
Peripheral Blood: Red Blood Cell Morphology 1016
••Protein Electrophoresis, Capillary Zone 1103
••Protein Electrophoresis, Serum 1104
••Protein Electrophoresis, Urine 1107
Protein, Quantitative, Urine 1108
Protein, Semiquantitative, Urine 1113
••Protein, Total, Serum 1114
•Urinalysis 1289
Viscosity, Serum or Plasma 1313

MONOCLONAL GAMMOPATHY OF UNKNOWN SIGNIFICANCE (MGUS) [273.1]
••Bone Marrow 296
Complete Blood Count 442
••Immunofixation Electrophoresis, Serum or Urine 768
•Immunoglobulin A 770
•Immunoglobulin G 775
Immunoglobulin M 779
••Protein Electrophoresis, Capillary Zone 1103
••Protein Electrophoresis, Serum 1104
•Protein Electrophoresis, Urine 1107
Protein, Total, Serum 1114

MONONUCLEOSIS (INFECTIOUS) [075] *see also* CYTOMEGALOVIRUS, LEUKEMOID REACTIONS, LYMPHADENOPATHY, TOXOPLASMOSIS
Alanine Aminotransferase, Serum 116
Alkaline Phosphatase, Serum 127
Aspartate Aminotransferase, Serum 216
Bilirubin, Total, Serum 265
Cold Agglutinin Titer 430
••Complete Blood Count 442
Cytomegalovirus Culture 495

MORQUIO SYNDROME [277.5]

M PROTEIN, SERUM *see* MONOCLONAL GAMMOPATHY, MONOCLONAL GAMMOPATHY OF UNKNOWN SIGNIFICANCE (MGUS)

MUCOPOLYSACCHARIDOSIS *see* MORQUIO SYNDROME

MUCORMYCOSIS *see* PHYCOMYCOSIS

MULTIPLE ENDOCRINE NEOPLASIA (MEN 1, MEN 2) [237.4] *see also* CARCINOID (INTESTINAL TRACT), CUSHING SYNDROME, GASTRINOMA, HYPERPARATHYROIDISM, HYPERPROLACTINEMIA/CHIARI-FROMMEL SYNDROME, INSULINOMA (BETA-CELL TUMOR), ISLET-CELL TUMORS, MEDULLARY CARCINOMA OF THYROID, PHEOCHROMOCYTOMA, PITUITARY/PITUITARY ADENOMA, ZOLLINGER-ELLISON SYNDROME

MULTIPLE SCLEROSIS [340] *see also* ALZHEIMER DISEASE, ATAXIA, DEMENTIA, NEUROSYPHILIS

MUMPS [072.9]

MUSCLE DISEASE [359.9] see also DERMATOMYOSITIS, MIXED CONNECTIVE TISSUE DISEASE (MCTD), MUSCULAR DYSTROPHY, MYASTHENIA GRAVIS, MYOSITIS

MUSCULAR DYSTROPHY [359.1] see also DUCHENNE MUSCULAR DYSTROPHY, MUSCLE DISEASE, MYOSITIS

MYASTHENIA GRAVIS [358.0] see also BOTULISM, GUILLAIN-BARRÉ SYNDROME, MUSCULAR DYSTROPHY, THYMOMA, TICK PARALYSIS

MYCOBACTERIAL INFECTION (ATYPICAL) [031.9] see also ACQUIRED IMMUNODEFICIENCY SYNDROME (AIDS), CAT SCRATCH DISEASE, LEPROSY, TUBERCULOSIS, TULAREMIA

MYCOBACTERIUM AVIUM-INTRACELLULARE (MAI) see MYCOBACTERIAL INFECTION (ATYPICAL)

(Continued)

MYELOMA [203.0] *(Continued)*

MYELOMENINGOCELE [741.9]

MYELOPROLIFERATIVE DISEASES *see* LEUKEMIA, MYELOFIBROSIS

MYOCARDIAL INFARCT [410.9] *see also* ANGINA, CHEST PAIN, CORONARY ARTERIAL DISEASE, CORONARY ARTERIAL DISEASE (RISK), HYPERLIPIDEMIA, SYPHILIS, VASCULITIS

MYOCARDITIS [429.0] *see also* ACQUIRED IMMUNODEFICIENCY SYNDROME (AIDS), ADENOVIRUS INFECTION, ARTHRITIS (RHEUMATOID), BRUCELLOSIS, CARDIOMYOPATHY, CHAGAS DISEASE, CHEST PAIN, CHURG-STRAUSS SYNDROME, CONGESTIVE HEART FAILURE, COXSACKIE VIRUS INFECTION, CYTOMEGALOVIRUS, DIPHTHERIA, EPSTEIN-BARR VIRUS, FUNGI, HEART, HERPESVIRUS INFECTION, INFECTIVE ENDOCARDITIS, LEGIONNAIRES' DISEASE, LEPTOSPIROSIS (WEIL SYNDROME), LYME DISEASE, MEASLES (RUBEOLA), MUMPS, PERICARDIAL EFFUSION/PERICARDITIS, POLYMYOSITIS, RELAPSING FEVER, RHEUMATIC FEVER, RICKETTSIAL INFECTION/ROCKY MOUNTAIN SPOTTED FEVER, RUBELLA, SARCOIDOSIS, SCLERODERMA, STREPTOCOCCAL INFECTION, SYPHILIS, SYSTEMIC LUPUS ERYTHEMATOSUS (SLE), TOXOPLASMOSIS, TRICHINOSIS, TYPHUS, VARICELLA-ZOSTER, WEGENER GRANULOMATOSIS

(Continued)

NEONATAL CHOLESTASIS WORK-UP [774.6] *(Continued)*

NEPHRITIS [583.9] *see also* AUTOIMMUNE DISEASES, NEPHROSIS/ NEPHROTIC SYNDROME, PURPURA (HENOCH-SCHÖNLEIN), SYSTEMIC LUPUS ERYTHEMATOSUS (SLE), VASCULITIS, WEGENER GRANULOMATOSIS

NEPHROLITHIASIS *see* KIDNEY STONE

NEPHROSIS/NEPHROTIC SYNDROME [581.9] *see also* AMYLOIDOSIS, ASCITES, DIABETIC NEPHROPATHY, GLOMERULONEPHRITIS, HEMOLYTIC-UREMIC SYNDROME, HYPOPROTEINEMIA, PURPURA (HENOCH-SCHÖNLEIN), RENAL FAILURE, RENAL VEIN THROMBOSIS, STREPTOCOCCAL INFECTION, SYSTEMIC LUPUS ERYTHEMATOSUS (SLE), THROMBOSIS, VASCULITIS

Key classical features of nephrotic syndrome include edema, hypoalbuminemia, and hypercholesterolemia as well as severe proteinuria with oval fat bodies. Important sequelae include sodium retention, thromboembolism, and infection.

(Continued)

NEUROPATHY [355.9] *(Continued)*

NEUROSYPHILIS [094.9] *see also* ALZHEIMER DISEASE, DEMENTIA, MULTIPLE SCLEROSIS, SYPHILIS

NIEMANN-PICK DISEASE [272.7] *see also* INBORN ERRORS OF METABOLISM, SPLENOMEGALY

NIPPLE DISCHARGE [611.79]

NITROPRUSSIDE POISONING [972.6]

NOCARDIOSIS [039.9] *see also* ABSCESS (BRAIN), ACTINOMYCOSIS, ALCOHOLISM, ALVEOLAR PROTEINOSIS, EMPYEMA, LYMPHADENOPATHY, LYMPHOMA, PNEUMONIA, TRANSPLANTATION, TUBERCULOSIS

NONALCOHOLIC STEATOHEPATITIS *see also* ALCOHOLISM, DIABETES MELLITUS, HEPATITIS, LIVER, METABOLIC SYNDROME, STEATOSIS (LIVER)

AST and ALT are generally increased threefold to fivefold and alkaline phosphatase is normal to moderately increased. ALT equals or is greater than AST, while the reverse is true in instances of alcoholic hepatitis. Obesity, diabetes mellitus, and hyperlipidemia are commonly associated. Other entities (eg, Wilson disease) can present as steatohepatitis.

NONTROPICAL SPRUE *see* CELIAC DISEASE

NUTRITIONAL STATUS *see* DIARRHEA, MALABSORPTION/MALDIGESTIVE DISEASES, MALNUTRITION, VOMITING

OBESITY [278.00]

OBSTRUCTIVE LUNG DISEASE [496] *see also* ALPHA₁-ANTITRYPSIN DEFICIENCY, BRONCHIECTASIS, BRONCHITIS, CHRONIC OBSTRUCTIVE PULMONARY DISEASE, CYSTIC FIBROSIS, HYPERCAPNIA

OCCIPITAL HORN SYNDROME

OLIGOSPERMIA *see* AZOOSPERMIA/OLIGOSPERMIA

OLIGURIA [788.5] *see also* DEHYDRATION/HYPOVOLEMIA, GLOMERULONEPHRITIS, LEPTOSPIROSIS (WEIL SYNDROME), POISONING, RENAL FAILURE, SHOCK, SYNDROME OF INAPPROPRIATE SECRETION OF ANTIDIURETIC HORMONE (SIADH), UREMIA, VASCULITIS

(Continued)

OLIGURIA [788.5] *(Continued)*

OOPHORITIS [614.2] *see also* PELVIC INFECTION/PELVIC INFLAMMATORY DISEASE (PID), SEXUALLY TRANSMITTED DISEASES

OPISTHORCHIASIS [121.0] *see also* PARASITIC INFESTATIONS

ORCHITIS [604.90] *see also* BRUCELLOSIS, GONORRHEA, LEPTOSPIROSIS (WEIL SYNDROME), LYMPHOCYTIC CHORIOMENINGITIS, MUMPS, PLEURODYNIA, PROSTATITIS, TUBERCULOSIS, VARICELLA-ZOSTER

OSTEITIS DEFORMANS *see* PAGET DISEASE OF BONE

OSTEITIS FIBROSA [733.29]

Osteitis fibrosa is the most common type of renal osteodystrophy.

OSTEOMALACIA [268.2] *see also* BONE DISEASE, HYPERPARATHYROIDISM, HYPOPHOSPHATASIA, METABOLIC BONE DISEASE, OSTEOPOROSIS, UREMIA

OSTEOMYELITIS [730.2] *see also* ASPERGILLOSIS, BONE DISEASE, BRUCELLOSIS, CANDIDIASIS, LEUKEMIA, NEUROBLASTOMA, Q FEVER, STREPTOCOCCAL INFECTION, TUBERCULOSIS

(Continued)

OVARIAN FUNCTION TESTS *(Continued)*

OXALURIA [592.9] *see also* CROHN DISEASE, ETHYLENE GLYCOL POISONING, KIDNEY STONE

PAGET DISEASE OF BONE [731.0] *see also* BONE DISEASE, METABOLIC BONE DISEASE

Isolated, marked increase of serum alkaline phosphatase, sometimes to very high levels.

PANCREATIC EXOCRINE FUNCTION TESTS/INSUFFICIENCY [577.8] *see also* ALCOHOLISM, CYSTIC FIBROSIS, MALABSORPTION/MALDIGESTIVE DISEASES, PANCREATITIS

PANCREATITIS [577.0] *see also* ALCOHOLISM, CARCINOMA (PANCREAS), CHOLECYSTITIS/CHOLEDOCHOLITHIASIS/CHOLELITHIASIS, CYSTIC FIBROSIS, CYST (PANCREATIC), DIABETES MELLITUS, HEMOLYTIC-UREMIC SYNDROME, HYPERGLYCEMIA, HYPERPARATHYROIDISM, HYPERTRIGLYCERIDEMIA, MALNUTRITION, OLIGURIA, PEPTIC ULCER

Increased ALT and alkaline phosphatase support the diagnosis of gallstone-associated pancreatitis, when the diagnosis of pancreatitis is established.

(Continued)

PARASITIC INFESTATIONS [136.9] *(Continued)*

PARATYPHOID INFECTION [002.9]

PARENTERAL NUTRITION (TOTAL)

PAROTITIS *see* ACTINOMYCOSIS, AMYLOIDOSIS, CAT SCRATCH DISEASE, HEAVY METAL/METAL POISONING, LEAD TOXICITY, MERCURY POISONING, MUMPS, SARCOIDOSIS, SJÖGREN SYNDROME, SYSTEMIC LUPUS ERYTHEMATOSUS (SLE), TOXOPLASMOSIS

PAROXYSMAL NOCTURNAL HEMOGLOBINURIA *see* HEMOGLOBINURIA (PAROXYSMAL NOCTURNAL) (PNH)

PARVOVIRUS B19 INFECTION

PASTURELLA MULTOCIDA [027.2]

PELVIC INFECTION/PELVIC INFLAMMATORY DISEASE (PID) [614.9] *see also* GONORRHEA, SEXUALLY TRANSMITTED DISEASES, SYPHILIS

PEMPHIGOID *see* BULLOUS SKIN DISEASE/PEMPHIGOID/PEMPHIGUS

PEMPHIGUS *see* BULLOUS SKIN DISEASE/PEMPHIGOID/PEMPHIGUS

PEPTIC ULCER [533.9] *see also* ANEMIA (IRON DEFICIENCY), BLEEDING (GASTROINTESTINAL), GASTRINOMA, GASTRITIS, G-CELL HYPERPLASIA, LIVER, MASTOCYTOSIS, PANCREATITIS, UREMIA, ZOLLINGER-ELLISON SYNDROME

PERICARDIAL EFFUSION/PERICARDITIS [423.9] *(Continued)*

PERICARDITIS (CONSTRICTIVE) [423.2] *see also* ARTHRITIS (RHEUMATOID), CARCINOMA (BREAST), CARCINOMA (LUNG), HISTOPLASMOSIS, LYMPHOMA, UREMIA

Other causes of constrictive pericarditis include prior surgery, trauma, and radiation therapy.

PERINAUD OCULOGLANDULAR SYNDROME *see also* CAT SCRATCH DISEASE, TULAREMIA

PERIODIC PARALYSIS, HYPERKALEMIC

An autosomal dominant, sudden increases of potassium cause muscular paralysis.

PERITONITIS [567.9] *see also* ASCITES, CIRRHOSIS, EFFUSIONS/ TRANSUDATES/EXUDATES, ENTEROCOLITIS (NECROTIZING), PERFORATED VISCUS, SEPTICEMIA, TUBERCULOSIS

PERNICIOUS ANEMIA *see* ANEMIA (MACROCYTIC/MEGALOBLASTIC)

PERTUSSIS [033.9]

PESTICIDES POISONING [989.4]

PETECHIA [782.7] *see also* BLEEDING, INFECTION (INTRAUTERINE), INFECTIVE ENDOCARDITIS, PLATELET FUNCTION TESTS, PURPURA, PURPURA (HENOCH-SCHÖNLEIN), VASCULITIS

(Continued)

PNEUMONIA [486] *(Continued)*

PNEUMONIA (*CANDIDA*) [112.4] *see also* FUNGI

PNEUMONIA (*PNEUMOCYSTIS CARINII*) *see* PNEUMOCYSTIS CARINII

PNEUMONIA (PRIMARY ATYPICAL) [486] *see also* CHLAMYDIA INFECTION, LEGIONNAIRES' DISEASE, *MYCOPLASMA* INFECTION, PNEUMONIA, RICKETTSIAL INFECTION/ROCKY MOUNTAIN SPOTTED FEVER

PNEUMONIA (VIRAL) [480.9] *see also* ADENOVIRUS INFECTION, CYTOMEGALOVIRUS, ENTEROVIRUS INFECTION, EPSTEIN-BARR VIRUS, HERPESVIRUS INFECTION, INFLUENZA, MEASLES (RUBEOLA), PNEUMONIA, RESPIRATORY SYNCYTIAL VIRUS INFECTION, RUBELLA, VARICELLA-ZOSTER

POISONING [988.8] *(Continued)*

POLIOMYELITIS [045.9] see also BOTULISM, ENTEROVIRUS INFECTION, TICK PARALYSIS, WEST NILE VIRUS INFECTION

POLYCYSTIC DISEASE (KIDNEY) [753.12]

POLYCYSTIC OVARY SYNDROME [256.4] see also ADENOMA (ADRENAL), CARCINOMA (ADRENAL), CONGENITAL ADRENAL HYPERPLASIA, CUSHING SYNDROME, HIRSUTISM, OVARIAN FUNCTION TESTS, VIRILIZATION

Manifestations include hirsutism, anovulation, oligomenorrhea or amenorrhea, often with obesity and not infrequently with diabetes mellitus.

POLYCYTHEMIA/POLYCYTHEMIA VERA [238.4] see also HYPERVISCOSITY, SPLENOMEGALY

(Continued)

POLYCYTHEMIA/POLYCYTHEMIA VERA [238.4] *(Continued)*

POLYDIPSIA [783.5] *see also* DIABETES INSIPIDUS, DIABETES MELLITUS, HYPONATREMIA, PHEOCHROMOCYTOMA, POLYURIA

POLYMYALGIA RHEUMATICA [725] *see also* ARTERITIS (TEMPORAL), MIXED CONNECTIVE TISSUE DISEASE (MCTD), MUSCLE DISEASE, SCLERODERMA, SYSTEMIC LUPUS ERYTHEMATOSUS (SLE)

POLYMYOSITIS [710.4] *see also* ARTHRITIS (RHEUMATOID), AUTOIMMUNE DISEASES, DERMATOMYOSITIS, FASCIITIS (EOSINOPHILIC), MIXED CONNECTIVE TISSUE DISEASE (MCTD), MUSCLE DISEASE, MUSCULAR DYSTROPHY, MYASTHENIA GRAVIS, MYOCARDITIS, MYOSITIS, SARCOIDOSIS, SCLERODERMA, SYSTEMIC LUPUS ERYTHEMATOSUS (SLE), VASCULITIS

POLYNEURITIS [356.9] *see also* GUILLAIN-BARRÉ SYNDROME

(Continued)

PRENATAL DIAGNOSIS [V28.8] *(Continued)*

PRIMARY BILIARY CIRRHOSIS *see* CIRRHOSIS (PRIMARY BILIARY)

PRIMARY EFFUSION LYMPHOMA *see also* EFFUSIONS/TRANSUDATES/ EXUDATES, KAPOSI SARCOMA

Human herpesvirus 8 is found in body-cavity-based B-cell lymphoma, primary effusion lymphoma. It is found as well in 70% to 90% of patients with Kaposi sarcoma.

PRIMARY SCLEROSING CHOLANGITIS *see* CHOLANGITIS (PRIMARY SCLEROSING)

PRIONS

PROCTITIS [569.49] *see also* COLITIS, COLITIS (ULCERATIVE), CROHN DISEASE

PROLACTINOMA *see also* HYPERPROLACTINEMIA/CHIARI-FROMMEL SYNDROME

PROPTOSIS [376.30] *see also* HYPERTHYROIDISM

PROSTATE *see* CARCINOMA (PROSTATE), PROSTATITIS

PROSTATITIS [601.9] *see also* CARCINOMA (PROSTATE), GONORRHEA, ORCHITIS

PSORIASIS [696.1] (Continued)

PURPURA (HENOCH-SCHÖNLEIN) [287.0] *see also* ARTHRITIS, BLEEDING (GASTROINTESTINAL), GLOMERULONEPHRITIS, NEPHRITIS, PETECHIA, PURPURA, VASCULITIS

The classic triad of arthritis, palpable purpura, and crampy abdominal pain may not always be found. Purpura, nephritis, and abdominal pain can be caused by Henoch-Schönlein purpura or by microscopic polyangiitis.

PYLORIC STENOSIS [537.0] *see also* BARTTER SYNDROME, DEHYDRATION/HYPOVOLEMIA, HYPOKALEMIA, TRISOMY, TURNER SYNDROME, VOMITING

PYRUVATE DEHYDROGENASE COMPLEX DEFICIENCY

Q FEVER [083.0] *see also* FEVER UNDETERMINED ORIGIN (FUO), LEGIONNAIRES' DISEASE, RICKETTSIAL INFECTION/ROCKY MOUNTAIN SPOTTED FEVER, TICK BITE

RABIES [071]

RAPE (ALLEGED) [V71.5]

RASH [782.1] *see also* DERMATITIS, LYME DISEASE, MEASLES (RUBEOLA), PURPURA, PURPURA (HENOCH-SCHÖNLEIN), RUBELLA, SCARLET FEVER

(Continued)

RASH [782.1] *(Continued)*

RAYNAUD PHENOMENON [443.0] *see also* ARTHRITIS, ARTHRITIS (RHEUMATOID), CIRRHOSIS (PRIMARY BILIARY), CREST SYNDROME ("LIMITED SCLERODERMA"), DERMATOMYOSITIS, ESOPHAGEAL DYSFUNCTION, LYMPHOMA, MACROGLOBULINEMIA OF WALDENSTRÖM, MIXED CONNECTIVE TISSUE DISEASE (MCTD), MYELOMA, SCLERODERMA, SYSTEMIC LUPUS ERYTHEMATOSUS (SLE)

Symptoms and signs of Raynaud type can also appear with carcinoma.

RECTAL BLEEDING [569.3] *see also* CARCINOMA (COLORECTAL), COLITIS, COLITIS (ULCERATIVE)

5-α REDUCTASE DEFICIENCY

RELAPSING FEVER [087.9]

Spirochetes may be seen in stained peripheral blood smears. Thrombocytopenia is common in louse-borne disease.

RENAL FAILURE [586] *see also* AMINOGLYCOSIDE TOXICITY, AMYLOIDOSIS, ATHEROSCLEROSIS, CYSTINOSIS/CYSTINURIA, DIABETES MELLITUS, ETHYLENE GLYCOL POISONING, GLOMERULONEPHRITIS, HEMOLYTIC-UREMIC SYNDROME, HYPERTENSION, KIDNEY STONE, MYELOMA, NEPHROSIS/NEPHROTIC SYNDROME, NEPHROTOXICITY, POISONING, POLYCYSTIC DISEASE (KIDNEY), RENAL VEIN THROMBOSIS, SCLERODERMA, SHOCK, SYSTEMIC LUPUS ERYTHEMATOSUS (SLE), TOLUENE POISONING, UREMIA, VASCULITIS, WEGENER GRANULOMATOSIS

RENAL INFARCT [593.81]

(Continued)

(Continued)

SEIZURES [780.3] *(Continued)*

SEMINOMA [186.9] *see also* CARCINOMA (TESTIS)

The ovarian equivalent is dysgerminoma. AFP and hCG are usually within normal limits, but are needed for differential diagnosis when only small biopsies are available.

SEPTICEMIA [038.9] *see also* ABSCESS (BRAIN), ABSCESS (LUNG), APPENDICITIS, BACTEREMIA, ENCEPHALITIS, FEVER UNDETERMINED ORIGIN (FUO), *HAEMOPHILUS INFLUENZAE* INFECTION, INFECTIVE ENDOCARDITIS, JAUNDICE, LEUKEMIA, LYMPHOMA, MENINGITIS, PERITONITIS, PNEUMONIA, REYE SYNDROME, STREPTOCOCCAL INFECTION

SERUM SICKNESS [999.5] *see also* ARTHRITIS, GLOMERULONEPHRITIS, IMMUNE COMPLEX DISEASE, SYSTEMIC LUPUS ERYTHEMATOSUS (SLE)

SEXUALLY TRANSMITTED DISEASES [099.9] *see also* ACQUIRED IMMUNODEFICIENCY SYNDROME (AIDS), *CHLAMYDIA* INFECTION, GONORRHEA, HEPATITIS, HERPESVIRUS INFECTION (GENITAL), LYMPHOGRANULOMA VENEREUM, NEUROSYPHILIS, SYPHILIS, UROGENITAL INFECTIONS

SHOCK [785.50] *see also* ADDISON DISEASE, DISSEMINATED INTRAVASCULAR COAGULATION (DIC), *HAEMOPHILUS INFLUENZAE* INFECTION, HYPOTENSION, MYOCARDIAL INFARCT, OLIGURIA, PHEOCHROMOCYTOMA, RENAL FAILURE, RENAL VEIN THROMBOSIS, SEPTICEMIA

(Continued)

SHOCK [785.50] *(Continued)*

SHWACHMAN-DIAMOND SYNDROME [288.0] *see also* ANEMIA, CYSTIC FIBROSIS, DIARRHEA, MALABSORPTION/MALDIGESTIVE DISEASES, PANCREATIC EXOCRINE FUNCTION TESTS/INSUFFICIENCY, THROMBOCYTOPENIA

SICKLE CELL ANEMIA *see* ANEMIA (SICKLE CELL AND VARIANT SICKLING HEMOGLOBINS)

SICKLING HEMOGLOBINS *see* ANEMIA (SICKLE CELL AND VARIANT SICKLING HEMOGLOBINS)

SIDEROBLASTIC ANEMIA *see* ANEMIA (SIDEROBLASTIC)

SIDEROSIS (TRANSFUSIONAL) *see also* HEMOCHROMATOSIS, HEMOSIDEROSIS

SINUSITIS [473.9]

SJÖGREN SYNDROME [710.2] *see also* ARTHRITIS (RHEUMATOID), AUTOIMMUNE DISEASES, CIRRHOSIS (PRIMARY BILIARY), LYMPHOMA, THYROIDITIS

SKIN LESION/SKIN ULCER [709.9] *see also* BULLOUS SKIN DISEASE/PEMPHIGOID/PEMPHIGUS, LANGERHANS' CELL HISTIOCYTOSIS, PSORIASIS, SYSTEMIC LUPUS ERYTHEMATOSUS (SLE)

SMALLPOX

SMOKE INHALATION *see* CARBON MONOXIDE POISONING, FIRE EXPOSURE

SOFT TISSUE SARCOMA *see* SARCOMA (SOFT TISSUE)
SPERM AGGLUTINATING ANTIBODIES *see* INFERTILITY (FEMALE),
    INFERTILITY (MALE)
SPHINGOLIPIDOSIS [272.7]

SPINA BIFIDA [741.9]

SPLENOMEGALY [789.2] *see also* ACQUIRED IMMUNODEFICIENCY
    SYNDROME (AIDS), AMYLOIDOSIS, ANEMIA (HEMOLYTIC), ANEMIA
    (SICKLE CELL AND VARIANT SICKLING HEMOGLOBINS), ANEMIA
    (SPHEROCYTOSIS), ARTHRITIS (RHEUMATOID), BRUCELLOSIS,
    CHAGAS DISEASE, CIRRHOSIS, CIRRHOSIS (PRIMARY BILIARY),
    CONGESTIVE HEART FAILURE, CYTOMEGALOVIRUS, GAUCHER
    DISEASE, HEMOLYTIC DISEASE OF THE NEWBORN, HEPATITIS,
    HEPATITIS (AUTOIMMUNE), HISTOPLASMOSIS, HODGKIN DISEASE,
    HYPERSPLENISM, INFECTIVE ENDOCARDITIS, LEISHMANIASIS,
    LEPTOSPIROSIS (WEIL SYNDROME), LEUKEMIA, LYMPHOMA,
    MACROGLOBULINEMIA OF WALDENSTRÖM, MALARIA,
    MONONUCLEOSIS (INFECTIOUS), MYELOFIBROSIS, NEUROBLASTOMA,
    NIEMANN-PICK DISEASE, POLYCYTHEMIA/POLYCYTHEMIA VERA,
    PORPHYRIA (CONGENITAL ERYTHROPOIETIC), PSITTACOSIS,
    SARCOIDOSIS, SCHISTOSOMIASIS, SEPTICEMIA, SYSTEMIC LUPUS
    ERYTHEMATOSUS (SLE), TANGIER DISEASE, THALASSEMIA,
    TOXOPLASMOSIS, TUBERCULOSIS, TYPHOID FEVER, UREMIA, WILSON
    DISEASE
See table in listing, Bilirubin, Total, Serum.

SPONDYLITIS (ANKYLOSING) [720.0] *see also* ARTHRITIS, ARTHRITIS
    (RHEUMATOID)

SPOROTRICHOSIS [117.1] *see also* FUNGI

SPRUE *see* CELIAC DISEASE, DIARRHEA, MALABSORPTION/MALDIGESTIVE
    DISEASES
STAPHYLOCOCCAL INFECTION [041.1] *see also* BRONCHITIS, EMPYEMA

(Continued)

STAPHYLOCOCCAL INFECTION [041.1] *(Continued)*

STARVATION [994.2] *see also* DIARRHEA, KETOSIS/KETONURIA/ KETOACIDOSIS, MALABSORPTION/MALDIGESTIVE DISEASES

STEATORRHEA *see* CIRRHOSIS (PRIMARY BILIARY), CROHN DISEASE, DIARRHEA, MALABSORPTION/MALDIGESTIVE DISEASES

STEATOSIS (LIVER) [571.8] *see also* ALCOHOLISM, DIABETES MELLITUS, LIVER, NONALCOHOLIC STEATOHEPATITIS

Steatosis is found with diabetes mellitus, ethanol abuse, and with obesity. Elevations of ALT and triglycerides are the most reliable markers. Serologic evaluation is recommended to rule out other entities.

STEIN-LEVENTHAL SYNDROME *see* POLYCYSTIC OVARY SYNDROME

STERILITY *see* INFERTILITY (FEMALE), INFERTILITY (MALE)

STIFF-PERSON SYNDROME [333.91]

STOMATITIS [528.0] *see also* PHARYNGITIS

STORAGE POOL DISEASE

STREPTOCOCCAL INFECTION [041.0] *see also* ARTHRITIS (INFECTIOUS/ INFLAMMATORY), BRONCHITIS, GLOMERULONEPHRITIS, INFECTIVE ENDOCARDITIS, MENINGITIS, OSTEOMYELITIS, PHARYNGITIS, PNEUMONIA, RHEUMATIC FEVER, URINARY TRACT INFECTION

STROKE *see* CEREBRAL INFARCTION/CEREBRAL THROMBOSIS/ CEREBROVASCULAR ACCIDENT (CVA)

STRONGYLOIDIASIS *see* DIARRHEA, PARASITIC INFESTATIONS

STUART-PROWER FACTOR *see* FACTOR X DEFICIENCY

SYNDROME X

SYNOVITIS *see* ARTHRITIS

SYPHILIS [097.9] *see also* ACQUIRED IMMUNODEFICIENCY SYNDROME (AIDS), GONORRHEA, HERPESVIRUS INFECTION (GENITAL), NEUROSYPHILIS, SEXUALLY TRANSMITTED DISEASES

SYSTEMIC LUPUS ERYTHEMATOSUS (SLE) [710.0] *see also* ANEMIA (HEMOLYTIC), ANTIPHOSPHOLIPID ANTIBODY SYNDROME, ARTHRITIS, ARTHRITIS (RHEUMATOID), AUTOIMMUNE DISEASES, CREST SYNDROME ("LIMITED SCLERODERMA"), DERMATOMYOSITIS, DISCOID LUPUS ERYTHEMATOSUS (DLE), LYMPHADENOPATHY, MIXED CONNECTIVE TISSUE DISEASE (MCTD), MYOSITIS, NEPHROSIS/ NEPHROTIC SYNDROME, RAYNAUD PHENOMENON, SCLERODERMA, SJÖGREN SYNDROME, SKIN LESION/SKIN ULCER, SPLENOMEGALY, VASCULITIS

(Continued)

TETANY [781.7] *(Continued)*

THALASSEMIA [282.4] *see also* ANEMIA, ANEMIA (HEMOLYTIC), HEMOGLOBINOPATHY

THROAT *see* PHARYNGITIS, RHEUMATIC FEVER, STREPTOCOCCAL INFECTION

THROMBASTHENIA *see* GLANZMANN DISEASE

THROMBOCYTHEMIA/THROMBOCYTOSIS [289.9; 238.7] *see also* AMYLOIDOSIS, LEUKEMIA, MYELOFIBROSIS, POLYCYTHEMIA/ POLYCYTHEMIA VERA, THROMBOCYTHEMIA/THROMBOCYTOSIS

THROMBOCYTOPATHY *see* BERNARD-SOULIER SYNDROME

THROMBOCYTOPENIA [287.5] *see also* BRUISING, CYTOMEGALOVIRUS, ECCHYMOSIS (SPONTANEOUS), EHRLICHIOSIS, GAUCHER DISEASE, GESTATIONAL THROMBOCYTOPENIA, HELLP SYNDROME, HEMOLYTIC-UREMIC SYNDROME, IDIOPATHIC THROMBOCYTOPENIC PURPURA (ITP), LEUKEMIA, MYELOFIBROSIS, PLATELET FUNCTION TESTS, PURPURA, SYSTEMIC LUPUS ERYTHEMATOSUS (SLE), TYPHOID FEVER, WILSON DISEASE, WISKOTT-ALDRICH SYNDROME

(Continued)

THROMBOSIS [453.9] *(Continued)*

THROMBOSIS (VENOUS) [453.8]

THROMBOTIC THROMBOCYTOPENIC PURPURA (TTP) [446.6] *see also* DISSEMINATED INTRAVASCULAR COAGULATION (DIC), HELLP SYNDROME, HEMOLYTIC-UREMIC SYNDROME, RENAL FAILURE

Predominantly affecting women, TTP is characterized by fever, renal failure with thrombocytopenia, and microangiopathic hemolytic anemia and CNS manifestations. See the Anemia Flowchart in the introduction to the Hematology chapter.

THYMOMA [212.6] *see also* HYPOGAMMAGLOBULINEMIA, MYASTHENIA GRAVIS

THYROIDITIS [245.9] *see also* HYPOTHYROIDISM, SJÖGREN SYNDROME

TRACHOMA [076.9] *(Continued)*

TRANSFUSION REACTION [999.6]

TRANSFUSION REACTION (FEBRILE) [999.8]

TRANSIENT ISCHEMIC ATTACK (TIA) [435.9]

TRANSMISSIBLE SPONGIFORM ENCEPHALOPATHIES

TRANSPLANTATION *see also* CANDIDIASIS, CRYPTOCOCCOSIS, CYTOMEGALOVIRUS, GRAFT-VERSUS-HOST DISEASE, LYMPHOMA, LYMPHOCYTES/LYMPHOPROLIFERATIVE DISORDERS, TOXOPLASMOSIS, TUBERCULOSIS

TRANSUDATES *see* EFFUSIONS/TRANSUDATES/EXUDATES

(Continued)

TUBERCULOSIS [011.9] *(Continued)*

TUBO-OVARIAN ABSCESS [614.2] *see also* GONORRHEA

TULAREMIA [021.9] *see also* CAT SCRATCH DISEASE, LYMPHOGRANULOMA VENEREUM, MYCOBACTERIAL INFECTION (ATYPICAL), SCROFULA, SEPTICEMIA, TICK BITE, TUBERCULOSIS

Tularemia is usually spread by tick bite or wild animal bite or scratch and is a cause of fever with a single prominent lymph node.

TUMOR PROGNOSIS/GRADE/STAGE

TURNER SYNDROME [758.6] *see also* AMENORRHEA, GROWTH HORMONE DEFICIENCY, HYPOGONADISM

TYPHOID FEVER [002.0] *see also* ENTERIC FEVER, SALMONELLOSIS

A bacterial disease caused by *Salmonella typhi*, typhoid is characterized by fever, abdominal pain, diarrhea, rose spots, splenomegaly, and prostration. Culture of blood and stool are indicated. Bone marrow culture is the most sensitive test. Leukopenia with increased bands, often decreased platelets, and anemia occur and DIC may develop.

TYPHUS [081.9]

TYROSINEMIA/TYROSINURIA [270.2] *see also* AMINOACIDURIAS

Viral Culture 1307
•Virus, Direct Detection by Fluorescent Antibody 1311

VIPOMA
•Pancreatic Polypeptide, Human, Serum or Plasma 998
••Vasoactive Intestinal Polypeptide, Plasma 1302

VIRAL INFECTION *see* ACQUIRED IMMUNODEFICIENCY SYNDROME
(AIDS), BITES (INSECT), COXSACKIE VIRUS INFECTION,
CYTOMEGALOVIRUS, DIARRHEA, HEPATITIS, HERPESVIRUS
INFECTION, MEASLES (RUBEOLA), MONONUCLEOSIS (INFECTIOUS),
MYOCARDITIS, PNEUMONIA, POLIOMYELITIS, SEXUALLY TRANSMITTED
DISEASES

VIRILIZATION [255.2] *see also* ADENOMA (ADRENAL), AMENORRHEA,
CARCINOMA (ADRENAL), CONGENITAL ADRENAL HYPERPLASIA,
CUSHING SYNDROME, HIRSUTISM, HYPOGONADISM, POLYCYSTIC
OVARY SYNDROME
Androstenedione, Serum. 158
•Cortisol, Free, Urine 459
Cortisol, Serum or Plasma 460
•Dehydroepiandrosterone and Dehydroepiandrosterone Sulfate, Serum or
Plasma. 506
Estradiol, Serum 553
Follicle Stimulating Hormone, Serum, Plasma, or Urine 609
Histopathology 733
17-Hydroxycorticosteroids, Urine 753
17-Hydroxyprogesterone, Whole Blood, Serum, or Plasma 755
Luteinizing Hormone, Blood or Urine 876
Prolactin, Serum 1094
•Testosterone, Total and Free, Serum or Plasma 1238

VISCOSITY *see* HYPERVISCOSITY
VITAMIN A DEFICIENCY [264.9] *see also* CIRRHOSIS (PRIMARY BILIARY),
CYSTIC FIBROSIS, MALABSORPTION/MALDIGESTIVE DISEASES
Vitamin A, Serum or Plasma 1314

VITAMIN B₆ DEFICIENCY [266.1]
Homocyst(e)ine, Plasma 741

VITAMIN B₁₂ DEFICIENCY *see* ANEMIA (MACROCYTIC/MEGALOBLASTIC)
VITAMIN D INTOXICATION *see* HYPERVITAMINOSIS D
VITAMIN K DEFICIENCY [269.0; 776.0]
Activated Partial Thromboplastin Time. 100
Coagulation Factor Assays 418
Complete Blood Count 442
Osteocalcin, Serum or Plasma 983
Osteocalcin (Undercarboxylated), Serum 984
Phenothiazines, Serum 1021
Phenytoin, Serum or Plasma. 1026
Plasma, Fresh Frozen. 1039
Plasma, Frozen, Donor Retested 1041
Primidone, Serum or Plasma 1091
••Prothrombin Time 1116

VITAMINS *see also* ALCOHOLISM, MALNUTRITION
Activated Partial Thromboplastin Time. 100
Antioxidant Concentrations, Plasma 192
•Ascorbic Acid, Serum or Plasma 215
Calcium, Serum 329
Cobalamin, Serum 424
Fibrinogen 583
Iron and Total Iron Binding Capacity/Transferrin, Serum 807
Methylmalonic Acid, Serum, Plasma, Urine, or Amniotic Fluid 908
Phosphorus, Urine 1032
Prothrombin Time 1116
Vitamin A, Serum or Plasma 1314
Vitamin B₆, Plasma or Serum 1315
Vitamin D, Serum 1318
Vitamin E, Serum or Plasma 1319

VOLUME EXPANSION *see* HYDRATION/OVERHYDRATION

VOMITING [787.03] *see also* ACIDOSIS/ALKALOSIS (ACID-BASE BALANCE),
BARTTER SYNDROME, HEPATITIS, HYPONATREMIA, INBORN ERRORS
OF METABOLISM, LEAD TOXICITY, MENINGITIS, PYLORIC STENOSIS,
RENAL FAILURE
Arsenic, Blood 209
Arsenic, Urine. 211
Bicarbonate, Blood 258
•Chloride, Serum, Plasma, or Blood 391
Creatinine, Serum or Plasma 474

(Continued)

VOMITING [787.03] *(Continued)*
- •• Electrolyte Panel, Serum 532
  - Ketone Bodies, Blood 816
  - Ketones, Urine 817
  - Osmolality, Serum 978
- • pH, Blood 1018
  - Phosphorus, Serum 1031
  - Porphobilinogen, Urine 1073
- • Potassium, Serum or Plasma 1078
- • Potassium, Urine 1080
  - Protein, Total, Serum 1114
  - Pseudocholinesterase, Serum 1122
  - Sodium, Serum or Plasma 1210
  - Sodium, Urine 1213
  - Specific Gravity, Urine 1216
  - Urea Nitrogen:Creatinine Ratio 1283
  - Urea Nitrogen, Serum or Plasma 1284
  - Urinalysis 1289

VON WILLEBRAND DISEASE [286.4] *see also* FACTOR VIII DEFICIENCY, HEMOPHILIA
- Activated Clotting Time 98
- Activated Partial Thromboplastin Time 100
- Antiplasmin 196
- Bleeding Time 270
- Coagulation Factor Assays 418
- Complete Blood Count 442
- Cryoprecipitate 481
- Factor VIII Concentrate 563
- • Platelet Aggregation 1045
- Prothrombin Time 1116
- •• von Willebrand Factor 1321

VULVOVAGINITIS [616.10] *see also* CANDIDIASIS, GONORRHEA, SEXUALLY TRANSMITTED DISEASES, *TRICHOMONAS* INFESTATION, UROGENITAL INFECTIONS
- •• Cervical/Vaginal Cytology 376
  - Fungal Culture, Biopsy or Body Fluid 619
  - Fungus Smear, Stain 626
  - Glucose, Postglucose Load, Plasma 647
- • Gram Stain 858
- •• Herpes Simplex Virus Culture and Antigen Detection 721
  - Herpesvirus Cytology 727
- • KOH Preparation 823
- •• *Neisseria gonorrhoeae* Culture and Smear 945
  - *Neisseria gonorrhoeae* Nucleic Acid Detection 947
- •• *Trichomonas* Preparation 1273

WARFARIN THERAPY
- Prothrombin Time 1116

WATER BALANCE *see* ACIDOSIS/ALKALOSIS (ACID-BASE BALANCE), DEHYDRATION/HYPOVOLEMIA

WATERY DIARRHEA-HYPOKALEMIA-HYPOCHLORHYDRIA (WDHH) SYNDROME
- •• Vasoactive Intestinal Polypeptide, Plasma 1302

WEGENER GRANULOMATOSIS [446.4] *see also* GLOMERULONEPHRITIS, NEPHRITIS, VASCULITIS
- •• Antineutrophil Cytoplasmic Antibody 187
  - Blood, Urine 281
  - Eosinophil Count 542
  - Glomerular Basement Membrane Antibody 638
- •• Histopathology 733
  - Lymph Node Biopsy 880
  - Protein Electrophoresis, Capillary Zone 1103
  - Protein Electrophoresis, Serum 1104
  - Urinalysis 1289

WEIGHT LOSS [783.2] *see also* ADDISON DISEASE, DIARRHEA, HEART, MALNUTRITION, METASTASIS, PITUITARY/PITUITARY ADENOMA, SYSTEMIC LUPUS ERYTHEMATOSUS (SLE), TUBERCULOSIS
- • Albumin, Serum 120
  - Drugs of Abuse Testing, Urine 525
- • Electrolyte Panel, Serum 532
- • Glucose, Fasting, Plasma 643
- • Glucose, Postglucose Load, Plasma 647
  - Glucose, Semiquantitative, Urine 650
  - Growth Hormone, Serum 662

WEIL DISEASE see LEPTOSPIROSIS (WEIL SYNDROME)

WERNER-MORRISON SYNDROME see also DEHYDRATION/HYPOVOLEMIA, DIARRHEA, MALABSORPTION/MALDIGESTIVE DISEASES

WEST NILE VIRUS INFECTION see also ENCEPHALITIS

WHIPPLE DISEASE [040.2]

WHOOPING COUGH see PERTUSSIS

WILMS TUMOR [189.0] see also NEUROBLASTOMA, POLYCYSTIC DISEASE (KIDNEY), RENAL VEIN THROMBOSIS

WILSON DISEASE [275.1] see also ANEMIA (HEMOLYTIC), CIRRHOSIS, DEMENTIA, FANCONI SYNDROME, JAUNDICE

Low serum copper and ceruloplasmin concentrations with high urinary copper are found. Low serum uric acid is secondary to renal tubular acidosis. Hepatic copper is >250 µg/g dry weight of liver. Serum alkaline phosphatase typically is low; bilirubin may be high; and AST, ALT mildly increased, AST > ALT. Look for Coombs-negative hemolytic anemia.

WISKOTT-ALDRICH SYNDROME [279.12] see also LEUKEMIA, LYMPHOMA, THROMBOCYTOPENIA

XANTHOMA [272.2]

(Continued)

Other books offered by
LEXI-COMP

PHARMACOGENOMICS HANDBOOK

Perfect Bound / Book Size: 4.25" x 7"
by Humma, Ellingrod, Kolesar

This exciting new title by Lexi-Comp, introduces
Pharmacogenomics to the forward-thinking healthcare
professional and student! It presents information concerning key
genetic variations that may influence drug disposition and/or
sensitivity. Brief introductions to fundamental concepts in
genetics and genomics are provided in order to bring the reader
up-to-date on these rapidly emerging sciences. This book
provides a foundation for all clinicians who will be called on to
integrate rapidly expanding genomic knowledge into the
management of drug therapy. A great introduction to
pharmacogenetic principles as well as a concise reference on key
polymorphisms known to influence drug response!

DRUG-INDUCED NUTRIENT DEPLETION HANDBOOK

Perfect Bound / Book Size: 4.25" x 7"
by Pelton, LaValle, Hawkins, Krinsky

A complete and up-to-date listing of all drugs known to deplete
the body of nutritional compounds.

This book is alphabetically organized and provides extensive
cross-referencing to related information in the various sections
of the book. Drug monographs identify the nutrients depleted
and provide cross-references to the nutrient monographs for
more detailed information on effects of depletion, biological
function & effect, side effects & toxicity, RDA, dosage range,
and dietary sources. this book also contains a studies & abstracts
section, a valuable appendix, and alphabetical &
pharmacological indexes.

NATURAL THERAPEUTICS POCKET GUIDE

Perfect Bound / Book Size: 4.375" x 8"
by Krinsky, LaValle, Hawkins, Pelton, Ashbrook Willis

Provides condition-specific information on common uses of
natural therapies. Each condition discussed includes the following:
review of condition, decision tree, list of commonly recommended
herbals, nutritional supplements, homeopathic remedies, lifestyle
modifications, and special considerations.

Provides herbal/nutritional/nutraceutical monographs with over 10
fields including references, reported uses, dosage, pharmacology,
toxicity, warnings & interactions, and cautions &
contraindications. The Appendix includes: drug-nutrient depletion,
herb-drug interactions, drug-nutrient interaction, herbal medicine
use in pediatrics, unsafe herbs, and reference of top herbals.

To order call toll free anywhere in the U.S.: 1-800-837-LEXI (5394)
Outside of the U.S. call: 330-650-6506 or online at www.lexi.com

Other books offered by
LEXI-COMP

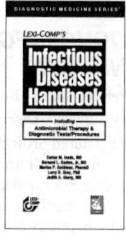

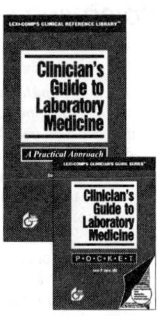

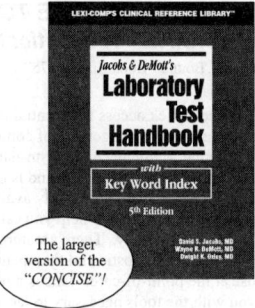
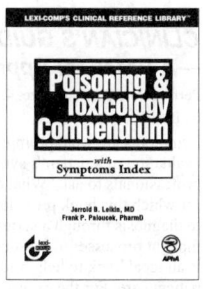

Other books offered by
LEXI-COMP

DRUG INFORMATION HANDBOOK FOR PSYCHIATRY

Perfect Bound / Book Size: 4.5" x 9"
by Fuller and Sajatovic

The source for comprehensive and clinically-relevant drug information for the mental health professional. Alphabetically arranged by generic and brand name for ease-of-use. There are up to 35 key fields of information including these unique fields: "Effect on Mental Status" and "Effect on Psychiatric Treatment." A special topics/issues section includes psychiatric assessment, major psychiatric disorders, major classes of psychotropic medications, psychiatric emergencies, special populations, enhanced patient education information section, and DSM-IV classification. Also contains a valuable appendix section, Pharmacologic Index, and Alphabetical Index.

RATING SCALES IN MENTAL HEALTH

Perfect Bound / Book Size: 5.5" x 8.5"
by Sajatovic and Ramirez

A basic guide to the rating scales in mental health, this is an ideal reference for psychiatrists, nurses, residents, psychologists, social workers, healthcare administrators, behavioral healthcare organizations, and outcome committees. It is designed to assist clinicians in determining the appropriate rating scale when assessing their client. A general concepts section provides text discussion on the use and history of rating scales, statistical evaluation, rating scale domains, and two clinical vignettes. Information on over 100 rating scales used in mental health is organized by condition. Appendix contains tables and charts in a quick reference format allowing clinicians to rapidly identify categories and characteristics of rating scales.

PSYCHOTROPIC DRUG INFORMATION HANDBOOK

Perfect Bound / Book Size: 4.375" x 8"
by Fuller and Sajatovic

This portable, yet comprehensive guide to psychotropic drugs provides healthcare professionals with detailed information on use, drug interactions, pregnancy risk factors, warnings/precautions, adverse reactions, mechanism of action, and contraindications. Alphabetically organized by brand and generic name, this concise handbook provides quick access to the information you need and includes patient education sheets on the psychotropic medications. It is the perfect pocket companion to the *Drug Information for Psychiatry.*

To order call toll free anywhere in the U.S.: 1-800-837-LEXI (5394)
Outside of the U.S. call: 330-650-6506 or online at www.lexi.com

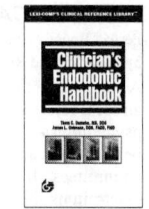

Other books offered by
LEXI-COMP

MANUAL OF CLINICAL PERIODONTICS

Spiral Bound / Book Size: 8.5" x 11"
by Serio and Hawley

A reference manual for diagnosis and treatment including sample treatment plans. It is organized by basic principles and is visually-cued with over 220 high quality color photos. The presentation is in a "question & answer" format. There are 12 chapters tabbed for easy access: 1) Problem-based Periodontal Diagnosis; 2) Anatomy, Histology, and Physiology; 3) Etiology and Disease Classification; 4) Assessment, Diagnosis, and Treatment Planning; 5) Prevention and Maintenance; 6) Nonsurgical Treatment; 7) Surgical Treatment: Principles; 8) Repair, Resection, and Regeneration; 9) Periodontal Plastic Surgery; 10) Periodontal Emergencies; 11) Implant Considerations; 12) Appendix

ORAL SOFT TISSUE DISEASES

Spiral Bound / Book Size: 8.5" x 11"
by Newland, Meiller, Wynn, Crossley

Designed for all dental professionals, a pictorial reference to assist in the diagnosis and management of oral soft tissue diseases, (over 160 photos). Easy-to-use, sections include: Diagnosis process: obtaining a history, examining the patient, establishing a differential diagnosis, selecting appropriate diagnostic tests, interpreting the results, etc.; white lesions; red lesions; blistering-sloughing lesions; ulcerated lesions; pigmented lesions; papillary lesions; soft tissue swelling (each lesion is illustrated with a color representative photograph); specific medications to treat oral soft tissue diseases; sample prescriptions; and special topics.

ORAL HARD TISSUE DISEASES

Spiral Bound / Book Size: 8.5" x 11"
by Newland

A reference manual for radiographic diagnosis, visually-cued with over 130 high quality radiographs is designed to require little more than visual recognition to make an accurate diagnosis. Each lesion is illustrated by one or more photographs depicting the typical radiographic features and common variations. There are 12 chapters tabbed for easy access: 1) Periapical Radiolucent Lesions; 2) Pericoronal Radiolucent Lesions; 3) Inter-Radicular Radiolucent Lesions; 4) Periodontal Radiolucent Lesions; 5) Radiolucent Lesions Not Associated With Teeth; 6) Radiolucent Lesions With Irregular Margins; 7) Periapical Radiopaque Lesions; 8) Periocoronal Radiopaque Lesions; 9) Inter-Radicular Radiopaque Lesions; 10) Radiopaque Lesions Not Associated With Teeth; 11) Radiopaque Lesions With Irregular Margins; 12) Selected Readings / Alphabetical Index

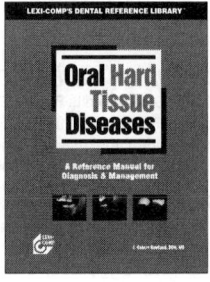

To order call toll free anywhere in the U.S.: 1-800-837-LEXI (5394)
Outside of the U.S. call: 330-650-6506 or online at www.lexi.com

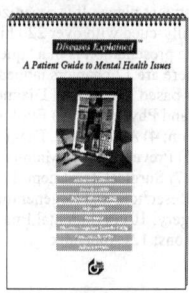
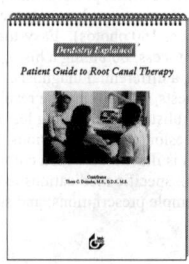
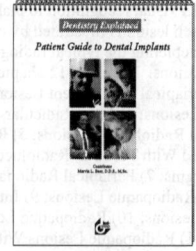

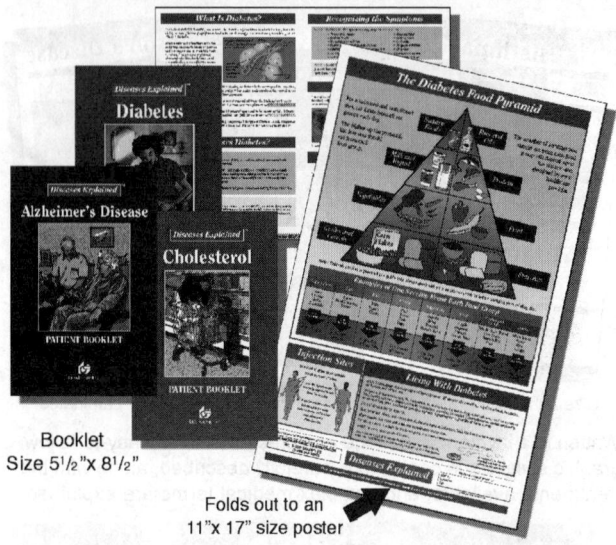

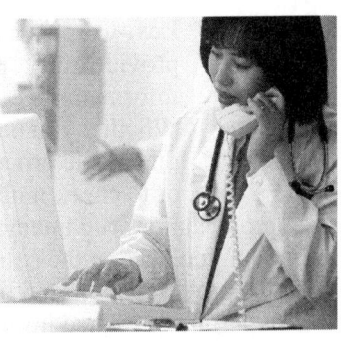

LEXI-COMP ON-HAND SOFTWARE LIBRARY
For Palm OS® and
Windows™ Powered Pocket PC Devices

Lexi-Comp's handheld software solutions provide quick, portable access to clinical information needed at the point-of-care. Whether you need laboratory test or diagnostic procedure information, to validate a dose, or to check multiple medications and natural products for drug interactions, Lexi-Comp has the information you need in the palm of your hand. Lexi-Comp also provides advanced linking technology to allow you to hyperlink to related information topics within a title or to the same topic in another title for more extensive information. No longer will you have to exit 5MCC to look up a dose in Lexi-Drugs or lab test information in Lexi-Diagnostic Medicine — seamlessly link between all databases to **save valuable time**.

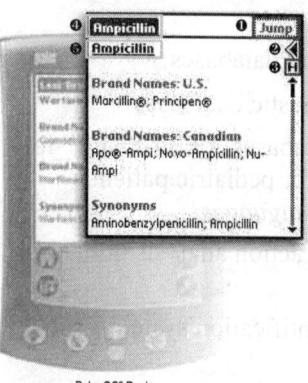

Palm OS® Device

New Navigational Tools:

❶ **"Jump"** provides a drop down list of available fields to easily navigate through information.

❷ **Back arrow** returns to the index from a monograph or to the "Installed Books" menu from the Index.

❸ **"H"** provides a linkable History to return to any of the last 12 Topics viewed during your session.

❹ **Title bar:** Tap the monograph or topic title bar to activate a menu to "Edit a Note" or return to the "Installed Books" menu.

❺ **Linking:** Link to another companion database by clicking the topic or monograph title link or within a database noted by various hyperlinked (colorized and underlined) text.

To order call toll free anywhere in the U.S.: 1-800-837-LEXI (5394)
Outside of the U.S. call: 330-650-6506 or online at www.lexi.com